THE SENSES:
A COMPREHENSIVE
REFERENCE

THE SENSES: A COMPREHENSIVE REFERENCE

Volume 4
OLFACTION AND TASTE

Volume Editors
Dr Stuart Firestein
Columbia University, New York, NY, USA

Dr Gary K. Beauchamp
Monell Chemical Senses Center, Philadelphia, PA, USA

Advisory Board

Dr Allan I. Basbaum
University of California, San Francisco, CA, USA

Dr Akimichi Kaneko
Keio University, Tokyo, Japan

Dr Gordon M. Shepherd
Yale University, New Haven, CT, USA

Dr Gerald Westheimer
University of California, Berkeley, CA, USA

ELSEVIER

AMSTERDAM BOSTON HEIDELBERG LONDON NEW YORK OXFORD
PARIS SAN DIEGO SAN FRANCISCO SINGAPORE SYDNEY TOKYO

Academic Press is an imprint of Elsevier
The Boulevard, Langford Lane, Kidlington, Oxford OX5 1GB, UK
525 B Street, Suite 1900, San Diego, CA 92101-4495, USA

First edition 2008

Notice
No responsibility is assumed by the publisher for any injury and/or damage to persons
or property as a matter of products liability, negligence or otherwise, or from any use
or operation of any methods, products, instructions or ideas contained in the material
herein, Because of rapid advances in the medical sciences, in particular, independent
verfication of diagnoses and drug dosages should be made

British Library Cataloguing in Publication Data
A catalogue record for this book is available from the British Library

Library of Congress Catalog Number: 2007939855

ISBN: 978-012-639482-5

For information on all Elsevier publications
visit our website at books.elsevier.com

Printed and bound in Canada

07 08 09 10 11 10 9 8 7 6 5 4 3 2 1

Contents

Contents of All Volumes

Contributors to All Volumes

B W Ache
University of Florida, Gainesville, FL, USA

P J Albrecht
Albany Medical College, Albany, NY, USA

J M Alexander
University of Wisconsin–Madison, Madison, WI, USA

T S Alioto
University of California, Berkeley, CA, USA

M Alvarez
Universidad Nacional Autónoma de México, México

B L Anderson
University of New South Wales, Sydney, NSW, Australia

D E Angelaki
Washington University School of Medicine, St. Louis, MO, USA

V Anseloni
University of Maryland Dental School, Baltimore, MD, USA

A V Apkarian
Northwestern University, Chicago, IL, USA

K M Armstrong
Stanford University School of Medicine, Stanford, CA, USA

K Bowmaker
University College London, London, UK

A A Bachmanov
Monell Chemical Senses Center, Philadelphia, PA, USA

C A Bagley
Johns Hopkins Hospital, Baltimore, MD, USA

R Bandler
University of Sydney, Sydney, NSW, Australia

L A Barlow
University of Colorado School of Medicine, Aurora, CO, USA

R Baron
Christian-Albrechts-Universität Kiel, Kiel, Germany

L M Bartoshuk
University of Florida, Gainesville, FL, USA

K I Baumann
University of Hamburg, Hamburg, Germany

G K Beauchamp
Monell Chemical Senses Center, Philadelphia, PA, USA

O Behrend
Humboldt-University, Berlin, Germany

K W Beisel
Creighton University, Omaha, NE, USA

F Benedetti
University of Turin Medical School, Turin, Italy

S Bensmaia
The Johns Hopkins University, Baltimore, MD, USA

D A Bereiter
University of Minnesota, Minneapolis, MN, USA

J Bergan
Stanford University School of Medicine, Stanford, CA, USA

I L Bernstein
University of Washington, Seattle, WA, USA

D M Berson
Brown University, Providence, RI, USA

T Berta
University of Lausanne, Lausanne, Switzerland

K Bielefeldt
University of Pittsburgh, Pittsburgh, PA, USA

L A Birder
University of Pittsburgh School of Medicine, Pittsburgh, PA, USA

F Birklein
University of Mainz, Mainz, Germany

J D Bohbot
Vanderbilt University, Nashville, TN, USA

R T Born
Harvard Medical School, Boston, MA, USA

J D Boughter Jr.
University of Tennessee Health Science Center, Memphis, TN, USA

S Bradesi
University of California, Los Angeles, CA, USA

R M Bradley
University of Michigan, Ann Arbor, MI, USA

A S Bregman
McGill University, Montreal, QC, Canada

K H Britten
University of California, Davis, CA, USA

M-C Broillet
University of Lausanne, Lausanne, Switzerland

S M Bromley
University of Pennsylvania, Philadelphia, PA, USA, UMDNJ-Robert Wood Johnson Medical School, Camden, NJ, USA

R M Burger
Lehigh University, Bethlehem, PA, USA

H Burton
Washington University School of Medicine, St. Louis, MO, USA

M R Byers
University of Washington, Seattle WA, USA

A Büschges
University of Cologne, Cologne, Germany

S W Cadden
University of Dundee, Dundee, UK

J N Campbell
Johns Hopkins University, Baltimore, MD, USA

J Caprio
Louisiana State University, Baton Rouge, LA, USA

C E Carr
University of Maryland, College Park, MD, USA

J Carroll
Medical College of Wisconsin, Milwaukee, WI, USA

E Carstens
University of California, Davis, CA, USA

M J Caterina
Johns Hopkins School of Medicine, Baltimore, MD, USA

B Cerf-Ducastel
San Diego State University, San Diego, CA, USA

F Cervero
McGill University, Montreal, QC, Canada

L M Chen
Vanderbilt University, Nashville, TN, USA

J Christensen-Dalsgaard
University of Southern Denmark, Odense, Denmark

T A Cleland
Cornell University, Ithaca, NY, USA

T J Coderre
McGill University, Montreal, QC, Canada

D Copenhagen
University of California San Francisco, CA, USA

R M Costanzo
Virginia Commonwealth University, Richmond, VA, USA

E Covey
University of Washington, Seattle, WA, USA

A D Craig
Barrow Neurological Institute, Phoenix, AZ, USA

W Cronin
University of Maryland, Baltimore, MD, USA

C Darian-Smith
Stanford University School of Medicine, Stanford, CA, USA

R Davis-Taber
Global Pharmaceutical Research and Development, Abbott Park, IL, USA

J W Dawson
Carleton University, Ottawa, ON, Canada

Y De Koninck
Centre de recherche Université Laval Robert-Giffard, Quebec, QC, Canada

V de Lafuente
Universidad Nacional Autónoma de México, México

I Decosterd
University of Lausanne, Lausanne, Switzerland

P H Delano
Universidad de Chile, Santiago, Chile

C D Derby
Georgia State University, Atlanta, GA, USA

S W G Derbyshire
University of Birmingham, Birmingham, UK

J A DeSimone
Virginia Commonwealth University, Richmond, VA, USA

J DeSimone
Virginia Commonwealth University, Richmond, VA, USA

M Devor
Hebrew University of Jerusalem, Jerusalem, Israel

R A DiCaprio
Ohio University, Athens, OH, USA

E Disbrow
University of California, San Francisco, CA, USA

J O Dostrovsky
University of Toronto, Toronto, ON, Canada

R L Doty
University of Pennsylvania, Philadelphia, PA, USA

A Dray
AstraZeneca Research and Development, Montreal, PQ, Canada

R Dubner
University of Maryland, Baltimore, MD, USA

G E DuBois
The Coca-Cola Company, Atlanta, GA, USA

B Duchaine
University College London, London, UK

V B Duffy
University of Connecticut, Storrs, CT, USA

J D Durrant
University of Pittsburgh, Pittsburgh, PA, USA

P L Edds-Walton
Parmly Hearing Institute, Chicago, IL, USA

E Eliav
UMDNJ-New Jersey Dental School, Newark, NJ, USA

M Ennis
University of Tennessee Health Science Center, Memphis, TN, USA

R S Erzurumlu
University of Maryland School of Medicine, Baltimore, MD, USA

R T Eskew Jr.
Northeastern University, Boston, MA, USA

T Euler
Max-Planck-Institute for Medical Research, Heidelberg, Germany

A Faurion
Neurobiologie Sensorielle, NOPA-NBS, INRA, Jouy en Josas, France

R R Fay
Loyola University Chicago, Chicago, IL, USA

D J Felleman
University of Texas Medical School, Houston, TX, USA

A S Feng
University of Illinois at Urbana-Champaign, Urbana, IL, USA

K M Fenn
University of Chicago, Chicago, IL, USA

R D Fernald
Stanford University, Stanford, CA, USA

J Ferraro
University of Kansas Medical Center, Kansas City, KS, USA

R B Fillingim
University of Florida College of Dentistry, Community Dentistry and Behavioral Science Gainesville, FL, USA

T E Finger
University of Colorado School of Medicine, Aurora, CO, USA

N B Finnerup
Aarhus University Hospital, Aarhus, Denmark

M F Fitzgerald
University College London, London, UK

J R Flanagan
Queen's University, Kingston, ON, Canada

H Flor
Central Institute of Mental Health, Mannheim, Germany

A Fontanini
Brandeis University, Waltham, MA, USA

D H Foster
University of Manchester, Manchester, UK

M E Frank
University of Connecticut Health Center, Farmington, CT, USA

M A Freed
University of Pennsylvania School of Medicine, Philadelphia, PA, USA

A S French
Dalhousie University, Halifax, NS, Canada

R Friedman
Vanderbilt University, Nashville, TN, USA

B Fritzsch
Creighton University, Omaha, NE, USA

M Frot
INSERM U879, Bron France

T Fukushima
The University of Tokyo School of Medicine, Tokyo, Japan

D N Furness
Keele University, Keele, UK

G Galizia
Universität Konstanz, Konstanz, Germany

J L Gallant
Helen Wills Neuroscience Institute, Berkeley, CA, USA

P D R Gamlin
University of Alabama at Birmingham, Birmingham, AL, USA

E P Gardner
Department of Physiology and Neuroscience, New York University School of Medicine, New York, NY, USA

G F Gebhart
University of Pittsburgh, Pittsburgh, PA, USA

C D Gilbert
The Rockefeller University, New York, NY, USA

D Rodriguez Gil
Yale University School of Medicine, New Haven, CT, USA

J I Glendinning
Barnard College, Columbia University, New York, NY, USA

P J Goadsby
University of California, San Francisco, CA, USA

P Gochee
University of Kansas Medical Center, Kansas City, KS, USA

M S Gold
University of Pittsburgh, Pittsburgh PA, USA

A W Goodwin
University of Melbourne, Parkville, Vic, Australia

J Gottlieb
Columbia University, New York, NY, USA

R H Gracely
University of Michigan Health System, VAMC, Ann Arbor, MI, USA

C A Greer
Yale University School of Medicine, New Haven, CT, USA

M Gridi-Papp
University of California, Los Angeles, CA, USA

M Grim
Charles University, Praha, Czech Republic

S E Grossman
Brandeis University, Waltham, MA, USA

B Grothe
Ludwig-Maximilians-University, Munich, Germany

M C Göpfert
University of Cologne, Cologne, Germany

T A Hackett
Vanderbilt University, Nashville, TN, USA

C M Hackney
University of Cambridge, Cambridge, UK

A Hajnal
Milton S. Hershey Medical Center, Hershey, PA, USA

Z Halata
University of Hamburg, Hamburg, Germany

R Hallworth
Creighton University, Omaha, NE, USA

R C Hardie
University of Cambridge, Cambridge, UK

K M Hargreaves
University of Texas Health Science Center, San Antonio, TX, USA

I A Harrington
Augustana College, Rock Island, IL, USA

J P Harris
University of California, San Diego, CA, USA

G J Hathway
University College London, London, UK

S E Hausselt
Max-Planck-Institute for Medical Research, Heidelberg, Germany

A Hayar
University of Arkansas for Medical Sciences, Little Rock, AR, USA

J E Hayes
Brown University, Providence, RI, USA

D He
Creighton University, Omaha, NE, USA

B Hedwig
University of Cambridge, Cambridge, UK

H E Heffner
University of Toledo, Toledo, OH, USA

R S Heffner
University of Toledo, Toledo, OH, USA

M M Heinricher
Oregon Health & Science University, Portland, OR, USA

A Hernández
Universidad Nacional Autónoma de México, México

A Hirsh
University of Florida, Gainesville, FL, USA

J R Holt
University of Virginia School of Medicine, Charlottesville, VA, USA

P Honore
Global Pharmaceutical Research and Development, Abbott Park, IL, USA

S S Hsiao
The Johns Hopkins University, Baltimore, MD, USA

J W Hu
University of Toronto, Toronto, ON, Canada

J Iglesias
Cuban Center for Neuroscience, Habana, Cuba

F Imamura
Yale University School of Medicine, New Haven, CT, USA

S L Ingram
Washington State University, Vancouver, WA, USA

J Isnard
Lyon I University and INSERM U879, Bron, France

G H Jacobs
University of California, Santa Barbara, CA, USA

W Jänig
Physiologisches Institut, Christian-Albrechts-Universität zu Kiel, Germany

W Jänig
Christian-Albrechts-Universität zu Kiel, Kiel, Germany

L Jasmin
Neurosurgery and Gene Therapeutics Research Institute, Los Angeles, CA, USA

T S Jensen
Aarhus University Hospital, Aarhus, Denmark

R S Johansson
Umeå University, Umeå, Sweden

S J St. John
Rollins College, Winter Park, FL, USA

B A Johnson
University of California, Irvine, CA, USA

B Johnson
UC Berkeley, Berkeley, CA, USA

J I Johnson
Michigan State University, East Lansing, MI, USA

J H Kaas
Vanderbilt University, Nashville, TN, USA

T Kamigaki
The University of Tokyo School of Medicine, Tokyo, Japan

E Kaplan
The Mount Sinai School of Medicine, New York, NY, USA

H Kasahara
The University of Tokyo School of Medicine, Tokyo, Japan

D B Katz
Brandeis University, Waltham, MA, USA

B J B Keats
Louisiana State University Health Sciences Center, New Orleans, LA, USA

K Keay
University of Sydney, Sydney, NSW, Australia

V Kefalov
Washington University School of Medicine, St. Louis, MO, USA

D R Ketten
Woods Hole Oceanographic Institution, Woods Hole, MA, USA

R M Khan
UC Berkeley, Berkeley, CA, USA

M C Killion
Etymotic Research Ltd., Elk Grove Village, IL, USA

J C Kinnamon
University of Denver, Denver, CO, USA

S C Kinnamon
Colorado State University, Fort Collins, CO, USA

K R Kluender
University of Wisconsin–Madison, Madison, WI, USA

E Knudsen
Stanford University School of Medicine, Stanford, CA, USA

T Kobayakawa
National Institute of Advanced Industrial Science and Technology (AIST), Tsukuba, Japan

H Komatsu
National Institute for Physiological Sciences, Okazaki, Japan

M Konishi
California Institute of Technology, Pasadena, CA, USA

H G Krapp
Imperial College London, London, UK

B Krekelberg
Rutgers University, Newark, NJ, USA

R F Krimm
University of Louisville School of Medicine, Louisville, KY, USA

L Krubitzer
University of California, Davis, CA, USA

T Kurahashi
Osaka University, Osaka, Japan

M Kössl
Johann Wolfgang Goethe Universität, Frankfurt/Main, Germany

S Lacey
Emory University School of Medicine, Atlanta, GA, USA

R Ladher
RIKEN Centre for Developmental Biology, Kobe, Japan

A K Lalwani
New York University School of Medicine, New York, NY, USA

G J Lavigne
Université de Montréal, Montreal, QC, Canada

H C Lawson
Johns Hopkins Hospital, Baltimore, MD, USA

D Le Bars
INSERM U-713, Paris, France

B B Lee
SUNY College of Optometry, New York, NY, USA

S Lee
Korea Institute of Science and Technology, Seoul, Korea

T Leinders-Zufall
University of Maryland School of Medicine, Baltimore, MD, USA

A Lelli
University of Virginia School of Medicine, Charlottesville, VA, USA

L Lemus
Universidad Nacional Autónoma de México, México

F A Lenz
Johns Hopkins Hospital, Baltimore, MD, USA

M Leon
University of California, Irvine, CA, USA

A R Light
University of Utah, Salt Lake City, UT, USA

D Lima
Universidade do Porto, Porto, Portugal

C Linster
Cornell University, Ithaca, NY, USA

W Li
The Rockefeller University, New York, NY, USA

P-M Lledo
Pasteur Institute, Paris, France

E R Loew
Cornell University, Ithaca, NY, USA

R Luna
Universidad Nacional Autónoma de México, México

D-G Luo
Johns Hopkins University School of Medicine, Baltimore, MD, USA

V Lyall
Virginia Commonwealth University, Richmond, VA, USA

H Machelska
Charité – Universitätsmedizin Berlin, Campus Benjamin Franklin, Berlin, Germany

E A Macpherson
University of Michigan, Ann Arbor, MI, USA

S F Maier
University of Colorado at Boulder, Boulder, CO, USA

H Maija
Helsinki University Hospital, Helsinki, Finland

P B Manis
The University of North Carolina at Chapel Hill, Chapel Hill, NC, USA

G A Manley
Technische Universität München, Garching, Germany

I Marc
Université Laval, Quebec City, QC, Canada

D Margoliash
University of Chicago, Chicago, IL, USA

R F Margolskee
Mount Sinai School of Medicine, New York, NY, USA

G R Martin
University of Birmingham, Birmingham, UK

S C Massey
University of Texas Medical School, Houston, TX, USA

F Mauguière
Lyon I University and INSERM U879, Bron, France

M Max
Mount Sinai School of Medicine, New York, NY, USA

B J May
The Johns Hopkins University School of Medicine, Baltimore, MD, USA

E A Mayer
University of California, Los Angeles, CA, USA

C H McCool
University of California, Davis, CA, USA

D H McDougal
University of Alabama at Birmingham, Birmingham, AL, USA

P A McGrath
The University of Toronto, Toronto, ON, Canada

E M McLachlan
Prince of Wales Medical Research Institute, Randwick, NSW, Australia

D G McLaren
University of Wisconsin, Madison, WI, USA

L M Mendell
State University of New York, Stony Brook, NY, USA

J A Mennella
Monell Chemical Senses Center, Philadelphia, PA, USA

S Mense
Institut für Anatomie und Zellbiologie, Universität Heidelberg, Heidelberg, Germany

W Meyerhof
German Institute of Human Nutrition Potsdam-Rehbruecke, Nuthetal, Germany

R A Meyer
Johns Hopkins University, Baltimore, MD, USA

H J Michalewski
University of California, Irvine, CA, USA

J C Middlebrooks
University of Michigan, Ann Arbor, MI, USA

E D Milligan
University of Colorado at Boulder, Boulder, CO, USA

Y Miyashita
The University of Tokyo School of Medicine, Tokyo, Japan

J S Mogil
McGill University, Montreal, QC, Canada

T Moore
Stanford University School of Medicine, Stanford, CA, USA

T Moser
University of Goettingen, Goettingen, Germany

V Nácher
Universidad Nacional Autónoma de México, México

P M Narins
University of California, Los Angeles, CA, USA

J Ngai
University of California, Berkeley, CA, USA

M A L Nicolelis
Duke University, Durham, NC, USA

R Norgren
Milton S. Hershey Medical Center, Hershey, PA, USA

P T Ohara
University of California, San Francisco, CA, USA

S Ohara
Johns Hopkins Hospital, Baltimore, MD, USA

K Okura
Tokushima Graduate School, Tokushima, Japan

D Oliver
Universität Freiburg, Freiburg, Germany

G A Orban
K.U. Leuven Medical School, Leuven, Belgium

D Osorio
University of Sussex, Brighton, UK

M H Ossipov
University of Arizona, Tucson, AZ, USA

C C Pack
McGill University School of Medicine, Montreal, PQ, Canada

G E Pickard
Colorado State University, Fort Collins, CO, USA

R J Pitts
Vanderbilt University, Nashville, TN, USA

G S Pollack
McGill University, Montreal, QC, Canada

A N Popper
University of Maryland, College Park, MD, USA

F Porreca
University of Arizona, Tucson, AZ, USA

C V Portfors
Washington State University, Vancouver, WA, USA

M Postma
University of Cambridge, Cambridge, UK

R J Prenger
University of California, Berkeley, CA, USA

T M Preuss
Emory University, Atlanta, GA, USA

D D Price
University of Florida, Gainesville, FL, USA

I Provencio
University of Virginia, Charlottesville, VA, USA

A C Puche
University of Maryland School of Medicine, Baltimore, MD, USA

S Puria
Stanford University, Stanford, CA, USA

H-X Qi
Vanderbilt University, Nashville, TN, USA

P Rainville
Université de Montréal, Montreal, QC, Canada

S N Raja
Johns Hopkins University, Baltimore, MD, USA

R Rajimehr
Massachusetts General Hospital, Charlestown, MA, USA

R L Reed
University of Florida, Gainesville, FL, USA

B E Reese
University of California, Santa Barbara, CA, USA

L Rela
Yale University School of Medicine, New Haven, CT, USA

K Ren
University of Maryland, Baltimore, MD, USA

B A Revill
Brandeis University, Waltham, MA, USA

J Reynolds
The Salk Institute for Biological Studies, San Diego, CA, USA

A Ribeiro-da-Silva
McGill University, Montreal, QC, Canada

F L Rice
Albany Medical College, Albany, NY, USA

F Rieke
University of Washington, Seattle, WA, USA

M Ringkamp
Johns Hopkins University, Baltimore, MD, USA

H L Rittner
Charité – Universitätsmedizin Berlin, Campus Benjamin Franklin, Berlin, Germany

D Robert
University of Bristol, Bristol, UK

W M Roberts
University of Oregon, Eugene, OR, USA

M E Robinson
University of Florida, Gainesville, FL, USA

L Robles
Universidad de Chile, Santiago, Chile

V Rodríguez
Cuban Center for Neuroscience, Habana, Cuba

I Rodriguez
University of Geneva, Geneva, Switzerland

A W Roe
Vanderbilt University, Nashville, TN, USA

E T Rolls
University of Oxford, Oxford, UK

R Romo
Universidad Nacional Autónoma de México, México

E W Rubel
University of Washington, Seattle, WA, USA

I Russell
University of Sussex, Brighton, UK

M A Rutherford
University of Oregon, Eugene, OR, USA

K Saito
University of Pennsylvania, Philadelphia, PA, USA

H Sakano
University of Tokyo, Tokyo, Japan

A N Salt
Washington University School of Medicine, St. Louis, MO, USA

J Sandkühler
Medical University of Vienna, Vienna, Austria

K Sathian
Emory University School of Medicine, Atlanta, GA, USA

R J Schafer
Stanford University School of Medicine, Stanford, CA, USA

S S Schiffman
Duke University Medical Center, Durham, NC, USA

M Schmelz
University of Heidelberg, Mannheim, Germany

J Schouenborg
Lund University, Lund, Sweden

B A Schulte
Medical University of South Carolina, Charleston, SC, USA

I Schwetz
Medical University, Graz, Austria

J E Schwob
Tufts University School of Medicine, Boston, MA, USA

V E Scott
Global Pharmaceutical Research and Development, Abbott Park, IL, USA

R V Shannon
House Ear Institute, Los Angeles, CA, USA

A Sharma
Columbia University, New York, NY, USA

L T Sharpe
University College London, London, UK

S M Sherman
The University of Chicago, Chicago, IL, USA

T Shimura
Osaka University, Osaka, Japan

J Siegel
Northwestern University, Evanston, IL, USA

C T Simons
Global Research and Development Center, Cincinnati, OH, USA

W Singer
Max Planck Institute for Brain Research, Frankfurt, Germany

D V Smith
The University of Tennessee College of Medicine, Memphis, TN, USA

M T Smith
John Hopkins Medical School, Baltimore, MD, USA

R G Smith
University of Pennsylvania, Philadelphia, PA, USA

J B Snow Jr.
University of Pennsylvania, Philadelphia, PA, USA

D J Snyder
Yale University, New Haven, CT, USA

N Sobel
UC Berkeley, Berkeley, CA, USA

P J Sollars
Colorado State University, Fort Collins, CO, USA

A C Spector
The Florida State University, Tallahassee, FL, USA

H Staecker
University of Kansas Medical Center, Kansas City, KS, USA

A Starr
University of California, Irvine, CA, USA

R Staud
University of Florida, Gainesville, FL, USA

E A Stauffer
University of Virginia School of Medicine, Charlottesville, VA, USA

G C Stecker
University of Washington, Seattle, WA, USA

C R Steele
Stanford University, Stanford, CA, USA

C Stein
Charité – Universitätsmedizin Berlin, Campus Benjamin Franklin, Berlin, Germany

L J Stein
Monell Chemical Senses Center, Philadelphia, PA, USA

A Stockman
University College London, London, UK

R Storms
Veterans Administration Medical Center, Kansas City, MO, USA

E Strettoi
Neuroscience Institute, Pisa, Italy

H Takeuchi
Osaka University, Osaka, Japan

E Thomson
Duke University, Durham, NC, USA

N Tian
Yale University, New Haven, CT, USA

D J Tollin
University of Colorado Health Sciences Center, Aurora, CO, USA

M Tominaga
National Institutes of Natural Sciences, Okazaki, Japan

R Tootell
Massachusetts General Hospital, Charlestown, MA, USA

K Touhara
The University of Tokyo, Chiba, Japan

S P Travers
The Ohio State University, Columbus, OH, USA

R D Treede
Johannes Gutenberg-University, Mainz, Germany

R D Treede
Ruprecht-Karls-University Heidelberg, Heidelberg, Germany

N F Troje
Queen's University, Kingston, ON, Canada

L O Trussell
Oregon Health and Science University, Portland, OR, USA

A Tsuboi
University of Tokyo, Tokyo, Japan

M J Valdés-Sosa
Cuban Center for Neuroscience, Habana, Cuba

D I Vaney
The University of Queensland, Brisbane, QLD, Australia

M Vater
Universität Potsdam, Golm, Germany

M Vorobyev
University of Queensland, Brisbane, QLD, Australia

E T Walters
University of Texas at Houston, Medical School, Houston, TX, USA

M E Warchol
Washington University School of Medicine, St. Louis, MO, USA

E Warrant
University of Lund, Lund, Sweden

W H Warren
Brown University, Providence, RI, USA

L R Watkins
University of Colorado at Boulder, Boulder, CO,USA

L A Werner
University of Washington, Seattle, WA, USA

U Wesselmann
The Johns Hopkins University School of Medicine, Baltimore, MD, USA

G Westheimer
University of California, Berkeley, CA, USA

K N Westlund
University of Texas Medical Branch, Galveston, TX, USA

H E Wheat
University of Melbourne, Parkville, Vic, Australia

M C Whitehead
University of California, San Diego, La Jolla, CA, USA

M C Whitman
Yale University School of Medicine, New Haven, CT, USA

M Wicklein
University College London, London, UK

M C Wiest
Duke University, Durham, NC, USA

J C Willer
INSERM U-731, Paris, France

M A Willis
Case Western Reserve University, Cleveland, OH, USA

W D Willis Jr
University of Texas Medical Branch, Galveston, TX, USA

J F Willott
University of South Florida, Tampa, FL

D A Wilson
University of Oklahoma, Norman, OK, USA

M Wilson
University of California, Davis, CA, USA

J M Wolfe
Brigham and Women's Hospital & Harvard Medical School, Cambridge, MA, USA

J N Wood
University College London, London, UK

H Wässle
Max-Planck-Institute for Brain Research, Frankfurt/Main, Germany

J E Yack
Carleton University, Ottawa, ON, Canada

T Yamamoto
Osaka University, Osaka, Japan

R Yang
University of Denver, Denver, CO, USA

K-W Yau
Johns Hopkins University School of Medicine, Baltimore, MD, USA

R P Yezierski
Comprehensive Center for Pain Research and The McKnight Brain Institute, University of Florida, Gainesville, FL, USA

W A Yost
Loyola University Chicago, Chicago, IL, USA

J M Young
Fred Hutchinson Cancer Research Center, Seattle, WA, USA

G Yovel
Tel Aviv University, Tel Aviv, Israel

A Zainos
Universidad Nacional Autónoma de México, México

H U Zeilhofer
University of Zurich, Zurich, Switzerland

D M Zeitler
New York University School of Medicine, New York, NY, USA

F G Zeng
University of California, Irvine, CA, USA

J-K Zubieta
University of Michigan, Ann Arbor, MI, USA

F Zufall
University of Maryland School of Medicine, Baltimore, MD, USA

L J Zwiebel
Vanderbilt University, Nashville, TN, USA

Introduction to Volume 4

This volume of the *The Senses: A Comprehensive Reference* provides a current review of the chemical senses of taste and smell. Historically, these were considered the minor senses. Descriptions of them were often combined in textbooks both because the natural stimuli are chemicals and because not much was known about them compared with vision and audition. Presumably, they were less studied since it did not seem so bad if humans lost their ability to smell or taste; blindness and deafness were much more serious concerns. However, one might justifiably argue that if all animal life were considered, these are the most important senses. Taste is devoted to a single overwhelmingly important function: it insures that an organism takes in appropriate nutrients and avoids poison. The sense of smell has more varied functions. It too is involved with recognizing food and motivating its intake but it also plays a critical role in monitoring the environment for danger and, perhaps most importantly, in regulating social and sexual activities. Thus, without these two senses most animals would neither eat nor mate! For humans, we have been learning more about the crucial roles taste and smell play in regulating food choice and intake, modulating social interactions, surveying the chemical environment, and providing pleasure. Loss or alteration of smell and taste are not trivial afflictions.

Fortunately, research in the chemical senses is no longer neglected. Indeed, the remarkable rate of progress may, at first blush, even make the idea of a handbook seem absurd. By the time you read these chapters there will be tens of new publications not covered in this book. But in fact it is all those papers that make a compendium like this useful, even if necessarily incomplete. Not long ago one could start out in the fields of olfaction or taste and come up to speed in the literature quite easily. Indeed, for some of us that was one of the attractions of the field. The advent of new techniques, their successful application to questions in chemical sensing, the attraction of investigators from other fields, suddenly transformed the chemical senses from the most mysterious to the most investigated of the sensory systems. Now it is critical to be able to read papers in molecular biology, anatomy, physiology, imaging, psychophysics, genetics, bioinformatics, genomics. . .

So the value of a handbook is as a quick but inclusive reference that will bring even senior investigators rapidly up to speed in an unfamiliar area. In this respect, the contributors to this volume have done an admirable job. Chapters cover all of the above topics in the context of specific systems in olfaction and taste, they cover historical literature (now anything published before 1995 it seems), and provide the kind of background that will facilitate appreciation of the up-to-date advances that appear monthly, if not weekly, in our dynamic field.

This handbook will also, we hope, serve new entrants in the field, especially students and postdoctoral fellows. Each chapter has extensive citations that are an excellent guide to the current literature, and will remain so for many years. The authors have endeavored not only to review the current state of the field, but also to identify important questions and remaining mysteries. Although there have been amazing advances in the past decade and a half, there are even more questions, and more interesting questions, than there were when the last edition of this Handbook appeared.

So, how about the next edition? What will it contain? Will it appear in print or only electromagnetically? What are the advances that will be chronicled in that edition? Who will the chapter authors be? This is a remarkable era in the chemical senses. Our bet is that it is only beginning. We thank the authors for their work to chronicle its current progress and to set the stage for future discoveries.

Gary K. Beauchamp

Dedication

David V. Smith (1943–2006)

David Smith was prominent among a cohort of taste physiologists born in the 1940s on three continents, who have collectively defined – or trained those who defined – gustatory activity in the central nervous system. They learned from the founders of our discipline: Yngve Zotterman, Carl Pfaffmann, Lloyd Beidler, and Masayasu Sato. Equipped with self-styled microelectrodes, they extended recordings from peripheral axons to the small, medial neurons of this ancient sense and taught us how taste selects from a perilous chemical environment to compose a healthy body.

Even as David made his way from his boyhood home in Memphis to study psychology in Knoxville forces were aligning in a competition that was to become the central motif of his career: labeled-line versus patterning. It was a binary that always concerned David, but never consumed him, as it did others of his era. It was our primary topic of professional conversation when David and I explored the South Pacific for a month in the early 1970s, and still the focus of a chapter we wrote three decades later. However voluminous the data, however sophisticated the analyses, they have never been sufficient to seal a victory, and now the issue, still unresolved if indeed a resolution exists, lies exhausted at the periphery of the field (see Chapter 4.17 A Perspective on Chemosensory Quality Coding by M. Frank)

David was always respectful of the coding arguments from both sides, as he was of those who made them. However, he could never divorce his thinking from the central discovery of gustatory electrophysiology – that taste cells are broadly tuned – and permit himself to favor a labeled-line strategy that would seem poorly suited to that finding. Thus, David remained an advocate of patterning even as he vowed unsuccessfully, three times in my presence, never to entertain the topic again.

David trained with Don McBurney in psychophysics, then with Pfaffmann in the electrophysiological techniques that became central to his life's work. He experimented on blowflies, frogs, mice, rats, rabbits, cats, and humans, but David's primary focus was on the hamster hindbrain. Over 30 years, David generated a body of data from the hamster that complemented each component that others had revealed in rats: anatomical connections, membrane qualities, coding principles, neurotransmitters, and centrifugal influences (see Chapter 4.12

Neurotransmitters in the Taste Pathway by R. Bradley). David's thinking was creative and original; his techniques precise; his analyses sophisticated and unbiased. He did not work in large groups – across all his publications David has a mean of fewer than 1.5 co-authors – but over time he worked with scores of colleagues, learning their techniques, sharing his, and always pressing for deeper understanding (see Chapter 4.15 Central Neural Processing of Taste Information by D. Smith and S. Travers).

His objective pursuit of information absent personal agendas made David a trusted colleague and leader. The Association for Chemoreception Sciences (AChemS; Minneapolis) elected him Executive Chairperson (now 'President') in 1985. David directed the neuroscience program at the University of Wyoming in the 1970s and 1980s, the Taste and Smell Center at Cincinnati in the 1980s and 1990s, served as Vice-Chair of Anatomy and Neurobiology at the University of Maryland School of Medicine in the 1990s and 2000s, then completed his life cycle, returning to Memphis as endowed Department Chair of Anatomy and Neurobiology at the Tennessee Health Science Center. In each role, David was fair, collegial, yet demanding, as he was as Executive Editor of *Chemical Senses*.

David was in the fullness of his personal and professional life, living in the city that had called him home, surrounded by appreciative colleagues and by his wife Michiko, whose devotion David requited. In my last chat with the healthy David in 2005, he expressed as much satisfaction with his life as his modesty would permit. It was all too brief.

Thomas R. Scott

4.01 Phylogeny of Chemical Sensitivity

B W Ache, University of Florida, Gainesville, FL, USA

J M Young, Fred Hutchinson Cancer Research Center, Seattle, WA, USA

4.01.1 Introduction

> Smells are surer than sounds and sights to make your heart-strings crack.
>
> Rudyard Kipling

It is difficult for humans, being the visually oriented creatures we are, to appreciate the fundamental importance of chemical sensitivity to life. The ability to detect and respond in an adaptive manner to chemical signals serves as the primary window to the sensory world for most species of animals. Chemical sensitivity is present even in the simplest of the extant life forms: bacteria, slime molds, and protozoans. Indeed, all living cells are irritable to chemicals. This predisposition of cells to be perturbed by chemicals led to the eventual evolution of specific receptor proteins to detect chemical signals and, ultimately, to specific chemosensory organs. The fundamental importance of chemical sensitivity to life and health was recognized by the award of the 2004 Nobel Prize in Physiology or Medicine to Drs. Linda Buck and Richard Axel for their pioneering discovery of olfactory receptor proteins and the understanding of olfactory organization these groundbreaking findings allowed.

Most of our detailed knowledge of chemical sensitivity comes from relatively few, but evolutionary distant species of animals that allow for phylogenetic comparison, including nematodes, insects, crustaceans, fish, frogs, mouse, and rat. Insights gained from rapid advances in molecular biology greatly facilitate this comparison, especially using the experimental model organisms *Caenorhabditis elegans*, *Drosophila*, zebrafish, and mice. Comparison across phylogenetically distant animals can be difficult, however. Even the most fundamental of distinctions such as that between smell and taste can blur when one considers, for example, that lobsters and fish have anatomically distinct olfactory systems that even sniff. Yet odors for these animals are sapid molecules in solution, not unlike the context in which tastants activate vertebrate taste receptor cells. The distinction between smell and taste is further obscured by the presence of taste receptor cells in fish that respond to the same molecules as fish olfactory receptor cells and with greater sensitivity than the olfactory receptor cells. Comparison across animals as phylogenetically distant as nematodes, flies, fish, and mice inevitably requires caution when even basic terminology such as olfaction and taste can be called into question. At the same time, understanding the phylogeny of chemical sensitivity can produce important new insight into how we detect smells and tastes. From the two examples just mentioned, for example, we learn that the fundamental difference between smell and taste must be based on something other than the signaling medium, the nature of the ligands, and the threshold of sensitivity – the descriptors traditionally used to distinguish these two chemosensory modalities in terrestrial vertebrates. We are challenged to understand what really is unique about the chemosensory modality we call olfaction. In addition to comparative studies leading to new questions and insights into chemical sensitivity, animal models provide the opportunity to perform experiments designed to answer those questions.

In this chapter, we take a broad view, trying to look through species-dependent differences in an attempt to reveal broad principles of chemical sensitivity. We conclude that, even though the details differ between phylogenetically diverse animals, general principles of chemical signal detection are shared by most species. Citations throughout are only representative and are not intended to be exhaustive.

4.01.2 Chemical Signals

4.01.2.1 Nature of Chemical Signals

Animals use chemical stimuli to communicate in an amazingly diverse array of behavioral contexts, many of which are hard to appreciate because the action of chemosensory cues is often subtle and difficult for humans to even envision. Chemical stimuli emanating from other species of organisms, known collectively as allelochemics (Whittaker, R. H. and Feeny, P. P., 1971), control prey localization, homing, symbiotic associations, territorial marking, predator deterrence and avoidance, metamorphosis and growth, and pollination, to name just a few examples. Other chemical stimuli of conspecific origin, known collectively as pheromones (Shorey, H. H., 1976), act far beyond their now well-known role as sex

attractants, a function which has been described in most animal phyla. For example, recognition pheromones denote the identity of individuals, social status, social group, and place; aggregation pheromones mediate feeding, sex, and aggression; dispersion pheromones maintain individual spacing and minimize predation; and reproductive pheromones trigger courtship displays and postures. In addition to these triggering functions, pheromones also serve priming functions in equally diverse contexts, in which the chemical stimulus additionally or alternately initiates longer term changes in the state of the recipient animal rather than just triggering immediate, overt responses (Vandenburg, J. G., 1983). The many, diverse behavioral roles chemical stimuli serve in phylogenetically diverse animals underscore the extreme importance of chemical sensitivity.

Many classes of molecules fall within the theoretical limits of molecular size and type for signal function, especially when one considers that these limits expand for chemical signals that can travel by bulk flow in aqueous media and that signals are not restricted to volatile molecules that can diffuse through air. The enormous information content inherent in having such a large number of potential signal molecules is one of the hallmarks of chemosensory communication. Given the great diversity of potential signal molecules, one might expect a given molecule to acquire signal function only occasionally in evolution. However, it is not uncommon to find that members of the same class of molecule or even the same molecular species have signal function in phylogenetically diverse animals. The role of purine nucleotides as chemosensory signals epitomizes this point. Guanosine 5'-monophosphate (GMP) synergizes the perceived intensity of monosodium glutamate in human taste, and methyl xanthenes, compounds that inhibit adenosine receptors, enhance sweet taste (Rifkin, B. and Bartoshuk, L. M., 1980). GMP and other purine nucleotides also increase the binding of glutamate to its receptors isolated from bovine taste papillae by several fold (Torii, K. and Cagan, R. H., 1980). Slime molds use adenosine 5'-monophosphate as an aggregating signal (Mato, J. M. et al., 1978), while adenosine 5'-triphosphate (ATP) in mammalian blood triggers gorging by blood-feeding insects (Friend, W. G. and Smith, J. J. B., 1982). Finally, lobsters recognize food sources in part based on the ratio of ATP to other adenosine nucleotides in their food (Zimmer-Faust, R. K. et al., 1988), and inosine 5'-monophosphate (IMP) activates gustatory receptors in marine flatfish (Mackie, A. M. and Adron, J. W., 1978). It is not clear whether such diversity of signaling roles for a single class of molecules reflects the early evolution and retention of specific receptor types, or convergent evolution of different receptor types to recognize especially biologically relevant ligands. Such questions can now be addressed as more chemoreceptor proteins are cloned and functionally expressed (see Section 4.01.3).

Chemical signals are rarely, if ever, single compounds in real-world situations. Rather, they are complex mixtures of compounds, where related signals can contain many of the same components in different ratios (Figure 1). Single chemical compounds can elicit physiological and behavioral responses and are used routinely in the laboratory to experimentally dissect chemosensory function. However, complete biological activity inevitably requires stimulation with complex, multicomponent

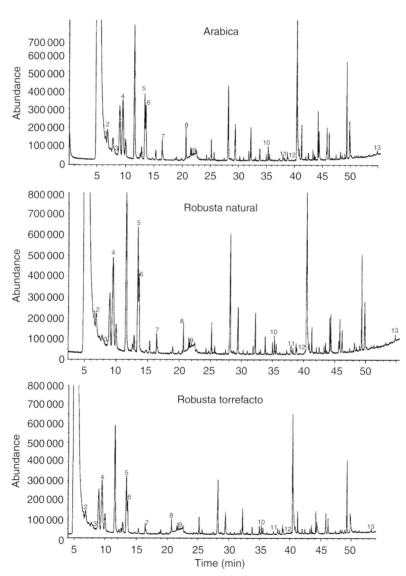

Figure 1 Natural odorants are complex, often highly redundant mixtures. Plots show GC–MS chromatograms of the aroma volatiles of three varieties of espresso coffee: Arabica, Robusta natural, and Robusta torrefacto. Ordinate, absorbance, measuring relative abundance. Abscissa, time in minutes. Similarly numbered peaks in each chromatogram reflect the same compound. Adapted with permission from Maeztu, L., Sanz, C., Andueza, S., Paz De Pena, M., Bello, J., and Cid, C. 2001. Characterization of espresso coffee aroma by static headspace GC-MS and sensory flavor profile. J. Agric. Food. Chem. 49, 5437–5444. Copyright 2001 American Chemical Society.

mixtures of chemical compounds. This generalization holds across phylogenetically diverse species. Relatively simple animals such as cnidarians, jellyfish and sea anemones, can capture and ingest their prey in response to single chemical compounds: glutathione elicits feeding in *Hydra* (Lenhoff, H. M. and Lindstedt, K. J., 1974). But even for these relatively simple animals, discharge of the prey-capturing organs, the cnidia, is regulated by at least two classes of receptors, one for amino and imino acids and another for N-acetylated sugars (Thorington, G. U. and Hessinger, D. A., 1988). Although insect pheromones were originally thought to be silver bullet odorants, as single chemical compounds can attract male to female moths in the field, natural insect pheromones also turn out to be mixtures. For example, the cabbage looper flies upwind in response to major components of the natural pheromone blend, but at least six other components of the female's pheromone gland are required to evoke the full behavioral response of male moths to virgin females (Bjostad, L. B. *et al.*, 1984). Indeed, it is often only subtle differences in the blend ratios of insect pheromones that keep sympatric insect species isolated. Mammals also use complex blends of compounds in pheromone secretions. Here, too, while single putative pheromone components can activate rodent vomeronasal (pheromone) chemoreceptor cells (Leinders-Zufall, T. *et al.*, 2000), it can be difficult to assign behavioral function to individual compounds due to the complexity of mammalian pheromonal secretions and the relatively small contribution of any one component. The scent secretions of tamarins, for example, lack marked differences in the qualitative composition of the marks of individuals of different gender or subspecies, suggesting that this level of behavioral discrimination must be encoded by subtle quantitative differences in the composition of scent mixtures (Belcher, A. M. *et al.*, 1988). Dissecting multicomponent stimulus blends can be daunting due to possible interactions among co-activated elements in the chemosensory pathway. When these interactions are inhibitory, for instance, removal of any one component from a mixture can result in expansion of the relative stimulus strength of the remaining components, masking the ability to detect the component's contribution to the stimulus capacity of the mixture.

4.01.2.2 Dynamics of Chemical Signals

Another hallmark of chemosensory signaling is that the signal is usually intermittent. Indeed, intermittency is fundamental to chemosensation, if not to sensory perception in general, by serving to offset adaptation and sharpen the onset of the signal (Dethier, V. G., 1987). Intermittency is an inherent property of the stimulus field or plume for many animals. Microbes and the smallest of eukaryotes that live within the viscous sublayer of the boundary zone of surfaces, and even those microorganisms that live suspended in the water column, experience stimulus spread that is dominated by diffusion and therefore concentration increases or decreases continuously. All larger animals, however, experience turbulent air or water flow; hence local currents and eddies dominate and perturb stimulus clouds emanating from point sources. Earlier models of odor plumes described the turbulent mixing of the chemical signal into the surrounding water or air as a diffusion process, producing homogeneous clouds with concentration that decreases with distance from the source. However, measurement of stimulus plumes on the scale at which they are encountered by most macroscopic animals in both water and air shows the plumes to be highly discontinuous (Koehl, M. A. R. *et al.*, 2001). As a result, chemoreceptors are only intermittently exposed to the stimulus as the animal moves through the medium or the medium moves over the animal (Figure 2). The specific parameters of the plume structure are medium-dependent and presumably contribute to shaping the dynamic aspects of chemical sensitivity

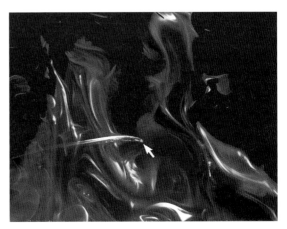

Figure 2 Odor plumes are highly discontinuous in space and time. Photograph of an odor field in a flume flowing from right to left past the antennule of a lobster. The odorant is a fluorescent dye illuminated by a laser sheet that penetrates the water mass in the plane of the organ. Note the scale of the discontinuities relative to the size of the olfactory organ, the tuft of sensilla on the distal end of the antennule (arrow). Unpublished photograph courtesy of M. Koehl.

in different animals. In addition to intermittency inherent in the stimulus field, animals often actively control the release of chemical signals in a pulsatile manner so as to enhance the intermittency of the stimulus field, further underscoring the fundamental importance of intermittency to chemical sensitivity. This phenomenon typically occurs in association with the release of pheromones, intraspecific chemical signals in which the sender and the receiver of the signal are coevolving. Intermittent release occurs in animals as phylogenetically diverse as slime molds (Mato, J. M. *et al.*, 1978) and insects (Christensen, T. A. *et al.*, 1994). In addition to intermittency inherent in the signal itself, active processes often intermittently gate access of the stimulus to the receptor organ, as considered in Section 4.01.5.2.

4.01.3 Chemoreceptor Proteins

Recognition of this diverse assortment of chemical signals is achieved largely using G-protein-coupled receptors (GPCRs), seven-transmembrane domain proteins that activate G-protein-based signaling cascades when activated by their ligands. Members of the GPCR superfamily recognize a wide range of both external and endogenous chemical signals including general odorants, pheromones, and tastants, as well as endogenous peptide, lipid, neurotransmitter, and nucleotide signals (Vassilatis, D. K. *et al.*, 2003). Several independent expansions of ancestral GPCRs have resulted in a number of sizeable, distantly related chemosensory gene families.

In 1991, Linda Buck and Richard Axel identified a large, diverse family of GPCRs expressed in the rat olfactory epithelium (Buck, L. and Axel, R., 1991) and proposed that they function as odorant receptors (commonly referred to as olfactory receptors, as distinct from a subsequently identified family proposed to be receptors for pheromonal odorants – but see below). This landmark discovery opened the door to the subsequent identification of a number of other chemosensory GPCR families, including insect olfactory and gustatory receptors (Clyne, P. J. *et al.*, 1999; Scott, K. *et al.*, 2001), nematode chemoreceptors (Troemel, E. R. *et al.*, 1995), the V1R and V2R families of mammalian pheromone receptors (Dulac, C. and Axel, R., 1995; Matsunami, H. and Buck, L. B, 1997), and the T1R and T2R families of mammalian taste receptors (Hoon, M. A. *et al.*, 1999; Chandrashekar, J. *et al.*, 2000). With the recent

availability of almost complete genome sequences for a variety of species, there has been an explosion in papers describing entire families of chemosensory GPCRs (Table 1). These vast quantities of sequence data allow systematic, in-depth studies of many aspects of chemosensory coding: evolution, receptor–ligand interactions, expression patterns, etc. The size of these chemosensory families varies greatly, from only three GPCRs in the entire yeast proteome (GPR1, a nutritional sensor, and the STE2 and STE3 pheromone receptors) and three receptors in the mammalian T1R family (Liao, J. and Schultz, P. G., 2003) to over 1000 functional members in the rodent olfactory receptor repertoires, comprising up to ~6% of all genes in the genome (Young, J. M. *et al.*, 2003; Rat Genome Sequencing Project Consortium, 2004).

4.01.3.1 Receptor Expression Patterns

In many cases, the function of these gene families was initially postulated based only on expression in the appropriate chemosensory organ. A common assumption has been that the olfactory epithelium and the odorant receptor family expressed therein recognize general (nonpheromonal) odorants and that the vomeronasal organ and its receptor families recognize pheromones. Recent data show that this may be an oversimplification: the vomeronasal organ appears to recognize some compounds which have no known pheromonal activity (Sam, M. *et al.*, 2001) and some pheromones can be recognized after ablation of the vomeronasal organ, perhaps by the nose (Brennan, P. A. and Keverne, E. B., 2004). Additional proof of chemosensory receptor function has been provided for some members of these gene families, for example, by demonstrating functional expression in heterologous cell lines (e.g., Krautwurst, D. *et al.*, 1998), knockout and transgenic rescue experiments (e.g., Mueller, K. L. *et al.*, 2005), or by correlating natural sequence variation with phenotypic variation in chemosensory abilities (e.g., Chandrashekar, J. *et al.*, 2000).

However, the expression pattern of many other genes in these families remains untested, and their functional assignment should be considered tentative as it is based only on sequence similarity. Some systematic studies have been performed to confirm chemosensory expression of subsets of these genes (e.g., Scott, K. *et al.*, 2001) but are not comprehensive. Indeed, expression of some chemosensory receptors has been demonstrated outside the tissue that would

Table 1   Size of selected chemosensory receptor gene families

	No. of genes[a]	No. of pseudogenes[a]	Reference
Olfactory receptors			
Human	388	414	(1)
Mouse	~1200	~300	(2)
Rat	~1430	~640	(3)
Dog	~1070	~230	(4)
Chicken	82	476	(5)
Frog	410	478	(5)
Zebrafish	102	35	(5)
V1R pheromone receptors			
Human	2	115	(6)
Mouse	165	165	(6)
Rat	106	110	(6)
Dog	8	54	(6)
Cow	32	41	(7)
Opossum	49	53	(7)
Zebrafish	1	0	(8)
V2R pheromone receptors			
Rat	~100 (total)		(9)
T1R taste receptors			
Human, mouse, rat	3	n.d.	(10)
T2R bitter taste receptors			
Human	25	11	(11)
Rat, mouse	36	7	(12)
Zebrafish	≥2	n.d.	(8)
Chicken	3	n.d.	(13)
Invertebrate chemoreceptors			
Fly olfactory	60	Few	(14)
Fly gustatory	68	Few	(14)
Mosquito olfactory	79	n.d.	(15)
Mosquito gustatory	76	n.d.	(15)
Nematode worm	~1100 (total)		(16)

[a]Most numbers represent lower bound estimates of gene numbers, as genome assemblies are incomplete, and draft assemblies can contain sequencing errors that mean some intact genes appear as pseudogenes. Many gene families have been reported in several studies; for brevity, a single representative report is cited (in general, more recent reports are chosen as they are based on best available genomic data); apologies are due to other authors.
n.d, not determined.
(1) Niimura Y. and Nei M. (2003); (2) Young J. M. *et al.* (2003); (3) Rat Genome Sequencing Project Consortium (2004); (4) Quignon P. *et al.* (2003); (5) Niimura Y. and Nei M. (2005); (6) Young J. M. *et al.* (2005); (7) Grus W. E. *et al.* (2005); (8) Pfister P. and Rodriguez I. (2005); (9) Herrada G. and Dulac C. (1997); (10) Hoon M. A. *et al.* (1999), Montmayeur J. P. *et al.* (2001), Liao J. and Schultz P. G. (2003); (11) Go Y. *et al.* (2005); (12) Wu S. V. *et al.* (2005); (13) Hillier L. W. *et al.* (2004); (14) Robertson H. M. *et al.* (2003); (15) Hill C. A. *et al.* (2002); (16) Robertson H. M. (2001).

be expected based on their family membership. For example, one human pheromone receptor gene appears to be expressed in the olfactory epithelium rather than the vomeronasal organ (Rodriguez, I. *et al.*, 2000), some mammalian olfactory receptors are expressed in nonolfactory tissues, principally the testis (Parmentier, M. *et al.*, 1992), and some bitter taste receptors are expressed in the nose (Finger, T. E. *et al.*, 2003) and gastrointestinal tract (Wu, S. V. *et al.*, 2002). It is possible that some of these reports may reflect promiscuous, irrelevant transcriptional activity, but a recent study provides convincing evidence

that the human olfactory receptor hOR17-4 is expressed in testis as well as the nose, responds to the chemical bourgeonal, and allows sperm to undergo chemotaxis toward bourgeonal sources (Spehr, M. *et al.*, 2003). Given their ability to detect a diverse range of ligands, it is perhaps unsurprising that chemosensory receptors have been co-opted to perform other functions.

Within a chemosensory organ, members of some chemosensory gene families exhibit a remarkable and mysterious expression pattern. For example, each neuron in the mammalian olfactory epithelium probably expresses only one allele of one of the thousand or more genes in the family (Chess, A. et al., 1994; Malnic, B. et al., 1999; Serizawa, S. et al., 2000), although it is still possible that this model is only partially correct (Mombaerts, P., 2004b). This one-neuron, one-gene expression regime ensures that responses to different ligands are segregated into different neurons, allowing fine discrimination between molecules. Axons of neurons expressing the same olfactory receptor converge at a limited number of locations in the olfactory bulb in the brain, thus integrating signals from functionally identical neurons (Mombaerts, P. et al., 1996). Vomeronasal neurons probably follow a similar expression regime (Rodriguez, I. et al., 1999), as do fruitfly olfactory neurons. Most fruitfly neurons each express two receptors: one, Or83b, that is broadly expressed in many olfactory neurons and a second receptor (or occasionally a pair of receptors) selected from among the rest of the gene family (Larsson, M. C. et al., 2004; Goldman, A. L. et al., 2005). The mechanism ensuring singular expression of olfactory receptors is still unclear, although negative feedback may be operating to ensure that once one receptor gene is chosen for expression, no other genes of the family are expressed (Serizawa, S. et al., 2003).

Other families of chemosensory GPCRs, however, are expressed under a less strictly controlled regime. Each mammalian bitter taste receptor cell expresses many or all of the ~30 members of the T2R bitter taste receptor family and is thus broadly tuned to detect a wide range of bitter tastants (Mueller, K. L. et al., 2005). Likewise, each of the nematode worm's 32 chemosensory neurons expresses multiple receptors (Troemel, E. R. et al., 1995). Very few cell types are therefore needed to detect a broad range of chemicals, with the corresponding disadvantage that discrimination between different ligands may be impossible without additional input from other chemosensory modalities. The specificity of neuronal tuning and wiring is discussed further in Section 4.01.6.2.

4.01.3.2 Receptor Structure

It is assumed that regions of the chemoreceptor proteins that vary between members of a family are involved in ligand specification and that regions that are conserved between family members are important for functions common to all members, such as signaling through the G protein (Buck, L. and Axel, R., 1991). Elucidating the three-dimensional structure of many of these GPCRs has

proved difficult, as they are transmembrane proteins and in some cases difficult to express in heterologous cell lines. However, distant similarity to rhodopsin, whose crystal structure is known, has allowed some groups to model the structure of a few olfactory receptors (e.g., Floriano, W. B. et al., 2000). Functional studies utilizing these models have been slow to appear, but one study shows that structural considerations can successfully predict which residues are important in determining ligand specificity (Katada, S. et al., 2005).

It has recently become clear that dimerization (both heterodimerization and homodimerization) is important for many GPCRs to function correctly. Dimerization can help achieve proper cell-surface expression, as well as being involved in signal transduction and in determining ligand specificity (Bouvier, M., 2001). The role of dimerization is still largely unknown for the chemosensory receptors, except for the mammalian T1R taste receptor family. This family has three members, T1R1, T1R2, and T1R3; cells expressing T1R1 and T1R3 (presumably functioning as a heterodimer) recognize L-amino acids (Nelson, G. et al., 2002), whereas cells expressing T1R2 and T1R3 recognize sweet tastes (Nelson, G. et al., 2001). The co-expression of olfactory receptor OR83b with other receptors in fruitflies (Larsson, M. C. et al., 2004) and of its ortholog in silk moths (Nakagawa, T. et al., 2005) suggests that these receptors may also function as heterodimers. Heterodimerization among members of the mammalian olfactory and pheromone receptor families is unlikely, as each neuron appears to express only a single gene of the family. There is, however, some evidence to suggest that these receptors dimerize with other GPCR or non-GPCR molecules (Loconto, J et al., 2003; Hague, C. et al., 2004; Saito, H. et al., 2004).

4.01.3.3 Receptor–Ligand Pairings

With the availability of vast numbers of chemosensory receptor sequences, it would be very useful to know which chemical ligand or ligands each receptor responds to. A few general principles emerge from what we know so far about receptor–ligand pairing: receptors may be broadly tuned (each olfactory receptor recognizes multiple odorants) or finely tuned to recognize a single ligand, like the pheromone receptors (Malnic, B. et al., 1999; Leinders-Zufall, T. et al., 2000). In addition, each ligand may activate multiple receptors, each to a different extent. Section 4.01.6.2 discusses these issues in more detail. Several different approaches have been taken to deciphering receptor–ligand pairings, including coupling expression in heterologous cell lines with functional assays (e.g., Krautwurst, D. *et al.*, 1998), correlating natural or engineered genetic variation in chemosensory receptors with phenotypic variation in chemosensory abilities (Montmayeur, J. P. *et al.*, 2001) and measuring ligand responsiveness of chemosensory neurons expressing particular receptors (Malnic, B. *et al.*, 1999; Bozza, T. *et al.*, 2002; Hallem, E. A. *et al.*, 2004). The success of these approaches has varied between the chemosensory receptor families (Mombaerts, P., 2004a). Many ligands have been identified for mammalian taste receptors and insect olfactory receptors. Unfortunately, heterologous expression (the approach which seems most promising for large-scale identification of receptor–ligand relationships) of many olfactory receptors has proved remarkably difficult. Three recent studies have identified proteins that interact with olfactory or pheromone receptors to facilitate cell-surface expression (Loconto, J. *et al.*, 2003; Hague, C. *et al.*, 2004; Saito, H. *et al.*, 2004). Use of these proteins may allow a much larger number of receptor–ligand pairs to be identified in the near future, findings that are likely to be of great interest and commercial value (Gilbert, A. N. and Firestein, S., 2002).

4.01.3.4   Receptor Gene Family Evolution

As well as greatly informing our knowledge of how chemicals are sensed, the chemosensory families provide fascinating case studies of gene family evolution. Although some chemosensory gene families have remained static over time (e.g., the mammalian T1R family has three members in both human and mouse, and each mouse gene clearly corresponds to a single human gene: Liao, J. and Schultz, P. G., 2003), others have changed greatly even between closely related species. A number of processes have shaped and continue to change these gene families: gene duplication increases the number of family members; accumulation of sequence changes in new gene duplicates creates novel ligand-binding capabilities; deletions and inactivating mutations cause gene loss; and gene conversion erases sequence divergence and/or creates new combinations of sequence variants (Sharon, D. *et al.*, 1999; Young, J. M. *et al.*, 2002). The net result of these ongoing processes is that different species can have very different functional receptor repertoires (Table 1). Figure 3 illustrates how some of these processes have shaped the

modestly sized human and mouse T2R bitter taste receptor families.

Differences between species in chemosensory receptor families may be driven by differences in the biological needs of the species – for example, some authors suggest that the loss of functional genes and accumulation of pseudogenes in the primate olfactory and perhaps pheromone receptor families were coincident with the acquisition of trichromatic vision (Zhang, J. and Webb, D. M., 2003; Gilad, Y. *et al.*, 2004), implying that our increased reliance on visual cues made chemosensation at least partially obsolete. For most single-copy genes in a genome, it is possible to identify a unique equivalent in other organisms, a fact that motivates the use of animal models to learn about human biology. The dramatic evolutionary change seen in some of these chemosensory families, together with the fact that even very minor changes in amino acid sequence can alter ligand-binding specificities (Katada, S. *et al.*, 2005), means that studies on individual chemoreceptors may not be particularly applicable across species.

The V1R pheromone receptor family provides perhaps the most striking example of evolutionary change in the chemosensory gene families. Recent studies (Grus, W. E. *et al.*, 2005; Young, J. M. *et al.*, 2005) have shown that rodents have over a hundred functional V1R genes, cows have 30 or more, and opossum at least 49, while dogs have only 8, and humans and chimps close to none. The lack of functional V1Rs in humans and chimps is in keeping with the finding that one of the downstream signaling components, $Trp2\beta$, is a pseudogene (Liman, E. R. *et al.*, 1999) and that the vomeronasal organ is absent or vestigial in these species (Trotier, D. *et al.*, 2000). Even between two closely related species, rat and mouse, V1R repertoires reveal dramatic species-specific expansions in some subfamilies, with as many as 23 mouse genes in a V1R subfamily that contains only one rat pseudogene (Grus, W. E. and Zhang, J., 2004; Young, J. M. *et al.*, 2005). It is possible that as two lineages separate during evolution, changes in pheromone receptor gene families establish or reinforce speciation barriers through their role in determining mating behaviors.

These genome-altering processes are ongoing, resulting in polymorphism between different individuals both in receptor sequences and in gene copy number. There are at least 26 human sequence polymorphisms where one variant encodes an apparently functional olfactory receptor and the other variant encodes a pseudogene (Menashe, I. *et al.*, 2003). The

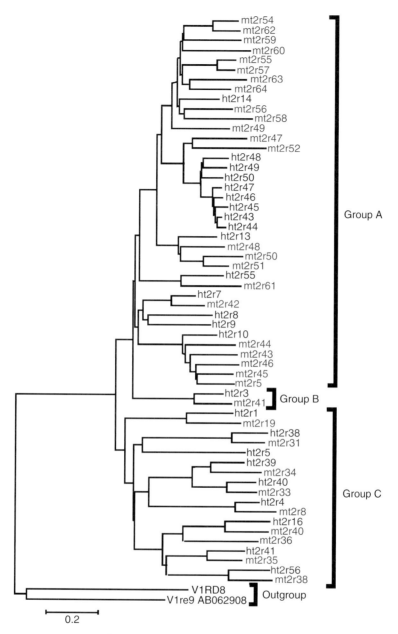

Figure 3 Chemosensory receptor gene families are evolving dynamically. Phylogenetic tree comparing 58 putatively functional human (blue) and mouse (red) bitter taste receptor genes showing that in some branches of the tree (groups B and C) many pairs of one-to-one human–mouse equivalents can be identified, suggesting that these genes have remained stable since the human and rodent lineages diverged. In other branches of the tree (group A), some genes have duplicated several times since the lineages diverged, resulting in clusters of very similar genes and loss of simple, one-to-one equivalence relationships. Stable genes may recognize tastants important to both species, whereas recently duplicated genes might be responsible for the differences between humans and rodents in their gustatory abilities. Adapted from Shi, P., Zhang, J., Yang, H., and Zhang, Y.-P. 2003. Adaptive diversification of bitter taste receptor genes in mammalian evolution. Mol. Biol. Evol. 205, 805–814, with permission from Oxford University Press.

functional implications of some individuals having many more functional olfactory receptors than others are not known, but could be significant. Many other sequence polymorphisms cause a single amino acid change whose functional consequences are currently unknown (e.g., Zhang, X. *et al.*, 2004). Copy number polymorphism of a group of human olfactory receptors has also been described. A recent duplication of a sequence block found near the ends of several human chromosomes results in different individuals having between 7 and 11 copies of one functional receptor and two pseudogenes, with many copies showing subtle differences in amino acid sequence (Trask, B. J. et al., 1998). Such variation can be difficult to detect; future studies will undoubtedly uncover additional examples of chemosensory receptor gene copy number polymorphism.

One can imagine two possible selective advantages to retaining duplicates of chemosensory receptors. Sequence changes accumulating in a new gene copy might result in novel ligand-binding specificities, providing a selective advantage to an individual who can now detect additional chemical signals. Alternatively, an increase in gene copy number of a given chemosensory receptor might increase the relative number of sensory cells choosing to express that receptor, resulting in increased sensitivity to the corresponding ligand. Evolutionary selective pressures favoring novel amino acid sequence variants are often described as positive selection: when sequence changes are not tolerated, selection is said to be purifying. Initial studies looking at orthologous receptor genes in different species, at paralogous genes within a species, and at polymorphisms between members of the same species are suggestive of positive selection acting on a subset of olfactory receptors (e.g., Hughes, A. L. and Hughes, M. K., 1993; Gilad, Y. et al., 2000), pheromone receptors (Emes, R. D. et al., 2004; Lane, R. P. et al., 2004), nematode chemosensory receptors (Thomas, J. H. et al., 2005), and mammalian bitter taste receptors (Shi, P. et al., 2003). It will be fascinating to learn whether the signs of positive selection observed in these genes correlate with the emergence of novel ligand-binding capabilities.

The functional consequences of changes in the chemosensory receptor families (both in copy number and in amino acid sequence) are mostly unclear at the moment, although some progress is being made in correlating sequence differences and changes in ligand responsiveness. A number of studies describe differential sensitivity to various odorants and tastants, for example, musk, isovaleric acid, and asparagus metabolites (Whissell-Buechy, D. and Amoore, J. E., 1973; Lison, M. et al., 1980; Pourtier, L. and Sicard, G., 1990), but the underlying sequence differences have been identified in only a few cases. Polymorphisms between mouse strains in the gene encoding T1R3 govern their sensitivity to several sweet tastants including saccharin and sucrose

(Montmayeur, J. P. et al., 2001). Variation in the mouse mT2R-5 bitter taste receptor affects cyclohex-imide sensitivity (Chandrashekar, J. et al., 2000), and variation in one of the human T2R genes determines whether the bitter compound phenylthiocarbamide can be tasted (Kim, U. K. et al., 2003).

4.01.4 Peripheral Chemoreceptor Cells

4.01.4.1 Receptor Cell Morphology

The evolution of multicellularity allowed for specialization of cellular function and the appearance of cells dedicated to the detection of chemical cues. Some chemosensory input is mediated by free nerve endings, as in the case of vertebrate trigeminal chemoreceptor neurons, but typically animals have evolved more specialized chemoreceptor cells. Chemoreceptor cells occur both internally, for example, in blood vessels and the gastrointestinal tract, where they monitor the internal milieu, and externally. The general morphology of specialized chemoreceptor cells, whether internal or external, is strikingly similar among species and systems, notwithstanding species-specific differences in their detailed morphology. With one main exception, chemoreceptor cells tend to be primary bipolar neurons in which the dendritic membrane terminates in a filamentous array generally assumed to increase the surface area for stimulus capture, and the axon extends without synapsing to the central nervous system (CNS). Comparing the morphology of olfactory receptor cells in vertebrates and insects and receptor cells in the nematode amphid organ demonstrates this point (Figure 4). Vertebrate taste cells, however, along with their probably evolutionarily related counterparts, solitary chemoreceptor cells (Finger, T. E. et al., 2003), are not bipolar neurons, but secondary-type sensory cells of epithelial origin that make synaptic contact with afferent nerve fibers in the periphery. These cells, too, terminate in filamentous extensions of the plasma membrane. The filamentous extensions of chemoreceptor cells can be either ciliary or microvillar in origin.

4.01.4.2 Receptor Cell Turnover

Chemosensory cells are necessarily exposed to the environment and therefore subject to environmental onslaught. Presumably in response to this stress, they have evolved the ability to turn over and replace

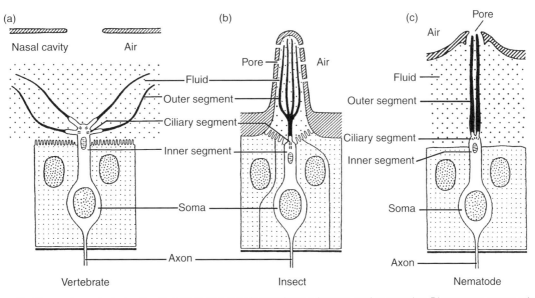

Figure 4 Comparison of primary chemoreceptor neurons in vertebrates, insects, and nematodes. Diagrams compare primary chemoreceptor neurons in the vertebrate olfactory epithelium (a), an insect olfactory sensillum (b), and the nematode amphid organ (c), all drawn to the same size. Note that all are small, bipolar neurons that terminate in a ciliary arbor and send their axon without branching to the CNS. Reproduced from Ache, B. W. 1991. Phylogeny of Smell and Taste. In: Smell and Taste in Health and Disease (*eds.* T. V. Getchell, *et al.*), pp. 3–18. Raven Press. © 1991 Lippincot Williams & Wilkins, with permission.

themselves throughout the life of the animal. Both olfactory and taste receptor cells in all vertebrates are characterized by well-established cycles of birth, maturation, and death (Graziadei, P. P. C. and Monti-Graziadei, G. A., 1978). Given that the olfactory receptor cells are neurons, this feature is remarkable, in that adult neurons generally are not considered to undergo neurogenesis. Receptor cell turnover is not limited to more complex animals, however. The same pulse-labeling technique used to document turnover in vertebrate olfactory receptor neurons shows that chemoreceptor neurons in the anterior tentacles (olfactory organs) of snails also turn over (Chase, R. and Rieling, J., 1986). Lobster olfactory receptor cells (aesthetascs) also turn over, moving from birth, to maturation, to senescence distally along the length of the olfactory organ as the animal grows and adds new segments to the olfactory organ (Steullet, P. *et al.*, 2000). While data are lacking for other species, functional constancy in three such phylogenetically diverse groups of animals argues that turnover is an adaptive property of specialized chemosensory cells that can be expected to occur more generally.

4.01.4.3 Signal Transduction

In addition to detecting the stimulus, chemosensitive cells need to transduce the signal by coupling it to one or more downstream effector molecules. Prokaryotes such as bacteria sense chemical stimuli such as sugars and amino acids differently than most eukaryotes, using transmembrane methyl-accepting chemotaxis proteins (MCPs) linked to a cytoplasmic histidine-aspartate phosphorylating (HAP) system (Wadhams, G. H. and Armitage, J. P., 2004). HAP systems rely on autophosphorylation of a histidine residue and subsequent transfer of a phosphoryl group to an aspartate residue on an associated regulatory protein to generate a response. As detailed in the previous section, eukaryotic chemoreceptor proteins are typically GPCRs that couple to downstream effectors through heteromeric GTP-binding proteins and intracellular second messengers. The second messenger typically targets ion channels that when activated alter the membrane potential of the cell and generate a graded, voltage-dependent response. In multicellular animals with specialized chemoreceptor cells, this leads to the generation of all-or-none electrical signals (action potentials or spikes) that propagate to the CNS with a frequency that is proportional to the magnitude of the graded change in membrane potential. Action potentials are generated within the sensory cell in the case of primary sensory receptor neurons or through synaptic coupling to a neuron in the case of secondary sensory receptor cells.

Two main intracellular signaling pathways are used in eukaryotic cells, utilizing cyclic nucleotides and phosphoinositide-derived signals. Both pathways seem to operate in a diverse range of species, with no clear evolutionary trend in the use of one signaling cascade over the other. Cyclic nucleotide signaling in chemosensory transduction is best understood in vertebrate olfactory receptor neurons (Figure 5(a)). The target of cyclic nucleotide signaling in these cells is the olfactory cyclic nucleotide-gated ion channel

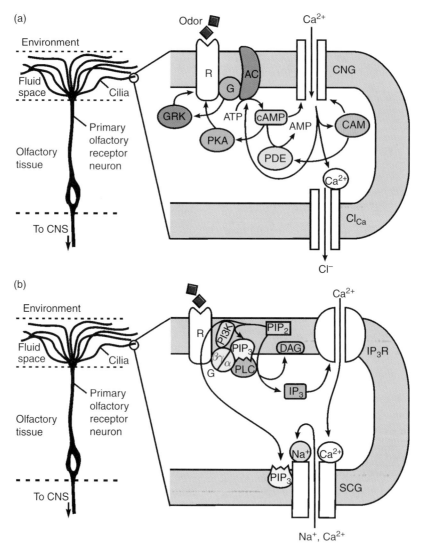

Figure 5 Two intracellular signaling cascades implicated in chemosensory transduction. (a) Diagram of cyclic nucleotide signaling in the transduction compartment (olfactory cilia) of vertebrate olfactory receptor neurons. Odorant molecules bind to a receptor protein (R) coupled to an olfactory specific G_s-protein (G) and activate a type III adenylyl cyclase (AC), increasing intracellular cAMP levels. cAMP targets an olfactory-specific cyclic-nucleotide gated ion channel (CNG), a nonselective cation channel that increases intracellular calcium and secondarily activates a calcium-activated chloride channel thought to carry the majority of the transduction current. Other regulatory pathways are also shown. (b) Diagram of phosphoinositide signaling as currently understood in the transduction compartment (outer dendrite) of lobster olfactory receptor neurons. Odorant molecules bind to a receptor protein (R) coupled to a G_q-protein and activate both phospholipase-C (PLC) and phosphoinositide 3-OH kinase (PI3K) to generate diacylglycerol (DAG) and inositol 1,4,5-trisphosphate (IP_3), and phosphatidylinositol 3,4,5-trisphosphate (PIP_3), respectively, from phosphatidylinositol 4,5-bisphosphate (PIP_2). PIP_3 in concert with release of extracellular calcium from a plasma membrane IP_3 receptor (IP_3R) targets a lobster homolog of a transient receptor potential channel, a nonselective cation channel that is modulated by both sodium and calcium (SGC) and is thought to carry the majority of the transduction current. Details of these pathways vary in other receptor cells and other species.

(Zufall, F. *et al.*, 1994), activation of which allows calcium entry into the cell that secondarily activates a calcium-activated chloride current in a two-step activation cascade. The latter current generates much of the excitatory receptor potential (Reisert, J. *et al.*, 2003). Cyclic nucleotide signaling also appears to operate in chemosensory transduction in slime molds (Newell, P. C. *et al.*, 1988), protozoans (*Paramecium*, Yang, W. Q. *et al.*, 1997), nematodes (*C. elegans*, Komatsu, H. *et al.*, 1999), and arthropods (lobster, Boekhoff, I. *et al.*, 1994). The role of phosphoinositide signaling in chemosensory transduction is less well established than that of cyclic nucleotide signaling and is perhaps best understood in mediating excitation of crustacean olfactory receptor cells (Figure 5(b)). There, the target of phosphoinositide signaling is a calcium-sensitive presumptive lobster homolog of the TRP family of ion channels (Bobkov, Y. V. and Ache, B. W., 2005). Odorants activate both the PLC- and the PI3K-mediated arms of this signaling cascade, allowing the channel to be targeted directly by 3-phosphoinositides and/or indirectly via gating of extracellular calcium from an associated plasma membrane InsP$_3$ receptor (Munger, S. *et al.*, 2000). Phosphoinositide signaling has also been implicated at least indirectly in chemosensory transduction in other, phylogenetically diverse species, including slime molds (Newell, P. C. *et al.*, 1988), nematodes (Colbert, H. A. *et al.*, 1997), insects (Boekhoff, I. *et al.*, 1990), fish (Bruch, R. C. and Teeter, J. H., 1989), mammalian vomeronasal receptor cells (Lucas, P. *et al.*, 2003), and receptor cells in the main olfactory epithelium of mammals (Spehr, M. *et al.*, 2002).

A question that has been the subject of some controversy, for example, in taste transduction in mammals (Zhang, Y. *et al.*, 2003), is whether individual receptor cells use both cyclic nucleotide and phosphoinositide signaling cascades for chemosensory transduction. There is no *a priori* need for a receptor cell to utilize multiple signaling cascades to encode the magnitude of receptor binding. However, when coupled to different receptors or to different sites on the same receptor in a ligand-specific manner, multiple signaling cascades can allow receptor cells to integrate responses to multiple chemical stimuli with potentially important consequences for quality coding (see Section 4.01.6.2). Many diverse species appear to utilize both signaling pathways, including slime molds (Newell, P. C. *et al.*, 1988), arthropods (lobster, Boekhoff, I. *et al.*, 1994), and mammalian olfactory (Spehr, M. *et al.*, 2002), and taste (Gilbertson, T. A. *et al.*, 2000) receptor cells,

suggesting that having both signaling pathways in the same cell may serve a fundamental role in chemosensory signaling, although one that is still being resolved.

Sensitivity to some chemicals can be mediated through non-GPCR-based pathways. Odor- or taste-gated ion channels have been proposed as a potential mechanism of chemosensory transduction in a small number of cases (e.g., Teeter, J. H. *et al.*, 1990), although ionotropic chemoreceptors do not appear to play a major role in chemosensory transduction. The best understood exceptions to the involvement of GPCRs in chemosensory signaling involve direct action of the stimulus on the cell without the need for cell-surface receptors. Ions such as Na$^+$ and protons serve as salt and sour stimuli, respectively, in mammalian taste. These ions are thought to permeate through steady-state ion channels and alter the membrane potential directly without the involvement of downstream signaling cascades (Gilbertson, T. A. *et al.*, 2000). While ions might act similarly in other species, this question has not received the same degree of attention as it has in mammalian taste receptor cells. It is known, however, that ammonia can penetrate the cell membrane of protozoans such as the ciliate, *Paramecium*, and alter chemotaxis through alkalization of the cytoplasm (Van Houten, J. L., 1998), suggesting that nonreceptor-mediated chemical sensitivity can work in parallel with receptor-mediated chemosensitivity in diverse species.

4.01.5 Perireceptor Processes

Activation and adaptation of chemoreceptor cells can be influenced by mechanical and biochemical events in the vicinity of the receptor cells. By modulating or filtering the chemical signal that actually reaches the site of transduction, these so-called perireceptor processes are essential components of chemosensory systems (Carr, W. E. *et al.*, 1990a). The importance of perireceptor processes in chemical sensitivity is still being appreciated, but these processes, too, appear to be conserved in diverse species, suggesting that they are fundamentally important in chemical signal detection.

4.01.5.1 Mechanical Processes

As noted earlier, chemoreceptor cells are seldom, if ever, continuously exposed to the stimulus, and the resulting intermittency of stimulation is fundamental to how they operate. In addition to the signal itself being intermittent, active processes often result in intermittent access of the stimulus to the receptor cells. Receptors within the taste papillae on the surface of the tongue, for example, are only intermittently contacted by the stimulus, the food bolus, when food is actively moved around the oral cavity by tongue movements and mastication. Perhaps the best known example of active intermittent sampling is sniffing in mammalian olfaction, a process that has an integral effect on odor processing in the olfactory bulb and lateral hypothalamus (Macrides, F. and Chorover, S. L., 1972) and that optimizes the perception of odor intensity in humans (Laing, D. G., 1985). A diverse range of active processes in other vertebrate and invertebrate species gate access of stimulus to their respective olfactory organs. Snakes flick their tongue, periodically bringing airborne volatiles to the vomeronasal organ (Kubie, J. and Halpern, M., 1975). Salamanders actively ventilate their olfactory and vomeronasal receptor cavities at 1–2 Hz (Kauer, J. S. and Shepherd, G. M., 1977). Octopus (Chase, R. and Wells, M. J., 1986) and cyclosmate fish like flounders (Doving, K. B. and Thommesen, G., 1977) actively pump water through their siphons or nasal chamber, respectively. Moths fan their wings to enhance air penetration through their olfactory sensillae (Loudon, C. and Koehl, M. A. R., 2000) and decapod crustaceans flick their olfactory organ (Schmitt, B. C. and Ache, B. W., 1979). The ubiquity of active gating suggests that time-locked intermittency is somehow fundamental to the recognition and discrimination of chemosensory signals. Indeed, the dynamic properties of downstream elements in the transduction cascade as well as the kinetics of synaptic interactions in the CNS may be tuned to such intermittency.

4.01.5.2 Biochemical Processes

The apical ends of chemoreceptor cells typically do not project into the environment directly, but instead into a fluid-filled compartment that in turn contacts the environment. This holds even for aquatic animals where desiccation and/or osmotic challenge would not necessarily be a problem. The composition of the fluid bathing the receptor cells is actively regulated by the organism and can contain enzymes, buffers, and other molecules capable of interacting with and potentially modifying the chemical signal (Pelosi, P., 1996). Best known of these are the so-called odorant-

binding proteins (OBPs), small soluble dimeric proteins that bind hydrophilic odorants. OBPs have been characterized in human olfactory mucus (Briand, L. et al., 2002), as well as in most terrestrial animals, including sheep, pigs, cows, rats, frogs, insects, snails (Pelosi, P. and Maida, R., 1990), and elephants (Lazar, J. et al., 2002). The common molecular properties and the occurrence of OBPs in such a phylogenetically diverse range of terrestrial animals suggest they played an important role in the terrestrialization of olfaction. OBPs in moth and rat are not homologous, suggesting the common molecular properties of at least these OBPs evolved convergently (Pevsner, J. et al., 1988). Given the homology of mammalian OBPs to the lipocalin superfamily of proteins (Tegoni, M. et al., 2000), it has often been assumed that OBPs serve to bind and transport hydrophobic ligands through the aqueous perireceptor environment. However, OBPs have also been proposed to serve other functions, including serving as molecular filters to specify and perhaps facilitate stimulus access to receptors (Vogt, R. G. et al., 1990). Another family of proteins, pheromone-binding proteins (PBPs), helps deliver volatile pheromone compounds. Some OBPs and PBPs are similar in sequence, suggesting that members of the OBP family may serve as molecular filters for pheromones rather than for general odors (Pelosi, P., 2001). However, the function of OBPs remains unclear for any type of odorant.

The perireceptor fluid also contains degradative enzymes that could deactivate the stimulus (Carr, W. E. S. et al., 1990b). These, too, occur in phylogenetically diverse species. Slime molds, for example, aggregate in response to the pulsatile release of cAMP from calling amoeboid cells. Both membrane-associated and soluble phosphodiesterases rapidly inactivate cAMP in the vicinity of the source, thereby sharpening the chemotaxic gradient (Janssens, P. M. W. and Van Haastert, P. J. M., 1987). An esterase (Vogt, R. G. et al., 1985) and an aldehyde oxidase (Rybczynski, R. et al., 1989) that rapidly degrade pheromones have been identified in the perireceptor fluid or lymph of insect olfactory receptor cells. Ectonucleotidases in the lobster olfactory organ progressively dephosphorylate adenosine nucleotides (feeding cues) into nonstimulatory adenosine (Trapido-Rosenthal, H. G. et al., 1987). Finally, the mammalian olfactory epithelium contains many catabolic enzymes that could potentially degrade various classes of odorants, including an olfactory-specific homolog of uridine diphosphate-glucuronosyl transferase (Dahl, A., 1988) and several isoforms of cytochrome P450 enzymes. Lower levels of cytochrome P450 are also found in the mammalian vomeronasal organ (Gu, J. *et al.*, 1999). It remains to be determined whether such degradative enzymes act on the signal *per se*, that is, whether they rapidly terminate the signal and therefore alter its dynamic properties or whether they serve to maintain the perireceptor space over longer intervals of time by minimizing the background levels of stimuli and/or removing potentially toxic environmental compounds.

4.01.6 Processing of Chemosensory Information

4.01.6.1 Central Neural Organization

Chemosensory organs are sometimes associated with a peripheral nerve net or plexus, as occurs, for example, in mollusks, resulting in an afferent signal that already reflects some degree of integration. Such peripheral chemosensory integration is best characterized in the vertebrate taste system, where groups of taste receptor cells, supporting cells, and their associated neural elements form a distinct end organ, the taste bud, with synaptic and ephaptic connections

suggestive of interactions among the cells reminiscent of those in the vertebrate retina (Huang, Y. J. *et al.*, 2005). Accordingly, afferent taste nerves transmit an integrated signal to the CNS. The more common organizational plan of chemosensory pathways, however, is for the primary chemosensory afferents to project without synapsing to the CNS, shifting the majority of signal processing to the brain. We are only just beginning to understand how the brain processes chemosensory information, but new molecular and imaging approaches to studying the CNS are poised to rapidly increase our level of understanding of the central neural substrates for molecular recognition (e.g., Zou, Z. *et al.*, 2001).

One system that has been particularly well studied in several, phylogenetically diverse animals is the first olfactory relay – the vertebrate olfactory bulb and its functional equivalent in arthropods, the insect antennal lobe, and the crustacean olfactory lobe. The rather striking analogy of the vertebrate olfactory bulb and the insect antennal lobe was noted by neuroanatomists as early as 1883 (Belloncei, G., 1883), even though some differences are apparent. The mammalian olfactory bulb is laminarly organized with neurons integral to the neuropil, the synaptic-containing region of the bulb (Figure 6(a)), while the

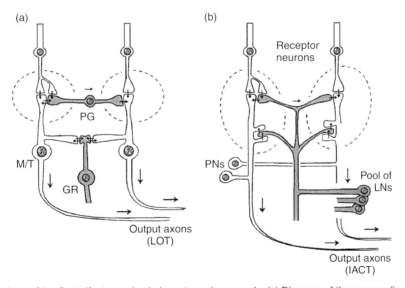

Figure 6 Comparison of the first olfactory relay in insects and mammals. (a) Diagram of the mammalian olfactory bulb showing receptor cells contacting mitral/tufted (M/T) and periglomerular (PG) cells in glomerularly organized neuropil (dashed circles), creating parallel output pathways in the lateral olfactory tract (LOT), transected by two levels of lateral inhibitory connections, one formed by the PG cells and the other by granular (GR) cells. (b) Diagram of the insect antennal lobe drawn in the same format as (a) showing essentially the same overall organization of the projection neurons (PNs) and various types of local interneurons (LNs). From Christensen, T. A. and White, J. 2000. Representation of Olfactory Information in the Brain. In: The Neurobiology of Taste and Smell, 2nd edn. (*eds.* T. E. Finger, W. L. Silver, and D. Restrepo), pp. 201–232. Wiley-Liss, Inc. Copyright 2000 John Wiley, Reprinted with permission of Wiley-Liss, Inc., a subsidiary of John Wiley & Sons, Inc.

insect antennal lobe lacks laminar organization, and its neurons are peripheral to the neuropil, with the synapses confined to the glomeruli (Figure 6(b)). To see the organizational similarity emerge, one has to look through these characteristic differences in the organization of vertebrate and invertebrate neuropil (Christensen, T. A. and White, J., 2000). Doing so shows that in both the olfactory bulb and the antennal lobe, the primary olfactory afferents strongly converge into glomerularly organized neuropil where they branch profusely and terminate on both projection neurons and local interneurons. As mentioned above, glomeruli are formed by the convergence of olfactory neurons that have chosen to express the same receptor and thus integrate signals from functionally identical neurons. Glomeruli in both structures contain the same complex, serial reciprocal synapses (Pinching, A. J. and Powell, T. P. S., 1971; Tolbert, L. P. and Hildebrand, J. G., 1981). The projection neurons take the output of one or a few (depending on the species) glomeruli directly to the next synaptic level, the olfactory cortex in mammals and the corpora pedunculata in arthropods. Local interneurons inherent to the neuropil create two levels of lateral connectivity across the afferent fiber-projection neuron throughput pathway. Interestingly, while most of the olfactory glomeruli tend to be morphologically uniform, one glomerulus in each structure, the modified glomerular complex in mammals (Teicher, M. H. *et al.*, 1980) and the macroglomerular complex in some insects (Matsumoto, S. G. and Hildebrand, J. G., 1981), is greatly enlarged and processes input from pheromone receptors. Many of these organizational features also occur in the first olfactory relay in terrestrial snails, the protocerebral lobe (Chase, R. and Tolloczko, B., 1986), further underscoring the potential utility of this organizational plan to olfactory signal detection.

The apparent conservation between species in the anatomical organization of the first olfactory relay suggests it has important functions in olfactory signal detection. Clearly, the synaptic connections within and among glomeruli are integral to the way olfactory information is processed at this level of the olfactory pathway, and molecular and imaging approaches are actively being used in both vertebrate and invertebrate models to decipher the role(s) of olfactory glomeruli in odor recognition (e.g., Bozza, T. *et al.*, 2004). While the functional role(s) of the first olfactory relay still remain to be understood, some conserved functional features can be identified. One

such conserved feature is presynaptic afferent inhibition (PAI), in which the terminals of the primary afferent fibers are contacted by inhibitory local interneurons. One distinct subpopulation of the local interneurons that laterally connect olfactory glomeruli mediates PAI in animals as diverse as mammals (Hayar, A. *et al.*, 2004) and lobsters (Wachowiak, M. *et al.*, 1997). There is functional evidence for PAI in lobsters, turtles, and rodents, and in both vertebrates and invertebrates, this involves paired neurotransmitters (Wachowiak, M. *et al.*, 2002). As is commonly the case in phylogenetic comparisons, the cellular mechanism differs even though the principal holds. PAI in lobsters reflects ionotropic receptor-mediated changes in membrane potential, while in the turtle, PAI reflects metabotropic receptor-mediated changes in intracellular calcium (Wachowiak, M. and Cohen, L. B., 1999).

4.01.6.2 Quality Coding

The question of how the molecular identity of a chemical stimulus is coded by the nervous system is fundamental to understanding chemical sensitivity. In spite of recent progress, however, the answer remains elusive for both smell and taste. The central argument is whether the identity or quality of the chemical signal is coded by labeled lines, in which the particular cell type that is activated specifies or labels the stimulus, or whether the quality of the stimulus is not inherent in any particular cell type but rather is represented by the pattern of activation across a population of cells with relatively broad overlapping response spectra in a combinatorial manner (Figure 7). Operationally, labeled line or combinatorial coding is often considered in reference to a particular level of a chemosensory pathway, e.g., the tuning or range of ligands that activate individual receptor cells, but the concept ultimately applies more to the overall coding strategy for the sensory modality in question. In that regard, labeled line and combinatorial coding are not mutually exclusive and could be integrated within a given level or across levels of a chemosensory pathway, as seen in the coding of odor mixtures in the lobster olfactory pathway (Derby, C. D., 2000). Both strategies appear to be used in a wide range of species. The choice of coding strategy may be more closely tied to the informational importance of the stimulus than the phylogenetic position of the species. Labeled line coding is expensive as it requires dedicated neural space and therefore may serve to detect stimuli of

Stimulus molecules   Cells

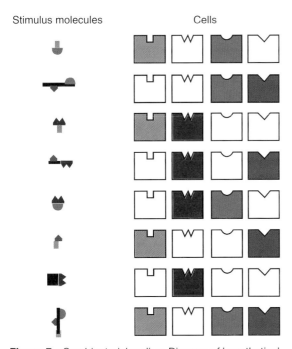

Figure 7 Combinatorial coding. Diagram of hypothetical stimulatory molecules with some shared as well as some different molecular features (left) and the different patterns of neural activity each molecule would elicit across a hypothetical population of four receptor cells each expressing one receptor protein tuned to a particular molecular feature (right). The cells that are activated by each molecule are colored. Note that many molecules activate multiple receptors, and each molecule generates a unique pattern of activity across the population of receptor cells. Reproduced from Malnic, B., Hirono, J., Sato, T., and Buck, L. B. 1999. Combinatorial receptor codes for odors. Cell 96, 713–723, with permission from Elsevier.

especially strong adaptive value, for example, the detection of salt in terrestrial mammals, or of pheromones in short-lived insects that hatch, mate, and die.

It is now generally agreed that odorants are coded in a combinatorial manner. This long-standing idea received strong support from recent evidence that individual mammalian olfactory receptor cells expressing a single, identified receptor protein can be activated by multiple different odorants and that individual odorants activate multiple receptor cells expressing different receptor proteins (Malnic, B. *et al.*, 1999). Similar findings for the olfactory receptors of insects (Hallem, E. A. *et al.*, 2004) and fish (Luu, P. *et al.*, 2004) suggest that combinatorial coding has been conserved in the evolution of olfaction. Combinatorial coding is also used in other chemosensory systems, as demonstrated by the hundreds of receptor proteins

expressed in the nematode *C. elegans*, which has only 32 chemoreceptor cells in the entire adult worm (Bargmann, C. I. and Kaplan, J. M., 1998).

Chemosensory receptor cells can have two (bipolar) modes of signaling, excitation and inhibition. Such flexibility adds to the combinatorial possibilities in chemosensory coding and has been observed in a phylogenetically diverse range of species. Odors inhibit as well as excite olfactory receptor cells in mollusks (squid, Lucero, M. T. *et al.*, 1992), arthropods (insects, lobsters, McClintock, T. S. and Ache, B. W., 1989; De Bruyne, M. *et al.*, 2001), fish (Kang, J. and Caprio, J., 1995), amphibians (frogs, Sanhueza, M. *et al.*, 2000), and mammals (rats, Duchamp-Viret, P. *et al.*, 1999). The cellular mechanism(s) that mediate this opponent input are still being explored, but as noted earlier, this phenomenon can be explained in at least some species by ligand-directed activation of opposing intracellular signaling pathways. Different chemical stimuli attract and repel even single-celled organisms such as *Paramecium* by polarizing the cell in opposite directions (Van Houten, J. L., 1998).

How tastants are coded has focused mainly on the mammalian gustatory system, where the ability to distinguish the so-called basic tastes – sweet, sour, bitter, salt, and amino acids – spawned the search for cell types coding for these qualities. Recent evidence suggests that bitter, sweet, and amino acid tastes are coded by labeled lines, at least at the level of the receptor cells (Zhang, Y. *et al.*, 2003; Mueller, K. L. *et al.*, 2005). It is possible, for example, to induce a sweet response, that is, attraction, to bitter stimuli that otherwise evoke aversion by genetically engineering a mouse to drive the expression of a bitter receptor in taste receptor cells that normally express sweet receptors. This finding implies that taste cells that normally express sweet taste receptors are hardwired to elicit attraction independently of what receptors or ligands are present. Consistent with this interpretation, some cells in the taste nerve as well as in the second gustatory relay in rodents, the parabranchial nucleus (PBN), are best for, that is, respond most strongly to, specific taste qualities and could be considered to function as labeled lines for those qualities (Smith, D. V., 1985). However, the labeled line principle may not exclusively hold. Smith and others found that no one cell type in the PBN can distinguish similar-tasting compounds and that other, more broadly tuned cells contribute to a combinatorial code for taste quality. In multivariate analyses, he found, for example, that sweet-best PBN neurons can distinguish sweet tasting compounds in a panel of tastants, but neither these nor the more broadly tuned cells by themselves were able to differentiate sweet from, say, bitter tasting compounds. Only both types of cells together contained sufficient information to differentiate sweet from bitter. These and other findings argue that, like odorants, tastants are coded in a combinatorial manner. Unfortunately, little additional insight into taste coding is available from other vertebrates or invertebrates, in large part due to the difficulty inherent in defining taste quality in nonmammalian species. In an extensive analysis of phytophagous caterpillar species with differing types of food preference, it was not possible to identify specific taste receptor cell types that could account for food preference, even in strict monophagous species, arguing that food preference was not based on a labeled line code (Dethier, V. G. and Crnjar, R. M., 1982). On the other hand, a distinct set of neurons in the main taste organ of flies, the labellum, exhibit potential hallmarks of a labeled line for sweet taste. These neurons express a receptor for trehalose, project to a distinct brain region, and mediate responsiveness to sweet (as defined by mammalian behavioral responses) compounds, while being unresponsive to bitter compounds (Thorne, N. et al., 2004). Whether tastants are coded using a common strategy or whether coding strategy is context- and/or species-dependent remains an open question.

4.01.7 Chemosensory-Mediated Behavior and Plasticity

As noted earlier, phylogenetically diverse animals use chemical stimuli to communicate in an amazingly diverse array of behavioral contexts. The ability of pheromones to convey remarkably specific information about gender, status, and individuality is just one particularly noteworthy example. Primate secretions are sufficiently complex to allow social information to be communicated by unique blends of the major and minor constituents of these secretions (Epple, G. et al., 1989). Chemosensory-mediated social recognition in mammals is better understood in mice, where genes of the major histocompatibility complex (MHC), the genes controlling immunological identity, have been known for some time to determine pheromone-mediated mating preference (Yamaguchi, M. et al., 1981; Yamazaki, K. et al., 1989). Recently, MHC class I peptides in the urine of mice, the peptides that mediate individual recognition in pregnancy block, were shown to

activate a subset of vomeronasal chemoreceptor cells (Leinders-Zufall, T. et al., 2004). Pheromones also mediate individual recognition in cnidarians – the jellyfish, corals, and sea anemones. These animals can reproduce asexually by budding and form large colonies or mats of genetically identical individuals. Clear borders between mats of such sea anemone individuals are actively maintained by intraspecific aggression (Francis, L., 1973). This behavior is triggered when distinct chemoreceptors on the tentacles contact foreign individuals in neighboring mats as the animals wave about (Lubbock, R., 1980), suggesting this is a chemosensory-mediated behavior. Thomas L. (1974) proposed the existence of a fundamental similarity between immunorecognition and chemical sensitivity, suggesting that the histocompatibility genes on which immunorecognition is based evolved to protect the integrity of the individual organism, for example from invasion of potential symbionts, by strategies such as imparting a characteristic odor to each individual. The presence of individual-specific pheromones in animals as phylogenetically diverse as coelenterates and mammals gives credence to his hypothesis.

All animals must adapt to changing environments, so it is not surprising that plasticity is a hallmark of chemosensory-mediated behavior. Behavioral plasticity encompasses a broad range of phenomena that do not necessarily share a common neural substrate. These phenomena range from simple alterations in levels of responsiveness such as sensitization and habituation to imprinting, to more complex forms of associative leaning. One type of plasticity of particular relevance to chemical sensitivity is food aversion learning, the long-term retention of experience gained from a single association between ingestion and subsequent illness that most humans have experienced at one time or another. This phenomenon is also well documented in other mammals, especially in coyotes and rats where it has been studied in the context of bait shyness (Garcia, J. et al., 1974). Food aversion also occurs in blue jays after eating toxic monarch butterflies. They subsequently seize but no longer consume monarchs due to the distinctively bitter taste of the toxic cardiac glycosides the butterflies sequester from the plants on which they feed (Brower, L. P. and Glazier, S. C., 1975). Food aversion even occurs in mollusks, where a single meal of carrot followed by exposure to quinidine sulfate, an irritant, causes the terrestrial slug, Limax, to reduce its preference for carrots (Gelperin, A., 1975). Interestingly, food aversion in slugs shows all the classical characteristics of food aversion in mammals, including rapid onset, the need for only a single pairing of the conditioned (CS) and unconditioned (US) stimulus, a long CS–US interval, persistence without reinforcement, association restricted to a specific CS, and enhancement through co-association with odor cues. Thus, even complex, associative chemosensory-mediated behavior, and possibly its underlying neural circuitry, can transcend broad species differences.

The early stages of olfactory processing are involved in experience-dependent behavioral modification in animals as phylogenetically diverse as mammals and mollusks. Early odor experience in rat pups, for example, dramatically enhances later responses of the animals to the familiar odor, a

phenomenon that is accompanied by specific physiological and morphological changes in the circuitry of the olfactory bulb (Coopersmith, R. and Leon, M., 1984). Similarly, the gastropod mollusk, *Limax*, can also be behaviorally conditioned to odors. Indeed, it is possible to condition the brain after it has been isolated from the animal and placed in a recording chamber using the same paradigm used to condition the intact animal (Gelperin, A., 1986), allowing direct access to the altered neural circuitry for biochemical and biophysical analyses. The network of olfactory interneurons in the protocerebral lobe of the slug shows coherent oscillations in local field potentials (LFPs) that can be modified by odor input, not unlike similar coherent LFP oscillations observed in insects and a wide array of vertebrates (Figure 8). The

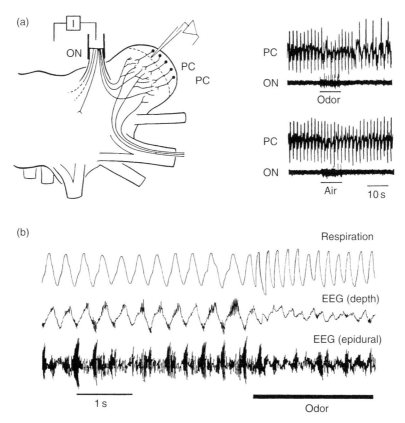

Figure 8 Dynamical structure of neural activity in the first olfactory relay implicated in odor discrimination and memory. (a) Left panel: diagram of the cerebral ganglion of a terrestrial slug showing a recording electrode in the protocerebral lobe (PC) and a stimulating electrode on the olfactory nerve (ON, shown here cut for electrical stimulation but normally intact). Right panel: recordings of neural oscillations from the preparation showing responses to potato odor (upper pair of traces) and clean air (lower pair of traces). Note the odor-induced changes in the waveform and frequency of the PC lobe oscillations. (b) Recordings from the rabbit olfactory bulb showing the 3–7 Hz respiratory rhythm (upper trace) and 40–80 Hz oscillatory EEGs recorded at depth (middle trace) and on the surface (bottom trace) of the bulb. Note the odor-induced changes in the waveform and frequency of the oscillatory field potentials. Reproduced from Ache, B. W. 1991. Phylogeny of Smell and Taste. In: Smell and Taste in Health and Disease (*eds*. T. V. Getchell, *et al*.), pp. 3–18. Raven Press. © 1991 Lippincot Williams & Wilkins, with permission.

presence of neural circuitry at the same synaptic level of the olfactory pathway with similar dynamical structure in animals as phylogenetically diverse as mollusks, insects, and mammals suggests that local circuit oscillations or the coherent neural activity that underlies them play a fundamental role in odor discrimination (Laurent, G., 2002). Odor memory has been shown to be associated with olfactory network dynamics in both insects (locusts, Stopfer, M. and Laurent, G., 1999) and mollusks (slugs, Gelperin, A., 1999). This research is leading to testable predictions as to how oscillatory dynamics may influence the acquisition and storage of odor memory.

4.01.8   Overview

In this chapter, we focus on features shared by different chemosensory systems and different species to reveal the guiding principles of chemical detection. Evolutionary conservation of these shared features presumably reflects the use of a common solution to the common need that all species have to interact with their chemical environment. The rather striking similarities seen in the organization of some of these chemosensory systems, especially olfaction, across a broad range of species suggest there is an optimal solution to the problem of detecting and discriminating chemosensory signals. Either this solution evolved relatively early and was subsequently retained or, more likely, different species convergently evolved the same or similar solutions to the chemosensory problem. As a result, the biological strategy for odor recognition, for example, should be worth emulating in designing biosensors to detect chemical compounds of importance in biomedicine, biosafety, and biodefense. Such biosensors need to detect multicomponent chemical signatures in the complex, dynamic chemical backgrounds that form real-world chemosensory environments. The utility of imitating this biological strategy to develop biosensors provides just one example of how one can learn from a broad, evolutionary approach to understanding chemical sensitivity.

The similarities seen between humans and other species in structural and functional elements of chemical sensing supports the continued use of animal models to investigate chemosensation. As noted by Dethier V. G. (1981), animal studies can tell us not only about those species but also about ourselves, provided we are careful to avoid the dual pitfalls of anthropomorphism and zoomorphism. Even knowing the extent to which chemical sensing in animals differs from that in humans cannot help but reveal something about ourselves.

Acknowledgments

B.W.A. gratefully acknowledges the contribution of his many students and postdoctoral associates over the years to the insight and ideas presented here. J.M.Y. thanks Barbara Trask and colleagues in the Trask laboratory for support, stimulating discussions, and explorations of the literature. The authors thank the National Institute of Deafness and Other Communication Disorders for support (DC05995 and DC04209). We also thank Ms M.L. Milstead for assistance with the illustrations and Ms J. Tulsian for assistance with the bibliography.

References

Ache, B. W. 1991. Phylogeny of Smell and Taste. In: Smell and Taste in Health and Disease (*eds*. T. V. Getchell, *et al.*), pp. 3–18. Raven Press.

Bargmann, C. I. and Kaplan, J. M. 1998. Signal transduction in the *Caenorhabditis elegans* nervous system. Annu. Rev. Neurosci. 21, 279–308.

Belcher, A. M., Epple, G., Kuderling, I., and Smith III A. B. 1988. Volatile components of scent material from the cotton-top tamarin (*Saguinus o. oedipus*): a chemical and behavioral study. J. Chem. Ecol. 14, 1367–1384.

Bjostad, L. B., Linn, C. E., Du, J. W., and Roelofs, W. L. 1984. Identification of new sex pheromone components in *Trichoplusia ni* predicted from biosynthetic precursors. J. Chem. Ecol. 10, 1309–1323.

Bobkov, Y. V. and Ache, B. W. 2005. Pharmacological properties and functional role of a TRP-related ion channel in lobster olfactory receptor neurons. J. Neurophysiol. 93, 1372–1380.

Boekhoff, I., Michel, W. C., Breer, H., and Ache, B. W. 1994. Single odors differentially stimulate dual second messenger pathways in lobster olfactory receptor cells. J. Neurosci. 14, 3304–3309.

Boekhoff, I., Raming, K., and Breer, H. 1990. Pheromone-induced stimulation of inositol-trisphophate formation in insect antennae is mediated by G-proteins. J. Comp. Physiol. B 160, 99–103.

Belloncei, G. 1883. Sur la structure et les rapports des lobes olfactifs dans les arthropods superieurs et les vertebras. Arch. Ital. Biol. 3, 191–196.

Bouvier, M. 2001. Oligomerization of G-protein-coupled transmitter receptors. Nat. Rev. Neurosci. 2, 274–286.

Bozza, T., Feinstein, P., Zheng, C., and Mombaerts, P. 2002. Odorant receptor expression defines functional units in the mouse olfactory system. J. Neurosci. 22, 3033–3043.

Bozza, T., McGann, J. P., Mombaerts, P., and Wachowiak, M. 2004. *In vivo* imaging of neuronal activity by targeted expression of a genetically encoded probe in the mouse. Neuron 42, 9–21.

Brennan, P. A. and Keverne, E. B. 2004. Something in the air? New insights into mammalian pheromones. Curr. Biol. 14, R81–R89.

Briand, L., Eliot, C., Nespoulous, C., Bezirard, V., Huet, J.-C., Henery, C., Blon, F., Trotier, D., and Pernollet, J.-C. 2002. Evidence of an odorant-binding protein in the human olfactory mucus: location, structural characterization, and odorant-binding properties. Biochemistry 41, 7241–7252.

Brower, L. P. and Glazier, S. C. 1975. Localization of heart poisons in the monarch butterfly. Science 208, 753–755.

Bruch, R. C. and Teeter, J. H. 1989. Second-Messenger Signaling Mechanisms in Olfaction. In: Chemical Senses: Vol. 1, Receptor Events and Transduction in Taste and Olfaction (eds. J. G. Brand, J. H. Teeter, R. H. Cagan, and M. R. Kare), pp. 283–298. Marcel Dekker.

Buck, L. and Axel, R. 1991. A novel multigene family may encode odorant receptors: a molecular basis for odor recognition. Cell 65, 175–187.

Carr, W. E., Gleeson, R. A., and Trapido-Rosenthal, H. G. 1990a. The role of perireceptor events in chemosensory processes. Trends Neurosci. 13, 212–215.

Carr, W. E. S., Trapido-Rosenthal, H. G., and Gleeson, R. A. 1990b. The role of degradative enzymes in chemosensory processes. Chem. Senses 15, 181–190.

Chandrashekar, J., Mueller, K. L., Hoon, M. A., Adler, E., Feng, L., Guo, W., Zuker, C. S., and Ryba, N. J. 2000. T2Rs function as bitter taste receptors. Cell 100, 703–711.

Chase, R. and Rieling, J. 1986. Autoradiographic evidence for receptor cell renewal in the olfactory epithelium of a snail. Brain Res. 384, 232–239.

Chase, R. and Tolloczko, B. 1986. Synaptic glomeruli in the olfactory system of a snail, Achatina fulica. Cell Tiss. Res. 246, 567–573.

Chase, R. and Wells, M. J. 1986 Chemotactic behavior in Octopus. J. Comp. Physiol. A 158, 375–381.

Chess, A., Simon, I., Cedar, H., and Axel, R. 1994. Allelic inactivation regulates olfactory receptor gene expression. Cell 78, 823–834.

Christensen, T. A. and White, J. 2000. Representation of Olfactory Information in the Brain. In: The Neurobiology of Taste and Smell, 2nd edn. (eds. T. E. Finger, W. L. Silver, and D. Restrepo), pp. 201–232. Wiley-Liss, Inc.

Christensen, T. A., Laskbrook, J. M., and Hildebrand, J. G. 1994. Neural activation of the sex pheromone gland in the moth Manduca sexta: real-time measurement of pheromone release. Physiol. Entomol. 19, 265–270.

Clyne, P. J., Warr, C. G., Freeman, M. R., Lessing, D., Kim, J., and Carlson, J. R. 1999. A novel family of divergent seven-transmembrane proteins: candidate odorant receptors in Drosophila. Neuron 22, 327–338.

Colbert, H. A., Smith, T. L., and Bargmann, C. I. 1997. OSM-9, a novel protein with structural similarity to channels is required for olfaction, mechanosensation, and olfactory adaptation in Caenorhabditis elegans. J. Neurosci. 17, 8259–8269.

Coopersmith, R. and Leon, M. 1984. Enhanced neural response to familiar olfactory cues. Science 225, 849–851.

Dahl, A. 1988. The Effect of Cytochrome P-450-Dependent Metabolism and Other Enzymes Activities on Olfaction. In: Molecular Neurobiology of the Olfactory System (eds. F. L. Margolis and T. V. Getchell), pp. 51–70. Plenum Press.

De Bruyne, M., Foster, K., and Carlson, J. 2001. Odor coding in the Drosophila antenna. Neuron 30, 537–552.

Derby, C. D. 2000. Learning from spiny lobsters about chemosensory coding of mixtures. Physiol. Behav. 69, 203–209.

Dethier, V. G. 1981. Fly, rat, and man: the continuing quest for an understanding of behavior. Proc. Am. Philos. Soc. 125, 460–466.

Dethier, V. G. 1987. Sniff, flick, and pulse: an appreciation of interruption. Proc. Am. Philos. Soc. 131, 159–176.

Dethier, V. G. and Crnjar, R. M. 1982. Candidate codes in the gustatory system of caterpillars. J. Gen. Physiol. 79, 549–569.

Doving, K. B. and Thommesen, G. 1977. Some Properties of the Fish Olfactory System. In: Olfaction and Taste, (eds. J. LeMagnen and P. MacLeod), Vol. 6, pp. 175–183. Information Retrieval.

Duchamp-Viret, P., Chaput, M., and Duchamp, A. 1999. Odor response properties of rat olfactory receptor neurons. Science 284, 2171–2174.

Dulac, C. and Axel, R. 1995. A novel family of genes encoding putative pheromone receptors in mammals. Cell 83, 195–206.

Emes, R. D., Beatson, S. A., Ponting, C. P., and Goodstadt, L. 2004. Evolution and comparative genomics of odorant- and pheromone-associated genes in rodents. Genome Res. 14, 591–602.

Epple, G., Belcher, A., Greenfield, K. L, Kuderling, I., Nordstrom, K., and Smith III A. B. 1989. Scent Mixtures Used as Social Signals in Two Primate Species: Saguinus fuscicollis and Saguinus o. oedipus. In: Perception of Complex Smells and Tastes (eds. D. G. Laing, W. S. Cain, R. L. McBride, and B. W. Ache), pp. 1–26. Academic Press.

Finger, T. E., Bottger, B., Hansen, A., Anderson, K. T., Alimohammadi, H., and Silver, W. L. 2003. Solitary chemoreceptor cells in the nasal cavity serve as sentinels of respiration. Proc. Natl. Acad. Sci. U. S. A. 100, 8981–8986.

Floriano, W. B., Vaidehi, N., Goddard, W. A., III, Singer, M. S., and Shepherd, G. M. 2000. Molecular mechanisms underlying differential odor responses of a mouse olfactory receptor. Proc. Natl. Acad. Sci. U. S. A. 97, 10712–10716.

Francis, L. 1973. Intraspecific aggression and its effect on the distribution of Anthopleura elegantissima and some related sea anemones. Biol. Bull. 144, 73–92.

Friend, W. G. and Smith, J. J. B. 1982. ATP analogs and other phosphate compounds as gorging stimulants for Rhodnius prolixus. J. Insect Physiol. 28, 371–376.

Garcia, J., Hankins, W. G., and Rusiniak, K. W. 1974. Behavioral regulation of the milieu interne in man and rat. Science 185, 824–831.

Gelperin, A. 1975. Rapid food-aversion learning by a terrestrial mollusk. Science 189, 567–570.

Gelperin, A. 1986. Complex associative learning in small neural networks. Trends Neurosci. 9, 323–368.

Gelperin, A. 1999. Oscillatory dynamics and information processing in olfactory systems. J. Exp. Biol. 202, 1855–1864.

Gilad, Y., Segre, D., Skorecki, K., Nachman, M. W., Lancet, D., and Sharon, D. 2000. Dichotomy of single-nucleotide polymorphism haplotypes in olfactory receptor genes and pseudogenes. Nat. Genet. 26, 221–224.

Gilad, Y., Wiebe, V., Przeworski, M., Lancet, D., and Paabo, S. 2004. Loss of olfactory receptor genes coincides with the acquisition of full trichromatic vision in primates. PLoS Biol. 2, E5.

Gilbert, A. N. and Firestein, S. 2002. Dollars and scents: commercial opportunities in olfaction and taste. Nat. Neurosci 5(Suppl. 1), 1043–1045.

Gilbertson, T. A., Damak, S., and Margolskee, R. F. 2000. The molecular physiology of taste transduction. Curr. Opin. Neurobiol. 10, 519–527.

Go, Y., Satta, Y., Takenaka, O., and Takahata, N. 2005. Lineage-specific loss of function of bitter taste receptor genes in humans and nonhuman primates. Genetics 170, 313–326.

Goldman, A. L., Van der Goes van Naters, W., Lessing, D., Warr, C. G., and Carlson, J. R. 2005. Coexpression of two functional odor receptors in one neuron. Neuron 45, 661–666.

Graziadei, P. P. C. and Monti-Graziadei, G. A. 1978. Continuous Nerve Cell Renewal in the Olfactory System. In: Handbook of Sensory Physiology (*ed*. M. Jacobson), Vol. 9, pp. 55–83. Springer.

Grus, W. E. and Zhang, J. 2004. Rapid turnover and species-specificity of vomeronasal pheromone receptor genes in mice and rats. Gene 340, 303–312.

Grus, W. E., Shi, P., Zhang, Y. P., and Zhang, J. 2005. Dramatic variation of the vomeronasal pheromone receptor gene repertoire among five orders of placental and marsupial mammals. Proc. Natl. Acad. Sci. U. S. A. 102, 5767–5772.

Gu, J., Dudley, C., Su, Tt., Spink, D. C., Zhang, Q.-Y., Moss, R. L., and Ding, X. 1999. Cytochrome P450 and steroid hydroxylase activity in mouse olfactory and vomeronasal mucose. Biochem. Biophys. Res. Comm. 266, 262–267.

Hague, C., Uberti, M. A., Chen, Z., Bush, C. F., Jones, S. V., Ressler, K. J., Hall, R. A., and Minneman, K. P. 2004. Olfactory receptor surface expression is driven by association with the beta2-adrenergic receptor. Proc. Natl. Acad. Sci. U. S. A. 101, 13672–13676.

Hallem, E. A., Ho, M. G., and Carlson, J. R. 2004. The molecular basis of odor coding in the *Drosophila* antenna. Cell 117, 965–979.

Hayar, A., Karnup, S., Ennis, M., and Shipley, M. T. 2004. External tufted cells: a major excitatory element that coordinates glomerular activity. J. Neurosci. 24, 6676–6685.

Herrada, G. and Dulac, C. 1997. A novel family of putative pheromone receptors in mammals with a topographically organized and sexually dimorphic distribution. Cell 90, 763–773.

Hill, C. A., Fox, A. N., Pitts, R. J., Kent, L. B., Tan, P. L., Chrystal, M. A., Cravchik, A., Collins, F. H., Robertson, H. M., and Zwiebel, L. J. 2002. G protein-coupled receptors in *Anopheles gambiae*. Science 298, 176–178.

Hillier, L. W., Miller, W., Birney, E., Warren, W., Hardison, R. C., Ponting, C. P., Bork, P., Burt, D. W., Groenen, M. A., Delany, M. E., *et al.* 2004. Sequence and comparative analysis of the chicken genome provide unique perspectives on vertebrate evolution. Nature 432, 695–716.

Hoon, M. A., Adler, E., Lindemeier, J., Battey, J. F., Ryba, N. J., and Zuker, C. S. 1999. Putative mammalian taste receptors: a class of taste-specific GPCRs with distinct topographic selectivity. Cell 96, 541–551.

Huang, Y. J., Maruyama, Y., Lu, K. S., Pereira, E., Plonsky, I., Baur, J. E., Wu, D., and Roper, S. D. 2005. Mouse taste buds use serotonin as a neurotransmitter. J. Neurosci. 25, 843–847.

Hughes, A. L. and Hughes, M. K. 1993. Adaptive evolution in the rat olfactory receptor gene family. J. Mol. Evol. 36, 249–254.

Janssens, P. M. W. and Van Haastert, P. J. M. 1987. Molecular basis of transmembrane signal transduction in *Dictyostelium discoideum*. Microbiol. Rev. 51, 396–418.

Kang, J. and Caprio, J. 1995. *In vivo* responses of single olfactory receptor neurons in the channel catfish, *Ictalurus punctatus*. J. Neurophysiol. 73, 172–177.

Katada, S., Hirokawa, T., Oka, Y., Suwa, M., and Touhara, K. 2005. Structural basis for a broad but selective ligand spectrum of a mouse olfactory receptor: mapping the odorant-binding site. J. Neurosci. 25, 1806–1815.

Kauer, J. S. and Shepherd, G. M. 1977 Analysis of the onset phase of olfactory bulb unit responses to odour pulses in the salamander. J. Physiol. (Lond.) 272, 495–516.

Kim, U. K., Jorgenson, E., Coon, H., Leppert, M., Risch, N., and Drayna, D. 2003. Positional cloning of the human quantitative trait locus underlying taste sensitivity to phenylthiocarbamide. Science 299, 1221–1225.

Koehl, M. A. R., Koseff, J. R., Crimaldi, J. P., McCay, M. G., Cooper, T., Wiley, M. B., and Moore, P. A. 2001. Lobster sniffing: antennule design and hydrodynamic filtering of information in an odor plume. Science 294, 1948–1951.

Komatsu, H., Jin, Y. H., L'Etoile, N., Mori, I., Bargmann, C. I., Akaike, N., and Ohshima, Y. 1999. Functional reconstitution of a heteromeric cyclic nucleotide-gated channel of *Caenorhabditis elegans* in cultured cells. Brain Res. 821, 160–168.

Krautwurst, D., Yau, K. W., and Reed, R. R. 1998. Identification of ligands for olfactory receptors by functional expression of a receptor library. Cell 95, 917–926.

Kubie, J. and Halpern, M. 1975. Laboratory observations of trailing behavior in garter snakes. J. Comp. Physiol. Psychol. 89, 667–674.

Laing, D. G. 1985. Optimum perception of odor intensity by humans. Physiol. Behav. 34, 569–574.

Lane, R. P., Young, J., Newman, T., and Trask, B. J. 2004. Species specificity in rodent pheromone receptor repertoires. Genome Res. 14, 603–608.

Larsson, M. C., Domingos, A. I., Jones, W. D., Chiappe, M. E., Amrein, H., and Vosshall, L. B. 2004. Or83b encodes a broadly expressed odorant receptor essential for *Drosophila* olfaction. Neuron 43, 703–714.

Laurent, G. 2002. Olfactory network dynamics and the coding of multidimensional signals. Nat. Rev. Neurosci. 3, 884–895.

Lazar, J., Greenwood, D. R., Rasmussen, L. E., and Prestwich, G. D. 2002. Molecular and functional characterization of an odorant binding protein of the Asian elephant, *Elephas maximus*: implications for the role of lipocalins in mammalian olfaction. Biochemistry 41, 11786–11794.

Lenhoff, H. M. and Lindstedt, K. J. 1974. Chemoreception in Aquatic Invertebrates with Special Emphasis on the Feeding Behavior of Coelenterates. In: Chemoreception in Marine Organisms (*eds*. P. T. Grant and A. M. Mackie), pp. 143–175. Academic Press.

Leinders-Zufall, T., Brennan, P., Widmayer, P., Prashanth-Chandramani, S., Maul-Pavicic, A., Jager, M., Li, X. H., Breer, H., Zufall, F., and Boehm, T. 2004. MHC class I peptides as chemosensory signals in the vomeronasal organ. Science 306, 1033–1037.

Leinders-Zufall, T., Lane, A. P., Puche, A. C., Ma, W., Novotny, M. V., Shipley, M. T., and Zufall, F. 2000. Ultrasensitive pheromone detection by mammalian vomeronasal neurons. Nature 405, 792–796.

Liao, J. and Schultz, P. G. 2003. Three sweet receptor genes are clustered in human chromosome 1. Mamm. Genome 14, 291–301.

Liman, E. R., Corey, D. P., and Dulac, C. 1999. TRP2: a candidate transduction channel for mammalian pheromone sensory signaling. Proc. Natl. Acad. Sci. U. S. A. 96, 5791–5796.

Lison, M., Blondheim, S. H., and Melmed, R. N. 1980. A polymorphism of the ability to smell urinary metabolites of asparagus. Br. Med. J. 281, 1676–1678.

Loconto, J., Papes, F., Chang, E., Stowers, L., Jones, E. P., Takada, T., Kumanovics, A., Lindahl, K. F., and Dulac, C. 2003. Functional expression of murine V2R pheromone receptors involves selective association with the M10 and M1 families of MHC class Ib molecules. Cell 112, 607–618.

Loudon, C. and Koehl, M. A. R. 2000. Sniffing by a silkworm moth: wing fanning enhances air penetration through and pheromone interception by antennae. J. Exp. Biol. 203, 2977–2990.

Lubbock, R. 1980. Clone-specific cellular recognition in a sea anemone. Proc. Natl. Acad. Sci. U. S. A. 77, 6667–6669.

Lucas, P., Ukhanov, K., Leinders-Zufall, T., and Zufall, F. 2003. A diacylglycerol-gated cation channel in vomeronasal neuron dendrites is impaired in *TRPC2* mutant mice: mechanism of pheromone transduction. Neuron 40, 551–561.

Lucero, M. T., Horrigan, F. T., and Gilly, W. F. 1992. Electrical responses to chemical-stimulation of squid olfactory receptor-cells. J. Exp. Biol. 162, 231–249.

Luu, P., Acher, F., Bertrand, H. O., Fan, J., and Ngai, J. 2004. Molecular determinants of ligand selectivity in a vertebrate odorant receptor. J. Neurosci. 24, 10128–10137.

Mackie, A. M. and Adron, J. W. 1978. Identification of inosine and inosine 5′-monophosphate as the gustatory feeding stimulants for the turbot, *Scophthalmus maximus*. Comp. Biochem. Physiol. A 60, 79–83.

Macrides, F. and Chorover, S. L. 1972. Olfactory bulb units: activity correlated with inhalation cycles and odor quality. Science 175, 84–87.

Maeztu, L., Sanz, C., Andueza, S., Paz De Pena, M., Bello, J., and Cid, C. 2001. Characterization of espresso coffee aroma by static headspace GC-MS and sensory flavor profile. J. Agric. Food. Chem. 49, 5437–5444.

Malnic, B., Hirono, J., Sato, T., and Buck, L. B. 1999. Combinatorial receptor codes for odors. Cell 96, 713–723.

Mato, J. M., Jastorff, B., Morr, M., and Konijn, T. M. 1978. A model for cyclic AMP-chemoreceptor interaction in *Dictyostelium discoideum*. Biochem. Biophys. Acta 544, 309–314.

Matsumoto, S. G. and Hildebrand, J. G. 1981. Olfactory mechanisms in the moth *Manduca sexta*: response characteristics and morphology of central neurons in the antennal lobes. Proc. R. Soc. Lond., B 213, 249–277.

Matsunami, H. and Buck, L. B. 1997. A multigene family encoding a diverse array of putative pheromone receptors in mammals. Cell 90, 775–784.

McClintock, T. S. and Ache, B. W. 1989. Hyperpolarizing receptor potentials in lobster olfactory receptor cells: implications for transduction and mixture suppression. Chem. Senses 14, 637–647.

Menashe, I., Man, O., Lancet, D., and Gilad, Y. 2003. Different noses for different people. Nat. Genet. 34, 143–144.

Mombaerts, P. 2004a. Genes and ligands for odorant, vomeronasal and taste receptors. Nat. Rev. Neurosci. 5, 263–278.

Mombaerts, P. 2004b. Odorant receptor gene choice in olfactory sensory neurons: the one receptor-one neuron hypothesis revisited. Curr. Opin. Neurobiol. 14, 31–36.

Mombaerts, P., Wang, F., Dulac, C., Chao, S. K., Nemes, A., Mendelsohn, M., Edmondson, J., and Axel, R. 1996. Visualizing an olfactory sensory map. Cell 87, 675–686.

Montmayeur, J. P., Liberles, S. D., Matsunami, H., and Buck, L. B. 2001. A candidate taste receptor gene near a sweet taste locus. Nat. Neurosci. 4, 492–498.

Mueller, K. L., Hoon, M. A., Erlenbach, I., Chandrashekar, J., Zuker, C. S., and Ryba, N. J. 2005. The receptors and coding logic for bitter taste. Nature 434, 225–229.

Munger, S., Gleeson, R. A., Aldrich, H. C., Rust, N. C., Ache, B. W., and Greenberg, R. M. 2000. Characterization of a phosphoinositide-mediated odor transduction pathway reveals plasma membrane localization of an inositol 1,4,5-trisphosphate receptor in lobster olfactory receptor neurons. J. Biol. Chem. 275, 20450–20457.

Newell, P. C., Europe-Finner, G. N., Small, N. V., and Liu, G. 1988. Inositol phosphates, G-proteins and *ras* genes involved in chemotactic signal transduction in *Dictyostelium discoideum*. J. Cell. Sci. 96, 669–673.

Nakagawa, T., Sakurai, T., Nishioka, T., and Touhara, K. 2005. Insect sex-pheromone signals mediated by specific combinations of olfactory receptors. Science 307, 1638–1642.

Nelson, G., Chandrashekar, J., Hoon, M. A., Feng, L., Zhao, G., Ryba, N. J., and Zuker, C. S. 2002. An amino-acid taste receptor. Nature 416, 199–202.

Nelson, G., Hoon, M. A., Chandrashekar, J., Zhang, Y., Ryba, N. J., and Zuker, C. S. 2001. Mammalian sweet taste receptors. Cell 106, 381–390.

Niimura, Y. and Nei, M. 2003. Evolution of olfactory receptor genes in the human genome. Proc. Natl. Acad. Sci. U. S. A. 100, 12235–12240.

Niimura, Y. and Nei, M. 2005. Evolutionary dynamics of olfactory receptor genes in fishes and tetrapods. Proc. Natl. Acad. Sci. U. S. A. 102, 6039–6044.

Parmentier, M., Libert, F., Schurmans, S., Schiffmann, S., Lefort, A., Eggerickx, D., Ledent, C., Mollereau, C., Gerard, C., Perret, J., *et al.* 1992. Expression of members of the putative olfactory receptor gene family in mammalian germ cells. Nature 355, 453–455.

Pelosi, P. 1996. Perireceptor events in olfaction. J. Neurobiol. 30, 3–19.

Pelosi, P. 2001. The role of perireceptor events in vertebrate olfaction. Cell Mol. Life Sci. 58, 503–509.

Pelosi, P. and Maida, R. 1990. Odorant-binding proteins in vertebrates and insects: similarities and possible common function. Chem. Senses 15, 205–215.

Pevsner, J., Reed, R. R., Feinstein, P. G., and Snyder, S. H. 1988. Molecular cloning of odorant-binding protein: member of a ligand carrier family. Science 241, 336–339.

Pfister, P. and Rodriguez, I. 2005. Olfactory expression of a single and highly variable V1r pheromone receptor-like gene in fish species. Proc. Natl. Acad. Sci. U. S. A. 102, 5489–5494.

Pinching, A. J. and Powell, T. P. S. 1971. The neuropil of the glomeruli of the olfactory bulb. J. Cell Sci. 9, 347–377.

Pourtier, L. and Sicard, G. 1990. Comparison of the sensitivity of C57BL/6J and AKR/J mice to airborne molecules of isovaleric acid and amyl acetate. Behav. Genet. 20, 499–509.

Quignon, P., Kirkness, E., Cadieu, E., Touleimat, N., Guyon, R., Renier, C., Hitte, C., Andre, C., Fraser, C., and Galibert, F. 2003. Comparison of the canine and human olfactory receptor gene repertoires. Genome Biol. 4, R80.

Rat Genome Sequencing Project Consortium. 2004. Genome sequence of the Brown Norway rat yields insights into mammalian evolution. Nature 428, 493–521.

Reisert, J., Bauer, P. J., Yau, K. W., and Frings, S. 2003. The Ca-activated Cl channel and its control in rat olfactory receptor neurons. J. Gen. Physiol. 122, 349–363.

Rifkin, B. and Bartoshuk, L. M. 1980. Taste synergism between monosodium glutamate and disodium 5′-guanylate. Physiol. Behav. 24, 1169–1172.

Robertson, H. M. 2001. Updating the str and srj (stl) families of chemoreceptors in *Caenorhabditis nematodes* reveals frequent gene movement within and between chromosomes. Chem. Senses 26, 151–159.

Robertson, H. M., Warr, C. G., and Carlson, J. R. 2003. Molecular evolution of the insect chemoreceptor gene superfamily in *Drosophila melanogaster*. Proc. Natl. Acad. Sci. U. S. A. 100 Suppl. 2, 14537–14542.

Rodriguez, I., Feinstein, P., and Mombaerts, P. 1999. Variable patterns of axonal projections of sensory neurons in the mouse vomeronasal system. Cell 97, 199–208.

Rodriguez, I., Greer, C. A., Mok, M. Y., and Mombaerts, P. 2000. A putative pheromone receptor gene expressed in human olfactory mucosa. Nat. Genet. 26, 18–19.

Rybczynski, R., Reagan, J., and Lerner, M. R. 1989. A pheromone-degrading aldehyde oxidase in the antennae of the moth *Manduca sexta*. J. Neurosci. 9, 1341–1353.

Saito, H., Kubota, M., Roberts, R. W., Chi, Q., and Matsunami, H. 2004. RTP family members induce functional expression of mammalian odorant receptors. Cell 119, 679–691.

Sam, M., Vora, S., Malnic, B., Ma, W., Novotny, M. V., and Buck, L. B. 2001. Odorants may arouse instinctive behaviors. Nature 412, 142.

Sanhueza, M., Schmachtenberg, O., and Bacigalupo, J. 2000. Excitation, inhibition, and suppression by odors in isolated toad and rat olfactory receptor neurons. J. Exp. Biol. 203, 253–262.

Schmitt, B. C. and Ache, B. W. 1979. Olfaction: responses of a decapod crustacean are enhanced by flicking. Science 205, 204–206.

Scott, K., Brady, R., Jr., Cravchik, A., Morozov, P., Rzhetsky, A., Zuker, C., and Axel, R. 2001. A chemosensory gene family encoding candidate gustatory and olfactory receptors in *Drosophila*. Cell 104, 661–673.

Serizawa, S., Ishii, T., Nakatani, H., Tsuboi, A., Nagawa, F., Asano, M., Sudo, K., Sakagami, J., Sakano, H., Ijiri, T., *et al.* 2000. Mutually exclusive expression of odorant receptor transgenes. Nat. Neurosci. 3, 687–693.

Serizawa, S., Miyamichi, K., Nakatani, H., Suzuki, M., Saito, M., Yoshihara, Y., and Sakano, H. 2003. Negative feedback regulation ensures the one receptor–one olfactory neuron rule in mouse. Science 302, 2088–2094.

Sharon, D., Glusman, G., Pilpel, Y., Khen, M., Gruetzner, F., Haaf, T., and Lancet, D. 1999. Primate evolution of an olfactory receptor cluster: diversification by gene conversion and recent emergence of pseudogenes. Genomics 61, 24–36.

Shi, P., Zhang, J., Yang, H., and Zhang, Y.-P. 2003. Adaptive diversification of bitter taste receptor genes in mammalian evolution. Mol. Biol. Evol. 205, 805–814.

Shorey, H. H. 1976. Animal Communication by Pheromones. Academic Press.

Smith, D. V. 1985. Brainstem Processing of Gustatory Information. In: Taste, Olfaction, and the Central Nervous System (*ed*. D. W. Pfaff), pp. 151–177. Rockefeller University Press.

Spehr, M., Wetzel, C. H., Hatt, H., and Ache, B. W. 2002. 3-Phosphoinositides modulate cyclic nucleotide signaling in olfactory receptor neurons. Neuron 33, 731–739.

Spehr, M., Gisselmann, G., Poplawski, A., Riffell, J. A., Wetzel, C. H., Zimmer, R. K., and Hatt, H. 2003. Identification of a testicular odorant receptor mediating human sperm chemotaxis. Science 299, 2054–2058.

Steullet, P., Cate, H. S., and Derby, C. D. 2000. A spatiotemporal wave of turnover and functional maturation of olfactory receptor neurons in the spiny lobster *Panulirus argus*. J. Neurosci. 20, 3282–3294.

Stopfer, M. and Laurent, G. 1999. Short-term memory in olfactory network dynamics. Nature 402, 664–668.

Teeter, J. H, Brand, J. G., and Kumazawa, T. 1990. A stimulus-activated conductance in isolated taste epithelial membranes. Biophys. J. 58, 253–259.

Teicher, M. H., Stewart, W. B., Kauer, J. S., and Shepherd, G. M. 1980. Suckling pheromone stimulation of a modified glomerular region in the developing rat olfactory bulb revealed by the 2-deoxy glucose method. Brain Res. 194, 530–535.

Tegoni, M., Pelosi, P., Vincent, F., Spinelli, S., Campanacci, V., Grolli, S., Ramoni, R., and Cambillau, C. 2000. Mammalian odorant binding proteins. Biochem. Biophys. Acta 1482, 229–240.

Thomas, J. H., Kelley, J. L., Robertson, H. M., Ly, K., and Swanson, W. J. 2005. Adaptive evolution in the SRZ chemoreceptor families of *Caenorhabditis elegans* and *Caenorhabditis briggsae*. Proc. Natl. Acad. Sci. U. S. A. 102, 4476–4481.

Thomas, L. 1974. The Lives of a Cell: Notes of a Biology Watcher. The Viking Press.

Thorne, N., Chromey, C., Bray, S., and Amrein, H. 2004. Taste perception and coding in *Drosophila*. Curr. Biol. 14, 1065–1079.

Thorington, G. U. and Hessinger, D. A. 1988. Control of cnidia discharge: I, Evidence for two classes of chemoreceptor. Biol. Bull. 174, 163–171.

Tolbert, L. P. and Hildebrand, J. G. 1981. Organization and synaptic ultrastructure of glomeruli in the antennal lobes of the moth *Manduca sexta*: a study using thin sections and freeze-fracture. Proc. R. Soc. Lond. B 213, 279–301.

Torii, K. and Cagan, R. H. 1980. Biochemical studies of taste sensation, IX, Enhancement of L-[3H] glutamate binding to bovine taste papillae by 5′-ribonucleotides. Biochem. Biophys. Acta 627, 313–323.

Trapido-Rosenthal, H. G., Carr, W. E. S., and Gleeson, R. A. 1987. Biochemistry of an olfactory purinergic system: dephosphorylation of excitatory nucleotides and uptake of adenosine. J. Neurochem. 49, 1174–1182.

Trask, B. J., Friedman, C., Martin-Gallardo, A., Rowen, L., Akinbami, C., Blankenship, J., Collins, C., Giorgi, D., Iadonato, S., Johnson, F., *et al.* 1998. Members of the olfactory receptor gene family are contained in large blocks of DNA duplicated polymorphically near the ends of human chromosomes. Hum. Mol. Genet. 7, 13–26.

Troemel, E. R., Chou, J. H., Dwyer, N. D., Colbert, H. A., and Bargmann, C. I. 1995. Divergent seven transmembrane receptors are candidate chemosensory receptors in *C. elegans*. Cell 83, 207–218.

Trotier, D., Eloit, C., Wassef, M., Talmain, G., Bensimon, J. L., Doving, K. B., and Ferrand, J. 2000. The vomeronasal cavity in adult humans. Chem. Senses 25, 369–380.

Vandenburg, J. G. 1983. Pheromones and Reproduction in Mammals. Academic Press.

Van Houten, J. L. 1998. Chemosensory transduction in *Paramecium*. Eur. J. Protistol. 34, 301–307.

Vassilatis, D. K., Hohmann, J. G., Zeng, H., Li, F., Ranchalis, J. E., Mortrud, M. T., Brown, A., Rodriguez, S. S., Weller, J. R., Wright, A. C., *et al.* 2003. The G protein-coupled receptor repertoires of human and mouse. Proc. Natl. Acad. Sci. U. S. A. 100, 4903–4908.

Vogt, R. G., Riddiford, L. M., and Prestwich, G. D. 1985. Kinetic measurements of a sex pheromone-degrading enzyme: the sensillar esterase of *Antheraea polyphemus*. Proc. Natl. Acad. Sci. U. S. A. 82, 8827–8831.

Vogt, R. G., Rybczynski, R., and Lerner, M. R. 1990. The Biochemistry of Odorant Reception and Transduction. In: Chemosensory Information Processing (*ed*. D. Schild), pp. 33–76. Springer.

Wachowiak, M. and Cohen, L. B. 1999. Presynaptic inhibition of primary olfactory afferents mediated by different mechanisms in lobster and turtle. J. Neurosci. 19, 8808–8817.

Wachowiak, M., Cohen, L. B., and Ache, B. W. 2002. Presynaptic inhibition of olfactory receptor neurons in crustaceans. Micro. Res. Tech. 58, 365–375.

Wachowiak, M., Diebel, C. E., and Ache, B. W. 1997. Local interneurons define functionally distinct regions within lobster olfactory glomeruli. J. Exp. Biol. 200, 989–1001.

Wadhams, G. H. and Armitage, J. P. 2004. Making sense of it all: bacterial chemotaxis. Nat. Rev. Mol. Cell Biol. 5, 1024–1037.

Whissell-Buechy, D. and Amoore, J. E. 1973. Odour-blindness to musk: simple recessive inheritance. Nature 242, 271–273.

Whittaker, R. H. and Feeny, P. P. 1971. Allelochemics: chemical interactions between species. Science 171, 757–770.

Wu, S. V., Chen, M. C., and Rozengurt, E. 2005. Genomic organization, expression and function of bitter taste receptors (T2R) in mouse and rat. Physiol. Genomics 22, 139–149.

Wu, S. V., Rozengurt, N., Yang, M., Young, S. H., Sinnett-Smith, J., and Rozengurt, E. 2002. Expression of bitter taste receptors of the T2R family in the gastrointestinal tract and enteroendocrine STC-1 cells. Proc. Natl. Acad. Sci. U. S. A. 99, 2392–2397.

Yamaguchi, M., Yamazaki, K., Beauchamp, G. K., Bard, J., Thomas, L., and Boyse, E. A. 1981. Distinctive urinary odors governed by the major histocompatibility locus of the mouse. Proc. Natl. Acad. Sci. U. S. A. 78, 5817–5820.

Yamazaki, K., Beauchamp, G. K., Bard, J., and Boyse, E. A. 1989. Sex-chromosomal odor types influence the maintenance of early pregnancy in mice. Proc. Natl. Acad. Sci. U. S. A. 86, 9399–9401.

Yang, W. Q., Braun, C., Plattner, H., Purvee, J., and Van Houten, J. L. 1997. Cyclic nucleotides in glutamate chemosensory signal transduction of *Paramecium*. J. Cell Sci. 110, 2567–2572.

Young, J. M., Friedman, C., Williams, E. M., Ross, J. A., Tonnes-Priddy, L., and Trask, B. J. 2002. Different evolutionary processes shaped the mouse and human olfactory receptor gene families. Hum. Mol. Genet. 11, 535–546.

Young, J. M., Shykind, B. M., Lane, R. P., Tonnes-Priddy, L., Ross, J. A., Walker, M., Williams, E. M., and Trask, B. J. 2003. Odorant receptor expressed sequence tags demonstrate olfactory expression of over 400 genes, extensive alternate splicing and unequal expression levels. Genome Biol. 4, R71.

Young, J. M., Kambere, M., Trask, B. J., and Lane, R. P. 2005. Divergent V1R repertoires in five species: amplification in rodents, decimation in primates, and a surprisingly small repertoire in dogs. Genome Res. 15, 231–240.

Zimmer-Faust, R. K., Gleeson, R. A., and Carr, W. E. S. 1988. The behavioral response of spiny lobsters to ATP: evidence for mediation by P2-like chemosensory receptors. Biol. Bull. 175, 167–174.

Zhang, J. and Webb, D. M. 2003. Evolutionary deterioration of the vomeronasal pheromone transduction pathway in catarrhine primates. Proc. Natl. Acad. Sci. U. S. A. 100, 8337–8341.

Zhang, X., Rodriguez, I., Mombaerts, P., and Firestein, S. 2004. Odorant and vomeronasal receptor genes in two mouse genome assemblies. Genomics 83, 802–811.

Zhang, Y., Hoon, M. A., Chandrashekar, J., Mueller, K. L., Cook, B., Wu, D., Zuker, C. S., and Ryba, N. J. P. 2003. Coding of sweet, bitter, and umami tastes: different receptor cells sharing similar signaling pathways. Cell 112, 293–301.

Zou, Z., Horowitz, L. F., Montmayeur, J. P., Snapper, S., and Buck, L. B. 2001. Genetic tracing reveals a stereotyped sensory map in the olfactory cortex. Nature 414, 173–179.

Zufall, F., Firestein, S., and Shepherd, G. M. 1994. Cyclic nucleotide-gated ion channels and sensory transduction in olfactory receptor neurons. Ann. Rev. Biophys. Biomol. Struct. 23, 577–607.

Further Reading

Hildebrand, J. G. and Shepherd, G. M. 1997. Mechanisms of olfactory discrimination: converging evidence for common principles across phyla. Ann. Rev. Neurosci. 20, 595–631.

Van Houten, J.l. 1994. Chemosensory transduction in eukaryotic microorganisms: trends for neuroscience? Trends Neurosci. 17, 62–71.

4.02 Chemistry of Gustatory Stimuli

G E DuBois, The Coca-Cola Company, Atlanta, GA, USA

J DeSimone and V Lyall, Virginia Commonwealth University, Richmond, VA, USA

Glossary

adaptation The phenomenon of taste desensitization which occurs on iterative exposure to a taste stimulus over a short period of time, a phenomenon believed to be a consequence of receptor phosphorylation by kinases specific for action on G-protein-coupled receptors.

concentration/response function A function which describes the relationship between tastant response (R) and stimulus concentration (C). Generally, C/R functions take the form of the law of mass action $R = R_m C/(k_d + C)$, where R_m is maximal response and k_d is the apparent receptor/tastant dissociation constant.

epithelial sodium channel A multi-subunit ion channel protein, usually referred to as ENaC, believed to be the principal initiator in the activation of salt-sensitive taste bud cells.

flavor profile A graphical description of the quality of the taste of a stimulus in terms of the relative intensities of multiple sensory attributes such as sweet, bitter, umami, sour, and salty.

G-protein Heterotrimeric guanosine triphosphate (GTP)-binding proteins constituted of α, β, and γ subunits which transduce the activation of G-protein-coupled receptors into cellular activation.

G-protein-coupled receptor A membrane protein receptor, usually referred to as a GPCR, which signals the presence of an extracellular ligand to an intracellular signaling protein known as a G protein, thus leading to cellular activation. GPCRs are constituted of extracellular ligand-binding domains and intracellular G-protein-binding domains connected by seven α-helical transmembrane segments.

gustatory stimuli Chemical compounds which initiate responses from taste bud cells and are perceived as sweet, bitter, umami, sour, or salty in quality of taste.

inositol trisphosphate receptor A protein receptor, present on intracellular calcium storage vesicles in sweet-, bitter-, and umami-sensitive taste bud cells, which is activated by its ligand inositol trisphosphate (IP_3) to promote the release of calcium ions into the cytoplasm, free calcium being a key element in the cascade of reactions leading to cellular excitation.

phospholipase $C_{\beta2}$ An enzyme present in sweet-, bitter-, and umami-sensitive taste bud cells which communicates receptor and G-protein activation by action on the membrane lipid phosphatidylinositol to produce inositol trisphosphate (IP_3) and diacylglycerol (DAG), where IP_3, and possibly DAG, are key elements in the cascade of reactions involved in excitation of taste bud cells.

polycystic-kidney-disease-like ion channel An ion channel protein of the transient receptor potential family thought to mediate the excitation of sour-sensitive taste bud cells.

T1R1/T1R3 The heterodimeric G-protein-coupled receptor constituted of the two proteins T1R1 and T1R3 which initiates the responses of umami-sensitive taste bud cells to monosodium glutamate and other savory stimulants.

T1R2/T1R3 The heterodimeric G-protein-coupled receptor constituted of the two proteins T1R2 and T1R3 which initiates the responses of sweet-sensitive taste bud cells to sweeteners.

T2R A member of the family of G-protein-coupled receptors (25 in humans) which initiates the responses of bitter-sensitive taste bud cells to bitterants.

temporal profile A graphical description of the temporal characteristics of a taste experience which includes quantification of the rate of taste onset and the rate of taste extinction.

transient receptor potential channel M5 An ion channel protein, usually referred to as TRPM5, which is activated by intracellular calcium and mediates the activation of sweet-, bitter-, and umami-sensitive taste bud cells through membrane depolarization.

transient receptor potential channel V1 An ion channel protein, usually referred to as TRPV1 and known to be the receptor for vanilloid trigeminal nerve stimulants including capsaicin, thought to work together with the epithelial sodium channel in excitation of salt-sensitive taste bud cells.

4.02.1 Introduction

All chemical compounds, whether organic, inorganic, or organometallic, exhibit taste and therefore are potential gustatory stimuli. This is a fundamental premise for the discussion to follow. In order to exhibit taste, however, compounds must be in solution and some compounds that appear to be tasteless are probably not really tasteless, but rather are insufficiently soluble to reach their thresholds for activation of any of the gustatory sensor systems. In this chapter, we undertake the challenge of developing a coherent picture of the relationship between chemical structure and biological activity for each of the common modalities of taste. Understanding the structure–activity relationship (SAR) within series of biologically active chemical compounds is the never-ending and elusive vision for

medicinal, flavor and fragrance, and agricultural chemists. The reason for the elusive nature of this vision is made clear by a quotation of Melvin Calvin, 1961, Nobel laureate in chemistry:

> It's no trick to get the right answer when you have all the data. The real creative trick is to get the right answer when you have only half the data in hand and half of it is wrong. And you don't know which half.

Thus, as we challenge ourselves to understand the chemistry of gustatory stimuli, it is important to recognize that some of the data that we are trying to make sense out of are misleading at best and, in some cases, just wrong. With this fact in mind, we proceed.

What is taste? In the lexicon of the consumer, taste is the total sensory experience that results from a superimposition of gustatory, olfactory, and chemesthetic sensory inputs to the central nervous system which occurs on sampling a food or beverage. And it can be argued that even the senses of vision and audition contribute significantly. In this Chapter, only gustatory stimuli are considered. Historically, it has been generally accepted that the sense of gustation is made up of the four modalities: sweet, bitter, sour, and salty. We now know that savory, often referred to by the Japanese word *umami*, is a fifth gustatory modality. In this Chapter, the relationship between chemical structure and gustatory activity, or gustatory SAR, for each of these five taste modalities is discussed. In each case, current understandings of the biochemical mechanisms are summarized first since without such a background, little sense can be made of SAR.

4.02.2 Sweet-Tasting Stimuli

Sweet taste is clearly a very important sensation as evidenced by the fact that wars have been fought and people enslaved over sugar, the prototypical sweet stimulus (Mintz, S. W., 1985). More sweeteners, principally sugar but also syrups derived from starch as well as a variety of noncaloric sweeteners, are added to foods and beverages than any other ingredient type. Human liking of sweet taste is innate as has been demonstrated by Steiner J. E. (1994). He studied the behavioral responses of newborns and clearly showed that they exhibit strong liking for sweettasting stimuli. As a consequence of the importance of sweet taste in the human diet, more research has been conducted on elucidating the biochemical pathways which mediate it and on the discovery and commercial development of noncaloric sweeteners than on bitter, umami, sour, and salty tastes combined. For this rationale, we discuss sweet-tasting stimuli first and in considerable detail.

4.02.2.1 Biochemistry of Sweet Taste

In spite of the importance of sweet taste to our daily enjoyment of life, until the late 1980s, the biochemical pathway which mediates it was largely unknown. Then, evidence began to accumulate that it must be G-protein-coupled receptor (GPCR) mediated. More specifically, sweet taste was thought to be the result of activation of several GPCRs since the findings of biochemical, electrophysiological, and psychophysical experiments could only be easily

explained by a plurality of receptors (Faurion, A., 1987; DuBois, G. E., 1997). And this expectation was supported by the fact that multiple subtypes of GPCRs commonly exist for other important signal molecules (e.g., acetylcholine, norepinephrine, dopamine, serotonin, etc.).

In the early 2000s, a breakthrough occurred dramatically improving our understanding of sweet taste. In 2001, a collaborative team from the laboratories of Charles Zuker (University of California, San Diego) and Nicholas Ryba (US National Institutes of Health) reported the discovery of the rat sweetener receptor (Nelson, G. *et al.*, 2001). In a functional assay, they showed that all substances that rats generalize to sucrose taste appear to be mediated by a single receptor which their observations suggested to be a heterodimer of two 7-transmembrane domain (TMD) proteins which they named T1R2 and T1R3. This heterodimeric sweetener receptor is usually written as T1R2/T1R3. Both T1R2 and T1R3 are members of the small family of class C GPCRs. The most studied members of the class C GPCRs are the 8 homodimeric metabotrophic glutamate, single heterodimeric gamma-aminobutyric acid type B, and single homodimeric extracellular calcium receptors which have recently been reviewed (Pin, J.-P. *et al.*, 2003). This rat receptor discovery was quickly followed by the report by Li X. *et al.* (2002) of Senomyx (La Jolla, CA) of parallel findings for the human system. Again, as in the rodent, the results were most consistent with human sweet taste initiation by the single heterodimeric receptor T1R2/T1R3. Heterologous cells (i.e., HEK-293 cells), in which both human T1R2 and human T1R3 were expressed, responded to all structural types of sweeteners tested in a manner consistent with expectation from sensory experiments. Thus, at the present time, there is a general consensus that the heterodimer T1R2/T1R3 is the sweetener receptor. Evidence has been presented for the mouse by the Zuker and Ryba laboratories that a T1R3-only receptor, perhaps a homodimer, is functional as a carbohydrate (CHO)-only sweetener receptor (Zhao, G. Q. *et al.*, 2003). However, no evidence has yet been reported on such a receptor in humans or other animal models. At about the same time as the pioneering work by the Zuker and Ryba laboratories, several other laboratories also identified the T1R3 component of the receptor (Bachmanov, A. A. *et al.*, 2001; Max, M. *et al.*, 2001; Montmayeur, J.-P. *et al.*, 2001; Sainz, E. *et al.*, 2001).

Class C GPCRs are unique in that they possess very large N-terminal Venus flytrap-like domains (VFDs). For the case of the metabotrophic glutamate receptor mGluR1, Kunishima N. *et al.* (2000) demonstrated that its VFD closes on binding glutamate, hence the analogy to a Venus flytrap. This precedent, and the fact that the sweetener receptor and the umami receptor, shown in parallel work by Li and co-workers to be the GPCR heterodimer T1R1/T1R3, contain the common subunit T1R3 lead to the expectation that sweeteners likely bind in the VFD of T1R2. Subsequent work by Li and co-workers (Xu, H. *et al.*, 2004) probed the fundamental question of sweetener-binding locus with the finding that, while some sweeteners do bind in the VFD of T1R2 (e.g., aspartame and neotame), at least one sweetener (i.e., cyclamate) does not, but rather binds within the 7-TMD of T1R3. The binding of cyclamate to the TMD of T1R3 was corroborated by Robert Margolskee and associates of Mt Sinai School of Medicine (Jiang, P. *et al.*, 2005) with site-directed mutagenesis studies which provided significant detail on the interactions of cyclamate with T1R3. Other work by Margolskee and associates on brazzein, a natural protein sweetener, showed that its locus of binding is in the cysteine-rich domain (CRD) of T1R3, a subunit of the protein which connects VFD and TMD domains (Jiang, P. *et al.*, 2004). The human sweetener receptor is the first Class C GPCR demonstrated to have multiple agonist-binding loci (orthosteric sites).

A topic of considerable controversy in the field of taste research has been that of taste quality coding. Taste bud cells (TBCs) are known to be innervated by nerve fibers of three gustatory nerves: the chorda tympani (CT), the glossopharyngeal, and the suprapetrosal nerves. Each of these nerves is a bundle of many individual fibers and some have argued that taste quality is coded by a cross-fiber pattern of activity and others have argued that individual fibers are specific for activity. Evidence for taste quality coding due to modality-specific fibers was provided by electrophysiological studies in chimpanzees by Goran Hellekant and Yuzo Ninomiya. They carried out single fiber recordings and demonstrated that some fibers responded only to sweeteners (Hellekant, G. and Ninomiya, Y., 1991) and other fibers responded only to bitterants (Hellekant, G. and Ninomiya, Y., 1994a; 1994b) leading them to conclude that taste quality is coded at the level of the TBC. In other words, they argued that individual TBCs are specific sensors for sweet, bitter, umami, sour, or salty. Further convincing evidence that taste

quality coding occurs at the level of the TBC comes from recent work of the Zuker and Ryba laboratories. Early in 2003, working with $PLC_{\beta2}$ knockout mice, they engineered mice in which they selectively rescued $PLC_{\beta2}$ function in bitter receptor expressing cells (Zhang, Y. *et al.*, 2003) and found that these mice responded normally to bitterants but still exhibited no responses to sweet or umami stimuli. Later, working with T_1R knockout mice that gave no responses to sweet or umami stimuli, they engineered mice in which human T1R2 function was added to T1R3-expressing TBCs (Zhao, G. Q. *et al.*, 2003) and found that these mice responded to aspartame, a compound sweet to humans but inactive in mice. They also selectively introduced an opioid receptor into sweet-sensing TBCs and observed that these mice now responded with attraction to opioid agonists. In summary, the evidence is now strong that taste quality is coded at the level of the TBC.

The recent identification of the sweetener receptor and elucidation of the mechanism of sweet taste coding was preceded by the discovery of specific G proteins that mediate sweet taste. In 1992, Robert Margolskee, then at Roche Institute of Molecular Biology, and associates reported that a G protein closely related to transducin is critical for sweet taste (McLaughlin, S. K. *et al.*, 1992). They called this G-protein gustducin but noted that low levels of transducin are also present in TBCs. And, in addition, very recently, Guy Servant and co-workers (Senomyx) obtained evidence showing that the sweetener receptor can also functionally interact with $G_{\alpha i(o)}$ proteins (Ozeck, M. *et al.*, 2004). $G_{i(2)}$ is a favored $G_{\alpha i(o)}$ candidate as it was shown to be the most highly expressed $G_{\alpha i(o)}$ in TBCs. Other key elements of the sweet taste transduction cascade were identified by Margolskee and associates (Pérez, C. A. *et al.*, 2002) and by the Zuker and Ryba team (Zhang, Y. *et al.*, 2003). They found that phospholipase $C_{\beta2}$ ($PLC_{\beta2}$), the inositol trisphosphate (IP_3) receptor (IP_3R), and the transient receptor potential (TRP) channel m5 (TRPM5) are key effector elements in sweet taste transduction. Thus, at this time, evidence exists for the participation of one receptor (T1R2/T1R3), as many as three G proteins (gustducin, transducin, and possibly $G_{i(2)}$), one enzyme ($PLC_{\beta2}$), a second messenger receptor (IP_3R), and an ion channel (TRPM5) in the activation of sweet-sensing TBCs. Recognition of (1) the relatively high concentrations of noncaloric sweeteners (i.e., millimolar levels) commonly employed in foods and beverages, (2) the fact that noncaloric sweeteners are commonly lipophilic, and (3) the fact that such lipophilic molecules are generally absorbed into cells led to speculation (DuBois, G. E., 1997) that sweeteners may initiate their activity at intracellular elements of the transduction cascade. Support for this idea is provided by the work of Michael Naim and associates (Hebrew University) who reported that some noncaloric sweeteners have the capability to directly activate G proteins (Naim, M. *et al.*, 1994). And further support is provided by recent work from the Naim group through studies showing that several noncaloric sweeteners are rapidly taken up into TBCs (Zubare-Samuelov, M. *et al.*, 2005). However, in view of the finding that every sweetener tested by Senomyx investigators activated HEK-293 cells in which the T1R2/T1R3 receptor is expressed, while otherwise identical control cells, lacking only the T1R2/T1R3 receptor, are unaffected, it remains to be demonstrated that any sweeteners do actually act at downstream elements in the sweet-sensing TBC activation cascade. Further, since we now know that bitter and umami taste stimuli initiate their sensations at the TBC level through the same effectors which mediate sweet taste signaling (i.e., gustducin, $PLC_{\beta 2}$, IP_3R, and TRPM5), if any sweeteners do initiate their activities at intracellular components of TBCs, then they should also simultaneously activate bitter- and umami-sensitive TBCs resulting in mixed modality (i.e., sweet/bitter/umami) compounds. And while sweet/bitter compounds are known, no compounds with sweet/umami or sweet/bitter/umami taste have been reported, thus leading to the overall conclusion that initiation of sweet taste activity at intracellular targets of TBCs seems unlikely. Nonetheless, since Naim and co-workers have demonstrated that small molecules including some sweeteners are readily taken up into TBCs, it is logical to expect that some such compounds may exist.

In summary, transduction of sweet taste is generally accepted as proceeding via activation of the heterodimeric sweetener receptor T1R2/T1R3 with subsequent activation of the G protein gustducin (and/or transducin and possibly $G_{i(2)}$). In this generally accepted transduction pathway, the gustducin $G_{\beta}G_{\gamma}$ subunit activates $PLC_{\beta 2}$ to act on membrane phosphatidylinositol to produce IP_3, which acts at its receptor IP_3R on intracellular Ca^{2+} storage sites to release Ca^{2+}. And finally, Ca^{2+} gates the TRPM5 ion channel enabling the inward flow of Na^+, depolarizing the sweet-sensing TBC, and initiating signaling to the CNS. Evidence

from the Zuker and Ryba laboratories argues that this gustducin $G_{\beta}G_{\gamma}$ pathway is the only sweet taste transduction pathway since mice in which the TRPM5 gene was partially deleted were observed to lack all behavioral and nerve responses to sweeteners (Zhang, Y. *et al.*, 2003). Earlier work, however, argues for a gustducin G_{α} pathway in sweet taste transduction, where G_{α} is the gustducin subunit which communicates with intracellular effectors (Varkevisser, B. and Kinnamon, S. C., 2000; Margolskee, R. F., 2002). And very recent work from the Margolskee laboratory (Damak, S. *et al.*, 2006), in which TRPM5 gene expression in the mouse was fully blocked, continues to argue for a sweet taste transduction pathway not mediated by TRPM5. In this work, the TRPM5 knockout mice continue to exhibit weak responses to sweeteners. And, in the electrophysiological component of this work, glossopharyngeal nerve responses were observed while the CT nerve responses were not, thus suggesting that transduction pathway may vary between TBCs innervated by the two different nerve systems. Thus, at this time, while we have a general consensus that the primary pathway for sweet taste transduction is the gustducin $G_{\beta}G_{\gamma}/PLC_{\beta 2}/IP_3R/$ TRPM5 pathway, it appears too early to conclude that it is the exclusive pathway.

4.02.2.2 Sweetener Structure–Activity Relationship

Since the early 1800s, chemists have discovered hundreds of synthetic and natural sweeteners and by the 1980s, at least 50 structural classes of sweet-tasting organic compounds were known. In an effort to make sense of this structural diversity, models were developed in an attempt to rationalize the relationship between structure and taste activity. These are pharmacophore models and the most well known of them is the so-called A-H/B model (Shallenberger, R. S. and Acree, T. E., 1967). In this model, it was hypothesized that all sweeteners contain H-bond donor and H-bond acceptor groups separated by not <2.5 or >4.0 Å. In effort to better explain empirical observations, many improved pharmacophore models have been developed since the A-H/B model (Kier, L. B., 1972; van der Heijden, A. *et al.*, 1985a; 1985b; Douglas, A. J. and Goodman, M. 1991; Rohse, H. and Belitz, H.-D., 1991; Temussi, P. A. *et al.*, 1991; Tinti, J.-M. and Nofre, C., 1991; Bassoli, A. *et al.*, 2002). An assumption implicit in all of these models is that sweetness is initiated following the binding of a sweetener to a single site (i.e., orthosteric site) on a single receptor. However, recent studies on the mapping of sweetener-binding loci to the sweetener receptor (Jiang, P. *et al.*, 2004; Xu, H. *et al.*, 2004) have demonstrated that the human sweetener receptor contains at least three orthosteric sites and thus the A-H/B model as well as all of the other pharmacophore models, while perhaps correct within structural classes of sweeteners, are not correct in the general sense. In summary, it appears that human sweetness is substantially, and perhaps exclusively, initiated by activation of a single receptor, the GPCR heterodimer T1R2/T1R3, at a plurality (i.e., ≥ 3) of orthosteric sites. And therefore ≥ 3 pharmacophore models are required to characterize the relationship between chemical structure and sweet taste. The first model consistent with this logic was developed at The NutraSweet Company for the aspartame orthosteric site (Culberson, J. C. and Walters, D. E., 1991). Later, an enhancement of this aspartame pharmacophore model, employing comparative molecular field analysis to provide quantitative predictive power, was developed at The Coca-Cola Company (D'Angelo, L. and Iacobucci, G., 1995).

Over the last 200 years, chemists have identified and characterized many, many sweet-tasting chemical compounds. Many of these compounds are synthetic but also many of them have been found in nature. The most common sweeteners, present in ordinary fruits and vegetables, are CHO sweeteners. However, health concerns related to obesity and diabetes led to the commercial development of noncaloric sweeteners found either in the chemist's laboratory or in nature. In Sections 4.02.2.2.1–4.02.2.2.4 of this discussion on sweeteners is given a representative sampling of chemical compounds found to exhibit sweet taste. Numerous reviews on sweetener SAR are available (e.g., Lee, C.-K., 1987; Marie, S. and Piggott, J. R., 1991; Shallenberger, R. S., 1993; O'Brien Nabors, L., 2001; Kim, N.-C. and Kinghorn, A. D., 2002). In this review, all commercially developed sweeteners are discussed. Chemical structures, empirical formulas, molecular weights, and, to the extent available, sweetness potencies (P) are provided. In Section 4.02.2.2.5 of this section on sweet-tasting stimuli, important information on differences in sensory characteristics of sweeteners is presented. As explained, P values for CHO sweeteners are constants, independent of sucrose reference concentration, while for synthetic and natural noncaloric sweeteners, they are variable with a strong dependence on sucrose reference concentration. Thus, P values for synthetic and natural noncaloric

sweeteners are given relative to a sucrose reference and are on a weight basis (P_w). P_w values can easily be converted to molar basis potencies (P_m) by multiplication by the factor $M_{compound}/M_{sucrose}$ (i.e., the ratio of the molecular weights of a compound and of sucrose). Sweetness potencies for the synthetic and natural noncaloric sweeteners are expressed as $P_w(X)$ where X is the sucrose reference concentration (% w/v). Ideally, all P_w values would be provided relative to a common sucrose reference, determined by a common protocol and be at a sucrose use level relevant to food and beverage applications (e.g., 10% sucrose). However, synthetic and natural noncaloric sweeteners have variable maximal responses that in many cases are much less than 10% sucrose sweetness equivalency. And oftentimes, literature sweetness potencies are reported without sucrose reference information. Such unreferenced sweetness potencies are reported here as $P_w(ur)$. As a consequence of these and other complexities, the comparison of sweetness potencies for sweeteners reported by different authors using different protocols is difficult at best.

4.02.2.2.1   Carbohydrate caloric, partially caloric and noncaloric sweeteners

CHOs are generally considered to be aldehyde- or ketone-containing polyhydroxylic compounds which most commonly exist in hemiacetal, acetal, hemiketal, or ketal forms. In nature, CHOs exist predominantly in polymeric form. Principal among the natural CHO polymers are starch and cellulose, both of which are glucose polymers, and chitin, a polymer of 2-acetamido-2-deoxyglucose. Polymeric CHOs are not sweet. At the same time, all low molecular weight CHOs are sweet-tasting compounds. Some low molecular weight CHOs are also bitter, but all are sweet with P_w values within a factor of 0.1–2 of that of sucrose. In the author's (GED) experience, reports in the literature to the contrary are incorrect. The most common of the CHO sweeteners is sucrose, a disaccharide constituted of glucose and fructose monosaccharide subunits. Sucrose is produced on a very large scale with 2005 world production forecast at 144.8×10^6 MT. In acidic foods and beverages, sucrose hydrolyzes to a 1/1 mixture of glucose and fructose as illustrated in Figure 1. This 1/1 glucose/fructose mixture is prepared on a commercial scale from sucrose and is called invert sugar. Other glucose/fructose mixtures are manufactured from starch on a very large scale. These products are known as high fructose starch syrups (HFSSs) and are available as

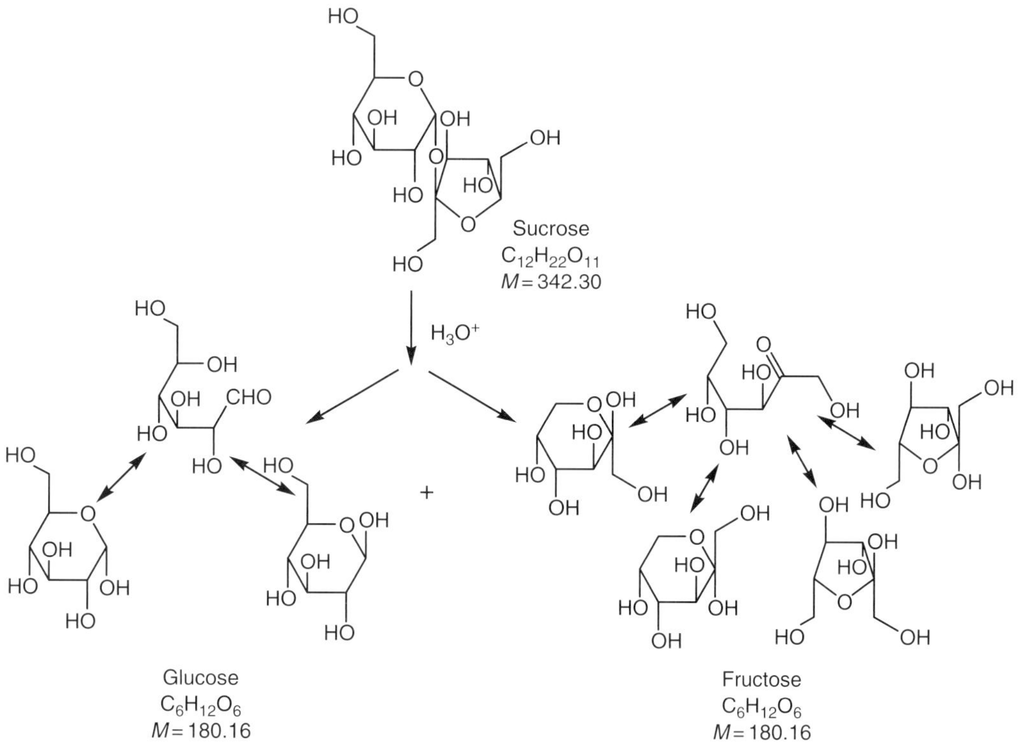

Figure 1 The hydrolysis of sucrose to glucose and fructose.

42% fructose (HFSS-42), 55% fructose (HFSS-55), and 90% fructose (HFSS-90). It is important to recognize that, while sucrose is a single chemical entity, glucose is a mixture of three tautomeric forms and fructose, a mixture of five tautomeric forms, the tautomeric forms of each monosaccharide being in rapid equilibrium with each other. And thus, in acidic foods or beverages sweetened with sucrose, the active sweetener system is a mixture of three compounds, with two of them present in eight tautomeric forms, all present simultaneously. Efforts have been made to determine the relative sweetness potencies of the different tautomeric forms of glucose and fructose. However, since these tautomers are in rapid equilibrium, the published sweetness potencies are highly speculative.

A great many other low molecular weight CHOs are known and all are sweet to humans. A sampling of mono-, di-, and oligosaccharides, intended to illustrate the diversity of sweet-tasting CHOs, is provided in Figures 2 and 3 along with their sweetness potencies where available. No attempt has been made to be comprehensive in this listing. The examples given are provided simply to illustrate the breadth of the SAR among CHO sweeteners. Some low molecular weight CHOs are reported to be bitter rather than sweet. As example, gentiobiose is reported to be bitter (Pfeilsticker, K. *et al.*, 1978). However, as a 25% solution, a commercial sample was observed to exhibit both sweet and bitter tastes (DuBois, G. E. *et al.*, 1994). In work on synthetic analogs of natural CHOs, many of them, even at high states of purity, exhibit bitterness (DuBois, G. E., 1997). However, in general, further purification resulted in purely sweet samples. It seems likely that the bitter tastes reported for many CHOs may be due to potent bitter contaminants.

The simplest monosaccharide CHOs, glycolaldehyde (**1**) and glyceraldehyde (**2**), cannot form monomeric cyclic acetals or ketals and are reported to exist primarily in dimeric form and thus their active pharmacophores are believed to be their dimeric hemiacetals. All of the higher molecular weight monosaccharide CHOs illustrated exist almost exclusively as cyclic hemiacetal or hemiketal tautomeric forms. And thus it is believed that their active pharmacophores are the cyclic tautomers. A very surprising finding among the monosaccharide CHOs is the total absence of enantiospecificity in some enantiomeric pairs. Thus Shallenberger R. S. *et al.* (1969) reported complete inability of human

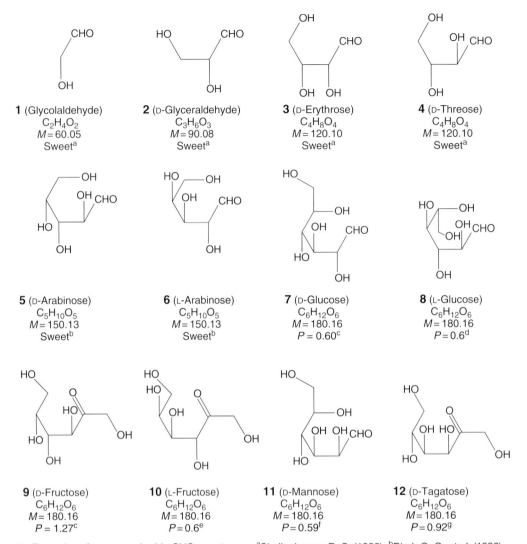

Figure 2 Examples of monosaccharide CHO sweeteners. [a]Shallenberger R. S. (1993); [b]Birch G. G. *et al.* (1996); [c]DuBois G. E. *et al.* (1991); [d]DuBois G. E. and Singer N. (1984); [e]DuBois G. E. (1995); [f]Schutz H. G. and Pilgrim F. J. (1957); and [g]Bertelsen H. *et al.* (2001).

subjects to differentiate D- and L-enantiomers for several sugars. Birch G. G. *et al.* (1996) corroborated the Shallenberger report for the case of D- and L-arabinose, **5** and **6**. And the author (GED) and Singer N. (1984) corroborated the Shallenberger report for the case of D- and L-glucose, **7** and **8**. In both the arabinose and the glucose enantiomeric pairs, identical tastes are observed when tasted at equal concentrations. In contrast, however, D- and L-fructose are quite different, with the natural D enantiomer being approximately twice as potent as the L enantiomer (DuBois, G. E., 1995). Given that CHO sweeteners are thought to activate the sweetener receptor T1R2/T1R3 through a binding interaction,

it is to be expected that enantiomers should differ in their binding effectiveness and thus the observed difference in potencies for the fructose enantiomers is expected. However, identical sweetness potencies within the D/L arabinose and D/L glucose enantiomeric pairs are very surprising.

It seems very unlikely that each enantiomer in a pair of CHO sweeteners should bind with equal affinity to the receptor and this reasoning led to the suggestion that CHOs as well as other highly hydrophilic compounds may activate the receptor via a nonbinding mechanism (DuBois, G. E., 1997). The binding sites for the natural agonists for family C GPCRs are known to be in the VFDs

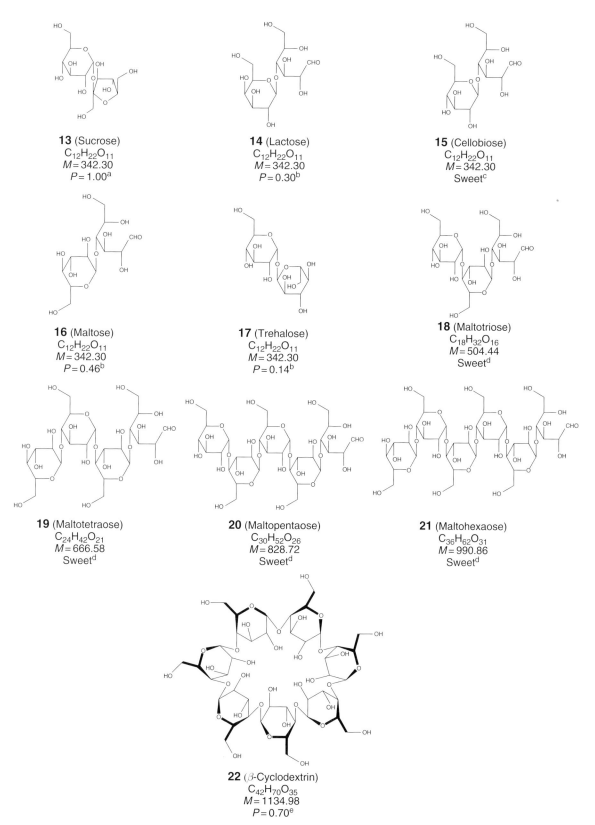

13 (Sucrose)
$C_{12}H_{22}O_{11}$
$M = 342.30$
$P = 1.00^a$

14 (Lactose)
$C_{12}H_{22}O_{11}$
$M = 342.30$
$P = 0.30^b$

15 (Cellobiose)
$C_{12}H_{22}O_{11}$
$M = 342.30$
Sweet^c

16 (Maltose)
$C_{12}H_{22}O_{11}$
$M = 342.30$
$P = 0.46^b$

17 (Trehalose)
$C_{12}H_{22}O_{11}$
$M = 342.30$
$P = 0.14^b$

18 (Maltotriose)
$C_{18}H_{32}O_{16}$
$M = 504.44$
Sweet^d

19 (Maltotetraose)
$C_{24}H_{42}O_{21}$
$M = 666.58$
Sweet^d

20 (Maltopentaose)
$C_{30}H_{52}O_{26}$
$M = 828.72$
Sweet^d

21 (Maltohexaose)
$C_{36}H_{62}O_{31}$
$M = 990.86$
Sweet^d

22 (β-Cyclodextrin)
$C_{42}H_{70}O_{35}$
$M = 1134.98$
$P = 0.70^e$

Figure 3   Examples of disaccharide and oligosaccharide CHO sweeteners. ^{a}DuBois G. E. *et al.* (1991); ^{b}Schutz H. G. and Pilgrim F. J. (1957); $^{c-e}$DuBois G. E. and D'Angelo L. L. (1996).

and these agonists are believed to activate the receptor by causing the VFD to adopt a closed conformation. And recent work on an analog of the GABA$_B$ family C GPCR, engineered to exist exclusively in a closed conformation (Kniazeff, J. *et al* 2004), shows that the closed conformation is fully active. The binding site for aspartame has been shown to be in the VFD of the T1R2 component of the T1R2/T1R3 receptor (Xu, H. *et al.*, 2004) and recent evidence (Nie, Y. *et al.*, 2005) suggests that sucrose binds in the VFDs of both T1R2 and T1R3. Thus, it appears that CHO sweeteners most likely modulate the equilibrium between the open (inactive) and closed (active) conformations of the VFD of T1R2 and perhaps T1R3 as well. If this is the case, it seems reasonable that the open/closed equilibria could be altered in two ways. The first possibility is that CHO sweeteners, like other family C GPCR agonists, bind to a site within the VFD. A second possibility, however, is that CHO sweeteners activate the receptor by a nonbinding mechanism where they simply alter the VFD open/closed equilibrium in favor of the closed conformation by a general solvent effect. Solution equilibria of all types are well known to be solvent dependent and a change in the open/closed equilibria for T1R2 and T1R3 to favor the closed forms that is promoted by a change in solvent from normal saliva to a concentrated CHO sweetener solution should not be a surprise. Thus, it could well be that CHO sweeteners showing no enantiospecificity (i.e., D/L arabinose and D/L glucose) may activate the receptor exclusively by a nonbinding mechanism and that CHO sweeteners which do show enantiospecificity (i.e., D/L fructose) activate the receptor by both binding and nonbinding mechanisms. If this nonbinding pathway for sweetener receptor activation does exist, then it is to be expected that all sweeteners acting by this pathway should exhibit similar levels of activity. And they do, as is made clear with the examples to follow.

The aldehyde or ketone moieties present in the CHO sweeteners in Figure 2 are not prerequisites for sweet taste. Simple alcohols as well as CHO reduction products, often referred to as polyols, are also sweet in taste. An exemplary list of polyol sweeteners is provided in Figure 4.

Within the last 20–30 years, due to the increased interest in reduced-calorie food and beverage products, substantial interest has been expended on the commercial development of reduced-calorie CHOs and polyols. Among naturally occurring CHOs and polyols already mentioned which are reduced in bioavailable energy content are the naturally occurring *meso*-erythritol and D-tagatose. In the United States, the FDA has accepted bioavailable calorie levels of 0.2 and 1.5 kcal g^{-1} for erythritol and D-tagatose, respectively, although, in Japan, an energy content for erythritol of 0.0 kcal g^{-1} is accepted. A number of semisynthetic reduced-calorie CHOs and polyols have been commercially developed. These compounds are listed in Figure 5.

A survey of the relationship between structure and activity for the CHO sweeteners shown above leads to the expectation that any polyhydroxylic organic compound may be sweet. In food applications, sucrose, the most common CHO sweetener, provides functionality beyond sweetness including freezing point depression (frozen desserts) and starch gelatinization point elevation (baked goods). These properties are colligative in nature and thus the ideal noncaloric CHO sweetener should have molecular weight comparable to that of sucrose. In addition, the quality of taste of CHO sweeteners was noted to be dependent on the ratio of hydroxyl groups to carbon atoms. This logic led the author (GED) and a team at The NutraSweet Company to the synthesis of a diverse group of approximately 100 CHO-like noncaloric sweeteners (DuBois, G. E. *et al.*, 1992). Compounds **35–38**, illustrated in Figure 6, are exemplary of this work and all are noncaloric and exhibit clean sweet tastes. At about the same time, Adam Mazur and co-workers (Procter & Gamble Company) reported the synthetic lactitol derivative **39** to be sweet and noncaloric.

4.02.2.2.2 *Synthetic noncaloric sweeteners*

The field of organic chemistry really had its beginning in the nineteenth century. Many new chemical reactions were discovered and novel chemicals synthesized. These new chemical compounds were characterized by techniques available at the time which included physical description (i.e., solid, liquid, crystalline form, etc.), color, boiling point, melting point, odor, and, of particular reference to this discussion, taste. The routine tasting of novel chemical compounds by chemists continued well into the twentieth century. And in this way, synthetic sweeteners were discovered among many structural classes of compounds. Examples of 21 structural classes of synthetic noncaloric sweeteners, along with chemical structures, molecular weights,

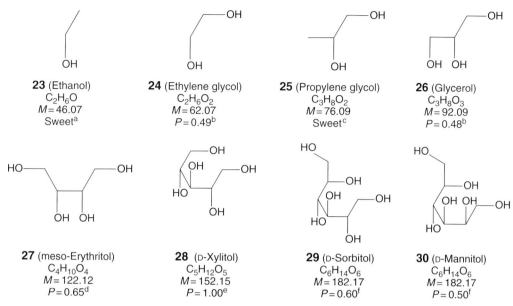

Figure 4 Examples of common polyol sweeteners. [a]Scinska A. *et al.* (2000); [b]Shallenberger R. S. (1993); [c,d]Embuscado M. E. and Patil S. K. (2001); [e]Olinger P. M. and Pepper T. (2001); and [f]Le A. S. and Bowe Mulderrig K. (2001).

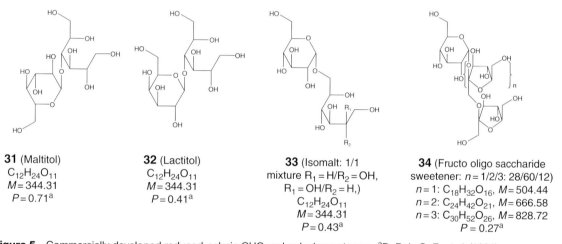

Figure 5 Commercially developed reduced-calorie CHO and polyol sweeteners. [a]DuBois G. E. *et al.* (1991).

empirical formulas, and sweetness potency information, are provided in Figures 7 and 8. Generic and common names are provided for sweeteners which have been commercially developed. The sweeteners are listed in the approximate order of their discovery. Additional information on the discoveries of each of these classes of sweeteners is as follows:

1. *Nitroanilines.* Muspratt J. and Hofmann A. (1846) reported sweet taste for *m*-nitroaniline (**40**) and this is the earliest report, of which the author (GED) is aware, of a compound more potent than common CHO sweeteners. Nearly a century

later, the very potent analog P-4000 (**41**) was reported by Blanksma J. J. and van der Weyden P. W. M. (1940) of the Rijks-Universiteit (Leiden, The Netherlands). For many years, P-4000 was the most potent sweetener known.

2. *N-Sulfonyl amides.* Saccharin (**42**) is the first sweetener of this structural type. The sweetness of saccharin was discovered by Constantine Fahlberg in the laboratory of Ira Remsen at Johns Hopkins University (Fahlberg, C. and Remsen, I., 1879) and was commercialized in the United States as the first product of the Monsanto Chemical Company. Much

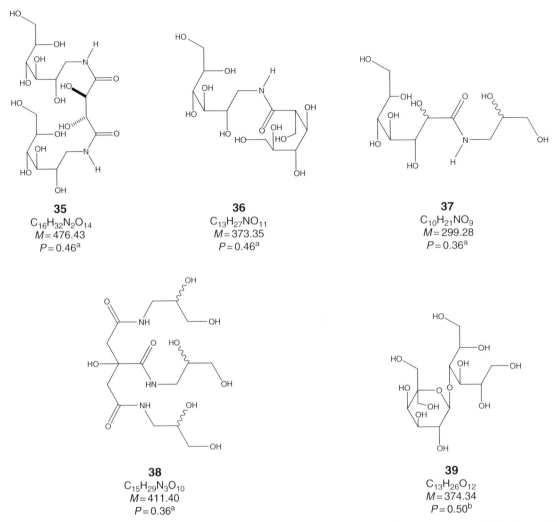

Figure 6   Synthetic noncaloric CHO-like sweeteners. [a]DuBois G. E. *et al.* (1992) and [b]Mazur A. W. and Mohlenkamp M. J. (1997).

later, the structurally similar oxathiazinone class of sweeteners was discovered by chemists Clauss K. and Jensen H. (1973) of Hoechst AG and acesulfame (**43**) selected for commercial development. Both saccharin (Na and Ca salts) and acesulfame (K salt) have broad approvals for use in foods and beverages. Also, both sweeteners exhibit significant bitter/ metallic off tastes and, for this reason, are almost always blended with other sweeteners.

3. *Aryl ureas.* Berlinerblau J. (1883), in the laboratory of R. Schmitt of Polytechnikum Dresden (Germany), reported the discovery of a series of sweet-tasting aryl ureas. Dulcin (**44**) is the best representative of this structural class of sweeteners which is the third structural class of synthetic

sweeteners to be reported. Dulcin was evaluated as a candidate for commercial development but, following the finding of significant toxicity in rat studies at 0.1% of the diet, was not further considered (Fitzhugh, O. G. *et al.*, 1951).

4. *Oximes.* Many oximes have been found to be sweet. The first example was reported by Furukawa S. and Tomizawa Z. (1920). This compound is the oxime (**45**) of perillaldehyde. It was reported to exhibit a sweetness potency of 3400 relative to 9% glucose. Since glucose is 0.6× sucrose in sweetness potency, its sucrose basis sweetness potency is calculated to be $P_w(5.4) = 3400$.

5. *2-Carboxyalkyl benzimidazoles.* Chatterjee B. (1929) of the University of Manchester (Manchester, UK)

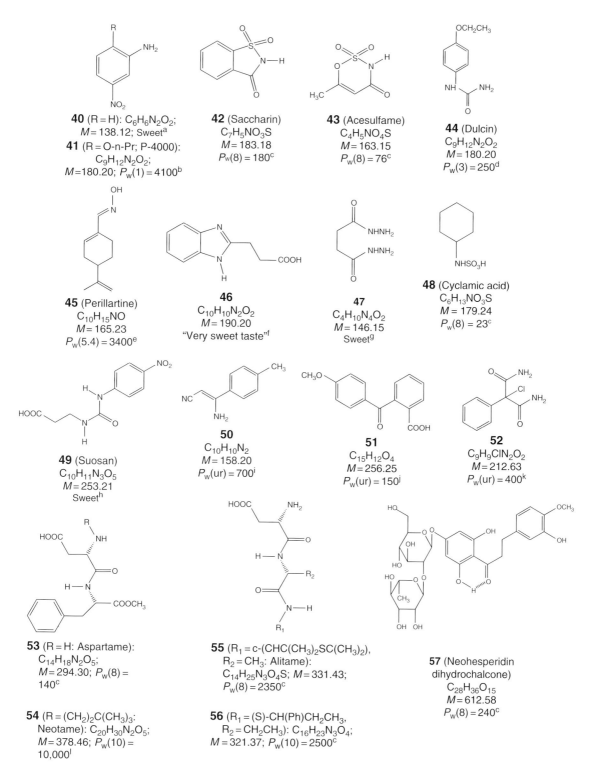

Figure 7 Synthetic noncaloric sweeteners and analogs discovered in the nineteenth and early twentieth centuries.
[a]Muspratt J. T. and Hofmann A. W. (1846); [b]Blanksma J. J. and van der Weyden P. W. M. (1940); [c]DuBois G. E. (2000); [d]Berlinerblau J. (1883); [e]Furukawa S. and Tomizawa Z. (1920); [f]Chatterjee B. (1929); [g]DeGraaf H. (1930); [h]Petersen S. and Müller E. (1948); [i]Dornow A. *et al.* (1949); [j]Runti C. and Galimberti S. (1957); [k]Runti C. and Collino F. (1964); and [l]Nofre C. and Tinti J.-M. (2000).

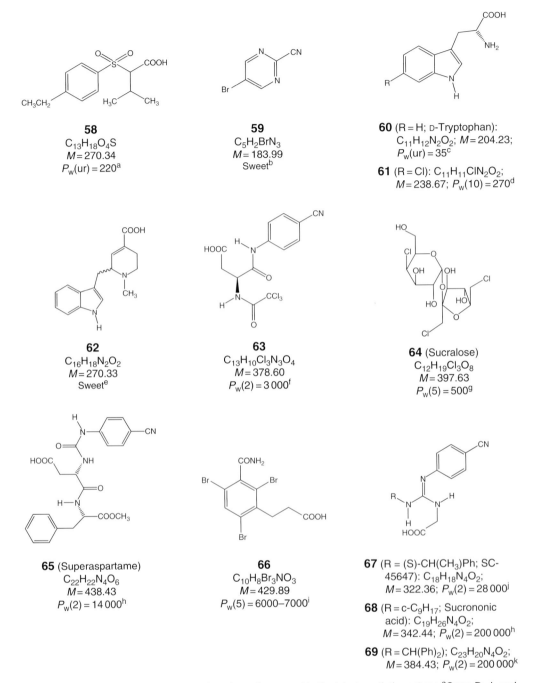

58
$C_{13}H_{18}O_4S$
$M = 270.34$
$P_w(ur) = 220^a$

59
$C_5H_2BrN_3$
$M = 183.99$
Sweet[b]

60 (R = H; D-Tryptophan):
$C_{11}H_{12}N_2O_2$; $M = 204.23$;
$P_w(ur) = 35^c$

61 (R = Cl): $C_{11}H_{11}ClN_2O_2$;
$M = 238.67$; $P_w(10) = 270^d$

62
$C_{16}H_{18}N_2O_2$
$M = 270.33$
Sweet[e]

63
$C_{13}H_{10}Cl_3N_3O_4$
$M = 378.60$
$P_w(2) = 3\,000^f$

64 (Sucralose)
$C_{12}H_{19}Cl_3O_8$
$M = 397.63$
$P_w(5) = 500^g$

65 (Superaspartame)
$C_{22}H_{22}N_4O_6$
$M = 438.43$
$P_w(2) = 14\,000^h$

66
$C_{10}H_8Br_3NO_3$
$M = 429.89$
$P_w(5) = 6000–7000^i$

67 (R = (S)-CH(CH₃)Ph; SC-
45647): $C_{18}H_{18}N_4O_2$;
$M = 322.36$; $P_w(2) = 28\,000^j$

68 (R = c-C₉H₁₇; Sucrononic
acid): $C_{19}H_{26}N_4O_2$;
$M = 342.44$; $P_w(2) = 200\,000^h$

69 (R = CH(Ph)₂); $C_{23}H_{20}N_4O_2$;
$M = 384.43$; $P_w(2) = 200\,000^k$

Figure 8   Synthetic noncaloric sweeteners and analogs discovered in the late twentieth century. [a]Cram D. J. and Ratajczak A. (1971); [b]Budesinsky Z. and Vavrina J. (1972); [c]Solms J. (1965); [d]Suarez T. *et al.* (1975); [e]Hofmann A. (1972); [f]Lapidus M. and Sweeney M. (1973); [g]Hough L. and Phadnis S. P. (1976); [h]Tinti J.-M. and Nofre C. (1991); [i]Gries H. and Mützel W. (1985); [j]Nofre C. *et al.* (1990); and [k]Nagarajan S. *et al.* (1992).

reported that members of this structural class of compounds show very sweet taste. Exemplary is 2-carboxyethylbenzimidazole (**46**).

6. *Hydrazides.* Simple carboxylic acid hydrazides were discovered to be sweet in taste by DeGraaf

H. (1930). Exemplary is the dihydrazide derivative of succinic acid (**47**).

7. *Sulfamates.* The discovery of sweet taste in this structural type of compounds was made by Michael Sveda in the laboratory of Professor L.

Audrieth at the University of Illinois (Audrieth, L. F. and Sveda, M., 1944). The best known sweetener of this structural class is *N*-cyclohexylsulfamic acid (**48**), commercially developed in the sodium and calcium salt forms which are well known as cyclamates. Cyclamates have been widely used in foods and beverages, usually in blends with other noncaloric sweeteners.

8. *Carboxyalkyl aryl ureas.* The discovery of sweet taste in this class of compounds was reported by Petersen S. and Müller E. (1948) of Farbenfabriken Bayer (Leverkusen, Germany). Later, a substantial number of sweet analogs were reported by Tinti J.-M. *et al.* (1982) of the Université Claude Bernard (Lyon, France).

9. *β-Amino cinnamonitriles.* Dornow A. *et al.* (1949) of Institut für Organische Chemie der Technischen Hochschule (Hanover, Germany) reported this structural class of sweeteners which is best exemplified by **50**.

10. *Carboxy benzophenones.* Runti C. and Galimberti S. (1957) of the University of Trieste reported this structural class of sweeteners which is best exemplified by **51**.

11. *Halogenated malonamides.* Runti C. and Collino F. (1964) of the University of Trieste reported this structural class of sweeteners which is best exemplified by **52**.

12. *Dipeptide esters and amides.* This is the structural class of noncaloric sweeteners of greatest commercial significance and is best exemplified by aspartame (**53**). The sweet taste of aspartame was serendipitously discovered in 1965 by James Schlatter in the laboratory of Robert Mazur of Searle Pharmaceutical Company (Mazur, R. H. *et al.*, 1969). Aspartame is unique among synthetic sweeteners as its metabolism leads only to natural amino acids and methanol, all of which are ingested in much higher amounts on consumption of common foods. Neotame (**54**), an analog of aspartame with substantially increased potency, was reported recently by Nofre C. and Tinti J.-M. (2000) of the Université Claude Bernard. It has also been commercially developed. Many dipeptide class sweeteners have been identified since the pioneering work by Mazur and co-workers. Included among such discoveries are the related dipeptide amides, exemplified by alitame (**55**) by Brennan T. M. and Hendrick M. E. (1983) of Pfizer Pharmaceutical Company, and **56** by Sweeny J.

G. *et al.* (1995) of The Coca-Cola Company. Alitame has also been commercially developed.

13. *Dihydrochalcones.* Horowitz R. M. and Gentili B. (1969), of the USDA Laboratory (Pasadena, CA), reported the discovery of a series of sweet-tasting dihydrochalcones. The discovery of sweet taste in this series of compounds was particularly unexpected since it was made in a study of bitter taste SAR within flavanoid glycosides which are well known to exhibit potent bitterness. Neohesperidin dihydrochalcone (**57**), the preferred compound in the series, has been commercially developed but has only seen limited use because of sweetness which is noticeably delayed in onset and which lingers very significantly.

14. *Sulfones.* The discovery of sweet taste in this class of compounds, exemplified by **58**, was made by Aleksander Ratajczak in the laboratory of Donald Cram at UCLA (Cram, D. J. and Ratajczak, A., 1971). This class of sweeteners has been described in detail by Polanski J. and Ratajczak A. (1993).

15. *Pyrimidines.* The discovery of potent sweet taste among substituted pyrimidines, exemplified by **59**, was reported by Budesinsky Z. and Vavrina J. of the Research Institute for Pharmacy and Biochemistry (Prague, Czechoslovakia) in 1972.

16. *D-Amino acids.* Many D-amino acids are sweet with the most potent being D-tryptophan (**60**) (Solms, J. *et al.*, 1965). Much more potent substituted analogs were reported by Suarez T. *et al.* (1975) of Eli Lilly Pharmaceutical Company including **61**. And, at about the same time, in a study of lysergic acid derivatives by Hofmann A. (1972) of Sandoz AG (Basel, Switzerland), **62** was reported to be twice as sweet as saccharin.

17. *N'-Acyl-aspartyl-anilides.* The discovery of this structural class of sweeteners, exemplified by **63**, which is clearly related to the carboxyalkyl aryl ureas, was reported by Lapidus M. and Sweeney M. (1973) of Wyeth Laboratories (Philadelphia, PA).

18. *Halogenated sugars.* The discovery of substantial elevation in sweetness potency of sucrose by halogen substitution of sugar hydroxyl groups was first reported by Hough L. and Phadnis S. P. (1976) of the University of London. The most well-known member of this structural class is sucralose (**64**) which has been commercially developed.

19. *Dipeptide ester/carboxyalkyl aryl urea hybrids.* Tinti J.-M. and Nofre C. (1991) of Université Claude Bernard (Lyon, France) reported the discovery

of highly potent hybrids of the dipeptide ester and carboxyalkyl aryl urea classes of sweeteners. Exemplary of such hybrid sweeteners is a compound that they named superaspartame (**65**). This finding strongly suggests that these two classes of sweeteners bind to the sweetener receptor at contiguous and overlapping sites.

20. *Halogenated carboxyalkyl benzamides.* Gries H. and Mützel W. (1985) of Schering Pharmaceutical Company AG (Berlin, Germany) reported the discovery of a highly potent class of substituted benzamides best exemplified by **66**.

21. *Guanidinoacetic acids.* In the mid-1980s, based on modeling made possible by their superaspartame discovery, Nofre C. *et al.* (1990) of the Université Claude Bernard (Lyon, France) discovered the guanidino-acetic acid class of sweeteners. This class is exemplified by SC-45647 (**67**) which has proven useful in mechanistic studies of sweet taste because of its activity in common animal models. It is further exemplified by the more potent sucrononic acid (**68**), also discovered by Nofre and co-workers and **69** discovered by Nagarajan S. *et al.* (1996) of The NutraSweet Company who were in collaboration with the Université Claude Bernard team.

4.02.2.2.3 *Natural noncaloric sweeteners*

As organic chemists discovered sweet taste in many novel synthetic compounds, they also elucidated the chemical structures of many sweet-tasting compounds found in nature. Natural product structure elucidation was a very difficult process up until the second half of the twentieth century. Although in very significant part, the difficulty in structure elucidation of natural noncaloric sweeteners was due to their structural complexity, the difficulty was also due to the absence of modern spectroscopic and chromatographic methods that now make such structure elucidations routine. What was impossible, or a very challenging task 50 years ago, can now usually be accomplished within a few days. Examples of noncaloric sweeteners found in nature are illustrated in Figures 9–12. Two research groups in particular have been active, beginning in the mid-1970s and continuing up until the mid-1990s, in the elucidation of the structures of natural noncaloric sweeteners. One was an American group led by A. Douglas Kinghorn of the University of Illinois at Chicago. And the second was a Japanese group led by Osamu Tanaka of Hiroshima University. Both groups made major contributions to our understanding of the diversity of potently sweet chemical compounds which exist in nature. The increase in structural complexity between the natural noncaloric sweeteners discussed here and the synthetic noncaloric sweeteners discussed above is noteworthy. A high percentage of the natural noncaloric sweeteners are glycosylated terpenoids or flavanoids, many times with unusual sugars, and still others are proteins. As a result of this structural complexity, economical preparation of natural noncaloric sweeteners by synthetic methods is quite unlikely. Examples of 21 structural classes of natural noncaloric sweeteners along with chemical structures, molecular weights, empirical formulas, and sweetness potency information are provided in Figures 9–12. Information on six protein sweeteners is provided below without structure. Common names are used for sweeteners which have been commercially developed or which, for other reasons, are best known by such names. The sweeteners are listed according to structural class (i.e., terpenoids, flavanoids, etc.). Additional information on the discoveries of each of these natural noncaloric sweeteners is as follows:

1. *Hernandulcin (70).* The only nonglycosylated sweet-tasting sesquiterpenoid identified in nature to date is hernandulcin. Kinghorn and associates (Compadre, C. M. *et al.*, 1987) reported the isolation of hernandulcin from the Mexican plant *Lippia dulcis* Trev. (Verbenaceae).

2. *Pine rosin diterpenoid (71).* One sweet-tasting nonglycosylated diterpenoid natural product has also been reported. Tahara A. *et al.* (1971) of The Institute of Physical and Chemical Research (Saitama-ken, Japan) reported elucidation of the structure of **71**, a potently sweet compound present in pine tree rosin.

3. *Mukurozioside (72).* A glycosylated sesquiterpenoid exhibiting sweet taste was first identified in *Sapindus mukurossi* Gaertn. (Sapindaceae) by Tanaka and co-workers (Kasai, R. *et al.*, 1986). It is present at high levels in the fruits of this plant and is responsible for its sweetness.

4. *Stevioside (73) and rebaudioside A (74).* A total of eight sweet-tasting glycosides of an entkaurene-type diterpenoid known as steviol have been isolated from the plant *Stevia rebaudiana* (Bertoni), indigenous to Paraguay. In the common variety of the plant, stevioside is the most abundant, and rebaudioside A, the second most abundant, steviol glycoside. Several groups have participated in elucidation of the structures of the steviol glycosides

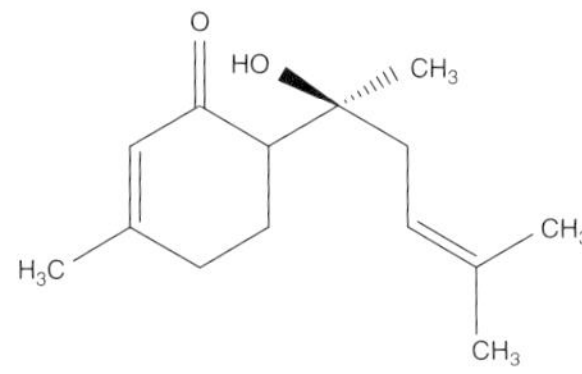

70 (Hernandulcin)
$C_{14}H_{22}O_2$
$M = 222.32$
$P_w(8.6) = 1800^a$

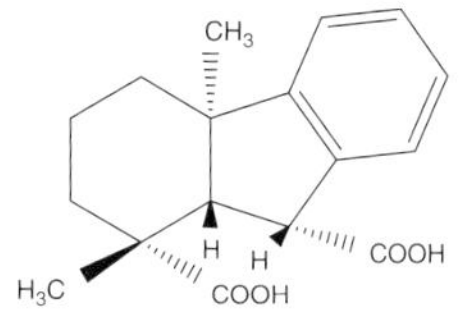

71
$C_{17}H_{20}O_4$
$M = 288.34$
$P_w(1) = 1600^b$

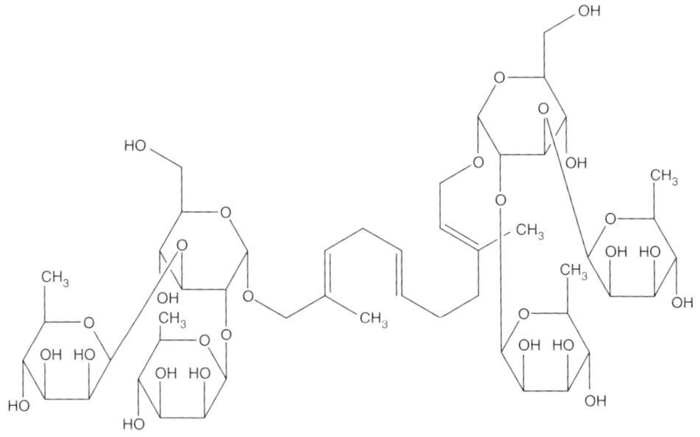

72 (Mukurozioside)
$C_{50}H_{84}O_{27}$
$M = 1117.19$
$P_w(2) = 1^c$

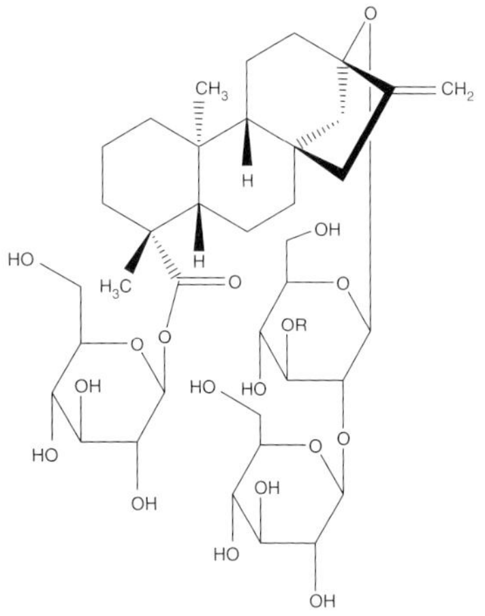

73 (R = H: Stevioside): $C_{38}H_{60}O_{18}$;
   $M = 804.87$; $P_w(5) = 120^d$

74 (R = β-D-Glu: Rebaudioside A): $C_{44}H_{70}O_{23}$;
   $M = 967.01$; $P_w(5) = 250^d$

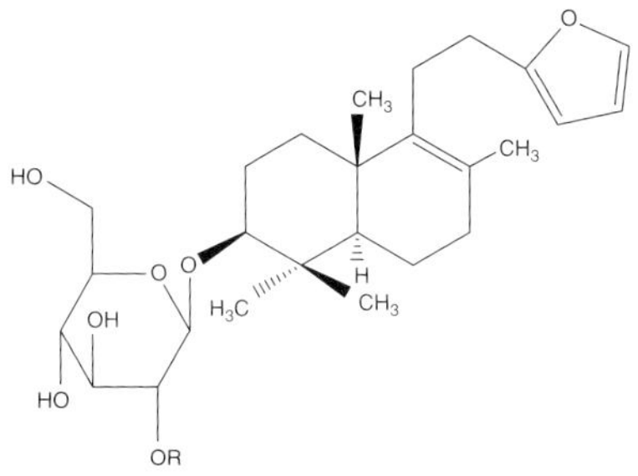

75 (R = β-D-Xyl: Baiyunoside): $C_{31}H_{48}O_{11}$; $M = 596.71$; $P_w(ur) = 500^e$

76   (R = α-L-Rham: Phlomisoside I): $C_{32}H_{50}O_{11}$; $M = 610.73$; $P_w(ur) = 500^e$

Figure 9   Sesquiterpenoid and diterpenoid natural noncaloric sweeteners. [a]Compadre C. M. *et al.* (1987); [b]Tahara A. *et al.* (1971); [c]Kasai R. *et al.* (1986); [d]DuBois G. E. (2000); and [e]Tanaka T. *et al.* (1983).

and these efforts have been reviewed (Crammer, B. and Ikan, R., 1987; Phillips, K. C., 1987).

5. *Baiyunoside (75) and phlomisoside I (76).* Two sweet-tasting labdane-type diterpenoid glycosides known as baiyunoside and phlomisoside were isolated from the Chinese plant *Phlomis betonicoides* Diels (Labiatae) by Tanaka T. *et al.* (1983; 1985).

6. *Glycyrrhizic acid (77).* The structure elucidation of the sweet-tasting triterpenoid glycoside known as glycyrrhizic acid, isolated from the European and central Asian shrub *Glycyrrhiza glabra* L. (Fabaceae), was completed by Lythgoe B. and Trippett S. (1950) of Cambridge University. The crude extract of this plant is known as licorice.

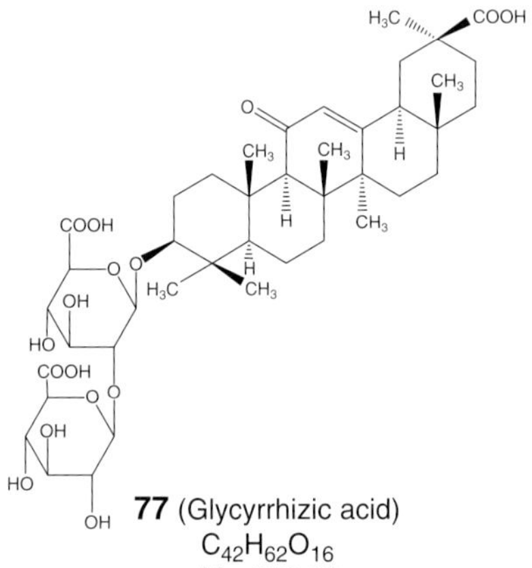

77 (Glycyrrhizic acid)
$C_{42}H_{62}O_{16}$
$M = 822.93$
$P_w(10) = 33^a$

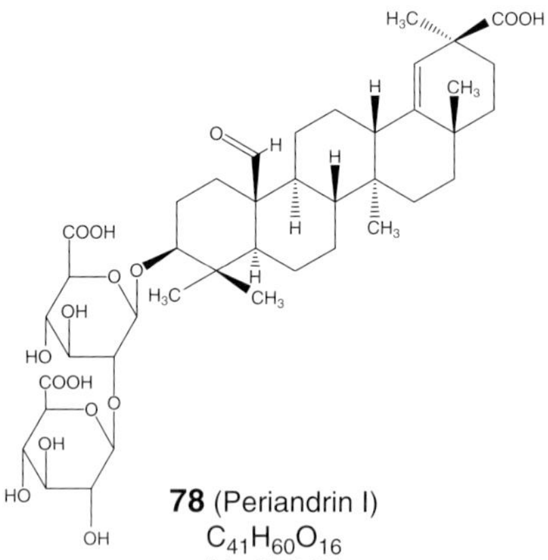

78 (Periandrin I)
$C_{41}H_{60}O_{16}$
$M = 808.91$
$P_w(0.9) = 90^b$

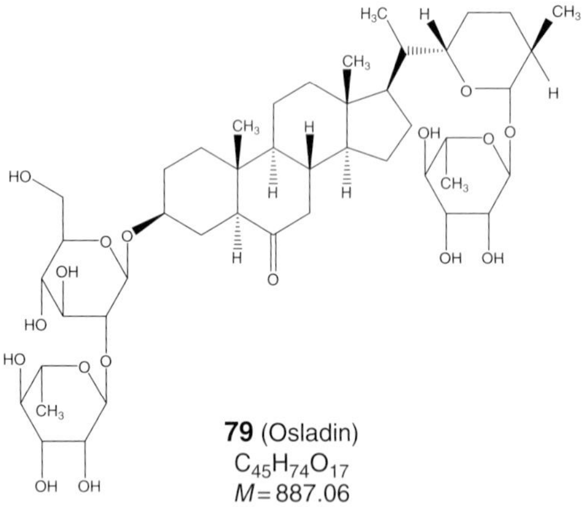

79 (Osladin)
$C_{45}H_{74}O_{17}$
$M = 887.06$
$P_w(ur) = 500^c$

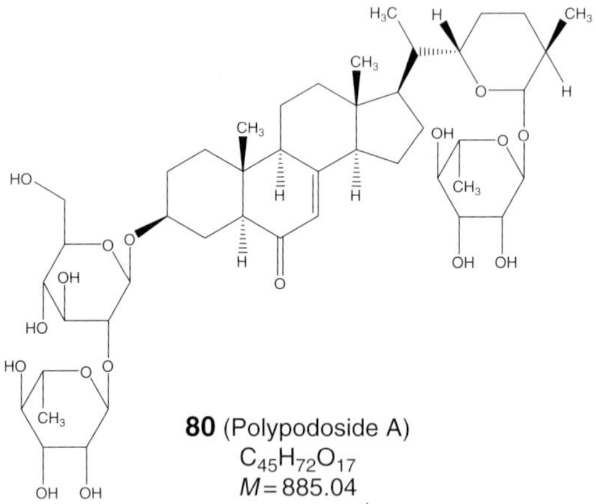

80 (Polypodoside A)
$C_{45}H_{72}O_{17}$
$M = 885.04$
$P_w(ur) = 600^d$

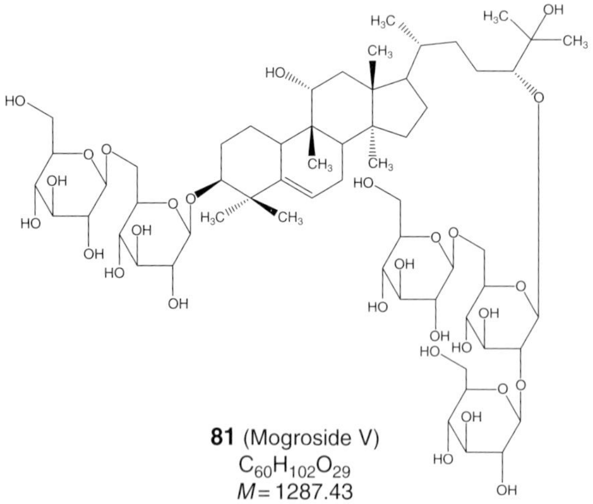

81 (Mogroside V)
$C_{60}H_{102}O_{29}$
$M = 1287.43$
$P_w(ur) = 250\text{-}425^e$

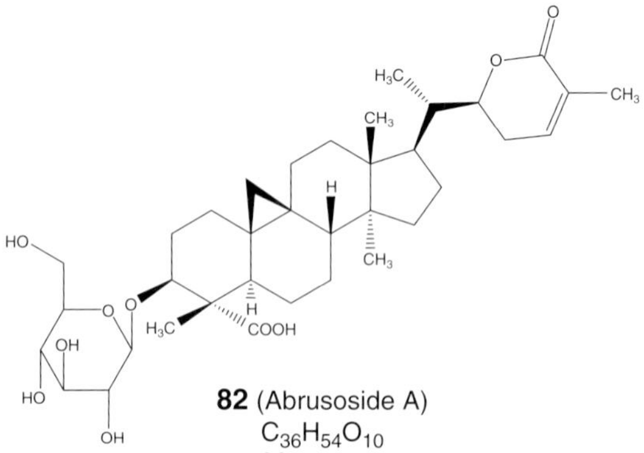

82 (Abrusoside A)
$C_{36}H_{54}O_{10}$
$M = 646.81$
$P_w(2) = 30^f$

Figure 10   (Continued)

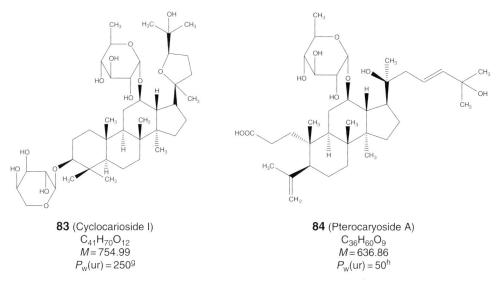

83 (Cyclocarioside I)
$C_{41}H_{70}O_{12}$
$M = 754.99$
$P_{w}(ur) = 250^{g}$

84 (Pterocaryoside A)
$C_{36}H_{60}O_{9}$
$M = 636.86$
$P_{w}(ur) = 50^{h}$

Figure 10 Triterpenoid natural noncaloric sweeteners. [a]DuBois G. E. (2000); [b]Hashimoto Y. *et al.* (1980); [c]Nishizawa M. and Yamada H. (1996); [d]Kim J. *et al.* (1988); [e]Kasai R. *et al.* (1989); [f]Choi Y.-H. *et al.* (1989); [g]Shu R. G. *et al.* (1995); and [h]Kennelly E. J. *et al.* (1995).

Glycyrrhizic acid is commercially available as the monoammonium salt and is characterized as exhibiting sweetness with a notable delay in onset and with a long linger. Glycyrrhizic acid and its salts have approval for use as flavors in the United States.

7. *Periandrin I (78).* The structure elucidation of the sweet-tasting triterpenoid glycosides known as the periandrins I–IV, isolated from the Brazilian plant *Periandra dulcis* Mart. (Leguminosae), was completed by Hashimoto Y. *et al.* (1980) of Kobe Women's College of Pharmacy in the early 1980s. The taste properties of the periandrins are similar to those of glycyrrhizic acid.

8. *Osladin (79).* The structure elucidation of the sweet-tasting triterpenoid glycoside known as osladin, isolated from the rhizomes of the fern *Polypodium vulgare* L. (Polypodiaceae), was only completed recently following an X-ray crystal structure determination by Nishizawa M. and Yamada H. (1996).

9. *Polypodoside A (80).* The structure elucidation of the sweet-tasting triterpenoid glycosides polypodosides A and B isolated from the North American fern *Polypodium glycyrrhiza* DC. Eaton (Polypodiaceae) was completed by Kinghorn and co-workers (Kim, J. *et al.*, 1988).

10. *Mogroside V (81).* The structure elucidation of the sweet-tasting cucurbitane-type triterpenoid glycosides mogrosides IV and V isolated from the Chinese medicinal plant known as Luo Han Guo

(*Siraitia grosvenorii* (Swingle) C. Jeffrey) was completed by Tanaka and co-workers in the late 1980s (Kasai, R. *et al.*, 1989; Matsumoto, K. *et al.*, 1990). Extracts of this plant are in common use in China and recently, a concentrate from the fruit of this plant has been commercially developed by Procter & Gamble for use in formulation of reduced-calorie beverages.

11. *Abrusoside A (82).* The structure elucidation of the sweet-tasting cycloartenol-type triterpenoid glycoside abrusoside A isolated from *Abrus precatorius* L. (Leguminosae) was completed by Kinghorn and co-workers (Choi, Y.-H. *et al.*, 1989).

12. *Cyclocarioside I (83).* The structure elucidation of the sweet-tasting dammarane-type triterpenoid glycoside cyclocarioside I, isolated from the leaves of the Chinese plant *Cyclocarya paliurus* Batal. Iljinsk (Juglandaceae) was completed by Shu R. G. *et al.* (1995) of Jiangxi College of Traditional Chinese Medicine. This plant is said to be used in China for the treatment of diabetes.

13. *Pterocaryoside A (84).* The structure elucidation of the sweet-tasting secodammarane-type triterpenoid glycosides pterocaryosides A and B, isolated from the leaves of *Pterocarya paliurus* Batal. (Juglandaceae) was reported by Kinghorn and co-workers (Kennelly, E. J. *et al.*, 1995). The leaves of this plant are used in China to sweeten foods.

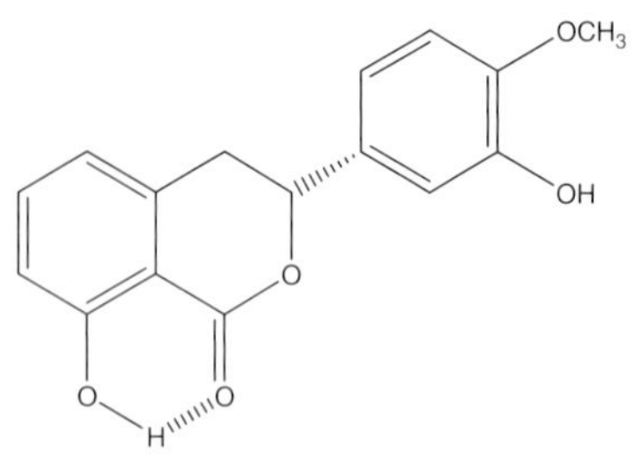

85 (Phyllodulcin)
$C_{16}H_{14}O_5$
$M = 286.28$
$P_w(3) = 400$[a]

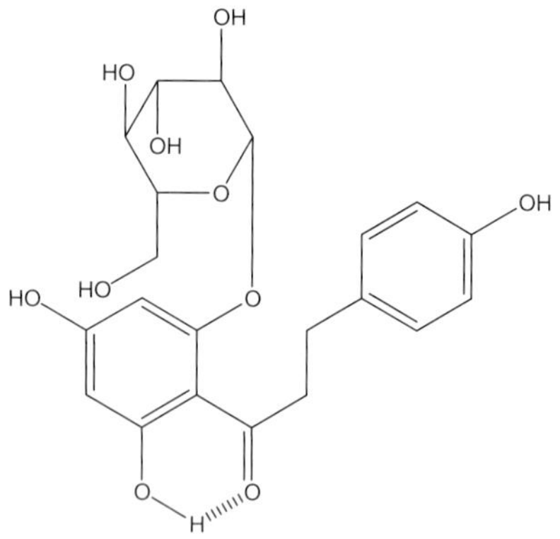

86 (Phloridzin)
$C_{21}H_{24}O_{10}$
$M = 436.41$
Sweet[b]

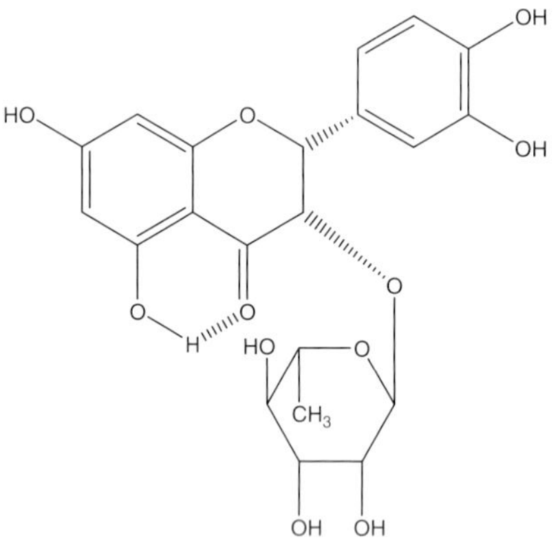

87 (Neoastilbin)
$C_{21}H_{22}O_{11}$
$M = 450.39$
Sweet[c]

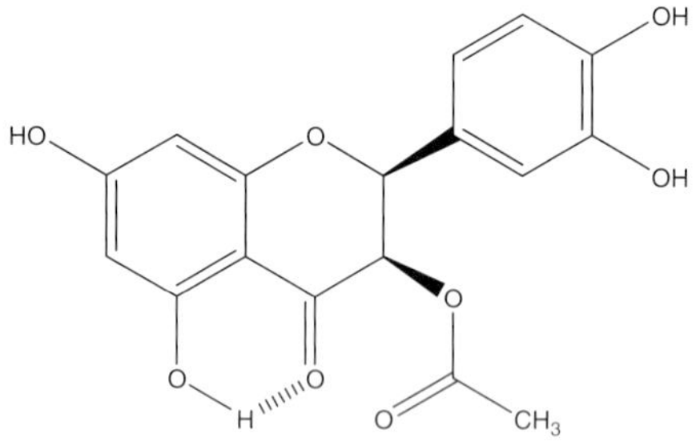

88
$C_{17}H_{14}O_8$
$M = 346.29$
$P_w(2) = 80$[d]

Figure 11 Naturally occuring dihydroisocoumarin, dihydrochalcone, and flavanone noncaloric sweeteners. [a]Arakawa H. and Nakazaki M. (1959); [b]Tanaka T. *et al.* (1982); [c]Kasai R. *et al.* (1988); and [d]Nanayakkara N. P. D. *et al.* (1988).

14. *Phyllodulcin (85).* The structure elucidation of the sweet-tasting 3,4-dihydroisocoumarin-type polyketide isolated from the leaves of *Hydrangea macrophylla* Seringe var. *thunbergii* (Siebold) Makino (Saxifragaceae) was completed by Arakawa H. and Nakazaki M. (1959) of Osaka City University. Extracts of the leaves of this plant are used in Japan to make a sweet tea at a festival known as Hamatsuri, a celebration of Buddha's birth.
15. *Phloridzin (86).* The sweet-tasting dihydrochalcone glycoside phloridzin is present in many parts of apple trees except for the mature fruit. Its structure was elucidated and first isolated from the leaves of *Symplocos lancitolia* Sieb. by Tanaka T. *et al.* (1982).
16. *Neoastilbin (87).* The flavanone glycoside neoastilbin is reported to be the sweet-tasting component in the leaves of *Engelhardtia chrysolepis* Hance (Juglandaceae), a tree which grows in a subtropical region of China. Its structure was elucidated by Tanaka and co-workers (Kasai, R. *et al.*, 1988).
17. *[2(R),3(R)]-Dihydroquercetin-3-O-acetate (88).* The structure elucidation of this sweet-tasting flavanone, isolated from *Tessaria dodoneifolia* (Hook

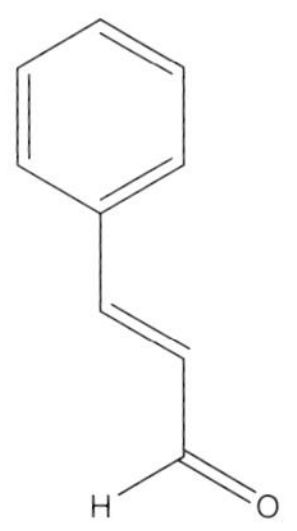

89 (Cinnamaldehyde)
C_9H_8O
$M = 132.16$
Sweet[a]

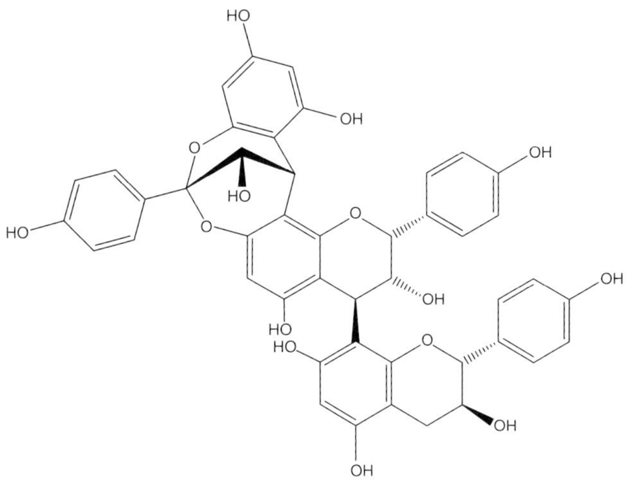

90 (Selligueain A)
$C_{45}H_{36}O_{15}$
$M = 816.76$
$P_w(2) = 35$[b]

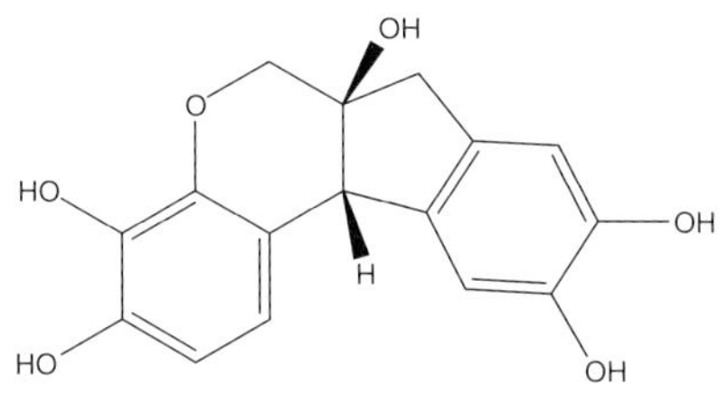

91 (Hematoxylin)
$C_{16}H_{14}O_6$
$M = 302.28$
$P_w(3) = 120$[c]

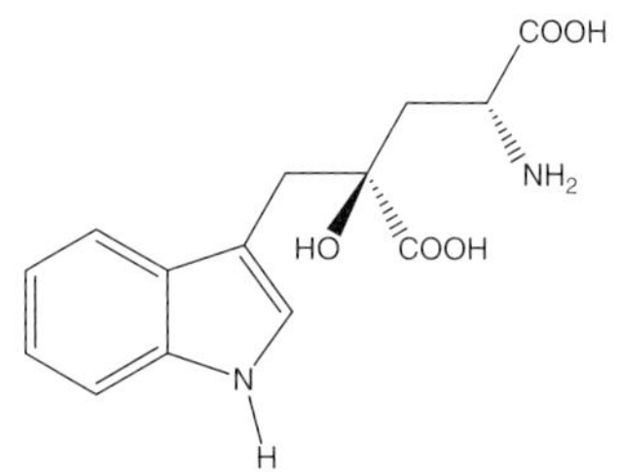

92 ((2R, 4R)-Monatin)
$C_{14}H_{16}N_2O_5$
$M = 292.29$
$P_w(5) = 2700$[d]

Figure 12 Phenylpropanoid, proanthocyanidin, benzo-[b]-indeno-[1,2-d]-pyran, and amino acid natural noncaloric sweeteners. [a]Hussain R. A. (1990); [b]Baek N.-I. et al. (1993); [c]Masuda H. et al. (1991); and [d]Bassoli A. et al. (2005).

and Arn.) Cabrera (Compositae), a plant indigenous to Paraguay, was reported by Kinghorn and co-workers (Nanayakkara, N. P. D. et al., 1988).

18. *Cinnamaldehyde (89).* trans-Cinnamaldehyde was identified as the sweet-tasting component in *Cinnamomum osmophloeum* Kanehira (Lauraceae) by Kinghorn and co-workers (Hussain, R. A. et al., 1990).

19. *Selligueain A (90).* The structure elucidation of the sweet-tasting proanthocyanidin selligueain A, following isolation from the rhizomes of the Indonesian fern *Selliguea feei* Bory (Polypodiaceae), was reported by Kinghorn and co-workers (Baek, N.-I. et al., 1993).

20. *Hematoxylin (91).* The benzo[b]indeno[1,2-d]pyran hematoxylin is reported to be the

sweet-tasting component in an extract of the heartwood of *Haematoxylon campechianum* L. (Leguminosaeae). Its structure was elucidated by Tanaka and co-workers (Masuda, H. et al., 1991). This compound had been used as a microbiological staining agent but was not previously recognized to be sweet.

21. *Monatin (92).* The sweet-tasting amino acid monatin was isolated from the roots of the African plant *Schlerochiton ilicifolius* A. Meeuse (Acanthaceae). The structure of monatin was reported by Louis Ackerman and co-workers of the University of Pretoria (Pretoria, South Africa) (Vleggaar, R. et al., 1992). In later work by Bassoli A. et al. (2005) of the University of Milan, it was found that all four stereoisomers are present in an extract

of the natural source, that all four stereoisomers exhibit sweetness, but that the most potent stereoisomer is the $2R/4R$ form.

22. *Thaumatin (93).* The sweet-tasting component in the berries of the West African plant *Thaumatococcus danielli* Benth, locally known as katempfe berries, was first isolated by van der Wel H. (1972) at Unilever and shown to be a protein. This protein called thaumatin was later shown to be a mixture of at least four similar 207 amino acid proteins. Thaumatin I, one of the two major forms, has $M = 22\,206$. The crystal structure of thaumatin was determined by Sung-Hou Kim and co-workers at the University of California at Berkeley (de Vos, A. M. *et al.*, 1985). Thaumatin exhibits potent sweetness $[P_w(10) = 1600]$ with a notable delay in onset and with a very pronounced sweetness linger (DuBois, G. E., 2000).

23. *Monellin (94).* The sweet-tasting component in the berries of the West African plant *Dioscoreophyllum cumminsii*, also known as the serendipity berry, was first isolated by Morris J. A. and Cagan R. H. (1972) at Monell Chemical Senses Center and shown to be a protein. Monellin consists of two separate peptide chains, the A chain of 45 and the B chain of 50 amino acid residues, which are associated noncovalently with each other and has $M = 11\,086$. And the crystal structure was determined by Sung-Hou Kim and co-workers at the University of California at Berkeley (Ogata, C. *et al.*, 1987). Like thaumatin, monellin exhibits potent sweetness $[P_w(\text{ur}) = 3000]$ with a notable delay in onset and with a very pronounced sweetness linger.

24. *Mabinlin (95).* The sweet-tasting component in the seeds of the Chinese plant *Capparis masaikai* Levl. (Capparidaceae) was shown to be a mixture of two proteins known as mabinlins I and II. In work by Masanori Kohmura and Yasuo Ariyoshi of Ajinomoto Company (1998), mabinlin II was determined to be composed of an A chain (33 amino acids) and a B chain (72 amino acids) which are connected via disulfide linkages. It was determined to have $M = 12\,441$ and sweetness recognition threshold of 0.1%. Since the threshold for sucrose sweetness recognition is *c.* 0.5%, it is only slightly more potent than sucrose.

25. *Brazzein (96).* The sweet-tasting component in the fruits of the African climbing vine *Pentadiplandra brazzeana* Baillon (Pentadiplandraceae) was demonstrated by Ming D. and Hellekant G.

(1994) of the University of Wisconsin to be a 54 amino acid residue protein with $M = 6473$ and $P_w(2) = 2000$.

26. *Miraculin (97).* A protein isolated from the fruits of the African plant *Richardella dulcifica* (Schum. et Thonn.) Baehni (Sapotaceae), often referred to as miracle fruit, is known as miraculin and was first shown to be a protein by Kurihara K. and Beidler L. (1968) of Florida State University. This protein, having $M = c.\ 24\,000$, is not sweet at all but, interestingly, causes sour solutions to taste sweet.

27. *Curculin (98).* The sweet-tasting component in the fruit of the Malaysian plant *Curculigo latifolia* was shown by Yoshie Kurihara and co-workers of Yokohama University to be a dimer of a 114 amino acid polypeptide $(M = 24\,734)$ (Yamashita, H. *et al.*, 1990). They named this protein curculin. Curculin, in addition to being a sweetener, causes sour solutions to taste sweet and also causes water to taste sweet. Curculin exhibits $P_w(6.8) = 550$.

4.02.2.2.4   *Sweet-tasting minerals*

Sweet taste is not limited to organic chemicals. Many inorganic salts are also known to exhibit sweet taste. Since the late 1800s, there have been many reports of sweet taste among inorganic salts and this topic has been reviewed by Shallenberger R. S. (1993). Sweet taste has been generally reported for alkali metal salts. As example, for NaCl, von Skramlik E. reported (1926) that its taste changes from tasteless at 9 mM, to slightly sweet at 10 mM, to sweet at 20 mM, to strongly sweet at 30 mM, to salty sweet at 40 mM to salty at 50 mM. In more recent work on the sweet taste of NaCl, Bartoshuk L. M. *et al.* (1978) of Yale University confirmed the presence of sweet taste but found that its observation is dependent on adaptation of the sensory system to water before tasting. Work in the author's laboratory (GED) is in general agreement with the reports of sweet taste for NaCl and other salts except that the sweetness observed is only weak before saltiness dominates the overall taste quality. Many other salts are known to exhibit sweet taste including beryllium (II) and lead (II) salts. Little analysis of the sweetness potencies of these salts appears to have been done, no doubt because of their significant toxicities. Rare earth salts, specifically $LaCl_3$, $TbCl_3$, and $EuCl_3$, have been observed to be potently sweet in the author's laboratory. All exhibit similar sweetness potencies $[P_w(5) = c.\ 100]$ but are disadvantaged in that they are strongly astringent as well as bitter in taste (DuBois, G. E., 2003).

4.02.2.3 Sensory Properties of Sweeteners

As is clear from the discussion above in II. A-D, many, many chemical compounds exhibit sweet taste. However, because of their natural occurrence in common fruits and vegetables, CHO sweeteners, and sucrose in particular, have become the consumer's standards for sweet taste. And it has been found that the synthetic and natural noncaloric sweeteners exhibit sweet tastes that differ from the tastes of CHO sweeteners in several ways. Principal differences include the following:

4.02.2.3.1 Concentration/response functions

Over the years, it has become common practice to define the sweetness potencies of sweeteners relative to sucrose as a reference. In doing so, the author (GED) *et al.* (1990) of The NutraSweet Company reported the finding that CHO sweetener potencies are independent of sucrose reference concentration while, in contrast, all synthetic and natural noncaloric sweeteners are strongly dependent on sucrose reference concentration. And it was demonstrated that the concentration/response (C/R) functions, where R is in units of percent sucrose equivalents, for all synthetic and natural noncaloric sweeteners are well modeled by the law of mass action $R = R_m C/(k_d + C)$, where R_m is the maximal response and k_d is the apparent sweetener/receptor dissociation constant. From these findings, the conclusion was drawn that all CHO sweeteners have a common R_m while all synthetic and natural noncaloric sweeteners have lower R_ms. Although other interpretations are possible, the data are consistent with the idea that all

noncaloric sweeteners are partial agonists relative to CHO sweeteners. Exemplary of C/R function behavior of noncaloric sweeteners are the C/R functions for sodium saccharin and sodium cyclamate which are illustrated in Figure 13.

4.02.2.3.2 Flavor profiles

In general, CHO sweeteners exhibit clean sweet tastes without bitter, sour, salty, or savory taste attributes. This is not generally the case, however, for synthetic and natural noncaloric sweeteners. In fact, clean sweet taste is the exception rather than the rule for noncaloric sweeteners. Bitter off taste is a common off taste among noncaloric sweeteners and is particularly noticeable in saccharin. It has now been demonstrated that the bitter off taste of saccharin is due to activation of specific bitterant receptors (Pronin, A. N. *et al.*, 2004). Exemplary of flavor profiles for noncaloric sweeteners are the flavor profiles for sodium saccharin and sodium cyclamate which are illustrated in Figure 14.

4.02.2.3.3 Temporal profiles

CHO sweeteners exhibit rapid onset of sweet taste and sweetness which dissipates relatively quickly. In general, all synthetic and natural noncaloric sweeteners exhibit sweet tastes which develop more slowly and which then dissipate more slowly, oftentimes very slowly. In a very recent report, Michael Naim and co-workers of Hebrew University speculate that the lingering sweetness character of many noncaloric sweeteners may be due to inhibition of a specific G-protein receptor kinase which is responsible for termination of receptor signaling (Zubare-Samuelov, M. *et al.*, 2005). Although this is a possible explanation, the cause of the atypical temporal profiles of noncaloric sweeteners remains to be determined with certainty.

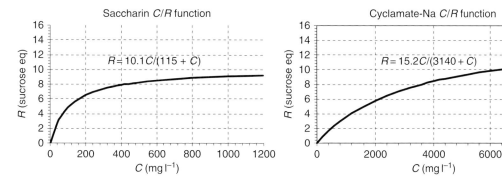

Figure 13 Concentration/response functions for sodium saccharin and sodium cyclamate.

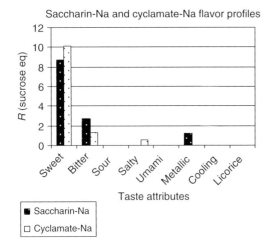

Figure 14 Comparison of flavor profiles of sodium saccharin and sodium cyclamate.

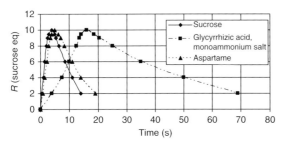

Figure 15 Comparison of temporal profiles of sucrose, aspartame, and monoammonium glycyrrhizinate.

Exemplary of temporal profiles for noncaloric sweeteners are the temporal profiles for sucrose, aspartame, and monoammonium glycyrrhizinate which are illustrated in Figure 15.

4.02.2.3.4 Adaptation

If a CHO sweetener is tasted iteratively over a short period of time, little change in sweetness intensity is noted. However, synthetic and natural noncaloric sweeteners exhibit significant decrements in perceived sweetness. The sensory system appears to desensitize to the noncaloric sweeteners to a much greater extent than is the case for CHO sweeteners. It is generally true that signaling from GPCRs is terminated by receptor phosphorylation by specific intracellular kinases. However, even if this is the case with the sweetener receptor, it is not clear as to why the desensitization following receptor activation with noncaloric sweeteners should exceed that promoted by CHO sweeteners. Exemplary of the desensitization differences between CHO sweeteners

and noncaloric sweeteners are the sensory panel estimates of sweetness intensity on iterative tasting of cola beverages sweetened with HFSS, aspartame, and an aryl urea type sweetener as is illustrated in Figure 16.

4.02.3 Bitter-Tasting Stimuli

It is generally accepted that the sense of bitter taste evolved as a mechanism whereby potentially toxic substances could be avoided. In fact, a great many natural and synthetic chemical compounds which exhibit toxicity are bitter in taste at some concentration. It is not apparent, however, that thresholds for bitter taste correlate well with thresholds for toxicity. And many perfectly safe naturally occurring compounds exhibit bitterness (e.g., caffeine and quinine) and provide desirable sensory attributes to foods and beverages. Still other bitter compounds (e.g., vitamins, minerals, and bioflavonoids) provide nutritional benefits. Another confounding factor is that some very toxic compounds (e.g., lead salts and beryllium salts) are not bitter and, to the contrary, are sweet. As a bottom line, bitter taste sensation, at best, may only be considered as a first line of defense against ingestion of toxic chemical compounds.

4.02.3.1 Biochemistry of Bitter Taste

A great deal of work has been done to elucidate the biochemical pathways that mediate the sense of bitter taste. And while a historical summary of this work is beyond the scope of this review, a few key discoveries will be pointed out preliminarily to a concise discussion of the current understanding of the biochemistry of bitter taste. Many reviews of the older literature are available and included among them are Belitz H. D. *et al.* (1985), Shallenberger R. S. (1993), Beauchamp G. K. (1994), and Roy G. (1997). In our view, the SAR for bitter-tasting compounds can only be appreciated after the mechanistic pathways that mediate bitter taste are understood.

Chemists have long been fascinated with the very high level of structural diversity among bitter-tasting chemical compounds. And, at least in part, because of this extraordinary diversity, Spielman A. I. *et al.*, (1992) had already concluded by the early 1990s that multiple mechanistic pathways must be involved. Also in the early 1990s, support was provided that bitter taste, at least for

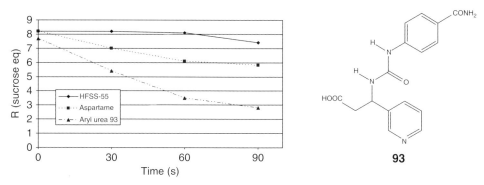

Figure 16 Comparison of the rates of sensory desensitization to HFSS-55, aspartame, and N-(4-carboxamidophenyl)-N'-(1-(3-pyridyl)-2-carboxy)-ethyl-urea (**93**) in a cola beverage system.

many compounds, must be mediated by one or more GPCRs. The support for GPCR involvement was the discovery of the G-protein gustducin and the demonstration that mice lacking gustducin could not detect common bitterants (Margolskee, R. L., 2004). The major breakthrough in understanding the biochemistry of bitter taste did not occur, however, until 2000. At this time, mouse and human genomic analysis led to the discovery of families of GPCRs referred to as T2Rs and these receptors were shown to be mediators of bitter taste signaling (Adler, E. *et al.*, 2000; Chandrashekar, J. *et al.*, 2000; Matsunami, H. *et al.*, 2000). More recently, comprehensive analysis of the human genome led to the conclusion that human bitter taste is mediated by a family of 25 T2Rs which are all expressed in the same subset of TBCs (Pronin, A. N. *et al.*, 2004). And, at the present time, substantial efforts are in place to identify the ligands which activate these 25 T2Rs, a process commonly referred to as deorphanization. Results to date include the finding (Bufe, B. *et al.*, 2002) that T2R16 mediates the activities of bitter glucopyranosides such as salicin and that T2R10 mediates the activity of strychnine. Additional findings (Pronin, A. N. *et al.*, 2004) include that T2R44 is activated by denatonium chloride and 6-nitrosaccharin and that T2R61 is activated by 6-nitrosaccharin. Extensive work is underway at this time to deorphanize all of the bitterant receptors. The finding that T2R44 responds to bitterants as diverse in structure as 6-nitrosaccharin and denatonium chloride, however, merits comment. It seems highly likely that bitter compounds will activate multiple receptors and the exception will be that a particular bitterant will activate a single T2R. And, in this sense, the bitter taste system

may be similar to the olfactory system where individual odorants typically activate multiple members of the set of 350–400 human olfactory receptors. The T2Rs and olfactory receptors are similar in structure, all being family A GPCRs, and are thought to bind their ligands at conserved agonist binding sites within the TMD. It should also be noted that while the primary structures of the T2Rs are known, the structures of the functional bitterant receptors are not. The principal complication is that many GPCRs exist in dimeric and perhaps even oligomeric forms (Breitwieser, G. E., 2004). If the bitterant receptors are dimeric, are they limited to homodimers or are they heterodimeric as well? If they are only homodimeric, then 25 T2Rs will enable formation of 25 $(T2R)_2$ receptors. If, however, all possible heterodimeric receptors form as well, the number of functional receptors will be 325. In view of the highly diverse relationship between chemical structure and bitter taste, such a large number of bitterant receptors would not be a surprise.

So, it appears that the dominant pathway for initiation of the bitter taste signaling process is activation of one or more T2Rs. The next question is how do activated T2Rs excite TBCs? As already noted, the G-protein gustducin is required for TBC responses to bitterants. And so, the question really is, how does activated gustducin lead to TBC excitation? The answer began to emerge in 2002 in work done in the Margolskee laboratory (Pérez, C. A. *et al.*, 2002). In differential cDNA screening experiments in mouse TBCs, they found the common expression of several transduction elements including the α subunit of gustducin ($G_{\alpha\text{gust}}$), the G-protein subunit $G_{\gamma 13}$, the enzyme phospholipase $C_{\beta 2}$ ($PLC_{\beta 2}$), the inositol 1,4,5-trisphosphate (IP_3) receptor type III (IP_3R3), and the TRP m5 ion channel (TRPM5). And they proposed that activated gustducin initiates a cascade of reactions as follows:

1. activation of $PLC_{\beta2}$ to convert phosphatidylinositol to IP_3,
2. action by IP_3 at its intracellular receptor IP_3R3 to promote release of Ca^{2+} from intracellular storage vesicles, and
3. Ca^{2+} binding to an intracellular site on the ion channel TRPM5 causing the channel to open allowing cations to enter the cell, thus causing the depolarization requisite for TBC excitation.

Strong support for this proposal was provided by a collaborative team from the Charles Zuker and Nicholds Ryba laboratories (Zhang, Y. *et al.*, 2003) that developed strains of mice in which expression of $PLC_{\beta2}$ and TRPM5, respectively, were knocked out. The strain lacking $PLC_{\beta2}$ was demonstrated to be incapable of responding to any of bitter, sweet, or umami taste stimuli. The same finding was made with the strain lacking TRPM5. This work demonstrated that responses to sweet, bitter, and umami stimuli are transduced by a common pathway and suggested that their signaling must be initiated by separate subsets of TBCs. This was demonstrated to be the case by developing a strain of mice from a $PLC_{\beta2}$ knockout strain in which $PLC_{\beta2}$ function was restored only in bitterant receptor expressing cells. These mice were found to have sensitivity to bitter stimuli but not to sweet or umami stimuli. Thus, it appears that coding of taste quality occurs at the cellular level. And therefore, some TBCs are sweetener sensors, some bitterant sensors and some umami sensors. This finding is consistent with work by Hellekant G. *et al.* (1997) showing that chimpanzee single CT nerve fibers emanating from taste buds show strong taste modality specificity. In general, fibers responding to the bitterants caffeine, quinine HCl, and denatonium benzoate were unresponsive to sucrose and fructose (sweet), citric acid (sour), NaCl (salty), and monosodium glutamate (umami). So, as a bottom line, some TBCs are bitterant-specific and they communicate with the brain via labeled lines so as to preserve their sensory quality coding.

Given that bitter taste may result from activation of any of 25, or even as many as 325, bitterant receptors in the bitter-sensitive subset of TBCs, it is obvious that a very broad bitterant SAR is to be expected. A finding that leads to the expectation of an even more diverse bitterant SAR is the observation that at least some bitter-tasting compounds readily cross the membranes of TBCs on a time scale consistent with the taste experience (Naim, M. *et al.*, 2002). Specifically the bitterants quinine and c-(Leu-Trp) were shown to readily accumulate at significant levels in TBCs within 1–2 s. As a result of these findings, Naim and co-workers speculated that many bitterants may initiate their effects at intracellular sites in the bitter-sensitive TBC activation cascade. In support of this idea, further work with TRPM5 (Damak, S. *et al.*, 2006) and $PLC_{\beta2}$ (Dotson *et al.*, 2005) knockout mice indicates that other sites could be involved in bitter signaling as these mice show some aversive responses at high concentrations of bitter compounds. In summary, for bitter taste, we have 25 bitterant receptors and at least 4 major intracellular components (i.e., gustducin, $PLC_{\beta2}$, IP_3R3, and TRPM5) that, in principle, may be sites of action for bitter-tasting chemical compounds. Without any knowledge of the SAR of bitterant compounds, the biochemistry of bitterant taste leads to the expectation of an extraordinarily diverse SAR.

As has already been noted, the bitterant SAR is an extraordinary diverse relationship as may be understood now based on the biochemistry of bitter taste. Volumes could be filled with the chemical structures of compounds reported to exhibit bitter taste. As a bottom line, it appears that bitter taste is the default taste. Thus as a general rule, if a compound's taste is not sweet, umami, sour, or salty, it will be bitter. The only factor in question is bitterness potency. Clearly, since bitterness is mediated by a family of membrane receptors, and probably by intracellular effectors as well, some bitterants will have high receptor affinities and some will not, thus resulting in a range of potencies. While this general rule is consistent with empirical observations on taste, poorly soluble compounds may appear to be exceptions. Clearly, in order to activate receptors, compounds must be in solution at sufficiently high concentration to reach their taste detection thresholds. And some apparently tasteless compounds are probably tasteless only because they are insufficiently soluble to initiate a threshold level of receptor activation. In such series of compounds, analogs of the poorly soluble compounds with higher solubility always exhibit taste and, almost always, it is bitter taste.

Interestingly, for some compounds thought to be purely sweet in taste, desensitization to sweet taste leads to bitterness perception (DuBois, G., 1985). It seems possible that signaling by sweet-sensitive

TBCs may be able to inhibit bitter-sensitive TBC signaling, an inhibition effect that is eliminated following desensitization of sweet-sensitive TBCs. Thus, for example, iterative tasting of a high concentration of aspartame results in aspartame exhibiting pronounced bitterness. Other highly hydrophilic sweeteners (e.g., sucrose, fructose, glucose) exhibit no bitterness at all even after attempted desensitization. Thus, it appears that some threshold hydrophobicity must be necessary for organic compounds to stimulate bitter-sensitive TBCs. Organic compounds that are too hydrophilic appear incapable of activating bitter-sensitive TBCs. Yet many inorganic compounds which are extremely hydrophilic are capable of activating bitter-sensitive TBCs. Many organic compounds are known which exhibit both sweet and bitter tastes. Typically such compounds exhibit only sweetness when evaluated at concentrations just above their taste thresholds, but at higher concentrations exhibit both sweet and bitter tastes. In such cases, it appears that sweet-sensitive TBCs are not able to inhibit activities of bitter-sensitive TBCs. It could be that the inhibitory effect of sweet-sensitive TBCs on bitter-sensitive cells only occurs when there is a wide difference between sweetness and bitterness potencies for compounds.

4.02.3.2 Bitterant Structure–Activity Relationship

An objective of this review is to summarize the SAR of bitter stimuli. However, given the extraordinary diversity of bitterant SAR, it would be of little value to provide a comprehensive listing of chemical compounds which have been reported to exhibit bitter taste. So, rather than attempt such a feat, a listing is provided of common bitter compounds that are either used in research or are present in common foods. Chemical structures, empirical formulas, and molecular weights are provided. Bitterant potencies are not generally available as is the case for sweetener potencies. Bitter taste has generally been a taste that food, beverage, and pharmaceutical manufacturers have tried to eliminate or minimize. And for this reason, there has not been a great deal of research in which bitterant potencies have been systematically determined as has been the case in the sweetener field. In some cases, bitter taste recognition thresholds (BRTs) have been reported as has been summarized by Glendinning J. I. (1994a; 1994b) and Belitz H.-D. and Wieser H. (1985).

A listing of ten common naturally occurring bitterants **94–103**, selected to illustrate the extraordinary diversity of bitterant SAR, with structures, molecular weights, empirical formulas, and BRT information provided in Figure 17, is as follows:

1. *Urea (94)*. Ureas in general are bitter in taste with the simplest member of the series exhibiting bitterness and bitterness potency increasing 100-fold on substitution with hydrophobic groups.
2. *Potassium chloride (95)*. KCl is commonly employed as a salt substitute by people suffering from hypertension. Its use is limited, however, by the presence of pronounced bitterness in addition to saltiness. Many other inorganic salts also exhibit bitter taste.
3. *Nicotine (96)*. Alkaloids, in general, are bitter in taste with nicotine being one of the simplest members of this class of compounds. It is the physiologically active and addictive component present in tobacco smoke.
4. *Caffeine (97)*. Caffeine is a fully methylated purine base present in many natural products including coffee beans, tea leaves and cola nuts. It, like other purine and pyrimidine bases, exhibits bitter taste.
5. *Catechin (98)*. Catechin is a flavanoid commonly found in tea.
6. *Quinine (99)*. Quinine is the most abundant alkaloid in the bark of the cinchona tree which is indigenous to South America. It was originally commercialized as an antimalarial but today is commonly employed as a bitterant in foods and beverages.
7. *Humulone (100)*. Humulone and related compounds are the bitter principles in hops which are key ingredients in beer.
8. *Limonin (101)*. Limonin is a member of the triterpenoid class of natural products and is the primary bitter principle in lemon and some other citrus fruit. It is a member of a group of natural products referred to as limonoids, all of which are bitter in taste.
9. *Naringin (102)*. Naringin is a member of a class of natural products known as flavanoid glycosides and is the principal bitter component present in the peel of grapefruit.
10. *Digitoxin (103)*. Digitoxin is a steroidal cardiac glycoside present in the plant *Digitalis purpurea* and is representative of a great many triterpenoid glycoside natural products, nearly all of which exhibit potent bitter taste.

A listing of ten bitterants, **104–113**, principally of synthetic origin, selected to further illustrate the

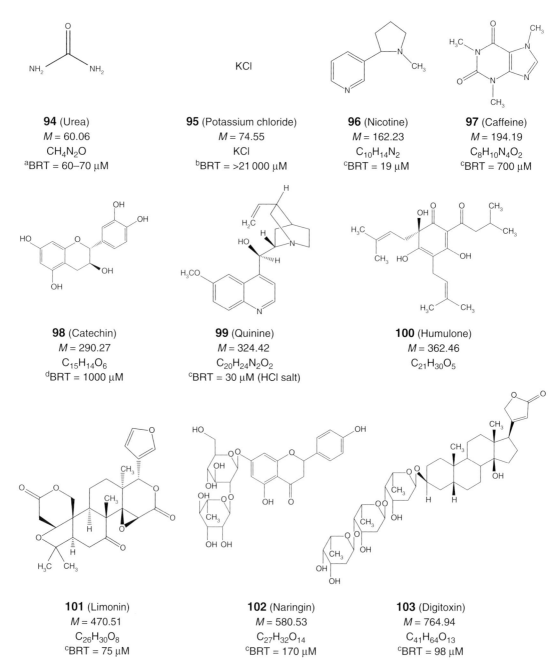

Figure 17 Chemical structures, molecular weights, empirical formulas, and bitterness recognition thresholds of some naturally occurring bitter-tasting chemical compounds. [a]Belitz H.-D. and Wieser H. (1985); [b]Henkin R. I. et al. (1963); [c]Glendinning J. I. (1994a; 1994b); and [d]Stark T. et al. (2005).

extraordinary diversity of bitterant SAR, with structures, molecular weights, empirical formulas, and BRT information provided in Figure 18, is as follows:

1. *Cyclohexylamine (104)*. Aliphatic amines in general are bitter-tasting compounds with bitterness potency increasing with the size of the aliphatic substituent.

Interestingly, on sulfamation (i.e., $R-NH_2 \rightarrow R-NHSO_3^- \, Na^+$) the bitter-tasting cyclohexylamine is converted into a sweet-tasting sulfamate salt commonly known as sodium cyclamate.

2. *Ethylpyrazine (105)*. Pyrazines are important flavor compounds exhibiting nutty and coffee-like aromatic characters. As a class of compounds,

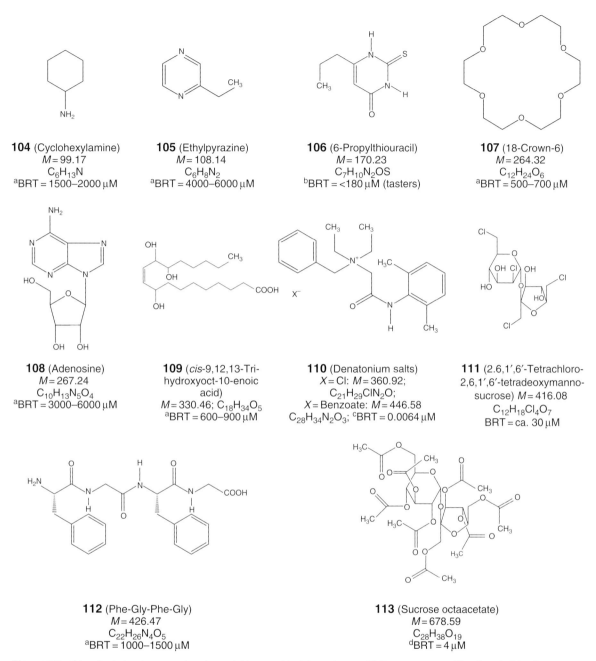

Figure 18 Chemical structures, molecular weights, empirical formulas, and bitterness recognition thresholds of some synthetic bitter-tasting chemical compounds. [a]Belitz H.-D. and Wieser H. (1985); [b]Chang W. I. et al. (2006); [c]Saroli A. (1986); and [d]Boughter J. D. and Whitney G. (1993).

however, they are generally bitter in taste with bitterness potency increasing with hydrophobicity.

3. *6-n-Propylthiouracil (PROP) (106)*. Thioureas are quite generally bitter in taste. PROP is a compound of substantial interest in that people exhibit a bimodal distribution in sensitivity to its bitterness with some individuals highly sensitive to its bitterness and some only sensitive to it at much higher concentrations. This difference in sensitivity has been demonstrated to have a genetic origin and is believed to be due to a mutation in T2R4 (Chandrashekar, J. et al., 2000).

4. *18-Crown-6 (107)*. Crown ethers are uniformly bitter in taste with bitterness potency increasing with hydrophobicity.

5. *Adenosine (108)*. Purine and pyrimidine nucleosides as well as their bases are generally bitter, where the deoxyribosides are more potently bitter than the ribosides and nucleotides are generally not bitter. As in other series of compounds, bitterness potency increases with hydrophobicity.

6. *cis-9,12,13-Tri-hydroxyoct-10-enoic acid (109)*. Hydroxylated fatty acids are common products of lipolysis and lipoxidation and are uniformly bitter. Such fat degradation products are responsible for the bitter off taste of many cereal and soy-based products.

7. *Denatonium salts (110)*. Quaternary ammonium salts in general are bitter in taste, although the most potent bitterants reported are denatonium salts as reported by Saroli A. (1986). Denatonium salts are commonly employed bitter stimulants employed in mechanistic studies in animal models.

8. *2,6,1′,6′-Tetrachloro-2,6,1′,6′-tetradeoxy-mannosucrose (111)*. Many halogenated deoxy sucrose derivatives are potently sweet compounds. Of particular note are 4,1′,6′-trichlorotrideoxy-galactosucrose (sucralose) and 4,6,1′,6′-tetrachloro-galactosucrose which are potent sweeteners. Interestingly, a few halogenated deoxy sucrose derivatives are potently bitter, with some such as **111** matching quinine in bitterness potency.

9. *(L)-Phenylalanyl-glycyl-(L)-phenylalanyl-glycine (112)*. Peptides, quite generally, are bitter as are hydrophobic amino acids. The exceptions are a few amino acids (e.g., glycine, D- and L-alanine, D-phenylalanine, and D-tryptophan) and a few peptide derivatives (e.g., aspartame) which are sweet in taste. And, as is the general trend, bitterness potency in such series of compounds increases with hydrophobicity. Many foods acquire bitter tastes due to the formation of bitter peptides on enzymatic breakdown of proteins. Tetrapeptide **112** is just one of the many bitter peptides which have been reported.

10. *Sucrose octaacetate (113)*. Acetylated sugars in general exhibit bitter taste, and sucrose octaacetate has been used extensively in the study of bitter taste in humans and animal model systems.

4.02.4 Umami-Tasting Stimuli

Until the close of the twentieth century, for many, taste was considered to be one of the five senses which may be subdivided into four basic tastes: sweet, sour, salty, and bitter. Even though this is the case, evidence began to accumulate for a fifth basic taste, savory taste. The first claim for savory taste was by Ikeda K. (1909) who isolated glutamic acid from a seaweed source and argued that its taste was unique. As the monosodium salt, the taste is savory and was given the Japanese word *umami* which means delicious. The literature on umami taste has been reviewed quite comprehensively by Kawamura Y. and Kare M. R. (1987) and more recently by Bellisle F. (1999).

4.02.4.1 Biochemistry of Umami Taste

At the beginning of the twenty-first century, conclusive evidence that umami taste is a fifth basic taste began to accumulate. First, Nirupa Chaudhari and associates (2000) at the University of Miami obtained evidence suggesting that the umami taste receptor is a truncated form of the metabotrophic glutamate receptor mGluR4. Shortly thereafter, a collaborative team from the laboratories of Charles Zuker and Nicholas Ryba reported discovery of a mouse amino acid receptor (Nelson, G. *et al.*, 2002). In a functional assay in HEK-293 cells, they showed that glutamic acid activity, as well as the activities of many other L-amino acids, is mediated by a single receptor, apparently a heterodimer of two 7-TMD proteins which they named T1R1 and T1R3, often written as T1R1/T1R3. Both T1R1 and T1R3 are members of the small family of class C GPCRs and T1R3 is the same 7-TMD protein present in the sweetener receptor T1R2/T1R3 already discussed. The rat amino acid receptor discovery was quickly followed by a report by Li X. *et al.* (2002) of Senomyx (La Jolla, CA) on parallel findings for the human system. HEK-293 cells, in which both human T1R1 and T1R3 were expressed, responded to L-glutamic acid in a manner consistent with expectation from sensory experiments. L-Aspartic acid was also active but all other amino acids tested were essentially inactive. Thus, while the rodent receptor is a general L-amino acid receptor, the human receptor is specific for glutamate and aspartate. In addition, inosine monophosphate (IMP) and other nucleotides, which appear synergistic with monosodium glutamate in taste tests, exhibited strong enhancements *in vitro*. On further analysis, it was learned that, in the absence of glutamate, these nucleotides are inactive. And therefore, they are not synergistic with glutamate. They are really enhancers of glutamate taste activity and likely function at the receptor as positive allosteric modulators. Positive and negative allosteric modulators of agonist activity are now known for many family C GPCRs as has been reviewed (Goudet, C. *et al.*, 2004).

Family C GPCRs are generally thought to bind their agonists within their large extracellular VFDs. Since the umami receptor is a heterodimeric receptor and since the umami and sweetener receptors contain the common subunit T1R3, it is thought, though not yet proven, that the glutamate binding site is most likely in the VFD of T1R1. Many other umami taste agonists have been identified as are reviewed below. Do they all bind in the glutamate binding site? In view of the finding that the closely related sweetener receptor T1R2/T1R3 binds agonists at no less than three sites, it seems likely that umami taste agonists also initiate their activities at a plurality of binding sites on T1R1/T1R3.

Sweet, bitter, and umami tastes are transduced by a common biochemical pathway. Experiments with gustducin and transducin knockout mice demonstrate that both of these G proteins mediate umami taste (He, W. *et al.*, 2004). Other work with PLC$_{\beta2}$ and TRPM5 knockout mice demonstrates the critical role for both of these effectors in umami taste (Zhang, Y. *et al.*, 2003). In summary, while the role of the truncated mGluR4 receptor is not yet clear, there is a general consensus that umami taste is primarily mediated by a receptor T1R1/T1R3 as well as several intracellular effectors (i.e., gustducin, PLC$_{\beta2}$, IP$_3$R, and TRPM5). It is also known that umami taste quality is coded at the cellular level, just as is the case for sweet and bitter taste qualities. Thus, within the taste bud, some TBCs are specific sensors for umami taste stimuli. They differ from the sweet- and bitter-sensitive TBCs in that they uniquely express the umami taste receptor T1R1/T1R3, while not expressing the sweetener receptor T1R2/T1R3 or bitterant T2R receptors.

4.02.4.2 Umami Tastant Structure–Activity Relationship

An objective of this review is to summarize the SAR of umami taste stimuli. Efforts aimed at discovery of novel umami tastants have not received the same intensity of focus as efforts to find novel sweeteners or to determine the structures of natural sweeteners or bitterants. Nonetheless, some umami tastant discovery work has been carried out in recent years and eight umami tastants **114–121**, with structures, molecular weights, empirical formulas, and umami taste potency information provided, are illustrated in Figure 19. Some di-, tri-, and tetra-peptides including Glu-Asp, Glu-Glu, Asp-Glu, Thr-Glu, Asp-Glu-Ser, Glu-Gly-Ser, Asp-Asp-Asp-Asp, and even the octapeptide Lys-Gly-Asp-Glu-Glu-Ser-Leu-Ala have been reported in the literature to exhibit umami taste. However, recent studies, as reviewed by Beksan E. *et al.* (2003), do not support the presence of umami taste in any of these compounds.

The purine nucleotides IMP (**122**), guanosine monophosphate (GMP (**123**)), and adenosine monophosphate (AMP (**124**)), as well as many synthetic analogs, have been reported to exhibit umami taste, as has been reviewed by Yamaguchi S. (1979). The chemical structures, empirical formulas, molecular weights, and relative umami taste activities are provided in Figure 20. In addition to exhibiting umami tastes themselves, these compounds are claimed to be strongly synergistic with monosodium glutamate. However, in the *in vitro* studies on the rat umami receptor carried out by Nelson G. *et al.* (2002) and on the human umami receptor carried out by Li X. *et al.* (2002), it was demonstrated that these purine nucleotides are inactive at the umami receptor if evaluated in the absence of glutamate or other amino acids with umami taste. Normally, in the *in vitro* receptor assays carried out in heterologous cell systems, the extracellular fluid contains low levels of amino acids. And, under these conditions, purine nucleotides were observed to exhibit activity. However, to ensure that the activity observed was due to the purine nucleotides rather than an enhancement effect of endogenous amino acids, the cell preparations were washed with amino acid free saline immediately before assay. Under these conditions, purine nucleotides give no responses while glutamate still exhibits its typical activity. Control cell preparations lacking the umami taste receptor are also unresponsive to purine nucleotides even without the cell washing step, thus providing evidence that the enhancement effect of purine nucleotides is receptor-mediated. In summary, it appears that purine nucleotides are not umami receptor agonists but are enhancers, presumably positive allosteric modulators, of glutamate and other amino acids. An observation consistent with

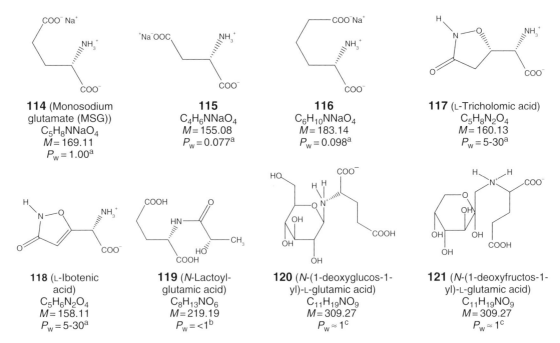

Figure 19 Chemical structures, molecular weights, empirical formulas, and umami taste potencies of some chemical compounds with umami taste. [a]Yamaguchi S. (1979); [b]Frérot E. and Escher S. D. (1997); and [c]Beksan E. *et al.* (2003).

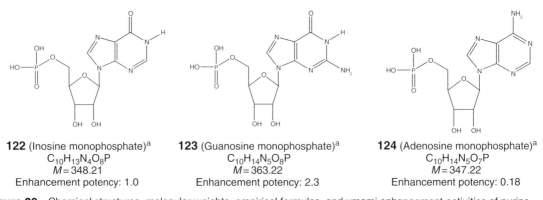

Figure 20 Chemical structures, molecular weights, empirical formulas, and umami enhancement activities of purine nucleotides. [a]Yamaguchi S. (1979).

this conclusion is that, while 0.1 mM IMP exhibits a weak umami taste, no increase in umami taste intensity is observed for 1.0 and 10.0 mM solutions (DuBois, G. E. and San Miguel, R. S., 2005). It appears that 0.1 mM IMP is sufficient to maximally enhance the umami taste of the glutamate endogenous to saliva and thus increased IMP levels are without effect. Since purine nucleotides are not umami taste agonists, the umami taste activities provided for the examples in Figure 20 are really metrics for their relative potencies as enhancers of the umami taste of umami agonists (i.e., glutamate and aspartate) in human saliva. It is noteworthy that

IMP, GMP, and AMP exhibit quite significant differences in activity as would be expected for a receptor-mediated effect.

4.02.5   Sour-Tasting Stimuli

Sour taste has always been accepted as one of the primary tastes and is very important to the enjoyment of many foods and beverages, especially those with sweet tastes. Here we provide a review of the current understanding of the cellular pathways which mediate sour taste followed by a review of knowledge on the taste of acids that are commonly used in foods and beverages.

4.02.5.1 Biochemistry of Sour Taste

There is no question that sour taste is a unique and primary taste, but yet, in spite of its importance to our daily enjoyment of foods and beverages, until quite recently, the biochemical pathways which mediate it were largely unknown. Then, evidence began to accumulate that it must be an ion channel-mediated activation pathway in specific subsets of TBCs. Sour taste is the response to acids. All acidic chemical compounds are sour in taste. Here we discuss pH homeostasis in the body and the role that sour taste sensation plays in it, the biochemical mechanisms for excitation of acid-sensitive TBCs, and the rationale for difference in sour taste for different acids.

pH homeostasis and sour taste. A unique characteristic of living organisms is that they maintain their internal milieu at constant pH. Even small deviations from the physiological pH can be life threatening for most organisms. Therefore, organisms have developed efficient cellular mechanisms to detect the presence of H^+ or acid equivalents in their immediate environment and to limit acid entry into their internal milieu. However, if some acid is absorbed into the internal milieu, it is immediately buffered and ultimately excreted. For the detection of acid in their environment most animal species rely on sour taste, and to a lesser extent on the sense of smell. Sour taste is elicited by acidic stimuli and is detected by a subset of specialized TBCs in the oral cavity referred to as acid-sensing TBCs. Many of the foods, juices, and beverages that humans consume are acidic in nature (Feldman, M. and Barnett, C.,1995). In humans, sour taste evokes an innate rejection response to extremely acidic or sour stimuli, limiting the *ad libitum* ingestion of acid (Beauchamp, G. K. *et al.*, 1991; Scott, T. R. and Plata-Salaman, C. R., 1991). In this regard, along with the lungs and kidneys, which are the primary organs for acid secretion, acid-sensing TBCs in the oral cavity, by restricting the intake of acid, function as part of a multiorgan system to maintain acid–base homeostasis.

Acid-sensing. Since most acids decrease extracellular pH (pH_o), it would seem reasonable to think that a decrease in the pH_o in the oral cavity should be the signal for sour taste transduction. However, if that

were the case, one would expect all acids at the same pH_o to taste equally sour. In our everyday experience and in animal model systems, different acids at same pH_o do not elicit the same neural responses or give the same perception of sourness (Beidler, L. M., 1967; Beidler, L. M. and Gross, G. W., 1971; Ogiso, K. *et al.*, 2000). In animal models it is well recognized that acetic acid at a fixed pH is always a more potent sour stimulus than an equivalent pH strong acid, such as HCl (Lyall, V. *et al.*, 2001). This indicates that a decrease in pH_o is not the signal for sour taste transduction. However, changes in intracellular pH (pH_i) of TBCs have been shown to correlate extremely well with the neural responses, indicating that it is the decrease in pH_i that is the proximate signal for sour taste transduction (Lyall, V. *et al.*, 2001; 2002a; 2002b; 2004; 2006).

Mechanisms for acid entry to TBCs. Acid transduction begins with the entry of H^+ or acid equivalents across the apical membrane of TBCs that causes a decrease in intracellular pH (pH_i) of acid-sensing TBCs. Acids exist in two forms: fully dissociated strong acids (e.g., HCl) and weak organic acids (e.g., acetic acid or citric acid) or dissolved CO_2, the precursor of carbonic acid. Accordingly, there are acid entry mechanisms for both strong and weak acids in the apical membrane of TBCs. For strong acids, acid entry involves H^+ flux through an apical proton channel that is blocked by divalent metal ions Zn^{2+} or Cd^{2+} (Lyall, V. *et al.*, unpublished observations). This channel is amiloride- and Ca^{2+}-insensitive but is activated by an increase in intracellular adenosine $3',5'$-cyclic monophosphate (cAMP) (Lyall, V. *et al.*, 2002a; 2004). However, there is some evidence to suggest that H^+-gated channels, such as the acid-sensing ion channel (ASIC) in the apical membrane and both ASIC (Ugawa S., *et al.*, 1998; Lin, W. *et al.*, 2002) and hyperpolarization-activated channels (HCN) (Stevens, D. R. *et al.*, 2001) in the basolateral membrane of TRCs, and TASK-2, a two pore domain K^+ channel (Lin, W. *et al.*, 2004; Richter, T. A. *et al.*, 2004a), may also play a role in sour taste transduction. These channels in the basolateral membrane could be activated if H^+ can cross tight junctions and decrease pH in the basolateral compartment. In mice, ASIC2 is not required for acid taste (Richter, T. A. *et al.*, 2004b) suggesting that at least in mice, ASIC channels do not play a major role in acid-sensing.

In contrast, weak organic acids do not seem to have apical membrane receptors. They enter TBCs across the apical membrane by passive diffusion as lipid-soluble undissociated neutral molecules. Once inside the cell they dissociate to generate H^+ intracellularly and decrease TBC pH_i (Lyall, V. *et al.*, 2001). For example, in the case of dissolved CO_2, the CO_2 molecules passively enter TBCs across the apical membrane and combine with H_2O in a reaction catalyzed by intracellular carbonic anhydrases to form carbonic acid (H_2CO_3), which then dissociates to yield $H^+ + HCO_3^-$.

Taste receptor cells also contain NPPB (**125**) sensitive stretch-activated Cl^- channels (Gilbertson, T. A., 2002). The presence of an NPPB-sensitive Cl^- channel activated by acid was demonstrated in mouse taste cells (Miyamoto, T., *et al.*, 1998), suggesting a role for Cl^- channels in acid taste transduction. In addition to Cl^- channels, there is also evidence that an NPPB-insensitive poorly selective cationic conductance in the apical receptive membranes of mouse TBCs may also be involved in sour taste transduction (Miyamoto, T., *et al.*, 1998). In summary, sour taste transduction is mediated via multiple pathways. Similarly, in the case of other acid-sensitive cells, such as the central chemosensitive neurons, a multiple factors model has been proposed for acid signaling (Putnam, R. W. *et al.*, 2004).

Neural responses to acidic stimuli. A decrease in TBC pH_i is responsible for the neural response profiles to acidic stimuli. Acidic stimuli elicit both CT and glossopharyngeal taste nerve responses that are composed of two components with distinct temporal characteristics. A rapid transient increase in the phasic neural response slowly declines to a quasi-steady state, defined as the tonic phase of the neural response. Treating rat tongue with specific membrane permeable blockers of carbonic anhydrases MK-417 (**126**) or MK-507 (**127**) inhibits both the phasic and tonic components of the CT response to CO_2 (Lyall, V. *et al.*, 2001; 2002b). This indicates that an acid-induced decrease in TBC pH_i is necessary to elicit both the phasic and the tonic components of the CT response to acid stimulation.

Following a decrease in pH_i, the proximate sour stimulus, the downstream transduction mechanisms for the phasic and tonic components of the CT response to acid stimulation follow distinct and separate pathways. An acid-induced decrease in pH_i of acid-sensing TBCs alters the cell cytoskeleton by shifting the F- to G-actin equilibrium toward G-actin resulting in cell shrinkage and by titrating the mean charge of the intracellular membrane-impermeant anions (which could include elements of the cytoskeleton) (Fraser, J. A. *et al.*, 2005). The pH_i-induced decrease in cell volume activates a flufenamic acid (**128**) sensitive shrinkage-activated

nonselective cation channel (SANSCC) in the basolateral membrane of TRCs. The entry of one or more monovalent cations through SANSCC depolarizes the receptor potential and is the basis for the phasic component of the CT response to acidic stimuli. Loading TBCs *in vivo* with the Ca^{2+}-chelator, BAPTA (**129**), did not affect the phasic part of the CT response to acids (Figure 21), indicating that phasic response is indifferent to changes in cytosolic Ca^{2+} (Lyall, V. *et al.*, 2006).

In acid-sensing TBCs a decrease in pH_i is followed by an increase in intracellular Ca^{2+} ($[Ca^{2+}]_i$) (Liu, L. and Simon, S. A., 2001a; 2001b; Richter, T. A. *et al.*, 2003). An increase in TBC $[Ca^{2+}]_i$ activates basolateral Na^+-H^+-exchanger-1 (NHE-1) and results in cell volume recovery and the neural adaptation to both strong and weak acid stimuli (Lyall, V. *et al.*, 2002a; 2004; Vinnikova, A. K. *et al.*, 2004; Lyall, V. *et al.*, 2006). Neural adaptation is related to the tonic part of the CT response to acidic stimuli (Figure 22). Therefore, a primary decrease in TBC pH_i followed by a secondary increase in $[Ca^{2+}]_i$, and the subsequent activation of basolateral NHE-1 are specific events related to the tonic part of the neural response only (Figure 23).

One unexpected outcome from work on the identity of the molecular taste receptors for sweet, bitter, and umami taste is that the molecular receptors for each of these modalities are segregated in separate subsets of receptor cells (Zhang, Y. *et al.*, 2003). Recent work on sour taste shows that this modality is also expressed uniquely in a separate

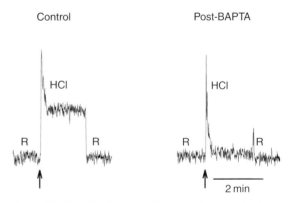

Figure 21 Loading fungiform TRCs with BAPTA *in vivo* eliminates tonic responses to HCl. Adapted from Lyall, V., Pasley, H., Phan, T-H. T., Mummalaneni, S., Heck, G. L., Vinnikova, A. K., and DeSimone, J. A. 2006. Intracellular pH modulates taste receptor volume and the phasic part of the chorda tympani response to acids. J. Gen. Physiol. 127, 15–34 with permission.

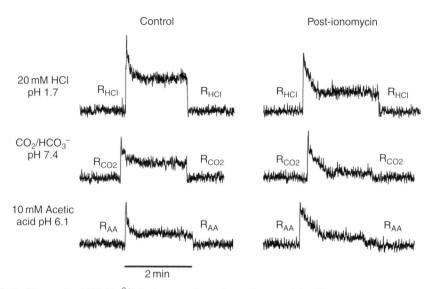

Figure 22 Effect of increasing TRC $[Ca^{2+}]_i$ by ionomycin on the tonic part of the CT response to acidic stimuli. CT responses were recorded before and after topical lingual application of 150 μM ionomycin (**130**). An increase in $[Ca^{2+}]_i$ increases the rate of adaptation to acidic stimuli. Adapted from Lyall, V., Heck, G. L., Vinnikova, A. K., Ghosh, S., Phan, T. H., Alam, R. I., Russell, O. F., Malik, S. A., Bigbee, J. W., and DeSimone, J. A. 2004. The mammalian amiloride-insensitive non-specific salt taste receptor is a vanilloid receptor-1 variant. J. Physiol. 558, 147–159 with permission.

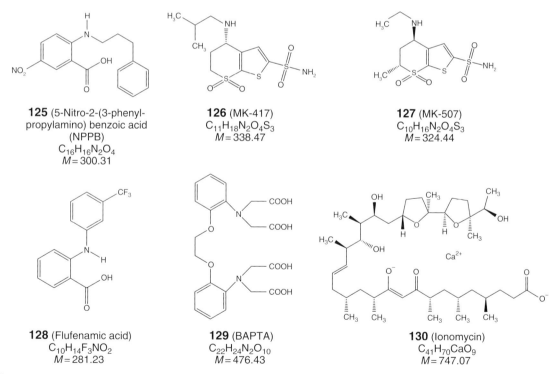

125 (5-Nitro-2-(3-phenyl-propylamino) benzoic acid (NPPB)
$C_{16}H_{16}N_2O_4$
$M = 300.31$

126 (MK-417)
$C_{11}H_{18}N_2O_4S_3$
$M = 338.47$

127 (MK-507)
$C_{10}H_{16}N_2O_4S_3$
$M = 324.44$

128 (Flufenamic acid)
$C_{10}H_{14}F_3NO_2$
$M = 281.23$

129 (BAPTA)
$C_{22}H_{24}N_2O_{10}$
$M = 476.43$

130 (Ionomycin)
$C_{41}H_{70}CaO_9$
$M = 747.07$

Figure 23 Molecular tools employed in the study of sour taste.

subset of taste receptor cells in the fungiform, circumvallate, and foliate papillae, and taste buds of the palate (Huang, A. L. *et al.*, 2006; Ishimaru, Y. *et al.*, 2006; Lopez-Jimenez, N. D. *et al.*, 2006). In the

taste buds of the circumvallate and foliate papillae, this subset of cells expresses two unique TRP proteins, designated PKD2L1 and PKD1L3, in each cell of the subset. In fungiform and palatal taste buds, only PKD2L1 is expressed in each of the cells of the subset. PKD2L1 and PKD1L3 localize preferentially in the apical regions of the taste receptor cells. When PKD2L1 and PKD1L3 are coexpressed in HEK cells they respond to the presence of citric acid and HCl by causing an increase in intracellular calcium concentration (Ishimaru, Y. *et al.*, 2006). It is unclear, however, if these proteins behave as calcium-conducting channels *in vivo*. That PKD2L1 containing cells are involved in sour taste is elegantly demonstrated in genetic ablation studies in which mice are engineered such that cells that would normally express PKD2L1 are destroyed by attenuated diphtheria toxin (Huang, A. L. *et al.*, 2006). In such mice, CT responses to acids were completely absent while responses to the other taste modalities remained intact. The fact that taste receptor cells lacking PKD2L1 fail to respond to weak and strong acids alike suggests that these proteins may be part of the intracellular pH-sensing system in sour taste receptor cells and in some central nervous system neurons where PKD2L1 is also expressed.

Neurotransmitter: TBCs detect the presence of acid in the oral cavity and elicit taste nerve responses. This implies that acid-induced changes in TBC pH_i, Ca^{2+}, cell shrinkage, and/or depolarization are communicated to the taste nerve fibers innervating TBCs via one or more neurotransmitters. Recent studies suggest the possible role for serotonin (**131**) (Kaya, N. *et al.*, 2004; Huang, Y. J. *et al.*, 2005) and ATP (**132**) (Finger, T. E., *et al.*, 2005). Serotonin, also known as 5-hydroxytryptamine (5-HT), is suggested to act on the 5-HT_3 receptor on the taste nerve fibers. However, the 5-HT_{3A} knockout mice did not show any differences

in their neural response to taste stimuli, including acids (Finger, T. E., *et al.*, 2005), indicating that serotonin may not be the taste neurotransmitter. In contrast to 5-HT_{3A} knockout mice, in a double knockout mouse model lacking ionotropic purinergic P2X receptor units $P2X_2$ and $P2X_3$, no neural taste responses in CT or glossopharyngeal nerve were observed for salty, bitter, sweet, umami, and acidic stimuli. This suggests that ATP may be the taste-specific neurotransmitter that transmits information from TBCs to taste nerves (Figure 24).

4.02.5.2 Acidulant Structure–Activity Relationship

As already noted, all acids exhibit sour taste. The proximate event for signaling by acid-sensing TBCs is a drop in intracellular pH. Strong mineral acids (e.g., HCl, H_2SO_4, H_3PO_4) reduce intracellular pH by the diffusion of hydronium ions through an apical proton channel. Organic acids (e.g., citric acid, malic acid, tartaric acid), however, while modulating intracellular pH by the hydronium ion/apical proton channel pathway to some degree, principally reduce intracellular pH following diffusion through the TBC membrane. This mechanistic rationale suggests that mineral acids at constant pH should exhibit equal sourness while the sournesses of organic acids should not be strongly correlated with pH but should be dependent on efficiencies of diffusion into the TBC. This expectation is consistent with observations. Exemplary listings of common mineral acids and organic acids, as well as their empirical formulas, molecular weights, and pK_as, are provided in Figures 25 and 26.

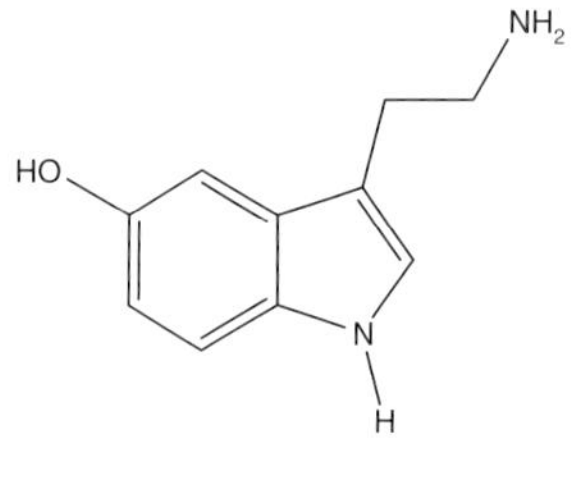

131 (Serotonin)
$C_{10}H_{12}N_2O$
$M = 176.22$

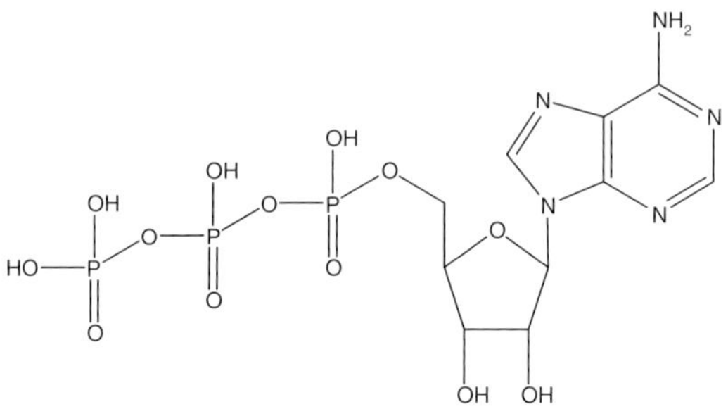

132 (Adenosine triphosphate (ATP))
$C_{10}H_{16}N_5O_{13}P_3$
$M = 507.18$

Figure 24 Possible taste neurotransmitters.

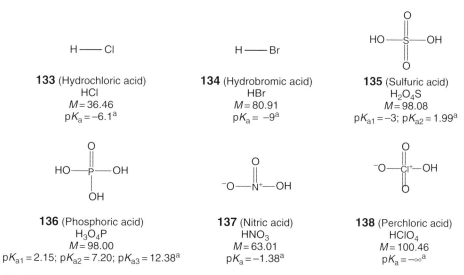

133 (Hydrochloric acid)
HCl
$M = 36.46$
$pK_a = -6.1$[a]

134 (Hydrobromic acid)
HBr
$M = 80.91$
$pK_a = -9$[a]

135 (Sulfuric acid)
H_2O_4S
$M = 98.08$
$pK_{a1} = -3$; $pK_{a2} = 1.99$[a]

136 (Phosphoric acid)
H_3O_4P
$M = 98.00$
$pK_{a1} = 2.15$; $pK_{a2} = 7.20$; $pK_{a3} = 12.38$[a]

137 (Nitric acid)
HNO_3
$M = 63.01$
$pK_a = -1.38$[a]

138 (Perchloric acid)
$HClO_4$
$M = 100.46$
$pK_a = -\infty$[a]

Figure 25 Chemical structures, empirical formulas, molecular weights, and pK_a constants for some common mineral acids.
[a]Dean J. A. (1985).

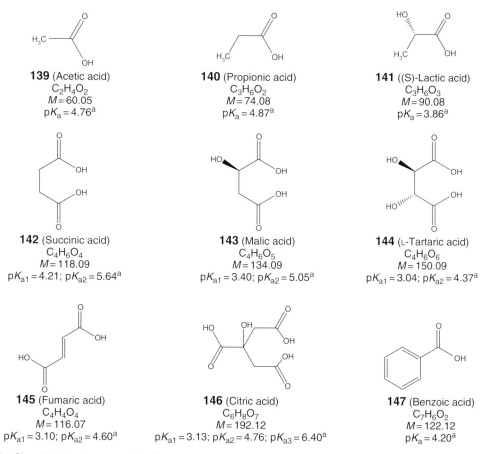

139 (Acetic acid)
$C_2H_4O_2$
$M = 60.05$
$pK_a = 4.76$[a]

140 (Propionic acid)
$C_3H_6O_2$
$M = 74.08$
$pK_a = 4.87$[a]

141 ((S)-Lactic acid)
$C_3H_6O_3$
$M = 90.08$
$pK_a = 3.86$[a]

142 (Succinic acid)
$C_4H_6O_4$
$M = 118.09$
$pK_{a1} = 4.21$; $pK_{a2} = 5.64$[a]

143 (Malic acid)
$C_4H_6O_5$
$M = 134.09$
$pK_{a1} = 3.40$; $pK_{a2} = 5.05$[a]

144 (L-Tartaric acid)
$C_4H_6O_6$
$M = 150.09$
$pK_{a1} = 3.04$; $pK_{a2} = 4.37$[a]

145 (Fumaric acid)
$C_4H_4O_4$
$M = 116.07$
$pK_{a1} = 3.10$; $pK_{a2} = 4.60$[a]

146 (Citric acid)
$C_6H_8O_7$
$M = 192.12$
$pK_{a1} = 3.13$; $pK_{a2} = 4.76$; $pK_{a3} = 6.40$[a]

147 (Benzoic acid)
$C_7H_6O_2$
$M = 122.12$
$pK_a = 4.20$[a]

Figure 26 Chemical structures, empirical formulas, molecular weights, and pK_a constants for some common organic acids.
[a]Dean J. A. (1985).

4.02.6   Salty-Tasting Stimuli

Salty taste has always been accepted as one of the primary tastes and is very important to the enjoyment of many foods, especially foods with umami tastes. In fact, until the recent identification of the umami receptor, many held the belief that umami taste was a superimposition of salty and sweet tastes. In umami taste, the presence of a cation (e.g., Na^+) is critical as without it, even the prototypical umami agonist monosodium glutamate does not exhibit umami taste. Here we provide a review of the current understanding of the biochemistry of salty taste followed by a review of observations on the taste of NaCl and a few other common salts.

4.02.6.1   Biochemistry of Salty Taste

Salty taste is a unique and primary taste and in spite of its importance to our daily enjoyment of foods, until quite recently, the biochemical pathways which mediate it were largely unknown. Then, evidence began to accumulate that it must be mediated by ion channels present in specific subsets of TBCs. Salty taste is the response to NaCl, the prototypical salty taste stimulus, as well as a number of other salts. Here we discuss the biological mechanisms which mediate salt homeostasis, the ion channels which mediate the excitation of salt-sensitive TBCs and the biochemical explanation for the differences in tastes between salts of differing composition.

Sodium balance and sodium appetite. Tight regulation of the ionic composition and osmotic pressure of the extracellular fluid compartment of terrestrial mammals is fundamental to their survival. This is accomplished, in part, through the agency of the regulatory hormone, aldosterone, which increases Na^+ transport across epithelial cells from the lumina of the renal collecting duct and colon (Rossier, B. C. *et al.*, 2002). Aldosterone action, therefore, reduces Na^+ excretion in urine and feces. On the input side, changes in Na^+ balance depend on the subject's eating behavior which in turn is influenced by the sense of taste and the subject's endocrine status as determined by current Na^+ balance (Denton, D. A., 1967). For example, in Na^+-replete rats, NaCl concentrations near isotonicity tend to be preferred over water while those in the hypertonic range are increasingly less palatable (Richter, C. P., 1936;

Fregly, M. J. and Rowland, N. E., 1986). Na^+-deprived and adrenalectomized rats, on the other hand, prefer even hypertonic NaCl solutions over water (Richter, C. P., 1936; Fregly, M. J. and Rowland, N. E., 1986; Scott, T. R. and Plata-Salaman, C. R., 1991). Increased salt preference has also been reported for human subjects taking diuretics on a Na^+-restricted diet (Beauchamp, G. K. *et al.*, 1990). Na^+-deficient wild rabbits taste and then readily consume Na^+ from salt licks while tasting and then rejecting nearby licks of nonsodium salts (Denten, D. A., 1967). These and other data indicate a connection between Na^+-specific taste and an animal's current interstitial fluid Na^+ levels and mineralocorticoid status.

Among the Na^+ transporters upregulated by aldosterone is the epithelial sodium channel (ENaC). ENaC is present in the membranes of various types of epithelial cells, including taste receptor cells of rats (Kretz, O. *et al.*, 1999; Lin, W. *et al.*, 1999). As such, ENaC would appear to be an excellent candidate for the Na^+-specific salt taste receptor. For the rat and other rodents, the preponderance of evidence indicates that this is the case.

Sodium taste and ENaC pharmacology. Na^+ transport across epithelia, including lingual epithelia, involves the passive flux of Na^+ from the luminal compartment (oral cavity for dorsal lingual epithelia) into the epithelial cells (for taste, receptor cells in taste buds and some surrounding nontaste epithelia) through apical membrane ENaC. Na^+ is then pumped across the basolateral membranes of the epithelial cells by the Na^+-K^+-ATPase (Verrey, F. *et al.*, 2000). Transepithelial Na^+ transport can be inhibited at the cell apical membrane by adding amiloride (**148**) or one of its analogs to the luminal solution, thereby reducing the Na^+ permeability of ENaC (Verrey, F. *et al.*, 2000). For the species in which transepithelial ion transport has been studied across the dorsal lingual epithelium, viz., dog (DeSimone, J. A. *et al.* 1984; Mierson, S. *et al.*, 1985; Simon, S. A. and Garvin, J. L., 1985), rat (Heck, G. L. *et al.*, 1984; Garvin, J. L. *et al.*, 1988; Mierson, S. *et al.*, 1988; Gilbertson, T. A. and Zhang, H., 1998), rabbit (Simon, S. A. *et al.*, 1986), and hamster (Gilbertson, T. A. and Zhang, H., 1998), the Na^+ transport paradigm appears to apply. The most compelling evidence that ENaC or some ENaC variants are indeed Na^+-specific salt taste receptor proteins is that the taste nerve response to NaCl is significantly inhibited by amiloride and its analogs in a variety of species. Taste responses to NaCl recorded in the afferent CT or in the nucleus of the solitary tract of rat (Schiffman, S. S. *et al.*, 1983; Heck, G. L. *et al.*, 1984; Brand, J. G. *et al.*, 1985; Hill, D. L. and Bour, T. C., 1985; Yoshii, K. *et al.*, 1986; Ninomiya, Y. and Funakoshi, M., 1988; Scott, T. R. and Giza, B. K., 1990; St. John, S. J. and Smith, D. V., 2000), hamster (Hettinger, T. P. and Frank, M. E., 1990), some mouse strains (Ninomiya, Y. *et al.*, 1989), and gerbil (Schiffman, S. S. *et al.*, 1990) are significantly inhibited by amiloride without effect on responses to stimuli of other taste modalities. Amiloride sensitivity is observed in single CT units of the chimpanzee that respond strongly to Na^+ and Li^+ salts, but not in units sensitive to both Na^+ and K^+ (Hellekant, G. *et al.*, 1997). Whole cell patch clamp studies on isolated rat and hamster taste buds show that amiloride blocks a Na^+ current across taste cell membranes, consistent with a role for ENaC in Na^+ taste reception (Avenet, P. and Lindemann, B., 1988; 1991; Gilbertson, T. A. *et al.*, 1993). Taken together the data are consistent with the conclusion that ENaC is the Na^+-specific salt taste receptor in many species including many rodents and nonhuman primates. In rats the amiloride-sensitive part of the response accounts for 75–80% of the total response to 100 mM NaCl(see Figure 23). Surprisingly, ENaC appears to play less of a role in human salt taste (Smith, D. V. and Ossebaard, C. A., 1995).

In rats, ENaC appears to be functional only in the taste buds innervated by the CT, that is, principally those in fungiform papillae. Behavioral studies convincingly show that the rat's ability to discriminate between NaCl and KCl depends on taste information from the CT and this ability is progressively lost with the NaCl solutions containing increasing amiloride concentrations (Spector, A. C. *et al.*, 1996). This behavioral change correlates perfectly with the amiloride concentration dependence of the diminution of the response to NaCl in Na^+-sensitive single units of the nucleus of the solitary tract. The cumulative molecular, immunocytochemical, electrophysiological, and behavioral evidence indicates, therefore, that ENaC is a Na^+-specific salt taste receptor, and most likely the only Na^+-specific receptor.

Amiloride-insensitive salt taste reception. As seen in Figure 27, after inhibition of the NaCl response by benzamil (**149**), a significant neural response remains. Analysis of single CT units in rats (Frank, M. E. *et al.*, 1983), hamsters (Frank, M. E. *et al.*, 1988), and chimpanzees (Hellekant, G. *et al.*, 1997) suggests the existence of a salt taste receptor type that is

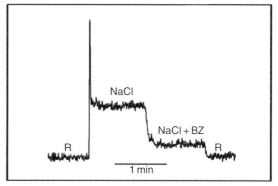

Figure 27 Integrated rat chorda tympani response to 100 mM NaCl followed by a response to 100 mM NaCl + 5 µM benzamil (**149;** Bz). R denotes the rinse solution (10 mM KCl). The NaCl response above baseline response (R) minus the NaCl + Bz response is the amiloride-sensitive part of the NaCl response (about 75% of the NaCl response).

cation-nonselective in addition to one that is Na^+-specific. Accordingly, one obvious candidate for such a taste receptor would be a nonselective cation channel. Unfortunately this fact alone does little to limit the number of possibilities, given the plethora of nonselective cation channels that have been described. One possibly fruitful approach is to seek pharmacological probes that might restrict the number of possible candidate receptor types, that is, an approach analogous to that leading to the discovery of ENaC as the Na^+-specific taste receptor through the action of its pharmacological inhibitor, amiloride. In this manner it has been shown that the amiloride-insensitive part of the NaCl CT response, as well as the responses to KCl, NH_4Cl, and $CaCl_2$, are enhanced by the vanilloids, resiniferatoxin (**150;** RTX), and capsaicin (**151**) (Lyall, V. *et al.*, x). At higher vanilloid concentrations the salt-evoked CT responses are inhibited. The amiloride-insensitive part of the NaCl CT response is completely inhibited by the vanilloid receptor-1 (VR1 or TRPV1) inhibitor, SB366971 (**152**) (Lyall, V. *et al.*, 2004). These results have led to the hypothesis that TRPV1, a nonselective cation channel, or more likely, one of its splice variants, is the nonselective cation salt taste receptor. Additional data in support of this are as follows (cf. Lyall, V. *et al.*, 2004):

1. The amiloride-insensitive part of the NaCl CT response increases with temperature and the effects of temperature and vanilloids are additive, similar to TRPV1.

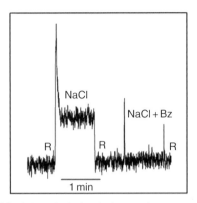

Figure 28 Integrated chorda tympani response to 100 mM NaCl and 100 mM NaCl + 5 µM benzamil (**148**; Bz) in a TRPV1 knockout mouse. Note there is no amiloride-insensitive response (i.e., no response to NaCl + Bz). Wildtype mice show an amiloride-insensitive NaCl response. This shows that the amiloride-insensitive response to NaCl depends on the expression of the TRPV1 gene.

2. Increases in temperature and vanilloid concentration, in turn, increase the response conductance (slope of the normalized CT response with voltage applied to the anterior lingual receptive field), that is, the taste receptor has ion channel characteristics *in situ*.
3. Vanilloids, external pH, and ATP lower the temperature threshold of the response.
4. A TRPV1 mRNA transcript common to several channels in the TRP receptor family was detected in rat fungiform TBCs (see also Liu, L. and Simon, S. A., 2001a; 2001b).
5. TRPV1 knockout mice lack an amiloride-insensitive NaCl CT response (see Figure 28) while control mice show an intact response.
6. Ethanol is an agonist of the amiloride-insensitive salt taste response (Lyall, V. *et al.*, 2005) similar to its agonist effect on cloned TRPV1 (Trevisani *et al.*, 2002).

These results suggest that the amiloride-insensitive nonselective cation salt taste receptor in the taste cells of fungiform papillae is a TRPV1 variant (TRPV1t). Two reasons indicating that TRPV1t differs from TRPV1 are as follows:

1. TRPV1t is constitutively active, while TRPV1 must be activated by temperature, pH, or vanilloids, and
2. Unlike TRPV1, TRPV1t is not affected by a drop in pH alone.

A vanilloid must be present for TRPV1t to show pH dependence. At present it is not known what structural differences between TRPV1t and TRPV1 account for these important functional differences. Also it is not known if TRPV1t accounts for the nonselective cation salt sensitivity in the glossopharyngeal nerve (i.e., whether or not TRPV1t is also present in circumvallate taste cells) (Figure 29).

Differences in taste for salts with common cations. Why is NaCl a more potent salty stimulus than Na_2SO_4 or Na_3PO_4? Based on the foregoing discussion on the involvement of ENaC and TRPV1 in human salty taste, these three sodium salts should be equivalent in salty taste at common Na^+ concentration. However, this is not the case. NaCl is the most potent of the Na^+ salts and the diminished potencies of other Na^+ salts have long puzzled investigators in gustation research. An explanation for this apparent paradox began to emerge in the 1980s with the work of Harry Harper at Rockefeller University (1987), the author (JD) and co-workers at Virginia Commonwealth University (Ye, Q. *et al.*, 1991), and Sid Simon and co-workers at Duke University (Elliot, E. J. and Simon, S. A. 1990; Simon, S. A. *et al.*, 1993). It was proposed that NaCl and other Na^+ salts are able to passively diffuse through the tight junctions between TBCs and that, for salts with small counterions (e.g., Cl^-), both cation and anion are able to diffuse through the tight junctions at comparable rates. On the other hand, it was recognized that, while this is the case for small anions, for Na^+ salts with large anions, the rates of diffusion for the anions would be lower than that for the Na^+ ion and that therefore, for such salts, a hyperpolarizing field potential would be imposed on salt-sensitive TBCs which would inhibit the diffusion of Na^+ ions through ion channels in the apical membrane. The net result of this differential ionic diffusion is that Na^+ salts with large anions are expected to be less potent in salty taste than are salts with small anions and this is exactly what is observed.

4.02.6.2 Salty Tastant Structure–Activity Relationship

As already discussed, salt (i.e., NaCl) is the prototypical salty stimulus. However, for reasons that are now clear based on an understanding of the biochemistry of salty taste, many other salts exhibit

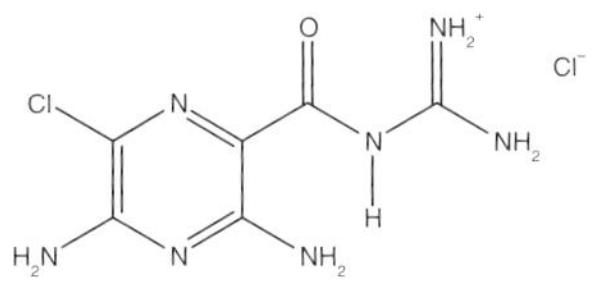

148 (Amiloride HCl)
$C_6H_9Cl_2N_7O$
$M = 266.09$

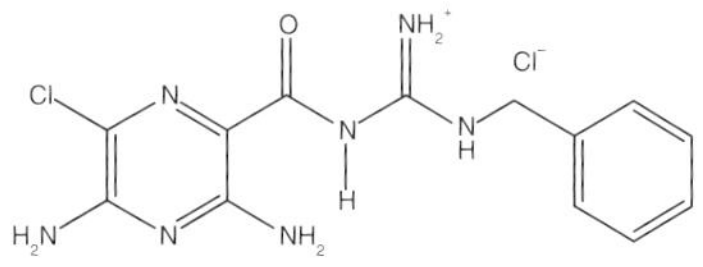

149 (Benzamil HCl)
$C_{13}H_{15}Cl_2N_7O$
$M = 356.21$

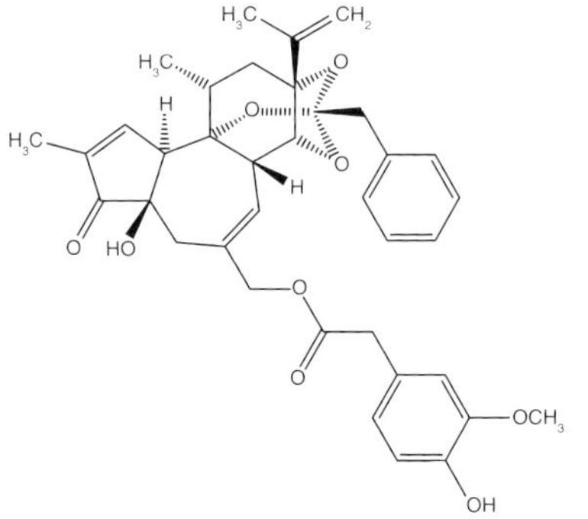

150 (Resiniferatoxin (RTX))
$C_{37}H_{40}O_9$
$M = 628.71$

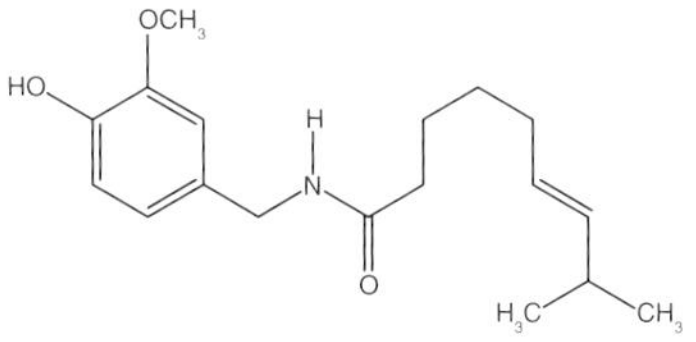

151 (Capsaicin)
$C_{18}H_{27}NO_3$
$M = 305.41$

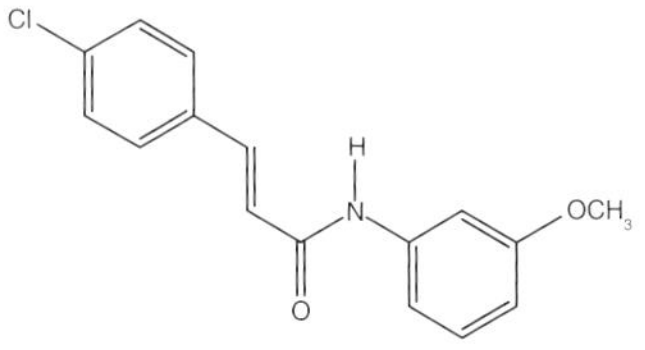

152 (SB360971)
$C_{16}H_{14}ClNO_2$
$M = 287.74$

Figure 29 Molecular tools employed in the study of salty taste.

salty taste. In general, salts exhibit multiple qualities of taste with sweet and bitter tastes being particularly common and even sourness observed for some (Shallenberger, R. S., 1993). Thus, as example, NaCl is perceived as sweet at low concentrations, salty and sweet at midrange concentrations, and only salty at higher concentrations. Bartoshuk L. M. *et al.* at Yale University (1978) noted that the observation of sweet taste for NaCl was protocol-dependent and required rinsing with water immediately before tasting the salt solution. KCl exhibits salty, sweet, and bitter qualities of taste. Thus, at low concentrations, KCl is weakly sweet and salty; at midrange concentrations, salty and weakly bitter; and at very high concentrations, only bitter. LiCl exhibits salty, sweet, and sour tastes. Thus, at low

concentrations, LiCl is weakly sweet and salty, and at high concentrations, salty and weakly sour. K_2SO_4 exhibits salty, sweet, sour, and bitter tastes. Thus, at low concentrations, K_2SO_4 is sweet and bitter; at midrange concentrations, sweet, salty, and bitter; and at high concentrations, salty, bitter, and sour. As a bottom line, the presence of salty taste is common in salts of many different types even though the purest salty taste is exhibited by NaCl. The saltiness potencies for a number of salts were determined by von Skramlik E. (1926) and have been reviewed by Beets M. G. J. (1978). These potencies, recalculated so as to place them on a scale where 1 M NaCl is defined as having a potency of 1.00, are given in Figure 30 along with empirical formulas and formula weights.

NaCl	NaBr	NaI	Na_2SO_4	$NaNO_3$
153	**154**	**155**	**156**	**157**
(Sodium chloride)	(Sodium bromide)	(Sodium iodide)	(Sodium sulfate)	(Sodium nitrate)
ClNa	BrNa	INa	Na_2O_4S	$NaNO_3$
FW = 58.44	FW = 102.89	FW = 149.89	FW = 142.04	FW = 84.99
$P_m(1) = 1.00$	$P_m(1) = 1.00$	$P_m(1) = 0.77$	$P_m(1) = 1.25$	$P_m(1) = 0.17$

$NaHCO_3$	LiCl	LiBr	LiI	$LiNO_3$
158	**159**	**160**	**161**	**162**
(Sodium bicarbonate)	(Lithium chloride)	(Lithium bromide)	(Lithium iodide)	(Lithium nitrate)
$CHNaO_3$	ClLi	BrNa	INa	$LiNO_3$
FW = 84.01	FW = 42.39	FW = 86.84	FW = 133.85	FW = 68.95
$P_m(1) = 0.21$	$P_m(1) = 0.44$	$P_m(1) = 0.79$	$P_m(1) = 0.57$	$P_m(1) = 0.23$

KCl	KBr	KI	K_2SO_4	KNO_3
163	**164**	**165**	**166**	**167**
(Potassium chloride)	(Potassium bromide)	(Potassium iodide)	(Potassium sulfate)	(Potassium nitrate)
ClK	BrK	IK	K_2O_4S	KNO_3
FW = 74.55	FW = 119.00	FW = 166.00	FW = 174.26	FW = 101.10
$P_m(1) = 1.36$	$P_m(1) = 1.16$	$P_m(1) = 0.54$	$P_m(1) = 0.26$	$P_m(1) = 0.14$

$KHCO_3$	NH_4Cl	NH_4Br	NH_4I	$(NH_4)_2SO_4$
168	**169**	**170**	**171**	**172**
(Potassium bicarbonate)	(Ammonium chloride)	(Ammonium bromide)	(Ammonium iodide)	(Ammonium sulfate)
$CHKO_3$	H_4ClN	H_4BrN	H_4IN	$H_8N_2O_4S$
FW = 100.12	FW = 53.49	FW = 97.94	FW = 144.94	FW = 132.14
$P_m(1) = 0.23$	$P_m(1) = 2.83$	$P_m(1) = 1.83$	$P_m(1) = 2.44$	$P_m(1) = 1.26$

NH_4NO_3	$CaCl_2$	$MgCl_2$	Mg_2SO_4	
173	**174**	**175**	**176**	
(Ammonium nitrate)	(Calcium chloride)	(Magnesium chloride)	(Magnesium sulfate)	
$H_4N_2O_3$	$CaCl_2$	Cl_2Mg	MgO_4S	
FW = 80.04	FW = 110.98	FW = 95.21	FW = 120.37	
$P_m(1) = 1.03$	$P_m(1) = 1.23$	$P_m(1) = 0.20$	$P_m(1) = 0.01$	

Figure 30 Empirical formulas, formula weights, and saltiness potencies of 24 common salts (Beets, M. G. J., 1978).

References

Arakawa, H. and Nakazaki, M. 1959. Absolute configuration of phyllodulcin. Chem. Ind. (Lond.), 671.

Audrieth, L. F. and Sveda, M. 1944. Preparation and properties of some N-substituted sulfamic acids. J. Org. Chem. 9, 89–101.

Avenet, P. and Lindemann, B. 1988. Amiloride-blockable sodium currents in isolated taste receptor cells. J. Membr. Biol. 105, 245–255.

Avenet, P. and Lindemann, B. 1991. Noninvasive recording of receptor cell action potentials and sustained currents from single taste buds maintained in the tongue: the response to mucosal NaCl and amiloride. J. Membr. Biol. 124, 33–41.

Bachmanov, A. A., Li, X., Reed, D. R., Ohmen, J. D., Li, S., Chen, Z., Tordoff, M. G., de Jong, P. J., Wu, C., West, D. B., Chatterjee, A., Ross, D. A., and Beauchamp, G. K. 2001. Positional cloning of the mouse saccharin preference (Sac) locus. Chem. Senses 26, 925–933.

Baek, N.-I., Chung, M.-S., Shamon, L., Kardono, L. B. S., Tsauri, S., Padmawinata, K., Pezzuto, J. M., Soejarto, D. D., and Kinghorn, A. D. 1993. Potential sweetening agents of plant origin. 29. Studies on Indonesian medicinal plants. 6.

Selligueain A, a novel highly sweet proanthocyanidin from the rhizomes of Selliguea feei. J. Nat. Prod. 56, 1532–1538.

Bartoshuk, L. M., Murphy, C., and Cleveland, C. T. 1978. Sweet taste of dilute NaCl: psychophysical evidence for a sweet stimulus. Physiol. Behav. 21, 609–613.

Bassoli, A., Borgonovo, G., Busnelli, G., Morini, G., and Drew, M. G. B. 2005. Monatin and its stereoisomers: chemoenzymatic synthesis and taste properties. Eur. J. Org. Chem 1652–1658.

Bassoli, A., Drew, M. G. B., Merlini, L., and Morini, G. 2002. General pseudoreceptor model for sweet compounds: a semiquantitative prediction of binding affinity for sweet-tasting molecules. J. Med. Chem. 45, 4402–4409.

Beauchamp, G. K. (ed.) 1994. Kirin International Symposium on Bitter Taste, Elsevier Science.

Beauchamp, G. K., Bertino, M., Burke, D., and Engelman, K. 1990. Experimental sodium depletion and salt taste in normal human volunteers. Am. J. Clin. Nutr. 51, 881–889.

Beauchamp, G. K, Cowart, B., and Schmidt, H. J. 1991. Development of Chemosensitivity and Preference. In: Smell and Taste in Health and Disease (ed. T. V. Getchell, R. L. Doty, L. M. Bartoshuk, and J. B. Snow), pp. 405–416. Raven Press.

Beets, M. G. J. 1978. Structure–Activity Relationships in Human Chemoreception. Applied Science Publishers.

Beidler, L. M. 1967. Anion Influences on Taste Receptor Responses. In: Olfaction and Taste II (*ed*. T. Hayashi), pp. 509–534. Pergamon Press.

Beidler, L. M. and Gross, G. W. 1971. The Nature of Taste Receptor Sites. In: Contributions to Sensory Physiology (*ed*. W. D. Neff), pp. 97–127. Academic Press.

Beksan, E., Schieberle, P., Robert, F., Blank, I., Fay, L. B., Schlictherle-Cerny, H., and Hofmann, T. 2003. Synthesis and sensory characterization of novel umami-tasting glutamate glycoconjugates. J. Agric. Food Chem. 51, 5428–5436.

Belitz, H.-D. and Wieser, H. 1985. Bitter compounds: occurrence and structure–activity relationships. Food Rev. Int. 1(2), 271–354.

Bellisle, F. 1999. Glutamate and the UMAMI taste: sensory, metabolic, nutritional and behavioural considerations. A review of the literature published in the last 10 years. Neurosci. Biobehav. Rev. 23, 423–438.

Berlinerblau, J. 1883. Über die einwirkung von chlorcyan auf ortho- und auf para-amidophenetol. Journal für Praktische Chemie 30, 97–115.

Bertelsen, H., Hansen, S. J., Laursen, R. S., Saunders, J., and Eriknauer, K. 2001. Tagatose. In: Alternative Sweeteners, 3rd edn (*ed*. L. O'BrienNabors), pp. 105–127. Marcel Dekker.

Birch, G. G., Parke, S., Siertsema, R., and Westwell, J. M. 1996. Specific volumes and sweet taste. Food Chem. 56(3), 223–230.

Blanksma, J. J. and van der Weyden, P. W. M. 1940. Relationship between taste and structure in some derivatives of meta-nitraniline. Recueil des Travaux Chimiques des Pays-Bas 59, 629–632.

Boughter, J. D., Jr. and Whitney, G. 1993. Human taste thresholds for sucrose octaacetate. Chem. Senses 18(4), 445–448.

Brand, J. G., Teeter, J. H., and Silver, W. L. 1985. Inhibition by amiloride of chorda tympani responses evoked by monovalent salts. Brain Res. 334, 207–214.

Breitwieser, G. E. 2004. G protein-coupled receptor oligomerization: implications for G protein activation and cell signaling. Circ. Res. 94, 17–27.

Brennan, T. M. and Hendrick, M. E., US Patent 4 411 925 (October 25, 1983) (To Pfizer, Inc).

Budesinsky, Z. and Vavrina, J. 1972. Nucleophilic substitutions in the 2-methylsulfonyl pyrimidine series. Collection Czechoslov. Chem. Commun. 37, 1721–1733.

Bufe, B., Hofmann, T., Krautwurst, D., Raguse, J. D., and Meyerhof, W. 2002. The human TAS2R16 receptor mediates bitter taste in response to α-glucopyranosides. Nature Genet. 32, 397–401.

Chandrashekar, J., Mueller, K. L., Hoon, M. A., Adler, E., Feng, L., Guo, W., Zuker, C. S., and Ryba, N. J. P. 2000. T2RT2RSsT2RSsT2RSs function as bitter taste receptors. Cell 100, 703–711.

Chang, W.-I., Chung, J.-W., Kim, Y.-K., Chung, S.-C., and Kho, H.-S. 2006. The relationship between phenylthiocarbamide (PTC) and 6-n-propylthiouracil (PROP) taster status and taste thresholds for sucrose and quinine. Arch Oral Bio 51, 427–432.

Chatterjee, B. 1929. Attempts to find new anti-malarials. Part IV. β-Benziminazolylethylamine and β-5(or 6)-ethoxybenziminazolylethylamine. J. Chem. Soc. 2965–2968.

Chaudhari, N., Landin, A. M., and Roper, S. D. 2000. A metabotropic glutamate receptor variant functions as a taste receptor. Nat Neurosci 3, 113–119.

Choi, Y.-H., Kinghorn, A. D., Shi, X., Zhang, H., and Teo, B. K. 1989. Abrusoside A: a new type of highly sweet triterpene glycoside. J. Chem. Soc., Chem. Commun. 887–888.

Clauss, K. and Jensen, H. 1973. Oxathiazinone dioxides, a novel group of sweetening agents. Angew. Chem. Int. Ed. Engl. 12, 869–876.

Compadre, C. M., Hussain, R. A., Lopez de Compadre, R. L., Pezzuto, J. M., and Kinghorn, A. D. 1987. The intensely sweet herb sesquiterpene hernandulcin: isolation, synthesis, characterization and preliminary safety evaluation. J. Agric. Food Chem., 35, 273–279.

Cram, D. J. and Ratajczak, A. 1971. 1-(Phenylsulfonyl)-cyclopropanecarboxylic acids. US Patent 3 598 868 (August 10, 1971; to The Regents of the University of California, Berkeley, CA, USA).

Crammer, B. and Ikan, R. 1987. Progress in the Chemistry and Properties of Rebaudiosides. In: Developments in Sweeteners-3 (*ed*. T. H. Grenby), pp. 45–64. Elsevier Applied Science.

Culberson, J. C. and Walters, D. E. 1991. Three-Dimensional Model for the Sweet Taste Receptor: Development and Use, Chapter 16. In: Sweeteners: Discovery, Molecular Design and Chemoreception (*ed*. D. E. Walters, F. T. Orthoefer, and G. E. DuBois), ACS Symposium Series 450, 214–223. American Chemical Society.

Damak, S., Rong, M., Yasumatsu, K., Kokrashvili, Z., Perez, C. A., Shigemura, N., Yoshida, R., Mosinger, B., Jr., Glendinning, J. I., Ninomiya, Y., and Margolskee, R. F. 2006. Trpm5 null mice respond to bitter, sweet and umami compounds. Chem. Senses 31, 253–264.

D'Angelo, L. L. and Iacobucci, G. A. 1995. QSAR studies of a group of structurally diverse sweeteners using CoMFA and design of a new series of aspartame analogs. *Presented at the 210th National Meeting of the American Chemical Society*, Chicago, IL, USA, August 20–24. Book of Abstracts, Pt 1, COMP-083, American Chemical Society.

Dean, J. A. (*ed*.). 1985. Lange's Handbook of Chemistry, 13th edn. McGraw-Hill Book Company.

DeGraaf, H. 1930. The relationship between taste and the constitution in the dihydrazides of dicarboxylic acids and their derivatives, Diss., Leiden, 138; Chem. Abstr. 24, 5723.

Denten, D. A. 1967. Salt Appetite. In: Handbook of Physiology, Section 6. Vol. 1 (*eds*. C. F. Code and W. Heidel), p. 433. American Physiological Society.

DeSimone, J. A., Heck, G. L., Mierson, S., and DeSimone, S. K. 1984. The active ion transport properties of canine lingual epithelia *in vitro*. Implications for gustatory transduction. J. Gen. Physiol. 83, 633–656.

Dornow, A., Kühlcke, I., and Baxmann, F. 1949. Über einige derivate der benzoylessigsäure, Chemische Berichte 82, 254–257.

Dotson, C. D., Roper, S. D., and Spector, A. C. 2005. PLC$_{\beta 2}$-Independent Behavioral Avoidance of Prototypical Bitter-Tasting Ligands. Chem. Senses 30, 593–600.

Douglas, A. J. and Goodman, M. 1991. Molecular Basis of Taste: A Stereoisomeric Approach, Chapter 10. In: Sweeteners: Discovery, Molecular Design and Chemoreception (*ed*. D. E. Walters, F. T. Orthoefer, and G. E. DuBois), pp. 128–142. ACS Symposium Series 450, American Chemical Society.

DuBois, G. E. 1985. Unpublished Results. G. D. Searle Pharmaceutical Company Skokie, IL, USA.

DuBois, G. E. 1995. Unpublished Results, The Coca-Cola Company, Atlanta, GA, USA.

DuBois, G. E. 1997. New Insights on the Coding of the Sweet Taste Message in Chemical Structure. In: Firmenich Jubilee Symposium Olfaction and Taste: A Century for the Senses 1895–1995 (*ed*. G. Salvadori), pp. 32–95, Allured Publishing Corp.

DuBois, G. E. 2000. Nonnutritive Sweeteners. In: Encyclopedia of Food Science & Technology (*ed*. Frederick J Francis), pp. 2245–2265. John Wiley & Sons,.

DuBois, G. E. 2003. Unpublished Results, The Coca-Cola Company, Atlanta, GA.

DuBois, G. E. and D'Angelo, L. L. 1996. Current thoughts on the workings of polyol sweeteners. *Oral Presentation, International Sweetener Symposium*, Jerusalem, Israel.

DuBois, G. E. and San Miguel, R. S. 2005. Unpublished Results, The Coca-Cola Company, Atlanta, GA, USA.

DuBois, G. E. and Singer, N. 1984. Unpublished Results, G. D. Searle Pharmaceutical Company, Skokie, IL, USA.

DuBois, G. E., D'Angelo, L. L., and King, G. A. 1994. Unpublished Results, The Coca-Cola Company, Atlanta, GA, USA.

DuBois, G. E., Walters, D. E., Schiffman, S. S., Warwick, Z. S., Booth, B. J., Pecore, S. D., Gibes, K., Carr, B. T., and Brands, L. M. 1991. Concentration–response relationships of sweeteners (*eds*. D. E. Walters, F. T. Orthoefer, and G. E. DuBois), pp. 261–276. ACS Symposium Series 450, American Chemical Society.

DuBois, G. E., Zhi, B., Roy, G. M., Stevens, S. Y., and Yalpani, M. 1992. New non-ionic polyol derivatives with sucrose-mimetic properties. J. Chem. Soc. Chem. Commun. 1604–1605.

Elliot, E. J. and Simon, S. A. 1990. The anion in salt taste: a possible role for paracellular pathways. Brain Res. 535(1), 9–17.

Embuscado, M. E. and Patil, S. K. 2001. Erythritol. In: Alternative Sweeteners, 3rd edn (*ed*. L. O'Brien Nabors), pp. 235–254. Marcel Dekker.

Fahlberg, C. and Remsen, I. 1879. Über die oxydation des orthotoluolsulfamids. Chem. Ber. 2, 469–473.

Faurion, A. 1987. Physiology of the Sweet Taste. In: Progress in Sensory Physiology, Vol. 8 (*eds*. H. Autrum, D. Ottoson, E. R. Perl, R. F. Schmidt, H. Suimazu, and W. D. Willis), pp. 129–201. Springer.

Feldman, M. and Barnett, C. 1995. Relationship between the acidity and osmolarity of popular beverages and reported postprandial heartburn. Gastroenterology 108, 125–131.

Finger, T. E., Danilova, V., Barrows, J., Bartel, D. L., Vigers, A. J., Stone, L., Hellekant, G., and Kinnamon, S. C. 2005. ATP signaling is crucial for communication from taste buds to gustatory nerves. Science 310, 1495–1499.

Fitzhugh, O. G., Nelson, A. A., and Frawley, J. P. 1951. A comparison of the chronic toxicities of synthetic sweetening agents. J. Am. Pharm. Assoc. 40, 583–586.

Frank, M. E., Bieber, S. L., and Smith, D. V. 1988. The organization of taste sensibilities in hamster chorda tympani nerve fibers. J. Gen. Physiol. 91, 861–896.

Frank, M. E., Contreras, R. J., and Hettinger, T. P. 1983. Nerve fibers sensitive to ionic taste stimuli in chorda tympani of the rat. J. Neurophysiol. 50, 941–960.

Fraser, J. A., Middlebrook, C. E., Usher-Smith, J. A., Schwiening, C. J., and Huang, C. L. 2005. The effect of intracellular acidification on the relationship between cell volume and membrane potential in amphibian skeletal muscle. J. Physiol. 563, 745–764.

Fregly, M. J. and Rowland, N. E. 1986. Hormonal and neural mechanisms of sodium appetite. NIPS 1, 51–54.

Frérot, E. and Escher, S. D. 1997. Flavoured products and method for preparing the same. WO 97/04667.

Furukawa, S. and Tomizawa, Z. 1920. J. Chem. Ind. Tokyo, 23, 342; Chem. Abstr. 1920, 14, 2839.

Garvin, J. L., Robb, R., and Simon, S. A. 1988. Spatial map of salts and saccharides on dog tongue. Am. J. Physiol. 255, R117–122.

Gilbertson, T. A. 2002. Hypoosmotic stimuli activate a chloride conductance in rat taste cells. Chem. Senses 27, 383–394.

Gilbertson, T. A. and Zhang, H. 1998. Self inhibition in amiloride-sensitive sodium channels in taste receptor cells. J. Gen. Physiol. 111, 667–677.

Gilbertson, T. A., Roper, S. D., and Kinnamon, S. C. 1993. Proton currents through amiloride-sensitive Na^+ channels in isolated hamster taste cells: enhancement by vasopressin and cAMP. Neuron 10, 931–942.

Glendinning, J. I. 1994a. Is the Bitter Rejection Response Always Adaptive? In: Kirin International Symposium on Bitter Taste (*ed*. G. K. Beauchamp), pp. 1217–1222. Elsevier Science.

Glendinning, J. I. 1994b. Is the bitter rejection response always adaptive? Physiol. Behav. 56(6), 1217–1227.

Goudet, C., Binet, V., Prezeau, L., and Pin, J.-P. 2004. Allosteric modulators of class-C G-protein-coupled receptors open new possibilities for therapeutic application. Drug Discov. Today: New Strategies 1(1), 125–133.

Gries, H. and Mützel, W. 1985. 3-Substituted 2,4,6-trihalogenated benzamides as sweetening agents, US Patent 4 522 839 (June 11, 1985).

Harper, H. W. 1987. A diffusion potential model of salt taste receptors. Ann. N. Y. Acad. Sci. 510, 349–351.

Hashimoto, Y., Ishizone, H., and Ogura, M. 1980. Periandrin II and IV, triterpenoid glycosides from *Periandra dulcis*. Phytochemistry 19, 2411–2415.

He, W., Yasumatsu, K., Varadarajan, V., Yamada, A., Lem, J., Ninomiya, Y., Margolskee, R. F., and Damak, S. 2004. Umami taste responses are mediated by alpha-transducin and alpha-gustducin. J. Neurosci. 24, 7674–7680.

Heck, G. L., Mierson, S., and DeSimone, J. A. 1984. Salt taste transduction occurs through an amiloride-sensitive sodium transport pathway. Science 223, 403–405.

Hellekant, G. and Ninomiya, Y. 1991. On the taste of umami in chimpanzee. Physiol. Behav. 49, 927–934.

Hellekant, G. and Ninomiya, Y. 1994a. Bitter Taste in Single Chorda Tympani Taste Fibers from Chimpanzee. In: Kirin International Symposium on Bitter Taste (*ed*. G. K. Beauchamp), pp. 1185–1188. Elsevier Science.

Hellekant, G. and Ninomiya, Y. 1994b. Bitter taste in single chorda tympani taste fibers from chimpanzee. Physiol. Behav. 56(6), 1185–1188.

Hellekant, G., Ninomiya, Y., and Danilova, V. 1997. Taste in chimpanzees II: single chorda tympani fibers. Physiol. Behav. 61, 829–841.

Henkin, R. I., Gill, J. R., Jr., and Bartter, F. C. 1963. Studies on taste thresholds in normal man and in patients with adrenal cortical insufficiency: the role of adrenal cortical steroids and of serum sodium concentration. J. Clinical Invest. 42, 727–735.

Hettinger, T. P. and Frank, M. E. 1990. Specificity of amiloride inhibition of hamster taste responses. Brain Res. 513, 24–34.

Hill, D. L. and Bour, T. C. 1985. Addition of functional amiloride-sensitive components to the receptor membrane: a possible mechanism for altered taste responses during development. Brain Res. 352, 310–313.

Hofmann, A. 1972. Ein neuer süssstoff aus der indolreihe. Helv. Chim. Acta 55, 2934–2940.

Horowitz, R. M. and Gentili, B. 1969. Taste and structure in phenolic glycosides. J. Agric. Food Chem. 17, 696–700.

Hough, L. and Phadnis, S. P. 1976. Enhancement in the sweetness of sucrose. Nature 263, 800.

Huang, A. L., Chen, X., Hoon, M. A., Chandrashekar, J., Guo, W., Trankner, D., Ryba, N. J., and Zuker, C. S. 2006. The cells and logic for mammalian sour taste detection. Nature 442, 934–938.

Huang, Y. J., Maruyama, Y., Lu, K. S., Pereira, E., Plonsky, I, Baur, J. E., Wu, D., and Roper, S. D. 2005. Using biosensors to detect the release of serotonin from taste buds during taste stimulation. Arch. Ital. Biol. 143, 87–96.

Hussain, R. A., Poveda, L. J., Pezzuto, J. M., Soejarto, D. D., and Kinghorn, A. D. 1990. Sweetening agents of plant origin: phenylpropanoid constituents of seven sweet-tasting plants. Econ. Bot. 44(2), 174–182.

Ikeda, K. 1909. On a new seasoning. J. Tokyo Chem. Soc. 30, 820–836.

Ishimaru, Y., Inada, H., Kubota, M., Zhuang, H., Tominaga, M., and Matsunami, H. 2006. Transient receptor potential family members PKD1L3 and PKD2L1 form a candidate sour taste receptor. Proc. Natl. Acad. Sci. U. S. A. 103, 12569–12574.

Jiang, P., Cui, M., Zhao, B., Snyder, L. A., Benard, B. M. J., Osman, R., Max, M., and Margolskee, R. F. 2005. Identification of the cyclamate interaction site within the transmembrane domain of the human sweet taste receptor subunit T_1R_3T1R3. J. Biol. Chem. 280, 34296–34305.

Jiang, P., Ji, Q., Liu, Z., Snyder, L. A., Bernard, M. J., Margolskee, R. F., and Max, M. 2004. The cysteine-rich region of T_1R_3T1R3 determines responses to intensely sweet proteins. J. Biol. Chem. 279, 45068–45075.

Kasai, R., Fujino, H., Kuzuki, T., Wong, W.-H., Goto, C., Yata, N., Tanaka, O., Yasuhara, F., and Yamaguchi, S. 1986. Acyclic sesquiterpene oligoglycosides from pericarps of *Sapindus mukurossi*. Phytochemistry 25, 871–876.

Kasai, R., Hirono, S., Chou, W.-H., Tanaka, O., and Chen, F.-H. 1988. Sweet dihydroflavanol rhamnoside from leaves of *Engelhardtia shrysolepis*, a Chinese folk medicine, Chem. Pharm. Bull. 36, 4167–4170.

Kasai, R., Nie, R.-L., Nashi, K., Ohtani, K., Zhou, J., Tao, G.-D., and Tanaka, O. 1989. Sweet cucurbitane glycosides from fruits of *Siraitia siamensis* (chi-zi luo-han-guo), a Chinese folk medicine. Agric. Biol. Chem., 53, 3347–3349.

Kawamura, Y. and Kare, M. R. (*eds.*). 1987. Umami: A Basic Taste, Marcel Dekker.

Kaya, N., Shen, T., Lu, S. G., Zhao, F. L., and Herness, S. 2004. A paracrine signaling role for serotonin in rat taste buds: expression and localization of serotonin receptor subtypes. Am. J. Physiol. Regul. Integr. Comp. Physiol. 286, R649–658.

Kennelly, E. J., Cai, L., Long, L., Shamon, L., Zaw, K., Zhou, B.-N., Pezzuto, J. M., and Kinghorn, A. D. 1995. Novel highly sweet secodammarane glycosides from *Pterocarya paliurus*. J. Agric. Food Chem. 43, 2602–2607.

Kier, L. B. 1972. Molecular theory of sweet taste. J. Pharm. Sci. 61, 1394–1397.

Kim, N.-C. and Kinghorn, A. D. 2002. Sweet-Tasting and Sweetness-Modifying Constituents of Plants. In: Studies in Natural Products Chemistry, Vol. 27 (*ed*. Atta-ur-Rahman), pp. 3–57, Elsevier Science B. V.

Kim, J., Pezzuto, J. M., Soejarto, D. D., Lang, F. A., and Kinghorn, A. D. 1988. Polypodoside A, an intensely sweet constituent of the rhizomes of polypodium glycyrrhiza<NET>. J. Nat. Prod. 51, 1166–1172.

Kniazeff, J., Saintot, P.-P., Goudet, C., Liu, J., Charnet, A., Guillon, G., and Pin, J.-P. 2004. Locking the $GABA_B$ G-protein-coupled receptor in its active state. J. Neurosci. 24(2), 370–377.

Kohmura, M. and Ariyoshi, Y. 1998. Chemical synthesis and characterization of the sweet protein mabinlin II. Biopolymers 46, 215–223.

Kretz, O., Barbry, P., Bock, R., and Lindemann, B. 1999. Differential expression of RNA and protein of the three pore-forming subunits of the amiloride-sensitive epithelial sodium channel in taste buds of the rat. J. Histochem. Cytochem. 47, 51–64.

Kunishima, N., Shimada, Y., Tsuji, Y., Sato, T., Yamamoto, M., Kumasaka, T., Nakanishi, S., Jingami, H., and Morikawa, K. 2000. Structural basis of glutamate recognition by a dimeric metabotrophic glutamate receptor. Nature 407, 971–977.

Kurihara, K. and Beidler, L. M. 1968. Taste-modifying protein from miracle fruit. Science 161, 1241–1243.

Lapidus, M. and Sweeney, M. 1973. L-4′-Cyano-3-(2,2,2-trifluoroacetamido)-succinanilic acid and related synthetic sweetening agents. J. Med. Chem. 16, 163–166.

Le, A. S. and Bowe Mulderrig, K. 2001. Sorbitol and Mannitol. In: Alternative Sweeteners, 3rd edn (*ed*. L. O'Brien Nabors), pp. 317–334. Marcel Dekker.

Lee, C.-K. 1987. The Chemistry and Biochemistry of the Sweetness of Sugars. In: Advances in Carbohydrate Chemistry and Biochemistry, Vol. 45 (*eds*. R. S. Tipson and D. Horton),pp. 199–351.. Academic Press.

Li, X., Staszewski, L., Xu, H., Durick, K., Zoller, M., and Adler, E. 2002. Human receptors for sweet and umami taste. PNAS 99, 4692–4696.

Lin, W., Burks, C. A., Hansen, D. R., Kinnamon, S. C., and Gilbertson, T. A. 2004. Taste receptor cells express pH-sensitive leak K^+ channels. J. Neurophysiol. 92, 2909–2919.

Lin, W., Finger, T. E., Rossier, B. C., and Kinnamon, S. C. 1999. Epithelial Na^+ channel subunits in rat taste cells: localization and regulation by aldosterone. J. Comp. Neurol. 405, 406–420.

Lin, W., Ogura, T., and Kinnamon, S. C. 2002. Acid-activated cation currents in rat vallate taste receptor cells. J. Neurophysiol. 88, 133–141.

Liu, L. and Simon, S. A. 2001a. Acidic stimuli activate two distinct pathways in taste receptor cells from rat fungiform papillae. Brain Res. 923, 58–70.

Lopez-Jimenez, N. D, Cavenagh, M. M, Sainz, E., Cruz-Ithier, M. A., Battey, J. F., and Sullivan, S. L. 2006. Two members of the TRPP family of ion channels, Pkd1l3 and Pkd2l1, are co-expressed in a subset of taste receptor cells. J. Neurochem. 98, 68–77.

Lyall, V., Alam, R. I., Malik, S. A., Phan, T-H. T., Vinnikova, A. K., Heck, J. L., and DeSimone, J. A. 2004. Basolateral Na^+-H^+ exchanger-1 in rat taste receptor cells is involved in neural adaptation to acidic stimuli. J. Physiol. 556, 159–173.

Lyall, V., Alam, R. I., Phan, D. Q., Ereso, G. L., Phan, T-H. T., Malik, S. A., Montrose, M. H., Chu, S., Heck, G. L., Feldman, G. M., and DeSimone, J. A. 2001. Decrease in rat taste receptor cell intracellular pH is the proximate stimulus in sour taste transduction. Am. J. Physiol. Cell Physiol. 281, C1005–C1013.

Lyall, V., Alam, R. I., Phan, T-H. T., Phan, D. Q., Heck, G. L., and DeSimone, J. A. 2002a. Excitation and adaptation in the detection of hydrogen ions by taste receptor cells: a role for cAMP and Ca^{2+}. J. Neurophysiol. 87, 399–408.

Lyall, V., Alam, R. I., Phan, T-H. T., Russell, O. F., Malik, S. A., Heck, G. L., and DeSimone, J. A. 2002b. Modulation of rat chorda tympani NaCl responses and intracellular Na^+ activity in polarized taste receptor cells by pH. J. Gen. Physiol. 120, 793–815.

Lyall, V., Heck, G. L., Phan, T. H., Mummalaneni, S., Malik, S. A., Vinnikova, A. K., and DeSimone, J. A. 2005. Ethanol modulates the VR-1 variant amiloride-insensitive salt taste receptor. II. Effect on chorda tympani salt responses. J. Gen. Physiol. 125, 587–600.

Lyall, V., Heck, G. L., Vinnikova, A. K., Ghosh, S., Phan, T. H., Alam, R. I., Russell, O. F., Malik, S. A., Bigbee, J. W., and DeSimone, J. A. 2004. The mammalian amiloride-insensitive non-specific salt taste receptor is a vanilloid receptor-1 variant. J. Physiol. 558, 147–159.

Lyall, V., Pasley, H., Phan, T-H. T., Mummalaneni, S., Heck, G. L., Vinnikova, A. K., and DeSimone, J. A. 2006. Intracellular pH modulates taste receptor volume and the phasic part of the chorda tympani response to acids. J. Gen. Physiol. 127, 15–34.

Lythgoe, B. and Trippett, S. 1950. The constitution of the disaccharide of glycyrrhinic acid. J. Chem. Soc., 1983–1990.

Margolskee, R. F. 2002. Molecular mechanisms of bitter and sweet taste transduction. J. Biol. Chem. 277, 1–4.

Margolskee, R. L. 2004. Insights into Taste Transduction and Coding from Molecular, Biochemical, and Transgenic

Studies. In: Challenges in Taste Chemistry and Biology, ACS Symposium Series 867 (eds. T. Hofmann, C. -T. Ho, and W. Pickenhagen), pp. 26–44. American Chemical Society.

Marie, S. and Piggott, J. R. (eds.) 1991. Handbook of Sweeteners. Blackie & Son.

Masuda, H., Ohtani, K., Mizutani, K., Ogawa, S., Kasai, R., and Tanaka, O. 1991. Chemical study on *Haematoxylon campechianum*: a sweet principle and new dibenz-[*b,d*]-oxocin derivatives. Chem. Pharm. Bull. 39, 1382–1384.

Matsumoto, K., Kasai, R., Ohtani, K., and Tanaka, O. 1990. Minor cucurbitane glycosides from fruits of *Siraitia grosvenori* (Cucurbitaceae). Chem. Pharm. Bull. 38, 2030–2032.

Matsunami, H., Montmayeur, J. P., and Buck, L. B. 2000. A family of candidate taste receptors in human and mouse. Nature 404, 601–604.

Max, M., Gopi Shanker, Y., Huang, L., Rong, M., Liu, Z., Campagne, F., Weinstein, H., Damak, S., and Margolskee, R. F. 2001. *Tas1r3*, encoding a new candidate taste receptor, is allelic to the sweet responsiveness locus *Sac*. Nat. Genet. 28, 58–63.

Mazur, A. W. and Mohlenkamp, M. J. 1997. Small, Non-nutritive Carbohydrates as Sucrose Substitutes for Foods. In: New Technologies for Healthy Foods & Nutraceuticals (ed. M. Yalpani), pp. 124–142. ATL Press.

Mazur, R. H., Schlatter, J. M., and Goldkamp, A. H. 1969. Structure–taste relationships of some dipeptides, J. Am. Chem. Soc. 91, 2684–2691.

McLaughlin, S. K., Mckinnon, P. J., and Margolskee, R. F. 1992. Gustducin is a taste-cell-specific G-protein closely related to the transducins, Nature 357, 563–569.

Mierson, S., Heck, G. L., DeSimone, S. K., Biber, T. U., and DeSimone, J. A. 1985. The identity of the current carriers in canine lingual epithelium *in vitro*. Biochim. Biophys. Acta. 816, 283–293.

Mierson, S., Welter, M. E., Gennings, C., and DeSimone, J. A. 1988. Lingual epithelium of spontaneously hypertensive rats has decreased short-circuit current in response to NaCl. Hypertension 11, 519–522.

Ming, D. and Hellekant, G. 1994. Brazzein, a new high-potency thermostable sweet protein from *Pentadiplandra brazzeana* B. FEBS Lett. 355, 106–108.

Mintz, S. W. 1985. Sweetness and Power: The Place of Sugar in Modern History. Penguin Books.

Miyamoto, T., Fujiyama, R., Okada, Y., and Sato, T. 1998. Sour transduction involves activation of NPPB-sensitive conductance in mouse taste cells. J. Neurophysiol. 80, 1852–1859.

Montmayeur, J.-P., Liberles, S. D., Matsunami, H., and Buck, L. B. 2001. A candidate taste receptor gene near a sweet taste locus. 2001. Nat. Neurosci. 4(5), 492–498.

Morris, J. A. and Cagan, R. H. 1972. Purification of monellin, sweet principle in *Dioscoreophyllum cuminsii*, Biochim. Biophys. Acta 261, 114–122.

Muspratt, J. T. and Hofmann, A. W. 1846. Über das nitranilin, ein neues zersetzungsprodukt des dinitrobenzols. Ann. Chem. 57, 201–224.

Nagarajan, S., Kellogg, M. S., DuBois, G. E., Williams, D. S., Gresk, C. J., and Markos, C. S. 1992. Understanding the mechanism of sweet taste: synthesis of tritium labeled guanidineacetic acids. J. Labelled Comp. Radiopharm. 31(8), 599–607.

Naim, M., Nir, S., Spielman, A. I., Noble, A. C., Peri, I., Rodin, S., and Samuelov-Zubare, M. 2002. Hypothesis of Receptor-Dependent and Receptor-Independent Mechanisms for Bitter and Sweet Taste Transduction: Implications for Slow Taste Onset and Lingering Aftertaste. In: Chemistry of Taste: Mechanisms, Behaviors and Mimics

(eds. P. Given and D. Paredes), pp. 1–17. ACS Symposium Series 825, American Chemical Society.

Naim, M., Seifert, R., Nürnberg, B., Grünbaum, L., and Schultz, G. 1994. Some taste substances are direct activators of G-proteins, Biochem. J. 297, 451–454.

Nanayakkara, N. P. D., Hussain, R. A., Pezzuto, J. M., Soejarto, D. D., and Kinghorn, A. D. 1988. An intensely sweet dihydroflavonol derivative based on a natural product lead compound, J. Med. Chem. 31, 1250–1253.

Nelson, G., Chandrashekar, J., Hoon, M. A., Feng, L., Zhao, G., Ryba, N. J. P., and Zuker, C. S. 2002. An amino-acid taste receptor. Nature 416, 199–202.

Nelson, G., Hoon, M. A., Chandrashekar, J., Zhang, Y., Ryba, N. J. P., and Zuker, C. S. 2001. Mammalian sweet taste receptors. Cell 106, 381–390.

Nie, Y., Vigues, S., Hobbs, J. R., Conn, Graeme, L., and Munger, S. D. 2005. Distinct contributions of T1R2T1R2 and T1R3T1R3 taste receptor subunits to the detection of sweet stimuli, Curr. Biol. 15(21), 1948–1952.

Ninomiya, Y. and Funakoshi, M. 1988. Amiloride inhibition of responses of rat single chorda tympani fibers to chemical and electrical tongue stimulations. Brain Res. 451, 319–325.

Ninomiya, Y., Sako, N., and Funakoshi, M. 1989. Strain differences in amiloride inhibition of NaCl responses in mice, *Mus musculus*. J. Comp. Physiol. 166, 1–5.

Nishizawa, M. and Yamada, H. 1996. Intensely Sweet Saponin Osladin: Synthetic and Structural Study. In: Saponins Used in Food and Agriculture (eds. G. R. Waller and K. Yamasaki), pp. 25–36. Plenum Press.

Nofre, C. and Tinti, J.-M. 2000. Neotame: discovery, properties and utility. Food Chem. 69(3), 245–257.

Nofre, C., Tinti, J.-M., and Ouar Chatzopoulos, F. 1990. Sweetening Agents, US Patent 4 921 939 (May 1, 1990), (to Universite Claude Bernard, Lyon, France).

O'Brien Nabors, L. (ed.) 2001. Alternative Sweeteners, 3rd edn, Marcel Dekker.

Ogata, C., Hatada, M., Tomlinson, G., Shin, W.-C., and Kim, S.-H. 1987. Crystal structure of the intensely sweet protein monellin, Nature (Lond.) 328, 739–742.

Ogiso, K., Shimizu, Y., Watanabe, K., and Tonasaki, K. 2000. Possible involvement of undissociated acid molecules in the acid response of the chorda tympani nerve of the rat. J. Neurophysiol. 83, 2776–2779.

Olinger, P. M. and Pepper, T. 2001. Xylitol. In: Alternative Sweeteners, 3rd edn (ed. L. O'Brien Nabors), pp. 335–365. Marcel Dekker.

Ozeck, M., Brust, P., Xu, H., and Servant, G. 2004. Receptors for bitter, sweet and umami taste couple to inhibitory G protein signaling pathways. Eur. J. Pharmacol. 489, 139–149.

Pérez, C. A., Huang, L., Rong, M., Kozak, J. A., Preuss, A. K., Zhang, H., Max, M., and Margolskee, R. F. 2002. Transient receptor potential channel expressed in taste receptor cells. Nat. Neurosci. 5(11), 1–8.

Petersen, S. and Müller, E. 1948. Über eine neue gruppe von süssstoffen. Chem. Ber. 81, 31–38.

Pfeilsticker, K., Ruffler, I., Engel, C., and Rehage, C. 1978. Relation between bitter taste and positive surface tension of pure substances in aqueous solutions. Lebensm. Wiss. Technol. 11(6), 323–329.

Phillips, K. C. 1987. Stevia: Steps in Developing a New Sweetener. In: Developments in Sweeteners, 3 (ed. T. H. Grenby), pp. 1–43. Elsevier Applied Science.

Pin, J.-P., Galvez, T., and Prezeau, L. 2003. Evolution, structure, and activation mechanism of family 3/C G-protein-coupled receptors. Pharmacol. Ther. 98, 325–354.

Polanski, J. and Ratajczak, A. 1993. A Structure–Taste Study of a New Class of Artificial Sweeteners, Arylsulphonylalkanoic Acids, Chapter 11. In: Sweet-Taste Chemoreception (eds. M. Mathlouthi, J. A. Kanters, and G. G. Birch), pp. 185–203. Elsevier Applied Science.

Pronin, A. N., Tang, H., Connor, J., and Keung, W. 2004. Identification of ligands for two bitter T2RT2RS receptors. Chem. Senses 29, 583–593.

Putnam, R. W., Filosa, J. A., and Ritucci, N. A. 2004. Cellular mechanisms involved in CO_2 and acid signaling in chemosensitive neurons. Am. J. Physiol. 287, C1493–C1526.

Richter, C. P. 1936. Increased salt appetite of adrenalectomized rats. Am. J. Physiol. 115, 155–161.

Richter, T. A., Caicedo, A., and Roper, S. D. 2003. Sour taste stimuli evoke Ca^{2+} and pH responses in mouse taste cells. J. Physiol. 547, 475–483.

Richter, T. A., Dvoryanchikov, G. A., Chaudhari, N., and Roper, S. D. 2004a. Acid-sensitive two-pore domain potassium (K_2P) channels in mouse taste buds. J. Neurophysiol. 92, 1928–1936.

Richter, T. A., Dvoryanchikov, G. A., Roper, S. D., and Chaudhari, N. 2004b. Acid-sensing ion channel-2 is not necessary for sour taste in mice. J. Neuroscience 24, 4088–4091.

Rohse, H. and Belitz, H.-D. 1991. Shape of Sweet Receptors Studied by Computer Modeling, Chapter 13. In: Sweeteners: Discovery, Molecular Design and Chemoreception (eds. D. E. Walters, F. T. Orthoefer, and G. E. DuBois), pp. 176–192. ACS Symposium Series 450, American Chemical Society.

Rossier, B. C., Pradervand, S., Schild, L., and Hummler, E. 2002. Epithelial sodium channel and the control of sodium balance: interaction between genetic and environmental factors. Annu. Rev. Physiol. 64, 877–897.

Roy, G. (ed.) 1997. Modifying Bitterness: Mechanism, Ingredients and Applications, Technomic Publishing AG.

Runti, C. and Collino, F. 1964. Relations between chemical constitution and sweet taste. Derivatives of p-methoxybenzoylbenzoic acid and of malonic acid diamides. XII, Ann. Chim. (Rome) 54, 431–440.

Runti, C. and Galimberti, S. 1957. The relations between chemical constitutions and sweet taste. Derivatives of p-methoxybenzoylbenzoic acid. Ann. Chim. (Rome) 47, 250–259.

Sainz, E., Korley, J. N., Battey, J. F., and Sullivan, S. L. 2001. Identification of a novel member of the T_1R family of putative taste receptors. J. Neurochem. 77, 896–903.

Saroli, A. 1986. Structure-activity relationship of bitter compounds related to denatonium chloride and dipeptide methyl esters. Zeitschrift für Lebensmittel-Untersuchung und Forschung 182, 118–120.

Schiffman, S. S., Gatlin, L. A., Sattely-Miller, E. A., Graham, B. G., Heiman, S. A., Stagner, W. C., and Erickson, R. P. 1994. The effect of sweeteners on bitter taste in young and elderly adults. Brain Res. Bull. 35(3), 189–204.

Schiffman, S. S., Lockhead, E., and Maes, F. W. 1983. Amiloride reduces the taste intensity of Na^+ and Li^+ salts and sweeteners. Proc. Natl. Acad. Sci. U. S. A. 80, 6136–6140.

Schiffman, S. S., Suggs, M. S., Cragoe, E. J., Jr., and Erickson, R. P. 1990. Inhibition of taste responses to Na^+ salts by epithelial Na^+ channel blockers in gerbil. Physiol. Behav. 47, 455–459.

Schutz, H. G. and Pilgrim, F. J. 1957. Sweetness of various compounds and its measurement. Food Res. 22, 206–213.

Scinska, A., Koros, E., Habrat, B., Kukwa, A., Kostowski, W., and Bienkowski, P. 2000. Bitter and sweet components of ethanol taste in humans. Drug and alcohol dependence 60(2), 199–206.

Scott, T. R. and Giza, B. K. 1990. Coding channels in the taste system of the rat. Science 249, 1585–1587.

Scott, T. R. and Plata-Salaman, C. R. 1991. Coding of Taste Quality. In: Smell and Taste in Health and Disease

(eds. T. V. Getchell, R. L. Doty, L. M. Bartoshuk, and J. B. Snow), pp. 345–368. Raven Press.

Shallenberger, R. S. 1993. Taste Chemistry. Blackie Academic & Professional.

Shallenberger, R. S. and Acree, T. E. 1967. Molecular Theory of Sweet Taste. Nature (Lond.) 216, 480–482.

Shallenberger, R. S., Acree, T. E., and Lee, C. Y. 1969. Sweet taste of D- and L-sugars and amino acids and the steric nature of their chemoreceptor site. Nature (Lond.) 221, 555–556.

Shu, R. G., Xu, C. R., and Li, L. N. 1995. Studies on the sweet principles from the leaves of Cyclocarya paliurus (Batal.) Iljinsk. Acta Pharmaceutica Sinica 30, 757–761.

Simon, S. A. and Garvin, J. L. 1985. Salt and acid studies on canine lingual epithelium. Am. J. Physiol. 249, C398–408.

Simon, S. A., Holland, V. F., Benos, D. J., and Zampighi, G. A. 1993. Transcellular and paracellular pathways in lingual epithelia and their influence in taste transduction. Microsc. Res. Tech. 26(3), 196–208.

Simon, S. A., Robb, R., and Garvin, J. L. 1986. Epithelial responses of rabbit tongues and their involvement in taste transduction. Am. J. Physiol. 251, R598–608.

von Skramlik, E. 1926. Handbuch der Physiologie der niederen sinne: Band I. Die physiologie des geruchs-und geschmacksinnes. G. Thieme.

Smith, D. V. and Ossebaard, C. A. 1995. Amiloride suppression of the taste intensity of sodium chloride: evidence from direct magnitude scaling. Physiol. Behav. 57, 773–777.

Solms, J. 1969. Taste of amino acids, peptides and proteins. Int. z. vitaminforsch. 39, 320–322.

Spector, A. C., Guagliardo, N. A., and St. John, S. J. 1996. Amiloride disrupts NaCl versus KCl discrimination performance: implications for salt taste coding in rats. J. Neurosci. 16, 8115–8122.

Spielman, A. I., Huque, T., Whitney, G., and Brand, J. G. 1992. The Diversity of Bitter Taste Mechanisms. In: Sensory Transduction (eds. D. Corey and S. D. Roper), pp. 307–324. Rockefeller University Press.

St. John, S. J. and Smith, D. V. 2000. Neural representation of salts in the rat solitary nucleus: brain stem correlates of taste discrimination. J. Neurophysiol. 84, 628–638.

Stark, T., Bareuther, S., and Hofmann, T. 2005. Sensory-guided decomposition of roasted cocoa nibs (Theobroma cacao) and structure determination of taste-active polyphenols. J. Agric. Food Chem. 53, 5407–5418.

Steiner, J. E. 1994. Behavior Manifestations Indicative of Hedonics and Intensity in Chemosensory Experience. In: Olfaction and Taste XI, Proceedings of the 11th International Symposium on Olfaction and Taste and of the 27th Japanese Symposium on Taste and Smell (eds. K. Kurihara, N. Suzuki, and H. J. Ogawa), pp. 284–287. Springer-Verlag.

Stevens, D. R., Seifert, R., Bufe, B., Muller, F., Kremmer, E., Gauss, R., Meyerhof, W., Kaupp, U. B., and Lindemann, B. 2001. Hyperpolarization-activated channels HCN1 and HCN4 mediate responses to sour stimuli. Nature 413, 631–635.

Suarez, T., Kornfeld, E. C., and Sheneman, J. M. 1975. Sweetening Agent. US Patent 3 899 592 (August 12, 1975).

Sweeny, J. G., D'Angelo, L. L., Ricks, E. A., and Iacobucci, G. A. 1995. Discovery and synthesis of a new series of high-potency L-aspartyl-D-α-aminoalkanoyl-(S)-α-alkylbenzylamide sweeteners. J. Agric. Food. Chem. 43(8), 1969–1976.

Tahara, A., Nakata, T., and Ohtsuka, Y. 1971. New type of compound with strong sweetness. Nature (Lond.), 233–619.

Tanaka, T., Kawamura, K., Kohda, H., Yamasaki, K., and Tanaka, O. 1982. Glycosides of the leaves of Symplocos spp. (Symplocaceae), Chem. Pharm. Bull. 30, 2421–2423.

Tanaka, T., Tanaka, O., Lin, Z.-W., and Zhou, J. 1985. Sweet and bitter glycosides of the Chinese plant drug, Bai-Yun-

Shen: revision of the assignment of the source plant and isolation of two new diterpene glycosides. Chem. Pharm. Bull. 33, 4275–4280.

Tanaka, T., Tanaka, O., Lin, Z.-W., Zhou, J., and Ageta, H. 1983. Sweet and bitter glycosides of the Chinese plant drug, Bai-Yun-Shen (Roots of *Salvia Digitaloides*). Chem. Pharm. Bull. 31, 780–783.

Temussi, P. A., Lelj, F., and Tancredi, T. 1991. Structure–Activity Relationship of Sweet Molecules, Chapter 11. In: Sweeteners: Discovery, Molecular Design and Chemoreception (*eds*. D. E. Walters, F. T. Orthoefer, and G. E. DuBois), pp. 143–161. ACS Symposium Series 450, American Chemical Society.

Tinti, J.-M. and Nofre, C. 1991. Why Does a Sweetener Taste Sweet? A New Model. In: Sweeteners: Discovery, Molecular Design and Chemoreception (*eds*. D. E. Walters, F. T. Orthoefer, and G. E. DuBois), pp. 206–213. ACS Symposium Series 450, American Chemical Society.

Tinti, J.-M., Nofre, C., and Peytavi, A.-M. 1982. Interaction of suosan with the sweet taste receptor. Z. Lebensm.-Unters. - Forsch. 175, 266–268.

Trevisani, M., Smart, D., Gunthorpe, M. J., Tognetto, M., Barbieri, M., Campi, B., Amadesi, S., Gray, J., Jerman, J. C., Brough, S. J., Owen, D., Smith, G. D., Randall, A. D., Harrison, S., Bianchi, A., Davis, J. B., and Geppetti, P. 2002. Ethanol elicits and potentiates nociceptor responses via the vanilloid receptor-1. Nat. Neurosci. 5, 546–551.

Ugawa, S., Minami, Y., Guo, W., Saishin, Y., Takatsuji, K., Yamamoto, T., Tohyama, M., and Shimada, S. 1998. Receptor that leaves a sour taste in the mouth. Nature 395, 555–556.

van der Heijden, A., van der Wel, H., and Peer, H. G. 1985a. Structure–activity relationships in sweeteners. I. Nitroanilines, sulphamates, oximes, isocoumarins and dipeptides, Chem. Senses 10, 57–72.

van der Heijden, A., van der Wel, H., and Peer, H. G. 1985b. Structure–activity relationships in sweeteners. II. Saccharins, acesulfames, chlorosugars, tryptophans and ureas. Chem. Senses 10, 73–88.

van der Wel, H. 1972. Isolation and characterization of thaumatin I and II, the sweet tasting proteins from *Thaumatococcus danielli* (Benth), Eur. J. Biochem. 31, 221–225.

Varkevisser, B. and Kinnamon, S. C. 2000. Sweet taste transduction in hamster: role of protein kinases. J. Neurophysiol. 83(5), 2526–2532.

Verrey, F., Hummler, E., Schild, L., and Rossier, B. C. 2000. Control of Na Transport by Aldosterone. In: The Kidney: Physiology and Pathophysiology, Vol. 1 (*eds*. D. W. Seldin and G. Giebisch), pp. 1441–1471. Lippincott.

Vinnikova, A. K., Alam, R. I., Malik, S. A., Ereso, G. L., Feldman, G. M., McCarty, J. M., Knepper, M. A., Heck, G. L., DeSimone, J. A., and Lyall, V. 2004. Na^+–H^+ exchange activity in taste receptor cells. J. Neurophysiol. 91, 1297–1313.

Vleggaar, R., Ackerman, L. G. J., and Steyn, P. S. 1992. Structure elucidation of monatin, a high-intensity sweetener

isolated from the plant *Schlerochiton ilicifolius*. J. Chem. Soc. Perkin Trans. I Organic and Bio-Organic Chemistry 3095–3098.

de Vos, A. M., Hatada, M., van der Wel, H., Krabbendam, H., Peerdemann, A. F., and Kim, S.-H. 1985. Three dimensional structure of thaumatin I, an intensely sweet protein, Proc. Natl. Acad. Sci. U. S. A. 82, 1406–1409.

Xu, H., Staszewski, L., Tang, H., Adler, E., Zoller, M., and Li, X. 2004. Different functional roles of T1R subunits in the heteromeric taste receptors. Proc. Natl. Acad. Sci. U. S. A. 101, 14258–14263.

Yamaguchi, S. 1979. The Umami Taste. In: (*ed*. J. C. Boudreau), pp. 33–51, ACS Symposium Series 115, American Chemical Society.

Yamashita, H., Theerasilp, S., Aiuchi, T., Nakaya, K., Nakamura, Y., and Kurihara, Y. 1990. Purification and complete amino acid sequence of a new type of sweet protein with taste-modifying activity, curculin. J. Biol. Chem. 265, 15770–15775.

Ye, Q., Heck, G. L., and DeSimone, J. A. 1991. The anion paradox in sodium taste reception: resolution by voltage clamp studies. Science 254, 724–726.

Yoshii, K., Kiyomoto, Y., and Kurihara, K. 1986. Taste receptor mechanism of salts in frog and rat. Comp. Biochem. Physiol. A. 85, 501–507.

Zhang, Y., Hoon, M. A., Chandrashekar, J., Mueller, K. L., Cook, B., Wu, D., Zuker, C. S., and Ryba, N. J. P. 2003. Coding of sweet, bitter, and umami tastes: different receptor cells sharing similar signaling pathways. Cell 112, 293–301.

Zhao, G. Q., Zhang, Y., Hoon, M. A., Chandrashekar, J., Erlenbach, I., Ryba, N. J. P., and Zuker, C. S. 2003. The receptors for mammalian sweet and umami taste. Cell 115, 255–266.

Zubare-Samuelov, M., Shaul, M. E., Peri, I., Aliluiko, A., Tirosh, O., and Naim, M. 2005. Inhibition of signal termination-related kinases by membrane-permeant bitter and sweet tastants: potential role in taste signal termination. Am. J. Physiol. Cell Physiol. 289, C483–C492.

Further Reading

Economic and Social Department, Food and Agriculture Organization of the United Nations (2004). Food Outlook No. 4, Sugar, http://www.fao.org//docrep/007/j3877e/j3877e/j3877e12.htm.

Hashimoto, Y., Ishizone, H., Saganuma, M., Ogura, M., Nakatsu, K., and Yoshioka, H. 1983. Periandrin I, a sweet triterpene glycoside from *Periandra dulcis*. Phytochemistry 22, 259–264.

Hashimoto, Y., Ohta, Y., Ishizone, H., Kuriyama, M., and Ogura, M. 1982. Periandri III, a novel sweet triterpene glycoside from *Periandra dulcis*. Phytochemistry 21, 2335–2337.

4.03 Insect Gustatory Systems

John I Glendinning, Barnard College, Columbia University, New York, NY, USA

4.03.1 Introduction

The sense of taste is vital to invertebrates. It helps them avoid noxious chemicals (White, P. R. and Chapman, R. F., 1990: Schoonhoven, L. M. *et al.*, 1992; Glendinning, J. I., 2002), select host plants and oviposition sites (Ma, W.-C. and Schoonhoven, L. M., 1973; Städler, E. *et al.*, 1995), and identify and discriminate food-related chemical stimuli (Dethier, V. G., 1976; Bernays, E. A. and Chapman, R. F., 1994; Glendinning, J. I. *et al.*, 2002). Taste also helps invertebrates coordinate digestion by stimulating salivation during meals (Friend, W. G. and Smith, J. J., 1971; Ribeiro, J. M. C. and Garcia, E. S., 1980; Watanabe, H. and Mizunami, M., 2006) and determining whether ingested food is directed to the midgut for digestion or the crop for storage (Schmidt, J. M. and Friend, W. G., 1991).

I will examine several features of taste function: (1) the organization of the peripheral and central gustatory system; (2) responses of taste cells to food-related stimuli; (3) central processing of gustatory input; (4) taste–mixture interactions; and (5) factors that modulate gustatory responses. I will limit discussion to insects because this is the invertebrate taxon whose gustatory system has been studied most extensively. Further, because this chapter focuses on the mechanisms by which taste modulates feeding of insects, I will not review comprehensively the literature on insect gustatory receptor (Gr) genes and associated signaling pathways. Detailed reviews of these topics can be found elsewhere (Clyne, P. J. *et al.*, 2000; Glendinning, J. I. *et al.*, 2000a; Ishimoto, H. *et al.*, 2000; Dahanukar, A. *et al.*, 2001; Scott, K. *et al.*, 2001; Robertson, H. M. *et al.*, 2003; Chyb, S., 2004; Hallem, E. A. *et al.*, 2006; Robertson, H. M. and Wanner, K. W., 2006).

4.03.2   Organization of the Peripheral Gustatory System

Insect taste cells are bipolar sensory neurons, which are housed in peg-shaped uniporous structures called basiconic sensilla (or taste sensilla) (Figure 1(a)). These taste sensilla can occur inside the oral cavity, on mouthpart appendages, and/or on the thorax, ovipositor, antennae, wing margins, and/or legs (Chapman, R. F., 1982; Hansen-Delkeskamp, E., 1992; Stocker, R. F., 1994; Schoonhoven, L. M. and van Loon, J. J. A., 2002; Shuichi Haupt, S., 2004). Owing to the external location of many taste sensilla, insects can taste foods simply by contacting them. The taste sensilla are innervated by the single, unbranched dendrites of two to six taste cells, which terminate near the distal tip of the sensillum (Figure 1(b)). The cell body of each cell is located

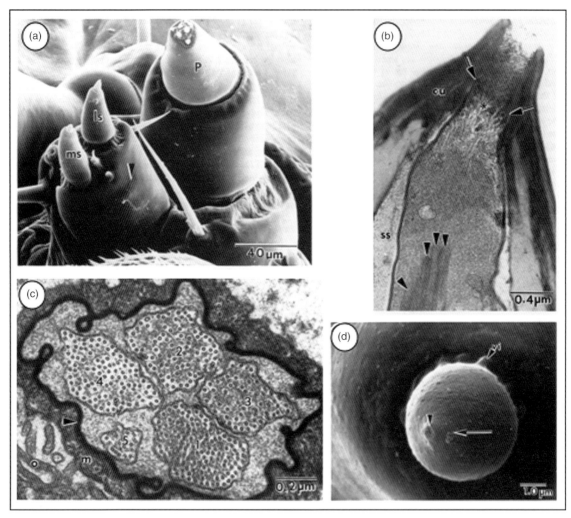

Figure 1   SEM micrographs of a taste sensillum from a caterpillar (*Mamestra configurata*, Lepidoptera: Noctuidae). (a) Left maxilla of caterpillar with lateral (*ls*) and medial (*ms*) styloconic sensilla, and a maxillary palp (*P*). (b) Longitudinal section of a lateral styloconic sensillum. At the top of the image, one can see the tip of the pore, which contains a plug of fenestrated fibrils (asterisk), and the dendritic sheath (arrows), which is fused with the cuticular wall of the sensillum. Four distal dendrites, each from different taste cells (arrowheads), can be seen near the bottom of the image; they are enclosed by the dendritic sheath. *cu*, cuticle; *ss*, sensillar sinus. (c) Proximal cross-section of a medial styloconic sensillum, showing four taste cells (*1–4*) and a mechanosensory neuron (*5*). All neurons are enclosed by a dendritic sheath (arrowhead), which is surrounded by a densely granulated intermediate sheath cell (m). *o*, outer sheath cell. (d) Apical view of the tip of a lateral styloconic sensillum with the terminal pore (arrowhead). These micrographs are from Shields, V. D. C. 1994. Ultrastructure of the uniporous sensilla on the galea of larval *Mamestra configurata* (Walker) (Lepidoptera: Noctuidae). Can. J. Zool. 72, 2016–2031.

near the base of the taste sensillum, and the axon extends directly to the central nervous system. In addition to taste cells, there is usually a single mechanosensory neuron in each taste sensillum (Figure 1(c)). The total number of taste sensilla per insect, and the distribution of these sensilla across the body surface, varies greatly among insect species (Chapman, R. F., 1982). Our ability to explore the membrane properties and intracellular signaling mechanisms of taste cells was increased by a recently developed method for isolating dissociated taste cells from pupal blowflies, *Phormia regina* (Murata, Y. *et al.*, 2006).

Taste transduction begins when a chemical stimulus encounters the pore at the tip of a taste sensillum, diffuses through a mucopolysaccharide substance in the pore, and then dissolves into the fluid surrounding the dendrites (Figures 1(b) and 1(d)). This fluid (or receptor lymph) constitutes the outer ionic milieu for the dendritic process of each taste cell. Once the chemical stimulus encounters the distal dendritic membrane, it is thought to interact with specific receptor proteins and/or ionic channels, leading to changes in membrane conductance (Morita, H., 1992). The changes induce an inward current that flows through the transduction ion channels and down to the base of the dendritic process, where current outflow causes depolarization. While the receptor lymph provides some of the driving electromotive force for this taste cell current, most of the force appears to come from a $+40$ to $+60$ mV potential difference between the receptor lymph space and the blood (or hemolymph), called the transepithelial voltage (Thurm, U. and Küppers, J., 1980; Kijima, H. *et al.*, 1995). The transepithelial voltage is established by electrogenic ion pumps located in the folded membrane of the tormogen and trichogen cells, which surround the taste cells.

4.03.3 Responses of Taste Cells to Food-Related Chemical Stimuli

Insect physiologists typically classify taste cells according to their best taste stimuli – that is, the stimuli that generate the strongest excitatory response. Accordingly, a taste cell that responds best to salts is called salt-sensitive, and a taste cell that responds best to compounds that humans describe as bitter is called bitter-sensitive. There are two caveats associated with this terminology. First, because most insect taste cells have been tested with only a limited number of taste stimuli, it is conceivable that the best stimulus for a taste cell has not yet been discovered. Second, the categorization of a taste cell as salt sensitive implies that stimulation of that taste cell generates a perception akin to saltiness in humans. At this point, we do not even know whether insects perceive distinct taste qualities like salty and sweet (see Section 4.03.5). For the sake of simplicity, I will use the conventional terminology for categorizing the taste cells, keeping in mind the above-stated caveats.

4.03.3.1 Bitter Compounds

Many of the compounds that taste bitter in humans also elicit an aversive response in insects (e.g., rapid cessation of feeding). Because virtually all naturally occurring poisons taste bitter to humans, the aversive response to these bitter compounds is thought to represent an evolved mechanism for limiting ingestion of toxic foods (Bate-Smith, E. C., 1972; Garcia, J. and Hankins, W. G., 1975; Brower, L. P., 1984; Brieskorn, C. H., 1990). The effectiveness of this taste-mediated poison detection system is limited by the fact that even though most poisons taste bitter, the opposite is not always true – that is, many compounds with a bitter taste are harmless (Bernays, E. A. and Chapman, R. F., 1987; Bernays, E. A., 1990; Rouseff, R. L., 1990; Bernays, E. A., 1991; Glendinning, J. I., 1994). Given that many nutritious plant tissues contain bitter compounds, insects (particularly herbivorous ones) would benefit from gustatory mechanisms that discriminate harmful and harmless bitter compounds.

Insects have several preingestive and postingestive mechanisms for detecting potentially toxic compounds in foods (Glendinning, J. I., 2002). The primary mechanism, however, is gustation. Insects have numerous taste cells that respond selectively to bitter compounds and elicit an aversive response when activated (Dethier, V. G., 1980; Schoonhoven, L. M. *et al.*, 1992). These bitter-sensitive taste cells have been discovered in Orthopterans (White, P. R. and Chapman, R. F., 1990; Chapman, R. F. *et al.*, 1991), Dipterans (Meunier, N. *et al.*, 2003), Lepidopterans (Schoonhoven, L. M. *et al.*, 1992), and some but not all Coleopterans (Mitchell, B. K., 1987; Messchemdorp, L. *et al.*, 1998). To illustrate the response properties of bitter-sensitive taste cells, I will feature the caterpillar of *Manduca sexta* (Lepidoptera: Sphingidae). This insect has a total of 62 taste cells distributed across five classes of bilaterally paired taste sensilla: one lateral styloconic, one medial styloconic, one epipharyngeal, and five maxillary palp sensilla (Figure 2(a)).

(a)

(b)

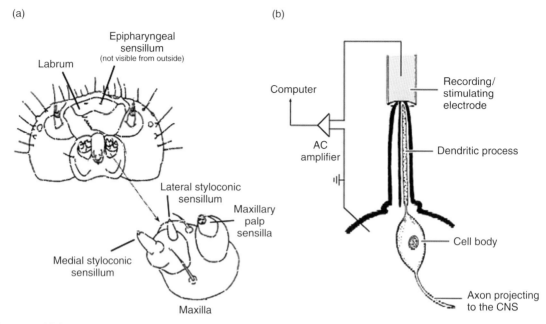

Figure 2 (a) Cartoon of the head of a *Manduca sexta* caterpillar, as viewed from below. An enlargement of the maxilla (indicated with an arrow) is provided to clarify the location of the medial and lateral styloconic sensilla. The epipharyngeal sensilla are located underneath the labrum, and thus are not visible in this diagram. (b) Illustration of the tip recording method, which was used to record excitatory responses of individual taste cells located within a taste sensillum. During a tip recording (Hodgson, E. S. *et al.*, 1955), the tip of a taste sensillum is inserted into the end of a glass recording/stimulating electrode, which is filled with a taste stimulus dissolved in an electrolyte solution (0.1 M KCl in deionized water). The taste stimulus solution diffuses through a pore in the tip of the sensillum and activates transduction mechanism(s) on the distal end of a taste cell's dendritic process; the electrode detects the ensuing action potentials. For clarity, only one taste cell is indicated. Note that the taste cell's axonal process projects directly to the central nervous system without synapsing. (a) This cartoon was adapted from Bernays, E. A. and Chapman, R. F. 1994. Host-Plant Selection by Phytophagous Insects. Chapman and Hall, their Figure 3.4.

Given that each taste sensillum contains one bitter-sensitive taste cell, *M. sexta* has a total of eight bilateral pairs of bitter-sensitive taste cells. Using the tip recording technique (Figure 2(b)), investigators have characterized the response of each bitter-sensitive taste cell to a structurally diverse range of substances that elicit immediate aversive responses: salicin, caffeine, aristolochic acid, *Canna* extract, and *Grindelia* extract (Figure 3). These recordings illustrate that the bitter-sensitive taste cell in the (1) epipharyngeal sensilla responds vigorously to salicin, caffeine, aristolochic acid, and *Canna* extract, (2) lateral styloconic sensilla responds vigorously to salicin, caffeine, aristolochic acid, and weakly to *Canna* extract, (3) medial styloconic sensilla responds vigorously to aristolochic acid and *Canna* extract, and weakly to caffeine and salicin, and (4) maxillary palp sensilla responds vigorously to *Grindelia* extract only (Glendinning, J. I. *et al.*, 2002). Behavioral studies have revealed that any bilateral pair of taste sensilla that responds vigorously to a particular bitter compound is also sufficient to mediate an aversive response to that compound (Glendinning,

J. I. *et al.*, 1999a; 2002; Glendinning, J. I., unpublished data). Taken together, these studies show that *M. sexta* possesses a heterogeneous population of bitter-sensitive taste cells – some with identical molecular receptive ranges (MRRs), others with partially overlapping MRRs, and yet others with unique MRRs. The adaptive significance of this taste cell diversity is that it facilitates the detection and discrimination of bitter taste stimuli.

To determine whether two bitter taste stimuli (e.g., caffeine and salicin) activate the same taste cell, one can examine the neural response of a taste cell to each compound, both alone and in binary mixture. In Figure 4(a), we show the response of the bitter-sensitive taste cell in the lateral styloconic sensillum to caffeine alone, salicin alone, and the binary mixture of both. Because the binary mixture caused the bitter-sensitive taste cell to generate approximately twice as many spikes as did caffeine or salicin alone, we infer that caffeine and salicin activated the same bitter-sensitive taste cell. In Figure 4b, we show the response of the bitter-sensitive taste cell in the lateral

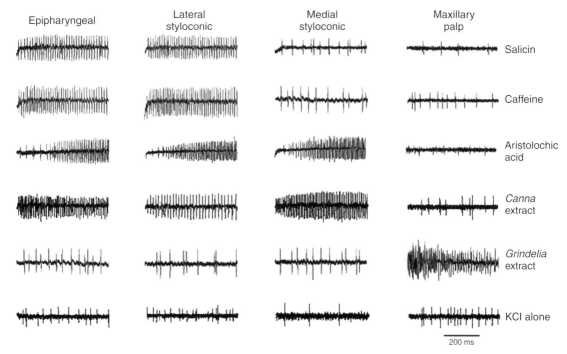

Figure 3 Typical excitatory responses of the four different classes of taste sensilla to five bitter taste stimuli: salicin, caffeine, aristolochic acid, *Canna* extract, and *Grindelia* extract (see Glendinning, J. I. *et al.*, 2002, for exact concentrations of each taste stimulus). The onset of stimulation occurred at the beginning of each trace, and each vertical line reflects the occurrence of an action potential. In most of the traces containing a large number of action potentials, there is a single taste cell (the bitter-sensitive taste cell) firing regularly, and another taste cell (the salt-sensitive taste cell) firing sporadically. The only exception is the multiunit response of the maxillary palp sensilla to the *Grindelia* extract, which probably contains action potentials from several bitter-sensitive taste cells. Because all bitter taste stimuli were presented in a 0.1 M KCl solution, we also show representative responses of each sensillum to the electrolyte solution alone for comparison (see bottom row of traces). Note that each bitter taste stimulus selectively activates bitter-sensitive taste cells in a subset of the taste sensilla (e.g., the *Grindelia* extract only stimulated bitter-sensitive taste cells in the maxillary palp). These traces are from Glendinning, J. I., Davis, A., and Ramaswamy, S. 2002. Contribution of different taste cells and signaling pathways to the discrimination of "bitter" taste stimuli by an insect. J. Neurosci. 22, 7281–7287.

styloconic sensillum to caffeine alone, *myo*-inositol alone, and the binary mixture of both. Because the response to the binary mixture consisted of two taste cells discharging out of phase with one another, and at the same rate as when each component was presented alone, we infer that each taste stimulus activated a different taste cell.

Several recent observations in adult *Drosophila* offer an explanation for why the bitter-sensitive taste cells in *Drosophila* (Meunier, N. *et al.*, 2003) and perhaps *M. sexta* (Figure 4) express distinct MRRs. The first observation is that the bitter-sensitive taste cells that express Gr66a (a presumed bitter taste receptor) respond to a diversity of bitter taste stimuli (Wang, Z. *et al.*, 2004; Marella, S. *et al.* 2006), and that silencing the Gr66a-expressing taste cells attenuates the aversive response to many bitter compounds (Wang, Z. *et al.*, 2004). The second observation is that knocking out the Gr66a receptor attenuates the responses of the bitter-sensitive taste

cells to caffeine and theophylline, but spares responses to most other bitter taste stimuli (Moon, S. J. *et al.*, 2006). Taken together, these observations indicate (1) that there are many different bitter taste receptors, (2) that each receptor has a relatively narrow and distinct MRR; and (3) that each class of bitter-sensitive taste cell expresses different combinations of the receptors.

A final gustatory mechanism for limiting ingestion of bitter and potentially toxic compounds is activated when insects encounter foods that contain mixtures of nutrients and bitter compounds. For instance, the labellar taste sensilla in some flies appear to lack bitter-sensitive taste cells *per se.* However, the excitatory response of the sugar-, salt-, and water-sensitive taste cells to their respective ligands is inhibited by the presence of bitter compounds (Morita, H., 1959; Dethier, V. G. and Bowdan, E., 1989; Meunier, N. *et al.*, 2003). This inhibition effectively eliminates the gustatory input that would normally stimulate feeding

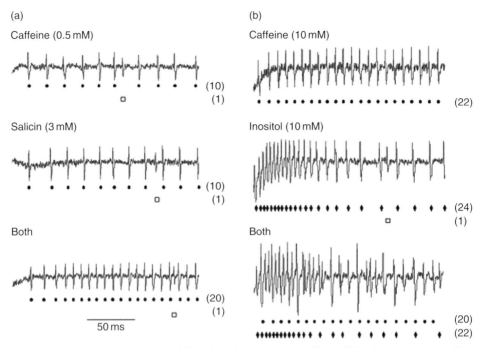

Figure 4 Excitatory responses of taste cells within a lateral styloconic sensillum of *Manduca sexta* caterpillars to single-component taste stimuli and binary mixtures. (a) Response of a bitter-sensitive taste cell to 0.5 mM caffeine alone, 3 mM salicin alone, and the binary mixture of both. Underneath each neural trace, we indicate the response of the bitter-sensitive taste cell (closed circle) and that of the salt-sensitive taste cell to the 0.1 mM KCl solvent (open square). We also indicate the total number of spikes assigned to each taste cell in parentheses. It is apparent that the bitter-sensitive taste cell generated twice as many spikes in response to the binary mixture than it did to either component alone. These traces are from Glendinning, J. I. and Hills, T. T. 1997. Electrophysiological evidence for two transduction pathways within a bittersensitive taste receptor. J. Neurophysiol. 78, 734–745. (b) Response of a bitter-sensitive taste cell to 10 mM caffeine, an inositol-sensitive taste cell to 10 mM *myo*-inositol alone, and both taste cells to the binary mixture of both taste stimuli. Underneath each neural trace, we indicate the response of the bitter-sensitive taste cell (closed circle), the response of the inositol-sensitive taste cell (closed diamonds), and the salt-sensitive taste cell (to the 0.1 mM KCl solvent) (open square). It is apparent that the bitter-sensitive taste cell and the inositol-sensitive taste cells each responded equally vigorously to their respective ligands, irrespective of whether they occurred alone or in binary mixture. Further, note that the response to the binary mixture contains two taste cells firing independently of one another; the unusually high-amplitude spikes reflect superpositions (i.e., instances when the two taste cells fired simultaneously). These traces are from Glendinning, J. I., Nelson, N., and Bernays, E. A. 2000b. How do inositol and glucose modulate feeding in *Manduca sexta* caterpillars? J. Exp. Biol. 203, 1299–1315.

(Dethier, V. G. and Bowdan, E., 1992). One limitation of this detection mechanism is that it is relatively nonselective – e.g., bitter stimuli that differ greatly in toxicity all produce a similar degree of inhibition (Dethier, V. G. and Bowdan, E., 1992).

4.03.3.2 Carbohydrates

Carbohydrates are strong feeding stimulants for many insects (Dethier, V. G., 1976; Bernays, E. A., 1985; Scheiner, R., 2004; Wang, Z. *et al.*, 2004). Most insects possess taste cells that respond selectively to saccharides (e.g., glucose, sucrose, fructose, and trehalose) and/or sugar alcohols (e.g., sorbitol, mannitol, and *myo*-inositol). Whereas Dipterans have carbohydrate-sensitive taste cells that respond both to saccharides and to sugar

alcohols, most Lepidopterans have separate taste cells for saccharides and sugar alcohols (Dethier, V. G., 1976; Schoonhoven, L. M. and van Loon, J. J. A., 2002).

The taste system of adult *Drosophila* is notable because it contains the only deorphaned sweet taste receptor in insects: Gr5a. Using a combination of molecular, physiological, and behavioral approaches, investigators demonstrated that Gr5a is encoded by the *Gr5a* gene, and that it is tuned specifically to trehalose, an abundant metabolic byproduct of yeast and an important food source for *Drosophila* (Tanimura, T. *et al.*, 1982; 1988; Dahanukar, A. *et al.*, 2001; Ueno, K. *et al.*, 2001; Chyb, S. *et al.*, 2003; Inomata, N. *et al.*, 2004). Although mutation of Gr5a attenuated the responsiveness of the sugar-sensitive taste cells to trehalose, it had no impact on responsiveness of the same taste cells to sucrose (Tanimura, T. *et al.*, 1982; 1988; Dahanukar, A. *et al.*, 2001; Ueno, K. *et al.*, 2001; Chyb, S. *et al.*, 2003; Inomata, N. *et al.*, 2004). In addition, two laboratories developed methods for selectively silencing the *Gr5a*-expressing taste cells in adult *Drosophila*; they did so by causing them to produce either tetanus (Thorne, N. *et al.*, 2004) or diphtheria (Wang, Z. *et al.*, 2004) toxin. Using the genetically modified flies, the laboratories demonstrated that flies expressing the toxins under the *Gr5a* driver showed an attenuated appetitive response to trehalose, but normal appetitive response to sucrose. Taken together, these molecular findings indicate that individual carbohydrate-sensitive taste cells express multiple taste receptors for sugars, each of which has a unique MRR. This inference is supported by electrophysiological studies showing that different classes of sensilla in *Drosophila* exhibit different response profiles for sugars (Hiroi, M. *et al.*, 2002).

Prior to the discovery of Gr5a, most work on sugar transduction focused on two hypothetical receptor sites: one that responds to pyranose sugars (e.g., glucose and sucrose) and another to furanose sugars (e.g., fructose) (Morita, H., 1992). The existence of these separate receptor sites was based on the following lines of evidence. Treating the sugar-sensitive taste cell with *p*-chloromercuribenzonate (a specific sulphydryl reagent) almost completely eliminated the response to pyranose sugars, but had no impact on the response to furanose sugars (Shimada, I. *et al.*, 1974). Conversely, treating the sugar-sensitive taste cell with amiloride abolished its response to furanose sugars, but not pyranose sugars (Liscia, A. *et al.*, 1997). Finally, Ozaki and her colleagues used affinity electrophoresis to isolate two putative taste receptor molecules; each of these proteins exhibited sugar-binding affinities and specificities that were consistent with separate pyranose- and furanose-binding sites (Ozaki, M. *et al.*, 1993). More work is needed to integrate these hypothetical receptor sites with our growing molecular understanding of insect gustation.

Based on sequence data from *Drosophila*, the 68 putative gustatory receptor genes (including *Gr5a*) are predicted to encode G-protein-coupled receptors (Robertson, H. M. *et al.*, 2003). In support of this prediction, several recent studies have reported that two G-protein subunits, Gsα and Gγ1, are both required for sugar detection in *Drosophila* (Ishimoto, H. *et al.*, 2005; Ueno, K. *et al.*, 2006). These G-protein subunits, in turn, appear to activate an IP$_3$ signaling pathway (Usui-Aoki, K. *et al.*, 2005). Other signaling molecules proposed to mediate sugar transduction include Gq-α (Seno, K. *et al.*, 2005b), nitric oxide (Schuppe, H. *et al.*, 2006), and calmodulin (Seno, K. *et al.*, 2005a).

4.03.3.3 Salts

Virtually all taste sensilla contain at least one salt-sensitive taste cell. Monovalent inorganic cations are the most effective stimulants for these taste cells (Dethier, V. G., 1976). Cations differ, however, in their stimulatory effectiveness (e.g., in the fleshfly, *Boettcherisca peregrina*, $Li^+ < Na^+ < Cs^+ < Rb^+ < K^+$) (den Otter, C. J., 1972). den Otter speculated that this order of stimulatory effectively does not vary as a function of ionic size or mobility because each cation interacts differentially with negative groups (probably phosphates) on the surface of the taste cells.

Flies show an appetitive response to low NaCl concentrations (5–10 mM) and aversive response to high NaCl concentrations ($\geq$500 mM) (Dethier, V. G., 1968; Hiroi, M. *et al.*, 2004). This concentration-dependent change in affective response to salts is thought to have evolved because low concentrations of salts satisfy nutritional needs, whereas higher concentrations threaten osmotic equilibrium. To explain the mechanistic basis of this behavior, Wang Z. *et al.* (2004) silenced the sugar-sensitive taste cells of *Drosophila* that express *Gr5a*. When these flies were offered a range of NaCl concentrations, they exhibited an attenuated appetitive response to the 5 mM concentration, but a normal aversive response to concentrations $\geq$500 mM. These latter findings, together with those of Hiroi M. *et al.* (2004), indicate that NaCl stimulates two different classes of taste cell – low concentrations stimulate a taste cell that elicits an appetitive response (presumably the so-called sugar-sensitive taste cell) and high concentrations stimulate a taste cell that elicits an aversive response (presumably a salt-sensitive taste cell).

In *Drosophila*, the aversive response to high (i.e., $\geq$500 mM) concentrations of NaCl is mediated in part by the *dpr* gene. This inference is based on behavioral studies with wildtype and *dpr*-knockout flies (Nakamura, M. *et al.*, 2002). The proboscis extension reflex was strongly inhibited by high NaCl concentrations in wildtype flies, but not in *dpr*-knockout flies. This *dpr* gene is thought to encode a protein with a single transmembrane domain and two Ig repeats, and to be expressed in a subset of taste cells (Nakamura, M. *et al.*, 2002).

An important, but unexplained, observation is that some insects can discriminate between different types of salts (Arms, K. *et al.*, 1974; Maes, F. W. and Bijpost,

S. C. A., 1979; Miyakawa, Y., 1981; 1982). In laboratory rats, the ability to discriminate Na^+ and K^+ salts appears to be mediated by differential input from Na^+-specific (amiloride-sensitive) and Na^+-nonspecific (amiloride-insensitive) cation channels (Spector, A. C. and Grill, H. J., 1992; Smith, D. V. *et al.*, 2000). The question of whether insects use a similar mechanism is unresolved. A recent study reported that *Drosophila* expresses a Na^+-nonspecific (amiloride-sensitive) cation channel (DEG/ENaC) in taste cells within the gustatory terminal organ (in larvae) and taste sensilla on the labellum, legs, and wing margins (in adults) (Liu, L. *et al.*, 2003). The inference that this channel is Na^+ nonspecific was based on the observation that disruption of the genes for DEG/ENaC (*ppk11* and *ppk19*) diminished behavioral and taste cell responsiveness to low (i.e., 10 mM) concentrations of both KCl and NaCl (Liu, L. *et al.*, 2003). Given that the gustatory response to 10 mM KCl in the genetically disrupted flies was still robust (i.e., $\sim$50% of that observed in wildtype flies), it is likely that *Drosophila* also has a DEG/ENaC-independent mechanism for detecting salts. Accordingly, differential input from the DEG/ENaC-dependent and DEG/ENaC-independent mechanisms could mediate the discrimination of Na^+ and K^+ salts.

4.03.3.4 Amino Acids

Amino acids stimulate feeding in many species of insect, but the behavioral responsiveness to any given amino acid varies considerably across species (Bernays, E. A., 1985; Albert, P. J. and Parisella, S., 1988; Hirao, T. and Ariai, N., 1990; Kim, J. H. and Mullin, C. A., 1998). A positive feeding response to amino acids may serve two adaptive functions. First, amino acids could be used as a proxy for protein content of foods, given that plant tissues with relatively high protein concentrations should, on average, have relatively high concentrations of free amino acids. If so, then amino acid-sensitive taste cells could help insects identify protein-rich plant tissues. Second, because some amino acids (e.g., phenylalanine) are considered limiting nutrients for plant-feeding insects (Bernays, E. A., 1982), amino acid-sensitive taste cells could help the insect maintain nutritional homeostasis.

In caterpillars, amino acids often stimulate specific taste cells (Dethier, V. G., 1973; Schoonhoven, L. M. and van Loon, J. J. A., 2002). For example, two closely related species of *Pieris* possess an amino acid-sensitive taste cell that shows concentration-dependent increases in response to ecologically relevant concentrations of 12 L-amino acids (Dethier, V. G., 1973; van Loon, J. J. A. and van Eeuwijk, F. A., 1989). Other species of caterpillar (e.g., *Grammia geneura*) detect amino acids with multiple taste cells (Bernays, E. A. and Chapman, E. A., 2001b). One of the amino acid-sensitive taste cells in *G. geneura* is unusual in that it responds to a diverse array of phagostimulants present in the insect's preferred food plants, including seven amino acids, three sugars, and a secondary plant compound (catalpol, an iridoid glycoside) (Bernays, E. A. *et al.*, 2000).

Dipterans use a variety of gustatory mechanisms for detecting amino acids. On the one hand, the tsetse fly (*Flossina fuscipes fuscipes*) has an amino acid-sensitive taste cell that responds selectively and vigorously 11 of the 20 L-amino acids present in animal proteins (van der Goes van Naters, W. M. and den Otter, C. J., 1998). On the other hand, the blowfly (*P. regina*), fleshfly (*B. peregrina*), and hoverfly (*Eristalis tenax*) all lack taste cells that respond to amino acids (Wolbarsht, M. L. and Hanson, F. E., 1965; Shiraishi, A. and Kuwabara, M., 1970; Goldrich, N. R., 1973; Wacht, S. *et al.*, 2000). Instead, these flies detect amino acids in more complex ways. For instance, when the labellar sensilla of blowflies and fleshflies were stimulated with each of 19 L-amino acids, the ensuing neural response fell into one of four categories. The amino acids (1) inhibited the excitatory response of all taste cells (aspartic acid, glutamic acid, histidine, arginine, and lysine); (2) stimulated the salt-sensitive taste cells (proline and hydroxyproline); (3) stimulated the sugar-sensitive taste cell (valine, leucine, isoleucine, methionine, phenylalanine, and tryptophan); or (4) failed to stimulate or inhibit any taste cell (glycine, alanine, serine, threonine, cysteine, and tyrosine) (Shiraishi, A. and Kuwabara, M., 1970; Goldrich, N. R., 1973). The behavioral significance of these neural responses is unclear because the same taste stimuli fail to elicit normal appetitive responses – that is, proboscis extension reflexes (Wolbarsht, M. L. and Hanson, F. E., 1965).

4.03.3.5 Acids

In humans, acids elicit the perception of sourness. Although organic acids tend to be more sour than inorganic acids (at the same pH), the intensity of sourness generally increases with proton concentration. Little is known about the behavioral responses of insects to low pH foods, however. There is one report showing that ascorbic acid stimulates feeding in *Pieris brassicae* caterpillars, but only when it is presented in binary mixture with sucrose (Ma, W.-C., 1972).

Organic acids occur in all plant tissues, and their concentrations can reach levels over 100 mM (e.g., citric acid from the Krebs cycle) (Harborne, J. B., 1973). Another nutritionally significant organic acid, ascorbic acid, occurs in plant tissues at concentrations of 1–10 mM (Schultz, J. C. and Lechowicz, M. J., 1986). The fact that ascorbic acid is an essential vitamin for most herbivorous insects may explain why caterpillars show an appetitive response to ascorbic acid when it co-occurs with sugars. On the other hand, the fact that acidic substances can be toxic and/or depress pH of the midgut may explain why many insects display an aversive response to highly acidic foods.

Ground beetles (Coleoptera, Carabidae) are the only known taxon of insects with taste cells that show pH-dependent changes in excitatory response. In these taste cells, the higher the pH, the higher the firing rate (Millius, M. *et al.*, 2006). For most other taxa of insects, acids appear to play a modulatory role – that is, alter the responsiveness of specific taste cells to their best stimuli. For example, at low concentrations, acids inhibit the responsiveness of salt-sensitive taste cells, whereas at higher concentrations, they inhibit the responsiveness of sugar- and inositol-sensitive taste cells and stimulate bitter-sensitive taste cells (Dethier, V. G. and Kuch, J. H., 1971; Bernays, E. A. *et al.*, 1998). These modulatory effects of acids are due to pH and not to specific effects of any particular acidic compound.

4.03.3.6 Miscellaneous Compounds

In addition to the taste stimuli described above, insect taste cells have been found to respond to chemicals not traditionally studied by taste physiologists. Below are four examples. First, the sugar-sensitive taste cells of blood-feeding insects (e.g., the fleshfly, *B. peregrina*) discharge in response to nucleotides in the blood of hosts (e.g., ADP and GDP) (Furuyama, A. *et al.*, 1999). Second, several herbivorous insects possess taste cells that respond to phytoecdysteroids in plant tissues (Ma, W.-C., 1969; Tanaka, Y. *et al.*, 1994; Marion-Poll, F. and Descoins, C., 2002; Calas, D. *et al.*, 2007). Stimulation of these taste cells elicits an aversive response. This response is thought to have evolved because phytoecdysteroids are found in about 6% of all land plant species, and their consumption can interfere with ecdysis. Third, some plant compounds, which are bitter and poisonous to mammals, actually

stimulate feeding in herbivorous insects. For example, several species of insect, which depend on pyrrolizidine alkaloids (PAs) for reproduction and chemical defense against predators, possess taste cells specifically tuned to PAs (Bernays, E. A. *et al.*, 2002; 2003). These PA-sensitive taste cells are thought to facilitate the location and consumption of PA-containing plants. Fourth, given the ease at which insects desiccate, it is not surprising that they have evolved taste cells tuned to water (Dethier, V. G., 1976). A recent study identified water-sensitive taste cells in adult *Drosophila* and showed that these taste cells project to a region of the subesophageal ganglion (SOG) that is distinct from that of the sugar- and bitter-sensitive taste cells (Inoshita, T. and Tanimura, T., 2006).

4.03.4 Organization of the Central Gustatory System

The SOG is considered the first relay center for taste processing because it is where the axons of most taste cells terminate (Kent, K. S. and Hildebrand, J. G., 1987; Mitchell, B. K. and Itagaki, H., 1992; Stocker, R. F., 1994; Mitchell, B. K. *et al.*, 1999; Thorne, N. *et al.*, 2004; Wang, Z. *et al.*, 2004). Exceptions to this pattern include instances where (1) taste cells in the legs of Orthopterans project to thoracic ganglia (Rogers, S. M. and Newland, P. L., 2002); and (2) a subpopulation of taste cells in the mouthpart sensilla of Lepidopterans project to the tritocerebrum (Kent, K. S. and Hildebrand, J. G., 1987; Jørgensen, K. *et al.*, 2006).

Several lines of evidence indicate that taste cells can directly activate feeding control circuits in the SOG. First, taste cell axons intermingle with (or occur in close apposition to) arborizations from the mouthpart motor neurons (Altman, J. S. and Kien, J., 1987). Second, some SOG interneurons have branches that occur in neuropil with both sensory and motor terminals (Altman, S. and Kien, J., 1987). Third, stimulating taste cells with plant extracts directly modulates the output of the mouthpart motor neurons (Griss, C. *et al.*, 1991). In addition to exerting effects within the SOG, afferent gustatory information appears to be relayed via the subesophageal-calycal tract to higher processing centers in the calyces of the mushroom body, which are important neuropils for learning and memory (Schröter, U. and Menzel, R., 2003).

To gain insight into the organization of the SOG, investigators initially used anatomical and cobalt-filling techniques, together with activity-dependent staining of single neurons, to determine the projection sites of axons from taste cells. These studies revealed that taste cells from different gustatory organs, different taste sensilla, or even the same sensillum can project to spatially distinct regions of the SOG (Kent, K. S. and Hildebrand, J. G., 1987; Shanbhag, S. R. and Singh, R. N., 1992; Stocker, R. F., 1994; Pollack, G. S. and Balakrishnan, R., 1997; Jørgensen, K. *et al.*, 2006).

Using molecular tools, investigators have been able to define more precisely the central projection patterns of taste cells. For example, work in *Drosophila* has shown that taste receptor expression can be used to characterize axonal projections from different classes of taste cell. Several investigators have shown that axonal projections to the SOG are segregated spatially according to both taste cell type (e.g.,

sugar sensitive versus bitter sensitive) and gustatory organ (proboscis, mouthparts, and legs) (Figure 5) (Thorne, N. *et al.*, 2004; Wang, Z. *et al.*, 2004; Inoshita, T. and Tanimura, T., 2006). These findings reveal that the SOG is organized according to a chemotopic and somatotopic organizational plan – i.e., the spatial pattern of activation reflects both the class of taste cell that was stimulated and the location of the taste stimulus on the body.

4.03.5 Central Processing of Gustatory Input

Mammals categorize taste stimuli according to three orthogonal dimensions: intensity, hedonics, and quality (Grill, H. J. and Berridge, K. C., 1985; Spector, A. C.,

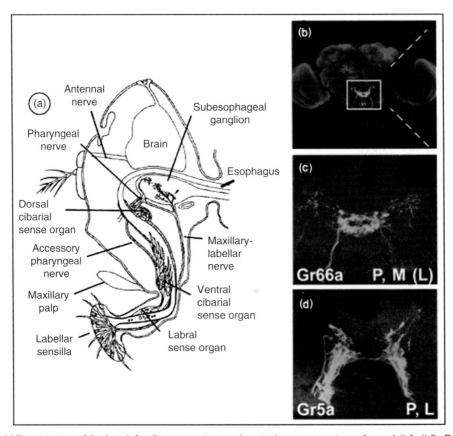

Figure 5 (a) Ultrastructure of the head, feeding apparatus, and central nervous system of an adult fruit fly *Drosophila melanogaster* (from Stocker, R. F. and Schorderet, M. 1981. Cobalt filling of sensory projections from internal and external mouthparts in *Drosophila*. Cell Tissue Res. 216, 513–523). (b) A whole-mount of the brain of *D. melanogaster*, showing axonal projects from taste cells (labeled with anti-GFP immunohistochemistry). The white box indicates the region of the SOG depicted in (c)–(d). (c–d) The green axons are from taste cells expressing a presumed bitter taste receptor, Gr66a (c), and the trehalose taste receptor, Gr5a (d). It is notable that the projection sites for taste cells expressing Gr66a are spatially distinct from those expressing Gr5a. (b)–(d) Are from Wang, Z., Singhvi, A., Kong, P., and Scott, K. 2004. Taste representations in the *Drosophila* brain. Cell 117, 981–991.

2000). Do insects use these three taste dimensions as well? Behavioral studies provide clear support for use of the intensity and hedonic dimensions. With regard to intensity, there are several reports of insects (1) exhibiting concentration-dependent changes in taste-mediated behavioral responsiveness to specific taste stimuli (Dethier, V. G., 1968; Bowdan, E., 1995; Glendinning, J. I. and Gonzalez, N. A., 1995; Glendinning, J. I. *et al.*, 1999a; Wang, Z. *et al.*, 2004), and (2) discriminating different intensities (i.e., concentrations) of salts in a classical conditioning paradigm (Maes, F. W. and Bijpost, S. C. A., 1979). With regard to hedonics, there are numerous reports of insects showing a taste-mediated preference or aversion to chemical stimuli (Harley, K. L. S. and Thorsteinson, A. J., 1967; Dethier, V. G., 1968; 1976; Bernays, E. A. and Chapman, R. F., 1994; Bowdan, E., 1995; Glendinning, J. I. and Gonzalez, N. A., 1995; Glendinning, J. I. *et al.*, 1999a; Wang, Z. *et al.*, 2004). The question of whether insects categorize taste stimuli into different qualities, however, has not been examined systematically.

To demonstrate that taste stimuli are grouped into different qualities (e.g., sweet, salty, sour, bitter, and umami), one cannot simply observe unconditioned behavioral responses. This is because taste stimuli with different taste qualities can elicit similar behavioral responses (Spector, A. C., 2000). For example, in many insects, high concentrations of bitter or salty taste stimuli elicit an aversive response. In the mammal literature, the most commonly used paradigm for determining whether two stimuli elicit distinct taste qualities is the conditioned taste aversion (CTA) (Frank, M. E. and Nowlis, G. H., 1989; Wiggins, L. L. *et al.*, 1989). In brief, one creates a CTA to a reference taste stimulus (e.g., quinine), and then asks whether the CTA generalizes to an isostimulatory concentration of another taste stimulus (e.g., NaCl). If the animal does not generalize the CTA, then one infers that the two taste stimuli have distinct taste qualities. On the other hand, if the animal does generalize the CTA, then one infers that the animal could not distinguish the two taste stimuli because they produce a common taste quality. While this CTA-generalization procedure has been used successfully in mammals, I am unaware of its application to insects. Given that insects are capable of forming CTAs (Gelperin, A., 1968; Lee, J. C. and Bernays, E. A., 1990), the CTA-generalization paradigm should provide a powerful tool for analyzing taste qualities in insects. If not, then an alternative approach would

be to generate a conditioned taste preference (CTP) to one taste stimulus (Gerber, B. *et al.*, 2004; Hendel, T. *et al.*, 2005.), and then test for generalization of the CTP to another taste stimulus.

The question of how gustatory input is processed is complicated by the observation that many taste-responsive neurons in the SOG respond to tactile input (Mitchell, B. K. and Itagaki, H., 1992; Rogers, S. M. and Simpson, S. J., 1999). This tactile input is derived from touch receptors within the taste sensilla, which generate action potentials when the taste sensilla are bent (e.g., while contacting food items). Even though tactile input from taste sensilla may provide critical proprioceptive feedback about the position of the chemosensilla relative to a food item, one might predict that the convergence of taste and tactile input on the same SOG interneurons would complicate taste processing. The insect may be able to reduce the contribution of tactile input, however, by limiting the strength of contact between the taste sensilla and food item (Rogers, S. M. and Simpson, S. J., 1999).

A recent study of *Drosophila* larvae revealed the power of applying molecular tools to the study of central taste processing (Melcher, C. and Pankratz, M. J., 2005). Using microarray analysis, the authors discovered a neuropeptide gene, *hugin* (*hug*), which is expressed in about 20 interneurons within the SOG. The axons of these interneurons project to the pharyngeal muscles, the ring gland (a central neuroendocrine organ), and the protocerebrum. On the other hand, the dendrites of the *hug*-expressing neurons receive input from taste cells and pharyngeal chemosensory cells. When the authors blocked synaptic transmission of the *hug*-expressing neurons, they altered the initiation of feeding in the larvae. Taken together, these observations suggest that the *hug*-expressing neurons are part of a neural circuit that modulates taste-mediated feeding responses.

4.03.5.1 Coding Mechanisms

The nature of the neural code for insect taste is unresolved. As in mammalian systems, there is debate over the relative importance of three different coding frameworks: labeled lines, ensemble (i.e., across fiber) patterns, and temporal patterns of spiking. The labeled-line framework hypothesizes that tastant identity is represented by neural activity in a limited subset of gustatory neurons. According to this model, activation of the sugar-sensitive taste cells would stimulate neural circuits within the SOG that initiate biting, whereas activation of a bitter-sensitive taste cell would inhibit the same neural circuits. Further, the behavioral response to a mixture of sugars and bitter taste stimuli should reflect the algebraic sum of the inputs from the carbohydrate- and bitter-sensitive taste cells, respectively (Schoonhoven, L. M. and Blom, F., 1988; Simmonds, M. S. J. and Blaney, W. M., 1991). A recent study (Marella, S. *et al.*, 2006) provided strong evidence that *Drosophila* uses labeled-line coding. The investigators genetically engineered two types of flies (e.g., A and B). Type A flies expressed a capsaicin receptor (VR1E600K) in the taste cells that express the Gr5a receptor (i.e., the sugar-sensitive taste cells), whereas type B flies expressed the same capsaicin receptor in taste cells that express the Gr66a receptor (i.e., the bitter-sensitive taste cells). Then, they showed that when offered a capsaicin-treated agar, the type A flies exhibited an appetitive response, while the type B flies exhibited an aversive response. This remarkable result shows that the nature of an insect's ingestive response to a chemical stimulus (i.e., appetitive or aversive) is not determined by the chemical itself; rather, it is determined by the specific class of taste cell (or labeled line) that is activated.

The ensemble coding framework hypothesizes that tastant identity is represented by the spatial pattern of activity across large populations of neurons (Schoonhoven, L. M. *et al.*, 1992; Smith, D. V. *et al.*, 2000). According to this framework, discriminable taste stimuli should each elicit different ensemble patterns (i.e., different magnitudes of discharge across the population of taste cells). Ensemble coding was evaluated in a study that sought to determine how *M. sexta* caterpillars discriminate the taste of three different host plants (Dethier, V. G. and Crnjar, R. M., 1982). After evaluating the three potential coding frameworks, the authors concluded that ensemble pattern provided the most parsimonious discrimination mechanism. Elsewhere, another study (Glendinning, J. I. *et al.*, 2002) demonstrated that *M. sexta* uses ensemble coding to discriminate between two aversive taste stimuli (salicin and *Grindelia* leaf extract), which activate different classes of bitter-sensitive taste cell.

While the labeled-line and ensemble coding frameworks focus on which neurons are activated within the gustatory neuraxis, the temporal coding framework focuses on how neurons are activated. The latter framework hypothesizes that tastant identity is represented by the precise temporal pattern of spiking within a neuron or across populations of neurons (Dethier, V. G. and Crnjar, R. M., 1982;

Katz, D. B. *et al.*, 2002; Katz, D. B., 2005). Support for this hypothesis comes from a recent study of *M. sexta* caterpillars (Glendinning, J. I. *et al.*, 2006). Several of the bitter-sensitive taste cells in this caterpillar generate a decelerating temporal pattern of spiking when stimulated by caffeine and salicin, and an accelerating temporal pattern of spiking when stimulated by aristolochic acid. The authors showed that the caterpillars use the distinct temporal patterns of spiking generated by salicin and aristolochic acid as a basis for discrimination.

It is important to emphasize two caveats associated with the coding frameworks discussed above. First, the three frameworks are not mutually exclusive. In fact, they probably function together in a complimentary manner within the same insect, as suggested by Dethier V. G. and Crnjar R. M. (1982). Second, given the profound effects of taste–mixture interactions, physiological state, and experience on responsiveness of individual taste cells to their respective ligands (see Sections 4.03.6 and 4.03.7), central processing of the peripheral signal is probably more dynamic than hitherto appreciated.

4.03.6 Taste–Mixture Interactions

Most studies of insect taste focus on single-component stimuli. While this approach offers many experimental advantages, it has limited ecological relevance because most foods contain complex mixtures of taste stimuli. Below I discuss our current (and admittedly, incomplete) understanding of taste–mixture interactions.

Four different types of taste–mixture interactions have been described in the peripheral taste system of insects. First, some compounds fail to produce any stimulatory effects on their own, but nevertheless inhibit the response of specific taste cells to their best stimulus (Dethier, V. G. and Bowdan, E., 1989; Bernays, E. A. and Chapman, R. F., 2000). Second, complex mixtures of taste stimuli can produce inhibitory effects in which the total number of impulses generated by the mixture is less than the algebraic sum of impulses generated by each of the mixture components alone (Ishikawa, S., 1967; Schoonhoven, L. M., 1978; Bernays, E. A. and Chapman, R. F., 2001a). Third, activation of one taste cell can inhibit the response of another taste cell to its own best stimulus (Mitchell, B. K., 1987; Blaney, W. M. and Simmonds, M. S. J., 1990; Chapman, R. F. *et al.*, 1991;

Schoonhoven, L. M. *et al.*, 1992; Shields, V. D. C. and Mitchell, B. K., 1995; Bernays, E. A. and Chapman, R. F., 2000). Fourth, taste mixtures can produce synergistic interactions – for example, simultaneous activation of two taste cells by their respective best stimuli can cause a single taste cell to discharge at a rate substantially higher than expected based on tests with each taste stimulus alone (Dethier, V. G., 1971; Dethier, V. G. and Kuch, J. H., 1971).

The mechanistic basis for peripheral mixture interactions is unclear. Some interactions may involve direct electrical communication between taste cells via gap junctions. In support of this suggestion, two studies have found what appear to be gap junctions between taste cells in flies (Isidoro, N. *et al.*, 1993) and thermo/hygrosensory neurons in a moth (Steinbrecht, R. A., 1989). However, another study failed to observe any evidence for gap junctions between taste cells in a caterpillar (Shields, V. D. C., 1994). A second possibility is that stimulation of one taste cell influences the responsiveness of others within the same taste sensillum through centrifugal feedback from the central nervous system. A third possibility is that activity in one taste cell may inhibit activity in a nearby cell through an ephaptic interaction (Jefferys, J. G., 1995; Bokil, H. *et al.*, 2001). A final possibility is that that a single taste cell could express two signaling pathways, and activation of one pathway inhibits (or enhances) the response of the other (Dethier, V. G. and Bowdan, E., 1989).

Taste–mixture interactions have also been observed in the central gustatory system. For instance, caffeine normally elicits a taste-mediated aversive response in *M. sexta* caterpillars (Glendinning, J. I. *et al.*, 1999a). However, when caffeine is presented in binary mixture with *myo*-inositol, the aversive response is completely attenuated (Glendinning, J. I. *et al.*, 2000b). Recordings from the peripheral taste system of this insect revealed that the bitter-sensitive taste cells exhibited the same response to caffeine, irrespective of whether the caffeine was presented alone or in binary mixture with *myo*-inositol; likewise, the inositol-sensitive taste cells exhibited the same response to *myo*-inositol, irrespective of whether the myo-inositol was presented alone or in binary mixture with caffeine (Glendinning, J. I. *et al.*, 2000b). The most parsimonious interpretation of these findings is that input from the inositol-sensitive taste cells inhibited the neural circuit in the central nervous system that mediates the aversive response to caffeine.

4.03.7 Factors that Modulate Gustatory Responsiveness

One of the most remarkable features of the insect taste system is its plasticity. To illustrate this point, I discuss below four factors known to increase (or decrease) the responsiveness of the gustatory system to specific taste stimuli.

4.03.7.1 Physical Changes in Taste Sensilla

In order for a chemical to stimulate a taste cell, it must first diffuse through the pore at the tip of a taste sensillum (see Figure 1). Although this terminal pore remains permanently open in many insect species, it opens and closes in Orthopterans (Bernays, E. A. *et al.*, 1972). The functional significance of terminal pore closure was revealed by a series of experiments in locusts, *Locusta migratoria* (Bernays, E. A. and Chapman, R. F., 1972). When the foregut of the locust becomes distended (e.g., after a meal), stretch receptors in the wall of the foregut are activated and send afferent signals to the brain via the posterior pharyngeal nerves and frontal connectives. These afferent signals stimulate the release of a diuretic hormone from the corpora cardiaca, which in turn cause pore closure in many of the taste sensilla. The functional effect of terminal pore closure is that it contributes to meal termination by eliminating phagostimulatory input from the peripheral taste system.

4.03.7.2 Nutrient Levels in Blood

Several investigators have discovered that changes in the concentration of specific nutrients in the blood (e.g., amino acids or trehalose) directly modulate the responsiveness of taste cells in grasshoppers and caterpillars to the same chemicals (Simmonds, M. S. J. *et al.*, 1992; Simpson, S. J. and Simpson, C. L., 1992; Bernays, E. A. *et al.*, 2004). In locusts, when blood levels of amino acids are low and sugars high, the responsiveness of the taste cells to amino acids is high and that to sugars is low (Simpson, S. J. and Simpson, C. L., 1992). In contrast, when blood levels of amino acids are high and sugars low, the responsiveness of the taste cells to amino acids is low and that to sugars is high. Because these modulatory effects of blood nutrients on taste cell responsiveness occur in the absence of neural or humoral links between the central nervous system and taste cells, they must be mediated peripherally. A recent study indicates that nutrient levels in the blood may modulate secretion of nitric oxide from glandular cells surrounding the taste cells, which in turn modulates the responsiveness of taste cells to their respective ligands (Schuppe, H. *et al.*, 2006).

4.03.7.3   Dietary Experience Effects

When insects are exposed repeatedly to the same taste stimulus (e.g., within and across successive meals), the responsiveness of their gustatory system to these stimuli can increase or decrease. The time scale over which these changes occur varies from seconds to days. Below, I discuss four ways that dietary experience modifies taste.

4.03.7.3.1   *Short-term sensory adaptation*
A common feature of all sensory receptor cells is that their firing rate diminishes over time in response to continuous stimulation (i.e., adapts), but recovers rapidly upon removal of the stimulus (i.e., disadapts). Adaptation and disadaptation occur over a period of milliseconds to seconds. To illustrate this process, we show adaptation curves for two taste cells in the lateral styloconic sensillum of *M. sexta* caterpillars (Figure 6). It is apparent that the response (as indicated by impulses per 100 ms) of the inositol-sensitive taste cell to 1 mM *myo*-inositol diminished by >50% over 500 ms, whereas that of the bitter-sensitive taste cell to

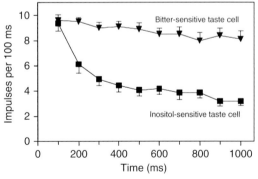

Figure 6 Adaptation rates of three different taste cells in the lateral styloconic sensillum of *Manduca sexta* caterpillars (as indicated by the number of spikes during each of 10 consecutive 100 ms bins). We show adaptation of the bitter-sensitive taste cell to 5 mM caffeine and the inositol-sensitive taste cell to 1 mM *myo*-inositol. Symbols indicate mean ±SE; *N* = 10 sensilla, each from different fifth instar caterpillars. These data are from Glendinning, J. I., Nelson, N., and Bernays, E. A. 2000b. How do inositol and glucose modulate feeding in *Manduca sexta* caterpillars? J. Exp. Biol. 203, 1299–1315.

5 mM caffeine diminished by <10% over the same time period.

The finding that bitter-sensitive taste cells are more resistant to adaptation than nutrient-sensitive taste cells has been reported for other insects as well (Schoonhoven, L. M., 1977). From a functional standpoint, this finding indicates that the relative amount of afferent input from different taste cells during a meal may not be constant, with the relative amount of input from the bitter-sensitive taste cells increasing over time. The extent to which these adaptation processes contribute to feeding during a meal is unclear, however, because insect gustatory organs do not remain in continuous contact with foods throughout a meal. Some gustatory organs make a series of drumming movements, resulting in intermittent contacts lasting only 100–200 ms (Blaney, W. M. and Duckett, A. M., 1975; Devitt, B. D. and Smith, J. J. B., 1985). These drumming movements could serve to minimize the effects of short-term adaptation in the taste cells and effectively increase the level of sensory input from taste cells over time (Blaney, W. M. and Duckett, A. M., 1975).

4.03.7.3.2   *Habituation*
When an animal is presented with a stimulus repeatedly across a sequence of trials, its responsiveness to that stimulus usually wanes (Hinde, R. A., 1970). This phenomenon is referred to as habituation if it meets several criteria – for example, it must be mediated centrally, recover spontaneously, generalize to related stimuli, and develop more readily in response to weak than intense stimuli (Thompson, R. F. and Spencer, W. A., 1966). For instance, fruit flies and bees normally exhibit a proboscis extension reflex for 5% sucrose, but this reflex extinguishes following 20 consecutive stimulations within a 5 min period (Duerr, J. S. and Quinn, W. G., 1982; Braun, G. and Bicker, G., 1992; Scheiner, R., 2004; Scheiner, R. *et al.*, 2004). This exposure-induced response decrement for sucrose was considered habituation because it met at least two of Thompson and Spencer's habituation criteria.

4.03.7.3.3   *Long-term adaptation*
When herbivorous insects are offered a diet containing an unpalatable but harmless compound as their only source of water and nutrients, they will usually sample the diet repeatedly. After ~24 h of such sampling, the insects eventually adapt their aversive response. This long-term adaptation phenomenon has been reported in several species of herbivorous

insect (Jermy, T. *et al.*, 1982; Simmonds, M. S. J. and Blaney, W. M., 1983; Szentesi, A. and Bernays, E. A., 1984; Blaney, W. M. and Simmonds, M. S. J., 1987; Usher, B. F. *et al.*, 1988; Glendinning, J. I. and Gonzalez, N. A., 1995; Bernays, E. A. *et al.*, 2004), but has been examined most extensively in caterpillars of *M. sexta* (Schoonhoven, L. M., 1969; 1978; Glendinning, J. I. *et al.*, 1999b; 2001a; 2001b; 2002). For example, 24 h of dietary exposure to a caffeine diet eliminates the aversive response to caffeine. The long-term adaptation phenomenon appears to be mediated peripherally because the dietary exposure regime profoundly desensitizes all of the bitter-sensitive taste cells to caffeine. Twenty-four hours of dietary exposure to a salicin diet also eliminates the aversive response to salicin. However, this latter adaptation phenomenon must be mediated centrally because there is no associated reduction in responsiveness of the bitter-sensitive taste cells. Taken together, these results indicate that the peripheral and central gustatory system of *M. sexta* each have independent mechanisms for overcoming the aversive response to harmless bitter compounds.

Because the long-term adaptation to salicin is mediated by the central gustatory system, and because it generalizes to related taste stimuli (Glendinning, J. I. *et al.*, 2001a; 2002), it can be viewed as a type of habituation. It is important to note, however, that the onset latency for the habituation to salicin (in *M. sexta*) is about 24 h, whereas that to sucrose (in flies and bees) is ≤5 min (Duerr, J. S. and Quinn, W. G., 1982; Braun, G. and Bicker, G., 1992; Scheiner, R., 2004; Scheiner, R. *et al.*, 2004). Further work is needed to determine whether these different onset latencies reflect (1) species differences, (2) greater difficulty habituating to aversive taste stimuli, and/or (3) the existence of multiple habituation mechanisms.

4.03.7.3.4 Induced food preferences

After feeding on a particular plant species for 1–2 days, many species of herbivorous insect induce a strong preference for that plant species (review in Szentesi, A. and Jermy, T., 1990). Despite the widespread occurrence of this phenomenon, the underlying physiological mechanisms are poorly understood. In all likelihood, a variety of mechanisms are activated, depending on the chemical constituents of the plant tissue and the insect species – e.g., induction of detoxification enzymes, changes in gustatory and/or olfactory responsiveness, and conditioned preferences (Bernays, E. A. and Weiss, M. R., 1996).

That taste contributes to induced food preferences was established recently in a study of *M. sexta* caterpillars, which are facultative specialists on plants in the family Solanaceae. When these caterpillars were reared on solanaceous foliage, they induced a strong and exclusive preference for the same foliage (Hanson, F. E. and Dethier, V. G., 1973; Yamamoto, R. T., 1974; de Boer, G. and Hanson, F. E., 1987; del Campo, M. L. and Renwick, J. A. A., 2000). A team of investigators discovered that indioside D (a steroidal glycoside) is one of the chemicals in the solanaceous foliage to which the caterpillars induce their preference (del Campo, M. L. *et al.*, 2001). These investigators also presented evidence that the taste cells in the caterpillar's styloconic sensilla become tuned to indioside D over the course of the induction process, and that this chemosensory tuning process is sufficient to explain the induced preference for solanaceous leaves (del Campo, M. L. *et al.*, 2001; del Campo, M. L. and Miles, C. I., 2003). According to the chemosensory tuning hypothesis, the taste cells sensitive to indioside D maintain their normal responsiveness, while those sensitive to other plant chemicals (e.g., sugars, salts, and bitter taste stimuli) become less responsive.

4.03.7.4 Health Status

A recent study demonstrated that peripheral gustatory sensitivity of two species of caterpillar changes in response to endoparasitic infection (Bernays, E. A. and Singer, M. S., 2005). More specifically, the investigators discovered that endoparasitic infections cause specific taste cells in the caterpillars to respond more vigorously to specific host plant compounds (i.e., PAs). Given that activation of the PA-sensitive taste cells stimulates increased feeding, it follows that an elevated responsiveness of the PA-sensitive taste cells would promote intake of PA-containing plant tissues. This sequence of events should benefit the caterpillars because PAs are toxic to the endoparasites and thus may help the caterpillars control (or even eliminate) the endoparasitic infection.

4.03.8 Future Directions

Despite all of the recent molecular advances in insect taste (review in Amrein, H. and Thorne, N., 2005), we still have a relatively simplistic understanding of how taste contributes to the control of feeding. This reflects a number of complicating factors. One is that investigators have distributed their efforts over a wide range of insect species. While this approach has offered important comparative insights, it has prevented a detailed analysis of the taste system in any one species. At this point, the insect species with the best-characterized taste system include adult blowflies (*P. regina*), adult *Drosophila melanogaster*, and larval *M. sexta*. By focusing future efforts on these species, investigators will be able to build a substantial body of knowledge, and ask even more sophisticated questions.

A second factor that has limited progress on insect taste is the widespread use of long-term feeding assays to assess taste function – for example, preference tests that span 2–6 h. Given that secondary plant compounds (and nutrients) can inhibit feeding in <5 min through postoral mechanisms (Reynolds, S. E., 1990; Timmins, W. A. and Reynolds, S. E., 1992; Glendinning, J. I. and Slansky Jr., F., 1994; Glendinning, J. I., 1996), it is likely that the results from feeding tests lasting ≥2 h will reflect contributions of both oral and postoral response mechanisms. The contribution of postoral mechanisms can confound the interpretation of behavioral assays that seek to isolate taste-mediated responses. Another problem with the use of long-term feeding assays is that they weaken attempts to link neural responses of the taste system to behavior. Because electrophysiological studies usually focus on the initial milliseconds (or seconds) of response, one cannot easily extrapolate these transient neural processes to behavioral processes that span 2–6 h. A more rational approach would be to determine the operative time frame for the feeding behavior of interest, and then to examine the neural responses that occur over that time frame. For instance, given that caterpillars and flies can discriminate taste stimuli in less than 1 s (Dethier, V. G., 1968; Dethier, V. G. and Crnjar, R. M., 1982), one should focus electrophysiological analyses of discrimination on the initial 1 s of neural response. To this end, investigators could use well-established techniques for measuring immediate taste-mediated responses – for example, the proboscis extension reflex in flies and bees (Dethier, V. G., 1976) or initial biting responses in caterpillars and beetles (Dethier, V. G. and Crnjar, R. M., 1982; Frazier, J. L., 1986; Sperling, J. L. and Mitchell, B. K., 1991; Glendinning, J. I. *et al.*, 2006).

A final factor that has hampered progress on insect taste is the absence of a framework for integrating taste into more generalized feeding control mechanisms. Such a framework has been developed for mammals (Spector, A. C., 2000) and has been of enormous value in helping understand how taste input modulates different aspects of the ingestive process. In mammals, taste processing has been grouped into three categories. Stimulus identification involves the discrimination of ecologically relevant taste stimuli, and the association of these taste stimuli with specific events or outcomes. These associative processes encompass both innate and conditioned responses. Ingestive motivation involves the central gustatory mechanisms that promote or inhibit the consumption of foods, and thereby represents the hedonic component of taste (i.e., whether an animal prefers or avoids a food). Digestive preparation involves the mechanisms by which taste activates feedforward mechanisms that aid in the digestive process. Application of this integrative framework to studies of insect taste would significantly enhance our ability to identify and analyze the central neuropils involved in different aspects of taste processing and better characterize phenotypic effects of genetic manipulations.

References

Albert, P. J. and Parisella, S. 1988. Feeding preferences of eastern spruce budworm larvae in two-choice tests with extracts of mature foliage and with pure amino acids. J. Chem. Ecol. 14, 1649–1656.

Altman, J. S. and Kien, J. 1987. Functional Organization of the Subesophageal Ganglion in Arthropods. In: Arthropod Brain: Its Evolution, Development, Structure and Functions (*ed*. A. P. Gupta), pp. 265–301. John-Wiley and Sons.

Amrein, H. and Thorne, N. 2005. Gustatory perception and behavior in *Drosophila melanogaster*. Curr. Biol. 15, R673–R684.

Arms, K., Feeny, P., and Lederhouse, R. C. 1974. Sodium: stimulus for puddling behavior by tiger swallowtail butterflies, *Papilio glaucus*. Science 185, 372–374.

Bate-Smith, E. C. 1972. Attractants and Repellents in Higher Animals. In: Phytochemical Ecology: Proceedings of the Phytochemical Society Symposium (*ed*. J. B. Harborne), pp. 45–56. Academic Press.

Bernays, E. A. 1982. The Insect on the Plant – a Closer Look. In: Proceedings of the 5th international symposium on insect-plant relationships (*eds*. J. H. Visser and A. K. Minks), pp. 3–17. Pudoc, Wageningen.

Bernays, E. A. 1985. Regulation of Feeding Behaviour. In: Comprehensive Insect Physiology, Biochemistry and Pharmacology, Vol. 4 (*eds*. G. A. Kerkut and L. I. Gilbert), pp. 1–32. Pergamon Press.

Bernays, E. A. 1990. Plant secondary compounds deterrent but not toxic to the grass specialist acridid *Locusta migratoria*: implications for the evolution of graminivory. Entomol. Exper. Appl. 54, 53–56.

Bernays, E. A. 1991. Relationship between deterrence and toxicity of plant secondary metabolites for the grasshopper *Schistocerca americana*. J. Chem. Ecol. 17, 2519–2526.

Bernays, E. A. and Chapman, R. F. 1972. The control of changes in peripheral sensilla associated with feeding in *Locusta migratoria* (L.). J. Exp. Biol. 57, 755–763.

Bernays, E. A. and Chapman, R. F. 1987. The Evolution of Deterrent Responses in Plant-Feeding Insects. In: Perspectives in Chemoreception and Behavior (eds. R. F. Chapman, E. A. Bernays, and J. G. Stoffolano, Jr.), pp. 159–173. Springer-Verlag.

Bernays, E. A. and Chapman, R. F. 1994. Host–Plant Selection by Phytophagous Insects. Chapman and Hall.

Bernays, E. A. and Chapman, R. F. 2000. A neurophysiological study of sensitivity to a feeding deterrent in two sister species of Heliothis with different diet breadths. J. Insect Physiol. 46, 905–912.

Bernays, E. A. and Chapman, R. F. 2001a. Taste cell responses in a polyphagous arctiid: towards a general pattern for caterpillars. J. Insect Physiol. 47, 1029–1043.

Bernays, E. A. and Chapman, R. F. 2001b. Electrophysiological responses of taste cells to nutrient mixtures in the polyphagous caterpillar of Grammia geneura. J. Comp. Physiol. A 187, 205–213.

Bernays, E. A. and Singer, M. S. 2005. Taste alteration and endoparasites. Nature 436, 476.

Bernays, E. A. and Weiss, M. R. 1996. Induced food preferences in caterpillars: the need to identify mechanisms. Entomol. Exper. Appl. 78, 1–8.

Bernays, E. A., Blaney, W. M., and Chapman, R. F. 1972. Changes in the chemoreceptor sensilla on the maxillary palps of Locusta migratoria in relation to feeding. J. Exp. Biol. 57, 745–753.

Bernays, E. A., Chapman, R. F., and Hartmann, T. 2002. A taste receptor neurone dedicated to the perception of pyrrolizidine alkaloids in the medial galeal sensillum of two polyphagous arctiid caterpillars. Physiol. Entomol. 27, 312–321.

Bernays, E. A., Chapman, R. F., Lamunyon, C. W., and Hartmann, T. 2003. Taste receptors for pyrrolizidine alkaloids in a monophagous caterpillar. J. Chem. Ecol. 29, 1709–1722.

Bernays, E. A., Chapman, R. F., and Singer, M. S. 2000. Sensitivity to chemically diverse phagostimulants in a single gustatory neuron of a polyphagous caterpillar. J. Comp. Physiol. A 186, 13–19.

Bernays, E. A., Chapman, R. F., and Singer, M. S. 2004. Changes in taste receptor cell sensitivity in a polyphagous caterpillar reflect carbohydrate but not protein imbalance. J. Comp. Physiol. A 190, 39–48.

Bernays, E. A., Glendinning, J. I., and Chapman, R. F. 1998. Plant acids modulate chemosensory responses in Manduca sexta larvae. Physiol. Entomol. 23, 193–201.

Blaney, W. M. and Duckett, A. M. 1975. The significance of palpation by the maxillary palps of Locusta migratoria (L.): an electrophysiological and behavioural study. J. Exp. Biol. 63, 701–712.

Blaney, W. M. and Simmonds, M. S. J. 1987. Experience: a Modifier of Neural and Behavioural Sensitivity. In: Insects–Plants (eds. V. Labeyrie, G. Fabres, and D. Lachaise), pp. 237–241. Dr. W. Junk Publishers.

Blaney, W. M. and Simmonds, M. S. J. 1990. A behavioural and electrophysiological study of the role of tarsal chemoreceptors in feeding by adults of Spodoptera, Heliothis virescens and Helicoverpa armigera. J. Insect Physiol. 36, 743–756.

Bokil, H., Laaris, N., Blinder, K., Ennis, M., and Keller, A. 2001. Ephaptic interactions in the mammalian olfactory system. J. Neurosci. 21, RC173.

Bowdan, E. 1995. The effects of a phagostimulant and a deterrent on the microstructure of feeding by Manduca sexta caterpillars. Entomologia Experimentalis et Applicata 77, 297–306.

Braun, G. and Bicker, G. 1992. Habituation of an appetitive reflex in the honeybee. J. Neurophysiol. 67, 588–598.

Brieskorn, C. H. 1990. Physiological and Therapeutic Aspects of Bitter Compounds. In: Bitterness in Foods and Beverages (ed. R. L. Rouseff), pp. 15–33. Elsevier.

Brower, L. P. 1984. Chemical Defense in Butterflies. In: The Biology of Butterflies (eds. R. I. Vane-Wright and P. R. Ackery), pp. 109–134. Academic Press.

Calas, D., Thiéry, D., and Marion-Poll, F. 2007. 20-Hydroxyecdysone deters oviposition and larval feeding in the European grapevine moth, Lobesia botrana. J. Chem. Ecol. 32, 2443–2454.

Chapman, R. F. 1982. Chemoreception: the significance of sensillum numbers. Adv. Insect Physiol. 16, 247–356.

Chapman, R. F., Ascoli-Christensen, A., and White, P. R. 1991. Sensory coding for feeding deterrence in the grasshopper Schistocerca americana. J. Exp. Biol. 158, 241–259.

Chyb, S. 2004. Drosophila gustatory receptors: from gene identification to functional expression. J. Insect Physiol. 50, 469–477.

Chyb, S., Dahanukar, A., and Carlson, J. R. 2003. Drosophila Gr5a encodes a taste receptor tuned to trehalose. Proc. Natl. Acad. Sci. U. S. A. 100(Suppl. 2), 14526–14530.

Clyne, P. J., Warr, C. G., and Carlson, J. R. 2000. Candidate taste receptors in Drosophila. Science 287, 1830–1834.

Dahanukar, A., Foster, K., van der Goes van Naters, W. M., and Carlson, J. R. 2001. A Gr receptor is required for response to the sugar trehalose in taste neurons of Drosophila. Nat. Neurosci. 4, 1182–1186.

de Boer, G. and Hanson, F. E. 1987. Differentiation of roles of chemosensory organs in food discrimination among host and non-host plants by larvae of the tobacco hornworm, Manduca sexta. Physiol. Entomol. 12, 387–398.

del Campo, M. L. and Miles, C. I. 2003. Chemosensory tuning to a host recognition cue in the facultative specialist larvae of the moth Manduca sexta. J. Exp. Biol. 206, 3979–3990.

del Campo, M. L. and Renwick, J. A. A. 2000. Induction of host specificity in larvae of Manduca sexta: chemical dependence controlling host recognition and developmental rate. Chemoecology 10, 115–121.

del Campo, M. L., Miles, C. I., Schroeder, F. C., Muellerk, C., Booker, R., and Renwick, J. A. 2001. Host recognition by the tobacco hornworm is mediated by a host plant compound. Nature 411, 186–189.

den Otter, C. J. 1972. Differential sensitivity of insect chemoreceptors to alkali cations. J. Insect Physiol. 18, 109–131.

Dethier, V. G. 1968. Chemosensory input and taste discrimination in the blowfly. Science 161, 389–391.

Dethier, V. G. 1971. A surfeit of stimuli: a paucity of receptors. Am. Sci. 59, 706–715.

Dethier, V. G. 1973. Electrophysiological studies of gustation in Lepidopterous larvae. II. Taste spectra in relation to food-plant discrimination. J. Comp. Physiol. 82, 103–134.

Dethier, V. G. 1976. The Hungry Fly: A Physiological Study of the Behavior Associated with Feeding. Harvard University Press.

Dethier, V. G. 1980. Evolution of receptor sensitivity to secondary plant substances with special reference to deterrents. Am. Nat. 115, 45–66.

Dethier, V. G. and Bowdan, E. 1989. The effect of alkaloids on the sugar receptors of the blowfly. Physiol. Entomol. 14, 127–136.

Dethier, V. G. and Bowdan, E. 1992. Effects of alkaloids on feeding by Phormia regina confirm the critical role of sensory inhibition. Physiol. Entomol. 17, 325–330.

Dethier, V. G. and Crnjar, R. M. 1982. Candidate codes in the gustatory system of caterpillars. J. Gen. Physiol. 79, 549–569.

Dethier, V. G. and Kuch, J. H. 1971. Electrophysiological studies of gustation in lepidopterous larvae. I. Comparative sensitivity to sugars, amino acids, and glycosides. Z. Vergl. Physiol. 72, 343–363.

Devitt, B. D. and Smith, J. J. B. 1985. Action of mouthparts during feeding in the dark-sided cutworm, *Euoxa messoria* (Lepidoptera: Noctuidae). Can. Entomol. 117, 343–349.

Duerr, J. S. and Quinn, W. G. 1982. Three *Drosophila* mutations that block associative learning also affect habituation and sensitization. Proc. Natl. Acad. Sci. U. S. A. 79, 3646–3650.

Frank, M. E. and Nowlis, G. H. 1989. Learned aversions and taste qualities in hamsters. Chem. Senses 14, 379–394.

Frazier, J. L. 1986. The Perception of Plant Allelochemicals that Inhibit Feeding. In: Molecular Aspects of Insect–Plant Interactions (*eds*. L. B. Brattsten and S. Ahmad), pp. 1–42. Plenum Press.

Friend, W. G. and Smith, J. J. 1971. Feeding in *Rhodnius prolixus*: mouthpart activity and salivation, and their correlation with changes of electrical resistance. J. Insect Physiol. 17, 233–243.

Furuyama, A., Koganezawa, M., and Shimida, I. 1999. Multiple receptor sites for nucleotide reception in the labellar taste receptor cells of the fleshfly *Boettcherisca peregrina*. J. Insect Physiol. 45, 249–255.

Garcia, J. and Hankins, W. G. 1975. The Evolution of Bitter and the Acquisition of Toxiphobia. In: Olfaction and Taste. V. Proceedings of the 5th International Symposium in Melbourne, Australia (*eds*. D. A. Denton and J. P. Coghlan), pp. 39–45. Academic Press.

Gelperin, A. 1968. Feeding behaviour of the praying mantis: a learned modification. Nature 217, 399–400.

Gerber, B., Scherer, S., Neuser, K., Michels, B., Hendel, T., Stocker, R. F., and Heisenberg, M. 2004. Visual learning in individually assayed *Drosophila* larvae. J. Exp. Biol. 207, 179–188.

Glendinning, J. I. 1994. Is the bitter rejection response always adaptive? Physiol. Behav. 56, 1217–1227.

Glendinning, J. I. 1996. Is chemosensory input essential for the rapid rejection of toxic foods? J. Exp. Biol. 199, 1523–1534.

Glendinning, J. I. 2002. How do herbivorous insects cope with noxious secondary plant compounds in their diet? Entomologia Experimentatlis et Applicata 104, 15–25.

Glendinning, J. I. and Gonzalez, N. A. 1995. Gustatory habituation to deterrent compounds in a grasshopper: concentration and compound specificity. Anim. Behav. 50, 915–927.

Glendinning, J. I. and Hills, T. T. 1997. Electrophysiological evidence for two transduction pathways within a bitter-sensitive taste receptor. J. Neurophysiol. 78, 734–745.

Glendinning, J. I. and Slansky, F., Jr. 1994. Interactions of allelochemicals with dietary constituents: effects on deterrency. Physiol. Entomol. 19, 173–186.

Glendinning, J. I., Brown, H., Capoor, M., Davis, A., Gbedemah, A., and Long, E. 2001b. A peripheral mechanism for behavioral adaptation to specific "bitter" taste stimuli in an insect. J. Neurosci. 21, 3688–3696.

Glendinning, J. I., Chaudhari, N., and Kinnamon, S. C. 2000a. Taste Transduction and Molecular Biology. In: The Neurobiology of Taste and Smell (*eds*. T. Finger, W. L. Silver, and D. Restrepo), 2nd edn., pp. 315–351. Wiley-Liss.

Glendinning, J. I., Davis, A., and Rai, M. 2006. Temporal coding mediates discrimination of "bitter" taste stimuli by an insect. J. Neurosci. 26, 8900–8908.

Glendinning, J. I., Davis, A., and Ramaswamy, S. 2002. Contribution of different taste cells and signaling pathways to the discrimination of "bitter" taste stimuli by an insect. J. Neurosci. 22, 7281–7287.

Glendinning, J. I., Domdom, S., and Long, E. 2001a. Selective adaptation to noxious foods by an insect. J. Exp. Biol. 204, 3355–3367.

Glendinning, J. I., Ensslen, S., Eisenberg, M. E., and Weiskopf, P. 1999b. Diet-induced plasticity in the taste system of an insect: localization to a single transduction pathway in an identified taste cell. J. Exp. Biol. 202, 2091–2102.

Glendinning, J. I., Nelson, N., and Bernays, E. A. 2000b. How do inositol and glucose modulate feeding in *Manduca sexta* caterpillars?. J. Exp. Biol. 203, 1299–1315.

Glendinning, J. I., Tarre, M., and Asaoka, K. 1999a. Contribution of different bitter-sensitive taste cells to feeding inhibition in a caterpillar (*Manduca sexta*). Behav. Neurosci. 113, 840–854.

Goldrich, N. R. 1973. Behavioral responses of *Phormia regina* (Meigen) to labellar stimulation with amino acids. J. Gen. Physiol. 61, 74–88.

Grill, H. J. and Berridge, K. C. 1985. Taste reactivity as a measure of the neural control of palatability. Prog. Psychobiol. Physiol. Psychol. 11, 1–61.

Griss, C., Simpson, S. J., Rohrbacher, J., and Rowell, C. H. F. 1991. Localization in the central nervous system of larval *Manduca sexta* (Lepidoptera: Sphingidae) of areas responsible for aspects of feeding behavior. J. Insect Physiol. 37, 477–482.

Hallem, E. A., Dahanukar, A., and Carlson, J. R. 2006. Insect odor and taste receptors. Annu. Rev. Entomol. 51, 113–135.

Hansen-Delkeskamp, E. 1992. Functional characterizaion of antennal contact chemoreceptors in the cockroach, *Periplaneta americana*. J. Insect Physiol. 38, 813–822.

Hanson, F. E. and Dethier, V. G. 1973. Role of gustation and olfaction in food plant discrimination in the tobacco hornworm, *Manduca sexta*. J. Insect Physiol. 19, 1019–1034.

Harborne, J. B. 1973. Phytochemical Methods. Chapman and Hall.

Harley, K. L. S. and Thorsteinson, A. J. 1967. The influence of plant chemicals on the feeding behavior, development, and survival of the two-striped grasshopper, *Melanoplus bivittatus* (Say), Acrididae: Orthoptera. Can. J. Zool. 45, 305–319.

Hendel, T., Michels, B., Neuser, K., Schipanski, A., Kaun, K., Sokolowski, M. B., Marohn, F., Michel, R., Heisenberg, M., and Gerber, B. 2005. The carrot, not the stick: appetitive rather than aversive gustatory stimuli support associative olfactory learning in individually assayed *Drosophila* larvae. J. Comp. Physiol. A 191, 265–278.

Hinde, R. A. 1970. Behavioural Habituation. In: Short-Term Changes in Neural Activity and Behaviour (*eds*. G. Horn and R. A. Hinde), pp. 3–40. Cambridge University Press.

Hirao, T. and Ariai, N. 1990. Gustatory and feeding responses to amino acids in the silkworm, *Bombyx mori*. Jpn. J. Appl. Entomol. Zool. 35, 73–76.

Hiroi, M., Marion-Poll, F., and Tanimura, T. 2002. Differentiated response to sugars among labellar chemosensilla in *Drosophila*. Zool. Sci. 19, 1009–1018.

Hiroi, M., Meunier, N., Marion-Poll, F., and Tanimura, T. 2004. Two antagonistic gustatory receptor neurons responding to sweet-salty and bitter taste in *Drosophila*. J. Neurobiol. 61, 333–342.

Hodgson, E. S., Lettvin, J. Y., and Roeder, K. D. 1955. Physiology of a primary chemoreceptor unit. Science 122, 417–418.

Inomata, N., Goto, H., Itoh, M., and Isono, K. 2004. A single amino-acid change of the gustatory receptor gene, *Gr5a*, has a major effect on trehalose sensitivity in a natural population of *Drosophila melanogaster*. Genetics 168, 1449–1458.

Inoshita, T. and Tanimura, T. 2006. Cellular identification of water gustatory receptor neurons and their central projection pattern in *Drosophila*. Proc. Natl. Acad. Sci. U. S. A. 103, 1094–1099.

Ishikawa, S. 1967. Maxillary Chemoreceptors in the Silkworm. In: Olfaction and Taste 2 (*ed*. T. Hayashi), pp. 761–777. Pergamon Press.

Ishimoto, H., Matsumoto, A., and Tanimura, T. 2000. Molecular identification of a taste receptor gene for trehalose in *Drosophila*. Science 289, 116–119.

Ishimoto, H., Takahashi, K., Ueda, R., and Tanimura, T. 2005. G-protein gamma subunit 1 is required for sugar reception in *Drosophila*. EMBO J. 24, 3259–3265.

Isidoro, N., Solinar, M., Baur, R., Roessingh, P., and Städler, E. 1993. Functional morphology of a tarsal sensillum of *Delia radicum* L (Diptera, Anthomyiidae) sensitive to important host-plant compounds. Int. J. Insect Morphol. Embryol. 39, 275–281.

Jefferys, J. G. 1995. Nonsynaptic modulation of neuronal activity in the brain: electric currents and extracellular ions. Physiol. Rev. 75, 689–723.

Jermy, T., Bernays, E. A., and Szentesi, A. 1982. The Effect of Repeated Exposure to Feeding Deterrents on Their Acceptability to Phytophagous Insects. In: Proceedings of the 5th International Symposium on Insect–Plant Relationships (eds. H. Visser and A. Minks), pp. 25–32. PUDOC.

Jørgensen, K., Kvello, P., Jørgen Almaas, T., and Mustaparta, H. 2006. Two closely located areas in the subesophageal ganglion and the tritocerebrum receive projections from gustatory receptor neurons located on the antennae and the proboscis in the moth *Heliothis virescens*. J. Comp. Neurol. 496, 121–134.

Katz, D. B. 2005. The many flavors of temporal coding in the gustatory cortex. Chem. Senses 30(Suppl. 1), 80–81.

Katz, D. B., Nicolelis, M. A., and Simon, S. A. 2002. Gustatory processing is dynamic and distributed. Curr. Opin. Neurobiol. 12, 448–454.

Kent, K. S. and Hildebrand, J. G. 1987. Cephalic sensory pathways in the central nervous system of larval *Manduca sexta* (Lepidoptera: Sphingidae). Philos. Trans. R. Soc. Lond. B Biol. Sci. 315, 1–36.

Kijima, H., Okada, Y., Oiki, S., Goshima, S., Nagata, N., and Kazawa, T. 1995. Free ion concentration in receptor lymph and a role of transepithelial voltage in the fly labellar taste receptor. J. Comp. Physiol. A 177, 123–133.

Kim, J. H. and Mullin, C. A. 1998. Structure-phagostimulatory relationships for amino acids in adult western corn rootworm, *Diabrotica virgifera virgifera*. J. Chem. Ecol. 24, 1499–1511.

Lee, J. C. and Bernays, E. A. 1990. Food tastes and toxic effects: Associative learning by the phytophagous grasshopper *Schistocerca americana* (Drury) (Orthoptera: Acrididae). Anim. Behav. 39, 163–173.

Liscia, A., Solari, P., Majone, R., Tomassini-Barbarossa, I., and Crnjar, R. 1997. Taste reception mechanisms in the blowfly: evidence of amiloride-sensitive and insensitive sites. Physiol. Behav. 62, 875–879.

Liu, L., Soren Leonard, A., Price, M. P., Johnson, W. A., and Welsh, M. J. 2003. Contribution of *Drosophlia* DEG/ENaC genes to salt taste. Neuron 39, 133–149.

Ma, W.-C. 1969. Some properties of gustation in the larvae of *Pieris brassicae*. Entomologia Experimentalis et Applicata 12, 584–590.

Ma, W.-C. 1972. Dynamics of feeding responses in *Pieris brassicae* Linn as a function of chemosensory input: a behavioral and electrophysiological study. Meded. Laudbouwhoge school Wageningen 72–11, 1–162.

Ma, W.-C. and Schoonhoven, L. M. 1973. Tarsal contact chemoreception hairs of the large white butterfly *Pieris brassicae* and their possible role in oviposition behavior. Entomologia Experimentatlis et Applicata 16, 343–357.

Maes, F. W. and Bijpost, S. C. A. 1979. Classical conditioning reveals discrimination of salt taste quality in the blowfly *Calliphora vicina*. J. Comp. Physiol. 133, 53–62.

Marella, S., Fischler, W., Kong, P., Asgarian, E., and Scott, K. 2006. Imaging taste responses in the fly brain reveals a functional map of taste category and behavior. Neuron 49, 285–295.

Marion-Poll, F. and Descoins, C. 2002. Taste detection of phytoecdysteriods in larvae of *Bombyx mori*, *Spodoptera littoralis* and *Ostrinia nubilalis*. J. Insect Physiol. 48, 467–476.

Melcher, C. and Pankratz, M. J. 2005. Candidate gustatory interneurons modulating feeding behavior in the *Drosophila* brain. PLoS Biol. 3, 1618–1628.

Messchemdorp, L., Smid, H. M., and Loon, J. J. A.v. 1998. The role of an epipharyngeal sensillum in the perception of feeding deterrents by *Leptinotarsa decemlineata* larvae. J. Comp. Physiol. A 183, 255–264.

Meunier, N., Marion-Poll, F., Rospars, J. P., and Tanimura, T. 2003. Peripheral coding of bitter taste in *Drosophila*. J. Neurobiol. 56, 139–152.

Millius, M., Merivee, E., Williams, I., Luik, A., Mänd, M., and Must, A. 2006. A new method for electrophysiological identification of antennal pH receptor cells in ground beetles: the example of *Pterostichus aethiops* (Panzer, 1796) (Coleoptera, Carabidae). J. Insect Physiol. 52, 960–967.

Mitchell, B. K. 1987. Interactions of alkaloids with galeal chemosensory cells of Colorado potato beetle. J. Chem. Ecol. 13, 2009–2022.

Mitchell, B. K. and Itagaki, H. 1992. Interneurons of the subesophageal ganglion of *Sarcophaga bullata* responding to gustatory and mechanosensory stimuli. J. Comp. Physiol. A 171, 213–230.

Mitchell, B. K., Itagaki, H., and Rivet, M.-P. 1999. Peripheral and central structures involved in insect gustation. Microsc. Res. Tech. 47, 401–415.

Miyakawa, Y. 1981. Bimodal response in a chemotactic behaviour of *Drosophila* larvae to monovalent salts. J. Insect Physiol. 27, 387–392.

Miyakawa, Y. 1982. Behavioural evidence for the existence of sugar, salt and amino acid taste receptor cells and some of their properties in *Drosophila* larvae. J. Insect Physiol. 28, 405–410.

Moon, S. J., Köttgen, M., Jiao, Y., Xu, H., and Montell, C. 2006. A taste receptor required for the caffeine response in vivo. Curr. Biol. 16, 1812–1817.

Morita, H. 1959. Initiation of spike potentials in contact chemosensory hairs of insects. III. D. C. stimulation and generator potential of labellar chemoreceptor of *Calliphora*. J. Cell Comp. Physiol. 54, 189–204.

Morita, H. 1992. Transduction process and impulse initiation in insect contact chemoreceptor. Zool. Sci. 9, 1–16.

Murata, Y., Ozaki, M., and Nakamura, T. 2006. Primary culture of gustatory receptor neurons from the Blowfly, *Phormia regina*. Chem. Senses 31, 497–504.

Nakamura, M., Baldwin, D., Hannaford, S., Palka, J., and Montell, C. 2002. Defective proboscus extension response (DPR), a member of the Ig superfamily required for the gustatory response to salt. J. Neurosci. 22, 3463–3472.

Ozaki, M., Amakawa, T., Ozaki, K., and Tokunaga, F. 1993. Two types of sugar-binding protein in the labellum of the fly. Putative taste receptor molecules for sweetness. J. Gen. Physiol. 102, 201–216.

Pollack, G. S. and Balakrishnan, R. 1997. Taste sensilla of flies: function, central neuronal projections, and development. Microsc. Res. Tech. 39, 532–546.

Reynolds, S. E. 1990. Feeding in Caterpillars: Maximizing or Optimizing Food Acquisition? In: Animal Nutrition and Transport Processes. I. Nutrition in Wild and Domestic Animals. Comparative Physiology, Vol. 5 (ed. J. Mellinger), pp. 106–118. Karger.

Ribeiro, J. M. C. and Garcia, E. S. 1980. The salivary and crop apyrase activity of *Rhodnius prolixus*. J. Insect Physiol. 26, 303–307.

Robertson, H. M. and Wanner, K. W. 2006. The chemoreceptor superfamily in the honey bee, *Apis mellifera*: expansion of the odorant, but not gustatory, receptor family. Genome Res. 16, 1395–1403.

Robertson, H. M., Warr, C. G., and Carlson, J. R. 2003. Molecular evolution of the insect chemoreceptor gene superfamily in *Drosophila melanogaster*. Proc. Natl. Acad. Sci. U. S. A. 100(Suppl. 2), 14537–14542.

Rogers, S. M. and Newland, P. L. 2002. Gustatory processing in thoracic local circuits of locusts. J. Neurosci. 22, 8324–8333.

Rogers, S. M. and Simpson, S. J. 1999. Chemo-discriminatory neurons in the sub-oesophageal ganglion of *Locusta migratoria*. Entomol. Exper. Appl. 91, 19–28.

Rouseff, R. L. 1990. Bitterness in Foods and Beverages. Elsevier Science Publishers.

Scheiner, R. 2004. Responsiveness to sucrose and habituation of the proboscis extension response in honey bees. J. Comp. Physiol. A 190, 727–733.

Scheiner, R., Sokolowski, M. B., and Erber, J. 2004. Activity of cGMP-dependent protein kinase (PKG) affects sucrose responsiveness and habituation in *Drosophila melanogaster*. Learn. Mem. 11, 303–311.

Schmidt, J. M. and Friend, W. G. 1991. Ingestion and diet destination in the mosquito *Culiseta inornata*: effects of carbohydrate configuration. J. Insect Physiol. 37, 817–828.

Schoonhoven, L. M. 1969. Sensitivity changes in some insect chemoreceptors and their effect on food selection behavior. Proc. Kon. Ned. Akad. Wet. Amsterdam. Ser. C 72, 491–498.

Schoonhoven, L. M. 1977. On the individuality of insect feeding behavior. Proceedings of the Koninklijke Nederlandse Akademie van Wetenschappen (Series C) 80, 341–350.

Schoonhoven, L. M. 1978. Long-term sensitivity changes in some insect taste receptors. Drug Res. 28, 2367.

Schoonhoven, L. M. and Blom, F. 1988. Chemoreception and feeding behavior in a caterpillar: towards a model of brain functioning in insects. Entomologia Experimentalis et Applicata 49, 123–129.

Schoonhoven, L. M. and van Loon, J. J. A. 2002. An inventory of taste in caterpillars: each species is its own key. Acta Zool. Acad. Sci. Hung. 48 (Suppl. 1), 215–263.

Schoonhoven, L. M., Blaney, W. M., and Simmonds, M. S. J. 1992. Sensory Coding of Feeding Deterrents in Phytophagous Insects. In: Insect–Plant Interactions, Vol. IV (ed. E. A. Bernays), pp. 59–79. CRC Press.

Schröter, U. and Menzel, R. 2003. A new ascending sensory tract to the calyces of the honeybee mushroom body, the subesophageal-caclcal tract. J. Comp. Neurol. 465, 168–178.

Schultz, J. C. and Lechowicz, M. J. 1986. Host–plant, larval age, and feeding behavior influence midgut pH in the gypsy moth (*Lymantria dispar*). Oecologia 71, 133–137.

Schuppe, H., Cuttle, M., and Newland, P. L. 2006. Nitric oxide modulates sodium taste via a cGMP-independent pathway. J. Neurobiol. 67, 219–232.

Scott, K., Brady, R., Jr., Cravchik, A., Morozov, P., Rzhetsky, A., Zuker, C., and Axel, R. 2001. A chemosensory gene family encoding candidate gustatory and olfactory receptors in *Drosophila*. Cell 104, 661–673.

Seno, K., Fujikawa, K., Nakamura, T., and Ozaki, M. 2005b. Gq-alpha subunit mediates receptor site-specific adaptation in the sugar taste receptor cell of the blowfly, *Phormia regina*. Neurosci. Lett. 377, 200–205.

Seno, K., Nakamura, T., and Ozaki, M. 2005a. Biochemical and physiological evidence that calmodulin is involved in the taste response of the sugar receptor cells of the blowfly, *Phormia regina*. Chem. Senses 30, 497–504.

Shanbhag, S. R. and Singh, R. N. 1992. Functional implications of the projections of neurons from individual labellar sensillum

of *Drosophila melanogaster* as revealed by neuronal-marker horseradish peroxidase. Cell Tissue Res. 267, 273–282.

Shields, V. D. C. 1994. Ultrastructure of the uniporous sensilla on the galea of larval *Mamestra configurata* (Walker) (Lepidoptera: Noctuidae). Can. J. Zool. 72, 2016–2031.

Shields, V. D. C. and Mitchell, B. K. 1995. The effect of phagostimulant mixtures on deterrent receptor(s) in two crucifer-feeding lepidopterous species. Philos. Trans. R. Soc. Lond. B 347, 459–464.

Shimada, I., Shiraishi, A., Kijima, H., and Morita, H. 1974. Separation of two receptor sites in a single labellar sugar receptor of the flesh-fly by treatment with *p*-chloromercuribenzoate. J. Insect Physiol. 20, 605–621.

Shiraishi, A. and Kuwabara, M. 1970. The effect of amino acids on the labellar hair chemosensory cells of the fly. J. Gen. Physiol. 56, 768–782.

Shuichi Haupt, S. 2004. Antennal sucrose perception in the honey bee (*Apis mellifera* L.): behaviour and electrophysiology. J. Comp. Physiol. A 190, 735–745.

Simmonds, M. S. J. and Blaney, W. M. 1983. Some Neurophysiological Effects of Azadirachtin on Lepidopterous Larvae and their Feeding Response. In: Proceedings of the Second International Neem Conference (*eds*. H. Schmutterer and K. R. S. Ascher), pp. 163–180. Deutsche Gesellschaft für Technische Zusammernarbeit (GTZ) GmbH.

Simmonds, M. S. J. and Blaney, W. M. 1991. Gustatory codes in lepidopterous larvae. Symposia Biologica Hungarica 39, 17–27.

Simmonds, M. S. J., Simpson, S. J., and Blaney, W. M. 1992. Dietary selection behaviour in *Spodoptera littoralis*: The effects of conditioning diet and conditioning period on neural responsiveness and selection behavior. J. Exp. Biol. 162, 73–90.

Simpson, S. J. and Simpson, C. L. 1992. Mechanisms controlling modulation by haemolymph amino acids of gustatory responsiveness in the locust. J. Exp. Biol. 168, 269–287.

Smith, D. V., St. John, S. J., and Boughter, J. D. 2000. Neuronal cell types and taste quality coding. Physiol. Behav. 69, 77–85.

Spector, A. C. 2000. Linking gustatory neurobiology to behavior in vertebrates. Neurosci. Biobehav. Rev. 24, 391–416.

Spector, A. C. and Grill, H. J. 1992. Salt taste discrimination after bilateral section of the chorda tympani or glossopharyngeal nerves. Am. J. Physiol. (Regul. Integr. Comp. Physiol.) 263, R169–R176.

Sperling, J. L. and Mitchell, B. K. 1991. A comparative study of host recognition and the sense of taste in *Leptinotarsa*. J. Exp. Biol. 157, 439–459.

Städler, E., Renwick, J. A. A., Radke, C. D., and Sachdev-Gupta, K. 1995. Tarsal contact chemoreceptor response to glucosinolates and cardenolides mediating ovipositioning in *Pieris rapae*. Physiol. Entomol. 20, 175–187.

Steinbrecht, R. A. 1989. The fine structure of thermohygrosensitive sensilla in the silkmoth *Bombyx mori*: receptor membrane structure and sensory cell contacts. Cell Tissue Res. 255, 49–57.

Stocker, R. F. 1994. The organization of the chemosensory system in *Drosophila melanogaster*: a review. Cell Tissue Res. 275, 3–26.

Stocker, R. F. and Schorderet, M. 1981. Cobalt filling of sensory projections from internal and external mouthparts in *Drosophila*. Cell Tissue Res. 216, 513–523.

Szentesi, A. and Bernays, E. A. 1984. A study of behavioural habituation to a feeding deterrent in nymphs of *Schistocerca gregaria*. Physiol. Entomol. 9, 329–340.

Szentesi, A. and Jermy, T. 1990. The role of experience in host plant choice by phytophagous insects. In: Insect–Plant

Interactions (*ed*. E. A. Bernays), Vol. 2, pp. 39–74. CRC Press.

Tanaka, Y., Asaoka, K., and Takeda, S. 1994. Different feeding and gustatory responses to ecdysone and 20-hydroxyecdysone by larvae of the silkworm, *Bombyx mori*. J. Chem. Ecol. 20, 125–133.

Tanimura, T., Isono, K., Takamura, T., and Shimada, I. 1982. Genetic dimorphism in the taste sensitivity to trehalose in *Drosophila melanogaster*. J. Comp. Physiol. 147, 433–437.

Tanimura, T., Isono, K., and Yamamoto, M. 1988. Taste sensitivity to trehalose and its alteration by gene dosage in *Drosophila melanogaster*. Genetics 119, 399–406.

Thompson, R. F. and Spencer, W. A. 1966. Habituation: a model phenomenon for the study of neuronal substrates of behavior. Psychol. Rev. 73, 16–43.

Thorne, N., Chromey, C., Bray, S., and Amrein, H. 2004. Taste perception and coding in *Drosophila*. Curr. Biol. 14, 1065–1079.

Thurm, U. and Küppers, J. 1980. Epithelial Potential of Insect Sensilla. In: Insect Biology of the Future (*eds*. M. Locke and D. Smith), pp. 735–763. Academic Press.

Timmins, W. A. and Reynolds, S. E. 1992. Physiological mechanisms underlying the control of meal size in *Manduca sexta* larvae. Phyisol. Entomol. 17, 81–89.

Ueno, K., Kohatsu, S., Clay, C., Forte, M., Isono, K., and Kidokoro, Y. 2006. Gsα is involved in sugar perception in *Drosophila melanogaster*. J. Neurosci. 26, 6143–6152.

Ueno, K., Ohta, M., Morita, H., Mikuni, Y., Nakajima, S., Yamamoto, K., and Isono, K. 2001. Trehalose sensitivity in *Drosophila* correlates with mutations in and expression of the gustatory receptor gene Gr5a. Curr. Biol. 11, 1451–1455.

Usher, B. F., Bernays, E. A., and Barbehenn, R. V. 1988. Antifeedant tests with larvae of *Pseudaletia unipuncta*: variability of behavioral response. Entomol. Exper. Appl. 48, 203–212.

Usui-Aoki, K., Matsumoto, K., Koganezawa, M., Kohatsu, S., Isono, K., Matsubayashi, H., and Yamamoto, M.-T. 2005. Targeted expression of IP3 sponge and IP3 dsRNA impairs sugar taste sensation in *Drosophila*. J. Neurogenet. 19, 123–141.

van der Goes van Naters, W. M. and den Otter, C. J. 1998. Amino acids as taste stimuli for tsetse flies. Physiol. Entomol. 23, 278–284.

van Loon, J. J. A. and van Eeuwijk, F. A. 1989. Chemoreception of amino acids in larvae of two species of *Pieris*. Physiol. Entomol. 14, 459–469.

Wacht, S., Lunau, K., and Hansen, K. 2000. Chemosensory control of pollen ingestion in the hoverfly *Eristalis tenax* by labellar taste hairs. J. Comp. Physiol. A 186, 193–203.

Wang, Z., Singhvi, A., Kong, P., and Scott, K. 2004. Taste representations in the *Drosophila* brain. Cell 117, 981–991.

Watanabe, H. and Mizunami, M. 2006. Classical conditioning of activities of salivary neurons in the cockroach. J. Exp. Biol. 209, 766–779.

White, P. R. and Chapman, R. F. 1990. Tarsal chemoreception in the polyphagous grasshopper *Schistocerca americana*: behavioural assays, sensilla distributions and electrophysiology. J. Comp. Physiol. A 167, 105–121.

Wiggins, L. L., Frank, R. A., and Smith, D. V. 1989. Generalization of learned taste aversions in rabbits: similarities among gustatory stimuli. Chem. Senses 14, 103–119.

Wolbarsht, M. L. and Hanson, F. E. 1965. Electrical and Behavioral Responses to Amino Acid Stimulation in the Blowfly. In: Olfaction and Taste, II (*ed*. T. Hayashi), pp. 749–760. Oxford.

Yamamoto, R. T. 1974. Induction of host plant specificity in the tobacco hornworm, *Manduca sexta*. J. Insect Physiol. 20, 641–650.

4.04 Aquatic Animal Models in the Study of Chemoreception

J Caprio, Louisiana State University, Baton Rouge, LA, USA

C D Derby, Georgia State University, Atlanta, GA, USA

Glossary

accessory lobes (ALs) The paired deutocerebral neuropils in crustaceans that are connected to the olfactory lobes through interneurons but do not receive direct input from primary chemoreceptor neurons.

accessory olfactory bulb (AOB) The initial portion of the vertebrate central nervous system that processes information arriving from the vomeronasal organ.

aesthetasc sensilla Also called olfactory sensilla, they are the cuticular sensory structures of crustaceans that are located on the distal end of the lateral flagellum of the antennule and contain only olfactory receptor neurons that project only to the olfactory lobes.

antennal lobes The paired deutocerebral neuropils of insects that process odor information arriving from olfactory receptor neurons on the antennae; analogous to the vertebrate olfactory bulbs and the crustacean olfactory lobes.

antennules The first pair of antennae of crustaceans that are composed of basal segments at the proximal end and two flagella (lateral and medial) at the distal end.

chemoreceptor neuron (CRN) First-order neurons that detect environmental chemicals and transmit that information via their axonal projections into the central nervous system.

deutocerebrum The midbrain of arthropods (positioned between the protocerebrum and the tritocerebrum), which receives sensory information from the antenna; in crustaceans, the deutocerebrum includes the olfactory lobes, ALs, and lateral antennular neuropils.

facial lobe (FL) The primary gustatory nucleus of the medulla in teleosts that receives input from the facial taste nerve (cranial nerve VII) that innervates taste buds on the external body surface, lips, and rostral oral cavity.

lateral antennular neuropils (LANs) The paired deutocerebral neuropils of crustaceans that receive input from the CRNs and mechanoreceptor neurons that innervate the nonaesthetasc sensilla on the lateral and medial flagella of the antennules.

median antennular neuropil (MAN) The unpaired deutocerebral neuropil of crustaceans that receives input from the CRNs and mechanoreceptor neurons that innervate the nonaesthetasc sensilla of the antennules.

nonolfactory chemosensilla Also called nonaesthetasc chemosensilla, they are a diverse assortment of cuticular sensory structures located on the antennule of crustaceans, which project into the LANs and MAN.

olfactory bulb (OB) The initial portion of the central nervous system in vertebrates that processes odor information arriving from olfactory receptor neurons.

olfactory lobe (OL) In teleosts, the portion of the forebrain that receives input from the olfactory tract, the axons of mitral cells, the output neurons of the OB. In crustaceans, the deutocerebral neuropil that receives input from the axons of olfactory receptor neurons of the aesthetasc sensilla of the lateral flagellum of the antennule.

olfactory receptor (OR) Molecules, typically G-protein-coupled receptors, located on CRNs, to which odorant molecules bind to initiate the olfactory transduction process.

olfactory receptor neuron (ORN) Primary receptor neurons whose dendrites contain the molecular ORs and whose axons project via the olfactory nerve into the olfactory region of the brain; in fish, the olfactory nerve is cranial nerve I projecting to the OB, and in crustaceans, it is a tract of the antennular nerve projecting to the OL.

perireception Events occurring around the olfactory and taste receptors that influence the chemical environment and the reception of chemical stimuli.

vagal lobe (VL) The primary gustatory nucleus of the medulla in teleosts that receives input from taste activity transmitted by the vagal taste nerve (cranial nerve X) that innervates taste buds within the oral cavity caudal to the first gill arch.

4.04.1 Introduction

The aim of this chapter is to provide a basic understanding of the chemosensory structures and mechanisms found in aquatic animals. Because of the vast literature on such an enormous topic, we have had to narrow considerably the scope of our review. First, we focus only on the major aquatic animal models in the study of chemoreception. We have selected two groups, representing the vertebrates and the invertebrates: the teleosts (bony fishes) and decapod crustaceans (spiny lobsters, clawed lobsters, crayfish, and crabs). Second, we focus on topics more aligned with biomedical than ecological issues: general structure of the chemosensory cells and their organization, and mechanisms of chemosensory transduction and the neural processing of chemical stimuli. Due to space limitations, topics such as pheromones (Stacey, N. E. and Sorensen, P. W., 2002; 2005) and the chemosensory basis of homing behavior (Døving, K. B., 1996; Døving, K. B. and Stabell, O. B., 2003) in fishes are not discussed. This chapter is also not an encyclopedic listing of how chemicals may influence various behaviors of aquatic organisms but focuses on chemosensory behaviors where there is better understanding of the neurophysiological substrate. Due to the necessary selection of research reports to include in this chapter from the numerous research articles that have allowed the field of aquatic chemoreception to progress to date, we take this opportunity to apologize to our coworkers whose work was not given the treatment that it deserves. Previously published excellent reviews in this field include Kleerekoper H. (1969), Hara T. J. (1982; 1992), Atema J. *et al.* (1989), Carr W. E. S. *et al.* (1990), Sandeman D. C. *et al.* (1992), Derby C. D. (2000), Weissburg M. J. (2000), Zimmer R. K. and Butman C. A. (2000), McClintock T. S. and Xu F. (2001), Ache B. W. (2002), Schachtner J. *et al.* (2005), Ache B. W. and

Young J. M. (2005), Moore P. A. and Berman D. A. (2005), and Atema J. and Steinbach M. A. (2007).

For terrestrial animals including mammals and insects, olfaction and taste are neatly distinguished based on the physical medium in which they operate. Olfaction is for volatile molecules that are delivered in air to the receptor epithelium, and taste is for water-soluble molecules that are delivered in an aqueous medium to the receptors. For fish and most crustaceans, which live in aquatic environments and whose chemosensory world is limited to water-soluble rather than to volatile chemical stimuli, this distinction between olfaction and taste is nonsensical.

Olfaction and taste are also distinguished within the vertebrates based on anatomical organization, regardless of the media in which the animals live. Olfaction is mediated by the olfactory nerve (cranial nerve (CN) I), and taste is mediated by modified epithelial (taste) cells innervated by CN VII (facial), IX (glossopharyngeal), or X (vagal), regardless of whether the animal is a human living in air, a fish living in water, or a frog that lives in both worlds. Yet this anatomical distinction, which works for vertebrates, is not instructive for the other $\sim 97\%$ of the animals – that is, the invertebrates.

A third way of distinguishing olfaction and taste is a functional one. Taste mediates more simple, reflexive behaviors – grabbing, biting, and swallowing (i.e., consummatory behavior), whereas olfaction mediates more complex behaviors, that is, search for chemicals from a distance, courtship behavior, and learning about odors (Atema, J. 1977; 1995); however, taste in some fish like catfishes mediates searching for distant food (i.e., appetitive behavior).

Another way to identify olfaction, at least for vertebrates and arthropods, is the organization of the first-order processing regions in the brains that receive the peripheral chemosensory input. This region – the olfactory bulb (OB) of vertebrates, the antennal lobe of insects, and the olfactory lobe of crustaceans – is organized into glomeruli, which contain the synapses between the olfactory receptor neurons (ORNs) and second-order neurons. Furthermore, there is an odotopic organization to these glomeruli, in which odorants of different chemical categories generate distinctive activation patterns across the glomeruli.

In this chapter, olfaction and taste in fish is distinguished according to the vertebrate-specific pattern of nerve innervation, which is consistent with the olfaction-has-glomeruli model. For crustaceans, the glomerular definition of olfaction, and thus their olfactory pathway, is defined as the aesthetasc–olfactory lobe pathway (as described below). However, distinguishing between olfaction and taste creates a binary categorization that does not adequately describe the diversity of chemical senses of any animal, including fish and crustaceans. For example, this categorization applied to fish combines the oropharyngeal pathway and the extraoral pathway, even though they differ in cranial innervation (IX/X versus VII) and function – the former controlling reflexive swallowing (consummatory behavior) and the latter for locating distant food and taking it into the mouth (appetitive behavior). In addition, this categorization does not include the solitary chemoreceptor cells, which individually are anatomically similar to taste cells, but they are not organized into bud-like structures and they synapse with taste, trigeminal, or spinal nerves (Kapoor, B. G. and Finger, T. E., 2003). For crustaceans, this binary classification results in identifying all nonaesthetasc chemoreceptive neurons (CRNs) as taste, even though they are extremely diverse in how they are packaged into functional units, how they connect to the central nervous system (CNS), their organization within the CNS, and the behaviors they mediate (see below). Combining them into a single taste category is too simplistic and restrictive, and thus, in this chapter they are termed nonaesthetasc or nonolfactory chemosensors.

4.04.2 Olfactory Transduction

4.04.2.1 Introduction

ORNs of teleosts and crustaceans have served as important models not only for understanding the specifics of transduction in these species but also for elucidating fundamental and general principles of olfactory transduction (Ache, B. W. and Zhainazarov, A. B., 1995; Bruch, R. C., 1996; Ache, B. W. 2002; Ache, B. W. and Young, J. M. 2005). Transduction processes that are well characterized and that we describe in this section include perireception (in crustaceans), molecular biology of receptor molecules (in fish), second messengers, ion channels, and kinases. Much has been learned about chemoreception in fishes from only a few animal models of the more than 25 000 extant teleost species. These model species include catfishes, zebrafish, goldfish, and salmonids. For crustaceans, the spiny lobster *Panulirus argus* and the clawed lobster *Homarus americanus* are the models of choice for studying olfactory transduction.

4.04.2.2 Fish

Genes for putative olfactory receptors (ORs) are identified from several teleosts (Ngai, J. *et al.*, 1993; Byrd, C. A. *et al.*, 1996; Cao, Y. *et al.*, 1998; Sun, H. *et al.*, 1999; Kondo, R. *et al.*, 2002). These ORs are G-protein-coupled receptors, although they are structurally highly dissimilar from the ORs of mammals and other vertebrates.

OR activation by odorants results in the formation of second messengers – cAMP and/or inositol 1,4,5-trisphosphate (IP_3)/diacylglycerol (DAG). These second messengers lead to the depolarization of ORNs via the gating of (1) cyclic nucleotide-gated (CNG) channels and subsequently Ca^{2+}-activated Cl^- channels (for the cAMP pathway), which amplifies the odor-activated signal (Schild, D. and Restrepo, D., 1998), (2) ciliary IP_3 channels (Bruch, R. C., 1996; Schild, D. and Restrepo, D., 1998), or (3) TRPC2 (transient receptor potential) channels (Sato, Y. *et al.*, 2005), leading in all three cases to increasing intracellular Ca^{2+}. Although excitatory conductances have only been described to date, ORNs of teleosts also exhibit odorant-induced suppression (Kang, J. and Caprio, J., 1995b), but little definitive information is known in fish about the responsible mechanisms. The cAMP pathway is the generally accepted excitatory mechanism for the ciliated ORNs (cORNs) of tetrapods, especially mammals (Gold, G. H., 1999), although recent evidence suggests that odor transduction in mammals likely also involves a cAMP-independent pathway (Lin, W. *et al.*, 2004). However, in microvillous receptor neurons of the vomeronasal system, a DAG transduction pathway was recently implicated (Lucas, P. *et al.*, 2003). Fish, however, do not possess a vomeronasal system, and both cORNs and microvillous ORNs (mORNs) are present in the main olfactory epithelium. Recent evidence from teleosts suggests that specific odorants activate ORs and the cAMP second messenger system in cORNs (Sato, K. and Suzuki, N., 2000; Hansen, A. *et al.*, 2003; Schmachtenberg, O. and Bacigalupo, J., 2004; Sato, Y. *et al.*, 2005), whereas other odorants activate members of the V2R gene family and the IP$_3$ (and possibly the diacyl glycerol) pathway of mORNs (Speca, D. J. *et al.*, 1999; Hansen, A. *et al.*, 2003; Sato, Y. *et al.*, 2005).

A recent molecular modeling study identified key amino acid residues that are involved in ligand binding and selectivity for the L-arginine OR (a V2R olfactory receptor) in goldfish (Luu, P. *et al.*, 2004). Identified was a key residue, methionine 389, in goldfish OR 5.24 that provides OR selectivity to basic amino acids, and another residue, lysine 386, in zebrafish that is critical for OR selectivity to acidic amino acids. Although neither the odorant nor receptor types are known for crypt ORNs, the transduction mechanism in channel catfish likely utilizes $G\alpha_o$ (Hansen, A. *et al.*, 2003), whereas in goldfish the same crypt ORN expresses both $G\alpha_o$ and $G\alpha_q$ (Hansen, A. *et al.*, 2004).

4.04.2.3 Crustaceans

The importance of perireceptor events, which influence the odorant environment around the receptors, has been well studied in spiny lobsters. Flicking of the olfactory organ leads to periodic loading and unloading of chemical stimuli around the densely packed aesthetascs (Schmitt, B. C. and Ache, B. W., 1979; Koehl, M. A. R. *et al.*, 2001). Lobsters, fish, and other aquatic animals do not appear to have odorant-binding proteins, whose function in terrestrial animals may include delivery of odorants to receptors or even be more directly central to receptor activation (Vogt, R. G., 2005; Xu, P. *et al.*, 2005). Lobsters, however, do have several perireceptor elements that are involved in inactivating or removing odorants from the receptor environment. These include ectoenzymes, such as ectonucleotidases, to convert the excitatory compounds AMP and ATP into non-excitatory adenosine, and transporters for odorants such as taurine, adenosine, and glutamate (Carr, W. E. S. *et al.*, 1990).

Despite the fact that genes for ORs are not yet identified in any crustacean, other lines of inquiry, including biochemical, molecular, and physiological, reveal much about the nature of these receptors and their role in transduction. Patch clamp (Hatt, H. and Ache, B. W., 1994), binding studies (Olson, K. S. *et al.*, 1992), and transmission electron microscopy (Blaustein, D. N. *et al.*, 1993) demonstrate odorant-binding receptors on the dendritic membrane of ORNs. Furthermore, biochemical analysis of receptor–ligand binding shows that high (nM to pM) and low (μM) affinity sites for taurine, AMP, glutamate, and arginine receptors on the ORN dendrites (Olson, K. S. *et al.*, 1992; Michel, W. C. *et al.*, 1993; Olson, K. S. and Derby, C. D., 1995).

These receptors are G-protein-coupled receptors, as demonstrated by physiological, biochemical, and molecular studies. Odor-activated physiological responses of ORNs are affected in predictable ways by blockers or activators of G proteins (Fadool, D. A. *et al.*, 1995). In addition, several G-protein subunits ($G\alpha_i$, $G\alpha_s$, $G\alpha_q$, and $G\beta$), phospholipase C-β (PLC-β), and a G-protein-coupled receptor kinase are identified from cDNA libraries of the antennules of lobsters (*Homarus*) (McClintock, T. S. and Xu, F., 2001). Furthermore, functional assays demonstrate a role for these molecules in either activation or desensitization of odorant responses (McClintock, T. S. and Xu, F., 2001).

Lobster ORNs have two distinct G-protein-coupled second messenger pathways (Ache, B. W., 2002). One of these is excitatory, mediated by phosphoinositol signaling. In this pathway, InsP$_3$ activates a cation-selective channel, leading to an inward conductance that normally depolarizes the cell. The other G-protein pathway is inhibitory, mediated by cyclic nucleotide signaling. In this pathway, cAMP activates an outward conductance, carried by K$^+$ through a CNG K$^+$ channel (Michel, W. C. and Ache, B. W., 1992), and suppresses a steady-state Cl$^-$ channel (Doolin, R. E. *et al.*, 2001; Doolin, R. E. and Ache, B. W., 2005a). Both excitatory and inhibitory pathways are present in the same ORN, as revealed by electrophysiological recordings from both individual cells (McClintock, T. S. and Ache, B. W., 1989; Michel, W. C. *et al.*, 1991) and from patches of ORN dendritic membranes (Hatt, H. and Ache, B. W., 1994).

A given odorant can be excitatory to some ORNs and inhibitory to others (Michel, W. C. *et al.*, 1991) and can activate either the inward or the outward conductance (Doolin, R. E. and Ache, B. W., 2005b). Furthermore, it appears that the two inhibitory conductances, Cl$^-$ and K$^+$, are not odor specific (Doolin, R. E. and Ache, B. W., 2005b). Thus, no odorant is exclusively excitatory or inhibitory across cells, and no receptor exclusively mediates excitation or inhibition.

These two second messenger-mediated pathways are the first step in odor activation, and this is followed by a second step that amplifies this signal and sets the gain of the ORN (Ache, B. W., 2002). This amplification involves a nonselective cation channel, which is a member of the TRP channel family (Zhainazarov, A. B. *et al.*, 2001; Bobkov, Y. V. and Ache, B. W., 2005) and may be functionally similar to the Ca^{2+}-activated Cl^- channel that participates in odor detection in mammals (Kleene, S. J., 1997).

These and other results lead to the conclusion that individual ORNs of spiny lobsters contain a rich diversity of transduction molecules. Together, these diverse intracellular pathways allow ORNs to do substantial processing and integration (Ache, B. W. *et al.*, 1998). In addition, ORN responsiveness can be modulated by gamma-aminobutyric acid ($GABA_A$) receptors on the soma (Zhainazarov, A. B., *et al.*, 1999) and by centrifugal synaptic inhibition at their axonal terminals (Wachowiak, M. *et al.*, 2002). An important question is, whether having ORNs with such complex integration properties is a feature of simpler animals (i.e., invertebrates), which have fewer ORNs than vertebrates, or whether vertebrate ORNs are similarly complex. Although this topic is hotly debated, recent evidence indicates that mammalian ORNs have both cAMP and IP_3 pathways that can interact in complex ways (Spehr, M. *et al.*, 2002; Zhainazarov, A. B. *et al.*, 2004).

4.04.2.4 Overview

Olfactory transduction in teleosts and crustaceans has many commonalities with each other and with other characterized ORNs. ORs are G-protein-coupled receptors linked via second messenger pathways to ion channels. A diversity of olfactory transduction cascades that lead to the activation of CNG, IP_3, and/or TRPC2 channels exist. Other channels are secondarily activated, which in turn amplify the signal. Single ORNs of teleosts and lobsters can have more than one second messenger pathway. In lobsters, these include specific excitatory and inhibitory pathways. For teleosts, dual excitatory ORN pathways are identified, but it is currently unknown whether odorant-induced suppression requires an additional inhibitory transduction pathway, or whether specific odorants can affect the resident pathways to result in suppression (Hallem, E. A. *et al.*, 2004). Multiple transduction cascades in

individual cells can enhance the diversity of odor responses generated by ORNs. Perireceptor events are important to aquatic animals – for example, ectoenzymes and transporters clean up the odor environment around the ORs – but other perireceptor processes not relevant to aquatic animals are absent – for example, no odorant-binding proteins.

4.04.3 Properties of the Peripheral Chemical Detectors

4.04.3.1 Introduction

Peripheral chemosensory systems have the task of extracting biologically important information about the type, amount, and temporal dynamics of environmental chemicals and sending this information to the CNS. In this section, we review the organization of peripheral olfactory systems of fish and crustaceans and describe how they perform these important tasks.

4.04.3.2 Fish

4.04.3.2.1 Specificity of single olfactory receptor neurons

Although numerous studies in fishes tested amino acids as odorant stimuli, few investigations focused primarily on the odorant specificities of individual ORNs (Mac Leod, N. K., 1976; Meredith, M. and Moulton, D. G., 1978; Meredith, M., 1981; Restrepo, D. *et al.*, 1990; Ivanova, T. T. and Caprio, J., 1993; Kang, J. and Caprio, J., 1995a; 1995b; 1997; Sato, K. and Suzuki, N., 2001; Friedrich, R. W. and Laurent, G., 2004). Electrophysiological studies in the channel catfish indicated that single ORNs had ongoing spontaneous activity ($\sim$5 spikes s^{-1}) that could be elevated or suppressed in response to amino acids (Kang, J. and Caprio, J., 1995b; 1997); furthermore, response suppression (25%) was the dominant response type (13% excited; 62% nonresponsive). A few single ORNs showed excitation to some amino acids and suppression to others. In an innovative study to reveal odor-induced spike activity of single ORNs whose responses were likely to have been shunted from an extracellular recording electrode due to the high ionic content of the pond water bathing the olfactory organ, the olfactory organ of bullhead catfish was bathed in water of low ionic content (Valentinčič, T. *et al.*, 2005). ORNs having low or no spontaneous activity were excited by specific amino acids. That excitation is the major type of response to amino acids in teleost ORNs was also reported for zebrafish where only 4% (3 of 76 ORNs) were suppressed by an amino acid (Friedrich, R. W. and Laurent, G., 2004). None of these reports provided evidence for the existence of specific neuron types that were excited by particular amino acids.

Current unpublished data (Nikonov A. A. and Caprio J.), however, indicate the existence of ORNs that are highly specific to basic and neutral amino acids, respectively, similar to those reported for group I neurons in the OB (see section 4.04.4.2.4) and FB (see Nikonov, A. A. and Caprio, J., 2007).

4.04.3.2.2 Correlation between olfactory receptor neuron type, transduction process, and odorant sensitivity

Several morphological types of ORNs occur in the olfactory epithelium of teleosts: ciliated (c), microvillous (m), and crypt (cp) (Figure 1). However, controversy exists as to whether these different cell types process specific types of odorants (see Eisthen, H. L., 2004, for a review). A whole-cell patch clamp study of isolated ORNs from rainbow trout indicated, however, that mORNs responded specifically to amino acids, whereas cORNs were more broadly tuned, responding to a wide variety of odorants including amino acids (Sato, K. and Suzuki, N., 2001). That mORNs also respond to amino acids was indicated via activity-dependent labeling by the ion channel permeant probe, agmatine, in zebrafish (Lipschitz, D. L. and Michel, W. C., 2002). Additional evidence to suggest that mORNs respond to amino acids was obtained in the goldfish where a V2R receptor for L-arginine was shown to couple to PLC and IP$_3$ production and be expressed in ORNs whose cell bodies were located superficially in the olfactory epithelium, a characteristic of mORNs (Speca, D. J. et al., 1999). Evidence that cORNs also respond to amino acids, as reported above for rainbow trout, comes from a perforated patch study of isolated cORNs in a marine fish, the Cabinza grunt (*Isacia conceptionis*). cORNs in this marine species responded to amino acids through the activation of CNG channels and calcium-activated conductances (Schmachtenberg, O. and Bacigalupo, J., 2004), confirming a similar report for cORNs in rainbow trout (Sato, K. and Suzuki, N., 2000).

The recent determination of the odotopic organization within the OB of a few species of teleosts also provided the pathway to link odorant type with receptor cell morphology. By placing DiI, a postmortem retrograde tracer, into select and discrete regions of the teleost OB shown to be excited by specific classes of odorants, a direct link between ORN morphology and the type of odorant detected was recently derived from studies in crucian carp, channel catfish, and goldfish. For crucian carp, the evidence suggests that mORNs respond to food odors (Hamdani, E. H. *et al.*, 2001a), whereas cORNs are involved in the detection of alarm substances (Hamdani, E. H. and Døving, K. B., 2002). For channel catfish, the combination of a variety of experimental procedures (light and electron microscopy, *in situ* hybridization, immunocytochemistry, electrophysiology, and pharmacology) provided the necessary information to link the specific anatomical type of ORN with the type of molecular OR, its specificity for odorant type, and its probable mechanism of transduction (Hansen, A. *et al.*, 2003). In channel catfish, bile salt odorants (possible socially relevant cues in teleosts; Sorensen, P. W. and Caprio, J., 1998), are detected primarily by ORs expressed within cORNs and transduced via the $G\alpha_{olf}$/cAMP cascade, whereas nucleotides are detected by V2Rs expressed in mORNs but transduced via an unknown (neither $G\alpha_{olf}$/cAMP nor $G\alpha_{q/11}$/PLC) cascade; however, recent evidence in zebrafish suggests that nucleotides possibly activate TRPC2 channels expressed in mORNs (Sato, Y. *et al.*, 2005). Amino acid odorants in channel catfish activate both cORNs (expressing ORs) and mORNs (expressing V2Rs) but via different transduction pathways in the two types of ORNs (Hansen, A. *et al.*, 2003). For goldfish, cORNs also express ORs and $G\alpha_{olf}$, and mORNs express V2R genes but are primarily immunoreactive for $G\alpha_o$; however, $G\alpha_{i-3}$ and $G\alpha_q$ are also indicated for mORNs that are shorter and that possess more stiff microvilli than other mORNs (Hansen, A. *et al.*, 2004). Independent pharmacological evidence in goldfish also indicated that bile salt odorants and some amino acids are detected via the cAMP pathway and that amino acids also utilize the IP$_3$ transduction pathway (Rolen, S. H. *et al.*, 2003), which is consistent with previous molecular investigations (Bruch, R. C., 1996).

Sex pheromones in teleosts are gonadal steroids and prostoglandins (Sorensen, P. W. and Caprio, J., 1998). Recent evidence in crucian carp indicate that axons of cp neurons project to the ventral OB where they synapse with mitral cells whose axons form the lateral bundle of the medial olfactory tract and mediate sexual behavior (Hamdani, E. H. and Døving, K. B., 2006; Lastein, S. *et al.*, 2006; Weltzien, F. *et al.*, 2003).

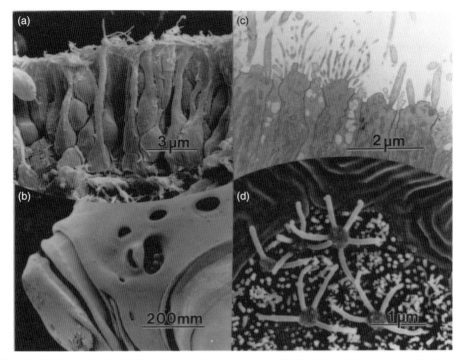

Figure 1   Teleost olfactory epithelia. (a) Scanning electron micrograph (SEM) of the adult sensory epithelium of a zebrafish, *Danio rerio*, showing some receptor cells with long dendrites, and some round receptor cells with short dendrites. (b) SEM of the olfactory organ of a zebrafish showing the organ, lateral line canals, and eye. (c) Transmission electron micrograph (TEM) of an apical longitudinal section of the olfactory epithelium of a goldfish, *Carassius auratus*, showing microvillous and ciliated olfactory receptor cells separated by supporting cells with vesicles. (d) SEM of the Indo-Pacific catfish, *Plotosus lineatus*, showing apical knobs of ciliated and microvillous receptor neurons. Reprinted from Sorensen, P. W. and Caprio, J. 1998. Chemoreception. In: The Physiology of Fishes (ed. D.H. Evans), 375–405. CRC Press LLC, by courtesy of Taylor and Francis Group, LLC. Courtesy: Eckart Zeiske and Anne Hanson.

4.04.3.3   Crustaceans

4.04.3.3.1   Organization

Being arthropods, crustaceans have an exoskeletal covering. Consequently, their chemosensory neurons are packaged into thin extensions of the cuticle, called setae or sensilla. A rich diversity of sensilla exists, which can be classified into olfactory and nonolfactory chemosensilla.

4.04.3.3.1.(i)   Olfactory sensilla   The olfactory sensilla, called aesthetascs, have been recognized in many crustacean taxa (Grünert, U. and Ache, B. W., 1988; Hallberg, E. *et al.*, 1992). Aesthetascs are distinguished by their peripheral structure and central connections. They are the only sensilla known to be innervated solely by chemosensory neurons and are typically richly innervated (hundreds of ORNs in some species such as spiny lobsters) (Derby, C. D. *et al.*, 2003). Aesthetascs are located on the paired first antennae (antennules) but only on the distal end of the lateral flagellum of the antennule (Figure 2). Chemical stimuli

pass through the porous cuticle of the aesthetascs and bind to receptor sites on the ORN dendrites. Each mature aesthetasc contains representatives of each of the functional types of ORNs, and thus, aesthetascs appear to be functional units of olfaction (Steullet, P. *et al.*, 2000b). The axons of the aesthetasc ORNs project ipsilaterally to the olfactory lobes (Schmidt, M. and Ache, B. W., 1996b) (see Section 4.04.4). Although aesthetascs are present in many crustacean taxa, they differ in their structure in functionally significant ways. For example, the organization of aesthetascs differs in aquatic, semiterrestrial, and terrestrial species (Ghiradella, H. *et al.*, 1968; Stensmyr, M. C. *et al.*, 2005). Aesthetasc sensilla contain not only sensory neurons but also supporting cells, which have functional overlap with supporting cells in insect and vertebrate olfactory organs. For example, supporting cells include (1) secretory cells associated with glands that produce a substance that likely coats the external surface of the aesthetasc cuticle (Schmidt, M. and Derby, C. D., 2005); (2) cells that secrete the sensillar cuticle; and (3) cells that produce ectoenzymes and

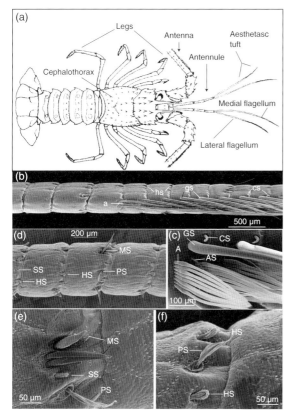

Figure 2 Peripheral chemosensory structures in spiny lobsters. (a) Drawing of a spiny lobster, *Panulirus argus*, showing the chemosensory organs. Each of the antennules (first antennae) has a medial flagellum and lateral flagellum. The sensory neurons are packaged into sensilla. The aesthetasc tuft, which contains aesthetasc sensilla and other associated sensilla (shown in the right half of b and in c), is located on the ventral side of the distal half of the lateral flagellum. Chemoreceptor neurons that innervate sensilla other than aesthetascs are called nonaesthetasc or nonolfactory chemoreceptor neurons. (b–f) Scanning electron micrographs of antennular sensilla. (b) The lateral flagellum at the boundary of the aesthetasc tuft. The aesthetasc tuft is to the right in this figure. The flagellum is divided into annuli, 12 of which are shown here. Each annulus in the aesthetasc region bears a similar complement of sensilla (shown in c), and each annulus in the nonaesthetasc region bears a similar complement of setae (shown in d–f). (c) Tuft region of the lateral flagellum. (d) Nontuft region of the lateral flagellum; the medial flagellum has similar types and organization of sensilla to this region. (e, f) High magnification views of nonaesthetasc regions of the lateral flagellum, showing four types of nonaesthetasc sensilla. A, aesthetasc sensillum; AS, asymmetric sensillum; CS, companion sensillum; GS, guard sensillum; HS, hooded sensillum; MS, medium-length simple sensillum; PS, plumose sensillum; SS, setuled sensillum. From figure 1 of Steullet, P., Dudar, O., Flavus, T., Zhou, M., and Derby, C. D. 2001. Selective ablation of antennular sensilla on the Caribbean spiny lobster *Panulirus argus* suggests that dual antennular chemosensory pathways mediate odorant activation of searching and localization of food. J. Exp. Biol. 204, 4259–4269, used with permission.

transporters that regulate the chemical environment of the fluid bathing the sensory cells (Carr, W. E. S. *et al.*, 1990; Gleeson, R. A. *et al.*, 1992).

4.04.3.3.1.(ii) Nonolfactory chemosensilla Nonolfactory sensilla are on all appendages and most of the body surface. Some of these, such as those on the mouthparts, are clearly gustatory in function, in that they control the ingestion of food (Derby, C. D. and Atema, J., 1982; Garm, A. *et al.*, 2005). Others are found on the legs, antennules, antennae, and general body surface (Derby, C. D., 1982; Schmidt, M. and Gnatzy, W., 1984; Cate, H. S. and Derby, C. D., 2001; 2002; Schmidt, M. and Derby, C. D., 2005), and consequently, these have many different functions. The nonolfactory chemosensilla are morphologically diverse, but they do have several common features. They are at least bimodally innervated, containing not only chemosensory neurons but also mechanosensory neurons (Schmidt, M. and Gnatzy, W., 1984; Cate, H. S. and Derby, C. D., 2002; Garm, A. *et al.*, 2005). Nonolfactory sensilla are more sparsely innervated than aesthetascs, typically with two or three mechanoreceptor neurons and fewer than 20 CRNs. Their cuticle is thick and not porous, and thus, chemicals enter through a terminal pore. The chemosensory neurons of these sensilla do not project to the olfactory lobe but rather to other neuropils in the CNS. This occurs even for nonolfactory antennular chemosensory neurons, which project ipsilaterally to the lateral antennular neuropils (LANs) and to the median antennular neuropil (MAN) (Schmidt, M. and Ache, B. W., 1996a) (see Section 4.04.4).

4.04.3.3.2 Sensitivity

4.04.3.3.2.(i) Response thresholds Olfactory and nonolfactory receptor neurons often differ in sensitivity. For *P. argus*, the sensitivity of antennular CRNs is much greater than for mouthpart CRNs (Thompson, H. A. and Ache, B. W., 1980; Derby, C. D. *et al.*, 1991; Garm, A. *et al.*, 2005). For example, thresholds for antennular CRNs (which include aesthetasc ORNs) are typically nanomolar and sometimes lower, whereas mouthpart CRNs have micromolar thresholds.

4.04.3.3.2.(ii) Tuning Crustaceans are sensitive to chemicals associated with most biologically important items, including food stimuli, social and sexual pheromones, settlement factors, and defensive compounds (Atema, J., 1988; Carr, W. E. S., 1988; Zimmer, R. K. and Butman, C. A., 2000; Kicklighter, C. E. *et al.*, 2005). Differences in sensitivity among species are partly attributable to the habitat in which they live. For example, for aquatic crustaceans, all sensors, including olfactory, are stimulated by water-soluble compounds. For terrestrial crustaceans, the antennules are sensitive to volatile chemicals, whereas the legs and mouthparts are sensitive to water-soluble stimuli (Rittschof, D. and Sutherland, J. P., 1986; Wellins, C. A. *et al.*, 1989; Stensmyr, M. C. *et al.*, 2005). Trophic level also accounts for interspecific differences in sensitivity. For example, carnivores such as *Panulirus* and *Homarus* have CRNs that respond to small, nitrogen-containing compounds – amino acids, amines, nucleotides, and peptides – that are prevalent in tissues of their animal prey and are relatively insensitive to carbohydrates and sugars (Zimmer-Faust, R. K. and Case, J. F., 1982a; Carr, W. E. S., 1988; Derby, C. D. and Atema, J., 1988; Zimmer-Faust, R. K., 1993). Herbivores and omnivores, such as fiddler crabs (Weissburg, M. J. and Zimmer-Faust, R. K., 1991), ghost crabs (Robertson, J. R. *et al.*, 1981; Trott, T. J. and Robertson, J. R., 1984), kelp crabs (Zimmer-Faust, R. K. and Case, J. F., 1982b), and crayfish (Ashby, E. A. and Larimer, J. L., 1965), often are sensitive to sugars common to plants, bacteria, and diatoms, as well as to some amino acids.

4.04.3.3.2.(iii) Specificity   The CRNs of crustaceans tend to be narrowly tuned, responding with high specificity to one compound or a set of structurally related compounds. Examples include CRNs of *H.americanus* tuned to predominantly to taurine, hydroxy-L-proline, or L-glutamate (Voigt, R. and Atema, J., 1992), and antennular CRNs of *P.argus* tuned primarily to taurine, glutamate, AMP, ATP, or ammonium (Derby, C. D. *et al.*, 1991; Derby, C. D., 2000). In the crayfish *Austropotamobius torrentium*, leg CRNs are less narrowly tuned but still have specificity to one of the following classes of compounds – amino acids, pyrimidines, or amines (Bauer, U. and Hatt, H., 1980; Bauer, U. *et al.*, 1981).

Within a species, there may be differences in specificity of CRNs in different organs. For example, in *P. argus*, individual antennular CRNs have narrower response profiles than mouthpart chemoreceptors (Garm, A. *et al.*, 2005).

Physiological studies of CRNs suggest that there are as many as a dozen types of molecular receptors that are heterogeneously distributed across the CRNs. Do single cells express a single receptor type, or more than one? It is often stated that ORNs of mammals and *Drosophila* express only one type of receptor, but this idea is being questioned in the face of new data (Mombaerts, P., 2004; Goldman, A. L.

et al., 2005), and it certainly is not true for CRNs in other animals such as *C. elegans* (Troemel, E. R., 1999). Although genes for chemoreceptors have not been identified in any crustacean, physiological studies strongly suggest that a given cell type has multiple receptor types, each coupled to either an excitatory or an inhibitory transduction cascade (see Section 4.04.2). However, the density of the receptor types on a CRN may vary, thus creating high response specificity (Cromarty, S. and Derby, C., 1997).

4.04.3.3.3   Temporal properties

Chemical stimuli in nature, even those released at a constant rate from a single odor source, are patchy and discontinuous. This is due to turbulence, which is ever-present though at different degrees in different environments. The temporal features of chemical stimuli influence the response pattern of a cell and consequently an animal's ability to find an odor source (Atema, J., 1995; Weissburg, M. J., 2000; Zimmer, R. K. and Butman, C. A., 2000; Koehl, M. A. R. *et al.*, 2001). Since the ability of CRNs to resolve temporal fluctuations determines the animal's ability to respond to them, it is important to identify these properties. The frequency following of CRNs is dependent on adaptation and disadaptation rates (Atema, J., 1995). Many crustacean CRNs have phasotonic response profiles, although this varies across cells (Borroni, P. F. and Atema, J., 1988; 1989; Weissburg, M. J. and Derby, C. D., 1995; Garm, A. *et al.*, 2005). CRNs on *Homarus* antennules can follow 4–5 odor pulses per second (Gomez, G. and Atema, J., 1996a; 1996b; Gomez, G. *et al.*, 1999). Since temporal variability of this magnitude is common in aquatic environments (see the above references), this ability to follow such pulses is obviously adaptive.

4.04.3.4   Processing Chemical Information by Single Peripheral Chemoreceptor Neurons of Crustaceans

To summarize our knowledge of neural coding by crustacean peripheral CRNs using a parallel construction with the section Processing Taste Information by Single Peripheral Taste Fibers by single peripheral taste fibers, we offer the following points:

1. Information about different classes of chemical stimuli can be transmitted to the CNS by different groups or types of CRNs; for example, the antennules, legs, and mouthparts of spiny lobsters and American lobsters have some CRNs most sensitive

to one or several amino acids, others most sensitive to one or several adenine nucleotides, and others most sensitive to ammonium ions (Voigt, R. and Atema, J., 1992; Derby, C. D., 2000; Garm, A. *et al.*, 2005).

2. The specificity of single CRNs is not limited to a particular class of chemical stimuli, although the strongest stimulus is usually several orders of magnitude more stimulatory than other stimuli; for example, taurine-best cells on antennules of spiny lobsters may also be sensitive to other amino acids, AMP, and ammonium, although the thresholds for taurine are 100–10 000 lower than for the other stimuli (Cromarty, S. and Derby, C., 1997).

3. Different CRNs having widely different tuning to members of the same class of chemicals can exist within the sensory appendage; for example, populations of CRNs with different amino acid specificities occur on the antennules, legs, and mouthparts of lobsters (Derby, C. D. *et al.*, 1991; Voigt, R. and Atema, J., 1992; Derby, C. D., 2000).

4. CRNs in different sensory appendages within the same species can have different chemical specificities; for example, in the American lobster, CRNs highly sensitive to hydroxy-L-proline are more abundant on the antennule than the legs or mouthparts (Voigt, R. and Atema, J., 1992).

5. The proportion and types of CRNs in different sensory appendages can vary within the same species. For example, cells with best responses to either hydroxy-L-proline, taurine, L-arginine, L-glutamate, betaine, and ammonium chloride occur in the antennules, legs, and mouthparts of the American lobster but in different proportions (Voigt, R. and Atema, J., 1992).

6. L-Amino acids are much more stimulatory than their respective D-isomers, but receptors for some D-isomers do exist (Michel, W. C. *et al.*, 1993).

7. Stimulus quantity appears to be coded by the frequency of action potentials generated. For individual CRNs, some have a dynamic concentration–response range of 5–6 log units of stimulus concentration, while others have more truncated concentration–response relationships. The absolute concentrations to which cells are responsive can differ widely (Derby, C. D. and Atema, J., 1988; Borroni, P. F. and Atema, J., 1988; 1989; Derby, C. D. *et al.*, 1991; Cromarty, S. and Derby, C., 1997).

8. Except for one report (Hatt, H., 1986), CRNs are not known to be sensitive to mechanical stimulation. However, they can respond to temperature and salinity changes (Schmidt, M., 1989).

4.04.3.5 Overview

Fish and crustaceans have a diversity of peripheral chemoreceptors. In teleosts, three types of ORNs have been identified: ciliated, microvillous, and crypt, which differ in their mode of transduction. The crustacean olfactory organ has many types of chemosensory neurons packaged into units called sensilla, but one of these sensillar types – the aesthetascs – is an olfactory sensillum based on its organization and central projections. The ability of fish and crustacean ORNs to detect and identify the type, amount, and temporal dynamics of chemical stimuli is well understood. ORNs generally have low spontaneous spiking activity, which is more usually increased but sometimes decreased by chemical stimulation. The population of ORNs on an olfactory organ responds to many chemicals, with small nitrogenous compounds, such as amino acids, amines, and nucleotides being the most effective. However, individual ORNs differ in their response specificities, which can be quite narrow. The response of ORNs increases in a concentration-dependent manner. Crustacean ORNs can follow pulses of chemicals up to at least 4–5 per second.

4.04.4 Organization of Central Olfactory Pathways

4.04.4.1 Introduction

The organization of the olfactory pathway of fishes and crustaceans has many elements common with that of other vertebrates, arthropods, and even other taxa. This section examines the connectivity between the olfactory periphery and CNS, the organization of the olfactory CNS, and how odor responses from the periphery are shaped by synaptic interactions within the olfactory CNS.

4.04.4.2 Fish

4.04.4.2.1 Organization of the olfactory bulb
Although the general anatomical organization of the olfactory system is similar across the vertebrates, the structure of the neural connections within the OB can be different between fish and mammals (reviewed in Dryer, L. and Graziadei, P. P. C., 1994). In contrast to mammals where a single mitral cell generally projects a single primary dendrite to a single large (50–200 μm in diameter), well-defined

glomerulus, mitral cells in the fish OB can project primary dendrites to several different, small (10–20 μm in diameter), and less well-defined glomeruli (Mori, K., 1995). However, the organization of the teleost OB may not be that radically different than in some mammals, since ~20% of the mitral cells in the main OB in rabbits possess multiple primary dendrites (Mori, K. *et al.*, 1983).

The anatomical organization in the teleost OB is similar to that observed in the accessory olfactory bulb (AOB) of the vomeronasal system in tetrapods where AOB mitral cells also have multiple apical dendrites, each of which innervates a small glomerulus (Satou, M., 1990; Takami, S. and Graziadei, P. P. C., 1991). This organization suggests that mitral cells in both the fish OB and the tetrapod AOB integrate odorant information from different types of ORNs (i.e., those expressing different molecular receptors) that provide input to the different glomeruli. A recent tracing study in the mouse, however, indicated that AOB mitral cells innervate glomeruli that only receive input from the same receptor type (i.e., a homotypic connectivity) (Del Punta, K. *et al.*, 2002). It is currently unknown whether such a homotypic connectivity is characteristic of the teleost OB (Satou, M. *et al.*, 1983; Riddle, D. R. and Oakley, B., 1992). Depending on the particular species of teleost, the anatomical organization within the OB can vary. For example, a recent investigation of the zebrafish OB indicated that the majority (~80%) of the mitral cells observed possessed a single primary apical dendrite (Fuller, C. L. *et al.*, 2005).

The dendritic field of a single mitral cell in fish can extend for more than 300–400 μm from the soma (Kosaka, T. and Hama, K., 1982b; Oka, Y., 1983) and terminate either in discrete tufts or in diffuse, brush-like ending (Kosaka, T. and Hama, K., 1982b; Riddle, D. R. and Oakley, B., 1992). However, in spite of this extensive dendritic field, the neuronal activities on one side of the OB are not influenced much by those in the opposite side (Satou, M., 1990), suggesting that there is little crossing of dendritic fields to opposite sides of the bulb. The recent evidence for odotopy within the OB of channel catfish (Nikonov, A. A. and Caprio, J., 2001) supports the functional isolation between right and left OB. In other aspects, mitral cells within fish lack basal dendrites, and thus, lateral interactions between mitral cells occur at the bases of their primary dendrites (Ichikawa, M., 1976; Kosaka, T. and Hama, K., 1982a; Oka, Y., 1983). Furthermore, axons of fish mitral cells arise not only from cell bodies but from thick dendrites (Dryer, L. and Graziadei, P. P. C., 1994). Ruffed neurons, a cell type not observed in

tetrapods, are also located within the mitral cell layer in teleosts. These cells synapse with granule cells but not with ORNs (Kosaka, T. and Hama, K., 1979; 1981). Tufted and periglomerular cells are apparently lacking in the teleost OB (Satou, M., 1990; Dryer, L. and Graziadei, P. P. C., 1994).

4.04.4.2.2   *Extrabulbar pathway*

Although the vast majority of axons of ORNs project to the OB, in some teleosts, a small subset ORN axons pass through the OB and project directly to the ventromedial telencephalon (Bazer, G. T. *et al.*, 1987; Honkanen, T. and Ekstrom, P., 1990; Szabo, T. *et al.*, 1991; Riddle, D. R. and Oakley, B., 1992; Becerra, M. *et al.*, 1994; Hofmann, M. H. and Meyer, D. L., 1995). Little functional information is known about this extrabulbar primary olfactory pathway (Fujita, I. *et al.*, 1991), but it is distinct from that of the terminal nerve (CN 0) whose axons also pass through the OB and connect to specific forebrain (FB) nuclei (Demski, L. S. and Northcutt, R. G., 1983).

4.04.4.2.3   *Odotopic representation of odorants in the teleost olfactory bulb*

Indication that the teleost OB is divided into different zones, each processing a specific class of biologically relevant odor, was initially indicated by Thommesen G. (1978), based on the recording of local field potentials in salmonids. This study showed that the lateral OB was highly responsive to amino acids, whereas the medial OB was more selective to bile acids/salts. Other evidence came from anatomical studies also in a salmonid (rainbow trout) that olfactory information arriving at the OB from the olfactory epithelium was likely sorted chemotopically (Riddle, D. R. *et al.*, 1993) and not topographically (Riddle, D. R. and Oakley, B., 1991). More direct evidence for the existence of an odotopic map in the teleost OB was derived from additional investigations of field potential in salmonids (Døving, K. B. *et al.*, 1980; Hara, T. J. and Zhang, C., 1996), calcium (Friedrich, R. W. and Korsching, S. I., 1997; Fuss, S. H. and Korsching, S. I., 2001) and voltage (Friedrich, R. W. and Korsching, S. I., 1998) imaging of olfactory nerve terminals within the OB of zebrafish, and a single-unit study in the OB of the channel catfish (Nikonov, A. A. and Caprio, J., 2001). The exceptional optical imaging study of odor responses to amino acids of olfactory receptor terminals within glomeruli of the lateral OB in zebrafish (Friedrich, R. W. and Korsching, S. I., 1997) provided the direct evidence that both odor identity and concentration are encoded by combinatorial glomerular activity patterns (Buck, L. B., 1996). Although some variations exist across the species of teleosts investigated, the general results of these studies indicate that food-related odors (i.e., amino acids and nucleotides) are mapped in OB regions (generally in the lateral OB) distinct from those of more socially relevant odors (bile salts, bile acids, and pheromones), which are mapped generally in the medial OB (Sorensen, P. W. *et al.*, 1991; Nikonov, A. A. and Caprio, J., 2001).

The medial–lateral distinction in odotopy within the teleost OB is consistent with mitral cell axons of the medial and lateral OB, respectively, projecting into the medial and lateral olfactory tracts (OTs) (Sheldon, R. E., 1912; Satou, M., 1990). The neuronal activities on one side of the fish OB are not influenced much by those on the opposite side, which is explained by limited dendritic fields of neurons in each part of the OB (Satou, M., 1990). In this respect, electrical stimulation experiments of distinct bundles of the OT in salmonids (Døving, K. B. *et al.*, 1980) and goldfish (Stacey, N. E. and Kyle, A. L., 1983; Demski, L. S. and Dulka, J. G., 1984) and behavioral investigations in crucian carp that had selected transections of different OT bundles (Hamdani, E. H. *et al.*, 2000; 2001a; 2001b; Weltzien, F. *et al.*, 2003) also support the previous evidence for a functional (food-related versus socially related) division between the medial and the lateral OB. This functional separation between types of odorants processed in medial and lateral OB is in sharp contrast to the two mirror-symmetric glomerular maps that exist in rodents (Lodovichi, C. *et al.*, 2003).

4.04.4.2.4   Odorant specificity of single olfactory bulb neurons

Electrophysiological investigations within the OB of the channel catfish (Nikonov, A. A. and Caprio, J., 2004) and zebrafish (Friedrich, R. W. and Laurent, G., 2004) provided information concerning the excitatory response spectrum to amino acids of single neurons (likely mitral cells) at the first synaptic level in the teleost olfactory system. Since the axons of ORNs expressing like ORs converge in the OB onto specific target glomeruli (Ressler, K. J. *et al.*, 1994; Vassar, R. *et al.*, 1994; Mombaerts, P. *et al.*, 1996; Strotmann, J. *et al.*, 2000), it is likely even in fish that the mitral cells whose apical dendrites innervate the respective glomeruli are tuned to odorants similarly as the ORNs that provide the input, but modified by intraglomerular circuitry (Yokoi, M. *et al.*, 1995; Sachse, S. and Galizia, G., 2002).

Two groups of neurons were recently identified in the OB of the channel catfish that were excited by amino acids. Group I OB neurons ($n = 91$ (37%)) were highly specific to the type of amino acid, that is, excited by acidic, basic, or neutral (those possessing short and long side chains, respectively) amino acids, and group II OB neurons ($n = 154$ (63%)) were also excited by a second amino acid but only at ≥10 times higher odorant concentration (Nikonov, A. A. and Caprio, J., 2004). Tested with an expanded series of amino acids, a subset of the group I OB units that were selective for neutral amino acids also distinguished neutral amino acids possessing long linear side chains from those with branched side chains. With increasing stimulus concentration, however, responses broadened such that single OB neurons responded selectively to neutral amino acids having either linear or branched side chains, but not to amino acids with either acidic or basic side chains. These results are consistent with the identification of relatively independent molecular ORs for acidic, basic, and neutral amino acids that were previously indicated in teleosts from electrophysiological cross-adaptation (Caprio, J. and Byrd, R. P., Jr. 1984; Caprio, J., *et al.* 1989; Sveinsson, T. and Hara, T. J., 1990; Michel, W. C. and Derbidge, D. S., 1997), biochemical binding (Cagan, R. H. and Zeiger, W. N., 1978; Brown, S. B. and Hara, T. J., 1981; Rhein, L. D. and Cagan, R. H., 1983; Rehnberg, B. G. and Schreck, C. B. 1986; Lo, Y. H. *et al.*, 1991), and calcium imaging (Friedrich, R. W. and Korsching, S. I., 1997; Fuss, S. H. and Korsching, S. I., 2001) studies. Relatively independent ORs for neutral amino acids with short and long side chains, respectively, were also reported (Caprio, J. and Byrd, R. P., Jr. 1984; Bruch, R. C. and Rulli, R. D., 1988; Sveinsson, T. and Hara, T. J., 1990).

It is rather intriguing that recent electrophysiological data obtained in zebrafish also provide evidence for OB unit selectivity to the type of amino acid prior to a reported declustering of odor representations that occurred over the first 800 ms of the response (Friedrich, R. W. and Laurent, G., 2001; Laurent, G. *et al.*, 2001). However, a similar effect was not observed in OB unit responses in channel catfish analyzed over a 3 s response period (Nikonov, A. A. and Caprio, J., 2004). Of 78 group I units analyzed from the OB of the channel catfish (i.e., those that were originally determined to be selectively responsive to only Met, Ala, Arg, or Glu over 3 s of response), 81% and 85% were similarly classified when analyzing the first and third seconds of the responses, respectively (Nikonov, A. A. and Caprio, J., 2004). A reanalysis (Caprio, J., unpublished data) was performed of zebrafish OB unit responses over the initial 400 ms of the response as reported in Figure 3 of Friedrich R. W. and Laurent

(a) Antennular olfactory pathway

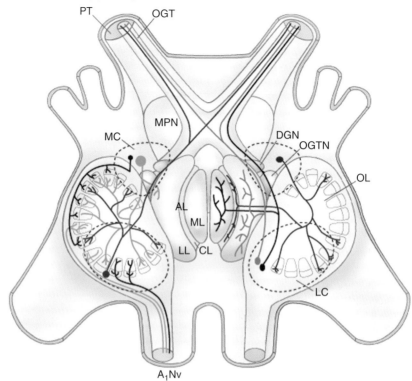

(b) Antennular nonolfactory chemosensory pathway

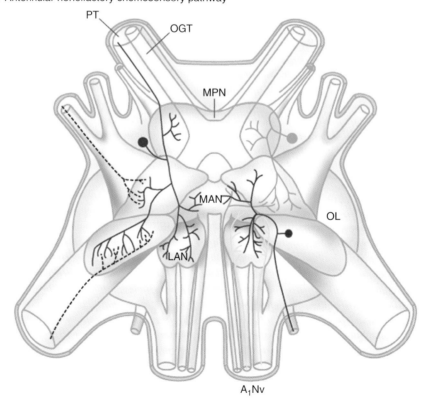

G. (2004) using the catfish scheme of identifying different groups of units by their selected excitatory responses to a specific type of amino acid. The results of this analysis suggested that 49 of the 58 units (84%) could be arranged into the same groups as had been reported for catfish – that is, those units that were preferentially excited by acidic, basic, short-chain neutral, and long-chain neutral amino acids. In addition, a fifth group emerged that was not detected in the channel catfish, those responsive to aromatic amino acids. Of these 49 zebrafish OB units, 63% were similarly classified when analyzing their responses over the first 1 s of the response, and 45% were similarly classified when the analysis time included 1.4 s of the response. These results suggest that there were not profound changes in responses of mitral cells in the zebrafish to the type of amino acid over response time and that the gradual dissolution of responses to types of amino acids over response time is possibly due to adaptational processes. For the zebrafish studies, however, the interpretation of the data is that responses of the population of mitral cells change over time in a stimulus-specific manner, thus providing a mechanism for the behavioral discrimination of individual amino acids (Friedrich, R. W. and Laurent, G., 2001). At present, the two different interpretations of the experimental results remain disputable.

Recently, the specificity of single OB units in the channel catfish to bile salts was reported (Rolen, S. H. and Caprio, J., 2007). OB neurons were identified that were excited selectively by taurine-conjugated bile salts and nonconjugated bile salts, respectively. A third type of OB neuron was rather nonselective and was excited by at least one member of each of three types of bile salts: taurine-conjugated, glycine-conjugated, and nonconjugated.

4.04.4.2.5 Olfactory forebrain

The OB odotopic maps of the aforementioned species of teleosts provide a blueprint for investigation of the organization of odor processing at the next ascending synaptic level in the olfactory system. The axons of the output neurons of the OB (i.e., mitral and ruffed cells) constitute the medial and lateral OTs, which terminate in the cerebral lobes of the olfactory FB. A critical question is whether an odor map exists in the FB, and if so, whether its organization is based on odorant structure as it is in the OB and the antennal lobe of insects. An alternate possibility is that an FB map is not based on odorant quality but on odorant function (e.g., feeding or social odorant cues), or even, that due to the apparent extensive overlap of medial and lateral OT projections, a chemotopic map may not exist in the FB.

Recent experiments performed to visualize FB neurons in mice (Zou, Z. et al., 2001; 2005) and mushroom body and lateral horn neurons of the protocerebrum in *Drosophila* (Wong, A. M. et al., 2002; Marin, E. C. et al.,

Figure 3 Antennular chemosensory pathways in the brain of spiny lobsters. (a) Summary diagram of the olfactory (aesthetasc) pathway, which is a ventral view of the brain. The olfactory lobe (OL) receives input from aesthetasc olfactory receptor neurons (ORNs) via the main antennular nerve (A_1Nv). ORNs typically terminate in a single glomerulus (blue), though there may be multiglomerular ORNs (black). The OL has two general types of interneurons: projection neurons (PNs), with their somata in the lateral somata cluster (LC); and local neurons (LNs), with their somata in the medial somata cluster (MC). Three types of OL interneurons are shown, distinguished according to color (red, black, and gray). In red is a PN with multiglomerular innervation in the OL and an axon that bifurcates and extends via the olfactory globular tracts (OGTs) to higher-order neuropils. Two types of OL LNs are shown (purple and green). The accessory lobe (AL) is organized into three regions: medial lobe (ML), central lobe (CL), and lateral lobe (LL). The AL receives no direct sensory input from receptor neurons, but it does receive input from the OL via LNs. AL LNs (purple), similar to those of the OL, have somata in the MC. The AL has PNs (black and gray), which are similar to PNs of the OL in having somata in the LC and bifurcated processes that exit the brain via the OGTs. The deutocerebral commissure neuropil (DCN) and olfactory globular tract neuropil (OGTN) are also innervated by processes from OL interneurons. (b) Summary diagram of the nonolfactory (nonaesthetasc) pathway, which is a dorsal view of the brain. Nonaesthetasc antennular chemoreceptor neurons (CRNs) and antennular mechanoreceptor neurons (yellow) project via the antennular nerve (A_1Nv) into the ipsilateral LAN. The LAN has two regions: a lateral lobe and a medial lobe. There is a general topotopy, in that receptor neurons from the antennular lateral flagellum innervate the lateral lobe of the LAN, and receptor neurons from the antennular medial flagellum innervate the medial lobe of the LAN. These receptor neuron terminals overlap with arborization of the antennular motor neurons (red). The MAN receives mechanoreceptor input via the A_1Nv from the statocyst (blue), an organ for maintaining equilibrium. PNs of the LAN leave the brain via the protocerebral tract (PT). The tegumentary neuropil (TN) and antennal neuropil (AnN) receive sensory input (dashed black) from the second antenna ($A_{II}Nv$). (a) is adapted from Schmidt, M. and Ache, B. W. 1996b. Processing of antennular input in the brain of the spiny lobster, *Panulirus argus*. II. The olfactory pathway. J. Comp. Physiol. A 178, 605–628. (b) is adapted from Schmidt, M. and Ache, B. W. 1996a. Processing of antennular input in the brain of the spiny lobster, *Panulirus argus*. I. Non-olfactory chemosensory and mechanosensory pathway of the lateral and median antennular neuropils. J. Comp. Physiol. A 178, 579–604. Courtesy: Amy Horner.

2002) that receive input from specific glomeruli of the OB (mice) or antennal lobe (*Drosophila*) provided a consistent picture of the general organization of OT projection patterns and suggested the possible logic for olfactory information processing in these brain structures. Common to both mice and *Drosophila* was a stereotypic projection of secondary olfactory projection neurons from specific glomerular modules of the OB or antennal lobe that terminate with considerable overlap in their multiple brain target areas. The recent findings that projection classes with similar axon terminal fields in the lateral horn in *Drosophila* tended to receive input from neighboring glomeruli (Marin, E. C. *et al.*, 2002) and that structurally similar odorants in the honeybee (Sachse, S. *et al.*, 1999), zebrafish (Friedrich, R. W. and Korsching, S. I., 1997), and rodents (Tsuboi, A. *et al.*, 1999; Malnic, B. *et al.*, 1999; Strotmann, J. *et al.*, 2000; Uchida, N. *et al.*, 2000) likely activate adjacent and overlapping glomeruli of the OB or antennal lobe suggest an odotopic organization in these central olfactory processing centers. However, recent studies indicated that this map in higher-order olfactory centers is different from that within the OB or antennal lobe (Zou, Z. *et al.*, 2001; 2005; Marin, E. C. *et al.*, 2002; Wong, A. M. *et al.*, 2002). Thus, in contrast to the OB, axonal arborizations in FB are diffuse and extensive, and projections from different OB glomeruli often overlap to varying degrees. This anatomical organizational pattern suggests that third-order neurons in the olfactory pathway integrate odor information arriving from multiple OB glomeruli, which possibly codes for odorant quality and which may also relate to the odor's behavioral significance (e.g., food or pheromones) (Johnson, D. M. G. *et al.*, 2000; Haberly, L. B., 2001; Wang, Y. *et al.*, 2001; Zou, Z. *et al.*, 2001; Marin, E. C. *et al.*, 2002; Wong, A. M. *et al.*, 2002).

For channel catfish, an FB odotopic map resembling somewhat that present in the OB was recently determined (Nikonov, A. A. *et al.*, 2005; Nikonov, A. A. and Caprio, J., 2007). Both the medial–lateral distinction between excitatory responses to bile salts and amino acids and the rostrocaudal distinction between excitatory responses to amino acids and nucleotides reflect a similar topographical organization within the OB of the same species. The amino acid-responsive terminal field appears homologous to the olfactory cortex and perhaps the olfactory tubercle, whereas the bile salt-responsive region, the medial terminal field, is possibly homologous to portions of the amygdala in amniote vertebrates (Wullimann, M. F. and Rink, E., 2002; Wullimann, M. F. and Mueller, T., 2004; Nikonov, A. A. *et al.*, submitted). Furthermore,

as suggested by the previous anatomical investigations of possible convergence within the FB of OT fibers emanating from different OB glomeruli, cell types not previously observed within the OB were evident. These included units that were excited by both amino acids and nucleotides and others excited by both neutral and basic amino acids; however, there was no evidence of the convergence of odor information arriving via the separate medial and lateral OTs. Thus, convergence occurred between OT fibers providing input of food-related odorants, but not between food-relevant and socially relevant chemical signals.

4.04.4.3   Crustaceans

4.04.4.3.1   *Organization of the olfactory lobes*

CRNs of the antennules are packaged into a diversity of types of sensilla (see Section 4.04.3). One of these types is specialized as the olfactory sensilla – the aesthetascs. The ORNs of the aesthetascs project ipsilaterally to the paired OL (Schmidt, M. and Ache, B. W., 1996b) (Figure 3). The OLs are thought to receive input almost exclusively from aesthetasc ORNs (Schmidt, M. and Ache, B. W., 1992; 1996b; Sandeman, D. C. *et al.*, 1992).

The OLs have a glomerular organization (Figure 3), generally similar to first-order olfactory neuropils of other animals – the antennal lobes of insects and the OBs of vertebrates (Hildebrand, J. G. and Shepherd, G. M., 1997; Strausfeld, N. J. and Hildebrand, J. G., 1999; Eisthen, H. L., 2002; Schachtner, J. *et al.*, 2005). The neuronal components of the OL glomeruli are also generally similar to that of the olfactory neuropils of vertebrates and insects. Besides the primary afferent input from the aesthetasc ORNs, there are projection interneurons (PNs), several classes of local interneurons (LNs), and centrifugal neurons. The location of the somata associated with these neurons is precise: the LNs have their somata in the medial cluster (cluster 9), and the PNs have their somata in the lateral cluster (cluster 10). The OL glomeruli have clear subdivisions: cap, subcap, and basal regions. These regions contain different neuronal types and synaptic interactions (Schmidt, M. and Ache, B. W., 1996b; Wachowiak, M. *et al.*, 1997). Most ORNs branch in only one glomerulus, primarily in the cap and subcap regions (Schmidt, M. and Ache, B. W., 1992). PNs have multiglomerular projections, with dense innervation in a few glomeruli and sparse innervation in many (Schmidt, M. and Ache, B. W., 1996b). The centrifugal fibers synapse on the somata of OL interneurons and are probably modulatory in function (Schmidt, M., 1997). The glomerular cap region is innervated by only two types of neurons: ORNs and multiglomerular GABAergic LNs. This implies that there are synapses between these neuronal types. This is probably the basis for the presynaptic inhibition from LNs onto ORNs (Wachowiak, M. *et al.*, 2002). Presynaptic inhibition in the OB also occurs in the vertebrates (Wachowiak, M. *et al.*, 2002).

Inhibition within the OL is mediated through at least two overlapping but functionally distinct inhibitory pathways (Wachowiak, M. and Ache, B. W., 1997). These are based on GABA and histamine, which shape the odor responses of OL units (Wachowiak, M. and Ache, B. W., 1998). These inhibitory circuits may have functional correlates with lateral inhibitory pathways in antennal lobe of insects and OB of vertebrates (Ache, B. W., 2002).

The number of OL glomeruli drastically varies across crustaceans, ranging from about 150 to 1300. The basis for this range is not obviously related to the number of aesthetascs, the animal's habitat, or phylogeny (Beltz, B. S. *et al.*, 2003).

Additionally, no clear sexually dimorphism in OLs is known, unlike has been described in some insects (e.g., the macroglomerular complex). This is in spite of the fact that sex pheromones are known in many crustaceans and their reception is mediated by aesthetascs (Gleeson, R. A., 1991; Kamio, M. *et al.*, 2002). It is possible that the OL possesses pheromone-specific glomeruli, but they are not morphologically distinct at a gross level.

Given findings that the glomerular organization in insects and vertebrates reflects odotopic mapping, it might be expected that the crustacean OL also has odotopic maps. There is indirect evidence for this, based on the fact that each aesthetasc expresses a diversity of ORNs (Steullet, P. *et al.*, 2000b) and the axons of ORNs from a single aesthetasc projects to many if not most glomeruli (Mellon, D. Jr. and Munger, S. D., 1990; Mellon, D. Jr. and Alones, V., 1993). The relationship between the number of ORNs and number of OL interneurons is instructive. For example, in spiny lobsters, each aesthetasc has ~300 ORNs (Derby, C. D. *et al.*, 2003). Aesthetascs appear to be functional units, with similar sets of ORNs in each aesthetasc (Steullet, P. *et al.*, 2000b); however, the number of different types of OR genes or ORNs is not known. Even assuming that there are 300 different types of ORs or ORNs, more than one type of ORN must innervate each glomerulus. However, direct tests of this hypothesis are lacking. Additionally, there is extensive convergence, given that there young adult spiny lobsters have about 300 000 ORNs per antennule projecting into 1200 glomeruli containing about 200 000 PNs and 100 000 LNs (Schmidt, M. and Ache, B. W., 1996b; Schachtner, J. *et al.*, 2005).

Closely associated with the OL, but lacking any primary sensory input, are the paired accessory lobes (ALs), olfactory globular tract neuropils, and deutocerebral commissure neuropils (Wachowiak, M. *et al.*, 1996; 1997; Sandeman, D. and Mellon, D., 2001; Ache, B. W., 2002; Schachtner, J. *et al.*, 2005) (Figure 3). The ALs are notable because of their large size in lobsters and crayfish and their glomerular organization. However, the AL glomeruli are smaller, more numerous, and of different synaptic organization than the OL glomeruli (Wachowiak, M. *et al.*, 1996; Schmidt, M. and Ache, B. W., 1996b). ALs have three regions – medial, central, and lateral – and each with specific synaptic connections (Wachowiak, M. *et al.*, 1996; Ache, B. W., 2002). For example, the medial and lateral regions have PNs, and the central region has interneurons connecting the AL to the OL as well as connection to the medial and lateral regions of the AL through LNs. The function of the AL is unknown, though its variance in size in different groups of crustaceans provides some grounds for speculation (Sandeman, D. C. *et al.*, 1993; Ache, B. W., 2002).

4.04.4.3.2 Organization of other chemosensory neuropils

All other antennular CRNs besides those in the aesthetascs, which include those on both the lateral and medial flagella of the antennules, project ipsilaterally to the paired LANs and to the unpaired MAN (Schmidt, M. and Ache, B. W., 1996a) (Figure 3). The LANs and MAN have a laminated organization, not glomerular. Unlike the OLs, the LANs and MAN also receive projections of mechanosensory neurons and antennular motor neurons. This organization is suggestive of topotopic maps, where space along the antennule is mapped topographically onto these neuropils (Schmidt, M. and Ache, B. W., 1993; 1996a; Schachtner, J. *et al.*, 2005); however, critical data are lacking.

4.04.4.3.3   Higher-order processing centers

The output interneurons from the OLs, ALs, and LANs project via the olfactory–globular tract (Figure 3) to a region of the protocerebrum called the terminal medulla and hemiellipsoid bodies, which are probably functionally equivalent to the mushroom bodies and lateral horn of insects (Schachtner, J. *et al.*, 2005). But the interneurons from the OLs, LANs, and ALs terminate in different regions of these neuropils, with phylogenetically diverse patterns of connection (Sullivan, J. M. and Beltz, B. S., 2004; 2005). The functional significance of these different projections and the activity of neurons in these higher-order neuropils are being explored (Sandeman, D. and Mellon, D., Jr. 2001; McKinzie, M. E. *et al.*, 2003).

4.04.4.3.4   Other chemoreceptor neuron processing centers

Virtually nothing is known about central processing of CRNs from the legs or mouthparts, except the identity of the regions into which their receptors project (e.g., Weissburg, M. J. *et al.*, 2001). While it is appreciated that the leg and mouthpart pathways control different behaviors than do the antennules (Derby, C. D., 2000; Horner, A. J. *et al.*, 2004; Garm, A. *et al.*, 2005), the mechanisms responsible for these differences remain largely unexplored.

4.04.4.4   Overview

The first-order olfactory processing center in both fishes and crustaceans, as with other vertebrates and arthropods, is glomerular in organization. These are the OB and AOB of fish, and the OL of crustaceans. The glomeruli are clusters of neuropil that contain the synapses between inputs (ORNs), outputs (mitral cells in fishes, projection neurons in crustaceans), and local interneurons. ORNs have ipsilateral projections, most being uniglomerular. The glomerular organization reflects an odotopy, with the odorants, at least food-related odorants, being represented across the glomeruli in a combinatorial fashion. Synaptic interactions within the OB and OL, which include inhibitory lateral connections, shape neuronal responses such that the output neurons of the OB and OL have different quality and temporal and response characteristics.

4.04.5   Neurogenesis and Turnover of Olfactory Neurons in Adult Crustaceans

Many crustaceans, fortunately including the model organisms for chemosensory research such as spiny lobsters, clawed lobsters, and crayfish, have indeterminate growth and add new ORNs to their olfactory organ and local and output interneurons to the OLs in the brain, throughout life including as adults (Steullet, P. *et al.*, 2000a; Harzsch, S., 2001; Derby, C. D. *et al.*, 2001a; Harrison, P. J. H. *et al.*, 2001a; Schmidt, M. 2001; Beltz, B. S. and Sandeman, D. C., 2003; Sandeman, R. and Sandeman, D., 2003).

Olfactory neurogenesis is highly flexible. Its rate can be modulated in adaptive ways by many variables. For example, following damage to the antennule, ORN neurogenesis increases dramatically so as to quantitatively compensate for the damage (Harrison, P. J. H. *et al.*, 2001a; 2003; 2004). As well, damage to the antennule modulates the rate of neurogenesis of olfactory interneurons (Sandeman, R. *et al.*, 1998; Hansen, A. and Schmidt, M., 2001; Sandeman, R. and Sandeman, D., 2003). Neurogenesis can also vary with internal factors, such as molt stage that is controlled by steroid hormones (Harrison, P. J. H. *et al.*, 2001a; 2001b), and environmental factors, such as circadian cycle (Goergen, E. M. *et al.*, 2002), season (Hansen, A. and Schmidt, M., 2004), environmental richness (Sandeman, R. and Sandeman, D., 2000), and social experience (Beltz, B. S. and Sandeman, D. C., 2003; Sandeman, R. and Sandeman, D., 2003; Song, C.-K. *et al.*, 2004).

In addition to continuous neurogenesis, olfactory neurons in the antennule, and to some extent in the brain, continuously turn over (Sandeman, R. and Sandeman, D. C., 1996; Steullet, P. *et al.*, 2000a; Schmidt, M., 2001; Harrison, P. J. H. *et al.*, 2001a; 2001b; 2003; 2004). The functional impact of this is that the olfactory organ completely turns over its ORNs after approximately four to five molts; the time for this depends on factors such as size of animal and environmental conditions, but for a young adult spiny lobster, this is approximately 1 year (Steullet, P. *et al.*, 2000a; Harrison, P. J. H. *et al.*, 2001a). The antennule is fully functional at any time, although the most proximal (newest) and distal (oldest) regions of aesthetascs have fewer aesthetascs and, at least in the case of the proximal (immature) aesthetascs, are not or less responsive to odors (Steullet, P. *et al.*, 2000a; 2000b). ORNs may be immature for several weeks or more after their birth (Steullet, P. *et al.*, 2000a; Harrison, P. J. H. *et al.*, 2001a), and olfactory interneurons may be immature (defined by typical expression of neurochemicals) for several months (Schmidt, M., 2001).

The fact that the site of proliferation of new ORNs is in a very small area of the olfactory organ gives experimental advantages of using crustacean models over others in exploring aspects of olfactory neurogenesis. For example, candidate molecules involved in ORN proliferation can be identified by using techniques such as representational difference analysis to identify transcripts that are enriched in the region of ORN proliferation compared to regions of mature ORNs. Initial analyses of this kind have identified interesting candidates (Stoss, T. D. *et al.*, 2004). These include (1) one whose expression levels also increase following damage, which also upregulates neurogenesis; (2) a member of the same protein family as follistatin, which is an antagonist of GDF 11, and GDF 11 is expressed by progenitors of mammalian ORNs and whose absence prevents proliferation of ORNs (Wu, H.-H. *et al.*, 2003); and (3) a growth factor of the PDGF/VEGF family. RT–PCR approaches (Chien, H. *et al.*, 2005) have also identified *splash*, a spiny lobster homologue of achaete-scute genes, and which is expressed in regions of ORN neurogenesis in the olfactory organ. Since the mammalian achaete-scute homologue, MASH, is involved in differentiation of ORNs in rodents, *splash* is an interesting initial candidate in the exploration of molecular mechanisms of control and modulation of olfactory neurogenesis in crustaceans.

4.04.6 Taste in Fish

4.04.6.1 Peripheral Taste Anatomy

Taste buds in teleosts are located generally on the lips and within the oral cavity, including high densities on the gill rakers (Jakubowski, M. and Whitear, M., 1990; Sorensen, P. W. and Caprio, J., 1998; Finger, T. E., and Simon, S. A., 2000; Hansen, A. and Reutter, K., 2004) (Figure 4). The buds can be elevated on epidermal hillocks, be flush with the surrounding epidermis, or sunken. Taste buds are also found on the external body surface, primarily on the face and in some species, such as catfishes, along the entire external body surface (Reutter, K., 1978). This increase in extraoral taste buds is thought not to result directly in an increase in sensitivity but in a greater ability to localize a taste source (Bardach, J. E. *et al.*, 1967). Taste bud

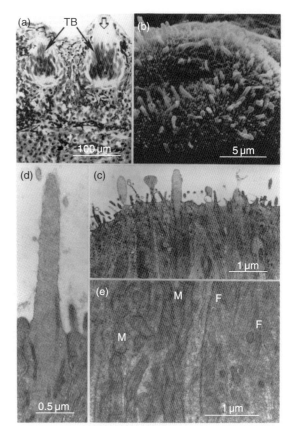

Figure 4 Teleost taste buds. (a) Light micrograph of a longitudinal section through two maxillary barbell taste buds of the bullhead catfish, *Ameiurus nebulosus*. TB indicates two taste buds and cells within; these taste buds are of the elevated type (unfilled arrow). At the lower portion of each taste bud, basal cells are observed. (b) Scanning electron micrograph (SEM) of a portion of the surface of a catfish (*Ameiurus*) taste bud showing numerous small and fewer large receptor microvilli. (c) Transmission electron micrograph (TEM) of the apical portion of an *Ameiurus* taste bud in longitudinal section. The apical portion of the light cells forms a single large receptor villus of a light cell and a small villus of an adjacent dark cell in longitudinal section from a *Silurus glanis* (European Wels catfish) taste bud. (e) Detail of the supranuclear region of a longitudinally sectioned *Silurus* taste bud that shows numerous mitochondria (M) within a light cell (left) and bundles of intermediate filaments (F) in a dark cell (right). Reprinted from Sorensen, P. W. and Caprio, J. 1998. Chemoreception. In: The Physiology of Fishes (ed. D.H. Evans), pp. 375–405. CRC Press LLC, by courtesy of Taylor and Francis Group, LLC. Courtesy: Klaus Reutter.

densities can vary tremendously depending on the species, the location of the receptive field, and the size of the specimen. In some fishes, such as minnows and certain cyprinids, up to ~300 taste buds mm^{-2} were estimated (Gomahr, H. *et al.*, 1992). Teleost taste buds contain up to ~100 cells that are generally divided into three types: light, dark, and basal cells (Reutter, K., 1978; Jakubowski, M. and Whitear, M., 1990). Light cells, considered the actual taste cell, possess a single large (0.5 μm thick, 1.5–3 μm long) club-shaped microvillus at its apical surface, whereas dark cells, which partially wrap light cells, have at their apical surface numerous small (0.1–0.2 μm thick, 0.5–1.0 μm long) microvilli and are considered supporting cells. However, dark cells in some species (e.g., ictalurid catfishes) form synapses with basal cells and may function also as taste cells (Reutter, K., 1978). Serotonin-rich basal cells, which resemble Merkel cells, number up to ~5 per taste bud, are oriented transversely to the longitudinal axis of the taste bud at the basal pole and form synaptic contacts with presumed taste cells. As in land vertebrates, taste cells undergo turnover. However, since teleosts are poikilothermic, the rate of turnover is dependent on the ambient water temperature. At temperatures of 14, 18, 22, and 30 °C, taste cells in channel catfish have an average life span of 40, 30, 15, and 12 days, respectively (Raderman-Little, R., 1979). Also, as in other vertebrates, taste buds, depending on their location, are innervated by branches of either the VII, IX, or X CN. In addition to taste cells located within taste buds, teleosts can possess solitary chemosensory cells (SCCs), which resemble taste receptor cells that reach densities in excess of $100\,000\,mm^{-2}$ and, depending on the location on the fish, form synaptic contact with different afferent nerve fibers (V, VII, and spinal) (Kotrschal, K., 1991; 1993; Finger, T. E., 1997; Hansen, A. and Reutter, K., 2004). Furthermore, there can be considerable overlap between input to the CNS of both SCC and gustatory neurons as in rocklings *Ciliata* and *Gaidropsarus* (Teleostei: Gadidae), or the SCC input via spinal nerves can be quite distinct from the taste system as in searobins (*Prionotus carolinus*) (Finger, T. E., 1982; Kotrschal, K. and Finger, T. E., 1996). These anatomical differences in central connections suggest that the SCCs of rocklings and searobins are not homologous. Physiological studies on SCCs are scant, but they suggest that SCCs in rocklings respond primarily to body mucus of heterospecific fish (Peters, R. C. *et al.*, 1991).

4.04.6.2   Taste Cell Physiology and Transduction

The major class of taste stimuli studied in a wide variety of fish species is amino acids, although additional gustatory stimuli investigated in particular

species include small peptides, nucleotides, bile salts and acids, aliphatic acids, quaternary ammonium compounds, and steroids (Marui, T. and Caprio, J., 1992; Hara, T. J., 1993). High-affinity taste receptor sites for L-alanine (dissociation constant (K_D) of 1.5 μM) and L-arginine (two affinity states: K_D values of 18 nM and 1.3 μM), respectively (Caprio, J., *et al.*, 1993), are located primarily on different taste cells within the same taste bud (Finger, T. E. *et al.*, 1996). The existence of independent gustatory receptor sites for L-alanine and L-arginine in channel catfish was determined previously from both electrophysiological and biochemical studies; a low-affinity receptor site for L-proline was also indicated (Caprio, J. *et al.*, 1993). Current evidence suggests that these three major classes of amino acid taste receptors in the channel catfish are coupled to activation of the taste receptor cells by two different mechanisms. At micromolar concentrations, L-alanine activates a G-protein-dependent increase in IP$_3$ or cAMP, but with a more rapid IP$_3$ production (Kalinoski, D. L. *et al.*, 1989). In contrast, both L-arginine and L-proline appear to be directly coupled to the activation of nonselective, but independent, cation channels (Kumazawa, T. *et al.*, 1998; Grosvenor, W. *et al.*, 2004). Further evidence of nonrandom expression of these amino acid taste receptors on taste cells is from electrophysiological studies of single taste fibers in the channel catfish, which has identified fiber types in the facial taste system that are highly responsive to L-alanine and L-arginine, respectively (Kohbara, J. *et al.*, 1992), and fiber types highly responsive to L-alanine, L-arginine, and L-proline, respectively, in the glossopharyngeal taste system (Ogawa, K. and Caprio, J., unpublished data).

Specific information concerning the molecular nature of taste receptors in vertebrates and the identification of their ligands has recently become available. As in mammals (Zhang, Y. *et al.*, 2003), fishes express two families of G-protein-coupled receptors, T1Rs and T2Rs in separate taste cells (Ishimaru, Y. *et al.*, 2005). T1Rs, members of class C GPCRs that possess an extensive N-terminus, exist as dimers. T1R dimers in fish detect amino acids (either T1R1/3 or multiple T1R2/3s), whereas T2Rs, which possess a short extracellular N-terminus and have not been shown to exist as dimers, detect bitter tastants (Oike, H. *et al.*, 2007).

4.04.6.3   Peripheral Nerve Taste Responses

Among the fishes tested electrophysiologically with amino acids, two groups were identified (Hara, T. J.

and Zielinski, B., 1989; Hara, T. J., 1993): (1) those whose facial taste system (those taste buds innervated by facial (CN VII) nerve fibers) responded to a number of different amino acids (i.e., wide response range) as demonstrated by channel catfish (Caprio, J., 1975; 1978) and (2) those whose facial taste systems were more selective (i.e., limited response range) as seen in salmonids (Hara, T. J. and Zielinski, B., 1989; Hara, T. J. *et al.*, 1999). Furthermore, multiunit recordings demonstrate that different branches of the facial nerve innervating different populations of taste buds located in different anatomical locations on the body of the fish have similar response properties and selectivities (Caprio, J., 1978; Davenport, C. J. and Caprio, J., 1982; Kanwal, J. S. *et al.*, 1987). The data are usually analyzed by comparing the magnitude of the integrated neural taste responses to the different chemicals at the same concentration. The relative magnitude of the integrated taste responses obtained even from different CNs can be similar as observed in rainbow trout (Kohbara, J. and Caprio, J., 2001). However, in other species such as channel catfish, taste responses between facial and glossopharyngeal/vagal nerves can be quite distinct (Kanwal, J. S. and Caprio, J., 1983; Kohbara, J. *et al.*, 1992). Facial taste thresholds for specific amino acids ranged generally between nano- and micromolar. Furthermore, depending on the species studied, taste thresholds of IX/X nerve fibers to specific amino acids can be either similar or higher than that determined for facial taste responses (Kanwal, J. S. and Caprio, J., 1983; Kohbara, J. and Caprio, J., 2001). Taste thresholds estimated from recordings from IX and/or X nerves in channel catfish were higher than those for VII, which is logical considering that high taste sensitivity of taste buds within the oropharyngeal cavity is not essential because of the higher stimulus concentrations present in the mouth once food intake occurs (Kanwal, J. S. and Caprio, J., 1987). The L-isomer of an amino acid was significantly more stimulatory than its D-isomer. However, a population of facial taste fibers in the sea catfish, *Arius felis*, was described that was more responsive to D- than to L-alanine (Michel, W. and Caprio, J., 1991) (see next section). In some fishes, sensitivity to a particular non-amino acid stimulus may even be greater than it is to amino acids. For example, the facial taste system of the rainbow trout to a bile salt, taurolithocholate, approaches picomolar concentrations, which is 4 log units lower than for L-proline, the most stimulatory amino acid (Hara, T. J. *et al.*, 1984; Yamashita, S. *et al.*, 2006).

4.04.6.4 Processing Taste Information by Single Peripheral Taste Fibers

Since taste receptor cells are secondary receptors, they must synapse with afferent nerve fibers (VII, IX, or X) to transmit gustatory information to the primary gustatory nucleus of the medulla. A single taste fiber may synapse with multiple taste cells, not all of which are necessarily located in the same taste bud. For example, in channel catfish, the number of taste buds innervated by a single taste fiber of the recurrent facial nerve that innervates taste buds on the flank increases with the size of the fish, from two taste buds per axon in small fish ($\sim$5 cm in length) to nearly 14 taste buds per axon profile in larger fish (37–40 cm) (Finger, T. E. *et al.*, 1991). Information on the processing of taste information by peripheral nerve fibers in teleosts is extremely limited and is based primarily on quantitative analyses of responses of facial taste fibers in only a few species. The data obtained with amino acid stimuli from these species, however, indicate that there is sufficient similarity in the tuning characteristics of gustatory neurons to justify the existence of different fiber types.

Although single-unit studies of the taste specificity of teleosts were accomplished in only a few species of the more than 25 000 that exist, some tentative generalizations can be drawn:

1. Taste information concerning different classes of taste stimuli can be transmitted to the medulla by different groups (i.e., types of facial taste fibers); for example, the facial taste system of Japanese puffer fish (*Fugu pardalis*) consists of different populations of taste fibers with specificities to amino acids, nucleotides, and an inorganic acid, respectively (Kiyohara, S. *et al.*, 1985).
2. The specificity of a single taste fiber is not limited to a particular class of chemical stimuli; for example, one taste fiber type in the yellowtail flounder is highly responsive to both a nucleotide (uridine-5′-monophosphate) and an amino acid (L-tryptophan) (Zeng, C. and Hidaka, I., 1990).
3. Different taste fiber types having widely different tuning to members of the same class of chemicals can exist within the same CN of a particular species; for example, populations of facial taste fibers with different amino acid specificities occur in channel catfish (Kohbara, J. *et al.*, 1992; Caprio, J., *et al.*, 1993), sea catfish (Michel, W. and Caprio, J., 1991), Japanese puffer (Kiyohara, S. *et al.*, 1985), and yellowtail (Zeng, C. and Hidaka, I., 1990).
4. Taste fibers of different CNs (VII versus IX and X) within the same species can have different chemical

specificities; for example, a population of taste fibers highly responsive to L-proline occurs for the IX but not VII taste system in channel catfish (Ogawa, K. and Caprio, J., unpublished data).

5. The proportion and types of taste fibers comprising different CNs can vary within the same species; for example, the population of L-alanine taste fibers within the IX taste nerve in channel catfish was reduced by 55% compared with VII.

6. In addition to the existence of populations of taste fibers that are highly responsive to particular L-amino acids, taste fibers for D-amino acids may also exist; for example, 38% of the single facial taste fibers analyzed from the sea catfish, *Arius felis*, were most responsive to D-alanine (Michel, W. and Caprio, J., 1991). However, this fiber type also responded to L-alanine, and the L-alanine fiber type responded to D-alanine. High concentrations of D-amino acids occur in the tissues of soft-bodied invertebrates (Preston, R. L. 1987), potential prey items of the sea catfish. In contrast, only 3% of the single facial taste fibers in the freshwater channel catfish were most responsive to D-alanine (Kohbara, J. *et al.*, 1992).

7. Irrespective of the of the chemical specificity of the particular taste fiber types occurring in a species, stimulus quantity appears to be coded by the frequency of action potentials generated. For individual taste fibers, some have a dynamic response range of 5–6 log units of stimulus concentration, whereas other fibers, even in the same fish, have more truncated dose–response relations (Michel, W. and Caprio, J., 1991; Kohbara, J. *et al.*, 1992).

8. Some peripheral taste fibers in a particular species may be bimodal, that is, they respond to both taste and mechanical stimuli as in catfish (Davenport, C. J. and Caprio, J., 1982; Ogawa, K. *et al.*, 1997), whereas in other species, such as the Japanese puffer, this does not appear to be the case (Kiyohara, S. *et al.*, 1985).

4.04.6.5   Taste and Tactile Input to the Central Nervous System (Medulla)

Chapter 10 in this volume and a previous report (Kanwal, J. S. and Finger, T. E., 1992) provide excellent reviews of the gustatory pathways within the CNS of teleosts. This section is thus limited to a brief summary of the topographical manner in which taste and tactile information is represented in the primary gustatory nucleus of the medulla of catfishes since much of the limited information on the

processing of taste information was derived from these teleosts known for their elaborate gustatory system. Limited electrophysiological information also exists concerning taste and tactile activity in higher-order, that is, pontine (Lamb, C. F. IV and Caprio, J., 1992) and diencephalic (Lamb, C. F. IV and Caprio, J., 1993), gustatory nuclei in catfish but is beyond the scope of this review.

Taste and tactile information (see point 8 above) arriving via VII and IX/X terminates in register and in a somatotopic and viscerotopic manner, respectively, within the special visceral sensory column of the rostral medulla that is equivalent to the gustatory portion of the nucleus of the solitary tract in mammals. Both taste and tactile information from the external body surface and rostral oral cavity in catfishes is processed within the facial lobe (FL) of the medulla and that within the more posterior oral cavity and gill rakers by the glossopharyngeal nucleus (possibly a small lobe in some species) and vagal lobe (VL) (Finger, T. E., 1976; Kanwal, J. S. and Caprio, J., 1987). There is, however, a disproportionate representation of structures (e.g., an enlarged volume of medulla for the representation of barbels and gill rakers), which is indicative of their relative functional importance for taste and tactile input. The FL, through its input from external body taste buds and its interconnections to the spinal cord and medullary reticular formation, is involved in localization of a stimulus and controlling swimming behavior (Kanwal, J. S. and Finger, T. E., 1997). As such, the FL map is well defined and precise. The glossopharyngeal and vagal systems, which receive input from taste buds within the oral cavity, in contrast connects with brainstem nuclei in the control of swallowing and oropharyngeal movements (Atema, J., 1971; Finger, T. E. and Morita, Y., 1985). The VL map is not so important for the localization of food and is thus more diffuse than that of the FL (Kanwal, J. S. and Caprio, J., 1988).

For teleosts that possess specialized taste and tactile structures, such as barbels, the special visceral column can be subdivided into two to three lobes (Kanwal, J. S. and Finger, T. E., 1992). Within each lobe a finer representation of body parts (e.g., barbels and gill arches) can be represented in lobules (Kanwal, J. S. and Caprio, J., 1987; Hayama, T. and Caprio, J., 1989; Kiyohara, S. and Caprio, J., 1996). For catfishes, taste and tactile inputs from each of the different barbels project to a different lobule within the FL (Kiyohara, S. *et al.*, 1986; Kanwal, J. S. and Caprio, J., 1988; Marui, T. *et al.*, 1988). In addition to the existence of peripheral bimodal (taste and tactile) fibers, mechanosensory input into the primary gustatory nucleus also occurs via mechanosensory-only facial (Davenport, C. J. and Caprio, J., 1982; Kiyohara, S. *et al.*, 1985) and trigeminal (Kiyohara, S. *et al.*, 1986) fibers. Medullary neurons responsive to taste stimuli (i.e., amino acids) were limited to smaller areas of the electrode tracks than neurons sensitive to touch alone (Marui, T., 1977; Marui, T. and Funakoshi, M., 1979; Marui, T. and Caprio, J., 1982; Kanwal, J. S. and Caprio, J., 1988). The overlap of taste and tactile maps is interesting from a neuroethological aspect as both types of input are involved in the detection and selective food ingestion in catfish and are likely processed simultaneously prior to activation of the respective motor neurons.

4.04.7 Processing of Mixtures

4.04.7.1 Introduction

In all natural environments, including aquatic, olfactory and gustatory receptor cells rarely encounter single pure chemicals as are usually presented in experimental studies. The composition of many biologically relevant mixtures in aquatic environments, particularly of food, is well characterized (Carr, W. E. S., 1988). Behavioral studies demonstrate the high potency of animal extracts and mixtures of amino acids to fishes and crustaceans. Additionally, however, the magnitude of the response to a chemical mixture is poorly predicted from knowledge of the responses to the individual components – revealing mixture interactions. These are either mixture suppression or mixture enhancement, in which the response to the mixture is less or greater, respectively, than expected. Obviously then, defining the expected response to a mixture becomes the critical issue. To do so for a CRN, it is critical to know several of its features: Does a CRN have both excitatory and inhibitory transduction pathways? Does a CRN have more than one type of receptor? If so, how are these coupled to the excitatory and/or inhibitory transduction pathways? Which mixture components interact with each receptor and transduction pathway? How do those components interact with the receptors – that is, are they agonists or antagonists, and is their action competitive or noncompetitive?

The more completely characterized is the cell, the more accurate is the predicted response to a mixture. In other words, mixture interactions are a consequence of an incomplete understanding of the above processes.

Fish and crustaceans, especially catfish and spiny lobsters, serve as model organisms to understand how mixtures are processed. These studies are built largely on electrophysiological and biochemical studies that have defined the diversity of receptor types and transduction pathways at the cellular level. Arguably, our understanding of how the sensory systems of animals detect, process, and ultimately lead to the discrimination and perception of mixtures is as complete for catfish and lobsters as it is for any animal. (And sadly, we still have a long way to go!) This section summarizes our understanding of mixture processing in these systems.

4.04.7.2 Fish

4.04.7.2.1 Introduction

Electrophysiological studies in teleosts attempted to predict the global response, that is, olfactory EOG (electro-oculography) and olfactory and gustatory integrated neural activity (Hidaka, I. *et al.*, 1976; Caprio, J., *et al.* 1989; Kohbara, J. and Caprio, J., 1996; Ogawa, K. and Caprio, J., 2000) to simple (binary and ternary) mixtures of amino acids. Additionally, a single olfactory electrophysiological investigation studied responses to complex mixtures (up to 10 components) of amino acids (Kang, J. and Caprio, J., 1991). The overall results of these investigations indicated that knowledge of the relative independence of the receptors for the component stimuli in a mixture, obtained from receptor binding and/or electrophysiological cross-adaptation experiments, was essential to successfully predict the magnitude of the electrophysiological response. Mixtures whose components showed little competition or cross-adaptation generated responses greater than those whose components were indicated to compete with a common receptor or with receptors having highly overlapping specificities. There was little evidence for mixture suppression. That mixture interactions are weak in the peripheral olfactory system in teleosts was also indicated in the zebrafish (Tabor, R. *et al.*, 2004). For olfaction, where single ORNs express one or at most a few ORs (Ngai, J. *et al.*, 1993; Mombaerts, P. 2004), the greater response to a mixture of amino acids was likely due to the activation of multiple populations of ORNs, each ORN within a specific population expressing one of the receptors for a particular component in the mixture. The greater response to mixtures observed in the peripheral taste system was also likely due to the activation of different receptors wherever expressed; for this effect observed in single taste fibers in the channel catfish (Ogawa, K. and Caprio, J., 1999) and Japanese puffer (Hidaka, I. *et al.*,

1976), the different receptors could theoretically be located on individual cells within a single taste bud or on different taste cells in different taste buds innervated by the single taste fiber. These studies also indicated for both olfactory and gustatory systems that the greater responses to mixtures are more likely due to the simultaneous activation of different receptors by the individual components of the mixture than by stimulus binding at the same total concentration to a single receptor. The greater response to a mixture whose components bind to relatively independent receptors than to a single or highly cross-reactive receptors may be the electrophysiological correlate of the behavioral observations that mixtures are often more stimulatory than individual components, even in cases when the concentration of the individual component is higher than the total concentration of the multiple-component stimuli within the mixture.

4.04.7.2.2   Responses of single units

4.04.7.2.2.(i)   Olfaction   There are few quantitative physiological studies of how single olfactory neurons in fishes respond to stimulus mixtures – primarily amino acids. For teleosts, only in the channel catfish was a quantitative study of single ORNs performed to stimulus mixtures (Kang, J. and Caprio, J., 1997), whereas responses of single OB neurons to stimulus mixtures were investigated in both the channel catfish (Kang, J. and Caprio, J., 1995b; 1997; Tabor, R. *et al.*, 2004) and zebrafish (Tabor, R. *et al.*, 2004). For channel catfish, the types of responses of single ORNs and OB neurons to binary mixtures comprising components that were either both excitatory, both suppressive, or both nonresponsive were similar (based on response type) to those of the components in 80% (for ORNs) and 82% (for OB neurons) of the trials. These results are comparable to a similar study in rats where 75% of the responses of single OB neurons to binary mixtures evoked similar response patterns as the components of the binary mixtures (Giraudet, P. *et al.*, 2002). Overall, these results indicate that profound mixture interactions are rare in simple mixtures where the components evoke similar responses.

For simple binary mixtures where the responses to the components were different (i.e., excitatory (E) and nonresponsive (N); suppressive (S) and N; E + S, respectively), responses of catfish OB neurons to the mixture were often similar in 94% of the trials to that for one of the components (Kang, J. and Caprio, J., 1995b). For example, in a binary mixture of an excitatory and a suppressive component, the response of an OB neuron to the mixture was either excitatory or suppressive in 94% of the trials – in only 6% of the trials was the mixture nonresponsive. For catfish ORNs, responses to the mixtures were similar to that for one of the components in 82% of the trials for E + S mixtures and 100% for both E + N and S + N mixtures (Kang, J. and Caprio, J., 1997). In a comparable study in the OB of the rat, units evoked a similar pattern of response, as did one of the components in the mixture in 83% of the trials (Giraudet, P. *et al.*, 2002). For zebrafish OB units, where binary mixtures of amino acids comprised an excitatory and a suppressive component ($n = 18$), the response to the binary mixture was dominated (in 85% of the cases) by the excitatory response, whereas the suppressive component occurred in the remaining 17% of the trials. In additional experiments, the component responses and those to the binary mixtures were complex, differing primarily in their temporal patterns (Tabor, R. *et al.*, 2004). In contrast, for OB units in channel catfish where excitatory and suppressive responses to the mixture components tested were equivalent (29%; 42% null), the response to the binary mixture was excitatory in 34%, suppressive in 41%, and null (no response significantly different from control) in 25% of the trials ($n = 32$) (Kang, J. and Caprio, J., 1995b). Mixture suppression or masking, where a component reduced or concealed the neuron's excitatory or inhibitory response to the other component, however, was observed with single ORN and OB neurons in catfish and single OB neurons in rat. This effect was observed when there was no significant response to a binary mixture composed of an effective stimulus (evoking either an E or S response) and a nonstimulatory component. Such mixture suppression occurred in 62% (for E + N components) and 68% (for S + N components) of these trials in catfish ORNs, 43% in catfish OB neurons, and 46% in rat OB neurons (when analyzing firing rates).

4.04.7.2.2.(ii)   Taste   Electrophysiological responses of integrated and single-unit facial taste responses to binary mixtures of amino acids in the channel catfish (Ogawa, K. and Caprio, J., 1999) were consistent with previous olfactory results obtained in the same species (see above). These reports collectively indicate that the magnitude of multiunit responses to binary and more complex olfactory and gustatory stimulus mixtures is greater if the component stimuli bind to relatively independent receptor sites than to the same or highly cross-reactive sites. The study also indicated that the greater taste activity observed to the mixture was not significantly different, whether recording the taste activity in a multifiber or a single-unit preparation. This finding is noteworthy in indicating that greater taste activity is not exclusively the result of the components in a binary mixture simultaneously activating different fiber types (i.e., having different chemical specificities); that is, the greater response also occurred by the activation of different taste receptors on taste cells innervated by the same single taste fiber. Also indicated was that the magnitude in enhanced taste activity could be significantly different across the different responding gustatory fiber types. In the Japanese puffer, *F. pardalis*, the amino acid derivative betaine (*N*-trimethyl glycine) enhanced taste activity of facial taste fibers to particular amino acids (Hidaka, I. *et al.*, 1976).

4.04.7.2.3 Behavioral discrimination

Fish are similar to humans in that components of binary mixtures are discriminable. Humans can identify up to approximately three components in either taste (Laing, D. G. *et al.*, 2002) or olfactory mixtures (Laing, D. G. and Francis, G. W., 1989). Whether a binary mixture of amino acids is detected by fishes as a unique odor or whether the qualities of the individual components are retained within the mixture was investigated in catfish (Valentinčič, T. *et al.*, 2000a). It was previously determined for catfish that the discrimination of chemicals is based on olfaction and not on taste (Valentinčič, T. *et al.*, 1994). Catfish conditioned to a binary mixture initially treat the mixture as the component eliciting the greater physiological response (based on the magnitude of the EOG in response to each component). This result is similar to that for humans, where binary odor mixtures are perceived as the more stimulatory component (Laing, D. G. and Willcox, M. E., 1983). Additional discrimination training in catfish, however, facilitates the discrimination of the less potent component in a binary mixture – that is, the binary mixture is no longer detected as its more stimulatory component. Thus, with enough training, catfish are able to elementally process mixtures (i.e., identify the separate components of the mixture).

4.04.7.3 Crustaceans

4.04.7.3.1 Electrophysiology

Electrophysiological studies of processing of mixtures by crustaceans have been performed exclusively using single cells rather than global responses such as EOGs. Thus, explanations are possible at the single-cell level as well as at the system level through analysis of populations of these single cells.

Single CRNs of crustaceans have multiple transduction pathways, including excitatory and inhibitory pathways and including more than one receptor type in the excitatory pathway (see Section 4.04.2). Thus, when trying to understand the response to a mixture, even a simple one such as a binary mixture, the magnitude and direction of the response (i.e., increase or decrease from spontaneous spiking rate) will depend on whether the components of a mixture activate the excitatory or the inhibitory pathway and on whether they activate the same or different receptor sites in a given pathway. As a consequence, responses to mixtures can be either more or less than the more excitatory component. The following are some examples.

If a CRN has both excitatory and inhibitory transduction pathways, and one chemical activates the excitatory pathway and another chemical activates the inhibitory pathway, then we would expect that a mixture of these two chemicals would generate a response less than excitatory chemical and greater than the inhibitory chemical. This has been observed from intracellular whole-cell patch clamp recording from lobster CRNs (Michel, W. C. and Ache, B. W., 1992). This type of effect can explain some cases of mixture suppression identified from extracellular measurement of spiking activity of single-unit recordings from lobster CRNs. These CRNs often have very low levels of spontaneously spiking activity – often much less than 1 spike per second – so that if a compound activates an inhibitory transduction pathway, this inhibition is not expressed as a reduction in spiking activity. If only spiking activity were being recorded, this would lead to an identification of mixture suppression. Yet knowing the transduction cascades activated by these components of a mixture leads to a predicted response that matches that observed.

If a CRN has two receptors types, each coupled to an excitatory transduction pathway as seems to be the case for many CRNs of crustaceans (see Section 4.04.2), then it is important to know how the components of a mixture interact with these receptor types. The predicted responses are different if the mixture components bind primarily to the same receptor or bind to different receptors. If the components compete for the same receptor, then whether they are competitive agonists or antagonists will influence the response generated by the mixture. The clearest demonstration of this is in a study by Cromarty S. and Derby C. (1997). This study examined responses of taurine-sensitive CRNs to binary mixtures of excitatory compounds, either competitive agonists (taurine and structural analogues of taurine) or noncompetitive agonists (taurine and structurally dissimilar molecules). Responses to binary mixtures of competitive agonists were significantly lower than responses to mixtures of noncompetitive agonists, but they were exactly that expected from a competitive model. Knowing how these compounds bind to receptors and how these receptors are coupled to transduction cascades allowed accurate predictions of responses to mixtures; without this, mixture interactions would be suggested to have occurred.

As described in the next section, behavioral studies suggest that spiny lobsters can use elemental processing to analyze mixtures, such that the mixture is perceived as a set of identifiable components (Livermore, A. et al., 1997). This behavioral finding has implications on how mixtures are processed by the peripheral olfactory system. It suggests that a mixture and its components should generate different neural response profiles and that the response profile of the mixture should have features in common with its components. In addition, if one component of a mixture dominates the perception of a mixture, then the neural profile generated by the mixture should be more similar to that of the dominant component. These ideas were tested using different blend ratios of binary mixtures (Steullet, P. and Derby, C. D., 1997). The results suggest that for the neural profiles generated in the peripheral olfactory system of spiny lobsters, the qualities of individual compounds are maintained when the compounds are mixed to form blends. This result suggests a neural explanation of the ability of spiny lobsters to elementally process an odor mixture.

4.04.7.3.2 Behavior

Mixture interactions have been examined at the behavioral level in some detail in several crustacean species, including the glass shrimp *Palaemonetes pugio* (Carr, W. E. S. and Derby, C. D., 1986a; 1986b), the Caribbean spiny lobster *P. argus* (Derby, C. D. et al., 1989; Daniel, P. C. and Derby, C. D., 1991), the California spiny lobster *Panulirus interruptus* (Zimmer-Faust, R. K. et al., 1984; Zimmer-Faust, R. K., 1987; 1993), and the American lobster *H. americanus* (Borroni, P. F. et al., 1986; Atema, J. et al., 1989). In these studies, typically the magnitude of the behavioral response (usually an appetitive response) to a mixture is compared to that of the responses to that mixture's components. The response magnitudes are then compared to each other and to the predicted responses calculated on a variety of models that fall into two classes: (1) those that assume the components interact with the same transduction pathways and (2) those that assume the components activate independent transduction pathways. The outcomes of these studies differ. For the glass shrimp, mixture enhancement or synergism, in which the mixture response was significantly greater than expected, was common. Compounds that by themselves had no activity, when mixed with each other or with compounds that were moderately active, generated very strong responses. Using an antennular flicking assay on the Caribbean spiny lobster and binary mixtures of compounds, mixture suppression was much more common, in which the response to a binary mixture was less than expected. In the California spiny lobster, adding a repellent compound (ammonium) to an attractive compound (ATP) predictably produced a response to the mixture that was less than that to the attractant alone (Zimmer-Faust, R. K. et al., 1984; Zimmer-Faust, R. K., 1987; 1993). From these results, it is clear that predicting responses to mixtures based only on responses to components is unreliable. We have argued earlier that with detailed knowledge of the transduction pathways within an ORN and the chemical stimuli that activate each of them, accurate predictions of mixture responses are possible. But of course for behavior, many neuronal levels exist between ORNs and motor output, including complex synaptic interactions. For example, some mixture interactions occur at the level of the CNS, independent of the ORNs (Derby, C. D. et al., 1985). Thus, generating a better understanding of processing of chemical information in the CNS is essential for understanding the behavior of animals to mixtures.

The ability of lobsters to discriminate among chemical stimuli is known from studies using aversive paradigms similar to those used on other animals. Such associative conditioning generates an aversive response to one chemical stimulus, even one previously attractive, and then animals can be examined for response generalization to other stimuli (Fine-Levy, J. B. et al., 1988). This procedure was used for several sets of stimuli that were also used in electrophysiological studies of mixture processing (see below). These include single compounds; binary mixtures that differ only in blend ratios, and their components; and complex mixtures of 29–35 components that mimic their natural food (crab, shrimp, oyster, and mullet). These results show that lobsters show the greatest generalization between stimuli that are closest in composition. For example, lobsters show much more generalization between crab mixture and shrimp mixture than between other mixtures, and this is highly correlated with the degree of similarity in the blend ratios of these mixtures (Fine-Levy, J. B. *et al.*, 1988; 1989; Fine-Levy, J. B. and Derby, C. D., 1991).

Behavioral studies also showed that spiny lobsters perceive chemical mixtures in different ways, depending on the salience of the components. They can learn them as a set of elemental cues if the salience of the components during learning is sufficiently high. In addition, they can also learn them as a configural cue (i.e., as a mixture-unique entity) if the salience of the individual components is not emphasized (Livermore, A. *et al.*, 1997; Derby, C. D., 2000).

4.04.7.4 Overview

Catfish and spiny lobsters are two of the more intensely studied animals regarding the neural processing and behavioral responses to chemical mixtures, and as such, they provide excellent case studies. Behavioral studies show that both catfish and spiny lobsters perceive the components of binary and even larger mixtures as discriminable elements, if the salience or relevance of those components is emphasized. This implies elemental processing. If the salience of the components is not emphasized, then mixtures are perceived as a configural (i.e., mixture-unique) cue. This leads to the conclusion that each discriminable component of a mixture should generate a unique neural profile, at least at early processing levels. Electrophysiological studies on catfish and lobsters support this conclusion. Individual studies identified examples of mixture interactions, in which the response to the mixture is different from that expected from the response to the mixture's components. However, the body of work on catfish and spiny lobsters demonstrates that the more completely characterized is a cell or a system, the more accurate is the predicted response to a mixture. If one knows the receptor types, the transduction cascades, how receptor types are coupled to those cascades, and which chemical stimuli interact with each receptor type and transduction cascade, then by using rigorous experimental procedures, responses to mixtures are highly predictable. A few studies on catfish and lobsters demonstrate this (Caprio, J., *et al.*, 1989; Kang, J. and Caprio, J., 1991; Cromarty, S. and Derby, C., 1997). The challenge is of course to build a detailed understanding of all of these molecular pathways. Since mixtures are the biologically relevant stimuli for animals, this is more than a worthy endeavor, as difficult as it is.

4.04.8 Behavioral Roles of Olfaction and Taste

4.04.8.1 Introduction

The chemical senses are involved to some degree in many behaviors of fishes and crustaceans. These include finding food, mates, habitat, and shelter, avoiding predators, social interactions, and homing. Although many studies investigated the chemosensory behavior of different species to chemical stimuli, the vast majority did not determine which chemosensory systems – olfaction, taste, or others – were primarily responsible for the behavioral results. Since both olfactory and taste systems for particular species of fishes and crustaceans can be highly sensitive to specific classes (e.g., amino acids) of chemical stimuli (Caprio, J., 1978; Goh, Y. and Tamura, T., 1980; Hara, T. J., 1994), without the appropriate controls, it can be confusing as to which system or systems are responsible for releasing a specific behavioral response.

4.04.8.2 Fish

Olfaction is involved in the detection and search for food and is uniquely implicated in the detection of pheromones and other chemicals that provide information concerning sex, social interactions, alarm, and homing. Taste, on the other hand, is also associated with food-related behaviors, but those that are more of a reflexive nature, such as ingestion, biting, and swallowing of food; for catfishes, taste also participates directly in the localization of the food source (Hara, T. J., 1993; Sorensen, P. W. and Caprio, J., 1998).

Typical feeding stimulants of fishes are amino acids, and both the olfactory and gustatory systems are generally highly stimulated by amino acids. However, the chemical specificity and the specific functions of the two chemosensory systems to these compounds can be quite distinct (Caprio, J., 1977; Goh, Y. and Tamura, T., 1980; Kohbara, J. *et al.*, 1992; Nikonov, A. A. and Caprio, J., 2004), and the feeding behaviors controlled by the two are not identical but are overlapping. For a generic fish (i.e., one without an elaborate extraoral taste system) such as rainbow trout, both olfactory (Hara, T. J., 1973) and gustatory (Marui, T. *et al.*, 1983) systems can be highly sensitive to amino acids. However, anosmic rainbow trout are incapable of initiating feeding in response to these chemical cues, in spite of a fully functioning taste system (Valentinčič, T. and Caprio, J., 1997). In contrast, for catfishes, which possess extensive extraoral and oral taste systems and a concomitant increase in the neural circuitry of the medulla facial and VLs to process gustatory information, both food search and ingestion are released in anosmic specimens (Bardach, J. E. *et al.*, 1967; Valentinčič, T. and Caprio, J., 1994). Olfaction in catfishes, however, is essential for the discrimination of different amino acid stimuli in a learning paradigm (Valentinčič, T. *et al.*, 1994; 2000b; Valentinčič, T., 2004).

4.04.8.3 Crustaceans

There are differences, as well as some overlap, in those behaviors controlled by olfaction (aesthetascs) and taste. In spiny lobsters, the antennules drive detection of and orientation to distant, attractive chemicals, including food-related chemicals and intraspecific chemicals such as aggregation pheromones and alarm pheromones (Derby, C. D. *et al.*, 2001b; Horner, A. J. *et al.*, 2004; Shabani, S. and Derby, C. D., 2006) and pheromones (aggregation cue and alarm cues) (Horner, A. J. and Derby, C. D., 2005; Shabani, S. and Derby, C. D., 2006). Interestingly, either the aesthetasc (olfactory) pathway alone or the nonaesthetasc antennular pathway alone is sufficient to mediate the animal's response to food odors (Horner, A. J. and Derby, C. D., 2005; Shabani, S. and Derby, C. D., 2006). In addition, both of these pathways can mediate discrimination of and learning about food-related chemicals, be they single compounds, simple mixtures, or complex mixtures. Thus, there is considerable redundancy between the aesthetasc and the nonaesthetasc antennular pathways. However, differences in the functions of the aesthetasc and nonaesthetasc antennular pathways have been identified as well. For example, only the aesthetasc pathway, and not the nonaesthetasc antennular pathway or extra-antennular chemosensory pathway, mediates attraction to aggregation pheromones or alarm pheromones (Horner, A. J. *et al.*, 2004; Horner, A. J. and Derby, C. D., 2005; Shabani, S. and Derby, C. D., 2006). "Aesthetases are also necessary for the response of clawed lobsters to social odors (Johnson, M. E. and Atema, J., 2005)". Thus, the aesthetascs uniquely possess detectors of aggregation and alarm cues.

It must be emphasized that although we have placed the nonaesthetasc antennular sensors in a single category, they differ not only structurally but also functionally. On the antennular flagella, there are nine different types of nonaesthetasc sensilla (Cate, H. S. and Derby, C. D., 2001). Although they all are bimodally innervated by chemosensory neurons and mechanosensory neurons that project to the LANs, they differ in setal structure and organization and in location on the antennule (Cate, H. S. and Derby, C. D., 2001; 2002). They also differ in function. For example, the asymmetric sensilla are uniquely responsible for mediating chemically activated antennular grooming behavior (Schmidt, M. and Derby, C. D., 2005).

Chemosensors on the legs of lobsters differ structurally and functionally from antennular sensory. In lobsters, they direct local searches to food chemicals, and when the leg touches the source of the chemicals, that item is grabbed and transferred to the mouthparts (Derby, C. D. and Atema, J., 1982).

Mouthpart chemoreceptors, with their high threshold and broad tuning (Garm, A. *et al.*, 2005), control whether food should be eaten (Derby, C. D. and Atema, J., 1982). Presumably they are checking the quality of the prey, such as determining whether feeding stimulants are present and deterrents are absent.

The hierarchy of control of behaviors by the different chemosensors as described above for lobsters may not apply completely to all crustaceans. For example, in the blue crab *Callinectes sapidus*, leg chemoreceptors can also play a role in orientation to distant cues (Keller, T. A. *et al.*, 2003).

Crustaceans demonstrate several types of learning involving their chemical senses, including habituation and aversive associative learning (Derby, C. D., 2000), one-trial flavor avoidance learning (hermit crab), and other important behaviors though less well-defined mechanistically (Caldwell, R. and Dingle, J., 1985; Wight, K. *et al.*, 1990). Aversive associative learning has been used to study spiny lobster's ability to discriminate chemical stimuli, including single-odorant compounds and different blend ratios of the same binary or complex mixtures (Derby, C. D., 2000). The associative task is learning to avoid a previously attractive food stimulus. Since this task is controlled by the antennules, expression of this learned task depends on the presence of fully functional antennules. But it does not require aesthetascs: either the aesthetasc (olfactory) pathway or the nonaesthetasc antennular pathway can mediate discrimination and learning about chemical stimuli (Steullet, P. *et al.*, 2001; 2002).

4.04.8.4 Overview

The chemical senses of fish and crustaceans control or modulate many of their behaviors. There are overlaps and differences in the behaviors that are mediated by the different chemosensory systems, including olfaction and taste. Both olfactory and nonolfactory systems can mediate orientation to distant food-related chemicals. Olfactory systems tend to control detection of pheromones and thus cues associated with intraspecific interactions. Some of the behaviors controlled by olfaction include learning about chemicals involved in discrimination. Taste and other nonolfactory systems are uniquely associated with chemically activated biting and swallowing of food and other more reflexive behaviors.

Acknowledgment

We also thank NIH and NSF for their support over the years, currently from NIH DC00312, NIH DC-03792, NSF IBN-0321444, and NSF IBN-0314970.

References

Ache, B. W. 2002. Crustaceans as Animal Models for Olfactory Research. In: Crustacean Experimental Systems in Neurobiology (*ed*. K. Wiese), pp. 189–199. Springer.

Ache, B. W. and Young, J. M. 2005. Olfaction: diverse species, conserved principles. Neuron 48, 417–430.

Ache, B. W. and Zhainazarov, A. B. 1995. Dual second-messenger pathways in olfactory transduction. Curr. Opin. Neurobiol. 5, 461–466.

Ache, B. W., Munger, S., and Zhainazarov, A. 1998. Organizational complexity in lobster olfactory receptor cells. Ann. N. Y. Acad. Sci. 855, 194–198.

Ashby, E. A. and Larimer, J. L. 1965. Modification of cardiac and respiratory rhythms in crayfish following carbohydrate chemoreception. J. Cell. Comp. Physiol. 65, 373–380.

Atema, J. 1971. Structures and functions of the sense of taste in the catfish, *Ictalurus natalis*. Brain Behav. Evol. 4, 273–294.

Atema, J. 1977. Functional Separation of Smell and Taste in Fish and Crustacea. In: Olfaction and Taste VI (*eds*. J. Le Magnen and P. Mac Leod), pp. 165–174. Information Retrieval Ltd

Atema, J. 1988. Distribution of Chemical Stimuli. In: Sensory Biology of Aquatic Animals (*eds*. J. Atema, R. R. Fay, A. N. Popper, and W. N. Tavolga), pp. 29–56. Springer.

Atema, J. 1995. Chemical signals in the marine environment: dispersal, detection, and temporal signal analysis. Proc. Natl. Acad. Sci. U. S. A. 92, 62–66.

Atema, J. and Steinbach, M. A. 2007. Chemical Communication in the Social Behavior of the lobster, *Homarus americanus*, and Other Decapod Crustacea. In: Ecology and Evolution of Social Behavior: Crustaceans as Model Systems (*eds*. E. Duffy and M. Thiel), Oxford University Press.

Atema, J., Borroni, P., Johnson, B., Voigt, R., and Handrich, L. 1989. Adaptation and Mixture Interactions in Chemoreceptor Cells: Mechanisms for Diversity and Contrast Enhancement. In: Perception of Complex Smells and Tastes (*eds*. D. G. Laing, W. S. Cain, R. L. McBride, and B. W. Ache), pp. 83–100. Academic Press.

Bardach, J. E., Todd, J. H., and Crickmer, R. 1967. Orientation by taste in fish of the genus *Ictalurus*. Science 155, 1276–1278.

Bauer, U. and Hatt, H. 1980. Demonstration of three different types of chemosensitive units in the crayfish claw using a computerized evaluation. Neurosci. Lett. 17, 209–214.

Bauer, U., Dudel, J., and Hatt, H. 1981. Characteristics of single chemoreceptive units sensitive to amino acids and related substances in the crayfish leg. J. Comp. Physiol. A 144, 67–74.

Bazer, G. T., Ebbesson, S. O. E., Reynolds, J. B., and Bailey, R. P. 1987. A cobalt-lysine study of primary olfactory projections in king salmon fry (*Oncorhynchus tshawytscha* Walbaum). Cell Tissue Res. 248, 499–503.

Becerra, M., Manso, M. J., Rodriguez-Moldes, I., and Anadón, R. 1994. Primary olfactory fibers project to the ventral telencephalon and preoptic region of trout (*Salmo trutta*): a developmental immunocytochemical study. J. Comp. Neurol. 342, 131–143.

Beltz, B. S. and Sandeman, D. C. 2003. Regulation of life-long neurogenesis in the decapod crustacean brain. Arthropod Struct. Dev. 32, 39–60.

Beltz, B. S., Kordas, K., Lee, M. M., Long, J. B., Benton, J. L., and Sandeman, D. C. 2003. Ecological, evolutionary, and functional correlates of sensilla number and glomerular density in the olfactory system of decapod crustaceans. J. Comp. Neurol. 455, 260–269.

Blaustein, D. N., Simmons, R. B., Burgess, M. F., Derby, C. D., Nishikawa, M., and Olson, K. S. 1993. Ultrastructural localization of 5'AMP odorant receptor sites on the dendrites of olfactory receptor neurons of the spiny lobster. J. Neurosci. 13, 2821–2828.

Bobkov, Y. V. and Ache, B. W. 2005. Pharmacological properties and functional role of a TRP-related ion channel in lobster olfactory receptor neurons. J. Neurophysiol. 93, 1372–1380.

Borroni, P. F. and Atema, J. 1988. Adaptation in chemoreceptor cells. I. Self-adapting backgrounds determine threshold and cause parallel shift of response function. J. Comp. Physiol. A 164, 67–74.

Borroni, P. F. and Atema, J. 1989. Adaptation in chemoreceptor cells. II. The effects of cross-adapting backgrounds depend on spectral tuning. J. Comp. Physiol. A 165, 669–677.

Borroni, P. F., Handrich, L. S., and Atema, J. 1986. The role of narrowly tuned taste cell populations in lobster (*Homarus americanus*) feeding behavior. Behav. Neurosci. 100, 206–212.

Brown, S. B. and Hara, T. J. 1981. Accumulation of chemostimulatory amino acids by a sedimentable fraction isolated from olfactory rosettes of rainbow trout (*Salmo gairdneri*). Biochem. Biophys. Acta 675, 149–162.

Bruch, R. C. 1996. Phosphoinositide second messengers in olfaction. Comp. Biochem. Physiol. 113B, 451–459.

Bruch, R. C. and Rulli, R. D. 1988. Ligand binding specificity of a neutral L-amino acid olfactory receptor. Comp. Biochem. Physiol. 91B, 535–540.

Buck, L. B. 1996. Information coding in the vertebrate olfactory system. Annu. Rev. Neurosci. 19, 517–544.

Byrd, C. A., Jones, J. T., Quattro, J. M., Rogers, M. E., Brunjes, P. C., and Vogt, R. G. 1996. Ontogeny of odorant receptor gene expression in zebrafish, *Danio rerio*. J. Neurobiol. 29, 445–458.

Cagan, R. H. and Zeiger, W. N. 1978. Biochemical studies of olfaction: binding specificity of radioactivity labeled stimuli to an isolated olfactory preparation from rainbow trout (*Salmo gairdneri*). Proc. Natl. Acad. Sci. U. S. A. 75, 4679–4683.

Caldwell, R. and Dingle, J. 1985. A test of individual recognition in the stomatopod *Gonodactylus festae*. Anim. Behav. 33, 101–106.

Cao, Y., Oh, B. C., and Stryer, L. 1998. Cloning and localization of two multigene receptor families in goldfish olfactory epithelium. Proc. Natl. Acad. Sci. U. S. A. 95, 11987–11992.

Caprio, J. 1975. High sensitivity of catfish taste receptors to amino acids. Comp. Biochem. Physiol. 52A, 247–251.

Caprio, J. 1977. Electrophysiological distinctions between the taste and smell of amino acids in catfish. Nature 266, 850–851.

Caprio, J. 1978. Olfaction and taste in the channel catfish: an electrophysiological study of the responses to amino acids and derivatives. J. Comp. Physiol. A 123, 357–371.

Caprio, J. and Byrd, R. P., Jr. 1984. Electrophysiological evidence for acidic, basic, and neutral amino acid olfactory receptor sites in the catfish. J. Gen. Physiol. 84, 403–422.

Caprio, J., Brand, J. G., Teeter, J. H., Valentinčič, T., Kalinoski, D. L., Kohbara, J., Kumazawa, T., and Wegert, S. 1993. The taste system of the channel catfish: from biophysics to behaviour. Trends Neurosci. 16, 192–197.

Caprio, J., Dudek, J., and Robinson, J. J., II 1989. Electro-olfactogram and multiunit olfactory receptor responses to binary and trinary mixtures of amino acids in the channel catfish, Ictalurus punctatus. J. Gen. Physiol. 93, 245–262.

Carr, W. E. S. 1988. The Molecular Nature of Chemical Stimuli in the Aquatic Environment. In: Sensory Biology of Aquatic Animals (eds. J. Atema, R. R. Fay, A. N. Popper, and W. N. Tavolga), pp. 3–27. Springer.

Carr, W. E. S. and Derby, C. D. 1986a. Behavioral chemoattractants for the shrimp, Palaemonetes pugio: identification of active components in food extracts and evidence of synergistic mixture interactions. Chem. Senses 11, 49–64.

Carr, W. E. S. and Derby, C. D. 1986b. Chemically stimulated feeding behavior in marine animals: importance of chemical mixtures and involvement of mixture interactions. J. Chem. Ecol. 12, 989–1011.

Carr, W. E. S., Gleeson, R. A., and Trapido-Rosenthal, H. G. 1990. The role of perireceptor events in chemosensory processes. Trends Neurosci. 13, 212–215.

Cate, H. S. and Derby, C. D. 2001. Morphology and distribution of setae on the antennules of the Caribbean spiny lobster Panulirus argus reveal new types of bimodal chemo-mechanosensilla. Cell Tissue Res. 304, 439–454.

Cate, H. S. and Derby, C. D. 2002. Ultrastructure and physiology of the hooded sensillum, a bimodal chemo-mechanosensillum of lobsters. J. Comp. Neurol. 442, 293–307.

Chien, H., Liu, H., Schmidt, M., Tai, P. C., and Derby, C. D. 2005. Molecular cloning and characterization of proneural genes in the olfactory organ of spiny lobsters. Chem. Senses 30, A91–A92.

Cromarty, S. and Derby, C. 1997. Multiple excitatory receptor types on individual olfactory neurons: implications for coding of mixtures in the spiny lobster. J. Comp. Physiol. A 180, 481–491.

Daniel, P. C. and Derby, C. D. 1991. Mixture suppression in behavior: the antennular flick response in the spiny lobster towards binary odorant mixtures. Physiol. Behav. 49, 591–601.

Davenport, C. J. and Caprio, J. 1982. Taste and tactile recordings from the ramus recurrens facialis innervating flank taste buds in the catfish. J. Comp. Physiol. A 147, 217–229.

Del Punta, K., Puche, A., Adams, N. C., Rodriguez, I., and Mombaerts, P. 2002. A divergent pattern of sensory axonal projections is rendered convergent by second-order neurons in the accessory olfactory bulb. Neuron 35, 1057–1066.

Demski, L. S. and Dulka, J. G. 1984. Functional-anatomical studies on sperm release evoked by electrical stimulation of the olfactory tract in goldfish. Brain Res. 291, 241–247.

Demski, L. S. and Northcutt, R. G. 1983. The terminal nerve: a new chemosensory system in vertebrates? Science 220, 435–437.

Derby, C. D. 1982. Structure and function of cuticular sensilla of the lobster Homarus americanus. J. Crust. Biol. 2, 1–21.

Derby, C. D. 2000. Learning from spiny lobsters about chemosensory coding of mixtures. Physiol. Behav. 69, 203–209.

Derby, C. D. and Atema, J. 1982. The function of chemo- and mechanoreceptors in lobster (Homarus americanus) feeding behaviour. J. Exp. Biol. 98, 317–327.

Derby, C. D. and Atema, J. 1988. Chemoreceptor Cells in Aquatic Invertebrates: Peripheral Mechanisms of Chemical Signal Processing in Decapod Crustaceans. In: Sensory Biology of Aquatic Animals (eds. J. Atema, R. R. Fay, A. N. Popper, and W. N. Tavolga), pp. 365–385. Springer.

Derby, C. D., Ache, B. W., and Kennel, E. W. 1985. Mixture suppression in olfaction: electrophysiological evaluation of the contribution of peripheral and central neural components. Chem. Senses 10, 301–316.

Derby, C. D., Cate, H. S., Steullet, P., and Harrison, P. J. H. 2003. Comparison of turnover in the olfactory organ of early juvenile stage and adult Caribbean spiny lobsters. Arthropod Struct. Dev. 31, 297–311.

Derby, C. D., Girardot, M.-N., and Daniel, P. C. 1991. Responses of olfactory receptor cells of spiny lobsters to binary mixtures. I. Intensity mixture interactions. J. Neurophysiol. 66, 112–130.

Derby, C. D., Girardot, M.-N., Daniel, P. C., and Fine-Levy, J. B. 1989. Olfactory Discrimination of Mixtures: Behavioral, Electrophysiological, and Theoretical Studies Using the Spiny Lobster Panulirus Argus. In: Perception of Complex Smells and Tastes (eds. D. G. Laing, W. S. Cain, R. L. McBride, and B. W. Ache), pp. 65–81. Academic Press.

Derby, C. D., Steullet, P., Cate, H. S., and Harrison, P. J. H. 2001a. A Compound Nose: Functional Organization and Development of Aesthetasc Sensilla. In: The Crustacean Nervous System (ed. K. Wiese), pp. 346–358. Springer.

Derby, C. D., Steullet, P., Horner, A. J., and Cate, H. S. 2001b. The sensory basis to feeding behavior in the Caribbean spiny lobster Panulirus Argus. Mar. Freshw. Res. 52, 1339–1350.

Doolin, R. E. and Ache, B. W. 2005a. Cyclic nucleotide signaling mediates an odorant-suppressible chloride conductance in lobster olfactory receptor neurons. Chem. Senses 30, 127–135.

Doolin, R. E. and Ache, B. W. 2005b. Specificity of odorant-evoked inhibition in lobster olfactory receptor neurons. Chem. Senses 30, 105–110.

Doolin, R. E., Zhainazarov, A. B., and Ache, B. W. 2001. An odorant-suppressed Cl^- conductance in lobster olfactory receptor cells. J. Comp. Physiol. A 187, 477–487.

Døving, K. B. 1996. Homing of Fish and the Nervous Pathways for Kin Recognition. In: Fish Pheromones. Origins and Mode of Action (eds. A. V. M. Canario and D. M. Power), pp. 98–109. University of Algarve.

Døving, K. B. and Stabell, O. B. 2003. Trails in Open Waters: Sensory Cues in Salmon Migration. In: Sensory Processing in the Aquatic Environment (eds. S. P. Collin and N. J. Marshall), pp. 39–52. Springer.

Døving, K. B., Selset, R., and Thommesen, G. 1980. Olfactory sensitivity to bile acids in salmonid fishes. Acta Physiol. Scand. 108, 123–131.

Dryer, L. and Graziadei, P. P. C. 1994. Mitral cell dendrites: a comparative approach. Anat. Embryol. (Berlin) 189, 91–106.

Eisthen, H. L. 2002. Why are olfactory systems of different animals so similar? Brain Behav. Evol. 59, 273–293.

Eisthen, H. L. 2004. The goldfish knows: olfactory receptor cell morphology predicts receptor gene expression. J. Comp. Neurol. 477, 341–346.

Fadool, D. A., Estey, S. J., and Ache, B. W. 1995. Evidence that a Gq-protein mediates excitatory odor transduction in lobster olfactory receptor neurons. Chem. Senses 20, 489–498.

Fine-Levy, J. B. and Derby, C. D. 1991. Effects of stimulus intensity and quality on discrimination of odorant mixtures by spiny lobsters in an associative learning paradigm. Physiol. Behav. 49, 1163–1168.

Fine-Levy, J. B., Daniel, P. C., Girardot, M., and Derby, C. D. 1989. Behavioral resolution of quality of odorant mixtures by spiny lobsters: differential aversive conditioning of olfactory responses. Chem. Senses 14, 503–524.

Fine-Levy, J. B., Girardot, M., Derby, C. D., and Daniel, P. C. 1988. Differential associative conditioning and olfactory discrimination in the spiny lobster *Panulirus argus*. Behav. Neural Biol. 49, 315–331.

Finger, T. E. 1976. Gustatory pathways in the bullhead catfish I. Connections of the anterior ganglion. J. Comp. Neurol. 165, 513–526.

Finger, T. E. 1982. Somatotopy in the representation of the pectoral fin and free fin rays in the spinal cord of the sea robin, *Prionotus carolinus*. Biol. Bull. 163, 154–161.

Finger, T. E. 1997. Evolution of taste and solitary chemoreceptor cell systems. Brain Behav. Evol. 50, 234–243.

Finger, T. E. and Morita, Y. 1985. Two gustatory systems: facial and vagal gustatory nuclei have different brainstem connections. Science 227, 776–778.

Finger, T. E. and Simon, S. A. 2000. Cell Biology of Taste Epithelium. In: The Neurobiology of Taste and Smell (eds. T. E. Finger, W. L. Silver, and D. Restrepo), pp. 287–314. Wiley-Liss Inc

Finger, T. E., Bryant, B. P., Kalinoski, D. L., Teeter, J. H., Bottger, B., Grosvenor, W., Cagan, R. H., and Brand, J. G. 1996. Differential localization of putative amino acid receptors in taste buds of the channel catfish *Ictalurus punctatus*. J. Comp. Neurol. 373, 129–138.

Finger, T. E., Drake, S. K., Kotrschal, K., Womble, M., and Dockstader, K. C. 1991. Postlarval growth of the peripheral gustatory system in the channel catfish, *Ictalurus punctatus*. J. Comp. Neurol. 314, 55–66.

Friedrich, R. W. and Korsching, S. I. 1997. Combinatorial and chemotopic odorant coding in the zebrafish olfactory bulb visualized by optical imaging. Neuron 18, 737–752.

Friedrich, R. W. and Korsching, S. I. 1998. Chemotopic, combinatorial, and noncombinatorial odorant representations in the olfactory bulb revealed using a voltage-sensitive axon tracer. J. Neurosci. 18, 9977–9988.

Friedrich, R. W. and Laurent, G. 2001. Dynamic optimization of odor representations by slow temporal patterning of mitral cell activity. Science 291, 889–894.

Friedrich, R. W. and Laurent, G. 2004. Dynamics of olfactory bulb input and output activity during odor stimulation in zebrafish. J. Neurophysiol. 91, 2658–2669.

Fujita, I., Sorensen, P. W., Stacey, N. E., and Hara, T. J. 1991. The olfactory system, not the terminal nerve, functions as the primary chemosensory pathway mediating responses to sex pheromones in male goldfish. Brain Behav. Evol. 38, 313–321.

Fuller, C. L., Ruthel, G., Warfield, K. L., Swenson, D. L., Basio, C. M., Aman, M. J., and Bavari, S. 2006. Mitral cells in the olfactory bulb of adult zebrafish (Danio rerio): morphology and distribution. J. Comp. Neurol. 499, 218–230.

Fuss, S. H. and Korsching, S. I. 2001. Odorant feature detection: activity mapping of structure response relationships in the zebrafish olfactory bulb. J. Neurosci. 21, 8396–8407.

Garm, A., Shabani, S., Hoeg, J. T., and Derby, C. D. 2005. Chemosensory neurons in the mouthparts of the spiny lobsters *Panulirus argus* and *Panulirus interruptus* (Crustacea: Decapoda). J. Exp. Mar. Biol. Ecol. 314, 175–186.

Ghiradella, H., Case, J. F., and Cronshaw, J. 1968. Structure of aesthetascs in selected marine and terrestrial decapods: chemoreceptor morphology and environment. Am. Zool. 8, 603–621.

Giraudet, P., Berthommier, F., and Chaput, M. 2002. Mitral cell temporal response patterns evoked by odor mixtures in the rat olfactory bulb. J. Neurophysiol. 88, 829–838.

Gleeson, R. A. 1991. Intrinsic Factors Mediating Pheromone Communication in the Blue Crab, *Callinectes sapidus*. In: Crustacean Sexual Biology (eds. R. T. Bauer and J. W. Martin), pp. 17–32. Columbia University Press.

Gleeson, R. A., Trapido-Rosenthal, H. G., McDowell, L. M., Aldrich, H. C., and Carr, W. E. S. 1992. Ecto-ATPase/ phosphatase activity in the olfactory sensilla of the spiny lobster, *Panulirus argus*: localization and characterization. Cell Tissue Res. 269, 439–445.

Goergen, E. M., Bagay, L. A., Rehm, K., Benton, J. L., and Beltz, B. S. 2002. Circadian control of neurogenesis. J. Neurobiol. 53, 90–95.

Goh, Y. and Tamura, T. 1980. Olfactory and gustatory responses to amino acids in two marine teleosts – red sea bream and mullet. Comp. Biochem. Physiol. 66C, 217–224.

Gold, G. H. 1999. Controversial issues in vertebrate olfactory transduction. Annu. Rev. Physiol. 61, 857–871.

Goldman, A. L., Van der Goes van Naters, W., Lessing, D., Warr, C. G., and Carlson, J. R. 2005. Coexpression of two functional odor receptors in one neuron. Neuron 45, 661–666.

Gomahr, H., Palzenberger, M., and Kotrschal, K. 1992. Density and distribution of external taste buds in cyprinids. Environ. Biol. Fish. 33, 125–134.

Gomez, G. and Atema, J. 1996a. Temporal resolution in olfaction: stimulus integration time of lobster chemoreceptor cells. J. Exp. Biol. 199, 1771–1779.

Gomez, G. and Atema, J. 1996b. Temporal resolution in olfaction. II. Time course of recovery from adaptation in lobster chemoreceptor cells. J. Neurophysiol. 76, 1340–1343.

Gomez, G., Voigt, R., and Atema, J. 1999. Temporal resolution in olfaction. III. Flicker fusion and concentration-dependent synchronization with stimulus pulse trains of antennular chemoreceptor cells in the American lobster. J. Comp. Physiol. A 185, 427–436.

Grosvenor, W., Kaulin, Y., Spielman, A. I., Bayley, D. L., Kalinoski, D. L., Teeter, J. H., and Brand, J. G. 2004. Biochemical enrichment and biophysical characterization of a taste receptor for L-arginine from the catfish, *Ictalurus punctatus*. BMC Neurosci. 5, 25p.

Grünert, U. and Ache, B. W. 1988. Ultrastructure of the aesthetasc (olfactory) sensilla of the spiny lobster, *Panulirus argus*. Cell Tissue Res. 251, 95–103.

Haberly, L. B. 2001. Parallel-distributed processing in olfactory cortex: new insights from morphological and physiological analysis of neuronal circuitry. Chem. Senses 26, 551–576.

Hallberg, E., Johansson, K. U. I., and Elofsson, R. 1992. The aesthetasc concept: structural variations of putative olfactory receptor cell complexes in Crustacea. J. Microsc. Res. Tech. 22, 325–335.

Hallem, E. A., Ho, M. G., and Carlson, J. R. 2004. The molecular basis of odor coding in the *Drosophila* antenna. Cell 117, 965–979.

Hamdani, E. H. and Døving, K. B. 2006. Specific projection of the sensory crypt cells in the olfactory system in crucian carp, *Carassius carassius*. Chem. Senses 31, 63–67.

Hamdani, E. H., Alexander, G., and Døving, K. B. 2001a. Projection of sensory neurons with microvilli to the lateral olfactory tract indicates their participation in feeding behaviour in crucian carp. Chem. Senses 26, 1139–1144.

Hamdani, E. H., Kasumyan, A., and Døving, K. B. 2001b. Is feeding behaviour in crucian carp mediated by the lateral olfactory tract? Chem. Senses 26, 1133–1138.

Hamdani, E. H., Stabell, O. B., Alexander, G., and Døving, K. B. 2000. Alarm reaction in the crucian carp is mediated by the medial bundle of the medial olfactory tract. Chem. Senses 25, 103–109.

Hansen, A. and Reutter, K. 2004. Chemosensory Systems in Fish: Structural, Functional and Ecological Aspects. In: The Senses of Fish: Adaptations for the Reception of Natural Stimuli (eds. G. von der Emde, J. Mogdans, and B. G. Kapoor), pp. 55–89. Narosa Publishing House Pvt. Ltd.

Hansen, A. and Schmidt, M. 2001. Neurogenesis in the central olfactory pathway of the adult shore crab Carcinus maenas is controlled by sensory afferents. J. Comp. Neurol. 441, 223–233.

Hansen, A. and Schmidt, M. 2004. Influence of season and environment on adult neurogenesis in the central olfactory pathway of the shore crab, Carcinus maenas. Brain Res. 1025, 85–97.

Hansen, A., Anderson, K., and Finger, T. E. 2004. Differential distribution of olfactory receptor neurons in goldfish: structural and molecular correlates. J. Comp. Neurol. 477, 347–359.

Hansen, A., Rolen, S. H., Anderson, K., Morita, Y., Caprio, J., and Finger, T. E. 2003. Correlation between olfactory receptor cell type and function in the channel catfish. J. Neurosci. 23, 9328–9339.

Hara, T. J. 1973. Olfactory responses to amino acids in rainbow trout, Salmo gairdneri. Comp. Biochem. Physiol. 44A, 407–416.

Hara, T. J. 1982. Chemoreception in Fishes. Elsevier.

Hara, T. J. 1992. Fish Chemoreception. Chapman & Hall.

Hara, T. J. 1993. Chemoreception. In: The Physiology of Fishes (ed. D. H. Evans), pp. 191–218. CRC Press.

Hara, T. J. 1994. Olfaction and gustation in fish: an overview. Acta Physiol. Scand. 152, 207–217.

Hara, T. J. and Zhang, C. 1996. Spatial projections to the olfactory bulb of functionally distinct and randomly distributed primary neurons in salmonid fishes. Neurosci. Res. 26, 65–74.

Hara, T. J. and Zielinski, B. 1989. Structural and functional development of the olfactory organ in teleosts. Trans. Am. Fish. Soc. 118, 183–194.

Hara, T. J., Carolfeld, J., and Kitamura, S. 1999. The variability of the gustatory sensibility in salmonids, with special reference to strain differences in rainbow trout, Oncorhynchus mykiss. Can. J. Fish. Aquat. Sci. 56, 13–24.

Hara, T. J., Macdonald, S., Evans, R. E., Marui, T., and Arai, S. 1984. Morpholine, Bile Acids and Skin Mucus as Possible Chemical Cues in Salmonid Homing: Electrophysiological Re-evaluation. In: Mechanisms of Migration in Fishes (eds. J. D. McCleave, G. P. Arnold, J. D. Dodson, and W. H. Neill), pp. 363–378. Plenum Publishing Corporation.

Harrison, P. J. H., Cate, H. S., and Derby, C. D. 2004. Localized ablation of olfactory receptor neurons induces both localized regeneration and widespread replacement of neurons in spiny lobsters. J. Comp. Neurol. 471, 72–84.

Harrison, P. J. H., Cate, H. S., Steullet, P., and Derby, C. D. 2001a. Structural plasticity in the olfactory system of adult spiny lobsters: postembryonic development permits life-long growth, turnover, and regeneration. Mar. Freshw. Res. 52, 1357–1365.

Harrison, P. J. H., Cate, H. S., Steullet, P., and Derby, C. D. 2003. Amputation-induced activity of progenitor cells leads to rapid regeneration of olfactory tissue in lobsters. J. Neurobiol. 55, 97–114.

Harrison, P. J. H., Cate, H. S., Swanson, E. S., and Derby, C. D. 2001b. Postembryonic proliferation in the spiny lobster antennular epithelium: rate of genesis of olfactory receptor neurons is dependent on molt stage. J. Neurobiol. 47, 51–66.

Harzsch, S. 2001. From Stem Cell to Structure: Neurogenesis in the CNS of Decapod Crustaceans. In: The Crustacean Nervous System (ed. K. Wiese), pp. 417–432. Springer.

Hatt, H. 1986. Responses of a bimodal neuron (chemo- and vibration-sensitive) on the walking legs of the crayfish. J. Comp. Physiol. A 159, 611–617.

Hatt, H. and Ache, B. W. 1994. Cyclic nucleotide- and inositol phosphate-gated ion channels in lobster olfactory receptor neurons. Proc. Natl. Acad. Sci. U. S. A. 91, 6264–6268.

Hayama, T. and Caprio, J. 1989. Lobule structure and somatotopic organization of the medullary facial lobe in the channel catfish Ictalurus punctatus. J. Comp. Neurol. 285, 9–17.

Hidaka, I., Nyu, N., and Kiyohara, S. 1976. Gustatory response in the puffer–IV effects of mixtures of amino acids and betaine. Bull. Fac. Fish. Mie Univ. 3, 17–28.

Hildebrand, J. G. and Shepherd, G. M. 1997. Mechanisms of olfactory discrimination: converging evidence for common principles across phyla. Annu. Rev. Neurosci. 20, 595–631.

Hofmann, M. H. and Meyer, D. L. 1995. The extrabulbar olfactory pathway: primary olfactory fibers bypassing the olfactory bulb in bony fishes? Brain Behav. Evol. 46, 378–388.

Honkanen, T. and Ekstrom, P. 1990. An immunocytochemical study of the olfactory projections in the three-spined stickleback, Gasterosteus aculeatus, L. J. Comp. Neurol. 292, 65–72.

Horner, A. J. and Derby, C. D. 2005. Chemical signals and chemosensory pathways involved in spiny lobster sheltering behavior. Chem. Senses 30, A97p.

Horner, A. J., Weissburg, M. J., and Derby, C. D. 2004. Dual antennular chemosensory pathways can mediate orientation by Caribbean spiny lobsters in naturalistic flow conditions. J. Exp. Biol. 207, 3785–3796.

Ishimaru, Y., Okada, S., Naito, H., Nagai, T., Yasuoka, A., Matsumoto, I., and Abe, K. 2005. Two families of candidate taste receptors in fishes. Mech. Develop. 122, 1310–1321.

Ivanova, T. T. and Caprio, J. 1993. Odorant receptor activated by amino acids in sensory neurons of the channel catfish Ictalurus punctatus. J. Gen. Physiol. 102, 1085–1105.

Jakubowski, M. and Whitear, M. 1990. Comparative morphology and cytology of taste buds in teleosts. Z. mikrosk.-anat. Forsch (Leipzig) 104, 529–560.

Johnson, M. E. and Atema, J. 2005. The olfactory pathway for individual recognition in the American lobster Homarus americanus. J. Exp. Biol. 208, 2865–2872.

Johnson, D. M. G., Illig, K. R., Behan, M., and Haberly, L. B. 2000. New features of connectivity in piriform cortex visualized by intracellular injection of pyramidal cells suggest that "primary" olfactory cortex functions like "association" cortex in other sensory systems. J. Neurosci. 20, 6974–6982.

Kalinoski, D. L., Huque, T., LaMorte, V. J., and Brand, J. G. 1989. Second-Messenger Events in Taste. In: Chemical Senses. Vol. 1: Receptor Events and Transduction in Taste and Olfaction (eds. J. G. Brand, J. H. Teeter, R. H. Cagan, and M. R. Kare), pp. 85–101. Marcel Dekker, Inc

Kamio, M., Matsunaga, S., and Fusetani, N. 2002. Copulation pheromone in the crab Telmessus cheiragonus (Brachyura: Decapoda). Mar. Ecol. Prog. Ser. 234, 183–190.

Kang, J. and Caprio, J. 1991. Electro-olfactogram and multiunit olfactory receptor responses to complex mixtures of amino acids in the channel catfish, Ictalurus punctatus. J. Gen. Physiol. 98, 699–721.

Kang, J. and Caprio, J. 1995a. Electrophysiological responses of single olfactory bulb neurons to binary mixtures of amino acids in the channel catfish, Ictalurus punctatus. J. Neurophysiol. 74, 1435–1443.

Kang, J. and Caprio, J. 1995b. In vivo responses of single olfactory receptor neurons in the channel catfish, Ictalurus punctatus. J. Neurophysiol. 73, 172–177.

Kang, J. and Caprio, J. 1997. In vivo responses of single olfactory receptor neurons of channel catfish to binary mixtures of amino acids. J. Neurophysiol. 77, 1–8.

Kanwal, J. S. and Caprio, J. 1983. An electrophysiological investigation of the oro-pharyngeal (IX-X) taste system in the channel catfish, *Ictalurus punctatus*. J. Comp. Physiol. A 150, 345–357.

Kanwal, J. S. and Caprio, J. 1987. Central projections of the glossopharyngeal and vagal nerves in the channel catfish, *Ictalurus punctatus*: clues to differential processing of visceral inputs. J. Comp. Neurol. 264, 216–230.

Kanwal, J. S. and Caprio, J. 1988. Overlapping taste and tactile maps of the oropharynx in the vagal lobe of the channel catfish, *Ictalurus punctatus*. J. Neurobiol. 19, 211–222.

Kanwal, J. S. and Finger, T. E. 1992. Central Representation and Projections of Gustatory Systems. In: Fish Chemoreception (ed. T. J. Hara), pp. 79–102. Chapman & Hall.

Kanwal, J. S. and Finger, T. E. 1997. Parallel medullary gustatospinal pathways in a catfish: possible neural substrates for taste-mediated food search. J. Neurosci. 17, 4873–4885.

Kanwal, J. S., Hidaka, I., and Caprio, J. 1987. Taste responses to amino acids from facial nerve branches innervating oral and extra-oral taste buds in the channel catfish, *Ictalurus punctatus*. Brain Res. 406, 105–112.

Kapoor, B. G. and Finger, T. E. 2003. Taste and Solitary Chemoreceptor Cells. In: Catfishes: Vol. 2 (eds. G. Arratia, B. G. Kapoor, M. Chardon, and R. Diogo), pp. 753–769. Science Publishers, Inc

Keller, T. A., Powell, I., and Weissburg, M. J. 2003. Role of appendages in chemically mediated orientation of blue crabs. Mar. Ecol. Prog. Ser. 261, 217–231.

Kicklighter, C. E., Shabani, S., Johnson, P. M., and Derby, C. D. 2005. Sea hares use novel antipredatory chemical defenses. Curr. Biol. 15, 549–554.

Kiyohara, S. and Caprio, J. 1996. Somatotopic organization of the facial lobe of the sea catfish *Arius felis* studied by transganglionic transport of horseradish peroxidase. J. Comp. Neurol. 368, 121–135.

Kiyohara, S., Hidaka, I., Kitoh, J., and Yamashita, S. 1985. Mechanical sensitivity of the facial nerve fibers innervating the anterior palate of the puffer, *Fugu pardalis*, and their central projection to the primary taste center. J. Comp. Physiol. A 157, 705–716.

Kiyohara, S., Houman, H., Yamashita, S., Caprio, J., and Marui, T. 1986. Morphological evidence for a direct projection of trigeminal nerve fibers to the primary gustatory center in the sea catfish, *Plotosus anguillaris*. Brain Res. 379, 353–357.

Kleene, S. J. 1997. High-gain, low-noise amplification in olfactory transduction. Biophys. J. 73, 1110–1117.

Kleerekoper, H. 1969. Olfaction in Fishes. Indiana University Press.

Koehl, M. A. R., Koseff, J. R., Crimaldi, J. P., McCay, M. G., Cooper, T., Wiley, M. B., and Moore, P. A. 2001. Lobster sniffing: antennule design and hydrodynamic filtering of information in an odor plume. Science 294, 1948–1951.

Kohbara, J. and Caprio, J. 1996. Taste responses to binary mixtures of amino acids in the sea catfish, *Arius felis*. Chem. Senses 21, 45–53.

Kohbara, J. and Caprio, J. 2001. Taste responses of the facial and glossopharyngeal nerves to amino acids in the rainbow trout. J. Fish Biol. 58, 1062–1072.

Kohbara, J., Michel, W., and Caprio, J. 1992. Responses of single facial taste fibers in the channel catfish, *Ictalurus punctatus*, to amino acids. J. Neurophysiol. 68, 1012–1026.

Kondo, R., Kaneko, S., Sun, H., Sakaizumi, M., and Chigusa, S. I. 2002. Diversification of olfactory receptor genes in the Japanese medaka fish, *Oryzias latipes*. Gene 282, 113–120.

Kosaka, T. and Hama, K. 1979. Ruffed cell: a new type of neuron with a distinctive initial unmyelinated portion of the axon in the olfactory bulb of the goldfish (*Carassius auratus*). J. Comp. Neurol. 186, 301–320.

Kosaka, T. and Hama, K. 1981. Ruffed cell: a new type of neuron with a distinctive initial unmyelinated portion of the axon in the olfactory bulb of the goldfish (*Carassius auratus*). III. Three-dimensional structure of the ruffed cell dendrite. J. Comp. Neurol. 201, 571–587.

Kosaka, T. and Hama, K. 1982a. Structure of the mitral cell in the olfactory bulb of the goldfish (*Carassius auratus*). J. Comp. Neurol. 212, 365–384.

Kosaka, T. and Hama, K. 1982b. Synaptic organization in the teleost olfactory bulb. J. Physiol. (Paris) 78, 707–719.

Kotrschal, K. 1991. Solitary chemosensory cells – taste, common chemical sense or what? Rev. Fish Biol. Fish. 1, 3–22.

Kotrschal, K. 1993. Sampling and behavioral evidence for mucus detection in a unique chemosensory organ: the anterior dorsal fin in rocklings. Zool. Jb. Physiol. 97, 47–67.

Kotrschal, K. and Finger, T. E. 1996. Secondary connections of the dorsal and ventral facial lobes in a teleost fish, the rockling (*Ciliata mustela*). J. Comp. Neurol. 370, 415–426.

Kumazawa, T., Brand, J. G., and Teeter, J. H. 1998. Amino acid-activated channels in the catfish taste system. Biophys. J. 75, 2757–2766.

Laing, D. G. and Francis, G. W. 1989. The capacity of humans to identify odors in mixtures. Physiol. Behav. 32, 809–814.

Laing, D. G. and Willcox, M. E. 1983. Perception of components in binary odour mixtures. Chem. Senses 7, 249–264.

Laing, D. G., Link, C., and Hutchinson, I. 2002. The limited capacity of humans to identify the components of taste mixtures and taste-odor mixtures. Perception 31, 617–635.

Lamb, C. F., IV and Caprio, J. 1992. Convergence of oral and extraoral information in the superior secondary gustatory nucleus of the channel catfish. Brain Res. 588, 201–211.

Lamb, C. F., IV and Caprio, J. 1993. Taste and tactile responsiveness of neurons in the posterior diencephalon of the channel catfish. J. Comp. Neurol. 337, 419–430.

Lastein, S., Hamdani, E. H., and Doving, K. B. 2006. Gender distinction in neural discrimination of sex pheromones in the olfactory bulb of crucian carp, *Carassius carassius*. Chem. Senses 31, 69–77.

Lin, W., Arellano, J., Slotnick, B., and Restrepo, D. 2004. Odors detected by mice deficient in cyclic nucleotide-gated channel subunit A2 stimulate the main olfactory system. J. Neurosci. 24, 3703–3710.

Lipschitz, D. L. and Michel, W. C. 2002. Amino acid odorants stimulate microvillar sensory neurons. Chem. Senses 27, 277–286.

Livermore, A., Hutson, M., Ngo, V., Hadjisimos, R., and Derby, C. D. 1997. Elemental and configural learning and the perception of odorant mixtures by the spiny lobster *Panulirus argus*. Physiol. Behav. 62, 169–174.

Lo, Y. H., Bradley, T. M., and Rhoads, D. E. 1991. L-Alanine binding sites and Na^+,K^+-ATPase in cilia and other membrane fractions from olfactory rosettes of Atlantic salmon. Comp. Biochem. Physiol. 98B, 121–126.

Lodovichi, C., Belluscio, L., and Katz, L. C. 2003. Functional topography of connections linking mirror-symmetric maps in the mouse olfactory bulb. Neuron 38, 265–276.

Lucas, P., Ukhanov, K., Leinders-Zufall, T., and Zufall, F. 2003. A diacylglycerol-gated cation channel in vomeronasal neuron dendrites is impaired in TRPC2 mutant mice: mechanism of pheromone transduction. Neuron 40, 551–561.

Luu, P., Achner, F., Bertrand, H.-O., Fan, J., and Ngai, J. 2004. Molecular determinants of ligand selectivity in a vertebrate odorant receptor. J. Neurosci. 24, 10128–10137.

Mac Leod, N. K. 1976. Spontaneous activity of single neurons in the olfactory bulb of the rainbow trout (*Salmo gairdneri*) and its modulation by olfactory stimulation with amino acids. Exp. Brain Res. 25, 267–278.

Malnic, B., Hirono, J., Sato, T., and Buck, L. 1999. Combinatorial receptor codes for odors. Cell 96, 712–723.

Marin, E. C., Jefferis, G. S. X. E., Komiyama, T., Zhu, H., and Luo, L. 2002. Representation of the glomerular olfactory map in the *Drosophila*. brain. Cell 109, 243–255.

Marui, T. 1977. Taste responses in the facial lobe of the carp, *Cyprinus carpio*. L. Brain Res. 130, 287–298.

Marui, T. and Caprio, J. 1982. Electrophysiological evidence for the topographical arrangement of taste and tactile neurons in the facial lobe of the channel catfish. Brain Res. 231, 185–190.

Marui, T. and Caprio, J. 1992. Teleost Gustation. In: Fish Chemoreception (*ed*. T. J. Hara), pp. 171–198. Chapman & Hall, Ltd

Marui, T. and Funakoshi, M. 1979. Tactile input to the facial lobe of the carp, *Cyprinus carpio* L. Brain Res. 177, 479–488.

Marui, T., Caprio, J., Kiyohara, S., and Kasahara, Y. 1988. Topographical organization of taste and tactile neurons in the facial lobe of the sea catfish, *Plotosus lineatus*. Brain Res. 446, 178–182.

Marui, T., Evans, R. E., Zielinski, B. S., and Hara, T. J. 1983. Gustatory responses of the rainbow trout (*Salmo gairdneri*) palate to amino acids and derivatives. J. Comp. Physiol. A 153, 423–433.

McClintock, T. S. and Ache, B. W. 1989. Hyperpolarizing receptor potentials in lobster olfactory receptor cells: implications for transduction and mixture suppression. Chem. Senses 14, 637–647.

McClintock, T. S. and Xu, F. 2001. Molecular Physiology of G-Proteins in Olfactory Transduction and CNS Neurotransmission in the Lobster. In: The Crustacean Nervous System (*ed*. K. Wiese), pp. 359–366. Springer.

McKinzie, M. E., Benton, J. L., Beltz, B. S., and Mellon, D. 2003. Parasol cells of the hemiellipsoid body in the crayfish *Procambarus clarkii*: dendritic branching patterns and functional implications. J. Comp. Neurol. 462, 168–179.

Mellon, D., Jr. and Alones, V. 1993. Cellular organization and growth-related plasticity of the crayfish olfactory midbrain. Microsc. Res. Tech. 24, 231–259.

Mellon, D., Jr. and Munger, S. D. 1990. Nontopographic projection of olfactory sensory neurons in the crayfish brain. J. Comp. Neurol. 296, 253–262.

Meredith, M. 1981. The analysis of response similarity in single neurons of the goldfish olfactory bulb using amino-acids as odor stimuli. Chem. Senses 6, 277–293.

Meredith, M. and Moulton, D. G. 1978. Patterned response to odor in single neurons of goldfish olfactory bulb: influence of odor quality and other stimulus parameters. J. Gen. Physiol. 71, 615–643.

Michel, W. C. and Ache, B. W. 1992. Cyclic nucleotides mediate an odor-evoked potassium conductance in lobster olfactory receptor cells. J. Neurosci. 12, 3979–3984.

Michel, W. and Caprio, J. 1991. Responses of single facial taste fibers in the sea catfish, *Arius felis*, to amino acids. J. Neurophysiol. 66, 247–260.

Michel, W. C. and Derbidge, D. S. 1997. Evidence of distinct amino acid and bile salt receptors in the olfactory system of the zebrafish, *Danio rerio*. Brain Res. 764, 179–187.

Michel, W. C., McClintock, T. S., and Ache, B. W. 1991. Inhibition of lobster olfactory receptor cells by an odor-activated potassium conductance. J. Neurophysiol. 65, 446–453.

Michel, W. C., Trapido-Rosenthal, H. G., Chao, E. T., and Wachowiak, M. 1993. Stereoselective detection of amino acids by lobster olfactory receptor neurons. J. Comp. Physiol. A 171, 705–712.

Mombaerts, P. 2004. Odorant receptor gene choice in olfactory sensory neurons: the one receptor-one neuron hypothesis revisited. Curr. Opin. Neurobiol. 14, 31–36.

Mombaerts, P., Wang, F., Dulac, C., Chao, S. K., Nemes, A., Mendelsohn, M., Edmondson, J., and Axel, R. 1996. Visualizing an olfactory sensory map. Cell 87, 675–686.

Moore, P. A. and Bergman, D. A. 2005. The smell of success and failure: the role of intrinsic and extrinsic chemical signals on the social behavior of crayfish. Integr. Comp. Biol. 45, 650–657.

Mori, K. 1995. Relation of chemical structure to specificity of response in olfactory glomeruli. Curr. Opin. Neurobiol. 5, 467–474.

Mori, K., Kishi, K., and Ojima, H. 1983. Distribution of dendrites of mitral, displaced mitral, tufted, and granule cells in the rabbit olfactory bulb. J. Comp. Neurol. 219, 339–355.

Ngai, J., Dowling, M. M., Buck, L., Axel, R., and Chess, A. 1993. The family of genes encoding odorant receptors in the channel catfish. Cell 72, 657–666.

Nikonov, A. A. and Caprio, J. 2001. Electrophysiological evidence for a chemotopy of biologically relevant odors in the olfactory bulb of the channel catfish. J. Neurophysiol. 86, 1869–1876.

Nikonov, A. A. and Caprio, J. 2004. Odorant specificity of single olfactory bulb neurons to amino acids in the channel catfish. J. Neurophysiol. 92, 123–134.

Nikonov, A. A. and Caprio, J. 2007. Responses of olfactory forebrain units to amino acids in the channel catfish. J. Neurophysiol. 97, 2490–2498.

Nikonov, A. A., Finger, T. E., and Caprio, J. 2005. Beyond the olfactory bulb: An odotopic map in forebrain. Proc. Natl. Acad. Sci. U.S.A 102, 18688–18693.

Ogawa, K. and Caprio, J. 1999. Facial taste responses of the channel catfish to binary mixtures of amino acids. J. Neurophysiol. 82, 564–569.

Ogawa, K. and Caprio, J. 2000. Glossopharyngeal taste responses of the channel catfish to binary mixtures of amino acids. Chem. Senses 25, 501–506.

Oike, H., Nagai, T., Furuyama, A., Okada, S., Aihara, Yoshiko, Ishimaru, Y, Marui, T., Matsumoto, I., Misaka, T., and Abe, K. 2007. Characterization of ligands for fish taste receptors. J. Neurosci. 27, 5584–5592.

Oka, Y. 1983. Golgi, electron microscopic and combined Golgi-electron microscopic studies of the mitral cells in the goldfish olfactory bulb. Neuroscience 8, 723–742.

Olson, K. S. and Derby, C. D. 1995. Inhibition of taurine 5´ AMP olfactory receptor sites of the spiny lobster *Panulirus argus* by odorant compounds and mixtures. J. Comp. Physiol. A 176, 527–540.

Olson, K. S., Trapido-Rosenthal, H. G., and Derby, C. D. 1992. Biochemical characterization of independent olfactory receptor sites for 5′-AMP and taurine in the spiny lobster. Brain Res. 583, 262–270.

Peters, R. C., Kotrschal, K., and Krautgartner, W. 1991. Solitary chemoreceptor cells of *Ciliata mustela* (Gadidae, Teleostei) are tuned to mucoid stimuli. Chem. Senses 16, 31–42.

Preston, R. L. 1987. Occurrence of D-amino acids in higher organisms: a survey of the distribution of D-amino acids in marine invertebrates. Comp. Biochem. Physiol. 87B, 55–62.

Raderman-Little, R. 1979. The effect of temperature on the turnover of taste bud cells in catfish. Cell. Tissue Kinet. 12, 269–280.

Rehnberg, B. G. and Schreck, C. B. 1986. The olfactory L-serine receptor in coho salmon: biochemical specificity and behavioral response. J. Comp. Physiol. A 159, 61–67.

Ressler, K. J., Sullivan, S. L., and Buck, L. B. 1994. Information coding in the olfactory system: evidence for a stereotyped and highly organized epitope map in the olfactory bulb. Cell 79, 1245–1255.

Restrepo, D., Miyamoto, T., Bryant, B. P., and Teeter, J. H. 1990. Odor stimuli trigger influx of calcium into olfactory neurons of the channel catfish. Science 249, 1166–1168.

Reutter, K. 1978. Taste organ in the bullhead (Teleostei). Adv. Anat. Embryol. Cell Biol. 55, 1–98.

Rhein, L. D. and Cagan, R. H. 1983. Biochemical studies of olfaction: binding specificity of odorants to cilia preparation from rainbow trout olfactory rosettes. J. Neurochem. 41, 569–577.

Riddle, D. R. and Oakley, B. 1991. Evaluation of projection patterns in the primary olfactory system of rainbow trout. J. Neurosci. 11, 3752–3762.

Riddle, D. R. and Oakley, B. 1992. Immunocytochemical identification of primary olfactory afferents in rainbow trout. J. Comp. Neurol. 324, 575–589.

Riddle, D. R., Wong, L. D., and Oakley, B. 1993. Lectin identification of olfactory receptor neuron subclasses with segregated central projections. J. Neurosci. 13, 3018–3033.

Rittschof, D. and Sutherland, J. P. 1986. Field studies on chemically mediated behavior in land hermit crabs: volatile and nonvolatile odors. J. Chem. Ecol. 12, 1273–1284.

Robertson, J. R., Fudge, J. A., and Vermeer, G. K. 1981. Chemical and live feeding stimulants of the sand fiddler crab, *Uca pugilator* (Bose). J. Exp. Mar. Biol. Ecol. 53, 47–64.

Rolen, S. H. and Caprio, J. 2007. Processing of bile salt odor information by single olfactory bulb neurons in the channel cat fish. J. Neurophysiol. 97, 4058–4068.

Sachse, S. and Galizia, G. 2002. Role of inhibition for temporal and spatial odor representation in olfactory output neurons: a calcium imaging study. J. Neurophysiol. 87, 1106–1117.

Sachse, S., Rappert, A., and Galizia, C. G. 1999. The spatial representation of chemical structures in the antennal lobe of honeybees: steps towards the olfactory code. Eur. J. Neurosci. 11, 3970–3982.

Sandeman, D. and Mellon, D., Jr. 2001. Olfactory Centers in the Brain of Freshwater Crayfish. In: The Crustacean Nervous System (*ed*. K. Wiese), pp. 386–404. Springer.

Sandeman, R. and Sandeman, D. C. 1996. Pre- and postembryonic development, growth and turnover of olfactory receptor neurones in crayfish antennules. J. Exp. Biol. 199, 2409–2418.

Sandeman, R. and Sandeman, D. 2000. "Impoverished" and "enriched" living conditions influence the proliferation and survival of neurons in crayfish brain. J. Neurobiol. 45, 215–226.

Sandeman, R. and Sandeman, D. 2003. Development, growth, and plasticity in the crayfish olfactory system. Microsc. Res. Tech. 60, 266–277.

Sandeman, R., Clarke, D., Sandeman, D., and Manly, M. 1998. Growth-related and antennular amputation-induced changes in the olfactory centers of crayfish brain. J. Neurosci. 18, 6195–6206.

Sandeman, D. C., Sandeman, R. E., Derby, C. D., and Schmidt, M. 1992. Morphology of the brain of crayfish, crabs, and spiny lobsters: a common nomenclature for homologous structures. Biol. Bull. 183, 304–326.

Sandeman, D. C., Scholtz, G., and Sandeman, R. E. 1993. Brain evolution in decapod Crustacea. J. Exp. Zool. 265, 112–133.

Sato, K. and Suzuki, N. 2000. The contribution of a Ca^{2+}-activated Cl^- conductance to amino-acid-induced inward current responses of ciliated olfactory neurons of the rainbow trout. J. Exp. Biol. 203, 253–262.

Sato, K. and Suzuki, N. 2001. Whole-cell response characteristics of ciliated and microvillous olfactory receptor neurons to amino acids, pheromone candidates and urine in rainbow trout. Chem. Senses 26, 1145–1156.

Sato, Y., Miyasaka, N., and Yoshihara, Y. 2005. Mutually exclusive glomerular innervation by two distinct types of olfactory sensory neurons revealed in transgenic zebrafish. J. Neurosci. 25, 4889–4897.

Satou, M. 1990. Synaptic organization, local neuronal circuitry, and functional segregation of the teleost olfactory bulb. Prog. Neurobiol. 34, 115–142.

Satou, M., Fujita, I., Ichikawa, M., Yamaguchi, K., and Ueda, K. 1983. Field potential and intracellular potential studies of the olfactory bulb in the carp: evidence for a functional separation of the olfactory bulb into lateral and medial subdivisions. J. Comp. Physiol. A 152, 319–333.

Schachtner, J., Schmidt, M., and Homberg, U. 2005. Organization and evolutionary trends of primary olfactory brain centers in Tetraconata (Crustacea + Hexapoda). Arthrop. Struct. Devel. 34, 257–299.

Schild, D. and Restrepo, D. 1998. Transduction mechanisms in vertebrate olfactory receptor cells. Physiol. Rev. 78, 429–458.

Schmachtenberg, O. and Bacigalupo, J. 2004. Olfactory transduction in ciliated receptor neurons of the Cabinza grunt, *Isacia conceptionis* (Teleostei: Haemulidae). Eur. J. Neurosci. 20, 3378–3386.

Schmidt, M. 1989. The hair-peg organs of the shore crab, *Carcinus maenas* (Crustacea, Decapoda): ultrastructure and functional properties of sensilla sensitive to changes in seawater concentration. Cell Tissue Res. 257, 609–621.

Schmidt, M. 1997. Distribution of presumptive chemosensory afferents with FMRFamide- or substance P-like immunoreactivity in decapod crustaceans. Brain. Res. 746, 71–84.

Schmidt, M. 2001. Neuronal differentiation and long-term survival of newly generated cells in the olfactory midbrain of the adult spiny lobster, *Panulirus argus*. J. Neurobiol. 48, 181–203.

Schmitt, B. C. and Ache, B. W. 1979. Olfaction: responses of a decapod crustacean are enhanced by flicking. Science 205, 204–206.

Schmidt, M. and Ache, B. W. 1992. Antennular projections to the midbrain of the spiny lobster. II. Sensory innervation of the olfactory lobe. J. Comp. Neurol. 318, 291–303.

Schmidt, M. and Ache, B. W. 1993. Antennular projections to the midbrain of the spiny lobster. III. Central arborizations of motoneurons. J. Comp. Neurol. 336, 583–594.

Schmidt, M. and Ache, B. W. 1996a. Processing of antennular input in the brain of the spiny lobster, *Panulirus argus*. I. Non-olfactory chemosensory and mechanosensory pathway of the lateral and median antennular neuropils. J. Comp. Physiol. A 178, 579–604.

Schmidt, M. and Ache, B. W. 1996b. Processing of antennular input in the brain of the spiny lobster, *Panulirus argus*. II. The olfactory pathway. J. Comp. Physiol. A 178, 605–628.

Schmidt, M. and Derby, C. D. 2005. Non-olfactory chemoreceptors in asymmetric setae activate antennular grooming behavior in the Caribbean spiny lobster *Panulirus argus*. J. Exp. Biol. 208, 233–248.

Schmidt, M. and Gnatzy, W. 1984. Are the funnel-canal organs the "campaniform sensilla" of the shore crab, *Carcinus maenas* (Decapoda, Crustacea)? II. Ultrastructure. Cell Tissue Res. 237, 81–93.

Shabani, S., Kamio, M., and Derby, C. D. 2006. Chemicals released by injured or disturbed conspecifics mediate defensive behaviors via the aesthetasc pathway in the spiny lobster *Panulirus argus*. Chem. Senses 31, A81–A82.

Sheldon, R. E. 1912. The olfactory tracts and centers in teleosts. J. Comp. Neurol. 22, 177–339.

Song, C.-K., Johnstone, L. M., Schmidt, M., Derby, C. D., and Edwards, D. H. 2004. Social experience decreases the rate of postembryonic neurogenesis in the olfactory system of juvenile *Procambarus clarkii*. Soc. Neurosci. Abstr. 754.12.

Sorensen, P. W. and Caprio, J. 1998. Chemoreception. In: The Physiology of Fishes (*ed*. D. H. Evans), pp. 375–405. CRC Press LLC.

Sorensen, P. W., Hara, T. J., and Stacey, N. E. 1991. Sex pheromones selectively stimulate the medial olfactory tracts of male goldfish. Brain Res. 558, 343–347.

Speca, D. J., Lin, D. M., Sorensen, P. W., Isacoff, E. Y., Ngai, J., and Dittman, A. 1999. Functional identification of a goldfish odorant receptor. Neuron 23, 487–498.

Spehr, M., Wetzel, C. H., Hatt, H., and Ache, B. W. 2002. 3-Phosphoinositides modulate cyclic nucleotide signaling in olfactory receptor neurons. Neuron 33, 731–739.

Stacey, N. E. and Kyle, A. L. 1983. Effects of olfactory tract lesions on sexual and feeding behavior in the goldfish. Physiol. Behav. 30, 621–628.

Stacey, N. E. and Sorensen, P. W. 2002. Fish Hormonal Pheromones. In: Hormones, Brain, and Behavior, Vol. 2 (eds. D. W. Pfaff, A. P. Arnold, S. Etgen, S. Fahrbach, and R. Rubin), pp. 375–435. Academic Press.

Stacey, N. E. and Sorensen, P. W. 2005. Hormones, Pheromones, and Reproductive Behaviors. In: Behaviour: Interactions with Fish Physiology (eds. K. A. Sloman, S. Balshine, and R. W. Wilson), Vol. 24 in Fish Physiology (series editors W. S. Hoar, D. J. Randall, and A.P. Farrell), Chap. 9, pp. 359–412. Academic Press, New York.

Stensmyr, M. C., Erland, S., Hallberg, E., Wallen, R., Greenaway, P., and Hansson, B. S. 2005. Insect-like olfactory adaptations in the terrestrial giant robber crab. Curr. Biol. 15, 116–121.

Steullet, P. and Derby, C. D. 1997. Coding of blend ratios of binary mixtures by olfactory neurons in the Florida spiny lobster, Panulirus argus. J. Comp. Physiol. A 180, 123–135.

Steullet, P., Cate, H. S., and Derby, C. D. 2000a. A spatiotemporal wave of turnover and functional maturation of olfactory receptor neurons in the spiny lobster Panulirus argus. J. Neurosci. 20, 3282–3294.

Steullet, P., Cate, H. S., Michel, W. C., and Derby, C. D. 2000b. Functional units of a compound nose: aesthetasc sensilla house similar populations of olfactory receptor neurons on the crustacean antennule. J. Comp. Neurol. 418, 270–280.

Steullet, P., Dudar, O., Flavus, T., Zhou, M., and Derby, C. D. 2001. Selective ablation of antennular sensilla on the Caribbean spiny lobster Panulirus argus suggests that dual antennular chemosensory pathways mediate odorant activation of searching and localization of food. J. Exp. Biol. 204, 4259–4269.

Steullet, P., Kruetzfeldt, D. R., Hamidani, G., Flavus, T., Ngo, V., and Derby, C. D. 2002. Dual parallel antennular chemosensory pathways mediate odor-associative learning and odor discrimination in the Caribbean spiny lobster Panulirus argus. J. Exp. Biol. 205, 851–867.

Stoss, T. D., Nickell, M. D., Hardin, H., Derby, C. D., and McClintock, T. S. 2004. Inducible transcript expressed by reactive epithelial cells at sites of olfactory sensory neuron proliferation. J. Neurobiol. 58, 355–368.

Strausfeld, N. J. and Hildebrand, J. G. 1999. Olfactory systems: common design, uncommon origins? Curr. Opin. Neurobiol. 9, 634–639.

Strotmann, J., Conzelmann, S., Beck, A., Feinstein, P., Breer, H., and Mombaerts, P. 2000. Local permutations in the glomerular array of the mouse olfactory bulb. J. Neurosci. 20, 6927–6938.

Sullivan, J. M. and Beltz, B. S. 2004. Evolutionary changes in the olfactory projection neuron pathways of eumalacostracan crustaceans. J. Comp. Neurol. 470, 25–38.

Sullivan, J. M. and Beltz, B. S. 2005. Integration and segregation of inputs to higher-order neuropils of the crayfish brain. J. Comp. Neurol. 481, 118–126.

Sun, H., Kondo, R., Shima, A., Naruse, K., Hori, H., and Chigusa, S. I. 1999. Evolutionary analysis of putative olfactory receptor genes of medaka fish, Oryzias latipes. Gene 231, 137–145.

Sveinsson, T. and Hara, T. J. 1990. Multiple olfactory receptors for amino acids in Arctic char (Salvelinus alpinus) evidenced by cross-adaptation experiments. Comp. Biochem. Physiol. 97A, 289–293.

Szabo, T., Blahser, S., Denizot, J. P., and Ravaille-Veron, M. 1991. Extensive primary olfactory projections beyond the olfactory bulb in teleost fish. Neurobiology 312, 555–560.

Tabor, R., Yaksi, E., Weislogel, J.-M., and Friedrich, R. W. 2004. Processing of odor mixtures in the zebrafish olfactory bulb. J. Neurosci. 24, 6611–6620.

Takami, S. and Graziadei, P. P. C. 1991. Light microscopic Golgi study of mitral/tufted cells in the accessory olfactory bulb of the adult rat. J. Comp. Neurol. 311, 65–83.

Thommesen, G. 1978. The spatial distribution of odour induced potentials in the olfactory bulb of char and trout (Salmonidae). Acta Physiol. Scand. 102, 205–217.

Thompson, H. A. and Ache, B. W. 1980. Threshold determination for olfactory receptors of the spiny lobster. Mar. Behav. Physiol. 7, 249–260.

Troemel, E. R. 1999. Chemosensory signaling in C. elegans. BioEssays 21, 1011–1020.

Trott, T. J. and Robertson, J. R. 1984. Chemical stimulants of cheliped flexion behavior by the western Atlantic ghost crab Ocypode quadrata (Fabricius). J. Exp. Mar. Biol. Ecol. 78, 237–252.

Tsuboi, A., Yoshihara, Y., Yamazaki, N., Kasai, H., Asai-Tsuboi, H., Komatsu, M., Serizawa, S., Ishii, T., Matsuda, Y., Nagawa, F., and Sakano, H. 1999. Olfactory neurons expressing closely linked and homologous odorant receptor genes tend to project their axons to neighboring glomeruli on the olfactory bulb. J. Neurosci. 19, 8409–8418.

Uchida, N., Takahashi, Y. K., Tanifuji, M., and Mori, K. 2000. Odor maps in the mammalian olfactory bulb: domain organization and odorant structural features. Nat. Neurosci. 3, 1035–1043.

Valentinčič, T. 2004. Taste and Olfactory Stimuli and Behavior in Fishes. In: The Senses of Fish (eds. G. von derEmde, J. Mogdans, and B. G. Kapoor), pp. 90–108. Narosa Publishing House.

Valentinčič, T. and Caprio, J. 1994. Consummatory feeding behavior to amino acids in intact and anosmic channel catfish Ictalurus punctatus. Physiol. Behav. 55, 857–863.

Valentinčič, T. and Caprio, J. 1997. Visual and chemical release of feeding behavior in adult rainbow trout. Chem. Senses 22, 375–382.

Valentinčič, T., Kralj, J., Stenovec, M., Koce, A., and Caprio, J. 2000a. The behavioral detection of binary mixtures of amino acids and their individual components by catfish. J. Exp. Biol. 203, 3307–3317.

Valentinčič, T., Metelko, J., Ota, D., Pirc, V., and Blejec, A. 2000b. Olfactory discrimination of amino acids in brown bullhead catfish. Chem. Senses 25, 21–29.

Valentinčič, T., Miklavc, P., Dolensek, J., and Plibersek, K. 2005. Correlations between olfactory discrimination, olfactory receptor neuron responses and chemotopy of amino acids in fishes. Chem. Senses 30(Suppl. 1), i312–i314.

Valentinčič, T., Wegert, S., and Caprio, J. 1994. Learned olfactory discrimination versus innate taste responses to amino acids in channel catfish, Ictalurus punctatus. Physiol. Behav. 55, 865–873.

Vassar, R., Chao, S. K., Sitcheran, R., Nuñez, J. M., Vosshall, L. B., and Axel, R. 1994. Topographic organization of sensory projections to the olfactory bulb. Cell 79, 981–991.

Vogt, R. G. 2005. Molecular Basis of Pheromone Detection in Insects. In: Comprehensive Insect Physiology, Biochemistry, Pharmacology and Molecular Biology. Vol. 3. Endocrinology (eds. L. I. Gilbert, K. Iatro, and S. Gill), pp. 753–804. Elsevier, London.

Voigt, R. and Atema, J. 1992. Tuning of chemoreceptor cells of the second antenna of the American lobster (Homarus *americanus*) with a comparison of four of its other chemoreceptor organs. J. Comp. Physiol. A 171, 673–683.

Wachowiak, M. and Ache, B. W. 1997. Dual inhibitory pathways mediated by GABA- and histaminergic interneurons in the lobster olfactory lobe. J. Comp. Physiol. A 180, 357–372.

Wachowiak, M. and Ache, B. W. 1998. Multiple inhibitory pathways shape odor-evoked responses in lobster olfactory projection neurons. J. Comp. Physiol. A 182, 425–434.

Wachowiak, M., Cohen, L. B., and Ache, B. W. 2002. Presynaptic inhibition of olfactory receptor neurons. Microsc. Res. Tech. 58, 365–375.

Wachowiak, M., Diebel, C. E., and Ache, B. W. 1996. Functional organization of olfactory processing in the accessory lobe of the spiny lobster. J. Comp. Physiol. A 178, 211–226.

Wachowiak, M., Diebel, C. E., and Ache, B. W. 1997. Local interneurons define functionally distinct regions within lobster olfactory glomeruli. J. Exp. Biol. 200, 989–1001.

Wang, Y., Wright, N. J. D., Guo, H.-F., Xie, Z., Svoboda, K., Malinow, R., Smith, D. P., and Zhong, Y. 2001. Genetic manipulation of the odor-evoked distributed neural activity in the *Drosophila* mushroom body. Neuron 29, 267–276.

Weissburg, M. J. 2000. The fluid dynamical context of chemosensory behavior. Biol. Bull. 198, 188–202.

Weissburg, M. J. and Derby, C. D. 1995. Regulation of sex-specific feeding behavior in fiddler crabs: physiological properties of chemoreceptor neurons in claws and legs of males and females. J. Comp. Physiol. A 176, 513–526.

Weissburg, M. J. and Zimmer-Faust, R. K. 1991. Ontogeny *versus* phylogeny in determining patterns of chemoreception: initial studies with fiddler crabs. Biol. Bull. 181, 205–215.

Weissburg, M. J., Derby, C. D., Johnson, O., McAlvin, B., and Moffett, J. M., Jr. 2001. Transsexual limb transplants in fiddler crabs and expression of novel sensory capabilities. J. Comp. Neurol. 440, 311–320.

Wellins, C. A., Rittschof, D., and Wachowiak, M. 1989. Location of volatile odor sources by ghost crab *Ocypode quadrata* (Fabricius). J. Chem. Ecol. 15, 1161–1169.

Weltzien, F., Hoglund, E., Hamdani, E. H., and Døving, K. B. 2003. Does the lateral bundle of the medial olfactory tract mediate reproductive behavior in male crucian carp? Chem. Senses 28, 293–300.

Wight, K., Francis, L., and Eldridge, D. 1990. Food aversion learning by the hermit crab *Pagurus granosimanus*. Biol. Bull. 178, 205–209.

Wong, A. M., Wang, J. W., and Axel, R. 2002. Spatial representation of the glomerular map in the *Drosophila* protocerebrum. Cell 109, 229–241.

Wu, H.-H., Ivkovic, S., Murray, R. C., Jaramillo, S., Lyons, K. M., Johnson, J. E., and Calof, A. L. 2003. Autoregulation of neurogenesis by GDF 11. Neuron 37, 197–207.

Wullimann, M. F. and Mueller, T. 2004. Teleostean and mammalian forebrains contrasted: evidence from genes to behavior. J. Comp. Neurol. 475, 143–162.

Yamashita, S., Yamada, T., and Hara, T. J. 2006. Gustatory responses to feeding- and non-feeding-stimulant chemicals, with an emphasis on amino acids, in rainbow trout. J. Fish Biol. 68, 783–800.

Xu, P., Atkinson, R., Jones, D. N., and Smith, D. P. 2005. Drosophila OBP LUSH is required for activity of pheromone-sensitive neurons. Neuron 45, 193–200.

Yokoi, M., Mori, K., and Nakanishi, S. 1995. Refinement of odor molecule tuning by dendrodendritic synaptic inhibition in the olfactory bulb. Proc. Natl. Acad. Sci. U. S. A. 3371–3375.

Zeng, C. and Hidaka, I. 1990. Single fiber responses in the palatine taste nerve of the yellowtail *Seriola quinqueradiata*. Nippon Suisan Gakkai Shi 56, 1611–1618.

Zhainazarov, A. B., Doolin, R. E., and Ache, B. W. 2001. Properties and Functional Role of a Sodium-Activated Nonselective Cation Channel in Lobster Olfactory Receptor Neurons. In: The Crustacean Nervous System (*ed.* K. Wiese), pp. 367–375. Springer.

Zhainazarov, A. B., Doolin, R. E., Hoegg, R., and Ache, B. W. 1999. GABA-mediated inhibition of primary olfactory receptor neurons. Biol. Signals Recept. 8, 348–359.

Zhang, Y., Hoon, M. A., Chandrashekar, J., Mueller, K. L., Cook, B., Wu, D., Zuker, C. S., and Ryba, N. J. 2003. Coding of sweet, bitter, and umami tastes: different receptor cells sharing similar signaling pathways. Cell 112, 293–301.

Zimmer, R. K. and Butman, C. A. 2000. Chemical signaling processes in the marine environment. Biol. Bull. 198, 168–187.

Zimmer-Faust, R. K. 1987. Crustacean chemical perception: towards a theory on optimal chemoreception. Biol. Bull. 172, 10–29.

Zimmer-Faust, R. K. 1993. ATP: a potent prey attractant evoking carnivory. Limnol. Oceanogr. 38, 1271–1275.

Zimmer-Faust, R. K. and Case, J. F. 1982a. Odors influencing foraging behavior of the California spiny lobster, *Panulirus interruptus*, and other decapod Crustacea. Mar. Behav. Physiol. 9, 35–58.

Zimmer-Faust, R. K. and Case, J. F. 1982b. Organization of food search in the kelp crab, *Pugettia producta* (Randall). J. Exp. Mar. Biol. Ecol. 57, 237–255.

Zimmer-Faust, R. K., Tyre, J. E., Michel, W. C., and Case, J. F. 1984. Chemical mediation of appetitive feeding in a marine decapod crustacean: the importance of suppression and synergism. Biol. Bull. 167, 339–353.

Zou, Z., Horowitz, L. F., Montmayeur, J.-P., Snapper, S., and Buck, L. B. 2001. Genetic tracing reveals a stereotyped sensory map in the olfactory cortex. Nature 414, 173–179.

Zou, Z., Li, F., and Buck, L. B. 2005. Odor maps in the olfactory cortex. Proc. Natl. Acad. Sci. U. S. A. 102, 7724–7729.

4.05 Ultrastructure of Taste Buds

J C Kinnamon and R Yang, University of Denver, Denver, CO, USA

Glossary

a-gustducin Alpha subunit of G protein-coupled receptor for bitter and sweet taste transduction pathways.

atypical mitochondria Large mitochondria with twisted-energized or swollen cristae found in Type II taste cells.

chorda tympani nerve Branch of the facial nerve (VIIth cranial) that innervates taste buds in the anterior two-thirds of the tongue.

circumvallate papillae Large, dome-shaped gustatory papillae located in the posterior one-third of the tongue; innervated by the glossopharyngeal nerve (IXth cranial).

facial nerve (Cranial nerve VII) Innervates taste buds fungiform and foliate papillae in the anterior two-thirds of the tongue.

foliate papillae Rows of gustatory papillae located on the sides of the posterior of the tongue; innervated by both the chorda tympani and glossopharyngeal nerve in rodents.

fungiform papillae Mushroom-shaped papillae located in the anterior two-thirds of the tongue; innervated by the chorda tympani nerve.

fusiform Spindle-shaped, with a wider center and tapered ends.

glossopharyngeal nerve Cranial nerve IX, which controls taste sensation in the posterior one-third of the tongue.

innervated Supplied with nerves.

intragemmal Located within the taste bud.

neurotropic Having an affinity for neural tissue, as in a neurotropic stain.

perigemmal Nerves that are found on the periphery of the taste bud.

pore substance Electron-dense material located in the taste pore of a taste bud.

subsurface cisternae Flat membranous stacks of smooth endoplasmic reticulum.

sustentacular taste cells Provides support or sustains.

umami Japanese word for delicious to describe the taste sensation when consuming amino acids such as glutamate.

vagus nerve Cranial nerve X, which innervated taste buds in the pharynx.

4.05.1   Introduction

Taste buds are onion-shaped end organs containing 50–150 cells that are specialized for detecting aqueous chemical stimuli. The taste bud is a dynamic system in which new taste cells (=taste bud cells) are continually being born, maturing, performing their sensory functions, eventually going into senescence, and ultimately dying – all within a period of from 10 days to 2 weeks (Beidler, L. and Smallman, R., 1965). However, recent evidence suggests that taste buds contain both long- and short-lived taste cells (Hamamichi, R. *et al.*, 2006). At the same time, the nervous system is believed to sprout new nerve processes and form synaptic connections with developing taste cells, while retracting synapses from old cells. Intragemmal taste cells are located within the taste bud proper, while perigemmal taste cells are located at the periphery of the taste bud. The goal of the present review is to provide the reader with a basic understanding of taste bud morphology.

Taste buds are surrounded by a relatively electron-dense, nongustatory epithelium replete with desmosomes, from which intermediate filaments project into the cytoplasm of the epithelial cells. The appearance of the nongustatory epithelial cells and the filamentous material is less organized near the basal region of the taste bud, making it more difficult to distinguish the basal region of the taste bud from the surrounding nongustatory epithelium. A typical mammalian taste bud is innervated by sensory nerve processes that enter at the base and ramify throughout the basal two-thirds of the taste bud. Nerve processes within a taste bud are termed intragemmal fibers. The intragemmal fibers originate from either the VII (facial), IX (glossopharyngeal), or the X (vagus) nerves (Figure 1). A small subset of taste cells form synapses onto the intragemmal nerve processes. Another set of nerve processes envelop but do not penetrate the taste bud; these are termed perigemmal nerves (however, see Nagy, J. *et al.*, 1982; Finger, T., 1986).

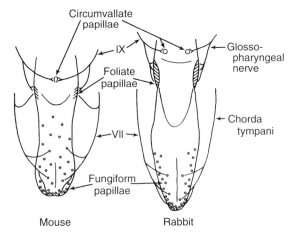

Figure 1   Innervation of rabbit and rat taste buds.

Although taste buds occur throughout the oral cavity, this review deals only with taste buds located on the foliate papillae of the rabbit and circumvallate taste buds of murines (rat and mouse). The rabbit foliate taste bud was selected because it was the first mammalian taste bud to be studied extensively and is the progenitor of the current system of taste cell classification. Taste buds of the rat and mouse were selected because these animals are the species of choice for physiological and molecular biology studies.

4.05.2   History

Taste bud structure has been studied for over a century, yet there is still considerable controversy concerning the classification of taste cells and an understanding of their functions (for reviews see Scalzi, H., 1967; Murray, R., 1973; Kinnamon, J., 1987; Roper, S., 1989; Reutter, K. and Witt, M., 1993; Miller, I., 1995; Finger, T., 1997; Pumplin, D. *et al.*, 1997; Finger, T. and Simon, S., 2000; Lindemann, B., 2001; Finger, T., 2005; Chandrashekar, J. *et al.*, 2006; Roper, S., 2006). The first descriptions of taste buds were in fish (Leydig, F., 1851). Mammalian taste buds were first studied by Schwalbe G. (1868) and Loven C. (1868). In the nineteenth century taste on cells were named, classified, and ascribed functions based on their staining properties, especially to neurotropic stains such as methylene blue (e.g., Wilson, J. and Edin, M., 1905). Electron microscopy was first used to characterize taste bud ultrastructure over five decades ago (Engstrom, H. and Rytzner, C., 1956a; 1965b). Since then there have been numerous studies on taste bud ultrastructure of mammalian and nonmammalian species (Table 1). Ultrastructural criteria have been used to classify taste cells into various types, including undifferentiated peripheral cells, type I (dark) cells, type II (light) cells, type III (receptor) cells, type IV (basal) cells, and intermediate cells. Traditionally, the morphological criteria used to classify taste cells have included:

1. electron density of the cytoplasm;
2. nuclear shape;
3. types and distribution of nuclear chromatin;
4. accumulations of specific organelles (e.g., polyribosomes, granular and smooth endoplasmic reticulum, and Golgi bodies);
5. number, types, and arrangement of apical microvilli; and
6. presence of synapses onto nerve processes.

Some newer criteria include the presence of subsurface cisternae (SSCs) of smooth endoplasmic reticulum and large, atypical mitochondria with tubular cristae (Royer, S. and Kinnamon, J., 1988; 1994; Clapp, T. *et al.*, 2004).

Table 1 Studies on taste bud ultrastructure of mammalian and nonmammalian species

Species	References
Axolotl	Fährmann W. (1967), Fährmann W. and Schuchardt E. (1967), Toyoshima K. *et al.* (1987), Wistuba J. *et al.* (1999).
Bird	Ganchrow D. *et al.* (1991; 1993; 1994; 1995; 1998), Stornelli M. *et al.* (2000), Witt M. *et al.* (2000).
Cat	Yoshie S. *et al.* (1997).
Cow	Tabata S. *et al.* (2003).
Dog	Baradi A. (1965), Kanazawa H. (1993).
Fish	Trujillo-Cenóz O. (1961), Hirata Y. (1966), Uga S. and Hama K. (1967), Welsch U. and Storch V. (1969), Kapoor B. *et al.* (1975), Grover-Johnson N. and Farbman A. (1976), Ovalle W. and Shinn S. (1977), Joyce E. and Chapman G. (1978), Reutter K. (1978; 1982; 1987; 1992), Fujimoto S. and Yamamoto K. (1980), Ono R. (1980), Toyoshima K. *et al.* (1984), Sbarbati A. *et al.* (1988a; 1998b; 1993; 2006), Witt M. and Reutter K. (1990), Reuter K. and Witt M. (1993, 2004), Boudriot F. and Reutter K. (2001).
Frog	Uga S. (1966), Uga S. and Hama K. (1967), Graziadei P. and DeHan R. (1971), Witt M. (1993), Witt M. *et al.* (2000).
Guinea-pig	Yoshie S. *et al.* (1990; 1991), Huang Y. *et al.* (2003).
Hamster	Miller R. and Chaudhry A. (1976).
Human	Graziadei P. (1969), Mattern C. *et al.* (1970), Takeda M. (1972), Paran N. *et al.* (1975), Azzali G. (1997), Witt M. and Reutter K. (1997), Witt M. *et al.* (2000).
Lamb	Sweazey R. *et al.* (1994).
Monkey	Murray R. and Murray A. (1960), Cottler-Fox M. *et al.* (1980, 1987), Ide C. and Munger B. (1980), Zahm D. and Munger B. (1983a; 1983b; 1985), Farbman A. *et al.* (1985).
Mouse	Mattern C. and Paran N. (1974), Takeda M. (1976; 1977), Akisaka T. and Oda M. (1978), Takeda M. and Kitao K. (1980), Takeda M. *et al.* (1981; 1985; 1988; 1989), Kinnamon J. *et al.* (1985; 1988), Delay R. *et al.* (1986), Royer S. and Kinnamon J. (1988; 1994), Esses B. *et al.* (1988), Kudoh M. (1988), Murray R. (1993), Sekerkova G. *et al.* (2005).
Mud puppy	Farbman A. and Yonkers J. (1971), Cummings T. *et al.* (1987), Delay R. and Roper S. (1988), Delay R. *et al.* (1993).
Rabbit	Trujillo-Cenóz O. (1957), De Lorenzo (1958; 1960; 1963), Iriki T. (1960), Nemetschek-Gansler H. and Ferner H. (1964), Murray R. and Murray A. (1967), Scalzi H. (1967), Jeppsson P. (1969), Murray R. 1969; 1973; 1986), Murray R. *et al.* (1969), Iwayama T. (1970), Fujimoto S. and Murray R. (1970), Sangiacomo C. (1970), Nada O. and Hirata K. (1975), Jahnke K. and Baur P. (1979), Toyoshima K. and Tandler B. (1989), Royer S. and Kinnamon J. (1991; 1994).
Rat	Farbman A. (1965a; 1965b; 1967), Gray E. and Watkins K. (1965), Beidler L. (1969), Uga S. (1969), Takeda M. and Hoshino T. (1975), Akisaka T. and Oda M. (1977; 1978), Graziadei P. and Graziadei M. (1978), Akisaka T. (1980), Toyoshima K. and Shimamura A. (1982), Kobayashi T. and Tomita H. (1986), Endo Y. (1988), Finger T. *et al.* (1990), Montavon P. *et al.* (1996), Yoshie S. *et al.* (1996), Sbarbati A. *et al.* (1998a; 1998b), Sakai H. *et al.* (1999), Yang R. *et al.* (2000a; 2000b; 2004; 2007), Yee C. *et al.* (2001), Seta *et al.* (2003), Clapp T. *et al.* (2004), Ma *et al.* (2007).
Squirrel	Popov V. *et al.* (1999).

4.05.3   A Classic Taste Bud: The Rabbit Foliate Taste Bud

The first ultrastructural studies on rabbit taste buds were carried out by Engstrom H. and Rytzner C. (1956a; 1956b) on foliate taste buds. They observed that taste cells had varying morphologies and staining properties, but found it difficult to classify the cells into different types. Because there appeared to be a continuum of cellular morphologies, Engstrom and Rytzner postulated that the different cell types represented stages of development of a single cell line. Since then the rabbit has been used extensively for ultrastructural studies on taste buds, most notably by Raymond Murray and coworkers (Murray, A., 1961; Murray, R. and Murray, A., 1966; 1967; Murray, R., 1969; Murray, R. *et al.*, 1969; Murray, R. and Murray, A., 1970; Murray, R., 1971; Murray, R. and Murray, A., 1971; Murray, R., 1973; Murray, R., 1986; Murray, R., 1993). The foliate papillae of a rabbit reside on both sides of the posterior tongue and contain numerous taste buds. These taste buds are innervated by branches of the glossopharyngeal nerve (IXth cranial nerve) (Murray, R. and Murray, A., 1971) as shown in Figure 1. This contrasts with the innervation of foliate papillae of the rat, which are innervated by branches of both the glossopharyngeal nerve and the chorda tympani branch of the facial nerve (VIIth cranial nerve) (Whiteside, B., 1927; Oakley, B., 1967).

As stated above, Engstrom H. and Rytzner C. (1956a; 1956b) did not classify rabbit foliate taste cells into specific types, postulating that the differing cellular morphologies represented stages of development of a single cell type. Later investigators began to classify rabbit taste cells into different types. Trujillo-Cenóz O. (1957) referred to neuroepithelial cells while De Lorenzo A. (1958; 1960; 1963) described receptor and sustentacular cells in rabbit foliate taste buds. Iriki T. (1960) classified taste cells from rabbit foliate taste buds as either gustatory or sustentacular. Murray A. (1961) observed two morphological cell types but eschewed the terms gustatory or sustentacular, instead proposing that both morphological cells types were gustatory in nature. Scalzi H. (1967), like Engstrom and Rytzner, supported the notion that there was most likely a single taste cell line in rabbit foliate taste buds, with cells differentiating from one cell type to another. This was reflected in the system of taste cell classification proposed by Scalzi, which included pregustatory, type Ia, type Ib, and type Ic taste cells. Murray R. and Murray A. (1967) described dark and light cells in rabbit foliate papillae, based primarily on the electron density of the cytoplasm. At that time, the light cell was believed to be the most likely candidate for being the gustatory receptor cell, but the dark cell was not completely excluded. Jeppsson P. (1969) adopted the dark and light cell nomenclature of Murray. In 1969 Murray and coworkers modified their system of classification for rabbit taste bud cells to include dark cells, light cells and introduced a new cell type – the type III or sensory cell (Murray, R., 1969; Murray, R. *et al.*, 1969; Murray, R., 1971). In addition, they adopted the type IV or basal cell proposed by Farbman A. (1965a) for rat taste buds. Following the nomenclature proposed by Farbman A. (1965a), Murray and coworkers referred to dark cells as type I cells, and light cells as type II cells. Iwayama T. (1970) and Fujimoto S. and Murray R. (1970) each carried out separate degeneration/regeneration studies on rabbit foliate taste buds and reached antipodal conclusions. Whereas Iwayama concluded that there was a single cell line that differentiated from dark (type I) to intermediate to light (type II) cells, Fujimoto S. and Murray R. (1970) concluded that the different cell types arose from separate cell lines. The same year Sangiacomo C. (1970) described type I, type II, and neurosecretory cells in rabbit foliate taste buds. Today, most investigators have adopted the nomenclature of Farbman A. (1965a) and Murray and co-workers (type I, type II, type III, and type IV) for most mammalian taste buds (see Murray, R., 1973).

4.05.3.1   General Morphological Features of Rabbit Foliate Taste Buds

A longitudinal section through a rabbit foliate taste bud reveals a broad, onion-shaped structure that rests on the basal lamina (Figure 2). The taste bud extends from the basal lamina to the surface of the lingual epithelium (Figure 2). The lamina propria underlying the basal lamina contains connective tissue, vasculature, and nerve processes. Rabbit foliate taste buds range from 30 to 70 µm in width and 50–60 µm in height (Murray, R., 1973; Royer, S. and Kinnamon, J., 1991). The intragemmal taste cells (=cells located within the taste bud proper) display a range of morphologies. Most of the taste cell nuclei are located in the bottom half to two-thirds of the taste bud. Taste cells are joined together at their apices by

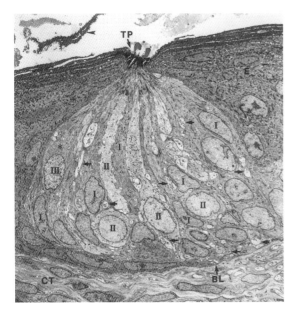

Figure 2 High-voltage electron micrograph (HVEM) of a longitudinal section through a rabbit foliate taste bud. The nucleus and part of the apical process (*) of a type III cell (III) appear at the left margin of this taste bud. Several nuclei and apical processes of type I and II cells are also present. Intragemmal nerve fibers are indicated by arrows. Two plumes of electron-dense pore substance extend above the taste pore (TP). Note the sloughing of the surface layer of the epithelium (E) at the upper left (arrowhead). BL, basal lamina; CT, connective tissue. ×2250. From Royer, S. and Kinnamon J. 1991. HVEM serial-section analysis of rabbit foliate taste buds: I. Type III cells and their synapses. J. Comp. Neurol. 306, 49–72.

tight junctions and by desmosomes elsewhere. Although there is considerable variability in cellular morphology within a taste bud, most taste cells are fusiform to spindle-shaped structures with centrally positioned nuclei. In favorable sections a few taste cells can be seen extending from the base of the taste bud to the taste pit, where apical microvilli from the taste cells project into a cavity below the epithelial surface. The taste pit opens into the oral cavity at the taste pore. The taste pore is typically filled with an electron-dense pore substance. It is not uncommon for foliate taste buds of the rabbit to possess multiple taste pores (Royer, S. and Kinnamon, J., 1991).

Prominent intercellular spaces have been described in taste buds (Nemetschek-Gansler, H. and Ferner, H., 1964; Baradi, A., 1965; Scalzi, H., 1967). Murray R. (1973) noted the high variability of intragemmal intercellular spaces and suggested these spaces were probably artifacts of specimen preparation. Our experience is similar; manipulation of

the osmolarity of the fixative can alter the size and shape of intercellular spaces within taste buds. Nerve processes ramify throughout the basal two-thirds of the taste bud. Synapses can be seen from type III cells onto nerve processes (Figure 5). We have used the following criteria to identify synapses (Royer, S. and Kinnamon, J., 1991):

1. a thickening of the presynaptic membrane, with a 15–20 nm cleft separating the parallel membranes;
2. clustering of vesicles adjacent to the membrane thickening; and
3. the appearance of this configuration in at least two consecutive 0.25 μm-thick sections.

When compared with taste cells of rodents, rabbit foliate taste cells are more distinctive and easier to classify into cell types. We have not observed any intermediate or transitional cell types in rabbit foliate taste buds.

4.05.3.2   Primary Foliate Taste Bud Cell Types

4.05.3.2.1   Type I cells

Type I cells (previously termed, dark cells), are slender cells with moderately expanded nuclear regions (Figures 2 and 3). No synaptic connections associated with type I cells have been observed. The type I cell has been referred to as being sustentacular by most investigators, implying that this cell type does not perform a sensory function, although one may wonder about the function of the apical microvilli of these cells. Both basal and apical processes are present, with the apical process terminating in approximately 30–40 long, slender, microvilli varying in length from 1 to 2 μm (Murray, R., 1973). In cross section, the microvilli possess hexagonally arranged electron-dense filamentous material (Scalzi, H., 1967), which is most likely actin. Unlike other cell types, type I cells also possess finger- or sheet-like cytoplasmic processes that separate and envelop type II cells, type III cells, and intragemmal nerve fibers in a Schwann cell-like manner (Murray, R., 1973; Royer, S. and Kinnamon, J., 1994). These processes can best be seen in transverse sections (Figure 3).

The cytoplasmic ground substance of a type I cell varies from intensely electron-dense to electron-lucent. The apical cytoplasm also contains varying amounts of granular and/or smooth endoplasmic reticulum. Perhaps the most distinguishing ultrastructural feature of the type I cell is the presence of numerous

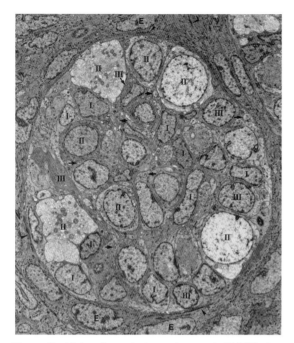

Figure 3   High-voltage electron micrograph (HVEM) of a transverse section through the nuclear region of a rabbit foliate taste bud. Cells of type I, type II, and type III are indicated. Several nerve fiber profiles are labeled with arrows. Note that the nuclei of three of the five type III cells in this taste bud lie along the right edge of the bud. The basal processes of the other two type III cells are located close to the left and top edges, respectively. One of these (at the center left) exhibits densely stained cytoplasm. The taste bud is surround by epithelial cells (E) containing numerous bundles of cytokeratin filaments (arrowheads). ×3900. From Royer, S. and Kinnamon J. 1991. HVEM serial-section analysis of rabbit foliate taste buds: I. Type III cells and their synapses. J. Comp. Neurol. 306, 49–72.

apical, electron-dense granules varying in size from 100 to 400 nm (Murray, R., 1973; Royer, S. and Kinnamon, J., 1994). The similarity of the contents of these granules to the pore substance found in the taste pore suggests that the pore substance is produced by the exocytosis of the apical granules in type I cells (Murray, R. and Murray, A., 1971).

The nuclei of type I cells are irregular in shape but generally are more slender when compared with the round-to-ovoid nuclei of type II cells. The nuclear matrix of a type I cell is considerably more electron-dense than the matrix of a type II or type III cell (Figure 3). Type I cell nuclei are more elongate than type II cell nuclei and contain patchy heterochromatin distributed throughout the nuclear matrix. Clumps of heterochromatin can also be seen adherent to the inner leaflet of the nuclear membrane.

4.05.3.2.2   Type II cells

Type II cells, previously termed, light cells, have a pronounced fusiform appearance, with a noticeably expanded nuclear region (Murray, R., 1973; Royer, S. and Kinnamon, J., 1994) (Figures 2 and 3). At low magnifications, type II cells are distinguished from type I and type III cells by the presence of large, round-to-ovoid nuclei in the type II cells. The nuclear matrix of a type II cell is more uniform in appearance, lacking much of the patchy heterochromatin present in the nuclei of type I cells.

The cytoplasmic ground substance of a type II cell is variable in appearance, but generally more electron-lucent than observed in a type I cell. Type II cells lack the enveloping, sheet-like cytoplasmic processes associated with type I cells, as well as the apical, dense granules characteristic of type I cells. The apical cytoplasm of type II cells often contains swollen or dilated cisternae of smooth endoplasmic reticulum and numerous mitochondria. The apical process of a type II cell terminates in several microvilli of generally uniform length, giving a brush-like appearance. In addition to conventional type II cells, we have observed a subset with large, round, lightly staining nuclei, and extremely electron-lucent, highly vesiculate cytoplasm, which we termed type II-VL (for very light) (Royer, S. and Kinnamon, J., 1994). These cells are primarily found at the periphery of the taste bud (e.g., Figure 3).

4.05.3.2.3   Type III cells

Early descriptions of cell types in rabbit foliate taste buds described only dark and light cells. A third cell type, termed the gustatory receptor, was described by Murray, R. *et al.*, in 1969. Type III cells are spindle-shaped with apical and basal processes and constitute approximately 5% of the cells within a rabbit foliate taste bud (5–15%: Murray, R. *et al.*, 1969; 5–7%: Royer, S. and Kinnamon, J., 1994). The type III cell is the only cell type in foliate taste buds of the rabbit that has been observed to form synaptic connections with nerve processes (Figures 4 and 5). Thus, the type III cell was considered to be the receptor cell (Murray, R., 1973). We found type III cells to be concentrated around the periphery of the taste bud (Figure 3; see also Figure 9: Royer, S. and Kinnamon, J., 1991). This contrasts with Murray's observation that type III cells are centrally placed (Murray, R., 1971). The apical region of the type III cell terminates in a single, blunt microvillus approximately 0.5–0.6 μm in diameter at its base and up to 4–5 μm in length (Royer, S. and Kinnamon, J., 1994). This contrasts

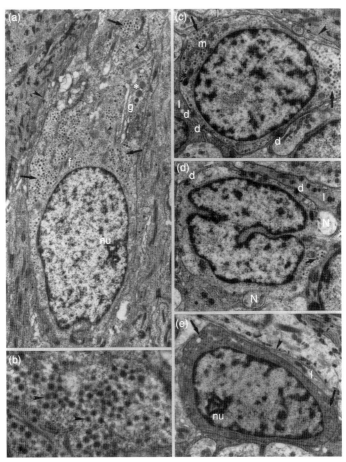

Figure 4 (a) Longitudinal section through the nucleus and apical process of a type III cell (also shown in Figure 2). The cytoplasm of the apical process contains abundant dense-cored vesicles (arrows), Golgi complex (g), lysosome-like dense bodies (*), granular endoplasmic reticulum (small arrowheads), and filaments (f). Nucleolus (nu). The margin of the taste bud indicated by the large arrowhead (cf. Figure 2). ×9600. (b) Higher magnification view of dense-cored vesicles located in the apical process of the type III cell shown in Figures 2 and a. The electron-lucent halo around the dense core is apparent in some of the vesicles (arrowhead). ×28 500. (c) Transverse section through the perikaryal region of a type III cell. A thin type I cell process (l) separates this cell from the border of the taste bud (arrowhead). The cytoplasm of the type III cell contains numerous dense-cored vesicles (arrows), several mitochondria (m), and a dense lysosome-like body (*). Desmosomes (d) attach the type III cell to adjacent type I cells. ×12 700. (d) Transverse section through the parikaryal region of a type III cell. The nuclear membrane displays two deep invaginations. Nerve fibers (N), desmosomes (d). ×11 300. (e) Transverse section through the perikaryal region of a type III cell located at the margin of a taste bud (arrowhead) and separated from surrounding cells by a thin type I cell process. Although the cytoplasm of this cell is quite electron-dense, groups of dense-cored vesicle can still be distinguished (arrows). ×13 100. From Royer, S. and Kinnamon J. 1991. HVEM serial-section analysis of rabbit foliate taste buds: I. Type III cells and their synapses. J. Comp. Neurol. 306, 49–72.

with the multiple microvilli at the apices of both type I and type II cells. In the upper third of the taste bud, type III cells are often separated from other cells by the cytoplasmic processes of type I cells, but may contact each other in the basal region. As with other taste cells, type III cells are joined to adjacent cells by desmosomes. Radiating from the desmosomes are numerous intermediate filaments that project into the cytoplasm.

The cytoplasm of a type III cell is variable, but usually moderately electron-dense (Royer, S. and Kinnamon, J., 1994). Scattered throughout the cytoplasm are mitochondria, granular and smooth endoplasmic reticulum, free ribosomes, and polyribosomes. The cytoplasmic ground substance of a small subset of type III cells is significantly more electron-dense (e.g., Figures 3 and 4). These cells have been termed dark type III cells (Murray R.,

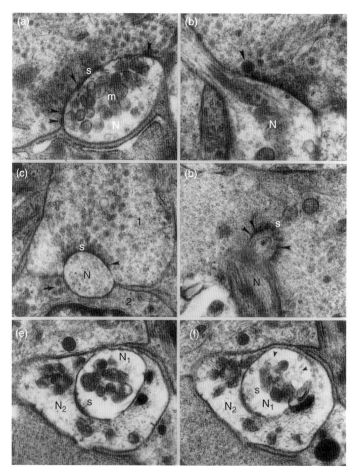

Figure 5 (a) Synapse (s) between a type III cell and a nerve process (N). Note the dense cloud of small, clear-cored vesicles as well as several large, dense-cored vesicles at the presynaptic site and presynaptic dense projections (arrowheads). The nerve fiber contains numerous mitochondria (m). ×36 900. (b) Synapse (s) from a type III cell onto an enlarged nerve process (N). Note the synaptic cleft and the very large dense-cored vesicle (arrowhead) adjacent to the synapse. ×44 600. (c) A small nerve fiber (N) lying between two adjacent type III cells (1 and 2) is postsynaptic to one of these cells (cell 1). Numerous dense-cored vesicles and presynaptic dense projections (arrowhead) are evident. The arrow points to a small group of dense-cored vesicles in cell 2 near its contact with the nerve fiber. ×29 900. (d) Synapse (s) from a type III cell onto a finger-like projection (*) of a nerve fiber (N). Presynaptic dense projections are indicated by arrowheads. ×37 900. (e, f) An axoaxonal synapse (s) between two nerve fibers. N_1 is presynaptic to N_2 and contains a cluster of clear vesicles adjacent to the presynaptic membrane. Dense material adheres to the postsynaptic membrane in N_2. Arrowheads in F indicate groups of clear vesicles in N_1. (e) ×32 600. (f) ×29 000. From Royer, S. and Kinnamon J. 1991. HVEM serial-section analysis of rabbit foliate taste buds: I. Type III cells and their synapses. J. Comp. Neurol. 306, 49–72.

and Murray, A., 1971). Murray R. (1986) speculated that dark type III cells might be developing or degenerating type III cells. A distinguishing feature of the type III cell is the presence of large, dense-cored vesicles (80–140 nm) with prominent electron-lucent halos (Figure 4). These vesicles are especially numerous in the cytoplasm surrounding the nucleus. These vesicles are distributed throughout the cytoplasm and should not be confused with the small, clear vesicles (40–60 nm) located at the presynaptic regions (see below).

The nuclei of type III cells are situated in the lower half of the taste bud. These nuclei are intermediate in appearance between the nuclei of type I and type II cells (Figures 2 and 3). Type III cell nuclei are ovoid to round in cross section, similar to type II cell nuclei, but elongate in longitudinal section like type I cell nuclei. The density of the nuclear matrix of a type III cell is intermediate between that of a type I and a type II cell. Like type I cells, the nuclei of type III cells possess prominent nuclear invaginations (Figures 3 and 4). Another similarity between the nuclei of type III cells and type I cells is the presence of clumps of heterochromatin scattered throughout the nuclear matrix, some of which are adherent to the nuclear membrane.

4.05.3.2.4 *Synaptic connections*

Type III cells are the only cells that have been observed to form synapses with intragemmal nerve fibers (Murray, R. and Murray, A., 1970; 1971; Murray, R., 1971; 1973; Royer, S. and Kinnamon, J., 1991; 1994). It is curious that both type I and type II cells possess apical microvilli – putative sites for sensory transduction – yet neither cell type forms conventional chemical synapses with nerve processes. We found that some type III cells had divergent synaptic connections with synapses from a single type III cell onto a maximum of six nerve processes. Convergence of input from two to three type III cells onto a single nerve process was also a common observation. Most synapses are located on the basolateral region of the type III cell. These synapses are usually asymmetrical with electron-dense presynaptic thickenings, often including presynaptic dense projections (Figure 5). Type III cell synapses in rabbit foliate taste buds are especially well developed when compared with synapses in taste buds of rodents. Round, clear vesicles (40–60 nm) of varying numbers are associated with the presynaptic active zone (Royer, S. and Kinnamon, J., 1991; 1994). Dense-cored vesicles can also be present at these synapses. In some instances, only a few dense-cored vesicles are interspersed among the clear vesicles. At other synapses dense-cored vesicles are the predominant vesicle type. Thus, there may be more than one type of synapse associated with type III cells, as well as more than one neurotransmitter/neuromodulator. On rare occasions we have observed neuroneuronal synapses in taste buds (e.g., Figures 5(e) and 5(f)).

4.05.4 Rodent Taste Buds

4.05.4.1 Innervation

Lingual taste buds in rodents are innervated by branches of the facial (VII cranial) and glossopharyngeal (IX cranial) nerves (Whiteside, B., 1927; Oakley, B., 1967; Beidler, L., 1969; Farbman, A. and Hellekant, G., 1978; Bradley, R. 1971; Whitehead, M. *et al.*, 1985) (Figure 1). Fungiform and anterior foliate taste buds of rodents are innervated by the chorda tympani branch of the VII cranial nerve. This differs from innervation of rabbit foliate taste buds, which

are innervated only by the glossopharyngeal (IX) nerve. Taste buds in the vallate and posterior foliate papillae of rodents are innervated by the fibers of the lingual–tonsillar branch of the glossopharyngeal cranial nerve (IX) (Whiteside, B., 1927; Oakley, B., 1967).

4.05.4.2 General Morphological Features

Recent evidence indicates that taste cells in rodents are epithelial in origin (Stone, L. and Finger, T., 1994; Stone, L. *et al.*, 1995). The classic study on taste cell turnover in the rat was carried out by Beidler L. and Smallman R. (1965), who found that taste cells were replaced approximately every 10 days to 2 weeks. Rodent taste buds are onion-shaped structures with spindle-shaped cells that span the entire length of the taste bud from the basal lamina up to the taste pore. Taste cell nuclei are usually located in the lower third of the taste bud. Based on their staining characteristics and ultrastructural features, taste cells are currently classified into type I (dark), II (light), type III, and type IV (basal) cells. The type I, type II, and type III cells extend their microvilli into the taste pore (Figure 7). The structural features of the microvilli are distinctive for each of the taste cell types (Figure 7). Electron-dense pore substance is present in the taste pore, although its function is unclear. Basal cells are found near the base of the bud, but these cells lack apical processes. Nerve fibers are abundant in the connective tissue underlying the taste buds and penetrate the taste bud at its base and course throughout the bud. Most nerve fibers terminate in the lower half of the taste bud, where they may receive synaptic contacts from taste cells.

4.05.4.3 Primary Taste Bud Cell Types

Farbman A. (1965a) introduced the nomenclature type I and type II taste cells and classified taste cells in rat taste buds into four types: peripheral, basal (type IV), type I (dark), and type II (light) cells (Figure 8). The type III cell was first described in the rat by Takeda M. and Hoshino T. (1975) and in circumvallate taste buds of the mouse by Takeda M. (1976). Type III cells were considered to be sensory cells because they were the only cell type forming synaptic contacts onto nerve processes in rodent taste buds (Takeda, M. and Hoshino, T., 1975; Takeda, M., 1976). Using ultrastructural criteria, we classified taste cells in mouse circumvallate taste buds into dark cells (type I), intermediate cells, light cells (type II), basal cells, and undifferentiated peripheral cells (Scalzi, H. 1967; Iwayama, T., 1970; Kinnamon, J. *et al.*, 1985; Delay, R. *et al.*, 1986; Kinnamon, J. *et al.*, 1988). Dark cells, intermediate cells, and light cells all appeared to form synaptic contacts with nerve processes (Kinnamon, J. *et al.*, 1985; 1988). Recent studies using immunofluorescence and immunoelectron microscopy with antisera markers for different taste cell types indicate that cells originally classified as intermediate cells in mouse circumvallate taste buds should properly be classified as type III cells (Yang, R. *et al.*, 2000a; Yee, C. *et al.*, 2001; Clapp, T. *et al.*, 2004; Yang, R. *et al.*, 2004). Thus, type III cells in rat circumvallate taste buds are believed to be the only cell type that forms classical synapses onto nerve processes (Yang, R. *et al.*, 2000b; Yee, C. *et al.*, 2001; Yang, R. *et al.*, 2004). Type II cells do not form classical synapses with nerve fibers. Instead, at the points of contact with nerve fibers, type II cells exhibit SSCs of smooth endoplasmic reticulum (Royer, S. and Kinnamon, J., 1988; Clapp, T. *et al.*, 2004), that suggest a functional contact with nerve processes.

4.05.4.3.1 *Type I cells*

Farbman A. (1965a) built upon the dark cell and light cell nomenclature of Murray and coworkers, referring to dark cells as type I cells and light cells as type II cells. Type I cells (Figures 6 and 8), are the most numerous taste cells (50–60%), usually located peripherally in the taste bud (Farbman, A. 1965a; Murray, R., 1973; Kinnamon, J. *et al.*, 1985; Delay, R. *et al.*, 1986). These cells are slender with electron-dense cytoplasm and elongate, electron-dense nuclei with, prominent, distinguishing nuclear invaginations. These cells span the entire length of the taste bud from the basal lamina to the taste pore (Figure 6). The apical end of a type I cell terminates in several long and slender microvilli (Figure 7). Perhaps the most distinguishing feature of type I cells involves the numerous dense-cored granules (100–400 nm in diameter) located in the apical cytoplasm (Figures 7 and 8). These dense-cored granules, which are conspicuously absent in type II cells, are present in the apical cytoplasm of type I cells in both mouse and rat foliate and circumvallate taste buds. This is not the case, however, with fungiform taste buds. Several investigators have reported that the apical cytoplasm of type I cells in fungiform taste buds lacks the dense-cored granules (Farbman, A. 1965a; Murray, R., 1973; Kinnamon, J. *et al.*, 1988; Royer, S. and Kinnamon, J.,

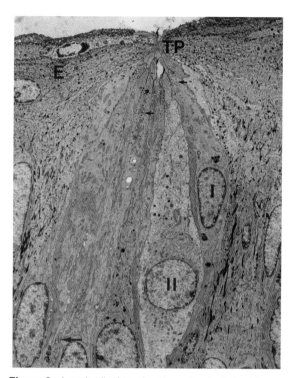

Figure 6 Longitudinal section through a taste bud from a rat circumvallate papilla. A type I cell (I) or dark cell contains electron-dense cytoplasm and a slender nucleus. Dense-cored granules (arrows) are present in the apical cytoplasm. A type II (II) or light cell has a large, circular nucleus and an electron-lucent cytoplasm. Taste pore (TP); nerve process (arrow); epithelium (E). ×4500. Adapted from Yang, R., Crowley, H., Rock, M. and Kinnamon, J. 2000a. Taste cells with synapses in rat circumvallate papillae display SNAP-25-like immunoreactivity. J. Comp. Neurol. 424, 205–215.

1991; Kinnamon, J. *et al.*, 1993; Reutter, K. and Witt, M., 1993; Miller, I., 1995). Other investigators have observed dense-cored granules in the type I cells of rat fungiform taste buds, but those granules were rod-shaped, only moderately electron-dense, and few in number. This contrasts with the circular, electron-dense apical vesicles observed in type I cells of circumvallate taste buds (Takeda, M. and Hoshino, T., 1975). These apical granules are believed to be secretory precursors of the dense substance in the taste pore (Murray, R. and Murray, A., 1967; Scalzi, H., 1967). The function of the dense pore substance is unknown, however, it may play an important role in facilitating and modifying the response of taste receptor cells to sapid stimuli (Murray, R. and Murray, A., 1970). Type I cells have irregular cytoplasmic processes that surround type II or type III taste cells. Thus, type I cells may function like

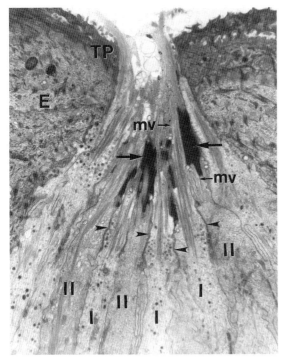

Figure 7 Longitudinal section through the taste pore region from a rat circumvallate taste bud. type I cells (I) are characterized by the presence of apical dense-cored granules (arrowheads) and several long, slender microvilli, whereas the apical microvilli of the type II cells (II) are brush-like, i.e., short and uniform in length. The most distinguishing ultrastructural feature of the type III cell (III) is the presence of a single, large, blunt microvillus that extends onto taste pore. The taste cells are joined by tight junctions (arrowheads). Microvilli (MV); pore substance (arrows); taste pore (TP); nongustatory epithelium (E). ×21 000.

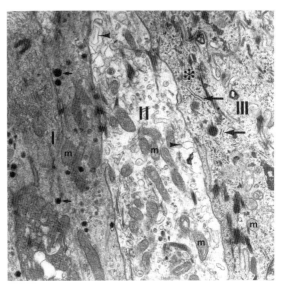

Figure 8 Longitudinal section showing the apical cytoplasm of three taste cells (I, II, and III) from a mouse taste bud. The electron-dense cytoplasm of the type I cell (I) contains numerous dense-cored granules (small arrows). The cytoplasm of the type I cell is rich in filaments and contains mitochondria and free ribosomes. The cytoplasm of the type II cell (II) is electron-lucent with numerous mitochondria (m) and swollen cisternae of agranular endoplasmic reticulum (arrowheads). Type III cells (III) exhibit electron density intermediate between that of type I (I) and type II (II) cells. The cytoplasm of type III cells contains numerous free ribosomes and rough endoplasmic reticulum. ×35 000.

glial cells in terms of isolation and nutrition (Reutter, K. and Witt, M., 1993; Pumplin, D. *et al.*, 1997). Although type I cells possess apical microvilli and are in direct contact with nerve processes, no synaptic contacts are believed to be present between type I cells and nerve processes (Farbman, A., 1965a; Murray, R., 1973; Reutter, K. and Witt, M., 1993). Although early studies suggested that type I cells might function as sensory receptors, it is generally accepted that they are restricted to a supportive or sustentacular role (Murray, R. and Murray, A., 1971; Murray, R., 1986; Lindemann, B., 1996; Lawton, D. *et al.*, 2000; Bartel, D. *et al.*, 2006).

The dark appearance of type I cell cytoplasm (Figure 8) results from an electron-dense ground substance, close packing of free ribosomes and poly-ribosomes, scattered cisternae of rough endoplasmic reticulum, and bundles of filamentous material surrounding the nucleus. The cytoplasm of type I cells also contains numerous mitochondria with densely staining matrices and lamellar cristae. These mitochondria are rod-shaped or elongate and randomly distributed throughout the cytoplasm. The Golgi bodies of type I cells typically consist of stacks of flattened cisternae, which are expanded only at their edges (Kinnamon, J. *et al.*, 1988).

The nuclei of type I cells are variable in profile. Compared with type II taste cells, the nuclei of type I cells are small, irregular, and electron-dense (Figure 6) (Kinnamon, J. *et al.*, 1985). The nuclei often contain one or more poorly defined nucleoli or aggregations of heterochromatin adhering to the inner leaflet of nuclear membrane (Kinnamon, J. *et al.*, 1988). The invaginations of type I cell nuclei are smaller than the deep prominent invaginations of type III cell nuclei.

4.05.4.3.2 Type II cells

The type II cell, initially termed light cell, constitutes approximately 15–30% of the cells in a taste bud

(Farbman, A., 1965a; Murray, R., 1973; Kinnamon, J. *et al.*, 1985; Delay, R. *et al.*, 1986) (Figures 6 and 8). Most type II cells extend the entire length of the taste bud from the basal lamina to the taste pore. A small subset of these cells, however, are located in the upper third of the taste bud and possess no apparent basal processes. These cells have a triangular profile in longitudinal sections. A typical type II cell possesses an electron-lucent cytoplasm and large circular or ovoid nuclei (Farbman, A., 1965a; Murray, R., 1973; Kinnamon, J. *et al.*, 1985; Delay, R. *et al.*, 1986; Kinnamon, J. *et al.*, 1988; Royer, S. and Kinnamon, J., 1988; Royer, S. and Kinnamon, J., 1991; Kinnamon, J. *et al.*, 1993; Royer, S. and Kinnamon, J., 1994; Pumplin, D. *et al.*, 1997). In contrast with the type I cell, the apical portion of a type II cell is wider than the type I cell and lacks the apical dense granules characteristic of the type I cells. The apical microvilli of a type II cell are numerous, short, and uniform in length, giving a brush-like appearance (Figure 7). Type II cells lack classical chemical synapses with nerve processes (Farbman, A., 1965a; Murray, R., 1973; Yee, C. *et al.*, 2001; Yang, R. *et al.*, 2004). However, other type II cell–nerve process contacts, such as SSCs of smooth endoplasmic reticulum and atypical mitochondria have been found at close appositions between taste cells and nerve processes (Farbman, A. 1965a; Fujimoto, S. and Murray, R., 1970; Murray, R., 1971; Takeda, M., 1976; Ide, C. and Munger, B., 1980; Royer, S. and. Kinnamon, J., 1988). Results from our laboratory suggest that SSCs are found exclusively at close appositions between type II cells and nerve processes in rat circumvallate taste buds (Clapp, T. *et al.*, 2004).

The cytoplasm of a type II cell contains an abundance of mitochondria, vesicles, and vacuoles (Farbman, A., 1965a) (Figure 6). The supranuclear region of a type II cell is rich in microchrondia, and more apically, the cytoplasm is replete with swollen vesicles of agranular endoplasmic reticulum. The cytoplasm also contains rough endoplasmic reticulum, free ribosmes, and intermediate filaments.

The nuclei of type II cells are large with a round-to-ovoid profiles. These nuclei are generally electron-lucent with a homogeneous nuclear matrix containing one to a few nucleoli (Kinnamon, J. *et al.*, 1985; Delay, R. *et al.*, 1986; Kinnamon, J. *et al.*, 1988; Reutter, K. and Witt, M., 1993) (Figure 6). In contrast with type I cell nuclei, the nuclei of type II cells generally lack invaginations and have fewer aggregations of heterochromatin and electron-dense material.

4.05.4.3.3　Type III cells

Type III cells (Figure 8) were first described in the foliate papillae of rabbits (Murray, R., 1973), but this cell type is present in other species as well, including the mouse (Takeda, M., 1976; 1977; Takeda, M. and Kitao, K., 1980; Takeda, M. *et al.*, 1981), rat (Takeda, M. and Hoshino, T., 1975; Yee, C. *et al.*, 2001; Yang, R. *et al.*, 2004), monkey (Farbman, A. *et al.*, 1985), and guinea-pig (Yoshie, S. *et al.*, 1990). Type III cells represent of the smallest proportion (5–15%) of taste cells in taste buds (Delay, R. *et al.*, 1986; Reutter, K. and Witt, M., 1993; Kinnamon, 1997). Type III cells are slender cells with a moderately electron-lucent cytoplasm. These cells span the entire length of taste buds from basal lamina to taste pore with a cytoplasm containing numerous ribosomes and cisternae of granular endoplasmic reticulum (Takeda, M. and Hoshino, T., 1975; Kinnamon, J. *et al.*, 1985; 1988; Reutter, K. and Witt, M., 1993; Kinnamon, J. 1987). The nuclei are intermediate in appearance between the nuclei of type I and type II cells (Takeda, M. and Hoshino, T., 1975; Yang, R. *et al.*, 2004; 2007). Type III cells in the mouse and rat taste buds lack the large dense-cored vesicles seen in the nuclear region of the cytoplasm of type III cells in rabbit foliate taste buds. The primary distinguishing features of type III cells in mouse and rat taste buds are as follows:

1. an elongate nucleus with large prominent invaginations;
2. a single, large, blunt microvillus that extends into the taste pore;
3. synapses onto nerve processes (Takeda, M. and Hoshino, T., 1975; Takeda, M., 1976; Royer, S. and Kinnamon, J., 1991); and
4. dense-cored vesicles (90–120 nm in diameter) and numerous clear vesicles (40–70 nm in diameter) are present at presynaptic sites (Kinnamon J. *et al.*, 1985; 1988; Yang, R. *et al.*, 2000a; 2004; 2007) (Figures 7, 9, and 10).

4.05.4.3.4　Type IV cells (basal cells)

Type IV cells lie in the basolateral region of taste buds. These cells are small, irregular, and similar to peripheral cells in morphology. but do not exend apical processes to the taste pore (Farbman, A., 1965a; 1965b; Reutter, K. and Witt, M., 1993). The most distinguishing feature of basal cells is the presence of intermediate filaments attached to the nuclear envelope (Delay, R. *et al.*, 1986). These cells

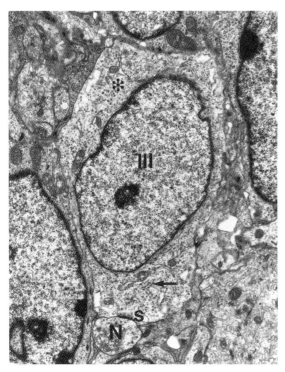

Figure 9 A low-magnification micrograph showing a mouse circumvallate type III taste cell (III) with a synapse (S) onto a nerve process. The cytoplasm of the taste cell contains numerous polyribosomes ribosomes (asterisk) and rough endoplasmic reticulum (arrow). The nucleus contains a small invagination and a prominent nucleolus. A group of small, clear synaptic vesicles are aggregated at the active zone of the synapse. The postsynaptic element, a nerve process (N) contains several mitochondria. ×15 000. Adapted from Kinnamon, J., Taylor, B., Delay, R. and Roper, S. 1985. Ultrastructure of mouse vallate taste buds. I. Taste cells and their associated synapses. J. Comp. Neurol. 235, 48–60.

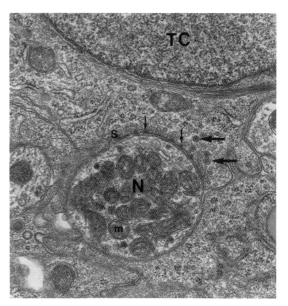

Figure 10 High magnification micrograph showing a macular synapse (S) from a mouse circumvallate taste cell (TC) onto a nerve process (N). Several small clear synaptic vesicles (40–70 nm) and a few large dense-cored vesicles (90–120 nm) (large arrows) are present at the presynaptic zone. Presynaptic dense projections (small arrows) can be seen at the synapse. The nerve process is replete with mitochondria (m), dense-cored vesicles and small clear vesicles. ×38 000. Adapted from Kinnamon, J., Sherman, T. and Roper, S. 1988. Ultrastructure of mouse vallate taste buds: III. Patterns of synaptic connectivity. J. Comp. Neurol. 270, 48–60.

are considered to be the precursor cells for other taste cell types in taste buds (Beidler, L. and Smallman, R., 1965; Conger, A. and Wells, M., 1969, Farbman, A., 1980, Stone *et al.*, 2002).

4.05.4.4 Taste Cell Synapses

De Lorenzo A. (1960) suggested that nerve fibers are in continual state of flux, forming synapses with new taste cells and retracting synapses from dying taste cells. Ultrastructural correlates of synapses in taste buds were first observed five decades ago (De Lorenzo, A. 1960; 1963; Gray, E. and Watkins, K., 1965; Hirata, Y., 1966), although Trujillo-Cenóz (1961) described contacts between taste cells and nerve fibers in catfish barbels. The presence or absence of synapses often has been used as a criterion

for sensory function in taste cells (Hirata, Y., 1966; Uga, S. and Hama, K., 1967; Murray, R., 1969; 1973; Takeda, M. and Hoshino, T., 1975; Takeda, M., 1976; Ide, C. and Munger, B., 1980; Zahm, D. and Munger, B., 1983a; 1983b; Kinnamon, J. *et al.*, 1985; 1988; 1993). Most investigators now consider the type III cell to be the chemoreceptor cell. In the mouse approximately 20% of the taste cells have synapses onto nerve processes in circumvallate taste buds (Kinnamon, J. *et al.*, 1985). Synapses in fungiform taste buds are fewer, but contain more vesicles than synapses in either circumvallate or foliate taste buds (Kinnamon, J. *et al.*, 1993). Virtually, all synapses in mammalian taste buds are from taste cells onto nerve fibers (Murray, R. and Murray, A., 1967; Scalzi, H., 1967; Beidler, L., 1969; Murray, R., 1969; Takeda, M. and Hoshino, T., 1975; Takeda, M., 1976; Miller, R. and Chaudhry, A., 1976; Takeda, M. *et al.*, 1981; Kinnamon, J. *et al.*, 1985; 1988; Royer, S. and Kinnamon, J., 1988; Kinnamon, J. *et al.*, 1993; Yang, R. *et al.*, 2000a; Yee, C. *et al.*, 2001; Yang, R. *et al.*, 2004). Taste cell synapses are generally located near the nuclear region of the presynaptic taste cell (Figure 9). The synaptic structure varies from macular to finger-like (Kinnamon, J. *et al.*, 1985; Yang, R. *et al.*, 2000a; 2004). The macular synapses are punctate and usually flat (Figure 10). A typical finger-like synapse, however, consists of a process from a postsynaptic nerve fiber protruding into an invagination of the presynaptic taste cell (Kinnamon, J. *et al.*, 1985). Synapses between intragemmal nerve processes are commonly observed during taste bud development (Kinnamon, J. *et al.*, 2005), but only rarely in taste buds of adult animals (however, see Graziadei, P. 1969; Royer, S. and Kinnamon, J., 1988; 1994). The function of neuroneuronal synapses in taste buds is not well understood.

4.05.4.5   Putative Nonsynaptic Connections

It is generally accepted that type II cells are important in the transduction of bitter, sweet, and umami taste, yet these cells do not form synapses onto nerve fibers. Other specialized contacts between taste cells and nerve processes have been observed that may be part of the transduction pathway, including SSCs of smooth endoplasmic reticulum and atypical mitochondria (Figures 11–13).

4.05.4.5.1   *Subsurface cisternae of smooth endoplasmic reticulum*

SSCs are narrow sacs of agranular endoplasmic reticulum that lie adjacent to the cytoplasmic leaflet of taste cell membranes at sites of close apposition with nerve processes (Figure 11) (Royer, S. and Kinnamon, J., 1988). The outer membranes of the cisternae are separated from the inner leaflet of the taste cell membrane by a narrow gap averaging 15 nm in width. The close apposition between the cisternae and cell membrane may extend for a distance up to 2–3 μm (Figure 11). SSCs have been reported in rat fungiform taste buds (Farbman, A., 1965a), rabbit foliate taste buds (Fujimoto, S. and Murray, R., 1970; Murray, R., 1971), mouse vallate and foliate (Takeda, M., 1976; Royer, S. and Kinnamon, J., 1988), and monkey laryngeal taste buds (Ide, C. and Munger, B., 1980). SSCs have been found at the close appositions between the plasma membrane of type II taste cells and nerve processes (Clapp, T. *et al.*, 2004). Although the function of SSCs is unclear, we speculate that the type II cells may communicate to the nerve processes via SSCs (Romanov, R. *et al.*, 2007).

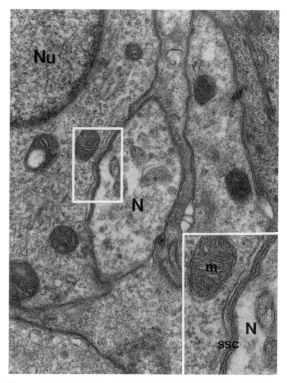

Figure 11   Electron micrograph showing a subsurface cisterna in close apposition to the junction of a taste cell (TC) with a nerve process (N). ×40 000. Inset: High magnification of the subsurface cisterna as it abuts the taste cell membrane. ×90 000. Adapted from Royer, S. and Kinnamon, J. 1988. Ultrastructure of mouse foliate taste buds: synaptic and nonsynaptic interactions between taste cells and nerve fibers. J. Comp. Neurol. 270, 11–24.

4.05.4.5.2   *Atypical mitochondria*

Most mitochondria in taste cells are elongate, slender structures with lamellar cristae rarely exceeding 0.3 μm in diameter (Royer, S. and Kinnamon, J., 1988) (Figure 12). At sites of close apposition between taste cells and nerve fibers, we occasionally observe unusually large, atypical microchondria (Figure 12). These atypical mitochondria are found only at the close appositions with nerve processes. The atypical mitochondria that we have observed have diameters two to three times larger and do not possess the lamellar or baffle cristae of conventional mitochondria. The cristae of the atypical mitochondria exhibit a twisted-energized or swollen twisted-energized configuration (Figure 12) (Korman, E. *et al.*, 1970; Williams, C. *et al.* 1970), presenting the impression of electron-dense sacs or tubules within the mitochondrion. The outer membranes of the mitochondria are separated from the taste cell membrane by a gap of approximately 20 nm. In favorable

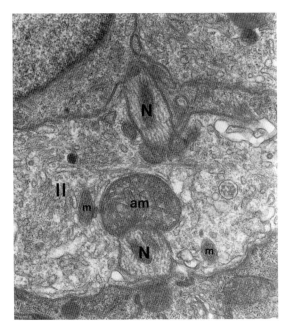

Figure 12 An atypical mitochondrion (AM) between a type II cell (II) and two nerve processes (N). The AM has a diameter two to three times larger than conventional mitochondria. The cristae of the atypical mitochondria exhibit a twisted-energized or swollen twisted-energized configuration. ×30 000. Adapted from Royer, S. and Kinnamon, J. 1988. Ultrastructure of mouse foliate taste buds: synaptic and nonsynaptic interactions between taste cells and nerve fibers. J. Comp. Neurol. 270, 11–24.

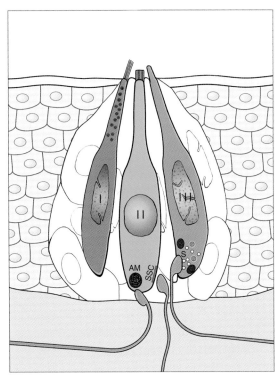

Figure 13 Diagram of a rat circumvallate taste bud showing a type I cell (I), a type II cell (II), and a type III cell (III). The type I cell has an electron-dense cytoplasm and a slender, elongate nucleus with invaginations and clumpy heterochromatin. The apical region of the type I cell contains dense granules and several long, slender microvilli. The type II cell is spindle-shaped with an electron-lucent cytoplasm. The nucleus of a type II cell is round to ovoid with a homogeneous nuclear matrix containing little heterochromatin. Subsurface cisternae (SSC) of smooth endoplasmic reticulum or atypical mitochondria (AM) are sometimes present at junctions with nerve processes. The apical cytoplasm of a type I cell often contains swollen cisternae of smooth endoplasmic reticulum. Several short, apical microvilli of uniform length give a brush-like appearance. The type III cell is slender with cytoplasm that is intermediate in electron density between a type I and type II cell. The slender nucleus contains patchy heterochromatin and pronounced invaginations. A single large, blunt, apical microvillus extends into the taste pore. Type III cells form synapses onto nerve processes. The presynaptic region of the type III cell contains numerous small, clear vesicles and a few large, dense-cored vesicles.

profiles, an electron-dense, wispy material can be seen connecting the mitochondrion to the inner leaflet of the taste cell plasma membrane. Because of their large size, the atypical mitochondria at taste cell–nerve fiber contacts would seem to be significant energy reservoirs and as such would be ready sources of ATP (Royer, S. and Kinnamon, J., 1988; Romanov, R. *et al.*, 2007). This is supported by recent results which demonstrate the importance of ATP as a neurotransmitter in taste buds (Finger, T., 2005).

4.05.5 Concluding Remarks

There is general agreement concerning the ultrastructure of circumvallate taste buds of rodents (Figure 13). The three most significant cell types are type I, type II, and type III cells. Features of type I (dark) cells: (1) slender cells with electron-dense cytoplasm, (2) elongate, electron-dense nuclei with prominent invaginations, (3) dense granules in the apical cytoplasm, and (4) several, long, slender microvilli projecting into the taste pore. Type I (dark) cells are believed to perform a glial function and to isolate other taste cells, but are not thought to be receptor cells. Features of type II (light) cells: (1) these cells are spindle-shaped with an expanded nuclear region and an electron-lucent cytoplasm, (2) type II cell have round-to-ovoid nuclei with a homogeneous, electron-lucent nuclear matrix,

(3) there are several short, apical microvilli of the same length, giving a brush-like appearance, and (4) recent electrophysiological and molecular biological studies indicate that type II cells are chemoreceptors. Although type II cells lack classical chemical synapses, they do possess specialized junctions with nerve processes involving SSCs of smooth endoplasmic reticulum or large, atypical mitochondria. Features of type III cells: (1) type III cells are slender and possess a moderately electron-lucent cytoplasm, (2) the nuclei are slender and contain prominent nuclear invaginations, (3) there is a single, blunt apical microvillus, and (4) type III cells have classical chemical synapses onto nerve processes with small, clear vesicles and large, dense-cored vesicles.

Our understanding of taste bud structure and function has increased greatly over the past 20 years. Knowledge of the ultrastructural features of a taste bud helps to understand the structural basis for taste function, but is not an explanation of taste cell function. Previously, investigators have made well-meaning attempts to infer taste cell function based solely on structural features. Several new approaches now make it possible to elucidate the function of taste cells and correlate this information with the structural features of taste cells. Immunohistochemistry and immunoelectron microscopy using markers for taste cells types and transduction pathways have provided significant indirect data that have enhanced our understanding of taste cell function. The use of the patch clamp recording technique has made it possible to characterize the electrophysiological properties of individual taste cells. Molecular and genetic approaches have made it possible to learn more about taste cell function by using transgenic and knockout mice. Recent studies have led to the identification of two families of G protein-coupled receptors (GPCRs) in taste cells: the T1Rs (T1R1, T1R2, and T1R3) (Hoon, M. *et al.*, 1999; Nelson, G. *et al.*, 2001; Sainz, E. *et al.*, 2001) and T2Rs (Adler, E. *et al.*, 2000). T1R1 is primarily expressed in the receptor cells of fungiform taste buds, and T1R2 is expressed in the receptor cells of circumvallate and foliate taste buds. T1R3, however, is broadly expressed in all lingual taste buds (Hoon, M. *et al.*, 1999; Sainz, E. *et al.*, 2001). T1R1 and T1R2 are expressed in separate taste cells and localize in the taste pore where the microvilli of the taste receptor cells exposed to taste stimuli. T1R1 and T1R3 are believed to function in the transduction of amino acid (umami) stimuli (Nelson, G. *et al.*, 2002; Li, X. *et al.*, 2002), whereas T1R2 and T1R3 transduce sweet

stimuli (Nelson, G. *et al.*, 2001; Li, X. *et al.*, 2002). The three T1Rs taste receptors are rarely coexpressed with α-gustducin in taste cells in circumvallate and foliate taste buds (Kim, M. *et al.*, 2003; Stone, L. *et al.*, 2007). T2Rs are exclusively expressed in taste receptor cells that contain gustducin, a G protein α subunit, implying that T2Rs are associated with bitter transduction (Adler, E. *et al.*, 2000; Chandrashekar, J. *et al.*, 2000; Matsunami, H. *et al.*, 2000). T1Rs and T2Rs are almost completely coexpress with TRMP5 and PLCβ2. Sweet, amino acid, and bitter reception are abolished inPLCβ2 knockout mice (Zhang, Y. *et al.*, 2003). T1Rs and T2Rs are believed to present in different subsets of type II taste cells. Most or all T2Rs are expressed in a particular subset of taste receptor cells that do not express T1Rs (Chandrashekar, J. *et al.*, 2000). Using combinations of the techniques described above it is now possible to truly begin to correlate structure with function in the taste cell and taste bud.

References

Adler, E., Hoon, M., Mueller, K., Chandrashekar, J., Ryba, N., and Zuker, C. 2000. A novel family of mammalian taste receptors. Cell 100, 693–702.

Akisaka, T. 1980. Morphological and functional aspect of the rat taste bud by means of electron microscope. J. Osaka Dent. Univ. 14, 1–28.

Akisaka, T. and Oda, M. 1977. The fine structural localization of adenosine triphosphatase activity on the taste bud in the fungiform papillae of the rat. Arch. Histol. Jpn. 41, 87–98.

Akisaka, T. and Oda, M. 1978. Taste buds in the vallate papillae of the rat studied with freeze-fracture preparation. Arch. Histol. Jpn. 41, 87–98.

Azzali, G. 1997. Ultrastructure and immunocytochemistry of gustatory cells in man. Ann. Anat. 179, 37–44.

Baradi, A. 1965. Intragemmal spaces in taste buds. Z. Zellforsch. 65, 313–318.

Bartel, D., Sullivan, S., Lavoie, E., Sevigny, J., and Finger, T. 2006. Nucleoside triphosphate diphosphohydrolase-2 is the ecto-ATPase of type I cells in taste buds. J. Comp. Neurol. 497, 1–12.

Beidler, L. and Smallman, R. 1965. Renewal of cells within taste buds. J. Cell. Biol. 27, 263–272.

Beidler, L. 1969. Innervation of Rat Fungiform Papilla. In: *Olfaction and Taste III* (ed. C. Pfaffmann), pp. 354–369. Rockefeller University Press.

Boudriot, F. and Reutter, K. 2001. Ultrastructure of the taste buds in the blind cave fish *Astyanax jordani* ("Anoptichthys") and the sighted river fish *Astyanax mexicanus* (Teleostei, Characidae). J. Comp. Neurol. 434, 428–444.

Bradley, R. 1971. Tongue Topography. In: *Handbook of Sensory Physiology* (ed. L. Beidler), pp. 1–30. Springer-Verlag.

Chandrashekar, J., Hoon, M., Ryba, N., and Zuker, C. 2006. The receptors and cells for mammalian taste. Nature 444, 288–294.

Chandrashekar, J., Mueller, K., Hoon, M., Adler, E., Feng, L., Guo, W., Zuker, C., and Ryba, N. 2000. T2Rs function as bitter taste receptors. Cell 100, 703–711.

Clapp, T., Yang, R., Stoick, C., Kinnamon, S., and Kinnamon, J. 2004. Morphologic characterization of rat taste receptor cells that express components of the phospholipase C signaling pathway. J. Comp. Neurol. 468, 311–321.

Conger, A. and Wells, M. 1969. Radiation and aging effect on taste structure and function. Radiat. Res. 37, 31–49.

Cottler-Fox, M., Arvidson, K., and Friberg, U. 1980. On the occurrence of dark and light cells in the fungiform taste buds. J. Dent. Res. 59, 191.

Cottler-Fox, M., Arvidson, K., Hammarlund, E., and Friberg, U. 1987. Fixation and occurrence of dark and light cells in taste buds of fungiform papillae. Scand. J. Dent. Res. 95, 417–427.

Cummings, T., Delay, R., and Roper, S. 1987. Ultrastructure of apical specializations of taste cells in the mudpuppy, *Necturus maculosus*. J. Comp. Neurol. 261, 604–615.

Delay, R. and Roper, S. 1988. Ultrastructure of taste cells and synapses in the mudpuppy *Necturus maculosus*. J. Comp. Neurol. 277, 268–280.

Delay, R., Kinnamon, J., and Roper, S. 1986. Ultrastructure of mouse vallate taste buds: II. Cell types and cell lineage. J. Comp. Neurol. 253, 242–252.

Delay, R., Taylor, R., and Roper, S. 1993. Merkel-like basal cells in *Necturus* taste buds contain serotonin. J. Comp. Neurol. 335, 606–613.

De Lorenzo, A. 1958. Electron microscopic observations on the taste buds of the rabbit. J. Biophys. Biochem. 4, 143–148.

De Lorenzo, A. 1960. Electron microscopy of the olfactory and gustatory pathways. Ann. Otol. Laryngol. 69, 410–420.

De Lorenzo, A. 1963. Studies on the Ultrastructure and Histophysiology of Cell Membranes, Nerve Fibers and Synaptic Junctions in Chemoreceptors. In: Olfaction and Taste (ed. Y. Zotterman), pp. 5–17. Pergamon.

Endo, Y. 1988. Exocytotic release of neurotransmitter substances from nerve endings in the taste buds of rat circumvallate papillae. Arch. Histol. Cytol. 51, 489–494.

Engstrom, H. and Rytzner, C. 1956a. The fine structure of taste buds and taste fibres. Ann. Otol. Rhinol. Laryngol. 65, 361–375.

Engstrom, H. and Rytzner, C. 1956b. The structure of taste buds. Acta Otolaryngol. 46, 361–367.

Esses, B., Jafek, B., Hommel, D., and Eller, P. 1988. Histological and ultrastructural changes of the murine taste bud following ionizing irradiation. Ear Nose Throat J. 67, 478–484.

Farbman, A. 1965a. Fine structure of the taste bud. J. Ultrastruct. Res. 12, 328–350.

Farbman, A. 1965b. Electron microscopy study of the developing taste bud in rat fungiform papilla. Dev. Biol. 11, 110–135.

Farbman, A. 1967. A particle–membrane complex in developing rat taste buds. J. Ultrastruct. Res. 19, 514–521.

Farbman, A. 1980. Renewal of taste bud cells in rat circumvallate papillae. Cell Tissue Kinet. 13, 349–357.

Farbman, A. and Hellekant, G. 1978. Quantitative analysis of the fiber population in rat chorda tympani nerves and fungiform papillae. Am. J. Anat. 153, 509–522.

Farbman, A. and Yonkers, J. 1971. Fine structure of the taste bud in the mud puppy, *Necturus maculosus*. Am. J. Anat. 131, 353–370.

Farbman, A., Hellekant, G., and Nelson, A. 1985. Structure of taste buds in foliate papillae of the rhesus monkey, *Macaca mulatta*. Am. J. Anat. 172, 41–56.

Fährmann, W. 1967. Licht- und elektronenmikroskopiche Befunde an den Geschmacksknospen der Axolotlzunge. Z. Mikr. Anat. Forsch. 77, 117–152.

Fährmann, W. and Schuchardt, E. 1967. Light and electron microscope observations on the taste buds of the axolotl tongue (German). Experientia 23, 657–659.

Finger, T. 1986. Peptide immunohistochemistry demonstrates multiple classes of perigemmal nerve fibers in the circumvallate papilla of the rat. Chem Senses 11, 135–144.

Finger, T. 1997. Evolution of taste and solitary chemoreceptor cell systems. Brain Behav. Evol. 50, 234–243.

Finger, T. 2005. Cell types and lineages in taste buds. Chem. Senses 30 (Suppl. 1), i54–i55.

Finger, T. and Simon, S. 2000. Cell Biology of Taste Epithelium. In: The Neurobiology of Taste and Smell (eds. T. Finger, W. Silver, and D. Restrepo), pp. 287–314. Wiley-Liss.

Finger, T., Womble, M., Kinnamon, J., and Ueda, T. 1990. Synapsin I-like immunoreactivity in nerve fibers associated with lingual taste buds of the rat. J. Comp. Neurol. 292, 283–290.

Fujimoto, S. and Murray, R. 1970. Fine structure of degeneration and regeneration in denervated rabbit vallate taste buds. Anat. Rec. 168, 393–414.

Fujimoto, S. and Yamamoto, K. 1980. Electron microscopy of terminal buds on the barbels of the silurid fish, *Corydoras paleatus*. Anat. Rec. 197, 133–141.

Ganchrow, D., Ganchrow, J., and Goldstein, R. 1991. Ultrastructure of palatal taste buds in the perihatching chick. Am. J. Anat. 192, 69–78.

Ganchrow, D., Ganchrow, J., Gross-Isseroff, R., and Kinnamon, J. 1995. Taste bud cell generation in the perihatching chick. Chem. Senses 20, 19–28.

Ganchrow, D., Ganchrow, J., Romano, R., and Kinnamon, J. 1994. Ontogenesis and taste bud cell turnover in the chicken. I. Gemmal cell renewal in the hatchling. J. Comp. Neurol. 345, 105–114.

Ganchrow, D., Ganchrow, J., Royer, S., Dovidpor, S., and Kinnamon, J. 1998. Identified taste bud cell proliferation in the perihatching chick. Chem. Senses 23, 333–341.

Ganchrow, J., Ganchrow, D., Royer, S., and Kinnamon, J. 1993. Aspects of vertebrate gustatory phylogeny: morphology and turnover of chick taste bud cells. Microsc. Res. Tech. 26, 6–19.

Gray, E. and Watkins, K. 1965. Electron microscopy of taste buds of the rat. Z. Zellforsch 66, 583–595.

Graziadei, P. 1969. The Ultrastructure of Vertebrate Taste Buds. In: Olfaction and Taste III (ed. C. Pfaffmann), pp. 315–330. Rockefeller University Press.

Graziadei, P. and DeHan, R. 1971. The ultrastructure of frogs' taste organs. Acta Anat. (Basel) 180, 563–603.

Graziadei, P. and Graziadei, G. 1978. Observations on the ultrastructure of ganglion cells in the circumvallate papilla of rat and mouse. Acta Anat. (Basel) 100, 289–305.

Grover-Johnson, N. and Farbman, A. 1976. Fine structure of taste buds in the barbel of the catfish, *Ictalurus punctatus*. Cell Tissue Res. 169, 395–403.

Hamamichi, R., Asano-Miyoshi, M., and Emori, Y. 2006. Taste bud contains both short-lived and long-lived cell populations. Neuroscience 141, 2129–2138.

Hirata, Y. 1966. Fine structure of the terminal buds on the barbels of some fishes. Arch. Histol. Jpn. 26, 507–523.

Hoon, M., Adler, E., Lindemeier, J., Battey, J., Ryba, N., and Zuker, C. 1999. Putative mammalian taste receptors: a class of taste-specific GPCRs with distinct topographic selectivity. Cell 96, 541–551.

Huang, Y., Wu, Y., and Lu, K. 2003. Immunoelectron microscopic studies on protein gene product 9.5 and calcitonin gene-related peptide in vallate taste cells and related nerves in the guinea pig. Microsc. Res. Tech. 62, 383–395.

Ide, C. and Munger, B. 1980. The cytologic composition of primate laryngeal chemosensory corpuscles. Am. J. Anat. 158, 193–209.

Iriki, T. 1960. Electron micrscopic observation of the taste buds of the rabbit. Acta Med. Univ. Kagoshima 2, 78–94.

Iwayama, T. 1970. Changes in the cell population of taste buds during degeneration and regeneration of their sensory innervation. Z. Zellforsch. 110, 487–495.

Jahnke, K. and Baur, P. 1979. Freeze-fracture study of taste bud pores in the foliate papillae of the rabbit. Cell Tissue Res. 200, 245–256.

Jeppsson, P. 1969. Studies on the structure and innervation of taste buds. An experimental and clinical investigation. Acta Otolaryngol. Suppl. 259, 50–99.

Joyce, E. and Chapman, G. 1978. Fine structure of the nasal barbel of the channel catfish, *Ictalurus punctatus*. J. Morphol. 158, 109–154.

Kanazawa, H. 1993. Fine structure of the canine taste bud with special reference to gustatory cell functions. Arch. Histol. Cytol. 56, 533–548.

Kapoor, B., Evans, H., and Pevzner, R. 1975. The gustatory system in fish. Adv. Mar. Biol. 13, 53–108.

Kim, M., Kusakabe, Y., Miura, H., Shindo, Y., Ninomiya, Y., and Hino, A. 2003. Regional expression patterns of taste receptors and gustducin in the mouse tongue. Biochem. Biophys. Res. Commun. 312, 500–506.

Kinnamon, J. 1987. Organization and Innervation of Taste Buds. In: Neurobiology of Taste and Smell (eds. T. Finger and W. Silver), pp. 277–297. John Wiley & Sons.

Kinnamon, J., Dunlap, M., and Yang, R. 2005. Synaptic connections in developing and adult rat taste buds. Chem. Senses 30(Suppl I), 60–61.

Kinnamon, J., Henzler, D., and Royer, S. 1993. HVEM ultrastructural analysis of mouse fungiform taste buds, cell types, and associated synapses. Microsc. Res. Tech. 26, 142–156.

Kinnamon, J., Sherman, T., and Roper, S. 1988. Ultrastructure of mouse vallate taste buds: III. Patterns of synaptic connectivity. J. Comp. Neurol. 270, 48–60.

Kinnamon, J., Taylor, B., Delay, R., and Roper, S. 1985. Ultrastructure of mouse vallate taste buds. I. Taste cells and their associated synapses. J. Comp. Neurol. 235, 48–60.

Kobayashi, T. and Tomita, H. 1986. Electron microscopic observation of vallate taste buds of zinc-deficient rats with taste disturbance. Auris Nasus Larynx 13(Suppl. 1), S25–S31.

Korman, E., Addink, A., Wakabayashi, T., and Green, D. 1970. A unified model of mitochondrial morphology. Bioenergetics 1, 9–32.

Kudoh, M. 1988. Ultrastructural and histochemical localization of acetylcholinesterase in the taste bud of mouse vallate papilla. Fukushima J. Med. Sci. 34, 27–44.

Lawton, D., Furness, D., Lindemann, B., and Hackney, C. 2000. Localization of the glutamate-aspartate transporter, GLAST, in rat taste buds. Eur. J. Neurosci. 12, 3163–3171.

Leydig, F. 1851. Ueber die aussere Haut einiger Susswasserfische. Z. Wiss. Zool. 3, 1–12.

Li, X., Staszewski, L., Xu, H., Durick, K., Zoller, M., and Adler, E. 2002. Human receptors for sweet and umami taste. Proc. Natl. Acad. Sci. U. S. A. 99, 4692–4696.

Lindemann, B. 1996. Chemoreception: tasting the sweet and bitter. Curr. Biol. 6, 1234–1237.

Lindemann, B. 2001. Receptors and transduction in taste. Nature 413, 219–225.

Loven, C. 1868. Beitrage zur Kenntnis vom Bau der Geschmackswarzchen der Zunge. Arch. Mikrosk. Anat. 4, 96–110.

Matsunami, H., Montmayeur, J., and Buck, L. 2000. A family of candidate taste receptors in human and mouse. Nature 404, 601–604.

Mattern, C. and Paran, N. 1974. Evidence of a contractile mechanism in the taste bud of the mouse fungiform papilla. Exp. Neurol. 44, 461–469.

Mattern, C., Daniel, W., and Henkin, R. 1970. The ultrastructure of the human circumvallate papilla. I. Cilia of the papillary crypt. Anat. Rec. 167, 175–181.

Ma, H., Yang, R., Thomas, S., and Kinnamon, J. 2007. Qualitative and quantitative differences between taste buds of the rat and mouse. BMC. Neurosci. 8, 5.

Miller, I. 1995. Anatomy of the Peripheral Taste System. In: Handbook of Olfaction and Gustation (ed. R. Doty), pp. 521–547. Marcel Dekker.

Miller, R. and Chaudhry, A. 1976. An ultrastructural study on the development of vallate taste buds of the golden Syrian hamster. Acta Anat. (Basel) 95, 190–206.

Montavon, P., Hellekant, G., and Farbman, A. 1996. Immunohistochemical, electrophysiological, and electron microscopical study of rat fungiform taste buds after regeneration of chorda tympani through the non-gustatory lingual nerve. J. Comp. Neurol. 367, 491–502.

Murray, A. 1961. Two gustatory types in rabbit taste buds. Anat. Rec. 139, 331.

Murray, R. 1969. Cell Types in Rabbit Taste Buds. In: Olfaction and Taste, III (ed. C. Pfaffmann), pp. 331–344. Rockefeller University Press.

Murray, R. 1971. Ultrastructure of Taste Receptors. In: Handbook of Sensory Physiology, Vol. IV (ed. L. M. Beidler), pp. 31–50. Springer-Verlag.

Murray, R. 1973. The Ultrastructure of Taste Buds. In: The Ultrastructure of Sensory Organs (ed. I. Friedmann), pp. 1–81. North-Holland.

Murray, R. 1986. The mammalian taste bud type III cell: A critical analysis. J. Ultrastruct. Molec. Struct. Res. 95, 175–188.

Murray, R. 1993. Cellular relations in mouse circumvallate taste buds. Microsc. Res. Tech. 26, 209–224.

Murray, R. and Murray, A. 1960. The fine structure of the taste buds of rhesus and cynomalgus monkeys. Anat. Rec. 138, 211–219.

Murray, R. and Murray, A. 1966. Fine Structure of Rabbit Taste Buds. In: Proceedings of the 6th International Congress on Electron Microscopy Vol. 2, (ed. R. Ueda), pp. 484–486. Maruzen.

Murray, R. and Murray, A. 1967. Fine structure of taste buds of rabbit foliate papillae. J. Ultrastruct. Res. 19, 327–353.

Murray, R. and Murray, A. 1970. The Anatomy and Ultrastructure of Taste Endings. In: Taste and Smell in Vertebrates (eds. G. Wolstenhome and J. Knight), pp. 3–30. J. & A. Churchill.

Murray, R. and Murray, A. 1971. Relations and possible significance of taste bud cells. Contrib. Sens. Physiol. 5, 47–95.

Murray, R., Murray, A., and Fujimoto, S. 1969. Fine structure of gustatory cells in rabbit taste buds. J. Ultrastruct. Res. 27, 444–461.

Nada, O. and Hirata, K. 1975. The occurrence of the cell type containing a specific monoamine in the taste bud of the rabbit's foliate papilla. Histochemistry 43, 237–240.

Nagy, J., Goedert, M., Hunt, S., and Bond, A. 1982. The nature of the substance P-containing nerve fibres in taste papillae of the rat tongue. Neuroscience 7, 3137–3151.

Nelson, G., Chandrashekar, J., Hoon, M., Feng, L., Zhao, G., Ryba, N., and Zuker, C. 2002. An amino-acid taste receptor. Nature 416, 199–202.

Nelson, G., Hoon, M., Chandrashekar, J., Zhang, Y., Ryba, N., and Zuker, C. 2001. Mammalian sweet taste receptors. Cell 106, 381–390.

Nemetschek-Gansler, H. and Ferner, H. 1964. Uber die ultrastruktur der geschmacksknospen. Z. Zellforsch 63, 155–178.

Oakley, B. 1967. Altered temperature and taste responses from cross-regenerated sensory nerves in the rat's tongue. J. Physiol. 188, 353–371.

Ono, R. 1980. Fine structure and distribution of epidermal projections associated with taste buds on the oral papillae in some loricariid catfishes Siluroidei: Loricariidae. J. Morphol. 164, 139–159.

Ovalle, W. and Shinn, S. 1977. Surface morphology of taste buds in catfish barbels. Cell Tissue Res. 178, 375–384.

Paran, N., Mattern, C., and Henkin, R. 1975. Ultrastructure of the taste bud of the human fungiform papilla. Cell Tissue Res. 161, 1–10.

Popov, V., Ignat'ev, D., and Lindemann, B. 1999. Ultrastructure of taste receptor cells in active and hibernating ground squirrels. J. Electron. Microsc. (Tokyo) 48, 957–969.

Pumplin, D., Yu, C., and Smith, D. 1997. Light and dark cells of rat vallate taste buds are morphologically distinct cell types. J. Comp. Neurol. 378, 389–410.

Reutter, K. 1978. Taste organ in the bullhead Teleostei. Anat. Embryol. Cell. Biol. 55, 1–98.

Reutter, K. 1982. Taste Organ in the Barbel of the Bullhead. In: Chemoreception in Fishes (ed. T. Hara), pp. 77–91. Elsevier.

Reutter, K. 1987. Specialized receptor villi and basal cells within the taste bud of the European silurid fish, *Silurus glanis* Teleostei. Ann. N. Y. Acad. Sci. 510, 570–573.

Reutter, K. 1992. Structure of the Peripheral Gustatory Organ, Represented by the Siluroid Fish *Plotosus lineatus* Thunberg. In: *Fish Chemoreception* (ed. T. Hara), pp. 60–78. Chapman & Hall.

Reutter, K. and Witt, M. 1993. Morphology of Vertebrate Taste Organs and Their Nerve Supply. In: *Mechanisms of Taste Transduction* (eds. S. Simon and S. Roper), pp. 29–82. CRC Press.

Reutter, K. and Witt, M. 2004. Are there efferent synapses in fish taste buds? J. Neurocytol. 33, 647–656.

Romanov, R., Rogachevskaja, O., Bystrova, M., Jiang, P., Margolskee, R., and Kolesnikov, S. 2007. Afferent neurotransmission mediated by hemichannels in mammalian taste cells. EMBO J. 26, 657–656.

Roper, S. 1989. The cell biology of vertebrate taste receptors. Annu. Rev. Neurosci. 12, 329–353.

Roper, S. 2006. Cell communication in taste buds. Cell Mol. Life Sci. 63, 1494–1500.

Royer, S. and Kinnamon, J. 1988. Ultrastructure of mouse foliate taste buds: synaptic and nonsynaptic interactions between taste cells and nerve fibers. J. Comp. Neurol. 270, 11–24.

Royer, S. and Kinnamon, J. 1991. HVEM serial-section analysis of rabbit foliate taste buds: I. Type III cells and their synapses. J. Comp. Neurol. 306, 49–72.

Royer, S. and Kinnamon, J. 1994. Application of serial sectioning and three-dimensional reconstruction to the study of taste bud ultrastructure and organization. Microsc. Res. Tech. 29, 381–407.

Sainz, E., Korley, J., Battey, J., and Sullivan, S. 2001. Identification of a novel member of the T1R family of putative taste receptors. J. Neurochem. 77, 896–903.

Sakai, H., Kaidoh, T., Morino, S., and Inoue, T. 1999. Three-dimensional architecture of the keratin filaments in epithelial cells surrounding taste buds in the rat circumvallate papilla. Arch. Histol. Cytol. 62, 375–381.

Sangiacomo, C. 1970. Neurosecretory cell types in normal taste bud. Experientia 26, 289–290.

Sbarbati, A., Benati, D., Crescimanno, C., and Osculati, F. 1998a. Membrane-bounded intratubular bodies in rat taste buds. J. Neurocytol. 27, 157–161.

Sbarbati, A., Franceschini, F., Zancanaro, C., Cecchini, T., Ciaroni, S., and Osculati, F. 1988b. The fine morphology of the basal cell in the frog's taste organ. J. Submicrosc. Cytol. Pathol. 20, 73–79.

Sbarbati, A., Merigo, F., Benati, D., Bernardi, P., Tizzano, M., Fabene, P., Crescimanno, C., and Osculati, F. 2006. Axon-like processes in type III cells of taste organs. Anat. Rec. A Discov. Mol. Cell Evol. Biol. 288, 276–279.

Sbarbati, A., Zancanaro, C., Ferrara, P., Franceschini, F., Accordini, C., and Osculati, F. 1993. Freeze-fracture characterization of cell types at the surface of the taste organ of the frog, *Rana esculenta*. J. Neurocytol. 22, 118–127.

Scalzi, H. 1967. The cytoarchitecture of gustatory receptors from the rabbit foliate papillae. Z. Zellforsch. 80, 413–435.

Schwalbe, G. 1868. Ueber die Geschmacksorgane der Saugethiere und des Menschen. Arch. Mikrosk. Anat. Entwickl. 4, 154–189.

Sekerkova, G., Freeman, D., Mugnaini, E., and Bartles, J. 2005. Espin cytoskeletal proteins in the sensory cells of rodent taste buds. J. Neurocytol. 34, 171–182.

Seta, Y., Seta, C., and Barlow, L. A. 2003. Notch-associated gene expression in embryonic and adult taste papillae and taste buds suggests a role in taste cell lineage decisions. J. Comp. Neurol. 464, 49–61.

Stone, L. and Finger, T. 1994. Mosaic analysis of the embryonic origin of taste buds. Chem. Senses 19, 725–735.

Stone, L., Barrows, J., Finger, T., and Kinnamon, S. 2007. Expression of T1Rs and gustducin in palatal taste buds of mice. Chem Senses. 32, 255–262.

Stone, L., Finger, T., Tam, P., and Tan, S. 1995. Taste receptor cells arise from local epithelium, not neurogenic ectoderm. Proc. Natl. Acad. Sci. U. S. A. 92, 1916–1920.

Stone, L., Tan, S., Tam, P., and Finger, T. 2002. Analysis of cell lineage relationships in taste buds. J. Neurosci. 22, 4522–29.

Stornelli, M., Lossi, L., and Giannessi, E. 2000. Localization, morphology and ultrastructure of taste buds in the domestic duck (*Cairina moschata* domestica L.) oral cavity. Ital. J. Anat. Embryol. 105, 179–188.

Sweazey, R., Edwards, C., and Kapp, B. 1994. Fine structure of taste buds located on the lamb epiglottis. Anat. Rec. 238, 517–527.

Tabata, S., Wada, A., Kobayashi, T., Nishimura, S., Muguruma, M., and Iwamoto, H. 2003. Bovine circumvallate taste buds: taste cell structure and immunoreactivity to alpha-gustducin. Anat. Rec. A Discov. Mol. Cell. Evol. Biol. 271, 217–224.

Takeda, M. 1972. Fine structure of developing taste buds in human fetal circumvallate papillae. Kaibogaku Zasshi. 47, 325–337.

Takeda, M. 1976. An electron microscopic study on the innervation in the taste buds of the mouse circumvallate papillae. Arch. Histol. Jpn. 39, 257–269.

Takeda, M. 1977. Uptake of 5-hydroxytryptophan by gustatory cells in the mouse taste bud. Arch. Histol. Jpn. 40, 243–250.

Takeda, M. and Hoshino, T. 1975. Fine structure of taste buds in the rat. Arch. Histol. Jpn. 37, 395–413.

Takeda, M. and Kitao, K. 1980. Effect of monoamines on the taste buds in the mouse. Cell Tissue Res. 210, 71–78.

Takeda, M., Obara, N., and Suzuki, Y. 1988. Intermediate filaments in mouse taste bud cells. Arch. Histol. Cytol. 51, 99–108.

Takeda, M., Obara, N., and Suzuki, Y. 1989. Cytoskeleton in the apical region of mouse taste bud cells. Shika Kiso Igakkai Zasshi 31, 317–323.

Takeda, M., Shishido, Y., Kitao, K., and Suzuki, Y. 1981. Biogenic monoamines in developing taste buds of mouse circumvallate papillae. Arch. Histol. Jpn. 44, 485–495.

Takeda, M., Suzuki, Y., and Shishido, Y. 1985. Effects of colchicine on the ultrastructure of mouse taste buds. Cell Tissue Res. 242, 409–416.

Toyoshima, K. and Shimamura, A. 1982. Virus-like particles in the taste buds of the circumvallate papilla of the rat. J. Electron Microsc. (Tokyo) 31, 273–275.

Toyoshima, K. and Tandler, B. 1989. Dense-cored vesicles and unusual lamellar bodies in type III gustatory cells in taste buds of rabbit foliate papillae. Acta Anat. (Basel) 135, 365–369.

Toyoshima, K., Miyamoto, K., and Shimamura, A. 1987. Fine structure of taste buds in the tongue, palatal mucosa and gill arch of the axolotl, *Ambystoma mexicanum*. Okajimas Folia Anat. Jpn. 64, 99–109.

Toyoshima, K., Nada, O., and Shimamura, A. 1984. Fine structure of monoamine-containing basal cells in the taste buds on the barbels of three species of teleosts. Cell Tissue Res. 235, 479–484.

Trujillo-Cenóz, O. 1957. Electron microscope study of the rabbit gustatory bud. Z. Zellforsch 46, 272–280.

Trujillo-Cenóz, O. 1961. Electron microscope observations on chemo- and mechano-receptor cells of fishes. Z. Zellforsch. 34, 654–676.

Uga, S. 1966. The fine structure of gustatory receptors and their synapses in frog's tongue. Symp. Cell Chem. 16, 75–85.

Uga, S. 1969. A study on the cytoarchitect of taste buds of rat cirumvallate papillae. Arch. Histol. Jpn. 31, 59–72.

Uga, S. and Hama, K. 1967. Electron microscopic studies on the region of the taste organ of carps and frogs. J. Electron Microsc. 16, 269–276.

Welsch, U. and Storch, V. 1969. Die Feinstruktur der Geschmacksknospen von Welsen [*Clarias batrachus* L. und *Kryptopterus bicirrhis* Cuvier et Valenciennes]. Z. Zellforsch. 100, 552–559.

Whiteside, B. 1927. Nerve overlap in the gustatory apparatus of the rat. J. Comp. Neurol. 44, 363–377.

Whitehead, M., Beeman, C., and Kinsella, B. 1985. Distribution of taste and general sensory nerve ending in fungiform papillae of the hamster. Am. J. Anat. 173, 185–202.

Williams, C., Vail, W., Harris, R., Caldwell, M., Green, D., and Valdivia, E. 1970. Conformational basis of energy transduction in membrane systems. VII. Configurational changes of mitochondria *in situ* and *in vitro*. Bioenergetics 1, 147–180.

Wilson, J. and Edin, M. 1905. The structure and function of the taste-buds in the larynx. Brain 28, 339–351.

Wistuba, J., Opolka, A., and Clemen, G. 1999. The epithelium of the tongue of *Ambystoma mexicanum*. Ultrastructural and histochemical aspects. Ann. Anat. 181, 523–536.

Witt, M. 1993. Ultrastructure of the taste disc in the red-bellied toad *Bombina orientalis* (Discoglossidae, Salientia). Cell Tissue Res. 272, 59–70.

Witt, M. and Reutter, K. 1990. Electron microscopic demonstration of lectin binding sites in the taste buds of the European catfish *Silurus glanis* (Teleostei). Histochemistry 94, 617–628.

Witt, M. and Reutter, K. 1997. Scanning electron microscopical studies of developing gustatory papillae in humans. Chem. Senses 22, 601–612.

Witt, M., Reutter, K., Ganchrow, D., and Ganchrow, J. 2000. Fingerprinting taste buds: intermediate filaments and their implication for taste bud formation. Philos. Trans. R. Soc. Lond. B Biol. Sci. 335, 1233–1237.

Yang, R., Crowley, H., Rock, M., and Kinnamon, J. 2000a. Taste cells with synapses in rat circumvallate papillae display SNAP-25-like immunoreactivity. J. Comp. Neurol. 424, 205–215.

Yang, R., Ma, H., Thomas, S., and Kinnamon, J. 2007. Immunocytochemical analysis of syntaxin-1 in rat circumvallate taste buds. J. Comp. Neurol., 502, 883–93.

Yang, R., Stoick, C., and Kinnamon, J. 2004. Synaptobrevin-2-like immunoreactivity is associated with vesicles at synapses in rat circumvallate taste buds. J. Comp. Neurol. 471, 59–71.

Yang, R., Tabata, S., Crowley, H., Margolskee, R., and Kinnamon, J. 2000b. Immunocytochemical localization of gustducin immunoreactivity in microvilli of type II taste cells of the rat. J. Comp. Neurol. 425, 139–151.

Yee, C., Yang, R., Böttger, B., Finger, T., and Kinnamon, J. 2001. "Type III" cells of rat taste buds: immunohistochemical and ultrastructural studies of neuron-specific enolase, protein gene product 9.5, and serotonin. J. Comp. Neurol. 440, 97–108.

Yoshie, S., Kanazawa, H., and Fujita, T. 1996. A possibility of efferent innervation of the gustatory cell in the rat circumvallate taste bud. Arch. Histol. Cytol. 59, 479–484.

Yoshie, S., Kanazawa, H., Nishida, A., and Fujita, T. 1997. Occurrence of subtypes of gustatory cells in cat circumvallate taste buds. Arch. Histol. Cytol. 60, 421–426.

Yoshie, S., Wakasugi, C., Teraki, Y., and Fujita, T. 1990. Fine structure of the taste bud in guinea pigs. I. Cell characterization and innervation patterns. Arch. Histol. Cytol. 53, 103–119.

Yoshie, S., Wakasugi, C., Teraki, Y., Iwanaga, T., and Fujita, T. 1991. Fine structure of the taste bud in guinea pigs. II. Localization of spot 35 protein, a cerebellar Purkinje cell-specific protein, as revealed by electron-microscopic immunocytochemistry. Arch. Histol. Cytol. 54, 113–118.

Zahm, D. and Munger, B. 1983a. Fetal development of primate chemosensory corpuscles. I. Synaptic relationships in late gestation. J. Comp. Neurol. 213, 146–162.

Zahm, D. and Munger, B. 1983b. Fetal development of primate chemosensory corpuscles. I. Synaptic relationships in early gestation. J. Comp. Neurol. 219, 36–50.

Zahm, D. and Munger, B. 1985. The innervation of the primate fungiform papilla – development, distribution and changes following selective ablation. Brain Res. Rev. 9, 147–186.

Zhang, Y., Hoon, M., Chandrashekar, J., Mueller, K., Cook, B., Wu, D., Zuker, C., and Ryba, N. 2003. Coding of sweet, bitter, and umami tastes: different receptor cells sharing similar signaling pathways. Cell 112, 293–301.

Further Reading

Beidler, L. and Smallman, R. 1963. Dynamics of taste cells. In: Olfaction and Taste (*ed.* Y. Zotterman), pp. 133–148. Pergamon Press.

Bigiani, A. and Roper, D. 1995. Estimation of the junctional resistance between electrically coupled receptor cells in *Necturus* taste buds. J. Gen. Physiol. 106, 705–725.

Finger, T., Danilova, V., Barrows, J., Bartel, D., Vigers, A., Stone, L., Hellekant, G., and Kinnamon, S. 2005. ATP signaling is crucial for communication from taste buds to gustatory nerves. Science 310, 1495–1499.

Fish, H., Malone, P., and Richter, C. 1944. The anatomy of the tongue of the domestic Norway rat. Anat. Rec. 89, 429–440.

Gilbertson, T., Damak, S., and Margolskee, R. 2000. The molecular physiology of taste transduction. Curr. Opin. Neurobiol. 10, 519–527.

Gomez-Ramos, P. and Rodriguez-Echandia, E. 1979. The fine structural effect of sialectomy on the taste bud cells in the rat. Tissue Cell 11, 19–29.

Kinnamon, S. and Margolskee, R. 1996. Mechanisms of taste transduction. Curr. Opinion Neurobiol. 6, 506–513.

Miller, I. and Bartoshuk, L. 1991. Taste Bud Distribution and Spatial Relationships. In: *Smell* and Taste in Health and Disease (*eds.* T. Getchell, R. Doty, L. Bartoshuk, and J. Snow), pp. 205–231. Raven Press.

Pumplin, D., Getschman, E., Boughter, J., Yu, C., and Smith, D. 1999. Differential expression of carbohydrate blood-group antigens on rat taste-bud cells: relation to the functional marker alpha-gustducin. J. Comp. Neurol. 415, 230–239.

Roper, S. 1992. The microphysiology of peripheral taste organs. J. Neurosci. 12, 1127–1134.

Travers, S. and Nicklas, K. 1990. Taste bud distribution in the rat pharynx and larynx. Anat. Rec. 227, 373–379.

Witt, M. and Reutter, K. 1996. Embryonic and early fetal development of human taste buds: a transmission electron microscopical study. Anat. Rec. 246, 507–523.

Yamasaki, H., Kubota, Y., Tagagi, H., and Tohyama, M. 1984. Immunoelectron-microscopic study on the fine structure of substance-P-containing fibers in the taste buds of the rat. J. Comp. Neurol. 227, 380–392.

Yang, J. and Roper, S. 1987. Dye-coupling in taste buds in the mudpuppy, *Necturus maculosus*. J. Neurosci. 7, 3561–3565.

Yoshii, K. 2005. Gap junctions among taste bud cells in mouse fungiform papillae. Chem. Senses 30(Suppl. i), 35–36.

4.06 Development of the Taste System

R F Krimm, University of Louisville School of Medicine, Louisville, KY, USA

L A Barlow, University of Colorado School of Medicine, Aurora, CO, USA

4.06.1 Introduction

At birth, all vertebrates possess a functional taste system. This sensory system is key for young animals to survive in a new world, in which they must avoid noxious and potentially lethal substances, yet ingest nutritionally valuable foods. From a developmental biologist's perspective, the taste system comprises four basic elements which arise in locations remote from one another and must connect accurately during embryogenesis. Specifically these components are: (1) taste buds arrayed within the mouth and pharynx; (2) sensory neurons residing within the cranial nerve ganglia which connect taste buds to (3) discrete gustatory regions of the hindbrain, which in turn connect to (4) upstream nuclei within the brain. This organization is already intricate from an anatomical perspective (see Chapter Gustatory Pathways in Fish and Mammals), but is truly a marvel of complexity when one considers its embryonic development. Taste buds develop *in situ* within the epithelia lining the oral and pharyngeal cavities, quite far from the developing brain. Likewise the sensory neurons destined to innervate taste buds also arise from neurogenic precursors outside of the brain. Finally, populations of hindbrain and higher order central neurons arise directly from specific regions of the forming neural tube. As we will review here, these elements each arise initially quite independently of, and at great distance from, the others, yet subsequently manage to connect to one another to produce a coordinated, fully functional taste system at birth (Figure 1).

In addition to these dynamic embryonic events, the taste system continues to develop postnatally. The cells within taste buds are constantly renewed throughout adult life, and the connections between the taste periphery and central targets remain plastic and modifiable by environmental influences. Despite this flexibility, we experience no alterations in our ability to taste.

In the past decade or so, studies of taste system development have been invigorated in large part by an expansion of experimental studies focused on embryonic development, as well as by use of transgenic mouse models in both embryonic and postnatal settings. These new approaches have provided new answers to major questions in the field: what are the molecular,

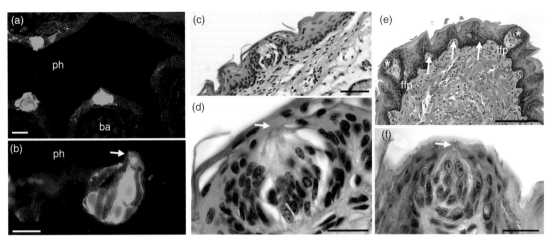

Figure 1 Depending upon the species or the location of taste buds within the oral cavity, taste buds can occur as discrete organs within the general epithelium or within more complex epithelial specializations. (a, b): Taste buds of juvenile axolotls (an aquatic salamander) are embedded directly in the oral and pharyngeal epithelia (ph; pharyngeal cavity). In these transverse sections through the pharyngeal region (dorsal is up), a subset of taste cells have been immunofluorescently stained (green) with antiserum against calretinin (Barlow, L. A. *et al.*, 1996). Note the characteristic onion shape of the taste organs and the narrow apical regions extruding into the pharyngeal cavity (arrow in b). (c, d): In mice, palatal taste buds are large and readily recognized as discrete organs within the epithelium. Again, these taste buds are found within the general epithelium, are onion-shaped (* in c), and narrow apically in the vicinity of the taste pore (arrow in d). (e, f): In contrast, taste buds on the mouse tongue are housed in elaborate papillae, including the fungiform papillae shown here (ffp). A single taste bud (* in e) sits in the apical papillary epithelium, while each papilla comprises an evaginated epithelium with a mesenchymal core. The fungiform papillae are interspersed with numerous, heavily keratinized filiform papillae (arrows in e). Despite their localization to fungiform papillae, the taste buds therein have the common onion shape and terminate apically in a narrow taste pore (arrow in f). Scale bars = 50 μm (a, c); 25 μm (b); 20 μm (d); 100 μm (e); 40 μm (f). (a, b, e, f) Transverse sections, dorsal is up. (c, d) Transverse sections, dorsal is down.

cellular, and tissue level interactions involved in taste organ development and patterning; how do sensory neurons come to innervate specifically both peripheral taste bud and central hindbrain targets; what types of cellular and molecular mechanisms maintain the mature taste system; and how well conserved are these developmental mechanisms in vertebrate evolution? These topics are thus the focus of our review.

4.06.2 The Taste Periphery: Of Buds and Bumps

To date, all vertebrate animals possess taste buds within the mouth and pharynx (although some fishes have external taste buds, see Gomahr, A. *et al.*, 1992; Reutter, K. and Witt, M., 1993; Hansen, A. *et al.*, 2002; Northcutt, R. G., 2005). These multicellular receptor organs are each composed of a mixed population of roughly 100 fusiform cells, which taper both apically and somewhat basally, to create an onion-shaped aggregate (Figure 2; and see Chapter Ultrastructure of Taste Buds). While this is the common organization for vertebrate taste

buds, there is an extremely wide variety in the morphology of epithelial structures associated with taste buds proper. In their most simplified state, taste buds occur within the general epithelium, and are not associated with any specialized epithelial appendages or structures (Figure 2). This is the case for aquatic salamanders, such as the axolotl (Fährmann, W., 1967; Toyoshima, K. *et al.*, 1987; Northcutt, R. G. *et al.*, 2000), where taste buds are readily detected in the oropharyngeal epithelium simply by their multicellularity and classic onion shape. In mammals, taste buds in the palate, pharynx, and larynx are also simple and onion shaped, residing directly within the epithelia (Mistretta, C. M., 1972; Belecky, T. L. and Smith, D. V., 1990; Zhang, C. and Oakley, B., 1996). However, in numerous instances, taste buds occur in the epithelium of specialized structures. For example, in catfish and zebrafish, individual taste buds are associated with taste hillocks (Gomahr, A. *et al.*, 1992; Reutter, K. and Witt, M., 1993; Hansen, A. *et al.*, 2002; Northcutt, R. G., 2005), which are slight epithelial elevations with small mesenchymal cores (Figure 2). On the tongues of mammals, taste buds are restricted to elaborate

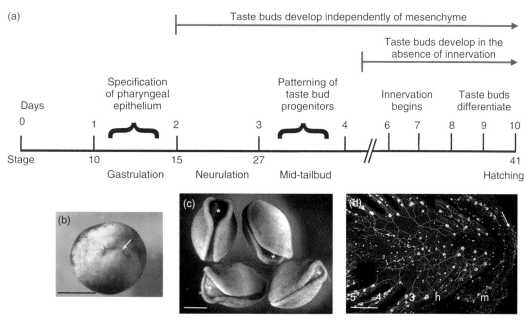

Figure 2 Time line of the experimental dissection of tissue level mechanisms governing taste bud development in axolotls. (a): Axolotl embryos develop from fertilization (day 0) to hatching (stage 41) over 10 days. Taste buds differentiate within a day of hatching, while the epithelium has already received innervation by 6 days. Nonetheless, taste buds develop in the absence of innervation (short red arrow; see text for full details). Further, taste buds develop in epithelium devoid of contact with cranial mesenchyme (long red arrow; see text). In fact, pharyngeal epithelium is specified by signals it receives during gastrulation and is patterned by tissue-intrinsic signaling during tailbud stages (horizontal brackets; see text). (b): At the onset of gastrulation, the dorsal blastopore lip is evident as a pigmented depression in the dorsal vegetal surface of an axolotl embryo (blue arrows; vegetal view). (c): During neurulation, the paired neural folds rise up from the dorsal surface of each axolotl embryo, and in this micrograph, the folds have approached the dorsal midline to fuse and form the neural tube. The anterior region (˚s) will give rise to the forebrain and midbrain, while the more posterior folds produce the hindbrain and spinal cord. (d): At hatching, both differentiated taste buds and their extensive innervation are evident in the internal pharyngeal surface of the lower jaw of axolotls. Using confocal microscopy on a larval axolotl head processed in whole mount for double immunofluorescence, taste buds (green) are revealed with antiserum against calretinin, and antiacetylated tubulin allows detection of nerves (red). m: mandible; h: hyoid arch; 3, 4, 5: third, fourth, and fifth branchial arches. Scale bars = 1 mm (b–d). Image acquired by Mark A. Parker.

structures, termed taste papillae (Mistretta, C. M., 1972). These appendages are composed of epithelial evaginations and/or invaginations, surrounding substantial mesenchymal cores (Figure 2). Taste buds reside in the epithelia of gustatory papillae, and taste cells within each bud extrude their apical microvilli into the lumen of a tightly arranged taste pore. The number of taste buds housed in these papillae can vary from 1 to 3 per papilla, as is the case for the smaller fungiform papillae in the anterior tongue, to several hundred per papillae as is the case for the large, posterior foliate and circumvallate papillae (Figure 2). In sum, while the morphology of taste buds is quite similar among vertebrates, ancillary epithelial structures associated with taste buds differ widely. This latter variability may reflect different selective pressures on taste epithelium, including

tasting in air versus in water, and the types of foods ingested by different animals.

Because aquatic salamanders have a taste bud organization uncomplicated by association with specialized epithelial structures, they perhaps can serve as a model for the most basic mechanisms driving the development of this system. Thus, that is where this review of new data on taste system development begins in earnest.

4.06.3 Development of the Taste Periphery: How Does Taste Bud Pattern Arise?

Recently, a number of studies providing insight into how taste buds develop have utilized either amphibian (axolotl) or mammalian (mouse and rat) embryos. And therefore, we will first review the insights into taste periphery development gleaned from studies of axolotls, then present the new findings in rodents that speak to these same issues.

4.06.3.1 Taste Bud Development in Axolotls

Axolotls are a species of aquatic salamander native to the high altitude lakes in the vicinity of Mexico City (Smith, H. M., 1969), and embryos of this species and those of its close relatives have been an important tool in studies of embryogenesis for nearly a century (e.g., Vogt, W., 1929; Mangold, O., 1931; Nieuwkoop, P. D. and Ubbels, G. A., 1972; Slack, J. M. and Forman, D., 1980; Armstrong, J. B. and Malacinski, G. M., 1989). In the last decade, we have taken advantage of the ease of microsurgical manipulation of these embryos at virtually any stage of development to explore tissue-level and cellular mechanisms regulating taste bud formation.

Although taste buds are first evident morphologically in the oral cavity as axolotl larvae hatch and begin to feed (Barlow, L. A. and Northcutt, R. G., 1995), the oral and pharyngeal epithelia are innervated by gustatory nerve fibers long before, during mid- to late embryonic stages. This sequence of events, innervation followed by taste bud differentiation, is common to all vertebrates examined to date (Landacre, F. L., 1907; Stone, L. S., 1940; Farbman, A. I., 1965; Bradley, R. M. and Stern, I. B., 1967; Reutter, K. and Witt, M., 1993; Whitehead, M. C. and Kachele, D. L., 1994; Barlow, L. A. *et al.*, 1996; Witt, M. and Reutter, K., 1996; Northcutt, R. G., 2005); and when combined with studies employing adult vertebrates demonstrating that maintenance of mature taste bud morphology is dependent upon intact innervation (e.g., Vintschgau, M. and Honigschmied, J., 1876; Torrey, T. W., 1934; Poritsky, R. and Singer, M., 1963; but see Whitehead, M. C. *et al.*, 1987), has led to a neural induction model for taste bud formation. Specifically, this model posits that during development, early arriving nerve fibers induce taste buds to form from an otherwise homogeneous local epithelium.

Over 10 years ago, expanding on ectopic grafting studies first published by Stone L. S. (1932; 1940), the neural induction model was tested using axolotl embryos and modern immunocytochemical methods to detect all nerve fibers (Barlow, L. A. *et al.*, 1996). The oropharyngeal epithelium of amphibian embryos,

destined to give rise to taste buds (Barlow, L. A. and Northcutt, R. G., 1995), was removed prior to innervation and cultured in isolation. Despite the complete absence of nerve contact, the explanted epithelia generated taste buds in appropriate densities and with a time course identical to that of intact embryos. Thus, at least in these amphibians, innervation is not required for taste bud formation.

Since taste buds are not induced by nerves, they might instead be induced by contact with cranial mesenchyme. This latter cell population contacts the oral epithelium during embryogenesis and is a known inducer of other tissues within the oral cavity, including teeth (Kollar, E. J., 1983; Lumsden, A. G. S., 1987; 1988) and cartilage (Graveson, A. C. and Armstrong, J. B., 1987). However, when oropharyngeal epithelia of early axolotl embryos are removed and cultured prior to and without subsequent contact with embryonic mesenchyme, taste buds still form in epithelial explants at the appropriate developmental stage, some 10 days later (Barlow, L. A., and Northcutt, R. G., 1997). Thus, taste bud development is independent of both nerves and cranial mesenchyme. Logically then, since no other tissues contact the oropharyngeal epithelium during embryogenesis, formation of taste buds is an intrinsic feature of this epithelium, one that is acquired early in embryonic development. In fact, this predisposition or specification (see Slack, J. M. W., 1991) to a taste bud-bearing fate is acquired during gastrulation and is due to signals from the embryonic notochord (Barlow, L. A., 2001). The implication here is that molecular signals from the notochord induce changes in the oropharyngeal epithelium such that the latter produces taste buds which differentiate 10 days later.

One unanticipated aspect of these results was how early the tissue-level interactions that set the stage for epithelium-intrinsic taste bud formation occurred – so much earlier than taste bud differentiation at the end of embryonic development. Given the large expanse of developmental time between the completion of gastrulation and taste bud differentiation 10 days later, how does this early intrinsic property transform into the patterned array of taste buds in the mouth? To address this question, we proposed that local cell–cell interactions were mediating pattern formation of presumed taste bud progenitor cells (Barlow, L. A. and Northcutt, R. G., 1998; Northcutt, R. G. and Barlow, L. A., 1998). To test if and when this was occurring, intercellular contacts were disrupted transiently in oropharyngeal epithelial explants at progressive developmental stages. These studies reveal a critical 12 h window, again long before taste bud differentiation and innervation of the epithelium, during which cell–cell communication is crucial for determining taste bud pattern when intercellular signals are interrupted during this phase, the total number and size of individual taste buds is dramatically increased (Parker, M. A. *et al.*, 2004). This result has led to our hypothesis that a population of distributed taste bud progenitor cells is induced at this earlier stage, which later goes on to divide and give rise to taste bud cells proper, a hypothesis that remains to be tested.

4.06.3.2 Taste Bud Development in Rodents

In broad overview, the development of taste buds in mice and rats is quite comparable to that of axolotls. Taste buds arise directly from the oral epithelium (Stone, L. M. *et al.*, 1995; 2002) and do not differentiate until perinatal stages, the precise timing depending upon their location. The sequence of taste bud formation has been best studied in the tongue because the future location of these organs can be discerned quite early in development via changes in the morphology of subsets of cells within the lingual epithelium. For example, in the anterior two-thirds of the tongue, fungiform papillae are first evident as two parasagittal rows of columnar epithelial foci (Mistretta, C. M., 1972; Farbman, A. I. and Mbiene, J.-P., 1991). These epithelial thickenings are termed placodes, which grow and invaginate to surround and create the mesenchymal core of each papilla. At birth, an onion-shaped aggregate of immature taste cells is present in the apex of each papilla, which subsequently differentiates fully into a mature taste bud sometime within the first postnatal week. Key to this designation of full differentiation is the postnatal formation of a definitive taste pore, which is the gold standard for a functional taste bud (Hosley, M. A. *et al.*, 1987; Farbman, A. I. and Mbiene, J.-P., 1991; Harada, S. *et al.*, 2000).

Intriguingly, lingual taste buds of the fungiform, foliate, and circumvallate papillae are not fully differentiated at birth in rodents (Mistretta, C. M., 1972; Oakley, B. *et al.*, 1991; Sbarbati, A. *et al.*, 1999), yet newborn mice and rats suckle almost immediately, inferring that they must possess a functional taste system. In fact, it is likely that the taste buds residing in the epithelium of the palate subserve this early requirement for sweet taste sensation in rodent pups (Harada, S. *et al.*, 1997; Sollars, S. I. and Hill, D. L., 2005). Palatal taste buds, as a population, differentiate prior to birth, in that over half

of them possess pores, compared to ∼15–20% of fungiform taste buds in newborns (Harada, S. *et al.*, 2000; Zhang, G. H. *et al.*, 2006). This result has been corroborated employing immunocytochemistry for a marker of a subtype of differentiated taste cells (Boughter, J. D. *et al.*, 1997; Yang, R. *et al.*, 2000). Using antiserum against alpha-gustducin, the majority of palatal taste buds have gustducin-immunoreactive taste cells at birth (El-Sharaby, A. *et al.*, 2001a; 2001b), while gustducin immunoreactivity is not detected in lingual taste buds until later, during the first postnatal week (El-Sharaby, A. *et al.*, 2001b; Zhang, G. H. *et al.*, 2006).

Regardless of location, and as in axolotls, however, taste bud differentiation occurs well after nerves contact the developing epithelium (Farbman, A. I., 1965; Farbman, A. I. and Mbiene, J.-P., 1991; Mbiene, J.-P. and Mistretta, C. M., 1997; Mbiene, J.-P., 2004; Sharaby, A. E. *et al.*, 2006), consistent with a neural induction model.

The neural induction hypothesis has now also been tested explicitly in rodents via lingual explant culture. Several groups have shown that the formation of fungiform placodes is independent of innervation. When lingual primordia are removed prior to placode formation and then cultured in the absence of innervation, these explants first develop placodes, which then begin papillary morphogenesis (Farbman, A. I. and Mbiene, J.-P., 1991; Mbiene, J.-P. *et al.*, 1997; Nosrat, C. A. *et al.*, 2001; Hall, J. M. *et al.*, 2003; Mistretta, C. M. *et al.*, 2003; Liu, H. X. *et al.*, 2004). Unfortunately, these cultures do not survive long enough to allow taste bud differentiation, and thus the precise relationship between placodes, papillae, and the late developing taste buds is unclear. While a number of studies have examined the impact of loss of sensory neurons via a neurotrophin knockout strategy in mice, this approach has shown primarily that innervation is required postnatally for taste bud maintenance rather than for testing the role of nerves in taste organ formation. These results will be addressed fully later on.

Currently, the accepted model for taste organ development is that placodes arise independently of innervation and, in turn, form rudimentary papillae in the absence of nerves. Shortly before birth, taste bud progenitor cells are organized within a small region of the papillary epithelium, and this latter process is thought to be nerve dependent (Farbman, A. I. and Mbiene, J.-P., 1991; Oakley, B. and Witt, M., 2004; Mistretta, C. M. and Liu, H. X., 2006). A counterview is that placodes actually represent taste bud progenitors (Mbiene, J.-P. and Roberts,

J. D., 2003). Thus the taste progenitor pool arises in a nerve-independent fashion, while full differentiation of these cells is again thought to rely on innervation; however, this early progenitor interpretation is controversial (see Oakley, B. and Witt, M., 2004).

Regardless of the identity of the taste placode cells, that is, progenitors for papillae or for taste buds, it is generally believed that taste papilla development is governed via epithelial–mesenchymal interactions. This assumption is based on similarities with other epithelial specializations, such as hair follicles, teeth, and feathers (Mistretta, C. M., 1998; Yu, M. *et al.*, 2002; Pispa, J. and Thesleff, I., 2003), yet until very recently, experimental tests of this idea were lacking. Through a series of elegant epithelial/mesenchymal recombination experiments employing embryonic lingual explants, Kim J. Y. *et al.* (2003) have shown that there are indeed signals between the epithelium and mesenchyme during early papilla morphogenesis. Moreover, the pattern of placode formation appears to reside within the lingual epithelium, while papillary morphogenesis requires general signals from mesenchyme. This latter finding suggests, as has been shown in axolotls (Parker, M. A. *et al.*, 2004), that taste organ pattern in rodent embryos is set up initially via cell–cell interactions intrinsic to the lingual epithelium. Consistent with a role for intercellular communication, disruption of gap junctional contacts in cultured mouse tongues results in alterations in taste placode pattern (Kim, J. Y. *et al.*, 2005).

4.06.3.3 Molecular Mechanisms of Taste Patterning

In addition to experimental tests of tissue and cellular interactions, there has been an expansion of molecular investigations of developing taste buds. A number of groups have shown that genes belonging to powerful developmental signaling pathways are expressed in the embryonic tongue and taste organs. The pathways investigated include: Sonic Hedgehog (Shh), Bone Morphogenetic Protein (BMP), Wnt/β-catenin, fibroblast growth factor (FGF), and Notch. All of these pathways are key regulators of numerous aspects of embryonic development, as well as of adult tissue homeostasis (Shh: Chuong, C. M. *et al.*, 2000; Marti, E. and Bovolenta, P., 2002; BMP: Tanabe, Y. and Jessell, T. M., 1996; Botchkarev, V. A. and Sharov, A. A., 2004; Wnt: Wodarz, A. and Nusse, R., 1998; Ciani, L. and Salinas, P. C., 2005; FGF:

Cornell, R. A. and Eisen, J. S., 2005; Dvorak, P. *et al.*, 2006; Notch: Blanpain, C. *et al.*, 2007).

In terms of taste organ patterning in rodent embryos, the Shh pathway is the most heavily studied to date. Shh is a secreted factor which is recognized by the transmembrane protein, Patched (Ptc) (see Athar, M. *et al.*, 2006). When Shh binds Ptc, this releases Smoothened (Smo), another transmembrane protein, from Ptc inhibition and activates an intracellular cascade, which in turn activates gene transcription. In particular, Shh signaling results in upregulation of Ptc and of the Gli1 transcription factor. Thus, Gli1 and Ptc expression are good indicators of active Shh signaling. In embryos, Shh ligand is present throughout the lingual epithelium as the tongue first forms from bilateral lingual swellings and becomes restricted to the columnar placodal epithelial cells as placodes arise. This focal Shh expression persists in papillae as they undergo morphogenesis (Bitgood, M. J. and McMahon, A. P., 1995; Hall, J. M. *et al.*, 1999; Jung, H. S. *et al.*, 1999; Mistretta, C. M. *et al.*, 2003; Liu, H. X. *et al.*, 2004). *Ptc* and *Gli1* are detected similarly throughout the lingual epithelium commencing with tongue formation but, unlike Shh, are also expressed in the subepithelial mesenchyme (Hall, J. M. *et al.*, 1999). Both epithelial and mesenchymal expression of *Ptc* and *Gli1* become progressively limited to the forming taste papillae, until both of these genes are detected in the epithelium in rings surrounding the Shh-expressing cells, as well as within the mesenchymal core of each papilla. In sum, these expression patterns are consistent in time and place with an important role for Shh signaling in taste organ development.

Expression of the BMP pathway has also been examined in the developing tongue of rodent embryos (Jung, H. S. *et al.*, 1999; Hall, J. M. *et al.*, 2003; Nie, X., 2005). BMPs comprise a small subfamily belonging to the large transforming growth factor-β (TGFβ) family of secreted factors (Kingsley, D. M., 1994). BMP ligands bind to heterodimeric BMP receptors which, through a series of transphosphorylations, result in activation of SMAD transcription factors, and thus of target gene transcription (Balemans, W. and Van Hul, W., 2002). BMP signaling is tightly regulated by a host of native, secreted antagonists, of which Noggin was the founding member (Zimmerman, L. B. *et al.*, 1996). When mRNA or protein for several BMP ligands is examined, low-level expression appears to commence throughout the epithelium and mesenchyme of the tongue rudiment as it forms (Jung, H. S. *et al.*, 1999; Zhou, Y. *et al.*, 2006). Expression, especially of *BMP4*, then consolidates within the developing placodal epithelium, coincident with Shh expression. In fact, there is a one-to-one correspondence between placodal cells expressing BMP4 and those expressing Shh (Hall, J. M. *et al.*, 2003). BMP4 protein is also detected within the subepithelial mesenchyme as papillae develop, whereas its mRNA is not, suggesting that the epithelial cells of the papillae are the source for the diffusible BMP4 found in the mesenchyme (Zhou, Y. *et al.*, 2006). Expression of most other genes of the BMP pathway has yet to be examined in the developing tongue, with the exception of Noggin, a secreted BMP2, 4, and 7 antagonist; Noggin-immunoreactivity is detected in a pattern similar to that described above for the BMP ligand (Zhou, Y. *et al.*, 2006). Like Shh, the timing and pattern of BMP expression is consistent with a role in taste organ formation and patterning.

Most recently, expression of the Wnt/β-catenin pathway has been explored in the developing taste periphery of mouse embryos. Wnts are secreted ligands, comprising 19 members to date (see Relevant Website section). The Wnt signal is transduced by several distinct, yet likely interacting, intracellular pathways (Moon, R. T. *et al.*, 2002; Bejsovec, A., 2005; Widelitz, R., 2005). Of these, the β-catenin pathway is the best studied (Wodarz, A. and Nusse, R., 1998). In the Wnt/β-catenin signaling cascade, Wnt ligands bind Frizzled receptors, which complex with intracellular proteins to inhibit the action of the GSK3β protein kinase. In the absence of Wnt, GSK3β phosphorylates β-catenin, thus targeting the latter for degradation. When Wnt is present, however, β-catenin is not degraded, and accumulates to sufficient levels to translocate to the nucleus and contribute to a transcriptional activating complex with TCF and LEF transcription factors. In this way, Wnt-responsive target genes are activated.

Wnt expression is evident in developing mouse tongues coincident with the appearance of taste placodes. In particular, *Wnt10b* is restricted to placodal epithelium, while *Wnt10a* is found broadly throughout the lingual epithelium as taste placodes form (Iwatsuki, K. *et al.*, 2007; Liu, F. *et al.*, 2007). Similarly, *Lef1*, a downstream transcription factor of the Wnt/β-catenin pathway is expressed in taste placodes and later in taste papilla epithelium (Iwatsuki, K. *et al.*, 2007). Via use of mice expressing a Wnt/β-catenin reporter transgene (TOPGAL transgenic strain; DasGupta, R. and Fuchs, E., 1999), the location of Wnt-responsive cells has also been determined. The TOPGAL gene reveals active Wnt/β-catenin signaling in the lingual epithelium

commencing as taste placodes form, and consistent with *Wnt10b* focal expression (Okubo, T. *et al.*, 2006; Iwatsuki, K. *et al.*, 2007; Liu, F. *et al.*, 2007). However, TOPGAL expression is also evident at these early stages in the intervening epithelial cells destined to give rise to filiform, that is, nongustatory, papillae of the lingual surface (Liu, F. *et al.*, 2007). Intriguingly, TOPGAL expression becomes increasingly restricted in the lingual epithelium, such that immediately prior to birth, it is detected only in immature taste buds in the apices of the fungiform papillae (Liu, F. *et al.*, 2007). Thus, Wnt/β-catenin signaling is implicated in taste organ formation, and possibly taste bud differentiation, by virtue of its spatiotemporal pattern of expression.

Compared with the extensive data on Shh, BMP, and recently Wnts, there is far less information on the expression of the FGF and Notch pathways in developing tongues. The FGF family comprises 22 genes to date, with numerous splice variants, which bind to FGF receptors (FGFRs 1–4; Itoh, N. and Ornitz, D. M., 2004). mRNA for several FGF ligands and a variety of FGFRs have been detected in the developing tongues of mice (Jung, H. S. *et al.*, 1999; Nie, X., 2005). *Fgf8*, which is known to function in early branchial arch patterning (Abu-Issa, R. *et al.*, 2002; Crump, J. G. *et al.*, 2004), is expressed in the lingual epithelium and subepithelial mesenchyme in the lateral lingual swellings prior to their fusion into the tongue rudiment (Jung, H. S. *et al.*, 1999). By the time the tongue proper has formed and taste placodes begin to appear, however, *Fgf8* expression is lost, suggesting that FGF8 functions in tongue formation, but does not have a direct role in taste placode formation.

The Notch pathway has numerous roles in cell fate decisions throughout embryonic development. Notch is a transmembrane receptor, which recognizes several transmembrane ligands, including Delta-like (Dll) and Jagged (or Serrate), in adjacent cells. When the Notch receptor is activated, gamma secretases cleave N, generating the N intracellular domain (N_{ICD}), which translocates to the nucleus and affects target gene transcription (Artavanis-Tsakonas, S., 1988; Cornell, R. A. and Eisen, J. S., 2005). In particular, in the developing nervous system and in neural stem cells, activated Notch signaling turns on transcription of Hes transcription factor genes and these cells assume a glial fate or remain undifferentiated; while cells in which Notch is not active transcribe neurogenic genes, such as Mash1, and assume a neural fate (Gaiano, N. and Fishell, G., 2002). In the developing tongue, Notch gene expression is not detected until well after Wnt,

Shh, and BMP expression have become focal within developing papillae (Seta, Y. *et al.*, 2003). As the papillae undergo morphogenesis, several Notch receptors and ligands are expressed broadly in the papillary epithelium. The transcription factors, *Mash1* and *Hes6*, by contrast, are expressed in small numbers of epithelial cells during papillary development (Seta, Y. *et al.*, 2003). Thus, while Shh, Wnt, and BMP expression are coincident with placode formation, Notch signaling appears to function later on, during papillary morphogenesis, and thus may regulate cell fate decisions within taste papillae.

4.06.3.4 Tests of Gene Function in Taste Organ Formation and Patterning

Recent tests of Wnt, Shh, and BMP function have revealed roles for each of these pathways in early taste organ development, as was predicted based upon the spatiotemporal expression of these genes. Perhaps the most complete studies have come from manipulation of Wnt/β-catenin, where the levels of signaling by this pathway have been altered genetically *in vivo* and via pharmacological manipulation *ex vivo* (Iwatsuki, K. *et al.*, 2007; Liu, F. *et al.*, 2007).

Using tissue-specific and inducible transgenic mouse lines, Wnt/β-catenin signaling was completely abolished within the lingual epithelium of mouse embryos, commencing just prior to the onset of taste placode formation. In these knockdown and knockout embryos, placodes are not evident, nor are evaginated papillae present at birth (Liu, F. *et al.*, 2007). Similarly, papillae are reduced, although not lost, in both Lef1$^{-/-}$ and Wnt10b$^{-/-}$ mice (Iwatsuki, K. *et al.*, 2007), implying that these two genes are not the only Wnt signaling genes involved in taste placode and papillary development. Conversely, when β-catenin is stabilized, that is, genetic alteration having rendered β-catenin resistant to degradation throughout the embryonic lingual epithelium, the entire lingual surface is transformed into taste papillae. Moreover, filiform papillae are absent, and taste buds within the taste papillae appear to have accelerated development (Liu, F. *et al.*, 2007).

Wnt/β-catenin signaling also appears to function as an initiator signal for taste placode formation, as it is upstream, at least genetically, of both Shh and BMPs. When Wnt/β-catenin signaling is lost, *Shh* is not expressed in the lingual epithelium, consistent with the observed loss of placodes. Similarly, *BMP4* is not expressed in the lingual epithelium of mouse embryos with epithelial loss of Wnt/β-catenin function (Liu, F.

et al., 2007). Moreover, in the gain-of-function genetic manipulation, as well as *ex vivo* when Wnt/β-catenin signaling is increased pharmacologically, *Shh* expression is increased, in that *Shh* placodal expression is expanded (Iwatsuki, K. *et al.*, 2007; Liu, F. *et al.*, 2007). Wnt/β-catenin is also upstream of Sox2, a transcription factor recently shown to be required for taste organ development (Okubo, T. *et al.*, 2006); elevation of Wnt/β-catenin signaling in cultured tongue explants increases Sox2-expressing taste placodes.

The role of Wnt/β-catenin as a positive regulator of Shh signaling, however, appears to be reciprocal, in that manipulation of Shh in cultured tongue explants alters expression of Wnt activity, as monitored by the TOPGAL reporter gene (Iwatsuki, K. *et al.*, 2007). Substantial data also implicate a key role for Shh in taste placode formation. When Shh is blocked in cultured tongue explants, taste placodes increase in number and size and occur closer together (Hall, J. M. *et al.*, 2003; Mistretta, C. M. *et al.*, 2003), while Shh ligand blocks papillary formation (Iwatsuki, K. *et al.*, 2007). These effects are clearly indirect, however, as blocking Shh causes placodal Shh expression to expand, as do taste papillae, while increasing Shh signaling *in vitro* causes loss of Shh expression just as taste papillae are lost. Moreover, Shh also regulates BMP4; Shh inhibition *in vitro* increases the number and size of BMP4-expressing placodes (Hall, J. M. *et al.*, 2003). Recently, BMPs have also been implicated in regulation of Shh in developing taste papillae (Zhou, Y. *et al.*, 2006). Again, using a culture approach, antagonizing BMP in lingual explants results in more and larger papillae, with commensurate increases in Shh expression, while exposure to BMPs reduces papillae and Shh expression. In sum, these functional data suggest that the molecular regulation of taste placode formation relies upon a network of interactions of at least the Wnt/β-catenin, Shh, and BMP pathways, which are likely integrated by additional intracellular and extracellular mechanisms.

4.06.4 Development of the Circumvallate Papilla, a Specialized Structure in Mammals

Until now, this review has focused on development of fungiform papillae and their resident taste buds, which arise as a distributed array on the anterior two-thirds of the tongue. However, mammals possess additional lingual taste buds housed in complex, and large, papillae at the posterior portion of the lingual tongue; these are the circumvallate and foliate papillae. In adult rodents, each of these structures possesses hundreds of taste buds embedded in the epithelia lining the trenches and folds of the circumvallate and foliate papillae, respectively. The circumvallate papilla in particular is known to have a number of additional anatomical specializations. von Ebner's glands (also known as posterior deep lingual glands) empty directly into the circumvallate trenches (Sbarbati, A. *et al.*, 2002; Lee, M. J. *et al.*, 2006; Suzuki, Y., 2006). Additionally, the mesenchymal core of the circumvallate is associated with a circumvallate ganglion (Graziadei, P. P. and Graziadei, G. A., 1978; Sbarbati, A. *et al.*, 2000). Thus, the development of the circumvallate papilla involves the coordinated formation of a complex taste organ, in concert with glandular and ganglionic components.

The circumvallate papilla of rodents begins to develop quite early, as the lateral lingual swellings of the tongue first fuse at the midline. At this time point, the epithelium of the future circumvallate has already begun to thicken (Jitpukdeebodintra, S. *et al.*, 2002). Shortly thereafter, it is recognizable as a broad epithelial thickening or placode, also visible on the dorsal surface of the tongue, coincident with the appearance of fungiform placodes (Paulson, R. B. *et al.*, 1985; AhPin, P. *et al.*, 1989; Wakisaka, S. *et al.*, 1996; Mbiene, J.-P. and Roberts, J. D., 2003). Subsequently, the lateral walls of the presumptive papillae invaginate and grow ventrally, surrounding a large mesenchymal core, which is invaded by neuroblasts of the papillary ganglion. Trench invagination continues, due to cell proliferation localized primarily to the growing tips of the trench epithelium (Lee, M. J. *et al.*, 2006), for the remainder of embryogenesis and into postnatal stages (Mistretta, C. M., 1972; Sbarbati, A. *et al.*, 2000; Jitpukdeebodintra, S. *et al.*, 2002). The trenches are thought to give rise to the von Ebner's glands via branching morphogenesis commencing late in embryonic development (Jitpukdeebodintra, S. *et al.*, 2002; Lee, M. J. *et al.*, 2006). The taste buds first differentiate within the circumvallate epithelium several days after birth (Mistretta, C. M., 1972; Oakley, B. *et al.*, 1991), although onion-shaped aggregates of cells considered to be immature taste buds are evident at birth in rodents (Mistretta, C. M., 1972).

Given the complex nature of the circumvallate papilla, are the molecular mechanisms governing its formation similar to those involved in fungiform development? At first glance, they may not be. Although both *Shh* and BMP4 are expressed in the developing circumvallate papilla (Hall, J. M. *et al.*, 1999; 2003; Lee, M. J. *et al.*, 2006), alterations in Shh signaling *in vitro* do not appear to affect circumvallate morphology or Shh expression therein (Hall, J. M. *et al.*, 2003; Mistretta, C. M. *et al.*, 2003; Liu, H. X. *et al.*, 2004). BMP2 and BMP4 do not appear to be expressed in the developing circumvallate, nor does manipulating BMP signaling in tongue explants appear to affect this papilla (Zhou, Y. *et al.*, 2006). Further, evidence of Wnt/β-catenin signaling was not observed by us in the developing circumvallate papilla (Liu, F. *et al.*, unpublished observation), suggesting that Wnt/β-catenin may not be active in this taste structure during embryogenesis. Thus, it appears that molecular mechanisms distinct from those involved in fungiform placode formation likely regulate circumvallate development. This dichotomy may be due to the increased complexity of this posterior taste papilla or may reflect an endodermal origin (Zhang, C. and Oakley, B., 1996).

Specifically, the oral cavity marks the boundary between the surface ectoderm and the foregut endoderm, although precisely where the border lies in rodents has been a topic of conjecture. One possibility, based on differential cytokeratin expression, is that the circumvallate lies in the endodermal epithelium, while the fungiform taste buds arise from ectoderm covering the anterior portion of the tongue (Zhang, C. and Oakley, B., 1996). Another molecular correlate of this supposition is that cytokeratin 14 (K14) is expressed in the epithelium of the anterior two-thirds of the embryonic tongue, encompassing the fungiform domain, yet is absent from the posterior third of the lingual epithelium where the circumvallate forms (Liu, F. *et al.*, 2007). This difference in K14 expression may indicate that each region arises from a different germ layer, fungiform from ectoderm, circumvallate from endoderm, or may simply reflect different molecular regulation of anterior versus posterior taste organs due to differences in their complexity.

4.06.5   Taste Bud Cell Lineage and Turnover

Postnatal taste buds comprise a heterogeneous population of approximately 100 cells, divided into roughly three cell types, I, II, and III, based upon ultrastructure and molecular markers (see Chapters Ultrastructure of Taste Buds and Taste Receptors). Briefly, type I cells are thought to function as glia, while types II and III are taste receptor cells. Type II cells express numerous bitter, sweet, and umami receptors and lack clear morphological synapses. In contrast, type III cells possess synapses and the transduction machinery for sour tastants. However, the intrigue from a developmental biologist's perspective in investigating mature taste buds is that the cells within each bud are continually replaced at a rapid pace, raising the questions of how taste cells are related to one another within a bud, and what is the nature of the cell lineage(s) that produces taste bud cells. Our current knowledge of taste cell turnover is limited primarily to birthdating studies for differentiated taste cell types.

Immature taste cells are thought to arise as a result of asymmetric division of between 7 and 13 putative stem or progenitor cells per taste bud (Beidler, L. M. and Smallman, R. L., 1965; Farbman, A. I., 1980; Stone, L. M. *et al.*, 2002). After ~3 days, taste cells differentiate, expressing appropriate cell type immuno-markers (Zhang, C. *et al.*, 1995; Cho, Y. K. *et al.*, 1998; Hirota, M. *et al.*, 2001). The relationship among cells of each bud is unclear, although type I cells likely represent a distinct lineage (Farbman, A. I., 1980). In terms of types II and III cells, either these are distinct cell lineages (Hamamichi, R. *et al.*, 2006), or a subset of type III cells may also give rise to type II cells (Miura, H. *et al.*, 2005; Ueda, K. *et al.*, 2006). Finally, taste cells are relatively short lived, as they die, via programmed cell death, on average 10 days after birth (Beidler, L. M. and Smallman, R. L., 1965; Farbman, A. I., 1980; Delay, R. J. *et al.*, 1986). However, it appears that not all taste cells have equivalent residence times in the taste bud. For example, in Farbman's tritiated thymidine birthdating study, he found that while the average lifespan for a likely type I or glial cell was 9 days, the persistence of type II receptor cells could be detected for as long as 25 days (Farbman, A. I., 1980). More recently, this finding has been confirmed using BrdU birthdating in combination with taste cell-type markers for types II and III (Hamamichi, R. *et al.*, 2006). A very short-lived population of cells is detected in taste buds with an average life span of 2 days, while a small proportion of type II and III cells are long-lived residents in the taste bud, detectable after 21 days. This perdurance of type II and III cells may reflect more elaborate differentiation and functional requirements for receptor cells compared with glial cells; the latter might be replaced on a shorter time frame. Alternatively, the very short lived cells may be immature taste cells of all three types and indicate that taste bud stem cells continually produce

an overabundance of immature cells to quickly accommodate injury or any errors in cell lineage decisions.

While it is well recognized that taste buds continually turn over, the molecular regulation of this process is unknown. Certainly a large number of regulatory genes and transcription factors are expressed in subsets of adult taste cells, implying that these factors function in taste cell turnover and lineage decisions within buds. For example, several embryonic signaling pathways persist in adult taste buds. *Shh* is expressed in a small number of rounded cells in the basal compartment of each taste bud, while the Shh receptor, *Ptc1*, localizes to cells surrounding the bud (Miura, H. *et al.*, 2001; 2003). Elements of the Notch pathway are also found in adult taste buds, including Notch receptors and ligands, as well as downstream transcription factors (Kusakabe, Y. *et al.*, 2002; Suzuki, Y. *et al.*, 2002; Miura, H. *et al.*, 2003; Seta, Y. *et al.*, 2003; Hamamichi, R. *et al.*, 2006; Seta, Y. *et al.*, 2006). In addition, a number of other signaling pathways not well described in embryonic taste organ development are present in adult taste buds, including the EGF pathway (McLaughlin, S. K., 2000; Sun, H. and Oakley, B., 2002) and the insulin-like growth factor (IGF) pathway (Suzuki, Y. *et al.*, 2005). Consistent also with a role in taste cell turnover, loss of expression of many of these gene products occurs when taste buds are denervated in nerve injury experiments, which are known to cause a transitory loss of taste buds, including *Shh* (Miura, H. *et al.*, 2004), *Mash1*(Seta, Y. *et al.*, 1999), and NeuroD, a transcription factor in the Notch pathway (Suzuki, Y. *et al.*, 2002). In each of these cases, as the nerve regenerates, taste buds reappear, as does the expression of these factors. However, these data are correlative. To elucidate the function of these gene pathways in taste cell turnover and to perhaps gain insight into the lineage relationships of taste cells within buds, inducible genetic manipulations will have to be employed.

Innervation is also known to regulate taste bud differentiation and maintenance, and this topic will be considered fully in a later section.

4.06.6   Development of Gustatory Innervation

The other major focus of this review is to discuss the development of the gustatory sensory neurons which contact taste buds in the lingual epithelium and convey taste information to the gustatory centers of the hindbrain. As mentioned in the introduction, the neurons that reside in the gustatory cranial ganglia arise embryonically from two neurogenic tissues that form both outside of the developing neural tube and distant from the oral epithelium. These two populations of developing sensory neurons must first locate one another and coalesce into a ganglion. They then send out both peripheral and central neurites to pathfind to the proper targets and make appropriate synaptic connections with taste buds and hindbrain neurons, respectively.

4.06.6.1 Development of the Sensory Neurons of the Gustatory Ganglia

Sensory neurons of the VII, IX, and X cranial nerve ganglia convey gustatory information from the mouth, pharynx, and larynx to the brain. Neurons of each of these ganglia arise from two discrete embryonic regions: the epibranchial placodes and a subset of the cranial neural crest (Narayanan, C. H. and Narayanan, Y., 1980; D'Amico-Martel, A. and Noden, D. M., 1983; Barlow, L. A. and Northcutt, R. G., 1995; Northcutt, R. G. and Brändle, K., 1995; Gross, J. B. *et al.*, 2003). Epibranchial placodes are a series of epithelial thickenings, which form in the ectoderm just dorsal to the branchial arches as the neural tube fuses. These placodes are neurogenic, in that they are mitotically active, producing immature neuroblasts which then migrate to assemble into portions of cranial ganglia VII, IX, and X (Webb, J. F. and Noden, D. M., 1993; Baker, C. V. and Bronner-Fraser, M., 2001; Begbie, J. and Graham, A., 2001a; Begbie, J. *et al.*, 2002; Schlosser, G., 2006). The cranial neural crest is a migratory population of cells, which gives rise to a broad variety of derivatives, including a portion of the sensory neurons and all of the glial cells of the VII, IX, and X cranial ganglia (Le Douarin, N. M. and Kalcheim, C., 1999).

Based primarily upon descriptive studies in catfish, it has been proposed that the epibranchial placodes give rise to the gustatory neurons of each ganglion. Catfish are rather unique with respect to the taste system; they have thousands of external taste buds distributed in the skin covering their head and trunk (Landacre, F. L., 1907; Atema, J., 1971; Reutter, K. *et al.*, 1974; Caprio, J. *et al.*, 1993). Regardless of location, these external taste buds are all innervated by a recurrent branch of the VII cranial nerve, and the ganglion for this nerve is greatly enlarged in catfish (Herrick, C. J., 1903; Kohbara, J. *et al.*, 1992). In

examining the developing epibranchial placode of the VII ganglion, Landacre found that it was also enlarged, and its size correlated with the size of the VII ganglion itself and with the vast number of taste buds this nerve innervates in catfish (Landacre, F. L., 1907; 1910). More recently, using grafting and transgenic technology in axolotls, we find that indeed epibranchial placodes give rise to sensory neurons that exclusively innervate taste buds (Harlow, D. and Barlow, L. A., in preparation). Moreover, neural crest-derived sensory neurons in these same ganglia provide only somatosensory innervation to the oral and pharyngeal epithelium, and never innervate taste buds. Thus despite coalescence of these two cell populations within the ganglia providing gustatory innervation to the oral cavity, the mature function of these neurons is determined by their embryonic origin.

Once these sensory neurons have come together to form the nascent cranial ganglia, these cells begin to send out processes both distally and proximally to pathfind to their peripheral taste bud and central nervous system (CNS) targets.

4.06.6.2 Development of Peripheral Taste Innervation

In rodents, fungiform taste buds and taste papillae are innervated by the chorda tympani (gustatory) branch of the VII nerve and the lingual (somatosensory) branch of the V nerve, while the circumvallate receives both gustatory and somatosensory innervation from the glossopharyngeal (IX) nerve. Fungiform innervation commences shortly after ganglion formation as the chorda tympani and lingual nerve fibers penetrate the base of the developing tongue rudiment (Mbiene, J.-P. and Mistretta, C. M., 1997; Scott, L. and Atkinson, M. E., 1998; Mbiene, J.-P., 2004), while the glossopharyngeal nerve enters the tongue 1 day later (Mbiene, J.-P. and Mistretta, C. M., 1997). Almost immediately after entering the tongue, the chorda tympani and lingual nerves, which have thus far traveled into the tongue as a unitary bundle, now defasciculate, and branches of each nerve extend toward the dorsal epithelial surface (Mbiene, J.-P., 2004). During this initial sensory innervation, the bulk of sensory fiber growth into the tongue in both rats and mice is from the chorda tympani nerve; growth of lingual fibers appears to follow the chorda tympani into the tongue (Rochlin, M. W. and Farbman, A. I., 1998; Mbiene, J.-P., 2004). As chorda tympani axons approach the epithelial surface they branch profusely. However, even 1 day prior to target innervation (at E13.5 in mice), only a few nerve bundles are specifically directed toward developing fungiform placodes (Mbiene, J.-P., 2004). In mice at this age, many nerve bundles have been described as having brush-like endings with numerous filopodia (Mbiene, J.-P., 2004). If so, these endings could be sampling the local environment to locate molecular cues that then guide the fiber bundles to individual fungiform placodes. Once chorda tympani fiber bundles are close to the epithelial surface they begin to grow toward taste placodes. Although lingual nerve bundles branch as they near the epithelial surface, growth of lingual fibers is not directed toward fungiform placodes (Mbiene, J.-P., 2004).

The initial progression of chorda tympani and lingual innervation in the tongue varies considerably among species, even between different rodents. In rats and mice, the chorda tympani is the first nerve to enter the tongue followed by the lingual fibers, while in hamsters, the lingual nerve is the first to enter the tongue and provides the bulk of its innervation (Whitehead, M. C. and Kachele, D. L., 1994). Also, unlike lingual innervation in rats and mice, the lingual nerve bundles in hamsters are actually directed toward fungiform papillae, with chorda tympani nerve bundles following (Whitehead, M. C. and Kachele, D. L., 1994).

By E14 in hamster, E14.5 in mouse, and E16 in rat, chorda tympani fibers have entered the core of the fungiform papillae and are clearly associated primarily with gustatory organs (Figure 3(a); Farbman, A. I. and Mbiene, J.-P., 1991; Whitehead, M. C. and Kachele, D. L., 1994; Mbiene, J.-P., 2004). By E16.5–E17 in rat and E14.5–E15 in mouse, chorda tympani fibers first penetrate the epithelium of individual taste papillae (Farbman, A. I., 1965; Farbman, A. I. and Mbiene, J.-P., 1991; Hall, J. M. *et al.*, 1999; Mbiene, J.-P. and Roberts, J. D., 2003). Initial innervation is sparse, but additional fibers continue to invade the epithelium at later embryonic ages. Synapse-like structures between gustatory afferents and fungiform taste buds are first observed several days after the initial innervation of fungiform papillae (Mbiene, J.-P. and Farbman, A. I., 1993). The sequence of events is comparable for the circumvallate papilla; glossopharyngeal fibers invade the circumvallate papilla core by mid-gestation and occasionally penetrate the epithelial surface (Figure 5(b); Mbiene, J.-P. and Mistretta, C. M., 1997). Circumvallate taste buds appear within the epithelial trenches and develop synapses during early postnatal development (Kinnamon, J. C. *et al.*, 2005). During this time, a robust neural plexus surrounds the

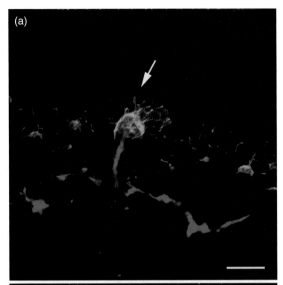

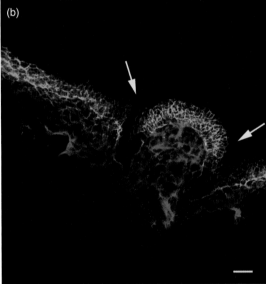

Figure 3 By E14.5 in the mouse, developing gustatory epithelia (anti-Shh; green) are associated with nerve fibers (anti-GAP43; red). Innervation has penetrated the epithelia of the developing fungiform papillae (a. arrow), although a neural bud is not yet present. The core of the developing circumvallatc papillae is filled with nerve fibers (b, between arrows). Scale bar = 20 μm.

taste bud; in addition to afferent synapses, this neural plexus contains neuroneuronal synapses. The role of this neural plexus is unclear, but it could be important for organizing initial taste bud–neuron connectivly.

Because chorda tympani nerve bundles are already associated with newly formed fungiform papillae by E14.5, the accuracy of initial targeting can be quantified at this age (Figure 4; Lopez, G. F. and Krimm, R. F., 2006b). This early targeting is incredibly accurate

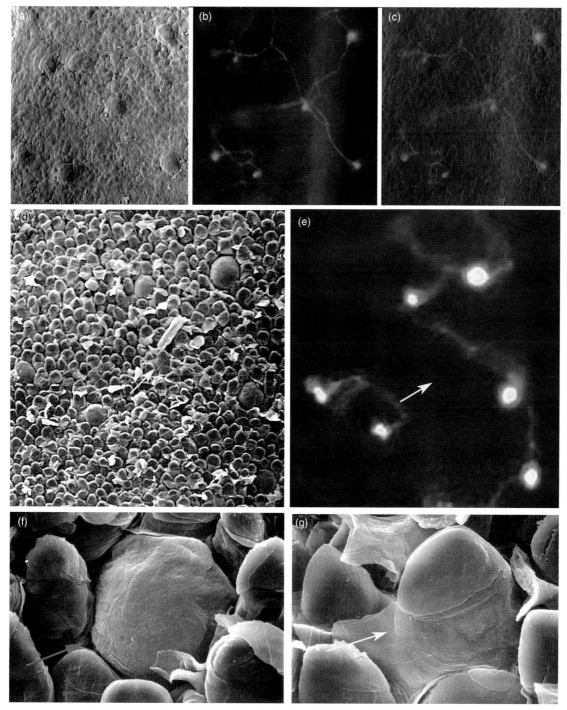

Figure 4   The locations of fungiform papillae (a,d) and innervation patterns (b,e) can be visualized in the same tongues by labeling first with DiI and then processing for scanning electron microscopy. Most fungiform papillae are innervated at both E14.5 (a,b,c) and E18.5 (d,e,f,g), although a few remain uninnervated. By E18.5, any uninnervated papillae (yellow arrow in d,g) have begun to adopt a morphology more similar to surround filiform papillae. Innervated papillae (red arrow in d and f) have a normal appearance.

in that most fungiform papillae are innervated and most fiber bundles are associated with a fungiform papilla. However, a few fungiform papillae are lacking innervation, and the chorda tympani innervates a few regions of the tongue that have no papillae. Between E14.5 and E18.5, most fungiform papillae previously lacking innervation now receive innervation, and any aberrant fiber bundles withdraw from nongustatory epithelium. Thus, a period of embryonic plasticity and rearrangement follows initial innervation.

Although the path gustatory axons take to their final destination is well described, axon guidance mechanisms are less clear. *In vitro*, initial axon outgrowth from gustatory ganglia requires addition of a neurotrophin to the culture media (Rao, H. *et al.*, 1997; Rochlin, M. W. *et al.*, 2000; Wiklund, P. and Ekstrom, P. A., 2000). Neurotrophins are a group of structurally and functionally related growth factors that exert a major influence on numerous aspects of sensory and sympathetic neuron development. The neurotrophins, brain-derived neurotrophic factors (BDNFs), neurotrophin-4 (NT4), and glial-derived neurotrophic factors (GDNFs) are capable of supporting neurite outgrowth of geniculate ganglion (VII) neurons, while neurotrophin-3 (NT3) and nerve growth factor (NGF) are not (Rao, H. *et al.*, 1997; Rochlin, M. W. *et al.*, 2000). Interestingly, removal of BDNF or NT4 does not disrupt the ability of chorda tympani axons to reach the tongue (Lopez, G. F. and Krimm, R. F., 2006a), indicating that neurotrophins are also functionally redundant for supporting initial axonal outgrowth *in vivo*.

Because the growth of chorda tympani axon bundles appears to follow predetermined pathways (Mbiene, J.-P. and Mistretta, C. M., 1997), the chorda tympani is probably guided by molecular cues in the environment. It is likely that multiple chemoattractive and chemorepulsive cues are required to direct gustatory axons from the geniculate (nVII) and petrosal (nIX) ganglia to the tongue and then to individual fungiform and circumvallate placodes (Tessier-Lavigne, M. and Goodman, C. S., 1996). For example, one factor produced by the tongue mesenchyme may encourage initial tongue innervation. A second factor produced by early gustatory epithelia may allow chorda tympani fibers to distinguish gustatory from surrounding epithelium. Finally, multiple chemorepulsive cues likely prevent chorda tympani axons from invading areas prematurely, from exiting the appropriate axonal pathway prematurely, and from innervating nongustatory regions of the epithelium.

One chemorepulsive factor important during both trigeminal and gustatory axon guidance is semaphorin 3A (Sema3A; Rochlin, M. W. and Farbman, A. I., 1998; Rochlin, M. W. *et al.*, 2000), which is expressed in developing tongue (Giger, R. *et al.*, 1996). Sema3A expression decreases medially to laterally across the tongue surface and prevents premature and aberrant growth of trigeminal and gustatory fibers into the tongue mid-region (Rochlin, M. W. and Farbman, A. I., 1998; Rochlin, M. W. *et al.*, 2000). In addition, as geniculate fibers near the epithelial surface, Sema3A may prevent premature penetration of the epithelium (Vilbig, R. *et al.*, 2004).

The lingual epithelium also produces chemoattractants (Gross, J. B. *et al.*, 2003; Vilbig, R. *et al.*, 2004). Axons of gustatory neurons isolated from the epibranchial placodes of axolotl are attracted to the pharyngeal endoderm (Gross, J. B. *et al.*, 2003). In addition, tongue explants from rats attract geniculate ganglion axons in culture (Vilbig, R. *et al.*, 2004). The chemoattractive factor emanating from the tongue epithelium has not yet been identified; however, it is probably not a neurotrophin because its chemoattractive properties are not blocked by the addition of high concentrations of neurotrophic factors to the culture media (Vilbig, R. *et al.*, 2004). This unidentified factor may be responsible for encouraging initial gustatory axon growth into the tongue and/or may direct chorda tympani branches to grow toward the dorsal tongue epithelia.

Although neurotrophins are not required to attract gustatory axons to the tongue, *in vitro* and *in vivo* studies support a chemotropic role for neurotrophins in the developing taste system. Growth cones reorient and grow toward a neurotrophin source (Ming, G. *et al.*, 1997; Paves, H. and Saarma, M., 1997; Ming, G. *et al.*, 1999; Tucker, K. L. *et al.*, 2001) and axons of the geniculate ganglion are attracted to beads containing BDNF (Rochlin, M. W. *et al.*, 2006). In the tongue, BDNF is located in gustatory epithelium while the surrounding somatosensory epithelium primarily expresses NT3 (Nosrat, C. A. and Olson, L., 1995; Nosrat, C. A. *et al.*, 1996; Nosrat, C. A., 1998; Fan, L. *et al.*, 2004). Thus, BDNF and NT3 are in the correct locations to direct gustatory fibers to fungiform papillae and somatosensory fibers to nongustatory epithelium. Consistent with these possible roles for BDNF and NT3, the disruption of the normal BDNF expression pattern *in vivo* prevents chorda tympani innervation of fungiform papillae (Ringstedt, T. *et al.*, 1999; Krimm, R. F. *et al.*, 2001; Lopez, G. F. and Krimm, R. F., 2006a). Specifically, overexpression of BDNF throughout the lingual epithelium, under the control of a keratin 14

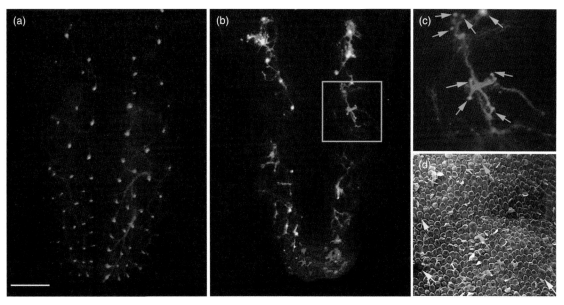

Figure 5 Innervation patterns are disrupted in BDNF-OE mice (b) compared with wild type mice (a). Fibers from the chorda tympani branch as they near the tongue surface in wild type mice (a) and innervate specific regions of the epithelium. In BDNF-OE mice (b), many thin processes are present near the epithelial surface; however, most fail to innervate fungiform papillae. The apparent disorganization of gustatory innervation can be seen more clearly in higher magnification of the tongue mid-region (c) along with scanning electron microscopy of the same region (d). While some of fungiform papillae appear to be innervated (blue arrows in c,d), some regions of the tongue epithelium are innervated by gustatory fibers even though no fungiform papillae are near by (green arrows in c). In addition, some fungiform papillae clearly remain uninnervated (yellow arrows in d). Scale bar in (a) = 500 μm and applies to (a) and (b).

promoter, disrupts the initial innervation of fungiform papillae and alters the normal morphology of gustatory sensory neurons (Figure 5; Krimm, R. F. *et al.*, 2001; Lopez, G. F. and Krimm, R. F., 2006a). BDNF over-expression also increases branching and attracts gustatory innervation to inappropriate regions of the taste epithelium. Finally, BDNF overexpression in both muscle and gustatory neurons disrupts chorda tympani target innervation (Ringstedt, T. *et al.*, 1999). Importantly, removal of functional BDNF also disrupts normal targeting (Lopez, G. F. *et al.*, 2006). In BDNF$^{-/-}$ mice, geniculate axons enter the tongue and grow toward the lingual epithelium normally, but once they near the epithelial surface, they grow wildly in the lamina propria and fail to penetrate the developing gustatory epithelia. Taken together, these findings indicate hat the chemoattractant properties of BDNF allow chorda tympani fibers to select and penetrate fungiform placodes as their appropriate target. It is still not clear whether NT3 is required to direct soma-tosensory afferents toward the nongustatory lingual epithelium, but NT3 expression patterns are consistent ith this possibility (Nosrat, C. A. and Olson, L., 1998).

4.06.6.3 Development of Projections of Gustatory Neurons to the Central Nervous System

The central axons of all cranial ganglion sensory neurons, including those of gustatory ganglia, project to the hindbrain. In axolotl, projections from early forming ganglia reach both peripheral and central targets earlier than later forming ganglia (Fritzsch, B. *et al.*, 2005). Each cranial ganglion forms its central and peripheral projections simultaneously, ruling out the possibility that innervation of peripheral targets directs central innervation and vice versa. Instead, the central and peripheral targets for these ganglia appear to be selected as the ganglia develop, with more ventrally located ganglia developing first and innervating central and peripheral targets that form early. More dorsally located ganglia develop later and innervate peripheral and central targets that develop later, thus suggesting that early molecular cues for ganglion differentiation may also influence the selection of central targets.

In rat CNS, geniculate ganglion axons begin to invade the first central nucleus of the gustatory system in the hindbrain, the nucleus of the solitary tract

(NTS), at E15 (Zhang, L. L. and Ashwell, K. W., 2001), the same age that peripheral fiber bundles arrive at fungiform papillae (Farbman, A. I. and Mbiene, J.-P., 1991). Thus, similar to axolotl, rat CNS projections from gustatory ganglia develop at the same time as peripheral projections. Consequently, peripheral connections with fungiform papillae probably do not control CNS targeting. Rather, as in axolotl, molecular cues for timing ganglia differentiation may also direct CNS targeting. Consistent with this idea, in mammals, the same transcription factor (e.g., Phox2b) regulates the development of geniculate and petrosal ganglia and their CNS target, the NTS, thus coordinating visceral sensory circuitry (Dauger, S. et al., 2003).

Centrally projecting taste afferents find their way to the hindbrain by following the path of migrating neuroglial neural crest from the hindbrain (Begbie, J. and Graham, A., 2001b). When the neural crest is ablated prior to migration, gustatory ganglion neurons fail to migrate internally from the epibranchial placodes. Central processes of these neurons form aberrant connections, frequently with adjacent ganglia, and fail to project toward the hindbrain. Thus, early developmental events such as the timing of ganglion differentiation, transcription factors, and migrating neural crest cells function together to regulate the initial growth of gustatory projections to the NTS.

4.06.6.4 Survival and Maintenance of Gustatory Neurons and Taste Buds

In addition to their demonstrated role in targeting of gustatory afferents, neurotrophins are also important for gustatory neuron survival. The traditional view of neurotrophic function is that sensory neuron targets regulate the number of surviving neurons by secreting limited amounts of neurotrophic compounds. BDNF may function as an important target-derived factor in the taste system, influencing neuronal survival after initial innervation of gustatory epithelium. BDNF mRNA is expressed in cells of the developing gustatory epithelium and taste buds (Nosrat, C. A. and Olson, L., 1995; Nosrat, C. A. et al., 1996), and BDNF expression begins prior to, and occurs independently of, gustatory innervation (Nosrat, C. A. and Olson, L., 1995; Nosrat, C. A. et al., 2001). The geniculate ganglion also produces BDNF (Schecterson, L. C. and Bothwell, M., 1992), which may influence neuronal survival through an autocrine or paracrine mechanism (Acheson, A. et al., 1995; Acheson, A. and Lindsay, R. M., 1996). The central target of geniculate and petrosal

neurons, the NTS, also produces BDNF, which may regulate survival of gustatory neurons (Acheson, A. et al., 1995).

Regardless of its source, BDNF is an important regulator of gustatory neuron number. $BDNF^{-/-}$ mice lose approximately half of their geniculate and nodose/petrosal ganglion neurons during development (Conover, J. C. et al., 1995; Liu, X. et al., 1995; Erickson, J. T. et al., 1996; Liebl, D. J. et al., 1997). Although both of these ganglia innervate taste and nontaste regions, many of the neurons lost in $BDNF^{-/-}$ mice probably innervate taste buds. Taste buds, in turn, require gustatory innervation for their continued maintenance (Hosley, M. A. et al., 1987; Nagato, T. et al., 1995; Sollars, S. I., 2005), and it is well established that fungiform and circumvallate taste buds are lost by birth in $BDNF^{-/-}$ mice (Nosrat, C. A. et al., 1997; Zhang, C. et al., 1997; Cooper, D. and Oakley, B., 1998; Oakley, B. et al., 1998; Mistretta, C. M. et al., 1999; Sun, H. and Oakley, B., 2002). The most likely explanation for these findings is that BDNF produced in either taste papillae, gustatory ganglia, or NTS is required for the survival of gustatory neurons, which in turn maintain taste buds. Therefore, the neurons lost in $BDNF^{-/-}$ mice are most likely gustatory neurons; however, retrograde labeling has not yet been used to determine the number of gustatory neurons that are lost in $BDNF^{-/-}$ mice.

Geniculate neurons are first lost in $BDNF^{-/-}$ mice between E12.5 and E14.5, indicating these neurons are first lost just before or as they innervate gustatory epithelia (Patel, A. V. and Krimm, R. F., 2005). However, geniculate neuron loss continues after E14.5, indicating the BDNF is required to support geniculate neuron survival throughout embryonic development.

BDNF is not the only neurotrophin that regulates taste system development. In addition, NT4 is expressed in tongue, taste buds, and geniculate ganglion neurons (Cho, T. T. and Farbman, A. I., 1999; Farbman, A. I. et al., 2004b; Takeda, M. et al., 2005a), and $NT4^{-/-}$ mice lose approximately half of their geniculate and nodose/petrosal ganglion neurons during development (Conover, J. C. et al., 1995; Liu, X. et al., 1995; Erickson, J. T. et al., 1996). Moreover, double $BDNF^{-/-}/NT4^{-/-}$ null mutants lose $\geq 90\%$ of geniculate and petrosal neurons (Conover, J. C. et al., 1995; Liu, X. et al., 1995; Erickson, J. T. et al., 1996), indicating that NT4- and BDNF-dependent neurons may represent two separate populations. Fungiform papillae are lost by birth in $NT4^{-/-}$ mice (Liebl, D. J. et al., 1999), but circumvallate taste buds are not, indicating that at least some of the NT4-dependent neurons in the geniculate ganglion are gustatory, while those in the petrosal ganglion are not. In NT4$^{-/-}$ mice, neurons are lost earlier than in BDNF$^{-/-}$ mice before E12.5, which is before target innervation (Patel, A. V. and Krimm, R. F., 2005). Also in NT4$^{-/-}$ mice, there is a separate wave of neuron loss between E14.5 and E16.5, distinct from neuronal loss in BDNF$^{-/-}$ mice. Thus, it is likely that NT4 plays two distinct roles at different stages of geniculate neuron development.

In addition to BDNF and NT4, NT3 may also play a regulatory role in gustatory neuron development. There is an abundance of NT3 in the tongue (Nosrat, C. A. *et al.*, 1996; Nosrat, C. A. and Olson, L., 1998; Fan, L. *et al.*, 2004) and ~45% of geniculate and nodose/petrosal neurons are lost in NT3$^{-/-}$ mice (Liebl, D. J. *et al.*, 1997). Oddly, while no fungiform papillae are lost in NT3$^{-/-}$ mice, double BDNF$^{-/-}$/NT3$^{-/-}$ null mutants lose more fungiform papillae than BDNF$^{-/-}$ mice (Nosrat, C. A. *et al.*, 2004), suggesting that NT3-supported trigeminal neurons maintain taste buds in the absence of BDNF. Alternatively, NT3 could function in conjunction with BDNF to support gustatory neurons, or NT3 could have a direct, nerve-independent influence on taste bud maintenance.

The neurotrophins, BDNF and NT4, function via the same receptors, TrkB, a member of the tyrosine kinase family, and p75, a pan-neurotrophin receptor. All mouse geniculate neurons express the TrkB receptor early in development (Yamout, A. *et al.*, 2005). While some geniculate neurons in rat continue to express TrkB postnatally and into adulthood, most express other neurotrophin receptors (Cho, T. T. and Farbman, A. I., 1999; Farbman, A. I. *et al.*, 2004a; 2004b). Expression of neurotrophin receptors in the geniculate ganglion may vary across rodent species; however, more likely, TrkB is downregulated postnatally. Mice lacking TrkB lose geniculate neurons by E12.5 (Fritzsch, B. *et al.*, 1997), and by E13.5, neuronal loss is greater in TrkB$^{-/-}$ mice than in NT4$^{-/-}$ or BDNF$^{-/-}$ mice, indicating that both neurotrophins regulate survival via TrkB. Because most geniculate and nodose/petrosal neurons are lost in TrkB$^{-/-}$ mice, taste buds may not receive gustatory innervation in TrkB$^{-/-}$ mice. Some fungiform taste buds are present at birth in TrkB$^{-/-}$ mice (Fritzsch, B. *et al.*, 1997); however, taste bud number has not been quantified, so how many remain is unclear.

Other factors are also required to maintain gustatory neurons and taste buds. Disruption of dystonin, a molecule important for cytoskeletal organization, causes a loss of petrosal ganglion neurons and a loss of circumvallate taste buds (Ichikawa, H. *et al.*, 2006). Although dystonin regulates neuronal cytoskeletal organization, only some subpopulations of neurons, primarily sensory neurons, are effected by its removal. There is a possible link between dystonin function and neuronal dependence on the neurotrophins NT3 and GDNF, in that the somatosensory neurons that are most sensitive to disruption of dystonin also tend to depend on either NT3 or GDNF for survival (Carlsten, J. A. *et al.*, 2001). Possible roles of NT3 in gustatory neuron development/maintenance have already been discussed. Currently, it is unclear whether GDNF plays a role in gustatory development. GDNF is expressed by the developing tongue (Nosrat, C. A. *et al.*, 1998) and adult taste buds (Takeda, M. *et al.*, 2004; 2005b) and geniculate neurons express GDNF receptors (Yamout, A. *et al.*, 2005). Furthermore, removal of GDNF reduces the number of petrosal ganglion neurons (Erickson, J. T. *et al.*, 1996). Thus a role for GDNF in gustatory system development or maintenance remains likely.

In addition to the known factors, an unidentified factor produced by gustatory neurons acts to maintain taste buds during development and into adulthood (Hosley, M. A. *et al.*, 1987; Oakley, B. *et al.*, 1990; Nagato, T. *et al.*, 1995; Sollars, S. I. *et al.*, 2002; Sollars, S. I., 2005). It is reasonable to hypothesize that this unknown factor is a growth factor, and growth factors typically function via tyrosine kinase receptors. With this possibility in mind, McLaughlin S. K. (2000) amplified both receptor and nonreceptor tyrosine kinases from a cDNA library to identify tyrosine kinases expressed in taste buds. The tyrosine kinase receptors, ErbB1 (the epithelial growth factor receptor, (Egfr)), ErbB2, ErbB3, and c-kit tyrosine kinase receptors are all preferentially expressed in taste buds (McLaughlin, S. K., 2000). ErbB2 typically functions as a coreceptor, but ErbB1 and ErbB3 have ligand-binding abilities. ErbB1/ErbB1 and ErbB1/ErbB2 heterodimers typically bind EGF, TGFα, and amphiregulin. Fungiform papillae and/or taste buds are lost and malformed in mice lacking Egfr/ErbB1 (Miettinen, P. J. *et al.*, 1995; Threadgill, D. W. *et al.*, 1995), but sensory innervation is unaffected (Sun, H. and Oakley, B., 2002). Thus, Egfr/ErbB1 signaling is important for the maintenance of taste buds.

One ligand of this receptor, EGF, is expressed in the tongue where it influences early fungiform placode development (Mistretta, C. M. *et al.*, 2005), indicating that taste bud loss in ErbB1$^{-/-}$ mice could be due to disruption of early placode development. Another intriguing possibility is that EGF in saliva maintains taste buds and fungiform papillae via ErbB1 (Morris-Wiman, J. *et al.*, 2000). Consistent with this interpretation, taste buds in circumvallate and foliate papillae are unaffected by either mutation of Egfr/ErbB1 or salivary gland removal (Morris-Wiman, J. *et al.*, 2000; Sun, H. and Oakley, B., 2002). Also, the addition of EGF to drinking water rescues the taste bud loss seen with salivary gland removal (Morris-Wiman, J. *et al.*, 2000). Lastly, it is possible that gustatory neurons produce an Erb ligand that is important for taste bud maintenance. This would be consistent with the role of other Erb ligands in the maintenance of sensory receptors (Hippenmeyer, S. *et al.*, 2002). Thus, while Erb receptor ligands may function to maintain taste buds, the exact function and source of these ligands is unclear. Equally elusive is the nature of the factor(s) produced by gustatory neurons that serves to support both taste bud differentiation during development and taste bud maintenance in adulthood.

4.06.7 Postnatal Plasticity in Taste Bud Innervation

Taste buds continue to develop postnatally, to turn over, and to reestablish connections with innervating neurons even into adulthood (Beidler, L. M. and Smallman, R. L., 1965). Changes in peripheral gustatory innervation patterns are related to, and may even regulate, changes in taste bud development. In adult rats, taste bud size is correlated with the number of geniculate ganglion neurons that innervate it (Krimm, R. F. and Hill, D. L., 1998b). This relationship develops over a prolonged period of postnatal development and is not observed until postnatal day 40 (PN40). Between PN10 and PN40, geniculate ganglion cells rearrange their peripheral connections (Krimm, R. F. and Hill, D. L., 2000), withdrawing innervation from certain taste buds, then innervating others. This constant rearrangement of peripheral connectivity does not appear to establish the relationship between taste bud size and geniculate neuron number. That is, individual ganglion neurons do not remove innervation from small taste buds in favor of larger taste buds. Instead, the number of neurons that innervate a given taste bud at PN10 predicts taste bud size at PN40. Thus, the number of neurons innervating a given taste bud is established early in development and taste buds grow to match the number of neurons innervating them.

Taste buds grow between PN10 and PN40 by adding taste cells (Hendricks, S. J. *et al.*, 2004). Taste

cells are constantly turning over, undergoing a continuous process of differentiation followed by death (Beidler, L. M. and Smallman, R. L., 1965). Postnatal differences in taste cell proliferation, differentiation, or life span could all increase the number of taste bud cells. The approximate 10-day life span of taste cells appears to remain constant during development (Hendricks, S. J. *et al.*, 2004) as does the proliferation rate of basal cells. However, the rate of taste cell addition accelerates. It is unclear what factor(s) controls the addition of new taste cells. Because taste buds increase in size in proportion to the amount of neural innervation they receive (Krimm, R. F. and Hill, D. L., 2000), the critical factor may be neuronal. Alternatively, factor(s) produced in fungiform papillae could regulate both the rate of taste cell addition and the number of neurons innervating the taste bud.

The number of geniculate neurons that typically innervate a fungiform taste bud varies considerably across rodent species. Rat taste buds are innervated by 3–14 ganglion neurons (Krimm, R. F. and Hill, D. L., 1998a; Shuler, M. G. *et al.*, 2004) while mouse taste buds are innervated by substantially fewer (2–7) (Zaidi, F. N. and Whitehead, M. C., 2006). Hamster fungiform taste buds are innervated by many more geniculate neurons (5–35) than either rat or mouse taste buds (Whitehead, M. C. *et al.*, 1999; Zaidi, F. N. and Whitehead, M. C., 2006). Differences in peripheral branching characteristics of ganglion cells (Zaidi, F. N. and Whitehead, M. C., 2006), rather than differences in the ratio of fungiform taste buds to geniculate neurons, appear to account for species variation. That is, geniculate neurons branch more and innervate more taste buds in hamsters than in mice (Zaidi, F. N. and Whitehead, M. C., 2006). It is unclear whether postnatal plasticity also diverges across species as a reflection of geniculate neuron branching.

4.06.8 Plasticity of Central Taste Connections

As discussed earlier, initial projection of gustatory axons to the NTS is dictated by early developmental cues, whereas further refinement occurs over a protracted developmental period after birth. The combined terminal field of the chorda tympani and the greater superficial petrosal (GSP) afferents steadily increases between PN7 and PN25 (Lasiter, P. S. and Bulcourf, B. B., 1995), illustrating that terminal field development is not complete until late postnatal stages. The terminal field of the chorda tympani nerve increases until PN15, then decreases in size (Sollars, S. I. *et al.*, 2006) before becoming adult-like at PN35. Postsynaptic neurons in the gustatory NTS also undergo remodeling between PN22 and PN28 (Liu, Y. S. *et al.*, 2000). For example, dendritic fields of salt-sensitive geniculate ganglion neurons decrease during this postnatal period. Because NTS connections are refined during late postnatal development, remodeling of these connections may be influenced by sensory experiences in the taste system.

In most sensory systems, including the taste system, blocking normal sensory experiences alters the development of central connections (Hubel, D. H. and Wiesel, T. N., 1970; Cummings, D. M. and Brunjes, P. C., 1997; Fox, K. *et al.*, 2002). Similarly, fungiform papillae/taste bud damage at PN2 stunts the terminal field expansion of the combined chorda tympani/GSP nerves in the NTS (Lasiter, P. L. and Kachele, D. L., 1990). The underlying mechanisms are unclear but may include injury-induced expression of numerous growth and/or immune factors or the loss of primary sensory afferents.

Developmental sodium restriction also influences later sensory experiences. Maternal sodium restriction during embryonic development specifically blocks postnatal development of chorda tympani nerve responses to sodium salts (Hill, D. L. and Przekop, P. R., 1988) and substantially alters the CNS projections of gustatory neurons. Chorda tympani and glossopharyngeal terminal fields within the rostral NTS are enlarged (King, C. T. and Hill, D. L., 1991; Krimm, R. F. and Hill, D. L., 1997; May, O. L. and Hill, D. L., 2006; Sollars, S. I. *et al.*, 2006) because connections are not refined and eliminated between PN25 and PN35, resulting in adult projections with an early developmental morphology (Sollars, S. I. *et al.*, 2006). Interestingly, unlike for the chorda tympani nerve, sodium responses in the GSP nerve are unaffected by sodium deprivation as is the size of the terminal field (Sollars, S. I. and Hill, D. L., 2000; May, O. L. and Hill, D. L., 2006). Alterations due to sodium restriction are not limited to the terminal fields of the affected primary afferents; postsynaptic neurons are also affected. Dendritic arbors of putative projection neurons in the gustatory NTS are longer in sodium-restricted rats than in controls (King, C. T. and Hill, D. L., 1993). The mechanisms through which sodium deprivation influences gustatory CNS development are not yet clear. Probably, the failure of the chorda tympani nerve to develop sodium sensitivity

reduces total postnatal chorda tympani activity, thus preventing postnatal refinement of the terminal field.

4.06.9   Conclusion

The general organization of the taste system is conserved across vertebrate species, from fish to humans: bipolar sensory neurons of the VII, IX, and X cranial ganglia convey taste information from aggregates of spindle-shaped cells, taste buds, located in the oral and pharyngeal epithelia to the first relay nucleus, the NTS, in the hindbrain. We propose that this conservation of anatomy is reflected in a conserved and basic plan for the development of elements of the taste system. Taste bud progenitors arise from the local epithelium and are induced and patterned via epithelium-intrinsic intercellular communication. While little is known specifically about the development of the NTS, it is clear that these cells arise from the hindbrain in all vertebrates. Gustatory neurons arise from epibranchial placodes and pathfind simultaneously to innervate both peripheral and central targets, allowing formation of a functional taste system.

Of course, there are variations in this theme, when one examines the details of development within several species. Chief among the differences encountered is the role of innervation in the differentiation of taste buds. In amphibians, taste buds differentiate fully in the complete absence of nerves, while full cytodifferentiation likely depends upon innervation in rodents. Additionally, the development of taste buds in many species, including mammals, is interwoven with the development of specialized epithelial structures, such as taste papillae. Thus, the interconnectedness of buds and papillae may obscure our ability to readily detect commonalities in taste organ developmental mechanisms across species.

However, by continuing to explore taste system development in a number of vertebrates, perhaps expanding to employ additional developmental models, such as zebrafish and chick, we will gain a clearer view of the basic mechanisms underlying development of this key sensory system.

References

Abu-Issa, R., Smyth, G., Smoak, I., Yamamura, K., and Meyers, E. N. 2002. Fgf8 is required for pharyngeal arch and cardiovascular development in the mouse. Development 129(19), 4613–4625.

Acheson, A. and Lindsay, R. M. 1996. Non target-derived roles of the neurotrophins. Philos. Trans. R. Soc. Lond. B Biol. Sci. 351(1338), 417–422.

Acheson, A., Conover, J. C., Fandl, J. P., DeChiara, T. M., Russell, M., Thadani, A., Squinto, S. P., Yancopoulos, G. D., and Lindsay, R. M. 1995. A BDNF autocrine loop in adult sensory neurons prevents cell death. Nature 374(6521), 450–453.

AhPin, P., Ellis, S., Arnott, C., and Kaufman, M. H. 1989. Prenatal development and innervation of the circumvallate papilla in the mouse. J. Anat. 162, 33–42.

Armstrong, J. B. and Malacinski, G. M. 1989. Developmental Biology of the Axolotl, p. 327. Oxford University Press.

Artavanis-Tsakonas, S. 1988. The molecular biology of the Notch locus and the fine tuning of differentiation in Drosophila. Trends Gen. 4, 95–100.

Atema, J. 1971. Structures and functions of the sense of taste in the catfish (Ictalurus natalis). Brain Behav. Evol. 4, 273–294.

Athar, M., Tang, X., Lee, J. L., Kopelovich, L., and Kim, A. L. 2006. Hedgehog signalling in skin development and cancer. Exp. Dermatol. 15(9), 667–677.

Baker, C. V. and Bronner-Fraser, M. 2001. Vertebrate cranial placodes I. embryonic induction. Dev. Biol. 232(1), 1–61.

Balemans, W. and Van Hul, W. 2002. Extracellular regulation of BMP signaling in vertebrates: a cocktail of modulators. Dev. Biol. 250(2), 231.

Barlow, L. A. 2001. Specification of pharyngeal endoderm is dependent on early signals from axial mesoderm. Development 128(22), 4573–4583.

Barlow, L. A. and Northcutt, R. G. 1995. Embryonic origin of amphibian taste buds. Dev. Biol. 169, 273–285.

Barlow, L. A. and Northcutt, R. G. 1997. Taste buds develop autonomously from endoderm without induction by cephalic neural crest or paraxial mesoderm. Development 124, 949–957.

Barlow, L. A. and Northcutt, R. G. 1998. The role of innervation in the development of taste buds: insights from studies of amphibian embryos. Ann. N. Y. Acad. Sci. 855, 58–69.

Barlow, L. A, Chien, C.-B., and Northcutt, R. G. 1996. Embryonic taste buds develop in the absence of innervation. Development 122, 1103–1111.

Begbie, J. and Graham, A. 2001a. The ectodermal placodes: a dysfunctional family. Philos. Trans. R. Soc. Lond. B Biol. Sci. 356(1414), 1655–1660.

Begbie, J. and Graham, A. 2001b. Integration between the epibranchial placodes and the hindbrain. Science 294(5542), 595–598.

Begbie, J., Ballivet, M., and Graham, A. 2002. Early steps in the production of sensory neurons by the neurogenic placodes. Mol. Cell Neurosci. 21(3), 502–511.

Beidler, L. M. and Smallman, R. L. 1965. Renewal of cells within taste buds. J. Cell Biol. 27, 263–272.

Bejsovec, A. 2005. Wnt pathway activation: new relations and locations. Cell 120(1), 11–14.

Belecky, T. L. and Smith, D. V. 1990. Postnatal development of palatal and laryngeal taste buds in the hamster. J. Comp. Neurol. 293, 646–654.

Bitgood, M. J. and McMahon, A. P. 1995. Hedgehog and Bmp genes are coexpressed at many diverse sites of cell–cell interaction in the mouse embryo. Dev. Biol. 172, 126–138.

Blanpain, C., Horsley, V., and Fuchs, E. 2007. Epithelial stem cells: turning over new leaves. Cell 128(3), 445–458.

Botchkarev, V. A. and Sharov, A. A. 2004. BMP signaling in the control of skin development and hair follicle growth. Differentiation 72(9-10), 512–526.

Boughter, J. D., Jr., Pumplin, D. W., Yu, C., Christy, R. C., and Smith, D. V. 1997. Differential expression of alpha-gustducin in taste bud populations of the rat and hamster. J. Neurosci. 17(8), 2852–2858.

Bradley, R. M. and Stern, I. B. 1967. The development of the human taste bud during the foetal period. J. Anat. 101, 743–752.

Caprio, J., Brand, J. G., Teeter, J. H., Valentincic, T., Kalinoski, D. L., Kohbara, J., Kumazawa, T., and Wegert, S. 1993. The taste system of the channel catfish: from biophysics to behavior. Trends Neurosci. 16, 192–197.

Carlsten, J. A., Kothary, R., and Wright, D. E. 2001. Glial cell line-derived neurotrophic factor-responsive and neurotrophin-3-responsive neurons require the cytoskeletal linker protein dystonin for postnatal survival. J. Comp. Neurol. 432(2), 155–168.

Cho, T. T. and Farbman, A. I. 1999. Neurotrophin receptors in the geniculate ganglion. Brain Res. Mol. Brain Res. 68(1–2), 1–13.

Cho, Y. K., Farbman, A. I., and Smith, D. V. 1998. The timing of alpha-gustducin expression during cell renewal in rat vallate taste buds. Chem. Senses 23(6), 735–742.

Chuong, C. M., Patel, N., Lin, J., Jung, H. S., and Widelitz, R. B. 2000. Sonic hedgehog signaling pathway in vertebrate epithelial appendage morphogenesis: perspectives in development and evolution. Cell Mol. Life Sci. 57(12), 1672–1681.

Ciani, L. and Salinas, P. C. 2005. WNTs in the vertebrate nervous system: from patterning to neuronal connectivity. Nat. Rev. Neurosci. 6(5), 351–362.

Conover, J. C., Erickson, J. T., Katz, D. M., Bianchi, L. M., Poueymirou, W. T., McClain, J., Pan, L., Helgren, M., Ip, N. Y., Boland, P., Friedman, B., Weigand, S., Vejsada, R., Kato, A., Dechiara, T. M., and Yancopoulos, G. D. 1995. Neuronal deficits, not involving motor neurons, in mice lacking BDNF and/or NT4. Nature 375(6528), 235–238.

Cooper, D. and Oakley, B. 1998. Functional redundancy and gustatory development in bdnf null mutant mice. Brain Res. Dev. Brain Res. 105(1), 79–84.

Cornell, R. A. and Eisen, J. S. 2005. Notch in the pathway: the roles of Notch signaling in neural crest development. Semin. Cell Dev. Biol. 16(6), 663–672.

Crump, J. G., Maves, L., Lawson, N. D., Weinstein, B. M., and Kimmel, C. B. 2004. An essential role for Fgfs in endodermal pouch formation influences later craniofacial skeletal patterning. Development 131(22), 5703–5716.

Cummings, D. M. and Brunjes, P. C. 1997. The effects of variable periods of functional deprivation on olfactory bulb development in rats. Exp. Neurol. 148(1), 360–366.

D'Amico-Martel, A. and Noden, D. M. 1983. Contributions of placodal and neural crest cells to avian cranial peripheral ganglia. Am. J. Anat. 166, 445–468.

DasGupta, R. and Fuchs, E. 1999. Multiple roles for activated LEF/TCF transcription complexes during hair follicle development and differentiation. Development 126(20), 4557–4568.

Dauger, S., Pattyn, A., Lofaso, F., Gaultier, C., Goridis, C., Gallego, J., and Brunet, J. F. 2003. Phox2b controls the development of peripheral chemoreceptors and afferent visceral pathways. Development 130(26), 6635–6642.

Delay, R. J., Kinnamon, J. C., and Roper, S. D. 1986. Ultrastructure of mouse vallate taste buds: II. Cell types and cell lineage. J. Comp. Neurol. 253(2), 242–252.

Dvorak, P., Dvorakova, D., and Hampl, A. 2006. Fibroblast growth factor signaling in embryonic and cancer stem cells. FEBS Lett. 580(12), 2869–2874.

El-Sharaby, A., Ueda, K., Kurisu, K., and Wakisaka, S. 2001a. Development and maturation of taste buds of the palatal epithelium of the rat: histological and immunohistochemical study. Anat. Rec. 263(3), 260–268.

El-Sharaby, A., Ueda, K., and Wakisaka, S. 2001b. Differentiation of the lingual and palatal gustatory epithelium of the rat as revealed by immunohistochemistry of alpha-gustducin. Arch. Histol. Cytol. 64(4), 401–409.

Erickson, J. T., Conover, J. C., Borday, V., Champagnat, J., Barbacid, M., Yancopoulos, G., and Katz, D. M. 1996. Mice lacking brain-derived neurotrophic factor exhibit visceral sensory neuron losses distinct from mice lacking NT4 and display a severe developmental deficit in control of breathing. J. Neurosci. 16(17), 5361–5371.

Fährmann, W. 1967. Licht- und electronenmikroskopische Untersuchungen an der Geschmacksknospe des neotenen Axolotls (Siredon mexicanum Shaw). Zeit Microsk Anat Forsch. 77, 117–152.

Fan, L., Girnius, S., and Oakley, B. 2004. Support of trigeminal sensory neurons by nonneuronal p75 neurotrophin receptors. Brain Res. Dev. Brain Res. 150(1), 23–39.

Farbman, A. I. 1965. Electron microscope study of the developing taste bud in rat fungiform papilla. Dev. Biol. 11, 110–135.

Farbman, A. I. 1980. Renewal of taste bud cells in rat circumvallate papillae. Cell Tissue Kinet. 13(4), 349–357.

Farbman, A. I. and Mbiene, J.-P. 1991. Early development and innervation of taste bud-bearing papillae on the rat tongue. J. Comp. Neurol. 304, 172–186.

Farbman, A. I., Brann, J. H., Rozenblat, A., Rochlin, M. W., Weiler, E., and Bhattacharyya, M. 2004a. Developmental expression of neurotrophin receptor genes in rat geniculate ganglion neurons. J. Neurocytol. 33(3), 331–343.

Farbman, A. I., Guagliardo, N., Sollars, S. I., and Hill, D. L. 2004b. Each sensory nerve arising from the geniculate ganglion expresses a unique fingerprint of neurotrophin and neurotrophin receptor genes. J. Neurosci Res. 78(5), 659–667.

Fox, K., Wallace, H., and Glazewski, S. 2002. Is there a thalamic component to experience-dependent cortical plasticity? Philos. Trans. R. Soc. Lond. B Biol. Sci. 357(1428), 1709–1715.

Fritzsch, B., Gregory, D., and Rosa-Molinar, E. 2005. The development of the hindbrain afferent projections in the axolotl: evidence for timing as a specific mechanism of afferent fiber sorting. Zoology (Jena) 108(4), 297–306.

Fritzsch, B., Sarai, P. A., Barbacid, M., and Silos-Santiago, I. 1997. Mice lacking the neurotrophin receptor trkB lose their specific afferent innervation but do develop taste buds. Int. J. Dev. Neurosci. 15, 563–576.

Gaiano, N. and Fishell, G. 2002. The role of notch in promoting glial and neural stem cell fates. Annu. Rev. Neurosci. 25, 471–490.

Giger, R., Wolfer, D. P., De Wit, G. M., and Verhaagen, J. 1996. Anatomy of rat semaphorin III/collapsin-1 mRNA expression and relationship to developing nerve tracts during neuroembryogenesis. J. Comp. Neurol. 375, 378–392.

Gomahr, A., Palzenberger, M., and Kotrschal, K. 1992. Density and distribution of external taste buds in cyprinids. Environ. Biol. Fishes 33, 125–134.

Graveson, A. C. and Armstrong, J. B. 1987. Differentiation of cartilage from cranial neural crest in the axolotl (Ambystoma mexicanum). Differentiation 35(1), 16–20.

Graziadei, P. P. and Graziadei, G. A. 1978. Observations on the ultrastructure of ganglion cells in the circumvallate papilla of rat and mouse. Acta Anat. (Basel) 100(3), 289–305.

Gross, J. B., Gottlieb, A. A., and Barlow, L. A. 2003. Gustatory neurons derived from epibranchial placodes are attracted to, and trophically supported by, taste bud-bearing endoderm in vitro. Dev. Biol. 264(2), 467–481.

Hall, J. M., Bell, M. L., and Finger, T. E. 2003. Disruption of sonic hedgehog signaling alters growth and patterning of lingual taste papillae. Dev. Biol. 255(2), 263–277.

Hall, J. M., Hooper, J. E., and Finger, T. E. 1999. Expression of sonic hedgehog, patched, and Gli1 in developing taste papillae of the mouse. J. Comp. Neurol. 406(2), 143–155.

Hamamichi, R., Asano-Miyoshi, M., and Emori, Y. 2006. Taste bud contains both short-lived and long-lived cell populations. Neuroscience 141(4), 2129–2138.

Hansen, A., Reutter, K., and Zeiske, E. 2002. Taste bud development in the zebrafish, Danio rerio. Dev. Dyn. 223(4), 483–496.

Harada, S., Yamaguchi, K., Kanemaru, N., and Kasahara, Y. 2000. Maturation of taste buds on the soft palate of the postnatal rat. Physiol. Behav. 68(3), 333–339.

Harada, S., Yamamoto, T., Yamaguchi, K., and Kasahara, Y. 1997. Different characteristics of gustatory responses between the greater superficial petrosal and chorda tympani nerves in the rat. Chem. Senses 22(2), 133–140.

Hendricks, S. J., Brunjes, P. C., and Hill, D. L. 2004. Taste bud cell dynamics during normal and sodium-restricted development. J. Comp. Neurol. 472(2), 173–182.

Herrick, C. J. 1903. On the phylogeny and morphological position of the terminal buds of fishes. J. Comp. Neurol. 13, 121–138.

Hill, D. L. and Przekop, P. R. 1988. Influences of dietary sodium on functional taste receptor development: a sensitive period. Science 241, 1826–1827.

Hippenmeyer, S., Shneider, N. A., Birchmeier, C., Burden, S. J., Jessell, T. M., and Arber, S. 2002. A role for neuregulin1 signaling in muscle spindle differentiation. Neuron 36(6), 1035–1049.

Hirota, M., Ito, T., Okudela, K., Kawabe, R., Hayashi, H., Yazawa, T., Fujita, K., and Kitamura, H. 2001. Expression of cyclin-dependent kinase inhibitors in taste buds of mouse and hamster. Tissue Cell 33(1), 25–32.

Hosley, M. A., Hughes, S. E., Morton, L. L., and Oakley, B. 1987. A sensitive period for the neural induction of taste buds. J. Neurosci. 7(7), 2075–2080.

Hubel, D. H. and Wiesel, T. N. 1970. The period of susceptibility to the physiological effects of unilateral eye closure in kittens. J. Physiol. 206(2), 419–436.

Ichikawa, H., De Repentigny, Y., Kothary, R., and Sugimoto, T. 2006. The survival of vagal and glossopharyngeal sensory neurons is dependent upon dystonin. Neuroscience 137(2), 531–536.

Itoh, N. and Ornitz, D. M. 2004. Evolution of the Fgf and Fgfr gene families. Trends Genet. 20(11), 563–569.

Iwatsuki, K., Liu, H. X., Gronder, A., Singer, M. A., Lane, T. F., Grosschedl, R., Mistretta, C. M., and Margolskee, R. F. 2007. Wnt signaling interacts with Shh to regulate taste papilla development. Proc. Natl Acad. Sci. U. S. A. 104(7), 2253–2258.

Jitpukdeebodintra, S., Chai, Y., and Snead, M. L. 2002. Developmental patterning of the circumvallate papilla. Int. J. Dev. Biol. 46(5), 755–763.

Jung, H. S., Oropeza, V., and Thesleff, I. 1999. Shh, Bmp-2, Bmp-4 and Fgf-8 are associated with initiation and patterning of mouse tongue papillae. Mech. Dev. 81(1–2), 179–182.

Kim, J. Y., Cho, S. W., Lee, M. J., Hwang, H. J., Lee, J. M., Lee, S. I., Muramatsu, T., Shimono, M., and Jung, H. S. 2005. Inhibition of connexin 43 alters Shh and Bmp-2 expression patterns in embryonic mouse tongue. Cell Tissue Res. 320(3), 409–415.

Kim, J. Y., Mochizuki, T., Akita, K., and Jung, H. S. 2003. Morphological evidence of the importance of epithelial tissue during mouse tongue development. Exp. Cell Res. 290(2), 217–226.

King, C. T. and Hill, D. L. 1991. Dietary sodium chloride deprivation throughout development selectively influences the terminal field organization of gustatory afferent fibers projecting to the rat nucleus of the solitary tract. J. Comp. Neurol. 303(1), 159–169.

King, C. T. and Hill, D. L. 1993. Neuroanatomical alterations in the rat nucleus of the solitary tract following early maternal

NaCl deprivation and subsequent NaCl repletion. J. Comp. Neurol. 333, 531–542.

Kingsley, D. M. 1994. The TGF-beta superfamily: new members, new receptors, and new genetic tests of function in different organisms. Genes Dev. 8(2), 133–146.

Kinnamon, J. C., Dunlap, M., and Yang, R. 2005. Synaptic connections in developing and adult rat taste buds. Chem. Senses 30(Suppl. 1), i60–i61.

Kohbara, J., Michel, W., and Caprio, J. 1992. Responses of single facial taste fibers in the channel catfish, *Ictalurus punctatus*, to amino acids. J. Neurophysiol. 68(4), 1012–1026.

Kollar, E. J. 1983. Epithelial–Mesenchymal Interactions in the Mammalian Integument: Tooth Development as a Model for Instructive Induction. In: Epithelial–Mesenchymal Interactions in Development (eds. R. H. Sawyer and J. F. Fallon), pp. 27–49. Praeger Publishers.

Krimm, R. F. and Hill, D. L. 1997. Early prenatal critical period for chorda tympani nerve terminal field development. J. Comp. Neurol. 378(2), 254–264.

Krimm, R. F. and Hill, D. L. 1998a. Innervation of single fungiform taste buds during development in rat. J. Comp. Neurol. 398(1), 13–24.

Krimm, R. F. and Hill, D. L. 1998b. Quantitative relationships between taste bud development and gustatory ganglion cells. Ann. N. Y. Acad. Sci. 855, 70–75.

Krimm, R. F. and Hill, D. L. 2000. Neuron/target matching between chorda tympani neurons and taste buds during postnatal rat development. J. Neurobiol. 43(1), 98–106.

Krimm, R. F., Miller, K. K., Kitzman, P. H., Davis, B. M., and Albers, K. M. 2001. Epithelial overexpression of BDNF or NT4 disrupts targeting of taste neurons that innervate the anterior tongue. Dev. Biol. 232(2), 508–521.

Kusakabe, Y., Miura, H., Hashimoto, R., Sugiyama, C., Ninomiya, Y., and Hino, A. 2002. The neural differentiation gene Mash-1 has a distinct pattern of expression from the taste reception-related genes gustducin and T1R2 in the taste buds. Chem. Senses 27(5), 445–451.

Landacre, F. L. 1907. On the place of origin and method of distribution of taste buds in *Ameiurus melas*. J. Comp. Neurol. 17, 1–66.

Landacre, F. L. 1910. The origin of the cranial ganglia in *Ameiurus*. J. Comp. Neurol. 20, 309–411.

Lasiter, P. S. and Bulcourf, B. B. 1995. Alterations in geniculate ganglion proteins following fungiform receptor damage. Dev. Brain Res. 89, 289–306.

Lasiter, P. L. and Kachele, D. L. 1990. Effects of early postnatal receptor damage on development of gustatory recipient zones within the nucleus of the solitary tract. Dev. Brain Res. 55, 57–71.

Le Douarin, N. M. and Kalcheim, C. 1999. The Neural Crest, p. 445. Cambridge University Press.

Lee, M. J., Kim, J. Y., Lee, S. I., Sasaki, H., Lunny, D. P., Lane, E. B., and Jung, H. S. 2006. Association of Shh and Ptc with keratin localization in the initiation of the formation of circumvallate papilla and von Ebner's gland. Cell Tissue Res. 325(2), 253–261.

Liebl, D. J., Mbiene, J. P., and Parada, L. F. 1999. NT4/5 mutant mice have deficiency in gustatory papillae and taste bud formation. Dev. Biol. 213(2), 378–389.

Liebl, D. J., Tessarollo, L., Palko, M. E., and Parada, L. F. 1997. Absence of sensory neurons before target innervation in brain-derived neurotrophic factor-, neurotrophin 3-, and TrkC-deficient embryonic mice. J. Neurosci. 17(23), 9113–9121.

Liu, F., Thirumangalathu, S., Gallant, N. M., Yang, S. H., Stoick-Cooper, C. L., Reddy, S. T., Andl, T., Taketo, M. M., Dlugosz, A. A., Moon, R. T., Barlow, L. A., and Millar, S. E.

2007. Wnt-beta-catenin signaling initiates taste papilla development. Nat. Genet. 39(1), 106–112.

Liu, H. X., Maccallum, D. K., Edwards, C., Gaffield, W., and Mistretta, C. M. 2004. Sonic hedgehog exerts distinct, stage-specific effects on tongue and taste papilla development. Dev. Biol. 276(2), 280–300.

Liu, X., Ernfors, P., Wu, H., and Jaenisch, R. 1995. Sensory but not motor neuron deficits in mice lacking NT4 and BDNF. Nature 375(6528), 238–241.

Liu, Y. S., Schweitzer, L., and Renehan, W. E. 2000. Development of salt responsive neurons in the nucleus of the solitary tract. J. Comp. Neurol. 425, 219–232.

Lopez, G. F. and Krimm, R. F. 2006a. Epithelial overexpression of BDNF and NT4 produces distinct gustatory axon morphologies that disrupt initial targeting. Dev. Biol. 292(2), 457–468.

Lopez, G. F. and Krimm, R. F. 2006b. Refinement of innervation accuracy following initial targeting of peripheral gustatory fibers. J. Neurobiol.

Lopez, G. F., Pa, V., and Krimm, R. F. 2006. The regulation of neural targeting in the developing gustatory system. Chem. Senses 154.

Lumsden, A. G. S. 1987. The Neural Crest Contribution to Tooth Development in the Mammalian Embryo. In: Developmental and Evolutionary Aspects of the Neural Crest (ed. P. F. A. Maderson), pp. 261–300. Wiley.

Lumsden, A. G. S. 1988. Spatial organization of the epithelium and the role of neural crest cells in the initiation of the mammalian tooth germ. Development 103(Suppl. 103), 155–169.

Mangold, O. 1931. Versuche zur Analyse der Entwicklung des Haftfadens beir Urodelen; ein Beispiel für die Inducktion artfrmeder Organe. Naturwissenschaften 45, 905–911.

Marti, E. and Bovolenta, P. 2002. Sonic hedgehog in CNS development: one signal, multiple outputs. Trends Neurosci. 25(2), 89–96.

May, O. L. and Hill, D. L. 2006. Gustatory terminal field organization and developmental plasticity in the nucleus of the solitary tract revealed through triple-fluorescence labeling. J. Comp. Neurol. 497(4), 658–669.

Mbiene, J.-P. 2004. Taste placodes are primary targets of geniculate but not trigeminal sensory axons in mouse developing tongue. J. Neurocytol. 33(6), 617–629.

Mbiene, J.-P. and Farbman, A. I. 1993. Evidence for stimulus access to taste cells and nerves during development: an electron microscopic study. Microsc. Res. Tech. 26(2), 94–105.

Mbiene, J.-P. and Mistretta, C. M. 1997. Initial innervation of embryonic rat tongue and developing taste papillae: nerves follow distinctive and spatially restricted pathways. Acta Anat. 160, 139–158.

Mbiene, J.-P. and Roberts, J. D. 2003. Distribution of keratin 8-containing cell clusters in mouse embryonic tongue: evidence for a prepattern for taste bud development. J. Comp. Neurol. 457(2), 111–122.

Mbiene, J.-P., MacCallum, D. K., and Mistretta, C. M. 1997. Organ cultures of embryonic rat tongue support tongue and gustatory papilla morphogenesis in vitro without intact sensory ganglia. J. Comp. Neurol. 377, 324–340.

McLaughlin, S. K. 2000. Erb and c-Kit receptors have distinctive patterns of expression in adult and developing taste papillae and taste buds. J. Neurosci. 20(15), 5679–5688.

Miettinen, P. J., Berger, J. E., Meneses, J., Phung, Y., Pedersen, R. A., Werb, Z., and Derynck, R. 1995. Epithelial immaturity and multiorgan failure in mice lacking epidermal growth factor receptor. Nature 376(6538), 337–341.

Ming, G., Lohof, A. M., and Zheng, J. Q. 1997. Acute morphogenic and chemotropic effects of neurotrophins on cultured embryonic *Xenopus* spinal neurons. J. Neurosci. 17(20), 7860–7871.

Ming, G., Song, H., Berninger, B., Inagaki, N., Tessier-Lavigne, M., and Poo, M. 1999. Phospholipase C-gamma and phosphoinositide 3-kinase mediate cytoplasmic signaling in nerve growth cone guidance. Neuron 23(1), 139–148.

Mistretta, C. M. 1972. Topographical and Histological Study of the Developing Rat Tongue, Palate and Taste Buds. In: Third Symposium on Oral Sensation and Perception the Mouth of the Infant (*ed.* J. F. Bosma), pp. 163–187. Charles C. Thomas.

Mistretta, C. M. 1998. The role of innervation in induction and differentiation of taste organs: introduction and background. Ann. N. Y. Acad. Sci. 855, 1–13.

Mistretta, C. M. and Liu, H. X. 2006. Development of fungiform papillae: patterned lingual gustatory organs. Arch. Histol. Cytol. 69(4), 199–208.

Mistretta, C. M., Goosens, K. A., Farinas, I., and Reichardt, L. F. 1999. Alterations in size, number, and morphology of gustatory papillae and taste buds in BDNF null mutant mice demonstrate neural dependence of developing taste organs. J. Comp. Neurol. 409(1), 13–24.

Mistretta, C. M., Grigaliunas, A., and Liu, H. X. 2005. Development of gustatory organs and innervating sensory ganglia. Chem. Senses 30(Suppl. 1), i52–i53.

Mistretta, C. M., Liu, H. X., Gaffield, W., and MacCallum, D. K. 2003. Cyclopamine and jervine in embryonic rat tongue cultures demonstrate a role for Shh signaling in taste papilla development and patterning: fungiform papillae double in number and form in novel locations in dorsal lingual epithelium. Dev. Biol. 254(1), 1–18.

Miura, H., Kato, H., Kusakabe, Y., Ninomiya, Y., and Hino, A. 2005. Temporal changes in NCAM immunoreactivity during taste cell differentiation and cell lineage relationships in taste buds. Chem Senses 30(4), 367–375.

Miura, H., Kato, H., Kusakabe, Y., Tagami, M., Miura-Ohnuma, J., Ninomiya, Y., and Hino, A. 2004. A strong nerve dependence of sonic hedgehog expression in basal cells in mouse taste bud and an autonomous transcriptional control of genes in differentiated taste cells. Chem. Senses 29(9), 823–831.

Miura, H., Kusakabe, Y., Kato, H., Miura-Ohnuma, J., Tagami, M., Ninomiya, Y., and Hino, A. 2003. Co-expression pattern of Shh with Prox1 and that of Nkx2.2 with Mash1 in mouse taste bud. Gene Expr. Patterns 3(4), 427–430.

Miura, H., Kusakabe, Y., Sugiyama, C., Kawamatsu, M., Ninomiya, Y., Motoyama, J., and Hino, A. 2001. Shh and Ptc are associated with taste bud maintenance in the adult mouse. Mech. Dev. 106(1–2), 143–145.

Moon, R. T., Bowerman, B., Boutros, M., and Perrimon, N. 2002. The promise and perils of Wnt signaling through beta-catenin. Science 296(5573), 1644–1646.

Morris-Wiman, J., Sego, R., Brinkley, L., and Dolce, C. 2000. The effects of sialoadenectomy and exogenous EGF on taste bud morphology and maintenance. Chem. Senses 25(1), 9–19.

Nagato, T., Matsumoto, K., Tanioka, H., Kodama, J., and Toh, H. 1995. Effect of denervation on morphogenesis of the rat fungiform papilla. Acta Anat. (Basel) 153(4), 301–309.

Narayanan, C. H. and Narayanan, Y. 1980. Neural crest and placodal contributions in the development of the glossopharyngeal–vagal complex in the chick. Anat. Rec. 196, 71–82.

Nie, X. 2005. Apoptosis, proliferation and gene expression patterns in mouse developing tongue. Anat. Embryol. (Berl.) 210(2), 125–132.

Nieuwkoop, P. D. and Ubbels, G. A. 1972. The formation of the mesoderm in Urodelean amphibians. IV. Qualitative evidence for the purely "ectodermal" origin of the entire mesoderm and of the pharyngeal endoderm. Wilhelm Roux's Arch. 169, 185–199.

Northcutt, R. G. 2005. Taste bud development in the channel catfish. J. Comp. Neurol. 482(1), 1–16.

Northcutt, R. G. and Barlow, L. A. 1998. Amphibians provide new insights into taste bud development. Trends Neurosci. 21(1), 38–42.

Northcutt, R. G. and Brändle, K. 1995. Development of branchiomeric and lateral line nerves in the axolotl. J. Comp. Neurol. 355, 427–454.

Northcutt, R. G., Barlow, L. A., Braun, C. B., and Catania, K. C. 2000. Distribution and innervation of taste buds in the axolotl. Brain Behav. Evol. 56(3), 123–145.

Nosrat, C. A. 1998. Neurotrophic factors in the tongue: expression patterns, biological activity, relation to innervation and studies of neurotrophin knockout mice. Ann. N. Y. Acad. Sci. 855, 28–49.

Nosrat, C. A. and Olson, L. 1995. Brain-derived neurotrophic factor mRNA is expressed in the developing taste bud-bearing tongue papillae of rat. J. Comp. Neurol. 360(4), 698–704.

Nosrat, C. A. and Olson, L. 1998. Changes in neurotrophin-3 messenger RNA expression patterns in the prenatal rat tongue suggest guidance of developing somatosensory nerves to their final targets. Cell Tissue Res. 292(3), 619–623.

Nosrat, C. A., Blomlof, J., ElShamy, W. M., Ernfors, P., and Olson, L. 1997. Lingual deficits in BDNF and NT3 mutant mice leading to gustatory and somatosensory disturbances, respectively. Development 124(7), 1333–1342.

Nosrat, C. A., Ebendal, T., and Olson, L. 1996. Differential expression of brain-derived neurotrophic factor and neurotrophin 3 mRNA in lingual papillae and taste buds indicates roles in gustatory and somatosensory innervation. J. Comp. Neurol. 376(4), 587–602.

Nosrat, C. A., MacCallum, D. K., and Mistretta, C. M. 1998. Early molecular events in gustatory papilla development are independent of nerve fibers. Chem. Senses. 23, 603–604.

Nosrat, C. A., MacCallum, D. K., and Mistretta, C. M. 2001. Distinctive spatiotemporal expression patterns for neurotrophins develop in gustatory papillae and lingual tissues in embryonic tongue organ cultures. Cell Tissue Res. 303(1), 35–45.

Nosrat, I. V., Agerman, K., Marinescu, A., Ernfors, P., and Nosrat, C. A. 2004. Lingual deficits in neurotrophin double knockout mice. J. Neurocytol. 33(6), 607–615.

Oakley, B. and Witt, M. 2004. Building sensory receptors on the tongue. J. Neurocytol. 33(6), 631–646.

Oakley, B., Brandemihl, A., Cooper, D., Lau, D., Lawton, A., and Zhang, C. 1998. The morphogenesis of mouse vallate gustatory epithelium and taste buds requires BDNF-dependent taste neurons. Brain Res. Dev. Brain Res. 105(1), 85–96.

Oakley, B., LaBelle, D. E., Riley, R. A., Wilson, K., and Wu, L.-H. 1991. The rate and locus of development of rat vallate taste buds. Dev. Brain Res. 58, 215–221.

Oakley, B., Wu, L. H., Lawton, A., and deSibour, C. 1990. Neural control of ectopic filiform spines in adult tongue. Neuroscience 36(3), 831–838.

Okubo, T., Pevny, L. H., and Hogan, B. L. 2006. Sox2 is required for development of taste bud sensory cells. Genes Dev. 20(19), 2654–2659.

Parker, M. A., Bell, M., and Barlow, L. A. 2004. Cell contact-dependent mechanisms specify taste bud number and size during a critical period early in embryonic development. Dev. Dyn. 230, 630–642.

Patel, A. V. and Krimm, R. F. 2005. BDNF and NT4 regulate geniculate ganglion neuron number at different developmental time points. Soc. Neurosci. Abst. Program # 252.17 (Online).

Paulson, R. B., Hayes, T. G., and Sucheston, M. E. 1985. Scanning electron microscope study of tongue development in the CD-1 mouse fetus. J. Craniofac. Genet. Dev. Biol. 5(1), 59–73.

Paves, H. and Saarma, M. 1997. Neurotrophins as *in vitro* growth cone guidance molecules for embryonic sensory neurons. Cell Tissue Res. 290(2), 285–297.

Pispa, J. and Thesleff, I. 2003. Mechanisms of ectodermal organogenesis. Dev. Biol. 262(2), 195–362.

Poritsky, R. and Singer, M. 1963. The fate of taste buds in tongue transplants to the orbit in the urodele *Triturus*. J. Exp. Zool. 153, 211–218.

Rao, H., Xu, Z., MacCallum, D. K., and Mistretta, C. M. 1997. BDNF and NGF differ in promoting neurite outgrowth from cultured embryonic rat geniculate and trigeminal ganglia. Chem. Senses 22, 775.

Reutter, K. and Witt, M. 1993. Morphology of Vertebrate Taste Organs and Their Nerve Supply. In: Mechanisms of Taste Transduction (eds. S. A. Simon and S. D. Roper), pp. 29–82. CRC Press.

Reutter, K., Breipohl, W., and Bijvank, G. J. 1974. Taste bud types in fishes II. Scanning electron microscopical investigations on *Xiphophorus helleri* Heckel (Poeciliidae, Cyprinodontiformes, Teleostei). Cell Tissue Res. 153, 151–165.

Ringstedt, T., Ibanez, C. F., and Nosrat, C. A. 1999. Role of brain-derived neurotrophic factor in target invasion in the gustatory system. J. Neurosci. 19(9), 3507–3518.

Rochlin, M. W. and Farbman, A. I. 1998. Trigeminal ganglion axons are repelled by their presumptive targets. J. Neurosci. 18(17), 6840–6852.

Rochlin, M. W., Egwiekhor, A., Vatterott, P., and Spec, A. 2006. BDNF attracts geniculate neurites, NT4 does not. Neurosci. Abstr. Program # 501.5 (Online).

Rochlin, M. W., O'Connor, R., Giger, R. J., Verhaagen, J., and Farbman, A. I. 2000. Comparison of neurotrophin and repellent sensitivities of early embryonic geniculate and trigeminal axons. J. Comp. Neurol. 422(4), 579–593.

Sbarbati, A., Crescimanno, C., Bernardi, P., Benati, D., Merigo, F., and Osculati, F. 2000. Postnatal development of the intrinsic nervous system in the circumvallate papilla–vonEbner gland complex. Histochem. J. 32(8), 483–488.

Sbarbati, A., Crescimanno, C., Bernardi, P., and Osculati, F. 1999. Alpha-gustducin-immunoreactive solitary chemosensory cells in the developing chemoreceptorial epithelium of the rat vallate papilla. Chem. Senses 24(5), 469–472.

Sbarbati, A., Merigo, F., Bernardi, P., Crescimanno, C., Benati, D., and Osculati, F. 2002. Ganglion cells and topographically related nerves in the vallate papilla/von Ebner gland complex. J. Histochem. Cytochem. 50(5), 709–718.

Schecterson, L. C. and Bothwell, M. 1992. Novel roles for neurotrophins are suggested by BDNF and NT-3 mRNA expression in developing neurons. Neuron 9(3), 449–463.

Schlosser, G. 2006. Induction and specification of cranial placodes. Dev. Biol. 294(2), 303–351.

Scott, L. and Atkinson, M. E. 1998. Target pioneering and early morphology of the murine chorda tympani. J. Anat. 192(Pt 1), 91–98.

Seta, Y., Seta, C., and Barlow, L. A. 2003. Notch-associated gene expression in embryonic and adult taste papillae and taste buds suggests a role in taste cell lineage decisions. J. Comp. Neurol. 464(1), 49–61.

Seta, Y., Stoick-Cooper, C. L., Toyono, T., Kataoka, S., Toyoshima, K., and Barlow, L. A. 2006. The bHLH transcription factors, Hes6 and Mash1, are expressed in distinct subsets of cells within adult mouse taste buds. Arch. Histol. Cytol. 69(3), 189–198.

Seta, Y., Toyono, T., Takeda, S., and Toyoshima, K. 1999. Expression of Mash1 in basal cells of rat circumvallate taste buds is dependent upon gustatory innervation. FEBS Lett. 444, 43–46.

Sharaby, A. E., Ueda, K., Honma, S., and Wakisaka, S. 2006. Initial innervation of the palatal gustatory epithelium in the rat as revealed by growth-associated protein-43 (GAP-43) immunohistochemistry. Arch. Histol. Cytol. 69(4), 257–272.

Shuler, M. G., Krimm, R. F., and Hill, D. L. 2004. Neuron/target plasticity in the peripheral gustatory system. J. Comp. Neurol. 472(2), 183–192.

Slack, J. M. W. 1991. The concepts of experimental Embryology In: From Egg to Embryo (eds. P. W. Barlow, D. Bray, P. B. Green, and J. M. W. Slack), pp. 9–33. Cambridge University Press

Slack, J. M. and Forman, D. 1980. An interaction between dorsal and ventral regions of the marginal zone in early amphibian embryos. J. Embryol. Exp. Morphol. 56, 283–299.

Smith, H. M. 1969. The Mexican axolotl: some misconceptions and problems. Bioscience 19(7), 593–597.

Sollars, S. I. 2005. Chorda tympani nerve transection at different developmental ages produces differential effects on taste bud volume and papillae morphology in the rat. J. Neurobiol. 64(3), 310–320.

Sollars, S. I. and Hill, D. L. 2000. Lack of functional and morphological susceptibility of the greater superficial petrosal nerve to developmental dietary sodium restriction. Chem. Senses 25(6), 719–727.

Sollars, S. I. and Hill, D. L. 2005. *In vivo* recordings from rat geniculate ganglia: taste response properties of individual greater superficial petrosal and chorda tympani neurones. J. Physiol. 564(Pt 3), 877–893.

Sollars, S. I., Smith, P. C., and Hill, D. L. 2002. Time course of morphological alterations of fungiform papillae and taste buds following chorda tympani transection in neonatal rats. J. Neurobiol. 51(3), 223–236.

Sollars, S. I., Walker, B. R., Thaw, A. K., and Hill, D. L. 2006. Age-related decrease of the chorda tympani nerve terminal field in the nucleus of the solitary tract is prevented by dietary sodium restriction during development. Neuroscience 137(4), 1229–1236.

Stone, L. M., Finger, T. E., Tam, P. P. L., and Tan, S.-S. 1995. Both ectoderm and endoderm give rise to taste buds in mice. Chem. Senses 20, 785–786.

Stone, L. M., Tan, S.-S., Tam, P. P. L., and Finger, T. E. 2002. Analysis of cell lineage relationships in taste buds. J. Neurosci. 22(11), 4522–4529.

Stone, L. S. 1932. Independence of taste organs with respect to their nerve fibers demonstrated in living salamanders. Proc. Soc. Exp. Biol. Med. 30, 1256–1257.

Stone, L. S. 1940. The origin and development of taste organs salamanders observed in the living condition. J. Exp. Zool. 83, 481–506.

Sun, H. and Oakley, B. 2002. Development of anterior gustatory epithelia in the palate and tongue requires epidermal growth factor receptor. Dev. Biol. 242(1), 31–43.

Suzuki, Y. 2006. Expression of IGFBPs in the developing mouse submandibular and von Ebner's glands. Anat. Embryol. (Berl.) 211(3), 189–196.

Suzuki, Y., Takeda, M., and Obara, N. 2002. Expression of NeuroD in the mouse taste buds. Cell Tissue Res. 307(3), 423–428.

Suzuki, Y., Takeda, M., Sakakura, Y., and Suzuki, N. 2005. Distinct expression pattern of insulin-like growth factor family in rodent taste buds. J. Comp. Neurol. 482(1), 74–84.

Takeda, M., Suzuki, Y., Obara, N., and Tsunekawa, H. 2005a. Immunohistochemical detection of neurotrophin-3 and -4, and their receptors in mouse taste bud cells. Arch. Histol. Cytol. 68(5), 393–403.

Takeda, M., Suzuki, Y., Obara, N., Uchida, N., and Kawakoshi, K. 2004. Expression of GDNF and GFR alpha 1 in mouse taste bud cells. J. Comp. Neurol. 479(1), 94–102.

Takeda, M., Suzuki, Y., Obara, N., Uchida, N., and Kawakoshi, K. 2005b. Expression of glial cell line-derived neurotrophic factor (GDNF) and GDNF family receptor alpha1 in mouse taste bud cells after denervation. Anat. Sci. Int. 80(2), 105–110.

Tanabe, Y. and Jessell, T. M. 1996. Diversity and pattern in the developing spinal cord. Science 274, 1115–1123.

Tessier-Lavigne, M. and Goodman, C. S. 1996. The molecular biology of axon guidance. Science 274, 1123–1133.

Threadgill, D. W., Dlugosz, A. A., Hansen, L. A., Tennenbaum, T., Lichti, U., Yee, D., LaMantia, C., Mourton, T., Herrup, K., Harris, R. C., Barnard, J. A., Yuspa, S. H., Coffey, R. J., and Magnuson, T. 1995. Targeted disruption of mouse EGF receptor: effect of genetic background on mutant phenotype. Science 269(5221), 230–234.

Torrey, T. W. 1934. The relation of taste buds to their nerve fibers. J. Comp. Neurol. 59, 203–220.

Toyoshima, K., Miyamoto, K., and Shimamura, A. 1987. Fine structure of taste buds in the tongue, palatal mucosa and gill arch of the axolotl, *Ambystoma mexicanum*. Okajimas Folia Anat. Jpn. 64, 99–110.

Tucker, K. L., Meyer, M., and Barde, Y. A. 2001. Neurotrophins are required for nerve growth during development. Nat. Neurosci. 4(1), 29–37.

Ueda, K., Ichimori, Y., Okada, H., Honma, S., and Wakisaka, S. 2006. Immunolocalization of SNARE proteins in both type II and type III cells of rat taste buds. Arch. Histol. Cytol. 69(4), 289–296.

Vilbig, R., Cosmano, J., Giger, R., and Rochlin, M. W. 2004. Distinct roles for Sema3A, Sema3F, and an unidentified trophic factor in controlling the advance of geniculate axons to gustatory lingual epithelium. J. Neurocytol. 33(6), 591–606.

Vintschgau, M. and Honigschmied, J. 1876. Nervus glossopharyngeus und Schmeckbecher. Arch. Gesamte Physiol. 14, 443–448.

Vogt, W. 1929. Gestaltungsanalyse am Amphibienkeim mit örtlicher Vitalfärbung. II. Gastrulation und Mesodermbildung bei Urodelen und Anuren. Wilhelm Roux's Arch. Entwicklungmech Org. 120, 385–706.

Wakisaka, S., Miyawaki, Y., Youn, S. H., Kato, J., and Kurisu, K. 1996. Protein gene-product 9.5 in developing mouse circumvallate papilla: comparison with neuron-specific enolase and calcitonin gene-related peptide. Anat. Embryol. (Berl.) 194(4), 365–372.

Webb, J. F. and Noden, D. M. 1993. Ectodermal placodes: contributions to the development of the vertebrate head. Am. Zool. 33(4), 434–447.

Whitehead, M. C. and Kachele, D. L. 1994. Development of fungiform papillae, taste buds, and their innervation in the hamster. J. Comp. Neurol. 340(4), 515–530.

Whitehead, M. C., Frank, M. E., Hettinger, T. P., Hou, L.-T., and Nah, H.-D. 1987. Persistence of taste buds in denervated fungiform papillae. Brain Res. 405, 192–195.

Whitehead, M. C., Ganchrow, J. R., Ganchrow, D., and Yao, B. 1999. Organization of geniculate and trigeminal ganglion cells innervating single fungiform taste papillae: a study with tetramethylrhodamine dextran amine labeling. Neuroscience 93(3), 931–941.

Widelitz, R. 2005. Wnt signaling through canonical and non-canonical pathways: recent progress. Growth Factors 23(2), 111–116.

Wiklund, P. and Ekstrom, P. A. 2000. Axonal outgrowth from adult mouse nodose ganglia *in vitro* is stimulated by neurotrophin-4 in a Trk receptor and mitogen-activated protein kinase-dependent way. J. Neurobiol. 45(3), 142–151.

Witt, M. and Reutter, K. 1996. Embryonic and early fetal development of human taste buds: a transmission electron microscopical study. Anat. Rec. 246, 507–523.

Wodarz, A. and Nusse, R. 1998. Mechanisms of Wnt signaling in development. Annu. Rev. Cell Dev. Biol. 14, 59–88.

Yamout, A., Spec, A., Cosmano, J., Kashyap, M., and Rochlin, M. W. 2005. Neurotrophic factor receptor expression and *in vitro* nerve growth of geniculate ganglion neurons that supply divergent nerves. Dev. Neurosci. 27(5), 288–298.

Yang, R., Tabata, S., Crowley, H. H., Margolskee, R. F., and Kinnamon, J. C. 2000. Ultrastructural localization of gustducin immunoreactivity in microvilli of type II taste cells in the rat. J. Comp. Neurol. 425(1), 139–151.

Yu, M., Wu, P., Widelitz, R. B., and Chuong, C. M. 2002. The morphogenesis of feathers. Nature 420(6913), 308–312.

Zaidi, F. N. and Whitehead, M. C. 2006. Discrete innervation of murine taste buds by peripheral taste neurons. J. Neurosci. 26(32), 8243–8253.

Zhang, C. and Oakley, B. 1996. The distribution and origin of keratin 20-containing taste buds in rat and human. Differentiation 61(2), 121–127.

Zhang, C., Brandemihl, A., Lau, D., Lawton, A., and Oakley, B. 1997. BDNF is required for the normal development of taste neurons *in vivo*. Neuroreport 8(4), 1013–1017.

Zhang, C., Cotter, M., Lawton, A., Oakley, B., Wong, L., and Zeng, Q. 1995. Keratin 18 is associated with a subset of older taste cells in the rat. Differentiation 59(3), 155–162.

Zhang, G. H., Deng, S. P., Li, L. L., and Li, H. T. 2006. Developmental change of alpha-gustducin expression in the mouse fungiform papilla. Anat. Embryol. (Berl.).

Zhang, L. L. and Ashwell, K. W. 2001. The development of cranial nerve and visceral afferents to the nucleus of the solitary tract in the rat. Anat. Embryol. (Berl.) 204(2), 135–151.

Zhou, Y., Liu, H. X., and Mistretta, C. M. 2006. Bone morphogenetic proteins and noggin: Inhibiting and inducing fungiform taste papilla development. Dev. Biol. 297(1), 198–213.

Zimmerman, L. B., De Jesus-Escobar, J. M., and Harland, R. M. 1996. The Spemann organizer signal noggin binds and inactivates bone morphogenetic protein 4. Cell 86(4), 599–606.

Relevant Website

http://www.stanford.edu – Stanford University.

4.07 The Sweet Taste of Childhood

J A Mennella, Monell Chemical Senses Center, Philadelphia, PA, USA

4.07.1 Introduction

Since the nineteenth century, sweet-tasting candies have been the first purchase children make with their own money (Woloson, W. A., 2002). The liking for all that tastes sweet is universal and a hallmark of development. Although strong evidence is lacking, it has been suggested that such preferences evolved to solve a basic nutritional problem of attracting children to sources of high energy during periods of maximal growth (Simmen, B. and Hladik, C. M., 1998; Beauchamp, G. K., 1999; Coldwell, S. E. and Oswald, T. K., 2004).

In this essay, the scientific literature that documents the innate and heightened preferences that children have for sweet-tasting foods and beverages when compared to adults will be reviewed. This review serves as a foundation for a discussion on the pain-reducing properties of sugars which, along with heightened sweet preferences, may be an identifying feature of childhood.

4.07.2 Ontogeny of Sweet Preferences

4.07.2.1 Fetal Life and Infancy

The preference for sweet tastes is present early in ontogeny and remarkably well conserved phylogenetically (Steiner, J. E. *et al.*, 2001). Although each measure has its limitations, the convergence of findings from the scientific literature suggests that from birth, infants are quite sensitive to and prefer sweet tastes. This ability to detect sweet tastes and, in turn, the efficacy of sweets to modulate a variety of behaviors may be evidenced as early as fetal life. It should be emphasized that although the response of human fetuses to sweet taste has never been directly investigated, indirect evidence from albeit methodologically flawed studies (see Mistretta, C. M. and Bradley, R. M., 1977), suggests that that prenatal monitoring of taste stimuli dissolved in amniotic fluid is possible during late gestation. That is, two studies reported increased fetal swallowing following the injection of sweet stimuli such as saccharin (De Snoo, K., 1937; Liley, A. W., 1972) into the amniotic fluid of pregnant women and decrease swallowing following the injection of a bitter-tasting substance (Liley, A. W., 1972).

Stronger evidence that the fetus can respond to taste stimuli comes from research on premature infants. To overcome methodological limitations and because premature infants often have immature suck–swallow coordination, innovative methods were developed that avoided the risk of fluid aspiration by embedding the taste substance in a nipple-shaped gelatin medium that released small amounts of the sweet substance when mouthed or sucked. Infants born preterm and tested between 33 and 40 weeks postconception produced stronger and more frequent sucking responses when offered the sucrose-sweetened nipple compared with a latex nipple (Maone, T. R. *et al.*, 1990).

After birth, the liking for sweet tastes is unequivocal. Newborns are sweet connoisseurs. They respond to even dilute sweet tastes, differentiate varying degrees of sweetness, and will consume more of a sugar solution when compared to water (Desor, J. A.

et al., 1973; Nowlis, G. H. and Kessen, W., 1976). When a sweet solution is placed in the oral cavity, the infant's face relaxes, resembling an expression of satisfaction which may be followed by a smile (Steiner, J. E., 1977; Rosenstein, D. and Oster, H., 1988). That the positive facial expressions elicited by sweet-tasting sugars are reflexlike is supported by the finding that single response components, such as tongue movements, can be reliably elicited in newborns by sweet tastes in a concentration-dependent manner (e.g., Nowlis, G. H. and Kessen, W., 1976; Weiffenbach, J. M., 1977). Moreover, neonates born with severe developmental malformations of the central nervous system (e.g., anencephalic infants) react to sweet tastes, as well as other tastants, like normal, term infants (Steiner, J. E., 1977).

Newborns also respond to sweet stimulation with changes in autonomic response (e.g., Crook, C. K. and Lipsitt, L. P., 1976; Lipsitt, L. P. *et al.*, 1976; Lipsitt, L. P., 1977; Haouari, N. *et al.*, 1995; Rao, M. *et al.*, 1997; Blass, E. M. and Watt, L. B., 1999), but the behavioral state of the infant modifies the direction of the change. For example, heart rate increased proportionally when increasing concentrations of sucrose were placed in the mouths of nonagitated infants (Ashmead, D. H. *et al.*, 1980). Conversely, heart rate decreased when sweet tastes were introduced to infants who were agitated, resulting in overall calmness (e.g., Blass, E. M. and Watt, L. B., 1999). Furthermore, electroencephalographic recordings of 2–3-day-old newborns revealed that infants respond to sucrose administration with an asymmetry of brain electrical activity, a response which is usually associated with hedonically positive emotional reactions or approach behavior (Fox, N. A. and Davidson, R. J., 1986).

4.07.2.2 Childhood and Adolescence

The preference for sweet taste is universal and evident among children around the world (e.g., Brazil (Tomita, N. E. *et al.*, 1999); France (Bellisle, F. *et al.*, 1990); Iraq (Jamel, H. A. *et al.*, 1997); Israel (Steiner, J. E. *et al.*, 1984); Mexico (Vazquez, M. *et al.*, 1982); The Netherlands (De Graaf, C. and Zandstra, E. H., 1999); and North America (Desor, J. A. *et al.*, 1975; Beauchamp, G. K. and Cowart, B. J., 1987; Liem, D. G. and Mennella, J. A., 2002). Both cross-sectional and longitudinal studies revealed that the preference for sweets remains heightened throughout childhood (Beauchamp, G. K. and Moran, M., 1984; Mennella,

J. A. *et al.*, 2005; Pepino, M. Y. and Mennella, J. A., 2005a) and early adolescence (Desor, J. A. *et al.*, 1975) and declines to adult levels during late adolescence (Desor, J. A. and Beauchamp, G. K., 1987).

The level of sucrose that children prefer, as measured in the laboratory, is significantly related to children's preferences for sweet-tasting foods such as cereals (Liem, D. G. and Mennella, J. A., 2002; Mennella, J. A. *et al.*, 2005) and beverages (Olson, C. M. and Gemmill, K. P., 1981; Mennella, J. A. *et al.*, 2005). Sweet taste can clearly drive consumption such that adding sugar to beverages such as Kool Aid™ (Beauchamp, G. K. and Moran, M., 1984) and solid foods such as spaghetti (Filer, L. J., 1978), ricotta, and jicama (Sullivan, S. A. and Birch, L. L., 1990) increased both liking and acceptance by children.

What causes the developmental shift in sweet preferences and consumption between childhood and adulthood remains a mystery, but this age-related decline in sweet preference has been observed in other mammals (Wurtman, J. J. and Wurtman, R. J., 1979; Bertino, M. and Wehmer, F., 1981; Marlin, N. A., 1983). Two hypotheses, not mutually exclusive, are presented here. First, children may be less sensitive to sweet tastes, thereby requiring larger amounts of sugars to obtain optimal sweetness. However, there is no strong evidence that the heightened sweet preference evidenced during infancy and childhood is due to decreased sensitivity to sweet tastes (see Cowart, B. J., 1981 for review). Whereas some investigators may report that children are more sensitive than adults when small regions of the anterior tongue are stimulated with sucrose (Stein, N. *et al.*, 1994), others found that taste sensitivity during childhood does not differ from that during adulthood (Cowart, B. J., 1981). Second, the heightened sweet preferences early in life may be linked to the growing child's need for calories (Drewnowski, A., 2000). Recent findings support this hypothesis. In brief, 11–15-year-old children were divided into sweet likers and dislikers based on their sucrose preferences (Coldwell, S. E. and Oswald, T. K., 2004). Although there were no differences between these two groups in measures such as sucrose detection thresholds, age, body mass index, percent of body fat, and pubertal development, growth rate was significantly lower in dislikers when compared to sweet likers suggesting that the age-related decline in sucrose preferences may be related to the cessation of growth.

4.07.2.3   Early Experiences and Contextual Learning about Sweets

Although the scientific evidence supports the hypothesis that sweet taste liking is innate and present at birth, if not earlier, the degree to which early experiences alter or modulate the development of sweet preferences later in life is largely unknown. Longitudinal studies revealed that babies who were routinely fed sweetened water (e.g., water sweetened with table sugar, Karo syrup or honey) during the first months of life exhibited a greater preference for sweetened water when tested at 6 months (Beauchamp, G. K. and Moran, M., 1982) and then again at 2 years of age (Beauchamp, G. K. and Moran, M., 1984) when compared to those who had little or no experience with sweetened water. Similarly, a more recent cross-sectional study on 6–10-year-old children revealed that such feeding practices may have longer-term effects on the preference for sweetened water than previously realized (Pepino, M. Y. and Mennella, J. A., 2005a).

Whether early experience with sweetened water modifies overall liking for sweetened foods is less uncertain. In fact, there is no compelling data suggesting that repeated exposure to sugar water results in a generalized heightened hedonic response to sweetness (Beauchamp, G. K. and Moran, M., 1984). Moreover, 4–7-year-old children whose mothers reported adding sugar to their foods on a routine basis were significantly more likely to prefer apple juices with added sugar and cereals with higher sugar contents than similar age children whose mothers reported never adding sugar to the foods at home (Liem, D. G. and Mennella, J. A., 2002). These data support the hypothesis that the context in which the taste experience occurred is an important factor and through familiarization, children develop a sense of what should, or should not, taste sweet (Beauchamp, G. K. and Cowart, B. J., 1985).

4.07.3   Pain-Reducing Properties of Sweet Tastes

4.07.3.1   Infancy

The pioneering research of Elliott Blass, Ronald Barr, and their colleagues revealed, more than a decade ago, that tasting a sweet substance modified crying behaviors and certain physiological responses in infants. A small amount of a sweet solution placed on the tongue of a crying newborn exerts a rapid, calming effect that persists for several minutes (Blass, E. M. and Hoffmeyer, L. B., 1991; Barr, R. G. et al., 1994). Sweet tastes can also blunt expressions of pain, and calm both preterm and full-term infants who have been subjected to painful events such as heel stick or circumcision (see Stevens, B. et al., 2004 for review). That sucrose attenuated a negative electroencephalographic response to a painful procedure (Fernandez, M. et al., 2003) suggests that sucrose blocks pain afferents which, in turn, diminish stress and cardiac changes. Because noncaloric sweet substances such as aspartame mimic the calming effects of sucrose (Barr, R. G. et al., 1999; Bucher, H. U. et al., 2000) and because the administration of sucrose by direct stomach loading is not effective (Ramenghi, L. A. et al., 1999), afferent signals from the mouth, rather than gastric or metabolic changes, appear to be responsible for the analgesic properties of sweet tastes.

Other flavors and orosensory components also reduce pain in infants and may act synergistically with sweet tastes. These include the flavor of mother's milk (Upadhyay, A. et al., 2004) and some of its constituents (i.e., fat, protein; Blass, E. M., 1997). Orotactile stimulation from a pacifier (Blass, E. M. and Hoffmeyer, L. B., 1991; Carbajal, R. et al., 1999), maternal skin-to-skin contact (Gray, L. et al., 2000), or being held before and during the painful procedure (Gormally, S. et al., 2001) also produce the analgesic effect, but the effects are transient when compared to when sweet taste is in the oral cavity (Smith, B. A. et al., 1990). Although the use of sterile sugar water to reduce pain in infants is now routine practice in several hospital nurseries because of its natural simplicity, viability and efficacy (Anand, K. J. S., 2001), we caution that more research is needed to determine sweet taste's efficacy in reducing pain after repeated administrations (Stevens, B. et al., 2004).

4.07.3.2   Childhood

The ability of sweet tastes to reduce pain continues during childhood as evidenced by the finding that the presence of sucrose, but not water, in the oral cavity delayed 8–11-year-old children's reporting of pain onset when undergoing a cold-induced pain stimulus test (Miller, A. et al., 1994; Pepino, M. Y. and Mennella, J. A., 2005b). Sucrose's efficacy in reducing pain is related to the hedonic value of sweet taste for the child (Pepino, M. Y. and Mennella, J. A., 2005b). The more children like sucrose, the better it works in increasing pain tolerance.

Less certain is whether sweet taste continues to be an analgesic during adulthood. While some studies suggested no effects (Miller, A. *et al.*, 1994; Mercer, M. E. and Holder, M. D., 1997), others reported increased tolerance to mechanical pain following consumption of palatable sweet food in women but not in men (Blass, E. *et al.*, 1987; Mercer, M. E. and Holder, M. D., 1997). From these observations, it can be postulated that if the analgesic response to intraoral sucrose is present during adulthood, it is less robust than that during childhood. The absence or diminution of an analgesic response to sucrose in adults is apparently not due to the lowered sucrose preference observed in adults overall (Miller, A. *et al.*, 1994) and is consistent with animal model studies that revealed that sweet-induced analgesia progressively declines during development and is absent by the third week of life in rats (Anseloni, V. C. *et al.*, 2002).

4.07.4 Concluding Remarks

The research discussed herein revealed that, as a group, children prefer significantly higher concentrations of sugars in foods and liquids than do adults. Nevertheless, there are striking individual and group differences in the levels of sweetness preferred (Desor, J. A. *et al.*, 1975; Beauchamp, G. K. and Moran, M., 1982; Pepino, M. Y. and Mennella, J. A., 2005a). The recent identification of some taste receptor genes (e.g., Drayna, D. *et al.*, 2003; Kim, U. K. *et al.*, 2003) has enabled us to explore the contribution of genetic variation (see McDaniel A. H. and Reed D. R., 2004). Alleles of the bitter-taste receptor *TAS2R38* gene explained much of the phenotypic variation in the sensitivity to the bitter compound, propylthiouracil, and partially explained individual differences in children's sweet preferences (Mennella, J. A. *et al.*, 2005; see also Keller, K. L. and Tepper, B. J., 2004). The future promises more research on how variation in other taste receptor genes, particularly those involved in sweet taste perception, contribute to individual differences in preferences and food habits. More knowledge about the factors that contribute to preferences for sweet-tasting foods and beverages in children, a generation who will struggle with obesity and diabetes, may suggest population-based strategies to overcome diet-induced disease and promote healthy eating habits.

Acknowledgments

Preparation of this manuscript was supported in part by NIH grants AA09523 and HD37119.

References

Anand, K. J. S. 2001. Consensus statement for the prevention and management of pain in the newborn. Arch. Pediatr. Adolesc. Med. 155, 173–180.

Anseloni, V. C., Weng, H. R., Terayama, R., Letizia, D., Davis, B. J., Ren, K., Dubner, R., and Ennis, M. 2002. Age-dependency of analgesia elicited by intraoral sucrose in acute and persistent pain models. Pain 97, 93–103.

Ashmead, D. H., Reilly, B. M., and Lipsitt, L. P. 1980. Neonates' heart rate, sucking rhythm, and sucking amplitude as a function of the sweet taste. J. Exp. Child Psychol. 29, 264–281.

Barr, R. G., Pantel, M. S., Young, S. N., Wright, J. H., Hendricks, L. A., and Gravel, R. 1999. The response of crying newborns to sucrose: is it a "sweetness" effect? Physiol. Behav. 66, 409–417.

Barr, R. G., Quek, V. S., Cousineau, D., Oberlander, T. F., Brian, J. A., and Young, S. N. 1994. Effects of intra-oral sucrose on crying, mouthing and hand–mouth contact in newborn and six-week-old infants. Dev. Med. Child Neurol. 36, 608–618.

Beauchamp, G. K. 1999. Factors Affecting Sweetness. In: Low-Calorie Sweeteners: Present and Future (ed. A. Corti), pp. 10–17. Karger. World Rev. Nutr. Diet.

Beauchamp, G. K. and Cowart, B. J. 1985. Congenital and experiential factors in the development of human flavor preferences. Appetite 6, 357–372.

Beauchamp, G. K. and Cowart, B. J. 1987. Development of Sweet Taste. In: Sweetness (ed. J. Dobbing), pp. 127–138. Springer.

Beauchamp, G. K. and Moran, M. 1982. Dietary experience and sweet taste preference in human infants. Appetite 3, 139–152.

Beauchamp, G. K. and Moran, M. 1984. Acceptance of sweet and salty tastes in 2-year-old children. Appetite 5, 291–305.

Bellisle, F., Dartois, A. M., Kleinknecht, C., and Broyer, M. 1990. Perceptions of and preferences for sweet taste in uremic children. J. Am. Diet. Assoc. 90, 951–954.

Bertino, M. and Wehmer, F. 1981. Dietary influence on the development of sucrose acceptability in rats. Dev. Psychobiol. 14, 19–28.

Blass, E., Fitzgerald, E., and Kehoe, P. 1987. Interactions between sucrose, pain and isolation distress. Pharmacol. Biochem. Behav. 26, 483–489.

Blass, E. M. 1997. Milk-induced hypoalgesia in human newborns. Pediatrics 99, 825–829.

Blass, E. M. and Hoffmeyer, L. B. 1991. Sucrose as an analgesic for newborn infants. Pediatrics 87, 215–218.

Blass, E. M. and Watt, L. B. 1999. Suckling- and sucrose-induced analgesia in human newborns. Pain 83, 611–623.

Bucher, H. U., Baumgartner, R., Bucher, N., Seiler, M., and Fauchere, J. C. 2000. Artificial sweetener reduces nociceptive reaction in term newborn infants. Early Hum. Dev. 59, 51–60.

Carbajal, R., Chauvet, X., Couderc, S., and Olivier-Martin, M. 1999. Randomised trial of analgesic effects of sucrose, glucose, and pacifiers in term neonates. BMJ 319, 1393–1397.

Coldwell, S. E. and Oswald, T. K. 2004. Growth-rate differs between children with adult-like versus child-like sugar preference. J. Dent. Res. 83 (Spec. Issue A), Abst. 0740.

Cowart, B. J. 1981. Development of taste perception in humans: sensitivity and preference throughout the life span. Psychol. Bull. 90, 43–73.

Crook, C. K. and Lipsitt, L. P. 1976. Neonatal nutritive sucking: effects of taste stimulation upon sucking rhythm and heart rate. Child Dev. 47, 518–522.

De Graaf, C. and Zandstra, E. H. 1999. Sweetness intensity and pleasantness in children, adolescents, and adults. Physiol. Behav. 67, 513–520.

De Snoo, K. 1937. Das trinkende kind im uterus [The drinking child in utero]. Monatsschr. Geburtsh Gynaekol. 105, 88–97.

Desor, J. A. and Beauchamp, G. K. 1987. Longitudinal changes in sweet preferences in humans. Physiol. Behav. 39, 639–641.

Desor, J. A., Greene, L. S., and Maller, O. 1975. Preferences for sweet and salty in 9- to 15-year-old and adult humans. Science 190, 686–687.

Desor, J. A., Maller, O., and Turner, R. E. 1973. Taste in acceptance of sugars by human infants. J. Comp. Phys. Psychol. 496–501.

Drayna, D., Coon, H., Kim, U. K., Elsner, T., Cromer, K., Otterud, B., Baird, L., Peiffer, A. P., and Leppert, M. 2003. Genetic analysis of a complex trait in the Utah Genetic Reference Project: a major locus for PTC taste ability on chromosome 7q and a secondary locus on chromosome 16p. Hum. Genet. 112, 567–572.

Drewnowski, A. 2000. Sensory control of energy density at different life stages. Proc. Nutr. Soc. 59, 239–244.

Fernandez, M., Blass, E. M., Hernandez-Reif, M., Field, T., Diego, M., and Sanders, C. 2003. Sucrose attenuates a negative electroencephalographic response to an aversive stimulus for newborns. J. Dev. Behav. Pediatr. 24, 261–266.

Filer, L. J. 1978. Studies of taste perception in infancy and childhood. Pediatr. Basics 5–9.

Fox, N. A. and Davidson, R. J. 1986. Taste-elicited changes in facial signs of emotion and the asymmetry of brain electrical activity in human newborns. Neuropsychologia 24, 417–422.

Gormally, S., Barr, R. G., Wertheim, L., Alkawaf, R., Calinoiu, N., and Young, S. N. 2001. Contact and nutrient caregiving effects on newborn infant pain responses. Dev. Med. Child Neurol. 43, 28–38.

Gray, L., Watt, L., and Blass, E. M. 2000. Skin-to-skin contact is analgesic in healthy newborns. Pediatrics 105, e14.

Haouari, N., Wood, C., Griffiths, G., and Levene, M. 1995. The analgesic effect of sucrose in full term infants: a randomised controlled trial. BMJ 310, 1498–1500.

Jamel, H. A., Sheiham, A., Watt, R. G., and Cowell, C. R. 1997. Sweet preference, consumption of sweet tea and dental caries; studies in urban and rural Iraqi populations. Int. Dent. J. 47, 213–217.

Keller, K. L. and Tepper, B. J. 2004. Inherited taste sensitivity to 6-n-propylthiouracil in diet and body weight in children. Obes. Res. 12, 904–912.

Kim, U. K., Jorgenson, E., Coon, H., Leppert, M., Risch, N., and Drayna, D. 2003. Positional cloning of the human quantitative trait locus underlying taste sensitivity to phenylthiocarbamide. Science 299, 1221–1225.

Liem, D. G. and Mennella, J. A. 2002. Sweet and sour preferences during childhood: role of early experiences. Dev. Psychobiol. 41, 388–395.

Liley, A. W. 1972. Disorders of Amniotic Fluid. In: Pathophysiology of Gestation, Vol. II, Fetal-placental Disorders (ed. N. S. Assali), pp. 157–206. Academic Press.

Lipsitt, L. P. 1977. Taste in Human Neonates: its Effects on Sucking and Heart Rate. In: Taste and Development: The Genesis of Sweet Preference (ed. J. M. Weiffenbach), pp. 125–142. U.S. Government Printing Office.

Lipsitt, L. P., Reilly, B. M., Butcher, M. J., and Greenwood, M. M. 1976. The stability and interrelationships of newborn sucking and heart rate. Dev. Psychobiol. 9, 305–310.

Maone, T. R., Mattes, R. D., Bernbaum, J. C., and Beauchamp, G. K. 1990. A new method for delivering a taste without fluids to preterm and term infants. Dev. Psychobiol. 23, 179–191.

Marlin, N. A. 1983. Early exposure to sugars influences the sugar preference of the adult rat. Physiol. Behav. 31, 619–623.

McDaniel, A. H. and Reed D. R. 2004. The Human Sweet Tooth and its Relationship to Obesity. In: Genomics and Proteomics in Nutrition (eds. N. Moustaid-Moussa and C. Berdanier), pp. 49–67. Dekker.

Mennella, J. A., Pepino, M. Y., and Reed, D. R. 2005. Genetic and environmental determinants of bitter perception and sweet preferences. Pediatrics 115, e216–e222.

Mercer, M. E. and Holder, M. D. 1997. Antinociceptive effects of palatable sweet ingesta on human responsivity to pressure pain. Physiol. Behav. 61, 311–318.

Miller, A., Barr, R. G., and Young, S. N. 1994. The cold pressor test in children: methodological aspects and the analgesic effect of intraoral sucrose. Pain 56, 175–183.

Mistretta, C. M. and Bradley, R. M. 1977. Taste in Utero: Theoretical Considerations. In: Taste and Development: The Genesis of Sweet Preference (ed. J. M. Weiffenbach), DHEW Publication No. NIH 77-1068, pp. 51–69. U.S. Government Printing Office.

Nowlis, G. H. and Kessen, W. 1976. Human newborns differentiate differing concentrations of sucrose and glucose. Science 191, 865–866.

Olson, C. M. and Gemmill, K. P. 1981. Association of sweet preference and food selection among four to five year old children. Ecol. Food Nutr. 11, 145–150.

Pepino, M. Y. and Mennella, J. A. 2005a. Factors contributing to individual differences in sucrose preference. Chem. Senses 30 (Suppl 1), i319–i320.

Pepino, M. Y. and Mennella, J. A. 2005b. Sucrose-induced analgesia is related to sweet preferences in children but not adults. Pain 119 , 210–218.

Ramenghi, L. A., Evans, D. J., and Levene, M. I. 1999. ''Sucrose analgesia'': absorptive mechanism or taste perception? Arch. Dis. Child Fetal Neonatal Ed. 80, F146–F147.

Rao, M., Blass, E. M., Brignol, M. M., Marino, L., and Glass, L. 1997. Reduced heat loss following sucrose ingestion in premature and normal human newborns. Early Hum. Dev. 48, 109–116.

Rosenstein, D. and Oster, H. 1988. Differential facial responses to four basic tastes in newborns. Child Dev. 59, 1555–1568.

Simmen, B. and Hladik, C. M. 1998. Sweet and bitter taste discrimination in primates: scaling effects across species. Folia Primatol. 69, 129–138.

Smith, B. A., Fillion, T. J. and Blass, E. M. 1990. Orally mediated sources of calming in 1 to 3-day-old human infants. Dev. Psychol. 26, 731–737.

Stein, N., Laing, D. G., and Hutchinson, I. 1994. Topographical differences in sweetness sensitivity in the peripheral gustatory system of adults and children. Brain Res. Dev. Brain Res. 82, 286–292.

Steiner, J. E. 1977. Facial expressions of the neonate infant indicating the hedonics of food-related chemical stimuli. In: Taste and Development: the Genesis of Sweet Preference (ed. J. M. Weiffenbach), pp. 173–188. U.S. Government Printing Office.

Steiner, J. E., Glaser, D., Hawilo, M. E., and Berridge, K. C. 2001. Comparative expression of hedonic impact: affective reactions to taste by human infants and other primates. Neurosci. Biobehav. Rev. 25, 53–74.

Steiner, J. E., Sgan-Cohen, H. D., and Nahas, J. 1984. Sweet preference and dental caries among Bedouin youth in Israel. Commun. Dent. Oral Epidemiol. 12, 386–389.

Stevens, B., Yamada, J., and Ohlsson, A. 2004. Sucrose for analgesia in newborn infants undergoing painful procedures. Cochrane Database Syst. Rev. Issue 3, CD001069.

Sullivan, S. A. and Birch, L. L. 1990. Pass the sugar, pass, the salt: experience dictates preference. Dev. Psychol. 26, 546–551.

Tomita, N. E., Nadanovsky, P., Vieira, A. L., and Lopes, E. S. 1999. Taste preference for sweets and caries prevalence in preschool children. Rev. Saude Publica 33, 542–546.

Upadhyay, A., Aggarwal, R., Narayan, S., Joshi, M., Paul, V. K., and Deorari, A. K. 2004. Analgesic effect of expressed breast milk in procedural pain in term neonates: a randomized, placebo-controlled, double-blind trial. Acta Paediatr. 93, 518–522.

Vazquez, M., Pearson, P. B., and Beauchamp, G. K. 1982. Flavor preferences in malnourished Mexican infants. Physiol. Behav. 28, 513–519.

Weiffenbach, J. M. 1977. Sensory mechanisms of the newborn's tongue. In: Taste and Development: the Genesis of Sweet Preference (*ed.* J. M. Weiffenbach), pp. 205–234. U.S. Government Printing Office.

Woloson, W. A. 2002. Refined Tastes: Sugar, Confectionery, and Consumers in Nineteenth-Century America. The Johns Hopkins University Press.

Wurtman, J. J. and Wurtman, R. J. 1979. Sucrose consumption early in life fails to modify the appetite of adult rats for sweet foods. Science 205, 321–322.

4.08 Taste Analgesia in Newborns

V Anseloni, University of Maryland Dental School, Baltimore, MD, USA

M Ennis, University of Tennessee Health Science Center, Memphis, TN, USA

Glossary

allodynia A painful response to a previously non-painful or subthreshold stimulus. Typically occurs after tissue injury and/or inflammation.

hyperalgesia A state characterized by enhanced sensation of pain. Typically occurs after tissue injury and/or inflammation.

mid-collicular transection A surgical procedure used to sever connections between the forebrain and the hindbrain.

von Frey filaments Filaments of various bending forces used to measure responses to mechanical cutaneous stimulation.

4.08.1 Taste Analgesia in Human Infants

Decades of studies have reported the analgesic and calming effects following intraoral infusion of sweet tastants (sucrose, glucose, fructose, aspartame), mother's milk, commercial formula, protein (provomin), fat (coconut, corn, and soy oil), and carbohydrates (polycose) on human infants (Blass, E. M. and Smith, B. A., 1992; Blass, E. M., 1997a; 1997c; Graillon, A. *et al.*, 1997; Blass, E. M. and Miller, L. W., 2001; Akman, I. *et al.*, 2002). Surprisingly, neither lactose nor colostrum has proven to produce analgesic or calming effects (Blass, E. M. and Smith, B. A., 1992; Blass, E. M., 1997a; 1997c; Blass, E. M. and Miller, L. W., 2001). Instead, the most potent taste analgesia is that triggered by sucrose.

The effects of sucrose have been well characterized in human infants. Analgesic and calming effects of taste analgesia are immediate and endure at least 5 min after sucrose termination (Blass, E. M. *et al.*, 1989; Smith, B. A. *et al.*, 1990; Blass, E. M. and Shah, A., 1995). When delivered repetitively (up to three times daily) over a 7-day period, there appears to be little or no tolerance to the analgesic effects of intraoral sucrose (Johnston, C. C. *et al.*, 2002). Intraoral infusion of microliter volumes (100–200 µl) of sucrose over a dose-range of 12–50% produce a flat dose–response relationship (Blass, E. M. *et al.*, 1989; Blass, E. M., 1997b; Overgaard, C. and Knudsen, A., 1999; Smith, B. A. *et al.*, 1990; Blass, E. M. and Hoffmeyer, L. B., 1991; Blass, E. M. and Smith, B. A., 1992).

Sucrose markedly diminishes crying in response to painful standard hospital procedures such as circumcision or blood collection via heel prick (Blass, E. M. and Hoffmeyer, L. B., 1991; Kaufman, G. E. *et al.*, 2002). In those studies, crying, grimacing, and heart rate have been the main assessments of nociception and calming in human newborns. The use of sucrose-coated pacifiers is currently recommended as a behavioral approach in the guidelines for managing pain in human newborns (Anand, K. J., 2001). However, despite the well-documented analgesic effects of sucrose, a clinically relevant question remains open for investigation: does repetitive administration of sucrose, especially in a neonatal intensive care unit (NICU) preterm infant, have long-term developmental consequences? A study by Johnston C. C. *et al.* (2002) reported that infants receiving intraoral sucrose up to three times daily for 1 week exhibited subsequent impairments in motor development and vigor. The cause of these deficits is unclear, but it is possible that the withdrawal of sucrose may render the infant more susceptible to subsequent pain.

Both preterm (Taddio, A. *et al.*, 2003) and full-term human infants of either cesarean section or vaginal deliveries readily exhibit analgesic and calming effects of sweet tastants. The age range of full term in those studies varied from few hours after birth (Blass, E. M. and Miller, L. W., 2001; Gray, L. *et al.*, 2002) to 2–6-week-old infants (Barr, R. G., *et al.*, 1994; Zeifman, D. *et al.*, 1996). Although rat studies have already unveiled the age-dependency of sucrose-induced analgesia, the age-dependency in human infants remains unclear. One major difference is that sucrose is effective in preterm human newborns, whereas in rats sucrose effects emerge between postnatal (P) day 2 and 3 (Anseloni, V. C. *et al.*, 2002). While the human developmental period over which intraoral sucrose is analgesic has not been systemically studied, it appears to be analgesic in 5–10-year-old children but not adults (Pepino, M. Y. and Mennella, J. A., 2005). Interestingly, this same study showed that analgesia elicited by sucrose correlates with sweet preferences in 5–10-year-old children but not adults. The more the children liked sucrose, the higher its analgesic efficacy. Therefore, the hedonic value of sucrose is reported to be a major component in triggering sucrose effects. It is unknown if there are gender-based differences in taste analgesia in human infants.

In agreement with animal studies discussed below, there is evidence that taste analgesia in human infants is mediated by central opioid mechanisms. Blass E. M. and Ciaramitaro V. (1994) reported that babies born to women who were maintained on methadone during pregnancy did not quiet to the taste of sucrose. This suggests an opioid mediation in the calming, and perhaps the analgesic, effects elicited by intraoral tastants. Taddio A. *et al.* (2003) reported that intraoral administration of sucrose in preterm infants does not lead to an increase in serum β-endorphin concentrations. However, synaptic release of opioid in the central nervous system (CNS) may not translate to increases in the periphery.

4.08.2   Taste Analgesia in Rodents

4.08.2.1   General Properties

Studies in rodents have revealed remarkable similarities of the properties of taste analgesia to those in human infants. As in human infants, intraoral infusion (20–140 µl) of dairy milk or half-and-half produces analgesia in P0 or P10 rats (Blass, E. M. *et al.*, 1991; Blass, E. M., 1997c). The analgesia develops rapidly (≤ 15 s) and persists for 0.5 min to at least 4 min after termination of infusion (Blass, E. M. *et al.*, 1991; Blass, E. M., 1997c). Milk infusions also reduced ultrasonic vocalizations, similar to its quieting effect in human infants (Blass, E. M. and Fitzgerald, E., 1988). Similar analgesic and quieting effects lasting up to 5 min were obtained by infusion of polycose or corn oil (Shide, D. J. and Blass, E. M., 1989).

Intraoral infusion (0.18 ml) of several sugars present in human and rat mother's milk were also shown to be analgesic in P10 rat pups (Blass, E. M. and Shide, D. J., 1994). Infusion of 0.22 M fructose, sucrose, or glucose reduced spontaneous ultrasonic vocalizations and increased paw withdrawal latencies to a conductive thermal stimulus. By contrast, lactose was ineffective. The rank-order of potencies was sucrose > glucose > fructose, paralleling findings in human infants. Testing of additional sugar concentrations (0.44 and 0.66 M) produced a flat dose–response relationship; that is, no additional effects were observed with higher concentrations. The reason for the flat dose–response function is unclear. Behavioral and electrophysiological studies demonstrate graded responses to these sugars over this concentration range in juvenile rodents. In P14–P20 hamsters, graded responses of the chorda tympani nerve are observed in response to sugars from 0.2 to 0.6 M, although these responses are of smaller amplitude than in adult animals (Hill, D. L., 1988). It is possible that the analgesic mechanisms (e.g., opioid release) activated by the sugars are saturated (i.e., ceiling effect) at 0.2 M and that a graded analgesic effect might be observed only with lower concentrations. Hill D. L. (1988) also reported that the chorda tympani responses to lactose (0.2–0.5 M) were essentially similar to those of fructose and glucose at identical concentrations. Extrapolating to rats, it is thus unclear why fructose and glucose produce analgesia, while lactose does not. One possibility is that central neural responses to lactose may be considerably less than the other sugars. Unfortunately, there is a relative paucity of information on the development of central gustatory neuronal responses to sugars during early postnatal development in rats.

Are the calming and analgesic effects of intraoral infusion explained by mechanical, somatosensory stimulation of tastant infusion? A number of studies have addressed this question, finding that intraoral infusion of distilled water, saline, or other salt stimuli do not reduce vocalizations and do not produce analgesia (Blass, E. M. *et al.*, 1987; Shide, D. J. and Blass, E. M., 1989; Blass, E. M. *et al.*, 1991; Blass, E. M. and Shide, D. J., 1994; Ren, K. *et al.*, 1997; Anseloni, V. C. *et al.*, 2002; 2005). Thus, the effects of intraoral infusion are not due to somatosensory stimulation from the cannula or fluid movement in the oral cavity, nor are they due to gustatory stimulation of other taste qualities (i.e., salt).

4.08.2.2 Oral Cavity versus Postingestional/Absorptive Mechanism?

An important question is if the quieting and analgesic effects of intraoral sucrose are mediated by stimulation of taste receptors in the oral cavity or to postingestional or absorptive mechanisms? The rapid onset (i.e., within seconds) of the effects of intraoral tastants suggest that they are triggered by oral cavity taste receptors. The contribution of postingestional mechanisms has not been directly examined in animal studies. However, this question was addressed in human infants. Ramenghi L. A. *et al.* (1999) found that while intraoral sucrose reduced crying and behavioral responses indicative of pain following heel lancing, direct stomach loading via a nasogastric tube was without effect. This suggests that stimulation of taste buds in the oral cavity by sucrose is necessary to induce analgesia.

Given that taste analgesia is due to stimulation of taste receptors in the oral cavity, it is of interest to determine which of the various taste-related cranial nerve branches are involved. Taste receptors are topographically distributed throughout the oral cavity and are innervated by three cranial nerves. The chorda tympani and the greater superficial petrosal branches of the facial nerve (VII) innervate the taste receptors of the anterior tongue and palate, respectively. In rats, receptors responsive to sweet stimuli are located primarily on the palate, consistent with observations that greater petrosal nerve fibers are more responsive to sugars than chorda tympani fibers. Taste receptors in the circumvallate and foliate papillae on the posterior tongue are innervated by the lingual branch of the glossopharyngeal nerve (IX), and those in the

oropharynx are innervated by superior laryngeal branch of the vagus nerve (X). Gustatory fibers of VII, IX, and X terminate in a rostral–caudal sequence in the rostral, gustatory portion of the nucleus tractus solitarius (rNTS; reviewed in Travers, S. P., 1993). Unfortunately, the contribution of these cranial nerves to taste analgesia has not been investigated. Additional studies are needed to clarify the roles of the gustatory cranial nerves in taste analgesia.

4.08.2.3 Neurochemical Mechanisms – Opioid Mediation

A number of rodent studies have investigated the neurochemical mechanisms underlying taste analgesia. The effects of oral sucrose are similar to those of opioid narcotics such as morphine. Morphine is analgesic, calming, and decreases vocalizations. Blass E. M. *et al.* (1991) reported that the analgesic effect of a single 20 μl intraoral bolus of milk was comparable in magnitude to that elicited by systemic injection of a moderate dose of the morphine (0.125 mg kg^{-1}). Other studies showed that the analgesic and quieting effects of intraoral milk and sugars were blocked by systemic injections of the nonselective opioid receptor antagonists naloxone or naltrexone (Blass, E. M. *et al.*, 1987; Blass, E. M. and Fitzgerald, E., 1988; Shide, D. J. and Blass, E. M., 1989; Blass, E. M. *et al.*, 1991). Taken together, these findings demonstrate that the calming and analgesic effects of intraoral fats and sugars are mediated by release of endogenous opioids.

Interestingly, additional studies showed that calming effects of intraoral tastants may also involve the gastrointestinal peptide, cholecystokinin (CCK). Calming effects, as measured by reductions in ultrasonic vocalizations, are elicited by systemic injection of CCK (Weller, A. and Blass, E. M., 1988). In the same study, CCK had no effect on pain thresholds. CCK receptor antagonists were subsequently found to block the calming effects of intraoral milk or corn oil, but not milk (Blass, E. M. and Shide, D. J., 1993; Weller, A. and Gispan, I. H., 2000). These studies indicate that the calming, but not the analgesic, effects of intraoral tastant involve a CCK component.

4.08.3 Outstanding Issues and Recent Developments

Sucrose-induced analgesia in rats and humans develops within seconds, lasts for minutes, is prevented by systemic injection of opiate receptor antagonists, and is compromised in human infants born to opiate-dependent mothers. Thus, the mechanisms appear to be remarkably well conserved phylogenetically and involve central opioid systems. A number of critical questions about taste analgesia remain unanswered. The rodent work summarized above was confined to the early postnatal period (P0–P10). Does taste analgesia exhibit a critical period in which intraoral tastants are effective? In other words, is this phenomenon developmentally transient? The pain-relieving effects of intraoral tastants were assayed in acute pain tests using primarily noxious thermal stimuli. Are intraoral tastant equally effective for other types of nociceptive stimuli, such as mechanical stimuli? Persistent inflammatory pain is a major consequence of tissue injury. Are intraoral tastants effective in attenuating hyperalgesia and allodynia associated with tissue injury? The neural circuits by which taste stimuli engage central opioid analgesic mechanisms are not known. The remainder of this chapter summarizes recent studies from our laboratories that have begun to address these questions.

4.08.3.1 Ontogeny of Taste-Analgesia and Efficacy for Inflammatory Pain

We investigated the effects of intraoral sucrose (7.5%) on nocifensive withdrawal responses to thermal and mechanical stimuli in naive rat pups and in pups in which Complete Freund's adjuvant (CFA) was injected in a fore- or hindpaw to produce inflammation (Anseloni, V. C. *et al.*, 2002). Pain thresholds were tested using conductive thermal and mechanical (von Frey filaments) stimuli from P0 (the day of birth) to P21. In noninflamed animals, for noxious thermal stimuli, sucrose-induced analgesia for the hindpaw emerged at P3, peaked at P7–P10, then progressively declined and was absent at P17 (Figure 1). For mechanical forepaw stimuli, sucrose-induced analgesia emerged, and was maximal at ~P10, then declined and was absent at P17. By contrast, maximal sucrose-induced analgesia for mechanical hindpaw stimuli was delayed (P13) compared to that for the forepaw, although it was also absent at P17. In inflamed animals, sucrose reduced hyperalgesia and allodynia assessed with mechanical stimuli (Figure 2). Sucrose-induced analgesia in inflamed animals was initially present at P3 for the forepaw and P13 for the hindpaw, and was absent by P17 for both limbs. Intraoral sucrose produced significantly greater effects on responses in fore- and

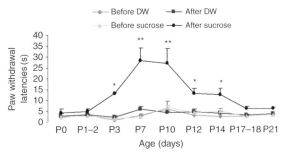

Figure 1 Effects of intraoral sucrose infusion of forepaw withdrawal latencies to thermal stimuli from P0 to P21. Infusion of distilled water (DW) did not influence paw withdrawal latencies ($P > 0.05$) and baseline withdrawal latencies were similar across P0–P21. Sucrose infusion did not significantly influence paw withdrawal latencies at P0–P2 or P17–P21. However, sucrose infusion significantly elevated withdrawal latencies from P3 to P14. Each time point includes a minimum of eight animals. Data are expressed as mean ± standard error of the mean paw withdrawal latency. *$P < 0.01$; **$P < 0.001$.

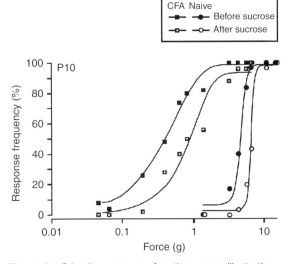

Figure 2 Stimulus–response function curves illustrating the effects of intraoral sucrose on response frequency of the forepaw to mechanical stimuli in noninflamed (naive) and inflamed (CFA) rats at P10; see text for details. Note the leftward shift in the stimulus–response function in inflamed rats compared to naive rats. After inflammation, rats respond to stimuli that are subthreshold in naive rats, indicating the development of inflammation-induced allodynia. There is also an increased response to suprathreshold stimuli, indicating the development of inflammation-induced hyperalgesia. Note that sucrose infusion (open symbols) diminished the response frequency, resulting in a significant rightward shift of the curves for both naive and inflamed rats ($P < 0.001$). Reprinted from Anseloni, V. C., Weng, H. R., Terayama, R., Letizia, D., Davis, B. J., Ren, K., Dubner, R., and Ennis, M. 2002. Age-dependency elicited by intraoral sucrose in acute and persistent pain models. Pain 97, 93–103, with permission from IASP.

hindpaws in inflamed rats than in naive rats indicating that it reduces hyperalgesia and allodynia beyond its effects on responses in naive animals.

The preceding study demonstrated that sucrose-induced analgesia is age-dependent and maximal at P10–P13, and is limited to the preweaning period (<P17) in rats. After P10–P13, the analgesic effect of intraoral sucrose progressively subsides as weaning approaches, and is absent at the approximate time of weaning (P17–P21). Extrapolating to humans, P7 in rats correlates to full-term human newborns and the range of P0–P20 in rats corresponds approximately to 24 weeks of gestation and the toddler stage, respectively, in human infants (Helmuth, L., 2000). Sucrose-induced analgesia in rats thus disappears at a neonatal age when locomotor activity is sufficiently developed for the animal to generate organized escape from noxious stimuli. The age-dependency of sucrose-induced analgesia and its differential maturation for the fore- and hindpaw may be due to developmental changes in endogenous analgesic mechanisms and developmental modulation of the interaction between gustatory and pain modulatory pathways (for review see Anseloni, V. C. et al., 2002). The findings above support the hypothesis that sucrose has a selective influence on analgesic mechanisms and that an enhanced sucrose effect takes place in hyperalgesic, inflamed animals as compared to naive animals. These results indicate that intraoral sucrose alleviates transient pain in response to thermal and mechanical stimuli, and also effectively reduces inflammatory hyperalgesia and allodynia.

The absence of an analgesic effect of sucrose infusion at P0–P2 was surprising given previous reports that intraoral milk infusion was analgesic in P0 pups delivered by cesarean section (Blass, E. M. et al., 1991). The reason for the difference between the two studies is unclear. Stress or changes in opioid mechanisms as a result of natural delivery may compromise the ability of milk and/or sucrose to produce analgesia for the first few postnatal days. Milk also contains a number of individual analgesic sugars and fats, and the combination of these, but not sucrose alone, may be sufficient to produce analgesia immediately after birth. Further studies are needed to address these questions.

4.08.3.2 Neural Circuitry

As reviewed above, the characteristics of sucrose-induced analgesia have been well established in rat pups and human infants. By contrast, nothing is known about the neurobiological substrate underlying this phenomenon. The ability of a small intraoral bolus of sucrose to trigger opiate-mediated analgesia suggests that there are connections from sweet taste-responsive brain neurons to the descending antinociceptive systems. In this regard, it is noteworthy that several gustatory areas have been reported to be involved in analgesia, including the parabrachial nucleus (Bester, H. et al., 1995), insular cortex (Burkey, A. R. et al., 1996), central nucleus of the amygdala (Kalivas, P. W. et al., 1982; Oliveras, J. L. and Besson, J. M., 1988), and lateral hypothalamus. Additionally, tract tracing studies have demonstrated that several gustatory areas project to two major brainstem antinociceptive sites involved in opiate-receptor-dependent analgesia: the midbrain periaqueductal gray (PAG) and the rostroventromedial medullary region that includes nucleus raphe magnus (RMG). For example, the parabrachial nucleus, insular cortex, central nucleus of the amygdala, and lateral hypothalamus densely innervate the PAG and some of these areas directly project to the RMG area (Behbehani, M. M. et al., 1988; Krukoff, T. L. et al., 1993; Behbehani, M. M., 1995; Hermann, D. M. et al., 1997). Thus, there is considerable overlap among, as well as numerous potential anatomical linkages between, components of gustatory pathway and sites known to be involved in centrally mediated analgesia.

To begin to assess these circuits, we used mid-collicular transections as well as immunohistochemistry for the protein product of the immediate early gene c-fos, to identify sites involved in the analgesic effect of sucrose in neonatal rats. Previous studies have shown that certain functions of the taste system do not require processing or feedback from higher-order gustatory structures. Adult animals exhibit appropriate motor and discriminative responses to taste stimuli following mid-collicular transections that isolate the parabrachial nucleus (PBN) from the taste thalamus and other forebrain gustatory structures. For example, rats will ingest palatable sweet taste solutions or milk and reject unpalatable bitter taste solutions following mid-collicular transections (Kornblith, C. L. and Hall, W. G., 1979; Grill, H. J. and Norgren, R., 1978; Grill, H. J. and Berridge, K., 1985). Thus, the discriminatory and basic motor aspects of taste behavior are retained in the brainstem caudal to the level of the PBN. Given these findings, we asked whether the neural circuitry necessary for mediating sucrose-induced analgesia in neonatal rats also resides within the brainstem and spinal cord. To assess this, mid-collicular transections that isolated the PBN, as well as more caudal structures, from mid- and forebrain components of the gustatory pathway were performed (Anseloni, V. C. et al., 2005). Also present posterior to the transection were several structures involved in antinociception, including the caudal PAG and RMG. We found that in the transected pups, the analgesic actions of intraoral sucrose infusion persisted. Moreover, sucrose-induced analgesia was enhanced compared to its actions in the sham pups. This suggested that descending projections from the mid- or forebrain may tonically facilitate central responses to noxious somatosensory input. Consistent with this, transection alone produced a decrease in withdrawal responses to noxious mechanical stimuli, resulting in behavioral hypoalgesia. However, a simple shift in baseline responses did not appear to account for the persistence of sucrose's analgesic effect in the transected pups. Transection alone increased the baseline withdrawal responses to roughly the same degree (30–40%) as did sucrose in the sham animals (40–50%). By contrast, sucrose produced a significantly greater increase (173%) following mid-collicular transection (Figure 3).

The preceding findings indicate that the neural circuitry underlying sucrose-induced analgesia resides within the brainstem and spinal cord. We next assessed c-fos activity elicited by brief intraoral infusion (90 μl in 90 s) of sucrose (0.2 M) or

ammonium chloride (a salt stimulus, 0.1 M) to identify candidate brainstem neurons involved in this circuitry (Anseloni, V. C. et al., 2005). Compared to control groups (intact, cannula, distilled water), both sucrose and ammonium chloride induced Fos expression in the rNTS, the first relay in the ascending gustatory pathway. Sucrose also elicited Fos expression in several brainstem areas associated with centrally mediated analgesia, including the PAG and the RMG. Taken together, these findings demonstrate that analgesia elicited by intraoral sucrose does not require involvement of the forebrain. Intraoral sucrose activates neurons in the PAG and RMG, two key brainstem sites critically involved in descending pain modulation.

Previous studies have shown that activation of the PAG and RMG produce analgesia and are two key brain substrates involved in the analgesic actions of opioids. RMG neurons are the main source of descending projections (via the dorsolateral funiculus) to the dorsal horn of the spinal cord. RMG stimulation, or morphine administration, inhibits dorsal horn neurons involved in pain transmission (Fields, H. L. et al., 1991; Behbehani, M. M., 1995). If, as suggested above by the study of Anseloni V. C. et al. (2005), intraoral sucrose produces analgesia via activation of RMG neurons, then intraoral sucrose should, similar to direct stimulation of RMG, inhibit spinal cord neuronal responses to nociceptive stimuli. Ren K. et al. (1997) found that intraoral sucrose markedly diminished Fos expression in the spinal cord elicited by inflammation of the forepaw. Thus, the effects of intraoral sucrose are similar to those of morphine, in that both blunt dorsal horn neuronal responses to noxious somatosensory input.

4.08.4 Future Directions

Despite our increased understanding of the neurobiological underpinnings and properties of taste analgesia in humans and rodents, significant questions remain. What mechanisms are responsible for the ontogeny and developmentally transient nature of this phenomenon? Where are the specific links where gustatory information accesses central analgesic circuits? What opioid neurotransmitters (e.g., enkephalin, dynorphin) and opiate receptor subtypes are involved, and where are they located? These will be difficult questions to answer, as opioids and opiate receptors are located in numerous components in both the ascending gustatory pathway and

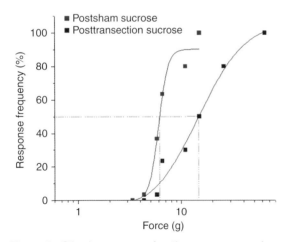

Figure 3 Stimulus–response function curves comparing the effects of intraoral sucrose on response frequency of the hindpaw to mechanical stimuli following sham and mid-collicular transection surgeries in P10–P12 rats. Sucrose after transection had an enhanced analgesic effect compared to sucrose in sham animals, as illustrated by the significant rightward shift in the stimulus–response curve ($P < 0.001$); see text for details.

descending antinociceptive circuits. Might analgesia elicited by nursing provide a natural neuroprotective function for the developing nervous system? The ability of natural stimuli to produce analgesia may have important consequences beyond transient or brief pain relief in developing infants. Preterm and certain high-risk infants undergo repeated clinical procedures which produce cutaneous inflammation accompanied by hyperalgesia lasting for several hours or even days (Anand, K. J., 1999). Recent studies have shown that tissue injury during the first few postnatal weeks produces changes in the long-term development of nociceptive circuits and pain thresholds that persist in adult animals (Ruda, M. A. *et al.*, 2000; Lidow, M. S. *et al.*, 2001; Ren, K. *et al.*, 2004). Thus, analgesia produced by natural stimuli during the preweaning period may protect against the deleterious effects of tissue injury and inflammation on the development and/or maturation of nociceptive circuits.

Acknowledgments

Supported by PHS grants DC03895 and NS41384.

References

Akman, I., Ozek, E., Bilgen, H., Ozdogan, T., and Cebeci, D. 2002. Sweet solutions and pacifiers for pain relief in newborn infants. J. Pain 3, 199–202.

Anand, K. J. 2001. International evidence-based group for neonatal pain. Consensus statement for the prevention and management of pain in the newborn. Arch. Pediatr. Adolesc. Med. 155, 173–180.

Anand, K. J., Coskun, V., Thrivikraman, K. V., Nemeroff, C. B., and Plotsky, P. M. 1999. Long-term behavioral effects of repetitive pain in neonatal rat pups. Physiol. Behav. 66, 627–637.

Anseloni, V. C., Ren, K., Dubner, R., and Ennis, M. 2005. A brainstem substrate for analgesia elicited by intraoral sucrose. Neuroscience 133, 231–243.

Anseloni, V. C., Weng, H. R., Terayama, R., Letizia, D., Davis, B. J., Ren, K., Dubner, R., and Ennis, M. 2002. Age-dependency elicited by intraoral sucrose in acute and persistent pain models. Pain 97, 93–103.

Barr, R. G., Quek, V. S., Cousineau, D., Oberlander, T. F., Brian, J. A., and Young, S. N. 1994. Effects of intra-oral sucrose on crying, mouthing and hand–mouth contact in newborn and six-week-old infants. Dev. Med. Child. Neurol. 36(7), 608–618.

Behbehani, M. M. 1995. Functional characteristics of the midbrain periaqueductal gray. Prog. Neurobiol. 46, 575–605.

Behbehani, M. M., Park, M. R., and Clement, M. E. 1988. Interactions between the lateral hypothalamus and the periaqueductal gray. J. Neurosci. 8, 2780–2787.

Bester, H., Menendez, L., Besson, J. M., and Bernard, J. F. 1995. Spino (trigemino) parabrachiohypothalamic pathway:

electrophysiological evidence for an involvement in pain processes. J. Neurophysiol. 73, 568–585.

Blass, E. M. 1997a. Infant formula quiets crying human newborns. J. Dev. Behav. Pediatr. 18, 162–165.

Blass, E. M. 1997b. Interactions between contact and chemosensory mechanisms in pain modulation in 10-day-old rats. Behav. Neurosci. 111, 147–154.

Blass, E. M. 1997c. Milk-induced hypoalgesia in human newborns. Pediatrics 99, 825–829.

Blass, E. M. and Ciaramitaro, V. 1994. A new look at some old mechanisms in human newborns: taste and tactile determinants of state, affect, and action. Monogr. Soc. Res. Child Dev. 59, 1–81.

Blass, E. M. and Fitzgerald, E. 1988. Milk-induced analgesia and comforting in 10-day-old rats: opioid mediation. Pharmacol. Biochem. Behav. 29, 9–13.

Blass, E. M. and Hoffmeyer, L. B. 1991. Sucrose as an analgesic for newborn infants. Pediatrics 87, 215–218.

Blass, E. M. and Miller, L. W. 2001. Effects of colostrum in newborn humans: dissociation between analgesic and cardiac effects. J. Dev. Behav. Pediatr. 22, 385–390.

Blass, E. M. and Shah, A. 1995. Pain-reducing properties of sucrose in human newborns. Chem. Senses 20, 29–35.

Blass, E. M. and Shide, D. J. 1993. Endogenous cholecystokinin reduces vocalizations in isolated 10-day-old rats. Behav. Neurosc. 107, 488–492.

Blass, E. M. and Shide, D. J. 1994. Some comparisons among the calming and pain-relieving effects of sucrose, glucose, fructose and lactose in infant rats. Chem. Senses 19, 239–249.

Blass, E. M. and Smith, B. A. 1992. Differential effects of sucrose, fructose, glucose, and lactose on crying in 1- to 3-day-old human infants: qualitative and quantitative considerations. Dev. Psych. 28, 804–810.

Blass, E. M., Fillion, T. J., Rochat, P., Hoffmeyer, L. B., and Metzger, M. A. 1989. Sensorimotor and motivational determinants of hand–mouth coordination in 1–3-day-old human infants. Dev. Psych. 25, 963–975.

Blass, E. M., Fitzgerald, E., and Kehoe, P. 1987. Interactions between sucrose, pain and isolation distress. Pharmacol. Biochem. Behav. 26, 483–489.

Blass, E. M., Jackson, A. M., and Smotherman, W. P. 1991. Milk-induced, opioid-mediated antinociception in rats at the time of cesarean delivery. Behav. Neurosci. 105, 677–686.

Burkey, A. R., Carstens, E., Wenniger, J. J., Tang, J., and Jasmin, L. 1996. An opioidergic cortical antinociception triggering site in the agranular insular cortex of the rat that contributes to morphine antinociception. J. Neurosci. 16, 6616–6623.

Fields, H. L., Heinricher, M. M., and Mason, P. 1991. Neurotransmitters in nociceptive medullary circuits. Ann. Rev. Neurosci. 14, 219–245.

Graillon, A., Barr, R. G., Young, S. N., Wright, J. H., and Hendricks, L. A. 1997. Differential response to intraoral sucrose, quinine and corn oil in crying human newborns. Physiol. Behav. 62, 317–325.

Gray, L., Miller, L. W., Philipp, B. L., and Blan, E. M. 2002. Breastfeeding is analgeric in healthy newborns. Pediatrics 109(4), 590–593.

Grill, H. J. and Berridge, K. 1985. Taste reactivity as a measure of neural control of palatability. Prog. Psychobiol. Physiolog. Psychol. 11, 1–61.

Grill, H. J. and Norgren, R. 1978. Chronically decerebrate rats demonstrate satiation but not bait shyness. Science 201, 267–269.

Helmuth, L. 2000. Early insult rewires pain circuits. Science 289, 521–522.

Hermann, D. M., Luppi, P. H., Peyron, C., Hinckel, P., and Jouvet, M. 1997. Afferent projections to the rat nuclei raphe magnus, raphe pallidus and reticularis gigantocellularis pars alpha demonstrated by iontophoretic application of choleratoxin (subunit b). J. Chem. Neuroanat. 13, 1–21.

Hill, D. L. 1988. Development of chorda tympani nerve taste responses in the hamster. J. Comp. Neurol. 268, 346–356.

Johnston, C. C., Filion, F., Snider, L., Majnemer, A., Limperopoulos, C., Walker, C. D., Veilleux, A., Pelausa, E., Cake, H., Stone, S., Sherrard, A., and Boyer, K. 2002. Routine sucrose analgesia during the first week of life in neonates younger than 31 weeks' postconceptional age. Pediatrics 110, 523–528.

Kalivas, P. W., Gau, B. A., Nemeroff, C. B., and Prange, A. J. 1982. Antinociception after microinjection of neurotensin into the central amygdaloid nucleus of the rat. Brain Res. 243, 279–286.

Kaufman, G. E., Cimo, S., Miller, L. W., and Blass, E. M. 2002. An evaluation of the effects of sucrose on neonatal pain with 2 commonly used circumcision methods. Am. J. Obstet. Gynecol. 186, 564–568.

Kornblith, C. L. and Hall, W. G. 1979. Brain transections selectively alter ingestion and behavioral activation in neonatal rats. J. Comp. Physiol. Psychol. 93, 1109–1117.

Krukoff, T. L., Harris, K. H., and Jhamandas, J. H. 1993. Efferent projections from the parabrachial nucleus demonstrated with the anterograde tracer *Phaseolus vulgaris* leucoagglutinin. Brain Res. Bull. 30, 163–172.

Lidow, M. S., Song, Z. M., and Ren, K. 2001. Long-term effects of short-lasting early local inflammatory insult. Neuroreport 12, 399–403.

Oliveras, J. L. and Besson, J. M. 1988. Stimulation-produced analgesia in animals: behavioral investigations. Prog. Brain Res. 77, 141–157.

Overgaard, C. and Knudsen, A. 1999. Pain-relieving effect of sucrose in newborns during heel prick. Biol. Neonate 75, 279–284.

Pepino, M. Y. and Mennella, J. A. 2005. Sucrose-induced analgesia is related to sweet preferences in children but not adults. Pain 119, 210–218.

Ramenghi, L. A., Evans, D. J., and Levene, M. I. 1999. "Sucrose analgesia": absorptive mechanism or taste perception? Arch. Dis. Child. Fetal Neonatal Ed. 80, F146–F147.

Ren, K., Anseloni, V., Zou, S. P., Wade, E. B., Novikova, S. I., Ennis, M., Traub, R. J., Gold, M. S., Dubner, R., and Lidow, M. S. 2004. Characterization of basal and re-inflammation-associated long-term alteration in pain responsivity following short-lasting neonatal local inflammatory insult. Pain 110, 588–596.

Ren, K., Blass, E. M., Zhou, Q., and Dubner, R. 1997. Suckling and sucrose ingestion suppress persistent hyperalgesia and spinal cord Fos expression after forepaw inflammation in infant rats. Proc. Natl. Acad. Sci. U. S. A. 94, 1471–1475.

Ruda, M. A., Ling, Q. D., Hohmann, A. G., Peng, Y. B., and Tachibana, T. 2000. Altered nociceptive neuronal circuits after neonatal peripheral inflammation. Science 289, 628–631.

Shide, D. J. and Blass, E. M. 1989. Opioidlike effects of intraoral infusions of corn oil and polycose on stress reactions in 10-day-old rats. Behav. Neurosci. 103, 1168–1175.

Smith, B. A., Fillion, T. J., and Blass, E. M. 1990. Orally mediated sources of calming in 1- to 3-day-old human infants. Dev. Psych. 26, 731–737.

Taddio, A., Shah, V., Shah, P., and Katz, J. 2003. Beta-endorphin concentration after administration of sucrose in preterm infants. Arch. Pediatr. Adolesc. Med. 157, 1071–1074.

Travers, S. P. 1993. Orosensory Processing in Neural Systems of the Nucleus of the Solitary Tract. In: Mechanisms of Taste Transduction (*eds*. S. A. Simon and S. D. Roper), pp. 339–394. CRC Press.

Weller, A. and Blass, E. M. 1988. Behavioral evidence for cholecystokinin-opiate interactions in neonatal rats. Am. J. Physiol. 255, R5901–R5907.

Weller, A. and Gispan, I. H. 2000. A cholecystokinin receptor antagonist blocks milk-induced but not maternal-contact-induced decrease of ultrasonic vocalization in rat pups. Dev. Psychobiol. 37, 35–43.

Zeifman, D., Delaney, S., and Blass, E. M. 1996. Sweet taste, looking, and calm in a 2- and 4-week-old infants: the eyes have it. Dev. Psych. 32, 1090–1099.

4.09 Taste Receptors

M Max, Mount Sinai School of Medicine, New York, NY, USA

W Meyerhof, German Institute of Human Nutrition Potsdam-Rehbruecke, Nuthetal, Germany

4.09.1 Introduction

Bitter taste has been considered to be part of a warning system designed by nature to prevent organisms from ingesting poisonous substances. Bitter transduction, therefore, appears to be associated with neural networks involved in avoidance behavior. In marked contrast, sweet and amino acid taste (also referred to as umami taste) are thought to act as calorie detectors and consequently elicit attractive behavioral patterns that promote intake. It turned out recently that in the periphery taste receptor cells form separate populations, one for each taste modality (Chandrashekar, J., 2006). It is apparently not important how a taste receptor cell population is stimulated to elicit the sensation of umami, sweet or bitter, it is only important that it is exited (Zhang, Y. *et al.*, 2003; Mueller, K. L. *et al.*, 2005). In this scenario, the taste receptors fulfill the task to convert a chemical structure into a biochemical response that is propagated to the cerebral cortex, where the activities of neurons reflect the cognitive perception. This chapter focuses on the receptors for sweet, umami, and bitter tastes. During the last ~7 years enormous progress has been in elucidating many of the receptor properties. But still the field is in dynamic expansion. The accumulated knowledge will be summarized with reference to the current problems and to identify future directions.

4.09.2 Bitter-Taste Receptors

4.09.2.1 Discovery of TAS2Rs

A number of reasonable assumptions provided the basis of cloning the bitter-taste receptor genes. First, the observation that the G-protein subunit α-gustducin plays a major role in bitter taste (McLaughlin, S. K. *et al.*, 1992; Wong, G. T. *et al.*, 1996a; 1996b; Ming, D. *et al.*, 1998; 1999; Ruiz-Avila, L. *et al.*, 2000) suggested that bitter receptors are G-protein-coupled receptors (GPCRs). Second, it has been hypothesized that a family of receptors must exist that allows detection of the numerous bitter compounds (Belitz, H. D. and Wieser, H., 1985). Third, it has been anticipated that the receptor genes are linked to chromosomal loci that have previously been demonstrated to influence the sensitivities for various bitter-tasting substances (Lush, I. E., 1981; 1984; 1986; Lush, I. E. and Holland, G., 1988). Searching the human genome data base ultimately led to the identification of the first members of a gene family encoding a new class of GPCRs termed T2Rs (Adler, E. *et al.*, 2000) or TRBs (Matsunami, H. *et al.*, 2000). With the gene sequences at hand several other groups subsequently isolated additional members of this gene family from humans, mice, rats, and various primate species (Bufe, B. *et al.*, 2002; Conte, C. *et al.*, 2002; 2003; Shi, P. *et al.*, 2003; Parry, C. M. *et al.*, 2004; Fischer, A. *et al.*, 2005; Go, Y. *et al.*, 2005). Today the complete repertoire of bitter receptor genes has been identified in humans, mice, rats, dogs, cows, opossums, chickens, frogs, and

several fish (Conte, C. *et al.*, 2002; 2003; Go, Y., 2006; Shi, P. and Zhang, J., 2006). Twenty-three of the 25 human *TAS2R* genes are present in two extended clusters on chromosomes 7q34-35 and 12p13.31-13.2, while one each is located on chromosome 5p15.31 and 7q31.32 (Adler, E. *et al.*, 2000; Matsunami, H. *et al.*, 2000; Bufe, B. *et al.*, 2004). In mice, all but two genes are found in two clusters on chromosome 6, the two exceptions being located on chromosomes 2 and 15. Due to the simultaneous cloning in different laboratories the nomenclature of bitter receptor genes is confusing. In this review article the authors use the nomenclature of Bufe B. *et al.* (2002), that has been approved by the human genome project nomenclature committee.

Like many other GPCRs, the coding regions of TAS2Rs are not interrupted by introns (Adler, E. *et al.*, 2000; Matsunami, H. *et al.*, 2000; Bufe, B. *et al.*, 2002; Conte, C. *et al.*, 2002; 2003) and likely fold into seven transmembrane regions with the corresponding intracellular and extracellular domains (Floriano, W. B. *et al.*, 2006; Miguet, L. *et al.*, 2006). TAS2Rs lack, however, any sequence relationship to other members of the GPCR superfamily, but do show sequence relationship among themselves. They also lack the typical signatures of other GPCRs, such as the DRY motif close to transmembrane domain (TMD) 3 and the two conserved cysteine residues in extracellular loops 2 and 3 thought to form disulphide bridges in other GPCRs. TAS2Rs also miss N-linked glycosylation consensus sequences in the N-terminal domain. TAS2Rs are characterized by short N- and C-terminal domains and a well conserved N-linked glyco-sylation consensus sequence in extracellular loop 2. They are therefore considered to form a separate class of GPCRs. Even within a species sequence identities amongst TAS2Rs vary greatly ranging from less then 20% for some members to more than 80% for others. The extracellular parts of the TAS2Rs and the parts of the transmembrane segments facing the extracellular space are less conserved than the intracellular domains and the parts of the transmembrane segments close to the cytoplasm. This is consistent with the idea that the former regions form heterogeneous binding pockets for the numerous bitter compounds, whereas the latter regions engage in G-protein activation that is likely less variable (Meyerhof, W., 2005). The C-terminal domain, which also shows little sequence conservation, could reflect differences in receptor trafficking and/or regulation.

4.09.2.2 Molecular Evolution of *TAS2R* Genes

The available TAS2R repertoires reveal different numbers of genes in humans, mice, rats, dogs, cows, opossums, chickens, frogs, and fish. Whereas chickens have only three genes, their number in mammals and amphibian is in the range of 20–50 (Go, Y., 2006; Shi, P. and Zhang, J., 2006). The high numbers of TAS2Rs in frogs are surprising as most bitter substances are of plant origin and frogs mostly feed on animals. It has been proposed that the TAS2R gene repertoires evolved from multiple ancestors in the lineage leading to teleosts and tetrapods by frequent gene duplications and deactivations with expansions in mammals and amphibia and contraction in chickens. Another characteristic of the TAS2R repertoires across species is a one-to-one orthology for some genes whereas others are lineage or species specific. The lineage specific TAS2Rs appear to be duplicated lately in evolution as their sequences are comparably well conserved and the genes are closely spaced on the chromosomes. These genes may be under positive selection driven by adaptation of species to changing nutritional habits during phylogenetic evolution and detect species specific bitter substances (Bufe, B. *et al.*, 2002; Shi, P. *et al.*, 2003). The *TAS2R* genes with a one-to-one orthology appear to be under more selective constraints and have been proposed to detect bitter compounds encountered by various species (Shi, P. *et al.*, 2003).

Evidence for reduced selective pressure of primate *TAS2R* genes compared with their rodent counterparts comes from the increasing number of pseudogenes from rodents to apes and human and from similar proportions of synonymous to nonsynonymous substitutions (Fischer, A. *et al.*, 2005; Go, Y. *et al.*, 2005). In the hominid lineage, the reduced functional constraints may have been arisen by diminished necessity to detect bitter substances due to decreased consumption of plant food, increased consumption of animal meat and detoxification of food through the controlled use of fire (Wang, X. *et al.*, 2004). Positive selection has also been shown to shape *TAS2R* genes. A sensitive allele of the human glucopyranoside receptor, TAS2R16, was generated during the Paleolithic and, with the migration of *Homo* out of Africa, spread all over the world while the original low sensitivity allele is only found in Africa (Soranzo, N. *et al.*, 2005). It has been proposed that increased sensitivity for frequent toxic glycosides allowed ancient humans to establish healthier diets with low contents of such plant secondary metabolites exerting a selective advantage in carriers of the sensitive allele. The low-sensitivity allele, restricted to Africa with a distribution similar to those of known anti-Malaria alleles, may have been associated with a higher intake of cyanogenic glycosides. A possible consequence could be chronic cyanide poisoning and sickle cell anemia. This provides a selective advantage to carriers of the low sensitivity allele as sickle cell anemia protects against malaria infection (Soranzo, N. *et al.*, 2005). In addition to loss of functional constraints and positive selection, balancing natural selection has been shown as an evolutionary force. It maintains functional nontaster alleles of the phenylthiocarbamide receptor, TAS2R38 (Kim, U. K. *et al.*, 2003; Bufe, B. *et al.*, 2005), in the human but not in the chimpanzee population (Wooding, S. *et al.*, 2004, Wooding, S. *et al.*, 2006).

4.09.2.3 Expression of *TAS2R* Genes in Oral and Extraoral Tissues

4.09.2.3.1 *Expression of TAS2R genes in oral tissues*

Fully consistent with their role as taste receptor genes, TAS2R messenger ribonucleic acids (mRNAs) have been found by reverse-transcriptase polymerase chain reaction (RT-PCR) and *in situ* hybridization in taste bud cells of all three types of lingual papillae in rodents and humans (Adler, E. *et al.*, 2000; Matsunami, H. *et al.*, 2000; Ueda, T. *et al.*, 2001; Bufe, B. *et al.*, 2002; Kim, M. R. *et al.*, 2003; Behrens, M. *et al.*, 2004; Kuhn, C. *et al.*, 2004; Bufe, B. *et al.*, 2005). TAS2R expression did not overlap with that of TAS1Rs supporting the finding that separate taste receptor cell populations exist for the different taste modalities (Zhang, Y. *et al.*, 2003; Mueller, K. L. *et al.*, 2005; Huang, A. L. *et al.*, 2006). From the number of cells labeled by *in situ* hybridization probes it has been estimated that some 15% of rodent taste bud cells express TAS2Rs (Adler, E. *et al.*, 2000). Based on a smaller set of data a similar estimate has been achieved in humans (Meyerhof, W. *et al.*, 2005). The proportion of TAS2R-positive cells in taste buds was similar for all types of taste papillae (Adler, E. *et al.*, 2000). Differences in TAS2R expression, however, were seen across papillae. Whereas almost all foliate and vallate taste buds contain TAS2R-positive cells, only less than 10% of the fungiform papillae do so (Adler, E. *et al.*, 2000). As gustatory information of fungiform papillae is largely transmitted by the chorda tympani nerve, while responses from foliate and vallate papillae are transmitted by the glossopharyngeal nerve the question arises whether or not the TAS2R expression pattern generates a chemotopic map. Although this question cannot be answered definitely, it is important to note that for some but not all bitter compounds glossopharyngeal responses differed from those of the chorda tympani (Dahl, M. *et al.*, 1997; Danilova, V. and Hellekant, G., 2003; 2004).

From *in situ* hybridization experiments in rodents with single and mixed TAS2R probes Adler E. *et al.* (2000) concluded that a large number of TAS2R genes, possibly the full repertoire, is expressed in the same subset of taste receptor cells that therefore form a uniform population of bitter-taste cells which could explain the uniformity of bitter taste. This conclusion was supported in phospholipase C-$\beta 2$ knockout mice that lost their responses to bitter stimuli. When phospholipase C-$\beta 2$ expression was rescued under the control of various *TAS2R* gene promoters, the animals acquired nerve and behavioral responses to many bitter stimuli (Mueller, K. L. *et al.*, 2005). However, this conclusion is challenged by another *in situ* hybridization study in mice that did not reveal the same degree of homogeneity amongst bitter-taste cells (Matsunami, H. *et al.*, 2000) and by physiological recordings in mouse taste tissues that revealed clear heterogeneity across bitter cells by demonstrating that taste cells responded only to one or two out of a collection of five different bitter compounds (Caicedo, A. and Roper, S. D., 2001). A nonuniform population of bitter receptor cells would be a prerequisite for mammals to discriminate among bitter stimuli. However, this ability has also been disputed (Whitney, G. and Harder, D. B., 1994; Dahl, M. *et al.*, 1997; Delwiche, J. F. *et al.*, 2001; Spector, A. C. and Kopka, S. L., 2002). Perhaps mammals categorize compounds into but can discriminate amongst them within taste modalities.

4.09.2.3.2 Expression of TAS2R genes in extraoral tissues

The recent observation that *TAS2R* gene expression and expression of taste-signaling molecules has been detected in extraoral tissues challenged the role of TAS2Rs as taste receptors. As long as the complete TAS2R repertoire has not been demonstrated to be present in taste tissues, some of them may not function as taste sensors but fulfill other specific tasks. Nevertheless, extraorally expressed TAS2Rs likely exert additional functions. Sites of extraoral TAS2R expression were the solitary chemoreceptor cells in the respiratory epithelium. Irrigation of the respiratory epithelium with bitter compounds elicited trigeminal responses and pronounced respiratory effects (Finger, T. E. *et al.*, 2003). These results indicate that, similar to their oral function to prevent intoxication, nasal bitter receptors could protect mammals from the aspiration of noxious substances. Other sites of TAS2R expression were gastrointestinal cells and the testes (Wu, S. V. *et al.*, 2002; Behrens, M. *et al.*, 2006) suggesting that they may be involved in metabolic and reproductive functions.

4.09.2.4 Functional Properties of Bitter Receptors

4.09.2.4.1 Tuning of TAS2Rs to their agonists

Most of the known properties of TAS2Rs have been uncovered in heterologous expression assays using mammalian and insect cell lines (Chandrashekar, J. *et al.*, 2000; Bufe, B. *et al.*, 2002; Sainz, E. *et al.*, 2007). Usually the recombinant receptors are expressed as chimeras containing N-terminally located domains of bovine rhodopsin (Chandrashekar, J. *et al.*, 2000; Sainz, E. *et al.*, 2007) or rat somatostatin type 3 receptor (Bufe, B. *et al.*, 2002) that serve as cell surface-targeting modules. For some receptors such targeting motifs can be omitted if auxiliary proteins of the receptor transport protein or receptor expression enhancing protein families are coexpressed (Saito, H. *et al.*, 2004; Behrens, M. *et al.*, 2006). Receptor activation is monitored by fluorometric detection of bitter compound-induced changes in cytosolic calcium concentrations in cell lines expressing the promiscuously coupling G-protein subunit Gα15 (Chandrashekar, J. *et al.*, 2000; Bufe, B. *et al.*, 2002) or chimeras of α-gustducin and Gα16 (Ueda, T. *et al.*, 2003; Behrens, M. *et al.*, 2004; Kuhn, C. *et al.*, 2004; Bufe, B. *et al.*, 2005). Alternative approaches rely on the incorporation of radiolabeled GTPγS into purified G-protein α-subunits added to membranes containing the recombinant TAS2Rs (Chandrashekar, J. *et al.*, 2000; Sainz, E. *et al.*, 2007).

With these assays, the cognate bitter compounds for a number of TAS2Rs have been identified (Table 1) and a number of important conclusions have been drawn from the experiments. First, the identified compounds taste bitter confirming the role of TAS2Rs as bitter receptors. Second, TAS2Rs are broadly tuned to detect numerous substances. Some of them are activated by compounds with common substructures. Human TAS2R16 for instance detects numerous β-glucopyranosides and hTAS2R38 various thioamides (Bufe, B.

Table 1   Cognate bitter compounds for TAS2Rs

TAS2R	Cognate compounds	Reference
hTAS2R4	Denatonium benzoate, PTC	Chandrashekar J. (2006)
mTAS2R5	Cycloheximide, lidocaine	Pronin A. N. *et al.* (2004); Chandrashekar J. (2006)
mTAS2R8	Denatonium benzoate	Chandrashekar J. (2006)
hTAS2R7	Atropine, chloroquine, papaverine, quinacrine, strychnine	Sainz E. *et al.* (2007)
hTAS2R10	Strychnine	Bufe B. *et al.* (2002)
hTAS2R14	Aristolochic acid, α-thujone, picrotin, naphtoic acid	Behrens M. *et al.* (2004); Sainz E. *et al.* (2007)
hTAS2R16	Numerous β-glucopyranosides	Bufe B. *et al.* (2002)
hTAS2R38	Numerous thioamides including PROP and PTC	Kim U. K. *et al.* (2003); Bufe B. *et al.* (2005)
hTAS2R43	Saccharin, acesulfame K, aristolochic acid, 6-nitrosaccharin	Kuhn C. *et al.* (2004)
hTAS2R44	Saccharin, acesulfame K, aristolochic acid	Kuhn C. *et al.* (2004); Pronin A. N. *et al.* (2004)
hTAS2R47	Denatonium benzoate, 6-nitrosaccharin PROP, 6-*n*-propylthiouracil; PTC, phenylthiocarbamide	Pronin A. N. *et al.* (2004)

et al., 2002; 2005). Human TAS2R7 and TAS2R14 are even more promiscuous and responded to chemically extremely diverse compounds (Behrens, M. *et al.*, 2004; Sainz, E. *et al.*, 2007). It has been proposed that during evolution different functional constraints shaped bitter-taste receptors to fulfill different tasks. More selective TAS2Rs could detect the prominent plant toxins in a familiar environment, whereas the promiscuous TAS2Rs may be important for accessing novel habitats (Behrens, M. *et al.*, 2004). Nevertheless, if the broad tuning is a property of all TAS2Rs, it would easily explain how mammals detect the numerous bitter substances in their environment with a small repertoire of bitter receptors. Third, it appears that the sensitivity of TAS2Rs for their agonists recorded *in vitro* corresponds reasonably well to the sensory performance of mice or human subjects (Chandrashekar, J. *et al.*, 2000; Bufe, B. *et al.*, 2002; Bufe, B. *et al.*, 2005) suggesting that the biochemical properties of the receptors determine perceptive performance. This assumption is impressively supported by experiments in transgenic mice. Wild-type mice are indifferent to β-glucopyranosides. However, if the human *TAS2R16* gene is expressed as a transgene in mice bitter receptor cells the animals acquire sensitivity toward these compounds with parameters similar to those seen in human subjects and hTAS2R16 receptor assays (Mueller, K. L. *et al.*, 2005). Fourth, although not examined specifically, it appears from Table 1 that some redundancy exists in TAS2R tuning. For instance, hTAS2R44 shares agonists with hTAS2R43 and hTAS2R14; hTAS2R10 with hTAS2R7; and hTAS2R4 with hTAS2R47. The physiological consequences, if there are any, are unknown. We may speculate here that, possibly, bitter

taste would have a backup system at its disposal in the case mutations silence TAS2Rs or that compounds activating several receptors would have greater bitterness than compounds with only one receptor or that compounds with multiple receptors could be detected over a wider concentration range than those with only one.

4.09.2.4.2   *Activation of TAS2Rs by agonists*

The broad tuning of TAS2Rs immediately raises the question of how these receptors accommodate their ligands. In contrast to neurotransmitter or peptide hormone receptors, which have affinities in the nanomolar or picomolar range for their receptors (Olias, G. *et al.*, 2004), TAS2Rs respond to micromolar and millimolar concentrations of their agonists (see references in Table 1). We may imagine that TAS2Rs have loosely fitting binding pockets in which the agonists are accommodated in a relatively unspecific fashion. Or, TAS2Rs may exhibit separate or overlapping highly specific binding sites for their agonists. Considerations about the structure-activities of bitter compounds at their TAS2Rs would support the latter conjecture. Human TAS2R16, for instance, specifically detects β-glucopyranosides but is not activated by β-galactopyranosides or α-glucopyranosides. Variations in the size and hydrophobicity of the aglycon influence critically the potencies of glucopyranosides at hTAS2R16. These data argue for specific interactions of molecular groups of the agonists with definite amino acid residues of the receptor. (Bufe, B. *et al.*, 2002). Analogous observations that have been made with hTAS2R38 support this conclusion (Bufe, B. *et al.*, 2005).

Unfortunately, the molecular architecture of TAS2Rs remained largely unknown. Only one report addressed this question experimentally by mutational analysis and provided limited insight into the interaction of the TAS2Rs with their agonists (Pronin, A. N. *et al.*, 2004). Analyses of domain swapping experiments and chimeras made of the closely related receptors, TAS2R43, which responded to 6-nitrosaccharin, and TAS2R44, which is sensitive to *N*-isopropyl-2-methyl-5-nitrobenzenesulfonamide (IMNB), suggested that amino acid residues responsible for selective receptor activation by IMNB are present in extracellular loop 1 but not 2. Both loops and other parts of the receptor, however, contain residues important for activation of hTAS2R43 by 6-nitrosaccharin. Further analysis uncovered the importance of residue G92 in the extracellular loop of hTAS2R44. Exchange of this residue for the corresponding asparagine of hTAS2R43 made the mutant receptor responsive to both ligands, whereas the reciprocal mutation in hTAS2R43 impaired the activation properties of this receptor. This residue was also of importance for denatonium activation of hTAS2R47, the corresponding mutants N92G and N92A being unresponsive. Also a tryptophan residue present at position 88 of these receptors and conserved in 21 out of 25 human TAS2Rs was critical for agonist activation of hTAS2R43, hTAS2R44, and hTAS2R47. Although these findings are of great importance and interest, they do not allow precise conclusions to be drawn about the binding motifs of TAS2Rs. The analyzed residues may be involved in ligand binding, activation mechanisms, or globally alter the structure of the receptors (Meyerhof, W., 2005).

4.09.2.4.3 G-protein coupling

There are ~20 different $G\alpha$ subunits falling into four major classes $G\alpha s$, $G\alpha i$, $G\alpha 12$, and $G\alpha q$ (Gudermann, T. *et al.*, 1997). α-Gustducin, a $G\alpha i$-type protein closely related to visual transducin, has originally been identified as a lingual G protein present in a subset of taste receptor cells and been suggested to transduce taste stimuli (McLaughlin, S. K. *et al.*, 1992). Biochemical, physiological, and genetic studies since then confirmed that α-gustducin is critically involved in bitter transduction (Wong, G. T. *et al.*, 1996a; 1996b; Ming, D. *et al.*, 1998; Ming, D. *et al.*, 1999; Wong, G. T. *et al.*, 1999; Caicedo, A. *et al.*, 2003). Direct proof was finally obtained by showing that agonist activation of

recombinant mT2R5 in insect cell membranes resulted in GDP/GTPγS exchange in exogenously added purified α-gustducin in the presence of purified β/γ G-protein subunits (Chandrashekar, J. *et al.*, 2000).

However, these results do not rule out the possibility that TAS2Rs may couple to other G-protein α-subunits. In fact, in human embryonic kidney cells previously transfected with DNA for mTAS2R5 that do not express α-gustducin, cycloheximide stimulation activated extracellular signal-regulated kinase activity in a pertussis toxin sensitive fashion, suggesting that this bitter receptor also couples to Gi/o-type G-proteins (Ozeck, M. *et al.*, 2004). This suggestion has been confirmed by reconstituting signal transduction of recombinant mTAS2R5, hTAS2R7, hTAS2R14, hTAS2R43, and hTAS2R47 in insect cell membranes in the presence of purified G-protein subunits (Sainz, E. *et al.*, 2007). None of the receptors coupled to $G\alpha s$ or $G\alpha q$. These experiments demonstrated that members of the TAS2R family are capable of signaling through multiple $G\alpha i$ subunits, including transducin, $G\alpha i$, and $G\alpha o$, but not through $G\alpha s$- and $G\alpha q$-type G-proteins. The findings are also in line with observations made in α-gustducin gene-targeted mice showing that bitter responses were clearly impaired but not abolished (Caicedo, A. *et al.*, 2003; Wong, G. T. *et al.*, 1996a) and with findings showing that transducin is present in taste tissue, partially rescues gustucin null mice, and functionally couples to bitter-taste receptors (He, H. T. *et al.*, 1990; Ruiz-Avila, L. *et al.*, 1995; Yang, H. *et al.*, 1999).

TAS2Rs activate phospholipase C-$\beta2$ but not through $G\alpha q$ suggesting that they do so via β/γ G-protein subunits (Sainz, E. *et al.*, 2007). Searches for such subunits that are involved in bitter taste identified $G\gamma 13$ which colocalizes with α-gustducin in taste receptor cells and physically interacts with $G\beta 1$, a subunit also expressed in taste cells (Huang, L. *et al.*, 1999). Blocking experiments with specific antibodies led to the proposal that the heterotrimeric G-protein may be composed of α-gustducin and $G\beta 1/\gamma 13$ or $G\beta 3/\gamma 13$ (Huang, L. *et al.*, 1999; Rossler, P. *et al.*, 2000). Reconstitution experiments with membranes containing recombinant receptors and purified G-protein subunits demonstrated that mTAS2R5, hTAS2R7, and hTAS2R47 indeed couple to $G\beta 1/\gamma 13$ but also to $G\beta 3/\gamma 13$ (Sainz, E. *et al.*, 2007). Together, the available data suggest that complex rules govern TAS2R-G-protein coupling and that coupling is influenced by the precise concentrations of subunits of all three classes and their individual affinities for each other and for the receptors.

4.09.2.5 Polymorphisms and Functional Heterogeneity of TAS2Rs

Genetically determined differences in bitter-taste sensitivity for various compounds have been seen in mice and mapped to specific chromosomal loci (Lush, I. E., 1981; 1984; 1986; Lush, I. E. and Holland, G., 1988). One of these loci encodes the bitter-taste receptor TAS2R5 of which strain specific functionally distinct variants exist that determine taste sensitivities for cycloheximide in strains of mice (Chandrashekar, J. et al., 2000). Thus, there is a molecular basis for explaining taste sensitivity ranging from genotype to receptor properties and sensory performance. In humans, inherited differences in bitter-taste sensitivity are known for some 70 years when it has been discovered that there are genetically determined pronounced individual differences in tasting phenylthiocarbamide (PTC; Blakeslee, A. F., 1931; 1932; Fox, A. L., 1932). PTC-tasters have been reported to taste PTC and chemically related compounds, such as propylthiouracil, at micromolar concentrations, whereas nontasters are taste-blind to these compounds even when they were exposed to them at several 100-fold higher concentrations (Bufe, B. et al., 2005). Eventually, positional cloning identified the *hTAS2R38* gene as a receptor for PTC and showed that three nonsynonymous single nucleotide polymorphisms, SNPs, define different hTAS2R38 haplotypes (Figure 1), of which two are frequent (Kim, U. K. et al., 2003). Depending on the amino acids present in the positions affected by the SNPs, the so-called PAV haplotype is associated with high sensitivity for PTC, whereas the AVI haplotype is linked to nontaster status (Kim, U. K. et al., 2003). A subsequent study then demonstrated that the different hTAS2R38 haplotypes encode functionally distinct hTAS2R38 variants and their sensitivities monitored *in vitro* determine perceptual performance of human subjects (Bufe, B. et al., 2005). Thus, also in humans we have a clear indication that the TAS2R genotype determines the functional properties of the encoded bitter receptor and, these in turn, are crucial for the sensory abilities of subjects.

A comprehensive evaluation of sequence and haplotype variation in the human *TAS2R* gene repertoire identified substantial coding sequence diversity. In a worldwide population sample of 55 individuals, more than four variant amino acid positions were found on average per *TAS2R* gene. Together, the 25 *hTAS2R* genes encode 151 different protein-coding haplotypes, yet some show translational stop codons (Kim, U. et al., 2005). It is reasonable to assume that

(a)

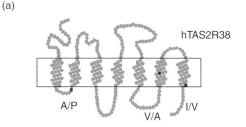

(b)

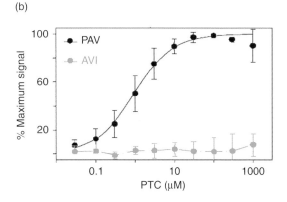

(c)

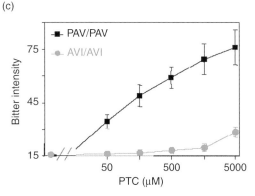

Figure 1 Molecular basis of tasting phenylthiocarbamide. (a) Schematic representation of TAS2R38. Each circle represents an amino acid residue. Amino acid residues in positions 49, 262, and 296 that are affected by single nucleotide polymorphisms are depicted in red and the specific amino acids present in these positions are indicated. (b) Postthreshold bitter intensity ratings of subject groups genotyped to be PAV or AVI homozygous. Please note the pronounced differences in taste sensitivity. (c) functional responses of hTAS2R38-PAV and hTAS2R38-AVI variants recorded in transfected cell lines. The differences in receptor activation closely resemble those seen in the taste experiment shown in (b). Data from Kim, M. R. et al. (2003) and Bufe, B. et al. (2005).

many of the variant receptors differ in their functional properties toward ligands by showing no responses, or altered ligand spectra, or modified sensitivities for agonists. This has already been shown for the alleles specifying hTAS2R16-N172 and hTAS2R16-K172 (Soranzo, N. *et al.*, 2005). We may also expect that the functional differences of TAS2Rs translate into perceptual differences in the population. This conclusion is supported by the observed differences in tasting chloramphenicol, strychnine, and quinine across subjects (Blakeslee, A. F., 1935; Sugino, Y. *et al.*, 2002).

4.09.2.6 Implications for Nutrition

It is still a matter of speculation to what extent, if at all, differences in taste sensitivities, particularly bitter sensitivity influences nutritional behavior. Food choice is determined by many parameters with the taste of food being crucial (Drewnowski, A., 1997; 2001). However, a mechanistic explanation is currently unavailable. It has been pointed out that bitter taste contributes to the enjoyment of food and beverages (Drewnowski, A., 2001). In contrast, moderate or strong bitter taste may lead to rejection of healthy compounds (Drewnowski, A. and Gomez-Carneros, C., 2000). Based on the finding that individual TAS2Rs mediate bitter taste for specific compounds and on the conjecture that TAS2R variants determine interindividual bitter-taste sensitivities, we have reason to expect that rejection of bitter substances occurs more easily in subjects carrying high-sensitivity haplotypes than in carriers of low-sensitivity haplotypes. Many studies have been carried out that aimed at associating PTC taster status or hTAS2R38 genotypes with intake behavior and risks for diseases (Drewnowski, A., 2004; Duffy, V. B. *et al.*, 2004b). Indeed such associations have been reported in some studies for preferences for vegetables, fruits, fats, alcohol, and sweeteners (Bartoshuk, L. M., 1979; Drewnowski, A. and Rock, C. L., 1995; Tepper, B. J. and Nurse, R. J., 1997) and risks for goiter, alcoholism, cancer, and depression (Harris, H. *et al.*, 1949; Facchini, F. *et al.*, 1990; Pelchat, M. L. and Danowski, S., 1992; Duffy, V. B. *et al.*, 2004a; Joiner, T. E., Jr. and Perez, M., 2004) while another study did not find such associations (Timpson, N. J. *et al.*, 2005). Also for the hTAS2R16-K172 low-sensitvity allele an association with risk for alcoholism has been detected (Hinrichs, A. L. *et al.*, 2006). Perhaps, all these studies may be biased by the fact that the food items monitored, i.e., vegetables or fruits in general, contain numerous bitter compounds while hTAS2R38 detects only thioamides. Future association studies should look into foods that are rich in hTAS2R38 agonists. To the authors' knowledge, to date only one preliminary study is available, which showed that PTC tasters avoided intake of plant-derived food containing possible hTAS2R38 agonists (Sandell, M. A. and Breslin, P. A., 2006).

4.09.3 Sweet-Taste Receptors

4.09.3.1 Tas1Rs

4.09.3.1.1 Tas1Rs belong to family C G-protein-coupled receptors

The receptors for sweet (Tas1R2 + Tas1R3 function as a sweet receptor) and umami (Tas1R1 + Tas1R3 also function as an umami receptor) taste qualities are GPCRs that, by their protein sequence and homology to other GPCRs, can be assigned to family C. This family has the typical 7TM spanning domain as seen in family A (rhodospinlike GPCRs) but also contains a large extracellular domain that is known to be the orthosteric endogenous ligand-binding domain. Other family members include the metabotropic glutamate binding receptors (mGluRs), calcium-sensing receptors, metabotropic gamma-aminobutyric acid (GABA) receptors and several putative odorant receptors. Although it is now thought that most GPCRs dimerize along their TM domains, family C receptors are known to be obligate dimers that dimerize along their extracellular domains as well as their TM domains. The Tas1 receptors are variously called Tas1R, Tas1R, and Tas1R. At the amino acid level Tas1Rs are 26–34% identical with each other, and $\sim$22–25% identical with mGluRs. The similarity of the Tas1Rs to mGluRs is much higher, ranging from 36% to 43%, making them members of GPCR family C. Family C GPCRs include the mGluRs, calcium-sensing receptors, metabotropic GABA receptors, and a variety of odorant (fish) and pheromone-sensing receptors. A truncated taste-expressed variant of mGluR4 (taste-mGluR4) is present in $\sim$40–70% of taste buds and may also function as an umami receptor (Chaudhari, N. *et al.*, 2000). Below we review what is known about the structure of family C receptors from crystal studies of mGluRs.

4.09.3.1.2 Structure–function studies of family C receptors' venus flytrap module

In addition to the heptihelical TMD common to GPCRs from family A, family C GPCRs have large extracellular region composed of two domains: the venus flytrap module (VFTM, $\sim$500 residues) and the cysteine-rich domain (CRD, $\sim$70 residues; Jingami, H. *et al.*, 2003). The VFTM is a two-lobed clamshell-like structure. The CRD lies between the VFTM and the heptahelical TMD. The CRD has nine highly conserved cysteines within its span of $\sim$70 residues. Based on extensive structure–function studies of family C receptors, the crystal structure of mGluR1, mGluR2, and mGluR3 (see below), and homology to periplasmic sugar and amino acid transporters, it is highly likely that the VFTM is the receptor's primary ligand-binding domain (Hermans, E. and Challiss, R. A., 2001). It is thought that ligand binding occurs independently to one or both monomers of each dimeric receptor. The ligand-binding site for the endogenous ligand lies in a cleft between the two prominent lobes, with the ligand fitting like a pearl inside a closed oyster. Family C GPCRs are obligate homodimers or heterodimers that dimerize via a discontinuous set of contacts in lobe 1 of the VFTM.

There are a number of structures of the VFTM (mGluR1) and VFTM + CRD (mGluR3) (Kunishima, N. *et al.*, 2000; Tsuchiya, D. *et al.*, 2002; Muto, T. *et al.*, 2007). The crystal structure of the homodimeric mGluR1–3 VFTM consists of two asymmetrical mirror-image monomers. Each VFTM monomer has two lobes that together form a clamshell. The two monomers dimerize by interacting along each lobe 1; the two lobes 2 have no part in the dimerization (Kunishima, N. *et al.*, 2000). Lobe 1 and lobe 2 of each monomer approach each other and their intersection forms the cleft that binds ligand. The ligand-bound, closed conformation of the monomer is the active form. Upon ligand binding and closure of the cleft, the dimer interface rotates upon itself, which brings the two lobes 2 within closer proximity. In mGluR1 the majority of ligand-contacting residues in lobe 1 make direct hydrogen bonds with glutamate, while those in lobe 2 make water-mediated hydrogen bonds and salt bridges (Jingami, H. *et al.*, 2003).

Tas1Rs and other family C GPCRs are expected to dimerize and to bind ligands in a manner similar to mGluR1. Comparing Tas1R and mGluR VFTMs (independently of the receptors' CRDs and TMDs) reveals amino acid identities of ~25% and similarities ~41%. The predicted secondary structures of helices and β-strands of the extracellular domains of Tas1Rs match that of the crystal structure of mGluR1 at nearly every point. The conformational change in family C GPCRs that takes place upon ligand binding appears to be transmitted from the VFTM and CRD to the TMD helical bundle, then onward to the cytoplasmic surface, where it contacts and activates the G protein.

4.09.3.1.3 Cysteine-rich domain structure

Recently a crystal structure of mGluR3 VFTM + CRD was published showing for the first time the structure of the CRD (Muto, T. *et al.*, 2007), which contains three β-sheets composed of two antiparallel β-strands. Eight of the nine conserved cysteines formed four structurally stabilizing disulphide bonds within the structure. A ninth cysteine forms a disulphide bond with a conserved cysteine at the bottom of lobe 2 of the same monomer VFTM. This structure is likely to hold for all the family C receptors except for the GABA receptors, which do not have the conserved cysteines in its linker domain. The CRD, with its connection to the motion of lobe 2 of the VFTM via the disulphide linkage is likely to be critical for transmittal of the conformational change from the ligand-binding VFTM to the TMD. The structure of the CRD suggests that there is a reorientation of the ends of the CRD from 104–126 Å to 46–90 Å apart between the resting and the active state of the VFTM. This suggests an activation mechanism that involves movements of the two TMDs relative to one another upon receptor activation (Muto, T. *et al.*, 2007). However, it is currently unknown how the VFTMs communicate with the TMDs and to what extent intra-TMD versus inter-TMD conformational change contributes to receptor activity.

4.09.3.1.4 Role of the transmembrane domain in family C G-protein-coupled receptors

A hallmark of GPCRs is that they have seven transmembrane helices constituting a TMD. Another key feature of GPCRs is that they mediate their intracellular actions via coupling to and activation of G proteins. Thus, activation of family A receptors by binding of ligand within the TMD is transmitted to the G protein by cytoplasmic loops adjacent to the TMD, whereas activation of family C receptors by ligand binding within the VFTM is transmitted to the TMD and then by cytoplasmic loops to the G protein. The solved crystal structures of bovine rhodopsin have been widely used as a structural templates to construct homology models of diverse GPCRs using comparative modeling techniques (Visiers, I. *et al.*, 2000; 2002; Malherbe, P. *et al.*, 2003; Miedlich, S. U. *et al.*, 2004; Petrel, C. *et al.*, 2004). The crystal structures of bovine rhodopsin are at present the only solved structures of any GPCRs that include the heptahelical TMD (Palczewski, K. *et al.*, 2000; Teller, D. C. *et al.*, 2001; Okada, T. *et al.*, 2002). The supposition that the general topology of the TMD of rhodopsin is shared by most or all other types of GPCRs, including family C receptors, is supported by the following two points. First, numerous studies indicate that allosteric binding sites of several family C receptors are within their TMDs. For example, several noncompetitive antagonists and agonists bind within the TMDs of mGluR1, mGluR5, and the calcium-sensing receptor (Pagano, A. *et al.*, 2000; Malherbe, P. *et al.*, 2003; Petrel, C. *et al.*, 2004). Second, although the overall sequence homology of TMDs between GPCRs in families A and C is low, key residues that define the TM helices are conserved in rhodopsin and in family C GPCRs.

4.09.3.2 Discovery of Tas1Rs

4.09.3.2.1 *Orphan receptors Tas1R1 and Tas1R2*

The taste receptors Tas1R1 and Tas1R2 were first identified in taste cell-specific libraries (Hoon, M. A. *et al.*, 1999). It was originally proposed that mTas1R2 might be a bitter-responsive receptor and that mTas1R1 might be a sweet-responsive taste receptor; the rationale being that Tas1R1 is expressed at the front of the tongue (which is sometimes thought to be more sensitive to sweet) and that Tas1R1 gene maps nearer to the *sac* locus at the distal end of mouse chromosome 4 (Hoon, M. A. *et al.*, 1999). While it is true that both Tas1R1 and Tas1R2 map to the distal end of mouse chromosome 4, refined mapping studies indicated that Tas1R1 and Tas1R2 are several centimorgans away from the *sac* locus, much too far to be the *sac* gene itself (Li, X. *et al.*, 2001).

The taste cell-specific expression of Tas1R1 and Tas1R2 suggested that both receptors might play roles in taste transduction. *In situ* hybridization showed that Tas1R1 and Tas1R2 have different expression patterns in rat taste tissue: rTas1R1 is localized to geschmackstreifen (taste stripe) and fungiform papillae, and rTas1R2 is localized primarily to circumvallate and foliate papillae. The significance of this regional expression pattern is unclear, since it does not correspond to any known pattern of tastant sensitivity in rodents (Lindemann, B., 1999; Sako, N. *et al.*, 2000). Prior to the discovery of Tas1R3, the function of the Tas1R1 and Tas1R2 orphan receptors could not be studied in heterologous cells because as was later shown, Tas1R1 and Tas1R2 form obligate heterodimers with Tas1R3 to function, hindering identification of their natural ligands.

4.09.3.2.2 *ID of possible umami receptors: truncated mGluR4, mGluR1*

A truncated taste-expressed variant of mGluR4 (taste-mGluR4) was proposed to function as an umami receptor. By *in situ* hybridization, this receptor is present in ~40–70% of the taste buds examined (Chaudhari, N. *et al.*, 2000). Taste mGluR4 displays the characteristics of an umami receptor, with sensitivity to glutamate in the millimolar range, consistent with concentrations known to elicit the sensation of umami. Heterologously expressed taste-mGluR4 shows two other hallmarks of the umami response: it is activated by L-AP4, which elicits a taste similar to L-glutamate in rats, and it is potentiated by 5'-ribonucleotides.

Upon identification of Tas1R3, it became possible to show that the combination of Tas1R1 + Tas1R3 gave rise to a receptor that was sensitive to a variety of L-amino acids including monosodium glutamate suggesting that this combination of Tas1Rs was an umami receptor (Li, X. *et al.*, 2002a; Zhao, G. Q. *et al.*, 2003).

4.09.3.2.3 *Identification of Sac locus GPCR Tas1R3 – Tas1R3 is Sac*

Inbred strains of mice vary widely in their behavioral and electrophysiological responses to saccharin, sucrose and other compounds that humans perceive as sweet. The *sac* locus was identified as the primary genetic determinant in mice of preference for saccharin and other sweet compounds, distinguishing taster strains (e.g., C57BL/6) from nontasters (e.g., DBA/2; Fuller, J. L., 1974; Lush, I. E., 1989; Lush, I. E. *et al.*, 1995; Blizard, D. A. *et al.*, 1999). To identify candidates for the *sac* gene we searched the sequence of the distal end of mouse chromosome 4 (to which *sac* maps) for genes that might encode GPCRs and other signal transduction elements. This search yielded Tas1R3, a previously unknown GPCR located within 20 kbp of the genetic marker closest to *sac* (Max, M. *et al.*, 2001). Tas1R3 is the third member of the Tas1R taste receptor family, and is most closely related to Tas1R1 and Tas1R2. *In situ* hybridization and immunohistochemistry showed that Tas1R3 is expressed selectively in a subset of taste receptor cells. Sequence analysis of mTas1R3 from eight independently derived strains of mice identified specific polymorphisms (T55A and I60T) that assort between taster and nontaster strains (Max, M. *et al.*, 2001). Other laboratories also identified Tas1R3 by RT-PCR or genomics-based approaches (Bachmanov, A. A. *et al.*, 2001; Kitagawa, M. *et al.*, 2001; Montmayeur, J. P. *et al.*, 2001; Nelson, G. *et al.*, 2001; Sainz, E. *et al.*, 2001). Transgenic expression of the taster form of Tas1R3 converted nontaster mice into tasters, demonstrating that Tas1R3 is indeed *sac* (Nelson, G. *et al.*, 2001; Damak, S., personal communication).

4.09.3.2.4 Are there other Tas1Rs?

A fourth related gene known as GPRC6A is related to Tas1R1-3 by 20–24% identity. It has been speculated that this gene might be a taste receptor gene by several groups; however, convincing expression in taste cells has not been found to date. Other studies identify this receptor as either an amino acid receptor or a second calcium-sensing receptor with important function in bone (Pi, M. *et al.*, 2005; Christiansen, B. *et al.*, 2006). The latter work is particularly compelling because there is extensive expression of this receptor in bone and it shows responses to calcium, calcium mimetics, and to an important bone peptide, osteocalcin. It remains to be seen if there is a role for this receptor in taste although as this juncture it appears unlikely.

4.09.3.3 Functional Identification of Tas1Rs

4.09.3.3.1 Tas1R2 + Tas1R3 is the primary sweet receptor

Damak S. *et al.* (2003) found that knockout mice lacking Tas1R3 showed no behavioral or nerve response to noncaloric sweeteners, and diminished responses to sugars and umami compounds. Zhao G. Q. *et al.* (2003) observed that knockout mice lacking both Tas1R2 and Tas1R3 had no responses to sweet compounds; mice lacking both Tas1R1 and Tas1R3 had no responses to umami compounds. Zhao G. Q. *et al.* (2003) also showed that mice expressing human Tas1R2 as a transgene were responsive to human-specific sweeteners such as aspartame and thaumatin that are detected by humans but not mice.

Calcium-imaging assays of HEK 293 cells transfected with Tas1R2, Tas1R3, and a reporter G-protein showed responses to a wide range of sweet ligands (Nelson, G. *et al.*, 2001; Li, X. *et al.*, 2002a; Jiang, P. *et al.*, 2004). If either Tas1R2 or Tas1R3 was omitted from the transfection no responses to sweet compounds were detected, leading to the conclusion that the combination of Tas1R2 + Tas1R3 is a broadly responding sweet receptor. In contrast, cells transfected with Tas1R1, Tas1R3, and the reporter G protein responded to amino acids, but not to sweeteners, indicating that the combination of Tas1R1 + Tas1R3 is an umami receptor (Nelson, G. *et al.*, 2001; Li, X. *et al.*, 2002b; Nelson, G. *et al.*, 2002). Apparently, Tas1R3 provides a common element necessary for the proper function of both sweet and umami receptors. Initially, it

was thought that ligand-binding determinants would be present only in Tas1R1 and Tas1R2, but not in Tas1R3; however, it appears that Tas1R3 also contributes to the ligand-binding process (see below). Although, heterologous expression of Tas1R1 or Tas1R2 alone yielded no responses to either sweet or umami compounds, expression of Tas1R3 alone yielded responses to trehalose, but not to other sweeteners (Ariyasu, T. *et al.*, 2003; Jiang, P., personal communication) although high concentrations of natural sweeteners appear to be sensed by homodimers of Tas1R3. Like other family C GPCRs, the Tas1Rs probably function as dimers (heterodimers, and in some cases homodimers).

HEK cells expressing hTas1R2 + hTas1R3 respond to all sweet-taste stimuli tested: sugars (sucrose, fructose, galactose, glucose, etc.); amino acids (glycine and D-tryptophan); sweet proteins (brazzein, monellin, and thaumatin); and synthetic sweeteners (cyclamate, saccharin, ace K, aspartame, dulcin, neotame, and sucralose; Nelson, G. *et al.*, 2001; Jiang, P. *et al.*, 2004; Xu, H. *et al.*, 2004). All of these responses are inhibited by the sweet-taste inhibitor lactisole, which acts on the TMD of Tas1R3 (Xu, H. *et al.*, 2004; Jiang, P. *et al.*, 2005a; Winnig, M. *et al.*, 2005). These results, along with the observation that human perception of sweet taste is inhibited by lactisole (Sclafani, A. and Perez, C., 1997). Schiffman S. S. *et al.* (1999) argue that Tas1R2 + Tas1R3 is the primary or the only sweet-taste receptor although that knockout Tas1R3 mice still respond to natural sweeteners suggests that there may be some alternative sugar-sensing mechanism that remains in the absence of the sweet receptor.

4.09.3.4 Ligand-Binding Sites on Tas1Rs

The sweet receptor appears to have multiple sites of interaction for sugars and sweet-tasting artificial compounds. The orthosteric site for sugars is likely to be the closed cleft of the VFTM Tas1R2, and perhaps the cleft of the VFTM of Tas1R3. Artificial sweeteners may also interact with the closed cleft of the VFTM of Tas1R2 in sites overlapping that of the orthosteric site for natural sugars as well as an allosteric site within the TMD of Tas1R3. We have proposed that the larger, noncarbohydrate-containing, naturally occurring peptide sweeteners found in some plants may interact with both the CRD of Tas1R3 and the VFTM of Tas1R2 (Jiang, P. *et al.*, 2004; Maillet, E., personal communication) although other groups have suggested that these proteins interact with the open cleft of both VFTMs (Morini, G. *et al.*, 2005; Walters, D. E. and Hellekant, G., 2006).

4.09.3.4.1 Orthosteric binding site within the VFTM of Tas1R2

Work from our laboratory indicates that the VFTM of Tas1R2 is one orthosteric binding site for natural sugars as well as forming a binding site for dipeptide and several other artificial sweeteners. We concluded this from studies where we mutated numerous amino acids that line the VFTM cleft and tested the resulting receptors for specific defects in the receptors ability to be activated by a variety of sweeteners (Maillet, E., personal communication). We have also employed receptors composed partly of human and partly of mouse sequences (chimeric receptors) to show which domains of the receptor are required for specific responses to sweeteners that only humans and other primates can taste (dipeptide sweeteners, cyclamate, brazzein, monellin, and others).

4.09.3.4.2 Activity assays map the sweet receptor aspartame interaction site to the VFTM of hTas1R2

Scientists in our laboratory assayed the effects of sweeteners on heterologously expressed mismatched receptors from mouse and human. While mTas1R2 + mTas1R3 responded to D-tryp, saccharin,

and sucrose, but not to the human-specific sweeteners (i.e., aspartame, monellin, and brazzein), the interspecies pairing hTas1R2 + mTas1R3 responded to aspartame and most other sweeteners, but not brazzein. These results indicate that responsiveness to aspartame requires the human form of Tas1R2, but does not require the human form of Tas1R3.

A chimeric receptor with the VFTM of human Tas1R2 and the rest from mouse Tas1R2 paired with mTas1R3 (i.e., hVFTM-CRD/mTMDT1R2 + mT1R3) gave responses to aspartame (Figure 2(a)). This indicates that responsiveness to aspartame requires human residues within the VFTM–CRD of Tas1R2, and not human residues within the TMD of Tas1R2. To localize the residues of hTas1R2 required for responsiveness to aspartame better we made additional mouse/human and human/mouse chimeras within the VFTM–CRD of TAS1R2 (Figure 2(b)) and coexpressed them with mTas1R3. The pairing of human/mouse chimeras h1-468/mT1R2 or h1-577/mT1R2 with mT1R3 gave normal responses to aspartame and neotame (Figure 2(b)). This indicated that responsiveness to aspartame and neotame requires human residues within residues 1–468 of the VFTM of hTas1R2 – the rest of Tas1R2 and all of Tas1R3 can be from mouse and still support responses to these human-specific dipeptide sweeteners. The pairing of mouse/

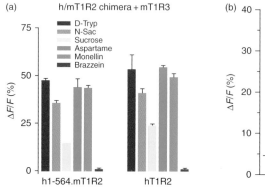 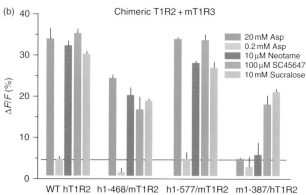

Figure 2 Sweet-receptor responsiveness to aspartame (Asp) and neotame requires human-specific residues within the venus flytrap module (VFTM) of TAS1R2. (a) HEK293E cells transiently transfected with the human/mouse chimera h.1–564.mTas1R2, mTas1R3, and Gα16-i3 were assayed for calcium mobilization in response to the indicated sweet ligands (see methods for details). h1–564.mTas1R2 Contains the extracellular domain of hTas1R2 and the transmembrane region of mTas1R2. Its response profile is comparable to that of hTas1R2. (b) Human/mouse and mouse/human chimeras of Tas1R2 were constructed and coexpressed in HEK cells along with mTas1R3 and Gα16-i3. SC45647 and sucralose served as positive controls for activity. By our naming convention h1-468/mTas1R2 contains residues 1–468 from hTas1R2 and the remainder from mTas1R2. Tas1R2 chimeras with residues 1–468 (VFTM) or 1–577 (VFTM plus part of cysteine-rich domain) from hTas1R2 were responsive to both aspartame and neotame. A Tas1R2 chimera containing mouse residues 1–387 (amino terminal two-thirds of VFTM) from mTas1R2 and the remainder from human Tas1R2 did not respond to aspartame or neotame. We conclude that specific human residues within the VFTM of hasS1R2 are required for responsiveness to aspartame and neotame. ΔF/F, the change of fluorescence at peak over the baseline fluorescence.

human chimera m1-387/hT1R2 with mTas1R3 gave no responses to aspartame or neotame (Figure 2(b)), indicating that one or more residues in the amino terminal two-thirds of the VFTM of hTAS1R2 are necessary for responsiveness toward aspartame and neotame. Based on mapping the interaction of aspartame and neotame to the VFTM of Tas1R2 and the overall homology of the Tas1Rs to mGlurR1, we reasoned that aspartame might bind within the cleft of the VFTM in a manner similar to that of glutamate in mGluR1.

To test this idea, scientists in our laboratory used homology modeling, automatic docking, chimeric and mutant receptors, and activity assays with heterologously expressed receptors to identify the potential binding site(s) within the VFTM of hTas1R2 for aspartame and neotame (Figure 3). We constructed models of Tas1R2 + Tas1R3 based upon the crystal structures of mGluR1 and used them to dock various sweeteners. For aspartame and neotame, our docking studies predicted that the NH^{3+} of aspartame and NH^{2+} of neotame interact through salt bridges with the negatively charged carbonyl groups of Tas1R2 Asp278 and Asp307, and through hydrogen bonds with a backbone oxygen atom of Ser303. The negatively charged carbonyl groups (COO−) of aspartame and neotame are proposed to interact through hydrogen bonds with the NH

backbone atoms of Tas1R2 Arg383 and Val384. The phenyl groups of aspartame and neotame are proposed to interact through pi-pi interactions with Tas1R2 Tyr215, and through hydrophobic interactions with Pro277. Morini G. et al. (2005) have also modeled the structure of the sweet receptor VFTMs (based on the structure of the VFTM of the mGluR1), and proposed that aspartame could bind within the VFTMs of both TAS1R2 and TAS1R3. The predictions of our model and theirs are in general agreement with regard to the interaction of aspartame with the VFTM of TAS1R2 (both models identified S165, E302, and Y215 as potential interactions), although, our model makes more extensive predictions of potential contacts for aspartame in the VFTM of TAS1R2. Furthermore, based on our activity assays with hTAS1R3 homodimers (unpublished data) we think it unlikely, although theoretically possible, that aspartame binds to the VFTM of TAS1R3.

4.09.3.4.3 Mutational analysis of the aspartame interaction site within the venus flytrap module of hTas1R2

Our models of docked aspartame and neotame suggested that there is a good possibility that the VFTM of Tas1R2 contains a cleft between its lobes 1 and 2 that could serve as a small molecule-binding site for sugars and/or dipeptides akin to the canonical glutamate binding site of mGluR1. Using our identification of residues within the presumptive small molecule binding cleft of hTas1R2 we carried out alanine scan mutagenesis to determine which, if any, of residues do indeed contribute to a binding pocket for aspartame and other dipeptide sweeteners. In a few cases of interest (e.g., D307) we followed up with more conservative replacements. We have tested these mutants for their responses to aspartame and neotame (and cyclamate – our control for overall efficacy of mutant receptors) (Figure 4). Some mutations had similar effects on responsiveness to both aspartame and cyclamate (e.g., R383A, D307A, E302A, Y215A, D142A, and I67A), suggesting that they had altered general responsiveness of the receptor, or altered coupling between the VFTM of Tas1R2 and the TMD of Tas1R3. Other mutations had relatively specific effects on the response to aspartame (e.g., D307N, D278A, S144A, and Y103A), suggesting that they might be part of the aspartame-binding site. These residues are implicated by our model as being part of the binding site for aspartame within the VFTM of hTas1R2.

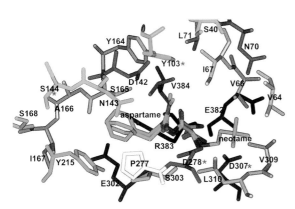

Figure 3 Model of cleft between lobes 1 and 2 of the venus flytrap module (VFTM) of hTas1R2 with docked aspartame and neotame. The metabotropic glutamate binding receptor (mGluR1; PDB:1EWK) crystal structure was used to model the VFTM of Tas1R2, then aspartame and neotame were individually docked onto the structure. Our model predicts that aspartame and neotame forms hydrogen bonds with the side chains of D278, D307, and S303, and makes hydrophobic interactions with the side chains of Y215, Y103, P277, and with the backbones of R383, V384, D142, and L279.

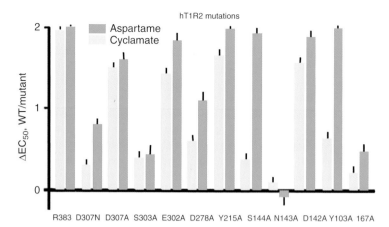

Figure 4 Effects of alanine scan mutations in the venus flytrap module (VFTM) of hTAS1R2 on responses to aspartame. Residues within a potential aspartame binding site of the VFTM of hTas1R2, predicted by homology with the metabotropic glutamate binding receptor binding site and by homology models of hTas1R2, were individually mutated to alanine (and in one case to asparagine (D307N)). Mutants were coexpressed with hTas1R3 and tested for responsiveness to aspartame and cyclamate. Some mutations had similar effects on responsiveness to both aspartame and cyclamate (e.g., R383A, D307A, E302A, Y215A, D142A, and I67A), suggesting that they altered general responsiveness of the receptor, or altered coupling between the VFTM of Tas1R2 and the TMD of TAS1R3. Other mutations had relatively specific effects on the response to aspartame (e.g., D307N, D278A, S144A, and Y103A), suggesting that they might be part of the aspartame binding site. ΔEC_{50}, the change in the effective concentration to produce 50% response; WT, wild type.

4.09.3.4.4   The expressed venus flytrap modules of both mTas1R2 and mTas1R3 bind sugars and sucralose

Work in Munger's laboratory, using direct physical binding assays, also shows that the expressed VFTMs of Tas1R2 and Tas1R3 react with sugars and sucralose as shown by changes in intrinsic fluorescence. The peak intrinsic tryptophan fluorescence of the mTas1R VFTM proteins was monitored and then the concentration–response relationship for this fluorescence was determined upon titration of VFTM with sweet ligands. Glucose, sucrose, and sucralose decreased the peak fluorescence intensity of mTas1R3VFTM$_{B6}$, exhibiting K_d values of 7.3 ± 0.7 mM, 2.9 ± 0.4 mM, and 0.9 ± 0.1 mM, respectively (Nie, Y. *et al.*, 2005), indicating binding of these ligands to the VFTM of mTas1R3. Ligand binding is known to stabilize a conformational change in the VFTMs of other class C GPCRs (Kunishima, N. *et al.*, 2000; Pin, J. P. *et al.*, 2003). A synchrotron radiation circular dichroism (SRCD) spectroscopy, assay was used to determine if Tas1R VFTMs undergo a ligand binding-dependent change in conformation. Addition of 5 mM glucose and sucrose were shown to produce large changes in the SRCD spectrum of the mTas1R2 and mTas1R3 VFTMs, although the changes seen differed between Tas1R2 and Tas1R3 VFTMs. These results demonstrate that the VFTMs of both

Tas1R3 and Tas1R2 undergo a conformational change upon ligand binding, supporting roles for each subunit in response to binding sweet ligands. Using these two techniques, Nie Y. *et al.* (2005) showed that a naturally occurring variant in mouse Tas1R3 VFTM, *sac* (I60T), which renders the mice less sensitive to sweeteners was shown to alter the function of the sweet-taste receptor in two important ways: (1) by decreasing affinity for sweet ligands and (2) by inhibiting the subunit's ability to undergo a conformational change upon ligand binding. The diminished sweet preference displayed by nontaster mice is likely due to the decreased ability of Tas1R3 to bind ligand and/or undergo a conformational change that may be involved in signal transduction (Nie, Y. *et al.*, 2005).

4.09.3.4.5   The cysteine-rich domain of Tas1R3

In 2004 we determined that the CRD of hTas1R3 contains residues critical for the sweet receptor's interaction with the sweet protein brazzein (Jiang, P. *et al.*, 2004). Brazzein is sweet to humans, but not to mice. Using mismatched pairs and chimeric receptors from mouse and human, we identified a small region of the CRD of hTAS1R3 that was required for the response to brazzein (residues 535–545).

Within this region of the CRD of TAS1R3 there are only five amino acids that differ between mTAS1R3 and hTAS1R3 (see figure 3a of Jiang, P. *et al.*, 2004). Starting with hTAS1R3 we individually mutated each of these five amino acids to their mouse counterparts (a loss-of-function experiment; see figure 3b of Jiang, P. *et al.*, 2004). In this way we identified hTAS1R3 A537 and F540 as contributing to brazzein's activity. To identify additional residues of the CRD of hTAS1R3 involved in responsiveness to brazzein we made a series of alanine-scan mutations within the CRD (Jiang, P. *et al.*, personal communication). Analysis of these mutants showed that D535A selectively abolished sweet receptor responsiveness to brazzein (Figure 5). The adjacent substitution, D534A, had no effect on sweet receptor responses to brazzein or the other sweeteners in our panel. Substituting glutamine for aspartate at position 535 (D535Q) also blocked responsiveness to brazzein (Jiang, P. *et al.*, personal communication, data not shown).

A new crystal structure of mGluR3 that includes the CRD region was recently published (Muto, T. *et al.*, 2007). We have constructed a homology model of the heterotrimer hTas1R2 + hTas1R3 based on this new structure (Figure 6). In this structure we see that D535, A537, and F540 lie on a single face of the CRD, making it an attractive site of interaction for a site on brazzein. In addition, evidence suggests that brazzein also makes contact with the VFTM of Tas1R2, suggesting that this protein sweetener acts by stabilizing an active conformation of the receptor by binding to multiple domains simultaneously.

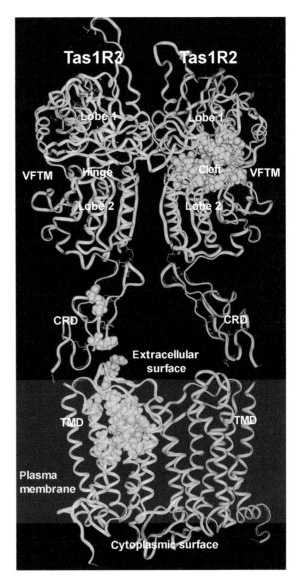

Figure 6 The homology model of all six domains of the sweet receptor (Tas1R2 + Tas1R3). The extracellular component of our model is based on the metabotropic glutamate binding receptor (mGluR3) venus flytrap module (VFTM) plus cysteine-rich domain (CRD) in the closed:closed configuration (2E4U). The two transmembrane domains are based on the crystal structure of rhodopsin (1HZX). The dimer interface between opposing helices 4 and 5 was chosen to best optimize the continuance of the CRD structure into that of TM helix 1 and is entirely speculative. The model is represented as a ribbon structure with only those residue side chains represented in Corey–Pauling–Koltun space filling representation. (CPK) that we have mutated and which alter receptor function in response to ligands. $\Delta F/F$, the change of fluorescence at peak over the baseline fluorescence.

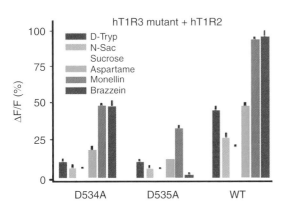

Figure 5 The D535A mutation in the cysteine-rich domain (CRD) of hTAS1R3 specifically abolishes sweet receptor responses to brazzein. Several residues within the CRD of hTas1R3 were individually mutated to alanine. Ala-scan mutants were coexpressed with hTas1R2 in HEK 293 cells along with a G-protein reporter, then tested for responsiveness to brazzein and a panel of sweeteners. D535A, but not D534A, showed a selective deficit in responsiveness to brazzein. D535 may participate in, or be close to the brazzein binding site in the CRD of hTas1R3.

Other workers (Morini, G. *et al.*, 2005; Walters, D. E. and Hellekant, G., 2006) have independently used molecular modeling to dock the protein sweeteners, brazzein and monellin, to the open cleft of both Tas1R2 and Tas1R3 using a docking strategy that relies heavily on electrostatics as its primary basis for fitting ligand to binding site. It seems unlikely, however, that these protein sweeteners could activate the receptor by interacting with it in an open cleft conformation given the likelihood that the sweet receptor's mode of activation is similar to that of the mGluRs and calcium-sensing receptor which are known to require the closed state of the VFTM for activity.

4.09.3.4.6 The transmembrane domain of Tas1R3

The transmembrane portion of the sweet receptor is composed of one TMD from Tas1R2 and one from Tas1R3. It is presumed that these TMDs dimerize as has recently been shown for family A receptors although the specific contact interface(s) have yet to be determined. Like family A receptors, family C receptors have intra-TMD binding sites for small molecules that can act as either agonists or antagonists or as enhancers. From a series of human/mouse chimeras of Tas1R3 it was determined that it is the TMD of hTas1R3 that is required for sweet receptor responsiveness to cyclamate and lactisole (Jiang, P. *et al.*, 2005b; summarized in Figure 7). More refined chimeras and the generation of mutants (human-to-mouse loss of function and mouse-to-human gain of function) led us to identify specific residues within TM helices 5 (A733) and 7 (L798), and extracellular loop 3 (R790) that were required for the receptor to interact with cyclamate and lactisole.

These findings suggested that both lactisole and cyclamate bind within the TMD of the sweet receptor in a manner similar to that seen with other ligands binding to family A GPCRs (e.g., catecholamines bound within the TMD of the β-adrenergic receptor, or retinal bound within the TMD of rhodopsin). An alignment of the TMD of hTas1R3 with that of rhodopsin was used to identify 17 residues forming a candidate ligand-binding pocket within the TMD of hTas1R3. Alanine was substituted for each of these 17 residues and tested for responsiveness to a variety of sweeteners and lactisole. This produced seven receptors that responded to no tested sweetener, two receptors that were reduced specifically in their response to cyclamate, one receptor that was specifically reduced in their response to lactisole, and three receptors that had altered responsiveness to both cyclamate and lactisole. These results suggest that the binding pockets for cyclamate and latisole overlap within the TMD of hTas1R3 (Figure 8).

4.09.4 Future Studies/Unanswered Questions

Many questions remain to be answered about the sweet receptor. Of primary interest is mapping the various binding sites for ligands that lead to its active

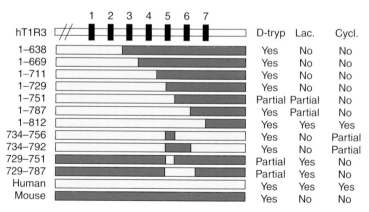

Figure 7 Schematic diagram showing responses of m/hTas1R3 chimeras to D-tryptophan, lactisole, and cyclamate. mTas1R3 portions are colored blue, hTas1R3 portions are yellow. Responsiveness of dimers of hTas1R2 + m/hTas1R3 chimeras to D-tryptophan (D-tryp), D-tryp + lactisole (Lac), or cyclamate (Cycl) are shown to the right. Chimeras containing mTas1R3 TM helix 5 did not respond to lactisole. The response to cyclamate by chimeras is more complicated and appears to require human residues in TM helices 5 and 7.

(a)

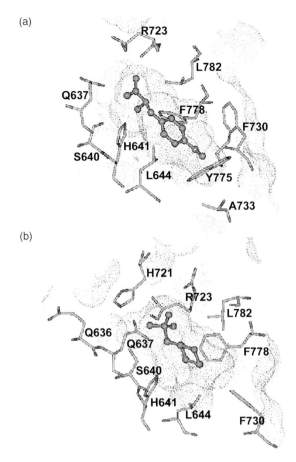

(b)

Figure 8 Models of the transmembrane domains (TMD) of the ligand-binding pocket of hTas1R3. (a) Lactisole docked within the TMD pocket. Lactisole's negatively charged propionic acid moiety is near H641 (TM helix 3). Lactisole's phenyl ring is predicted to form pi–pi interactions with F778 (TM helix 6). Lactisole's methoxyl group extends into a hydrophobic pocket and faces A733 (TM helix 5). (b) Cyclamate docked within the TMD pocket. The sulfamate group of cyclamate extends toward the top part of the pocket formed by TM helix 3 and extracellular loop 2. Residues Q637, S640, H641, H721, and R723 are in proximity to form ionic or hydrogen bonds with cyclamate's sulfamate group. Residues L644, F730, and F778 are predicted to form a hydrophobic pocket around cyclamate's cyclohexanyl group. Note that the binding pockets for the two ligands overlap but are not identical.

conformation. That there are over 50 molecules that taste sweet to humans that vary in structure from small molecule natural sugars, artificial sweeteners of multiple classes and even small proteins as well as at least one antagonist, suggests a great complexity to the receptor. Our work and that of others has demonstrated that there are multiple binding sites in at least

four domains (VFTMs of Tas1R2 and Tas1R3 and the CRD and TMD of Tas1R3) and with overlapping binding sites for various ligands in each. New work from Jay Slack at Givaudan (personal communication, data not shown) suggests that the TMD of Tas1R2 may also be a binding site for perillartine (an aldoxime-type sweetener). An understanding of binding sites for sweeteners might well lead to a rational design for artificial sweeteners with better taste profiles. Binding-site studies along with investigations that use them to understand domain:domain interactions that lead to receptor activation may eventually give us the ability to rationally design better more potent sweeteners. Of even more interest, we might discover ways to enhance the sweetness of natural sugars with compounds that have no taste of their own.

An active area of investigation, not only of the sweet receptor but of all family C receptors is the conformational changes that the receptors undergo that leads to activation of G protein. Of particular interest is whether the monomer, the dimer or some larger grouping of receptors provides maximal activity in response to sweeteners. Of interest in this regard is the observation that many artificial sweeteners, while more potent than natural sugars, are none-the-less, at best, only partial agonists. Studies investigating additivity or synergy between sweeteners that bind to different domains of the receptor will be particularly important in our understanding of signal transmission within the sweet receptor.

Some of the knowledge gathered for sweet-taste receptors is still lacking for bitter-taste receptors. This applies in particular to the molecular architecture of TAS2Rs. Complex mutational analyses along with molecular modeling approaches are required. Clearly for a complete understanding of bitter-taste coding the cognate bitter substances for all TAS2Rs need to be identified, although this will be a never-ending project in light of the numerous bitter compounds and the extensive species differences. In humans, it emerges that there are pronounced variations among *TAS2R* genes, a fact that is expected to affect eating behavior. Uncovering the underlying mechanisms is scientifically interesting and of great potential medical and social importance for eating disorders, obesity, and their consequences. In this context, pharmacological treatment of taste in patients suffering from eating disorders with compounds developed by aid of *in vitro* receptor assay could be a therapeutic option. Another important development in the field would be the elucidation of how bitter receptor cells communicate with the nervous system and how bitter transmission proceeds and is processed in relation to the other taste modalities. With some optimism we can await the future which will us bring exciting discoveries.

References

Adler, E., Hoon, M. A., Mueller, K. L., Chandrashekar, J., Ryba, N. J., and Zuker, C. S. 2000. A novel family of mammalian taste receptors [see comments]. Cell 100, 693–702.

Ariyasu, T., Matsumoto, S., Kyono, F., Hanaya, T., Arai, S., Ikeda, M., and Kurimoto, M. 2003. Taste receptor T1R3 is an essential molecule for the cellular recognition of the disaccharide trehalose. In Vitro Cell Dev. Biol. Anim. 39, 80–88.

Bachmanov, A. A., Li, X., Reed, D. R., Ohmen, J. D., Li, S., Chen, Z., Tordoff, M. G., de Jong, P. J., Wu, C., West, D. B., Chatterjee, A., Ross, D. A., and Beauchamp, G. K. 2001. Positional cloning of the mouse saccharin preference (Sac) locus. Chem. Senses 26, 925–933.

Bartoshuk, L. M. 1979. Bitter taste of saccharin related to the genetic ability to taste the bitter substance 6-*n*-propylthiouracil. Science 205, 934–935.

Behrens, M., Bartelt, J., Reichling, C., Winnig, M., Kuhn, C., and Meyerhof, W. 2006. Members of RTP and REEP gene families influence functional bitter taste receptor expression. J. Biol. Chem. 281, 20650–20659.

Behrens, M., Brockhoff, A., Kuhn, C., Bufe, B., Winnig, M., and Meyerhof, W. 2004. The human taste receptor hTAS2R14 responds to a variety of different bitter compounds. Biochem. Biophys. Res. Commun. 319, 479–485.

Belitz, H. D. and Wieser, H. 1985. Bitter compounds: occurrence and structure–activity relationship. Food Rev. Int. 1, 271–354.

Blakeslee, A. F. 1931. Genetics of sensory thresholds: taste for pheny thio carbamide. Science 74, 607.

Blakeslee, A. F. 1932. Genetics of sensory thresholds: taste for pheny thio carbamide. Proc. Natl. Acad. Sci. U. S. A. 18, 120–130.

Blakeslee, A. F. 1935. A dinner demonstration of threshold differences in taste and smell. Science 81, 504–507.

Blizard, D. A., Kotlus, B., and Frank, M. E. 1999. Quantitative trait loci associated with short-term intake of sucrose, saccharin and quinine solutions in laboratory mice. Chem. Senses 24, 373–385.

Bufe, B., Breslin, P. A., Kuhn, C., Reed, D. R., Tharp, C. D., Slack, J. P., Kim, U. K., Drayna, D., and Meyerhof, W. 2005. The molecular basis of individual differences in phenylthiocarbamide and propylthiouracil bitterness perception. Curr. Biol. 15, 322–327.

Bufe, B., Hofmann, T., Krautwurst, D., Raguse, J. D., and Meyerhof, W. 2002. The human TAS2R16 receptor mediates bitter taste in response to beta-glucopyranosides. Nat. Genet. 32, 397–401.

Bufe, B., Schöley-Pohl, E., Krautwurst, D., Hofmann, T., and Meyerhof, W. 2004. Identification of human bitter taste receptors. Am. Chem. Soc. 45–59.

Caicedo, A. and Roper, S. D. 2001. Taste receptor cells that discriminate between bitter stimuli. Science 291, 1557–1560.

Caicedo, A., Pereira, E., Margolskee, R. F., and Roper, S. D. 2003. Role of the G-protein subunit alpha-gustducin in taste cell responses to bitter stimuli. J, Neurosci. 23, 9947–9952.

Chandrashekar, J. 2006. The receptors and cells for mammalian taste. Nature 444, 288–294.

Chandrashekar, J., Mueller, K. L., Hoon, M. A., Adler, E., Feng, L., Guo, W., Zuker, C. S., and Ryba, N. J. 2000. T2Rs function as bitter taste receptors. Cell 100, 703–711.

Chaudhari, N., Landin, A. M., and Roper, S. D. 2000. A metabotropic glutamate receptor variant functions as a taste receptor. Nat. Neurosci. 3, 113–119.

Christiansen, B., Wellendorph, P., and Brauner-Osborne, H. 2006. Known regulators of nitric oxide synthase and arginase are agonists at the human G-protein-coupled receptor GPRC6A. Br. J. Pharmacol. 147, 855–863.

Conte, C., Ebeling, M., Marcuz, A., Nef, P., and Andres-Barquin, P. J. 2002. Identification and characterization of human taste receptor genes belonging to the TAS2R family. Cytogenet. Genome Res. 98, 45–53.

Conte, C., Ebeling, M., Marcuz, A., Nef, P., and Andres-Barquin, P. J. 2003. Evolutionary relationships of the Tas2r receptor gene families in mouse and human. Physiol. Genomics 14, 73–82.

Dahl, M., Erickson, R. P., and Simon, S. A. 1997. Neural responses to bitter compounds in rats. Brain Res. 756, 22–34.

Damak, S., Rong, M., Yasumatsu, K., Kokrashvili, Z., Varadarajan, V., Zou, S., Jiang, P., Ninomiya, Y., and Margolskee, R. F. 2003. Detection of sweet and umami taste in the absence of taste receptor T1r3. Science 301, 850–853.

Danilova, V. and Hellekant, G. 2003. Comparison of the responses of the chorda tympani and glossopharyngeal nerves to taste stimuli in C57BL/6J mice. BMC Neurosci. 4, 5.

Danilova, V. and Hellekant, G. 2004. Sense of taste in a New World monkey, the common marmoset. II. Link between behavior and nerve activity. J. Neurophysiol. 92, 1067–1076.

Delwiche, J. F., Buletic, Z., and Breslin, P. A. 2001. Covariation in individuals' sensitivities to bitter compounds: evidence supporting multiple receptor/transduction mechanisms. Percept. Psychophys. 63, 761–776.

Drewnowski, A. 1997. Taste preferences and food intake. Annu. Rev. Nutr. 17, 237–253.

Drewnowski, A. 2001. The science and complexity of bitter taste. Nutr. Rev. 59, 163–169.

Drewnowski, A. 2004. 6-n-Propylthiouracil Sensitivity, Food Choices, and Food Consumption. In: Genetic Variation in Taste Sensitivity (eds. J. Presscott and B. J. Tepper), pp. 179–194. Marcel.

Drewnowski, A. and Gomez-Carneros, C. 2000. Bitter taste, phytonutrients, and the consumer: a review. Am. J. Clin. Nutr. 72, 1424–1435.

Drewnowski, A. and Rock, C. L. 1995. The influence of genetic taste markers on food acceptance. Am. J. Clin. Nutr. 62, 506–511.

Duffy, V. B., Davidson, A. C., Kidd, J. R., Kidd, K. K., Speed, W. C., Pakstis, A. J., Reed, D. R., Snyder, D. J., and Bartoshuk, L. M. 2004a. Bitter receptor gene (TAS2R38), 6-*n*-propylthiouracil (PROP) bitterness and alcohol intake. Alcohol Clin. Exp. Res. 28, 1629–1637.

Duffy, V. B., Lucchina, L. A., and Bartoshuk, L. M. 2004b. Genetic Variation in Taste: Potential Biomarker for Cardiovascular Disease Risk? In: Genetic Variation in Taste Sensitivity (eds. J. Prescott and B. J. Tepper), pp. 195–228. Marcel.

Facchini, F., Abbati, A., and Campagnoni, S. 1990. Possible relations between sensitivity to phenylthiocarbamide and goiter. Hum. Biol. 62, 545–552.

Finger, T. E., Bottger, B., Hansen, A., Anderson, K. T., Alimohammadi, H., and Silver, W. L. 2003. Solitary chemoreceptor cells in the nasal cavity serve as sentinels of respiration. Proc. Natl. Acad. Sci. U. S. A. 100, 8981–8986.

Fischer, A., Gilad, Y., Man, O., and Paabo, S. 2005. Evolution of bitter taste receptors in humans and apes. Mol. Biol. Evol. 22, 432–436.

Floriano, W. B., Hall, S., Vaidehi, N., Kim, U., Drayna, D., and Goddard, W. A., 3rd. 2006. Modeling the human PTC bitter-taste receptor interactions with bitter tastants. J. Mol. Model. 12, 931–941.

Fox, A. L. 1932. The Relationship between chemical constitution and taste. Proc. Nat. Acad. Sci. U. S. A. 18, 115–120.

Fuller, J. L. 1974. Single-locus control of saccharin preference in mice. J. Hered. 65, 33–36.

Go, Y. 2006. Lineage-specific expansions and contractions of the bitter taste receptor gene repertoire in vertebrates. Mol. Biol. Evol. 23, 964–972.

Go, Y., Satta, Y., Takenaka, O., and Takahata, N. 2005. Lineage-specific loss of function of bitter taste receptor genes in humans and non-human primates. Genetics 170, 313–326.

Gudermann, T., Schoneberg, T., and Schultz, G. 1997. Functional and structural complexity of signal transduction via G-protein-coupled receptors. Annu. Rev. Neurosci. 20, 399–427.

Harris, H., Kalmus, H., and Trotter, W. R. 1949. Taste sensitivity to phenylthiourea in goitre and diabetes. Lancet 2, 1038.

He, H. T., Rens-Domiano, S., Martin, J. M., Law, S. F., Borislow, S., Woolkalis, M., Manning, D., and Reisine, T. 1990. Solubilization of active somatostatin receptors from rat brain. Mol. Pharmacol. 37, 614–621.

Hermans, E. and Challiss, R. A. 2001. Structural, signalling and regulatory properties of the group I metabotropic glutamate receptors: prototypic family C G-protein-coupled receptors. Biochem. J. 359, 465–484.

Hinrichs, A. L., Wang, J. C., Bufe, B., Kwon, J. M., Budde, J., Allen, R., Bertelsen, S., Evans, W., Dick, D., Rice, J., Foroud, T., Nurnberger, J., Tischfield, J. A., Kuperman, S., Crowe, R., Hesselbrock, V., Schuckit, M., Almasy, L., Porjesz, B., Edenberg, H. J., Begleiter, H., Meyerhof, W., Bierut, L. J., and Goate, A. M. 2006. Functional variant in a bitter-taste receptor (hTAS2R16) influences risk of alcohol dependence. Am. J. Hum. Genet. 78, 103–111.

Hoon, M. A., Adler, E., Lindemeier, J., Battey, J. F., Ryba, N. J., and Zuker, C. S. 1999. Putative mammalian taste receptors: a class of taste-specific GPCRs with distinct topographic selectivity. Cell 96, 541–551.

Huang, A. L., Chen, X., Hoon, M. A., Chandrashekar, J., Guo, W., Trankner, D., Ryba, N. J., and Zuker, C. S. 2006. The cells and logic for mammalian sour taste detection. Nature 442, 934–938.

Huang, L., Shanker, Y. G., Dubauskaite, J., Zheng, J. Z., Yan, W., Rosenzweig, S., Spielman, A. I., Max, M., and Margolskee, R. F. 1999. Ggamma13 colocalizes with gustducin in taste receptor cells and mediates IP3 responses to bitter denatonium. Nat. Neurosci. 2, 1055–1062.

Jiang, P., Cui, M., Zhao, B., Liu, Z., Snyder, L. A., Benard, L. M., Osman, R., Margolskee, R. F., and Max, M. 2005a. Lactisole interacts with the transmembrane domains of human T1R3 to inhibit sweet taste. J. Biol. Chem. 280, 15238–15246.

Jiang, P., Cui, M., Zhao, B., Snyder, L. A., Benard, L. M., Osman, R., Max, M., and Margolskee, R. F. 2005b. Identification of the cyclamate interaction site within the transmembrane domain of the human sweet taste receptor subunit T1R3. J. Biol. Chem. 280, 34296–34305.

Jiang, P., Ji, Q., Liu, Z., Snyder, L. A., Benard, L. M., Margolskee, R. F., and Max, M. 2004. The cysteine-rich region of T1R3 determines responses to intensely sweet proteins. J. Biol. Chem. 279, 45068–45075.

Jingami, H., Nakanishi, S., and Morikawa, K. 2003. Structure of the metabotropic glutamate receptor. Curr. Opin. Neurobiol. 13, 271–278.

Joiner, T. E., Jr. and Perez, M. 2004. Phenylthiocarbamide tasting and family history of depression, revisited: low rates of depression in families of supertasters. Psychiatry Res. 126, 83–87.

Kim, M. R., Kusakabe, Y., Miura, H., Shindo, Y., Ninomiya, Y., and Hino, A. 2003. Regional expression patterns of taste receptors and gustducin in the mouse tongue. Biochem. Biophys. Res. Commun. 312, 500–506.

Kim, U., Wooding, S., Ricci, D., Jorde, L. B., and Drayna, D. 2005. Worldwide haplotype diversity and coding sequence variation at human bitter taste receptor loci. Hum. Mutat. 26, 199–204.

Kim, U. K., Jorgenson, E., Coon, H., Leppert, M., Risch, N., and Drayna, D. 2003. Positional cloning of the human quantitative trait locus underlying taste sensitivity to phenylthiocarbamide. Science 299, 1221–1225.

Kitagawa, M., Kusakabe, Y., Miura, H., Ninomiya, Y., and Hino, A. 2001. Molecular genetic identification of a candidate receptor gene for sweet taste. Biochem. Biophys. Res. Commun. 283, 236–242.

Kuhn, C., Bufe, B., Winnig, M., Hofmann, T., Frank, O., Behrens, M., Lewtschenko, T., Slack, J. P., Ward, C. D., and Meyerhof, W. 2004. Bitter taste receptors for saccharin and acesulfame K. J. Neurosci. 24, 10260–10265.

Kunishima, N., Shimada, Y., Tsuji, Y., Sato, T., Yamamoto, M., Kumasaka, T., Nakanishi, S., Jingami, H., and Morikawa, K. 2000. Structural basis of glutamate recognition by a dimeric metabotropic glutamate receptor. Nature 407, 971–977.

Li, X., Inoue, M., Reed, D. R., Huque, T., Puchalski, R. B., Tordoff, M. G., Ninomiya, Y., Beauchamp, G. K., and Bachmanov, A. A. 2001. High-resolution genetic mapping of the saccharin preference locus (Sac) and the putative sweet taste receptor (T1R1) gene (Gpr70) to mouse distal Chromosome 4. Mamm. Genome 12, 13–16.

Li, X., Staszewski, L., Xu, H., Durick, K., Zoller, M., and Adler, E. 2002a. Human receptors for sweet and umami taste. Proc. Natl. Acad. Sci. U. S. A. 99, 4692–4696.

Li, X., Staszewski, L., Xu, H., Durick, K., Zoller, M., and Adler, E. 2002b. Human receptors for sweet and umami taste. Proc. Natl. Acad. Sci. U. S. A. 99, 4692–4696.

Lindemann, B. 1999. Receptor seeks ligand: on the way to cloning the molecular receptors for sweet and bitter taste. Nat. Med. 5, 381–382.

Lush, I. E. 1981. The genetics of tasting in mice. I. Sucrose octaacetate. Genet. Res. 38, 93–95.

Lush, I. E. 1984. The genetics of tasting in mice. III. Quinine. Genet. Res. 44, 151–160.

Lush, I. E. 1986. The genetics of tasting in mice. IV. The acetates of raffinose, galactose and beta-lactose. Genet. Res. 47, 117–123.

Lush, I. E. 1989. The genetics of tasting in mice. VI. Saccharin, acesulfame, dulcin and sucrose. Genet. Res. 53, 95–99.

Lush, I. E. and Holland, G. 1988. The genetics of tasting in mice. V. Glycine and cycloheximide. Genet. Res. 52, 207–212.

Lush, I. E., Hornigold, N., King, P., and Stoye, J. P. 1995. The genetics of tasting in mice. VII. Glycine revisited, and the chromosomal location of Sac and Soa. Genet. Res. 66, 167–174.

Malherbe, P., Kratochwil, N., Zenner, M. T., Piussi, J., Diener, C., Kratzeisen, C., Fischer, C., and Porter, R. H. 2003. Mutational analysis and molecular modeling of the binding pocket of the metabotropic glutamate 5 receptor negative modulator 2-methyl-6-(phenylethynyl)-pyridine. Mol. Pharmacol. 64, 823–832.

Matsunami, H., Montmayeur, J. P., and Buck, L. B. 2000. A family of candidate taste receptors in human and mouse [see comments]. Nature 404, 601–604.

Max, M., Shanker, Y. G., Huang, L., Rong, M., Liu, Z., Campagne, F., Weinstein, H., Damak, S., and

Margolskee, R. F. 2001. Tas1r3, encoding a new candidate taste receptor, is allelic to the sweet responsiveness locus Sac. Nat. Genet. 28, 58–63.

McLaughlin, S. K., McKinnon, P. J., and Margolskee, R. F. 1992. Gustducin is a taste-cell-specific G protein closely related to the transducins. Nature 357, 563–569.

Meyerhof, W. 2005. Elucidation of mammalian bitter taste. Rev. Physiol. Biochem. Pharmacol. 154, 37–72.

Meyerhof, W., Behrens, M., Brockhoff, A., Bufe, B., and Kuhn, C. 2005. Human bitter taste perception. Chem. Senses 30 Suppl 1, i14–i15.

Miedlich, S. U., Gama, L., Seuwen, K., Wolf, R. M., and Breitwieser, G. E. 2004. Homology modeling of the transmembrane domain of the human calcium sensing receptor and localization of an allosteric binding site. J. Biol. Chem. 279, 7254–7263.

Miguet, L., Zhang, Z., and Grigorov, M. G. 2006. Computational studies of ligand-receptor interactions in bitter taste receptors. J. Recept. Signal Transduct. Res. 26, 611–630.

Ming, D., Ninomiya, Y., and Margolskee, R. F. 1999. Blocking taste receptor activation of gustducin inhibits gustatory responses to bitter compounds. Proc. Natl. Acad. Sci. U. S. A. 96, 9903–9908.

Ming, D., Ruiz-Avila, L., and Margolskee, R. F. 1998. Characterization and solubilization of bitter-responsive receptors that couple to gustducin. Proc. Natl. Acad. Sci. U. S. A. 95, 8933–8938.

Montmayeur, J. P., Liberles, S. D., Matsunami, H., and Buck, L. B. 2001. A candidate taste receptor gene near a sweet taste locus. Nat. Neurosci. 4, 492–498.

Morini, G., Bassoli, A., and Temussi, P. A. 2005. From small sweeteners to sweet proteins: anatomy of the binding sites of the human T1R2–T1R3 receptor. J. Med. Chem. 48, 5520–5529.

Mueller, K. L., Hoon, M. A., Erlenbach, I., Chandrshekar, J., Zuker, C. S., and Ryba, N. J. P. 2005. The receptors and coding logic for bitter taste. Nature 434, 225–229.

Muto, T., Tsuchiya, D., Morikawa, K., and Jingami, H. 2007. Structures of the extracellular regions of the group II/III metabotropic glutamate receptors. Proc. Natl. Acad. Sci. U. S. A. 104, 3759–3764.

Nelson, G., Chandrashekar, J., Hoon, M. A., Feng, L., Zhao, G., Ryba, N. J., and Zuker, C. S. 2002. An amino-acid taste receptor. Nature 416, 199–202.

Nelson, G., Hoon, M. A., Chandrashekar, J., Zhang, Y., Ryba, N. J., and Zuker, C. S. 2001. Mammalian sweet taste receptors. Cell 106, 381–390.

Nie, Y., Vigues, S., Hobbs, J. R., Conn, G. L., and Munger, S. D. 2005. Distinct contributions of T1R2 and T1R3 taste receptor subunits to the detection of sweet stimuli. Curr. Biol. 15, 1948–1952.

Okada, T., Fujiyoshi, Y., Silow, M., Navarro, J., Landau, E. M., and Shichida, Y. 2002. Functional role of internal water molecules in rhodopsin revealed by X-ray crystallography. Proc. Natl. Acad. Sci. U. S. A. 99, 5982–5987.

Olias, G., Viollet, C., Kusserow, H., Epelbaum, J., and Meyerhof, W. 2004. Regulation and function of somatostatin receptors. J. Neurochem. 89, 1057–1091.

Ozeck, M., Brust, P., Xu, H., and Servant, G. 2004. Receptors for bitter, sweet and umami taste couple to inhibitory G protein signaling pathways. Eur. J. Pharmacol. 489, 139–149.

Pagano, A., Ruegg, D., Litschig, S., Stoehr, N., Stierlin, C., Heinrich, M., Floersheim, P., Prezeau, L., Carroll, F., Pin, J. P., Cambria, A., Vranesic, I., Flor, P. J., Gasparini, F., and Kuhn, R. 2000. The non-competitive antagonists 2-methyl-6-(phenylethynyl)pyridine and 7-hydroxyiminocyclopropan[b]chromen-1a-carboxylic acid ethyl ester interact with overlapping binding pockets in the transmembrane region of group I metabotropic glutamate receptors. J. Biol. Chem. 275, 33750–33758.

Palczewski, K., Kumasaka, T., Hori, T., Behnke, C. A., Motoshima, H., Fox, B. A., Le Trong, I., Teller, D. C., Okada, T., Stenkamp, R. E., Yamamoto, M., and Miyano, M. 2000. Crystal structure of rhodopsin: A G protein-coupled receptor. Science 289, 739–745.

Parry, C. M., Erkner, A., and le Coutre, J. 2004. Divergence of T2R chemosensory receptor families in humans, bonobos, and chimpanzees. Proc. Natl. Acad. Sci. U. S. A. 101, 14830–14834.

Pelchat, M. L. and Danowski, S. 1992. A possible genetic association between PROP-tasting and alcoholism. Physiol. Behav. 51, 1261–1266.

Petrel, C., Kessler, A., Dauban, P., Dodd, R. H., Rognan, D., and Ruat, M. 2004. Positive and negative allosteric modulators of the Ca^{2+}-sensing receptor interact within overlapping but not identical binding sites in the transmembrane domain. J. Biol. Chem. 279, 18990–18997.

Pi, M., Faber, P., Ekema, G., Jackson, P. D., Ting, A., Wang, N., Fontilla-Poole, M., Mays, R. W., Brunden, K. R., Harrington, J. J., and Quarles, L. D. 2005. Identification of a novel extracellular cation-sensing G-protein-coupled receptor. J. Biol. Chem. 280, 40201–40209.

Pin, J. P., Galvez, T., and Prezeau, L. 2003. Evolution, structure, and activation mechanism of family 3/C G-protein-coupled receptors. Pharmacol. Ther. 98, 325–354.

Pronin, A. N., Tang, H., Connor, J., and Keung, W. 2004. Identification of Ligands for Two Human Bitter T2R Receptors. Chem. Senses 29, 583–593.

Rossler, P., Boekhoff, I., Tareilus, E., Beck, S., Breer, H., and Freitag, J. 2000. G protein betagamma complexes in circumvallate taste cells involved in bitter transduction. Chem. Senses 25, 413–421.

Ruiz-Avila, L., McLaughlin, S. K., Wildman, D., McKinnon, P. J., Robichon, A., Spickofsky, N., and Margolskee, R. F. 1995. Coupling of bitter receptor to phosphodiesterase through transducin in taste receptor cells. Nature 376, 80–85.

Ruiz-Avila, L., Ming, D., and Margolskee, R. F. 2000. An In vitro assay useful to determine the potency of several bitter compounds. Chem. Senses 25, 361–368.

Sainz, E., Cavenagh, M. M., Gutierrez, J., Battey, J. F., Northup, J. K., and Sullivan, S. L. 2007. Functional characterization of human bitter taste receptors. Biochem. J. 403, 537–543.

Sainz, E., Korley, J. N., Battey, J. F., and Sullivan, S. L. 2001. Identification of a novel member of the T1R family of putative taste receptors. J. Neurochem. 77, 896–903.

Saito, H., Kubota, M., Roberts, R. W., Chi, Q., and Matsunami, H. 2004. RTP family members induce functional expression of mammalian odorant receptors. Cell 119, 679–691.

Sako, N., Harada, S., and Yamamoto, T. 2000. Gustatory information of umami substances in three major taste nerves. Physiol. Behav. 71, 193–198.

Sandell, M. A. and Breslin, P. A. 2006. Variability in a taste-receptor gene determines whether we taste toxins in food. Curr. Biol. 16, R792–794.

Schiffman, S. S., Booth, B. J., Sattely-Miller, E. A., Graham, B. G., and Gibes, K. M. 1999. Selective inhibition of sweetness by the sodium salt of $+/-$2-(4-methoxyphenoxy)propanoic acid. Chem Senses 24, 439–447.

Sclafani, A. and Perez, C. 1997. Cypha [propionic acid, 2-(4-methoxyphenol) salt] inhibits sweet taste in humans, but not in rats. Physiol. Behav. 61, 25–29.

Shi, P. and Zhang, J. 2006. Contrasting modes of evolution between vertebrate sweet/umami receptor genes and bitter receptor genes. Mol. Biol. Evol. 23, 292–300.

Shi, P., Zhang, J., Yang, H., and Zhang, Y. P. 2003. Adaptive diversification of bitter taste receptor genes in Mammalian evolution. Mol. Biol. Evol. 20, 805–814.

Soranzo, N., Bufe, B., Sabeti, P. C., Wilson, J. F., Weale, M. E., Marguerie, R., Meyerhof, W., and Goldstein, D. B. 2005. Positive selection on a high-sensitivity allele of the human bitter-taste receptor TAS2R16. Curr. Biol. 15, 1257–1265.

Spector, A. C. and Kopka, S. L. 2002. Rats fail to discriminate quinine from denatonium: implications for the neural coding of bitter-tasting compounds. J. Neurosci. 22, 1937–1941.

Sugino, Y., Umemoto, A., and Mizutani, S. 2002. Insensitivity to the bitter taste of chloramphenicol: an autosomal recessive trait. Genes Genet. Syst. 77, 59–62.

Teller, D. C., Okada, T., Behnke, C. A., Palczewski, K., and Stenkamp, R. E. 2001. Advances in determination of a high-resolution three-dimensional structure of rhodopsin, a model of G-protein-coupled receptors (GPCRs). Biochemistry 40, 7761–7772.

Tepper, B. J. and Nurse, R. J. 1997. Fat perception is related to PROP taster status. Physiol. Behav. 61, 949–954.

Timpson, N. J., Christensen, M., Lawlor, D. A., Gaunt, T. R., Day, I. N., and Smith, G. D. 2005. TAS2R38 (phenylthiocarbamide) haplotypes, coronary heart disease traits, and eating behavior in the British Women's Heart and Health Study. Am. J. Clin. Nutr. 81, 1005–1011.

Tsuchiya, D., Kunishima, N., Kamiya, N., Jingami, H., and Morikawa, K. 2002. Structural views of the ligand-binding cores of a metabotropic glutamate receptor complexed with an antagonist and both glutamate and Gd^{3+}. Proc. Natl. Acad. Sci. U. S. A. 99, 2660–2665.

Ueda, T., Ugawa, S., Ishida, Y., Shibata, Y., Murakami, S., and Shimada, S. 2001. Identification of coding single-nucleotide polymorphisms in human taste receptor genes involving bitter tasting. Biochem. Biophys. Res. Commun. 285, 147–151.

Ueda, T., Ugawa, S., Yamamura, H., Imaizumi, Y., and Shimada, S. 2003. Functional interaction between T2R taste receptors and G-protein α-subunits expressed in taste receptor cells. J. Neurosci. 23, 7376–7380.

Visiers, I., Ballesteros, J. A., and Weinstein, H. 2002. Three-dimensional representations of G protein-coupled receptor structures and mechanisms. Methods Enzymol. 343, 329–371.

Visiers, I., Braunheim, B. B., and Weinstein, H. 2000. Prokink: a protocol for numerical evaluation of helix distortions by proline. Protein Eng. 13, 603–606.

Walters, D. E. and Hellekant, G. 2006. Interactions of the sweet protein brazzein with the sweet taste receptor. J. Agric. Food Chem. 54, 10129–10133.

Wang, X., Thomas, S. D., and Zhang, J. 2004. Relaxation of selective constraint and loss of function in the evolution of human bitter taste receptor genes. Hum. Mol. Genet. 13, 2671–2678.

Whitney, G. and Harder, D. B. 1994. Genetics of bitter perception in mice. Physiol. Behav. 56, 1141–1147.

Winnig, M., Bufe, B., and Meyerhof, W. 2005. Valine 738 and lysine 735 in the fifth transmembrane domain of rTas1r3 mediate insensitivity towards lactisole of the rat sweet taste receptor. BMC Neurosci 6, 22.

Wong, G. T., Gannon, K. S., and Margolskee, R. F. 1996a. Transduction of bitter and sweet taste by gustducin. Nature 381, 796–800.

Wong, G. T., Ruiz-Avila, L., and Margolskee, R. F. 1999. Directing gene expression to gustducin-positive taste receptor cells. J. Neurosci. 19, 5802–5809.

Wong, G. T., Ruiz-Avila, L., Ming, D., Gannon, K. S., and Margolskee, R. F. 1996b. Biochemical and transgenic analysis of gustducin's role in bitter and sweet transduction. Cold Spring Harb. Symp. Quant. Biol. 61, 173–184.

Wooding, S., Bufe, B., Grassi, C., Howard, M. T., Stone, A. C., Vazquez, M., Dunn, D. M., Meyerhof, W., Weiss, R. B., and Bamshad, M. J. 2006. Independent evolution of bitter-taste sensitivity in humans and chimpanzees. Nature 440, 930–934.

Wooding, S., Kim, U. K., Bamshad, M. J., Larsen, J., Jorde, L. B., and Drayna, D. 2004. Natural selection and molecular evolution in PTC, a bitter-taste receptor gene. Am. J. Hum. Genet. 74, 637–646.

Wu, S. V., Rozengurt, N., Yang, M., Young, S. H., Sinnett-Smith, J., and Rozengurt, E. 2002. Expression of bitter taste receptors of the T2R family in the gastrointestinal tract and enteroendocrine STC-1 cells. Proc. Natl. Acad. Sci. U. S. A. 99, 2392–2397.

Xu, H., Staszewski, L., Tang, H., Adler, E., Zoller, M., and Li, X. 2004. Different functional roles of T1R subunits in the heteromeric taste receptors. Proc. Natl. Acad. Sci. U. S. A. 101, 14258–14263.

Yang, H., Wanner, I. B., Roper, S. D., and Chaudhari, N. 1999. An optimized method for in situ hybridization with signal amplification that allows the detection of rare mRNAs. J. Histochem. Cytochem 47, 431–446.

Zhang, Y., Hoon, M. A., Chandrashekar, J., Mueller, K. L., Cook, B., Wu, D., Zuker, C. S., and Ryba, N. J. 2003. Coding of sweet, bitter, and umami tastes. Different receptor cells sharing similar signaling pathways. Cell 112, 293–301.

Zhao, G. Q., Zhang, Y., Hoon, M. A., Chandrashekar, J., Erlenbach, I., Ryba, N. J., and Zuker, C. S. 2003. The receptors for mammalian sweet and umami taste. Cell 115, 255–266.

4.10 Taste Transduction

S C Kinnamon, Colorado State University, Fort Collins, CO, USA

R F Margolskee, Mount Sinai School of Medicine, New York, NY, USA

4.10.1 Introduction

During the past decade considerable progress has been made in revealing the basic mechanisms of taste transduction and signaling. This explosion of information is due in large part to the sequencing of the rodent and human genomes, allowing identification of the taste receptors, downstream signaling molecules, and effector ion channels involved in taste transduction. Further, these discoveries have led to the production of gene-targeted and transgenic mice, where the role of individual signaling proteins can be assessed *in vivo*. Here we review recent progress in quality-specific transduction, focusing primarily on mammalian taste cells. In addition, we discuss the identity and potential roles of neurotransmitters in taste cells and how taste information is communicated to afferent nerve fibers. Detailed information about G-protein-coupled taste receptors (GPCRs) is covered in Chapter Taste Receptors.

4.10.1.1 Taste Qualities

The sense of taste affords organisms the ability to discriminate nutritionally important substances from potentially harmful ones. The gustatory system detects and discriminates nonvolatile stimuli representing five basic taste qualities – sweet, salty, sour, bitter, and umami. Sweet, salty, and umami stimuli are generally appetitive, because they represent chemicals required for proper nutrition. These include sugars for energy production, sodium (Na^+) for ionic homeostasis, and amino acids as a source for protein synthesis. Bitter and sour stimuli are naturally aversive and warn against ingestion of potentially toxic or poisonous substances.

Recently obtained evidence suggests that fat and water may each constitute additional primary taste qualities. In the case of fat, it is well established that mammals prefer lipid enriched foods. However, until recently, this preference for fat was presumed to be due only to its pleasant textural qualities (mouth feel). New data suggest that specific fatty acid receptors and fatty acid-modulated ion channels are expressed in taste receptor cells, where they may be involved in the detection of fatty acids as taste stimuli (for review, Laugerette, F. *et al.*, 2007). Similarly, water-permeable ion channels were recently found on the apical membrane of taste cells, suggesting they may be the source of the gustatory neural response to water (Watson, K. J. *et al.*, 2007). However, whether water represents a specific taste

or whether it is involved in volume regulation of taste cells is not known.

4.10.1.2 Taste Buds, Taste Cell Types, and Afferent Innervation

Taste buds are the transducing elements of gustatory sensation. A typical taste bud is an onion shaped end organ comprising 50–100 spindle shaped cells that extend from the basal lamina to the surface of the tongue. The apical portion of each taste cell contains microvilli, which protrude through an opening in the epithelium called the taste pore. The basolateral membrane makes synaptic contacts with afferent nerve fibers, which project taste information to the brain. Four types of cells can be distinguished based on morphological and functional criteria. Type I cells, also known as dark cells, make up about 50% of the cells in a taste bud. These cells are generally considered to have a support function, and are not likely to be involved directly in transducing taste stimuli, although their apical membranes do reach the taste pore. The membranes of type I cells wrap around other cells in a glial-like fashion, and they express transporters and enzymes for uptake and inactivation of transmitters (Lawton, D. M. *et al.*, 2000; Bartel, D. L. *et al.*, 2006; Figure 1.) Type II cells, also known as light cells, make up about 30% of the taste bud. These cells express the taste receptors and downstream signaling elements implicated in responses to bitter, sweet, and umami taste stimuli. These cells associate with afferent nerve fibers, but do not form definitive synapses with pre- and postsynaptic specializations (Clapp, T. R. *et al.*, 2004; 2006). Type III cells, also known as intermediate cells, make up about 20% of cells in the

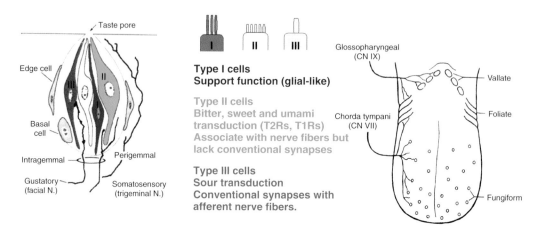

Figure 1 Cell types and innervation of taste buds. Type II taste cells are coded green, while Type III taste cells are coded red in subsequent figures. (Modified, with permission, from Finger and Simon, 2000).

taste bud. Recent evidence suggests that type III cells are the transducers of acidic (sour) stimuli (Richter, T. A. *et al.*, 2003; Kataoka, S. *et al.*, 2007). These cells form prominent synapses with afferent nerve fibers and may also be involved in cell-to-cell signaling in the taste bud (Herness, S. *et al.*, 2005; Roper, S. D., 2006). Proliferative basal cells are also found in the basolateral regions of the taste bud. These spherical cells divide to form new taste cells, which turn over every 10 days to 2 weeks. Ultrastructural features of these cell types are reviewed in Chapter Ultrastructure of Taste Buds.

Lingual taste buds are housed in connective tissue papillae. The anterior two-thirds of the tongue contains fungiform papillae, each of which contain one or two taste buds. These taste buds are innervated by the chorda tympani branch of the facial nerve (CN VII). The posterior region of the tongue contains the circumvallate and foliate papillae, each of which contain hundreds of taste buds. These taste buds are innervated primarily by the glossopharyngeal nerve (CN IX), but the anterior foliate taste buds are innervated by a branch of the chorda tympani nerve. Taste buds are also located on the soft palate, where they are innervated by the greater superficial petrosal nerve, a branch of CN VII, and on the larynx and esophagus, where they are innervated by a branch of the vagus nerve (CN X).

4.10.1.3 Taste Receptor Cell Properties

Simple taste stimuli, such as salts and acids, act on apically located ion channels, while more complex stimuli, such as sugars and plant alkaloids, bind to GPCRs and activate intracellular second messenger signaling cascades. In all cases, transduction appears to result in membrane depolarization and action potential initiation. Action potentials in taste cells were first observed over two decades ago, using intracellular recordings from mudpuppy taste cells (Roper, S., 1983). Action potentials are now known to be a general property of type II and type III rodent taste cells and they are generated regularly in response to most taste stimuli (Avenet, P. and Lindemann, B., 1991; Gilbertson, T. A. *et al.*, 1992; Cummings, T. A. *et al.*, 1993; Yoshida, R. *et al.*, 2006). Underlying action potential generation are tetrodotoxin-sensitive Na^+ channels (Herness, M. S. and Sun, X. D., 1995), a variety of potassium (K^+) channels (Chen, Y. *et al.*, 1996; Liu, L. *et al.*, 2005), and voltage-gated Ca^{2+} channels (Behe, P. *et al.*, 1990). Recent data obtained from specific taste cell types

indicated that while both type II and type III cells express voltage-gated Na^+ and K^+ channels, only type III taste cells express voltage-gated Ca^{2+} channels (Medler, K. F. *et al.*, 2003; Clapp, T. R. *et al.*, 2006; DeFazio, R. A. *et al.*, 2006). The role of action potentials in the transduction process has been unclear, since taste cells are short receptor cells that should be capable of transmitter release without requiring action potentials. Recent data using cell attached recordings from a polarized, semi intact preparation show that the breadth of tuning of individual taste cells in response to taste stimuli is nearly identical to the breadth of tuning of single afferent nerve fibers, suggesting that action potentials may be required for transmitter release (Yoshida, R. *et al.*, 2006).

4.10.2 Quality Specific Transduction

During the past 10–15 years several candidate taste transduction elements have been identified in taste receptor cells. Validation of candidates as confirmed taste transduction elements requires their presence in taste cells along with the demonstration of their involvement, from *in vivo* and/or *in vitro* studies, in specific taste transduction pathways underlying particular taste qualities.

4.10.2.1 Salt Taste

Ingestion of salts, particularly NaCl, is critical for the maintenance of ionic homeostasis. The cation of the salt is the main component detected, and several cations, including Na^+, Li^+, K^+, Ca^{2+}, and NH_4^+ contribute to salty taste perception. When salts are present in sufficient concentration in the oral cavity, cations permeate ionic channels in the apical membrane of the taste cells, causing cationic influx and membrane depolarization. Monovalent cations also permeate the tight junctions at the apex of the taste bud, where they interact with the basolateral membrane of taste cells and generate a hyperpolarizing field potential (Ye, Q. *et al.*, 1991). This paracellular shunt pathway reduces the magnitude of the receptor potential and attenuates the perception of saltiness. The magnitude of the paracellular shunt is strongly influenced by the anion of the salt. Small anions such as chloride (Cl^-) diffuse readily into the paracellular space, where they blunt the hyperpolarizing field potential produced by the cations. In contrast, large anions such as gluconate and acetate fail to permeate the paracellular pathway. Thus, salts with large anions are perceived to be less salty than equivalent concentrations of salts with small anions (Elliott, E. J. and Simon, S. A., 1990; Ye, Q. et al., 1994).

4.10.2.1.1 Amiloride-sensitive versus amiloride-insensitive mechanisms

Salt taste transduction mechanisms vary in different species, depending on the ionic composition of the diet (Boudreau, J. C. et al., 1985). Herbivores and other mammals on a low Na^+ diet have a Na^+-specific appetite, mediated by highly Na^+-specific channels that are inhibited by the diuretic drug amiloride. Amiloride-sensitive Na^+ channels were originally proposed as a salt transduction mechanism by John DeSimone and colleagues, who observed that the chorda tympani response to NaCl in rats is sensitive to amiloride (Heck, G. L. et al., 1984). Whole cell patch-clamp studies have confirmed the presence of amiloride-sensitive Na^+ currents in hamster (Gilbertson, T. A. et al., 1992; 1993; Gilbertson, T. A. and Fontenot, D. T., 1998), rat (Doolin, R. E. and Gilbertson, T. A., 1996), and mouse (Miyamoto, T. et al., 2001) fungiform taste cells. However, other mechanisms for salt transduction exist, because about 20% of the chorda tympani response and the entire glossopharyngeal response to NaCl is amiloride-insensitive in rodents (Ninomiya, Y. et al., 1989; Gannon, K. S. and Contreras, R. J., 1995; Ninomiya, Y., 1998). Amiloride-insensitive mechanisms appear to predominate in species with adequate

dietary Na^+ levels, including humans, although both mechanisms exist to some extent in most species. While amiloride-sensitive mechanisms are highly Na^+ selective, amiloride-insensitive mechanisms do not discriminate between cations. Indeed, when the chorda tympani nerve is transected, or when amiloride is present, rats lose the ability to discriminate Na^+ from K^+ in behavioral discrimination assays (Spector, A. C. et al., 1996). Amiloride-sensitive and -insensitive mechanisms are likely distributed in different taste cells, because NaCl responses in single chorda tympani nerve fibers are either inhibited completely by amiloride, or unaffected by amiloride (Figure 2). Fibers that are inhibited by amiloride are narrowly tuned to NaCl (i.e., N fibers), while amiloride-insensitive fibers are broadly tuned to several salts as well as acids (i.e., H fibers; for a recent review, see Spector, A. C. and Travers, S. P., 2005).

4.10.2.1.2 ENaC – the amiloride-sensitive salt taste receptor

The epithelial Na^+ channel ENaC is now believed to be the Na^+-specific, amiloride-sensitive salt taste receptor. ENaC was first cloned from colon epithelium, but is expressed widely in many Na^+-transporting epithelial tissues (for review, see Garty, H. and Palmer, L. G., 1997). The functional channel is composed of three homologous subunits, α, β, and γ. Expression studies in heterologous cells suggest that all three subunits are required for normal function (Canessa, C. M. et al., 1994). In taste cells,

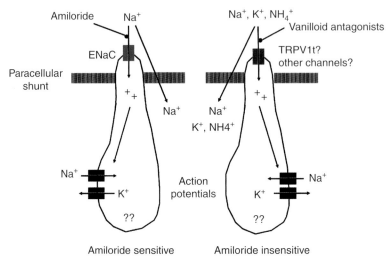

Figure 2 Salt transduction. The epithelial Na^+ channel ENaC is the amiloride-sensitive, Na^+ specific salt taste receptor. Other apical channels, including a variant of the capsaicin receptor TRPV1, mediate the amiloride-insensitive salt response, which is non-selective for cations. These mechanisms appear to be expressed in different populations of taste cells, but in neither case is it clear which taste cell type (II or III) mediates the response.

molecular and immunocytochemical studies indicate that α-, β-, and γ-ENaC are expressed in a large fraction of rat taste cells in all papilla. However, the density of expression of β- and γ-ENaC is significantly lower in foliate and circumvallate taste cells compared to fungiform taste cells (Kretz, O. *et al.*, 1999; Lin, W. *et al.*, 1999). This differential expression is consistent with the presence of functional amiloride-sensitive Na$^+$ currents in fungiform, but not in vallate or foliate taste cells (Doolin, R. E. and Gilbertson, T. A., 1996). ENaC in taste cells shares electrophysiological and pharmacological properties with ENaC in other tissues, including a high sensitivity to amiloride, high selectivity for Na$^+$ over K$^+$, significant permeability to protons, and self-inhibition by extracellular Na$^+$ (Gilbertson, T. A. *et al.*, 1993; Doolin, R. E. and Gilbertson, T. A., 1996; Gilbertson, T. A. and Zhang, H., 1998). Further, the same hormones that regulate Na$^+$ transport in epithelial cells regulate ENaC expression in taste cells. Both aldosterone and vasopressin increase the number of functional channels in the apical membrane (Herness, M. S., 1992; Gilbertson, T. A. *et al.*, 1993; Lin, W. *et al.*, 1999), while atrial natriuretic peptide (ANP) and oxytocin decrease them (Lindemann, B. *et al.*, 1999). Vasopressin acts via the V$_2$ receptor to increase the density of existing channels, probably via cAMP-dependent phosphorylation, while aldosterone likely acts to increase transcription of the channel, as it does in other tissues. Interestingly, aldosterone treatment of rats causes upregulation of both β- and γ-ENaC in taste cells, resulting in a larger density of amiloride-sensitive currents in fungiform taste cells. In vallate and foliate taste cells, which are normally unresponsive to amiloride, aldosterone treatment induces a small amiloride-sensitive Na$^+$ current in about half of the taste cells (Lin, W. *et al.*, 1999).

4.10.2.1.3 *Amiloride-insensitive mechanisms*

In contrast to the amiloride-sensitive mechanism, much less is known about the basis of amiloride-insensitive salt mechanisms, and multiple mechanisms are likely to contribute. The amiloride-insensitive mechanism was originally believed to be solely a consequence of the paracellular shunt pathway. Na$^+$ and other monovalent cations penetrate the tight junctions and come into contract with the basolateral membrane of taste cells. However, amiloride is impermeable and therefore inaccessible to basolateral Na$^+$ channels. Although ENaC channels are restricted to the apical

membrane in most transporting epithelia, some studies have suggested basolateral amiloride-sensitive Na$^+$ channels are present in taste cells. Ussing chamber studies in rat showed that amiloride inhibited a Na$^+$ current in the basolateral regions of taste epithelia, but these channels were less sensitive to amiloride than were apical channels ($K_i = 52\,\mu M$ compared to $0.2\,\mu M$), and less highly selective for Na$^+$ (Mierson, S. *et al.*, 1996). Whether this mechanism contributes significantly to the amiloride-insensitive Na$^+$ response is not clear. The concentration of Na$^+$ in the extracellular space near taste cells is at least 140 mM, and it is difficult to conceive how salt taste stimuli could achieve sufficient concentrations in the paracellular space to create a driving force for Na$^+$ to enter the cells.

Recent studies suggest that, in addition to the paracellular pathway, apical nonselective cation channels may contribute to the amiloride-insensitive mechanism. Evidence for apical nonselective cation channels was first obtained from whole cell recordings in polarized mouse taste cells, where apical NaCl stimulation was shown to elicit both amiloride-sensitive and amiloride-insensitive inward currents (Miyamoto, T. *et al.*, 2001). Further evidence for apical nonselective cation channels was obtained by the use of a novel pharmacological probe, cetylpyridinium chloride (CPC). When applied to the rat tongue, CPC selectively modulates the amiloride-insensitive component of the chorda tympani response to NaCl as well as responses to KCl and NH$_4$Cl, suggesting that the conductance may be a nonselective cation channel (Lyall, V. *et al.*, 2001). CPC is chemically related to the vanilloid family of modulators, which suggested that the channel may be related to the vanilloid family of TRP channels, the TRPVs. Indeed, a variant of TRPV1, called TRPV1t, is expressed in fungiform taste receptor cells (Lyall, V. *et al.*, 2004b). Several lines of evidence suggested a role of TRPV1t in the amiloride-insensitive salt response. First, other vanilloid modulators, including resiniferatoxin and capsaicin, selectively modulate the amiloride-insensitive chorda tympani response, and the response is temperature-sensitive, as is characteristic of TRPV channels. Further, blockers of TRPV1, including capsazepine and SB-366791, inhibit the amiloride-insensitive response. Finally, TRPV1 knockout mice appear to be lack the amiloride-insensitive chorda tympani response to NaCl (Lyall, V. *et al.*, 2004b). However, a more detailed analysis of chorda tympani responses in the knockout mice revealed that although the tonic portion of the amiloride-insensitive salt response is missing, some phasic response remains in the knockout (Treesukosol, Y. *et al.*, 2007). Further, behavioral studies revealed nearly identical detection thresholds for NaCl in knockout and wild-type mice, even in the presence of amiloride (Ruiz, C. *et al.*, 2006; Treesukosol, Y. *et al.*, 2007). Two bottle preference tests showed that TRPV1 knockout mice prefer NaCl at concentrations normally avoided by wild-type mice, suggesting that TRPV1t may play a role in the avoidance of high concentrations of salt (Ruiz, C. *et al.*, 2006), but this must be confirmed with brief access tests to rule out post ingestive effects. What these behavioral data imply is that additional mechanisms must contribute to the amiloride-insensitive salt response. Further studies are needed, particularly on circumvallate and foliate taste buds, where the amiloride-insensitive response predominates.

4.10.2.2 Sour Taste

Acidic stimuli elicit sour taste, a naturally aversive taste quality that likely evolved to prevent ingestion of strong acids, spoiled foods, and unripe fruit. As is the case with salt transduction, sour taste transduction mechanisms likely involve both apical and paracellular pathways, and mechanisms vary considerably in different species. Generally, however, weak organic acids produce larger nerve responses and taste more sour than strong highly dissociated acids at equivalent pH (for recent review, see DeSimone, J. A. and Lyall, V., 2006). Thus, extracellular protons are unlikely to be the effective stimulus in sour transduction.

4.10.2.2.1 Intracellular versus extracellular pH as a stimulus

A preponderance of data suggests that intracellular pH (pH_i), is the proximate stimulus mediating sour taste transduction. Similar concentrations of buffered and unbuffered acids produce equivalent chorda tympani nerve responses and equivalent decreases in pH_i, even though their pH_o levels are vastly different. Direct evidence for the role of intracellular pH in sour transduction was obtained recently by exposing the rat tongue to nigericin, a K^+-H^+ exchanger that equilibrates pH_i with pH_o when high extracellular K^+ is present. Under these conditions, intracellular pH can be controlled rather precisely simply by varying the pH of extracellular solution. In the presence of nigericin and K^+, the threshold for eliciting a chorda tympani response decreases from pH 4.5 to pH 6.5 (Lyall, V. *et al.*, 2007). It is proposed that neutral acids,

such as citric or acetic acid readily permeate the plasma membrane of taste cells in their undissociated form. Once inside the cell they dissociate, resulting in a decreased pH_i (Lyall, V. *et al.*, 2001). CO_2 also readily crosses the membrane, where it is converted intracellularly to H_2CO_3 by carbonic anhydrase. Strong acids also decrease intracellular pH, likely by permeation of proton channels, but lower pH values are required for equivalent effects. In addition to proton permeable channels, several proton-modulated ion channels are expressed in taste cells and these may also contribute to sour transduction, by depolarizing taste cells and activating voltage-gated Ca^{2+} channels. The identity of channels proposed to be involved in sour taste is summarized below.

4.10.2.2.2 Proton-permeable ion channels

In mice and rats, evidence exists for two types of apical proton channels. The chorda tympani response to HCl, but not to weak acids or CO_2, is decreased by several proton channel blockers, including Zn^{2+}, Cd^{2+} and diethylpolycarbonate (DEPC; DeSimone, J. A. *et al.*, 2007). The pharmacology of the response suggests that the proton influx may occur via the NADPH oxidase-linked proton channel, gp91^{phox}. This hypothesis was recently tested with gp91^{phox} knockout mice, in which approximately 60% of the chorda tympani response to HCl was diminished. The remaining response was insensitive to specific modulators of gp91^{phox}, but was enhanced by membrane permeant analogs of cAMP (Lyall, V. *et al.*, 2002; DeSimone, J. A. *et al.*, 2007). The identity of the cAMP-modulated proton channel is not known.

The epithelial Na channel ENaC is also permeable to H^+ in the absence of extracellular Na^+. In hamsters, where salivary Na^+ is very low, ENaC appears to contribute to sour transduction (Gilbertson, T. A. *et al.*, 1992; 1993; Gilbertson, D. M. and Gilbertson, T. A., 1994). However, amiloride has no effect on acid responses in mice (Miyamoto, T. *et al.*, 2000; Richter, T. A. *et al.*, 2003) or rats (DeSimone, J. A. *et al.*, 1995), suggesting that this mechanism may be unique to hamsters.

4.10.2.2.3 Proton-gated ion channels

A number of proton-gated cation channels are expressed in taste cells and several have been suggested to serve as sour receptors or modulators. However, none is known definitively to contribute to sour transduction *in vivo*. Acid-sensing ion channels (ASICs), which are proton-gated cation channels that mediate pain perception accompanying tissue acidosis

(Waldmann, R. *et al.*, 1999), have been proposed as sour receptors. ASIC2a and 2b are co-expressed in many rat taste cells (Ugawa, S. *et al.*, 1998; 2003), and when heterologously expressed, the channels mediate a transient, amiloride-sensitive inward current in response to a decrease in extracellular pH. In a subset of isolated rat taste cells, pH 4.5 similarly elicits a transient inward current that is partially sensitive to high concentrations of amiloride (Lin, W. *et al.*, 2002). However, ASIC2 knockout mice have normal behavioral and electrophysiological responses to low pH, suggesting that the acid-gated inward current is likely mediated by a different mechanism, at least in mice (Kinnamon, S. C. *et al.*, 2000; Richter, T. A. *et al.*, 2004b). Again, species differences may be involved here, since much less ASIC2 mRNA is present in mouse taste cells compared to those of rat (Richter, T. A. *et al.*, 2004b).

The hyperpolarization-activated channels HCN1 and HCN4 are expressed in a subset of rat taste cells. These channels were suggested to serve as sour receptors, because, when expressed in heterologous cells, they are activated by low external pH (Stevens, D. R. *et al.*, 2001). However, Cs^+, a specific blocker of HCN channels, has no effect on acid-induced Ca^{2+} responses in mouse taste cells (Richter, T. A. *et al.*, 2003), suggesting that HCN channels may play other roles in the taste bud.

In polarized mouse taste cells, apically applied citric acid elicits an inward current that is sensitive to NPPB, a Cl^- channel blocker (Miyamoto, T. *et al.*, 1998). These data suggest that an acid-gated Cl^- channel may be involved in sour transduction. Two acid-sensitive Cl^- channels have been identified in taste cells, CIC-4 and its splice variant, CIC-4A (Huang, L. *et al.*, 2005). Whether either of these channels is directly involved in acid detection remains to be determined.

4.10.2.2.4 Proton-blocked K^+ channels

Most K^+ channels are inhibited by a decrease in pH_o. K^+ channel block could conceivably contribute to acid-induced depolarization, if the K^+ channels were localized to the apical membrane and conductive at the resting potential. In mudpuppy, voltage-gated K^+ channels are restricted to the apical membrane, where they are blocked directly by protons to depolarize the membrane (Kinnamon, S. C. *et al.*, 1988; Roper, S. D. and McBride, D. W., Jr., 1989). However, in mammals, several studies have indicated that K^+ channels appear to be absent from the apical membrane, at least in fungiform taste cells (Miyamoto, T. *et al.*, 1996; Furue, H. and Yoshii, K., 1997; Miyamoto, T. *et al.*, 1998). Thus, this mechanism is unlikely to contribute

to sour transduction in mammals. Basolateral K^+ channels could contribute to sour transduction, if they are inhibited by a decrease in intracellular pH. Several members of the two-pore domain K^+ channels (K_{2P} or KCNK) are expressed in rat taste cells, including TWIK 1, TASK 1-2, TREK 1-2, and TRAAK 1 (Lin, W. *et al.*, 2004; Richter, T. A. *et al.*, 2004a). Some of these channels can be modulated by intracellular acidification, however whether they are located in the taste cells that communicate sour information to the nervous system is unknown.

4.10.2.2.5 PKD1L3 and PKD2L1

Recently, two members of the TRPP family of transient receptor potential (TRP) channels, PKD1L3 and PKD2L1, were identified in taste cells by expression screening. These channels are expressed in a specific subset of taste cells, distinct from the type II taste cells involved in bitter, sweet, and umami transduction (Huang, A. L. *et al.*, 2006; Ishimaru, Y. *et al.*, 2006; LopezJimenez, N. D. *et al.*, 2006). When coexpressed in heterologous cells, PKD2L1 and PKD1L3 respond to low pH_o with an increase in intracellular Ca^{2+}. The increase in intracellular Ca^{2+} occurs with a delay, suggesting that intracellular, rather than extracellular pH may be mediating the effect (Ishimaru, Y. *et al.*, 2006). Recent immunocytochemical studies using antibodies directed against PKD2L1 show that although the channel is located in the apical third of the taste bud, most channels appear to be located basolateral to the tight junctions, where they would be more likely to respond to changes in intracellular pH. Further, the channel is expressed exclusively in the type III taste cells, those cells that express voltage-gated Ca^{2+} channels (Kataoka, S. *et al.*, 2007). These data agree well with Richter T. A. *et al.* (2003), who show that apically applied acids elicit a decrease in intracellular pH in most taste cells, but only those taste cells with voltage-gated Ca^{2+} channels show a concomitant increase in intracellular Ca^{2+} (Richter, T. A. *et al.*, 2003). These data suggest that the taste cells with voltage-gated Ca^{2+} channels are the cells that communicate sour taste information to the nervous system.

Further evidence for a role of PKD channels in sour taste came from molecular studies of the Zuker laboratory. Targeted expression of diphtheria toxin (DTA) was used to ablate those taste cells which expressed PKD2L1. These mice completely lack chorda tympani responses to all acids, both weak and strong, but have normal responses to other taste stimuli (Huang, A. L. *et al.*, 2006). These data suggest that the same taste cells transduce both strong and weak acids, and that the sour transducing taste cells are likely to be a unique subset of taste cells. However, PKD channel knockouts will be needed to determine if any of the PKD proteins are involved in sour transduction (Figure 3).

4.10.2.2.6 Downstream signaling effectors in sour taste

The DeSimone laboratory has combined single cell imaging with chorda tympani nerve recording to investigate sour transduction downstream of the decrease in intracellular pH (for review, DeSimone, J. A. and Lyall, V., 2006). They have found that the phasic part of the chorda tympani response to both strong and weak acids is mediated by different mechanisms than is the tonic portion of the response. The phasic part of the response is associated with a rapid, pH_i-mediated decrease in taste cell volume, caused by a change in the equilibrium between F-

actin and G-actin. This in turn activates a flufenamic acid-sensitive shrinkage-activated cation channel on the basolateral membrane. Blocking either the conversion of F-actin to G-actin, or the cation channel completely suppresses the phasic portion of the chorda tympani, with no effect on the tonic portion of the response (Lyall, V. *et al.*, 2006). In contrast, the tonic portion of the response appears to be Ca^{2+}-dependent. It is likely that the decrease in pH_i activates a cation channel, possibly PKD2L1, which results in membrane depolarization and activation of voltage-gated Ca^{2+} channels, mediating transmitter release. The increase in intracellular Ca^{2+} not only mediates the tonic part of the chorda tympani response to acids but also causes adaptation to acids, by activating a basolateral Na^+-H^+ exchanger, NHE1 (Lyall, V. *et al.*, 2004a; Vinnikova, A. K. *et al.*, 2004). This exchanger likely attenuates the decrease in pH_i, which would cause adaptation of the inward induced by acid stimulation. Indeed, acid stimulation induces a rapidly adapting current in taste cells (Lin, W. *et al.*, 2002), but further studies will be required to determine if the adaptation phase is dependent on NHE1.

4.10.2.3 Transduction Mechanisms Common to Bitter, Sweet, and Umami Taste Qualities

Although different GPCRs have been shown to underlie the initial detection of bitter, sweet and umami tastants (see Chapter Taste Receptors), there are several transduction elements that are common to pathways underlying all three taste qualities. As is typical of signaling pathways in sensory and nonsensory systems, specific GPCRs couple through their associated G proteins to initiate signaling cascades involving effector enzymes, second messengers and second messenger-activated/inhibited ion channels and other end targets.

4.10.2.3.1 G proteins

The first taste signaling element identified was α-gustducin, an α-transducin-like G protein α-subunit selectively expressed in $\sim$25% of taste cells, specifically of type II (McLaughlin, S. K. *et al.*, 1992; Boughter, J. D., Jr. *et al.*, 1997). In *in vitro* assays gustducin heterotrimers couple to isolated native bitter receptors (Ruiz-Avila, L. *et al.*, 2000) and heterologously expressed bitter-responsive T2R receptors (Chandrashekar, J. *et al.*, 2000). Heterotrimers containing α-gustducin chimeras

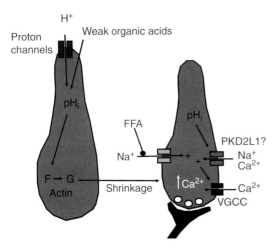

Figure 3 Sour transduction. Proton channels mediate the response to strong acids, while weak organic acids permeate the cell membrane in the undissociated state. Both mechanisms are present in the same taste cells, and both lead to decreases in intracellular pH, which is the proximate stimulus for sour transduction. The decrease in pHi leads to a change in the equilibrium between F and G actin, causing the taste cell to shrink in volume. The shrinkage activates a flufenamic-sensitive cation channel, which elicits the phasic portion of the afferent nerve response. The tonic portion of the nerve response is also mediated by the decrease in pHi, which activates a proton-gated non-selective cation channel, possibly PKD2L1. This causes membrane depolarization and activation of voltage-gated Ca^{2+} channels, presumably in Type III taste cells, since these are the only taste cells that express voltage gated Ca^{2+} channels.

capable of promoting calcium mobilization (to monitor receptor activity) also couple to heterologously expressed sweet-responsive T1R2 + T1R3 receptors and umami-responsive T1R1 + T1R3 receptors (Nelson, G. et al., 2001; Li, X. et al., 2002; Nelson, G. et al., 2002). Studies of α-gustducin knockout mice have demonstrated gustducin's key role in vivo in bitter, sweet and umami taste transduction (Wong, G. T. et al., 1996; Ming, D. et al., 1998; Ruiz-Avila, L. et al., 2001; Ruiz, C. J. et al., 2003; Glendinning, J. I. et al., 2005; Danilova, V. et al., 2006). That α-gustducin knockout mice have markedly reduced, but not entirely absent, behavioral and nerve responses to the bitter, sweet and umami compounds suggests that other taste cell-expressed G proteins may couple to these GPCRs normally or in the absence of gustducin. Consistent with this inference is the observation that transgenic expression of a dominant-negative version of α-gustducin in the α-gustducin knockout background further diminished taste responses to bitter and sweet (Ruiz-Avila, L. et al., 2001).

In addition to α-gustducin the following G α-subunits are known to be expressed in taste cells: Gαs, Gαq, Gα14, Gα15, Gαi2, Gαi3, α-transducin (McLaughlin, S. K. et al., 1992; Ruiz-Avila, L. et al., 1995; Boughter, J. D., Jr. et al., 1997; Kusakabe, Y. et al., 1998). Heterologously expressed mT2R5 selectively coupled to α-gustducin over the other G proteins examined (αi, αs, αo, or αq) (Chandrashekar, J. et al., 2000). Although α-transducin knockouts had no effect on taste responses, double knockouts lacking α-transducin and α-gustducin showed that α-transducin is involved in umami taste, but not in bitter or sweet (He, W. et al., 2004).

Upon activation, heterotrimeric G proteins such as gustducin dissociate into α and $\beta\gamma$ components, each capable of carrying a signal to one or more downstream effectors. Gβ1, Gβ3, and Gγ13 are expressed with α-gustducin and Gαi2 in type II taste cells (Huang, L. et al., 1999; Clapp, T. R. et al., 2001). α-Gustducin expressing taste cells are a subset of those which express Gγ13 and Gαi2 (Huang, L. et al., 1999; Adler, E. et al., 2000; Matsunami, H. et al., 2000; Clapp, T. R. et al., 2001). The T2R bitter receptors are only expressed in a subset of gustducin-positive type II taste cells (Adler, E. et al., 2000; Matsunami, H. et al., 2000; Kim, M. R. et al., 2003) and likely couple preferentially to gustducin over any other G proteins in those same taste cells. Marked regional differences in coexpression of gustducin and taste receptors have been noted: in posterior fields (i.e., circumvallate papillae) gustducin was typically found along with T2Rs, while in anterior fields (fungiform papillae and palatal taste buds) gustducin is more often seen coexpressed with T1Rs (Huang, L. et al., 1999; Adler, E. et al., 2000; Matsunami, H. et al., 2000; Clapp, T. R. et al., 2001; Stone, L. M. et al., 2007). This pattern is consistent with gustducin's involvement in responses to bitter, sweet, and umami taste stimuli.

4.10.2.3.2 Effector enzymes

Phospholipase β2 (PLCβ2), phospholipase A2, cNMP phosphodiesterase (PDE), and adenylyl cyclase (AC) type 8 have been identified in type II taste cells as candidate effector enzymes to mediate G protein-dependent taste responses (Ruiz-Avila, L. et al., 1995; Rossler, P. et al., 1998; Oike, H. et al., 2006a; Trubey, K. R. et al., 2006). α-Gustducin and α-transducin activate retinal and taste PDEs (Ruiz-Avila, L. et al., 1995), whereas gustducin's $\beta\gamma$ complex activates taste PLCβ2 (Huang, L. et al., 1999; Matsunami, H. et al., 2000; Yan, W. et al., 2001). In in vitro assays antibodies against α-gustducin, but not those against gustducin's β subunits, blocked stimulation of taste tissue PDE activity by native bitter taste receptors (Yan, W. et al., 2001). Conversely, antibodies against Gβ3 or Gγ13 blocked the IP$_3$ response of taste tissue receptors to bitters, but had no effect on cNMP levels (Huang, L. et al., 1999; Yan, W. et al., 2001). Thus, heterotrimeric gustducin interacts with two different effectors to effect two responses in taste cells: 1. hydrolysis of cNMPs via α-subunit activation of PDE, and 2. generation of IP$_3$ and DAG via $\beta\gamma$ activation of PLCβ2. The subsequent steps in the PLCβ2 pathway are activation of IP$_3$ receptors (type III IP$_3$ receptors are selectively expressed in the same type II taste cells that express PLCβ2 (Clapp, T. R. et al., 2001) and release of Ca^{2+} from internal stores followed by activation of the calcium-activated cation channel Trpm5 (see below). Arachidonic acid generated by taste cell-expressed phospholipase A2 (Oike, H. et al., 2006a) may activate Trpm5 (Oike, H. et al., 2006b). The subsequent steps in the cNMP pathway are presently unclear: decreased cNMPs may act on protein kinases which in turn may regulate taste cell ion channel activity, or cNMP levels may regulate directly the activity of cNMP-gated and cNMP-inhibited ion channels expressed in taste cells (Kolesnikov, S. S. and Margolskee, R. F., 1995; Misaka, T. et al., 1997; 1998; 1999).

Both PLCβ2 and AC have been implicated in sweet taste transduction (reviewed in (Kinnamon, S.

C. and Margolskee, R. F., 1996; Lindemann, B., 1996; Herness, M. S. and Gilbertson, T. A., 1999). In both biochemical and physiological studies taste responses to sugars differ from those to artificial sweeteners. In biochemical studies with taste tissue, sucrose elicits an elevation of cAMP, while certain artificial sweeteners lead to the generation of IP$_3$ (reviewed in Kinnamon, S. C. and Margolskee, R. F., 1996; Lindemann, B., 1996; Herness, M. S. and Gilbertson, T. A., 1999). In physiological studies with rat taste cells, sucrose led to Ca^{2+} influx, while the artificial sweeteners saccharin and SC45647 elevated Ca^{2+} via release from internal stores (Bernhardt, S. J. et al., 1996). Although a single heterodimeric sweet receptor (T1R2 + T1R3) mediates responses to both natural and artificial sweeteners (see Chapter Taste Receptors), it appears that activation by different types of sweet compounds elicits differential activation of downstream effectors, presumably by promoting activation of one or another G protein coupled to the sweet receptor.

4.10.2.3.3 PLCβ2 knockout mice

PLCβ2 knockout mice were reported to lack all nerve and behavioral responses to bitter, sweet and umami compounds (Zhang, Y. et al., 2003; Damak, S. et al., 2006). In a subsequent behavioral study it was shown that these knockout mice displayed aversive responses to high concentrations of the bitter compounds quinine and denatonium (Dotson, C. D. et al., 2005). Although

the PLCβ2 knockout mice showed no appetitive lick responses to sucrose at even the highest concentrations, their aversive responses to high concentrations of bitters were shifted to about a tenfold higher threshold (Dotson, C. D. et al., 2005). This indicates that at least for bitters there are PLCβ2-independent mechanisms that come to bear at high concentrations.

4.10.2.3.4 Target channels

Trpm5 is a transient receptor potential ion channel subtype that is selectively expressed in type II taste cells along with α-gustducin, Gγ13 and PLCβ2 (Clapp, T. R. et al., 2001; Perez, C. A. et al., 2002). Trpm5 is cation selective and in heterologous expression systems can be activated by IP$_3$ or intracellular Ca^{2+} (Hofmann, T. et al., 2003; Liu, D. and Liman, E. R., 2003; Prawitt, D. et al., 2003) or arachidonic acid (Oike, H. et al., 2006b). Recent studies with caged IP$_3$ and caged Ca^{2+} show that native Trpm5 in excised taste cell patches is gated directly by Ca^{2+} (Zhang, Y. et al., 2007). Similarly to PLCβ2 knockout mice, Trpm5 knockout mice have pronounced deficits in their nerve and behavioral responses to bitter, sweet and umami compounds (Zhang, Y. et al., 2003; Damak, S. et al., 2006). However, Trpm5-independent transduction mechanisms contributing to bitter, sweet and umami were revealed by the reduced but not absent nerve and behavioral responses observed with Trpm5 knockout mice (Damak, S. et al., 2006; Figure 4). It was noted that the effects of knocking

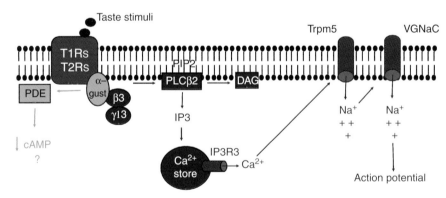

Figure 4 Transduction of bitter, sweet, and umami taste stimuli. Although different receptors are involved (T2Rs for bitter and T1Rs for sweet and umami stimuli), all utilize similar downstream signaling effectors in Type II taste cells. Receptor binding leads to activation of a heterotrimeric G protein, gustducin. The α subunit activates phosphodiesterase (PDE) to decrease intracellular cAMP, while the $\beta\gamma$ partners stimulate phospholipase Cβ2 (PLCβ2) to produce the second messengers IP3 and diacylglycerol (DAG). The IP3 binds to the Type III IP3 receptor (IP3R3), causing release of Ca^{2+} from intracellular stores and subsequent Ca^{2+}-dependent activation of a monovalent cation channel, TRPM5. This causes membrane depolarization, activation of voltage-gated Na$^+$ channels (VGNaC), and action potentials, which are believed to be required for transmitter release. Knockout of PLCβ2, TRPM5, or α-gustducin compromises responses to bitter, sweet, and umami taste stimuli. The role of the decrease in intracellular cAMP is not known, but it may regulate phosphorylation of downstream signaling effectors or target cyclic nucleotide modulated ion channels.

out Trpm5 varied with the taste field and type of taste quality examined: with bitter compounds, glossopharyngeal nerve responses were greatly diminished while chorda tympani nerve responses were normal; with umami compounds, glossopharyngeal nerve responses were intact while chorda tympani responses were greatly diminished. That target channels in addition to Trpm5 may exist in taste cells comes from the detection of another Ca^{2+}-activated cation conductance in excised taste cell patches from Trpm5 knockout mice (Zhang, Y. *et al.*, 2007). Thus Trpm5 is the primary target channel for the taste cell phosphoinositide signaling pathway: activation of $PLC\beta2$ by $G\gamma13$ would generate IP_3 to activate IP_3RIII, released intracellular Ca^{2+} would activate Trpm5 leading to cation influx and taste cell depolarization. In addition there may be other targets of this pathway as well as other target channels in other pathways.

4.10.2.4 Transduction Elements Specific to Particular Taste Qualities

Although several elements are common to the transduction of bitter, sweet, and umami taste qualities, the only transduction element, aside from the receptors themselves, that is involved in one but not all three taste qualities is α-transducin. As noted above, α-transducin knockouts had a specific effect on umami responses, however, this was only revealed in the α-gustducin knockout background – single knockouts lacking only α-transducin had no effect on bitter sweet or umami (He, W. *et al.*, 2004). That the closely related G protein α-subunits, α-transducin and α-gustducin are not interchangeable in taste signaling comes from studies in which α-transducin was added back as a transgene in α-gustducin knockout mice (He, W. *et al.*, 2002).

That umami transduction might employ unique transduction elements not in common with those of the other GPCR-dependent taste qualities might stem from the multiplicity of receptors and channels implicated in umami. As is further discussed in Chapter Taste Receptors both GPCRs and ion channels have been implicated as umami receptors. Amongst the GPCRs implicated in umami responses are ionotropic glutamate receptors (Lin, W. and Kinnamon, S. C., 1999; Caicedo, A. *et al.*, 2000b), metabotropic glutamate receptors (mGluRs) (Monastyrskaia, K. *et al.*, 1999; Chaudhari, N. *et al.*, 2000), and the heterodimer of T1R1 and T1R3 (Nelson, G. *et al.*, 2001; Li, X. *et al.*, 2002).

4.10.3 Other Transduction Mechanisms

4.10.3.1 Fat Taste

Preference for fatty foods is ubiquitous among mammals. This is not surprising, since lipids represent a significant source of energy and have appealing orosensory qualities (for recent review, see Laugerette, F. *et al.*, 2007). However, whether fat is actually detected as a taste stimulus has been controversial. If fat is detected as a taste stimulus, long chain fatty acids appear to mediate the effect. Rodents prefer long chain fatty acids to either triglycerides or medium-chain fatty acids in preference assays (Fukuwatari, T. *et al.*, 1997) and conditioned taste aversion tests show that rats can specifically detect long chain fatty acids (McCormack, D. N. *et al.*, 2006). Long chain fatty acids are produced from triglycerides by the action of lingual lipase, an enzyme that has been identified in the secretion of von Ebner's glands, which bathe the circumvallate taste buds. In mice, inhibition of this enzyme reduced preference for triglycerides, but not for free fatty acids (Kawai, T. and Fushiki, T., 2003).

4.10.3.1.1 Fatty acid modulation of taste cells

The first evidence for a direct effect of fatty acids was obtained from whole-cell recordings showing fatty acids block delayed rectifier K^+ channels in rat taste cells (Gilbertson, T. A. *et al.*, 1997). These channels are blocked directly by long chain polyunsaturated fatty acids in fungiform taste buds, and by monounsaturated fatty acids in vallate and foliate taste buds. The fatty acid inhibition of K^+ channels depolarizes taste cells and is hypothesized to modulate sensitivity to conventional taste stimuli. In support of this model, low concentrations of linoleic acid augmented preference for subthreshold concentrations of saccharin in two bottle preference tests (Gilbertson, T. A. *et al.*, 2005). Further, licking rates of rats to suprathreshold concentrations of prototypical sweet, salty, sour, and bitter compounds were altered by addition of linoleic and oleic acids to the taste stimuli. Licking rates to sweet stimuli were augmented by the fatty acids, while licking rates to bitter, sour, and salty stimuli were decreased (Pittman, D. W. *et al.*, 2006). Similar psychophysical studies in humans, however, have failed to show a significant effect of fatty acids on either detection thresholds or intensity ratings of taste stimuli (Mattes, R. D., 2007), suggesting fatty acid modulation of taste may be less important in humans.

4.10.3.1.2 Fatty acid taste receptors

Evidence was obtained recently for a fatty acid transporter that may serve as a specific taste receptor for fat (Fukuwatari, T. *et al.*, 1997; Laugerette, F. *et al.*, 2005). CD36, a lipid-binding protein, is expressed in the apical region of taste cells in both mice and rats. It is highly expressed in circumvallate papillae, but less so in foliate and fungiform papillae. The role of CD36 as a fatty acid transducer is supported by its predicted structure. The protein contains a large extracellular hydrophobic pocket and a C terminal sequence that associates with Src kinases, suggesting a role in intracellular signaling (Laugerette, F. *et al.*, 2005). More definitive evidence for its role in fat sensing came from studies in CD36 knockout mice. These mice lack preference for long chain fatty acids, but have normal preference for other taste stimuli. Further, these mice lack a lipid-induced cephalic phase response normally present in wild-type mice (Laugerette, F. *et al.*, 2005). These data provide the strongest evidence to date that fat may be the sixth primary taste. However, further work is necessary to determine how CD36 activation is converted into a taste signal, which taste cells detect fat, and how fat information is encoded in the nervous system with respect to other taste qualities. Also, it will be important to determine if CD36 is expressed in human taste cells, where it could serve as a substrate to modulate fat intake.

Some evidence also exists for GPCRs for fat. GPR 120 is a fatty acid receptor expressed in enteroendocrine cells of the colon. mRNA for GPR 120 was recently found in circumvallate taste buds, and immunocytochemistry showed intense labeling in the apical region of a subset of rat taste cells (Matsumura, S. *et al.*, 2007). Thus, its expression pattern is consistent with a potential role in fat sensing, but definitive evidence must await knockout of the gene.

4.10.3.2 Water Taste

Although water is not classically considered to have a taste, it is well established that gustatory nerve fibers respond to water applied to the tongue (for review, see Gilbertson, T. A. *et al.*, 2006). The mechanisms involved in the water response are only now beginning to be understood. Gilbertson T. A. (2002) showed that hypo-osmotic solutions lead to an increase in cell capacitance accompanied by activation of a Cl^- conductance. This conductance is similar physiologically and pharmacologically to the well studied swelling-activated Cl^- current observed in other cell types (Jentsch, T. J., 1996). Such a current likely plays a role in regulating taste cell volume in response to varying osmotic taste stimuli. Recently, the water-permeable ion channels involved in the hypo-osmotic swelling were identified (Watson, K. J. *et al.*, 2007). Several aquaporins (AQPs) are expressed in taste cells, including AQP1 and AQP2, which label predominately the basolateral membrane, and AQP5, which labels both apical and basolateral membranes. Whole-cell recording showed that a specific AQP inhibitor significantly reduces hypo-osmotic-induced Cl^- currents in taste cells, suggesting that AQPs are the source of the water influx. Further studies are needed to determine how this leads to the well-characterized water response in afferent nerve fibers.

4.10.4 Signaling to Afferent Nerve Fibers and Adjacent Taste Cells

Taste cells communicate sensory information to afferent nerve fibers as well as to adjacent cells within taste buds (for review, see Herness, S. *et al.*, 2005; Roper, S. D., 2006). Many transmitters have been proposed to contribute to chemical communication in the taste bud, including serotonin, adenosine triphosphate (ATP), glutamate, acetylcholine, norepinephrine, gamma-aminobutyric acid (GABA), and various peptides. Indeed, evidence exists for expression of each of these neuroactive compounds or their receptors in taste buds, but to date only two, ATP and serotonin, have met all the criteria required to be considered as a taste cell transmitter or modulator. By analogy to the nervous system, these criteria include: (1) presence in the presynaptic cell, (2) release upon stimulation, (3) activation of the post-synaptic cell, usually through specific receptors, and (4) mechanism for termination of transmitter action.

4.10.4.1 Adenosine Triphosphate as a Transmitter

ATP is now believed to be the primary transmitter that links taste cells to activation of gustatory afferent fibers (Finger, T. E. *et al.*, 2005). The first clue that ATP might serve as a taste cell transmitter was the observation that the ionotropic purinergic receptors P2X2 and P2X3 are abundantly expressed on gustatory afferent terminals (Bo, X. *et al.*, 1999). Definitive evidence for a role of P2X2 and P2X3 was obtained from P2X2/P2X3 double knockout mice, which lack behavioral and afferent nerve responses to most taste stimuli (Finger, T. E. *et al.*, 2005). Further, taste stimuli evoke release of ATP from taste cells (Finger, T. E. *et al.*, 2005; Huang, L. *et al.*, 2007; Romanov, R. A. *et al.*, 2007), and the released ATP is likely degraded by NTPDase2, an ecto ATPase abundantly expressed on the membranes of type I (support) taste cells (Bartel, D. L. *et al.*, 2006). Thus, ATP appears to fit all the criteria required to serve as a transmitter in the taste bud. The released ATP also binds to G-protein-coupled P2Y receptors on taste cell membranes (Kataoka, S. *et al.*, 2004; Bystrova, M. F. *et al.*, 2006; Huang, L. *et al.*, 2007), suggesting that ATP may also serve as an autocrine or paracrine modulator in the taste bud. Recent studies using sensitive biosensor cells expressing purinergic receptors have shown that either taste stimuli (Huang, L. *et al.*, 2007) or membrane depolarization (Romanov, R. A. *et al.*, 2007) can evoke ATP release from single type II taste cells, likely via gap junction hemichannels composed of pannexin 1 or connexin 43. Release can be evoked in the absence of Ca^{2+} (Romanov, R. A. *et al.*, 2007), which may explain the curious observation that although type II taste cells generate action potentials, they lack voltage-gated Ca^{2+} channels and conventional synapses with the nervous system (Clapp, T. R. *et al.*, 2004; 2006; DeFazio, R. A. *et al.*, 2006).

4.10.4.2 Serotonin and Other Neuroactive Modulators

Serotonin (5-HT) was originally believed to be the primary taste cell transmitter, since it is widely expressed in taste buds of a variety of species from fish to mammals (Kim, D. J. and Roper, S. D., 1995). In nonmammalian vertebrates serotonin is expressed in Merkel-like basal cells, where it appears to serve as a paracrine modulator in the taste bud (Delay, R. J. *et al.*, 1997). In mammalian taste cells, it is expressed in a subpopulation of type III taste cells (Yee, C. L. *et al.*, 2001). Recent studies show that serotonin is released from single type III cells in response to membrane depolarization and from intact taste buds in response to bitter taste stimuli, presumably in

response to ATP released from type II taste cells (Huang, L. *et al.*, 2007). Enzymes for serotonin biosynthesis (DeFazio, R. A. *et al.*, 2006; Ortiz-Alvarado, R. *et al.*, 2006; Seta, Y. *et al.*, 2007) and reuptake (Ren, Y. *et al.*, 1999) are expressed in taste buds, as well as specific serotonin receptors. Both $5\text{-}HT_{1A}$ and $5\text{-}HT_{3}$ were amplified by RT-PCR from RNA extracts of taste buds (Kaya, N. *et al.*, 2004). Immunocytochemical labeling showed that $5\text{-}HT_{1A}$ is expressed on the membranes of nonserotonergic taste cells, where it likely mediates the effects of serotonin as a paracrine neuromodulator (Figure 5). $5\text{-}HT_{1A}$ receptor agonists modulate taste cells by inhibiting calcium-activated K^{+} currents and voltage-dependent Na^{+} currents, suggesting that serotonin affects the membrane excitability of taste cells (Herness, M. S. and Chen, Y., 1997; Herness, S. and Chen, Y., 2000). In contrast, taste cell membranes lack expression of $5\text{-}HT_{3}$ receptors, suggesting it may be present on afferent nerve terminals (Kaya,

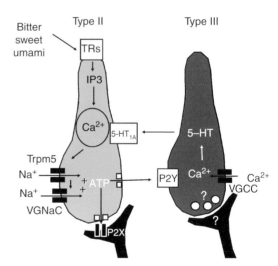

Figure 5 Proposed role of ATP and serotonin in afferent signaling. Bitter, sweet, and umami taste stimuli evoke ATP release from Type II taste cells. The ATP subsequently binds to the ionotropic purinergic receptors P2X2 and P2X3, which are abundantly expressed on afferent nerve fibers. Recent data suggest that the release of ATP is non-vesicular, involving voltage- and Ca^{2+}-gated gap junction hemichannels. ATP also binds to G protein-coupled purinergic receptors (P2Y) on the membranes of Type III taste cells, causing them to release serotonin (5-HT). The 5-HT likely binds to $5\text{-}HT_{1A}$ receptors on Type II taste cells, modulating their sensitivity to sweet and bitter taste stimuli. A major unanswered question is the identity and role of the transmitter at the prominent synapses between Type III taste cells and nerve fibers, but it is not believed to be 5-HT, because 5-HT3 knockout mice have normal behavioral responses to taste stimuli.

N. *et al.*, 2004). However, mice lacking 5-HT$_3$ receptors have normal behavioral responses to taste stimuli (Finger, T. E. *et al.*, 2005), indicating that serotonin is not likely to be involved in direct activation of afferent fibers. Recent behavioral data suggest a role of serotonin in modulation of taste thresholds in humans. Serotonin reuptake inhibitors, which increase the life time of serotonin in the synaptic cleft, enhance sensitivity to sweet and bitter taste stimuli in healthy adult human subjects (Heath, T. P. *et al.*, 2006). Further studies will be required to determine if 5-HT$_{1A}$ receptors play a role in this modulation.

Norepinephrine (noradrenaline) and its receptors are expressed in subsets of taste cells, suggesting a role in cell-to-cell communication. α- and β-adrenoceptor agonists inhibit outward K$^+$ currents in taste cells, and multiple α- and β-adrenoceptors are expressed in lingual epithelium (Herness, S. *et al.*, 2002). However, which taste cell types are involved, and whether norepinephrine is an autocrine or paracrine modulator has not been determined. As is the case with serotonin, norepinephrine reuptake inhibitors enhance taste sensitivity in humans, suggesting a role for norepinephrine in modulating human taste thresholds (Heath, T. P. *et al.*, 2006).

Several neuropeptides are expressed in taste buds, including cholecystokinin (CCK), vasoactive intestinal peptide (VIP), and neuropeptide Y (NPY; Shen, T. *et al.*, 2005; Zhao, F. L. *et al.*, 2005). These peptides are all members of the brain-gastrointestinal tract family of peptides, which modulate feeding circuits in the brain. Thus, it is of interest that they are also expressed in taste receptor cells, the gateway to food intake. All three peptides are largely coexpressed with gustducin in circumvallate taste buds, suggesting that they are expressed primarily in bitter sensitive taste cells. Interestingly, NPY and CCK both modulate an inwardly rectifying K$^+$ current in taste cells, but in opposite directions – NPY enhances the current to hyperpolarize taste cells, while CCK inhibits it, resulting in taste cell depolarization.

4.10.4.3 Other Candidate Transmitters

While it is clear that ATP released from type II cells activates purinergic receptors on gustatory nerve fibers, the identity of the transmitter at the prominent synapse between type III taste cells and nerve fibers remains enigmatic. Investigators have failed to detect ATP release from type III taste cells (Huang, L. *et al.*, 2007; Romanov, R. A. *et al.*, 2007), and serotonin is unlikely to directly activate nerve fibers, as discussed above. These synapses are normally associated with clusters of clear vesicles (see Chapter Ultrastructure of Taste Buds), suggesting they may contain a conventional transmitter. Glutamate has long been proposed as a transmitter in taste buds, due to the expression of the glial glutamate transporter GLAST in type I taste cells (Lawton, D. M. *et al.*, 2000), and the expression of ionotropic glutamate receptors on the basolateral membranes of taste cells (Caicedo, A. *et al.*, 2000a). Another conventional transmitter candidate is acetylcholine, which modulates taste cells via muscarinic receptors (Ogura, T., 2002; Ogura, T. and Lin, W., 2005). However, the source of acetylcholine is still unclear. Recently, GAD67, involved in the synthesis of GABA, was shown to be expressed exclusively in type III cells (DeFazio, R. A. *et al.*, 2006), suggesting GABA may play a role at this synapse. Further studies are necessary to determine if type III taste cells release any of these potential transmitter candidates.

References

Adler, E., Hoon, M. A., Mueller, K. L., Chandrashekar, J., Ryba, N. J., and Zuker, C. S. 2000. A novel family of mammalian taste receptors. Cell 100, 693–702.

Avenet, P. And and Lindemann, B. 1991. Noninvasive recording of receptor cell action potentials and sustained currents from single taste buds maintained in the tongue: the response to mucosal NaCl and amiloride. J. Membr. Biol. 124, 33–41.

Bartel, D. L., Sullivan, S. L., Lavoie, E. G., Sevigny, J., and Finger, T. E. 2006. Nucleoside triphosphate diphosphohydrolase-2 is the ecto-ATPase of type I cells in taste buds. J. Comp. Neurol. 497, 1–12.

Behe, P., DeSimone, J. A., Avenet, P., and Lindemann, B. 1990. Membrane currents in taste cells of the rat fungiform papilla. Evidence for two types of Ca currents and inhibition of K currents by saccharin. J. Gen. Physiol. 96, 1061–1084.

Bernhardt, S. J., Naim, M., Zehavi, U., and Lindemann, B. 1996. Changes in IP3 and cytosolic Ca^{2+} in response to sugars and non-sugar sweeteners in transduction of sweet taste in the rat. J. Physiol. 490, 325–336.

Bo, X., Alavi, A., Xiang, Z., Oglesby, I., Ford, A., and Burnstock, G. 1999. Localization of ATP-gated P2X2 and P2X3 receptor immunoreactive nerves in rat taste buds. Neuroreport 10, 1107–1111.

Boudreau, J. C., Sivakumar, L., Do, L. T., White, T. D., Oravec, J., and Hoang, N. K. 1985. Neurophysiology of geniculate ganglion (facial nerve) taste systems: species comparisons. Chem. Senses 10, 89–127.

Boughter, J. D., Jr., Pumplin, D. W., Yu, C., Christy, R. C., and Smith, D. V. 1997. Differential expression of alpha-gustducin in taste bud populations of the rat and hamster. J. Neurosci. 17, 2852–2858.

Bystrova, M. F., Yatzenko, Y. E., Fedorov, I. V., Rogachevskaja, O. A., and Kolesnikov, S. S. 2006. P2Y isoforms operative in mouse taste cells. Cell Tissue Res. 323, 377–382.

Caicedo, A., Jafri, M. S., and Roper, S. D. 2000a. *In situ* Ca^{2+} imaging reveals neurotransmitter receptors for glutamate in taste receptor cells. J. Neurosci. 20, 7978–7985.

Caicedo, A., Kim, K. N., and Roper, S. D. 2000b. Glutamate-induced cobalt uptake reveals non-NMDA receptors in rat taste cells. J. Comp. Neurol. 417, 315–324.

Canessa, C. M., Schild, L., Buell, G., Thorens, B., Gautschi, I., Horisberger, J. D., and Rossier, B. C. 1994. Amiloride-sensitive epithelial Na^+ channel is made of three homologous subunits. Nature 367, 463–467.

Chaudhari, N., Landin, A. M., and Roper, S. D. 2000. A metabotropic glutamate receptor variant functions as a taste receptor. Nat. Neurosci. 3, 113–119.

Chandrashekar, J., Mueller, K. L., Hoon, M. A., Adler, E., Feng, L., Guo, W., Zuker, C. S., and Ryba, N. J. 2000. T2Rs function as bitter taste receptors. Cell 100, 703–711.

Chen, Y., Sun, X. D., and Herness, S. 1996. Characteristics of action potentials and their underlying outward currents in rat taste receptor cells. J. Neurophysiol. 75, 820–831.

Clapp, T. R., Medler, K. F., Damak, S., Margolskee, R. F., and Kinnamon, S. C. 2006. Mouse taste cells with G protein-coupled taste receptors lack voltage-gated calcium channels and SNAP-25. BMC Biol. 4, 7.

Clapp, T. R., Stone, L. M., Margolskee, R. F., and Kinnamon, S. C. 2001. Immunocytochemical evidence for co-expression of Type III IP3 receptor with signaling components of bitter taste transduction. BMC Neurosci. 2, 6.

Clapp, T. R., Yang, R., Stoick, C. L., Kinnamon, S. C., and Kinnamon, J. C. 2004. Morphologic characterization of rat taste receptor cells that express components of the phospholipase C signaling pathway. J. Comp. Neurol. 468, 311–321.

Cummings, T. A., Powell, J., and Kinnamon, S. C. 1993. Sweet taste transduction in hamster taste cells: evidence for the role of cyclic nucleotides. J. Neurophysiol. 70, 2326–2336.

Damak, S., Rong, M., Yasumatsu, K., Kokrashvili, Z., Perez, C. A., Shigemura, N., Yoshida, R., Mosinger, B., Jr., Glendinning, J. I., Ninomiya, Y., and Margolskee, R. F. 2006. Trpm5 null mice respond to bitter, sweet, and umami compounds. Chem. Senses 31, 253–264.

Danilova, V., Damak, S., Margolskee, R. F., and Hellekant, G. 2006. Taste responses to sweet stimuli in α-gustducin knockout and wild-type mice. Chem. Senses 31, 573–580.

DeFazio, R. A., Dvoryanchikov, G., Maruyama, Y., Kim, J. W., Pereira, E., Roper, S. D., and Chaudhari, N. 2006. Separate populations of receptor cells and presynaptic cells in mouse taste buds. J. Neurosci. 26, 3971–3980.

Delay, R. J., Kinnamon, S. C., and Roper, S. D. 1997. Serotonin modulates voltage-dependent calcium current in Necturus taste cells. J. Neurophysiol. 77, 2515–2524.

DeSimone, J. A. and Lyall, V. 2006. Taste receptors in the gastrointestinal tract III. Salty and sour taste: sensing of sodium and protons by the tongue. Am. J. Physiol. Gastrointest. Liver Physiol. 291, G1005–1010.

DeSimone, J. A., Callaham, E. M., and Heck, G. L. 1995. Chorda tympani taste response of rat to hydrochloric acid subject to voltage-clamped lingual receptive field. Am. J. Physiol. 268, C1295–1300.

DeSimone, J. A., Phan, T. H., Heck, G. L., Mummalaneni, S., Sturz, G. R., and Lyall, V. 2007. Proton flux through NADPH oxidase-linked H^+ channel (gp91^{phox}) is involved in eliciting chorda tympani (CT) taste nerve responses to strong acids. Chem. Senses Abstract. (in press).

Doolin, R. E. and Gilbertson, T. A. 1996. Distribution and characterization of functional amiloride-sensitive sodium channels in rat tongue. J. Gen. Physiol. 107, 545–554.

Dotson, C. D., Roper, S. D., and Spector, A. C. 2005. PLCbeta2-independent behavioral avoidance of prototypical bitter-tasting ligands. Chem. Senses 30, 593–600.

Elliott, E. J. and Simon, S. A. 1990. The anion in salt taste: a possible role for paracellular pathways. Brain Res. 535, 9–17.

Finger, T. E., Danilova, V., Barrows, J., Bartel, D. L., Vigers, A. J., Stone, L., Hellekant, G., and Kinnamon, S. C. 2005. ATP signaling is crucial for communication from taste buds to gustatory nerves. Science 310, 1495–1499.

Fukuwatari, T., Kawada, T., Tsuruta, M., Hiraoka, T., Iwanaga, T., Sugimoto, E., and Fushiki, T. 1997. Expression of the putative membrane fatty acid transporter (FAT) in taste buds of the circumvallate papillae in rats. FEBS Lett. 414, 461–464.

Furue, H. and Yoshii, K. 1997. In situ tight-seal recordings of taste substance-elicited action currents and voltage-gated Ba currents from single taste bud cells in the peeled epithelium of mouse tongue. Brain Res. 776, 133–139.

Gannon, K. S. and Contreras, R. J. 1995. Sodium intake linked to amiloride-sensitive gustatory transduction in C57BL/6J and 129/J mice. Physiol. Behav. 57, 231–239.

Garty, H. and Palmer, L. G. 1997. Epithelial sodium channels: function, structure, and regulation. Physiol. Rev. 77, 359–396.

Gilbertson, D. M. and Gilbertson, T. A. 1994. Amiloride reduces the aversiveness of acids in preference tests. Physiol. Behav. 56, 649–654.

Gilbertson, T. A. 2002. Hypoosmotic stimuli activate a chloride conductance in rat taste cells. Chem. Senses 27, 383–394.

Gilbertson, T. A. and Fontenot, D. T. 1998. Distribution of amiloride-sensitive sodium channels in the oral cavity of the hamster. Chem. Senses 23, 495–499.

Gilbertson, T. A. and Zhang, H. 1998. Self-inhibition in amiloride-sensitive sodium channels in taste receptor cells. J. Gen. Physiol. 111, 667–677.

Gilbertson, T. A., Avenet, P., Kinnamon, S. C., and Roper, S. D. 1992. Proton currents through amiloride-sensitive Na channels in hamster taste cells. Role in acid transduction. J. Gen. Physiol. 100, 803–824.

Gilbertson, T. A., Baquero, A. F., and Spray-Watson, K. J. 2006. Water taste: the importance of osmotic sensing in the oral cavity. J. Water Health 4(Suppl 1), 35–40.

Gilbertson, T. A., Fontenot, D. T., Liu, L., Zhang, H., and Monroe, W. T. 1997. Fatty acid modulation of K^+ channels in taste receptor cells: gustatory cues for dietary fat. Am. J. Physiol. 272, C1203–C1210.

Gilbertson, T. A., Liu, L., Kim, I., Burks, C. A., and Hansen, D. R. 2005. Fatty acid responses in taste cells from obesity-prone and -resistant rats. Physiol. Behav. 86, 681–690.

Gilbertson, T. A., Roper, S. D., and Kinnamon, S. C. 1993. Proton currents through amiloride-sensitive Na^+ channels in isolated hamster taste cells: enhancement by vasopressin and cAMP. Neuron 10, 931–942.

Glendinning, J. I., Bloom, L. D., Onishi, M., Zheng, K. H., Damak, S., Margolskee, R. F., and Spector, A. C. 2005. Contribution of alpha-gustducin to taste-guided licking responses of mice. Chem. Senses 30, 299–316.

He, W., Danilova, V., Zou, S., Hellekant, G., Max, M., Margolskee, R. F., and Damak, S. 2002. Partial rescue of taste responses of alpha-gustducin null mice by transgenic expression of alpha-transducin. Chem. Senses 27, 719–727.

He, W., Yasumatsu, K., Varadarajan, V., Yamada, A., Lem, J., Ninomiya, Y., Margolskee, R. F., and Damak, S. 2004. Umami taste responses are mediated by alpha-transducin and alpha-gustducin. J. Neurosci. 24, 7674–7680.

Heath, T. P., Melichar, J. K., Nutt, D. J., and Donaldson, L. F. 2006. Human taste thresholds are modulated by serotonin and noradrenaline. J. Neurosci. 26, 12664–12671.

Heck, G. L., Mierson, S., and DeSimone, J. A. 1984. Salt taste transduction occurs through an amiloride-sensitive sodium transport pathway. Science 223, 403–405.

Herness, M. S. 1992. Aldosterone increases the amiloride-sensitivity of the rat gustatory neural response to NaCl. Comp. Biochem. Physiol. Comp. Physiol. 103, 269–273.

Herness, M. S. and Chen, Y. 2000. Serotonergic agonists inhibit calcium-activated potassium and voltage-dependent sodium currents in rat taste receptor cells. J. Membr. Biol. 173, 127–138.

Herness, M. S. and Gilbertson, T. A. 1999. Cellular mechanisms of taste transduction. Annu. Rev. Physiol. 61, 873–900.

Herness, M. S. and Sun, X. D. 1995. Voltage-dependent sodium currents recorded from dissociated rat taste cells. J. Membr. Biol. 146, 73–84.

Herness, S. and Chen, Y. 1997. Serotonin inhibits calcium-activated K^+ current in rat taste receptor cells. Neuroreport 8, 3257–3261.

Herness, S., Zhao, F. L., Kaya, N., Lu, S. G., Shen, T., and Sun, X. D. 2002. Adrenergic signalling between rat taste receptor cells. J. Physiol. 543, 601–614.

Herness, S., Zhao, F. L., Kaya, N., Shen, T., Lu, S. G., and Cao, Y. 2005. Communication routes within the taste bud by neurotransmitters and neuropeptides. Chem. Senses 30(Suppl 1), i37–i38.

Hofmann, T., Chubanov, V., Gudermann, T., and Montell, C. 2003. TRPM5 is a voltage-modulated and Ca(2+)-activated monovalent selective cation channel. Curr. Biol. 13, 1153–1158.

Huang, A. L., Chen, X., Hoon, M. A., Chandrashekar, J., Guo, W., Trankner, D., Ryba, N. J., and Zuker, C. S. 2006. The cells and logic for mammalian sour taste detection. Nature 442, 934–938.

Huang, L., Cao, J., Wang, H., Vo, L. A., and Brand, J. G. 2005. Identification and functional characterization of a voltage-gated chloride channel and its novel splice variant in taste bud cells. J. Biol. Chem. 280, 36150–36157.

Huang, L., Shanker, Y. G., Dubauskaite, J., Zheng, J. Z., Yan, W., Rosenzweig, S., Spielman, A. I., Max, M., and Margolskee, R. F. 1999. Ggamma13 colocalizes with gustducin in taste receptor cells and mediates IP3 responses to bitter denatonium. Nat. Neurosci. 2, 1055–1062.

Huang, Y. J., Maruyama, Y., Dvoryanchikov, G., Pereira, E., Chaudhari, N., and Roper, S. D. 2007. The role of pannexin 1 hemichannels in ATP release and cell–cell communication in mouse taste buds. Proc. Natl. Acad. Sci. U. S. A. 104, 6436–6441.

Ishimaru, Y., Inada, H., Kubota, M., Zhuang, H., Tominaga, M., and Matsunami, H. 2006. Transient receptor potential family members PKD1L3 and PKD2L1 form a candidate sour taste receptor. Proc. Natl. Acad. Sci. U. S. A. 103, 12569–12574.

Jentsch, T. J. 1996. Chloride channels: a molecular perspective. Curr. Opin. Neurobiol. 6, 303–310.

Kataoka, S., Hansen, A., Matsunami, H., and Finger, T. E. 2007. The sour taste receptor PKD2L1 is expressed by Type III taste cells in the mouse. Chem. Senses Abst. In press.

Kataoka, S., Toyono, T., Seta, Y., Ogura, T., and Toyoshima, K. 2004. Expression of P2Y1 receptors in rat taste buds. Histochem. Cell Biol. 121, 419–426.

Kawai, T. and Fushiki, T. 2003. Importance of lipolysis in oral cavity for orosensory detection of fat. Am. J. Physiol. Regul. Integr. Comp. Physiol. 285, R447–R454.

Kaya, N., Shen, T., Lu, S. G., Zhao, F. L., and Herness, S. 2004. A paracrine signaling role for serotonin in rat taste buds: expression and localization of serotonin receptor subtypes. Am. J. Physiol. Regul. Integr. Comp. Physiol. 286, R649–R658.

Kim, D. J. and Roper, S. D. 1995. Localization of serotonin in taste buds: a comparative study in four vertebrates. J. Comp. Neurol. 353, 364–370.

Kim, M. R., Kusakabe, Y., Miura, H., Shindo, Y., Ninomiya, Y., and Hino, A. 2003. Regional expression patterns of taste receptors and gustducin in the mouse tongue. Biochem. Biophys. Res. Commun. 312, 500–506.

Kinnamon, S. C. and Margolskee, R. F. 1996. Mechanisms of taste transduction. Curr. Opin. Neurobiol. 6, 506–513.

Kinnamon, S. C., Dionne, V. E., and Beam, K. G. 1988. Apical localization of K^+ channels in taste cells provides the basis for sour taste transduction. Proc. Natl. Acad. Sci. U. S. A. 85, 7023–7027.

Kinnamon, S. C., Price, M. P., Stone, L. M., Lin, W., and Welsh, M. J. 2000. The acid-sensing ion channel BNC1 is not required for sour taste transduction; 2000; Brighton, UK. p. 80.

Kolesnikov, S. S. and Margolskee, R. F. 1995. A cyclic-nucleotide-suppressible conductance activated by transducin in taste cells. Nature 376, 85–88.

Kretz, O., Barbry, P., Bock, R., and Lindemann, B. 1999. Differential expression of RNA and protein of the three pore-forming subunits of the amiloride-sensitive epithelial sodium channel in taste buds of the rat. J. Histochem. Cytochem. 47, 51–64.

Kusakabe, Y., Yamaguchi, E., Tanemura, K., Kameyama, K., Chiba, N., Arai, S., Emori, Y., and Abe, K. 1998. Identification of two α-subunit species of GTP-binding proteins, Gα15 and Gαq, expressed in rat taste buds. Biochim. Biophys. Acta 1403, 265–272.

Laugerette, F., Gaillard, D., Passilly-Degrace, P., Niot, I., and Besnard, P. 2007. Do we taste fat? Biochimie 89, 265–269.

Laugerette, F., Passilly-Degrace, P., Patris, B., Niot, I., Febbraio, M., Montmayeur, J. P., and Besnard, P. 2005. CD36 involvement in orosensory detection of dietary lipids, spontaneous fat preference, and digestive secretions. J. Clin. Invest. 115, 3177–3184.

Lawton, D. M., Furness, D. N., Lindemann, B., and Hackney, C. M. 2000. Localization of the glutamate–aspartate transporter, GLAST, in rat taste buds. Eur. J. Neurosci. 12, 3163–3171.

Li, X., Staszewski, L., Xu, H., Durick, K., Zoller, M., and Adler, E. 2002. Human receptors for sweet and umami taste. Proc. Natl. Acad. Sci. U. S. A. 99, 4692–4696.

Lin, W. and Kinnamon, S. C. 1999. Physiological evidence for ionotropic and metabotropic glutamate receptors in rat taste cells. J. Neurophysiol. 82, 2061–2069.

Lin, W., Burks, C. A., Hansen, D. R., Kinnamon, S. C., and Gilbertson, T. A. 2004. Taste receptor cells express pH-sensitive leak K^+ channels. J. Neurophysiol. 92, 2909–2919.

Lin, W., Finger, T. E., Rossier, B. C., and Kinnamon, S. C. 1999. Epithelial Na^+ channel subunits in rat taste cells: localization and regulation by aldosterone. J. Comp. Neurol. 405, 406–420.

Lin, W., Ogura, T., and Kinnamon, S. C. 2002. Acid-activated cation currents in rat vallate taste receptor cells. J. Neurophysiol. 88, 133–141.

Lindemann, B. 1996. Taste reception. Physiol. Rev. 76, 718–766.

Lindemann, B., Gilbertson, T. A., and Kinnamon, S. C. 1999. Amiloride-sensitive sodium channels in taste. Curr. Topics Membr. 47, 315–336.

Liu, D. and Liman, E. R. 2003. Intracellular Ca^{2+} and the phospholipid PIP2 regulate the taste transduction ion channel TRPM5. Proc. Natl. Acad. Sci. U. S. A. 100, 15160–15165.

Liu, L., Hansen, D. R., Kim, I., and Gilbertson, T. A. 2005. Expression and characterization of delayed rectifying K^+ channels in anterior rat taste buds. Am. J. Physiol. Cell Physiol. 289, C868–C880.

LopezJimenez, N. D., Cavenagh, M. M., Sainz, E., Cruz-Ithier, M. A., Battey, J. F., and Sullivan, S. L. 2006. Two members of the TRPP family of ion channels, Pkd1l3 and Pkd2l1, are co-expressed in a subset of taste receptor cells. J. Neurochem. 98, 68–77.

Lyall, V., Alam, R. I., Malik, S. A., Phan, T. H., Vinnikova, A. K., Heck, G. L., and DeSimone, J. A. 2004a. Basolateral Na^+-H^+ exchanger-1 in rat taste receptor cells is involved in neural adaptation to acidic stimuli. J. Physiol. 556, 159–173.

Lyall, V., Alam, R. I., Phan, D. Q., Ereso, G. L., Phan, T. H., Malik, S. A., Montrose, M. H., Chu, S., Heck, G. L., Feldman, G. M., and DeSimone, J. A. 2001. Decrease in rat taste receptor cell intracellular pH is the proximate stimulus in sour taste transduction. Am. J. Physiol. Cell Physiol. 281, C1005–C1013.

Lyall, V., Alam, R. I., Phan, T. H., Phan, D. Q., Heck, G. L., and DeSimone, J. A. 2002. Excitation and adaptation in the detection of hydrogen ions by taste receptor cells: a role for cAMP and Ca^{2+}. J. Neurophysiol. 87, 399–408.

Lyall, V., Heck, G. L., Vinnikova, A. K., Ghosh, S., Phan, T. H., Alam, R. I., Russell, O. F., Malik, S. A., Bigbee, J. W., and DeSimone, J. A. 2004b. The mammalian amiloride-insensitive non-specific salt taste receptor is a vanilloid receptor-1 variant. J. Physiol. 558, 147–159.

Lyall, V., Pasley, H., Phan, T. H., Mummalaneni, S., Heck, G. L., Vinnikova, A. K., and DeSimone, J. A. 2006. Intracellular pH modulates taste receptor cell volume and the phasic part of the chorda tympani response to acids. J. Gen. Physiol. 127, 15–34.

Lyall, V., Sturz, G. R., Phan, T. T., Heck, G. L., Mummalaneni, S., and DeSimone, J. A. 2007. Nigericin shifts the pH threshold for the chorda tympani (CT) taste nerve response from 4.5 to 6.5. Chem. Senses Abstract. (in press).

Matsumura, S., Mizushige, T., Yoneda, T., Iwanaga, T., Tsuzuki, S., Inoue, K., and Fushiki, T. 2007. GPR expression in the rat taste bud relating to fatty acid sensing. Biomed. Res. 28, 49–55.

Matsunami, H., Montmayeur, J. P., and Buck, L. B. 2000. A family of candidate taste receptors in human and mouse. Nature 404, 601–604.

Mattes, R. D. 2007. Effects of linoleic acid on sweet, sour, salty and bitter taste thresholds and intensity ratings of adults. Am. J. Physiol. Gastrointest. Liver Physiol. 292, G 1243–G 1248.

McCormack, D. N., Clyburn, V. L., and Pittman, D. W. 2006. Detection of free fatty acids following a conditioned taste aversion in rats. Physiol. Behav. 87, 582–594.

McLaughlin, S. K., McKinnon, P. J., and Margolskee, R. F. 1992. Gustducin is a taste-cell-specific G protein closely related to the transducins. Nature 357, 563–569.

Medler, K. F., Margolskee, R. F., and Kinnamon, S. C. 2003. Electrophysiological characterization of voltage-gated currents in defined taste cell types of mice. J. Neurosci. 23, 2608–2617.

Mierson, S., Olson, M. M., and Tietz, A. E. 1996. Basolateral amiloride-sensitive Na^+ transport pathway in rat tongue epithelium. J. Neurophysiol. 76, 1297–1309.

Ming, D., Ruiz-Avila, L., and Margolskee, R. F. 1998. Characterization and solubilization of bitter-responsive receptors that couple to gustducin. Proc. Natl. Acad. Sci. U. S. A. 95, 8933–8938.

Misaka, T., Ishimaru, Y., Iwabuchi, K., Kusakabe, Y., Arai, S., Emori, Y., and Abe, K. 1999. A gustatory cyclic nucleotide-gated channels CNGgust, is expressed in the retina. Neuroreport 10, 743–746.

Misaka, T., Kusakabe, Y., Emori, Y., Arai, S., and Abe, K. 1998. Molecular cloning and taste bud-specific expression of a novel cyclic nucleotide-gated channel. Ann. N. Y. Acad. Sci. 855, 150–159.

Misaka, T., Kusakabe, Y., Emori, Y., Gonoi, T., Arai, S., and Abe, K. 1997. Taste buds have a cyclic nucleotide-activated channel, CNGgust. J. Biol. Chem. 272, 22623–22629.

Miyamoto, T., Fujiyama, R., Okada, Y., and Sato, T. 1998. Sour transduction involves activation of NPPB-sensitive

conductance in mouse taste cells. J. Neurophysiol. 80, 1852–1859.

Miyamoto, T., Fujiyama, R., Okada, Y., and Sato, T. 2000. Acid and salt responses in mouse taste cells. Prog. Neurobiol. 62, 135–157.

Miyamoto, T., Miyazaki, T., Fujiyama, R., Okada, Y., and Sato, T. 2001. Differential transduction mechanisms underlying NaCl- and KCl-induced responses in mouse taste cells. Chem. Senses 26, 67–77.

Miyamoto, T., Miyazaki, T., Okada, Y., and Sato, T. 1996. Whole-cell recording from non-dissociated taste cells in mouse taste bud. J. Neurosci. Methods 64, 245–252.

Monastyrskaia, K., Lundstrom, K., Plahl, D., Acuna, G., Schweitzer, C., Malherbe, P., and Mutel, V. 1999. Effect of the umami peptides on the ligand binding and function of rat mGlu4a receptor might implicate this receptor in the monosodium glutamate taste transduction. Br. J. Pharmacol. 128, 1027–1034.

Nelson, G., Chandrashekar, J., Hoon, M. A., Feng, L., Zhao, G., Ryba, N. J., and Zuker, C. S. 2002. An amino-acid taste receptor. Nature 416, 199–202.

Nelson, G., Hoon, M. A., Chandrashekar, J., Zhang, Y., Ryba, N. J., and Zuker, C. S. 2001. Mammalian sweet taste receptors. Cell 106, 381–390.

Ninomiya, Y. 1998. Reinnervation of cross-regenerated gustatory nerve fibers into amiloride-sensitive and amiloride-insensitive taste receptor cells. Proc. Natl. Acad. Sci. U. S. A. 95, 5347–5350.

Ninomiya, Y., Sako, N., and Funakoshi, M. 1989. Strain differences in amiloride inhibition of NaCl responses in mice, Mus musculus. J. Comp. Physiol. A 166, 1–5.

Ogura, T. 2002. Acetylcholine increases intracellular Ca^{2+} in taste cells via activation of muscarinic receptors. J. Neurophysiol. 87, 2643–2649.

Ogura, T. and Lin, W. 2005. Acetylcholine and acetylcholine receptors in taste receptor cells. Chem. Senses 30 (Suppl 1), i41.

Oike, H., Matsumoto, I., and Abe, K. 2006a. Group IIA phospholipase A(2) is coexpressed with SNAP-25 in mature taste receptor cells of rat circumvallate papillae. J. Comp. Neurol. 494, 876–886.

Oike, H., Wakamori, M., Mori, Y., Nakanishi, H., Taguchi, R., Misaka, T., Matsumoto, I., and Abe, K. 2006b. Arachidonic acid can function as a signaling modulator by activating the TRPM5 cation channel in taste receptor cells. Biochim. Biophys. Acta 1761, 1078–1084.

Ortiz-Alvarado, R., Guzman-Quevedo, O., Mercado-Camargo, R., Haertle, T., Vignes, C., and Bolanos-Jimenez, F. 2006. Expression of tryptophan hydroxylase in developing mouse taste papillae. FEBS Lett. 580, 5371–5376.

Perez, C. A., Huang, L., Rong, M., Kozak, J. A., Preuss, A. K., Zhang, H., Max, M., and Margolskee, R. F. 2002. A transient receptor potential channel expressed in taste receptor cells. Nat. Neurosci. 5, 1169–1176.

Pittman, D. W., Labban, C. E., Anderson, A. A., and O'Connor, H. E. 2006. Linoleic and oleic acids alter the licking responses to sweet, salt, sour, and bitter tastants in rats. Chem. Senses 31, 835–843.

Prawitt, D., Monteilh-Zoller, M. K., Brixel, L., Spangenberg, C., Zabel, B., Fleig, A., and Penner, R. 2003. TRPM5 is a transient Ca^{2+}-activated cation channel responding to rapid changes in $[Ca]_i^{2+}$. Proc. Natl. Acad. Sci. U. S. A. 100, 15166–15171.

Ren, Y., Shimada, K., Shirai, Y., Fujimiya, M., and Saito, M. 1999. Immunocytochemical localization of serotonin and serotonin transporter (SET) in taste buds of rat. Brain Res. Mol. Brain Res. 74, 221–224.

Richter, T. A., Caicedo, A., and Roper, S. D. 2003. Sour taste stimuli evoke Ca^{2+} and pH responses in mouse taste cells. J. Physiol. 547, 475–483.

Richter, T. A., Dvoryanchikov, G. A., Chaudhari, N., and Roper, S. D. 2004a. Acid-sensitive two-pore domain potassium (K2P) channels in mouse taste buds. J. Neurophysiol. 92, 1928–1936.

Richter, T. A., Dvoryanchikov, G. A., Roper, S. D., and Chaudhari, N. 2004b. Acid-sensing ion channel-2 is not necessary for sour taste in mice. J. Neurosci. 24, 4088–4091.

Romanov, R. A., Rogachevskaja, O. A., Bystrova, M. F., Jiang, P., Margolskee, R. F., and Kolesnikov, S. S. 2007. Afferent neurotransmission mediated by hemichannels in mammalian taste cells. Embo J. 26, 657–667.

Roper, S. 1983. Regenerative impulses in taste cells. Science 220, 1311–1312.

Roper, S. D. 2006. Cell communication in taste buds. Cell Mol. Life Sci. 63, 1494–1500.

Roper, S. D. and McBride, D. W., Jr. 1989. Distribution of ion channels on taste cells and its relationship to chemosensory transduction. J. Membr. Biol. 109, 29–39.

Rossler, P., Kroner, C., Freitag, J., Noe, J., and Breer, H. 1998. Identification of a phospholipase C beta subtype in rat taste cells. Eur. J. Cell. Biol. 77, 253–261.

Ruiz, C., Gutknecht, S., Delay, E., and Kinnamon, S. 2006. Detection of NaCl and KCl in TRPV1 knockout mice. Chem. Senses 31, 813–820.

Ruiz, C. J., Wray, K., Delay, E., Margolskee, R. F., and Kinnamon, S. C. 2003. Behavioral evidence for a role of alpha-gustducin in glutamate taste. Chem. Senses 28, 573–579.

Ruiz-Avila, L., McLaughlin, S. K., Wildman, D., McKinnon, P. J., Robichon, A., Spickofsky, N., and Margolskee, R. F. 1995. Coupling of bitter receptor to phosphodiesterase through transducin in taste receptor cells. Nature 376, 80–85.

Ruiz-Avila, L., Ming, D., and Margolskee, R. F. 2000. An *in vitro* assay useful to determine the potency of several bitter compounds. Chem. Senses 25, 361–368.

Ruiz-Avila, L., Wong, G. T., Damak, S., and Margolskee, R. F. 2001. Dominant loss of responsiveness to sweet and bitter compounds caused by a single mutation in alpha-gustducin. Proc. Natl. Acad. Sci. U. S. A. 98, 8868–8873.

Seta, Y., Kataoka, S., Toyono, T., and Toyoshima, K. 2007. Immunohistochemical localization of aromatic L: -amino acid decarboxylase in mouse taste buds and developing taste papillae. Histochem. Cell Biol. 127, 415–422.

Shen, T., Kaya, N., Zhao, F. L., Lu, S. G., Cao, Y., and Herness, S. 2005. Co-expression patterns of the neuropeptides vasoactive intestinal peptide and cholecystokinin with the transduction molecules alpha-gustducin and T1R2 in rat taste receptor cells. Neuroscience 130, 229–238.

Spector, A. C. and Travers, S. P. 2005. The representation of taste quality in the mammalian nervous system. Behav. Cogn. Neurosci. Rev. 4, 143–191.

Spector, A. C., Guagliardo, N. A., and St John, S. J. 1996. Amiloride disrupts NaCl versus KCl discrimination performance: implications for salt taste coding in rats. J. Neurosci. 16, 8115–8122.

Stevens, D. R., Seifert, R., Bufe, B., Muller, F., Kremmer, E., Gauss, R., Meyerhof, W., Kaupp, U. B., and Lindemann, B. 2001. Hyperpolarization-activated channels HCN1 and HCN4 mediate responses to sour stimuli. Nature 413, 631–635.

Stone, L. M., Barrows, J., Finger, T. E., and Kinnamon, S. C. 2007. Expression of T1Rs and gustducin in palatal taste buds of mice. Chem. Senses 32, 255–262.

Treesukosol, Y., Lyall, V., Heck, G. L., Desimone, J. A., and Spector, A. C. 2007. A psychophysical and electrophysiological analysis of salt taste in Trpv1 null mice. Am. J. Physiol. Regul. Integr. Comp. Physiol. 292, R 1799–1809.

Trubey, K. R., Culpepper, S., Maruyama, Y., Kinnamon, S. C., and Chaudhari, N. 2006. Tastants evoke cAMP signal in taste buds that is independent of calcium signaling. Am. J. Physiol. Cell. Physiol. 291, C237–244.

Ugawa, S., Minami, Y., Guo, W., Saishin, Y., Takatsuji, K., Yamamoto, T., Tohyama, M., and Shimada, S. 1998. Receptor that leaves a sour taste in the mouth. Nature 395, 555–556.

Ugawa, S., Yamamoto, T., Ueda, T., Ishida, Y., Inagaki, A., Nishigaki, M., and Shimada, S. 2003. Amiloride-insensitive currents of the acid-sensing ion channel-2a (ASIC2a)/ASIC2b heteromeric sour-taste receptor channel. J. Neurosci. 23, 3616–3622.

Vinnikova, A. K., Alam, R. I., Malik, S. A., Ereso, G. L., Feldman, G. M., McCarty, J. M., Knepper, M. A., Heck, G. L., DeSimone, J. A., and Lyall, V. 2004. Na$^+$-H$^+$ exchange activity in taste receptor cells. J. Neurophysiol. 91, 1297–1313.

Waldmann, R., Champigny, G., Lingueglia, E., De Weille, J. R., Heurteaux, C., and Lazdunski, M. 1999. H(+)-gated cation channels. Ann. N. Y. Acad. Sci. 868, 67–76.

Watson, K. J., Kim, I., Baquero, A. F., Burks, C. A., Liu, L., and Gilbertson, T. A. 2007. Expression of aquaporin water channels in rat taste buds. Chem. Senses. 32, 411–421.

Wong, G. T., Gannon, K. S., and Margolskee, R. F. 1996. Transduction of bitter and sweet taste by gustducin. Nature 381, 796–800.

Yan, W., Sunavala, G., Rosenzweig, S., Dasso, M., Brand, J. G., and Spielman, A. I. 2001. Bitter taste transduced by PLC-beta(2)-dependent rise in IP(3) and alpha-gustducin-dependent fall in cyclic nucleotides. Am. J. Physiol. Cell Physiol. 280, C742–C751.

Ye, Q., Heck, G. L., and DeSimone, J. A. 1991. The anion paradox in sodium taste reception: resolution by voltage-clamp studies. Science 254, 724–726.

Ye, Q., Heck, G. L., and DeSimone, J. A. 1994. Effects of voltage perturbation of the lingual receptive field on chorda tympani responses to Na$^+$ and K$^+$ salts in the rat: implications for gustatory transduction. J. Gen. Physiol. 104, 885–907.

Yee, C. L., Yang, R., Bottger, B., Finger, T. E., and Kinnamon, J. C. 2001. "Type III" cells of rat taste buds: immunohistochemical and ultrastructural studies of neuron-specific enolase, protein gene product 9.5, and serotonin. J. Comp. Neurol. 440, 97–108.

Yoshida, R., Yasumatsu, K., Shigemura, N., and Ninomiya, Y. 2006. Coding channels for taste perception: information transmission from taste cells to gustatory nerve fibers. Arch. Histol. Cytol. 69, 233–242.

Zhang, Y., Hoon, M. A., Chandrashekar, J., Mueller, K. L., Cook, B., Wu, D., Zuker, C. S., and Ryba, N. J. 2003. Coding of sweet, bitter, and umami tastes: different receptor cells sharing similar signaling pathways. Cell 112, 293–301.

Zhang, Z., Zhao, Z., Margolskee, R. F., and Liman, E. R. 2007. The transduction channel TRPM5 is gated by intracellular calcium in taste cells. J. Neurosci. 27, 5777–5786.

Zhao, F. L., Shen, T., Kaya, N., Lu, S. G., Cao, Y., and Herness, S. 2005. Expression, physiological action, and coexpression patterns of neuropeptide Y in rat taste-bud cells. Proc. Natl. Acad. Sci. U. S. A. 102, 11100–11105.

4.11 Gustatory Pathways in Fish and Mammals

M C Whitehead, University of California, San Diego, La Jolla, CA, USA

T E Finger, University of Colorado School of Medicine, Aurora, CO, USA

Glossary

alar plate A sensory region located dorsally in the embryonic spinal cord and hindbrain.

barbels Whisker-like, taste bud-bearing organs located near the mouth in some fishes, for example catfish.

basal plate A motor region located ventrally in the embryonic spinal cord and hindbrain.

bauplan Building plan: In taxonomy, the common properties of the members of a systematic group of animals.

brachium conjunctivum A white matter landmark in the pons consisting of axon bundles exiting the cerebellum.

Golgi stain A specialized stain showing cell bodies, dendrites and axons of a small, random subset of neurons in a histological section.

lemniscal pathway A sensory pathway that ascends from the spinal cord or brainstem, through the thalamus, to the cerebral cortex, that is from first-order to second-order to third-order nuclei, etc.

limbic Brain regions and functions concerned with visceral processes, e.g., the emotional or homeostatic status of the organism.

Nissl stain A routine stain showing all neuronal cell bodies in a histological section.

nucleus ambiguus A group of motor neurons that innervate pharyngeal muscles.

parabrachial nucleus A group of neurons surrounding the brachium conjunctivum in the dorsal pons.

primary sensory nucleus A brain region first activated by peripheral inputs; the neuronal group in the medulla receiving direct connections from peripheral nerves.

second-order sensory nucleus A brain region, located rostral to the medulla, that receives input from a primary sensory nucleus.

somatotopic A spatial representation that maps the sensory surface of an organism within the nervous system.

telencephalic Related to the forebrain.

teleost A group of bony, ray-finned, fishes with movable upper jaws.

third-order sensory nucleus A brain region that receives input from a second-order nucleus.

4.11.1 Introduction and Overview

The sense of taste in all complex animals begins with specialized chemoreceptors located in and around the mouthparts. These receptor organs are crucial in determining the palatability of potential food items and in orchestrating the appropriate ingestive or avoidance reactions. The organization of the neural centers activated by the receptors is the topic of this chapter, which presents a comparative perspective. Comparison of invertebrate and vertebrate systems may be instructive in terms of recognizing general principles of processing or organization, but data from one group is not likely to be directly applicable to the other group. For example, taste receptor organs in invertebrates involve sensory neurons with axons directly extending into the central nervous system. In the vertebrates, taste receptor organs (taste buds) consist of modified epithelial cells, not neurons. The taste bud cells connect with the primary taste neurons which lie in remote cranial ganglia, and it is the central processes of these ganglion cells that convey the taste message into the central nervous system. Thus, the taste system in invertebrates, for example, leech, fruit fly, is analogous, but not homologous to the taste system in vertebrates although commonalities do exist, probably through convergence (Finger, T. E., 2006). In contrast, the taste system in the different vertebrates is homologous (derived from a common ancestral condition), and so similarities in organization and mechanism are likely. Many such homologies are apparent when comparing the taste systems of different groups within vertebrates.

The general structure and organization of the taste system in vertebrates suggest that this system evolved in the earliest vertebrates and therefore is part of the common bauplan of all vertebrates. Taste buds are present in all vertebrates including lampreys, elasmobranches, bony fishes, amphibia, and amniotes including mammals. Comparison of the organization of the taste systems in various vertebrates contributes to an understanding of which features are part of the common organizational plan and which features may be derived for particular species or groups. This chapter first describes the general organizational plan for the central representation of taste within the brains of vertebrates. Then it examines in more detail the neuroanatomical organization of the taste system in two well-studied vertebrate groups: teleost fish and mammals, especially rodents.

4.11.1.1 Functional Organization of the Taste System in Vertebrates

Taste buds serve as sentinels for the digestive tract. They are positioned to allow the animal to assess the chemical composition of potential food items before they are given entrance to the alimentary canal. Not surprisingly then, stimulation of taste buds can evoke either ingestive or rejection reflexes. Yet taste does more than to merely serve as a go/no-go switch for foods. The sense of taste offers a conscious awareness of the composition of foods and is used to preferentially ingest particular substances according to past experience. Concomitant with a role in feeding, the sense of taste is involved in triggering digestive reflexes appropriate to the items ingested.

The roles for the gustatory system can be divided into three broad functional domains: (1) local brainstem pathways – reflexive action involving oropharyngeal muscles and digestive processes, (2) lemniscal pathways – mediating conscious awareness of what is being eaten, and (3) basal forebrain (limbic) pathways – visceral functions relating to hunger status, homeostasis, and drive state. Accordingly, it is not surprising that taste inputs to the brain ultimately influence centers associated with each of these functions, respectively: (1) oromotor and visceromotor nuclei of the brainstem, (2) a lemniscal system conveying information to specific gustatory areas of the forebrain, for example, gustatory cortex, and (3) basal forebrain regions, for example, the hypothalamus and amygdala, areas that are interconnected with the gustatory brainstem and typically associated with drive states, emotions, and homeostasis.

4.11.1.1.1 Primary sensory nuclei

In all vertebrates, taste information is conveyed to the central nervous system via three cranial nerves: the facial (CN VII), glossopharyngeal (CN IX), and vagus (CN X) nerves. The taste fibers in these nerves project to a restricted portion of the visceral sensory column of the medulla, an area situated dorsolateral to the sulcus limitans, a groove in the lateral wall of the ventricle. In the embryo, this sulcus separates alar plate from basal plate, and therefore separates sensory nuclei dorsally, from motor nuclei ventrally. In the medulla, the embryonic neural tube spreads apart dorsally to form the fourth ventricle so that the alar plate comes to lie dorsolateral of the basal plate. The visceral sensory column lies along the medioventral margin of the alar plate, where it is adjacent to the basal plate (Figure 1).

The viscerosensory column (Herrick, C. J., 1922) is, however, not a single homogeneous structure any

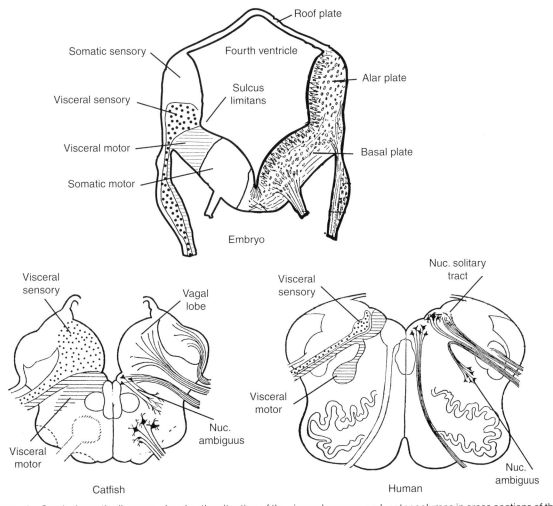

Figure 1 Semischematic diagrams showing the situation of the visceral sensory and motor columns in cross sections of the medulla in the embryo (above), the adult catfish (lower left) and adult human (lower right). In all situations, the visceral sensory column lies immediately dorsal to the sulcus limitans, while the visceral motor column lies immediately ventral. This same relationship holds in the adult brain of all vertebrate species. Adapted from Herrick, C. J. 1922. An Introduction to Neurology. W.B. Saunders Co.

more than the dorsolaterally adjacent somatic sensory column is a unitary structure. The somatic sensory column, for example, contains special somatic components: cochlear nuclei and vestibular nuclei, as well as general somatic components, i.e., trigeminal nuclei. Similarly, the visceral sensory column contains special visceral components for taste, and general visceral components for interoceptive functions carried on the vagus nerve including those mediated by digestive, cardiac, and respiratory afferent fibers. The taste portion of the visceral sensory column lies rostral and lateral to the general visceral components. In amniote vertebrates, the visceral sensory column is represented by the nucleus of the solitary tract (NST), containing separable gustatory and general visceral subdivisions. In fishes, the visceral sensory column is grossly divisible into lobes or lobules which protrude dorsally into the fourth ventricle. These lobules form clear anatomical boundaries between the gustatory and general visceral components of the nuclear complex. The gustatory lobes in the medulla in fishes are homologous to the gustatory portion of the NST in other vertebrates.

The gustatory and general visceral portions of the visceral sensory complex differ not only in terms of inputs but also in terms of connectivity with the rest of the brain. The general visceral components of the visceral complex maintain intimate reflex interconnections with the subjacent dorsal motor nucleus of the vagus, a preganglionic parasympathetic nucleus (Rogers, R. C. and McCann, M. J., 1993) and areas of the reticular formation that influence autonomic outflow (Ross, C. A. *et al.*, 1985). In addition, the general visceral subnuclei are extensively interconnected with limbic structures such as hypothalamus and amygdala (Ricardo, J. A. and Koh, E. T., 1978). The gustatory nuclei not only have interconnections with these limbic areas but possess unique connections as well. The reflex connections of the gustatory subnuclei involve oromotor centers and salivatory nuclei much more so than they do the dorsal motor nucleus of the vagus (Cunningham, E. T. and Sawchenko, P. E., 2000). In addition, the gustatory nuclei give rise to a unique lemniscal system conveying gustatory information to the thalamus and thence to a primary gustatory region of the cerebrum (e.g., gustatory cortex in mammals).

4.11.1.1.2 Local reflex pathway

Activation of the gustatory system will result in oropharyngeal movements (whether they be ingestion and swallowing or gagging and rejection) as well as reflex salivation and activation of the GI tract. The gustatory nuclei have reflex connections ending on oromotor nuclei (trigeminal motor nuc., nuc. ambiguus, hypoglossal nuc.) and in the adjacent premotor reticular formation. The more rostral gustatory nuclei tend to have more connections with reticular formation and anterior oromotor centers, while more caudal gustatory nuclei have more substantial connections with oropharyngeal motoneurons of the nucleus ambiguus. Homologous reflex pathways in fish and mammals will be described.

4.11.1.1.3 The gustatory lemniscus

Like other special senses, the gustatory system maintains a chain of connections within the brain with an ultimate target in the telencephalon. A labeled-line sensory pathway that ascends from the medulla to the thalamus and finally to cerebral cortex is known as a lemniscus. In mammals, the ultimate target of the gustatory lemniscus is primary gustatory cortex. In nonmammalian vertebrates, the telencephalic target is a restricted area within the dorsal telencephalon.

The gustatory lemniscus starts from the primary gustatory nuclei of the medulla to reach a specific thalamic relay nucleus. The ascending fibers proceed along two routes. Some percentage of the fibers more in some species (e.g., monkey) fewer in others (e.g., rat) ascend through the brainstem directly to reach the ipsilateral taste nucleus of the thalamus. Other fibers from the primary sensory nuclei ascend only as far as the ipsilateral secondary viscerosensory nucleus of the dorsal pons (i.e., pontine taste area (Norgren, R. and Leonard, C. M., 1973), parabrachial nuc. (Norgren, R. and Pfaffmann, C., 1975), in mammals, superior secondary gustatory nuc. (Herrick, C. J., 1905) in fish. The secondary viscerosensory nucleus of the pons, like the primary viscerosensory column of the medulla, contains distinct gustatory and general visceral components. The gustatory component of the secondary viscerosensory nucleus projects to the ipsilateral thalamic taste nucleus that is targeted by the direct ascending gustatory lemniscal system. The thalamic gustatory nucleus relays taste information to the telencephalic gustatory target which, in mammals, is gustatory cortex.

4.11.1.2 Visceral–Limbic Connections

The gustatory system, consistent with its visceral sensory nature, has extensive interconnection with the portions of the limbic system involved in regulation of homeostatic functions and drive states. The gustatory modality is crucial in the selection of particular foods. The choice of preferred foods changes according to an animal's physiological state (specific appetites, e.g., sodium appetite) and past experience. For example, a sodium-deprived rat will specifically seek out sodium salts in preference to other salts. Similarly, an animal that has fallen ill after ingesting a novel food will develop a strong aversion to the taste associated with that food (conditioned taste aversion). Both the specific appetites and conditioned taste aversions require coordination of limbic areas (hypothalamus and amygdala) with incoming gustatory information. Hence it is not surprising to find that ascending gustatory information reaches these limbic areas as well as the thalamocortical system.

4.11.1.3   The Gustatory Sense as a Visceral Modality

The gustatory sense has a unique location at the boundary between the external environment and the internal milieu. In some vertebrates (and even young children!), the gustatory system is used as an exteroceptive sense to gather information about the external world. But in all vertebrates, the gustatory modality is used to monitor intake of substances into the alimentary tract and to protect the airways from foreign substances, for example, laryngal taste buds which line the airway side of the larynx detect food or liquid that might be inhaled (Dickman, J. D. and Smith, D. V., 1988). The farther posterior a taste bud lies in the oropharynx, the more it serves as a visceral sensory organ; the more anterior a taste bud lies (e.g., tip of the tongue; barbels and lips of fishes), the more it is used as an exteroceptive special sense.

The most posterior taste buds are innervated by the vagus nerve, a mixed nerve containing some fibers which convey taste from these buds, while other fibers convey multiple general visceral modalities from interoceptive chemoreceptive end organs. These general visceral modalities, for example, information on blood gases, cardiac functions, upper GI tract functions, project into caudal portions of the brainstem viscerosensory complex (nuc. solitary tract). These inputs are relayed to brainstem visceromotor centers, to the pontine viscerosensory relay (parabrachial complex), and to the limbic areas of the forebrain including hypothalamus and amygdala. These limbic connections are highly reminiscent of the limbic relays for taste information, which they

parallel to a large extent; their organization reflects the visceral nature of the gustatory system.

4.11.2   Gustatory Centers in Fish: A Comparative Viewpoint

This section will describe how taste information is represented in the brains of nonmammalian vertebrates and especially teleost (bony) fishes. Understanding the organization of the gustatory system across vertebrates is important in two ways. First, it allows us to abstract general organizational principles that are evolutionarily conserved and which are likely to underlie the central organization of taste systems in mammals including humans. Second, the gustatory system is comparatively large in many fishes, and so the central taste nuclei represent a much larger portion of the brain, thus making the system more amenable to study. The basic connectivity of the brain is preserved through evolution, so understanding these pathways in fish will shed light on the basic organization of these systems in mammals.

What are the major features of the central gustatory system that can be discerned in these comparative studies? First, the primary principle of organization of the primary visceral nuclei, including the taste nuclei, is a somatotopic, or organotopic, representation. In the taste nuclei of fishes, a fine-grained topography is a recurrent theme across species, with particular sensory appendages or specializations being represented in elaborate corresponding central structures. This detailed somatotopic arrangement is mirrored in the primary general visceral nuclei in both fish and amniote vertebrates (Katz, D. M. and Karten, H. J., 1983; Altschuler, S. M. *et al.*, 1989) where different organs of the digestive and respiratory systems target specific subnuclei of the primary visceral sensory complex. Second, reflex systems arising from the different parts of the gustatory nuclear complex are different according to the end organ or part of the oral cavity innervated. Accordingly, facial taste information, which arises from more distal sensory fields, has connectivity more related to movements of the body, head, or tongue, whereas vagal taste information, arising from the posterior part of the oral cavity, has connectivity with motor systems driving swallowing. Third, despite the elaborate detail of the primary gustatory nuclei, the second-order nuclei do not mirror this somatotopic encoding of the information. The implication is that either the somatotopic information is encoded in some other fashion as it is relayed to higher centers or else the somatotopic information is utilized for processing in the primary sensory nuclei but is less important higher in the neuraxis. Nonetheless, despite the loss of the fine-grain somatotopy (Marui, T., 1981), the secondary nucleus does maintain the gross separation of facial and vagal gustatory information (Morita, Y. *et al.*, 1980) as well as gustatory and general visceral functions (Finger, T. E. and Kanwal, J. S., 1992).

4.11.2.1 Primary Gustatory Nuclei

A great deal of diversity exists in terms of the relative size and complexity of the taste system in the 20 000 species of teleost fishes. Some, for example, zebrafish, have a relatively undistinguished gustatory system encompassing both intraoral and external taste buds. Other fishes, including catfishes (order: siluriformes), many carps (family: cyprinidae) including goldfish, goatfish (family: Mullidae), and *Heterotis*, an unusual African fish unrelated to the mainstream teleosts, have elaborate taste bud arrays and central specializations. Because of the diverse species exhibiting such gustatory specializations, the elaboration of the system must have occurred independently several times during evolution. Despite these independent enlargements of the system in some species, numerous similarities still exist, and these must point to the underlying organizational scheme that formed early in vertebrate evolution. Such traits are likely to be consistent throughout all groups from fishes to amniote vertebrates including humans.

In all vertebrates, taste inputs arrive on three cranial nerves: CN VII (facial), CN IX (glossopharyngeal), and CN X (vagus). The facial nerve innervates taste buds situated farthest from the esophagus. In mammals, these are on the tip of the tongue, but in most fishes, taste buds occur on the external epithelium as well as in the oral cavity. Those taste buds lying outside of the oral cavity are always innervated by branches of the facial nerve. In fishes with specialized taste appendages (barbels), for example, catfish (Atema, J., 1971) and goatfish (Kiyohara, S. *et al.*, 2002), there is a concomitant relative enlargement of the facial taste system. Conversely, other fish species exhibit a relative enlargement and specialization of the vagal taste system which innervates elaborations of palatal or branchial taste organs (Braford, M. R. J., 1986; Finger, T. E., 1988).

As in other vertebrates, the three taste nerves enter the medulla to terminate in the viscerosensory column of nuclei extending along the wall of the fourth ventricle from near the obex to the level of the pontomedullary junction. The facial nerve terminates most rostrally in this column while the vagus nerve ends most caudally; the glossopharyngeal zone occupies an intermediate position. Even in fishes lacking an elaborately derived gustatory system, this overall organization is obvious. The three taste nerves terminate in serial order: facial terminals most anterior, glossopharyngeal terminals in the middle, and vagal taste inputs most posterior of the three. In all fishes, the general visceral components of the vagus nerve (e.g., innervating gut) terminate still more posteriorly in the viscerosensory column (Morita, Y. and Finger, T. E., 1985b; 1987).

Because the facial nerve taste system and the vagal nerve taste system innervate different populations of taste buds and terminate in clearly different lobes of the medulla, straightforward lesion experiments have been feasible to determine functional and behavioral differences between these systems. Atema J. (1971) ablated either the facial or vagal lobe in catfish to show that the facial taste system is involved in food localization, while the vagal taste system is necessary for ingestion and swallowing. The reflex connections arising from each of the gustatory lobes are commensurate with the behaviors and movements driven by each of these subsystems. The facial lobe is connected to premotor reticular formation that is involved in orientation to external food sources (Finger, T. E. and Morita, Y., 1985; Morita, Y. and Finger, T. E., 1985a); the vagal lobe has a tight reflex connection to the nuc. ambiguus motoneurons necessary for swallowing of identified food items.

4.11.2.1.1 Facial nerve specializations

The peripheral gustatory organs are especially numerous in those species with elaborate external gustatory appendages used to find food in the environment. Such species include catfishes, goatfishes, and gadids (cods and their relatives). These species are not closely related, emphasizing that gustatory specializations have evolved numerous times during fish phylogeny. Nonetheless, in each of these forms, the elaboration of the external, facially innervated taste periphery results in elaboration of the rostral part of the gustatory sensory column. In fish with an elaborate gustatory periphery, the rostral part of the viscerosensory column bulges medially and dorsally into the fourth ventricle to form an obvious facial lobe (Figure 2).

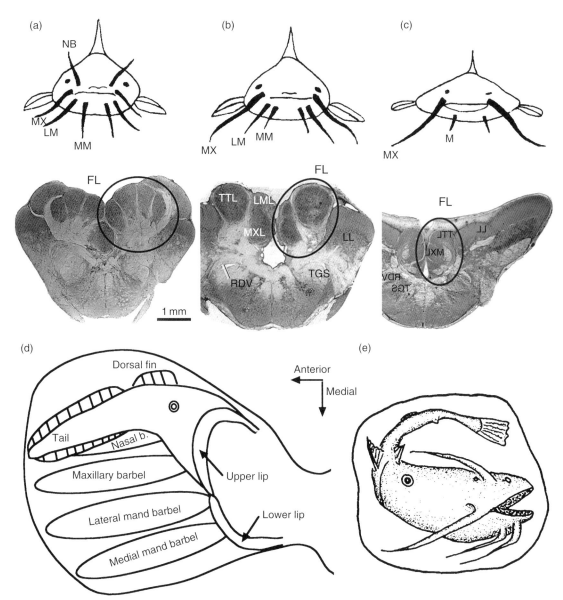

Figure 2 (a–c) The relationship between the number of barbels and the number of lobules in the facial lobe can be appreciated in these matched front views of the fish and cross sections of the rostral medulla in three different species of catfishes. (a) *Plotosus lineatus* has four barbels on each side and four lobules to the facial lobe. (b) *Arius felis* has three barbels and three lobules. (c) *Silurus asotus* has only two barbels and two lobules. (d, e) Semischematic diagrams of the piscunculus in the facial lobes of two species of catfish. A piscunculus is a distorted map of the receptor surface in the brain similar to the somatosensory homunculus drawn for the representation of touch in human somatosensory cortex. (d) The detailed somatotopic organization of the facial lobe of *Plotosus lineatus* can be seen in this figure. The seemingly disproportionate size of the barbels is related to the high density of taste buds on these organs. Representation in the facial lobe is in proportion to amount of taste input and not physical size of the receptor organ. (e) A more fanciful piscunculus projected onto the facial lobe of the bullhead catfish, *Ictalurus nebulosus*. (a, c) Parts of this figure were reprinted from Kiyohara, S. and Caprio, J. 1996. Somatotopic organization of the facial lobe of the sea catfish *Arius felis* studied by transganglionic transport of horseradish peroxidase. J. Comp. Neurol. 368, 121–135, with permission of Wiley Press. (d) Reprinted with permission from Kiyohara, S. and Tsukahara, J. 2005. Barber Taste System in Catfish and Goatfish. In: Fish chemosenser (*eds*. K. Reutter and B. G. Kapoor), pp. 175–211. Science Publishers. (e) Reprinted with permission from Wiley Press from Finger, T. E. 1978. Gustatory pathways in the bullhead catfish. II. Facial lobe connections. J. Comp. Neurol. 180, 691–705.

The structure and complexity of the facial lobe in each species match well the peripheral elaborations present on the animal. In catfishes, the number of lobules present exactly corresponds to the number of peripheral barbels the fish has (Kiyohara, S. and Caprio, J., 1996). In goatfish, the elongate single barbel is represented in a recurved tubular neuropil that maintains a precise somatotopic map of the peripheral end organ (Kiyohara, S. *et al.*, 2002). In codfish, with a simple single barblet, the facial lobe is present but is not particularly elaborate.

4.11.2.1.2   *Vagus nerve specializations*
Some fishes have an elaborate intra-oral food-sorting apparatus innervated by the vagus nerve. In carps, this mechanism involves the gill arches and palate

(Sibbing, F. A. *et al.*, 1986; Osse, J. W. *et al.*, 1997). In the unrelated Osteoglossomorph, *Heterotis,* the food sorting is accomplished in a specialization of the gill chamber (Braford, M. R. J., 1986). In both cases, the vagal taste information enters an elaborate, laminated vagal lobe. In goldfish (and probably *Heterotis*), the taste periphery is represented in a fine-grain orototopic fashion, matching the structure of the peripheral end organ: spiraled in *Heterotis*, a hemisphere in goldfish (Morita, Y. and Finger, T. E., 1985b) (Figure 3).

4.11.2.2   Brainstem Reflex Connections

The gustatory lobes in fishes give rise to local connections to adjacent nuclei and reticular formation

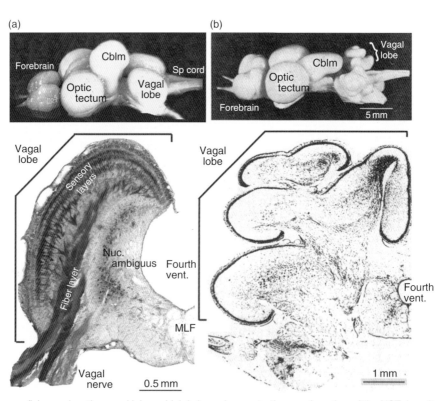

Figure 3   In some fish species, the vagal lobe, which is homologous to the vagal portion of the NST, is quite elaborate and may constitute nearly 20% of the entire brain mass. (a) In the common goldfish, *Carassius auratus*, the vagal lobe is a large, laminated hemisphere protruding from the dorsolateral surface of the medulla. Above: Lateral view of the brain. The large size of the vagal lobe can be appreciated in relation to the rest of the brain. Below: Cross section of the medulla through the vagal lobe, showing the left half of the brain (medial to the right). The laminated nature of the vagal lobe in goldfish is readily apparent in this preparation in which the vagus nerve was filled with the tracer horseradish peroxidase (HRP). The sensory layers lie along the outer half of the lobe overlying the fiber layer. The sensory layers are homologous to the vagal taste part of the NST. The motoneurons that innervate the palate, part of the nuc. ambiguus (visceral motor column) lie deeper in the lobe and are retrogradely labeled by HRP in this preparation. (b) In the African fish, *Heterotis niloticus*, the vagal lobe assumes a spiral shape matching a spiraled food-sorting organ of the branchial cavity (Braford, M. R. J., 1986). Above: The vagal lobe corkscrews dorsolaterally outward from the medulla. Below: Cross section of the medulla through the vagal lobe showing the left side of the brain. Three turns of the spiral can be seen in section. The superficial layer of small closely packed cells is similar to the superficial layer of the vagal lobe in goldfish (left).

areas of the medulla as well as ascending and descending connections (Herrick, C. J., 1905; Finger, T. E., 1978; Finger, T. E. and Morita, Y., 1985). The distinct nature of these reflex connections is more evident in these fishes than it is in mammals, but the same principle obtains. Reflex connections from each division of the visceral sensory column are appropriate for the particular organ or system represented (Figure 4). Thus, for example, general visceral sensory nuclei (GVM) have reflex connections with the general visceral motor nuclei (DMNX) that drive the autonomic nervous system (Altschuler, S. M. *et al*, 1989; Goehler, L. E. and Finger, T. E., 1992; Rogers,

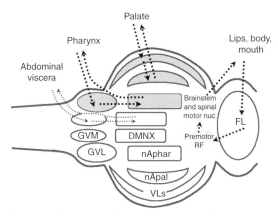

Figure 4 Schematic diagram of a dorsal view of the medulla of a goldfish, anterior to the right. The different reflex systems for the four visceral subsystems are illustrated. The abdominal viscera are innervated by the general visceral branch of the vagus nerve which projects centrally to the medial division of the general visceral nucleus (GVM) and commissural nucleus of Cajal. As in mammals, these maintain an intimate reflex system with the dorsal motor nucleus of the vagus (DMNX). Three semi-independent taste reflex systems also can be identified (shaded in gray): (1) pharyngeal, (2) palatal and caudal oral cavity, and (3) taste buds on the lips and external body surface. Each is connected to a different motor subsystem. The pharyngeal sensory nucleus (GVL) has reflex connections to the pharyngeal motor pool (nAphar). The sensory layers of the vagal lobe (VLs) have a topographically organized reflex connection with the pool of nuc. ambiguus neurons that extend into the vagal lobe motor layer (nApal). The facial lobe gustatory system, which drives complex movements of the body and head for orientation to external food sources, has connections to the premotor reticular formation. DMNX, dorsal motor nucleus of the vagus; FL, facial lobe; GVL, general visceral nucleus, lateral component; GVM, general visceral nucleus medial component; nApal, palatal portion of the nucleus ambiguus (=motor layers of the vagal lobe); nAphar, pharyngeal portion of nucleus ambiguus; VLs, sensory layers of the vagal lobe. Adapted from Goehler L. E. and Finger T. E. (1992) and Kanwal J. S. and Finger T. E. (1997).

R. C. and McCann, M. J., 1993). Within the brainstem gustatory complex itself, distinct reflex channels are evident according to the particular peripheral organ innervated. Thus the subnucleus representing the pharynx (GVL) maintains reflex connections with the motoneurons that drive the pharyngeal musculature (nAphar). Similarly, the areas representing the palate (VLs) have reflex connections with motoneurons driving the palate (nApal – the motor layer of the vagal lobe). Finally, the areas representing taste buds on the lips and external body surface are connected to the subjacent reticular formation which provides coordinated output to the motor nuclei necessary to drive movements of the body and head (Kanwal, J. S. and Finger, T. E., 1997).

4.11.2.3 Second-Order Gustatory Nuclei

The principal target for the ascending secondary gustatory fiber systems is the superior secondary gustatory nucleus (Herrick, C. J., 1905) which is homologous to the pontine taste area or parabrachial nuclear complex in mammals (Norgren, R. and Leonard, C. M., 1973; Norgren, R. and Pfaffmann, C., 1975). In fishes, as in mammals, the superior secondary gustatory nucleus lies in the dorsal pons adjacent to the superior cerebellar peduncle, or brachium conjunctivum. Whereas in mammals, the gustatory portion of the parabrachial complex is a small, poorly differentiated region; in fishes with large gustatory systems, the superior secondary gustatory nucleus is a large well-differentiated nucleus clearly divided into three or more compartments (Finger, T. E., 1978; Morita, Y. *et al.*, 1980; Finger, T. E. and Kanwal, J. S., 1992). The facial lobe (taste) projects to the most medial subnucleus; the vagal lobe (taste) projects laterally, and the commissural nucleus of Cajal and associated general visceral nuclei project to a superior secondary general visceral nucleus, ventral and anterior of the secondary gustatory nuclei (Morita, Y. *et al.*, 1980; Finger, T. E. and Kanwal, J. S., 1992; Yoshimoto, M. *et al.*, 1998). The secondary gustatory and general visceral nuclei exhibit distinct histochemical differences as well as the differences in connectivity (Finger, T. E. and Kanwal, J. S., 1992). A substantial bilateral projection exists so that the superior secondary gustatory nucleus receives information from both sides of the body and mouth (Figure 5).

Although the superior secondary gustatory nucleus in fishes is often very large and well differentiated, the fine topography so evident in the primary gustatory centers appears largely lost in the

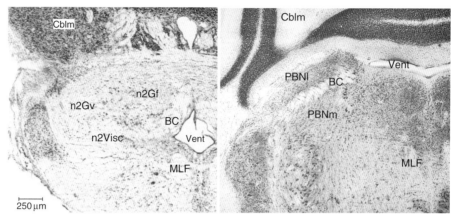

Figure 5   The secondary gustatory nuclei of the pons in (left) a catfish (*Ictalurus punctatus*) and (right) a mouse shown at the same magnification. In fishes, this nucleus is the superior secondary gustatory nucleus (n2G) whereas in mammals, it is called the parabrachial nucleus (PBN) or pontine taste area. In these species, as in all other vertebrates, the secondary gustatory nuclei lie adjacent or surrounding the Brachium conjunctivum (BC) which carries the ascending output from the cerebellum. In fishes, the n2G is divisible into a general visceral zone (n2visc) and a gustatory zone, itself divisible into a facial taste relay (n2Gf) and a vagal taste relay (n2Gv). The parabrachial complex also contains several subdivisions, the most prominent being a lateral subnucleus (PBNl) dorsolateral to the BC and a medial subnucleus (PBNm) ventral to the BC. BC, brachium conjunctivum; Cblm, cerebellum; MLF, medial longitudinal fasciculus; vent, ventricle (rostral end of fourth ventricle).

second-order nucleus. Receptive fields are large rather than discrete (Marui, T., 1981; Lamb, C. F. and Caprio, J., 1992), and small injections of tracer into the primary gustatory nuclei label widespread terminals throughout large areas of the secondary gustatory complex.

4.11.2.4   Ascending Fiber System

As mentioned above, ascending gustatory information targets two different sorts of nuclei in the forebrain. The first is a specific taste lemniscal pathway with an ultimate telencephalic target. The second is a more diffuse ascending system targeting a variety of limbic structures including the hypothalamus and amygdala.

In mammals, the telencephalic target of the gustatory lemniscus is gustatory cortex; the thalamic relay is VPMpc (parvocellular subnucleus of the ventroposteriomedial nucleus of the thalamus). Forebrain organization in fish is still being debated, but the clearest formulation names the third-order thalamic gustatory target as the preglomerular tertiary gustatory nucleus (Wullimann, M. F., 1988; Rink, E. and Wullimann, M. F., 1998; Yoshimoto, M. *et al.*, 1998). This tertiary gustatory nucleus relays taste information to a caudal, medial portion of the dorsal forebrain (dDm) (Kanwal, J. S. *et al.*, 1988; Yoshimoto, M. *et al.*, 1998). Based on connectivity

and physiology, the gustatory part of dDm appears homologous to gustatory cortex.

The secondary gustatory nucleus also gives rise to a more diffuse ascending fiber system. In mammals, the system extends from the parabrachial nucleus (PBN) and terminates in various limbic structures including hypothalamus and amygdala (Nomura, S. *et al.*, 1979; Halsell, C. B., 1992). Likewise in fishes, the superior secondary gustatory nucleus projects broadly to limbic forebrain areas including the hypothalamus and probable amygdalar equivalent zones (Lamb, C. F. and Finger, T. E., 1996; Yoshimoto, M. *et al.*, 1998; Folgueira, M. *et al.*, 2003).

4.11.3   Gustatory Pathways in Mammals

This section describes the taste and general viscerosensory systems in mammals and highlights the many features of the organization of these systems in mammals which are comparable to those described for fish in the last section.

4.11.3.1   Primary Sensory Nucleus

The central sensory nucleus that receives taste input from the periphery, in mammals, is the NST. The nucleus is an extensive viscerosensory cell grouping that is small in the dorsoventral and mediolateral planes, but long rostrocaudally. In fact, the NST extends throughout nearly the entire length of the medulla, from the level of the cochlear nucleus rostrally, where it is positioned somewhat laterally beneath the vestibular nuclei, to the spinal-medullary junction caudally, where it is positioned on the midline below the area postrema. The NST has, embedded laterally within it, the histologically distinct, rostrocaudally running solitary tract, the structure that gives the nucleus its name.

4.11.3.1.1 Primary afferent inputs

The solitary tract contains all the axons of the cranial nerves that provide input to the NST from the periphery. Thus, afferent axons from cranial nerves VII, IX, and X all travel in the tract and give off terminal branches that synapse in the NST. These inputs, which include both taste (special viscerosensory) and general viscerosensory axons (e.g., from the gastrointestinal organs), are arranged topographically in the nucleus (Figure 6). This topography is primarily

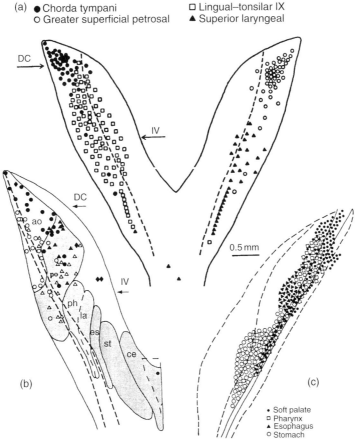

Figure 6 Horizontal depiction of the entire nucleus of the solitary tract in rat showing the topography of primary afferent inputs and of responses to oral and alimentary tract stimulation. Rostral is at the top of the figures. (a) Distribution of terminals of the peripheral gustatory nerves. Chorda tympani and greater superficial petrosal terminals, both of the seventh nerve, overlap and are located rostral to the ninth (lingual-tonsilar IX) nerve and the tenth (superior laryngeal) nerve. (b) Distribution of electrophysiological responses to oral stimulation. Taste responses from the anterior oral cavity (ao), closed circles; taste responses from the posterior oral cavity (po), closed triangles; taste responses from the whole mouth, closed diamonds; responses to mechanical stimulation of the anterior oral cavity, open circles; responses to mechanical stimulation of the posterior oral cavity, open triangles. Caudal shaded areas indicate primary afferent terminals from the pharynx (ph), larynx (la), esophagus (es), stomach (st), and cecum (ce). (c) Differential, but overlapping terminations from the soft palate, closed circles; pharynx, open squares; esophagus, closed triangles; and stomach, open circles. The dashed line in (a) represents a parcellation of the NST into medial and lateral regions. DC and IV arrows indicate, respectively, the level of the dorsal cochlear nucleus, and union of the NST with the fourth ventrical medially. Reproduced from Norgren, R. 1995. Gustatory System. In: The Rat Nervous System (ed. G. Paxinos), pp. 751–771. Academic Press, with permission from Elsevier, as adapted from Travers, S. P. 1993. Orosensory Processing in Neural Systems of the Nucleus of Solitary Tract. In: Mechanisms of Taste Transduction (eds. S. A. Simon and S. D. Roper), pp. 339–424. CRC press.

in the rostrocaudal plane reflecting the different medullary entry points of the NST's cranial nerve inputs, from the rostral-most VII to the caudal-most X. Moreover, the pattern of gustatory inputs defines an oropharyngeal topography within the NST wherein the taste buds of different regions of the oropharyngeal cavity are represented centrally in a somatotopic arrangement. Cranial nerve VII-innervated buds are on the anterior tongue and on the apposing palate, the central endings of the nerve branches innervating these two fields are largely coincident and located at the rostral pole of the NST. Cranial nerve IX-innervated buds are on the posterior tongue, the nerve's gustatory endings are overlapping but immediately caudal to those of CN VII (Whitehead, M. C., 1988a; 1988b in hamster; May, O. L. and Hill, D. L., 2006, in rat). Cranial nerve X-innervated buds are farthest posterior (on the larynx), the nerve's endings are located farthest caudal in the gustatory NST.

Primary afferent inputs are also arranged with a gross modality-specific topography in that gustatory axons synapse in rostral levels of the NST, while general visceral ones synapse caudally. This rostrocaudal arrangement is related to the differential representation of these two broad functional categories of axon types in VII, IX, and X. Thus, for example, VII which contains taste but no general visceral axons synapses rostrally, X with some taste, but predominant general viscerosensory axons synapses caudally. At any particular rostrocaudal level of the NST, primary afferent endings are concentrated in some cell groups of the nucleus, light or absent in others, indicating that there are subdivisions within the nucleus that can be defined on the basis of cytoarchitecture and inputs (Figure 7). Taste primary afferent synapses, in particular, are concentrated in a centrally located cell grouping in the rostral half of the NST, termed the rostral central subdivision. They synapse to a lesser extent also in the rostral lateral subdivision. These two subdivisions, and others involved in higher order, lemniscal connections of the taste pathways, are characterized by differences in the densities, sizes, and shapes of constituent neurons. The functional segregation within the NST further extends to the general visceral inputs themselves, with different visceral organs being represented in different longitudinal domains (i.e., subdivisions) within the NST (Katz, D. M. and Karten, H. J., 1983; Altschuler, S. M. *et al.*, 1989).

4.11.3.1.2 Subdivisions and cell types

Cells of the NST, generally, are smaller than those that surround the nucleus, for example, in the vestibular nuclei above, or the reticular formation below. This feature and the nucleus' felt-like neuropil, evident in silver stains and as differentially heterochromatic Nissl staining, allow the boundaries of the NST to be readily drawn (Whitehead, M. C., 1988a; 1988b). Cell groupings within the NST give the nucleus a heterogeneous cytological organization (Figure 8). A number of investigators have subdivided the nucleus based on routine cell staining with the Nissl method. Most studies recognize medial and lateral differences in cell packing and have simply divided the nucleus at every rostrocaudal level into two parts, medial and lateral (see, e.g., Torvik, A., 1956 in rat; Beckstead, R. M. and Norgren, R., 1979 in monkey; Hamilton, R. B. and Norgren, R., 1984 in rat). In hamster, additional staining methods, most notably, the Golgi stain which shows the dendrites and axons of cells, provided evidence for a third area, the central subdivision, that is cytologically distinct and interposed between the medial and lateral subdivisions (Whitehead, M. C., 1988a; 1988b). Subsequent studies in rat have similarly identified the rostral central subdivision as the prominent zone of neural activation by gustatory inputs from the periphery (e.g., Harrer, M. I. and Travers, S. P., 1996; Travers, S. P. and Hu, H., 2000). This central area coincides with parts of the lateral and medial subdivisions of earlier parcellation schemes (Torvik, A., 1956; Beckstead, R. M. and Norgren, R., 1979).

The central subdivision of the NST is distinguished by cells that are relatively small and densely packed and that exhibit three types of dendritic morphologies (Figure 9). Elongate cells have ovoid cell bodies and dendrites that extend in the mediolateral plane. Tufted neurons have dendrites or dendritic branches with one predominant orientation, usually dorsally or ventrally. Stellate cells have rounded or polygonal cell bodies and dendrites that radiate in all directions. A variety of small stellate cell with lumpy, irregular dendrites and a very short, locally arborizing axons, i.e., a local circuit neuron, is also present. The dendrites of all three cell types bear spiny or pedunculated appendages.

Correlating the cytoarchitectonic subdivisions with the locations of taste nerve endings, labeled by application of tracers to gustatory branches of the facial (CN VII), glossopharyngeal (CN IX), and vagus (CN X) nerves, established that the taste primary afferent termination zone coincides with the

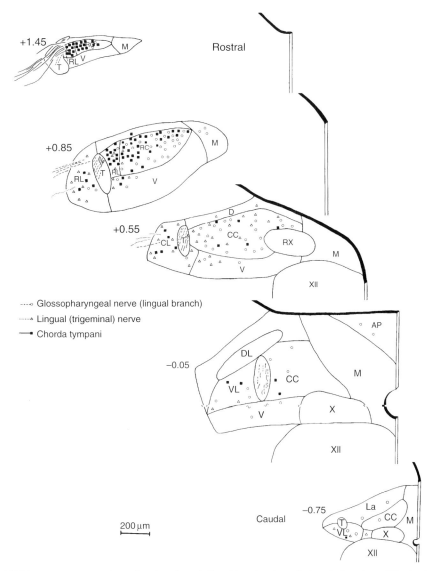

Figure 7 The NST in transverse sections showing the relationship of oral cavity inputs to cytoarchitectonic subdivisions in hamster (see Figure 8 for abbreviations). Taste afferent terminals from the anterior tongue (chorda tympani, closed squares) are concentrated in the rostral central subdivision; terminals from the posterior tongue (glossopharyngeal nerve, lingual branch, open circles) overlap but are densest more caudal than chorda ones in the rostral central subdivision. Nontaste, trigeminal inputs from the anterior tongue (lingual nerve, open triangles) overlap taste inputs but are concentrated in the rostral lateral subdivision. Numbers at left indicate distance (in mm) of each section from the obex. Reproduced from Whitehead M. C. (1988a; 1988b).

rostral central subdivision (Figure 7). The rostral lateral subdivision also received many taste endings, with fewer and fewer endings in the caudal central and caudal lateral subdivisions, respectively. Considering the entire extent and outlines of the NST, the distribution of all taste endings is to rostral and lateral regions of the nuclear complex.

The synaptic organization of taste inputs within the rostral central subdivision has been defined for the hamster (Whitehead, M. C., 1986; Brining, S. K. and Smith, S. K., 1996). Taste afferent endings, labeled with the marker horseradish peroxidase, were found to synapse *via* round vesicle-containing endings and asymmetrical membrane specializations primarily with small caliber dendrites and dendritic spines (Figure 10). Synapses of this type are likely excitatory. The taste afferent endings synapse in association with other, presumably nonprimary, endings. Some of

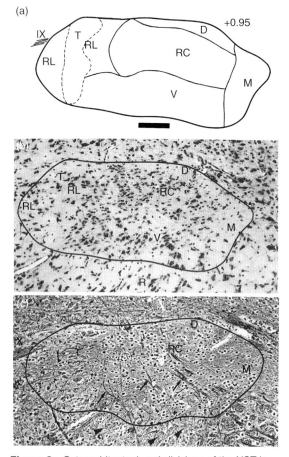

Figure 8 Cytoarchitectonic subdivisions of the NST in transverse sections at the level of the glossopharyngeal sensory nerve root in hamster (level similar to the second from the top in Figure 7). (a) Drawing of the outlines of subdivisions in the Nissl-stained section (b) The rostral central subdivision (RC) is characterized by smaller, densely packed, round, oval, and elongate cells; the medial subdivision (M) is sparsely cellular; the ventral subdivision (V) has larger somewhat clustered cells; the rostral lateral subdivision is moderately cellular and has the solitary tract (T) embedded within it. (c) Reduced silver-stained section adjacent to that in (b) showing the felt-like neuropil of the NST and thick fibers of passage arcing from the vestibular nuclei dorsal to the NST through the rostral central (RC) and ventral (V) subdivisions. Number at upper right indicates distance (in mm) of the sections from the obex. Reproduced from Whitehead M. C. (1988a; 1988b).

these nonprimary endings contain flat vesicles and engage in symmetrical synaptic junctions indicative of inhibitory synapses. The presence of presumed inhibitory endings is consistent with the identification of GABA-ergic cells and endings in the rostral NST (Lasiter, P. S. and Kachele, D. L., 1988; King, M. S., 2003) and similar small neurons in the rostral central

subdivision that have locally arborizing axons, that is, inhibitory local circuit neurons. The complex synaptic milieu of taste endings on second-order neuronal dendrites, distant from their cell bodies, allows for modulation of taste-elicted activity in the afferent endings themselves or in the second-order neurons by other inputs, including interneuronal ones.

Other subdivisions border the rostral central subdivision (see, e.g., Figure 8). They differ in the extent to which they are involved in taste-related pathways (Whitehead, M. C., 1988a; 1988b). Thus, there are ventral and medial subdivisions bordering the rostral central subdivision which receive sparse taste afferent input (Figure 7). Similarly, caudally in the NST, subdivisions include caudal central, ventral medial, and ventrolateral which receive sparse taste input, but receive other inputs including general visceral primary afferent axonal endings from internal organs (Figure 6). These caudal areas of the NST project to the rostral areas and allow for interoceptive (e.g., gastrointestinal) and oral (taste) interactions within the NST (Karimnamazi, H. et al., 2002).

4.11.3.2   Brainstem Reflex Connections

Taste stimulation elicits highly stereotyped oromotor behaviors that effect ingestion of palatable compounds (e.g., licking of sweet tastants) and rejection of aversive tastes (e.g., tongue protrusion and gaping to eject bitter, potentially poisonous, compounds) (Grill, H. J. and Norgren, R., 1978a; 1978b). These motor behaviors result from activation and modulation of local gustatory sensory–motor reflex pathways involving short distance connections confined within the medulla. These local, intramedullary pathways and the neurons they engage have, to some level of detail, been defined anatomically and functionally. The rostral (gustatory) NST projects heavily to the nearby, subjacent reticular formation (Travers, J. B., 1988; Beckman, M. E. and Whitehead, M. E., 1991) (Figures 11 and 12). The reticular formation, in turn, projects to motor nuclei involved in oromotor behaviors (Travers, J. B. and Norgren, R., 1983) (Figures 12 and 13). These include the hypoglossal nucleus for tongue movements, the trigeminal motor nucleus for jaw movements, the facial motor nucleus for facial movements, and nucleus ambiguus for pharyngeal movements associated with swallowing. Though less well studied because of the small size of the NST and the difficulty of mapping short connections, there are intranuclear connections that ultimately influence reticular and oromotor neuronal responses to tastes. These intranuclear

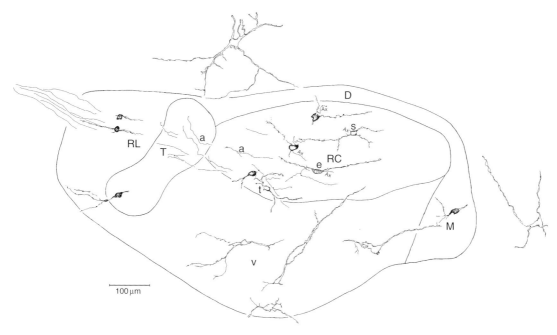

Figure 9 Cell types and axons (Golgi-Cox method) of the rostral NST in hamster at the level of the glossopharyngeal nerve root (fibers at upper left) (level similar to that in Figure 8). Primary afferent axons of the nerve root and solitary tract (a) parallel the dendrites of elongate cells in the rostral lateral subdivision and in the rostral central subdivision (e). A stellate cell (s) also has mediolaterally oriented dendrites; t, tufted neuron. Cells of the ventral subdivision are larger with longer dendrites than those of other subdivisions. Larger still, with more elaborate dendritic branching, are the cells dorsal (vestibular) and medial to the NST which are drawn for comparison. (Abbreviations of subdivisions as in Figure 8.) Reproduced from Whitehead M. C. (1988a; 1988b).

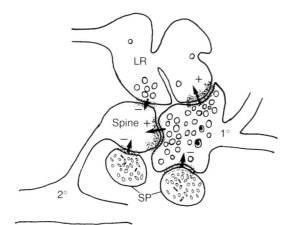

Figure 10 A synaptic circuit for taste (facial, including chorda tympani) primary afferent endings in the rostral central subdivision of the NST in hamster. Taste endings (1°) synapse on dendritic spines of second-order neurons (2°) and on other (possibly dendritic) endings (LR), which in turn influence transmission by means of LR (dendro?)-dendritic synapse. Transmission through the taste axon-second-order neuron circuit is also modulated pre- and postsynaptically by SP axonal endings with flat vesicles and symmetrical synaptic junctions. The signs (+) and (−) indicate possible excitatory and inhibitory influences. Reproduced from Whitehead, M.C. 1986. Anatomy of the gustatory system in the hamster: synaptology of facial afferent terminals in the solitary nucleus. J. Comp. Neurol. 244, 72–85.

connections appear to differentially involve the cytoarchitectonic subdivisions.

Small neuroanatomical tracer injections confined to the rostral NST and heavily involving the rostral central subdivision demonstrate that there are intra-NST projections from this taste-afferent recipient region to ventral and medial portions of the NST at more caudal levels. Thus, the medial and ventral subdivisions at intermediate rostrocaudal levels of the nucleus receive heavy input from the rostral central NST. Outside the NST, heavy axonal projections terminate in the parvicellular medullary reticular formation immediately ventral to the NST (Beckman, M. E. and Whitehead, M. C., 1991) (Figures 11 and 12). The injection sites yielding these local projections, while small, invariably include the ventral subdivision, so it is difficult to resolve all the possible short distance connections. Nevertheless, the evidence to date is consistent with a pathway whereby neurons in the rostral central subdivision are activated by taste-elicited impulses arriving from primary afferent synapses (Figure 12). In turn, many of these first-order central neurons project to other regions of the NST, including to the ventral subdivision. The ventral subdivision (and, perhaps, the rostral central

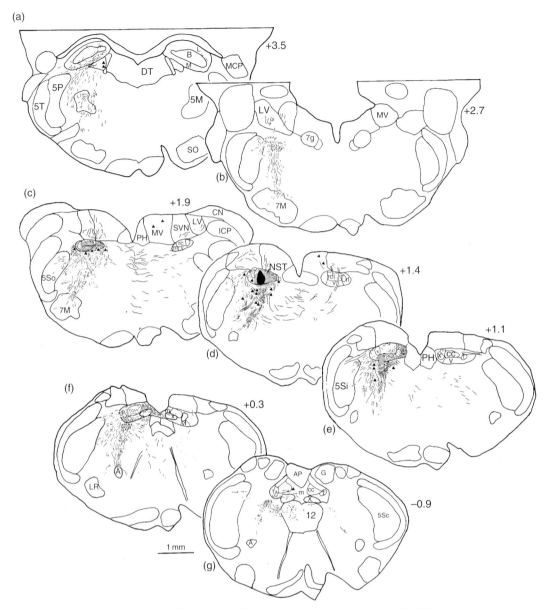

Figure 11 Ascending (lemniscal) projections and local (reflex) projections from the rostral NST in hamster seen after an injection of neuroanatomical tracer (HRP) into the gustatory domain (center of injection site at same level as in Figure 8, includes rostral central and ventral subdivisions). Letters identifying nuclei are shown on the right side of the brain in order to best see labeled endings (dots) on the right. Ascending axons from the NST terminate rostrally in the medial PBN (B-M) and sparsely in the trigeminal motor nucleus (5M). Local and caudally directed labeled axons terminate in the reticular formation (RF) ventral to the NST and lateral to the hypoglossal nucleus (12) and, sparsely, within the hypoglossal, vagal (X) and facial (7M) motor nuclei. Nucleus ambiguus (A). Intra-NST label extends caudally to the level of the area postrema (AP) and is concentrated in the medial (m) and ventral (v) subdivisions. Triangles indicate retrogradely labeled neuronal cell bodies. Numbers at right indicate distance (in mm) of each section from the obex. Reproduced from Beckman, M. E. and Whitehead, M. C. 1991. Intramedullary connections of the nucleus of the solitary tract in the hamster. Brain Res. 557, 265–279, with permission from Elsevier.

subdivision, directly) projects heavily to the parvo-cellular reticular formation (Beckman, M. E. and Whitehead, M. C., 1991) which, in turn, projects heavily to all the oromotor nuclei (Travers, J. B. and Norgren, R., 1983; Travers, J. B., 1988).

Combining Fos labeling during taste stimulation with neuroanatomical retrograde labeling of cells that project to the parvocellular reticular formation revealed a pattern consistent with a role for ventrally projecting neurons and of the ventral NST

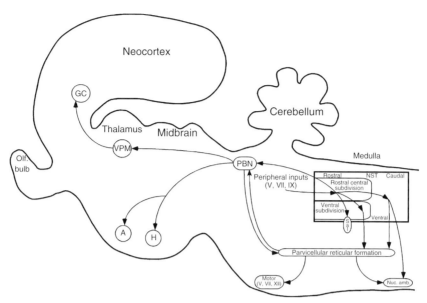

Figure 12 Local and ascending pathways of the mammalian central gustatory system diagrammed on a sagittal view of the brain. The local (reflex) pathway involves cells in the rostral central subdivision of the NST that receive input from primary afferent peripheral fibers. These cells can project to the ventral subdivision of the NST either directly or via the caudal NST. The ventral subdivision, in turn, projects to the reticular formation. Finally, the parvicellular reticular formation projects to nucleus ambiguus and the oral motor nuclei (motor V, VII, XII). Local projections also involve the salivatory nuclei (Sal). The major ascending (lemniscal) pathway involves cells of the rostral central subdivision that project to the parabrachial nucleus of the pons (PBN) which, in turn, projects to the parvocellular ventroposteromedial nucleus of the thalamus (VPM). This gustatory part of the thalamus projects to agranular insular cortex (GC). A second ascending pathway involves a projection from the PBN to the basal forebrain (principally to the lateral hypothalamus (H), amygdala (A), and bed nucleus of the stria terminalis (not shown).

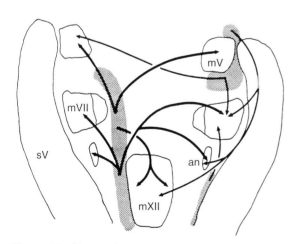

Figure 13 Diagram (horizontal plane) of the regions of the reticular formation (in gray cross-hatching) that project to the oral motor nuclei in rat. The reticular-motor projection neurons are located throughout the length of the medulla, including the area below the NST as in Figure 11. Projections are to nucleus ambiguus (an), the hypoglossal nucleus (mXII), the facial (mVII), and trigeminal (mV) motor nuclei bilaterally. Reproduced from Travers, J. B., and Norgren, R. 1983. Afferent projections to the oral motor nuclei in the rat. J. Comp. Neurol. 220, 280–298.

subdivision in influencing premotor reticular formation neurons (Travers, S. P. and Hu, H., 2000). Electrical stimulation of the second-order gustatory nucleus (the PBN, in rat) activates cells in the ventral NST and elicits oromotor behaviors, illustrating an influence of descending impulses from the second-order taste center on the medullary reflex pathway (King, M. S. *et al.*, 2003).

4.11.3.2.1 Topography of taste quality

Functional evidence that the rostral central and lateral subdivisions of the rostral NST are principally involved in taste information processing derives from experiments seeking to determine whether taste quality is represented topographically in the NST. Electrophysiological recordings, either single or multiunit methods, have not, despite many investigations, revealed a consistent or clear-cut chemotopic organization (e.g., Pfaffmann, C. *et al.*, 1961; Scott, T. R. *et al.*, 1986; Mistretta, C. M., 1988). Nevertheless, reconstructions of sites of taste-activated neurons in relation to subdivisions of the NST localize most of them, in hamster, to the rostral central and rostral lateral subdivisions (McPheeters, M. *et al*, 1990;

Whitehead, M. C. *et al.*, 1993). More recently, anatomical localization of NST cells activated by particular taste stimuli was accomplished with immunocytochemistry for Fos, the protein product of the c-fos gene. Fos-positive cells attributable to taste stimulation were differentially concentrated in the rostral central subdivision (Harrer, M. I. and Travers, S. P., 1996, in rat). Although no straightforward chemotopography was evident and sucrose and quinine activated cells were intermingled, the latter were concentrated medially within the central subdivision evidence for a crude chemotopography as has been suggested in earlier electrophysiological studies (Halpern, B. P. and Nelson, L. M. 1965) (Figure 14). This crude chemotopy may reflect the differential distribution of different taste receptors within the oral cavity (Chandrashekar, J. *et al.*, 2006) making it difficult to distinguish a chemotopic map from an orotopic one.

4.11.3.3 Lemniscal Pathway

The lemniscal pathway in the mammalian gustatory system extends from the NST through the thalamus to the cerebral cortex for perception (Figure 12). In rodents, but not primates, all information ascending to the thalamocortical system undergoes an obligatory synaptic interruption in the PBN of the pons. The PBN is also a site of synaptic interruption in the pathway leading to the basal forebrain (see Section 4.11.3.4).

The projection from the NST to the PBN is massive and involves axons originating from cells in rostral and caudal levels of the medullary nucleus (Norgren, R. and Leonard, C. M., 1973; Norgren, R., 1978; Beckstead, R. M. *et al.*, 1980; Travers, J. B., 1988). Thus, both taste and general viscerosensory impulses project to the pons. The pontine representations of taste and nontaste (e.g., gastric) signals are separated among the subdivisions of the PBN (Karimnamazi, H. *et al.*, 2002). The gustatory PBN projection neurons of the NST have been identified and are predominantly located in the rostral central subdivision, the same region receiving densest taste inputs from the periphery (Whitehead, M. C., 1990). Specific cell types in the rostral central subdivision that project to the PBN, in hamster, are the elongate and stellate cells. These projection neurons, back-

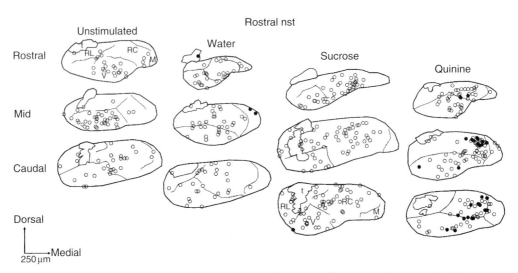

Figure 14 Distribution of cells immuno-positive for Fos elicited by two qualitatively and hedonically distinct taste stimuli, sucrose and quinine, within the rostral nucleus of the solitary tract of rat. Three anteroposterior levels in the transverse plane are shown for one animal from each of four stimulus conditions. Thin lines depict boundaries of cytoarchitectonic subdivisions (abbreviations as in Figure 8). Filled and open circles indicate intensely and moderately stained nuclei, respectively, of Fos-positive cells. Taste-stimulated animals had more numerous Fos-positive cells than the water and unstimulated animals, particularly in the rostral central subdivision. Within the central subdivision, Fos-positive neurons following quinine stimulation were concentrated medially; positive neurons following sucrose stimulation were distributed more evenly along the mediolateral axis. Despite their differential distribution, sucrose- and quinine-activated neurons were intermingled. Reproduced from Harrer, M. I. and Travers, S .P. 1996. Topographic organization of Fos-like immunoreactivity in the rostral nucleus of the solitary tract evoked by gustatory stimulation with sucrose and quinine. Brain Res. 711, 125–137, with permission from Elsevier.

filled with neuroanatomical markers, have been analyzed with the electron microscope to establish that they receive primary-like synaptic inputs (Whitehead, M. C., 1993). Therefore, the NST–PBN gustatory projection neurons correspond to first-order central neurons that are contacted directly by taste afferent endings from the periphery.

4.11.3.3.1 Second-order (parabrachial) nucleus

The PBN is a complex of different cytoarchitectonic subdivisions that surround the brachium conjunctivum (Fulwiler, C. E. and Saper, C. B., 1984; Halsell, C. B. and Frank, M. E., 1989). These subdivisions are generally grouped into either the lateral division, dorsal and lateral to the brachium conjunctivum and bordering the external surface of the pons, or the medial division, medial to the brachium. Gustatory-responsive neurons have been localized predominantly in the medial division, while general viscerosensory neurons have been localized in the lateral division. Anatomical studies support this functional parcellation: the medial PBN receives input from the rostral, gustatory NST; the lateral PBN receives input from the caudal, viscerosensory NST. In particular, the medial PBN, predominantly its caudal and ventral parts, receives projections from cells of the rostral central and rostral lateral subdivisions of the NST. The lateral PBN, predominantly its intermediate and rostral–dorsal parts, receives projections from the caudal central and medial subdivisions of the NST (Whitehead, M. C., 1990). PBN projection cells of the rostral NST, that is, gustatory cells of the ascending viscerosensory system, are more numerous than PBN projection cells of the caudal NST, that is, general viscerosensory cells. The pontine projections are predominantly ipsilateral, although a small minority of NST neurons project to the contralateral PBN. The most prominent subdivision within the PBN receiving input from the gustatory NST is the medial subdivision, especially at caudal levels of the nucleus. Gustatory projections also extend, but lightly, into the medial part of the lateral subdivision. Within the gustatory PBN neurons responsive to anterior and posterior regions of the oral cavity are located in different cytoarchitectonic subdivisions, evidence for at least a gross orotopy at this second-order level of the ascending taste system in mammals (Halsell, C. B. and Travers, S. P., 1997), just as in fish (see Section 4.11.3.2).

4.11.3.3.2 Third-order (thalamic) nucleus and gustatory cortex

In contrast with other lemniscal sensory systems in which the thalamus receives input directly from the medulla, in the taste system, excepting for Old World monkeys (Hamilton, R. B. et al., 1985), the thalamus receives input from second-order central neurons in the PBN (Figure 12). The PBN-thalamus projection terminates in a small medial-most sliver of the ventroposteromedial thalamic nucleus (VPM) (Norgren, R. and Leonard, C. M., 1973). The VPM generally processes information about somatosensation of the face relayed rostrally from first-order central neurons in the trigeminal sensory nucleus of the brainstem. The gustatory-specific thalamic nucleus, the VPMgus parvocellularis (pc), is the medial-most part of the VPM. Thus, functions are topographically represented; third-order taste neurons lie just medial to second-order general sensory neurons processing, for example, touch and temperature information from the face, mouth, and tongue. Indeed, taste and lingual touch neurons occur side by side in the VPM. The VPMgus pc projects to gustatory neocortex (Figure 12).

Gustatory neocortex lies in the insular region, dorsal to the rhinal fissure. This region, identified as taste-responsive (Kosar, E. et al., 1986 in rat, Wehby, R. G. and London, J. A., 1995 in hamster), is characterized cytoarchitectonically as being agranular or dysgranular, that is, lacking a well-defined layer IV (Kosar, E. et al., 1986) As in thalamus, somatotopy is evident in the cortical taste area. Taste neurons are located in a region bordering (ventral to) somatosensory primary cortical neurons representing the tongue, which in turn are just ventral to mouth and face representations (Norgren, R. and Wolf, G., 1975).

4.11.3.4 Basal Forebrain and Descending Pathways

In addition to the thalamocortical gustatory pathway, there are dense ascending taste projections to the basal forebrain (Figure 12). Axons from cells in the NST and PBN terminate heavily in the lateral hypothalamus, the amygdala, and the bed nucleus of the stria terminalis (Saper, C. B., 1982; Halsell, C. B., 1992; Nakashima, M. et al., 2000; Pritchard, T. C. et al., 2000). These limbic projections likely mediate autonomic functions related to activation of the taste system and hedonic aspects of feeding. These limbic regions, together with the pontine and

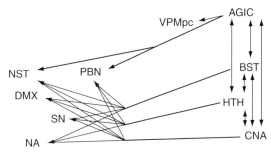

Figure 15 Descending pathways of the mammalian gustatory system schematically represented in the sagittal plane (rostral is at the right). Lines with single arrowheads indicate major long, descending forebrain projections to the hindbrain. Forebrain regions include agranular insular (gustatory) cortex (AGIC), the bed nucleus of the stria terminalis (BST), the hypothalamus (HTH), and the central nucleus of the amygdala (CNA). Hindbrain nuclei that receive terminals from the forebrain gustatory-related areas are the NST, the PBN, the dorsal motor vagal nucleus (DMX), the salivatory nuclei (SN) and nucleus ambiguus (NA). Double arrows indicate local forebrain reciprocal interconnections. Reproduced from Norgren, R. 1985. Taste and the autonomic nervous system. Chem. Senses 10, 143–161, by permission of Oxford University Press.

thalamocortical areas, are reciprocally interconnected with each other (Figure 15).

Descending axons from forebrain, taste-related areas also synapse in the NST (Figure 15). Projections from gustatory cortex to the taste region of the NST are particularly dense and terminate heavily in the rostral central subdivision, the same subnuclear zone that contains primary afferent synapse and PBN projection neurons (Whitehead, M. C. *et al.*, 2000). Thus, descending impulses from cortex are positioned to influence the first central taste area. In addition, there are descending connections from gustatory cortex to the amygdala, the hypothalamus, and the PBN (Reep, R. L. and Winans, S. S., 1982). These areas, therefore, interconnect both the gustatory NST and the gustatory neocortex. The interconnections are quite precise and related to gustatory functions (Karimnamazi, H. and Travers, J. B., 1998; Cho, Y. K. *et al.*, 2003). Thus, gustatory cortex receives projections from the lateral hypothalamus, the same region that sends descending axons to the rostral NST. Cortical connections with the PBN are reciprocal and most heavily involve the medial PBN, the major termination site for axons ascending from the rostral NST, and a site originating projections back to the rostral NST. The gustatory cortex projects heavily to the central amygdaloid nucleus, a

site of origin of amygdalofugal axons that project to the rostral NST. The anterior and basolateral amygdala are also reciprocally connected with gustatory cortex (Reep, R. L. and Winans, S. S., 1982; Whitehead, M. C. *et al.*, 2000). By these connections, the forebrain, particularly the limbic system, has access to both taste information and regions involved in autonomic outflow, for example, for energy and hydromineral regulation (Norgren, R., 1985; Grill, H. J. and Kaplan, J. M., 2002).

4.11.4 Conclusion

In summary, the general organization of the central gustatory pathways is similar in mammals and teleost fishes. This implies that this organizational scheme evolved early in vertebrate phylogeny. Not only are the broad patterns of connectivity similar between these disparate groups but also the details of connectivity and neurotransmitters. For example, glutamate acting on ionotropic glutamate receptors is the neurotransmitter at the first synapse of the system within the medullary gustatory nuclei (Bradley, R. M. *et al.*, 1996; Smeraski, C. A. *et al.*, 1996; Li, C. S. and Smith, D. V., 1997; Smeraski, C. A. *et al.*, 1998; 1999). Different cell populations in the primary taste relay give rise to the reflex connections and the ascending fiber systems (Finger, T. E., 1978; Whitehead, M. C., 1990; Halsell, C. B. *et al.*, 1996; Gill, C. F. *et al.*, 1999). In addition, neurons of the secondary gustatory nucleus can either project widely to both lemniscal and limbic targets or may show more restricted projections to the lemniscal nuclei only (Nomura, S. *et al.*, 1979; Halsell, C. B., 1992; Lamb, C. F. and Finger, T. E., 1996). Taken together, these findings indicate that the basic organization of the central gustatory system is phylogenetically old and stable through evolution. This stability may reflect the crucial nature of this system for survival.

References

Altschuler, S. M., Bao, X. M., Bieger, D., Hopkins, D. A., and Miselis, R. R. 1989. Viscerotopic representation of the upper alimentary tract in the rat: sensory ganglia and nuclei of the solitary and spinal trigeminal tracts. J. Comp. Neurol. 283, 248–268.

Atema, J. 1971. Structures and functions of the sense of taste in the catfish (*Ictalurus natalis*). Brain Behav. Evol. 4, 273–294.

Beckman, M. E. and Whitehead, M. C. 1991. Intramedullary connections of the nucleus of the solitary tract in the hamster. Brain Res. 557, 265–279.

Beckstead, R. M. and Norgren, R. 1979. An autoradiographic examination of the central distribution of the trigeminal, facial, glossopharyngeal, and vagal nerves in the monkey. J. Comp. Neurol. 184, 455–472.

Beckstead, R. M., Morse, J. R., and Norgren, R., 1980. The nucleus of the solitary tract in the monkey: projections to the thalamus and brain stem nuclei. J. Comp. Neurol. 190, 259–282.

Bradley, R. M., King, M. S., Wang, L., and Shu, X. 1996. Neurotransmitter and neuromodulator activity in the gustatory zone of the nucleus tractus solitarius. Chem. Senses 21, 377–385.

Braford, M. R. J. 1986. *De Gustibus Non Est Disputandem*: a spiral center for taste in the brain of the teleost fish, *Heterotis niloticus*. Science 232, 489–491.

Brining, S. K. and Smith, D. V. 1996. Distribution and synaptology of glossopharyngeal afferent nerve terminals in the nucleus of the solitary tract of the hamster. J. Comp. Neurol. 365, 556–574.

Chandrashekar, J., Hoon, M. A., Ryba, N. J., and Zuker, C. S. 2006. The receptors and cells for mammalian taste. Nature 444, 288–294.

Cho, Y. K., Li, C. S., and Smith, D. V. 2003. Descending influences from the lateral hypothalamus and amygdala converge onto medullary taste neurons. Chem. Senses 28, 155–171.

Cunningham, E. T. and Sawchenko, P. E., Jr. 2000. Dorsal medullary pathways subserving oromotor reflexes in the rat: implications for the central neural control of swallowing. J. Comp. Neurol. 417, 448–466.

Dickman, J. D. and Smith, D. V. 1988. Response properties of fibers in the hamster superior laryngeal nerve. Brain Res. 45, 25–38.

Finger, T. E. 1978. Gustatory pathways in the bullhead catfish. II. Facial lobe connections. J. Comp. Neurol. 180, 691–705.

Finger, T. E. 1988. Sensorimotor mapping and oropharyngeal reflexes in goldfish, *Carassius auratus*. Brain Behav. Evol. 31, 17–24.

Finger, T. E. 2006. Evolution of Taste. In: Evolution of the Nervous System: Non-mammalian Vertebrates (eds. J. Kaas, G. Striedter, T. Bullock, T. Preuss, J. Rubenstein, and L. Krubitzer), pp. 423–441. Elsevier.

Finger, T. E. and Kanwal, J. S. 1992. Ascending general visceral pathways within the brainstems of two teleost fishes: *Ictalurus punctatus* and *Carassius auratus*. J. Comp. Neurol. 320, 509–520.

Finger, T. E. and Morita, Y. 1985. Two gustatory systems: facial and vagal gustatory nuclei have different brainstem connections. Science 227, 776–778.

Folgueira, M., Anadon, R., and Yanez, J. 2003. Experimental study of the connections of the gustatory system in the rainbow trout, *Oncorhynchus mykiss*. J. Comp. Neurol. 465, 604–619.

Fulwiler, C. E. and Saper, C. B. 1984. Subnuclear organization of the efferent connections of the parabrachial nucleus in the rat. Brain Res. Rev. 7, 229–259.

Gill, C. F., Madden, J. M., Roberts, B. P., Evans, L. D., and King, M. S. 1999. A subpopulation of neurons in the rat rostral nucleus of the solitary tract that project to the parabrachial nucleus express glutamate-like immunoreactivity. Brain Res. 821, 251–262.

Goehler, L. E. and Finger, T. E. 1992. Functional organization of vagal reflex systems in the brain stem of the goldfish, *Carassius auratus*. J. Comp. Neurol. 319, 463–478.

Grill, H. J. and Kaplan, J. M. 2002. The neuroanatomical axis for control of energy balance. Front Neuroendocrinol. 23, 2–40.

Grill, H. J. and Norgren, R. 1978a. The taste reactivity test. I. Mimetic responses to gustatory stimuli in neurologically normal rats. Brain Res. 143, 263–79.

Grill, H. J. and Norgren, R. 1978b. The taste reactivity test. II. Mimetic responses to gustatory stimuli in chronic thalamic and chronic decerebrate rats. Brain Res. 143, 281–97.

Halpern, B. P. and Nelson, L. M. 1965. Bulbar gustatory responses to anterior and to posterior tongue stimulation in the rat. Am. J. Physiol. 209, 105–110.

Halsell, C. B. 1992. Organization of parabrachial nucleus efferents to the thalamus and amygdala in the golden hamster. J. Comp. Neurol. 317, 57–78.

Halsell, C. B. and Frank, M. E. 1989. Cytoarchitecture of the pontine taste area in the golden hamster. Chem. Senses 14, 707.

Halsell, C. B. and Travers, S. P. 1997. Anterior and posterior oral cavity responsive neurons are differentially distributed among parabrachial subnuclei in rat. J. Neurophysiol. 78, 920–38.

Halsell, C. B., Travers, S. P., and Travers, J. B. 1996. Ascending and descending projections from the rostral nucleus of the solitary tract originate from separate neuronal populations. Neuroscience 72, 185–197.

Hamilton, R. B. and Norgren, R. 1984. Central projections of gustatory nerves in the rat. J. Comp. Neurol. 222, 560–577.

Hamilton, R. B., Pritchard, T. C., and Norgren, R. 1985. Projections to the gustatory thalamus of new world primates. Soc. Neurosci. Abstr. 11, 1259.

Harrer, M. I. and Travers, S. P. 1996. Topographic organization of Fos-like immunoreactivity in the rostral nucleus of the solitary tract evoked by gustatory stimulation with sucrose and quinine. Brain Res. 711, 125–137.

Herrick, C. J. 1905. The central gustatory paths in the brains of bony fishes. J. Comp. Neurol. 15, 375–456.

Herrick, C. J. 1922. An Introduction to Neurology. W.B. Saunders Co.

Kanwal, J. S. and Finger, T. E. 1997. Parallel medullary gustatospinal pathways in a catfish: possible neural substrates for taste-mediated food search. J. Neurosci. 17, 4873–4885.

Kanwal, J. S., Finger, T. E., and Caprio, J. 1988. Forebrain connections of the gustatory system in ictalurid catfishes. J. Comp. Neurol. 278, 353–376.

Karimnamazi, H. and Travers, J. B. 1998. Differential projections from gustatory responsive regions of the parabrachial nucleus to the medulla and forebrain. Brain Res. 813, 283–302.

Karimnamazi, H., Travers, S. P., and Travers, J. B. 2002. Oral and gastric input to the parabrachial nucleus of the rat. Brain Res. 957, 193–206.

Katz, D. M. and Karten, H. J. 1983. Visceral representation within the nucleus of the tractus solitarius in the pigeon, *Columba livia*. J. Comp. Neurol. 218, 42–73.

King, M. S. 2003. Distribution of immunoreactive GABA and glutamate receptors in the gustatory portion of the nucleus of the solitary tract in rat. Brain Res. Bull. 60, 241–54.

King, M. S., Smith, M. T., Jani, K. P., and King, C. T. 2003. The locations of Fos-immunoreactive neurons following electrical stimulation of the gustatory parabrachial nucleus in conscious rats. Soc. Neurosci. Abstr. 594. 8.

Kiyohara, S. and Caprio, J. 1996. Somatotopic organization of the facial lobe of the sea catfish *Arius felis* studied by transganglionic transport of horseradish peroxidase. J. Comp. Neurol. 368, 121–135.

Kiyohara, S. and Tsukahara, J. 2005. Barbel Taste System in Catfish and Goatfish. In: Fish chemosenser (eds. K. Reutter and B. G. Kapoor), pp. 175–211. Science Publishers.

Kiyohara, S., Sakata, Y., Yoshitomi, T., and Tsukahara, J. 2002. The 'goatee' of goatfish: innervation of taste buds in the barbels and their representation in the brain. Proc. R. Soc. B. Biol. Sci. 269, 1773–1780.

Kosar, E., Grill, H. J., and Norgren, R. 1986. Gustatory cortex in the rat. I. Physiological properties and cytoarchitecture. Brain Res. 379, 329–241.

Lamb, C. F. and Caprio, J. 1992. Convergence of oral and extraoral information in the superior secondary gustatory nucleus of the channel catfish. Brain Res. 588, 201–211.

Lamb, C. F. and Finger, T. E. 1996. Axonal projection patterns of neurons in the secondary gustatory nucleus of channel catfish. J. Comp. Neurol. 365, 585–593.

Lasiter, P. S. and Kachele, D. L. 1988. Organization of GABA and GABA-transaminase containing neurons in the gustatory zone of the nucleus of the solitary tract. Brain Res. Bull. 21, 623–636.

Li, C. S. and Smith, D. V. 1997. Glutamate receptor antagonists block gustatory afferent input to the nucleus of the solitary tract. J. Neurophysiol. 77, 1514–1525.

Marui, T. 1981. Taste responses in the superior secondary gustatory nucleus of the carp, *Cyprinus carpio* L. Brain Res. 217, 59–68.

May, O. L. and Hill, D. L. 2006. Gustatory terminal field organization and developmental plasticity in the nucleus of the solitary tract revealed through triple-fluorescence labeling. J. Comp. Neurol. 497, 658–669.

McPheeters, M., Hettinger, T. P., Nuding, S. C., Savoy, L. D., Whitehead, M. C., and Frank, M. E. 1990. Taste-responsive neurons and their locations in the solitary nucleus of the hamster. Neuroscience 34, 745–758.

Mistretta, C. M. 1988. How the anterior tongue and taste responses are mapped onto the NST in fetal and perinatal sheep. Chem. Senses 13, 719–720.

Morita, Y. and Finger, T. E. 1985a. Reflex connections of the facial and vagal gustatory systems in the brainstem of the bullhead catfish, *Ictalurus nebulosus*. J. Comp. Neurol. 231, 547–558.

Morita, Y. and Finger, T. E. 1985b. Topographic and laminar organization of the vagal gustatory system in the goldfish, *Carassius auratus*. J. Comp. Neurol. 238, 187–201.

Morita, Y. and Finger, T. E. 1987. Topographic representation of the sensory and motor roots of the vagus nerve in the medulla of goldfish, *Carassius auratus*. J. Comp. Neurol. 264, 231–249.

Morita, Y., Ito, H., and Masai, H. 1980. Central gustatory paths in the crucian carp, *Carassius carassius*. J. Comp. Neurol. 191, 119–132.

Nakashima, M., Uemura, M., Yasui, K., Ozaki, H. S., Tabata, S., and Taen, A. 2000. An anterograde and retrograde tract-tracing study on the projections from the thalamic gustatory area in the rat: distribution of neurons projecting to the insular cortex and amygdaloid complex. Neurosci. Res. 36, 297–309.

Nomura, S., Mizuno, N., Itoh, K., Matsuda, K., Sugimoto, T., and Nakamura, Y. 1979. Localization of parabrachial nucleus neurons projecting to the thalamus or the amygdala in the cat using horseradish peroxidase. Exp. Neurol. 64, 375–385.

Norgren, R. 1978. Projections from the nucleus of the solitary tract in the rat. Neuroscience 3, 207–218.

Norgren, R. 1985. Taste and the autonomic nervous system. Chem. Senses 10, 143–161.

Norgren, R. 1995. Gustatory System. In: The Rat Nervous System (*ed*. G. Paxinos), pp. 751–771. Academic Press.

Norgren, R. and Leonard, C. M. 1973. Ascending central gustatory pathways. J. Comp. Neurol. 150, 217–237.

Norgren, R. and Pfaffmann, C. 1975. The pontine taste area in the rat. Brain Res. 91, 99–117.

Norgren, R. and Wolf, G. 1975. Projections of thalamic gustatory and lingual areas in the rat. Brain Res. 92, 123–129.

Osse, J. W., Sibbing, F. A., and van den Boogaart, J. G. 1997. Intra-oral food manipulation of carp and other cyprinids:

adaptations and limitations. Acta Physiol. Scand. Suppl. 638, 47–57.

Pfaffmann, C., Erickson, R. O., Frommer, G. P., and Halpern, B. P. 1961. Gustatory Discharges in the Rat Medulla and Thalamus. In: Sensory Communication (*ed*. W. A. Rosenblith), pp. 455–473. Wiley.

Pritchard, T. C., Hamilton, R. B., and Norgren, R. 2000. Projections of the parabrachial nucleus in the old world monkey. Exp. Neurol. 165, 101–117.

Reep, R. L. and Winans, S. S. 1982. Afferent connections of dorsal and ventral agranular insular cortex in the hamster. Neuroscience 7, 1265–1288.

Ricardo, J. A. and Koh, E. T. 1978. Anatomical evidence of direct projections from the nucleus of the solitary tract to the hypothalamus, amygdala, and other forebrain structures in the rat. Brain Res. 153, 1–26.

Rink, E. and Wullimann, M. F. 1998. Some forebrain connections of the gustatory system in the goldfish *Carassius auratus* visualized by separate DiI application to the hypothalamic inferior lobe and the torus lateralis. J. Comp. Neurol. 394, 152–170.

Rogers, R. C. and McCann, M. J. 1993. Intramedullary connections of the gastric region in the solitary nucleus: a biocytin histochemical tracing study in the rat. J. Auton. Nerv. Syst. 42(2), 119–130.

Ross, C. A., Ruggiero, D. A., and Reis, D. J. 1985. Projections from the nucleus tractus solitarii to the rostral ventrolateral medulla. J. Comp. Neurol. 242, 511–534.

Saper, C. B. 1982. Convergence of autonomic and limbic connections in the insular cortex of the rat. J. Comp. Neurol. 210, 163–173.

Scott, T. R., Yazley, S., Sienkiewicz, J., and Rolls, E. T. 1986. Gustatory responses in the nucleus tractus solitarius of the alert cynomolgus monkey. J. Neurophysiol. 55, 182–200.

Sibbing, F. A., Osse, J. M. W., and Terlouw, A. 1986. Food handling in the carp (*Cyrinus carpio*): its movement patterns, mechanisms, and limitations. J. Zool. 210, 161–203.

Smeraski, C. A., Dunwiddie, T. V., Diao, L., and Finger, T. E. 1998. Excitatory amino acid neurotransmission in the primary gustatory nucleus of the goldfish *Carassius auratus*. Ann. N. Y. Acad. Sci. 855, 442–449.

Smeraski, C. A., Dunwiddie, T. V., Diao, L., and Finger, T. E. 1999. NMDA and non-NMDA receptors mediate responses in the primary gustatory nucleus in goldfish. Chem. Senses 24, 37–46.

Smeraski, C. A., Dunwiddie, T. V., Diao, L. H., Magnusson, K. R., and Finger, T. E. 1996. Glutamate receptors mediate synaptic responses in the primary gustatory nucleus in goldfish. Soc. Neurosci. Abstr. 22.

Torvik, A. 1956. Afferent connections to the sensory trigeminal nuclei, the nucleus of the solitary tract and adjacent structure. J. Comp. Neurol. 106, 51–132.

Travers, J. B. 1988. Efferent projections from the anterior nucleus of the solitary tract of the hamster. Brain Res. 457, 1–11.

Travers, S. P. 1993. Orosensory Processing in Neural Systems of the Nucleus of Solitary Tract. In: Mechanisms of Taste Transduction (*eds*. S. A. Simon and S. D. Roper), pp. 339–424. CRC press.

Travers, J. B. and Norgren, R. 1983. Afferent projections to the oral motor nuclei in the rat. J. Comp. Neurol., 220, 280–298.

Travers, S. P. and Hu, H. 2000. Extranuclear projections of rNST neurons expressing gustatory-elicited Fos. J. Comp Neurol. 427, 124–138.

Wehby, R. G. and London, J. A. 1995. Cyto-, myelo-, and chemoarchitecture of the insular cortex of the Syrian golden hamster. AChems. Abstr. 17, 251.

Whitehead, M. C. 1986. Anatomy of the gustatory system in the hamster: synaptology of facial afferent terminals in the solitary nucleus. J. Comp. Neurol. 244, 72–85.

Whitehead, M. C. 1988a. Neuroanatomy of the gustatory system. Gerodontics 4, 239–243.

Whitehead, M. C. 1988b. Neuronal architecture of the nucleus of the solitary tract in the hamster. J. Comp. Neurol., 276, 547–572.

Whitehead, M. C. 1990. Subdivisions and neuron types of the nucleus of the solitary tract that project to the parabrachial nucleus in the hamster. J. Comp. Neurol. 301, 554–574.

Whitehead, M. C. 1993. Distribution of synapses on identified cell types in a gustatory subdivision of the nucleus of the solitary tract. J. Comp. Neurol. 332, 326–340.

Whitehead, M. C., Bergula, A., and Holliday, K. 2000. Forebrain projections to the rostral nucleus of the solitary tract in the hamster. J. Comp. Neurol. 422, 429–447.

Whitehead, M. C., McPheeters, M., Savoy, L. D., and Frank, M. E. 1993. Morphological types of neurons located at taste-responsive sites in the solitary nucleus of the hamster. Microsc. Res. Tech. 26, 245–359.

Wullimann, M. F. 1988. The tertiary gustatory center in sunfishes is not nucleus glomerulosus. Neurosci. Lett. 86, 6–10.

Yoshimoto, M., Albert, J. S., Sawai, N., Shimizu, M., Yamamoto, N., and Ito, H. 1998. Telencephalic ascending gustatory system in a cichlid fish, *Oreochromis* (Tilapia) *niloticus*. J. Comp. Neurol. 392, 209–226.

Further Reading

Mistretta, C. M. 1998. How the anterior tongue and taste responses are mapped onto the NST in fetal and perinatal sheep. Chem. Senses. 13, 719–720.

4.12 Neurotransmitters in the Taste Pathway

R M Bradley, University of Michigan, Ann Arbor, MI, USA

4.12.1 Introduction

In general there are two groups of investigators studying the nucleus of the solitary tract (NST). The largest group investigates the caudal nucleus that is involved in control of the digestive, cardiovascular, and respiratory systems. A much smaller cadre of investigators study the rostral nucleus (rNST) that forms the first central relay in the central taste pathway. Gustatory, tactile, and thermal information is relayed from the oral cavity via the chorda tympani and greater superficial petrosal branches of the facial nerve and the lingual branch of the glossopharyngeal nerve to the rNST, sometimes referred to as the gustatory NST. From there information either passes to rostral brain areas or to brainstem motor systems involved in facial and masticatory movements or the secretion of saliva.

Synaptic processing is involved at all stages of the pathway and investigators of both the rostral and caudal NST have extensively studied the synapses using anatomical, neurophysiological, and immunocytochemical techniques. The present review is restricted to the current knowledge of the neurotransmitters involved in the taste pathway which has been the focus of reviews (Bradley, R. M. *et al.*, 1996; Bradley, R. M. and Grabauskas, G., 1998; Smith, D. V. *et al.*, 1998). Neurotransmitters involved in the caudal NST have been extensively reviewed (Van Giersbergen, P. L. M. *et al.*, 1992; Andresen, M. C. and Kunze, D. L., 1994; Lawrence, A. J. and Jarrott, B., 1996) and whereas there are similarities between both the rostral and caudal divisions of the NST there are significant differences.

While these differences often reflect true differences in neurotransmitter expression, they may also be due to the fact that investigators studying a particular neurotransmitter restrict the investigation to either the caudal or rostral division. Moreover it is not always possible in investigations using coronal sections to determine how far rostral in the NST the sections are studied (e.g., see Ambalavanar, R. *et al.*, 1998; Saha, S. *et al.*, 2001) even though the title of the reports implies that the whole NST is being investigated. When investigators use horizontal sections of the entire extent of the NST differences in neurotransmitter distribution become evident (see Davis, B. J. and Jang, T., 1988).

4.12.2 Neurotransmitters in the Sensory Ganglia Involved in Transmission of Taste Information

Gustatory information derived from taste receptor cells is conveyed to the first central relay in the taste pathway via afferent sensory fibers. The cell bodies of these fibers are collected in peripheral sensory ganglia which include the geniculate (GG, facial nerve VII), petrosal (PG, glossopharyngeal IX), and nodose ganglia (NG, vagus nerve X). The neurotransmitter composition of the PG and NG has been studied using immunocytochemical techniques and these ganglia contain glutamate, substance P (SP), tyrosine hydroxylase, vasoactive intestinal polypeptide, calcitonin gene-related peptide, galanin, and aspartate

(Helke, C. J. and Hill, K. M., 1988; Helke. C. J. and Niederer, A. J., 1990; Czyzyk-Kreska, M. F. *et al.*, 1991; Helke, C. J. and Rabchevsky, A., 1991; Ichikawa, H. *et al.*, 1991; Finley, J. C. W. *et al.*, 1992; Okada, J. and Miura, M., 1992). No similar immunocytochemical survey has been undertaken for the GG. However, because afferent sensory neurons with cell bodies in the PG and GG supply a heterogeneous population of receptors, retrograde tracing techniques have been used to isolate the neurons that innervate taste buds for detailed investigation. Neurons in the GG and PG supplying the tongue, fluorescently labeled with either Fluorogold or true blue, were immunoreacted for glutamate receptors (Caicedo, A. *et al.*, 2004) or acutely dissociated and analyzed with patch-clamp recording (King, M. S. and Bradley, R. M., 2000; Koga, T. and Bradley, R. M., 2000). The pharmacological properties of neurons of the NG innervating laryngeal taste buds have not been determined.

There were significant differences in the response of PG and GG gustatory neurons to the application of acetylcholine (ACh), serotonin (5-HT), SP, and gamma-aminobutyric acid (GABA). Whereas PG neurons responded to ACh, serotonin, SP, and GABA, GG neurons only responded to SP and GABA (Koga, T. and Bradley, R. M., 2000; Figure 1).

Gustatory neurons of the GG have been examined for their expression and response to glutamate receptor agonists. Ionotropic glutamate receptors are divided into two major groups based on their sensitivity to the agonists *N*-methyl-D-aspartate (NMDA) and α-amino-3-hydroxy-5-methyl-4-isoxazoleproprionic acid (AMPA). Dissociated GG taste neurons responded to application of both agonists (King, M. S. and Bradley, R. M., 2000; Figure 2). By using antibodies to NMDA and AMPA receptor subunits, Caicedo A. *et al.* (2004) were able to demonstrate that gustatory neurons in the GG were strongly immunoreactive for GluR2/3, GluR4, and NR1 subunits suggesting that putative AMPA receptors are mainly composed of GluR3 and GluR4 subunits while NMDA receptors consist of NR1 receptor subunits.

4.12.3 Neurotransmitters in the Rostral Nucleus of the Solitary Tract Involved in Processing of Taste Information

Both anatomical and electrophysiological techniques have been used to define the neurotransmitters and neuromodulators and their receptors within the rNST.

Glutamate and GABA have been shown to play a role in synaptic transmission in the rNST. Glutamate is the afferent excitatory transmitter released from gustatory afferent terminals (Wang, L. and Bradley, R. M., 1995; Li, C. S. and Smith, D. V., 1997) and is contained in neurons of the rNST (Sweazey, R. D., 1995). GABA immunoreactivity is present in the rNST, primarily contained in small ovoid neurons (Lasiter, P. S. and Kachele, D. L., 1988; Davis, B. J., 1993; Sweazey, R. D., 1996; Leonard, N. L. *et al.*, 1999). Electrophysiological recordings in brain slices indicate that neuron of the rNST express both AMPA and NMDA receptors as well as $GABA_A$ and $GABA_B$ receptors (Liu, H. *et al.*, 1993; Wang, L. and Bradley, R. M., 1993; Bradley, R. M. *et al.*, 1996; Li, C. S. and Smith, D. V., 1997; Figure 2). The brainstem slice experiments all suggest that rNST neurons are under tonic GABAergic inhibition. This was directly tested *in vivo* on taste-responsive neurons in which the spontaneous activity was suppressed in a concentration-dependent manner by microinjection of GABA (Smith, D. V. and Li, C. S., 1998).

The distribution of glutamate and GABA receptors in the rNST is not uniform. AMPA and NMDA receptors are concentrated in the rostral central division of the rNST which receives most of the primary afferent input (Whitehead, M. C., 1988). In contrast, GABA is more uniformly distributed throughout the rNST (King, M. S., 2003). These differences in distribution of the glutamate and GABA receptors are thought to represent their different roles in gustatory processing, glutamate being involved in excitatory input from afferent taste fibers, and GABA mediating generalized tonic inhibition of the whole rNST.

Study of rNST synaptic potentials reveals that they consist of a complex mixture of excitatory and inhibitory potentials. The synaptic potentials result from excitatory afferent input mediated by glutamate and inhibitory transmission is mediated primarily at $GABA_A$ synapses (Grabauskas, G. and Bradley, R. M., 1996). By eliminating the excitatory component with glutamate receptor blockers the inhibitory component can be isolated, allowing the study of pure inhibitory postsynaptic potentials (IPSPs) in rNST. In addition, instead of using single-shock stimuli to investigate synaptic potentials, trains of stimuli (tetanic stimulation) that mimic the frequency of afferent input to the rNST evoked by gustatory stimulation reveals additional characteristics of inhibitory activity in the rNST. Tetanic stimulation at frequencies of 10–30 Hz resulted in sustained hyperpolarizing IPSPs that could be blocked by the $GABA_A$

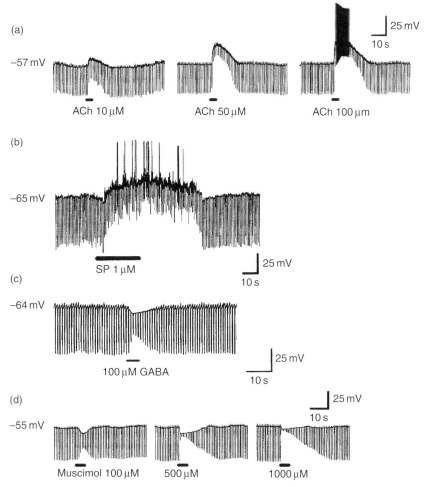

Figure 1 Voltage responses recorded from petrosal (PG) and geniculate ganglion (GG) neurons innervating the tongue to hyperpolarizing injections of −100 pA, 100 ms current pulses. (a) Membrane responses of a PG neuron to a 5 s application (bar) of increasing concentrations of acetylcholine (ACh). The ACh application induced membrane depolarization accompanied by a decrease in input resistance. The response is dose-dependent and returns to control levels within 30 s after termination of application. The depolarization resulting from the highest concentration of ACh initiated a burst of action potentials. (b) Membrane response of a PG neuron to a 5 s application (bar) of substance P (SP). In this neuron SP application depolarized the neuron and initiated the production of action potentials. (c) Membrane response of a GG neuron to a 5 s application (bar) of gamma-aminobutyric acid (GABA). In this neuron GABA application hyperpolarized the neuron with a decrease in input resistance. (d) Membrane responses of a PG neuron to a 5 s application (bar) of increasing concentrations of the $GABA_A$ receptor agonist muscimol. The mucimol application induced dose-dependent membrane hyperpolarization accompanied by a decrease in input resistance. Reproduced from Koga T. and Bradley R. M. 2000. Biophysical properties and responses to neurotransmitters of petrosal and geniculate ganglion neurons innervating the tongue. J. Neurophysiol. 84, 1404–1413, used with permission from the American Physiological Society.

antagonist bicuculline. Because afferent sensory information to the rNST consists of relatively high-frequency spike trains, these short-term changes in inhibitory synaptic activity could potentially play a key role in taste processing by facilitating synaptic integration (Grabauskas, G. and Bradley, R. M., 1998).

A single tetanic stimulation results in potentiation of single stimulus shock-evoked IPSPs for a considerable time. This potentiation can last over 30 min (Figure 3).

Tetanic stimulation results in activation of all postsynaptic $GABA_A$ receptors and induces long-lasting changes in the presynaptic GABAergic neuron. These long-lasting changes of the presynaptic neuron facilitate the release of GABA during single stimulus shock and, as a consequence, more postsynaptic receptors are activated during a single stimulus shock. Long-term potentiation of inhibition in the gustatory relay nucleus has the capability to profoundly alter transmission of

(a)

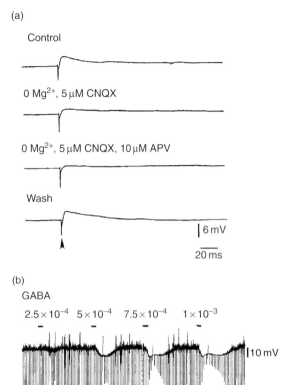

Control

0 Mg^{2+}, 5 μM CNQX

0 Mg^{2+}, 5 μM CNQX, 10 μM APV

Wash

6 mV

20 ms

(b)

GABA

2.5×10^{-4} 5×10^{-4} 7.5×10^{-4} 1×10^{-3}

10 mV

100 pA

1 min

Figure 2 (a) Effect of glutamate receptor antagonists on excitatory postsynaptic potential (EPSP) recorded from a rostal nucleus of the solitary tract (rNST) neuron in a brain slice preparation superfused with physiological saline that did not contain magnesium. The EPSPs were initiated (arrowhead) by electrical stimulation of the solitary tract. Mean of 10 EPSPs under control conditions and after application of CNQX (an α-amino-3-hydroxy-5-methyl-4-isoxazoleproprionic acid (AMPA)/kinate glutamate receptor antagonist) alone and CNQX and APV (an N-methyl-D-aspartate (NMDA) glutamate receptor blocker). CNQX reduced the EPSP leaving a CNQX-resistant component. The CNQX resistant component was blocked by application of APV. (b) Voltage responses recorded from an rNST neuron in a brain slice preparation to hyperpolarizing injections of −100 pA, 100 ms current pulses. The neuron was exposed to increasing 5 s (bar) concentrations of gamma-aminobutyric acid (GABA) interspersed with control saline rinses. GABA hyperpolarizes the neuron in a concentration-dependent manner accompanied by a decrease in input resistance. (a) Reproduced from Wang L. and Bradley R. M. 1995. *In vitro* study of afferent synaptic transmission in the rostral gustatory zone of the rat nucleus of the solitary tract. Brain Res. 702, 188–198, with permission from Elsevier. (b) Reproduced from Wang L. and Bradley R. M. 1993. Influence of GABA on neurons of the gustatory zone of the rat nucleus of the solitary tract. Brain Res. 616, 144–153, with permission from Elsevier.

taste information and therefore may be of significance in the mechanism of taste-related learning phenomena (Grabauskas, G. and Bradley, R. M., 1999).

The postnatal development of inhibition in rNST has also been examined (Grabauskas, G. and Bradley, R. M., 2001). All IPSPs recorded in postnatal rats were hyperpolarizing, but the rise and decay time constants of the single stimulus shock-evoked IPSPs decreased, and the inhibition response–concentration function to the GABA antagonist bicuculline shifted to the left during development. In postnatal 0 (P0)–P14, but not older animals, the IPSPs had a bicuculline insensitive component that was sensitive to block by picrotoxin, suggesting a transient expression of GABA$_C$ receptors. These developmental changes in inhibitory synaptic activity may operate to shape synaptic activity in early development of the rat gustatory system during a time of maturation of taste preferences and aversions (Grabauskas, G. and Bradley, R. M., 2001).

Other neurotransmitters and neuropeptides have been identified in the gustatory NST including SP, enkephalin, and catecholamines (Davis, B. J. and Jang, T., 1988; Davis, B. J. and Kream, R. M., 1993; Davis, B. J. and Smith, H. M., 1999). SP immunoreactivity is found within the glossopharyngeal nerve (Cuello, A. C. and Kanazawa, I., 1978; Helke, C. J. and Hill, K. M., 1988) as well as within the rNST (Cuello, A. C. and Kanazawa, I., 1978; Jacquin, T. *et al.*, 1989). Application of SP to neurons of the rNST in an *in vitro* brain slice preparation depolarized the majority of the neurons studied (68%) in a dose-dependent manner (King, M. S. *et al.*, 1993). The excitatory effect of SP was the result of direct postsynaptic action on the neurons (King, M. S. *et al.*, 1993). Similar results were reported when SP was microinjected onto rNST neurons in an *in vivo* preparation (Davis, B. J. and Smith, D. V., 1997). In this study, SP excited 48% of the neurons and enhanced the response of the rNST neurons to stimulation of the tongue with taste stimuli. In both the *in vitro* and *in vivo* studies a small number of neurons were inhibited by SP application demonstrated by membrane hyperpolarization or suppression of taste responses, respectively. Thus, SP effectively modulates the activity of gustatory neurons in rNST.

Opioids also modulate the activity of gustatory neurons in rNST (Li, C. S. *et al.*, 2003). During extracellular recording from rNST neurons, met-enkephalin was microinjected resulting in the suppression of taste initiated responses in 24% of the recorded neurons. The opioid antagonist naltrexone effectively blocked the inhibitory effect

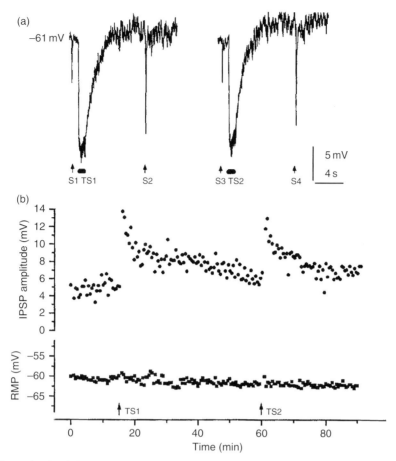

Figure 3 Effect of tetanic stimulation on inhibitory postsynaptic potentials (IPSPs) recorded from a rostal nucleus of the solitary tract (rNST) neuron in a brain slice preparation. IPSPs were isolated by chronic application of glutamate receptor blockers and elicited by electrical stimulation of the rNST. (a) After eliciting a control IPSP (S1) a tetanic stimulation (TS1) evoked potentiation of the amplitude of a subsequent IPSP (S2). Forty-five minutes after the initial potentiation the IPSP returns to control levels (S3). A second tetanic stimulation (TS2) results in potentiation again (S4). (b) Tetanic stimulation (TS1) potentiates the IPSP amplitude (●) for a considerable length of time. Tetanic stimulation does not change the resting potential (■). Once the IPSP returns to control levels a second tetanic stimulation (TS2) results in potentiation again. RMP, resting membrane potential. Reproduced from Grabauskas G. and Bradley R. M. 1999. Potentiation of GABAergic synaptic transmission in the rostral nucleus of the solitary tract. Neuroscience 94, 1173–1182, with permission from Elsevier.

of met-enkephalin suggesting that opiates maintain a tonic inhibitory action on a subpopulation of taste responsive neurons in rNST.

4.12.4 Neurotransmitters in the Parabrachial Nucleus Involved in Processing of Taste Information

In rodents and lagomorphs, gustatory information processed at the brainstem level ascends and relays in the parabrachial nucleus (PBN) in the dorsal pons (Norgren, R., 1978; Travers, J. B., 1988; Whitehead, M. C., 1990). Taste-responsive PBN neurons also

have descending connections to the rNST as well as to medullary motor nuclei (Karimnamazi, H. and Travers, J. B., 1998). These connections imply that the PBN plays a significant role in taste processing by transmitting information in both a rostral and caudal direction.

The PBN has been dived into a number of major regions (Fulwiler, C. E. and Saper, C. B., 1984) with the central medial (CM), ventrolateral (VL), and waist area (W) being important in taste processing because these areas have been shown to receive input from the rNST and neurons in these areas respond to taste stimuli flowed over the tongue (Norgren, R. and Pfaffmann, C., 1975). Synaptic input from the rNST to the PBN is mostly excitatory and mediated by glutamate since this neurotransmitter has been shown to be contained in a subset of rNST neurons that project to the PBN (Chamberlin, N. L. and Saper, C. B., 1995; Gill, C. F. *et al.*, 1999). Glutamate receptors have been shown to be present throughout the PBN but neurons in the W express AMPA (Chamberlin, N. L. and Saper, C. B., 1995; Guthmann, A. and Herbert, H., 1999c), NMDA receptors (Zidichouski, J. A. *et al.*, 1996; Guthman, A. and Herbert, H., 1999b) as well as metabotropic glutamate receptors (Zidichouski, J. A. *et al.*, 1996; Guthman, A. and Herbert, H., 1999a).

Investigators have also demonstrated that PBN neurons receive inhibitory input from the NST (Granata, A. R. and Kitai, S. T., 1989; Jhamandas, J. H. and Harris, K. H., 1992; Suemori, K. *et al.*, 1994). Based on the results of injections of GABA agonists and antagonists into the PBN this inhibition has been demonstrated to be GABAergic (Christie, M. J. and North, R. A., 1988; Mollace, V. *et al.*, 1991). The effect of GABA was studied on gustatory PBN neurons *in vitro* indicating that GABA functions as an inhibitory neurotransmitter in the gustatory PBN, mediated principally by GABA$_A$ receptors (Kobashi, M. and Bradley, R. M., 1998). At present no specific information is available on other neurotransmitters active on neurons in the gustatory PBN or on other synaptic relays in the ascending gustatory pathway.

4.12.5 Role of Neurotransmitters in Brainstem Taste-Initiated Reflex Activity

4.12.5.1 Gusto-Salivary Reflex Activity

Saliva is essential in the initial stages of taste transduction, acting as a solvent for taste stimuli, as a transport medium of the dissolved stimuli, and a possible source of ions that pass through taste receptor apical ion channels to depolarize or hyperpolarize the taste cell (Bradley, R. M., 1991). In concert, stimulation of taste receptors reflexly initiates saliva flow (Matsuo, R., 1999a). This reflex pathway involves afferent input to the rNST that then synapses with parasympathetic secretomotor neurons innervating the salivary glands. The parasympathetic neurons are collected in a column of brainstem neurons termed salivatory nuclei that are closely associated with the medial border of the rNST. Dendrites of the salivatory neurons penetrate into the rNST (Kim, M. *et al.*, 2004) facilitating synaptic contacts.

Recordings from identified neurons of the salivatory nuclei have demonstrated that the postsynaptic potentials initiated by stimulation of the solitary tract or adjacent neuropil are a mixture of excitation and inhibition (Mitoh, Y. *et al.*, 2004; Bradley, R. M. *et al.*, 2005). Application of synaptic agonists and antagonists have revealed that the excitatory component is mediated by NMDA and AMPA receptors and that inhibition results from stimulation of GABA$_A$ receptors (Mitoh, Y. *et al.*, 2004; Suwabe, T. and Bradley, R. M., 2005). Glycinergic receptors have been demonstrated on some salivatory neurons also (Mitoh, Y. *et al.*, 2004). Somata and dendrites of salivatory neurons receive symmetric synapses from glutamatergic, GABA, and glycine immunoreactive varicosities (Kobayashi, M. *et al.*, 1997). Salivatory nucleus neuron somata were shown to be strongly positive for the glutamate receptor subtypes NR1, NR2A, and either unlabeled or weakly positive for other glutamate receptor subtypes indicating that postsynaptic activity is mediated principally by NMDA receptors (Kim, M. *et al.*, 2005; Figure 4). Based on the neurophysiological measures these excitatory and inhibitory synapses are either monosynaptic connections between primary afferent fibers or involve polysynaptic connections via neurons in the rNST (Suwabe, T. and Bradley, R. M., 2005).

The salivary reflex is also influenced by descending input from higher brain centers (Matsuo, R., 1999b) and a number of neuropeptides have been hypothesized to be involved in this descending modulation of secretion as demonstrated by strong immunoreactivity for SP, serotonin, neuropeptide Y, somatostatin, tyrosine hydroxylase, vasoactive intestinal peptide, and calcitonin gene-related peptide (Nemoto, T. *et al.*, 1995). Of these neuropeptides, SP and serotonin have been shown to excite salivatory neurons in a concentration-dependent manner (Suwabe, T. and Bradley, R. M., 2007).

4.12.5.2 Synaptic Effects of Neuromodulators Known to have Profound Effects on Feeding Behavior on rNST Neurons

Injection of the benzodiazepine (BZ) receptor agonist diazepam into the fourth ventricle enhances food palatability and intake (Berridge, K. C. and Peciña, S., 1995), and it has been hypothesized that this phenomenon results from the action of diazepam on rNST neurons. BZs facilitate GABA-induced IPSPs

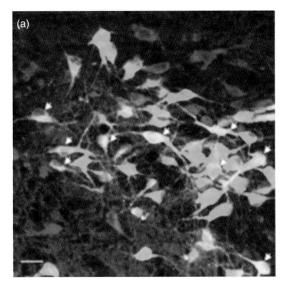
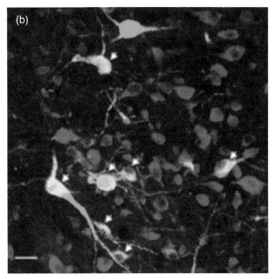

Figure 4 Glutamate receptor immunoreactivity of identified neurons of the salivatory nucleus. Parasympathetic secretomotor neurons of the inferior salivatory nucleus were identified by retrograde transport of a fluorescent marker. Horizontal sections of the medulla were subsequently immunoreacted for glutamate receptor subunits tagged with a fluorescent label. The resulting sections were examined using confocal microscopy. (a) Brainstem neurons labeled with antibodies to NR1 receptors (red) and retrogradely labeled neurons of the inferior salivatory nucleus (green). Doubled labels neurons appear yellow and are indicated by white arrows. (b) Brainstem neurons labeled with antibodies to NR2A receptors (red) and retrogradely labeled neurons of the inferior salivatory nucleus (green). Doubled labels neurons appear yellow and are indicated by white arrows. Note that many neurons in the brainstem are strongly labeled with the receptor antibody, but not all the salivatory neurons express the glutamate receptor. Scale bar = 25 μm.

in postsynaptic neurons that have $GABA_A$ receptors by modulating chloride channel activity. In brain slices maintained at body temperature, pharmacologically isolated IPSPs are diazepam sensitive demonstrating that rNST neurons express $GABA_A$ receptors with BZ binding sites. It is possible, therefore, that application of BZ into the fourth ventricle influences food intake by potentiation of GABAergic neurotransmission in the rNST (Grabauskas, G. and Bradley, R. M., 2000).

Cholecystokinin (CCK), a peptide secreted by gastrointestinal tract-lining cells, plays an important role in digestion. CCK has also been shown to cause satiety by acting centrally (Baile, C. A. *et al.*, 1986) and neurons immunoreactive for CCK have been located in the NST (Kubota, Y. *et al.*, 1983) and the PBN, including the W (Kubota, Y. *et al.*, 1983; Block, C. H. and Hoffman, G., 1987; Sutin, E. L. and Jacobowitz, D. M., 1988; Herbert, H. and Saper, C. B., 1990; Födor, M. *et al.*, 1992). Thus, CCK is present in the NST and taste-responsive area of the PBN suggesting that this neuropeptide may have a role in taste processing. While investigators using electrophysiological methods have demonstrated that CCK can modulate the excitatory response initiated by

vagal stimulation (Saleh, T. M. and Cechetto, D. F., 1996) there are no similar studies on taste-responsive neurons in the PBN.

4.12.6 Conclusions

The essence of neural processing involves synaptic interactions and synaptic interactions involve neurotransmitters. Investigators studying taste processing have only recently turned their attention to the role of neurotransmitters at each relay in the taste pathway. In the review by Travers J. B. *et al.* (1987) on gustatory neural processing in the hindbrain, neurotransmitters are not even mentioned. Since that time a number of investigators have demonstrated both anatomically and physiologically the role neurotransmission plays in taste processing. They have determined that glutamate is the excitatory transmitter between afferent taste fibers and neurons of the rNST. They have demonstrated that inhibition mediated principally by $GABA_A$ receptors is an important factor in modulating the afferent input and that neurons of the rNST are under tonic GABA inhibition. Moreover, a form of long-term inhibition must also be factored into gustatory processing by rNST circuits. It is important to also remember that afferent information is distributed to a number of destinations after arriving at the brainstem and that each of these circuits may use different neurotransmitters. Other neurotransmitters are also important as well, and some of these have been characterized, particularly SP. Besides neurotransmitters active in the ascending pathway other neurotransmitters are found in the descending pathways from more rostral brain areas. While much more information remains to be acquired it is evident from the current state of knowledge that the apparent complexity of synaptic interactions at all levels of the taste pathway should be taken into account when the subject of central taste coding is discussed.

References

Ambalavanar, R., Ludlow, C. L., Wenthold, R. J., Tanaka, Y., Damirjian, M., and Petralia, R. S. 1998. Glutamate receptor subunits in the nucleus of the tractus solitarius and other regions of the medulla oblongata in the cat. J. Comp. Neurol. 402, 75–92.

Andresen, M. C. and Kunze, D. L. 1994. Nucleus tractus solitarius – gateway to neural circulatory control. Annu. Rev. Physiol. 56, 93–116.

Baile, C. A., McLaughlin, C. L., and Della-Fera, M. A. 1986. Role of cholecystokinin and opioid peptides in control of food intake. Physiol. Rev. 66, 172–234.

Berridge, K. C. and Peciña, S. 1995. Benzodiazepines, appetite, and taste palatability. Neurosci. Biobehav. Rev. 19, 121–131.

Block, C. H. and Hoffman, G. 1987. Neuropeptides and monoamine components of the parabrachial pontine complex. Peptides 8, 267–283.

Bradley, R. M. 1991. Salivary Secretion. In: Smell and Taste in Health and Disease (eds. T. V. Getchel, R. L. Doty, L. M. Bartoshuk, and J. B. J. Snow), pp. 127–144. Raven Press.

Bradley, R. M. and Grabauskas, G. 1998. Neural Circuits for Taste – Excitation, Inhibition, and Synaptic Plasticity in the Rostral Gustatory Zone of the Nucleus of the Solitary Tract. In: Olfaction and Taste XII (ed. C. Murphy), pp. 467–474. New York Academy of Sciences.

Bradley, R. M., Fukami, H., and Suwabe, T. 2005. Neurobiology of the gustatory-salivary reflex. Chem. Senses 30, i70–i71.

Bradley, R. M., King, M. S., Wang, L., and Shu, X. Q. 1996. Neurotransmitter and neuromodulatory activity in the gustatory zone of the nucleus tractus solitarius. Chem. Senses 21, 377–385.

Chamberlin, N. L. and Saper, C. B. 1995. Differential distribution of AMPA-selective glutamate receptor subunits in the parabrachial nucleus of the rat. Neuroscience 68, 435–443.

Caicedo, A., Zucchi, B., Pereira, E., and Roper, S. D. 2004. Rat gustatory neurons in the geniculate ganglion express glutamate receptor subunits. Chem. Senses 29, 463–471.

Christie, M. J. and North, R. A. 1988. Agonist at m-opioid, M_2-muscarinic and $GABA_B$-receptors increase the same potassium conductances in rat lateral parabrachial neurones. Br. J. Pharmacol. 95, 896–902.

Cuello, A. C. and Kanazawa, I. 1978. The distribution of substance P immunoreactive fibers in the rat central nervous system. J. Comp. Neurol. 178, 129–156.

Czyzyk-Kreska, M. F., Bayliss, D. A., Seroogy, K. B., and Millhorn, D. E. 1991. Gene expression for peptides in neurons of the petrosal and nodose ganglia in rat. Exp. Brain Res. 83, 411–418.

Davis, B. J. 1993. GABA-like immunoreactivity in the gustatory zone of the nucleus of the solitary tract in the hamster: light and electron microscopic studies. Brain Res. Bull. 30, 69–77.

Davis, B. J. and Jang, T. 1988. Tyrosine hydroxylase-like and dopamine beta-hydroxylase-like immunoreactivity in the gustatory zone of the nucleus of the solitary tract in the hamster: light- and electron-microscopic studies. Neuroscience 27, 949–964.

Davis, B. J. and Kream, R. M. 1993. Distribution of tachykinin- and opioid-expressing neurons in the hamster solitary nucleus: An immuno- and in situ hybridization histochemical study. Brain Res. 616, 6–16.

Davis, B. J. and Smith, D. V. 1997. Substance P modulates taste responses in the nucleus of the solitary tract of the hamster. Neuroreport 8, 1723–1727.

Davis, B. J. and Smith, H. M. 1999. Neurokinin-1 receptor immunoreactivity in the nucleus of the solitary tract in the hamster. Neuroreport 10, 1003–1006.

Finley, J. C. W., Polak, J., and Katz, D. M. 1992. Transmitter diversity in carotid body afferent neurons: dopaminergic and peptidergic phenotypes. Neuroscience 51, 973–987.

Födor, M., Gorcs, T. J., and Plakovits, M. 1992. Immunohistochemical study on the distribution of neuropeptides within the pontine tegmentum-particularly the parabrachial nucleus and the locus coeruleus of the human brain. Neuroscience 46, 891–908.

Fulwiler, C. E. and Saper, C. B. 1984. Subnuclear organization of the efferent connections of the parabrachial nucleus in the rat. Brain Res. Rev. 7, 229–256.

Gill, C. F., Madden, J. M., Roberts, B. P., Evans, L. D., and King, M. S. 1999. A subpopulation of neurons in the rat rostral nucleus of the solitary tract that project to the parabrachial nucleus express glutamate-like immunoreactivity. Brain Res. 821, 251–262.

Grabauskas, G. and Bradley, R. M. 1996. Synaptic interactions due to convergent input from gustatory afferent fibers in the rostral nucleus of the solitary tract. J. Neurophysiol. 76, 2919–2927.

Grabauskas, G. and Bradley, R. M. 1998. Tetanic stimulation induces short-term potentiation of inhibitory synaptic activity in the rostral nucleus of the solitary tract. J. Neurophysiol. 79, 595–604.

Grabauskas, G. and Bradley, R. M. 1999. Potentiation of GABAergic synaptic transmission in the rostral nucleus of the solitary tract. Neuroscience 94, 1173–1182.

Grabauskas, G. and Bradley, R. M. 2000. Effect of diazepam on inhibitory postsynaptic potentials in nucleus of the solitary tract of neonatal rats is temperature dependent. Chem. Senses 25, 678.

Grabauskas, G. and Bradley, R. M. 2001. Postnatal development of inhibitory synaptic transmission in the rostral nucleus of the solitary tract. J. Neurophysiol. 85, 2203–2212.

Granata, A. R. and Kitai, S. T. 1989. Intracellular study of nucleus parabrachialis and nucleus tractus solitarii interconnections. Brain Res. 492, 281–292.

Guthmann, A. and Herbert, H. 1999a. Distribution of metabotropic glutamate receptor subunits in the parabrachial and kölliker-fuse nuclei of the rat. Neuroscience 89, 873–881.

Guthmann, A. and Herbert, H. 1999b. Expression on N-methyl-D-aspartate receptor subunits in the rat parabrachial and kölliker-fuse nuclei of the rat and in selected ponuomedullary brainstem nuclei. J. Comp. Neurol. 415, 501–517.

Guthmann, A. and Herbert, H. 1999c. *In situ* hybridization analysis of flip/flop splice variants of AMPA-type glutamate receptor subunits in the rat parabrachial and kölliker-fuse nuclei. Mol. Brain Res. 74, 145–157.

Helke, C. J. and Hill, K. M. 1988. Immunohistochemical study of neuropeptides in vagal and glossopharyngeal afferent neurons in the rat. Neuroscience 26, 539–551.

Helke, C. J. and Niederer, A. J. 1990. Studies on the coexistance of substance P with other putative neurotransmitters in the nodose and petrosal ganglia. Synapse 5, 144–151.

Helke, C. J. and Rabchevsky, A. 1991. Axotomy alters putative neurotransmitters in visceral sensory neurons of the nodose and petrosal ganglia. Brain Res. 551, 44–51.

Herbert, H. and Saper, C. B. 1990. Cholecystokinin-, galanin-, and corticotropin-releasing factor-like immunoreactive projections from the nucleus of the solitary tract to the parabrachial nucleus in the rat. J. Comp. Neurol. 293, 581–598.

Ichikawa, H., Jacobowitz, D. M., Winsky, L., and Helke, C. J. 1991. Calretinin-immunoreactivity in vagal and glossopharyngeal sensory neurons of the rat. Distribution and coexistence with putative transmitter agents. Brain Res. 557, 316–321.

Jacquin, T., Denavit-Saubié, M., and Champagnat, J. 1989. Substance P and serotonin mutually reverse their excitatory effects in the rat nucleus tractus solitarius. Brain Res. 502, 214–222.

Jhamandas, J. H. and Harris, K. H. 1992. Influence of nucleus tractus solitarius stimulation and baroreceptor activation on rat parabrachial neurons. Brain Res. Bull. 28, 565–571.

Karimnamazi, H. and Travers, J. B. 1998. Differential projections from gustatory responsive regions of the parabrachial nucleus to the medulla and forebrain. Brain Res. 813, 283–302.

Kim, M., Chiego, D. J., and Bradley, R. M. 2005. Characterization of the gusto-salivary reflex: ionotropic glutamate receptor expression in preganglionic neurons of the rat inferior salivatory nucleus. Program No. 281. 4, Abstract viewer/Itinerary Planner. Society for Neuroscience.

King, M. S. 2003. Distribution of immunoreactive GABA and glutamate receptors in the gustatory portion of the nucleus of the solitary tract in rat. Brain Res. Bull. 60, 241–254.

King, M. S. and Bradley, R. M. 2000. Biophysical properties and responses to glutamate receptor agonists of identified subpopulations of rat geniculate ganglion neurons. Brain Res. 866, 237–246.

King, M. S., Wang, L., and Bradley, R. M. 1993. Substance P excites neurons in the gustatory zone of the rat nucleus tractus solitarius. Brain Res. 619, 120–130.

Kobashi, M. and Bradley, R. M. 1998. Effects of GABA on neurons of the gustatory and visceral zones of the parabrachial nucleus in rats. Brain Res. 799, 323–328.

Kobayashi, M., Nemoto, T., Nagata, H., Konno, A., and Chiba, T. 1997. Immunohistochemical studies on glutamatergic, GABAergic and glycinergic axon viscosities presynaptic to parasympathetic preganglionic neurons in the superior salivatory nucleus of the rat. Brain Res. 766, 72–82.

Koga, T. and Bradley, R. M. 2000. Biophysical properties and responses to neurotransmitters of petrosal and geniculate ganglion neurons innervating the tongue. J. Neurophysiol. 84, 1404–1413.

Kubota, Y., Inagaki, S., Shiosaka, S., Cho, H. J., Tateishi, K., Hasimura, E., Hamaoka, T., and Tohyama, M. 1983. The distribution of cholecystokinin octapeptide-like structures in the lower brain stem of the rat: an immunohistochemical analysis. Neuroscience 9, 587–604.

Lasiter, P. S. and Kachele, D. L. 1988. Organization of GABA and GABA-transaminase containing neurons in the gustatory zone of the nucleus of the solitary tract. Brain Res. Bull. 21, 623–636.

Lawrence, A. J. and Jarrott, B. 1996. Neurochemical modulation of cardiovascular control in the nucleus tractus solitarius. Prog. Neurobiol. 48, 21–53.

Leonard, N. L., Renahan, W. E., and Schweitzer, L. 1999. Structure and function of gustatory neurons in the nucleus of the solitary tract. IV. The morphology and synaptology of GABA-immunoreactive terminals. Neuroscience 92, 151–162.

Li, C. S. and Smith, D. V. 1997. Glutamate receptor antagonists block gustatory afferent input to the nucleus of the solitary tract. J. Neurophysiol. 77, 1514–1525.

Li, C. S., Davis, B. J., and Smith, D. V. 2003. Opioid modulation of taste responses in the nucleus of the solitary tract. Brain Res. 965, 21–34.

Liu, H., Behbehani, M. M., and Smith, D. V. 1993. The influence of GABA on cells in the gustatory region of the hamster solitary nucleus. Chem. Senses 18, 285–305.

Matsuo, R. 1999a. Interrelation of Taste and Saliva. In: Neural Mechanisms of Salivary Gland Secretion (*eds*. J. R. Garrett, J. Ekström, and L. C. Anderson), pp. 185–195. Karger.

Matsuo, R. 1999b. Central Connections for Salivary Innervation and Efferent Impulse Formation. In: Neural Mechanisms of Salivary Secretion (*eds*. J. R. Garrett, J. Ekström, and L. C. Anderson), pp. 26–43. Karger.

Mitoh, Y., Funahashi, M., Kobashi, M., and Matsuo, R. 2004. Excitatory and inhibitory postsynaptic currents of the superior salivatory nucleus innervating the salivary glands and tongue in the rat. Brain Res. 999, 62–72.

Mollace, V., De Francesco, E. A., Fersini, G., and Nistico, G. 1991. Age-dependent changes in cardiovascular responses induced by muscimol infused into the nucleus tractus solitarii and nucleus parabrachialis in rats. Br. J. Pharmacol. 103, 1802–1806.

Nemoto, T., Konno, A., and Chiba, T. 1995. Synaptic contact of neuropeptide- and amine-containing axons on parasympathetic preganglionic neurons in the superior salivatory nucleus of the rat. Brain Res. 685, 33–45.

Norgren, R. 1978. Projections from the nucleus of the solitary tract in the rat. Neuroscience 3, 207–218.

Norgren, R. and Pfaffmann, C. 1975. The pontine taste area in the rat. Brain Res. 91, 99–117.

Okada, J. and Miura, M. 1992. Transmitter substances contained in the petrosal ganglion cells determined by a double-labeling method in the rat. Neurosci. Lett. 146, 33–36.

Saha, S., Sieghart, W., Fritschy, J. M., McWilliam, P. N., and Batten, T. F. 2001. γ-Aminobutyric acid receptor ($GABA_A$) subunits in rat nucleus tractus solitarii (NTS) revealed by polymerase chain reaction (PCR) and immunohistochemistry. Mol. Cell. Neurosci. 17, 241–257.

Saleh, T. M. and Cechetto, D. F. 1996. Peptide changes in the parabrachial nucleus following cervical vagal stimulation. J. Comp. Neurol. 366, 390–405.

Smith, D. V. and Li, C. S. 1998. Tonic GABAergic inhibition of taste-responsive neurons in the nucleus of the solitary tract. Chem. Senses 23, 159–169.

Smith, D. V., Li, C. S., and Davis, B. J. 1998. Excitatory and Inhibitory Modulation of Taste Responses in the Hamster Brainstem. In: Olfaction and Taste XII (*ed*. C. Murphy), pp. 450–456. New York Academy of Sciences.

Suemori, K., Kobashi, M., and Adachi, A. 1994. Effects of gastric distension and electrical stimulation of dorsomedial medulla on neurons in parabrachial nucleus of rats. J. Auton. Nerv. Syst. 48, 221–229.

Sutin, E. L. and Jacobowitz, D. M. 1988. Immunocytochemical localization of peptides and other neurochemicals in the rat laterodorsal nucleus and adjacent area. J. Comp. Neurol. 270, 243–270.

Suwabe, T. and Bradley, R. M. 2005. Characterization of the gusto-salivary reflex: excitatory postsynaptic potentials recorded from identified rat inferior salivatory neurons. Program No. 281. 3, Abstract viewer/Itinerary Planner. Society for Neuroscience.

Suwabe, T. and Bradley, R. M. 2007. Effects of 5-hydroxytryptamine and substance P on neurons of the inferior salivatory nucleus. J. Neurophysiol. 97, 2605–2611.

Sweazey, R. D. 1995. Distribution of aspartate and glutamate in the nucleus of the solitary tract of the lamb. Exp. Brain Res. 105, 241–253.

Sweazey, R. D. 1996. Distribution of GABA and glycine in the lamb nucleus of the solitary tract. Brain Res. 737, 275–286.

Travers, J. B. 1988. Efferent projections from the anterior nucleus of the solitary tract of the hamster. Brain Res. 457, 1–11.

Travers, J. B., Travers, S. P., and Norgren, R. 1987. Gustatory neural processing in the hindbrain. Annu. Rev. Neurosci. 10, 595–632.

Van Giersbergen, P. L. M., Palkovits, M., and de Jong, W. 1992. Involvement of neurotransmitters in the nucleus tractus solitarii in cardiovascular regulation. Physiol. Rev. 72, 789–824.

Wang, L. and Bradley, R. M. 1993. Influence of GABA on neurons of the gustatory zone of the rat nucleus of the solitary tract. Brain Res. 616, 144–153.

Wang, L. and Bradley, R. M. 1995. In vitro study of afferent synaptic transmission in the rostral gustatory zone of the rat nucleus of the solitary tract. Brain Res. 702, 188–198.

Whitehead, M. C. 1988. Neuronal architecture of the nucleus of the solitary tract in the hamster. J. Comp. Neurol. 276, 547–572.

Whitehead, M. C. 1990. Subdivisions and neuron types of the nucleus of the solitary tract that project to the parabrachial nucleus in the hamster. J. Comp. Neurol. 301, 554–574.

Zidichouski, J. A., Easaw, J. C., and Jhamandas, J. H. 1996. Glutamate receptor subtypes mediate excitatory synaptic responses of rat lateral parabrachial neurons. Am. J. Physiol. Heart Circ. Physiol. 270, H1557–H1567.

4.13 Functional Magnetic Resonance Imaging Study of Taste

A Faurion, Neurobiologie Sensorielle, NOPA-NBS, INRA, Jouy en Josas, France

T Kobayakawa, National Institute of Advanced Industrial Science and Technology (AIST), Tsukuba, Japan

B Cerf-Ducastel, San Diego State University, San Diego, CA, USA

Glossary

chemotopy Similar to tonotopy or somatotopy. For taste, still an hypothesis: the different chemicals would elicit an activation in a given localization of the primary projection area in the cortex.

EGM Electrogustometry. The application of a very low current on the tongue moves the saliva cations toward the microvillosities of the taste pores; detection thresholds are easily and precisely measured varying the current.

finger span A technique for perceived intensity evaluation. The forefinger is attached to the cursor of a potentiometer, while the thumb is fixed at the end of the resistor: the distance between both fingers tips, a reflect of the articulation opening, measures the perceived intensity.

fMRI Functional magnetic resonance imaging, a brain imaging technique.

MEG Magnetoencephalography, a brain imaging technique.

NMR Nuclear magnetic resonance.

TEP Tomography by emission of positrons, a brain imaging technique.

T–I Time–intensity. The perception profile is continuously measured during the stimulation or even after the stimulation has ended, during a persisting aftertaste.

4.13.1 History of Functional Magnetic Resonance Imaging Applied to the Study of Task-Dependent Cerebral Activation

Roy C. W. and Sherrington C. S. (1890) were the first to suggest that an increase of the local blood flow may signal an increased neuronal activity, which is now possibly visualized using nuclear magnetic resonance (NMR) imaging (Lauterbur, P. C., 1973). Neural activity results in an increased blood flow with a relative decrease in deoxyhemoglobin and an increase in oxyhemoglobin. Many efforts were developed to map brain tissue and brain activity (Le Bihan, D. *et al.*, 1986; Belliveau *et al.*, 1991; Kwong, K. K. *et al.*, 1992; Ogawa, S. *et al.*, 1992), leading to the first observation of an increased oxygen content in the visual area under stimulation. Functional magnetic resonance imaging (fMRI) uses the magnetic resonance of protons in living tissues. Spins carried by molecules are labeled by gradient pulses. During the relaxation time after a first pulse, a second one is applied which will result in a net phase shift. A relative decrease in deoxyhemoglobin (blood oxygenated level-dependent effect, BOLD) is reflected as an increase in the relaxation time, hence, as an increase of the magnetic resonance signal. The signal is an indirect reflect of the neural activity. Fundamental research in fMRI imaging is still under development to record signals that should be as close as possible from the neuronal activity.

After stimulating the spins, echoes are collected to form an image of the moving spins. With single-shot brain images that typically last between 25 and 100 ms (for example in echo-planar imaging, EPI), the technique allows powerful statistics and minimizes motion artifacts, offering a high spatial resolution of 1 mm or better. New statistical developments now care for naturally different amplitudes of the signal for repeated stimuli (in event-related experiments). The most important breakthrough brought by fMRI is that averaging data across brains is not necessary as in positron emission tomography (PET), thus allowing interpretation of individual brain activity and the study of interindividual differences (functional interindividual differences). After a brilliant start with visual, counting, finger tapping, speech and memory tasks, chemoreception was a new challenge with the difficulty of mastering stimulation from a distance. A whole body magnet puts the subject's mouth at a minimum of 1.60 m from the outside of the machine, and every apparatus including magnetic parts must be localized outside of the shielded room, some 8 m apart.

But, as for PET, the temporal resolution is limited by the timing of the physiological event actually recorded, which is not the electrical activity of neurons but the delayed hemodynamic response, between 5 and 8 s. The temporal data acquisition is *de facto* averaged with this time constant. For a better time resolution, magnetoencephalography (MEG), directly recording the brain electrical activity, is a complementary choice. Magnetic fields are measured using SQUID sensors on the whole scalp. The signal emitted by active regions of the brain is collected in each sensor, sampled, and filtered. A number of trials (30–40) are summated. The magnitude of the neuromagnetic field at each sensor is compared to a magnitude calculated at the same sensor based on a theoretical model by the Grynszpan–Geselowitz equation (Grynszpan, F. and Geselowitz, D. B., 1973). The difference observed at each sensor results from the neuronal activity, the localization of which is to be found by mathematically resolving the inverse problem. The number of activated foci responsible for the signal is arbitrarily determined by the experimenter, and their localization, at first putative, is confirmed by the minimization algorithm. The advantage of the technique is that the recorded signal is directly generated by the electrical activity of neurones with a temporal precision of 1 ms.

Associating both fMRI and MEG techniques enables researchers to look at firing foci in individual brains with the precision of 1 ms and 1 mm, provided the experimental paradigm records enough psychophysical data for the interpretation of the function of the task-activated foci. Not only similarities but also differences of results may be informative as both techniques look at physiological events through different time windows.

4.13.2 Mapping Brain Activity in Response to Taste Stimulation

The pioneering work imaging brain activation for taste was from Kinomura S. with PET in 1994 showing thalamus, insular cortex, anterior cingulate gyrus, parahippocampal gyrus, lingual gyrus, caudate nucleus, and temporal gyrus activation (Kinomura, S. *et al.*, 1994). The same year, we started experimenting with fMRI in the 3T of the Service Frédéric Joliot (CEA) at Orsay, Director D. Le Bihan. We developed a stimulating method in order to suppress somatic information, sensory adaptation, and head movements consecutive to swallowing, using a continuous flow alternating water and stimulus in an on–off paradigm (Cerf, B. *et al.*, 1996; 1998). Small bolus (100 µl) was repetitively sent to the subject's mouth, through 2 m long silicone tubes every 3 s to minimize adaptation (pulsatile stimulation). This procedure ensured a negligible increase of the volume of fluid in the mouth with respect to saliva. Subjects were allowed to swallow *ad libitum*; hence no mouth movement could be correlated with the stimulation paradigm. The time–intensity (T–I) perception profile, which was recorded with the finger-span technique using a linear potentiometer, was used as a template to correlate to the NMR signal of each pixel or voxel. This template allowed visualizing more numerous activated clusters of pixels (+400%) than the convolution (Bandettini, P. A., *et al.*, 1993) of the stimulation paradigm waveform with a standard 6 s time-constant filter. For example, the correlation between the NMR

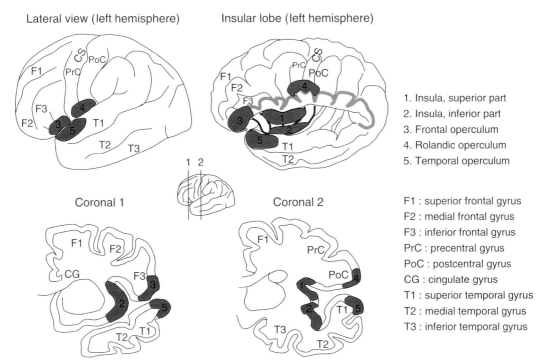

Lateral view (left hemisphere)   Insular lobe (left hemisphere)

1. Insula, superior part
2. Insula, inferior part
3. Frontal operculum
4. Rolandic operculum
5. Temporal operculum

Coronal 1   Coronal 2

F1 : superior frontal gyrus
F2 : medial frontal gyrus
F3 : inferior frontal gyrus
PrC : precentral gyrus
PoC : postcentral gyrus
CG : cingulate gyrus
T1 : superior temporal gyrus
T2 : medial temporal gyrus
T3 : inferior temporal gyrus

Figure 1 Brain areas activated by taste stimuli, observed with fMRI in human subjects. From Cerf, B. 1998. Doctoral dissertation, Paris.

signal and the T–I perception profile was above 0.6 in 30% of cases (thresholded at 0.4) and above 0.7 in 10% of cases instead of below 0.6 in standard conditions (Cerf, B. *et al.*, 1996; Van de Moortele, P.-F. *et al.*, 1997). The explanation comes from the waveform of recorded profiles which confirmed a slow rate of rise of the taste perception in these conditions (usually above 6 s time constant), which could vary with the stimulus and also with the subject who however individually exhibited a high reproducibility across sessions. Introducing the T–I perception measurement in the magnet enabled finding out pure taste activation that was barely detected otherwise (so-called taste is usually associated with somatosensory and retro-olfactory converging information). Hence, associating quantitative psychophysical evaluation to functional brain imaging is an important factor of success and an important factor for interpreting results. Besides, it provides a control of the stimulation. This recording of perception could also be run off-line, out of the magnet, for control sessions.

Using these conditions, and single subject analysis rather than group analysis, taste areas were localized (Figure 1) in various parts of the insula and in frontal, temporal, rolandic opercula, including pre- and post-central gyri (Cerf, B. *et al.*, 1998; Faurion, A. *et al.*,

1999; Cerf-Ducastel, B. *et al.*, 2001) as expected from electrophysiological recordings in the monkey (Ruch, T. C. and Patton, H. D., 1946; Bagshaw, M. H. and Pribram, K. H., 1953; Benjamin, R. M. and Burton, H., 1968; Ogawa, H. *et al.*, 1985; Pritchard, T. C. *et al.*, 1986; Yaxley, S. *et al.*, 1990; Scott, T. R. *et al.*, 1991; Ogawa, H., 1994; Kadohisa, M. *et al.*, 2005) or clinical observations in humans (Bornstein, W. S., 1940; Lewey, F. H., 1943; Penfield, W. and Faulk, M. E., 1955; Motta, T. A., 1959; Hausser-Hauw, C. and Bancaud, J., 1987). Other localizations of activation of lower occurrence were not all mentioned at that time. More interestingly, a bilateral projection was found at the superior part of insula, whereas a small location in the anterior inferior insular area was found activated in the left hemisphere for strict right-handers and in the right hemisphere for left-handers (Faurion, A. *et al.*, 1999).

4.13.3 The Insular/Opercular Area and the Role of Anterior Inferior Insula

The activations due to pure taste and somatosensory or trigeminal oral stimuli, such as strong acid or potassium aluminate, were studied comparatively in the same subjects and experimental sessions. They were shown to appear in the same regions of the brain of 12 right-handed subjects (Cerf-Ducastel, B. *et al.*, 2001). However, it was possible to make out both senses on the basis of the coactivation of areas across experiments, such as the rolandic operculum for somatosensory stimuli and the anterior inferior insula for taste. Factor analysis of the table of activated brain areas across experiments showed that the right and left rolandic operculum were discriminated from all other regions when experiments included somato-sensory stimuli, whereas the right and left anterior inferior insula were discriminated from all other regions when experiments included only pure taste. It appears that multidimensional analysis sorted out the somatosensory projection area for the tongue for strong acid and astringent stimuli as an internal control of the experiment and, on the contrary, the anterior inferior insula appeared taste specific. For taste stimuli, the anterior inferior insular area was not only found again activated in the left hemisphere for right-handers but also statistically coactivated with the left angular gyrus (Cerf-Ducastel, B. *et al.*, 2001), a structure that is known to be related to comprehension of language in the dominant hemisphere of these right-handers. Furthermore, it was shown with MEG (Kobayakawa, T. *et al.*, in preparation) that this very region was the latest one to be activated, between 800 and 1400 ms for saccharin, after all other regions and after the most prominent brain activation had been recorded between 200 and 500 ms. This set of results suggests an integrative function for the inferior insula with special functions related to semantics for taste in the dominant hemisphere of right-handers, and an initial activation in the bilateral upper part of insula. A controversy remains concerning the first location of activation in this upper part of the insula, either anterior for some authors or posterior for other ones. For studying initial activation, MEG is a preferred technique, *vide infra*: timing of activation.

A much clearer picture of brain activation was obtained looking at the spatiotemporal pattern of activation. The time course of brain image formation was different for experiments including NaCl and saccharin in MEG studies, as expected from electrophysiological recordings of taste nerves in rodents. Moreover, both the temporal dimension given by the apparition of different dipoles in various places successively in the brain and the movement of each dipole we can follow during a few dozens of milliseconds helped visualizing the cross-talk activity between different brain areas and within each area during the first second of stimulation. In many

behavioral experiments, this period is fairly sufficient to identify the signification of the stimulus.

Using electrogustometric stimulation at low current intensities for taste which excludes tingling sensation, in right-handed subjects, Barry M. A. *et al.* (2001) could not show a simple relationship between the side of the stimulus applied and the side where the activation was found in insular and opercular areas at individual level. However, on the group analysis, they found a predominant right activation in the frontal opercular area. Finally, in this study, taste appears to project more to the right hemisphere than lingual touch. May be the group analysis averaged different results in different subjects, or a more complex data treatment is already performed prior to this projection. By comparison, and except in the inferior anterior insula, we observed an alternated right–left prevalence depending on anteriority, with equal left and right projection in the anterior insula/opercular area at $20 < y < 30$ and a more pronounced right insular activation at $0 < y < 10$ (less anterior), which may also look subject dependent. The electrogustometric (EGM) stimulation at low current is an interesting technique to confirm that actual taste and not somatosensory responses were found in this study in the postcentral gyrus: taste (saliva cations being applied iontophoretically) elicited activation at a more ventral location of the postcentral gyrus than in Pardo J. V. *et al.* (1997) where both taste and somatosensory senses were recruited. A recent clinical study tended to show a predominantly ipsilateral projection for taste and a more prominent contralateral projection for lingual touch information (Berlucchi, G. *et al.*, 2004). Taking together their results and ours, Barry *et al.*, looking more precisely at insula, which they divided into nine parts (i.e., anterior, middle, posterior x superior, central, inferior), concluded that taste-evoked activation in the superior part seems to predominate in the right-hand side, that activation in the central part appears bilateral and that activation in the inferior part is in the dominant left hemisphere of right-handers.

4.13.4 Timing of Activation: Magnetoencephalography Experiments

MEG is the best technique to locate the first temporal activation (primary taste area). Kobayakawa T. *et al.* (1996; 1999; 2005) showed an initial activation in the most posterior superior region of insula and rolandic operculum at onset latencies below 100 ms for salt and about 120–140 ms for saccharin. A complementary argument to assess this area was the first one to be activated came from EEG simultaneously recorded (Mizoguchi, C. *et al.*, 2002). Using electrogustometric stimulation, Yamamoto C. *et al.* (2003) showed, with MEG, a more anterior insular early projection, close to frontal operculum as shown in electrophysiological studies in primates. These discrepancies might refer to different experimental conditions.

4.13.5   Intensity Coding, Absence of Chemotopy

In the insula, Small D. M. *et al.* (2003) with fMRI and Kobayakawa T. *et al.* (2005) with MEG found indication of the coding of intensity of taste. O'Doherty J. P. *et al.* (2002) and De Araujo I. E. T. *et al.* (2003) confirmed, with fMRI, the existence of unimodal neuronal activity for taste in the anterior superior insula ($y = 3$ to 14) and frontal operculum.

O'Doherty J. *et al.* (2001) looked for topography of taste quality, but they found no trace of different localization of activation for glucose and salt: neither group analysis nor single subject analysis did provide any evidence of chemical topography in either insula or opercula. This result is not unexpected and matches well with (1) the indication of great interindividual differences of quality perception by subjects: any averaging of eventually different chemotopies would fail showing different locations for different taste qualities and with (2) the existence a continuum of numerous tastes, each stimulus being coded by a different neuronal pattern of activation, which is its signature.

4.13.6   Imagination of Taste versus Perception of Taste

Kobayashi M. *et al.* (2004) showed that the left insula was activated by gustatory imagery tasks and that the middle and superior frontal gyri which are occasionally found activated in taste experiments seemed to play an important role in the generation of hallucination of taste (top-down information, evoked imagery without the stimulus) rather than in taste perception due to the stimulus (bottom-up information).

4.13.7   Learning by Familiarization

Faurion A. *et al.* (1998; 2002) showed plasticity on repetitive imaging of subjects submitted to a protocol of familiarization to novel tastes: the number of pixels significantly changed in the insular/opercular area between the first fMRI session in which the subject tasted the novel stimuli for the first time and the second fMRI session which occurred after a heavy psychophysics session in which subjects evaluated the intensity of the novel stimuli about 120 times. The number of pixels did not change any more in the third fMRI session run after 10 more psychophysics sessions. When looking for a psychophysical parameter correlated with this pixel number variation, it was shown that the group of experiments showing an increase of the pixel number was the same as the group of experiments showing an increase of the hedonic evaluation with familiarization and that the group of experiments showing a decrease of the pixel number was the same as the group of experiments showing a decrease of the hedonic evaluation with familiarization. This result pointed to some relationship in the insular/opercular area between the evolution of the number of activated pixels and the direction of the evolution of the hedonic assessment in process of familiarization to foods, a situation that frequently occurs in everyday life for taste.

4.13.8   Orbitofrontal Cortex, an Integrative Area Contributing to Hedonic Valence and Food Reward

In the literature, the orbitofrontal cortex (OFC) is thought to be the region where the reward value of the food is expressed. If they did not find chemotopy in the insula, O'Doherty J. *et al.* (2001) showed that considering subjects taken individually, the areas activated for glucose (pleasant) and for salt (slightly unpleasant) were separated and only slightly overlapping. Averaging subjects tended to blur these interindividual differences. Hence the OFC responds to affectively positive and affectively negative taste stimuli in topographically different locations.

Small D. M. *et al.* (2003) found pleasant/aversive value in the anterior part of insula extending to the OFC. Similarly, De Araujo I. E. T. *et al.* (2003) mapped three distinct functions in the continuous region extending from the far anterior insula toward and including the orbitofrontal cortex. They looked at taste and olfactory convergence in relation with pleasantness of flavor. If unimodal neurones for taste were found in the anterior superior insula and frontal operculum, taste and olfaction seem to converge at the far anterior (inferior) insula where it joins the OFC (MNI coordinates: 45, 15, −9), in a region even more caudal than the caudal OFC (ventral forebrain) and in the nearby cingulate gyrus, which seems to respond to the mere association of both sensations. Small D. M. *et al.* (2004) proposed that the OFC together with the insula and the anterior cingulate gyrus constitute the key components of the network underlying flavor perception, which is in agreement with electrophysiological recordings: this region is known in the primate to receive olfactory input and secondary taste projection from the primary taste area.

In the lateral anterior OFC, a positive intermodality synergistic interaction appeared as the activation in this region seems to be more pronounced for the stimulation with the mixture of sucrose taste and strawberry odor (flavor) than by the sum of the activation to each stimulus separately.

In a more medial part of the anterior OFC appeared information relevant to taste and smell subjectively matching together (judgment of congruence between taste and smell). This region was also shown to respond to pleasant odors (Zald, D. H. and Pardo, J. V., 1997) and pleasant touch (Rolls, E. T. *et al.*, 2003).

Finally, a region responding to anticipation of pleasant and unpleasant taste was found only in the OFC where regions responding to expectation of reward and reward receipt were separated (O'Doherty, J. P. *et al.*, 2002), whereas other areas in striatum (putamen, nucleus accumbens) and amygdala responded to both. The OFC appears as the key structure for efficient and rapid stimulus reinforcement of an association learning between visual or olfactory stimulus and a positive or negative reinforcer such as taste or touch (Rolls, E. T., 2004).

For the sake of comparison, Cerf, Kobayakawa, and Faurion together showed a complex long-lasting integration process including semantics and taste in the dominant hemisphere in the anterior inferior insula (-45 ± 4; 7 ± 11; -3 ± 5 and 39 ± 10; 28 ± 6; -5 ± 3), a region that is nearby the frontal operculum and the caudal OFC, and this region may be the same as, or very close to, areas that were shown to be activated by taste (38; 20; −4; −32; 22; 0) in Kringelbach M. L. *et al.* (2004), by sucrose and strawberry odor (−31; 22; 0) in de Araujo I. E. T. *et al.* (2003), by glucose (−33; 17; 2) in O'Doherty J. *et al.* (2001), and unpleasant taste (−39; 27; −3) in Small D. M. *et al.* (2003).

4.13.9 Amygdala Involved in Hedonic Valence and Memorization

In amygdala, Small D. M. *et al.* (2003) found intensity of taste, as well as aversive value. O'Doherty J. *et al.* (2001) suggested that amygdala may signal as well aversive taste and affectively positive taste. Slightly different was the result of Zald D. H. *et al.* (2002) showing that the amygdala would respond preferentially to the aversive quinine and poorly to nonaversive stimulus such as sucrose. O'Doherty J. P. *et al.* (2002) showed that the OFC and not the amygdala responded to changes in pleasantness associated with eating (satiety effect). The authors propose that flavor novelty might be expressed in the amygdala on the basis of match/mismatch association of odor and flavor. These results are confirmed by Small D. M. *et al.* (2003). It could be hypothesized that in amygdala, novelty triggers some memorization mechanisms and recall memories with cultural clues. Activation for unpleasant taste may be more pronounced than for mild pleasant taste as unpleasant taste is more impressive and probably triggers more actively the memorization circuits. Zald D. H. (2003), in a very comprehensive review, points to the fact that the amplitude of pleasantness and unpleasantness for stimuli presented is usually not equivalent, a factor that may obscure the role of the amygdala with respect to positive hedonic valence stimuli. The author stresses the importance of dissociating the arousal value from the emotional value of the stimulus presented. Another interesting point for amygdala is that it seems to be activated for low-intensity subliminal stimuli which do not elicit conscious perception. Based on animal literature, a direct fast connection between thalamus and amygdala may be the source of behavioral responses related to nonperceived subliminal though emotionally relevant stimuli.

4.13.10 Other Areas

The anterior cingulate is known to be related to emotion. De Araujo I. E. and Rolls E. T. (2004) showed an activation in the anterior cingulum for fat independent of viscosity and for glucose, suggesting a response to the hedonic positive somatosensory aspect of fat. Small D. M. *et al.* (2004) found also activation in the anterior cingulate for congruent odor taste combination, activation depending on previous experience with taste/smell combinations.

Also the precentral gyrus at various z coordinates is often activated at various z coordinates during taste stimulation; its contribution to taste the function of which in taste is still to be determined.

Other locations were found in various studies which were at first not considered as they were either seldom or not expected from previous electrophysiological studies. As their roles were not easily interpreted, they could have been taken as putative artifacts. However, they repetitively appear in recently published studies, the number of which, although still low, now significantly increases (from one in 1994 to more than 10 in 2004) (e.g., the ventral and medial part of the thalamus, the lingual gyrus in the visual area, and the cerebellum). More numerous studies might confirm the validity of these activated areas as relevant to the taste stimulus.

Finally, Kringelbach M. L. *et al.* (2004) showed that not only the anterior insula, the frontal operculum, and the caudal OFC do respond to taste, but that taste activation is also found in the dorsolateral prefrontal cortex at a rather high level ($z = 36$). This region is already known to be implicated in response selection, working memory, attention, and integration (i.e., executive control).

4.13.11 Conclusive Remarks

So far, one of the key points for interpreting images remains that the examined areas benefit from being identified by eye recognition in anatomical images, coordinates being only indicative as they can be different from one imaging technique to another one and from one subject to another one (anatomical differences). Second, different imaging techniques (PET, fMRI, or MEG) or different stimulation protocols may identify different locations responding to taste in a given area; for example, MEG studies (Kobayakawa, T. *et al.*, 1996; 1999) did not find middle insula activation and PET studies find also more posterior activation in insula (Frey, S. and Petrides, M., 1999; Small, D. M. *et al.*, 1999) than fMRI experiments do, which also show anterior superior insular activation. The different time windows of the techniques resulting in different

temporal averaging might be responsible for focusing on different regions, or the tasks given to subjects were different and accounted for different responses.

Complementary studies with MEG and fMRI associating both high temporal and spatial resolutions allow observing the dynamics of activation in the brain in response to a given tastant, with dipoles flashing as questions and answers in and toward various regions alternately, a cross-talk pattern which is specific of the quality and contextual properties of the stimulus. Such results confirm (1) the specific signature of each single tastant and (2) a widespread representation of taste information across various regions of the brain. Reliable tools are now available for mapping the individual coding of taste sensation and its relevant cultural signification, for example, for thoroughly investigating innate and acquired interindividual differences in taste perception.

References

Bagshaw, M. H. and Pribram, K. H. 1953. Cortical organization in gustation (*Macaca mulatta*). J. Neurophysiol. 16, 499–508.

Bandettini, P. A., Jesmanowicz, A., Wong, E. C., and Hyde, J. S. 1993. Processing strategies for functional MRI of the human brain. Magn. Reson. Med. 30, 161–173.

Barry, M. A., Gatenby, J. C., Zeiger, J. D., and Gore, J. C. 2001. Hemispheric dominance of cortical activity evoked by focal electrogustometry stimuli. Chem. Senses 26, 471–782.

Belliveau, J. W., Kennedy, D. N., McKinstry, R. C., Buchbinder, B. R., Weisskoff, R. M., Cohen, M. S., Verea, J. M., Brady, T. J., and Rosen, B. R. 1991. Functional mapping of the human visual cortex by magnetic resonance imaging. Science 254, 716–719.

Benjamin, R. M. and Burton, H. 1968. Projection of taste nerve afferents to anterior opercular-insular cortex in squirrel monkey (*Saimiri sciureus*). Brain Res. 7, 221–231.

Berlucchi, G., Moro, V., Guerrini, C., and Aglioti, S. M. 2004. Dissociation between taste and tactile extinction on the tongue after right brain damage. Neuropsychologia 42, 1007–1016.

Bornstein, W. S. 1940. Cortical representation of taste in man and monkey. II. The localization of the cortical taste area in man and a method of measuring impairment of taste in man. Yale J. Biol. Med. 13, 133–156.

Cerf, B., Mac Leod, P., Van de Moortele, P.-F., Le Bihan, D., and Faurion, A. 1998. Functional lateralization of human gustatory cortex related to handedness disclosed by MRI study. Ann. N. Y. Acad. Sci. 855, 575–578.

Cerf, B., Van de Moortele, P.-F., Giacomini, E., Mac Leod, P., Faurion, A., and Le Bihan, D. 1996. Correlation of perception to temporal variations of fMRI signal: a taste study. Proc. Int. Soc. Magn. Reson. Med. (ISMRM), 280.

Cerf-Ducastel, B., Van de Moortele, P.-F., Mac Leod, P., Le Bihan, D., and Faurion, A. 2001. Interaction of gustatory and lingual somatosensory perceptions at the cortical level in the human: a functional magnetic resonance imaging study. Chem. Senses 26, 371–383.

De Araujo, I. E. and Rolls, E. T. 2004. Representation in the human brain of food texture and oral fat. J. Neurosci. 24(12), 3086–3093.

De Araujo, I. E. T., Rolls, E. T., Kringelbach, M. L., McGlone, F., and Phillips, N. 2003. Taste-olfactory convergence, and the representation of the pleasantness of flavour, in the human brain, Eur. J. Neurosci. 18, 2059–2068.

Faurion, A., Cerf, B., Le Bihan, D., and Pillias, A. M. 1998. fMRI study of taste cortical areas in humans (activations observed in relation to the hedonic and semantic status of the stimulus). Ann. N. Y. Acad. Sci. 585, 535–545.

Faurion, A., Cerf, B., Van de Moortele, P.-F., Lobel, E., Mac Leod, P., and Le Bihan, D. 1999. Human taste cortical areas studied with functional magnetic resonance imaging: evidence of functional lateralization related to handedness. Neurosci. Lett. 277, 189–192.

Faurion, A. B., Cerf, A. M., Pillias, N., and Boireau, N. 2002. Increased Taste Sensitivity by Familiarization to "Novel" Stimuli. Psychophysics, fMRI and Electrophysiological Techniques Suggest Modulations at Peripheral and Central Levels. In: Olfaction, Taste and Cognition (eds. C. Rouby, B. Schaal, D. Dubois, R. Gervais, and A. Holley), pp. 350–366. Cambridge University Press.

Frey, S. and Petrides, M. 1999. Re-examination of the human taste region: a positron emission tomography. Eur. J. Neurosci. 11, 2985–2988.

Grynszpan, F. and Geselowitz, D. B. 1973. Model studies of the magnetocardiogram. Biophys. J. 13(9), 911–925.

Hausser-Hauw, C. and Bancaud, J. 1987. Gustatory hallucinations in epileptic seizures. Electrophysiological, clinical and anatomical correlates. Brain 110, 339–359.

Kadohisa, M., Rolls, E. T., and Verhagen, J. V. 2005. Neuronal representations of stimuli in the mouth: The primate insular taste cortex, orbitofrontal cortex and amygdala. Chem. Senses 30, 1–19.

Kinomura, S., Kawashima, K., Yamada, S., Ono, M., Itoh, S., Yoshioka, T., Yamaguchi, H., Matsui, H., Itoh, R., Goto, T., Fujiwara, K., Satoh, K., and Fukuda, H. 1994. Functional anatomy of taste perception in the human brain studied with positron emission tomography, regional cerebral blood flow, taste stimulation, insular cortex, cingulate cortex, parahippocampal gyrus, human brain. Brain Res. 659, 263–266.

Kobayakawa, T., Endo, H., Ayabe-Kanamura, S., Kumagai, T., Yamaguchi, Y., Kikuchi, Y., Takeda, T., Saito, S., and Ogawa, H. 1996. The primary gustatory area in human cerebral cortex studied by magnetoencephalography. Neurosci. Lett. 212, 155–158.

Kobayakawa, T., Ogawa, H., Kaneda, H., Ayabe-Kanamura, S., Endo, H., and Saito, S. 1999. Spatio-temporal analysis of cortical activity evoked by gustatory stimulation in human. Chem. Senses 24, 201–209.

Kobayakawa, T., Wakita, M., Saito, S., Gotow, N., Sakai, N., and Ogawa, H. 2005. Location of the primary gustatory area in humans and its properties studied by magnetoencephalography. Chem. Senses 30(Suppl.1), i226–i227.

Kobayashi, M., Takeda, M., Hattori, N., Fukunaga, M., Sasabe, T., Inoue, N., Nagai, Y., Sawada, T., Sadato, N., and Watanabe, Y. 2004. Functional imaging of gustatory perception and imagery: "top-down" processing of gustatory signals. Neuroimage 23(4), 1271–1282.

Kringelbach, M. L., de Araujo, I. E. T., and Rolls, E. T. 2004. Taste-related activity in the human dorsolateral prefrontal cortex. Neuroimage 21, 781–788.

Kwong, K. K., et al. 1992. Dynamic magnetic resonance imaging of human brain activity during primary sensory stimulation. Proc. Natl. Acad. Sci. U. S. A. 89, 5675–5679.

Lauterbur, P. C. 1973. Image formation by induced local interaction. Examples employing nuclear magetic resonance imaging Nature 241, 190–191.

Le Bihan, D., et al. 1986. MR imaging of intravoxel incoherent motions: application to diffusion and perfusion in neurologic disorders. Radiology 161, 401–407.

Lewey, F. H. 1943. Aura of taste preceding convulsions associated with a lesion of the parietal operculum: report of a case. Arch. Neurol. Psychiatr. 50, 575–578.

Mizoguchi, C., Kobayakawa, T., Saito, S., and Ogawa, H. 2002. Gustatory evoked cortical activity in humans studied by simultaneous EEG and MEG recording. Chem. Senses. 27(7), 629–634.

Motta, T. A. 1959. G. I centri corticali del gusto. Bulletino delle Scienze Mediche 131, 480–493.

O'Doherty, J. P., Deichmann, R., Critchley, H. D., and Dolan, R. J. 2002. Neural responses during anticipation of a primary taste reward. Neuron 33(5), 815–826.

O'Doherty, J., Rolls, E. T., Francis, S., Bowtell, R., and McGlone, F. 2001. Representation of pleasant and aversive taste in the human brain. J. Neurophysiol. 85, 1315–1321.

Ogawa, H. 1994. Gustatory cortex of primates: anatomy and physiology. Neurosci. Res. 20, 1–13.

Ogawa, H., Ito, S. I., and Nomura, T. 1985. Two distinct projection areas from tongue nerves in the frontal operculum of macaque monkeys as revealed with evoked potential mapping. Neurosci. Res. 2, 447–459.

Ogawa, S., et al. 1992. Intrinsic signal changes accompanying sensory stimulation: functional brain mapping with magnetic resonance imaging. Proc. Natl. Acad. Sci. U. S. A. 89, 5951–5955.

Pardo, J. V., Wood, T. D., Costello, P. A., Pardo, P. J., and Lee, J. T. 1997. PET study of the localization and laterality of lingual somatosensory processing in humans. Neurosci. Lett. 234, 23–26.

Penfield, W. and Faulk, M. E. 1955. The insula. Further observations on its function. Brain 78(4), 445–470.

Pritchard, T. C., Hamilton, R. B., Morse, J. R., and Norgren, R. 1986. Projections of thalamic gustatory and lingual areas in the monkey, Macaca fascicularis. J. Comp. Neurol. 244, 213–228.

Rolls, E. T. 2004. The functions of the orbitofrontal cortex. Brain Cogn. 55(1), 11–29. Review.

Rolls, E. T., O'Doherty, J., Kringelbach, M. L., Francis, S., Bowtell, R., and McGlone, F. 2003. Representations of pleasant and painful touch in the human orbitofrontal and cingulate cortices. Cereb. Cortex 13(3), 308–17.

Roy, C. W. and Sherrington, C. S. 1890. On the regulation of the blood supply of the brain. J. Physiol. 11, 85–108.

Ruch, T. C. and Patton, H. D. 1946. The relation of the deep opercular cortex to taste. Fed. Proc. 5, 89–90.

Scott, T. R., Plata-Salaman, C. R., Smith, V. L., and Giza, B. K. 1991. Gustatory neural coding in the monkey cortex: stimulus intensity. J. Neurophysiol. 65(1), 76–86.

Small, D. M., Gregory, M. D., Mak, Y. E., Gitelman, D., Mesulam, M. M., and Parish, T. 2003. Dissociation of neural representation of intensity and affective valuation in human gustation. Neuron 39, 701–711.

Small, D. M., Voss, J., Mak, Y. E., Simmons, K. B., Parrish, T., and Gitelman, D. 2004. Experience-dependent neural integration of taste and smell in the human brain. J. Neurophysiol. 92(3), 1892–1903.

Small, D. M., Zald, D. H., Jones-Gotman, M., Zatorre, R. J., Pardo, J. V., Frey, S., and Petrides, M. 1999. Human cortical gustatory areas: a review of functional neuroimaging data. Neuroreport 10, 7–14.

Van de Moortele, P.-F., Cerf, B., Lobel, E., Paradis, A.-L., and Faurion, A. 1997. Latencies in fMRI time-series: effect of slice acquisition order and perception. NMR Biomed. 10, 230–236.

Yamamoto, C., Takehara, Morikawa, K., Nakagawa, S., Yamaguchi, M., Iwaki, S., Tonoike, M., and Yamamoto, T. 2003. Magnetoencephalographic study of cortical activity evoked by electrogustatory stimuli. Chem. Senses 28, 245–251.

Yaxley, S., Rolls, E. T., and Sienkiewicz, Z. J. 1990. Gustatory responses of single neurons in the insula of the macaque monkey. J. Neurophysiol. 63(4), 689–700.

Zald, D. H. 2003. The human amygdala and the emotional evaluation of sensory stimuli. Brain Res. Brain Res. Rev. 41(1), 88–123.

Zald, D. H. and Pardo, J. V. 1997. Emotion, olfaction, and the human amygdala: amygdala activation during aversive olfactory stimulation. Proc. Natl. Acad. Sci. U. S. A. 94(8), 4119–4124.

Zald, D. H., Hagen, M. C., and Pardo, J. V. 2002. Neural correlates of tasting concentrated quinine and sugar solutions. J. Neurophysiol. 87, 1067–1075.

Relevant Website

http://www.madic.org/index.php – For technical aspects.

4.14 Amiloride-Sensitive Ion Channels

J A DeSimone and V Lyall, Virginia Commonwealth University, Richmond, VA, USA

Glossary

pS Picosiemens, a conductance unit (10^{-12} siemens).

rENaC Rat ENaC, ENaC cloned from rat tissue.

TRPV1t Taste variant of TRPV1, transient receptor potential cation channel, subfamily V, member 1

4.14.1 Epithelial Na$^+$ Channel

4.14.1.1 Epithelial Na$^+$ Channel Function

Sodium (Na$^+$) absorption across many epithelial tissues depends on the epithelial Na$^+$ channel (ENaC), an apical membrane integral protein. Net absorption of Na$^+$ from the luminal side of the tissue to the submucosal side can occur when the luminal Na$^+$ concentration is equal to or less than the interstitial fluid Na$^+$ concentration, that is, Na$^+$ transport across epithelial cells is overall active, that is, coupled to metabolic energy sources (Ussing, H. H. and Zerahn, K., 1951; Verrey, F. *et al.*, 2000). However, the initial influx of Na$^+$ from lumen to cell occurs by passive electrodiffusion down both the Na$^+$ concentration gradient and the apical membrane electrical potential gradient. Cell Na$^+$ is then actively extruded across the basolateral membrane through sodium–potassium (Na$^+$-K$^+$) pumps (Horisberger, J. D., 2004). Net Na$^+$ transport can be pharmacologically blocked by either

adding ouabain to the basolateral side to inhibit the Na$^+$-K$^+$ pumps or with amiloride on the apical side to reduce the Na$^+$ conductance of ENaC (Verrey, F. *et al.*, 2000). Before the structure of ENaC was determined, the potent effect of amiloride as a Na$^+$-transport blocker proved to be a reliable indication of the channel's presence, hence its much cited earlier name: amiloride-sensitive epithelial Na$^+$ channel. Transepithelial Na$^+$ transport can be monitored *in vitro* in a variety of tissues as the major component of the short-circuit current (I_{sc}) across the epithelium mounted as a tissue sheet between a pair of Ussing chambers allowing solutions with ionic composition approximating extracellular fluid to bathe both apical and basolateral tissue surfaces (Ussing, H. H. and Zerahn, K., 1951). If Na$^+$ enters the epithelial cells exclusively through apical membrane ENaC, then amiloride or its analogs will significantly inhibit the I_{sc} (Will, P. C. *et al.*, 1985). If Na$^+$ is the major ion actively absorbed, as in the case of amphibian urinary bladder, then I_{sc} will be reduced nearly to zero (Palmer, L. G. *et al.*, 1982).

Mammalian tissues, for which the presence of ENaC has either been directly demonstrated or inferred from electrophysiological, pharmacological, or immunocytochemical studies include: the distal convoluted tubule and collecting tubules of the kidney, colon, trachea, sweat and salivary glands, and dorsal lingual epithelium including taste receptor cells (Garty, H. and Palmer, L. G., 1997). The single channel biophysical properties of Na^+-transporting ion channels obtained from patch clamp recordings from rat cortical collecting tubules and the A6 cell line from *Xenopus* kidney correspond well with those of cloned ENaC expressed in *Xenopus* oocytes (Canessa, C. M. *et al.*, 1994). Based on such comparisons the consensus opinion is that the Na^+-transporting properties of most epithelia derive from ENaC. First, ion transport in epithelia and through cloned ENaC is selective for Na^+ over K^+ with a Na^+:K^+ permeability ratio in excess of 20 (Garty, H. and Palmer, L. G., 1997; Rossier, B. C. *et al.*, 2002). Epithelial and cloned ENaC demonstrate a unitary conductance of 5 pS, an open probability of 0.5, and a dissociation constant for amiloride binding of 0.1–1.0 μM (Garty, H. and Palmer, L. G., 1997). The aldosterone-responsive epithelia of the kidney and colon are critical in the control of blood volume, blood pressure, and Na^+ balance. Consequently, in humans loss-of-function mutations in ENaC result in pseudohypoaldosteronism type 1, a severe salt wasting disease, and gain-of-function mutations cause pseudoaldosteronism, a form of salt-sensitive hypertension (Rossier, B. C. *et al.*, 2002). Na^+ consumption, in which Na^+ taste plays a role, is a factor in overall Na^+ balance. All three ENaC subunits are expressed in rat taste buds (Kretz, O. *et al.*, 1999). It is perhaps not surprising, therefore, that in mammals utilizing ENaC as a Na^+-specific taste receptor, that ENaC levels in taste cells are also modulated by aldosterone (Lin, W. *et al.*, 1999). The sensory consequence of this is that the amiloride-sensitive component of the chorda tympani taste nerve response to sodium chloride (NaCl) in rats increases following aldosterone treatment (Herness, M. S., 1992).

4.14.1.2   Epithelial Na^+ Channel Structure

ENaC is an integral membrane protein composed of three subunits, designated α, β, and γ, each of which is comprised of about 650–700 amino acids

(Canessa, C. M. *et al.*, 1994). Evidence suggests the native channel consists of two α, one β, and one γ subunit (Firsov, D. *et al.*, 1998; Kosari, F. *et al.*, 1998; Figure 1).

The essential Na^+-conducting properties of ENaC are found only in the α subunit. While the β and γ subunits have no intrinsic channel function, their coexpression with the α subunit enhanced its amiloride-sensitive conductance by more than 100-fold (Canessa, C. M. *et al.*, 1994). Each subunit has just two membrane-spanning domains. Both the N- and C-termini are relatively short and located on the cytoplasmic side while the extracellular domain consists of a loop, containing nearly 70% of each subunit's amino acids, connecting the membrane-spanning hydrophobic segments (Benos, D. J. and Stanton, B. A., 1999). Evidence suggests that the major amiloride binding sites in rat ENaC involve a six amino acid residue track (α-278-283, WYRFHY), within the extracellular domain of α-rENaC (Kieber-Emmons, T. *et al.*, 1995). Histidine at α-282 appears to provide an essential electropositive moiety because mutagenesis of it to negative aspartic acid resulted in a 40-fold increase in the binding dissociation constant between amiloride and ENaC, while substitution with positive arginine decreased the dissociation constant by sixfold (Lin, C. *et al.*, 1994). The extracellular domain may also bind an accessory protein that is the site of action of serine proteases. The action of channel activating protease 1 (CAP1, prostasin), an endogenous serine protease, increases the Na^+ flux through ENaC, an effect that can be reproduced with trypsin using ENaC expressed in *Xenopus* oocytes (Chraïbi, A. *et al.*, 1998) and in rat taste receptors (see Section 4.14.2.7). Deletion of the N-terminal region of each subunit eliminates channel activity suggesting that it

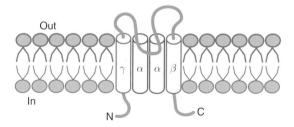

Figure 1   Model of epithelial Na^+ channel illustrating the stoichiometry among the subunits. Each subunit has two transmembrane domains (not shown) with C- and N-terminal regions in the cytoplasmic domain as shown. Reprinted, with permission, from the Annual Review of Physiology, Volume 64 © 2002 by Annual Reviews.

is important in regulating channel gating and proper assembly of functional ENaC from its subunits (Chalfant, M. L. *et al.*, 1999a). Point substitutions of serine for glycine in the N-terminus at G95 of (α-rENaC) rat ENaC, G37 of β-rENaC, or G40 of γ-rENaC significantly decrease ENaC conductance by decreasing the channel open probability (Grunder, S. *et al.*, 1997). The significance of the serine for glycine substitution at G37 of β-rENaC is evident from the fact that it is responsible for pseudohypoaldosteronism type 1. The second membrane-spanning region (M2) of each subunit and part of the pre-M1 domain contain the channel selectivity filter, sites responsible for conductance inhibition by divalent cations, and further sites involved in amiloride inhibition (Benos, D. J. and Stanton, B. A., 1999). The C-terminal region of each subunit is the most variable, and probably accounts for species and tissue differences in some of ENaC's regulatory properties, such as sensitivity to cyclic AMP (cAMP). In airway cells (Kunzelmann, K. *et al.*, 2000), renal collecting duct cells (Handler, J. S. and Orloff, J., 1981), and taste cells (Gilbertson, T. A. *et al.*, 1993; see Figure 1) cAMP stimulates amiloride-sensitive Na^+ transport through ENaC, while in mammalian colon and urinary bladder cAMP has no effect (Garty, H. and Palmer, L. G., 1997). Point mutations in a proline-rich region in the β- and γ-subunit C-termini of ENaC are sufficient to produce the gain-of-function disease, pseudoaldosteronism. This proline-rich region is important for normal internalization of ENaC from the plasma membrane and its absence results in overexpression of ENaC and a gain-of-function disease resulting in severe hypertension.

4.14.1.3   Epithelial Na^+ Channel Regulation

Salt and water balance in terrestrial mammals is tightly regulated by the hormone, aldosterone which exerts stimulatory effects on ENaC, Na^+-K^+ pumps, and various membrane ion exchangers in cells of the renal collecting duct and colon (Rossier, B. C. *et al.*, 2002). Aldosterone-sensitive Na^+ transport occurs in the duct cells of salivary and sweat glands and in the taste cells (Herness, M. S., 1992; Lin, W. *et al.*, 1999). Following a latent period of 0.5–1.0 h, the aldosterone early response, which lasts from 1.5 to 3.0 h, consists of activation of preexisting Na^+ channels and pumps. The late response (at 6–24 h) is characterized by *de novo* synthesis of ENaC and Na^+-K^+ pumps. In rat colon (Lingueglia, E. *et al.*, 1994) and rat taste cells (Lin, W. *et al.*, 1999) aldosterone increases the

expression of β- and γ-ENaC without effect on α-ENaC. A second hormone, vasopressin, also increases Na^+ transport through its effects on ENaC activity. The effect is well studied in amphibian skin, urinary bladder, and amphibian-derived A6 cells (Li, J. H. Y. *et al.*, 1982). In mammals vasopressin acts on renal collecting duct cells and on taste cells (Gilbertson, T. A. *et al.*, 1993). Vasopressin binds to the membrane receptor, V_2, which in turn activates adenylyl cyclase resulting in an increased cell cAMP concentration. When cell cAMP levels are elevated pharmacologically in vasopressin-responsive epithelia, including taste cells (Gilbertson, T. A. *et al.*, 1993), the effect mimics vasopressin. cAMP increases the Na^+ conductance of epithelia by increasing the apical membrane density of ENaC without changing the channel open time probability (Morris, R. G. and Schafer, J. A., 2002). The flux of Na^+ through ENaC is observed to be locally downregulated over a range of increasing apical Na^+ concentrations, a process referred to as Na^+ self-inhibition (Lindemann, B., 1984). Na^+ self-inhibition is one of many processes by which cells resist excessive increase in osmotic pressure. This self-regulatory process has been demonstrated in rat taste cells showing amiloride sensitivity (Gilbertson, T. A. and Zhang, H., 1998). Na^+ self-inhibition can itself be inhibited by a range of compounds including: sulfhydryl reagents, organic cations, cadmium (Cd^{2+}), and zinc (Zn^{2+}). An inhibitory effect of *p*-hydroxymercuribenzoate on Na^+ self-inhibition has been shown in rat fungiform taste cells (Gilbertson, T. A. and Zhang, H., 1998). The inhibitory effect of Zn^{2+} on Na^+ self-inhibition can be demonstrated by an enhanced rat chorda tympani response to NaCl in the presence of Zn^{2+} (see Section 4.14.2.8). Intracellular calcium (Ca^{2+}) and intracellular pH exert potent regulatory effects on ENaC under physiological conditions. An increase in intracellular Ca^{2+} inhibits the Na^+ flux through ENaC (Ishikawa, T. *et al.*, 1998) as does a decrease in intracellular pH (Palmer, L. G. and Frindt, G., 1987; Chalfant, M. L. *et al.*, 1999b). These inhibitory effects of intracellular Ca^{2+} and pH can be demonstrated in the taste system in recordings of the rat chorda tympani response to NaCl (see Sections 4.14.2.3 and 4.14.2.4). In ENaC-expressing oocytes, an increase in the extracellular osmotic pressure due to nonelectrolytes such as mannitol at hyperosmotic concentrations, will also increase the Na^+ conductance of ENaC (Ji, H. L. *et al.*, 1998). This effect of hyperosmotic mannitol can be observed under physiological conditions in recordings of the NaCl chorda tympani response (Lyall, V. *et al.*, 1999; see Section 4.14.2.5). The temperature dependence of the ENaC conductance is unusual in that it is greater at lower temperature than at higher (Chraïbi, A. and Horisberger, J. D., 2002). This appears to be due to a decrease in the channel open probability with increasing temperature. This accounts for the striking inhibition of the amiloride-sensitive part of the chorda tympani response to NaCl with increasing temperature (see Section 4.14.2.6). Proteases can eliminate the inhibitory effect of increased temperature and also can block Na^+ self-inhibition (Chraïbi, A. and Horisberger, J. D., 2002), which indicates that the temperature effect is connected with the mechanism of Na^+ self-inhibition and that proteases modulate ENaC conductance by blocking that mechanism.

4.14.2 Amiloride-Sensitive Salt Taste Receptor

4.14.2.1 Hormonal Regulation of Epithelial Na$^+$ Channel and Salt Taste

In rats, where ENaC plays a major role in the animal's behavior toward Na^+ salts versus non-Na^+ salts (Spector, A. C. *et al.*, 1996), data support a role for both aldosterone and vasopressin in taste cell function. Injection of Sprague-Dawley rats with aldosterone increased apical immunoreactivity to β- and γ-ENaC, increased the number of amiloride-sensitive fungiform taste cells, and enhanced the magnitude of the Na^+ current (Lin, W. *et al.*, 1999). The sensory consequence of this is that following aldosterone injection the amiloride-sensitive component of the rat chorda tympani response to NaCl was significantly increased (Herness, M. S., 1992). Vasopressin and its associated second messenger, cAMP, increased the amiloride-sensitive Na^+ current in hamster fungiform taste cells (Gilbertson, T. A. *et al.*, 1993), suggesting a role for vasopressin in taste nerve activity toward Na^+ salts. It should be emphasized that these hormonal influences are likely to be less of a factor in human salt taste because the amiloride-sensitive NaCl-taste response in humans represents the smaller component of the two salt taste-sensing mechanisms (Schiffman, S. S. *et al.*, 1983; Smith, D. V. and Ossebaard, C. A., 1995; Lyall, V. *et al.*, 2004a).

4.14.2.2 Regulation of Epithelial Na$^+$ Channel by cAMP and Salt Taste

While no data are available on the action of vasopressin on the chorda tympani response to NaCl, topical application of 8-(4-chlorophenylthio)-cAMP (CPT-cAMP), a membrane permeable analog of cAMP, to the rat tongue significantly enhances the response to NaCl as seen in Figure 2(a) (Russell, O. F. *et al.*, 2002). The increased response is amiloride-sensitive (data not shown) indicating that cAMP increased ENaC activity, which is consistent with its well-studied action in Na^+-transporting epithelia.

The effect of cAMP on the chorda tympani response to NaCl also suggests a possible mechanism for taste mixture interaction at the cellular level. If a subset of ENaC-containing taste cells is also responsive to tastants that modulate intracellular cAMP, it is clear that tastant-evoked modulation of cAMP can also affect cell responsiveness to NaCl as part of a stimulus mixture.

4.14.2.3 Regulation of Epithelial Na$^+$ Channel by Intracellular pH and Salt Taste

The decrease in the Na^+ conductance of ENaC with decreasing intracellular pH, observed in a variety of Na^+-transporting epithelia and in ENaC-expressing oocytes, also occurs in ENaC-containing taste cells (Lyall, V. *et al.*, 2002). A decrease in intracellular pH in rat taste cells decreases the Na^+ flux across the apical membranes and this pH-induced modulation of ENaC conductance in taste cells significantly affects the chorda tympani response to NaCl (Lyall, V. *et al.*, 2002; Figure 2(b)). Figure 2(b) shows the effect of extracellular pH on the neural response to 100 mM NaCl. Data show, however, that the pH regulatory site is intracellular, and that extreme extracellular pH variation (e.g., pH 2 or 10.3) produces intracellular pH changes of only a few tenths of a unit (Lyall, V. *et al.*, 2002). These changes are, however, sufficient to produce a strong effect on the Na^+ conductivity of ENaC and, therefore, account for the mixture suppression of ENaC-based salt taste by low pH and its mixture enhancement at high pH.

4.14.2.4 Regulation of Epithelial Na$^+$ Channel by Intracellular Ca^{2+} and Salt Taste

An increase in intracellular Ca^{2+} due to an increase in TRPM5 cation conductance occurs in taste cells during transduction for the sweet, bitter, and

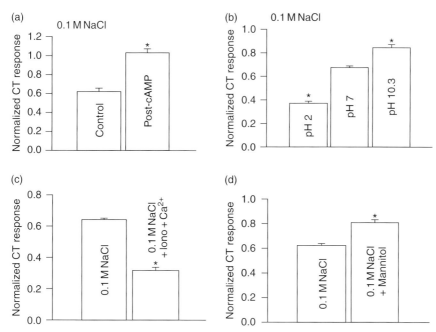

Figure 2 (a) Tonic rat chorda tympani (CT) response to 0.1 M sodium chloride (NaCl) normalized to the response to 0.3 M ammonium chloride (NH$_4$Cl) before (control) and following treatment of the tongue with chlorophenylthio-cAMP (CPT-cAMP) for 20 min. The increase in response due to cyclic AMP (cAMP) is consistent with an enhancement of the Na$^+$ flux through epithelial Na$^+$ channel (ENaC; cf. Russell, O. F. *et al.*, 2002). (b) Normalized CT response to 0.1 M NaCl as a function of solution pH. Results illustrate that the inhibitory effect of decreasing pH on Na$^+$ transport by ENaC is reflected in the neural response to NaCl (cf. Lyall, V. *et al.*, 2002). (c) Normalized CT response to 0.1 M NaCl before (control, contains 10 mM calcium chloride (CaCl$_2$)) and following treatment of the rat tongue with 150 µM ionomycin (iono) plus 10 mM CaCl$_2$. Increasing the intracellular Ca^{2+} concentration of ENaC-containing taste cells causes a decrease in the neural response to NaCl, which is consistent with the observed decrease in Na$^+$ conductance of ENaC due to increasing cytosolic Ca^{2+} (cf. Russell, O. F. *et al.*, 2002). (d) Normalized CT response to 0.1 M NaCl (control) and 0.1 M NaCl plus 1 M mannitol. The effect of hyperosmotic mannitol on the neural response is consistent with the observed increase in the Na$^+$ conductance of ENaC due to hyperosmotic mannitol. Each bar represents the mean ± standard error ($n = 3$; cf. Lyall, V. *et al.*, 1999). (b) Reproduced from The Journal of General Physiology, 2002, 120: 793–815. Copyright 2002 The Rockefeller University Press. (d) Reproduced from Lyall, V., Heck, G. L., DeSimone, J. A., and Feldman, G. M. 1999. Effects of osmolarity on taste receptor cell size and function. Am. J. Physiol. 277, C800–C813, used with permission from the American Physiological Society.

umami taste qualities (Zhang, Y. *et al.*, 2003). Sour taste also involves an increase in intracellular Ca^{2+} that both sustains transduction (Liu, L. and Simon, S. A., 2001; Richter, T. A. *et al.*, 2003) and serves as a potent stimulus for the activation of the Na$^+$-H$^+$ exchanger isoform 1, the major cellular sour taste adaptation mechanism (Lyall, V. *et al.*, 2004b; Vinnikova, A. K. *et al.*, 2004). As already discussed, ENaC conductance is reduced by an increase in intracellular Ca^{2+}. As Figure 2(c) shows, increasing taste cell intracellular Ca^{2+} using the Ca^{2+} ionophore, ionomycin, significantly suppresses the chorda tympani response to NaCl (Russell, O. F. *et al.*, 2002). Thus, any subset of cells expressing both ENaC and a second transduction or adaptation mechanism resulting in an increase in intracellular Ca^{2+}, will lead to a suppression of the cells responsiveness to NaCl.

4.14.2.5 Regulation of Epithelial Na$^+$ Channel by Osmolarity and Salt Taste

As mentioned above (Section 4.14.1.3), hyperosmotic mannitol increases ENaC conductance. On this basis one would expect hyperosmotic mannitol to increase the chorda tympani response to NaCl and the increased response to be abolished by amiloride significantly. This is in fact what is observed (Lyall, V. *et al.*, 1999; Figure 2(d)). Thus, ENaC-based salt taste reception is enhanced in a hyperosmotic stimulus mixture. This osmotic effect accounts for some of the taste properties in rats and hamsters of the short-chain polysaccharide, polycose. During polycose preparation low levels of salt are added which, due to the osmotic effect of the saccharide, add a detectable salt taste component to polycose solutions (Lyall, V. *et al.*, 1999). This may account, in part, for the fact that polycose stimulates salt-sensitive units in the rat nucleus of the solitary tract and the strong behavioral preference rats show for polycose solutions (Giza, B. K. *et al.*, 1991).

4.14.2.6   Regulation of Epithelial Na$^+$ Channel by Temperature and Salt Taste

In oocytes expressing ENaC, the steady-state amiloride-sensitive current due to 100 mM NaCl at 36 °C was less than 50% of the current at 27 °C (Chraïbi, A. and Horisberger, J. D., 2002). This unusual inhibitory effect of increasing temperature on ENaC conductance is reflected in the amiloride-sensitive part of the chorda tympani response to NaCl. As Figure 3(c) shows the amiloride-sensitive part of the chorda tympani response to 100 mM NaCl decreased about 50% from 30.0 to 36.8 °C and rapidly decreased to zero at 43.0 °C. By contrast the amiloride-insensitive

part of the NaCl response increased with temperature reaching a maximum at 43.0 °C and then declined slowly with temperature. This temperature dependence is one of many properties that indicate that the amiloride-insensitive NaCl response is mediated through a variant of the vanilloid receptor 1 (TRPV1t; Lyall, V. *et al.*, 2004a). It is clear that most of the taste sensitivity to salt at temperatures above 37 °C is not due to ENaC but rather to TRPV1t.

4.14.2.7   Regulation of Epithelial Na$^+$ Channel by Proteases and Salt Taste

ENaC conductance is modulated by endogenous proteases, an effect that can be mimicked using low concentrations of exogenous trypsin (Chraïbi, A. *et al.*, 1998). Trypsin, as might be anticipated, has an enhancing effect on the chorda tympani response to NaCl. As seen in Figure 3(a) the effect is specific for

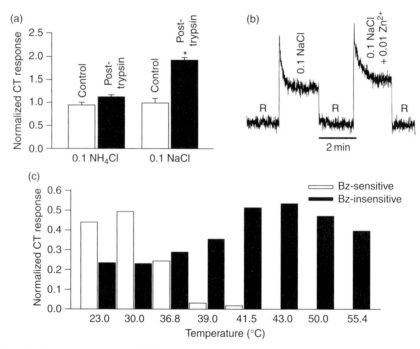

Figure 3 (a) Normalized rat chorda tympani (CT) responses to 0.1 M ammonium chloride (NH$_4$Cl) and 0.1 M sodium chloride (NaCl). For each salt white bars indicate control responses and black bars indicate responses following treatment of the rat tongue with 20 μg ml^{-1} trypsin for 20 min. Note trypsin had no effect on the response to NH$_4$Cl, but increased the tonic response to NaCl by 100%. This is consistent with the observed enhancing effect of trypsin and other proteases on the Na$^+$ conductance of epithelial Na$^+$ channel (ENaC). Each bar represents the mean ± standard error ($n = 3$). (b) Chorda tympani response to 30 mM NaCl relative to a baseline response in 10 mM potassium chloride (KCl) followed by a response to 30 mM NaCl plus 10 mM zinc chloride (ZnCl$_2$). The 10 mM ZnCl$_2$ does not give a neural response above baseline. The increase in the response to NaCl by Zn^{2+} is consistent with the observed inhibitory action of Zn^{2+} on the Na$^+$ self-inhibition property of ENaC. (c) The amiloride-sensitive component of the chorda tympani response to 0.1 M NaCl (white bars) and the amiloride-insensitive component (black bars) as a function of temperature. Note that the amiloride-sensitive component rapidly diminishes above 30 °C, a property consistent with the observed temperature dependence of ENaC. In contrast the amiloride-insensitive component increases with temperature to a maximum at 43 °C and then slowly diminishes with increasing temperature.

responses to NaCl which is consistent with a mechanism involving ENaC and protease blockage of Na^+ self-inhibition.

4.14.2.8 Na^+ Self-Inhibition of Epithelial Na^+ Channel and Salt Taste

A wide variety of substances interfere with Na^+ self-inhibition of ENaC. Zn^{2+} concentrations too low to give a chorda tympani response in the rat, can nonetheless significantly enhance the response to NaCl. Figure 3(b) shows the chorda tympani response to 30 mM NaCl relative to a baseline in 10 mM KCl. A subsequent stimulation with 30 mM NaCl plus 10 mM $ZnCl_2$ shows a significantly enhanced response. This is consistent with Zn^{2+}-induced blockage of Na^+ self-inhibition that is in turn consistent with the effect of temperature and trypsin shown in Figures 3(c) and 3(a), respectively.

4.14.3 Summary and Conclusions

The demonstration of ENaC in rat fungiform taste receptor cells, the close conformity of the rat amiloride-sensitive neural response to NaCl, and the properties of ENaC expressed in oocytes, and the results of rat salt taste behavioral studies leave no doubt that ENaC is the Na^+-specific salt taste receptor located in the anterior tongue. All three ENaC subunits are found in fungiform taste cells in the anterior tongue, but only α-ENaC is found in abundance in the circumvallate taste cells of the posterior tongue. This may explain why amiloride sensitivity is found in the chorda tympani response to NaCl (anterior tongue), but not in the glossopharyngeal nerve response (posterior tongue) (Kretz, O. *et al.*, 1999). It should be noted, however, that treatment of rats with aldosterone induced an amiloride-sensitive Na^+ current in cells of the circumvallate papillae that normally do not have amiloride sensitivity (Lin, W. *et al.*, 1999). This suggests that perhaps aldosterone increases the targeting of ENaC subunits to the apical membrane of taste cells. A definitive answer requires, however, the cloning of each ENaC subunit from both fungiform and circumvallate taste receptor cells, a task not yet completed. Until that is accomplished it remains unclear whether the ENaC subunits in taste cells share 100% homology with complementary subunits in colon or kidney collecting tubules. While ENaC is a major salt taste receptor in rodents and other mammals, it is much less

important in humans. For that reason the amiloride-insensitive salt taste receptor is beginning to receive more attention especially now that some progress has been made in its characterization (Lyall, V. *et al.*, 2004a).

References

Benos, D. J. and Stanton, B. A. 1999. Functional domains within the degenerin/epithelial sodium channel (Deg/ENaC) superfamily of ion channels. J. Physiol. 520, 631–644.

Canessa, C. M., Schild, L., Buell, G., Thorens, B., Gautschi, I., Horisberger, J. D., and Rossier, B. C. 1994. Amiloride-sensitive epithelial Na^+ channel is made of three homologous subunits. Nature 367, 463–467.

Chalfant, M. L., Denton, J. S., Berdiev, B. K., Ismailov, I. I., Benos, D. J., and Stanton, B. A. 1999b. Intracellular H^+ regulates the α subunit of ENaC, the epithelial Na^+ channel. Am. J. Physiol. 276, C477–C486.

Chalfant, M. L., Denton, J. S., Langloh, A. L., Karlson, K. H., Loffing, J., Benos, D. J., and Stanton, B. A. 1999a. The NH_2 terminus of the epithelial sodium channel contains an endocytic motif. J. Biol. Chem. 274, 32889–32896.

Chraïbi, A. and Horisberger, J. D. 2002. Na self inhibition of human epithelial Na channel: temperature dependence and effect of extracellular proteases. J. Gen. Physiol. 120, 133–145.

Chraïbi, A., Vallet, V., Firsov, D., Hess, S. K., and Horisberger, J. D. 1998. Protease modulation of the activity of the epithelial sodium channel expressed in *Xenopus* oocytes. J. Gen. Physiol. 111, 127–138.

Firsov, D., Gautschi, I., Merillat, A. M., Rossier, B. C., and Schild, L. 1998. The heterotetrameric architecture of the epithelial sodium channel (ENaC). EMBO J. 17, 344–352.

Garty, H. and Palmer, L. G. 1997. Epithelial sodium channels: function, structure, and regulation. Physiol. Rev. 77, 359–396.

Gilbertson, T. A. and Zhang, H. 1998. Self-inhibition in amiloride-sensitive sodium channels in taste receptor cells. J. Gen. Physiol. 111, 667–677.

Gilbertson, T. A., Roper, S. D., and Kinnamon, S. C. 1993. Proton currents through amiloride-sensitive Na^+ channels in isolated hamster taste cells: enhancement by vasopressin and cAMP. Neuron 10, 931–942.

Giza, B. K., Scott, T. R., Sclafani, A., and Antonucci, R. F. 1991. Polysaccharides as taste stimuli: their effect in the nucleus tractus solitarius of the rat. Brain Res. 555, 1–9.

Grunder, S., Firsov, D., Chang, S. S., Jaeger, N. F., Gautschi, I., Schild, L., Lifton, R. P., and Rossier, B. C. 1997. A mutation causing pseudohypoaldosteronism type 1 identifies a conserved glycine that is involved in the gating of the epithelial sodium channel. EMBO J. 16, 899–907.

Handler, J. S. and Orloff, J. 1981. Antidiuretic hormone. Annu. Rev. Physiol. 43, 611–624.

Herness, M. S. 1992. Aldosterone increases the amiloride-sensitivity of the rat gustatory neural response to NaCl. Comp. Biochem. Physiol. 103A, 269–273.

Horisberger, J. D. 2004. Recent insights into the structure and the mechanism of the sodium pump. Physiology 19, 377–387.

Ishikawa, T., Marunaka, Y., and Rotin, D. 1998. Electrophysiological characterization of the rat epithelial Na^+ channel (rENaC) expressed in MDCK cells. Effects of Na^+ and Ca^{2+}. J. Gen. Physiol. 111, 825–846.

Ji, H. L., Fuller, C. M., and Benos, D. J. 1998. Osmotic pressure regulates $\alpha\beta\gamma$-rENaC expressed in *Xenopus* oocytes. Am. J. Physiol. 275, C1182–C1190.

Kieber-Emmons, T., Lin, C., Prammer, K. V., Villalobos, A., Kosari, F., and Kleyman, T. R. 1995. Defining topological similarities among ion transport proteins with anti-amiloride antibodies. Kidney Int. 48, 956–964.

Kosari, F., Sheng, S., Li, J., Mak, D. O., Foskett, J. K., and Kleyman, T. R. 1998. Subunit stoichiometry of the epithelial sodium channel. J. Biol. Chem. 273, 13469–13474.

Kretz, O., Barbry, P., Bock, R., and Lindemann, B. 1999. Differential expression of RNA and protein of the three pore-forming subunits of the amiloride-sensitive epithelial sodium channel in taste buds of the rat. J. Histochem. Cytochem. 47, 51–64.

Kunzelmann, K., Schreiber, R., Nitschke, R., and Mall, M. 2000. Control of epithelial Na$^+$ conductance by the cystic fibrosis transmembrane conductance regulator. Pfluegers Arch. 440, 193–201.

Li, J. H., Palmer, L. G., Edelman, I. S., and Lindemann, B. 1982. The role of sodium-channel density in the natriferic response of the toad urinary bladder to an antidiuretic hormone. J. Membr. Biol. 64, 77–89.

Lin, C., Kieber-Emmons, T., Villalobos, A. P., Foster, M. H., Wahlgren, C., and Kleyman, T. R. 1994. Topology of an amiloride-binding protein. J. Biol. Chem. 269, 2805–2813.

Lin, W., Finger, T. E., Rossier, B. C., and Kinnamon, S. C. 1999. Epithelial Na$^+$ channel subunits in rat taste cells: localization and regulation by aldosterone. J. Comp. Neurol. 405, 406–420.

Lindemann, B. 1984. Fluctuation analysis of sodium channels in epithelia. Annu. Rev. Physiol. 46, 497–515.

Lingueglia, E., Renard, S., Waldmann, R., Voilley, N., Champigny, G., Plass, H., Lazdunski, M., and Barbry, P. 1994. Different homologous subunits of the amiloride-sensitive Na$^+$ channel are differently regulated by aldosterone. J. Biol. Chem. 269, 13736–13739.

Liu, L. and Simon, S. A. 2001. Acidic stimuli activate two distinct pathways in taste receptor cells from rat fungiform papillae. Brain Res. 923, 58–70.

Lyall, V., Alam, R. I., Malik, S. A., Phan, T. H., Vinnikova, A. K., Heck, G. L., and DeSimone, J. A. 2004b. Basolateral Na$^+$-H$^+$ exchanger-1 in rat taste receptor cells is involved in neural adaptation to acidic stimuli. J. Physiol. 556, 159–173.

Lyall, V., Alam, R. I., Phan, T. H., Russell, O. F., Malik, S. A., Heck, G. L., and DeSimone, J. A. 2002. J. Gen. Physiol. 120, 793–815.

Lyall, V., Heck, G. L., DeSimone, J. A., and Feldman, G. M. 1999. Effects of osmolarity on taste receptor cell size and function. Am. J. Physiol. 277, C800–C813.

Lyall, V., Heck, G. L., Vinnikova, A. K., Ghosh, S., Phan, T. H., Alam, R. I., Russell, O. F., Malik, S. A., Bigbee, J. W., and DeSimone, J. A. 2004a. The mammalian amiloride-insensitive non-specific salt taste receptor is a vanilloid receptor-1 variant. J. Physiol. 558, 147–159.

Morris, R. G. and Schafer, J. A. 2002. cAMP increases density of ENaC subunits in the apical membrane of MDCK cells in direct proportion to amiloride-sensitive Na$^+$ transport. J. Gen. Physiol. 120, 71–85.

Palmer, L. G. and Frindt, G. 1987. Effects of cell Ca and pH on Na channels from rat cortical collecting tubule. Am. J. Physiol. 253, F333–F339.

Palmer, L. G., Li, J. H., Lindemann, B., and Edelman, I. S. 1982. Aldosterone control of the density of sodium channels in the toad urinary bladder. J. Membr. Biol. 64, 91–102.

Richter, T. A., Caicedo, A., and Roper, S. D. 2003. Sour taste stimuli evoke Ca^{2+} and pH responses in mouse taste cells. J. Physiol. 547, 475–483.

Rossier, B. C., Pradervand, S., Schild, L., and Hummler, E. 2002. Epithelial sodium channel and the control of sodium balance: interaction between genetic and environmental factors. Annu. Rev. Physiol. 64, 877–897.

Russell, O. F., Phan, T. T., Alam, R. I., Heck, G. L., Lyall, V., and DeSimone, J. A. 2002. Acute regulation of rat NaCl taste responses by cAMP and calcium. Chem. Senses 27, 661–671.

Schiffman, S. S., Lockhead, E., and Maes, F. W. 1983. Amiloride reduces the taste intensity of Na$^+$ and Li$^+$ salts and sweeteners. Proc. Natl. Acad. Sci. U. S. A. 80, 6136–6140.

Smith, D. V. and Ossebaard, C. A. 1995. Amiloride suppression of the taste intensity of sodium chloride: evidence from direct magnitude scaling. Physiol. Behav. 57, 773–777.

Spector, A. C., Guagliardo, N. A., and St. John, S. J. 1996. Amiloride disrupts NaCl versus KCl discrimination performance: implications for salt taste coding in rats. J. Neurosci. 16, 8115–8122.

Ussing, H. H. and Zerahn, K. 1951. Active transport of sodium as the source of electric current in the short-circuited frog skin. Acta Physiol. Scand. 23, 110–127.

Verrey, F., Hummler, E., Schild, L., and Rossier, B. C. 2000. Control of Na Transport by Aldosterone. In: (*eds.* D. W. Seldin and G. Giebisch),The Kidney: Physiology and Pathophysiology Vol. 1, pp. 1441–1471.. Lippincott.

Vinnikova, A. K., Alam, R. I., Malik, S. A., Ereso, G. L., Feldman, G. M., McCarty, J. M., Knepper, M. A., Heck, G. L., DeSimone, J. A., and Lyall, V. 2004. Na$^+$-H$^+$ exchange activity in taste receptor cells. J. Neurophysiol. 91, 1297–1313.

Will, P. C., Cortright, R. N., DeLisle, R. C., Douglas, J. G., and Hopfer, U. 1985. Regulation of amiloride-sensitive electrogenic sodium transport in the rat colon by steroid hormones. Am. J. Physiol. 248, G124–G132.

Zhang, Y., Hoon, M. A., Chandrashekar, J., Mueller, K. L., Cook, B., Wu, D., Zuker, C. S., and Ryba, N. J. 2003. Coding of sweet, bitter, and umami tastes: different receptor cells sharing similar signaling pathways. Cell 112, 293–301.

4.15 Central Neural Processing of Taste Information

D V Smith[†], The University of Tennessee College of Medicine, Memphis, TN, USA

S P Travers, The Ohio State University, Columbus, OH, USA

4.15.1 Introduction

The sense of taste has a critical role throughout the animal kingdom because it detects and discriminates key chemicals in the animal and plant matter, which constitutes potential foodstuff. Diet varies among species and the gustatory systems of different animals exhibit important specializations. Nevertheless, mammalian species, the topic of the present chapter, possess the ubiquitous ability to sense nutritive carbohydrates and amino acids, as well as electrolytes such as sodium. Similarly, mammals distinguish potential corrosive or toxic substances including acids and alkyloids. These different groups of stimuli are described by human subjects as having distinctive qualities: most carbohydrates are sweet, acids are sour, alkaloids and other toxins are bitter, and sodium is salty; other electrolytes have mixed tastes. Amino acid taste is less well understood, but in humans, L-glutamate seems to have a special potency for activating a selective amino acid receptor (Li, X. *et al.*, 2002b) and is often assumed to have a unique taste, best known as umami. In rodents, a larger variety of L-amino acids appear to share a common transduction mechanism (Nelson, G. *et al.*, 2002), though at least glutamate is probably closely aligned perceptually with carbohydrates (Yamamoto, T. *et al.*, 1991; Stapleton, J. R. *et al.*, 1999).

Making distinctions between tastants is vital for maintaining a diet that promotes homeostasis and avoids harmful substances. The structure and function of the G-protein-coupled receptors and ion channels with the requisite specificity for this discrimination have been described in elegant detail in the last decade (reviewed in Chandrashekar, J. *et al.*, 2006). Agreement is mounting that these receptors are distributed in a relatively segregated manner (Nelson, G. *et al.*, 2001; Huang, A. L. *et al.*, 2006), such that different taste receptor cells are specialized for detecting particular classes of chemicals (Yoshida, R. *et al.*, 2006). Similarly, a notable degree of specificity is maintained

†Deceased

in the individual gustatory nerve fibers that innervate taste receptor cells, and carry this information into the brain (e.g., Frank, M., 1973; Frank, M. E. *et al.*, 1983; Frank, M. E., 1991). Thus, much of the peripheral code for taste quality is arguably sparse with just a few different groups of receptor cells and their innervating fibers comprising labeled lines. However, there is not unanimous agreement on this coding scheme, and there are important caveats particularly with regard to the code for sodium versus acids, as many peripheral nerve fibers are broadly responsive across these classes of chemicals (e.g., Frank, M., 1973; Frank, M. E. *et al.*, 1983; Frank, M. E., 1991, #241). Further, in the central nervous system, convergence of primary afferents (Ogawa, H. *et al.*, 1982; Travers, S. P. *et al.*, 1986; Sweazey, R. D. and Smith, D. V., 1987; Vogt, M. B. and Mistretta, C. M., 1990) and complex integrative processing within central circuits, gives rise to neurons with heterogeneous tuning curves, and although some neurons remain narrowly tuned (e.g., Nishijo, H. and Norgren, R., 1990; Geran, L. C. and Travers, S. P., 2006); many are broad (e.g., Nishijo, H. and Norgren, R., 1990; Di Lorenzo, P. M. and Victor, J. D., 2003; Lemon, C. H. and Smith, D. V., 2005; Geran, L. C. and Travers, S. P., 2006), as summarized in Spector A. C. and Travers S. P. (2005). As a consequence, whether any aspect of central coding employs dedicated pathways, or whether central coding relies entirely upon comparison across neurons is still a topic of considerable debate (see Chapter A Perspective on Chemosensory Quality Coding). Furthermore, there is growing interest in the possibility of a contribution of temporal coding to making distinctions among tastes (see Chapter Neural Ensembles in Taste Coding).

Overlaid on any scheme for taste quality coding are additional, equally important aspects of gustatory processing. Foods possess not only taste, but also texture, thermal, and olfactory components. These multiple characteristics are integrated largely through central circuits to produce flavor. In addition, though some taste preferences and aversions are innate, they are not hardwired. The central nervous system plays a key role in providing the requisite plasticity to meet changing homeostatic states and environmental demands. Thus, the desirability of even the most delectable dessert is suppressed by satiety. Limited specific mechanisms appear to be present to accommodate particular changes in homeostatic state. Satiety has a more pronounced impact on palatable, for example, sweet substances (Foster, L. A. *et al.*, 1996), and a sodium deficit leads to an increased preference for this specific

ion (Contreras, R. and Hatton, G., 1975). However, deficits in other micro- or macronutrients, for example, vitamins or essential amino acids, are not associated with such predetermined changes in gustatory responsiveness. Instead, organisms learn to associate the flavor of a particular food with the correction of a deficit (Markison, S. *et al.*, 1999; 2000). Learning is also essential to assure that an animal will avoid contaminated food sources that do not have an innately aversive taste (Garcia, J. *et al.*, 1970).

This chapter presents data on the functional characteristics of central gustatory neurons. We consider the topographic organization of the system, as well as several key features of neurons including their tuning curves, receptive field organization, and inputs from other modalities. In addition, the plasticity of these circuits as evidenced by the impact of homeostatic state, learning, neurotransmitters and neuromodulators, and interactions between different regions of the gustatory neuraxis, is discussed. The study of the cellular basis of these features had lagged behind a systems level analysis; but significant information is now available at the level of the first-order relay, the rostral nucleus of the solitary tract (rNST). Some key ideas regarding taste quality coding are presented in the subsections following the current chapter. For a fuller perspective, the reader is referred to several recent reviews that present divergent opinions (Spector, A. C. and Travers, S. P., 2005; Chandrashekar, J. *et al.*, 2006; Jones, L. M. *et al.*, 2006; Simon, S. A. *et al.*, 2006; Hallock, R. M. and Di Lorenzo, P. M., 2006).

A schematic diagram of gustatory system organization highlighting those regions and connections discussed in this review is presented in Figure 1. Further details regarding anatomical organization can be found in the chapter in this volume by Whitehead and Finger. Classic decerebration studies by Grill H. J. and Norgren R. (1978b) established important hierarchical features of the system. When transections are made just rostral to the midbrain, the existing brainstem circuitry still possesses impressive regulatory capacity. Decerebrate rats do not voluntarily eat or drink but if fluids are introduced into their mouths, stereotyped oral behaviors, virtually identical to those observed in intact animals, assure that sweet substances are ingested and that bitter substances are rejected. Similar behaviors are observed in anencephalic human infants, establishing phylogenetic generality (Steiner, J. E., 1973). Moreover, this restricted component of the system possesses some degree of modifiability by homeostatic state, as filling the stomach with liquid food shortens bouts of licking

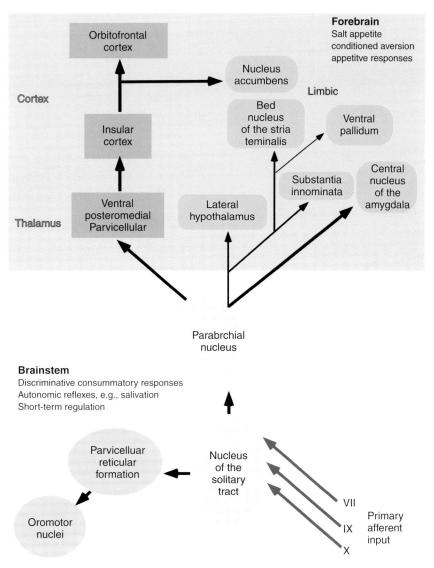

Figure 1   Schematic diagram depicting primary afferent input, and the major, direct connections of the gustatory system in rodents. Line thickness gives a rough indication of the strength of a given connection. Primate gustatory pathways are similar, despite some notable differences that are discussed in the text. The diagram also indicates two main functional divisions within the system: brainstem and forebrain, with each division associated with different behavioral repertoires. Importantly, however, there are extensive reciprocal connections between the forebrain and brainstem (see Figure 2), indicating important interactions between these two major divisions. Likewise, the gustatory forebrain can be loosely divided into thalamocortical and limbic divisions, which are differentially associated with perceptual and motivational function; however, again there are extensive anatomical connections between these networks (not shown), forming a substrate for functional interactions.

elicited by intraoral infusions of sucrose, just as it does in normal rats (Grill, H. J. and Norgren, R., 1978a). However, without the forebrain, critical aspects of modifiability are lost. Thus, longer-term changes in homeostatic state, such as that induced by food (Seeley, R. J. *et al.*, 1994; Kaplan, J. M. *et al.*, 2000) or sodium (Grill, H. J. *et al.*, 1986) deprivation cause modifications in the intake of different tasting substances in the intact, but not decerebrate rat. Finally,

the effects of learning are not expressed in decerebrate animals. In intact animals, when a novel taste is paired with gastrointestinal malaise, the animal subsequently rejects that taste, a phenomenon known as conditioned taste aversion (Garcia, J. *et al.*, 1970); decerebrate rats do not exhibit this switch in behavior. Although the system is arranged hierarchically, this is not to say that the components of the system operate in isolation to perform these different functions.

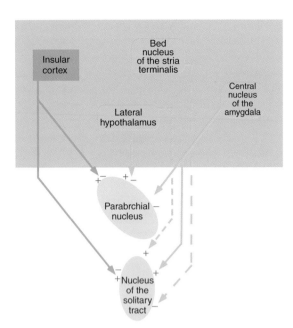

Figure 2   Major, direct, descending pathways from forebrain to brainstem, which have been documented both anatomically and physiologically. Note that all of these regions also receive ascending gustatory input, as shown in Figure 1. The most prevalent effects of electrical stimulation or glutamate activation of each forebrain structure on gustatory responses are also shown. Some effects are mixed, some largely excitatory, and others largely inhibitory.

Instead, the connections shown in Figure 2 comprise a substrate indicative of a high degree of interaction between the forebrain and hindbrain, and there are a multitude of other interactions, particularly between various regions of the forebrain.

4.15.2   Properties of Brainstem Gustatory Neurons

Taste information is processed through two brainstem nuclei on its way to the forebrain (Figure 1). Gustatory axons from the VII, IX, and X cranial nerves terminate first within the rNST, from which axons arise to project predominantly to the parabrachial nuclei (PbN) of the pons, but also to synaptic targets within the medulla, including the subadjacent reticular formation and indirectly and in some cases directly to oral motor nuclei. Cells within the nucleus of the solitary tract (NST) and PbN are also targeted by descending axons from all gustatory-responsive areas of the forebrain, providing a substrate for the modulation of taste activity in response to physiological need and prior experience. Attempts to understand the neural substrate for gustatory

processing have addressed the morphology and physiology of cells within these brainstem nuclei.

4.15.2.1   Neuron Types in the Nucleus of the Solitary Tract and Parabrachial Nuclei

Progress in our understanding of many sensory or motor systems has come from a thorough description of the cells comprising the networks involved. In the gustatory brainstem, there have been numerous attempts to categorize neurons on the basis of their morphologies, intrinsic membrane properties, and taste responsiveness.

4.15.2.1.1   Morphological cell types
Neurons within the NST have been examined using a variety of anatomical techniques in an attempt to classify them on the basis of their morphologies. Purely anatomical studies have shown several prominent cell types, including multipolar (stellate), elongate (fusiform), and small ovoid cells. Several different parameters have been used to classify these cells. One of the earliest schemes grouped hamster NST cells according to their size and the invagination of their nuclei, resulting in four groups: large and small cells with or without nuclear invaginations (Davis, B. J. and Jang, T., 1986). In Nissl-stained material, Whitehead M. C. (1988) described three major cell types in the hamster NST: elongate, ovoid, and round, based on the shape of the soma, which he quantified by measuring each cell's form factor. Form factor is a measure of roundness, given by the formula $ff = 4\pi\text{area}/\text{perimeter}^2$, ranging from 1.0 for a perfectly round cell to 0.0 for a straight line. Elongate cells were those with $ff < 0.7$, ovoid with ff between 0.7 and 0.9, and round with $ff > 0.9$. Elongate cells were more common in the rostral lateral subdivision of the NST, whereas those cells within the ventral subdivision were typically larger cells. Throughout the NST, ovoid cells were the most common. Lasiter P. S. and Kachele D. L. (1988) described similar cell types in the rat NST and further demonstrated that the ovoid cells had smaller cross-sectional areas than other cell types.

Later studies used Golgi impregnation to examine cells in the hamster NST. Davis B. J. and Jang T. (1988a) described two cell types:

1. large, fusiform cells with two or three relatively unbranched primary dendrites and
2. smaller, stellate cells with more numerous and more highly branched dendrites.

Using Golgi methods (Whitehead, M. C., 1988) also described elongate and stellate cells and another class he termed tufted cells, which were large neurons with several dendrites, one of which exhibited many branches. There are some discrepancies between these studies in what constitutes large versus small or medium neurons and a thorough description of these neuronal classification studies and their limitations was provided in a later review article (Schweitzer, L. *et al.*, 1995).

Immunocytochemical studies have demonstrated that many of the small ovoid neurons express the inhibitory neurotransmitter gamma-aminobutyric acid (GABA), suggesting that these numerous cells comprise a network of inhibitory interneurons within the NST (Lasiter, P. S. and Kachele, D. L., 1988). In addition, tract-tracing experiments have demonstrated that the majority of NST neurons projecting to the PbN are elongate and stellate cells (Whitehead, M. C., 1990). Further, some of these larger neurons expressed tyrosine hydroxylase or dopamine-β-hydroxylase (Davis, B. J. and Jang, T., 1988b). Overall, the idea that there are numerous small ovoid cells serving as interneurons and elongate cells and large multipolar cells comprising the PbN projection neurons within the NST is fairly well accepted. That scheme was used by King M. S. and Bradley R. M. (1994) in their attempt to relate the biophysical properties of rat NST neurons to their morphology. These investigators grouped rat NST neurons that were labeled with biocytin during *in vitro* slice recording experiments into ovoid, elongate and multipolar cells, although they admitted that such a straightforward classification was not always easy. Nevertheless, the 58 neurons that they reconstructed were divided into 9 elongate, 21 multipolar, and 28 ovoid cells. These cell types, however, were not clearly related to the biophysical properties of the neurons (see below). A similar attempt to correlate electrophysiology and morphology was made by Renehan W. E. *et al.* (1994; 1996), who labeled rat NST cells with biocytin *in vivo* after recording their responses to gustatory stimulation. Although there was some indication that cells responding to quinine were smaller than those responding to sucrose, NaCl or HCl and that some cells had greater or lesser dendritic dimensions, there was little or no relationship between gustatory responsiveness and cell classification into elongate, multipolar, or ovoid neurons. These authors actually used a hierarchical cluster analysis to classify rat NST neurons based on their morphologies, but the

resultant six classifications did not relate clearly to taste responsiveness either. Thus, cells in the NST are classifiable into groups based on morphology, as is common in many areas of the brain, but so far the functional significance of such a classification is not apparent.

4.15.2.1.2 Cell types based on biophysical properties

With the advent of brain slice techniques, it has become possible to characterize many of the intrinsic membrane properties of brainstem gustatory neurons, particularly those in the rostral portion of the NST. In pioneering work on rat medullary slices, Bradley R. M. and Sweazey R. D. (1990; 1992) demonstrated four different groups of NST neurons based on their responses to injected current (Figure 3). Three groups of cells (groups I–III) exhibited regular firing in whole-cell recordings in response to a 1200 ms, 100 pA depolarizing current injection. These groups were separable on the basis of the effects of a preceding hyperpolarization on this firing pattern produced by a 150 ms, 200 pA hyperpolarizing pulse. The firing of group I cells was severely disrupted by hyperpolarization, whereas those of group III were completely unaffected by hyperpolarizing the membrane. Group II cells showed regular firing to a depolarizing current step after an initial delay produced by the preceding hyperpolarization; this delay was produced by activation of a potassium A current (I_{KA}) (Tell, F. and Bradley, R. M., 1994) as it was this current that was blocked by 4-aminopyridine. Cells of group IV showed only a brief, rapid burst of action potentials to depolarization, which was variably affected by prior hyperpolarization. These different groups of neurons would be expected to respond differently to excitatory and inhibitory influences in the processing of gustatory information.

In a more recent study of rat NST neurons, a group of cells was identified that exhibited a depolarizing response to hyperpolarization, known as I_h current (Uteshev, V. V. and Smith, D. V., 2006). In response to hyperpolarizing current steps, these cells exhibit a nonohmic response in which the membrane potential depolarizes during the hyperpolarization, due to hyperpolarization-activated cation channels. This current is blocked by 1 mM CsCl applied to the bath. These cells were largely a different subset than those showing I_A current and were differentially distributed within the rNST. Cells with I_A current were

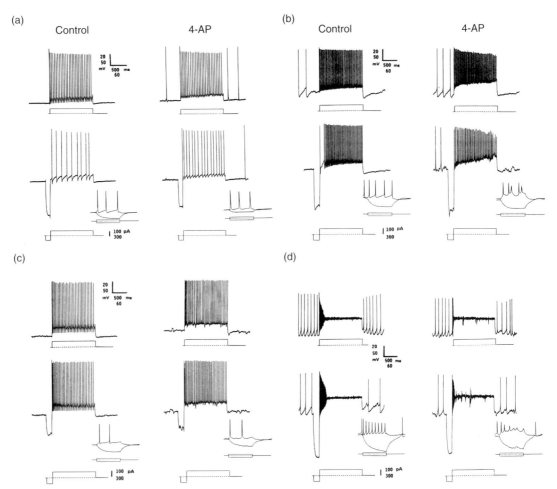

Figure 3   Types of rostral nucleus of the solitary tract (rNST) neurons defined based on membrane properties demonstrated in *in vitro* preparations. Types I–IV are shown in panels (a)–(d), respectively. Groups I–III exhibit repetitive firing throughout a depolarizing current pulse; group IV only a brief train of action potentials. Groups I–IV differ according to the effect of a preceding hyperpolarizing current pulse. Group III neurons are little affected. However, type I neurons respond with a more irregular and reduced firing and type II with a delay in the initiation of the response. Although not present in this example, a preceding hyperpolarizing pulse often shortens the burst of action potentials exhibited by type IV neurons. The response delay observed in type III neurons is eliminated by 4-aminopyridine (4-AP), indicating that a hyperpolarization elicited A-current is responsible for the delay. From Tell, F. and Bradley, R. M. 1994. Whole-cell analysis of ionic currents underlying the firing pattern of neurons in the gustatory zone of the nucleus tractus solitarii. J. Neurophysiol. 71, 479–492, with permission.

found in the medial half of the NST, where the majority of neurons projecting to the PbN are located (Halsell, C. B. *et al.*, 1996). On the other hand, those neurons expressing I_h were predominantly located in the lateral half (Uteshev, V. V. and Smith, D. V., 2006), where most of the cells project into oral motor circuits through the subadjacent reticular formation (Halsell, C. B. *et al.*, 1996). Although NST neurons can be grouped according to their responses to injected current (Bradley, R. M. and Sweazey, R. D., 1992) or by the presence of I_A or I_h current

(Uteshev, V. V. and Smith, D. V., 2006), there does not appear to by any clear relationship between such classifications and cell morphology (King, M. S. and Bradley, R. M., 1994; Uteshev, V. V. and Smith, D. V., 2006).

4.15.2.1.3   Cell types based on gustatory responsiveness

A direct way to understand gustatory processing by the brainstem would be to categorize neurons on the basis of their taste responsiveness. Such an effort has been made by almost all investigators working in this area, but the most pervasive of these classifications are the best-stimulus categories first described by Frank M. (1973) for taste-responsive fibers of the chorda tympani nerve. Chorda tympani fibers were classified into groups (S-, N-, H-, or Q-best) based on which of four basic taste stimuli: sucrose, NaCl, HCl, or quinine were most effective when presented to the anterior tongue at midrange concentrations.

A best-stimulus categorization based simply on the basis of identifying the most effective among four stimuli is subject to criticism because it is nearly a foregone conclusion that at least one stimulus will be nominally more effective than the others. However, what is more compelling is that, within a best-stimulus group, the cells are relatively similar in their overall pattern of responsiveness. Thus, when the four stimuli were ordered along the abscissa from most to least preferred (i.e., S, N, H, and Q), chorda tympani fibers peaked at a single point, with profiles indicative of S-, N-, or H-best fibers. Q-best fibers (except for one) were not evident in the hamster chorda tympani nerve, but are common in the glossopharyngeal nerve of both hamsters (Hanamori, T. *et al.*, 1988) and rats (Frank, M. E., 1991). In fact, this similarlity between members of a group holds up to the challenge of applying more stimuli and to categorizing response profiles in a more quantitative fashion; for example, by using hierarchical cluster analysis, which is a numerical taxonomic procedure (e.g., Rodieck, R. W. and Brening, R. K., 1982). When such an analysis is applied to the responses of hamster chorda tympani fibers to 13 stimuli, three groups of neurons emerge (Frank, M. E. *et al.*, 1988), which almost without exception correspond to the S-, N-, and H-best classifications. This analysis demonstrates that neurons within a group are more similar to one another than to members of other groups and provides a quantitative basis for classifying neurons in this system. In a series of experiments on the hamster brainstem, Smith and his co-workers showed that a best-stimulus classification worked well to categorize neurons of the NST (Travers, J. B. and Smith, D. V., 1979) and PbN (Van Buskirk, R. L. and Smith, D. V., 1981). Further, multivariate analyses of the response profiles of these cells, even including responses from as many as 18 stimuli, resulted in classifications that were, to a large extent, predicted from the best stimulus (Smith, D. V. *et al.*, 1983). Similar analyses of rat NST (Nakamura, K. and Norgren, R., 1993; Lemon, C. H. and Smith, D. V., 2005; Geran, L. C. and Travers, S. P., 2006) or PbN (Nishijo, H. and

Norgren, R., 1990) neurons also show such neural groups, but only if taste stimuli are applied to receptors responsive to sweet stimuli. Earlier studies in rats, in which stimuli were not applied to the palate, did not present clear evidence for neural groups based on hierarchical clustering (Woolston, D. C. and Erickson, R. P., 1979) and, in fact, argued against the existence of functional neural groups. Examples from a recent study that included neurons responsive to inputs from the anterior tongue, palate, and posterior tongue are depicted in Figure 4. As a result of including neurons with inputs from the posterior foliate papillae, this study described groups of cells with optimal responses to bitter stimuli as well as to sweeteners (and amino acids), sodium, and acids. Thus, cells within the gustatory brainstem are readily classified into groups based on similarities and differences in their responses to taste stimulation.

When a classification scheme can be supported by biological parameters, it increases the confidence one places in such a categorization. An example of such support is the funneling of input from amiloride-sensitive epithelial sodium channels (ENaCs) on the taste receptor cells into a single class of neurons in the peripheral and central nervous system. Fibers of the chorda tympani nerve that respond best to NaCl exhibit responses to NaCl that are blockable in a dose-dependent fashion by lingual application of the diuretic drug amiloride (Ninomiya, Y. and Funakoshi, M., 1988; Hettinger, T. P. and Frank, M. E., 1990). The substantial response to NaCl in HCl-best fibers is much less affected by amiloride treatment. This segregation of amiloride-sensitive input to the NaCl- but not the HCl-best neurons is maintained in the NST (Scott, T. R. and Giza, B. K., 1990; Giza, B. K. and Scott, T. R., 1991). Application of amiloride to the tongue disrupts performance of rats on a two-lever operant task in which they have been taught to discriminate NaCl from KCl, implicating amiloride-sensitive NaCl-best neurons as an important contributor to the rat's ability to discriminate these two salts (Spector, A. C. *et al.*, 1996; see also Hill, D. L. *et al.*, 1990).

Another systematic relationship between chemosensitivity and a biological parameter has been described for neurons optimally responsive to bitter tasting stimuli. In a study including electrical stimulation of the PbN to test for antidromic invasion, Cho Y. K. *et al.* (2002a) found a substantial group of quinine-best neurons based on responses of the anterior tongue to equally effective concentrations of the basic stimuli: 0.032 M

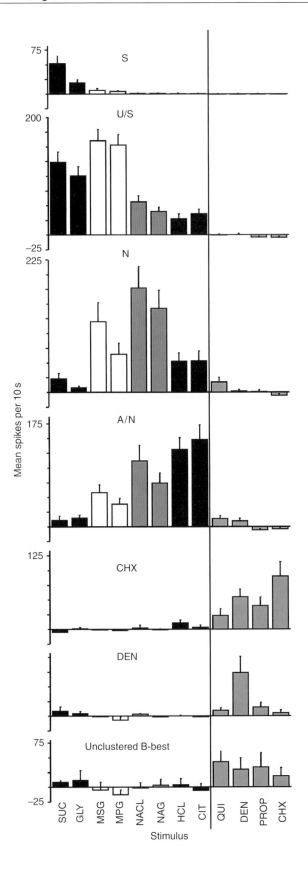

sucrose, 0.032 M NaCl, 0.0032 M citric acid, and 0.032 M QHCl (quinine-HCl). When this strong concentration of QHCl is used, QHCl-best neurons are readily evident in the hamster NST following stimulation of the anterior tongue. Interestingly, all 23 QHCl-best neurons in this sample that projected axons to the PbN had conduction velocities that were less than 25% of those of the other neurons. In other words, this best-stimulus classification identified a subset of small, slowly conducting neurons. This finding corresponds well with those of Renehan W. E. *et al.* (1995) , who found that neurons responding specifically to QHCl in the rat NST were significantly smaller than other cells, and of Geran L. C. and Travers S. P. (2006) who also found that bitter-best rat NST neurons had very slow conduction velocities.

Overall, numerous studies of the responsiveness of central gustatory neurons have described groups of cells based on their best stimulus. This classification scheme corresponds well to those categories derived from more quantitative techniques such as hierarchical cluster analysis, even those based on responses to many different stimuli. To some extent, because of the relatively broader tuning of central taste neurons in comparison to peripheral gustatory fibers, the correspondence between a best-stimulus classification and a hierarchical cluster solution is not quite as tight as for peripheral fibers (Smith, D. V. *et al.*, 1983). It is also clear that changes in relative stimulus concentrations can result in changes in best-stimulus classification, as concentration–response functions for different stimuli are not always parallel (e.g., see Smith, D. V. and Travers, J. B., 1979). Nevertheless, although to some extent dependent on stimulus concentration, receptive field stimulated, and species, best-stimulus groups exhibit remarkable

consistencies between laboratories and the use of such a classification scheme encompasses an important organizing principle in this system.

4.15.2.2 Neural Processing Through the Brainstem

Taste fibers traveling in the chorda tympani and greater superficial petrosal branches of the VII nerve, the lingual–tonsillar branch of the IX nerve, and the internal branch of the superior laryngeal nerve send their axons into the NST in an orderly rostral to caudal distribution (Contreras, R. J. *et al.*, 1982; Hamilton, R. B. and Norgren, R., 1984). However, the terminations of these nerves also exhibit significant overlap, giving rise to convergence between axons within the same nerve (Vogt, M. B. and Mistretta, C. M., 1990) and between inputs from completely separate receptor fields (Travers, S. P. *et al.*, 1986; Travers, S. P. and Norgren, R., 1991).

4.15.2.2.1 Breadth of tuning of nucleus of the solitary tract and parabrachial nuclei neurons

One parameter that is important in every sensory system is the tuning profile of afferent neurons. In taste, early studies described this feature by tabulating the number of stimuli among those representing the basic taste qualities that evoked activity in a recorded neuron; that is, the breadth of tuning of gustatory fibers and central neurons was simply described in terms of how many stimuli produced responses (e.g., Frank, M. and Pfaffmann, C., 1969; Perrotto, R. S. and Scott, T. R., 1976). However, such an approach depends heavily on the definition of a response, which varies across laboratories. In

---

Figure 4 Examples of central neuron types (rat nucleus of the solitary tract (NST)) based on chemosensitive response profiles. Types were identified using cluster analysis, and generally corresponded to a best-stimulus classification. The response types resemble those in other studies, though additional features are apparent, based on including neurons responsive to the posterior tongue, as well as including L-glutamate in the stimulus array. Two neuron types were robustly responsive to sucrose including a small group of posterior tongue responsive neurons with a relatively specific response to sucrose (S neurons), and a larger group comprised mainly of anterior oral cavity responsive neurons. The latter group also responded well to glycine, and in fact responded nominally better to glutamate (U/S neurons). Two groups responded best to sodium salts (N neurons) or acids. As is common, acid-best neurons were also robustly responsive to sodium salts, and thus named A/N neurons. Finally, there were neurons with posterior tongue receptive fields optimally and relatively specifically responsive to bitter-tasting chemicals. The multiple types of bitter-best neurons provide preliminary evidence for heterogeneous processing of different (ionic vs. nonionic) bitter tastants. SUC, 300 mM sucrose; GLY, 300 mM glycine; MSG, 30 mM monosodium glutamate + 3 mM inosine monophosphate; MPG, 30 mM monopotassium L-glutamate + 3 mM inosine monophosphate; NACL, 100 mM NaCl; NAG, 100 mM NaGluconate; HCL, 10 mM HCl; CIT, 30 mM citric acid; QUI, 3 mM quinine hydrochloride; DEN, 10 mM denatonium benzoate; PROP, 7 mM propylthiouracil; CHX, 0.01 mM cycloheximide. From Geran, L. C. and Travers, S. P. 2006. Single neurons in the nucleus of the solitary tract respond selectively to bitter taste stimuli. J. Neurophysiol. 96, 2513–2527, with permission.

1979, Smith D. V. and Travers J. B. introduced the entropy measure as a way to quantify the breadth of tuning of gustatory neurons. This measure is given by the equation:

$$H = -K \sum p_i (\log p_i), \quad i = 1, n$$

where H is the breadth of responsiveness, K is a scaling constant, and p_i is the proportional response to each of n stimuli. The p_i's for each cell are derived by converting the neural response profile for that cell to a proportional profile, the response to each stimulus being expressed as a proportion of the total response to all four stimuli. This manipulation maintains the relative response to all stimuli and does not depend upon a response criterion. Therefore, a cell might be considered broadly responsive if it responds well to its best stimulus and strongly to one other or weakly to two or three others. An $H = 1.0$ indicates a cell that is equally responsive to all four stimuli, whereas an $H = 0.0$ indicates a cell responding exclusively to one stimulus of the four. This measure has seen extensive use over the past 25 years in the study of both peripheral and central gustatory neurons (see Spector, A. C. and Travers, S. P., 2005 for a thorough review of this literature). Recently, Spector A. C. and Travers S. P. (2005) have proposed an additional measure of response tuning, which they termed the noise-to-signal measure, which takes into account the relative magnitudes of the responses to the most and second most effective stimuli. Using this approach, a cell with a large second response would be considered more broadly tuned than one with responses to all three other stimuli if their responses were smaller. Both approaches together can provide a good description of the breadth of tuning of gustatory neurons.

Comparing the breadth of tuning across levels of the gustatory system is difficult, even with a standard way of measuring response breadth. Different laboratories often use different stimuli and concentrations, and also study neurons receiving input from different regions of the mouth, all of which can impact how broadly tuned the cells appear. For example, Figure 4 depicted bitter-best neurons with relatively narrow tuning to bitter stimuli and posterior tongue receptive fields. However, another recent study, which concentrated on the anterior mouth-responsive portion of NST, and which used different concentrations of electrolytes and different bitter tastants, reported that bitter-responsive neurons responded nearly equally to salty and sour stimuli (Lemon, C. H. and Smith, D. V., 2005) (Figure 5).

To avoid such complications, a direct comparison of hamster chorda tympani, NST, and PbN responses was made using the same stimuli and concentrations. The results demonstrated that cells within the brainstem are more broadly tuned than peripheral gustatory axons (Van Buskirk, R. L. and Smith, D, V., 1981). Among chorda tympani fibers, the sucrose-best neurons are the most narrowly tuned (mean $H = 0.386$, compared to 0.587 for NaCl-best and 0.668 for HCl-best neurons). At the level of the NST, sucrose-best neurons are significantly more broadly responsive ($H = 0.586$) and this breadth increases again at the level of the PbN ($H = 0.743$). Thus, the one neuron type in the periphery which stands out for its narrow tuning becomes systematically more broadly tuned as information converges onto neurons in the medulla and pons of the hamster. In fact, at the level of the pons, sucrose-best neurons are more broadly tuned than the broadly tuned HCl-best fibers of the chorda tympani nerve. There is a wealth of data on several species and several levels of the nervous system about the breadth of responsiveness of gustatory neurons (summarized in Spector, A. C. and Travers, S. P., 2005). Overall, many cells in the central nervous system are relatively broadly responsive to stimuli representing different taste qualities; however, some neurons with more specific tuning profiles can also be observed.

4.15.2.2.2 Convergence, orotopy, and oral somatosensory responsiveness

Fibers of the chorda tympani nerve typically innervate taste cells in multiple fungiform papillae, although the sensitivities of each branch appear to be highly similar, suggesting that taste axons are guided to make connections to particular receptors (Oakley, B. *et al.*, 1979; see also Ninomiya, Y., 1998). Thus convergence of separate axonal branches within the tongue does not appreciably increase the breadth of a neuron's responsiveness. On the other hand, when recordings are made from neurons in the NST, separate receptive fields are seen to converge (Vogt, M. B. and Mistretta, C. M., 1990), resulting in greater breadth of responsiveness. Further, axons from completely separate receptor populations have been shown to converge on NST neurons in the rat (Ogawa, H. and Hayama, T., 1984), particularly between the palate and anterior tongue (Travers, S. P. *et al.*, 1986; Travers, S. P. and Norgren, R., 1991). These neurons often responded best to NaCl applied to the anterior tongue and sucrose applied to the

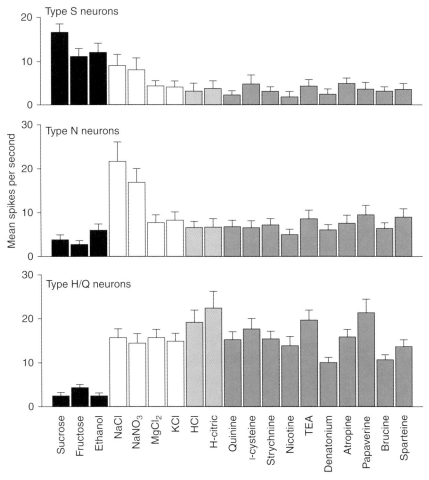

Figure 5 Further examples of rat nucleus of the solitary tract (NST) neuron types based on chemosensitive response profiles. The N neurons strongly resemble the N neurons in Figure 4, and the S neurons are similar to the S/U neurons in Figure 4. However, neurons with selective responses to bitter tastants were not apparent. Instead, neurons with optimal responses to acids responded nearly as well to salts and bitter stimuli. Stimuli: 320 mM sucrose, 500 mM fructose, 40% ethanol, 10 mM NaCl, 8 mM NaNO$_3$, 3 mM HCl, 5 mM citric acid, 10 mM MgCl$_2$, 5 mM KCl, 15 mM denatonium benzoate, 3 mM L-cysteine, 1 mM nicotine, 5 mM strychnine, 40 mM tetraethylammonium, 30 mM atropine, 5 mM brucine, 30 mM papaverine, 9 mM sparteine, 7 mM quinine hydrochloride. From Lemon, C. H. and Smith, D. V. 2005. Neural representation of bitter taste in the nucleus of the solitary tract. J. Neurophysiol. 94, 3719–3729 with permission.

palate. Convergence from different parts of the oral cavity onto cells in the hamster NST has also been described (Sweazey, R. D. and Smith, D. V., 1987). Convergence of peripheral axons onto neurons of the NST is probably responsible for the increased breadth of responsiveness seen in NST neurons over peripheral fibers. Although there are little or no data on whether sucrose-best fibers of the chorda tympani nerve exhibit amiloride-sensitive responses to NaCl, responses to NaCl in sucrose-best neurons in the NST are blocked by amiloride (Boughter, J. D., Jr. and Smith, D. V., 1998; St. John, S. J. and Smith, D. V., 2000), probably reflecting convergence between peripheral NaCl-best and sucrose-

best chorda tympani axons onto cells of the NST. At the level of the PbN, in addition to an increase in the breadth of tuning of sucrose-best neurons (Van Buskirk, R. L. and Smith, D .V., 1981), there is direct evidence that inputs from the anterior and posterior oral cavity converge onto single neurons (Halsell, C. B. and Travers, S. P., 1997).

As a result of the diverse anatomical distribution of taste buds across the tongue, palate, and oropharyngeal and laryngeal epithelia and their innervation by three different cranial nerves, the central representation of gustatory input is arranged in an orotopic fashion (Contreras, R. J. et al., 1982; Hamilton, R. B. and Norgren, R., 1984; Travers, S. P. and Norgren, R.,

1995). Within the medulla, afferent neurons of the VII nerve, including those within the chorda tympani and the greater superficial petrosal nerves, terminate in the most rostral pole of the NST. Overlapping this distribution but extending more caudally is the input from the IX nerve and this distribution is overlapped and extended more caudally still by afferent terminals of the gustatory axons of the X nerve. Consequently, the oral cavity is roughly represented spatially within the NST, with input from the anterior tongue and palate most rostral and that from the epiglottis most caudal within the nucleus. In rodents there are some differences in responsiveness to different stimuli across the various receptor fields, most notably a greater sensitivity to quinine and other bitter stimuli on the posterior tongue. This difference is reflected in the responsiveness of cells in different regions of the NST (Sweazey, R. D. and Smith, D. V., 1987; Geran, L. C. and Travers, S. P., 2006). In the PbN, gustatory input to the external lateral–inner subdivision is exclusively from the posterior oral cavity, whereas both posterior and anterior oral inputs are comingled in the central medial and ventral lateral subdivisions (Halsell, C. B. and Travers, S. P., 1997). In fact, orotopy is observed all the way to the cortex, where input from the VII and IX nerves is still somewhat segregated (e.g., Benjamin, R. M. and Akert, K., 1959). Thus input from the posterior and anterior oral cavities is partially segregated at all levels of the gustatory system. Functionally, gustatory information carried in the VII nerve, but not the IX, appears to be critical for taste discrimination (Spector, A. C. and Grill, H. J., 1992; Spector, A. C. et al., 1997; St. John, S. J. and Spector, A. C., 1998), whereas aversive taste reactivity (e.g., gapes) depends upon IX nerve input (Travers, J. B. et al., 1987; Grill, H. J. et al., 1992). These data suggest that gustatory input carried by different cranial nerves may subserve somewhat different functional roles, as is seen quite clearly in fish, where the VII nerve is involved in food seeking and the IX in ingestion (Bardach, J. E. et al., 1967; Atema, J., 1971). Although the degree of anatomical overlap between cranial nerve inputs in mammals makes such a relationship more difficult to discern, a similar functional dichotomy between VII and IX nerve gustatory function may exist.

In addition to processing gustatory information, the rNST also receives notable inputs from oral somatosensory receptors. Somatosensory input arising from the anterior mouth reaches the nucleus though projections of oral divisions of the trigeminal nerve; and that from the posterior mouth through the IX nerve (Beckstead, R. M. and Norgren, R., 1979; Contreras, R. J. et al., 1982; Hamilton, R. B. and Norgren, R., 1984). Neurons that are unresponsive to taste, but instead responsive to nonnociceptive mechanical stimulation of the oral mucosa are also arranged in an orderly orotopic fashion, laterally adjacent to the location of taste-responsive neurons (Travers, S. P. and Norgren, R., 1995). In addition, a number of gustatory-responsive neurons can also be activated by mechanical stimulation of the mouth, with such responses stronger and more prevalent in posterior-responsive neurons receiving inputs from the IX nerve (Travers, S. P. and Norgren, R., 1995).

4.15.2.2.3 Modulation of gustatory responsiveness

Gustatory neurons in the rodent PbN send projections to two distinct forebrain systems: thalamocortical and limbic (Figure 1). Double-labeling experiments have shown that some PbN neurons project to the ventral posterior medial nucleus, parvicellularis (VPMpc) of the thalamus, whereas others project to the central nucleus of the amygdala (CeA) (Voshart, K. and van der Kooy, D., 1981). From the VPMpc, gustatory information is transmitted to the gustatory insular cortex. PbN neurons also project to the lateral hypothalamus, bed nucleus of the stria terminalis, and the substantia innominata (Norgren, R., 1974; 1976). Although the precise route may be different (see below), taste information in the primate also reaches limbic, as well as thalamocortical forebrain sites. These various areas of the limbic forebrain pathway have been implicated in food and water regulation, sodium appetite, taste aversion learning, and the response to stress, suggesting a substrate through which taste activity interfaces with motivated behavior. Both the insular cortex and many of these limbic areas have feedback projections to the NST and PBN (Figure 2) (van der Kooy, D. et al., 1984; Veening, J. G. et al., 1984; Moga, M. M. et al., 1990b; Halsell, C. B., 1998; Whitehead, M. C. et al., 2000), providing a mechanism by which higher-level processing can impact taste responses at the initial stages of the pathway.

4.15.2.2.3.(i) Physiological alterations The first indications of a modulatory effect of the forebrain on taste activity came from studies on decerebrate rats, which demonstrated that taste responses in brainstem nuclei were changed by severing the connections with the forebrain (Hayama, T. et al., 1985; Di Lorenzo, P. M., 1988; Mark, G. P. et al., 1988). Further experiments

using reversible lesions of the insular cortex with procaine demonstrated both increased and decreased responding to taste stimulation in the PbN (Di Lorenzo, P.M. and Monroe, S., 1990) as well as the NST (Di Lorenzo, P. M. and Monroe, S., 1995). These data demonstrated that the responsiveness of gustatory neurons in the brainstem are not simply the result of the integration of afferent inputs, but are modifiable by descending circuits. Several kinds of experiments have illustrated that physiological and experiential factors may modulate the responsiveness of taste neurons in the brainstem; descending modulation presumably makes a key contribution to these influences, though local circuits most likely play a role as well.

4.15.2.2.3.(ii) Homeostatic mechanisms: feeding and appetite

Gustatory input is critical to the initiation and control of ingestive behavior, guiding the choice of nutritive foods, and promoting the avoidance of toxic substances. A number of studies have shown that manipulations that alter an animal's state of satiety can have an influence on the responses of gustatory neurons in the brainstem. For example, hyperglycemia produced by intravenous infusion of glucose decreased responses to orally applied glucose in cells of the rat NST by an average of 43% (Giza, B. K. and Scott, T. R., 1983) Responses to NaCl and HCl were considerably less affected (20% and 16%, respectively) by blood glucose levels and those to QHCl (3%) not at all. Similarly, rats decreased their preference for glucose following an intravenous glucose load, which had no effect on their behavioral responses to quinine (Giza, B. K. and Scott, T. R., 1987). Therefore, when blood sugar levels are elevated, gustatory responses in the brain are decreased and the intake of glucose is concomitantly reduced. Intravenous infusion of other satiety factors, including insulin and pancreatic glucagon, also decreased NST multiunit responses to glucose, although cholecystokinin (CCK) had no effect on taste activity (Giza, B. K. et al., 1992). Similarly, gastric distension has been shown to decrease the responsiveness of NST neurons to taste stimulation (Glenn, J. F. and Erickson, R. P., 1976). These factors, all of which can be signals contributing to satiety, appear to decrease gustatory sensitivity to nutritive taste stimuli.

It is generally believed that in the short-term regulation of feeding behavior, the sensory components of food play a dominant role (Smith, G. P., 1996). In sham-feeding experiments in which orosensory stimulation is isolated from any postingestive effects, fat infusions into the duodenum decrease food intake (e.g., Greenberg, D. et al., 1990; Foster, L. A. et al., 1996). Electrophysiological evidence indicates that intraduodenal infusion of lipid reversibly decreases the responsiveness of PbN taste neurons to sucrose and somewhat to NaCl, with no effect on responses to acid or quinine (Figure 6, Hajnal, A. et al., 1999).

The effects were also differential for neuron type, indicating a further specialization for this modulation. Sucrose responses were most affected in sucrose-best neurons, and in particular those with the most specific responses to sucrose. These effects were rapid, beginning within 5–10 min, and reversible within about 40 min after the beginning of a 10 min lipid infusion. These data demonstrate that fat introduced into the duodenum has a specific effect on taste mechanisms, providing for decreased taste sensitivity to stimuli that signal nutritional value.

Another area in which nutritional need results in altered gustatory responsiveness is sodium appetite. Normal rats behaviorally choose to ingest NaCl at concentrations that maintain sodium homeostasis. Thus an isotonic concentration of about 0.15 M is avidly consumed, whereas lower concentrations are less preferred and higher concentrations become increasingly avoided; indeed rats can voluntarily mix NaCl and water to maintain isotonic levels of intake (Rinaman, L. et al., 1993). However, when plasma sodium concentration declines, a condition known as hyponatremia, there are two mechanisms for returning sodium to normal levels. One is to conserve sodium through activation of the renin–antiotensin–aldosterone system to increase sodium resorption and the other is to increase the ingestion of sodium by creating a specific sodium appetite (see McCaughey, S. A. and Scott, T. R., 1998). Like most other mammals, rats that are made sodium deficient will exhibit a sodium appetite that is characterized by an increase in the intake of sodium salts, such as NaCl (Jalowiec, J. E. and Stricker, E. M., 1970; Contreras, R. and Hatton, G., 1975), which is ingested at concentrations that are normally behaviorally neutral or avoided. Investigation of the responsiveness of gustatory neurons following sodium deficiency demonstrates that the responsiveness to NaCl is decreased, particularly in those neurons most responsive to sodium salts (i.e., those with amiloride-sensitive input) in both peripheral chorda tympani axons (Contreras, R. J., 1977; Contreras, R. J. and Frank, M., 1979) and in neurons of the NST (Jacobs, K. M. et al., 1988) and PbN (Shimura, T. et al., 1997a). These results have been interpreted to

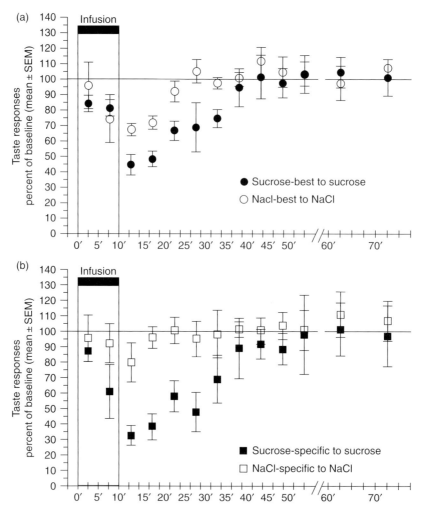

Figure 6 Effects of intraduodenal lipid infusion on taste responses in parabrachial nuclei (PbN) neurons. (a) Mean responses to sucrose in sucrose-best neurons, and mean responses to NaCl in NaCl-best neurons before and at different times after the infusion. Both the responses to sucrose and NaCl were suppressed rapidly, but the sucrose response was much more profoundly affected. (b) The sucrose effect was more pronounced in the subset of sucrose-best neurons that were most selectively tuned to this tastant; the opposite was true for the suppression of the NaCl response; lipids infusion was less effective in suppressing the NaCl response of NaCl-specific neurons. From Hajnal, A., Takenouchi, K., and Norgren, R. 1999. Effect of intraduodenal lipid on parabrachial gustatory coding in awake rats. J. Neurosci. 19, 7182–7190, with permission.

mean that a sodium-deprived rat would require greater ingestion of NaCl to establish the same level of sensory input as a normal rat (Contreras, R. J. and Frank, M., 1979). Whether or not this decrease in the sodium responsiveness is actually required for the expression of sodium appetite is unclear, however, since varying the way in which sodium appetite is induced can actually lead to opposite changes; that is, increases in brainstem taste responses (Tamura, R. and Norgren, R., 1997), even though the animal still consumes more sodium. Thus, the relationship between changes in taste responsiveness and changes in ingestive behavior do not appear simple.

4.15.2.2.3.(iii) Learning: conditioned preferences and aversions Palatable taste stimuli that are associated with gastrointestinal illness become aversive, a learning phenomenon known as conditioned taste aversion (CTA), first described by Garcia J. *et al.* (1970) . When rats are made ill following exposure to a novel stimulus, they associate the resulting illness with the taste of the stimulus and thereafter are averse to it and to other stimuli with a similar taste (Nachman, M., 1963; Halpern, B. P. and Tapper, D. N., 1971). Such behavior has survival value for the animal by helping it to avoid the ingestion of food that contains toxic substances. The generalization of CTA has been used as a psychophysical tool to assess the similarity in quality among gustatory stimuli in experimental animal models (e.g., Smith, D. V. and Travers, J. B., 1979; Smith, D. V. *et al.*, 1979; Nowlis, G. H. *et al.*, 1980), as animals will avoid other stimuli in proportion to the perceived similarity to the conditioned stimulus (CS) (Smith, D. V. and Theodore, R. M., 1984). Several studies have shown that such conditioning alters the responsiveness of gustatory neurons, generally by making the CS a somewhat more effective stimulus. For example, Chang F. C. and Scott T. R. (1984) demonstrated that a CTA to saccharin increased the responsiveness to the CS in those neurons most responsive to sweet substances. Similarly, conditioning an aversion to NaCl resulted in an increased response to the CS in NaCl-best neurons of the rat PbN (Shimura, T. *et al.*, 1997b). Such an increase is restricted to those neurons receiving amiloride-sensitive sodium input, but this increase in responsiveness was not evident in decerebrate animals (Tokita, K. *et al.*, 2004). Thus, as shown above, sodium need leads to a decreased neural response to NaCl and increased consumption, whereas CTA leads to an increased neural response to NaCl and decreased consumption. On the other hand, when the CS was a normally nonpreferred HCl solution, the effects of CTA were the opposite: PbN responses to the CS were reduced (Shimura, T. *et al.*, 2002). When rats are conditioned to prefer a $MgCl_2$ or citric acid solution, there were slight shifts in neural responses in the rat NST, making the CS more distinct from other nonpreferred solutions, but these effects were much less robust than the effects of CTA (Giza, B. K. *et al.*, 1997).

4.15.2.2.4 Neurotransmitters and neuromodulators in the nucleus of the solitary tract

Recently, significant progress has been made in identifying the neurotransmitters and neuromodulators along the gustatory pathway, and defining their roles in gustatory information processing. In the brainstem, most is known about the NST.

4.15.2.2.4.(i) Excitatory amino acids

Much of the excitatory transmission in the central nervous system is mediated by excitatory amino acids (EAAs), such as glutamate (Mayer, M. L. and Westbrook, G. L., 1987). In the visceral portions of the NST, several EAA agonists had been shown to elicit cardiovascular responses when injected into the medial NST, mimicking the effects of stimulation of baroreceptor afferent fibers, and these responses are blocked by both *N*-methyl-D-aspartic acid (NMDA) and non-NMDA antagonists (Guyenet, P. G. *et al.*, 1987; Kubo, T. and Kihara, M., 1988a; 1988b). Synaptic transmission between primary vagal afferent fibers and second-order cells of the caudal NST appears to be mediated by glutamate (Drewe, J. A. *et al.*, 1990; Glaum, S. R. *et al.*, 1992; Vardhan, A. *et al.*, 1993; Andresen, M. and Yang, M., 1994). In studies done *in vitro* on rat brainstem slices, the responses of cells in the gustatory region of the rNST to electrical stimulation of the solitary tract are reduced or blocked by antagonists of both NMDA and α-amino-3-hydroxy-5-methyl-4-isoxazolepropionic acid (AMPA)/kainate receptors (Wang, L. and Bradley, R. M., 1995), suggesting that both NMDA and non-NMDA receptors may mediate transmission between primary gustatory fibers and second-order neurons of the NST. The direct involvement of EAA receptors in the transmission of gustatory information was shown in an *in vivo* experiment, in which gustatory responses in individual cells of the hamster NST were blocked by EAA receptor antagonists (Li, C. S. and Smith, D. V., 1997). Microinjection of kynurenic acid, a broad-spectrum EAA antagonist, into the vicinity of taste-responsive neurons of the NST completely and reversibly blocked their activity. In these *in vivo* experiments, the AMPA/kainate receptor antagonist 6-cyano-7-nitroquinoxaline-2,3-dione (CNQX) also completely and reversibly blocked these responses, although the NMDA antagonist DL-2-amino-5-phosphonovalerate (APV) was less effective. All gustatory responses in the NST were blocked by CNQX, regardless of the type of cell or nature of the gustatory stimulus (Li, C. S. and Smith, D. V., 1997). Because the *in vitro* experiments had shown colocalization of NMDA and non-NMDA receptors on many of the same NST neurons (Wang, L. and Bradley, R. M., 1995), it is likely that the NMDA receptors play a secondary role which is not easily revealed *in vivo*.

4.15.2.2.4.(ii) Gamma-aminobutyric acid, tonic inhibition

There are many small ovoid neurons within the NST that express the inhibitory neurotransmitter GABA in both rats (Lasiter, P. S. and Kachele, D. L., 1988) and hamsters (Davis, B. J., 1993). Slice experiments on the rNST had suggested that cells within the gustatory NST could be inhibited by GABA and excited by the $GABA_A$ receptor antagonist, bicuculline methodide (BICM), suggesting a tonic GABAergic inhibitory network within this region (Liu, H. *et al.*, 1993; Wang, L. and Bradley, R. M., 1993). When GABA was microinjected into the hamster NST *in vivo*, taste-responsive neurons showed a dose-dependent inhibition of spontaneous and evoked activity and their responses to taste stimulation were enhanced in a dose-dependent manner by BICM; 63% of the taste-responsive cells in the NST were modulated by GABAergic mechanisms in these experiments (Smith, D. V. and Li, C. S., 1998). The responses of all tastants were enhanced by BICM administration, showing that particular neurons, but not particular stimuli, are modulated by GABAergic inhibitory circuits. The result of the removal of GABAergic inhibition was an increase in the breadth of responsiveness of NST neurons to gustatory stimulation as the cells became more responsive to all stimuli, similar to the effects on tuning seen in other sensory systems following BICM (Kyriazi, H. T. *et al.*, 1996; Lazareva, N. A. *et al.*, 1997).

One source of inhibition of NST gustatory neurons arises from the gustatory cortex (GC), which sends descending connections to the NST (van der Kooy, D. *et al.*, 1984). Stimulation of GC results in both excitation and inhibition of the activity of taste neurons in the NST (Di Lorenzo, P. M. and Monroe, S., 1995). Both electrical stimulation of the GC and excitation produced by the glutamate analog DL-homocysteic acid (DLH) produced either excitation or inhibition in about 34% of taste-responsive cells in the hamster NST (Smith, D. V. and Li, C., 2000). The inhibition produced by GC stimulation was blocked by infusion of BICM into the vicinity of the recorded cell in the NST. Of the eight cells inhibited by GC stimulation, seven were most responsive to NaCl, suggesting that cortical-evoked inhibition of NST activity may be somewhat stimulus-specific (Smith, D. V. and Li, C., 2000), although the effects of GC stimulation on the responses to NaCl themselves were not tested in this study. Gustatory neurons in both the PbN (Kobashi, M. and Bradley, R. M., 1998) and insular cortex (Ogawa, H. *et al.*, 1998) are also modulated by GABAergic mechanisms.

4.15.2.2.4.(iii) Peptides: substance P and opioids

Immunocytochemical studies show that there are substance P (SP)-containing neurons and fiber terminals within the rNST (South, E. H. and Ritter, R. C., 1986; Davis, B. J. and Kream, R. M., 1993). In experiments on *in vitro* brainstem slices in rats (King M. S. *et al.*, 1993), SP transiently depolarized 80 of 117 (68%) rNTS neurons in a dose-dependent manner. Submicromolar concentrations of SP had potent excitatory effects, and the half-maximal response occurred at 0.6 μM. The depolarizing effect of SP was accompanied by an increase in input resistance

in 81% of the responsive neurons and persisted in low Ca^{2+} and high Mg^{2+} saline as well as in the presence of tetrodotoxin (TTX), suggesting direct postsynaptic action of SP on the recorded neurons. SP also hyperpolarized four neurons (4%) and had no effect on 33 cells (28%). When gustatory neurons were recorded *in vivo* from hamster NST, about 48% were excited by microinjection of SP into the nucleus, whereas 9% were inhibited (Davis, B. J. and Smith, D. V., 1997). In this experiment, only NaCl was used as a taste stimulus, so it is as yet unclear whether SP produces differential effects according to stimulus. However, the responses to NaCl were clearly enhanced by SP.

Endogenous opiates have been implicated in gustatory palatability (e.g., Rideout, H. J. and Parker, L. A., 1996) and earlier studies had shown that systemic administration of morphine resulted in inhibition of taste-evoked multiunit activity in the rat PbN (Hermann, G. E. and Novin, D., 1980) Injection of naltrexone (NLTX), a nonselective opioid receptor antagonist, into the NST blocks feeding induced by neuropeptide Y administration into the paraventricular nucleus (Kotz, C. M. *et al.*, 1997), implicating an opioid-mediated influence within the NST. Further, μ-opioid receptors are expressed by rNST neurons that receive synaptic input from neurons in the central nucleus of the amygdala (Pickel, V. M. and Colago, E. E., 1999), an area known to provide modulatory input to the gustatory NST (Li, C. S. *et al.*, 2002a). Within the gustatory portion of the rNST, there are both nerve fibers and cell bodies expressing methionine enkephalin (met-ENK) immunoreactivity (Davis, B. J. and Kream, R. M., 1993). In neurons recorded *in vivo* from the hamster NST, microinjection of met-ENK reversibly reduced the responses of 39 of 165 (23.6%) taste-responsive neurons in a dose-dependent manner (Li, C. S. *et al.*, 2003). This inhibition was blocked by NLTX, although NLTX alone had no effect, suggesting that opioid inhibition in these neurons is not tonic. Thus, along with GABAergic mechanisms, opioids exert an inhibitory effect on gustatory processing in the NST.

4.15.2.2.4.(iv) Other neurotransmitters: dopamine, acetylcholine, and nitric oxide

Less is known about the function of other neurotransmitters in the gustatory NST, although there is some evidence for dopaminergic, cholinergic, and nitrergic mechanisms. Within the gustatory portion of the hamster NST, neurons expressing tyrosine hydroxylase and dopamine-β-hydroxylase have been described (Davis, B. J. and Jang, T., 1988b). In electron micrographic studies,

(Davis, B. J., 1998) demonstrated that NST neurons expressing tyrosine hydroxylase received synaptic input directly from chorda tympani nerve axons, suggesting that these neurons are second-order gustatory cells. The rNST also contains neurons that express nitric oxide synthase (NOS), the synthetic enzyme for nitric oxide and stain for NADPH diaphorase (NADPHd), a marker for NOS (Ruggiero, D. A. *et al.*, 1996; Travers, S. P. and Travers, J. B., 2006). In fact, about 15–20% of the neurons that express Fos-like immunoreactivity in response to gustatory stimulation with quinine or citric acid are double-labeled for NADPHd suggesting that nitric oxide may be a transmitter in a subset of taste cells. Further, many of these Fos neurons intermingle with NOS fibers, implying that nitric oxide could modulate taste responses (Travers, S. P. and Travers, J. B., 2006). However, to date, there are no functional data showing a role for dopamine or nitric oxide in gustatory processing. On the other hand, application of acetylcholine (ACh) to cells recorded in rat brain slices through the rNST show that many of these neurons are modulated by ACh, mediated through both nicotinic and muscarinic ACh receptors (Uteshev, V. V and Smith, D. V., 2006). Nicotinic responses, which were always excitatory, were mediated by either $\alpha 7$ or non-$\alpha 7$ nicotinic AChRs and comprised about 66% (50/76) of the cells examined. Responses mediated by muscarinic AChRs were seen in 42% (31/73) neurons tested. These responses were inhibitory near the resting potential, resulting in a block of evoked spikes. Extensive acetylcholinesterase (AChE) immunostaining is closely associated with the terminal fields of both the chorda tympani and glossopharyngeal nerves in the rNST (Barry, M. A. *et al.*, 2001) and there are neurons expressing choline acetyltransferease (ChAT) in the medial portion of the rNST (Ruggiero, D. A. *et al.*, 1990). These latter neurons are part of the parasympathetic preganglionic neurons of the inferior salivatory nucleus (Contreras, R. J. *et al.*, 1980), which could conceivably modulate taste activity as a consequence of parasympathetic activity. There are as yet no data on cholinergic modulation of taste responses *in vivo*, nor is it clear whether all of the cholinergic input to this area is derived from these parasympathetic neurons.

4.15.2.2.5 Descending pathways from forebrain targets: modulation of nucleus of the solitary tract and parabrachial nuclei

As delineated above, decerebration has been shown to modulate brainstem gustatory activity, indicating that the responsiveness of brainstem gustatory neurons reflects not only sensory input from the taste receptors, but descending influences from forebrain circuits. There are direct descending projections from the GC, the lateral hypothalamus, the central nucleus of the amygdala, and the bed nucleus of the stria terminalis to both the NST and PbN gustatory regions (van der Kooy, D. *et al.*, 1984; Moga, M. M. *et al.*, 1990b; Allen, G. V. *et al.*, 1991; Halsell, C. B., 1998). A series of recent experiments on both rats and hamsters has demonstrated that each of these forebrain targets of the gustatory system provides modulatory control over taste-responsive neurons in the NST and PbN (Figure 2).

4.15.2.2.5.(i) Gustatory cortex   When the activity of GC was blocked by local infusion of procaine, both increases and decreases in the taste responses were recorded from the rat NST (Di Lorenzo, P. M. and Monroe, S., 1995) and PbN (Di Lorenzo, P. M., 1990), showing that these cells can be either facilitated or inhibited by cortical influences. Similar results were seen as a result of electrical or chemical stimulation of the GC in hamsters (Smith, D. V. and Li, C., 2000). Of 50 neurons recorded from the hamster NST, 17 (34%) were modulated by ipsilateral cortical stimulation. About half of these (9/17) were excited and half (8/17) were inhibited. Although the excitatory effects were distributed across different cell types, the inhibitory effects were significantly more common in NST cells responding best to NaCl. In a more recent experiment, in which stimulating electrodes were implanted bilaterally in the GC, a slightly greater influence of the contralateral cortex was noted: 16 of 50 cells (32%) were modulated ipsilaterally, whereas 20 of 50 (40%) responded to contralateral stimulation (Smith, D. V. *et al.*, 2004). Eleven of these neurons received converging input from both sides of the cortex, so that among the 50 cells recorded, 25 (50%) were modulated by one or both sides. A similar experiment on the rat showed that taste-responsive neurons in the PbN are also modulated by GC stimulation, with both inhibitory and excitatory responses (Lundy, R. F., Jr. and Norgren, R., 2004a). Both of these stimulation studies are compatible with the earlier studies in which the cortex was anesthetized with procaine, which resulted in both increased and decreased taste responses in PbN and NST (Di Lorenzo, P. M., 1990).

4.15.2.2.5.(ii) Lateral hypothalamus   Cells within the lateral hypothalamus respond to taste stimuli

applied to the oral cavity (Norgren, R., 1970; Yamamoto, T. *et al.*, 1989a) or alter their activity during food ingestion (Sasaki, K. *et al.*, 1984; Ono, T. *et al.*, 1986). There are descending projections from the lateral hypothalamus to both the PbN and NST (van der Kooy, D. *et al.*, 1984; Moga, M. M. *et al.*, 1990a). Earlier studies had suggested that stimulation of the lateral hypothalamus enhanced the responses of rat NST neurons evoked by chorda tympani nerve stimulation (Bereiter, D. *et al.*, 1980) or by electrical stimulation of the tongue (Matsuo, R. *et al.*, 1984). Electrical and chemical stimulation of the lateral hypothalamus in hamsters produced orthodromic modulation of half (49/99) of NST taste-responsive neurons (Cho, Y. K. *et al.*, 2002b) (Grossman, S. P. *et al.*, 1978; Frank, R. A. and Stutz, R. M., 1982). Most of these effects (44/99) were excitatory, as shown in the example in Figure 7.

Thus about half the taste-responsive cells of the NST are modulated (mostly excited) by the lateral hypothalamus. Stimulation of this region of the brain induces feeding behavior and, conversely, lateral hypothalamic lesions reduce food intake (Grossman, S. P. *et al.*, 1978; Frank, R. A. and Stutz, R. M., 1982), and tthese effects have been shown to interact with taste-guided behavior (Murzi, E. *et al.*, 1986; Touzani, K. *et al.*, 1990; Conover, K. L. and Shizgal, P., 1994). Thus, it is interesting to speculate that the lateral hypothalamus could enhance the signal-to-noise ratio in NST taste neurons during bouts of feeding (Cho, Y. K. *et al.*, 2002b). Lateral hypothalamic stimulation has also been shown to modulate taste activity in neurons of the rat PbN, although the excitatory and inhibitory influences were equal (Lundy, R. F., Jr. and Norgren, R., 2004a).

4.15.2.2.5.(iii) Central nucleus of the amygdala

The central nucleus of the amygdala contains neurons that respond differentially to hedonically positive and negative taste stimuli (Nishijo, H. *et al.*, 1998) and both the central nucleus and basolateral amygdala are involved in CTA learning (Yamamoto, T. *et al.*, 1994). The central nucleus of the amygdala provides a direct descending projection to both the NST and PbN (Norgren, R., 1976; Veening, J. G. *et al.*, 1984; Halsell, C. B., 1998). When stimulating electrodes were implanted bilaterally in the hamster central nucleus, 36 of 109 NST taste-responsive neurons (33%) were orthodromically modulated; 33 of these cells were excited and 3 were inhibited. Interestingly, those neurons that were modulated by the central nucleus of the amygdala had a significantly smaller response to taste stimuli than those that were not modulated. However, like the modulation of cells by the lateral hypothalamus, there was no specific effect on any one stimulus or cell type within the NST. On the other hand, in the PbN, Lundy R. F., Jr. and Norgren R. (2001) reported that stimulation of the central nucleus of the amygdala in the rat produced mostly suppressive effects, with an overall effect of enhancing the salience of the response to NaCl, suggesting that descending control from the amygdala could play a role in the mechanisms of sodium appetite. Stimulation of the central nucleus of the amygdala also produced more inhibition than excitation in another experiment (Tokita, K. *et al.*, 2004), in which it was shown that alterations in PbN neuronal responses to NaCl following CTA are eliminated by decerebration. These authors argued that descending influences from the central nucleus are important for the gustatory changes seen in the PbN following conditioning. Thus, one of the roles of the descending influence of the central nucleus of the amygdala on both the PbN and NST may be to produce alterations in brainstem responding following CTA, such as changes in taste reactivity seen, for example, to sucrose following aversive conditioning (Grill, H. J., 1985).

In both the NST (Cho, Y. K. *et al.*, 2003) and PbN (Lundy, R. F., Jr. and Norgren, R., 2004a), taste-responsive cells are often modulated by more than one forebrain structure. For example, of 113 cells in the hamster NST that were modulated by stimulation of either the lateral hypothalamus or central nucleus of the amygdala, 52 were responsive to stimulation of both sites. In the rat PbN, 60 neurons were tested for their responses to stimulation of the GC, central nucleus of the amygdale and lateral hypothalamus and 53 of them were modulated; 37 of them responded to more than one of these sites, with 17 neurons influenced by two sites and 20 by all three. In both the NST and PbN studies, the influence of different sites was most often similar, although there were cells that were excited by one site and inhibited by another. In a recent study of the effects of CeA and LH stimulation on neurons of the hamster PbN, most cells of the PbN were antidromically activated by stimulation of these sites (Li, C. S. *et al.*, 2005), which are synaptic targets of output neurons of the PbN (Norgren, R., 1976). Of 101 neurons tested in this study, 83 were antidromically driven by either the lateral hypothalamus or central nucleus of the amygdala and 57 of these by both, indicating a direct projection from the PbN to both of these areas. Only a small number of cells were

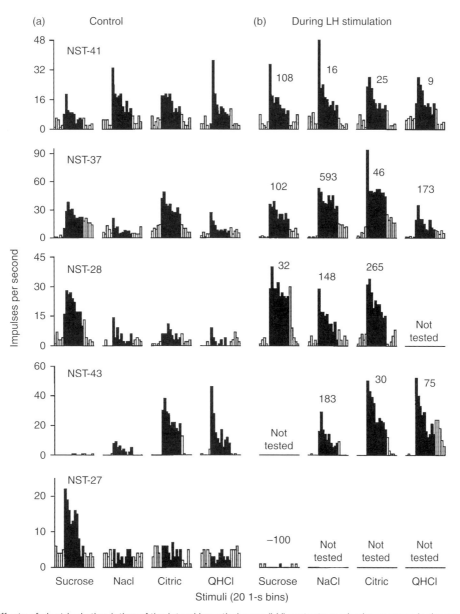

Figure 7 Effects of electrical stimulation of the lateral hypothalamus (LH) on taste-evoked responses in the rostral nucleus of the solitary tract (NST). Left: Gustatory responses in five neurons in the absence of LH stimulation. Right: Taste-evoked activity of these same neurons during concurrent stimulation (100 Hz, 0.2 ms pulses) using an intensity that was just subthreshold for altering spontaneous activity. As was typical for the entire population, the taste responses of most (4/5) of these neurons were amplified by activation of the LH. From Cho, Y. K. *et al.*, 2002b. Taste responses of neurons of the hamster solitary nucleus are enhanced by lateral hypothalamic stimulation. J. Neurophysiol. 87, 1981–1992, with permission.

orthodromically modulated, with five cells excited and eight inhibited by one of these forebrain areas. Clearly, there are complex descending influences on the gustatory responses of cells in both the NST and PbN.

4.15.2.2.5.(iv) Bed nucleus of the stria terminalis

In addition to the ventral projection of the PbN to the lateral hypothalamus and central nucleus of the amygdala, there is also a projection to the dorsolateral bed nucleus of the stria terminalis (Norgren, R., 1976), although to date there are no electrophysiological data showing gustatory responsiveness in this area. Nevertheless, the bed nucleus sends a descending projection to both the NST and the PbN (van der Kooy, D. *et al.*, 1984; Moga, M. M. *et al.*, 1990b; Whitehead, M. C. *et al.*, 2000). Stimulation of the bed nucleus of the stria terminalis inhibited 29 of 101 hamster NST taste neurons, whereas only 7 were excited. All cell types were affected, although the number of NaCl-best neurons were fewer and citric acid-best greater among the modulated cells than expected by chance. Although we know that various subnuclei of the bed nucleus of the stria terminalis are implicated in a number of neural systems, including those involved in responses to stress (Morley, J. E. and Levine, A. S., 1980; Herman, J. P. and Cullinan, W. E., 1997; Cecchi, M. *et al.*, 2002), and in motivation, reward, and drug addition (Walker, J. R. *et al.*, 2000; Macey, D. J. *et al.*, 2003), how it plays a role in taste processing is not clear. Descending feedback from this region could conceivably be involved in the effects of stress on eating behavior (Morley, J. E. and Levine, A. S., 1980; Fullerton, D. T. *et al.*, 1985; Walker, J. R. *et al.*, 2000; Hagan, M. M. *et al.*, 2003; Macey, D. J. *et al.*, 2003).

These kinds of stimulation experiments have demonstrated extensive centrifugal modulation of brainstem gustatory activity, in both the NST and PbN. Essentially, every forebrain target of the gustatory system, including the GC, lateral hypothalamus, central nucleus of the amygdala, and bed nucleus of the stria terminalis exerts a descending modulatory influence on the processing of taste information. The hypothalamus and amygdala have predominantly an excitatory effect on the NST, whereas the cortex and bed nucleus produce significant inhibition of medullary taste responses. This extensive neural substrate no doubt underlies the modulation of taste activity by physiological and experiential factors and further research should be directed toward determining how these pathways are engaged by alterations in blood glucose, gastric distension, CTA, and other physiological conditions known to alter taste activity.

4.15.3 Gustatory Responsiveness of Forebrain Neurons

Although we know a great deal about the responsiveness of brainstem gustatory neurons, there are also data on gustatory areas beyond the pons, including the gustatory thalamic area, the taste cortex, and some regions of the limbic forebrain, including the lateral hypothalamus, central nucleus of the amygdala, nucleus accumbens, and ventral pallidum. The types of cells seen in brainstem areas, that is, sucrose-best, NaCl-best, etc., are also evident in more rostral nuclei. However, cells that appear to be tuned to

stimulus palatability more than taste quality *per se* are also evident, especially in the limbic regions. Whereas the data on brainstem taste responses are almost exclusively from the rodent, and with notable exceptions (Nakamura, K. and Norgren, R., 1991; Nishijo, H. and Norgren, R., 1991; Nishijo, H. *et al.*, 1991; Nakamura, K. and Norgren, R., 1993), mostly from anesthetized preparations, much of the forebrain data is from primates, and regardless of species, frequently studies utilize awake, behaving preparations. Studies based on functional imaging in humans further provide a novel perspective for studying forebrain gustatory function. Due to space constraints, we will concentrate on cortical and limbic processing of taste information.

4.15.3.1 Taste Cortex

4.15.3.1.1 Organization and relationship to other sensory modalities

As discussed in Chapter Gustatory Pathways in Fish and Mammals, primary GC in both rodents and primates is located in insular cortex (Pritchard, T. C. *et al.*, 1986; Kosar, E. *et al.*, 1986a; 1986b). In primates, the gustatory insular cortex is buried in the Sylvian fissure but the primary gustatory area extends from the insula onto the opercular cortex (Pritchard, T. C. *et al.*, 1986). In both rodents and primates, gustatory signals reach the insular cortex via a relay in the most medial part of the ventroposteromedial nucleus of the thalamus (VPMpc). However, in rodents taste information reaches the thalamus only after making a synapse in the parabrachial nucleus (Norgren, R. and Leonard, C. M., 1971; 1973); in primates there is a direct rNST–thalamic pathway (Beckstead, R. M. *et al.*, 1980). Although a gustatory relay in the primate PBN was initially considered unlikely, recent human imaging studies hint that a pontine relay may be a possibility in these species as well (Small, D. M. *et al.*, 2003; Topolovec, J. C. *et al.*, 2004). However, because the rostral pole of the monkey NST does not project to the PBN (Beckstead, R. M. *et al.*, 1980), nor are anterior tongue taste responses apparent in the pons (Pritchard, T. C. *et al.*, 2000), any gustatory PBN representation in primates is likely to arise from posterior tongue or pharyngeal taste buds (discussed in Travers, S. P., 1993).

In primates, taste-responsive neurons are also located in the orbitofrontal cortex and this area has typically been considered to constitute secondary GC (Baylis, L. L. *et al.*, 1995); taste-related activity has recently been reported in an analogous region in rodents as well (Gutierrez, R. *et al.*, 2006). Further, emerging data, including human imaging studies, suggest that there is a high degree of complexity in cortical processing such that multiple hierarchical gustatory relays may be located within both the insular and orbitofrontal cortex (discussed in Pritchard, T. C. and Norgren, R., 2004). In addition to confirming the involvement of insular/opercular and orbitofrontal cortex in representing gustatory information in humans, imaging studies implicate other cortical areas, most consistently the cingulate, but also the posterior parietal or prefrontal, in processing some aspects of the gustatory signal (e.g., Small, D. M. *et al.*, 2003; 2004).

Cortical taste representation is complex with regard to its gustatory orotopic organization and relationship to the representation of innocuous somatosensation, visceral sensation, pain, and olfactory signals. Consistent with lower levels of the neuraxis, at least in subprimates, taste inputs to the cortex are arranged orotopically. Based on electrically stimulating the chorda tympani and glossopharyngeal nerves and recording evoked potentials (Benjamin, R. M. and Pfaffmann, C., 1955; Yamamoto, T. and Kawamura, Y., 1975; Yamamoto, T. *et al.*, 1980) or single units (Hanamori, T. *et al.*, 1997), information from the anterior oral cavity is located more anteriorly; that from the posterior oral cavity more posteriorly. A recent experiment using intrinsic optical imaging suggests that taste quality may also have a systematic topographic organization along the anterior–posterior axis in cortex. Although not exhibiting complete segregation, activity elicited by sweet and salty tastants (sucrose and NaCl) was located more anteriorly than that produced by stimulation with sour or bitter stimuli (citric acid and quinine) (Accolla, R. *et al.*, 2007).

4.15.3.1.1.(i) Taste–visceral relationships Gustatory orotopy is actually part of a more extensive orderly topography in insular cortex, again mirroring the brainstem and thalamic organizations. The representation of inputs arising from the postoral gastrointestinal tract and other visceral organs extends posteriorly from the posterior oral cavity gustatory representation (Cechetto, D. F. and Saper, C. B., 1987; Squire, L. R. *et al.*, 2003). The degree of segregation or overlap between taste signals and visceral signals has not been thoroughly explored, but some degree of convergence onto single neurons in rodents seems likely (Yamamoto, T. *et al.*, 1989b; Hanamori, T.

et al., 1998a; 1998b; but see Cechetto, D. F. and Saper, C. B., 1987). In nonhuman primates, single-unit responses in the insula evoked by visceral stimulation have mostly been assessed indirectly, by noting how satiety affects gustatory responses. In the insular cortex, feeding a monkey to satiety does not alter the magnitude of gustatory responses (Yaxley, S. *et al.*, 1988), suggesting segregation between gustatory and visceral signals in this region. However, this segregation in the primary GC seems to be a major difference between monkeys and rodents, as in rodents, taste and visceral signals converge even at brainstem levels (e.g., Giza, B. K. *et al.*, 1992; Hajnal, A. *et al.*, 1999; Baird, J. P. *et al.*, 2001). In monkeys, convergence of taste and visceral signals has been reported to be a higher-level function first observed in the orbitofrontal cortex (Rolls, E. T. *et al.*, 1989; Pritchard, T. C. *et al.*, 2005), and is reflected in a decline of taste responses coincident with the onset of satiety. Interestingly, the decline in gustatory responsiveness is limited to the stimulus used to induce satiation, and this neural change parallels a specific change in behavior, a phenomenon termed sensory-specific satiety (Rolls, B. J. *et al.*, 1981; 1986). Human imaging studies further demonstrate that gustatory-evoked activity in the orbitofrontal cortex is altered by satiety (Kringelbach, M. L. *et al.*, 2003), although, in contrast to the monkey, insular responses also seem to be affected (Kringelbach, M. L. *et al.*, 2003; Smeets, P. A. *et al.*, 2006; Uher, R. *et al.*, 2006).

4.15.3.1.1.(ii) Taste–somatosensory relationships As with lower levels of the neuraxis, gustatory responses in primary taste cortex have a proximate relationship with the representation of oral somatosensory, as well as visceral sensory inputs. The primary somatosensory cortex (S1) is comprised of somatosensory-specific neurons. In primates, S1 is located in Brodmann's areas 3a, 3b, 1, and 2. S1 in rodents is in a homologous location. In both species, S1 is medial to the primary gustatory representation in insular cortex. The somatotopy in S1 is arranged such that the oral representation is most lateral and abuts the gustatory-responsive zone (Squire, L. R. *et al.*, 2003). The extent to which there is a segregated region, comprised only of taste-responsive neurons, remains an unanswered question. There is evidence for an orderly medial-to-lateral progression of oral tactile, thermal, and gustatory responses, suggestive of a certain degree of segregation (Kosar, E. *et al.*, 1986a). However, other data indicate that cells responsive to oral mechanical stimulation are intermingled with taste cells in this region. Further, a large subset of gustatory cortical neurons also respond to oral mechanical stimuli (Katz, D. B. *et al.*, 2001; Ogawa, H. and Wang, X. D., 2002; Wang, X. and Ogawa, H., 2002). In fact, taste cells with oral mechanical RF's have also been reported to receive additional inputs from remote regions of the glaborous skin of the body (Wang, X. and Ogawa, H., 2002). Thus it is tempting to conclude that nearly all cortical taste neurons also have a tactile input. However, as recently pointed out by Lundy R. F., Jr. and Norgren R. (2004b), only a few studies have actually tested responsiveness of insular neurons to both modalities, particularly in a quantitative fashion. Thus, though convergence seems pervasive, the relative magnitudes of taste and somatosensory responses within individual neurons are unclear (Wang, X. and Ogawa, H., 2002). Furthermore, since some converging somatosensory responses appear nociceptive since they require high pressures or pinch (Ogawa, H. and Wang, X. D., 2002), a more thorough examination is required to rule out nonspecific effects, such as arousal. Nevertheless, the apparent multimodal nature of many neurons in primary taste cortex raises the question of how these neurons are able to clearly code the presence of a taste apart from somatosensory stimuli. A proposed solution to this conundrum is the interesting suggestion that the different modalities are coded by temporally distinct components of the response; that is, the initial 0.5 s of the response codes somatosensory events and the subsequent response codes gustatory-specific information (Katz, D. B. *et al.*, 2001).

A novel hypothesis regarding the relationship of gustatory to somatosensory cortex was recently suggested by Kaas J. H. (2005) . Observations in primates have revealed an ever-expanding region for somatosensory cortex, indicating that it is much more extensive than evident in the original descriptions. Neurophysiological mapping studies indicate that multiple, higher-level somatosensory maps extend laterally from areas 3a, 3b, and 1 into the insula. These regions receive spinal and extraoral trigeminal, in addition to an intraoral representation, and arguably encompass what has classically been considered primary GC (e.g., Jain, N. *et al.*, 2001). These findings prompted Kaas J. H. to propose the provocative idea that there may be no pure primary GC, but instead, that the insular taste area is an integral part of the intraoral representations embedded in higher-order somatosensory zones. He further speculates that this may account for the common observation that single-unit studies observe only a small fraction of neurons in the insular GC that actually respond to taste stimulation (e.g., Scott, T.

R. *et al.*, 1986; Plata-Salaman, C. R. *et al.*, 1996; Wang, X. and Ogawa, H., 2002). Although such a radical restructuring of the concept of primary GC is intriguing, further experiments are required for verification. Another possibility is that it is simply difficult to precisely locate the relatively small region receiving direct inputs from the gustatory thalamus (Pritchard, T. C. *et al.*, 1986), making it equally plausible that there is a small, more purely taste core adjacent to a larger area with mixed inputs.

The complexity of the insular region is also highlighted by discoveries about the nociceptive pathway. Neuroanatomical (Jasmin, L. *et al.*, 2004), neurophysiological (Ito, S. I., 1998; Hanamori, T. *et al.*, 1998b; Gauriau, C. and Bernard, J. F., 2004), and functional imaging studies (Petrovic, P. *et al.*, 2002; Bingel, U. *et al.*, 2003; Schweinhardt, P. *et al.*, 2006) in humans have identified the insular cortex as a major locus for processing pain information. These signals reach the insula via multiple pathways, including a spinoparabrachial pathway that relays through the ventroposteromedial and nearby regions of the thalamus (Menendez, L. *et al.*, 1996; Bester, H. *et al.*, 1999; Krout, K. E. and Loewy, A. D., 2000; Jasmin, L. *et al.*, 2004), and in primates through an additional evolutionarily newer direct spinal–thalamic–insular pathway that synapses in a posteriorly adjacent distinct nucleus in the thalamus, the posterior portion of the ventromedial nucleus (Craig, A. D., 2004). In fact, these pathways appear to carry not only nociceptive signals, but also thermal signals (Craig, A. D. *et al.*, 2000; Craig, A. D., 2003a; 2003b). The representation of pain, thermal, visceral, and taste information in the insular cortex has led to the suggestion that it is a region unified for processing the homeostatic state of the body (Craig, A. D., 2003a; 2003b).

4.15.3.1.1.(iii) Taste–olfactory relationships

Because the everyday appreciation of food and drink does not rely on taste alone, but rather on the combination of taste with oral somatosensory and olfactory sensations, it is especially important to understand where these two types of chemical signals converge centrally. The first point of convergence appears to be in the forebrain, but neither the initial site of convergence nor the responsible pathway(s) is entirely clear. Furthermore, species differences are likely. In the rodent, anatomical data suggest that olfactory information can impact the insular cortex via a least two routes, a direct projection from the olfactory bulb (Shipley, M. T. and Geinisman, Y., 1984), or a connection from the endopyriform nucleus (Behan, M. and Haberly, L. B., 1999; Fu, W. *et al.*, 2004). However, whether the insular area described in these studies is actually gustatory is not certain, nor is it clear that olfactory signals actually converge on a significant proportion of taste cells in rodent insular cortex. Other sites of forebrain convergence are also possible. A recent optical imaging/recording brain slice study demonstrated that signals from the pyriform (olfactory) and dysgranular (probably taste) insular cortex both reach the endopyriform nucleus, and converge on single cells (Fu, W. *et al.*, 2004). In addition, another forebrain region, the amygdalopyriform transition area, seems to receive relatively direct olfactory and gustatory information over a different route (Santiago, A. C. and Shammah-Lagnado, S. J., 2005). In this case, the olfactory inputs arise from neurons in the pyriform cortex, whereas putative gustatory inputs arise from the parabrachial nucleus. That the parabrachial input is gustatory seems quite likely because the cells of origin depicted in photomicrographs are centered in precisely the same part of the waist region where electrophysiological studies routinely report taste responses (e.g., Norgren, R. and Pfaffmann, C., 1975). There has not been a systematic neurophysiological investigation of gustatory–olfactory convergence in the rodent forebrain, although one report did document a few such convergent cells in insular cortex (Yamamoto, T. *et al.*, 1989b). In contrast, in the monkey, responses to both taste and olfactory stimulation have been recorded from significant populations of single neurons in the lateral orbitofrontal cortex (Rolls, E. T. and Baylis, L. L., 1994), although such convergence has not been observed in the insula. However, it may be premature to rule out taste–olfactory convergence in the insula. Indeed, as summarized by Small, D. M. and Prescott, J. (2005), human imaging studies reveal that olfactory stimuli activate the same insular and orbitofrontal regions as do tastants, although the obverse is not true for primary olfactory cortex. Imaging studies also indicate other cortical regions where taste and smell interact or overlap, with the anterior cingulate being particularly notable. Importantly, the human imaging studies point to critical integrative features of this convergence; that is, synergy between taste and olfactory signals occurs prominently with retronasal but not orthonasal olfactory signals (Small, D. M. *et al.*, 2005) and further requires that the taste and olfactory signals are congruent (Figure 8). Thus a

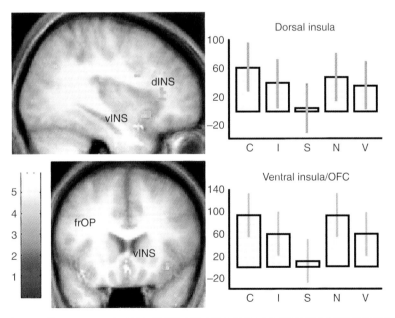

Figure 8 Functional magnetic resonance imaging (fMRI) activation of the insular region following presentation of a sweet stimulus (0.18 M sucrose) mixed with vanilla. Both the dorsal insula (primary gustatory cortex) and the ventral insula, a higher-order processing region, are activated in a supra-additive fashion by this congruent, familiar mixture. The graphs on the right depict these results quantitatively and compare the sucrose–vanilla congruent mixture (C), with an incongruent mixture (NaCl–vanilla, I), and with each of the components alone. S, sucrose; N, NaCl; V, vanilla. (a) From Small, D. M., Voss, J., Mak, Y. E., Simmons, K .B., Parrish, T., and Gitelman, D. 2004. Experience-dependent neural integration of taste and smell in the human brain. J. Neurophysiol. 92, 1892–1903, with permission.

familiar olfactory–taste pair, vanilla–sweet, yielded superadditive functional magnetic resonance imaging (fMRI) responses in the insula and several other forebrain regions but an unfamiliar, incongruent pair, vanilla–salty, did not (Small, D. M. *et al.*, 2004).

4.15.3.1.2 Gustatory cell types and processing

4.15.3.1.2.(i) Response profiles and firing rates

Compared to other sensory systems, the transformation of gustatory information by cortical circuits at first glance seems subtle. Some differences have been reported, for example, a minority of insular neurons in the rat appeared to code palatability rather than quality (Yamamoto, T. *et al.*, 1989b). Nevertheless, in both rodents (Yamamoto, T. *et al.*, 1985; 1989b; Ogawa, H. *et al.*, 1992) and primates (Yaxley, S. *et al.*, 1990), many studies of primary GC have yielded reports of best-stimulus neurons quite similar to those in peripheral nerves and brainstem. The same is true for neurons in caudolateral (secondary) GC (Baylis, L. L. and Rolls, E. T., 1991) though, as noted above, more modulation by satiety and interaction with other modalities probably occurs in this higher-order cortex. On the face of it, the lack of an obvious transformation at the cortical level is not due to using restricted stimulus arrays, or suppressed cortical activity in anesthetized preparations. A significant number of these studies used large numbers of tastants (e.g., Smith-Swintosky, V. L. *et al.*, 1991; Plata-Salaman, C. R. *et al.*, 1992; 1993; Scott, T. R. *et al.*, 1994; Plata-Salaman, C. R. *et al.*, 1995; Scott, T. R. *et al.*, 1999), included taste mixtures (Miyaoka, Y. and Pritchard, T. C., 1996; Plata-Salaman, C. R. *et al.*, 1996), and many were carried out in awake preparations.

Though the relative response profiles appear similar to those at lower levels of the neuraxis, it is notable that lower response rates are often characteristic of cortical taste cells. In addition, in contrast to the concentrated populations of taste neurons found in the brainstem, gustatory cortical neurons often appear to be intermingled with a larger population of nongustatory cells (e.g., Scott, T. R. *et al.*, 1986; Plata-Salaman, C. R. *et al.*, 1996; Wang, X. and Ogawa, H., 2002). The reason for this apparent sparseness is not known. As discussed above, Kaas J. H. (2005) suggests that one possibility is that multiple gustatory cortical regions are intertwined with secondary somatosensory cortex; that is, there is no pure GC. Another possibility is suggested by the findings of Pritchard T. C. *et al.* (2005). These investigators

recently explored a different cortical area in monkey, the medial orbitofrontal cortex, a region implicated in taste function by anatomical experiments. In this previously unexplored area, the proportion of taste cells was markedly higher than in many other cortical studies. A careful reconstruction of recording sites showed that the incidence of gustatory cells was highly dependent upon the exact recording location; the concentrated core of taste-responsive neurons was very restricted. This suggests that some apparent sparseness in other studies could be due to the inherent technical difficulties in precisely targeting gustatory cortical regions, particularly in chronic recording studies. Interestingly, this study hints at another possible reason for sparseness since 24% of the neurons did not respond significantly to any of the four standard tastants but did respond to another (but unspecified) taste stimulus. This finding emphasizes that much remains to be learned about what constitutes the optimal stimulus configuration for cortical cells. Thus, cortical neurons may require more specific taste stimuli, combinations of taste stimuli or gustatory stimulation coincident with inputs from other modalities. For example, some taste cells may require particular combinations of learned taste and olfactory stimuli, as suggested by the human imaging studies (Small, D. M. *et al.*, 2004), or more speculatively, other learned combinations such as mixtures of different tastes, or of specific taste/texture combinations.

4.15.3.1.2.(ii) Breadth of tuning

As discussed above, central neurons are, on average, more broadly tuned than those in the periphery. However, there are divergent reports concerning whether there is a systematic trend for tuning to broaden (Stapleton, J. R. *et al.*, 2006), sharpen (Scott, T. R. *et al.*, 1986; Rolls, E. T. *et al.*, 1990), or remain unchanged in cortex, compared to the brainstem (summarized in Spector, A. C. and Travers, S. P., 2005). At present, it is difficult to resolve this issue, in part, due to the variety of animal models (rodent, primate), types of preparation (acute anesthetized, chronic awake), recording procedures (single-unit sharp electrode; microwire bundles), and analytic techniques (summed activity, time course) used in the cortical experiments. Chronic, awake preparations arguably provide the most powerful tool for revealing the full range of responsiveness of central sensory neurons. This is particularly true for the cortex, because of the profound anesthetic suppression of sensory responses at this level of the nervous system (Sloan, T. B., 1998), as well as the active behaviors sometimes required for revealing cortical response properties (e.g., Krupa, D. J. *et al.*, 2004). However, even when consideration is limited to awake preparations, varied tuning curves are apparent between studies. For example, although both groups of investigators recorded from a similar cortical region with microwire bundles, Stapleton, J. R. *et al.* (2006) reported an *H* value of 0.97, indicating that different stimuli evoked responses of roughly equal magnitude, whereas Yamamoto T. *et al.* (1989b) reported an average *H* value of 0.54, indicating much more specificity. In fact, although they tend to be more consistent, varied tuning curves can be observed within studies; in a typical instance, Pritchard T. C. *et al.* (2005) reported *H* values that ranged from 0.28 to 0.98. Thus, one conclusion emphasized in another recent review is that central tuning curves are inherently heterogeneous, including both narrowly and broadly tuned neurons best suited to subserve distinct functions, such as perception, motivation, and learning (Spector, A. C. and Travers, S. P., 2005).

4.15.3.1.2.(iii) *Response time course, behavioral context*   Other more recent perspectives for devising a taste code involve alternate analyses which take time course and behavioral context into account (Katz, D. B. *et al.*, 2001; 2002; Hallock, R. M. and Di Lorenzo, P. M., 2006; Stapleton, J. R. *et al.*, 2006). In one experiment that analyzed just the initial 150 ms of the response, firing rates alone indicated almost no stimulus specificity, but when time course was considered, more specificity emerged (Stapleton, J. R. *et al.*, 2006). Another report of insular neurons not only observed very broadly tuned cells, but further found that the tuning curves of many neurons changed during a single experimental session, in parallel with decreases in the animal's attentional state (Fontanini, A. and Katz, D. B., 2006). Thus, unlike most previous studies, these recent reports provide little evidence for the classic best stimulus categories of neurons based on differential response magnitudes. Instead, they focus on the complex dynamic features of gustatory cortical responses and the key influence of behavioral state. It is important to point out, however, that the differences apparent in these more recent studies are probably due not only to differences in the type of analysis or behavioral paradigm, but also from differences in sampled populations. Although earlier studies similarly used awake animals, they also typically tested each stimulus on just a few trials and used simple response measures of summed activity; thus, taste responses were defined

on the basis of relatively obvious changes in response rate (e.g., Scott, T. R. *et al.*, 1986; Yamamoto, T. *et al.*, 1989b; Rolls, E. T. *et al.*, 1990; Pritchard, T. C. *et al.*, 2005). In contrast, many current studies that take time course into account also tend to perform many repeated trials, a combination that adds a great deal of statistical power for identifying variations in neuronal firing. Such approaches extend the taste-responsive population to neurons that formerly would not have been considered to be gustatory (Katz, D. B. *et al.*, 2001). To what extent they include those neurons with more obvious responses is not entirely clear. Furthermore, the advantage conferred by this increased statistical power makes it important to demonstrate that these relatively subtle, but statistically demonstrable changes in time course or magnitude are behaviorally meaningful, as well as to attend to potential contributions from nearby oral thermal and tactile neurons. Nevertheless, it is clear that studies in behaving animals that use more complex behavioral paradigms and more powerful analytic techniques have great potential.

A particularly intriguing study has shed additional light on the function of rodent orbitofrontal cortex as it relates to rewarding tastes and voluntary ingestive behavior (Gutierrez, R. *et al.*, 2006; Travers, J. B., 2006). Neurons in this area of the cortex exhibited clear modulations of activity predictive of the beginning, occurrence, or end of a licking bout (Figure 9), and were more subtly modified depending on whether the rat was drinking sucrose or water. Thus, similar to what has been suggested for primate (Rolls, E. T. and Scott, T. R., 2003), rodent orbitofrontal cortex appears to monitor the reward value of tastes, and play a direct role in controlling reward-seeking behavior. Indeed, temporary inactivation of the rat orbitofrontal cortex with a $GABA_A$ agonist, muscimol, altered the microstructure of licking behavior. Overall intake remained relatively stable, but ingestion occurred in shorter bouts, indicative of a disruption of appetitive behavior (Gutierrez, R. *et al.*, 2006).

4.15.3.1.3   *Neurotransmitters and neuromodulators in the insular cortex*
4.15.3.1.3.(i) *Excitatory amino acids*   In GC, a large variety of neurotransmitters, modulators, and their receptors comprise a rich substrate for processing sensory information. Taste responses in the majority of rat insular cortex neurons are dependent on glutamatergic input acting via non-NMDA, as well as NMDA receptors (Otawa, S. *et al.*, 1995), implying that

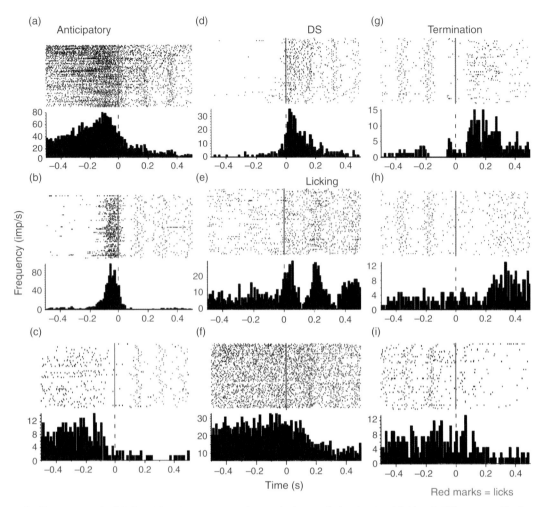

Figure 9 Responses of orbitofrontal cortex neurons prior to, during, and after bouts of licking 0.4 M sucrose. Each graph shows a different neuron, and depicts raster displays for multiple trials above a summed perievent histogram. The red dashes indicate licks. Each neuron exhibits a unique response pattern, but each column shows a given type based on whether the neuron alters its firing rate prior to a licking bout (Anticipatory), during the licking bout/stimulus (DS), or after the termination of the bout (Termination). From Gutierrez, R., Carmena, J. M., Nicolelis, M. A., and Simon, S. A. 2006. Orbitofrontal ensemble activity monitors licking and distinguishes among natural rewards. J. Neurophysiol. 95, 119–133, with permission.

glutamate is a key transmitter in the thalamocortical projection. Interestingly, in distinct contrast to the NST, NMDA receptor antagonists were as potent at reducing taste responses as non-NMDA antagonists, perhaps reflecting the greater plasticity inherent in cortical regions (Otawa, S. *et al.*, 1995). Ogawa and his group further demonstrated that both spontaneous and taste-evoked activity in the insular cortex is modulated by inhibitory amino acids, ACh, and peptides (Ogawa, H. *et al.*, 1998; 2000; Hasegawa, K. and Ogawa, H., 2007). These neurotransmitter systems likely play a major role in taste processing as 40–60% of insular cells were sensitive to iontophoresis of their antagonists and agonists.

4.15.3.1.3.(ii) Gamma-aminobutyric acid Similar to the NST, gustatory cortical neurons appear to be under tonic inhibition (Ogawa, H. *et al.*, 1998). Iontophoresis of the $GABA_A$ antagonist, bicuculline, onto insular cortical neurons yielded increases in spontaneous rate in 69% of the taste neurons sampled. Interestingly, only 51% of the entire population of insular cortical neurons (taste and nontaste combined) exhibited such an effect, suggesting that gustatory neurons may be subject to a greater degree of tonic inhibition than surrounding cells. Bicuculline exerted an effect on the taste responses themselves in a somewhat smaller (38%) but still significant proportion of neurons. As expected, taste responses usually increased following bicuculline application; however, a more surprising outcome was that the increases were stimulus specific. These differential effects produced marked alterations in response profiles, including changes of best-stimulus and breadth of tuning (Figure 10).

Similar to the effects of bicuculline in the somatosensory cortex (Dykes, R. W. *et al.*, 1984; Tremere, L. *et al.*, 2001), receptive field size of gustatory neurons also increased when GABA$_A$ inhibition was removed. Before drug application most neurons had receptive fields restricted to the anterior tongue, but after bicuculline, inputs from the posterior tongue and palate became apparent. A particularly dramatic outcome was the recruitment of previously silent neurons that lacked both spontaneous and evoked activity; after bicuculline, these neurons became responsive to taste stimuli.

4.15.3.1.3.(iii)   Acetylcholine   ACh also affected taste responses in a substantial proportion (63%) of insular neurons but spontaneous rate was affected less often (17%; Hasegawa, K. and Ogawa, H., 2007). In contrast to bicuculline, iontophoresis of ACh both increased and decreased gustatory responses, though the action was usually in the same direction for different stimuli in a given cell. Nevertheless, within cells, responses to different

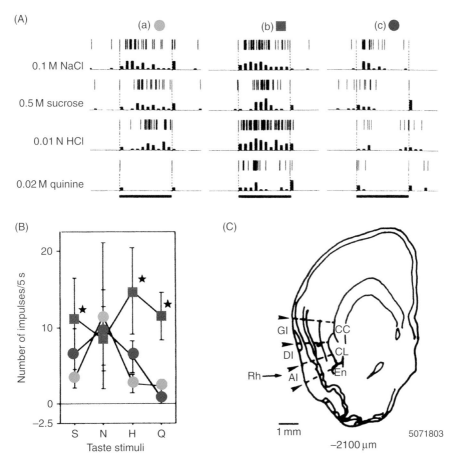

Figure 10   Example of an insular cortical neuron in which taste responses were increased in a stimulus-specific manner by the gamma-aminobutyric acid (GABA$_A$) antagonist, bicuculline. (A) Perievent histograms of a single trial (a) before, (b) during, and (c) after iontophoresis of bicuculline. (B) Mean responses for 5 s summed activity before (dark blue circles), during (red squares), and after (light blue circles). Before application of the GABA$_A$ antagonist, the neuron responded most robustly to NaCl but weakly to quinine. Following bicuculline, all of the responses, except that to NaCl, were significantly increased (asterisks), resulting in a best-stimulus change from NaCl to HCl, and an increase in breadth of tuning. (C) Location of the recorded cell in dysgranular insular (DI) cortex adapted from Ogawa, H., Hasegawa, K., Otawa, S., and Ikeda, I. 1998. GABAergic inhibition and modifications of taste responses in the cortical taste area in rats. Neurosci. Res. 32, 85–95, with permission.

stimuli were affected to varying extents, producing a net change in the chemosensitive profile, rather than a simple change in the overall level of responsiveness. Indeed the best stimulus changed in nearly half (19/43) of the cells. In contrast to other sensory cortical areas where ACh seems to sharpen tuning profiles (e.g., Sillito, A. M. and Kemp, J. A., 1983), ACh effects in taste cortex were more complex, and could give rise to either increases or decreases in neuronal selectivity. In about 2/3 of the cases, atropine antagonized the actions of ACh, suggesting a muscarinic basis; however, the remaining third of the effects were presumably nicotinic.

4.15.3.1.3.(iv) Peptides Consistent with the presence of SP and calcitonin gene-related peptide (CGRP) fibers in insular cortex, (Mantyh, P. W. and Hunt, S. P., 1984; Yasui, Y. *et al.*, 1989; Yamamoto, T. *et al.*, 1990), both peptides altered spontaneous rate, as well as taste responses in gustatory cells (Ogawa, H. *et al.*, 2000). Spontaneous rate was affected in 40% of neurons and the effects were largely (~86%) facilitory, consistent with the known cellular actions of the NK1 and CGRP receptors (Strand, F., 1999). Somewhat surprisingly, although these peptides also altered taste responses in a large proportion of insular neurons (~50%), the effects were mainly suppressive; that is, opposite to the effects on spontaneous rate, suggestive of an indirect effect through inhibitory interneurons. Furthermore, these peptides modulated responses to various taste stimuli to different degrees within individual neurons. Thus, for the three neurotransmitter systems evaluated so far, each is associated with unique net effects on spontaneous and evoked activity. However, all produce stimulus-specific alterations, which give rise to changes in response profiles in individual neurons. These diverse actions seem well suited to provide the substrate for cortical plasticity, such as those instances that have been observed during alterations of attentional state and homeostasis discussed above.

4.15.3.1.3.(v) Neurochemical substrate for conditioned taste aversion Analysis of the insular substrate important for conditioned taste aversion provides additional insights into the functions of the various neurotransmitters in taste cortex. Insular infusions of muscarinic antagonists blunt the acquisition of a conditioned taste aversion, but only if they are given prior to taste exposure, suggesting that that ACh impacts critical aspects of processing the taste

signal (Ferreira, G. *et al.*, 2002). However, the effect is not simply on taste perception since infusions prior to retrieval are ineffective (Naor, C. and Dudai, Y., 1996; Berman, D. E. *et al.*, 2000). Interestingly, novel taste stimuli are much more potent in establishing an aversion, and microdialysis experiments have revealed that novel but not familiar taste stimuli elicit ACh release in the insular cortex (Miranda, M. I. *et al.*, 2000). There appears to be a complex relationship between stimulus novelty and muscarinic action in the insula, however, as muscarinic antagonists interfere only with the attenuation of neophobia, but do not abolish neophobia itself (reviewed in Bermudez-Rattoni, F. *et al.*, 2004). NMDA antagonists also blunt the acquisition of a conditioned taste aversion (Rosenblum, K. *et al.*, 1997; Gutierrez, H. *et al.*, 1999; Ferreira, G. *et al.*, 2002), but unlike muscarinic agonists, have effects when injected either before or after the presentation of the taste stimulus (Ferreira, G. *et al.*, 2002). Beta adrenergic antagonists also interfere with acquisition of a conditioned taste aversion, and unlike NMDA or muscarinic receptors, interfere with extinction once the aversion is established (Berman, D. E. and Dudai, Y., 2001). Thus multiple cortical transmitter systems have specific effects on this learned behavior. Although some of these effects may be due to short-term modulations of taste or visceral responses and second messenger systems, it is clear that long-term retention of conditioned taste aversion requires protein synthesis in insular cortex (Rosenblum, K. *et al.*, 1993). The immediate-early gene, c-fos, is a transcription factor that is differentially upregulated in the insular cortex upon exposure to novel taste stimuli (Koh, M. T. and Bernstein, I. L., 2005) and appears to serve as a critical intermediary, at least for protein synthesis required for conditioned taste aversion retention (Yasoshima, Y. *et al.*, 2006). The downstream targets of this gene are unknown, but an earlier report suggested that the amount of CGRP in insular cortex was higher after exposure to aversive compared to preferred stimuli, including a previously preferred stimulus, to which a conditioned taste aversion had been given (Yamamoto, T. *et al.*, 1990). Although this discussion focuses on the role of insular cortex in conditioned taste aversion, this does not imply that the cortex is the sole substrate for this behavior. Instead, interactions between cortical (e.g., Schafe, G. E. and Bernstein, I. L., 1998), subcortical (e.g., Schafe, G. E. and Bernstein, I. L., 1996), and brainstem sites (e.g., Reilly, S. *et al.*, 1993) are critical.

4.15.3.2 Limbic Forebrain

Structures in the limbic forebrain and hypothalamus are intimately involved in motivated behaviors, including ingestion. To maintain homeostasis, a complex neural network must integrate a wide variety of internal and external signals with past experience. Although the final consummatory sequences of eating and drinking are characterized by relatively fixed, species-specific patterns of oromotor movements, the appetitive behaviors necessary to reach this endpoint must be flexible in order to accommodate changing environmental demands. The ventral forebrain receives all the relevant information to accomplish this task, including gustatory signals.

The highly interconnected nature of forebrain circuits suggests that taste has the potential to influence a large number of limbic and hypothalamic structures. As shown in Figure 1, in the rodent, anatomical tracing experiments have demonstrated direct projections from the taste-responsive regions in the parabrachial nucleus to the central nucleus of the amygdala and the bed nucleus of the stria terminialis, as well to a contiguous ventral region including the lateral hypothalamus, substantia innominata, fundus striatum, and ventral pallidum (Norgren, R. and Leonard, C. M., 1973; Norgren, R., 1974; 1976; Bernard, J. F. et al., 1993; Alden, M. et al., 1994; Karimnamazi, H. and Travers, J. B., 1998). In the primate, gustatory information reaches many of these same structures, but the most direct route is likely through connections from the GC (discussed in Pritchard, T. C. and Norgren, R., 2004). In addition to this anatomical evidence, antidromic activation has been used to verify that some of the projections to the rodent lateral hypothalamus, substantia innominata, and amygdala do carry taste information (Norgren, R., 1974; 1976); this has not yet been demonstrated for the other sites. Besides this direct pathway from the PBN, there are a number of potential indirect pathways to these regions, and others. Of particular note is the projection from the insular cortex to the nucleus accumbens (Reynolds, S. M. and Zahm, D. S., 2005), a structure with a prominent role in motivational processes. However, the insular–accumbens projection does not appear to be from the primary taste cortex, that is, it is not from the dysgranular region, which receives a direct projection from the gustatory thalamus (Kosar, E. et al., 1986b), nor the surrounding granular cortex, where taste responses have also been recorded (Ogawa, H. et al., 1992), but rather from the ventrally adjacent agranular insular cortex (Reynolds, S. M. and Zahm, D. S., 2005). However, the dysgranular insular cortex projects to the agranular (Shi, C. J. and Cassell, M. D., 1998), providing a potential pathway for taste information to reach nucleus accumbens.

Taste responses in the limbic forebrain have been studied most frequently in the lateral hypothalamus (Burton, M. J. et al., 1976; Ono, T. et al., 1986; Rolls, E. T. et al., 1986; Schwartzbaum, J. S., 1988) and amygdala, with an emphasis on the central nucleus (Yan, J. and Scott, T. R., 1996; Nishijo, H. et al., 1998). A more recent focus has been on the nucleus accumbens (Roitman, M. F. et al., 2005; Taha, S. A. and Fields, H. L., 2005) and ventral pallidum (Tindell, A. J. et al., 2004; 2006). Virtually all of the recording preparations assessing gustatory responsiveness in the limbic forebrain utilize awake, behaving animals, presumably because such responses would be difficult to detect in anesthetized preparations (but see Azuma, S. et al., 1984). Ample evidence for a taste influence on the limbic forebrain is provided from a score of reports of presumptive taste responses observed in the context of food or fluid rewards obtained in operant tasks. Often, these same cells are influenced by other unconditioned sensory stimuli like olfaction, or other aspects of the task, including the onset or offset of the conditioned stimulus or operant responding (e.g., Rolls, E. T. et al., 1979; Sanghera, M. K. et al., 1979; Schwartzbaum, J. S., 1988; Yamamoto, T. et al., 1989a; Oomura, Y. et al., 1991; Williams, G. V. et al., 1993; Tindell, A. J. et al., 2004). There is less information on the detailed nature of gustatory processing in these circuits. In the lateral hypothalamus and central nucleus of the amygdala, it is clear that taste influences just a limited proportion of encountered neurons (Schwartzbaum, J. S., 1988; Yan, J. and Scott, T. R., 1996; Nishijo, H. et al., 2000), suggesting that they are a specialized population. In the nucleus accumbens and ventral pallidum, a larger proportion of neurons seems to be active during taste stimulation (Tindell, A. J. et al., 2004; Roitman, M. F. et al., 2005; Taha, S. A. and Fields, H. L., 2005; Tindell, A. J. et al., 2006), but since different testing paradigms have been used in the various areas, direct comparisons are difficult. Compared to what is known about the thalamocortical pathway, gustatory-modulated neurons in all of the limbic-related regions appear to have a wider variety of nongustatory inputs (Oomura, Y. et al., 1991; Karadi, Z. et al., 1992; Nishijo, H. et al., 1998), and frequently exhibit marked decrements, as well as increments in firing, in response to taste stimuli (Schwartzbaum, J. S., 1988; Oomura, Y. et al., 1991; Roitman, M. F. et al., 2005). Although single neurons in the lateral hypothalamus (Yan, J. and Scott, T. R., 1996) and amygdala (Scott, T. R. *et al.*, 1993; Nishijo, H. *et al.*, 1998), respond differentially to different tastants, they do not appear to code for quality, *per se*. A systematic comparison of amygdalar and PBN taste responses in the rat instead suggested that the distinctions that amygdalar neurons make are more simply related to binary palatability decisions than those in the PBN, which make additional distinctions between different tastants (Nishijo, H. *et al.*, 1998); a detailed study in the monkey came to a similar conclusion (Scott, T. R. *et al.*, 1993). The pervasive influence of satiety on taste responses in the lateral hypothalamus and amygdala suppots a primary role for these cells in coding palatability (Burton, M. J. *et al.*, 1976; Rolls, E. T. *et al.*, 1986; Yan, J. and Scott, T. R., 1996).

Likewise, palatability appears to be a critical stimulus dimension for neurons in the nucleus accumbens and ventral pallidum. Previous studies had suggested that nucleus accumbens neurons responded to rewarding stimuli, including sweet tastes but the apparent gustatory responses had been observed during operant tasks (e.g., Nicola, S. M. *et al.*, 2004), making it unclear whether they were were innate or instead required learning and/or a more complex behavioral context. In addition, little attention had been paid to stimuli at the opposite end of the spectrum. A recent study that utilized stimulus delivery via intraoral cannulas indeed demonstrated that the hedonically negative taste stimulus, quinine, as well as hedonically positive sucrose, could drive accumbens neurons and that the responses were apparent when the animal initially encountered the gustatory stimuli; that is, learning was not required (Roitman, M. F. *et al.*, 2005). Further, different neurons responded to sucrose and quinine. Over the course of several trials, these taste cells in accumbens also developed distinctive responses to the conditioned visual and auditory stimuli that signaled tastant delivery. Similar to the unconditioned responses to the tastants themselves, neurons that responded to the sucrose cue were mostly separate from those responsive to the quinine cue although the relationships between the conditioned and unconditioned responses were complicated. This study also suggested that, for the most part, sucrose- and quinine-responsive neurons responded differentially, that is, with response decrements and increments, respectively. Another description of accumbens taste responses provides a different perspective on the significance of the excitatory and inhibitory responses (Taha, S. A. and Fields, H. L., 2005). These investigators observed accumbens activity while water-deprived rats licked water and varying concentrations of sucrose. Similar to what Roitman and coworkers observed, most responses to sucrose were inhibitory. However, these inhibitory responses appeared most directly related to the licking behavior, *per se*. Decrements did not vary as a function of sucrose concentration, and licking to water or even a dry spout resulted in similar decreases in firing rate. However, in agreement with the other report, these investigators also observed a small population of sucrose-driven responses, which exhibited positive concentration–response functions. Furthermore, the excitatory responses were modulated by context, since a given concentration of sucrose elicited larger responses when paired with water, than with a higher concentration of sugar. Such a contrast paradigm is well-known to change the reward value of a gustatory stimulus (Flaherty, C. F. *et al.*, 1994).

The ventral pallidum, a target of accumbens projections (reviewed in Zahm, D. S., 2000) also contains gustatory-responsive neurons whose firing rates are modulated by the hedonic nature of the stimulus (Tindell, A. J. *et al.*, 2004; 2006). An experiment which tracked pallidal taste responses before, during, and after induction of a salt appetite is particularly illuminating in this regard (Tindell, A. J. *et al.*, 2006) (Figure 11).

Prior to and following sodium deprivation, ventral pallidal neurons responded more robustly to a highly preferred concentration of sucrose (0.5 M) than to a highly aversive, hypertonic (1.5 M) concentration of sodium chloride. Following deprivation, this same concentration of salt elicited parallel increases in positively hedonic behaviors and firing in ventral pallidal neurons. In fact, firing rates to sodium were now equivalent to those elicited by sucrose. Interestingly, deprivation also caused pallidal neurons to become more broadly tuned, at least with regard to the limited number of taste qualities tested. When neurons were classified as a function of responsiveness to sucrose, salt, and water, the largest proportion of neurons responded to one of the stimuli prior to deprivation, in contrast to all three during the deprived state. These results provide a rather dramatic example of gustatory malleability in the face of changing homeostatic demands.

4.15.4 Conclusions

The critical role of the gustatory system in guiding feeding behavior is accomplished by a network characterized by a high degree of discriminative and

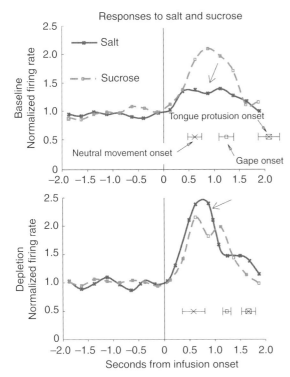

Figure 11 Mean firing rate of ventral pallidum neurons during infusions of 0.5 M sucrose (dotted line) and 1.5 M NaCl (solid line) during a normal homeostatic state (top graph) and during sodium deprivation (bottom graph). In the normal state, palatable sucrose induces a brief, twofold increase in firing rate but NaCl elicits a much smaller response. Under the sodium deprivation condition, when NaCl is as palatable as sucrose, both stimuli elicit similar increases in the activity of ventral pallidal neurons. The symbols beneath the neural responses indicate the onset of the oromotor responses elicited by these stimuli. The onset of the neural response precedes licking or gaping, but is not tightly coupled to these behaviors. From Tindell, A. J., Smith, K. S., Pecina, S., Berridge, K. C. and Aldridge, J. W. 2006. Ventral pallidum firing codes hedonic reward: when a bad taste turns good. J. Neurophysiol. 96, 2399–2409, with permission.

integrative capacity, and an equally high degree of plasticity necessary to meet shifting environmental and homeostatic demands. The operations of this system have become increasingly clarified, but key questions remain unanswered. First, the central code(s) for taste quality remains an area of active investigation (see Chapters Neural Ensembles in Taste Coding and A Perspective on Chemosensory Quality Coding). The broad tuning of many central neurons is suggestive of an ensemble code, but its precise nature, including the size and dynamic aspects of the necessary ensemble, are just now coming to light. In addition, there is a subset of more narrowly tuned central neurons. It is unclear whether

such cells simply comprise part of an ensemble code, or instead are a specialized subpopulation that uses a sparse code to accomplish particular functions. It has become increasingly apparent that single central taste neurons also can respond to olfactory and somatosensory stimuli. The likely function of such convergence is to give rise to the compound sensation of flavor. Human imaging studies further suggest that integration between modalities not only requires special combinations of stimuli, but also relies upon learning. However, the way in which single gustatory neurons combine these multiple aspects of a compound flavor stimulus signals requires much more scrutiny. Similarly, the list of manipulations that can modify taste responses, including changes in homeostatic state, learning, modulation by neurotransmitters, and activiation of other neural structures, is continuing to grow. The challenge is to be able to define how these changes relate mechanistically to one another, and which changes in neural responsiveness comprise the critical alternations that drive adaptive behavior.

References

Accolla, R., Bathellier, B., Petersen, C. C. H., and Carleton, A. 2007. Differential spatial representation of taste modalities in the rat gustatory cortex. J. Neurosci. 27, 1396–1404.

Alden, M., Besson, J. M., and Bernard, J. F. 1994. Organization of the efferent projections from the pontine parabrachial area to the bed nucleus of the stria terminalis and neighboring regions: a PHA-L study in the rat. J. Comp. Neurol. 341, 289–314.

Allen, G. V., Saper, C. B., Hurley, K. M., and Cechetto, D. F. 1991. Organization of visceral and limbic connections in the insular cortex of the rat. J. Comp. Neurol. 311, 1–16.

Andresen, M. and Yang, M. 1994. Excitatory Amino Acid Receptors and Afferent Synaptic Transmission in The Nucleus Tractus Solitarius. In: Nucleus of the Solitary Tract (ed. I. Barraco), pp. 187–192. CRC Press.

Atema, J. 1971. Structures and functions of the sense of taste in the catfish (Ictalurus natalis). Brain Behav. Evol. 4, 273–294.

Azuma, S., Yamamoto, T., and Kawamura, Y. 1984. Studies on gustatory responses of amygdaloid neurons in rats. Exp. Brain Res. 56, 12–22.

Baird, J. P., Travers, S. P., and Travers, J. B. 2001. Integration of gastric distension and gustatory responses in the parabrachial nucleus. Am. J. Physiol. Regul. Integr. Comp. Physiol. 281, R1581–R1593.

Bardach, J. E., Todd, J. H., and Crickmer, R. 1967. Orientation by taste in fish of the genus Ictalurus. Science 155, 1276–1278.

Barry, M. A., Haglund, S., and Savoy, L. D. 2001. Association of extracellular acetylcholinesterase with gustatory nerve terminal fibers in the nucleus of the solitary tract. Brain Res. 921, 12–20.

Baylis, L. L. and Rolls, E. T. 1991. Responses of neurons in the primate taste cortex to glutamate. Physiol. Behav. 49, 973–979.

Baylis, L. L., Rolls, E. T., and Baylis, G. C. 1995. Afferent connections of the caudolateral orbitofrontal cortex taste area of the primate. Neuroscience 64, 801–812.

Beckstead, R. M. and Norgren, R. 1979. An autoradiographic examination of the central distribution of the trigeminal, facial, glossopharyngeal, and vagal nerves in the monkey. J. Comp. Neurol. 184, 455–472.

Beckstead, R. M., Morse, J. R., and Norgren, R. 1980. The nucleus of the solitary tract in the monkey: projections to the thalamus and brain stem nuclei. J. Comp. Neurol. 190, 259–282.

Behan, M. and Haberly, L. B. 1999. Intrinsic and efferent connections of the endopiriform nucleus in rat. J. Comp. Neurol. 408, 532–548.

Benjamin, R. M. and Akert, K. 1959. Cortical and thalmic areas involved in taste discrimination in the albino rat. J. Comp. Neurol. 111, 231–259.

Benjamin, R. M. and Pfaffmann, C. 1955. Cortical localization of taste in albino rat. J. Neurophysiol. 18, 56–64.

Bereiter, D., Berthoud, H. R., and Jeanrenaud, B. 1980. Hypothalamic input to brain stem neurons responsive to oropharyngeal stimulation. Exp. Brain Res. 39, 33–39.

Berman, D. E. and Dudai, Y. 2001. Memory extinction, learning anew, and learning the new: dissociations in the molecular machinery of learning in cortex. Science 291, 2417–2419.

Berman, D. E., Hazvi, S., Neduva, V., and Dudai, Y. 2000. The role of identified neurotransmitter systems in the response of insular cortex to unfamiliar taste: activation of ERK1-2 and formation of a memory trace. J. Neurosci. 20, 7017–7023.

Bermudez-Rattoni, F., Ramirez-Lugo, L., Gutierrez, R., and Miranda, M. I. 2004. Molecular signals into the insular cortex and amygdala during aversive gustatory memory formation. Cell Mol. Neurobiol. 24, 25–36.

Bernard, J. F., Alden, M., and Besson, J. M. 1993. The organization of the efferent projections from the pontine parabrachial area to the amygdaloid complex: a *Phaseolus vulgaris* leucoagglutinin (PHA-L) study in the rat. J. Comp. Neurol. 329, 201–229.

Bester, H., Bourgeais, L., Villanueva, L., Besson, J. M., and Bernard, J. F. 1999. Differential projections to the intralaminar and gustatory thalamus from the parabrachial area: a PHA-L study in the rat. J. Comp. Neurol. 405, 421–449.

Bingel, U., Quante, M., Knab, R., Bromm, B., Weiller, C., and Buchel, C. 2003. Single trial fMRI reveals significant contralateral bias in responses to laser pain within thalamus and somatosensory cortices. Neuroimage 18, 740–748.

Boughter, J. D., Jr and Smith, D. V. 1998. Amiloride blocks acid responses in NaCl-best gustatory neurons of the hamster solitary nucleus. J. Neurophysiol. 80, 1362–1372.

Bradley, R. M. and Sweazey, R. D. 1990. *In vitro* intracellular recordings from gustatory neurons in the rat solitary nucleus. Brain Res. 508, 168–171.

Bradley, R. M. and Sweazey, R. D. 1992. Separation of neuron types in the gustatory zone of the nucleus tractus solitarii on the basis of intrinsic firing properties. J. Neurophysiol. 67, 1659–1668.

Burton, M. J., Rolls, E. T., and Mora, F. 1976. Effects of hunger on the responses of neurons in the lateral hypothalamus to the sight and taste of food. Exp. Neurol. 51, 668–677.

Cecchi, M., Khoshbouei, H., Javors, M., and Morilak, D. A. 2002. Modulatory effects of norepinephrine in the lateral bed nucleus of the stria terminalis on behavioral and neuroendocrine responses to acute stress. Neuroscience 112, 13–21.

Cechetto, D. F. and Saper, C. B. 1987. Evidence for a viscerotopic sensory representation in the cortex and thalamus in the rat. J. Comp. Neurol. 262, 27–45.

Chandrashekar, J., Hoon, M. A., Ryba, N. J., and Zuker, C. S. 2006. The receptors and cells for mammalian taste. Nature 444, 288–294.

Chang, F. C. and Scott, T. R. 1984. Conditioned taste aversions modify neural responses in the rat nucleus tractus solitarius. J. Neurosci. 4, 1850–1862.

Cho, Y. K., Li, C. S., and Smith, D. V. 2002a. Gustatory projections from the nucleus of the solitary tract to the parabrachial nuclei in the hamster. Chem. Senses 27, 81–90.

Cho, Y. K., Li, C. S., and Smith, D. V. 2002b. Taste responses of neurons of the hamster solitary nucleus are enhanced by lateral hypothalamic stimulation. J. Neurophysiol. 87, 1981–1992.

Cho, Y. K., Li, C. S., and Smith, D. V. 2003. Descending influences from the lateral hypothalamus and amygdala converge onto medullary taste neurons. Chem. Senses 28, 155–171.

Conover, K. L. and Shizgal, P. 1994. Competition and summation between rewarding effects of sucrose and lateral hypothalamic stimulation in the rat. Behav. Neurosci. 108, 537–548.

Contreras, R. J. 1977. Changes in gustatory nerve discharges with sodium deficiency: a single unit analysis. Brain Res. 121, 373–378.

Contreras, R. J. and Frank, M. 1979. Sodium deprivation alters neural responses to gustatory stimuli. J. Gen. Physiol. 73, 569–594.

Contreras, R. and Hatton, G. 1975. Gustatory adaptation as an explanation for dietary-induced sodium appetite. Physiol. Behav. 15, 569–576.

Contreras, R. J., Beckstead, R. M., and Norgren, R. 1982. The central projections of the trigeminal, facial, glossopharyngeal and vagus nerves: an autoradiographic study in the rat. J. Auton. Nerv. Syst. 6, 303–322.

Contreras, R. J., Gomez, M. M., and Norgren, R. 1980. Central origins of cranial nerve parasympathetic neurons in the rat. J. Comp. Neurol. 190, 373–394.

Craig, A. D. 2003a. Interoception: the sense of the physiological condition of the body. Curr. Opin. Neurobiol. 13, 500–505.

Craig, A. D. 2003b. Pain mechanisms: labeled lines versus convergence in central processing. Annu. Rev. Neurosci. 26, 1–30.

Craig, A. D. 2004. Distribution of trigeminothalamic and spinothalamic lamina I terminations in the macaque monkey. J. Comp. Neurol. 477, 119–148.

Craig, A. D., Chen, K., Bandy, D., and Reiman, E. M. 2000. Thermosensory activation of insular cortex. Nat. Neurosci. 3, 184–190.

Davis, B. J. 1993. GABA-like immunoreactivity in the gustatory zone of the nucleus of the solitary tract in the hamster: light and electron microscopic studies. Brain Res. Bull. 30, 69–77.

Davis, B. J. 1998. Synaptic relationships between the chorda tympani and tyrosine hydroxylase-immunoreactive dendritic processes in the gustatory zone of the nucleus of the solitary tract in the hamster. J. Comp. Neurol. 392, 78–91.

Davis, B. J. and Jang, T. 1986. The gustatory zone of the nucleus of the solitary tract in the hamster: light microscopic morphometric studies. Chem. Senses 11, 213–228.

Davis, B. J. and Jang, T. 1988a. A Golgi analysis of the gustatory zone of the nucleus of the solitary tract in the adult hamster. J. Comp. Neurol. 278, 388–396.

Davis, B. J. and Jang, T. 1988b. Tyrosine hydroxylase-like and dopamine beta-hydroxylase-like immunoreactivity in the gustatory zone of the nucleus of the solitary tract in the hamster: light- and electron-microscopic studies. Neuroscience 27, 949–964.

Davis, B. J. and Kream, R. M. 1993. Distribution of tachykinin- and opioid-expressing neurons in the hamster solitary nucleus: an immuno- and in situ hybridization histochemical study. Brain Res. 616, 6–16.

Davis, B. J. and Smith, D. V. 1997. Substance P modulates taste responses in the nucleus of the solitary tract of the hamster. Neuroreport 8, 1723–1727.

Di Lorenzo, P. M. 1988. Taste responses in the parabrachial pons of decerebrate rats. J. Neurophysiol. 59, 1871–1887.

Di Lorenzo, P. M. 1990. Corticofugal influence on taste responses in the parabrachial pons of the rat. Brain Res. 530, 73–84.

Di Lorenzo, P. M. and Monroe, S. 1990. Taste responses in the parabrachial pons of ovariectomized rats. Brain Res. Bull. 25, 741–748.

Di Lorenzo, P. M. and Monroe, S. 1995. Corticofugal influence on taste responses in the nucleus of the solitary tract in the rat. J. Neurophysiol. 74, 258–272.

Di Lorenzo, P. M. and Victor, J. D. 2003. Taste response variability and temporal coding in the nucleus of the solitary tract of the rat. J. Neurophysiol. 90, 1418–1431.

Drewe, J. A., Miles, R., and Kunze, D. L. 1990. Excitatory amino acid receptors of guinea pig medial nucleus tractus solitarius neurons. Am. J. Physiol. 259, H1389–H1395.

Dykes, R. W., Landry, P., Metherate, R., and Hicks, T. P. 1984. Functional role of GABA in cat primary somatosensory cortex: shaping receptive fields of cortical neurons. J. Neurophysiol. 52, 1066–1093.

Ferreira, G., Gutierrez, R., De La Cruz, V., and Bermudez-Rattoni, F. 2002. Differential involvement of cortical muscarinic and NMDA receptors in short- and long-term taste aversion memory. Eur. J. Neurosci. 16, 1139–1145.

Flaherty, C. F., Turovsky, J., and Krauss, K. L. 1994. Relative hedonic value modulates anticipatory contrast. Physiol. Behav. 55, 1047–1054.

Fontanini, A. and Katz, D. B. 2006. State-dependent modulation of time-varying gustatory responses. J. Neurophysiol. 96, 3183–3193.

Foster, L. A., Nakamura, K., Greenberg, D., and Norgren, R. 1996. Intestinal fat differentially suppresses sham feeding of different gustatory stimuli. Am. J. Physiol. 270, R1122–R1125.

Frank, M. 1973. An analysis of hamster afferent taste nerve response functions. J. Gen. Physiol. 61, 588–618.

Frank, M. and Pfaffmann, C. 1969. Taste nerve fibers: a random distribution of sensitivities to four tastes. Science 164, 1183–1185.

Frank, M. E. 1991. Taste-responsive neurons of the glossopharyngeal nerve of the rat. J. Neurophysiol. 65, 1452–1463.

Frank, R. A. and Stutz, R. M. 1982. Behavioral changes induced by basolateral amygdala self-stimulation. Physiol. Behav. 28, 661–665.

Frank, M. E., Bieber, S. L., and Smith, D. V. 1988. The organization of taste sensibilities in hamster chorda tympani nerve fibers. J. Gen. Physiol. 91, 861–896.

Frank, M. E., Contreras, R. J., and Hettinger, T. P. 1983. Nerve fibers sensitive to ionic taste stimuli in chorda tympani of the rat. J. Neurophysiol. 50, 941–960.

Fu, W., Sugai, T., Yoshimura, H., and Onoda, N. 2004. Convergence of olfactory and gustatory connections onto the endopiriform nucleus in the rat. Neuroscience 126, 1033–1041.

Fullerton, D. T., Getto, C. J., Swift, W. J., and Carlson, I. H. 1985. Sugar, opioids and binge eating. Brain Res. Bull. 14, 673–680.

Garcia, J., Kovner, R., and Green, K. 1970. Cue properties vs palatability of flavors in avoidance learning. Psychon. Sci. 20, 313–314.

Gauriau, C. and Bernard, J. F. 2004. Posterior triangular thalamic neurons convey nociceptive messages to the secondary somatosensory and insular cortices in the rat. J. Neurosci. 24, 752–761.

Geran, L. C. and Travers, S. P. 2006. Single neurons in the nucleus of the solitary tract respond selectively to bitter taste stimuli. J. Neurophysiol. 96, 2513–2527.

Giza, B. K. and Scott, T. R. 1983. Blood glucose selectively affects taste-evoked activity in rat nucleus tractus solitarius. Physiol. Behav. 31, 643–650.

Giza, B. K. and Scott, T. R. 1987. Blood glucose level affects perceived sweetness intensity in rats. Physiol. Behav. 41, 459–464.

Giza, B. K. and Scott, T. R. 1991. The effect of amiloride on taste-evoked activity in the nucleus tractus solitarius of the rat. Brain Res. 550, 247–256.

Giza, B. K., Ackroff, K., McCaughey, S. A., Sclafani, A., and Scott, T. R. 1997. Preference conditioning alters taste responses in the nucleus of the solitary tract of the rat. Am. J. Physiol. 273, R1230–R1240.

Giza, B. K., Scott, T. R., and Vanderweele, D. A. 1992. Administration of satiety factors and gustatory responsiveness in the nucleus tractus solitarius of the rat. Brain Res. Bull. 28, 637–639.

Glaum, S. R., Slater, N. T., Rossi, D. J., and Miller, R. J. 1992. Role of metabotropic glutamate (ACPD) receptors at the parallel fiber-Purkinje cell synapse. J. Neurophysiol. 68, 1453–1462.

Glenn, J. F. and Erickson, R. P. 1976. Gastric modulation of gustatory afferent activity. Physiol. Behav. 16, 561–568.

Greenberg, D., Smith, G. P., and Gibbs, J. 1990. Intraduodenal infusions of fats elicit satiety in sham-feeding rats. Am. J. Physiol. 259, R110–R118.

Grill, H. J. 1985. Introduction: physiological mechanisms in conditioned taste aversions. Ann. N. Y. Acad. Sci. 443, 67–88.

Grill, H. J. and Norgren, R. 1978a. Chronically decerebrate rats demonstrate satiation but not bait shyness. Science 201, 267–269.

Grill, H. J. and Norgren, R. 1978b. The taste reactivity test. II. Mimetic responses to gustatory stimuli in chronic thalamic and chronic decerebrate rats. Brain Res. 143, 281–297.

Grill, H. J., Schulkin, J., and Flynn, F. W. 1986. Sodium homeostasis in chronic decerebrate rats. Behav. Neurosci. 100, 536–543.

Grill, H. J., Schwartz, G. J., and Travers, J. B. 1992. The contribution of gustatory nerve input to oral motor behavior and intake-based preference. I. Effects of chorda tympani or glossopharyngeal nerve section in the rat. Brain Res. 573, 95–104.

Grossman, S. P., Dacey, D., Halaris, A. E., Collier, T., and Routtenberg, A. 1978. Aphagia and adipsia after preferential destruction of nerve cell bodies in hypothalamus. Science 202, 537–539.

Gutierrez, R., Carmena, J. M., Nicolelis, M. A., and Simon, S. A. 2006. Orbitofrontal ensemble activity monitors licking and distinguishes among natural rewards. J. Neurophysiol. 95, 119–133.

Gutierrez, H., Hernandez-Echeagaray, E., Ramirez-Amaya, V., and Bermudez-Rattoni, F. 1999. Blockade of N-methyl-D-aspartate receptors in the insular cortex disrupts taste aversion and spatial memory formation. Neuroscience 89, 751–758.

Guyenet, P. G., Filtz, T. M., and Donaldson, S. R. 1987. Role of excitatory amino acids in rat vagal and sympathetic baroreflexes. Brain Res. 407, 272–284.

Hagan, M. M., Chandler, P. C., Wauford, P. K., Rybak, R. J., and Oswald, K. D. 2003. The role of palatable food and hunger as trigger factors in an animal model of stress induced binge eating. Int. J. Eat Disord. 34, 183–197.

Hajnal, A., Takenouchi, K., and Norgren, R. 1999. Effect of intraduodenal lipid on parabrachial gustatory coding in awake rats. J. Neurosci. 19, 7182–7190.

Hallock, R. M. and Di Lorenzo, P. M. 2006. Temporal coding in the gustatory system. Neurosci. Biobehav. Rev. 30, 1145–1160.

Halpern, B. P. and Tapper, D. N. 1971. Taste stimuli: quality coding time. Science 171, 1256–1258.

Halsell, C. B. 1998. Differential distribution of amygdaloid input across rostral solitary nucleus subdivisions in rat. Ann. N. Y. Acad. Sci. 855, 482–485.

Halsell, C. B. and Travers, S. P. 1997. Anterior and posterior oral cavity responsive neurons are differentially distributed among parabrachial subnuclei in rat. J. Neurophysiol. 78, 920–938.

Halsell, C. B., Travers, S. P., and Travers, J. B. 1996. Ascending and descending projections from the rostral nucleus of the solitary tract originate from separate neuronal populations. Neuroscience 72, 185–197.

Hamilton, R. B. and Norgren, R. 1984. Central projections of gustatory nerves in the rat. J. Comp. Neurol. 222, 560–577.

Hanamori, T., Kunitake, T., Kato, K., and Kannan, H. 1997. Convergence of afferent inputs from the chorda tympani, lingual–tonsillar and pharyngeal branches of the glossopharyngeal nerve, and superior laryngeal nerve on the neurons in the insular cortex in rats. Brain Res. 763, 267–270.

Hanamori, T., Kunitake, T., Kato, K., and Kannan, H. 1998a. Neurons in the posterior insular cortex are responsive to gustatory stimulation of the pharyngolarynx, baroreceptor and chemoreceptor stimulation, and tail pinch in rats. Brain Res. 785, 97–106.

Hanamori, T., Kunitake, T., Kato, K., and Kannan, H. 1998b. Responses of neurons in the insular cortex to gustatory, visceral, and nociceptive stimuli in rats. J. Neurophysiol. 79, 2535–2545.

Hanamori, T., Miller, I. J., Jr, and Smith, D. V. 1988. Gustatory responsiveness of fibers in the hamster glossopharyngeal nerve. J. Neurophysiol. 60, 478–498.

Hasegawa, K. and Ogawa, H. 2007. Effects of acetylcholine on coding of taste information in the primary gustatory cortex in rats. Exp Brain Res. 179, 97–109

Hayama, T., Ito, S., and Ogawa, H. 1985. Responses of solitary tract nucleus neurons to taste and mechanical stimulations of the oral cavity in decerebrate rats. Exp. Brain Res. 60, 235–242.

Herman, J. P. and Cullinan, W. E. 1997. Neurocircuitry of stress: central control of the hypothalamo-pituitary-adrenocortical axis. Trends Neurosci. 20, 78–84.

Hermann, G. E. and Novin, D. 1980. Morphine inhibition of parabrachial taste units reversed by naloxone. Brain Res. Bull. 5(Suppl. 4), 169–173.

Hettinger, T. P. and Frank, M. E. 1990. Specificity of amiloride inhibition of hamster taste responses. Brain Res. 513, 24–34.

Hill, D. L., Formaker, B. K., and White, K. S. 1990. Perceptual characteristics of the amiloride-suppressed sodium chloride taste response in the rat. Behav. Neurosci. 104, 734–741.

Huang, A. L., Chen, X., Hoon, M. A., Chandrashekar, J., Guo, W., Trankner, D., Ryba, N. J., and Zuker, C. S. 2006. The cells and logic for mammalian sour taste detection. Nature 442, 934–938.

Ito, S. I. 1998. Possible representation of somatic pain in the rat insular visceral sensory cortex: a field potential study. Neurosci. Lett. 241, 171–174.

Jacobs, K. M., Mark, G. P., and Scott, T. R. 1988. Taste responses in the nucleus tractus solitarius of sodium-deprived rats. J. Physiol. (Lond.) 406, 393–410.

Jain, N., Qi, H. X., Catania, K. C., and Kaas, J. H. 2001. Anatomic correlates of the face and oral cavity representations in the somatosensory cortical area 3b of monkeys. J. Comp. Neurol. 429, 455–468.

Jalowiec, J. E. and Stricker, E. M. 1970. Restoration of body fluid balance following acute sodium deficiency in rats. J. Comp. Physiol. Psychol. 70, 94–102.

Jasmin, L., Burkey, A. R., Granato, A., and Ohara, P. T. 2004. Rostral agranular insular cortex and pain areas of the central nervous system: a tract-tracing study in the rat. J. Comp. Neurol. 468, 425–440.

Jones, L. M., Fontanini, A., and Katz, D. B. 2006. Gustatory processing: a dynamic systems approach. Curr. Opin. Neurobiol. 16, 420–428.

Kaas, J. H. 2005. The future of mapping sensory cortex in primates: three of many remaining issues. Philos. Trans. R. Soc. Lond. B. Biol. Sci. 360, 653–664.

Kaplan, J. M., Roitman, M., and Grill, H. J. 2000. Food deprivation does not potentiate glucose taste reactivity responses of chronic decerebrate rats. Brain Res. 870, 102–108.

Karadi, Z., Oomura, Y., Nishino, H., Scott, T. R., Lenard, L., and Aou, S. 1992. Responses of lateral hypothalamic glucose-sensitive and glucose- insensitive neurons to chemical stimuli in behaving rhesus monkeys. J. Neurophysiol. 67, 389–400.

Karimnamazi, H. and Travers, J. B. 1998. Differential projections from gustatory responsive regions of the parabrachial nucleus to the medulla and forebrain. Brain Res. 813, 283–302.

Katz, D. B., Simon, S. A., and Nicolelis, M. A. 2001. Dynamic and multimodal responses of gustatory cortical neurons in awake rats. J. Neurosci. 21, 4478–4489.

Katz, D. B., Simon, S. A., and Nicolelis, M. A. 2002. Taste-specific neuronal ensembles in the gustatory cortex of awake rats. J. Neurosci. 22, 1850–1857.

King, M. S. and Bradley, R. M. 1994. Relationship between structure and function of neurons in the rat rostral nucleus tractus solitarii. J. Comp. Neurol. 344, 50–64.

King, M. S., Wang, L., and Bradley, R. M. 1993. Substance P excites neurons in the gustatory zone of the rat nucleus tractus solitarius. Brain Res. 619, 120–130.

Kobashi, M. and Bradley, R. M. 1998. Effects of GABA on neurons of the gustatory and visceral zones of the parabrachial nucleus in rats. Brain Res. 799, 323–328.

Koh, M. T. and Bernstein, I. L. 2005. Mapping conditioned taste aversion associations using c-Fos reveals a dynamic role for insular cortex. Behav. Neurosci. 119, 388–398.

van der Kooy, D., Koda, L. Y., McGinty, J. F, Gerfen, C. R., and Bloom, F. E. 1984. The organization of projections from the cortex, amygdala, and hypothalamus to the nucleus of the solitary tract in rat. J. Comp. Neurol. 224, 1–24.

Kosar, E., Grill, H. J., and Norgren, R. 1986a. Gustatory cortex in the rat. I. Physiological properties and cytoarchitecture. Brain Res. 379, 329–341.

Kosar, E., Grill, H. J., and Norgren, R. 1986b. Gustatory cortex in the rat. II. Thalamocortical projections. Brain Res. 379, 342–352.

Kotz, C. M., Billington, C. J., and Levine, A. S. 1997. Opioids in the nucleus of the solitary tract are involved in feeding in the rat. Am. J. Physiol. 272, R1028–R1032.

Kringelbach, M. L., O'Doherty, J., Rolls, E. T., and Andrews, C. 2003. Activation of the human orbitofrontal cortex to a liquid food stimulus is correlated with its subjective pleasantness. Cereb. Cortex 13, 1064–1071.

Krout, K. E. and Loewy, A. D. 2000. Parabrachial nucleus projections to midline and intralaminar thalamic nuclei of the rat. J. Comp. Neurol. 428, 475–494.

Krupa, D. J., Wiest, M. C., Shuler, M. G., Laubach, M., and Nicolelis, M. A. 2004. Layer-specific somatosensory cortical activation during active tactile discrimination. Science 304, 1989–1992.

Kubo, T. and Kihara, M. 1988a. Evidence for gamma-aminobutyric acid receptor-mediated modulation of the aortic baroreceptor reflex in the nucleus tractus solitarii of the rat. Neurosci. Lett. 89, 156–160.

Kubo, T. and Kihara, M. 1988b. N-methyl-D-aspartate receptors mediate tonic vasodepressor control in the caudal ventrolateral medulla of the rat. Brain Res. 451, 366–370.

Kyriazi, H. T., Carvell, G. E., Brumberg, J. C., and Simons, D. J. 1996. Quantitative effects of GABA and bicuculline methiodide on receptive field properties of neurons in real and simulated whisker barrels. J. Neurophysiol. 75, 547–560.

Lasiter, P. S. and Kachele, D. L. 1988. Organization of GABA and GABA-transaminase containing neurons in the gustatory zone of the nucleus of the solitary tract. Brain Res. Bull. 21, 623–636.

Lazareva, N. A., Shevelev, I. A., Eysel, U. T., and Sharaev, G. A. 1997. Bicuculline and orientation tuning of neurons of the visual cortex. Neurophysiology 27, 42–49.

Lemon, C. H. and Smith, D. V. 2005. Neural representation of bitter taste in the nucleus of the solitary tract. J. Neurophysiol. 94, 3719–3729.

Li, C. S. and Smith, D. V. 1997. Glutamate receptor antagonists block gustatory afferent input to the nucleus of the solitary tract. J. Neurophysiol. 77, 1514–1525.

Li, C. S., Cho, Y. K., and Smith, D. V. 2002a. Taste responses of neurons in the hamster solitary nucleus are modulated by the central nucleus of the amygdala. J. Neurophysiol. 88, 2979–2992.

Li, C. S., Cho, Y. K., and Smith, D. V. 2005. Modulation of parabrachial taste neurons by electrical and chemical stimulation of the lateral hypothalamus and amygdala. J. Neurophysiol. 93, 1183–1196.

Li, C. S., Davis, B. J., and Smith, D. V. 2003. Opioid modulation of taste responses in the nucleus of the solitary tract. Brain Res. 965, 21–34.

Li, X., Staszewski, L., Xu, H., Durick, K., Zoller, M., and Adler, E. 2002b. Human receptors for sweet and umami taste. Proc. Natl. Acad. Sci. U. S. A. 99, 4692–4696.

Liu, H., Behbehani, M. M., and Smith, D. V. 1993. The influence of GABA on cells in the gustatory region of the hamster solitary nucleus. Chem. Senses 18, 285–305.

Lundy, R. F., Jr. and Norgren, R. 2001. Pontine gustatory activity is altered by electrical stimulation in the central nucleus of the amygdala. J. Neurophysiol. 85, 770–783.

Lundy, R. F., Jr. and Norgren, R. 2004a. Activity in the hypothalamus, amygdala, and cortex generates bilateral and convergent modulation of pontine gustatory neurons. J. Neurophysiol. 91, 1143–1157.

Lundy, R. F., Jr. and Norgren, R. 2004b. Gustatory System. In: The Rat Nervous System (ed. G. Paxinos), pp. 890–911. Elsevier Academic Press.

Macey, D. J., Smith, H. R., Nader, M. A., and Porrino, L. J. 2003. Chronic cocaine self-administration upregulates the norepinephrine transporter and alters functional activity in the bed nucleus of the stria terminalis of the rhesus monkey. J. Neurosci. 23, 12–16.

Mantyh, P. W. and Hunt, S. P. 1984. Neuropeptides are present in projection neurones at all levels in visceral and taste pathways: from periphery to sensory cortex. Brain Res. 299, 297–312.

Mark, G. P., Scott, T. R., Chang, F. C., and Grill, H. J. 1988. Taste responses in the nucleus tractus solitarius of the chronic decerebrate rat. Brain Res. 443, 137–148.

Markison, S., Gietzen, D. W., and Spector, A. C. 1999. Essential amino acid deficiency enhances long-term intake but not short-term licking of the required nutrient. J. Nutr. 129, 1604–1612.

Markison, S., Thompson, B. L., Smith, J. C., and Spector, A. C. 2000. Time course and pattern of compensatory ingestive behavioral adjustments to lysine deficiency in rats. J. Nutr. 130, 1320–1328.

Matsuo, R., Shimizu, N., and Kusano, K. 1984. Lateral hypothalamic modulation of oral sensory afferent activity in nucleus tractus solitarius neurons of rats. J. Neurosci. 4, 1201–1207.

Mayer, M. L. and Westbrook, G. L. 1987. The physiology of excitatory amino acids in the vertebrate central nervous system. Prog. Neurobiol. 28, 197–276.

McCaughey, S. A. and Scott, T. R. 1998. The taste of sodium. Neurosci. Biobehav. Rev. 22, 663–676.

Menendez, L., Bester, H., Besson, J. M., and Bernard, J. F. 1996. Parabrachial area: electrophysiological evidence for an involvement in cold nociception. J. Neurophysiol. 75, 2099–2116.

Miranda, M. I., Ramirez-Lugo, L., and Bermudez-Rattoni, F. 2000. Cortical cholinergic activity is related to the novelty of the stimulus. Brain Res. 882, 230–235.

Miyaoka, Y. and Pritchard, T. C. 1996. Responses of primate cortical neurons to unitary and binary taste stimuli. J. Neurophysiol. 75, 396–411.

Moga, M. M., Herbert, H., Hurley, K. M., Yasui, Y., Gray, T. S., and Saper, C. B. 1990a. Organization of cortical, basal forebrain, and hypothalamic afferents to the parabrachial nucleus in the rat. J. Comp. Neurol. 295, 624–661.

Moga, M. M., Saper, C. B., and Gray, T. S. 1990b. Neuropeptide organization of the hypothalamic projection to the parabrachial nucleus in the rat. J. Comp. Neurol. 295, 662–682.

Morley, J. E. and Levine, A. S. 1980. Stress-induced eating is mediated through endogenous opiates. Science 209, 1259–1261.

Murzi, E., Hernandez, L., and Baptista, T. 1986. Lateral hypothalamic sites eliciting eating affect medullary taste neurons in rats. Physiol. Behav. 36, 829–834.

Nachman, M. 1963. Learned aversion to the taste of lithium chloride and generalization to other salts. J. Comp. Physiol. Psychol. 56, 343–349.

Nakamura, K. and Norgren, R. 1991. Gustatory responses of neurons in the nucleus of the solitary tract of behaving rats. J. Neurophysiol. 66, 1232–1248.

Nakamura, K. and Norgren, R. 1993. Taste responses of neurons in the nucleus of the solitary tract of awake rats: an extended stimulus array. J. Neurophysiol. 70, 879–891.

Naor, C. and Dudai, Y. 1996. Transient impairment of cholinergic function in the rat insular cortex disrupts the encoding of taste in conditioned taste aversion. Behav. Brain Res. 79, 61–67.

Nelson, G., Chandrashekar, J., Hoon, M. A., Feng, L., Zhao, G., Ryba, N. J., and Zuker, C. S. 2002. An amino-acid taste receptor. Nature 416, 199–202.

Nelson, G., Hoon, M. A., Chandrashekar, J., Zhang, Y., Ryba, N. J., and Zuker, C. S. 2001. Mammalian sweet taste receptors. Cell 106, 381–390.

Nicola, S. M., Yun, I. A., Wakabayashi, K. T., and Fields, H. L. 2004. Firing of nucleus accumbens neurons during the consummatory phase of a discriminative stimulus task depends on previous reward predictive cues. J. Neurophysiol. 91, 1866–1882.

Ninomiya, Y. 1998. Reinnervation of cross-regenerated gustatory nerve fibers into amiloride-sensitive and amiloride-insensitive taste receptor cells. Proc. Natl. Acad. Sci. U. S. A. 95, 5347–5350.

Ninomiya, Y. and Funakoshi, M. 1988. Amiloride inhibition of responses of rat single chorda tympani fibers to chemical and electrical tongue stimulations. Brain Res. 451, 319–325.

Nishijo, H. and Norgren, R. 1990. Responses from parabrachial gustatory neurons in behaving rats. J. Neurophysiol. 63, 707–724.

Nishijo, H. and Norgren, R. 1991. Parabrachial gustatory neural activity during licking by rats. J. Neurophysiol. 66, 974–985.

Nishijo, H., Ono, T., and Norgren, R. 1991. Parabrachial gustatory neural responses to monosodium glutamate ingested by awake rats. Physiol. Behav. 49, 965–971.

Nishijo, H., Ono, T., Uwano, T., Kondoh, T., and Torii, K. 2000. Hypothalamic and amygdalar neuronal responses to various tastant solutions during ingestive behavior in rats. J. Nutr. 130, 954S–959S.

Nishijo, H., Uwano, T., Tamura, R., and Ono, T. 1998. Gustatory and multimodal neuronal responses in the amygdala during licking and discrimination of sensory stimuli in awake rats. J. Neurophysiol. 79, 21–36.

Norgren, R. 1970. Gustatory responses in the hypothalamus. Brain Res. 21, 63–77.

Norgren, R. 1974. Gustatory afferents to ventral forebrain. Brain Res. 81, 285–295.

Norgren, R. 1976. Taste pathways to hypothalamus and amygdala. J. Comp. Neurol. 166, 17–30.

Norgren, R. and Leonard, C. M. 1971. Taste pathways in rat brainstem. Science 173, 1136–1139.

Norgren, R. and Leonard, C. M. 1973. Ascending central gustatory pathways. J. Comp. Neurol. 150, 217–237.

Norgren, R. and Pfaffmann, C. 1975. The pontine taste area in the rat. Brain Res. 91, 99–117.

Nowlis, G. H., Frank, M. E., and Pfaffmann, C. 1980. Specificity of acquired aversions to taste qualities in hamsters and rats. J. Comp. Physiol. Psychol. 94, 932–942.

Oakley, B., Jones, L. B., and Kaliszewski, J. M. 1979. Taste responses of the gerbil IXth nerve. Chem. Senses 4, 79–87.

Ogawa, H. and Hayama, T. 1984. Receptive fields of solitario-parabrachial relay neurons responsive to natural stimulation of the oral cavity in rats. Exp. Brain Res. 54, 359–366.

Ogawa, H. and Wang, X. D. 2002. Neurons in the cortical taste area receive nociceptive inputs from the whole body as well as the oral cavity in the rat. Neurosci. Lett. 322, 87–90.

Ogawa, H., Hasegawa, K., and Murayama, N. 1992. Difference in taste quality coding between two cortical taste areas, granular and dysgranular insular areas, in rats. Exp. Brain Res. 91, 415–424.

Ogawa, H., Hasegawa, K., and Nakamura, T. 2000. Action of calcitonin gene-related peptide (CGRP) and substance P on neurons in the insular cortex and the modulation of taste responses in the rat. Chem. Senses 25, 351–359.

Ogawa, H., Hasegawa, K., Otawa, S., and Ikeda, I. 1998. GABAergic inhibition and modifications of taste responses in the cortical taste area in rats. Neurosci. Res. 32, 85–95.

Ogawa, H., Hayama, T., and Ito, S. 1982. Convergence of input from tongue and palate to the parabrachial nucleus neurons of rats. Neurosci. Lett. 28, 9–14.

Ono, T., Sasaki, K., Nishino, H., Fukuda, M., and Shibata, R. 1986. Feeding and diurnal related activity of lateral hypothalamic neurons in freely behaving rats. Brain Res. 373, 92–102.

Oomura, Y., Nishino, H., Karadi, Z., Aou, S., and Scott, T. R. 1991. Taste and olfactory modulation of feeding related neurons in behaving monkey. Physiol. Behav. 49, 943–950.

Otawa, S., Takagi, K., and Ogawa, H. 1995. NMDA and non-NMDA receptors mediate taste afferent inputs to cortical taste neurons in rats. Exp. Brain Res. 106, 391–402.

Perrotto, R. S. and Scott, T. R. 1976. Gustatory neural coding in the pons. Brain Res. 110, 283–300.

Petrovic, P., Kalso, E., Petersson, K. M., and Ingvar, M. 2002. Placebo and opioid analgesia – imaging a shared neuronal network. Science 295, 1737–1740.

Pickel, V. M. and Colago, E. E. 1999. Presence of mu-opioid receptors in targets of efferent projections from the central nucleus of the amygdala to the nucleus of the solitary tract. Synapse 33, 141–152.

Plata-Salaman, C. R., Scott, T. R., and Smith-Swintosky, V. L. 1992. Gustatory neural coding in the monkey cortex: L-amino acids. J. Neurophysiol. 67, 1552–1561.

Plata-Salaman, C. R., Scott, T. R., and Smith-Swintosky, V. L. 1993. Gustatory neural coding in the monkey cortex: the quality of sweetness. J. Neurophysiol. 69, 482–493.

Plata-Salaman, C. R., Scott, T. R., and Smith-Swintosky, V. L. 1995. Gustatory neural coding in the monkey cortex: acid stimuli. J. Neurophysiol. 74, 556–564.

Plata-Salaman, C. R., Smith-Swintosky, V. L., and Scott, T. R. 1996. Gustatory neural coding in the monkey cortex: mixtures. J. Neurophysiol. 75, 2369–2379.

Pritchard, T. C. and Norgren, R. 2004. Gustatory System. In: The Human Nervous System. Chap. 31 (eds. G. Paxinos and J. Mai), Elsevier.

Pritchard, T. C., Edwards, E. M., Smith, C. A., Hilgert, K. G., Gavlick, A. M., Maryniak, T. D., Schwartz, G. J., and Scott, T. R. 2005. Gustatory neural responses in the medial orbitofrontal cortex of the old world monkey. J. Neurosci. 25, 6047–6056.

Pritchard, T. C., Hamilton, R. B., Morse, J. R., and Norgren, R. 1986. Projections of thalamic gustatory and lingual areas in the monkey, Macaca fascicularis. J. Comp. Neurol. 244, 213–228.

Pritchard, T. C., Hamilton, R. B., and Norgren, R. 2000. Projections of the parabrachial nucleus in the old world monkey. Exp. Neurol. 165, 101–117.

Reilly, S., Grigson, P. S., and Norgren, R. 1993. Parabrachial nucleus lesions and conditioned taste aversion: evidence supporting an associative deficit. Behav. Neurosci. 107, 1005–1017.

Renehan, W. E., Jin, Z., Zhang, X., and Schweitzer, L. 1994. Structure and function of gustatory neurons in the nucleus of the solitary tract. I. A classification of neurons based on morphological features. J. Comp. Neurol. 347, 531–544.

Renehan, W. E., Jin, Z., Zhang, X., and Schweitzer, L. 1996. Structure and function of gustatory neurons in the nucleus of the solitary tract: II. Relationships between neuronal morphology and physiology. J. Comp. Neurol. 367, 205–221.

Renehan, W. E., Zhang, X., Beierwaltes, W. H., and Fogel, R. 1995. Neurons in the dorsal motor nucleus of the vagus may integrate vagal and spinal information from the GI tract. Am. J. Physiol. 268, G780–G790.

Reynolds, S. M. and Zahm, D. S. 2005. Specificity in the projections of prefrontal and insular cortex to ventral striatopallidum and the extended amygdala. J. Neurosci. 25, 11757–11767.

Rideout, H. J. and Parker, L. A. 1996. Morphine enhancement of sucrose palatability: analysis by the taste reactivity test. Pharmacol. Biochem. Behav. 53, 731–734.

Rinaman, L., Verbalis, J. G., Stricker, E. M., and Hoffman, G. E. 1993. Distribution and neurochemical phenotypes of caudal medullary neurons activated to express cFos following peripheral administration of cholecystokinin. J. Comp. Neurol. 338, 475–490.

Rodieck, R. W. and Brening, R. K. 1982. On classifying retinal ganglion cells by numerical methods. Brain Behav. Evol. 21, 42–46.

Roitman, M. F., Wheeler, R. A., and Carelli, R. M. 2005. Nucleus accumbens neurons are innately tuned for rewarding and aversive taste stimuli, encode their predictors, and are linked to motor output. Neuron 45, 587–597.

Rolls, E. T. and Baylis, L. L. 1994. Gustatory, olfactory, and visual convergence within the primate orbitofrontal cortex. J. Neurosci. 14, 5437–5452.

Rolls, E. T. and Scott, T. R. 2003. Central Taste Anatomy and Physiology. In: Handbook of Olfaction and Gustation. Chap. 33 (ed. R. L. Doty), Marcel Dekker.

Rolls, E. T., Murzi, E., Yaxley, S., Thorpe, S. J., and Simpson, S. J. 1986. Sensory-specific satiety: food-specific reduction in responsiveness of ventral forebrain neurons after feeding in the monkey. Brain Res. 368, 79–86.

Rolls, B. J., Rolls, E. T., Rowe, E. A., and Sweeney, K. 1981. Sensory specific satiety in man. Physiol. Behav. 27, 137–142.

Rolls, E. T., Sanghera, M. K., and Roper-Hall, A. 1979. The latency of activation of neurones in the lateral hypothalamus and substantia innominata during feeding in the monkey. Brain Res. 164, 121–135.

Rolls, E. T., Sienkiewicz, Z. J., and Yaxley, S. 1989. Hunger modulates the responses to gustatory stimuli of single neurons in the caudolateral orbitofrontal cortex of the macaque monkey. Eur. J. Neurosci. 1, 53–60.

Rolls, E. T., Yaxley, S., and Sienkiewicz, Z. J. 1990. Gustatory responses of single neurons in the caudolateral orbitofrontal cortex of the macaque monkey. J. Neurophysiol. 64, 1055–1066.

Rosenblum, K., Berman, D. E., Hazvi, S., Lamprecht, R., and Dudai, Y. 1997. NMDA receptor and the tyrosine phosphorylation of its 2B subunit in taste learning in the rat insular cortex. J. Neurosci. 17, 5129–5135.

Rosenblum, K., Meiri, N., and Dudai, Y. 1993. Taste memory: the role of protein synthesis in gustatory cortex. Behav. Neural Biol. 59, 49–56.

Ruggiero, D. A., Giuliano, R., Anwar, M., Stornetta, R., and Reis, D. J. 1990. Anatomical substrates of cholinergic–autonomic regulation in the rat. J. Comp. Neurol. 292, 1–53.

Ruggiero, D. A., Mtui, E. P., Otake, K., and Anwar, M. 1996. Central and primary visceral afferents to nucleus tractus solitarii may generate nitric oxide as a membrane-permeant neuronal messenger. J. Comp. Neurol. 364, 51–67.

Sanghera, M. K., Rolls, E. T., and Roper-Hall, A. 1979. Visual responses of neurons in the dorsolateral amygdala of the alert monkey. Exp. Neurol. 63, 610–626.

Santiago, A. C. and Shammah-Lagnado, S. J. 2005. Afferent connections of the amygdalopiriform transition area in the rat. J. Comp. Neurol. 489, 349–371.

Sasaki, K., Ono, T., Muramoto, K., Nishino, H., and Fukuda, M. 1984. The effects of feeding and rewarding brain stimulation on lateral hypothalamic unit activity in freely moving rats. Brain Res. 322, 201–211.

Schafe, G. E. and Bernstein, I. L. 1996. Forebrain contribution to the induction of a brainstem correlate of conditioned taste aversion: I. The amygdala. Brain Res. 741, 109–116.

Schafe, G. E. and Bernstein, I. L. 1998. Forebrain contribution to the induction of a brainstem correlate of conditioned taste aversion. II. Insular (gustatory) cortex. Brain Res. 800, 40–47.

Schwartzbaum, J. S. 1988. Electrophysiology of taste, feeding and reward in lateral hypothalamus of rabbit. Physiol. Behav. 44, 507–526.

Schweinhardt, P., Glynn, C., Brooks, J., McQuay, H., Jack, T., Chessell, I., Bountra, C., and Tracey, I. 2006. An fMRI study of cerebral processing of brush-evoked allodynia in neuropathic pain patients. Neuroimage 32, 256–265.

Schweitzer, L., Jin, Z., Zhang, X., and Renehan, W. E. 1995. Cell types in the rostral nucleus of the solitary tract. Brain Res. Brain Res. Rev. 20, 185–195.

Scott, T. R. and Giza, B. K. 1990. Coding channels in the taste system of the rat. Science 249, 1585–1587.

Scott, T. R., Giza, B. K., and Yan, J. 1999. Gustatory neural coding in the cortex of the alert cynomolgus macaque: the quality of bitterness. J. Neurophysiol. 81, 60–71.

Scott, T. R., Karadi, Z., Oomura, Y., Nishino, H., Plata-Salaman, C. R., Lenard, L., Giza, B. K., and Aou, S. 1993. Gustatory neural coding in the amygdala of the alert macaque monkey. J. Neurophysiol. 69, 1810–1820.

Scott, T. R., Plata-Salaman, C. R., and Smith-Swintosky, V. L. 1994. Gustatory neural coding in the monkey cortex: the quality of saltiness. J. Neurophysiol. 71, 1692–1701.

Scott, T. R., Yaxley, S., Sienkiewicz, Z. J., and Rolls, E. T. 1986. Gustatory responses in the frontal opercular cortex of the alert cynomolgus monkey. J. Neurophysiol. 56, 876–890.

Seeley, R. J., Grill, H. J., and Kaplan, J. M. 1994. Neurological dissociation of gastrointestinal and metabolic contributions to meal size control. Behav. Neurosci. 108, 347–352.

Shi, C. J. and Cassell, M. D. 1998. Cortical, thalamic, and amygdaloid connections of the anterior and posterior insular cortices. J. Comp. Neurol. 399, 440–468.

Shimura, T., Komori, M., and Yamamoto, T. 1997a. Acute sodium deficiency reduces gustatory responsiveness to NaCl in the parabrachial nucleus of rats. Neurosci. Lett. 236, 33–36.

Shimura, T., Tanaka, H., and Yamamoto, T. 1997b. Salient responsiveness of parabrachial neurons to the conditioned stimulus after the acquisition of taste aversion learning in rats. Neuroscience 81, 239–247.

Shimura, T., Tokita, K., and Yamamoto, T. 2002. Parabrachial unit activities after the acquisition of conditioned taste aversion to a non-preferred HCl solution in rats. Chem. Senses 27, 153–158.

Shipley, M. T. and Geinisman, Y. 1984. Anatomical evidence for convergence of olfactory, gustatory, and visceral afferent pathways in mouse cerebral cortex. Brain Res. Bull. 12, 221–226.

Sillito, A. M. and Kemp, J. A. 1983. Cholinergic modulation of the functional organization of the cat visual cortex. Brain Res. 289, 143–155.

Simon, S. A., de Araujo, I. E., Gutierrez, R., and Nicolelis, M. A. 2006. The neural mechanisms of gustation: a distributed processing code. Nat. Rev. Neurosci. 7, 890–901.

Sloan, T. B. 1998. Anesthetic effects on electrophysiologic recordings. J. Clin. Neurophysiol. 15, 217–226.

Small, D. M. and Prescott, J. 2005. Odor/taste integration and the perception of flavor. Exp. Brain Res. 166, 345–357.

Small, D. M., Gerber, J. C., Mak, Y. E., and Hummel, T. 2005. Differential neural responses evoked by orthonasal versus retronasal odorant perception in humans. Neuron 47, 593–605.

Small, D. M., Gregory, M. D., Mak, Y. E., Gitelman, D., Mesulam, M. M., and Parrish, T. 2003. Dissociation of neural representation of intensity and affective valuation in human gustation. Neuron 39, 701–711.

Small, D. M., Voss, J., Mak, Y. E., Simmons, K. B., Parrish, T., and Gitelman, D. 2004. Experience-dependent neural integration of taste and smell in the human brain. J. Neurophysiol. 92, 1892–1903.

Smeets, P. A., de Graaf, C., Stafleu, A., van Osch, M. J., Nievelstein, R. A., and van der Grond, J. 2006. Effect of satiety on brain activation during chocolate tasting in men and women. Am. J. Clin. Nutr. 83, 1297–1305.

Smith, G. P. 1996. The direct and indirect controls of meal size. Neurosci Biobehav. Rev. 20, 41–46.

Smith, D. V. and Li, C. S. 1998. Tonic GABAergic inhibition of taste-responsive neurons in the nucleus of the solitary tract. Chem. Senses 23, 159–169.

Smith, D. V. and Li, C. 2000. GABA-mediated corticofugal inhibition of taste-responsive neurons in the nucleus of the solitary tract [In Process Citation]. Brain Res. 858, 408–415.

Smith, D. V. and Theodore, R. M. 1984. Conditioned taste aversions: generalization to taste mixtures. Physiol. Behav. 32, 983–989.

Smith, D. V. and Travers, J. B. 1979. A metric for the breadth of tuning of gustatory neurons. Chem. Senses 4, 215–229.

Smith, D. V., Li, C. S., and Cho, Y. C. 2004. Forebrain modulation of brainstem gustatory processing. Chem. Senses 30(Suppl. 1), i176–i177.

Smith, D. V., Travers, J. B., and Van Buskirk, R. L. 1979. Brainstem correlates of gustatory similarity in the hamster. Brain Res. Bull. 4, 359–372.

Smith, D. V., Van Buskirk, R. L., Travers, J. B., and Bieber, S. L. 1983. Gustatory neuron types in hamster brain stem. J. Neurophysiol. 50, 522–540.

Smith-Swintosky, V. L., Plata-Salaman, C. R., and Scott, T. R. 1991. Gustatory neural coding in the monkey cortex: stimulus quality. J. Neurophysiol. 66, 1156–1165.

South, E. H. and Ritter, R. C. 1986. Substance P-containing trigeminal sensory neurons project to the nucleus of the solitary tract. Brain Res. 372, 283–289.

Spector, A. C. and Grill, H. J. 1992. Salt taste discrimination after bilateral section of the chorda tympani or glossopharyngeal nerves. Am. J. Physiol. 263, R169–R176.

Spector, A. C. and Travers, S. P. 2005. The representation of taste quality in the mammalian nervous system. Behav. Cogn. Neurosci. Rev. 4, 143–191.

Spector, A. C., Markison, S., St. John, S. J., and Garcea, M. 1997. Sucrose vs. maltose taste discrimination by rats depends on the input of the seventh cranial nerve. Am. J. Physiol. 272, R1210–R1218.

Spector, A. C., Redman, R., and Garcea, M. 1996. The consequences of gustatory nerve transection on taste-guided licking of sucrose and maltose in the rat. Behav. Neurosci. 110, 1096–1109.

Squire, L. R., Bloom, F. E., McConnell, S. K., Roberts, J. L., Spitzer, N. C., and Zigmond, M. J. 2003. Fundamental Neuroscience, Academic Press.

St. John, S. J. and Smith, D. V. 2000. Neural representation of salts in the rat solitary nucleus: brain stem correlates of taste discrimination. J. Neurophysiol. 84, 628–638.

St. John, S. J. and Spector, A. C. 1998. Behavioral discrimination between quinine and KCl is dependent on input from the seventh cranial nerve: implications for the functional roles of the gustatory nerves in rats. J. Neurosci. 18, 4353–4362.

Stapleton, J. R., Lavine, M. L., Wolpert, R. L., Nicolelis, M. A., and Simon, S. A. 2006. Rapid taste responses in the gustatory cortex during licking. J. Neurosci. 26, 4126–4138.

Stapleton, J. R., Roper, S. D., and Delay, E. R. 1999. The taste of monosodium glutamate (MSG), L-aspartic acid, and N-methyl-D-aspartate (NMDA) in rats: are NMDA receptors involved in MSG taste? Chem. Senses 24, 449–457.

Steiner, J. E. 1973. The gustofacial response: observation on normal and anencephalic newborn infants. Symp. Oral Sens. Percept. 4, 254–278.

Strand, F. 1999. Neuropeptides: Regulators of Physiological Processes. MIT Press.

Sweazey, R. D. and Smith, D. V. 1987. Convergence onto hamster medullary taste neurons. Brain Res. 408, 173–184.

Taha, S. A. and Fields, H. L. 2005. Encoding of palatability and appetitive behaviors by distinct neuronal populations in the nucleus accumbens. J. Neurosci. 25, 1193–1202.

Tamura, R. and Norgren, R. 1997. Repeated sodium depletion affects gustatory neural responses in the nucleus of the solitary tract of rats. Am. J. Physiol. 273, R1381–R1391.

Tell, F. and Bradley, R. M. 1994. Whole-cell analysis of ionic currents underlying the firing pattern of neurons in the gustatory zone of the nucleus tractus solitarii. J. Neurophysiol. 71, 479–492.

Tindell, A. J., Berridge, K. C., and Aldridge, J. W. 2004. Ventral pallidal representation of pavlovian cues and reward: population and rate codes. J. Neurosci. 24, 1058–1069.

Tindell, A. J., Smith, K. S., Pecina, S., Berridge, K. C., and Aldridge, J. W. 2006. Ventral pallidum firing codes hedonic reward: when a bad taste turns good. J. Neurophysiol. 96, 2399–2409.

Tokita, K., Karadi, Z., Shimura, T., and Yamamoto, T. 2004. Centrifugal inputs modulate taste aversion learning associated parabrachial neuronal activities. J. Neurophysiol. 92, 265–279.

Topolovec, J. C., Gati, J. S., Menon, R. S., Shoemaker, J. K., and Cechetto, D. F. 2004. Human cardiovascular and gustatory brainstem sites observed by functional magnetic resonance imaging. J. Comp. Neurol. 471, 446–461.

Touzani, K., Ferssiwi, A., and Velley, L. 1990. Localization of lateral hypothalamic neurons projecting to the medial part of the parabrachial area of the rat. Neurosci. Lett. 114, 17–21.

Travers, J. B. 2006. Organization and expression of reward in the rodent orbitofrontal cortex. Focus on "orbitofrontal ensemble activity monitors licking and distinguishes among natural rewards. J. Neurophysiol. 95, 14–15.

Travers, S. P. (ed.) 1993. Orosensory Processing in Neural Systems of the Nucleus of the Solitary Tract pp. 339–394. CRC Press.

Travers, S. P. and Norgren, R. 1991. Coding the sweet taste in the nucleus of the solitary tract: differential roles for anterior tongue and nasoincisor duct gustatory receptors in the rat. J. Neurophysiol. 65, 1372–1380.

Travers, S. P. and Norgren, R. 1995. Organization of orosensory responses in the nucleus of the solitary tract of rat. J. Neurophysiol. 73, 2144–2162.

Travers, S. P. and Travers, J. B. 2006. Taste-evoked Fos expression in nitrergic neurons in the nucleus of the solitary tract and reticular formation of the rat. J. Comp. Neurol. 500, 746–760.

Travers, S. P., Pfaffmann, C., and Norgren, R. 1986. Convergence of lingual and palatal gustatory neural activity in the nucleus of the solitary tract. Brain Res. 365, 305–320.

Travers, J. B. and Smith, D. V. 1979. Gustatory sensitivities in neurons of the hamster nucleus tractus solitarius. Sens. Processes 3, 1–26.

Travers, J. B., Grill, H. J., and Norgren, R. 1987. The effects of glossopharyngeal and chorda tympani nerve cuts on the ingestion and rejection of sapid stimuli: an electromyographic analysis in the rat. Behav. Brain Res. 25, 233–246.

Tremere, L., Hicks, T. P., and Rasmusson, D. D. 2001. Expansion of receptive fields in raccoon somatosensory cortex in vivo by GABA(A) receptor antagonism: implications for cortical reorganization. Exp. Brain Res. 136, 447–455.

Uher, R., Treasure, J., Heining, M., Brammer, M. J., and Campbell, I. C. 2006. Cerebral processing of food-related stimuli: effects of fasting and gender. Behav. Brain Res. 169, 111–119.

Uteshev, V. V. and Smith, D. V. 2006. Cholinergic modulation of neurons in the gustatory region of the nucleus of the solitary tract. Brain Res. 1084, 38–53.

Van Buskirk, R. L. and Smith, D. V. 1981. Taste sensitivity of hamster parabrachial pontine neurons. J. Neurophysiol. 45, 144–171.

Vardhan, A., Kachroo, A., and Sapru, H. N. 1993. Excitatory amino acid receptors in the nucleus tractus solitarius mediate the responses to the stimulation of cardio-pulmonary vagal afferent C fiber endings. Brain Res. 618, 23–31.

Veening, J. G., Swanson, L. W., and Sawchenko, P. E. 1984. The organization of projections from the central nucleus of the amygdala to brainstem sites involved in central autonomic regulation: a combined retrograde transport-immunohistochemical study. Brain Res. 303, 337–357.

Vogt, M. B. and Mistretta, C. M. 1990. Convergence in mammalian nucleus of solitary tract during development and functional differentiation of salt taste circuits. J. Neurosci. 10, 3148–3157.

Voshart, K. and van der Kooy, D. 1981. The organization of the efferent projections of the parabrachial nucleus of the forebrain in the rat: a retrograde fluorescent double-labeling study. Brain Res. 212, 271–286.

Walker, J. R., Ahmed, S. H., Gracy, K. N., and Koob, G. F. 2000. Microinjections of an opiate receptor antagonist into the bed nucleus of the stria terminalis suppress heroin self-administration in dependent rats. Brain Res. 854, 85–92.

Wang, L. and Bradley, R. M. 1993. Influence of GABA on neurons of the gustatory zone of the rat nucleus of the solitary tract. Brain Res. 616, 144–153.

Wang, L. and Bradley, R. M. 1995. *In vitro* study of afferent synaptic transmission in the rostral gustatory zone of the rat nucleus of the solitary tract. Brain Res. 702, 188–198.

Wang, X. and Ogawa, H. 2002. Columnar organization of mechanoreceptive neurons in the cortical taste area in the rat. Exp. Brain Res. 147, 114–123.

Whitehead, M. C. 1988. Neuronal architecture of the nucleus of the solitary tract in the hamster. J. Comp. Neurol. 276, 547–572.

Whitehead, M. C. 1990. Subdivisions and neuron types of the nucleus of the solitary tract that project to the parabrachial nucleus in the hamster. J. Comp. Neurol. 301, 554–574.

Whitehead, M. C., Bergula, A., and Holliday, K. 2000. Forebrain projections to the rostral nucleus of the solitary tract in the hamster. J. Comp. Neurol. 422, 429–447.

Williams, G. V., Rolls, E. T., Leonard, C. M., and Stern, C. 1993. Neuronal responses in the ventral striatum of the behaving macaque. Behav. Brain Res. 55, 243–252.

Woolston, D. C. and Erickson, R. P. 1979. Concept of neuron types in gustation in the rat. J. Neurophysiol. 42, 1390–1409.

Yamamoto, T. and Kawamura, Y. 1975. Cortical responses to electrical and gustatory stimuli in the rabbit. Brain Res. 94, 447–463.

Yamamoto, T., Matsuo, R., Fujimoto, Y., Fukunaga, I., Miyasaka, A., and Imoto, T. 1991. Electrophysiological and behavioral studies on the taste of umami substances in the rat. Physiol. Behav. 49, 919–925.

Yamamoto, T., Matsuo, R., Ichikawa, H., Wakisaka, S., Akai, M., Imai, Y., Yonehara, N., and Inoki, R. 1990. Aversive taste stimuli increase CGRP levels in the gustatory insular cortex of the rat. Neurosci. Lett. 112, 167–172.

Yamamoto, T., Matsuo, R., and Kawamura, Y. 1980. Localization of cortical gustatory area in rats and its role in taste discrimination. J. Neurophysiol. 44, 440–455.

Yamamoto, T., Matsuo, R., Kiyomitsu, Y., and Kitamura, R. 1989a. Response properties of lateral hypothalamic neurons during ingestive behavior with special reference to licking of various taste solutions. Brain Res. 481, 286–297.

Yamamoto, T., Matsuo, R., Kiyomitsu, Y., and Kitamura, R. 1989b. Taste responses of cortical neurons in freely ingesting rats. J. Neurophysiol. 61, 1244–1258.

Yamamoto, T., Shimura, T., Sako, N., Yasoshima, Y., and Sakai, N. 1994. Neural substrates for conditioned taste aversion in the rat. Behav. Brain Res. 65, 123–137.

Yamamoto, T., Yuyama, N., Kato, T., and Kawamura, Y. 1985. Gustatory responses of cortical neurons in rats. II. Information processing of taste quality. J. Neurophysiol. 53, 1356–1369.

Yan, J. and Scott, T. R. 1996. The effect of satiety on responses of gustatory neurons in the amygdala of alert cynomolgus macaques. Brain Res. 740, 193–200.

Yasoshima, Y., Sako, N., Senba, E., and Yamamoto, T. 2006. Acute suppression, but not chronic genetic deficiency, of c-fos gene expression impairs long-term memory in aversive taste learning. Proc. Natl. Acad. Sci. U. S. A. 103, 7106–7111.

Yasui, Y., Saper, C. B., and Cechetto, D. F. 1989. Calcitonin gene-related peptide immunoreactivity in the visceral sensory cortex, thalamus, and related pathways in the rat. J. Comp. Neurol. 290, 487–501.

Yaxley, S., Rolls, E. T., and Sienkiewicz, Z. J. 1988. The responsiveness of neurons in the insular gustatory cortex of the macaque monkey is independent of hunger. Physiol. Behav. 42, 223–229.

Yaxley, S., Rolls, E. T., and Sienkiewicz, Z. J. 1990. Gustatory responses of single neurons in the insula of the macaque monkey. J. Neurophysiol. 63, 689–700.

Yoshida, R., Shigemura, N., Sanematsu, K., Yasumatsu, K., Ishizuka, S., and Ninomiya, Y. 2006. Taste responsiveness of fungiform taste cells with action potentials. J. Neurophysiol. 96, 3088–3095.

Zahm, D. S. 2000. An integrative neuroanatomical perspective on some subcortical substrates of adaptive responding with emphasis on the nucleus accumbens. Neurosci. Biobehav. Rev. 24, 85–105.

4.16 Neural Ensembles in Taste Coding

A Fontanini, S E Grossman, B A Revill, and D B Katz, Brandeis University, Waltham, MA, USA

Glossary

coherence The state of sticking together, here meaning neurons firing (or changing firing) at the same times.

context A set of variables having to do with the environment, either internal or external.

ensembles Groups of neurons recorded simultaneously.

forebrain The rostral-most portion of the brain, including the thalamus, hypothalamus, basal ganglia, limbic system, and of course cerebral cortex.

hierarchy Classification for a system containing a series of levels, each of which does a particular job or set of jobs.

interactions Activity in one element (here, a neuron or brain region) having an impact on activity in another.

pattern completion A property of a neural network architecture, whereby the act of setting a small subset of units into a state reflecting a particular global pattern causes, through time, the rest of the network to complete this global pattern.

receptive field The range of stimuli to which a neuron responds.

taste space A theoretical organization of the relative similarities between the stimuli within a particular modality.

top-down Interregional interactions reflecting the influence of a higher level of a hierarchy on a lower level of the same hierarchy.

4.16.1 Introduction

Two main theories of taste coding, the labeled-line (LL) and across-neuron pattern (ANP) hypotheses, dominate research in gustation. The two theories are similar, in that both explain taste coding in terms of activity in populations of neurons. They differ in only one regard: according to the LL theory, a particular subgroup of neurons communicates information for a particular taste, and therefore a reduction in firing among one subgroup (say, sucrose-best neurons) codes a reduction in that taste (sweetness); according to the ANP theory, the entire population of taste-responsive neurons participates in all codes, and therefore a reduction in the response of sucrose-best neurons is expected to code both a reduction in sweetness and an increase in some other

taste quality (which is coded, in part, by specifically low firing rates in sucrose-best neurons).

It is not our intent to pursue these distinctions here. Excellent pro-LL (Scott, K., 2004) and pro-ANP (Smith, D. V. and St. John, S. J., 1999) reviews have been written in the last 10 years. Instead, we will discuss a basic aspect of neural population function that is largely lacking from debates over taste coding. In this essay, we will suggest that coding in distributed neural populations is intrinsically interactive, and that future advances in our theories of gustatory population coding will therefore require an accounting of such interactions. We will present data demonstrating that such interactions occur in sensory systems, including the gustatory system. This will lead to a discussion of the spatial and temporal structures that such interactions introduce into population activity, and finally to our suggestion that dynamic population codes may be best thought of not as coding stimuli themselves, but rather as driving the transformation of sensory-related activity into action-related activity. Guided by seminal theorizing about perception and action (Gibson, J. J., 1966; Erickson, R., 1984) and by data from taste and other sensory systems, we will argue that the time has come to think about gustatory population coding in a way that is orthogonal to the LL–ANP debate.

4.16.2 Population Coding and Neural Interactions

While the LL and ANP hypotheses both refer to activity in populations of neurons, neither suggests a specific role for interactions between the neurons. Such interactions, embodied at the simplest level by concepts such as lateral or reciprocal inhibition and excitation, would in fact introduce complications into the decoding of a LL or ANP code, because such interactions cause neural activity to be modulated through time (see below). In fact, any purely spatial coding scheme (such as both LL and ANP) proposed to function in a network in which neurons interact must include specification of the operative time period during which the spatial code will be polled for content.

By comparison, most conceptions of neural population function explicitly rely on convergence and/or feedback between nodes (see, for instance, Nagai, T. *et al.*, 1992; van Vreeswijk, C. and Sompolinsky, H., 1998; Nagai, T., 2000; Masuda, N. and Aihara, K., 2003). The earliest neural network models, founded on basic neuroscientific principles but developed before researchers had the means to do ensemble electrophysiology, suggested that population coding relies on information transfer among an entire set of neural elements (McClelland, J. L. and Rumelhart, D. E., 1981). No individual group of neurons in these models (save the input and output nodes) is profitably described as coding any particular stimulus. Instead, processing is a function of the interactions between neurons. Input to such models – even LL input – is transformed by interconnected neural networks into dynamic patterns in which individual neural elements seldom code stimulus attributes independently of their neighbors (see, for instance, Lumer, E. D. *et al.*, 1997; Rabinovich, M. I. *et al.*, 2000; Sporns, O. *et al.*, 2000).

Data collected in the light of such neural network modeling has suggested that real brain systems engage in exactly this sort of population coding. A variety of vertebrate and invertebrate sensory, cognitive, and motor systems may make explicit use of interactions between neurons (see below). These interactions introduce considerable complexity into single-neuron behavior and do processing work in neural systems. And while most of this work has been done in the visual, auditory, somatosensory, and olfactory systems, it now appears that the same can be said of the gustatory system – neuronal interactions occur both within and between brain regions in the taste neuroaxis. We will now briefly discuss those anatomical and physiological data.

Interactive processing within single brain regions can be revealed in cross-correlogram (CCG) and cross-coherence peaks, which indicate that one neuron produces action potentials in a consistent temporal relationship to those of another (Brody, C. D., 1999). When that correlated activity is specific to particular stimuli, it suggests that neural coding may involve population interactions. Such coding has been observed to occur in the somatosensory (Roy, S. and Alloway, K. D., 1999), auditory (e.g., Eggermont, J. J., 1994; DeCharms, R. C. and Merzenich, M. M., 1996; Eggermont, J. J., 2000), visual (e.g., Brosch, M. *et al.*, 1997; Lampl, I. *et al.*, 1999; Bretzner, F. *et al.*, 2001; Yoshimura, Y. *et al.*, 2005), and olfactory (e.g., Wehr, M. and Laurent, G., 1996; Christensen, T. A. *et al.*, 2003) systems, and similar findings have been reported in the frontal cortex of monkeys performing a GO NO-GO task (Vaadia, E. *et al.*, 1995) and in motor cortex (Hatsopoulos, N. G. *et al.*, 2003).

In the taste system as well, pairs of cortical neurons produce correlated spike patterns during the presentation of particular subsets of tastes (Yokota, T. *et al.*, 1996; Nakamura, T. and Ogawa, H., 1997; Yokota, T. and Satoh, T., 2001; Katz, D. B. *et al.*, 2002b). Taste administration recruits taste-specific but overlapping neuronal ensembles, including some neurons that are broadly tuned and some that according to classic single-neuron analyses cannot even be identified as taste responsive, that is, neurons with flat or unremarkable responses to taste administration may still be involved in taste-specific cross-correlations (Katz, D. B. *et al.*, 2002b). Taste-specific cross-correlations have also been observed in the nucleus of the solitary tract (NTS, Adachi, M. *et al.*, 1989) and in the pontine parabrachial nuclei (PbN, Yamada, S. *et al.*, 1990; Adachi, M., 1991). Patterns of neural interactions appear to provide an additional source of taste-related information not inherent in single-unit spike trains, suggesting that taste neurons are embedded in interactive ensembles.

Of course, these intraregional ensembles are themselves embedded in larger interactive networks. Various dye-labeling techniques have demonstrated the existence of reciprocal connections among taste-responsive nuclei in brainstem, thalamic, limbic, and cortical regions. Just as injections of horse-radish per-oxidase into PbN have revealed two parallel ascending taste pathways in rodents – a parabrachio-thalamo-cortical and parabrachio-amygdaloid path-way (Halsell, C. B., 1992) – similar methods have revealed descending pathways back to the brainstem (van der Kooy, D. *et al.*, 1984; Huang, T. *et al.*, 2003). Reciprocal connections have been shown to exist between various pairs of taste regions, including gustatory cortex (GC) and thalamus (Nakashima, M. *et al.*, 2000), GC and amygdala (McDonald, A. J. and Jackson, T. R., 1987), and amygdala and PbN (Takeuchi, Y. *et al.*, 1982). These data suggest that taste information is processed in a distributed system of information exchange involving ascending and descending pathways.

Interregional connectivity in the taste system is both convergent and functional (Di Lorenzo, P. M. and Monroe, S., 1997), just as it is in other systems (e.g., Kay, L. M. *et al.*, 1996; Alonso, J. M. *et al.*, 2001; Alloway, K. D. and Roy, S. A., 2002; Villalobos, M. E. *et al.*, 2005). For example, stimulation of GC, amygdala, or hypothalamus modifies single-neuron PbN responses to lingual application of the four basic tastes (Lundy, R. F. and Norgren, R., 2004; Li, C. S. *et al.*, 2005). A large percentage of the PbN taste neurons that receive any sort of feedback receive it from at least two forebrain sites, suggesting that multiple feedback loops may converge upon single brainstem taste neurons. Modulation of neurons in NTS, the very first central taste relay, via manipulation of both central amygdala and lateral hypothalamus, produces similar results (Cho, Y. K. *et al.*, 2003).

The specific function of feedback from forebrain to brainstem has been suggested to be a sharpening of receptive field (i.e., forebrain stimulation typically reduces the number of stimuli to which brainstem neurons respond, see Lundy, R. F. and Norgren, R., 2004). It might be tempting to conclude that the ultimately sharpened responses are evidence for a LL code, but in fact these data compellingly demonstrate that feed-forward projections are relatively broadly tuned (reflecting either broadly receptive transduction mechanisms or very early mixing of pathways), and that the apparent labeling of brainstem neurons represents an interactive network effect, in which the broad responses are tuned up by forebrain populations.

4.16.3 Functional Implications of Interactive Population Coding

The existence of interactive ensembles has strong implications for neural network function. Specifically, within- and between-region interactions affect neural activity by introducing structure – both spatial and temporal – into spontaneous and evoked neural activity. Through such imposition of structure, interneuronal interactions imbue neural activity with contextual and behavioral specificity. We will now discuss studies demonstrating that these phenomena are prominent in many neural systems and, to the extent that they have been studied, in the taste system as well.

Classical analysis of sensory neural coding relies on the tacit assumption that spontaneous, prestimulus neural activity is random and uncorrelated – noise with which stimulus-evoked signals must compete. In populations of interacting neural elements, however, spontaneous activity is far from random. Hallmark phenomena of interactive processing, such as pattern completion (Rumelhart, D. E. and McClelland, J. L., 1986), ensure that spontaneous activity can drive a network into preferred coherent states.

Little work on this topic has yet been done in the taste neuroaxis, but visual cortex has been shown to spontaneously attain spatially coherent global states (Arieli, A. *et al.*, 1996). Images of intrinsic V1 signals, keyed to the spontaneous firing of individual neurons, are nearly identical to the maps produced when those neurons' best stimuli are presented (Tsodyks, M. *et al.*, 1999; Kenet, T. *et al.*, 2003). Analogously, motor cortical networks have been suggested to attain states, referred to as preshapes, that predict population codes for particular movements well in advance of such movements (Bastian, A. *et al.*, 1998; 2003). Cortex can produce best-stimulus responses both in the presence and absence of these stimuli.

The complementary result has also been shown: cortex produces surprisingly variable responses to static stimuli. In fact, the specific spatial structure of cortical responses is less dependent on the exact physical stimulus than on the cognitive context, that is, the animal's interpretation of the stimulus at a particular moment. In the visual system, for example, attention modulates the magnitude of sensory activity, enhancing responses to the attended object and suppressing responses to unattended ones (e.g., Fischer, B. and Boch, R., 1985; Reynolds, J. H. and Chelazzi, L., 2004). In situations in which a stimulus is bistable (e.g., the figure-ground illusion), meanwhile, cortical neurons shift firing rates in direct relation to changes in perception, despite the fact that the actual sensory stimulus is static (Leopold, D. A. and Logothetis, N. K., 1996; Otterpohl, J. R. *et al.*, 2000). Cognitive states also influence the correlations between neurons responding to simultaneously presented stimuli, modifying spatial patterns of neural interactions in cortical regions (Hatsopoulos, N. G. *et al.*, 1998; 2003), and thereby tagging the ensembles recruited to code particular real-world objects; the visual responses to disconnected line segments that are perceived to be part of a single occluded rod fire in synchrony, for instance, while responses to identical line segments that do not present the percept of an occluded rod do not (Singer, W., 1993; see also Harris, K. D., 2005).

In the taste system, as well, attentional variables affect cortical coding: when a rat ceases to pay attention to taste stimuli, the responses of 40% of the neurons in taste cortex suddenly change their receptive fields (Fontanini, A. and Katz, D. B., 2006). These changes are not random, but rather represent an interpretable modulation of perceptual taste space – specifically, an increase in the salience of the palatability dimension. Changes in orofacial responses to the tastes (Grill, H. J. and Norgren, R., 1978; Berridge, K. C., 2000) confirm this interpretation (Fontanini, A. and Katz, D. B., 2006). Such placing of coherent responses into a meaningful context has been suggested to have its source in feedback from higher neural centers (Engel, A. K. *et al.*, 2001; Buffalo, E. A. *et al.*, 2005). In other words, inclusion of multiple brain regions in the coding population places the neural code into the motivational and cognitive context, such that sensory responses reflect the meaning of stimuli, and not simply their physical makeup.

While it has not yet been shown that taste networks organize into analogous spatial coherence (but see Yoshimura, H. *et al.*, 2004), it is becoming clear that neural responses to taste administration, like those to other stimuli (e.g., Golomb, D. *et al.*, 1994; Vaadia, E. *et al.*, 1995; MacLeod, K, and Laurent, G.,

1996; Seidemann, E. *et al.*, 1996; Compte, A. *et al.*, 2000; Kirkland, K. L. *et al.*, 2000; Bazhenov, M. *et al.*, 2001; Miller, P. *et al.*, 2005), do have temporal structure. Such temporal structure has been extensively described in both brainstem (Di Lorenzo, P. M. and Schwartzbaum, J. S., 1982; Erickson, R. P. *et al.*, 1994; Di Lorenzo, P. M. and Victor, J. D., 2003) and cortex (Katz, D. B. *et al.*, 2001; see also Tabuchi, E. *et al.*, 2002). The use of chronic recordings in active, tasting rats has allowed us to observe not only slower dynamics but also oscillations (Fontanini, A. and Katz, D. B., 2005) in taste cortex, a region rife with inhibitory cross-talk (Ogawa, H. *et al.*, 1998). These cortical dynamics have been directly linked to network functioning (Katz, D. B. *et al.*, 2002b), a finding that is consistent with paired-pulse studies showing that two identical inputs, separated by just long enough to allow the processing of information from the first input to begin, cause reliably distinct patterns of response in NTS neurons (Lemon, C. H. and Di Lorenzo, P. M., 2002; Di Lorenzo, P. M. *et al.*, 2003). Gustatory responses are clearly dynamic, as predicted by interactive population models (but not by either the LL or ANP model).

Analogous to the above findings on spatial coherence, perceptual relevance emerges in the temporal structure of cortical taste responses, much as it does in other systems (Sugase, Y. *et al.*, 1999; Friedrich, R. W. and Laurent, G., 2001). While data from primates have suggested that only prefrontal cortical activity is affected by important changes in state (i.e., satiety, Rolls, E. T. *et al.*, 1989), rat primary cortical responses clearly contain both sensory and cognitive components. These responses can be divided in three epochs, each of which reflects a particular stage of gustatory processing – somatosensation, chemosensation, and palatability (Katz, D. B. *et al.*, 2001). Attention-related changes in stimulus palatability are preferentially expressed in changes in late-epoch coding (Fontanini, A. and Katz, D. B., 2006), a fact that reinforces the conclusion that the three epochs represent genuine temporal coding, rather than trivially reflecting the outcome of processing.

The evolution of activity during these three epochs could be the result of interactions/reverberations between the gustatory system and other high-order areas known to code palatability, such as the amygdala (Nishijo, H. *et al.*, 1998). It has also recently been shown that even later aspects of cortical responses in rats change as a taste becomes familiar (Bahar, A. S. *et al.*, 2004). Thus, cognitive processes are apparent both in the spatial and temporal structures of sensory responses, as one expects when examining a system functioning via interacting populations of neurons.

4.16.4 The Purposes of Neural Interactions in Taste

It is clear that the behavior of neurons in primary GC must be interpreted in terms of context – both the physical context of the networks (of neurons and brain areas) into which the neuron is connected and the cognitive context of the task. Primary GC is not a passive receiver, but rather a dynamic processor of information constantly engaged in a behaviorally dependent interplay with other regions; taste neurons in cortex interact with populations of neurons throughout the taste neuroaxis in the process of stimulus coding. A fundamental issue that remains to be addressed, however, is the definition of the term coding. Put another way, the question is: what kind of processes are interacting populations of taste neurons involved in?

It has been suggested that neural ensembles may interact for the purposes of binding and/or effective signal transmission, collapsing neurons into a functional, synchronous ensemble in order that this ensemble may code a context-embedded percept. Intraregional connectivity is thought to allow flexible assembly of a broad range of possible ensembles, while longer-range connections modulate that assembly according to global (cognitive) states (Brosch, M. *et al.*, 1997). This hypothesis has been further extended to consider time-varying responses, wherein distinct subpopulations of neurons fire synchronously at different times during the response (Vaadia, E. *et al.*, 1995; Wehr, M. and Laurent, G., 1996).

Synchrony is potentially powerful, because synchronous input can be decoded by coincidence-detecting readout cells whose firing represents a synthetic and sparsened version of the input. Perhaps the most well-studied example comes from the insect olfactory system. Kenyon cells in the mushroom body of the locust and fly receive complex, time-varying inputs from populations of projecting neurons in the antenna lobe (Laurent, G. *et al.*, 2001; Lei, H. *et al.*, 2004; Wilson, R. I. *et al.*, 2004) and respond to this very complex input with relatively few – and very reliable – action potentials following the arrival of synchronous inputs (Perez-Orive, J. *et al.*, 2002). The attractiveness of the synchronization hypothesis lies in the ease with which it

fits into an overall vision of the brain as a coding–decoding device: every area codes the input according to some rules and feeds its output to a higher level area that synthesizes it into a sparser representation. In this framework the brain is treated as a linear hierarchical system, where the output of a level is sent to the next, and where the highest level decodes a fully contextualized percept.

The evidence collected thus far, however, does not provide strong support for such a model of gustatory function. As described above, information traversing the gustatory system is hierarchical only to a first approximation – for every ascending pathway between brainstem and forebrain there is matching feedback. In the context of this recurrent organization it is hard to identify the putative decoding zone in which convergence and sparsening might occur. Recordings confirm that the taste selectivity of neural responses does not increase by a great deal as one ascends through the system (Yamamoto, T. *et al.*, 1985; Ogawa, H. *et al.*, 1990; Nishijo, H. *et al.*, 1998), and there is not much evidence for synchrony or oscillations during gustatory processing. In fact, while oscillations occur prominently in gustatory cortex, they specifically occur when rats are not engaged in taste processing (Fontanini, A. and Katz, D. B., 2005). Significant CCGs between pairs of neurons separated by more than 100 μm, meanwhile, appear to reflect simultaneous changes of firing rate, rather than synchronous firing of action potentials (Katz, D. B. *et al.*, 2002b).

Alternative theories avoid the coding–decoding issue entirely, simply by suggesting that processing of a sensory stimulus can be accomplished in the absence of coding of the sensory stimulus. Several researchers and philosophers (see, for instance, Varela, F. J. *et al.*, 1991; van Gelder, T., 1992; Eggermont, J. J., 1998; Engel, A. K. *et al.*, 2001), including more than one chemosensory scientist (Freeman, W. J. and Skarda, C. A., 1994; Halpern, B. P., 2000), have suggested that coding and representation may be constructs with more relevance to computer function than to brain function. Such a notion is *prima facie* attractive, because all extant concepts of taste coding run into trouble over the fact of the wide range of response latencies produced in gustatory behavior – the fact that some taste responses can be produced as little as 200 ms following taste administration, while others take well over 1 s (Halpern, B. P., 2005). Furthermore, it makes sense from an ecological prospective, in that the function of gustation is to provide information for the crucial decision of ingesting or rejecting food (Gibson, J. J., 1966). Taste is therefore intimately linked to orofacial motor behavior – a linkage that is explicit in the fact that the brainstem taste relays are almost directly connected into orofacial motor-neuron pools (Travers, J. B. and Norgren, R., 1983; Travers, J. B. *et al.*, 2000). It may be reasonably argued that the job of populations of neurons in the gustatory neuroaxis is not to code tastes at all, but rather to transform taste input into motor output. In such a scheme, the spatial hierarchy in which each successive brain region contains more highly processed information is replaced by a temporal hierarchy in which successive time points contain more highly processed information.

This, then, would be the function of interacting neural populations – to transform input, through poststimulus time, into a form adequate for driving behavior. Stimulus-related input does not get represented by neural firing (or even by neural synchrony) for any particular finite time, but rather sets in motion a dynamic process of population interaction (Harris, K. D., 2005). This is consistent not only with the recent data on temporal coding and cross-correlations in taste (Katz, D. B. *et al.*, 2002a), but also with data from other systems showing: (1) action-oriented responses in early sensory relays (Kay, L. M. and Laurent, G., 1999, Shuler, M. G. and Bear, M. F., 2006); (2) emergence of perceptual information through time (Sugase, Y. *et al.*, 1999); and (3) intriguing new data suggesting that visual input merely perturbs active processes already underway in V1 (Fiser, J. *et al.*, 2004).

4.16.5 Conclusions

These are necessarily speculative musings, because the study of interactive population coding in taste cortex and the larger taste neuroaxis is still in its infancy. We do not write this review in an attempt to resolve the debate between LL and ANP theories – both theories have their adherents, both can be thought to receive support from data, and as of now neither can be disproved in the central nervous system. Our suggestion is that we will need a more explicit account of interacting populations, perhaps adapted from research in other systems, if we are to provide the necessary framework for a complete understanding of the functioning of taste networks. These interactions exist in the taste system, as they do in other, more extensively examined systems and in realistic neural models. The extant theory and data suggest that they may be central to sensory function – function that may conceivably have more to do with transforming input smoothly into motor output than with actual coding of the stimulus *per se*. It will likely be years before these issues can be completely resolved.

References

Adachi, M. 1991. The mechanism of taste quality discrimination in rat pontine parabrachial nucleus. Aichi Gakuin Daigaku Shigakkai Shi 29, 283–299.

Adachi, M., Ohshima, T., Yamada, S., and Satoh, T. 1989. Cross-correlation analysis of taste neuron pairs in rat solitary tract nucleus. J. Neurophysiol. 62, 501–509.

Alloway, K. D. and Roy, S. A. 2002. Conditional cross-correlation analysis of thalamocortical neurotransmission. Behav. Brain Res. 135, 191–196.

Alonso, J. M., Usrey, W. M., and Reid, R. C. 2001. Rules of connectivity between geniculate cells and simple cells in cat primary visual cortex. J. Neurosci. 21, 4002–4015.

Arieli, A., Sterkin, A., Grinvald, A., and Aertsen, A. 1996. Dynamics of ongoing activity: explanation of the large variability in evoked cortical responses. Science 273, 1868–1871.

Bahar, A. S., Dudai, Y., and Ahissar, E. 2004. Neural signature of taste familiarity in the gustatory cortex of the freely behaving rat. J. Neurophysiol. 92, 3298–3308.

Bastian, A., Riehle, A., Erlhagen, W., and Schoner, G. 1998. Prior information preshapes the population representation of movement direction in motor cortex. Neuroreport 9, 315–319.

Bastian, A., Schoner, G., and Riehle, A. 2003. Preshaping and continuous evolution of motor cortical representations during movement preparation. Eur. J. Neurosci. 18, 2047–2058.

Bazhenov, M., Stopfer, M., Rabinovich, M., Abarbanel, H. D., Sejnowski, T. J., and Laurent, G. 2001. Model of cellular and network mechanisms for odor-evoked temporal patterning in the locust antennal lobe. Neuron 30, 569–581.

Berridge, K. C. 2000. Measuring hedonic impact in animals and infants: microstructure of affective taste reactivity patterns. Neurosci. Biobehav. Rev. 24, 173–198.

Bretzner, F., Aitoubah, J., Shumikhina, S., Tan, Y. F., and Molotchnikoff, S. 2001. Modulation of the synchronization between cells in visual cortex by contextual targets. Eur. J. Neurosci. 14, 1539–1554.

Brody, C. D. 1999. Disambiguating different covariation types. Neural Comput. 11, 1527–1535.

Brosch, M., Bauer, R., and Eckhorn, R. 1997. Stimulus-dependent modulations of correlated high-frequency oscillations in cat visual cortex. Cereb. Cortex 7, 70–76.

Buffalo, E. A., Bertini, G., Ungerleider, L. G., and Desimone, R. 2005. Impaired filtering of distracter stimuli by TE neurons following V4 and TEO lesions in macaques. Cereb. Cortex 15, 141–151.

Cho, Y. K., Li, C. S., and Smith, D. V. 2003. Descending influences from the lateral hypothalamus and amygdala converge onto medullary taste neurons. Chem. Senses 28, 155–171.

Christensen, T. A., Lei, H., and Hildebrand, J. G. 2003. Coordination of central odor representations through transient, non-oscillatory synchronization of glomerular output neurons. Proc. Natl. Acad. Sci. U. S. A. 100, 11076–11081.

Compte, A., Brunel, N., Goldman-Rakic, P. S., and Wang, X. J. 2000. Synaptic mechanisms and network dynamics underlying spatial working memory in a cortical network model. Cereb. Cortex 10, 910–923.

DeCharms, R. C. and Merzenich, M. M. 1996. Primary cortical representation of sounds by the coordination of action-potential timing. Nature 381, 610–613.

Di Lorenzo, P. M. and Monroe, S. 1997. Transfer of information about taste from the nucleus of the solitary tract to the parabrachial nucleus of the pons. Brain Res. 763, 167–181.

Di Lorenzo, P. M. and Schwartzbaum, J. S. 1982. Coding of gustatory information in the pontine parabrachial nuclei of the rabbit: temporal patterns of neural response. Brain Res. 251, 245–257.

Di Lorenzo, P. M. and Victor, J. D. 2003. Taste response variability and temporal coding in the nucleus of the solitary tract of the rat. J. Neurophysiol. 90, 1418–1431.

Di Lorenzo, P. M., Lemon, C. H., and Reich, C. G. 2003. Dynamic coding of taste stimuli in the brainstem: effects of brief pulses of taste stimuli on subsequent taste responses. J. Neurosci. 23, 8893–8902.

Eggermont, J. J. 1994. Neural interaction in cat primary auditory cortex II. Effects of sound stimulation. J. Neurophysiol. 71, 246–270.

Eggermont, J. J. 1998. Is there a neural code? Neurosci. Biobehav. Rev. 22, 355–370.

Eggermont, J. J. 2000. Sound-induced synchronization of neural activity between and within three auditory cortical areas. J. Neurophysiol. 83, 2708–2722.

Engel, A. K., Fries, P., and Singer, W. 2001. Dynamic predictions: oscillations and synchrony in top-down processing. Nat. Rev. Neurosci. 2, 704–716.

Erickson, R. 1984. On the neural bases of behavior. Am. Sci. 72, 233–241.

Erickson, R. P., Di Lorenzo, P. M., and Woodbury, M. A. 1994. Classification of taste responses in brain stem: membership in fuzzy sets. J. Neurophysiol. 71, 2139–2150.

Fischer, B. and Boch, R. 1985. Peripheral attention versus central fixation: modulation of the visual activity of prelunate cortical cells of the rhesus monkey. Brain Res. 345, 111–123.

Fiser, J., Chiu, C., and Weliky, M. 2004. Small modulation of ongoing cortical dynamics by sensory input during natural vision. Nature 431, 573–578.

Fontanini, A. and Katz, D. B. 2005. 7 to 12 Hz activity in rat gustatory cortex reflects disengagement from a fluid self-administration task. J. Neurophysiol. 93, 2832–2840.

Fontanini, A. and Katz, D. B. 2006. State-dependent modulation of time-varying gustatory responses. J. Neurophysiol. 96, 3183–3193.

Freeman, W. J. and Skarda, C. A. 1994. Representations: Who Needs Them? In: Brain Organization and Memory: Cells, Systems, and Circuits (eds. J. L. McGaugh, N. Weinberger, and G. Lynch), pp. 375–380. Oxford University Press.

Friedrich, R. W. and Laurent, G. 2001. Dynamic optimization of odor representations by slow temporal patterning of mitral cell activity. Science 291, 889–894.

van Gelder, T. 1992. What might cognition be if not computation? J. Philo. 91, 345–381.

Gibson, J. J. 1966. The Senses Considered as Perceptual Systems. Houghton Mifflin.

Golomb, D., Wang, X. J., and Rinzel, J. 1994. Synchronization properties of spindle oscillations in a thalamic reticular nucleus model. J. Neurophysiol. 72, 1109–1126.

Grill, H. J. and Norgren, R. 1978. The taste reactivity test. I. Mimetic responses to gustatory stimuli in neurologically normal rats. Brain Res. 143, 263–279.

Halpern, B. P. 2000. Sensory coding, decoding, and representations. Unnecessary and troublesome constructs? Physiol. Behav. 69, 115–118.

Halpern, B. P. 2005. Temporal characteristics of human taste judgements as calibrations for gustatory event-related potentials and gustatory magnetoencephalographs. Chem. Senses 30(Suppl. 1), i228–i229.

Halsell, C. B. 1992. Organization of parabrachial nucleus efferents to the thalamus and amygdala in the golden hamster. J .Comp. Neurol. 317, 57–78.

Harris, K. D. 2005. Neural signatures of cell assembly organization. Nat. Rev. Neurosci. 6, 399–407.

Hatsopoulos, N. G., Ojakangas, C. L., Paninski, L., and Donoghue, J. P. 1998. Information about movement direction obtained from synchronous activity of motor cortical neurons. Proc. Natl. Acad. Sci. U. S. A. 95, 15706–15711.

Hatsopoulos, N. G., Paninski, L., and Donoghue, J. P. 2003. Sequential movement representations based on correlated neuronal activity. Exp. Brain. Res. 149, 478–486.

Huang, T., Yan, J., and Kang, Y. 2003. Role of the central amygdaloid nucleus in shaping the discharge of gustatory neurons in the rat parabrachial nucleus. Brain Res. Bull. 61, 443–452.

Katz, D. B., Nicolelis, M. A., and Simon, S. A. 2002a. Gustatory processing is dynamic and distributed. Curr. Opin. Neurobiol. 12(4), 448–454.

Katz, D. B., Simon, S. A., and Nicolelis, M. A. 2001. Dynamic and multimodal responses of gustatory cortical neurons in awake rats. J. Neurosci. 21, 4478–4489.

Katz, D. B., Simon, S. A., and Nicolelis, M. A. 2002b. Taste-specific neuronal ensembles in the gustatory cortex of awake rats. J. Neurosci. 22, 1850–1857.

Kay, L. M. and Laurent, G. 1999. Odor- and context-dependent modulation of mitral cell activity in behaving rats. Nat. Neurosci. 2, 1003–1009.

Kay, L. M., Lancaster, L. R., and Freeman, W. J. 1996. Reafference and attractors in the olfactory system during odor recognition. Int. J. Neural Syst. 7, 489–495.

Kenet, T., Bibitchkov, D., Tsodyks, M., Grinvald, A., and Arieli, A. 2003. Spontaneously emerging cortical representations of visual attributes. Nature 425, 954–956.

Kirkland, K. L., Sillito, A. M., Jones, H. E., West, D. C., and Gerstein, G. L. 2000. Oscillations and long-lasting correlations in a model of the lateral geniculate nucleus and visual cortex. J. Neurophysiol. 84, 1863–1868.

van der Kooy, D., Koda, L. Y., McGinty, J. F., Gerfen, C. R., and Bloom, F. E. 1984. The organization of projections from the cortex, amygdala, and hypothalamus to the nucleus of the solitary tract in rat. J. Comp. Neurol. 224, 1–24.

Lampl, I., Reichova, I., and Ferster, D. 1999. Synchronous membrane potential fluctuations in neurons of the cat visual cortex. Neuron 22, 361–374.

Laurent, G., Stopfer, M., Friedrich, R. W., Rabinovich, M. I., Volkovskii, A., and Abarbanel, H. D. 2001. Odor encoding as an active, dynamical process: experiments, computation, and theory. Annu. Rev. Neurosci. 24, 263–297.

Lei, H., Christensen, T. A., and Hildebrand, J. G. 2004. Spatial and temporal organization of ensemble representations for different odor classes in the moth antennal lobe. J. Neurosci. 24, 11108–11119.

Lemon, C. H. and Di Lorenzo, P. M. 2002. Effects of electrical stimulation of the chorda tympani nerve on taste responses in the nucleus of the solitary tract. J. Neurophysiol. 88, 2477–2489.

Leopold, D. A. and Logothetis, N. K. 1996. Activity changes in early visual cortex reflect monkeys' percepts during binocular rivalry. Nature 379, 549–553.

Li, C. S., Cho, Y. K., and Smith, D. V. 2005. Modulation of parabrachial taste neurons by electrical and chemical stimulation of the lateral hypothalamus and amygdala. J. Neurophysiol. 93, 1183–1196.

Lumer, E. D., Edelman, G. M., and Tononi, G. 1997. Neural dynamics in a model of the thalamocortical system. I. Layers, loops and the emergence of fast synchronous rhythms. Cereb. Cortex 7, 207–227.

Lundy, R. F., Jr. and Norgren, R. 2004. Activity in the hypothalamus, amygdala, and cortex generates bilateral and convergent modulation of pontine gustatory neurons. J. Neurophysiol. 91, 1143–1157.

MacLeod, K. and Laurent, G. 1996. Distinct mechanisms for synchronization and temporal patterning of odor-encoding neural assemblies. Science 274, 976–979.

Masuda, N. and Aihara, K. 2003. Duality of rate coding and temporal coding in multilayered feedforward networks. Neural Comput. 15, 103–125.

McClelland, J. L. and Rumelhart, D. E. 1981. An interactive activation model of context effects in letter perception. Part 1. An account of basic findings. Psychol. Rev. 88, 375–407.

McDonald, A. J. and Jackson, T. R. 1987. Amygdaloid connections with posterior insular and temporal cortical areas in the rat. J. Comp. Neurol. 262, 59–77.

Miller, P., Brody, C. D., Romo, R., and Wang, X. J. 2005. A recurrent network model of somatosensory parametric working memory in the prefrontal cortex. Cereb. Cortex 15, 679.

Nagai, T. 2000. Artificial neural networks estimate the contribution of taste neurons to coding. Physiol. Behav. 69, 107–113.

Nagai, T., Yamamoto, T., Katayama, H., Adachi, M., and Aihara, K. 1992. A novel method to analyse response patterns of taste neurons by artificial neural networks. Neuroreport 3, 745–748.

Nakamura, T. and Ogawa, H. 1997. Neural interaction between cortical taste neurons in rats: a cross-correlation analysis. Chem. Senses 22, 517–528.

Nakashima, M., Uemura, M., Yasui, K., Ozaki, H. S., Tabata, S., and Taen, A. 2000. An anterograde and retrograde tract-tracing study on the projections from the thalamic gustatory area in the rat: distribution of neurons projecting to the insular cortex and amygdaloid complex. Neurosci. Res. 36, 297–309.

Nishijo, H., Uwano, T., Tamura, R., and Ono, T. 1998. Gustatory and multimodal neuronal responses in the amygdala during licking and discrimination of sensory stimuli in awake rats. J. Neurophysiol. 79, 21–36.

Ogawa, H., Hasegawa, K., Otawa, S., and Ikeda, I. 1998. GABAergic inhibition and modifications of taste responses in the cortical taste area in rats. Neurosci. Res. 32, 85–95.

Ogawa, H., Ito, S., Murayama, N., and Hasegawa, K. 1990. Taste area in granular and dysgranular insular cortices in the rat identified by stimulation of the entire oral cavity. Neurosci. Res. 9, 196–201.

Otterpohl, J. R., Haynes, J. D., Emmert-Streib, F., Vetter, G., and Pawelzik, K. 2000. Extracting the dynamics of perceptual switching from 'noisy' behaviour: an application of hidden Markov modelling to pecking data from pigeons. J. Physiol. Paris 94, 555–567.

Perez-Orive, J., Mazor, O., Turner, G. C., Cassenaer, S., Wilson, R. I., and Laurent, G. 2002. Oscillations and sparsening of odor representations in the mushroom body. Science 297, 359–365.

Rabinovich, M. I., Huerta, R., Volkovskii, A., Abarbanel, H. D., Stopfer, M., and Laurent, G. 2000. Dynamical coding of sensory information with competitive networks. J. Physiol. Paris 94, 465–471.

Reynolds, J. H. and Chelazzi, L. 2004. Attentional modulation of visual processing. Annu. Rev. Neurosci. 27, 611–647.

Rolls, E. T., Sienkiewicz, Z. J., and Yaxley, S. 1989. Hunger modulates the responses to gustatory stimuli of single neurons in the caudolateral orbitofrontal cortex of the macaque monkey. Eur. J. Neurosci. 1, 53–60.

Roy, S. and Alloway, K. D. 1999. Synchronization of local neural networks in the somatosensory cortex: a comparison of stationary and moving stimuli. J. Neurophysiol. 81, 999–1013.

Rumelhart, D. E. and McClelland, J. L. 1986. Parallel Distributed Processing: Explorations in the Microstructure of Cognition, MIT Press.

Scott, K. 2004. The sweet and the bitter of mammalian taste. Curr. Opin. Neurobiol. 14, 423–427.

Seidemann, E., Meilijson, I., Abeles, M., Bergman, H., and Vaadia, E. 1996. Simultaneously recorded single units in the frontal cortex go through sequences of discrete and stable states in monkeys performing a delayed localization task. J. Neurosci. 16, 752–768.

Shuler, M. G. and Bear, M. F. 2006. Reward timing in the Primary Visual Cortex. Science 311, 1606–1609.

Singer, W. 1993. Synchronization of cortical activity and its putative role in information processing and learning. Annu. Rev. Physiol. 55, 349–374.

Smith, D. V. and St. John, S. J. 1999. Neural coding of gustatory information. Curr. Opin. Neurobiol. 9, 427–435.

Sporns, O., Tononi, G., and Edelman, G. M. 2000. Connectivity and complexity: the relationship between neuroanatomy and brain dynamics. Neural Netw. 13, 909–922.

Sugase, Y., Yamane, S., Ueno, S., and Kawano, K. 1999. Global and fine information coded by single neurons in the temporal visual cortex. Nature 400, 869–873.

Tabuchi, E., Yokawa, T., Mallick, H., Inubushi, T., Kondoh, T., Ono, T., and Torii, K. 2002. Spatio-temporal dynamics of brain activated regions during drinking behavior in rats. Brain Res. 951, 270–279.

Takeuchi, Y., McLean, J. H., and Hopkins, D. A. 1982. Reciprocal connections between the amygdala and parabrachial nuclei: ultrastructural demonstration by degeneration and axonal transport of horseradish peroxidase in the cat. Brain Res. 239, 583–588.

Travers, J. B. and Norgren, R. 1983. Afferent projections to the oral motor nuclei in the rat. J. Comp. Neurol. 220, 280–298.

Travers, J. B., DiNardo, L. A., and Karimnamazi, H. 2000. Medullary reticular formation activity during ingestion and rejection in the awake rat. Exp. Brain Res. 130, 78–92.

Tsodyks, M., Kenet, T., Grinvald, A., and Arieli, A. 1999. Linking spontaneous activity of single cortical neurons and the underlying functional architecture. Science 286, 1943–1946.

Vaadia, E., Haalman, I., Abeles, M., Bergman, H., Prut, Y., Slovin, H., and Aertsen, A. 1995. Dynamics of neuronal interactions in monkey cortex in relation to behavioural events. Nature 373, 515–518.

Varela, F. J., Thompson, E. T., and Rosch, E. 1991. The Embodied Mind: Cognitive Science and Human Experience. MIT Press.

Villalobos, M. E., Mizuno, A., Dahl, B. C., Kemmotsu, N., and Muller, R. A. 2005. Reduced functional connectivity between V1 and inferior frontal cortex associated with visuomotor performance in autism. Neuroimage 25, 916–925.

van Vreeswijk, C. and Sompolinsky, H. 1998. Chaotic balanced state in a model of cortical circuits. Neural Comput. 10, 1321–1371.

Wehr, M. and Laurent, G. 1996. Odour encoding by temporal sequences of firing in oscillating neural assemblies (w/ commentary). Nature 384, 162–166.

Wilson, R. I., Turner, G. C., and Laurent, G. 2004. Transformation of olfactory representations in the *Drosophila* antennal lobe. Science 303, 366–370.

Yamada, S., Ohshima, T., Oda, H., Adachi, M., and Satoh, T. 1990. Synchronized discharge of taste neurons recorded simultaneously in rat parabrachial nucleus. J. Neurophysiol. 63, 294–302.

Yamamoto, T., Yuyama, N., Kato, T., and Kawamura, Y. 1985. Gustatory responses of cortical neurons in rats. II. Information processing of taste quality. J. Neurophysiol. 53, 1356–1369.

Yokota, T. and Satoh, T. 2001. Three-dimensional estimation of the distribution and size of putative functional units in rat gustatory cortex as assessed from the inter-neuronal distance between two neurons with correlative activity. Brain Res. Bull. 54, 575–584.

Yokota, T., Eguchi, K., and Satoh, T. 1996. Correlated discharges of two neurons in rat gustatory cortex during gustatory stimulation. Neurosci. Lett. 209, 204–206.

Yoshimura, Y., Dantzker, J. L., and Callaway, E. M. 2005. Excitatory cortical neurons form fine-scale functional networks. Nature 433, 868–873.

Yoshimura, H., Sugai, T., Fukuda, M., Segami, N., and Onoda, N. 2004. Cortical spatial aspects of optical intrinsic signals in response to sucrose and NaCl stimuli. Neuroreport 15, 17–20.

4.17 A Perspective on Chemosensory Quality Coding

M E Frank, University of Connecticut Health Center, Farmington, CT, USA

Glossary

Dedicated sensory neurons They are activated by single or multiple chemosensory receptors to directly and separately code each chemosensory quality.

Discrete stimuli They are separate entities that may share parts and sizes. Chemosensory stimuli are examples; each chemosensory receptor has a discrete molecular domain.

Dynamic chemosensory coding It promotes brief identification of the strong and novel in a complex chemical mixture by combining mixture suppression and rapid sensory adaptation.

Hedonic valence It is positive when a chemosensory stimulus is perceived as good; negative when a chemosensory stimulus is perceived as bad.

Odor notes They are distinct perceptual olfactory qualities; there are hundreds, grouped as object-like (e.g., flower, fruit, animal) or process-like (e.g., spoiled, excreted, burned).

Perceptual continua They are constructed for color and pitch by red-green and blue-yellow opponent cells for each retinal position and multi-cell frequency maps, respectively.

Quality constancy This occurs if a chemosensory quality remains the same in different contexts, allowing chemosensory stimulus recognition.

Quality tuning They may be broad or fine, depending on how stimuli are biologically detected and species-selective perceptual requirements. An example of broad tuning is the tripartite cellular code for color carried by three pigments, each found in a dedicated retinal cone. An example of fine receptor tuning is the best-frequency cellular code for pitch carried by hundreds of cochlear hair cells.

Somatotopic neural maps These represent distinct touch, pain, warm, and cold senses side by side for each bodily place: for example, on legs, arms, head, nose, mouth, and tongue.

Sound space This is coded in interaural time differences but visual space is coded by positions in images projected on the retina.

Spectral stimuli They are continuous and uni-dimensional; examples are the visible spectrum of electromagnetic waves and the audible spectrum of airborne pressure waves.

Taste qualities They are distinct perceptual gustatory qualities; there are few; primarily sweet, salty, sour, and bitter.

4.17.1 The Sweet–Salty–Sour–Bitter Decision

In humans, the essence of most tastes is described by a handful of distinct qualities. There are hundreds of odor qualities. The logic of chemosensory coding must take into account the hundreds of different olfactory receptors (ORs) (Axel, R., 2005; Buck, L. B., 2005) compared to the many fewer receptors implicated in taste reception (Chandrashekar, J. *et al.*, 2006), both of which are now being characterized for a multiple mammalian species.

For tastes, besides quality, that is, the sweet–salty–sour–bitter decision, there is a good–bad decision (Nowlis, G. H. and Frank, M. E., 1981), as there is for smells (Anderson, A. K. *et al.*, 2003). The gustatory and olfactory systems code hedonic attributes that are associated with the water-soluble and airborne chemicals that they, respectively, detect. The good–bad decision about tastes and smells can be changed with experience (Cain, W. S. and Johnson, F., 1978; Palmerino, C. C. *et al.*, 1980). In contrast, the sweet–salty–sour–bitter and odor note decision, although it may be enhanced or dulled via intensity shifts (Contreras, R. J. and Lundy, R. F. 2000), appears unchanged and recoverable. The neural basis for sweet–salty–sour–bitter decisions, like odor-quality decisions, persists intact throughout the brain. This is consistent with taste-quality (Frank, M. E. *et al.*, 2003) and odor-quality (Goyert, H. F. *et al.*, 2007) constancy. The constancies are necessary and sufficient for chemosensory stimulus recognition, and instruct us on the logic of chemosensory coding in the central nervous system (CNS).

4.17.2 Stimulus and Perceptual Domains

It would simplify the study of sensory systems if all systems detected stimuli and coded perceptions in one way (Erickson, R. P., 1985). However, by their nature, different kinds of stimuli do not fall along comparable stimulus dimensions (Frank, M. E. and Hettinger, T. P., 2005). It may be convenient to use well-studied senses such as visual color or auditory pitch as model systems, but is it productive? Color and pitch are spectral senses in which stimuli fall on convenient physical continua that can be sampled with a frequency sweep. Sampling chemosensory stimuli, in contrast, is discrete; there is no single stimulus dimension to sweep across.

4.17.2.1 Nonchemosensory

4.17.2.1.1 Pitch
Pitches, like tastes and smells, neither define external space, like color, nor identify stimulation sites, like the somatotopic body senses. The stimulus domain of pitch is temporal, a spectrum of molecular pressure waves. In the cochlea, these waves are transformed to a cellular continuum. The waves localize according to frequency from the apex to the base of the basilar membrane (Békésy, G. von, 1970). Pitch is detected mechanically by multiple individual receptor cells (inner hair cells) located at multiple points along the vibrating membrane. The activated cells create a neuronal continuum in cranial nerve (CN) VIII that is transmitted to the neurons in the brainstem cochlear nucleus. CN VIII fibers mostly connect one to one to inner hair cells, a connectivity that sets the fibers' quite specific best-frequency tuning (Goldstein, E. B., 2007). This is fine tuning.

Thus, a temporal stimulus domain is translated into an intercellular domain to code pitch in the auditory nerve. In the CNS, frequency is mapped multiple times from cochlear nucleus to auditory cortex, and, as a consequence, pitch information is present in the form of maps everywhere in the auditory CNS. Taste quality information is also found throughout the gustatory CNS but does not appear to be in the form of sweet–salty–sour–bitter maps; rather, tastes and other oral senses originating from the same regions may be aligned (Chapter 4.15; McPheeters, M. *et al.*, 1990; Lundy, R. F. and Norgren, R., 2004; Kaas, J. H., 2005). Odor-quality signals detected by the subsets of olfactory sensory neurons (OSNs), each exclusively expressing a single OR variant, converge onto a few particular glomeruli in the olfactory bulb; but this stereotyped OR-based representation is neither somatotopic nor recapitulated elsewhere in the brain (Buck, L. B., 2005).

4.17.2.1.2 Color
The most popular model used to instruct studies of chemosensory systems is color vision (Erickson, R. P., 1985; Shepherd, G. M., 2005). Colors are parts of every discrete spot of every image in the visual field. In that sense, color, unlike pitch, is a spectral percept distributed in external space. Color stimuli are electromagnetic waves within a continuous visible spectrum and spectral sensitivity

is species-dependent (Glosmann, M. and Anhelt, P. K., 2002). The continuum of visible frequencies is partitioned by a few chemical pigment chromophores. In humans each chromophore is located exclusively in one of the three subsets of cone receptor cells in the retina. The three human cones contain all our color information. Each cone pigment absorbs light across the entire visible spectrum, with absorbance peaking at different frequencies. This is a broad tuning, characteristic of pigment absorbance.

Thus, a visible spectral wavelength continuum is reduced to a tripartite cellular code present in separate cells: so-called red (L), green (M), and blue (S) cones. Within the retina, the three cone spectra are combined to yield blue-yellow (L + M) and red-green opponent retinal ganglion cells associated with every spot on the retina, and these two opponents are represented in neurons distributed throughout the visual system from thalamus to visual cortex (Goldstein, E. B., 2007). Exactly how the continuity of colors that we perceive localized to each spot in our visual experience is reconstructed from absorbance differences of several pigments by the brain is unknown (Lennie, P., 2000). In the gustatory system there are peripheral and central neurons that respond antagonistically to good–bad stimulus pairs. (Chapter 4.15; McPheeters, M. et al., 1990; Formaker, B. K. et al., 1997).

4.17.2.2 Chemosensory

In contrast to the spectral senses of color and pitch, chemosensory stimulus detection in mammals involves discrete stimulus–receptor biochemical interactions, reactions with specified chemosensory molecular ranges (Bargman, C. L., 2006). Not needing to first convert spectral information to chemical or mechanical information in order to activate cells, the chemical senses coding logic can be more direct and represents phyletically early sensing systems. Candidate G protein-coupled receptors (GPCRs) have been discovered for (1) sweet taste; (2) mostly unidentified, species-specific, bitter tastes (Chandrashekar, J., et al., 2006); and (3) multiple, mostly unidentified, species-specific odors (Axel, R., 2005; Buck, L. B., 2005). In nearly all the cases, the environmental ligands that match the receptors are unknown.

4.17.2.2.1 Taste

The perceptual simplicity of taste has facilitated the identification of a single heterodimeric GPCR taste

receptor (TR) for sweet compounds (Tas1r2–Tas1r3), which has a molecular range coincident with compounds that taste sweet to humans (Li, X. et al., 2002). The molecular range is species-selective. Tas1r2 is not expressed in cats (Li, X. et al., 2005) and, consequently, cats do not detect sweet stimuli. Just as each OR is expressed in a dedicated set of OSNs, the single sweet TR is found in a subset of taste receptor cells (TRCs) likely dedicated for signaling sweet stimuli (Chandrashekar, J. et al., 2006).

An alternate arrangement has been implicated for bitter taste. Multiple candidate bitter receptors have been discovered, but all 30 T2Rs may be expressed in each cell of a subset of TRCs. Similar arrangements have been seen for nematode chemosensory receptors (Bargman, C. L., 2006). Each T2R receptor of a given species would detect distinct dangerous chemicals, possibly produced by other competing species. (Hettinger, T. P. et al., 2007). In contrast to sweet, no single TR could service the chemical diversity of bitters with appropriate specificity. However, this tentative bitter coding logic fails to account for the fact that all bitter stimuli do not necessarily elicit a unitary sensation. Rodent chorda tympani and glossopharyngeal nerves, both of which may innervate TRCs containing T2Rs, carry distinct information about aversive stimuli to the brain (St. John, S. J. and Spector, A. C., 1998; King, C. T. et al., 2000) and code for several distinct aversive perceptions (Frank, M. E. et al., 2004; Hettinger, T. P. et al., 2007) that have perhaps been reduced to a single bitter in humans.

4.17.2.2.2 Smell

In comparison to TRCs, identification of the function of particular dedicated OSNs has been a struggle, likely for the same reason that odor qualities have not been satisfactorily classified (Cain, W. S., 1978): there are so many odors and many may be sensed only by a few related species (Niimura, Y. and Nei, M., 2006). Just like tastes, odors are not distinguished merely as feature detectors of stimulus functional groups, shapes, or sizes. Rather, TRs and ORs likely evolved to make distinctions among particular whole molecules important for survival, some sharing functional groups, and with as much specificity as required for recognition (Bargman, C. L., 2006; Mori, T. et al., 2006; Goyert, H. F. et al., 2007).

Thus, like the body senses (McBurney, D. H. and Gent, J. F., 1979), there is a distinctly possible direct coupling of TR or OR with a discrete chemosensory quality, taste, or odor. Deconstruction of stimulus dimensions and their reconstruction into perceptual dimensions by the brain (Buck, L. B., 2005) may be appropriate for continuous spectral senses partitioned by a few receptor cell types like color. However, deconstruction and subsequent reconstruction (Turin, L., 2006) may be unnecessary for multiple discrete chemosensory domains.

4.17.3 Processing of Chemosensory Quality Signals

Although processing can potentially occur among the TRCs within taste buds (Kaya, N. *et al.*, 2004; Roper, S. D., 2006), recent findings strongly suggest that perceptual taste categories are isomorphic with TR specificity (Chandrashekar, J. *et al.*, 2006). This is consistent with modular specificities defined by single taste-fiber neural responses transmitted to the CNS (Frank, M. E., 2000; Frank, M. E. *et al.*, 2005).

4.17.3.1   Mixtures

Sweet–salty–sour–bitter is a set of distinct taste perceptions that is not transformed in stimulus mixtures (Frank, M. E. *et al.*, 2003). Figure 1 demonstrates this with the specificity of intake suppression driven by conditioned taste aversions, which do not alter the sweet–salty–sour–bitter decision (Hettinger, T. P. *et al.*, 2007). The taste of NaCl is recognized by hamsters in the binary mixtures of NaCl and sucrose, or NaCl and quinine, or the ternary mixture of NaCl and sucrose and quinine. The mixture percepts contain salt-like components. There are specific adjustments in intensity, however. One example is the suppression of quinine recognition in the presence of NaCl (Figure 1); but it can be recovered by an increase in quinine concentration (Nowlis, G. H. and Frank, M. E., 1981). There is also a mixture suppression of intensity in ternary mixtures (e.g., sucrose results in Figure 1). Mixture suppression, seen in humans (Bartoshuk, L. M., 1975), focuses mixture perception on the stronger mixture components.

In taste, mixture processing begins in the taste bud (Formaker, B. K. *et al.*, 1997), perhaps between the multiple TRCs containing cells expressing receptors for distinct qualities, and may continue within the brainstem nucleus of the solitary tract, which, like the olfactory bulb, contains ample inhibitory interneurons within its circuitry (Chapter 4.15; Bradley, R. M., 2006). In chemosensory mixtures, stronger stimuli suppress the weaker one more than the weaker

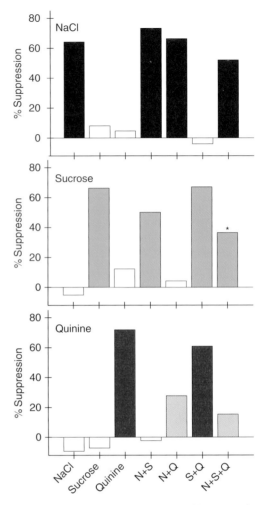

Figure 1   Recognition of components of binary and ternary taste mixtures by hamsters. The percentage suppression of intake measures stimulus recognition. NaCl (100 mM) was recognized (blue bars) in mixtures containing NaCl (N). Sucrose (100 mM) was recognized (green bars) in mixtures containing sucrose (S). Quinine (1 mM, Q) was recognized (red bars) in sucrose–quinine mixtures but not in NaCl–quinine mixtures (pink bars). *Sucrose recognition was weaker for the ternary S+N+Q mixture than for sucrose alone. Data are from Frank, M. E., Formaker, B. K., and Hettinger, T. P. 2003. Taste responses to mixtures: analytic processing of taste quality. Behav. Neurosci. 117, 228–235.

suppress the stronger (Laing, D. G. *et al.*, 1984). In this context, there is also an important temporal process to consider, rapid sensory adaptation.

4.17.3.2   Chemosensory Quality Constancy

Quality tuning of brainstem and forebrain taste neurons appear, on average, to broaden when compared to tuning of receptors or taste nerve fibers (Chapter 4.15). Explanations of how tuning profiles change at different levels of the nervous system should lead to a greater understanding of the roles of identified CNS neurons with distinct inputs and projections (Bradley, R. M., 2006). Likely some CNS transformations serve to stabilize a compound's taste or odor quality in a complex labile chemical environment. Stimuli that we usually experience are not pure or equally strong but are combinations of multiple compounds that wax and wane in time. In both taste and smell, identification of mixture components falls off rapidly with decreases in intensity and with increases in the number of components (Laing, D. H. *et al.*, 1984; Livermore, A. and Laing, D. H. 1998; Laing, D. H. *et al.*, 2002). In odor mixtures, rapid adaptation and mixture suppression determine which characteristic odors will be identified at a given moment (Goyert, H. F. *et al.*, 2007). Unlike taste for which taste buds are potential sites of TRC interaction, OSN interactions are initiated in the complex layered olfactory bulb via well-defined inhibitory circuitry (Yokoi, M. *et al.*, 1995) before they are transmitted to higher central levels.

Figure 2 shows correct identification of four characteristic odors by people. Each odor is recognized more often if other mixture components are sniffed for a few seconds right before (solid color bars) and less often if the test component itself is sniffed right before (diagonal striped bars). Instead of cross-adapting, the other odors release mixture suppression among the signals for the four compounds. For example, the odor of vanillin, labeled vanilla, is recognized more frequently in two-, three-, and four-component mixtures when the other mixture components, isopropyl alcohol, menthol (labeled mint), and phenethyl alcohol (labeled rose), are sniffed first. Vanillin recognition is poorer after it is sampled. The dynamic coding mechanisms of mixture suppression and rapid adaptation promote recognition of characteristic odors of newly introduced and stronger chemosensory stimuli. Chemosensory circuits may have evolved to focus our attention on the recognition of chemicals that we need to know are there in the environment or the mouth: serious new opportunities for, or dangers to, our survival.

4.17.4   Conclusion

Stimuli that we usually encounter are not pure chemicals but combinations of varying amounts of multiple compounds that come and go. In such shifting contexts, a few taste or odor qualities are identified at a time. This dynamic chemosensory coding (Goyert, H. F. *et al.*, 2007) warrants further study in peripheral taste neurons, which transmit signals to the CNS after they are processed by taste buds, and in mitral and tufted cells, which transmit the processed signals from the olfactory bulb to the olfactory cortex. The taste or odor perceived is typically characteristic of the pure component, a chemosensory quality constancy: the objective of quality coding. The Pied Beauty described in G. M. Hopkins' poem (Nowlis, G. H. and Frank, M. E., 1981) for chemosensory perceptions may not be appreciated all at once, except in our memories.

Acknowledgment

This cameo is supported by NIH grant DC004099.

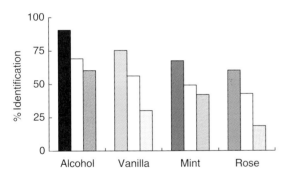

Figure 2 Identification of components in two-, three-, and four-component mixtures. The stimuli (labels) were 1 M isopropyl alcohol (alcohol), 5 mM vanillin (vanilla), 1 mM *l*-menthol (mint), and 0.5 mM phenethyl alcohol (rose). Correct component identification was greatest following sniffing other mixture components (solid color bars) for a few seconds and poorest following sniffing stimuli containing the indicated component (striped color bars). White bars are controls. Data are from Goyert, H. F., Frank, M. E., Gent, J. F., and Hettinger, T. P. 2007. Characteristic component odors emerge from mixtures after selective adaptation. Brain Res. Bull. 72, 1–9.

References

Anderson, A. K., Christoff, K., Stappen, I., Panitz, D., Ghahremani, D. G., Glover, G., Gabrieli, J. D., and Sobel, N. 2003. Dissociated neural representations of intensity and valence in human olfaction. Nature Neurosci. 6, 196–202.

Axel, R. 2005. Scents and sensibility: a molecular logic of olfactory perception (Nobel lecture). Angew. Chem. Int. Ed. 44, 6111–6127.

Bargman, C. I. 2006. Comparative chemosensation from receptors to ecology. Nature 444, 295–301.

Bartoshuk, L. M. 1975. Taste mixtures: is mixture suppression related to compression? Physiol. Behav. 14, 643–649.

Békésy, G. von. 1970. Travelling waves as frequency analysers in the cochlea. Nature 225, 1207–1209.

Bradley, R. M. 2006. rNST Circuits. In: The Role of the Nucleus of the Solitary Tract in Gustatory Processing (ed. R. M. Bradley), Chapter 7, pp. 137–151. CRC Press.

Buck, L. B. 2005. Unraveling the sense of smell (Nobel lecture). Angew. Chem. Int. Ed. 44, 6128–6140.

Cain, W. S. 1978. History of Research on Smell. In: Handbook of Perception, Vol. VIA, Tasting and Smelling (eds. E. C. Carterette and M. P. Friedman), pp. 197–229. Academic Press.

Cain, W. S. and Johnson, F., Jr. 1978. Lability of odor pleasantness: influence of mere exposure. Perception 7, 459–465.

Chandrashekar, J., Hoon, M. A., Ryba, N. J., and Zuker, C. S. 2006. The receptors and cells for mammalian taste. Nature 444, 288–294.

Contreras, R. J. and Lundy, R. F., Jr. 2000. Gustatory neuron types in the periphery: a functional perspective. Physiol. Behav. 69, 41–52.

Erickson, R. P. 1985. Definitions: A Matter of Taste. In: Taste, Olfaction, and the Central Nervous System (ed. D. W. Pfaff), pp. 129–150. Rockefeller University Press.

Formaker, B. K., MacKinnon, B. I., Hettinger, T. P., and Frank, M. E. 1997. Opponent effects of quinine and sucrose on single fiber taste responses of the chorda tympani nerve. Brain Res. 772, 239–242.

Frank, M. E. 2000. Neuron types, receptors, behavior, and taste quality. Physiol. Behav. 69, 53–62.

Frank, M. E. and Hettinger, T. P. 2005. What the tongue tells the brain about taste. Chem. Senses 30(Suppl. 1), i68–i69.

Frank, M. E., Bouverat, B. P., MacKinnon, B. I., and Hettinger, T. P. 2004. The distinctiveness of ionic and nonionic bitter stimuli. Physiol. Behav. 80, 421–431.

Frank, M. E., Formaker, B. K., and Hettinger, T. P. 2005. Peripheral gustatory processing of sweet stimuli by golden hamsters. Brain Res. Bull. 66, 70–84.

Frank, M. E., Formaker, B. K., and Hettinger, T. P. 2003. Taste responses to mixtures: analytic processing of taste quality. Behav. Neurosci. 117, 228–235.

Glosmann, M. and Anhelt, P. K. 2002. A mouse-like retinal cone in the Syrian hamster: S opsin coexpressed with M opsin in a common cone photoreceptor. Brain Res. 929, 139–146.

Goldstein, E. B. 2007. Sensation and Perception, 7th edn. Thomson Wadsworth.

Goyert, H. F., Frank, M. E., Gent, J. F., and Hettinger, T. P. 2007. Characteristic component odors emerge from mixtures after selective adaptation. Brain Res. Bull. 72, 1–9.

Hettinger, T. P., Formaker, B. K., and Frank, M. E. 2007. Cycloheximide: no ordinary bitter stimulus. Behav. Brain Res. 180, 4–17.

Kaas, J. H. 2005. The future or mapping sensory cortex in primates: three of the remaining issues. Philos. Trans. R. Soc. Lond. B 360, 653–664.

Kaya, N., Shen, T., Lu, S. G., Zhao, F. L., and Herness, S. 2004. A paracrine signaling role for serotonin in rat taste buds: expression and localization of serotonin receptor subtypes. Am. J. Physiol. Regul. Integr. Comp. Physiol. 286, R649–R658.

King, C. T., Garcea, M., and Spector, A. C. 2000. Glossopharyngeal nerve regeneration is essential for the complete recovery of quinine-stimulated oromotor rejection behaviors and central patterns of neuronal activity in the nucleus of the solitary tract in the rat. J. Neurosci. 20, 8426–8434.

Laing, D. G., Link, C., Jinks, A. L., and Hutchinson, I. 2002. The limited capacity of humans to identify the components of taste mixtures and taste–odour mixtures. Perception 31, 617–635.

Laing, D. G., Panhuber, H., Willcox, M. E., and Pittman, E. A. 1984. Quality and intensity of binary odor mixtures. Physiol. Behav. 33, 309–319.

Lennie, P. 2000. Color Vision. In: Principles of Neural Science, 4th edn. (eds. E. R. Kandel, J. H. Schwartz, and T. M. Jessell), pp. 572–589. McGraw-Hill.

Li, X., Li, W., Wang, H., Cao, J., Maehashi, K., Huang, L., Bachmanov, A. A., Reed, D. R., Legrand-Defretin, V., Beauchamp, G. K., and Brand, J. G. 2005. Pseudogenization of a sweet-receptor gene accounts for cats' indifference toward sugar. PloS Genet. 1, 27–35.

Li, X., Staszewski, L., Xu, H., Durick, K., Zoller, M., and Adler, E. 2002. Human receptors for sweet and umami taste. Proc. Natl. Acad. Sci. U. S. A. 99, 4692–4696.

Livermore, A. and Laing, D. G. 1998. The influence of odor type on the discrimination and identification of odorants in multicomponent odor mixtures. Physiol. Behav. 65, 311–320.

Lundy, R. F., Jr. and Norgren, R. 2004. Gustatory System. In: The Rat Nervous System (ed. G. Paxinos), pp. 891–921. Elsevier Academic Press.

McBurney, D. H. and Gent, J. F. 1979. On the nature of taste qualities. Psychol. Bull. 86, 151–167.

McPheeters, M., Hettinger, T. P., Nuding, S. C., Savoy, L. D., Whitehead, M. C., and Frank, M. E. 1990. Taste-responsive neurons and their locations in the solitary nucleus of the hamster. Neuroscience 34, 745–758.

Mori, K., Takahashi, Y. K., Igarashi, K. M., and Yamaguchi, M. 2006. Maps of odorant molecular features in the mammalian olfactory bulb. Physiol. Rev. 86, 409–433.

Niimura, Y. and Nei, M. 2006. Evolutionary dynamics of olfactory and other chemosensory receptor genes in vertebrates. J. Hum. Genet. 51, 505–517.

Nowlis, G. H. and Frank, M. E. 1981. Quality Coding in Gustatory Systems of Rats and Hamsters. In: Perception of Behavioral Chemicals (ed. D. M. Norris), pp. 59–80. Elsevier/North Holland Biomedical Press.

Palmerino, C. C., Rusiniak, K. W., and Garcia, J. 1980. Flavor-illness aversions: the peculiar roles of odor and taste in memory for poison. Science 208, 753–755.

Roper, S. D. 2006. Cell communication in taste buds. Cell Mol. Life Sci. 63, 1494–1500.

Shepherd, G. M. 2005. Outline of a theory of olfactory processing and its relevance to humans. Chem. Senses 30(Suppl. 1), i3–i5.

St. John, S. J. and Spector, A. C. 1998. Behavioral discrimination between quinine and KCl is dependent on input-from the seventh cranial nerve: implications for the functional roles of the gustatory nerves in rats. J. Neurosci. 18, 4353–4362.

Turin, L. 2006. The Secret of Scent: Adventures in Perfume and the Science of Smell. HarperCollins.

Yokoi, M., Mori, K., and Nakanishi, S. 1995. Refinement of odor molecule tuning by dendrodendritic synaptic inhibition in the olfactory bulb. Proc. Natl. Acad. Sci. U. S. A. 92, 3371–3375.

4.18 Oral Chemesthesis and Taste

C T Simons, Global Research and Development Center, Cincinnati, OH, USA

E Carstens, University of California, Davis, CA, USA

Glossary

adaptation Reduction in the response magnitude of a sensory neuron during a maintained stimulus.

chemesthesis Sensations of a burning, stinging, tingling, or anesthetic quality elicited by chemical stimulation of the mucosa or skin.

desensitization (tachyphylaxis) Reduction in response to reapplication of a stimulus. Self-desensitization refers to a reduced response to reapplication of the same stimulus, and cross-desensitization is a reduction in the expected magnitude of a response to one stimulus when preceded by a different conditioning stimulus.

masking Inability or reduced ability to detect a stimulus in the presence of another stimulus, usually of a higher intensity.

nociceptor Sensory receptor innervating skin, mucosa, or other tissues that respond to noxious (potentially or overtly damaging) stimulation.

noxious Of sufficient intensity to threaten or overtly damage tissue.

pain An unpleasant sensory and emotional experience associated with actual or potential tissue damage, or described in terms of such damage.

sensitization Progressive increase in perceived sensory intensity during repeated application of a fixed stimulus (also refers to increased response of a sensory neuron following a noxious conditioning stimulus).

stimulus-induced recovery After desensitization, repeated reapplication of the stimulus induces a progressive increase in response that eventually reaches the same level as before desensitization.

TRP Transient receptor potential class of ion channels, some of which respond to chemicals or temperature changes.

Oral somatosensation provides important information regarding the texture, temperature, and pungency associated with oral stimuli. These stimuli are detected, respectively, by the presence of mechanosensitive, thermosensitive, and nociceptive neurons that are dispersed throughout the oral epithelium (Nagy, J. I. *et al.*, 1982; Whitehead, M. C. and Frank, M. E., 1983; Ito, J. and Oyagi, S., 1995). Whereas mechanosensitive neurons have receptor endings specialized to detect specific mechanically evoked sensations (e.g., vibration), thermosensitive neurons and nociceptors are thinly (Aδ-fibers) or unmyelineated (C-fibers) and are believed to have free nerve endings that respond to non-noxious heat and cold or various types of noxious stimuli including those of thermal, mechanical, or chemical origin (Dubner, R. and Bennett, G. J., 1983; Sessle, B. J., 1987; Willis, W. D. and Westlund, K. N., 1997; Millan, M. J., 1999). Although nociceptors are specifically tuned to convey sensations of pain, it is currently unclear whether nociceptors responsive to a particular class of noxious stimuli exist (e.g., chemonociceptors) (Schmidt, R. et al., 1995). Sensitivity to chemical irritants serves to protect organisms against exposure to potentially harmful compounds. This specialized sense, defined in the early 1900s (Parker, G. H., 1912) as the common chemical sense, is mediated by chemonociceptive nerve fibers and is separate from the system that provides chemosensitivty to taste and smell molecules (although at high concentrations, salts and acids can evoke sensations of pain; Carstens, E. et al., 1998; Dessirier, J. M. et al., 2000a; 2001b). More recently, objections to the term common chemical sense has resulted in the term chemesthesis being used to describe the sensations resulting from chemical activation of somatosensory pathways (Green, B. G. et al., 1990). Chemesthesis refers to the ability to detect and respond to potentially noxious chemical stimuli across the entire body, not just in the oral or nasal mucosa. Whereas mucosal membranes (such as the cornea and oral and nasal epithelium) are generally more sensitive to chemical irritants, the mechanisms that subserve the detection of noxious chemicals are likely the same as those occurring in skin. The important feature which differentiates oral chemesthesis from generalized chemesthesis relates to its contribution to flavor. The pungency associated with horseradish, the burn of hot chili peppers, and the tingle induced by carbonated beverages all result from the activation of neural pathways sensitive to specific chemicals. In fact, along with taste and smell, oral chemesthesis is often considered to be one of the three main contributors to food and beverage flavor profiles. As such, increasing attention has been paid to oral chemesthesis and its impact on taste and flavor perception.

4.18.1 Oral Chemesthesis: Anatomy

The detection of oral irritants is achieved primarily by the activation of the trigeminal nerve, although the glossopharyngeal and vagus nerves also participate (Figure 1). The trigeminal nerve, as the name suggests, is comprised of three divisions that innervate the whole of the orofacial region. The mandibular and maxillary divisions are responsible for conveying somatosensory information from the oral cavity, whereas the ophthalmic division, via the ciliary and ethmoid nerves, innervates the cornea and nasal epithelium, respectively. The mandibular branch is the largest of the three divisions and innervates the anterior tongue and mandibular

dental pulp. Unlike the maxillary and ophthalmic divisions, the mandibular division is a mixed nerve and contains both afferent and efferent fibers (Shankland, W. E., 2001b). Of particular importance for detecting oral irritants, trigeminal innervation of the anterior tongue occurs via the lingual nerve (Sostman, A. L. and Simon, S. A., 1991; Wang, Y. et al., 1993). Although nociceptive fibers are found across the entire lingual surface, they terminate most densely in close proximity to taste buds (Nagy, J. I. et al., 1982; Finger, T., 1986; Whitehead, M. C. and Kachele, D. L., 1994; Whitehead, M. C. et al., 1999). This so-called perigemmal pattern of innervation is different from that of gustatory fibers which make intragemmal (within taste bud) synaptic contact with taste receptor cells. The midfacial region is innervated by the maxillary division of the trigeminal nerve. The maxillary division is a sensory nerve comprised exclusively of afferent fibers (Shankland, W. E., 2001a). The maxillary branch innervates many important structures for the detection of oral irritants including the soft and hard palate, the tonsils, and the nasopharyngeal mucosa (Shankland, W. E., 2001a). Although not involved in oral chemesthesis, the ophthalmic division of the trigeminal nerve is important in nasal irritation (Silver, W. L. and Moulton, D. G., 1982). The ethmoid nerve conveys afferent signals from the nasal mucosa, and sensitivity to nasal irritants has been linked to sick-building syndrome (Shusterman, D., 2003) and other disorders including seasonal allergic rhinitis (Shusterman, D., 2002).

In addition to the trigeminal nerve, the glossopharyngeal and vagus nerves also contribute to oral chemosensitivity by responding to oral irritants, as well as tastants, delivered to the posterior oral cavity. Specifically, the glossopharyngeal nerve responds to chemoirritants delivered to the posterior tongue (Barry, M. A. et al., 1993; Hayakawa, T. et al., 2005) as well as the soft palate (Hayakawa, T. et al., 2005) and upper oropharynx (Hayakawa, T. et al., 2005). The vagus nerve is activated by irritants delivered to the lower pharynx and the epiglottis (Boucher, Y. et al., 2003).

Primary afferent somatosensory fibers from the oral cavity and throat project, in a topographical manner, to central relays in the brainstem. Trigeminal afferents carry information to the brainstem trigeminal complex where there appears to be some segregation in the processing of different types of somatosensory inputs. Nociceptive and thermally sensitive fibers project to superficial layers (I–II) of

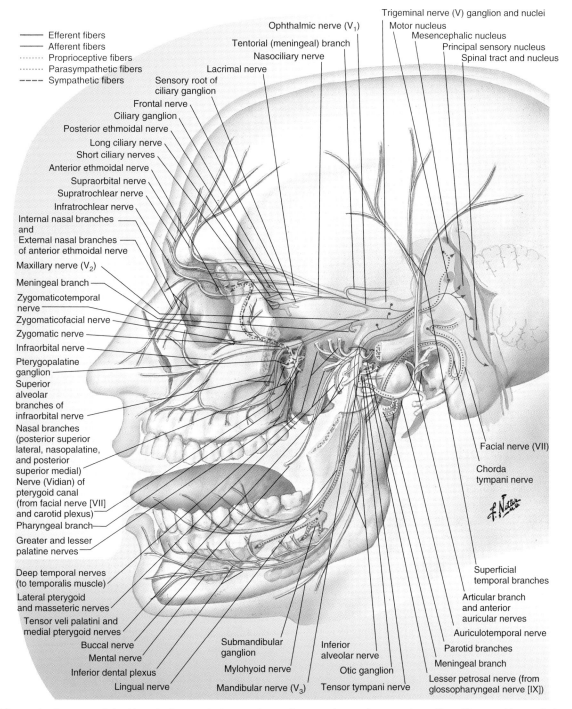

Figure 1 Anatomy of the trigeminal nerve, its innervation territory, and central connections. From Netter, with permission.

the trigeminal complex, whereas innocuous low-threshold mechanical stimuli activate neurons in superficial as well as deeper (III–IV) trigeminal layers (Hu, J. W. *et al.*, 2005). As the majority of studies investigating oral chemesthesis have focused on lingual sensitivity, more is known regarding the topographical organization of lingual inputs compared to other nerve inputs (however, see Rentmeister-Bryant, H. and Green, B. G. 1997; Green, B. G. 1998). In particular, it has been shown that lingual nociceptors project to the superficial layers of the trigeminal subnucleus caudalis (Vc) (Hayashi, H., 1985; Jacquin, M. F. *et al.*, 1986; Komai, M. and Bryant, B. P., 1993; Strassman, A. M. and Vos, B. P., 1993; Coimbra, F. and Coimbra, A., 1994). There they form synaptic connections with second-order neurons involved in relaying nociceptive information to higher centers (Kruger, L. and Michel, F., 1962; Yokota, T., 1975; Amano, N. *et al.*, 1986; Strassman, A. M. and Vos, B. P., 1993; Carstens, E. *et al.*, 1995; Raboisson, P. *et al.*, 1995; Carstens, E. *et al.*, 1998; Simons, C. T. *et al.*, 1999; Dessirier, J. M. *et al.*, 2000b; Simons, C. T. *et al.*, 2000; Carstens, E. *et al.*, 2002a). Although Vc has been the most extensively studied (for review see Sessle, B. J. and Greenwood, L. F., 1976; Hu, J. W. *et al.*, 1981; Dubner, R. and Bennett, G. J., 1983), the more rostral trigeminal subnuclei interpolaris and oralis are also involved with processing and relaying oral pain information (Greenwood, L. F. and Sessle, B. J., 1976; Sessle, B. J. and Greenwood, 1976; Nord, S. G. and Young, R. F., 1979; Hu, J. W. *et al.*, 1981; Hayashi, H. *et al.*, 1984; Hu, J. W. and Sessle, B. J., 1984; Hayashi, H. and Tabata, T., 1989; Dallel, R. *et al.*, 1990; Jacquin, M. F. and Rhoades, R. W., 1990; Raboisson, P. *et al.*, 1991; Hu, J. W. *et al.*, 1992; Ohya, A., 1992; Ohya, A. *et al.*, 1993; Dallel, R. *et al.*, 1996).

Oral somatosensory information conveyed by non-trigeminal nerve fibers also project topographically to brainstem nuclei. Both, the glossopharyneal nerve and the vagus nerve project to the nucleus of the solitary tract (NTS). *c-fos* studies of nicotine-induced throat irritation have shown that vagal afferents project to the most caudal aspects of the NTS (Boucher, Y. *et al.*, 2003). The glossopharyngeal nerve, on the other hand, projects more rostrally (Hamilton, R. B. and Norgren, R., 1984). Importantly, although both nerves encode the quality and intensity of oral somatosensory and taste stimuli, the processing of such information is segregated and distributed to different parts of the NTS. Taste processing occurs in rostral medial NTS, whereas somatosensory processing occurs more laterally (Barry, M. A. *et al.*, 1993). Interestingly, collateral branches from trigeminal afferents also project to rostromedial NTS, providing a substrate by which trigeminal–taste interactions might occur (Jacquin, M. F. *et al.*, 1982; Whitehead, M. C. and Frank, M. E., 1983; Marfurt, C. F. and Rajchert, D. M. 1991; Boucher, Y. *et al.*, 2003).

Two classes of second-order nociceptive neurons reside in Vc and NTS. The majority tend to be of the wide-dynamic-range (WDR) type and respond, in graded fashion, to both innocuous and noxious somatosensory stimuli (Dubner, R. and Bennett, G. J., 1983; Sessle, B. J., 1987). The existence of WDR cells is possible due to the convergence of nociceptive and non-nociceptive peripheral afferents onto second-order cells. As the stimulus intensity increases from innocuous to noxious, the second-order cell integrates the afferent volleys first from non-nocicepive cells and then from nociceptors as well. Thus, an innocuous stimulus elicits a low level of activity from WDR cells relative to a noxious stimulus that will evoke larger responses. The other class of second-order nociceptor includes those that respond exclusively to high-intensity, noxious stimuli. These nociceptive-specific (NS) cells receive input exclusively from peripheral C- or Aδ- fibers (Villanueva, L. *et al.*, 1988). Electrophysiolgical studies have revealed that relative to the numbers of WDR-type cells, NS cells tend to be less prevalent, at least in Vc (see, e.g., Carstens, E. *et al.*, 1998). Nociceptors sensitive to chemical irritants exist in both classes of cells (Hu, J. W., 1990; Carstens, E. *et al.*, 1998; Simons, C. T. *et al.*, 1999; Dessirier, J. M. *et al.*, 2000b). Interestingly, both neuron types exhibit similar response characteristics to chemical stimuli. Specifically, they tend to be broadly tuned, with a given Vc neuron capable of responding to a variety of noxious chemicals (Carstens, E. *et al.*, 1998; Simons, C. T. *et al.*, 1999; Dessirier, J. M. *et al.*, 2000b) (Figure 2). This suggests that either peripheral nociceptors express receptor mechanisms to a broad range of irritant chemicals or, alternatively nociceptors in the periphery are specifically tuned, but convergence of these neurons in Vc broadens the tuning characteristic of central neurons. The former explanation is most parsimonious with peripheral fiber and trigeminal ganglion studies in which a single nociceptive fiber or ganglion cell responds to multiple chemical irritants (Wang, Y. *et al.*, 1993; Bryant, B. P. and Moore, P. A., 1995; Liu, L. and Simon, S. A., 1996a; 1996c; Liu, L. *et al.*, 1997; Liu, L. and Simon, S. A., 1997; 1998; Steen, K. H. *et al.*, 1999).

From the first central relay in the trigeminal complex or NTS, oral somatosensory information remains segregated as it is conveyed to third-order sites. In humans, the dorsal column pathway transmits tactile information from layers III–IV of the trigeminal complex to the contralateral ventroposteriomedial (VPM) thalamus (Kandel, E. R. *et al.*, 2000). The anterolateral tract conveys thermal and nociceptive signals from superficial layers to a more lateral thalamic region, the contralateral

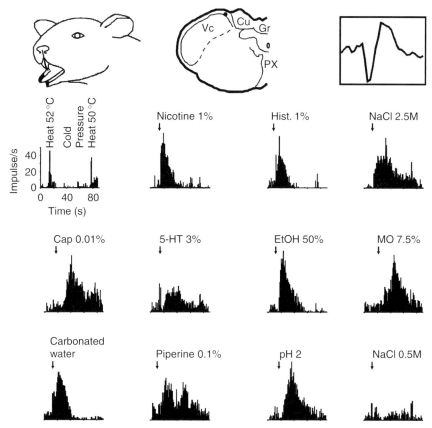

Figure 2 Nonselective responses of Vc neuron to different oral irritants. The figurines in the upper row show, from left to right, a drawing of the rat's face with the receptive field of this neuron on the tongue and chin (black), a cross-section of the medulla showing the recording site (dot) in the superficial part of dorsomedial Vc, and the action potential waveform. The lower three rows show peristimulus–time histograms (PSTHs, bin width 1 s) of responses to various stimuli. Calibrations given in left-hand PSTH, second row. Cap, capsaicin; Cu, cunate n.; EtOH, ethanol; Gr, gracile n.; Hist., histamine; 5-HT, serotonin; MO, mustard oil; pH2, HCl at pH of 2. PX, pyramidal decussation; Vc, trigeminal subnucleus caudalis. Reproduced from Carstens, E., Iodi Carstens, M., Dessirier, J. M., O'Mahony, M., Simons, C. T., Sudo, M., and Sudo, S. 2002. It hurts so good: oral irritation by spices and carbonated drinks and the underlying neural mechanisms. Food Qual. Pref. 13, 431–443, with permission from Elsevier; adapted from Carstens, E., Kuenzler, N., and Handwerker, H. O. 1998. Activation of neurons in rat trigeminal subnucleus caudalis by different irritant chemicals applied to oral or ocular mucosa. J. Neurophysiol. 80, 4654–4692.

ventroposteriolateral (VPL) thalamus (Kandel, E. R. *et al.*, 2000). In rodents, where many of the functional anatomical studies have been completed, ascending somatosensory information has an obligatory relay in the contralateral pontine parabrachial nucleus (Feil, K. and Herbert, H., 1995). From the parabrachial nucleus, somatosensory projections are to the appropriate regions of the ventroposterior thalamus.

Following thalamic processing, ascending anterolateral tract and dorsal column pathways project to the somatosensory (SI) cortex, an area found in the postcentral gyrus of the anterior parietal lobe. Somatosensory projections from the thalamus are organized such that a representation of the body is mapped onto the cortex (Kandel, E. R. *et al.*, 2000).

This map, known as a homunculus, is arranged so that body regions having greater sensory acuity occupy a larger region of the SI cortex. Thus, oral structures (lips, teeth, tongue, and pharynx) occupy a relatively large region of lateral SI cortex compared to legs and trunk, which occupy a smaller region of medial SI cortex. Projections from SI are directed toward secondary cortical areas in the parietal lobe (LaMotte, R. H. and Mountcastle, V. B., 1979) and tertiary areas in the insula and orbitofrontal cortex (Verhagen, J. V. *et al.*, 2003; Kadohisa, M. *et al.*, 2004; 2005). Oral somatosensory inputs into the insula and orbitofrontal cortex are of particular interest to taste and flavor interactions. The primary gustatory cortex is found in the insula (Yaxley, S. *et al.*, 1990; Small, D. M.

et al., 1999; Stapleton, J. R. *et al.*, 2006), and the orbitofrontal cortex receives converging taste, smell, somatosensory, and visual information to encode stimulus valence (Rolls, E. T. and Baylis, L. L., 1994; Critchley, H. D. and Rolls, E. T., 1996).

4.18.2 Oral Chemesthesis: Physiology

The role of chemonociceptive neurons in the oral cavity is to signal the presence of stimuli that are potentially or overtly damaging prior to ingestion. However, not all stimuli that activate nociceptors are damaging. For example, capsaicin, the pungent principle in chili peppers, binds to the vanilloid-1 receptor (TRPV1 or VR-1) expressed in nociceptors (Caterina, M. J. *et al.*, 1997; Tominaga, M. *et al.*, 1998; Caterina, M. J. *et al.*, 1999; Nagy, I. and Rang, H. P., 1999; Caterina, M. J. *et al.*, 2000) to elicit a burning sensation that, in some cuisines, is highly sought after. Capsaicin, however, is relatively benign and TRPV1 is ordinarily present in nociceptors to detect the presence of noxious thermal or acidic stimuli (Tominaga, M. *et al.*, 1998; Cesare, P. *et al.*, 1999; Nagy, I. and Rang, H. P., 1999; Jordt, S. E. *et al.*, 2000).

The process of detecting chemesthetic stimuli is initiated at the receptor level. In the oral cavity, trigeminal ganglion neurons express a variety of receptors that are activated by the presence of chemical irritants (Wang, Y. *et al.*, 1993; Bryant, B. P. and Moore, P. A., 1995; Liu, L. and Simon, S. A., 1996a; 1996c; 1997; Liu, L. *et al.*, 1997; Liu, L. and Simon, S. A., 1998; Steen, K. H. *et al.*, 1999). A variety of methods, including patch-clamp techniques and calcium imaging, have been used to study the response characteristics of trigeminal ganglion neurons to oral irritants, and in the last decade, a wealth of information has been generated regarding the receptor and cellular mechanisms underlying irritant chemical detection. The great majority of known chemoirritant receptors are of the transient receptor potential (TRP) superfamily (for review see Clapham, D. E., 2003; Wang, H. and Woolf, C. J., 2005) of cation channels. All mammalian TRP channels are 6-transmembrane polypeptides that assemble as tetramers to form cation channels (Clapham, D. E., 2003). To date, TRP channels have been shown to transduce and discriminate a large variety of sensations ranging from taste (sweet, bitter, and umami) (Perez, C. A. *et al.*, 2002; Zhang, Y. *et al.*, 2003) to temperature (Caterina, M. J. *et al.*, 1997; 1999; McKemy, D. D. *et al.*, 2002; Peier, A. M. *et al.*, 2002b; Smith, G. D. *et al.*, 2002; Xu, H. *et al.*, 2002), pressure (Mizuno, A. *et al.*, 2003; Suzuki, M. *et al.*, 2003), and noxious chemicals (Caterina, M. J. *et al.*, 1997; Tominaga, M. *et al.*, 1998; Trevisani, M. *et al.*, 2002; Yang, B. H. *et al.*, 2003; Bandell, M. *et al.*, 2004; Jordt, S. E. *et al.*, 2004; Bautista, D. M. *et al.*, 2005; Macpherson, L. J. *et al.*, 2005; Moqrich, A. *et al.*, 2005). Interestingly, most of the chemosensitive TRP channels identified thus far are also responsive to other noxious stimuli. For instance, the first channel to be identified as chemosensitive, TRPV1, has been shown to be activated by noxious heat ($>43\,°C$) in addition to the irritants capsaicin, acids (Caterina, M. J. *et al.*, 1997; Tominaga, M. *et al.*, 1998), ethanol (Trevisani, M. *et al.*, 2002), eugenol (Yang, B. H. *et al.*, 2003), and camphor (Xu, H. *et al.*, 2005) (see Table 1).

The chemosensitive TRP channels come from three different subfamilies. The TRPV family is characterized primarily by its thermosensitvity (Clapham, D. E., 2003). Of the six members in this subfamily, four appear to be thermosensitive and three express chemosensitivity (Table 1). In addition to TRPV1's well-known sensitivity to noxious heat, capsaicin (Caterina, M. J. *et al.*, 1997), acidic stimuli (Tominaga, M. *et al.*, 1998), and ethanol (Trevisani, M. *et al.*, 2002), TRPV3, responds to innocuous warmth ($>30\,°C$) (Peier, A. M.

Table 1 Thermo- and chemosensitive transient receptor potentials (TRPs)

	Temperature sensitivity	*Chemical sensitivity*
TRPV1 (VR1)	$>43\,°C$	Capsaicin, acid, ethanol, eugenol camphor
TRPV2 (VRL-1)	$>52\,°C$	
TRPV3	$>30\,°C$	Camphor, carvacol, eugenol, thymol
TRPV4	$>24\,°C$	Acid
TRPM8	$<25\,°C$	Menthol, icilin
TRPA1	$<18\,°C$	Allyl isothiocyanate (mustard oil), cinnamic aldehyde, methyl salicylate (wintergreen oil), eugenol (clove oil), gingerol (ginger), allicin (garlic)

et al., 2002b; Smith, G. D. *et al.*, 2002; Xu *et al.*, 2002) as well as camphor (Moqrich, A. *et al.*, 2005), carvacol, eugenol, and thymol (Xu, H. *et al.*, 2006). TRPV4 is responsive to pressure (Mizuno, A. *et al.*, 2003; Suzuki, M. *et al.*, 2003), innocuous warming (>24 °C) (Guler, A. D. *et al.*, 2002), and low pH (Suzuki, M. *et al.*, 2003).

Like the TRPV subfamily, the TRPM and TRPA receptor subfamilies display sensitivity to multiple stimuli (Table 1). In regard to chemesthesis, TRPM8 and TRPA1 (ANKTM1) are well characterized. TRPM8 has been shown to respond to cold (<25 °C) as well as the cooling agents menthol and icillin (McKemy, D. D. *et al.*, 2002; Peier, A. M. *et al.*, 2002a; Andersson, D. A. *et al.*, 2004; Chuang, H. H. *et al.*, 2004). At high concentrations, menthol has been shown to evoke activity in nociceptive neurons (Xing, H. *et al.*, 2006) and elicit sensations of both cooling and irritation when applied to humans (Green, B. G. 1992; Cliff, M. A. and Green, B. G., 1994; Dessirier, J. M. *et al.*, 2001a). Initial studies indicated that TRPA1 was responsive to noxious cold (<18 °C) (Story, G. M. *et al.*, 2003), although this has been challenged (Jordt, S. E. *et al.*, 2004) and it was recently reported that knockout mice lacking TRPA1 do not exhibit any deficit in responsiveness to noxious cold (Bautista, D. M. *et al.*, 2006). Subsequently, TRPA1 has been shown to additionally respond to pungent compounds including mustard oil, cinnamic aldehyde, wintergreen oil, clove oil, ginger, and garlic (Bandell, M. *et al.*, 2004; Jordt, S. E. *et al.*, 2004; Bautista, D. M. *et al.*, 2005; Macpherson, L. J. *et al.*, 2005) (see Table 1).

Additional receptor families have been implicated for the detection of other oral irritants. For instance, in addition to TRPV1, acid-sensitive ion channels (ASICs) have been shown to be present in trigeminal ganglion cells (Olson, T. H. *et al.*, 1998) and respond to low pH (Lingueglia, E. *et al.*, 1997; Waldmann, R. *et al.*, 1997; Price, M. P. *et al.*, 2001). Similarly, nicotinic acetylcholinergic receptors signal the presence of nicotine irritation (Liu, L. and Simon, S. A. 1996d; Dessirier, J. M. *et al.*, 1998; Liu, L. *et al.*, 1998; Marubio, L. M. *et al.*, 1999).

Electrophysiological recordings elicited by lingual application of chemesthetic stimuli have been made along the entire neural pathway. Remarkably, there is great correspondence between the patterns of activity generated at different levels of the neuraxis. Unlike other systems, activity generated in the periphery seems to be faithfully transmitted to higher brain levels, thus providing a solid foundation for studying the neural correlates of irritant sensation.

Lingual nerve recordings have been made in response to application of a variety of chemesthetic stimuli including acids (Sostman, A. L. and Simon, S. A., 1991; Komai, M. and Bryant, B. P., 1993), vanilloids (Wang, Y. *et al.*, 1993), menthol (Wang, Y. *et al.*, 1993; Lundy, R. F., Jr. and Contreras, R. J., 1994; 1995), ethanol (Danilova, V. and Hellekant, G. 2002), and astringent (Shiffman, S. S., *et al.*, 1992) and tingling (Bryant, B. P. and Mezine, I., 1999) compounds. The difficulty in isolating single C- and Aδ-fibers makes this technique extremely labor intensive and, consequently, whole-nerve recordings are often made. Despite these technical limitations, when combined with patch-clamp and calcium-imaging studies of trigeminal ganglion cells, important information has been gathered detailing cellular processes and coding mechanisms. Indeed, increasing numbers of studies suggest that the processing of chemoirritant information remains segregated from the processing of non-nociceptive chemically evoked sensations. For instance, chemical irritants (e.g., capsaicin, pentanoic acid, and mustard oil) tend to activate the same population of chemosensitive trigeminal cells, whereas menthol, when delivered at a low, nonirritating dose, activates a different population (Jordt, S. E. *et al.*, 2004; Abe, J. *et al.*, 2005; Kobayashi, K. *et al.*, 2005). Some of the overlap in chemical sensitivity is the result of a specific receptor being activated by multiple irritants as is the case with TRPV1 which is activated by vanilloids (Caterina, M. J. *et al.*, 1997), protons (Tominaga, M. *et al.*, 1998; Jordt, S. E. *et al.*, 2000), and ethanol (Trevisani, M. *et al.*, 2002). Additionally, multiple studies have shown individual trigeminal ganglion cells or fibers to respond to a diverse sample set of chemical irritants, suggesting that multiple receptor mechanisms are coexpressed in the same cell. The fact that TRPM8 is expressed in a population of neurons separate from those expressing TRPV1 or TRPA1 (Jordt, S. E. *et al.*, 2004; Abe, J. *et al.*, 2005; Kobayashi, K. *et al.*, 2005) provides a mechanism by which chemically evoked irritant and thermal sensations are kept separate. If the pungency associated with high menthol concentrations can be attributed to activation of TRPA1, then a chemesthetic model supporting the separate processing of irritant and thermal stimuli can be developed. However, to date, rodent TRPA1 has not been shown to be menthol sensitive.

Centrally, *c-fos* immunohistochemistry has been used to show that similar populations of neurons in superficial Vc are activated by lingual application of a variety of chemical irritants including nicotine, capsaicin, piperine, histamine, acids, and high salt concentrations (Carstens, E. *et al.*, 1995; Simons, C. T. *et al.*, 1999; Dessirier, J. M. *et al.*, 2000; Carstens, E. *et al.*, 2001; Sudo, S. *et al.*, 2003). However, immuno-histochemical studies are static representations of neural activity and cannot show whether a given neuron responds to multiple chemical stimuli. Thus, electrophysiological recordings of Vc neural activity are necessary to delineate the response properties of chemosensitive neurons. To date, extracellular recordings suggest that Vc neurons respond to multiple classes of chemoirritants (Carstens, E. *et al.*, 1998; Simons, C. T. *et al.*, 1999; Dessirier, J. M. *et al.*, 2000; Sudo, S. *et al.*, 2002; Simons, C. T. *et al.*, 2003c; Sudo, S. *et al.*, 2003; Simons, C. T. *et al.*, 2004) (Figure 2). Recent studies indicate that neurons in superficial laminae of Vc that respond to noxious tongue cooling also respond to menthol, and that the large majority of these additionally respond to noxious heat, capsaicin, mustard oil, and other irritant chemicals (Carstens, E. *et al.*, 2005) (Figure 3). This multiple sensitivity is true for WDR and NS cell types. Thus, although in the periphery, chemoirritation may be coded along labeled lines, at the level of the first central synapse, this specificity appears to be lost.

Few studies have looked at the chemosensitivity of neurons residing in thalamus or somatosensory cortex. Injections of capsaicin under the skin of the lip in rat increased spontaneous firing rate of thalamocortical neurons within 15 min and resulted in reorganization of VPM and SI whisker receptive fields within 1 h (Katz, D. B. *et al.*, 1999). These findings are in contrast to studies at the brainstem and spinal level wherein capsaicin induces immediate increases in neuronal firing (e.g., Carstens, E. *et al.*, 1998; Dessirier, J. M. *et al.*, 2000), and the time course for capsaicin-induced reorganization is significantly shorter (Simone, D. A. *et al.*, 1991; Pettit, M. J. and Schwark, H. D., 1996; Chiang, C. Y. *et al.*, 1997). Capsaicin has been shown to activate neurons in higher cortical levels including the orbitofrontal cortex and amygdala (Rolls, E. T. *et al.*, 2003; Kadohisa, M. *et al.*, 2004; 2005). Of the population of macaque orbitofrontal neurons responding to taste, temperature, or viscosity, nearly 16% also responded to capsaicin (Kadohisa, M. *et al.*, 2004). Notably, the response characteristics of neurons in orbitofrontal cortex activated by thermal stimuli were different from those activated by capsaicin, indicating that the representation of capsaicin in this region is different from other somatosensory stimuli such as temperature.

Perceptually, chemesthetic sensations elicited by various irritant chemicals are often verbalized using the same descriptor. Typically, these sensations are described as burning or pungent regardless of whether the stimulus is, for example, capsaicin or menthol. However, it is unknown whether these chemically diverse compounds all elicit sensations that are qualitatively identical or, alternatively, whether the compounds evoke qualitatively different sensations that are described as similar due to a limited vocabulary. Neurophysiological evidence has shown that a single nociceptor can express multiple receptor subtypes and respond to a variety of chemoirritants. However, subjects can easily discriminate the irritant

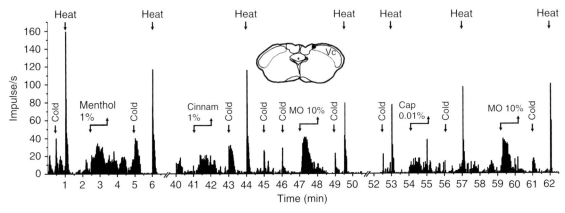

Figure 3 Representative individual example of cold- and menthol-responsive Vc unit that additionally responded to noxious heat and other irritant chemicals. Shown are peristimulus–time histograms (PSTHs; bin width 1 s) of activity. Arrows indicate time of application of noxious cold (0–3 °C water) or heat (53 °C water) stimuli, and bars with downward and upward arrows indicate start and end of chemical application. Cinnam: cinnamaldehyde; MO, mustard oil; Cap, capsaicin. Inset shows recording site (dot).

sensation evoked from different chemical compounds. Certainly, temporal cues assist subjects in this discrimination task. For instance, some compounds, such as allyisothiocyanate, have a quick onset and relatively short duration, whereas others have a slower onset (nicotine) or longer duration (capsaicin). Further studies are needed to delineate whether chemesthetic submodalities exist. Further advances in neuroimaging techniques might allow investigators to determine the neural underpinnings that allow subjects to differentiate chemical irritants.

4.18.3 Chemoirritation: Temporal Phenomena

Most sensory systems maintain sensitivity to conditions of repeated or prolonged stimulation by a process called adaptation. Adaptation is a specialized cellular response that results in a decreased number of evoked action potentials despite a maintained stimulus. Cellular adaptation results in the loss of the conscious perception of a constant stimulus. Unique to the chemesthetic sense, irritant stimuli often evoke patterns of sensation that do not conform to the classic models of adaptation. Indeed the sensation elicited by irritant stimuli can increase or decrease with repeated stimulation depending upon chemical compound and the interstimulus interval (ISI).

Sensitization is the phenomenon encountered when a chemesthetic stimulus elicits an irritant sensation of progressively increasing magnitude when applied repeatedly to the tongue (Stevens, D. A. and Lawless, H. T., 1987; Green, B. G., 1989; Cliff, M. A. and Green, B. G., 1996; Prescott, J. and Stevenson, R. J., 1996b; Dessirier, J. M. *et al.*, 1997; Dessirier, J. M. *et al.*, 1999; Prescott, J. and Swain-Campbell, N., 2000). The most well-studied sensitizing irritant is the vanilloid capsaicin. Numerous investigators have noted that when capsaicin is applied to the oral mucosa at ISIs of less than 3 min, the perceived irritant sensation grows (Figure 4(a), left). However, following a short hiatus of greater than 3 min, a condition of self-desensitization (Figure 4, right) develops where further capsaicin applications are perceived as less irritating (Green, B. G., 1989; Geppetti, P. *et al.*, 1993; Cliff, M. A. and Green, B. G., 1998; Green, B. G., 1998; Green, B. G. and Rentmeister-Bryant, H., 1998). With continued application, the perceived irritation builds up again to eventually reach the same level obtained prior to

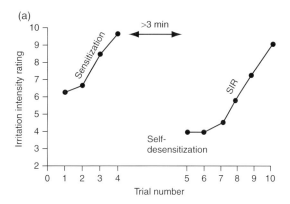

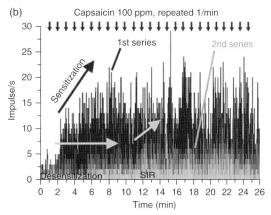

Figure 4 Capsaicin sensitization, desensitization, and stimulus-induced recovery. (a) Graph plots human psychophysical intensity ratings of irritation elicited by repeated application of capsaicin (10 ppm) to the tongue at 1-min interstimulus intervals. (b) Example of response of Vc neuron to repeated application of capsaicin (at arrows). Black histogram shows the response to initial series of capsaicin applications and gray histogram shows the response to second series of applications 30 min after the end of the first series. Note the progressive increase in response to the first series of applications (sensitization), the absence of response to the initial capsaicin applications in the second series (desensitization) with a delayed increased (stimulus-induced recovery = SIR). (a) Data adapted from Green, B. G. 1996. Rapid recovery from capsaicin desensitization during recurrent stimulation. Pain 68, 245–253. (b) Adapted from Dessirier, J. M., Simons, C. T., Sudo, M., Sudo, S., and Carstens, E. 2000b. Sensitization, desensitization and stimulus-induced recovery of trigeminal neuronal responses to oral capsaicin and nicotine. J. Neurophysiol. 84, 1851–1862.

desensitization (Figure 4(a), right); this condition has been termed stimulus-induced recovery (SIR) (Green, B. G., 1996; Green, B. G. and Rentmeister-Bryant, H., 1998). Other vanilloids, including piperine (Stevens, D. A. and Lawless, H. T., 1987; Green, B. G., 1996; Dessirier, J. M. *et al.*, 1999), acids (Green, B. G., 1996; Dessirier, J. M. *et al.*, 2000a), and high concentrations of salt (Dessirier, J. M. *et al.*, 2001b), elicit patterns of irritation similar to those evoked by capsaicin. Allylisothiocyanate (Simons, C. T. *et al.*, 2003b), nicotine (Dessirier, J. M. *et al.*, 1999), menthol (Cliff, M. A. and Green, B. G., 1994; 1996; Green, B. G. and McAuliffe, B. L., 2000; Dessirier, J. M. *et al.*, 2001a), zingerone (Prescott, J. and Stevenson, R. J.,

1996a), cinnamaldehyde (Prescott, J. and Swain-Campbell, N., 2000), and ethanol (Prescott, J. and Swain-Campbell, N., 2000), however, show a pronounced desensitization with repeated application. Figure 5(a) shows desensitization with menthol. Furthermore, reapplication of the same chemical

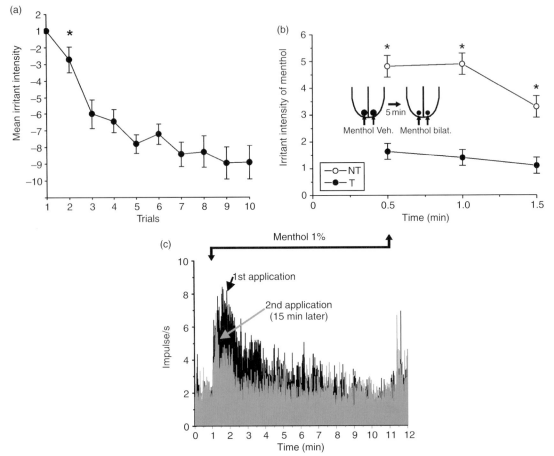

Figure 5 Menthol desensitization. (a) Mean irritant intensity reported 25 s following stimulus onset as a function of trials of menthol (19.2 mM) delivered sequentially at 1 min intervals. Intensity ratings were made using a bipolar category scale in which the magnitude of irritation elicited by the initial stimulus is rated 1; increasingly negative numbers indicate weaker intensity compared to the first trial. Error bars: SE *: significantly different from initial rating ($P < 0.05$, ANOVA). (b) Menthol self-desensitization, assessed using half-tongue two-alternative forced-choice procedure. Menthol was applied to one side of the tongue and vehicle to the other (left-hand inset showing figurine of tongue), followed 5 min later by bilateral application of menthol (right-hand inset). Graph plots the mean irritant intensity for the menthol-treated (T, filled circles) and nontreated (NT, open circles) side of the tongue vs. time following bilateral application of menthol. Note the significant self-desensitization. Error bars: SE Asterisks indicate significant difference between T and NT ($P < 0.05$, t-test). (c) Desensitization of responses of Vc neurons to menthol. Black peristimulus–time histogram (PSTH, bin width 1 s) shows averaged responses of 14 Vc neurons to the first application of menthol to the dorsal tongue for 10 min by constant flow (downward and upward arrows denote onset and cessation of menthol application). Note the initial increase in response followed by a decline to baseline levels (desensitization) despite continued menthol, similar to the psychophysical ratings (a). Gray PSTH shows averaged response of the same group of Vc neurons to a second application of menthol 15 min later. Note that the response was smaller compared to the first menthol application (self-desensitization), also similar to the psychophysical findings (b). (a) From Dessirier, J. M., O'Mahony, M., and Carstens, E. 2001a. Oral irritant properties of menthol: sensitizing and desensitizing effects of repeated application and cross-desensitization to nicotine. Physiol. Behav. 73, 25–36. (b) From Dessirier, J. M., O'Mahony, M., and Carstens, E. 2001a. Oral irritant properties of menthol: sensitizing and desensitizing effects of repeated application and cross-desensitization to nicotine. Physiol. Behav. 73, 25–36.

following a hiatus elicits weaker irritation. An example of this self-desensitization with menthol is shown in Figure 5(b).

In addition to sensitization, desensitization, and SIR, many compounds cross-interact to influence another irritant's temporal properties. Information from these studies sheds important light on the transduction and coding mechanisms underlying chemesthesis. For instance, following a hiatus from capsaicin application, the irritant sensation evoked by either piperine (Green, B. G., 1996), nicotine (Dessirier, J. M. *et al.*, 1997), mustard oil (Simons, C. T. *et al.*, 2003b), citric acid (Dessirier, J. M. *et al.*, 2000a), or carbonated water (Dessirier, J. M. *et al.*, 2001c) is reduced. Similar results have been obtained with menthol and nicotine (Dessirier, J. M. *et al.*, 2001a). This phenomenon has been termed cross-desensitization. Cross-sensitization and cross-SIR have been noted for NaCl–capsaicin (Dessirier, J. M. *et al.*, 2001b) and citric acid–capsaicin (Dessirier, J. M. *et al.*, 2000a), respectively. When the temporal chemesthetic properties of an irritant are influenced by the application of a different irritant, several conclusions can be drawn. For instance, the results can be used to suggest that the two irritants share a common receptor mechanism as has been shown with capsaicin and piperine or ethanol. Similarly, for irritant compounds that activate different receptors, like capsaicin and mustard oil or nicotine and menthol, cross-interactions suggest that the same nociceptive cells or neural pathways are involved in the processing of the chemesthetic information.

The different patterns of sensation (i.e., sensitization, desensitization, and SIR) most likely reflect competing cellular mechanisms that differentially dominate during the continued application of the chemical irritant. To date, there is no evidence that sensitization occurs within a single isolated trigeminal ganglion cell. In fact, with repeated application of capsaicin, tachyphylaxis has been observed whereby the size of the evoked inward current decreases (Cholewinski, A. *et al.*, 1993; Liu, L. and Simon, S. A. 1996b; Koplas, P. A. *et al.*, 1997). Thus, the mechanism proposed to mediate sensitization is one of spatial summation (Carstens, E. *et al.*, 2002). In this model (Figure 6), a chemoirritant applied to the tongue is believed to diffuse into the lingual tissue where it accesses previously quiescent nociceptive cells. As long as the pool of irritant stimuli is maintained, sensitization is believed to continue until the rate of diffusion is equaled by the rate at which the stimulus is broken down or absorbed into the blood stream. When the diffusion rate equals the absorption rate, the perceived irritant is hypothesized to plateau and when the absorption rate is greater than the diffusion rate, mechanisms subserving tachyphylaxis dominate, and a state of desensitization results.

Electrophysiological studies in rat Vc neurons have indicated that a remarkable correlation exists between their chemical responsivity and the sensations perceived in human psychophysical studies. Capsaicin, when applied with a short ISI to the rat tongue, elicits activity in Vc neurons that progressively increases over time to reach a plateau approximately 7 min following stimulus administration (Figure 4(b)) (Dessirier, J. M. *et al.*, 2000b). When capsaicin is reapplied following a

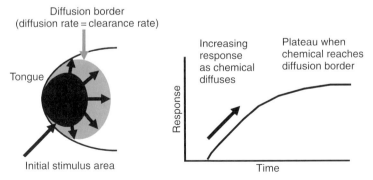

Figure 6 Spatial recruitment of lingual nociceptors to explain sensitization. Left-hand drawing shows initial area (black) of chemical stimulus applied to the tongue. With repeated application, the area increases to a maximum diffusion border (gray) where the rate of chemical diffusion equals the rate of clearance. As the chemical diffuses, it progressively excites more nociceptors by spatial recruitment, causing a progressive rise in sensation (right-hand graph) that plateaus as the chemical reaches its diffusion border.

hiatus of 15 min, evoked activity is significantly delayed; however, further administration causes the firing rate to eventually increase (Figure 4(b)). These data are suggestive of desensitization and SIR, respectively. Conversely, menthol and other chemicals elicit a desensitizing pattern of irritation in human psychophysical studies (see above) which is also reflected in the response pattern of Vc neurons. When applied to the tongue of anesthetized rats, menthol (Carstens, E. *et al.*, 2005), nicotine (Dessirier, J. M. *et al.*, 2000b), and allylisothiocyanate (Simons, C. T. *et al.*, 2004) have been shown to evoke activity in Vc neurons that decreases with continued application. An example with menthol is shown in Figure 5(c) (black histogram). Note that when menthol was reapplied after a 15 min hiatus, it elicited a weaker response (Figure 5(c), gray histogram) consistent with human psychophysical studies (Figure 5(b)).

4.18.4 Irritant–Taste Interactions

Foods containing irritant chemicals continue to gain in popularity in the United States and elsewhere. However, despite the increasing consumption of spicy foods, few studies have looked at irritation–taste interactions. From a nutritional perspective, taste is a major contributor to diet selection (Scott, T. R. and Verhagen, J. V., 2000). Thus, alterations in perceptions of taste resulting from the presence of irritant stimuli could impact nutritional status. Moreover, the existence of irritant–taste interactions could alter the way sensory methodologies are utilized within the food industry. Many irritant chemicals evoke sensitizing or desensitizing patterns of irritation (see above). If these chemicals similarly modulate sensations of taste, new precautions or methodologies must be developed to ensure that the sensory data obtained from spicy products are useful.

Once thought to be tasteless, vanilloids (capsaicin, piperine, and zingerone), as well as menthol, have recently been shown to evoke a bitter taste from a subpopulation of individuals when applied directly to taste papillae (Green, B. G. and Schullery, M. T., 2003; Green, B. G. and Hayes, J. E., 2003; 2004). For those subjects sensitive to the taste quality of capsaicin and zingerone, the intensity appears to be stronger when applied to circumavallate as opposed to fungiform papillae. The mechanism appears to be vanilloid activation of taste receptor cells. However, whether these compounds activate TRPV1 or a particular receptor from the T2R bitter family is still unclear.

Despite the recent observations that most, if not all, irritants evoke a sensation of taste, anecdotal evidence suggests that these same pungent chemicals modify the gustatory qualities of many foods (Prescott, J. *et al.*, 1993). Subjects often report not liking pungent spices (e.g., chili peppers) because the irritant quality blocks or masks the normal taste quality of the food they want to enjoy. A number of psychophysical studies have attempted to quantify the effects reported anecdotally. Most investigations examined the impact of vanilloids (capsaicin and piperine) on taste perception (Szolcsányi, J., 1977; Lawless, H. T. and Stevens, D. A., 1984; Lawless, H. T. *et al.*, 1985; Cowart, B. J., 1987; Prescott, J. *et al.*, 1993; Karrer, T. and Bartoshuk, L., 1995; Prescott, J. and Stevenson, R. J., 1995; Stevenson, R. J. and Prescott, J., 1997, Simons, C. T. *et al.*, 2002); however, CO_2 has also been addressed (Cometto-Muñiz, J. E. *et al.*, 1987; Cowart, B. J., 1998).

Initial studies reported that recognition thresholds of taste stimuli were not affected by rinsing with aqueous solutions of capsaicin (Szolcsányi, J., 1977). However, recognition thresholds are notoriously inconsistent and often show great intraindividual variability (O'Mahony, M. *et al.*, 1976). At suprathreshold concentrations of taste stimuli, the effects of capsaicin appear to be variable. Following this initial study, Lawless H. T. and Stevens D. A. (1984) noted modest-to-significant reductions in the perceived intensity of sour, bitter, and sweet stimuli following a whole-mouth capsaicin rinse. The effect of piperine was more pronounced with a significant suppression noted of all taste qualities. Similarly, evaluated as sucrose– or NaCl–capsaicin mixtures, perceived sweet intensity was decreased, whereas there was minimal impact on perceived saltiness (Prescott, J. *et al.*, 1993). These findings were confirmed in a subsequent study (Stevenson, R. J. and Prescott, J., 1997). Most recently, capsaicin pretreatment was found to suppress the perceived intensity of sucrose-induced sweetness, quinine-induced bitterness, and glutamate-induced umami (Simons, C. T. *et al.*, 2002). It is tempting to speculate that across these studies, the taste sensations that consistently show suppression when capsaicin irritation is present are those that typically involve G-protein-coupled receptors (GPCRs); however, it should be emphasized that quinine is thought to have a mode of action that is independent of GPCRs (Chen, Y. and Herness, M. S., 1997).

Some of this interstudy variability can be attributed to the method of capsaicin/tastant delivery as evidence suggests that pretreating the oral cavity with irritants prior to evaluating tastes results in a greater apparent reduction in taste intensity compared to rinsing with irritant–tastant mixtures (Cowart, B. J., 1987). Moreover, confounding influences of solution concentrations and temperature have not always been controlled. However, despite these methodological differences, pretreating the oral cavity with capsaicin or piperine appears to consistently suppress the intensity of perceived sweetness.

Several studies also investigated the effect of capsaicin desensitization on taste perception (Gilmore, M. M. and Green, B. G., 1993; Karrer, T. and Bartoshuk, L., 1995; Simons, C. T. *et al.*, 2002). Capsaicin desensitization reduced the sensation elicited by citric acid and NaCl, but this effect was minimized when controlling for context effects (Gilmore, M. M. and Green, B. G., 1993). In fact, when controlling for context effect, only the sensations elicited by the highest levels of NaCl or citric acid were reduced. At high concentrations, both citric acid and NaCl evoke irritant sensations that are difficult to separate from the taste sensations. The reduction in sensation reported at the high tastant concentrations, therefore, might be due to capsaicin's ability to cross-desensitize the irritation elicited by NaCl or citric acid as opposed to desensitizing taste sensation *per se*. Similar results were found using a different methodology (Simons, C. T. *et al.*, 2002). Self-desensitization evoked by low capsaicin concentrations (1.5 or ppm) had no effect on the taste sensations elicited by sucrose, quinine, monosodium glutamate, NaCl, or citric acid. However, at high capsaicin concentrations (100 ppm or more), self-desensitization led to a reduction in the perceived bitter intensity elicited by propthiouracil or quinine as well as the saltiness and sourness of NaCl and citric acid solutions (Karrer, T. and Bartoshuk, L., 1995). However, at lower capsaicin concentrations (10 ppm), these same authors saw no effect of capsaicin desensitization on perceived intensity of taste solutions (Karrer, T. and Bartoshuk, L., 1995).

The presence of CO_2 appears to variably modulate the perceived intensity of taste solutions. When added to aqueous solutions of sucrose, citric acid, NaCl, or quinine, CO_2 was found to slightly increase the perceived intensity of salty and sour sensations (Cometto-Muñiz, J. E. *et al.*, 1987). However, these results were not interpreted as a CO_2 enhancement of saltiness or sourness, but instead attributed to subject

confusion between the sensation elicited by low concentrations of NaCl or citric acid with the pungent sensation elicited by CO_2. In a subsequent experiment (Cowart, B. J., 1998), carbonation was found to have no effect on total taste intensity of model taste solutions, but increased the intensity of the sweet and salty subcomponents of the stimulus. On the surface, it is surprising that CO_2 would increase the perceived intensity of some tastants, whereas vanilloids tend to have a suppressive effect. However, CO_2, being an acid, is sour and would therefore be expected to increase the perceived intensity of some taste sensations.

Despite the psychophysical evidence for irritant–taste interactions, it is not clearly understood where they occur. Clearly, in the case of vanilloid-evoked taste sensations (Green, B. G. and Schullery, M. T., 2003; Green, B. G. and Hayes, J. E., 2003; 2004), the compound is acting to excite taste receptor cells. However, in the case of irritant modulation of taste sensations, there are a number of potential sites and mechanisms. For instance, it is possible that irritants and tastants interact at the cognitive level as has been proposed for gustation and olfaction. Given the distinct anatomical separation of the olfactory and gustatory neural pathways, this scenario makes sense. However, in the case of oral irritants and tastants, the two chemical modalities stimulate neurons located within the same oral tissue, most notably the tongue. This provides a substrate by which chemicals (i.e., substance P, calcitonin gene-related peptide (CGRP)) released locally from trigeminal nerve endings in response to a noxious stimulus might modulate the activity of taste receptor cells or gustatory neurons (Wang, Y. *et al.*, 1995; Osada, K. *et al.*, 1997; Simon, S. A. *et al.*, 2003). Therefore, it is possible that intermodality communication occurs peripherally. Other potential sites of interaction also exist. Trigeminal primary afferent neurons project not only to brainstem nuclei involved in processing nociceptive information but also send collateral branches to the gustatory NTS (Beckstead, R. M. and Norgren, R., 1979). Similarly, second-order projections from Vc have also been identified in the NTS (Menétrey, D. and Basbaum, A. I., 1987). Therefore, oral irritant and taste information might converge onto second-order neurons residing in the taste nucleus. Nociceptive input could, therefore, directly modulate NTS gustatory neurons or, alternatively, excite inhibitory gamma-amino-*n*-butyric acid (GABA)-ergic or opioidergic interneurons that have been identified within this area (Lynch, W. C.

et al., 1985; Davis, B. J., 1993; Davis, B. J. and Kream, R. M., 1993). Either mechanism of central interaction could suppress the activity of gustatory cells resulting in alterations to taste perception.

4.18.4.1 Peripheral Interactions

Several investigations have addressed the possibility that trigeminal and gustatory interactions occur in the periphery. In particular, three studies have looked directly at the consequence of lingual nerve stimulation on the response of chorda tympani fibers (Wang, Y. *et al.*, 1995; Osada, K. *et al.*, 1997; Simon, S. A. *et al.*, 2003). In the first (Wang, Y. *et al.*, 1995), chorda tympani responses to NaCl were recorded before, during, and after electrical stimulation of the lingual nerve. It was found that the NaCl-induced activity in these fibers was significantly depressed during simultaneous lingual nerve stimulation (Figure 7). The effect was presumed to be due to the release of neuroactive peptides from the peripheral terminals of trigeminal nociceptive nerve endings that directly affected the responses of gustatory neurons to NaCl. Anatomical studies have shown that taste receptor cells or perigemmal neurons express receptors for a variety of neuropeptides including substance P (Chang, G. Q. *et al.*, 1996) and CGRP (van Rossum, D. *et al.*, 1997), both of which are released from C-fiber terminals following electrical stimulation or via an axon reflex (Holzer, P., 1998).

In the second study (Osada, K. *et al.*, 1997), chorda tympani responses to NaCl or KCl were recorded in the presence or absence of capsaicin. The majority of responses to NaCl and KCl were depressed when capsaicin was presented simultaneously with the gustatory stimulus (Figure 8). Interestingly, in a subpopulation of neurons, capsaicin treatment significantly enhanced the NaCl or KCl response. As in the previous study, the attenuated gustatory response was attributed to the presence of neuroactive peptides released in response to capsaicin.

However, with the completion of more recent studies, alternative explanations are possible. For instance, studies suggest that the amiloride-insensitive pathway mediating NaCl sensitivity involves a TRPV1 variant (Lyall, V. *et al.*, 2004; 2005) that responds to the TRPV1 agonists resiniferatoxin and capsaicin. As such, the possibility exists that simultaneous capsaicin application may directly gate the TRPV1 variant to evoke a heightened NaCl or KCl response. Similarly, if the TRPV1 variant has altered desensitization kinetics, capsaicin application might be speculated to result in attenuated NaCl responses. In addition to the potential modulation of the TRPV1 variant, recent studies confirm that capsaicin inhibits voltage-sensitive ion channels expressed in taste receptor cells (Park, K. *et al.*, 2003, Costa, R. M. *et al.*, 2005). The conductance of outwardly rectifying K-channels was found to be completely inhibited in the presence of 500 nM capsaicin (Park, K. *et al.*, 2003), suggesting a mechanism by which capsaicin could enhance gustatory responses of taste receptor cells. In a similar set of experiments, 30 μM capsaicin was found to inhibit both the peak inward current and the peak outward current by 20% and 30%, respectively (Costa, R. M. *et al.*, 2005). Attenuation of the voltage-sensitive inward current would result in decreased sensitivity to any gustatory stimulus. Interestingly, support for a non-TRPV1-mediated mechanism was observed in a behavioral experiment where the preference for sucrose exhibited by TRPV1-knockout mice was abolished in the presence of capsaicin (Costa, R. M. *et al.*, 2005).

Substance P and CGRP are released from the peripheral terminals of nociceptive nerve endings in response to an activating noxious stimulus (Holzer, P., 1998). The close proximity of nociceptive free nerve endings to taste receptor cells suggests that peptides released might modulate gustatory signaling. To specifically address the direct effect of neuropeptides on gustatory signaling, Simon S. A. *et al.* (2003) recorded chorda tympani taste responses in the presence and

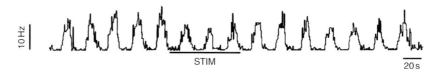

Figure 7 Peripheral trigeminal modulation of taste nerve activity. Shown are responses (in PSTH format) of a single chorda tympani fiber to repeated application of 0.1 M NaCl to the tongue, without and during electrical stimulation of the lingual nerve (0.5 ms biphasic pulses at 15 Hz and 5 V). Note the depressant effect of lingual nerve stimulation on tastant-evoked responses. From Wang, Y., Erickson, R. P., and Simon, S. A. 1995. Modulation of rat chorda tympani nerve activity by lingual nerve stimulation. J. Neurophysiol. 73, 1468–1483, figure 8B, with permission.

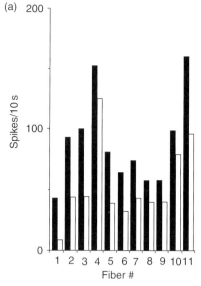

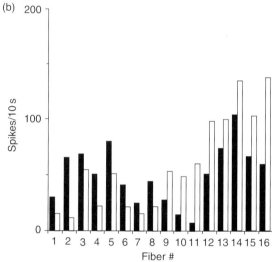

Figure 8 Capsaicin modulation of taste nerve responses. Graphs plot the responses of single N-type (a: responsive to sodium salts) and E-type (b: broadly responsive to cations) chorda tympani fibers. Filled bars show responses to 100 mM NaCl. Open bars show responses to 100 mM NaCl plus 100 ppm capsaicin. Note that responses of N-type fibers were depressed in the presence of capsaicin, while responses of some E-type fibers were depressed and others enhanced in the presence of capsaicin. From Osada, K., Komai, M., Bryant, B. P., Suzuki, H., Goto, A., Tsunoda, K., Kimura, S., and Furukawa, Y. 1997. Capsaicin modifies responses of rat chorda tympani nerve fibers to NaCl. Chem. Senses 22, 249–255, figure 4, with permission.

absence of substance P and CGRP. Whole-nerve recordings in response to lingual NaCl, HCl, quinine HCl, and sucrose were made from the chorda tympani of rats. Simultaneously, substance P or CGRP was infused into the ipsilateral lingual artery. Infusion of CGRP reduced the chorda tympani response to NaCl and HCl to about 80% of preinfusion levels (Figure 9). Importantly, these effects were shown to be independent of any vasoactive or thermal change induced by neuropeptide infusion. Conversely, substance P injection had no noticeable effect on NaCl or HCl when injected at concentrations that did not induce vasodilation or increased lingual temperature. In addition, neither substance P nor CGRP had any apparent effect on sucrose- or quinine-evoked responses. However, for these two taste responses, methodological limitations were thought to have mitigated the reliable detection of possible neuropeptide effects.

Results from central gustatory recordings corroborate the existence of peripheral interactions between the trigeminal and gustatory systems. Responses to sucrose, quinine HCl, monosodium glutamate, NaCl, and citric acid recorded from taste cells in the NTS were shown to be significantly suppressed following prolonged (7 min) lingual capsaicin (100 μM) application (Figure 10) (Simons, C. T. et al., 2003a). Across all tastants, capsaicin suppressed the evoked response to approximately 57% of precapsaicin levels. Interestingly, the same level of capsaicin suppression was observed after bilateral trigeminal ganglionectomy, indicating that an intact trigeminal pathway is not required for capsaicin attenuation of taste responses (Simons, C. T. et al., 2003a).

Anatomical studies further support a peripheral effect of capsaicin (Simons, C. T. et al., 2003a). In addition to the direct effects of neuroactive peptides mentioned previously, substance P and CGRP have potent vasoactive properties that induce changes in the permeability of the vascular wall leading to a localized inflammatory response including plasma extravasation and edema. Tests for plasma extravasation typically involve the systemic injection of dye into the bloodstream followed by a noxious stimulus. When plasma extravasation occurs, dye that has leaked out of the vasculature is found to have stained surrounding tissue. Following lingual capsaicin application, plasma extravasation was found to be localized to the intragemmal space of taste papillae (Figure 11), and this effect was seen in both intact and ganglionectomized animals. This finding provides additional evidence that neuropeptides act in close proximity to taste receptor cells and suggests that edema localized to the taste bud may also play a role in capsaicin-induced taste suppression.

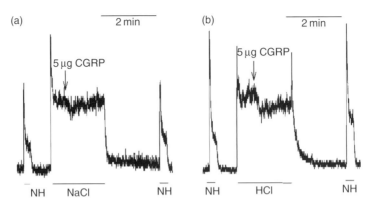

Figure 9 Depressant effects of calcitonin gene-related peptide (CGRP) on chorda tympani fiber responses to tastants. Shown are the effects of lingual artery perfusion of 5 μg CGRP in 100 μl KH buffer (arrows). Responses to 0.1 M NH_4Cl bracket each trial. (a) CGRP (at arrow) decreased the tonic responses to 0.1 M NaCl. (b) CGRP (arrow) decreased response to 10 mM HCl. From Simon, S. A., Liu, L., and Erickson, R. P. 2003. Neuropeptides modulate rat chorda tympani responses. Am. J. Physiol. (Regul. Integr. Comp. Physiol.) 284, R1494–R1505, figure 7.

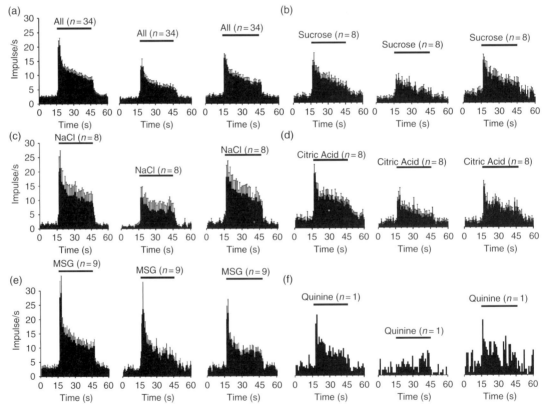

Figure 10 Suppression of tastant-evoked responses of nucleus of the solitary tract (NTS) units by oral capsaicin. Each triad of panels shows averaged peristimulus–time histograms (PSTHs) of NTS unit responses to the indicated tastant, before capsaicin (left PSTHs), immediately after cessation of capsaicin (middle PSTHs), and 12 min after cessation of capsaicin (right PSTHs). Horizontal bars indicate 30 s duration of tastant stimulus. Error bars indicate SEM. (a) Data from all 34 units were pooled to show averaged responses to the best tastant. Note suppression of response (middle panel; to 57%) immediately after capsaicin with recovery (right panel). (b) Sucrose. Note suppression (middle panel; to 63%) immediately after capsaicin with recovery (right panel). (c) NaCl. Note suppression (middle panel; to 44%) immediately after capsaicin with recovery (right panel). (d), Citric acid. Note suppression (middle panel; to 50%) immediately after capsaicin with partial recovery (right panel). (e) Monosodium glutamate (MSG). Note suppression (middle panel; to 73%) immediately after capsaicin with little or no recovery (right panel). (f), Quinine; individual example showing that the response of the unit to quinine (left panel) was depressed immediately after capsaicin (middle panel) with recovery (right panel). From Simons, C. T., Boucher, Y., and Carstens, E. 2003a. Suppression of central taste transmission by oral capsaicin. J. Neurosci. 23, 978–985, figure 2, with permission.

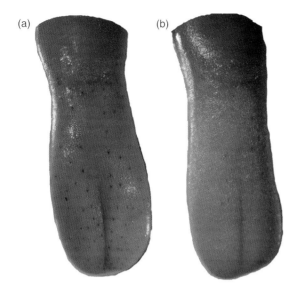

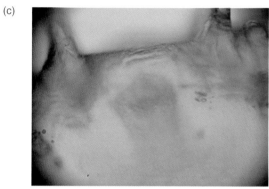

Figure 11 Capsaicin-induced plasma extravasation. (a) Photograph of rat tongue treated previously with topical capsaicin. Note accumulation of blue dye within taste papillae across the lingual surface. (b) Photograph of rat tongue pretreated with ethanol. Note absence of punctuate dye extravasation. (c), Photomicrograph of section through fungiform papillae (100× magnification). The black arrowhead indicates the dorsal lingual surface, whereas the white arrowhead shows a filliform papilla. Note that blue dye accumulates within the taste bud, with little or none in surrounding tissues. From Simons, C. T., Boucher, Y., and Carstens, E. 2003a. Suppression of central taste transmission by oral capsaicin. J. Neurosci. 23, 978–985, figure 6, with permission.

4.18.4.2 Central Interactions – Nucleus of the Solitary Tract

Several investigators have found that second-order neurons located in the caudal portion of the gustatory nucleus respond not only to sapid stimuli but also to touch and temperature (Ogawa, H. and Hayama, T., 1984; Ogawa, H. *et al.*, 1984; 1988; Halsell, C. B. *et al.*, 1993; Travers, S. P. and Norgren, R., 1995). Although

the chorda tympani nerve is well known to respond to tastants as well as to mechanical and thermal stimuli (Ogawa, H. *et al.*, 1968; Robinson, P. P., 1988), these findings nevertheless give rise to the possibility that gustatory and somatosensory (including nociceptive) information converge onto cells located within the gustatory NTS. Anatomic evidence additionally provides support. Primary afferent trigeminal fibers have been shown to send collateral branches into the NTS (Beckstead, R. M. and Norgren, R., 1979), and second-order neurons originating in Vc similarly project onto the same region (Davis, B. J. and Kream, R. M., 1993). Moreover, trigeminal afferent fibers projecting to the NTS contain the neuromodulator substance P (South, E. H. and Ritter, R. C., 1986), and dense neurokinin 1 (substance P receptor) labeling has been observed in the same dorsomedial and dorsolateral subnuclei (Davis, B. J and Smith, D. V., 1997). Physiological findings confirm that substance P has a modulatory effect on gustatory cellular activity. Whole-cell patch-clamp recordings in rat brainstem slices showed the intrinsic activity of NTS units was influenced by the application of substance P (0.1–4 μM) to the bathing medium (King, M. S. *et al.*, 1993). Indeed, substance P induced a transient ($\sim$7 min) depolarization in 70% of the tested neurons that was accompanied by an increase in the input resistance. In contrast, only 4% of neurons exhibited a hyperpolarization following substance P exposure. These findings suggest that substance P has primarily an excitatory effect that is mediated by the closure of ion channels. Similar results were obtained from whole animal recordings where microinjection of substance P into the hamster NTS was found to modulate the responses to electrical or NaCl stimulation of the tongue (Davis, B. J and Smith, D. V. 1997). In 60% of the neurons tested, substance P injection increased the activity evoked by electrical stimulation. In the remaining 40%, substance P substantially reduced (to approximately 50%) the electrically evoked response. Correspondingly, nearly 50% of the NaCl-evoked responses were increased following injection of substance P into the NTS. However, in contrast to the electrical stimulus, only 9% of NaCl-evoked responses were inhibited by substance P. Thus, substance P appears to be a potent and varied modulator of central gustatory cells, although it is still speculative this effect is mediated by trigeminal activity.

Dense innervation of the NTS by inhibitory GABAergic and opioidergic neurons (Lynch, W. C.

et al., 1985; Davis, B. J, 1993; Davis, B. J. and Kream, R. M., 1993) suggests another mechanism of inhibitory modulation. Noxious stimuli directed at the oral cavity could activate trigeminal nociceptors that further excite inhibitory interneurons to directly suppress tastant-evoked activity in NTS.

Functional studies directly testing the hypothesis that trigeminal activity modulates central gustatory responses have recently been completed (Boucher, Y. *et al.*, 2003; Simons, C. T. *et al.*, in press). Single-cell gustatory responses were recorded from rats before and after trigeminal ganglionectomy (Boucher, Y. *et al.*, 2003). Ganglioectomy *per se* did not modulate taste responses, suggesting that trigeminally mediated excitatory or inhibitory processes are not tonically active. However, with simultaneous electrical activation of the central cut end of the mandibular nerve, 64% of gustatory responses were modulated relative to prestimulation levels. Indeed, approximately 21% of the gustatory responses were inhibited by simultaneous mandibular nerve stimulation, whereas 43% of the responses were enhanced (Boucher, Y. *et al.*, 2003). Consistent with a peripheral mode of suppression, central gustatory responses were attenuated when the peripheral cut end of the mandibular nerve was stimulated.

Nicotine activation of trigeminal afferents has a similar effect (Simons, C. T. *et al.*, in press). Topical application of nicotine to the tongue for a period of 4 min suppressed taste responses recorded in NTS. This effect was dose dependent. Whereas low nicotine concentrations (0.84 mM) had no effect, higher doses (8.4 mM or 600 mM) suppressed taste-evoked responses to 54% and 27%, respectively. The attenuation was nearly equal for all taste qualities. Mecamylamine was shown to reduce the nicotine's suppressive effect of taste responses, suggesting the involvement of nicotinic cholinergic receptors. Interestingly, following bilateral ganglionectomy, nicotine application had no effect on taste responses, indicating that nicotine suppression of taste occurs centrally and is mediated through the trigeminal system. This is in marked contrast to capsaicin which was shown to have a peripheral mechanism of suppression.

4.18.4.3 Cognitive Interactions

In addition to the possible peripheral and central mechanisms discussed above, high-level interactions occurring in the cortex might contribute to the effect of oral irritants on taste perception. The convergence of somatosensory and taste pathways into the insular and orbitofrontal cortex provides the neural substrate for these interactions to occur (Cerf-Ducastel, B. *et al.*, 2001). Several studies have identified neurons in the orbitofrontal cortex that respond to gustatory as well as to somatosensory stimuli (Rolls, E. T. *et al.*, 1999; 2003; Verhagen, J. V. *et al.*, 2003; Kadohisa, M. *et al.*, 2004; 2005). In particular, neurons were identified that responded to taste but also to the texture (Rolls, E. T. *et al.*, 1999; 2003; Verhagen, J. V. *et al.*, 2003) and temperature (Kadohisa, M. *et al.*, 2004; 2005) of a stimulus as well as the presence of capsaicin (Kadohisa, M. *et al.*, 2004; 2005). These studies indicate that areas involved in high-level stimulus processing can be modulated by a variety of multimodal inputs to impact both perception and palatability.

There are several possible ways that higher-order somatosensory–taste interactions can result in altered perceptual responses. It is well known that attention is a primary determinant of perception (Marks, L. E. and Wheeler, M. E., 1998; Driver, J. and Frackowiak, R. S., 2001). Attending to a stimulus brings it into the consciousness where it can be evaluated and processed. It is plausible, under some paradigms, that the presence of oral irritation draws attention away from the gustatory stimulus such that it is perceived as less intense. Similarly, context effects have often been shown to impact perceived stimulus intensity. Any evaluation of stimulus strength is made in the context of other stimuli. When a given stimulus is evaluated in the presence of a stronger stimulus, intensity of the first stimulus is often reported as much less than when that same stimulus is evaluated alone or in the presence of a weaker stimulus. Changes to intensity judgments due to context have been explained by Parducci's range–frequency theory (Parducci, A., 1965). Two studies assessed the context effect in regard to the mechanism by which trigeminal irritants modulate taste perceptions (Gilmore, M. M. and Green, B. G., 1993; Simons, C. T. *et al.*, 2002).

In one study (Simons, C. T. *et al.*, 2002), context effect was minimized by using a half-tongue methodology in which capsaicin was applied unilaterally but taste intensity was evaluated bilaterally. In this condition, context effect would be expected to affect perceived taste on both sides equally not just the side receiving capsaicin pretreatment (Simons, C. T. *et al.*, 2002). Moreover, context effect is a generalized phenomenon and would not be expected to impact only specific taste qualities. However, these results were not observed. In fact, capsaicin irritation suppressed only the perceived intensity of sweetness, bitterness, and umami sensations. Another study noted that context effect plays a role in the suppression of salt and sour tastes following zingerone desensitization as well as playing lesser role following capsaicin desensitization (Gilmore, M. M. and Green, B. G., 1993). Thus, psychological mechanisms including attention and context effect may contribute to the chemesthetic suppression of taste observed both anecdotally and experimentally, and this is a factor that should be controlled for.

Masking is another mechanism by which the presence of oral irritants might impact the perception of gustatory stimuli. Masking is a phenomenon that occurs when the presence of one stimulus obscures the presence of another. In most cases, masking can be achieved simply by increasing the intensity of a second stimulus. In the food and beverage industry, off-notes related to a particular ingredient can be masked by increasing the concentration of a more desirable ingredient. Thus, the presence of a bitter stimulus is often masked by increasing the level of sucrose or other palatable substance. There are limitations as to how effective masking can be. For instance, it is often difficult to mask very intense stimuli. Additionally, often masking agents have a concentration at which they are most effective. Going beyond that level can result in additional undesirable attributes. Capsaicin might serve as a masking agent. The perception of an irritating sensation could mask the sensation evoked by gustatory stimuli. The convergence of taste and somatosensory information in the orbitofrontal cortex (Kadohisa, M. *et al.*, 2004; 2005) provides a mechanism by which a chemoirritants can modulate gustatory information. However, as with context effects, masking is a general phenomena, and it would not be expected to impact only select taste qualities.

4.18.5 Perspectives

The study of the chemical senses is beginning to flourish. In particular, chemesthesis research is attracting the attention of a greater number of investigators. Light is beginning to be shed on questions that were unresolved just a few years ago. However, to further the knowledge of chemesthesis, a multidisciplinary approach using genetic, molecular, physiological, and perceptual methods will be needed. Unlike many sensory systems, chemesthetic information appears to be transmitted relatively

faithfully from the periphery to higher centers of the brain. This provides an ideal opportunity to link perceptual endpoints to the neural and molecular correlates that drive them. As the era of molecular biology continues to unfold, undoubtedly additional receptors will be identified that contribute to chemesthetic perception. It will be imperative to understand how these receptors are distributed across trigeminal neurons and how they contribute to the sensations elicited by chemical stimuli. In addition, further studies are needed to delineate whether qualitative differences exist between the sensations elicited by different chemoirritants. The cross-modal interactions occurring between chemesthesis and taste are deserving of more attention. It is clear that multiple mechanisms contribute to the ability of oral irritants to suppress gustatory processing. However, to date, only a few of the many chemesthetic stimuli have been evaluated for effects on taste. It will be interesting to determine whether taste suppression is a phenomenon common to all oral irritants or, alternatively, if only select compounds have this capacity.

References

Abe, J., Hosokawa, H., Okazawa, M., Kandachi, M., Sawada, Y., Yamanaka, K., Matsumura, K., and Kobayashi, S. 2005. TRPM8 protein localization in trigeminal ganglion and taste papillae. Brain Res. Mol. Brain Res. 136, 91–98.

Amano, N., Hu, J. W., and Sessle, B. J. 1986. Responses of neurons in feline trigeminal subnucleus caudalis (medullary dorsal horn) to cutaneous, intraoral, and muscle afferent stimuli. J. Neurophysiol. 55, 227–243.

Andersson, D. A., Chase, H. W., and Bevan, S. 2004. TRPM8 activation by menthol, icilin, and cold is differentially modulated by intracellular pH. J. Neurosci. 24, 5364–5369.

Bandell, M., Story, G. M., Hwang, S. W., Viswanath, V., Eid, S. R., Petrus, M. J., Earley, T. J., and Patapoutian, A. 2004. Noxious cold ion channel TRPA1 is activated by pungent compounds and bradykinin. Neuron 41, 849–857.

Barry, M. A., Halsell, C. B., and Whitehead, M. C. 1993. Organization of the nucleus of the solitary tract in the hamster: acetylcholinesterase, NADH dehydrogenase, and cytochrome oxidase histochemistry. Microsc. Res. Tech. 26, 231–244.

Bautista, D. M., Jordt, S.-V., Nikai, T., Tsuruda, P. R., Read, A. J., Poblete, J., Yamoah, E. N., Basbaum, A. I., and Julius, D. 2006. TRPA1 mediates the inflammatory actions of environmental irritants and proalgesic agents. Cell 124(6), 1269–1282.

Bautista, D. M., Movahed, P., Hinman, A., Axelsson, H. E., Sterner, O., Hogestatt, E. D., Julius, D., Jordt, S. E., and Zygmunt, P. M. 2005. Pungent products from garlic activate the sensory ion channel TRPA1. Proc. Natl. Acad. Sci. U. S. A. 102, 12248–12252.

Beckstead, R. M. and Norgren, R. 1979. An autoradiographic examination of the central distribution of the trigeminal, facial, glossopharyngeal, and vagal nerves in the monkey. J. Comp. Neurol 184, 455–472.

Boucher, Y., Simons, C. T., Cuellar, J. M., Jung, S. W., Carstens, M. I., and Carstens, E. 2003. Activation of brain stem neurons by irritant chemical stimulation of the throat assessed by c-fos immunohistochemistry. Exp. Brain Res. 148, 211–218.

Bryant, B. P. and Mezine, I. 1999. Alkylamides that produce tingling paresthesia activate tactile and thermal trigeminal neurons. Brain Res. 842, 452–460.

Bryant, B. P. and Moore, P. A. 1995. Factors affecting the sensitivity of the lingual trigeminal nerve to acids. Am. J. Physiol. 268, R58–65.

Carstens, E., Iodi Carstens, M., Dessirier, J. M., O'Mahony, M., Simons, C. T., Sudo, M., and Sudo, S. 2002. It hurts so good: oral irritation by spices and carbonated drinks and the underlying neural mechanisms. Food Qual. Pref. 13, 431–443.

Carstens, E., Kuenzler, N., and Handwerker, H. O. 1998. Activation of neurons in rat trigeminal subnucleus caudalis by different irritant chemicals applied to oral or ocular mucosa. J. Neurophysiol. 80, 4654–4692.

Carstens, E., Merrill, A. W., Zanotto, K., and Iodi Carstens, M. 2005 Cold- and menthol-sensitive neurons in superficial trigeminal subnucleus caudalis (Vc). Program No. 747.4. 2005 Abstract Viewer/Itinerary Planner. Society for Neuroscience.

Carstens, E., Saxe, I., and Ralph, R. 1995. Brainstem neurons expressing c-Fos immunoreactivity following irritant chemical stimulation of the rat's tongue. Neuroscience 69, 939–953.

Carstens, E., Simons, C. T., Dessirier, J.-M., Iodi-Carstens, M., and Jinks, S. L. 2001. Role of neuronal nicotinic acetylcholine receptors in the activation of neurons in the trigeminal subnucleus caudalis by nicotine delivered to the oral mucosa. Exp. Brain Res. 132, 375–383.

Caterina, M. J., Leffler, A., Malmberg, A. B., Martin, W. J., Trafton, J., Petersen-Zeitz, K. R., Koltzenburg, M., Basbaum, A. I., and Julius, D. 2000. Impaired nociception and pain sensation in mice lacking the capsaicin receptor [see comments]. Science 288, 306–313.

Caterina, M. J., Rosen, T. A., Tominaga, M., Brake, A. J., and Julius, D. 1999. A capsaicin-receptor homologue with a high threshold for noxious heat. Nature 398, 436–441.

Caterina, M. J., Schumacher, M. A., Tominaga, M., Rosen, T. A., Levine, J. D., and Julius, D. 1997. The capsaicin receptor: a heat-activated ion channel in the pain pathway [see comments]. Nature 389, 816–824.

Cerf-Ducastel, B., Van de Moortele, P. F., MacLeod, P., Le Bihan, D., and Faurion, A. 2001. Interaction of gustatory and lingual somatosensory perceptions at the cortical level in the human: a functional magnetic resonance imaging study. Chem. Senses 26, 371–383.

Cesare, P., Moriondo, A., Vellani, V., and McNaughton, P. A. 1999. Ion channels gated by heat. Proc. Natl. Acad. Sci. U.S.A. 96, 7658–7563.

Chang, G. Q., Vigna, S. R., and Simon, S. A. 1996. Localization of substance P NK-1 receptors in rat tongue. Regul. Pept. 63, 85–89.

Chen, Y. and Herness, M. S. 1997. Electrophysiological actions of quinine on voltage-dependent currents in dissociated rat taste cells. Pflugers Arch. 434, 215–226.

Chiang, C. Y., Hu, J. W., and Sessle, B. J. 1997. NMDA receptor involvement in neuroplastic changes induced by neonatal capsaicin treatment in trigeminal nociceptive neurons. J. Neurophysiol. 78, 2799–2803.

Cholewinski, A., Burgess, G. M., and Bevan, S. 1993. The role of calcium in capsaicin-induced desensitization in rat cultured dorsal root ganglion neurons. Neuroscience 55, 1015–1023.

Chuang, H. H., Neuhausser, W. M., and Julius, D. 2004. The super-cooling agent icilin reveals a mechanism of coincidence detection by a temperature-sensitive TRP channel. Neuron 43, 859–869.

Clapham, D. E. 2003. TRP channels as cellular sensors. Nature 426, 517–524.

Cliff, M. A. and Green, B. G. 1994. Sensory irritation and coolness produced by menthol: evidence for selective desensitization of irritation. Physiol. Behav. 56, 1021–1029.

Cliff, M. A. and Green, B. G. 1996. Sensitization and desensitization to capsaicin and menthol in the oral cavity: interactions and individual differences. Physiol. Behav. 59, 487–494.

Coimbra, F. and Coimbra, A. 1994. Dental noxious input reaches the subnucleus caudalis of the trigeminal complex in the rat, as shown by c-fos expression upon thermal or mechanical stimulation. Neurosci. Lett. 173, 201–204.

Cometto-Muñiz, J. E., García-Medina, M. R., Calviño, A. M., and Noriega, G. 1987. Interactions between CO_2 oral pungency and taste. Perception 16, 629–640.

Costa, R. M., Liu, L., Nicolelis, M. A., and Simon, S. A. 2005. Gustatory Effects of capsaicin that are independent of TRPV1 receptors. Chem. Senses 30, i198–i200.

Cowart, B. J. 1987. Oral chemical irritation: does it reduce perceived taste intensity? Chem. Senses 12, 467–479.

Cowart, B. J. 1998. The addition of CO_2 to traditional taste solutions alters taste quality. Chem. Senses 23, 397–402.

Critchley, H. D. and Rolls, E. T. 1996. Hunger and satiety modify the responses of olfactory and visual neurons in the primate orbitofrontal cortex. J. Neurophysiol. 75, 1673–1686.

Dallel, R., Luccarini, P., Molat, J. L., and Woda, A. 1996. Effects of systemic morphine on the activity of convergent neurons of spinal trigeminal nucleus oralis in the rat. Eur. J. Pharmacol. 314, 19–25.

Dallel, R., Raboisson, P., Woda, A., and Sessle, B. J. 1990. Properties of nociceptive and non-nociceptive neurons in trigeminal subnucleus oralis of the rat. Brain Res. 521, 95–106.

Danilova, V. and Hellekant, G. 2002. Oral sensation of ethanol in a primate model III: responses in the lingual branch of the trigeminal nerve of Macaca mulatta. Alcohol 26, 3–16.

Davis, B. J. 1993. GABA-like immunoreactivity in the gustatory zone of the nucleus of the solitary tract in the hamster: light and electron microscopic studies. Brain Res. Bull. 30, 69–77.

Davis, B. J. and Kream, R. M. 1993. Distribution of tachykinin- and opioid-expressing neurons in the hamster solitary nucleus: an immuno- and in situ hybridization histochemical study. Brain Res. 616, 6–16.

Davis, B. J. and Smith, D. V. 1997. Substance P modulates taste responses in the nucleus of the solitary tract of the hamster. Neuroreport 8, 1723–1727.

Dessirier, J. M., Nguyen, N., Sieffermann, J. M., Carstens, E., and O'Mahony, M. 1999. Oral irritant properties of piperine and nicotine: psychophysical evidence for asymmetrical desensitization effects. Chem. Senses 24, 405–413.

Dessirier, J. M., O'Mahony, M., and Carstens, E. 1997. Oral irritant effects of nicotine: psychophysical evidence for decreased sensation following repeated application and lack of cross-desensitization to capsaicin. Chem. Senses 22, 483–492.

Dessirier, J. M., O'Mahony, M., and Carstens, E. 2001a. Oral irritant properties of menthol: sensitizing and desensitizing effects of repeated application and cross-desensitization to nicotine. Physiol. Behav. 73, 25–36.

Dessirier, J. M., O'Mahony, M., Iodi-Carstens, M., and Carstens, E. 2000a. Sensory properties of citric acid: psychophysical evidence for sensitization, self-desensitization, cross-desensitization and cross-stimulus-induced recovery following capsaicin. Chem. Senses 25, 769–780.

Dessirier, J. M., O'Mahony, M., Iodi-Carstens, M., Yao, E., and Carstens, E. 2001b. Oral irritation by sodium chloride: sensitization, self-desensitization, and cross-sensitization to capsaicin. Physiol. Behav. 72, 317–324.

Dessirier, J. M., O'Mahony, M., Sieffermann, J. M., and Carstens, E. 1998. Mecamylamine inhibits nicotine but not capsaicin irritation on the tongue: psychophysical evidence that nicotine and capsaicin activate separate molecular receptors. Neurosci. Lett. 240, 65–68.

Dessirier, J. M., Simons, C. T., O'Mahony, M., and Carstens, E. 2001c. The oral sensation of carbonated water: cross-desensitization by capsaicin and potentiation by amiloride. Chem. Senses 26, 639–643.

Dessirier, J. M., Simons, C. T., Sudo, M., Sudo, S., and Carstens, E. 2000b. Sensitization, desensitization and stimulus-induced recovery of trigeminal neuronal responses to oral capsaicin and nicotine. J. Neurophysiol. 84, 1851–1862.

Driver, J. and Frackowiak, R. S. 2001. Neurobiological measures of human selective attention. Neuropsychologia 39, 1257–1262.

Dubner, R. and Bennett, G. J. 1983. Spinal and trigeminal mechanisms of nociception. Annu. Rev. Neurosci. 6, 381–418.

Feil, K. and Herbert, H. 1995. Topographic organization of spinal and trigeminal somatosensory pathways to the rat parabrachial and Kolliker–Fuse nuclei. J. Comp. Neurol. 353, 506–528.

Finger, T. 1986. Peptide immunohistochemistry demonstrates multiple classes of perigemmal nerve fibers in the circumvallate papilla of the rat. Chem. Senses 11, 135–144.

Geppetti, P., Tramontana, M., Del Bianco, E., and Fusco, B. M. 1993. Capsaicin-desensitization to the human nasal mucosa selectively reduces pain evoked by citric acid. Br. J. Clin. Pharmacol. 35, 178–183.

Gilmore, M. M. and Green, B. G. 1993. Sensory irritation and taste produced by NaCl and citric acid: effects of capsaicin desensitization. Chem. Senses 18, 257–272.

Green, B. G. 1989. Capsaicin sensitization and desensitization on the tongue produced by brief exposures to a low concentration. Neurosci. Lett. 107, 173–178.

Green, B. G. 1992. The sensory effects of l-menthol on human skin. Somatosens. Mot. Res. 9, 235–244.

Green, B. G. 1996. Rapid recovery from capsaicin desensitization during recurrent stimulation. Pain 68, 245–253.

Green, B. G. 1998. Capsaicin desensitization and stimulus-induced recovery on facial compared to lingual skin. Physiol. Behav. 65, 517–523.

Green, B. G. and Hayes, J. E. 2003. Capsaicin as a probe of the relationship between bitter taste and chemesthesis. Physiol. Behav. 79, 811–821.

Green, B. G. and Hayes, J. E. 2004. Individual differences in perception of bitterness from capsaicin, piperine and zingerone. Chem. Senses 29, 53–60.

Green, B. G. and McAuliffe, B. L. 2000. Menthol desensitization of capsaicin irritation. Evidence of a short-term anti-nociceptive effect. Physiol. Behav. 68, 631–639.

Green, B. G. and Rentmeister-Bryant, H. 1998. Temporal characteristics of capsaicin desensitization and stimulus-induced recovery in the oral cavity. Physiol. Behav. 65, 141–149.

Green, B. G. and Schullery, M. T. 2003. Stimulation of bitterness by capsaicin and menthol: differences between lingual areas innervated by the glossopharyngeal and chorda tympani nerves. Chem. Senses 28, 45–55.

Green, B. G., Mason, J. R., and Kare, M. R. 1990. General Discussion. In: Chemical Senses, Irritation, (eds. B. G. Green, J. R., Mason, and M. R. Kare), Vol. 2. pp. 325–340. Dekker.

Greenwood, L. F. and Sessle, B. J. 1976. Inputs to trigeminal brain stem neurones from facial, oral, tooth pulp and pharyngolaryngeal tissues: II. Role of trigeminal nucleus caudalis in modulating responses to innocuous and noxious stimuli. Brain Res. 117, 227–238.

Guler, A. D., Lee, H., Iida, T., Shimizu, I., Tominaga, M., and Caterina, M. 2002. Heat-evoked activation of the ion channel, TRPV4. J. Neurosci. 22, 6408–6414.

Halsell, C. B., Travers, J. B., and Travers, S. P. 1993. Gustatory and tactile stimulation of the posterior tongue activate overlapping but distinctive regions within the nucleus of the solitary tract. Brain Res. 632, 161–173.

Hamilton, R. B. and Norgren, R. 1984. Central projections of gustatory nerves in the rat. J. Comp. Neurol. 222, 560–577.

Hayakawa, T., Maeda, S., Tanaka, K., and Seki, M. 2005. Fine structural survey of the intermediate subnucleus of the nucleus tractus solitarii and its glossopharyngeal afferent terminals. Anat. Embryol. (Berl.) 210, 235–244.

Hayashi, H. 1985. Morphology of terminations of small and large myelinated trigeminal primary afferent fibers in the cat. J. Comp. Neurol. 240, 71–89.

Hayashi, H. and Tabata, T. 1989. Physiological properties of sensory trigeminal neurons projecting to mesencephalic parabrachial area in the cat. J. Neurophysiol. 61, 1153–1160.

Hayashi, H., Sumino, R., and Sessle, B. J. 1984. Functional organization of trigeminal subnucleus interpolaris: nociceptive and innocuous afferent inputs, projections to thalamus, cerebellum, and spinal cord, and descending modulation from periaqueductal gray. J. Neurophysiol. 51, 890–905.

Holzer, P. 1998. Neurogenic vasodilatation and plasma leakage in the skin. General Pharmacol. 30, 5–11.

Hu, J. W. 1990. Response properties of nociceptive and non-nociceptive neurons in the rat's trigeminal subnucleus caudalis (medullary dorsal horn) related to cutaneous and deep craniofacial afferent stimulation and modulation by diffuse noxious inhibitory controls. Pain 41, 331–345.

Hu, J. W. and Sessle, B. J. 1984. Comparison of responses of cutaneous nociceptive and nonnociceptive brain stem neurons in trigeminal subnucleus caudalis (medullary dorsal horn) and subnucleus oralis to natural and electrical stimulation of tooth pulp. J. Neurophysiol. 52, 39–53.

Hu, J. W., Dostrovsky, J. O., and Sessle, B. J. 1981. Functional properties of neurons in cat trigeminal subnucleus caudalis (medullary dorsal horn). I. Responses to oral-facial noxious and nonnoxious stimuli and projections to thalamus and subnucleus oralis. J. Neurophysiol. 45, 173–192.

Hu, J. W., Sessle, B. J., Raboisson, P., Dallel, R., and Woda, A. 1992. Stimulation of craniofacial muscle afferents induces prolonged facilitatory effects in trigeminal nociceptive brain-stem neurones. Pain 48, 53–60.

Hu, J. W., Sun, K. Q., Vernon, H., and Sessle, B. J. 2005. Craniofacial inputs to upper cervical dorsal horn: implications for somatosensory information processing. Brain Res. 1044, 93–106.

Ito, J. and Oyagi, S. 1995. Localization of trigeminal sensory innervation to the tongue and pharynx of the cat using horseradish peroxidase as tracer. Eur. Arch. Otorhinolaryngol. 252, 172–175.

Jacquin, M. F. and Rhoades, R. W. 1990. Cell structure and response properties in the trigeminal subnucleus oralis. Somatosens. Mot. Res. 7, 265–288.

Jacquin, M. F., Renehan, W. E., Mooney, R. D., and Rhoades, R. W. 1986. Structure–function relationships in rat medullary and cervical dorsal horns. I. Trigeminal primary afferents. J. Neurophysiol. 55, 1153–1186.

Jacquin, M. F., Semba, K., Rhoades, R. W., and Egger, M. D. 1982. Trigeminal primary afferents project bilaterally to dorsal horn and ipsilaterally to cerebellum, reticular formation, and cuneate, solitary, supratrigeminal and vagal nuclei. Brain Res. 246, 285–291.

Jordt, S. E., Bautista, D. M., Chuang, H. H., McKemy, D. D., Zygmunt, P. M., Hogestatt, E. D., Meng, I. D., and Julius, D. 2004. Mustard oils and cannabinoids excite sensory nerve fibres through the TRP channel ANKTM1. Nature 427, 260–265.

Jordt, S. E., Tominaga, M., and Julius, D. 2000. Acid potentiation of the capsaicin receptor determined by a key extracellular site. Proc. Natl. Acad. Sci. U. S. A. 97, 8134–8139.

Kadohisa, M., Rolls, E. T., and Verhagen, J. V. 2004. Orbitofrontal cortex: neuronal representation of oral temperature and capsaicin in addition to taste and texture. Neuroscience 127, 207–221.

Kadohisa, M., Rolls, E. T., and Verhagen, J. V. 2005. Neuronal representations of stimuli in the mouth: the primate insular taste cortex, orbitofrontal cortex and amygdala. Chem. Senses 30, 401–419.

Kandel, E. R., Schwartz, J. H., and Jessell, T. A. 2000. Principles of Neural Science, 4th edn. McGraw-Hill.

Karrer, T. and Bartoshuk, L. 1995. Effects of capsaicin desensitization on taste in humans. Physiol. Behav. 57, 421–429.

Katz, D. B., Simon, S. A., Moody, A., and Nicolelis, M. A. 1999. Simultaneous reorganization in thalamocortical ensembles evolves over several hours after perioral capsaicin injections. J. Neurophysiol. 82, 963–977.

King, M. S., Wang, L., and Bradley, R. M. 1993. Substance P excites neurons in the gustatory zone of the rat nucleus tractus solitarius. Brain Res. 619, 120–130.

Kobayashi, K., Fukuoka, T., Obata, K., Yamanaka, H., Dai, Y., Tokunaga, A., and Noguchi, K. 2005. Distinct expression of TRPM8, TRPA1, and TRPV1 mRNAs in rat primary afferent neurons with adelta/c-fibers and colocalization with trk receptors. J. Comp. Neurol. 493, 596–606.

Komai, M. and Bryant, B. P. 1993. Acetazolamide specifically inhibits lingual trigeminal nerve responses to carbon dioxide. Brain Res. 612, 122–129.

Koplas, P. A., Rosenberg, R. L., and Oxford, G. S. 1997. The role of calcium in the desensitization of capsaicin responses in rat dorsal root ganglion neurons. J. Neurosci. 17, 3525–3537.

Kruger, L. and Michel, F. 1962. A single neuron analysis of buccal cavity representation in the sensory trigeminal complex of the cat. Arch. Oral Biol. 7, 491–503.

LaMotte, R. H. and Mountcastle, V. B. 1979. Disorders in somesthesis following lesions of parietal lobe. J. Neurophysiol. 42, 400–419.

Lawless, H. T. and Stevens, D. A. 1984. Effects of oral chemical irritation on taste. Physiol. Behav. 32, 995–998.

Lawless, H. T., Rozin, P., and Shenker, J. 1985. Effect of oral capsaicin on gustatory, olfactory and irritant sensations and flavor identification in humans who regularly or rarely consume chili pepper. Chem. Senses 10, 579–589.

Lingueglia, E., de Weille, J. R., Bassilana, F., Heurteaux, C., Sakai, H., Waldmann, R., and Lazdunski, M. 1997. A modulatory subunit of acid sensing ion channels in brain and dorsal root ganglion cells. J. Biol. Chem. 272, 29778–29783.

Liu, L. and Simon, S. A. 1996a. Capsaicin and nicotine both activate a subset of rat trigeminal ganglion neurons. Am. J. Physiol. 270, C1807–C1814.

Liu, L. and Simon, S. A. 1996b. Capsaicin-induced currents with distinct desensitization and Ca2+ dependence in rat trigeminal ganglion cells. J. Neurophysiol. 75, 1503–1514.

Liu, L. and Simon, S. A. 1996c. Similarities and differences in the currents activated by capsaicin, piperine, and zingerone in rat trigeminal ganglion cells. J. Neurophysiol. 76, 1858–1869.

Liu, L. and Simon, S. A. 1996d. Capsaicin and nicotine both activate a subset of rat trigeminal ganglion neurons. Am. J. Physiol. (Cell. Physiol.) 270, C1807–C1814.

Liu, L. and Simon, S. A. 1997. Capsazepine, a vanilloid receptor antagonist, inhibits nicotinic acetylcholine receptors in rat trigeminal ganglia. Neurosci. Lett. 228, 29–32.

Liu, L. and Simon, S. A. 1998. The influence of removing extracellular Ca^{2+} in the desensitization responses to capsaicin, zingerone and olvanil in rat trigeminal ganglion neurons. Brain Res. 809, 246–252.

Liu, L., Chang, G. Q., Jiao, Y. Q., and Simon, S. A. 1998. Neuronal nicotinic acetylcholine receptors in rat trigeminal ganglia. Brain Res. 809, 238–245.

Liu, L., Lo, Y., Chen, I., and Simon, S. A. 1997. The responses of rat trigeminal ganglion neurons to capsaicin and two nonpungent vanilloid receptor agonists, olvanil and glyceryl nonamide. J. Neurosci. 17, 4101–4111.

Lundy, R. F., Jr. and Contreras, R. J. 1994. Neural responses of thermal-sensitive lingual fibers to brief menthol stimulation. Brain Res. 641, 208–216.

Lundy, R. F., Jr. and Contreras, R. J. 1995. Tongue adaptation temperature influences lingual nerve responses to thermal and menthol stimulation. Brain Res. 676, 169–177.

Lyall, V., Heck, G. L., Vinnikova, A. K., Ghosh, S., Phan, T. H., Alam, R. I., Russell, O. F., Malik, S. A., Bigbee, J. W., and DeSimone, J. A. 2004. The mammalian amiloride-insensitive non-specific salt taste receptor is a vanilloid receptor-1 variant. J. Physiol. 558, 147–159.

Lyall, V., Heck, G. L., Vinnikova, A. K., Ghosh, S., Phan, T. H., and Desimone, J. 2005. A novel vanilloid receptor-1 (VR-1) variant mammalian salt taste receptor. Chem. Senses 30, i42–i43.

Lynch, W. C., Watt, J., Krall, S., and Paden, C. M. 1985. Autoradiographic localization of kappa opiate receptors in CNS taste and feeding areas. Pharmacol. Biochem. Behav. 22, 699–705.

Macpherson, L. J., Geierstanger, B. H., Viswanath, V., Bandell, M., Eid, S. R., Hwang, S., and Patapoutian, A. 2005. The pungency of garlic: activation of TRPA1 and TRPV1 in response to allicin. Curr. Biol. 15, 929–934.

Marfurt, C. F. and Rajchert, D. M. 1991. Trigeminal primary afferent projections to ''non-trigeminal'' areas of the rat central nervous system. J. Comp. Neurol. 303, 489–511.

Marks, L. E. and Wheeler, M. E. 1998. Attention and detectability of weak taste stimuli. Chem. Senses 23, 19–29.

Marubio, L. M., del Mar Arroyo-Jimenez, M., Cordero-Erausquin, M., Lena, C., Le Novere, N., de Kerchove d'Exaerde, A., Huchet, M., Damaj, M. I., and Changeux, J. P. 1999. Reduced antinociception in mice lacking neuronal nicotinic receptor subunits. Nature 398, 805–810.

McKemy, D. D., Neuhausser, W. M., and Julius, D. 2002. Identification of a cold receptor reveals a general role for TRP channels in thermosensation. Nature 416, 52–58.

Menétrey, D. and Basbaum, A. I. 1987. Spinal and trigeminal projections to the nucleus of the solitary tract: a possible substrate for somatovisceral and viscerovisceral reflex activation. J. Comp. Neurol. 255, 439–450.

Millan, M. J. 1999. The induction of pain: an integrative review. Prog. Neurobiol. 57, 1–164.

Mizuno, A., Matsumoto, N., Imai, M., and Suzuki, M. 2003. Impaired osmotic sensation in mice lacking TRPV4. Am. J. Physiol. (Cell. Physiol.) 285, C96–C101.

Moqrich, A., Hwang, S. W., Earley, T. J., Petrus, M. J., Murray, A. N., Spencer, K. S., Andahazy, M., Story, G. M., and Patapoutian, A. 2005. Impaired thermosensation in mice lacking TRPV3, a heat and camphor sensor in the skin. Science 307, 1468–1472.

Nagy, I. and Rang, H. P. 1999. Similarities and differences between the responses of rat sensory neurons to noxious heat and capsaicin. J. Neurosci. 19, 10647–10655.

Nagy, J. I., Goedert, M., Hunt, S. P., and Bond, A. 1982. The nature of the substance P-containing nerve fibres in taste papillae of the rat tongue. Neuroscience 7, 3137–3151.

Nord, S. G. and Young, R. F. 1979. Effects of chronic descending tractotomy on the response patterns of neurons in the trigeminal nuclei principalis and oralis. Exp. Neurol. 65, 355–372.

Ogawa, H. and Hayama, T. 1984. Receptive fields of solitario-parabrachial relay neurons responsive to natural stimulation of the oral cavity in rats. Exp. Brain Res. 54, 359–366.

Ogawa, H., Hayama, T., and Yamashita, Y. 1988. Thermal sensitivity of neurons in a rostral part of the rat solitary tract nucleus. Brain Res. 454, 321–331.

Ogawa, H., Imoto, T., and Hayama, T. 1984. Responsiveness of solitario-parabrachial relay neurons to taste and mechanical stimulation applied to the oral cavity in rats. Exp. Brain Res. 54, 349–358.

Ogawa, H., Sato, M., and Yamashita, S. 1968. Multiple sensitivity of chorda typani fibres of the rat and hamster to gustatory and thermal stimuli. J. Physiol. 199, 223–240.

Ohya, A. 1992. Responses of trigeminal subnucleus interpolaris neurons to afferent inputs from deep oral structures. Brain Res. Bull. 29, 773–781.

Ohya, A., Tsuruoka, M., Imai, E., Fukunaga, H., Shinya, A., Furuya, R., Kawawa, T., and Matsui, Y. 1993. Thalamic- and cerebellar-projecting interpolaris neuron responses to afferent inputs. Brain Res. Bull. 32, 615–621.

Olson, T. H., Riedl, M. S., Vulchanova, L., Ortiz-Gonzalez, X. R., and Elde, R. 1998. An acid sensing ion channel (ASIC) localizes to small primary afferent neurons in rats. Neuroreport 9, 1109–1113.

O'Mahony, M., Kingsley, L., Harji, A., and Davies, M. 1976. What sensation signals the salt taste threshold? Chem. Senses Flavor 2, 177–188.

Osada, K., Komai, M., Bryant, B. P., Suzuki, H., Goto, A., Tsunoda, K., Kimura, S., and Furukawa, Y. 1997. Capsaicin modifies responses of rat chorda typani nerve fibers to NaCl. Chem. Senses 22, 249–255.

Parducci, A. 1965. Category judgment: a range-frequency model. Psychol. Rev. 72, 407–418.

Park, K., Brown, P. D., Kim, Y. B., and Kim, J. S. 2003. Capsaicin modulates K+ currents from dissociated rat taste receptor cells. Brain Res. 962, 135–143.

Parker, G. H. 1912. The relation of smell, taste and the common chemical sense in vertebrates. J. Acad. Natl. Sci. 15, 221–234.

Peier, A. M., Moqrich, A., Hergarden, A. C., Reeve, A. J., Andersson, D. A., Story, G. M., Earley, T. J., Dragoni, I., McIntyre, P., Bevan, S., and Patapoutian, A. 2002a. A TRP channel that senses cold stimuli and menthol. Cell 108, 705–715.

Peier, A. M., Reeve, A. J., Andersson, D. A., Moqrich, A., Earley, T. J., Hergarden, A. C., Story, G. M., Colley, S., Hogenesch, J. B., McIntyre, P., Bevan, S., and Patapoutian, A. 2002b. A heat-sensitive TRP channel expressed in keratinocytes. Science 296, 2046–2049.

Perez, C. A., Huang, L., Rong, M., Kozak, J. A., Preuss, A. K., Zhang, H., Max, M., and Margolskee, R. F. 2002. A transient receptor potential channel expressed in taste receptor cells. Nat. Neurosci. 5, 1169–1176.

Pettit, M. J. and Schwark, H. D. 1996. Capsaicin-induced rapid receptive field reorganization in cuneate neurons. J. Neurophysiol. 75, 1117–1125.

Prescott, J. and Stevenson, R. J. 1995. Effects of oral chemical irritation on tastes and flavors in frequent and infrequent users of chili. Physiol. Behav. 58, 1117–1127.

Prescott, J. and Stevenson, R. J. 1996a. Desensitization to oral zingerone irritation: effects of stimulus parameters. Physiol. Behav. 60, 1473–1480.

Prescott, J. and Stevenson, R. J. 1996b. Psychophysical responses to single and multiple presentations of the oral irritant zingerone: relationship to frequency of chili consumption. Physiol. Behav. 60, 617–624.

Prescott, J. and Swain-Campbell, N. 2000. Responses to repeated oral irritation by capsaicin, cinnamaldehyde and ethanol in PROP tasters and non-tasters. Chem. Senses 25, 239–246.

Prescott, J., Allen, S., and Stephens, L. 1993. Interactions between oral chemical irritation, taste and temperature. Chem. Senses 18, 389–404.

Price, M. P., McIlwrath, S. L., Xie, J., Cheng, C., Qiao, J., Tarr, D. E., Sluka, K. A., Brennan, T. J., Lewin, G. R., and Welsh, M. J. 2001. The DRASIC cation channel contributes to the detection of cutaneous touch and acid stimuli in mice. Neuron 32, 1071–1083.

Raboisson, P., Bourdiol, P., Dallel, R., Clavelou, P., and Woda, A. 1991. Responses of trigeminal subnucleus oralis nociceptive neurones to subcutaneous formalin in the rat. Neurosci. Lett. 125, 179–182.

Raboisson, P., Dallel, R., Clavelou, P., Sessle, B. J., and Woda, A. 1995. Effects of subcutaneous formalin on the activity of trigeminal brain stem nociceptive neurones in the rat. J. Neurophysiol. 73, 496–505.

Rentmeister-Bryant, H. and Green, B. G. 1997. Perceived irritation during ingestion of capsaicin or piperine: comparison of trigeminal and non-trigeminal areas. Chem. Senses 22, 257–266.

Robinson, P. P. 1988. The characteristics and regional distribution of afferent fibres in the chorda typani of the cat. J. Physiol. 406, 345–357.

Rolls, E. T. and Baylis, L. L. 1994. Gustatory, olfactory, and visual convergence within the primate orbitofrontal cortex. J. Neurosci. 14, 5437–5452.

Rolls, E. T., Critchley, H. D., Browning, A. S., Hernadi, I., and Lenard, L. 1999. Responses to the sensory properties of fat of neurons in the primate orbitofrontal cortex. J. Neurosci. 19, 1532–1540.

Rolls, E. T., Verhagen, J. V., and Kadohisa, M. 2003. Representations of the texture of food in the primate orbitofrontal cortex: neurons responding to viscosity, grittiness, and capsaicin. J. Neurophysiol. 90, 3711–3724.

van Rossum, D., Hanisch, U. K., and Quirion, R. 1997. Neuroanatomical localization, pharmacological characterization and functions of CGRP, related peptides and their receptors. Neurosci. Biobehav. Rev. 21, 649–678.

Schiffman, S. S., Suggs, M. S., Sostmon, A. C., and Simon, S. A. 1992. Chorda typani and lingual nerve responses to astringent compounds in rodents. Physiol. Behav. 51, 55–63.

Schmidt, R., Schmelz, M., Forster, C., Ringkamp, M., Torebjork, E., and Handwerker, H. 1995. Novel classes of responsive and unresponsive C nociceptors in human skin. J. Neurosci. 15, 333–341.

Scott, T. R. and Verhagen, J. V. 2000. Taste as a factor in the management of nutrition. Nutrition 16, 874–885.

Sessle, B. J. 1987. The neurobiology of facial and dental pain: present knowledge, future directions. J. Dent. Res. 66, 962–981.

Sessle, B. J. and Greenwood, L. F. 1976. Inputs to trigeminal brain stem neurones from facial, oral, tooth pulp and pharyngolaryngeal tissues: I. Responses to innocuous and noxious stimuli. Brain Res. 117, 211–226.

Shankland, W. E., II. 2001a. The trigeminal nerve. Part iii: the maxillary division. Cranio 19, 78–83.

Shankland, W. E., II. 2001b. The trigeminal nerve. Part IV: the mandibular division. Cranio 19, 153–161.

Shusterman, D. 2002. Review of the upper airway, including olfaction, as mediator of symptoms. Environ. Health Perspect. 110, 649–653.

Shusterman, D. 2003. Toxicology of nasal irritants. Curr. Allergy Asthma Rep. 3, 258–265.

Silver, W. L. and Moulton, D. G. 1982. Chemosensitivity of rat nasal trigeminal receptors. Physiol. Behav. 28, 927–931.

Simon, S. A., Liu, L., and Erickson, R. P. 2003. Neuropeptides modulate rat chorda tympani responses. Am. J. Physiol. (Regul. Integr. Comp. Physiol.) 284, R1494–R1505.

Simone, D. A., Sorkin, L. S., Oh, U., Chung, J. M., Owens, C., LaMotte, R. H., and Willis, W. D. 1991. Neurogenic hyperalgesia: central neural correlates in responses of spinothalamic tract neurons. J. Neurophysiol. 66, 228–246.

Simons, C. T., Boucher, Y., and Carstens, E. 2003a. Suppression of central taste transmission by oral capsaicin. J. Neurosci. 23, 978–985.

Simons, C. T., Boucher, Y., Iodi Carstens, M., and Carstens, E. (2006) Nicotine suppression of gustatory neurons in the nucleus of the solitary tract. J. Neurophysiol. 96, 1877–1886.

Simons, C. T., Carstens, M. I., and Carstens, E. 2003b. Oral irritation by mustard oil: self-desensitization and cross-desensitization with capsaicin. Chem. Senses 28, 459–465.

Simons, C. T., Dessirier, J. M., Carstens, M. I., O'Mahony, M., and Carstens, E. 1999. Neurobiological and psychophysical mechanisms underlying the oral sensation produced by carbonated water. J. Neurosci. 19, 8134–8144.

Simons, C. T., Dessirier, J.-M., Carstens, M. I., O'Mahony, M., and Cartens, E. 2000. The Tingling Sensation of Carbonated Drinks is Mediated by a Carbonnic Anhydrase-Dependent Excitation of Trigeminal Nociceptive Neurons. In: Proceedings of the 9th World Congress on Pain, Vol. 16. (eds. M. Devor, M. C. Rowbotham, and Z. Wiesenfeld-Hallin). IASP Press.

Simons, C. T., O'Mahony, M., and Carstens, E. 2002. Taste suppression following lingual capsaicin pre-treatment in humans. Chem. Senses 27, 353–365.

Simons, C. T., Sudo, S., Sudo, M., and Carstens, E. 2003c. Mecamylamine reduces nicotine cross-desensitization of trigeminal caudalis neuronal responses to oral chemical irritation. Brain Res. 991, 249–253.

Simons, C. T., Sudo, S., Sudo, M., and Carstens, E. 2004. Mustard oil has differential effects on the response of trigeminal caudalis neurons to heat and acidity. Pain 110, 64–71.

Small, D. M., Zald, D. H., Jones-Gotman, M., Zatorre, R. J., Pardo, J. V., Frey, S., and Petrides, M. 1999. Human cortical gustatory areas: a review of functional neuroimaging data. Neuroreport 10, 7–14.

Smith, G. D., Gunthorpe, M. J., Kelsell, R. E., Hayes, P. D., Reilly, P., Facer, P., Wright, J. E., Jerman, J. C., Walhin, J. P., Ooi, L., Egerton, J., Charles, K. J., Smart, D., Randall, A. D., Anand, P., and Davis, J. B. 2002. TRPV3 is a temperature-sensitive vanilloid receptor-like protein. Nature 418, 186–190.

Sostman, A. L. and Simon, S. A. 1991. Trigeminal nerve responses in the rat elicited by chemical stimulation of the tongue. Arch. Oral Biol. 36, 95–102.

South, E. H. and Ritter, R. C. 1986. Substance P-containing trigeminal sensory neurons project to the nucleus of the solitary tract. Brain Res. 372, 283–289.

Stapleton, J. R., Lavine, M. L., Wolpert, R. L., Nicolelis, M. A., and Simon, S. A. 2006. Rapid taste responses in the gustatory cortex during licking. J. Neurosci. 26, 4126–4138.

Steen, K. H., Wegner, H., and Reeh, P. W. 1999. The pH response of rat cutaneous nociceptors correlates with extracellular [Na+] and is increased under amiloride. Eur. J. Neurosci. 11, 2783–2792.

Stevens, D. A. and Lawless, H. T. 1987. Enhancement of responses to sequential presentation of oral chemical irritants. Physiol. Behav. 39, 63–65.

Stevenson, R. J. and Prescott, J. 1997. Judgments of chemosensory mixtures in memory. Acta Psychol. 95, 195–214.

Story, G. M., Peier, A. M., Reeve, A. J., Eid, S. R., Mosbacher, J., Hricik, T. R., Earley, T. J., Hergarden, A. C., Andersson, D. A., Hwang, S. W., McIntyre, P., Jegla, T., Bevan, S., and Patapoutian, A. 2003. ANKTM1, a TRP-like channel expressed in nociceptive neurons, is activated by cold temperatures. Cell 112, 819–829.

Strassman, A. M. and Vos, B. P. 1993. Somatotopic and laminar organization of fos-like immunoreactivity in the medullary and upper cervical dorsal horn induced by noxious facial stimulation in the rat. J. Comp. Neurol. 331, 495–516.

Sudo, S., Sudo, M., Simons, C. T., Dessirier, J. M., and Carstens, E. 2002. Sensitization of trigeminal caudalis neuronal responses to intraoral acid and salt stimuli and desensitization by nicotine. Pain 98, 277–286.

Sudo, S., Sudo, M., Simons, C. T., Dessirier, J. M., Iodi Carstens, M., and Carstens, E. 2003. Activation of neurons in trigeminal caudalis by noxious oral acidic or salt stimuli is not reduced by amiloride. Brain Res. 969, 237–243.

Suzuki, M., Mizuno, A., Kodaira, K., and Imai, M. 2003. Impaired pressure sensation in mice lacking TRPV4. J. Biol. Chem. 278, 22664–22668.

Szolcsányi, J. 1977. A pharmacological approach to elucidation of the role of different nerve fibres and receptor endings in mediation of pain. J. Physiol. 73, 251–259.

Tominaga, M., Caterina, M. J., Malmberg, A. B., Rosen, T. A., Gilbert, H., Skinner, K., Raumann, B. E., Basbaum, A. I., and Julius, D. 1998. The cloned capsaicin receptor integrates multiple pain-producing stimuli. Neuron 21, 531–543.

Travers, S. P. and Norgren, R. 1995. Organization of orosensory responses in the nucleus of the solitary tract of rat. J. Neurophysiol. 73, 2144–2162.

Trevisani, M., Smart, D., Gunthorpe, M. J., Tognetto, M., Barbieri, M., Campi, B., Amadesi, S., Gray, J., Jerman, J. C., Brough, S. J., Owen, D., Smith, G. D., Randall, A. D., Harrison, S., Bianchi, A., Davis, J. B., and Geppetti, P. 2002. Ethanol elicits and potentiates nociceptor responses via the vanilloid receptor-1. Nat. Neurosci. 5, 546–551.

Verhagen, J. V., Rolls, E. T., and Kadohisa, M. 2003. Neurons in the primate orbitofrontal cortex respond to fat texture independently of viscosity. J. Neurophysiol. 90, 1514–1525.

Villanueva, L., Bouhassira, D., Bing, Z., and Le Bars, D. 1988. Convergence of heterotopic nociceptive information onto subnucleus reticularis dorsalis neurons in the rat medulla. J. Neurophysiol. 60, 980–1009.

Waldmann, R., Champigny, G., Bassilana, F., Heurteaux, C., and Lazdunski, M. 1997. A proton-gated cation channel involved in acid-sensing. Nature 386, 173–177.

Wang, H. and Woolf, C. J. 2005. Pain TRPs. Neuron 46, 9–12.

Wang, Y., Erickson, R. P., and Simon, S. A. 1993. Selectivity of lingual nerve fibers to chemical stimuli. J. Gen. Physiol. 101, 843–866.

Wang, Y., Erickson, R. P., and Simon, S. A. 1995. Modulation of rat chorda tympani nerve activity by lingual nerve stimulation. J. Neurophysiol. 73, 1468–1483.

Whitehead, M. C. and Frank, M. E. 1983. Anatomy of the gustatory system in the hamster: central projections of the chorda tympani and the lingual nerve. J. Comp. Neurol. 220, 378–395.

Whitehead, M. C. and Kachele, D. L. 1994. Development of fungiform papillae, taste buds, and their innervation in the hamster. J. Comp. Neurol. 340, 515–530.

Whitehead, M. C., Ganchrow, J. R., Ganchrow, D., and Yao, B. 1999. Organization of geniculate and trigeminal ganglion cells innervating single fungiform taste papillae: a study with tetramethylrhodamine dextran amine labeling. Neuroscience 93, 931–941.

Willis, W. D. and Westlund, K. N. 1997. Neuroanatomy of the pain system and of the pathways that modulate pain. J. Clin. Neurophysiol. 14, 2–31.

Xing, H., Ling, J., Chen, M., and Gu, J. G. 2006. Chemical and cold sensitivity of two distinct populations of TRPM8-expressing somatosensory neurons. J. Neurophysiol. 95, 1221–1230.

Xu, H., Blair, N. T., and Clapham, D. E. 2005. Camphor activates and strongly desensitizes the transient receptor potential vanilloid subtype 1 channel in a vanilloid-independent mechanism. J. Neurosci. 25, 8924–8937.

Xu, H., Delling, M., Jun, J. C., and Clapham, D. E. 2006. Oregano, thyme and clove-derived flavors and skin sensitizers activate specific TRP channels. Nat. Neurosci. 9, 628–635.

Xu, H., Ramsey, I. S., Kotecha, S. A., Moran, M. M., Chong, J. A., Lawson, D., Ge, P., Lilly, J., Silos-Santiago, I., Xie, Y., DiStefano, P. S., Curtis, R., and Clapham, D. E. 2002. TRPV3 is a calcium-permeable temperature-sensitive cation channel. Nature 418, 181–186.

Yang, B. H., Piao, Z. G., Kim, Y. B., Lee, C. H., Lee, J. K., Park, K., Kim, J. S., and Oh, S. B. 2003. Activation of vanilloid receptor 1 (VR1) by eugenol. J. Dent. Res. 82, 781–785.

Yaxley, S., Rolls, E. T., and Sienkiewicz, Z. J. 1990. Gustatory responses of single neurons in the insula of the macaque monkey. J. Neurophysiol. 63, 689–700.

Yokota, T. 1975. Excitation of units in marginal rim of trigeminal subnucleus caudalis elicited by tooth pulp stimulation. Brain Res. 95, 154–158.

Zhang, Y., Hoon, M. A., Chandrashekar, J., Mueller, K. L., Cook, B., Wu, D., Zuker, C. S., and Ryba, N. J. 2003. Coding of sweet, bitter, and umami tastes: different receptor cells sharing similar signaling pathways. Cell 112, 293–301.

Further Reading

Brand, G. and Jacquot, L. 2002. Sensitization and desensitization to allyl isothiocyanate (mustard oil) in the nasal cavity. Chem. Senses 27, 593–598.

Carstens, E., Simons, C. T., Dessirier, J.-M., Sudo, M., Sudo, S., Iodi Carstens, M., and Jinks, S. L. 2002. Activation of Neurons in Trigeminal Subnucleus Caudalis (Vc) by Irritant Chemical Stimulation: Extracellular Single-Unit Recording and c-fos Immunohistochemical Methods. In: Methods in Chemosensory Research (eds. S. A. Simon and M. A. L. Nicolelis), pp. 267–292. CRC Press.

Dessirier, J. M., Simons, C. T., Carstens, M. I., O'Mahony, M., and Carstens, E. 2000. Psychophysical and neurobiological evidence that the oral sensation elicited by carbonated water is of chemogenic origin. Chem. Senses 25, 277–284.

4.19 Genetics and Evolution of Taste

J D Boughter, Jr., University of Tennessee Health Science Center, Memphis, TN, USA

A A Bachmanov, Monell Chemical Senses Center, Philadelphia, PA, USA

Glossary

complex trait A trait influenced by multiple factors, including effects of genes, environment, and their interactions.

congenic strain A strain that is identical to one of the progenitor inbred strains (inbred partner or background strain) except for a particular locus of interest that originates from the other progenitor strain (donor strain). Traditional congenic strains are produced through a series of at least 10 lineal backcrosses. Marker-assisted selective breeding of congenic strains (speed congenics) requires a smaller number of backcross generations.

conserved synteny Correspondence in order of chromosomal locations of orthologous genes in different species.

inbred strain A strain that has been bred using brother–sister matings for at least 20 consecutive generations. Same-sex individuals within an inbred strain are considered to be ∼99% identical genetically.

haplotype Haploid genotype; a combination of linked alleles at an individual chromosome; may refer to a set of linked genes or to a set of sequence variants within a gene.

locus A location in the genome. It may correspond to a gene, an anonymous DNA segment (marker), or a genomic region linked to a phenotypical variation (qualitative or quantitative trait locus).

ortholog Genes or proteins in different species that have similar sequences, which suggests that they have originated from a common ancestral gene.

paralog Genes in the same species with similar sequence, which suggests that they originated from a common ancestral gene.

phenotype The observable characteristics of an organism, as determined by both genetic makeup and environmental influences.

recombinant inbred strains A set of independently derived strains, all of which originate from the same two inbred progenitor strains. The two inbred progenitor strains are outcrossed to produce F_1 hybrids, which are intercrossed to produce F_2 hybrids, followed by consecutive generations of brother–sister matings. After 20 generations of brother–sister mating, mice within each recombinant inbred strain are genetically nearly identical.

segregating cross A cross between inbred strains, in which alleles of polymorphic markers or genes segregate among subgroups of the cross with different genotypes. Commonly produced segregating crosses are hybrids of the second filial (F_2) or a backcross (N_2) generation.

4.19.1 Taste Is a Complex Trait

Behavioral genetic approaches have been instrumental in the discovery of genes that influence taste phenotypes. The investigation of genetic bases of taste began early in the twentieth century with the description of human differences in sensitivity to the bitter-tasting chemical phenylthiocarbamide (PTC) (Fox, A. L., 1931; 1932). In the years following this discovery, behavioral genetic approaches conducted predominately in mice identified considerable variation in taste responses among inbred strains. Chromosomal mapping of genetic loci influencing preferences for sweet- and bitter-tasting stimuli led to the identification of families of G-protein-coupled receptors (GPCR) for sweet and bitter taste. However, the search for genes influencing taste function is not necessarily limited to receptors. Situated at the beginning of the alimentary canal, the gustatory system is a key player in homeostatic systems dealing with both fluid and food intake. In fact, taste and feeding are linked in terms of reciprocal connectivity between brain regions, and taste detection has been shown to be heavily modulated by postingestive feedback, including neural and hormonal factors (e.g., Scott, T. R., 2001; Bloomgarden, Z. T., 2006). Additionally, evidence indicates that taste sensitivity is modifiable by environmental factors such as diet and experience (Duffy, V. B. and Bartoshuk, L. M., 2000; Tichansky, D. S. *et al.*, 2006). This interactivity implies that taste sensitivity is a complex trait, and as such taste-related behavior phenotypes are likely to be determined by multiple genes and factors (Figure 1). This review will focus primarily on the study of rodent taste behavior genetics, as human genetic approaches are reviewed elsewhere (Chapter Propylthiouracil (PROP) Taste). However, an evolutionary perspective on taste receptor genes will include both vertebrate and invertebrate species.

Traditionally, the five major taste qualities have been described using terms that reflect human perceptual experience (sweet, salty, bitter, sour, and umami). Although there are many studies indicating that the mechanisms underlying perception of different taste qualities are similar in humans and nonhuman species, one should use caution when applying human descriptors to animal sensory function. It is perhaps more accurate to describe taste quality perception by nonhuman animals using chemical names of taste stimuli (e.g., sodium taste or sucrose-like taste), but for brevity the human descriptors are used throughout this review.

4.19.2 Genetic Approaches to Taste Behavior

4.19.2.1 Assessing Taste Behavior

Taste ability includes multiple aspects such as quality, intensity, and hedonic value. In addition, taste behavior may be influenced by nongustatory sensory factors such as olfaction and somatosensation. Early studies of human genetic variation of taste sensitivity concentrated on either responses to a single concentration of a stimulus (i.e., tasters vs. nontasters of PTC) or measurement of detection thresholds (Harris, H. and Kalmus, H., 1949; Wooding, S., 2006). More recent studies have utilized psychophysical methods such as suprathreshold magnitude estimation (Bartoshuk, L. M. *et al.*, 1994) to evaluate additional aspects of taste perception. In rodents, the majority of behavioral genetic studies have been conducted using 48 h two-bottle preference test, where an animal is presented with two bottles in the home cage, one containing water, and one containing a taste solution. After 24 h, the position of the bottles is switched to control for side preference (Richter, C., 1932). In these tests, the taste quality and intensity of the stimulus are inferred based

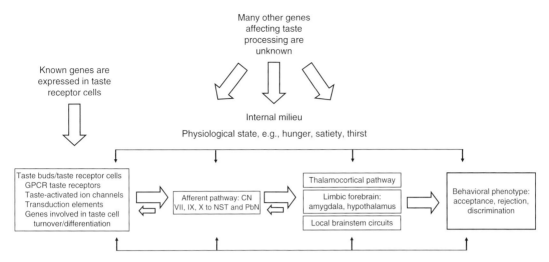

Figure 1 Components of the taste system and its associated genes. The search for the genetic basis of taste has primarily focused on taste receptors and other components of the taste signaling pathway expressed in peripheral taste receptor cells, and their relationship to taste behavior phenotypes. Other genes must play a role in complex interactions between peripheral mechanisms, brain taste centers, and behavioral outcomes. All of these steps in taste processing are subject to modulation by factors associated with the internal and external milieu, including physiological systems involved in food and fluid intake, and by other, nongenetic factors such as diet, learning, and exteroceptory cues like odors or visual stimuli. CN, cranial nerve; NST, nucleus of the solitary tract; PbN, parabrachial nucleus.

on level of avoidance or preference for the solution bottle. Such tests are relatively simple to conduct, and offer a considerable advantage in that one may conduct simultaneous testing of large groups of animals that are required for the genetic analysis. Tests involving multiple stimulus bottles are also useful (Tordoff, M. G. and Bachmanov, A. A., 2003). However, behavior over this relatively long time frame may be influenced not only by an animal's perception of a stimulus, but also by postingestive consequences. Taste-salient, brief-access tests minimize postingestive cues by restricting sampling times to short durations (<30 s). Such tests have been recently modified for use in high-throughput screening of taste function in mouse populations (Boughter, J. D., Jr. *et al.*, 2002; Glendinning, J. I. *et al.*, 2002). Behavioral test methods for animals are reviewed extensively in Chapter Behavioral Analysis of Taste Function in Rodent Models.

4.19.2.2 Behavioral Genetic Techniques

The goal of behavioral genetics is to understand how genes influence individual differences. Usually, a starting point for discerning genetic effects on behavior in rodents is the use of inbred strains. After 20 generations of inbreeding, same-sex individuals within a given strain are considered to be ~99% identical genetically. When individuals from multiple inbred strains are compared on a particular behavior or phenotype, the mean difference among these strains means is considered to be due to genetic factors. Commonly used strains in biomedical research include C57BL/6 (B6), DBA, 129, and BALB. Indeed, early studies with inbred mice indicated substantial strain differences in solution intake for various taste qualities (McClearn, G. E. and Rodgers, D., 1959; 1961; Hoshishima, K. *et al.*, 1962). Since then, inbred strains have been extensively characterized with regards to taste ability using not only intake tests, but also taste-salient behavioral tasks, taste nerve and central nervous system (CNS) physiology, and anatomical methods (e.g., Ninomiya, Y. *et al.*, 1984; Shingai, T. and Beidler, L. M., 1985; Ninomiya, Y. and Funakoshi, M., 1989; Kachele, D. L. and Lasiter, P. S., 1990; Lush, I. E., 1991; Eylam, S. and Spector, A. C., 2004; Inoue, M. *et al.*, 2004a; Boughter, J. D., Jr. *et al.*, 2005; McCaughey, S. A., 2007). Inbred strains of rats have not been used as commonly, although rat strains differ in salt and sweet taste responses (Sollars, S. I. *et al.*, 1990; Goodwin, F. L. and Amit, Z., 1998).

Differences between strains serve as the starting point for genetic mapping studies. Using Mendelian crosses of parental strains (i.e., F_1 and F_2 hybrids, and backcrosses) the investigator may estimate the relative size and number of genetic factors contributing to the behavior (Figure 2). Individuals from a segregating generation such as F_2 can be both phenotyped with respect to taste behavior, and genotyped using a set of genetic markers such as single nucleotide polymorphisms (SNPs). Significant association between phenotype and marker results in the identification of qualitative or quantitative trait loci (QTL) that pinpoint regions of the genome where genes influencing the trait of interest reside. For example, QTL studies conducted in an F_2 cross of B6 (high sweetener preferring) and 129 (low preferring)

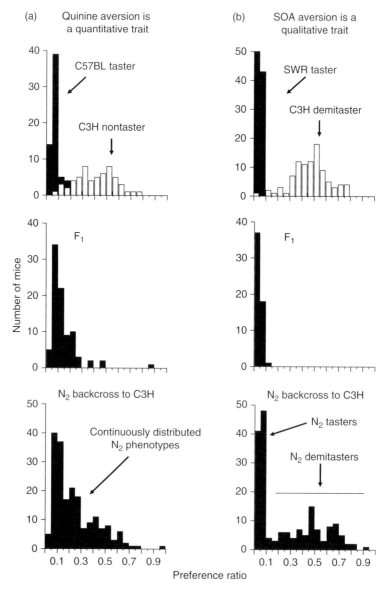

Figure 2 Examples of quantitative and qualitative traits for bitter taste aversion. Graphs show individual distributions of preference ratios for quinine (a) or sucrose octaacetate (SOA) (b) solutions. For quinine (a), there is overlap in the distributions between C57BL/6 (taster) and C3H/HeB (nontaster) progenitor inbred strains. An F_1 generation consists primarily of phenotypic tasters, whereas the N_2 backcross to C3H is comprised of a continuous distribution of phenotypes ranging from taster to nontaster. This continuous distribution is consistent with control by more than one quantitative trait locus. For SOA (b), there is almost no overlap in the distributions of SWR (taster) and C3H (demitaster) progenitor inbred strains. The F_1 distribution is comprised of phenotypic tasters. The N_2 backcross distribution is bimodal, consisting of about half tasters, half demitasters. Unlike the case for quinine, this distribution is consistent with expectations from a single locus model. SOA aversion is therefore a qualitative trait. Data from Boughter J. D. Jr., *et al.* (1992) and Boughter J. D. Jr., and Whitney G. (1995).

mouse strains demonstrate loci linked to both behavioral and physiological responses to sweeteners (Inoue, M. *et al.*, 2004b). Recombinant inbred (RI) strains (Figure 3) have also been used to identify QTLs influencing behavioral or neural phenotypes, including the frequently used BXD set, derived from B6 and DBA/2 mice (Peirce, J. L. *et al.*, 2004). Notably, BXDs have been used to map QTLs influencing sweet and bitter taste in mice (Lush, I. E. and Holland, G., 1988; Nelson, T. M. *et al.*, 2005). Finally, congenic strains (Figure 4) have been used extensively in taste behavioral genetic research (Whitney, G. *et al.*, 1989; Boughter, J. D., Jr. and Whitney, G., 1995; Bachmanov, A. A. *et al.*, 2001a; Li, X. *et al.*, 2001). A congenic strain is one that is identical to a

progenitor inbred strain except for the substitution of a particular locus of interest (Flaherty, L., 1981). This is accomplished through a series of (at least 10) lineal backcrosses. The newly created mice are used to evaluate an allelic variant against a particular strain background. For example, 129.B6 congenics (created from 129P3 and B6 mice) were used to localize the *Sac* locus influencing sweetener preference in mice (Li, X. *et al.*, 2001).

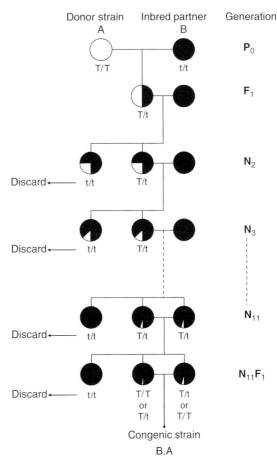

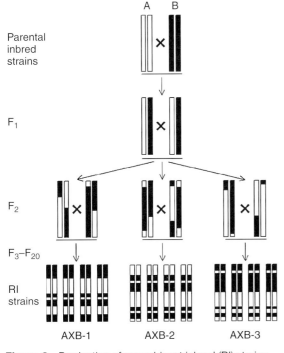

Figure 3 Production of recombinant inbred (RI) strains. A pair of homologous autosomes is shown in different generations of RI strains production. Genetic material of two hypothetical progenitor inbred strains, A and B, is shown in white and black respectively. Recombinant chromosomes (combining parts from both progenitor strains) are first observed in the F_2 generation. Randomly chosen F_2 mice are mated, and their offspring is maintained by brother–sister mating for at least another 18 generations. In the subsequent generations of breeding, additional chromosomal recombinations occur, and mice of each RI strain become genetically more similar. After 20 generations of brother–sister mating, mice within each RI strain are genetically nearly identical (illustrated by chromosomal pairs of four mice from thee hypothetical RI strains).

Figure 4 Backcross protocol for congenic strain development involving transfer of an autosomal dominant taster allele T from a donor strain (strain A; white circle) onto the genomic background of an inbred partner (strain B; black circle) with a recessive nontaster allele t. F_1 heterozygote offspring from the parent strains are mated to mice from the partner strain. Only phenotypic taster offspring (T/t) from subsequent generations are selected for backcrossing to the partner strain. With each backcross generation, one-half of the remaining genetic material from the donor strain is lost, except for material closely linked on the chromosome bearing the locus of interest. By generation N_{11}, the introgressed chromosome segment is isolated on a ~99% genomic background from the partner strain. Tasters from generation N_{11} and subsequent generations are sib intercrossed.

4.19.3   Bitter Taste

A large and diverse array of molecules evoke the sensation of bitterness, and the ability to detect and avoid these stimuli is assumed to have evolved as a mechanism to prevent ingestion of toxins (Spielman, A. I. *et al.*, 1992). In fact, a strong correlation exists in the animal kingdom between bitter sensitivity and tolerance to toxic compounds (Glendinning, J. I., 1994). It has long been appreciated that there is substantial variation among mouse strains in regard to intake and aversion of bitter-tasting compounds. The discovery of genetic loci influencing bitter sensitivity led to the identification of a family of GPCR genes, dubbed *Tas2r*s. These receptors are characterized by a short N-terminus with a potential transmembrane ligand-binding domain. To date, about 25 human and 36 mouse bitter taste receptor genes have been identified, although only a handful have known ligands.

4.19.3.1   Phenylthiocarbamide Taste in Man and Mouse

Variation in taste sensitivity for the bitter goitrogenic compound PTC is a classic example of an inherited trait in humans. It has been the source of much genetic and sensory research since the 1930s (see Chapter Propylthiouracil (PROP) Taste). It is now known that up to 85% of the variation in PTC sensitivity among humans is accounted for by allelic variation of a single bitter taste receptor gene on chromosome 7q, dubbed *TAS2R38* (Kim, U. K. *et al.*, 2003; Kim, U. K. and Drayna, D., 2005). Spurred on by the early discovery of human variation in PTC sensitivity, investigators sought a parallel finding in animal models. There was some evidence for a PTC dimorphism in chimpanzees (Fischer, R. A. *et al.*, 1939), suggesting that taste polymorphisms were not restricted to humans. Richter C. and Clisby K. H. (1941) reported individual variation in PTC avoidance thresholds of rats, although the genetic analysis of this was not pursued. Mice from a BALB strain displayed greater aversion of PTC in two-bottle tests than mice from other strains, including B6, although this difference required 10–12 consecutive days of testing to develop (Klein, T. W. and DeFries, J. C., 1970; Whitney, G. and Harder, D. B., 1986a). Several intercross and backcross generations created from BALB and B6 strains yielded segregation ratios consistent with phenotypic control by a single autosomal locus, with a dominant taster allele. It was suggested

that this effect reflects other variables in addition to sensory ability, such as sensitivity to postingestive effects of PTC, which is extremely toxic to mice. A direct comparison of intake tests and brief-access procedures demonstrated that differential avoidance of PTC among inbred strains depends on the amount of stimulus consumed, and not necessarily on immediate orosensory or taste cues (Nelson, T. M. *et al.*, 2003). Furthermore, PTC has been shown to be nearly as effective as LiCl in serving as the unconditioned stimulus in a conditioned taste aversion (CTA) experiment (St. John, S. J. *et al.*, 2005). Although sensitivity to PTC and the chemically related compound 6-*n*-propylthiouracil (PROP) is highly correlated among human subjects, aversions to these compounds do not correlate among mice (Barnicot, N. *et al.*, 1951; Whitney, G. and Harder, D. B., 1994). In summary, an ortholog to the human PTC receptor *TAS2R38* has not yet been described in mice, a finding perhaps not surprising in light of species-specific differences in the bitter taste gene family (Shi, P. *et al.*, 2003).

4.19.3.2   Genetics of Taste Responses to Other Bitter Compounds in Mice

The genetic basis of differences in taste sensitivity to other bitter compounds has been studied throughout the last several decades. Inbred mouse strains have been found to differ in aversion to a large number of both toxic and nontoxic bitter stimuli, including (but not limited to) quinine, strychnine, acetylated sugars, $MgCl_2$, cycloheximide, denatonium benzoate, isohumulone, quinine, and sucrose octaacetate (SOA) (Warren, R. P. and Lewis, R. C., 1970; Lush, I. E., 1981; 1982; 1984; 1986; Lush, I. E. and Holland, G., 1988; Whitney, G. and Harder, D. B., 1994; Bachmanov, A. A. *et al.*, 1996a; Boughter, J. D., Jr. and Whitney, G., 1998; Nelson, T. M. *et al.*, 2003). Aversion of the acetylated sugar SOA among mice is predominantly due to variation at a single qualitative trait locus, *Soa*, although SOA aversion may also be influenced by additional loci and environment. Aversion ratios among segregating generations of mice were consistent with variation at a single autosomal locus (*Soa*), with the taster allele dominant over either a demitaster (partial sensitivity) or nontaster allele (Figure 2; Lush, I. E., 1981; Whitney, G. and Harder, D. B., 1986b; Harder, D. B. *et al.*, 1992; Boughter, J. D., Jr. and Whitney, G., 1995). Results of both taste-salient brief-access behavioral tests and peripheral taste nerve physiology yielded concentration-by-concentration similarity with two-bottle data and confirmed that the strain difference was gustatory in nature (Inoue, M. *et al.*, 2001a; Boughter, J. D., Jr. *et al.*, 2002). However, at least one other study has found evidence for polygenic control of SOA aversion (Le Roy, I. *et al.*, 1999). The aversion phenotype is subject to environmental modulation: unlike control tasters, taster mice raised on drinking water adulterated with a normally avoided concentration of SOA display only minimal avoidance of SOA in a two-bottle test with water (Harder, D. B. *et al.*, 1989). Unlike SOA avoidance, aversion to quinine and PROP in two-bottle tests has been demonstrated to be under polygenic control in segregating crosses or RI strains (Figure 2; Boughter, J. D., Jr. *et al.*, 1992; Bachmanov, A. A. *et al.*, 1996a; Harder, D. B. and Whitney, G., 1998). Interestingly, strain differences in intake of bitter-tasting drugs such as nicotine have been shown (Robinson, S. F. *et al.*, 1996; Butt, C. M. *et al.*, 2005), but it is not known whether these differences are related to taste or postingestive effects of these drugs.

4.19.3.3 Bitter Taste Loci and Receptors

Using a segregating cross of mice, Capeless C. G. *et al.* (1992) mapped a single locus influencing SOA sensitivity, *Soa*, to a region on distal mouse chromosome 6. This locus overlapped with the *Prp* genes encoding salivary proteins (Bachmanov, A. A. *et al.*, 2001a); although these genes were considered as candidates for the *Soa* locus, they have been shown not to influence SOA taste (Harder, D. B. *et al.*, 2000). Studies with RI and congenic strains linked aversion to a number of other bitter compounds, including quinine, cycloheximide, and PROP, to this same region (Lush, I. E. and Holland, 1988; Lush, I. E. *et al.*, 1995; Boughter, J. D., Jr. and Whitney, G., 1998; Blizard, D. A. *et al.*, 1999). Subsequent genome mining studies discovered a cluster of GPCR bitter taste receptor genes (*Tas2rs*) at this location (Adler, E. *et al.*, 2000; Chandrashekar, J. *et al.*, 2000; Matsunami, H. *et al.*, 2000). Recent characterization of the *Tas2r* receptor gene family in mice indicates 36 genes distributed into two clusters on distal chromosome 6 and one on chromosome 15, including 24 genes at the *Soa-Qui* locus that collectively comprise a large haplotype for bitter taste sensitivity (Wu, S. V. *et al.*, 2005; Nelson, T. M. *et al.*, 2005; Shi, P. *et al.*, 2003). It is likely that allelic variation of these receptors is responsible for phenotypical variation in sensitivity to different bitter tastants linked to the *Soa-Qui* locus.

Analysis of *Tas2r* sequences in strains of quinine taster (B6) and nontaster (DBA/2) mice reveals considerable variation across the *Tas2r* genes (Nelson, T. M. *et al.*, 2005). Among bitter receptors in the mouse, only two have been deorphanized thus far with respect to ligand, one responding exclusively to cycloheximide (encoded by the *Tas2r105* gene), the other to PROP and denatonium (encoded by *Tas2r108*) (Chandrashekar, J. *et al.*, 2000; Conte, C. *et al.*, 2006; Mueller, K. L. *et al.*, 2005). In contrast, several human T2R receptors (aside from T2R38) have been characterized with respect to ligand binding, including T2R16 which responds rather narrowly to a class of bitter chemicals known as β-glucopyranosides, T2R14 that responds to a variety of bitter stimuli, and T2R43 and T2R44 that respond to the bitter aftertaste of saccharin (Bufe, B. *et al.*, 2002; Behrens, M. *et al.*, 2004; Kuhn, C. *et al.*, 2004).

4.19.3.4 Evolution of the T2R Receptors

The study of taste receptor gene evolution has provided key insights into the dynamic process of adaptation of different species to their feeding ecology and environment. Invertebrates such as insects possess aversive responses to bitter-tasting or deterrents chemicals (Glendinning, J. I. and Hills, T. T., 1997; Glendinning, J. I. *et al.*, 1998). A *Drosophila Gr66a* taste receptor responsible for aversive response to caffeine (Moon, S. J. *et al.*, 2006), a compound that tastes bitter to humans, is unrelated to vertebrate T2R genes. This suggests that receptors for detection of aversive taste evolved independently in vertebrate and invertebrate species.

Fish and chickens have a small number of the T2R genes (ten or less; Table 1) and no pseudogenes. Mammals and frogs have much larger numbers of the T2R genes (21–64) and a high proportion of T2R pseudogenes (15–44%). An analysis of relatedness of the T2R genes in different species suggests a complex evolution of this gene family. The common ancestor of tetrapods and teleost fish already had multiple T2Rs. This initial expansion in the common ancestor of tetrapods was followed by additional independent expansions in frogs and mammals and contraction in birds (Go, Y., 2006; Shi, P. and Zhang, J., 2006).

These evolutionary changes likely reflect adaptation to different diets. It appears that the number of functional T2R genes in mammals correlates with levels of toxic compounds in different types of food. Toxic bitter compounds are more common in plant than in animal tissues (Glendinning, J. I., 1994). Correspondingly, omnivores (potentially exposed to the largest number of toxic compounds) have the

Table 1   Numbers of T1R and T2R genes in vertebrate

Species	T1R genes			T2R genes	
	T1R1 genes	T1R2 genes	T1R3 genes	Functional and putatively functional genes	Pseudogenes
Human (*Homo sapiens*)	1	1	1	28	16
Mouse (*Mus musculus*)	1	1	1	36	7
Rat (*Rattus norvegicus*)	1	1	1	37	7
Cow (*Bos taurus*)	Unknown	Unknown	Unknown	19	15
Dog (*Canis familiaris*)	1	1	1	16	5
Cat (*Felis catus*)[a]	1	0	1	Unknown	Unknown
Opossum (*Monodelphis domestica*)	1	1	1	29	5
Chicken (*Gallus gallus*)	1	0	1	3	0
Frog (*Xenopus tropicals*)	0	0	0	52	12
Fugu fish (*Takifugu rubripes*)	1	2	1	4	0
Puffer fish (*Tetraodon nigroviridis*)	1	3	1	6	0
Zebra fish (*Danio rerio*)	1	2	1	10	0
Medaka fish (*Oryzias latipes*)	1	3	1	0	0

[a]Cat *Tas1r2* and *Tas1r3* data from Li X. *et al.* (2005); cat Tas1r1 is based on a cat genome sequence (GenBank accession AANG01000989) with a corresponding predicted protein 90% identical to dog T1R1.
Data from Li X. *et al.* (2005), Shi P. and Zhang J. (2006), Liao J. and Schultz P.G. (2003), Shi P. *et al.* (2003), Go Y. *et al.* (2005), Li X. *et al.* (2002a; 2002b; 2002c), Hoon M. A. *et al.* (1999), Bachmanov A. A. *et al.* (2001), Lagerstrom M. C. *et al.* (2006), Adler E. *et al.* (2000), Ishimaru Y. *et al.* (2005), Go Y. (2006), Conte C. *et al.* (2002; 2003a; 2003b), Wang X. *et al.* (2004), Wu S. U. *et al.* (2005), Bjarnadottir T. K. *et al.* (2005; 2006), Fischer R. A. *et al.* (2005), and Parry C. M. *et al.* (2004). When different sources indicated different numbers of genes, the highest estimate is shown.

largest number of T2R genes, carnivores (potentially exposed to the smallest number of toxic compounds) seem to have the smallest number of T2R genes, and herbivores are intermediate (Shi, P. and Zhang, J., 2006) (see also Table 1).

The repertoire of the T2R genes is highly conserved between the mouse and rat (Wu, S. V. *et al.*, 2005) and among primate species, including humans (Fischer, A. *et al.*, 2005; Go, Y. *et al.*, 2005; Parry, C. M. *et al.*, 2004). Primates have a smaller number of intact T2R genes and a higher number and proportion of pseudogenes compared with rodents (Conte, C. *et al.*, 2003a; Go, Y. *et al.*, 2005) (Table 1). Comparison of T2R sequences in rodents and primates showed that several major groups of these genes were present prior to divergence of their lineages (Conte, C. *et al.*, 2003a). Consistent with this, mouse and human T2R genes reside in chromosomal regions with conserved synteny, indicating that major duplication events occurred prior to the separation of primates and rodents (Bufe, B. *et al.*, 2002). Approximately 1/4 of the T2R genes have 1:1 mouse–human orthology, nearly 1/4 showed expansion in humans, and ~1/2 showed expansion in mice (Shi, P. *et al.*, 2003; Bjarnadottir, T. K. *et al.*, 2005). T2R gene families in rodents and primates appear to have undergone a complex combination of local and interchromosomal duplications, deletions, pseudogenization, and

positive selection driving divergence of paralogous T2Rs (Conte, C. *et al.*, 2003b; Shi, P. *et al.*, 2003; Liman, E. R., 2006; Shi, P. and Zhang, J., 2006). The role of tandem gene duplication in evolution of the T2R genes is supported by clustering of these genes in a few chromosomal regions (Fredriksson, R. *et al.*, 2003; Shi, P. *et al.*, 2003).

Current literature provides conflicting views on evolution of the T2R genes in primates. Several studies suggested reduced selective constraint and neutral evolution of T2R genes in primates compared with rodents (Wang, X. *et al.*, 2004; Fischer, A. *et al.*, 2005; Go, Y. *et al.*, 2005). These conclusions were based on analyses of rates of synonymous and nonsynonymous nucleotide substitutions, higher number of pseudogenes in primates compared with rodents, and allelic segregation of pseudogenes in humans (Conte, C. *et al.*, 2003b; Kim, U. K. *et al.*, 2004; Wang, X. *et al.*, 2004; Fischer, A. *et al.*, 2005; Go, Y. *et al.*, 2005; Meyerhof, W., 2005). On the other hand, several other studies have detected signs of natural selection of the T2R genes in humans, as is likely the case with other species (Shi, P. *et al.*, 2003; Wooding, S. *et al.*, 2004; Soranzo, N. *et al.*, 2005; Shi, P. and Zhang, J., 2006). For example, signs of positive selection were found for human *TAS2R16* gene, which encodes a receptor for β-glucopyranosides. Its ancestral allele associated with lower sensitivity to β-glucopyrano-sides is under positive selection only in central Africa, while in the rest of the world the evolutionarily derived allele associated with an increased sensitivity to β-glucopyranosides is under positive selection. It was suggested that that the global pattern of allelic variation of the *TAS2R16* gene depends on selective pressures of protection against malaria in Africa and protection against toxins in malaria-free zones (Soranzo, N. *et al.*, 2005). Taster and nontaster alleles of the *TAS2R38* gene encoding a PTC taste receptor are maintained in different human populations under balancing selection. It was hypothesized that the *TAS2R38* nontaster allele encodes a functional receptor responsive to toxic bitter substances other than PTC, so that *TAS2R38* heterozygotes gain a fitness advantage because of their ability to perceive and avoid a larger number of bitter toxins compared with homozygotes (Wooding, S. *et al.*, 2004).

Evidence for both neutral evolution and natural selection of T2R genes in primates may reflect adaptation to their feeding ecology. If primates are exposed to a smaller number of toxins than their ancestors were, this may result in loss of selective pressure for some *TAS2R* genes. However, receptors involved in detection of toxins to which primates are still exposed would remain under stabilizing, balancing, or positive selection.

4.19.4   Sweet and Umami Taste

Sweetness is evoked by a diversity of chemicals, including simple carbohydrates, and in most species the sweet taste is associated with appetitive consummatory behavior (Bachmanov, A. A., 2008). Umami is a unique taste quality, described by humans as meaty or brothy, evoked by the amino acid glutamate (e.g., monosodium glutamate or MSG) and some other L-type amino acids and purine 5′-nucleotides (Lindemann, B., 2000) These two distinct taste qualities are linked because their function depends on shared GPCR proteins from the T1R family. In the mouse, three T1R-encoding genes, *Tas1r1*, *Tas1r2*, and *Tas1r3*, are located on chromosome 4 (Hoon, M. A. *et al.*, 1999; Kitagawa, M. *et al.*, 2001; Max, M. *et al.*, 2001; Montmayeur, J. P. *et al.*, 2001; Nelson, G. *et al.*, 2001; Sainz, E. *et al.*, 2001; Reed, D. R. *et al.*, 2004). The three corresponding receptor proteins, T1R1, T1R2, and T1R3, comprise a small family of class C GPCRs. This class includes also the metabotropic glutamate and extracellular calcium-sensing receptors, all of which have a large N-terminal

extracellular domain. When T1R3 is coexpressed in a heterologous system with T1R2, it functions as a broad-spectrum sweet taste receptor (Nelson, G. *et al.*, 2001; Li, X. *et al.*, 2002b; Nelson, G. *et al.*, 2002). In contrast, a heterodimer of T1R1 and T1R3 proteins functions as an umami taste receptor in humans, and is more broadly tuned in rodents to respond to L-amino acids (Li, X. *et al.*, 2002c; Nelson, G. *et al.*, 2002; Damak, S. *et al.*, 2003; Zhao, G. Q. *et al.*, 2003).

4.19.4.1   Strain Differences in Responses to Sweet-Tasting Stimuli

Taste sensitivity to compounds classified as tasting sweet, such as sugars, certain amino acids, and artificial sweeteners, varies among human individuals and there is evidence this may be determined genetically (e.g., Greene, L. S. *et al.*, 1975; Gent, J. F. and Bartoshuk, L. M, 1983; Reed, D. R. *et al.*, 1997). Genetic variation in sweet taste responsiveness exists also among inbred mouse strains. In early studies, C57BL mice displayed a significantly greater avidity for sucrose than did DBA/2 mice (McClearn, G. E. and Rodgers, D., 1959; 1962). Fuller J. L. (1974) noted that DBA/2 mice were indifferent to certain concentrations of sodium saccharin in two-bottle tests, whereas B6 mice strongly preferred it. Segregation ratios in a cross of these strains were reported to be consistent with a single locus, two-allele model. An autosomal gene (*Sac*) was proposed to explain this variation, with a dominant *Sac*^{b} allele in the B6 strain and a recessive *Sac*^{d} allele in the DBA/2 strain. Fuller proposed that this genetic variation affected a hedonic-based response rather than sensory threshold. Subsequent studies by Lush I. E. (1989) suggested that the *Sac* locus also influenced preference to other synthetic sweeteners, including acesulfame and dulcin. Consistent with this, earlier reports had noted that saccharin and sucrose intakes correlated among inbred strains (Pelz, W. E. *et al.*, 1973; Ramirez, I. and Fuller, J. L., 1976). Thus, early studies were supportive of a common genetic basis and single taste receptor underlying sweet taste in mice.

A detailed analysis of sweet taste behavior has been conducted with mice from the B6 and 129 strains. Compared with 129 mice, B6 mice display a higher preference for a large number of sweeteners, including sugars (sucrose and maltose) sweet-tasting amino acids (glycine, D-phenylalanine, D-tryptophan, L-proline, and L-glutamine), and several but not all noncaloric sweeteners (saccharin, acesulfame, dulcin, sucralose, and SC-45647) (Lush, I. E., 1989; Belknap, J. K. *et al.*, 1993; Capeless, C. G. and Whitney, G., 1995; Lush, I. E.

et al., 1995; Bachmanov, A. A. *et al.*, 2001d). This phenotypic difference is not due to a generalized difference in taste responsiveness or differences in appetite, but is rather specific to variation in sweet taste responsiveness (Bachmanov, A. A. *et al.*, 1996b; 2001c). Differences between these strains in preference for a sweet-tasting amino acid, glycine (Bachmanov, A. A. *et al.*, 2001d) appeared to depend on mechanisms distinct from those affecting responses to many other sweeteners. Both B6 and 129 mice generalized CTA between glycine and several other sweeteners, demonstrating that they perceive the sucrose-like taste of glycine. Thus, the lack of a strong glycine preference by 129 mice cannot be explained by their inability to perceive its sweetness (Manita, S. *et al.*, 2006). Despite differences in glycine intakes and preferences, chorda tympani (CT) responses to glycine are similar in mice from both strains (Inoue, M. *et al.*, 2001b). Neither behavioral nor neural responses to glycine are influenced by the *Tas1r3* genotype (Eylam, S. and Spector, A. C., 2004; Inoue, M. *et al.*, 2004b), suggesting that variation in taste responses to glycine depends on other genes.

4.19.4.2 Mapping and Identification of Sweet Taste Receptor Genes

Linkage studies conducted with BXD RI strains, or in crosses with B6 and DBA/2 or 129 strains, detected at least one to two loci on mouse chromosome 4 influencing sweetener preferences (Lush, I. E., 1989; Belknap, J. K. *et al.*, 1992; Phillips, T. J. *et al.*, 1994; Lush, I. E. *et al.*, 1995; Blizard, D. A. *et al.*, 1999). Bachmanov A. A. *et al.* (1997) reported the existence of two distinct loci on chromosome 4 (including *Sac*) in a B6 by 129 cross that collectively accounted for more than 50% of the genetic variability in sucrose intake; these loci were also shown to affect the magnitude and threshold of the CT response to sucrose. A subsequent positional cloning study demonstrated that the *Sac* locus in the subtelomeric region of chromosome 4 corresponded to the *Tas1r3* taste receptor gene (Figure 5) (Bachmanov, A. A. *et al.*, 2001b; Li, X. *et al.*, 2002a).

Polymorphisms of *Tas1r3* have been linked to sweet taste responses attributed to allelic variants of the *Sac* locus (Figure 5). A comprehensive quantitative analysis of *Tas1r3* sequence variants associated with saccharin preference using 30 inbred mouse strains yielded association at a haplotype including the amino acid substitution of isoleucine to threonine at position 60 (Ile60Thr) (Reed, D. R. *et al.*, 2004). A subsequent *in vitro* study showed that a corresponding site-directed

mutation changes binding affinity of the T1R3 protein to several sweeteners (Nie, Y. *et al.*, 2005).

The T1R2 + T1R3 sweet receptor responds to a broad variety of ligands in both humans and rodents (although there are species differences), including sugars, artificial sweeteners, and sweet-tasting amino acids (Nelson, G. *et al.*, 2001; Li, X. *et al.*, 2002b; Nelson, G. *et al.*, 2002). There is also evidence that T1R3 functions as a low-affinity sugar receptor alone, probably as a homodimer (Zhao, G. Q. *et al.*, 2003).

Aside from the *Tas1r* family, there may be other genes that affect sweet taste. Multigenic inheritance of sweetener preferences was shown in a number of studies (Ramirez, I. and Fuller, J. L., 1976; Phillips, T. J. *et al.*, 1994; Capeless, C. G. and Whitney, G., 1995; Lush, I. E. *et al.*, 1995; Bachmanov, A. A. *et al.*, 1996a). Mice lacking T1R3 have detection thresholds for sucrose identical to wild types, and can discriminate among sweet and savory stimuli (Delay, E. R. *et al.*, 2006). In F_2 hybrids produced from B6 and 129 strains, intake and neural response of the sweet amino acid glycine was not linked to *Tas1r3* (Inoue, M. *et al.*, 2004b), suggesting the possibility of a non-T1R mechanism. One of the genetic loci affecting sweet taste responses is *dpa* (D-phenylalanine aversion), which affects ability of mice to generalize CTA between D-phenylalanine and sucrose, inferring that *dpa* affects ability to detect sweetness of D-phenylalanine. The *dpa* locus also affects responses of sweet-sensitive fibers of the CT nerve to D-phenylalanine. B6 mice carry a dominant allele of *dpa* that determines an ability to recognize the sweetness of D-phenylalanine, whereas BALB/c mice carry a recessive *dpa* allele conferring inability to detect D-phenylalanine sweetness. The *dpa* locus was mapped to proximal chromosome 4, a region distinct from the subtelomeric chromosome 4 harboring the *Tas1r* genes (Ninomiya, Y. *et al.*, 1987; 1991; 1992; Shigemura, N. *et al.*, 2005). A locus on proximal chromosome 4, in the *dpa* region, was found to be suggestively linked to consumption of, and CT responses to, sucrose (Bachmanov, A. A. *et al.*, 1997). An epistatic interaction between effects on sucrose intake of this locus and the *Tas1r3* locus suggests that these two loci may encode interacting components of sweet taste transduction (Bachmanov, A. A. *et al.*, 1997).

4.19.4.3 Umami Taste Genetics

Little is known about genetic variation in umami taste responses. Humans appear to differ in perception of glutamate taste (Lugaz, O. *et al.*, 2002), but it is not

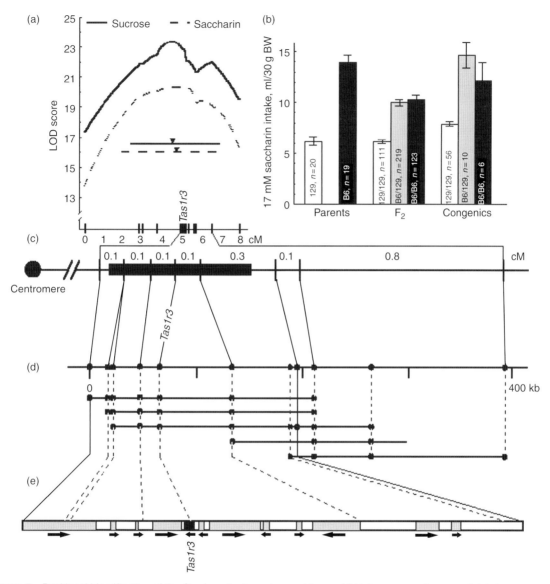

Figure 5 Positional identification of the *Sac* (saccharin preference) locus. (a) Linkage map of mouse distal chromosome 4 based on data from the B6 × 129 F_2 intercross. The *X*-axis shows distances between markers in recombination units (cM). The *Y*-axis shows the logarithm of the odds ratio (LOD) scores for sucrose and saccharin consumption. The LOD score peaks (indicated by black triangles) and confidence intervals (solid horizontal line for sucrose, 4.5 cM, and dotted horizontal line for saccharin, 5.3 cM) define the genomic region of the *Sac* locus. (b) Average daily 17 mM saccharin consumption by mice from parental 129 and B6 strains (left), B6 × 129 F_2 hybrids (center), and congenic 129.B6-*Sac* mice (right) in 96 h two-bottle tests with water (means ± SE). *Tas1r3* genotypes of the F_2 and congenic mice and mouse numbers are indicated on the bars. Differences between parental strains and among the F_2 and congenic genotypes were significant ($p < 0.0001$, ANOVA). F_2 and congenic B6 homozygotes and heterozygotes for *Tas1r3* did not differ from each other, and had higher saccharin intakes compared with 129 homozygotes ($p < 0.0001$, posthoc tests). (c) Linkage map of the *Sac*-containing region defined based on the size of the donor fragment in the 129.B6-*Sac* congenic strain (black box). Distances between markers were estimated based on the B6 × 129 F_2 intercross (see panel a). (d) A contig of bacterial artificial chromosome (BAC) clones and physical map of the *Sac* region. BAC clones are represented by horizontal lines. Dots indicate marker content of the BAC clones. (e) Genes within the *Sac*-containing interval. Filled areas indicate predicted genes. Arrows indicate the predicted direction of transcription. Reproduced from Bachmanov, A. A., Reed, D. R., Li, X., and Beauchamp, G. K. 2002. Genetics of sweet taste preferences. Pure Appl. Chem. 74, 1135–1140, with permission from IUPAC.

known whether this variation has a genetic basis. Genetically determined differences in responses to umami taste stimuli were shown in mice: in two-bottle tests, B6 mice display higher consumption of MSG and inosine monophosphate than do mice from the 129 strain (Bachmanov, A. A. *et al.*, 2000). These strain differences seem to be specific for umami stimuli and are not associated with variation in sweet or salty tastes (Bachmanov, A. A. *et al.*, 1996a; 1998; 2000).

A heterodimer of the T1R1 and T1R3 proteins is thought to function as a receptor for umami taste stimuli. Consistent with this, disruption of the *Tas1r1* and/or *Tas1r3* genes diminishes behavioral or neural responses of knockout mice to umami taste stimuli (Nelson, G. *et al.*, 2002; Damak, S. *et al.*, 2003; Zhao, G. Q. *et al.*, 2003). However, natural allelic variation of the *Tas1r3* gene does not affect umami taste responses in mice: an analysis of the F_2 hybrids between the B6 and 129 inbred mouse strains has shown that the *Tas1r3* allelic variants are not associated with behavioral or neural taste responses to umami stimuli (Inoue, M. *et al.*, 2004b). Therefore, although the T1R3 receptor is involved in transduction of umami taste, the B6/129 sequence variants affecting its sensitivity to sweeteners do not affect its sensitivity to umami compounds. Recent studies of umami taste transduction indicate the existence of more than one receptor responding to this class of stimuli (Chaudhari, N. *et al.*, 2000; Maruyama, Y. *et al.*, 2006); allelic variation of these genes might underlie genetic differences in umami taste.

4.19.4.4 Evolution of T1R Receptors

Numbers of the T1R genes in different vertebrate species range from complete absence in the frog to five in the puffer fish and medaka fish (Table 1). With the exception of the frog, all vertebrates have single T1R1 and T1R3 genes, but the number of the T1R2 genes varies: the chicken and Felidae species (domestic cat, tiger, and cheetah; Li, X. *et al.*, 2005) lack functional T1R2 genes, but several fish species have two to three T1R2 genes. The vertebrate T1R receptors are not found in invertebrates (Bjarnadottir, T. K. *et al.*, 2005) and are not related to a Drosophila taste receptor for a sugar trehalose encoded by the *Gr5a* gene (Dahanukar, A. *et al.*, 2001; Ueno, K. *et al.*, 2001; Chyb, S. *et al.*, 2003; Inomata, N. *et al.*, 2004; Isono, K. *et al.*, 2005; Hallem, E. A. *et al.*, 2006).

These data allow us to establish changes in the T1R repertoire during evolution. Although many vertebrate and invertebrate animals detect the taste of sugars and avidly consume them, receptors for sugars evolved independently in these two lineages. Most vertebrates have three clades of T1R genes, suggesting that all three evolved before separation of tetrapods and teleosts (Shi, P. and Zhang, J., 2006). However, another group concluded from a similar analysis that fish and mammalian T1R1 and T1R3 genes are orthologs, but fish and mammalian T1R2 evolved independently (Ishimaru, Y. *et al.*, 2005). It appears that after separation of tetrapods and teleosts, duplication of the T1R2 gene occurred in the evolution of different fish species. Several Felidae species (Li, X. *et al.*, 2005) and the chicken (Lagerstrom, M. C. *et al.*, 2006; Shi, P. and Zhang, J., 2006) lost the T1R2 gene, and the tongueless western clawed frog lost all three T1R genes. Thus, pseudogenization of the T1R2 gene occurred multiple times in evolution.

Loss of the T1R2 gene in cats and chickens must result in the absence of the T1R2 + T1R3 sweet taste receptor, which corresponds to the lack of taste responses to sweeteners in these species (Kare, M. R., 1961; Halpern, B. P., 1962; Bartoshuk, L. M *et al.*, 1975; Beauchamp, G. K. *et al.*, 1977; Ganchrow, J. R. *et al.*, 1990). Interestingly, some birds recognize sugar taste (Matson, K. D. *et al.*, 2000; 2001), suggesting that they may have a functional T1R2. The T1R2 pseudogenization and lack of sweet taste responsiveness in cats are probably results of these animals being obligate carnivores that do not seek sugars in their food, and thus do not have a selective advantage of having a functional sweet taste receptor that recognizes sugars. Dogs, which are carnivores from the Canidae family, have a functional T1R2 gene structure (Li, X. *et al.*, 2005; Shi, P. and Zhang, J., 2006) and are attracted to sugars (Grace, J. and Russek, M., 1969). Thus, loss of the T1R2 gene in cats and chickens may be a consequence of their feeding behavior that does not require a sweet taste receptor for proper food choice. However, a reverse causative relationship cannot be excluded, when a loss-of-function mutation in the T1R2 gene resulted in loss of sweet taste sensation, which in turn altered feeding behavior of these animals.

The role of natural selection in evolution of the T1R genes was shown in several studies involving comparisons of the paralogous genes and analyses of their within-species variation. Sequence divergence among the paralogous T1R genes appears to be governed by positive selection (Shi, P. and Zhang, J., 2006). Most of the inferred positively selected sites are located in the N-terminal extracellular domains of the T1R proteins (Shi, P. and Zhang, J., 2006), which participate in ligand binding (Xu, H. *et al.*, 2004; Jiang, P. *et al.*, 2005a, Nie, Y. *et al.*, 2005). Analysis of within-species *TAS1R* sequence variation among humans suggested that *TAS1R* variants have come to their current frequencies under positive natural selection during population growth, which implies that the coding sequence variants affect receptor function (Kim, U. K. *et al.*, 2006).

Results of these studies suggest that the evolutionary changes of the T1R receptors are likely to affect their ligand binding properties. This is consistent with currently available data on species differences in ligand specificity of the T1R1 + T1R3 and T1R2 + T1R3 receptors (Nelson, G. *et al.*, 2001; 2002; Li, X. *et al.*, 2002c; Winnig, G. *et al.*, 2005). Several studies have shown that sequence variation between the T1R orthologs underlies species differences in responses to several amino acids, sweeteners and sweet taste blockers. T1R variants that determine species differences in ligand binding were found in the cysteine-rich region of the extracellular N-terminal domain (Jiang, P. *et al.*, 2004) and in the transmembrane domains (Jiang, P. *et al.*, 2005a; 2005b; Winnig, G. *et al.*, 2005).

Although ligands for the T1R receptors have been experimentally confirmed only for humans and rodents, it is likely that their orthologs in other species have similar ligand specificities. This would make sense because sugars and amino acids, the main natural T1R ligands, are important nutrients for animals of many different species. This explains the lower diversity and slower evolution of the T1R genes compared with the T2R genes involved in bitter taste (Shi, P. and Zhang, J., 2006). Nevertheless, variation in the T1R genes among species may reflect species differences in diets, for example the amino acid composition of food or presence of sweet proteins in some tropical fruits.

4.19.5   Salt and Sour Taste

Ionic taste stimuli such as H^+ or Na^+ are thought to be transduced primarily via apically located ion channels. A large portion of NaCl or salt taste is thought to be dependent on the influx of sodium through epithelial sodium channels, members of the degenerin/ENaC superfamily of ion channels. ENaC is blocked by the diuretic compound amiloride. Taste receptor cells (TRCs) express Na^+-selective channels that are also blocked by amiloride and share a number of other physiological properties with the ENaCs (for a review, see Boughter, J. D., Jr. and Gilbertson, T. A., 1999; Gilbertson, T. A. *et al.*, 2000). Entry of the sodium ion

through these channels depolarizes the taste cell. However, the specific identity of these channels, as well as the mechanisms by which salty taste of non-sodium salts is transduced, are not well understood (Chandrashekar, J. *et al.*, 2006). Sour-tasting (acid) stimuli also depolarize TRCs either by activating cation channels, or by direct permeation of protons through ion channels, including ENaC (Gilbertson, T. A. *et al.*, 1992; Gilbertson, T. A. and Boughter, J. D., Jr., 2003). Additionally, transient receptor potential (TRP) channels have been shown to be involved in acid taste (Huang, A. L. *et al.*, 2006; Ishimaru, Y. *et al.*, 2006).

Comparatively few studies have reported genetic variation affecting taste responses to salty or sour substances. Inbred strain differences have been reported for NaCl preferences in both mice (Hoshishima, K. *et al.*, 1962; Ninomiya, Y. *et al.*, 1989; Beauchamp, G. K., 1990; Lush, I. E., 1991; Bachmanov, A. A. *et al.*, 2002; Tordoff, M. G. *et al.*, 2007) and rats (Midkiff, E. E. *et al.*, 1985). Among commonly used strains, B6 and BALB are NaCl avoiding, and their CT response to NaCl is amiloride sensitive, whereas DBA/2, C3H/He, and 129 mice are NaCl-tolerant and their CT response to NaCl is amiloride-insensitive (Ninomiya, Y. *et al.*, 1989; Gannon, K. and Contreras, R. J., 1993; Ninomiya, Y., 1996). Experiments using isolated taste buds indicated that the strain difference might be due to a difference in the density of functional amiloride-sensitive Na^+ channels on taste cells (Miyamoto, T. *et al.*, 1999). Although B6 and DBA/2 mice differ in terms of the amiloride sensitivity of their CT response to NaCl (Ninomiya, Y. *et al.*, 1989), these strains do not differ in detection or discrimination of salts, and these behaviors are attenuated by amiloride in a similar fashion in both strains (Eylam, S. and Spector, A. C., 2003; 2005). More studies are needed to determine whether there is a relationship between amiloride sensitivity of neural responses to NaCl, and NaCl preference. The genetic variation in voluntary NaCl consumption may also depend on a variety of other factors in addition to peripheral taste responsiveness.

4.19.6   Taste and Other Complex Phenotypes

4.19.6.1   Taste and Diet

Taste sensitivity is associated with food choice and intake, and taste preferences are a predictor of propensity to consume sweet or high-fat foods (e.g., Bartoshuk, L., 1980; Drewnowski, A. *et al.*, 1985; Anliker, J. A. *et al.*, 1991; Drewnowski, A. J. *et al.*, 1998; Salbe, A. D. *et al.*,

2004). At least some of this association is underlied by allelic variation of taste receptor genes (Mennella, J. A. *et al.*, 2005; Dinehart, M. E. *et al.*, 2006; Reed, D. R. and McDaniel, A. H., 2006; Sandell, M. A. and Breslin, P. A., 2006). Unexpectedly, haplotypes at the *TAS2R38* gene encoding the PTC receptor were significantly related to children's to preference for sucrose and sweet-tasting foods and beverages: Those possessing taster alleles tended to be sweet preferrers; the authors speculate that the PTC gene may be in linkage disequilibrium with sweet taste genes (Mennella, J. A. *et al.*, 2005). Other studies have shown that *TAS2R38* genotype or PROP taster status affects perception of bitterness of vegetables, such as broccoli, turnips, or Brussels sprouts, and hence may influence dietary selection (Dinehart, M. E. *et al.*, 2006; Sandell, M. A. and Breslin, P. A., 2006). It is furthermore possible that there is a relationship between genetic determinates of taste sensitivity and obesity (Bartoshuk, L. M *et al.*, 2006).

There is a strong relationship between taste and feeding (Scott, T. R., 2001), and future taste-related gene discovery may focus on the interaction of taste processing with homeostatic systems involved in the regulation of feeding, appetite, fluid balance, reward, and drug administration.

4.19.6.2   Taste and Drug Intake

It has long been appreciated that a consistent predictor of alcohol (ethanol) drinking in rodents is avidity for consumption of sweet-tasting compounds. Positive correlations between ethanol and sweetener intake have been noted in multiple strains of rats and mice, and a common genetic basis is strongly supported by a robust association of these phenotypes in both selected lines and segregating populations (McClearn, G. E. and Rodgers, D., 1959; Belknap, J. K. *et al.*, 1993; Overstreet, D. H. *et al.*, 1993; Bachmanov, A. A. *et al.*, 1996b; Dess, N. K. *et al.*, 1998; Stewart, R. B. *et al.*, 2003). There is also substantial evidence for an association between alcohol drinking and sweet taste in humans (Yamamoto, M. E. *et al.*, 1991; Kampov-Polevoy, A. *et al.*, 1997; 2003; 2004). *Tas1r3* knockout mice do not prefer ethanol unlike their wild-type controls, strongly arguing for a direct role of the sweet taste receptor in mediating ethanol intake (Blednov, Y. A. *et al.*, 2007). In rats, ethanol applied to the tongue selectively activates a subpopulation of sweet-responsive taste neurons in the CNS, and this activity is blocked by lingual application of gurmarin, a peripheral antagonist of sweet taste (Lemon, C. H. *et al.*, 2004).

There is also an association of bitter taste sensitivity with alcohol, caffeine and nicotine use (Hall, M. J. *et al.*, 1975; Mela, D. J., 1989; Mela, D. J. *et al.*, 1992; Mattes, R. D., 1994; Enoch, M. A. *et al.*, 2001; Snedecor, S. M. *et al.*, 2006). This association may be due to individual variation in bitter taste receptors. For example, subjects possessing nontaster haplotypes at the *TAS2R38* locus rate alcoholic beverages as less bitter, and more preferred (Duffy, V. B. *et al.*, 2004; Lanier, S. A. *et al.*, 2005). Similarly, there is a greater percentage of bitter-insensitive individuals among smokers than nonsmokers, suggesting these nontasters may be at risk for nicotine addiction (Enoch, M. A. *et al.*, 2001). Because there is a strong relationship between smoking and alcohol use in humans (Pomerleau, C. S. *et al.*, 2004) it is reasonable to assume taste may be a common factor in both addictions. However, oral somatosensory factors may also play a role because both nicotine and ethanol are trigeminal irritants (e.g., Carstens, E. *et al.*, 1998). In addition, causative relationships between taste and drug intake may be more complex, as drug intake may influence taste sensitivity. There is evidence that long-term smoking or oral tobacco use may contribute to a decrement in overall taste acuity (Sato, K. *et al.*, 2002; Khan, G. J. *et al.*, 2003), and taste detection thresholds for caffeine are elevated in habitual caffeine drinkers relative to nonusers (Mela, D. J. *et al.*, 1992).

4.19.7   Conclusions

Historically, behavioral genetic studies of taste and intake behaviors have been instrumental in the eventual discovery of G-protein-coupled taste receptors. In particular, the discovery of the PTC polymorphism in humans stimulated research into the genetics basis of taste behavior. In rodents, strain differences in responsiveness to bitter- and sweet-tasting stimuli, primarily in two-bottle tests, provided linkages to chromosome regions later found to harbor taste receptor genes. However, the balance of evidence suggests that taste-related behaviors are under polygenic control, and other genes have yet to be elucidated. Within- and between-species differences in the T1R and T2R genes are associated with variation in taste-related behaviors. The evolution of the T1R and T2R gene families reflects adaptation strategies of different species to their chemical environments and feeding ecology.

References

Adler, E., Hoon, M. A., Mueller, K. L., Chandrashekar, J., Ryba, N. J. P., and Zuker, C. S. 2000. A novel family of mammalian taste receptors. Cell 100, 693–702.

Anliker, J. A., Bartoshuk, L., Ferris, A. M., and Hooks, L. D. 1991. Children's food preferences and genetic sensitivity to the bitter taste of 6-n-propylthiouracil (PROP). Am. J. Clin. Nutr. 54, 316–320.

Bachmanov, A. A. 2008. Genetic Architecture of Sweet Taste. In: Sweetness and Sweeteners: Biology, Chemistry and Psychophysics. ACS Symposium Series (eds. D. K. Weerasinghe and G. E. Dubois), Oxford University Press.

Bachmanov, A. A., Beauchamp, G. K., and Tordoff, M. G. 2002. Voluntary consumption of NaCl, KCl, $CaCl_2$, and NH_4Cl solutions by 28 mouse strains. Behav. Genet. 32, 445–457.

Bachmanov, A. A., Li, X., Li, S., Neira, M., Beauchamp, G. K., and Azen, E. A. 2001a. High-resolution genetic mapping of the sucrose octaacetate taste aversion (Soa) locus on mouse Chromosome 6. Mamm. Genome 12, 695–699.

Bachmanov, A. A., Li, X., Reed, D. R., Ohmen, J. D., Li, S., Chen, Z., Tordoff, M. G., De Jong, P. J., Wu, C., West, D. B., Chatterjee, A., Ross, D. A., and Beauchamp, G. K. 2001b. Positional cloning of the mouse saccharin preference (Sac) locus. Chem. Senses 26, 925–933.

Bachmanov, A. A., Reed, D. R., Li, X., and Beauchamp, G. K. 2002. Genetics of sweet taste preferences. Pure Appl. Chem. 74, 1135–1140.

Bachmanov, A. A., Reed, D. R., Ninomiya, Y., Inoue, M., Tordoff, M. G., Price, R. A., and Beauchamp, G. K. 1997. Sucrose consumption in mice: major influence of two genetic loci affecting peripheral sensory responses. Mamm. Genome 8, 545–548.

Bachmanov, A. A., Reed, D. R., Tordoff, M. G., Price, R. A., and Beauchamp, G. K. 1996a. Intake of ethanol, sodium chloride, sucrose, citric acid, and quinine hydrochloride solutions by mice: a genetic analysis. Behav. Genet. 26, 563–573.

Bachmanov, A. A., Reed, D. R., Tordoff, M. G., Price, R. A., and Beauchamp, G. K. 2001c. Nutrient preference and diet-induced adiposity in C57BL/6ByJ and 129P3/J mice. Physiol. Behav. 72, 603–613.

Bachmanov, A. A., Tordoff, M. G., and Beauchamp, G. K. 1996b. Ethanol consumption and taste preferences in C57BL/6ByJ and 129/J mice. Alcoholism: Clin. Exp. Res. 20, 201–206.

Bachmanov, A. A., Tordoff, M. G., and Beauchamp, G. K. 1998. Voluntary sodium chloride consumption by mice: Differences among five inbred strains. Behav. Genet. 28, 117–124.

Bachmanov, A. A., Tordoff, M. G., and Beauchamp, G. K. 2000. Intake of umami-tasting solutions by mice: a genetic analysis. J. Nutr. 130, 935S–941S.

Bachmanov, A. A., Tordoff, M. G., and Beauchamp, G. K. 2001d. Sweetener preference of C57BL/6ByJ and 129P3/J mice. Chem. Senses 26, 905–913.

Barnicot, N., Harris, H., and Kalmus, H. 1951. Taste thresholds of further eighteen compounds and their correlation with PTC thresholds. Ann. Eugenics 16, 119–128.

Bartoshuk, L. 1980. Influence of chemoreception and psychologic state on food selection. Int. J. Obesity 4, 351–355.

Bartoshuk, L. M., Duffy, V. B., Hayes, J. E., Moskowitz, H. R., and Snyder, D. J. 2006. Psychophysics of sweet and fat perception in obesity: problems, solutions and new perspectives. Phil. Trans. R. Soc. Lond. B Biol. Sci. 361, 1137–1148.

Bartoshuk, L. M., Duffy, V. B., and Miller, I. J. 1994. PTC/PROP tasting: anatomy, psychophysics, and sex effects. Physiol. Behav. 56, 1165–1171.

Bartoshuk, L. M., Jacobs, H. L., Nichols, T. L., Hoff, L. A., and Ryckman, J. J. 1975. Taste rejection of nonnutritive sweeteners in cats. J. Comp. Physiol. Psychol. 89, 971–975.

Beauchamp, G. K. 1990. Genetic control over salt preference in inbred strains of mice. Chem. Senses 15, 551.

Beauchamp, G. K., Maller, O., and Rogers, J. G. 1977. Flavor preferences in cats (Felis catus and Panthera sp.). J. Comp. Physiol. Psychol. 91, 1118–1127.

Behrens, M., Brockhoff, A., Kuhn, C., Bufe, B., Winnig, M., and Meyerhof, W. 2004. The human taste receptor hTAS2R14 responds to a variety of different bitter compounds. Biochem. Biophys. Res. Commun. 319, 479–485.

Belknap, J. K., Crabbe, J. C., Plomin, R., Mcclearn, G. E., Sampson, K. E., O'Toole, L. A., and Gora-Maslak, G. 1992. Single-locus control of saccharin intake in BXD/Ty recombinant inbred (RI) mice: some methodological implications for RI strain analysis. Behav. Genet. 22, 81–100.

Belknap, J. K., Crabbe, J. C., and Young, E. R. 1993. Voluntary consumption of ethanol in 15 inbred mouse strains. Psychopharmacology (Berl.) 112, 503–510.

Bjarnadottir, T. K., Fredriksson, R., and Schioth, H. B. 2005. The gene repertoire and the common evolutionary history of glutamate, pheromone (V2R), taste(1) and other related G protein-coupled receptors. Gene 362, 70–84.

Blednov, Y. A., Walker, D., Martinez, M., Levine, M., Damak, S., and Margolskee, R. F. 2007. Perception of sweet taste is important for voluntary alcohol consumption in mice. Genes Brain Behav. doi: 10.1111/j.1601-183X.2007.00309.X.

Blizard, D. A., Kotlus, B., and Frank, M. E. 1999. Quantitative trait loci associated with short-term intake of sucrose, saccharin and quinine solutions in laboratory mice. Chem. Senses 24, 373–385.

Bloomgarden, Z. T. 2006. Gut hormones and related concepts. Diabetes Care 29, 2319–2324.

Boughter, J. D., Jr. and Gilbertson, T. A. 1999. From channels to behavior: an integrative model of NaCl taste. Neuron 22, 213–215.

Boughter, J. D., Jr. and Whitney, G. 1995. C3.SW-Soa^a heterozygous congenic taster mice. Behav. Genet. 25, 233–237.

Boughter, J. D., Jr. and Whitney, G. 1998. Behavioral specificity of the bitter taste gene Soa. Physiol. Behav. 63, 101–108.

Boughter, J. D., Jr., Harder, D. B., Capeless, C. G., and Whitney, G. 1992. Polygenic determination of quinine sensitivity among mice. Chem. Senses 17, 427–434.

Boughter, J. D., Jr., Raghow, S., Nelson, T. M., and Munger, S. D. 2005. Inbred mouse strains C57BL/6J and DBA/2J vary in sensitivity to a subset of bitter stimuli. BMC Genet. 6, 36.

Boughter, J. D., Jr., St. John, S. J., Noel, D. T., Ndubuizu, O., and Smith, D. V. 2002. A brief-access test for bitter taste in mice. Chem. Senses 27, 133–142.

Bufe, B., Hofmann, T., Krautwurst, D., Raguse, J. D., and Meyerhof, W. 2002. The human TAS2R16 receptor mediates bitter taste in response to beta-glucopyranosides. Nat. Genet. 32, 397–401.

Butt, C. M., King, N. M., Hutton, S. R., Collins, A. C., and Stitzel, J. A. 2005. Modulation of nicotine but not ethanol preference by the mouse Chrna4 A529T polymorphism. Behav. Neurosci. 119, 26–37.

Capeless, C. G. and Whitney, G. 1995. The genetic basis of preference for sweet substances among inbred strains of mice: preference ratio phenotypes and the alleles of the Sac and dpa loci. Chem. Senses 20, 291–298.

Capeless, C. G., Whitney, G., and Azen, E. A. 1992. Chromosome mapping of Soa, a gene influencing gustatory sensitivity to sucrose octaacetate in mice. Behav. Genet. 22, 655–663.

Carstens, E., Kuenzler, N., and Handwerker, H. O. 1998. Activation of neurons in rat trigeminal subnucleus caudalis by different irritant chemicals applied to oral or ocular mucosa. J. Neurophysiol. 80, 465–492.

Chandrashekar, J., Hoon, M. A., Ryba, N. J., and Zuker, C. S. 2006. The receptors and cells for mammalian taste. Nature. 444, 288–294.

Chandrashekar, J., Mueller, K. L., Hoon, M. A., Adler, E., Feng, L., Guo, W., Zuker, C. S., and Ryba, J. P. 2000. T2Rs function as bitter taste receptors. Cell 100, 703–711.

Chaudhari, N., Landin, A. M., and Roper, S. D. 2000. A metabotropic glutamate receptor variant functions as a taste receptor. Nat. Neurosci. 3, 113–119.

Chyb, S., Dahanukar, A., Wickens, A., and Carlson, J. R. 2003. Drosophila Gr5a encodes a taste receptor tuned to trehalose. Proc. Natl. Acad. Sci. U. S. A. 100(Suppl. 2), 14526–14530.

Conte, C., Ebeling, M., Marcuz, A., and Andres-Barquin, P. J. 2003a. Identification of the T2R repertoire of taste receptor genes in the rat genome sequence. Genome Lett. 2, 155–161.

Conte, C., Ebeling, M., Marcuz, A, Nef, P., and Andres-Barquin, P. J. 2002. Identification and characterization of human taste receptor genes belonging to the TAS2R family. Gene Mapp. Clon. Seq. 98, 45–53.

Conte, C., Ebeling, M., Marcuz, A., Nef, P., and Andres-Barquin, P. J. 2003b. Evolutionary relationships of the Tas2r receptor gene families in mouse and human. Physiol. Genomics 14, 73–82.

Conte, C., Guarin, E., Marcuz, A., and Andres-Barquin, P. J. 2006. Functional expression of mammalian bitter taste receptors in *Caenorhabditis elegans*. Biochimie. 88, 801–806.

Dahanukar, A., Foster, K., Van der Goes Van Naters, W. M., and Carlson, J. R. 2001. A Gr receptor is required for response to the sugar trehalose in taste neurons of Drosophila. Nat. Neurosci. 4, 1182–1186.

Damak, S., Rong, M., Yasumatsu, K., Kokrashvili, Z., Varadarajan, V., Zou, S., Jiang, P., Ninomiya, Y., and Margolskee, R. F. 2003. Detection of sweet and umami taste in the absence of taste receptor T1r3. Science 301, 850–853.

Delay, E. R., Hernandez, N. P., Bromley, K., and Margolskee, R. F. 2006. Sucrose and monosodium glutamate taste thresholds and discrimination ability of T1R3 knockout mice. Chem. Senses. 31, 351–357.

Dess, N. K., Badia-Elder, N. E., Thiele, T. E., Kiefer, S. W., and Blizard, D. A. 1998. Ethanol consumption in rats selectively bred for differential saccharin intake. Alcohol 16, 275–278.

Dinehart, M. E., Hayes, J. E., Bartoshuk, L. M., Lanier, S. L., and Duffy, V. B. 2006. Bitter taste markers explain variability in vegetable sweetness, bitterness, and intake. Physiol. Behav. 87, 304–313.

Drewnowski, A., Henderson, S. A., and Barratt-Fornell, A. 1998. Genetic sensitivity to 6-*n*-propylthiouracil and sensory responses to sugar and fat mixtures. Physiol. Behav. 63, 771–777.

Drewnowski, A. J., Brunzell, J. D., Sande, K., Iverius, P. H., and Greenwood, M. R. 1985. Sweet tooth reconsidered: taste responsiveness in human obesity. Physiol. Behav. 35, 617–622.

Duffy, V. B. and Bartoshuk, L. M. 2000. Food acceptance and genetic variation in taste. J. Am. Dietetic Assoc. 100, 647–655.

Duffy, V. B., Davidson, A. C., Kidd, J. R., Kidd, K. K., Speed, W. C., Pakstis, A. J., Reed, D. R., Snyder, D. J., and Bartoshuk, L. M. 2004. Bitter receptor gene (TAS2R38), 6-*n*-propylthiouracil (PROP) bitterness and alcohol intake. Alcoholism: Clin. Exp. Res. 28, 1629–1637.

Enoch, M. A., Harris, C. R., and Goldman, D. 2001. Does a reduced sensitivity to bitter taste increase the risk of becoming nicotine addicted? Addictive Behav. 26, 399–404.

Eylam, S. and Spector, A. C. 2003. Oral amiloride treatment decreases taste sensitivity to sodium salts in C57BL/6J and DBA/2J mice. Chem. Senses 28, 447–458.

Eylam, S. and Spector, A. C. 2004. Stimulus processing of glycine is dissociable from that of sucrose and glucose based on behaviorally measured taste signal detection in Sac 'taster' and 'non-taster' mice. Chem. Senses 29, 639–649.

Eylam, S. and Spector, A. C. 2005. Taste discrimination between NaCl and KCl is disrupted by amiloride in inbred mice with amiloride-insensitive chorda tympani nerves. Am. J. Physiol. Regul. Integr. Comp. Physiol. 288, R1361–R1368.

Fischer, A., Gilad, Y., Man, O., and Paabo, S. 2005. Evolution of bitter taste receptors in humans and apes. Mol. Biol. Evol. 22, 432–436.

Fischer, R. A., Ford, E. B., and Huxley, J. 1939. Taste testing the anthropoid apes. Nature 144, 750.

Flaherty, L. 1981. Congenic Strains. In: The Mouse in Biomedical Research. Vol. 1. History, Genetics, and Wild Mice (*eds.* H. L. Foster, J. D. Small, and J. G. Fox), pp. 215–222. Academic Press.

Fox, A. L. 1931. Six in ten "tasteblind" to bitter chemical. Sci. News Lett. 9, 249.

Fox, A. L. 1932. The relationship between chemical constituion and taste. Proc. Natl. Acad. Sci. U. S. A. 18, 115–120.

Fredriksson, R., Lagerstrom, M. C., Lundin, L. G., and Schioth, H. B. 2003. The G-protein-coupled receptors in the human genome form five main families. Phylogenetic analysis, paralogon groups, and fingerprints. Mol. Pharmacol. 63, 1256–1272.

Fuller, J. L. 1974. Single-locus control of saccharin preference in mice. J. Heredity 65, 33–36.

Ganchrow, J. R., Steiner, J. E., and Bartana, A. 1990. Behavioral reactions to gustatory stimuli in young chicks (*Gallus gallus domesticus*). Dev. Psychobiol. 23, 103–117.

Gannon, K. and Contreras, R. J. 1993. Sodium intake linked to amiloride-sensitive gustatory transduction in C57BL/6J and 129/J mice. Physiol. Behav. 57, 231–239.

Gent, J. F. and Bartoshuk, L. M. 1983. Sweetness of sucrose, neohesperidin dihydrochalcone, and saccharin is related to genetic ability to taste the bitter substance 6-*n*-propylthiouracil. Chem. Senses 7, 265–272.

Gilbertson, T. A. and Boughter, J. D., Jr. 2003. Taste transduction: appetizing times in gustation. NeuroReport 14, 905–911.

Gilbertson, T. A., Avenet, P., Kinnamon, S. C., and Roper, S. D. 1992. Proton currents through amiloride-sensitive Na channels in hamster taste cells: role in acid transduction. J. Gen. Physiol. 100, 803–824.

Gilbertson, T. A., Damak, S., and Margolskee, R. F. 2000. The molecular physiology of taste transduction. Curr. Opin. Neurobiol. 10, 519–527.

Glendinning, J. I. 1994. Is the bitter rejection response always adaptive? Physiol. Behav. 56, 1217–1227.

Glendinning, J. I. and Hills, T. T. 1997. Electrophysiological evidence for two transduction pathways within a bitter-sensitive taste receptor. J. Neurophysiol. 78, 734–745.

Glendinning, J. I., Gresack, J., and Spector, A. C. 2002. A high-throughput screening procedure for identifying mice with aberrant taste and oromotor function. Chem. Senses 27, 461–474.

Glendinning, J. I., Valcic, S., and Timmermann, B. N. 1998. Maxillary palps can mediate taste rejection of plant allelochemicals by caterpillars. J. Comp. Physiol. A 183, 35–43.

Go, Y. 2006. Lineage-specific expansions and contractions of the bitter taste receptor gene repertoire in vertebrates. Mol. Biol. Evol. 23, 964–972.

Go, Y., Satta, Y., Takenaka, O., and Takahata, N. 2005. Lineage-specific loss of function of bitter taste receptor genes in humans and nonhuman primates. Genetics 170, 313–1326.

Goodwin, F. L. and Amit, Z. 1998. Do taste factors contribute to the mediation of ethanol intake? Ethanol and saccharin–quinine intake in three rat strains. Alcoholism: Clin. Exp. Res. 22, 837–844.

Grace, J. and Russek, M. 1969. The influence of previous experience on the taste behavior of dogs toward sucrose and saccharin. Physiol. Behav. 4, 553–558.

Greene, L. S., Desor, J. A., and Maller, O. 1975. Heredity and experience: their relative importance in the development of taste preference in man. J. Comp. Physiol. Psychol. 89, 279–284.

Hall, M. J., Bartoshuk, L. M., Cain, W. S., and Stevens, J. C. 1975. PTC taste blindness and the taste of caffeine. Nature 253, 442–443.

Hallem, E. A., Dahanukar, A., and Carlson, J. R. 2006. Insect odor and taste receptors. Annu. Rev. Entomol. 51, 113–135.

Halpern, B. P. 1962. Gustatory nerve responses in the chicken. Am. J. Physiol. 203, 541–544.

Harder, D. B. and Whitney, G. 1998. A common polygenic basis for quinine and PROP avoidance in mice. Chem. Senses 23, 327–332.

Harder, D. B., Azen, E. A., and Whitney, G. 2000. Sucrose octaacetate avoidance in nontaster mice is not enhanced by two type-A Prp transgenes from taster mice. Chem. Senses 25, 39–45.

Harder, D. B., Capeless, C. G., Maggio, J. C., Boughter, J. D., Gannon, K. S., Whitney, G., and Azen, E. A. 1992. Intermediate sucrose octa-acetate sensitivity suggests a third allele at mouse bitter taste locus *Soa* and *Soa-Rua* identity. Chem. Senses 17, 391–401.

Harder, D. B., Maggio, J. C., and Whitney, G. 1989. Assessing gustatory detection capabilities using preference procedures. Chem. Senses 14, 547–564.

Harris, H. and Kalmus, H. 1949. The measurement of taste sensitivity to phenylthiourea (P.T.C.). Ann. Eugenics 15, 24–31.

Hoon, M. A., Adler, E., Lindemeier, J., Battey, J. F., Ryba, N. J., and Zuker, C. S. 1999. Putative mammalian taste receptors: a class of taste-specific GPCRs with distinct topographic selectivity. Cell 96, 541–551.

Hoshishima, K., Yokoyama, S., and Seto, K. 1962. Taste sensitivity in various strains of mice. Am. J. Physiol. 202, 1200–1204.

Huang, A. L., Chen, X., Hoon, M. A., Chandrashekar, J., Guo, W., Trankner, D., Ryba, N. J., and Zuker, C. S. 2006. The cells and logic for mammalian sour taste detection. Nature 442, 934–938.

Inomata, N., Goto, H., Itoh, M., and Isono, K. 2004. A single-amino-acid change of the gustatory receptor gene, Gr5a, has a major effect on trehalose sensitivity in a natural population of *Drosophila melanogaster*. Genetics 167, 1749–1758.

Inoue, M., Beauchamp, G. K., and Bachmanov, A. A. 2004a. Gustatory neural responses to umami taste stimuli in C57BL/6ByJ and 129P3/J mice. Chem. Senses 29, 789–795.

Inoue, M., Li, X., Mccaughey, S. A., Beauchamp, G. K., and Bachmanov, A. A. 2001a. Soa genotype selectively affects mouse gustatory neural responses to sucrose octaacetate. Physiol. Genomics 5, 181–186.

Inoue, M., McCaughey, S. A., Bachmanov, A. A., and Beauchamp, G. K. 2001b. Whole nerve chorda tympani responses to sweeteners in C57BL/6ByJ and 129P3/J mice. Chem. Senses 26, 915–923.

Inoue, M., Reed, D. R., Li, X., Tordoff, M. G., Beauchamp, G. K., and Bachmanov, A. A. 2004b. Allelic variation of the *Tas1r3* taste receptor gene selectively affects behavioral and neural taste responses to sweeteners in the F_2 hybrids between C57BL/6ByJ and 129P3/J mice. J. Neurosci. 24, 2296–2303.

Ishimaru, Y., Inada, H., Kubota, M., Zhuang, H., Tominaga, M., and Matsunami, H. 2006. Transient receptor potential family members PKD1L3 and PKD2L1 form a candidate sour taste receptor. Proc. Natl. Acad. Sci. U. S. A. 103, 12569–12574.

Ishimaru, Y., Okada, S., Naito, H., Nagai, T., Yasuoka, A., Matsumoto, I., and Abe, K. 2005. Two families of candidate taste receptors in fishes. Mech. Dev. 122, 1310–1321.

Isono, K., Morita, H., Kohatsu, S., Ueno, K., Matsubayashi, H., and Yamamoto, M. T. 2005. Trehalose sensitivity of the gustatory receptor neurons expressing wild-type, mutant and ectopic Gr5a in Drosophila. Chem. Senses 30(Suppl.1), i275–i276.

Jiang, P., Cui, M., Ji, Q., Snyder, L., Liu, Z., Benard, L., Margolskee, R. F., Osman, R., and Max, M. 2005a. Molecular mechanisms of sweet receptor function. Chem. Senses 30(Suppl. 1), i17–i18.

Jiang, P., Cui, M., Zhao, B., Snyder, L. A., Benard, L. M., Osman, R., Max, M., and Margolskee, R. F. 2005b. Identification of the cyclamate interaction site within the transmembrane domain of the human sweet taste receptor subunit T1R3. J. Biol. Chem. 280, 34296–34305.

Jiang, P., Ji, Q., Liu, Z., Snyder, L. A., Benard, L. M., Margolskee, R. F., and Max, M. 2004. The cysteine-rich region of T1R3 determines responses to intensely sweet proteins. J. Biol. Chem. 279, 45068–45075.

Kachele, D. L. and Lasiter, P. S. 1990. Murine strain differences in taste responsivity and organization of the rostral nucleus of the solitary tract. Brain Res. Bull. 24, 239–247.

Kalmus, H. 1971. Genetics of Taste. In: Handbook of Sensory Physiology (*ed*. L. M. Beidler), pp. 165–179. Springer.

Kampov-Polevoy, A. B., Eick, C., Boland, G., Khalitov, E., and Crews, F. T. 2004. Sweet liking, novelty seeking, and gender predict alcoholic status. Alcoholism: Clin. Exp. Res. 28, 1291–1298.

Kampov-Polevoy, A., Garbutt, J. C., and Janowsky, D. 1997. Evidence of preference for a high-concentration sucrose solution in alcoholic men. Am. J. Psychiatry 154, 269–270.

Kampov-Polevoy, A. B., Garbutt, J. C., and Khalitov, E. 2003. Family history of alcoholism and response to sweets. Alcoholism: Clin. Exp. Res. 27, 1743–1749.

Kare, M. R. 1961. Comparative Aspects of the Sense of Taste. In: Physiological and Behavioral Aspects of Taste (*eds*. M. R. Kare and B. P. Halpern). The University of Chicago Press.

Khan, G. J., Mehmood, R., Salah Ud, D., and Ihtesham Ul, H. 2003. Effects of long-term use of tobacco on taste receptors and salivary secretion. J. Ayub Med. College Abbottabad 15, 15, 37–39.

Kim, U. K. and Drayna, D. 2005. Genetics of individual differences in bitter taste perception: lessons from the PTC gene. Clin. Genet. 67, 275–280.

Kim, U. K., Breslin, P. A., Reed, D., and Drayna, D. 2004. Genetics of human taste perception. J. Dental Res. 83, 448–453.

Kim, U. K., Jorgenson, E., Coon, H., Leppert, M., Risch, N., and Drayna, D. 2003. Positional cloning of the human quantitative trait locus underlying taste sensitivity to phenylthiocarbamide. Science 299, 1221–1225.

Kim, U. K., Wooding, S., Riaz, N., Jorde, L. B., and Drayna, D. 2006. Variation in the human TAS1R taste receptor genes. Chem. Senses 31, 599–611.

Kitagawa, M., Kusakabe, Y., Miura, H., Ninomiya, Y., and Hino, A. 2001. Molecular genetic identification of a candidate receptor gene for sweet taste. Biochem. Biophys. Res. Commun. 283, 236–242.

Klein, T. W. and Defries, J. C. 1970. Similar polymorphism of taste sensitivity to PTC in mice and men. Nature 225, 555–557.

Kuhn, C., Bufe, B., Winnig, M., Hofmann, T., Frank, O., Behrens, M., Lewtschenko, T., Slack, J. P., Ward, C. D., and Meyerhof, W. 2004. Bitter taste receptors for saccharin and acesulfame K. J. Neurosci. 24, 10260–10265.

Lagerstrom, M. C., Hellstrom, A. R., Gloriam, D. E., Larsson, T. P., Schioth, H. B., and Fredriksson, R. 2006. The G protein-coupled receptor subset of the chicken genome. PLoS Comput. Biol. 2, e54.

Lanier, S. A., Hayes, J. E., and Duffy, V. B. 2005. Sweet and bitter tastes of alcoholic beverages mediate alcohol intake in of-age undergraduates. Physiol. Behav. 83, 821–831.

Le Roy, I., Pager, J., and Roubertoux, P. L. 1999. Genetic dissection of gustatory sensitivity to bitterness (sucrose octaacetate) in mice. C. R. Acad. Sci. Série III 322, 831–836.

Lemon, C. H., Brasser, S. M., and Smith, D. V. 2004. Alcohol activates a sucrose-responsive gustatory neural pathway. J. Neurophysiol. 92, 536–544.

Li, X., Bachmanov, A. A., Li, S., Chen, Z., Tordoff, M. G., Beauchamp, G. K., De Jong, P. J., Wu, C., Chen, L., West, D. B., Ross, D. A., Ohmen, J. D., and Reed, D. R. 2002a. Genetic, physical, and comparative map of the subtelomeric region of mouse Chromosome 4. Mamm. Genome 13, 5–19.

Li, X., Inoue, M., Reed, D. R., Huque, T., Puchalski, R. B., Tordoff, M. G., Ninomiya, Y., Beauchamp, G. K., and Bachmanov, A. A. 2001. High-resolution genetic mapping of the saccharin preference locus (Sac) and the putative sweet taste receptor (T1R1) gene (Gpr70) to mouse distal Chromosome 4. Mamm. Genome 12, 13–16.

Li, X., Li, W., Wang, H., Cao, J., Maehashi, K., Huang, L., Bachmanov, A. A., Reed, D. R., Legrand-Defretin, V., Beauchamp, G. K., and Brand, J. G. 2005. Pseudogenization of a sweet-receptor gene accounts for cats' indifference toward sugar. PLoS Genet. 1, 27–35.

Li, X., Staszewski, L., Xu, H., Durick, K., Zoller, M., and Adler, E. 2002b. Human receptors for sweet and umami taste. Proc. Natl. Acad. Sci. U. S. A. 99, 4692–4696.

Liao, J. and Schultz, P. G. 2003. Three sweet receptor genes are clustered in human chromosome 1. Mamm. Genome 14, 291–301.

Liman, E. R. 2006. Use it or lose it: molecular evolution of sensory signaling in primates. Pflugers Arch. 453, 125–131.

Lindemann, B. 2000. A taste for umami. Nat. Neurosci. 3, 99–100.

Lugaz, O., Pillias, A. M., and Faurion, A. 2002. A new specific ageusia: some humans cannot taste L-glutamate. Chem. Senses 27, 105–115.

Lush, I. E. 1981. The genetics of tasting in mice. I. Sucrose octaacetate. Genet. Res. 38, 93–95.

Lush, I. E. 1982. The genetics of tasting in mice. II. Strychnine. Chem. Senses 7, 93–98.

Lush, I. E. 1984. The genetics of tasting in mice. III. Quinine. Genet. Res. 44, 151–160.

Lush, I. E. 1986. The genetics of tasting in mice. IV. The acetates of raffinose, galactose and beta-lactose. Genet. Res. 47, 117–123.

Lush, I. E. 1989. The genetics of tasting in mice. VI. Saccharin, acesulfame, dulcin and sucrose. Genet. Res. 53, 95–99.

Lush, I. E. 1991. The Genetics of Bitterness, Sweetness, and Saltiness in Strains of Mice. In: Genetics of Perception and Communication (eds. C. J. Wysocki and M. R. Kare). Dekker.

Lush, I. E. and Holland, G. 1988. The genetics of tasting in mice. V. Glycine and cycloheximide. Genet. Res. 52, 207–212.

Lush, I. E., Hornigold, N., King, P., and Stoye, J. P. 1995. The genetics of tasting in mice. VII. Glycine revisited, and the chromosomal location of Sac and Soa. Genet. Res. 66, 167–174.

Manita, S., Bachmanov, A. A., Li, X., Beauchamp, G. K., and Inoue, M. 2006. Is glycine "sweet" to mice? Mouse strain differences in perception of glycine taste. Chem. Senses 31, 785–793.

Maruyama, Y., Pereira, E., Margolskee, R. F., Chaudhari, N., and Roper, S. D. 2006. Umami responses in mouse taste cells indicate more than one receptor. J. Neurosci. 26, 2227–2234.

Matson, K. D., Millam, J. R., and Klasing, K. C. 2000. Taste threshold determination and side-preference in captive cockatiels (Nymphicus hollandicus). Appl. Anim. Behav. Sci. 69, 313–326.

Matson, K. D., Millam, J. R., and Klasing, K. C. 2001. Thresholds for sweet, salt, and sour taste stimuli in cockatiels (Nymphicus hollandicus). Zoo Biol. 20, 1–13.

Matsunami, H., Montmayeur, J. P., and Buck, L. B. 2000. A family of candidate taste receptors in human and mouse. Nature 404, 601–604.

Mattes, R. D. 1994. Influences on acceptance of bitter foods and beverages. Physiol. Behav. 56, 1229–1236.

Max, M., Shanker, Y. G., Huang, L., Rong, M., Liu, Z., Campagne, F., Weinstein, H., Damak, S., and Margolskee, R. F. 2001. Tas1r3, encoding a new candidate taste receptor, is allelic to the sweet responsiveness locus Sac. Nat. Genet. 28, 58–63.

McCaughey, S. A. 2007. Taste-evoked responses to sweeteners in the nucleus of the solitary tract differ between C57BL/6ByJ and 129P3/J mice. J. Neurosci. 27, 35–45.

McClearn, G. E. and Rodgers, D. 1961. Genetic factors in alcohol preference of laboratory mice. J. Comp. Physiol. Psychol. 54, 116–119.

McClearn, G. E. and Rodgers, D. A. 1959. Differences in alcohol preference among inbred strains of mice. Q. J. Stud. Alcohol 20, 691–695.

Mela, D. J. 1989. Gustatory function and dietary habits in users and nonusers of smokeless tobacco. Am. J. Clin. Nutr. 49, 482–489.

Mela, D. J., Mattes, R. D., Tanimura, S., and Garcia-Medina, M. R. 1992. Relationships between ingestion and gustatory perception of caffeine. Pharmacol. Biochem. Behav. 43, 513–521.

Mennella, J. A., Pepino, M. Y., and Reed, D. R. 2005. Genetic and environmental determinants of bitter perception and sweet preferences. Pediatrics 115, e216–222.

Meyerhof, W. 2005. Elucidation of mammalian bitter taste. Rev. Physiol. Biochem. Pharmacol. 154, 37–72.

Midkiff, E. E., Bernstein, I. L., Fitts, D. A., and Simpson, J. B. 1985. Salt preference and salt appetite of Fischer-344 rats. Chem. Senses 10, 446.

Miyamoto, T., Fujiyama, R., Okada, Y., and Sato, T. 1999. Strain difference in amiloride-sensitivity of salt-induced responses in mouse non-dissociated taste cells. Neurosci. Lett. 277, 13–16.

Montmayeur, J. P., Liberles, S. D., Matsunami, H., and Buck, L. B. 2001. A candidate taste receptor gene near a sweet taste locus. Nat. Neurosci. 4, 492–498.

Moon, S. J., Kottgen, M., Jiao, Y., Xu, H., and Montell, C. 2006. A taste receptor required for the caffeine response in vivo. Curr. Biol. 16, 1812–1817.

Mueller, K. L., Hoon, M. A., Erlenbach, I., Chandrashekar, J., Zuker, C. S., and Ryba, N. J. P. 2005. The receptors and coding logic for bitter taste. Nature 434, 225–229.

Nelson, G., Chandrashekar, J., Hoon, M. A., Feng, L., Zhao, G., Ryba, N. J., and Zuker, C. S. 2002. An amino-acid taste receptor. Nature 416, 199–202.

Nelson, G., Hoon, M. A., Chandrashekar, J., Zhang, Y., Ryba, N. J., and Zuker, C. S. 2001. Mammalian sweet taste receptors. Cell 106, 381–390.

Nelson, T. M., Munger, S. D., and Boughter, J. D., Jr. 2003. Taste sensitivities to PROP and PTC vary independently in mice. Chem. Senses 28, 695–704.

Nelson, T. M., Munger, S. D., and Boughter, J. D., Jr. 2005. Haplotypes at the Tas2r locus on distal chromosome 6 vary with quinine taste sensitivity in inbred mice. BMC Genet. 6, 32.

Nie, Y., Vigues, S., Hobbs, J. R., Conn, G. L., and Munger, S. D. 2005. Distinct contributions of T1R2 and T1R3 taste receptor subunits to the detection of sweet stimuli. Curr. Biol. 15, 1948–1952.

Ninomiya, Y. 1996. Salt taste responses of mouse chorda tympani neurons: evidence for existence of two different amiloride-sensitive receptor components for NaCl with different temperature dependencies. J. Neurophysiol. 76, 3550–3554.

Ninomiya, Y. and Funakoshi, M. 1989. Behavioural discrimination between glutamate and the four basic taste substances in mice. Comp. Biochem. Physiol. 92, 365–370.

Ninomiya, Y., Higashi, T., Mizukoshi, T., and Funakoshi, M. 1987. Genetics of the ability to perceive sweetness of D-phenylalanine in mice. Ann. N. Y. Acad. Sci. 510, 527–529.

Ninomiya, Y., Mizukoshi, T., Higashi, T., Katsukawa, H., and Funakoshi, M. 1984. Gustatory neural responses in three different strains of mice. Brain Res. 302, 305–314.

Ninomiya, Y., Nomura, T., and Katsukawa, H. 1992. Genetically variable taste sensitivity to D-amino acids in mice. Brain Res. 596, 349–352. Dekker.

Ninomiya, Y., Sako, N., and Funakoshi, M. 1989. Strain differences in amiloride inhibition of NaCl responses in mice, *Mus musculus*. J. Comp. Physiol. 166, 1–5.

Ninomiya, Y., Sako, N., Katsukawa, H., and Funakoshi, M. 1991. Taste Receptor Mechanisms Influenced by a Gene on Chromosome 4 in Mice. In: Genetics of Perception and Communication (eds. C. J. Wysocki and M. R. Kare), pp. 267–278.

Overstreet, D. H., Kampov-Polevoy, A. B., Rezvani, A. H., Murelle, L., Halikas, J. A., and Janowsky, D. S. 1993. Saccharin intake predicts ethanol intake in genetically heterogeneous rats as well as different rat strains. Alcoholism: Clin. Exp. Res. 17, 366–369.

Parry, C. M., Erkner, A., and Le Coutre, J. 2004. Divergence of T2R chemosensory receptor families in humans, bonobos, and chimpanzees. Proc. Natl. Acad. Sci. U. S. A. 101, 14830–14834.

Peirce, J. L., Lu, L., Gu, J., Silver, L. M., and Williams, R. W. 2004. A new set of BXD recombinant inbred lines from advanced intercross populations in mice. BMC Genet. 5, 7.

Pelz, W. E., Whitney, G., and Smith, J. C. 1973. Genetic influences on saccharin preference of mice. Physiol. Behav. 10, 263–265.

Phillips, T. J., Crabbe, J. C., Metten, P., and Belknap, J. K. 1994. Localization of genes affecting alcohol drinking in mice. Alcoholism: Clin. Exp. Res. 18, 931–941.

Pomerleau, C. S., Marks, J. L., Pomerleau, O. F., and Snedecor, S. M. 2004. Relationship between early experiences with tobacco and early experiences with alcohol. Addictive Behav. 29, 1245–1251.

Ramirez, I. and Fuller, J. L. 1976. Genetic influence on water and sweetened water consumption in mice. Physiol. Behav. 16, 163–168.

Reed, D. R., Bachmanov, A. A., Beauchamp, G. K., Tordoff, M. G., and Price, R. A. 1997. Heritable variation in food preferences and their contribution to obesity. Behav. Genet. 27, 373–387.

Reed, D. R., Li, S., Li, X., Huang, L., Tordoff, M. G., Starling-Roney, R., Taniguchi, K., West, D. B., Ohmen, J. D., Beauchamp, G. K., and Bachmanov, A. A. 2004. Polymorphisms in the taste receptor gene (*Tas1r3*) region are associated with saccharin preference in 30 mouse strains. J. Neurosci. 24, 938–946.

Reed, D. R. and McDaniel, A. H. 2006. The human sweet tooth. BMC Oral Health 6(Suppl. 1), S17.

Richter, C. 1932. Mineral appetite of parathyroidectomized rats. Am. J. Med. Sci. 9, 9.

Richter, T. A. and Clisby, K. H. 1941. Phenylthiocarbamide taste thresholds of rats and human beings. Am. J. Physiol. 134, 157–164.

Robinson, S. F., Marks, M. J., and Collins, A. C. 1996. Inbred mouse strains vary in oral self-selection of nicotine. Psychopharmacology (Berl.) 124, 332–339.

Rodgers, D. A. and McClearn, G. E. 1962. Mouse strain differences in preference for various concentrations of alcohol. Q. J. Stud. Alcohol 23, 26–33.

Sainz, E., Korley, J. N., Battey, J. F., and Sullivan, S. L. 2001. Identification of a novel member of the T1R family of putative taste receptors. J. Neurochem. 77, 896–903.

Salbe, A. D., Delparigi, A., Pratley, R. E., Drewnowski, A., and Tataranni, P. A. 2004. Taste preferences and body weight changes in an obesity-prone population. Am. J. Clin. Nutr. 79, 372–378.

Sandell, M. A. and Breslin, P. A. 2006. Variability in a taste-receptor gene determines whether we taste toxins in food. Current Biol. 16, R792–R794.

Sato, K., Endo, S., and Tomita, H. 2002. Sensitivity of three loci on the tongue and soft palate to four basic tastes in smokers and non-smokers. Acta Otolaryngol. 2002, 74–82.

Scott, T. R. 2001. The role of taste in feeding. Appetite 37, 111–113.

Shi, P. and Zhang, J. 2006. Contrasting modes of evolution between vertebrate sweet/umami receptor genes and bitter receptor genes. Mol. Biol. Evol. 23, 292–300.

Shi, P., Zhang, J., Yang, H., and Zhang, Y. P. 2003. Adaptive diversification of bitter taste receptor genes in mammalian evolution. Mol. Biol. Evol. 20, 805–814.

Shigemura, N., Yasumatsu, K., Yoshida, R., Sako, N., Katsukawa, H., Nakashima, K., Imoto, T., and Ninomiya, Y. 2005. The role of the dpa locus in mice. Chem. Senses 30(Suppl. 1), i84–i85.

Shingai, T. and Beidler, L. M. 1985. Interstrain differences in bitter taste responses in mice. Chem. Senses 10, 51–55.

Snedecor, S. M., Pomerleau, C. S., Mehringer, A. M., Ninowski, R., and Pomerleau, O. F. 2006. Differences in smoking-related variables based on phenylthiocarbamide "taster" status. Addictive Behav. 31, 2309–2312.

Sollars, S. I., Midkiff, E. E., and Bernstein, I. L. 1990. Genetic transmission of NaCl aversion in the Fisher-344 rat. Chem. Senses 15, 521–527.

Soranzo, N., Bufe, B., Sabeti, P. C., Wilson, J. F., Weale, M. E., Marguerie, R., Meyerhof, W., and Goldstein, D. B. 2005. Positive selection on a high-sensitivity allele of the human bitter-taste receptor TAS2R16. Curr. Biol. 15, 1257–1265.

Spielman, A. I., Huque, T., Whitney, G., and Brand, J. G. 1992. The diversity of bitter taste signal transduction mechanisms. Soc. Gen. Physiol. Ser. 47, 307–324.

St. John, S. J., Pour, L., and Boughter, J. D., Jr. 2005. Phenylthiocarbamide produces conditioned taste aversions in mice. Chem. Senses 30, 377–382.

Stewart, R. B., Bice, P., Foroud, T., Lumeng, L., Li, T. K., and Carr, L. G. 2003. Correlation of saccharin and ethanol intake in the F2 progeny of HAD2 and LAD2 crosses. Alcoholism: Clin. Exp. Res. 27, 49A.

Tichansky, D. S., Boughter, J. D., Jr., and Madan, A. K. 2006. Taste change after laparoscopic Roux-en-Y gastric bypass and laparoscopic adjustable gastric banding. Surg. Obes. Relat. Dis. 2, 440–444.

Tordoff, M. G. and Bachmanov, A. A. 2003. Mouse taste preference tests: why only two bottles? Chem. Senses 28, 315–324.

Tordoff, M. G., Bachmanov, A. A., and Reed, D. R. 2007. Forty mouse strain survey of water and sodium intake. Physiol. Behav. Apr. 1 (Epub ahead of print).

Ueno, K., Ohta, M., Morita, H., Mikuni, Y., Nakajima, S., Yamamoto, K., and Isono, K. 2001. Trehalose sensitivity in

Drosophila correlates with mutations in and expression of the gustatory receptor gene Gr5a. Curr. Biol. 11, 1451–1455.

Wang, X., Thomas, S. D., and Zhang, J. 2004. Relaxation of selective constraint and loss of function in the evolution of human bitter taste receptor genes. Hum. Mol. Genet. 13, 2671–2678.

Warren, R. P. and Lewis, R. C. 1970. Taste polymorphism in mice involving a bitter sugar derivative. Nature 227, 77–78.

Whitney, G. and Harder, D. B. 1986a. PTC preference among laboratory mice: understanding of a previously "unreplicated" report. Behav. Genet. 16, 409–416.

Whitney, G. and Harder, D. B. 1986b. Single locus control of sucrose octaacetate tasting among mice. Behav. Genet. 16, 559–574.

Whitney, G. and Harder, D. B. 1994. Genetics of bitter perception in mice. Physiol. Behav. 56, 1141–1147.

Whitney, G., Harder, D. B., and Gannon, K. S. 1989. The B6.SW bilineal congenic sucrose octaacetate (SOA)-taster mice. Behav. Genet. 19, 409–416.

Winnig, M., Bufe, B., and Meyerhof, W. 2005. Valine 738 and lysine 735 in the fifth transmembrane domain of rTas1r3 mediate insensitivity towards lactisole of the rat sweet taste receptor. BMC Neurosci. 6, 22.

Wooding, S. 2006. Phenylthiocarbamide: a 75-year adventure in genetics and natural selection. Genetics 172, 2015–2023.

Wooding, S., Kim, U. K., Bamshad, M. J., Larsen, J., Jorde, L. B., and Drayna, D. 2004. Natural selection and molecular evolution in PTC, a bitter-taste receptor gene. Am. J. Hum. Genet. 74, 637–646.

Wu, S. V., Chen, M. C., and Rozengurt, E. 2005. Genomic organization, expression, and function of bitter taste receptors (T2R) in mouse and rat. Physiol. Genomics 22, 139–149.

Xu, H., Staszewski, L., Tang, H., Adler, E., Zoller, M., and Li, X. 2004. Different functional roles of T1R subunits in the heteromeric taste receptors. Proc. Natl. Acad. Sci. U. S. A. 101, 14258–14263.

Yamamoto, M. E., Block, G. D., and Ishii, E. 1991. Food patterns among adolescents: influence of alcohol consumption. Alcoholism: Clin. Exp. Res. 15, 359.

Zhao, G. Q., Zhang, Y., Hoon, M. A., Chandrashekar, J., Erlenbach, I., Ryba, N. J., and Zuker, C. S. 2003. The receptors for mammalian sweet and umami taste. Cell 115, 255–266.

Relevant Websites

http://www.ensembl.org – Ensemble browser of sequenced genomes.

http://www.ncbi.nlm.nih.gov/gquery – Entrez (National Center for Biotechnology Information cross-database search).

http://www.ncbi.nlm.nih.gov/Genbank – GenBank.

http://gdbwww.gdb.org/gdb – Human Genome Database.

http://www.monell.org/MMTPP – Monell Mouse Taste Phenotyping Project.

http://www.informatics.jax.org – Mouse Genome Database.

http://www.jax.org/phenome – Mouse Phenome Database.

http://www.ncbi.nlm.nih.gov/mapview – National Center for Biotechnology Information browser of sequenced genomes.

http://rgd.mcw.edu – Rat Genome Database.

http://genome.ucsc.edu – University of California Santa Cruz browser of sequenced genomes.

4.20 Propylthiouracil (PROP) Taste

D J Snyder, Yale University, New Haven, CT, USA

V B Duffy, University of Connecticut, Storrs, CT, USA

J E Hayes, Brown University, Providence, RI, USA

L M Bartoshuk, University of Florida, Gainesville, FL, USA

Glossary

category scale A method of rating perceived intensity in which subjects place ratings in categories (e.g., a 9-point scale, in which 1 = very weak, 5 = medium, and 9 = very strong).

chorda tympani A branch of the facial nerve (cranial nerve VII) that carries taste information from the mobile, anterior two-thirds of the tongue.

flavor The combined sensations of taste, oral somatosensation, and retronasal olfaction.

gLMS General labeled magnitude scale, a method of rating perceived intensity. The scale is a numbered line running from 0 = no sensation to 100 = strongest imaginable sensation of any kind with intermediate labels: 1.4 = barely detectable, 6 = weak, 16 = moderate, 35 = strong, and 53 = very strong. This spacing of labels gives the scale ratio properties, that is, a rating of 40 denotes a perceived intensity twice as strong as a rating of 20.

magnitude matching A method of rating-perceived intensity that permits valid comparisons across groups. This technique depends on the ability of subjects to match the intensities of different types of sensations. Subjects rate the perceived intensities of stimuli of interest relative to an unrelated standard. By assuming that variability in the intensity of the standard is equivalent across two or more groups, magnitude matching permits group comparisons of the stimulus of interest.

orthonasal olfaction Odor sensations arising from environmental stimuli that are sniffed.

PROP 6-n-Propylthiouracil, a compound that binds to the T2R38 receptor and to which some individuals are taste blind.

PTC Phenylthiocarbamide, a compound that binds to the T2R38 receptor and to which some individuals are taste blind.

retronasal olfaction Odor sensations arising from foods in the mouth; these are pumped into the nasal cavity via the rear of the throat during chewing and swallowing.

supertaster An individual who experiences the most intense oral sensations (i.e., taste, oral burn, oral touch) and the most intense retronasal olfactory sensations.
tastant A substance that evokes a taste sensation.

VAS Visual analog scale, a method of rating-perceived intensity in which subjects mark a line labeled with the minimum and maximum intensities possible for a given sensory experience.

4.20.1 Discovery of Taste Blindness

Taste blindness was discovered by accident in 1931. Arthur Fox, a chemist working for DuPont de Nemours, synthesized some phenylthiocarbamide (PTC), and when he attempted to put the compound into a bottle, some of it escaped. One of his colleagues, Dr. C. R. Noller, an eminent organic chemist, noticed a bitter taste; Fox did not. Testing co-workers (Fox, A. L., 1931) revealed that the majority tasted PTC as bitter. When Fox and a prominent geneticist of the day asked 2550 attendees at a meeting of the American Association for the Advancement of Science to vote on the quality of taste from PTC crystals (Blakeslee, A. F. and Fox, A. L., 1932), they confirmed that most tasted bitter (tasters) but a substantial number (28%) did not (nontasters). When *The Journal of Heredity* published these data, the editor had small pieces of filter paper impregnated with PTC inserted into the journal. Readers could purchase these PTC papers which facilitated testing of a variety of groups (e.g., the Dionne quintuplets were all tasters; Ford, N. and Mason, A. D., 1941). Family studies (Snyder, L. H., 1931) suggested that nontasting was recessive.

The informal testing of that era was displaced by more quantitative threshold methodology (Harris, H. and Kalmus, H., 1949) that produced the now-familiar bimodal distribution of PTC thresholds. Evaluation of a series of compounds showed that those containing the N–C=S group (including PTC and 6-*n*-propylthiouracil (PROP)) produced the bimodal distribution. Studies in this early era focused on assessing percentages of tasters by race, sex, and various disease states.

Fischer brought new insights to taste blindness. He stopped using PTC and switched to PROP because it lacked the sulfurous odor of PTC. In addition, he added an assessment for quinine; he reasoned that PROP thresholds assessed a specific sensitivity while quinine thresholds assessed a general ability to taste (Fischer, R., 1967). He began to consider the implications of sensory variation for behavior relevant to health; he found associations between taste thresholds (PROP and quinine) and food preferences, alcohol abuse, cigarette smoking, and body weight.

Most recently, Drayna's group found that the gene for PTC/PROP tasting is located on chromosome 7 (Kim, U. K. *et al.*, 2003). Those individuals with nontaster thresholds (PROP threshold > 0.0002 M) were homozygous for the recessive allele; those with taster thresholds (PROP threshold < 0.0001 M) were either homozygous for the dominant allele or were heterozygous. The human genome project has resulted in the discovery of a variety of other taste-related genes as well (e.g., Mombaerts, P., 2004). In the future, the elucidation of ligands for the receptors expressed by those genes and the development of more accurate methods for oral sensory genotype/phenotype analysis will make possible more detailed studies of the associations between oral sensation and health.

4.20.2 New Psychophysical Methods and the Discovery of Supertasters

4.20.2.1 Visual Analog Scales and Category Scales: Invalid for Across-Group Psychophysical Comparisons

Fischer lacked the psychophysical tools we have today, but his speculations on health outcomes associated with genetic differences in taste foreshadowed the results that are now emerging. With the new tools that modern scaling studies provide, the magnitude of the associations between sensation and behavior can finally be quantified.

The search for associations between diet, smoking, alcohol intake, and the ability to taste PROP requires comparisons between the sensations experienced by nontasters and tasters. Unfortunately, many of the studies searching for these associations used labeled scales (e.g., visual analog scale (VAS) and category scales like the Natick 9-point scale or the labeled magnitude scale (Green, B. G. *et al.*, 1993)). Comparisons across nontasters and tasters cannot be made with these scales. That is, these scales are valid for making within-subject comparisons or making comparisons across groups where the group members are randomly selected, but they are invalid for making across-group comparisons. The explanation will be briefly reviewed here but the interested reader is referred to more complete treatments of the issue (Bartoshuk, L. M. 2002; Bartoshuk, L. M. *et al.*, 2002; Bartoshuk, L. M. *et al.*, 2004a; 2005b; Snyder, D. J. *et al.*, 2004; Bartoshuk, L. M. *et al.*, 2006b).

VAS and category scales are usually anchored in terms of the minimum and maximum of the sensation to be measured. For example, a VAS used to assess sweetness was a line labeled "not at all" at one end and "extremely" at the other end (Salbe, A. D. *et al.*, 2004); similarly, a 9-point category scale for sweetness was labeled 1 = "not at all" and 9 = "extremely" (Drewnowski, A. *et al.*, 1997). Intermediate values on category scales may be labeled as well; another 9-point category scale was labeled 1 = "none," 3 = "slight," 5 = "moderate," 7 = "strong," and 9 = "extreme" (Kamen, J., 1959).

Unfortunately, the intensity labels on these scales do not denote the same perceived intensities to all. Suppose that groups of nontasters and tasters were presented with a series of sucrose solutions varying in concentration from near threshold to very concentrated and asked to select one they would describe as strong. In this situation, nontasters and tasters would select similar concentrations. However, if they were then asked to don earphones and set the loudness of a tone to match the sweetness of the sucrose they had selected, the two groups would behave differently. The nontasters would match the sweetness they experienced to a lower tone intensity than would the tasters. Since we have no reason to believe that nontasters and tasters hear differently, this result shows that nontasters experienced less sweetness than did the tasters even though both groups described the sweetness as strong (e.g., see Bartoshuk, L. M. and Snyder, D. J. 2004 or Snyder, D. J. *et al.*, 2004 for the data on which this example was based).

In other words, intensity descriptors are relative. The mental intensity ruler that we apply to sensations behaves as if it were printed on elastic; this ruler expands or contracts with the sensory domain of interest, the physiological capacity of that domain, and the individual's experience. The relativity of intensity descriptors was captured in a dramatic quote from Stevens S. S. (1958, p. 633): "Mice may be called large or small, and so may elephants, and it is quite understandable when someone says it was a large mouse that ran up the trunk of the small elephant."

Invalid comparisons have been discussed periodically throughout the history of scale development (e.g., Aitken, R. C. B. 1969; Narens, L. and Luce, R. D. 1983; Biernat, M. and Manis, M. 1994; Birnbaum, M. H. 1999). These errors most often result in the failure to see real differences, but occasionally they will produce erroneous differences that appear to go in the wrong direction (e.g., see the results for KCl taste in Schifferstein, H. N. J. and Frijters, J. E. R., 1991 and NaCl taste in Drewnowski, A. *et al.*, 1997). Most of the early failures to find differences between nontasters and tasters can be attributed to this measurement error.

4.20.2.2 Magnitude Matching

The sweetness experiment above demonstrates a serious problem with sensory measurement, but it also suggests a solution. When two groups apply the same intensity descriptor to a given sensory experience, this does not mean that the two groups are experiencing the same perceived intensity. In order to reveal differences in perceived intensities between two groups, members of those groups should be asked to compare the sensation of interest to an unrelated standard. There will be variation in the perception of the standard within each of the two groups, but if the standard is unrelated to the sensation of interest, that variation will be roughly the same for both groups. The use of an unrelated standard led to the formalization of the method of magnitude matching (Hall, M. J. *et al.*, 1975; Marks, L. E. and Stevens, J. C., 1980; Stevens, J. C. and Marks, L. E., 1980; Marks, L. E. *et al.*, 1988). Magnitude matching is feasible using any ratio method (e.g., magnitude estimation (Stevens, S. S., 1969), general labeled magnitude scale (Bartoshuk, L. M. *et al.*, 2004b), general VAS (Bartoshuk, L. M. *et al.*, 2005b)) that permits assessments of both sensations of interest and independent standards on the same scale. Recent work shows that these standards need not be actual stimuli, but may include remembered sensations (e.g., the brightest light ever seen, the loudest sound ever heard, the strongest sensation of any kind ever experienced) (Bartoshuk, L. M. *et al.*, 2005b).

4.20.2.3 Supertasting, Oral Burn, and Oral Touch

As data collected with magnitude matching accumulated, significant differences among tasters became apparent. A subgroup of tasters were found to experience not only the greatest bitterness from PROP but also the greatest taste intensities from all other tastants as well. These supertasters also experienced the most intense burn from oral irritants (e.g., chili peppers) as well as the most intense oral tactile sensations (e.g., fats) (Duffy, V. B. *et al.*, 1996; Tepper, B. J. and Nurse, R. J. 1997; Prutkin, J. M. *et al.*, 2000; Hayes, J. E. *et al.*, 2007).

Insights from Miller's anatomical studies help to explain these associations. Miller, I. J. and Whitney G. (1989) discovered a relationship between taste intensity and taste bud density in mice, and they suggested that a similar relationship could explain human variation in taste intensity. They were correct; the density of fungiform papillae (i.e., structures that contain taste buds on the anterior tongue) shows great variability in humans. Moreover, the taste buds on fungiform papillae are surrounded by nerve fibers that mediate oral burn/pain sensations, and other areas on fungiform papillae are innervated by fibers that mediate oral touch sensations (e.g., see Bartoshuk, L. M. *et al.*, 2006a for a discussion). Thus, the more fungiform papillae an individual has, the more intense the experiences of taste, oral burn, and oral touch.

4.20.2.4 Pathology

Inhibition among brain regions mediating oral sensation also contributes to individual differences (e.g., Bartoshuk, L. M. *et al.*, 2005c). While widespread oral sensory damage compromises perceived taste intensity, damage to a localized part of this system can release inhibition in other parts, producing paradoxical increases in perceived taste intensity. Further, the presence of central interactions between the taste and trigeminal systems means that taste damage may intensify nontaste oral sensations. In short, oral sensory variation is the joint result of genetic, anatomical and pathological variation. One of the major goals of modern research into associations between sensory experience and health is to assess the degree

of variance explained by each of these sources and their interactions (e.g., Dinehart, M. E. *et al.*, 2006).

Oral sensory pathology cannot be assessed without testing function at individual nerve fields in the mouth, mainly because disinhibition maintains the constancy of whole-mouth perception, thus masking local pathology. We assess regional oral sensation with a spatial test using scaling methodology that includes an unrelated standard (e.g., Dinehart, M. E. *et al.*, 2006).

4.20.2.5 How Are Supertasters Defined?

As our understanding of taste blindness has evolved, so too has our ability to assess oral sensory function accurately. The discovery of taste blindness involved a binary question: does PTC taste bitter or not (Fox, A. L., 1931)? Threshold testing (Fischer, R. and Griffin, F., 1964) subsequently led to quantitative criteria to classify subjects as nontasters or tasters of PROP, while the development of suprathreshold scaling methods subdivided tasters into supertasters and medium tasters (Bartoshuk, L. M. *et al.*, 1994). PROP supertasters proved to be supertasters in general – that is, they experience the most intense tastes from non-PROP tastants – and super-perceivers of oral burn and oral touch. Recent advances in anatomical and genetic assessment have added further complexity, but for historical and practical reasons, we believe that psychophysical criteria are the most important determinants of oral sensory function; after all, taster status should reflect how people actually experience taste.

We initially suspected that taste blindness arose from a single gene (i.e., supertasters = homozygous dominant, medium tasters = heterozygous, nontasters = recessive), but the discovery that fungiform papilla density correlates with suprathreshold PROP bitterness (Bartoshuk, L. M. *et al.*, 1994) suggested another possibility: supertasters of PROP might be tasters (homozygous or heterozygous) bearing the largest number of fungiform papilla. Fortunately, the identification of a gene (*T2R38*) encoding a PTC/PROP taste receptor (Kim, U. K. *et al.*, 2003) permitted further resolution. The most common alleles for this gene are distinguished by specific amino acid differences at crucial positions – Ala49Pro, Val262Ala, and Ile296Val – with the two most common haplotypes noted as PAV and AVI. Nontasters carry two recessive alleles (AVI/AVI), while tasters are heterozygous (PAV/AVI) or carry two dominant alleles (PAV/PAV). Genotype analysis alone, however, is insufficient for classification, mainly because *T2R38* genotypes associate with threshold measures of PTC/PROP bitterness (Kim, U. K. *et al.*, 2003), but not with suprathreshold measures. Individuals expressing PAV/PAV perceive slightly more suprathreshold PROP bitterness than do those expressing PAV/AVI, but the difference is too small to account for psychophysical measures of supertasting (Duffy, V. B. *et al.*, 2004a).

Figure 1 demonstrates the utility of using only suprathreshold perceived intensity to define taster status. Threshold measures (left panel) divide subjects into the two groups identified in early studies (i.e., nontasters and tasters). But a different picture emerges when both of these groups are subdivided by suprathreshold PROP intensity (right panel). Consistent with our previous work, we separated subjects with taster thresholds (i.e., PAV/PAV or PAV/AVI) into those who perceive the least bitterness (i.e., medium tasters) and those who perceive the most (i.e., supertasters), resulting in greater group differences across a wider intensity range. When we separated subjects with nontaster thresholds (i.e., AVI/AVI) in the same fashion, we found that some of these individuals produce a suprathreshold function that classifies them as medium tasters. In other words, genetic analysis of *T2R38* fails to identify supertasters and includes among nontasters those who experience considerable bitterness from PROP.

Why is this? Recently, Hayes and Duffy (Hayes, J. E., 2007) showed that fungiform papilla density is largely independent of *T2R38* genotype. We previously reported (Bartoshuk, L. M. *et al.*, 1994) that nontasters tend to have the fewest fungiform papillae, but that finding was an artifact caused by classifying subjects by both threshold and suprathreshold PROP bitterness. In fact, we excluded subjects with discordant results, such as those with nontaster thresholds who tasted PROP as fairly bitter, but we now know that those AVI/AVI individuals with the most fungiform papillae taste PROP as fairly bitter. These individuals are supertasters for many oral stimuli, but not for those that bind to T2R38 (e.g., PTC/ PROP). *T2R38* variation fails to account for differences in perceived intensity because it fails to account for differences in oral anatomy.

The term supertaster was originally applied to those individuals who perceived the most intense taste sensations from a variety of taste stimuli, including PROP. We continue to support this view; taster status should reflect responses to multiple oral sensory cues. Because many oral sensations are influenced by anatomy, it might seem that anatomy is more accurate than PROP bitterness for classification, but pathology complicates the picture because damage to the taste system can alter oral perception while leaving fungiform papillae intact. In truth, anatomy, genetics and pathology help us to determine the underlying basis of oral sensation, but psychophysical assessment of perceived intensity remains the critical measure of oral sensory function – and the only way to identify supertasters. PROP

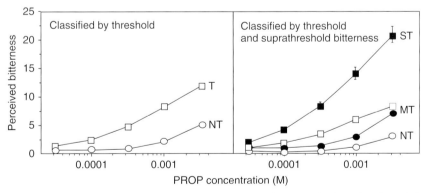

Figure 1 Bitterness of PROP (±SE) assessed with magnitude matching (i.e., magnitude estimation relative to a tone standard). The left panel shows the PROP functions for individuals with PROP thresholds > 0.0002 M (i.e., nontasters, NT, indicated by circles) and those with thresholds < 0.0001 M (i.e., tasters, T, indicated by squares). PROP thresholds strongly reflect different *T2R38* genotypes: NTs express AVI/AVI; Ts express PAV/AVI or PAV/PAV. The right panel shows the same groups subdivided into two groups each: those who tasted the least bitterness from PROP (open symbols) and those who tasted the most bitterness from PROP (filled symbols). The resulting intermediate function (i.e., medium tasters, MT) reflects a combination of PAV/PAV, PAV/AVI, and AVI/AVI genotypes. Data from Bartoshuk, L. M. 2000. Comparing sensory experiences across individuals: recent psychophysical advances illuminate genetic variation in taste perception. Chem. Senses 25, 447–460.

bitterness may provide only one picture of oral sensation, but it is quite a good picture because it represents the combined influences of *T2R38* genotype, fungiform papilla density, and disease. Oral sensation may be influenced by additional factors with uncertain impact (e.g., hormones, trophic factors, other taste/tactile genes), but PROP bitterness has proven valuable in identifying relationships between taste intensity, dietary choice, and long-term health outcomes (see below). That said, there is nothing intrinsically special about PTC/PROP except that it is a probe stimulus – one that elicits highly variable responses that often show strong agreement with other oral stimuli.

Because the functions of PROP bitterness across concentration diverge (right panel), the most accurate way to sort individuals based on PROP bitterness is to use the highest practical concentration; consistent with Figure 1, we typically use a 0.0032 M solution. An alternative to testing with PROP solutions is to use impregnated filter papers, similar to those used by Blakeslee A. F. and Fox A. L. (1932). To produce the most intense bitter taste possible with a paper test, we soak small pieces of filter paper (3 cm diameter, Whatman #1) in a saturated PROP solution (e.g., Bartoshuk, L. M. *et al.*, 2002), yielding ~1.6 mg PROP per paper. These PROP papers produce similar variability in perceived bitterness as a solution of 0.0032 M PROP.

4.20.2.6 Supertasters and Flavor

When food is consumed, the sensations we know as flavor result from mixtures of taste and oral somatosensory cues, as well as olfactory sensations produced when chewing and swallowing force food odors behind the palate and into the nasal cavity. These odor sensations are called retronasal to distinguish them from environmental odors that are sniffed, or orthonasal olfaction. Although both types of odor cues stimulate the same olfactory receptors high in the nasal cavity, retronasal olfaction is perceptually localized to the mouth and orthonasal olfaction to the nose; the two are processed in related but distinct brain regions (Small, D. M. and Jones-Gotman, M., 2001). For many years, the localization of retronasal olfaction to the mouth was thought to be mediated by oral touch (Hollingworth, H. L. and Poffenberger, A. T., 1917), but we now know that taste plays a critical role as well. If the chorda tympani taste nerve is anesthetized on one side of the mouth, retronasal olfactory cues are less intense and appear to arise

from the unanesthetized side of the mouth (Fast, K. *et al.*, 2000; Snyder, D. J. *et al.*, 2001). This association explains a curious clinical phenomenon: some patients with compromised taste and oral touch report a normal ability to smell, yet they claim that flavors are diminished or absent. The brain appears to rely on the simultaneous presence of multiple cues to determine how to process incoming olfactory information – odor + sniffing = orthonasal olfaction; odor + taste + oral touch + oromotor activity = retronasal olfaction. In patients with normal smell but damaged innervation of the oral cavity, food-related odors may fail to be discriminated as retronasal because the concomitant oral cues that identify odor input as coming from the mouth are absent or reduced.

Connections between taste and retronasal olfaction suggest that perceptual differences associated with oral sensory genetics, anatomy, and pathology should extend to flavor perception; not surprisingly, supertasters perceive more intense flavors than do nontasters (Duffy, V. B. *et al.*, 2003; Bartoshuk, L. M. *et al.*, 2005a). This finding helps to explain interesting curiosities (e.g., chefs are more likely to be supertasters than chance would predict; Bartoshuk, L. M. *et al.*, 2004a), but more importantly, it contributes to robust differences in food behavior.

4.20.3 Implications of Oral Sensory Variation for Nutrition and Health

Food preferences influence food intake. Because diet is an important risk factor for a variety of chronic diseases, we have found that oral sensory variation is associated with several of these diseases. The methodological advances we have described enabled the discovery of these associations.

4.20.3.1 Cancer Risk

Links between PTC/PROP tasting and cancer first appeared some years ago (e.g., Milunicová, A. *et al.*, 1969), but more recent studies have specifically targeted oral sensory variation as a potential contributor to cancer risk. The bitter tastes of phytonutrients with cancer-preventive properties can be a barrier to their intake (Drewnowski, A. and Gomez-Carneros, C., 2000). PROP tasting is associated with reduced vegetable intake (Basson, M. D. *et al.*, 2005; Dinehart, M. E. *et al.*, 2006; Bell, K. I. and Tepper, B. J., 2006; Sandell, M. A. and Breslin, P. A., 2006). Those who taste PROP as most bitter reported the lowest preferences for traditionally disliked vegetables (e.g., Brussels sprouts, kale) because they taste more bitter and less sweet (Dinehart, M. E. *et al.*, 2006). One of the risk factors for colon cancer is low vegetable intake (McCullough, M. L. *et al.*, 2003), and data from routine colonoscopies show that men who tasted PROP as most bitter had the most colonic polyps (Basson, M. D. *et al.*, 2005).

4.20.3.2   Cardiovascular Disease Risk

Body mass (BMI) is an important risk factor for cardiovascular disease. The first data linking BMI to PROP tasting came from Fischer (Fischer, R. *et al.*, 1966), who used thresholds for PTC and quinine to show that thinner individuals tended to be tasters. Recent work supports this early observation and suggests that those who taste less bitterness from PROP and other compounds have dietary preferences that support increased energy intake (e.g., higher preferences for fats, sweets, alcohol). For example, overweight (but not obese) middle-aged and elderly individuals tend to find PROP less bitter (Lucchina, L. *et al.*, 1995; Tepper, B. J. and Ullrich, N. V., 2001; Duffy, V. B. *et al.*, 2004b; Duffy, V. B., 2004; Goldstein, G. L. *et al.*, 2005).

Taste pathology appears to interact with genetic variability to affect BMI over time. Diminished bitter taste on the anterior tongue suggests damage to the chorda tympani taste nerve; both are associated with higher BMI (Lanier, S. A. *et al.*, 2005a). Severe middle ear infections may permanently damage the chorda tympani nerve as it traverses the middle ear, and prior histories of such infection are associated with increased BMI, especially in PROP supertasters (Tanasescu, M. *et al.*, 2000; Snyder, D. J. *et al.*, 2003; Bartoshuk, L. M. *et al.*, 2006a).

4.20.3.3   Alcohol Intake

Suggestions that PTC/PROP nontasters are most likely to abuse alcohol date back many years (e.g., Driscoll, K. E. *et al.*, 2006). The sensory properties of alcohol vary with PROP status: supertasters perceive the most intense bitterness and oral burn from alcoholic beverages, and so they like and consume them least (Bartoshuk, L. M., 1993; Duffy, V. B. *et al.*, 2004a; 2004c). Similar to the results observed with vegetables, supertasters find alcoholic beverages more bitter and less sweet than do nontasters (Lanier, S. A. *et al.*, 2005b).

4.20.3.4   Smoking

As with so many health associations related to PTC/PROP, nontasting was first associated with tobacco use by Fischer (Fischer, R. *et al.*, 1963). Modern studies support these early observations (e.g., Snedecor, C. M. *et al.*, 2006), suggesting that supertasters smoke the least (Snyder, D. J. *et al.*, 2005).

4.20.3.5   Future Interventions?

Taste blindness appears to explain some of the widespread variation in food choice. Accordingly, dietary recommendations that account for differences in oral sensation may promote food enjoyment and healthier dietary habits. While supertasters may benefit from oral sensory hindrances to overconsuming alcohol, sweets, and fats, they may have less desire to consume recommended intakes of vegetables. Manipulations that reduce vegetable bitterness (e.g., bitterness inhibition; Ming, D. *et al.*, 1999) may increase preferences for these important foods. Meanwhile, nontasters experience fewer negative sensory experiences from vegetables, but may benefit from dietary counseling to curb alcohol and carbohydrate intake, consume healthful dietary fats, and balance energy intake with physical activity. In short, understanding the behavioral implications of oral sensory variation permits us to discover effective interventions for improved health.

References

Aitken, R. C. B. 1969. Measurement of feelings using visual analogue scales. Proc. R. Soc. Med. 62, 989–993.

Bartoshuk, L. M. 1993. The biological basis of food perception and acceptance. Food Qual. Preference 4, 21–32.

Bartoshuk, L. M. 2000. Comparing sensory experiences across individuals: recent psychophysical advances illuminate genetic variation in taste perception. Chem. Senses 25, 447–460.

Bartoshuk, L. M. 2002. Self reports and across-group comparisons: a way out of the box. APS Obs. 15(7), 26–28.

Bartoshuk, L. M. and Snyder, D. J. 2004. Psychophysical Measurement of Human Taste Experience. In: Handbook of Behavioral Neurobiology (*eds*. E. Stricker and S. Woods), pp. 89–107. Plenum.

Bartoshuk, L. M., Christensen, C. M., Duffy, V., Sheridan, K., Small, D. M., and Synder, D. 2005a. PROP and retronasal olfaction. Chem. Senses 30, A236.

Bartoshuk, L. M., Duffy, V. B., Chapo, A. K., Fast, K., Yiee, J. H., and Hoffman, H. J. 2004a. From psychophysics to the clinic: missteps and advances. Food Qual. Pref. 15, 617–632.

Bartoshuk, L. M., Duffy, V. B., Green, B. G., Hoffman, H. J., Ko, C.-W., Lucchina, L. A., Marks, L. E., Snyder, D. J., and Weiffenbach, J. 2004b. Valid across-group comparisons with labeled scales: the gLMS vs magnitude matching. Physiol. Behav. 82, 109–114.

Bartoshuk, L. M., Duffy, V. B., Fast, K., Green, B. G., Prutkin, J. M., and Snyder, D. J. 2002. Labeled scales (e.g., category, Likert, VAS) and invalid across-group comparisons. What we have learned from genetic variation in taste. Food Qual. Pref. 14, 125–138.

Bartoshuk, L. M., Duffy, V. B., Hayes, J. E., Moskowitz, H., and Snyder, D. J. 2006a. Psychophysics of sweet and fat perception in obesity: problems, solutions and new perspectives. Philos. Trans. R. Soc. Biol. Sci. 361, 1137–1148.

Bartoshuk, L. M., Duffy, V. B., and Miller, I. J. 1994. PTC/PROP tasting: anatomy, psychophysics, and sex effects. Physiol. Behav. 56, 1165–1171.

Bartoshuk, L. M., Fast, K., and Snyder, D. 2005b. Differences in our sensory worlds: Invalid comparisons with labeled scales. Curr. Dir. Psychol. Sci. 14, 122–125.

Bartoshuk, L. M., Snyder, D. J., and Duffy, V. B. 2006b. Hedonic gLMS: valid comparisons for food liking/disliking across obesity, age, sex and PROP status. Chem. Senses 31, A50.

Bartoshuk, L. M., Snyder, D. J., Grushka, M., Berger, A. M., Duffy, V. B., and Kveton, J. F. 2005c. Taste damage: previously unsuspected consequences. Chem. Senses 30(Suppl. 1), (suppl. 1) i218–i219.

Basson, M. D., Bartoshuk, L. M., Dichello, S. Z. L. P., Weiffenbach, J., and Duffy, V. B. 2005. Association between 6-n-propylthiouracil (PROP) bitterness and colonic neoplasms. Dig. Dis. Sci. 50, 483–489.

Bell, K. I. and Tepper, B. J. 2006. Short-term vegetable intake by young children classified by 6-n-propylthiouracil bitter-taste phenotype. Am. J. Clin. Nutr. 84, 245–251.

Biernat, M. and Manis, M. 1994. Shifting standards and stereotype-based judgements. J. Person. Soc. Psychol. 66, 5–20.

Birnbaum, M. H. 1999. How to show that 9>221: collect judgements in a between-subjects design. Psychol. Meth. 4, 243–249.

Blakeslee, A. F. and Fox, A. L. 1932. Our different taste worlds. J. Hered. 23, 97–107.

Dinehart, M. E., Hayes, J. E., Bartoshuk, L. M., Lanier, S. L., and Duffy, V. B. 2006. Bitter taste markers explain variability in vegetable sweetness, bitterness, and intake. Physiol. Behav. 87, 304–313.

Drewnowski, A. and Gomez-Carneros, C. 2000. Bitter taste, phytonutrients, and the consumer: a review. Am. J. Clin. Nutr. 72, 1424–1435.

Drewnowski, A., Henderson, S. A., Shore, A. B., and Barratt-Fornell, A. 1997. Nontasters, tasters, and supertasters of 6-n-propylthiouracil (PROP) and hedonic response to sweet. Physiol. Behav. 62, 649–655.

Driscoll, K. E., Perez, M., Cukrowicz, K. C., Butler, M., and Joiner, T. E. 2006. Associations of phenythiocarbamide tasting to alcohol problems and family history of alcoholism differ by gender. Psychiatry Res. 143, 21–27.

Duffy, V. B. 2004. Associations between oral sensation, dietary behaviors and risk of cardiovascular disease (CVD). Appetite 43, 5–9.

Duffy, V. B., Bartoshuk, L. M., Lucchina, L. A., Snyder, D. J., and Tym, A. 1996. Supertasters of PROP (6-n-propylthiouracil) rate the highest creaminess to high-fat milk products. Chem. Senses 21, 598.

Duffy, V. B., Chapo, A. K., Hutchins, H. L., and Bartoshuk, L. M. 2003. Retronasal olfactory intensity: associations with taste. Chem. Senses 28, A33.

Duffy, V. B., Davidson, A. C., Kidd, J. R., Kidd, K. K., Speed, W. C., and Pakstis, A. J. 2004a. Bitter receptor gene (TAS2R38), 6-n-propylthiouracil (PROP) bitterness and alcohol intake. Alcohol. Clin. Exp. Res. 28, 1629–1637.

Duffy, V. B., Lucchina, L. A., and Bartoshuk, L. M. 2004b. Genetic Variation in Taste: Potential Biomarker for Cardiovascular Disease Risk? In: Genetic Variations in Taste Sensitivity: Measurement, Significance and Implications (eds. J. Prescott and B. J. Tepper), pp. 195–228. Dekker.

Duffy, V. B., Peterson, J. M., and Bartoshuk, L. M. 2004c. Associations between taste genetics, oral sensation and alcohol intake. Physiol. Behav. 82, 435–445.

Fast, K., Tie, K., Bartoshuk, L. M., Kveton, J. F., and Duffy, V. B. 2000. Unilateral anesthesia of the chorda tympani nerve suggests taste may localize retronasal olfaction. Chem. Senses 25, 614–615.

Fischer, R. 1967. Genetics and Gustatory Chemoreception in Man and Other Primates. In: The Chem. Senses and Nutrition (eds. M. R. Kare and O. Maller), pp. 621–681. John Hopkins Press.

Fischer, R. and Griffin, F. 1964. Pharmacogenetic aspects of gustation. Drug Res. 14, 673–686.

Fischer, R., Griffin, F., and Kaplan, A. R. 1963. Taste thresholds, cigarette smoking, and food dislikes. Med. Exp. 9, 151–167.

Fischer, R., Griffin, F., and Rockey, M. 1966. Gustatory chemoreception in man: multidisciplinary aspects and perspectives. Perspect. Biol. Med. 9, 549–577.

Ford, N. and Mason, A. D. 1941. Taste reactions of the Dionne quintuplets. J. Hered. 10, 365–368.

Fox, A. L. 1931. Six in ten "tasteblind" to bitter chemical. Sci. News Lett. 9, 249.

Goldstein, G. L., Daun, H., and Tepper, B. J. 2005. Adiposity in middle-aged women is associated with genetic taste blindness to 6-n-propylthiouracil. Obes. Res. 13, 1017–1023.

Green, B. G., Shaffer, G. S., and Gilmore, M. M. 1993. A semantically-labeled magnitude scale of oral sensation with apparent ratio properties. Chem. Senses 18, 683–702.

Hall, M. J., Bartoshuk, L. M., Cain, W. S., and Stevens, J. C. 1975. PTC taste blindness and the taste of caffeine. Nature 253, 442–443.

Harris, H. and Kalmus, H. 1949. The measurement of taste sensitivity to phenylthiourea (P.T.C.). Ann. Eugen. 15, 24–31.

Hayes, J. E. 2007. Statistical modeling of taste markers, food preference and intake, and anthropometrics. PhD dissertation, University of Connecticut.

Hayes, J. E., Bartoshuk, L. M., Kidd, J. R., and Duffy, V. B. 2007. TAS2R38 genotype, fungiform papillae and suprathreshold taste response. Chem. Senses 32 (in press).

Hollingworth, H. L. and Poffenberger, A. T. 1917. The Sense of Taste. Moffat, Yard and Company.

Kamen, J. 1959. Interaction of sucrose and calcium cyclamate on perceived intensity of sweetness. J. Food Sci. 24, 279–282.

Kim, U. K., Jorgenson, E., Coon, H., Leppert, M., Risch, N., and Drayna, D. 2003. Positional cloning of the human quantitative trait locus underlying taste sensitivity to phenylthiocarbamide. Science 299, 1221–1225.

Lanier, S. A., Hutchins, H., and Duffy, V. B. 2005a. Taste and dietary predictors of central adipositiy in adult females. Chem. Senses 30, A34.

Lanier, S. A., Hayes, J. E., and Duffy, V. B. 2005b. Sweet and bitter tastes of alcoholic beverages mediate alcohol intake in of-age undergraduates. Physiol. Behav. 83, 821–831.

Lucchina, L., Bartoshuk, L. M., Duffy, V. B., Marks, L. E., and Ferris, A. M. 1995. 6-n-Propylthiouracil perception affects nutritional status of independent-living older females. Chem. Senses 20, 735.

Marks, L. E. and Stevens, J. C. 1980. Measuring Sensation in the Aged. In: Aging in the 1980's: Psychological Issues (ed. L. W. Poon), pp. 592–598. American Psychological Association.

Marks, L. E., Stevens, J. C., Bartoshuk, L. M., Gent, J. G., Rifkin, B., and Stone, V. K. 1988. Magnitude matching: the measurement of taste and smell. Chem. Senses 13, 63–87.

McCullough, M. L., Robertson, A. S., Chao, A., Jacobs, E. J., Stampfer, M. J., Jacobs, D. R., Diver, W. R., Calle, E. E., and Thun, M. J. 2003. A prospective study of whole grains, fruits, vegetables and colon cancer risk. Cancer Causes Control 14, 959–970.

Miller, I. J. and Whitney, G. 1989. Sucrose octaacetate-taster mice have more vallate taste buds than non-tasters. Neurosci. Lett. 100, 271–275.

Milunicová, A., Jandová, A., Laurová, L., Novotná, J., and Skoda, V. 1969. Hereditary blood and serum types, PTC test and level of the fifth fraction of serum lactatedehydrogenase in females with gyneological cancer (II. Communication). Neoplasma 16, 311–316.

Ming, D., Ninomiya, Y., and Margolskee, R. F. 1999. Blocking taste receptor activation of gustducin inhibits gustatory responses to bitter compounds. Proc. Natl. Acad. Sci. U. S. A. 17, 9903–9908.

Mombaerts, P. 2004. Genes and ligands for odorant, vomeronasal and taste receptors. Nat. Rev. Neurosci. 5, 263–278.

Narens, L. and Luce, R. D. 1983. How we may have been misled into believing in the interpersonal comparability of utility. Theor. Decis. 15, 247–260.

Prutkin, J. M., Duffy, V. B., Etter, L., Fast, K., Gardner, E., and Lucchina, L. A. 2000. Genetic variation and inferences about perceived taste intensity in mice and men. Physiol. Behav. 69, 161–173.

Salbe, A. D., DelParigi, A., Pratley, R. E., Drewnowski, A., and Tataranni, P. A. 2004. Taste preferences and body weight changes in an obesity-prone population. Am. J. Clin. Nutr. 79, 372–378.

Sandell, M. A. and Breslin, P. A. 2006. Variability in a taste-receptor gene determines whether we taste toxins in food. Curr. Biol. 16, R792–R794.

Schifferstein, H. N. J. and Frijters, J. E. R. 1991. The perception of the taste of KCl, NaCl, and quinine HCl is not related to PROP-sensitivity. Chem. Senses 16, 303–317.

Small, D. M. and Jones-Gotman, M. 2001. Neural substrates of taste/smell interactions and flavour in the human brain. Chem. Senses 26, 1034.

Snedecor, C. M., Pomerleau, C. S., Mehringer, A. M., Ninowski, R., and Pomerleau, O. F. 2006. Differences in smoking-related variables based on phenylthiocarbamide 'taster' status. Addict. Behav. 31, 2309–2312.

Snyder, L. H. 1931. Inherited taste deficiency. Science 74, 151–152.

Snyder, D. J., Davidson, A. C., Kidd, J. R., Kidd, K. K., Speed, W. C., Pakstis, A. J., Cubells, J. F., O'Malley, S. S., and Bartoshuk, L. M. 2005. Oral sensation influence tobacco use: genetic and psychophysical evidence. Presented at the Annual Meetings of the Society for Research on Nicotine and Tobacco, Prague, Czech Republic.

Snyder, D. J., Duffy, V. B., Chapo, A. K., Cobbett, L. E., and Bartoshuk, L. M. 2003. Childhood taste damage modulates obesity risk: effects on fat perception and preference. Obes. Res. 11(Suppl), A147.

Snyder, D. J., Dwivedi, N., Mramor, A., Bartoshuk, L. M., and Duffy, V. B. 2001. Taste and touch may contribute to the localization of retronasal olfaction: unilateral and bilateral anesthesia of cranial nerves V/VII. Paper presented at the Society of Neuroscience Abstract, San Diego, CA, USA.

Snyder, D. J., Fast, K., and Bartoshuk, L. M. 2004. Valid comparisons of suprathreshold stimuli. J. Conscious. Stud. 11, 40–57.

Stevens, J. C. and Marks, L. E. 1980. Cross-modality matching functions generated by magnitude estimation. Percept. Psychophys. 27, 379–389.

Stevens, S. S. 1969. Sensory scales of taste intensity. Percept. Psychophys. 6, 302–308.

Stevens, S. S. 1958. Adaptation-level vs the relativity of judgment. Am. J. Psychol. 71, 633–646.

Tanasescu, M., Ferris, A. M., Himmelgreen, D. A., Rodriguez, N., and Perez-Escamilla, R. 2000. Biobehavioral factors are associated with obesity in Puerto Rican children. J. Nutrit. 130, 1734–1742.

Tepper, B. J. and Nurse, R. J. 1997. Fat perception is related to PROP taster status. Physiol. Behav. 61, 949–954.

Tepper, B. J. and Ullrich, N. V. 2001. Influence of genetic taste sensitivity to 6-n-propylthiouracil (PROP), dietary restraint and disinhibition on body mass index in middle-aged women. Physiol. Behav. 75, 305–312.

Further Reading

Hayes, J. and Duffy, V. B. 2007. Revisiting sucrose-fat mixtures: sweetness and creaminess vary with phenotypic markers of oral sensation. Chem. Senses, in press.

Snyder, D. J., Prescott, J., and Bartoshuk, L. M. 2006. Modern Psychophysics and the Assessment of Human Oral Sensation. In: Taste and Smell: An Update (eds. T. Hummel and A. Welge-Lüssen). Karger.

Trock, B. J., Lanza, E., and Greenwald, P. 1990. High fiber diet and colon cancer: a critical review. Prog. Clin. Biol. Res. 346, 145–157.

4.21 Salt Taste

G K Beauchamp and L J Stein, Monell Chemical Senses Center, Philadelphia, PA, USA

Glossary

KCl Potassium chloride.
LiCl Lithium chloride.

Na Sodium.
NaCl Sodium chloride; salt.

Sodium chloride (NaCl: herein referred to as salt unless otherwise noted) has played a central role in human society throughout history. As a highly valued commodity, wars have been fought to control salt access and availability, and it has served as a major basis for taxation for millennia (see Gandhi, M., 1930 for example). People have been paid in salt (salaries) and salt has been transported, sometimes requiring considerable effort and expense, across long distances (reviews: Denton, D., 1982; Kurlansky, M., 2002).

Until recently, a major problem for many people has been to obtain sufficient salt, both for themselves and also for their domestic animals, especially herbivorous ones that cannot obtain sufficient quantities of sodium (Na) from most plants. Times have changed, however, and the salt problem in much of today's world is not one of insufficiency, but of excess.

In this essay, we do not address health implications of high salt consumption; these are covered in numerous reviews and commentaries (e.g., Taubes, G., 1998; Sacks, F. M. *et al.*, 2001; Alderman, M. H., 2002; Hooper, L. *et al.*, 2002; De Wardener, H. E. *et al.*, 2004; Karppanen, H. and Mervaalo, E., 2006). The consensus is that intake of too much salt – exactly what constitutes too much for any given individual likely depends on many factors – is unhealthy for at least some proportion of the population. Consequently, achieving a more thorough understanding of salt taste, including how it is detected, the role it plays in food palatability, and how salt intake may be modified, carries public health implications. A better understanding of all aspects of salt taste is thus a necessity.

In what follows, we discuss four topics. First, we focus on how salt in the mouth evokes the sensation of saltiness. As will be shown, our knowledge of this area has advanced greatly over the past years but remains incomplete. Second, we discuss why people seem to like salt so much. This section considers both innate and acquired factors. In the third part, we consider the varied reasons for adding salt to human foods. Although these sections primarily address human salt taste and consumption, issues relevant to other animals will sometimes be raised. The final section provides an overview of strategies that might be employed to address the public health mandate to reduce salt intake.

4.21.1 What Makes Salt Taste Salty?

NaCl and LiCl (lithium chloride) are the only two substances that taste almost purely salty, lacking virtually any bitter, sour, sweet, or umami quality. Of the sodium salts, several, such as Na-acetate, are predominantly salty but weakly so, while most others taste salty and something else, usually bitter. A few other compounds are salty with additional taste characteristics. For example, KCl is quite salty but also bitter, particularly so to some people. A small number of amino acids have a salty taste component. Nevertheless, it is striking how few salty compounds exist, and of these, how even fewer are purely salty. This paucity of salty stimuli contrasts starkly with bitter and even sweet taste, for each of which there are numerous and diverse molecular stimulants. These observations suggest a highly specific mechanism for salty taste detection. And since the primary ion in salty taste is thought to be Na (Bartoshuk, L. M., 1980), this mechanism must be specific to sodium.

A major advance in our understanding of salt taste mechanisms was made independently in the early 1980s by Schiffman S. and colleagues (Schiffman, S. S. *et al.*, 1983) and DeSimone J. and colleagues (DeSimone, J. A. *et al.*, 1981). These investigators hypothesized that the Na taste recognition mechanism might be the same or very similar to mechanisms for sodium recognition in other parts of the body, particularly the kidneys. Following this logic, dissociated Na dissolved in saliva was postulated to activate a Na-specific ion channel. Over the ensuing 25 years, this important insight has received a substantial body of support from biophysical, physiological, and sensory studies that employed mainly animal models such as rats, mice, and dogs. Particularly telling support came from studies demonstrating that a major component of salt taste transduction is inhibited by amiloride, a diuretic that blocks sodium channels. However, most of these studies also indicated the additional existence of a nonamiloride-suppressible component to salt taste transduction (see, e.g., Geran, L. C. and Spector, A. C., 2000). The nature of this second element remains obscure, although it might involve a nonspecific vanilloid receptor-1 analog (Lyall, V. *et al.*, 2004).

The sodium ion channel scenario is elegant in its simplicity and intuitive appeal. In particular, it can account for the specificity of Na as the main component of saltiness perception. Moreover, sodium ion channels have been implicated in the salt taste perception of animal models ranging from sheep to rats to flies (e.g., Liu, L. *et al.*, 2003). Nevertheless, the hypothesis is not completely compelling. Although some of the original data from human psychophysical studies showed that amiloride reduced human perception of salty taste, subsequent studies have been inconsistent. Indeed, in an extremely thorough review of this literature, Halpern concluded that amiloride has little or no effect on human perception of NaCl saltiness (Halpern, B. P., 1998). These human data are quite puzzling in the face of the overwhelming evidence for an amiloride effect to reduce salt taste perception in other animal models.

Perhaps this should not be so surprising; species differences are common for other taste transduction mechanisms, as they are in many other biological areas. Yet given the fundamental nature of salt taste, along with the widespread evidence for involvement of a sodium channel in so many species, it seems implausible that humans would possess a completely novel mechanism for detecting NaCl. Perhaps the channel is sufficiently altered in humans that amiloride is ineffective; this would account for the fact that some human sodium channels are not amiloride blockable. What is needed to fully understand human salt taste is a comprehensive explanation of the molecular mechanisms that underlie its detection. While this is not yet available, it is likely to be forthcoming quite soon.

As noted above, salt taste transduction also includes an amiloride-insensitive component, which again is not well understood (Halpern, B. P., 1998). Activation of this component, which is not specific to sodium, may not be perceived as fully or even partially salty. Nevertheless, identification of this mechanism and its function will be necessary for a full understanding of the molecular basis for salt taste.

4.21.2 Why Do We Like the Taste of Salt?

Attempts to answer this complicated question are generally addressed at two levels: one involving true physiological salt need and the second involving need-free salt intake.

4.21.2.1 Liking for Salt: Sodium Need

It is widely assumed that the very existence of a mechanism to specifically detect Na is due to the special role Na plays in animal biology. This rationale has been thoroughly reviewed elsewhere (Denton, D., 1982) and includes the importance of Na in nerve conduction, energy release, and acid–base balance. Moreover, since Na is not evenly distributed in nature and is present in very low concentrations in many plants, it is assumed that a specific means to detect its presence would be highly adaptive. These theoretical arguments are supported by the widespread existence across species of mechanisms for detecting salt, by salt taste's apparent status as one of a small group of basic tastes in many organisms, and by the motivating effects of salt taste, particularly when Na is scarce (McBurney, D. H. and Gent, J. F., 1979).

This last point requires elaboration. A tremendous number of observational and experimental studies (again reviewed with great scholarship by Denton, D., 1982; for a more recent review see Schulkin, J., 1991) demonstrate that a sodium-deficient animal, especially an herbivore or omnivore, will go to great lengths to obtain Na, using salty taste to detect sources of Na. This need-based search for and consumption of salt has provided a model system that demonstrates innate and specific mechanisms for detecting Na and has identified some of the hormonal substrates that underlie need-based salt appetite. This vast area of research is difficult to summarize, but apparently Na depletion elicits changes in hormones and central neural system organization that over time alter avidity for the taste of salt. Indeed, the taste quality elicited by salt may change as a function of sodium depletion and hence need (for review, see McCaughey, S. A. and Scott, T. R., 1998). A peripheral component may also contribute to increased avidity for Na in the face of need, as Na depletion may alter salt taste transduction mechanisms (Contreras, R. J. and Frank, M., 1979) in addition to central mechanisms regulating the avidity for salt. Put simply, Na depletion seems to make salt taste both better and different in animal models such as rodents and sheep.

Once again, species differences are apparent. As might be anticipated based on a meat diet naturally rich in sodium, carnivores may not respond physiologically or behaviorally to Na depletion in the same ways that omnivores and herbivores do. Perhaps carnivores such as dogs and cats never experience the consequences of Na depletion. As meat-eating animals, they either ingest sufficient food to survive, consequently ensuring adequate Na, or they die of starvation.

A more puzzling species difference again involves humans. The evidence that Na depletion in adult humans stimulates a specific, intense desire for the taste of salt is weak (Beauchamp, G. K., 1991). Although some human experimental studies indicate that Na loss is followed by an elevation in salt liking, this does not truly mimic the responses of animal models, where extreme avidity follows depletion. A partial explanation may be that human experimental studies cannot induce the same degree of depletion routinely employed in animal model studies.

In one area, species differences may not exist. When Na depletion occurs very early in human life as a result of varied clinical conditions, a profound elevation in the desire for salt occurs, similar to that seen in experimental rodent models. For obvious reasons, there are no experimental studies of early Na depletion in humans; however, the weight of clinical evidence (Beauchamp, G. K. et al., 1991) indicates that such heightened avidity is remarkably persistent. This observation is mirrored in some animal model studies (e.g., Thaw, A. K. et al., 2000), demonstrating a permanent elevation of salt taste avidity following early Na depletion. The relative roles of central and peripheral changes in modulating this effect remain unknown.

4.21.2.2 Liking for Salt: Need-Free Consumption

Although it makes some evolutionary sense that many species would develop mechanisms to enable detection of environmental Na and facilitate ingestion under conditions of depletion, why do they consume it avidly when not deficient? More specifically, why do humans in almost all developed societies, who are probably never truly sodium deficient, so like their food with substantial added salt?

Two explanations have been proffered. The first suggests that herbivores and some omnivores have an innate liking for the taste of salt. According to this rationale, it is to an organism's benefit to consume salt when it is available. There is no cost for this behavior, with the exception of some adult-onset diseases, as excess sodium is easily removed by the kidneys. Furthermore, it has been hypothesized that high Na intake may reflect evolutionary pressure to protect against sudden dehydration (Fessler, D. M. T., 2003). A problem with this line of argument centers on the fact that sodium cannot be stored. What then is the advantage of consuming salt in excess whenever it is available? Perhaps, this behavior allows animals learn its location so as to be able to quickly locate it in preparation, from an evolutionary perspective, for later deficits. Indeed, sodium-deprived rats do use learned associations between environmental cues and salt taste to guide them to locations where they have previously ingested salt (Stouffer, E. M. and White, N. M., 2005).

The second explanation is that individual exposure to salt, particularly early exposure, results in a heightened desire or avidity (e.g., Smirga, M. *et al.*, 2002). This is a commonly-held belief among nutritionists, dieticians, physicians, and others (e.g., MacGregor, G. A. and de Wardener, H. E., 1998). But this explanation raises yet another question. Why should a taste experienced in infancy or even adulthood take on heightened value? This is often called taste imprinting but, unlike classic imprinting where the animal learns important characteristics of its parent, the adaptive value of taste imprinting is obscure, particularly for omnivorous species like humans or rats.

Experimental studies in animal models and humans provide several examples of effects of early dietary exposure to salt on subsequent perception and preference. As noted previously, sodium depletion early in life may cause an elevation of salt preference. The opposite – high sodium consumption early in life – also has been reported to have modest effects in heightening later preferences (Contreras, R. J. and Kosten, T., 1983). Furthermore, amniotic fluid, saliva, and milk all contain sodium, consumption of which could establish or heighten a preference, and many studies with rodent models, particularly rats, demonstrate that salt solutions around isotonicity are avidly consumed on first presentation.

As with most arguments pitting innateness versus learning, this dichotomy is almost assuredly false. Liking for salt and salty foods is surely the result of an interaction between innate and acquired information. Human infants at birth are either indifferent to salt or reject it, particularly at hypertonic concentrations (Beauchamp, G. K. *et al.*, 1986). By approximately 4–6 months of age, infants show a preference (relative to plain water) for low-to-moderate saline solutions around the level of isotonicity. This age-related hedonic shift may represent in part the maturation of an innate transductive or neurological mechanism. Some rodent studies have shown the ability to detect salty taste matures postnatally (for review, see Hill, D.L. and Mistretta, C.M., 1990); this may also be the case for humans. Prior exposure to foods containing salt may modulate this developmental change (e.g., Harris, G. and Booth, D. A., 1985). We have recently found (Stein, L. J. *et al.*, in preparation) that infants fed starchy foods between months 2 and 6 of life show heightened salt preferences relative to infants not exposed to these foods. Since one characteristic accompanying ingestion of starchy foods is likely a substantial exposure to salt, it is possible that this difference represents effects of prior exposure to salt. Perhaps, salt exposure during a critical period of receptive element maturation permanently alters peripheral and/or central structures and is thereby particularly potent in establishing childhood and perhaps even adult habits of intake. Alternatively, the heightened salt preference could reflect a shift in response to increased salt in the diet, similar to what is observed in adults (Bertino, M. *et al.*, 1986; see also discussion below). This is surely an area that should be investigated since it has major public health implications.

Although liking for salt in foods by children and adults is virtually universal, there are substantial individual differences. The degree to which genetic differences in salt taste perception could underlie these differences is unknown, but as the human salt taste mechanism(s) becomes better characterized, the influence of genetic variation will simultaneously become more clear.

In addition to probable effects of early exposure, environmental factors play a major role in determining salt avidity and consumption. The existence of societies where salt intake is very low has been interpreted to indicate that higher intake, such as that of Western societies, is the result of habit and commercial interests (e.g., MacGregor, G. A. and de Wardener, H. E., 1998). However, in most of these societies, low salt intake reflects low availability; wherever salt is inexpensive and widely available, it is noteworthy that most cultures consume roughly the same amount (see data in Intersalt Cooperative Research Group, 1988). This observation leads one to wonder whether there remains an unidentified (physiological?) explanation for human salt consumption. This question was raised many years ago (Bunge, G., 1902; Kaunitz, H., 1956) and remains worthy of consideration, particularly in light of calls for substantial population-wide reduction in salt consumption.

4.21.3 What Are the Functions of Salt in Food?

In the foregoing, it has been implicitly assumed that the proximate reason humans avidly consume Na is because we like the experience of the salty taste. This is undoubtedly part of the reason for modern human salt consumption, but surely not the only one, for salt plays many other roles in human food and nutrition.

Virtually all the salt people consume is in or on food; very few individuals eat pure salt outside that

Table 1   Number of recipes for which salt is called for in La Cuisine de France (Turgeon, C., 1964)

Food group	Number of recipes	Number calling for salt	Number calling for other salty ingredients	% calling for salt
Vegetables	142	131	11	100
Soups	51	51	0	100
Sauces	26	26	0	100
Starches (potato, rice, pasta, etc.)	59	57	2	100
Pastry[a]	136	70	10	58

[a]Many of the pastry recipes are for frosting-like items that are mainly sugar.
Note: Not tabulated are entrees, eggs, fish, poultry and game, meat, salads, desserts, and wine. Many of these use or are served with the items in the table.

context. In foods, salt has several technical functions (Man, C. M. D., 2007). It is an excellent preservative, preventing bacterial growth. It also serves structural functions during cooking and food processing by interacting with other ingredients to affect the texture and color of food. In addition to salt's functional benefits, it has a very significant additional strong point for both industry and the home cook: it is cheap. Consequently, any acceptable salt alternative must not only taste good and function as well as salt, but must also be inexpensive.

In addition to these functional features that make salt so important in foods, salt plays a sensory role over and above making food taste salty. Consider again bread and bread products. The salt in these makes up the single largest source of sodium consumed in the US diet; yet, we typically do not perceive bread as salty. Bread without salt tastes insipid or bad; when salt is added, the flavor is enhanced without making the bread taste salty. Thus salt, when added to food, must have functions other than adding saltiness. What are they?

One supposition is that the sense of taste imparts not only a quality, but also a sensation of quantity, amount, or bulk. It is difficult to know how to experimentally approach this concept; something similar to mouthfeel might be involved. It is as if taste stimuli (umami and sweet perhaps having some of the same characteristics as salt) might also interact with tactile fibers or other chemosensory structures to give a sense that the food has substance.

If this idea of mouthfeel is hard to encompass, yet another function of salt is better established. Sodium is an outstanding inhibitor of certain bitter-tasting compounds (Breslin, P. A. S. and Beauchamp, G. K., 1995). We have shown (Breslin, P. A. S. and Beauchamp, G. K., 1997) that a sodium salt added to a mixture of bitter and sweet suppresses the bitterness,

thereby releasing the sweet from inhibition. The overall sensory effect in the three-component model system is to enhance sweetness. We suggest that this can help explain why salt is added to so many foods when the intent is not to add saltiness: the salt suppresses off tastes and enhances good ones. Indeed, in a survey of recipes from one traditional French cook book (Turgeon, C., 1964) (Table 1), we found that most included either salt or a condiment-like olives that is high in salt. Clearly, the value of salt to food is demonstrated by its frequent use both by the home cook and by the multinational food companies.

4.21.4   What Can Be Done to Reduce Salt Consumption?

If salt reduction is the goal, what reasonable strategies are possible? There appear to be two approaches to reducing sodium consumption. In both salt is removed from foods; in the first, it is claimed that people would adjust to or at least tolerate the changes in taste (if the nontaste functions could be accounted for by other means), whereas in the second, an alternative taste-active substance is added to offset the loss of salt.

Consider the first strategy, which assumes that excess intake (relative to known physiological need) is due to habit and that habits can be changed. Clinical and anecdotal literature (see Beauchamp, G. K., 1991) reveal that when people undertake a low-sodium diet, the immediate response is misery. However, the lower sodium diet eventually becomes accepted, and in fact, foods containing the previous amount of salt are perceived as too salty (e.g., Stefansson, V., 1946). Many years ago, we provided experimental evidence to support this observation

(Bertino, M. *et al.*, 1983). After assuming a diet with about a 50% reduction in sodium content for 2–3 months, volunteers came to prefer their food with lower salt levels. They acclimated to this diet. We also showed (Bertino, M. *et al.*, 1986) that preference could be moved in the other direction: when people were put on a higher salt diet, they began to like more concentrated salt in their foods.

This evidence supports the view that if somehow we were to all undertake a low sodium diet, it would soon come to be the norm and no one would notice or feel deprived. But is this really possible? One widely recommended strategy is for food suppliers to gradually reduce the salt in foods so that the change is not noticed. Indeed, it appears that one can reduce the salt in some foods as much as 25% without consumers being aware (Girgis, S. *et al.*, 2003). But if people do not notice the reduction, will their taste preference change? And without a change in preference, can significant reductions really be obtained? Another issue that has not been experimentally addressed is whether one could lower the expected level of salt in one segment of the diet (e.g., commercial soups) and not the remainder of the foods. Would people come to prefer a lowered salt level in the single food if the overall levels were not correspondingly lowered or would they just choose not to eat as much soup (or add salt to it)? I am aware of no research that bears on this point. Finally, how would one deal with the nontaste functions of salt discussed above?

In summary, there is reason to believe that sodium intake could be successfully lowered through a unified dietary reduction in salt (assuming that the drive to consume salt in excess of need referred to in the previous section does not exist), but many practical difficulties need to be addressed.

Diabetics should not consume sugar in excess. Until fairly recently, this meant that they could not have sweeteners in their diet. Then came the introduction of saccharin and other alternative sweeteners. Could there be a saccharin or aspartame or sucralose for salt? Unlike the many molecules that function as alternative sweeteners, very few purely salty compounds have been identified. Indeed, given what we know about the receptor mechanisms for sweet and salty taste, it is quite likely that few or no such compounds will be found. Sweet taste recognition involves a classical seven-transmembrane G-coupled receptor system. It acts similarly to a lock and key, and apparently any molecule having the key's features will taste sweet (see Nelson, G.

et al., 2001; reviews: Margolskee, R. R., 2002; Kim, U.-K. *et al.*, 2004). In marked contrast, as discussed above, the high specificity of the sodium channel involved in salt taste transduction may mandate that no sodium-free salt taste substitutes exist.

Amiloride studies inform us that it is possible to modify the intensity of saltiness (at least in animal models). This leads to the question: If saltiness can be inhibited, can it also be enhanced? Might it be possible to find a compound or compounds that, when added to a reduced sodium food, make that food taste acceptably salty? Some amino acids and amino acid salts have this property, thus demonstrating proof of principle (Tamura, M. *et al.*, 1989; Lee, T. D., 1992; Riha III, W. E. *et al.*, 1997). What is needed is a concerted effort, most efficiently using the (yet unidentified) human salt receptor as a screening device, to search for such molecules. Of course, if one is found, many practical problems will arise. Is it safe? Does it have the other functions of salt, for example the bitter blocking activity? Is it sufficiently inexpensive? Will people accept a new chemical in their food? In spite of these problems, the search for a salt enhancer is worthwhile and has promise.

4.21.5 Final Comment

There is still much to learn about how salt taste is perceived and what factors are responsible for preferences for salt in foods. Given the apparent role of excess salt intake in human disease, primarily essential hypertension, it is imperative that this gap in our knowledge be closed.

References

Alderman, M. H. 2002. Salt, blood pressure and health: a cautionary tale. Int. J. Epidemiol. 31, 311–315.

Bartoshuk, L. M. 1980. Sensory Analysis of the Taste of NaCl. In: Biological and Behavioral Aspects of Salt Intake (*eds.* M. R. Kare, M. J. Fregly, and R. A. Bernard), pp. 83–98. Academic Press.

Beauchamp, G. K. 1991. Salt Preference in Humans. In: Encyclopedia of Human Biology, Vol. 6, p. 715. Academic Press.

Beauchamp, G. K., Bertino, M., and Engelman, K. 1991. Human Salt Appetite. In: Chemical Senses, Appetite and Nutrition (*eds.* M. I. Friedman, M. G. Tordoff, and M. R. Kare), pp. 85–107. Dekker.

Beauchamp, G. K., Cowart, B. J., and Moran, M. 1986. Developmental changes in salt acceptability in human infants. Dev. Psychobiol. 19, 75–83.

Bertino, M., Beauchamp, G. K., and Engelman, K. 1983. Long-term reduction in dietary sodium alters the taste of salt. Am. J. Clin. Nutr. 36, 1134–1144.

Bertino, M., Beauchamp, G. K., and Engelman, K. 1986. Increasing dietary salt alters salt taste. Physiol. Behav. 38, 203–213.

Breslin, P. A. S. and Beauchamp, G. K. 1995. Suppression of bitterness by sodium: variation among bitter stimuli. Chem. Senses 20, 609–623.

Breslin, P. A. S. and Beauchamp, G. K. 1997. Sodium salts potentiate flavour by suppressing bitterness. Nature 287, 563.

Bunge, G. 1902. Textbook of Physiological and Pathological Chemistry, 2nd edn. Blakison.

Contreras, R. J. and Frank, M. 1979. Sodium deprivation alters neural responses to gustatory stimuli. J. Gen. Physiol. 73, 569–594.

Contreras, R. J. and Kosten, T. 1983. Prenatal and early postnatal sodium chloride intake modifies the solution preferences of adult rats. J. Nutr. 113, 1051–1062.

Denton, D. 1982. The Hunger for Salt. Springer-Verlag, Inc.

DeSimone, J. A., Heck, G. L., and DeSimone, S. K. 1981. Active ion transport in dog tongue: a possible role in taste. Science 214, 1039–1041.

De Wardener, H. E., He, F. J., and MacGregor, G. A. 2004. Plasma sodium and hypertension. Kidney Int. 66, 2454–2466.

Fessler, D. M. T. 2003. An evolutionary explanation of the plasticity of salt preferences: prophylaxis against sudden dehydration. Med. Hypotheses 61(3), 412–415.

Gandhi, M. 1930. Monograph on Common Salt. Federation of Indian Chambers of Commerce and Industry.

Geran, L. C. and Spector, A. C. 2000. Sodium taste detectability in rats is independent of anion size: the psychophysical characteristics of the transcellular sodium taste transduction pathway. Behav. Neurosci. 114(6), 1229–1238.

Girgis, S., Prescott, N. B., Prendergast, J., Dumbrell, S., Turner, C., and Woodward, M. 2003. A one-quarter reduction in the salt content of bread can be made without detection. Eur. J. Clin. Nutr. 57(4), 616–620.

Halpern, B. P. 1998. Amiloride and vertebrate gustatory responses to NaCl. Neurosci.. Biobehav. Rev. 23, 5–47.

Harris, G. and Booth, D. A. 1985. Sodium preference in food and previous dietary exposure in 6-month-old infants. IRCS Med. Sci. 13, 1178–1179.

Hill, D. L. and Mistretta, C. M. 1990. Developmental neurobiology of salt taste sensation. Trends Neurosci. 13, 188–195.

Hooper, L., Bartlett, C., Smith, G. D., and Ebrahim, S. 2002. Systematic review of long term effects of advice to reduce dietary salt in adults. BMJ 325, 1–9.

Intersalt Cooperative Research Group. 1988. An international study of electrolyte excretion and blood pressure: Results for 24 hour urinary sodium and potassium excretion. Br. Med. J. 297, 319–328.

Karppanen, H. and Mervaala, E. 2006. Sodium intake and hypertension. Prog. Cardiovase. Dis. 49(2), 59–75.

Kaunitz, H. 1956. Causes and consequences of salt consumption. Nature 178, 1141–1144.

Kim, U.-K., Breslin, P. A. S., and Drayna, D. 2004. Genetics of human taste perception. J. Dent. Res. 83(6), 448–453.

Kurlansky, M. 2002. Salt: A World History, 484 pp. Walker and Company.

Lee, T. D., inventor; Kraft General Foods Inc., assignee. Seasoned food product with a salt enhancer. US Patent 5 176 934. April 16, 1992.

Liu, L., Leonard, A. S., Motto, D. G., Feller, M. A., Price, M. P., Johnson, W. A., and Welsh, M. J. 2003. Contribution of drosophila DEG/ENaC genes to salt taste. Neuron 39(1), 133–146.

Lyall, V., Heck, G. L., Ninnikova, A. K., Shosh, S., Phan, T-H. T., Alan, R. I., Russell, O. F., Makik, S. A., Bigbee, J. W., and DeSimone, J. A., 2004. The mammalian amiloride-insensitive non-specific salt taste receptor is a vanilloid receptor-1 variant. J. Physiol. 558(1), 147–159.

MacGregor, G. A. and de Wardener, H. E. 1998. Salt, Diet and Health, 227 pp. Cambridge University Press.

Man, C. M. D. 2007. Technological Functions of Salt in Food Products. In: Reducing Salt in Foods: Practical Strategies (eds. D. Kilcast and F. Angus), pp. 157–173. Woodhead Publishing and CRC Press.

Margolskee, R. R. 2002. Molecular mechanisms of bitter and sweet taste transduction. J. Biol. Chem. 277(1), 1–4.

McBurney, D. H. and Gent, J. F. 1979. On the nature of taste qualities. Psychol. Bull. 86, 151–167.

McCaughey, S. A. and Scott, T. R. 1998. The taste of sodium. Neurosci. Biobehav. Rev. 22, 663–676.

Nelson, G., Hoon, M. A., Chandrashekar, J., Zhang, Y., Ryba, N. J. P., and Zuker, C. S. 2001. Mammalian sweet taste receptors. Cell 106, 381–390.

Riha, III,W. E., Brand, J. G., and Breslin, P. A. S. 1997. Salty taste enhancement with amino acids. Chem. Senses 22, 778.

Sacks, F. M., Svetkey, P., Vollmer, W. M., Appel, L., Bray, G. A., Harsha, D., Obarzanek, E., Conlin, P. R., Miller, III, E. R., Simons-Morton, D. G., Karanja, N., and Lin, P.-H. 2001. Effects on blood pressure of reduced dietary sodium and the dietary approaches to stop hypertension (DASH) diet. 2001. N. Engl. J. Med. 344(1), 3–10.

Schiffman, S. S., Lockhead, E., and Maes, F. W. 1983. Amiloride reduces the taste intensity of Na and Li salts and sweeteners. Proc. Natl. Acad. Sci. U. S. A. 80, 6136–6140.

Schulkin, J. 1991. Sodium Hunger: The Search for a Salty Taste. Cambridge University Press.

Smriga, M., Kameishi, M., and Torii, K. 2002. Brief exposure to NaCl during early postnatal development enchances adult intake of sweet and salty compounds. Neuroreport. 13(18), 2565–2569.

Stefansson, V. 1946. Not by Bread Alone. Macmillan.

Stouffer, E. M. and White, N. M. 2005. A latent cue preference based on sodium depletion in rats. Learn. Mem. 12, 549–552.

Tamura, M., Seki, T., Yoshihiro, K., Tada, M., Kikuchi, E., and Okai, H. 1989. An enhancing effect on the saltiness of sodium chloride of added amino acids and their esters. Agric. Biol. Chem. 53, 1625–1633.

Taubes, G. 1998. The political history of salt. Science 281, 898–907.

Thaw, A. K., Frankmann, S., and Hill, D. L. 2000. Behavioral taste responses of developmentally NaCl-restricted rats to various concentrations of NaCl. Behav. Neurosci. 114, 437–441.

Turgeon, C. 1964. La Cuisine de France, Mapie, the Countess de Toulouse-Lautrec, Edited and translated by Charlotte Turgeon, 763 pp. The Orion Press.

Further Reading

Appel, L. J., Moore, T. J., Obarzanek, E., Vollmer, W. M., Svetkey, L. P., Sacks, F. M., Bray, G. A., Vogt, T. M., Cutler, J. A., Windhauser, M. M., Lin, P.-H, and Karanja, N. 1997. A clinical trial of the effects of dietary patterns on blood pressure. N. Engl. J. Med. 336(16), 1117–1124.

Beauchamp, G. K. and Cowart, B. 1990. Preference for high salt concentrations among children. Dev. Psychol. 26, 539–545.

Beauchamp, G. K., Bertino, M., Burke, D., and Engelman, K. 1990. Experimental sodium depletion and salt taste in normal human volunteers. Am. J. Clin. Nutr. 51, 881–889.

He, F. J. and MacGregor, G. A. 2003. How far should salt intake be reduced? Hypertension 42, 1093–1099.

Zinner, S. H., McGarvey, S. T., Lipsitt, L. P., and Rosner, B. 2002. Neonatal blood pressure and salt taste responsiveness. Hypertension 40, 280–285.

4.22 Behavioral Analysis of Taste Function in Rodent Models

S J St. John, Rollins College, Winter Park, FL, USA

A C Spector, The Florida State University, Tallahassee, FL, USA

4.22.1 Introduction

Researchers studying sensation and perception face the daunting task of making inferences about the internal experience of the subject without the benefit of direct measurement. In humans, perception is sometimes inferred through verbal report. This is not possible in nonhuman animals, from which the primary neurobiological database derives. Consequently, perceptual processes must be assessed through other behavioral means. This poses significant methodological and conceptual challenges, but is an absolutely essential component in any attempt to understand the neural principles underlying perception. Indeed, the strategic use of well-designed and complementary behavioral procedures, coupled with theoretically relevant manipulations of the nervous system, can lead to rich insights into the relatively concealed workings of our own sensory machinery.

Comparisons of the consequences of neural manipulations on the performance of animals tested on diverse behavioral tasks have historically provided powerful breakthroughs in understanding sensory function. A vivid example of this was the use of multiple visual tasks to better define the blindness caused by tectal lesions in hamsters (Schneider, G. E., 1969). Following such lesions, hamsters located food only by touch and smell, and did not orient their heads to motion, but did learn to associate a visual cue with the availability of water. In contrast, animals with lesions of the primary visual cortex located food by sight and oriented their heads in response to moving stimuli, but did not learn to associate a visual cue with water if the stimulus was sufficiently complex. This study, and others like it, highlights the danger of misinterpreting functional deficits, or lack thereof, following neural manipulations, if the scope of the behavioral analysis is limited. In other words, if only a single behavioral task had been employed to assess the effects of the lesion, different conclusions may have been reached regarding the functional use of sight. Without question, such experiments helped refine the concept of blindness and have contributed greatly to our view of the brain as a parallel processor.

In this vein, the purpose of this chapter is to discuss some basic interpretive properties of standard behavioral tasks that have been, and can be, used to infer the taste capabilities of nonhuman animal subjects and to provide some examples of how several of these paradigms have been successfully used in efforts to reveal the functional organization of the gustatory system.

4.22.2   Domains of Taste Function

Taste function can be heuristically divided into at least three domains (see Pfaffmann, C. *et al.*, 1979; Norgren, R., 1985; Scott, T. R. and Mark, G. P., 1986; Spector, A. C., 2000). The sensory-discriminative domain refers to the processes that allow animals to discriminate between taste compounds. This function can be subdivided into intensity, providing information regarding the amount of the stimulus,

and quality, providing information about the identity of the stimulus. The motivational domain refers to processes that promote or discourage ingestion of chemical substances. This function can be subdivided into an appetitive/avoidance component, which refers to actions that bring the animal closer to, or take the animal further from, the stimulus, and a consummatory/rejection component, which refers to reflex-like oromotor movements that lead to the ingestion or expulsion of stimuli contacting taste receptors (Craig, W., 1918; see also Grill, H. J. *et al.*, 1987; Berridge, K. C., 2004). Finally, the physiological domain refers to the ability of taste stimuli to trigger autonomic responses such as salivation (e.g., Pavlov, I. P., 1902).

In mammals, taste buds are distributed throughout the oral cavity in distinct fields each innervated by a different branch of three cranial nerves (Bradley, R. M., 1971; Miller, I. J., Jr., 1977; Travers, S. P. and Nicklas, K., 1990). In the central nervous system, gustatory projections form two divergent pathways as they ascend from the caudal brainstem to the forebrain – a thalamocortical pathway, characteristic of many sensory systems, and a ventral forebrain pathway, involving the amygdala, hypothalamus, and other structures (see Norgren, R., 1992; Travers, S. P., 1993; Norgren, R., 1995). Thus, the anatomical organization of the gustatory system invites speculation regarding the functional significance of these neural circuits. One effective strategy that has been used to uncover the relationship between nervous system structures and gustatory function is to alter the flow of information along these pathways and assess the consequences on taste-related behavioral tasks. Several decades of such experiments have taught us that a single behavioral measure is not sufficient to capture the full impact of a given manipulation on gustatory function. Just as blindness can be a relative term, so can aguesia. Competence or impairment on one behavioral task involving taste stimuli does not necessarily predict the same level of performance on another after a given experimental manipulation of the nervous system.

Some examples of such functional dissociations can be found in the following pages. The remainder of Section 4.22.3 will briefly review some of the methodological and interpretive properties of various behavioral paradigms used to assess gustatory processes in animal models, and Section 4.22.4 will illustrate the successful use of such procedures in neurobiological studies.

4.22.3 Behavioral Assessment of Taste Function

The assessment of taste function in animals requires careful consideration of the methodological features of the behavioral procedures used. Performance in some behavioral paradigms reflects the outcome of processes involving more than one functional domain (Table 1). These behavioral paradigms differ in some fundamental ways.

Procedures that involve the measurement of intake over prolonged periods of stimulus presentation (e.g., two-bottle preference test) are governed partly by the affective potency of the stimulus, but are also vulnerable to the influence of postingestive events (e.g., McCleary, R. A., 1953; Shuford, E., Jr., 1959; Mook, D. G., 1963; Rabe, E. F. and Corbit, J. D., 1973; Weingarten, H. P. and Watson, S. D., 1982; Nissenbaum, J. W. and Sclafani, A., 1987a). Intake tests involving the use of the sham drinking preparation (in which the ingested content flows out of an open gastric cannula) provide greater confidence that taste processing – or at least orosensory processing – guides behavior (e.g., Davis, J. D. and Campbell, C. S., 1973; Weingarten, H. P. and Watson, S. D., 1982; Schneider, L. H. *et al.*, 1986; Nissenbaum, J. W. and Sclafani, A., 1987a). Similarly, the presentation of fluids in brief-access trials, which minimize the influence of postingestive receptor systems during the response, can be used to assess the motivational characteristics (i.e., affective potency) of taste stimuli (see Young, P. T., 1952; Young, P. T. and Madsen, C. H., Jr., 1963; Grill, H. J. and Berridge, K. C., 1985). These procedures involve both an appetitive/avoidance component and a consummatory/rejection component. Other tasks allow investigators to assess the appetitive/avoidance component (e.g., a progressive ratio procedure) independent of the consummatory/rejection component and vice versa (e.g., an oromotor taste reactivity procedure).

In still other procedures, taste compounds are used as discriminative stimuli in operant-conditioning procedures or as conditioned stimuli in classical conditioning paradigms (e.g., Koh, S. D. and Teitelbaum, P., 1961; Morrison, G. R. and Norrison, W. 1966; Morrison, G. R., 1967; Slotnick, B. M., 1982; Spector, A. C. *et al.*, 1990b; Stapleton, J. R. *et al.*, 2002). These tasks provide a means by which sensory-discriminative taste function can be assessed separately from the affective properties of a stimulus. For example, hydrochloric acid or citric acid, quinine, and high concentrations of sodium chloride (NaCl), are all avoided by rats in intake and brief-access taste tests, but rats can be trained to discriminate these compounds in operant or classical conditioning paradigms (Morrison, G. R., 1967; Nowlis, G. H. *et al.*, 1980).

Regardless of the behavioral procedure employed, the potential for nongustatory cues associated with other chemosensory receptor systems must be considered. Perhaps the most vexing of these signals arises from stimulation of the trigeminal and olfactory systems, particularly at higher stimulus concentrations. For example, the viscosity of sugar solutions or the irritant properties of salts and acids (e.g., Gilmore, M. M. and Green, B. G., 1993) produce trigeminal sensations that could potentially affect behavioral responses. Although many prototypical taste compounds have very low vapor pressures and are thus suboptimal olfactory stimuli, there is evidence that rats can detect the smell of some taste solutions (Miller, S. D. and Erickson, R. P., 1966; Van Buskirk, R. L., 1981; Rhinehart-Doty, J. A. *et al.*, 1994; Capaldi, E. D. *et al.*, 2004). In contrast, at least with certain types of stimulus delivery designs (e.g., Spector, A. C. *et al.*, 1990a), it is clear that odor cues, if present, are insufficient at maintaining normal performance in some taste-related tasks when certain gustatory nerves are transected – a neural manipulation unlikely to have any direct effect on the olfactory system (e.g., Spector, A. C. *et al.*, 1990b; Slotnick *et al.*, 1991; Spector, A. C. *et al.*, 1996b; St. John, S. J. and Spector, A. C., 1996; St. John, S. J. and Spector, A. C., 1998). In any event, such concerns often necessitate empirical assays of the contribution of extragustatory sensory cues to behavior.

4.22.3.1 Two-Bottle Preference Test

4.22.3.1.1 Procedure

In the two-bottle preference test, animals are presented with two bottles usually placed on their cages for ~24 h. One bottle contains a taste solution and the other contains water (or some other taste solution). The relative intake of the two solutions is calculated and then the bottles are refilled, the positions reversed, and the test repeated. The repetition of the test with the bottles reversed serves to offset the appearance of a solution preference that may instead reflect a position preference. The determination of preference is based on the 48 h values. If the animals are unfamiliar with having access to two bottles or to bottles of the type used for the experiment, it is advisable to present habituation trials consisting of only water in advance of the experimental tests.

In most applications of this procedure, it is the outcome of the behavioral response, intake, which is

Table 1 Functional features of behavioral taste tests

Behavioral procedure	Taste quality measure	Intensity measure	Hedonic measure	Appetitive/ avoidance component	Consummatory/ reflex component	Postingestive influence	Deprivation necessary
Unconditioned two-bottle preference test	No[a]	Yes	Yes	Yes	Yes[b]	Yes[c]	Variable[d,e]
Unconditioned brief-access taste test	No[a]	Yes	Yes	Yes	Yes[f]	Minimized	Variable[d]
Unconditioned oromotor taste reactivity test	No[a]	Yes	Yes	No	Yes	Minimized	No
Unconditioned sham drinking test	No[a]	Yes	Yes	Yes	Yes[b]	No[g]	Variable[d]
Progressive ratio taste reinforcer efficacy test	No[a]	Yes	Yes	Yes	No	Minimized	Variable[d]
Conditioned taste preference/aversion: two-bottle preference test	Yes	Yes	Yes	Yes	Yes[b]	Yes[c]	Variable[e]
Conditioned taste preference/aversion: brief-access taste test	Yes	Yes	Yes	Yes	Yes[f]	Minimized	Yes
Conditioned taste preference/aversion: oromotor taste reactivity test	Yes	Yes	Yes	No	Yes	Minimized	No
Conditioned taste preference/aversion: progressive ratio test	Yes	Yes	Yes	Yes	No	Minimized	Yes
Operantly conditioned taste detection	No	Yes	No	No	No	Minimized	Yes
Operantly conditioned taste intensity discrimination	No	Yes	No	No	No	Minimized	Yes
Operantly conditioned taste discrimination	Yes	No	No[h]	No	No	Minimized	Yes

[a] Because of the specificity of responses to sodium salts in sodium-depleted animals, this test could be considered as a sensory measure of taste quality under certain conditions.

[b] Though not explicitly measured. If this test is coupled with a licking pattern analysis, then burst of licks could be considered as reflecting a strong consummatory component.

[c] The extent depends upon the duration of the test.

[d] Deprivation is not always required for unconditionally reinforcing stimuli (like sucrose).

[e] Deprivation is not required for long-duration tests (24 h, overnight, etc.).

[f] Though not explicitly measured. Trials initiated would be considered a pure appetitive measure, but the licks taken on a trial in some sense represents both appetitive and consummatory responses.

[g] Or virtually none. There is some evidence of some small amounts of nutrients passing into the duodenum during sham drinking.

[h] There may be some stimulus pairs (e.g., sucrose vs. quinine) where hedonically based taste discriminations could be possible.

measured. However, in some cases, investigators have measured licking behavior itself and have thus been able to reconstruct the pattern of drinking (e.g., LeMagnen, J. and Devos, M., 1980; Smith, J. C. and Foster, D. F., 1980; Spector, A. C. and Smith, J. C., 1984; Davis, J. D., 1989; Davis, J. D. and Smith, G. P., 1992; Davis, J. D., 1996; Spector, A. C. and St. John, S. J., 1998). When this is done in rodents, it is clear that drinking occurs in bursts of licks separated by pauses. A collection of bursts followed by a long pause could be construed as an ingestive bout or a meal if the solution is caloric. The number of bursts represents a relatively pure assessment of the appetitive behavior (i.e., initiations of drinking) and the size of a burst is heavily influenced by consummatory processes (e.g., swallowing). Thus, more refined measurement of behavior during a two-bottle preference test can provide meaningful information regarding the effects of taste on the components of motivated action.

4.22.3.1.2 Strengths

The two-bottle preference test has a long history, and thus a large body of normative data exists using this procedure. It is also relatively inexpensive, requires no customized equipment, and does not involve complicated training protocols for either animals or the staff. It is also usually amenable to testing a large number of subjects simultaneously.

4.22.3.1.3 Limitations

This test has been used to assess unconditioned and conditioned taste preferences and aversions. When used to measure unconditioned responses to taste compounds it does not provide any explicit information about the discriminative qualitative features of the stimulus (i.e., an animal might avoid both hydrochloric acid and quinine relative to water, but a similar level of avoidance does not preclude that that animal could discriminate these tastes from one another). The two-bottle preference test can provide some information about intensity, but it is difficult to discern the intensity of the stimulus independent of its hedonic characteristics because it is the latter that help drive the intake. For example, Eylam S. and Spector A. C. (2002) found that the NaCl detection thresholds measured with an operant-conditioning procedure (discussed below) were well below the concentration at which the same mice started to reliably avoid the stimulus in a two-bottle preference test. Thus, although mice could detect low concentrations of NaCl, they apparently displayed no consistent preference or aversion for them (Figure 1). In contrast, detection thresholds for

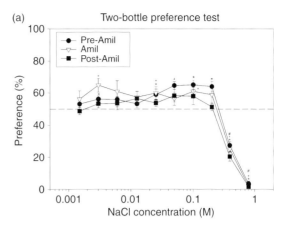

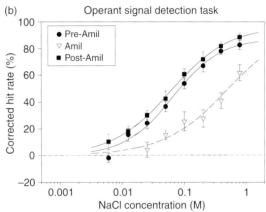

Figure 1 (a) Mean ($\pm$ standard error of the mean) percentage NaCl intake by C57BL/6J mice ($n = 7$) in a 48 h two-bottle preference test in which ascending concentrations of NaCl were pitted against water in three series of tests with (Amil, open triangles) and without (preAmil, closed circles; post-Amil, closed squares) 100 μM amiloride serving as the solvent. Significant differences from 50%: * for pre-Amil, + for Amil, and # for post-Amil. These mice had been previously tested for detection thresholds in an operant signal detection task (see (b)). (b) Mean ($\pm$ standard error of the mean) percentage hits on NaCl trials corrected for false alarms on water trials in an operant signal detection task as a function of NaCl concentration. The mice were tested in three phases of sessions. In the first (preAmil, filled circles) and third phase (post-Amil, filled squares), the NaCl was dissolved in water. In the second phase of sessions (Amil, open triangles), the NaCl was dissolved in 100 μM amiloride. In comparing (a) and (b) note that the preference-avoidance function is relatively flat until the concentration exceeds 0.2 M, whereas the dynamic range of responsiveness in the operant signal detection task occurs at lower concentrations. Also, amiloride had little effect on NaCl preference in these mice, but had a striking effect on NaCl detectability in the operant task. Adapted with permission from Eylam, S. and Spector, A. C. The effect of amiloride on operantly conditioned performance in an NaCl taste detection task and NaCl preference in C57BL/6J mice. Behav. Neurosci. 116, 149–159, 2002, Copyright 2002, American Psychological Association.

sucrose and glucose in mice correspond reasonably well with preference thresholds reported in the literature in mice (Bachmanov, A. A. *et al.*, 2001) likely because the affective value of these stimuli grows sharply once they are detected (see Eylam, S. and Spector, A. C., 2004). It is prudent to use the term preference threshold to indicate the concentration that subjects treat differently than water in two-bottle preference tests, and reserve the term detection threshold for concentrations discriminated from water in assays (like operant-conditioning procedures) where behavior is not dependent on hedonic evaluations.

As mentioned above, one principal shortcoming of the long-term two-bottle preference test is that post-ingestive events can influence intake and thus can complicate the interpretation of taste effects. For example, St. John S. J. *et al.* (2005) recently demonstrated that the dramatic increased avoidance of phenylthiocarbamide with ascending concentration in certain mouse strains was most likely due to the formation of a conditioned aversion to that compound over successive intake tests, rather than the expression of unconditioned avoidance due to its taste.

4.22.3.2 Brief-Access Taste Test

4.22.3.2.1 *Procedure*

In the brief-access taste test, animals are presented with very brief access to taste solutions, on the order of a few seconds, during which licking responses are measured (e.g., Young, P. T. and Trafton, C. L., 1964; Davis, J. D., 1973; Yamamoto, T. and Asai, K., 1986; Krimm, R. F. *et al.*, 1987; Smith, J. C. *et al.*, 1992; Breslin, P. A. *et al.*, 1993b; Spector, A. C. *et al.*, 1993a; 1993b; O'Keefe, G. B. *et al.*, 1994; St. John, S. J. *et al.*, 1994; Contreras, R. J. *et al.*, 1995; Markison, S. *et al.*, 1995; Spector, A. C. *et al.*, 1996b; Boughter, J. D. *et al.*, 2002; Glendinning, J. I. *et al.*, 2002; Zhang, Y. *et al.*, 2003; Dotson, C. D. and Spector, A. C., 2004). Many trials are presented during a session and responses to a variety of different solutions varying in concentration and/or solute can be tested. For normally preferred compounds, the test can be conducted in nondeprived animals, but for normally avoided compounds, the test must be conducted in fluid-deprived animals so that the subjects are highly motivated to sample the stimuli in search of water.

4.22.3.2.2 *Strengths*

Because immediate responses to small volumes of taste stimuli are measured, this test minimizes the influence of postingestive events on the behavior. Furthermore,

the brief-access taste test can measure changes in responsiveness past the point at which preference tests reach asymptote as the concentration is raised. Because many trials are presented in a single session, full concentration–response functions can be obtained in a single test session. Random presentation minimizes order effects which are sometimes present in concentration series of longer-duration intake tests. In general, very orderly concentration–response functions are derived in the brief-access taste test, although the dynamic range of responsiveness is biased toward the higher concentrations (i.e., as with the two-bottle preference test, the concentration at which the lick rate differs from water may or may not be closely related to the detection threshold).

4.22.3.2.3 *Limitations*

First, like the two-bottle preference test, the brief-access taste procedure relies on the hedonic characteristics of the stimulus to drive responses and thus does not provide explicit information regarding the discriminative qualitative features of a taste compound. For example, animals will avoid licking both NaCl and quinine in a concentration-dependent manner even though they can discriminate between these stimuli in several other tasks (e.g., Morrison, G. R., 1967; Nowlis, G. H. *et al.*, 1980; Spector, A. C. *et al.*, 1993a; Cauthon, R. *et al.*, 1994; St. John, S. J. *et al.*, 1994; Spector, A. C., 1995b; Boughter, J. D. *et al.*, 2002; Glendinning, J. I. *et al.*, 2002; Geran, L. C. *et al.*, 2004).

Second, although concentration–response functions using this method are highly reliable within a given context, the context can affect the outcome. For example, the concentration–response function can be shifted to the left when an animal is tested with only one concentration per session pitted against water, rather than randomly presenting all concentrations on each day of testing (Glendinning, J. I. *et al.*, 2005).

Third, the brief-access taste test provides a measure of the affective potency of a taste stimulus, but does not clearly differentiate between appetitive/avoidance and consummatory/rejection responses. The former behavior is associated with the animal approaching the drinking spout and the latter behavior is reflexively triggered by the contact of the oral receptors with the stimulus. Although it may be possible to reconstruct the pattern of licking behavior during a trial to try to distinguish these processes, the very brief duration of the observation period (usually 5–10 s) places limitations on such an analysis. In contrast, trials initiated could be considered a measure of appetitive behavior, but this datum is also under the influence of the postingestive properties of the stimulus and the physiological state of the animal. For example, it is likely that the more fluid-deprived an animal is, the more trials it will initiate.

Finally, and unlike a long-term preference test, the brief-access taste test must be conducted differently for normally preferred and normally avoided stimuli. Animals will not initiate trials to nonpreferred stimuli unless they are water-deprived; in contrast, water-deprived rats will lick all concentrations of preferred substances at a maximal rate (a ceiling effect). Depending on the overall goals of the research, the physiological state of testing may constrain interpretation.

4.22.3.3   Oromotor Taste Reactivity Test

4.22.3.3.1   Procedure

Chronic indwelling cannulae can be surgically implanted into the oral cavity through which taste solutions can be infused at a rate and pattern under the control of the experimenter. Oral motor and somatic responses can then be monitored (Grill, H. J. and Norgren, R., 1978c; 1978d; Grill, H. J. et al., 1987; Spector, A. C. et al., 1988; Brining, S. K. et al., 1991; Berridge, K. C., 1996). Animals can swallow or expel the infused fluid. Ingestive responses elicited by normally preferred stimuli include tongue protrusions and low-amplitude mouth movements. Aversive responses elicited by normally avoided stimuli include gapes, chin rubs, head shakes, and forelimb flails. Generally the responses are videotaped or measured with electromyographic (EMG) electrodes placed in relevant oromotor muscles (e.g., Travers, J. et al., 2000).

4.22.3.3.2   Strengths

Orofacial responses are triggered by the stimulus contacting the oral receptors and no appetitive action is necessary for stimulus delivery making the test useful for assessing consummatory/rejection behavior almost purely. In addition, because the experimenter initiates stimulation directly, there is no need for food or water deprivation to promote stimulus sampling.

The test is typically very short, on the order of a minute, precluding any substantial postingestive influence.

4.22.3.3.3   Limitations

As with the other tests described thus far, taste reactivity is dependent on the affective properties of the stimulus, and thus does not indicate the relative discriminability of stimuli or provide high-fidelity

information about detection thresholds. Unlike most behavioral assays, this procedure requires a surgical procedure, albeit one that rodents tolerate well. Also, unlike most behavioral assays, the dependent variable cannot be automatically counted (e.g., by a lickometer circuit) unless EMG electrodes are surgically implanted (see Travers, J. B. et al., 1997). Thus, time and training are required of the experimenter to score videotapes frame by frame.

4.22.3.4   Progressive Ratio Taste Reinforcer Efficacy Test

4.22.3.4.1   Procedure

In the progressive ratio taste reinforcer efficacy test, animals are trained to execute a specific operant response, such as lever pressing or dry-spout licking, to receive a small volume of a taste solution. Initially, the number of responses required to produce the taste reinforcer is low, but the requirement is progressively raised as the session continues until the animal reaches a breakpoint and ceases to respond further (Hodos, W., 1961; Hodos, W. and Kalman, G., 1963; Reilly, S., 1999; Reilly, S. and Trifunovic, R., 1999; Sclafani, A. and Ackroff, K., 2003). The breakpoint is thought to reflect the efficacy of the reinforcer and thus can be construed as a measure of the affective potency of a taste stimulus.

4.22.3.4.2   Strengths

Once trained, deprivation states are unnecessary for normally preferred stimuli (e.g., sucrose), but water deprivation is required to test normally avoided stimuli (e.g., quinine). In the latter case, the motivation to rehydrate is pitted against the aversive characteristics of the taste stimulus. Depending on the step-size schedule, the actual total amount of reinforcer delivered can be kept relatively small, thus attenuating the influence of postingestive factors and satiation. Just as the oromotor taste reactivity test is a pure measure of consummatory/rejection responses, the progressive ratio is a pure measure of appetitive behavior. This is because the behavior, which leads to the presentation of the taste reinforcer, is produced without the stimulus contacting the receptors.

4.22.3.4.3   Limitations

This procedure has the same limitations as some of the other procedures discussed in that the responses are driven by the affective evaluation of the stimulus.

4.22.3.5 Conditioned Taste Aversion

4.22.3.5.1 Procedure

Typically, when ingestion of a specific taste stimulus is followed by injection of a viscerally aversive agent (e.g., lithium chloride (LiCl); see Garcia, J. *et al.*, 1955; Nachman, M. and Ashe, J. H., 1973; Domjan, M., 1980; Barker, L. M. *et al.*, 1999), often after only a single conditioning trial (depending on the dose of the aversive agent), animals will display robust avoidance and rejection behavior in response to the conditioned taste stimulus. Because animals will generalize their conditioned responses to some taste compounds but not others, this procedure has been useful for making inferences about the degree of qualitative similarity between taste stimuli (e.g., Nachman, M., 1963; Tapper, D. N. and Halpern, B. P., 1968; Nowlis, G. H. *et al.*, 1980; Smith, D. V. and Theodore, R. M., 1984).

4.22.3.5.2 Strengths

To the extent that taste aversion generalization paradigms provide information about the sensory-discriminative features of taste solutions, the taste aversion methodology is technically simpler than operant discrimination paradigms. In addition, training is usually effective after a single conditioning trial. If a gustometer is used to measure conditioned avoidance responses, a variety of test compounds can be included in the stimulus array and generalization profiles can be obtained in a single test session.

4.22.3.5.3 Limitations

From a practical perspective, one must be concerned with the possibility of extinction effects that can occur with continued testing that involve the conditioned stimulus or compounds that bear some qualitative similarity to it. This places limits on the number of test sessions and test stimuli that can be included in the design. From a conceptual perspective, conditioned taste aversions are functionally complex measures in that the quality, intensity, hedonics, and even physiological responses (e.g., Berridge, K. *et al.*, 1981) are all potentially involved. These procedures are designed to change the hedonic properties of a specific stimulus identified by its quality. Like all conditioning processes the concentration of the taste stimuli must be considered because generalization can take place along the intensity dimension (Hagstrom, E. C. and Pfaffmann, C., 1959; Tapper, D. N. and Halpern, B. P., 1968; Nowlis, G. H., 1974; Scott, T.

R. and Giza, B. K., 1987; Spector, A. C. and Grill, H. J., 1988). Accordingly, a given experimental manipulation of the gustatory system can affect behavioral outcomes in this paradigm by influencing function in one of several of these domains.

Second, it must be remembered that generalization and discrimination, while related, are distinct psychological concepts. For example, taste aversions to sucrose generalize to maltose (e.g., Nissenbaum, J. W. and Sclafani, A., 1987b; Spector, A. C. and Grill, H. J., 1988) and glutamate (Heyer, B. R. *et al.*, 2003), but sucrose can also be discriminated from these compounds on the basis of gustatory cues (Spector, A. C. *et al.*, 1997; Stapleton, J. R. *et al.*, 2002; Heyer, B. R. *et al.*, 2004).

4.22.3.6 Operantly Conditioned Taste Detection and Discrimination

4.22.3.6.1 Procedure

Operantly conditioned taste detection and discrimination tasks have a well-established history of use in the psychophysical examination of a variety of sensory systems in animals (see Blough, D. and Blough, P., 1977; Stebbins, W. C. and Berkley, M. A., 1990). In the study of gustatory processes, generally a taste stimulus is used as a cue for responding in a specific way (see Carr, W. J., 1952; Harriman, A. E. and MacLeod, R. B., 1953; Koh, S. D. and Teitelbaum, P., 1961; Morrison, G. R. and Norrison, W., 1966; Morrison, G. R., 1967; Shaber, G. S. *et al.*, 1970; Slotnick, B. M., 1982; Brosvic, G. M. and Slotnick, B. M., 1986; Spector, A. C. *et al.*, 1990b; Willner, P. *et al.*, 1990; Thaw, A. K. and Smith, J. C., 1992; St. John, S. J. *et al.*, 1997; Stapleton, J. R. *et al.*, 2002; Spector, A. C., 2003). If the response is correctly executed the behavior can be reinforced (e.g., the animal is rewarded); if the response is incorrectly executed, the animal can be punished. For instance, a thirsty rat can be trained to press a left-hand lever if a five-lick stimulus sample is NaCl and a right-hand lever if the sample is water. A correct response can lead to access to 20 licks of water (which is rewarding because the animal is on a water-restriction schedule) and an incorrect response can lead to a 30 s time-out, further delaying the opportunity to receive water. The concentration of the NaCl stimulus can be varied across trials until it reaches a low enough value that the animal does not differentially respond between the taste and water stimulus; thus a detection threshold can be measured. If the water stimulus in the prior example is replaced with a second taste stimulus (e.g., potassium chloride (KCl)), then the procedure can be used to measure the degree of discriminability between the two taste compounds being tested.

4.22.3.6.2 Strengths

These procedures possess some of the same benefits as the brief-access taste test in that small stimulus samples are used and immediate responses are measured curtailing the influence of postingestive events on responses. In addition, the taste compounds serve as signals for other events (reward or punishment) and, thus, the responses are under the control of the reinforcement contingencies established and not the inherent motivational characteristics of the sapid stimuli.

4.22.3.6.3 Limitations

Under ideal circumstances, the strength of this procedure is that it allows a relatively pure assay of the sensory-discriminative aspects of the tastant. However, animals working for reinforcement may use whatever cues are available. If all of the concentrations of one stimulus are more reinforcing or more aversive than all of the concentrations of the comparison stimulus, then discrimination might arguably take place along a hedonic dimension. For example, if rats were trained to discriminate sucrose from quinine, it is conceivable that such a discrimination could be guided by the hedonic valence of the stimuli as opposed to their qualitative characteristics. In general, because the hedonic potency of a taste stimulus often covaries with its concentration, and several concentrations of each stimulus are presented within a session, there is likely overlap in the affective valence of the discriminated compounds such that the hedonic properties of the stimuli are rendered irrelevant cues. Nevertheless, there may be some stimulus pairs (e.g., sucrose vs. quinine) where hedonically based taste discriminations could be possible.

An additional consideration in the use of discrimination procedures is that it remains possible that animals might use taste cues that are not strictly related to the perceptual quality of the compounds to guide their performance. For example, perhaps compounds evoking similar qualities might vary in their kinetics of reaching or activating receptors such that they differ in the rise and/or decay times of the sensation. Potentially, that information could be used by the animal to competently discriminate between the two stimuli. Although such a possibility remains speculative, it is difficult to entirely dismiss. Because

animals potentially have a variety of taste cues not strictly related to the quality of the stimulus per se to guide performance, a failure for an animal to discriminate between two compounds is an interpretively powerful result, provided that learning and motivational factors can be discounted.

4.22.4 Use of Multiple Behavioral Assays Can Provide Rich Insight into the Organization of the Gustatory System

As noted in the introduction, the use of multiple behavioral assays is critical to gain a comprehensive understanding of sensory function; no single behavioral procedure provides a gold standard test. A given assay may reveal taste deficits after a particular experimental manipulation, but other assays may indicate normal function. This is because each behavioral paradigm is influenced by various aspects of taste processing as summarized in Table 1. An experimental manipulation may affect one aspect of taste function (e.g., quality) but not another (e.g., hedonic), and whether performance in a task is affected will depend on the degree to which the behavior in the assay is dependent on the relevant function. Some examples are provided below.

4.22.4.1 Peripheral Mechanisms of Salt Taste

Decades ago, Pfaffmann and Beidler (Pfaffmann, C., 1941; Beidler, L. M., 1953; 1954; Pfaffmann, C., 1955) showed that the chorda tympani (CT) nerve of the rat was exceptionally responsive to sodium salts. Thus, although this nerve innervating the anterior tongue transmits afferent information from a minority of taste buds (in the rat, less than 15% of the total), the electrophysiological data promoted the rational expectation that transection of the CT nerve would severely impair taste sensitivity to salt. Pfaffmann C. (1952) tested this hypothesis by using the two-bottle preference test and comparing salt preference in intact rats, rats with CT nerve transection, and rats with combined transection of both the CT and glossopharyngeal (GL, innervates posterior tongue) nerves (denervating >70% of the total taste buds). To his surprise, none of these transections had an effect on salt preference; an outcome reported by others using similar behavioral paradigms (Richter, C. P., 1939; Akaike, N. *et al.*, 1965; Vance, W. B., 1967;

Grill, H. J. and Schwartz, G. J., 1992; Grill, H. J. *et al.*, 1992). Although some or all of this maintained responsiveness in the neurotomized rats may have been due to the influence of the NaCl on postlaryngeal receptors, a similar lack of effect on salt responsiveness has been observed even when a brief-access taste test, which minimizes the contribution of postingestive events, was used as an assay (Cauthon, R. *et al.*, 1994).

Adulteration of NaCl solutions with the epithelial sodium channel blocker amiloride also has little or no effect on NaCl preference measured in a two-bottle preference test in either C57BL/6 (B6) mice (Figure 1) or F-344 rats (Chappell, J. P. *et al.*, 1998; Eylam, S. and Spector, A. C., 2002). Like the lack of effect of CT nerve transection on NaCl preference, the ineffectiveness of amiloride to alter ingestive behavior in this test was unexpected because the drug notably suppresses CT and greater superficial petrosal (GSP) responses to NaCl in most rodents (e.g., Heck, G. I. *et al.*, 1984; Brand, J. G. *et al.*, 1985; DeSimone, J. A. and Ferrell, F., 1985; Herness, M. S., 1987; Ninomiya, Y. *et al.*, 1989; Hettinger, T. P. and Frank, M. E., 1990; Simon, S. A. *et al.*, 1993; Ye, Q. *et al.*, 1993; Doolin, R. E. and Gilbertson, T. A., 1996; Sollars, S. I. and Hill, D. L., 1998; Lindemann, B. *et al.*, 1999).

One seemingly straightforward conclusion consistent with these behavioral data is that the CT and the epithelial sodium channels in taste receptor cells are not critical in the maintenance of taste sensitivity to NaCl, but such a verdict would be premature. Responsiveness in both the two-bottle preference test and the brief-access taste test are driven by the hedonic characteristics of the stimulus. Moreover, the former procedure is vulnerable to postingestive events and the latter procedure does not have optimal sensitivity in the low concentration range. Neither procedure provides explicit information on the qualitative characteristics of the taste stimulus. Thus, to conclude that rodents had normal salt taste perception following these manipulations in the peripheral gustatory system would be akin to concluding that hamsters are visually normal after visual cortex lesions based on the results of one or only a few behavioral assays.

As it turns out, suppression of the amiloride-sensitive salt transduction pathway or removal of the afferent input from the CT has severe consequences on sodium taste discriminability. When operant conditioning procedures are used to train rats to respond differentially to NaCl and water and the concentration of the stimulus is varied, the resulting detection threshold is shifted 1–2 orders of magnitude by transection of the CT (Spector, A. C. *et al.*, 1990b; Slotnick, B. M. *et al.*, 1991; Kopka, S. L. and Spector, A. C., 2001). When the nerve regenerates, the NaCl detection threshold returns to normal (Kopka, S. L. and Spector, A. C., 2001). Likewise, stimulus adulteration with amiloride, which appears to be tasteless to rats and mice (Markison, S. and Spector, A. C., 1995; Eylam, S. *et al.*, 2003), raises detection thresholds for NaCl and sodium gluconate, but not KCl, measured in operantly trained rats and mice (Geran, L. C. *et al.*, 1999; Geran, L. C. and Spector, A. C., 2000a; 2000b; Eylam, S. and Spector, A. C., 2002; 2003). It is important to note, however, that while CT transection raises the NaCl detection threshold measured in operant conditioning procedures, when a similar task was used to assess NaCl difference thresholds (i.e., just noticeable differences between NaCl concentrations) at mid-range concentrations, CT (or GL) transection had little effect (Colbert, C. L. *et al.*, 2004). Apparently, the input from the remaining gustatory nerves is sufficient in maintaining the ability of rats to make normal intensity discriminations within the NaCl concentration range that is detectable after CT transection.

Thus, amiloride treatment and CT transection decrease the sensitivity of some rodents to NaCl. But do these manipulations have any effect on the qualitative nature of NaCl taste? Rats can be trained to discriminate between salts in operant-conditioning paradigms. In these designs, concentration is varied so that intensity cues are irrelevant. Interestingly, performance of rats in an NaCl versus KCl operant discrimination task is significantly impaired by CT transection, but is unaffected by GL transection (Spector, A. C. and Grill, H. J., 1992; St. John, S. J. *et al.*, 1997; Kopka, S. L. *et al.*, 2000). Combined transection of the CT and GSP, removing all of the taste input of cranial nerve VII, drops performance on either an NaCl versus NH_4Cl or on an NH_4Cl versus KCl taste discrimination task to chance levels, but GL transection is without effect (Geran, L. C. and Spector, A. C., 2002; Geran, L. C. *et al.*, 2002b).

Stimulus adulteration with amiloride also leads to chance performance on a NaCl versus KCl discrimination task in rats and mice (Spector, A. C. *et al.*, 1996a; Eylam, S. and Spector, A. C., 2005). In these experiments, animals had many more errors on NaCl trials than on KCl trials suggesting that when the stimuli are mixed with amiloride, NaCl tastes more similar to KCl than to unadulterated NaCl. In support of this conclusion, taste aversions conditioned to

NaCl mixed with amiloride generalize to nonsodium salts in a brief-access taste test (Hill, D. L. *et al.*, 1990). In contrast, when unadulterated NaCl is used as a conditioned stimulus, rats (and other rodents) will specifically avoid NaCl and LiCl but not nonsodium salts (Nowlis, G. H. *et al.*, 1980; Hill, D. L. *et al.*, 1990; see also Nachman, M., 1963).

Another way in which CT transection and amiloride treatment have been shown to affect the taste quality of NaCl is through procedures that take advantage of the physiological and behavioral mechanisms that apparently evolved to protect herbivores and some omnivores from dangerous deficiencies in body sodium. When rats are acutely depleted of sodium, hormonal mechanisms are engaged that facilitate the reabsorption of sodium and trigger neural circuits that potentiate appetitive and consummatory behavior promoting the search for and ingestion of sodium salts (see Denton, D., 1982; Schulkin, J., 1991). Sodium-depleted rats will display increased intake of NaCl in brief-access, short- and long-term drinking tests (Richter, C. P., 1936; Handal, P. J., 1965; Jalowiec, J. E. and Stricker, E. M., 1970; Wolf, G., 1982; Epstein, A. N., 1984; Bernstein I. L. and Hennessy C. J., 1987; Rowland, N. E. and Fregly, M. J., 1992; Breslin, P. A. *et al.*, 1993a; O'Keefe, G. B. *et al.*, 1994; Brot, M. D. *et al.*, 2000), will enhance their licking response for sodium salts over nonsodium salts in brief-access taste tests (Breslin, P. A. *et al.*, 1993b; 1995; Markison, S. *et al.*, 1995; Geran, L. C. and Spector, A. C., 2004), and will show increased ingestive oral motor consummatory responses to NaCl (Berridge, K. C. *et al.*, 1984; Grill, H. J. and Berridge, K. C., 1985; Grill, H. J. and Bernstein, I. L., 1988). The appetite is specific for salts with a sodium cation; the only exception is lithium which is treated as if it is sodium. Transection of the CT compromises, and amiloride treatment eliminates, the vigor and cation-specificity of depletion-induced sodium appetite in rats (Bernstein I. L. and Hennessy C. J., 1987; McCutcheon, N. B., 1991; Sollars, S. I. and Bernstein, I. L., 1992; Breslin, P. A. *et al.*, 1993b; Sollars, S. I. and Bernstein, I. L., 1994; Breslin, P. A. *et al.*, 1995; Markison, S. *et al.*, 1995; Frankmann, S. P. *et al.*, 1996; Roitman, M. F. and Bernstein, I. L., 1999; Brot, M. D. *et al.*, 2000; Geran, L. C. and Spector, A. C., 2004).

Collectively, the findings from these various behavioral procedures in rats indicate that absolute sensitivity to NaCl, as well as the perceived quality of this salt stimulus, is partially dependent on the input

of the CT nerve that likely derives from the activation of the amiloride-sensitive salt transduction pathway. This is consistent with the fact that the rat CT has narrowly tuned fibers that respond to NaCl and LiCl but do not respond well to other salts or acids (Frank, M. E. *et al.*, 1983; Ninomiya, Y. and Funakoshi, M., 1988; Lundy, R. F., Jr. and Contreras, R. J., 1999; Sollars, S. I. and Hill, D. L., 2005). Moreover, the sodium responsiveness of these fibers is significantly suppressed by lingual amiloride treatment (Ninomiya, Y. and Funakoshi, M., 1988). There are, however, other sodium-responsive fibers in the CT and in other gustatory nerves that are not affected by the application of amiloride on their receptor fields (Ninomiya, Y. and Funakoshi, M., 1988; Formaker, B. K. and Hill, D. L., 1991; Kitada, Y. *et al.*, 1998; Sollars, S. I. and Hill, D. L., 2005). It would appear that these amiloride-insensitive fibers are sufficient at providing input that can maintain normal unconditioned avoidance responses to salt. This is presumably one reason why the two-bottle preference test and the brief-access taste test, when used to assess unconditioned responses to NaCl, reveal normal function after CT transection or amiloride treatment. Thus, many of the procedures listed in Table 1 have nicely complemented each other and have collectively provided insight into the principles and peripheral mechanisms associated with salt taste in the rodent model.

4.22.4.2 Use of Behavioral Procedures to Study the Functional Neuroanatomy of the Central Gustatory System

Some of the behavioral procedures in Table 1 have been used to assess taste function after production of experimental lesions in central gustatory structures. Interpretation of the behavioral effects following manipulations of the brain is always challenging because any resulting impairments might have a variety of functional origins. Furthermore, any behavioral deficit following a lesion may not be due to the disruption of processing in the damaged brain site, but might be due to the interruption of information flow to brain regions anatomically downstream. Nevertheless, the complementary use of different behavioral procedures combined with theoretically relevant anatomical manipulations of the brain has offered insight into the functional organization of the central gustatory system in at least the rat model. Some examples are provided below.

4.22.4.2.1 Taste reactivity in decerebrate rats

Gustatory signals from the periphery are transmitted to the rostral one-third of the nucleus of the solitary tract (NST) by branches of cranial nerves VII, IX, and X. In rodents, the main ascending gustatory pathway begins with the projections of the second-order neurons in the NST to third-order taste-responsive neurons in the parabrachial nucleus (PBN). The projections of the third-order neurons in the PBN form two pathways destined for forebrain sites. The first terminates in the parvicellular subdivision of the ventral posteromedial nucleus of the thalamus, which, in turn, projects to the insular cortex. The second pathway terminates in several ventral forebrain regions including the lateral hypothalamus, the central nucleus of the amygdala, and the substantia innominata (see Norgren, R. and Leonard, C. M., 1973; Norgren, R., 1974; 1976; Hamilton, R. B. and Norgren, R., 1984; Halsell, C. B., 1992; Travers, S. P., 1993; Norgren, R., 1995).

About 30 years ago, Grill H. J. and Norgren R. (1978d) made a remarkable discovery that began to shed light on the functional significance of hindbrain and forebrain gustatory circuits. They developed the taste reactivity test procedure (see Section 4.22.3.3) to test the responsiveness of supracollicular decerebrate rats to prototypical taste stimuli (Grill, H. J. and Norgren, R., 1978c). In this neural preparation, the forebrain is neurally isolated from the rest of the nervous system. Animals that survive this procedure do not spontaneously eat and drink and have impairments in body temperature regulation, but they can nonetheless be chronically maintained by gavage feeding and close control of environmental conditions (Grill, H. J. and Norgren, R., 1978b). Despite the fact that these animals are severely compromised behaviorally, they display relatively normal concentration-dependent oral motor responses to taste stimuli (Grill, H. J. and Norgren, R., 1978d). Intraoral infusions of quinine elicit gapes and intraoral infusions of sucrose elicit tongue protrusions and mouth movements. These findings provide convincing evidence that the neural circuits of the caudal brainstem are sufficient to support taste-elicited oral motor reflexes.

In intact rats, taste reactivity can be modified by learning. If the intraoral delivery of a palatable fluid is followed by visceral malaise, animals will display aversive oral motor responses to this normally palatable sugar stimulus when tested on subsequent occasions (e.g., Grill, H. J. and Norgren, R., 1978a;

Grill, H. J. and Berridge, K. C., 1985; Spector, A. C. et al., 1988). Grill H. J. and Norgren R. (1978a) went onto show that even though decerebrate rats display relatively normal unconditioned taste reactivity, these animals do not change their response profile even after several pairings of the taste stimulus with a malaise-inducing LiCl injection. Regardless of the functional basis for the impairment, it is clear that the forebrain is necessary for this particular form of taste-related learning to be expressed. The taste reactivity paradigm is the only type of taste-related behavioral measure from Table 1 that can be used in the examination of decerebrate rats. However, its application in the assessment of both unconditioned and conditioned responses to taste stimuli led to fundamental insights concerning the functional significance of caudal brainstem versus forebrain gustatory circuits.

4.22.4.2.2 Taste function after lesions in the gustatory nuclei

One limitation of the decerebration strategy is that it involves a very crude and expansive neural manipulation affecting many different sensory, motor, and autonomic systems. Thus, while functions that survive this substantial neural insult are conceptually meaningful, not much can be said of behaviors that are impaired or eliminated. The interpretative significance of any lesion-induced behavioral deficit depends heavily on the anatomical specificity of the neural manipulation. For close to 50 years, researchers have targeted brain sites along the gustatory pathway and have assessed the degree of behavioral competence following the production of lesions. The early work generally involved large lesions and almost the exclusive use of intake tests to assess taste function (Patton, H. D. et al., 1944; Benjamin, R. M., 1955a; 1955b; Benjamin, R. M. and Akert, K., 1958; Oakley, B. and Pfaffmann, C., 1962; Blomquist, A. J. and Antem, A., 1967; Loullis, C. C. et al., 1978; Braun, J. J. et al., 1982; Hill, D. L. and Almli, C. R., 1983; Lasiter, P. S., 1985; see also, Spector, A. C., 1995a; Reilly, S., 1998). Later investigations involved more anatomically precise lesions often placed with electrophysiological guidance and incorporated more diverse behavioral paradigms (but see Oakley, B. and Pfaffmann, C., 1962; Oakley, B., 1965). Rather than review the entire spectrum of results from such studies, it is worth noting a few findings that highlight the interpretive power of comprehensive behavioral analyses.

The consequences of lesions directed toward the second central relay in the rodent gustatory system, the PBN, provide an interpretive complement to the findings from experiments involving supracollicular transections described in Section 4.22.4.2.1. Overall, a PBN lesion is considerably less drastic than decerebration, but at the same time is anatomically more severe in terms of interrupting gustatory neural circuits. Decerebration presumably leaves intact the NST and PBN as well as connections between these areas and medullary nuclei, including those involved in orofacial motor and physiological reflexes (see Travers, J. B. *et al.*, 1999). Lesions in the gustatory zone of the dorsal pons, in contrast, should limit gustatory processing to the NST and its local connections, to the extent that the PBN is an obligate relay for forebrain taste projections.

From that perspective, it is noteworthy that electrophysiologically guided lesions in the PBN have relatively little effect on the intake of a variety of prototypical taste stimuli by rats (in contrast to the complete suppression of appetitive behavior in the decerebrate). Nonetheless, intake is not normal following this manipulation; investigators have reported overconsumption of several prototypical taste stimuli (e.g., Flynn, F. W. *et al.*, 1991b). Converging evidence for an alteration in taste processing comes from brief-access taste testing (see Section 4.22.3.2). Substantial changes in behavioral responses to quinine, sucrose, and NaCl have been reported in rats with bilateral PBN lesions (Spector, A. C. *et al.*, 1993a; 1995) collectively suggesting that rats are made hypoguesic by this lesion. The simplest explanation is that PBN lesion reduces the perceived intensity of gustatory stimuli, but these data also suggest that hedonic processing and possibly stimulus identification remain intact. For example, although the effective concentration range is shifted to higher concentrations, rats with PBN lesion nonetheless avoid quinine and NaCl in a concentration-dependent manner, and increase licking of sucrose in a concentration-dependent manner.

The simple explanation of hypoguesia (weakened taste intensity) is complicated by another study which demonstrated that detection thresholds measured using operant procedures sometimes were not altered following PBN lesion (Spector, A. C. *et al.*, 1995). That is, individual animals with PBN lesion showed elevated detection thresholds in some cases and not in others. Importantly, all of these animals failed to express a learned taste aversion to either sucrose or NaCl after multiple conditioning trials, a finding that served as functional confirmation of the PBN lesion. Similar individual animal variability has also been reported in operant discrimination procedures; rats with PBN lesion that fail to form conditioned taste aversions are in some cases unable to discriminate NaCl from KCl in an operantly trained discrimination task, but in other cases show near-normal discrimination performance (Spector, A. C., 1995a). Whether the individual differences are related to prelesion training differences, the exact locus and extent of the lesion, or differences in the plasticity of pre-PBN circuitry is unknown. What is clear from this work is that rats with lesions centered in the taste-responsive zone of the dorsal pons are not aguesic, and that their gustatory competence can, in some cases, emulate that of intact rats depending on the task.

Importantly, this residual competence is informative in interpretation of other effects of PBN lesion. As noted, a hallmark of PBN lesion is that, like decerebration, this neural manipulation prevents the formation of conditioned taste aversions (DiLorenzo, P. M., 1988; Ivanova, S. F. and Bures, J., 1990; Flynn, F. W. *et al.*, 1991a; Yamamoto, T. and Fujimoto, Y., 1991; Spector, A. C. *et al.*, 1992; Reilly, S. *et al.*, 1993; Scalera, G. *et al.*, 1995; Spector, A. C. *et al.*, 1995; Grigson, P. S. *et al.*, 1997; 1998; Sclafani, A. *et al.*, 2001) Such a deficit would be expected if PBN lesions rendered the rat aguesic or severely hypoguesic to taste solutions; likewise, rats would not be able to express taste aversions if PBN lesions prevented the identification of a specific taste solution. However, work cited above converges on the conclusion that although some deficits in detection thresholds, suprathreshold sensitivity, and taste discriminability can be seen, those deficits do not always occur and are not severe enough to explain the deficits in taste aversion acquisition. Indeed, if the taste aversion is conditioned presurgically, rats with PBN lesions display normal avoidance of the conditioned taste stimulus, but are unable to learn an aversion to a novel sapid compound (Grigson, P. S. *et al.*, 1997). Clearly, the use of multiple behavioral tests has provided considerable interpretive power in diagnosing the role of the PBN in taste aversion learning and leads to the conclusion that the taste-responsive portion of this nucleus must remain intact for taste signals to be conditionally associated with visceral malaise (see Spector, A. C. *et al.*, 1992; Spector, A. C., 1995a; Norgren, R. and Grigson, P. S., 1996; Grigson, P. S. *et al.*, 1997; 1998).

Likewise, PBN lesions eliminate the expression of a salt appetite (Flynn, F. W. *et al.*, 1991a; Scalera, G. *et al.*, 1995), impair the formation of a conditioned flavor preference (Sclafani, A. *et al.*, 2001), and prevent the avoidance of diets deficient in amino acids or protein (Fromentin, G. *et al.*, 2000). Against the backdrop of the behavioral work discussed so far, the interpretation of the cause of these deficits would not rely on the animal failing to taste NaCl, the conditioned stimulus, or the deficient diet. Rather, like with taste aversion, the PBN or its downstream processing targets appear to be indispensable for behaviors that require the integration of taste and viscerosensory signals.

4.22.5 Final Comments

This chapter has provided an overview of the behavioral techniques that have been used to assess taste function in animal, primarily rodent, models. We have attempted to show how these various paradigms can complement each other leading to a more comprehensive interpretation of the effects of experimental manipulations on the gustatory system, be they anatomical, pharmacological, or genetic. Greater coordination in the application of behavioral, neurobiological, and genetic techniques, all of which are evolving rapidly, should ultimately lead to a better understanding of the neural mechanisms underlying taste function and provide a logical bridge potentially connecting findings based on animal models to the study of human gustation.

References

Akaike, N., Hiji, Y., and Yamada, K. 1965. Taste preference and aversion in rats following denervation of the chorda tympani and IXth nerve. Kumamoto Med. J. 18, 108–109.

Bachmanov, A. A., Tordoff, M. G., and Beauchamp, G. K. 2001. Sweetener preference of C57BL/6ByJ and 129P3/J mice. Chem. Senses 26, 905–913.

Barker, L. M., Best, M. R., and Domjan, M. 1999. Learning Mechanisms in Food Selection. Baylor University Press.

Beidler, L. M. 1953. Properties of chemoreceptors of tongue of rat. J. Neurophysiol. 16, 595–607.

Beidler, L. M. 1954. A theory of taste stimulation. J. Gen. Physiol. 38, 133–139.

Benjamin, R. M. 1955a. Cortical taste mechanisms studied by two different test procedures. J. Comp. Physiol. Psychol. 48, 119–122.

Benjamin, R. M. 1955b. The effect of fluid deprivation on taste deficits following cortical lesions. J. Comp. Physiol. Psychol. 48, 502–505.

Benjamin, R. M. and Akert, K. 1958. Cortical and thalamic areas involved in taste discrimination in the albino rat. J. Comp. Neurol. 111, 231–260.

Bernstein I. L. and Hennessy C. J. 1987. Amiloride-sensitive sodium channels and expression of sodium appetite in rats. Am. J. Physiol. Regul. Integr. Comp. Physiol. 253, R371–R374.

Berridge, K., Grill, H. J., and Norgren, R. 1981. Relation of consummatory responses and preabsorptive insulin release to palatability and learned taste aversions. J. Comp. Physiol. Psychol. 95, 363–382.

Berridge, K. C. 1996. Food reward: brain substrates of wanting and liking. Neurosci. Biobehav. Rev. 20, 1–25.

Berridge, K. C. 2004. Motivation concepts in behavioral neuroscience. Physiol. Behav. 81, 179–209.

Berridge, K. C., Flynn, F. W., Schulkin, J., and Grill, H. J. 1984. Sodium depletion enhances salt palatability in rats. Behav. Neurosci. 98, 652–660.

Blomquist, A. J. and Antem, A. 1967. Gustatory deficits produced by medullary lesions in the white rat. J. Comp. Physiol. Psychol. 63, 439–443.

Blough, D. and Blough, P. 1977. Animal Psychophysics. In: Handbook of Operant Behavior (eds. W. K. Honig and J. E. R. Straddon), pp. 514–539. Prentice-Hall.

Boughter, J. D., St. John, S. J., Noel, D. T., Ndubuizu, O., and Smith, D. V. 2002. A brief-access test for bitter taste in mice. Chem. Senses 27, 133–142.

Bradley, R. M. 1971. Tongue Topography. In: Taste (ed. Beidler, L. M.), pp. 1–30. Springer.

Brand, J. G., Teeter, J. H., and Silver, W. L. 1985. Inhibition by amiloride of chorda tympani responses evoked by monovalent salts. Brain Res. 334, 207–214.

Braun, J. J., Lasiter, P. S., and Kiefer, S. W. 1982. The gustatory neocortex of the rat. Physiol. Psychol. 10, 13–45.

Breslin, P. A., Kaplan, J. M., Spector, A. C., Zambito, C. M., and Grill, H. J. 1993a. Lick rate analysis of sodium taste-state combinations. Am. J. Physiol. 264, R312–R318.

Breslin, P. A., Spector, A. C., and Grill, H. J. 1993b. Chorda tympani section decreases the cation specificity of depletion-induced sodium appetite in rats. Am. J. Physiol. 264, R319–R323.

Breslin, P. A., Spector, A. C., and Grill, H. J. 1995. Sodium specificity of salt appetite in Fischer-344 and Wistar rats is impaired by chorda tympani nerve transection. Am. J. Physiol. 269, R350–R356.

Brining, S. K., Belecky, T. L., and Smith, D. V. 1991. Taste reactivity in the hamster. Physiol. Behav. 49, 1265–1272.

Brosvic, G. M. and Slotnick, B. M. 1986. Absolute and intensity-difference taste thresholds in the rat: evaluation of an automated multi-channel gustometer. Physiol. Behav. 38, 711–717.

Brot, M. D., Watson, C. H., and Bernstein, I. L. 2000. Amiloride-sensitive signals and NaCl preference and appetite: a lick-rate analysis. Am. J. Physiol. Regul. Integr. Comp. Physiol. 279, R1403–R1411.

Capaldi, E. D., Hunter, M. J., and Privitera, G. J. 2004. Odor of taste stimuli in conditioned "taste" aversion learning. Behav. Neurosci. 118, 1400–1408.

Carr, W. J. 1952. The effect of adrenalectomy upon the NaCl taste threshold in rat. J. Comp. Physiol. Psychol. 45, 377–380.

Cauthon, R., Garcea, M., and Spector, A. C. 1994. Taste-guided unconditioned licking to suprathreshold sodium chloride solutions is unaffected by selective lingual denervation. Chem. Senses 19, 452.

Chappell, J. P., St. John, S. J., and Spector, A. C. 1998. Amiloride does not alter NaCl avoidance in Fischer-344 rats. Chem. Senses 23, 151–157.

Colbert, C. L., Garcea, M., and Spector, A. C. 2004. Effects of selective lingual gustatory deafferentation on suprathreshold taste intensity discrimination of NaCl in rats. Behav. Neurosci. 118, 1409–1417.

Contreras, R. J., Carson, C. A., and Pierce, C. E. 1995. A novel psychophysical procedure for bitter taste assessment in rats. Chem. Senses 20, 305–312.

Craig, W. 1918. Appetites and aversions as constituents of instincts. Biol. Bull. 34, 91–107.

Davis, J. D. 1973. The effectiveness of some sugars in stimulating licking behavior in the rat. Physiol. Behav. 11, 39–45.

Davis, J. D. 1989. The Microstructure of Ingestive Behavior. In: The Psychobiology of Human Eating Disorders (eds. L. H. Schneider, S. J. Cooper, and K. A. Halmi), pp. 106–121. Annals of the New York Academy of Sciences.

Davis, J. D. 1996. Deterministic and probabilistic control of the behavior of rats ingesting liquid diets. Am. J. Physiol. Regul. Integr. Comp. Physiol. 270, R793–R800.

Davis, J. D. and Campbell, C. S. 1973. Peripheral control of meal size in the rat: effect of sham feeding on meal size and drinking rate. J. Comp. Physiol. Psychol. 83, 379–387.

Davis, J. D. and Smith, G. P. 1992. Analysis of the microstructure of the rhythmic tongue movements of rats ingesting maltose and sucrose solutions. Behav. Neurosci. 106, 217–228.

Denton, D. 1982. The Hunger for Salt. Springer.

DeSimone, J. A. and Ferrell, F. 1985. Analysis of amiloride inhibition of chorda tympani taste response of rat to NaCl. Am. J. Physiol. Regul. Integr. Comp. Physiol. 249, R52–R61.

DiLorenzo, P. M. 1988. Long-delay learning in rats with parabrachial pontine lesions. Chem. Senses 13, 219–229.

Domjan, M. 1980. Ingestional Aversion Learning: Unique and General Processes. In: Advances in the Study of Behavior, vol. 2 (eds. J. Rosenblatt, R. Hinde, C. Beer, and M. Busnel), pp. 215–336. Academic Press.

Doolin, R. E. and Gilbertson, T. A. 1996. Distribution and characterization of functional amiloride-sensitive sodium channels in rat tongue. J. Gen. Physiol. 107, 545–554.

Dotson, C. D. and Spector, A. C. 2004. The relative affective potency of glycine, L-serine and sucrose as assessed by a brief-access taste test in inbred strains of mice. Chem. Senses 29, 489–498.

Epstein, A. N. 1984. The dependence of the salt appetite of the rat on the hormonal consequences of sodium deficiency. J. Physiol. 79, 494–498.

Eylam, S. and Spector, A. C. 2002. The effect of amiloride on operantly conditioned performance in an NaCl taste detection task and NaCl preference in C57BL/6J mice. Behav. Neurosci. 116, 149–159.

Eylam, S. and Spector, A. C. 2003. Oral amiloride treatment decreases taste sensitivity to sodium salts in C57BL/6J and DBA/2J mice. Chem. Senses 28, 447–458.

Eylam, S. and Spector, A. C. 2004. Stimulus processing of glycine is dissociable from that of sucrose and glucose based on behaviorally measured taste signal detection in Sac 'taster' and 'non-taster' mice. Chem. Senses 29, 639–649.

Eylam, S. and Spector, A. C. 2005. Taste discrimination between NaCl and KCl is disrupted by amiloride in inbred mice with amiloride-insensitive chorda tympani nerves. Am. J. Physiol. Regul. Integr. Comp. Physiol. 288, R1361–R1368.

Eylam, S., Tracy, T., Garcea, M., and Spector, A. C. 2003. Amiloride is an ineffective conditioned stimulus in taste aversion learning in C57BL/6J and DBA/2J mice. Chem. Senses 28, 681–689.

Flynn, F. W., Grill, H. J., Schulkin, J., and Norgren, R. 1991a. Central gustatory lesions: II. Effects on sodium appetite, taste aversion learning, and feeding behaviors. Behav. Neurosci. 105, 944–954.

Flynn, F. W., Grill, H. J., Schwartz, G. J., and Norgren, R. 1991b. Central gustatory lesions: I. Preference and taste reactivity tests. Behav. Neurosci. 105, 933–943.

Formaker, B. K. and Hill, D. L. 1991. Lack of amiloride sensitivity in SHR and WKY glossopharyngeal taste responses to NaCl. Physiol. Behav. 50, 765–769.

Frank, M. E., Contreras, R. J., and Hettinger, T. P. 1983. Nerve fibers sensitive to ionic taste stimuli in chorda tympani of the rat. J. Neurophysiol. 50, 941–960.

Frankmann, S. P., Sollars, S. I., and Bernstein, I. L. 1996. Sodium appetite in the sham-drinking rat after chorda tympani nerve transection. Am. J. Physiol. Regul. Integr. Comp. Physiol. 271, R339–R345.

Fromentin, G., Feurte, S., Nicolaidis, S., and Norgren, R. 2000. Parabrachial lesions disrupt responses of rats to amino acid devoid diets, to protein-free diets, but not to high-protein diets. Physiol. Behav. 70, 381–389.

Garcia, J., Kimeldorf, D. J., and Koelling, R. A. 1955. Conditioned aversion to saccharin resulting from exposure to gamma radiation. Science 122, 157–158.

Geran, L. C. and Spector, A. C. 2000a. Amiloride increases sodium chloride taste detection threshold in rats. Behav. Neurosci. 114, 623–634.

Geran, L. C. and Spector, A. C. 2000b. Sodium taste detectability in rats is independent of anion size: the psychophysical characteristics of the transcellular sodium taste transduction pathway. Behav. Neurosci. 114, 1229–1238.

Geran, L. C. and Spector, A. C. 2002. Glossopharyngeal nerve transection does not compromise chloride salt discrimination in the rat. Chem. Senses 27, A78.

Geran, L. C. and Spector, A. C. 2004. Anion size does not compromise sodium recognition by rats after acute sodium depletion. Behav. Neurosci. 118, 178–183.

Geran, L. C., Garcea, M., and Spector, A. C. 2004. Nerve regeneration-induced recovery of quinine avoidance after complete gustatory deafferentation of the tongue. Am. J. Physiol. Regul. Integr. Comp. Physiol. 287, R1235–R1243.

Geran, L. C., Garcea, M., and Spector, A. C. 2002. Transecting the gustatory branches of the facial nerve impairs NH(4)Cl vs. KCl discrimination in rats. Am. J. Physiol. Regul. Integr. Comp. Physiol. 283, R739–R747.

Geran, L. C., Guagliardo, N. A., and Spector, A. C. 1999. Chorda tympani nerve transection, but not amiloride, increases the KCl taste detection threshold in rats. Behav. Neurosci. 113, 185–195.

Gilmore, M. M. and Green, B. G. 1993. Sensory irritation and taste produced by NaCl and citric acid: effects of capsaicin desensitization. Chem. Senses 18, 257–272.

Glendinning, J. I., Bloom, L. D., Onishi, M., Zheng, K. H., Damak, S., Margolskee, R. F., and Spector, A. C. 2005. Contribution of α-gurtducin to taste-guided licking responses of mice. Chem. Senses 30, 299–316.

Glendinning, J. I., Gresack, J., and Spector, A. C. 2002. A high-throughput screening procedure for identifying mice with aberrant taste and oromotor function. Chem. Senses 27, 461–474.

Grigson, P. S., Reilly, S., Shimura, T., and Norgren, R. 1998. Ibotenic acid lesions of the parabrachial nucleus and conditioned taste aversion: further evidence for an associative deficit in rats. Behav. Neurosci. 112, 160–171.

Grigson, P. S., Shimura, T., and Norgren, R. 1997. Brainstem lesions and gustatory function. 3. The role of the nucleus of the solitary tract and the parabrachial nucleus in retention of a conditioned taste aversion in rats. Behav. Neurosci. 111, 180–187.

Grill, H. J. and Bernstein, I. L. 1988. Strain differences in taste reactivity to NaCl. Am. J. Physiol. Regul. Integr. Comp. Physiol. 255, R424–R430.

Grill, H. J. and Berridge, K. C. 1985. Taste reactivity as a measure of the neural control of palatability. Prog. Psychobiol. Physiol. Psychol. 11, 1–61.

Grill, H. J. and Norgren, R. 1978a. Chronic decerebrate rats demonstrated satiation, but not bait shyness. Science 201, 267–269.

Grill, H. J. and Norgren, R. 1978b. Neurological tests and behavioral deficits in chronic thalamic and chronic decerebrate rats. Brain Res. 143, 299–312.

Grill, H. J. and Norgren, R. 1978c. The taste reactivity test. I. Mimetic responses to gustatory stimuli in neurologically normal rats. Brain Res. 143, 263–279.

Grill, H. J. and Norgren, R. 1978d. The taste reactivity test. II. Mimetic responses to gustatory stimuli in chronic thalamic and chronic decerebrate rats. Brain Res. 143, 281–297.

Grill, H. J. and Schwartz, G. J. 1992. The contribution of gustatory nerve input to oral motor behavior and intake-based preference. II. Effects of combined chorda tympani and glossopharyngeal nerve section in the rat. Brain Res. 573, 105–113.

Grill, H. J., Schwartz, G. J., and Travers, J. B. 1992. The contribution of gustatory nerve input to oral motor behavior and intake-based preference. I. Effects of chorda tympani or glossopharyngeal nerve section in the rat. Brain Res. 573, 95–104.

Grill, H. J., Spector, A. C., Schwartz, G. J., Kaplan, J. M., and Flynn, F. W. 1987. Evaluating Taste Effects on Ingestive Behavior. In: Techniques in the Behavioral and Neural Sciences, vol. 1: Feeding and Drinking (eds. F. Toates and N. Rowland), pp. 151–188. Elsevier.

Hagstrom, E. C. and Pfaffmann, C. 1959. The relative taste effectiveness of different sugars for the rat. J. Comp. Physiol. Psychol. 59, 259–262.

Halsell, C. B. 1992. Organization of parabrachial nucleus efferents to the thalamus and amygdala in the golden hamster. J. Comp. Neurol. 317, 57–78.

Hamilton, R. B. and Norgren, R. 1984. Central projections of gustatory nerves in the rat. J. Comp. Neurol. 222, 560–577.

Handal, P. J. 1965. Immediate acceptance of sodium salts by sodium deficient rats. Psychon. Sci. 3, 315–316.

Harriman, A. E. and MacLeod, R. B. 1953. Discriminative thresholds of salt for normal and adrenalectomized rats. Am. J. Psychol. 66, 465–471.

Heck, G. I., Mierson, S., and DeSimone, J. A. 1984. Salt taste transduction occurs through an amiloride-sensitive sodium transport pathway. Science 223, 403–405.

Herness, M. S. 1987. Effect of amiloride on bulk flow and iontophoretic taste stimuli. J. Comp. Physiol. A 160, 281–288.

Hettinger, T. P. and Frank, M. E. 1990. Specificity of amiloride inhibition of hamster taste responses. Brain Res. 513, 24–34.

Heyer, B. R., Taylor-Burds, C. C., Mitzelfelt, J. D., and Delay, E. R. 2004. Monosodium glutamate and sweet taste: discrimination between the tastes of sweet stimuli and glutamate in rats. Chem. Senses 29, 721–729.

Heyer, B. R., Taylor-Burds, C. C., Tran, L. H., and Delay, E. R. 2003. Monosodium glutamate and sweet taste: generalization of conditioned taste aversion between glutamate and sweet stimuli in rats. Chem. Senses 28, 631–641.

Hill, D. L. and Almli, C. R. 1983. Parabrachial nuclei damage in infant rats produces residual deficits in gustatory preferences/aversions and sodium appetite. Dev. Psychobiol. 16, 519–533.

Hill, D. L., Formaker, B. K., and White, K. S. 1990. Perceptual characteristics of the amiloride-suppressed sodium chloride taste response in the rat. Behav. Neurosci. 104, 734–741.

Hodos, W. 1961. Progressive ratio as a measure of reward strength. Science 134, 943–944.

Hodos, W. and Kalman, G. 1963. Effects of increment size and reinforcer volume on progressive ratio performance. J. Exp. Anal. Behav. 6, 387–392.

Ivanova, S. F. and Bures, J. 1990. Acquisition of conditioned taste aversion in rats is prevented by tetrodotoxin blockade of a small midbrain region centered around the parabrachial nuclei. Physiol. Behav. 48, 543–549.

Jalowiec, J. E. and Stricker, E. M. 1970. Restoration of body fluid balance following acute sodium deficiency in rats. J. Comp. Physiol. Psychol. 70, 94–102.

Kitada, Y., Mitoh, Y., and Hill, D. L. 1998. Salt taste responses of the IXth nerve in Sprague-Dawley rats: lack of sensitivity to amiloride. Physiol. Behav. 63, 945–949.

Koh, S. D. and Teitelbaum, P. 1961. Absolute behavioral taste thresholds in the rat. J. Comp. Physiol. Psychol. 54, 223–229.

Kopka, S. L. and Spector, A. C. 2001. Functional recovery of taste sensitivity to sodium chloride depends on regeneration of the chorda tympani nerve after transection in the rat. Behav. Neurosci. 115, 1073–1085.

Kopka, S. L., Geran, L. C., and Spector, A. C. 2000. Functional status of the regenerated chorda tympani nerve as assessed in a salt discrimination task. Am. J. Physiol. Regul. Integr. Comp. Physiol. 278, R720–R731.

Krimm, R. F., Nejad, M. S., Smith, J. C., Miller, I. J., Jr., and Beidler, L. M. 1987. The effect of bilateral sectioning of the chorda tympani and the greater superficial petrosal nerves on the sweet taste in the rat. Physiol. Behav. 41, 495–501.

Lasiter, P. S. 1985. Thalamocortical relations in taste aversion learning: II. Involvement of the medial ventrobasal thalamic complex in taste aversion learning. Behav. Neurosci. 99, 477–495.

LeMagnen, J. and Devos, M. 1980. Parameters of the meal pattern in rats: their assessment and physiological significance. Neurosci. Biobehav. Rev. 4, 1–11.

Lindemann, B., Gilbertson, T. A., and Kinnamon, S. C. 1999. Amiloride-Sensitive Sodium Channels in Taste. Academic Press.

Loullis, C. C., Wayner, M. J., and Jolicoeur, F. B. 1978. Thalamic taste nuclei lesions and taste aversion. Physiol. Behav. 20, 653–655.

Lundy, R. F., Jr. and Contreras, R. J. 1999. Gustatory neuron types in rat geniculate ganglion. J. Neurophysiol. 82, 2970–2988.

Markison, S. and Spector, A. C. 1995. Amiloride is a poor conditioned stimulus in taste aversion. Chem. Senses 20, 739–740.

Markison, S., St. John, S. J., and Spector, A. C. 1995. Glossopharyngeal nerve transection does not compromise the specificity of taste-guided sodium appetite in rats. Am. J. Physiol. Regul. Integr. Comp. Physiol. 269, R215–R221.

McCleary, R. A. 1953. Taste and post-ingestion factors in specific-hunger behavior. J. Comp. Physiol. Psychol. 46, 411–420.

McCutcheon, N. B. 1991. Sodium deficient rats are unmotivated by sodium chloride solutions mixed with the sodium channel blocker amiloride. Behav. Neurosci. 105, 764–766.

Miller, I. J., Jr. 1977. Gustatory Receptors of the Palate. In: Food Intake and Chemical Senses (eds. Y. Katsuki, M. Sato, S. Takagi, and Y. Oomura), pp. 173–186. University of Tokyo Press.

Miller, S. D. and Erickson, R. P. 1966. The odor of taste solutions. Physiol. Behav. 1, 145–146.

Mook, D. G. 1963. Oral and postingestional determinants of the intake of various solutions in rats with esophageal fistulas. J. Comp. Physiol. Psychol. 56, 645–659.

Morrison, G. R. 1967. Behavioral response patterns to salt stimuli in the rat. Can. J. Psychol. 21, 141–152.

Morrison, G. R. and Norrison, W. 1966. Taste detection in the rat. Can. J. Psychol. 20, 208–217.

Nachman, M. 1963. Learned aversion to the taste of lithium chloride and generalization to other salts. J. Comp. Physiol. Psychol. 56, 343–349.

Nachman, M. and Ashe, J. H. 1973. Learned taste aversions in rats as a function of dosage, concentration, and route of administration of LiCl. Physiol. Behav. 10, 73–78.

Ninomiya, Y. and Funakoshi, M. 1988. Amiloride inhibition of responses of rat single chorda tympani fibers to chemical and electrical tongue stimulations. Brain Res. 451, 319–325.

Ninomiya, Y., Sako, N., and Funakoshi, M. 1989. Strain differences in amiloride inhibition of NaCl responses in mice, Mus Musculus. J. Comp. Physiol. A 166, 1–5.

Nissenbaum, J. W. and Sclafani, A. 1987a. Sham-feeding response of rats to polycose and sucrose. Neurosci. Biobehav. Rev. 11, 215–222.

Nissenbaum, J. W. and Sclafani, A. 1987b. Qualitative differences in polysaccharide and sugar tastes in the rat: a two-carbohydrate taste model. Neurosci. Biobehav. Rev. 11, 187–196.

Norgren, R. 1974. Gustatory afferents to ventral forebrain. Brain Res. 81, 285–295.

Norgren, R. 1976. Taste pathway to hypothalamus and amygdala. J. Comp. Neurol. 166, 17–30.

Norgren, R. 1985. Taste and the autonomic nervous system. Chem. Senses 10, 143–161.

Norgren, R. 1992. The central gustatory and visceral afferent systems arising from the nucleus of the solitary tract. Adv. Physiol. Sci. 16, 359–366.

Norgren, R. 1995. Gustatory System. In: The Rat Nervous System (ed. G. Paxinos), pp. 751–771. Academic Press.

Norgren, R. and Grigson, P. S. 1996. The Role of the Central Gustatory System in Learned Taste Aversions. In: Perception, Memory and Emotion: Frontiers in Neuroscience (eds. T. Ono, B. McNaughton, S. Molotchnikoff, E. Rolls, and H. Nishijo), pp. 479–497. Pergamon.

Norgren, R. and Leonard, C. M. 1973. Ascending central gustatory pathways. J. Comp. Neurol. 150, 217–238.

Nowlis, G. H. 1974. Conditioned stimulus intensity and acquired alimentary aversions in the rat. J. Comp. Physiol. Psychol. 86, 1173–1184.

Nowlis, G. H., Frank, M. E., and Pfaffmann, C. 1980. Specificity of acquired aversions to taste qualities in hamsters and rat. J. Comp. Physiol. Psychol. 94, 932–942.

O'Keefe, G. B., Schumm, J., and Smith, J. C. 1994. Loss of sensitivity to low concentrations of NaCl following bilateral chorda tympani nerve sections in rats. Chem. Senses 19, 169–184.

Oakley, B. 1965. Impaired operant behavior following lesions of the thalamic taste nucleus. J. Comp. Physiol. Psychol. 59, 202–210.

Oakley, B. and Pfaffmann, C. 1962. Electrophysiologically monitored lesions in the gustatory thalamic relay of the albino rat. J. Comp. Physiol. Psychol. 55, 155–160.

Patton, H. D., Ruch, T. C., and Walker, A. E. 1944. Experimental hypoguesia from Horsley-Clarke lesions of the thalamus in Macaca mulatta. J. Neurophysiol. 7, 171–184.

Pavlov, I. P. 1902. The Work of the Digestive Glands. Charles Griffin.

Pfaffmann, C. 1941. Gustatory afferent impulses. J. Cell Comp. Physiol. 17, 243–258.

Pfaffmann, C. 1952. Taste preference and aversion following lingual denervation. J. Comp. Physiol. Psychol. 45, 393–400.

Pfaffmann, C. 1955. Gustatory nerve impulses in rat, cat, and rabbit. J. Neurophysiol. 18, 429–440.

Pfaffmann, C., Frank, M., and Norgren, R. 1979. Neural mechanisms and behavioral aspects of taste. Ann. Rev. Psychol. 30, 283–325.

Rabe, E. F. and Corbit, J. D. 1973. Postingestional control of sodium chloride solution drinking in the rat. J. Comp. Physiol. Psychol. 84, 268–274.

Reilly, S. 1998. The role of the gustatory thalamus in taste-guided behavior. Neurosci. Biobehav. Rev. 22, 883–901.

Reilly, S. 1999. Reinforcement value of gustatory stimuli determined by progressive ratio performance. Pharmacol. Biochem. Behav. 63, 301–311.

Reilly, S. and Trifunovic, R. 1999. Progressive ratio performance in rats with gustatory thalamus lesions. Behav. Neurosci. 113, 1008–1019.

Reilly, S., Grigson, P. S., and Norgren, R. 1993. Parabrachial nucleus lesions and conditioned taste aversion: evidence supporting an associative deficit. Behav. Neurosci. 107, 1005–1017.

Rhinehart-Doty, J. A., Schumm, J., Smith, J. C., and Smith, G. P. 1994. A non-taste cue of sucrose in short-term taste tests in rats. Chem. Senses 19, 425–431.

Richter, C. P. 1936. Increased salt appetite in adrenalectomized rats. Am. J. Physiol. Regul. Integr. Comp. Physiol. 115, 155–161.

Richter, C. P. 1939. Transmission of taste sensation in animals. Trans. Am. Neurol. Assoc. 65, 49–50.

Roitman, M. F. and Bernstein, I. L. 1999. Amiloride-sensitive sodium signals and salt appetite: multiple gustatory pathways. Am. J. Physiol. 276, R1732–R1738.

Rowland, N. E. and Fregly, M. J. 1992. Repletion of acute sodium deficit in rats drinking either low or high concentrations of sodium chloride solution. Am. J. Physiol. Regul. Integr. Comp. Physiol. 262, R419–R425.

Scalera, G., Spector, A. C., and Norgren, R. 1995. Excitotoxic lesions of the parabrachial nuclei prevent conditioned taste aversions and sodium appetite in rats. Behav. Neurosci. 109, 997–1008.

Schneider, G. E. 1969. Two visual systems. Science 163, 895–902.

Schneider, L. H., Gibbs, J., and Smith, G. P. 1986. D-2 selective receptor antagonists suppress sucrose sham feeding in the rat. Brain Res. Bull. 17, 605–611.

Schulkin, J. 1991. Sodium Hunger: the Search for a Salty Taste. Cambridge University Press.

Sclafani, A. and Ackroff, K. 2003. Reinforcement value of sucrose measured by progressive ratio operant licking in the rat. Physiol. Behav. 79, 663–670.

Sclafani, A., Azzara, A. V., Touzani, K., Grigson, P. S., and Norgren, R. 2001. Parabrachial nucleus lesions block taste and attenuate flavor preference and aversion conditioning in rats. Behav. Neurosci. 115, 920–933.

Scott, T. R. and Giza, B. K. 1987. A measure of taste intensity discrimination in the rat through conditioned taste aversion. Physiol. Behav. 41, 315–320.

Scott, T. R. and Mark, G. P. 1986. Feeding and taste. Prog. Neurobiol. 27, 293–317.

Shaber, G. S., Brent, R. L., and Rumsey, J. A. 1970. Conditioned suppression taste thresholds in the rat. J. Comp. Physiol. Psychol. 73, 193–201.

Shuford, E., Jr. 1959. Palatability and osmotic pressure of glucose and sucrose solutions as determinants of intake. J. Comp. Physiol. Psychol. 52, 150–153.

Simon, S. A., Holland, V. F., Benos, D. J., and Zampighi, G. A. 1993. Transcellular and paracellular pathways in lingual epithelia and their influence in taste transduction. Microsc. Res. Tech. 26, 196–208.

Slotnick, B. M. 1982. Sodium chloride detection threshold in the rat determined using a simple operant taste discrimination task. Physiol. Behav. 28, 707–710.

Slotnick, B. M., Sheelar, S., and Rentmeister-Bryant, H. 1991. Transection of the chorda tympani and insertion of ear pins for stereotaxic surgery: equivalent effects on taste sensitivity. Physiol. Behav. 50, 1123–1127.

Smith, D. V. and Theodore, R. M. 1984. Conditioned taste aversions: generalization to taste mixtures. Physiol. Behav. 32, 983–989.

Smith, J. C. and Foster, D. F. 1980. Some determinants of intake of glucose + saccharin solutions. Physiol. Behav. 25, 127–133.

Smith, J. C., Davis, J. D., and O'Keefe, G. B. 1992. Lack of an order effect in brief contact taste tests with closely spaced test trials. Physiol. Behav. 52, 1107–1111.

Sollars, S. I. and Bernstein, I. L. 1992. Sodium appetite after transection of the chorda tympani nerve in Wistar and Fischer 344 rats. Behav. Neurosci. 106, 1023–1027.

Sollars, S. I. and Bernstein, I. L. 1994. Gustatory deafferentation and desalivation: effects of NaCl preference of Fischer 344 rats. Am. J. Physiol. Regul. Integr. Comp. Physiol. 266, R510–R517.

Sollars, S. I. and Hill, D. L. 1998. Taste responses in the greater superficial petrosal nerve: substantial sodium salt and amiloride sensitivities demonstrated in two rat strains. Behav. Neurosci. 112, 991–1000.

Sollars, S. I. and Hill, D. L. 2005. *In vivo* recordings from geniculate ganglia: taste response properties of individual greater superficial petrosal and chorda tympani neurones. J. Physiol. 564, 877–893.

Spector, A. C. 1995a. Gustatory function in the parabrachial nuclei: implications from lesion studies in rats. Rev. Neurosci. 6, 143–175.

Spector, A. C. 1995b. Gustatory parabrachial lesions disrupt taste-guided quinine responsiveness in rats. Behav. Neurosci. 109, 79–90.

Spector, A. C. 2000. Linking gustatory neurobiology to behavior in vertebrates. Neurosci. Biobehav. Rev. 24, 391–416.

Spector, A. C. 2003. Psychophysical Evaluation of Taste Function in Non-Human Mammals. In: Handbook of Olfaction and Gustation (*ed*. R. L. Doty), pp. 861–879. Marcel Dekker.

Spector, A. C. and Grill, H. J. 1988. Differences in the taste quality of maltose and sucrose in rats: issues involving the generalization of conditioned taste aversions. Chem. Senses 13, 95–113.

Spector, A. C. and Grill, H. J. 1992. Salt taste discrimination after bilateral section of the chorda tympani or glossopharyngeal nerves. Am. J. Physiol. Regul. Integr. Comp. Physiol. 263, R169–R176.

Spector, A. C. and Smith, J. C. 1984. A detailed analysis of sucrose drinking in the rat. Physiol. Behav. 33, 127–136.

Spector, A. C. and St. John, S. J. 1998. Role of taste in the microstructure of quinine ingestion by rats. Am. J. Physiol. 274, R1687–R1703.

Spector, A. C., Andrews-Labenski, J., and Letterio, F. C. 1990a. A new gustometer for psychophysical taste testing in the rat. Physiol. Behav. 47, 795–803.

Spector, A. C., Breslin, P., and Grill, H. J. 1988. Taste reactivity as a dependent measure of the rapid formation of conditioned taste aversion: a tool for the neural analysis of taste-visceral associations. Behav. Neurosci. 102, 942–952.

Spector, A. C., Grill, H. J., and Norgren, R. 1993a. Concentration-dependent licking of sucrose and sodium chloride in rats with parabrachial gustatory lesions. Physiol. Behav. 53, 277–283.

Spector, A. C., Guagliardo, N. A., and St. John, S. J. 1996a. Amiloride disrupts NaCl versus KCl discrimination performance: implications for salt taste coding in rats. J. Neurosci. 16, 8115–8122.

Spector, A. C., Markison, S., St. John, S. J., and Garcea, M. 1997. Sucrose vs. maltose taste discrimination by rats depends on the input of the seventh cranial nerve. Am. J. Physiol. Regul. Integr. Comp. Physiol. 41, R1210–R1228.

Spector, A. C., Norgren, R., and Grill, H. J. 1992. Parabrachial gustatory lesions impair taste aversion learning in rats. Behav. Neurosci. 106, 147–161.

Spector, A. C., Redman, R., and Garcea, M. 1996b. The consequences of gustatory nerve transection on taste-guided licking of sucrose and maltose in the rat. Behav. Neurosci. 110, 1096–1109.

Spector, A. C., Scalera, G., Grill, H. J., and Norgren, R. 1995. Gustatory detection thresholds after parabrachial nuclei lesions in rats. Behav. Neurosci. 109, 939–954.

Spector, A. C., Schwartz, G. J., and Grill, H. J. 1990b. Chemospecific deficits in taste detection after selective gustatory deafferentation in rats. Am. J. Physiol. Regul. Integr. Comp. Physiol. 258, R820–R826.

Spector, A. C., Travers, S. P., and Norgren, R. 1993b. Taste receptors on the anterior tongue and nasoincisor ducts of rats contribute synergistically to behavioral responses to sucrose. Behav. Neurosci. 107, 694–702.

St. John, S. J. and Smith, D. V. 2000. Neural representation of salts in the rat solitary nucleus: brain stem correlates of taste discrimination. J. Neurophysiol. 84, 628–638.

St. John, S. J. and Spector, A. C. 1996. Combined glossopharyngeal and chorda tympani nerve transection elevates quinine detection thresholds in rats (*Rattus norvegicus*). Behav. Neurosci. 110, 1456–1468.

St. John, S. J. and Spector, A. C. 1998. Behavioral discrimination between quinine and KCl is dependent on input from the seventh cranial nerve: implications for the functional roles of the gustatory nerves in rats. J. Neurosci. 18, 4353–4362.

St. John, S. J., Garcea, M., and Spector, A. C. 1994. Combined, but not single, gustatory nerve transection substantially alters taste-guided licking behavior to quinine in rats. Behav. Neurosci. 108, 131–140.

St. John, S. J., Markison, S., Guagliardo, N. A., Hackenberg, T. D., and Spector, A. C. 1997. Chorda tympani transection and selective desalivation differentially disrupt two-lever salt discrimination performance in rats. Behav. Neurosci. 111, 450–459.

St. John, S. J., Pour, L., and Boughter, J. D., Jr. 2005. Phenylthiocarbamide produces conditioned taste aversions in mice. Chem. Senses 30, 377–382.

Stapleton, J. R., Luellig, M., Roper, S. D., and Delay, E. R. 2002. Discrimination between the tastes of sucrose and monosodium glutamate in rats. Chem. Senses 27, 375–382.

Stebbins, W. C. and Berkley, M. A. 1990. Comparative Perception: Complex Signals. John Wiley & Sons.

Tapper, D. N. and Halpern, B. P. 1968. Taste stimuli: a behavioral categorization. Science 161, 708–710.

Thaw, A. K. and Smith, J. C. 1992. Conditioned suppression as a method of detecting taste thresholds in the rat. Chem. Senses 17, 211–223.

Travers, J. B., DiNardo, L. A., and Karimnamazi, H. 1997. Motor and premotor mechanisms of licking. Neurosci. Biobehav. Rev. 21, 631–647.

Travers, J. B., Urbanek, K., and Grill, H. J. 1999. Fos-like immunoreactivity in the brain stem following oral quinine stimulation in decerebrate rats. Am. J. Physiol. Regul. Integr. Comp. Physiol. 277, R384–R394.

Travers, J., DiNardo, L., and Karimnamazi, H. 2000. Medullary reticular formation activity during ingestion and rejection in the awake rat. Exp. Brain Res. 130, 78–92.

Travers, S. P. 1993. Orosensory Processing in Neural Systems of the Nucleus of the Solitary Tract. In: Mechanisms of Taste Transduction (*eds*. S. A. Simon and S. D. Roper), pp. 339–393. CRC Press.

Travers, S. P. and Nicklas, K. 1990. Taste bud distribution in the rat pharynx and larynx. Anatom. Rec. 227, 373–379.

Van Buskirk, R. L. 1981. The role of odor in the maintenance of flavor aversion. Physiol. Behav. 27, 189–193.

Vance, W. B. 1967. Hypogeusia and taste preference behavior in the rat. Life Sci. 6, 743–748.

Weingarten, H. P. and Watson, S. D. 1982. Sham feeding as a procedure for assessing the influence of diet palatability on food intake. Physiol. Behav. 28, 401–407.

Willner, P., Papp, M., Phillips, G., Maleeh, M., and Muscat, R. 1990. Pimozide does not impair sweetness discrimination. Psychopharmacology 102, 278–282.

Wolf, G. 1982. Refined salt appetite methodology for rats demonstrated by assessing sex differences. J. Comp. Physiol. Psychol. 96, 1016–1021.

Yamamoto, T. and Asai, K. 1986. Effects of gustatory deafferentation on ingestion of taste solutions as seen by licking behavior in rats. Physiol. Behav. 37, 299–305.

Yamamoto, T. and Fujimoto, Y. 1991. Brain mechanisms of taste aversion learning in the rat. Brain Res. Bull. 27, 403–406.

Ye, Q., Heck, G. L., and DeSimone, J. A. 1993. Voltage dependence of the rat chorda tympani response to Na^+ salts: implications for the functional organization of taste receptor cells. J. Neurophysiol. 70, 167–178.

Young, P. T. 1952. The role of hedonic processes in the organization of behavior. Psychol. Rev. 59, 249–257.

Young, P. T. and Madsen, C. H., Jr. 1963. Individual isohedons in sucrose-sodium chloride and sucrose-saccharin gustatory areas. J. Comp. Physiol. Psychol. 56, 903–909.

Young, P. T. and Trafton, C. L. 1964. Activity contour maps as related to preference in four gustatory stimulus areas of the rat. J. Comp. Physiol. Psychol. 58, 68–75.

Zhang, Y., Hoon, M. A., Chandrashekar, J., Mueller, K. L., Cook, B., Wu, D., Zuker, C. S., and Ryba, N. J. 2003. Coding of sweet, bitter, and umami tastes: different receptor cells sharing similar signaling pathways. Cell 112, 293–301.

4.23 Flavor Aversion Learning

I L Bernstein, University of Washington, Seattle, WA, USA

Glossary

c-Fos protein Product of immediate early gene c-fos, regulates the induction of downstream target genes as a transcription factor.

conditioned stimulus (CS) A neutral stimulus which comes to be associated, after training, with an unconditioned stimulus.

emetic Causing nausea and/or vomiting.

latent inhibition A process by which preexposure to a stimulus without consequence retards the learning of subsequent conditioned associations with that stimulus.

N-methyl-D-aspartate (NMDA) receptors One class of glutamate receptors which potentiates synapses in several neural pathways and is

believed to be importantly involved in neural plasticity.

phosphorylation The addition of a phosphate to a protein or a small molecule which can switch on or off cellular regulatory events.

protein kinase A cAMP-dependent protein kinase, also known as protein kinase A (PKA), refers to a family of enzymes whose activity is dependent on the level of cyclic AMP (cAMP) in the cell.

unconditioned stimulus (US) A stimulus used in classical conditioning which produces an unconditioned response without any learning.

Flavor cues such as taste, odor, and texture, are critical to food selection. These cues contribute to the palatability of food but they also play a protective role by discouraging the consumption of dangerous edibles. Flavor aversion refers to the avoidance of foods or fluids based on these cues. Aversions can be innate or they can be learned. Innate aversions, as exemplified by bitter avoidance, do not require prior experience, probably evolved to deter consumption of toxic chemicals that could be detected by their taste, and tend to be shared by members of a species. Learned aversions are those that are based on experience, such as consumption of a food prior to a bout of illness or discomfort. In such cases, illness can be a consequence of ingestion or coincidentally associated with it, such as a food eaten just before the onset of a bout of gastroenteritis (stomach flu). Among humans learned aversions tend to

be idiosyncratic; that is, individuals differ from one another with respect to the foods that happen to become associated with the malaise.

The topic of this chapter is the sort of aversions which are acquired; those commonly referred to as learned taste aversions. A brief description of the distinctive characteristics of this learning, a review of what is known about the neural mechanisms that underlie it, and a summary of some clinical implications will be provided.

4.23.1 General Description

Animals and humans learn to avoid foods which may have made them sick, and they do this with remarkable facility. This learning is a form of associative conditioning in which an individual comes to avoid a taste (conditioned stimulus (CS)) which has previously been paired with a treatment that produces transient illness (unconditioned stimulus (US)). Taste aversion learning is unusually robust and has clear adaptive advantages, particularly for omnivorous species which must select from a range of nutrients, some of which may be toxic. However, conditioned taste aversions (CTAs) can also be maladaptive, as in the case where safe, nutritious foods are avoided because of a chance association with symptoms of nausea or malaise. In cases such as gastroenteritis or cancer chemotherapy, the symptoms are not causally related to the food in question but the association and consequent distaste develops anyway (Bernstein, I. L. and Borson, S., 1986).

Several features of taste aversion learning are striking when they are compared to other learning paradigms (Riley, A. L., 2005). CTAs occur rapidly and potently under conditions that could not support other types of learning. Virtually every other conditioning paradigm requires close temporal proximity between presentation of CS and US. CTAs are routinely acquired after delays of several minutes or an hour between exposure to CS taste and US illness (Garcia, J. et al., 1966). When one considers that long delay learning of taste aversions is often displayed after only one conditioning trial, the unusual potency of this learning is quite impressive. The requirement for temporal contiguity, with an optimal range being 500 ms to 2 s is so common a feature of associate learning paradigms that it has been used to build models of the types of cellular signaling processes that might underlie plasticity (Abrams, T. W. and Kandel, E. R., 1988). In this regard taste aversion learning is clearly an anomaly. This raises the question of how neural signaling in CTA circuits bridge a CS–US intervals of minutes and hours (Koh, M. T. et al., 2003). The degree to which similar or unique mechanisms of neural plasticity are recruited and/or modified for this task remains to be determined.

The robustness of food aversion learning is also underscored by the findings of Roll D. L. and Smith J. C. (1972) that rats will acquire aversions even when they are completely anesthetized during and after the time when the US is administered. These findings, replicated and extended by Bermúdez-Rattoni F. et al. (1988), imply that these associations can be formed while unconscious.

In the laboratory, taste aversion learning has most often been studied in rats. This is likely to have been a fortuitous choice since rats have proven to be remarkably good at taste aversion learning. This may be due

to two characteristics of this species; first, they are omnivores and opportunistic foragers which means that in their natural environment they frequently sample new foods and are thus at risk for encountering toxins. Second, they are unable to vomit which means that once they eat something toxic they cannot dispose of it easily. For these reasons, this species may have been under particularly strong selective pressure to develop a robust and efficient taste aversion conditioning mechanism as a means of toxin avoidance. But rats are certainly not unique in this ability. Taste aversion learning has been demonstrated in an enormous range of animal species, from the very simple, such as the garden slug, *limax* (Sahley, C. L. et al., 1981), to the complex, such as humans (Garb, J. L. and Stunkard, A. J., 1974; Bernstein, I. L., 1978; Logue, A. W. et al., 1981; Midkiff, E. E. and Bernstein, I. L., 1985; Bernstein, I. L. and Borson, S., 1986).

4.23.2 Conditioned Taste Aversion: The Conditioned Stimulus

4.23.2.1 Salience of Taste Cues

Typical CTA training consists of the presentation of a palatable taste solution, such as dilute saccharin, followed by the injection of a nonlethal toxin. In untrained rats saccharin is highly preferred and consumed with enthusiasm. This makes the behavior of the conditioned animals particularly striking, because they avoid the saccharin and appear to actually be disgusted by it.

The earliest characterizations of taste aversion learning emphasized the ease with which taste cues were associated with subsequent illness, and the relative difficulty of associating cues in other sensory modalities (e.g., sounds and sights) with illness (Garcia, J. et al., 1974). This makes sense from an adaptive perspective as foods, and the sensations associated with their ingestion, are the most likely culprits when unpleasant gastrointestinal consequences arise. This left open the question of the salience of other food-related sensations, especially olfaction. When we commonly refer to the flavors of foods and fluids, we conflate taste and odor cues so that taste aversions may actually be flavor aversions, aversions to both the taste and odor of a food. In the laboratory, the contribution of these two modalities can be assessed independently. In rats, odor cues, presented without tastes, are much less effective than taste cues as CSs in aversion conditioning (Palmerino, C. C. et al., 1980). In contrast, if an odor is conditioned in compound with a taste a phenomenon referred to as taste potentiation of odor is observed. That it, after conditioning the compound (taste plus odor), when the odor is tested in isolation, it is clear that it has become a highly potent cue for avoidance. Under most normal circumstances, then, odors may be important cues in aversion conditioning. Furthermore, it should be noted that reliance on odor cues allows animals to actually avoid ingestion of a toxic or suspicious substance, whereas reliance on taste cues require some sampling and that can be risky.

4.23.2.2 Salience of Novel Tastes

The robust and rapid CTA acquisition which has been described relies heavily on a CS taste characteristic that has not been mentioned yet, that is, the novelty of the taste. Aversions arise extremely rapidly if a CS food or taste is novel, but if an organism has had prior, safe exposure to the food, conditioning is significantly retarded (Revusky, S. H. and Bedarf, E. W., 1967; Garcia, J. et al., 1974) or prevented completely (Koh, M. T. and Bernstein, I. L., 2005). In the laboratory, one or two safe exposures to a taste prior to conditioning can dramatically attenuate learning.

This characteristic of taste aversion learning is dramatic but not unique. An analogous phenomenon has been described in other associative conditioning paradigms and the general process whereby CS pre-exposure retards conditioning is referred to as latent inhibition (Lubow, R. E., 1973). CTA acquisition shares additional characteristics with other associative learning paradigms. Conditioning strength is positively correlated with CS intensity, US intensity, and number of CS–US pairings. CTAs display generalization and, although very durable, CTAs are subject to extinction (Riley, A. L., 2005).

4.23.3 Conditioned Taste Aversions: The Unconditioned Stimulus

Laboratory studies of taste aversion learning commonly involve administration of a drug, such as lithium chloride (LiCl), as the US. LiCl, and many other US drugs, are known to act as emetics. This might lead one to assume that if a treatment is capable of acting as a US in taste aversion conditioning it probably causes nausea, or at least produces unpleasant symptoms. This is almost certainly not the case. Instead, the CTA literature yields a remarkably varied list of drugs and other treatments which are effective USs in conditioning (Riley, A. L., 2005).

Careful examination of these treatments does not reveal particular feature(s) that they have in common (Gamzu, E. et al., 1985). Although emetic drugs are highly effective as USs, many drugs that are not emetic can be effective (Grant, V. L., 1987; Goudie, 1979). Even more puzzling, some effective drugs do not appear to be particularly aversive. A striking demonstration of this is the fact that drugs which humans and animals will self-administer (Hunt, T. and Amit, Z., 1987) can support CTAs. So the claim that taste aversion conditioning can serve as a behavioral assay of aversiveness or toxicity should be met with skepticism.

Although it is clear that nausea is not necessary for the development of food aversions, there is compelling evidence that nausea may play a unique and particularly potent role in this learning. Evidence both from humans and rats (Pelchat, M. L. and Rozin, P., 1982; Pelchat, M. et al., 1983) suggests that the prototypical learned food aversion, where the food becomes genuinely distasteful, generally is associated with symptoms of nausea. One demonstration of this point emerged from a survey undertaken by Pelchat M. L. and Rozin P. (1982) of people who suffered from food allergies. When allergic symptoms included nausea, subjects reported not only that they avoided the food, but that they actually disliked the taste. In contrast, when the allergic response included symptoms such as mouth sores or hives, subjects avoided the food but reported no change in the food's hedonic rating. In other words these subjects avoided eating the food because it would cause unpleasant symptoms, not because they actually found it distasteful. Thus, Pelchat M. L. and Rozin P. (1982) argue that nausea is not only sufficient for the establishment of a taste aversion but is actually necessary for the development of hedonic shifts or distaste toward foods. It is this shift in the hedonic or incentive value of a taste that, some believe, constitutes a genuine learned taste aversion.

4.23.3.1 Neural Mediation of Food Aversion Learning

When humans develop aversions as a result of the coincidental association between consumption of a food and chemotherapy or symptoms of gastroenteritis, the aversions appear to defy cognition. That is, strong aversions arise despite a person's awareness that the target food was not actually the cause of their illness. These findings suggest that food aversion learning is based on relatively primitive associative mechanisms perhaps involving neural integration occurring in lower brain areas such as the brainstem.

Taste and visceral information converge in the brain stem, within the nucleus of the solitary tract (NTS) although there is little or no overlap between the regions responsive to gustatory and visceral input (Travers, J. B. *et al.*, 1987). Gustatory neurons from the facial, glossopharyngeal, and vagus nerves terminate in the rostral portion of the NTS. The NTS would, therefore, appear to be a critical part of the pathway mediating taste aversion learning. However, Grigson P. S. *et al.* (1997a) found that rats with extensive lesions of the rostral (gustatory) zone of the NTS demonstrated normal taste aversion learning. Since reception of gustatory information would appear to be obligatory to taste aversion learning, one might assume the NTS lesions, though extensive, did not completely eliminate incoming gustatory signals and that a degraded signal is sufficient for the learning.

In contrast to the rather limited role indicated for NTS, the next site in the ascending gustatory and visceral relay, the pontine parabrachial nucleus (PBN), appears to be crucial for taste aversion acquisition. Rats with lesions of the PBN are unable to acquire a CTA (Reilly, S. *et al.*, 1993). This deficit appears to be quite specific. It does not appear to be due to an inability to taste the CS or experience the US but rather to an inability to associate the two stimuli at the time of conditioning. Also, if conditioning occurs prior to lesioning, PBN lesions do not disrupt CTA expression (Grigson, P. S. *et al.*, 1997b). Thus, the PBN appears necessary for taste aversion acquisition but not expression.

Although taste and visceral information converge within the brainstem and pons, these regions do not appear to be sufficient for the integration needed for taste aversion learning. The chronic decerebrate rat has been shown to be incapable of forming the associations necessary for taste aversion learning (Grill, H. J. and Norgren, R., 1978) which suggests that integration of taste and visceral inputs within the brainstem of the rat is insufficient to support the acquisition or expression of taste aversions. A variety of studies indicate that forebrain structures within the ascending gustatory projection are involved in taste aversion learning (Chambers, K. C., 1990; Yamamoto, T. and Fujimoto, Y., 1991). Two regions of particular interest are the amygdala and gustatory (insular) cortex.

Studies examining the effects of amygdala lesions on taste aversion learning have been somewhat inconsistent. Some of the inconsistency may be attributable to differences in lesion method as well as details of the conditioning protocol (Schafe, G. E. *et al.*, 1998; Morris, R. *et al.*, 1999). Overall, a role for the amygdala in CTA acquisition and expression has been supported by lesion studies as well as studies using methods which lack the interpretational limitations of the lesion approach (Schafe, G. E. and Bernstein, I. L., 1996; Koh, M. T. *et al.*, 2003; Koh, M. T. and Bernstein, I. L., 2005). For example, inhibition of immediate early gene expression and protein kinase A signaling within the amygdala has been demonstrated to interfere with CTA acquisition (Lamprecht, R. and Dudai, Y., 1996; Koh, M. T. *et al.*, 2002).

Like the amygdala, the insular cortex (IC) has strong reciprocal connections with NTS and PBN (Saper, C. B., 1982) and lesion studies have been quite consistent in demonstrating attenuation of taste aversion learning after destruction of this area (Lasiter, P. S. and Glanzman, D. L., 1982; 1985; Bermúdez-Rattoni, F. and McGaugh, J. L., 1991). Furthermore, reversible interference with cholinergic function or blockade of *N*-methyl-D-aspartate (NMDA) receptors in rat IC disrupts taste aversion learning (Naor, C. and Dudai, Y., 1996), and protein tyrosine phosphorylation is altered in the IC following training (Rosenblum, K. *et al.*, 1995). Combining IC lesions or inactivation with staining for c-Fos protein also provides evidence for a dynamic role for the IC in CTA learning. Collectively, results support an important role for this region in taste memory and taste aversion learning.

Thus, lesion studies and new molecular techniques, as well as neuroanatomical approaches, have begun to define the pathways involved in processing taste and visceral signals, and in forming the durable associations between them which are the basis of a CTA. The cellular mechanisms which underlie these associations have also begun to be identified. A recent review provides an intriguing theoretical perspective on these processes (Bermúdez-Rattoni, F., 2004).

4.23.4 Food Aversions in Humans

Food aversion learning in humans has been examined principally through the use of survey methods (Garb, J. L. and Stunkard, A. J., 1974; Logue, A. W. *et al.*, 1981; Midkiff, E. E. and Bernstein, I. L., 1985). Characteristics of food aversions reported by humans are generally similar to those observed in laboratory studies of rats. Acquisition often occurs in a single trial, frequently with long delays interposed between tasting the food and experiencing illness. Aversions were more likely to develop to less familiar as well as less preferred foods. Extinction was reported to be more effective at reducing aversions than forgetting. This raises the question of what role this powerful and primitive learning plays in human food choice.

The extent to which food aversions, learned as a consequence of the association of specific foods with gastrointestinal symptoms, contribute to overall food choice is not simple to determine. Some who have evaluated this issue have concluded that the role of aversion conditioning is minor. They base this on the following reasoning. Many studies have shown that a majority of people surveyed report having experienced the development of a food aversion at some time in their life. However, if the number of people reporting existing food aversions and the number of foods they avoid due to these experiences are calculated, this number is small relative to the large number of foods people seem to dislike which leaves a vast number of food aversions unaccounted for in terms of their cause (Rozin, P. and Vollmecke, T. A., 1986).

Another line of reasoning leads to a different conclusion. There are good reasons to believe that verbal reports of learned food aversions underestimate the actual number of such aversions. Memory researchers distinguish between implicit and explicit memories (Squire, L. R. *et al.*, 1993). Explicit memories are those you are aware of consciously while implicit memories are those you have but are not consciously aware of. Since classical conditioning is more likely to generate implicit than explicit memories, most food aversion learning experiences in humans may form implicit, not explicit, memories. Research indicates that implicit memory formation is less susceptible to anesthesia (Andrade, J., 1995) which might explain the resistance of food aversion learning to anesthesia (Roll, D. L. and Smith, J. C., 1972; Bermúdez-Rattoni, F. *et al.*, 1988). Thus, if all but the most severe learned food aversions form implicit not explicit memories, subjects' recall of taste aversion experiences may not be the most sensitive way to assess aversion conditioning. If this is true then conditioned food aversions may affect food choices and hedonic responses to foods unconsciously and such conditioning may play a significant role in human food choice but this role would be quite difficult to assess methodologically. Also some people who are chronically ill, prone to motion sickness, or otherwise vulnerable to experiencing unpleasant gastrointestinal symptoms may be at a much higher risk of developing these aversions and having them limit their acceptable food choices (Bernstein, I. L. and Borson, S., 1986).

4.23.5 Clinical Implications

As previously noted, learned taste aversions often arise when the consumption of a food or drink is followed by nausea or gastrointestinal malaise. Sometimes the illness is actually caused by the food or beverage consumed, but it is frequently the case that the food and illness are only coincidentally associated. An example of aversions in which there is indeed a causal relationship between CS and US is the common occurrence of aversions to specific alcoholic beverages, such as bourbon or tequila, which were consumed in excess and which led to nausea and vomiting. Aversions to specific alcoholic beverages frequently appear in surveys done with college students, for obvious reasons (Logue, A. W. *et al.*, 1981; Midkiff, E. E. and Bernstein, I. L., 1985). Another situation in which food aversions develop is in patients receiving cancer chemotherapy (Bernstein, I. L., 1978). Drugs used to treat cancer, while frequently effective in halting the growth of tumor cells, can have severe side effects including nausea and vomiting. Cancer patients often develop aversions to specific foods they consume before these treatments, despite their recognition that their chemotherapy and not their lunch was the cause of these symptoms (Bernstein, I. L. and Borson, S., 1986). Recognition that cancer patients are at risk for the development of learned food aversions and that this can affect their appetite and nutritional status has led to interventions which can prevent the development of these aversions (Broberg, D. J. and Bernstein, I. L., 1987).

The rapid acquisition and robustness of learned taste aversions have made them an attractive tool in therapeutic efforts to modify problem behaviors such as alcoholism and smoking. In point of fact, a natural consequence of excessive alcohol consumption is illness, including severe nausea and vomiting. As noted above, these symptoms are effective in inducing aversions to specific alcoholic beverages in many college students. Do alcoholics fail to acquire such aversions? If they acquire such aversions, how do the aversions subsequently affect their drinking? A survey of alcoholics in a treatment facility indicated that alcoholics do acquire taste aversions to alcoholic beverages as a consequence of overconsumption. However, when a taste aversion to a specific alcoholic beverage was acquired, alcoholics avoided that particular beverage, choosing some other alcoholic beverages instead (Logue, A. W. *et al.*, 1983). Chemical aversion therapy involves the use of drugs, such as emetine, to induce nausea and vomiting after alcohol consumption. This form of therapy has been incorporated into a range of alcoholism treatment packages. Available outcome data from a number of these programs indicate that the effectiveness of their treatment packages in promoting abstinence is quite good. However, the contribution of chemical aversion therapy, *per se*, to the effectiveness of the entire treatment package has not been explicitly evaluated and therefore it remains unknown. Aversion therapy has also been used as an aid to smoking cessation. This involves a rapid smoking procedure which conforms to the taste aversion learning paradigm in that it involves pairing the flavor cues associated with smoking cigarettes with aversive symptoms such as nausea and headache. This procedure has been reported to be effective in producing aversions to the taste and smell of cigarettes and to improve abstinence rates in smoking cessation programs.

References

Abrams, T. W. and Kandel, E. R. 1988. Is contiguity detection in classical conditioning a system or a cellular property? Learning in Aplysia suggests a possible molecular site. Trends Neurosci. 11, 128–135.

Andrade, J. 1995. Learning during anaesthesia: a review. Br. J. Psychol. 86, 479–506.

Bermúdez-Rattoni, F. 2004. Molecular mechanisms of taste-recognition memory. Nat. Rev. Neurosci. 5, 209–217.

Bermúdez-Rattoni, F. and McGaugh, J. L. 1991. Insular cortex and amygdala lesions differentially affect acquisition on inhibitory avoidance and conditioned taste aversion. Brain Res. 549, 165–170.

Bermúdez-Rattoni, F., Forthman, D. L., Sanchez, M. A., Perez, J. L., and Garcia, J. 1988. Odor and taste aversions conditioned in anesthetized rats. Behav. Neurosci. 102, 726–732.

Bernstein, I. L. 1978. Learned taste aversions in children receiving chemotherapy. Science 200, 1302–1303.

Bernstein, I. L. and Borson, S. 1986. Learned food aversion: a component of anorexia syndromes. Psychol. Rev. 93, 462–472.

Broberg, D. J. and Bernstein, I. L. 1987. Candy as a scapegoat in the prevention of food aversions in children receiving chemotherapy. Cancer 60, 2344–2347.

Chambers, K. C. 1990. A neural model for conditioned taste aversions. Annu. Rev. Neurosci. 13, 373–385.

Gamzu, E., Vincent, G., and Boff, E. 1985. A pharmacological perspective of drugs used in establishing conditioned food aversions. Ann. N. Y. Acad. Sci. 443, 231–249.

Garb, J. L. and Stunkard, A. J. 1974. Taste aversions in man. Am. J. Psychiatry 131, 1204–1207.

Garcia, J., Ervin, R. R., and Koelling, R. A. 1966. Learning with prolonged delay of reinforcement. Psychon. Sci. 5, 121–122.

Garcia, J., Hankins, W. G., and Rusiniak, K. W. 1974. Behavioral regulation of the milieu interne in man and rat. Science 185, 824–831.

Goudie, A. J. 1979. Aversine stimulus properties of drugs. Neuropharmacology 18, 971–979.

Grant, V. L. 1987. Do conditioned taste aversions result from activation of emetic mechanisms? Psychopharmacology 93, 405–415.

Grigson, P. S., Shimura, T., and Norgren, R. 1997a. Brainstem lesions and gustatory function: II. The role of the nucleus of the solitary tract in NA+ appetite, conditioned taste aversion and conditioned odor aversion in rats. Behav. Neurosci. 111, 169–179.

Grigson, P. S., Shimura, T., and Norgren, R. 1997b. Brainstem lesions and gustatory function: III. The role of the nucleus of the solitary tract and the parabrachial nucleus in retention of a conditioned taste aversion in rats. Behav. Neurosci. 111, 180–187.

Grill, H. J. and Norgren, R. 1978. Chronically decerebrate rats demonstrate satiation but not bait shyness. Science 201, 267–269.

Hunt, T. and Amit, Z. 1987. Conditioned taste aversion induced by self-administered drugs: paradox revisited. Neurosci. Biobehav. Rev. 11, 107–130.

Koh, M. T. and Bernstein, I. L. 2005. Mapping conditioned taste aversion associations using c-fos reveals a dynamic role for insular cortex. Behav. Neurosci. 119, 388–398.

Koh, M. T., Thiele, T. E., and Bernstein, I. L. 2002. Inhibition of protein kinase A activity interferes with long-term, but not short-term, memory of conditioned taste aversions. Behav. Neurosci. 116, 1070–1074.

Koh, M. T., Wilkins, E. E., and Bernstein, I. L. 2003. Novel taste elevates c-fos expression in the central amygdala and insular cortex: implication for taste aversion learning. Behav. Neurosci. 117, 1416–1422.

Lamprecht, R. and Dudai, Y. 1996. Transient expression of c-fos in rat amygdala during training is required for encoding conditioned taste aversion memory. Learn. Memory 3, 31–41.

Lasiter, P. S. and Glanzman, D. L. 1982. Cortical substrates of taste aversion learning: dorsal prepiriform (insular) lesions disrupt taste aversion learning. J. Comp. Physiol. Psychol. 96, 376–392.

Lasiter, P. S. and Glanzman, D. L. 1985. Cortical substrates of taste aversion learning: Involvement of dorsolateral amygdaloid nuclei and temporal neocortex in taste aversion learning. Behav. Neurosci. 99, 257–276.

Logue, A. W., Logue, K. R., and Strauss, K. E. 1983. The acquisition of taste aversions in humans with eating and drinking disorders. Behav. Res. Ther. 21, 275–289.

Logue, A. W., Ophir, I., and Strauss, K. E. 1981. The acquisition of taste aversions in humans. Behav. Res. Ther. 19, 319–333.

Lubow, R. E. 1973. Latent inhibition. Psychol. Bull. 79, 398–407.

Midkiff, E. E. and Bernstein, I. L. 1985. Targets of learned food aversions in humans. Physiol. Behav. 34, 839–841.

Morris, R., Frey, S., Kasambira, T., and Petrides, M. 1999. Ibotenic acid lesions of the basolateral, but not the central, amygdala interfere with conditioned taste aversion: evidence from a combined behavioral and anatomical tract-tracing investigation. Behav. Neurosci. 113, 291–302.

Naor, C. and Dudai, Y. 1996. Transient impairment of cholinergic function in the rat insular cortex disrupts the encoding of taste in conditioned taste aversion. Behav. Brain Res. 79, 61–67.

Palmerino, C. C., Rusiniak, K. W., and Garcia J. 1980. Flavor illness aversions: the peculiar role of odor and taste in memory for poison. Science 208, 753–755.

Pelchat, M. L. and Rozin, P. 1982. The special role of nausea in the acquisition of food dislikes by humans. Appetite 3, 341–351.

Pelchat, M., Grill, H. J., Rozin, P., and Jacobs, J. 1983. Quality of acquired responses to tastes by *Rattus norvegicus* depends upon type of associated discomfort. J. Comp. Physiol. Psychol. 97, 140–153.

Reilly, S., Grigson, P. S., and Norgren, R. 1993. Parabrachial nucleus lesions and conditioned taste aversion: evidence supporting an associative deficit. Behav. Neurosci. 107, 1005–1017.

Revusky, S. H. and Bedarf, E. W. 1967. Association of illness with prior ingestion of novel foods. Science 155, 219–220.

Riley, A. L. 2005. Conditioned Taste Aversion: An Annotated Bibliography. http://ctalearning.com.

Roll, D. L. and Smith, J. C. 1972. Conditioned Taste Aversion in Anesthetized Rats. In: Biological Boundaries of Learning (*eds*. M. E. P. Seligman and J. L. Hager), pp. 98–102. Appleton-Century-Crofts.

Rosenblum, K., Schul, R., Meiri, N., Hadari, Y. R., Zick, Y., and Dudai, Y. 1995. Modulation of protein tyrosine phosphorylation in rat insular cortex after conditioned taste aversion training. Proc. Natl. Acad. Sci. U. S. A. 92, 1157–1161.

Rozin, P. and Vollmecke, T. A. 1986. Food likes and dislikes. Annu. Rev. Nutr. 6, 433–456.

Sahley, C. L., Gelperin, A., and Rudy, J. 1981. One-trial associative learning modifies food odor preferences of a terrestrial mollusc. Proc. Natl. Acad. Sci. U. S. A. 78, 640–642.

Saper, C. B. 1982. Reciprocal parabrachial-cortical connections in the rat. Brain Res. 52, 91–97.

Schafe, G. E. and Bernstein, I. L. 1996. Forebrain contribution to the induction of a brainstem correlate of conditioned taste aversion: I. The amygdala. Brain Res. 741, 109–116.

Schafe, G. E., Thiele, T. E., and Bernstein, I. L. 1998. Conditioning method dramatically alters the role of amygdala in taste aversion learning. Learn. Memory 15, 481–492.

Squire, L. R., Knowlton, B., and Musen, G. 1993. The structure and organization of memory. Annu. Rev. Psychol. 44, 453–495.

Travers, J. B., Travers, S. P., and Norgren, R. 1987. Gustatory neural processing in the hindbrain. Annu. Rev. Neurosci. 10, 595–632.

Yamamoto, T. and Fujimoto, Y. 1991. Brain mechanisms of taste aversion learning in the rat. Brain Res. Bull. 27, 403–406.

Further Reading

Bernstein, I. L. 1991. Flavor Aversion. In: Smell and Taste in Health and Disease (*eds*. T. V. Getchell, R. L. Doty, L. M. Bartoshuk, and J. B. Snow), pp. 417–428. Raven Press.

Yamamoto, T., Shimura, T., Sako, N., Yasoshima, Y., and Sakai, N. 1994. Neural substrates for conditioned taste aversion in the rat. Behav. Brain Res. 65, 123–137.

4.24 Roles of Taste in Feeding and Reward

T Yamamoto and T Shimura, Osaka University, Osaka, Japan

Glossary

conditioned taste aversion (CTA) Avoidance and aversion to a substance after an association learning between the ingestion of the substance (CS) and subsequent malaise (US) with gastrointestinal disorders and nausea. Animals remember the taste for a long time and reject its ingestion at subsequent exposures. CTA has unique characteristics such that strong and long-lasting memory can be acquired by a single pairing of the CS and US even with a long (up to several hours) CS–US interval.

When we take a bite of our favorite food, we experience a variety of sensations, and almost simultaneously we will smile and perceive it to be delicious. We will then be motivated to have more bites, the jaws and tongue move rhythmically with salivary secretion and active gastrointestinal functions to ingest the food. Eventually, the ingestive behavior finishes with the satisfaction of feeling full. Sometimes if something is palatable, we will overeat. Some of our favorite foods may have been innately determined, e.g., cakes and chocolates with innately preferred sweet tastes, but others are acquired after good experiences, or as a more specific example, on the basis of association learning between taste perception and nutritive postingestional effects. Conversely, even favorite foods can become aversive and avoided as a consequence of an unpleasant experience including postingestional malaise. Such taste-mediated behaviors can be commonly observed in everyone. In this chapter, we review the literature regarding these different aspects of pleasantness and rewarding eating behavior and discuss the underlying anatomy, neural substrates, and chemical mediators in the brain, learning and memory processes, and finally the activation patterns in the human brain.

4.24.1   Central Pathways Involved in Hedonic Evaluation

4.24.1.1   Inputs from the Gustatory Pathway to Reward System and Feeding Center

Central gustatory pathways have been well studied and documented in monkeys (Rolls, E. T., 2004) and rodents, especially rats (Norgren, R., 1995). Figure 1 shows a schematic diagram of some of the gustatory pathways in rats. Branches of the facial, glossopharyngeal, and vagus nerves, which synapse with receptor cells in the taste buds, convey taste messages to the first relay nucleus, the rostral part of the nucleus tractus solitarius (NTS). The second relay nucleus for ascending taste inputs is the parabrachial nucleus (PBN) of the pons. The third relay nucleus is the medial parvocellular component of the ventrobasal complex of the thalamus (VPMpc). This thalamic nucleus projects to the gustatory cortex (GC), the cortical gustatory area in the insular cortex (IC). Other ascending projections from the PBN are to the lateral hypothalamus (LH) (not shown in Figure 1), the central nucleus of the amygdala, and the bed nucleus of the stria terminalis (not shown in Figure 1). In monkeys, however, ascending fibers of neurons in the gustatory area of the NTS directly reach the VPMpc, bypassing the PBN (Beckstead, R. M. et al., 1980). It is known that general visceral inputs also project in a similar fashion to the brain regions in parallel with the gustatory projections described above.

The neural pathway of the brain reward system has also been studied and documented (Berridge, K. C. and Robinson, T. E., 1998; Wise, R. A., 2002 for reviews). As shown in Figure 1 and described in more detail in the following sections in this chapter, the essential components are the ventral tegmental area (VTA) in the midbrain, the origin of the mesolimbic dopamine system, the nucleus accumbens (Acc), an essential interface between motivation (e.g., palatability) and action (e.g., feeding), and the ventral pallidum (VP) situated between the Acc and LH, known as the feeding center.

When we consider the roles of taste in feeding and reward, we have to make clear how the taste system interacts with the reward and feeding systems. However, this has not yet been well clarified. The amygdala (AMY), the prefrontal cortex (PFC) including the ventrolateral (or anterior sulcal) and dorsomedial cortices, and IC are the candidates for the interfaces between the two systems. The gustatory insular cortex sends axons to the PFC (Saper, C. B., 1982; Shi, C.-J. and Cassell, M. D., 1998), and the dorsomedial PFC neurons actually respond to gustatory stimuli (Lukáts, B. et al., 2002; Karádi, Z. et al., 2005). Among other structures, the PFC is interconnected with the feeding-related subcortical areas such as the basal forebrain (Divac, I. et al., 1978), amygdala (Pérez-Jaranay, J. M. and Vives, F., 1991), LH (Kita, H. and Oomura, Y., 1981), VTA (Divac, I. et al., 1978; Kosobud, A. E. et al., 1994), and Acc (Brog, J. S. et al., 1993). Behavioral studies have shown that the PFC is associated with various mechanisms in the central control of feeding, for example, lesions of the dorsomedial PFC result in finickiness (Kolb, B. and Nonneman, A. J., 1975) and impairment of conditioned taste aversion (CTA) (Hernádi, I. et al., 2000; Karádi, Z. et al., 2005), while lesion and electrical stimulation of the ventrolateral PFC induce feeding disturbances (Kolb, B. and Nonneman, A. J., 1975; Brandes, J. S. and Johnson, A. K., 1978) and feeding (Bielajew, C. and Trzcinska, M., 1994), respectively.

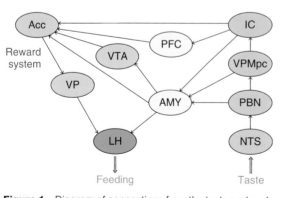

Figure 1   Diagram of connections from the taste system to reward system and feeding center. NTS, nucleus of tractus solitarius; PBN, parabrachial nucleus; VPMpc, parvocellular part of the ventralis posteromedial thalamic nucleus; IC, insular cortex; PFC, prefrontal cortex; AMY, amygdala; VTA, ventral tegmental area; Acc, nucleus of accumbens; VP, ventral pallidum; LH, lateral hypothalamic area. Note that this diagram does not include all of the connections.

4.24.1.2   Central Representation of Flavor

It is widely accepted that flavor is the integrated sensation of taste and odor, and often the somatic sensations of texture and temperature are included (Bartoshuk, L. M. and Beauchamp, G. K., 1994 for a review). Flavor contributes significantly to hedonic evaluation of food and feeding behavior. The brain sites involved in flavor evaluation are thought to be the regions where taste and other sensory inputs are integrated. In rodents, although neurons in the NTS, PBN, and VPMpc are more or less responsive to taste, tactile, and thermal stimulation of the tongue, more intense overlapping activation including olfactory and visceral inputs are in regions of the GC (Shipley, M. T. and Geinisman, Y., 1984; Yamamoto, T. *et al.*, 1989a), amygdala (Azuma, S. *et al.*, 1984; Nishijo, H. *et al.*, 1998c), and hypothalamus (Yamamoto, T. *et al.*, 1989b). In monkeys, single-neuron recording studies have identified neurons responsive to both taste and smell in the insula/operculum, the primary GC (Scott, T. R. and Plata-Salaman, C. R., 1999 for a review), and orbitofrontal cortex (Rolls, E. T. and Baylis, L. L., 1994). Some neurons in the monkey orbitofrontal cortex receive visual (Rolls, E. T. and Baylis, L. L., 1994) and somatosensory inputs related to fat-like texture (Rolls, E. T. *et al.*, 1999) and viscosity of the texture (Rolls, E. T. *et al.*, 2003) in the mouth. Neurons in the amygdala also respond to different sensory inputs (Nishijo, H., *et al.*, 1988a; 1988b). In a recent neuroimaging study in humans, Small D. M. *et al.* (2004) suggest that the insula, orbitofrontal cortex, and anterior cingulate cortex are key components of the network underlying flavor perception and that taste–smell integration is dependent on the mode of olfactory delivery, the orthonasal or retronasal routes, and previous experience with taste and smell combinations.

4.24.1.3 Taste-Mediated Reflexes

Taste stimuli evoke a variety of autonomic and hormonal reflex responses including salivary secretion, increases in gastric motility and gastric acid, exocrine pancreatic secretion, and endocrine secretions (Pavlov, I. P., 1902; Janowitz, H. D. *et al.*, 1950; Novis, B. H. *et al.*, 1971; Fischer, U. *et al.*, 1972; Steffens, A. B., 1976; Nicolaidis, S., 1977; Ohara, I. *et al.*, 1988; Niijima, A., 1991; Ohara, I. *et al.*, 1996). In addition to these responses, taste stimuli elicit somatic motor responses involving mainly lingual and oral-facial muscles (Steiner, J. E., 1973; Grill, H. J. and Norgren, R. 1978a). Cephalic phase insulin release is a reflexly induced secretion elicited by oral stimulation with taste stimuli especially including sweet- (Steffens, A. B., 1976; Berthoud, H. R. *et al.*, 1980) and umami-tasting (Niijima, A. *et al.*, 1990) substances. Such a taste quality-specific reflex can also be seen for salivary secretion, as shown below.

4.24.1.4 Salivary Secretion

Salivary secretion occurs when sapid foods are put in the mouth, termed the gustatory-salivary reflex. The volume and chemical composition of saliva reflexly induced by taste stimulation are markedly different depending on the quality (Baxter, H., 1933; Kerr, A. C., 1961; Newbrun, E., 1962; Funakoshi, M. and Kawamura, Y., 1967; Gantt, W. H., 1973) and nature of the gustatory stimulus. Dry bread induces a copious serous saliva and meat elicits a small amount of mucous saliva (Pavlov, I. P., 1902). Gjorstrup P. (1980) found that sweet stimuli regularly produced a large increase in the amylase secretion from the parotid gland of conscious rabbits, salty stimuli had a small effect, and sour and bitter stimuli had a negligibly small effect.

It is generally known that the most copious salivary secretion is induced by sour (or acid) stimulation (Kerr, A. C., 1961; Emmelin, N. and Holmberg, J., 1967; Funakoshi, M. and Kawamura, Y., 1967) and that the amount of secretion increases with the stimulus concentration (Feller, R. P. *et al.*, 1965; Chauncey, H. H. and Feller, R. P., 1967). The gustatory-salivary reflex has been studied in anesthetized decerebrate animals (Yamamoto, T. and Kawamura, Y., 1977; Kawamura, Y. and Yamamoto, T., 1978; Matsuo, R. and Yamamoto, Y., 1989; Matsuo, R. *et al.*, 1989; 2001) where taste neurons in the lower brainstem (NTS and/or PBN) and the salivatory nuclei are involved. These studies have shown that the magnitude of response of the gustatory afferent neurons to taste stimulation is correlated with the volume of salivary secretion regardless of the taste quality, except for copious secretions to strong acids (Kawamura, Y. and Yamamoto, T., 1978) and bitter tastes accompanying rejection behaviors (Matsuo, R. *et al.*, 2001). According to Matsuo R. *et al.* (2001), aversive taste information from the PBN reaches the salivatory secretory center via the reticular formation ventral to the PBN with a slight contribution of the connection between the NTS and the salivatory nuclei.

Since the amount of salivary secretion elicited by taste stimulation is markedly reduced after lesions of the cortical taste area in conscious dogs (Funakoshi, M. *et al.*, 1972), higher central nervous system structures may also be involved in the modulation of activity of the reflex arc in conscious animals.

4.24.1.5 Somatic Motor Responses

Animals as well as humans accept and ingest sweet and slightly salty substances (e.g., sucrose and NaCl), while they reject very sour and bitter substances (e.g., HCl and quinine), accompanied by particular oral-facial movements. The possibility that such acceptance or rejection behaviors based on the hedonics of taste are determined in the brainstem without higher cortical involvement was first noticed by Miller F. R. and Sherrington C. R. (1916).

Effects of taste stimuli on tongue motility have been reported in human neonates. For example, Weiffenbach J. M. and Thach B. T. (1973) reported that a drop of glucose solution, but not water, to the tongue elicited lateral tongue movement toward the site of the sapid stimulation. Nowlis G. H. (1973) has pointed out that human newborn infants are capable of eliciting differential tongue movements depending on the quality of gustatory stimuli. Moreover, Steiner J. E. (1973) observed that human newborn infants showed active and differential oral-facial expression based on the hedonic dimension of gustatory stimuli. The responses to sugar (sweet taste) and quinine (bitter taste) were characterized by expressions of pleasure (or relaxation and acceptance) and displeasure (or rejection), respectively. He concluded that the neural substrate of such hedonic behavior is located at the brainstem level, because essentially identical features of facial expression were observed in anencephalic and hydroanencephalic neonates.

Taste-elicited behavioral displays were also investigated in rats (Grill, H. J. and Norgren, R., 1978a; 1978b) and newborn rabbit pups (Ganchrow, J. R. et al., 1979). Grill H. J. and Norgren R. (1978a) analyzed the pattern of tongue, jaw, and body movements in normal and decerebrate rats in response to an injection of a small amount of taste solution into the oral cavity and showed that sucrose, NaCl, and HCl solutions elicited a response sequence beginning with rhythmic mouth movements followed by rhythmic tongue protrusions and then lateral tongue movements, while quinine induced a response beginning with gaping followed by various characteristic body responses. These behavioral experiments on animals again show that taste-elicited behaviors can be grouped into acceptance and rejection based on hedonics of tastes, and these behaviors depend only on the lower brainstem.

When taste solutions were infused into the mouth, c-*fos* was strongly expressed in the lateral and intermediate zones of the medullary reticular formation, which includes premotor neurons of masticatory, facial and tongue motor systems (DiNardo, L. A. and Travers, J. B., 1994; Travers, J. B. *et al.*, 1999). This area

receives inputs from the taste areas of the NTS (Shammah-Lagnado, S. J. et al., 1992; Halsell, C. B. et al., 1996) and PBN (Herbert, H. et al., 1990; Karimnamazi, H. and Travers, J. B., 1998). These findings suggest that taste information reaches the motor nuclei through the medullary reticular formation to generate well-organized taste-elicited motor reactions.

4.24.2 Substances Related to Palatability

This section describes several neurochemicals that appear to be related to taste hedonics, especially to palatability. We discuss the possible involvement of dopamine with palatability in the next section.

4.24.2.1 Benzodiazepine

Benzodiazepines such as chlordiazepoxide and diazepam are widely prescribed for the treatment of anxiety disorders. A considerable body of evidence has indicated that benzodiazepine agonists, which facilitate the opening of Cl^- channels in response to $GABA_A$ receptor activation, promote food intake (Berridge, K. C. and Peciña, S., 1995 for a review). Although in most studies, this hyperphagic effect has been shown in rodents, benzodiazepines also enhances food intake in rhesus monkeys (Foltin, R. W. *et al.*, 1985), baboons (Foltin, R. W. *et al.*, 1989), and humans (Haney, M. *et al.*, 1997).

Because of their well-known anxiolytic and sedative actions, the facilitatory effect of benzodiazepines on food intake was once interpreted as an indirect consequence (Poschel, B. P. H., 1971). However, the following results suggest that benzodiazepine-induced hyperphagia is not due to secondary effects of benzodiazepine on anxiety or arousal. First, benzodiazepine agonists increase food intake in rats under nonstressful conditions of the home cage (McLaughlin, C. L. and Baile, C. A., 1979; Cole, S. O., 1983). Second, while sedative effects are rapidly reduced by repeated treatments with benzodiazepine, tolerance to the hyperphagic effects slowly develops (Wise, R. A. and Dawson, V., 1974). Third, while some partial agonists of benzodiazepine (Ro16-6028, Ro17-1812, Ro23-0364) facilitate food intake, these drugs do not produce a strong sedative effect (Cooper, S. J. and Green, A. E., 1993). Other types of benzodiazepines (CGS 9895 and CGS 9896) diminish behavioral expressions of anxiety and arousal without facilitating food intake (Cooper, S. J. and Yerbury, R. E., 1986). Thus, the benzodiazepine neural systems that modulate anxiety are thought to be separable from those that increase feeding.

In earlier studies, the hyperphagic effect of benzodiazepines was considered to be due to an anti-satiety effect (Margules, D. L. and Stein, L., 1967) or to a hunger-inducing effect (Wise, R. A. and Dawson, V., 1974). Several lines of evidence, however, strongly suggest that benzodiazepines specifically enhance the hedonic palatability of food (Cooper, S. J. and Estall, L. B., 1985; Berridge, K. C. and Peciña, S., 1995). For example, systemic administration of benzodiazepines selectively increase the intake of preferred foods such as cookies, sweetened foods, and saccharin solution, but not the intake of regular chow or plain water (Cooper, S. J. and McClelland, A., 1980; Cooper, S. J., 1987; Cooper, S. J. and Yerbury, R. E., 1988; Parker, L. A., 1991).

Since these studies used the volume (g or ml) of food or fluid consumed as a measure for pharmacological modulation of feeding, some factors other than intraoral taste stimulation might have affected the results. Cooper S. J. *et al.* (1988) reported that a benzodiazepine agonist facilitated, and an antagonist reduced, sucrose sham feeding in the gastric-fistulated rat, suggesting that benzodiazepine-induced hyperphagia is not due to postingestional factors.

Although the hedonic aspect of the taste of food and fluid is one of the most important factors that affect feeding, it is difficult to determine the perceived palatability in animals from the volume of food or fluid consumed. Grill H. J. and Norgren R. (1978a) developed a more direct method to quantify palatability (termed taste reactivity), which measures a characteristic set of orofacial and somatic responses that rats display when taste solutions are infused directly into their mouths. For example, sucrose solution elicits an ingestive taste reactivity pattern of tongue protrusions, paw licking, and mouth movements, while bitter quinine solution elicits an aversive taste reactivity pattern such as gaping, chin rubbing, and paw pushing. Mildly aversive taste solutions elicit predominantly passive drips. Using the taste reactivity test, Berridge K. C. and Treit D. (1986) have showed that systemic injection of chlordiazepoxide, a benzodiazepine agonist, selectively facilitates positive hedonic responses to sucrose infusions into the mouth. In contrast, chlordiazepoxide has little or no effect on aversive reactions elicited by quinine. Since such a selective enhancement of hedonic reactivity patterns by benzodiazepine agonists is blocked by benzodiazepine antagonist Ro 15-1788 (Treit, D. *et al.*, 1987), the effects of benzodiazepines on taste palatability are benzodiazepine receptor specific. Similar enhancements of positive palatability to taste stimuli induced by benzodiazepines have been reported by several studies using the taste reactivity test (Gray, R. W. and Cooper, S. J., 1995; Söderpalm, A. H. and Hansen, S., 1998). In addition to taste reactivity, Higgs S. and Cooper S. J. (1997; 1998a) showed that benzodiazepines increased bouts of licking for highly palatable solutions by a microstructural analysis of licking behavior in a brief contact test in the rat.

Aside from these behavioral and pharmacological evaluations, it is important to identify the site(s) in the central nervous system on which benzodiazepines act to enhance palatability. The distribution of benzodiazepine receptors is relatively greater in the forebrain than in the lower brainstem (Richards, J. G. and Mohler, H., 1984). Nevertheless, recent evidence strongly suggests that benzodiazepines act on the particular receptors in the lower brainstem and enhance taste palatability. In an attempt to determine a possible site of benzodiazepine action, Berridge K. C. (1988) examined taste reactivity in chronic decerebrate rats that are known to demonstrate normal affective taste reactivity (Grill, H. J. and Norgren, R., 1978b). Systemic benzodiazepine injection facilitated the positive taste reactivity in response to normally preferred taste solutions, indicating that the midbrain and/or hindbrain seem to contain the minimum benzodiazepine receptors and neural circuitry required to enhance positive hedonic reactions to taste. The notion is supported by the result that in normal rats, benzodiazepine is more effective at eliciting feeding (Higgs, S. and Cooper, S. J., 1996a) and enhancing positive hedonic reactions to oral sucrose (Peciña, S. and Berridge, K. C., 1996) when microinjections are made in the fourth ventricle (lower brainstem) than in the lateral ventricle (forebrain). Furthermore, direct administration of the benzodiazepine agonist midazolam into the PBN increases the consumption of palatable food (Higgs, S. and Cooper, S. J., 1996b) and positive taste reactivity (Söderpalm, A. H. and Berridge, K. C., 2000), suggesting that the PBN is the most probable region on which benzodiazepines act to influence food intake. Because the PBN is the second gustatory relay, it is likely that benzodiazepines act on specific receptors in the PBN to modify taste information so as to enhance taste palatability and facilitate consumption of food.

Some results suggest that the enhancement of palatability induced by intra-PBN injection of benzodiazepine is mediated by $GABA_A$ receptors in the PBN. It is well known that benzodiazepines act on the $GABA_A$ receptor to increase its affinity for GABA. In fact, GABA functions as an inhibitory neurotransmitter in both gustatory and visceral part of the PBN, mediated in part, by $GABA_A$ receptors (Kobashi, M. and Bradley, R. M., 1998). Shimura T. et al. (2004) found that systemic injection of midazolam increased the intake of a sucrose solution in wildtype mice but not in mice deficient in the 65-kDa isoform of glutamate decarboxylase (GAD65), a GABA synthetic enzyme in the central nervous system. The results indirectly suggest that GAD65-generated GABA is necessary for benzodiazepines to enhance taste palatability in the PBN. Nevertheless, greater understanding of the physiological significance of benzodiazepines for taste palatability awaits further investigation.

4.24.2.2   Opioids

Various endogenous opioids and their receptor subtypes are found in diverse brain regions, particularly within sites involved in emotion, pain and stress, and endocrine function and feeding (Mansour, A. et al., 1987). Long before the discovery of endogenous opioids, it had been demonstrated that systemic morphine administrations increase food intake in rats (Flowers, S. H. et al., 1929; Martin, W. R. et al., 1963). Since the demonstrations that opioid antagonists reduce food intake at appropriate dosages (Holtzman, S. G., 1974; Brown, D. R. and Holtzman, S. G., 1979), many studies have examined the involvement of opioids in the regulation of food intake (Glass, M. J. et al., 1999 for a review). In general, opioid agonists facilitate and antagonists diminish food consumption. Although precise mechanisms are still unclear, evidence supports the hypothesis that opioids modulate food and taste palatability, leading to increased food intake (Cooper, S. J. and Kirkham, T. C., 1993 for a review). One reason for this hypothesis is that the hyperphagic effect is selective to normally preferred food and tastes. For example, the preference for sweet solutions over water is abolished by systemic administration of opioid antagonist naloxone in rats (Le Magnen, J. et al., 1980). Naltrexone, an opioid antagonist, does not modify the energy intake in control rats receiving ordinary chow but suppresses hyperphagia induced by a highly palatable diet (Apfelbaum, M. and

Mandenoff, A., 1981). Both naloxone and naltrexone abolish the preference for a highly palatable saccharin solution in water-deprived rats due to a selective decrease in the consumption of the saccharin solution, with no effect on water intake (Cooper, S. J., 1983).

Studies with the taste reactivity test show that systemic morphine administrations selectively increase positive hedonic reactions to a mixture of sucrose and quinine solution without enhancing aversive reactions (Doyle, T. G. et al., 1993), suggesting that opioid-induced feeding is at least partly mediated by increased palatability of food. Since opioid receptor antagonists reduce sucrose sham feeding in the gastric-fistulated rat (Kirkham, T. C. and Cooper, S. J., 1988), intraoral but not postingestional factors seem to affect opioid-induced modification of feeding. In addition, using a microstructural analysis of licking pattern for sucrose or fat emulsion, Higgs S. and Cooper S. J. (1998b) showed that palatability is modulated by opioids.

In fact, blood levels of β-endorphin levels increase after the consumption of palatable sweeteners and decrease following the intake of aversive quinine in thirsty rats (Yamamoto, T. et al., 2000). β-Endorphin levels in the cerebrospinal fluid, on the other hand, increase after water consumption by thirsty rats. The difference in β-endorphin levels between blood and cerebrospinal fluid suggests that two opiate systems separately function during the intake of fluid.

Experimental work in humans also supports the notion that the modification of feeding produced by opioid agonists and antagonists is mediated by a hedonic shift in the palatability of food and fluid. Fantino M. et al. (1986) showed that naltrexone produces a significant decrease on a subjective pleasure scale for sweetened solution in normal human subjects. Naloxone reduces taste preferences for sweet high-fat foods such as cookies or chocolate in both binge eaters and controls (Drewnowski, A. et al., 1992).

Intracerebroventricular as well as systemic administration of opioid agonists facilitates food intake (Gosnell, B. A., 1987). Moreover, the opioid antagonist naloxone causes a reduction in food intake when injected intracerebroventricularly at a dose that is ineffective when administered peripherally (Gosnell, B. A., 1987). Morphine administrations into the lateral ventricle increase both positive hedonic taste reactivity patterns and feeding (Peciña, S. and Berridge, K. C., 1995). These results favor the notion that opioids act centrally to modify food intake.

Although opioids are believed to affect multiple feeding systems (Glass, M. J. *et al.*, 1999), recent evidence strongly suggests that the nucleus accumbens and its associated circuitry are primarily involved in palatability-dependent feeding (Kelley, A. E. *et al.*, 2002 for a review). We describe several lines of evidence that support this hypothesis in the next section.

The hypothalamus is known to be an important region of opioid action for food intake. Microinjections of β-endorphin into the ventromedial nucleus (Grandison, L. and Guidotti, A., 1977) or the paraventricular nucleus (McLean, S. and Hoebel, B. G., 1983) of the hypothalamus facilitates food intake in satiated rats. On the contrary, localized injections of naloxone into the paraventricular nucleus of the hypothalamus significantly reduces food intake (Woods, J. S. and Leibowitz, A. F., 1985). However, these studies suggest that opioids in the hypothalamus are involved in energy needs rather than the palatability of food.

The gustatory relays in the lower brainstem also appear to be sites on which opioids act to facilitate feeding. Infusion of the selective mu-opioid receptor agonist [d-Ala2,N-Me-Phe4,Gly5-ol]enkephalin (DAMGO) into the lateral PBN increases food intake and naloxone antagonizes DAMGO-induced feeding (Wilson, J. D. *et al.*, 2003) Injection of DAMGO into the NTS, the first gustatory relay, stimulates food intake, whereas injection of either the delta or kappa receptor agonist has no effect (Kotz, C. M. *et al.*, 1997). Microinfusion of methionine-enkephalin into the NTS suppresses both spontaneous and evoked activity of more than 20% of taste responsive neurons (Li, C. S. *et al.*, 2003). These effects are blockable by naltrexone, which alone is without effect, suggesting that opioids play some role in processing of taste information in the NTS.

4.24.2.3 Cannabinoid

Recent evidence suggests that the endocannabinoid system is involved in the control of food intake mainly by modulating food and taste palatability (Cooper, S. J., 2004 for a review). The psychoactive effects of marijuana are largely attributable to a constituent called THC (delta-9-tetrahydrocannabinol). Various cannabinoids which are chemicals of the same class as THC may also be psychoactive (Kephalas, T. A. *et al.*, 1976). Since the discovery of a specific receptor in the brain for cannabinoids and endogenous ligands (Devane, W. A. *et al.*, 1992),

increasing attention has been directed to possible behavioral functions of endocannabinoids, including a role in feeding. Although earlier studies had suggested that marijuana facilitates appetite and food intake in humans (Abel, E. L., 1971; Foltin, R. W. *et al.*, 1986), a hyperphagic effect had not been demonstrated clearly in animal studies until recently. Several recent studies have shown that endocannabinoids can facilitate food intake under certain conditions. For example, both the exogenous cannabinoid THC and the endocannabinoid arachidonoyl ethanolamide (anandamide) stimulate eating in presatiated rats, with a marked reduction in latency to feed (Williams, C. M. and Kirkham, T. C., 2002). Microstructural analyses of licking reveal that THC and anandamide significantly increase both the total number of licks and bout duration (Higgs, S. *et al.*, 2003). These results suggest that cannabinoids promote eating by increasing the incentive value of food.

Microinjection of 2-arachidonoyl glycerol (2-AG) into the nucleus accumbens shell robustly increases food intake (Kirkham, T. C. *et al.*, 2002). Since subcutaneous administration of a cannabinoid receptor antagonist blocked the hyperphagic response, cannabinoids in the nucleus accumbens appear to be involved in the facilitation of food intake. Microinjection of the cannabinoid receptor agonist CP55,940 into the fourth, but not lateral, ventricle increases consumption of highly palatable sweetened condensed milk (Miller, C. C. *et al.*, 2004), suggesting that cannabinoids can act on the hindbrain to enhance food intake. Nevertheless, our knowledge about behavioral mechanisms of cannabinoid-induced hyperphagia is still quite limited. Further studies are needed to conclude that endogenous cannabinoids directly affect palatability leading to food intake.

4.24.2.4 Substances Related to Negative Hedonics

While studies of substances related to negative hedonics of taste are quite limited, Yamamoto T. *et al.* (1990) showed that the calcitonin gene-related peptide (CGRP)-like immunoreactivity levels in the gustatory insular cortex were increased by aversive taste stimuli such as quinine and even by palatable stimuli such as NaCl and sucrose, if animals had been trained to avoid them by conditioning. The results suggest that CGRP in the GC is implicated in rejection or avoidance behaviors to aversive taste stimuli.

Recent evidence suggests that diazepam-binding inhibitor (DBI), the only known endogenous ligand of the benzodiazepine receptor, is related to negative hedonics of taste. DBI-like peptide is released by aversive quinine stimuli (Manabe, Y. *et al.*, 2000). Injection of a DBI fragment into the fourth ventricle suppressed the intake of 5% sucrose solution and water and increased the aversive taste reactivity to NaCl in mice (Manabe, Y. *et al.*, 2001). In contrast with the facilitatory effect of exogenous benzodiazepine agonists for positive palatability, the possible role of the endogenous DBI for negative hedonics is of interest.

Histamine, which is widely distributed in the mammalian central nervous system as a neurotransmitter or neuromodulator, also seems to be related to negative hedonics of taste. Intraoral infusions of quinine-HCl increase hypothalamic histamine release, but infusions of sucrose or saccharin produced a decrease in histamine levels, suggesting that histamine plays a role in the behavioral expression to aversive taste stimuli (Treesukosol, Y. *et al.*, 2005).

4.24.3   Role of the Reward System in Taste-Mediated Behavior

4.24.3.1   What is the Reward System?

Rats (Olds, J. and Milner, P. M., 1954), humans (Bishop, M. P. *et al.*, 1963), and many other species will perform a response, such as pressing a lever, in order to obtain direct electrical stimulation to specific sites in their own brains. Originally, Olds and Milner argued that the specific brain sites that mediate self-stimulation are those that normally mediate the pleasurable effects of natural rewards (e.g., food, water, and sex). Although there have been numerous debates for a relatively long period, the current consensus seems to be that the circuits mediating intracranial self-stimulation phenomena are reward circuits. Similarly, it is well documented that the brain reward systems are neural substrates for the rewarding effects of addictive drugs. The mesolimbic dopamine system, arising from dopaminergic neurons in the VTA of the midbrain to terminate in the nucleus accumbens, has been shown to play a particularly important role in natural as well as drug rewards (Wise, R. A., 2002 for a review).

Many studies have shown that activation of dopamine systems is triggered in animals by encounters with food, sex, drugs of abuse, and intracranial self-stimulation (Berridge, K. C. and Robinson, T. E.,

1998 for a review). In humans, a functional MRI study also reported that presentation of rewards such as cocaine to cocaine-dependent subjects modulated activity in dopamine target sites including the nucleus accumbens (Breiter, H. C. *et al.*, 1997).

4.24.3.2   The Role of Dopamine in Food Reward

It is well known that dopamine affects feeding behavior (Blackburn, J. R. *et al.*, 1992 for a review). Systemic administration of pimozide, a dopamine receptor antagonist, attenuates lever pressing and running for food reward in hungry rats (Wise, R. A. *et al.*, 1978). Pimozide also suppresses the consumption of a saccharin-glucose solution in a dose-related manner (Xenakis, S. and Sclafani, A., 1981; Geary, N. and Smith, G. P., 1985). These early studies favor the view that dopamine receptor blockade reduces the rewarding impact of palatable stimuli, including food (Wise, R. A., 1982). However, taste reactivity analyses have shown that dopamine is less likely involved in modifying palatability. Berridge and his colleagues (Berridge, K. C. *et al.*, 1989; Berridge, K. C., 1996) injected 6-hydroxydopamine (6-OHDA), a selective neurotoxin for dopamine, bilaterally into the ascending dopaminergic bundle, at a point in the LH where fibers join together from midbrain A9 and A10 sites on their way to the neostriatum and nucleus accumbens. In these rats, dopamine was depleted by 95–99% in both the neostriatum and nucleus accumbens and rats became aphagic. However, positive hedonic reactions to sucrose were not suppressed relative to control levels even in these aphagic rats. Aversive reactions to bitter solutions were also unaltered by these selective dopaminergic lesions.

Electrophysiological experiments also suggest that dopamine neurons in the mesolimbic system are less related to the evaluation of taste palatability. Dopamine neurons in the midbrain including the VTA of the monkey show short, phasic activation in a rather homogenous fashion after the presentation of liquid and food rewards, and visual or auditory stimuli that predict reward, suggesting that these dopamine neurons preferentially report rewarding events (Schultz, W. and Romo, R., 1990). However, these dopamine neurons appear to encode a discrepancy between the prediction of reward (a reward prediction error) rather than reward per se (Ljungberg, T. *et al.*, 1992). Recently, we have found that more than half of the neurons recorded in the

VTA of freely behaving rats change their firing during licking of liquid reward (Shimura, T. *et al.*, 2005). Since there is no difference in the responsiveness to taste solutions and water, these neurons may be involved in fluid reward rather than the hedonic impact of taste.

Shimura T. *et al.* (2002a) showed evidence that the mesolimbic dopaminergic system interacts with the benzodiazepine and/or opioid systems to exhibit the normal intake pattern for palatable fluid. Electrolytic or 6-OHDA lesions of the VTA suppress the consumption of a preferred sucrose or NaCl solution without influencing the intake of other tastes or water. Systemic injections of the benzodiazepine agonist midazolam significantly facilitates the consumption of a preferred sucrose in sham control but not in VTA-lesioned rats. Midazolam does not affect the consumption of an aversive quinine solution in either lesioned or sham animals. Moreover, while systemic administration of morphine, an opioid agonist, selectively increases the consumption of sucrose solution in control rats without affecting the consumption of quinine solution, the intake of both sucrose and quinine solutions remains unchanged by morphine injections in VTA-lesioned rats. These results suggest that the dopaminergic mediation is required for the normal expression of both benzodiazepine- and opioid-induced overconsumption of palatable fluid. According to the current concept that food reward contains distinguishable functional components, liking (palatability) and wanting (incentive motivation) (Berridge, K. C., 1996), the mesolimbic dopamine system seems to mediate wanting rather than liking for food and fluid reward. In fact, hyperdopaminergic dopamine transporter-knockdown mutant mice have higher wanting for a sweet reward in a runway task (Peciña, S. *et al.*, 2003). But sucrose taste fails to elicit higher orofacial liking reactions from mutant mice in an affective taste reactivity test. On the other hand, since sucrose sham feeding increases dopamine levels in the nucleus accumbens in a concentration-dependent manner (Hajnal, A. *et al.*, 2004), the mesolimbic dopamine system seems to be implicated in a reward effect of taste palatability.

4.24.3.3   The Nucleus Accumbens

The nucleus accumbens has long been considered to be an essential interface between motivation and action (Mogenson, G. J. *et al.*, 1980). Neuroanatomical and histochemical studies have revealed that the accumbens is composed of two major subregions, the core, and the shell extending medially, ventrally, and laterally around the core. The functions of core and shell subregions are thought to be different on the basis of distinctive anatomical profiles (Heimer, L. *et al.*, 1991). Considerable evidence has indicated that the nucleus accumbens is critically involved in palatability-induced feeding behavior. The hyperphagic effect of opioids described in the previous section has been shown to be most prominent when opioids are injected into the nucleus accumbens, especially into the shell subdivision. For example, microinjections of DAMGO induce a robust, dose-dependent increase in food intake that is blocked by coadministration of naltrexone (Bakshi, V. P. and Kelley, A. E., 1993). It is noted that the hyperphagic effects of DAMGO are selective to highly palatable taste stimuli such as a high-fat diet (Zhang, M. *et al.*, 1998; Zhang, M. and Kelley, A. E., 2000), sucrose solution (Zhang, M. and Kelley, A. E., 1997), saccharin, NaCl, and ethanol solutions (Zhang, M. and Kelley, A. E., 2002). In addition, morphine microinjections into the nucleus accumbens shell not only facilitates feeding but also selectively increases positive hedonic patterns of behavioral affective reaction elicited by oral sucrose, using the taste reactivity test of hedonic palatability (Peciña, S. and Berridge, K. C., 2000).

The nucleus accumbens receives afferent inputs from corticolimbic and thalamic structures, which are primarily mediated by excitatory amino acids (Robinson, T. G. and Beart, P. M., 1988). Microinjections of 6,7-dinitroquinoxaline-2,3-dione (DNQX), a glutamatergic α-amino-3-hydroxy-5-methylisoxazole-4-propionic acid (AMPA) receptor antagonist, but not the NMDA antagonist 2-amino-5-phosphonopentanoic acid (AP-5), into the medial shell subregion resulted in a pronounced feeding response (Maldonado-Irizarry, C. S. *et al.*, 1995). DNQX significantly facilitates intake of solid and liquid food, but does not significantly affect water intake or gnawing behavior, suggesting that DNQX is acting on a system specifically involved with the regulation of food intake (Stratford, T. R. *et al.*, 1998). Moreover, the feeding response is completely inhibited by concurrent infusion of the $GABA_A$ agonist muscimol into the LH, a major projection area of the accumbens shell (Maldonado-Irizarry, C. S. *et al.*, 1995). These findings demonstrate a selective role for non-NMDA receptors in the nucleus accumbens shell in ingestive behavior and suggest an important functional link between the nucleus accumbens and the LH.

The majority of cells projecting from the accumbens shell are medium spiny neurons that use GABA as a neurotransmitter (Meredith, G. E. *et al.*, 1993). Acute inhibition of cells in the accumbens shell by administration of the $GABA_A$ receptor agonist muscimol or the $GABA_B$ receptor agonist baclofen elicits intense, dose-related feeding without altering water intake (Stratford, T. R. and Kelley, A. E., 1997). Moreover, injections of muscimol into the accumbens shell greatly increase the number of cells exhibiting Fos-like immunoreactivity in the LH (Stratford, T. R. and Kelley, A. E., 1999). Immunohistochemical results show that intra-accumbens shell muscimol treatment increases the percentage of orexin/hypocretin-containing neurons expressing Fos in the lateral hypothalamic area (Baldo, B. A. *et al.*, 2004). The same treatment fails to increase Fos expression in melanin concentrating hormone-containing neurons in the LH. These results suggest that extensive reciprocal interaction between the nucleus accumbens shell and LH regulates feeding behavior.

The VP is a main output target of the GABAergic neuron in the nucleus accumbens. From the VP, GABAergic efferents project to the LH (Groenewegen, H. J. *et al.*, 1993). Thus, the VP is anatomically interposed between neurons in the nucleus accumbens and those in the LH that are known to be intimately related to feeding behavior. Blockade of $GABA_A$ receptors in the VP with bicuculline elicits a strong feeding response in satiated rats without affecting water intake (Stratford, T. R. *et al.*, 1999). Recently, we have found that microinjection of bicuculline into the VP enhances the intake of a preferred saccharin but not quinine solution and water in dehydrated rats, suggesting that the overconsumption produced by GABA blockade in the VP is specific to palatable tastes (Shimura, T. *et al.*, 2006).

Compared to evidence concerning the efferent connections of the nucleus accumbens that may mediate food intake, the role of afferent inputs that convey taste and visceral information to the accumbens are poorly understood. Some anatomical evidence shows that the accumbens receives taste and visceral information through direct input from the NTS in the medulla (Ricardo, J. A. and Koh, E. T., 1978; Saper, C. B., 1982). The accumbens also receives taste information from the insular GC (Brog, J. S. *et al.*, 1993). The amygdala is likely an important source of information about taste and visceral functions to the nucleus accumbens. Taste information received by the central nucleus of the amygdala via the NTS

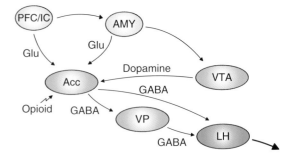

Figure 2   Diagram of reward system responsible for palatability-induced feeding. Glu, glutamate. Other abbreviations are the same as those in Figure 1.

and PBN projects caudally to the VTA, then arrives at the nucleus accumbens. The insular GC – basolateral amygdala – accumbens pathway is also suggested in rats (McDonald, A. J. and Jackson, T. R., 1987; McDonald, A. J. *et al.*, 1999). Because of these afferent as well as efferent connections, the nucleus accumbens appears to be well suited to regulate food intake. Although increasing evidence suggests that the nucleus accumbens shell has a key role for palatability-induced consumption as described above, the precise mechanism for processing of taste palatability in the accumbens is not yet understood.

The role of the reward system in palatability-induced feeding is summarized in Figure 2. Taste relays in the lower brainstem are responsible for the palatability shift mediated primarily by benzodiazepines, at least in rodents. Converging corticolimbic input of taste and visceral information to the nucleus accumbens shell is likely involved in mediation of palatability-induced food intake. The GABAergic efferents from the nucleus accumbens shell project to the LH, partly via the VP, so as to regulate the intake of food that is hedonically desirable. Opioid mechanisms in the ventral striatum, especially in the nucleus accumbens, are critically involved in the enhancement of taste palatability. The dopaminergic projection arising from the VTA to the nucleus accumbens is implicated in wanting rather than liking for food and liquid reward.

4.24.4   Role of Taste in Feeding

4.24.4.1   Feeding Behavior to Palatable and Aversive Foods

Substances and liquids with sweet taste, umami taste, and salty taste of hypotonic and isotonic sodium salts are palatable and preferred, while those with sour taste and bitter taste are aversive and are rejected. Such taste-mediated behaviors have been shown by organisms through the long history of evolution so that sweetness is a signal for energy sources; umami, a signal for protein; saltiness, a signal for minerals; sourness, for rotten food; bitterness, for poisons. These relationships are innately common among different species of animals, that is, human newborn babies show ingestive behavior with mild facial expressions to sweeteners and rejective behavior with aversive expressions to acids and bitter substances, whereas rats show the characteristic oral-lingual movements and body reactions as described in the previous section of this chapter and show a very similar preference–aversion pattern of intake to different sapid solutions as do humans (Figure 3)

The neural substrates for taste-mediated ingestion are basically located in the brainstem without the involvement of higher centers of the gustatory system since similar hedonic reactions to taste stimuli can be seen in anencephalic or hydroanencephalic human infants (Steiner, J. E., 1973) and decerebrate rats (Grill, H. J. and Norgren, R., 1978b). The possible existence of anatomical segregation of taste-responsive neurons in both the qualitative and hedonic aspects of taste at the brainstem level (Yamamoto, T. *et al.*, 1994a) may serve an efficient basis of eliciting quality-specific and hedonic-oriented gustatory reflexes. Other lines of evidence through c-*fos* immunohistological and

electrophysiological studies suggest that there are dual separate pathways for sensory and hedonic aspects of taste in each station of the ascending gustatory route including the NTS, PBN, and VPMpc, and in the GC (Sewards, T. V., 2004, for a review).

It is also necessary to add that higher brain centers which may not be directly involved in the processing of gustatory quality discrimination play important roles in evaluation of hedonic values of foods and tastes. As is described in detail in the following sections of this chapter, such regions include the amygdala, orbitofrontal cortex, cingulate cortex, reward system, and hypothalamus.

Besides the innate preference or aversion behaviors, nutrients or minerals that are essential for living organisms will become very palatable, preferred, and craved when those substances are deficient. The well-known and documented instances are salt appetite and amino acid deficiency. When animals are deficient in body sodium, they ingest excessive amounts of salt solutions at high concentrations which are normally avoided. They also increase their intake of lower salt concentrations. The mean response of neurons to NaCl is reduced in the NTS of sodium-deprived rats (Jacobs, K. M. *et al.*, 1988; Nakamura, K. and Norgren, R., 1995). It has further been demonstrated that sodium depletion causes not only lowered responses to NaCl but also an alteration in the neural representation of NaCl (Jacobs, K. M.

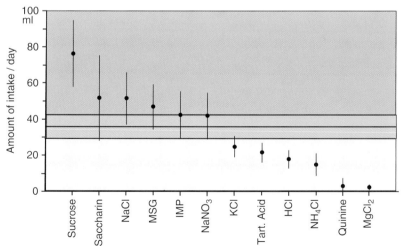

Figure 3 Amount of fluid intake/day for 12 different taste stimuli with the single bottle method in rats. Each value with a vertical line and a horizontal bar with upper and lower horizontal bars (water intake, blue belt) indicates means ± SEs (*n* = 15). The chemicals on the abscissa are arranged from left to right in the order of most to least preferred. If the amount of intake of a stimulus is over the blue belt (within the pink area), the stimulus is preferable, and if within the yellow area, the stimulus is aversive. Adapted from Yamamoto, T., Yuyama, N., Kato, T., and Kawamura, Y. 1985b Gustatory responses of cortical neurons in rats. III. Neural and behavioral measures compared. J. Neurophysiol. 53, 1370–1386.

et al., 1988). As for PBN neurons (Shimura, T. *et al.*, 1997a), taste responses to 0.3 and 0.5 M solutions of NaCl were significantly lower in the sodium-deprived rats than in controls, and correlation coefficients of responses among sodium salts and those between 0.5 M NaCl and sweeteners were larger in the sodium-deprived rats than in controls. These results suggest that PBN neurons in sodium-deprived rats are critically implicated in induction of quantitative and qualitative changes of gustatory sense to sodium salts. Rats display a preference for a particular amino acid when they are deprived of the amino acid. For example, when rats were fed with a lysine-deficient diet, the order of preference was lysine > saline > monosodium glutamate > glycine > threonine > water > arginine > histidine, while in control rats, arginine > saline > monosodium glutamate > glycine > water > threonine > histidine > lysine (Tabuchi, E. *et al.*, 1991). These investigators found that some neurons in the LH responded specifically to lysine when the animals were lysine deficient, suggesting that the LH is involved in mediating the preference for deficient amino acids.

4.24.4.2 Learning and Memory Related to Feeding Behavior

Feeding behavior can be modified by learning through gustatory experience associated with pleasant or unpleasant consequences.

4.24.4.3 Preference Learning

There are two types of learned taste preference: (1) attenuation of neophobia and (2) conditioned taste preference (CTP). When an animal ingests a harmless new substance or liquid, it shows neophobia (see the left in Figure 4) and it increases the consumption at subsequent exposures after learning that the substance is safe to consume (see the upper-right in Figure 4). Through this process of the attenuation of neophobia (or learned safety), foods can be classified as familiar-safe (Nachman, M. and Jones, D. R., 1974; Domjan, M., 1976). Although each part of the gustatory pathway may be concerned with neophobia, since lesions of only one of these parts attenuate neophobia (Yamamoto, T. *et al.*, 1995), a recent study (Bahar, A. *et al.*, 2004) shows that the GC plays an important role in the recognition of whether the taste is familiar or novel. Gutiérrez R. *et al.* (2003) show that cortical muscarinic receptors are important during a 2- to 4-h continuous period in which noxious

consequences of food ingestion are absent for the formation and consolidation of learned safety.

When ingestion of a neutral or mildly aversive food is associated with good postingestive visceral sensation, those foods become hedonically positive, or rewarding and preferred. This phenomenon is referred to as CTP (Fanselow, G. and Birk, J., 1982). CTP is attained gradually but more strongly via a flavor conditioning process. Previous studies demonstrated that rats could learn to prefer a taste solution paired with intragastric nutrient infusion (Elizalde, G. and Sclafani, A., 1990), opiate administration (Mucha, R. F. and Herz, A., 1986), and intracranial self-stimulation (Olds, J. and Milner, P. M., 1954). Nutrients can have positive postingestive actions that influence food selection and increase consumption (Sclafani, A., 2004). Giza B. K. *et al.* (1997) showed that the consumption of even an innately aversive $MgCl_2$ solution could be preferred by pairing with a nutrient, polycose, in rats. The rewarding properties of food that promote eating and influence food choice result from the central integration of orosensory and viscerosensory stimuli. The central neural mechanism of association of taste with postingestive food reward is not fully understood. The PBN and LH may play some important roles in flavor preference learning (Sclafani, A. *et al.*, 2001; Touzani, K. and Sclafani, A., 2001). The central dopamine system may also have a critical role in flavor learning (Azzara, A. V. *et al.*, 2001), whereas the involvement of opioid systems is less certain (Azzara, A. V. *et al.*, 2000).

4.24.4.4 Aversion Learning

When ingestion of a substance is followed by malaise such as gastrointestinal discomfort and nausea, associative learning between the ingested substance and internal consequences is quickly established; animals remember the taste for a long time and reject its ingestion on subsequent exposure (see the lower-right in Figure 4). This phenomenon is called a CTA or taste aversion learning.

CTA has the following characteristics, which are distinguished from classical conditioning (Bures, J. *et al.*, 1988): (1) Strong and long-lasting CTAs to novel taste stimuli can be established in a single learning procedure, that is, after one pairing of conditioned stimulus (CS) and unconditioned stimulus (US). (2) Successful CTAs can develop to the CS after delays of as long as 4–12 h between exposure to the CS and delivery of the US. CTA can be considered as a kind of

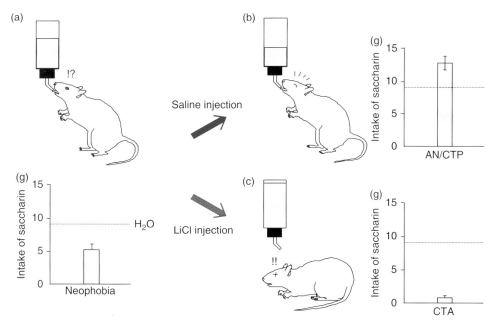

Figure 4 Schematic drawing of neophobia, attenuation of neophobia (AN), conditioned taste preference (CTP), and conditioned taste aversion (CTA). The magnitude of saccharin intake in each graph is taken from Yamamoto, T., Fujimoto, Y., Shimura, T., and Sakai, N. 1995. Conditioned taste aversion in rats with excitotoxic brain lesions. Neurosci. Res. 22, 31–49.

fear learning and serves as a defense mechanism of the organism to avoid ingestion of harmful substances. CTA is easily acquired in humans as well, for example, our survey (unpublished) for Osaka University students shows that a significant proportion of their food dislikes is based on CTA acquired before the middle grade of the elementary school (or below 10 years old). After the acquisition of CTA to an artificial sweetener, saccharin, this sweet and palatable substance still tastes sweet but changes to an aversive stimulus. Taste quality may not change, but the hedonic aspect changes drastically. This fact may indicate the existence of separate neural representations of the sensory aspects and the hedonic aspects of taste in the gustatory pathway (Sewards, T. V., 2004 for a review). Although the neural substrate and the molecular mechanisms of CTA have recently been well studied (Yamamoto, T. *et al.*, 1994b; Bures, J. *et al.*, 1998; Yamamoto, T. *et al.*, 1998; Bermúdez-Rattoni, F., 2004; Yamamoto, T., 2007 for reviews), a detailed discussion of the contribution of the different brain regions to formation of the CTA is outside the scope of this review except for the following few topics.

Electrophysiological studies of unit activity in GC neurons in freely ingesting and anesthetized rats have revealed that the GC processes multiple aspects of taste-related responses including sensory and

hedonic responses (Yamamoto, T. *et al.*, 1989a), convergent responses (Yamamoto, T. *et al.*, 1988; 1989a; Hanamori, T. *et al.*, 1998), temporal responses (Katz, D. B. *et al.*, 2001), anticipation (Yamamoto, T. *et al.*, 1988), and familiarity (Bahar, A. *et al.*, 2004). These response characteristics of GC neurons may not only contribute to the acquisition and retention of CTA but may be actively involved in the formation of CTA as revealed by occurrence of short-term and long-term potentiated responses to the CS after pairing its ingestion with the US (LiCl) (Yamamoto, T. *et al.*, 1989a; Yasoshima, Y. and Yamamoto, T. 1998). The amygdala is one of the most important areas in the integration of the CS and US. Yamamoto T. *et al.* (1994b), Yamamoto T. *et al.* (1995), and Yasoshima Y. *et al.* (2006) found that the basolateral nucleus was more important in the acquisition and retention of CTA in comparison with the central nucleus of the amygdala. Electrical stimulation of the central nucleus enhanced the responsiveness of the PBN neurons that responded best to the CS (Tokita, K. *et al.*, 2004). Taking these results together with those from other literature, it is plausible that the basolateral amygdala, which is involved in the formation of fear learning (Maren, S. and Fanselow, M. S., 1996 for a review), is important in the hedonic shift from palatable to aversive, and that the central nucleus, which is known to receive taste inputs together with other sensory inputs, contributes to the enhancement of gustatory responses to the CS (Shimura, T. *et al.*, 1997b; 2002b; Tokita, K. *et al.*, 2004), which facilitates detecting and avoiding the harmful substance.

4.24.4.5 Orexigenic Substances

Palatability is one of the most important factors that regulates food and fluid intake. Animals usually prefer innately palatable sweet-tasting substances, and their consumption often exceeds the need for homeostatic repletion. The effects of palatable food on feeding behavior may be related to functions of the newly identified chemical mediators that regulate appetite, for example, orexins (orexin-A and -B), melanin-concentrating hormone (MCH), agouti-related protein (AgRP), ghrelin, and neuropeptideY (NPY), all of which are known to stimulate feeding and thus, to function as orexigenic peptides (Inui, A., 2000; Schwartz, M. W. *et al.*, 2000). These peptides are produced in the hypothalamus and are well studied in terms of feeding, energy homeostasis, and obesity. Among these peptides, Furudono Y. *et al.* (2006) have suggested that orexin-A, NPY, and MCH are involved, in this order of effectiveness, in palatability-induced ingestion on the bases of the following. Administration of orexin-A into the lateral ventricle in rats increased both water and saccharin intake compared to vehicle administration. Similarly, administration of MCH increased the intake of both fluids. However, NPY injections enhanced only the intake of saccharin. These results suggest that orexin-A, MCH, and NPY are related to the facilitation of drinking. On the other hand, ghrelin and AgRP had no effects on fluid intake. When the changes in mRNA levels for orexin, NPY, and MCH in the hypothalamus were examined after drinking saccharin or water, saccharin, but not water, facilitated an increase in the orexin and NPY mRNA levels. MCH mRNA levels did not differ between saccharin and water intake and remained stable. Increased ingestion may be closely related to enhanced digestive functions. To address this issue, Kobashi M. *et al.* (2003) examined the effects of intraventricular injections of orexin-A on gastric motility. They measured the motility changes as alterations in intragastric pressure, by means of two balloons in the proximal and distal parts of the stomach (Kobashi, M. *et al.*, 2000). The proximal stomach is known to function as a reservoir, whereas the distal stomach's main function is to stir and drain

gastric contents. Orexin-A induced strong rhythmical gastric motility in the distal stomach, while the vehicle had no effect. Since the function of the distal stomach is to stir and discharge gastric contents, gastric motility induced by orexin-A may facilitate digestive function. In the proximal stomach, orexin-A induced relaxation, while the vehicle had no effect. It is suggested that the relaxation of the proximal stomach enables it to accept more food, and the facilitation of phasic contractions of the distal stomach accelerates draining of the increased gastric contents. Thus, it is plausible that the orexigenic action of orexin-A accompanies the activation of digestive function. These results may answer a common question about why you can eat palatable sweet-tasting foods faster and in larger amount than neutral or unpalatable foods.

4.24.5 Brain Imaging Study on Hedonic Evaluation in Humans

When we wish to analyze the role of taste in feeding and reward in humans, the most suitable method now available is the noninvasive brain imaging techniques such as functional magnetic resonance imaging (fMRI), positron emission tomography (PET), and magnetoencephalography (MEG).

A single chemical solution placed into the mouth can elicit both sensory (or qualitative) perception and hedonic (preferred or aversive) feeling, for example, sucrose is sweet and palatable, while 1 M NaCl is salty and aversive and quinine hydrochloride is bitter and aversive. Therefore, we can analyze sensory and/ or hedonic aspects when these taste stimuli are used.

A number of researchers have identified the primary taste area (PTA) in the opercular-insular cortex by using fMRI (Faurion, A. *et al.*, 1999; O'Doherty, J. *et al.*, 2001; Schoenfeld, M. A. *et al.*, 2004), PET (Kinomura, S. *et al.*, 1994; Small, D. M. *et al.*, 1999 for a review), and MEG (Kobayakawa, T. *et al.*, 1996; Yamamoto, C. *et al.*, 2003; 2006). One of the issues in the PTA concerning taste quality representation is the existence of chemotopy, that is, topographic arrangement of taste responsiveness to each of the five basic tastes: sweet, umami, salty, sour, and bitter. In spite of the limitations of analysis due to the anatomical position and small size of the PTA, Schoenfeld M. A. *et al.* (2004) used the fMRI method and showed the possible existence of chemotopy with a high interindividual variability, although with some considerable overlap. They reported that

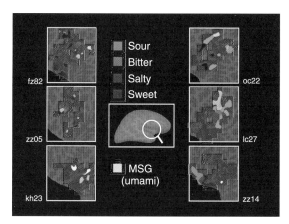

Figure 5 The functional magnetic resonance imaging (fMRI) analysis of the topography of hemodynamic activity elicited by citric acid (sour), caffeine (bitter), NaCl (salty), sucrose (sweet), and monosodium glutamate (umami) on the flattened cortical surfaces of the insular/opercular cortex (encircled area in the center of the figure) of the right hemisphere of six human subjects. Each panel corresponding to each subject shows a composite pattern of responses to multiple stimulations with each of the five taste stimuli denoted by the different colors. Note the different patterns of hemodynamic activity for the five tastes with considerable overlap and a high interindividual topographical variability. Taste-specific patterns, however, were stable over time in each subject. From Schoenfeld, M. A., Neuer, G., Tempelmann, C., Schüßler, K., Noesselt, T., Hopf, J. M., and Heinze, H. J. 2004. Functional magnetic resonance tomography correlates of taste perception in the human primary taste cortex. Neuroscience 127, 347–353.

the taste-specific patterns as shown in Figure 5 were stable over time in each subject. The existence of chemotopy and implication of its importance in taste quality discrimination have been implicated in the rat GC (Yamamoto, T. *et al.*, 1985a; 1985b; 1989a; Yoshimura, H. *et al.*, 2005; Accolla, R. *et al.*, 2007) and suggested in the monkey PTA (Scott, T. R. *et al.*, 1986).

The PTA is involved not only in receiving and processing of taste inputs for qualitative and quantitative discrimination of taste stimuli but also in gustatory imagery by recalling gustatory memories. According to Kobayashi M. *et al.* (2004) in fMRI study, there is an asymmetrical topography of activation in the insula: the left insula is predominantly activated by gustatory imagery tasks, while the right insula, by actual gustatory stimulation in the mouth.

Recent fMRI and PET studies have shown that human amygdala responds specifically to negative emotion, especially to facial expressions of anger, fear, or sorrow (Hyman, S. E., 1988). There is a report,

however, that the left amygdala is activated by a facial expression of fear and right amygdala is activated by an expression of happiness (Morris, J. S. *et al.*, 1996). Unpleasant odor stimulation also activates the amygdala. Zald D. H. and Pardo J. V. (1997) showed that strong unpleasant odor of sulfate compounds activated the bilateral amygdala and left orbitofrontal cortex. There are high correlations among the left amygdala, left orbitofrontal cortex, and perceived aversive feeding, indicating strong functional interactions between the amygdala and orbitofrontal cortex and important roles in emotional evaluation of odor.

Concerning the activation of the amygdala by taste stimulation in humans, Zald D. H. *et al.* (1998) reported in a PET study that 1 M NaCl, a strong and aversive taste stimulus, activated the right amygdala and neutral water and palatable chocolate induced no response in the amygdala, suggesting that the amygdala is responsive to aversive stimuli. Further, Zald D. H. *et al.* (2002) showed in a PET study that bitter and aversive quinine (0.02 M), but not sweet and palatable sucrose (30%), induced big responses in the left amygdala. Both quinine and sucrose excited the right posterior part of the orbitofrontal cortex, but the left inferior frontal pole and the anterior part of the orbitofrontal cortex were activated only by quinine stimulation. On the other hand, O'Doherty J. *et al.* (2001) found in fMRI study that both pleasant (1 M glucose) and unpleasant (0.1 M NaCl) taste stimuli evoked responses in the amygdala and orbitofrontal cortex, but the areas activated were different between the two stimuli. Thus, the activated areas are not consistent among researchers.

As already described, a flavor, taste-odor integration is very important in the sensory evaluation of foods. Small D. M. *et al.* (1997) in a PET study found that mismatched combinations such as salty and strawberry, sour and coffee, sweet and soy sauce, bitter and grapefruit, in contrast to the matched combinations such as salty and soy sauce, bitter and coffee, sour and grapefruit, sweet and strawberry, induced stronger activation in the left amygdala and the bilateral forebrain basal area. From these results, they suggest that these areas are important in processing of neophobic or unpleasant taste stimuli. In a further study using fMRI technique, Small D. M. *et al.* (2004) observed superadditive responses during the perception of the congruent flavor (vanilla/sweet) compared with the sum of its constituents in the anterior cingulate cortex, dorsal insular, anterior ventral insular extending into the caudal orbitofrontal cortex, frontal operculum, ventral lateral PFC, and posterior parietal cortex. Taken together with previous findings in the literature (O'Doherty, J. *et al.*, 2001; Small, D. M. *et al.*, 2001; Zald, D. H. *et al.*, 2002), their results suggest that the insular, orbitofrontal, and anterior cingulate cortices are key components of the network underlying flavor perception.

Acknowledgments

The preparation of the manuscript for this chapter and some of the work reported herein was supported in part by the Grant-in Aid for 21st Century COE program.

References

Abel, E. L. 1971. Effects of marijuana on the solution of anagrams, memory and appetite. Nature 231, 260–261.

Accolla, R., Bathellier, B., Petersen, C. C. H., and Carleton, A. 2007. Differential spatial representation of taste modalities in the rat gustatory cortex. J. Neurosci. 27, 1396–1404.

Apfelbaum, M. and Mandenoff, A. 1981. Naltrexone suppresses hyperphagia induced in the rat by a highly palatable diet. Pharmacol. Biochem. Behav. 15, 89–91.

Azuma, S., Yamamoto, T., and Kawamura, Y. 1984. Studies on gustatory responses of amygdaloid neurons in rats. Exp. Brain Res. 56, 12–22.

Azzara, A. V., Bodnar, R. J., Delamater, A. R., and Sclafani, A. 2000. Naltrexone fails to block the acquisition or expression of a flavor preference conditioned by intragastric carbohydrate infusions. Pharmacol. Biochem. Behav. 67, 545–557.

Azzara, A. V., Bodnar, R. J., Delamater, A. R., and Sclafani, A. 2001. D_1 but not D_2 dopamine receptor antagonism blocks the acquisition of a flavor preference conditioned by intragastric carbohydrate infusions. Pharmacol. Biochem. Behav. 68, 709–720.

Bahar, A., Dudai, Y., and Ahissar, E. 2004. Neural signature of taste familiarity in the gustatory cortex of the freely behaving rat. J. Neurophysiol. 92, 3298–3308.

Bakshi, V. P. and Kelley, A. E. 1993. Feeding induced by opioid stimulation of the ventral striatum: role of opiate receptor subtypes. J. Pharmacol. Exp. Ther. 265, 1253–1260.

Baldo, B. A., Gual-Bonilla, L., Sijapati, K., Daniel, R. A., Landry, C. F., and Kelley, A. E. 2004. Activation of a subpopulation of orexin/hypocretin-containing hypothalamic neurons by GABAA receptor-mediated inhibition of the nucleus accumbens shell, but not by exposure to a novel environment. Eur. J. Pharmacol. 19, 376–386.

Bartoshuk, L. M. and Beauchamp, G. K. 1994. Chemical senses. Annu. Rev. Psychol. 45, 419–449.

Baxter, H. 1933. Variations in the inorganic constituents of mixed and parotid gland saliva activated by reflex stimulation in the dog. J. Biol. Chem. 102, 203–217.

Beckstead, R. M., Morse, J. R., and Norgren, R. 1980. The nucleus of the solitary tract in the monkey: projections to the thalamus and brain stem nuclei. J. Comp. Neurol. 190, 259–282.

Bermúdez-Rattoni, F. 2004. Molecular mechanisms of taste-recognition memory. Nat. Rev. Neurosci. 5, 209–217.

Berridge, K. C. 1988. Brainstem systems mediate the enhancement of palatability by chlordiazepoxide. Brain Res. 447, 262–268.

Berridge, K. C. 1996. Food reward: brain substrates of wanting and liking. Neurosci. Biobehav. Rev. 20, 1–25.

Berridge, K. C. and Peciña, S. 1995. Benzodiazepines, appetite, and taste palatability. Neurosci. Biobehav. Rev. 19, 121–131.

Berridge, K. C. and Robinson, T. E. 1998. What is the role of dopamine in reward: hedonic impact, reward learning, or incentive salience? Brain Res. Rev. 28, 309–369.

Berridge, K. C. and Treit, D. 1986. Chlordiazepoxide directly enhances positive ingestive reactions in rats. Pharmacol. Biochem. Behav. 24, 217–221.

Berridge, K. C., Venier, I. L., and Robinson. T. E. 1989. Taste reactivity analysis of 6-hydroxydopamine-induced aphagia: implications for arousal and anhedonia hypotheses of dopamine function. Behav. Neurosci. 103, 36–45.

Berthoud, H. R., Trimble, E. R., Siegel, E. G., Bereiter, D. A., and Jeanrenaud, B. 1980. Cephalic-phase insulin secretion in normal and pancreatic islet-transplanted rats. Am. J. Physiol. 238, E336–E340.

Bielajew, C. and Trzcinska, M. 1994. Characteristics of stimulation-induced feeding sites in the sulcal prefrontal cortex. Behav. Brain Res. 61, 29–35.

Bishop, M. P., Elder, S. T., and Heath, R. G. 1963. Intracranial self-stimulation in man. Science 140, 394–396.

Blackburn, J. R., Pfaus, J. G., and Phillips, A. G. 1992. Dopamine functions in appetitive and defensive behaviours. Prog. Neurobiol. 39, 247–279.

Brandes, J. S. and Johnson, A. K. 1978. Recovery of feeding in rats following frontal neocortical ablations. Physiol. Behav. 20, 763–770.

Breiter, H. C., Gollub, R. L., Weisskoff, R. M., Kennedy, D. N., Makris, N., Berke, J. D., Goodman, J. M., Kantor, H. L., Gastfriend, D. R., Riorden, J. P., Mathew, R. T., Rosen, B. R., and Hyman, S. E. 1997. Acute effects of cocaine on human brain activity and emotion. Neuron 19, 591–611.

Brog, J. S., Salyapongse, A., Deutch, A. Y., and Zahm, D. S. 1993. The patterns of afferent innervation of the core and shell in the "accumbens" part of the rat ventral striatum: immunohistochemical detection of retrogradely transported fluoro-gold. J. Comp. Neurol. 338, 255–278.

Brown, D. R. and Holtzman, S. G. 1979. Suppression of deprivation-induced food and water intake in rats and mice by naloxone. Pharmacol. Biochem. Behav. 11, 567–573.

Bures, J., Bermúdez-Rattoni, F., and Yamamoto, T. 1998. Conditioned Taste Aversion: Memory of a Special Kind, pp. 1–178. Oxford University Press.

Chauncey, H. H. and Feller, R. P. 1967. Measurement of Human Gustatory Perception Using the Parotid Gland Secretion Rate. In: Olfaction and Taste (*ed.* T. Hayashi), Vol. II, pp. 265–280. Pergamon.

Cole, S. O. 1983. Combined effects of chlordiazepoxide treatment and food deprivation on concurrent measures of feeding and activity. Pharmacol. Biochem. Behav. 18, 369–372.

Cooper, S. J. 1983. Effects of opiate agonists and antagonists on fluid intake and saccharin choice in the rat. Neuropharmacology 22, 323–328.

Cooper, S. J. 1987. Chlordiazepoxide-induced selection of saccharin-flavoured food in the food-deprived rat. Physiol. Behav. 41, 539–542.

Cooper, S. J. 2004. Endocannabinoids and food consumption: comparisons with benzodiazepine and opioid palatability-dependent appetite. Eur. J. Pharmacol. 500, 37–49.

Cooper, S. J. and Estall, L. B. 1985. Behavioural pharmacology of food, water and salt intake in relation to drug actions at benzodiazepine receptors. Neurosci. Biobehav. Rev. 9, 5–19.

Cooper, S. J. and Green, A. E. 1993. The benzodiazepine receptor partial agonists, bretazenil (Ro 16-6028) and Ro 17-1812, affect saccharin preference and quinine aversion in the rat. Behav. Pharmacol. 4, 81–85.

Cooper, S. J. and Kirkham, T. C. 1993. Opioid Mechanisms in the Control of Food Consumption and Taste Preferences. In: Handbook of Experimental Pharmacology. Opioids II, Vol. 104/II (ed. A. Herz), pp. 239–262. Springer-Verlag.

Cooper, S. J. and McClelland, A. 1980. Effects of chlordiazepoxide, food familiarization, and prior shock experience on food choice in rats. Pharmacol. Biochem. Behav. 12, 23–28.

Cooper, S. J. and Yerbury, R. E. 1986. Benzodiazepine-induced hyperphagia: stereospecificity and antagonism by pyrazoloquinolines, CGS 9895 and CGS 9896. Psychopharmacology (Berlin) 89, 462–466.

Cooper, S. J. and Yerbury, R. E. 1988. Clonazepam selectively increases saccharin ingestion in a two-choice test. Brain Res. 456, 173–176.

Cooper, S. J., van der Hoek, G., and Kirkham, T. C. 1988. Bi-directional changes in sham feeding in the rat produced by benzodiazepine receptor ligands. Physiol. Behav. 42, 211–216.

Devane, W. A., Hanus, L., Breuer, A., Pertwee, R. G., Stevenson, L. A., Griffin, G., Gibson, D., Mandelbaum, A., Etinger, A., and Mechoulam, R. 1992. Isolation and structure of a brain constituent that binds to the cannabinoid receptor. Science 258, 1946–1949.

DiNardo, L. A. and Travers, J. B. 1994. Hypoglossal neural activity during ingestion and rejection in the awake rat. J. Neurophysiol. 72, 1181–1191.

Divac, I., Kosmal, A., Bjorklund, A., and Lindvall, O. 1978. Subcortical projections to the prefrontal cortex in the rat as revealed by the horseradish peroxidase technique. Neuroscience 3, 785–796.

Domjan, M. 1976. Determinants of the enhancement of flavored-water intake by prior exposure. J. Exp. Psychol. Anim. Behav. Process 2, 17–27.

Doyle, T. G., Berridge, K. C., and Gosnell, B. A. 1993. Morphine enhances hedonic taste palatability in rats. Pharmacol. Biochem. Behav. 46, 745–749.

Drewnowski, A., Krahn, D. D., Demitrack, M. A., Nairn, K., and Gosnell, B. A. 1992. Taste responses and preferences for sweet high-fat foods: evidence for opioid involvement. Physiol. Behav. 51, 371–379.

Elizalde, G. and Sclafani, A. 1990. Flavor preferences conditioned by intragastric Polycose infusions: a detailed analysis using an electronic esophagus preparation. Physiol. Behav. 47, 63–77.

Emmelin, N. and Holmberg, J. 1967. Impulse frequency in secretory nerves of salivary glands. J. Physiol. Lond. 191, 205–214.

Fanselow, G. and Birk, J. 1982. Flavor–flavor associations induce hedonic shifts in taste preference. Anim. Learn. Behav. 10, 223–228.

Fantino, M., Hosotte, J., and Apfelbaum, M. 1986. An opioid antagonist, naltrexone, reduces preference for sucrose in humans. Am. J. Physiol. 251, R91–R96.

Faurion, A., Cerf, B., Van De Moortele, P. F., Lobel, E., Mac Leod, P., and Le Bihan, D. 1999. Human taste cortical areas studied with functional magnetic resonance imaging: evidence of functional lateralization related to handedness. Neurosci. Lett. 277, 189–192.

Feller, R. P., Sharon, I. M., Chauncey, H. H., and Shannon, I. L. 1965. Gustatory perception of sour, sweet, and salt mixtures using parotid gland flow rate. J. Appl. Physiol. 20, 1341–1344.

Fischer, U., Hommel, H., Ziegler, M., and Michael, R. 1972. The mechanism of secretion after oral glucose administration. I. Multiphasic course of insulin mobilization after oral

administration of glucose in conscious dogs. Diabetologia 8, 104–110.

Flowers, S. H., Dunham, E. S., and Barbour, H. G. 1929. Addiction edema, and withdrawal edema in morphinized rats. Proc. Soc. Exp. Biol. Med. 26, 572–574.

Foltin, R. W., Brady, J. V., and Fischman, M. V. 1986. Behavioral analysis of marijuana effects on food intake in humans. Pharmacol. Biochem. Behav. 25, 577–582.

Foltin, R. W., Ellis, S., and Schuster, C. R. 1985. Specific antagonism by Ro15-1788 of benzodiazepine-induced increases in food intake in rhesus monkeys. Pharmacol. Biochem. Behav. 23, 249–252.

Foltin, R. W., Fischman, M. W., and Byrne, M. F. 1989. Food intake in baboons: effects of diazepam. Psychopharmacology (Berl.) 97, 443–447.

Funakoshi, M. and Kawamura, Y. 1967. Relations Between Taste Qualities and Parotid Gland Secretion Rate. In: Olfaction and Taste II (ed. T. Hayashi), pp. 281–287. Pergamon Press.

Funakoshi, M., Kasahara, Y., Yamamoto, T., and Kawamura, Y. 1972. Taste Coding and Central Perception. In: Olfaction and Taste IV (ed. D. Schneider), pp. 336–342. Wissenschaftlich Verlagsgesellschaft MBH.

Furudono, Y., Ando, C., Yamamoto, C., Kobashi, M., and Yamamoto, T. 2006. Involvement of specific orexigenic neuropeptides in sweetener-induced overconsumption in rats. Behav. Brain Res. 175, 241–248.

Ganchrow, J. R., Oppenheimer, M., and Steiner, J. E. 1979. Behavioral displays to gustatory stimuli in newborn rabbit pups. Chem. Senses Flavor 4, 49–61.

Gantt, W. H. 1973. The nervous secretion of saliva. Quantitative studies in the natural unconditioned reflex secretion of parotid saliva. Am. J. Physiol. 119, 493–507.

Geary, N. and Smith, G. P. 1985. Pimozide decreases the positive reinforcing effect of sham fed sucrose in the rat. Pharmacol. Biochem. Behav. 22, 787–790.

Giza, B. K., Ackroff, K., McCaughey, S. A., Sclafani, A., and Scott, T. R. 1997. Preference conditioning alters taste responses in the nucleus of the solitary tract of the rat. Am. J. Physiol. 273, R1230–R1240.

Gjorstrup, P. 1980. Taste and chewing as stimuli for the secretion of amylase from the parotid gland of the rabbit. Acta. Physiol. Scand. 110, 295–301.

Glass, M. J., Billington, C. J., and Levine, A. S. 1999. Opioids and food intake: distributed functional neural pathway? Neuropeptides 33, 360–368.

Gosnell, B. A. 1987. Central structures involved in opioid-induced feeding. Fed. Proc. 46, 163–167.

Grandison, L. and Guidotti, A. 1977. Stimulation of food intake by muscimol and β-endorphin. Neuropharmacology 16, 533–536.

Gray, R. W. and Cooper, S. J. 1995. Benzodiazepines and palatability: taste reactivity in normal ingestion. Physiol. Behav. 58, 853–859.

Grill, H. J. and Norgren, R. 1978a. The taste reactivity test. I. Mimetic responses to gustatory stimuli in neurologically normal rats. Brain Res. 143, 263–279.

Grill, H. J. and Norgren, R. 1978b. The taste reactivity test. II. Mimetic responses to gustatory stimuli in chronic thalamic and chronic decerebrate rats. Brain Res. 143, 281–297.

Groenewegen, H. J., Berendse, H. W., and Haber, S. N. 1993. Organization of the output the ventral striatopallidal system in the rat: ventral pallidal efferents. Neuroscience 57, 113–142.

Gutiérrez, R., Téllez, L. A., and Bermúdez-Rattoni, F. 2003. Blockade of cortical muscarinic but not NMDA receptors prevents a novel taste from becoming familiar. Eur. J. Neurosci. 17, 1556–1562.

Hajnal, A., Smith, G. P., and Norgren, R. 2004. Oral sucrose stimulation increases accumbens dopamine in the rat. Am. J. Physiol. Regul. Integr. Comp. Physiol. 286, R31–R37.

Halsell, C. B., Travers, S. P., and Travers, J. B. 1996. Ascending and descending projections from the rostral nucleus of the solitary tract originate from separate neuronal populations. Neuroscience 72, 185–197.

Hanamori, T., Kunitake, T., Kato, K., and Kannan, H. 1998. Responses of neurons in the insular cortex to gustatory, visceral, and nociceptive stimuli in rats. J. Neurophysiol. 79, 2535–2545.

Haney, M., Comer, S. D., Fischman, M. W., and Foltin, R. W. 1997. Alprazolam increases food intake in humans. Psychopharmacology (Berl.) 132, 311–314.

Heimer, L., Zahm, D. S., Churchill, L., Kalivas, P. W., and Wohltmann, C. 1991. Specificity in the projection patterns of accumbal core and shell in the rat. Neuroscience 41, 89–125.

Herbert, H., Moga, M. M., and Saper, C. B. 1990. Connections of the parabrachial nucleus with the nucleus of the solitary tract and the medullary reticular formation in the rat. J. Comp. Neurol. 293, 540–580.

Hernádi, I., Karádi, Z., Vígh, J., Petykó, Z., Egyed, R., Berta, B., and Lénárd, L. 2000. Alterations of conditioned taste aversion after microiontophoretically applied neurotoxins in the medial prefrontal cortex of the rat. Brain Res. Bull. 53, 751–758.

Higgs, S. and Cooper, S. J. 1996a. Increased food intake following injection of the benzodiazepine receptor agonist midazolam into the IVth ventricle. Pharmacol. Biochem. Behav. 55, 81–86.

Higgs, S. and Cooper, S. J. 1996b. Hyperphagia induced by direct administration of midazolam into the parabrachial nucleus of the rat. Eur. J. Pharmacol. 313, 1–9.

Higgs, S. and Cooper, S. J. 1997. Midazolam-induced rapid changes in licking behaviour: evidence for involvement of endogenous opioid peptides. Psychopharmacology (Berl.) 131, 278–286.

Higgs, S. and Cooper, S. J. 1998a. Effects of benzodiazepine receptor ligands on the ingestion of sucrose, Intralipid, and maltodextrin: an investigation using a microstructural analysis of licking behavior in a brief contact test. Behav. Neurosci. 112, 447–457.

Higgs, S. and Cooper, S. J. 1998b. Evidence for early opioid modulation of licking responses to sucrose and Intralipid: a microstructural analysis in the rat. Psychopharmacology (Berl.) 139, 342–355.

Higgs, S., Williams, C. M., and Kirkham, T. C. 2003. Cannabinoid influences on palatability: microstructural analysis of sucrose drinking after Δ9-tetrahydrocannabinol, anandamide, 2-arachidonoyl glycerol and SR141716. Psychopharmacology (Berlin) 165, 370–377.

Holtzman, S. G. 1974. Behavioral effects of separate and combined administration of naloxone and d-amphetamine. J. Pharmacol. Exp. Ther. 189, 51–60.

Hyman, S. E. 1988. Neurobiology: a new image for fear and emotion. Nature 393, 417–418.

Inui, A. 2000. Transgenic approach to the study of body weight regulation. Pharmacol. Rev. 52, 35–62.

Jacobs, K. M., Mark, G. P., and Scott, T. R. 1988. Taste responses in the nucleus tractus solitarius of sodium-deprived rats. J. Physiol. (Lond.) 406, 393–410.

Janowitz, H. D., Hollander, F., Orringer, P., Levy, M. H., Winkelstein, A., Kaufman, M. R., and Margolin, S. G. 1950. A quantitative study of the gastric secretory response to sham feeding in a human subject. Gastroenterology 16, 104–116.

Karádi, Z., Lukáts, B., Papp, S.z., Egyed, R., Lénárd, L., and Takács, G. 2005. Involvement of forebrain glucose-monitoring neurons in taste information processing:

electrophysiological and behavioral studies. Chem. Senses 30, i168–i169.

Karimnamazi, H. and Travers, J. B. 1998. Differential projections from gustatory responsive regions of the parabrachial nucleus to the medulla and forebrain. Brain Res. 813, 283–302.

Katz, D. B., Simon, S. A., and Nicolelis, M. A. L. 2001. Dynamic and multimodal responses of gustatory cortical neurons in awake rats. J. Neurosci. 21, 4478–4489.

Kawamura, Y. and Yamamoto, T. 1978. Studies on neural mechanisms of the gustatory-salivary reflex in rabbits. J. Physiol. (Lond.) 285, 35–47.

Kelley, A. E., Bakshi, V. P., Haber, S. N., Steininger, T. L., Will, M. J., and Zhang, M. 2002. Opioid modulation of taste hedonics within the ventral striatum. Physiol. Behav. 76, 365–377.

Kephalas, T. A., Kiburis, J., Michael, C. M., Miras, C. J., and Papadakis, D. P. 1976. Some Aspects of Cannabis Smoke Chemistry. In: Marihuana: Chemistry, Biochemistry, and Cellular Affects (ed. G. G. Nahas), pp. 39–49. Springer-Verlag.

Kerr, A. C. 1961. The Physiological Regulation of Salivary Secretions in Man, Pergamon Press.

Kinomura, S., Kawashima, R., Yamada, K., Ono, S., Itoh, M., Yoshioka, S., Yamaguchi, T., Matsui, H., Miyazawa, H., Itoh, H., Goto, R., Fujiwara, T., Satoh, K., and Fukuda, H. 1994. Functional anatomy of taste perception in the human brain studied with positron emission tomography. Brain Res. 659, 263–266.

Kirkham, T. C. and Cooper, S. J. 1988. Attenuation of sham feeding by naloxone is stereospecific: evidence for opioid mediation of orosensory reward. Physiol. Behav. 43, 845–847.

Kirkham, T. C., Williams, C. M., Fezza, F., and Di Marzo, V. 2002. Endocannabinoid levels in rat limbic forebrain and hypothalamus in relation to fasting, feeding and satiation: stimulation of eating by 2-arachidonoyl glycerol. Br. J. Pharmacol. 136, 550–557.

Kita, H. and Oomura, Y. 1981. Reciprocal connections between the lateral hypothalamus and the frontal cortex in the rat: Electrophysiological and anatomical observations. Brain Res. 213, 1–16.

Kobashi, M. and Bradley, R. M. 1998. Effects of GABA on neurons of the gustatory and visceral zones of the parabrachial nucleus in rats. Brain Res. 799, 323–328.

Kobashi, M., Furudono, Y., Matsuo, R., and Yamamoto, T. 2003. Central orexin facilitates gastric relaxation and contractility in rats. Neurosci. Lett. 332, 171–174.

Kobashi, M., Mizutani, M., and Matsuo, R. 2000. Water stimulation of the posterior oral cavity induces inhibition of gastric motility. Am. J. Physiol. 279, R778–R785.

Kobayakawa, T., Endo, H., Ayabe-Kanamura, S., Kumagai, T., Yamaguchi, Y., Kikuchi, Y., Takeda, T., Saito, S., and Ogawa, H. 1996. The primary gustatory area in human cerebral cortex studied by magnetoencephalography. Neurosci. Lett. 212, 155–158.

Kobayashi, M., Takeda, M., Hattori, N., Fukunaga, M., Sasabe, T., Inoue, N., Nagai, Y., Sawada, T., Sadato, N., and Watanabe, Y. 2004. Functional imaging of gustatory perception and imagery: top-down processing of gustatory signals. Neuroimage 23, 1271–1282.

Kolb, B. and Nonneman, A. J. 1975. Prefrontal cortex and the regulation of food intake in the rat. J. Comp. Physiol. Psychol. 88, 806–815.

Kosobud, A. E., Harris, G. C., and Chapin, J. K. 1994. Behavioral associations of neuronal activity in the ventral tegmental area of the rat. J. Neurosci. 14, 7117–7129.

Kotz, C. M., Billington, C. J., and Levine, A. S. 1997. Opioids in the nucleus of the solitary tract are involved in feeding in the rat. Am. J. Physiol. 272, R1028–R1032.

Le Magnen, J., Marfaing-Jallat, P., Miceli, D., and Devos, M. 1980. Pain modulating and reward systems: a single brain mechanism? Pharmacol. Biochem. Behav. 12, 729–733.

Li, C. S., Davis, B. J., and Smith, D. V. 2003. Opioid modulation of taste responses in the NTS. Brain Res. 965, 21–34.

Ljungberg, T., Apicella, P., and Schultz, W. 1992. Responses of monkey dopamine neurons during learning of behavioral reactions. J. Neurophysiol. 67, 145–163.

Lukáts, B., Papp, S.z., Szalay, C.s., Gode, J., Lénárd, L., and Karádi, Z. 2002. Gustatory neurons in the nucleus accumbens and the mediodorsal prefrontal cortex of the rat. Acta Physiol. Hung. 89, 250.

Maldonado-Irizarry, C. S., Swanson, C. J., and Kelley, A. E. 1995. Glutamate receptors in the nucleus accumbens shell control feeding behavior via the lateral hypothalamus. J. Neurosci. 15, 6779–6788.

Manabe, Y., Kuroda, K., Imaizumi, M., Inoue, K., Sako, N., Yamamoto, T., Fushiki, T., and Hanai, K. 2000. Diazepam-binding inhibitor-like activity in rat cerebrospinal fluid after stimulation by an aversive quinine taste. Chem. Senses 25, 739–746.

Manabe, Y., Toyoda, T., Kuroda, K., Imaizumi, M., Yamamoto, T., and Fushiki, T. 2001. Effect of diazepam binding inhibitor (DBI) on the fluid intake, preference and the taste reactivity in mice. Behav. Brain Res. 126, 197–204.

Mansour, A., Khachaturian, H., Lewis, M. E., Akil, H., and Watson, S. J. 1987. Autoradiographic differentiation of mu, delta, and kappa opioid receptors in the rat forebrain and midbrain. J. Neurosci. 7, 2445–2464.

Maren, S. and Fanselow, M. S. 1996. The amygdala and fear conditioning: has the nut been cracked? Neuron 16, 237–240.

Margules, D. L. and Stein, L. 1967. Neuroleptics vs. Tranquilizers: Evidence from Animals of Mode and Site of Action. In: Neuropsychopharmacology (ed. H. Brill), pp. 74–80. Excerpta Medica Foundation.

Martin, W. R., Wikler, A., Eades, C. G., and Persor, F. T. 1963. Tolerance to and physical dependence on morphine in rats. Psychopharmacologia 4, 247–260.

Matsuo, R. and Yamamoto, T. 1989. Gustatory-salivary reflex: neural activity of sympathetic and parasymathetic fibers innervating the submandibular gland of the hamster. J. Auton. Nerv. Syst. 26, 187–197.

Matsuo, R., Yamamoto, T., Yoshitaka, K., and Morimoto, T. 1989. Neural substrates for reflex salivation induced by taste, mechanical, and thermal stimulation of the oral region in decerebrate rats. Jpn. J. Physiol. 39, 349–357.

Matsuo, R., Yamauchi, Y., Kobashi, M., Funahashi, M., Mitoh, Y., and Adachi, A. 2001. Role of parabrachial nucleus in submandibular salivary secretion induced by bitter taste stimulation in rats. Auton. Neurosci. 88, 61–73.

McDonald, A. J. and Jackson, T. R. 1987. Amygdaloid connections with posterior insular and temporal cortical areas in the rat. J. Comp. Neurol. 262, 59–77.

McDonald, A. J., Shammah-Lagnado, S. J., Shi, C., and Davis, M. 1999. Cortical afferents to the extended amygdala. Ann. N. Y. Acad. Sci. 877, 309–338.

McLaughlin, C. L. and Baile, C. A. 1979. Cholecystokinin, amphetamine and diazepam and feeding in lean and obese Zucker rats. Pharmacol. Biochem. Behav. 10, 87–93.

McLean, S. and Hoebel, B. G. 1983. Feeding induced by opiates injected into the paraventricular hypothalamus. Peptides 4, 287–292.

Meredith, G. E., Pennartz, C. M., and Groenewegen, H. J. 1993. The cellular framework for chemical signaling in the nucleus accumbens. Prog. Brain Res. 99, 3–24.

Miller, C. C., Murray, T. F., Freeman, K. G., and Edwards, G. L. 2004. Cannabinoid agonist, CP 55,940, facilitates intake of palatable foods when injected into the hindbrain. Physiol. Behav. 80, 611–616.

Miller, F. R. and Sherrington, C. R. 1916. Some observations on the buccopharygeal stage of reflex deglutition the cat. Q. J. Exp. Physiol. 9, 147–186.

Mogenson, G. J., Jones, D. L., and Yim, C. Y. 1980. From motivation to action: functional interface between limbic system and the motor system. Prog. Neurobiol. 14, 69–97.

Morris, J. S., Frith, C. D., Perrett, D. I., Rowland, D., Young, A. W., Calder, A. J., and Dolan, R. J. 1996. A differential neural response in the human amygdala to fearful and happy facial expressions. Nature 383, 812–815.

Mucha, R. F. and Herz, A. 1986. Preference conditioning produced by opioid active and inactive isomers of levorphanol and morphine in rat. Life Sci. 38, 241–249.

Nachman, M. and Jones, D. R. 1974. Learned taste aversions over long delays in rats: the role of learned safety. J. Comp. Physiol. Psychol. 86, 949–956.

Nakamura, K. and Norgren, R. 1995. Sodium-deficient diet reduces gustatory activity in the nucleus of the solitary tract of behaving rats. Am. J. Physiol. 269: R647–R661.

Newbrun, E. 1962. Observations on the amylase content and flow rate of human saliva following gustatory stimulation. J. Dent. Res. 41, 459–465.

Nicolaidis, S. 1977. Sensory-Neuroendocrine Reflexes and Their Anticipatory and Optimizing Role on Metabolism. In: The Chemical Senses and Nutrition (eds. M. R. Kare and O. Maller), pp. 123–140. Academic Press.

Niijima, A. 1991. Effects of oral and intestinal stimulation with umami substance on gastric vagus activity. Physiol. Behav. 49, 1025–1028.

Niijima, A., Togiyama, T., and Adachi, A. 1990. Cephalic-phase insulin release induced by taste stimulus of monosodium glutamate (umami taste). Physiol. Behav. 48, 905–908.

Nishijo, H., Ono, T., and Nishino, H. 1988a. Topographic distribution of modality-specific amygdalar neuron in alert monkey. J. Neurosci. 8, 3556–3569.

Nishijo, H., Ono, T., and Nishino, H. 1988b. Single neuron responses in amygdala of alert monkey during complex sensory stimulation with affective significance. J. Neurosci. 8, 3570–3583.

Nishijo, H., Uwano, T., Tamura, R., and Ono, T. 1998c. Gustatory and multimodal neuronal responses in the amygdala during licking and discrimination of sensory stimuli in awake rats. J. Neurophysiol. 79, 21–36.

Norgren, R. 1995. Gustatory System. In: The Rat Nervous System (ed. G. Paxinos), pp. 751–771. Academic Press.

Novis, B. H., Bank, S., and Marks, I. N. 1971. The cephalic phase of pancreatic secretion in man. Scand. J. Gastroenterol. 6, 417–421.

Nowlis, G. H. 1973. Taste-Elicited Tongue Movements in Human Newborn Infants: An Approach to Palatability. In: 4th Symposium on Oral Sensation and Perception (ed. J. F. Bosma), pp. 292–303. US Government Printing Office.

O'Doherty, J., Rolls, E. T., Francis, S., Bowtell, R., and McGlone, F. 2001. Representation of pleasant and aversive taste in the human brain. J. Neurophysiol. 85, 1315–1321.

Ohara, I., Naruse, M. M., and Itokawa, Y. 1996. The effect of palatability and feeding conditions on digestive functions in rats. J. Am. Coll. Nutr. 15, 186–191.

Ohara, I., Otsuka, S., and Yugari, Y. 1988. Cephalic phase response of pancreatic exocrine secretion in conscious dogs. Am. J. Physiol. 254, G416–G423.

Olds, J. and Milner, P. M. 1954. Positive reinforcement produced by electrical stimulation of septal area and other regions of rat brain. J. Comp. Physiol. Psychol. 47, 419–427.

Parker, L. A. 1991. Chlordiazepoxide nonspecifically enhances consumption of saccharin solution. Pharmacol. Biochem. Behav. 38, 375–377.

Pavlov, I. P. 1902. The Work of the Digestive Glands. Charles Griffin.

Peciña, S. and Berridge, K. C. 1995. Central enhancement of taste pleasure by intraventricular morphine. Neurobiology Bp 3, 269–280.

Peciña, S. and Berridge, K. C. 1996. Brainstem mediates diazepam enhancement of palatability and feeding: microinjections into forth ventricle versus lateral ventricle. Brain Res. 727, 22–30.

Peciña, S. and Berridge, K. C. 2000. Opioid site in nucleus accumbens shell mediates eating and hedonic 'liking' for food: map based on microinjection Fos plumes. Brain Res. 863, 71–86.

Peciña, S., Cagniard, B., Berridge, K. C., Aldridge, J. W., and Zhuang, X. 2003. Hyperdopaminergic mutant mice have higher "wanting" but not 'liking' for sweet rewards. J. Neurosci. 23, 9395–9402.

Pérez-Jaranay, J. M. and Vives, F. 1991. Electrophysiological study of the response of medial prefrontal cortex neurons to stimulation of the basolateral nucleus of the amygdala in the rat. Brain Res. 564, 97–101.

Poschel, B. P. H. 1971. A simple and specific screen for benzodiazepine-like drugs. Psychopharmacologia 19, 193–198.

Ricardo, J. A. and Koh, E. T. 1978. Anatomical evidence of direct projections from the nucleus of the solitary tract to the hypothalamus, amygdala, and other forebrain structures in the rat. Brain Res. 153, 1–26.

Richards, J. G. and Mohler, H. 1984. Benzodiazepine receptors. Neuropharmacology 23, 233–242.

Robinson, T. G. and Beart, P. M. 1988. Excitant amino acid projections from rat amygdala and thalamus to nucleus accumbens. Brain Res. Bull. 20, 467–471.

Rolls, E. T. 2004. The functions of the orbitofrontal cortex. Brain Cogn. 55, 11–29.

Rolls, E. T. and Baylis, L. L. 1994. Gustatory, olfactory, and visual convergence within the primate orbitofrontal cortex. J. Neurosci. 14, 5437–5452.

Rolls, E. T., Critchley, H. D., Browning, A. S., Hernadi, I., and Lenard, L. 1999. Responses to the sensory properties of fat of neurons in the primate orbitofrontal cortex. J. Neurosci. 19, 1532–1540.

Rolls, E. T., Verhagen, J. V., and Kadohisa, M. 2003. Representations of the texture of food in the primate orbitofrontal cortex: neurons responding to viscosity, grittiness, and capsaicin. J. Neurophysiol. 90, 3711–3724.

Saper, C. B. 1982. Convergence of autonomic and limbic connections in the insular cortex of the rat. J. Comp. Neurol. 210, 163–173.

Schoenfeld, M. A., Neuer, G., Tempelmann, C., Schüßler, K., Noesselt, T., Hopf, J. M., and Heinze, H. J. 2004. Functional magnetic resonance tomography correlates of taste perception in the human primary taste cortex. Neuroscience 127, 347–353.

Schultz, W. and Romo, R. 1990. Dopamine neurons of the monkey midbrain: contingencies of responses to stimuli eliciting immediate behavioral reactions. J. Neurophysiol. 63, 607–624.

Schwartz, M. W., Woods, S. C., Porte, D., Jr., Sleeley, R. J., and Baskin, D. G. 2000. Central nervous system control of food intake. Nature 404, 661–671.

Sclafani, A. 2004. Oral and postoral determinants of food reward. Physiol. Behav. 81, 773–779.

Sclafani, A., Azzara, A. V., Touzani, K., Grigson, P. S., and Norgren, R. 2001. Parabrachial nucleus lesions block taste and attenuate flavor preference and aversion conditioning in rats. Behav. Neurosci. 115, 920–933.

Scott, T. R. and Plata-Salaman, C. R. 1999. Taste in the monkey cortex. Physiol. Behav. 67, 489–511.

Scott, T. R., Yaxley, S., Siekiewicz, Z. J., and Rolls, E. T. 1986. Gustatory responses in the frontal opercular cortex of the alert cynomolgus monkey. J. Neurophysiol. 56, 876–890.

Sewards, T. V. 2004. Dual separate pathways for sensory and hedonic aspects of taste. Brain Res. Bull. 62, 271–283.

Shammah-Lagnado, S. J., Costa, M. S. M. O., and Ricardo, J. A. 1992. Afferent connections of the parvocellular reticular formation: a horseradish peroxidase study in the rat. Neuroscience 50, 403–425.

Shi, C.-J. and Cassell, M. D. 1998. Cortical, thalamic, and amygdaloid connections of the anterior and posterior insular cortices. J. Comp. Neurol. 399, 440–468.

Shimura, T., Imaoka, H., Okazaki, Y., Kanamori, Y., Fushiki, T., and Yamamoto, T. 2005. Involvement of the mesolimbic system in palatability-induced ingestion. Chem. Senses 30, i188–i189.

Shimura, T., Imaoka, H., and Yamamoto, T. 2006. Neurochemical modulation of ingestive behavior in the ventral pallidum. Eur. J. Neurosci. 23, 1596–1604.

Shimura, T., Kamada, Y., and Yamamoto, T. 2002a. Ventral tegmental lesions reduce overconsumption of normally preferred taste fluid in rats. Behav. Brain Res. 134, 123–130.

Shimura, T., Komori, M., and Yamamoto, T. 1997a. Acute sodium deficiency reduces gustatory responsiveness to NaCl in the parabrachial nucleus of rats. Neurosci. Lett. 236, 33–36.

Shimura, T., Tanaka, H., and Yamamoto, T. 1997b. Salient responsiveness of parabrachial neurons to the conditioned stimulus after the acquisition of taste aversion learning in rats. Neuroscience 81, 239–247.

Shimura, T., Tokita, K., and Yamamoto, T. 2002b. Parabrachial unit activities after the acquisition of conditioned taste aversion to a nonpreferred HCl solution in rats. Chem. Senses 27, 153–158.

Shimura, T., Watanabe, U., Yanagawa, Y., and Yamamoto, T. 2004. Altered taste function in mice deficient in the 65-kDa isoform of glutamate decarboxylase. Neurosci. Lett. 356, 171–174.

Shipley, M. T. and Geinisman, Y. 1984. Anatomical evidence for convergence of olfactory, gustatory, and visceral afferent pathways in mouse cerebral cortex. Brain Res. Bull. 12, 221–226.

Small, D. M., Jones-Gotman, M., Zatorre, R. J., Petrides, M., and Evance, A. C. 1997. Flavor processing: more than the sum of its part. Neuroreport 8, 3913–3917.

Small, D. M., Voss, J., Mak, Y. E., Simmons, K. B., Parrish, T., and Gitelman, D. 2004. Experience-dependent neural integration of taste and smell in the human brain. J. Neurophysiol. 92, 1892–1903.

Small, D. M., Zald, D. H., Jones-Gotman, M., Zatorre, R. J., Pardo, J. V., Stephen, F., and Petrides, M. 1999. Human cortical gustatory areas: a review of functional neuroimaging data. Neuroreport 10, 7–14.

Small, D. M., Zatorre, R. J., Dagher, A., Evans, A. C., and Jones-Gotman, M. 2001. Changes in brain activity related to eating chocolate: from pleasure to aversion. Brain 124, 1720–1733.

Söderpalm, A. H. and Berridge, K. C. 2000. The hedonic impact and intake of food are increased by midazolam microinjection in the parabrachial nucleus. Brain Res. 877, 288–297.

Söderpalm, A. H. and Hansen, S. 1998. Benzodiazepines enhance the consumption and palatability of alcohol in the rat. Psychopharmacology (Berl.) 137, 215–222.

Steffens, A. B. 1976. Influence of the oral cavity on insulin release in the rat. Am. J. Physiol. 230, 1411–1415.

Steiner, J. E. 1973. The Gustofacial Response: Observation on Normal and Anencephalic Newborn Infants. In: 4th Symposium on Oral Sensation and Perception (*ed*. J. F. Bosma), pp. 254–278. US Government Printing Office.

Stratford, T. R. and Kelley, A. E. 1997. GABA in the nucleus accumbens shell participates in the central regulation of feeding behavior. J. Neurosci. 17, 4434–4440.

Stratford, T. R. and Kelley, A. E. 1999. Evidence of a functional relationship between the nucleus accumbens shell and lateral hypothalamus subserving the control of feeding behavior. J. Neurosci. 19, 11040–11048.

Stratford, T. R., Kelley, A. E., and Simansky, K. J. 1999. Blockade of GABAA receptors in the medial ventral pallidum elicits feeding in satiated rats. Brain Res. 825, 199–203.

Stratford, T. R., Swanson, C. J., and Kelley, A. 1998. Specific changes in food intake elicited by blockade or activation of glutamate receptors in the nucleus accumbens shell. Behav. Brain Res. 93, 43–50.

Tabuchi, E., Ono, T., Nishijo, H., and Torii, K. 1991. Amino acid and NaCl appetite, and LHA neuron responses of lysine-deficient rat. Physiol. Behav. 49, 951–964.

Tokita, K., Karádi, Z., Shimura, T., and Yamamoto, T. 2004. Centrifugal inputs modulate taste aversion learning associated parabrachial neuronal activities. J. Neurophysiol. 92, 265–279.

Touzani, K. and Sclafani, A. 2001. Conditioned flavor preference and aversion: role of the lateral hypothalamus. Behav. Neurosci. 115, 84–93.

Travers, J. B., Urbanek, K., and Grill, H. J. 1999. Fos-like immunoreactivity in the brain stem following oral quinine stimulation in decerebrate rats. Am. J. Physiol. 277, R384–R394.

Treesukosol, Y., Ishizuka, T., Yamamoto, C., Senda, K., Tsutsumi, S., Yamatodani, A., and Yamamoto, T. 2005. Hypothalamic histamine release by taste stimuli in freely moving rats: possible implication of palatability. Behav. Brain Res. 164, 67–72.

Treit, D., Berridge, K. C., and Schultz, C. E. 1987. The direct enhancement of positive palatability by chlordiazepoxide is antagonized by Ro 15-1788 and CGS 8216. Pharmacol. Biochem. Behav. 26, 709–714.

Weiffenbach, J. M. and Thach, B. T. 1973. Elicited Tongue Movements: Touch and Taste in the Mouth of the Neonate. In: 4th Symposium on Oral Sensation and Perception (*ed*. J. F. Bosma), pp. 232–243. US Government Printing Office.

Williams, C. M. and Kirkham, T. C. 2002. Observational analysis of feeding induced by Δ9-THC and anandamide. Physiol. Behav. 76, 241–250.

Wilson, J. D., Nicklous, D. M., Aloyo, V. J., and Simansky, K. J. 2003. An orexigenic role for mu-opioid receptors in the lateral parabrachial nucleus. Am. J. Physiol. Regul. Integr. Comp. Physiol. 285, R1055–R1065.

Wise, R. A. 1982. Neuroleptics and operant behavior: the anhedonia hypothesis. Behav. Brain Sci. 5, 39–87.

Wise, R. A. 2002. Brain reward circuitry: insights from unsensed incentives. Neuron 36, 229–240.

Wise, R. A. and Dawson, V. 1974. Diazepam-induced eating and lever pressing for food in sated rats. J. Comp. Physiol. Psychol. 86, 930–941.

Wise, R. A., Spindler, J., deWit, H., and Gerber, G. J. 1978. Neuroleptic-induced "anhedonia" in rats: pimozide blocks reward quality of food. Science 201, 262–264.

Woods, J. S. and Leibowitz, S. F. 1985. Hypothalamic sites sensitive to morphine and naloxone: effects on feeding behavior. Pharmacol. Biochem. Behav. 23, 431–438.

Xenakis, S. and Sclafani, A. 1981. The effects of pimozide on the consumption of a palatable saccharin-glucose solution in the rat. Pharmacol. Biochem. Behav. 15, 435–442.

Yamamoto, C., Nagai, H., Takahashi, K., Nakagawa, S., Yamaguchi, M., Tonoike, M., and Yamamoto, T. 2006.

Cortical representation of taste modifying action of miracle fruit in humans. Neuroimage 33, 1145–1151.

Yamamoto, C., Takehara, S., Morikawa, K., Nakagawa, S., Yamaguchi, M., Iwaki, S., Tonoike, M., and Yamamoto, T. 2003. Magenetoencephalographic study of cortical activity evoked by electrogustatory stimuli. Chem. Senses 28, 245–251.

Yamamoto, T. 2007. Brain regions responsible for the expression of conditioned taste aversion in rats. Chem. Senses 32, 105–109.

Yamamoto, T. and Kawamura, Y. 1977. Responses of the submandibular secretory nerve to taste stimuli. Brain Res. 130, 152–155.

Yamamoto, T., Fujimoto, Y., Shimura, T., and Sakai, N. 1995. Conditioned taste aversion in rats with excitotoxic brain lesions. Neurosci. Res. 22, 31–49.

Yamamoto, T., Matsuo, R., Ichikawa, H., Wakisaka, S., Akai, M., Imai, Y., Yonehara, N., and Inoki, R. 1990. Aversive taste stimuli increase CGRP levels in the gustatory insular cortex of the rat. Neurosci. Lett. 112, 167–172.

Yamamoto, T., Matsuo, R., Kiyomitsu, Y., and Kitamura, R. 1988. Sensory inputs from the oral region to the cerebral cortex in behaving rats: An analysis of unit responses in cortical somatosensory and taste areas during ingestive behavior. J. Neurophysiol. 60, 1303–1321.

Yamamoto, T., Matsuo, R., Kiyomitsu, Y., and Kitamura, R. 1989a. Taste responses of cortical neurons in freely ingesting rats. J. Neurophysiol. 61, 1244–1258.

Yamamoto, T., Matsuo, R., Kiyomitsu, Y., and Kitamura, R. 1989b. Response properties of lateral hypothalamic neurons during ingestive behavior with special reference to licking of various taste solutions. Brain Res. 481, 286–297.

Yamamoto, T., Nagai, T., Shimura, T., and Yasoshima, Y. 1998. Roles of chemical mediators in the taste system. Jpn. J. Pharmacol. 76, 325–348.

Yamamoto, T., Sako, N., and Maeda, S. 2000. Effects of taste stimulation on β-endorphin levels in rat cerebrospinal fluid and plasma. Physiol. Behav. 69, 345–350.

Yamamoto, T., Shimura, T., Sakai, N., and Ozaki, N. 1994a. Representation of hedonics and quality of taste stimuli in the parabrachial nucleus of the rat. Physiol. Behav. 56, 1197–1202.

Yamamoto, T., Shimura, T., Sako, N., Yasoshima, Y., and Sakai, N. 1994b. Neural substrates for conditioned taste aversion in the rat. Behav. Brain Res. 65, 123–137.

Yamamoto, T., Yuyama, N., Kato, T., and Kawamura, Y. 1985a. Gustatory responses of cortical neurons in rats. II. Information processing of taste quality. J. Neurophysiol. 53, 1356–1369.

Yamamoto, T., Yuyama, N., Kato, T., and Kawamura, Y. 1985b. Gustatory responses of cortical neurons in rats. III. Neural and behavioral measures compared. J. Neurophysiol. 53, 1370–1386.

Yasoshima, Y. and Yamamoto, T. 1998. Short-term and long-term excitability changes of the insular cortical neurons after the acquisition of taste aversion learning in behaving rats. Neuroscience 284, 1–5.

Yasoshima, Y., Scott, T. R., and Yamamoto, T. 2006. Memory-dependent c-Fos expression in the nucleus accumbens and extended amygdala following the expression of a conditioned taste aversive in the rat. Neuroscience 141, 35–45.

Yoshimura, H., Sugai, T., Segami, N., and Onoda, N. 2005. Chemotopic arrangement for taste quality discrimination in the cortical taste area. Chem. Senses 30, i164–i165.

Zald, D. H. and Pardo, J. V. 1997. Emotion, olfaction, and the human amygdala: Amygdala activation during aversive olfactory stimulation. Proc. Natl. Acad. Sci. U. S. A. 94, 4119–4124.

Zald, D. H., Hagen, M. C., and Pardo, J. V. 2002. Neural correlates of tasting concentrated quinine and sugar solutions. J. Neurophysiol. 87, 1068–1075.

Zald, D. H., Lee, J. T., Fluegel, K. W., and Pardo, J. V. 1998. Aversive gustatory stimulation activates limbic circuits in humans. Brain 121, 1143–1154.

Zhang, M. and Kelley, A. E. 1997. Opiate agonists microinjected into the nucleus accumbens enhance sucrose drinking in rats. Psychopharmacology (Berl.) 132, 350–360.

Zhang, M. and Kelley, A. E. 2000. Enhanced intake of high-fat food following striatal mu-opioid stimulation: microinjection mapping and fos expression. Neuroscience 99, 267–277.

Zhang, M. and Kelley, A. E. 2002. Intake of saccharin, salt, and ethanol solutions is increased by infusion of a mu opioid agonist into the nucleus accumbens. Psychopharmacology (Berl.) 159, 415–423.

Zhang, M., Gosnell, B. A., and Kelley, A. E. 1998. Intake of high-fat food is selectively enhanced by mu opioid receptor stimulation within the nucleus accumbens. J. Pharmacol. Exp. Ther. 285, 908–914.

Further Reading

Ikuno, H. and Sakaguchi, T. 1990. Gastric vagal functional distribution in the secretion of gastric acid produced by sweet taste. Brain Res. Bull. 25, 429–431.

Yamamoto, T. 1994. A Neural Model for Taste Aversion Learning. In: Olfaction and Taste XI (eds. K. Kurihara, N. Suzuki, and H, Ogawa), pp. 471–474. Springer.

Relevant Website

http://www.ctalearning.com – A. L. Riley and K. B. Freeman. Conditioned Taste Aversion: An Annotated Bibliography.

4.25 Dopamine Release by Sucrose

A Hajnal and R Norgren, Milton S. Hershey Medical Center, Hershey, PA, USA

4.25.1 Ingestive Behavior and Reward

Sucrose is a prototypical reward in animal behavior studies. It is a simple compound, can be delivered easily in water, and sustains reliable performance without food or water restriction. Under ordinary circumstances, humans report sucrose as sweet and pleasurable. The generality of this experience allows scientists to infer that similar subjective events occur in other animals. In behavioral terms, reward is defined operationally as an event – a stimulus – that increases the probability of a contiguous response being emitted again (Thorndike, E., 1911). When studying the neural bases for reward, however, operational definitions are less satisfactory because the goal is to understand the so-called intervening variable, that is, what humans term pleasure.

The normal behavioral response to sucrose is ingestion. It is organized in the caudal brainstem because the entire sequence appears in anencephalic infants and chronically decerebrate rats neither of which have functional forebrains (Steiner, J. E., 1973; Grill, H. J. and Norgren, R., 1978a). In decerebrate rats, ingestion and rejection are near normal but largely stimulus bound (Grill, H. J. and Norgren, R., 1978b). Although short-term plasticity can occur, behavior in decerebrates cannot be altered permanently by experience (Grigson, P. S. *et al.*, 1997). The forebrain is necessary for learning and memory to influence consummatory behavior. This influence can be manifested in the consummatory behavior itself, such as with a learned preference or aversion (Sclafani, A., 1997). More frequently, the influence of learning appears in the appetitive phase, as when a deprived animal returns to a place where it had previously encountered food (Craig, W., 1918). Thus, the forebrain is necessary for the full expression of the motivation and reward systems that activate and direct the appetitive behavior needed to bring the animal into contact with the consummatory stimuli (Norgren, R. *et al.*, 2003).

In short, the neural circuits needed for discriminative ingestion and rejection behavior are complete in the brainstem. The systems that animate adaptive behavior in the real world – learning, memory, motivation, and reward – require the forebrain. Therefore, in order to investigate the neural basis for sucrose reward, we must identify an index for this function that depends on the forebrain and is not directly related to behavior. Although far from perfect, the best candidate available is dopamine release, specifically in the nucleus accumbens (NAcc; Ikemoto, S. and Panksepp, J., 1999; Wise, R. A., 2005).

4.25.2 Dopamine and Reward

The neural substrate for reward remains a challenge because the concept itself is illusive. Despite this limitation, extensive research has been devoted to the neural bases of reward. As a result, a network of nuclei and neurotransmitters has been identified that influence reward functions, even if the nature of that influence is far from settled. Among the most studied of these putative reward substrates is the mesolimbic or mesoaccumbens dopamine system. Experimental manipulations of this system disrupt behavior guided by both natural and non-natural rewards (Smith, G. P., 1995; Wise, R. A., 2002). Conversely, both natural and drug rewards activate this pathway (Smith, G. P., 1995; Wise, R. A., 2002; 2005) which originates from the ventral tegmental area (VTA) of the brainstem (Oades, R. D. and Halliday, G. M., 1987). Although its role in reward is debated (Berridge, K. C. and Robinson, T. E., 1998; Cannon, C. M. and Palmiter, R. D., 2003), no one denies that dopamine release occurs in the mesoaccumbens dopamine system when normally preferred stimuli are presented to an animal. The contention arises over interpreting the significance of the release. Does it signify a broader phenomenon, such as arousal, attention, learning, or reward, or a more restricted category, such as uncertainty, novelty, expected value, or incentive (Di Chiara, G., 1998; Ikemoto, S. and Panksepp, J., 1999; Salamone, J. D. *et al.*, 2003; Schultz, W., 2005; Wise, R. A., 2005)? In what follows, and particularly in our own work, we assume only that dopamine release tracks the incentive valence of reward, that it provides a forebrain index of sensory events that the animal will work to obtain.

The presentation or even the anticipation of food stimulates dopamine neurons in the VTA and dopamine release preferentially in the shell of the NAcc (Church, W. H. *et al.*, 1987; Hernandez, L. and Hoebel, B. G., 1988a; 1988b; Radhakishun, F. S. *et al.*, 1988; Yoshida, M. *et al.*, 1992; Inoue, K. *et al.*, 1993; Westerink, B. H. *et al.*, 1994; Wilson, C. *et al.*, 1995). In addition to the NAcc, food intake results in dopamine release in other terminal regions of the VTA projections such in the prefrontal cortex (Hernandez, L. and Hoebel, B. G., 1990; Feenstra, M. G. and Botterblom, M. H., 1996) and the central nucleus of the amygdala (Hajnal, A. and Lenard, L., 1997). Although imaging methods have been useful, primarily in human studies (Wang, G. J. *et al.*, 2002; McClure, S. M. *et al.*, 2004; Rolls, E. T., 2005), the majority of available data derive from *in vivo* microdialysis studies, a method that permits direct measurement of central dopamine release and metabolism during behavior (Hoebel, B. G. *et al.*, 1989; Westerink, B. H., 1995). Dialysis typically requires 5–20 min samples to detect significant changes in dopamine flux. Advances in voltammetry have decreased the detection threshold during ingestive behavior into the millisecond range (Kiyatkin, E. A. and Gratton, A., 1994; Roitman, M. F. *et al.*, 2004).

4.25.3 Dopamine and Sucrose Preference

During ingestion dopamine release in the NAcc is a function of the oral stimulus, previous experience, and motivational state (Bassareo, V. and Di Chiara, G., 1997; 1999; Norgren, R. *et al.*, 2003; Smith, G. P., 2004). Most relevant feeding experiments were performed under deprivation condition that by itself augments the incentive value of external stimuli (Carr, K. D., 1996; Pothos, E. N., 2001). Deprivation also prevents habituation of accumbens dopamine release (i.e., latent inhibition) after repeated exposure to ingestive stimuli (Bassareo, V. and Di Chiara, G., 1999). Nevertheless, rats fed *ad libitum* on standard food still exhibit an increase of NAcc dopamine when given a more preferred diet (Ahn, S. and Phillips, A. G., 1999; Cenci, M. A. *et al.*, 1992; Mitchell, J. B. and Gratton, A., 1992; Wilson, C. *et al.*, 1995). In fact, highly preferred food itself reliably elicits an increase in the extracellular dopamine in the NAcc (Mark, G. P. *et al.*, 1991; Mitchell, J. B. and Gratton, A., 1992; Hajnal, A. and Norgren, R., 2001) and also increases somatodendritic dopamine release in the VTA, where the mesoaccumbens pathway originates (Zhang, H. *et al.*, 1994). Moreover, repeated access to palatable sucrose solution consistently results in dopamine release in the NAcc over several days (Hajnal, A. and Norgren, R., 2001; 2002; Rada, P. *et al.*, 2005), an effect that can be reproduced even when the postabsorptive effects are eliminated using sham feeding preparation (Hajnal, A. *et al.*, 2004a). This observation supports the notion that dopamine tracks the hedonic value of an (oro)sensory stimulus.

Another indication that dopamine influences feeding behavior in general and sucrose preference in particular comes from studies that interfere with the postsynaptic receptors for this neurotransmitter. Systemic injection of D1 and D2 dopamine receptor antagonists reduced both real and sham feeding of sucrose (Geary, N. and Smith, G. P., 1985; Smith, G. P. and Schneider, L. H., 1988; Schneider, L. H., 1989; Cooper, S. J. *et al.*, 1993; Duong, A. and Weingarten, H. P., 1993; Hsiao, S. and Smith, G. P., 1995). Learned preference for sucrose also is decreased with administration of dopamine receptor antagonists (Yu, W. Z. *et al.*, 2000a; 2000b). Furthermore, dopamine-deficient (DD) mice can eat but are not motivated to do so (Szczypka, M. S. *et al.*, 1999b), suggesting that DD mice are unable to integrate and process the neuronal signals necessary to stimulate and maintain feeding. In fact, dopamine replacement re-establishes preference for sucrose in DD mice (Szczypka, M. S. *et al.*, 1999a). Lesions to the VTA also have been shown to reduce sucrose preference (Shimura, T. *et al.*, 2002). These studies demonstrate that dopamine and its receptors are important for initiating and maintaining sucrose ingestion, but they do not reveal where in the brain (or the body) these actions take place.

The opposite relationship, that is, that sucrose intake can be increased by pharmacological manipulation, provides more specific information. In this case, nomifensine, a blocker of dopamine uptake, infused into the accumbens by reverse microdialysis increased dopamine in the accumbens and increased sucrose intake (Hajnal, A. and Norgren, R., 2001). Although nomifensine also affects other monoamine transporters, low doses of the dopamine receptor antagonists SCH 23390 and sulpiride dialyzed together with the reuptake blocker into the accumbens abolished both the increased dopamine and the increased sucrose licking (Hajnal, A. and Norgren, R., 2001).

Less specific evidence for accumbens dopamine modulating reward arises from the cross-sensitization effects of sucrose and the indirect dopamine agonist amphetamine. Compared with NaCl-treated controls, rats previously sensitized to amphetamine increased their locomotor activity after brief access to sucrose (Avena, N. M. and Hoebel, B. G., 2003a). Similarly, rats receiving restricted sucrose access exhibited higher locomotor activity when subsequently injected with amphetamine (Avena, N. M. and Hoebel, B. G., 2003b; Hajnal, A. *et al.*, 2004b). In addition, rats that spontaneously ingested more sucrose showed increased amphetamine-induced dopamine overflow in the caudal NAcc relative to those that licked less sucrose (Sills, T. L. and Vaccarino, F. J., 1996). These studies imply a reciprocal relationship between dopamine flux and ingestion of sucrose. Dopamine tracks sucrose intake even if the causal direction is less obvious.

Direct measures of accumbens dopamine during sucrose ingestion are surprisingly scarce. Oddly, indirect measures are the norm. In these experiments, accumbens dopamine is more often measured using a conditioned stimulus that was paired with sucrose or some other normally preferred chemical such as saccharin. In naive rats licking a saccharin solution causes an increase in accumbens dopamine, measured with microdialysis (Mark, G. P. *et al.*, 1991; Grigson, P. S. *et al.*, 2004). Fructose has similar effects in rats (Hajnal, A., personal communication). Both saccharin and fructose taste sweet to humans, are preferred by rats (Sclafani, A. and Nissenbaum, J. W., 1985; Sclafani, A. and Mann, S., 1987; Sclafani, A. *et al.*, 1998), and exert reinforcing effects when used as conditioned stimuli (Sclafani, A. and Ackroff, K., 1994). Microdialysis experiments with

unconditioned sucrose, however, have been done only recently in our laboratory.

In the first experiment, we established the effect of sucrose ingestion on dopamine in the NAcc. Rats were placed on a 16 h water deprivation schedule and trained to lick either water or 0.3 M sucrose for 20 min each morning for about 1 week. Then the microdialysis probes were inserted into the caudomedial NAcc and left in place for 3 days. The fluid access schedule was continued as before, that is, each rat had access to water and sucrose at least once. Although deprived overnight (∼18 h), ingesting water failed to alter dopamine levels. Licking 0.3 M sucrose for 20 min, however, produced a 300% increase in dopamine overflow in the NAcc (Hajnal, A. and Norgren, R., 2001). The effect was impressive, but linking it directly to the reward value of sucrose was not that straightforward. The rats ingested considerably more sucrose solution than water and, even in 20 min, the sucrose would have produced metabolic feedback to which the brain was sensible. To control for these possibly confounding effects, we did two further experiments using essentially the same paradigm.

First, we trained rats to sham feed sucrose – the rats lick normally from a spout, but the fluid drains from a chronic gastric fistula. Then they were allowed to sham lick three different concentrations of sucrose, while accumbens dopamine release was assessed by microdialysis. Even without significant metabolic consequences, accumbens dopamine increased as a function of sucrose concentration. In a third set of rats, we again varied concentration, but clamped total intake during the 20 min sampling period to 75% of what each rat normally ingested of the weakest solution. Even with ingested volume controlled, accumbens dopamine release rose as a function of increasing sucrose concentration (Hajnal, A. *et al.*, 2004a). If we operationally define hedonic value by relative preference, then sucrose reward varies directly with its concentration because, in brief access tests, rats prefer stronger solutions to weaker ones (Davis, J. D., 1973; Spector, A. C. *et al.*, 1993; Contreras, R. J. *et al.*, 1995). Because accumbens dopamine release also varies directly with sucrose concentration, we can take these values as a forebrain index of gustatory reward. Although the exact relationship between dopamine and reward remains undetermined, this effect is sufficient to propose NAcc dopamine as a tool to track capacity of taste stimuli to engage motivational systems.

4.25.4 Central Pathways of Sucrose Reward

Intraoral intake of sucrose increases dopamine in the nucleus of the solitary tract (Bednar, I. *et al.*, 1994), the first central relay of the gustatory system (Lundy, R. F., Jr. and Norgren, R., 2004b). The NAcc receives axonal projections from the caudal part of the nucleus of the solitary tract, but the taste neurons are in the rostral half of the nucleus (Zagon, A. *et al.*, 1994). The accumbens also sends reciprocal projections to the nucleus of the solitary tract (Brog, J. S. *et al.*, 1993; Stratford, T. R. and Kelley, A. E., 1999) via a circuit that includes the parabrachial nucleus (PBN; Usuda, I. *et al.*, 1998). Given its distribution, the involvement of this direct loop in gustatory functions is conceivable, but unlikely. In rodents, the major rostral projection of gustatory neurons is from the anterior aspect of the nucleus of the solitary tract to the caudomedial PBN. From the PBN, the central gustatory system bifurcates, one arm forming a standard thalamocortical axis, the other distributing widely in the limbic forebrain (Lundy, R. F., Jr. and Norgren, R., 2004a). The PBN, which includes the second central gustatory relay, projects densely to the central nucleus of the amygdala, the lateral hypothalamus, and the bed nucleus of the stria terminalis (Norgren, R., 1976; 1978), all of which send axons to the NAcc shell (Kirouac, G. J. and Ganguly, P. K., 1995). The thalamocortical gustatory system can also reach NAcc via substantial connections to the central nucleus of the amygdala, lateral hypothalamus, and the prefrontal cortex (Kosar, E. *et al.*, 1986; Shi, C. J. and Cassell, M. D., 1998; Lundy, R. F., Jr. and Norgren, R., 2004b).

Because all these areas figure prominently in theories of motivation and reward, we decided to determine which forebrain taste pathway supported the release of dopamine in the NAcc during sucrose licking (Pfaffmann, C. *et al.*, 1977). After bilateral lesions of the thalamic taste relay, the accumbens dopamine response during sucrose ingestion was identical to that of sham-operated controls. After similar damage in the PBN, sucrose licking elicited less than one-third of the NAcc dopamine as the controls (Figure 1; Hajnal, A. and Norgren, R., 2005).

We performed another, similar experiment using immunohistochemical staining of Fos, the nuclear phosphoprotein product of the early-immediate gene c-Fos (Mungardee, S. S. *et al.*, 2004). After having similar, bilateral lesions the PBN or the thalamic

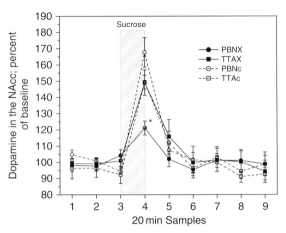

Figure 1 Extracellular levels of dopamine in the nucleus accumbens (NAcc), expressed as a percentage of mean baseline ($\pm$ standard error of the mean) before, during, and after a 20 min presentation of 0.3 M sucrose (bracket). The rats received ibotenic-acid microlesions in the parabrachial nucleus PBN (PBNX), the thalamic taste area (TTAX), or control surgeries using physiological NaCl injections to the same areas (PBNc, TTAc). Asterisk denotes statistically significant difference based on *post hoc* comparisons between groups ($^{*}P < 0.01$). Adapted from Hajnal, A. and Norgren, R. (2005) Taste pathways that mediate accumbens dopamine release by sapid sucrose. Physiol. Behav. 84, 363–369, with permission from Elsevier.

taste relay, rats were allowed to sham-feed sucrose or water for 1 h. Their brains were then cut and stained for the Fos protein. In the controls and the rats that received thalamic lesions (Figure 2), sham-licking sucrose produced more Fos in the shell of the NAcc than did ingestion of water. In contrast, lesions of the gustatory PBN reduced the overall level of Fos staining and eliminated the differential effect of licking sucrose (Figure 2). These complimentary data represent the first demonstration that the affective character of a sensory stimulus might separate from the thalamocortical stream as early as the second central synapse.

Knowing the pathways through which a normally preferred taste influences a forebrain index of reward says little about the sensory signal that elicits the effect. Sucrose appears to be inherently rewarding. Nevertheless, a full gastrointestinal tract can reduce sugar intake to near zero, implying that the sensory signal no longer is preferred. Conversely, during a negative sodium balance, a normally rejected 0.5 M NaCl solution is avidly consumed. In fact, for most stimuli, reward value is imposed either through learning or through the physiological state of the animal. This could be accomplished by altering the

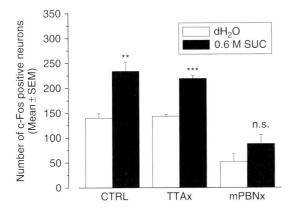

Figure 2 Number of c-Fos positive neurons in the shell of the nucleus accumbens after ibotenic acid microlesions to the thalamic taste area (TTAx) or the medial PBN (mPBNx) in rats exposed to sham drinking of 0.6 M sucrose (SUC) or water (dH$_2$O). Asterisks denote results from *post hoc* analysis (**, $P < 0.01$; ***, $P < 0.001$; n.s., not significant; SEM, standard error of the mean).

sensory message itself, before it engages the motivation and reward systems, by gating the message within those systems, or both.

We have examined the former proposition because the convergence of visceral afferent, gustatory, and forebrain centrifugal systems in the brainstem relays provides the anatomical framework for such an interaction. Alterations in the response properties of taste neurons in the nucleus of the solitary tract and in the PBN are coincident with behavioral changes in hedonic value (Norgren, R. *et al.*, 2003). For instance, intraduodenal lipid infusions, which reduces intake of sucrose (Foster, L. A. *et al.*, 1996), also decreased the response of PBN sucrose-best cells to sucrose by 55%, but was without effect on sucrose-evoked responses in other cell types (Hajnal, A. *et al.*, 1999). Similar if less specific modulation of taste processing also can be produced by inducing a negative sodium balance or by activating descending projections from reciprocally connected forebrain regions such as the lateral hypothalamus, central amygdala, and the gustatory cortex (Contreras, R. J., 1977; Nakamura, K. and Norgren, R., 1995; Lundy, R. F., Jr. and Norgren, R., 2001; Cho, Y. K. *et al.*, 2002; 2003; Tamura, R. and Norgren, R., 2003; Lundy, R. F., Jr. and Norgren, R., 2004a; Li, C. S. *et al.*, 2005). These observations demonstrate that both physiological and forebrain modulation occur before gustatory neural gustatory activity reaches the mesoaccumbens dopamine system. Whether or not this modulation influences the capacity of tastes to

release dopamine, much less their hedonic tone, remains to be determined.

4.25.5 Chronic Effects of Sucrose Ingestion

So far we have discussed only the acute effects of oral sucrose stimulation. In addition to short-term dopamine activation, ingestion of preferred foods, such as glucose or sucrose may result in a lasting neuroadaptation within the mesoaccumbens dopamine system (Colantuoni, C. *et al.*, 2001; Avena, N. M. and Hoebel, B. G., 2003b). We observed that, after as few as 6 days, scheduled, daily, 20 min access to 0.3 M sucrose results in upregulation of the dopamine membrane transporter and downregulation of the D2 dopamine receptors (Bello, N. T. *et al.*, 2002; 2003). Although the function of this neuroadaptation is not clear, accumulating evidence supports the hypothesis that dopamine plays an important role metabolic regulation and energy balance, including in human obesity (Figlewicz, D. P. *et al.*, 1994; 1996; 1998; Patterson, T. A. *et al.*, 1998; Wang, J. *et al.*, 2001; Wang, G. J. *et al.*, 2002; Wang, J. *et al.*, 2002). Chronic food restriction reduces basal dopamine levels in rats and increases consummatory responses to both food and drug rewards (Pothos, E. N. *et al.*, 1995). Similar manipulations also increase bingeing on palatable meals (Hagan, M. M. and Moss, D. E., 1997). Finally, we demonstrated that blocking dopamine reuptake in NAcc of deprived rats accentuates sucrose intake (Hajnal, A. and Norgren, R., 2001). Taken together, these data indicate that dopamine and the NAcc contribute to the control of feeding behavior in part through influencing preference.

4.25.6 Sucrose Reward without Dopamine

It is important to remember that we use accumbens dopamine flux as an index of reward, not as its equivalent. The most dramatic proof of this distinction comes from genetically engineered mice that never produce dopamine, but exhibit near normal sucrose preferences (Cannon, C. M. and Palmiter, R. D., 2003). Opioids are clearly involved with affective responses and, within the NAcc, they may be responsible for the hedonic modulation of taste information (Kelley, A. E. *et al.*, 2002). Opioid antagonists reduce intake of palatable foods in real (Cooper, S. J.,

1982; 1983; Cooper, S. J. *et al.*, 1985; Lynch, W. C., 1986) and sham feeding (Kirkham, T. C. and Cooper, S. J., 1988; Kirkham, T. C., 1990). They also decrease oral motor responses to sucrose in the taste reactivity test (Parker, L. A. *et al.*, 1992). Conversely, when given intermittent access to sugar solutions, rats exhibit increased binding of μ-1 opioid receptors in the NAcc (Colantuoni, C. *et al.*, 2001). Similar evidence exists for serotonin, neuropeptide Y, cholecystokinin, leptin, and acetylcholine, among others (Waldbillig, R. J. and Bartness, T. J., 1982; Gosnell, B. A. and Hsiao, S., 1984; Montgomery, A. M. and Burton, M. J., 1986; Leibowitz, S. F. *et al.*, 1992; Eckel, L. A. and Ossenkopp, K. P., 1994; De Jonghe, B. C. *et al.*, 2005; Hajnal, A. *et al.*, 2005; Kelley, A. E. *et al.*, 2005). All of these factors ultimately may affect the activation of the mesoaccumbens dopamine system, but the relationships are not simple.

Even if the final common path for all these other neurotransmitters does include dopamine, many of them also influence the code for gustatory afferent activity more directly, even at the level of the taste buds. Specifically, neuropeptide Y, cholecystokinin, acetylcholine, and serotonin receptors all have been identified on taste receptor cells (Herness, S. and Chen, Y., 1997; Herness, S. *et al.*, 2002; Huang, Y. J. *et al.*, 2005; Ogura, T. and Lin, W., 2005; Zhao, F. L. *et al.*, 2005). Insulin and leptin receptors also occur on taste receptor cells (Shigemura, N. *et al.*, 2003; Suzuki, Y. *et al.*, 2005). In fact, leptin receptors occur exclusively on sweet taste receptor cells. The exaggerated sweet preference seen in some obese animal models might be related to a malfunction in these leptin receptors (Kawai, K. *et al.*, 2000; Ninomiya, Y. *et al.*, 2002; Shigemura, N. *et al.*, 2004). In normal mice, administration of leptin suppresses single fiber responses in the chorda tympani and glossopharyngeal nerves to sucrose and saccharin but not to other stimuli (Kawai, K. *et al.*, 2000). These direct peripheral effects on the taste system illustrate the multiple systems that can influence gustatory coding. All of these neurotransmitters and peptides play important parts in the control of ingestive behavior and energy balance. Thus, they are obvious candidates for mediating the hedonic tone of taste whose biologic role is to arbitrate ingestion and rejection behavior.

4.25.7   Summary

Sucrose ingestion releases dopamine within the NAcc. This response is concentration dependent, but does not require gastrointestinal or metabolic feedback. This central response arises from the oral gustatory properties of sucrose. Specifically, the accumbens response appears to reflect the hedonic character of sweet stimuli in general. When saccharin is made aversive by association with visceral malaise, this stimulus switches from increasing extracellular dopamine to inhibiting its release (Mark, G. P. *et al.*, 1991). While sucrose licking does increase dopamine release, the obverse also is true. Increasing dopamine availability in the NAcc increases intake of sucrose solutions and blocking D1 dopamine receptors inhibits it (Hajnal, A. and Norgren, R., 2001).

Despite these close relationships, a direct causal link between taste and reward in the accumbens is unlikely. Sucrose intake is not compromised by dopaminergic lesions and is near normal in dopamine knockout mice. Thus, dopamine activation is related to the orosensory properties of sucrose and probably to its hedonic value, but that increase is not the equivalent of sucrose reward. In fact, accumbens dopamine may not be necessary for the expression of sucrose preference, but it does track the hedonic value underlying the behavior. If true, this assertion does not apply just to sucrose or even to gustatory stimuli but probably to rewards in general (Wightman, R. M. and Robinson, D. L., 2002). Normal biological rewards are elicited by external events, consummatory stimuli, of which sucrose is a prime exemplar (Craig, W., 1918). The reward value, however, is not determined by the stimulus, but depends on the immediate and remembered circumstances of the animal. The hedonic valence is then added to the afferent sensory neural activity resulting from the stimulus somewhere between the peripheral receptors and behavior. Determining how this comes about remains a central goal of neuroscience.

References

Ahn, S. and Phillips, A. G. 1999. Dopaminergic correlates of sensory-specific satiety in the medial prefrontal cortex and nucleus accumbens of the rat. J. Neurosci. 19, RC29.

Avena, N. M. and Hoebel, B. G. 2003a. A diet promoting sugar dependency causes behavioral cross-sensitization to a low dose of amphetamine. Neuroscience 122, 17–20.

Avena, N. M. and Hoebel, B. G. 2003b. Amphetamine-sensitized rats show sugar-induced hyperactivity (cross-sensitization) and sugar hyperphagia. Pharmacol. Biochem. Behav. 74, 635–639.

Bassareo, V. and Di Chiara, G. 1997. Differential influence of associative and nonassociative learning mechanisms on the responsiveness of prefrontal and accumbal dopamine transmission to food stimuli in rats fed *ad libitum*. J. Neurosci. 17, 851–861.

Bassareo, V. and Di Chiara, G. 1999. Modulation of feeding-induced activation of mesolimbic dopamine transmission by appetitive stimuli and its relation to motivational state. Eur. J. Neurosci. 11, 4389–4397.

Bednar, I., Qian, M., Qureshi, G. A., Kallstrom, L., Johnson, A. E., Carrer, H., and Sodersten, P. 1994. Glutamate inhibits ingestive behaviour. J. Neuroendocrinol. 6, 403–408.

Bello, N. T., Lucas, L. R., and Hajnal, A. 2002. Repeated sucrose access influences dopamine D2 receptor density in the striatum. Neuroreport 13, 1575–1578.

Bello, N. T., Sweigart, K. L., Lakoski, J. M., Norgren, R., and Hajnal, A. 2003. Restricted feeding with scheduled sucrose access results in an upregulation of the rat dopamine transporter. Am. J. Physiol. Regul. Integr. Comp. Physiol. 284, R1260–R1268.

Berridge, K. C. and Robinson, T. E. 1998. What is the role of dopamine in reward: hedonic impact, reward learning, or incentive salience? Brain Res. Brain Res. Rev. 28, 309–369.

Brog, J. S., Salyapongse, A., Deutch, A. Y., and Zahm, D. S. 1993. The patterns of afferent innervation of the core and shell in the "accumbens" part of the rat ventral striatum: immunohistochemical detection of retrogradely transported fluoro-gold. J. Comp. Neurol. 338, 255–278.

Cannon, C. M. and Palmiter, R. D. 2003. Reward without dopamine. J. Neurosci. 23, 10827–10831.

Carr, K. D. 1996. Feeding, drug abuse, and the sensitization of reward by metabolic need. Neurochem. Res. 21, 1455–1467.

Cenci, M. A., Kalen, P., Mandel, R. J., and Bjorklund, A. 1992. Regional differences in the regulation of dopamine and noradrenaline release in medial frontal cortex, nucleus accumbens and caudate-putamen: a microdialysis study in the rat. Brain Res. 581, 217–228.

Cho, Y. K., Li, C. S., and Smith, D. V. 2002. Taste responses of neurons of the hamster solitary nucleus are enhanced by lateral hypothalamic stimulation. J. Neurophysiol. 87, 1981–1992.

Cho, Y. K., Li, C. S., and Smith, D. V. 2003. Descending influences from the lateral hypothalamus and amygdala converge onto medullary taste neurons. Chem. Senses 28, 155–171.

Church, W. H., Justice, J. B., Jr., and Neill, D. B. 1987. Detecting behaviorally relevant changes in extracellular dopamine with microdialysis. Brain Res. 412, 397–399.

Colantuoni, C., Schwenker, J., McCarthy, J., Rada, P., Ladenheim, B., Cadet, J. L., Schwartz, G. J., Moran, T. H., and Hoebel, B. G. 2001. Excessive sugar intake alters binding to dopamine and mu-opioid receptors in the brain. Neuroreport 12, 3549–3552.

Contreras, R. J. 1977. Changes in gustatory nerve discharges with sodium deficiency: a single unit analysis. Brain Res. 121, 373–378.

Contreras, R. J., Carson, C. A., and Pierce, C. E. 1995. A novel psychophysical procedure for bitter taste assessment in rats. Chem. Senses 20, 305–312.

Cooper, S. J. 1982. Palatability-induced drinking after administration of morphine, naltrexone and diazepam in the non-deprived rat. Subst. Alcohol Actions Misuse 3, 259–266.

Cooper, S. J. 1983. Effects of opiate agonists and antagonists on fluid intake and saccharin choice in the rat. Neuropharmacology 22, 323–328.

Cooper, S. J., Francis, J., and Barber, D. J. 1993. Selective dopamine D-1 receptor agonists, SK&F 38393 and CY 208-243 reduce sucrose sham-feeding in the rat. Neuropharmacology 32, 101–102.

Cooper, S. J., Jackson, A., Morgan, R., and Carter, R. 1985. Evidence for opiate receptor involvement in the consumption of a high palatability diet in nondeprived rats. Neuropeptides 5, 345–348.

Craig, W. 1918. Appetites and aversions as constituents of instincts. Biol. Bull. 91–107.

Davis, J. D. 1973. The effectiveness of some sugars in stimulating licking behavior in the rat. Physiol. Behav. 11, 39–45.

De Jonghe, B. C., Hajnal, A., and Covasa, M. 2005. Increased oral and decreased intestinal sensitivity to sucrose in obese, prediabetic CCK-A receptor-deficient OLETF rats. Am. J. Physiol. Regul. Integr. Comp. Physiol. 288, R292–R300.

Di Chiara, G. 1998. A motivational learning hypothesis of the role of mesolimbic dopamine in compulsive drug use. J. Psychopharmacol. 12, 54–67.

Duong, A. and Weingarten, H. P. 1993. Dopamine antagonists act on central, but not peripheral, receptors to inhibit sham and real feeding. Physiol. Behav. 54, 449–454.

Eckel, L. A. and Ossenkopp, K. P. 1994. Cholecystokinin reduces sucrose palatability in rats: evidence in support of a satiety effect. Am. J. Physiol. 267, R1496–R1502.

Feenstra, M. G. and Botterblom, M. H. 1996. Rapid sampling of extracellular dopamine in the rat prefrontal cortex during food consumption, handling and exposure to novelty. Brain Res. 742, 17–24.

Figlewicz, D. P., Brot, M. D., McCall, A. L., and Szot, P. 1996. Diabetes causes differential changes in CNS noradrenergic and dopaminergic neurons in the rat: a molecular study. Brain Res. 736, 54–60.

Figlewicz, D. P., Patterson, T. A., Johnson, L. B., Zavosh, A., Israel, P. A., and Szot, P. 1998. Dopamine transporter mRNA is increased in the CNS of Zucker fatty (fa/fa) rats. Brain Res. Bull. 46, 199–202.

Figlewicz, D. P., Szot, P., Chavez, M., Woods, S. C., and Veith, R. C. 1994. Intraventricular insulin increases dopamine transporter mRNA in rat VTA/substantia nigra. Brain Res. 644, 331–334.

Foster, L. A., Nakamura, K., Greenberg, D., and Norgren, R. 1996. Intestinal fat differentially suppresses sham feeding of different gustatory stimuli. Am. J. Physiol. 270, R1122–R1125.

Geary, N. and Smith, G. P. 1985. Pimozide decreases the positive reinforcing effect of sham fed sucrose in the rat. Pharmacol. Biochem. Behav. 22, 787–790.

Gosnell, B. A. and Hsiao, S. 1984. Effects of cholecystokinin on taste preference and sensitivity in rats. Behav. Neurosci. 98, 452–460.

Grigson, P. S., Acharya, N. K., and Hajnal, A. 2004. A single saccharin-morphine pairings leads to a conditioned reduction in CS intake and accumbens dopamine. Program No. 119.17, 2004 Abstract Viewer/Itinerary Planner. Society for Neuroscience (online).

Grigson, P. S., Kaplan, J. M., Roitman, M. F., Norgren, R., and Grill, H. J. 1997. Reward comparison in chronic decerebrate rats. Am. J. Physiol. 273, R479–R486.

Grill, H. J. and Norgren, R. 1978a. The taste reactivity test. II. Mimetic responses to gustatory stimuli in chronic thalamic and chronic decerebrate rats. Brain Res. 143, 281–297.

Grill, H. J. and Norgren, R. 1978b. Chronically decerebrate rats demonstrate satiation but not bait shyness. Science 201, 267–269.

Hagan, M. M. and Moss, D. E. 1997. Persistence of binge-eating patterns after a history of restriction with intermittent bouts of refeeding on palatable food in rats: implications for bulimia nervosa. Int. J. Eat. Disord. 22, 411–420.

Hajnal, A. and Lenard, L. 1997. Feeding-related dopamine in the amygdala of freely moving rats. Neuroreport 8, 2817–2820.

Hajnal, A. and Norgren, R. 2001. Accumbens dopamine mechanisms in sucrose intake. Brain Res. 904, 76–84.

Hajnal, A. and Norgren, R. 2002. Repeated access to sucrose augments dopamine turnover in the nucleus accumbens. Neuroreport 13, 2213–2216.

Hajnal, A. and Norgren, R. 2005. Taste pathways that mediate accumbens dopamine release by sapid sucrose. Physiol. Behav. 84, 363–369.

Hajnal, A., Covasa, M., Acharya, N. K., and Bello, N. T. 2004b. Altered dopamine functions in obese CCK-A receptor deficient (OLETF) rats: reduced motor activity and responsiveness to amphetamine and chronic sucrose feeding, lower dopamine transporter binding. Program No. 75.19, 2004 Abstract Viewer/Itinerary Planner. Society for Neuroscience (online).

Hajnal, A., Covasa, M., and Bello, N. T. 2005. Altered taste sensitivity in obese, pre-diabetic OLETF rats lacking CCK-1 receptors. Am. J. Physiol. Regul. Integr. Comp. Physiol. 289, R1675–R1686.

Hajnal, A., Smith, G. P., and Norgren, R. 2004a. Oral sucrose stimulation increases accumbens dopamine in the rat. Am. J. Physiol. Regul. Integr. Comp. Physiol. 286, R31–R37.

Hajnal, A., Takenouchi, K., and Norgren, R. 1999. Effect of intraduodenal lipid on parabrachial gustatory coding in awake rats. J. Neurosci. 19, 7182–7190.

Hernandez, L. and Hoebel, B. G. 1988a. Feeding and hypothalamic stimulation increase dopamine turnover in the accumbens. Physiol. Behav. 44, 599–606.

Hernandez, L. and Hoebel, B. G. 1988b. Food reward and cocaine increase extracellular dopamine in the nucleus accumbens as measured by microdialysis. Life Sci. 42, 1705–1712.

Hernandez, L. and Hoebel, B. G. 1990. Feeding can enhance dopamine turnover in the prefrontal cortex. Brain Res. Bull. 25, 975–979.

Herness, S. and Chen, Y. 1997. Serotonin inhibits calcium-activated K$^+$ current in rat taste receptor cells. Neuroreport 8, 3257–3261.

Herness, S., Zhao, F. L., Lu, S. G., Kaya, N., and Shen, T. 2002. Expression and physiological actions of cholecystokinin in rat taste receptor cells. J. Neurosci. 22, 10018–10029.

Hoebel, B. G., Hernandez, L., Schwartz, D. H., Mark, G. P., and Hunter, G. A. 1989. Microdialysis studies of brain norepinephrine, serotonin, and dopamine release during ingestive behavior. Theoretical and clinical implications. Ann. N. Y. Acad. Sci. 575, 171–191; discussion 192–173.

Hsiao, S. and Smith, G. P. 1995. Raclopride reduces sucrose preference in rats. Pharmacol. Biochem. Behav. 50, 121–125.

Huang, Y. J., Maruyama, Y., Lu, K. S., Pereira, E., Plonsky, I., Baur, J. E., Wu, D., and Roper, S. D. 2005. Mouse taste buds use serotonin as a neurotransmitter. J. Neurosci. 25, 843–847.

Ikemoto, S. and Panksepp, J. 1999. The role of nucleus accumbens dopamine in motivated behavior: a unifying interpretation with special reference to reward-seeking. Brain Res. Brain Res. Rev. 31, 6–41.

Inoue, K., Kiriike, N., Okuno, M., Ito, H., Fujisaki, Y., Matsui, T., and Kawakita, Y. 1993. Scheduled feeding caused activation of dopamine metabolism in the striatum of rats. Physiol. Behav. 53, 177–181.

Kawai, K., Sugimoto, K., Nakashima, K., Miura, H., and Ninomiya, Y. 2000. Leptin as a modulator of sweet taste sensitivities in mice. Proc. Natl. Acad. Sci. U. S. A. 97, 11044–11049.

Kelley, A. E., Baldo, B. A., Pratt, W. E., and Will, M. J. 2005. Corticostriatal-hypothalamic circuitry and food motivation: integration of energy, action and reward. Physiol. Behav. 86, 773–795.

Kelley, A. E., Bakshi, V. P., Haber, S. N., Steininger, T. L., Will, M. J., and Zhang, M. 2002. Opioid modulation of taste hedonics within the ventral striatum. Physiol. Behav. 76, 365–377.

Kirkham, T. C. 1990. Enhanced anorectic potency of naloxone in rats sham feeding 30% sucrose: reversal by repeated naloxone administration. Physiol. Behav. 47, 419–426.

Kirkham, T. C. and Cooper, S. J. 1988. Attenuation of sham feeding by naloxone is stereospecific: evidence for opioid mediation of orosensory reward. Physiol. Behav. 43, 845–847.

Kirouac, G. J. and Ganguly, P. K. 1995. Topographical organization in the nucleus accumbens of afferents from the basolateral amygdala and efferents to the lateral hypothalamus. Neuroscience 67, 625–630.

Kiyatkin, E. A. and Gratton, A. 1994. Electrochemical monitoring of extracellular dopamine in nucleus accumbens of rats lever-pressing for food. Brain Res. 652, 225–234.

Kosar, E., Grill, H. J., and Norgren, R. 1986. Gustatory cortex in the rat. II. Thalamocortical projections. Brain Res. 379, 342–352.

Leibowitz, S. F., Xuereb, M., and Kim, T. 1992. Blockade of natural and neuropeptide Y-induced carbohydrate feeding by a receptor antagonist PYX-2. Neuroreport 3, 1023–1026.

Li, C. S., Cho, Y. K., and Smith, D. V. 2005. Modulation of parabrachial taste neurons by electrical and chemical stimulation of the lateral hypothalamus and amygdala. J. Neurophysiol. 93, 1183–1196.

Lundy, R. F., Jr. and Norgren, R. 2001. Pontine gustatory activity is altered by electrical stimulation in the central nucleus of the amygdala. J. Neurophysiol. 85, 770–783.

Lundy, R. F., Jr. and Norgren, R. 2004a. Activity in the hypothalamus, amygdala, and cortex generates bilateral and convergent modulation of pontine gustatory neurons. J. Neurophysiol. 91, 1143–1157.

Lundy, R. F., Jr. and Norgren, R. 2004b. Gustatory System. In: The Rat Nervous System (ed. G. Paxinos), pp. 891–921. Elsevier.

Lynch, W. C. 1986. Opiate blockade inhibits saccharin intake and blocks normal preference acquisition. Pharmacol. Biochem. Behav. 24, 833–836.

Mark, G. P., Blander, D. S., and Hoebel, B. G. 1991. A conditioned stimulus decreases extracellular dopamine in the nucleus accumbens after the development of a learned taste aversion. Brain Res. 551, 308–310.

McClure, S. M., York, M. K., and Montague, P. R. 2004. The neural substrates of reward processing in humans: the modern role of fMRI. Neuroscientist 10, 260–268.

Mitchell, J. B. and Gratton, A. 1992. Partial dopamine depletion of the prefrontal cortex leads to enhanced mesolimbic dopamine release elicited by repeated exposure to naturally reinforcing stimuli. J. Neurosci. 12, 3609–3618.

Montgomery, A. M. and Burton, M. J. 1986. Effects of peripheral 5-HT on consumption of flavoured solutions. Psychopharmacology (Berl.) 88, 262–266.

Mungardee, S. S., Lundy, R. F., Jr., Caloiero, V. G., and Norgren, R. 2004. Forebrain c-Fos expression following sham exposure to sucrose after central gustatory lesions: a quantitative study, Society for Neuroscience. Abstract Viewer/Itenary Planner. Washington, DC.

Nakamura, K. and Norgren, R. 1995. Sodium-deficient diet reduces gustatory activity in the nucleus of the solitary tract of behaving rats. Am. J. Physiol. 269, R647–R661.

Ninomiya, Y., Shigemura, N., Yasumatsu, K., Ohta, R., Sugimoto, K., Nakashima, K., and Lindemann, B. 2002. Leptin and sweet taste. Vitam. Horm. 64, 221–248.

Norgren, R. 1976. Taste pathways to hypothalamus and amygdala. J. Comp. Neurol. 166, 17–30.

Norgren, R. 1978. Projections from the nucleus of the solitary tract in the rat. Neuroscience 3, 207–218.

Norgren, R., Grigson, P. S., Hajnal, A., and Lundy, R. F., Jr. 2003. Motivational Modulation of Taste. In: Cognition and Emotion in the Brain (eds. T. Ono, G. Matsumoto, R. R. Llinas, A. Berthoz, R. Norgren, H. Nishijo, and R. Tamura), pp. 319–334. Elsevier.

Oades, R. D. and Halliday, G. M. 1987. Ventral tegmental (A10) system: neurobiology. 1. Anatomy and connectivity. Brain Res. 434, 117–154.

Ogura, T. and Lin, W. 2005. Acetylcholine and acetylcholine receptors in taste receptor cells. Chem. Senses 30 Suppl 1, i41.

Parker, L. A., Maier, S., Rennie, M., and Crebolder, J. 1992. Morphine- and naltrexone-induced modification of palatability: analysis by the taste reactivity test. Behav. Neurosci. 106, 999–1010.

Patterson, T. A., Brot, M. D., Zavosh, A., Schenk, J. O., Szot, P., and Figlewicz, D. P. 1998. Food deprivation decreases mRNA and activity of the rat dopamine transporter. Neuroendocrinology 68, 11–20.

Pfaffmann, C., Norgren, R., and Grill, H. J. 1977. Sensory affect and motivation. Ann. N. Y. Acad. Sci. 290, 18–34.

Pothos, E. N. 2001. The effects of extreme nutritional conditions on the neurochemistry of reward and addiction. Acta Astronaut. 49, 391–397.

Pothos, E. N., Creese, I., and Hoebel, B. G. 1995. Restricted eating with weight loss selectively decreases extracellular dopamine in the nucleus accumbens and alters dopamine response to amphetamine, morphine, and food intake. J. Neurosci. 15, 6640–6650.

Rada, P., Avena, N. M., and Hoebel, B. G. 2005. Daily bingeing on sugar repeatedly releases dopamine in the accumbens shell. Neuroscience 134, 737–744.

Radhakishun, F. S., van Ree, J. M., and Westerink, B. H. 1988. Scheduled eating increases dopamine release in the nucleus accumbens of food-deprived rats as assessed with on-line brain dialysis. Neurosci. Lett. 85, 351–356.

Roitman, M. F., Stuber, G. D., Phillips, P. E., Wightman, R. M., and Carelli, R. M. 2004. Dopamine operates as a subsecond modulator of food seeking. J. Neurosci. 24, 1265–1271.

Rolls, E. T. 2005. Taste, olfactory, and food texture processing in the brain, and the control of food intake. Physiol. Behav. 85, 45–56.

Salamone, J. D., Correa, M., Mingote, S., and Weber, S. M. 2003. Nucleus accumbens dopamine and the regulation of effort in food-seeking behavior: implications for studies of natural motivation, psychiatry, and drug abuse. J. Pharmacol. Exp. Ther. 305, 1–8.

Schneider, L. H. 1989. Orosensory self-stimulation by sucrose involves brain dopaminergic mechanisms. Ann. N. Y. Acad. Sci. 575, 307–319.

Schultz, W. 2006. Behavioral theories and the neurophysiology of reward. Annu. Rev. Psychol. 57, 87–115.

Sclafani, A. 1997. Learned controls of ingestive behaviour. Appetite 29, 153–158.

Sclafani, A. and Ackroff, K. 1994. Glucose- and fructose-conditioned flavor preferences in rats: taste versus postingestive conditioning. Physiol. Behav. 56, 399–405.

Sclafani, A. and Mann, S. 1987. Carbohydrate taste preferences in rats: glucose, sucrose, maltose, fructose and polycose compared. Physiol. Behav. 40, 563–568.

Sclafani, A. and Nissenbaum, J. W. 1985. On the role of the mouth and gut in the control of saccharin and sugar intake: a reexamination of the sham-feeding preparation. Brain Res. Bull. 14, 569–576.

Sclafani, A., Thompson, B., and Smith, J. C. 1998. The rat's acceptance and preference for sucrose, maltodextrin, and saccharin solutions and mixtures. Physiol. Behav. 63, 499–503.

Shi, C. J. and Cassell, M. D. 1998. Cortical, thalamic, and amygdaloid connections of the anterior and posterior insular cortices. J. Comp. Neurol. 399, 440–468.

Shigemura, N., Miura, H., Kusakabe, Y., Hino, A., and Ninomiya, Y. 2003. Expression of leptin receptor (Ob-R) isoforms and signal transducers and activators of transcription (STATs) mRNAs in the mouse taste buds. Arch. Histol. Cytol. 66, 253–260.

Shigemura, N., Ohta, R., Kusakabe, Y., Miura, H., Hino, A., Koyano, K., Nakashima, K., and Ninomiya, Y. 2004. Leptin modulates behavioral responses to sweet substances by influencing peripheral taste structures. Endocrinology 145, 839–847.

Shimura, T., Kamada, Y., and Yamamoto, T. 2002. Ventral tegmental lesions reduce overconsumption of normally preferred taste fluid in rats. Behav. Brain Res. 134, 123–130.

Sills, T. L. and Vaccarino, F. J. 1996. Individual differences in sugar consumption following systemic or intraaccumbens administration of low doses of amphetamine in nondeprived rats. Pharmacol. Biochem. Behav. 54, 665–670.

Smith, G. P. 1995. Dopamine and Food Reward. In: Progress in Psychobiology and Physiological Psychology (eds. S. Fluharty, A. Morrion, J. Sprague, and E. Stellar), pp. 83–144. Academic Press.

Smith, G. P. 2004. Accumbens dopamine mediates the rewarding effect of orosensory stimulation by sucrose. Appetite 43, 11–13.

Smith, G. P. and Schneider, L. H. 1988. Relationships between mesolimbic dopamine function and eating behavior. Ann. N. Y. Acad. Sci. 537, 254–261.

Spector, A. C., Grill, H. J., and Norgren, R. 1993. Concentration-dependent licking of sucrose and sodium chloride in rats with parabrachial gustatory lesions. Physiol. Behav. 53, 277–283.

Steiner, J. E. 1973. The gustofacial response: observation on normal and anencephalic newborn infants. Symp. Oral Sens. Percept. 254–278.

Stratford, T. R. and Kelley, A. E. 1999. Evidence of a functional relationship between the nucleus accumbens shell and lateral hypothalamus subserving the control of feeding behavior. J. Neurosci. 19, 11040–11048.

Suzuki, Y., Takeda, M., Sakakura, Y., and Suzuki, N. 2005. Distinct expression pattern of insulin-like growth factor family in rodent taste buds. J. Comp. Neurol. 482, 74–84.

Szczypka, M. S., Mandel, R. J., Donahue, B. A., Snyder, R. O., Leff, S. E., and Palmiter, R. D. 1999a. Viral gene delivery selectively restores feeding and prevents lethality of dopamine-deficient mice. Neuron 22, 167–178.

Szczypka, M. S., Rainey, M. A., Kim, D. S., Alaynick, W. A., Marck, B. T., Matsumoto, A. M., and Palmiter, R. D. 1999b. Feeding behavior in dopamine-deficient mice. Proc. Natl. Acad. Sci. U. S. A. 96, 12138–12143.

Tamura, R. and Norgren, R. 2003. Intracranial renin alters gustatory neural responses in the nucleus of the solitary tract of rats. Am. J. Physiol. Regul. Integr. Comp. Physiol. 284, R1108–R1118.

Thorndike, E. 1911. Animal Intelligence: Experimental Studies. MacMillan.

Usuda, I., Tanaka, K., and Chiba, T. 1998. Efferent projections of the nucleus accumbens in the rat with special reference to subdivision of the nucleus: biotinylated dextran amine study. Brain Res. 797, 73–93.

Waldbillig, R. J. and Bartness, T. J. 1982. The suppression of sucrose intake by cholecystokinin is scaled according to the magnitude of the orosensory control over feeding. Physiol. Behav. 28, 591–595.

Wang, G. J., Volkow, N. D., and Fowler, J. S. 2002. The role of dopamine in motivation for food in humans: implications for obesity. Expert Opin. Ther. Targets 6, 601–609.

Wang, J., Liu, Z. L., and Chen, B. 2001. Association study of dopamine D2, D3 receptor gene polymorphisms with motor fluctuations in PD. Neurology 56, 1757–1759.

Westerink, B. H. 1995. Brain microdialysis and its application for the study of animal behaviour. Behav. Brain Res. 70, 103–124.

Westerink, B. H., Teisman, A., and de Vries, J. B. 1994. Increase in dopamine release from the nucleus accumbens in response to feeding: a model to study interactions between drugs and naturally activated dopaminergic neurons in the rat brain. Naunyn Schmiedebergs Arch. Pharmacol. 349, 230–235.

Wightman, R. M. and Robinson, D. L. 2002. Transient changes in mesolimbic dopamine and their association with 'reward'. J. Neurochem. 82, 721–735.

Wilson, C., Nomikos, G. G., Collu, M., and Fibiger, H. C. 1995. Dopaminergic correlates of motivated behavior: importance of drive. J. Neurosci. 15, 5169–5178.

Wise, R. A. 2002. Brain reward circuitry: insights from unsensed incentives. Neuron 36, 229–240.

Wise, R. A. 2005. Forebrain substrates of reward and motivation. J. Comp. Neurol. 493, 115–121.

Yoshida, M., Yokoo, H., Mizoguchi, K., Kawahara, H., Tsuda, A., Nishikawa, T., and Tanaka, M. 1992. Eating and drinking cause increased dopamine release in the nucleus accumbens and ventral tegmental area in the rat: measurement by *in vivo* microdialysis. Neurosci. Lett. 139, 73–76.

Yu, W. Z., Silva, R. M., Sclafani, A., Delamater, A. R., and Bodnar, R. J. 2000a. Role of D(1) and D(2) dopamine receptors in the acquisition and expression of flavor-preference conditioning in sham-feeding rats. Pharmacol. Biochem. Behav. 67, 537–544.

Yu, W. Z., Silva, R. M., Sclafani, A., Delamater, A. R., and Bodnar, R. J. 2000b. Pharmacology of flavor preference conditioning in sham-feeding rats: effects of dopamine receptor antagonists. Pharmacol. Biochem. Behav. 65, 635–647.

Zagon, A., Totterdell, S., and Jones, R. S. 1994. Direct projections from the ventrolateral medulla oblongata to the limbic forebrain: anterograde and retrograde tract-tracing studies in the rat. J. Comp. Neurol. 340, 445–468.

Zhang, H., Kiyatkin, E. A., and Stein, E. A. 1994. Behavioral and Pharmacological modulation of ventral tegmental dendritic dopamine release. Brain Res. 656, 59–70.

Zhao, F. L., Shen, T., Kaya, N., Lu, S. G., Cao, Y., and Herness, S. 2005. Expression, physiological action, and coexpression patterns of neuropeptide Y in rat taste-bud cells. Proc. Natl. Acad. Sci. U. S. A. 102, 11100–11105.

4.26 The Representation of Flavor in the Brain

E T Rolls, University of Oxford, Oxford, UK

Glossary

flavor A sensation produced by the combination of the taste, smell, and texture of food. **reward** A stimulus such as food for which an animal will work (see Rolls, E. T., 2005a).

sensory-specific satiety A reduction in the pleasantness of and appetite for a food that has been eaten to satiety in a meal which is partly specific to the food eaten in a meal.

umami The fifth taste, produced by the stimuli monosodium glutamate and 5'-ribonucleotides such as inosine monophosphate and guanosine monophosphate.

4.26.1 Taste Processing in the Primate Cortex

4.26.1.1 Pathways

A diagram of the taste and related olfactory, somatosensory, and visual pathways in primates is shown in Figure 1. Of particular interest is that in primates there is a direct projection from the rostral part of the nucleus of the solitary tract (NTS) to the taste thalamus and thus to the primary taste cortex in the frontal operculum and adjoining insula, with no pontine taste area and associated subcortical projections as in rodents (Norgren, R., 1984; Pritchard, T. C. et al., 1986). This emphasis on cortical processing of taste in primates may be related to the great development of the cerebral cortex in primates, and the advantage of using extensive and similar cortical analysis of inputs from every sensory modality before the analyzed representations from each modality are brought together in multimodal regions to form representations of flavor.

4.26.1.2 The Primary and Secondary Taste Cortex

The primary taste cortex is in the rostral insula and adjacent frontal operculum (Pritchard, T. C. et al., 1986). A secondary cortical taste area in primates was discovered by Rolls E. T. et al. (1990) in the caudolateral orbitofrontal cortex, extending several millimeters in front of the primary taste cortex, and also extending to more medial parts of the orbitofrontal cortex (Rolls, E. T. and Baylis, L. L., 1994).

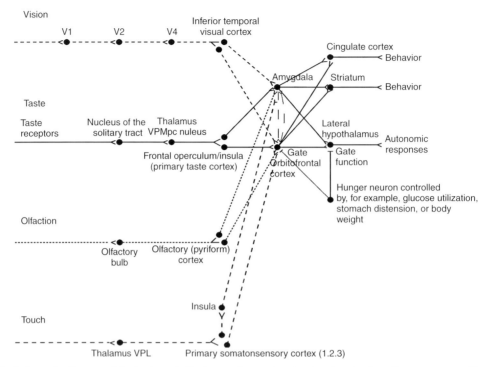

Figure 1 Schematic diagram of the taste and olfactory pathways in primates showing how they converge with each other and with visual pathways. The gate functions shown refer to the finding that the responses of taste neurons in the orbitofrontal cortex and the lateral hypothalamus are modulated by hunger. VPMpc, ventralposteromedial thalamic nucleus; V1, V2, V4, visual cortical areas.

4.26.1.3   Five Prototypical Tastes, Including Umami

In the primary and secondary taste cortex, there are many neurons that respond best to each of the four classical prototypical tastes sweet, salt, bitter and sour (Scott, T. R. *et al.*, 1986; Yaxley, S. *et al.*, 1988; Rolls, E. T. *et al.*, 1990; Rolls, E. T., 1997; Rolls, E. T. and Scott, T. R., 2003), but also there are many neurons that respond best to umami tastants such as glutamate (which is present in many natural foods such as tomatoes, mushrooms and milk; Baylis, L. L. and Rolls, E. T., 1991) and inosine monophosphate (which is present in meat and some fish such as tuna; Rolls, E. T. *et al.*, 1996a).

4.26.2   The Pleasantness of the Taste of Food

The modulation of the reward value of a sensory stimulus such as the taste of food by motivational state, for example hunger, is one important way in which motivational behavior is controlled (Rolls, E. T., 2005a). The subjective correlate of this modulation is that food tastes pleasant when hungry, and tastes hedonically neutral

when it has been eaten to satiety. We have found that the modulation of taste-evoked signals by motivation is not a property found in early stages of the primate gustatory system including the nucleus of the solitary tract (Yaxley, S. *et al.*, 1985) and the primary taste cortex (frontal opercular, Rolls, E. T. *et al.*, 1988; insular, Yaxley, S. *et al.*, 1988). In contrast, in the secondary taste cortex, in the caudolateral part of the orbitofrontal cortex, the responses of neurons to the taste of glucose decreases to zero while the monkey is fed glucose to satiety (Rolls, E. T. *et al.*, 1989). It is an important principle that the identity of a taste, and its intensity, are represented separately (in the primary taste cortex) from its pleasantness (in the secondary taste cortex) (Rolls, E. T., 2005a). Thus it is possible to represent what a taste is, and to learn about it, even when we are not hungry.

4.26.3   The Representation of Flavor: Convergence of Olfactory and Taste Inputs

Neuronal responses in the primate primary taste cortex are not driven by olfactory inputs during normal taste/smell tests (Verhagen, J. V. *et al.*, 2004).

However, we found (Rolls, E. T. and Baylis, L. L., 1994) that in the orbitofrontal cortex taste areas, of 112 single neurons which responded to any of these modalities, many were unimodal (taste 34%, olfactory 13%, visual 21%), but were found in close proximity to each other. Some single neurons showed convergence, responding for example to taste and visual inputs (13%), taste and olfactory inputs (13%), and olfactory and visual inputs (5%). Some of these multimodal single neurons had corresponding sensitivities in the two modalities, in that they responded best to sweet tastes (e.g., 1M glucose), and responded more in a visual discrimination task to the visual stimulus which signified sweet fruit juice than to that which signified saline; or responded to sweet taste, and in an olfactory discrimination task to fruit odor (see Figure 2). The different types of neurons (unimodal in different modalities, and multimodal) were frequently found close to one another in tracks made into this region, consistent with the hypothesis that the multimodal representations are actually being formed from unimodal inputs to this region.

It thus appears to be in these orbitofrontal cortex areas that flavor representations are built, where flavor is taken to mean a representation which is evoked best by a combination of gustatory and olfactory input.

The primate amygdala has neurons that combine representations of taste and oral texture (Kadohisa, M. *et al.*, 2005b), and visual stimuli (Wilson, F. A. W. and Rolls, E. T., 2005), and olfactory inputs also reach the amygdala, but less is known about olfactory-taste association learning in the primate amygdala.

4.26.4 The Rules Underlying the Formation of Flavor Representations in the Primate Cortex

Critchley H. D. and Rolls E. T. (1996c) showed that 35% of orbitofrontal cortex olfactory neurons categorized odors based on their taste association in an olfactory-to-taste discrimination task. Rolls E. T. *et al.* (1996a) found that 68% of orbitofrontal cortex odor-responsive neurons modified their responses in some way following changes in the taste reward associations of the odorants during olfactory-taste discrimination learning and its reversal. (In an olfactory discrimination experiment, if a lick response to one odor, the S+, is made a drop of glucose taste reward is obtained; if incorrectly a lick response is made to another odor, the S−, a drop of aversive saline is obtained. At some time in the experiment, the contingency between the odor and the taste is reversed, and when the meaning of the two odors alters, so does the behavior. It is of interest to investigate in which parts of the olfactory system the neurons show reversal, for where they do, it can be concluded that the neuronal response to the odor depends on the taste with which it is associated, and does not depend primarily on the physicochemical structure of the odor). An example of a neuron showing olfactory-to-taste reversal is shown in Figure 3. These findings demonstrate directly a coding principle in primate olfaction whereby the responses of some orbitofrontal cortex olfactory neurons are modified by and depend upon the taste with which the

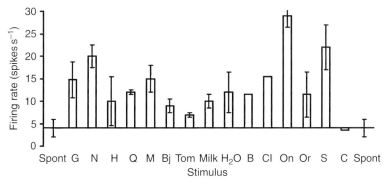

Figure 2 Olfactory to taste convergence on to a single neuron in the macaque orbitofrontal cortex. Spont, spontaneous; G, 1 M glucose; N, 0.1 M NaCl; H, 0.01 M hycrochloride; Q, 0.001 M quinine hycrochloride; M, 0.1 M monosodium glutamate; Bj, 20% blackcurrant juice; Tom, tomato juice; B, banana odor; Cl, clove oil odor; On, onion odor; Or, orange odor; S, salmon odor; C, control no-odor presentation. The mean responses ± standard error of the means are shown. The neuron responded best to the savory tastes of NaCl and monosodium glutamate and to the consonant odors of onion and salmon. Adapted from Rolls, E. T. and Baylis, L. L. 1994. Gustatory, olfactory, and visual convergence within the primate orbitofrontal cortex. J. Neurosci. 14, 5437–5452.

(a)

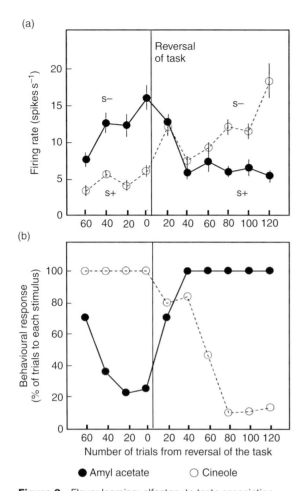

(b)

● Amyl acetate   ○ Cineole

Figure 3 Flavor learning: olfactory to taste association reversal by an orbitofrontal cortex neuron. (a) The activity of a single orbitofrontal olfactory neuron during the performance of a two-odor olfactory discrimination task and its reversal is shown. Each point represents the mean poststimulus activity of the neuron in a 500 ms period on approximately 10 trials of the different odorants. The standard errors of these responses are shown. The odorants were amyl acetate (closed circle; initially s−) and cineole (open circles; initially s+). After 80 trials of the task the reward associations of the stimuli were reversed. This neuron reversed its responses to the odorants following the task reversal. (b) The behavioral responses of the monkey during the performance of the olfactory discrimination task. The number of lick responses to each odorant is plotted as a percentage of the number of trials to that odorant in a block of 20 trials of the task. Adapted from Rolls, E. T., Critchley, H., Mason, R., and Wakeman, E. A. 1996. Orbitofrontal cortex neurons: role in olfactory and visual association learning. J. Neurophysiol. 75, 1970–1981.

odor is associated (Rolls, E. T., 2001; 2002a; 2002b; Deco, G. and Rolls, E. T., 2005b).

Some neurons in the same orbitofrontal cortex region are influenced by the sight as well as by the

taste of food, and this is a learned convergence (Thorpe, S. J. *et al.*, 1983; Rolls, E. T. *et al.*, 1996b). As a result of this learning, food choice based on the visual–taste association can be made, and potentially the sight of food can influence its perceived taste. It is of interest that the olfactory-taste association learning is less complete, and much slower, than the modifications found for orbitofrontal visual neurons during visual–taste learning and its reversal (Rolls, E. T. *et al.*, 1996b). This relative inflexibility of olfactory responses is consistent with the need for some stability in odor–taste associations to facilitate the formation and perception of flavors.

Some orbitofrontal cortex olfactory neurons do not code in relation to the taste with which the odor is associated (Critchley, H. D. and Rolls, E. T., 1996c) so that there is also a taste-independent representation of odor in this region.

4.26.5 The Responses of Orbitofrontal Cortex Taste and Olfactory Neurons to the Texture, and Temperature of Food

The texture of food may be considered as a component of the flavor of food. The texture of food, including its viscosity, influences some neurons in the primary taste cortex that have taste responses (Verhagen, J. V. *et al.*, 2004). These texture inputs also thereby influence neurons in the primate orbitofrontal cortex (Rolls, E. T. *et al.*, 2003b), which thus becomes a region where flavor representations can be influenced by the taste, smell, sight and texture of food. Some of the orbitofrontal cortex neurons with texture-related responses encode parametrically the viscosity of food in the mouth (shown using a methyl cellulose series in the range 1–10 000 centiPoise) (see Figure 3), and others independently encode the particulate quality of food in the mouth, produced quantitatively for example by adding 20–100 μm microspheres to methyl cellulose (Rolls, E. T. *et al.*, 2003b). Other neurons respond to water, and others to the somatosensory stimuli astringency as exemplified by tannic acid (Critchley, H. D. and Rolls, E. T., 1996a), and to capsaicin (Rolls, E. T. *et al.*, 2003b; Kadohisa, M. *et al.*, 2004; 2005a).

Texture in the mouth is an important indicator of whether fat is present in a food, which is important not only as a high value energy source, but also as a potential source of essential fatty acids. In the orbitofrontal cortex, Rolls E. T. *et al.* (1999) have found a population of neurons that responds when fat is in the mouth. The fat-related responses of these neurons are produced at least in part by the texture of the food rather than by chemical receptors sensitive to certain chemicals, in that such neurons typically respond not only to foods such as cream and milk containing fat, but also to paraffin oil (which is a pure hydrocarbon) and to silicone oil $(Si(CH_3)O_2)_n)$. Moreover, the texture channel through which these fat-sensitive neurons are activated are separate from viscosity sensitive channels, in that the responses of these neurons cannot be predicted by the viscosity of the oral stimuli (Verhagen, J. V. et al., 2003). Some of the fat-related neurons do though have convergent inputs from the chemical senses, in that in addition to taste inputs, some of these neurons respond to the odor associated with a fat, such as the odor of cream (Rolls, E. T. et al., 1999). Feeding to satiety with fat (e.g., cream) decreases the responses of these neurons to zero on the food eaten to satiety, but if the neuron receives a taste input from for example glucose taste, that is not decreased by feeding to satiety with cream. Thus there is a representation of the macronutrient fat in this brain area, and the activation produced by fat is reduced by eating fat to satiety.

In addition, we have shown recently (Kadohisa, M. et al., 2004; 2005a) that some neurons in the orbitofrontal cortex reflect the temperature of substances in the mouth, and that this temperature information is represented independently of other sensory inputs by some neurons, and in combination with taste or texture by other neurons.

4.26.6 The Representation of the Pleasantness of Flavor in the Brain: Olfactory and Visual Sensory-Specific Satiety, and Their Representation in the Primate Orbitofrontal Cortex

In the orbitofrontal cortex, it is found that the decreases in the responsiveness of the neurons are relatively specific to the food with which the monkey has been fed to satiety. For example, in seven experiments in which the monkey was fed glucose solution, neuronal responsiveness decreased to the taste of the glucose but not to the taste of blackcurrant juice. Conversely, in two experiments in which the monkey was fed to satiety with fruit juice, the responses of the neurons decreased to fruit juice but not to glucose (Rolls, E. T. et al., 1989).

It has also been possible to investigate whether the olfactory representation in the orbitofrontal cortex is affected by hunger, and thus whether the pleasantness of odor is represented in the orbitofrontal cortex. In satiety experiments, Critchley H. D. and Rolls E. T. (1996b) showed that the responses of some olfactory neurons to a food odor are decreased during feeding to satiety with a food (e.g., fruit juice) containing that odor. In particular, seven of nine olfactory neurons that were responsive to the odors of foods, such as blackcurrant juice, were found to decrease their responses to the odor of the satiating food. The decrease was typically at least partly specific to the odor of the food that had been eaten to satiety, potentially providing part of the basis for sensory-specific satiety. It was also found for eight of nine neurons that had selective responses to the sight of food that they demonstrated a sensory-specific reduction in their visual responses to foods following satiation. These findings show that the olfactory and visual representations of food, as well as the taste representation of food, in the primate orbitofrontal cortex are modulated by hunger. Usually a component related to sensory-specific satiety can be demonstrated. It is thus the orbitofrontal cortex which computes sensory-specific satiety, and it is in areas such as this and the areas that receive from it that neuronal activity may be related to whether a food tastes pleasant, and to whether the food should be eaten (see Scott, T. R. et al., 1995; Critchley, H. D. and Rolls, E. T., 1996c; Rolls, E. T. and Rolls, J. H., 1997; Rolls, E. T., 1999; 2000a; 2000b; Rolls, E. T. and Scott, T. R., 2003; Rolls, E. T., 2005a; 2005b; 2006).

The enhanced eating when a variety of foods is available, as a result of the operation of sensory-specific satiety, may have been advantageous in evolution in ensuring that different foods with important different nutrients were consumed, but today in humans, when a wide variety of foods is readily available, it may be a factor that can lead to overeating and obesity (Rolls, E. T., 2005a).

4.26.7 Functional Neuroimaging Studies in Humans

4.26.7.1 Taste

In humans it has been shown in neuroimaging studies using functional magnetic resonance imaging (fMRI) that taste activates an area of the anterior insula/frontal operculum, which is probably the primary taste cortex, and part of the orbitofrontal cortex, which is probably the secondary taste cortex (Francis, S. *et al.*, 1999; Small, D. M. *et al.*, 1999; O'Doherty, J. *et al.*, 2001; de Araujo, I. E. T. *et al.*, 2003b; Faurion, A. *et al.*, 2005). Another study has recently shown that umami taste stimuli, including monosodium glutamate, activate similar cortical regions of the human taste system to those activated by a prototypical taste stimulus, glucose (de Araujo, I. E. T. *et al.*, 2003a). A part of the rostral anterior cingulate cortex (ACC) was also activated in this study, as it is in many studies by taste, odor, flavor, and oral texture (Rolls, E. T., 2007). O'Doherty J. *et al.* (2001) showed that the human amygdala was as much activated by the affectively pleasant taste of glucose as by the affectively negative taste of NaCl, and thus provided evidence that the human amygdala is not especially involved in processing aversive as compared to rewarding stimuli.

4.26.7.2   Odor

In humans, in addition to activation of the pyriform (olfactory) cortex (Zald, D. H. and Pardo, J. V., 1997; Sobel, N. *et al.*, 2000; Poellinger, A. *et al.*, 2001), there is strong and consistent activation of the orbitofrontal cortex by olfactory stimuli (Zatorre, R. J. *et al.*, 1992; Francis, S. *et al.*, 1999). In an investigation of where the pleasantness of olfactory stimuli might be represented in humans, O'Doherty J. *et al* (2000) showed that the activation of an area of the orbitofrontal cortex to banana odor was decreased (relative to a control vanilla odor) after bananas were eaten to satiety. Thus activity in a part of the human orbitofrontal cortex olfactory area is related to sensory-specific satiety, and this is one brain region where the pleasantness of odor is represented. Flavor sensory-specific satiety is also represented in the human orbitofrontal cortex, in that when a whole food (either chocolate milk, or tomato juice) is eaten to satiety activation to the flavor of the food eaten to satiety, but not to the other flavor, decreases in the orbitofrontal cortex but not in the primary taste cortex (Kringelbach, M. L. *et al.*, 2003; see Figure 4).

An important issue is whether there are separate regions of the brain discriminable with fMRI that represent pleasant and unpleasant odors. To investigate this, we measured the brain activations produced by three pleasant and three unpleasant odors. The pleasant odors chosen were linalyl acetate (floral, sweet), geranyl acetate (floral), and alpha-ionone (woody, slightly food-related). (Chiral substances were used as racemates.) The unpleasant odors chosen were hexanoic acid, octanol, and isovaleric acid. We found that they activated dissociable parts of the human brain (Rolls, E. T. *et al.*, 2003a). Pleasant but not unpleasant odors were found to activate a medial region of the rostral orbitofrontal cortex. Further, there was a correlation between the subjective pleasantness ratings of the six odors given during the investigation with activation of a medial region of the rostral orbitofrontal cortex. In contrast, a correlation between the subjective unpleasantness ratings of the six odors was found in regions of the left and more lateral orbitofrontal cortex. Activation was also found in the ACC, with a middle part of the anterior cingulate activated by both pleasant and unpleasant odors, and a more anterior part of the ACC showing a correlation with the subjective pleasantness ratings of the odors. Activation in primary olfactory cortical areas was not correlated with the pleasantness of the odor, but was correlated with the intensity (Rolls, E. T. *et al.*, 2003a).

4.26.7.3   Olfactory-Taste Convergence to Represent Flavor

To investigate where in the human brain interactions between taste and odor stimuli may be realized to implement flavor, we performed an event-related fMRI study with sucrose and monosodium glutamate taste, and strawberry and methional (chicken) odors, delivered unimodally or in different combinations (de Araujo, I. E. T. *et al.*, 2003c). The brain regions that were activated by both taste and smell included parts of the caudal orbitofrontal cortex, amygdala, insular cortex and adjoining areas, and ACC. It was shown that a small part of the anterior (putatively agranular) insula responds to unimodal taste and to unimodal olfactory stimuli; and that a part of the anterior frontal operculum is a unimodal taste area (putatively primary taste cortex) not activated by olfactory stimuli. Activations to combined olfactory and taste stimuli where there was little or no activation to either alone (providing positive evidence for interactions between the olfactory and taste inputs) were found in a lateral anterior part of the orbitofrontal cortex. Correlations with consonance ratings for the smell and taste combinations, and for their pleasantness, were found in a medial anterior part of the orbitofrontal cortex (see Figure 5). These results provide evidence on the neural substrate for the convergence of taste and olfactory stimuli to produce flavor in humans, and where the pleasantness of flavor is represented in the human brain (de Araujo, I. E. T. *et al.*, 2003c).

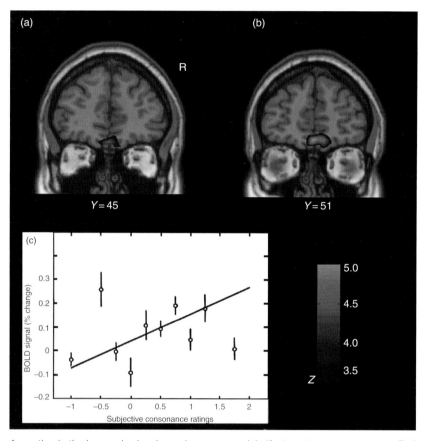

Figure 4   Flavor formation in the human brain, shown by cross-modal olfactory–taste convergence. Brain areas where activations were correlated with the subjective ratings for stimulus (taste–odor) consonance and pleasantness. (a) A second-level, random effects analysis based on individual contrasts (the consonance ratings being the only effect of interest) revealed a significant activation in a medial part of the anterior orbitofrontal cortex. (b) Random effects analysis based on the pleasantness ratings showed a significant cluster of activation located in a (nearby) medial part of the anterior orbitofrontal cortex. The images were thresholded at $P < 0.0001$ for illustration. (c) The relation between the blood oxygen level-dependent (BOLD) signal from the cluster of voxels in the medial orbitofrontal cortex shown in (a) and the subjective consonance ratings. The analyses shown included all the stimuli included in this investigation. The means and standard errors of the mean across subjects are shown, together with the regression line, for which $r = 0.52$. Adapted from de Araujo, I. E. T., Rolls, E. T., Kringelbach, M. L., McGlone, F., and Phillips, N. 2003c. Taste-olfactory convergence, and the representation of the pleasantness of flavour, in the human brain. Eur. J. Neurosci. 18, 2374–2390.

We have also investigated how the flavor of savory foods is produced. Umami taste is produced by glutamate acting on a fifth taste system. However, glutamate presented alone as a taste stimulus is not highly pleasant, and does not act synergistically with other tastes (sweet, salt, bitter and sour). McCabe C. and Rolls E. T. (2007) showed that when glutamate is given in combination with a consonant, savory, odor (vegetable), the resulting flavor can be much more pleasant. Moreover, we showed using functional brain imaging with fMRI that the glutamate and savory odor combination produced much greater activation of the medial orbitofrontal cortex and pregenual cingulate cortex than the sum of the activations by the taste and olfactory

components presented separately. Supralinear effects were much less (and significantly less) evident for sodium chloride and vegetable odor. Further, activations in these brain regions were correlated with the pleasantness, consonance of the taste and olfactory components, and the fullness of the flavor, of the stimuli. We thus proposed that glutamate acts by the nonlinear effects it can produce when combined with a consonant odor (McCabe, C. and Rolls, E. T., 2007). I therefore propose the concept that umami can be thought of as a rich and delicious flavor that is produced by a combination of glutamate taste and a consonant savory odor. Glutamate is thus a flavor enhancer because of the way that it can combine supralinearly with consonant odors.

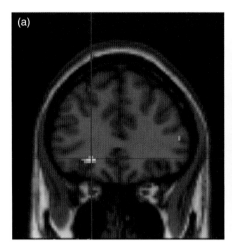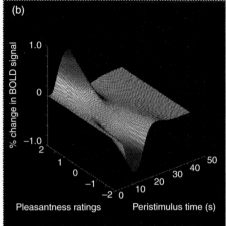

Figure 5   Areas of the human orbitofrontal cortex with activations correlating with pleasantness ratings for flavor in a sensory-specific satiety design. (a) Coronal section through the region of the orbitofrontal cortex from the random effects group analysis showing the peak in the left orbitofrontal cortex (Talairach co-ordinates $X, Y, Z = -22, 34, -8$, z-score $= 4.06$), in which the blood oxygen level-dependent (BOLD) signal in the voxels shown in yellow was significantly correlated with the subjects' subjective pleasantness ratings of the foods throughout an experiment in which the subjects were hungry and found the food pleasant, and were then fed to satiety with the food, after which the pleasantness of the food decreased to neutral or slightly unpleasant. The design was a sensory-specific satiety design, and the pleasantness of the food not eaten in the meal, and the BOLD activation in the orbitofrontal cortex, were not altered by eating the other food to satiety. The two foods were tomato juice and chocolate milk. (b) Plot of the magnitude of the fitted hemodynamic response from a representative single subject against the subjective pleasantness ratings (on a scale from -2 to $+2$) and peristimulus time in seconds. Adapted from Kringelbach, M. L., O'Doherty, J., Rolls, E. T., and Andrews, C. 2003. Activation of the human orbitofrontal cortex to a liquid food stimulus is correlated with its subjective pleasantness. Cereb. Cortex 13, 1064–1071.

4.26.7.4   Cognitive Influences on Olfactory and Flavor Processing

In line with the neuronal convergence of visual and taste inputs on to single neurons in the orbitofrontal cortex (see above), it is found that the sight of food can influence its perceived flavor, and an fMRI correlate of this has been reported (Osterbauer, R. A. *et al.*, 2005). An example of this interaction is that if a white wine is colored red, then adjectives used to describe the flavor of red wine are used to describe the flavor of the white wine.

To investigate how cognitive this influence could be, de Araujo I. E. T. *et al.* (2005) delivered a standard test odor (isovaleric acid with added cheese flavor), but paired it on some trials with a (visually presented) word label Cheddar cheese, and on other trials with the word label body odor. It was found that the word label produced a large modulation of the olfactory activation in the secondary olfactory cortex in the orbitofrontal cortex (with also some modulation in the amygdala). Moreover, the activations in the orbitofrontal cortex were correlated with the pleasantness ratings given by the subjects of the test odor, which were influenced by the word label. In a

control with clean air from the olfactometer, the word labels had much less influence, so that their effect was in particular to modulate the activations being produced in the secondary olfactory cortex by an olfactory stimulus. Thus cognitive influences from the linguistic level, the level of words, can reach down into the secondary olfactory cortex in the orbitofrontal cortex, and modulate their representation of olfactory stimuli (de Araujo, I. E. T. *et al.*, 2005). The mechanism is probably similar to the way in which top-down attentional processes operate, that is by a biased competition mechanism (Rolls, E. T. and Deco, G., 2002; Deco, G. and Rolls, E. T., 2003; 2004; 2005a; 2005c).

It is thus becoming possible to start to understand not only where flavor is represented in the brain, but also how the affective value of smell, taste and flavor are represented, how these representations are influenced by visual stimuli and cognitive states, and how these representations fit into a wider picture of the brain processes underlying the affective or hedonic value of stimuli. This in turn helps to advance understanding of the neural basis of appetite, the control of food intake, and emotion, and their disorders (Rolls, E. T., 2005a).

Acknowledgments

Some of the research from the author's laboratory was supported by the Medical Research Council.

References

Baylis, L. L. and Rolls, E. T. 1991. Responses of neurons in the primate taste cortex to glutamate. Physiol. Behav. 49, 973–979.

Critchley, H. D. and Rolls, E. T. 1996a. Hunger and satiety modify the responses of olfactory and visual neurons in the primate orbitofrontal cortex. J. Neurophysiol. 75, 1673–1686.

Critchley, H. D. and Rolls, E. T. 1996b. Responses of primate taste cortex neurons to the astringent tastant tannic acid. Chem. Senses 21, 135–145.

Critchley, H. D. and Rolls, E. T. 1996c. Olfactory neuronal responses in the primate orbitofrontal cortex: analysis in an olfactory discrimination task. J. Neurophysiol. 75, 1659–1672.

de Araujo, I. E. T., Kringelbach, M. L., and Rolls, E. T. 2003a. Taste-olfactory convergence, and the representation of the pleasantness of flavour, in the human brain. Eur. J. Neurosci. 18, 2374–2390.

de Araujo, I. E. T., Kringelbach, M. L., Rolls, E. T., and Hobden, P. 2003b. The representation of umami taste in the human brain. J. Neurophysiol. 90, 313–319.

de Araujo, I. E. T., Kringelbach, M. L., Rolls, E. T., and McGlone, F. 2003c. Human cortical responses to water in the mouth, and the effects of thirst. J. Neurophysiol. 90, 1865–1876.

de Araujo, I. E. T., Rolls, E. T., Kringelbach, M. L., McGlone, F., and Phillips, N. 2003c. Taste-olfactory convergence, and the representation of the pleasantness of flavour, in the human brain. Eur. J. Neurosci. 18, 2374–2390.

de Araujo, I. E. T., Rolls, E. T., Velazco, M. I., Margot, C., and Cayeux, I. 2005. Cognitive modulation of olfactory processing. Neuron 46, 671–679.

Deco, G. and Rolls, E. T. 2003. Attention and working memory: a dynamical model of neuronal activity in the prefrontal cortex. Eur. J. Neurosci. 18, 2374–2390.

Deco, G. and Rolls, E. T. 2004. A neurodynamical cortical model of visual attention and invariant object recognition. Vision Res. 44, 621–644.

Deco, G. and Rolls, E. T. 2005a. Attention, short-term memory, and action selection: a unifying theory. Prog. Neurobiol. 76, 236–256.

Deco, G. and Rolls, E. T. 2005b. Synaptic and spiking dynamics underlying reward reversal in orbitofrontal cortex. Cerebral Cortex 15, 15–30.

Deco, G. and Rolls, E. T. 2005c. Neurodynamics of biased competition and co-operation for attention: a model with spiking neurons. J. Neurophysiol. 94, 295–313.

Faurion, A., Kobayakawa, T., and Cerf-Ducastel, B. 2005. Cerebral imaging in taste. Chem. Senses 30, i230–i231.

Francis, S., Rolls, E. T., Bowtell, R., McGlone, F., O'Doherty, J., Browning, A., Clare, S., and Smith, E. 1999. The representation of pleasant touch in the brain and its relationship with taste and olfactory areas. Neuroreport 10, 453–459.

Kadohisa, M., Rolls, E. T., and Verhagen, J. V. 2004. Orbitofrontal cortex neuronal representation of temperature and capsaicin in the mouth. Neuroscience 127, 207–221.

Kadohisa, M., Rolls, E. T., and Verhagen, J. V. 2005a. Neuronal representations of stimuli in the mouth: the primate insular taste cortex, orbitofrontal cortex, and amygdala. Chem. Senses 30, 401–409.

Kadohisa, M., Rolls, E. T., and Verhagen, J. V. 2005b. The primate amygdala: neuronal representations of the viscosity, fat texture, grittiness and taste of foods. Neuroscience 132, 33–48.

Kringelbach, M. L., O'Doherty, J., Rolls, E. T., and Andrews, C. 2003. Activation of the human orbitofrontal cortex to a liquid food stimulus is correlated with its subjective pleasantness. Cereb. Cortex 13, 1064–1071.

McCabe, C. and Rolls, E. T. 2007. Umami: a delicious flavor formed by convergence of taste and olfactory pathways in the human brain. Eur. J. Neurosci. 25, 1855–1864.

Norgren, R. 1984. Central Neural Mechanisms of Taste. In: Handbook of Physiology – The Nervous System III Sensory Processes 1 (ed. I. Darien-Smith), pp. 1087–1128. American Physiological Society.

O'Doherty, J., Rolls, E. T., Francis, S., Bowtell, R., McGlone, F., Kobal, G., Renner, B., and Ahne, G. 2000. Sensory-specific satiety related olfactory activation of the human orbitofrontal cortex. Neuroreport 11, 893–897.

O'Doherty, J., Rolls, E. T., Francis, S., Bowtell, R., and McGlone, F. 2001. The representation of pleasant and aversive taste in the human brain. J. Neurophysiol. 85, 1315–1321.

Osterbauer, R. A., Matthews, P. M., Jenkinson, M., Beckmann, C. F., Hansen, P. C., and Calvert, G. A. 2005. The color of scents: chromatic stimuli modulate odor responses in the human brain. J. Neurophysiol. 93, 3434–3441.

Poellinger, A., Thomas, R., Lio, P., Lee, A., Makris, N., Rosen, B. R., and Kwong, K. K. 2001. Activation and habituation in olfaction – an fMRI study. Neuroimage 13, 547–560.

Pritchard, T. C., Hamilton, R. B., Morse, J. R., and Norgren, R. 1986. Projections of thalamic gustatory and lingual areas in the monkey, Macaca fascicularis. J. Comp. Neurol. 244, 213–228.

Rolls, E. T. 1997. Taste and olfactory processing in the brain and its relation to the control of eating. Crit. Rev. Neurobiol. 11, 263–287.

Rolls, E. T. 1999. The Brain and Emotion. Oxford University Press.

Rolls, E. T. 2000a. The orbitofrontal cortex and reward. Cereb. Cortex 10, 284–294.

Rolls, E. T. 2000b. Taste, Olfactory, Visual and Somatosensory Representations of the Sensory Properties of Foods in the Brain, and their Relation to the Control of Food Intake. In: Neural and Metabolic Control of Macronutrient Intake (eds. H. R. Berthoud and R. J. Seeley), pp. 247–262. CRC Press.

Rolls, E. T. 2001. The rules of formation of the olfactory representations found in the orbitofrontal cortex olfactory areas in primates. Chem. Senses 26, 595–604.

Rolls, E. T. 2002a. The Cortical Representation of Taste and Smell. In: Olfaction, Taste and Cognition (eds. G. Rouby, B. Schaal, D. Dubois, R. Gervais, and A. Holley), pp. 367–388. Cambridge University Press.

Rolls, E. T. 2002b. The Functions of the Orbitofrontal Cortex. In: Principles of Frontal Lobe Function (eds. D. T. Stuss and R. T. Knight), pp. 354–375. Oxford University Press.

Rolls, E. T. 2005a. Emotion Explained. Oxford University Press.

Rolls, E. T. 2005b. Taste, olfactory, and food texture processing in the brain, and the control of food intake. Physiol. Behav. 85, 45–56.

Rolls, E. T. 2006. Brain mechanisms underlying flavour and appetite. Philos. Trans. R. Soc. B 361, 1123–1136.

Rolls, E. T. 2007. The Anterior and Midcingulate Cortices and Reward. In: Cingulate Neurobiology and Disease, Vol. 1,

Infrastructure, Diagnosis and Treatment (*ed*. B. Vogt), Chapter 8. Oxford University Press.

Rolls, E. T. and Baylis, L. L. 1994. Gustatory, olfactory, and visual convergence within the primate orbitofrontal cortex. J. Neurosci. 14, 5437–5452.

Rolls, E. T. and Deco, G. 2002. Computational Neuroscience of Vision. Oxford University Press.

Rolls, E. T. and Rolls, J. H. 1997. Olfactory sensory-specific satiety in humans. Physiol. Behav. 61, 461–473.

Rolls, E. T. and Scott, T. R. 2003. Central Taste Anatomy and Neurophysiology. In: Handbook of Olfaction and Gustation, 2nd edn. (*ed*. R. L. Doty), pp. 679–705. Dekker.

Rolls, E. T., Critchley, H. D., and Treves, A. 1996b. The representation of olfactory information in the primate orbitofrontal cortex. J. Neurophysiol. 75, 1982–1996.

Rolls, E. T., Critchley, H. D., Browning, A. S., Hernadi, A., and Lenard, L. 1999. Responses to the sensory properties of fat of neurons in the primate orbitofrontal cortex. J. Neurosci. 19, 1532–1540.

Rolls, E. T., Critchley, H., Wakeman, E. A., and Mason, R. 1996a. Responses of neurons in the primate taste cortex to the glutamate ion and to inosine 5′-monophosphate. Physiol. Behav. 59, 991–1000.

Rolls, E. T., Kringelbach, M. L., and de Araujo, I. E. T. 2003a. Different representations of pleasant and unpleasant odors in the human brain. Eur. J. Neurosci. 18, 695–703.

Rolls, E. T., Scott, T. R., Sienkiewicz, Z. J., and Yaxley, S. 1988. The responsiveness of neurones in the frontal opercular gustatory cortex of the macaque monkey is independent of hunger. J. Physiol. 397, 1–12.

Rolls, E. T., Sienkiewicz, Z. J., and Yaxley, S. 1989. Hunger modulates the responses to gustatory stimuli of single neurons in the caudolateral orbitofrontal cortex of the macaque monkey. Eur. J. Neurosci. 1, 53–60.

Rolls, E. T., Verhagen, J. V., and Kadohisa, M. 2003b. Representations of the texture of food in the primate orbitofrontal cortex: neurons responding to viscosity, grittiness and capsaicin. J. Neurophysiol. 90, 3711–3724.

Rolls, E. T., Yaxley, S., and Sienkiewicz, Z. J. 1990. Gustatory responses of single neurons in the caudolateral orbitofrontal cortex of the macaque monkey. J. Neurophysiol. 64, 1055–1066.

Scott, T. R., Yan, J., and Rolls, E. T. 1995. Brain mechanisms of satiety and taste in macaques. Neurobiology 3, 281–292.

Scott, T. R., Yaxley, S., Sienkiewicz, Z. J., and Rolls, E. T. 1986. Gustatory responses in the frontal opercular cortex of the alert cynomolgus monkey. J. Neurophysiol. 56, 876–890.

Small, D. M., Zald, D. H., Jones-Gotman, M., Zatorre, R. J., Pardo, J. V., Frey, S., and Petrides, M. 1999. Human cortical gustatory areas: a review of functional neuroimaging data. Neuroreport 10, 7–14.

Sobel, N., Prabkakaran, V., Zhao, Z., Desmond, J. E., Glover, G. H., Sullivan, E. V., and Gabrieli, J. D. E. 2000. Time course of odorant-induced activation in the human primary olfactory cortex. J. Neurophysiol. 83, 537–551.

Thorpe, S. J., Rolls, E. T., and Maddison, S. 1983. Neuronal activity in the orbitofrontal cortex of the behaving monkey. Exp. Brain Res. 49, 93–115.

Verhagen, J. V., Kadohisa, M., and Rolls, E. T. 2004. The primate insular/opercular taste cortex: neuronal representations of the viscosity, fat texture, grittiness and taste of foods in the mouth. J. Neurophysiol. 92, 1685–1699.

Verhagen, J. V., Rolls, E. T., and Kadohisa, M. 2003. Neurons in the primate orbitofrontal cortex respond to fat texture independently of viscosity. J. Neurophysiol. 90, 1514–1525.

Wilson, F. A. W. and Rolls, E. T. 2005. The primate amygdala and reinforcement: a dissociation between rule-based and associatively-mediated memory revealed in amygdala neuronal activity. Neuroscience 133, 1061–1072.

Yaxley, S., Rolls, E. T., and Sienkiewicz, Z. J. 1988. The responsiveness of neurons in the insular gustatory cortex of the macaque monkey is independent of hunger. Physiol. Behav. 42, 223–229.

Yaxley, S., Rolls, E. T., Sienkiewicz, Z. J., and Scott, T. R. 1985. Satiety does not affect gustatory activity in the nucleus of the solitary tract of the alert monkey. Brain Res. 347, 85–93.

Zald, D. H. and Pardo, J. V. 1997. Emotion, olfaction, and the human amygdala: amygdala activation during aversive olfactory stimulation. Proc. Natl. Acad. Sci. U. S. A. 94, 4119–4124.

Zatorre, R. J., Jones-Gotman, M., Evans, A. C., and Meyer, E. 1992. Functional localization of human olfactory cortex. Nature 360, 339–340.

Further Reading

Aggleton, J. P. (*ed*). 2000. The Amygdala 2nd. Edn., Oxford University Press.

Zald, D. H. and Rauch, S. L. (*eds.*) 2006. The Orbitofrontal Cortex. Oxford University Press.

4.27 The Aging Gustatory System

S S Schiffman, Duke University Medical Center, Durham, NC, USA

4.27.1 Introduction

Demographic shifts in the global population are occurring rapidly with escalation in both the number and the percentage of elderly persons (Cohen, J. E., 2003; Administration on Aging, 2004). Over 400 million people worldwide are currently 65 years of age or greater (Davies, A. M., 1989; Cohen, J. E., 2003), and this number is expected to exceed 1.5 billion by 2050 (Cohen, J. E., 2003). A significant proportion of this population will have taste impairments that can potentially alter the health and well-being of the elderly. While disorders of taste can occur throughout the lifespan, they are far more prevalent in an older population. Diagnostic terms for taste disorders include ageusia (absence of taste), hypogeusia (diminished sensitivity of taste), and dysgeusia (distortion of normal taste). Research studies and clinical reports indicate that hypogeusia and dysgeusia commonly occur in older individuals (Schiffman, S. S., 1993; 1997) while ageusia is relatively rare (Pribitkin, E. *et al.*, 2003). The aim of this chapter is to provide an overview of gustatory perception in older individuals. Topics to be covered include changes of taste perception at thresholds levels, changes of taste perception at suprathreshold levels, medications and medical conditions associated with taste alterations, as well as taste losses in the absence of drugs and disease.

An intact sense of taste is vital for persons of all ages. Taste is a phylogenetically primitive sense that serves both as a precautionary mechanism and as the initial internal analyzer for materials ingested into the oral cavity. It enables us to distinguish between nutritive chemicals (such as sugars) and potentially harmful compounds (such as bitter toxic alkaloids).

Failure to detect the bitter taste warning from alkaloid poisons at the gateway to the stomach and intestines can have deadly consequences. Overall, taste sensations serve as a quality check for the identity and safety of food in the mouth (i.e., should we swallow the contents in our mouths or spit them out?). In addition to their role in identity and safety, taste sensations (along with smell and trigeminal sensations) contribute to the digestion of food by triggering salivary, gastric, pancreatic, and intestinal secretions that prepare the body for the absorption of nutrients (Giduck, S. A. *et al.*, 1987; Schiffman, S. S., and Warwick, Z. S., 1992; Teff, K. L. and Engelman, K., 1996). Gustatory signals additionally serve as indicators of the nutritional value of a food from a learned association of the food's taste with its postingestive effects (Warwick, Z. S. and Schiffman, S. S., 1991; Drewnowski, A. *et al.*, 1995; Stubbs, R. J. and Whybrow, S., 2004). This learned association of taste signals with metabolic consequences enables food intake and meal size to be modulated in anticipation of nutritional needs. Thus, alterations in the ability to taste are not merely an inconvenience but rather can impact survival, general health, and quality of life.

The qualitative range of taste has not been fully established. Traditionally, much taste research has presumed the existence of only four (or possibly five) independent taste qualities, the four so-called basic or primary tastes (sweet, sour, salty, and bitter) and a fifth quality, the taste of glutamate salts called umami. However, data are accumulating that suggest other sensory qualities such as fatty (Gilbertson, T. A. *et al.*, 1997), metallic (Schiffman, S.S., 2000), starchy/polysaccharide (Sclafani, A., 2004), chalky (Schiffman, S. S., and Erickson, R. P., 1993), and astringent (Schiffman, S. S., *et al.*, 1992) are carried by taste nerves. Thus, while the terms sweet, sour, salty, and bitter may be familiar and have linguistic relevance, they do not describe the entire range of tastes perceived by humans and other animals.

The ability to perceive the taste signals of specific sensory qualities (such as salty or sweet) are especially important for subgroups of the elderly population such as those with hypertension who must comply with salt-restricted diets or those with diabetes who must reduce their consumption of sugar (Schiffman, S. S., 1997). Reduced sensitivity to the oral component of fats in the elderly can increase their risk of cardiovascular disease, diabetes, hypertension, and other conditions in which high fat intake is contraindicated. Taste losses along with other physiological and psychological factors also contribute to the observed decline in food intake in many older persons (Morley, J. E., 2001). Overall, taste disorders that impair or distort the taste of palatable as well as unpalatable substances can increase the risk of numerous adverse safety and health-related issues for an elderly population.

4.27.2   Changes of Taste at Threshold Levels

4.27.2.1   Taste Thresholds in a General Elderly Population

An overview of research studies indicates that the detection threshold (DT) and recognition threshold (RT) for simple basic tastes (sweet, sour, salty, and bitter) are moderately elevated in older persons compared to a younger cohort (e.g., Richter, C. P. and Campbell, K. H., 1940; Harris, H. and Kalmus, H., 1949; Bourlière, F. *et al.*, 1958; Hinchcliff, R. 1958; Cooper, R. M. *et al.*, 1959; Kalmus, H. and Trotter, W. R., 1962; Glanville, E. V. *et al.*, 1964; Smith, S. E. and Davies, P. D., 1973; Grzegorczyk, P. B. *et al.*, 1979; Murphy, C., 1979; Schiffman, S.S. *et al.*, 1979; Dye, C. J. and Koziatek, D. A., 1981; Schiffman, S. S. *et al.*, 1981; Moore, L. M. *et al.*, 1982; Weiffenbach, J. M. *et al.*, 1982; Schiffman, S. S. *et al.*, 1990; 1991; Schiffman, S. S., 1993; Schiffman, S. S. *et al.*, 1994; Stevens, J. C. *et al.*, 1995; Stevens, J. C., 1996; Stevens J. C. and Traverzo, A., 1997; Schiffman, S. S. *et al.*, 1998; Mojet, J. *et al.*, 2001; Yamauchi, Y. *et al.*, 2002; Fukunaga, A. *et al.*, 2005). An elevated DT indicates that the elderly require the presence of more molecules (or ions) for a sensation to be perceived compared to younger persons – that is, the elderly have reduced absolute sensitivity. An elevated RT signifies that a greater concentration of a tastant is required before it can be correctly identified. Furthermore, thresholds are higher for a compound in a mixture than in a simple aqueous solution (Stevens, J. C. and Traverzo, A., 1997), that is, other ingredients in a mixture can mask significant food components such as NaCl, making it harder to detect and recognize salt in food than alone in a simple aqueous solution.

Tables 1–9 show the results of studies that compare mean thresholds for the elderly with those of the young for a range of compounds including sodium salts with different anions (Schiffman, S. S. *et al.*, 1990), bitter compounds (Schiffman, S. S. *et al.*, 1994), sweeteners (Schiffman, S. S. *et al.*, 1981), acids

Table 1   Mean detection and recognition thresholds for elderly and young subjects for sodium salts with different anions

	Detection thresholds			Recognition thresholds for saltiness		
	Elderly, E (M)	Young, Y (M)	E/Y	Elderly, E (M)	Young, Y (M)	E/Y
MSG (monosodium glutamate, $\lambda = 26.0$)[a]	0.00638	0.00126	5.06	0.0091	0.00207	4.41
Na acetate ($\lambda = 40.9$)	0.0190	0.00242	7.84	0.0229	0.00952	2.41
Na ascorbate ($\lambda = 39.7$)	0.0250	0.00404	6.19	0.0265	0.00809	3.28
Na carbonate ($\lambda = 44.5$)	0.00829	0.00218	3.79	0.0234	0.00425	5.51
Na chloride ($\lambda = 76.3$)	0.01850	0.00238	7.76	0.0227	0.00815	2.79
Na citrate ($\lambda = 210.6$)	0.0130	0.000531	24.5	0.0187	0.00190	9.84
Na phosphate monobasic ($\lambda = 33.0$)	0.0160	0.00307	5.21	0.0253	0.01140	2.22
Na succinate ($\lambda = 117.6$)	0.0138	0.000854	16.2	0.0167	0.00217	7.71
Na sulfate ($\lambda = 160.0$)	0.0283	0.000981	28.8	0.0349	0.00322	10.86
Na tartrate ($\lambda = 119.2$)	0.0159	0.00151	10.5	0.0277	0.00295	9.39

[a]λ is the molar conductivity of the anion at infinite dilution, 25 °C, ohm^{-1} cm^2 mol^{-1}.
Data are from Schiffman, S. S., Crumbliss, A. L., Warwick, Z. S., and Graham, B. G. 1990. Thresholds for sodium salts in young and elderly subjects: correlation with molar conductivity of anion. Chem. Senses 15, 671–678.

Table 2   Mean detection and recognition thresholds for bitter compounds in elderly and young subjects

	Detection thresholds			Recognition thresholds		
	Elderly (E)	Young (Y)	E/Y	Elderly (E)	Young (Y)	E/Y
Caffeine ($\log P = -0.097$)[a]	1.99 mM	1.30 mM	1.53	6.74 mM	1.87 mM	3.60
Denatonium benzoate ($\log P = -0.088$)	0.0323 µM	0.0115 µM	2.81	0.0387 µM	0.0123 µM	3.14
KNO$_3$ ($\log P = -4.0$)	32.7 mM	1.91 mM	17.1	271 mM	5.97 mM	45.4
MgCl$_2$ ($\log P = -3.026$)	5.20 mM	1.02 mM	5.10	21.8 mM	20.3 mM	1.07
MgNO$_3$ ($\log P = -4.0$)	33.3 mM	1.40 mM	23.8	191 mM	14.8 mM	12.9
MgSO$_4$ ($\log P = -4.0$)	6.08 mM	0.323 mM	18.8	14.8 mM	2.59 mM	5.71
Naringin ($\log P = -0.114$)	0.138 mM	0.0427 mM	3.23	0.195 mM	0.0561 mM	3.48
Phenylthiocarbamide ($\log P = +0.743$)	1.26 mM	0.591 mM	2.13	1.74 mM	1.21 mM	1.44
Quinine HCl ($\log P = +1.082$)	8.07 µM	3.99 µM	2.02	12.3 µM	4.75 µM	2.59
Quinine sulfate ($\log P = +1.036$)	8.75 µM	2.04 µM	4.29	12.3 µM	2.53 µM	4.86
Sucrose octaacetate ($\log P = +1.010$)	5.32 µM	3.89 µM	1.37	22.8 µM	5.30 µM	4.30
Urea ($\log P = -2.190$)	0.116 M	0.103 M	1.12	0.245 M	0.134 M	1.83

[a]Logarithm of the 1-octanol/water partition coefficient.
Data are from Schiffman, S. S., Gatlin, L. A., Frey, A. E., Heiman, S. A., Stagner, W. C., and Cooper, D. C. 1994. Taste perception of bitter compounds in young and elderly persons: relation to lipophilicity of bitter compounds. Neurobiol. Aging 15, 743–750.

(Schiffman, S. S., 1993), astringent compounds (Schiffman, S. S. et al., 1989), amino acids (Schiffman, S. S. et al., 1979) including glutamate salts (Schiffman, S. S. et al., 1991), fats (Schiffman, S. S., et al., 1998), and gums (Schiffman, S. S. et al., 1989). The subjects in these studies were elderly individuals who took an average of 3.4 medications but who otherwise led active, normal lives. For DTs, the ratio of DT (elderly)/DT (young) revealed that DTs in the elderly were higher by the following amounts: 11.6 times higher for sodium salts; 2.7 times higher for sweeteners; 4.3 times higher for acids; 2.8 for astringent compounds; 7.0 times higher for bitter compounds; 2.5 times higher for amino acids; 5.0 times higher for glutamate salts;

3.1 times higher for fats/oils; and 3.7 times higher for polysaccharides/gums. The average loss in detection across this qualitative range of compounds is 4.74. However, the degree of loss within a group (such as sodium salts) was not uniform but rather varied with the chemical structure of the compounds tested. For RTs, the ratio of RT (elderly)/RT (young) revealed that RTs in the elderly were higher by the following amounts: 5.8 times higher for sodium salts; 2.1 times higher for sweeteners; 6.8 times higher for acids with sour tastes; 3.0 for astringent compounds; 7.5 times higher for bitter compounds; and 3.0 times higher for polysaccharides/gums. The average loss in recognition across these taste qualities is 4.7.

Table 3   Mean detection and recognition thresholds for sweeteners in elderly and young subjects

	Detection thresholds			Recognition thresholds		
	Elderly (E)	Young (Y)	E/Y	Elderly (E)	Young (Y)	E/Y
Acesulfame-K	74.7 µM	44.4 µM	1.68	239 µM	161 µM	1.48
Aspartame	91.3 µM	22.4 µM	4.07	124 µM	44.9 µM	2.76
Calcium cyclamate	0.412 mM	0.266 mM	1.55	1.69 mM	1.33 mM	1.27
Fructose	10.1 mM	4.39 mM	2.30	26.4 mM	16.6 mM	1.59
Monellin	0.0913 µM	0.0195 µM	4.67		0.0676 µM	
Neohesperidin dihydrochalcone	4.60 µM	2.20 µM	2.09		5.28 µM	
Rebaudioside	13.0 µM	4.61 µM	2.82		13.6 µM	
Sodium saccharin	42.4 µM	14.7 µM	2.88	137 µM	49.7 µM	2.76
Stevioside	16.0 µM	5.31 µM	3.02		23.7 µM	
Thaumatin	0.133 µM	0.0716 µM	1.86		0.201 µM	
D-Tryptophan	0.322 mM	0.109 mM	2.95	1.45 mM	0.546 mM	2.66

Data are from Schiffman, S. S., Lindley, M. G., Clark, T. B., and Makino, C. 1981. Molecular mechanism of sweet taste: relationship of hydrogen bonding to taste sensitivity for both young and elderly. Neurobiol. Aging 2, 173–185.

Table 4   Mean detection and recognition thresholds for acids in elderly and young subjects

	Detection thresholds			Recognition thresholds for sourness		
Stimulus	Elderly, E (mM)	Young, Y (mM)	E/Y	Elderly, E (mM)	Young, Y (mM)	E/Y
Acetic acid	0.273	0.106	2.58	0.819	0.294	2.79
Ascorbic acid	0.725	0.281	2.58	2.190	0.396	5.53
Citric acid	0.375	0.0498	7.53	0.816	0.131	6.23
Glutamic acid	0.463	0.0920	5.03	1.500	0.309	4.85
Hydrochloric acid	0.200	0.0179	11.17	0.477	0.0226	21.11
Succinic acid	0.188	0.132	1.42	1.330	0.174	7.64
Sulfuric acid	0.100	0.0468	2.14	0.170	0.0468	3.63
Tartaric acid	0.163	0.0864	1.89	0.297	0.131	2.27

Data are from Schiffman, S. S. 1993. Perception of taste and smell in elderly persons. Crit. Rev. Food Sci. Nutr. 33, 17–26.

Table 5   Mean detection and recognition thresholds for acidic/astringent compounds in elderly and young subjects

	Detection thresholds			Recognition thresholds for astringency		
Stimulus	Elderly, E (mM)	Young, Y (mM)	E/Y	Elderly, E (mM)	Young, Y (mM)	E/Y
Gallic acid	0.780	0.250	3.12	2.07	1.10	1.88
Tartaric acid	0.220	0.0549	4.01	0.324	0.0689	4.70
Tannic acid	0.072	0.0271	2.66	0.295	0.0528	5.59
Catechin	1.48	1.18	1.25	2.500	1.56	1.60
Ammonium alum	0.172	0.0780	2.21	0.487	0.244	2.00
Potassium alum	0.454	0.120	3.78	1.380	0.723	1.91

4.27.2.1.1   Sodium salts

Table 1 compares the taste DTs and RTs for 10 sodium salts in young and elderly persons (see Schiffman, S. S. et al., 1990). The DTs of these salts were highly correlated with the molar conductivity of the anion. Molar conductivity (λ) is a measure of the electrical charge carried by the anion per unit time. Furthermore, age-related decrements as determined by the ratio DT (elderly)/DT (young) were greatest for anions with the largest molar conductivity (Na sulfate, Na tartrate, Na citrate, and Na succinate). That is, age-related losses in sensitivity to sodium salts were greatest for anions with the highest charge mobility. RTs are also given in Table 1 for those subjects who did perceive saltiness; however, some subjects did not recognize a salty

component for some of these sodium salts at concentrations 64 times higher than the DT.

4.27.2.1.2 Bitter compounds

Table 2 compares the DTs and RTs for 12 bitter compounds in young and elderly subjects (see Schiffman, S. S. et al., 1994). Subjects included both tasters and nontasters of the bitter compound phenylthiocarbamide (PTC). A strong relationship between bitter threshold values and the logarithm of the octanol/water partition coefficient ($\log P$) was found for both young and elderly subjects. The logarithm of the 1-octanol/water partition coefficient ($\log P$) is a measure of the lipophilicity of the bitter compound. The greatest losses at the threshold level for the elderly were for the least lipophilic compounds, that is, $MgNO_3$, $MgSO_4$, and KNO_3. There were no differences in DTs or RTs for tasters and nontasters of PTC with the exception of PTC itself.

4.27.2.1.3 Sweet compounds

Table 3 compares the taste DTs and RTs for 11 sweeteners in young and elderly persons. Sweeteners with the highest DTs are natural sugars for both age groups (see Schiffman, S. S. et al., 1981). The fact that high concentrations are needed for the detection of sugars may have had survival value for past generations who needed to choose only the most caloric foods in a natural environment. Sweeteners with the lowest DTs (i.e., low concentrations are

Table 6 Mean detection thresholds for amino acids in elderly and young subjects

Stimulus	Elderly (E)	Young (Y)	E/Y
L-Alanine	19.5 mM	16.2 mM	1.20
L-Arginine	1.12 mM	1.20 mM	0.93
L-Arginine HCl	2.39 mM	1.23 mM	1.94
L-Asparagine	9.33 mM	1.62 mM	5.75
L-Aspartic acid	0.501 mM	0.182 mM	2.75
L-Cysteine	0.390 mM	0.0630 mM	6.19
L-Cysteine HCl	20.0 μM	16.0 μM	1.25
L-Glutamic acid	0.100 mM	0.0630 mM	1.59
L-Glutamine	26.9 mM	9.77 mM	2.75
L-Glycine	0.0617 M	0.0309 M	2.00
L-Histidine	6.45 mM	1.23 mM	5.24
L-Histidine HCl	0.389 mM	0.0794 mM	4.90
L-Isoleucine	12.0 mM	7.41 mM	1.62
L-Leucine	12.9 mM	6.45 mM	2.00
L-Lysine	2.24 mM	0.708 mM	3.16
L-Lysine HCl	2.09 mM	0.447 mM	4.68
L-Methionine	2.63 mM	3.72 mM	0.71
L-Phenylalanine	19.1 mM	6.61 mM	2.89
L-Proline	0.0372 M	0.0151 M	2.46
L-Serine	0.0263 M	0.0209 M	1.26
L-Threonine	0.020 M	0.0257 M	0.78
L-Tryptophan	2.88 mM	2.29 mM	1.26
L-Valine	0.0115 M	0.00416 M	2.76

Data are from Schiffman, S. S., Hornack, K., and Reilly, D. 1979. Increased taste thresholds of amino acids with age. Am. J. Clin. Nutr. 32, 1622–1627.

Table 7 Mean detection thresholds for glutamate salts with and without the taste enhancer inosine 5′-monophosphate (IMP) in elderly and young subjects

Stimulus	Elderly (E)	Young (Y)	E/Y
Sodium glutamate	2.83 mM	0.902 mM	3.14
Sodium glutamate with 0.1 mM IMP	0.888 mM	0.113 mM	7.86
Sodium glutamate with 1 mM IMP	0.145 mM	0.0480 mM	3.02
Potassium glutamate	7.69 mM	0.902 mM	8.53
Potassium glutamate with 0.1 mM IMP	0.549 mM	0.106 mM	5.18
Potassium glutamate with 1 mM IMP	0.0928 mM	0.0108 mM	8.59
Ammonium glutamate	4.26 mM	1.08 mM	3.94
Ammonium Glutamate with 0.1 mM IMP	0.458 mM	0.139 mM	3.29
Ammonium glutamate with 1 mM IMP	0.129 mM	0.0343 mM	3.76
Calcium diglutamate	1.09 mM	0.292 mM	3.73
Calcium diglutamate with 0.1 mM IMP	0.327 mM	0.0606 mM	5.40
Calcium diglutamate with 1 mM IMP	0.0692 mM	0.0190 mM	3.64
Magnesium diglutamate	1.86 mM	0.253 mM	7.35
Magnesium diglutamate with 0.1 mM IMP	0.289 mM	0.0421 mM	6.86
Magnesium diglutamate with 1 mM IMP	0.0452 mM	0.0257 mM	1.76
IMP	1.99 mM	0.430 mM	4.63

Data are from Schiffman, S. S., Frey, A. E., Luboski, J. A., Foster, M. A., and Erickson, R. P. 1991. Taste of glutamate salts in young and elderly subjects: role of inosine 5′-monophosphate and ions. Physiol. Behav. 49, 843–854.

Table 8   Mean detection thresholds (%) for three oils (MCT, soybean, and mineral) in four emulsifiers (acacia gum, Emplex, Tween 80, and Na caseinate) in oil-in-water emulsion

Oil	Emulsifier	Elderly subjects (%)	Young subjects (%)	Ratio, E/Y
MCT[a]	Acacia	10.1	2.85	3.54
Soybean	Acacia	12.9	4.02	3.20
Mineral	Acacia	9.77	4.43	2.20
MCT	Emplex	25.0	3.93	6.37
Soybean	Emplex	14.9	6.52	2.28
Mineral	Emplex	20.0	8.85	2.26
MCT	Tween 80	19.3	5.35	3.60
Soybean	Tween 80	17.7	5.85	3.02
Mineral	Tween 80	19.9	5.77	3.44
MCT	Na caseinate	13.6	6.18	2.20
Soybean	Na caseinate	13.0	5.35	2.43
Mineral	Na caseinate	13.4	4.27	3.13

[a]Medium-chain triglycerides.
Data are from Schiffman, S. S., Graham, B. G., Sattely-Miller, E. A., and Warwick, Z. S. 1998. Orosensory perception of dietary fat. Curr. Dir. Psychol. Sci. 7, 137–143.

Table 9   Mean detection and recognition thresholds for five polysaccharides/gums

Stimulus	Detection thresholds			Recognition thresholds for thickness		
	Elderly, E (%)	Young, Y (%)	E/Y	Elderly, E (%)	Young, Y (%)	E/Y
Acacia gum	1.02	0.644	1.58	3.12	1.44	2.17
Guar gum	0.116	0.057	2.04	0.42	0.22	1.91
Locust bean gum	0.43	0.061	7.05	0.74	0.22	3.36
Xanthan gum	0.238	0.0396	6.01	0.41	0.071	5.77
Algin	0.115	0.0605	1.90	0.26	0.15	1.73

Data from Schiffman, S. S., Rascoe, D., and Garcia, R. A. 1989. The effects of age and race on thresholds and magnitude estimates of edible gums. Chem. Senses 14, 743.

detected) for both age groups were those compounds with the greatest possible sites capable of intermolecular hydrogen bonding with the sweetener taste receptor. The RTs for the large sweetener molecules such as monellin and thaumatin were extremely variable in the elderly, so means are not given.

4.27.2.1.4   Acids

Table 4 compares the DTs and RTs for eight acids with sour tastes in young and elderly persons (Schiffman, S. S., 1993). The loss in sensitivity with age for HCl, the acid with the lowest molecular weight, was significantly greater than losses for the organic acids.

4.27.2.1.5   Astringent compounds

DTs and RTs for six astringent compounds are given in Table 5. DTs for astringent compounds are in the same general range as that for the acids in Table 4. This is most likely due to the fact that all of these

astringent compounds have pH values below 3.5 at the concentrations tested. RTs for astringency were up to six times higher than the DTs for astringent compounds in young subjects and four times higher for elderly subjects. Each of these astringent compounds was reported to have sour and bitter tastes that emerged at concentrations intermediate between the DT and RT for astringency. Astringent compounds are found in natural food products and are important sensory components of many beverages such as wine, tea, and fruit juices (Bate-Smith, E. C., 1954; Robichaud, J. L. and Noble, A. C., 1990). Astringency is associated with a drying or a puckering sensation on the tongue and in the oral cavity. A portion of this sensation appears to be transmitted through gustatory nerves because patients with severed trigeminal nerves report drying or puckering sensations on the tongue (the trigeminal nerve innervates the buccal cavity of the mouth and the first 2/3 of the tongue) (Schiffman, S. S., unpublished clinical tests).

4.27.2.1.6 Amino acids

Table 6 compares the taste DTs for 19 L-amino acids and 4 monohydrochloride derivatives in young and elderly persons (see Schiffman, S. S. *et al.*, 1979). The ratio DT (elderly)/DT (young) for 20 of the 23 amino acid compounds was >1, which was highly statistically significant. There were no major differences in the degree of loss with compound type although age-related losses tended to be higher for two amino acids with side chains containing basic groups (L-histidine and L-lysine) and their monohydrochloride derivatives.

4.27.2.1.7 Glutamate salts

Table 7 compares the DTs for five glutamate salts (sodium glutamate, potassium glutamate, ammonium glutamate, calcium diglutamate, and magnesium diglutamate) in young and elderly persons (see Schiffman, S. S. *et al.*, 1991). The effect of inosine 5′-monophosphate (IMP), a taste enhancer, on taste thresholds of glutamate compounds at two concentrations (0.1 and 1.0 mM IMP) was also investigated. While 0.1 mM IMP lowered thresholds for all the five salts in young but not older subjects, 1 mM IMP lowered thresholds in both young and elderly groups.

4.27.2.1.8 Fats (emulsified oils)

Recent data suggest that fats and oils have taste components because sensations from these stimuli can be detected on the lateral posterior sides of the tongue without tongue movement (Schiffman, S. S. *et al.*, 1998), and electrophysiological data indicate that fatty acids activate taste nerves (Gilbertson, T. A. *et al.*, 1997). Table 8 compares the mean DTs for three different oils in four different emulsifiers in young and elderly subjects. The oils were refined, bleached, deodorized soybean oil (long-chain triglyceride, LCT); medium-chain triglyceride (MCT) oil; and light mineral oil. These fats are LCTs, MCTs, and a mixture of liquid hydrocarbons from petroleum, respectively. Oil-in-water emulsions were made with each oil using one of the four different emulsifiers: Polysorbate 80 (Tween® 80), sodium stearoyl lactylate (Emplex), sodium caseinate, and acacia gum. Statistical analysis revealed significant main effects for age (elderly had higher thresholds) and emulsifier. The low DTs of the oils emulsified in acacia gum (which suspends fat globules in a matrix rather than acting as a surfactant) accounted for the main effect of emulsifier. There were no significant differences in the thresholds with or without nose clips. The mean DT of the oils across all conditions for young subjects was 5.30% oil (v/v) while mean DTs for the elderly was 15.8% oil (v/v). The lowest mean DT was obtained for MCT oil in acacia gum for young subjects (2.85%). The highest mean DT was obtained for MCT oil in Emplex for elderly subjects (25.0%).

4.27.2.1.9 Complex polysaccharides/gums

Gums are complex polysaccharides used as thickening and gelling agents in many foods such as mayonnaise, ice cream, syrups, cheese, and dietetic foods (Klose, R. E. and Glicksman, M., 1972). A portion of the oral sensation from polysaccharides in humans is transmitted through gustatory nerves because patients with severed trigeminal nerves can differentiate between water alone and gums in water when dropped on the tongue (Schiffman, S. S., unpublished clinical tests). Animal data also suggest that polysaccharides may have a taste (Sclafani, A. 2004). DTs and RTs for gums commonly added to foods are given in Table 9 (from Schiffman, S. S. *et al.* 1989). The ratio of DT (elderly)/DT (young) varied with the gum structure from 1.58 for acacia gum to 7.05 for locust bean gum. On the descriptor scales, subjects not only commented on the thickness but also used adjectives such as starchy.

4.27.2.1.10 Electric taste

Taste thresholds for electric stimulation of the tongue have also been compared in young and elderly subjects. Like thresholds for chemicals, these thresholds tend to be elevated with increasing age (Krarup, B. 1958; Hughes, G. 1969; Nakazato, M. *et al.*, 2002; Terada, T. *et al.*, 2004; Kettaneh, A. *et al.* 2005).

4.27.2.2 Nonuniformity of Taste Loss

It can be seen from the above studies that different stimuli have varying degrees of loss depending upon their chemical structure. This fact is reemphasized by a study of 65 elderly (aged 65–85 years) who were taking no medications other than vitamins. Correlations of taste DTs and RTs for NaCl, quinine HCl (QHCl), and sucrose with one another for unmedicated elderly are given in Table 10. DTs and RTs for a specific tastant were highly correlated with one another: QHCl DT–QHCl RT ($r = 0.69$), NaCl DT–NaCl RT ($r = 0.64$), and sucrose DT–sucrose RT ($r = 0.5$). RTs for NaCl, sucrose, and QHCl were also significantly correlated with one

Table 10   Correlation between thresholds with the same and different taste qualities

	NaCl DT	NaCl RT	Sucrose DT	Sucrose RT	QHCl DT	QHCl RT
NaCl DT	1.00	0.64, $p < 0.0001$	0.12, $p = 0.1886$	0.19, $p = 0.0304$	0.14, $p = 0.1228$	0.15, $p = 0.0861$
NaCl RT		1.00	0.13, $p = 0.1569$	0.30, $p = 0.0006$	0.17, $p = 0.0616$	0.27, $p = 0.0023$
Sucrose DT			1.00	0.50, $p < 0.0001$	0.08, $p = 0.3920$	0.06, $p = 0.5056$
Sucrose RT				1.00	0.19, $p = 0.0352$	0.26, $p = 0.0029$
QHCl DT					1.00	0.69, $p < 0.0001$
QHCl RT						1.00

another but to a lesser degree than the DTs and RTs for a specific tastant: NaCl RT–sucrose RT ($r = 0.30$), NaCl RT–QHCl RT ($r = 0.27$), and sucrose RT–QHCl RT ($r = 0.26$). All other taste intercorrelations among taste measures in Table 10 were 0.19 or less. Thus, if a person has a loss for the salty taste of NaCl, this does not necessarily mean that the person has a loss for other compounds.

4.27.2.3   Taste Thresholds and Nutritional Value

An interesting trend in threshold concentration levels in Tables 1–9 for both young and elderly is that compounds with high caloric or nutritional value such as sugars, fats, and amino acids tend to have higher DTs than certain noxious bitter compounds that can be detected in minute amounts. One possible explanation for the higher DTs for sugars, amino acids, and fats is that this allows only the high-caloric foods to be perceived in a natural setting rather than foods with low nutritional value. That is, too much taste at low concentrations could inhibit the intake of adequate calories.

4.27.3   Suprathreshold Taste Recognition and Identification

4.27.3.1   Magnitude Estimation

Suprathreshold taste studies using magnitude estimation (which relates the perceived intensity to concentration) show decrements in perceived intensity in the elderly (see Schiffman, S. S., 1997, for review). The slope of the psychophysical function that relates perceived intensity and concentration declines with age for a broad range of stimuli (Schiffman, S. S. and Clark, T. B. 1980; Schiffman, S. S., et al., 1981; Cowart, B. J., 1983; Schiffman, S. S., et al., 1994; Mojet, J. et al., 2003). Cowart B. J. (1983) found that the ratios of slope (young)/slope

(elderly) for basic tastes were as follows: sucrose, 1.39; citric acid, 1.30; NaCl, 1.25, and quinine sulfate, 1.73. Schiffman S. S. and Clark T. B. (1980) reported that the slopes of the perceived intensity/concentration curves for 23 amino acids for elderly subjects were always flatter than the slopes for young subjects. The mean ratio of slope (young)/slope (elderly) for all the 23 amino acids was 2.55. However, the ratio of the slopes varied greatly with the amino acids. The ratios for two amino acids with side chain containing acidic groups, L-aspartic acid and L-glutamic acid, were 10.9 and 4.5, respectively, while the ratios for L-lysine and L-proline were only 1.10 and 1.28, respectively.

Schiffman S. S. et al. (1981) reported that the mean ratio of slope (young)/slope (elderly) for ten sweeteners (all of those in Table 3 except monellin) was 2.06 with ratios greater than 1.0 for all sweeteners. Although the variability in the slopes for sweeteners was not as great as for amino acids, there was non-uniformity in the flattening of the slopes, which suggests that receptor site compositions for sweeteners can be altered by the aging process. The greatest depression in slope with age was found for thaumatin, rebaudioside, and neohesperidin dihydrochalcone, which are relatively large sweetener molecules with more possible sites for hydrogen bonding with the receptor. Schiffman S. S. et al. (1994) also used magnitude estimation to compare slopes for a set of bitter compounds (caffeine, denatonium benzoate, $MgCl_2$, $MgNO_3$, quinine HCl, quinine sulfate, and sucrose octa-acetate) in young and elderly subjects. Both tasters and nontasters of the bitter compound PTC were included as subjects in the study. The mean ratio of slope (young)/slope (elderly) for tasters was 1.76 and the mean ratio of slope (young)/slope (elderly) for nontasters was 1.19. This suggests that suprathreshold loss can be slightly greater for tasters than for nontasters. Mojet J. et al. (2003) also reported that the slopes of the psychophysical functions for the sweet, bitter, and umami tastants in water were flatter in the elderly than in the young. As a group these studies indicate that suprathreshold taste intensity is reduced in elderly persons, and there is evidence for a differential decline based on the chemical structure. The relative contribution of normal aging and other health factors to this decline is not known, however, because medications were not controlled in these studies.

4.27.3.2 Difference Thresholds

Losses also occur in the ability to discriminate between different suprathreshold intensities of the same stimulus. Gilmore M. M. and Murphy C. (1989) found that while young subjects generally needed a 34% difference in concentration to perceive a perceptible difference in the bitterness of caffeine, the elderly required an increment of 74% increase. Schiffman S. S. (1993) reported that while young subjects generally needed only a 6–12% difference in a moderately intense concentration of NaCl to perceive a change, the elderly required an increment of 25% to distinguish a difference in saltiness. Nordin S. *et al.* (2003) also found that difference

thresholds for citric acid and NaCl were larger in older individuals.

4.27.3.3 Qualitative Perception in the Elderly

In order to determine if there are any age-related differences in the qualitative taste perception, young and elderly subjects were asked to rate 20 tastants. Each tastant was evaluated on five different adjectives (sweet, sour, salty, bitter, and other) for a total of ten points distributed among the five qualities (Schiffman, S. S., 2000). Subjects were also asked to judge whether each stimulus gave the impression of a single taste (singular) or gave the impression that it existed in the form of two or more component tastes (nonsingular). The mean ratings for sweet, salty, sour, bitter, and other for each tastant in young and elderly subjects along with the percentage of subjects who perceived the tastant as singular are given in Table 11. Several findings emerged from the study. First, neither young nor elderly categorized tastants into four discrete groups; rather the majority of tastes were found to have an other component. That is,

Table 11 Mean quality and singularity ratings of 20 tastants by 12 young and 12 elderly subjects[a]

	Sweet		Salty		Sour		Bitter		Other		Singular (%)	
	Y	E	Y	E	Y	E	Y	E	Y	E	Y	E
Acetic acid (0.1 M)	0.8	1.7	1.1	0.8	4.8	6.2	3.1	0.2	0.2	1.1	58.3	25.0
L-Arginine (0.5 M)	1.0	0	0.2	0.8	1.3	0	2.4	0.9	5.1	8.3	66.7	58.3
Citric acid (0.01 M)	0.2	0	0.8	0.5	4.8	5.1	3.2	3.1	1.0	1.3	58.3	25.0
L-Cysteine HCl (0.005 M)	0.9	0	0.9	0.8	4.0	3.2	0.9	2.3	3.3	3.7	25.0	41.7
Fructose (0.8 M)	9.3	8.9	0.1	0	0	0	0.1	0	0.5	1.1	83.3	83.3
Glycine (0.5 M)	5.4	7.7	0.5	0.1	0.3	0.3	0.3	1.4	3.5	0.5	58.3	50.0
HCl (0.01 M)	0.1	0.3	1.7	0.4	5.5	5.7	2.4	1.2	0.3	2.4	33.3	58.3
KCl (0.3 M)	0	0.4	5.5	6.1	2.0	0.8	1.4	2.7	1.1	0	25	25.0
K_2SO_4 (0.3 M)	0.3	1.8	1.4	0.5	2.3	0.5	3.1	2.8	2.9	4.4	25	8.3
$MgCl_2$ (0.1 M)	0.8	1	1.7	0.8	1.3	0	4.8	2.6	1.4	5.6	16.7	50.0
Na glutamate (0.05 M)	0.7	1.1	1.1	0.5	0.6	0	1.0	0.7	6.6	7.7	50	66.7
Na acetate (0.85 M)	0.3	0	6.3	3.6	1.1	3.5	1.1	0.4	1.2	2.5	16.7	50.0
NaCl (0.2 M)	0.3	0.1	7.8	6.8	0.9	1.5	0.8	1.1	0.2	0.5	58.3	50.0
Na_2CO_3 (0.1 M)	0.2	0	0.7	1.5	1.2	1.3	5.5	3.3	2.4	3.9	25	33.3
Na saccharin (0.0025 M)	7.4	8.2	0.1	0.4	0.8	0	0.7	0.4	1.0	1.0	33.3	50.0
NH_4Cl (0.1 M)	0	0	3.6	3.3	1.6	1.2	4.5	5.0	0.3	0.5	8.3	41.7
Quinine HCl (0.001 M)	0	0	0	0	0.4	0	9.3	9.1	0.3	0.9	91.7	16.7
Succinic acid (0.01 M)	0.3	0.8	0.8	2.8	5.5	4.0	1.6	1.1	1.8	1.3	41.7	33.3
Sucrose (0.8 M)	8.9	9.0	0	0	0	0.8	0	0	1.1	0.2	75.0	91.7
Urea (1.0 M)	0.3	0	0.8	0.3	0.4	0	6.2	4.8	2.3	4.9	58.3	75.0
Totals	37.2	41.0	35.1	30.0	38.8	34.1	52.4	43.1	36.5	51.8	908.2	933.3

[a]Subjects rated the taste of each compound on five different scales (sweet, sour, salty, bitter, and other) for a total of 10 points distributed in any way among the five categories. Subjects also judged whether each stimulus was singular (gave the impression of a single taste) or nonsingular (gave the impression that it existed in the form of two or more component tastes).

neither young nor elderly subjects, while wearing nose plugs, limited their ratings of the 20 tastants to sweet, sour, salty, and bitter alone. Second, the magnitude of the total (sum) ratings on other for elderly subjects was greater than for any of the four so-called primary tastes, that is, these stimuli clearly had qualities other than sweet, salty, sour, or bitter. The higher number of off-tastes perceived by the elderly were most likely the result of medications and medical conditions (see Schiffman, S. S., 2007). Third, stimuli that are considered to be primary or basic tastes such as NaCl (salty), urea (bitter), and citric acid (sour) were not found to be any more singular (which one should expect if a taste is basic or unique) than other stimuli for both age groups although the degree of singularity depended on the stimulus.

4.27.3.4　Other Suprathreshold Findings in the Elderly

Murphy C. (1986) found that the elderly tend to find salt and sugar more pleasant at higher concentrations that do younger subjects, which is consistent with the loss of sensitivity. Warwick Z. S. and Schiffman S. S. (1990) found that the fat content in sugar/fat and salt/fat mixtures was unrelated to pleasantness ratings for elderly subjects, which is consistent with the fact that elderly have elevated thresholds for oils. Byrd E. and Gertman S. (1959) found no significant age-related decreases in the ability to identify suprathreshold concentrations of sweet (sucrose), sour (citric acid), salty (NaCl), and bitter (quinine HCl) in water but there was a consistent trend toward diminished positive identification with age. However, the elderly were clearly less able to discriminate between the taste qualities in real food products (Mojet, J. et al., 2004). Schiffman S. (1979) also found that the elderly (while blindfolded) have reduced ability to identify blended foods on the basis of taste (and smell).

4.27.4　Medications Associated with Taste Alterations

Medications can play a major role in taste losses and distortions (Schiffman, S. S. et al., 2002; Schiffman, S. S., 2007) but the majority of taste changes attributed to specific drugs are based on clinical reports rather than on experimental studies. Oral sensory complaints including taste problems

are more prevalent in elderly who use medications (Schiffman, S. S., 1997; Schiffman, S. S., and Zervakis, J., 2002; Nagler, R. M. and Hershkovich, O., 2005). Table 12 lists hundreds of medications that have been clinically associated with complaints such as loss of taste, altered taste, and metallic taste. Most major drug classes have been implicated in taste disturbances. However, because quantitative assessments of large populations taking specific medications have not been performed, it is not yet known whether clinically reported taste complaints (e.g., case reports) are indeed gustatory or olfactory in nature, or if they are actually caused by medications or the medical conditions they are designed to treat. Furthermore, according to the Physicians' Desk Reference (2000–2005), adverse taste complaints from medications afflict only a minority of patients taking a drug but the actual prevalence of medication-induced losses awaits determination by experimental testing procedures.

Table 12　Medications associated clinically and experimentally with taste loss

AIDS- and HIV-related therapeutic drugs
Didanosine
Indinavir
Lamivudine
Nelfinavir
Nevirapine
Pyrimethamine
Ritonavir
Saquinavir
Stavudine
Zalcitabine
Zidovudine

Amebicides and anthelmintics
Metronidazole
Niclosamide
Niridazole

Anesthetics
Benzocaine (ethyl aminobenzoate)
Dibucaine hydrochloride
Euprocin
Lidocaine
Procaine hydrochloride
Propofol
Tropacocaine

Anticholesteremics and antilipidemics
Atorvastatin calcium
Cholestyramine
Clofibrate

(*Continued*)

Table 12   (Continued)

Fluvastatin sodium
Gemfibrozil
Lovastatin
Pravastatin sodium
Probucol
Simvastatin

Anticoagulants
Phenindione
Warfarin sodium

Antihistamines
Chlorpheniramine maleate
Loratadine
Terfenadine and pseudoephedrine

Antimicrobials
Amphotericin B
Ampicillin
Atovaquone
Aztreonam
Bleomycin
Carbenicillin indanyl sodium
Cefamandole
Cefpodoxime proxetil
Ceftriaxone sodium
Cefuroxime axetil
Cinoxacin
Ciprofloxacin
Clarithromycin
Clindamycin phosphate
Clofazimine
Dapsone
Enoxacin
Ethambutol hydrochloride
Griseofulvin
Imipenem–cilastatin sodium
Lincomycin HCl
Lomefloxacin HCl
Mezlocillin sodium
Norfloxacin
Ofloxacin
Pentamidine isethionate
Piperacillin and tazobactam sodium
Pyrimenthamine
Rifabutin
Sulfamethoxazole
Trimethoprim
Tetracyclines
Ticarcillin disodium and clavulanate potassium
Tyrothricin

Antiproliferative, including immunosuppressive agents
Aldesleukin
Azathioprine
Carmustine
Cisplatin
Carboplatin
Doxorubicin and methotrexate
Fluorouracil

Interferon α-2a (recombinant)
Interferon α-2b (recombinant)
Vincristine sulfate

Antirheumatic, antiarthritic, analgesic, antipyretic, and anti-inflammatory
Auranofin
Aurothioglucose
Benoxaprofen
Butorphanol tartrate
Choline magnesium trisalicylate
Colchicine
Dexamethasone
Diclofenac potassium/diclofenac sodium
Dimethyl sulfoxide
Etodolac
Fenoprofen calcium
Flurbiprofen
Gold,
Gold sodium thiomalate
Hydrocortisone
Hydromorphone HCl
Ibuprofen
Ketoprofen
Ketorolac tromethamine
Morphine sulfate
Nabumetone
Nalbuphine HCl
Oxaprozin
D-Penicillamine and penicillamine
Pentazocine lactate
Phenylbutazone
Piroxicam
Salicylates
Sulindac
Sumatriptan succinate
5-Thiopyridoxine

Antispasmodics, irritable bowel syndrome
Dicylomine HCl
Oxybutynin chloride
Phenobarbital + hyoscyamine SO4 + atropine
SO4 + scopolamine hydrobromide

Antithyroid agents
Carbimazole
Methimazole
Methylthiouracil
Propylthiouracil
Thiouracil

Antiulcerative
Clidinium bromide
Famotidine
Glycopyrrolate
Hyoscyamine sulfate
Mesalamine
Misoprostol
Omeprazole
Propantheline bromide
Sulfasalazine

(Continued)

Table 12   (Continued)

Antiviral
Acyclovir
Foscarnet sodium
Idoxuridine
Interferon α-n3
Interferon β-1b
Rimantadine HCl

Agents for dental hygiene
Sodium fluoride
Sodium lauryl sulfate
Chlorhexidine digluconate mouthrinses

Bronchodilators and antiasthmatic drugs
Albuterol sulfate
Beclomethasone dipropionate
Bitolterol mesylate
Cromolyn sodium
Ephedrine HCl + phenobarbitol + potassium
iodide + theophylline calcium salicylate
Flunisolide
Metaproterenol sulfate
Nedocromil
Pirbuterol acetate inhalation aerosol
Terbutaline sulfate

*Cardiovascular medications including diuretics,
antiarrhythmic, antihypertensive, and antifibrillatory agents*
Acetazolamide
Adenosine
Amiodarone HCl
Amiloride and its analogs
Amlodipine besylate
Benazepril HCl + hydrochlorothiazide
Bepridil HCl
Betaxolol HCl
Bisoprolol fumarate and bisoprolol fumarate with
hydrochlorothiazide
Captopril and captopril/hydrochlorothiazide
Clonidine
Diazoxide
Diltiazem
Doxazosin mesylate
Enalapril and derivatives
Esmolol HCl
Ethacrynic acid
Flecainide acetate
Fosinopril sodium
Guanfacine HCl
Hydrochlorothiazide
Labetalol HCl
Metolazone
Mexiletine HCl
Moricizine HCl
Nifedipine
Procainamide HCl
Propafenone HCl
Propranolol HCl
Ramipril
Spironolactone
Tocainide HCl

Hyper- and hypoglycemic drugs
Diazoxide
Glipizide
Phenformin and derivatives

Hypnotics and sedatives
Estazolam
Flurazepam HCl
Midazolam HCl
Prochlorperazine
Promethazine HCl
Quazepam
Triazolam
Zolpidem tartrate
Zoplicone

*Muscle relaxants and drugs for treatment of Parkinson's
disease*
Baclofen
Chlormezanone
Cyclobenzaprine HCl
Dantrolene sodium
Levodopa
Methocarbamol
Pergolide mesylate
Selegiline HCl

Psychopharmacologic including antiepileptic
Alprazolam
Amitriptyline HCl
Amoxapine
Buspirone HCl
Carbamazepine
Chlordiazepoxide + amitriptyline HCl
Clomipramine HCl
Clozapine
Desipramine HCl
Doxepin HCl
Felbamate
Fluoxetine HCl
Imipramine HCl and imipramine pamoate
Lithium carbonate
Maprotiline HCl
Nortriptyline HCl
Paroxetine HCl
Perphenazine–amitriptyline HCl
Phenytoin
Pimozide
Protriptyline HCl
Psilocybin
Risperidone
Sertraline HCl
Trazodone HCl
Trifluoperazine HCl
Trimipramine maleate
Venlafaxine HCl

Sympathomemetic drugs
Amphetamine
Benzphetamine HCl
Dextroamphetamine sulfate
Fenfluramine HCl
Mazindol

(*Continued*)

Table 12   (Continued)

Methamphetamine HCl
Phendimetrazine tartrate
Phentermine resin and phentermine HCl

Vasodilators
Bamifylline hydrochloride
Diltiazem
Dipyridamole
Isosorbide mononitrate
Nitroglycerin patch
Oxyfedrine

Others (indication)
Allopurinol (reduces serum and urinary uric acid)
Antihemophilic factor (recombinant) (clotting factor –
hemophilia)
Antithrombin III (human) (antithrombin III deficiency)
Calcitonin (Paget's disease, hypercalcemia, osteoporosis)
Cyclosporine (prevent rejection of liver, heart, and kidney
transplants)
Etidronate (hypercalcemia, antipsoriatic)
Etretinate (antipsoriatic)
Germine monoacetate (Eaton–Lambert syndrome)
Granisetron HCl (antiemetic/antinauseant)
Histamine phosphate (control for allergic skin testing)
Iohexol (diagnostic imaging product)
Iron sorbitex (hematinic)
Leuprolide acetate (inhibits gonadotropin secretion/
prostatic cancer)
Levamisole HCl (immunomodulator – restores depressed
immune function)
Mesna (detoxifying agent)
Methazolamide (carbonic anhydrase inhibitor)
Methylergonovine maleate (prevents postpartum
hemorrhage)
Midodrine (hypotension)
Nicotine (smoking cessation)
Nicotine polacrilex (smoking cessation)
Oxycodone HCl (pain)
Pentoxifylline (blood viscosity modulator)
Potassium iodide (expectorant)
Sermorelin acetate (diagnostic)
Succimer (lead poisoning)
Terbinafine
Ursodiol (gall stone dissolution)
Vitamin D/calcitriol (hypocalcemia)
Vitamin K1/phytonadione (coagulation disorders)

See Schiffman S. S. and Zervakis J. (2002) for more details.

Currently, little is known about the sites of action or cascade of cellular events by which medications induce taste complaints. However, medications can impact taste perception at several levels of the nervous system including the peripheral receptors, chemosensory neural pathways, and/or the brainstem and brain. Medications can induce a taste at receptor sites that are localized on either the apical or the basolateral membranes of taste cells. The plasma concentrations of some drugs (e.g., saquinavir) are greater than the taste thresholds and thus can activate receptors on the blood side of taste cells (Schiffman, S. S. *et al.*, 1999). For most drugs, however, the salivary or plasma concentrations of drugs tend to be lower than the taste threshold values (see Schiffman, S. S. *et al.*, 2002). Yet, drugs or their metabolites can accumulate in taste buds over time to reach concentrations that are greater than taste DTs. Drugs can also alter taste transduction mechanisms or can permeate the blood–brain barrier to interfere with taste signaling at the neural levels.

4.27.5   Medical Conditions Associated with Taste Complaints

A vast range of medical conditions has been reported to affect the sense of taste (Table 13), and clinical studies of wasting patients indicate that taste losses are especially severe (Schiffman, S. S., 1983; Schiffman, S. S., and Wedral, E., 1996). However, in most of these studies, patients who were evaluated for taste perception were also being treated with medications for their medical condition, so it is impossible to determine the relative contribution of disease states and medications to taste losses. Cancer is an example of a chronic medical condition in which patients are especially vulnerable to taste disorders (Schiffman, S. S. and Graham, B. G., 2000). Taste changes occur both in untreated cancer patients (Brewin, T. B., 1980; Ovesen, L. *et al.*, 1991) and those treated with chemotherapy (Nielsen, S. S. *et al.*, 1980; Fetting, J. *et al.*, 1985; Lindley, C. *et al.*, 1996) and radiation (Conger, A. D., 1973).

Cancer disproportionately impacts the elderly (Cohen, H. J., 1998) with 60% of all malignant tumors diagnosed in persons aged 65 years and above. Furthermore, 69% of all cancer deaths occur in this age group. Table 14 gives an overview of the taste changes that have been measured in cancer patients using various psychophysical testing techniques. The data suggest that cancer and its treatment impair the ability to detect the presence of basic tastes, reduce the perceived intensity of suprathreshold concentrations of tastants, and interfere with the ability to discriminate and identify tastes and smells. Threshold losses have also been detected using electrogustometry (Berteretche, M. V. *et al.*, 2004). Table 15 gives an overview of the food aversions, complaints, and altered preferences reported by cancer patients.

Table 13 Medical conditions associated clinically and experimentally with taste loss

Nervous system disorders
Alzheimer's disease
Bell's palsy
Damage to chorda tympani
Guillain–Barre syndrome
Familial dysautonomia
Head trauma
Multiple sclerosis
Raeder's paratrigeminal syndrome
Tumors and lesions

Nutritional disorders
Cancer
Chronic renal failure
Liver disease including cirrhosis
Niacin (vitamin B$_3$) deficiency
Thermal burn
Zinc deficiency

Endocrine disorders
Adrenal cortical insufficiency
Congenital adrenal hyperplasia
Cretinism
Cushing's syndrome
Panhypopituitarism
Hypothyroidism
Diabetes mellitus
Gonadal dysgenesis (Turner's syndrome)
Pseudohypoparathyroidism

Local disorders
Facial hypoplasia
Glossitis and other oral disorders
Leprosy
Oral Crohn's disease
Radiation therapy
Sjögren's syndrome

Other
Amyloidosis and sarcoidosis
Cystic fibrosis
High altitude
Hypertension
Influenza-like infections
Laryngectomy
Major depressive disorder

See Schiffman S. S. and Zervakis J. (2002) for more details.

These data suggest that at least half of the cancer patients may have impaired taste functioning at some point during the course of their disease and treatment (e.g., DeWys, W. D. and Walters, K., 1975). The duration of losses can last from several weeks to 6 months or more (Conger, A. D., 1973; Mossman, K. *et al.*, 1982; Ophir, D. *et al.*, 1988; Maes, A. *et al.*, 2002).

The causes of altered taste perception in cancer are not well understood but metabolic changes induced by the presence of a neoplasm as well as damage to the sensory receptors by radio- and chemotherapies are likely involved. Radiation therapy and chemotherapy not only affect the turnover of taste and smell receptors but can also alter the anatomical integrity of the taste bud. Oral complications of cancer such as infections (fungal, viral, bacterial), ulcers, drug-induced stomatitis, and dry mouth may also play a role. In a recent study, tasting of the bitterness of 6-*n*-propylthiouracil (PROP) bitterness was also associated with the development of colonic neoplasms (Basson, M. D., *et al.*, 2005). The ability to taste the bitterness of PROP is genetically determined. The presumed reason for this association was that persons who perceive the bitter taste PROP would be more likely to avoid foods such as vegetables that are high in antioxidants.

4.27.6 Cause of Taste Losses in the Absence of Drugs and Disease

The cause of taste changes in the elderly in the absence of medications and disease is not fully understood. Early studies with large sample sizes suggested that losses in the number of papillae and/or taste buds occurred in the elderly (Arey, L. B. *et al.*, 1935; Mochizuki, Y., 1937; 1939; Moses, S. W. *et al.*, 1967). However, later studies in a smaller number of older individuals suggested that such losses with age are minimal (Arvidson, A., 1979; Miller, I. J., Jr., 1986; Bradley, R. M., 1988; Miller, I. J., Jr., 1988). These later studies are difficult to reconcile with earlier studies and with more recent psychophysical studies that show striking localized deficits (i.e., regional losses) in taste perception over different areas of the tongue in older individuals (Bartoshuk, L. M. *et al.*, 1987; Matsuda, T. and Doty, R. L., 1995). Furthermore, studies of the lingual surface have found entirely flat areas on the surface of the tongue without papillae in old age (Kobayashi, K. *et al.*, 2001).

Reduced renewal of taste cells may account for some age-related losses. While taste cells normally replicate every 10–10.5 days (Beidler, L. M. and Smallman, R. L., 1965), delayed cell renewal and highly vacuolated cytoplasm have been reported to occur in taste buds of aged mice (Fukunaga, A., 2005). Decreased levels of estrogen and testosterone in the elderly may play a role in reduced cell renewal in taste buds since animal models indicate that estrogen and testosterone increase mitotic activity (Eartly, H. *et al.*, 1951; Zhang, Z. *et al.*, 1998). Endocrine factors

Table 14   Alterations of taste performance in cancer patients

Taste loss	Type of cancer	Effect of therapy	Reference
Elevated detection and recognition thresholds for NaCl (salt), sucrose (sweet), HCl (sour), and urea (bitter) prior to radiotherapy; salty, sweet, and bitter further impaired by radiotherapy	Various malignant neoplasms	Radiotherapy further impaired taste loss	Bolze M. S. et al. (1982)
Elevated recognition thresholds for sucrose (sweet), HCl (sour), quinine HCl (bitter) during radiotherapy; recovery by 120 days	Oropharyngeal cancers	Radiotherapy	Conger A. D. (1973)
Elevated detection and recognition thresholds, especially for bitter and salt thresholds during radiotherapy	Head and neck	Radiotherapy	Mossman K. L. and Henkin R. I. (1978)
Elevated NaCl (salty) recognition thresholds	Breast and colon	Prior to treatment	Carson J. A. and Gormican A. (1977)
Elevated recognition threshold for hydrochloric acid (sour); individual differences in bitter and sweet threshold changes	Lung	Prior to therapy	Williams L. R. and Cohen M. H. (1978)
Elevated taste recognition thresholds for NaCl, sucrose, quinine sulfate, picric acid; thresholds returned to normal 6 weeks post-treatment	Oropharyngeal	During and after radiotherapy	Kalmus H. and Farnsworth D. (1959)
Thresholds for NaCl (salt), tartaric acid (sour), sucrose (sweet), and quinine (bitter) elevated by radiation and chemotherapy; recovery was not complete by 1 year	Oral squamous cell carcinoma	Radiation and chemotherapy	Tomita Y. and Osaki T. (1990)
Elevated glucose recognition threshold	Various malignant neoplasms	During chemotherapy	Bruera E. et al. (1984)
Significant increase in electrical taste detection threshold; no change in smell threshold	Lung, ovary, and breast	Increase in untreated patients; thresholds decreased only in patients who responded to chemotherapy (after 2 to 3 months)	Ovesen L. et al. (1991)
Significant decrease in recognition threshold for urea (bitter)	Gastrointestinal		Hall J. C. et al. (1980)
Loss of ability to discriminate between different concentrations of salt, sweet, sour, and bitter	Melanoma	During nine courses of chemotherapy	Mulder N. H. et al. (1983)
Significant reduction in smell identification in patients with estrogen receptor; positive breast cancer	Breast cancer	Mixed sample (treated and untreated)	Lehrer S. et al. (1985)

including castration and thyroidectomy have long been known to depress cell proliferation in other tissues such as small-bowel epithelium (Hopper, A. F. et al., 1972; Williamson, R. C., 1978). There may be some shrinkage of taste projection areas in the brainstem because images of the cross section of the upper part of the medulla oblongata suggest slight reductions in size with age and tooth loss in some individuals (Yamamoto, T. et al., 2005). Decreased performance on psychophysical taste tests in the elderly can also be related to structural changes in the hippocampus and amygdala (Scheibel, M. E. and Scheibel, A. B., 1975; Tomlinson, B. E. and Henderson, G., 1976). Damage to the amygdala and the hippocampus has been shown to interfere with registering and valuing behavior (Pribram, K. H. 1971). Environmental pollutants can also impact taste receptors and/or projections to the central nervous system (Schiffman, S. S. and Nagle, H. T., 1992). Hyposalivation and low perceived salivary flow have also been reported to lower satisfaction with tasting foods (Ikebe, K. et al., 2002).

Table 15   Taste complaints and altered food preferences in cancer patients

Sensory loss	Type of cancer	Effect of therapy	Reference
Food aversions and cravings	Various malignant neoplasms	Radiotherapy	Brewin T. B. (1982)
Reduced palatability of high-protein foods, cereals, and sweets in patients with taste aversions	Various	Treated and untreated	Vickers Z. M. et al. (1981)
All food tasted nauseating, greasy or rancid; wine tasted metallic; water tasted salty	Oropharyngeal	Developed during first two weeks of radiotherapy	Kalmus H. and Farnsworth D. (1959)
Patients developed aversions to sweets, meats, caffeinated beverages, high fat and greasy foods during therapy	Breast and lung	Prior to and during chemotherapy	Mattes R. D. et al. (1987)
Symptom of reduced appetite correlates with elevated recognition threshold for sucrose (sweet); meat aversion correlates with lowered thresholds for urea (bitter)	Various		DeWys W. D. and Walters K. (1975)
Highly varied hedonic responses to beverages containing five suprathreshold concentrations of citric acid (in lemonade), NaCl (in unsalted tomato juice), urea (in tonic water), and sucrose (in cherry drink); anorectics preferred lower sweetness levels than nonanorectics; yet sweet foods constituted a greater percentage of their daily caloric intake	Upper gastrointestinal and lung	Patients on chemotherapy had less distinct preference for any of the five concentrations of sucrose, particularly high levels	Trant A. S. et al. (1982)
Percentage of patients reporting taste problems increased from 18% prior to radiation to over 80% during the 5th week of radiation; foods with abnormal taste included high-protein foods (meat, eggs, dairy), fruits, vegetables, sweet, breads, cereal, coffee, and tea	Head and neck cancer	During radiotherapy	Chencharick J. D. and Mossman K. L. (1983)
Complaints of metallic, bitter, or decreased taste; distorted sweet taste	Breast and lung	Chemotherapy – cisplatin	Rhodes V. A. et al. (1994)

The mechanism for some taste losses may also involve neurodegeneration that can accompany cytokine-mediated inflammatory processes (see Viviani, B. et al., 2004, for description). Inflammatory processes involving cytokines and inflammation mediators with subsequent cell death have been implicated in age-related losses in olfaction (Ge, Y. et al., 2002; Conley, D. B. et al., 2003; Kern, R. C. et al., 2004; Raviv, J. R. and Kern, R. C., 2004). Similarly, taste alterations subsequent to infection as well as losses in normal aging may also be caused by the neurotoxic effects of cytokines and mediators. It is noteworthy that impairments in taste perception are associated with the presence of *Candida albicans* in the oral cavity (Sakashita, S. et al., 2004). A general decline in immune functioning, including cytokine dysregulation, is a hallmark of aging (Wilson, C. J. et al., 2002). Thus, inflammatory

processes that induce neural damage and apoptosis in the taste system may account for reduced sensitivity from normal aging as well as from infections. Environmental pollutants may play a role in the inflammatory processes (see Schiffman, S. S. and Nagle, H. T., 1992).

4.27.7   Final Comment

Losses in taste perception occur in the elderly with elevated DTs and RTs as well as decrements in perceived suprathreshold intensity. While loss in taste sensitivity occurs during normal aging in the absence of disease or medications, losses can be significantly exacerbated by medical conditions and drugs. Dysgeusia increases in persons taking medications (e.g., cardiac medications), but the relative contribution of medications and medical conditions to these taste distortions is not fully understood.

Acknowledgments

This research was supported by a grant to Dr. Susan Schiffman and Duke University from the National Institute on Aging AG00443.

References

Administration on Aging AoA 2004. A Profile of Older Americans: 2004, pp. 1–18. U.S. Department of Health and Human Services, http://www.aoa.gov/prof/Statistics/profile/2004/2004profile.pdf, accessed 21 December 2005.

Arey, L. B., Tremaine, M. J., and Monzingo, F. L. 1935. The numerical and topographical relations of taste buds to human circumvallate papillae throughout the life span. Anat. Rec. 64, 9–25.

Arvidson, K. 1979. Location and variation in number of taste buds in human fungiform papillae. Scand. J. Dent. Res. 87, 435–442.

Bartoshuk, L. M., Desnoyers, S., Hudson, C., Marks, L., O'Brien, M., Catalanotto, F., Gent, J., Williams, D., and Ostrum, K. M. 1987. Tasting on localized areas. Ann. N. Y. Acad. Sci. 510, 166–168.

Basson, M. D., Bartoshuk, L. M., Dichello, S. Z., Panzini, L., Weiffenbach, J. M., and Duffy, V. B. 2005. Association between 6-n-propylthiouracil (PROP) bitterness and colonic neoplasms. Dig. Dis. Sci. 50(3), 483–489.

Bate-Smith, E. C. 1954. Astringency in foods. Food 23, 124–135.

Beidler, L. M. and Smallman, R. L. 1965. Renewal of cells within taste buds. J. Cell Biol. 27, 263–272.

Berteretche, M. V., Dalix, A. M., d'Ornano, A. M., Bellisle, F., Khayat, D., and Faurion, A. 2004. Decreased taste sensitivity in cancer patients under chemotherapy. Support Care Cancer 12(8), 571–576.

Bolze, M. S., Fosmire, G. J., Stryker, J. A., Chung, C. K., and Flipse, B. G. 1982. Taste acuity, plasma zinc levels, and weight loss during radiotherapy: a study of relationships. Radiology 144, 163–169.

Bourlière, F., Cendron, H., and Rapaport, A. 1958. Modification avec l'age des seuils gustatifs de perception et de reconnaissance aux saveurs salée et sucrée, chez l'homme. Gerontologia 2, 104–112.

Bradley, R. M. 1988. Effects of aging on the anatomy and neurophysiology of taste. Gerodontics 4, 244–248.

Brewin, T. B. 1980. Can a tumour cause the same appetite perversion or taste change as a pregnancy? Lancet 2, 907–908.

Brewin, T. B. 1982. Appetite perversions and taste changes triggered or abolished by radiotherapy. Clin. Radiol. 33, 471–475.

Bruera, E., Carraro, S., Roca, E., Cedaro, L., and Chacon, R. 1984. Association between malnutrition and caloric intake, emesis, psychological depression, glucose taste, and tumor mass. Cancer Treat. Rep. 68, 873–876.

Byrd, E. and Gertman, S. 1959. Taste sensitivity in ageing persons. Geriatrics 14, 381–384.

Carson, J. A. and Gormican, A. 1977. Taste acuity and food attitudes of selected patients with cancer. J. Am. Diet Assoc. 70, 361–365.

Chencharick, J. D. and Mossman, K. L. 1983. Nutritional consequences of the radiotherapy of head and neck cancer. Cancer 51, 811–815.

Cohen, H. J. 1998. Cancer and Aging: Overview. In: American Society of Clinical Oncology Education Book (ed. M. C. Perry), pp. 223–226. American Society of Clinical Oncology.

Cohen, J. E. 2003. Human population: the next half century. Science 302, 1172–1175.

Conger, A. D. 1973. Loss and recovery of taste acuity in patients irradiated to the oral cavity. Radiat. Res. 53, 338–347.

Conley, D. B., Robinson, A. M., Shinners, M. J., and Kern, R. C. 2003. Age-related olfactory dysfunction: cellular and molecular characterization in the rat. Am. J. Rhinol. 17(3), 169–175.

Cooper, R. M., Bilash, I., and Zubek, J. P. 1959. The effect of age on taste sensitivity. J. Gerontol. 14, 56–58.

Cowart, B. J. 1983. Direct scaling of the intensity of basic tastes: a life span study. Paper presented at the Annual Meeting of the Association of Chemoreception Sciences, Sarasota, FL.

Davies, A. M. 1989. Older populations, aging individuals and health for all. World Health Forum 10, 299–306; discussion 306–321.

DeWys, W. D. and Walters, K. 1975. Abnormalities of taste sensation in cancer patients. Cancer 36, 1888–1896.

Drewnowski, A., Krahn, D. D., Demitrack, M. A., Nairn, K., and Gosnell, B. A. 1995. Naloxone, an opiate blocker, reduces the consumption of sweet high-fat foods in obese and lean female binge eaters. Am. J. Clin. Nutr. 61, 1206–1212.

Dye, C. J. and Koziatek, D. A. 1981. Age and diabetes effects on threshold and hedonic perception of sucrose solutions. J. Gerontol. 36, 310–315.

Eartly, H., Grad, B., and Leblond, C. P. 1951. The antagonistic relationship between testosterone and thyroxine in maintaining the epidermis of the male rat. Endocrinology 49(6), 677–686.

Fetting, J., Wilcox, P. M., Sheidler, V. R., Enterline, J. P., Donehower, R. C., and Grochow, L. B. 1985. Tastes associated with parenteral chemotherapy for breast cancer. Cancer Treat. Rep. 69, 1249–1251.

Fukunaga, A. 2005. Age-related changes in renewal of taste bud cells and expression of taste cell-specific proteins in mice. Kokubyo Gakkai Zasshi, Japan 72, 84–89.

Fukunaga, A., Uematsu, H., and Sugimoto, K. 2005. Influences of aging on taste perception and oral somatic sensation. J. Gerontol. A Biol. Sci. Med. Sci. 60(1), 109–113.

Ge, Y., Tsukatani, T., Nishimura, T., Furukawa, M., and Miwa, T. 2002. Cell death of olfactory receptor neurons in a rat with nasosinusitis infected artificially with Staphylococcus. Chem. Senses 27(6), 521–527.

Giduck, S. A., Threatte, R. M., and Kare, M. R. 1987. Cephalic reflexes: their role in digestion and possible roles in absorption and metabolism. J. Nutr. 117, 1191–1196.

Gilbertson, T. A., Fontenot, D. T., Liu, L., Zhang, H., and Monroe, W. T. 1997. Fatty acid modulation of K^+ channels in taste receptor cells: gustatory cues for dietary fat. Am. J. Physiol. 272, C1203–C1210.

Gilmore, M. M. and Murphy, C. 1989. Aging is associated with increased Weber ratios for caffeine, but not for sucrose. Percept. Psychophys. 46(6), 555–559.

Glanville, E. V., Kaplan, A. R., and Fischer, R. 1964. Age, sex, and taste sensitivity. J. Gerontol. 19, 474–478.

Grzegorczyk, P. B., Jones, S. W., and Mistretta, C. M. 1979. Age-related differences in salt taste acuity. J. Gerontol. 34, 834–840.

Hall, J. C., Staniland, J. R., and Giles, G. R. 1980. Altered taste thresholds in gastro-intestinal cancer. Clin. Oncol. 6, 137–142.

Harris, H. and Kalmus, H. 1949. The measurement of taste sensitivity to phenylthiourea (P.T.C.). Ann. Eugen. 15, 24–31.

Hinchcliff, R. 1958. Clinical quantitative gustometry. Acta Otolaryngol. 49, 453–466.

Hopper, A. F., Rose, P. M., and Wannemacher, R. W., Jr. 1972. Cell population changes in the intestinal mucosa of protein-depleted or starved rats. II. Changes in cellular migration rates. J. Cell Biol. 53(1), 225–230.

Hughes, G. 1969. Changes in taste sensitivity with advancing age. Gerontol. Clin. 11, 224–230.

Ikebe, K., Sajima, H., Kobayashi, S., Hata, K., Morii, K., Nokubi, T., and Ettinger, R. L. 2002. Association of salivary flow rate with oral function in a sample of community-dwelling older adults in Japan. Oral Surg. Oral Med. Oral Pathol. 94(2), 184–190.

Kalmus, H. and Farnsworth, D. 1959. Impairment and recovery of taste following irradiation of the oropharynx. J. Laryngol. Otol. 73, 180–182.

Kalmus, H. and Trotter, W. R. 1962. Direct assessment of the effect of age on P.T.C. sensitivity. Ann. Hum. Genet. 26, 145–149.

Kern, R. C., Conley, D. B., Haines, G. K., 3rd, and Robinson, A. M. 2004. Treatment of olfactory dysfunction. II. Studies with minocycline. Laryngoscope 114(12), 2200–2204.

Kettaneh, A., Paries, J., Stirnemann, J., Steichen, O., Eclache, V., Fain, O., and Thomas, M. 2005. Clinical and biological features associated with taste loss in internal medicine patients. A cross-sectional study of 100 cases.Appetite 44(2), 163–169.

Klose, R. E. and Glicksman, M. 1972. Gums. In: Handbook of Food Additives, 2nd edn. (ed. T. E. Furia), pp. 295–359. The Chemical Rubber.

Kobayashi, K., Kumakura, M., Yoshimura, K., and Shindo, J. 2001. Stereo-structural study of the lingual papillae and their connective tissue cores in relation to ageing changes in the human tongue. Ital. J. Anat. Embryol. 106(2 Suppl. 1), 305–311.

Krarup, B. 1958. Electro-gustometry: a method for clinical taste examinations. Acta Otolaryngol. (Stockh.) 49, 294–305.

Lehrer, S., Levine, E., and Bloomer, W. D. 1985. Abnormally diminished sense of smell in women with oestrogen receptor positive breast cancer. Lancet 2, 333.

Lindley, C., Lowder, D., Sauls, A., McCune, J., Sawyer, W., and Eatmon, T. 1996. Patient perception of the impact and magnitude of the side-effects of chemotherapy: the Coates study revisited (Meeting abstract). Proc. Annu. Meet. Am. Soc. Clin. Oncol 15, A1652.

Maes, A., Huygh, I., Weltens, C., Vandevelde, G., Delaere, P., Evers, G., and Van den Bogaert, W. 2002. De Gustibus: time scale of loss and recovery of tastes caused by radiotherapy. Radiother. Oncol. 63(2), 195–201.

Matsuda, T. and Doty, R. L. 1995. Regional taste sensitivity to NaCl: relationship to subject age, tongue locus and area of stimulation. Chem. Senses 20(3), 283–290.

Mattes, R. D., Arnold, C., and Boraas, M. 1987. Management of learned food aversions in cancer patients receiving chemotherapy. Cancer Treat. Rep. 71, 1071–1078.

Miller, I. J., Jr. 1986. Variation in human fungiform taste bud densities among regions and subjects. Anat. Rec. 216(4), 474–482.

Miller, I. J., Jr. 1988. Human taste bud density across adult age groups. J. Gerontol. 43(1), B26–B30.

Mochizuki, Y. 1937. An observation on the numerical and topographical relations of the taste buds to circumvallate papillae of Japanese. Okajimas Folia Anat. Jpn 15, 595–608.

Mochizuki, Y. 1939. Studies on the papillae foliata of Japanese. II. The number of taste buds.Okajimas Folia Anat. Jpn 18, 355–369.

Mojet, J., Christ-Hazelhof, E., and Heidema, J. 2001. Taste perception with age: generic or specific losses in threshold sensitivity to the five basic tastes? Chem. Senses 26(7), 845–860.

Mojet, J., Heidema, J., and Christ-Hazelhof, E. 2003. Taste perception with age: generic or specific losses in supra-threshold intensities of five taste qualities? Chem. Senses 28(5), 397–413.

Mojet, J., Heidema, J., and Christ-Hazelhof, E. 2004. Effect of concentration on taste–taste interactions in foods for elderly and young subjects. Chem. Senses 29(8), 671–681.

Moore, L. M., Nielsen, C. R., and Mistretta, C. M. 1982. Sucrose taste thresholds: age-related differences. J. Gerontol. 37, 64–69.

Morley, J. E. 2001. Decreased food intake with aging. J. Gerontol. A Biol. Sci. Med. Sci. 56 Spec. No. 2, 81–88.

Moses, S. W., Rotem, Y., Jagoda, N., Talmor, N., Eichhorn, F., and Levin, S. 1967. A clinical, genetic and biochemical study of familial dysautonomia in Israel. Isr. J. Med. Sci. 3, 358–371.

Mossman, K. L. and Henkin, R. I. 1978. Radiation-induced changes in taste acuity in cancer patients. Int. J. Radiat. Oncol. Biol. Phys. 4, 663–670.

Mossman, K., Shatzman, A., and Chencharick, J. 1982. Long-term effects of radiotherapy on taste and salivary function in man. Int. J. Radiat. Oncol. Biol. Phys. 8, 991–997.

Mulder, N. H., Smit, J. M., Kreumer, W. M., Bouman, J., Sleijfer, D. T., Veeger, W., and Schraffordt Koops, H. 1983. Effect of chemotherapy on taste sensation in patients with disseminated malignant melanoma. Oncology 40, 36–38.

Murphy, C. 1979. The Effect of Age on Taste Sensitivity. In: Special Senses in Aging: A Current Biological Assessment (eds. S. S. Han and D. H. Coons), pp. 21–33. University of Michigan Institute of Gerontology.

Murphy, C. 1986. Taste and Smell in the Elderly. In: Clinical Measurement of Taste and Smell (eds. H. L. Meiselman and R. S. Rivlin), pp. 343–372. MacMillan Publishing.

Nagler, R. M. and Hershkovich, O. 2005. Relationships between age, drugs, oral sensorial complaints and salivary profile. Arch. Oral Biol. 50(1), 7–16.

Nakazato, M., Endo, S., Yoshimura, I., and Tomita, H. 2002. Influence of aging on electrogustometry thresholds. Acta Otolaryngol. Suppl. 546, 16–26.

Nielsen, S. S., Theologides, A., and Vickers, Z. M. 1980. Influence of food odors on food aversions and preferences in patients with cancer. Am. J. Clin. Nutr. 33, 2253–2261.

Nordin, S., Razani, L. J., Markison, S., and Murphy, C. 2003. Age-associated increases in intensity discrimination for taste. Exp. Aging Res. 29(3), 371–381.

Ophir, D., Guterman, A., and Gross-Isseroff, R. 1988. Changes in smell acuity induced by radiation exposure of the olfactory mucosa. Arch. Otolaryngol. Head Neck Surg. 114, 853–855.

Ovesen, L., Srensen, M., Hannibal, J., and Allingstrup, L. 1991. Electrical taste detection thresholds and chemical smell detection thresholds in patients with cancer. Cancer 68, 2260–2265.

Pribitkin, E., Rosenthal, M. D., and Cowart, B. J. 2003. Prevalence and causes of severe taste loss in a chemosensory clinic population. Ann. Otol. Rhinol. Laryngol. 112(11), 971–978.

Pribram, K. H. 1971. Languages of the Brain, p. 345. Prentice-Hall.

Raviv, J. R. and Kern, R. C. 2004. Chronic sinusitis and olfactory dysfunction. Otolaryngol. Clin. North Am. 37(6), 1143–1157, v–vi.

Rhodes, V. A., McDaniel, R. W., Hanson, B., Markway, E., and Johnson, M. 1994. Sensory perception of patients on selected antineoplastic chemotherapy protocols. Cancer Nurs. 17, 45–51.

Richter, C. P. and Campbell, K. H. 1940. Sucrose taste thresholds of rats and humans. Am. J. Physiol. 128, 291–297.

Robichaud, J. L. and Noble, A. C. 1990. Astringency and bitterness of selected phenolics in wine. J. Sci. Food Agric. 53, 343–353.

Sakashita, S., Takayama, K., Nishioka, K., and Katoh, T. 2004. Taste disorders in healthy "carriers" and "non-carriers" of *Candida albicans* and in patients with candidosis of the tongue. J. Dermatol. 31(11), 890–897.

Scheibel, M. E. and Scheibel, A. B. 1975. Structural Changes in the Aging Brain. In: Aging: Clinical, Morphological, and Neurochemical Aspects in the Aging Central Nervous System (*eds*. H. Brody, D. Harman, and J. Ordy), Vol. 1, pp. 11–37. Raven Press.

Schiffman, S. 1979. Changes in Taste and Smell with Age: Psychophysical Aspects. In: Sensory Systems and Communication in the Elderly, Vol. 10, Aging (*eds*. J. M. Ordy and K. Brizzee), pp. 227–246. Raven Press

Schiffman, S. S. 1983. Taste and smell in disease. N. Engl. J. Med. 308, 1275–1279, 1337–1343.

Schiffman, S. S. 1993. Perception of taste and smell in elderly persons. Crit. Rev. Food Sci. Nutr. 33, 17–26.

Schiffman, S. S. 1997. Taste and smell losses in normal aging and disease. JAMA 278, 1357–1362.

Schiffman, S. S. 2000. Taste quality and neural coding: implications from psychophysics and neurophysiology. Physiol. Behav. 69, 147–159.

Schiffman, S. S. 2007. Critical illness and changes in sensory perception. Proc. Nutr. Soc. (in press).

Schiffman, S. S. and Clark, T. B. 1980. Magnitude estimates of amino acids for young and elderly subjects. Neurobiol. Aging 1, 81–91.

Schiffman, S. S. and Erickson, R. P. 1993. Psychophysics: Insights into Transduction Mechanisms and Neural Coding. In: Mechanisms of Taste Transduction (*eds*. S. A. Simon and S. D. Roper), pp. 395–424. CRC Press.

Schiffman, S. S. and Graham, B. G. 2000. Taste and smell perception affect appetite and immunity in the elderly. Eur. J. Clin. Nutr. 54(Suppl. 3), S54–S63.

Schiffman, S. S. and Nagle, H. T. 1992. Effect of environmental pollutants on taste and smell. Otolaryngol. Head Neck Surg. 106, 693–700.

Schiffman, S. S. and Warwick, Z. S. 1992. The Biology of Taste and Food Intake. In: The Science of Food Regulation: Food Intake, Taste, Nutrient Partitioning, and Energy Expenditure (*eds*. G. A. Bray and D. H. Ryan), Vol. 2, pp. 293–312. Pennington Center Nutrition Series Louisiana State University Press.

Schiffman, S. S. and Wedral, E. 1996. Contribution of taste and smell losses to the wasting syndrome. Age Nutr. 7, 106–120.

Schiffman, S. S. and Zervakis, J. 2002. Taste and smell perception in the elderly: effect of medications and disease. Adv. Food Nutr. Res. 44, 248–346.

Schiffman, S. S., Crumbliss, A. L., Warwick, Z. S., and Graham, B. G. 1990. Thresholds for sodium salts in young and elderly subjects: correlation with molar conductivity of anion. Chem. Senses 15, 671–678.

Schiffman, S. S., Frey, A. E., Luboski, J. A., Foster, M. A., and Erickson, R. P. 1991. Taste of glutamate salts in young and elderly subjects: role of inosine 5'-monophosphate and ions. Physiol. Behav. 49, 843–854.

Schiffman, S. S., Gatlin, L. A., Frey, A. E., Heiman, S. A., Stagner, W. C., and Cooper, D. C. 1994. Taste perception of bitter compounds in young and elderly persons: relation to lipophilicity of bitter compounds. Neurobiol. Aging 15, 743–750.

Schiffman, S. S., Graham, B. G., Sattely-Miller, E. A., and Warwick, Z. S. 1998. Orosensory perception of dietary fat. Curr. Dir. Psychol. Sci. 7, 137–143.

Schiffman, S. S., Hornack, K., and Reilly, D. 1979. Increased taste thresholds of amino acids with age. Am. J. Clin. Nutr. 32, 1622–1627.

Schiffman, S. S., Lindley, M. G., Clark, T. B., and Makino, C. 1981. Molecular mechanism of sweet taste: relationship of hydrogen bonding to taste sensitivity for both young and elderly. Neurobiol. Aging 2, 173–185.

Schiffman, S. S., Rascoe, D., and Garcia, R. A. 1989. The effects of age and race on thresholds and magnitude estimates of edible gums. Chem. Senses 14, 743.

Schiffman, S. S., Suggs, M. S., Sostman, A. L., and Simon, S. A. 1992. Chorda tympani and lingual nerve responses to astringent compounds in rodents. Physiol. Behav. 51, 55–63.

Schiffman, S. S., Zervakis, J., Graham, B. G., and Westhall, H. L. 2002. Age-Related Chemosensory Losses: Effect of Medications. In: Chemistry of Taste (*eds*. P. Givens and D. Paredes), pp. 94–108. American Chemical Society.

Schiffman, S. S., Zervakis, J., Heffron, S., and Heald, A. E. 1999. Effect of protease inhibitors on the sense of taste. Nutrition 15, 767–772.

Sclafani, A. 2004. The sixth taste? Appetite 43, 1–3.

Smith, S. E. and Davies, P. D. 1973. Quinine taste thresholds: a family study and a twin study. Ann. Hum. Genet. 37, 227–232.

Sternlieb, I. and Scheinberg, I. H. 1964. Penicillamine therapy for hepatolenticular degeneration. JAMA 189, 748–754.

Stevens, J. C. 1996. Detection of tastes in mixture with other tastes: issues of masking and aging. Chem. Senses 21(2), 211–221.

Stevens, J. C. and Traverzo, A. 1997. Detection of a target taste in a complex masker. Chem. Senses 22(5), 529–534.

Stevens, J. C., Cruz, L. A., Hoffman, J. M., and Patterson, M. Q. 1995. Taste sensitivity and aging: high incidence of decline revealed by repeated threshold measures. Chem. Senses 20(4), 451–459.

Stubbs, R. J. and Whybrow, S. 2004. Energy density, diet composition and palatability: influences on overall food energy intake in humans. Physiol. Behav. 81(5), 755–764.

Teff, K. L. and Engelman, K. 1996. Palatability and dietary restraint: effect on cephalic phase insulin release in women. Physiol. Behav. 60, 567–573.

Terada, T., Sone, M., Tsuji, K., Mishiro, Y., and Sakagami, M. 2004. Taste function in elderly patients with unilateral middle ear disease. Acta Otolaryngol. Suppl. 553, 113–116.

Tomita, Y. and Osaki, T. 1990. Gustatory impairment and salivary gland pathophysiology in relation to oral cancer treatment. Int. J. Oral Maxillofac. Surg. 19, 299–304.

Tomlinson, B. E. and Henderson, G. 1976. Some Quantitative Cerebral Findings in Normal and Demented Old People. In: Neurobiology of Aging (*eds*. R. D. Terry and S. Gershon), Vol. 3 , pp. 183–204. Raven Press.

Trant, A. S., Serin, J., and Douglass, H. O. 1982. Is taste related to anorexia in cancer patients? Am. J. Clin. Nutr. 36, 45–58.

Vickers, Z. M., Nielsen, S. S., and Theologides, A. 1981. Food preferences of patients with cancer. J. Am. Diet. Assoc. 79, 441–445.

Viviani, B., Bartesaghi, S., Corsini, E., Galli, C. L., and Marinovich, M. 2004. Cytokines role in neurodegenerative events. Toxicol. Lett 149(1–3), 85–89.

Warwick, Z. S. and Schiffman, S. S. 1990. Sensory evaluations of fat-sucrose and fat-salt mixtures: relationship to age and weight status. Physiol. Behav. 48, 633–636.

Warwick, Z. S. and Schiffman, S. S. 1991. Flavor–calorie relationships: effect on weight gain in rats. Physiol. Behav. 50, 465–470.

Weiffenbach, J. M., Baum, B. J., and Burghauser, R. 1982. Taste thresholds: quality specific variation with human aging. J. Gerontol. 37, 372–377.

Williams, L. R. and Cohen, M. H. 1978. Altered taste thresholds in lung cancer. Am. J. Clin. Nutr. 31, 122–125.

Williamson, R. C. 1978. Intestinal adaptation. Mechanisms of control. N. Engl. J. Med. 298(26), 1444–1450.

Wilson, C. J., Finch, C. E., and Cohen, H. J. 2002. Cytokines and cognition – the case for a head-to-toe inflammatory paradigm. J. Am. Geriatr. Soc. 50(12), 2041–2056.

Yamamoto, T., Shibata, M., Goto, N., Ezure, H., Ito, J., and Suzuki, M. 2005. Morphometric evaluation of the human tractus solitarius. Okajimas Folia Anat. Jpn 82(1), 5–8.

Yamauchi, Y., Endo, S., and Yoshimura, I. 2002. A new whole-mouth gustatory test procedure. II. Effects of aging, gender and smoking. Acta Otolaryngol. Suppl. 546, 49–59.

Zhang, Z., Laping, J., Glasser, S., Day, P., and Mulholland, J. 1998. Mediators of estradiol-stimulated mitosis in the rat uterine luminal epithelium. Endocrinology 139(3), 961–966.

4.28 Signal Transduction in the Olfactory Receptor Cell

H Takeuchi and T Kurahashi, Osaka University, Osaka, Japan

4.28.1 Outline of the Olfactory Signal Transduction

During the past decades, a huge amount of knowledge has been accumulated in the research filed regarding the molecular mechanisms mediating olfactory energy conversion. The outline of this signal transduction is summarized as follows (see Figures 2(a) and 6). In the genomic DNA, the olfactory receptor cells (ORCs) prepare multiple types of genes encoding seven-transmembrane-domain receptors. Individual receptor cells select only one type of receptor protein gene for the functional expression on their ciliary surface. The receptor proteins are coupled to the olfaction-specific G protein (termed G$_{olf}$) that activates type-III adenylyl cyclase (AC). This enzyme converts cytoplasmic ATP to cyclic AMP (cAMP). These sequential chemical reactions finally lead to the openings of ion channels that underlie cell excitation (transduction channel). Although the natural ligands for olfaction have large diversity exceeding 100 000 varieties, the signal thus converges into only one type of second messenger, cAMP, through the enzymatic cascade equipped in the sensory cilia. Another parallel pathway utilizing phospholipase C (PLC) – InsP$_3$ – may coexist for the chemical sensation for particular species of animals and/or odorants.

4.28.2 Receptor Potential as an Initial Membrane Excitation

As in the general neurons, excitation of the ORCs is the change in the electrical potential across the plasma membrane. The resting membrane potential is set at -50 to -70 mV. Odorant stimulation induces depolarization (change in the membrane potential to the positive direction; Figure 1).

The key molecular element that converts the chemical energy into the biological signal is the transduction channel located in the ciliary membrane. Openings of transduction channels cause a change in the membrane potential depending on the input resistance of the ORC membrane. Usually, the input resistance of the ORC is >5 GΩ, and therefore, only 1 pA of current can induce >5 mV of change in the membrane potential. This initial depolarization is called the receptor potential. The time course of the receptor potential is roughly the same as the time course of opening and closing of the transduction channel. This potential change is characterized by its graded property; information about the strength of the stimulants is represented as a magnitude of the depolarization (Figure 1). Since this potential is slow and graded, it cannot conduct to the brain along the thin and long axon; its amplitude is exponentially reduced with the longitudinal

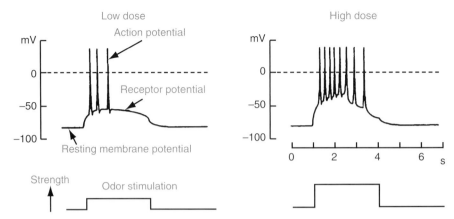

Figure 1 Membrane excitation of the single ORC to low and high doses of odorant stimuli.

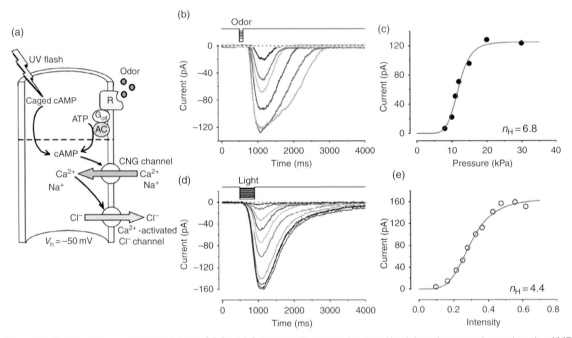

Figure 2 Dose–response relation of single ORCs. (a) Olfactory cilia were stimulated both by odorant and cytoplasmic cAMP emitted by the photolysis of caged cAMP. (b) Odorant-induced response. (c) Dose–response relation. (d) cAMP-induced response. Different cell from (b). (e) Intensity–response relation. Adapted from Takeuchi, H. and Kurahashi, T. 2005. Mechanism of signal amplification in the olfactory sensory cilia. J. Neurosci. 25(48), 11084–11091.

distance from the generation site (cilia). In order to transmit the odorant information to the brain, ORCs induce self-regenerating action potentials (spikes) that do not show signal decay during its conduction. In the action potentials, firing frequencies represent the information about the strength of the stimulants.

The transduction current is unique, in that the magnitude of the excitation shows a very high cooperativity; a small change in the odorant dose is reflected to a big change in the response amplitude. When the inward current responses recorded under the whole-cell voltage-clamp condition are plotted versus

stimulus strength, the relation is fitted by the Hill equation (eqn [1]) with very high Hill coefficient (n_H),

$$I = I_{\max} \frac{C^{n_H}}{C^{n_H} + K_{1/2}^{n_H}} \qquad [1]$$

where I is the current, C is the dose of the ligand, and $K_{1/2}$ is the half-maximum dose. Hill coefficient is an index to express cooperativity and is about 5–6 (Figure 2). If there is no cooperativity, this number becomes 1. Since the same n_H value can be obtained when the cytoplasmic cAMP concentration ($[cAMP]_i$) is increased by the photolysis of caged cAMP, the high cooperativity is concluded to be achieved after the cAMP production (see below for the detailed mechanisms).

4.28.3 Differences of Single Olfactory Receptor Cells in Functional Ligand Affinities

As described in earlier chapters, genomic DNA contains a large family of genes encoding the odorant receptors, and individual ORCs choose only one type of gene for functional expression. Therefore, investigating the spectrum of single ORCs is equivalent to the survey of the functional spectrum of single odorant receptor in native cells. Since the number of ligand is over 100 000 and since the number of the odorant receptor is 100–1000, it is obvious that the binding of odorants to the odorant receptor is one to many.

When single ORCs were stimulated by two different species of odorants, ORCs show a variety of responsiveness, including cells responding to either types of odorants, cells responding to both, and cells responding to neither (Figure 3). The presence of cells responding to different types of odorants indicates that the single odorant receptor can bind to many different odorants. One may think that the ligand–receptor interaction is an independent event (i.e., random bindings of ligands into the receptors). It seems likely, however, that in the ORCs, the odorant receptors have preferential recognition for certain spectrum of odorants. When the responsiveness to both cineole and lilial stimuli are examined in single receptor cells, there is a positive dependence in the conditional probabilities (therefore the positive correlations, Figure 3(c)); cells responding to cineole are observed in cells responding to lilial more frequently than in cells insensitive to lilial, vice versa cells responding to lilial are observed in cells responding to cineole more frequently than in cells insensitive to cineole (Figure 3(d)). There may be some correlation between the preference of the odorant receptor and the quality of the smell. At this point, unfortunately, there is no systematic study in the receptor preference of ligands and needs to be clarified in future.

4.28.4 Can Olfactory Receptor Cells Detect a Single Odorant Molecule?

In vision, the rod photoreceptor cell can detect a signal from the quantum unit of light, a single photon. Following this knowledge, scientists have been looking for the unitary events in the olfactory response; namely, a response induced by a single odorant molecule. In 1995, Menini and co-workers reported that the response of single ORCs to a low concentration of odorant (c. 1 µM, detecting-threshold dose) was very noisy. The noise component was observed as bump-like events, and they suggested that these bump responses may express unitary events caused by the hitting of single odorant molecules to the single olfactory receptor proteins. The amplitude of the events was approximately 1 pA. Immediately after their observations, however, Lowe and Gold showed that application of cAMP into the cytoplasm of the ORCs causes responses with very similar profile as odorant responses that have bump-like current

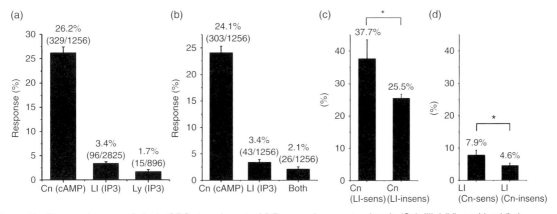

Figure 3 Responsiveness of single ORCs to odorants. (a) Responsiveness to cineole (Cn), lilial (Ll), and lyral (Ly). (b) Responsiveness of single ORCs when cells were stimulated by both Cn and Ll. (c and d) Conditional probabilities derived from the experiment of (b) Adapted from Takeuchi, H., Imanaka, Y., Hirono, J., and Kurahashi, T. 2003. Cross-adaptation between olfactory responses induced by two subgroups of odorant molecules. J. Gen. Physiol. 122, 255–264.

fluctuations. Since there is no odorants in the experiments that employed the cAMP introduction, bump-like current responses are not attributable to the odorant bindings to the receptors. Their interpretation for the bump response is that the olfactory signal transduction is intrinsically noisy and that the basal noise is clipped off by a strong nonlinear amplification established by the sequential openings of two types of ionic channels: cyclic nucleotide-gated (CNG) and Cl^- channels (see Section 4.28.9).

Just very recently, Bhandawat and co-workers showed that unitary events for olfactory transduction were observed only when the Ca ions were removed from the surrounding media. In this specific condition, the amplitude of the unitary events was 0.9 pA. At the physiological condition containing Ca^{2+}, however, unitary events became undetectable and could be estimated only with a noise analysis by variance/mean–mean relationship and were estimated to be 0.01 pA. This value is very small; even smaller than the basal fluctuations by the intrinsic noise. It thus seems unlikely that the unitary events for the single odorant detection are observed as an olfactory response. One may feel this feature in the ORC quite puzzling, in that rod photoreceptors can detect single photon. However, even among the photoreceptor cell, cone-type photoreceptor cells are not able to detect single photon. In this sense, it is rather appropriate to consider that the rod photoreceptor cell is specifically adapted to have a unique and strong amplification system. As for the signal amplification in the ORC, a recent work by Takeuchi and Kurahashi has directly shown that the enzymatic processes of olfactory sensory cilia have low amplification (see Section 4.28.9).

4.28.5 Enzyme – Second Messenger: cyclic AMP and InsP$_3$

For a long period of time, it has been believed that two independent pathways utilizing either AC-cAMP system or the PLC-InsP3 system mediate the olfactory signal transduction. In other words, odorant species are divided into two subgroups: one is called the cAMP odorant and another InsP$_3$ odorant (for history and evidence, see Schild, D. and Restrepo, D., 1998). There have been several approaches that have tried to identify the second messenger mediating olfactory transduction. Gold and his collaborators used gene-targeting methods to find out if electro-olfactogram (EOG) responses are affected in mice that lacks molecular elements constituting the cAMP pathway (CNG channels; G_{olf}; AC type III). EOG responses to InsP$_3$ odorants were abolished or reduced in animals that lacked any of those elemental proteins. In addition, Zufall's group showed that the inhibitor of AC reduced the response induced by InsP$_3$ odorants. In a recent work by Takeuchi and co-workers, the effects of cytoplasmic cNMPs in sensory cilia were directly examined in cells that retained their response abilities to odorants (including InsP$_3$ odorants). They showed that responses induced by both cAMP and InsP$_3$ odorants resemble each other and show almost perfect symmetrical cross-adaptation. The degree of cross-adaptation between the cAMP odorant and the InsP$_3$ odorant matched perfectly, which cannot be explained simply by coincidences in parallel systems. Furthermore, they showed that the response induced by the cytoplasmic cAMP is changed depending on the dose of InsP$_3$ odorants (Figure 4). The additive

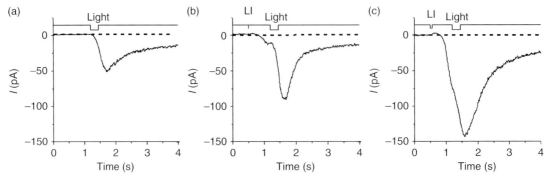

Figure 4 Summation of cAMP-induced current on the lilial (LI)-induced current in the single ORC. Cytoplasmic cAMP was jumped by a flash photolysis of caged cAMP. Lilial stimuli were applied from the puffer pipette to the cilia. Note that the amplitude of cAMP response (light) is affected by the presence of the lilial. Takeuchi, H. and Kurahashi, T. 2003. Identification of second messenger mediating signal transduction in the olfactory receptor cell. J. Gen. Physiol. 122(5), 557–567.

effects could be explained by assuming that InsP$_3$ odorants do increase the cytoplasmic cAMP. All together, it is highly likely that the main stream of olfactory signal transduction is uniformly mediated by cAMP for a wide variety of odorants through a universal second messenger.

Just very recently, however, Restrepo's group showed that mice lacking their CNG channels could still have abilities of responding to particular types of odorants. This result indicates that there is a pathway that is independent from the cAMP-mediated pathway. However, EOG responses in mutant mice were all reduced in comparison to those observed in the wild type. Nonlinear summation observed in the InsP$_3$ odorant response (see Figure 4) does indicate the increase of cAMP, but does not completely rule out the possibilities for the some fractions responses to be evoked in the InsP$_3$-mediated pathway. Coupled with the fact that ORCs in some particular animals (e.g., lobster) use InsP$_3$ as a second messenger, certain fraction of the responses may be mediated by the InsP$_3$-dependent pathway.

4.28.6 Cyclic Nucleotide-Gated Channel

The discovery of cationic channel that is directly gated by cytoplasmic cyclic nucleotide was first made on a work applied to the rod photoreceptor cell. This finding actually ended the long-standing question of what kind of substance (second messenger) transmits the information from the disk membrane (where photopigments are located) to the plasma membrane (where ionic channels mediating cell excitation are present). This historical finding provided a big impact not only on the research on the photoreceptor cell, but also for the research on other sensory and neuronal signal transduction system. Nakamura T. and Gold G.H. (1987) applied the inside-out patch clamp method onto the ciliary membrane of the toad ORC and confirmed electrophysiologically the presence of homologous channel in the ORC. Following the electrophysiological evidence, the gene that encodes the CNG channel was identified. The amino acid sequences of the cloned CNG channels have high identity between rod and ORCs. Both express six membrane-spanning domains and one pore region (Figure 5). The ionic channel is cation-selective, and this is well consistent with the fact that the odorant-activated channels are cation-selective when measured under the Ca^{2+}-free condition (therefore no involvements of Cl$^-$ component).

It is speculated now that the native olfactory channel consists of tetramer using three alpha subunits (CNGA2 and CNGA4) and one beta subunit CNGB1b (stoichiometry 2:1:1 CNGA2:CNGA4:CNGB1b). Among them, CNGA4 subunit seems to be essential to establish Ca^{2+}-calmodulin (Ca-CAM) effect that regulates the olfactory adaptation (see Section 4.28.10).

In the amphibian, CNG channels express a strong localization to the ciliary membrane where olfactory transduction takes place. The density of the channels are as high as 1000 µm^{-2}. In the rodents, however, the distribution of the CNG channels seems to be relatively spread. At this point, there is no physiological explanation why CNG channels are strongly polarized in the amphibian and why they are spread over the cell in the rodent.

Single events for the CNG current can be directly observed when the divalent cations were

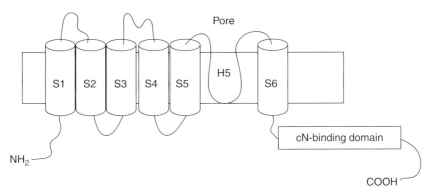

Figure 5 Structure of the subunit of the CNG channel. Both N- and C-terminals are located in the cytoplasmic side. There are six transmembrane domains (S1–S6) and one loop region (H5).

removed from the media. The size of unitary current is about $1\,pA$ when the membrane potential is held at $+50\,mV$. Open–close transition is about the order of $100\,ms$. In the presence of physiological Ca^{2+} and Mg^{2+}, however, such unitary conductance cannot be observed, because of a flickering-block of channels by divalent cations. This makes the channel event extremely small. Furthermore, the channel density is extremely high. These features allow the summed transduction current to express very high S/N ratio.

4.28.7 Ca^{2+}-Activated Cl^- Channel

Kleene S. and Gesteland R. C. (1991) first reported that olfactory sensory cilia contained Cl^- channels that were sensitive to cytoplasmic Ca^{2+}. Since CNG channels have a very high permeability to Ca^{2+} ($P_{Ca}/P_{Na} = 6.5$), opening of the CNG channel allows Ca^{2+} to enter into the ciliary cytoplasm during the odorant response, which leads to the activation of Ca^{2+}-activated Cl^- channels located in the cilia. Initially, the real function of the Cl^- channel was not identified, and a likely possibility assumed at that point was that Cl^- channel may be responsible for the hyperpolarization of the cell; generally, in neurons, Cl^- channels are functional for generating inhibitory responses. Kurahashi T. and Yau K. W. (1993) demonstrated, however, that in the newt ORCs, the transduction inward current actually consists of two ionic components, cationic and anionic currents. Surprisingly, the Cl^- current was inward, which causes a membrane depolarization. In 1995, Ache's group showed that cytoplasmic Cl^- concentration in the ORC was high enough to induce depolarizing response. Recent works revealed that Cl ions are up-taken into the cell with an Na^+–K^+–$2Cl^-$ co-transporter called NKCC1.

As for the physiological roles of this unusual Cl^- channel, there are two possible molecular functions proposed for the ORC. First, asymmetrical ion flow achieved by the both cations and anions may serve a stability of the inward current under unstable ionic conditions which seems to be caused quite frequently in the olfactory organ that is directly exposed to the external environment. Second, addition of the sequential ion channel causes a boosting of the total current, providing a signal amplification. Especially with both CNG and Cl^- currents having high cooperativities (Hill coefficient of about 2 for each), the final current expresses

extremely high nonlinear amplification (cooperativity of 5–6; see Figure 1).

Ca^{2+}-activated Cl^- channel has a unitary conductance of $0.8\,pS$ and the density 70 channels μm^{-2}, when measured with the noise analysis method (by Kleene). Similar to the CNG channels, this channel has small unitary event and the high density. Again these features establish a signal transduction with high S/N ratio. At this point, unfortunately, the molecular structure of this Cl^- channel is not yet identified.

4.28.8 Ca^{2+}-Activated K^+ Channel (Possibility for the Inhibitory Response)

Potassium channel that is directly gated by cytoplasmic Ca^{2+} is commonly distributed to the neuronal system including the ORC. Bacigalupo's group has shown that the olfactory sensory cilia contain this type of channel with inside-out preparation and the immuno-cyto-chemistry. With the same mechanism as the Ca^{2+}-activated Cl^- channel, it is very likely that this K^+ channel is open when the CNG channel allows Ca^{2+} to enter the ciliary cytoplasm.

However, the physiological role of this K^+ channel is still under speculation. One of the most intriguing possibilities is that the opening of K^+ channel may induce cell hyperpolarization and therefore cause inhibitory responses in the ORC activities. Although the actual equilibrium potential for K^+ at the sensory cilia is not identified in the ORC, common sense of the neuronal system is that K^+ channel is inhibitory. Since very early stage of electrophysiological works on ORCs, it has been reported that certain types of odorants and ORCs express inhibitory responses. These cells may express a large number of K^+ channels on their ciliary plasma membrane.

4.28.9   Signal Amplification: Comparison with the Phototransduction

For a long period of time, researches on olfactory transduction system have been limited on rather qualitative descriptions because of experimental difficulties accompanying the use of diverse ligands, multiple receptors, and the fine structure of the sensory cilia. By overcoming those experimental difficulties, a recent work by Takeuchi and Kurahashi was conducted to understand the mechanism of signal amplification in the olfactory transduction system. cAMP dynamics was inversely estimated from the activities of CNG channels. As a result, cAMP-production rate was in an order of 2×10^4 s^{-1} per cilium at the maximum odorant stimulus. The cAMP molecules must be produced from many odorant receptors on the ciliary surface (immunostaining against the receptor protein generally shows their presences to be homogeneous, spanning the entire cilia). It is therefore suggested that the odorant receptor triggers very small activities of enzymatic reaction.

It has been pointed out that the olfactory transduction system is very homologous as the phototransduction system, in that both systems are mediated by receptor – G protein – effecter enzyme and the CNG channel. In the rod photoreceptor cell, light-activated protein reactions have been fairly well described. Activation of single rhodopsin by a single photon triggers 10^2–10^3 molecular changes in transducin–phosphodiesterase (PDE) cascade, which leads to the breakdown of 10^4–10^5 cGMP molecules in a second (Figure 6). This number is much bigger than that observed in the ORC, as is obvious in a comparison between extreme opposite situations, maximum versus unitary rates.

As described before, Bhandawat *et al.* showed that the olfactory unitary event, presumably evoked by a single odorant molecule, could be very small. They considered that such small unitary event was attributable to that the lifetime of individual odorant receptor molecules was very short (i.e., 1 ms). As comparison, in rods, a single photon switches the status of single rhodopsin molecule into an active form that lasts for >1 s. During its long lifetime, many transducin–PDE molecules are activated, which leads to the generation of the single photon response. Since the excitation of the single rhodopsin by a single photon is by chance and very instantaneous, rods may have to acquire a high amplification at the receptor–enzyme level. In case of chemical sensations, however, stimulants can be staying around until they are removed.

Nevertheless, it may be still puzzling that the olfactory enzymes use very low signal amplification. In olfaction, however, activation of G-protein-coupled receptor (GPCR) produces cAMP, instead of the breakdown of cGMP in the rod. Therefore, the low amplification in olfactory enzymes would be an efficient way to avoid loss of ATP. Apparently, signal transduction with a small number of molecules is achieved by the fine ciliary structure that has high surface–volume ratio in which even a small change in the absolute number of molecules is reflected as a big change in the concentration. In addition, the ORC has a unique and strong nonlinear amplification detecting a slight change in the odorant dose, which is regulated by Ca^{2+} that flows through the CNG channel; cytoplasmic increase of Ca^{2+} in turn activates excitatory Cl$^-$ current to boost the net

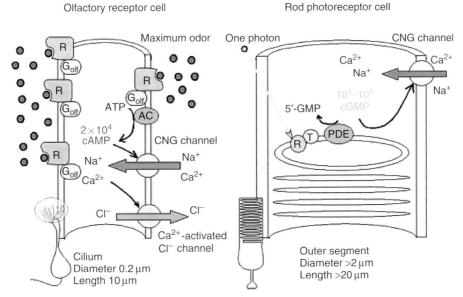

Figure 6 Mechanisms of olfactory transduction and phototransduction.

transduction current. Thus, the sequential openings of two ion channels establish a high nonlinear amplification in olfaction, utilizing Ca^{2+} as a third messenger that is not present in the phototransduction system.

4.28.10 Adaptation

It has been known that the olfactory sensation becomes gradually smaller during the long exposure to the odorant environment (Figure 7(a)). Also, when the odorant pulses are delivered twice, the secondary response becomes smaller than the initial response. The reduced response becomes recovered as the inter-interval is prolonged (Figure 8).

These phenomena are generally called as fatigue, desensitization, or adaptation. One may use these words to the same concept when the response amplitude is gradually reduced as time processed. However, the fundamental concepts are thought to be different.

Among them, the word adaptation is used especially when the dose–response relation is shifted to

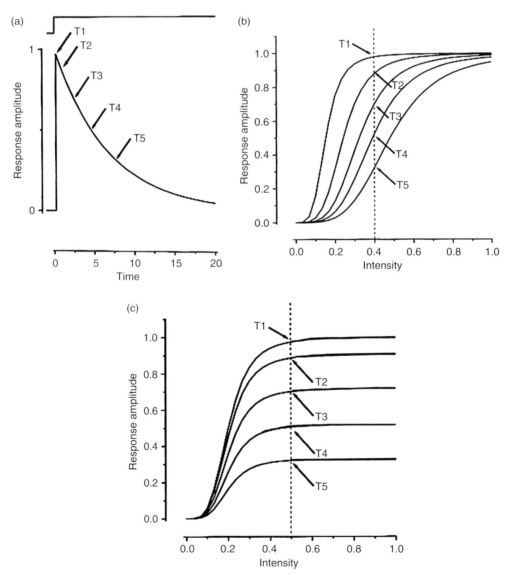

Figure 7 Response decay explained by either the change in $K_{1/2}$ (b) or change in I_{max} (c). (a) Time course of the odorant-induced current. (b) Gradual change of the dose–response relation accompanying the increase of $K_{1/2}$. (c) Gradual change of the dose–response relation accompanying the reduction of I_{max}.

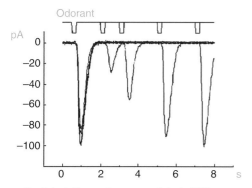

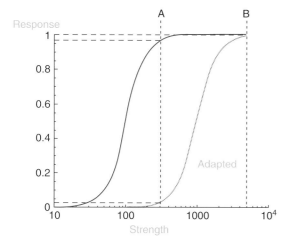

Figure 8 Adaptation and recovery of single ORC responses. Note that the amplitude of the secondary responses is reduced in size. This current reduction is attributable to the change in the dose–response relation as in Figure 9. As the inter-interval is increased, the current amplitude returns to the initial level. Complete recovery is observed after c. 10 s.

Figure 9 Contribution of sensory adaptation on establishing a wide dynamic range. Dose–response relation is shifted in order to expand the dynamic range. Under the standard condition (red curve), the ORC cannot discriminate the change in the stimulus strength between A and B. In the adapted state (blue, dashed line), A and B can be discriminated as a difference in the response amplitude.

cover a wider range for the stimulus strength. As illustrated in Figure 9, shift of the dose–response curve (change in $K_{1/2}$) can establish two competitive functions; signal amplification and wide dynamic range. In contrast, desensitization or fatigue during a long step includes all decaying phenomena. For instance, change in the I_{max} can also express the same phenomenon, but in this case, the dynamic range is not changed (Figures 7(b) and 7(c)).

The ORC itself has an ability of adaptation. This sensory adaptation is achieved in the ciliary molecular circuit that utilizes the Ca^{2+} movement through the CNG channel. Influx of Ca^{2+} causes an increase in the cytoplasmic Ca^{2+} concentration. Cytoplasmic Ca^{2+} binds to the CAM that has already attached to the CAM-binging domain locating N-terminal cytoplasmic chain of the CNG channel. This causes reduction of the ligand affinity leading to the change of the total sensitivity. In contrast, mechanisms causing olfactory fatigue are still not yet well identified.

Ca ions that flowed through the CNG channels play crucial roles in the ORCs as described (activation of Cl^- and K^+ channel, regulation for adaptation). This Ca^{2+} must be excluded from the cilia. Unfortunately, at this moment, there is little information about the Ca^{2+} extrusion system. Electrophysiological data show that Ca^{2+} extrusion mechanism in the olfactory sensory cilia is dependent on the presence of external Na^+. As a homologous biological system, photoreceptor outer segment is equipped with Na/Ca/K exchanger that regulates cytoplasmic Ca^{2+} level. There is a possibility that the same molecular element is present in the olfactory sensory cilia (for detail, see Matthews, H. R. and Reisert, J., 2003).

4.28.11 Functional Roles of Molecular Elements in the Olfactory Cilia

As described, the molecular network within the sensory cilia establishes an elegant and sophisticated signal processing. Active time courses of the molecular elements were investigated by the series of works done by Takeuchi H. and Kurahashi T. (2002; 2003; 2005), in which odorant-induced response and cAMP-induced response were directly compared in the living olfactory cilia (Figure 10).

Very short lifetime of the active odorant receptor is extended while the signal is transmitted to G_{olf} and AC (Figure 11). cAMP molecules survives longer than the active time of AC, until they are hydrolyzed by the activity of PDE. In addition, cytoplasmic Ca^{2+} may be kept for some period of time, until Ca^{2+}-binding proteins trap them or until they are excluded. These time courses of cytoplasmic factors will further extend the active time for the signals.

During the active lifetime of cAMP, CNG channels are kept open, which allows Ca^{2+} to enter into the cytoplasm. This Ca^{2+} opens Ca^{2+}-activated Cl^- channels. At this stage, the signal amplitude is boosted in a nonlinear manner. A long step of odorant

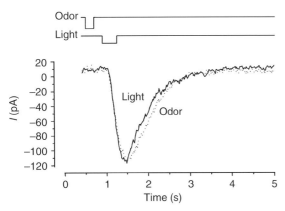

Figure 10 Estimation of odor-activated AC activity by a response-clamp experiment. Cytoplasmic cyclic nucleotide (cGMP having the same effect as cAMP) was increased by photolysis of cytoplasmic caged cGMP. The light condition was adjusted to induce an identical waveform of response as the odorant response. In this experimental condition, the period of light exposure is comparable to the activation of odorant-induced AC activity. Note that the light exposure is delayed from the odorant stimulation and that the period is longer. Also, the current response lasts longer than the light exposure. This indicates that cytoplasmic elements (cAMP and/or Ca^{2+}) remain longer than the AC activity. From Takeuchi, H. and Kurahashi, T. 2002. Photolysis of caged cyclic AMP in the ciliary cytoplasm of the newt olfactory receptor cell. J. Physiol. (Lond.) 541(Pt 3), 825–833.

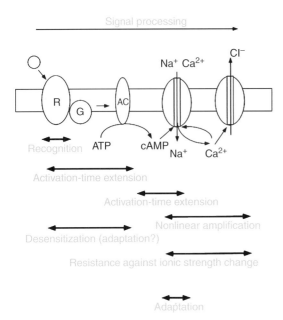

Figure 11 Functional roles of olfactory molecular elements.

stimulation causes gradual increase of cAMP production rate, and $[cAMP]_i$ increases superlinearly with time for over a second. Therefore, the rising phase of the response becomes very rapid and instantaneous.

The combination of cation and anion components is also important for the resistance against the change in the external ionic strength. Having both ion channels together, any change in the cationic current accompanied with the change in the NaCl concentration can be compensated by the inverse stream of the anions.

Adaptation is regulated by the Ca^{2+}-feedback to the CNG channel where very high nonlinear amplification is established. In addition, there is a possibility that desensitization is also regulated at the upstream point from the AC. These molecular roles are summarized in Figure 11.

4.28.12 Resting Membrane Potential and Spike Discharges

As has been generally known for neurons, the olfactory cell is equipped with an excitatory membrane having several types of voltage-gated ion channel. Most of them are distributed in the interstitial area (dendrosomatic regions) of the ORC. As described, the resting potential of the ORC is about -50 to $-70\,mV$. This value is close to the equilibrium potential for the K ions, and it is therefore likely that the resting potential is determined by the K-dominant component. In case of rodents, an ionic channel that is activated by the hyperpolarization (I_h channel) seems to be involved in the determination of the resting potential. Since I_h channels have permeability to both Na^+ and K^+, the resting potential of these cells may be a bit depolarized. This may provide an important function that the resting potential is closed to the spike-firing threshold. However, such I_h channels are not observed in amphibian. Because of this, amphibian ORCs have very high input resistance ($>5\,G\Omega$). Such high input resistance would be remarkable especially at the hyperpolarizing voltage region. It may be possible to think that inhibitory responses (presumably evoked by the Ca^{2+}-activated K^+ channels) are dominant for the amphibian.

For the spike discharges, voltage-dependent Na^+ and Ca^{2+} channels have been shown to be involved in their generations. Essentially, Na^+ channel is a major element that is needed for the spike generation, and Ca^{2+} channel (especially T-type) has an ability of lowering the threshold. The falling phase is shaped by the action of K^+ channel (delayed rectifier and transient K^+ channels) (Table 1).

Table 1   Ionic channels equipped in the olfactory receptor cell

	Functions
Transduction channels	
CNG channel	Initiation of membrane excitation
Ca^{2+}-activated Cl^- channel	Boosting the transduction current
Ca^{2+}-activated K^+ channel	Inhibitory response (?)
$InsP_3$-gated channel	Parallel transduction pathway (?)
Voltage-dependent channels	
Na^+ channel	Initiation of spikes
L-type Ca^{2+} channel	Ca^{2+} uptake, generation of spike
T-type Ca^{2+} channel	Lowering the spike threshold
Delayed K^+ channel	Sharpen the spike falling phase
Transient K^+ channel (A channel)	Sharpen the spike falling phase
	Repetitive spikes (re-activation of Na)
Ca^{2+}-activated K^+ channel	Resting membrane potential
	Spike falling phase
Hyperpolarization-activated cation (h) channel (rodents)	Resting membrane potential

4.28.13   Modulation of Membrane Excitability of the Olfactory Receptor Cell

As described, the ORC is the entrance for the olfactory sensation. It is likely that this step is controlled by a number of endocrine factors. First observation was that adrenalin changed spike initiation patterns by modulating activities of voltage-gated Na^+ and Ca^{2+} channels via an ATP-dependent protein kinase. Besides, dopamine, Ach, serotonin, and LHRH have been shown to modulate the activities of ORCs (for details; see bibliography by Narusuye, K. *et al.*, 2003 and related articles). These findings raise a possibility that the olfactory input is changed depending on the physical and emotional status. In the psychological works, there are cumulative data showing that the olfactory sensation is strongly influenced by the body condition. It is highly expected that in near future the modulatory effects by endocrine factors on ORCs are correlated with the human olfactory sensations.

References

Kleene, S. and Gesteland, R. C. 1991. Calcium-activated chloride conductance in frog olfactory cilia. J. Neurosci. 11, 3624–3629.

Kurahashi, T. and Yau, K. W. 1993. Co-existence of cationic and chloride components in odorant-induced current of vertebrate olfactory receptor cells. Nature 363, 71–74.

Matthews, H. R. and Reisert, J. 2003. Calcium, the two-faced messenger of olfactory transduction and adaptation. Curr. Opin. Neurobiol. 13(4), 469–475.

Nakamura, T. and Gold, G. H. 1987. A cyclic nucleotide-gated conductance in olfactory receptor cilia. Nature 325, 442–444.

Narusuye, K., Kawai, F., and Miyachi, E. 2003. Spike encoding of olfactory receptor cells. Neurosci. Res. 46(4), 407–413.

Schild, D. and Restrepo, D. 1998. Transduction mechanisms in vertebrate olfactory receptor cells. Physiol. Rev. 78(2), 429–466.

Takeuchi, H. and Kurahashi, T. 2002. Photolysis of caged cyclic AMP in the ciliary cytoplasm of the newt olfactory receptor cell. J Physiol. 541(Pt 3), 825–833.

Takeuchi, H. and Kurahashi, T. 2003. Identification of second messenger mediating signal transduction in the olfactory receptor cell. J. Gen. Physiol. 122(5), 557–567.

Takeuchi, H. and Kurahashi, T. 2005. Mechanism of signal amplification in the olfactory sensory cilia. J. Neurosci. 25(48), 11084–11091.

Takeuchi, H., Imanaka, Y., Hirono, J., and Kurahashi, T. 2003. Cross-adaptation between olfactory responses induced by two subgroups of odorant molecules. J. Gen. Physiol. 122, 255–264.

Further Reading

Bhandawat, V., Reisert, J., and Yau, K. W. 2005. Elementary response of olfactory receptor neurons to odorants. Science 308, 1931–1934.

Buck, L. and Axel, R. 1991. A novel multigene family may encode odorant receptors: a molecular basis for odor recognition. Cell 65, 175–187.

Delgado, R., Saavedra, M. V., Schmachtenberg, O., Sierralta, J., and Bacigalupo, J. 2003. Presence of Ca^{2+}-dependent K^+ channels in chemosensory cilia support a role in odor transduction. J. Neurophysiol. 90, 2022–2028.

Kurahashi, T. and Menini, A. 1997. Mechanism of odorant adaptation in the olfactory receptor cell. Nature 385, 725–729.

Zhainazarov, A. B. and Ache, B. W. 1995. Odor-induced currents in Xenopus olfactory receptor cells measured with perforated-patch recording. J. Neurosci. 74, 479–483.

4.29 Olfactory Cyclic Nucleotide-Gated Ion Channels

M-C Broillet, University of Lausanne, Lausanne, Switzerland

4.29.1 Introduction

Nakamura T. and Gold G. H. (1987) using excised patches of cilia membrane from olfactory neurons recorded macroscopic ionic currents directly activated by cAMP. The ion channel responsible for this conductance has since then been cloned and identified in several species (Kaupp, U. B. and Seifert, R., 2002). Cyclic nucleotide-gated (CNG) channels form a family of ion channels that are structurally related to voltage-gated channels (Figure 1). They require the binding of at least two cyclic nucleotide molecules for activation (Zufall, F. *et al.*, 1991; Biskup, C. *et al.*, 2007). Although they have recently been identified in an assortment of cell types and tissues, they are most prevalent in the peripheral sensory receptor cells of the visual and olfactory systems (Kaupp, U. B. and Seifert, R., 2002; Pifferi, S. *et al.*, 2006). In olfactory neurons, these channels can be activated by either cAMP ($K_d = 20\,\mu M$) or cGMP ($K_d = 5\,\mu M$), although it is generally believed that under normal physiological conditions it is a rise in intracellular cAMP that is responsible for channel activation (Firestein, S. and Zufall, F., 1994; Shepherd, G. M., 1994). A high density of CNG channels is present on the ciliary membrane of olfactory neurons. These channels are selective for cations, and their activation leads to cell membrane depolarization. Native CNG channels are known to be permeable to Ca^{2+} ions, but because the Ca^{2+} ions bind to sites on their way through the pore they also act to block monovalent currents (Frings, S. *et al.*, 1995; Zagotta, W. N. and Siegelbaum, S. A., 1996).

4.29.2 Phylogeny

In vertebrates, six members of the CNG channel gene family have been identified. These genes are grouped according to sequence similarity into two subtypes, CNGA and CNGB (Bradley, J. *et al.*, 2001a). Additional genes coding for CNG channels have been cloned from *Drosophila melanogaster* and *Caenorhabditis elegans*. The phylogenetic relationship of these channels is shown in Figure 2.

The first cDNA clone for a subunit of a CNG channel, CNGA1 (previously called α1, CNG1 or RCNC1) has been isolated from bovine retina (Kaupp, U. B. *et al.*, 1989). CNGA1 is expressed in rod photoreceptors and produces functional channels that are gated by cGMP when expressed exogenously either in *Xenopus* oocytes or in a human embryonic kidney cell line (HEK293). Mutations in CNGA1 in humans cause an autosomal recessive form of retinitis pigmentosa, a degenerative form of blindness (Dryja, T. P. *et al.*, 1995). Later, the second subunit of the rod channel, CNGB1 (previously called β_1 CNG4 or RCNC2), has been isolated and cloned (Chen, T.-Y. *et al.*, 1993; Korschen, H. G. *et al.*, 1995). CNGB1 subunits expressed alone do not produce functional CNG channels, but coexpression of CNGA1 and CNGB1 subunits yields heteromeric channels with permeation, modulation, pharmacology, and cyclic-nucleotide specificity similar to that of native channels (Chen, T.-Y. *et al.*, 1993; Korschen, H. G. *et al.*, 1995). CNG channels form tetrameric

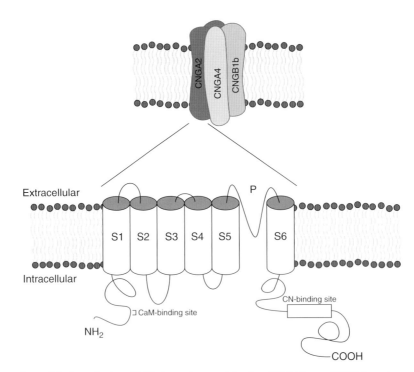

Figure 1 Tetrameric model of an olfactory CNG channel comprising two CNGA2, one CNGA4, and one CNGB1b subunits. A schematic representation of the two-dimensional architecture of one CNG channel subunit is also shown. S1–S6 are the putative transmembrane domains and P is the putative pore region. The cyclic nucleotide (CN)-binding site is defined by homology to the sequences of cAMP and cGMP binding proteins. The position of the Ca^{2+}-calmodulin (CaM) binding site is indicated for an olfactory CNGA2 subunit.

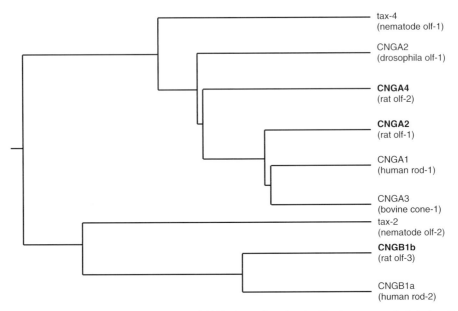

Figure 2 Phylogenetic tree of the olfactory and retinal CNG channels subunits. The tree was calculated on the basis of the sequence alignments with the transmembrane domains and the cyclic nucleotide binding site of the respective subunits.

molecules (Varnum, M. D. *et al.*, 1995; Gordon, S. E. and Zagotta, W. N., 1995b; Liu, D. T. *et al.*, 1996) with a stoichiometry of three CNGA1 subunits and one CNGB1 subunit (Weitz, D. *et al.*, 2002; Zheng, J. *et al.*, 2002; Zhong, H. *et al.*, 2002).

The CNG channels from cone photoreceptors are composed of two other types of subunits, CNGA3 (previously called α2, CNG3 or CCNC1) and CNGB3 (previously called β2, CNG6 or CCNC2) (Bonigk, W. *et al.*, 1993; Gerstner, A. *et al.*, 2000) with a stoichiometry of two CNGA3 and two CNGB3 (Peng, C. *et al.*, 2004). CNGA3 subunits, but not CNGB3 subunits, form functional channels when expressed in heterologous systems. Mutations in human CNGA3 and CNGB3 have been linked to complete achromatopsia (also referred to as rod mono-chromacy or total color blindness), a rare, autosomal recessive inherited and congenital disorder character-ized by the complete inability to discriminate between colors (Kohl, S. *et al.*, 1998; Sundin, O. H. *et al.*, 2000; Wissinger, B. *et al.*, 2001).

Native olfactory CNG channels are constructed from three different but highly homologous subunits (Figure 1), the CNGA2 (previously called α3, CNG2 or OCNC1) (Dhallan, R. S. *et al.*, 1990), the CNGA4 (previously called α4, CNG5 or OCNC2) (Bradley, J. *et al.*, 1994; Liman, E. R. and Buck, L. B., 1994), and the CNGB1b (previously called β1, CNG4.3 or RCNC2) (Sautter, A. *et al.*, 1998; Bonigk, W. *et al.*, 1999) subunits. The stoichiometry of olfactory CNG channels has been identified using fluorescently tagged subunits and fluorescence resonance energy transfer (Zheng, J. and Zagotta, W. N., 2004). They are tetrameric molecules composed of two CNGA2, one CNGA4 and one CNGB1b subunits. The CNGA2 subunits form cAMP-activated channels when heterologously expressed in mammalian cells (Dhallan, R. S. *et al.*, 1990). This conductance is less sensitive to cAMP than the native rat channel and shows very little rectification in the absence of divalent cations, whereas the native CNG conductance is out-ward rectifying in the presence of physiological calcium concentration (Frings, S. *et al.*, 1992; Bonigk, W. *et al.*, 1999). The second olfactory subunit cloned (CNGA4) is 52% identical to the CNGA2 (Bradley, J. *et al.*, 1994; Liman, E. R. and Buck, L. B., 1994). The expression of CNGA4 alone in *Xenopus* oocytes or its transfection into HEK293 cells shows no activation by cyclic nucleotides (Bradley, J. *et al.*, 1994; Liman, E. R. and Buck, L. B., 1994). When the CNGA2 and the CNGA4 are coexpressed in mammalian cells (Bradley, J. *et al.*, 1994) or *Xenopus* oocytes (Liman, E. R. and

Buck, L. B., 1994), an outward rectifying cation con-ductance exhibiting a cAMP sensitivity closer to the native CNG channel is observed. The third subunit of the olfactory CNG channel, the CNGB1b (Sautter, A. *et al.*, 1998; Bonigk, W. *et al.*, 1999), is a spliced form of the rod photoreceptor CNGB1a subunit. When this subunit is coexpressed with the CNGA2 and CNGA4 subunit in HEK293 cells, it creates channels with sensitivity for cAMP, discrimination between Na^+ and K^+, single-channel conductance, kinetics of open–closed transitions and the presence of subcon-ductance states similar to the native olfactory channel.

Two extra CNG channel subunits have been iden-tified in *D. melanogaster*. The first one is expressed in the antennae and in the visual system, suggesting that CNG channels may be involved in the transduction of light in invertebrates (Baumann, A. *et al.*, 1994) and the second, a putative CNG-like channel subunit, is expressed in the brain (Miyazu, M. *et al.*, 2000). Two CNG channel subunits, Tax-2 and Tax-4, have also been cloned in *C. elegans* (Coburn, C. M. and Bargmann, C. I., 1996; Komatsu, H. *et al.*, 1996); they are required for chemosensation, thermosensation, and normal axon outgrowth of sensory neurons.

4.29.3 Structure

CNG channels are composed of four subunits around a centrally located pore (Gordon, S. E. and Zagotta, W. N., 1995b; Liu, D. T. *et al.*, 1996; Varnum, M. D. and Zagotta, W. N., 1996). Each subunit contains six transmembrane segments, a reentrant P-loop and intracellular amino- and carboxy-terminal regions (Figure 1) (Kaupp, U. B. *et al.*, 1989; Molday, R. S. *et al.*, 1991; Wohlfart, P. *et al.*, 1992; Liu, D. T. *et al.*, 1996). The P-loop and the S6 segment line the ion-conducting pore. The carboxy-terminal region contains a cyclic nucleotide (CN)-binding site and a region connecting it to the S6 segment (the C-linker region). The amino-terminal region and the region following the CN-binding domain have specialized functions for each of the CNG channel subtypes. A series of elegant experiments with cloned channels and chimeric constructs have identified several regions as well as specific residues distributed, in particular, throughout the approximately 500 amino acids of the CNGA1 (retinal) or the CNGA2 (olfactory) subunit proteins that play a key role in channel regulation (Matulef, K. and Zagotta, W. N., 2003). The characteristics of these regions will be discussed below.

4.29.3.1 Pore

The pore region of the channel controls both the single-channel conductance and the pore diameter of the channel (Goulding, E. H. *et al.*, 1993). The determination of how ions permeate the channel and how the ionic selectivity occurs is of great physiological importance because ion permeation is responsible for generating membrane depolarization which is critical for electrical signaling, but also the CNG channels are permeable to Ca^{2+} which is an important element in the activation of intracellular targets. Balasubramanian S. *et al.* (1997) have demonstrated that the permeation properties of the olfactory CNG channels are significantly different from those of photoreceptor CNG channels. Their results further indicate that Na^+ currents through these channels do not obey the independence principle and show saturation kinetics with K(m)s in the range of 100–150 mM. They also display a lack of voltage dependence of conductance in asymmetric solutions that suggests that ion-binding sites are situated midway along the channel pore.

Wells G. B. and Tanaka J. C. (1997) have developed a two-site, Eyring rate theory model of ionic permeation for CNG channels. The parameters of the model are optimized by simultaneously fitting current–voltage data sets from excised photoreceptor patches in electrolyte solutions containing one or more of the following ions: Na^+, Ca^{2+}, Mg^{2+}, and K^+. The model accounts well for the shape of the IV relations, the binding affinity for Na^+, the reversal potential values with single-sided additions of calcium or magnesium and biionic KCl, and the $K_{1/2}$ and voltage dependence for divalent block from the cytoplasmic side of the channel. The differences between the predicted $K_{1/2}$s for extracellular block by Ca^{2+} and Mg^{2+} and the values obtained from heterologous expression of only the CNGA1 subunit of the channel suggest that the CNGB1 subunit or a cell-specific factor affects the interaction of divalent cations at the external but not the internal face of the channel. The model predicts concentration-dependent permeability ratios with single-sided addition of calcium and magnesium and anomalous mole fraction effects under a limited set of conditions for both monovalent and divalent cations. Calcium and magnesium are predicted to carry 21 and 10%, respectively, of the total current in the retinal rod cell at −60 mV.

In addition to permeating the CNG channel, Ca^{2+} also profoundly blocks the current flow carried by monovalent cations through the channels (Zagotta, W. N. and Siegelbaum, S. A., 1996) similar to the behavior observed in voltage-activated Ca^{2+}-channels (Almers, W. and McCleskey, E. W., 1984). Supposedly, this behavior implies the high-affinity binding of Ca^{2+} to a single acidic amino acid residue located in the pore of the channel (E363 for the rod CNG channel and E333 for the catfish olfactory CNG channel) (Root, M. J. and MacKinnon, R., 1993). These authors have also implicated this particular glutamate residue as being important in the external rapid proton block of CNG channels, another characteristic that the CNG channels share with Ca^{2+} channels. Gavazzo P. *et al.* (1997) have examined the modulation by internal protons of the bovine CNGA1 and CNGA2 subunits. Increasing internal proton concentrations causes a partial blockage of the single-channel current, consistent with protonation of a single acidic site with a pK1 of 4.5–4.7, both in rod and in olfactory CNG channels. Channel gating properties are also affected by internal protons. The open probability at low cyclic nucleotide concentrations is greatly increased by lowering pH$_i$, and the increase is larger when channels are activated by cAMP than by cGMP. Therefore, internal protons affect both channel permeation and gating properties, causing a reduction in single-channel current and an increase in open probability. These effects are likely to be caused by different titratable groups on the channel.

Amino acid conservation among a wide range of P-loop-containing channels suggests that the structures of the KcsA, a bacterial potassium channel (Doyle, D. A. *et al.*, 1998), and of the MthK, another channel (Jiang, Y. *et al.*, 2002a; 2002b), may serve as general models for the closed and open state conformations of CNG channels. Experimental evidence using cysteine mutations suggests that the cytoplasmic opening of the CNG channel pore is narrow when channels are closed and widens when channels open (Flynn, G. E. and Zagotta, W. N., 2001). Cysteine-scanning mutagenesis studies suggest that the pore helix, near the selectivity filter, undergoes a conformational change during channel activation (Becchetti, A. *et al.*, 1999; Liu, J. and Siegelbaum, S. A., 2000). Recently, site-directed mutagenesis and inside-out patch-clamp recordings (Qu, W. *et al.*, 2006) have been used to investigate ion permeation and selectivity of the rat CNGA2 channels expressed in HEK293 cells. A single point mutation of the negatively charged P-loop glutamate (E342) to either a positively charged lysine or arginine has resulted in functional channels, which consistently respond to cGMP, although the currents are generally extremely small. These results show that it is predominantly the charge of the E342 residue in the P-loop that controls the cation–anion selectivity of this channel. These results also have potential implications for the determinants of anion–cation selectivity in the large family of P-loop-containing channels.

4.29.3.2 Cyclic Nucleotide-Binding Domain

Since the demonstration by Fesenko E. E. *et al.* (1985) and Nakamura T. and Gold G. H. *et al.* (1987) that CNG channels could be directly activated by intracellular cyclic nucleotides, a highly conserved stretch of approximately 120 amino acids homologous to the CN-binding domains of other proteins like the cAMP or cGMP protein kinases has been identified. It consists of a short amino-terminal α helix preceding an eight-stranded anti-parallel β roll that is followed by two α helices (Zagotta, W. N. and Siegelbaum, S. A., 1996). A structural model of the retinal CNG channel CN-binding site has also been constructed (Kumar, V. D. and Weber, I. T., 1992).

The retinal and olfactory CNG channels share a high degree of sequence similarity (over 80% amino acids identity) in the CN-binding sites, but they have very different cyclic nucleotide selectivities. cGMP is a more potent and effective agonist of the retinal channel (Fesenko, E. E. *et al.*, 1985) while cAMP and cGMP have very similar effects on the olfactory channel (Zufall, F. *et al.*, 1994). Varnum M. D. *et al.* (1995) have investigated the molecular mechanism for ligand discrimination of CNG channels. They have found that the retinal photoreceptors and olfactory neurons CNG channels are differentially activated by ligands that varied only in their purine ring structure. The nucleotide selectivity of the retinal bovine CNG channel (cGMP > cIMP $\gg$ cAMP) is significantly altered by neutralization of a single aspartic acid residue (D604) in the CN-binding site (cGMP $\geq$ cAMP > cIMP). Substitution by a nonpolar residue at this position inverts agonist selectivity (cAMP $\gg$ cIMP $\geq$ cGMP). These effects result from an alteration in the relative ability of the agonists to promote the allosteric conformational change associated with channel activation, not from a modification in their initial binding affinity. These authors propose a general mechanism for guanine nucleotide discrimination, in common with that observed in high-affinity GTP-binding proteins,

involving the formation of a pair of hydrogen bonds between the aspartic acid side chain and N1 and N2 of the guanine ring. This amino acid residue (D604) appears to play a critical role in the selective activation of the retinal CNG channel by cGMP. The presence of a methionine at this position in CNGA4 is sufficient to explain the altered ligand specificity of the native olfactory channel (Shapiro, M. S. and Zagotta, W. N., 2000).

The functional effects of each ligand-binding event have always been difficult to assess because ligands continuously bind and unbind at each site. Furthermore, in retinal rod photoreceptors, the low cytoplasmic concentration of cyclic GMP means that CNG channels exist primarily in partially liganded states. Ruiz M. L. and Karpen J. W. (1997) have studied single CNG channel behavior with the use of a photo-affinity analogue of cGMP that tethered cGMP moieties covalently to their binding sites to show that single retinal CNG channels could be effectively locked in four distinct ligand-bound states. Their results indicate that CNG channels open more than they would spontaneously when two ligands are bound (approximately 1% of the maximum current), significantly more with three ligands bound (approximately 33%), and open maximally with four ligands bound. In each ligand-bound state, channels open to two or three different conductance states. These findings place strong constraints on the activation mechanism of CNG channels.

This issue has been investigated further by Liu D. T. *et al.* (1998). These authors observe the effects of individual binding events on channel activation by studying CNG channels containing one, two, three, or four functional CN-binding sites. They find that the binding of a single ligand significantly increases channel opening, although four ligands are required for full channel activation. Their data are inconsistent with models in which the four subunits activate in a single concerted step (Monod-Wyman-Changeux model) or in four independent steps (Hodgkin-Huxley model). Instead, the four subunits of the channel may associate and activate as two independent channel dimers.

A molecular mechanism for the conformational changes that occur in the ligand-binding domain during channel activation has been proposed (Varnum, M. D. *et al.*, 1995). In this mechanism, the CN initially binds to the closed channel primarily by interactions between the β roll and the ribose and cyclic phosphate of the CN. CN binding is followed by a conformational change in the CN-binding site that is coupled to the opening of the pore. This conformational change in the CN-binding site is proposed to involve a relative movement of the C-helices toward the β rolls of each subunit, allowing the D604 residues to interact with the purine rings of the bound CN. This interaction could provide a significant portion of the energy required to drive an otherwise unfavorable opening conformational change. This mechanism is supported by additional experiments. Mutation in the β roll of R559, the residue that forms the primary salt bridge with the phosphate of the CN, dramatically inhibits the initial binding of ligand, decreasing the apparent affinity of the channel for cAMP and cGMP (Tibbs, G. R. *et al.*, 1998). Furthermore, cysteine modification of C505 in the β roll primarily affects the initial binding of cGMP, whereas modification of an introduced cysteine in the C-helix, G597C, primarily affects the agonist potency (Matulef, K. *et al.*, 1999). It has been shown that cysteine residues in the C-helix produce an intersubunit disulfide bond primarily when the channel is closed (Matulef, K. and Zagotta, W., 2002; Mazzolini, M. *et al.*, 2002), suggesting that the C-helices might be nearer to each other or more flexible in the closed state of the channel and separate upon opening. It is still not known how the binding of the ligands to the channel subunits is translated into channel opening. The activation of olfactory CNG channels has been studied further by photolysis-induced jumps of cGMP or cAMP (Nache, V. *et al.*, 2005) and found to be highly cooperative. A model containing three highly cooperative binding sites with a ligand affinity high–low–high described their experimental data most adequately. Recently, Biskup C. *et al.* (2007) have studied homomeric CNGA2 channels in inside-out membrane patches by simultaneously determining channel activation and ligand binding, using the fluorescent cGMP analogue 8-DY547-cGMP as the ligand. They have shown that four ligands bind to the channels and that there is significant interaction between the binding sites. Among the binding steps, the second is most critical for channel opening: its association constant is three orders of magnitude smaller than the others and it triggers a switch from a mostly closed to a maximally open state. These results contribute to unraveling the role of the subunits in the cooperative mechanism of CNGA2 channel activation and could be of general relevance for the activation of other ion channels and receptors.

4.29.3.3 C-Linker region

The CN-binding site is connected to the last transmembrane segment of the channel by a chain of approximately 90 amino acids known as the C-linker region. Experimental evidences suggest that the N-terminal region of the channel and the C-linker region influence the apparent agonist affinity and efficacy with which the cyclic nucleotides open the channel. The C-linker region has been implicated by several studies as being critical in the gating reaction that leads to channel activation subsequent to CN binding.

As a first attempt toward understanding the channel-gating process, Gordon S. E. and Zagotta W. N. (1995a) have studied the mechanism of potentiation of expressed rod CNG channels by Ni^{2+}. They have found that coordination binding of Ni^{2+} between histidine residues (H420) on adjacent channel subunits occurs when the channels are open. Mutation of H420 to lysine completely eliminates the potentiation by Ni^{2+} but does not markedly alter the apparent cGMP affinity of the channel, indicating that the introduction of positive charge at the Ni^{2+}-binding site is not sufficient to produce potentiation. Deletion or mutation of most of the other histidines present in the channel do not diminish Ni^{2+} potentiation. These authors have also examined the role of subunit interactions in Ni^{2+} potentiation by generating heteromultimeric channels using dimers of the rod CNG channel. Injection of single heterodimers (wt/H420Q or H420Q/wt) in which one subunit contains H420 and the other does not resulted in channels that are not potentiated by Ni^{2+}. However, coinjection of both heterodimers into *Xenopus* oocytes results in channels that exhibit potentiation. The H420 residues probably occurs predominantly in nonadjacent subunits when each heterodimer is injected individually, but when the two heterodimers are coinjected, the H420 residues could occur in adjacent subunits as well. Their results suggest that the mechanism of Ni^{2+} potentiation involves intersubunit coordination of Ni^{2+} by H420. Based on the preferential binding of Ni^{2+} to open channels, they suggest that alignment of H420 residues of neighboring subunits into the Ni^{2+}-coordinating position may be associated with channel opening. These authors have also identified the corresponding histidine residue on olfactory CNGA2 subunits (H396) to have an opposite effect, an inhibition, on Ni^{2+} binding (Gordon, S. E. and Zagotta, W. N., 1995c). Thus, this particular C-linker region of the channel probably undergoes a movement during the opening transition (Gordon, S. E. and Zagotta, W. N., 1995a; 1995b). Indeed, a histidine scan of the region just below the S6 in CNGA1 channels has found that histidines introduced at positions 416 and 420 caused Ni^{2+} to stabilize the open state (Johnson, J. P., Jr. and Zagotta, W. N., 2001). The state dependence of Ni^{2+} coordination suggests a model in which a translation and clockwise rotation of this region relative to the central axis of the pore are involved in channel activation.

The cone and olfactory CNG channels also differ considerably in cyclic nucleotide affinity and efficacy. Zong X. *et al.* (1998) have found that three amino acids in the C-linker region are major determinants of gating in these CNG channels. Indeed, the replacement of three amino acids in the cone C-linker by the corresponding amino acids of the olfactory channel (I439V, D481A, and D494S) profoundly enhances the cAMP efficacy and increases the affinities for cAMP and cGMP. Unlike the wild-type cone channel, the mutated channel exhibits similar (olfactory) single-channel kinetics for both cGMP and cAMP, explaining the increase in cAMP efficacy. Therefore, the identified amino acids appear to be major determinants of CNG channel gating.

Among the other key amino acid residues of the C-terminal end of the channel are a series of intracellularly located cysteine residues in or near the CN-binding site (Biel, M. *et al.*, 1996). These residues are thought to affect the gating reaction either through subunit–subunit interactions or within single subunits (Varnum, M. D. and Zagotta, W. N., 1996; Gordon, S. E. *et al.*, 1997). At least one of those cysteines has been proposed as a site which undergoes redox modulation by reactive nitrogen species *via* S-nitrosylation. Indeed, biochemical experiments allowed Broillet M.-C. and Firestein S. (1996) to find that the cysteine in position C460 in the CNGA2 subunit of the rat is the critical residue in the reaction that leads to channel activation by nitric oxide (NO). This particular cysteine residue is located within the C-linker region just N-terminal to the CN-binding site.

Brown L. A. *et al.* (1998) have probed the structural changes that occur during channel activation by CN using SH-modifying reagents on the bovine CNGA1 subunit. Treatment with these reagents dramatically potentiates the channel's response to both cAMP and cGMP. This potentiation is abolished by conversion of the cysteine residue C481 to a nonreactive alanine residue. Potentiation occurs more rapidly in the presence of saturating cGMP, indicating that this region

of the channel is more accessible when the channel is open. C481 is located in the C-linker region between the S6 transmembrane domain and CN-binding site and corresponds to the C460 residue of the olfactory channel which is the NO target site. These results suggest that this region of the channel undergoes significant movement during the activation process and is critical for coupling ligand binding to pore opening. Brown L. A. *et al.* (1998) also claim that potentiation, however, is not mediated by the recently reported interaction between the amino- and carboxy-terminal regions of the CNGA1 subunit because deletion of the entire amino-terminal domain had little effect on potentiation by SH-modifying reagents.

Other parts of the C-linker have also been found to affect the allosteric opening transition. The *C. elegans* TAX-4 channel has a much higher cyclic nucleotide efficacy and sensitivity than the bovine CNGA1 channel (Komatsu, H. *et al.*, 1996; Paoletti, P. *et al.*, 1999). These differences are largely due to three residues in the C-linker, R460, I465, and N466 (numbers correspond to CNGA1 channels) (Paoletti, P. *et al.*, 1999). In addition, differences between gating of CNGA3 and CNGA2 channels have been attributed to three amino acids in the C-linker (I415, D457, and D470) (Zong, X. *et al.*, 1998). Also, protons bind to H468 in CNGA1 channels, causing a potentiation similar to that caused by Ni^{2+} (Gordon, S. E. *et al.*, 1996). Another residue in the C-linker near the beginning of the CNBD, C481, undergoes state-dependent modification by the cysteine-modifying reagents *N*-ethylmaleimide and methanethiosulfonate-ethyltrimethylammonium (Gordon, S. E. *et al.*, 1997; Brown, L. A. *et al.*, 1998). A fluorophore attached to this site has also been shown to undergo state-dependent quenching (Zheng, J. and Zagotta, W. N., 2000). These studies suggest that the entire C-linker region may be involved in the allosteric opening transition.

4.29.3.4 Amino-Terminal Domain

The amino-terminal region of some CNG channels has been found to affect the allosteric opening transition. CNGA2 channels have a lower free energy of opening compared with that of CNGA1 channels (Goulding, E. H. *et al.*, 1994; Liu, M. *et al.*, 1994; Gordon, S. E. *et al.*, 1995b; Fodor, A. A. *et al.*, 1997). Replacing the amino-terminal domain of CNGA1 with that of CNGA2 decreases the free energy of opening, whereas replacing the amino-terminal domain of CNGA2 with that of CNGA1 increases the free energy of opening (Goulding, E. H. *et al.*, 1994; Gordon, S. E. *et al.*, 1995b). Single-channel analysis shows that the CNGA2 amino terminus stabilizes the open state, causing a dramatic increase in the open probability for partial agonists such as cAMP (Sunderman, E. R. and Zagotta, W. N., 1999). Deleting part of the CNGA2 amino-terminal region decreases the open probability for cAMP and the apparent affinity for cGMP, suggesting that this region has an autoexcitatory effect on channel gating (Liu, M. *et al.*, 1994).

Using polypeptides expressed in bacteria, the amino- and carboxy-terminal regions of CNGA2 have been found to interact directly (Varnum, M. D. and Zagotta, W. N., 1997). Furthermore, this interaction can be blocked by the addition of Ca^{2+}-calmodulin, not by the addition of either Ca^{2+} or calmodulin alone (Varnum, M. D. and Zagotta, W. N., 1997) leading to a proposed mechanism for Ca^{2+}-calmodulin modulation in which the CNGA2 amino-terminal domain has an autoexcitatory effect by interacting with the CNGA2 carboxy-terminal domain, and Ca^{2+}-calmodulin inhibits CNGA2 channels by preventing this interaction.

4.29.4 Physiology

One important model system to study the different functions of CNG channels is the olfactory system (Firestein, S. and Zufall, F., 1994). The remarkable capacity to discriminate among a wide range of odor molecules begins at the level of the olfactory sensory neurons (OSNs). These particular neurons perform the complex task of converting the chemical information contained in the odor molecules into changes in membrane potential (Shepherd, G. M., 1994). The different steps of the transduction cascade (Figure 3) can be summarized as follows: when a receptor molecule is occupied by an odorant, it activates a specific GTP-binding protein (G_{olf}), which modulates the activity of an adenylyl cyclase (AC type III), an enzyme producing the second messenger cAMP. cAMP directly activates a CNG channel representing the first step in the generation of the electrical response. Ca^{2+} entering *via* the CNG channel gates a Ca^{2+}-activated Cl^- channel that contributes substantially to the sensory response. Ca^{2+} also binds to calmodulin lowering the ligand sensitivity of the CNG channels. Ca^{2+}-calmodulin also stimulates the activity of a phosphodiesterase reducing the concentration of cAMP. Ca^{2+} is finally extruded by an Na^+/Ca^{2+} exchanger. This cascade of events results in the cell membrane shifting its resting potential from -65 to -45 mV. This depolarization spreads by passive current flow through the dendrite to the soma where it activates voltage-gated Na^+ channels, initiating impulse generation. The combination of Na^+ current with voltage-dependent K^+ currents and with a small

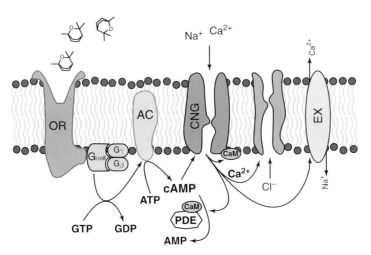

Figure 3 The olfactory transduction cascade. The binding of an odorant molecule (cineole is represented here) to the odorant receptor (OR) leads to the interaction of the receptor to a GTP-binding protein (G_{olf}). This interaction in turn leads to the release of the GTP-coupled alpha subunit of the G protein, which then stimulates the adenylyl cyclase (AC) to produce elevated levels of cAMP. The increase of cAMP opens cyclic nucleotide-gated channels (CNG) causing an odorant-induced inward current carried by Na^+ and Ca^{2+} ions and an alteration of the membrane potential. Ca^{2+} entering *via* the CNG channel gates a Ca^{2+}-activated Cl^- channel that contributes substantially to the sensory response. Ca^{2+} also binds to calmodulin (CaM) lowering the ligand sensitivity of the CNG channels. Ca^{2+}-calmodulin also stimulates the activity of a phosphodiesterase (PDE) reducing the concentration of cAMP. Ca^{2+} is finally extruded by a Na^+/Ca^{2+} exchanger (EX).

Ca^{2+} current acts to produce one or more action potentials that can propagate *via* the axon to the olfactory bulb of the brain.

In OSNs, odors do not only trigger the signaling cascade but also generate various mechanisms to fine-tune the odor-induced current, including a low-selective odor inhibition of the olfactory signal. This wide-range olfactory inhibition is taking place at the level of the CNG channels (Delay, R. and Restrepo, D., 2004; Chen, T. Y. *et al.*, 2006). Interestingly, the inhibitory effect is small in the homomeric CNGA2 channels but larger in channels consisting all olfactory native subunits. The potency of the suppression on the cloned CNG channel appears to be smaller than that previously shown in native olfactory neurons, suggesting that CNG channels switch on and off the olfactory signaling pathway and that the on and off signals may both be amplified by the subsequent olfactory signaling steps.

Nonmotile cilia on OSNs compartmentalize signaling molecules, including odorant receptors and CNG channels, allowing for efficient, spatially confined responses to sensory stimuli. Indeed, CNG channels have been found to have a particular spatial distribution along the ciliary length of OSNs (Flannery, R. J. *et al.*, 2006). The proximal cilia segment, which (in frog) is the first 20% of the cilium, appears to express a small fraction of the CNG channels, whereas the distal segment contains the majority, mostly clustered in one region. Movement of proteins within the cilia is governed by intraflagellar transport, a process that facilitates bidirectional movement along microtubules. Work in *C. elegans* has established that heterotrimeric and homodimeric kinesin-2 family members play a critical role in anterograde transport. Recently, it has been shown that the ciliary targeting of olfactory CNG channels requires the presence of both the CNGB1b channel subunit and the kinesin-2 motor protein, KIF17 (Flannery, R. J. *et al.*, 2006). A critical carboxy-terminal motif (RVxP) on CNGB1b is required for ciliary trafficking of olfactory CNG channels. These results have been further confirmed by the invalidation of the CNGB1 protein leading to olfactory dysfunction and subciliary CNG channel trapping (Michalakis, S. *et al.*, 2006). Indeed, in CNGB1-/- mice, the CNGA2/CNGA4 channels are targeted to the plasma membrane of olfactory knobs but fail to be trafficked into olfactory cilia. Interestingly, a similar trafficking defect can be observed in mice deficient for the CNGA4 subunit. These results demonstrate that CNGB1 has a dual function *in vivo*. First, it gives the olfactory CNG channel a variety of modulatory and biophysical properties tailored to the specific requirements of olfactory transduction. Second, together with the CNGA4 subunit, CNGB1 is needed for ciliary targeting of the olfactory CNG channel.

Olfactory axons are able to elongate and establish synaptic contacts with target neurons in the olfactory bulb throughout adulthood. Axonal pathfinding and target recognition are critical steps in the formation of specific axonal connections in the developing nervous system. These processes could be controlled by local cues, either molecular or structural, that develop in the region of the axon tract. Recently, cyclic nucleotides have been receiving much attention as important regulators in determining whether environmental cues are functioning as attractive or repulsive for growth cone behavior (Song, H. J. *et al.*, 1997; Hopker, V. H. *et al.*, 1999; Song, H. J. and Poo, M. M., 1999). The effects of cyclic nucleotides on growth cone behavior may be partially mediated by CNG channels. CNG channel subunits have been identified in chemosensory neurons in *C. elegans* and shown to be important in the growth and targeting of axons (Coburn, C. M. and Bargmann, C. I., 1996). Active growth cones have been shown to exhibit Ca^{2+} waves which may spread into the neurite (Gu, X. and Spitzer, N. C., 1995; Spitzer, N. C. *et al.*, 1995). These effects are dependent on extracellular calcium, but none of the voltage-gated Ca^{2+} calcium channels expressed in growth cones seem to be involved (Gomez, T. M. *et al.*, 1995; Gottmann, K. and Lux, H. D., 1995). It is possible that CNG channels play an important function in the regulation of Ca^{2+} in the growth cone. Besides the effects of cyclic nucleotides on growth cone behavior, NO has also been shown to significantly affect growth cone behavior. Cheung W. S. *et al.* (2000) showed that NO can stabilize growth cone filopodia, similar to previous findings with 8-Br-cAMP. Van Wagenen S. and Rehder V. (1999) have extended these findings by showing that the response of the growth cone to NO is mediated via a cascade implying cGMP and entry of external Ca^{2+}. These effects of NO on growth cone behavior may also be mediated by CNG channels. Thus, Kafitz K. W. *et al.* (2000) have shown that cyclic nucleotides and NO activate a calcium entry mechanism in olfactory receptor growth cones with properties similar to the CNG channels. Since CNG channels, especially the NO-gated homomeric CNGA4 channels, are highly permeable to Ca^{2+} (Broillet, M.-C. and Firestein, S., 1997) this may provide a direct mechanism through which NO can regulate calcium levels and behavior of growth cones. In summary, CNG channels present at the membrane of the olfactory growth cones might play a key role in OSNs development and turnover.

4.29.5   Modulation

CNG channels are modulated by cellular factors, including phosphorylation enzymes (Gordon, S. E. et al., 1992; Molokanova, E. et al., 1997; Muller, F. et al., 1998), transition metal divalent cations (Ildefonse, M. et al., 1992; Karpen, J. W. et al., 1993; Gordon, S. E. and Zagotta, W. N., 1995a), lipids such as diacylglycerol (Gordon, S. E. et al., 1995a; Crary, J. I. et al., 2000a; 2000b), endogenous Ca^{2+}-binding proteins (Gordon, S. E. and Zagotta, W. N., 1995b) (Balasubramanian, S. et al., 1996; Rebrik, T. I. and Korenbrot, J. I., 1998) and, in particular, Ca^{2+}-calmodulin (Hsu, Y.-T. and Molday, R. S., 1993; Chen, T.-Y. and Yau, K.-W. 1994; Gordon, S. E. and Zagotta, W. N., 1995b).

When exposed to short repetitive pulses of odors, OSNs rapidly adapt to the stimulus by decreasing their responsiveness in a Ca^{2+}-dependent manner (Kurahashi, T. and Shibuya, T., 1990). Kurahashi T. and Menini A. (1997) have investigated the localization of the principal molecular mechanism for adaptation in the olfactory transduction process and have shown that both the cAMP- and the odor-induced responses have similar adaptation properties, indicating that the entire adaptation process takes place after the production of cAMP and may be mediated by Ca^{2+}-calmodulin-dependent inhibition of the olfactory CNG channel. Experiments on olfactory adaptation performed in knockout mice of the CNGA4 subunit have confirmed that the molecular mechanism for adaptation is localized at the channel level, most likely through CNG channel inhibition by Ca^{2+}-calmodulin (Munger, S. D. et al., 2001). Moreover, comparing properties of native channels with heterologously expressed heteromeric channels, the modulatory subunits CNGA4 and CNGB1b have been shown to be responsible for the physiological modulation of the olfactory CNG channel by Ca^{2+}-calmodulin. Bradley J. et al. (2001a; 2001b; 2004) and Munger S. D. et al. (2001) measured, in excised patches containing native heteromeric olfactory CNG channels, a fast current inhibition upon addition of Ca^{2+}-calmodulin that persisted for several

seconds also after calmodulin was removed in Ca^{2+}-free solution. CNGA4 indeed has calmodulin binding sites. CNGA4 has a IQ-type calmodulin binding site located at the C-terminal region, while CNGB1b has a similar IQ-type site located at the N-terminal region and a basic amphiphilic α-helix (Baa) motif in the C-terminal region. It has been shown that the IQ-type sites are necessary and sufficient for Ca^{2+}-calmodulin channel inhibition, whereas the Baa-type site is not necessary (Bradley, J. et al., 2004; 2005). However, the molecular mechanism by which the binding of Ca^{2+}-calmodulin decreases the ligand sensitivity of the olfactory channel is not yet understood.

Multiple studies of native CNG channels have shown that the addition of micromolar concentrations of intracellular Ca^{2+} is able to decrease the channel sensitivity to cGMP or cAMP, probably by activating a Ca^{2+}-responsive endogenous factor already pre-associated with the channel (Kramer, R. H. and Siegelbaum, S. A., 1992; Lynch, J. W. and Lindemann, B., 1994; Gordon, S. E. et al., 1995b; Balasubramanian, S. et al., 1996; Rebrik, T. I. and Korenbrot, J. I. 1998; Kleene, S. J., 1999; Bradley, J. et al., 2004). Bradley J. et al. (2004) have shown that Ca^{2+}-free calmodulin, called apocalmodulin, is able to bind to the heterologously expressed heteromeric olfactory CNG channels even in the absence of Ca^{2+}. Moreover, when Ca^{2+} concentration rises above 100 nM, Ca^{2+} can rapidly modulate the CNG channel sensitivity by directly binding to the pre-associated calmodulin. Furthermore, it has been suggested (Bradley, J. et al., 2004) that, in native channels also, the pre-associated endogenous factor can be apocalmodulin, although a demonstration is still missing. Since Ca^{2+} enters into the olfactory cilia through the CNG channel itself, the pre-associated Ca^{2+}-responsive factor can provide a very fast feedback modulation at the channel level. It has been further verified that the CNGA4 channel subunit plays a central role in Ca^{2+}-calmodulin-dependent odor adaptation (Bradley, J. et al., 2001b; Kelliher, K. R. et al., 2003). Recently, phosphatidylinositol-3,4,5-trisphosphate (PIP3) has been shown to inhibit the activation of the olfactory CNG channels binding to the amino-terminal region of the channel occluding the action of Ca^{2+}-calmodulin (Brady, J. D. et al., 2006).

In addition to their activation by cyclic nucleotides, NO-generating compounds can directly open the olfactory CNG channels through a redox reaction that results in the S-nitrosylation of a free SH group on a cysteine residue (Broillet, M.-C. and

Firestein, S., 1996). This posttranslational modification, comparable to phosphorylation, can modulate the activity of different proteins such as the NMDA receptor, caspases, and others. The NO target site on the CNG channel has been identified by mutating the four candidate intracellular cysteine residues Cys-460, Cys-484, Cys-520, and Cys-552 of the rat CNGA2 subunit into serine residues. All mutant channels continue to be activated by cyclic nucleotides, but only one of them, the C460S mutant channel, exhibits a total loss of NO sensitivity (Broillet, M.-C., 2000). This result is further supported by a similar lack of NO sensitivity that is found for a natural mutant of this precise cysteine residue, the *D. melanogaster* CNG channel. Cys-460 is located in the C-linker region of the channel known to be important in channel gating. Kinetic analyses suggest that at least two of these Cys-460 residues on different channel subunits are involved in the activation by NO. These results show that one single cysteine residue is responsible for NO sensitivity but that several channel subunits need to be activated for channel opening by NO. The corresponding cysteine on the CNGA4 subunit is the cysteine 352. This subunit cannot form homomeric channels that can be activated by cyclic nucleotides, but they can be activated by NO. The properties of these NO-gated channels reveal striking differences with those of either the native rat olfactory CNG channels or the homomeric CNGA2 channels such as the presence of open channel noise and a very different calcium permeability (Broillet, M.-C. and Firestein, S., 1997). These homomeric CNGA4 channels may be expressed at different cellular locations such as the immature olfactory neurons or the vomeronasal neurons and fulfill different physiological roles than the heteromeric CNG channels.

Recently, patch-clamp experiments have revealed that many OSNs respond not only to odors, but also to mechanical stimuli delivered by pressure ejections of odor-free Ringer solution (Grosmaitre, X. *et al.*, 2007). The mechanical responses correlate directly with the pressure intensity and show several properties similar to those induced by odors, including onset latency, reversal potential and adaptation to repeated stimulation. Blocking adenylyl cyclase III or knocking out the CNGA2 subunit eliminates both the odor and the mechanical responses, suggesting that both are mediated by a shared cAMP/CNG channel cascade. This mechanosensitivity enhances the firing frequency of individual neurons when they are weakly stimulated by odors and most likely drives the rhythmic activity (theta oscillation) in the olfactory bulb to synchronize with respiration.

4.29.6 Pharmacology

Although there is a formidable array of specific blockers for sodium, calcium, and potassium channels, specific blockers for CNG channels are scarce. A number of pharmacological agents including L-*cis*-diltiazem (Haynes, L. W., 1992), tetracaine (Fodor, A. A. *et al.*, 1997), pimozide (Nicol, G. D., 1993), and LY-58558 (Leinders-Zufall, T. and Zufall, F., 1995) have been reported to block the current through CNG channels, but these agents have significant drawbacks. First, they exert their effect on the cytoplasmic face of the channel, making their utility questionable in more intact preparations. Second, these blockers have multiple targets, making results difficult to interpret. Recently, Brown R. L. *et al.* (1999) have found a peptide toxin purified from the venom of the Australian King Brown snake, which they have named pseudechetoxin (PsTx). When applied to the extracellular face of membrane patches containing rat CNGA2 channels, PsTx blocks the cGMP-dependent current with a K_i of 5 nM. The block is independent of voltage and requires only a single molecule of toxin. PsTx also blocks CNG channels composed of bovine CNGA1 subunits with high affinity (100 nM), but it is less effective on the heteromeric version of the rod channel (K_i approximately 3 μM). PsTs acts as a pore blocker occluding the entrance to the channel by forming high-affinity contacts with the pore turret (Brown, R. L. *et al.*, 1999). This toxin is structurally classified as a cysteine-rich secretory protein and exhibits structural features that are quite distinct from those of other known small peptidic channel blockers (Yamazaki, Y. *et al.*, 2002). It has now been crystallized and analyzed using X-ray diffraction. PsTx could to be a valuable pharmacological tool for studies on the structure and physiology of CNG channels.

4.29.7 Conclusion

The activation of CNG channels is a complex process comprising the initial ligand binding and a consecutive allosteric transition from a closed to an open configuration. The recent studies on CNG channels have demonstrated the importance of different regions of the protein in activation and modulation of this class of ion channels.

The direct activation by cyclic nucleotides is not only controlled at the CN-binding site where cyclic nucleotide recognition and channel activation take place but particular amino acid residues located elsewhere on the protein also modulate the affinity and efficacy of cAMP or cGMP. Allosteric changes in the CN-binding site, movement of certain regions of the channel molecule, like the C-linker region, seem to be of fundamental importance in the control of the gating mechanisms of the channel. The pore region is responsible for ionic selectivity and for the control of physiological Ca^{2+} permeation, whereas the amino-terminal and the carboxy-terminal regions of the channel take part in channel modulation.

CNG channels first identified in rods and olfactory receptors are now distributed throughout the different cells of the body and may have important roles such as cell motility, secretion, development, and neural plasticity. Indeed, CNG channels have been found in the central nervous system (Kingston, P. A. et al., 1996; Bradley, J. et al., 1997; Wei, J. Y. et al., 1998) and have been implicated in processes as diverse as synaptic modulation, central communication, plasticity, and axon outgrowth in animals ranging from the nematode to mammals (Coburn, C. M. and Bargmann, C. I., 1996; Zufall, F. et al., 1997). CNG channel subunits, in particular the olfactory CNG channel subunits have been identified in the brain (Bradley, J. et al., 1997; Kingston, P. A. et al., 1999; Podda, M. V. et al., 2005; Tetreault, M. L. et al., 2006). Specific subsets of neurons such as the CA1 and CA3 neurons of the hippocampus express CNG channel subunits (Kuzmiski, J. B. and MacVicar, B. A., 2001) suggesting that these channels have a particular function in the central nervous system that is related specifically to certain cell types, rather than being of a general housekeeping nature (Zufall, F. et al., 1997; Wei, J. Y. et al., 1998). Ca^{2+} imaging studies have found that a rise in intracellular Ca^{2+} in hippocampal neurons could result from elevated intracellular cyclic nucleotide concentrations, suggesting that CNG channels play a role in the synaptic plasticity underlying learning and memory (Leinders-Zufall, T. et al., 1995; Bradley, J. et al., 1997; Kingston, P. A. et al., 1999). Indeed, long-term potentiation is attenuated in mutant mice lacking CNGA2 (Parent, A. et al., 1998).

A CNG channel, similar to the olfactory CNGA2, is also expressed throughout the heart of the mouse, but its function remains unclear (Ruiz, M. L. et al., 1996). As sensors of cyclic nucleotide concentrations and conduits for Ca^{2+} entry, CNG channels may play a role in regulating the heart rate and contraction. The cloning and first functional characterization of a plant CNG channel occurred from the *Arabidopsis* cDNA (Leng, Q. et al., 1999). This channel could, like its animal counterparts, translate stimulus-induced changes in cytosolic cyclic nucleotides into altered cell membrane potential and/or cation flux as part of signal cascade pathway(s) that remain to be identified. Future work will reveal new functions for CNG channels as well as more information about their structure–functions and modulation.

Acknowledgments

I gratefully thank Olivier Randin for assistance in producing the figures. M.-C. B. is supported by grants from the Fonds National Suisse de la Recherche and the Leenaards Foundation.

References

Almers, W. and McCleskey, E. W. 1984. Non-selective conductance in calcium channels of frog muscle: calcium selectivity in a single-file pore. J. Physiol. 353, 585–608.

Balasubramanian, S., Lynch, J. W., and Barry, P. H. 1996. Calcium-dependent modulation of the agonist affinity of the mammalian olfactory cyclic nucleotide-gated channel by calmodulin and a novel endogenous factor. Membr. Biol. 152, 13–23.

Balasubramanian, S., Lynch, J. W., and Barry, P. H. 1997. Concentration dependence of sodium permeation and sodium ion interactions in the cyclic AMP-gated channels of mammalian olfactory receptor neurons. J. Membr. Biol. 159(1), 41–52.

Baumann, A., Frings, S., Godde, M., Seifert, R., and Kaupp, U. B. 1994. Primary structure and functional expression of a *Drosophila* cyclic nucleotide-gated channel present in eyes and antennae. EMBO J. 13, 5040–5050.

Becchetti, A., Gamel, K., and Torre, V. 1999. Cyclic nucleotide-gated channels. Pore topology studied through the accessibility of reporter cysteines. J. Gen. Physiol. 114, 377–392.

Biel, M., Zong, X., and Hofmann, F. 1996. Cyclic nucleotide-gated cation channels: molecular diversity, structure and cellular functions. Trends Cardiovasc. Med. 6, 274–280.

Biskup, C., Kusch, J., Schulz, E., Nache, V., Schwede, F., Lehmann, F., Hagen, V., and Benndorf, K. 2007. Relating ligand binding to activation gating in CNGA2 channels. Nature 446, 440–443.

Bonigk, W., Altenhofen, W., Muller, F., Dose, A., Illing, M., Molday, R. S., and Kaupp, U. B. 1993. Rod and cone photoreceptor cells express distinct genes for cGMP-gated channels. Neuron 10, 865–877.

Bonigk, W., Bradley, J., Muller, F., Sesti, F., Boekhoff, I., Ronnett, G. V., Kaupp, U. B., and Frings, S. 1999. The native rat olfactory cyclic nucleotide-gated channel is composed of three distinct subunits. J. Neurosci. 19, 5332–5347.

Bradley, J., Bonigk, W., Yau, K. W., and Frings, S. 2004. Calmodulin permanently associates with rat olfactory CNG channels under native conditions. Nat. Neurosci. 7, 705–710.

Bradley, J., Frings, S., Yau, K. W., and Reed, R. 2001a. Nomenclature for ion channel subunits. Science 294, 2095–2096.

Bradley, J., Li, J., Davidson, N., Lester, H. A., and Zinn, K. 1994. Heteromeric olfactory cyclic nucleotide-gated channels: a subunit that confers increased sensitivity to cAMP. Proc. Natl. Acad. Sci. U. S. A. 91, 8890–8894.

Bradley, J., Reisert, J., and Frings, S. 2005. Regulation of cyclic nucleotide-gated channels. Curr. Opin. Neurobiol. 15, 343–349.

Bradley, J., Reuter, D., and Frings, S. 2001b. Facilitation of calmodulin-mediated odor adaptation by cAMP-gated channel subunits. Science 294, 2176–2178.

Bradley, J., Zhang, Y., Bakin, R., Lester, H. A., Ronnett, G. V., and Zinn, K. 1997. Functional expression of the heteromeric ''olfactory'' cyclic nucleotide-gated channel in the hippocampus: a potential effector of synaptic plasticity in brain neurons. J. Neurosci. 17, 1993–2005.

Brady, J. D., Rich, E. D., Martens, J. R., Karpen, J. W., Varnum, M. D., and Brown, R. L. 2006. Interplay between PIP3 and calmodulin regulation of olfactory cyclic nucleotide-gated channels. Proc. Natl. Acad. Sci. U. S. A. 103, 15635–15640.

Broillet, M.-C. 2000. A single intracellular cysteine residue is responsible for the activation of the olfactory cyclic nucleotide-gated channel by NO. J. Biol. Chem. 275, 15135–15141.

Broillet, M.-C. and Firestein, S. 1996. Direct activation of the olfactory cyclic nucleotide-gated channel through modification of sulfhydryl groups by NO compounds. Neuron 16, 377–385.

Broillet, M.-C. and Firestein, S. 1997. Beta-subunits of the olfactory cyclic nucleotide-gated channel form a nitric oxide activated Ca^{2+} channel. Neuron. 18, 951–958.

Brown, L. A., Snow, S. D., and Haley, T. L. 1998. Movement of gating machinery during the activation of rod cyclic nucleotide-gated channels. Biophys. J. 75, 825–833.

Brown, R. L., Haley, T. L., West, K. A., and Crabb, J. W. 1999. Pseudechetoxin: a peptide blocker of cyclic nucleotide-gated ion channels. Proc. Natl. Acad. Sci. U. S. A. 96, 754–759.

Chen, T.-Y. and Yau, K.-W. 1994. Direct modulation by Ca^{2+}-calmodulin of cyclic nucleotide-activated channel of rat olfactory receptor neurons. Nature 368, 545–548.

Chen, T.-Y., Peng, Y.-W., Dhallan, R. S., Ahamed, B., Reed, R. R., and Yau, K.-W. 1993. A new subunit of the cyclic nucleotide-gated cation channel in retinal rods. Nature 362, 764–767.

Chen, T. Y., Takeuchi, H., and Kurahashi, T. 2006. Odorant inhibition of the olfactory cyclic nucleotide-gated channel with a native molecular assembly. J. Gen. Physiol. 128, 365–371.

Cheung, W. S., Bhan, I., and Lipton, S. A. 2000. Nitric oxide (NO) stabilizes whereas nitrosonium (NO+) enhances filopodial outgrowth by rat retinal ganglion cells in vitro. Brain Res. 868, 1–13.

Coburn, C. M. and Bargmann, C. I. 1996. A putative cyclic nucleotide-gated channel is required for sensory development and function in C. elegans. Neuron 17, 695–706.

Crary, J. I., Dean, D. M., Maroof, F., and Zimmerman, A. L. 2000a. Mutation of a single residue in the S2–S3 loop of CNG channels alters the gating properties and sensitivity to inhibitors. J. Gen. Physiol. 116, 769–780.

Crary, J. I., Dean, D. M., Nguitragool, W., Kurshan, P. T., and Zimmerman, A. L. 2000b. Mechanism of inhibition of cyclic nucleotide-gated ion channels by diacylglycerol. J. Gen. Physiol. 116, 755–768.

Delay, R. and Restrepo, D. 2004. Odorant responses of dual polarity are mediated by cAMP in mouse olfactory sensory neurons. J. Neurophysiol. 92, 1312–1319.

Dhallan, R. S., Yau, K. W., Schrader, K. A., and Reed, R. R. 1990. Primary structure and functional expression of a cyclic nucleotide-activated channel from olfactory neurons. Nature 347, 184–187.

Doyle, D. A., Morais Cabral, J., Pfuetzner, R. A., Kuo, A., Gulbis, J. M., Cohen, S. L., Chait, B. T., and MacKinnon, R. 1998. The structure of the potassium channel: molecular basis of K^+ conduction and selectivity. Science 280, 69–77.

Dryja, T. P., Finn, J. T., Peng, Y. W., McGee, T. L., Berson, E. L., and Yau, K. W. 1995. Mutations in the gene encoding the alpha subunit of the rod cGMP-gated channel in autosomal recessive retinitis pigmentosa. Proc. Natl. Acad. Sci. U. S. A. 92, 10177–10181.

Fesenko, E. E., Kolesnikov, S. S., and Lyubarsky, A. L. 1985. Induction by cyclic GMP of cationic conductance in plasma membrane of retinal rod outer segment. Nature 313, 310–313.

Firestein, S. and Zufall, F. 1994. The cyclic nucleotide-gated channel of olfactory receptor neurons. Semin. Cell Biol. 5, 39–46.

Flannery, R. J., French, D. A., and Kleene, S. J. 2006. Clustering of cyclic nucleotide-gated channels in olfactory cilia. Biophys. J. 91, 179–188.

Flynn, G. E. and Zagotta, W. N. 2001. Conformational changes in S6 coupled to the opening of cyclic nucleotide-gated channels. Neuron 30, 689–698.

Fodor, A. A., Black, K. D., and Zagotta, W. N. 1997. Tetracaine reports a conformational change in the pore of cyclic nucleotide-gated channels. J. Gen. Physiol. 110, 591–600.

Frings, S., Lynch, J. W., and Lindemann, B. 1992. Properties of cyclic nucleotide-gated channels mediating olfactory transduction. J. Gen. Physiol. 100, 45–67.

Frings, S., Seifert, R., Godde, M., and Kaupp, U. B. 1995. Profoundly different calcium permeation and blockage determine the specific function of distinct cyclic nucleotide-gated channels. Neuron 15, 169–179.

Gavazzo, P., Picco, C., and Menini, A. 1997. Mechanism of modulation by internal protons of cyclic nucleotide-gated channels cloned from sensory cells. Proc. R. Soc. Lond. B. Biol. Sci. 264(1385), 1157–1165.

Gerstner, A., Zong, X., Hofmann, F., and Biel, M. 2000. Molecular cloning and functional characterization of a new modulatory cyclic nucleotide-gated subunit from mouse retina. J. Neurosci. 20, 1324–1332.

Gomez, T. M., Snow, D. M., and Letourneau, P. C. 1995. Characterization of spontaneous calcium transients in nerve growth cones and their effect on growth cone migration. Neuron 14, 1233–1246.

Gordon, S. E., Brautigan, D. L., and Zimmerman, A. L. 1992. Protein phosphatases modulate the apparent agonist affinity of the light-regulated ion channel in retinal rods. Neuron 9, 739–748.

Gordon, S. E., Downing-Park, J., Tam, B., and Zimmerman, A. L. 1995a. Diacylglycerol analogs inhibit the rod cGMP-gated channel by a phosphorylation-independent mechanism. Biophys. J. 69, 409–417.

Gordon, S. E., Downing-Park, J., and Zimmerman, A. L. 1995b. Modulation of the cGMP-gated ion channel in frog rods by calmodulin and an endogenous inhibitory factor. J. Physiol. 486(Pt 3), 533–546.

Gordon, S. E., Oakley, J. C., Varnum, M. D., and Zagotta, W. N. 1996. Altered ligand specificity by protonation in the ligand binding domain of cyclic nucleotide-gated channels. Biochemistry 35, 3994–4001.

Gordon, S. E., Varnum, M. D., and Zagotta, W. N. 1997. Direct interaction between amino- and carboxyl-terminal domains of cyclic nucleotide-gated channels. Neuron 19, 431–441.

Gordon, S. E. and Zagotta, W. N. 1995a. A histidine residue associated with the gate of the cyclic nucleotide-activated channels in rod photoreceptors. Neuron 14, 177–183.

Gordon, S. E. and Zagotta, W. N. 1995b. Localization of regions affecting an allosteric transition in cyclic nucleotide-activated channels. Neuron 14, 857–864.

Gordon, S. E. and Zagotta, W. N. 1995c. Subunit interactions in coordination of Ni^{2+} in cyclic nucleotide-gated channels. Proc. Natl. Acad. Sci. U. S. A. 92, 10222–10226.

Gottmann, K. and Lux, H. D. 1995. Growth cone calcium ion channels: properties, clustering, and functional roles. Perspect. Dev. Neurobiol. 2, 371–377.

Goulding, E. H., Tibbs, G. R., Liu, D., and Siegelbaum, S. A. 1993. Role of H5 domain in determining pore diameter and ion permeation through cyclic nucleotide-gated channels. Nature 61–64.

Goulding, E. H., Tibbs, G. R., and Siegelbaum, S. A. 1994. Molecular mechanism of cyclic nucleotide-gated channel activation. Nature 372, 369–374.

Grosmaitre, X., Santarelli, L. C., Tan, J., Luo, M., and Ma, M. 2007. Dual functions of mammalian olfactory sensory neurons as odor detectors and mechanical sensors. Nat. Neurosci. 10, 348–354.

Gu, X. and Spitzer, N. C. 1995. Distinct aspects of neuronal differentiation encoded by frequency of spontaneous Ca^{2+} transients. Nature 375, 784–787.

Haynes, L. W. 1992. Block of the cyclic GMP-gated channel of vertebrate rod and cone photoreceptors by *l-cis*-diltiazem. J. Gen. Physiol. 100, 783–801.

Hopker, V. H., Shewan, D., Tessier-Lavigne, M., Poo, M., and Holt, C. 1999. Growth-cone attraction to netrin-1 is converted to repulsion by laminin-1. Nature 401, 69–73.

Hsu, Y.-T. and Molday, R. S. 1993. Modulation of the cGMP-gated channel of rod photoreceptor cells by calmodulin. Nature 361, 76–79.

Ildefonse, M., Crouzy, S., and Bennett, N. 1992. Gating of retinal rod cation channel by different nucleotides: comparative study of unitary currents. J. Membr. Biol. 130, 91–104.

Jiang, Y., Lee, A., Chen, J., Cadene, M., Chait, B. T., and MacKinnon, R. 2002a. Crystal structure and mechanism of a calcium-gated potassium channel. Nature 417, 515–522.

Jiang, Y., Lee, A., Chen, J., Cadene, M., Chait, B. T., and MacKinnon, R. 2002b. The open pore conformation of potassium channels. Nature 417, 523–526.

Johnson, J. P., Jr. and Zagotta, W. N. 2001. Rotational movement during cyclic nucleotide-gated channel opening. Nature 412, 917–921.

Kafitz, K. W., Leinders-Zufall, T., Zufall, F., and Greer, C. A. 2000. Cyclic GMP evoked calcium transients in olfactory receptor cell growth cones. Neuroreport 11, 677–681.

Karpen, J. W., Brown, R. L., Stryer, L., and Baylor, D. A. 1993. Interactions between divalent cations and the gating machinery of cyclic GMP-activated channels in salamander retinal rods. J. Gen. Physiol. 101, 1–25.

Kaupp, U. B. and Seifert, R. 2002. Cyclic nucleotide-gated ion channels. Physiol. Rev. 82, 769–824.

Kaupp, U. B., Niidome, T., Tanabe, T., Terada, S., Bonigk, W., Stuhmer, W., Cook, N. J., Kangawa, K., Matsuo, H., Hirose, T., Miyata, T., and Numa, S. 1989. Primary structure and functional expression from complementary DNA of the rod photoreceptor cyclic GMP-gated channel. Nature 342, 762–766.

Kelliher, K. R., Ziesmann, J., Munger, S. D., Reed, R. R., and Zufall, F. 2003. Importance of the CNGA4 channel gene for odor discrimination and adaptation in behaving mice. Proc. Natl. Acad. Sci. U. S. A. 100, 4299–4304.

Kingston, P. A., Zufall, F., and Barnstable, C. J. 1996. Rat hippocampal neurons express genes for both rod retinal and olfactory cyclic nucleotide-gated channels: novel targets for cAMP/cGMP function. Proc. Natl. Acad. Sci. U. S. A. 93, 10440–10445.

Kingston, P. A., Zufall, F., and Barnstable, C. J. 1999. Widespread expression of olfactory cyclic nucleotide-gated channel genes in rat brain: implications for neuronal signalling. Synapse 32, 1–12.

Kleene, S. J. 1999. Both external and internal calcium reduce the sensitivity of the olfactory cyclic-nucleotide-gated channel to cAMP. J. Neurophysiol. 81, 2675–2682.

Kohl, S., Marx, T., Giddings, I., Jagle, H., Jacobson, S. G., Apfelstedt-Sylla, E., Zrenner, E., Sharpe, L. T., and Wissinger, B. 1998. Total colourblindness is caused by mutations in the gene encoding the alpha-subunit of the cone photoreceptor cGMP-gated cation channel. Nat. Genet. 19, 257–259.

Komatsu, H., Mori, I., Rhee, J.-S., Akaike, N., and Ohshima, Y. 1996. Mutations in a cyclic nucleotide-gated channel lead to abnormal thermosensation and chemosensation in *C. elegans*. Neuron 17, 707–718.

Korschen, H. G., Illing, M., Seifert, R., Sesti, F., Williams, A., Gotzes, S., Colville, C., Muller, F., Dose, A., Godde, M., *et al.* 1995. A 240 kDa protein represents the complete beta subunit of the cyclic nucleotide-gated channel from rod photoreceptor. Neuron 15, 627–636.

Kramer, R. H. and Siegelbaum, S. A. 1992. Intracellular Ca^{2+} regulates the sensitivity of cyclic nucleotide-gated channels in olfactory receptor neurons. Neuron 9, 897–906.

Kumar, V. D. and Weber, I. T. 1992. Molecular model of the cyclic GMP-binding domain of the cyclic GMP-gated ion channel. Biochemistry 31, 4643–4649.

Kurahashi, T. and Menini, A. 1997. Mechanism of odorant adaptation in the olfactory receptor cell. Nature 385, 725–729.

Kurahashi, T. and Shibuya, T. 1990. Ca^{2+}-dependent adaptive properties in the solitary olfactory receptor cells of the newt. Br. Res. 515, 261–268.

Kuzmiski, J. B. and MacVicar, B. A. 2001. Cyclic nucleotide-gated channels contribute to the cholinergic plateau potential in hippocampal CA1 pyramidal neurons. J. Neurosci. 21, 8707–8714.

Leinders-Zufall, T. and Zufall, F. 1995. Block of cyclic nucleotide-gated channels in salamander olfactory receptor neurons by the guanylyl cyclase inhibitor LY83583. J. Neurophysiol. 74, 2759–2762.

Leinders-Zufall, T., Rosenboom, H., Barnstable, C. J., Shepherd, G. M., and Zufall, F. 1995. A calcium-permeable cGMP-activated cation conductance in hippocampal neurons. Neuroreport 6, 45–49.

Leng, Q., Mercier, R. W., Yao, W., and Berkowitz, G. A. 1999. Cloning and first functional characterization of a plant cyclic nucleotide-gated cation channel. Plant Physiol. 121, 753–761.

Liman, E. R. and Buck, L. B. 1994. A second subunit of the olfactory cyclic nucleotide-gated channel confers high sensitivity to cAMP. Neuron 13, 611–621.

Liu, J. and Siegelbaum, S. A. 2000. Change of pore helix conformational state upon opening of cyclic nucleotide-gated channels. Neuron 28, 899–909.

Liu, D. T., Tibbs, G. R., Paoletti, P., and Siegelbaum, S. A. 1998. Constraining ligand-binding site stoichiometry suggests that a cyclic nucleotide-gated channel is composed of two functional dimers. Neuron 21, 235–248.

Liu, D. T., Tibbs, G. R., and Siegelbaum, S. A. 1996. Subunit stoichiometry of cyclic nucleotide-gated channels and effects of subunit order on channel function. Neuron 16, 983–990.

Liu, M., Chen, T., Ahamed, B., Li, J., and Yau, K. W. 1994. Calcium-calmodulin modulation of the olfactory cyclic nucleotide-gated cation channel. Science 266, 1348–1354.

Lynch, J. W. and Lindemann, B. 1994. Cyclic nucleotide-gated channels of rat olfactory receptor cells: divalent cations control the sensitivity to cAMP. J. Gen. Physiol. 103, 87–106.

Matulef, K. and Zagotta, W. 2002. Multimerization of the ligand binding domains of cyclic nucleotide-gated channels. Neuron 36, 93–103.

Matulef, K. and Zagotta, W. N. 2003. Cyclic nucleotide-gated ion channels. Annu. Rev. Cell Dev. Biol. 19, 23–44.

Matulef, K., Flynn, G. E., and Zagotta, W. N. 1999. Molecular rearrangements in the ligand-binding domain of cyclic nucleotide-gated channels. Neuron 24, 443–452.

Mazzolini, M., Punta, M., and Torre, V. 2002. Movement of the C-helix during the gating of cyclic nucleotide-gated channels. Biophys. J. 83, 3283–3295.

Michalakis, S., Reisert, J., Geiger, H., Wetzel, C., Zong, X., Bradley, J., Spehr, M., Huttl, S., Gerstner, A., Pfeifer, A., Hatt, H., Yau, K. W., and Biel, M. 2006. Loss of CNGB1 protein leads to olfactory dysfunction and subciliary cyclic nucleotide-gated channel trapping. J. Biol. Chem. 281, 35156–35166.

Miyazu, M., Tanimura, T., and Sokabe, M. 2000. Molecular cloning and characterization of a putative cyclic nucleotide-gated channel from Drosophila melanogaster. Insect Mol. Biol. 9, 283–292.

Molday, R. S., Molday, L. L., Dose, A., Clark-Lewis, I., Illing, M., Cook, N. J., Eismann, E., and Kaupp, U. B. 1991. The cGMP-gated channel of the rod photoreceptor cell characterization and orientation of the amino terminus. J. Biol. Chem. 266, 21917–21922.

Molokanova, E., Trivedi, B., Savchenko, A., and Kramer, R. H. 1997. Modulation of rod photoreceptor cyclic nucleotide-gated channels by tyrosine phosphorylation. J. Neurosci. 17, 9068–9076.

Muller, F., Bonigk, W., Sesti, F., and Frings, S. 1998. Phosphorylation of mammalian olfactory cyclic nucleotide-gated channels increases ligand sensitivity. J. Neurosci. 18(1), 164–173.

Munger, S. D., Lane, A. P., Zhong, H., Leinders-Zufall, T., Yau, K. W., Zufall, F., and Reed, R. R. 2001. Central role of the CNGA4 channel subunit in Ca^{2+}-calmodulin-dependent odor adaptation. Science 294, 2172–2175.

Nache, V., Schulz, E., Zimmer, T., Kusch, J., Biskup, C., Koopmann, R., Hagen, V., and Benndorf, K. 2005. Activation of olfactory-type cyclic nucleotide-gated channels is highly cooperative. J. Physiol. 569, 91–102.

Nakamura, T. and Gold, G. H. 1987. A cyclic nucleotide-gated conductance in olfactory receptor cilia. Nature 325, 442–444.

Nicol, G. D. 1993. The calcium channel antagonist, pimozide, blocks the cyclic GMP-activated current in rod photoreceptors. J. Pharmacol. Exp. Ther. 265, 626–632.

Paoletti, P., Young, E. C., and Siegelbaum, S. A. 1999. C-Linker of cyclic nucleotide-gated channels controls coupling of ligand binding to channel gating. J. Gen. Physiol. 113, 17–33.

Parent, A., Schrader, K., Munger, S. D., Reed, R. R., Linden, D. J., and Ronnett, G. V. 1998. Synaptic transmission and hippocampal long-term potentiation in olfactory cyclic nucleotide-gated channel type 1 null mouse. Am. Physiol. Soc. 79, 3295–3301.

Peng, C., Rich, E. D., and Varnum, M. D. 2004. Subunit configuration of heteromeric cone cyclic nucleotide-gated channels. Neuron 42, 401–410.

Pifferi, S., Boccaccio, A., and Menini, A. 2006. Cyclic nucleotide-gated ion channels in sensory transduction. FEBS Lett. 580, 2853–2859.

Podda, M. V., Marcocci, M. E., Del Carlo, B., Palamara, A. T., Azzena, G. B., and Grassi, C. 2005. Expression of cyclic nucleotide-gated channels in the rat medial vestibular nucleus. Neuroreport 16, 1939–1943.

Qu, W., Moorhouse, A. J., Chandra, M., Pierce, K. D., Lewis, T. M., and Barry, P. H. 2006. A single P-loop glutamate point mutation to either lysine or arginine switches the cation–anion selectivity of the CNGA2 channel. J. Gen. Physiol. 127, 375–389.

Rebrik, T. I. and Korenbrot, J. I. 1998. In intact cone photoreceptors, a Ca^{2+}-dependent, diffusible factor modulates the cGMP-gated ion channels differently than in rods. J. Gen. Physiol. 112, 537–548.

Root, M. J. and MacKinnon, R. 1993. Identification of an external divalent cation-binding site in the pore of a cGMP-activated channel. Neuron 11, 459–466.

Ruiz, M. L. and Karpen, J. W. 1997. Single cyclic nucleotide-gated channels locked in different ligand-bound states. Nature 389, 389–392.

Ruiz, M. L., London, B., and Nadal-Ginard, B. 1996. Cloning and characterization of an olfactory cyclic nucleotide-gated channel expressed in mouse heart. J. Mol. Cell Cardiol. 28, 1453–1461.

Sautter, A., Zong, X., Hofmann, F., and Biel, M. 1998. An isoform of the rod photoreceptor cyclic nucleotide-gated channel b subunit expressed in olfactory neurons. Proc. Natl. Acad. Sci. U. S. A. 95, 4696–4701.

Shapiro, M. S. and Zagotta, W. N. 2000. Structural basis for ligand selectivity of heteromeric olfactory cyclic nucleotide-gated channels. Biophys. J. 78, 2307–2320.

Shepherd, G. M. 1994. Discrimination of molecular signals by the olfactory receptor neuron. Neuron 13, 771–790.

Song, H. J. and Poo, M. M. 1999. Signal transduction underlying growth cone guidance by diffusible factors. Curr. Opin. Neurobiol. 9, 355–363.

Song, H. J., Ming, G. L., and Poo, M. M. 1997. cAMP-induced switching in turning direction of nerve growth cones. Nature 388, 275–279.

Spitzer, N. C., Olson, E., and Gu, X. 1995. Spontaneous calcium transients regulate neuronal plasticity in developing neurons. J. Neurobiol. 26, 316–324.

Sunderman, E. R. and Zagotta, W. N. 1999. Sequence of events underlying the allosteric transition of rod cyclic nucleotide-gated channels. J. Gen. Physiol. 113, 621–640.

Sundin, O. H., Yang, J. M., Li, Y., Zhu, D., Hurd, J. N., Mitchell, T. N., Silva, E. D., and Maumenee, I. H. 2000. Genetic basis of total colourblindness among the Pingelapese islanders. Nat. Genet. 25, 289–293.

Tetreault, M. L., Henry, D., Horrigan, D. M., Matthews, G., and Zimmerman, A. L. 2006. Characterization of a novel cyclic nucleotide-gated channel from zebrafish brain. Biochem. Biophys. Res. Commun. 348, 441–449.

Tibbs, G. R., Liu, D. T., Leypold, B. G., and Siegelbaum, S. A. 1998. A state-independent interaction between ligand and a conserved arginine residue in cyclic nucleotide-gated channels reveals a functional polarity of the cyclic nucleotide binding site. J. Biol. Chem. 273, 4497–4505.

Van Wagenen, S. and Rehder, V. 1999. Regulation of neuronal growth cone filopodia by nitric oxide. J. Neurobiol. 39, 168–185.

Varnum, M. D. and Zagotta, W. N. 1996. Subunit interactions in the activation of cyclic nucleotide-gated ion channels. Biophys. J. 70, 2667–2679.

Varnum, M. D. and Zagotta, W. N. 1997. Interdomain interactions underlying activation of cyclic nucleotide-gated channels. Science 278, 110–113.

Varnum, M. D., Black, K. D., and Zagotta, W. N. 1995. Molecular mechanism for ligand discrimination of cyclic nucleotide-gated channels. Neuron. 15, 619–625.

Wei, J. Y., Roy, D. S., Leconte, L., and Barnstable, C. J. 1998. Molecular and pharmacological analysis of cyclic nucleotide-gated channel function in the central nervous system. Prog. Neurobiol. 56, 37–64.

Weitz, D., Ficek, N., Kremmer, E., Bauer, P. J., and Kaupp, U. B. 2002. Subunit stoichiometry of the CNG channel of rod photoreceptors. Neuron 36, 881–889.

Wells, G. B. and Tanaka, J. C. 1997. Ion selectivity predictions from a two-site permeation model for the cyclic nucleotide-gated channel of retinal rod cells. Biophys. J. 72(1), 127–140.

Wissinger, B., Gamer, D., Jagle, H., Giorda, R., Marx, T., Mayer, S., Tippmann, S., Broghammer, M., Jurklies, B., Rosenberg, T., Jacobson, S. G., Sener, E. C., Tatlipinar, S., Hoyng, C. B., Castellan, C., Bitoun, P., Andreasson, S., Rudolph, G., Kellner, U., Lorenz, B., Wolff, G., Verellen-Dumoulin, C., Schwartz, M., Cremers, F. P., Apfelstedt-Sylla, E., Zrenner, E., Salati, R., Sharpe, L. T., and Kohl, S. 2001. CNGA3 mutations in hereditary cone photoreceptor disorders. Am. J. Hum. Genet. 69, 722–737.

Wohlfart, P., Haase, W., Molday, R. S., and Cook, N. J. 1992. Antibodies against synthetic peptides used to determine the topology and site of glycosylation of the cGMP-gated channel from bovine rod photoreceptors. J. Biol. Chem. 267, 644–648.

Yamazaki, Y., Brown, R. L., and Morita, T. 2002. Purification and cloning of toxins from elapid venoms that target cyclic nucleotide-gated ion channels. Biochemistry 41, 11331–11337.

Zagotta, W. N. and Siegelbaum, S. A. 1996. Structure and function of cyclic nucleotide-gated channels. Annu. Rev. Neurosci. 19, 235–263.

Zheng, J. and Zagotta, W. N. 2000. Gating rearrangements in cyclic nucleotide-gated channels revealed by patch-clamp fluorometry. Neuron 28, 369–374.

Zheng, J. and Zagotta, W. N. 2004. Stoichiometry and assembly of olfactory cyclic nucleotide-gated channels. Neuron 42, 411–421.

Zheng, J., Trudeau, M. C., and Zagotta, W. N. 2002. Rod cyclic nucleotide-gated channels have a stoichiometry of three CNGA1 subunits and one CNGB1 subunit. Neuron 36, 891–896.

Zhong, H., Molday, L. L., Molday, R. S., and Yau, K. W. 2002. The heteromeric cyclic nucleotide-gated channel adopts a 3A:1B stoichiometry. Nature 420, 193–198.

Zong, X., Zucker, H., Hofmann, F., and Biel, M. 1998. Three amino acids in the C-linker are major determinants of gating in cyclic nucleotide-gated channels. EMBO J. 17, 353–362.

Zufall, F., Firestein, S., and Shepherd, G. M. 1991. Analysis of single cyclic-nucleotide gated channels in olfactory receptor cells. J. Neurosci. 11, 3573–3580.

Zufall, F., Firestein, S., and Shepherd, G. M. 1994. Cyclic nucleotide-gated ion channels and sensory transduction in olfactory receptor neurons. Annu. Rev. Biophys. Biomol. Struct. 23, 577–607.

Zufall, F., Shepherd, G. M., and Barnstable, C. J. 1997. Cyclic nucleotide-gated channels as regulators of CNS development and plasticity. Curr. Opin. Neurobiol. 7, 404–412.

4.30 Structure, Expression, and Function of Olfactory Receptors

K Touhara, The University of Tokyo, Chiba, Japan

4.30.1 Introduction

Titus Lucretius Carus, a Roman poet and philosopher, proposed in his book *On the Nature of Things* (50 BC) that a variety of odors exist because each odorant possesses a unique structure (Lucretius, T., 1995). In the mid-twentieth century, this concept was established as the stereospecific receptor theory, which provides an explanation for the molecular mechanisms underlying the remarkable olfactory sensing system (Amoore, J. E., 1963). The receptor theory postulates that there are receptor sites for odorants and that odor perception occurs only when the structure of an odorant molecule and the binding site match. Other theories including vibrational theory, puncturing theory, radiational theory, and absorption theory have been proposed to explain odor perception, but they remain unproven.

Based on the evidence for the involvement of G protein-mediated signaling and ion channels in olfactory signal transduction (Pace, U. *et al.*, 1985; Sklar, P. B. *et al.*, 1986; Nakamura, T. and Gold, G. H., 1987; Breer, H. *et al.*, 1990; Dhallan, R. S. *et al.*, 1990), Buck and Axel hypothesized that receptor proteins for odorants must be expressed in the olfactory epithelium and must belong to a family of seven-transmembrane G-protein-coupled receptors (GPCRs) (Buck, L. B., 2004). State-of-the-art molecular biology techniques allowed them to identify a multigene family that appears to encode receptors for odorants (Buck, L. and Axel, R., 1991). The proteins were named olfactory or odorant receptors (ORs). The former name was given because the proteins are expressed in olfactory neurons and play a crucial role in olfaction, and the latter because the proteins bind odorant molecules; however, strictly saying, neither terminology is correct because ORs have since been found in tissues other than the olfactory epithelium and because ORs have functions in addition to acting as odorant sensors (Spehr, M. *et al.*, 2006a). Nonetheless, the main role of the OR family is to provide a molecular basis for recognition and discrimination of thousands of odorant molecules in the olfactory sensory system.

4.30.2 Structure

4.30.2.1 Protein Structure

OR proteins are classified as members of the GPCR superfamily because of the presence of structural features common to GPCRs (Buck, L. and Axel, R., 1991; Strader, C. D. *et al.*, 1994) and because they couple to and activate heterotrimeric G proteins (Krautwurst, D. *et al.*, 1998; Kajiya, K. *et al.*, 2001). ORs possess seven hydrophobic transmembrane domains, a disulfide bond between the conserved cysteines in the extracellular loops, a conserved glycosylation site in the N-terminal region (Katada, S. *et al.*, 2004), and several amino acid sequences that are conserved in the OR family (Zozulya, S. *et al.*, 2001; Zhang, X. and Firestein, S., 2002) (Figure 1). The OR consensus sequences reside

within the cytoplasmic side of each transmembrane regions and include PMYFFL (transmembrane domain [TM]2), MAYDRYVAIC (TM3), KAFSTC (TM6), and PMLNPXXY (TM7). Although these consensus sequences vary somewhat between species, they have been widely used to retrieve OR genes from genome. In addition, extensive motif analysis has revealed more than 80 specific short motifs, some of which constitute signature sequences for the OR subfamily of a certain species or have implications for the function and evolution the ORs (Liu, A. H. *et al.*, 2003). These conserved motifs likely contribute to the proper folding of ORs in the plasma membrane so that ORs can bind odorants and couple to appropriate G proteins. The transmembrane regions, on the other hand, help form the odorant-binding pocket (Figure 1). The sequences in the binding pocket are relatively variable, allowing ORs to bind a wide spectrum of odorant molecules (Singer, M. S. *et al.*, 1996; Pilpel, Y. and Lancet, D., 1999; Floriano, W. B. *et al.*, 2000; Singer, M. S., 2000; Floriano, W. B. *et al.*, 2004; Man, O. *et al.*, 2004; Katada, S. *et al.*, 2005). Overall, the three-dimensional structure of the OR is probably similar to that of rhodopsin, which has been determined by a combination of computer modeling, site-directed mutation, and

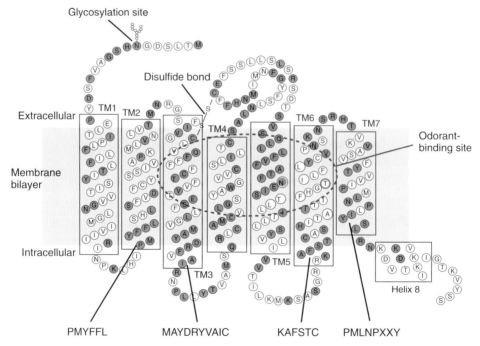

Figure 1 Structure and topology of an olfactory receptor (OR) protein. Conserved sequence motifs, glycosylation site, and disulfide bond in the OR family are shown. The odorant-binding site is in the space formed by TM3-6 (dotted circle). Red, variable amino acids in the OR family; blue, conserved amino acids in the OR family. TM, transmembrane domain.

X-ray crystallography and which is the only GPCR structure that has been solved so far (see Odorant-Binding Site) (Katada, S. *et al.*, 2005).

4.30.2.2   Olfactory Receptor Genes

It has been estimated that the OR family includes at least several hundred members in the rat (Buck, L. and Axel, R., 1991). In the last decade or so, the genome sequence projects have enabled the comprehensive analysis of the OR gene family and have revealed the genomic structure and distribution of the OR genes from various organisms (Mombaerts, P., 2004). The most recent data for vertebrate ORs are shown in Table 1. In mammals, the OR repertoire comprises 800–1500 members, whereas in fish, the OR family has only approximately 100 members (Niimura, Y. and Nei, M., 2005c). Thus, expansion of the OR gene family occurred when animals shifted

their living space from an aquatic to a terrestrial environment.

Based on phylogenetic analysis, the mammalian OR genes can be classified into two different groups: class I and class II (Glusman, G. *et al.*, 2001; Zozulya, S. *et al.*, 2001; Zhang, X. and Firestein, S., 2002) (Figure 2). This classification is based on the finding that frog (*Xenopus laevis*) has two different classes of OR genes: one (class I) that is similar to fish OR genes and a second (class II) that is similar to mammalian OR genes (Freitag, J. *et al.*, 1995). For example, the numbers of intact class I OR genes in human and mouse are 57 and 115 of 388 and 1037 total intact OR genes, respectively (Niimura, Y. and Nei, M., 2005c). Later studies revealed, however, that mammalian class I genes formed a distinct clade from fish OR genes that can be further divided into nine subgroups (Niimura, Y. and Nei, M., 2005c) (Figure 2). The amphibian OR repertoire turns out to be similar to both fish and mammal ORs,

Table 1   Functional and nonfunctional OR genes in various vertebrate species

	Organism	Total genes	Intact genes	Pseudogene (%)	Reference[a]
Mammal	Human	802	388	52	Niimura Y. and Nei M. (2005a; 2005b; 2005c)
	Mouse	1391	1037	25	Niimura Y. and Nei M. (2005a; 2005b; 2005c)
	Rat	1493	1202	20	Quignon P. *et al.* (2005)
	Dog	1094	872	20	Quignon P. *et al.* (2005)
Bird	Chicken[b]	554	78	86	Niimura Y. and Nei M. (2005a; 2005b; 2005c)
Amphibian	*X. tropicalis*	888	410	54	Niimura Y. and Nei M. (2005a; 2005b; 2005c)
Fish	Zebrafish	133	98	26	Niimura Y. and Nei M. (2005a; 2005b; 2005c)
	Puffer fish	94	40	57	Niimura Y. and Nei M. (2005a; 2005b; 2005c)

[a]Note that there have been many reports on the estimates of the numbers of olfactory receptor (OR) genes, but the most recent ones are shown here. The numbers will be subject to change slightly with time in the future.
[b]The actual number of chicken ORs appears to be much larger due to numerous short contings in the draft genome.

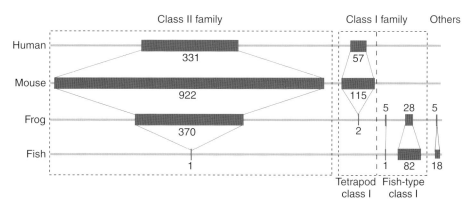

Figure 2   Numbers of functional olfactory receptor (OR) genes belonging to different groups in four species. The class II family has originated from one ancestral gene in fish. The class I family can be divided into two classes; tetrapod class I and fish-type class I.

suggesting that there are at least three evolutional lineages (i.e., fish, amphibian, and mammal) in the vertebrate OR gene family (Niimura, Y. and Nei, M., 2005c).

Relatively few OR genes have been found in insects. For example, there are 62 OR genes expressed in the antenna and the maxillary palps in *Drosophila melanogaster* (Clyne, P. J. *et al.*, 1999; Gao, Q. and Chess, A., 1999; Vosshall, L. B. *et al.*, 1999). Similar numbers of OR genes have been found in the genome of the malaria mosquito (Fox, A. N. *et al.*, 2002; Hill, C. A. *et al.*, 2002) and from several moths (Krieger, J. *et al.*, 2002; 2003; Nakagawa, T. *et al.*, 2005). The insect ORs have the following unique features: (1) they lack obvious homology with vertebrate ORs or other GPCRs (Robertson, H. M. *et al.*, 2003); (2) there do not appear to be pseudogenes (Robertson, H. M. *et al.*, 2003); (3) they are coexpressed and heterodimerize with the unique OR, which is highly conserved across diverse insect species, also known as the Or83b family (Larsson, M. C. *et al.*, 2004; Nakagawa, T. *et al.*, 2005; Neuhaus, E. M. *et al.*, 2005); and (4) their membrane topology is distinct from conventional GPCRs so that their N-termini are located intracellularly (Benton, R. *et al.*, 2006).

4.30.2.3 Olfactory Receptor Pseudogenes

A significant portion of the OR gene family has been pseudogenized in vertebrates. Hominoids possess a high pseudogene content (~50%), whereas only ~20% of OR genes are pseudogenes in mouse and dog and 25–35% are pseudogenes in primates (Gilad, Y. *et al.*, 2004; Niimura, Y. and Nei, Y., 2005c; Quignon, P. *et al.*, 2005) (Table 1). The ORs are the most changeable gene family and still undergo rapid molecular evolution by tandem gene duplication and pseudogenization in the lineage of each animal species. For example, the mouse lineage acquired a few hundred new OR genes after the human–mouse divergence, indicating that both gene expansion in the mouse lineage and gene loss in the human lineage contribute to the differences in the numbers of functional ORs (Niimura, Y. and Nei, M., 2005b). The OR gene dynamics could also be pressured by the state of the environment and the type of sense that is utilized for social and sexual behavior. The fraction of pseudogenes in the OR family has increased during evolution in the order of rodents, monkeys, and humans, suggesting that the reduced sense of smell correlates with the loss of functional OR

genes. Indeed, in whale and dolphin, in which the auditory system is dominant, 70–80% of OR genes appear to be pseudogenes (Y. Go, personal communication).

4.30.2.4 Genomic Structure

The OR genes are encoded by an exon that contains an intronless coding region of ~1 kb along with one to four upstream exons (Lane, R. P. *et al.*, 2001; Young, J. M. and Trask, B. J., 2002). The OR genes form genomic clusters and are distributed widely in chromosomes; however, the distribution is not even, and almost half of the mouse OR genes are located on chromosomes 2 and 7, which have orthologous relationships with chromosome 11 in human (Niimura, Y. and Nei, M., 2005a). The majority of the OR gene family has multiple transcriptional isoforms (Asai, H. *et al.*, 1996; Lane, R. P. *et al.*, 2001; Volz, A. *et al.*, 2003; Fukuda, N. and Touhara, K., 2006). Further, OR genes are highly polymorphic as suggested for human leukocyte antigen-linked OR genes (Ehlers, A. *et al.*, 2000; Eklund, A. C. *et al.*, 2000) and dog OR genes (Tacher, S. *et al.*, 2005), which may account for individual differences in the ability to perceive odor and may sometimes cause specific anosmia.

4.30.3 Expression

4.30.3.1 Single-Receptor Gene Choice

Each olfactory sensory neuron expresses just one of the 1000 ORs (Serizawa, S. *et al.*, 2004; Shykind, B. M., 2005). The one-neuron one-receptor rule has been confirmed by a variety of techniques, namely, *in situ* hybridization (Ngai, J. *et al.*, 1993; Ressler, K. J. *et al.*, 1993; Vassar, R. *et al.*, 1993), single-cell reverse transcriptase-polymerase chanin reaction (RT-PCR) analysis (Malnic, B. *et al.*, 1999; Touhara, K. *et al.*, 1999), and transgenic experiments (Qasba, P. and Reed, R. R., 1998; Serizawa, S. *et al.*, 2000; Vassalli, A. *et al.*, 2002; Serizawa, S. *et al.*, 2003). Further, the selected OR appears to be transcribed from only one of two alleles, either the maternal or paternal allele (Chess, A. *et al.*, 1994; Ishii, T. *et al.*, 2001). This mutually exclusive expression pattern is preserved even between transgenes and endogenous copies, supporting the stochastic model (Qasba, P. and Reed, R. R., 1998; Serizawa, S. *et al.*, 2000; Vassalli, A. *et al.*, 2002; Serizawa, S. *et al.*, 2003). Indeed, the single OR choice is observed from a minigene that contains a few kilobases upstream of the transcription start site, implicating the involvement of cis-regulatory elements (Qasba, P. and Reed, R. R., 1998; Vassalli, A. *et al.*, 2002; Oka, Y. *et al.*, 2006).

The expression of one cluster of OR genes including MOR28 in transgenic mice requires a long upstream region of ~100 kb containing an enhancer element called the H-region (Serizawa, S. *et al.*, 2000; 2003). This locus control region, H, appears to associate with multiple OR promoters and act as a cis-acting enhancer element (Lomvardas, S. *et al.*, 2006) (Figure 3). In olfactory neurons, one of the two H-region alleles is methylated, suggesting that the H-enhancer is functionally monoallelic, therefore

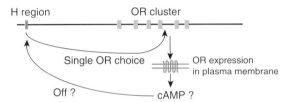

Figure 3 A model for olfactory receptor (OR) gene expression. The H-region containing an enhancer element associates with multiple OR promoters on different chromosomes to activate OR gene expression. The expressed functional OR proteins transmit inhibitory signals to block the action of the H-enhancer, resulting in the expression of only one functional OR in a sensory neuron.

ensuring selection of a single receptor (Lomvardas, S. *et al.*, 2006).

In addition to positive regulation of OR gene expression by cis-regulatory elements, negative feedback regulation must occur to suppress the expression of additional receptors (Serizawa, S. *et al.*, 2003; Lewcock, J. W. and Reed, R. R., 2004; Shykind, B. M. *et al.*, 2004). It appears that functional OR proteins but not mRNA play an inhibitory role in preventing further activation of other OR genes, although the molecular mechanism remains to be elucidated (Figure 3).

4.30.3.2 Spatial Expression Pattern

In mouse and rat, *in situ* hybridization experiments using different OR probes have suggested that each OR gene is expressed only in a certain region of the olfactory epithelium (Nef, P. *et al.*, 1992; Strotmann, J. *et al.*, 1992; Ressler, K. J. *et al.*, 1993; Vassar, R. *et al.*, 1993; Strotmann, J. *et al.*, 1994). Thus, the epithelium can be divided into at least four spatial zones expressing different sets of OR genes (Sullivan, S. L. *et al.*, 1996) (Figure 4). The distinct spatial expression pattern appears to be conserved among the vertebrate species, although the expression of fish ORs is not as clearly segregated as that of mammalian ORs (Ngai, J. *et al.*, 1993; Weth, F. *et al.*, 1996). Initially, the

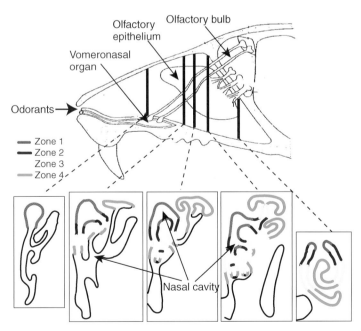

Figure 4 Anatomic structure of nasal cavity and olfactory receptor (OR) expression zones in the olfactory epithelium. The spatial expression zones are depicted in four different colors for clarity, but OR genes are expressed in a unique and distinct zone in a continuous and overlapping manner.

spatial zones observed in rodents were thought to be sharply segregated, but careful *in situ* hybridization studies revealed that each OR gene is expressed in a unique and distinct zone in a continuous and overlapping manner (Norlin, E. M. *et al.*, 2001; Iwema, C. L. *et al.*, 2004; Miyamichi, K. *et al.*, 2005). Therefore, multiple overlapping bands are formed along the dorsomedial/ventrolateral axis, although the boundary of the most dorsal region corresponding to the olfactory cell adhesion molecule (OCAM)-negative neuronal layer was maintained.

There are some exceptions to the zonal expression rule. First, some OR genes are expressed in the septal organ, which is an isolated island of olfactory sensory neurons located on the nasal septum near the entrance of the nasopharynx (Kaluza, J. F. *et al.*, 2004; Tian, H. and Ma, M., 2004). Second, the OR37 subfamily shows a patch-like distribution in clustered populations of neurons in central turbinates (Strotmann, J. *et al.*, 1999). Although there are some promoter motifs that might control the expression domain in the nasal cavity, the regulatory factors that determine the positional information remain to be identified.

4.30.3.3 Temporal Expression Pattern

In rodents, OR genes start to be expressed during the embryonic stages. The onset of expression in mice is at embryonic day (E) 12–14 as determined by *in situ* hybridization for several ORs (Strotmann, J. *et al.*, 1995; Sullivan, S. L. *et al.*, 1995). Microarray analysis of the developmental course of OR gene expression revealed that the number of transcribed OR genes remains at a relatively low level until birth (100–200 ORs), and a large number of OR genes are detectable only after birth (Zhang, X. *et al.*, 2004). At 2–3 weeks after birth, the number of detected OR genes reached a peak (~500 ORs) (Zhang, X. *et al.*, 2004). Recently, specific antibodies against ORs have become available (Barnea, G. *et al.*, 2004; Strotmann, J. *et al.*, 2004), and they have revealed that OR proteins are present in dendritic knobs of olfactory sensory neurons as early as E12 and before the initiation of ciliogenesis (Schwarzenbacher, K. *et al.*, 2005). Because the olfactory cilia are formed at E13, the OR proteins appear to migrate into developing cilia, and strong labeling by the antibody is observed in the expanded tips of the cilia. Up to E18, the number of cilia per neuron continues to increase, during which OR immunostaining is clearly visible.

4.30.3.4 Ectopic Expression

OR expression is not restricted to olfactory sensory neurons. Expression of ORs in nonolfactory tissues was first observed in the male germ line (Parmentier, M. *et al.*, 1992). Since then, ectopic OR transcripts have been found in a variety of tissues including the spleen and insulin-secreting cells (Blache, P. *et al.*, 1998), lingual epithelial cells of the tongue (Durzynski, L. *et al.*, 2005; Gaudin, J. C. *et al.*, 2006), ganglia of the autonomic nervous system (Weber, M. *et al.*, 2002), pyramidal neurons in the cerebral cortex of the brain (Otaki, J. M. *et al.*, 2004), the colon (Yuan, T. T. *et al.*, 2001), myocardial cells in developing heart (Drutel, G. *et al.*, 1995; Ferrand, N. *et al.*, 1999), the prostate gland (Yuan, T. T. *et al.*, 2001), and erythroid cells (Feingold, E. A. *et al.*, 1999). Polyclonal antibodies against an OR expressed in dog sperm bind specifically to the midpiece of the flagellum of mature sperm (Vanderhaeghen, P. *et al.*, 1993; Walensky, L. D. *et al.*, 1995). In other cases, the transcripts have been detected mainly by RT-PCR, and therefore, the expression level might be too low to be considered functional. Indeed, an RNase protection assay and microarray analysis estimated that a total of 50–70 OR genes are expressed in mouse testis (Vanderhaeghen, P. *et al.*, 1997a; 1997b; Zhang, X. *et al.*, 2004), whereas a careful *in situ* hybridization study revealed that only up to 10–20 OR genes were actually detected at a significant level (Fukuda, N. and Touhara, K., 2006) (Figure 5(a)). The testicular ORs are expressed in a subset of the seminiferous

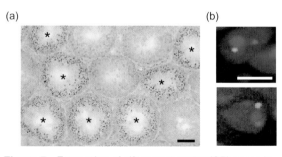

(a)　　　　　　　　　　(b)

Figure 5 Expression of olfactory receptor (OR) genes in mouse testis. (a) *In situ* hybridization showing expression of an OR in a subset of the seminiferous tubules (asterisks). Scale bar = 0.1 mm. (b) Double-*in situ* hybridization of ORs showing two signals derived from two ORs in nuclei of round spermatids. Scale bar = 10 μm. Adapted from Fukuda, N. and Touhara, K. 2006. Developmental expression patterns of testicular olfactory receptor genes during mouse spermatogenesis. Genes Cells 11, 71–81, with permission from Blackwell Publishing.

tubules, suggesting that they are expressed at specific stages during spermatogenesis (Fukuda, N. and Touhara, K., 2006). Phylogenetic analysis suggests that there is no unique characteristic sequence similarity among the testicular OR genes (Vanderhaeghen, P. *et al.*, 1997b; Zhang, X. *et al.*, 2004; Fukuda, N. and Touhara, K., 2006). Double-*in situ* hybridization experiments showed that, in contrast to olfactory neurons, more than one OR could be expressed in a single spermatogenic cell, indicating that the transcriptional regulation of OR genes in testis is distinct from that in the olfactory system (Fukuda, N. and Touhara, K., 2006) (Figure 5(b)).

4.30.4 Function I: An Odorant Sensor

4.30.4.1 General Considerations for Functional Expression

Understanding of OR function has progressed slowly due to a lack of appropriate heterologous systems for expressing and assaying odorant responses (McClintock, T. S. and Sammeta, N., 2003). Therefore, olfactory neurons themselves were first targeted as an OR expression system (Zhao, H. *et al.*, 1998) because they should possess the appropriate cellular machinery for expressing ORs and transmitting odorant signals. Adenovirus-mediated gene transfer was used to overexpress ectopic ORs in the olfactory epithelium, and the ligand responses were measured either by electro-olfactography (Zhao, H. *et al.*, 1998) or by Ca^{2+} imaging of infected neurons (Touhara, K. *et al.*, 1999) (Figure 6(a)). In addition, a gene-targeting approach was utilized to tag defined OR-expressing

olfactory neurons with green fluorescent proteins, after which the responses of the fluorescent cells to the cognate odorants were recorded (Bozza, T. *et al.*, 2002; Grosmaitre, X. *et al.*, 2006). Although this homologous *in vivo* expression system gives reliable odorant responses, high-throughput ligand screening necessitates establishment of heterologous expression systems.

Several attempts have been made to achieve functional expression of ORs on the cell surface in heterologous systems. In some cases, adding an N-terminal leader sequence from another GPCR resulted in a limited expression of functional ORs in the plasma membrane and in a successful odorant response in a heterologous system such as HEK293 cells (Krautwurst, D. *et al.*, 1998; Wetzel, C. H. *et al.*, 1999; Kajiya, K. *et al.*, 2001). Additional studies have shown that glycosylation of the N-terminus of ORs is required for proper translocation to the plasma membrane (Katada, S. *et al.*, 2004). In addition, the one-transmembrane protein RTP1, which has been referred to as an OR chaperone, appears to enhance cell-surface expression of ORs, and many ORs have been deorphanized by coexpressing them with RTP1 (Saito, H. *et al.*, 2004). In addition, Ric8B, a putative guanine nucleotide exchange factor for $G\alpha olf$, promotes efficient signal transduction of ORs (Von Dannecker, L. E. *et al.*, 2006). Thus, introduction of all the factors required for OR expression present in olfactory neurons is expected to solve the problems of heterologous expression. In addition, a method for maintaining cell-surface localization is needed because the ORs undergo continuous internalization and recycling in heterologous cells (Jacquier, V. *et al.*, 2006). Thus, the level of surface expression appears

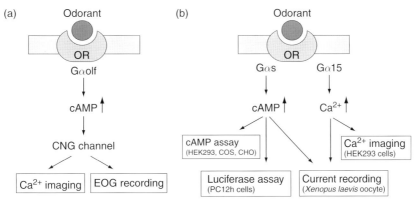

Figure 6 Odorant response assays for an olfactory receptor (OR). (a) Homologous expression of an OR in olfactory neurons. Odorant responses are monitored by Ca^{2+} imaging or electro-olfactogram (EOG). CNG, cyclic nucleotide gated. (b) Heterologous expression of an OR in mammalian cell lines or *Xenopus laevis* oocytes. Odorant responses are measured by $G\alpha s$-cAMP or $G\alpha 15$-Ca^{2+} signaling.

to be critical for OR assays. Finally, it should be noted that there are dramatic differences even between HEK293 cells from different laboratories.

4.30.4.2 Odorant Response Assays

Ca^{2+} imaging has been one of the most common and reliable methods for detecting activation of orphan GPCRs in high-throughput ligand screening. The Ca^{2+} imaging assay, however, is limited to GPCRs that couple to $G\alpha q$-type G-proteins. For GPCRs that couple to unknown G proteins, the G-protein α subunit, $G\alpha 15$, is often used to force the signaling to an inositol phosphate-mediated cascade because of its promiscuity for GPCRs. This strategy has been successfully applied to ORs that couple to the olfactory $G\alpha s$-type G-protein, $G\alpha olf$, which mediates increases in cAMP in olfactory neurons. Thus, in HEK293 cells coexpressing $G\alpha 15$ and ORs, Ca^{2+} responses are observed when the cells are stimulated with their ligands (Krautwurst, D. *et al.*, 1998; Kajiya, K. *et al.*, 2001; Touhara, K. *et al.*, 2006) (Figures 6(b) and 7).

Without coexpression of $G\alpha 15$, ORs activate endogenous $G\alpha s$ upon ligand stimulation in various mammalian cell lines (e.g., HEK293, COS-7, and CHO-K1 cells) (Kajiya, K. *et al.*, 2001; Katada, S. *et al.*, 2003). Odorant-induced cAMP increases have been measured using an enzyme-linked immunoassay (Touhara, K. *et al.*, 2006) (Figure 6(b)). Alternatively, a luciferase-reporter assay system using the zif268 promoter allows luminescent detection of cAMP increases upon stimulation with an odorant (Katada, S. *et al.*,

2003). For example, in PC12h cells cotransfected with an OR and the luciferase-reporter construct, odorant stimulation produces dose-dependent increases in luciferase activity (Katada, S. *et al.*, 2003).

The *X. laevis* oocyte is another good heterologous expression system for ORs (Figure 6(b)). Ca^{2+} increases via an endogenous Ca^{2+}-dependent Cl^- channel can be electrophysiologically detected by coexpressing $G\alpha 15$, and $G\alpha s$-mediated cAMP increases can be detected by coexpressing a cAMP-dependent Cl^- channel, CFTR (Katada, S. *et al.*, 2003; Abaffy, T. *et al.*, 2006; Touhara, K. *et al.*, 2006). The whole-cell voltage-clamp method allows for measurement of whole-cell ion currents flowing through the membrane via these ion channels. The current induced by an odorant stimulus can be further amplified by measuring the conductance at $+60\,mV$ with depolarization step pulses (Touhara, K. *et al.*, 2006).

To elucidate the structural features of an OR at the atomic level, it is necessary to develop a system for overexpressing and crystallizing the OR. This is a problem not only for ORs but also other GPCRs because of the difficulty in handling seven-transmembrane proteins. In this regard, yeast has been proposed as a good overexpression system for future structural analyses of ORs (Pajot-Augy, E. *et al.*, 2003).

4.30.4.3 Olfactory Receptor Pharmacology

The best characterized OR is a mouse OR, mOR-EG (MOR174-9), which was originally isolated from a single eugenol-responsive neuron by Ca^{2+} imaging

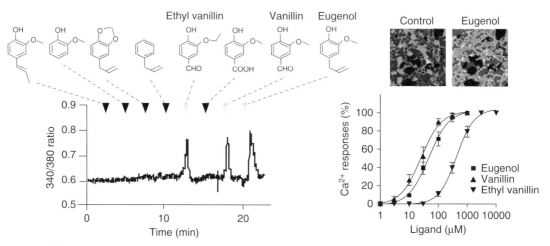

Figure 7 Ca^{2+} responses and dose–response profiles for various ligands in HEK293 cells expressing mOR-EG and $G\alpha 15$. mOR-EG responds to eugenol, vanillin, and ethyl vanillin with different EC_{50} values. Red cells indicate high levels of intracellular Ca^{2+} and blue cells represent the basal levels in pseudocolored images.

and single-cell RT-PCR techniques (Kajiya, K. *et al.*, 2001). The mOR-EG recognizes 22 odorants that share certain molecular determinants, with EC_{50} values ranging from a few to several hundred micromolar (Katada, S. *et al.*, 2005) (Figures 7 and 8). The structure–activity relationship studies for mOR-EG suggested that ORs have a broad but selective molecular receptive range, and the selectivity is determined by shape, size, functional group, and length of a ligand. This pharmacological concept is generally common to many ORs as has been shown by studies on rat I7 (Araneda, R. C. *et al.*, 2000) and other deorphanized ORs (Wetzel, C. H. *et al.*, 1999; Bozza, T. *et al.*, 2002; Gaillard, I. *et al.*, 2002; Levasseur, G. *et al.*, 2003; Spehr, M. *et al.*, 2003; Matarazzo, V. *et al.*, 2005; Shirokova, E. *et al.*, 2005); however, there seem to be some ORs whose ligand spectra are relatively narrow. The ligand specificity appears to be determined by the environment of the binding site in each OR. Considering that there are a thousand ORs, each of which has a unique ligand spectrum, and that each odorant is recognized by multiple ORs, it is reasonable to conclude that odorant information is coded by a combination of ORs activated by the odorant (Malnic, B. *et al.*, 1999; Kajiya, K. *et al.*, 2001; Touhara, K., 2002) (Figure 9).

As is the case of ligands for other GPCRs, odorants function not only as agonists for some ORs but also as antagonists for others, suggesting that the interaction between ORs and odorants in the mixture can be more complex than expected (Figure 10). For example, the eugenol response to mOR-EG was potently blocked by some structurally related odorants such as methyl isoeugenol and isosafrol (Oka, Y. *et al.*, 2004a; 2004b). The antagonism was observed in mOR-EG-expressing olfactory neurons using an intact olfactory epithelium slice (Omura, M. *et al.*, 2003; Oka, Y. *et al.*, 2004b). Furthermore, rat I7 has been shown to be antagonized by citral (Araneda, R. C. *et al.*, 2000), and undecanal was found to be an inhibitor of hOR17-4 (Spehr, M. *et al.*, 2003). Thus, the antagonism between odorants may contribute significantly to the formation of the receptor code for an odorant mixture in the olfactory system. In reality, however, this type of interaction may be rare because few odorants act as antagonists. Indeed, optical recording in the olfactory bulb shows that a receptor code for a natural odor such as coffee or cumin is basically the sum of the responses to its individual components (Lin da, Y. *et al.*, 2006). Quantitative recording using a Ca^{2+} imaging technique, however, demonstrated that a partial inhibition of the response by antagonists occurs in the OR-defined glomerulus, which may contribute to a change in odor perception (Y. Oka and K. Touhara, unpublished observation).

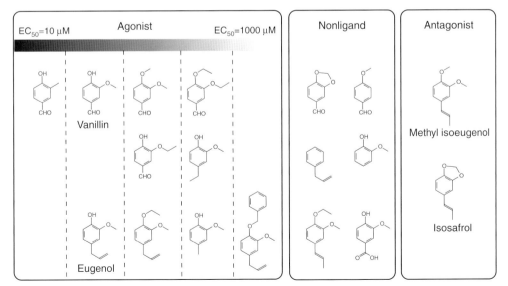

Figure 8 Structure of agonists and antagonists of mOR-EG. The compounds on the left-most side exhibit the lowest EC_{50} values. Adapted from Touhara, K., Katada, S., Nakagawa, T., and Oka, Y. 2006. Ligand screening of olfactory receptors. In: *G Protein-Coupled Receptors: Structure, Function, and Ligand Screening* (eds. T. Haga and S. Takeda), pp. 85–109. CRC press.

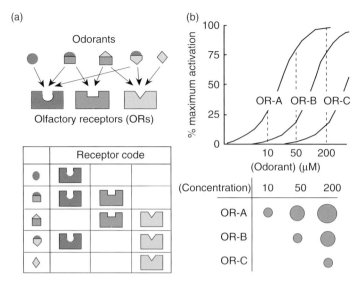

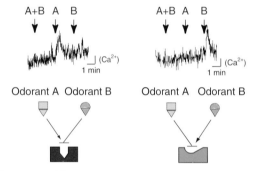

Figure 9 A combinatorial receptor code for an odorant. (a) A receptor code for an odorant is determined by a combination of activated olfactory receptors (ORs). (b) An odorant is recognized by a subset of ORs that have different affinities, thereby resulting in different receptor codes with increasing odorant concentrations.

Figure 10 Olfactory receptor antagonism between odorants. Ca^{2+} response profiles of olfactory neurons that respond to odorant A or B that is inhibited by odorant B or A, respectively (see Oka, Y. *et al.*, 2004 for more detail).

4.30.4.4 Odorant-Binding Site

Multiple alignment analysis of the OR superfamily identified highly conserved and variable regions that are probably involved in structural organization and in ligand recognition, respectively (Buck, L. and Axel, R., 1991; Zhang, X. and Firestein, S., 2002). Some of the earlier attempts such as correlated mutation analysis (Singer, M. S. *et al.*, 1996) and Fourier analysis of multiple OR sequences (Pilpel, Y. and Lancet, D., 1999; Man, O. *et al.*, 2004) further supported the hypothesis that variable residues are responsible for odorant binding. Alternatively, when cognate ligands for ORs are known, odorant-sensing residues have been predicted using a combination of

computer modeling and docking simulation. A total of 10–20 residues in TM3, 4, 5, and 6 are predicted to be involved in ligand recognition and responsible for odorant specificity of ORs based on the coordinates of bacteriorhodopsin, a non-GPCR seven-transmembrane helix protein (Singer, M. S. and Shepherd, G. M., 1994), a later rhodopsin X-ray structure (Floriano, W. B. *et al.*, 2000; Singer, M. S., 2000), a recent computationally determined three-dimensional structure predicted using only the primary sequence (Vaidehi, N. *et al.*, 2002; Floriano, W. B. *et al.*, 2004; Hall, S. E. *et al.*, 2004; Hummel, P. *et al.*, 2005), and computer modeling and calculation of the lowest ligand-binding energy for several ORs.

A systematic experimental approach to identify the odorant-binding site was taken for mOR-EG (Katada, S. *et al.*, 2005). To define the odorant-binding environment, functional analysis was carried out on a series of site-directed mutants along with parallel ligand docking simulations. Most of the critical residues involved in odorant recognition, and therefore sensitive to mutation, were hydrophobic, and the binding pocket was formed by TM3, TM5, and TM6 (Katada, S. *et al.*, 2005) (Figure 11). The spatial location of the binding pocket is similar to that for other biogenic GPCRs, but unlike other GPCRs, which tend to form multiple electrostatic interactions with their ligands, hydrophobic amino acids appear to play critical roles in odorant recognition by ORs. The accuracy of the binding model was confirmed by

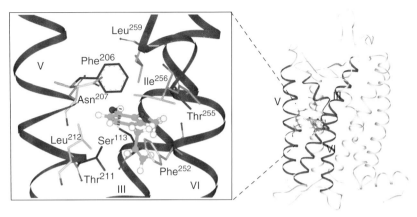

Figure 11 A model for binding of eugenol to mOR-EG and the orientation of amino acids in the binding site. Eugenol is shown in green, and TMIII, V, and VI are shown as red ribbons. Adapted from Katada, S., Hirokawa, T., Oka, Y., Suwa, M., and Touhara, K. 2005. Structural basis for a broad but selective ligand spectrum of a mouse olfactory receptor: mapping the odorant-binding site. J. Neurosci. 25, 1806–1815, Copyright 2005 by the Society of Neuroscience, with permission.

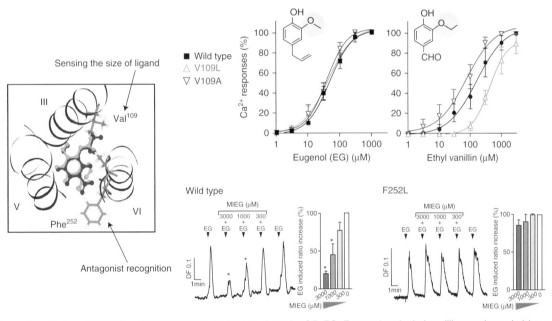

Figure 12 Roles of Val109 and Phe252 in the binding pocket of mOR-EG. Eugenol and ethyl vanillin are shown in blue and pink, respectively, in top view of the bound mOR-EG structures. The mutational analyses show that Val109 senses the size of the ligand, while Phe252 is involved in antagonist recognition (see Katada, S. et al., 2005 for more detail). Adapted from Katada, S., Hirokawa, T., Oka, Y., Suwa, M., and Touhara, K. 2005. Structural basis for a broad but selective ligand spectrum of a mouse olfactory receptor: mapping the odorant-binding site. J. Neurosci. 25, 1806–1815, Copyright 2005 by the Society of Neuroscience, with permission.

the fact that single amino acid changes caused predictable changes in agonist and antagonist specificity (Katada, S. *et al.*, 2005) (Figure 12). Thus, the stereospecific receptor theory was proven by the functional identification of the odorant-binding site, and it appears that ligand information is transduced by the three-dimensional configuration of the binding pocket and its specific odorant ligand.

4.30.4.5 Odorant Sensitivity and Specificity

The sensitivity of ORs to odorants in heterologous expression systems is relatively high compared to that in the olfactory system; specifically, the threshold concentrations for the best ligand in each OR range from a few to several hundred micromolar (Mombaerts, P., 2004). This is not simply due to problems of the heterologous expression system because the sensitivity of olfactory sensory neurons expressing the same OR is similar (Oka, Y. *et al.*, 2004b; 2006) (Figure 13(a)). The odorant sensitivity of glomeruli innervated by olfactory neurons expressing a defined OR, however, is approximately 1000-fold higher than that of peripheral neurons (Oka, Y. *et al.*, 2006) (Figure 13(a)). In addition, the specificity of the in vivo odorant response in an OR-defined glomerulus is different from that suggested by the *in vitro* pharmacology in a heterologous system (Oka, Y. *et al.*, 2006) (Figure 13(b)). The higher apparent sensitivity and specificity to odorants *in vivo* seems to be due to a contribution of the olfactory mucosa, which provides a site for efficiently concentrating odorants and carrying them to the receptor site. An important caveat to this is that the ligand specificity of ORs in heterologous OR expression systems may not always reflect the specificity observed in the olfactory bulb *in vivo*.

4.30.5 Function II: Others

4.30.5.1 Axon Guidance

Olfactory sensory neurons expressing a given OR project their axons to defined glomeruli in the olfactory bulb (Ressler, K. J. *et al.*, 1994; Vassar, R. *et al.*, 1994; Mombaerts, P. *et al.*, 1996) (Figure 14). Receptor deletion or swapping experiments suggest that the OR protein itself is instructive for the guidance of olfactory neuron axons (Wang, F. *et al.*, 1998). Indeed, the OR proteins are present in olfactory axonal terminals (Barnea, G. *et al.*, 2004; Strotmann, J. *et al.*, 2004) as well as in cells of the cribriform mesenchyme, which are involved in fasciculation and sorting of outgrowing axons (Schwarzenbacher, K. *et al.*, 2006). A point mutation in the OR protein results in the formation of novel glomeruli, indicating that the OR amino acid sequence encodes information about axonal identity (Feinstein, P. and Mombaerts, P., 2004). The OR, however, is not

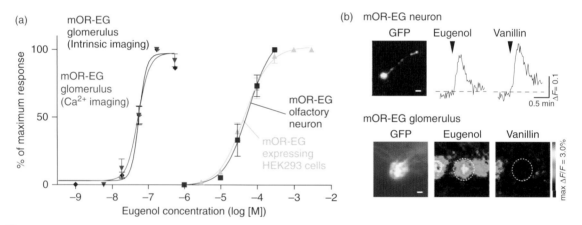

Figure 13 Different odorant sensitivity and specificity between olfactory neurons expressing mOR-EG and the corresponding mOR-EG glomerulus. (a) Dose–response analysis of mOR-EG *in vivo* and *in vitro*. The mOR-EG glomerulus exhibits higher sensitivity compared to the response properties determined in HEK293 cells or in mOR-EG-expressing neurons. (b) Odorant responses in mOR-EG-expressing olfactory neurons (upper panel) and in mOR-EG glomerulus (lower panel). The mOR-EG-expressing neurons showed responses to vanillin, whereas the mOR-EG glomerulus did not respond to vanillin (see Oka, Y. *et al.*, 2006 for more detail).

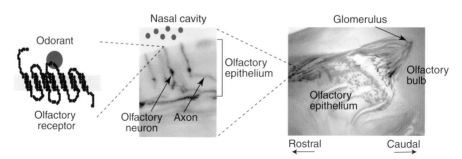

Figure 14 Fluorescent images of the olfactory epithelium and the olfactory bulb of gene-targeted mice expressing LacZ in mOR-EG-expressing olfactory neurons.

the only player in axonal convergence; membrane receptors such as the ephrin receptor and the neuropilin may be coordinately involved in determining the sensory map in the olfactory bulb (Pasterkamp, R. J. *et al.*, 1999; Cutforth, T. *et al.*, 2003). In addition, the instructive role in axon guidance is not restricted to an OR. Also, *β2*-adrenergic receptor can substitute for an OR in glomerular formation, suggesting that GPCRs have common functions (Feinstein, P. *et al.*, 2004). Both ORs and the *β2*-adrenergic receptor couple to the stimulatory G proteins, Gαs, which may account for the common functions of these receptors. In agreement with this, OR-mediated Gαs signaling has been shown to be one of the positional determinants (Imai, T. *et al.*, 2006). Thus, the OR proteins expressed on the olfactory neuronal cilia and dendrites recognize odorants from the external environment, whereas the same OR proteins localized on axon termini appear to function as key mediators of proper axon guidance.

4.30.5.2 Chemosensor

Analogous to axon guidance, wherein olfactory neurons are attracted and directed to a defined glomerulus by the ORs, testicular ORs are likely involved in proper navigation of sperm toward the egg in mammals (Spehr, M. *et al.*, 2006a). In sperm, human OR17-4 recognizes bourgeonal, a floral aldehyde odorant, resulting in induction of chemotaxis and chemokinesis (Spehr, M. *et al.*, 2003). Similarly, MOR23 expressed in mouse sperm recognizes the floral aldehyde odorant lyral, which regulates sperm motility (Fukuda, N. *et al.*, 2004) (Figure 15). Since both bourgeonal and lyral are synthetic fragrances, identification of the endogenous ligands for these ORs is necessary to demonstrate the biological relevance of sperm ORs. Multiple testicular ORs are expressed by various types of spermatids and spermatocytes in different developmental stages in mice, suggesting that some ORs are involved in spermatogenesis rather than in sperm physiology (Fukuda, N. and Touhara, K., 2006) (Figure 5). In any event, ORs may function as chemosensors in nonolfactory tissues.

4.30.5.3 Pheromone Detector

Pheromones are substances that are utilized for intraspecies communication (Karlson, P. and Luscher, M., 1959; Stowers, L. and Marton, T. F., 2005; Touhara, K., in press). In mice, pheromones are detected by two olfactory systems: the main olfactory system and the vomeronasal system (Halpern, M. and Martinez-

Marcos, A., 2003; Baxi, K. N. *et al.*, 2006; Spehr, M. *et al.*, 2006b). Thus, the ORs in the main olfactory system recognize both odorants and volatile pheromones. In fish, ORs seem to play a role in recognizing putative pheromones such as bile acid, prostaglandin, and steroid derivatives (Friedrich, R. W. and Korsching, S. I., 1998; Sorensen, P. W. and Caprio, J., 1998; Luu, P. *et al.*, 2004). Similarly, some insect ORs are utilized as sex pheromone receptors (Nakagawa, T. *et al.*, 2005). In contrast to mammalian ORs, which function as monomers, formation of heterodimers is required for the function of ORs as pheromone sensors in insects (Nakagawa, T. *et al.*, 2005) (Figure 16). During evolution, each species

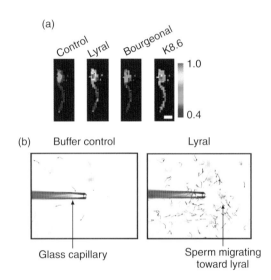

Figure 15 Responses of mouse sperm to lyral, a ligand for MOR23. (a) Pseudocolored image of Ca^{2+} levels in mature mouse sperm responding to lyral and K8.6. (b) Images of mouse sperm that show accumulation around the tip of a glass microcapillary containing lyral. Adapted from Fukuda, N., Yomogida, K., Okabe, M., and Touhara, K. 2004. Functional characterization of a mouse testicular olfactory receptor and its role in chemosensing and in regulation of sperm motility. J. Cell. Sci. 117, 5835–5845.

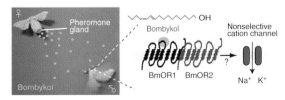

Figure 16 Responses of male silk moths to sex pheromone, bombykol, produced by female moths. Bombykol is recognized by a hetrodimeric complex of two olfactory receptors (ORs), BmOR1 and BmOR2, and elicits nonselective cation channel activity (see Nakagawa, T. *et al.*, 2005 for more detail).

selected an appropriate OR as a pheromone sensor, and the sensor has acquired unique signal transduction mechanisms, which provide high sensitivity and specificity for the target pheromone.

4.30.6 Remaining Problems

The discovery of the OR superfamily and subsequent experimental evidence validated the stereospecific receptor theory, which states that each functional group of an odorant is a molecular determinant for the odorant's interaction with an OR. The fact that the olfactory neuronal circuits follow the one-neuron one-receptor rule and OR-instructed axonal convergence suggests that a combinatorial receptor code for each odorant is transmitted to the olfactory bulb without the information being mixed.

Several questions remain to be answered. For example, how are OR proteins transported to the cilia? What is the mechanism mediating the negative feedback of OR expression? How is the zonal expression pattern formed? Is the presence of olfactory mucus crucial for the olfactory sensitivity and specificity? Is there an endogenous ligand for ORs expressed in nonolfactory tissues? Is the vibrational theory completely dead? How are small molecules such as CO_2 and H_2S recognized? What is the three-dimensional structure of an OR? What are the molecular dynamics of an OR upon ligand binding? Is there a relationship between OR polymorphism and anosmia? What are the differences in neuronal circuits between odorants with physiological effects and those without? Some of these questions will have been explored by the time that this review is in press, but there are always remaining questions because olfaction is the most mysterious of the senses.

Acknowledgments

I would like to thank the members of Touhara laboratory for providing data. This work was supported in part by grants from the Ministry of Education, Science, Sports, and Culture (MEXT), the Japan Society for the Promotion of Science (JSPS), and the Program for Promotion of Basic Research Activities for Innovative Biosciences (PROBRAIN) in Japan.

References

Abaffy, T., Matsunami, H., and Luetje, C. W. 2006. Functional analysis of a mammalian odorant receptor subfamily. J. Neurochem. 97, 1506–1518.

Amoore, J. E. 1963. Stereochemical theory of olfaction. Nature 198, 271–272.

Araneda, R. C., Kini, A. D., and Firestein, S. 2000. The molecular receptive range of an odorant receptor. Nat. Neurosci. 3, 1248–1255.

Asai, H., Kasai, H., Matsuda, Y., Yamazaki, N., Nagawa, F., Sakano, H., and Tsuboi, A. 1996. Genomic structure and transcription of a murine odorant receptor gene: differential initiation of transcription in the olfactory and testicular cells. Biochem. Biophys. Res. Commun. 221, 240–247.

Barnea, G., O'Donnell, S., Mancia, F., Sun, X., Nemes, A., Mendelsohn, M., and Axel, R. 2004. Odorant receptors on axon termini in the brain. Science 304, 1468.

Baxi, K. N., Dorries, K. M., and Eisthen, H. L. 2006. Is the vomeronasal system really specialized for detecting pheromones? Trends Neurosci. 29, 1–7.

Benton, R., Sachse, S., Michnick, S. W., and Vosshall, L. B. 2006. Atypical membrane topology and heteromeric function of Drosophila odorant receptors in vivo. PLoS Biol. 4, e20.

Blache, P., Gros, L., Salazar, G., and Bataille, D. 1998. Cloning and tissue distribution of a new rat olfactory receptor-like (OL2). Biochem. Biophys. Res. Commun. 242, 669–672.

Bozza, T., Feinstein, P., Zheng, C., and Mombaerts, P. 2002. Odorant receptor expression defines functional units in the mouse olfactory system. J. Neurosci. 22, 3033–3043.

Breer, H., Boekhoff, I., and Tareilus, E. 1990. Rapid kinetics of second messenger formation in olfactory transduction. Nature 345, 65–68.

Buck, L. B. 2004. The search for odorant receptors. Cell 116, S117–S119, 111 p following S119.

Buck, L. and Axel, R. 1991. A novel multigene family may encode odorant receptors: a molecular basis for odor recognition. Cell 65, 175–187.

Chess, A., Simon, I., Cedar, H., and Axel, R. 1994. Allelic inactivation regulates olfactory receptor gene expression. Cell 78, 823–834.

Clyne, P. J., Warr, C. G., Freeman, M. R., Lessing, D., Kim, J., and Carlson, J. R. 1999. A novel family of divergent seven-transmembrane proteins: candidate odorant receptors in Drosophila. Neuron 22, 327–338.

Cutforth, T., Moring, L., Mendelsohn, M., Nemes, A., Shah, N. M., Kim, M. M., Frisen, J., and Axel, R. 2003. Axonal ephrin-As and odorant receptors: coordinate determination of the olfactory sensory map. Cell 114, 311–322.

Dhallan, R. S., Yau, K. W., Schrader, K. A., and Reed, R. R. 1990. Primary structure and functional expression of a cyclic nucleotide-activated channel from olfactory neurons. Nature 347, 184–187.

Drutel, G., Arrang, J. M., Diaz, J., Wisnewsky, C., Schwartz, K., and Schwartz, J. C. 1995. Cloning of OL1, a putative olfactory receptor and its expression in the developing rat heart. Receptors Channels 3, 33–40.

Durzynski, L., Gaudin, J. C., Myga, M., Szydlowski, J., Gozdzicka-Jozefiak, A., and Haertle, T. 2005. Olfactory-like receptor cDNAs are present in human lingual cDNA libraries. Biochem. Biophys. Res. Commun. 333, 264–272.

Ehlers, A., Beck, S., Forbes, S. A., Trowsdale, J., Volz, A., Younger, R., and Ziegler, A. 2000. MHC-linked olfactory receptor loci exhibit polymorphism and contribute to extended HLA/OR-haplotypes. Genome Res. 10, 1968–1978.

Eklund, A. C., Belchak, M. M., Lapidos, K., Raha-Chowdhury, R., and Ober, C. 2000. Polymorphisms in the

HLA-linked olfactory receptor genes in the Hutterites. Hum. Immunol. 61, 711–717.

Feingold, E. A., Penny, L. A., Nienhuis, A. W., and Forget, B. G. 1999. An olfactory receptor gene is located in the extended human beta-globin gene cluster and is expressed in erythroid cells. Genomics 61, 15–23.

Feinstein, P. and Mombaerts, P. 2004. A contextual model for axonal sorting into glomeruli in the mouse olfactory system. Cell 117, 817–831.

Feinstein, P., Bozza, T., Rodriguez, I., Vassalli, A., and Mombaerts, P. 2004. Axon guidance of mouse olfactory sensory neurons by odorant receptors and the beta2 adrenergic receptor. Cell 117, 833–846.

Ferrand, N., Pessah, M., Frayon, S., Marais, J., and Garel, J. M. 1999. Olfactory receptors, Golf alpha and adenylyl cyclase mRNA expressions in the rat heart during ontogenic development. J. Mol. Cell. Cardiol. 31, 1137–1142.

Floriano, W. B., Vaidehi, N., and Goddard, W. A., III. 2004. Making sense of olfaction through predictions of the 3-D structure and function of olfactory receptors. Chem. Senses 29, 269–290.

Floriano, W. B., Vaidehi, N., Goddard, W. A., III. Singer, M. S., and Shepherd, G. M. 2000. Molecular mechanisms underlying differential odor responses of a mouse olfactory receptor. Proc. Natl. Acad. Sci. U.S.A. 97, 10712–10716.

Fox, A. N., Pitts, R. J., and Zwiebel, L. J. 2002. A cluster of candidate odorant receptors from the malaria vector mosquito, Anopheles gambiae. Chem. Senses 27, 453–459.

Freitag, J., Krieger, J., Strotmann, J., and Breer, H. 1995. Two classes of olfactory receptors in Xenopus laevis. Neuron 15, 1383–1392.

Friedrich, R. W. and Korsching, S. I. 1998. Chemotopic, combinatorial, and noncombinatorial odorant representations in the olfactory bulb revealed using a voltage-sensitive axon tracer. J. Neurosci. 18, 9977–9988.

Fukuda, N. and Touhara, K. 2006. Developmental expression patterns of testicular olfactory receptor genes during mouse spermatogenesis. Genes Cells 11, 71–81.

Fukuda, N., Yomogida, K., Okabe, M., and Touhara, K. 2004. Functional characterization of a mouse testicular olfactory receptor and its role in chemosensing and in regulation of sperm motility. J. Cell. Sci. 117, 5835–5845.

Gaillard, I., Rouquier, S., Pin, J. P., Mollard, P., Richard, S., Barnabe, C., Demaille, J., and Giorgi, D. 2002. A single olfactory receptor specifically binds a set of odorant molecules. Eur. J. Neurosci. 15, 409–418.

Gao, Q. and Chess, A. 1999. Identification of candidate Drosophila olfactory receptors from genomic DNA sequence. Genomics 60, 31–39.

Gaudin, J. C., Breuils, L., and Haertle, T. 2006. Mouse orthologs of human olfactory-like receptors expressed in the tongue. Gene 381, 42–48.

Gilad, Y., Wiebe, V., Przeworski, M., Lancet, D., and Paabo, S. 2004. Loss of olfactory receptor genes coincides with the acquisition of full trichromatic vision in primates. PLoS Biol. 2, E5.

Glusman, G., Yanai, I., Rubin, I., and Lancet, D. 2001. The complete human olfactory subgenome. Genome Res. 11, 685–702.

Grosmaitre, X., Vassalli, A., Mombaerts, P., Shepherd, G. M., and Ma, M. 2006. Odorant responses of olfactory sensory neurons expressing the odorant receptor MOR23: a patch clamp analysis in gene-targeted mice. Proc. Natl. Acad. Sci. U. S. A. 103, 1970–1975.

Hall, S. E., Floriano, W. B., Vaidehi, N., and Goddard, W. A., III. 2004. Predicted 3-D structures for mouse I7 and rat I7 olfactory receptors and comparison of predicted odor recognition profiles with experiment. Chem. Senses 29, 595–616.

Halpern, M. and Martinez-Marcos, A. 2003. Structure and function of the vomeronasal system: an update. Prog. Neurobiol. 70, 245–318.

Hill, C. A., Fox, A. N., Pitts, R. J., Kent, L. B., Tan, P. L., Chrystal, M. A., Cravchik, A., Collins, F. H., Robertson, H. M., and Zwiebel, L. J. 2002. G protein-coupled receptors in Anopheles gambiae. Science 298, 176–178.

Hummel, P., Vaidehi, N., Floriano, W. B., Hall, S. E., and Goddard, W. A., III. 2005. Test of the binding threshold hypothesis for olfactory receptors: explanation of the differential binding of ketones to the mouse and human orthologs of olfactory receptor 912-93. Protein Sci. 14, 703–710.

Imai, T., Suzuki, M., and Sakano, H. 2006. Roles of G protein-mediated cAMP signals in positioning glomeruli in the mouse olfactory bulb. Science 314, 657–661.

Ishii, T., Serizawa, S., Kohda, A., Nakatani, H., Shiroishi, T., Okumura, K., Iwakura, Y., Nagawa, F., Tsuboi, A., and Sakano, H. 2001. Monoallelic expression of the odourant receptor gene and axonal projection of olfactory sensory neurones. Genes Cells 6, 71–78.

Iwema, C. L., Fang, H., Kurtz, D. B., Youngentob, S. L., and Schwob, J. E. 2004. Odorant receptor expression patterns are restored in lesion-recovered rat olfactory epithelium. J. Neurosci. 24, 356–369.

Jacquier, V., Prummer, M., Segura, J. M., Pick, H., and Vogel, H. 2006. Visualizing odorant receptor trafficking in living cells down to the single-molecule level. Proc. Natl. Acad. Sci. U.S.A.

Kajiya, K., Inaki, K., Tanaka, M., Haga, T., Kataoka, H., and Touhara, K. 2001. Molecular bases of odor discrimination: Reconstitution of olfactory receptors that recognize overlapping sets of odorants. J. Neurosci. 21, 6018–6025.

Kaluza, J. F., Gussing, F., Bohm, S., Breer, H., and Strotmann, J. 2004. Olfactory receptors in the mouse septal organ. J. Neurosci. Res. 76, 442–452.

Karlson, P. and Luscher, M. 1959. Pheromones: a new term for a class of biologically active substances. Nature 183, 55–56.

Katada, S., Hirokawa, T., Oka, Y., Suwa, M., and Touhara, K. 2005. Structural basis for a broad but selective ligand spectrum of a mouse olfactory receptor: mapping the odorant-binding site. J. Neurosci. 25, 1806–1815.

Katada, S., Nakagawa, T., Kataoka, H., and Touhara, K. 2003. Odorant response assays for a heterologously expressed olfactory receptor. Biochem. Biophys. Res. Commun. 305, 964–969.

Katada, S., Tanaka, M., and Touhara, K. 2004. Structural determinants for membrane trafficking and G protein selectivity of a mouse olfactory receptor. J. Neurochem. 90, 1453–1463.

Krautwurst, D., Yau, K. W., and Reed, R. R. 1998. Identification of ligands for olfactory receptors by functional expression of a receptor library. Cell 95, 917–926.

Krieger, J., Klink, O., Mohl, C., Raming, K., and Breer, H. 2003. A candidate olfactory receptor subtype highly conserved across different insect orders. J. Comp. Physiol. A Neuroethol. Sens. Neural Behav. Physiol. 189, 519–526.

Krieger, J., Raming, K., Dewer, Y. M., Bette, S., Conzelmann, S., and Breer, H. 2002. A divergent gene family encoding candidate olfactory receptors of the moth Heliothis virescens. Eur. J. Neurosci. 16, 619–628.

Lane, R. P., Cutforth, T., Young, J., Athanasiou, M., Friedman, C., Rowen, L., Evans, G., Axel, R., Hood, L., and Trask, B. J. 2001. Genomic analysis of orthologous mouse and human olfactory receptor loci. Proc. Natl. Acad. Sci. U.S.A. 98, 7390–7395.

Larsson, M. C., Domingos, A. I., Jones, W. D., Chiappe, M. E., Amrein, H., and Vosshall, L. B. 2004. Or83b encodes a broadly expressed odorant receptor essential for Drosophila olfaction. Neuron 43, 703–714.

Levasseur, G., Persuy, M. A., Grebert, D., Remy, J. J., Salesse, R., and Pajot-Augy, E. 2003. Ligand-specific dose-response of heterologously expressed olfactory receptors. Eur. J. Biochem. 270, 2905–2912.

Lewcock, J. W. and Reed, R. R. 2004. A feedback mechanism regulates monoallelic odorant receptor expression. Proc. Natl. Acad. Sci. U.S.A. 101, 1069–1074.

Lin da, Y., Shea, S. D., and Katz, L. C. 2006. Representation of natural stimuli in the rodent main olfactory bulb. Neuron 50, 937–949.

Liu, A. H., Zhang, X., Stolovitzky, G. A., Califano, A., and Firestein, S. J. 2003. Motif-based construction of a functional map for mammalian olfactory receptors. Genomics 81, 443–456.

Lomvardas, S., Barnea, G., Pisapia, D. J., Mendelsohn, M., Kirkland, J., and Axel, R. 2006. Interchromosomal interactions and olfactory receptor choice. Cell 126, 403–413.

Lucretius, L. 1995. *On the Nature of Things: De rerum natura.* The Johns Hopkins Univ. Pr., Anthony M. Esolen, transl. Baltimore.

Luu, P., Acher, F., Bertrand, H. O., Fan, J., and Ngai, J. 2004. Molecular determinants of ligand selectivity in a vertebrate odorant receptor. J. Neurosci. 24, 10128–10137.

Malnic, B., Hirono, J., Sato, T., and Buck, L. B. 1999. Combinatorial receptor codes for odors. Cell 96, 713–723.

Man, O., Gilad, Y., and Lancet, D. 2004. Prediction of the odorant binding site of olfactory receptor proteins by human-mouse comparisons. Protein Sci. 13, 240–254.

Matarazzo, V., Clot-Faybesse, O., Marcet, B., Guiraudie-Capraz, G., Atanasova, B., Devauchelle, G., Cerutti, M., Etievant, P., and Ronin, C. 2005. Functional characterization of two human olfactory receptors expressed in the baculovirus Sf9 insect cell system. Chem. Senses 30, 195–207.

McClintock, T. S. and Sammeta, N. 2003. Trafficking prerogatives of olfactory receptors. Neuroreport 14, 1547–1552.

Miyamichi, K., Serizawa, S., Kimura, H. M., and Sakano, H. 2005. Continuous and overlapping expression domains of odorant receptor genes in the olfactory epithelium determine the dorsal/ventral positioning of glomeruli in the olfactory bulb. J. Neurosci. 25, 3586–3592.

Mombaerts, P. 2004. Genes and ligands for odorant, vomeronasal and taste receptors. Nat. Rev. Neurosci. 5, 263–278.

Mombaerts, P., Wang, F., Dulac, C., Chao, S. K., Nemes, A., Mendelsohn, M., Edmondson, J., and Axel, R. 1996. Visualizing an olfactory sensory map. Cell 87, 675–686.

Nakagawa, T., Sakurai, T., Nishioka, T., and Touhara, K. 2005. Insect sex-pheromone signals mediated by specific combinations of olfactory receptors. Science 307, 1638–1642.

Nakamura, T. and Gold, G. H. 1987. A cyclic nucleotide-gated conductance in olfactory receptor cilia. Nature 325, 442–444.

Nef, P., Hermans-Borgmeyer, I., Artieres-Pin, H., Beasley, L., Dionne, V. E., and Heinemann, S. F. 1992. Spatial pattern of receptor expression in the olfactory epithelium. Proc. Natl. Acad. Sci. U. S. A. 89, 8948–8952.

Neuhaus, E. M., Gisselmann, G., Zhang, W., Dooley, R., Stortkuhl, K., and Hatt, H. 2005. Odorant receptor heterodimerization in the olfactory system of *Drosophila melanogaster*. Nat. Neurosci. 8, 15–17.

Ngai, J., Chess, A., Dowling, M. M., Necles, N., Macagno, E. R., and Axel, R. 1993. Coding of olfactory information: topography of odorant receptor expression in the catfish olfactory epithelium. Cell 72, 667–680.

Niimura, Y. and Nei, M. 2005a. Comparative evolutionary analysis of olfactory receptor gene clusters between humans and mice. Gene 346, 13–21.

Niimura, Y. and Nei, M. 2005b. Evolutionary changes of the number of olfactory receptor genes in the human and mouse lineages. Gene 346, 23–28.

Niimura, Y. and Nei, M. 2005c. Evolutionary dynamics of olfactory receptor genes in fishes and tetrapods. Proc. Natl. Acad. Sci. U.S.A. 102, 6039–6044.

Norlin, E. M., Alenius, M., Gussing, F., Hagglund, M., Vedin, V., and Bohm, S. 2001. Evidence for gradients of gene expression correlating with zonal topography of the olfactory sensory map. Mol. Cell. Neurosci. 18, 283–295.

Oka, Y., Katada, S., Omura, M., Suwa, M., Yoshihara, Y., and Touhara, K. 2006. Functional odorant-receptor map in the mouse olfactory bulb: in vivo sensitivity and specificity. Neuron 52, 857–869.

Oka, Y., Nakamura, A., Watanabe, H., and Touhara, K. 2004a. An odorant derivative as an antagonist for an olfactory receptor. Chem Senses 29, 815–822.

Oka, Y., Omura, M., Kataoka, H., and Touhara, K. 2004b. Olfactory receptor antagonism between odorants. EMBO J. 23, 120–126.

Omura, M., Sekine, H., Shimizu, T., Kataoka, H., and Touhara, K. 2003. In situ Ca^{2+} imaging of odor responses in a coronal olfactory epithelium slice. Neuroreport 14, 1123–1127.

Otaki, J. M., Yamamoto, H., and Firestein, S. 2004. Odorant receptor expression in the mouse cerebral cortex. J. Neurobiol 58, 315–327.

Pace, U., Hanski, E., Salomon, Y., and Lancet, D. 1985. Odorant-sensitive adenylate cyclase may mediate olfactory reception. Nature 316, 255–258.

Pajot-Augy, E., Crowe, M., Levasseur, G., Salesse, R., and Connerton, I. 2003. Engineered yeasts as reporter systems for odorant detection. J. Recept. Signal Transduct. Res. 23, 155–171.

Parmentier, M., Libert, F., Schurmans, S., Schiffmann, S., Lefort, A., Eggerickx, D., Ledent, C., Mollereau, C., Gerard, C., Perret, J. et al. 1992. Expression of members of the putative olfactory receptor gene family in mammalian germ cells. Nature 355, 453–455.

Pasterkamp, R. J., Ruitenberg, M. J., and Verhaagen, J. 1999. Semaphorins and their receptors in olfactory axon guidance. Cell. Mol. Biol. (Noisy-le-grand) 45, 763–779.

Pilpel, Y. and Lancet, D. 1999. The variable and conserved interfaces of modeled olfactory receptor proteins. Protein Sci. 8, 969–977.

Qasba, P. and Reed, R. R. 1998. Tissue and zonal-specific expression of an olfactory receptor transgene. J. Neurosci. 18, 227–236.

Quignon, P., Giraud, M., Rimbault, M., Lavigne, P., Tacher, S., Morin, E., Retout, E., Valin, A. S., Lindblad-Toh, K., Nicolas, J., and Galibert, F. 2005. The dog and rat olfactory receptor repertoires. Genome Biol. 6, R83.

Ressler, K. J., Sullivan, S. L., and Buck, L. B. 1993. A zonal organization of odorant receptor gene expression in the olfactory epithelium. Cell 73, 597–609.

Ressler, K. J., Sullivan, S. L., and Buck, L. B. 1994. Information coding in the olfactory system: evidence for a stereotyped and highly organized epitope map in the olfactory bulb. Cell 79, 1245–1255.

Robertson, H. M., Warr, C. G., and Carlson, J. R. 2003. Molecular evolution of the insect chemoreceptor gene superfamily in *Drosophila melanogaster*. Proc. Natl. Acad. Sci. U. S. A. 100 (Suppl. 2), 14537–14542.

Saito, H., Kubota, M., Roberts, R. W., Chi, Q., and Matsunami, H. 2004. RTP family members induce functional expression of mammalian odorant receptors. Cell 119, 679–691.

Schwarzenbacher, K., Fleischer, J., and Breer, H. 2005. Formation and maturation of olfactory cilia monitored by odorant receptor-specific antibodies. Histochem. Cell Biol. 123, 419–428.

Schwarzenbacher, K., Fleischer, J., and Breer, H. 2006. Odorant receptor proteins in olfactory axons and in cells of the cribriform mesenchyme may contribute to fasciculation and sorting of nerve fibers. Cell Tissue Res. 323, 211–219.

Serizawa, S., Ishii, T., Nakatani, H., Tsuboi, A., Nagawa, F., Asano, M., Sudo, K., Sakagami, J., Sakano, H., Ijiri, T., Matsuda, Y., Suzuki, M., Yamamori, T., and Iwakura, Y. 2000. Mutually exclusive expression of odorant receptor transgenes. Nat. Neurosci. 3, 687–693.

Serizawa, S., Miyamichi, K., Nakatani, H., Suzuki, M., Saito, M., Yoshihara, Y., and Sakano, H. 2003. Negative feedback regulation ensures the one receptor–one olfactory neuron rule in mouse. Science 302, 2088–2094.

Serizawa, S., Miyamichi, K., and Sakano, H. 2004. One neuron–one receptor rule in the mouse olfactory system. Trends Genet. 20, 648–653.

Shirokova, E., Schmiedeberg, K., Bedner, P., Niessen, H., Willecke, K., Raguse, J. D., Meyerhof, W., and Krautwurst, D. 2005. Identification of specific ligands for orphan olfactory receptors. G protein-dependent agonism and antagonism of odorants. J. Biol. Chem. 280, 11807–11815.

Shykind, B. M. 2005. Regulation of odorant receptors: one allele at a time. Hum. Mol. Genet. 14 (Spec No. 1), R33–R39.

Shykind, B. M., Rohani, S. C., O'Donnell, S., Nemes, A., Mendelsohn, M., Sun, Y., Axel, R., and Barnea, G. 2004. Gene switching and the stability of odorant receptor gene choice. Cell 117, 801–815.

Singer, M. S. 2000. Analysis of the molecular basis for octanal interactions in the expressed rat 17 olfactory receptor. Chem. Senses 25, 155–165.

Singer, M. S. and Shepherd, G. M. 1994. Molecular modeling of ligand–receptor interactions in the OR5 olfactory receptor. Neuroreport 5, 1297–1300.

Singer, M. S., Weisinger-Lewin, Y., Lancet, D., and Shepherd, G. M. 1996. Positive selection moments identify potential functional residues in human olfactory receptors. Receptors Channels 4, 141–147.

Sklar, P. B., Anholt, R. R., and Snyder, S. H. 1986. The odorant-sensitive adenylate cyclase of olfactory receptor cells. Differential stimulation by distinct classes of odorants. J. Biol. Chem. 261, 15538–15543.

Sorensen, P. W. and Caprio, J. 1998. Chemoreception. In: The Physiology of Fishes (ed. D. H. Evans), pp. 375–405. CRC Press.

Spehr, M., Gisselmann, G., Poplawski, A., Riffell, J. A., Wetzel, C. H., Zimmer, R. K., and Hatt, H. 2003. Identification of a testicular odorant receptor mediating human sperm chemotaxis. Science 299, 2054–2058.

Spehr, M., Schwane, K., Riffell, J. A., Zimmer, R. K., and Hatt, H. 2006a. Odorant receptors and olfactory-like signaling mechanisms in mammalian sperm. Mol. Cell. Endocrinol. 250, 128–136.

Spehr, M., Spehr, J., Ukhanov, K., Kelliher, K. R., Leinders-Zufall, T., and Zufall, F. 2006b. Parallel processing of social signals by the mammalian main and accessory olfactory systems. Cell. Mol. Life Sci. 63, 1476–1484.

Stowers, L. and Marton, T. F. 2005. What is a pheromone? Mammalian pheromones reconsidered. Neuron 46, 699–702.

Strader, C. D., Fong, T. M., Tota, M. R., and Underwood, D. 1994. Structure and function of G protein-coupled receptors. Annu. Rev. Biochem. 63, 101–132.

Strotmann, J., Hoppe, R., Conzelmann, S., Feinstein, P., Mombaerts, P., and Breer, H. 1999. Small subfamily of olfactory receptor genes: structural features, expression pattern and genomic organization. Gene 236, 281–291.

Strotmann, J., Levai, O., Fleischer, J., Schwarzenbacher, K., and Breer, H. 2004. Olfactory receptor proteins in axonal processes of chemosensory neurons. J. Neurosci. 24, 7754–7761.

Strotmann, J., Wanner, I., Helfrich, T., Beck, A., and Breer, H. 1994. Rostro-caudal patterning of receptor-expressing olfactory neurones in the rat nasal cavity. Cell Tissue Res. 278, 11–20.

Strotmann, J., Wanner, I., Helfrich, T., and Breer, H. 1995. Receptor expression in olfactory neurons during rat development: in situ hybridization studies. Eur. J. Neurosci. 7, 492–500.

Strotmann, J., Wanner, I., Krieger, J., Raming, K., and Breer, H. 1992. Expression of odorant receptors in spatially restricted subsets of chemosensory neurones. Neuroreport 3, 1053–1056.

Sullivan, S. L., Adamson, M. C., Ressler, K. J., Kozak, C. A., and Buck, L. B. 1996. The chromosomal distribution of mouse odorant receptor genes. Proc. Natl. Acad. Sci. U.S.A. 93, 884–888.

Sullivan, S. L., Bohm, S., Ressler, K. J., Horowitz, L. F., and Buck, L. B. 1995. Target-independent pattern specification in the olfactory epithelium. Neuron 15, 779–789.

Tacher, S., Quignon, P., Rimbault, M., Dreano, S., Andre, C., and Galibert, F. 2005. Olfactory receptor sequence polymorphism within and between breeds of dogs. J. Hered. 96, 812–816.

Tian, H. and Ma, M. 2004. Molecular organization of the olfactory septal organ. J. Neurosci. 24, 8383–8390.

Touhara, K. 2002. Odor discrimination by G protein-coupled olfactory receptors. Microsc. Res. Tech. 58, 135–141.

Touhara, K. Molecular biology of peptide pheromone production and reception in mice. Adv. Genet. (in press).

Touhara, K., Katada, S., Nakagawa, T., and Oka, Y. 2006. Ligand screening of olfactory receptors. In: G Protein-Coupled Receptors: Structure, Function, and Ligand Screening (eds. T. Haga and S. Takeda), pp. 85–109. CRC press.

Touhara, K., Sengoku, S., Inaki, K., Tsuboi, A., Hirono, J., Sato, T., Sakano, H., and Haga, T. 1999. Functional identification and reconstitution of an odorant receptor in single olfactory neurons. Proc. Natl. Acad. Sci. U. S. A. 96, 4040–4045.

Vaidehi, N., Floriano, W. B., Trabanino, R., Hall, S. E., Freddolino, P., Choi, E. J., Zamanakos, G., and Goddard, W. A., III. 2002. Prediction of structure and function of G protein-coupled receptors. Proc. Natl. Acad. Sci. U. S. A. 99, 12622–12627.

Vanderhaeghen, P., Schurmans, S., Vassart, G., and Parmentier, M. 1993. Olfactory receptors are displayed on dog mature sperm cells. J. Cell Biol. 123, 1441–1452.

Vanderhaeghen, P., Schurmans, S., Vassart, G., and Parmentier, M. 1997a. Molecular cloning and chromosomal mapping of olfactory receptor genes expressed in the male germ line: evidence for their wide distribution in the human genome. Biochem. Biophys. Res. Commun. 237, 283–287.

Vanderhaeghen, P., Schurmans, S., Vassart, G., and Parmentier, M. 1997b. Specific repertoire of olfactory receptor genes in the male germ cells of several mammalian species. Genomics 39, 239–246.

Vassalli, A., Rothman, A., Feinstein, P., Zapotocky, M., and Mombaerts, P. 2002. Minigenes impart odorant receptor-specific axon guidance in the olfactory bulb. Neuron 35, 681–696.

Vassar, R., Chao, S. K., Sitcheran, R., Nunez, J. M., Vosshall, L. B., and Axel, R. 1994. Topographic organization of sensory projections to the olfactory bulb. Cell 79, 981–991.

Vassar, R., Ngai, J., and Axel, R. 1993. Spatial segregation of odorant receptor expression in the mammalian olfactory epithelium. Cell 74, 309–318.

Volz, A., Ehlers, A., Younger, R., Forbes, S., Trowsdale, J., Schnorr, D., Beck, S., and Ziegler, A. 2003. Complex transcription and splicing of odorant receptor genes. J. Biol. Chem. 278, 19691–19701.

Von Dannecker, L. E., Mercadante, A. F., and Malnic, B. 2006. Ric-8B promotes functional expression of odorant receptors. Proc. Natl. Acad. Sci. U. S. A. 103, 9310–9314.

Vosshall, L. B., Amrein, H., Morozov, P. S., Rzhetsky, A., and Axel, R. 1999. A spatial map of olfactory receptor expression in the Drosophila antenna. Cell 96, 725–736.

Walensky, L. D., Roskams, A. J., Lefkowitz, R. J., Snyder, S. H., and Ronnett, G. V. 1995. Odorant receptors and desensitization proteins colocalize in mammalian sperm. Mol. Med. 1, 130–141.

Wang, F., Nemes, A., Mendelsohn, M., and Axel, R. 1998. Odorant receptors govern the formation of a precise topographic map. Cell 93, 47–60.

Weber, M., Pehl, U., Breer, H., and Strotmann, J. 2002. Olfactory receptor expressed in ganglia of the autonomic nervous system. J. Neurosci. Res. 68, 176–184.

Weth, F., Nadler, W., and Korsching, S. 1996. Nested expression domains for odorant receptors in zebrafish olfactory epithelium. Proc. Natl. Acad. Sci. U. S. A. 93, 13321–13326.

Wetzel, C. H., Oles, M., Wellerdieck, C., Kuczkowiak, M., Gisselmann, G., and Hatt, H. 1999. Specificity and sensitivity of a human olfactory receptor functionally expressed in human embryonic kidney 293 cells and *Xenopus laevis* oocytes. J. Neurosci. 19, 7426–7433.

Young, J. M. and Trask, B. J. 2002. The sense of smell: genomics of vertebrate odorant receptors. Hum. Mol. Genet. 11, 1153–1160.

Young, J. M., Shykind, B. M., Lane, R. P., Tonnes-Priddy, L., Ross, J. A., Walker, M., Williams, E. M., and Trask, B. J. 2003. Odorant receptor expressed sequence tags demonstrate olfactory expression of over 400 genes, extensive alternate splicing and unequal expression levels. Genome Biol. 4, R71.

Yuan, T. T., Toy, P., McClary, J. A., Lin, R. J., Miyamoto, N. G., and Kretschmer, P. J. 2001. Cloning and genetic characterization of an evolutionarily conserved human olfactory receptor that is differentially expressed across species. Gene 278, 41–51.

Zhang, X. and Firestein, S. 2002. The olfactory receptor gene superfamily of the mouse. Nat. Neurosci. 5, 124–133.

Zhang, X., Rogers, M., Tian, H., Zou, D. J., Liu, J., Ma, M., Shepherd, G. M., and Firestein, S. J. 2004. High-throughput microarray detection of olfactory receptor gene expression in the mouse. Proc. Natl. Acad. Sci. U.S.A. 101, 14168–14173.

Zhao, H., Ivic, L., Otaki, J. M., Hashimoto, M., Mikoshiba, K., and Firestein, S. 1998. Functional expression of a mammalian odorant receptor. Science 279, 237–242.

Zozulya, S., Echeverri, F., and Nguyen, T. 2001. The human olfactory receptor repertoire. Genome Biol. 2, RESEARCH0018.

4.31 Regulation of Expression of Odorant Receptor Genes

A Tsuboi and H Sakano, University of Tokyo, Tokyo, Japan

Glossary

ISH In situ hybridization
LCR Locus control region
OB Olfactory bulb

OE Olfactory epithelium
OR Odorant receptor
OSN Olfactory sensory neuron

4.31.1 Organization and Structure of the Odorant Receptor Genes

A multigene family encoding odorant receptor (OR) molecules was first identified in rat, which consists of hundreds of related OR genes (Buck, L. and Axel, R., 1991). In humans, only ~350 of the 800 OR genes have an intact open reading frame and are assumed to be functional (Glusman, G. et al., 2001; Malnic, B. et al., 2004). In the mouse, of ~1400 OR sequences, some 20% are pseudogenes (Young, J. M. et al., 2002; Zhang, X. and Firestein, S., 2002; Godfrey, P. A. et al., 2004; Zhang, X. et al., 2004a). The mouse repertoire of ~1100 functional OR genes is the largest multigene family that encompasses ~5% of all genes. Mouse ORs are grouped into ~230 subfamilies, based on the amino acid identities of >40% (Zhang, X. and Firestein, S., 2002). Vertebrate OR genes are divided phylogenetically into two different classes, I and II. The class I genes resemble the fish OR genes, while the class II genes are unique to terrestrial vertebrates. The mouse OR genes are clustered at ~50 different loci spread over almost all chromosomes. Each cluster contains 1–189 OR genes, with an average intergenic distance of ~25 kb. Most OR genes are composed of two to five exons, where usually one exon encodes the entire protein. At least two-thirds of OR genes exhibit multiple transcriptional variants with alternative isoforms of noncoding exons (Young, J. M. et al., 2003). Comparative analysis of the mouse and human OR genes revealed that coding homologies within the clusters are accounted for by recent gene duplication as well as gene conversion among the coding sequences (Nagawa, F. et al., 2002). To maintain the integrity of the domain structure of the olfactory sensory map, gene conversion may have played an active role in maintaining the sequence similarities between the paralogs of the OR genes.

4.31.2 Zone-Specific Expression of Odorant Receptor Genes in the Olfactory Epithelium

In rodents, the olfactory epithelium (OE) can be divided into at least four distinct zones, based on the expression patterns of OR genes (Ressler, K. J. *et al.*, 1993; Vassar, R. *et al.*, 1993; Strotmann, J. *et al.*, 1994a; 1994b). The most dorsomedial zone of the OE is defined as zone 1, according to the nomenclature of Sullivan S. L. *et al.* (1996). Zone 1 olfactory sensory neurons (OSNs) are positive for olfactory medium-chain acyl-CoA synthetase (O-MACS) (Oka, Y. *et al.*, 2003), but negative for olfactory cell adhesion molecule (OCAM) (Yoshihara, Y. *et al.*, 1997). Conversely, nonzone 1 OSNs are negative for O-MACS, but positive for OCAM.

4.31.2.1 Class I Odorant Receptors

The class I ORs were first identified in fishes (Ngai, J. *et al.*, 1993) and later found in amphibians (Freitag, J. *et al.*, 1995). They are assumed to recognize water-soluble odorous ligands. Class I ORs are also found in mammals, including human (Glusman, G. *et al.*, 2001; Malnic, B. *et al.*, 2004) and mouse (Zhang, X. and Firestein, S., 2002; Young, J. M. *et al.*, 2002; Zhang, X. *et al.*, 2004a; Godfrey, P. A. *et al.*, 2004). In the mouse, there are over 150 class I OR genes, classified into 42 subfamilies (Zhang, X. and Firestein, S., 2002; Zhang, X. *et al.*, 2004a). They are all clustered at a single locus on the mouse chromosome 7. A high-throughput microarray and *in situ* hybridization analyses revealed that almost all class I OR genes are expressed in zone 1 (Zhang, X. *et al.*, 2004b; Tsuboi, A. *et al.* 2006). It is possible that they are coordinately regulated by a common locus control region (LCR) and/or promoter sequences.

4.31.2.2 Class II Odorant Receptors

In contrast to class I, class II OR genes account for a large share of the mammalian OR repertoire. ISH analyses with various OR gene probes demonstrated that each OR gene is expressed in a restricted area in the OE. It has been reported that OSNs expressing a given OR gene are confined to one of the four OE zones but are randomly distributed within the zone (Ressler, K. J. *et al.*, 1993; Vassar, R. *et al.*, 1993; Strotmann, J. *et al.*, 1994a; 1994b). This appears to be true for zone 1-specific class I genes (Tsuboi, A.

et al., 2006). However, for most class II OR gene, the expression area is not always confined to one of the four conventional zones (Norlin, E. M. *et al.*, 2001; Iwema, C. L. *et al.*, 2004; Miyamichi, K. *et al.*, 2005). Expression areas appear to be specific to each OR gene and are arranged in an overlapping and continuous manner in the OE. It was demonstrated by DiI retrograde staining experiments that the dorsal/ventral arrangement of glomeruli in the olfactory bulb (OB) is well correlated with the expression areas of corresponding ORs along the dorsomedial/ventrolateral (DM/VL) axis in the OE (Miyamichi, K. *et al.*, 2005). How is the area-specific expression regulated for the OR genes? It has been assumed that zone-specific transcription factors for the OR genes are responsible for the regulation. However, if the expression areas of various ORs are arranged in a continuous and overlapping manner, we may have to consider an alternative mechanism, one that detects a gradient or the relative location in the OE for each OR gene along the DM/VL axis. If this is the case, the choice of the OR genes may be more restricted by the location of the OSN in the OE than what has been thought.

4.31.3 Single Odorant Receptor Expression in Each Olfactory Sensory Neuron

OSN axons expressing the same OR converge to a specific set of glomeruli in the OB. In the mammalian olfactory system, single OR gene choice and OR-instructed axonal projection are fundamental to the conversion of olfactory signals to a topographic map in the OB. Quantitative analysis of ISH led to the suggestion that each OSN expresses a limited member of OR gene, possibly one (Ngai, J. *et al.*, 1993; Ressler, K. J. *et al.*, 1993; Vassar, R. *et al.*, 1993). Subsequent cDNA analyses of a single OSN revealed that only one OR is expressed in each cell (Malnic, B. *et al.*, 1999). The first evidence for the one glomerulus–one receptor rule came from ISH of OB slices detecting OR mRNA at the axon termini within a glomerulus (Ressler, K. J. *et al.*, 1994; Vassar, R. *et al.*, 1994; Tsuboi, A. *et al.*, 1999; 2006). Each OR probe detected a pair of glomeruli, one glomerulus on the lateral and the other on the medial side of the OB. Subsequent studies using the marker-tagged OR gene allowed visualization of individual axons (Mombaerts, P. *et al.*, 1996; Wang, F. *et al.*, 1998). In such animals, all OSN axons expressing a given OR converge on a specific set of glomeruli. Thus, odorant stimuli that have activated OSNs in the OE are converted to a topographic map of activated glomeruli, that is, an odor map in the OB (Uchida, N. *et al.*, 2000).

4.31.4 Allelic Exclusion of Odorant Receptor Genes

Autosomal genes come in two alleles, the paternal and maternal, and most of them are expressed biallelically. However, some are expressed monoallelically, for example, the antigen receptor genes and a few interleukin genes. Allelic exclusion was also observed in the OR gene system by using PCR to detect the polymorphisms in the transcripts from the paternal and maternal alleles of a particular OR gene (Chess, A. *et al.*, 1994). Gene-targeted mice, in which both alleles were differently tagged, have demonstrated the monoallelic expression of the OR gene (Strotmann, J. *et al.*, 2000): among thousands of cells examined, no OSN expressed both markers, simultaneously. Furthermore, the nuclei of OSNs were analyzed by fluorescent *in situ* hybridization (FISH) using OR gene probes (Ishii, T. *et al.*, 2001). RNA- and DNA-FISH were performed to detect the transcriptionally active site and the chromosomal location of the genomic sequence, respectively. In this experiment, DNA-FISH detected two genomic loci in the nuclei, one for the paternal and the other for the maternal allele, while RNA-FISH detected the primary transcripts at only one of the two genomic sites. Together, these results confirmed the monoallelic expression of the OR gene. Monoallelic expression of OR molecules appears to be quite important, not only in recognizing the ligand, but also in projecting axons to a specific site in the OB. If two different OR molecules were expressed from both alleles, this would cause confusion in the bundling and pathfinding of OSN axons.

Exclusion has also been found between the endogenous and the transgenic OR gene alleles, and even among the tandemly linked transgenic alleles with the same coding and promoter sequences (Serizawa, S. *et al.*, 2000). These studies demonstrate that under any circumstance, no more than one OR allele can be expressed in each neuron. Such an unusual mode of gene expression, monoallelic and mutually exclusive, has previously been shown only for the antigen receptor genes of the immune system.

4.31.5 Single Odorant Receptor Gene Expression

4.31.5.1 Positive Regulation

What kind of mechanisms would be possible to maintain the expression of only a single OR gene in each OSN? On the basis of previous studies on other multigene families, three activation mechanisms have been considered for the choice and activation of OR genes (Kratz, E. *et al.*, 2002):

1. DNA recombination, which brings a promoter and the enhancer region into close proximity;
2. gene conversion, which transfers a copy of the gene into the expression cassette; and
3. LCR, which interacts with only one promoter site.

Irreversible DNA changes, that is, recombination and gene conversion, have been attractive explanations for single OR gene expression because of the many parallels between the immune and the olfactory systems. However, two groups have independently cloned mice from postmitotic OSN nuclei, and found that the mice showed no irreversible DNA changes in the OR genes (Eggan, K. *et al.*, 2004; Li, J. *et al.*, 2004). Furthermore, these mice did not exhibit monoclonal expression of the OR in the donor OSN. Thus, the DNA rearrangement models for OR gene activation were finally excluded. Because the gene translocation models appeared unlikely, the third possibility was explored, a *cis*-acting DNA region, for example, LCR, that might regulate the expression of the OR genes. Sequence comparison of the mouse and human genomes revealed a 2 kb homology (H) region far upstream of the *MOR28* cluster (Figure 1) (Nagawa, F. *et al.*, 2002). Deletion of the H region abolished the expression of all OR gene members in the cluster on a yeast artificial chromosome (YAC) transgene (Serizawa, S. *et al.*, 2003). Furthermore, attachment of the H region to the truncated transgenic constructs lacking the upstream sequence restored the expression of the OR gene cluster (Serizawa, S. *et al.*, 2003). Interestingly, when the H region was relocated closer to the cluster, the number of OSNs expressing the proximal OR gene was greatly increased. This probably explains why the H region is located far from the OR gene cluster in the chromosome. If the H region were located too close to the cluster, the proximal OR gene would be activated frequently and the downstream genes would rarely be chosen. These results indicate that the H region is a *cis*-acting LCR that activates the *MOR28* cluster, although it is yet to be studied whether the LCR is commonly

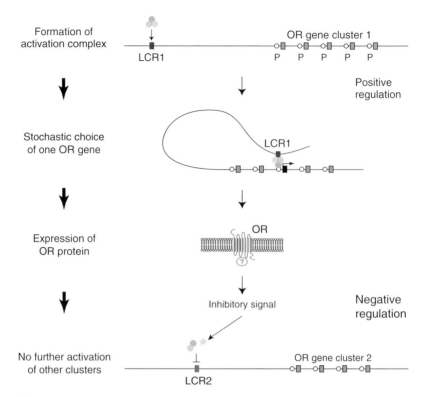

Figure 1   A model for the single odorant receptor (OR) gene expression. It is assumed that the activation complex formed in the locus control region (LCR) stochastically chooses one promoter (P) site by random collision, activating one particular OR gene member within the cluster (positive regulation). Once a functional gene is expressed, the OR molecules may transmit inhibitory signals to block the further activation of additional OR genes or clusters (negative regulation). Stochastic activation of an OR gene and negative feedback regulation by the OR gene product, together, can ensure the maintenance of the one receptor–one neuron rule in the mammalian olfactory system.

found in other OR gene clusters. In some transgenic experiments, a short stretch of DNA containing only the exons of the OR gene and its promoter are sufficient for the transgene to be expressed (Qasba, P. and Reed, R. R., 1998; Vassalli, A. *et al.*, 2002; Lewcock, J. W. and Reed, R. R., 2004). How does this happen? It is possible that these short DNA segments contain an enhancer or have integrated into a transcriptionally active region of chromatin, nearby another enhancer or LCR. It appears that two regulatory regions, LCR and promoter, cooperate in expressing the OR genes. It is possible that the chromatin-remodeling complex formed in the LCR activates the OR gene cluster, and that the LCR–promoter association initiates the transcription of the targeted OR gene (Figure 1).

4.31.5.2   Negative Regulation

It is important to ask how the expression of one particular OR gene is maintained. It appears that the activation complex formed in the LCR interacts

with only one promoter site in the OR gene cluster (Figure 1), as has been reported for other gene systems, for example, the human visual pigment genes (Wang, Y. *et al.*, 1999; Smallwood, P. M. *et al.*, 2002) and globin genes (Li, Q. *et al.*, 1999). However, this would not preclude the activation of a second OR from the other allele or other OR gene clusters. To achieve mutually exclusive expression, negative feedback regulation may be needed to prohibit the further activation of other OR genes or OR gene clusters. It is known that a substantial number of pseudogenes are present in the mammalian OR gene families; ~20% of total OR genes in mouse (Young, J. M. *et al.*, 2002; Zhang, X. and Firestein, S., 2002; Godfrey, P. A. *et al.*, 2004; Zhang, X. *et al.*, 2004a) and ~60% in human are nonfunctional (Glusman, G. *et al.*, 2001; Malnic, B. *et al.*, 2004). In the scenario where the activated LCR has committed to a pseudogene, another LCR must undergo a similar process to ensure the activation of an intact OR gene. Once a functional OR gene has been chosen and expressed, this trial must be halted. It has been proposed that the functional OR proteins may prevent further activation of other OR genes in the cell (Figure 2). Coexpression analysis of OSNs with the naturally occurring frameshift mutants supported the feedback mechanism (Serizawa, S. *et al.*, 2003) (Figure 2). Activation of secondary OR genes was also seen with transgenic OR genes lacking either the entire coding region (Serizawa, S. *et al.*, 2003; Feinstein, P. *et al.*, 2004; Lewcock, J. W. and Reed, R. R., 2004; Shykind, B. M. *et al.*, 2004) or the start codon (Lewcock, J. W. and Reed, R. R., 2004) (Figure 2). These observations suggest that the OR gene product – not mRNA but protein – has a regulatory role in preventing the secondary activation of other OR genes (Figure 2). How is this negative regulation achieved? In the antigen receptor genes, further V(D)J joining ceases as soon as a functional gene is generated, whereas rearrangement continues when the joining is unsuccessful (i.e., out-of-frame). For the Ig heavy-chain genes, the pre-B-cell receptor provides an inhibitory signal via spleen tyrosine kinase (Syk) (Schweighoffer, E. *et al.*, 2003) to

preclude further rearrangement in the heavy-chain locus, possibly by changes in chromatin structure (Karasuyama, H. *et al.*, 1996; Chowdhury, D. and Sen, R., 2004; Goldmit, M. and Bergman, Y., 2004; Mostoslavsky, R. *et al.*, 2004). An analogous feedback regulation is postulated for OR gene expression; however, the exact nature of inhibitory signals and target molecules is yet to be discovered. In the feedback model, attempts to activate functional OR genes can continue in two classes of OSNs: those that have not yet activated any OR gene and those that have activated OR pseudogenes. This prevents a large fraction of developing OSNs from becoming receptorless. Other possible mechanisms have been proposed to ensure the monogenic expression of the OR gene. Among them, the oligogenic model assumes that the OR gene activation occurs stochastically in each OSN according to a Poisson distribution (Mombaerts, P., 2004). In this model, it is assumed that OSNs that do not express an OR cannot survive, and that those expressing two or more OR genes are actively eliminated (negative selection), possibly because of the confusion in bundling and pathfinding.

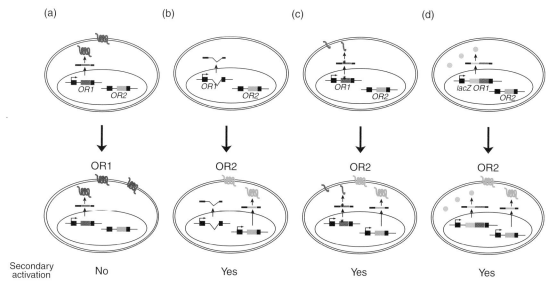

Figure 2   Inhibition of secondary odorant receptor (OR) gene activation by functional OR proteins. (a) Functional OR. When the activated OR gene (*OR1*) expresses functional OR proteins, activation of other OR genes (*OR2*) is inhibited. (b) Coding deletion (no OR). When the entire coding region is deleted, the activated mutant gene (*OR1*) cannot suppress the secondary activation of other OR genes (*OR2*). (c) Frameshift OR (short peptide). From the frameshifted pseudogenes, the full-length mRNA is synthesized. However, such pseudogenes express only short abnormal peptides because of premature stop codons created by frameshift mutations, and permit secondary activation of other OR genes (*OR2*). (d) The mRNA-alone construct (no OR). From this *lacZ–OR1* fusion gene, bicistronic mRNA is synthesized. Owing to the absence of the initiation codon in *OR1*, this mRNA produces only β-galactosidase (β-Gal). Because no OR1 protein is expressed, the secondary activation of other OR genes (*OR2*) is permitted.

4.31.6   *Cis*-Elements and *Trans*-Factors for Odorant Receptor Gene Expression

What kind of regulatory sequences and protein factors are needed for the single OR gene choice and zone-specific expression? Using full-length cDNA and 5′-RACE (rapid amplification of cDNA ends) analyses, conserved promoter sequences have been searched for in the OR gene system. However, no common promoter motifs were found for the 18 OR genes belonging to the *P2* cluster (Lane, R. P. *et al.*, 2001). Recently, based on the expression of transgenic minigenes, putative homeodomain, and Olf-1/EBF (O/E)-like binding sequences were identified in the immediately upstream regions of mouse, rat, and human OR genes including *MOR23*, *M71* (Vassalli, A. *et al.*, 2002), and *mOR37* (Hoppe, P. *et al.*, 2003). The homeodomain sequence was found to interact with a LIM-homeodomain protein, Lhx2 (Hoppe, P. *et al.*, 2003; Hirota, J. and Mombaerts, P., 2004), whereas the O/E-like motif is known to be recognized by O/E-1, -2, -3, and -4 helix–loop–helix transcription factors (Wang, S. S. *et al.*, 1997; 2002). Furthermore, the minimal promoter region for the *M71* gene expression has been defined with the transgenic minigene containing the homeodomain and O/E-like binding sequences (Rothman, A. *et al.*, 2005). Double mutations in both sites abolished the transgene expression. It has been proposed that the homeodomain and O/E sequences regulate the probability of *M71* gene choice differentially across the OE.

Recently, two competing gene loci in naive T helper (Th) cells, one for interleukin (IL)-4 on chromosome 11 and the other for γ interferon on chromosome 10, associate with each other via LCR in the IL-4 locus (Spilianakis, C. G. *et al.*, 2005). Such interchromosomal associations may not be restricted to the Th lymophokine system, but may instead underlie the general phenomenon of alternative gene regulation of multiple genetic loci, for example, the OR gene system. Because there are more than 50 different OR gene clusters scattered on many different chromosomes, the system needs to limit the number of activated clusters, ideally to one, to reduce the likelihood of the simultaneous activation of multiple OR genes. If the OR gene LCR can act interchromosomally like the IL-4 LCR, it would eliminate the need for every OR gene cluster to carry its own LCR (Lomvardas, S. *et al.*, 2006). Although it is yet to be clarified how many LCRs are present in the OR gene system, *trans*-acting LCR can contribute to the mutually exclusive activation of OR gene clusters scattered on many different chromosomes. Spatial organization of active chromatin structures and their cross talk within the nucleus may play key roles in regulating the coordinate expression of competing gene loci on different chromosomes.

4.31.7   Conclusion

In the mammalian olfactory system, each OSN activates only one functional OR gene in a monoallelic and mutually exclusive manner. Such unique expression forms the genetic basis for the OR-instructed axonal projection of OSNs to the OB (Imai, T. *et al.*, 2006; Serizawa, S. *et al.*, 2006). Thus, the one neuron–one receptor rule is essential for the conversion of olfactory signals received in the OE into a topographical map in the OB. One example of LCR has been reported for the mouse OR gene cluster which contains *MOR28*. Like the LCR in other gene systems, physical interaction between the LCR and the promoter probably enables the expression of only one OR gene within the cluster. This model is attractive because it reduces the likelihood of the simultaneous activation of two different OR genes, from a probability among ~1400 genes to that among ~50 loci. Recent transgenic experiments demonstrated that the mutant OR genes lacking either the entire coding sequence or the start codon can permit a second OR gene to be expressed. An inhibitory role has been postulated for the functional OR proteins to prevent the further activation of other OR genes, although the exact mechanism has yet to be elucidated. Stochastic activation of an OR gene by LCR and negative feedback regulation by the OR gene product, together, might ensure the maintenance of the one neuron–one receptor rule in the mammalian olfactory system.

References

Buck, L. and Axel, R. 1991. A novel multigene family may encode odorant receptors: a molecular basis for odor recognition. Cell 65, 175–187.

Chess, A. *et al.* 1994. Allelic inactivation regulates olfactory receptor gene expression. Cell 78, 823–834.

Chowdhury, D. and Sen, R. 2004. Mechanisms for feedback inhibition of the immunoglobulin heavy chain locus. Curr. Opin. Immunol. 16, 235–240.

Eggan, K. *et al.* 2004. Mice cloned from olfactory sensory neurons. Nature 428, 44–49.

Feinstein, P. *et al.* 2004. Axon guidance of mouse olfactory sensory neurons by odorant receptors and the beta2 adrenergic receptor. Cell 117, 833–846.

Freitag, J. *et al.* 1995. Two classes of olfactory receptors in *Xenopus laevis*. Neuron 15, 1383–1392.

Glusman, G. *et al.* 2001. The complete human olfactory subgenome. Genome Res. 11, 685–702.

Godfrey, P. A. *et al.* 2004. The mouse olfactory receptor gene family. Proc. Natl. Acad. Sci. U. S. A. 101, 2156–2161.

Goldmit, M. and Bergman, Y. 2004. Monoallelic gene expression: a repertoire of recurrent themes. Immunol. Rev. 200, 197–214.

Hirota, J. and Mombaerts, P. 2004. The LIM-homeodomain protein Lhx2 is required for complete development of mouse olfactory sensory neurons. Proc. Natl. Acad. Sci. U. S. A. 101, 8751–8755.

Hoppe, P. *et al.* 2003. The clustered olfactory receptor gene family 262: genomic organization, promotor elements, and interacting transcription factors. Genome Res. 13, 2674–2685.

Imai, T. *et al.* 2006. Odorant receptor-derived cAMP signals direct axonal targeting. Science 314, 657–661.

Ishii, T. *et al.* 2001. Monoallelic expression of the odourant receptor gene and axonal projection of olfactory sensory neurones. Genes Cells 6, 71–78.

Iwema, C. L. *et al.* 2004. Odorant receptor expression patterns are restored in lesion-recovered rat olfactory epithelium. J. Neurosci. 24, 356–369.

Karasuyama, H. *et al.* 1996. Surrogate light chain in B cell development. Adv. Immunol. 63, 1–41.

Kratz, E. *et al.* 2002. Odorant receptor gene regulation: implications from genomic organization. Trends Genet. 18, 29–34.

Lane, R. P. *et al.* 2001. Genomic analysis of orthologous mouse and human olfactory receptor loci. Proc. Natl. Acad. Sci. U. S. A. 98, 7390–7395.

Lewcock, J. W. and Reed, R. R. 2004. A feedback mechanism regulates monoallelic odorant receptor expression. Proc. Natl. Acad. Sci. U. S. A. 101, 1069–1074.

Li, Q. *et al.* 1999. Locus control regions: coming of age at a decade plus. Trends Genet. 15, 403–408.

Li, J. *et al.* 2004. Odorant receptor gene choice is reset by nuclear transfer from mouse olfactory sensory. Nature 428, 393–399.

Lomvardas, S. *et al.* 2006. Interchromosomal interactions and olfactory recepter choice. Cell 126, 403–413.

Malnic, B. *et al.* 1999. Combinatorial receptor codes for odors. Cell 96, 713–723.

Malnic, B. *et al.* 2004. The human olfactory receptor gene family. Proc. Natl. Acad. Sci. U. S. A. 101, 2584–2589.

Miyamichi, K. *et al.* 2005. Continuous and overlapping expression domains of odorant receptor genes in the olfactory epithelium determine the dorsal/ventral positioning of glomeruli in the olfactory bulb. J. Neurosci. 25, 3586–3592.

Mombaerts, P. 2004. Odorant receptor gene choice in olfactory sensory neurons: the one receptor–one neuron hypothesis revisited. Curr. Opin. Neurobiol. 14, 31–36.

Mombaerts, P. *et al.* 1996. Visualizing an olfactory sensory map. Cell 87, 675–686.

Mostoslavsky, R. *et al.* 2004. The lingering enigma of the allelic exclusion mechanism. Cell 118, 539–544.

Nagawa, F. *et al.* 2002. Genomic analysis of the murine odorant receptor MOR28 cluster: a possible role of gene conversion in maintaining the olfactory map. Gene 292, 73–80.

Ngai, J. *et al.* 1993. The family of genes encoding odorant receptors in the channel catfish. Cell 72, 657–666.

Norlin, E. M. *et al.* 2001. Evidence for gradients of gene expression correlating with zonal topography of the olfactory sensory map. Mol. Cell. Neurosci. 18, 283–295.

Oka, Y. *et al.* 2003. O-MACS, a novel member of the medium-chain acyl-CoA synthetase family, specifically expressed in the olfactory epithelium in a zone-specific manner. Eur. J. Biochem. 270, 1995–2004.

Qasba, P. and Reed, R. R. 1998. Tissue and zonal-specific expression of an olfactory receptor transgene. J. Neurosci. 18, 227–236.

Ressler, K. J. *et al.* 1993. A zonal organization of odorant receptor gene expression in the olfactory epithelium. Cell 73, 597–609.

Ressler, K. J. *et al.* 1994. Information coding in the olfactory system: evidence for a stereotyped and highly organized epitope map in the olfactory bulb. Cell 79, 1245–1255.

Rothman, A. *et al.* 2005. The promoter of the mouse odorant receptor gene *M71*. Mol. Cell. Neurosci. 28, 535–546.

Schweighoffer, E. *et al.* 2003. Unexpected requirement for ZAP-70 in pre-B cell development and allelic exclusion. Immunity 18, 523–533.

Serizawa, S. *et al.* 2000. Mutually exclusive expression of odorant receptor transgenes. Nat. Neurosci. 3, 687–693.

Serizawa, S. *et al.* 2003. Negative feedback regulation ensures the one receptor–one olfactory neuron rule in mouse. Science 302, 2088–2094.

Serizawa, S. *et al.* 2006. A neuronal identity code for the odorant receptor specific and activity dependent axon sorting. Cell 127, 1057–1069.

Shykind, B. M. *et al.* 2004. Gene switching and the stability of odorant receptor gene choice. Cell 117, 801–815.

Smallwood, P. M. *et al.* 2002. Role of a locus control region in the mutually exclusive expression of human red and green cone pigment genes. Proc. Natl. Acad. Sci. U. S. A. 99, 1008–1011.

Spilianakis, C. G. *et al.* 2005. Interchromosomal associations between alternatively expressed loci. Nature 435, 637–645.

Strotmann, J. *et al.* 1994a. Rostrocaudal patterning of receptor-expressing olfactory neurones in the rat nasal cavity. Cell Tissue Res. 278, 11–20.

Strotmann, J. *et al.* 1994b. Olfactory neurons expressing distinct odorant receptor subtypes are spatially segregated in the nasal epithelium. Cell Tissue Res. 276, 429–438.

Strotmann, J. *et al.* 2000. Local permutations in the glomerular array of the mouse olfactory bulb. J. Neurosci. 20, 6927–6938.

Sullivan, S. L. *et al.* 1996. The chromosomal distribution of mouse odorant receptor genes. Proc. Natl. Acad. Sci. U. S. A. 93, 884–888.

Tsuboi, A. *et al.* 1999. Olfactory neurons expressing closely linked and homologous odorant receptor genes tend to project their axons to neighboring glomeruli on the olfactory bulb. J. Neurosci. 19, 8409–8418.

Tsuboi, A. *et al.* 2006. Olfactory sensory neurons expressing class I odorant receptors converge their axons on an anotero-dorsal domain of the olfactory bulb in the mouse. Eur. J. Neurosci. 23, 1436–1444.

Uchida, N. *et al.* 2000. Odor mps in the mammalian olfactory bulb: domain organization and odorant structural features. Nat. Neurosci. 3, 1035–1043.

Vassalli, A. *et al.* 2002. Minigenes impart odorant receptor-specific axon guidance in the olfactory bulb. Neuron 35, 681–696.

Vassar, R. *et al.* 1993. Spatial segregation of odorant receptor expression in the mammalian olfactory epithelium. Cell 74, 309–318.

Vassar, R. *et al.* 1994. Topographic organization of sensory projections to the olfactory bulb. Cell 79, 981–991.

Wang, S. S. *et al.* 1997. The characterization of the Olf-1/EBF-like HLH transcription factor family: implications in olfactory gene regulation and neuronal development. J. Neurosci. 17, 4149–4158.

Wang, F. *et al.* 1998. Odorant receptors govern the formation of a precise topographic map. Cell 93, 47–60.

Wang, Y. *et al.* 1999. Mutually exclusive expression of human red and green visual pigment-reporter transgenes occurs at high frequency in murine cone photoreceptors. Proc. Natl. Acad. Sci. U. S. A. 96, 5251–5256.

Wang, S. S. *et al.* 2002. Cloning of a novel Olf-1/EBF-like gene, O/E-4, by degenerate oligo-based direct selection. Mol. Cell. Neurosci. 20, 404–414.

Yoshihara, Y. *et al.* 1997. OCAM: A new member of the neural cell adhesion molecule family related to zone-to-zone projection of olfactory and vomeronasal axons. J. Neurosci. 17, 5830–5842.

Young, J. M. *et al.* 2002. Different evolutionary processes shaped the mouse and human olfactory receptor gene families. Hum. Mol. Genet. 11, 535–546.

Young, J. M. *et al.* 2003. Odorant receptor expressed sequence tags demonstrate olfactory expression of over 400 genes, extensive alternative splicing and unequal expression levels. Genome Biol. 4, R71.

Zhang, X. and Firestein, S. 2002. The olfactory receptor gene superfamily of the mouse. Nat. Neurosci. 5, 124–133.

Zhang, X. *et al.* 2004a. Odorant and vomeronasal receptor genes in two mouse genome assemblies. Genomics 83, 802–811.

Zhang, X. *et al.* 2004b. High-throughput microarray detection of olfactory receptor gene expression in the mouse. Proc. Natl. Acad. Sci. U. S. A. 101, 14168–14173.

4.32 Genomics of Odor Receptors in Zebrafish

J Ngai and T S Alioto, University of California, Berkeley, CA, USA

4.32.1 Introduction

The *OR* gene superfamily is the largest multigene superfamily described in mammalian genomes. Surveys of the completed mouse genome sequence have indicated the existence of about 1068 potential intact *OR* genes (comprising at least 228 subfamilies) and 334 pseudogenes (Zhang, X. and Firestein, S., 2002; Zhang, X. *et al.*, 2004). In humans, there are ~350 intact *OR* genes and ~400 pseudogenes (Glusman, G. *et al.*, 2001; Zozulya, S. *et al.*, 2001; Niimura, Y. and Nei, M., 2003; Malnic, B. *et al.*, 2004). By way of contrast, molecular cloning and genomic DNA blot hybridizations in fish species suggest an *OR* repertoire size approximately five- to tenfold smaller than that of mammalian species (Ngai, J. *et al.*, 1993; Barth, A. L. *et al.*, 1997).

An understanding of how vertebrate olfactory receptor repertoires evolved can be gained from comparing the properties and organization of genes from divergent vertebrate species. In this regard, the zebrafish, *Danio rerio*, provides a useful model for comparative genomics studies, owing to intensive efforts to sequence this species' genome as well as a large research community using the zebrafish as a model experimental organism. In early studies, approximately 30 zebrafish *OR* sequences were identified previously using polymerase chain reaction (PCR) and homology-based techniques (Barth, A. L. *et al.*, 1996; Byrd, C. A. *et al.*, 1996; Weth, F. *et al.*, 1996; Barth, A. L. *et al.*, 1997; Vogt, R. G. *et al.*, 1997; Dugas, J. C. and Ngai, J., 2001). Although a number of phylogenetic reconstructions have been made (Freitag, J. *et al.*, 1995; 1998; 1999; Glusman, G. *et al.*, 2000; Dugas, J. C. and Ngai, J., 2001; Aparicio, S. *et al.*, 2002; Zhang, X. and Firestein, S., 2002; Irie-Kushiyama, S. *et al.*,

2004; Zhang, X. *et al.*, 2004), a more accurate view of the OR superfamily's evolutionary history would be facilitated by comparisons between genomic datasets that include a more complete representation of member genes from each species. To this end, genome database mining on the zebrafish genome sequence provided by the Sanger Institute *D. rerio* Sequencing Project has allowed a description of the complete *OR* gene repertoire in this species (Alioto, T. S. and Ngai, J., 2005; Niimura, Y. and Nei, M., 2005). An analysis of these receptor sequences and comparison with other vertebrate OR repertoires provide insights into the evolution of this chemosensory receptor superfamily in vertebrates.

4.32.2 The Zebrafish OR Repertoire

Iterative BLAST searches of the third (Zv3) and fourth (Zv4) draft zebrafish genome assemblies were conducted using previously identified ORs as query sequences (Alioto, T. S. and Ngai, J., 2005) (see also Niimura, Y. and Nei, M., 2005). This search identified 143 intact *OR* genes (136 with no disruptions), 7 partial genes, 10 pseudogenes greater than 700 bp in length, and 15 gene fragments shorter than 700 bp (Table 1). The total *OR* gene count is a conservative estimate of the true size of the OR repertoire, with between 78% (136/175) and 86% ([143 + 7]/175) of identifiable OR sequences consisting of potentially functional *OR* genes.

OR families are typically defined as monophyletic groups, with members that share greater than 40% amino acid identity, whereas subfamily members share greater than 60% amino acid identity (Lancet, D. and Ben-Arie, N., 1993). Using these operational

Table 1   Summary of identified teleost OR genes

	Zebrafish	*Fugu*	*Tetraodon*
Intact genes[a]	143 (7)	44 (3)	42 (6)
Partial genes[b]	7	9	4
Pseudogenes[c]	10	4	11
Total	160	57	57

[a]A gene is operationally defined as intact if it possesses a full-length OR protein coding sequence with no more than one disruption. For each species, the number of genes with a single disruption is listed in parentheses; the remainder contain no disruptions.
[b]A sequence encoding ≤275 contiguous amino acids, missing specific features characteristic of ORs (see the text for details), or missing start or stop codons is classified as a partial gene.
[c]A sequence is defined as a pseudogene if it is a partial gene with one or more disruption or a full-length gene with two or more disruptions.
From Alioto, T.S. and Ngai, J. 2005. The odorant receptor repertoire of teleost fish. BMC Genomics 6–173.

definitions, zebrafish *OR* genes were classified into families and subfamilies by reconstructing their phylogeny using neighbor-joining and maximum likelihood algorithms (Alioto, T. S. and Ngai, J., 2005). Clades of *OR* genes with less than ~40% and less than ~60% interbranch amino acid identity were used to group genes into distinct families and subfamilies, respectively. The average percent identity between families is approximately 25%, while the maximum observed percent identity between any two ORs of different families is 39% (Alioto, T. S. and Ngai, J., 2005).

4.32.3   Genomic Distribution of Zebrafish OR Genes

Previous studies have demonstrated that *OR* genes are clustered in vertebrate genomes (Ben-Arie, N. *et al.*, 1994; Dugas, J. C. and Ngai, J., 2001; Young, J. M. *et al.*, 2002; Zhang, X. and Firestein, S., 2002; Zhang, X. *et al.*, 2004). In mammalian genomes, *OR* genes are distributed widely, residing on 18 chromosomes in the mouse (Zhang, X. and Firestein, S., 2002) and 21 chromosomes in humans (Glusman, G. *et al.*, 2001; Malnic, B. *et al.*, 2004). From the zebrafish Zv3 and Zv4 assemblies, 119 of the identified zebrafish *OR* genes are distributed in five major clusters containing between 14 and 31 genes each (Alioto, T. S. and Ngai, J., 2005). There are two clusters on chromosome 15, two on chromosome 21, one on chromosome 10, several small clusters on chromosomes 8, 14, and 17, and in a few cases, genes exist as singletons (Figure 1). Genomic locations were assignable for ~80% of the *OR* genes identified, with 29 remaining unassigned (Alioto, T. S. and Ngai, J.,

2005). Subfamilies are largely contiguous, and subfamily members usually share the same transcriptional orientation, suggesting tandem duplication as a mechanism of expansion within a subfamily (Dugas, J. C. and Ngai, J., 2001; Kratz, E. *et al.*, 2002). The tight clustering of subfamily members in the zebrafish genome differs from the organization seen in mammalian genomes, which exhibit a higher degree of dispersion of related *OR* genes (Ben-Arie, N. *et al.*, 1994; Dugas, J. C. and Ngai, J., 2001; Young, J. M. *et al.*, 2002; Zhang, X. and Firestein, S., 2002; Zhang, X. *et al.*, 2004).

4.32.4   Phylogeny of Zebrafish OR Genes

A phylogenetic tree of the 143 intact *OR* genes and four full-length pseudogenes was constructed using a neighbor joining algorithm (Figure 2(a); Alioto, T. S. and Ngai, J., 2005). From this information, it is possible to discern the relationships between the OR genes and infer their evolutionary origins. Based on this analysis, zebrafish ORs could be classified into eight families (≥40% intrafamily sequence identity) and 40 subfamilies (≥60% intrasubfamily sequence identity). Most of the gene families contain between 12 and 40 genes each; the two smallest families, family A and family B, contain six genes and one gene each, respectively. The intrasubfamily identity threshold was lowered for three subfamilies, OR102, OR115, and OR125, to generate monophyletic clades. High bootstrap support (Figure 2(a)) justifies these classifications, with all subfamilies exhibiting bootstrap scores of 100%. In addition, the topology of the phylogenetic tree shown in Figure 2 is supported by an independent phylogenetic reconstruction using a maximum likelihood algorithm (Alioto, T. S. and Ngai, J., 2005).

4.32.5   Comparison of the Zebrafish, Mouse, and Pufferfish OR Repertoires

To gain additional insight into how the *OR* gene superfamily evolved in vertebrates, the zebrafish ORs were aligned to the set of OR sequences identified in the mouse genome (Alioto, T. S. and Ngai, J., 2005). Mouse ORs can be classified into two groups, class I and class II, each showing on average greater than 40% intragroup sequence identity (Zhang, X. and Firestein, S., 2002). Based on their greater similarity to the limited number of fish *OR* genes previously identified, class I genes from amphibians and mammals have been

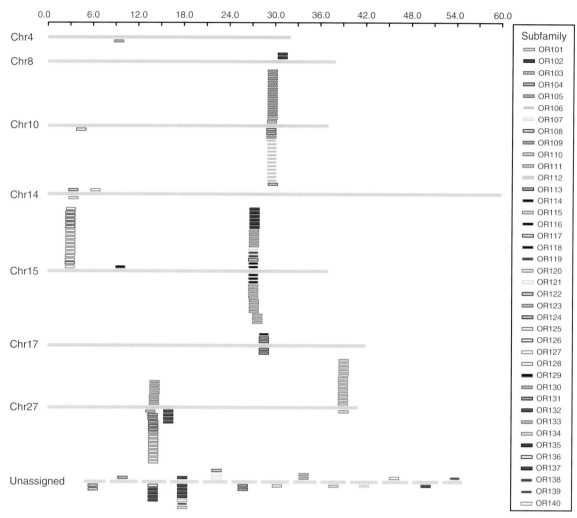

Figure 1 Chromosomal distribution of zebrafish *OR* genes. The majority of OR genes are organized in large clusters at only a few loci in the zebrafish genome. OR genes are depicted as boxes above (plus strand) or below (minus strand) a line representing each chromosome that encodes ORs. Genes are color coded according to subfamily. From Alioto, T.S. and Ngai, J. 2005. The odorant receptor repertoire of teleost fish. BMC Genomics 6, 173.

referred to as fish-like (Freitag, J. *et al.*, 1995; Glusman, G. *et al.*, 2001; Zhang, X. and Firestein, S., 2002). However, an analysis of the complete set of zebrafish *OR* genes indicates that this view cannot be generalized to the entire fish OR repertoire (Alioto, T. S. and Ngai, J., 2005). Mammalian class I and class II genes can in fact be grouped more closely with only two of eight approximately equidistantly related zebrafish families; class I genes show close similarity to only a small subset of zebrafish *OR* genes (*OR112-1*, *OR113-1*, *OR113-2*, and *OR114-1*, which together comprise family A), and one zebrafish gene (*OR101-1*, comprising the single member family B) clusters together with mammalian class II genes (Figure 2(b)). Overall, mouse class I exhibits similar average pairwise identity to the zebrafish

families (27.3 ± 4.8% identity (mean ± standard deviation); range: 17–32%) as mouse class II (27.7 ± 5.5%; range: 18–38%); the difference in mean values is not significant in a two-tailed *t*-test ($p = 0.89$).

A comparison with other teleost *OR* genes further reveals that six of the eight zebrafish OR families overlap with families from two pufferfish species, fugu (*Takifugu rubripes*) and tetraodon (*Tetraodon nigroviridis*) (Table 1 and Figure 2(c); Alioto, T. S. and Ngai, J., 2005). Families B and G do not appear to be present in the two pufferfish genomes. Interestingly, family H is clearly the most divergent family of teleost ORs (Figure 2(c); Alioto, T. S. and Ngai, J., 2005). The location of the outgroup melanocortin receptor between family H and the other families supports the

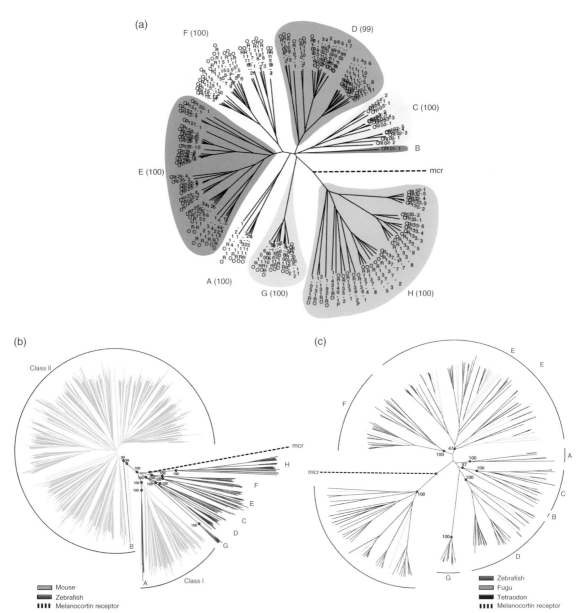

Figure 2   Phylogeny of zebrafish and other vertebrate OR families. (a) Phylogeny of zebrafish receptors. A neighbor-joining tree was constructed based on an alignment of the predicted amino acid sequences of 143 intact genes and four full-length pseudogenes identified from the zebrafish genome. *OR* genes are named by subfamily and colored by family. The eight gene families are labeled A–H. The zebrafish melanocortin receptor branch (dotted line labeled mcr) indicates the root of the tree. Bootstrap scores for each family are indicated in parentheses. (b) Phylogenetic relationship among zebrafish and mouse odorant receptors. The following sets of genes were aligned and used to construct a tree by neighbor joining: the mouse odorant receptors, mORs; the subset of 136 intact zebrafish ORs with no disruptions (highlighted in red); and mouse melanocortin receptors (mcr). Note the presence of *OR* gene subfamilies *OR112*, *OR113*, and *OR114* (family A) within the class I clade and zebrafish *OR101-1* (family B) within the Class II clade. Bootstrap scores corresponding to selected nodes are indicated. (c) Phylogeny of the complete OR repertoires of zebrafish, fugu and tetraodon. One hundred thirty-six nondisrupted genes from zebrafish, 42 nondisrupted genes from fugu, and 42 nondisrupted genes from tetraodon were used in this neighbor-joining analysis. Families are labeled A–H and correspond to the zebrafish families shown in panel (a). Bootstrap scores corresponding to selected nodes are indicated. From Alioto, T.S. and Ngai, J. 2005. The odorant receptor repertoire of teleost fish. BMC Genomics 6, 173.

conclusion that this family is the result of a very ancient gene duplication event. Based on the degree of divergence from other *OR* gene families, it is possible that the genes comprising family H may not in fact encode bona fide odorant receptors. However, family H forms a cluster distinct from non-OR G-protein-coupled receptors (GPCRs) in a phylogenetic tree comprising mouse and zebrafish ORs together with a set of 199 non-OR type I (rhodopsin class) mouse GPCRs (Figure 3). Thus, we consider the family H sequences operationally as *OR* genes. More generally, this phylogenetic reconstruction based on OR and non-OR GPCRs reveals that the ORs as a group are distinct from the other type I GPCRs.

4.32.6 Conserved Motifs in Predicted OR Protein Sequences

Previous studies of vertebrate ORs have identified a number of conserved sequence motifs characteristic of these receptors (Pilpel, Y. and Lancet, D., 1999;

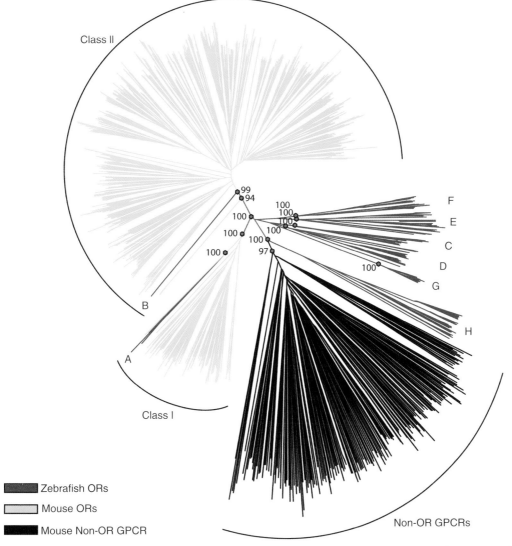

Figure 3 Phylogeny of zebrafish and mouse ORs rooted by mouse non-OR GPCRs. The following sets of genes were aligned: the mouse odorant receptors, mORs; the subset of 136 intact zebrafish ORs with no disruptions; and 199 mouse non-odorant rhodopsin-like GPCRs downloaded from GPCRDB (G-protein-coupled receptor database). A phylogenetic tree was then generated by neighbor joining. The expanded representation of non-OR GPCRs in this phylogeny more clearly demonstrates the segregation of OR and non-OR GPCRs as well as the approximately equal distance of zebrafish families C–H from the mouse class I and class II OR genes. From Alioto, T.S. and Ngai, J. 2005. The odorant receptor repertoire of teleost fish. BMC Genomics 6, 173.

Zozulya, S. *et al.*, 2001; Zhang, X. and Firestein, S., 2002). These include the following: an N-linked glycosylation site NX[TS]X in the N-terminal domain; the motif MA[FY][DE]RYVAIC located at the third transmembrane domain (TM3)/second intracellular loop (IC2) junction which is thought to interact with G-proteins (specifically G$_{olf}$); three conserved cysteine residues in the second extracellular loop (EC2) thought to partake in disulfide bonding; and the motif KAFSTCXSH in IC3 containing an intracellular cysteine conserved in GPCRs and potential phosphorylation sites. These motifs are conserved in all the zebrafish *OR* families, with the exception of

Family H, in which only the MAYDRYVAIC motif is conserved. This sequence conservation is illustrated by a sequence logo generated from the alignment of predicted full-length zebrafish OR coding sequences (Figure 4(a); Alioto, T. S. and Ngai, J., 2005). In this representation, the relative frequency with which an amino acid appears at a given position is reflected by the height of its one-letter amino acid code in the logo, with the total height at a given position proportional to the level of sequence conservation. Interestingly, when compared to the sequence logo representing the alignment of mouse class I and class II ORs (Figure 4(b)), the zebrafish OR logo shows lower conservation

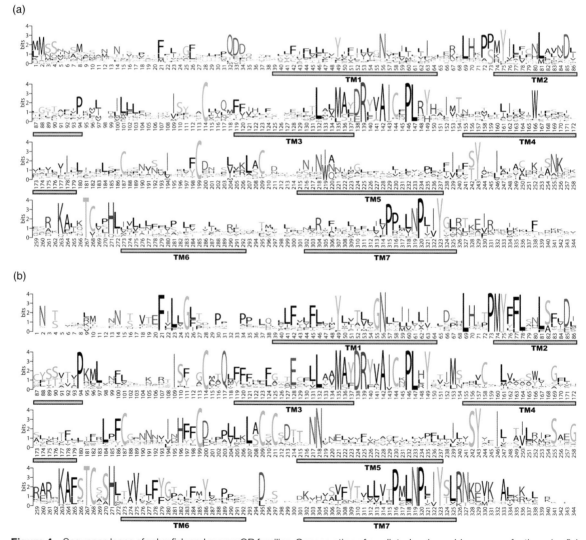

Figure 4   Sequence logos of zebrafish and mouse OR families. Conservation of predicted amino acid sequence for the zebrafish (a) and mouse (b) OR repertoires is shown graphically (see the text). *Y*-axis, information content. *X*-axis, residue position. For this analysis, positions with gaps in more than 95% of sequences, as well as poorly aligned N- and C-terminal sequences, were removed. Positions in the species-specific logos are identical according to this alignment. From Alioto, T.S. and Ngai, J. 2005. The odorant receptor repertoire of teleost fish. BMC Genomics 6, 173.

amongst the predicted zebrafish receptor sequences (reflected by the more numerous and shorter letters at individual positions in the logo), revealing the greater diversity within the zebrafish versus mouse OR superfamily (Alioto, T. S. and Ngai, J., 2005).

4.32.7 Evolution of the Vertebrate OR Gene Repertoire

The characterization of the complete *OR* repertoires from both fish and mammalian species allows an informed analysis of *OR* gene evolution in the vertebrate lineage. From the comparative analysis of mammalian and teleost OR sequences, a model for OR gene evolution in vertebrates can be proposed (Alioto, T. S. and Ngai, J., 2005): *OR* genes in present-day vertebrates likely descended from eight ancestral *OR* genes (or gene families) that existed at the time of the split between ray-finned and lobe-finned fish (the ancestors of teleosts and tetrapods, respectively) approximately 450 million years ago (Hedges, S. B., 2002). A phylogenetic reconstruction based on mouse and zebrafish ORs and 199 mouse non-OR GPCRs (Figure 3) indicates that the ORs form a group distinct from all other type I GPCRs, possibly reflecting a very ancient duplication event(s) and/or rapid divergence of the ORs in the evolution of type I GPCRs. The estimate of ancestral *OR* gene number is based on the identification of eight *OR* gene families in teleosts, two of which show somewhat higher similarity to the two *OR* gene families in mammals. The grouping of zebrafish and pufferfish *OR* genes into common families indicates that the gene duplication events that gave rise to the major OR families probably occurred prior to the speciation of teleosts. In addition, the greater similarities between zebrafish family A and mouse class I, and between zebrafish family B and mouse class II, infer that the ancestral genes for these families existed before the tetrapodon/teleost split. Our model therefore suggests a history during which ancestral genes or gene families were selectively lost during the evolution of the different vertebrate lineages (Figure 5). Of the ancestral families, zebrafish retained eight families, fugu and tetraodon retained six families, and mammals retained two families.

It is also possible that the four to six gene families unique to teleosts descended from family A/class I and/or family B/class II ancestral genes, after the tetrapodon/teleost split. Such a scenario seems unlikely, however, considering the roughly equivalent

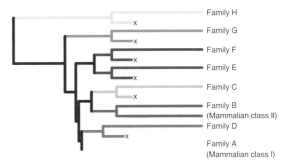

Figure 5   Model for vertebrate *OR* gene family evolution. We hypothesize the existence of a minimum of eight ancestral *OR* genes or gene families at the time of the last common ancestor between the tetrapodon and fish lineages. The phylogeny of OR families can be explained by differential family and subfamily expansion and gene death (X's on mouse lineages). Mammalian class I and class II represent a large expansion of two of these major OR families, each of which also encompasses zebrafish *OR* genes. Branch lengths not drawn to scale.

degree of divergence exhibited between seven of the eight teleost gene families (including family A and family B). Moreover, amphibian and avian OR genes can be grouped into six of the eight identified OR families (family A/class I, family B/class II, family C, family E, family F, and family H), further implicating the presence of common ancestral genes for these families prior to the tetrapodon/teleost split (Niimura, Y. and Nei, M., 2005).

Mechanisms of gene or family loss in a particular vertebrate lineage may have involved a number of processes, for example, gene conversion, pseudogenization of all genes in a family, unequal crossover recombination events during meiosis, or larger chromosomal rearrangements. From the available data, it is not possible to infer the precise order and rate of *OR* gene family expansion and contraction, or speciation events. Nonetheless, six of the retained *OR* gene families were subject to a substantial net expansion and diversification in zebrafish (and to a lesser degree in the pufferfish), while the other two ancestors gave rise to the present-day mammalian class I and class II ORs as well as to a small number of zebrafish genes. We hypothesize that relaxed selective pressure on a subset of the ancestral tetrapodon OR repertoire led to the loss of major *OR* gene families in the mammalian lineage. The expansion within the two remaining gene families was likely driven by the adaptation to the terrestrial odorous environment. Thus, different selective pressures found in the aquatic and terrestrial environments led to different sizes and shapes of the OR repertoires of fish and mammals.

It is generally thought that the diversity of OR sequences – as represented in the number of receptor families – underlies the diversity of chemical structures or odor space that can be detected by an organism's olfactory system. Thus, with approximately six to eight *OR* gene families retained over evolutionary time (vs. two in mammals), fish may be capable of detecting a larger diversity of chemical structures than mammals. However, the larger total number of OR sequences in mammals ($\sim$1000 vs. $\sim$100 in fish) presumably allows a finer discrimination amongst the compounds that are detected by the mammalian olfactory system.

References

Alioto, T. S. and Ngai, J. 2005. The odorant receptor repertoire of teleost fish. BMC Genomics 6, 173.

Aparicio, S., Chapman, J., Stupka, E., Putnam, N., Chia, J. M., Dehal, P., Christoffels, A., Rash, S., Hoon, S., Smit, A., *et al.* 2002. Whole-genome shotgun assembly and analysis of the genome of *Fugu rubripes*. Science 297, 1301–1310.

Barth, A. L., Dugas, J. C., and Ngai, J. 1997. Noncoordinate expression of odorant receptor genes tightly linked in the zebrafish genome. Neuron 19, 359–369.

Barth, A. L., Justice, N. J., and Ngai, J. 1996. Asynchronous onset of odorant receptor expression in the developing zebrafish olfactory system. Neuron 16, 23–34.

Ben-Arie, N., Lancet, D., Taylor, C., Khen, M., Walker, N., Ledbetter, D. H., Carrozzo, R., Patel, K., Sheer, D., Lehrach, H., and North, M. A. 1994. Olfactory receptor gene cluster on human chromosome 17: possible duplication of an ancestral receptor repertoire. Hum. Mol. Genet. 3, 229–235.

Byrd, C. A., Jones, J. T., Quattro, J. M., Rogers, M. E., Brunjes, P. C., and Vogt, R. G. 1996. Ontogeny of odorant receptor gene expression in zebrafish, *Danio rerio*. J. Neurobiol. 29, 445–458.

Dugas, J. C. and Ngai, J. 2001. Analysis and characterization of an odorant receptor gene cluster in the zebrafish genome. Genomics 71, 53–65.

Freitag, J., Beck, A., Ludwig, G., von Buchholtz, L., and Breer, H. 1999. On the origin of the olfactory receptor family: receptor genes of the jawless fish (*Lampetra fluviatilis*). Gene 226, 165–174.

Freitag, J., Krieger, J., Strotmann, J., and Breer, H. 1995. Two classes of olfactory receptors in *Xenopus laevis*. Neuron 15, 1383–1392.

Freitag, J., Ludwig, G., Andreini, I., Rossler, P., and Breer, H. 1998. Olfactory receptors in aquatic and terrestrial vertebrates. J. Comp. Physiol. [A] 183, 635–650.

Glusman, G., Sosinsky, A., Ben-Asher, E., Avidan, N., Sonkin, D., Bahar, A., Rosenthal, A., Clifton, S., Roe, B.,

Ferraz, C., *et al.* 2000. Sequence, structure, and evolution of a complete human olfactory receptor gene cluster. Genomics 63, 227–245.

Glusman, G., Yanai, I., Rubin, I., and Lancet, D. 2001. The complete human olfactory subgenome. Genome Res. 11, 685–702.

Hedges, S. B. 2002. The origin and evolution of model organisms. Nat. Rev. Genet. 3, 838–849.

Irie-Kushiyama, S., Asano-Miyoshi, M., Suda, T., Abe, K., and Emori, Y. 2004. Identification of 24 genes and two pseudogenes coding for olfactory receptors in Japanese loach, classified into four subfamilies: a putative evolutionary process for fish olfactory receptor genes by comprehensive phylogenetic analysis. Gene 325, 123–135.

Kratz, E., Dugas, J. C., and Ngai, J. 2002. Odorant receptor gene regulation: implications from genomic organization. Trends Genet. 18, 29–34.

Lancet, D. and Ben-Arie, N. 1993. Olfactory receptors. Curr. Biol. 3, 668–674.

Malnic, B., Godfrey, P. A., and Buck, L. B. 2004. The human olfactory receptor gene family. Proc. Natl. Acad. Sci. U. S. A. 101, 2584–2589.

Ngai, J., Dowling, M. M., Buck, L., Axel, R., and Chess, A. 1993. The family of genes encoding odorant receptors in the channel catfish. Cell 72, 657–666.

Niimura, Y. and Nei, M. 2003. Evolution of olfactory receptor genes in the human genome. Proc. Natl. Acad. Sci. U. S. A. 100, 12235–12240.

Niimura, Y. and Nei, M. 2005. Evolutionary dynamics of olfactory receptor genes in fishes and tetrapods. Proc. Natl. Acad. Sci. U. S. A. 102, 6039–6044.

Pilpel, Y. and Lancet, D. 1999. The variable and conserved interfaces of modeled olfactory receptor proteins. Protein Sci. 8, 969–977.

Vogt, R. G., Lindsay, S. M., Byrd, C. A., and Sun, M. 1997. Spatial patterns of olfactory neurons expressing specific odor receptor genes in 48-hour-old embryos of zebrafish *Danio rerio*. J. Exp. Biol. 200 (Pt 3), 433–443.

Weth, F., Nadler, W., and Korsching, S. 1996. Nested expression domains for odorant receptors in zebrafish olfactory epithelium. Proc. Natl. Acad. Sci. U. S. A. 93, 13,321–13,326.

Young, J. M., Friedman, C., Williams, E. M., Ross, J. A., Tonnes-Priddy, L., and Trask, B. J. 2002. Different evolutionary processes shaped the mouse and human olfactory receptor gene families. Hum. Mol. Genet. 11, 535–546.

Zhang, X. and Firestein, S. 2002. The olfactory receptor gene superfamily of the mouse. Nat. Neurosci. 5, 124–133.

Zhang, X., Rodriguez, I., Mombaerts, P., and Firestein, S. 2004. Odorant and vomeronasal receptor genes in two mouse genome assemblies. Genomics 83, 802–811.

Zozulya, S., Echeverri, F., and Nguyen, T. 2001. The human olfactory receptor repertoire. Genome Biol. 2, RESEARCH0018.

Relevant Website

http://www.gpcr.org/7tm – G-protein-coupled receptor database.

4.33 Genomics of Invertebrate Olfaction

J D Bohbot, R J Pitts, and L J Zwiebel, Vanderbilt University, Nashville, TN, USA

Glossary

gene A segment of DNA that represents a fundamental unit of heredity.

gene cluster Set of two or more related genes physically located nearby in the genome and encoding similar proteins.

genome The complete genetic content of an organism.

G-protein-coupled receptor (GPCR) Superfamily of seven transmembrane proteins functioning in signal transduction cascades including odorant/taste receptors, photoreceptors, dopamine receptors, acetylcholine receptors, and opioid receptors.

insect Arthropods with three pairs of legs and three body parts.

invertebrate Heterogenous cluster of animals based on their lack of internal skeleton including protozoa, annelids, echinoderms, mollusks, and arthropods.

locus Specific chromosomal location of a gene in the genome.

nematode Commonly known as roundworms based on their worm-like appearance.

odorant Chemical compound stimulating the olfactory system.

odorant-binding protein Small water-soluble proteins thought to bind odorants.

odorant receptor Subclass of membrane proteins belonging to the GPCR superfamily displaying affinity to various odorants.

olfaction Sense of smell.

ortholog Two genes in two separate species that derived from a common ancestral gene by speciation.

sensillum Morphological unit consisting of sensory neurons and accessory cells located on the surface of sensory tissues.

signal transduction Cellular process consisting of converting a signal into another. It is supported by a cascade of molecular components including receptors, enzymes, and ion channels.

synteny Describes the conserved arrangement of genes between related species.

4.33.1 Introduction

The diversity of the animal lineages and their adaptations to various environments are a testament to the formidable natural creativity of life. Within the context of the requirement for mobility, constraints such as gravity have been addressed differently across phyla: while some organisms escape immobility through the use of 2, 4, or 1000 legs, others choose to fly or crawl. In a similar context we may ask the question: how have the challenges of olfaction been answered in phylogenetically diverse phyla?

Olfaction is of fundamental importance in the life cycle and life histories of invertebrates such as insects, crustaceans, or round worms and has been described as the chemosensory modality dedicated to the detection of small airborne chemicals called odorants. Several observations challenge this definition. Aquatic animals use anatomically and molecularly similar chemosensory systems to detect low molecular weight, water-soluble compounds, blurring the distinction between olfaction and gustation. More compelling is the fact that, in general terms, the biochemistry of chemical recognition (chemosensation), that is, the initial physical interactions between the organism olfactory machinery and the environmental stimulus, occurs in a uniformly water-soluble medium. Accordingly, although the distinctive characteristics of gustatory and olfactory processes remain to be determined, the olfactory system must achieve four fundamental requirements that may best be summarized as: speed, tuning breadth, discrimination, and sensitivity. These elements represent the fact that within a few milliseconds, the organism must recognize and discriminate low-concentration odorants amongst a potentially unlimited set of qualitatively different chemical stimuli.

These requirements are even more salient among invertebrates that typically rely on olfaction as their main sensory modality. Experimental expediency and the relative simplicity of invertebrates compared to vertebrates have fostered the study of model organisms such as insects, crustaceans, and nematodes to advance our understanding of the underlying molecular and cellular mechanisms of olfaction. On the surface, the organization and the physiology of the chemosensory systems of arthropods and nematodes vary widely. Are these differences solely a reflection of genetic variability or do they derive from conserved principles? From the early 1980s to late 1990s, several olfactory components had been identified by biochemical and genetic extraction methods. However, the bulk of olfactory genes remained largely uncharacterized especially in the case of highly divergent gene families for which a homology-based approach proved inadequate. The availability of whole genome sequence of *Caenorhabditis elegans* and *Drosophila melanogaster* would change the equation forever (Consortium, C. E. S., 1998; Adams, M. D. *et al.*, 2000) to allow the analysis of large amounts of data and to facilitate the identification of olfactory gene families (Table 1). Importantly, many candidate olfactory genes are still to be functionally validated while others with unknown function remain unidentified. In this context, it is noteworthy that much of the work on invertebrate olfaction has been driven by visual and olfactory system studies in mammals.

What are olfactory genes and how many are there in any given organism? What are their functions? How and when do these gene families interact to generate complex phenotypes? The genomic era is still in its infancy, but it has already opened tremendous opportunities to study these questions. As more genomes become available, comparative genomics will likely reveal how chromosomal organization and rearrangements affect the nascent divergence of phylogenetically related species. Of particular interest is the analysis of orthologous genes across diverse taxa. Although at present, the olfactory genomics of invertebrates is limited to a handful of phylogenetically distant organisms, entire orthologous gene families have nevertheless been identified facilitating the elucidation of function using reverse genetic tools or by functional inference in several instances.

Table 1   Known gene families implicated in olfactory primary coding in *Anopheles gambiae*, *Drosophila melanogaster*, and *Caenorhabditis elegans*

		D. melanogaster	*An. gambiae*	*C. elegans*	References
Receptor genes	*Obp*	33/1	29	ND	a, b
	Csp	4/1*	6	ND	c, d
	Or/Cr	62/37	79/2	1699/1	e, f, g, h, i, j
Gα subunit genes		6/1	6/1	21/5	k, l, m, n, o
Effector genes	*Tm-Gc*	12	ND	28/2	p, q, r
	Tm-Ac	6/1	ND	4	s, t
	Plc	3/1	ND	1/1*	u, v
	Pde	2/1	ND	ND	w, x
Ion channel genes	*Cng*	4/1	ND	6/2	y, z, a', b'
	K channels (eag)	2/1	ND	2	c'
	Trp	4	ND	11/2	d', e', f'
RGS genes	*Arr*	4/2	4/2	2/1	g', h'
	Grk	2	2	2/1	i', j'
	Ip3k	2/1	ND	ND	v
	Itpr	1/1	ND	ND	k'
	nPkc	1	ND	2/2	l'
	Calcineurin	ND	ND	1/1	m'
	Pkg	ND	ND	2/1	n'
	Dgk	ND	ND	5/2	o'
	RdgB	1/1	ND	ND	p'

[a]Hekmat-Scafe D. S. *et al.* (2002); [b]Xu, P. X. *et al.* (2003); Xu, P. *et al.* (2005); [c]Wanner K. W. *et al.* (2004); [d]McKenna M. P. *et al.* (1994); [e]Hill C. A. *et al.* (2002); [f]Vosshall L. B. *et al.* (2000); [g]Hallem E. A. and Carlson J. R. (2006); [h]Robertson H. M. and Thomas J. H. (2006); [i]Sengupta P. *et al.* (1996); [j]Hallem E. A. *et al.* (2004b); [k]Jansen G. *et al.* (1999); [l]Lans H. *et al.* (2004); [m]Zwaal R. R. *et al.* (1997); [n]Rutzler M. R. *et al.* (2006); [o]Roayaie K. *et al.* (1998); [p]Fitzpatrick D. A. *et al.* (2006); [q]L'etoile N. D. and Bargmann C. I. (2000); [r]Birnby D. A. *et al.* (2000); [s]Martin F. *et al.* (2001); [t]Korswagen H. C. *et al.* (1998); [u]Riesgo-Escovar J. *et al.* (1995); [v]Gomez-Diaz C. *et al.* (2006); [w]Gomez-Diaz C. *et al.* (2004); [x]Martin F. *et al.* (2001); [y]Baumann A. *et al.* (1994); [z]Komatsu H. *et al.* (1999); [a']Coburn C. M. and Bargmann, (1996); [b']Kaupp U. P. and Seifert R. (2002); [c']Dubin A. E. *et al.* (1998); [d']Stortkuhl K. F. *et al.* (1999); [e']Tobin D. *et al.* (2002); [f']Colbert H. A. and Bargmann C. I. (1995); [g']Palmitessa A. *et al.* (2005); [h']Merrill C. E. *et al.* (2003; 2005); [i']Cassil J. A. *et al.* (1991); [j']Fukuto H. S. *et al.* (2004); [k']Deshpande M. *et al.* (2000); [l']Okochi Y. *et al.* (2005); [m']Kuhara A. *et al.* (2002); [n']L'etoile N. D. *et al.* (2002); [o']Matsuki M. *et al.* (2006); [p']Riesgo-Escovar J. R. *et al.* (1994).
Numbers on the left indicate genes with putative olfactory functions and numbers on the right indicate functionally characterized genes. Genes with putative olfactory function are indicated by an asterisk* and unidentified family are indicated by ND (nondetermined). RGS, regulator of G protein signaling; Tm, transmembrane.

This chapter will discuss the current state of knowledge of gene families involved in odor reception, olfactory transduction, and adaptation. It will mainly draw from the study of three best-described invertebrate genomes: *D. melanogaster*, *Anopheles gambiae*, and *C. elegans*. We have included a brief comparative description of the cellular and the molecular participants in the invertebrate olfactory organs. A more detailed discussion of invertebrate olfactory systems is described in chapter 22. We structured our discussion to mimic the successive stages of primary olfactory sensing and adaptation, beginning with odor recognition, olfactory transduction and ending with regulation of primary sensory processing, as we review the organization, structure, regulation, and function of olfactory genes in insects and nematodes.

4.33.2 An Overview of Invertebrate Chemosensory Organs and Their Molecular Components

In insects, olfaction is mediated by thousands of primary olfactory receptor neurons (ORNs) segregated by groups of two or more in small olfactory organs called sensilla (Figure 1). The dendrites of these neurons project in perforated hollow cuticular structures of various morphological types and are mainly arrayed on the antennal surface. In *C. elegans*, volatile chemicals are detected by five pairs (AWA, AWB, AWC, ASH, and ADL) of chemoreceptor neurons (CRNs) located in the main sensory organs known as the amphid sensilla (Bargmann, C. I. *et al.*, 1993). These organs sit on each side of the buccal cavity and communicate with the external environment via a prominent pore. Each CRN responds to many different compounds (Bargmann, C. I. and Horvitz, H. R., 1991). Despite overall morphological similarity, it is becoming more and more apparent that morphological

differences result from ecological adaptations rather than based on olfactory-related requirements. In the case of the terrestrial giant robber crab, olfactory receptors (aesthetascs) display morphological and

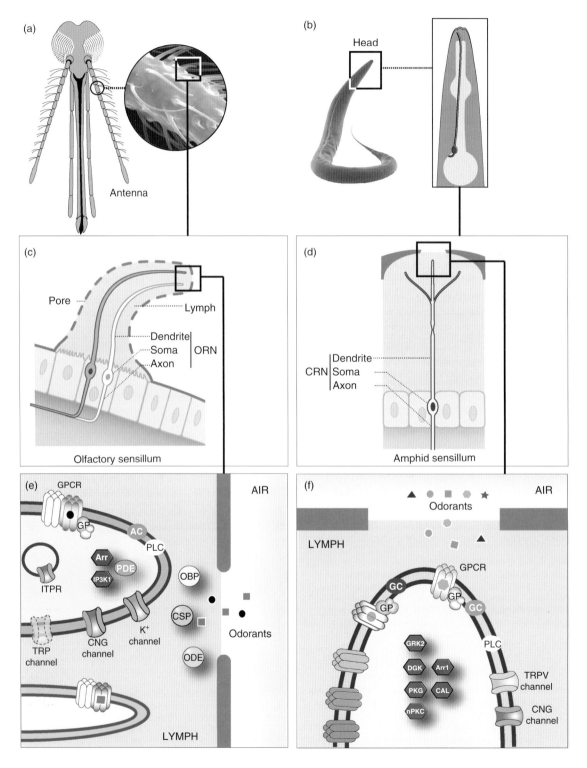

physiological attributes intermediate between marine crustaceans and insects (Stensmyr, M. C. *et al.*, 2005). Drosophila olfactory sensilla exhibit over a dozen morphological subtypes, which do not seemingly relate to olfactory function (Shanbhag, S. R. *et al.*, 1999; Stocker, R. F., 2001).

Despite the diverse morphologies, the partitioned nature of these structures remains invariant between insects, crustaceans, and nematodes and represents a defining hallmark of invertebrates. Thus, it is not surprising that similar adaptive features are reflected by common cellular and molecular strategies. Olfactory sensilla are highly specialized cell clusters encased in the olfactory epithelium such that the ORNs are separated and protected from the environment by an aqueous fluid on its apical side and physiologically supported by accessory cells on its basal side. The physical and biochemical demands imposed by this system require an odorant to cross the extracellular sensillum lymph to reach the molecular receptor located on the surface of the dendritic membrane in order to activate the intracellular transduction machinery. Furthermore, another defining specificity is that insect odorant receptors (ORs) almost certainly require the association with a heterodimer partner homolog (Benton, R. *et al.*, 2006) for proper activation of intracellular downstream effectors such as a heterotrimeric G protein (GP) (Laue, M. *et al.*, 1997) and a cyclic nucleotide-gated (CNG) ion channel (Baumann, A. *et al.*, 1994; Dubin, A. E. and Harris, G. L., 1997).

Studies in cockroach, Drosophila, and lobsters implicate both a GP-coupled adenylyl cyclase (AC) and a GP-coupled phospholipase C (PLC) transduction cascades in olfactory reception (Boekhoff, I. *et al.*, 1990a; 1990b; Boekhoff, I. *et al.*, 1994; Riesgo-Escovar, J. *et al.*, 1995; Deshpande, M. *et al.*, 2000; Martin, F. *et al.*, 2001). Putative targets of these effector genes

include a CNG channel and a cyclic nucleotide-modulated K^+ channel (Baumann, A. *et al.*, 1994; Dubin, A. E. *et al.*, 1998). Signal termination appears to be mediated by several enzymes including a cyclic adenosine monophosphate (cAMP)-phosphodiesterase (cAMP-PDE), an inositol 1,4,5-triphosphate kinase 1 (IP3K1), and two sensory arrestins (ARRs) (Martin, F. *et al.*, 2001; Merrill, C. E. *et al.*, 2002; 2003; Gomez-Diaz, C. *et al.*, 2004; Merrill, C. E. *et al.*, 2005; Gomez-Diaz, C. *et al.*, 2006). Regulation of GP signaling requires a transient receptor potential (TRP) Ca^{2+} channel (Bergamasco, C. and Bazzicalupo, P., 2006) and an inositol 1,4,5-triphosphate receptor (ITPR).

Rapid ORN activation depends on cellular and molecular mechanisms that facilitate the transport and the clearance of biologically active/toxic compounds. Although G-protein-coupled receptors (GPCRs), also known as seven-transmembrane receptors, and their GP partners are invariably used as ORs across diverse phyla, there is significant variation as to how these receptors interact with their molecular environment. Within invertebrate lineages, the main difference between arthropods and nematodes consists of divergent molecular strategies upstream of the GPCRs within the fluid compartment of the sensillum. Insect ORs interact with an extracellular perireceptor pathway that potentially includes various odorant-binding protein (OBP) members, possibly chemosensory proteins (CSPs) and odorant-degrading enzymes (ODEs).

ODEs comprise members of the cytochrome P450s, gluthatione-*S*-transferase and carboxylesterases supergene families that have often been implicated in olfactory signaling based on inferential expression patterns. In particular, in mosquitoes and other insects, ODEs play an important role in insecticide metabolism/resistance (Ranson, H. *et al.*, 2002), and several reports have suggested a possible

Figure 1 Comparison of the olfactory systems of arthropods and nematodes. (a)The antenna of a mosquito and (b) the head of *Caenorhabditis elegans* are magnified in (c) and (d), respectively, to show the cellular organization of insect olfactory sensillum and amphid sensillum. Two odorant receptor neurons (ORNs) and only one of the chemoreceptor neuron (CRN) are shown for clarity purposes. Close up of the dendritic termini are shown in (e) and (f) to show the various molecular components involved in olfactory sensing in both invertebrate systems. The molecular components of odor reception include the bona fide odorant-activated G-protein-coupled receptors (GPCRs). An additional role in odor coding has been proposed for the insect odorant binding proteins (OBPs) and chemosensory proteins (CSPs). Heterotrimeric G proteins (GPs) activate a variety of intracellular signal transduction pathways. Two main intracellular signaling routes are mediated by a phospholipase C (PLC) and adenylyl/guanylyl cyclase (AC and GC). These effector proteins target various receptors and ion channels such as an inositol triphosphate receptor (ITPR) and transient receptor potential (TRP), cyclic nucleotide-gated (CNG) potassium (K^+) channels. Regulation of GP/GPCR signaling is mediated by arrestins (ARR) and G protein-coupled receptor kinases (GRK). Calcineurin (CAL), protein kinases G and C (PKG, nPKC), and diacyl glycerol kinases (DGKs) modulate primary sensing by targeting various components of the transductory cascade. Picture of the *C. elegans* was kindly provided by Z. Altun and D. Hall.

role for ODEs in xenobiotic/odorant clearance based solely on their expression in olfactory tissues (Hovemann, B. T. *et al.*, 1997; Rogers, M. E. *et al.*, 1999; Maibeche-Coisne, M. *et al.*, 2005; Lycett, G. J. *et al.*, 2006). Owing to this paucity of specific data tying these genes to olfactory signaling, an in-depth analysis of ODEs superfamilies has not been included in this chapter. A review of ODEs is available from Vogt R. G. (2003).

For expediency we will focus on the chemosensory gene families in *C. elegans* restricted to the AWA, AWB and AWC neurons, which together mediate chemotaxis to a range of volatile attractants. A detailed review of the full chemical sensitivity in *C. elegans* is available from Bergamasco C. and Bazzicalupo P. (2006). Interestingly, no OBPs, CSPs, or ODEs have been described in nematodes (Rubin, G. M. *et al.*, 2000). The transduction pathways of *C. elegans* and insects are generally similar as they both use GPCRs and GPs (Zwaal, R. R. *et al.*, 1997; Roayaie, K. *et al.*, 1998; Lans, H. *et al.*, 2004) to target, in some cases, homologous ion channels (Colbert, H. A. and Bargmann, C. I., 1995; Tobin, D. *et al.*, 2002). However, *C. elegans* seems to favor cyclic guanosine monophosphate (cGMP) over cAMP as a second messenger (Birnby, D. A. *et al.*, 2000; L'etoile, N. D. and Bargmann, C. I., 2000). Moreover, the regulation of GP signaling in *C. elegans* involves several protein kinases (L'etoile, N. D. *et al.*, 2002; Fukuto, H. S. *et al.*, 2004; Okochi, Y. *et al.*, 2005; Matsuki, M. *et al.*, 2006), a single ARR and calcineurin (Kuhara, A. *et al.*, 2002; Palmitessa, A. *et al.*, 2005).

4.33.3 Odor Recognition

Upon entering the sensillum, an odorant is presumed to transverse the sensillum lymph in order to activate the OR. While this process is the initial component of odor recognition in arthropods and worms, the precise mechanisms and functional roles of molecular players such as OBPs and CSPs remain unclear. Nevertheless, these gene families have been proposed to play significant roles in the general process of perireception, that is, interaction of an odorant with molecular components preceding OR activation. Indeed, *Ors*, *Obps*, and *Csps* are ancient and diverse gene families, which are likely a reflection of their evolutionary value in olfactory signal transduction.

4.33.3.1 Insect Odorant-Binding Proteins

The OBP designation derives from their initial discovery in the silk moth *Antheraea polyphemus* where the first pheromone-binding protein (*Apol*PBP) was shown to bind *in vitro* one of the female sex pheromone component in the aqueous lymph bathing the pheromone-sensitive neuron (Vogt, R. G. and Riddiford, L. M., 1981). Membership in this gene family is contingent on the following four requirements: antennal-specific expression, putative odor-binding capability, overall size of approximately 14 kDa and the presence of six conserved cysteine residues at characteristic positions (six-cysteine motif). OBPs of the lipocalin family have also been described in higher vertebrates but despite their common name with their insect counterparts, both families do not share any similarity at the sequence and at the structural levels suggesting convergent evolution in both lineages (Pelosi, P., 1994; Tegoni, M. *et al.*, 2000). With the growing number of OBP reports in other insect orders, it became apparent that all criteria could not always be met and OBP membership relied primarily on circumstantial functional evidences in the face of relaxed sequence homology. The identification of multiple OBPs in several insect species and the description of a new subfamily of OBPs called general odorant-binding proteins (GOBPs) in Lepidoptera provided good support that OBPs were part of a multigene family with an intrinsic role in perireception (Vogt, R. G. *et al.*, 1989; 1991; Robertson, H. M. *et al.*, 1999). This notion would be strengthened by the identification of *Obp* genes in a wide range of insect species including Coleoptera (Nikonov, A. A. *et al.*, 2002), Hymenoptera (Briand, L. *et al.*, 2001), Diptera (Kim, M. S. *et al.*, 1998), Orthoptera (Picone, D. *et al.*, 2001), Dictyoptera (Riviere, S. *et al.*, 2003), and Heteroptera (Vogt, R. G. *et al.*, 1999). Approximately 20 years after the initial *Apol*PBP characterization, the availability of *D. melanogaster* and *An. gambiae* genomes resulted in an explosion of putative *Obps* when over 50 genes were described based on similar structural features but not necessarily based on similar functional properties (Hekmat-Scafe, D. S. *et al.*, 2002; Xu, P. X. *et al.*, 2003).

4.33.3.1.1 *Comparative evolutionary genomics*

The genomes of *D. melanogaster* and *An. gambiae* contain 49 and 57 potential *Obp* genes, respectively (Hekmat-Scafe, D. S. *et al.*, 2002; Xu, P. X. *et al.*, 2003).

Within this classification, the six-cysteine motif, as a structural standard, has not always been completely adhered to, and *Obp* classification has relied on more broadly based sequence homology strategies using different variations of the BLAST algorithm (Altschul, S. F. *et al.*, 1990). Alignment of deduced OBP sequences indicates a diverse family of proteins with low-average amino acid identity within a given species of approximately 15–20% range (from 5% to 100% and from 9.2% to 62.6% range in *An. gambiae* and *D. melanogaster*, respectively) (Galindo, K. and Smith, D. P., 2001). OBP protein size typically ranges from 14 to 35 kDa and most have a potential peptide signal at their N-terminus. Moreover, this heterogeneous group encompasses several variations of the typical *Obp* gene group. In *D. melanogaster*, the *Obp* gene family includes 29 classic, 14 plus-C, and 6 minus-C genes. In *An. gambiae*, there are 29 typical, 12 plus-C, and 16 atypical *Obp* genes (see description below). The classic *Obp* genes are predicted to encode short 14 kDa proteins with six stereotypically placed cysteine residues and represent the archetype to which all other *Obp* types are compared. Most *Obp* genes include up to three introns that are distributed between the nine conserved positions shown in Figure 2. Generally, the first intron is located immediately downstream of the DNA encoding the putative peptide signal, which is presumably necessary for the extracellular expression of the mature protein. In the majority of cases (>80%), these introns occur between codons (phase 0) (Hekmat-Scafe, D. S. *et al.*, 2002). Classical *Obps* are predicted to encode six α-helical proteins and to form a globular compact structure reinforced by the presence of three disulfide bridges as shown by the structural studies on the pheromone-binding protein (PBP) of *Bombyx mori* (*Bmor*PBP) (Sandler, B. H. *et al.*, 2000).

Although these genes are distributed throughout the genome, most of them are cytogenetically clustered (Figure 3). In the case of *D. melanogaster*, more than half of the *Obp* genes (57%) are located on chromosome *2R*, where 14 *Obp* genes are clustered within a 825-kb stretch that also encompasses OR gene *Or56a* (Galindo, K. and Smith, D. P., 2001; Graham, L. A. and Davies, P. L., 2002; Hekmat-Scafe, D. S. *et al.*, 2002). In contrast, chromosome 2L of *D. melanogaster* carries only one *Obp* gene. *An. gambiae Obp* gene clusters are more dispersed and characteristically smaller. In both species these gene clusters tend to be localized at the tips of chromosome and rather than being organized in tandem arrays, typically occur in forward and backward orientations, suggesting that this family is rapidly evolving via multiple duplication and rearrangement events.

Many sequence analyses have been done on OBPs mainly to understand the structural relationships between members of this highly divergent gene family. All OBPs described to date have been restricted to insects within the Neoptera superorder. They include Lepidoptera, Diptera, Hymenoptera, Coleoptera (holometabolous insects), and one Hemiptera (hemimetabolous insects). Phylogenetic analyses have shown that three subfamilies seem to be monophyletic (i.e., subgroups of OBPs belonging to one insect lineage). The PBP/GOBP subfamily is restricted to the Lepidoptera (Figure 4). Their

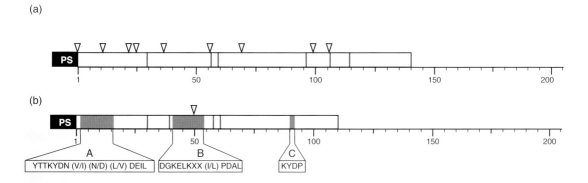

Figure 2 Gene structure and protein motifs of the *Obp* and *Csp* gene families. (a) Odorant-binding protein (OBP): The intron locations (arrowheads) and cysteine residues (vertical red bars) are shown relative to a scale of the average protein size in amino acids. Data from Hekmat-Scafe, D. S., Scafe, C. R., Mckinney, A. J., and Tanouye, M. A. 2002. Genome-wide analysis of the odorant-binding protein gene family in *Drosophila melanogaster*. Genome Res. 12, 1357–1369. (b) Chemosensory protein (CSP): Conserved amino acid motifs A, B, and C are shaded in gray. Data from Wanner, K. W., Willis, L. G., Theilmann, D. A., Isman, M. B., Feng, Q., and Plettner, E. 2004. Analysis of the insect os-d-like gene family. J. Chem. Ecol. 30, 889–911.

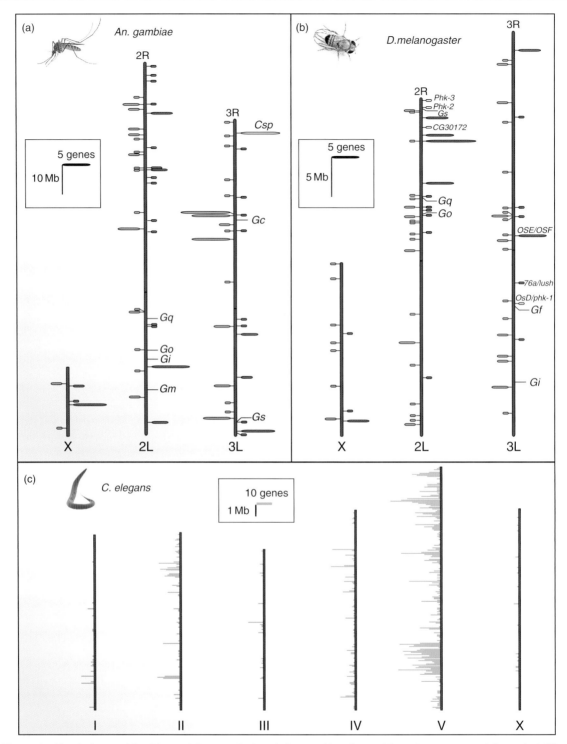

Figure 3 Physical map of *Or, Obp,* and *Cr* genes in *Anopheles gambiae, Drosophila melanogaster,* and *Caenorhabditis elegans.* Genomic locations of *Or* (blue), *Csps* (yellow), and *Obps* (red) are listed on each chromosome of (a) *An. gambiae* and (b) *D. melanogaster* (data from Hill, C. A. *et al.*, 2002; Robertson, H. M. *et al.*, 2003; Xu, P. X. *et al.*, 2003; Hekmat-Scafe, D. S. *et al.*, 2002; Galindo, K. and Smith, D. P., 2001; Zhou, J. J. *et al.*, 2004; Biessmann, H. *et al.*, 2005; Wanner, K. W. *et al.*, 2004). (c) Positions in the entire *C. elegans* genome of all analyzed members of chemoreceptor (*Cr*) families. Adapted from Robertson, H. M. and Thomas, J. H. The putative chemoreceptor families of *C. elegans* (January 6, 2006), WormBook, ed. The *C. elegans* Research Community, WormBook, doi/10.1895/wormbook.1.66.1, http://www.wormbook.org. Picture of the *C. elegans* was kindly provided by Z. Altun and D. Hall.

Odorant-binding protein

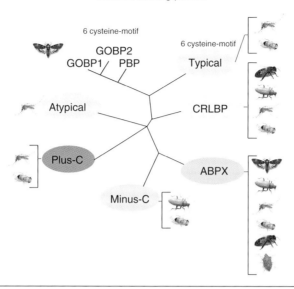

Odorant receptor

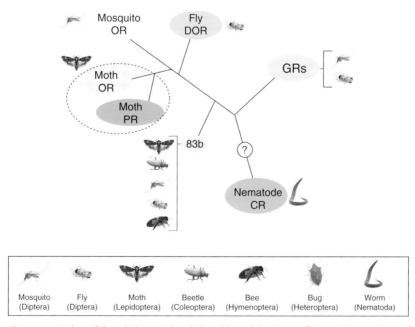

Figure 4 Stylized representation of the phylogenetic relationships of the insect *Obp* and invertebrate *Or* gene families. While the phylogenetic trees of odorant-binding proteins (OBPs) and odorant receptors (ORs) proteins are meant to display the various subfamilies (colored ovals), the phylogenetic relationships represented here do not represent true phylogenetic distances.

absence from other insect lineages suggests that this family arose within the Lepidoptera. This subfamily is subdivided into three subgroups based on sequence homology: PBP, GOBP1, and GOBP2 groups.

PBPs are highly divergent (20–95% amino acid identity) while within one GOBP group the percentage of amino acid increases to 90%. In the case of the *Manduca sexta* PBP1–GOBP2 cluster, gene duplication is one of the mechanisms that is likely to be responsible for generating molecular diversity (Vogt, R. G. *et al.*, 2002). Here, it was shown that both genes are approximately 2 kb apart, share identical exon-intron boundaries and are oriented in the same direction. As in the case of *D. melanogaster*, evolutionarily related genes seemingly share similar genomic features such as close cytogenetic position (gene clustering) and identical exon–intron boundaries. Although, it is important to know that these common features do not necessarily mean similar expression pattern and/or functional relationships.

Plus-C *Obps* have been identified only in Dipterans and thus must have arisen prior to the mosquito/fly divergence ~250 million years ago (mya). In Diptera, there are approximately a dozen Plus-C subfamily members per genome that are distributed in a few clusters (Zhou, J. J. *et al.*, 2004). The Plus-C group contains several divergent features from the classical *Obps*. Most notably a unique exon/intron usage not present in other *Obp* subgroups. In addition, Plus-C proteins average length 250 AA in length including 12 cysteine residues. Eleven Plus-C *Obps* have been described in *Drosophila pseudoobscura* (Zhou, J. J. *et al.*, 2004) and each of them has a strong counterpart in *D. melanogaster* as suggested by phylogenetic analysis. Significantly, the clustering pattern and synteny between these two Drosophilids is also extremely conserved for animals that diverged approximately 25–55 mya (Richards, S. *et al.*, 2005), and this may be indicative of similar function across all Drosophilids. This genomic organization suggests that many of these genes arose via duplication and evolved rapidly. There is only partial sequence similarity between Plus-C OBPs in these two species (overall 20%), and the majority has been shown to be expressed in the antenna and possess a peptide signal (Zhou, J. J. *et al.*, 2004; Biessmann, H. *et al.*, 2005).

Minus-C *Obp* genes have also been identified in several insect orders including Diptera and Coleoptera (Figure 4). Most *D. melanogaster* Minus-C genes (four out of six) are clustered at the tip of chromosome 3R, exhibit very low sequence similarity (~20% amino acid identity), and lack two of the six canonical cysteine residues, usually C2 and C5. Interestingly, non-Diptera Minus-C *Obps* also lack cysteines C2 and C5. Among them, *Tenebrio molitor* B1, B2, and Thp12 are hemolymph proteins (Paesen, G. C. and Happ, G. M., 1995; Rothemund, S. *et al.*, 1999; Graham, L. A. and Davies, P. L., 2002) suggesting that their *Drosophila* homologs may also have nonolfactory functions. However, the other two *Drosophila* Minus-C proteins each contain

all six cysteine residues, are expressed in the antennae (Galindo, K. and Smith, D. P., 2001) raising the possibility that olfactory function may be dependent upon the retention of a set of particular structural motifs.

Atypical *Obps* are more similar to the classic *Obp* genes but distinct from the plus-C class, and thus far have been found only in *An. gambiae* where 16 genes are predicted, and it is likely that this subfamily arose after the fly/mosquito divergence (Xu, P. X. *et al.*, 2003). Indeed homology searches have failed to identify atypical *Obps* in *Drosophila*. While atypical *Obps* display the six-cysteine hallmark of classic *Obps*, they display a characteristically elongated C-terminus containing up to seven additional cysteine residues located downstream of the six-cysteine motif. Furthermore, atypical *Obp* genes lack introns in their predicted coding regions. Two clusters of seven and three genes are located on chromosome X and 3L, respectively, with the remaining genes being scattered throughout the genome. As is typical for other *Obp* classes these genes share a low 20.2% amino acid identity (Hekmat-Scafe, D. S. *et al.*, 2002).

The antennal-binding protein family (ABPX) (Krieger, J. *et al.*, 1996) encompasses *Obps* from several insect orders (Figure 4). They were first identified in *B. mori* where they displayed little homology to the PBP/GOBP subfamily. To date, this subfamily includes several Drosophila and mosquito genes including olfactory-specific proteins E and F (*OS-E*, *OS-F*), *Ag-Obp1*, *Ag-Obp2*, *Ag-Obp3*, and *Ag-Obp17*, the *Lygus* antennal protein (LAP) (Vogt, R. G. *et al.*, 1999), several beetle *Obps* (Wojtasek, H. and Leal, W. S., 1999), and a single queen bee *Obp* (Danty, E. *et al.*, 1999). Lastly, a separate but related class of OBP-like proteins known as the chemical sense-related lipophilic ligand-binding proteins (CRLBP) were first identified in the contact chemoreceptor sensilla of the blowfly *Phormia regina* (Ozaki, M. *et al.*, 1995; Tsuchihara, K. *et al.*, 2005). Other CRLBP members have been identified in Coleoptera (Wojtasek, H. and Leal, W. S., 1999), Diptera, and in the honeybee *Apis mellifera* (Danty, E. *et al.*, 1999).

4.33.3.1.2 Functional genomics of Obps

Obps or *Obp*-like genes have been identified in vertebrates and insects. Although both OBP families are phylogenetically unrelated, both taxa have developed similar mechanisms at the perireceptor level to facilitate or to modulate the interaction between ORs and their ligands. It is therefore striking that no OBPs have been described in *C. elegans* suggesting that the nematode chemosensory system has evolved without the requirement of *Obps* for its signal transduction pathway. This raises the possibility as to whether OBPs are required for olfactory functions in all cases. What is the evidence supporting such function for this gene family? And, do these proteins possess *bone fide* odorant-binding capabilities?

The assumption of their olfactory involvement was for more than 20 years circumstantial and relied on *in vitro* experiments that were rather difficult to interpret. Evidences largely drew from the differential pattern of *Obp* expression that was distinguished by high levels of antennal transcripts. Indeed, the central dogma of OBP expression studies consistently maintained that moth PBPs were preferentially expressed in male pheromone-sensitive trichoid sensilla whereas GOBPs were expressed in the general plant-volatile-sensitive basiconic sensilla of both sexes (Vogt, R. G. *et al.*, 1991). This situation changed when the expression of PBPs and GOBPs was reported in male and female adult moths and of GOBPs in larvae albeit with different expression patterns (Vogt, R. G. *et al.*, 2002). Furthermore, immunocytochemistry experiments were able to localize their exact expression to the sensillum lymph bathing the ORN (Steinbrecht, R. A., 1998).

The crystal structure of *B. mori* PBP and its natural bound ligand, bombykol, provided a structural model to investigate the nature of this molecular relationship and the potential consequences on OR activation by such a complex (Sandler, B. H. *et al.*, 2000). Five additional OBP structures have been resolved from different insect orders including *Dmel*-LUSH (Kruse, S. W. *et al.*, 2003), cockroach *Leucophaea maderae* PBP (Lartigue, A. *et al.*, 2003), the honeybee *A. mellifera* ASP1 (Lartigue, A. *et al.*, 2004), *A. polyphemus* PBP (Mohanty, S. *et al.*, 2004), and *An. gambiae* OBP1 (Wogulis, M. *et al.*, 2006). In all cases a similar potential of hydrogen (PH)-dependent mechanism seems likely to be responsible for the release of the ligand.

Definitive evidence for olfactory function was provided by the characterization of the only mutant defective for OBP expression: the *Drosophila OBP76a* mutant, also referred to as *lush* (Figure 5). *Lush* mutants display an abnormal attraction to toxic levels of ethanol (Kim, M. S. *et al.*, 1998) and a loss of sensitivity to the aggregation of pheromone 11-*cis*-vaccenyl acetate (VA) (Xu, P. *et al.*, 2005). These two compounds are detected by two different subsets of trichoid sensilla also corresponding to *lush* expression. Transgenic rescue and the introduction of LUSH protein into the recording pipette of VA-sensitive neuron in mutant flies were shown to

restore the electrophysiological response to this compound. More importantly, using alternative putative carrier such as DmelOBP83a (OSF) or bovine serum albumin (BSA) did not restore the wild-type phenotype demonstrating that *lush* is specifically required for the transduction pathway of VA.

If *Obp* expression in olfactory sensilla can be used to infer olfactory function, then their expression in nonolfactory tissues requires the assumption of alternative functions. In *Drosophila*, where Galindo K. and Smith D. P. (2001) studied the expression pattern of over 30 *Obps* by LacZ promoter fusions, four classes were distinguished based on their expression pattern: olfactory, taste, olfactory and taste, and nonchemosensory. The last class includes the Drosophila *Obp19d* (*pbprp2*) whose expression is quite ubiquitous and is expressed in olfactory sensilla coeloconic, in various taste organs, and in the epidermis (Park, S. K. *et al.*, 2000; Galindo, K. and Smith, D. P., 2001; Shanbhag, S. R. *et al.*, 2001). Accordingly, a scavenger function has been proposed for this class of *Obps* (Park, S. K. *et al.*, 2000; Shanbhag, S. R. *et al.*, 2001).

4.33.3.2 Insect Chemosensory Proteins

The *Csp* gene family also encodes a group of small, water-soluble proteins presumed to act as carriers for small hydrophobic compounds. CSPs are smaller ($\sim$110 AA) than OBPs and have four cysteine residues whose positions are also strictly conserved. The crystal structure of the noctuid *Mamestra brassicae* CSP-A6 has six α-helices surrounding a hydrophobic binding pocket and two disulfide bridges in a folding pattern different from that of OBPs (Lartigue, A. *et al.*, 2002). The first member of this family (A10) was identified in *D. melanogaster* and due to its antennal expression was denoted as olfactory-specific D or OS-D (McKenna, M. P. *et al.*, 1994). Subsequent members of this family were referred as OS-D like, pherokine (Phk), or sensory appendage protein (SAP) based on their expression in various sensory structures. CSPs have been identified in many insect orders including Hymenoptera, Diptera, Lepidoptera, Dictyoptera, Orthoptera, Hemiptera, and Phasmatodea. The hallmark of the encoded protein family includes four invariant cysteine residues at conserved positions $CX_6CX_{18}CX_2$. There are two classes of CSPs. The first class contains three conserved amino acid motifs: an N-terminal motif A [YTTKYDN(V/I)(N/D)(L/V)DEIL], a central motif B [DGKELKXX(I/L)PDAL], and a C-terminal motif C [KYDP] (Figure 2). Proteins that diverge from this pattern belong

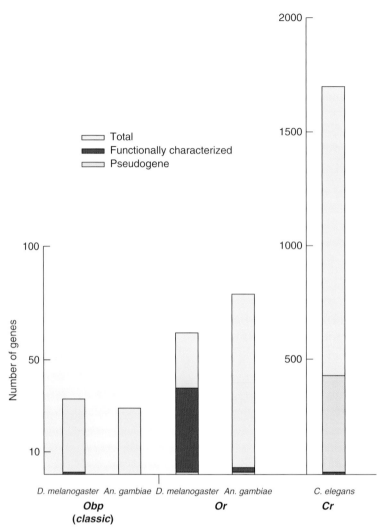

Figure 5   Known gene families in insects and nematodes. (a) *Or* and *Obp* gene repertoires in *Anopheles gambiae*, *Drosophila melanogaster*, and *Caenorhabditis elegans*. Green, total number of genes; blue, number of pseudogenes; and brown, functionally characterized genes.

to a second class. The conservation of sequence, structural motifs, intron site, and genomic clustering in the insect Neoptera infraclass (insect with wing folding) suggest an ancient origin predating the Neoptera–Paleoptera divergence. See Wanner K. W. *et al.* (2004) for a detailed review.

4.33.3.2.1   *Genomic organization and gene structure of Csps*

How do *Csp* repertoires compare to *Obp* repertoires? The genome of *D. melanogaster*, *D. pseudoobscura*, *A. mellifera*, and *An. gambiae* have 4, 4, 6, and 6 *Csp* genes, respectively (Wanner, K. W. *et al.*, 2004). Generally, CSPs are more conserved than OBPs with an average of 50% amino acid

identity within and between species and typically have 0 or 1 introns (Figure 2). When present, the intron splice site is strictly conserved one nucleotide past an invariant lysine residue (phase 1 intron). All *An. gambiae Csp* genes are clustered within a 120 kb stretch on chromosome 3R, while in *Drosophila*, Os-D is isolated on chromosome 3L and three other *Csps* are located within ∼1 Mb of each other on chromosome 2R. Synteny is highly conserved between *D. melanogaster* and *D. pseudoobscura*.

4.33.3.2.2   *Functional genomics of Csps*

The function of the *Csp* genes has been inferred exclusively based on indirect evidence such as expression patterns and structural studies. Contrary to *Obps*, *Csps* are, for the most part, expressed in several chemosensory and nonchemosensory tissues. *In situ* hybridization and *in vitro* binding studies of the moth *M. brassicae Csp* (*MbraCsp*) genes and their encoded proteins correlate their expression to the antenna and proboscis tissues. CSPs were also found in the pheromone gland, which is lacking olfactory organs (Nagnan-Le Meillour, P. *et al.*, 2000; Jacquin-Joly, E. *et al.*, 2001). In locusts different subclasses of *Csps* are associated with different tissues including the gustatory sensilla chaetica, which are distributed throughout the body and in the subcuticular space between the antennal epithelium and the cuticle (Angeli, S. *et al.*, 1999; Jin, X. *et al.*, 2005). Interestingly, in locust sensilla chaetica, CSPs are restricted to the noninnervated sensillum lymph compartment suggesting that they act outside of chemosensory processes.

The Drosophila genome (Figure 3) contains four *Csp* genes and encapsulates the multifunctional reality of the *Csp* gene family. *DmelOS-D/pherokine-1* expression is associated with the coeloconic sensilla on the antenna. Pherokine-2 (Phk-2 also known as ejaculatory bulb protein III PEBIII) is expressed solely in male ejaculatory bulbs, which contains VA. During copulation, VA is transferred to the female and has an antiaphrodisiac effect on male courtship (Brieger, G. and Butterworth, F. M., 1970). Interestingly, Phk-2 shares 50% identity with *M. brassicae* CSP (MbraOBP2) that binds VA *in vitro* (Bohbot, J. *et al.*, 1998). Phk-2 is also 53% identical to the *Periplaneta americana* protein p10, whose expression is significantly enhanced during leg regeneration of juvenile cockroaches and is also expressed in other tissues of juveniles and adults (antenna and head) (Kitabayashi, A. N. *et al.*, 1998). Both Phk-2 and Phk-3 expression are enhanced following viral and bacterial infection, respectively (Sabatier, L. *et al.*, 2003). The authors suggested that pherokines, including OSD, present a defense mechanism against microorganism entry regions such as the sensillum where the lymph and the environment are in contact.

In conclusion, the ubiquitous expression pattern of CSPs, their specific mode of regulation, and broad ligand recognition capabilities (Nagnan-Le Meillour, P. *et al.*, 2000; Jacquin-Joly, E. *et al.*, 2001; Lartigue, A. *et al.*, 2002) are consistent with the suggestion that this gene family is involved in the transport of hydrophobic ligands in various physiological pathways.

4.33.3.3 Odorant Receptors

The most numerous and well-studied chemosensory genes are the ORs, followed distantly by the gustatory

receptors (GRs). While the ORs are a clear subclass of chemoreceptors in insects (Figure 4) (Hill, C. A. *et al.*, 2002; Robertson, H. M. *et al.*, 2003; Bohbot, J., *et al.*, 2007; Nene, V., *et al.*, 2007) the distinction between olfaction and gustation, and thus ORs and GRs, is less clear in the nematode *C. elegans* (Bergamasco, C. and Bazzicalupo, P., 2006). For the purpose of this chapter, the *C. elegans* chemosensory receptors will be considered as olfactory genes, although their relationship to insect ORs is ambiguous at best (Figure 4). As a class, ORs dominate the invertebrate GPCR landscape, accounting for one-quarter to one-half of all GPCRs in insect genomes (Hill, C. A. *et al.*, 2002; Bohbot *et al.*, in preparation) and perhaps more than half of all GPCRs in *C. elegans* (Ono, Y. *et al.*, 2005; Robertson, H. M. and Thomas, J. H., 2006).

As a whole, the invertebrate ORs are an extremely divergent family of genes, often displaying very low similarity among members within the same species and with few apparent orthologues between species (Robertson, H. M. *et al.*, 2003; Ache, B. W. and Young, J. W., 2005; Robertson, H. M. and Thomas, J. H., 2006). While the vertebrate ORs are also characteristically divergent they show much more similarity than their invertebrate counterparts (Mombaerts, P., 2004). Furthermore, vertebrate ORs lack introns while the invertebrate *Or* genes contain introns, making their annotations considerably less straightforward. Intron gain and loss in invertebrate ORs is evident and has been used as one indicator of their phylogenetic relationships (Robertson, H. M., 1998; 2000; Robertson, H. M. *et al.*, 2003). As is the case for vertebrate *Or* genes, invertebrate OR loci are often clustered, ranging from just a few tandem genes in *D. melanogaster* (Robertson, H. M. *et al.*, 2003) to an enormous proliferation of loci on *C. elegans* chromosome V (Figure 3) (Robertson, H. M. and Thomas, J. H., 2006). In many cases, tandem or clustered groups of genes are paralogous, representing recent duplications of receptors. Another distinction between the vertebrate and invertebrate OR families is in the number of apparent pseudogenes. Vertebrate genomes contain as many as 50% *Or* pseudogenes (Mombaerts, P., 2004; Ache, B. W. and Young, J. M., 2005), and as many as one-third of the chemoreceptors have been reported as pseudogenes in the N2 strain of *C. elegans* (Robertson, H. M., 1998; 2000; Robertson, H. M. and Thomas, J. H., 2006). In contrast, insects families retain very few *Or* pseudogenes (Hill, C. A. *et al.*, 2002; Robertson, H. M. *et al.*, 2003; Ache, B. W. and Young, J. M., 2005; Bohbot *et al.*, in preparation). Together with the widespread occurrence of single nucleotide polymorphisms in the insect and the nematode ORs (Hill, C. A. *et al.*, 2002; Robertson, H. M. and Thomas, J. H., 2006; Bohbot *et al.*, in preparation), their overall divergence, intron flux, and duplications suggest that the invertebrate ORs are very rapidly evolving gene families. This rapid evolution may have arisen out of the need for adapting to changing environments, mating specificity, and otherwise differing life histories of divergent species (Robertson, H. M. *et al.*, 2003; Robertson, H. M. and Thomas, J. H., 2006).

4.33.3.3.1 Insect odorant receptors

The identification of the first vertebrate ORs (Buck, L. and Axel, R., 1991) led to a major effort by several groups interested in identifying insect ORs. Nearly a decade later, the release of the *D. melanogaster* genome facilitated the identification of a family of ORs from *D. melanogaster* (Clyne, P. J. *et al.*, 1999; Gao, Q. and Chess, A., 1999; Vosshall, L. B. *et al.*, 1999). Since then a significant body of work has been put forth describing insect *Or* genes, their patterns of expression, functions, and implications for odor coding (Robertson, H. M. *et al.*, 2003; Hallem, E. *et al.*, 2004a; Hallem, E. A. *et al.*, 2004b; Jacquin-Joly, E. and Merlin, C., 2004; Ache, B. W. and Young, J. M., 2005; Dahanukar, A. *et al.*, 2005; Jefferis, G. S. X. E., 2005; Rutzler, M. and Zwiebel, L., 2005). The completion of the genome of the malaria mosquito, *An. gambiae* (Holt, R. A. *et al.*, 2002), led to the description of a second complete insect OR gene family and to the comparison of *An. gambiae* and *D. melanogaster* OR repertoires (Fox, A. N. *et al.*, 2001; Hill, C. A. *et al.*, 2002). Multiple ORs have now been identified in species from at least four other insect orders including Diptera, Lepidoptera, Hymenoptera (Krieger, J. *et al.*, 2003), and Coleoptera (reviewed in Hallem, E. A. *et al.*, 2006). Furthermore, a single OR is highly conserved across insect orders, widely coexpressed with other ORs, and required for their function (Krieger, J. *et al.*, 2003; Larsson, M. C. *et al.*, 2004; Pitts, R. J. *et al.*, 2004; Jones, W. D. *et al.*, 2005; Xia, Y. and Zwiebel, L. J., 2006) (Figure 4). It is apparent that with several insect genome projects nearly completed or about to be initiated, the future of comparative genomics is certain to yield new and exciting information.

4.33.3.3.1.(i) Drosophila odorant receptors

When homology-based cloning efforts repeatedly failed, the first wave of insect OR identification was the direct result of a novel approach for bioinformatics-based searches of the *D. melanogaster* genome project

(Clyne, P. J. *et al.*, 1999). The complete set of *D. melanogaster* 60 *Or* genes (*DOrs*) encoding 62 Drosophila olfactory receptor (DOR) peptides in the Drosophila genome (Robertson, H. M. *et al.*, 2003) followed soon after (Vosshall, L. B. *et al.*, 1999; Vosshall, L. B., 2000), launching a revolution in insect olfactory research, most of which will be reviewed in other chapters in this volume. DORs are a very divergent family, most sharing less than 20% identity, with limited subfamilies having >40% identity (Vosshall, L. B., 2000). In contrast, their vertebrate counterparts generally display a much higher level of similarity of at least 40% among mouse ORs, for example (Zhang, X. and Firestein, S., 2002).

DOrs display interesting gene structures, some of which are conserved among animal ORs. For example, *DOrs* are dispersed as single genes throughout the Drosophila genome (Figure 3) (Robertson, H. M. *et al.*, 2003). Those genes that are grouped occur as doublets or triplets and generally are most similar to one another, suggesting that they represent gene duplication events (Robertson, H. M. *et al.*, 2003). The *DOr19a* and *DOr19b* genes, which are tightly linked, inverted relative to one another, and whose peptide translations are >90% identical, are an example of one such recent duplication (Robertson, H. M. *et al.*, 2003). In contrast to the widely dispersed locations of *DOrs*, *Ors* from other insects and *C. elegans* are often found in moderate to large clusters (Figure 3) (Hill, C. A. *et al.*, 2002; Robertson, H. M. and Thomas, J. H., 2006), and significant clustering is also common for vertebrate *Or* genes (Mombaerts, P., 2004). Two *DOrs* apparently produce multiple proteins by alternate splicing (Robertson, H. M. *et al.*, 2003). *DOr46a* and *DOr69a* each produce a second 'b' peptide, *DOr46a* by the splicing of a single C-terminal exon encoding the seventh transmembrane (TM) and *DOr69a* by the splicing of two C-terminal exons encoding the sixth and seventh TMs, to unique N-terminal exons (Robertson, H. M. *et al.*, 2003). Alternative splicing is a feature common to insect GRs (Clyne, P. J. *et al.*, 2000; Hill, C. A. *et al.*, 2002) and has been documented at low frequency in vertebrate ORs (Young, J. *et al.*, 2003), which lack introns (Mombaerts, P., 2004). In contrast, *DOrs* contain between one and nine introns with typically four of the C-terminal introns likely to be ancestral in the *DOr* lineage, two of which are shared with *DGrs*, reflecting their common derivation (Figure 6) (Robertson, H. M. *et al.*, 2003).

DOrs are expressed in adult antennae, maxillary palps, and in the dorsal organ of larvae (Clyne, P. J. *et al.*, 1999; Gao, Q. and Chess, A., 1999; Vosshall, L. B.

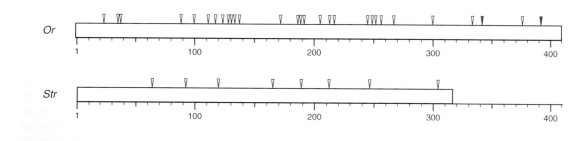

Figure 6 Locations of introns in the *Or/Str* genes in *Drosophila* and *Caenorhabditis elegans*. The intron locations (arrowheads) are shown relative to a scale of the average receptor size in amino acids. Red introns indicate ancient introns. Data from Robertson H. M. *et al.* (2003) and Robertson H. M. (1998).

et al., 1999; Vosshall, L. B. *et al.*, 2000). Their expression is limited to specific sensillar types in stereotypic regions of the antennae, and their patterns and projections are conserved from individual to individual (Vosshall, L. B. *et al.*, 2000). While generally little is known about the regulation of *DOr* expression, one important example stands out. The *D. melanogaster* mutant, *abnormal chemosensory jump 6* (*acj6*), was isolated based on its reduced jumping response to odor stimuli (McKenna, M. P. *et al.*, 1989), as well as defects of olfactory physiology (Ayer, R. K. and Carlson, J., 1991). *Acj6* encodes a Pitl Octl Unc-86 (POU) domain transcription factor that is expressed in ORNs and directly effects the expression of several *DOrs* (Clyne, P. J. *et al.*, 1999). The regulation of *Or* expression level and location is likely to play a significant role in establishing olfactory sensitivity. Future studies will undoubtedly reveal the *cis*-acting DNA elements and *trans*-acting protein factors that are involved in *Or* gene regulation and that whether these elements are conserved among invertebrates.

Functional studies of invertebrate ORs will be more thoroughly discussed in other chapters in this volume. But as they relate to receptor gene similarities the following points are worth mentioning. More than half of the *DOrs* have been functionally characterized (Figure 5) (Dobritsa, A. A. *et al.*, 2003; Hallem, E. *et al.*, 2004a; Goldman, A. L. *et al.*, 2005; Kreher, S. A. *et al.*, 2005) and most receptors appear broadly tuned to many different odors across different molecular classes (Hallem, E. *et al.*, 2004a). Surprisingly, phylogenetically related receptors are not necessarily tuned to related compounds (Hallem, E. *et al.*, 2004a). For example, DOR9a and DOR47a, while forming a monophyletic clade in a phylogenetic analysis, are tuned to multiple, nonoverlapping odorants (Hallem, E. *et al.*, 2004a). It is too early to comment as to whether this complex relationship between ligand specificity and OR homology may prove to be a general trend for invertebrate ORs.

4.33.3.3.1.(ii) *Anopheles odorant receptors*

Soon after the identification of *DOrs*, candidate *Or* genes were identified with homology-based approaches from *An. gambiae* (*AgGPRors*) (Fox, A. N. *et al.*, 2001; 2002). The completion of the *An. gambiae* genome sequence (Holt, R. A. *et al.*, 2002) facilitated the characterization of 79 *AgGPRors*, representing the second complete set of insect *Ors*. Like their *DOr* counterparts, the *AgGPRors* are a very diverse group of genes, the vast majority share less than 20% identity at the amino acid level, although 28 pairwise identities are higher than 70% (Fox, A. N. *et al.*, 2001; Fox, A. N. *et al.*, 2002; Hill, C. A. *et al.*, 2002). Indeed, several AgGPRors are so similar (>90% identity for proteins encoded by *AgGPRors75,76,78* and *AgGPRors77,79*) as to suggest that they are either very recent duplications or alternatively alleles of the same gene that has been erroneously ascribed separate loci in the annotation of a mixed haplotype genome.

Phylogenetic comparison of the *DmOrs* and *AgGPRors* illustrates the extreme divergence between insect OR that has become the hallmark of these gene families. Most groupings with high bootstrap support are made up of receptors from the same species while only a few highly divergent ORs from both species are grouped together (Hill, C. A. *et al.*, 2002).

AgGPROR7 is the sequence as well as the functional orthologue of DOR83b and represents a unique OR subfamily (see below), the two sharing a remarkable 78% identity and ∼90% similarity (Pitts, R. J. *et al.*, 2004). The next best examples are AgGPRor2 and DmOr43a, which share just 36% identity and ∼70% similarity (Fox, A. N. *et al.*, 2001; Hill, C. A. *et al.*, 2002). Interestingly, AgGPRor2 and DmOr43a do not appear to share odorant ligand specificity (Storkuhl, K. F. and Kettler, R., 2001; Wetzel, C. H. *et al.*, 2001; Hallem, E. *et al.*, 2004a; Hallem, E. A. *et al.*, 2004b).

The *AgGPRors* are dispersed on all three chromosomes, although there are several tightly linked loci with as many as nine genes (Figure 3) (Hill, C. A. *et al.*, 2002). In each case the clustered genes are very similar to one another and are generally transcribed in the same direction, indicating that they have likely arisen as a result of multiple duplication events (Hill, C. A. *et al.*, 2002). Like the *DOrs*, all *AgGPRors* contain introns (Fox, A. N. *et al.*, 2001; 2002; Hill, C. A. *et al.*, 2002; Pitts, R. J. *et al.*, 2004), and many *AgGPRors* share one or more 3′ ancestral introns of *DOrs*, supporting a common, distant derivation (Pitts, R. J. *et al.*, 2004; Zwiebel lab, unpublished data).

To date, two *AgGPRors* have been functionally characterized (Hallem, E. A. *et al.*, 2004b). One of those genes, *AgGPRor1*, displays enhanced female expression that is downregulated after blood feeding (Fox, A. N. *et al.*, 2001) and is sensitive to a component of human sweat (Hallem, E. A. *et al.*, 2004b). This is an intriguing finding and may ultimately lead to the design of strategies to interfere in Anopheles, if not mosquito-specific, in the olfactory-driven behaviors, such as host-seeking, in these disease vectors.

4.33.3.3.1.(iii) Lepidopteran odorant receptors

The first Lepidopteran candidate *Ors* (*HR1–HR9*) was identified and cloned from the tobacco budworm moth, *Heliothis virescens*, as a result of screening a proprietary genome project (Krieger, J. *et al.*, 2002). Twelve more *Ors* were subsequently identified, many of which have been classified as candidate pheromone receptors (PRs) and are discussed below (Krieger, J. *et al.*, 2004). While this analysis is presumed to represent only a subset of the OR repertoire from *H. virescens*, we can make several inferences albeit limited in their scope from this work. As noted for other insect ORs, the putative HRs have a canonical seven-TM structure, but are nevertheless a very divergent group of genes whose conceptual

translations share very low sequence identity with *D. melanogaster* ORs and GRs (Krieger, J. *et al.*, 2002). Intraspecifically, HRs share low identity among the nonpheromone receptor candidates, generally 8–15%, the highest identities being 52% and 34% between HR7 and HR9 (Krieger, J. *et al.*, 2002) and HR7 and HR12, respectively (Krieger, J. *et al.*, 2004). Although genomic DNA fragments were identified from the *H. virescens* genome, no data regarding introns or map positions were presented (Krieger, J. *et al.*, 2002), and the fact that the *H. virescens* genome is a nonpublic resource has made genomics comparisons with other insect ORs impossible at present. All candidate receptors examined by RT-PCR were expressed in antennae with a subset expressed in other chemosensory tissues (Krieger, J. *et al.*, 2002; 2004). *In situ* hybridization revealed that two receptors, *HR6* and *HR8*, are expressed in distinct groups of antennal neurons (Krieger, J. *et al.*, 2002). Interestingly, it has been suggested that a subset of the identified *H. virescens* *ORs* might, in fact, encode functional GRs. Specifically, HRs1, 4, and 5 are expressed in tissues with potential gustatory function (Krieger, J. *et al.*, 2002) and form a monophyletic clade distinct from other HRs that shares their most significant cross-species similarity with *D. melanogaster* GRs (Krieger, J. *et al.*, 2004).

Eleven candidate PRs have now been identified from two moth species: six from *H. virescens* (Krieger, J. *et al.*, 2004; 2005) and five from the silkmoth, *B. mori* (Sakurai, T. *et al.*, 2004; Krieger, J. *et al.*, 2005; Nakagawa, T. *et al.*, 2005). Currently little is known about the genomics of these receptors beyond their coding sequences, peptide alignments, and phylogenies; accordingly, any speculation as to potential paralogy and orthology among these receptors must therefore await the completion of a more extended set of characterizations. Only the genomic structure of *BmOr-1* has been presented (Sakurai, T. *et al.*, 2004). The PR conceptual translations have low sequence similarity to ORs from other insect species and they form their own monophyletic clade within the limited moth receptor family (Krieger, J. *et al.*, 2004; Sakurai, T. *et al.*, 2004; Krieger, J. *et al.*, 2005). Their similarity to one another ranges from 35% to 70% identity within a species to greater than 40% identity between BmOR-3 and HRs 4, 11, 14, 15, and 16 (Krieger, J. *et al.*, 2005). The PRs are also longer, on average, than other insect ORs averaging 430 AA in *H. virescens* and 425 AA in *B. mori* (Krieger, J. *et al.*, 2005) as compared to an average of ∼400 AA for *D. melanogaster* and *An. gambiae* ORs, respectively (Hill, C. A. *et al.*, 2002; Robertson, H. M. *et al.*, 2003).

A limited set of expression analyses and functional characterizations have been carried out for PRs. The expression of five of the six candidate *H. virescens* PRs and four of the five candidate *B. mori* PRs are either enhanced in or specific to male antennae, which is consistent with a putative role in pheromone sensing (Krieger, J. *et al.*, 2004; Sakurai, T. *et al.*, 2004; Krieger, *et al.*, 2005; Nakagawa, T. *et al.*, 2005; Gohl, T. and Krieger, J. R., 2006). Furthermore, an antibody against HR13 labels dendrites in the sensilla trichodea type 1, which are found only on male antennae and most of which are sensitive to the major component of the *H. virescens* female pheromone (Gohl, T. and Krieger, J. R., 2006). The silkmoth receptor BmOR-1 detects bombykol, the sole female sex pheromone, as shown by ectopic expression in female antennae, as well as heterologous expression in *Xenopus laevis* oocytes (Sakurai, T. *et al.*, 2004; Nakagawa, T. *et al.*, 2005). In these latter studies, while BmOR-1 alone induced a significant response to bombykol in oocytes, coexpression of the 83b subfamily member BmOR-2 greatly increased both the fraction of responsive oocytes and the magnitude of the individual responses (Nakagawa, T. *et al.*, 2005) leading to the suggestion, subsequently verified by *in situ* hybridization, that *BmOr-1* and *BmOr-3* neurons coexpress *BmOr2* (Nakagawa, T. *et al.*, 2005). In contrast, another study also used *in situ* hybridization to localize these receptors and drew the opposite conclusion, stating that neither *BmOr-1* nor *BmOr-3* colocalize with *BmOr-2* (Krieger, J. *et al.*, 2005). This striking paradox must await further investigation for its resolution and will undoubtedly be important, given the implication that 83b subfamily members may not be required for pheromone sensing in insects, but instead function only in general odorant perception.

4.33.3.3.1.(iv) *DOr83b family* From its initial identification and subsequent expression analysis, *DOr83b* gene was singled out as a potentially extraordinary OR (Vosshall, L. B. *et al.*, 2000). It is much longer than the average DOR, having an extended loop between the fourth and fifth putative transmembrane regions and, more strikingly, is expressed in nearly all antennae and maxillary palp neurons in *D. melanogaster* (Vosshall, L. B. *et al.*, 2000). The subsequent identification of *AgGPRor7* in *An. gambiae*, *AaGPRor7* in *Ae. Aegypti*, and *HR2* in *H. virescens* demonstrated that *DOr83b* is also extremely well conserved in other insects and represents an important receptor subfamily (Krieger, J. *et al.*, 2002; Melo,

A. C. *et al.*, 2004; Pitts, R. J. *et al.*, 2004). In addition, conserved intron position between *DOr83b* and *AgGPRor7* as well as microsynteny with respect to neighboring genome regions also suggested true orthology, the first of its kind among insect ORs (Hill, C. A. *et al.*, 2002). Krieger J. *et al.* (2003) identified member genes from species representing multiple insect orders including Lepidoptera (*B. mori* and *Antheraea pernyi*), Diptera (*Calliphora erythrocephala*), Hymenoptera (*A. mellifera*) and Coleoptera (*T. molitor*), demonstrating the broad conservation of the 83b family.

The *DOr83b* subgroup is unique to insects and forms an ancestral monophyletic clade within the larger insect OR family, which interestingly is more similar than other ORs to GRs (Figure 4) (Robertson, H. M. *et al.*, 2003). The exceptional degree of sequence conservation and expression characteristics raised the obvious hypothesis that the *DOr83b* family might represent a nonconventional OR subfamily that is crucial to insect olfactory signal transduction (Vosshall, L. B., 2000). Indeed, several studies have since demonstrated its functional significance in olfaction. Initially, *DOr83b* knockdowns carried out in *D. melanogaster* using either reverse genetics or RNA interference (Larsson, M. C. *et al.*, 2004; Neuhaus, E. M. *et al.*, 2005) revealed an absolute requirement for *DOr83b* in normal olfactory responses. Furthermore, DOR83b was shown necessary and sufficient for the proper localization and retention of other conventional DOrs to the dendritic membrane (Larsson, M. C. *et al.*, 2004; Benton, R. *et al.*, 2006). Testing this hypothesis even further, transgenic expression of *DOr83b* orthologues from three different insect species in the antennae of *DOr83b*-mutant flies restored odorant sensitivity as well as the correct localization of conventional *DOrs* to ORN dendrites, providing direct evidence of their functional conservation (Jones, W. D. *et al.*, 2005). While heterologous expression studies revealed that members of the DOr83b family dimerize with conventional ORs and thereby enhance odorant response sensitivity (Wetzel, C. H. *et al.*, 2001; Sakurai, T. *et al.*, 2004; Neuhaus, E. M. *et al.*, 2005), DOR83b does not independently confer odorant sensitivity (Dobritsa, A. A. *et al.*, 2003; Benton, R. *et al.*, 2006). Lastly, recent studies suggest that DOr83b and other DOrs may have an unexpectedly inverted conformation in the dendritic membrane, with their N-termini located intracellularly and their C-termini located extracellularly (Benton, R. *et al.*, 2006). This potentially unique conformation, along with the apparent lack of an 83b subfamily member of nonconventional ORs, further distinguishes the molecular components of insect olfactory signal transduction from its counterparts in both vertebrate and *C. elegans*. It will now be interesting to determine how far back in the hexapod lineage this receptor can be identified. We may indeed find that the principles of olfaction elucidated in insect model systems are conserved in older hexapods and even in other arthropod lineages, giving us insight into the evolution of olfaction in general, and potentially providing opportunities for broad-based control strategies that rely on olfaction for their implementation.

4.33.3.3.2 Caenorhabditis elegans chemoreceptors

Four years after the identification of vertebrate ORs (Buck, L. and Axel, R., 1991), the first candidate invertebrate chemoreceptors (*CeCrs*) were identified through bioinformatic methods in the soil nematode, *C. elegans* (Troemel, E. R. *et al.*, 1995). The expression of a subset of *CeCrs* was observed in chemosensitive neurons, consistent with the proposed chemoreceptor functions (Troemel, E. R. *et al.*, 1995). Since then, more than 20 gene families composed of nearly 1300 intact genes, and over 400 apparent pseudogenes, accounting for roughly 7% of all *C. elegans* genes, have been described (Robertson, H. M. and Thomas, J. H., 2006). By convention, *CeCr* gene family names usually begin with the initials *sr* for serpentine receptor, the exception being the *str* family, which stands for seventransmembrane domain (Troemel, E. R. *et al.*, 1997).

The *CeCr* genes share many of the characteristics common to insect ORs inasmuch as they are very divergent, both within and between families (Robertson, H. M. and Thomas, J. H., 2006). While their topology places them squarely within the GPCR superfamily, their phylogenetic relationships to insect ORs have not been fully explored, making it difficult to address whether they share a common chemoreceptor lineage or were derived independently (Figure 4). What is clear is that the *CeCrs* cluster on a massive scale in the *C. elegans* genome (Figure 3). As a rule, closely related genes tend to be located near each other (Robertson, H. M., 2000; 2001; Chen, N. *et al.*, 2005). Similar to other invertebrate *Ors*, *CeCrs* contain small introns (Robertson, H. M., 1998; 2000). One distinguishing overall characteristic of the *CeCr* gene products is their small size (Figure 6). As we have already noted above, insect ORs have an average length of about 400 AA. However, the *CeCr* peptides are generally predicted to

have lengths of 350 AA or less, resulting in smaller loop regions between predicted transmembrane domains (Troemel, E. R. *et al.*, 1995; Robertson, H. M., 1998; Robertson, H. M., 2000; Chen, N. *et al.*, 2005; Robertson, H. M. and Thomas, J. H., 2006 #3625). Finally, the small number of chemosensory neurons (<40) relative to the large number of *CeCrs* implies that multiple chemoreceptor proteins must be expressed in single sensory neurons (Figure 1) (Bergamasco, C. and Bazzicalupo, P., 2006), a feature that is not shared with insect ORs that are thought to express a single conventional *Or* (Ache, B. W. and Young, J. M., 2005).

Although not all *CeCrs* are expected to act as traditional ORs, sensitive to volatile odorants, we will consider them together as one large superfamily with the understanding that future studies will begin to elucidate the spectrum of *CeCr* functions. A recent review summarizes the size and the diversity of the *CeCr* superfamily in this way, "This large genetic investment might result from an extreme dependence on chemosensory abilities in the absence of visual and auditory systems. On the basis of genetic complexity, this system probably constitutes the bulk of signal transduction that occurs in *C. elegans*" (Robertson, H. M. and Thomas, J. H., 2006). Furthermore, the completion of the *Caenorhabditis briggsae* genome allowed comparative studies of candidate chemoreceptors between *C. elegans* and *C. briggsae*, species that have diverged 80–110 mya (Stein, L. D. *et al.*, 2003) and revealed extreme divergence between the two *Cr* repertoires resulting from frequent amplifications of *CeCrs* relative to their *C. briggsae* counterparts (Robertson, H. M. and Thomas, J. H., 2006).

4.33.3.3.2.(i) Str/srh gene families

The phylogenetically related *str* and *srh* genes are the largest individual *CeCr* families together comprising about 600 genes (Robertson, H. M. and Thomas, J. H., 2006). Phylogenetic studies of *str/srh* genes within *C. elegans* and between *C. elegans* and *C. briggsae* suggest that these receptors are in a constant state of change driven by the processes of gene duplication, deletion, intron gain and loss, and movement within the respective genome (Robertson, H. M., 1998; 2000; 2001). Interestingly, many of the nearly 200 apparent *str* and *srh* pseudogenes in the N2 strain encode full-length *CeCr* proteins in other nematode strains (Stewart, M. K. *et al.*, 2005). This finding and the discovery of several other nucleotide polymorphisms indicate extensive gene diversity in wild *C. elegans* populations

(Stewart, M. K. *et al.*, 2005). Incredibly, more than 80% of the *str* and *srh* genes are concentrated on chromosome V (Robertson, H. M., 1998; 2000; 2001; Robertson, H. M. and Thomas, J. H., 2006). This intriguing theme is repeated for most other *CeCr* loci (Figure 3) (Ono, Y. *et al.*, 2005; Robertson, H. M. and Thomas, J. H., 2006).

4.33.3.3.2.(ii) Sra/srab gene families

The *sra* and *srab* are two of the smaller *CeCr* families, but important in the amount of attention they have received. The *sra* genes were among the first candidate *CeCrs* identified (Troemel, E. R. *et al.*, 1995). This *sra* family is composed of 32 genes and 7 pseudogenes (Robertson, H. M. and Thomas, J. H., 2006). Twenty-two *srab* genes and five pseudogenes were recently identified based on their similarity to the *sra* genes (Chen, N. *et al.*, 2005). Both *sra* and *srab* families have an average translation of about 330 AA (Robertson, H. M. and Thomas, J. H., 2006). *Sra* genes are concentrated on chromosomes I and II, while the *srab* genes are almost all located on chromosome V and nearly all *srab* family members lack an *sra*-defining protein motif, and phylogenetic analysis clearly segregates them into a distinct subfamily (Chen, N. *et al.*, 2005). There are small numbers of orthologues between *sra* and *srab* genes in *C. elegans* and *C. briggsae* based on sequence alignments and genome synteny (Chen, N. *et al.*, 2005). Those *C. elegans/C. briggsae* gene pairs that are existent are substantially diverged and are not considered to be true orthologues (Robertson, H. M. and Thomas, J. H., 2006). Furthermore, the *sra* and *srab* gene families are significantly expanded in *C. elegans* relative to *C. briggsae* due to apparent duplication events in *C. elegans* (Chen, N. *et al.*, 2005).

4.33.3.3.2.(iii) CeCr function

Genetic studies have supported a chemoreceptive role for selective *CeCr* genes. Certainly, the most compelling example of this is the characterization of *odr-10*, an *str* family member that encodes a receptor mediating the olfactory response to diacetyl (Sengupta, P. *et al.*, 1996). *Odr-10* remains the sole *CeCr* with an identified ligand (Figure 5).

In the absence of additional functional studies, numerous investigations have examined *CeCr* expression patterns as a means of supporting their classification as chemoreceptors. For example, four *sra* genes were expressed in chemosensitive amphid neurons in the head (Troemel, E. R. *et al.*, 1995) while expression of three *srab* genes was observed in tail chemosensitive neurons, two of those also showing expression in head chemosensory neurons (Chen, N. *et al.*, 2005). *Str-2* is expressed only in AWC neurons (Colosimo, M. E. *et al.*, 2003) and *odr-10* is expressed in AWA neurons (Sengupta, P. *et al.*, 1996). Furthermore, while a summary of the data from several promoter: Green Fluorescent Protein (GFP) expression studies illustrates that most *CeCrs* (48/63) are expressed in sensory tissues while many others (15/63) are not (Robertson, H. M. and Thomas, J. H., 2006). Another study identified a promoter element, common to numerous *srb* and *sri* genes, that is capable of ectopically driving *odr-10* expression in the ADL neuron (McCarroll, S. A. *et al.*, 2005). Finally, a microarray-based study determined that five *CeCrs* were among the most prevalent genes expressed in the RNA derived from the AWB neuron, a result confirmed by promoter:GFP experiments (Colosimo, M. E. *et al.*, 2004).

While many of the experiments described above support a chemoreceptive function for *CeCrs*, the supposition that all *CeCr* gene products act either as volatile ORs or soluble tastant receptors remains tenuous at best. Multiple *CeCrs* from different families, as suggested either by their similarities to other genes with known function, expression patterns, or mutant phenotypes, might well have other functions. For example, the *srw* family shares its origins with FMRFamide and other peptide receptors (Robertson, H. M. and Thomas, J. H., 2006). Although a proposed role for the *srw* gene as environmental peptide receptors has merit, some of these 145 genes may operate in other capacities, perhaps as cell–cell communication receptors. In fact, at least one *CeCr* does not seem to function as a chemoreceptor at all. *sra-11* is expressed in interneurons downstream of sensory neurons and an *sra-11* null mutation disrupts olfactory imprinting, a specific type of olfactory-driven memory (Remy, J.-J. and Hobert, O., 2005) for at least two different odorants suggesting that *sra-11* has a general role in the process. While it seems likely that the majority of *CeCrs* do indeed act as receptors of environmental stimuli, much more study is required to fully appreciate the multiplicity of receptive roles, both external and internal, that this large superfamily may or may not carry out.

4.33.4 Signaling Cascade

For better or worse, the identification of genes involved in olfactory transduction in invertebrates has historically been restricted to the molecular context derived from the knowledge acquired in the vertebrate visual and olfactory systems. These pathways have been elegantly studied and extensively reviewed elsewhere (Hildebrand, J. G. and Shepherd, G. M., 1997), and molecular, cellular, and genetic studies indicate that, for the most part, similar gene families are required in invertebrate olfactory transduction. Even so, it is essential to appreciate that several studies elaborated in this chapter have pointedly demonstrated that insect olfactory systems display several unique characteristics that argue against a wholesale acceptance of vertebrate signal transduction paradigms. Indeed, the novel function and conformation of both nonconventional and conventional ORs in Drosophila has fostered the suggestion that insect chemosensory transduction processes may in fact involve non-GPCR mechanisms (Benton, R. *et al.*, 2006).

4.33.4.1 Guanine Nucleotide-Binding Proteins and Effector Genes

4.33.4.1.1 Insects

GPs are a heterotrimeric assemblage of an alpha (α), beta (β) and gamma (γ) subunits as ancient as the eukaryote divergence (Jones, A. M. and Assmann, S. M., 2004). GPs may best be considered as molecular switches: located in the cell membrane, they modulate the activity of a variety of intracellular effector molecules once activated by GPCRs. Although the exact mechanism of GPs activation remains ambiguous, odorant-induced conformational change of the GPCR is likely to be required. The Gα subunit in particular seems to be interacting directly with the OR since both visual and olfactory systems rely on GPCRs (Grobner, G. *et al.*, 2000; Vogt, R. G., 2006).

Six Gα genes have been identified in the genomes of *D. melanogaster* and *An. gambiae* (Figure 3). *An. gambiae* G$_q\alpha$ (*agq*), *ago*, *agi*, and *agm* are located within a large 24-Mb segment on chromosome 2L while *ags* and *agc* are located on 3L and 3R, respectively. Phylogenetic analysis supports that *ago* and *agi* derive from a common ancestor as suggested by their relative genomic proximity (~3 Mb) (Rutzler, M. R. *et al.*, 2006). Drosophila Gα genes (*dg*) are located ~2 Mb apart on 2R and 3L with *dgo* and *dgq*. Importantly, splice variants of G$_q\alpha$ are components of the phototransduction pathway and are also expressed in chemosensory cells (Scott, K. *et al.*, 1995; Talluri, S. *et al.*, 1995). According to immunolocalization studies, G$_q\alpha$ protein is associated to the dendrite of *B. mori*

and *An. gambiae* ORN (Laue, M. *et al.*, 1997; Rutzler, M. R. *et al.*, 2006), and functional evidence for G$_q\alpha$ requirement in olfactory transduction has been provided by RNAi mediated silencing experiments in the Drosophila antenna (Kalidas, S. and Smith, D. P., 2002). In these studies, altered olfactory responses were odor-dependent suggesting that there are more than one type of olfactory GP or, indeed, that alternative transduction mechanisms coexist.

In fact, molecular and electrophysiological approaches provide support for the hypothesis that two transduction pathways may be active in invertebrate olfactory transduction. One depends on the second messenger 1,4,5 inositol triphosphate (IP3) and the other depends on cAMP. In the lobster *Panulirus argus*, both cAMP and IP3 pathways are activated in response to stimulation with complex odor mixtures suggesting that both an AC and a PLC are targeted (Boekhoff, I. *et al.*, 1994).

A PLCβ is almost certainly involved in the olfactory transduction pathway of insects. In Drosophila, olfactory and visual physiology is dependent on the activity of the *norpA* PLC gene, which acts downstream of GPs (Riesgo-Escovar, J. *et al.*, 1995) although *norpA* activity in olfactory tissues is thus far restricted to the maxillary palp. The transient increase of IP3, a breakdown product of PIP2 by PLC, following exposure to odorants also supports a GP-mediated transduction pathway in insects (Breer, H. *et al.*, 1990). In Drosophila, recovery from olfactory adaptation is mediated by a phosphatidylinositol transfer protein gene dubbed *retinal degeneration B* (*Rdgb*), (Riesgo-Escovar, J. R. *et al.*, 1994). *Rdgb* mutants display delayed return to the resting potential following exposure to odorants. *Rdgb* encodes a six-transmembrane domain protein, which contains Ca$^+$ binding domains involved in IP3 formation. In addition to being expressed in the eye, this gene is present in the antenna, labellum, and maxillary palp where it is thought to play a role in membrane metabolism.

Mutations of the Drosophila rutabaga (*rut*) and dunce (*Dun*) genes affect the cAMP transduction pathway (Martin, F. *et al.*, 2001; Gomez-Diaz, C. *et al.*, 2004). *Rut* and *Dun* encode an AC and a cAMP-PDE, respectively. Both genes express in the antenna, and the mutants exhibit defective electrophysiological and behavioral olfactory responses coherent with the initial stage of odorant transduction. Both genes are members of small families: six ACs and two cAMP-PDE have been identified in the fly genome.

4.33.4.1.2 Caenorhabditis elegans

Twenty-one *Gα* genes are in the genome of *C. elegans* (Jansen, G. *et al.*, 1999; Bergamasco, C. and Bazzicalupo, P., 2006). Two clusters of five and nine genes are located on chromosomes I and V, respectively. Phylogenetic analysis shows that there is one orthologue for each of the four vertebrate *Gα* subunits (*Gα$_{o/i}$, Gα$_s$, Gα$_q$*, and *Gα$_{12}$*) sharing over 68% amino acid identity (Roayaie, K. *et al.*, 1998). *C. elegans* GPs are widely expressed in the nervous system but six *Gα* subunits are expressed in the two pairs of AWA and AWC neurons that mediate odorant attraction (Bargmann, C. I. *et al.*, 1993; Troemel, E. R. *et al.*, 1995; 1997; Jansen, G. *et al.*, 1999). Based on GFP fusion constructs and immunocytological studies, five Gα proteins (GPA-2, GPA-3, GPA-5, GPA-13, and ODR-3) are localized to the ciliated dendrites suggesting that they presumably interact with GPCRs in a complex signaling network (Jansen, G. *et al.*, 1999; Lans, H. *et al.*, 2004).

In *C. elegans*, GP targets are likely to include AC and guanylyl cyclase (GC), cGMP-PDE and phospholipases. While in vertebrate, olfaction CNG channels activation is mediated by an AC/cAMP pathway, this role is performed in *C. elegans* by GCs. Two transmembrane GP-sensitive GCs (tGCs) are clearly required for chemosensory transduction in *C. elegans*. Both ODR-1 and DAF-11 (abnormal *da*uer *f*ormation) are located in the cilia of various chemosensory neurons including the olfactory neuron AWC (Birnby, D. A. *et al.*, 2000; L'etoile, N. D. and Bargmann, C. I., 2000). At least 26 additional transmembrane and 7 soluble GC genes have been identified in the *C. elegans* genome, suggesting that this organism favors cGMP over cAMP second messengers to activate downstream ion channels. Phylogenetic analyses show that 21 of the tGCs form a lineage-specific expansion that correlates with the proximity of these genes on the chromosomes (Fitzpatrick, D. A. *et al.*, 2006).

4.33.4.2 Ion Channels

CNG and TRP ion channels function in many vertebrate sensory transduction pathways. *Drosophila* and *C. elegans* each have four and six CNG channel genes, respectively, and at least two reports suggest that ion channel genes are active in Drosophila olfaction. One gene encodes an apparent vertebrate CNG Ca$^+$ channel homolog (*Dmel*CNG) and the gene *ether a go-go* (*eag*) encodes a putative cyclic nucleotide-modulated K$^+$ channel (Baumann, A. *et al.*, 1994; Dubin, A. E. *et al.*, 1998). *Dmel*CNG is expressed in the antennal and eye tissues and physiological studies suggest that this is a Ca$^+$ channel modulated by cGMP (Baumann, A. *et al.*, 1994). Mutant *eag* flies exhibit defective olfactory phenotype in a subset of olfactory sensilla (Dubin, A. E. *et al.*, 1998). Although not related to *Dmel*CNG, the *eag* gene also requires cyclic nucleotides for activity and contains a consensus cyclic nucleotide-binding site. No other ion channels have been identified as possible downstream target of effector genes in primary olfactory sensing. However, there is one report suggesting that a TRP channel is involved in olfactory adaptation in the developing Drosophila antenna (Stortkuhl, K. F. *et al.*, 1999), but this gene could not be detected in mature adult antenna. Lastly, an ITPR is expressed strongly in this tissue and is required for the maintenance of olfactory adaptation (Deshpande, M. *et al.*, 2000).

The *C. elegans* CNG consists of the α-subunit TAX-2 (abnormal chemo*tax*is) and the β-subunit TAX-4 involved in thermosensation and chemosensation (Coburn, C. M. and Bargmann, C. I., 1996). TAX-2 and TAX-4 are most similar to the vertebrate α and β subunits of the rod photoreceptors. Heterologous expression of TAX-2/TAX-4 has shown that this CNG channel is preferably activated by cGMP over cAMP and following its activation opens to Ca$^+$ currents (Komatsu, H. *et al.*, 1999) and loss of function mutants in either gene-display-defective chemotaxis to a subset of volatile compounds in the AWC and AWB neurons (Coburn, C. M. and Bargmann, C. I., 1996).

In contrast to the TAX-2/TAX-4 CNG channel that is active in AWB and AWC neurons, most chemosensory neurons in *C. elegans* use a TRP (vanilloid-related subfamily)-related ion channel for primary sensation and adaptation (Colbert, H. A. and Bargmann, C. I., 1995) whose closest vertebrate homologue is the nociceptive capsaicin receptor (Minke, B. and Cook, B., 2002). The *C. elegans* genome harbors five *trp* genes including *osm-9* (abnormal *osm*otic avoidance) and four *ocr* genes (*osm-9/c*apsaicin *r*eceptor-related). Combinations of OSM-9 and OCR subunits determine the function of the resulting TRP channels (Tobin, D. *et al.*, 2002). *Osm-9 lf* mutants in the AWA and ASH neurons exhibit strong chemosensory defects caused by the suppression of calcium currents after stimulation with a wide range of odorants. This suggests that most, if not all, transductory signaling converges on this ion channel. In the AWA and ASH neurons, OSM-9 requires OCR-2 for signal transduction and both are located on the ciliated end of the dendrite (for review, see Bergamasco, C. and Bazzicalupo, P., 2006). In the AWC neuron, OSM-9 alone is involved in olfactory adaptation, negatively modulating olfactory responses to prolonged odor exposure (Colbert, H. A. and Bargmann, C. I., 1995).

4.33.5 Regulators of Signaling

4.33.5.1 Arrestins and G-Protein-Coupled Receptor Kinases

In mammalian systems, signal termination is mediated by two small gene families: GPCR kinases (GRKs) and ARRs. GRKs phosphorylate activated GPCRs, which are in turn targeted by ARRs, which disrupts the GPCR–GP complex (Krupnick, J. G. et al., 1997). This promotes several events including clathrin-mediated receptor internalization/degradation and other downstream pathways (for reviews, see Lefkowitz, R. J. and Shenoy, S. K., 2005; Reiter, E. and Lefkowitz, R. J., 2006).

Thus far, two putative Drosophila (*DmGrk1* and *DmGrk2*) and two *C. elegans* GRKs (*Ce-Grk-1* and *Ce-Grk-2*) have been identified (Cassill, J. A. et al., 1991; Fukuto, H. S. et al., 2004; Reiter, E. and Lefkowitz, R. J., 2006). Not surprisingly, invertebrate *Grk* genes share considerable homology with their vertebrate counterparts. *Ce*-GRK1 and *Ce*-GRK-2 are 56% and 66% identical with human GRK5 and GRK3, respectively. *DmGrk1/2* and *Ce-Grk2* are broadly expressed throughout development and in various tissues including the CNS suggesting that they may function in multiple developmental and neuronal pathways. While the function of *Grk* genes in Drosophila olfaction has not been established, *Ce-grk2* (*lf*) mutants exhibit a significant chemosensory defect consisting of a loss of response to octanol (Fukuto, H. S. et al., 2004).

As is the case for vertebrates, four *arrestin* genes have been identified in the genome of *D. melanogaster* and *An. gambiae* (*DmArr1-4* and *AgArr1-4*, respectively). *DmArr1* and *DmArr2* are expressed in photoreceptors and ORNs (Merrill, C. E. et al., 2002). Consistent with a role in olfactory signal transduction, hypomorphic alleles of both *DmArr1* and *DmArr2* display significant reductions in olfactory responses to various odorants, and *DmArr1* larvae show defective chemotaxis in behavioral experiments (Merrill, C. E. et al., 2005). Interestingly, these phenotypes are variable depending on odorants stimuli suggesting that in insect ORNs, *Arr1* and *Arr2* may participate in two independent desensitization pathways.

There are four subclasses of insect ARRs based on sequence similarity. Insects *Arr1* and *Arr2* encode approximately 400 AA proteins. They are referred to as sensory ARRs because of their broad expression in the olfactory and visual tissues. However, while the ARR1 primary amino acid sequence contains all the conserved domains of vertebrate ARRs, ARR2 lacks the C-terminal clathrin binding domain (Merrill, C. E. et al., 2002; 2003). *Arr3* genes on the other hand encode considerably larger proteins and include the *D. melanogaster Kurtz* (*DmKrz*), a gene expressed throughout development in a multitude of tissues including the nervous system (Roman, G. et al., 2000) and mutations in which larval are lethal. *Arr3* genes exhibit most of the canonical motifs found in vertebrate nonvisual ARRs. The fourth subclass of ARR genes (*Arr4* alleles) is significantly different from the other types, hence their annotation as atypical.

In contrast, the *C. elegans* genome apparently encodes a single ARR (*Ce-Arr1*) gene, which shares significant sequence homology to vertebrate nonvisual ARRs (65% similarity) and is most similar to the nonsensory ARR3/KURTZ ARR subtypes with 70% amino acid similarity, including well-characterized motifs known to interact with various molecular partners (Palmitessa, A. et al., 2005). However, in contrast to *Kurtz*, *Ce-Arr1* mutants in *C. elegans* are viable and develop normally. Interestingly, while null mutants fail to adapt the following odorant stimulation, adaptation is restored in the C-terminal-truncated *Ce-Arr1* rescue mutant albeit with reduced recovery from adaptation capabilities demonstrating a dual role for *Ce-Arr1* in adaptation and recovery from adaptation in the AWC neuron. Although expressed broadly throughout the nervous system, *Ce-Arr1* is more highly expressed within the AWA, AWB, AWC, ASH, and ADL chemosensory neurons, where it overlaps with the expression of sensory GPs as described above. Palmitessa A. et al. (2005) determined that desensitization and adaptation require the N-terminal region to prevent the phosphorylated GPCR from interacting with the GP, while recovery from adaptation depends on the various components of the endocytic machinery and is mediated by the C-terminal domain.

4.33.5.2 The Caenorhabditis elegans Calcineurin TAX-6

The vertebrate calcineurin is a Ca^+/calmodulin (CaM)-dependent protein phophatase that acts as a positive regulator of calcium-dependent signals in various biological pathways. The CaM and the B-subunit are required to activate the catalytic activity of the A-subunit. Ce-TAX-6 is 77% identical to the A-subunit (Kuhara, A. *et al.*, 2002). A putative homolog of the B-subunit has been identified in the worm genome and share 80% amino acid identity with its vertebrate counterpart. *Tax-6* mutants display several phenotypes indicating the pleiotropic nature of this gene. In the AWC neuron, *tax-6* mutants display diminished olfactory responsiveness reflected by enhanced adaptation when exposed to various odorants. The mechanism underlying this chemosensory is unknown but it requires the presence of a functional TRPV OSM-9 channel.

4.33.5.3 Other Protein Kinases

In addition to the core components of olfactory signal transduction thus far discussed, there is good evidence that implicates an ever increasing set of protein kinase-associated pathways in invertebrate chemosensory processes. Indeed, in further support of IP3-based second messenger pathways in insect olfactory transduction cascades is the identification of the Drosophila *IP3 kinase1* gene (*IP3K1*) (Gomez-Diaz, C. *et al.*, 2006). IP3K1 mediates IP3 degradation and is expressed in multiple locations including the olfactory appendages and the head. Importantly, gain-of-function *IP3K* Drosophila mutants exhibit diminished olfactory sensitivity due to increased recovery times (Gomez-Diaz, C. *et al.*, 2006).

In *C. elegans*, the signal termination in the AWC olfactory neuron is characterized by short- and long-term adaptation components that are mediated by a cGMP-dependent kinase (EGL-4) (L'etoile, N. D. *et al.*, 2002). EGL-4 and mammal protein kinase G (PKG) share similar functional domains including cGMP-binding motifs and a serine/threonine kinase domain (L'etoile, N. D. *et al.*, 2002). The latter domain has been shown to regulate the TAX-2/TAX-4 CNG channel by targeting a specific phosphorylation site on the TAX-2 subunit. A longer adaptation mechanism seems to be mediated by EGL-4 nuclear translocation and by regulation of gene expression. This mechanism is present in mammalian systems and is assumed by PKG (Gudi, T. *et al.*, 1997).

Expression and mutant analyses show that the protein kinase C (PKC) genes *ttx-4* and *tpa-1* are involved in multiple sensory pathways including olfaction (Okochi, Y. *et al.*, 2005). These genes encode protein homologs to the human novel PKC (over 50% amino acid identity) and Drosophila PKC (60% amino acid identity), and both have Di Acyl Glycerol-binding domains. The chemotaxis defects observed in *ttx-4* and *tpa-1 lf* mutants are rescued by an exogenous DAG analog implicating these two genes in the regulation of the second messenger pathway pertaining to olfactory sensation.

Lastly, five DAG kinase genes (*dgk*) have been identified in the *C. elegans* genome of which *dgk-1* and *dgk-3* act redundantly to reduce DAG levels and have been shown to participate in olfactory adaptation (Matsuki, M. *et al.*, 2006). The effects of DAG and other genes known to participate in DAG-mediated pathways provide strong support that phospholipase-based effectors are active in this complex regulatory network. Still, the jury is out as to the precise mechanisms underlying olfactory signaling within the invertebrate systems under study. At this point there is a growing body of circumstantial evidence but, if we may take the courtroom analogy one step further, no smoking gun to provide hard proof implicating any specific pathway(s).

4.33.6 Conclusion and Perspectives

The study of the olfactory subgenomes of insects and worms indicate that olfactory organs use similar gene families for the detection and conversion of chemical signals. Whether these shared molecular strategies are the result of evolutionary convergence as a response to a common chemical environment and/or a reflection of shared ancestry is difficult to discern. The large *Obp/Or* gene families described in this chapter display elements of similar genomic dynamics. For example, the majority of genes in a given family within a given genome occur in clusters and their phylogenetic relations surely reflect sequence homology that often correlates with cytogenetic position implying that local gene expansion has resulted from intrachromosomal gene duplications. It can also be reasonably argued that sequence homology between vertebrate and invertebrate gene families, in particular in the components of signal transduction, provides strong support for shared ancestry. Although important, this maxim is not so evident in the case of the chemosensory GPCRs and OBPs, which may represent two genetic interfaces between the ORN and the odors. The repertoires of these gene families are highly divergent even within species with only few orthologues shared between species. This type of comparison clearly indicates that this family displays lineage-specific expansions and reductions indicating ongoing genetic turn over and rapid evolution.

All GPCRs are presumed to have seven-transmembrane domains and interact with highly conserved or analogous transductory elements. The identification of which is the result of both directed research based on the mammalian olfactory and visual systems and also through discovery-based approaches epitomized by the use of forward genetic screens. Whereas both worms and insects use GPCRs as the initial recognition step, insects have developed a singular mechanism requiring the combination of a generic nonconventional GPCR and an odorant-specific conventional OR. While it appears likely that this unique adaptation is profound and may be specific to arthropods, the functional implications of this mechanism within these taxa remain unknown. In addition, the proposed reverse topology of insect ORs potentially makes this system truly novel as it adds a level a complexity in the molecular architecture of olfactory transduction. Interestingly, despite all these structural differences and sequence divergence, GPCRs of vertebrate and invertebrate interact with conserved GPs.

An equally striking difference between insects and worms resides in the divergent molecular strategies observed upstream of the chemosensory GPCRs. While insects have a moderate number of *Ors* and *Obps*, worms do not have *Obps* but have a much larger *Or* repertoire. It has been proposed that OBPs contribute to the odor coding by interacting with the ORs. But as we pointed earlier, there are very few reports suggesting a strict requirement of these proteins for proper olfactory function. Also, the number of authentic *C. elegans* chemoreceptors are probably smaller as indicated by a recent research.

Overall, it is certain that the number of genes involved in olfactory pathways will increase as more genetic screens and reverse genetic tools are applied to invertebrate systems. The most challenging task lays ahead in forging an understanding of the regulation and the network of epistatic interactions among olfactory genes that serve to provide the basis for the downstream processing of olfactory information that ultimately results in learning, memory, and which form the basis for critical behavioral decisions.

References

Ache, B. W. and Young, J. M. 2005. Olfaction: diverse species, conserved principles. Neuron 48, 417–430.

Adams, M. D., Celniker, S. E., Holt, R. A., Evans, C. A., Gocayne, J. D., Amanatides, P. G., Scherer, S. E., Li, P. W., Hoskins, R. A., Galle, R. F., George, R. A., Lewis, S. E., Richards, S., Ashburner, M., Henderson, S. N., Sutton, G. G., Wortman, J. R., Yandell, M. D., Zhang, Q., Chen, L. X., Brandon, R. C., Rogers, Y. H., Blazej, R. G., Champe, M., Pfeiffer, B. D., Wan, K. H., Doyle, C., Baxter, E. G., Helt, G., Nelson, C. R., Gabor, G. L., Abril, J. F., Agbayani, A., An, H. J., Andrews-Pfannkoch, C., Baldwin, D., Ballew, R. M., Basu, A., Baxendale, J., Bayraktaroglu, L., Beasley, E. M., Beeson, K. Y., Benos, P. V., Berman, B. P., Bhandari, D., Bolshakov, S., Borkova, D., Botchan, M. R., Bouck, J., Brokstein, P., Brottier, P., Burtis, K. C., Busam, D. A., Butler, H., Cadieu, E., Center, A., Chandra, I., Cherry, J. M., Cawley, S., Dahlke, C., Davenport, L. B., Davies, P., De Pablos, B., Delcher, A., Deng, Z., Mays, A. D., Dew, I., Dietz, S. M., Dodson, K., Doup, L. E., Downes, M., Dugan-Rocha, S., Dunkov, B. C., Dunn, P., Durbin, K. J., Evangelista, C. C., Ferraz, C., Ferriera, S., Fleischmann, W., Fosler, C., Gabrielian, A. E., Garg, N. S., Gelbart, W. M., Glasser, K., Glodek, A., Gong, F., Gorrell, J. H., Gu, Z., Guan, P., Harris, M., Harris, N. L., Harvey, D., Heiman, T. J., Hernandez, J. R., Houck, J., Hostin, D., Houston, K. A., Howland, T. J., Wei, M. H., Ibegwam, C., et al. 2000. The genome sequence of *Drosophila melanogaster*. Science 287, 2185–2195.

Altschul, S. F., Gish, W., Miller, W., Myers, E. W., and Lipman, D. J. 1990. Basic local alignment search tool. J. Mol. Biol. 215, 403–410.

Angeli, S., Ceron, F., Scaloni, A., Monti, M., Monteforti, G., Minnocci, A., Petacchi, R., and Pelosi, P. 1999. Purification, structural characterization, cloning and immunocytochemical localization of chemoreception proteins from *Schistocerca gregaria*. Eur. J. Biochem. 262, 745–754.

Ayer, R. K., Jr. and Carlson, J. 1991. Acj6: a gene affecting olfactory physiology and behavior in *Drosophila*. PNAS 88, 5467–5471.

Bargmann, C. I. and Horvitz, H. R. 1991. Chemosensory neurons with overlapping functions direct chemotaxis to multiple chemicals in *C. elegans*. Neuron 7, 729–742.

Bargmann, C. I., Hartwieg, E., and Horvitz, H. R. 1993. Odorant-selective genes and neurons mediate olfaction in *C. elegans*. Cell 74, 515–527.

Baumann, A., Frings, S., Godde, M., Seifert, R., and Kaupp, U. B. 1994. Primary structure and functional expression of a *Drosophila* cyclic nucleotide-gated channel present in eyes and antennae. Embo J. 13, 5040–5050.

Benton, R., Sachse, S., Michnick, S. W., and Vosshall, L. B. 2006. Atypical membrane topology and heteromeric function of *Drosophila* odorant receptors *in vivo*. PLoS Biol. 4, e20.

Bergamasco, C. and Bazzicalupo, P. 2006. Signaling in the chemosensory systems: chemical sensitivity in *Caenorhabditis elegans*. Cell Mol. Life Sci. 63, 1510–1522.

Biessmann, H., Nguyen, Q. K., Le, D., and Walter, M. F. 2005. Microarray-based survey of a subset of putative olfactory genes in the mosquito *Anopheles gambiae*. Insect Mol. Biol. 14, 575–589.

Birnby, D. A., Link, E. M., Vowels, J. J., Tian, H., Colacurcio, P. L., and Thomas, J. H. 2000. A transmembrane guanylyl cyclase (daf-11) and hsp90 (daf-21) regulate a common set of chemosensory behaviors in *Caenorhabditis elegans*. Genetics 155, 85–104.

Boekhoff, I., Michel, W. C., Breer, H., and Ache, B. W. 1994. Single odors differentially stimulate dual second messenger pathways in lobster olfactory receptor cells. J. Neurosci. 14, 3304–3309.

Boekhoff, I., Raming, K., and Breer, H. 1990a. Pheromone-induced stimulation of inositol-trisphosphate formation in insect antennae is mediated by G-proteins. J. Comp. Physiol. B 160, 99–103.

Boekhoff, I., Strotmann, J., Raming, K., Tareilus, E., and Breer, H. 1990b. Odorant-sensitive phospholipase C in insect antennae. Cell Signal 2, 49–56.

Bohbot, J., Pitts, R. J., Kwon, H.-W. and Rützler, M. 2007. Molecular characterization of the *Aedes aegypti* odorant receptor gene family. Insect Mol. Biol. (accepted).

Bohbot, J., Sobrio, F., Lucas, P., and Nagnan-Le Meillour, P. 1998. Functional characterization of a new class of odorant-binding proteins in the moth *Mamestra brassicae*. Biochem. Biophys. Res. Commun. 253, 489–494.

Breer, H., Boekhoff, I., and Tareilus, E. 1990. Rapid kinetics of second messenger formation in olfactory transduction. Nature 345, 65–68.

Briand, L., Nespoulous, C., Huet, J. C., Takahashi, M., and Pernollet, J. C. 2001. Ligand binding and physico-chemical properties of asp2, a recombinant odorant-binding protein from honeybee (*Apis mellifera* l.). Eur. J. Biochem. 268, 752–760.

Brieger, G. and Butterworth, F. M. 1970. *Drosophila melanogaster*: identity of male lipid in reproductive system. Science 167, 1262.

Buck, L. and Axel, R. 1991. A novel mulitgene family may encode odorant receptors: a molecular basis for odor recognition. Cell 65, 175–187.

Cassill, J. A., Whitney, M., Joazeiro, C. A., Becker, A., and Zuker, C. S. 1991. Isolation of *Drosophila* genes encoding G protein-coupled receptor kinases. Proc. Natl. Acad. Sci. U. S. A. 88, 11067–11070.

Chen, N., Pai, S., Zhao, Z., Mah, A., Newbury, R., Johnsen, R. C., Altun, Z., Moerman, D. G., Baillie, D. L., and Stein, L. D. 2005. Identification of a nematode chemosensory gene family. PNAS 102, 146–151.

Clyne, P. J., Warr, C. G., and Carlson, J. R. 2000. Candidate taste receptors in *Drosophila*. Science 287, 1830–1834.

Clyne, P. J., Warr, C. G., Freeman, M. R., Lessing, D., Kim, J., and Carlson, J. R. 1999. A novel family of divergent seven-transmembrane proteins: candidate odorant receptors in *Drosophila*. Neuron 22, 327–338.

Coburn, C. M. and Bargmann, C. I. 1996. A putative cyclic nucleotide-gated channel is required for sensory development and function in *C. elegans*. Neuron 17, 695–706.

Colbert, H. A. and Bargmann, C. I. 1995. Odorant-specific adaptation pathways generate olfactory plasticity in *C. elegans*. Neuron 14, 803–812.

Colosimo, M. E., Brown, A., Mukhopadhyay, S., Gabel, C., Lanjuin, A. E., Samuel, A. D. T., and Sengupta, P. 2004. Identification of thermosensory and olfactory neuron-specific genes via expression profiling of single neuron types. Curr. Biol. 14, 2245–2251.

Colosimo, M. E., Tran, S., and Sengupta, P. 2003. The divergent orphan nuclear receptor odr-7 regulates olfactory neuron gene expression via multiple mechanisms in *C. elegans*. Genetics 165, 1779–1791.

Consortium, C. E. S. 1998. Genome sequence of the nematode *C. elegans*: a platform for investigating biology. Science 282, 2012–2018.

Dahanukar, A., Hallem, E. A., and Carlson, J. R. 2005. Insect chemoreception. Curr. Opin. Neurobiol. 15, 423–430.

Danty, E., Briand, L., Michard-Vanhee, C., Perez, V., Arnold, G., Gaudemer, O., Huet, D., Huet, J. C., Ouali, C., Masson, C., and Pernollet, J. C. 1999. Cloning and expression of a queen pheromone-binding protein in the honeybee: an olfactory-specific, developmentally regulated protein. J. Neurosci. 19, 7468–7475.

Deshpande, M., Venkatesh, K., Rodrigues, V., and Hasan, G. 2000. The inositol 1,4,5-trisphosphate receptor is required for maintenance of olfactory adaptation in *Drosophila* antennae. J. Neurobiol. 43, 282–288.

Dobritsa, A. A., Van Der Goes Van Naters, W., Warr, C. G., Steinbrecht, R. A., and Carlson, J. R. 2003. Integrating the molecular and cellular basis of odor coding in the *Drosophila* antenna. Neuron 37, 827–841.

Dubin, A. E. and Harris, G. L. 1997. Voltage-activated and odor-modulated conductances in olfactory neurons of *Drosophila melanogaster*. J. Neurobiol. 32, 123–137.

Dubin, A. E., Liles, M. M., and Harris, G. L. 1998. The K^+ channel gene ether a go–go is required for the transduction of a subset of odorants in adult *Drosophila melanogaster*. J. Neurosci. 18, 5603–5613.

Fitzpatrick, D. A., O'halloran, D. M., and Burnell, A. M. 2006. Multiple lineage specific expansions within the guanylyl cyclase gene family. BMC Evol. Biol. 6, 26.

Fox, A. N., Pitts, R. J., Robertson, H. M., Carlson, J. R., and Zwiebel, L. J. 2001. Candidate odorant receptors from the malaria vector mosquito *Anopheles gambiae* and evidence of down-regulation in response to blood feeding. Proc. Natl. Acad. Sci. U. S. A. 98, 14693–14697.

Fox, A. N., Pitts, R. J., and Zwiebel, L. J. 2002. A cluster of candidate odorant receptors from the malaria vector mosquito, *Anopheles gambiae*. Chem. Senses 27, 453–459.

Fukuto, H. S., Ferkey, D. M., Apicella, A. J., Lans, H., Sharmeen, T., Chen, W., Lefkowitz, R. J., Jansen, G., Schafer, W. R., and Hart, A. C. 2004. G protein-coupled receptor kinase function is essential for chemosensation in *C. elegans*. Neuron 42, 581–593.

Galindo, K. and Smith, D. P. 2001. A large family of divergent *Drosophila* odorant-binding proteins expressed in gustatory and olfactory sensilla. Genetics 159, 1059–1072.

Gao, Q. and Chess, A. 1999. Identification of candidate *Drosophila* olfactory receptors from genomic DNA sequence. Genomics 60, 31–39.

Gohl, T. and Krieger, J. R. 2006. Immunolocalization of a candidate pheromone receptor in the antenna of the male moth, *Heliothis virescens*. Invertebr. Neurosci. 6, 13–21.

Goldman, A. L., Van Der Goes Van Naters, W., Lessing, D., Warr, C. G., and Carlson, J. R. 2005. Coexpression of two functional odor receptors in one neuron. Neuron 45, 661–666.

Gomez-Diaz, C., Martin, F., and Alcorta, E. 2004. The camp transduction cascade mediates olfactory reception in *Drosophila melanogaster*. Behav. Genet. 34, 395–406.

Gomez-Diaz, C., Martin, F., and Alcorta, E. 2006. The inositol 1,4,5-triphosphate kinase1 gene affects olfactory reception in *Drosophila melanogaster*. Behav. Genet. 36, 309–321.

Graham, L. A. and Davies, P. L. 2002. The odorant-binding proteins of *Drosophila melanogaster*: annotation and characterization of a divergent gene family. Gene 292, 43–55.

Grobner, G., Burnett, I. J., Glaubitz, C., Choi, G., Mason, A. J., and Watts, A. 2000. Observations of light-induced structural changes of retinal within rhodopsin. Nature 405, 810–813.

Gudi, T., Lohmann, S. M., and Pilz, R. B. 1997. Regulation of gene expression by cyclic GMP-dependent protein kinase requires nuclear translocation of the kinase: identification of a nuclear localization signal. Mol. Cell Biol. 17, 5244–5254.

Hallem, E., Ho, M. G., and Carlson, J. R. 2004a. The molecular basis of odor coding in the *Drosophila* antenna. Cell 117, 965–979.

Hallem, E. A. and Carlson, J. R. 2006. Coding of odors by a receptor repertoire. Cell 125, 143–160.

Hallem, E. A., Dahanukar, A., and Carlson, J. R. 2006. Insect odor and taste receptors. Annu. Rev. Entomol. 51, 113–135.

Hallem, E. A., Nicole Fox, A., Zwiebel, L. J., and Carlson, J. R. 2004b. Olfaction: mosquito receptor for human-sweat odorant. Nature 427, 212–213.

Hekmat-Scafe, D. S., Scafe, C. R., Mckinney, A. J., and Tanouye, M. A. 2002. Genome-wide analysis of the odorant-binding protein gene family in *Drosophila melanogaster*. Genome Res. 12, 1357–1369.

Hildebrand, J. G. and Shepherd, G. M. 1997. Mechanisms of olfactory discrimination: converging evidence for common principles across phyla. Annu. Rev. Neurosci. 20, 595–631.

Hill, C. A., Fox, A. N., Pitts, R. J., Kent, L. B., Tan, P. L., Chrystal, M. A., Cravchik, A., Collins, F. H., Robertson, H. M., and Zwiebel, L. J. 2002. G protein-coupled receptors in *Anopheles gambiae*. Science 298, 176–178.

Holt, R. A., Subramanian, G. M., Halpern, A., Sutton, G. G., Charlab, R., Nusskern, D. R., Wincker, P., Clark, A. G., Ribeiro, J. M., Wides, R., Salzberg, S. L., Loftus, B., Yandell, M., Majoros, W. H., Rusch, D. B., Lai, Z., Kraft, C. L., Abril, J. F., Anthouard, V., Arensburger, P., Atkinson, P. W., Baden, H., De Berardinis, V., Baldwin, D., Benes, V., Biedler, J., Blass, C., Bolanos, R., Boscus, D., Barnstead, M., Cai, S., Center, A., Chatuverdi, K., Christophides, G. K., Chrystal, M. A., Clamp, M., Cravchik, A., Curwen, V., Dana, A., Delcher, A., Dew, I., Evans, C. A., Flanigan, M., Grundschober-Freimoser, A., Friedli, L., Gu, Z., Guan, P., Guigo, R., Hillenmeyer, M. E., Hladun, S. L., Hogan, J. R., Hong, Y. S., Hoover, J., Jaillon, O., Ke, Z., Kodira, C., Kokoza, E., Koutsos, A., Letunic, I., Levitsky, A., Liang, Y., Lin, J. J., Lobo, N. F., Lopez, J. R., Malek, J. A., Mcintosh, T. C., Meister, S., Miller, J., Mobarry, C., Mongin, E., Murphy, S. D., O'brochta, D. A., Pfannkoch, C., Qi, R., Regier, M. A., Remington, K., Shao, H., Sharakhova, M. V., Sitter, C. D., Shetty, J., Smith, T. J., Strong, R., Sun, J., Thomasova, D., Ton, L. Q., Topalis, P., Tu, Z., Unger, M. F., Walenz, B., Wang, A., Wang, J., Wang, M., Wang, X., Woodford, K. J., Wortman, J. R., Wu, M., Yao, A., Zdobnov, E. M., Zhang, H., Zhao, Q., *et al.* 2002. The genome sequence of the malaria mosquito *Anopheles gambiae*. Science 298, 129–149.

Hovemann, B. T., Sehlmeyer, F., and Malz, J. 1997. *Drosophila melanogaster* NADPH-cytochrome p450 oxidoreductase: pronounced expression in antennae may be related to odorant clearance. Gene 189, 213–219.

Jacquin-Joly, E. and Merlin, C. 2004. Insect olfactory receptors: contributions of molecular biology to chemical ecology. J. Chem. Ecol. 30, 2359–2397.

Jacquin-Joly, E., Vogt, R. G., Francois, M. C., and Nagnan-Le Meillour, P. 2001. Functional and expression pattern analysis of chemosensory proteins expressed in antennae and pheromonal gland of *Mamestra brassicae*. Chem. Senses 26, 833–844.

Jansen, G., Thijssen, K. L., Werner, P., Van Der Horst, M., Hazendonk, E., and Plasterk, R. H. 1999. The complete family of genes encoding g proteins of *Caenorhabditis elegans*. Nat. Genet. 21, 414–419.

Jefferis, G. S. X. E. 2005. Insect olfaction: a map of smell in the brain. Curr. Biol. 15, R668–R670.

Jin, X., Brandazza, A., Navarrini, A., Ban, L., Zhang, S., Steinbrecht, R. A., Zhang, L., and Pelosi, P. 2005. Expression and immunolocalisation of odorant-binding and chemosensory proteins in locusts. Cell Mol. Life Sci. 62, 1156–1166.

Jones, A. M. and Assmann, S. M. 2004. Plants: the latest model system for G-protein research. EMBO Rep. 5, 572–578.

Jones, W. D., Nguyen, T. A., Kloss, B., Lee, K. J., and Vosshall, L. B. 2005. Functional conservation of an insect odorant receptor gene across 250 million years of evolution. Curr. Biol. 15, R119–R121.

Kalidas, S. and Smith, D. P. 2002. Novel genomic cdna hybrids produce effective RNA interference in adult *Drosophila*. Neuron 33, 177–184.

Kaupp, U. B. and Seifert, R. 2002. Cyclic nucleotide-gated ion channels. Physiol. Rev. 82, 769–824.

Kim, M. S., Repp, A., and Smith, D. P. 1998. Lush odorant-binding protein mediates chemosensory responses to alcohols in *Drosophila melanogaster*. Genetics 150, 711–721.

Kitabayashi, A. N., Arai, T., Kubo, T., and Natori, S. 1998. Molecular cloning of cdna for p10, a novel protein that increases in the regenerating legs of *Periplaneta americana* (American cockroach). Insect Biochem. Mol. Biol. 28, 785–790.

Komatsu, H., Jin, Y. H., L'etoile, N., Mori, I., Bargmann, C. I., Akaike, N., and Ohshima, Y. 1999. Functional reconstitution of a heteromeric cyclic nucleotide-gated channel of *Caenorhabditis elegans* in cultured cells. Brain Res. 821, 160–168.

Korswagen, H. C., van der Linden, A. M., and Plasterk, R. H. 1998. G protein hyperactivation of the *Caenorhabditis elegans* adenylyl cyclase SGS-1 induces neuronal degeneration. EMBO J. 17, 5059–5065.

Kreher, S. A., Kwon, J. Y., and Carlson, J. R. 2005. The molecular basis of odor coding in the *Drosophila* larva. Neuron 46, 445–456.

Krieger, J., Große-Wilde, E., Gohl, T., and Breer, H. 2005. Candidate pheromone receptors of the silkmoth *Bombyx mori*. Eur. J. Neurosci. 21, 2167–2176.

Krieger, J., Grosse-Wilde, E., Gohl, T., Dewer, Y. M., Raming, K., and Breer, H. 2004. Genes encoding candidate pheromone receptors in a moth (*Heliothis virescens*). Proc. Natl. Acad. Sci. U. S. A. 101, 11845–11850.

Krieger, J., Klink, O., Mohl, C., Raming, K., and Breer, H. 2003. A candidate olfactory receptor subtype highly conserved across different insect orders. J. Comp. Physiol. A Neuroethol. Sens. Neural Behav. Physiol. 189, 519–526.

Krieger, J., Raming, K., Dewer, Y. M., Bette, S., Conzelmann, S., and Breer, H. 2002. A divergent gene family encoding candidate olfactory receptors of the moth *Heliothis virescens*. Eur. J. Neurosci. 16, 619–628.

Krieger, J., Von Nickisch-Rosenegk, E., Mameli, M., Pelosi, P., and Breer, H. 1996. Binding proteins from the antennae of *Bombyx mori*. Insect Biochem. Mol. Biol. 26, 297–307.

Krupnick, J. G., Gurevich, V. V., and Benovic, J. L. 1997. Mechanism of quenching of phototransduction. Binding competition between arrestin and transducin for phosphorhodopsin. J. Biol. Chem. 272, 18125–18131.

Kruse, S. W., Zhao, R., Smith, D. P., and Jones, D. N. 2003. Structure of a specific alcohol-binding site defined by the odorant binding protein lush from *Drosophila melanogaster*. Nat. Struct. Biol. 10, 694–700.

Kuhara, A., Inada, H., Katsura, I., and Mori, I. 2002. Negative regulation and gain control of sensory neurons by the *C. elegans* calcineurin tax-6. Neuron 33, 751–763.

L'etoile, N. D. and Bargmann, C. I. 2000. Olfaction and odor discrimination are mediated by the *C. elegans* guanylyl cyclase odr-1. Neuron 25, 575–586.

L'etoile, N. D., Coburn, C. M., Eastham, J., Kistler, A., Gallegos, G., and Bargmann, C. I. 2002. The cyclic GMP-dependent protein kinase EGL-4 regulates olfactory adaptation in *C. elegans*. Neuron 36, 1079–1089.

Lans, H., Rademakers, S., and Jansen, G. 2004. A network of stimulatory and inhibitory Galpha-subunits regulates olfaction in *Caenorhabditis elegans*. Genetics 167, 1677–1687.

Larsson, M. C., Domingos, A. I., Jones, W. D., Chiappe, M. E., Amrein, H., and Vosshall, L. B. 2004. Or83b encodes a broadly expressed odorant receptor essential for *Drosophila* olfaction. Neuron 43, 703–714.

Lartigue, A., Campanacci, V., Roussel, A., Larsson, A. M., Jones, T. A., Tegoni, M., and Cambillau, C. 2002. X-ray structure and ligand binding study of a moth chemosensory protein. J. Biol. Chem. 277, 32094–32098.

Lartigue, A., Gruez, A., Briand, L., Blon, F., Bezirard, V., Walsh, M., Pernollet, J. C., Tegoni, M., and Cambillau, C. 2004. Sulfur single-wavelength anomalous diffraction crystal structure of a pheromone-binding protein from the honeybee Apis mellifera l. J. Biol. Chem. 279, 4459–4464.

Lartigue, A., Gruez, A., Spinelli, S., Riviere, S., Brossut, R., Tegoni, M., and Cambillau, C. 2003. The crystal structure of a cockroach pheromone-binding protein suggests a new ligand binding and release mechanism. J. Biol. Chem. 278, 30213–30218.

Laue, M., Maida, R., and Redkozubov, A. 1997. G-protein activation, identification and immunolocalization in pheromone-sensitive sensilla trichodea of moths. Cell Tissue Res. 288, 149–158.

Lefkowitz, R. J. and Shenoy, S. K. 2005. Transduction of receptor signals by beta-arrestins. Science 308, 512–517.

Lycett, G. J., Mclaughlin, L. A., Ranson, H., Hemingway, J., Kafatos, F. C., Loukeris, T. G., and Paine, M. J. 2006. Anopheles gambiae p450 reductase is highly expressed in oenocytes and in vivo knockdown increases permethrin susceptibility. Insect Mol. Biol. 15, 321–327.

Maibeche-Coisne, M., Merlin, C., Francois, M. C., Porcheron, P., and Jacquin-Joly, E. 2005. P450 and p450 reductase cdnas from the moth Mamestra brassicae: cloning and expression patterns in male antennae. Gene 346, 195–203.

Martin, F., Charro, M. J., and Alcorta, E. 2001. Mutations affecting the camp transduction pathway modify olfaction in Drosophila. J. Comp. Physiol. A 187, 359–370.

Matsuki, M., Kunitomo, H., and Iino, Y. 2006. Goalpha regulates olfactory adaptation by antagonizing Gqalpha-DAG signaling in Caenorhabditis elegans. Proc. Natl. Acad. Sci. U. S. A. 103, 1112–1117.

Mccarroll, S. A., Li, H., and Bargmann, C. I. 2005. Identification of transcriptional regulatory elements in chemosensory receptor genes by probabilistic segmentation. Curr. Biol. 15, 347–352.

Mckenna, M. P., Hekmat Scafe, D. S., Gaines, P., and Carlson, J. R. 1994. Putative Drosophila pheromone-binding proteins expressed in a subregion of the olfactory system. J. Biol. Chem. 269, 16340–16347.

Mckenna, M., Monte, P., Helfand, S. L., Woodard, C., and Carlson, J. 1989. A simple chemosensory response in Drosophila and the isolation of acj mutants in which it is affected. PNAS 86, 8118–8122.

Melo, A. C., Rutzler, M., Pitts, R. J., and Zwiebel, L. J. 2004. Identification of a chemosensory receptor from the yellow fever mosquito, Aedes aegypti, that is highly conserved and expressed in olfactory and gustatory organs. Chem. Senses 29, 403–410.

Merrill, C. E., Pitts, R. J., and Zwiebel, L. J. 2003. Molecular characterization of arrestin family members in the malaria vector mosquito, Anopheles gambiae. Insect Mol. Biol. 12, 641–650.

Merrill, C. E., Riesgo-Escovar, J., Pitts, R. J., Kafatos, F. C., Carlson, J. R., and Zwiebel, L. J. 2002. Visual arrestins in olfactory pathways of Drosophila and the malaria vector mosquito Anopheles gambiae. Proc. Natl. Acad. Sci. U. S. A. 99, 1633–1638.

Merrill, C. E., Sherertz, T. M., Walker, W. B., and Zwiebel, L. J. 2005. Odorant-specific requirements for arrestin function in Drosophila olfaction. J. Neurobiol. 63, 15–28.

Minke, B. and Cook, B. 2002. Trp channel proteins and signal transduction. Physiol. Rev. 82, 429–472.

Mohanty, S., Zubkov, S., and Gronenborn, A. M. 2004. The solution NMR structure of Antheraea polyphemus PBP provides new insight into pheromone recognition by pheromone-binding proteins. J. Mol. Biol. 337, 443–451.

Mombaerts, P. 2004. Genes and ligands for odorant, vomeronasal and taste receptors. Nat. Rev. Neurosci. 5, 263–278.

Nagnan-Le Meillour, P., Cain, A. H., Jacquin-Joly, E., Francois, M. C., Ramachandran, S., Maida, R., and Steinbrecht, R. A. 2000. Chemosensory proteins from the proboscis of Mamestra brassicae. Chem. Senses 25, 541–553.

Nakagawa, T., Sakurai, T., Nishioka, T., and Touhara, K. 2005. Insect sex-pheromone signals mediated by specific combinations of olfactory receptors. Science 307, 1638–1642.

Nene, V., Wortman, J. R., Lawson, D., Haas, B., Kodira, C., Tu, Z. J., Loftus, B., Xi, Z., Megy, K., Grabherr, M., Ren, Q., Zdobnov, E. M., Lobo, N. F., Campbell, K. S., Brown, S. E., Bonaldo, M. F., Zhu, J., Sinkins, S. P., Hogenkamp, D. G., Amedeo, P., Arensburger, P., Atkinson, P. W., Bidwell, S., Biedler, J., Birney, E., Bruggner, R. V., Costas, J., Coy, M. R., Crabtree, J., Crawford, M., Debruyn, B., Decaprio, D., Eiglmeier, K., Eisenstadt, E., El-Dorry, H., Gelbart, WM., Gomes, S. L., Hammond, M., Hannick, L. I., Hogan, J. R., Holmes, M. H., Jaffe, D., Johnston, J. S., Kennedy, R. C., Koo, H., Kravitz, S., Kriventseva, E. V., Kulp, D., Labutti, K., Lee, E., Li, S., Lovin, D. D., Mao, C., Mauceli, E., Menck, C. F., Miller, J. R., Montgomery, P., Mori, A., Nascimento, A. L., Naveira, H. F., Nusbaum, C., O'leary, S., Orvis, J., Pertea, M., Quesneville, H., Reidenbach, K. R., Rogers, Y. H., Roth, C. W., Schneider, J. R., Schatz, M., Shumway, M., Stanke, M., Stinson, E. O., Tubio, J. M., Vanzee, J. P., Verjovski-Almeida, S., Werner, D., White, O., Wyder, S., Zeng, Q., Zhao, Q., Zhao, Y., Hill, C. A., Raikhel, A. S., Soares, M. B., Knudson, D. L., Lee, N. H., Galagan, J., Salzberg, S. L., Paulsen, I. T., Dimopoulos, G., Collins, F. H., Birren, B., Fraser-Liggett, C. M., and Severson, D. W. 2007. Science 316, 1718–1723.

Neuhaus, E. M., Gisselmann, G., Zhang, W., Dooley, R., Stortkuhl, K., and Hatt, H. 2005. Odorant receptor heterodimerization in the olfactory system of Drosophila melanogaster. Nat. Neurosci. 8, 15–17.

Nikonov, A. A., Peng, G., Tsurupa, G., and Leal, W. S. 2002. Unisex pheromone detectors and pheromone-binding proteins in scarab beetles. Chem. Senses 27, 495–504.

Okochi, Y., Kimura, K. D., Ohta, A., and Mori, I. 2005. Diverse regulation of sensory signaling by C. elegans nPKC-epsilon/eta TTX-4. Embo J. 24, 2127–2137.

Ono, Y., Fujibuchi, W., and Suwa, M. 2005. Automatic gene collection system for genome-scale overview of G-protein coupled receptors in eukaryotes. Gene 364, 63–73.

Ozaki, M., Morisaki, K., Idei, W., Ozaki, K., and Tokunaga, F. 1995. A putative lipophilic stimulant carrier protein commonly found in the taste and olfactory systems. A unique member of the pheromone-binding protein superfamily. Eur. J. Biochem. 230, 298–308.

Paesen, G. C. and Happ, G. M. 1995. The B proteins secreted by the tubular accessory sex glands of the male mealworm beetle, Tenebrio molitor, have sequence similarity to moth pheromone-binding proteins. Insect Biochem. Mol. Biol. 25, 401–408.

Palmitessa, A., Hess, H. A., Bany, I. A., Kim, Y. M., Koelle, M. R., and Benovic, J. L. 2005. Caenorhabditus elegans arrestin regulates neural G protein signaling and olfactory adaptation and recovery. J. Biol. Chem. 280, 24649–24662.

Park, S. K., Shanbhag, S. R., Wang, Q., Hasan, G., Steinbrecht, R. A., and Pikielny, C. W. 2000. Expression patterns of two putative odorant-binding proteins in the olfactory organs of Drosophila melanogaster have different implications for their functions. Cell Tissue Res. 300, 181–192.

Pelosi, P. 1994. Odorant-binding proteins. Crit. Rev. Biochem. Mol. Biol. 29, 199–228.

Picone, D., Crescenzi, O., Angeli, S., Marchese, S., Brandazza, A., Ferrara, L., Pelosi, P., and Scaloni, A. 2001. Bacterial expression and conformational analysis of a chemosensory protein from *Schistocerca gregaria*. Eur. J. Biochem. 268, 4794–4801.

Pitts, R. J., Fox, A. N., and Zwiebel, L. J. 2004. A highly conserved candidate chemoreceptor expressed in both olfactory and gustatory tissues in the malaria vector, *Anopheles gambiae*. Proc. Natl. Acad. Sci. U. S. A. 101, 5058–5063.

Ranson, H., Claudianos, C., Ortelli, F., Abgrall, C., Hemingway, J., Sharakhova, M. V., Unger, M. F., Collins, F. H., and Feyereisen, R. 2002. Evolution of supergene families associated with insecticide resistance. Sci. 298, 179–181.

Reiter, E. and Lefkowitz, R. J. 2006. GRKs and beta-arrestins: roles in receptor silencing, trafficking and signaling. Trends Endocrinol. Metab. 17, 159–165.

Remy, J.-J. and Hobert, O. 2005. An interneuronal chemoreceptor required for olfactory imprinting in *C. elegans*. Science 309, 787–790.

Richards, S., Liu, Y., Bettencourt, B. R., Hradecky, P., Letovsky, S., Nielsen, R., Thornton, K., Hubisz, M. J., Chen, R., Meisel, R. P., Couronne, O., Hua, S., Smith, M. A., Zhang, P., Liu, J., Bussemaker, H. J., Van Batenburg, M. F., Howells, S. L., Scherer, S. E., Sodergren, E., Matthews, B. B., Crosby, M. A., Schroeder, A. J., Ortiz-Barrientos, D., Rives, C. M., Metzker, M. L., Muzny, D. M., Scott, G., Steffen, D., Wheeler, D. A., Worley, K. C., Havlak, P., Durbin, K. J., Egan, A., Gill, R., Hume, J., Morgan, M. B., Miner, G., Hamilton, C., Huang, Y., Waldron, L., Verduzco, D., Clerc-Blankenburg, K. P., Dubchak, I., Noor, M. A., Anderson, W., White, K. P., Clark, A. G., Schaeffer, S. W., Gelbart, W., Weinstock, G. M., and Gibbs, R. A. 2005. Comparative genome sequencing of *Drosophila pseudoobscura*: chromosomal, gene, and *cis*-element evolution. Genome Res. 15, 1–18.

Riesgo-Escovar, J., Raha, D., and Carlson, J. R. 1995. Requirement for a phospholipase C in odor response: overlap between olfaction and vision in *Drosophila*. Proc. Natl. Acad. Sci. U. S. A. 92, 2864–2868.

Riesgo-Escovar, J. R., Woodard, C., and Carlson, J. R. 1994. Olfactory physiology in the *Drosophila* maxillary palp requires the visual system gene *rdgB*. J. Comp. Physiol. A 175, 687–693.

Riviere, S., Lartigue, A., Quennedey, B., Campanacci, V., Farine, J. P., Tegoni, M., Cambillau, C., and Brossut, R. 2003. A pheromone-binding protein from the cockroach *Leucophaea maderae*: cloning, expression and pheromone binding. Biochem. J. 371, 573–579.

Roayaie, K., Crump, J. G., Sagasti, A., and Bargmann, C. I. 1998. The G alpha protein ODR-3 mediates olfactory and nociceptive function and controls cilium morphogenesis in *C. elegans* olfactory neurons. Neuron 20, 55–67.

Robertson, H. M. 1998. Two large families of chemoreceptor genes in the nematodes *Caenorhabditis elegans* and *Caenorhabditis briggsae* reveal extensive gene duplication, diversification, movement, and intron loss. Genome Res. 8, 449–463.

Robertson, H. M. 2000. The large srh family of chemoreceptor genes in *Caenorhabditis* nematodes reveals processes of genome evolution involving large duplications and deletions and intron gains and losses. Genome Res. 10, 192–203.

Robertson, H. M. 2001. Updating the str and srj (stl) families of chemoreceptors in *Caenorhabditis* nematodes reveals frequent gene movement within and between chromosomes. Chem. Senses 26, 151–159.

Robertson, H. M. and Thomas, J. H. 2006. The putative chemoreceptor families of *C. elegans*, http://www.wormbook.org, doi/10.1895/wormbook.1.66.1.

Robertson, H. M., Martos, R., Sears, C. R., Todres, E. Z., Walden, K. K., and Nardi, J. B. 1999. Diversity of odourant binding proteins revealed by an expressed sequence tag project on male *Manduca sexta* moth antennae. Insect Mol. Biol. 8, 501–518.

Robertson, H. M., Warr, C. G., and Carlson, J. R. 2003. Molecular evolution of the insect chemoreceptor gene superfamily in *Drosophila melanogaster*. Proc. Natl. Acad. Sci. U. S. A. 100(Suppl. 2), 14537–14542.

Rogers, M. E., Jani, M. K., and Vogt, R. G. 1999. An olfactory-specific glutathione-*S*-transferase in the sphinx moth *Manduca sexta*. J. Exp. Biol. 202, 1625–1637.

Roman, G., He, J., and Davis, R. L. 2000. Kurtz, a novel nonvisual arrestin, is an essential neural gene in *Drosophila*. Genetics 155, 1281–1295.

Rothemund, S., Liou, Y. C., Davies, P. L., Krause, E., and Sonnichsen, F. D. 1999. A new class of hexahelical insect proteins revealed as putative carriers of small hydrophobic ligands. Structure 7, 1325–1332.

Rubin, G. M., Yandell, M. D., Wortman, J. R., Gabor Miklos, G. L., Nelson, C. R., Hariharan, I. K., Fortini, M. E., Li, P. W., Apweiler, R., Fleischmann, W., Cherry, J. M., Henikoff, S., Skupski, M. P., Misra, S., Ashburner, M., Birney, E., Boguski, M. S., Brody, T., Brokstein, P., Celniker, S. E., Chervitz, S. A., Coates, D., Cravchik, A., Gabrielian, A., Galle, R. F., Gelbart, W. M., George, R. A., Goldstein, L. S., Gong, F., Guan, P., Harris, N. L., Hay, B. A., Hoskins, R. A., Li, J., Li, Z., Hynes, R. O., Jones, S. J., Kuehl, P. M., Lemaitre, B., Littleton, J. T., Morrison, D. K., Mungall, C., O'farrell, P. H., Pickeral, O. K., Shue, C., Vosshall, L. B., Zhang, J., Zhao, Q., Zheng, X. H., and Lewis, S. 2000. Comparative genomics of the eukaryotes. Science 287, 2204–2215.

Rutzler, M. and Zwiebel, L. 2005. Molecular biology of insect olfaction: recent progress and conceptual models. J. Comp. Physiol. A Neuroethol. Sens. Neural Behav. Physiol., 1–14.

Rutzler, M. R., Lu, T., and Zwiebel, L. J. 2006. The ga encoding gene family of the malaria vector mosquito *Anopheles gambiae*: expression analysis and immunolocalization of agaq and agao in female antennae. J. Comp. Neurol., 499, 533–545.

Sabatier, L., Jouanguy, E., Dostert, C., Zachary, D., Dimarcq, J. L., Bulet, P., and Imler, J. L. 2003. Pherokine-2 and -3. Eur. J. Biochem. 270, 3398–3407.

Sakurai, T., Nakagawa, T., Mitsuno, H., Mori, H., Endo, Y., Tanoue, S., Yasukochi, Y., Touhara, K., and Nishioka, T. 2004. Identification and functional characterization of a sex pheromone receptor in the silkmoth *Bombyx mori*. Proc. Natl. Acad. Sci. U. S. A. 101, 16653–16658.

Sandler, B. H., Nikonova, L., Leal, W. S., and Clardy, J. 2000. Sexual attraction in the silkworm moth: structure of the pheromone-binding-protein-bombykol complex. Chem. Biol. 7, 143–151.

Scott, K., Becker, A., Sun, Y., Hardy, R., and Zuker, C. 1995. Gq alpha protein function *in vivo*: genetic dissection of its role in photoreceptor cell physiology. Neuron 15, 919–927.

Sengupta, P., Chou, J. H., and Bargmann, C. I. 1996. Odr-10 encodes a seven transmembrane domain olfactory receptor required for responses to the odorant diacetyl. Cell 84, 899–909.

Shanbhag, S. R., Hekmat-Scafe, D., Kim, M. S., Park, S. K., Carlson, J. R., Pikielny, C., Smith, D. P., and Steinbrecht, R. A. 2001. Expression mosaic of odorant-binding proteins in *Drosophila* olfactory organs. Microsc. Res. Tech. 55, 297–306.

Shanbhag, S. R., Müller, B., and Steinbrech, R. A. 1999. Atlas of olfactory organs of *Drosophila melanogaster*. 1.Types, external organization, innervation and distribution of olfactory sensilla. Int. J. Insect Morphol. Embryol. 28, 377–397.

Stein, L. D., Bao, Z., Blasiar, D., Blumenthal, T., Brent, M. R., Chen, N., Chinwalla, A., Clarke, L., Clee, C., Coghlan, A., Coulson, A., Eustachio, P., Fitch, D. H. A., Fulton, L. A., Fulton, R. E., Griffiths-Jones, S., Harris, T. W., Hillier, L. W., Kamath, R., Kuwabara, P. E., Mardis, E. R., Marra, M. A., Miner, T. L., Minx, P., Mullikin, J. C., Plumb, R. W., Rogers, J., Schein, J. E., Sohrmann, M., Spieth, J., Stajich, J. E., Wei, C., Willey, D., Wilson, R. K., Durbin, R., and Waterston, R. H. 2003. The genome sequence of *Caenorhabditis briggsae*: a platform for comparative genomics. PLoS Biol. 1, e45.

Steinbrecht, R. A. 1998. Odorant-binding proteins: expression and function. Ann. N. Y. Acad. Sci. 855, 323–332.

Stensmyr, M. C., Erland, S., Hallberg, E., Wallen, R., Greenaway, P., and Hansson, B. S. 2005. Insect-like olfactory adaptations in the terrestrial giant robber crab. Curr. Biol. 15, 116–121.

Stewart, M. K., Clark, N. L., Merrihew, G., Galloway, E. M., and Thomas, J. H. 2005. High genetic diversity in the chemoreceptor superfamily of *Caenorhabditis elegans*. Genetics 169, 1985–1996.

Stocker, R. F. 2001. *Drosophila* as a focus in olfactory research: mapping of olfactory sensilla by fine structure, odor specificity, odorant receptor expression, and central connectivity. Microsc. Res. Tech. 55, 284–296.

Storkuhl, K. F. and Kettler, R. 2001. Functional analysis of an olfactory receptor in *Drosophila melanogaster*. PNAS 98, 9381–9385.

Stortkuhl, K. F., Hovemann, B. T., and Carlson, J. R. 1999. Olfactory adaptation depends on the trp Ca^{2+} channel in *Drosophila*. J. Neurosci. 19, 4839–4846.

Talluri, S., Bhatt, A., and Smith, D. P. 1995. Identification of a *Drosophila* G protein alpha subunit (dGq alpha-3) expressed in chemosensory cells and central neurons. Proc. Natl. Acad. Sci. U. S. A. 92, 11475–11479.

Tegoni, M., Pelosi, P., Vincent, F., Spinelli, S., Campanacci, V., Grolli, S., Ramoni, R., and Cambillau, C. 2000. Mammalian odorant binding proteins. Biochim. Biophys. Acta 1482, 229–240.

Tobin, D., Madsen, D., Kahn-Kirby, A., Peckol, E., Moulder, G., Barstead, R., Maricq, A., and Bargmann, C. 2002. Combinatorial expression of TRPV channel proteins defines their sensory functions and subcellular localization in *C. elegans* neurons. Neuron 35, 307–318.

Troemel, E. R., Chou, J. H., Dwyer, N. D., Colbert, H. A., and Bargmann, C. I. 1995. Divergent seven transmembrane receptors are candidate chemosensory receptors in *C. elegans*. Cell 83, 207–218.

Troemel, E. R., Kimmel, B. E., and Bargmann, C. I. 1997. Reprogramming chemotaxis responses: sensory neurons define olfactory preferences in *C. elegans*. Cell 91, 161–169.

Tsuchihara, K., Fujikawa, K., Ishiguro, M., Yamada, T., Tada, C., Ozaki, K., and Ozaki, M. 2005. An odorant-binding protein facilitates odorant transfer from air to hydrophilic surroundings in the blowfly. Chem. Senses 30, 559–564.

Vogt, R. G. 2003. Insect Pheromone Biochemistry and Molecular Biology. Elsevier Academic Press.

Vogt, R. G. 2006. How sensitive is a nose? Sci. STKE, pe8.

Vogt, R. G. and Riddiford, L. M. 1981. Pheromone binding and inactivation by moth antennae. Nature 293, 161–163.

Vogt, R. G., Callahan, F. E., Rogers, M. E., and Dickens, J. C. 1999. Odorant binding protein diversity and distribution among the insect orders, as indicated by LAP, an OBP-related protein of the true bug *Lygus lineolaris* (Hemiptera, Heteroptera). Chem. Senses 24, 481–495.

Vogt, R. G., Kohne, A. C., Dubnau, J. T., and Prestwich, G. D. 1989. Expression of pheromone binding proteins during antennal development in the gypsy moth *Lymantria dispar*. J. Neurosci. 9, 3332–3346.

Vogt, R. G., Prestwich, G. D., and Lerner, M. R. 1991. Odorant-binding-protein subfamilies associate with distinct classes of olfactory receptor neurons in insects. J. Neurobiol. 22, 74–84.

Vogt, R. G., Rogers, M. E., Franco, M. D., and Sun, M. 2002. A comparative study of odorant binding protein genes: differential expression of the PBP1–GOBP2 gene cluster in *Manduca sexta* (Lepidoptera) and the organization of OBP genes in *Drosophila melanogaster* (Diptera). J. Exp. Biol. 205, 719–744.

Vosshall, L. B. 2000. Olfaction in *Drosophila*. Curr. Opin. Neurobiol. 10, 498–503.

Vosshall, L. B., Amrein, H., Morozov, P. S., Rzhetsky, A., and Axel, R. 1999. A spatial map of olfactory receptor expression in the *Drosophila* antenna. Cell 96, 725–736.

Vosshall, L. B., Wong, A. M., and Axel, R. 2000. An olfactory sensory map in the fly brain. Cell 102, 147–159.

Wanner, K. W., Willis, L. G., Theilmann, D. A., Isman, M. B., Feng, Q., and Plettner, E. 2004. Analysis of the insect os-d-like gene family. J. Chem. Ecol. 30, 889–911.

Wetzel, C. H., Behrendt, H., Gisselmann, G., Storkuhl, K. F., Hovemann, B., and Hatt, H. 2001. Functional expression and characterization of a *Drosophila* odorant receptor in a heterologous cell system. PNAS 98, 9377–9380.

Wogulis, M., Morgan, T., Ishida, Y., Leal, W. S., and Wilson, D. K. 2006. The crystal structure of an odorant binding protein from *Anopheles gambiae*: evidence for a common ligand release mechanism. Biochem. Biophys. Res. Commun. 339, 157–164.

Wojtasek, H. and Leal, W. S. 1999. Degradation of an alkaloid pheromone from the pale-brown chafer, *Phyllopertha diversa* (Coleoptera: Scarabaeidae), by an insect olfactory cytochrome p450. FEBS Lett. 458, 333–336.

Xia, Y. and Zwiebel, L. J. 2006. Identification and characterization of an odorant receptor from the West Nile Virus mosquito, *Culex quinquefasciatus.* Insect Biochem. Mol. Biol. 36, 169–176.

Xu, P., Atkinson, R., Jones, D. N., and Smith, D. P. 2005. *Drosophila* OBP LUSH is required for activity of pheromone-sensitive neurons. Neuron 45, 193–200.

Xu, P. X., Zwiebel, L. J., and Smith, D. P. 2003. Identification of a distinct family of genes encoding atypical odorant-binding proteins in the malaria vector mosquito, *Anopheles gambiae*. Insect Mol. Biol. 12, 549–560.

Young, J., Shykind, B., Lane, R., Tonnes-Priddy, L., Ross, J., Walker, M., Williams, E., and Trask, B. 2003. Odorant receptor expressed sequence tags demonstrate olfactory expression of over 400 genes, extensive alternate splicing and unequal expression levels, http://genomebiology.Com/2003/4/11/r71. Genome Biol. 4, R71.

Zhang, X. and Firestein, S. 2002. The olfactory receptor gene superfamily of the mouse. Nat. Neurosci. 5, 124–133.

Zhou, J. J., Huang, W., Zhang, G. A., Pickett, J. A., and Field, L. M. 2004. "Plus-c" odorant-binding protein genes in two *Drosophila* species and the malaria mosquito *Anopheles gambiae*. Gene 327, 117–129.

Zwaal, R. R., Mendel, J. E., Sternberg, P. W., and Plasterk, R. H. 1997. Two neuronal G proteins are involved in chemosensation of the *Caenorhabditis elegans* Dauer-inducing pheromone. Genetics 145, 715–727.

4.34 Regeneration of the Olfactory Epithelium

J E Schwob, Tufts University School of Medicine, Boston, MA, USA

R M Costanzo, Virginia Commonwealth University, Richmond, VA, USA

Glossary

BHLH Basic helix–loop–helix transcription factor family.

CDKI Cyclin-dependent kinase inhibitor.

FACS Fluorescence-activated cell sorter.

GBC Globose basal cell.

HBC Horizontal basal cell.

MeBr Methyl bromide.

OB Olfactory bulb.

OE Olfactory epithelium.

ON Olfactory nerve.

OR Odorant receptor.

4.34.1 Introduction and Early Literature

The capacity of the olfactory epithelium (OE), especially its population of sensory neurons, to recover after injury, reconnect with the central nervous system (CNS), and reestablish function has been known from the first half of the twentieth century (Nagahara, Y., 1940). In addition to experimental studies on the consequences of transection of the olfactory nerve (ON) along its course from the OE to the olfactory bulb (OB), an extensive literature documents sensory recovery in humans in the large cohort of children and young adults who were treated by irrigation with zinc sulfate as a means of polio prophylaxis. Subsequent experimental investigations demonstrated anatomical reconstitution of the epithelium following zinc sulfate treatment in a variety of experimental species, including primates (Schultz, E. W., 1941; Schultz, E. W. and Gebhardt, S. P., 1942; Schultz, E. W., 1960; Mulvaney, B. D. and Heist, H. E., 1971; Matulionis, D. H., 1975; 1976). Studies performed by Graziadei, Moulton, and colleagues in the 1960s and 1970s demonstrated that the epithelium of laboratory animals harbored a population of rapidly proliferating cells that functioned as neuronal progenitors and gave rise to neurons (Moulton, D. G. *et al.*, 1970; Graziadei, P. P. and Metcalf, J. F., 1971; Graziadei, P. P., 1973; Moulton, D. G., 1974; Graziadei, P. P. C. and Monti Graziadei, G. A., 1978; Graziadei, P. P. and Monti Graziadei, G. A., 1979). This remarkable and seminal discovery invigorated the cellular and molecular investigations of olfactory neurogenesis, which have carried on to this day.

From a teleological perspective, the persistence of neurogenesis in the normal adult OE and the anatomical and functional recovery following injury to the epithelium or the nerve are highly adaptive. The epithelium is particularly vulnerable to any number of toxins, infectious agents, or other forms of insult, given the direct contact of the neural elements with the ambient, airborne environment. Moreover, the neurons and their axons form a direct conduit from the outer world into the CNS that may be used as a route of entry to the brain by pathogens, such as neurotropic viruses (Twomey, J. A. *et al.*, 1979; Monath, T. P. *et al.*, 1983; Stroop, W. G. *et al.*, 1984; Barnett, E. M. *et al.*, 1993; Barnett, E. M. and Perlman, S., 1993). Thus, a propensity for the neurons to die with little provocation may provide a survival advantage

in the short term for the organism by preventing the transmission of pathogens, but requires some capacity for compensatory reconstitution of the system after the exposure period has passed.

The intent of this review is to summarize the current state of our knowledge regarding cellular renewal in the normal OE, the enhanced production of neurons and nonneural cell types after injury, reinnervation of the bulb following epithelial reconstitution, and restoration of sensory function subsequent to anatomical recovery. We will not consider the vomeronasal system.

For heuristic purposes, our discussion will focus first on the epithelium and its capacity for reconstitution and then turn to the ON and its innervation of the bulb.

4.34.2 Structure of the Olfactory Mucosa: Layers and Compartments

The olfactory mucosa forms a continuous sheet lining the posterodorsal part of the nasal septum, the tips of the opposing nasal turbinates, and much of the turbinate shafts and cul-de-sacs at the same level (Moulton, D. G. and Beidler, L. M., 1967; Grazaiadei, P. P. C., 1971). In addition, a small patch of olfactory mucosa, termed the nasal septal organ (NSO) (of Masera), detaches from this sheet at an early embryological stage and comes to reside just dorsal to the opening of the nasopharyngeal duct (Bojsen-Moller, F., 1975; Weiler, E. and Farbman, A. I., 2003); the NSO is, for all intents and purposes, identical to the main OE in morphology, molecular phenotype, and function, and will not be considered separately (Taniguchi, K. *et al.*, 1993).

The olfactory mucosa consists of a pseudostratified epithelium and a lamina propria that rests directly on the bony and cartilagenous skeleton of the nose (de Lorenzo, A. J., 1957; Andres, K. H., 1966; Moulton, D. G. and Beidler, L. M., 1967; Grazaiadei, P. P. C., 1971; Graziadei, P. P. and Monti Graziadei, G. A., 1979). The epithelium contains a handful of cell types: neurons, supporting cells and basal cells (Figure 1(a)). Each of these categories may be further subdivided.

4.34.2.1 Sensory Neurons

The population of neurons consists of both mature, ciliated, olfactory marker protein (OMP)-expressing and immature, GAP-43 (+) cells that have yet to

(a)

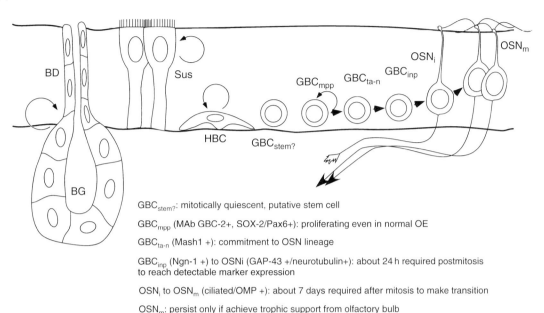

GBC$_{stem?}$: mitotically quiescent, putative stem cell

GBC$_{mpp}$ (MAb GBC-2+, SOX-2/Pax6+): proliferating even in normal OE

GBC$_{ta-n}$ (Mash1 +): commitment to OSN lineage

GBC$_{inp}$ (Ngn-1 +) to OSNi (GAP-43 +/neurotubulin+): about 24 h required postmitosis to reach detectable marker expression

OSN$_i$ to OSN$_m$ (ciliated/OMP +): about 7 days required after mitosis to make transition

OSN$_m$: persist only if achieve trophic support from olfactory bulb

(b)

GBC$_{mpp}$ (MAb GBC-2+, SOX-2/Pax6+): give rise ultimately to OSNs, GBCs, Sus, HBCs, duct/gland cells and ciliated columnar respiratory epithelial cells in the context of the MeBr-lesioned OE..

GBC$_{sus}$ (Hes1+): appear immediately after lesion and then differentiates into Sus cell(s)

GBC$_{ta-n}$ (Mash1+): committment to the neuronal lineage is delayed until 2 days after lesion

GBC$_{inp}$ (Ngn-1+, NeuroD+): do not return until 3 days after lesion, 1 day prior to reappearance of OSN$_i$ (GAP-43+/neurotubulin+)

Figure 1 The processes of cellular renewal differ between normal and lesioned olfactory epithelium. (a) In normal epithelium or in epithelium in which the population of neurons has been depleted as a consequence of acute damage to olfactory nerve or bulb or where neuronal turnover is accelerated for neurons born following bulb ablation, the various cell types arise via self-replacement or, in the case of the sensory neurons, from the mitosis of globose basal cells (GBC). (b) In contrast to the foregoing, following methyl bromide exposure, more promiscuous lineage relationships emerge: GBCs will give rise to neurons and themselves (a form of self-renewal), of course, but also to horizontal basal cells (HBCs), sustentacular (Sus) cells, Bowman gland (BG), and duct (BD) cells, and even respiratory epithelial (RE) cells. In essence, GBCs are demonstrably as potent in the context of the nasal epithelium as the cells of the olfactory placode. BD cells also evince a greater than normal potency and contribute to the replacement of the Sus cell population. The GBC population is heterogeneous with respect to functional phenotype, and the various kinds can be distinguished on the basis of their expression of member of various transcription factor families.

elaborate cilia (Verhaagen, J. *et al.*, 1989; Meiri, K. F. *et al.*, 1991). The population of typical olfactory neurons can be subdivided further on the basis of which single odorant receptor (OR) is expressed from among the family of about 1300 OR genes (Buck, L. and Axel, R., 1991; Zhang, X. and Firestein, S., 2002). That choice is constrained by a neuron's position along the dorsomedial-to-ventrolateral axis of the epithelium. Along that axis, each OR is limited to a narrow band that reaches through the anteroposterior extent of the OE (Ressler, K. J. *et al.*, 1993; Vassar, R. *et al.*, 1993). Although the spatial organization has been described as sharply bounded and zonal, more recent studies suggest that each OR-defined band overlaps partially with a number of other ones (Iwema, C. L. *et al.*, 2004; Miyamichi, K. *et al.*, 2005). As a consequence, the organization of OR expression across epithelial space is better described as a very large number of bands that gradually shift in position along that dorsomedial-to-ventrolateral axis, rather than a very small number of zones. From a developmental perspective, the organization in shifting, sharply bounded bands is as easily explained as the four-zone model; for example, a cell's response to a gradient of a signaling molecule can generate shifting bands as easily as a discrete number of stripes (Wolpert, L., 1989). Other features of the neuronal phenotype are also position-dependent including connectivity and expression of certain cell-surface markers, and may also be sharply bounded (Mori, K. *et al.*, 1985; Schwob, J. E. and Gottlieb, D. I., 1986; 1988; Schwarting, G. A. and Crandall, J. E., 1991; Schwarting, G. A. *et al.*, 1992; Yoshihara, Y. *et al.*, 1997).

In addition to the foregoing, atypical olfactory neurons are concentrated in (but not limited to) the epithelium lining the cul-de-sacs of the posterior nasal cavity, which differ from the majority. The sensory transduction apparatus is based on guanylyl cyclase expression rather than adenylyl cyclase; they lack OMP; finally, they project to the necklace glomeruli at the posterior edge of the glomerular sheet (Shinoda, K. *et al.*, 1989; 1993, Juilfs, D. M. *et al.*, 1997; Ring, G. *et al.*, 1997). The neurons elaborate thin unmyelinated axons that gather at the basal lamina under the arch-like process of the horizontal basal cells (HBCs) where the axons first come into contact with the glia of the ON.

4.34.2.2 Supporting Cells

The microvillar-capped supporting cells whose nuclei reside at the apex of the epithelium can be subdivided into sustentacular (Sus) and microvillar cells. Sus cells contain abundant smooth endoplasmic reticulum, while microvillar cells do not. Both express cytokeratins 8 and 18, but otherwise have different biochemical profiles: Sus cells stain with MAbs SUS-1 and SUS-4, while microvillar cells label with the MAb 1A-6, the lectin Iβ-4 from *Bandeiraea simplicifolia* (BS-I), and express ankyrin. Furthermore, other characteristics can be used to differentiate the two types of microvillar cells (Hempstead, J. L. and Morgan, J. I., 1985a; Goldstein, B. J. and Schwob, J. E., 1996; Asan, E. and Drenckhahn, D., 2005).

4.34.2.3 Basal Cells

The population of basal cells can be subdivided on the basis of their relationship to the basal lamina. Horizontal (or dark) basal cells (HBCs) attach directly to the basal lamina via desmosomes, express cytokeratins 5 and 14, the epidermal growth factor-receptor (EGF-R), intercellular adhesion molecule (ICAM), integrins, and a carbohydrate moiety recognized by the lectin BS-I (Graziadei, P. P. and Monti Graziadei, G. A., 1979; Holbrook, E. H. *et al.*, 1995; Carter, L. A. *et al.*, 2004).

Globose (or light) basal cells (GBCs) are situated at a remove from the basal lamina, and are distributed nonuniformly across the tangential extent of the epithelium (Graziadei, P. P. and Monti Graziadei, G. A., 1979; Holbrook, E. H. *et al.*, 1995; Loo, A. T. *et al.*, 1996). Conventionally, GBCs have been identified and defined on exclusionary grounds; cells have been classified as GBCs if morphologic simplicity and molecular phenotype indicate that they are not HBCs, not neurons, and not duct cells. A few positive markers have been generated, which are not strictly limited to the GBC population (Goldstein, B. J. and Schwob, J .E., 1996; Goldstein, B. J. *et al.*, 1997; Jang, W. *et al.*, 2003). For example, several monoclonal antibodies (MAbs) label GBCs along with neurons in normal epithelium, but not the other cell types. One of these MAbs – GBC-3 – binds to the 34 kDa, that is, immature, form of the laminin-receptor protein (iLRP) (Jang, W. *et al.*, 2007).

In addition, subsets of GBCs are labeled by a variety of TFs of the basic helix–loop–helix (bHLH), LHX, Msx, Pax, and SOX families, as well as others (Cau, E. *et al.*, 1997; 2000, Norlin, E. M. *et al.*, 2001; Cau, E. *et al.*, 2002; Suzuki, Y. *et al.*, 2003; Hirota, J. and Mombaerts, P., 2004; Kolterud, A. *et al.*, 2004; Guo, Z. *et al.*, 2005b). Thus, it appears that the population of GBCs is markedly heterogeneous on biochemical grounds that likely correspond with differences in their capacities as progenitor cells.

A third type of basal cell has also been noted on ultrastructural grounds; it extends a foot process, contacts the basal lamina via an adherens junction, and is more irregularly shaped than most other GBCs (Graziadei, P. P. and Monti Graziadei, G. A., 1979; Holbrook, E. H. *et al.*, 1995). Some authors have considered these cells as a transition stage between HBCs and GBCs (or vice versa) (Graziadei, P. P. and Monti Graziadei, G. A., 1979), but no evidence other than morphological can be adduced for this notion. Because, in fact, the third type cells attach by adherens and not desmosomal junctions, there are really no data, morphologic or otherwise, to indicate intermediacy of this third type of basal cell or to draw a lineage relationship between GBCs and HBCs. However, little else is known about this type.

4.34.2.4 Bowman's Glands/Ducts

The final elements in the epithelium are Bowman's ducts, which extend from the Bowman's glands in the lamina propria to the apical surface where the mucus formed by the glands is discharged. The glands themselves are found within the lamina propria. The ducts and gland cells share some phenotypic characteristics with Sus cells, including a common reactivity with several MAbs and high expression of cytochrome P450s (Hempstead, J. L. and Morgan, J. I., 1985a; 1985b; Chen, Y. *et al.*, 1992; Huard, J. M. *et al.*, 1998).

4.34.2.5 Cells of the Lamina Propria

Other elements in the lamina propria include fibroblasts, migrating lymphocytes, and tissue macrophages. There may be other cell types within the lamina propria; for example, cells express retinoic acid (RA)-synthesizing enzymes at differential levels as a function of position along a dorsomedial-to-ventrolateral axis have not yet been fully characterized as to type. Whether fibroblastic or not, they may be critical for the reassembly of the system after lesion.

4.34.2.6 Fascicles of the Olfactory Nerve

The fascicles of the olfactory and vomeronasal nerve also course through the lamina propria. At a macroscopic level, the nerve begins as small bundles of axons that are evident where the basal lamina is breached.

During their course posteriorward, the bundles gradually sink more deeply into the lamina propria where they obliquely contact, fuse with, and form larger fascicles of the nerve. The microscopic organization of the fascicles differs markedly from that of other peripheral nerves. The glia of the ON fascicles, termed olfactory ensheathing cells (OECs), are found within the perineurial sheath, which is formed by what have been termed ON fibroblasts (de Lorenzo, A. J., 1957; Andres, K. H., 1966; Barber, P. C. and Lindsay, R. M., 1982; Field, P. *et al.*, 2003; Li, Y. *et al.*, 2005). The perikarya of the OECs are found plastered against the edge of the nerve or of the endoneurial septae that subdivide the larger fascicles, while their processes interdigitate among the olfactory axons. In contrast to the mature spinal nerve in which single unmyelinated axons are segregated each within its own individual trough (Peters, A. *et al.*, 1991), collections of olfactory axons are in intimate contact with each other and isolated from the environment by the sheet-like processes of the OECs (de Lorenzo, A. J., 1957; Andres, K. H., 1966; Barber, P. C. and Lindsay, R. M., 1982; Field, P. *et al.*, 2003; Li, Y. *et al.*, 2005). In that respect, the relationship of bundles of olfactory axons and OEC glia in the adult is like that of the Schwann cell and axons of growing spinal nerves in the embryo, only writ large (Peters, A. *et al.*, 1991). The unique relationship between olfactory axons and glia begins within the pocket formed between the overarching processes of the HBCs and the basal lamina where the axons exit the epithelium and is maintained up to the entry of the axons into the neuropil of the olfactory glomeruli (Holbrook, E. H. *et al.*, 1995; Li, Y. *et al.*, 2005).

4.34.3 Ongoing Olfactory Neurogenesis

The intrinsic capacity of the OE to produce neurons throughout life was implicit in the demonstrations of structural and functional recovery by the olfactory system following experimental injury. However, a more direct understanding of the basis for ongoing neurogenesis came from the discoveries of the 1960s and 1970s by Graziadei, Moulton, and their respective colleagues that cells near the base of the OE in adult rodents incorporate labeled analogues of thymidine after pulse administration and that the thymidine-labeled cells can be chased with time and watched as they differentiate into neurons and then eventually disappear (Figure 1(a)) (Moulton, D. G. *et al.*, 1970; Graziadei, P. P. and Metcalf, J. F.,

1971; Graziadei, P. P., 1973; Moulton, D. G., 1974; Graziadei, P. P.C. and Monti Graziadei, G. A., 1978; Graziadei, P. P. and Monti Graziadei, G. A., 1979). During their cellular differentiation, the newborn neurons first extend a dendrite and express the axonal growth protein GAP-43 (Verhaagen, J. *et al.*, 1989; Meiri, K. F. *et al.*, 1991; Schwob, J. E. *et al.*, 1995). Within about a week of their birth the neurons have elaborated cilia, turned off GAP-43, and turned on OMP (Miragall, F. and Monti Graziadei, G. A., 1982; Schwob, J. E. *et al.*, 1995). OR expression lags the onset of neuronal differentiation by a few days, but is seen before OMP is expressed (Iwema, C. L. and Schwob, J. E., 2003).

Of the two cell types in the basal zone of the OE, tracing of cell lineage by use of a replication-incompetent, retrovirally derived vector (RRVV) or by dye injection into basal cells both show that neurons derive from GBCs (Schwartz Levey, M. *et al.*, 1992; Caggiano, M. *et al.*, 1994; Schwob, J. E. *et al.*, 1994a). The lineage analyses correspond with the demonstration that there is a parallel increase in GBC proliferation and neuronal production occasioned by damage to the ON and OB (Schwartz Levey, M. *et al.*, 1991). Subsequent analyses of the sequential expression of TFs that activate neurogenesis (i.e., Mash1, Ngn1, and NeuroD1) by GBCs during development and after lesion, provide a further indication that GBCs are the proximate precursors of neurons in the OE (Guillemot, F. *et al.*, 1993; Gordon, M. K. *et al.*, 1995; Cau, E. *et al.*, 1997; 2002; Manglapus, G. L. *et al.*, 2004).

Does GBC proliferation and neuron production persist throughout life? In most humans and experimental animals, the answer is yes. However, in humans, biopsies do show areas of epithelium that remain olfactory in character (i.e., containing microvillar-capped Sus cells) but are aneuronal, suggesting that the neurogenic capacity of the olfactory stem and progenitor cells can be exhausted (Holbrook, E. L. *et al.*, 2005). Interestingly, the same neurogenic exhaustion is observed in rodents in which an oncogene is expressed under the control of the OMP promoter (Largent, B. L. *et al.*, 1993); as the expression of an oncogene in a terminally differentiated cell often initiates programmed cell death (Al-Ubaidi, M. R. *et al.*, 1997; Feddersen, R. M. *et al.*, 1997), neuronal turnover may be grossly accelerated, and from an early stage, in these transgenic mice.

At the time the original data were obtained indicating adult neurogenesis, only the olfactory neurons were believed to be subject to replacement. Subsequently, the birth of other neuronal populations in adults, including the interneurons of the OB, was demonstrated elsewhere in the adult nervous system (Kaplan, M. S. and Hinds, J. W., 1977; Kaplan, M. S. *et al.*, 1985; Lois, C. and Alvarez-Buylla, A., 1993; 1994; Menezes, J. R. *et al.*, 1995). Nonetheless, the rate of production as a percentage of the established population appears higher in the OE than anywhere else, and more subject to regulation on the basis of need.

The original observations on basal cell proliferation were interpreted as an indication that sensory neurons within the OE had a fixed, finite lifespan, like epithelial cells of gut or skin (Moulton, D. G. *et al.*, 1970; Graziadei, P. P. and Metcalf, J. F., 1971; Graziadei, P. P., 1973; Moulton, D. G., 1974; 1975; Graziadei, P. P. C. and Monti Graziadei, G. A., 1978; Graziadei, P. P. and Monti Graziadei, G. A., 1979). Moreover, the rate of proliferation was used as a means of estimating that the neurons lived for about a month, on average. However, the pulse–chase data cannot be interpreted as providing evidence for the notion that the turnover of mature, synaptically connected olfactory sensory neurons (OSNs) occurs automatically at the end of a fixed (relatively) brief lifespan. Neither the original nor subsequent data preclude the notion that turnover might take place within a reserve population of unconnected, relatively immature neurons. In fact, neurons undergoing programmed cell death or apoptosis are found within both the immature and OMP-positive population in the normal adult OE (Holcomb, J. D. *et al.*, 1995; Deckner, M. L. *et al.*, 1997); in that vein it is also important to note that the onset of OMP expression is NOT by itself an indication of synaptic connectivity. Subsequent demonstrations have also weakened the association between ongoing neurogenesis and any prescribed replacement of mature neurons in the protected environment experienced by laboratory-housed animals. For example, the lifespan of newly born OSNs from animals that are in a pathogen-free, filtered air environment may be as long as a year (Hinds, J. W. *et al.*, 1984). Moreover, when the preexisting neurons are tagged by retrograde transport of a stable marker from the bulb, the labeled OSNs persist for a period of time that is substantially longer than a month (Mackay-Sim, A. and Kittel, P. W., 1991b); unfortunately, these data do not permit a determination of whether lifespan is indefinite once a neuron becomes established. Finally, a mechanism exists that allows the system to regulate survival of newly generated neurons on the basis of their connectivity status, which would be required if lifespan is not inherently limited. In more specific terms, OSNs exhibit a trophic dependency on the OB for their prolonged survival, such that lifespan is abbreviated for neurons born in the absence of the bulb (Monti Graziadei, G. A., 1983; Biffo, S. *et al.*, 1992; Schwob, J. E. *et al.*, 1992; Carr, V. M. and Farbman, A. I., 1993; Mackay-Sim, A. and Chuah, M. L., 2000). Trophic dependency is a means by which the system might eliminate those newly born neurons that fail to make any or the appropriate synaptic connections (Purves, D., 1988). Thereby the size of the sensory population may be regulated in accordance with the status of the target, which might, itself, be modified by sensory activity and odorant saliency, among other influences. Given these provisos, it is fair to say that the definitive experiments have not been done that would allow us to determine the lifespan of preexisting sensory neurons in the laboratory environment or any other milieu.

4.34.4 Response to Injury – Reconstitution of the Epithelium

The primary olfactory system has been subjected to three general forms of experimental injury: direct toxic injury to the OE; traumatic injury to the ON, OB, or both; and viral infection following intranasal inoculation.

4.34.4.1 Toxic Injury to the Olfactory Epithelium

Damage to the OE ensues in some cases by direct contact of a toxin with the epithelium; the two most widely used of these are zinc-containing salts, in particular zinc sulfate, delivered by intranasal infusion (Smith, C. G., 1938; Schultz, E. W., 1960; Matulionis, D. H., 1975; 1976; Harding, J. W. *et al.*, 1978; Cancalon, P., 1982; Burd, G. D., 1993), and MeBr, for which passive inhalation is the route of exposure (Hurtt, M. E. *et al.*, 1987; 1988; Schwob, J. E. *et al.*, 1995). In other cases, some toxins damage or destroy the OE after systemic administration; among these are 3-methyl indole (skatol), dichlobenil, and methimazole (Schwob, J. E. *et al.*, 1994b; Brittebo, E. B., 1995; Genter, M. B. *et al.*, 1995; 1996; Slotnick, B. *et al.*, 2001). The mechanism by which zinc sulfate causes relatively selective damage to the OE after intranasal administration is not known in any detail, although speculation has centered on the protein-coagulative properties of the salt (Schultz, E. W. and Gebhardt, S. P., 1942). MeBr, which effect is absolutely selective for damage

to the OE within the nose, is apparently converted to an active free radical toxin via biotransformation enzymes in the epithelium, most likely one or more members of the cytochrome P450 family (Reed, C. J. *et al.*, 1995). Other features of the olfactory toxicity of MeBr that reflect the need for activation include the relative insensitivity of younger animals, the strain dependence of efficacy in mice, and the absolute refractoriness to lesion under circumstances when the biotransformation systems are substantially diminished relative to normal (e.g., the OE cannot be lesioned again by a subsequent exposure to toxin during the immediate post-MeBr lesion period when cytochrome P450 enzymes are far below normal levels) (Schwob, J. E., unpublished results). Other systemically acting toxins, for example, skatol, may also exert their effect after enzymatic activation by that system.

The immediate effects of lesion have been studied in some detail for both zinc sulfate and MeBr. However, much of the zinc literature predates the availability and use of cell-type-specific reagents to study reconstitution of the OE. Accordingly, the timing and events of epithelial recovery is better known from the studies of MeBr lesions published more recently (Figure 1(b)). In virtually all cases where a direct comparison can be made, the findings with MeBr are highly analogous to those seen following zinc sulfate.

Inhalation of the toxin in MeBr-sensitive animals leads to a rapid loss (sloughing) of all neurons and supporting cells in over 95% of the OE within 24 h, and damage to those populations in the remainder of the tissue (Schwob, J. E. *et al.*, 1995). In addition, the Bowman's glands in the lamina propria are damaged but incompletely. The GBC, HBC, and duct cell populations also sustain some loss. In particular, HBCs disappear from the ventral OE in lesioned rats (but not mice) by the third day after exposure. In contrast, agents like methimazole take several days to produce maximal damage, which makes study of the initial events during epithelial regeneration difficult due to the concurrence of cell loss and replacement (Brittebo, E. B., 1995; Genter, M. B. *et al.*, 1995; Bergstrom, U. *et al.*, 2003). Within 2 days after MeBr, the remaining cells undergo a quick upsurge in proliferation. The first neurons to reappear do so at about 4 days after lesion, and then mature, that is, express OMP, around the middle of the second week after lesion, at roughly the same time as the first returning axons reinnervate glomeruli in the bulb. Likewise, the Sus cells begin to reappear within a few days, and reform a distinct layer at the apex of the epithelium by 10 days after lesion. By 6–8 weeks after MeBr, the majority of the OE has stabilized to a state that is indistinguishable from normal, as judged by the rate of GBC proliferation, the relative proportion of immature-to-mature OSNs, and the restoration of normal or near-normal levels of expression of cytochrome P450 enzymes (Schwob, J. E. *et al.*, 1995).

Depending on its severity, the initial lesion may preclude the recovery of the lesioned epithelium as olfactory, and the events underlying failure may model the pathophysiology of some forms of human olfactory dysfunction (Schwob, J. E. *et al.*, 1994b; Jang, W. *et al.*, 2003). For example, MeBr lesion can be enhanced such that all GBCs are eliminated from patches of the OE; the GBC-depleted areas reconstitute as respiratory epithelium (a form of metaplasia) (Jang, W. *et al.*, 2003). The same kind of metaplasia, is observed following zinc sulfate infusion or a combination of skatol injection and MeBr inhalation, and is often more extensive than with MeBr (Smith, C. G., 1938; Burd, G. D., 1993; Schwob, J. E. *et al.*, 1994b).

4.34.4.2 Epithelial Response to Nerve Transection or Olfactory Bulb Ablation

In contrast to the effects of and response to directly acting toxins, cell loss following nerve or bulb injury is limited to the OSNs. Nerve transection or bulb ablation causes retrograde degeneration of the neurons via a caspase-mediated process of programmed cell death (or apoptosis) within the space of a few days (Harding, J. *et al.*, 1977; Monti Graziadei, G. A. and Graziadei, P. P., 1979; Simmons, P. A. *et al.*, 1981; Costanzo, R. M. and Graziadei, P. P., 1983; Monti Graziadei, G. A., 1983; Verhaagen, J. *et al.*, 1990; Schwartz Levey, M. *et al.*, 1991; Carr, V. M. and Farbman, A. I., 1992; Gordon, M. K. *et al.*, 1995; Holcomb, J. D. *et al.*, 1995; Calof, A. L. *et al.*, 1998; Cowan, C. M. *et al.*, 2001; Cowan, C. M. and Roskams, A. J., 2002). Neuronal loss causes the upregulation of GBC proliferation only; the feedback inhibition from OSNs onto the mitotic activity of GBCs in the normal OE, which is apparently mediated by bone morphogenetic proteins (BMPs) 4 and 7 and by growth differentiation factor (GDF)11, is released (Gordon, M. K. *et al.*, 1995; Shou, J. *et al.*, 1999; 2000; Wu, H. H. *et al.*, 2003). Moreover, cytokines, including leukemia inhibitory factor (LIF), that are released by infiltrating macrophages may also play a role in the activation of GBCs in response to neuronal

degeneration (Nan, B. *et al.*, 2001; Bauer, S. *et al.*, 2003). As a consequence, expansive proliferation occurs within the GBC population, followed by an increased rate of neuronal production (Figure 1(a)) (Monti Graziadei, G. A. and Graziadei, P. P., 1979; Schwartz Levey, M. *et al.*, 1991). Basal cell proliferation and neuron production remains elevated over the long term in bulb-ablated animals, and perhaps nerve transected ones as well (Schwob, J. E. *et al.*, 1992). Following ablation, the abbreviated lifespan caused by the loss of trophic support from the bulb is such that the neuronal population is dominated by immature OSNs, instead of mature ones, while the size of the population always remains less than that of normal OE, despite accelerated neurogenesis (Verhaagen, J. *et al.*, 1990; Biffo, S. *et al.*, 1992; Schwob, J. E. *et al.*, 1992; Konzelmann, S. *et al.*, 1998). In contrast, nerve transection permits substantial though incomplete reinnervation of the bulb as well as some mistargeting; as a consequence, the population of OSNs recovers more fully toward normal than after bulb ablation (Monti Graziadei, G. A. and Graziadei, P. P., 1979; Graziadei, P. P. and Monti Graziadei, G. A., 1980; Costanzo, R. M., 1984; 1985; Koster, N. L. and Costanzo, R. M., 1996; Yee, K. K. and Costanzo, R. M., 1998; Costanzo, R. M., 2000; Christensen, M. D. *et al.*, 2001; Costanzo, R. M., 2005).

Other cell populations in the OE show a response to the lesion, although it is not mitotic. For example, the levels of cytochrome P450 enzymes decrease in Sus cells (Walters, E. *et al.*, 1992). In addition, Sus cells become active phagocytes (Suzuki, Y. *et al.*, 1995; 1996).

4.34.4.3 Effects of Viral Infection on the Olfactory Epithelium

The damage to the epithelium caused by viral infection is a much more frequent antecedent of olfactory dysfunction in humans than either toxin exposure or head trauma (for which nerve transection and/or bulb ablation is generally held to be analogous) (Doty, R. L., 1979; Moran, D. T. *et al.* 1992; Seiden, A. M., 2004; Holbrook, E. H. and Leopold, D. A., 2006). Biopsies of the human mucosa and specimens taken at autopsy indicate that the replacement of olfactory by respiratory epithelium is not a rare event in the human mucosa (Nakashima, T. *et al.*, 1984; Nakashima, T. *et al.*, 1985). Indeed, loss of neurogenic capacity due to respiratory metaplasia or stem cell exhaustion can become complete, leading to anaxonic, fibrotic ON fascicles (Holbrook, E. H. *et al.*, 2005). (Collagen deposition and fibrosis of ON fascicles is also observed in Macaque, but not in rodents (Graziadei, P. P. *et al.*, 1980).) Experimental models that duplicate the changes observed following viral infections in humans are rare and mechanistically obscure. However, it is worth noting that certain neurotropic viruses can gain entry to the CNS by infection of OSNs and anterograde transport to, and release from, the terminals of the ON, without killing the infected sensory neurons (Barthold, S. W., 1988; Schwob, J. E. *et al.*, 2001). To the extent that they cause downstream damage to the OB, the OE can come to resemble that which is harvested from bulbectomized animals.

4.34.5 Stem and Progenitor Cells of the Adult Olfactory Epithelium

The unique capacity of the OE to reconstitute after injury, for example, following inhalation of MeBr, reflects and requires the persistence of one or more types of neurocompetent stem cells, from which spring the downstream progenitors that eventually give rise to the various missing cell types. Of necessity, neurocompetent stem cells must be found within one or more of the three cell types that are spared following MeBr lesion: GBCs, HBCs, or duct cells. Of equal necessity, candidate olfactory stems must satisfy a certain operational definition that emerged from studies of cellular renewal in tissues like skin and gut, as follows. Tissue stem cells are totipotent (with respect to the tissue in question), infinitely self-renewing, and mitotically quiescent when the tissue is undamaged (Hall, P. A. and Watt, F. M., 1989; Rao, M. S., 1999; Watt, F. M., 2001; Alonso, L. and Fuchs, E., 2003; Pevny, L. and Rao, M. S., 2003). In the conventional model, progenitors of progressively more limited differentiative potency or capacity for self-renewal or both arise from and are downstream of the stems. Most investigators identify a kind of lineage-restricted progenitor that retains a capacity for expansive proliferation and call it a transit-amplifying (TA) cell (Potten, C. S. and Loeffler, M., 1990; Potten, C. S., 1998; Watt, F. M., 2001; Alonso, L. and Fuchs, E., 2003). Further downstream are the cells that are charged with giving rise to the differentiated tissue elements, and are called immediate precursors, for example, immediate neuronal precursors (INPs) in the OE.

4.34.5.1 Some Global Basal Cells Are Multipotent Progenitors

Our current understanding of the players and processes underlying cellular renewal in the olfactory system is based on RRVV lineage tracing experiments (Huard, J. M. *et al.*, 1998), *in vivo* transplantation studies using purified populations of cells (Goldstein, B. J. *et al.*, 1998; Chen, X. *et al.*, 2004), on gene expression analysis following lesion or in the embryonic placode (Manglapus, G. L. *et al.*, 2004), on the consequences of gene mutation (Cau, E. *et al.*, 1997; Fisher, A. and Caudy, M., 1998; Cau, E. *et al.*, 2000; 2002; Murray, R. C. *et al.*, 2003), and on *in vitro* studies (Shou, J. *et al.*, 1999; Satoh, M. and Yoshida, T., 2000; Carter, L. A. *et al.*, 2004). All of these data indicate that some cells within the population of GBCs are capable of a broad multipotency in the context of the MeBr-lesioned OE; hence, the GBCs may ultimately be responsible for the reconstitution of much of the OE after lesion and not merely the neuronal population (Figures 1(b) and 2). In aggregate, GBCs give rise to GBCs (a form of self-renewal), neurons, Sus cells, microvillar cells, HBCs, duct and gland cells, and even respiratory epithelial cells. Not all cell types are found in all clones – individual clones from transplanted GBCs or from RRVV-labeled progenitors may contain multiple cell types, or may consist of neurons or Sus cells alone – but there is a rough correlation between large clones, a greater number of cell types, and the presence of GBCs (i.e., a form of self-renewal). Hence, the differentiative potency of GBCs in the MeBr-lesioned OE seems to be as broad as that of the olfactory placodal cells, which are the embryonic precursors for all of the nasal lining. It is important to note that some multipotent cells present in the normal epithelium are active in the mitotic cycle as shown by infectability with RRVV *ex vivo* prior to transplantation into a host. Thus, the decision between neuropotency and multipotency is one that is made on a continuous basis in the normal OE.

4.34.5.2 Globose Basal Cells Are the Only Population for Which Functional Heterogeneity Has Been Demonstrated

Among the GBC population of both embryonic and adult OE are subsets that can be distinguished one from the next on the basis of their expression of TFs of the bHLH family of TFs, which in turn appear to correlate with difference in their potency as precursors (Cau, E. *et al.*, 1997; 2000; 2002; Manglapus, G. L. *et al.*, 2004). In particular, observations on mice bearing engineered mutations of the bHLH family members as well as analyses of TF expression in developing and regenerating OE have lead to a conceptualization of flow through the GBC population that acknowledges the diversity of precursor types (Figure 2). At the origin are the stem cells and broadly multipotent GBCs (GBCs$_{MPP}$) capable of giving rise to all cell types present in the OE as well as the respiratory epithelial cells normally found elsewhere in the nasal lining; it is likely that some, possibly most, broadly potent GBCs are not stems. Downstream of them are cells en route to terminally differentiated neurons only; among the downstream elements committed to neurogenesis are transit-amplifying GBCs (GBCs$_{TA}$) and INPs. Based on a comparison with TF markers, GBCs of all three functional types – MPPs, TAs, and INPs – are actively proliferating in the normal OE. In the usual formulation, the cellular flow in this model is depicted as unidirectional, and from less committed to more (Figure 2). However, as a greater appreciation for the plasticity of stems and progenitors has emerged, a case can be made for reevaluating the conventional notion of unidirectional flow.

Some of the molecular correlates of multipotency among the GBCs can be adduced (Figure 2). GBCs$_{MPP}$ express antigens that are recognized by MAbs GBC-1, GBC-2, and GBC-3 (the latter recognizing a carbohydrate moiety on iLRP) based on the comparison of timing of expression after MeBr lesion, the outcome of RRVV infection, and the use of these antibodies for fluorescence-activated cell sorter (FACS) (Goldstein, B. J. and Schwob, J. E., 1996; Goldstein, B. J. *et al.*, 1997; Huard, J. M. *et al.*, 1998; Jang, W. *et al.*, 2003; Chen, X. *et al.*, 2004; Jang, W. *et al.*, 2007). However, none of the three differentiates GBC$_{MPP}$ from GBCs committed to making neurons selectively.

The timing of bHLH TF expression relative to recovery of the OE after MeBr lesion (Manglapus, G. L. *et al.*, 2004) and the pattern of gene expression and epithelial composition in Mash1 (−/−) mutant animals, (Guillemot, F. *et al.*, 1993; Cau, E. *et al.*, 1997) help narrow and contrast the molecular profile of the MPPs and other progenitors. These data suggest that the GBCs$_{MPP}$ do not express Mash1, nor the TFs NeuroD and Ngn1, which are expressed after Mash1 in the embryonic and in the regenerating OE. GBCs$_{TA}$ that are downstream of the GBCs$_{MPP}$ and

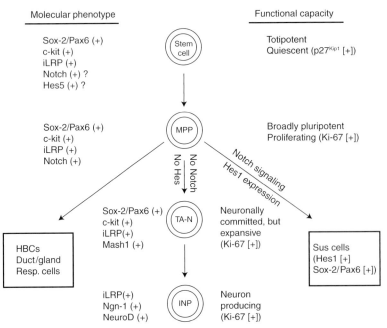

Figure 2 Functional classification and categorization of globose basal cells (GBCs) correlates with molecular phenotype. Central to differences in progenitor capacity seem to be members of the basic helix–loop–helix family of transcription factors (Hes, Mash, Ngn, and NeuroD) as well as the transcription factors Sox-2 and Pax6. For the latter factors, the pattern of expression in normal epithelium and during recovery from methyl bromide (MeBr) lesion suggests that they repress neuronal differentiation, which they do in other parts of the nervous system, a function which is required up to the final progenitor cell stage in the epithelium.

committed to a neuronal fate express Mash1, based on an analysis of Mash1 mutant mice and of the timing of its expression relative to lesion and embryonic development. Other GBCS that are further downstream and act as immediate neuronal precursors or producers (GBCs$_{INP}$) express NeuroD and/or Ngn1. All three functional types are present and mitotically active in the normal OE.

The same kind of analysis eliminates the bHLH TF Hes1 as a marker for, or critical to, the function of GBCs$_{MPP}$. Although *Hes1* expression appears in a substantial number of GBCs within 1 day of MeBr lesion, quickly thereafter the *Hes1* (+) cells shift apicalward and differentiate into Sus cells (Manglapus, G. L. *et al.*, 2004). The differentiation of *Hes1* (+) GBCs (GBCs$_{Sus}$) into Sus cells fits with the specific expression of Hes1 by Sus cells in OE of normal or bulbectomized animals; in these settings the GBC population makes only neurons. Thus, *Hes1* expression by GBCs does not indicate multipotency *per se* but seems to convey commitment to a Sus cell lineage. Parenthetically, *Hes1* is not needed to make Sus cells when expression of the *Mash1* gene is eliminated by genetic recombination, as supporting cells of the *Mash1* mutant mice do not express *Hes1* (Cau, E. *et al.*, 2000). These data fit with a demonstrated role for Hes1 protein, namely the suppression of *Mash1* expression; in the absence of *Mash1*, no such repression is needed.

However, the expression patterns of the TFs SOX-2 and Pax6 – high-mobility group (HMG) and paired domain TFs, respectively, that frequently cooperate in regulating gene transcription – indicate that GBCs$_{MPP}$ can be identified as GBCs that express these two TFs, and are not marked by expression of Mash1, or HBC, or Sus cell markers (cell types which also express the two TFs). In more specific terms, all GBCs in the OE from the immediate postlesion period express SOX-2 and Pax6 (Guo, Z. *et al.*, 2005a; 2005b); at this time the vast majority of GBCs are proliferating and either multipotent or making Sus cells. In their joint expression of SOX-2 and Pax6, GBCs$_{MPP}$ resemble the cells of the olfactory placode/pit (Grindley, J. C. *et al.*, 1995), which are the embryonic founders for the OE (as well as all other parts of the lining of the nasal cavity). Furthermore, stem cells of the embryonic CNS also seem to express both SOX-2 and Pax6 (Gotz, M. *et al.*, 1998; Marquardt, T. *et al.*, 2001; Chi, N. and Epstein, J. A., 2002; Avilion, A. A. *et al.*, 2003; Graham, V. *et al.*, 2003; Ellis, P. *et al.*, 2004; Brazel, C. Y. *et al.*, 2005; Englund, C. *et al.*, 2005).

In the final analysis, despite satisfying the multipotency criterion, it is unlikely that the actively proliferating GBCs$_{MPP}$ are stems, but rather they are most likely downstream of the bona fide stem cells.

4.34.5.3 Quiescent Globose Basal Cells

Among the GBC population are ones that express another characteristic of stem cells, namely prolonged mitotic quiescence. Hinds (Hinds, J. W. *et al.*, 1984) and then MacKay-Sim (Mackay-Sim, A. and Kittel, P., 1991a) recognized a population of basal cells that retained thymidine label for a prolonged period of time, but did not determine whether the label-retaining basal cells were HBCs or GBCs or both. Recent work extends the earlier observations and shows that some GBCs in normal OE express the cyclin-dependent kinase inhibitor (CDKI) p27^{Kip1}, and some retain BrdU label for periods of a month or more (Chen, X., 2003). MeBr lesion causes them to reenter the mitotic cycle with the consequence that p27 (+) cells disappear during the ramping up of proliferation during the acute postlesion period. Beginning 4 days after lesion, some GBCs return to quiescence; p27 (+) GBCs reappear and BrdU label begins to be retained by some GBCs. The kinetic characteristics of this population are like those of stems in other tissues, for example, the cornea.

4.34.5.4 Are Some Cells Among the Globose Basal Cells Tissue Stem Cells of the Olfactory Epithelium?

Of the characteristics of stem cells, some GBCs satisfy the totipotency criterion (with regard to the OE), including regeneration of GBCs, which implies at least a limited capacity for self-renewal. The data also show that some GBCs are quiescent for a long period of time. Finally, loss of GBCs results in a failure of neurogenesis and respiratory metaplasia. All of these are necessary characteristics for identifying stem cells, but are not sufficient. Unfortunately, we do not yet have the kinds of assays that allow us to define extensive self-renewal, as has been done for CNS stem cells, among others. Nonetheless, the findings gathered to date are highly suggestive that mitotically quiescent, SOX-2/Pax6 (+) GBCs are true tissue stem cells.

4.34.5.5 Progenitor Capacity of Other Epithelial Cell Types

The other stem cell candidates in the OE are the HBCs and duct/gland cells (Mulvaney, B. D. and Heist, H. E., 1971; Graziadei, P. P. and Monti Graziadei, G. A., 1979). The duct/gland cells do not seem to have any degree of neuropotency. When MeBr exposure is adjusted to cause a more severe lesion of the OE destroying all of the HBCs and GBCs in an area, the damaged OE is replaced by respiratory epithelium, even though gland cells proliferate and erupt superficial to the basal lamina (Jang, W. et al., 2003). Nor do transplanted duct cells do anything other than give rise to Sus/duct cells (Chen, X. et al., 2004). Although there is some evidence for expression of neuronal markers by HBCs in vitro (Satoh, M. and Yoshida, T., 2000; Carter, L. A. et al., 2004), transplantation is not effective in generating other cell types. In the lesioned OE of mice, the HBCs do not engraft (perhaps because they are preserved and proliferate throughout OE in contrast to their disappearance from ventral OE as a consequence of toxin exposure in rats) (Chen, X. et al., 2004). In MeBr-lesioned rats, HBCs labeled by the uptake of fluorescent latex microspheres (beads) do engraft but do not seem to do anything except make other HBCs. However, morphological analysis, and rare RRVV-labeled clones suggest that the HBCs can generate Sus cells as well as themselves in the context of the MeBr-lesioned OE. By no means can the extant in vivo data be considered definitive. Thus, we cannot yet discard the hypothesis that neurocompetent stem cells might also be found among the HBCs.

4.34.6 Reinervation of the Olfactory Bulb

Anecdotal reports of functional recovery in humans following either epithelial lesion or nerve transection imply some degree of reinnervation of the bulb by the newly born neurons of the reconstituted epithelium. In the context of the piecemeal turnover of the OSN population during normal life, that nascent innervation of the bulb by newly born neurons is remarkably precise as shown intuitively by the perceptual stability of olfaction in humans and the maintenance of precise connectivity in lab animals, both of which are the norm in adult life (Figure 3(a)). What then of the wholesale loss and reformation of the axonal projection onto the bulb following peripheral lesion? What do axon growth and synaptogenesis in that setting tell us of mechanisms regulating innervation of the bulb and their action following lesion?

4.34.6.1 Normal Organization of the Axonal Projection from the Olfactory Epithelium onto the Olfactory Bulb

The projection of the OE onto the OB is organized at three levels: At its coarsest, the organization of the projection is rhinotopic, in which there is a rough correspondence between regions of the OE and the part of the glomerular layer that they target: dorsal OE to dorsomedial OB, lateral OE to lateral and dorsolateral OB, etc. (Astic, L. and Saucier, D., 1986; Saucier, D. and Astic, L., 1986; Astic, L. et al., 1987; Schoenfeld, T. A. et al., 1994). The boundaries between one target region in the bulb and the next are probably not sharp. The graded, shifting projection at the boundary is demonstrated by the distribution of axons from biochemically distinct populations of neurons, that is, ventrolateral, OCAM/mamFasII (+) versus dorsomedial, OCAM/mamFasII (−), which are sharply delimited in the OE but whose axons intermingle in selected glomeruli at the margins between groups of wholly positive or wholly negative glomeruli (Mori, K. et al., 1985; Schwob, J. E. and Gottlieb, D. I., 1986). As a consequence, the axon terminal field of the neuropil of some of these border glomeruli exhibit intermediate levels of immunohistochemical staining with the anti-OCAM/mamFasII MAb RB-8.

The next more precise level of the projection – that is, onto groups of glomeruli – reflects the identity of the OR expressed by the neurons in question. Thus, neurons in the unlesioned adult OE expressing a particular OR, for example, as shown for P2, OR37A, and OR37C, will target one or two glomeruli within a roughly 250 μm-diameter patch consisting of about 30 or so potential target glomeruli (Strotmann, J. et al., 2000; Schaefer, M. L. et al., 2001). The target patch is found in a consistent location from animal to animal relative to defined landmarks of the bulb. For convenience, that level of specificity can be termed receptotopic. Within that patch, location is inconsistent; for example, the glomerulus innervated by OR37A is found in all possible relations to the one innervated by OR37C in the same patch: anterior, posterior, medial, or lateral to the OR37C glomerulus. Thus, a patch is shared as a target by multiple different kinds of OR-defined sensory neurons.

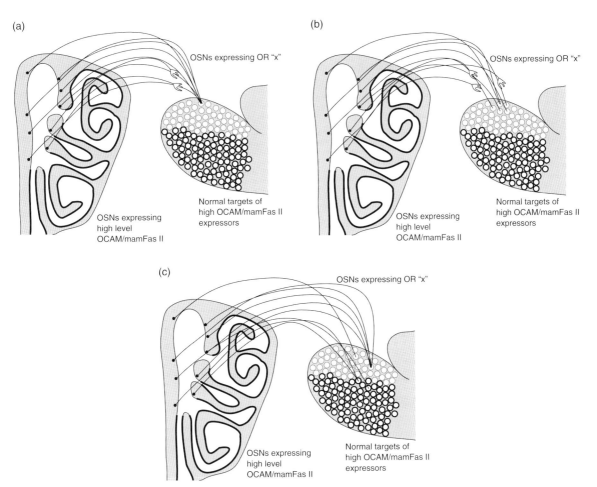

Figure 3 The accuracy of reinnervation of the olfactory bulb by newly born sensory neurons depends on the extent of neuron loss, and on the occurrence of damage to the olfactory nerve. (a) Glomerular targeting in the peripheral olfactory projection of normal animals, i.e., during piecemeal turnover of the neuronal population is highly accurate, and maintains preexisting glomerulotopy. The same outcome is observed when one odorant receptor (OR)-defined group of sensory neurons is ablated and then regenerated. The thick dark line in the cartoon of the nose on the right indicates the part of the olfactory epithelium in which neurons (OSNs) express OCAM/mamFas II at a high level. The glomeruli outlined with the thick dark line represent the normal targets for the high OCAM/mamFas II-expressing neurons. (b) Like the normal, following wholesale neuronal turnover caused by exposure to methyl bromide (MeBr), rhinotopy is restored, and the newly born neurons target approximately the same area within the glomerular sheet, reestablishing receptotopy. However, in contrast with normal, no longer do all of the group of OR-defined neurons converge onto a single glomerulus. (c) Following transection of the olfactory nerve, rhinotopy is distorted – as indicated by the low OCAM/mamFas II-expressing OSNs that innervate what are normally targets for high expressors – and neither targeting to the usual region of the glomerular layer nor convergence onto a single glomerulus is restored.

At its most precise, the projection onto the normal OB is characterized by the convergence of like fibers, that is, those expressing the same OR, most often onto a single glomerulus within the patch, which they seem to fill homogeneously and from which other OR-defined axonal types are apparently excluded (Mombaerts, P. *et al.*, 1996; Wang, F. *et al.*, 1998; Potter, S. M. *et al.*, 2001; Treloar, H. B. *et al.*, 2002; Feinstein, P. and Mombaerts, P., 2004; Feinstein, P. *et al.*, 2004). Studies during the normal

development of the projection suggest that unlike fibers intermingle initially within the common patch, but then the active coalescence of like fibers and their progressive segregation from unlike fibers comes to account for both the remarkable convergence of like fibers and the final position of the target glomerulus within the patch (Conzelmann, S. *et al.*, 2001). For convenience, the level of specificity implied by the convergence of like fibers onto a single glomerulus can be termed glomerulotopic.

4.34.6.2 Organization of the Projection Following Recovery from Injury to the Periphery

To what extent are the three levels of specificity that characterize the projection from OE onto OB in normal animals restored following reconstitution of the periphery and reinnervation of the bulb? Two settings may be contrasted: one in which the sensory neuron population undergoes wholesale turnover, for example, the cycle of destruction and reconstitution initiated by a general olfactotoxic agent (Schwob, J. E. *et al.*, 1999; Iwema, C. L. and Schwob, J. E., 2001; St John, J. A. and Key, B., 2003; Carr, V. M. *et al.*, 2004); the second, in which a specific OR-defined set of neurons has been eliminated through genetic means, for example, selective killing of P2 neurons via P2- and tetracycline-regulated expression of diptheria toxin (Gogos, J. A. *et al.*, 2000). In the case of MeBr lesion, the newly born neurons begin to fill the glomerular layer by 7 days after lesion, begin to reenter glomeruli near the end of the second week, fill them by the end of the third week and have stabilized by the end of the eighth week after lesion (Schwob, J. E. *et al.*, 1999).

The first issue that must be addressed following wholesale turnover of the sensory neuronal population is the fidelity with which the number and distribution of the various OR-defined classes of OSNs is restored across the epithelial sheet. Recent experiments utilized unilateral naris closure during the period of MeBr exposure (which has the effect of protecting the closed side from the effects of the gas almost completely), long postlesion survivals, and *in situ* hybridization (ISH) using multiple OR probes to compare lesioned-recovered and unlesioned OE (Iwema, C. L. *et al.*, 2004). Remarkably, the spatial patterning of OR expression was equivalent on the two sides, and the numbers of neurons labeled with the probe were also very similar in the lesioned-recovered epithelium as compared to the unlesioned. The extent of recovery is all the more striking given the partial depletion of the GBC population by the lesion and the degree of respiratory metaplasia in the ventrolateral areas of the epithelium. *A priori*, one can imagine two mechanisms by which position across the epithelial sheet comes to constrain OR expression: either the stem cells have a memory of place which is responsible for spatially imposed restrictions as the neurons are regenerated, or environmental cues derived from the lamina propria drive OR selection, at least after lesion from which effects the lamina cells would be protected. Of the two, the damage to the GBC population, which almost certainly must vary from animal to animal with respect to the OR-defined boundaries, suggests that cues external to the stem–progenitor-differentiating neuron hold sway and drive OR choice.

Despite the impressive degree to which the periphery is restored, a severe epithelial lesion is accompanied by targeting errors during reinnervation of the bulb. Projection errors can be detected at a very gross level as a reduction in the overall extent of reinnervation. Thus, when the severity of the initial lesion is enhanced, such that the recovery of the lesioned epithelium is incomplete and reduced to about four-fifths of its normal surface area, glomeruli at and near the posterior margins of the glomerular sheet remain denervated permanently (Schwob, J. E. *et al.*, 1999). Given that the reduction in the innervating population is not selective to a particular OR-defined type, the failure to reinnervate the posterior bulb indicates errors in receptotopic targeting. Interestingly, that failure in the anteroposterior dimension does not seem to interfere with the establishment of rhinotopic territories. Retrograde tract tracing experiments following selective labeling of the dorsomedial versus ventrolateral glomerular layer suggest little or no breakdown in the gross rhinotopic targeting onto the bulb (Schwob and Hamlin, unpublished results). That return to normalcy is confirmed by using the OCAM/mamFasII-specific MAb RB-8 to define the respective projections of the corresponding epithelial territories onto the bulb of rats following recovery from unilateral epithelial lesion.

The analyses of reinnervation by specific types of OSNs confirm that interpretation. Following peripheral lesion of P2-IRES-tauLacZ strain of transgenic mice with MeBr or with dichlobenil, the P2 fibers no longer converge to a single glomerulus on each surface of the bulb (Figure 3(b)) (Iwema, C. L. and Schwob, J. E., 2001; St. John, J. A. and Key, B., 2003). Instead, multiple glomeruli in the vicinity of the original one become innervated, but the P2 fibers in them are now only one part of a mix of axons of different types, in other words, the P2 fibers are only some of the OMP (+) axons innervating those glomeruli. Indeed, the P2 fibers tend to be plastered at the edges of the glomerular neuropil where they are more densely clustered than usual. The projection of other neuron types onto the bulb is also degraded relative to the unlesioned projection; like P2 neurons they come to innervate more than the normal number of glomeruli (Carr, V. M. *et al.*, 2004). Thus, the capacity to target a specific region or target patch within the glomerular layer is maintained, at least at anterior levels of the bulb, but no longer do the fibers converge properly and selectively.

In contrast, the projection to the P2 glomeruli is restored to its normal uniglomerular pattern following selective ablation by genetic means of all but a very small percentage of the extant P2 neurons, and their subsequent regeneration (Gogos, J. A. *et al.*, 2000). However, the environment experienced by the nascent P2 axons is quite different between global lesions and the selective ablation of one and only one population in two respects. First, following selective ablation the new P2 axons grow through ON fascicles that are very like normal having lost only about 0.1% of their fibers (less than the number that degenerate due to neuronal turnover – whether immature or not). The status of the ON itself seems to play some role during reinnervation judging by the loss of regional targeting during recovery from nerve transection. Second, the capacity to return to a specific region within the glomerular layer (true of the reinnervating axons with either selective or global destruction of the neurons) may reflect the fasciculation of like axons – not the P2 axons in this case, but all of the other types of axons that are targeting the area. In this interpretation, newly born, non-P2 neurons that target adjacent glomeruli may track or fasciculate along preexisting like axons to their proper target, that is, being actively directed away from the unoccupied P2 glomerulus. In contrast, the P2 neurons are not directed away from the P2 glomerulus, and would, therefore, have an advantage over unlike fibers in targeting it during their regeneration. In the free-for-all that would occur in the glomerular layer when preexisting axons are eliminated and a coherent wave of nascent fibers reach the bulb, it is only too easy to imagine that P2 or other axons would innervate multiple glomeruli and become stably ensconced in them. Some experimental evidence can be adduced for the importance of the preexisting fibers and the specificity of reinnervation. Neurons innervating atypical, necklace glomeruli at the posterior margin of the bulb are relatively less severely affected by MeBr lesion than the vast majority innervating the typical glomeruli, as a consequence of their concentration at the cribriform plate in areas relatively lacking cytochrome P450-expressing supporting cells. Thus, one can adjust the lesion conditions to spare about 50% of the preexisting neurons; in this case, the regenerated axons that target the necklace glomeruli end up reinnervating their original ones (which can be identified on the basis of position at the margin of the glomerular layer) (Ring, G. *et al.*, 1997). In contrast, lesions that destroy upward of 80% of the preexisting necklace neurons lead to mistargeting of necklace neurons into nonnecklace glomeruli.

4.34.6.3 Disorganization of the Projection Following Olfactory Nerve Transection

Although the projection of epithelium onto bulb is restored at the level of rhinotopy and receptotopy following MeBr lesion or other injuries to the periphery, ON transection results in significant changes to the ON projection patterns (Figure 3(c)). Following a complete transection of the ON, reinnervation of the bulb does not restore the normal topographical mapping of projections onto the OB (Christensen, M. D. *et al.*, 2001). The physical disruption of these nerves fibers and their perineural sheaths presents a challenge to regenerating nerve fibers as they grow back and reestablish connections in the bulb. The absence of guidance molecules present during early development as well as other factors may leave the regenerating axons with little or no cues to find their normal targets. ON transection experiments with P2-ITL mice show that regenerating axons often project widely across the OB during recovery, losing receptotopic specificity as shown by their innervation of multiple glomerular patches and losing rhinotopic specificity to a minor degree (Figure 3(c)) (Costanzo, R. M., 2000; 2005). Some of the regenerated axons projected diffusely while others innervated only a single locus in the bulb. In addition, while P2 nerve fibers continue to converge onto individual glomeruli, these glomeruli often received a mixed innervation that includes additional OR types. These results demonstrate that normal mapping of OR information onto the bulb is more substantially disrupted following recovery from nerve transection than after peripheral lesion. The remapping of odorant projections in the OB elicited by nerve transection potentially explains why animals must relearn how to perform odor discrimination tasks after recovery. Nonetheless, the rewired OB apparently provides sufficient spatial resolution of OR mapping to restore odor discrimination function.

4.34.6.4 Functional Recovery Following Epithelial Lesion or Nerve Transection

Despite the degradation in mapping of the OE onto the bulb after recovery from either peripheral injury or nerve lesion, there is a surprising degree of recovery in sensory function following the reconstitution of the epithelium and reinnervation of the bulb. In the case of MeBr lesion, rats trained on a five-odorant identification task – a task on which they normally perform at 90% correct or better – show a rapid return to criterion performance when retesting begins at 2 months postlesion when the regenerated projection has stabilized (Youngentob, S. L. and Schwob, J. E., 1997). However, multidimensional scaling analysis of the patterns in the stimulus–response matrix across the animals indicates that the lesioned-recovered animals do perceive and process the test odorants differently than unlesioned animals tested after equivalent rest periods. In other words, the lesioned-recovered animals make different kinds of mistakes, that is, confusions, when performing the task.

Restoration of odor-mediated behavior following recovery from ON transection has been reported in a variety of species including the pigeon (Oley, N. *et al.*, 1975; Walker, J. C. *et al.*, 1981), goldfish (von Rekowski, C. and Zippel, H. P., 1993), mouse (Harding, J. W. and Wright, J. W., 1979), and hamster (Yee, K. K. and Costanzo, R. M., 1995). There is a direct correlation between the time course of OB reinnervation and recovery of odor detection and discrimination function. These studies raise an important question; does the perception of an odor change after recovery from nerve transection, when normal receptotopic specificity is lost? To address this question Yee K. K. and Costanzo R. M. (1998) studied performance on an odor discrimination tasks before and 40 days after ON transection in hamster. Animals were trained to discriminate between two unfamiliar odors, strawberry, and cinnamon, before surgery and then retested after 40 days of recovery. At this time point, the OB is reinnervated by a large number of newly regenerated ON fibers. Animals receiving a sham surgical procedure that leaves the ONs intact were able to perform the odor discrimination task 40 days after sham surgery, without food reinforcement. Those animals that received the nerve transection procedure performed at chance levels. However, if food reinforcement was given when the correct odorant was selected, recovery animals were able to relearn the odor discrimination task within a few days. When animals were tested on day 40 with completely new odors, baby powder, and coffee, both the sham and the reinnervated animals were able to learn the odor discrimination task within a few days. These results suggest that although there are likely to be alterations in the way nerve fibers reconnect to the OB following nerve transaction, after recovery these animals are capable of making odor discriminations

but they must relearn the task. It is likely that the original perception of the odors as cinnamon and strawberry changed and that the animals must learn how to reinterpret these odors using the new mapping of ORs onto the bulb. Remarkably, the olfactory system has the capacity to restore its connections to the OB after injury, and even when significant changes occur to the OR mapping the animal can learn to use the new maps and recover olfactory function.

4.34.6.5 Summary

Aberrant connections are formed in the peripheral olfactory system following recovery from peripheral lesion or nerve transection. Rhinotopy and receptotopy – but not glomerultopy – are restored following epithelial lesion. In contrast, rhinotopy is somewhat compromised while receptotopy and glomerulotopy are severely disrupted following nerve transection. Nonetheless, neither form of lesion precludes sensory function following reconstitution of the epithelium and reinnervation of the bulb. However, the errors – which are of greater magnitude following nerve lesion – do change the encoding of odorants by the OB. Despite the reinnervation of the bulb, for these lesioned animals, as for humans presenting with olfactory dysfunction, a rose may no longer smell the same as a rose.

4.34.7 Conclusions and Unanswered Questions

The foregoing provides a summary of what we know of the recovery of the peripheral olfactory apparatus – the epithelium and its projection onto the bulb – following either reversible destruction of the epithelium or transection of the ON. From the phenomenological perspective, we now know the events that occur during epithelial reconstitution in substantial detail and those underlying reinnervation of the bulb somewhat less. However, our knowledge of cellular and molecular mechanism is still very far from complete. An integrated picture of the signals and transduction cascades that combine to regulate the choice point faced by the multipotent stem/progenitors is lacking and will remain obscure until certain conditions are met. Among them we need reagents and protocols that can be used to identify, isolate and characterize olfactory stem cells, and have them recapitulate the differentiation *in vitro* that they accomplish so effortlessly *in vivo*.

Once the stem cells are in hand, subsidiary questions of their therapeutic usefulness in the OE; for example, in replacing stem cells whose destruction or exhaustion underlies olfactory dysfunction, or in other settings, including their differentiation into OECs and implantation into injured spinal cord (Li, Y. *et al.*, 1997), can be more profitably addressed. Likewise, we have little in-depth understanding of the mechanisms for maintaining the precision of axon targeting in the adult, which goes somewhat awry following epithelial lesion and is substantially disordered by injury to theON. If the sparing of preexisting fibers does promote accurate reinnervation, how are like-axons recognized and how do they form a substrate for growth? The recovery of olfactory function despite the scrambling of connections raises the specter of reorganization in central olfactory processing and leads to questions of whether sensory activity during the reinnervation process might impact that recovery. As our understanding of basic mechanism deepens it is reasonable to expect that useful therapies will begin to emerge.

4.34.8   Note Added in Proof

Subsequent to the preparation of this chapter a paper was published describing the contributions of HBCs to the repair of the OE following injury by MeBr or another direct acting toxin (Leung, C. T. *et al.*, 2007). Their data report the use of a genetic lineage tracing technique in which cells that derive from HBCs are marked by the expression of LacZ due to the activation of Cre recombinase selectively in cells that make cytokeratin 5. In the epithelium of normal or bulbectomized mice, HBCs give rise only to HBCs throughout the life of the animal. However, after direct epithelial injury HBCs contribute to the roughly the same cell types as GBCs after transplantation, namely HBCs and GBCs themselves, neurons, sustentacular cells, and duct/gland cells; GBCs can give rise to respiratory epithelial cells, but HBCs have not yet been shown to give rise to them. Thus, it would appear that the HBCs, despite their relatively late emergence during the perinatal period in the development of the OE, serve as an additional reserve population with extensive multipotency *in vivo* and *in vitro* that is called upon when GBCs have been destroyed (a characteristic of the MeBr-lesioned mouse to a much greater extent than in lesioned rats). The persistence of two cell populations with the multipotency that is characteristic of tissue stem cells is reminiscent of other neurogenic

populations, including that of the subventricular zone in the adult forebrain in which both type B and type C cells have the characteristics of neural stem cells. Furthermore, the situation in the olfactory epithelium may be a reflection of a more primitive character than that of the differentiated CNS. The GBCs share characteristics of the cells that compose the olfactory placode and the early stages of its invagination, including the expression of Pax6, Sox2, and, somewhat later, Mash1, before the HBCs emerge around the time of birth. The sequence may be conceptualized as if the primitive neuroepithelial cells of the neural plate and early tube are retained even after radial glia and subsequent glial forms of the subventricular zone emerge to serve as neurocompetent stem cells.

References

Alonso, L. and Fuchs, E. 2003. Stem cells of the skin epithelium. Proc. Natl. Acad. Sci. U. S. A. 100(Suppl. 1), 11830–11835.

Andres, K. H. 1966. The fine structure of the olfactory region of macrosmatic animals. Z. Zellforsch. Mikrosk. Anat. 69, 140–154.

Asan, E. and Drenckhahn, D. 2005. Immunocytochemical characterization of two types of microvillar cells in rodent olfactory epithelium. Histochem. Cell Biol. 123, 157–168.

Astic, L. and Saucier, D. 1986. Anatomical mapping of the neuroepithelial projection to the olfactory bulb in the rat. Brain Res. Bull. 16, 445–454.

Astic, L., Saucier, D., and Holley, A. 1987. Topographical relationships between olfactory receptor cells and glomerular foci in the rat olfactory bulb. Brain Res. 424, 144–152.

Avilion, A. A., Nicolis, S. K., Pevny, L. H., Perez, L., Vivian, N., and Lovell-Badge, R. 2003. Multipotent cell lineages in early mouse development depend on SOX2 function. Genes Dev. 17, 126–140.

Barber, P. C. and Lindsay, R. M. 1982. Schwann cells of the olfactory nerves contain glial fibrillary acidic protein and resemble astrocytes. Neuroscience 7, 3077–3090.

Barnett, E. M. and Perlman, S. 1993. The olfactory nerve and not the trigeminal nerve is the major site of CNS entry for mouse hepatitis virus, strain JHM. Virology 194, 185–191.

Barnett, E. M., Cassell, M. D., and Perlman, S. 1993. Two neurotropic viruses, herpes simplex virus type 1 and mouse hepatitis virus, spread along different neural pathways from the main olfactory bulb. Neuroscience 57, 1007–1025.

Barthold, S. W. 1988. Olfactory neural pathway in mouse hepatitis virus nasoencephalitis. Acta Neuropathol. (Berl.) 76, 502–506.

Bauer, S., Rasika, S., Han, J., Mauduit, C., Raccurt, M., Morel, G., Jourdan, F., Benahmed, M., Moyse, E., and Patterson, P. H. 2003. Leukemia inhibitory factor is a key signal for injury-induced neurogenesis in the adult mouse olfactory epithelium. J. Neurosci. 23, 1792–1803.

Bergstrom, U., Giovanetti, A., Piras, E., and Brittebo, E. B. 2003. Methimazole-induced damage in the olfactory mucosa: effects on ultrastructure and glutathione levels. Toxicol. Pathol. 31, 379–387.

Biffo, S., Pognetto, M. S., Di Cantogno, L. V., Perroteau, I., and Fasolo, A. 1992. Bulbectomy enhances neurogenesis and cell turnover of primary olfactory neurons but does not abolish carnosine expression. Eur. J. Neurosci. 4, 1398–1406.

Bojsen-Moller, F. 1975. Demonstration of terminalis, olfactory, trigeminal and perivascular nerves in the rat nasal septum. J. Comp. Neurol. 159, 245–256.

Brazel, C. Y., Limke, T. L., Osborne, J. K., Miura, T., Cai, J., Pevny, L., and Rao, M. S. 2005. Sox2 expression defines a heterogeneous population of neurosphere-forming cells in the adult murine brain. Aging Cell 4, 197–207.

Brittebo, E. B. 1995. Metabolism-dependent toxicity of methimazole in the olfactory nasal mucosa. Pharmacol. Toxicol. 76, 76–79.

Buck, L. and Axel, R. 1991. A novel multigene family may encode odorant receptors: a molecular basis for odor recognition. Cell 65, 175–187.

Burd, G. D. 1993. Morphological study of the effects of intranasal zinc sulfate irrigation on the mouse olfactory epithelium and olfactory bulb. Microsc. Res. Tech. 24, 195–213.

Caggiano, M., Kauer, J. S., and Hunter, D. D. 1994. Globose basal cells are neuronal progenitors in the olfactory epithelium: a lineage analysis using a replication-incompetent retrovirus. Neuron 13, 339–352.

Calof, A. L., Rim, P. C., Askins, K. J., Mumm, J. S., Gordon, M. K., Iannuzzelli, P., and Shou, J. 1998. Factors regulating neurogenesis and programmed cell death in mouse olfactory epithelium. Ann. N. Y. Acad. Sci. 855, 226–229.

Cancalon, P. 1982. Degeneration and regeneration of olfactory cells induced by $ZnSO_4$ and other chemicals. Tissue Cell 14, 717–733.

Carr, V. M. and Farbman, A. I. 1992. Ablation of the olfactory bulb up-regulates the rate of neurogenesis and induces precocious cell death in olfactory epithelium. Exp. Neurol. 115, 55–59.

Carr, V. M. and Farbman, A. I. 1993. The dynamics of cell death in the olfactory epithelium. Exp. Neurol. 124, 308–314.

Carr, V. M., Ring, G., Youngentob, S. L., Schwob, J. E., and Farbman, A. I. 2004. Altered epithelial density and expansion of bulbar projections of a discrete HSP70 immunoreactive subpopulation of rat olfactory receptor neurons in reconstituting olfactory epithelium following exposure to methyl bromide. J. Comp. Neurol. 469, 475–493.

Carter, L. A., MacDonald, J. L., and Roskams, A. J. 2004. Olfactory horizontal basal cells demonstrate a conserved multipotent progenitor phenotype. J. Neurosci. 24, 5670–5683.

Cau, E., Casarosa, S., and Guillemot, F. 2002. Mash1 and Ngn1 control distinct steps of determination and differentiation in the olfactory sensory neuron lineage. Development 129, 1871–1880.

Cau, E., Gradwohl, G., Casarosa, S., Kageyama, R., and Guillemot, F. 2000. Hes genes regulate sequential stages of neurogenesis in the olfactory epithelium. Development 127, 2323–2332.

Cau, E., Gradwohl, G., Fode, C., and Guillemot, F. 1997. Mash1 activates a cascade of bHLH regulators in olfactory neuron progenitors. Development 124, 1611–1621.

Chen, X. 2003. Functional Capacity of Progenitor Cells in the Olfactory Epithelium. PhD thesis, Program in Cell, Molecular, and Developmental Biology. Tufts University

Chen, X., Fang, H., and Schwob, J. E. 2004. Multipotency of purified, transplanted globose basal cells in olfactory epithelium. J. Comp. Neurol. 469, 457–474.

Chen, Y., Getchell, M. L., Ding, X., and Getchell, T. V. 1992. Immunolocalization of two cytochrome P450 isozymes in rat nasal chemosensory tissue. Neuroreport 3, 749–752.

Chi, N. and Epstein, J. A. 2002. Getting your Pax straight: pax proteins in development and disease. Trends Genet. 18, 41–47.

Christensen, M. D., Holbrook, E. H., Costanzo, R. M., and Schwob, J. E. 2001. Rhinotopy is disrupted during the re-innervation of the olfactory bulb that follows transection of the olfactory nerve. Chem. Senses 26, 359–369.

Conzelmann, S., Malun, D., Breer, H., and Strotmann, J. 2001. Brain targeting and glomerulus formation of two olfactory neuron populations expressing related receptor types. Eur. J. Neurosci. 14, 1623–1632.

Costanzo, R. M. 1984. Comparison of neurogenesis and cell replacement in the hamster olfactory system with and without a target (olfactory bulb). Brain Res. 307, 295–301.

Costanzo, R. M. 1985. Neural regeneration and functional reconnection following olfactory nerve transection in hamster. Brain Res. 361, 258–266.

Costanzo, R. M. 2000. Rewiring the olfactory bulb: changes in odor maps following recovery from nerve transection. Chem. Senses 25, 199–205.

Costanzo, R. M. 2005. Regeneration and rewiring the olfactory bulb. Chem. Senses 30(Suppl. 1), 133–134.

Costanzo, R. M. and Graziadei, P. P. 1983. A quantitative analysis of changes in the olfactory epithelium following bulbectomy in hamster. J. Comp. Neurol. 215, 370–381.

Cowan, C. M. and Roskams, A. J. 2002. Apoptosis in the mature and developing olfactory neuroepithelium. Microsc. Res. Tech. 58, 204–215.

Cowan, C. M., Thai, J., Krajewski, S., Reed, J. C., Nicholson, D. W., Kaufmann, S. H., and Roskams, A. J. 2001. Caspases 3 and 9 send a pro-apoptotic signal from synapse to cell body in olfactory receptor neurons. J. Neurosci. 21, 7099–7109.

Deckner, M. L., Risling, M., and Frisen, J. 1997. Apoptotic death of olfactory sensory neurons in the adult rat. Exp. Neurol. 143, 132–140.

Doty, R. L. 1979. A review of olfactory dysfunctions in man. Am. J. Otolaryngol. 1, 57–79.

Ellis, P., Fagan, B. M., Magness, S. T., Hutton, S., Taranova, O., Hayashi, S., McMahon, A., Rao, M., and Pevny, L. 2004. SOX2, a persistent marker for multipotential neural stem cells derived from embryonic stem cells, the embryo or the adult. Dev. Neurosci. 26, 148–165.

Englund, C., Fink, A., Lau, C., Pham, D., Daza, R. A., Bulfone, A., Kowalczyk, T., and Hevner, R. F. 2005. Pax6, Tbr2, and Tbr1 are expressed sequentially by radial glia, intermediate progenitor cells, and postmitotic neurons in developing neocortex. J. Neurosci. 25, 247–251.

Feddersen, R. M., Yunis, W. S., O'Donnell, M. A., Ebner, T. J., Shen, L., Iadecola, C., Orr, H. T., and Clark, H. B. 1997. Susceptibility to cell death induced by mutant SV40 T-antigen correlates with Purkinje neuron functional development. Mol. Cell. Neurosci. 9, 42–62.

Feinstein, P. and Mombaerts, P. 2004. A contextual model for axonal sorting into glomeruli in the mouse olfactory system. Cell 117, 817–831.

Feinstein, P., Bozza, T., Rodriguez, I., Vassalli, A., and Mombaerts, P. 2004. Axon guidance of mouse olfactory sensory neurons by odorant receptors and the beta2 adrenergic receptor. Cell 117, 833–846.

Field, P., Li, Y., and Raisman, G. 2003. Ensheathment of the olfactory nerves in the adult rat. J. Neurocytol. 32, 317–324.

Fisher, A. and Caudy, M. 1998. The function of hairy-related bHLH repressor proteins in cell fate decisions. BioEssays 20, 298–306.

Genter, M. B., Deamer, N. J., Blake, B. L., Wesley, D. S., and Levi, P. E. 1995. Olfactory toxicity of methimazole: dose–response and structure–activity studies and characterization of flavin-containing monooxygenase activity in the Long–Evans rat olfactory mucosa. Toxicol. Pathol. 23, 477–486.

Genter, M. B., Owens, D. M., Carlone, H. B., and Crofton, K. M. 1996. Characterization of olfactory deficits in the rat following administration of 2,6-dichlorobenzonitrile (dichlobenil), 3,3′-iminodipropionitrile, or methimazole. Fundam. Appl. Toxicol. 29, 71–77.

Gogos, J. A., Osborne, J., Nemes, A., Mendelsohn, M., and Axel, R. 2000. Genetic ablation and restoration of the olfactory topographic map. Cell 103, 609–620.

Goldstein, B. J. and Schwob, J. E. 1996. Analysis of the globose basal cell compartment in rat olfactory epithelium using GBC-1, a new monoclonal antibody against globose basal cells. J. Neurosci. 16, 4005–4016.

Goldstein, B. J., Fang, H., Youngentob, S. L., and Schwob, J. E. 1998. Transplantation of multipotent progenitors from the adult olfactory epithelium. Neuroreport 9, 1611–1617.

Goldstein, B. J., Wolozin, B. L., and Schwob, J. E. 1997. FGF2 suppresses neuronogenesis of a cell line derived from rat olfactory epithelium. J. Neurobiol. 33, 411–428.

Gordon, M. K., Mumm, J. S., Davis, R. A., Holcomb, J. D., and Calof, A. L. 1995. Dynamics of MASH1 expression *in vitro* and *in vivo* suggest a non-stem cell site of MASH1 action in the olfactory receptor neuron lineage. Mol. Cell Neurosci. 6, 363–379.

Gotz, M., Stoykova, A., and Gruss, P. 1998. Pax6 controls radial glia differentiation in the cerebral cortex. Neuron 21, 1031–1044.

Graham, V., Khudyakov, J., Ellis, P., and Pevny, L. 2003. SOX2 functions to maintain neural progenitor identity. Neuron 39, 749–765.

Grazaiadei, P. P. C. 1971. The Olfactory Mucosa of Vertebrates. In: Chemical Senses. Part I Olfaction (*ed.* L. M. Beidler), pp. 27–58. Springer-Verlag.

Graziadei, P. P. 1973. Cell dynamics in the olfactory mucosa. Tissue Cell 5, 113–131.

Graziadei, P. P. and Metcalf, J. F. 1971. Autoradiographic and ultrastructural observations on the frog's olfactory mucosa. Z. Zellforsch. Mikrosk. Anat. 116, 305–318.

Graziadei, P. P. C. and Monti Graziadei, G. A. 1978. Continuous Nerve Cell Renewal in the Olfactory System. In: Handbook of Sensory Physiology (*ed.* M. Jacobson), Vol. IX, pp. 55–82. Springer.

Graziadei, P. P. and Monti Graziadei, G. A. 1979. Neurogenesis and neuron regeneration in the olfactory system of mammals. I. Morphological aspects of differentiation and structural organization of the olfactory sensory neurons. J. Neurocytol. 8, 1–18.

Graziadei, P. P. and Monti Graziadei, G. A. 1980. Neurogenesis and neuron regeneration in the olfactory system of mammals. III. Deafferentation and reinnervation of the olfactory bulb following section of the fila olfactoria in rat. J. Neurocytol. 9, 145–162.

Graziadei, P. P., Karlan, M. S., Graziadei, G. A., and Bernstein, J. J. 1980. Neurogenesis of sensory neurons in the primate olfactory system after section of the fila olfactoria. Brain Res. 186, 289–300.

Grindley, J. C., Davidson, D. R., and Hill, R. E. 1995. The role of Pax-6 in eye and nasal development. Development 121, 1433–1442.

Guillemot, F., Lo, L. C., Johnson, J. E., Auerbach, A., Anderson, D. J., and Joyner, A. L. 1993. Mammalian achaete-scute homolog 1 is required for the early development of olfactory and autonomic neurons. Cell 75, 463–476.

Guo, Z., Manglapus, G. L., Harris, M. E., and Schwob, J. E. 2005a. Expression of Pax6 in olfactory epithelium. Chem. Senses 30, A186.

Guo, Z., Manglapus, G. L., and Schwob, J. E. 2005b. Characterization of Sox2 expression in olfactory epithelium. Soc. Neurosci. Abstr. 31, Program No. 478.477.

Hall, P. A. and Watt, F. M. 1989. Stem cells: the generation and maintenance of cellular diversity. Development 106, 619–633.

Harding, J. W. and Wright, J. W. 1979. Reversible effects of olfactory nerve section on behavior and biochemistry in mice. Brain Res. Bull. 4, 17–22.

Harding, J. W., Getchell, T. V., and Margolis, F. L. 1978. Denervation of the primary olfactory pathway in mice. V. Long-term effect of intranasal ZnSO$_4$ irrigation on behavior, biochemistry and morphology. Brain Res. 140, 271–285.

Harding, J., Graziadei, P. P., Monti Graziadei, G. A., and Margolis, F. L. 1977. Denervation in the primary olfactory pathway of mice. IV. Biochemical and morphological evidence for neuronal replacement following nerve section. Brain Res. 132, 11–28.

Hempstead, J. L. and Morgan, J. I. 1985a. A panel of monoclonal antibodies to the rat olfactory epithelium. J. Neurosci. 5, 438–449.

Hempstead, J. L. and Morgan, J. I. 1985b. Monoclonal antibodies reveal novel aspects of the biochemistry and organization of olfactory neurons following unilateral olfactory bulbectomy. J. Neurosci. 5, 2382–2387.

Hinds, J. W., Hinds, P. L., and McNelly, N. A. 1984. An autoradiographic study of the mouse olfactory epithelium: evidence for long-lived receptors. Anat. Rec. 210, 375–383.

Hirota, J. and Mombaerts, P. 2004. The LIM-homeodomain protein Lhx2 is required for complete development of mouse olfactory sensory neurons. Proc. Natl. Acad. Sci. U. S. A. 101, 8751–8755.

Holbrook, E. H. and Leopold, D. A. 2006. An updated review of clinical olfaction. Curr. Opin. Otolaryngol. Head. Neck. Surg. 14, 23–28.

Holbrook, E. H., Leopold, D. A., and Schwob, J. E. 2005. Abnormalities of axon growth in human olfactory mucosa. Laryngoscope 115, 2144–2154.

Holbrook, E. H., Szumowski, K. E., and Schwob, J. E. 1995. An immunochemical, ultrastructural, and developmental characterization of the horizontal basal cells of rat olfactory epithelium. J. Comp. Neurol. 363, 129–146.

Holcomb, J. D., Mumm, J. S., and Calof, A. L. 1995. Apoptosis in the neuronal lineage of the mouse olfactory epithelium: regulation *in vivo* and *in vitro*. Dev. Biol. 172, 307–323.

Huard, J. M., Youngentob, S. L., Goldstein, B. J., Luskin, M. B., and Schwob, J. E. 1998. Adult olfactory epithelium contains multipotent progenitors that give rise to neurons and non-neural cells. J. Comp. Neurol. 400, 469–486.

Hurtt, M. E., Morgan, K. T., and Working, P. K. 1987. Histopathology of acute toxic responses in selected tissues from rats exposed by inhalation to methyl bromide. Fundam. Appl. Toxicol. 9, 352–365.

Hurtt, M. E., Thomas, D. A., Working, P. K., Monticello, T. M., and Morgan, K. T. 1988. Degeneration and regeneration of the olfactory epithelium following inhalation exposure to methyl bromide: pathology, cell kinetics, and olfactory function. Toxicol. Appl. Pharmacol. 94, 311–328.

Iwema, C. L. and Schwob, J. E. 2001. P2 olfactory sensory neurons do not converge appropriately in the olfactory bulb following peripheral lesion. Soc. Neurosci. Abstr. 27, in press.

Iwema, C. L. and Schwob, J. E. 2003. Odorant receptor expression as a function of neuronal maturity in the adult rodent olfactory system. J. Comp. Neurol. 459, 209–222.

Iwema, C. L., Fang, H., Kurtz, D. B., Youngentob, S. L., and Schwob, J. E. 2004. Odorant receptor expression patterns are restored in lesion-recovered rat olfactory epithelium. J. Neurosci. 24, 356–369.

Jang, W., Kim, K. P., and Schwob, J. E. 2007. Non-integrin laminin receptor precursor protein is expressed on olfactory stem and progenitor cells. J. Comp. Neurol. 502, 367–381.

Jang, W., Youngentob, S. L., and Schwob, J. E. 2003. Globose basal cells are required for reconstitution of olfactory epithelium after methyl bromide lesion. J. Comp. Neurol. 460, 123–140.

Juilfs, D. M., Fulle, H. J., Zhao, A. Z., Houslay, M. D., Garbers, D. L., and Beavo, J. A. 1997. A subset of olfactory neurons that selectively express cGMP-stimulated phosphodiesterase (PDE2) and guanylyl cyclase-D define a unique olfactory signal transduction pathway. Proc. Natl. Acad. Sci. U. S. A. 94, 3388–3395.

Kaplan, M. S. and Hinds, J. W. 1977. Neurogenesis in the adult rat: electron microscopic analysis of light radioautographs. Science 197, 1092–1094.

Kaplan, M. S., McNelly, N. A., and Hinds, J. W. 1985. Population dynamics of adult-formed granule neurons of the rat olfactory bulb. J. Comp. Neurol. 239, 117–125.

Kolterud, A., Alenius, M., Carlsson, L., and Bohm, S. 2004. The Lim homeobox gene Lhx2 is required for olfactory sensory neuron identity. Development 131, 5319–5326.

Konzelmann, S., Saucier, D., Strotmann, J., Breer, H., and Astic, L. 1998. Decline and recovery of olfactory receptor expression following unilateral bulbectomy. Cell Tissue Res. 294, 421–430.

Koster, N. L. and Costanzo, R. M. 1996. Electrophysiological characterization of the olfactory bulb during recovery from sensory deafferentation. Brain Res. 724, 117–120.

Largent, B. L., Sosnowski, R. G., and Reed, R. R. 1993. Directed expression of an oncogene to the olfactory neuronal lineage in transgenic mice. J. Neurosci. 13, 300–312.

Leung, C. T., Coulombe, P. A., and Reed, R. R. 2007. Contribution of olfactory neural stem cells to tissue maintenance and regeneration. Nat. Neurosci. 10, 720–726.

Li, Y., Field, P. M., and Raisman, G. 1997. Repair of adult rat corticospinal tract by transplants of olfactory ensheathing cells. Science 277, 2000–2002.

Li, Y., Field, P. M., and Raisman, G. 2005. Olfactory ensheathing cells and olfactory nerve fibroblasts maintain continuous open channels for regrowth of olfactory nerve fibres. Glia 52, 245–251.

Lois, C. and Alvarez-Buylla, A. 1993. Proliferating subventricular zone cells in the adult mammalian forebrain can differentiate into neurons and glia. Proc. Natl. Acad. Sci. U. S. A. 90, 2074–2077.

Lois, C. and Alvarez-Buylla, A. 1994. Long-distance neuronal migration in the adult mammalian brain. Science 264, 1145–1148.

Loo, A. T., Youngentob, S. L., Kent, P. F., and Schwob, J. E. 1996. The aging olfactory epithelium: neurogenesis, response to damage, and odorant-induced activity. Int. J. Dev. Neurosci. 14, 881–900.

de Lorenzo, A. J. 1957. Electron microscopic observations of the olfactory mucosa and olfactory nerve. J. Biophys. Biochem. Cytol. 3, 839–850.

Mackay-Sim, A. and Chuah, M. I. 2000. Neurotrophic factors in the primary olfactory pathway. Prog. Neurobiol. 62, 527–559.

Mackay-Sim, A. and Kittel, P. 1991a. Cell dynamics in the adult mouse olfactory epithelium: a quantitative autoradiographic study. J. Neurosci. 11, 979–984.

Mackay-Sim, A. and Kittel, P. W. 1991b. On the life span of olfactory receptor neurons. Eur. J. Neurosci. 3, 209–215.

Manglapus, G. L., Youngentob, S. L., and Schwob, J. E. 2004. Expression patterns of basic helix–loop–helix transcription factors define subsets of olfactory progenitor cells. J. Comp. Neurol. 479, 216–233.

Marquardt, T., Ashery-Padan, R., Andrejewski, N., Scardigli, R., Guillemot, F., and Gruss, P. 2001. Pax6 is required for the multipotent state of retinal progenitor cells. Cell 105, 43–55.

Matulionis, D. H. 1975. Ultrastructural study of mouse olfactory epithelium following destruction by $ZnSO_4$ and its subsequent regeneration. Am. J. Anat. 142, 67–89.

Matulionis, D. H. 1976. Light and electron microscopic study of the degeneration and early regeneration of olfactory epithelium in the mouse. Am. J. Anat. 145, 79–99.

Meiri, K. F., Bickerstaff, L. E., and Schwob, J. E. 1991. Monoclonal antibodies show that kinase C phosphorylation of GAP-43 during axonogenesis is both spatially and temporally restricted in vivo. J. Cell Biol. 112, 991–1005.

Menezes, J. R., Smith, C. M., Nelson, K. C., and Luskin, M. B. 1995. The division of neuronal progenitor cells during migration in the neonatal mammalian forebrain. Mol. Cell Neurosci. 6, 496–508.

Miragall, F. and Monti Graziadei, G. A. 1982. Experimental studies on the olfactory marker protein. II. Appearance of the olfactory marker protein during differentiation of the olfactory sensory neurons of mouse: an immunohistochemical and autoradiographic study. Brain Res. 239, 245–250.

Miyamichi, K., Serizawa, S., Kimura, H. M., and Sakano, H. 2005. Continuous and overlapping expression domains of odorant receptor genes in the olfactory epithelium determine the dorsal/ventral positioning of glomeruli in the olfactory bulb. J. Neurosci. 25, 3586–3592.

Mombaerts, P., Wang, F., Dulac, C., Chao, S. K., Nemes, A., Mendelsohn, M., Edmondson, J., and Axel, R. 1996. Visualizing an olfactory sensory map. Cell 87, 675–686.

Monath, T. P., Cropp, C. B., and Harrison, A. K. 1983. Mode of entry of a neurotropic arbovirus into the central nervous system. Reinvestigation of an old controversy. Lab. Invest. 48, 399–410.

Monti Graziadei, G. A. 1983. Experimental studies on the olfactory marker protein. III. The olfactory marker protein in the olfactory neuroepithelium lacking connections with the forebrain. Brain Res. 262, 303–308.

Monti Graziadei, G. A. and Graziadei, P. P. 1979. Neurogenesis and neuron regeneration in the olfactory system of mammals. II. Degeneration and reconstitution of the olfactory sensory neurons after axotomy. J. Neurocytol. 8, 197–213.

Moran, D. T., Jafek, B. W., Eller, P. M., and Rowley, J. C., III. 1992. Ultrastructural histopathology of human olfactory dysfunction. Microsc. Res. Tech. 23, 103–110.

Mori, K., Fujita, S. C., Imamura, K., and Obata, K. 1985. Immunohistochemical study of subclasses of olfactory nerve fibers and their projections to the olfactory bulb in the rabbit. J. Comp. Neurol. 242, 214–229.

Moulton, D. G. 1974. Dynamics of cell populations in the olfactory epithelium. Ann. N. Y. Acad. Sci. 237, 52–61.

Moulton, D. G. 1975. Cell Renewal in The Olfactory Epithelium. In: Olfaction and Taste, V (eds. D. A. Denton and J. P. Coghlan), pp. 111–114. Academic Press.

Moulton, D. G. and Beidler, L. M. 1967. Structure and function in the peripheral olfactory system. Physiol. Rev. 47, 1–52.

Moulton, D. G., Celebi, G., and Fink, R. P. 1970. Olfaction in Mammals – Two Aspects: Proliferation of Cells in The Olfactory Epithelium and Sensitivity to Odours. In: Ciba Foundation Symposium on Taste and Smell in Vertebrates (eds. G. E. W. Wolstenholme and J. Knight), pp. 227–250. Churchill.

Mulvaney, B. D. and Heist, H. E. 1971. Regeneration of rabbit olfactory epithelium. Am. J. Anat. 131, 241–252.

Murray, R. C., Navi, D., Fesenko, J., Lander, A. D., and Calof, A. L. 2003. Widespread defects in the primary olfactory pathway caused by loss of Mash1 function. J. Neurosci. 23, 1769–1780.

Nagahara, Y. 1940. Experimentelle Studien uber die histologiischen Veranderungen des Geruchsorgan nach der Olfactoriusdurschneidung. Beitrage zur Kenntnis des feineren Baus des Geruchsorgans. Jpn. J. Med. Sci. V. Pathol. 5, 165–169.

Nakashima, T., Kimmelman, C. P., Snow, J. B., and Jr 1984. Structure of human fetal and adult olfactory neuroepithelium. Arch. Otolaryngol. 110, 641–646.

Nakashima, T., Kimmelman, C. P., Snow, J. B., and Jr 1985. Olfactory marker protein in the human olfactory pathway. Arch. Otolaryngol. 111, 294–297.

Nan, B., Getchell, M. L., Partin, J. V., and Getchell, T. V. 2001. Leukemia inhibitory factor, interleukin-6, and their receptors are expressed transiently in the olfactory mucosa after target ablation. J. Comp. Neurol. 435, 60–77.

Norlin, E. M., Alenius, M., Gussing, F., Hagglund, M., Vedin, V., and Bohm, S. 2001. Evidence for gradients of gene expression correlating with zonal topography of the olfactory sensory map. Mol. Cell Neurosci. 18, 283–295.

Oley, N., DeHan, R. S., Tucker, D., Smith, J. C., and Graziadei, P. P. 1975. Recovery of structure and function following transection of the primary olfactory nerves in pigeons. J. Comp. Physiol. Psychol. 88, 477–495.

Peters, A., Palay, S. L., and Webster, H. D. 1991. The Fine Structure of the Nervous System: Neurons and Their Supporting Cells. Oxford University Press.

Pevny, L. and Rao, M. S. 2003. The stem-cell menagerie. Trends Neurosci. 26, 351–359.

Potten, C. S. 1998. Stem cells in gastrointestinal epithelium: numbers, characteristics and death. Philos. Trans. R. Soc. Lond. B Biol. Sci. 353, 821–830.

Potten, C. S. and Loeffler, M. 1990. Stem cells: attributes, cycles, spirals, pitfalls and uncertainties. Lessons for and from the crypt. Development 110, 1001–1020.

Potter, S. M., Zheng, C., Koos, D. S., Feinstein, P., Fraser, S. E., and Mombaerts, P. 2001. Structure and emergence of specific olfactory glomeruli in the mouse. J. Neurosci. 21, 9713–9723.

Purves, D. 1988. Body and Brain. A Trophic Theory of Neural Connections. Harvard University Press.

Rao, M. S. 1999. Multipotent and restricted precursors in the central nervous system. Anat. Rec. (New Anat.) 257, 137–148.

Reed, C. J., Gaskell, B. A., Banger, K. K., and Lock, E. A. 1995. Olfactory toxicity of methyl iodide in the rat. Arch. Toxicol. 70, 51–56.

von Rekowski, C. and Zippel, H. P. 1993. In goldfish the qualitative discriminative ability for odors rapidly returns after bilateral nerve axotomy and lateral olfactory tract transection. Brain Res. 618, 338–340.

Ressler, K. J., Sullivan, S. L., and Buck, L. B. 1993. A zonal organization of odorant receptor gene expression in the olfactory epithelium. Cell 73, 597–609.

Ring, G., Mezza, R. C., and Schwob, J. E. 1997. Immunohistochemical identification of discrete subsets of rat olfactory neurons and the glomeruli that they innervate. J. Comp. Neurol. 388, 415–434.

Satoh, M. and Yoshida, T. 2000. Expression of neural properties in olfactory cytokeratin-positive basal cell line. Brain Res. Dev. Brain Res. 121, 219–222.

Saucier, D. and Astic, L. 1986. Analysis of the topographical organization of olfactory epithelium projections in the rat. Brain Res. Bull. 16, 455–462.

Schaefer, M. L., Finger, T. E., and Restrepo, D. 2001. Variability of position of the P2 glomerulus within a map of the mouse olfactory bulb. J. Comp. Neurol. 436, 351–362.

Schoenfeld, T. A., Clancy, A. N., Forbes, W. B., and Macrides, F. 1994. The spatial organization of the peripheral olfactory system of the hamster. Part I: Receptor neuron projections to the main olfactory bulb. Brain Res. Bull. 34, 183–210.

Schultz, E. W. 1941. Regeneration of olfactory cells. Proc. Soc. Exp. Biol. Med. 46, 41–43.

Schultz, E. W. 1960. Repair of the olfactory mucosa with special reference to regeneration of olfactory cells (sensory neurons). Am. J. Pathol. 37, 1–19.

Schultz, E. W. and Gebhardt, S. P. 1942. Studies on chemical prophylaxis of experimental poliomyelitis. J. Infect. Dis. 70, 7–50.

Schwarting, G. A. and Crandall, J. E. 1991. Subsets of olfactory and vomeronasal sensory epithelial cells and axons revealed by monoclonal antibodies to carbohydrate antigens. Brain Res. 547, 239–248.

Schwarting, G. A., Deutsch, G., Gattey, D. M., and Crandall, J. E. 1992. Glycoconjugates are stage- and position-specific cell surface molecules in the developing olfactory system. 1. The CC1 immunoreactive glycolipid defines a rostrocaudal gradient in the rat vomeronasal system. J. Neurobiol. 23, 120–129.

Schwartz Levey, M., Chikaraishi, D. M., and Kauer, J. S. 1991. Characterization of potential precursor populations in the mouse olfactory epithelium using immunocytochemistry and autoradiography. J. Neurosci. 11, 3556–3564.

Schwartz Levey, M., Cinelli, A. R., and Kauer, J. S. 1992. Intracellular injection of vital dyes into single cells in the salamander olfactory epithelium. Neurosci. Lett. 140, 265–269.

Schwob, J. E. and Gottlieb, D. I. 1986. The primary olfactory projection has two chemically distinct zones. J. Neurosci. 6, 3393–3404.

Schwob, J. E. and Gottlieb, D. I. 1988. Purification and characterization of an antigen that is spatially segregated in the primary olfactory projection. J. Neurosci. 8, 3470–3480.

Schwob, J. E., Huard, J. M., Luskin, M. B., and Youngentob, S. L. 1994a. Retroviral lineage studies of the rat olfactory epithelium. Chem. Senses 19, 671–682.

Schwob, J. E., Saha, S., Youngentob, S. L., and Jubelt, B. 2001. Intranasal inoculation with the olfactory bulb line variant of mouse hepatitis virus causes extensive destruction of the olfactory bulb and accelerated turnover of neurons in the olfactory epithelium of mice. Chem. Senses 26, 937–952.

Schwob, J. E., Szumowski, K. E., and Stasky, A. A. 1992. Olfactory sensory neurons are trophically dependent on the olfactory bulb for their prolonged survival. J. Neurosci. 12, 3896–3919.

Schwob, J. E., Youngentob, S. L., and Meiri, K. F. 1994b. On the formation of neuromata in the primary olfactory projection. J. Comp. Neurol. 340, 361–380.

Schwob, J. E., Youngentob, S. L., and Mezza, R. C. 1995. Reconstitution of the rat olfactory epithelium after methyl bromide-induced lesion. J. Comp. Neurol. 359, 15–37.

Schwob, J. E., Youngentob, S. L., Ring, G., Iwema, C. L., and Mezza, R. C. 1999. Reinnervation of the rat olfactory bulb after methyl bromide-induced lesion: timing and extent of reinnervation. J. Comp. Neurol. 412, 439–457.

Seiden, A. M. 2004. Postviral olfactory loss. Otolaryngol. Clin. North. Am. 37, 1159–1166.

Shinoda, K., Ohtsuki, T., Nagano, M., and Okumura, T. 1993. A possible functional necklace formed by placental antigen X-P2-immunoreactive and intensely acetylcholinesterase-reactive (PAX/IAE) glomerular complexes in the rat olfactory bulb. Brain Res. 618, 160–166.

Shinoda, K., Shiotani, Y., and Osawa, Y. 1989. Necklace olfactory glomeruli form unique components of the rat primary olfactory system. J. Comp. Neurol. 284, 362–373.

Shou, J., Murray, R. C., Rim, P. C., and Calof, A. L. 2000. Opposing effects of bone morphogenetic proteins on neuron production and survival in the olfactory receptor neuron lineage. Development 127, 5403–5413.

Shou, J., Rim, P. C., and Calof, A. L. 1999. BMPs inhibit neurogenesis by a mechanism involving degradation of a transcription factor. Nat. Neurosci. 2, 339–345.

Simmons, P. A., Rafols, J. A., and Getchell, T. V. 1981. Ultrastructural changes in olfactory receptor neurons following olfactory nerve section. J. Comp. Neurol. 197, 237–257.

Slotnick, B., Bodyak, N., and Davis, B. J. 2001. Olfactory marker protein immunohistochemistry and the anterograde transport of horseradish peroxidase as indices of damage to the olfactory epithelium. Chem. Senses 26, 605–610.

Smith, C. G. 1938. Changes in the olfactory mucosa and the olfactory nerves following intranasal treatment with one per cent zinc sulphate. Can. Med. Assoc. J. 39, 138–140.

St John, J. A. and Key, B. 2003. Axon mis-targeting in the olfactory bulb during regeneration of olfactory neuroepithelium. Chem. Senses 28, 773–779.

Stroop, W. G., Rock, D. L., and Fraser, N. W. 1984. Localization of herpes simplex virus in the trigeminal and olfactory systems of the mouse central nervous system during acute and latent infections by *in situ* hybridization. Lab. Invest. 51, 27–38.

Strotmann, J., Conzelmann, S., Beck, A., Feinstein, P., Breer, H., and Mombaerts, P. 2000. Local permutations in the glomerular array of the mouse olfactory bulb. J. Neurosci. 20, 6927–6938.

Suzuki, Y., Mizoguchi, I., Nishiyama, H., Takeda, M., and Obara, N. 2003. Expression of Hes6 and NeuroD in the olfactory epithelium, vomeronasal organ and non-sensory patches. Chem. Senses 28, 197–205.

Suzuki, Y., Schafer, J., and Farbman, A. I. 1995. Phagocytic cells in the rat olfactory epithelium after bulbectomy. Exp. Neurol. 136, 225–233.

Suzuki, Y., Takeda, M., and Farbman, A. I. 1996. Supporting cells as phagocytes in the olfactory epithelium after bulbectomy. J. Comp. Neurol. 376, 509–517.

Taniguchi, K., Arai, T., and Ogawa, K. 1993. Fine structure of the septal olfactory organ of Masera and its associated gland in the golden hamster. J. Vet. Med. Sci. 55, 107–116.

Treloar, H. B., Feinstein, P., Mombaerts, P., and Greer, C. A. 2002. Specificity of glomerular targeting by olfactory sensory axons. J. Neurosci. 22, 2469–2477.

Twomey, J. A., Barker, C. M., Robinson, G., and Howell, D. A. 1979. Olfactory mucosa in herpes simplex encephalitis. J. Neurol. Neurosurg. Psychiatr. 42, 983–987.

Vassar, R., Ngai, J., and Axel, R. 1993. Spatial segregation of odorant receptor expression in the mammalian olfactory epithelium. Cell 74, 309–318.

Verhaagen, J., Oestreicher, A. B., Gispen, W. H., and Margolis, F. L. 1989. The expression of the growth associated protein B50/GAP43 in the olfactory system of neonatal and adult rats. J. Neurosci. 9, 683–691.

Verhaagen, J., Oestreicher, A. B., Grillo, M., Khew-Goodall, Y. S., Gispen, W. H., and Margolis, F. L. 1990. Neuroplasticity in the olfactory system: differential effects of central and peripheral lesions of the primary olfactory pathway on the expression of B-50/GAP43 and the olfactory marker protein. J. Neurosci. Res. 26, 31–44.

Walker, J. C., Smith, J. C., Rashotte, M. E., and Switzer, R. C. 1981. Behavioral and Anatomical Study of Olfactory Nerve Reconstitution in Pigeons. In: Olfaction and Taste VII (*ed*. H. Van der Starre), p. 409. Information Retrieval Ltd.

Walters, E., Buchheit, K., and Maruniak, J. A. 1992. Receptor neuron losses result in decreased cytochrome P-450 immunoreactivity in associated non-neuronal cells of mouse olfactory mucosa. J. Neurosci. Res. 33, 103–111.

Wang, F., Nemes, A., Mendelsohn, M., and Axel, R. 1998. Odorant receptors govern the formation of a precise topographic map. Cell 93, 47–60.

Watt, F. M. 2001. Stem cell fate and patterning in mammalian epidermis. Curr. Opin. Genet. Dev. 11, 410–417.

Weiler, E. and Farbman, A. I. 2003. The septal organ of the rat during postnatal development. Chem. Senses 28, 581–593.

Wolpert, L. 1989. Positional information revisited. Development 107(Suppl.), 3–12.

Wu, H. H., Ivkovic, S., Murray, R. C., Jaramillo, S., Lyons, K. M., Johnson, J. E., and Calof, A. L. 2003. Autoregulation of Neurogenesis by GDF11. Neuron 37, 197–207.

Yee, K. K. and Costanzo, R. M. 1995. Restoration of olfactory mediated behavior after olfactory bulb deafferentation. Physiol. Behav. 58, 959–968.

Yee, K. K. and Costanzo, R. M. 1998. Changes in odor quality discrimination following recovery from olfactory nerve transection. Chem. Senses 23, 513–519.

Yoshihara, Y., Kawasaki, M., Tamada, A., Fujita, H., Hayashi, H., Kagamiyama, H., and Mori, K. 1997. OCAM: A new member of the neural cell adhesion molecule family related to zone-to-zone projection of olfactory and vomeronasal axons. J. Neurosci. 17, 5830–5842.

Youngentob, S. L. and Schwob, J. E. 1997. Changes in odorant quality perception following methyl bromide induced lesions of the olfactory epithelium. Chem. Senses 22, 830–831.

Zhang, X. and Firestein, S. 2002. The olfactory receptor gene superfamily of the mouse. Nat. Neurosci. 5, 124–133.

4.35 Regeneration in the Olfactory Bulb

P-M Lledo, Pasteur Institute, Paris, France

Glossary

adult neurogenesis The entire set of events leading to the production of new neurons in the adult brain, from precursor cell division to functionally integrated survival.

commitment Engaging in a program leading to differentiation. For a stem cell, this means exit from self-renewal.

green fluorescent protein (GFP) An autofluorescent protein that was originally isolated from the jellyfish *Aequorea victoria*. It can be genetically conjugated with proteins to make them fluorescent. The most widely used mutant, enhanced GFP (eGFP), has an emission maximum at 510 nm.

lentivirus vector A retrovirus vector derived from human immunodeficiency virus (HIV), which infects both dividing and nondividing cells.

multipotent Can form multiple lineages that constitute an entire tissue or tissues. Example: hematopoietic stem cells.

neurogenic niche Cellular microenvironment providing support and stimuli necessary to sustain self-renewal.

oligopotent Able to form two or more lineages within a tissue. Example: a neural stem cell that can create a subset of neurons in the brain.

pluripotent Able to form all the body's cell lineages, including germ cells, and some or even all extraembryonic cell types. Example: embryonic stem cells.

potency The range of commitment options available to a cell.

precursor CNS stem cells and all progenitors are generally referred to as precursor cells.

progenitor A mitotic cell with a fast dividing cell cycle that retains the ability to proliferate and to give rise to terminally differentiated cells, but is not capable of indefinite self-renewal.

stem cell A cell that can continuously produce unaltered daughters and also has the ability to produce daughter cells that have different, more restricted properties.

stem cells plasticity Unproven notion that tissue stem cells may broaden potency in response to physiological demands or insults.

totipotent Sufficient to form entire organism. Totipotency is seen in zygote and plant meristem cells; not demonstrated for any vertebrate stem cell.

transit-amplifying cell Proliferative stem cell progeny fated for differentiation. Initially may not be committed and may retain self-renewal.

two-photon microscopy Microscopy in which fluorochromes are stimulated, not by a single high-energy photon, but quasi-simultaneously by two photons of low energy, greatly reducing both light scattering and photodynamic sample damage.

unipotent Forms a single lineage. Example: spermatogonial stem cells.

4.35.1 Introduction

Modern neurobiology has shown that complex biological systems, such as the vertebrate brain, are constructed from similar, repeated units. In the central nervous system (CNS), these units, or modules, are composed of networks with opposite effects – excitation and inhibition. Excitation was initially assumed to play the dominant role in information processing, whereas inhibition was thought to prevent excessive activity. Local inhibitory neurons appeared to be simple elements that regulated excitatory neuronal activity. Thus, as in the Chinese Yin and Yang philosophy, the activity of a large network of principal excitatory (glutamatergic) neurons was thought to be purposefully balanced by local circuit inhibitory neurons.

The γ-aminobutyric acid (GABA)-releasing neuron, also referred to as the local circuit inhibitory interneuron, is one of the most studied types of interneuron. Recent investigations have indicated that the initial concept of interneuron is an oversimplification, requiring readjustment to accommodate cell types and functions that would not strictly fit the definition cited above. For instance, combined anatomical, electrophysiological, and pharmacological approaches have recently provided new insight into the roles of interneurons at several levels. At the cellular level, these cells govern action potential generation, firing pattern, precise timing of the discharge of individual principal cells in relation to the emergent behavior of the entire cell assembly, membrane potential oscillations, and dendritic calcium spikes. At the network level, inhibitory interneurons are considered to play an important role in controlling synaptic strength and plasticity and the generation of large-scale synchronous oscillatory activity.

Interneurons have generated considerable research interest, but studies of their function are limited by their tremendous heterogeneity in morphology and connectivity, the principal neurons being much more uniform in their appearance. This limitation has tended to push attempts to characterize interneuron subtypes into the domain of neurochemical content, based on calcium-binding proteins or neuromodulators. However, neurochemically identical cells may have surprisingly different functional properties, further complicating classification. Furthermore, as inhibitory interneurons exert pleiotropic effects on brain microcircuits, any characterization based solely on function is almost certain to prove problematic.

Thus, a classification of interneurons can only be considered valid if it takes into account morphological, biochemical, and physiological characteristics. This criterion is satisfied for the first central relay of sensory information about odor, the olfactory bulb. This chapter provides an overview of the constitution of the bulbar interneuron population, current knowledge concerning the functions of these cells, their replacement and the possible role of adult neurogenesis in perception and memory. Due to space constraints, this review focuses exclusively on the mammalian olfactory bulb.

4.35.2 The Mammalian Olfactory Bulb Network

The main projection neurons of the bulb, the mitral cells, are located in a single lamina, the mitral cell layer (Figure 1). The primary dendrite of the mitral cell, extending vertically from the soma, contacts a single glomerulus, at which massive interactions with periglomerular interneurons and olfactory nerve terminals occur. Periglomerular interneurons are heterogeneous, including both GABAergic and dopaminergic subtypes. In contrast, mitral cell secondary dendrites radiate horizontally, for up to 1 μm, almost entirely spanning the olfactory bulb (Mori, K. *et al.*, 1983). In the external plexiform layer, they interact with inhibitory axonless interneurons – the granule cells – which constitute the largest cellular population within the bulb. The projection neurons engage with this extensive local circuit through dendrodritic synaptic contacts. This reciprocal circuit provides the basis for the reliable, spatially localized, recurrent inhibition of mitral cells (Rall, W. *et al.*, 1966). The synaptic depolarization of mitral cells, driven by persistent excitatory input from the sensory neurons, triggers glutamate release from the dendrites of these cells, resulting in the depolarization of interneuron dendrites and spines. This, in turn, elicits the direct release of GABA into the mitral cell. Thus, in the olfactory bulb, the quality of the odor stimulus is initially encoded by a specific combination of activated mitral cells, as a function of GABAergic inhibition. As a single granule cell is believed to contact a number of output neurons (Shepherd, G. M. *et al.*, 2004), the reciprocal inhibitory synaptic connection helps to synchronize mitral cell activity (Lagier, S. *et al.*, 2004).

Following activation of the glomeruli, the network of output neurons and local interneurons adds a second

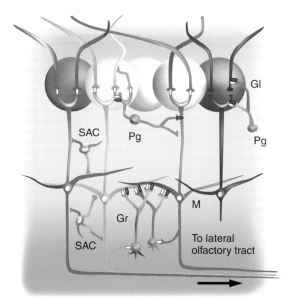

Figure 1 Wiring of the olfactory bulb. Each sensory neuron expresses only one of the 1000 odorant receptor genes and the axons from all cells expressing that particular receptor converge onto one or a few glomeruli (Gl) in the olfactory bulb. The nearly 2000 glomeruli in the rat olfactory bulb are spherical knots of neuropil, about 50–100 μm in diameter, which contain the incoming axons of sensory neurons and the apical dendrites of the projection neuron of the olfactory bulb, the mitral cell (M). Mitral axons leaving the olfactory bulb project widely to higher brain structures including the piriform cortex, hippocampus, and amygdala. Processing of the olfactory message into the bulb occurs through three populations of interneurons: periglomerular cells (Pg), granule cells (Gr), and short axon cells (SAC).

combinatorial layer to the representation of olfactory information. Contacts mediated by bulbar output neurons and local interneurons may contribute to the global reformatting of odor representations, in the form of a stimulus-dependent, temporal redistribution of activity across the olfactory bulb. In this context, lateral inhibition can be seen as a means of sharpening odor representation by a neuronal population (see below). Inhibitory circuits with different topologies are likely to form, depending on the sensory neuron afferents stimulated and the local interneuron connections activated following stimulation. Two combinatorial encoders therefore seem to be involved in information processing (Lledo, P.-M. *et al.*, 2005). The first is the olfactory receptor repertoire expressed by the ensemble of sensory neurons, which transduces receptor activation patterns into a glomerular odor image (Korsching, S., 2002). The secondary encoder is the intricate interneuron network, which extracts higher-order features from the odor images, converting

them into timing relationships across the firing output neurons (Laurent, G., 2002). Three assumptions can be made, based on the combination of these two encoders of the olfactory bulb. First, the neural synchronization of glomerular outputs may enhance the representation of a complex olfactory stimulus by integrating the different signal streams activated by the odor into a unified olfactory image in the sensory cortex. Second, the activation of neuronal assemblies in the bulb modulates groups of neurons in the forebrain, bringing into play an olfactory signal-decoding mechanism. It has been suggested that the olfactory forebrain uses coincidence detector neurons to identify the various patterns of activity in the bulb (Mori, K. *et al.*, 1999). In this model, synchronized oscillations are seen as a means by which signals from different olfactory neurons are temporally connected for detection. Finally, the neurons in the olfactory bulb that relay activity to the rest of the brain interact extensively with each other, both directly and through a network of coupled interneurons. The relay neurons can engage in highly synchronous oscillatory activity, thereby forming assemblies that conveying information about a particular odor blend. As described below, the continuous turnover of interneurons plays an important role in adjusting the spatiotemporal processes of the olfactory bulb to the level of complexity and novelty of odors encountered by the animal.

4.35.3 Building Neuronal Networks during Adulthood

Postdevelopmental neurogenesis is conserved across evolutionary boundaries, from crustaceans to higher vertebrates, such as birds, rodents, and primates, including humans. The extent of postnatal neurogenesis decreases with increasing brain complexity, with adult neurogenesis in lower vertebrates, such as lizards, providing a massive additional supply of neurons capable of regenerating entire areas of the brain, whereas mammalian neurogenesis is restricted to only a few regions of the brain and to the replacement of newborn neurons. The degree of postdevelopmental neurogenesis in a given species may depend on a trade-off between the benefits accruing from the creation of new neurons and the problems generated by these neurons in terms of their integration into the network circuitry. In more complex brains such as those of mammals, the problems of newborn neuron integration seem to outweigh the potential benefits and adult neurogenesis in normal conditions is probably confined to just two regions: the subgranular zone of the hippocampal dentate gyrus and the subventricular zone (SVZ), which contributes interneurons to the olfactory bulb (Alvarez-Buylla, A. and Garcia-Verdugo, J. M., 2002). There is some evidence to suggest that constitutive neurogenesis occurs in other regions of the adult brain, but this remains to be demonstrated (Rakic, P., 2002).

4.35.3.1 Producing New Neurons

Astrocytes in the adult SVZ, which lines the border between the striatum and the lateral ventricle to form a neurogenic niche, act as slow-dividing adult neural stem cells with high potency, generating transit-amplifying cells before transforming into neuroblast precursors (Doetsch, F., 2003) (Figure 2). These neuroblasts proceed toward the olfactory bulb along an intricate path of migration known as the rostral migratory stream (RMS). More than 30 000 neuroblasts leave the rodent SVZ and enter the RMS each day (Alvarez-Buylla, A. and Lim, D. A., 2004). In the RMS, they are not guided by radial glia, but instead migrate tangentially in chains through tubular structures formed by specialized astrocytes. These newborn cells then detach from the chains and migrate radially into the olfactory bulb, where they mature into olfactory inhibitory interneurons of two main types – granule cells and periglomerular cells – and integrate into networks. Both these cell types make only local contacts, directly or indirectly modulating the processing of sensory information by the mitral cells.

4.35.3.2 Maturation and Functional Integration

The use of replication-incompetent lentiviruses to transduce stem cells in the SVZ and label them with enhanced green fluorescent protein (eGFP) has made it possible to characterize the commitment of adult stem cells. The presence of GFP has made possible both morphological and electrophysiological characterization of the newly formed olfactory bulb neurons during their migration and differentiation (Belluzzi, O. *et al.*, 2003; Carleton, A. *et al.*, 2003). The morphology of newly generated cells becomes more complex in the first few weeks after their generation. Granule cells form more elaborate dendrites extending into the external plexiform layer of the bulb, and become fully mature morphologically within about 2 weeks. In contrast, periglomerular cells may take about 4 weeks to develop their full dendritic and axonal morphology (Belluzzi, O. *et al.*, 2003); a period of time similar to that required for the maturation of hippocampal cells generated during adulthood (Figure 3). During this maturation, immature bulbar interneurons express transient marker proteins, including TUJ1 and TUC4 (Lledo, P.-M. *et al.*, 2006). Functionally, as in the adult hippocampus, the first synapses made with new cells are GABAergic, and these contacts are followed by glutamatergic inputs several days later. The developmental sequence of voltage-dependent currents and synaptic connections differentiates between newly formed periglomerular and granule cells (and between bulbar and hippocampal granule cells). In periglomerular cells, maturation of the voltage-dependent sodium current, and consequently the capacity of the newly generated cells to fire action

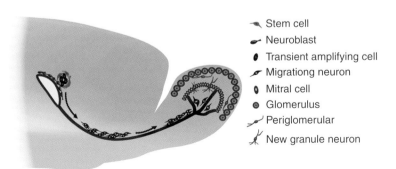

Figure 2 The subventricular zone (SVZ)–olfactory bulb pathway constitutes one of the two neurogenic niches in the adult forebrain. Mauve arrows indicate the tangential migration of neuroblasts from the lateral ventricle (filled in blue) toward the olfactory bulb. Neurogenesis refers to the replacement of bulbar local interneurons in the olfactory bulb.

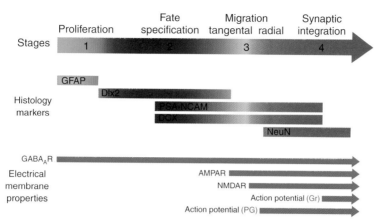

Figure 3 Postmitotic maturation of newborn neurons in the adult forebrain. The schema shows the functional stages through which adult-born neuron progress before becoming fully mature. This simplified state of affairs comes with a number of caveats, since the maturation of individual adult-born neurons has yet to be followed, maturation stages are likely to reflect a continuum rather than discrete steps, and maturation time points are likely to vary considerably from cell to cell. For this last reason, ages here represent the earliest age at which particular properties have been observed in a given cell type. Note also that maturation certainly does not stop at stage 4: arbor complexity and spine density can increase up to 4 months after the birth of a newborn neuron. It is noteworthy that receptor expression precedes synaptic activation, and γ-aminobutyric acid (GABA)-mediated input precedes glutamatergic input. Olfactory bulb granule cells show atypical maturation in that they only fire sodium-based action potentials after they have begun to receive synaptic input.

potentials, seems to precede the appearance of synaptic contacts, whereas the opposite pattern is observed in granule cells (Belluzzi, O. *et al.*, 2003; Carleton, A. *et al.*, 2003). The later maturation of granule cell excitability may prevent the preexisting circuitry in the adult from being disrupted during the insertion of newly formed cells. Alternatively, as even mature granule cells lack an axon, it may result purely from morphological differences between granule and periglomerular neurons. Further studies are required to determine whether newly generated neurons form functional synapses with their downstream target neurons and release appropriate neurotransmitters, which would unequivocally demonstrate their integration into adult networks.

Below, I summarize the contributions of early work in the field to the current framework for understanding the role of newborn interneurons in the adult olfactory bulb network. I will also review the more recent findings that have dramatically reshaped our view of the integration of newly formed neurons into adult networks. At first glance, this neurogenesis may be seen as a putative response to the death and turnover of the primary olfactory receptor neurons that occurs throughout the life of the animal. However, as we will see below, adult neurogenesis also endows the bulb with unique properties.

4.35.4 Functions of Newly Formed Interneurons in the Olfactory Bulb

Ongoing neurogenesis and migration in adults have been extensively documented, but their functional consequences remain unclear. As the largest neuronal population in the olfactory bulb consists of granule cells, which outnumber the projection neurons by at least 100:1 (Shepherd, G. M. *et al.*, 2004), and one granule cell contacts several mitral cells, with each mitral cell, in turn, contacting a large number of pyramidal cells from the pyriform cortex, the integration of transient newly formed neurons into an upstream position in the olfactory bulb is ideal for amplifying the effects of neurogenesis throughout the olfactory pathway. If the continuous generation of interneurons throughout the animal's life is necessary for olfaction, then changes in migration processes – resulting in changes in the recruitment of new interneurons – should affect olfactory processing. Studies of transgenic mice with low levels of adult neurogenesis have been carried out to test this hypothesis. For example, NCAM-mutant mice have been studied as these mice display lower than normal levels of neuroblast migration. The quantification of bulbar neurogenesis revealed 40% fewer newly formed interneurons in adult knockout mice than in the wild type.

This impairment of neurogenesis was accompanied by changes in odor discrimination (Gheusi, G. *et al.*, 2000), revealing a specific role for newly formed interneurons in the downstream coding of olfactory information. This observation was supported by observations in another transgenic line with impaired GABA$_A$ receptor-mediated synaptic inhibition (Nusser, Z. *et al.*, 2001) and by theoretical analyses (Cecchi, G. A. *et al.*, 2001). Based on these data, it appears that a critical level of inhibition, mediated by the activation of GABA$_A$ receptors located on the secondary dendrites of bulbar projection neurons, is required for olfactory processing. In addition, as the olfactory bulb is also involved in consolidating processes associated with long-term odor memory, changes in the number of GABAergic interneurons may regulate both olfactory perception and memory.

This issue has been investigated, by studies aiming to determine whether a change in the number of newly formed bulbar interneurons alters olfactory memory. In these experiments, animals were reared in a complex olfactory environment. Exposure to enriched sensory surroundings has many positive effects on brain structure and function, increasing the number of dendritic branches and spines, enlarging synapses, increasing the number of glial cells and improving performance in tests of spatial memory (Rosenzweig, M. R. and Bennett, E. L., 1996). Mice reared in enriched olfactory environments produce twice as many new interneurons as other mice. This increase is specific to the olfactory bulb, with no changes observed in the other neurogenic zone, the dentate gyrus of the adult hippocampus. Animals with higher levels of neurogenesis retain learned olfactory information for longer periods of time than controls (Rochefort, C. *et al.*, 2002). In particular, animals reared in enriched olfactory environments recognize familiar odors in more readily and in a more sustained manner than animals reared in standard conditions. Although the potential consequences of an enriched environment for synaptic efficacy and/or mitral cell activity were not specifically explored, these results are consistent with the existence of a correlation between the size of the newly formed interneuron population and odor memory.

4.35.5 Properties Conferred by Neurons Generated in Adulthood

Although recent estimates suggest that many more neurons are produced during adulthood than was initially thought, the estimated rate of neuron production in adulthood remains lower than that during development. Consequently, if the neurons generated during adulthood have similar functional properties to those generated in early life, then adult neurogenesis probably has an insignificant impact. In contrast, if adult-generated neurons have unique properties, giving them a greater impact than more mature neurons, then their continual integration into the functional circuitry would be expected to have major effects. Recent observations of both the hippocampus and the olfactory bulb are consistent with this second hypothesis. Indeed, newly formed neurons seem to display robust plasticity that cannot be inhibited by GABA, which is not the case for mature neurons. As new neurons are structurally plastic (Figure 4), they may well be very sensitive to changes in the animal's environment and life experiences. This suggests that adult neurogenesis is functionally important, as it results in a continual influx of neurons that are, at least temporarily, immature, with unique physiological properties.

As described above, bulbar interneurons play an essential role in shaping the olfactory information that reaches the olfactory cortex. As the cells generated in the adult SVZ differentiate exclusively into bulbar interneurons, they probably contribute to essential aspects of olfactory processing. Experimental studies and modeling have provided evidence that newly formed granule cells play a role in olfactory discrimination, but the more general functions of these cells remain to be characterized. Adult neurogenesis is unlikely to be of immediate functional

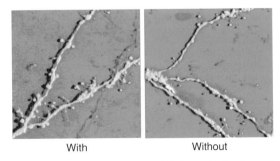

With   Without

Figure 4 The adult neurogenesis brings plastic elements to preexisting neuronal networks. Images of green fluorescent protein (GFP)-labeled newborn interneurons taken with a two-photon microscope. The newborn interneurons developed in active (left panel) or inoperative (right panel) olfactory bulb networks. Note reduced spine density for newborn interneurons maturing in the quiescent olfactory bulb.

benefit, because it takes several days to generate a new, functionally integrated interneuron (Figure 3). Cells generated during adulthood develop neurites within a few weeks. It is clear that the new connections cannot constitute a response to the particular functional event triggering neurogenesis, because these connections may well not form until after the trigger event has ended. Olfactory bulb neurogenesis should therefore be considered a long-term, as opposed to a short-term adjustment of the bulbar circuitry to a higher level of experience governed by olfaction.

4.35.6   Adult Neurogenesis: A Neural Basis for Experience-Induced Plasticity

It seems reasonable to assume that the mind and brain benefit from active living; this hypothesis is supported by several observations, including the results of epidemiological studies. It has also been clearly established that both physical and intellectual activities have a positive influence on the incidence of neurodegenerative disorders and cognitive decline. The continuous neurogenesis observed in the olfactory region of adult animals provides us with a unique model for studying the effects of sensory-driven activity and its benefits. In the olfactory bulb, the role of environment-induced activity can be explored not only during the neonatal period, when the developing structures are refined and stabilized, but also in adulthood, when more delicate regulation occurs. Several studies have demonstrated that changes in the level of sensory activity do not affect the proliferation and tangential migration of neuroblasts. Instead, they seem to regulate the number of newly generated neurons reaching the olfactory bulb and the level of cell death. Thus, sensory-driven activity may play an important role in several steps of adult neurogenesis (Figure 5). This, in turn, would greatly modify the number of new interneurons, thereby altering the inhibitory–excitatory balance in the bulbar neuronal networks. There is clear evidence that adult neurogenesis can be regulated, but most of the data on adult mammalian neurogenesis have been obtained from laboratory animals, kept under artificial conditions very different from their natural habitat. Even the odor-enriched environments used in experiments are arguably much poorer in terms of sensory stimulation than conditions in the wild.

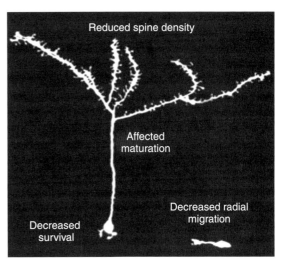

Figure 5   Role of sensory activity in shaping the adult neurogenesis. Using sensory deprivation paradigm, it has been demonstrated that odor-induced activity is required for normal radial migration of neuroblasts, their survival and maturation and the maintenance of spine density.

4.35.7   Ethological Relevance of Adult Neurogenesis

Although it is clear that the adult olfactory bulb integrates neurons throughout life, the reason for this remains unclear. What are the new neurons used for and how can an existing functional neural network integrate and even actively recruit new neurons? Current theories concerning cognition are based on the assumption that the adult brain is a stable network in terms of the number of neurons. According to this scheme, structural neural plasticity occurs only at the level of synapses, dendrites, and neurites. The demonstration of adult bulbar neurogenesis – the generation of new interneurons from resident neuronal stem cells and their integration into the bulbar circuit – has called this assumption into question. In light of results on the activity-dependent regulation of adult bulbar neurogenesis, some conclusions concerning why and how new neurons contribute to bulbar function can be made. I suggest that neurogenesis does not simply add neurons to enhance memory, but that it also strategically inserts new cells with unique properties (Figure 3). The multiple stages in the maturation of new neurons might provide a more extended substrate for plasticity. The newly formed neurons may provide the olfactory bulb with a large repository of plastic cells, enabling it to respond more effectively to changes in the environment and to life experiences (Figure 5). For instance, new neurons may respond preferentially to hormones, learning, and other forms of experience-induced activity. These aspects have been well documented in various biologically relevant models described briefly below.

Female mice form a memory of the pheromones of the male they mate with, a phenomenon known as the Bruce effect. This memory leads to disruption of the pregnancy when the female is exposed to the pheromones of other males (Bruce, H. M., 1959). It is formed during mating and requires a period of about 3–4 h of exposure to the male's pheromones. Brennan P. A. *et al.* (1995) used a microdialysis approach to explore changes in a range of neurotransmitters that occur when female mice form a memory of their stud male. The main change in the accessory olfactory bulb is a decrease in glutamate/ GABA ratio following memory formation in response to the mating male's pheromones. The increase in GABA release reflects a higher inhibitory tone at the reciprocal synapses. Brennan P. A. *et al.* (1995) interpreted this mechanism as a way of reducing excitation from mitral cells, which mediate the input from the mating male's pheromones, prevent the blockage of pregnancy. In contrast, exposure to a different male would activate a different subpopulation of mitral cells, not previously subjected to an increase in inhibitory control, thereby eliciting central pregnancy-blocking mechanisms. These processes are not restricted to the formation of a memory of the mating male's pheromones. Similar events have been observed in the olfactory bulb of ewes during the development of a specific ability to recognize their own lambs, enabling them to reject other lambs that attempt to suckle. After parturition, an increase in glutamate and GABA release occurs in the ewe's olfactory bulb in response to odors from her own lamb, but not to odors from other lambs (Kendrick, K. M. *et al.*, 1992). Once the ewe has established a selective bond with its lamb, the ratio of glutamate to GABA decreases in response to the lamb's odor, suggesting, once again, the existence of changes in the gain of reciprocal synapses between mitral and granule cells.

These examples illustrate changes in the excitatory–inhibitory balance in the olfactory bulb occurring during the establishment of specific social preferences. Similar changes have recently been shown to occur in circumstances involving olfactory learning in a broader context. For instance, the use of a conditioning procedure in which adult mice were trained, using odor, to dig through wood shavings to obtain a buried piece of sugar, has produced interesting findings. Following conditioning, the presentation of the odor previously associated with the reward induces an increase in both glutamate and GABA levels in the olfactory bulb and a decrease in the balance between excitatory and inhibitory transmitters. In contrast, no change is observed if animals are exposed to a nonconditioning odor, indicating that the increase in inhibition occurs only for relevant odors. These findings highlight the significance of the changes in the excitatory–inhibitory ratio that occur in the olfactory bulb following learning. The neurochemistry of inhibition enhancement is reminiscent of the upregulation of the permanent supply of bulbar interneurons correlated with a stronger memory in odor-enriched mice. Together, these findings indicate that adult bulbar neurogenesis may be involved in many natural types of behavior yet to be explored.

4.35.8 Concluding Remarks

We still do not know why there are new neurons in the adult brain, but research into the regulation of adult neurogenesis, and an increasing number of studies investigating the function of the new cells, have led to two conclusions. First, bulbar neurogenesis is affected by relevant behavior, including behavior specifically related to bulbar function, in particular. Second, changes in the rate or extent of neurogenesis have an effect on subsequent behavior. Thus, behavior may change the structure of the olfactory bulb, and changes in structure can subsequently modify, or at least affect, subsequent behavior. The correlation between activity and adult neurogenesis suggests that new neurons are involved general aspects of bulbar function, probably sustaining the ability of the adult network to adapt information processing to relevant ethological needs.

The sense of smell has played an essential role in mammalian evolution. Odor representation is dynamic and highly complex, and may require unique plasticity mechanisms. Neurogenesis, migration, and the replacement of new neurons in the olfactory bulb are probably involved in this adaptive mechanism. As it takes time for new neurons to mature and become integrated into synapses, adult neurogenesis probably contributes to slow, long-term adjustments of the bulbar circuitry, rather than to fast and acute plastic changes. However, new neurons may also be involved in faster mechanisms. Several new bulbar interneurons die within a month of maturation. Recent results show that newly formed granule cells maintain synaptic contacts during this period of rapid cell death. The selective elimination of neurons may make it possible to modify the circuitry more rapidly. The ability to mold the integration of new neurons and to eliminate older cells continually, without depleting the neuronal population, may provide neuronal networks with adaptive potential unequalled by synaptic plasticity alone.

In this brief review, I have tackled two of the main problems relating to the existence of newborn neurons in the adult olfactory bulb. Both are fundamental issues. First, I have discussed the role of adult neurogenesis in the context of bulbar function. Assuming that the bulb processes odor information before relaying it to the olfactory cortex, I hypothesize that adult neurogenesis allows the olfactory bulb to adjust the degree of processing appropriately. Second, the existence of a pool of juvenile neurons, enabling the system to adapt to future similar situations, raises the possibility that adult neurogenesis acts *post hoc* to provide a structural basis for brain plasticity. Research on adult neurogenesis in the olfactory bulb is thus not only interesting in itself, but also provides new avenues of exploration to increase our understanding of adult brain plasticity.

Acknowledgments

I thank Cecile Moreau for help with illustrations. This work was supported by the Fédération pour la Recherche sur le Cerveau, the Fondation pour la Recherche Médicale and the Agence Nationale de la Recherche (ANR-05-Neur-028-01).

References

Alvarez-Buylla, A. and Garcia-Verdugo, J. M. 2002. Neurogenesis in adult subventricular zone. J. Neurosci. 22, 629–634.

Alvarez-Buylla, A. and Lim, D. A. 2004. For the long run: maintaining germinal niches in the adult brain. Neuron 41, 683–686.

Belluzzi, O., Benedusi, M., Ackman, J., and LoTurco, J. J. 2003. Electrophysiological differentiation of new neurons in the olfactory bulb. J. Neurosci. 23, 10411–10418.

Brennan, P. A., Kendrick, K. M., and Keverne, K. B. 1995. Neurotransmitter release in the accessory olfactory bulb during and after the formation of an olfactory memory in mice. Neuroscience 69, 1075–1086.

Bruce, H. M. 1959. An exteroceptive block to pregnancy in the mouse. Nature 184, 105.

Carleton, A., Petreanu, L. T., Lansford, R., Alvarez-Buylla, A., and Lledo, P. M. 2003. Becoming a new neuron in the adult olfactory bulb. Nat. Neurosci. 6, 507–518.

Cecchi, G. A., Petreanu, L. T., Alvarez-Buylla, A., and Magnasco, M. O. 2001. Unsupervised learning in a model of adult neurogenesis. J. Comp. Neurosci. 11, 175–182.

Doetsch, F. 2003. The glial identity of neural stem cells. Nat. Neurosci. 6, 1127–1134.

Gheusi, G., Cremer, H., McLean, H., Chazal, G., Vincent, J. D., and Lledo, P. M. 2000. Importance of newly generated neurons in the adult olfactory bulb for odor discrimination. Proc. Natl. Acad. Sci. U. S. A. 97, 1823–1828.

Kendrick, K. M., Levy, F., and Keverne, E. B. 1992. Changes in the sensory processing of olfactory signals induced by birth in sheep. Science 256, 833–836.

Korsching, S. 2002. Olfactory maps and odor images. Curr. Opin. Neurobiol. 12, 387–392.

Lagier, S., Carleton, A., and Lledo, P. M. 2004. Interplay between local GABAergic interneurons and relay neurons generates gamma oscillations in the rat olfactory bulb. J. Neurosci. 24, 4382–4392.

Laurent, G. 2002. Olfactory network dynamics and the coding of multidimensional signals. Nat. Rev. Neurosci. 3, 884–895.

Lledo, P.-M., Gheusi, G., and Vincent, J. D. 2005. Information processing in the mammalian olfactory system. Physiol. Rev. 85, 281–317.

Lledo, P.-M., Grubb, M., and Alonso, M. 2006. Adult neurogenesis and functional plasticity in neuronal circuits. Nat. Neurosci. Rev. 7, 179–193.

Mori, K., Kishi, K., and Ojima, H. 1983. Distribution of dendrites of mitral, displaced mitral, tufted, and granule cells in the rabbit olfactory bulb. J. Comp. Neurol. 219, 339–355.

Mori, K., Nagao, H., and Yoshihara, Y. 1999. The olfactory bulb: coding and processing of odor molecule information. Science 286, 711–715.

Nusser, Z., Kay, L. M., Laurent, G., Homanics, G. E., and Mody, I. 2001. Disruption of GABAA receptors on GABAergic interneurons leads to increased oscillatory power in the olfactory bulb network. J. Neurophysiol. 86, 2823–2833.

Rakic, P. 2002. Adult neurogenesis in mammals: an identity crisis. J. Neurosci. 22, 614–618.

Rall, W., Shepherd, G. M., Reese, T. S., and Brightman, M. W. 1966. Dendrodendritic synaptic pathway for inhibition in the olfactory bulb. Exp. Neurol. 14, 44–56.

Rochefort, C., Gheusi, G., Vincent, J. D., and Lledo, P. M. 2002. Enriched odor exposure increases the number of newborn neurons in the adult olfactory bulb and improves odor memory. J. Neurosci. 22, 2679–2689.

Rosenzweig, M. R. and Bennett, E. L. 1996. Psychobiology of plasticity: effects of training and experience on brain and behavior. Behav. Brain Res. 78, 57–65.

Shepherd, G. M., Chen, W. R., and Greer, C. A. 2004. Olfactory Bulb. In: The Synaptic Organization of the Brain, 5th edn. (ed. G. M. Shepherd), pp. 165–216. Oxford University Press.

Further Reading

Kashiwadani, H., Sasaki, Y. F., Uchida, N., and Mori, K. 1999. Synchronized oscillatory discharges of mitral/tufted cells with different molecular receptive ranges in the rabbit olfactory bulb. J. Neurophysiol. 82, 1786–1792.

Petreanu, L. and Alvarez-Buylla, A. 2002. Maturation and death of adult-born olfactory bulb granule neurons: role of olfaction. J. Neurosci. 22, 6106–6113.

Rall, W. and Shepherd, G. M. 1968. Theoretical reconstruction of field potentials and dendrodendritic synaptic interactions in olfactory bulb. J. Neurophysiol. 31, 884–915.

Relevant Websites

http://www.apuche.org – Olfactory Image Archive at University of Maryland, Baltimore (USA).

http://leonserver.bio.uci.edu – Olfactory maps at University of California, Irvine (USA).

http://www.pasteur.fr – Perception and Memory Laboratory at Pasteur Institute (France).

http://www.uchsc.edu – Restrepo Laboratory at the University of Colorado Health Center (USA).

http://senselab.med.yale.edu – The SenseLab Project at Yale University (USA).

4.36 Architecture of the Olfactory Bulb

C A Greer, M C Whitman, L Rela, F Imamura, and D Rodriguez Gil, Yale University School of Medicine, New Haven, CT, USA

Glossary

cilia Nonmotile extensions of the olfactory sensory neurons within the lumen of the nasal cavity. The cilia are suspended in a layer of mucous, express odor receptors on their surface, and increase significantly the surface area of the olfactory sensory neurons.

dendrodendritic synapse Synapses occurring between two dendrites or between a dendrite and a dendritic spine. In the olfactory bulb dendrodendritic synapses form between mitral/tufted cells and granule cells in the external plexiform layer and between mitral/tufted cells and peri/juxtaglomerular cells in the glomerular layer.

gap junctions Intercellular channels that allow the diffusion of ions and metabolites and constitute the structural basis of electrical synapses. They may be homotypic, composed of identical hemichannels (or connexons) provided by each of the connected cells, or heterotypic, composed of different connexons. In vertebrates, each connexon is normally a hexamer of connexins.

glomerulus Spherical area of neuropil where olfactory sensory neuron axons terminate and make synaptic contact with mitral, tufted, and periglomerular cells. Glomeruli are molecularly homogeneous in that all of the sensory axons terminating in one glomerulus arise from sensory neurons expressing the same odor receptor.

granule cell A major interneuron in the olfactory bulb. Granule cells are anaxonic, they lack axons, and both input and output synapses occur through specialized dendritic spines found on their apical dendrites within the external plexiform layer. Granule cells establish these dendrodendritic synapses with the secondary dendrites of mitral and tufted cells. Input synapses, from centrifugal fibers and axon collaterals of mitral and tufted cells, also occur at the cell bodies and basal dendrites in the granule cell layer. New granule cells continue to be generated in the adult and migrate into the olfactory bulb via the rostral migratory stream.

lateral olfactory tract A myelinated tract of axons from mitral and tufted cells extending from the olfactory bulb to the cortex. As axons exit the lateral olfactory tract they terminate in layer Ia of cortex, where they synapse on the apical dendrites of cortical pyramidal neurons.

mitral cell Major projection neuron from olfactory bulb to cortex. The cell bodies comprise a monolayer in the adult olfactory bulb. The apical dendrite arborizes in a single glomerulus while the lateral dendrites extend out in the external plexiform layer perpendicular to the radial axis of the olfactory bulb. The mitral cell axons distribute broadly in piriform cortex.

olfactory ensheathing cells Glia that encompass fascicles of olfactory sensory neuron axons in the olfactory nerve and in the olfactory nerve layer of the olfactory bulb. The ensheathing cells are important in the ongoing growth of new axons from the epithelium to the olfactory bulb. Due to their ability to support axon outgrowth in the olfactory system, the ensheathing cells have also found application in spinal cord injury where transplants of the ensheathing cells can improve axon regeneration.

olfactory nerve A collection of olfactory sensory neuron axon fascicles, surrounded by the processes of olfactory ensheathing cells, that extends from the olfactory epithelium to the olfactory bulb. As the nerve merges with the olfactory bulb, it becomes the olfactory nerve layer.

olfactory sensory neuron A specialized receptor found in the olfactory epithelium. The apical dendrite extends in the lumen of the nasal cavity where multiple cilia extend. The olfactory sensory neurons each express only 1 of 1000 different odor receptors in rodents. The olfactory sensory neurons are derived from a population of basal cells that appear to provide for ongoing replacement of the sensory neurons throughout life.

peri/juxtaglomerular cells These surround the glomeruli of the olfactory bulb and extend dendritic processes into the glomeruli where some receive input from the olfactory sensory neuron axons and most participate in dendrodendritic synapses with the apical dendrites of mitral and tufted cells. Axons from the peri/juxtaglomerular cells can extend considerable distances in the glomerular layer to mediate interactions with distant glomeruli. Peri/juxtaglomerular cells can be further divided into multiple subpopulations based on their molecular phenotype including the expression of neurotransmitters and calcium-binding proteins. Like granule cells, peri/juxtaglomerular cells continue to be generated in the adult.

rostral migratory stream A well-defined glial lined tract via which neuroblasts from the subventricular zone migrate into the olfactory bulb.

sustentacular cells Nonsensory supporting cells that surround the olfactory sensory neurons in the olfactory epithelium. The sustentacular cells may contribute to the removal of odorants or other molecules from the nasal cavity.

tufted cell Includes three subpopulations of olfactory bulb neurons. Deep and middle tufted cells, named for the positions of the cell bodies in the external plexiform layer, are organized and project to higher regions in a manner similar to that of mitral cells. External tufted cells are proximal to the glomeruli and mediate intrabulbar connections.

4.36.1 Introduction

The olfactory bulb is the first relay in the processing of volatile odor information in the central nervous system. Odor processing begins in the olfactory epithelium (OE), where the sensory neurons are located, and then proceeds in a hierarchical manner into the olfactory bulb, followed by primary olfactory (piriform) cortex. Tertiary cortical projections are widespread throughout the neuroaxis. As has been noted previously, the fundamental principles of organization and information flow in the olfactory system are broadly conserved across species as well as phyla.

The volatile signals processed by the olfactory system are used widely in regulating feeding behavior and preferences, avoidance, territorial marking, recognition of self versus nonself, emotional states, among others. A comprehensive review of the behavioral significance of the olfactory system is beyond the scope of this chapter, as is a fully integrated description of hierarchical organization of the pathway. The focus here will be on the structural and synaptic organization of the olfactory bulb, with reference to the OE and primary olfactory cortex as needed to help interpret general architectural principles found in the olfactory bulb. Because the organization of olfactory systems, and the olfactory bulb in particular, is widely conserved, much of the information presented here will be based on studies carried out in rodents. As will be reviewed below, the olfactory bulb is laminated, like an onion, and includes a minimum of seven primary layers: (1) the olfactory nerve layer (ONL); (2) the glomerular layer; (3) the external plexiform layer (EPL); (4) the mitral cell layer; (5) the internal plexiform layer (IPL); (6) the granule cell layer; and (7) the subependymal layer or rostral migratory stream. Each of the layers is distinguished by its cellular composition and/or synaptic organization. We will begin with a brief review of the OE because it gives rise to the axons that will constitute the outermost layer of the olfactory bulb.

We apologize to those whose work has not received appropriate attention in this chapter due to space limitations. For further details on aspects of the organization of the olfactory bulb discussed here, readers are referred to several recent excellent and comprehensive reviews (Shepherd, G. M. *et al.*, 2004; Lledo, P. M. *et al.*, 2006; Wachowiak, M. and Shipley, M. T., 2006).

4.36.2 Nasal Cavity

4.36.2.1 Olfactory Epithelium

Odor processing begins in the OE where odorants, ligands, bind with subpopulations of olfactory sensory neurons (OSNs). The epithelium is pseudostratified and includes three primary populations of cells: the OSNs, sustentacular or supporting cells, and a basal cell population that generates new sensory neurons throughout life (Graziadei, P. P. C. and Monti-Graziadei, G. A., 1978; Newman, M. P. *et al.*, 2000). The sensory neurons are highly polarized, with an apical dendrite extending to the epithelial surface where it enlarges to include multiple cilia and a thin axon ($\sim$0.2 μm in diameter) extending from the basal pole (Cuschieri, A. and Bannister, L. H., 1975). Deep to the basal lamina, the axons coalesce into fascicles that ascend and become the olfactory nerve. As the axons exit the epithelium, they are accompanied by a specialized glia cell, the olfactory ensheathing cell/glia (OEC) (Doucette, J. R., 1984). While the axons remain unmyelinated, processes from the OECs do surround the fascicles of the olfactory nerve and interdigitate among the axons. Diverse lines of evidence suggest that the OECs play an important role in the ability of the olfactory nerve to sustain continued axon growth as new sensory neurons are generated (Kafitz, K. W. and Greer, C. A., 1999).

Since the identification in 1991 (Buck, L. and Axel, R., 1991) of the large family of genes ($\sim$1000 in rodents) coding for odor receptors (ORs), the tools of molecular biology have provided for unique insights into the organization of the OE and nerve. The OSNs are believed to express only one functional OR (Chess, A. *et al.*, 1994; Serizawa, S. *et al.*, 2003), and the application of embryonic stem cell engineering has allowed the introduction of markers, either LacZ or GFP, downstream from the promoters of specific ORs (Mombaerts, P. *et al.*, 1996). Thus, the full repertoire of sensory neurons expressing a specific OR can be visualized in their entirety, from their cell bodies in the OE to the terminal portions of their axons in the olfactory bulb. Although axons of the OSNs expressing the same OR coalesce onto one or two glomeruli in the olfactory bulb (typically one glomerulus on each side of the olfactory bulb), the cell bodies are stochastically distributed within a limited region/zone of the OE (Ressler, K. J. *et al.*, 1993; Vassar, R. *et al.*, 1993; Miyamichi, K. *et al.*, 2005) where they neighbor with OSNs expressing different ORs. Initially, it was thought that the OE was delimited into four distinct zones, each of which contained a specific subset of ORs (Ressler, K. J. *et al.*, 1993; Vassar, R. *et al.*, 1993). The most recent evidence, however, is that while each OR has a very limited band of expression within the OE, these are contiguous and overlapping (Iwema, C. L. *et al.*, 2004; Miyamichi, K. *et al.*, 2005). The heterogeneous mixture of molecularly defined OSNs continues as their axons exit the OE; axon fascicles within the OE but deep to the basal lamina are initially composed of heterogeneous populations of axons and show no evidence of topographical sorting.

The olfactory nerve exits the OE as the fascicles begin to enter the cranial cavity via a series of small foramen in the cribiform plate of the ethmoid bone. As the nerve contacts the rostral and rostral–ventral

quadrants of the olfactory bulb, a complex process of defasciculation begins and the outermost layer of the olfactory bulb, the ONL, is formed (Au, W. W. *et al.*, 2002) (Figure 1).

4.36.3   Olfactory Bulb – Laminar Organization

4.36.3.1   Olfactory Nerve Layer

The OECs that were associated with the OSN axons as they exited the OE continue to define the organization of fascicles in the ONL. The OECs in the outermost olfactory nerve layer (ONLo) express the low-affinity nerve growth factor receptor p75 (Au, W. W. *et al.*,

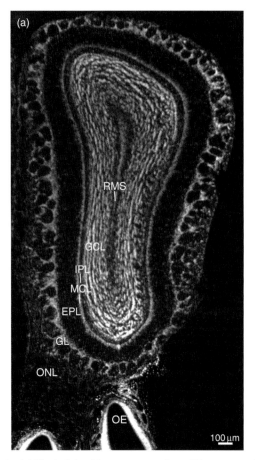

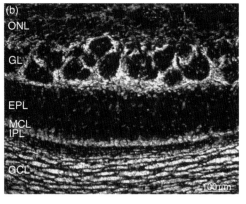

Figure 1   (a) Nissl-stained transverse section of the mouse olfactory bulb showing the laminar organization. The dorsal recess of the nasal cavity and olfactory epithelium are seen below. The spherical glomeruli delineated by periglomerular cells, the paucity of cells within the external plexiform layer, and the well-defined mitral cell layer and deeper granule cell layer are distinctive features of the olfactory bulb. Note that at this level, the laminae are continuous around the circumference of the olfactory bulb. (b) Higher magnification image of the Nissl-stained olfactory bulb. Note the differences in the sizes of the cells surrounding the glomeruli, periglomerular versus juxtaglomerular, and the large cell bodies in the mitral cell layer. The granule cells are distributed in clusters, islets, between which axons from both mitral/tufted cells and centrifugal axons travel. ONL, olfactory nerve layer; GL, glomerular layer; EPL, external plexiform layer; MCL, mitral cell layer; IPL, inner plexiform layer; GR, granule cell layer; RMS, rostral migratory stream; OE, olfactory epithelium. Calibration bars, 100 μm.

2002), while those in the inner nerve layer (ONLi) express the transcription factor TROY (Hisaoka, T. *et al.*, 2004). In parallel, the organization of fascicles differs in the ONLo and ONLi (Au, W. W. *et al.*, 2002). In the former, fascicles defined by the processes of the OECs are small, well defined, and tend to have axons with aligned trajectories. In the ONLi, the fascicles are larger and the trajectories of axons are often mixed. This has led to the suggestion that axons in the ONLo are *en route* while those in the ONLi are reorganizing into molecularly homogeneous fascicles that will coalesce into the same glomerulus. Throughout their length, the OSN axons remain fairly stable with a diameter of ~0.2 μm, but as they enter the ONLi and draw proximal to the glomerular layer, they increase somewhat in diameter to ~0.3 μm. Similar increases are seen elsewhere in the nervous system as axons begin to approach a target region of arborization and synapse formation. OSN axons do not branch within the ONL; arborization of the axons occurs after they enter a glomerulus (Klenoff, J. R. and Greer, C. A., 1998) (Figure 2).

4.36.3.2 Glomerular Layer

The glomeruli of the olfactory bulb, like glomeruli elsewhere in the nervous system, are compact areas of neuropil and contain the first synapses from the OSN axons. The glomeruli are encapsulated by glia cells, some of which extend processes into the neuropil (Kasowski, H. J. *et al.*, 1999; Wachowiak, M. and Shipley, M. T., 2006 for detailed review). In addition, glomeruli are delineated by a large population of peri- and juxtaglomerular cells.

The OSN axons penetrate a glomerulus at the top, predominately from the ONLi, although many examples are found of single axons or very small fascicles of two to five axons that enter from a variety of tangents (Treloar, H. B. *et al.*, 2002) including the EPL (see below). Within the glomerulus, each axon arborizes to generate 10–12 branches, each of which ends in a terminal bouton (Klenoff, J. R. and Greer, C. A., 1998). The OSN axons within the glomerulus also characteristically have varicosities along their length. Both the varicosities and terminal boutons establish synapses with the dendrites of the projection neurons, mitral and tufted cells, and interneurons, the peri/juxtaglomerular cells (Kasowski, H. J. *et al.*, 1999).

A hallmark of glomerular organization is the specificity with which a subset of axons coalesce within a specific glomerulus. As noted above, while all of the OSNs expressing a specific OR are broadly distributed

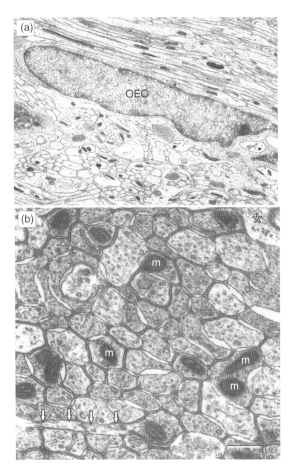

Figure 2 Electron micrographs from the olfactory nerve layer of the olfactory bulb. (a) Low magnification showing both longitudinally and transversely sectioned olfactory sensory neuron axons in the nerve layer. The nucleus (OEC) and some of the surrounding cytoplasm of an olfactory ensheathing cell is in the center of the image. Note that processes extending from the ensheathing cells surround bundles of the axons to form fascicles. (b) Higher magnification of a subset of olfactory sensory neuron axons. Ensheathing cell process (* and arrows) distribute among the axons. The fine structure of the axons includes mitochondria (m) as well as micro- and neurotubules. There is some variability in cross-sectional diameter, but the mean diameter of the olfactory sensory axons approximates 0.2 μm. OEC, olfactory ensheathing cell; m, mitochondria. Calibration bar (shown in b), 1.5 μm in (a) and 0.2 μm in (b).

across a restricted zone in the epithelium, after their axons reach the olfactory bulb, they defasciculate and reorganize into fascicles targeted for a specific region of the bulb (i.e., caudal–medial). The refinement of fascicles into more homogeneous collections of axons that express the same OR occurs when the axons are proximal to the glomerulus, predominately within the ONLi. While early in development there is evidence of heterogeneous OSN axons terminating in the same glomerulus (Zou, D. J. *et al.*, 2004), in the adult all of the axons synapsing within a glomerulus express the same OR (Treloar, H. B. *et al.*, 2002). The specificity and potential mechanisms that influence the coalescence of molecularly homogeneous OSN axons are discussed elsewhere in this volume but may include functional activity as well as adhesion mediated via the expression of ORs in both the axon and growth cone (Feinstein, P. and Mombaerts, P., 2004; Imai, T. *et al.*, 2006; Serizawa, S. *et al.*, 2006).

The interior of the glomerulus contains three primary elements: (1) the arborized axons of the OSNs; (2) the arborized dendritic processes of the projection neurons and glomerular layer neurons; and (3) the processes of glia cells. The interior of the glomerulus is free of cell bodies; rather, both the glia cell and peri/juxtaglomerular cell bodies help to define the perimeter of the glomerulus. At the ultrastructural level, two primary compartments can be identified within the glomerulus (Kasowski, H. J. *et al.*, 1999; Schoppa, N. E. and Westbrook, G. L., 2001). The first contains bundles of dendrites with no or few axons. The second contains numerous axonal boutons and scattered dendritic processes. As will be noted below, local circuit dendrodendritic synapses occur predominately within the dendritic compartment of the glomerulus, while the axodendritic synapses from the OSNs are found predominately within the axonal compartment. Between the intraglomerular compartments, the thin processes of glia cells contribute to segregation and isolation of synaptic circuits. The organization of synapses within the glomerulus will be discussed further below.

4.36.3.3 External Plexiform Layer

The EPL contains scattered populations of cell bodies, including tufted cells and short axon cells, but it is predominately an area of complex neuropil. The apical dendritic processes of projection neurons extend unbranched through the EPL as they target glomeruli where they arborize. In contrast, the secondary dendritic process of the projection neurons extend laterally very broadly in the EPL, in some cases through two-third of the circumference of the olfactory bulb (Mori, K. *et al.*, 1981). The secondary dendrites exhibit some branching, but it appears limited to no more than third or fourth order (Macrides, F. and Schneider, S. P., 1982; Mori, K. *et al.*, 1983; Orona, E. *et al.*, 1984). Also arborizing here are the dendritic processes of the main population of interneurons in the olfactory bulb, the granule cells. The spiny

dendrites of these interneurons establish reciprocal dendrodendritic synapses with the secondary dendrites of the projection neurons.

4.36.3.4 Mitral Cell Layer

The mitral cell layer includes the somata of the primary population of projection neurons in the olfactory bulb, the mitral cells. The mitral cells constitute a monolayer in the olfactory bulb, or a sheet that is roughly the thickness of a single mitral cell soma. Additionally, scattered throughout the mitral cell layer are the somata of numerous granule cells. In fact, granule cells make up the majority of cells with somata in the mitral cell layer (Frazier, L. L. and Brunjes, P. C., 1988; Imamura, F. *et al.*, 2006).

4.36.3.5 Internal Plexiform Layer

The IPL includes the axons of projection neurons that will later coalesce to form the lateral olfactory tract, projecting to olfactory cortical targets. The IPL also includes the axons of intrabulbar cells from the juxtaglomerular region that appear to contact homotypic regions on the opposite side of the olfactory bulb (Liu, W. L. and Shipley, M. T., 1994; Belluscio, L. *et al.*, 2002).

4.36.3.6 Granule Cell Layer

The granule cell layer includes organized islets of granule cell somata whose dendrites project into the EPL. Coursing between the islets of granule cells are the axons from the projection neurons that are moving toward the caudal–ventral–lateral aspect of the olfactory bulb where they join to form the lateral olfactory tract. As will be discussed further below, the granule cell layer can be divided into superficial, deep, and intermediate granule cell populations.

4.36.3.7 Subependymal Zone – Rostral Migratory Stream

The innermost region of the olfactory bulb is alternatively identified as a residual of the ventricular extension present during early development, the subependymal zone, or as the terminal region of the rostral migratory stream, the path of migrating neuroblasts, which form new olfactory bulb interneurons in the adult. The rostral migratory stream appears well defined and compact at the caudal pole of the olfactory bulb, as it extends from the olfactory

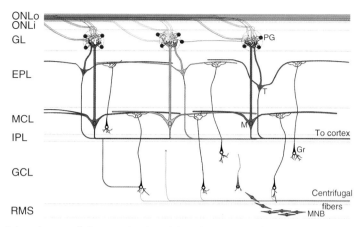

Figure 3 Schematic of the primary cellular organization of the olfactory bulb. Subpopulations of axons, shown in magenta, green, and blue, innervate specific glomeruli based on odor receptor expression. The apical dendritic processes of the projection neurons, mitral, and tufted cells arborize within the glomerulus, as do the dendritic processes of the periglomerular cells. The other primary population of interneurons, granule cells, have their somata located in the granule cell layer and an apical spiny dendrite that arborizes in the external plexiform layer. Axons of periglomerular cells mediate interglomerular actions, while centrifugal axons distribute across the olfactory bulb laminae. ONLo, outer olfactory nerve layer; ONLi, inner olfactory nerve layer; GL, glomerular layer; EPL, external plexiform layer; MCL, mitral cell layer; IPL, internal plexiform layer; GCL, granule cell layer; RMS, rostral migratory stream; PG, periglomerular cell; T, tufted cell; M, mitral cell, Gr, granule cell; MNB, migrating neuroblast.

peduncle, but becomes widely dispersed and indistinct in the rostral pole of the olfactory bulb (Figure 3).

4.36.4 Olfactory Bulb – Cell Populations

4.36.4.1 Projection Neurons

4.36.4.1.1 Mitral cells

Mitral cells are the most prominent population of projection neurons. Their cell bodies, 20–25 μm in diameter, are located in a monolayer, the mitral cell layer. A single apical dendrite (there are some exceptions to this rule) extends from the mitral cell to a single glomerulus where it arborizes extensively including fifth- and sixth-order branches (Price, J. L. and Powell, T. P., 1970c). The numbers are somewhat variable, but approximately 25 mitral cells terminate in each of the 2000 glomeruli found in the mouse olfactory bulb (Royet, J. P. *et al.*, 1998). Also extending from the mitral cell are secondary dendrites that remain restricted to the EPL and can continue for up to two-third of the circumference of the olfactory bulb (Mori, K. *et al.*, 1981). Of interest, the secondary dendrites of mitral cells appear to remain restricted to the deepest portion of the EPL, proximal to the mitral cell layer (Mori, K. *et al.*, 1983; Orona, E. *et al.*, 1984). An axon extends from the basal pole of the mitral cell and

joins other axons in the deeper IPL. Myelination occurs quickly, most likely within the IPL. Collateral branches of mitral cell axons appear to extend up into the EPL (Ramon, Y. and Cajal, S., 1911), though the evidence for this is sometimes conflicting (Kishi, K. *et al.*, 1984; Orona, E. *et al.*, 1984). The major axon branches continue on toward the caudal–ventral–lateral aspect of the olfactory bulb where they form the lateral olfactory tract. The axons of the mitral cells distribute broadly throughout olfactory cortical structures.

4.36.4.1.2 Tufted cells

The second population of projection neurons are tufted cells whose diameter is only slightly less than that of mitral cells, approximately 20 μm. Tufted cells are located both proximal to the mitral cell layer (deep tufted cells) and midway in the EPL (middle tufted cells). There is also a population of external tufted cells that are included in the juxtaglomerular population of cells (Pinching, A. J. and Powell, T. P., 1971c), but these appear to have their axons restricted to intrabulbar circuits (Liu, W. L. and Shipley, M. T., 1994; Belluscio, L. *et al.*, 2002). The apical dendrites of the middle and deep tufted cells behave similarly to those of mitral cells – extensive arborization within a single glomerulus. The secondary dendrites, however, differ in that they remain restricted to the most superficial portion of the EPL, proximal to the glomerular layer (Macrides, F. and Schneider, S. P., 1982; Mori, K. *et al.*, 1983; Orona, E. *et al.*, 1984). The differential distribution of the secondary dendrites of mitral and tufted cells, in the deep and superficial EPL, respectively, coupled with subpopulations of granule cells discussed below, lends credibility to the notion that the mitral and tufted cells may mediate parallel circuits for processing odor information. The axons of the middle and deep tufted cells, like those of the mitral cells, are myelinated, collect initially in the IPL, and then coalesce with the formation of the lateral olfactory tract. Their cortical targets appear to differ from those of mitral cells, with the tufted cells restricted to the most rostral regions of olfactory cortex and the more medial olfactory tubercle (Haberly, L. B. and Price, J. L., 1977).

4.36.4.2 Interneurons

As has been reviewed by a number of authors, there is an almost bewildering array of interneurons found in the olfactory bulb (Schneider, S. P. and Macrides, F., 1978; Shipley, M. T. and Ennis, M., 1996; Shepherd, G. M. *et al.*, 2004; Wachowiak, M. and Shipley, M. T., 2006). Those discussed below represent the major populations whose connectivity and functional properties are best understood.

4.36.4.2.1 *Periglomerular cells*

The largest population of glomerular layer interneurons are the periglomerular cells. As suggested by their name, the cell bodies, 6–8 µm in diameter, surround and define individual glomeruli. They have brushy and highly branched dendritic processes, often including spine-like extensions. The dendritic process typically arborizes in a single glomerulus, although exceptions in which two or more glomeruli are innervated by separate arbors from a single cell are sometimes seen (Pinching, A. J. and Powell, T. P., 1971c). The axons of periglomerular cells can extend for three to six glomeruli distant from the somata and often branch (Pinching, A. J. and Powell, T. P., 1972; Wachowiak, M. and Shipley, M. T., 2006 for review). The target regions of the periglomerular axons are not always evident, but appear to include extraglomerular apical trunks of projection neurons as well as other periglomerular cells (Pinching, A. J. and Powell, T. P., 1971b). In recent years, it has become increasingly clear that the periglomerular cells may be further subdivided into several subpopulations based on their molecular phenotype and synaptic

connections (Kosaka, K. and Kosaka, T., 2005a; Whitman , M. C. and Greer, C. A., 2007).

4.36.4.2.2 *External tufted cells*

External tufted cells, found predominately at the glomerular layer, EPL border, are slightly larger than the periglomerular cells, approximately 12–15 µm in diameter, with a cell body that is generally elliptical. An apical dendritic process runs parallel to the glomerular layer until it enters and arborizes in a single glomerulus. A secondary dendrite, much shorter than those encountered in mitral or the deep and middle tufted cells (Macrides, F. and Schneider, S. P., 1982), runs parallel and proximal to the glomerular layer. The axons of the external tufted cells extend through the EPL and into the IPL where they travel to a homotypic site on the opposite side of the olfactory bulb (Liu, W. L. and Shipley, M. T., 1994). The external tufted cells may provide a functional link between the two complementary regions in each half of the olfactory bulb that receive the same odor information (Belluscio, L. *et al.*, 2002).

4.36.4.2.3 *Short axon cells*

Short axon cells in the glomerular layer, located proximal to the nerve layer, are the rarest glomerular layer interneurons. They are slightly larger than periglomerular cells, 8–12 µm in diameter, and have a dendritic arbor that appears to surround the glomeruli, rather than densely innervating them as is seen with the periglomerular cells. Their axons, like those of the periglomerular cells, remain restricted to the juxtaglomerular region. Recent evidence suggests that this may contribute to circuits that include the external tufted cells and periglomerular cells of distant glomeruli.

4.36.4.2.4 *Van Gehuchten cells*

Short axon cells, Van Gehuchten cells (Schneider, S. P. and Macrides, F., 1978), are the predominate interneurons found in the EPL, and even these are relatively rare. They have a limited dendritic arbor within the EPL and axonal branches that distribute within both the EPL and mitral cell layer where they are thought to contact tufted and mitral cells, respectively.

4.36.4.2.5 *Blanes cells*

Blanes cells, first described at the end of the nineteenth century (Blanes, T., 1898), have remained largely unknown until recent data suggested they may interact with olfactory bulb granule cells (Pressler, R. T. and Strowbridge, B. W., 2006). Blanes cells are found throughout the granule cell layer of the olfactory bulb. They have an extensive arborization of spiny dendrites that remain restricted to the granule cell layer. Blanes cell axons remain local with the granule cell layer and are believed to contact the basal dendrites or somata of granule cells. With a cell diameter of approximately 15 μm, Blanes cells are easily distinguished from granule cells.

4.36.4.2.6 Granule cells

The granule cell layer is distinctive due to the arrangement of granule cell somata into rows or islets. Granule cells are small, 8–10 μm in diameter. They have a well-developed apical dendrite that extends into the EPL where it arborizes and becomes heavily invested with dendritic spines. A smaller less arborized basal dendrite is also present that remains restricted to the granule cell layer. Of perhaps greatest interest, the olfactory bulb granule cells are anaxonic; they do not develop an axon (Price, J. L. and Powell, T. P., 1970a). All of the synaptic input and output functions occur via the dendrites. The dendritic spines, also known as gemmules, of the granule cell are somewhat unusual in both receiving and making synapses. Contiguous compartments within the spine are specialized for efferent or afferent function (Rall, W. et al., 1966; Price, J. L. and Powell, T. P., 1970b). In addition, the spine necks are often quite long (>5 μm), and the major axes of the spine head can exceed 2 μm × 5 μm (Woolf, T. B. et al., 1991a; 1991b). The distribution/density of spines is highest throughout the arbor in the EPL, with only an occasional spine seen in the ascending apical dendritic trunk (Price, J. L. and Powell, T. P., 1970a). Granule cells can be divided into at least three subpopulations: (1) those whose cell bodies are located deep within the granule cell layer and whose apical dendritic arbor is limited to the deep EPL, proximal to the mitral cell layer; (2) those whose cell bodies are located more superficially in the granule cell layer and whose dendrites extend unbranched through the deep EPL until they reach the superficial EPL where they arborize extensively; and (3) those whose apical dendrites arborize within both the superficial and deep EPL (Mori, K. et al., 1983; Orona, E. et al., 1983; Greer, C. A., 1987; Imamura, F. et al., 2006). Recalling the differential distribution of the secondary dendrites of mitral and tufted cells (Mori, K. et al., 1983), the organization of the granule cell dendrites suggest that deep granule

cells may interact preferentially with mitral cell dendrites while the superficial granule cells interact preferentially with tufted cell dendrites. Other populations of interneurons, such as Blanes cells, Henson's cells and short axon Golgi cells are also found in the granule cell layer but occur at a low frequency and are not as well characterized (Price, J. L. and Powell, T. P., 1970c; Schneider, S. P. and Macrides, F., 1978).

4.36.4.3 Glia

Beginning in the nerve layer, a distinct population of glia are identified, the OEC/glia. These have thin processes that wrap fascicles of OSN axons, beginning deep to the basal lamina in the OE and continuing up to the olfactory bulb glomeruli (Doucette, R., 1991; 1993; Au, W. W. et al., 2002). Several lines of evidence suggest that the olfactory ensheathing glia may contribute to the ability of the OSN axons to successfully extend and innervate glomeruli throughout life (Doucette, R., 1990). Of further interest, the olfactory ensheathing glia have captured the interest of laboratories working on problems such as spinal cord injury where transplants of glia purified from the olfactory nerve appear to promote and improve recovery following injury (Ramon-Cueto, A. et al., 1998; Bartolomei, J. C. and Greer, C. A., 2000).

The shells of the glomeruli are formed, in part, by astrocytes that surround and delineate the surface (Bailey, M. S. and Shipley, M. T., 1993; Chang, C. Y. et al., 2003). Processes extending from their cell bodies distribute within the glomeruli and contribute to the segregation of the axonal and dendritic synaptic compartments (Chao, T. I. et al., 1997).

The EPL, mitral cell layer, and granule cell layer have small populations of astrocytes. It is of interest to point out that many of the synapses within the EPL lack the tight glia wrapping seen elsewhere in the nervous system (Bailey, M. S. and Shipley, M. T., 1993). The rostral migratory stream, however, particularly at the more caudal pole of the olfactory bulb, is defined in part by a compact population of astrocytes that surround and closely interdigitate with the migrating neuroblasts. Much remains to be learned about the properties of this population of glia, but they do appear to contribute to both the migratory dynamics of the neuroblasts and their topographically restricted path (Lois, C. et al., 1996) (Figure 4).

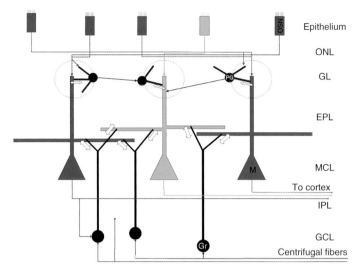

Figure 4 Schematic illustrating primary synaptic interactions in the olfactory bulb. After sorting into specific glomeruli, the axons of the olfactory sensory neurons make excitatory synapses onto the apical dendrites of the projection neurons, mitral, and tufted cells, as well as onto the periglomerular cell dendrites. Intraglomerular circuits also include reciprocal dendrodendritic synapses between the mitral/tufted cell dendrites and the periglomerular cell dendrites. Periglomerular cells can also influence activity in neighboring glomeruli by axodendritic synapses onto mitral/tufted cells and other periglomerular cells. In the external plexiform layer, spiny granule cell dendrites establish reciprocal dendrodendritic synapses with the secondary or lateral dendrites of mitral/tufted cells. Centrifugal axons terminate heavily in the granule cell layer as well as in the external plexiform and glomerular layers of the olfactory bulb. OSN, olfactory sensory neuron; PG, periglomerular cell, M, mitral cell; Gr, granule cell. Arrows indicate the polarity of the synapses – unidirectional or bidirectional.

4.36.5 Olfactory Bulb – Synaptic Organization

4.36.5.1 Glomerular Layer

Our models of synaptic organization in the glomerular layer are becoming increasingly complex as new tools are applied that distinguish between subpopulations of juxtaglomerular neurons and the distribution of their axons and dendritic processes (Chen, W. R. and Shepherd, G. M., 2006; Wachowiak, M. and Shipley, M. T., 2006). Many questions regarding the specificity of synapses within and between glomeruli remain unanswered. The synaptic organization discussed below reflects our current understanding of the basic model for which most information has been compiled.

As was noted earlier, the neuropil of the glomeruli can be divided into two distinct compartments: (1) the axonal compartment which contains the terminal boutons of the OSN axons and scattered dendritic processes; and (2) the dendritic compartment which contains bundles of closely apposed dendrites. This segregation is the foundation for the isolation of primary afferent versus local circuit synapses in the glomerulus (Schild, D. and Riedel, H., 1992; Kasowski, H. J. *et al.*, 1999) (Figure 5).

The axons of the OSNs do not branch until they innervate a glomerulus. As they enter the glomerulus, primarily at the boundary of the ONL and glomerular layer, the axons are tightly bundled but within the glomerular neuropil they separate and arborize (Pinching, A. J. and Powell, T. P., 1971a). Each axon establishes only a limited number of synapses; usually no more than 10–15 Gray Type I asymmetrical synapses with the dendrites of projection neurons or interneurons. The OSN axon specializations include both *en passant* varicosities and terminal boutons. Both specializations are filled with many small spherical vesicles, characteristic of glutamate, and often appear quite electron dense. Single dendritic processes are often seen surrounded by multiple OSN axon boutons, each of which is making a synapse. Although the distribution of the terminal boutons and *en passant* specialization of the axons is spatially restricted within the glomerulus, it is not known if their synapses converge onto the dendrites of the same cell or diverge and broadly synapse with the dendrites of many cells. Of interest, the axodendritic synapses from the OSN axons are found predominately toward the thinner tips/branches of the dendrites within the glomerulus (Pinching, A. J. and Powell, T. P., 1971a; Shepherd, G. M., 1972).

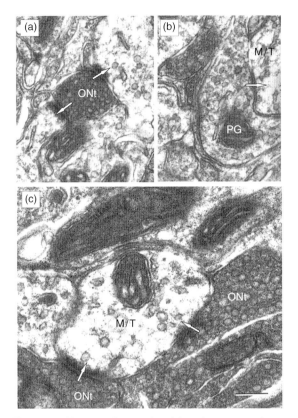

Figure 5   Electron micrographs of the primary synaptic appositions in the glomeruli of the olfactory bulb. In (a) and (c) olfactory sensory axon terminal boutons (ONt) are seen making asymmetrical synapses onto pale dendrites tentatively identified as mitral/tufted (M/T). The ONt is characterized by its electron dense appearance and clustering many small spherical vesicles adjacent to the presynaptic membrane. The target dendrite is recognized by the asymmetrical thickening of the postsynaptic membrane. In (b) a periglomerular cell (PG) is seen making a symmetrical synapse on an M/T dendrite. The PG dendrite has slightly elliptically shaped vesicles, and the pre- and postsynaptic membrane are symmetrical. Arrows indicate the polarity of the synapses. ONt, olfactory sensory axon terminal bouton; PG, periglomerular cell dendrite; M/T, mitral/tufted cell dendrite. Calibration bar (shown in c), 160 nm in (a) and 200 nm in (b) and (c).

Within the dendritic compartment of the glomerulus, the tightly bundled dendrites of the projection neurons and periglomerular cells establish reciprocal dendrodendritic synapses. The excitatory Gray Type I dendritic synapse is characterized by a small collection of spherical vesicles closely apposed to the dendritic membrane of a mitral or tufted cell. The apposed process, a periglomerular cell dendrite, has a thick postsynaptic specialization. Although the reciprocal synapse from the periglomerular cell to the mitral/tufted cell is occasionally seen in the same plane of section, it is much more common for them to be separated and recognized as reciprocal connections only following serial sectioning and reconstruction. The periglomerular synapse can be found in both spine-like processes or in the shafts of periglomerular cell dendrites. The inhibitory Gray Type II synapse from the periglomerular cell dendrite to the mitral/tufted cell dendrite includes a collection of vesicles, often elliptical in shape and adjacent to the presynaptic membrane. The specialization on the postsynaptic membrane is a thickening symmetrical to that seen on the presynaptic membrane. In contrast to the location of the primary afferent axodendritic synapses on the finer terminal branches of the dendrites, the Gray Type II synapses are most often found on larger trunks and at branch points (Pinching, A. J. and Powell, T. P., 1971a; Shepherd, G. M., 1972). This is consistent with the spatial distribution of excitatory and inhibitory synapses described in cortex and suggests that the inhibitory synapse can function as a gatekeeper that modulates excitatory inputs.

Recent evidence suggests that external tufted cells and glomerular layer short axon cells may also contribute to dynamic regulation of center-surround inhibition among glomeruli (Aungst, J. L. *et al.*, 2003). The short axon cells are believed to influence glomeruli up to 500 μm distant. Activation of the short axon cells occurs via a direct input of the OSN axons onto external tufted cells which, in turn, excite the short axon cells. While the full details of this and other glomerular circuits remain to be established, it is clear that the glomerular layer includes the circuitry for complex processing and integration of odor stimuli.

The terminals of centrifugal axons are also found in both the glomerular neuropil and juxtaglomerular space. These arise from several sources including the anterior olfactory nucleus, horizontal limb of the diagonal band, the raphe nucleus, and the locus coeruleus. While it is evident that these circuits help to modify intraglomerular information processing, their full nature has yet to be resolved (e.g., Wachowiak, M. and Shipley, M. T., 2006).

4.36.5.2   External Plexiform Layer

The EPL is the site of the most well-studied reciprocal dendrodendritic synapse in the mammalian nervous system: the mitral/tufted granule cell dendritic spine synapse (Hirata, Y., 1964; Andres, K. H., 1965; Rall, W. *et al.*, 1966; Landis, D. M. *et al.*, 1974;

Jackowski, A. *et al.*, 1978). The basic features are similar to those described about for the glomerular local circuits but, in the EPL the reciprocal synapse can often be seen in a single plane since the specialization is restricted to the granule cell spine (gemmule). The secondary dendrites of mitral/tufted cells can be quite large (>4 µm in cross section), and the clusters of spherical vesicles associated with the presynaptic specialization of the Gray Type I synapse often appear quite isolated. At the level of the electron microscope, there are no obvious structural determinants that regulate the clustering of vesicles, although recent evidence shows that vesicle-associated proteins regulating clustering, mobilization, and recycling are present in these presynaptic dendrites (Greer laboratory, unpublished observations). The apposed granule cell spine has

an asymmetric thickening immediately opposite the clustering of vesicles within the mitral/tufted cell dendrite (Figure 6).

The granule cell spine, in addition to the asymmetric postsynaptic membrane specialization, appears somewhat unique. It is filled with a large cluster of elliptical vesicles that are closely apposed to the symmetrical membrane thickenings characteristic of the Gray Type II synapse. Often the transition in the structure of the membrane between the asymmetrical postsynaptic thickening and the presynaptic symmetrical thickening is abrupt; the two specializations do not always appear to be separated by nonspecialized membrane leading to speculation about the independence of these two synapses. The granule cell dendritic spine, as was noted earlier, often has a very thin and long neck. Nevertheless, mitochondria can be

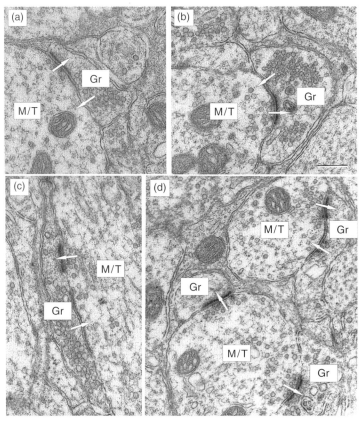

Figure 6 Electron micrographs of the reciprocal dendrodendritic synapses in the external plexiform layer of the olfactory bulb. In (a)–(d), examples of reciprocal dendrodendritic synapses between mitral/tufted (M/T) dendrites and granule cell spines (Gr) are shown. The asymmetrical synapse from the M/T to the Gr is characterized by a smaller collection of vesicles adjacent to the presynaptic membrane of the M/T dendrite and an asymmetrical thick postsynaptic specialization on the Gr membrane. The reciprocal symmetrical synapse is characterized by a larger collection of elliptical or heterogeneous vesicles and a symmetrical thickening of the pre- and postsynaptic membranes. Arrows indicate the polarity of the synapses. M/T, mitral/tufted cell dendrite; Gr, granule cell spine. Calibration bar (shown in b), 200 nm.

found extending from the parent dendrite into the spine head, as does smooth endoplasmic reticulum (Cameron, H. A. *et al.*, 1991; Woolf, T. B. *et al.*, 1991a). Reports of free and polyribosome clusters are also common. As was found for the mitral/tufted cell dendrites, the granule cell spines exhibit an array of vesicle-associated proteins consistent with the role as a presynaptic compartment. Given the immediate proximity of the reciprocal synapses found in the EPL, one can argue that this is the smallest microcircuit found in the nervous system.

4.36.5.3 Mitral Cell Layer

The ascending dendrites of granule cells also establish somatodendritic synapses with mitral cell bodies in the mitral cell layer. The features of these appear indistinguishable from those described above for the dendrodendritic synapses. Although it seems reasonable that tufted cell bodies may also be involved in somatodendritic synapses with granule cells in the EPL, none have been described.

4.36.5.4 Internal Plexiform Layer

Immunohistochemical and electron microscopy studies suggest that the collaterals of external tufted cells make synapses onto the ascending dendrites of granule cells within the IPL (Liu, W. L. and Shipley, M. T., 1994). More recent studies suggest that these connections are topographically organized so that the glomeruli on the opposite sides of the olfactory bulb that receive input from the same OR are synaptically linked via the external tufted cell circuit (Belluscio, L. *et al.*, 2002; Lodovici, C. *et al.*, 2003; Marks, C. A. *et al.*, 2006).

4.36.5.5 Granule Cell Layer

The basal dendrites of the granule cells in the granule cell layer receive synaptic input from a variety of sources (Price, J. L. and Powell, T. P., 1970a). Axon collaterals from mitral and tufted cells form asymmetric synapses on granule cell basal dendrites as do axons from the anterior olfactory nucleus, anterior commissure, olfactory cortex, horizontal limb of the diagonal band (Carson, K. A., 1984; Shipley, M. T. and Adamek, G. D., 1984), the locus coeruleus (McLean, J. H. *et al.*, 1989; McLean, J. H. and Shipley, M. T., 1991), and the raphe nucleus (McLean, J. H. and Shipley, M. T., 1987). While their specific functions and role in modulating

information processing in the olfactory bulb are not entirely clear, it is evident that the synaptic actions of these circuits will activate the granule cells and, in turn, their dendrodendritic circuits within the EPL. Thus, as is typical of most sensory systems, centrifugal and feedback circuits contribute to the flow of information through the olfactory bulb.

4.36.5.6 Electrical Synapses in the Olfactory Bulb

Gap junctions were long known to be the structural correlates of electrical coupling between cells (Rash, J. E. *et al.*, 1998); however, accumulating evidence that electrical coupling can confer unique properties to neural circuits has lately increased the interest in their distribution and properties (Rela, L. and Szczupak, L., 2004). As in other regions of the brain, gap junctions have been found in the olfactory bulb. In the ONL, gap junctions are abundant in the olfactory ensheathing glia, but seem to be absent from OSN axons (Rash, J. E. *et al.*, 2005). Within glomeruli, gap junctions are found between mitral/tufted dendrites, as well as between mitral/tufted cell dendrites and periglomerular cell dendrites, perhaps complementing the reciprocal dendrodendritic synapses (Pinching, A. J. and Powell, T. P., 1971b; Christie, J. M. *et al.*, 2005; Kosaka, T. and Kosaka, K., 2005b; Rash, J. E. *et al.*, 2005). Consistent with an interaction between chemical and electrical synapses, in the EPL examples of mixed synapses that include both gap junctions and conventional dendrodendritic synapses are reported between mitral/tufted cell secondary dendrites and granule cell spines. In the mitral cell layer, somatodendritic synapses between mitral and granule cells may also involve a combination of chemical and electrical components, as suggested by the identification of structures compatible with gap junctions in mitral/tufted cell perikarya (Miragall, F. *et al.*, 1996). Finally, gap junction coupling between granule cell bodies represents a plausible mechanism to synchronize the activity of clusters of these inhibitory interneurons (Reyher, C. K. *et al.*, 1991; Gibson, J. R. *et al.*, 1999; Galarreta, M. and Hestrin, S., 2001).

4.36.6 Lateral Olfactory Tract and Olfactory Cortex

The lateral olfactory tract forms at the caudal–lateral–ventral aspect of the olfactory bulb with the coalescence of mitral and tufted cell axons following their passage through the IPL and islets of granule cells. Because the axons are myelinated, in gross dissection the tract appears white. As the tract proceeds caudally, it moves to a more ventral position. Within the tract, the myelinated profiles of the projection neuron axons have a bimodal distribution for cross-sectional diameter. The larger profiles ($\sim$1.42 µm) are believed to arise from mitral cells while the smaller profiles ($\sim$1.14 µm) are axons from the middle and deep tufted cells (Price, J. L. and Sprich, W. W., 1975; Bartolomei, J. C. and Greer, C. A., 2000). In cortex, the axons exit the lateral olfactory tract and enter layer Ia where they make Gray Type I excitatory synapses with the spiny apical dendrites of both layer II and layer III pyramidal cells (Westrum, L. E., 1975; Wouterlood, F. G. and Nederlof, J., 1983; Schwob, J. E. and Price, J. L., 1984). In cortex, the distribution of the projection neuron axons is complex and remains somewhat controversial (Wilson, D. A. et al., 2006). One line of evidence from functional studies suggest that the axons have terminal fields that maintain the molecular specificity of OR input first established at the level of the glomerulus (Zou, Z. et al., 2001; 2005). This is also consistent with recent analyses using pseudorabies virus to characterize olfactory bulb – cortical topography (Willhite, D. C. et al., 2006). Earlier retrograde tracing, however, suggested that axon collaterals of single mitral cells diverged broadly throughout the rostral–caudal extent of cortex (Haberly, L. B. and Price, J. L., 1977; Scott, J. W. et al., 1980; Luskin, M. B. and Price, J. L., 1982; Zou, Z. et al., 2001; Fukushima, N. et al., 2002). Further discussion of the synaptic organization of piriform cortex is beyond the scope of this chapter other than to comment that it is complex and includes feedback circuits both within cortex as well as centrifugals to the ipsilateral olfactory bulb and to the contralateral olfactory pathway via the anterior commissure (de Olmos, J. et al., 1978; Haberly, L. B. and Price, J. L., 1978a; 1978b; Luskin, M. B. and Price, J. L., 1983a; 1983b). New studies will help to clarify the complex relationship between the olfactory bulb and cortex including the organization of the cortically derived centrifugal axons that return to the bulb.

4.36.7 Adult Neurogenesis and the Olfactory Bulb

While it has been recognized for some time that the OSNs turn over and are replaced by a population of basal cells in the OE, the magnitude of adult neurogenesis of olfactory bulb neurons has been recognized relatively recently (Altman, J., 1969; Kaplan, M. S. and Hinds, J. W., 1977; Lois, C. and Alvarez-Buylla, A., 1994). Several excellent reviews are available that summarize our current understanding of neurogenesis within the subventricular zone, the migration of neuroblasts via the rostral migratory stream into the olfactory bulb, and the ensuing differentiation of the migrating cells into granule and periglomerular cells (Alvarez-Buylla, A. and Garcia-Verdugo, J. M., 2002; Lennington, J. B. et al., 2003; Doetsch, F. and Hen, R., 2005; Lledo, P. M. and Saghatelyan, A., 2005; Lledo, P. M. et al., 2006). In the context of the plasticity of olfactory bulb synaptic circuits, many questions remain about the mechanisms influencing cellular loss and replacement among these populations of interneurons and the determinants of newly established synapses (Whitman, M. C. and Greer, C. A., 2007). It may be that the ongoing turnover of these populations of interneurons is a solution to the dynamic and transient nature of odor stimuli – an approach to nervous system plasticity that may function in concert with other mechanisms of synaptic plasticity including long-term potentiation and long-term depression (Figure 7).

4.36.8 Summary

The molecular, cellular, and synaptic organization of the olfactory bulb continue to pose significant challenges in understanding the dynamic nature of this structure and how the organization of the system may contribute to odor coding/processing. The advent of powerful molecular tools toward the end of the twentieth century has resulted in significant new insights into the complex wiring of the olfactory bulb, but many questions remain unanswered. Among the challenges that remain are understanding the developmental events that underlie the specificity of afferent axon convergence as well as the extraordinary precision of local synaptic circuits in the olfactory bulb and their ongoing remodeling.

Acknowledgments

Work from the author's laboratory has been generously supported by grants from NIH-NIDCD and NIH-NIA. The authors express the appreciation to other members of the laboratory for helpful discussions.

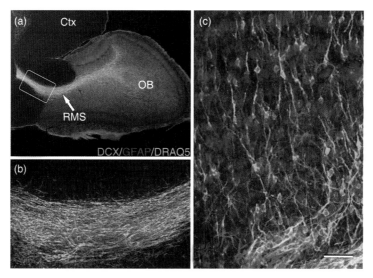

Figure 7 The rostral migratory stream and migrating neurons in the olfactory bulb. In (a), a sagittal section of the rostral brain and olfactory bulb shows the rostral migratory stream (RMS) as it travels from the subventricular zone, through the olfactory peduncle, and into the olfactory bulb. Doublecortin (DCX) staining shows the migrating neuroblasts while staining for glial fibrillary acitic protein (GFAP) shows the glial cells that surround the RMS. DRAQ5 identifies nuclei. In (b), the boxed area in (a) is shown at higher magnification. The density of GFAP-positive glia surrounding the chains of DCX-positive migrating neuroblasts is more apparent. In (c), the DCX-positive cells are exiting the RMS (bottom of image) and beginning to either differentiate as granule cells or migrate to the glomerular layer and differentiate into periglomerular cells. Ctx, cortex; RMS, rostral migratory stream; OB, olfactory bulb. Calibration bar (shown in c), 300 µm in (a); 100 µm in (b); and 40 µm in (c).

References

Altman, J. 1969. Autoradiographic and histological studies of postnatal neurogenesis. IV. Cell proliferation and migration in the anterior forebrain, with special reference to persisting neurogenesis in the olfactory bulb. J. Comp. Neurol. 137, 433–457.

Alvarez-Buylla, A. and Garcia-Verdugo, J. M. 2002. Neurogenesis in adult subventricular zone. J. Neurosci. 22, 629–634.

Andres, K. H. 1965. The fine structure of the olfactory bulb in rats with special reference to the synaptic connections. Z. Zellforsch. Mikrosk. Anat. 65, 530–561.

Au, W. W., Treloar, H. B., and Greer, C. A. 2002. Sublaminar organization of the mouse olfactory bulb nerve layer. J. Comp. Neurol. 446, 68–80.

Aungst, J. L., Heyward, P. M., Puche, A. C., Karnup, S. V., Hayar, A., Szabo, G., and Shipley, M. T. 2003. Centre-surround inhibition among olfactory bulb glomeruli. Nature 426, 623–629.

Bailey, M. S. and Shipley, M. T. 1993. Astrocyte subtypes in the rat olfactory bulb: morphological heterogeneity and differential laminar distribution. J. Comp. Neurol. 328, 501–526.

Bartolomei, J. C. and Greer, C. A. 2000. Olfactory ensheathing cells: bridging the gap in spinal cord injury. Neurosurgery 47, 1057–1069.

Belluscio, L., Lodovichi, C., Feinstein, P., Mombaerts, P., and Katz, L. C. 2002. Odorant receptors instruct functional circuitry in the mouse olfactory bulb. Nature 419, 296–300.

Blanes, T. 1898. Sobre algunos puntos dudoses de la estructuradel bulbo olfatorio. Revta Trimest. Microgr. 3, 99–127.

Buck, L. and Axel, R. 1991. A novel multigene family may encode odorant receptors: a molecular basis for odor recognition. Cell 65, 175–187.

Cameron, H. A., Kaliszewski, C. K., and Greer, C. A. 1991. Organization of mitochondria in olfactory bulb granule cell dendritic spines. Synapse 8, 107–118.

Carson, K. A. 1984. Quantitative localization of neurons projecting to the mouse main olfactory bulb. Brain Res. Bull. 12, 629–634.

Chang, C. Y., Chien, H. F., Jiangshieh, Y. F., and Wu, C. H. 2003. Microglia in the olfactory bulb of rats during postnatal development and olfactory nerve injury with zinc sulfate: a lectin labeling and ultrastrucutural study. Neurosci. Res. 45, 325–333.

Chao, T. I., Kasa, P., and Wolff, J. R. 1997. Distribution of astroglia in glomeruli of the rat main olfactory bulb: exclusion from the sensory subcompartment of neuropil. J. Comp. Neurol. 388, 191–210.

Chen, W. R. and Shepherd, G. M. 2006. The olfactory glomerulus: a cortical module with specific functions. J. Neurocytol. 34, 353–360.

Chess, A., Simon, I., Cedar, H., and Axel, R. 1994. Allelic inactivation regulates olfactory receptor gene expression. Cell 78, 823–834.

Christie, J. M., Bark, C., Hormuzdi, S. G., Helbig, I., Monyer, H., and Westbrook, G. L. 2005. Connexin36 mediates spike synchrony in olfactory bulb glomeruli. Neuron 46, 761–772.

Cuschieri, A. and Bannister, L. H. 1975. The development of the olfactory mucosa in the mouse: electron microscopy. J. Anat. 119, 471–498.

de Olmos, J., Hardy, H., and Heimer, L. 1978. The afferent connections of the main and the accessory olfactory bulb formations in the rat: an experimental HRP-study. J. Comp. Neurol. 181, 213–244.

Doetsch, F. and Hen, R. 2005. Young and excitable: the function of new neurons in the adult mammalian brain. Curr. Opin. Neurobiol. 15, 121–128.

Doucette, J. R. 1984. The glial cells in the nerve fiber layer of the rat olfactory bulb. Anat. Rec. 210, 385–391.

Doucette, R. 1990. Glial influences on axonal growth in the primary olfactory system. Glia 3, 433–449.

Doucette, R. 1991. PNS-CNS transitional zone of the first cranial nerve. J. Comp. Neurol. 312, 451–466.

Doucette, R. 1993. Glial cells in the nerve fiber layer of the main olfactory bulb of embryonic and adult mammals. Microsc. Res. Tech. 24, 113–130.

Feinstein, P. and Mombaerts, P. 2004. A contextual model for axonal sorting into glomeruli in the mouse olfactory system. Cell 117, 817–831.

Frazier, L. L. and Brunjes, P. C. 1988. Unilateral odor deprivation: early postnatal changes in olfactory bulb cell density and number. J. Comp. Neurol. 269, 355–370.

Fukushima, N., Oikawa, S., Yokouchi, K., Kawagishi, K., and Moriizumi, T. 2002. The minimum number of neurons in the central olfactory pathway in relation to its function: a retrograde fiber tracing study. Chem. Senses 27, 1–6.

Galarreta, M. and Hestrin, S. 2001. Electrical synapses between GABA-releasing interneurons. Nat. Rev. Neurosci. 2, 425–433.

Gibson, J. R., Beierlein, M., and Connors, B. W. 1999. Two networks of electrically coupled inhibitory neurons in neocortex. Nature 402, 75–79.

Graziadei, P. P. C. and Monti-Graziadei, G. A. 1978. Continuous Nerve Cell Renewal in the Olfactory System. In: Handbook Sensory Physiology. Development of Sensory Systems (ed. M. Jacobson), Vol. IX, pp. 55–83. Springer.

Greer, C. A. 1987. Golgi analyses of dendritic organization among denervated olfactory bulb granule cells. J. Comp. Neurol. 257, 442–452.

Haberly, L. B. and Price, J. L. 1977. The axonal projection patterns of the mitral and tufted cells of the olfactory bulb in the rat. Brain Res. 129, 152–157.

Haberly, L. B. and Price, J. L. 1978a. Association and commissural fiber systems of the olfactory cortex of the rat. II. Systems originating in the olfactory peduncle. J. Comp. Neurol. 181, 781–807.

Haberly, L. B. and Price, J. L. 1978b. Association and commissural fiber systems of the olfactory cortex of the rat. J. Comp. Neurol. 178, 711–740.

Hirata, Y. 1964. Some observations on the fine structure of the synapses in the olfactory bulb of the mouse, with particular reference to the atypical synaptic configurations. Arch. Histol. Jpn. 24, 293–302.

Hisaoka, T., Morikawa, Y., Kitamura, T., and Senba, E. 2004. Expression of a member of tumor necrosis factor receptor superfamily, TROY, in the developing olfactory system. Glia 45, 313–324.

Imai, T., Suzuki, M., and Sakano, H. 2006. Odorant receptor-derived cAMP signals direct axonal targeting. Science 314, 657–661.

Imamura, F., Nagao, H., Naritsuka, H., Murata, Y., Taniguchi, H., and Mori, K. 2006. A leucine-rich repeat membrane protein, 5T4, is expressed by a subtype of granule cells with dendritic arbors in specific strata of the mouse olfactory bulb. J. Comp. Neurol. 495, 754–768.

Iwema, C. L., Fang, H., Kurtz, D. B., Youngentob, S. L., and Schwob, J. E. 2004. Odorant receptor expression patterns are restored in lesion-recovered rat olfactory epithelium. J. Neurosci. 24, 356–369.

Jackowski, A., Parnavelas, J. G., and Lieberman, A. R. 1978. The reciprocal synapse in the external plexiform layer of the mammalian olfactory bulb. Brain Res. 159, 17–28.

Kafitz, K. W. and Greer, C. A. 1999. Olfactory ensheathing cells promote neurite extension from embryonic olfactory receptor cells in vitro. Glia 25, 99–110.

Kaplan, M. S. and Hinds, J. W. 1977. Neurogenesis in the adult rat: electron microscopic analysis of light radioautographs. Science 197, 1092–1094.

Kasowski, H. J., Kim, H., and Greer, C. A. 1999. Compartmental organization of the olfactory bulb glomerulus. J. Comp. Neurol. 407, 261–274.

Kishi, K., Mori, K., and Ojima, H. 1984. Distribution of local axon collaterals of mitral, displaced mitral, and tufted cells in the rabbit olfactory bulb. J. Comp. Neurol. 225, 511–526.

Klenoff, J. R. and Greer, C. A. 1998. Postnatal development of olfactory receptor cell axonal arbors. J. Comp. Neurol. 390, 256–267.

Kosaka, K. and Kosaka, T. 2005a. Synaptic organization of the glomerulus in the main olfactory bulb: compartments of the glomerulus and heterogeneity of the periglomerular cells. Anat. Sci. Int. 80, 80–90.

Kosaka, T. and Kosaka, K. 2005b. Intraglomerular dendritic link connected by gap junctions and chemical synapses in the mouse main olfactory bulb: electron microscopic serial section analyses. Neuroscience 131, 611–625.

Landis, D. M., Reese, T. S., and Raviola, E. 1974. Differences in membrane structure between excitatory and inhibitory components of the reciprocal synapse in the olfactory bulb. J. Comp. Neurol. 155, 67–91.

Lennington, J. B., Yang, Z., and Conover, J. C. 2003. Neural stem cells and the regulation of adult neurogenesis. Reprod. Biol. Endocrinol. 1, 99.

Liu, W. L. and Shipley, M. T. 1994. Intrabulbar associational system in the rat olfactory bulb comprises cholecystokinin-containing tufted cells that synapse onto the dendrites of GABAergic granule cells. J. Comp. Neurol. 346, 541–558.

Lledo, P. M. and Saghatelyan, A. 2005. Integrating new neurons into the adult olfactory bulb: joining the network, life–death decisions, and the effects of sensory experience. Trends Neurosci. 28, 248–254.

Lledo, P. M., Alonso, M., and Grubb, M. S. 2006. Adult neurogenesis and functional plasticity in neuronal circuits. Nat. Rev. Neurosci. 7, 179–193.

Lodovichi, C., Belluscio, L., and Katz, L. C. 2003. Functional topography of connections linking mirror-symmetric maps in the mouse olfactory bulb. Neuron 38, 265–276.

Lois, C. and Alvarez-Buylla, A. 1994. Long-distance neuronal migration in the adult mammalian brain. Science 264, 1145–1148.

Lois, C., Garcia-Verdugo, J. M., and Alvarez-Buylla, A. 1996. Chain migration of neuronal precursors. Science 271, 978–981.

Luskin, M. B. and Price, J. L. 1982. The distribution of axon collaterals from the olfactory bulb and the nucleus of the horizontal limb of the diagonal band to the olfactory cortex, demonstrated by double retrograde labeling techniques. J. Comp. Neurol. 209, 249–263.

Luskin, M. B. and Price, J. L. 1983a. The topographic organization of associational fibers of the olfactory system in the rat, including centrifugal fibers to the olfactory bulb. J. Comp. Neurol. 216, 264–291.

Luskin, M. B. and Price, J. L. 1983b. The laminar distribution of intracortical fibers originating in the olfactory cortex of the rat. J. Comp. Neurol. 216, 292–302.

Macrides, F. and Schneider, S. P. 1982. Laminar organization of mitral and tufted cells in the main olfactory bulb of the adult hamster. J. Comp. Neurol. 208, 419–430.

Marks, C. A., Cheng, K., Cummings, D. M., and Belluscio, L. 2006. Activity-dependent plasticity in the olfactory intrabulbar map. J. Neurosci. 26, 11257–11266.

McLean, J. H. and Shipley, M. T. 1987. Serotonergic afferents to the rat olfactory bulb. I. Origins and laminar specificity of serotonergic inputs in the adult rat. J. Neurosci. 7, 3016–3028.

McLean, J. H. and Shipley, M. T. 1991. Postnatal development of the noradrenergic projection from locus coeruleus to the olfactory bulb in the rat. J. Comp. Neurol. 304, 467–477.

McLean, J. H., Shipley, M. T., Nickell, W. T., Aston-Jones, G., and Reyher, C. K. 1989. Chemoanatomical organization of the noradrenergic input from locus coeruleus to the olfactory bulb of the adult rat. J. Comp. Neurol. 285, 339–349.

Miragall, F., Simburger, E., and Dermietzel, R. 1996. Mitral and tufted cells of the mouse olfactory bulb possess gap junctions and express connexin43 mRNA. Neurosci. Lett. 216, 199–202.

Miyamichi, K., Serizawa, S., Kimura, H. M., and Sakano, H. 2005. Continuous and overlapping expression domains of odorant receptor genes in the olfactory epithelium determine the dorsal/ventral positioning of glomeruli in the olfactory bulb. J. Neurosci. 25, 3586–3592.

Mombaerts, P., Wang, F., Dulac, C., Chao, S. K., Nemes, A., Mendelsohn, M., Edmondson, J., and Axel, R. 1996. Visualizing an olfactory sensory map. Cell 87, 675–686.

Mori, K., Kishi, K., and Ojima, H. 1983. Distribution of dendrites of mitral, displaced mitral, tufted, and granule cells in the rabbit olfactory bulb. J. Comp. Neurol. 219, 339–355.

Mori, K., Nowycky, M. C., and Shepherd, G. M. 1981. Electrophysiological analysis of mitral cells in the isolated turtle olfactory bulb. J. Physiol. 314, 281–294.

Newman, M. P., Feron, F., and Mackay-Sim, A. 2000. Growth factor regulation of neurogenesis in adult olfactory epithelium. Neuroscience 99, 343–350.

Orona, E., Rainer, E. C., and Scott, J. W. 1984. Dendritic and axonal organization of mitral and tufted cells in the rat olfactory bulb. J. Comp. Neurol. 226, 346–356.

Orona, E., Scott, J. W., and Rainer, E. C. 1983. Different granule cell populations innervate superficial and deep regions of the external plexiform layer in rat olfactory bulb. J. Comp. Neurol. 217, 227–237.

Pinching, A. J. and Powell, T. P. 1971a. The neuropil of the glomeruli of the olfactory bulb. J. Cell. Sci. 9, 347–377.

Pinching, A. J. and Powell, T. P. 1971b. The neuropil of the periglomerular region of the olfactory bulb. J. Cell. Sci. 9, 379–409.

Pinching, A. J. and Powell, T. P. 1971c. The neuron types of the glomerular layer of the olfactory bulb. J. Cell. Sci. 9, 305–345.

Pinching, A. J. and Powell, T. P. 1972. Experimental studies on the axons intrinsic to the glomerular layer of the olfactory bulb. J. Cell. Sci. 10, 637–655.

Pressler, R. T. and Strowbridge, B. W. 2006. Blanes cells mediate persistent feedforward inhibition onto granule cells in the olfactory bulb. Neuron 49, 889–904.

Price, J. L. and Powell, T. P. 1970a. The morphology of the granule cells of the olfactory bulb. J. Cell Sci. 7, 91–123.

Price, J. L. and Powell, T. P. 1970b. The synaptology of the granule cells of the olfactory bulb. J. Cell Sci. 7, 125–155.

Price, J. L. and Powell, T. P. 1970c. The mitral and short axon cells of the olfactory bulb. J. Cell Sci. 7, 631–651.

Price, J. L. and Sprich, W. W. 1975. Observations on the lateral olfactory tract of the rat. J. Comp. Neurol. 162, 321–336.

Rall, W., Shepherd, G. M., Reese, T. S., and Brightman, M. W. 1966. Dendrodendritic synaptic pathway for inhibition in the olfactory bulb. Exp. Neurol. 14, 44–56.

Ramon-Cueto, A., Plant, G. W., Avila, J., and Bunge, M. B. 1998. Long-distance axonal regeneration in the transected adult rat spinal cord is promoted by olfactory ensheathing glia transplants. J. Neurosci. 18, 3803–3815.

Ramón y Cajal, S. 1911. Histologie du Système Nerveux de l'Homme et des Vertébrés. Paris, Maloine.

Rash, J. E., Davidson, K. G., Kamasawa, N., Yasumura, T., Kamasawa, M., Zhang, C., Michaels, R., Restrepo, D., Ottersen, O. P., Olson, C. O., and Nagy, J. I. 2005. Ultrastructural localization of connexins (Cx36, Cx43, Cx45), glutamate receptors and aquaporin-4 in rodent olfactory mucosa, olfactory nerve and olfactory bulb. J. Neurocytol. 34, 307–341.

Rash, J. E., Yasumura, T., and Dudek, F. E. 1998. Ultrastructure, histological distribution, and freeze-fracture immunocytochemistry of gap junctions in rat brain and spinal cord. Cell Biol. Int. 22, 731–749.

Reyher, C. K., Lubke, J., Larsen, W. J., Hendrix, G. M., Shipley, M. T., and Baumgarten, H. G. 1991. Olfactory bulb granule cell aggregates: morphological evidence for interperikaryal electrotonic coupling via gap junctions. J. Neurosci. 11, 1485–1495.

Rela, L. and Szczupak, L. 2004. Gap junctions: their importance for the dynamics of neural circuits. Mol. Neurobiol. 30, 341–357.

Ressler, K. J., Sullivan, S. L., and Buck, L. B. 1993. A zonal organization of odorant receptor gene expression in the olfactory epithelium. Cell 73, 597–609.

Royet, J. P., Distel, H., Hudson, R., and Gervais, R. 1998. A re-estimation of the number of glomeruli and mitral cells in the olfactory bulb of rabbit. Brain Res. 788, 35–42.

Schild, D. and Riedel, H. 1992. Significance of glomerular compartmentalization for olfactory coding. Biophys. J. 61, 704–715.

Schneider, S. P. and Macrides, F. 1978. Laminar distributions of internuerons in the main olfactory bulb of the adult hamster. Brain Res. Bull. 3, 73–82.

Schoppa, N. E. and Westbrook, G. L. 2001. Glomerulus-specific synchronization of mitral cells in the olfactory bulb. Neuron 31, 639–651.

Schwob, J. E. and Price, J. L. 1984. The development of lamination of afferent fibers to the olfactory cortex in rats, with additional observations in the adult. J. Comp. Neurol. 223, 203–222.

Scott, J. W., McBride, R. L., and Schneider, S. P. 1980. The organization of projections from the olfactory bulb to the piriform cortex and olfactory tubercle in the rat. J. Comp. Neurol. 194, 519–534.

Serizawa, S., Miyamichi, K., Nakatani, H., Suzuki, M., Saito, M., Yoshihara, Y., and Sakano, H. 2003. Negative feedback regulation ensures the one receptor-one olfactory neuron rule in mouse. Science 302, 2088–2094.

Serizawa, S., Miyamichi, K., Takeuchi, H., Yamagishi, Y., Suzuki, M., and Sakano, H. 2006. A neuronal identity code for the odorant receptor-specific and activity-dependent axon sorting. Cell 127, 1057–1069.

Shepherd, G. M. 1972. Synaptic organization of the mammalian olfactory bulb. Physiol. Rev. 52, 864–917.

Shepherd, G. M., Chen, W., and Greer, C. A. 2004. Olfactory Bulb. In: Synaptic Organization of the Brain (ed. G. M. Shepherd), pp. 165–216. Oxford University Press.

Shipley, M. T. and Adamek, G. D. 1984. The connections of the mouse olfactory bulb: a study using orthograde and retrograde transport of wheat germ agglutinin conjugated to horseradish peroxidase. Brain Res. Bull. 12, 669–688.

Shipley, M. T. and Ennis, M. 1996. Functional organization of olfactory system. J. Neurobiol. 30, 123–176.

Treloar, H. B., Feinstein, P., Mombaerts, P., and Greer, C. A. 2002. Specificity of glomerular targeting by olfactory sensory axons. J. Neurosci. 22, 2469–2477.

Vassar, R., Ngai, J., and Axel, R. 1993. Spatial segregation of odorant receptor expression in the mammalian olfactory epithelium. Cell 74, 309–318.

Wachowiak, M. and Shipley, M. T. 2006. Coding and synaptic processing of sensory information in the glomerular layer of the olfactory bulb. Semin. Cell Dev. Biol. 17, 411–423.

Westrum, L. E. 1975. Electron microscopy of synaptic structures in olfactory cortex of early postnatal rats. J. Neurocytol. 4, 713–732.

Whitman, M. C. and Greer, C. A. 2007. Adult-generated neurons exhibit diverse developmental fates. Dev. Neurobiol. 67, 1079–1093.

Willhite, D. C., Nguyen, K. T., Masurkar, A. V., Greer, C. A., Shepherd, G. M., and Chen, W. R. 2006. Viral tracing identifies distributed columnar organization in the olfactory bulb. Proc. Natl. Acad. Sci. U. S. A. 103, 12592–12597.

Wilson, D. A., Kadohisa, M., and Fletcher, M. L. 2006. Cortical contributions to olfaction: plasticity and perception. Semin. Cell Dev. Biol. 17, 462–470.

Woolf, T. B., Shepherd, G. M., and Greer, C. A. 1991a. Serial reconstructions of granule cell spines in the mammalian olfactory bulb. Synapse 7, 181–192.

Woolf, T. B., Shepherd, G. M., and Greer, C. A. 1991b. Local information processing in dendritic trees: subsets of spines in granule cells of the mammalian olfactory bulb. J. Neurosci. 11, 1837–1854.

Wouterlood, F. G. and Nederlof, J. 1983. Terminations of olfactory afferents on layer II and III neurons in the entorhinal area: degeneration-Golgi-electron microscopic study in the rat. Neurosci. Lett. 36, 105–110.

Zou, D. J., Feinstein, P., Rivers, A. L., Mathews, G. A., Kim, A., Greer, C. A., Mombaerts, P., and Firestein, S. 2004. Postnatal refinement of peripheral olfactory projections. Science 304, 1976–1979.

Zou, Z., Li, F., and Buck, L. B. 2005. Odor maps in the olfactory cortex. Proc. Natl. Acad. Sci. U. S. A. 102, 7724–7729.

Zou, Z., Horowitz, L. F., Montmayeur, J. P., Snapper, S., and Buck, L. B. 2001. Genetic tracing reveals a stereotyped sensory map in the olfactory cortex. Nature 414, 173–179.

4.37 Physiology of the Main Olfactory Bulb

M Ennis, University of Tennessee Health Science Center, Memphis, TN, USA

A Hayar, University of Arkansas for Medical Sciences, Little Rock, AR, USA

Glossary

apamin A blocker of SK-type potassium channels.
AP5 An NMDA receptor antagonist.
CNQX An AMPA receptor antagonist.
field potentials Extracellularly recorded voltages representing summed currents from a population of neurons.

gabazine A GABA$_A$ receptor antagonist.
ZD 7288 A blocker of Ih channels.
4-AP A blocker of A-type potassium channels (I_A).

4.37.1 Overview of Main Olfactory Bulb Circuitry

4.37.1.1 Projections of Olfactory Receptor Neurons to Main Olfactory Bulb

As discussed elsewhere (Chapter Signal Transduction in the Olfactory Receptor Cell), odors are transduced by olfactory receptor neurons (ORNs), giving rise to action potentials. The action potentials propagate along the axons of ORNs – the olfactory nerve (ON) fibers. These fibers form bundles which then collect as groups of fascicles, pass through the cribriform plate, and synapse in the main olfactory bulb (MOB). Within the MOB, ON axons terminate and synapse with neural elements in the glomerular layer (GL). Based on anatomical considerations (high packing density of unmyelinated axons), it has been speculated that ephaptic interactions (current spread through the extracellular space) might synchronize ON fibers. Computational modeling studies (Bokil, H. *et al.*, 2001) suggest that spikes in a single ON axon evoke spikes in adjacent axons, thus leading to synchronous firing of a large number of axons. These findings suggest that ephaptic interactions among neighboring axons may synchronize spikes among ON fibers converging on the same glomerulus.

ORNs expressing the same odorant receptor project to one or two glomeruli located on the medial and/or lateral side of each MOB (Figure 1; Ressler, K. J. *et al.*, 1993; 1994; Vassar, R. *et al.*, 1994; Mombaerts, P. *et al.*, 1996; Wang, F. *et al.*, 1998; Potter, S. M. *et al.*, 2001; Treloar, H. B. *et al.*, 2002). Studies in transgenic animals showed that this projection pattern is topographically fixed across animals. That is, the same glomeruli identified in different mice receive inputs from the same restricted population of ORNs bearing the same receptor (Mombaerts, P. *et al.*, 1996; Wang, F. *et al.*, 1998; Potter, S. M. *et al.*, 2001; Treloar, H. B. *et al.*, 2002). Rough calculations confirm an approximately 1:2 ratio between the number of different types of receptors (~1000) and the total number of glomeruli (~1800) in mice.

ORNs utilize glutamate as their primary neurotransmitter (Sassoe-Pognetto, M. *et al.*, 1993). Carnosine, a soluble dipeptide, is uniquely expressed in high concentrations in mammalian ORNs, and it is present in ON axon terminals in the GL (Ferriero, D. and Margolis, F. L., 1975; Margolis, F. L., 1980; Biffo, S. *et al.*, 1990). Carnosine colocalizes with glutamate in the ON axon terminals (Sassoe-Pognetto, M. *et al.*, 1993), and it satisfies criteria for neurotransmitter candidacy, including (1) carnosine synthetic and degradative enzymes are present in ORNs; (2) the peptide is released by depolarization in a Ca^{2+}-dependent manner in ON synaptosomes; (3) high-affinity binding sites for carnosine are present in the GL (Ferriero, D. and Margolis, F. L., 1975; Margolis, F. L., 1980; Burd, G. D. *et al.*, 1982; Rochel, S. and Margolis, F. L., 1982; Margolis, F. L. *et al.*, 1983; Margolis, F. L. and Grillo, M., 1984; Margolis, F. L. *et al.*, 1985; 1987; Biffo, S. *et al.*, 1990). Zinc and copper are also present in high concentrations in ON axon terminals (Biffo, S. *et al.*, 1990). The potential neuromodulatory roles of carnosine, zinc, and copper are discussed below.

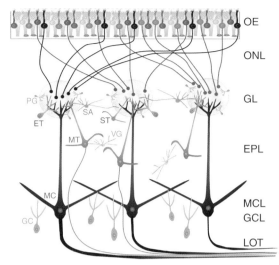

Figure 1 Schematic illustration of the basic circuitry of the main olfactory bulb (MOB) including the projections of olfactory receptor neurons (ORNs) from the olfactory epithelium (OE) to the glomerular layer (GL). Note that ORNs expressing different odorant receptor genes (shown as blue, red, or green cells) are interspersed and widely distributed, yet the axons of ORNs expressing the same odorant receptor gene converge onto the same glomerulus (or pairs of medial and lateral glomeruli) in the GL. Axons of ORNs travel in the olfactory nerve layer (ONL) and synapse in the GL on the dendrites of mitral cells (MC), tufted cells (external tufted cell, ET; middle tufted cell, MT), and generic juxtaglomerular (JG) neurons, which include periglomerular (PG) cells, ET cells, and short axon (SA) cells. SA cells interconnect different glomeruli. There are serial and reciprocal synapses between the apical dendrites of mitral/tufted cells and the processes of JG neurons. Superficial tufted (ST) cells are located in the superficial EPL or at the GL–EPL border. The lateral dendrites of mitral/tufted cells form serial and reciprocal synapses with the apical dendrites of granule cells (GC) in the EPL. The majority of GCs are concentrated in the GC layer (GCL) but many lie within the MCL. The axons of mitral/tufted cells project locally to GCs (not shown) and also to primary olfactory cortex via the lateral olfactory tract (LOT). The bulb also contains other populations of interneurons neurons, including the Van Gehuchten (VG) cells within the EPL.

4.37.1.2 MOB Circuitry

The MOB in rodents is situated at the rostral pole of the cranial cavity, and it is connected to the frontal cortex by a slender peduncle. The bulb can be thought of as a cylinder composed of distinct layers or laminae that are concentrically organized. These layers, from superficial to deep, are the ON layer (ONL), GL, external plexiform layer (EPL), mitral cell layer (MCL), internal plexiform layer (IPL), granule cell layer (GCL), and the ependymal layer (Figure 1). The ONL consists of ON axons and glial cells (Cajal, R. S. Y., 1911a; 1911b; Pinching, A. J. and

Powell, T. P., 1971b; Doucette, R., 1989). Deep to the ONL, the GL is comprised of neutropil-rich ovoid structures – the glomeruli – each of which is surrounded by a shell of small neurons and glia. Within the glomeruli, ON axons form synapses with mitral and tufted cells, as well as with the intrinsic neurons of the GL – the juxtaglomerular (JG) cells. Adjacent glomeruli are somewhat isolated from each other by astrocytes residing in the glomerular shell (Bailey, M. S. and Shipley, M. T., 1993). The EPL lies beneath or deep to the glomeruli, and it primarily consists of dense neuropil formed by the dendrites of mitral cells and GCs that ascend from the MCL and GCL, respectively. The EPL also contains several subtypes of tufted cells and intrinsic interneurons. The dominant feature of the EPL is nevertheless the extensive dendrodendritic synapses between mitral/tufted cells and GCs. Deep to the EPL, the MCL is a thin layer that contains the somata of mitral cells, as well as numerous GCs (Cajal, R. S. Y., 1911a; 1911b). Together with tufted cells, mitral cells are the major class of output cells of the bulb. They extend a single apical dendrite into the GL, where it arborizes extensively throughout much of a single glomerulus (Figure 1). The apical dendrites are synaptically contacted by ON terminals (Price, J. L. and Powell, T. P. S., 1970a; Shepherd, G. M., 1972). The secondary or lateral dendrites of mitral cells ramify in the EPL where they form dendrodendritic synapses with dendrites of GCs. Mitral/tufted cell axons terminate within the bulb in the IPL and GCL (Mori, K. *et al.*, 1983; Price, J. L. and Powell, T. P. S., 1970c), or exit the MOB and innervate a number of olfactory-related brain regions collectively known as the primary olfactory cortex (POC). Deep to the MCL, the IPL is the relatively thin layer with a low density of cells. The GCL is the deepest neuronal layer in the bulb, and it contains the largest number of cells. Most of the neurons of the GCL are the GCs, but there are also small numbers of Golgi cells, Cajal cells, and Blanes cells. The GCs are inhibitory GABAergic cells that form dendrodendritic synapses with mitral/tufted cells in the EPL.

4.37.2 Neurophysiology of the Glomerular Layer

4.37.2.1 Neurophysiological Properties of Glomerular Layer Neurons

The neurons of the GL are classified into three cell types, which include: (1) periglomerular (PG) cells,

(2) external tufted (ET) cells, and (3) short axon (SA) cells (Golgi, C. 1875; Van Gehuchten, L. E. and Martin, A., 1891; Blanes, T., 1898; Cajal, R. S. Y., 1911a; 1911b; Pinching, A. J. and Powell, T. P., 1971a; 1971b; 1971c; 1972b; 1972c). Collectively, the intrinsic neurons of the GL are referred to as JG cells. The term JG is also used here with regard to cited studies in which the subtype of glomerular neuron was not specified. The morphology and features of these cells are only briefly reviewed here as more detailed descriptions are available (Hayar, A. *et al.*, 2004a; 2004b).

4.37.2.1.1 ET cells

These are relatively large (10–15 μm) cells that are dispersed in the JG regions surrounding/deep to the glomeruli (Figures 1 and 2). Most have one apical dendrite that arborizes extensively throughout one glomerulus (Pinching, A. J. and Powell, T. P., 1971a; Hayar, A. *et al.*, 2004a; 2004b). Rarely, ET cells have two or three apical dendrites that ramify in different glomeruli. Most ET cells have secondary or lateral dendrites that extend in the superficial EPL. Some ET cells have axons that appear to synapse with PG cells or SA cells, or more infrequently project out of MOB (Pinching, A. J. and Powell, T. P., 1971a). ET cells are somewhat similar to tufted cells of the EPL and to mitral cells, but growing evidence suggests that all tufted cell subtypes exhibit distinct anatomical and physiological properties.

The most distinctive physiological feature of ET cells *in vitro* is their spontaneous rhythmical bursting (Hayar, A. *et al.*, 2004a; 2004b; 2005) (Figure 2). JG cells with burst characteristics have been reported *in vivo* (Getchell, T. V. and Shepherd, G. M., 1975; Wellis, D. P. and Scott, J. W., 1990). However, because of the difficulty of recording small JG neurons *in vivo*, the identity of these cells and the basis of their bursting behavior remained unknown. The rhythmical burst-firing mode was characteristic of morphologically confirmed ET neurons (Hayar, A. *et al.*, 2004a). By contrast, PG and SA cells do not spontaneously generate spike bursts nor can they be induced to do so by intracellular current injections (Figure 2). Each ET cell bursts at its own characteristic frequency. As a population, ET cell burst frequencies range from ∼1 to 8 Hz, with a mean of 3.3 ± 0.18 bursts/s. This range overlaps with the theta frequency range (2–12 Hz) prominent in oscillatory neural activity in the rodent olfactory network (see Oscillations and Synchrony in Main Olfactory Bulb); the theta range includes components related to low frequency (1–3 Hz) passive

sniffing as well as a higher frequency component (5–10 Hz) characteristic of active investigative sniffing (Adrian, E. D., 1950; Welker, W. I., 1964; Macrides, F. *et al.*, 1982; Eeckman, F. H. and Freeman, W. J., 1990; Kay, L. M. and Laurent, G., 1999; Kay, L. M., 2003). ET cells also receive spontaneous bursts of inhibitory postsynaptic currents (IPSCs) from PG cells (Hayar, A. *et al.*, 2005). ON stimulation evokes an excitatory postsynaptic current (EPSC) in ET cells that is followed by IPSC bursts (Hayar, A. *et al.*, 2005). Both the spontaneous and ON-evoked IPSCs in ET cells are driven primarily by activation of AMPA receptors.

Several lines of evidence indicate that bursting is an intrinsic property of ET cells. First, bursting deteriorates rapidly after establishment of whole-cell recording mode. This pronounced rundown of bursting may explain the low reported incidence of spontaneous bursting in JG cells in some whole-cell recording studies (Bardoni, R. *et al.*, 1995; Puopolo, M. and Belluzzi, O. 1996; McQuiston, A. R. and Katz, L. C., 2001). The rundown of bursting could be due to intracellular dialysis of an intracellular messenger important to maintain spontaneous activity (Alreja, M. and Aghajanian, G. K., 1995). Additional evidence in support of the intrinsic mechanism for bursting are the findings that: (1) burst frequency is voltage dependent and (2) bursting persists in blockers of glutamate and $GABA_A$ receptors. This eliminates the possibility that bursting is driven by glutamatergic input from the ON, glutamatergic dendrodendritic interactions among ET cells and/or mitral cells, or by disinhibition (Aroniadou-Anderjaska, V. *et al.*, 1999b; Isaacson, J. S., 1999; Carlson, G. C. *et al.*, 2000; Friedman, D. and Strowbridge, B. W., 2000; Salin, P. A. *et al.*, 2001; Schoppa, N. E. and Westbrook, G. L., 2001). Moreover, spontaneous bursting was not blocked by Cd^{2+}, which suppresses Ca^{2+}-dependent neurotransmitter release, ruling out the potential involvement of other neurotransmitters. Depolarizing current injections evoke in ET cells a low-threshold Ca^{2+} spike (LTS) that was eliminated by the Ca^{2+} channel blockers, Cd^{2+} and Ni^{2+} (McQuiston, A. R. and Katz, L. C., 2001; Hayar, A. *et al.*, 2004a). This suggested that the LTS might generate ET cell bursting. However, this is unlikely for several reasons. First, bursting in ET cells persisted after the LTS was blocked. Second, ET cell bursting was abolished by extracellular TTX or by intracellular QX-314, whereas the LTS persisted. Third, the activation threshold of the LTS (−38 mV) was approximately 15 mV more depolarized than the membrane potential from which bursting arises, on average −53 mV. Ca^{2+} may modulate bursting as

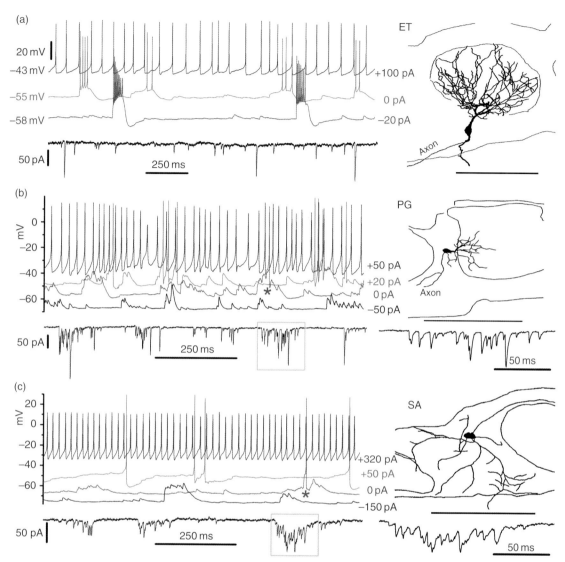

Figure 2 External tufted (ET) cells spontaneously generate rhythmic spike bursts, whereas periglomerular (PG) and short axon (SA) cells receive spontaneous bursts of excitatory synaptic input. Panels (a), (b), and (c) show the typical electrophysiological and morphological features of an ET, a PG, and a SA cell, respectively. The ET cell has highly branched tufted dendrites that ramify throughout a single glomerulus. The PG cell has relatively small soma and one to three relatively thick primary dendritic shafts that give rise to thinner branches ramifying within a subregion of a single glomerulus. The SA cell has several poorly branched dendrites that extended into or between two and four glomeruli. Scale bars below drawings, 100 μm. Current clamp recordings show resting spontaneous activity and effects of current injections. At rest (0 pA), the ET cell generates spontaneous bursts of spikes, whereas the PG and SA cells receive spontaneous bursts of EPSPs and generates spikes infrequently (asterisks). Bottom traces in a, b, and c show voltage-clamp recordings (holding potential = −60 mV) of spontaneous EPSCs in the same cells; regions enclosed in the box in b and c are shown at faster timescale at right. Note the bursting pattern of EPSCs in the PG and SA cells. Adapted from Hayar, A., Karnup, S., Ennis, M., and Shipley, M. T. 2004b. External tufted cells: a major excitatory element that coordinates glomerular activity. J. Neurosci. 24, 6676–6685, with permission from The Society for Neuroscience.

Ca^{2+} channel blockers prolonged burst duration (Hayar, A. *et al.*, 2004a). Ca^{2+} channel blockers also increased the interburst interval, but this could be due to a charge screening effect. Moreover, ET cells have a hyperpolarization-activated cation conductance (Ih)

that is prominent at resting membrane potential (Figure 3). Ih current was found in all JG cells (Cadetti, L. and Belluzzi, O., 2001). However, other results (Hayar, A., unpublished observations) indicate that Ih is very strong in ET cells compared to PG and

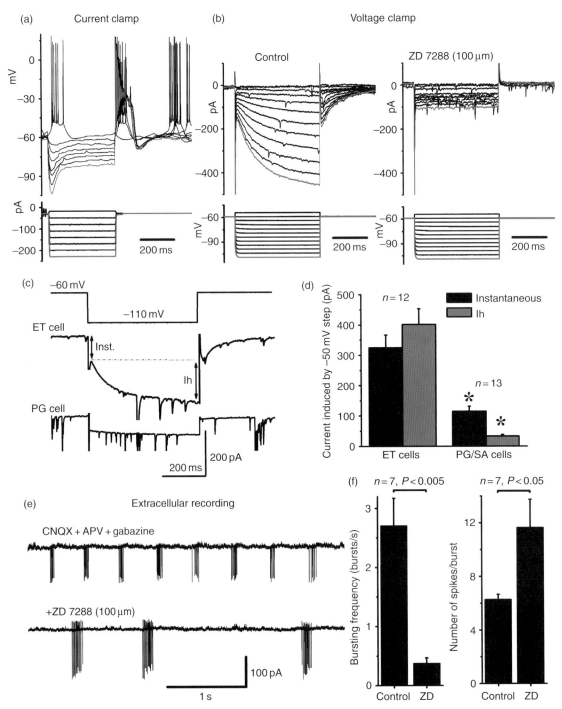

Figure 3 External tufted (ET) cells have a prominent Ih current. (a) Responses of an ET to hyperpolarizing and depolarizing current steps (bottom) in current clamp mode. Note the sag induced by membrane hyperpolarization, indicative of the presence of Ih current. After releasing the cell from the hyperpolarizing current, a rebound burst of spikes occurred. (b) Response of the same cell to membrane voltage steps (bottom) in voltage-clamp mode. A time-dependent inward current (Ih) was produced with hyperpolarizing voltage steps ($<-70\,mV$) followed by a rebound smaller inward current, which were blocked by the selective Ih current blocker, ZD 7288. (c) A periglomerular (PG) cell exhibits a much smaller Ih current (in response to a hyperpolarizing step from -60 to $-110\,mV$, voltage-clamp mode) than an ET cell. (d) Group data show that ET cells have significantly larger Ih current than PG and short axon (SA) cells. The instantaneous current (Inst) reflects the membrane conductance which is larger in ET cells, indicating that PG/SA cells have on average a relative much higher input resistance. (e) The bursting frequency of an extracellularly recorded spontaneously bursting ET cell (in the presence of CNQX, AP5, and gabazine) decreased in response to application of the Ih blocker ZD 7288. However, the number of spikes/burst increased as shown in the grouped data in (f).

SA cells and that it contributes to tonic depolarization of ET cells because the specific Ih blocker, ZD 7288, significantly reduced the bursting frequency (Figure 3). Ih participates in burst timing and rhythmicity in other brain structures (DiFrancesco, D. 1993; Lüthi, A. and McCormick, D. A. 1998; Pape, H. C. 1996) and is modulated by Ca^{2+}. Thus, Ih and Ca^{2+} may play a role in burst termination and/or setting the interburst interval.

Rhythmic bursting, but not the LTS, was blocked by extracellular TTX and by intracellular dialysis with QX-314. These results suggest that, as for many neocortical neurons (Brumberg, J. C. et al., 2000), burst generation in ET cells requires slowly inactivating Na^+ channels, that is, a persistent Na^+ current. Indeed, all ET cells tested had a prominent TTX-sensitive, persistent Na^+ current. The characteristics of this current are well suited to trigger spontaneous bursting in ET cells. The inward current activates near $-60\,mV$, slightly hyperpolarized to the mean ET cell resting potential ($-52\,mV$). Thus, it is reasonable to conjecture that as this persistent Na^+ current slowly depolarizes the ET cell membrane, additional Na^+ channels are activated further depolarizing the membrane to the threshold for action potential generation. This current maintains a level of depolarization sufficient to generate additional action potentials until another mechanism(s), possibly involving Ca^{2+} (see above), terminates the burst by transiently hyperpolarizing the membrane below the activation threshold for the persistent Na^+ current. As the membrane repolarizes, the persistent Na^+ current is re-engaged, and the burst cycle is repeated.

4.37.2.1.2 PG cells

These cells are the most numerous cells in the GL, and they are thought to be inhibitory in nature. They are small (5–8 µm), spherical or ovoid, and they are distributed in the PG regions surrounding the glomeruli (Figures 1 and 2). Their dendrites are typically restricted to a small subregion of a glomerulus (Pinching, A. J. and Powell, T. P., 1971a; Hayar, A. et al., 2004a; 2004b). Their dendrites receive asymmetrical (morphologically excitatory) synapses from ET and mitral/tufted cell dendrites; only a relatively small subpopulation of PG cells receives synapses from ON terminals. Some of the mitral/tufted cell to PG cell synapses are paired with reciprocal symmetrical (morphologically inhibitory) synapses from PG cells back onto the parent mitral/tufted cell dendrites (Pinching, A. J. and Powell, T. P., 1971b;

Toida, K. et al., 1998; Kasowski, H. J. et al., 1999; Toida, K. et al., 2000). Physiological recordings indicate that PG cells also receive monosynaptic dendrodendritic excitatory input from ET cells (Hayar, A. et al., 2004b). Axons of PG cells are rare but have been reported to extend over distances equivalent to four to five glomeruli (Blanes, T., 1898; Pinching, A. J. and Powell, T. P., 1971a; Hayar, A. et al., 2004a). They appear to form symmetrical (morphologically inhibitory) synapses onto mitral/tufted cell dendrites and onto ET cells and other JG cells (Pinching, A. J. and Powell, T. P., 1971c). Most PG cells are GABAergic (Ribak, C. E. et al., 1977; Mugnaini, E. et al., 1984; Kosaka, T. et al., 1985; 1987a; 1987b; 1987c; Kosaka, K. et al., 1987; 1988). The GL also contains the largest population of dopamine (DA)-containing cells in the brain. The majority of DA-containing cells are PG cells, but some ET cells are also dopaminergic (Halász, N. et al., 1977; Davis, B. J. and Macrides, F., 1983; Halász, N. et al., 1985; Gall, C. et al., 1986; McLean, J. H. and Shipley, M. T., 1988). About 80% of the DA cells contain GABA, which corresponds to about 50% of GABAergic cells (Kosaka, T. et al., 1985; Gall, C. M. et al., 1987; Kosaka, T. et al., 1987b; Kosaka, K. et al., 1987; 1988; 1995). Therefore, the DA PG neurons are regarded as a subpopulation of GABAergic neurons. Some PG cell subtypes also stain for thyrotropin-releasing hormone (TRH), enkephalin, nicotinamide adenine dinucleotide phosphate (NADPH)-diaphorase, neuropeptide-Y (NPY), somatostatin, and vasoactive intestinal polypeptide (VIP) (Bogan, N. et al., 1982; Davis, B. J. et al., 1982; Gall, C. et al., 1986; Scott, J. W. et al., 1987; Merchenthaler, I. et al., 1988; Tsuruo, Y. et al., 1988; Sanides-Kohlrausch, C. and Wahle, P., 1990a; Davis, B. J., 1991; Alonso, J. R. et al., 1993; Kosaka, K. et al., 1995; 1998; Kosaka, T. et al., 1987c).

The spontaneous activity patterns of PG and SA cells differ markedly from ET cells. PG and SA cells in vitro have relatively low levels of spontaneous spike activity and lack the capacity to generate spike bursts in response to depolarizing currents (Figure 2) (Hayar, A. et al., 2004a; 2004b). In slices, spikes in PG cells are driven primarily by spontaneous glutamatergic excitatory postsynaptic potentials (EPSPs). However, because of their heterogeneous neurochemical characteristics, PG cells are also likely to be functionally heterogeneous. For example, some studies have reported that different PG cell subtypes appear to exhibit different K^+

conductances (Puopolo, M. and Belluzzi, O. 1998) as well as different firing behaviors (McQuiston, A. R. and Katz, L. C., 2001). An estimated 10% of PG neurons in adulthood are positive for tyrosine hydroxylase (TH) (McLean, J. H. and Shipley, M. T., 1988; Kratskin, I. and Belluzzi, O., 2003), the rate-limiting enzyme for DA synthesis. Dopaminergic neurons have been recently identified *in vitro* using a transgenic mouse strain harboring an eGFP (enhanced green fluorescent protein) reporter construct under the promoter of TH. The most prominent feature of these cells when recorded in acutely dissociated cell culture preparations was the presence of regular spontaneous spiking at $\sim$8 Hz. In these cells, five main voltage-dependent conductances were identified (Pignatelli, A. *et al.*, 2005): the two having largest amplitude were a fast transient Na^+ current and a delayed rectifier K^+ current. In addition, they have three smaller inward currents, sustained by Na^+ ions (persistent type) and by Ca^{2+} ions (low-voltage-activated (LVA) and high-voltage-activated (HVA)). The pacemaking activity was shown to be supported by the interplay of the persistent Na^+ current and of a T-type Ca^{2+} current. Transgenic mice in which catecholaminergic neurons expressed human placental alkaline phosphatase (PLAP) on the outer surface of the plasma membrane were also used to identify dissociated dopaminergic PG neurons (Puopolo, M. *et al.*, 2005). Dopaminergic PG cells spontaneously generated action potentials in a rhythmic fashion with an average frequency of 8 Hz. It was found that substantial Ca^{2+} current and TTX-sensitive Na^+ current flow during the interspike depolarization. These results show that dopaminergic PG cells have intrinsic pacemaking activity, supporting the possibility that they can maintain a tonic release of DA to modulate the sensitivity of the olfactory system during odor detection.

4.37.2.1.3 SA cells

These cells are roughly the same size (8–12 μm) as ET cells. They are distinguished by multiple dendrites that seem to harvest information from multiple glomeruli (Figures 1 and 2) (Hayar, A. *et al.*, 2004b). The dendrites may receive synaptic inputs from ET and mitral/tufted cell dendrites (Hayar, A. *et al.*, 2004b), tufted cell axon collaterals, or other SA cells. SA cells have axons that can extend up to 1–2 mm within the GL (Aungst, J. L. *et al.*, 2003). The axons appear to synapse primarily onto the dendrites of PG cells (Pinching, A. J. and Powell, T. P., 1971a). They do not receive direct ON input (Pinching, A. J.

and Powell, T. P., 1971c; Hayar, A. *et al.*, 2004b). Their resting spontaneous activity patterns in MOB slices appear to be very similar to that of PG cells (Hayar, A. *et al.*, 2004b).

4.37.2.2 Electrophysiology of Intraglomerular Circuitry

4.37.2.2.1 Excitatory systems in the GL

4.37.2.2.1.(i) ON glutamatergic synaptic input to JG and mitral/tufted cells All studies to date have identified glutamate as the major transmitter at ON synapses onto the dendrites of JG neurons, as well as those of mitral/tufted cells in the GL. Release of glutamate from ON terminals is controlled by N- and P/Q-type Ca^{2+} channels (Isaacson, J. S. and Strowbridge, B. W., 1998; Murphy, G. J. *et al.*, 2004). Each glomerulus contains the apical dendritic tufts of about 20 mitral cells, 200 tufted cells, and 1500–2000 JG cells (reviewed in Ennis, M. *et al.*, 2007). A variety of studies demonstrated that sensory transmission from ON axon terminals to these dendrites is mediated by glutamate acting primarily at alpha-amino-3-hydroxy-5-methyl-4-isoxazole propionic acid (AMPA) and *N*-methyl D-aspartate (NMDA) ionotropic glutamate receptor subtypes (Bardoni, R. *et al.*, 1996; Ennis, M. *et al.*, 1996; Aroniadou-Anderjaska, V. *et al.*, 1997; Chen, W. R. and Shepherd, G. M., 1997; Keller, A. *et al.*, 1998; Aroniadou-Anderjaska, V. *et al.*, 1999a; Ennis, M. *et al.*, 2001). Thus, JG and mitral/tufted cell responses to ON input are excitatory (Figures 4 and 5). In addition, metabotropic glutamate receptors (mGluRs) are expressed by nearly all MOB neurons (see Ennis, M. *et al.*, 2007 for review). Comparatively less is known about the role of mGluRs at ON synapses. Electrophysiological and Ca^{2+}-imaging studies have reported that ON stimulation evokes an mGluR1-sensitive synaptic component in $\sim$40% of mitral cells in normal physiological conditions in the slice (De Saint Jan, D. and Westbrook, G. L., 2005; Ennis, M. *et al.*, 2006; Yuan, Q. and Knopfel, T., 2006). In the presence of ionotropic glutamate receptor antagonists, ON-evoked, mGluR1-mediated synaptic responses were markedly enhanced by inhibition of glutamate transporters (e.g., TBOA) (Figure 5). Thus, activation of mGluR1 is tightly regulated by glutamate uptake mechanisms. mGluR1-mediated responses were maximal with bursts of 50–100 Hz ON spikes, similar to frequencies of odor-induced spikes in ORNs *in vivo* (Duchamp-Viret, P. *et al.* 1999). Activation of mGluRs has also been suggested

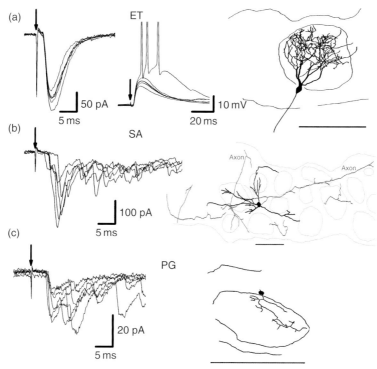

Figure 4 Olfactory nerve (ON)-evoked synaptic responses in juxtaglomerular (JG) cell subtypes. Panels a–c show six superimposed voltage-clamp traces (holding potential = −60 mV in this and subsequent panels) of ON-evoked EPSCs in external tufted (ET), short axon (SA), and periglomerular (PG) cells, respectively. Arrows indicate time of stimulation. Cell morphology is shown at right. Scale bars, 100 μm. ON stimulation produced short, constant latency EPSCs in the ET cell (a, left panel), indicative of monosynaptic input. In current clamp, ON stimulation evoked a constant, short latency EPSPs of variable amplitude (a, middle panel). The largest EPSPs triggered the generation of a burst of action potentials (truncated for clarity). Panels b and c show that ON stimulation evoked longer, variable latency EPSCs bursts in SA (b) and PG (c) cells, indicative of di- or polysynaptic responses. Adapted from Hayar, A. *et al.*, 2004, J. Neurosci. 24, 6676–6685., with permission from The Society for Neuroscience.

to produce long-term depression at ON to mitral cell synapses (Mutoh, H. *et al.*, 2005).

All ET cells receive monosynaptic ON input, which appears to be primarily mediated by AMPA and NMDA receptors in normal physiological conditions *in vitro* (Figure 4) (Hayar, A. *et al.*, 2004a; 2004b; 2005). If the amplitude of an ON-evoked EPSP reaches threshold for spike generation, a spike burst is always triggered in ET cells (Hayar, A. *et al*, 2004a). Further increases in ON stimulation strength produce the same all-or-none burst. Thus, ET cells receive monosynaptic ON input that is converted into an all-or-none spike burst. Accordingly, ET cells amplify suprathreshold sensory input at the first stage of synaptic transfer in the olfactory system. ET cells also readily entrain to rhythmic ON input delivered at theta frequencies (5–10 Hz) characteristic of investigative sniffing in rodents. As noted above, the intrinsic spontaneous bursting rate of ET cells ranges from 1 to 8 bursts/s.

This might suggest that ET cells with different spontaneous bursting rates may be preferentially entrained by repetitive sensory input at or near their intrinsic bursting frequency. This turns out not to be the case as ET cells, irrespective of their intrinsic spontaneous bursting frequencies, can be entrained by repetitive, 5–10 Hz ON input (Hayar, A. *et al.*, 2004a).

The majority of PG cells exhibit longer latency, prolonged bursts of EPSP/Cs in response to ON stimulation than ET cells (Figure 4). The long, variable latency of these responses is indicative of di- or polysynaptic ON input. Thus, most SA and PG cells do not appear to receive direct ON input, perhaps because their dendrites ramify in glomerular compartments devoid of ON terminals. Only 20% of PG cells had short, relatively constant latency EPSCs following ON stimulation, and in some of these cells, the short latency synaptic response was followed by a delayed burst of EPSP/Cs. SA cells also

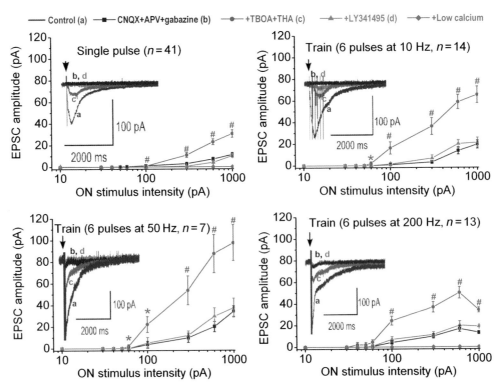

Figure 5 Olfactory nerve (ON)-evoked EPSCs elicited in the presence of glutamate uptake inhibitors (TBOA and THA) are dependent on stimulation intensity and frequency. The four panels show plots of the peak amplitude of ON-evoked EPSCs as a function of ON stimulation intensity (10–1000 μA); each panel shows data for a different ON stimulation frequency. Insets in each graph show traces of ON-evoked EPSCs in different pharmacological conditions; traces are averages of five ON-evoked EPSCs. Colored lines correspond to the pharmacological conditions indicated at the top; note that control data are not plotted on the line graphs. TBOA–THA significantly increases the ON-evoked EPSCs at a threshold intensity of 60–100 μA for all frequencies. Residual responses remaining in the presence of LY341495 and CNQX, APV, and gabazine were abolished by low Ca^{2+}-ACSF, as shown for the single-pulse and 200 Hz stimulation frequencies. $^*P < 0.05$, #$P < 0.001$ vs. CNQX, APV, and gabazine. Adapted from Ennis, M., Zhu, M., Heinbockel, T., and Hayar, A. 2006. Olfactory nerve-evoked metabotropic glutamate receptor-mediated responses in olfactory bulb mitral cells. J. Neurophysiol. 95, 2233–2241, with permission from The American Physiological Society.

responded to ON stimulation with long, variable latency, prolonged bursts of EPSP/Cs and never exhibited responses consistent with monosynaptic ON input (Figure 4). These findings indicate that (1) SA cells and most PG cells lack direct ON input and (2) ET cells along with mitral/tufted cells are the major postsynaptic targets of ON inputs.

Since most (~80%) PG cells, and all SA cells, lack monosynaptic ON input, this suggests that most PG cells are functionally associated with glomerular compartments lacking ON terminals. Anatomical studies have revealed that the glomeruli have a bicompartmental organization. Each glomerulus has several interdigitating compartments, one of which is rich in ON terminals and the second of which is devoid of ON input (Kosaka, K. *et al.*, 1997; Kasowski, H. J. *et al.*, 1999). Calbindin-positive JG neurons extend their dendrites

only into glomerular compartments devoid of ON terminals, suggesting that they do not receive direct sensory innervation (Toida, K. *et al.*, 1998; 2000). It is reasonable to speculate, therefore, that the SA and PG cells that do not receive direct input (Hayar, A. *et al.*, 2004b) may correspond to these calbindin-positive JG neurons. These PG cells might provide localized inhibition to mitral/tufted cell dendrites, or other nearby JG cells. Dopaminergic and GABAergic PG cells presynaptically inhibit ON terminals (Hsia, A. Y. *et al.*, 1999; Wachowiak, M. and Cohen, L. B., 1999; Aroniadou-Anderjaska, V. *et al.*, 2000; Berkowicz, D. A. and Trombley, P. Q., 2000; Ennis, M. *et al.*, 2001; Murphy, G. J. *et al.*, 2005; Wachowiak, M. *et al.*, 2005). Since at least 20% of PG cells receive monosynaptic ON input, it is possible that these cells primarily mediate this presynaptic inhibition of the ON. In contrast to PG cells, SA cells extend dendrites and axons across multiple glomeruli and are thought to be involved in interglomerular functions, such as center-surround inhibition of neighboring glomeruli (Aungst, J. L. *et al.*, 2003).

4.37.2.2.1.(ii) Spillover of dendritically released glutamate

As noted above, the apical dendrites of ET and mitral/tufted cells that ramify in the glomeruli release glutamate. Although the dendrites of these cell types do not form synapses with each other, the release of glutamate from the apical dendrites has been reported to produce nonsynaptically mediated excitation (i.e., glutamate spillover) of the parent cell releasing glutamate (self- or autoexcitation) or neighboring cells (Figure 6). Such excitation is facilitated by removal of Mg^{2+} from the extracellular media, which enhances the activation of NMDA receptors. Alternatively, spillover-mediated excitatory responses can be facilitated in the presence of physiological levels of extracellular Mg^{2+} by blockade of $GABA_A$ receptors (Salin, P. A. *et al.*, 2001; Friedman, D. and Strowbridge, B. W., 2000). Under such conditions, intracellular depolarization of individual mitral cells produces long-duration, NMDA receptor-dependent excitation of the same cell or adjacent mitral cells (Aroniadou-Anderjaska, V. *et al.*, 1999b; Isaacson, J. S., 1999; Friedman, D. and Strowbridge, B. W., 2000; Salin, P. A. *et al.*, 2001; Christie, J. M. and Westbrook, G. L., 2006). ET cells have also been reported to exhibit such self-excitation (Murphy, G. J. *et al.*, 2005). The mitral cell self-excitation can be blocked by intracellular Ca^{2+} chelation or blockade of voltage-gated Ca^{2+} channels (Friedman, D. and Strowbridge, B. W., 2000; Salin, P. A. *et al.*, 2001). The self-excitation responses are graded and increase with the number of spikes or depolarizing pulses applied to the mitral cell (Friedman, D. and Strowbridge, B. W., 2000; Salin, P. A. *et al.*, 2001). It has been suggested that the NMDA autoreceptors may serve to increase the firing frequency of mitral cells during prolonged discharges (Friedman, D. and Strowbridge, B. W., 2000). Such excitation can also be evoked by antidromic activation of mitral cells (Figure 6) (Chen, W. R. and Shepherd, G. M., 1997; Aroniadou-Anderjaska, V. *et al.*, 1999b). This glutamate spillover-mediated, NMDA receptor-dependent excitation appears to occur chiefly among the lateral dendrites of mitral cells, while an AMPA receptor-dependent spillover occurs among the apical dendrites of mitral/tufted cells (Aroniadou-Anderjaska,

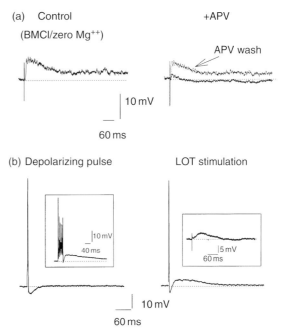

Figure 6 Mitral cells generate an NMDA receptor-dependent EPSP in response to antidromic stimulation of neighboring mitral/tufted cells. Panels a and b show recordings from two different cells in Mg^{2+}-free medium and bicuculline (10 μM). (a) Antidromic activation of mitral/tufted cells by lateral olfactory tract (LOT) stimulation, subthreshold for the recorded neuron, evoked an EPSP that was reversibly blocked by bath-applied APV (50 μM). (b) Mitral cell shown did not produce a prolonged depolarization in response to a 5 ms intracellular current pulse. However, a prolonged depolarization was evoked by longer current pulses (40 ms, shown in inset) or LOT stimulation with or without (inset) an antidromic spike. Traces are averages of five to ten sweeps. Reprinted from V. Aroniadou-Anderjaska, V., Ennis, M., and Shipley, M. T. 1999b. Dendrodendritic recurrent excitation in mitral cells of the rat olfactory bulb. J. Neurophysiol. 82, 489–494, with permission from The American Physiological Society.

V. *et al.*, 1999b; Salin, P. A. *et al.*, 2001; Schoppa, N. E. and Westbrook, G. L., 2001; 2002; Christie, J. M. and Westbrook, G. L., 2006). Electrotonic coupling among mitral cell dendrites in the same glomerulus facilitates spillover responses mediated by AMPA receptors (Schoppa, N. E. and Westbrook, G. L., 2002; Christie, J. M., and Westbrook, G. L., 2006).

Other glomerulus-specific glutamate-mediated excitatory interactions among mitral cells have been reported. Reciprocal glutamate-mediated excitation was reported between closely adjacent mitral cells whose apical dendrites extended into the same glomerulus (Urban, N. N. and Sakmann, B., 2002). The latency for the mitral to mitral cell EPSPs was surprisingly short, in line with monosynaptic mediation, despite the fact that mitral cells do not form anatomical synapses with each other. The EPSP was primarily mediated by AMPA receptors and originated within the GL, suggesting that they are generated in the apical dendrites of mitral cells. In contrast to the preceding studies, Carlson G. C. *et al.* (2000) reported that ON stimulation and antidromic activation of multiple mitral/tufted cells, but never activation of single mitral cells, elicited long-lasting depolarizations (LLDs) in mitral cells. The LLDs were all-or-none in nature, required activation of AMPA receptors, and originated in the glomeruli. Further, spontaneous LLDs were synchronous among mitral cells associated with the same, but not different glomeruli. The LLDs appear to be a network phenomenon, presumably reflecting recurrent, intraglomerular glutamate release from an ensemble of mitral/tufted cell apical dendrites. Recently, Karnup S. V. *et al.* (2006) described field potentials that are generated spontaneously in the GL in olfactory bulb slices (Figure 7). These spontaneous glomerular local field potentials (sGLFPs) had variable shape and amplitude and occurred at irregular intervals. They are similar to LLDs in mitral cells (Carlson,

G. C. *et al.*, 2000) in that they are mediated mostly by AMPA/kainate receptors and are enhanced during blockade of GABA$_A$ receptors. They persisted after the removal of the MCL, indicating that they are predominantly generated from by GL neurons. Nevertheless, they were correlated with mitral cell postsynaptic LLDs, suggesting the existence of a common generator (Figure 7). The observation that sGLFPs precede LLDs in mitral cells suggests that mitral cells are followers rather than initiators of population events. The most likely scenario is that mitral cells are activated when a critical number of glutamate-releasing JG neurons (presumably ET cells) are activated. Such a triggering mechanism may be produced by synchronous spike bursts in ET cells of the same glomerulus (Hayar, A. *et al.*, 2004a; 2004b). The role of spillover-mediated excitation among mitral/tufted cells is discussed below in Synchrony.

4.37.2.2.1.(iii) ET and mitral/tufted cell dendrodendritic input to PG and SA cells ET and mitral/tufted cells form glutamatergic dendrodendritic synapses with PG and SA cells. This topic is

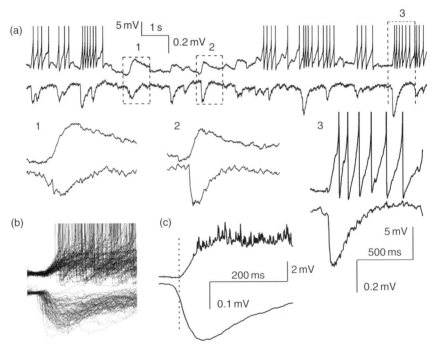

Figure 7 Glomerular and mitral cell long-lasting depolarizations (LLDs). Simultaneous recordings of spontaneous LLDs (sLLDs) in a mitral cell and spontaneous glomerular layer field potentials (sGLFPs) from the glomerulus containing the mitral cell tuft. (a) Dual sweep of intracellular (upper trace) and field activity (lower trace). Spikes of the mitral cell are truncated. Insets with an expanded timescale demonstrate variability of sGLFP locked to sLLD and different kinetics of sGLFP as compared with sLLD. (b) Overlapping sweeps with sLLD/sGLFP pairs ($n = 80$). Note 10-fold difference between smallest and biggest sGLFP indicating different amount of cells involved in successive sGLFP in a given glomerulus. (c) Averaged sLLD and sGLFP ($n = 80$) with the reference point set at the sLLD onset show 9.8 ms delay of sLLD upon initiation of sGLFP.

discussed below in *ET and mitral/tufted cell dendroden-dritic interactions with PG and SA cells.*

4.37.2.2.2   Inhibitory systems in the GL

4.37.2.2.2.(i)   Presynaptic inhibition of ON terminals

4.37.2.2.2.(i).(a)   DA and D2 receptors   The GL contains several hundred thousand DA neurons – PG neurons – but the MOB receives no known extrinsic DA input. In mammals, D1 receptor mRNA is expressed in the GL and GCL (Coronas, V. *et al.*, 1997), while immunocytochemical localization of D1 receptors is faint and primarily in the GCL (Levey, A. I. *et al.*, 1993). By contrast, D1-like binding is present at very low levels in all layers of the MOB with the exception of the ONL (Mansour, A. *et al.*, 1990; Nickell, W. T. *et al.*, 1991; Coronas, V. *et al.*, 1997). The functional significance of D1 receptors in MOB remains unclear. By contrast, only the ONL and GL have high densities of D2 receptors in rats and mice (Mansour, A. *et al.*, 1990; Nickell, W. T. *et al.*, 1991; Levey, A. I. *et al.*, 1993; Coronas, V. *et al.*, 1997; Koster, N. L. *et al.*, 1999) and to a lesser extent in guinea pigs, but not in cats or monkeys (Camps, M. *et al.*, 1990). In the GL, the JG neurons express D2 receptors (Mansour, A. *et al.*, 1990). Some immuno-cytochemical labeling for D2 receptors, as well as *in situ* hybridization, has been reported in the GCL and EPL; however D2-binding sites are consistently restricted to the ONL and GL (Mansour, A. *et al.*, 1990; Levey, A. I. *et al.*, 1993; Coronas, V. *et al.*, 1997). Other anatomical evidence indicates that most, if not all, of the D2 receptors in the GL occur on ON axon terminals. ORNs express D2 receptors and bulbectomy, a manipulation that causes retrograde degeneration of ORNs, eliminates D2 receptor mRNA in the olfactory epithelium (Koster, N. L. *et al.*, 1999). Taken together, these findings indicate that DA released from JG neurons may presynaptically modulate ON terminals via activation of D2 receptors.

In agreement with this, DA and D2 receptor agonists reduced spontaneous and ON-evoked activity in mitral and JC cells (Figure 8), as well as odor-evoked activity in the GL and odor detection performance, in a variety of species (Nowycky, M. C. *et al.*, 1983; Doty, R. L. and Risser, J. M., 1989; Sallaz, M. and Jourdan, F., 1992; Hsia, A. Y. *et al.*, 1999; Wachowiak, M. and Cohen, L. B., 1999; Berkowicz, D. A. and Trombley, P. Q., 2000; Ennis, M. *et al.*, 2001). These effects are mediated by presynaptic suppression of glutamate release from ON terminals via inhibition of Ca^{2+} influx (Wachowiak, M. and Cohen, L. B., 1999). In a similar manner, DA and D2 receptor agonists suppressed spontaneous and ON-evoked activity in JG cells but had no effect on mitral to JG cell transmission (Ennis, M. *et al.*, 2001). The inhibitory effects of DA were abolished in D2 receptor-knockout mice (Ennis, M. *et al.*, 2001). Presynaptic inhibition of ON terminals by DA (and GABA, see below) provides a mechanism for increasing the range of concentrations that can be processed by MOB neurons: as activity increases in ON terminals, DA JG cells are more strongly excited. This, in turn, provides negative feedback onto ON terminals, reducing the release of glutamate. Such a scheme would effectively increase the dynamic range of information transfer from ORNs to MOB neurons. Interestingly, systemic administration of D2 receptor agonists has been reported to prevent odorant-specific 2-dexy-glucose patterns in MOB and to reduce odorant detectability (Doty, R. L. and Risser, J. M., 1989; Sallaz, M. and Jourdan, F., 1992). Related to this question of how DA participates in odor processing is the degree to which these receptors are tonically active *in vivo*? If, for example, ON terminals are tonically inhibited by DA via D2 receptors, this might serve to filter out weak signals (noise). This might sharpen the spatial pattern of active glomeruli and facilitate detection of predominant odors. There is experimental support for this possibility. Blockade of D2-like receptors by systemic administration of spiperone increased the number of mitral cells that responded to single or multiple odorants (Wilson, D. A. and Sullivan, R. M., 1995). One interpretation of this study is that reduced D2 presynaptic inhibition of ON terminals increases the odor responsiveness of mitral cells but does so at the cost of reduced odorant discrimination.

4.37.2.2.2.(i).(b)   GABA and GABA_B receptors $GABA_B$ receptors play a presynaptic inhibitory role similar to that just described for D2 receptors. As noted above, GABAergic PG cells represent a large population of GL interneurons. In the rat MOB, the glomeruli have the highest concentration of $GABA_B$ receptors as determined by radioligand binding (Bowery, N. G. *et al.*, 1987; Chu, D. C. M. *et al.*, 1990) and by immunohistochemical localization of $GABA_B$ receptor subunits (Margeta-Mitrovic, M. *et al.*, 1999). EM immunohistochemistry revealed that the dense labeling in the GL is due to the presence of $GABA_B$ receptors on ON terminals and on the somata of PG cells (Bonino, M. *et al.*, 1999). A variety of imaging

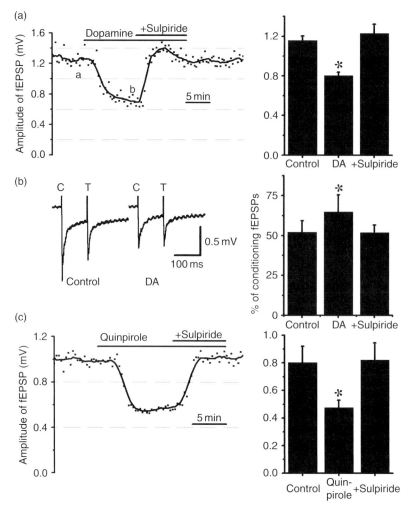

Figure 8 Dopamine (DA) suppresses the olfactory nerve (ON)-evoked field excitatory postsynaptic potential (fEPSP) recorded in the rat glomerular layer (GL). (a) Line graph from a typical experiment showing the suppression of the GL fEPSP by bath application of DA (40 μM). Values represent the peak amplitude ± SE of the fEPSP. DA-induced suppression (30.7%) was fully reversed by the D2 antagonist sulpiride (100 μM). Group data at the right from similar experiments show that DA significantly suppresses ($n = 5$, $^*P = 0.005$) the ON-evoked fEPSP, an effect fully reversed by sulpiride ($n = 4$). (b) Records show responses to paired-pulse stimulation of the ON (100-ms interstimulus intervals) before (Control) and during application of DA (DA); control and DA records correspond to time points a and b indicated in panel a. In control artificial cerebrospinal fluid (ACSF), paired ON shocks produced pronounced paired-pulse depression of the test fEPSP. DA preferentially suppressed the conditioning shock fEPSP and decreased paired-pulse depression. Traces are averages of five sweeps. Bar graph of group data showing the change in paired-pulse responses to ON stimulation; data are expressed as the amplitude of the test fEPSP as a percentage of the conditioning fEPSP. Note that DA significantly reduced paired-pulse depression ($n = 4$, $^*P = 0.01$). (c) Line graph showing the suppression of the GL fEPSP by bath application of quinpirole (100 μM). Sulpiride (100 μM) fully reversed the quinpirole-induced suppression. Group data for similar experiments are shown in the bar graphs to the right ($n = 10$, $^*P < 0.0001$). Reprinted from Ennis, M., Zhou, F. M., Ciombor, K. J., Aroniadou-Anderjaska, V., Hayar, A., Borrelli, E., Zimmer, L. A., Margolis, F., and Shipley, M. T. 2001. Dopamine D2 receptor-mediated presynaptic inhibition of olfactory nerve terminals. J. Neurophysiol. 86, 2986–2997, with permission from The American Physiological Society.

and electrophysiological studies have provided solid evidence that GABA released from PG neurons presynaptically inhibits glutamate release from ON terminals via activation of these GABA_B receptors (Figure 9) (Keller, A. *et al.*, 1998; Wachowiak, M. and Cohen, L. B., 1999; Aroniadou-Anderjaska, V. *et al.*, 2000; Palouzier-Paulignan, B. *et al.*, 2002; Murphy, G. J. *et al.*, 2005). The presynaptic inhibition of glutamate release is mediated by the suppression of Ca^{2+} influx into ON terminals (Wachowiak, M. *et al.*,

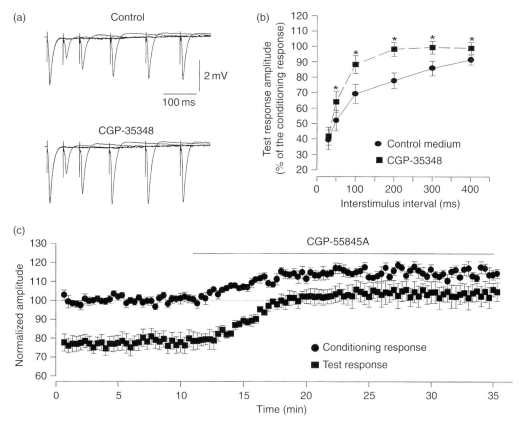

Figure 9 GABA$_B$ receptors on olfactory nerve (ON)terminals are activated both tonically and in response to ON stimulation. GABA$_B$ antagonists increase both conditioning and test responses of mitral/tufted cells to paired-pulse stimulation of the ON, while reducing paired-pulse depression (PPD) of test responses. In all experiments, AP5 is included in the medium. (a) Responses to paired-pulse stimulation of the ON (interstimulus intervals 50, 100, 200, 300, and 400 ms) before and after application of the GABA$_B$ antagonist CGP-35348 (1 mM). The GABA$_B$ antagonist increased the amplitude of the conditioning response and reduced or blocked PPD. Each trace is an average of five sweeps, and five traces are superimposed. (b) Group data ($n = 10$) of the effects of CGP-35348 (500 μM to 1 mM) on PPD of the ON-evoked glomerular field EPSP. The reduction of PPD was statistically significant (*$P < 0.05$) at interstimulus intervals from 50 to 400 ms. (c) Time course of the effects of CGP-55845A (10 μM) on conditioning and test responses (interstimulus interval 100 ms). Group data from six slices. Reprinted from Aroniadou-Anderjaska, V., Zhou, F. M., Priest, C. A., Ennis, M., and Shipley, M. T. 2000. Tonic and synaptically evoked presynaptic inhibition of sensory input to the rat olfactory bulb via GABA(B) heteroreceptors. J. Neurophysiol. 84, 1194–1203, with permission from The American Physiological Society.

2005). Stimulation of individual PG cells has been reported to inhibit, via GABA$_B$ receptors, ON input onto the stimulated cell (Murphy, G. J. *et al.*, 2005). The MOB also contains the highest levels of the putative inhibitory transmitter taurine, exceeding concentrations of GABA and glutamate (Collins, G. G., 1974; Margolis, F. L., 1974; Banay-Schwartz, M. *et al.*, 1989a; 1989b; Ross, C. D. *et al.*, 1995; Kamisaki, Y. *et al.*, 1996). Taurine is found in ON axons, in various neurons, and in astrocytes (Kratskin, I. L. *et al.*, 2000; Kratskin, I. and Belluzzi, O., 2002; Pow, D. V. *et al.*, 2002). In the ON terminals and some postsynaptic dendrites, taurine is colocalized with glutamate (Didier, A. *et al.*, 1994). Observation of spontaneous

taurine release from MOB synaptosomes suggests that taurine may be abundantly released (Kamisaki, Y. *et al.*, 1996). In electrophysiological recordings, taurine directly activated presynaptic GABA$_B$ receptors and inhibited ON terminals, and it also induced Cl$^-$ currents in mitral/tufted cells. Surprisingly, taurine had no direct effect on PG cells (Belluzzi, O. *et al.*, 2004).

4.37.2.2.2.(i).(c) ET and mitral/tufted cell dendro-dendritic interactions with PG and SA cells As noted above, ET and mitral/tufted cells form excitatory glutamatergic dendrodendritic synapses with PG and SA cells in the glomeruli. PG cells, and perhaps SA cells, in turn, form inhibitory GABAergic dendrodendritic synapses with ET and mitral/tufted cells. Although the ET and mitral/tufted cell input is excitatory, this topic is considered here for simplicity. Compared to the dendrodendritic synapses between mitral/tufted and GCs in the EPL, far less is known about the properties of the dendrodendritic synapses in the GL. PG and SA cells were reported to receive spontaneous bursts of glutamatergic EPSCs (Hayar, A. *et al.*, 2004b). Since SA cells and most PG cells do not receive direct ON input, this input must arise from other excitatory elements within the glomeruli. Prime candidates include the apical dendrites of ET cells and mitral tufted cells. Consistent with this, stimulation of the MCL or lateral olfactory tract (LOT) has been used to produce antidromic spikes which propagate into, and trigger glutamate release from, mitral/tufted cell apical dendrites. Such stimulation was reported to excite JG cells in the GL (Ennis, M. *et al.*, 2001). More recently, spontaneous spike bursts or direct depolarization of individual ET cells to elicit spike bursts were reported to trigger EPSC bursts in PG and SA cells mediated by activation of AMPA receptors (Figure 10) (Hayar, A. *et al.*, 2004b). The latency of the evoked EPSCs (0.85 ms) was indicative of monosynaptic input to PG cells from ET cells. Additional studies demonstrated that this ET cell to PG/SA cell excitatory transmission is intraglomerular as it was only observed between cells associated with the same glomerulus. Stimulation of an individual ET cell was reported to produce large EPSCs and Ca^{2+} spikes in PG cells (Murphy, G. J. *et al.*, 2005). The NMDA receptor antagonist 2-amino-5-phosphono-valeric acid (APV) had minimal effects on the EPSP but abolished the Ca^{2+} spike. The same study reported that stimulation of a single ET cell excites two to seven unidentified JG cells within the same glomerulus, suggesting that individual ET cells synapse with multiple PG cells.

Single spikes in ET cells are relatively ineffective in triggering GABA release from PG cells (Murphy, G. J. *et al.*, 2005). Multiple spikes, leading to LVA Ca^{2+} currents, are much more effective in releasing GABA from PG cells (Murphy, G. J. *et al.*, 2005). These findings suggest that GABA release from PG cells may be preferentially triggered when these cells receive strong synaptic input. ET cells, individually and collectively, provide strong input to PG cells. Spike bursts in individual ET cells robustly activate PG cells. Additionally, individual ET cells appear to activate, on average, five PG cells (Murphy, G. J. *et al.*, 2005). Since ET cells spontaneously generate synchronous spike bursts and provide convergent input to a multiple PG cells, they are likely to drive tonic release of GABA from PG cells. By contrast, dendrodendritic input from mitral cells, which do not spontaneously burst *in vitro*, may play a larger role in driving GABA release following sensory input.

PG cells contain GABA, and thus activation of these cells will cause dendrodendritic inhibition of ET and mitral/tufted cell dendrites in the glomeruli. PG cell axons also form symmetrical synapses onto the mitral/tufted cell dendrites and onto PG and SA cells (Pinching, A. J. and Powell, T. P., 1971c). Physiological studies support the notion that PG cells receive excitatory input from mitral/tufted cells and ET cells, and in return, make feedforward and feedback inhibitory GABAergic synapses onto these cells (Shepherd, G. M. and Greer, C. A., 1998; Hayar, A. *et al.*, 2004b). Such inhibition is thought to be primarily mediated by $GABA_A$ receptors. ET cells exhibit spontaneous bursts of IPSCs, indicating that these are tonically inhibited by GABAergic inputs (Figure 11) (Hayar, A. *et al.*, 2005). Spontaneous or ON-evoked EPSCs in ET cells are followed by bursts of IPSCs that are synchronous among ET cells of the same glomerulus. Thus, inhibitory GABAergic input from PG cells may coordinate the activity of ET cells associated with the same glomerulus. Intracellular depolarization of ET cells leads to GABAergic inhibition due to activation of dendrodendritic synapses with PG cells (Murphy, G. J. *et al.*, 2005). Using this paradigm, the feedback inhibition produced by ET cell depolarization was reduced by nimodopine, a blocker of L-type Ca^{2+} channels; nimodipine, however, did not reduce GABA release from PG cells. Subsequent experiments, using paired recordings of PG cells, suggested that GABA exocytosis from these cells is governed primarily by activation of HVA, P/Q-type Ca^{2+} channels. The L-type antagonist nimodipine did not directly alter GABA exocytosis from PG cells, leading to the conclusion that activation of LVA currents can facilitate GABA release from PG cells, but these channels are not directly coupled to GABA exocytosis. These studies and others (Smith, T. C. and Jahr, C. E., 2002) also indicate that PG cells, under certain circumstances, can release GABA onto themselves, and perhaps neighboring PG cells. GABA has been reported to depolarize PG cells at their resting potential, probably due to elevated intracellular chloride concentrations (Siklos, L. *et al.*, 1995; Smith, T. C. and Jahr, C. E., 2002). It was suggested

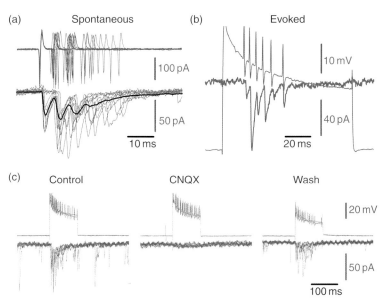

Figure 10 External tufted (ET) cells provide monoexcitatory synaptic input to periglomerular (PG) and short axon (SA) cells. (a) Dual recordings from an ET cell in cell-attached mode (blue traces) and a PG/SA cell in whole-cell voltage-clamp mode (red traces). Left panel (Spontaneous) shows 10 superimposed spontaneous spike burst-triggered traces (triggered on the first spike in the burst); black line is the average of 135 similar traces. Note that spontaneous action currents (blue trace) are accompanied by phase-locked busts of EPSCs (red trace). (b) (Evoked) shows an example of a burst of EPSCs (red trace) evoked by a burst of spikes elicited by extracellular injection of a positive current pulse (700 pA, 100 ms). Note close correspondence between evoked spikes in the ET cell and EPSCs in the SA/PG cell. (d) Recordings from the same two cells shown in panels a and b before (Control), during, and after (Wash) application of CNQX (10 μM). Note that bursts of EPSCs (red traces) in the SA/PG cell elicited by current injection-evoked spikes (blue traces) in the ET cell are completely blocked by CNQX (middle) in a reversible manner (right). CNQX also reversibly abolished spontaneous EPSCs. All panels show five superimposed sweeps. Adapted from Hayar, A., Karnup, S., Ennis, M., and Shipley, M. T. 2004b. External tufted cells: a major excitatory element that coordinates glomerular activity. J. Neurosci. 24, 6676–6685, with permission from The Society for Neuroscience.

that GABA inhibits PG cells by activating a chloride conductance that reduces the neuronal input resistance and shunts excitatory inputs. GABA released from PG cells was recently shown to inhibit other PG cells in the same glomerulus via GABA$_A$ receptors (Murphy, G. J. *et al.*, 2005). Because PG cell dendrites ramify within a restricted portion of a glomerulus, their functional interactions are presumably localized to microdomains of the extensive mitral/tufted cell dendritic arbors, or to nearby JG cells (Kasowski, H. J. *et al.*, 1999).

4.37.2.2.2.(ii) Neuromodulation in the GL
Neuromodulatory systems in the GL include carnosine and certain heavy metals (copper and zinc), and neuromodulatory inputs from centrifugal afferents. Centrifugal inputs are discussed below (see Neurophysiology of Neuromodulatory Inputs to Main Olfactory Bulb). Carnosine, a dipeptide synthesized by ORNs, is localized in ON terminals in the GL and fulfills many criteria for neurotransmitter candidacy. However, no direct postsynaptic actions of carnosine have been revealed to date (MacLeod, N. K. and Straughan, D. W., 1979; Nicoll, R. A. and Alger, B. E., 1980; Frosch, M. P. and Dichter, M. A., 1982; Trombley, P. Q. *et al.*, 1998). Carnosine did not affect currents evoked by glutamate, GABA, or glycine in cultured MOB neurons (Trombley, P. Q. *et al.*, 1998). Because carnosine is a chelator of both zinc and copper, it has been suggested that it might modulate transmission at ON synapses by regulating zinc and copper. Depending on the concentration, both zinc and copper can augment or block responses mediated by NMDA and GABA receptors. Both zinc and copper inhibit *N*-methyl D-aspartate (NMDA) and gamma-aminobutyric acid (GABA) receptor-mediated currents and synaptic transmission in MOB neurons (Trombley, P. Q. and Shepherd, G. M., 1996; Trombley, P. Q. *et al.*, 1998). Carnosine prevented the actions of copper and reduced the effects of zinc. These results suggest that carnosine may indirectly influence glutamate actions on MOB

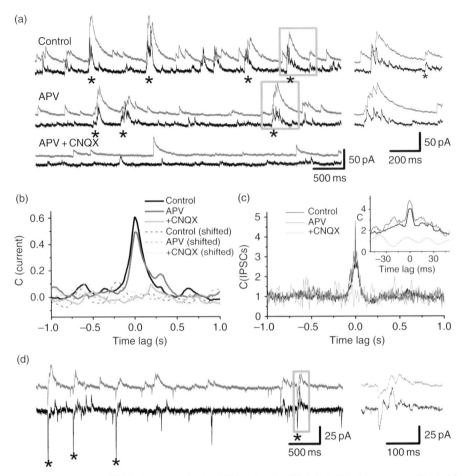

Figure 11 Synchronous bursts of IPSCs in external tufted (ET) cell pairs. All data in this figure were obtained from the same ET cell pair. (a) Simultaneous whole-cell voltage-clamp recordings (holding potential = 0 mV) from two ET cells (gray and black traces, respectively) show synchronous IPSC bursts (asterisks) before and during application of AP5 (50 μM). Note that although AP5 reduced burst frequency, bursts remained synchronous in the two cells. Additional application of CNQX eliminated the IPSC burst synchrony. Areas highlighted in rectangles are shown at faster timescale at right. (b) Membrane current cross-correlograms (50 s recording samples, 2 ms bins) in control, APV, and after additional application of CNQX. Note the significant peak at zero lag time in control and APV. There was no significant correlation after application of CNQX. The 99.73% confidence limit was determined by cross-correlating traces shifted by 5 s (dashed traces). (c) Cross-correlogram of the IPSC trains (5 min recording samples, 1 ms bins) show a significant narrow peak (see inset) at zero time lag, indicating synchronous IPSCs in the two recorded cells. The narrow peak was superimposed on a broader peak corresponding to the longer duration correlated IPSC bursts. CNQX abolished correlation and synchrony of IPSCs. (d) Simultaneous whole-cell voltage-clamp recordings (holding potential = −30 mV) from the same two ET cells showing synchronous EPSCs followed by IPSC bursts; recording highlighted in rectangle is shown at faster timescale at right. Recordings obtained before the application of synaptic blockers (CNQX, AP5, gabazine) in (a). Adapted from Hayar, A., Shipley, M. T., and Ennis, M. 2005. Olfactory bulb external tufted cells are synchronized by multiple intraglomerular mechanisms. J. Neurosci. 25, 8197–8208, with permission from The Society for Neuroscience.

neurons by modulating the effects of synaptically released zinc and copper. What roles might carnosine, zinc, and copper play in olfactory processing? At present, there are no clear answers to this question. Is carnosine, like other peptide transmitters, preferentially released by high frequencies of ON activity? If so, carnosine may be preferentially released during intense odor stimulation. Zinc is known to be preferentially released during high-frequency neural activity. There are neurotoxic effects of zinc and copper, and Trombley (Horning, M. S. et al., 2000) has speculated that carnosine, by preventing the actions of these metals, may serve an important neuroprotective function, perhaps to protect MOB neurons during intense or high-frequency activity.

4.37.2.3 Electrophysiology of Interglomerular Circuitry

Interglomerular interactions are poorly understood. As noted above, SA cells frequently have multiple dendrites (three to five) that extend over several (two to four) glomeruli (Aungst, J. L. *et al.*, 2003). Thus, these cells may receive odor information from multiple glomeruli. These cells do not receive direct ON input, but instead appear to receive olfactory input indirectly from ET and mitral/tufted cells (Hayar, A. *et al.*, 2004b). In turn, SA cells give rise to an axon that extends for considerable distances (~0.85 mm) in the GL (Aungst, J. L. *et al.*, 2003). Imaging studies suggest that the SA cells may be involved in lateral inhibition between glomeruli. Stimulation of the GL was reported to produce monosynaptic excitation of most ET and PG cells neurons in neighboring glomeruli mediated by activation of ionotropic glutamate receptors (Aungst, J. L. *et al.*, 2003). This suggests the intriguing possibility that SA cells are excitatory. Further, a prepulse to one glomerulus inhibited ON-evoked excitation of mitral cells in another glomerulus. It was concluded that the net functional effect of SA cell activation is to excite PG cells in other glomeruli, causing the release of GABA and inhibition of mitral cell responses to ON input. Taken together, these findings suggest that SA cells represent a major neural element for interglomerular interactions and that these cells may provide a substrate for center-surround inhibition – a common mechanism to enhance stimulus contrast in sensory circuits.

4.37.3 Neurophysiology of External Plexiform Layer

4.37.3.1 Tufted Cells

Tufted cells are the most numerous cells of the EPL. Three subclasses of tufted cells are recognized based on location in the EPL: superficial, middle, and deep tufted cells (Cajal, R. S. Y., 1890; Pinching, A. J. and Powell, T. P., 1971a; Macrides, F. and Schneider, S. P., 1982; Orona, E. *et al.*, 1984). Like ET cells and mitral cells, tufted cells of the mammalian EPL have one (or rarely several) apical dendrite(s) ending in a glomerular tuft (Cajal, R. S. Y., 1890). Like mitral cells, their apical dendrites receive ON input, and they form reciprocal dendrodendritic synapses with the dendrites of PG and SA cells. Their lateral dendrites ramify in the EPL and form dendrodendritic

synapses with GCs and other inhibitory interneurons within the EPL. The neurophysiology of dendrodendritic interactions between mitral/tufted cells and GCs in the EPL is discussed below. Axonal projections of middle and deep tufted cells are similar, but not identical, to those of mitral cells (Schoenfeld, T. A. and Macrides, F., 1984; Schoenfeld, T. A. *et al.*, 1985). The local axon collaterals of tufted cells course mainly in the IPL. The projections of tufted cells beyond the MOB terminate densely in the anterior olfactory nucleus and, to a lesser extent, in other rostral olfactory cortical structures (Schoenfeld, T. A. *et al.*, 1985; Scott, J. W., 1986). Few tufted cell axons project into more caudal POC regions.

Tufted cells utilize glutamate as their principle transmitter (Liu, C. J. *et al.*, 1989; Christie, J. M. *et al.*, 2001), but they also stain for a number of other substances, including cholecystokinin (CCK). *In vivo* electrophysiological recordings have shown that the sensitivity of tufted cell subtypes to ON stimulation is correlated with the depth of the lateral dendrites within the EPL. The more superficial tufted cells are more easily excited by ON stimulation than the deeper tufted cells (Schneider, S. P. and Scott, J. W., 1983; Wellis, D. P. *et al.*, 1989; Ezeh, P. I. *et al.*, 1993). In response to odor stimulation, superficially located tufted cells also exhibit more prolonged excitation than mitral cells (Luo, M. and Katz, L., 2001). More recent studies have reported that tufted cells can be excited for up to several minutes following odorant presentation (Luo, M. and Katz, L., 2001). Thus, anatomically distinct tufted cell subtypes appear to be functionally distinct. In slices, tufted cells have been observed to exhibit spontaneous and ON-evoked bursting activity somewhat similar to that of ET cells (Hamilton, K. A. *et al.*, 2005). The extent to which such bursting varies among different tufted cell subtypes, as well as the mechanisms generating bursting, is unknown.

Many superficial tufted cells contain CCK (Seroogy, K. B. *et al.*, 1985), and their axons give rise to a reciprocal network, the intrabulbar association system (IAS) that connects lateral and medial regions of each MOB (Schoenfeld, T. A. *et al.*, 1985). The axons of the IAS travel to the opposite side of the bulb and terminate on the apical dendrites of GCs coursing through the IPL en route to the EPL (Liu, W. L. and Shipley, M. T., 1994). The IAS projection exhibits a high degree of point-to-point specificity. The IAS projections from tufted cells associated with a single glomerulus preferentially target the IPL deep to the second glomerulus on the other side of MOB

(Belluscio, L. *et al.*, 2002; Lodovichi, C. *et al.*, 2003). Thus, the IAS appears to modulate the activities of subsets of MOB neurons receiving input from ORNs expressing the same receptor, on opposite sides of the MOB. CCK typically is excitatory, and therefore the IAS may depolarize GCs.

4.37.3.2 EPL Interneurons

The EPL contains anatomically and neurochemically heterogeneous subtypes of intrinsic interneurons (for review, see Ennis, M. *et al.*, 2007). Many EPL interneurons stain for GABA and are therefore presumed to be inhibitory (Mugnaini, E. *et al.*, 1984; Gall, C. M. *et al.*, 1987; Kosaka, T. *et al.*, 1987c; Ohm, T. G. *et al.*, 1990). EPL interneuron subtypes are neurochemically diverse and stain for a number of neurotransmitter markers, including NADPH diaphorase (Scott, J. W. *et al.*, 1987; Villalba, R. M. *et al.*, 1989; Alonso, J. R. *et al.*, 1995), substance P (Baker, H., 1986; Wahle, P. *et al.*, 1990), enkephalin (Bogan, N. *et al.*, 1982; Davis, B. J. *et al.*, 1982), NPY (Gall, C. *et al.*, 1986; Scott, J. W. *et al.*, 1987; Sanides-Kohlrausch, C. and Wahle, P., 1990a), neurotensin and somatostatin (Matsutani, S. *et al.*, 1988), and VIP (Gall, C. *et al.*, 1986; López-Mascaraque, L. *et al.*, 1989; Sanides-Kohlrausch, C. and Wahle, P., 1990b; Nakajima, T. *et al.*, 1996). The role of these transmitters in the function of EPL interneurons is not known. EPL interneurons are thought to interact chiefly with mitral/tufted cell dendrites (López-Mascaraque, L. *et al.*, 1990; Nagayama, S. *et al.*, 2004) and possibly GC dendrites (Schneider, S. P. and Macrides, F., 1978). Combined morphological and electrophysiological analyses showed that GABAergic EPL interneurons with highly varicose dendrites are excited polysynaptically by ON stimulation, most likely via input from nearby mitral/tufted cells (Hamilton, K. A. *et al.*, 2005). The interneurons exhibit high levels of spontaneous glutamatergic synaptic activity mediated by AMPA/kainate receptors, consistent with GluR1 AMPA receptor subunit staining in these cells (Petralia, R. S. and Wenthold, R. J., 1992; Giustetto, M. *et al.*, 1997; Montague, A. A. and Greer, C. A., 1999; Hamilton, K. A. and Coppola, D. M., 2003). Many EPL interneurons have dendrites that spanned several adjacent glomeruli, suggesting that they may provide localized inhibition of mitral/tufted cells that are topographically related to overlying glomeruli (Hamilton, K. A. *et al.*, 2005).

4.37.4 Neurophysiology of Mitral Cells

4.37.4.1 Anatomical Features

Deep to the EPL is the MCL, a thin layer that contains the somata of mitral cells as well as numerous GCs (Cajal, R. S. Y., 1911a; 1911b). In fact, there are ~40,000 mitral cells (Meisami, E., 1989) and ~100,000 GCs in the MCL (Frazier, L. L. and Brunjes, P. C., 1988). Thus, mitral cells comprise less than one-half of the cells in the MCL. Together with tufted cells, mitral cells are the major class of output cells of the bulb. As noted above, mitral cells extend a single apical dendrite into one glomerulus. There are about 25 mitral cells (and 50 tufted cells) associated with a single glomerulus (Cajal, R. S. Y., 1911a; 1911b; Allison, A., 1953). The apical dendrites are synaptically contacted by ON terminals (Price, J. L. and Powell, T. P. S., 1970a; Shepherd, G. M., 1972). The secondary or lateral dendrites of mitral cells ramify in the EPL where they may extend up to 1–2 mm. These lateral dendrites participate in dendrodendritic synapses with dendrites of GCs, as reviewed below (see Dendrodendritic Transmission Between Mitral/Tufted Cells and Granule Cells). Mitral cells have been subdivided into two classes, Type I and Type II, based on extension of dendrites into the deep or superficial parts of the EPL, respectively (Orona, E. *et al.*, 1984). Type II mitral cells are more easily excited by ON stimulation than Type I mitral cells (Schneider, S. P. and Scott, J. W., 1983; Wellis, D. P. *et al.*, 1989; Ezeh, P. I. *et al.*, 1993). Mitral cell lateral dendrites receive centrifugal axon inputs and inputs from EPL interneurons (Jackowski, A. *et al.*, 1978; Toida, K. *et al.*, 1996). Mitral cells give off axon collaterals, which terminate within the bulb, in the IPL and GCL (Price, J. L. and Powell, T. P. S., 1970c; Mori, K. *et al.*, 1983), or exit the MOB and innervate a number of olfactory-related brain regions collectively known as the POC. Mitral (and tufted) cells are glutamatergic, and the neuropeptide corticotropin-releasing factor (CRF) has been demonstrated in mitral and some tufted cells. Substance P has been detected in approximately one-half of mitral cells by in situ hybridization (Warden, M. K. and Young, W. S., 1988), but they do not stain for substance P immunocytochemistry (Inagaki, S. *et al.*, 1982; Shults, C. W. *et al.*, 1984; Baker, H., 1986).

4.37.4.2 Spontaneous Discharge and Intrinsic Membrane Properties

In vivo extracellular recording studies in anesthetized animals indicate that most mitral cells fire at fairly high spontaneous rates (6–30 Hz; mean rate, ~18 Hz) (Chaput, M. and Holley, A., 1979; Chaput, M., 1983; Yu, G.-Z. *et al.*, 1993; Jiang, M. R. *et al.*, 1996). They also exhibit periodic, long-duration (several minutes) tonic increases and decreases in firing rates that may be related to anesthesia (Yu, G.-Z. *et al.*, 1993; Jiang, M. R. *et al.*, 1996); mitral cells are less responsive to ON during the periods of reduced spontaneous activity. In awake rats, mitral cells firing rates range from 1 to 33 Hz with a mean rate of 12 Hz (Kay, L. M. and Laurent, G., 1999). A detailed consideration of mitral cell odor responses, receptive fields, and plasticity is beyond the scope of this chapter, and readers are referred to recent work in this area (Kashiwadani, H. *et al.*, 1999; Kay, L. M. and Laurent, G., 1999; Luo, M. and Katz, L., 2001; Fletcher, M. L. and Wilson, D. A., 2003; Nagayama, S. *et al.*, 2004; Mori, K. *et al.*, 2006).

Mitral cell spontaneous firing patterns and odor responses *in vivo* are modulated as a function of the respiratory cycle. Recent patch clamping studies in anesthetized rodents has provided new information on properties of mitral spontaneous and odor-evoked discharge. Mitral cells exhibit subthreshold membrane potential oscillations or action potentials synchronized to or modulated by respiration, which occurs in the theta frequency range, 1–12 Hz (Philpot, B. D. *et al.*, 1997; Chalansonnet, M. and Chaput, M. A., 1998; Kay, L. M. and Laurent, G., 1999; Charpak, S. *et al.*, 2001; Luo, M. and Katz, L., 2001; Debarbieux, F. *et al.*, 2003; Margrie, T. W. and Schaefer, A. T., 2003). The peak depolarization of the membrane potential oscillation coincides with the peak of the inhalation phase (Margrie, T. W. and Schaefer, A. T., 2003). The rhythmic membrane potential oscillations are blocked by glutamate receptor antagonists, demonstrating that they are synaptically mediated (Margrie, T. W. and Schaefer, A. T., 2003). Combined intrinsic imaging of glomerular odor responses and intracellular recordings from mitral cells suggest that mitral cells recorded below odor responsive glomeruli are excited by the odor, whereas those cells distant from the activated glomeruli show no response or are inhibited (Luo, M. and Katz, L., 2001). In responses to excitatory odors, mitral cells typically generate EPSPs and spikes that are synchronized to the inspiratory phase of respiration which occurs in the theta frequency range (Charpak, S. *et al.*, 2001; Margrie, T. W. *et al.*, 2001; Cang, J. and Isaacson, J. S., 2003; Debarbieux, F. *et al.*, 2003; Margrie, T. W. and Schaefer, A. T., 2003). EPSP amplitudes vary proportionately with the strength of the odor stimulus, and spikes are often launched from the rising phase of odor-evoked EPSPs (Cang, J. and Isaacson, J. S., 2003; Margrie, T. W. and Schaefer, A. T., 2003). The latency of the first spike following odor stimuli is inversely proportional to the number of spikes per respiratory cycle such that shorter latencies are observed when spike bursts are elicited (Cang, J. and Isaacson, J. S., 2003; Margrie, T. W. and Schaefer, A. T., 2003). This suggests that the initial spike latency may provide information about the strength of sensory input; that is, the concentration of odors. Sufficiently strong odor stimulation triggers a spike burst, with spikes in the burst occurring within the β–γ frequency range (20–60 Hz) (Debarbieux, F. *et al.*, 2003). The interburst spike frequency is constant across bursts that vary in the number of spikes (Margrie, T. W. and Schaefer, A. T., 2003). This suggests that the instantaneous spike frequency probably is not modulated by network-driven factors. Mitral cells also exhibit inhibitory responses to odors (Luo, M. and Katz, L., 2001; Margrie, T. W. *et al.*, 2001; Cang, J. and Isaacson, J. S., 2003). The strength of odor-evoked inhibitory postsynaptic potentials (IPSPs) does not vary with the concentration of the odor stimulus (Cang, J. and Isaacson, J. S., 2003). However, the IPSP amplitudes are larger during the initial respiratory cycle compared to subsequent cycles in the presence of an odorant.

In slices, mitral cell spontaneous discharge is more modest (1–6 Hz, mean, 3 Hz), presumably because tonic sensory input is absent (Ennis, M. *et al.*, 1996, Ciombor, K. J. *et al.*, 1999; Heyward, P. M. *et al.*, 2001). Mitral cell spontaneous firing is intrinsically generated and persists in the presence of blockers of ionotropic glutamate and GABA receptors (Ennis, M. *et al.*, 1996; Heyward, P. M. *et al.*, 2001). Patch clamp recordings have revealed that mitral cells exhibit intrinsic membrane bistability (Heyward, P. M. *et al.*, 2001; Heinbockel, T. *et al.*, 2004). They spontaneously alternate between a perithreshold upstate membrane potential and a more hyperpolarized (approximately −10 mV) downstate membrane potential (Figure 13). Bistability is typically encountered in mitral cells recorded in the depth of the slice (>70 μm from the surface) and may be absent in superficially located cells. Spontaneous or ON-evoked spikes are readily launched from the upstate, and a spike afterhyperpolarization or a hyperpolarization induced by an IPSP or current injection of sufficient amplitude resets the cell to the downstate.

In the downstate, more robust ON input is necessary to trigger spikes. The upstate is also characterized by high-frequency (10–50 Hz) subthreshold membrane potential oscillations that appear to be mediated by a balance between opposing K^+ currents and TTX-sensitive Na^+ currents (Desmaisons, D. *et al.*, 1999; Heyward, P. M. *et al.*, 2001; Balu, R. *et al.*, 2004). Inhibitory synaptic input (e.g., IPSPs) can reset the phase of the oscillations and the timing of mitral cell spikes (Desmaisons, D. *et al.*, 1999). The downstate appears to be maintained by slowly inactivating outward currents (e.g., K^+ currents), and the transition from the downstate to the upstate is due to inactivation of the outward currents and activation of regenerative voltage-dependent inward currents (e.g., TTX-sensitive, persistent Na^+ currents). The transition from the upstate to the downstate may reflect voltage-dependent inactivation of inward currents and activation of outward currents initiated by action potentials. Other aspects of mitral cell spontaneous activity are discussed below in Oscillations and Synchrony in Main Olfactory Bulb.

Further studies indicated that mitral cell-evoked activity is regulated by interactions between subthreshold TTX-sensitive Na^+ currents and by 4-amonipyridine (4-AP)-sensitive K^+ currents, such as transient outward or I_D currents (Balu, R. *et al.*, 2004). In response to sustained depolarization pulses, mitral cells generate intermittent spike clusters at 20–40 Hz, with clusters occurring with variable timing at theta frequencies (1–5 Hz). Brief repolarizations to recover inactivated Na^+ currents during sustained depolarizing pulses eliminated spike variability. Brief depolarizing pulses or simulated EPSPs mimicking rhythmic ON input during sniffing elicited highly precise spike clusters. The first EPSP frequently failed to elicit spikes, but decreased the spike threshold for subsequent EPSPs. 4-AP caused spikes to be triggered by the first EPSP but impaired precise spike timing to subsequent EPSPs. 4-AP also abolished intermittent or clustered firing during sustained depolarizations. Taken together, these findings suggest that intrinsic properties of mitral cells yield variable spiking responses to sustained depolarizations but allow temporally precise or phase-locked spiking responses to brief phasic input. Based on this, it has been suggested that mitral cells may be functionally analogous to high-pass filters, preferentially responding to phasic events that occur at theta frequencies (Balu, R. *et al.*, 2004).

Mitral cells also exhibit afterhyperpolarizations mediated by Ca^{2+}-dependent K^+ conductances

(Maher, B. J. and Westbrook, G. L., 2005). Step depolarization of mitral cells to $+10\,mV$ elicits outward currents that are attenuated by Cd^{2+} or the SK K^+ channel antagonist apamin. This current was reported to be absent in GCs. The SK current could be activated by Ca^{2+} influx via NMDA receptors or voltage-dependent Ca^{2+} channels. Apamin application increased the depolarization-evoked firing frequency of mitral cells. Other studies indicate that mitral cells lack, or have very weak, H-type currents (Djurisic, M. *et al.*, 2004). Mitral cells also express Kv1.3, a rapidly activating, moderately slow inactivating type of voltage-gated K^+ channels (Fadool, D. A. *et al.*, 2004). Mitral cells from Kv1.3-knockout mice have more depolarized resting potentials and smaller, broader spikes in comparison with those from wildtype mice.

4.37.4.3 Dendritic Spike Propagation

In vivo, odors have been reported to elicit fast prepotentials in mitral cells thought to represent truncated spikes generated in the apical dendrites (see Mori, K., 1987 for review). Similar fast prepotentials have been observed in slices (Chen, W. R. and Shepherd, G. M., 1997). More recent findings in MOB slices demonstrate that both the apical and lateral dendrites of mitral cells actively propagate action potentials. Spike elicited at the soma backpropagate nondecrementally along the apical dendrite and vice versa (Bischofberger, J. and Jonas, P., 1997; Chen, W. R. *et al.*, 1997; Isaacson, J. S. and Strowbridge, B. W., 1998; Debarbieux, F. *et al.*, 2003). The ability of the apical dendrite to generate spikes is due to the presence of TTX-sensitive Na^+ channels (Bischofberger, J. and Jonas, P., 1997). These studies suggest that with sufficiently strong ON input, spikes can be initiated in the apical dendrite and conducted to the soma without decrement. However, the site of spike initiation can be controlled by inhibitory inputs to the soma. With leak levels of ON stimulation, spikes are preferentially triggered at the soma, but this can be blocked by IPSPs at the soma or proximal segments of the lateral dendrite (Chen, W. R. *et al.*, 1997; Djurisic, M. *et al.*, 2004). With stronger levels of ON input, inhibitory inputs to the soma shifted the spike initiation site to the apical dendrite. ON-evoked EPSPs by contrast are decremental, and they decrease in amplitude by ~30% over a 300-µm distance in the apical dendrite (Djurisic, M. *et al.*, 2004).

Similar studies demonstrate actively propagating Na^+ spikes in mitral cell lateral dendrites

(Margrie, T. W. *et al.*, 2001; Xiong, W. and Chen, W. R., 2002). This suggests that spikes in single mitral cells, via activation of mitral to granule synapses, could conceivably inhibit mitral cells at considerable distances throughout the MOB. However, whether the spike propagation is nondecremental along the full length of the lateral dendrite is unclear as propagation has been reported to be attenuating (Margrie, T. W. *et al.*, 2001; Lowe, G., 2002; Christie, J. M. and Westbrook, G. L., 2003) or nonattenuating (Margrie, T. W. *et al.*, 2001; Xiong, W. and Chen, W. R., 2002; Debarbieux, F. *et al.*, 2003). Spike attenuation was reported to be caused by activation of A-type K^+ channels in the lateral dendrites (Christie, J. M. and Westbrook, G. L., 2003). Regenerative Ca^{2+} currents do not appear to play a role in dendritic spikes (Charpak, S. *et al.*, 2001; Margrie, T. W. *et al.*, 2001; Xiong, W. and Chen, W. R., 2002; Christie, J. M. and Westbrook, G. L., 2003). Local synaptic inhibitory input (GABAergic IPSPs) can block propagation of spikes in the lateral dendrite (Margrie, T. W. *et al.*, 2001; Lowe, G., 2002; Xiong, W. and Chen, W. R., 2002). Such inhibition is more influential when it occurs at the soma and/or proximal dendrites. Thus, inhibitory input to the soma and lateral dendrites may spatially constrain the range and magnitude of lateral inhibition (see also Dendrodendritic Inhibition).

4.37.4.4 Modulation by mGluRs and DA

In dissociated cultured rat and frog MOB neuronal preparations, Group I mGluRs increased Ca^{2+} release from internal stores in mitral/tufted cells as well as in MOB interneurons (Geiling, H. and Schild, D., 1996; Carlson, G. C. *et al.*, 1997), or it depolarized and increased the frequency of miniature excitatory postsynaptic currents in mitral cells (Schoppa, N. E. and Westbrook, G. L., 1997). Other studies indicate that activation of Group III mGluRs with AP4 inhibits Ca^{2+} currents in mitral cells and presynaptically decreases mitral cell to GC synaptic transmission (Trombley, P. Q. and Westbrook, G. L., 1992). More recent studies in rat and mouse MOB slices demonstrate that activation of mGluR1 directly depolarizes and increases the firing of MCs and that these effects persist in the presence of blockers of fast synaptic transmission (Figure 12) (Heinbockel, T. *et al.*, 2004). The same study showed that mGluR1 induces a voltage-dependent inward current consisting of multiple components. mGluR1 antagonists also altered mitral cell membrane potential bistability, increasing

the duration of the upstates and downstates, and substantially attenuated ON-evoked spikes (Figure 13). These findings suggest that endogenous glutamate tonically modulates MC excitability and responsiveness to ON input via activation of mGluR1. Although DA has no direct effect on resting membrane properties of mitral cells, pharmacological activation of D2 receptors was found to reduce glutamate release from mitral cells onto olfactory bulb interneurons in culture via inhibition of N and/or P/Q HVA Ca^{2+} channels (Davila, N. G. *et al.*, 2003).

4.37.5 Neurophysiology of Neurons in the Granule Cell Layer

4.37.5.1 Neuron Types of the GCL

GCs are axon-less cells with small cell bodies that are mostly tightly packed into row-like aggregates of three to nine somata in the GCL (Reyher, C. K. *et al.*, 1991). More superficially located GCs are also found mixed with mitral cell bodies within the MCL. Most GCs have an apical dendrite that ramifies within the EPL and shorter basal dendrites that ramify within the GCL. A differential sublaminar distribution has been observed for the dendrites of superficial and deep GCs. The apical dendrites of superficial GCs have very dense spines, and they terminate within both the superficial and deep portions of the EPL. By contrast, the apical dendrites of deeper GCs terminate preferentially within the deep EPL (Orona, E. *et al.*, 1983; Greer, C. A., 1987). Most GCs contain GABA (Ribak, C. E. *et al.*, 1977), but some contain enkephalin (Bogan, N. *et al.*, 1982; Davis, B. J. *et al.*, 1982). GC apical dendrites receive asymmetrical synapses from, and make symmetrical synapses onto, mitral/tufted cells. GCs also receive asymmetrical synapses from a variety of centrifugal afferents, including inputs from neuromodulatory transmitter systems (see Neurophysiology of Neuromodulatory Inputs to Main Olfactory Bulb) and POC (see Neurophysiology of Primary Olfactory Cortical Inputs to Main Olfactory Bulb) (Price, J. L. and Powell, T. P. S., 1970c). Centrifugal fibers arise from neurons in POC (e.g., piriform cortex (PC) and anterior olfactory nucleus) and comprise the bulk of synaptic contacts onto GC somata and proximal dendrites within the GCL (Price, J. L. and Powell, T. P. S., 1970c). GCs also receive synapses from the collateral branches of mitral and tufted cell axons, as well as inputs from

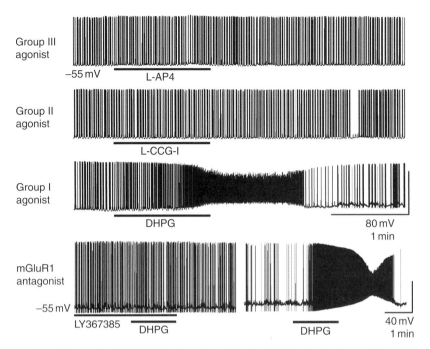

Figure 12 The group I mGluR agonist (RS)-3,5-dihydroxyphenylglycine (DHPG) depolarizes and increases the firing of mitral cells. Mitral cells were activated by group I, but not group II or III, mGluR agonists. Bath application of the group III mGluR agonist L(+)-2-amino-4-phosphonobutyric acid (AP4; 100 μM) or the group II mGluR agonist (2S,3S,4S)-CCG/(2S,1'S,2'S)-2-(carboxycyclopropyl)glycine (L-CCG-I; 20 μM) did not alter the membrane potential or firing rate of this mouse mitral cell (MC) in current clamp recordings. The same cell was activated by the selective group I agonist DHPG. The selective mGluR1 antagonist ({alpha}S)-{alpha}-amino-{alpha}-[(1S,2S)-2-carboxycyclopropyl]-9H-xanthine-9-propanoic acid (LY367385, 100 μM) blocked the actions of the group I mGluR agonist DHPG (50 μM) in another mouse MC. After washout (>15 min), reapplication of DHPG robustly depolarized and increased MC spontaneous discharge. All experiments were performed in the presence of blockers of fast synaptic transmission: CNQX (10 μM), APV (50 μM), and gabazine (5 μM). Reprinted from Heinbockel, T., Heyward, P., Conquet, F., and Ennis, M. 2004. Regulation of main olfactory bulb mitral cell excitability by metabotropic glutamate receptor mGluR1. J. Neurophysiol. 92, 3085–3096, with permission from The American Physiological Society.

Golgi, Cajal, and Blanes cells (Price, J. L. and Powell, T. P. S., 1970c).

The IPL and GCL also contain several interneuron subtypes (short-axon cells; Golgi, Cajal, and Blanes cells) that have dendrites and axons that ramify within the EPL, MCL, and GCL (Price, J. L. and Powell, T. P. S., 1970b; Schneider, S. P. and Macrides, F., 1978; Cajal, R. S. Y., 1890; Van Gehuchten, L. E. and Martin, A., 1891; Blanes, T., 1898; López-Mascaraque, L. *et al.*, 1990). These cells stain for a number of transmitters, including GABA, VIP, NPY, enkephalin, and soma-tostatin (see Ennis, M. *et al.*, 2007 for review). With the exception of Blanes cells, very little is known about the functions of the deep interneurons, but all are presumed to be inhibitory. Blanes cells have numerous dendrites emerging from all sides of the soma. The axons from these cells can extend consid-erable distances, but they typically remain within the GCL (Cajal, R. S. Y., 1911a; 1911b; Pressler, R. T. and Strowbridge, B. W., 2006). The anatomical

features of other interneurons in the IPL/GCL are reviewed elsewhere (Ennis, M. *et al.*, 2007).

4.37.5.2 Neurophysiology of GCs

In vivo and in slice preparations, GCs spontaneous spiking is relatively infrequent, probably due to their relatively hyperpolarized resting potential (−65 to −75 mV) and appears to be driven primarily by spontaneous glutamatergic input (Wellis, D. P. and Scott, J. W., 1990; Cang, J. and Isaacson, J. S., 2003; Zelles, T. *et al.*, 2006; Heinbockel, T. *et al.*, 2007). Synaptic input or direct depolarization can elicit Na^+ and Ca^{2+} spikes in GCs (Halabisky, B. *et al.*, 2000; Pinato, G. and Midtgaard, J., 2003; 2005; Egger, V. *et al.*, 2005). By contrast to mitral cells, odor-evoked responses in GCs attenuate rapidly after the first respiratory cycle (Cang, J. and Isaacson, J. S., 2003).

GCs express Ca^{2+} currents which activate at appro-ximately −60 mV and peak at 0–5 mV (Chen, W. R.

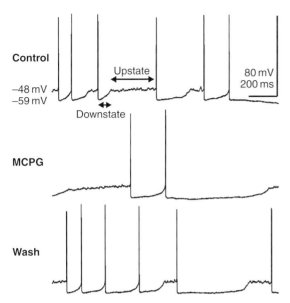

Figure 13 Activation and blockade of mGluRs modulates mitral cell (MC) spontaneous activity and membrane bistability. Upper trace: Current clamp recording showing membrane bistability in a rat MC. Middle trace: Bath application of MCPG (500 μM) reduced the firing frequency of the MC and prolonged the upstates and downstates. Bottom trace: the effects of MCPG were reversible on washout. Reprinted from Heinbockel, T., Heyward, P., Conquet, F., and Ennis, M. 2004. Regulation of main olfactory bulb mitral cell excitability by metabotropic glutamate receptor mGluR1. J. Neurophysiol. 92, 3085–3096, with permission from The American Physiological Society.

et al., 2000; Isaacson, J. S. and Vitten, H., 2003). These currents involve both LVA T-type Ca^{2+} currents and HVA Ca^{2+} currents (Isaacson, J. S. and Vitten, H., 2003). T-type currents can be activated by depolarization subthreshold for spike initiation and have been linked to GABA release from GCs (Egger, V. *et al.*, 2003). Small subthreshold EPSPs seem to produce Ca^{2+} transients restricted to dendritic spines (Egger, V. *et al.*, 2005). These Ca^{2+} transients appear to involve several sources, including LVA and HVA Ca^{2+} channels, NMDA receptors, and Ca^{2+}-induced Ca^{2+} release (Egger, V. *et al*, 2005). Sufficiently large EPSPs can trigger an all-or-none LTS that propagates throughout the GC dendrites (Egger, V. *et al.*, 2005). LTSs in GC soma have been described in amphibians (Pinato, G. and Midtgaard, J., 2003; 2005).

GC activity is strongly regulated by several K$^+$ conductances. They have a strong transient A-type K$^+$ current, I$_A$ (Schoppa, N. E. and Westbrook, G. L., 1999). I$_A$ in these cells activates at approximately −44 mV, at or near the threshold for spike generation (−47 mV). Many I$_A$ channels are only half-inactivated at GC resting membrane potential, and thus are available to affect depolarizing inputs (Schoppa, N. E. and Westbrook, G. L., 1999). Blockade of I$_A$ with 4-AP decreases the lag or delay for evoked spikes. The I$_A$ channels are expressed in the distal dendrites of GCs where they tend to counter brief depolarizing synaptic inputs. BK-type or high-conductance Ca^{2+}-activated K$^+$ currents involved in spike repolarization of afterhyperpolarization are engaged in GCs by strong depolarization or activation of NMDA receptors (Isaacson, J. S. and Murphy, G. J. 2001). Anatomically, gap junctions have been reported among GCs (Reyher, C. K. *et al.*, 1991), although electrophysiological studies found no evidence for electrotonic coupling (Schoppa, N. E., 2006). GCs in the GCL are potently activated by mGluR1 agonists, which depolarize and increase the firing rate of these cells (Heinbockel, T. *et al.*, 2007). By contrast, mGluR1 antagonists reduce mitral/tufted cell-evoked excitation of GCs.

4.37.5.3 Neurophysiology of Blanes Cells

Blanes cells are neurophysiologically distinct from GCs (Pressler, R. T. and Strowbridge, B. W., 2006). They have lower input resistance, more depolarized resting potential, and more hyperpolarized spike threshold than GCs. Stimulation of Blanes cells elicits monosynaptic GABAergic IPSPs in GCs, but Blanes cells do not receive reciprocal input from GCs. These cells generate prominent afterdepolarizations (ADPs) after action potentials that are triggered by Ca^{2+} influx through non-L-type channels. ADPs were suppressed by flufenamic acid, a blocker of nonspecific cation currents (I$_{CAN}$). Taken together, these findings suggest that the ADP is mediated by a Ca^{2+}-dependent I$_{CAN}$. Brief depolarization triggers unusually prolonged (up to 44 min) persistent spiking in Blanes cells. Persistent firing could also be triggered by excitatory synaptic input elicited by stimulation in the GCL or the GL. Stimuli that evoked persistent firing in Blanes cells also produced prolonged barrages of IPSCs in GCs. These findings indicate that Blanes cells play a major role in prolonged modulation of the excitability state of GCs, which in turn would impact on the activity of mitral/tufted output neurons. Blanes cells preferentially target the basal dendrites and somata of GCs, and like GCs, Blanes cells may receive centrifugal feedback projections from POC. If so, then direct activation of GCs by POC inputs may be followed by prolonged Blanes cell-mediated inhibition of GCs. Thus, the activity state of Blanes cells may provide a gate on excitatory feedback projections from POC to GCs, which in turn, would modulate GC inhibition of mitral/tufted cells.

4.37.6 Dendrodendritic Transmission Between Mitral/Tufted Cells and Granule Cells

4.37.6.1 Overview of Mitral/Tufted Cell–GC Dendrodendritic Interactions

The majority of synapses of the EPL are between the lateral dendrites of mitral/tufted cells and the dendrites of GCs. Most of the synapses are: (1) asymmetrical synapses from the mitral/tufted cell lateral dendrites onto GC dendrites, and (2) symmetrical synapses from the spines (gemmules) of GC dendrites onto the mitral/tufted cell lateral dendrites (Price, J. L. and Powell, T. P. S., 1970b; 1970c; 1970d). These synapses are mostly reciprocal (Hirata, Y., 1964; Rall, W. *et al.*, 1966; Price, J. L. and Powell, T. P. S., 1970d; Woolf, T. B. *et al.*, 1991) and therefore occur in roughly equal proportion (Jackowski, A. *et al.*, 1978). One mitral/tufted cell can therefore receive feedback inhibition as well as lateral inhibition from GCs that are excited by other mitral/tufted cells.

4.37.6.2 Excitatory Transmission from Mitral/Tufted Cells to GCs

Spike-evoked Ca^{2+} transients in mitral cells, and presumably glutamate release, is abolished by Cd^{2+} but not Ni^{2+}, indicating that it requires activation of HVA Ca^{2+} channels (Isaacson, J. S. and Strowbridge, B. W., 1998). Intracellular Ca^{2+} buffering in mitral cells indicates that the Ca^{2+} channels that trigger dendritic glutamate release are located nearby active release sites (Isaacson, J. S. and Strowbridge, B. W., 1998). In normal or Mg^{2+}-free extracellular media, stimulation of mitral/tufted cells evokes dual-component EPSCs in GCs, consisting of a fast AMPA receptor component and a slow NMDA receptor component (Isaacson, J. S. and Strowbridge, B. W., 1998; Schoppa, N. E. *et al.*, 1998; Aroniadou-Anderjaska, V. *et al.*, 1999a; Chen, W. R. *et al.*, 2000; Isaacson, J. S., 2001). AMPA receptors in cultured GCs desensitize rapidly and have little or no Ca^{2+} permeability (Blakemore, L. J. and Trombley, P. Q., 2003), suggesting that GC AMPA receptors

must include GluR2 subunits, which regulate Ca^{2+} permeability (Jardemark, K. *et al.*, 1997). The AMPA receptor synaptic component evoked by mitral/tufted cell input is less effective than the NMDA receptor component in evoking spikes in GCs, especially in Mg^{2+}-free conditions (Schoppa, N. E. *et al.*, 1998). As noted above, block of I_A enhances the ability of the AMPA receptor component to trigger spikes in GCs (Schoppa, N. E. and Westbrook, G. L., 1999). Paired-pulse stimulation of mitral cells can produce either facilitation or depression of the GC excitatory response to the second pulse, but on average leads to paired-pulse facilitation (Dietz, S. B. and Murthy, V. N., 2005). With repetitive stimulation trains, mitral to GC responses were suppressed less than GC to mitral cell inhibitory responses. Mitral/tufted cell to GC transmission may be modulated by mGluRs and monoaminergic transmitters (see Section Neurophysiology of Neuromodulatory Inputs to Main Olfactory Bulb). Activation of Group III mGluR was reported presynaptically to decrease mitral cell to GC synaptic transmission (Trombley, P. Q. and Westbrook, G. L., 1992).

4.37.6.3 Inhibitory Transmission from GCs to Mitral/Tufted Cells

Activation of GCs evokes IPSPs/IPSCs in mitral cells mediated by activation of $GABA_A$ receptors (Chen, W. R. *et al.*, 2000; Isaacson, J. S. and Vitten, H., 2003; Dietz, S. B. and Murthy, V. N., 2005). Paired-pulse or repetitive activation of GCs typically produces paired-pulse inhibition of mitral cell synaptic responses; that is, the response to the second or subsequent pulses are smaller than that to the first (Isaacson, J. S. and Vitten, H., 2003; Dietz, S. B. and Murthy, V. N., 2005). Pharmacological activation of $GABA_B$ receptors on GCs has been reported to reduce GABA release from these cells via inhibition of HVA Ca^{2+} currents (Isaacson, J. S. and Vitten, H., 2003).

4.37.6.4 Dendrodendritic Inhibition

4.37.6.4.1 Self-inhibition

Intracellular stimulation of single mitral cells in the presence or absence of TTX results in a dendrodentrically mediated feedback IPSPs/IPSCs, that is, self-inhibition or feedback inhibition (Isaacson, J. S. and Strowbridge, B. W., 1998; Schoppa, N. E. *et al.*, 1998; Chen, W. R. *et al.*, 2000; Halabisky, B. *et al.*, 2000; Dietz, S. B. and Murthy, V. N., 2005). Subsequent studies demonstrated that TTX has opposite effects on self-inhibition that depend upon the strength of the mitral cell depolarizing pulse; that is, the amount of glutamate released by the mitral cell (Halabisky, B. *et al.*, 2000). Short-pulse (2–3 ms) self-inhibition is reduced, whereas long-pulse (>25 ms) inhibition is enhanced, by TTX. The feedback inhibition is of long duration (1–2 s) and consists of a flurry of individual IPSPs/IPSCs. The slow kinetics suggest that the feedback IPSPs/IPSCs is mediated by asynchronous GABA release from multiple GCs (Schoppa, N. E. *et al.*, 1998). In normal extracellular levels of Mg^{2+}, the self-inhibition is reduced to similar levels by AMPA or NMDA receptor antagonism and is abolished when antagonists to both receptors are applied (Isaacson, J. S. and Strowbridge, B. W., 1998). Mg^{2+}-free conditions enhance self-inhibition and cause it to be dominated by activation of NMDA receptors; under these conditions, AMPA receptors have only a minor effect of the magnitude of self-inhibition (Isaacson, J. S. and Strowbridge, B. W., 1998; Schoppa, N. E. *et al.*, 1998; Chen, W. R. *et al.*, 2000; Isaacson, J. S., 2001; Halabisky, B. and Strowbridge, B. W., 2003). Similar findings were obtained for tufted cell self-inhibition (Christie, J. M. *et al.*, 2001). As will be discussed below, lateral inhibition has been reported to be strongly regulated by the I_A current in GCs. Self-inhibition has been reported to be unaffected (Halabisky, B. *et al.*, 2000) or enhanced (Schoppa, N. E. and Westbrook, G. L., 1999; Isaacson, J. S., 2001) by blockers of I_A. Self-inhibition is unaffected by the L-type Ca^{2+} channel antagonist (nifedipine) or T- or R-type Ca^{2+} channel antagonists (50 μM Cd^{2+}) but is markedly attenuated by the P/Q- and N-type antagonist ω-conotoxin MVIIC (Isaacson, J. S. and Strowbridge, B. W., 1998). Self-inhibition is reduced by $GABA_B$ receptor agonists, presumably by reducing HVA currents in GCs (Isaacson, J. S. and Vitten, H., 2003). It is also reduced by mGluR antagonists (Heinbockel, T. *et al.*, 2007).

4.37.6.4.2 Lateral inhibition

Stimulation of one mitral cell in the presence of TTX results in a dendrodendritically mediated mitral-granule-mitral cell IPSPs/IPSCs, that is, feedforward or lateral inhibition (Isaacson, J. S. and Strowbridge, B. W., 1998; Urban, N. N. and Sakmann, B., 2002). Like self-inhibition, lateral inhibition in Mg^{2+}-free conditions is completely abolished by NMDA receptor antagonists; in normal extracellular media, AMPA receptors seem to play a more important role when the lateral inhibition is evoked by weaker stimulation (Isaacson, J. S. and Strowbridge, B. W., 1998; Schoppa, N. E. *et al.*, 1998). The spatial extent of lateral inhibition among mitral cells has been reported to be greater than that for tufted cells. Tufted cell lateral inhibition is limited to several glomerular widths (<400 μm) as opposed to 750 μm for mitral cells (Christie, J. M. *et al.*, 2001). The weaker lateral inhibition of tufted cells could merely be due to the fact that their lateral dendrites are relatively short (Mori, K. *et al.*, 1983; Orona, E. *et al.*, 1984). The I_A current strongly regulates the duration of lateral inhibition and the role of AMPA receptors. 4-AP reduces the duration of lateral inhibition and increases the contribution of AMPA receptors. As noted above, 4-AP blocks I_A, which normally occludes or counters the brief AMPA receptor-mediated depolarization and spiking in GCs. I_A blockade disinhibits the AMPA receptor-mediated synaptic component, reduces the latency of evoked spikes, and therefore leads to more rapid and synchronous GABA release from GCs (Schoppa, N. E. and Westbrook, G. L., 1999). Consequently, 4-AP reduces the contribution of the slow kinetics of the NMDA receptor-mediated synaptic component to GC GABA release. The results were somewhat different for self-inhibition, in which the amplitude of early IPSC component was markedly increased, as was the total duration of the IPSC. Blockade of SK currents in mitral cells enhances lateral inhibition (Maher, B. J. and Westbrook, G. L., 2005).

4.37.6.4.3 Role of Ca^{2+} influx through N-methyl D-aspartate receptors and voltage-dependent Ca^{2+} channels

Later studies demonstrated that Ca^{2+} influx through the NMDA receptor can directly trigger GABA release from GCs (Halabisky, B. *et al.*, 2000; Chen, W. R. *et al.*, 2000; Isaacson, J. S., 2001). Thus, NMDA applied focally into the EPL in the presence of the HVA Ca^{2+} channel blocker, Cd^{2+}, elicited IPSCs in mitral cells. Similar results were obtained in Mg^{2+}-free conditions when flash photolysis was used to increase intracellular Ca^{2+} in the mitral cell (Chen, W. R. *et al.*, 2000; Isaacson, J. S., 2001); the photolysis-induced IPSC could be blocked by NMDA receptor antagonists or Cd^{2+} (Chen, W. R. *et al.*, 2000; Isaacson, J. S., 2001). The NMDA-evoked IPSC did not depend on Ca^{2+} release from internal stores as it was unaffected by thapsigargin (Halabisky, B. *et al.*, 2000; Chen, W. R. *et al.*, 2000). However, IPSCs elicited by NMDA or KCl application in the GCL were blocked by Cd^{2+} (Halabisky, B. *et al.*, 2000). These findings suggest that Ca^{2+} influx via NMDA receptors near dendro-dendritic synapses, presumably on GC spines, is directly coupled to GABA release. However, depolarization at more remote sites (GC somata/proximal dendrites) induces Ca^{2+} influx via HVA Ca^{2+} channels.

4.37.6.4.4 Local vs. global modes of dendrodendritic inhibition

Current views are that weak stimulation producing small synaptic responses in GCs, and thus Ca^{2+} influx into isolated GCs spines, triggers GABA release and inhibition of mitral cell dendrites that are synaptically coupled to those spines. Such inhibition would tend to be spatially localized. Both NMDA receptors and voltage-dependent Ca^{2+} channels can provide the Ca^{2+} influx necessary for GABA release from spines. By contrast, stronger levels of mitral/tufted cell input that generate actively propagating Na^+ spikes would produce more global or widespread inhibition of mitral/tufted cells (Chen, W. R. *et al.*, 2000; Egger, V. *et al.*, 2003). As noted above, sufficiently large EPSPs subthreshold for Na^+ spikes can trigger an all-or-none LTS that propagates throughout the GC dendrites (Egger, V. *et al.*, 2005). Thus, the LTS is an additional candidate for global lateral inhibition.

4.37.6.4.5 Temporal modulation of dendrodendritic inhibition

In slices, paired-pulse mitral cell stimulation experiments show that self-inhibition elicited by the second pulse was markedly reduced and recovered with a time constant of ~6 s (Dietz, S. B. and Murthy, V. N., 2005). This finding may, in part, account for the decrement in GABAergic synaptic input to mitral cells across respiratory cycles during odor stimulation (Cang, J. and Isaacson, J. S., 2003). The observation that GC odor responses subside across sniffs may also lead to reduced dendrodendritic inhibition during respiration. Thus, the strength of dendrodendritic inhibition is likely to be temporally modulated with respect to respiration. Other studies indicate that higher frequency local γ-oscillatory activity generated by mitral cell–GC interactions may also modulate GABAergic inhibition. Local γ-frequency stimulation in the GCL or ELP can facilitate mitral cell feedback and lateral inhibition by relieving the Mg^{2+} block of NMDA receptors on GCs (Halabisky, B. and Strowbridge, B. W., 2003).

4.37.7 Neurophysiology of New, Adult-Born Neurons

The deepest layer in the MOB is the subependymal layer (also called subventricular zone), which is a cell-poor region lining the ventricle (if present) in adults. Most MOB interneurons originate postnatally from progenitor cells within this layer (Hinds, J. W., 1968; Altman, J., 1969; Bayer, S. A., 1983). In adults, interneurons (primarily GCs and PG cells) are continually generated from these progenitor cells, and their offspring generated en route migrates to the MOB within the rostral migratory stream (RMS; Luskin, M. B., 1993; Lois, C. and Alvarez-Buylla, A., 1994; Smith, C. and Lushkin, M. B., 1998; Wichterle, H. *et al.*, 2001). Two subsets of new interneurons have been identified, which express either TH, which is required for DA synthesis, or Ca^{2+} calmodulin-dependent protein kinase IV during migration (Baker, H. *et al.*, 2001). Both the progenitors and new interneurons express functional $GABA_A$ receptors. Electrophysiological studies indicate that the new interneurons subsequently express functional AMPA receptors, then NMDA receptors, before they exhibit spiking activity, responses to ON stimulation, and spontaneous glutamatergic EPSCs (Belluzzi, O. *et al.*, 2003; Carleton, A. *et al.*, 2003). These electrophysiological studies provide evidence that the new interneurons become functionally integrated into the MOB circuitry. Neural cell-adhesion protein-deficient mice, which exhibit defective migration of new interneurons into the GCL and a reduced MOB size, have been shown to exhibit impaired odor discrimination (Gheusi, G. *et al.*, 2000). Although threshold detection and short-term odor memory were normal in these mice, in normal mice both olfactory memory and survival of the new interneurons were improved following rearing in an odor-enriched environment (Rochefort, C. *et al.*, 2002). Thus, the neurons of the subependymal layer and RMS appear to be important both to MOB development and to certain olfactory functions during adulthood.

4.37.8 Neurophysiology of Primary Olfactory Cortical Inputs to Main Olfactory Bulb

Extrinsic afferent input to MOB, also referred to as centrifugal fibers, can be subdivided into two classes: (1) inputs arising from nonolfactory, so-called neuromodulatory transmitter systems including ACh, norepinephrine (NE), and 5-HT (these neuromodulatory inputs are discussed below in Neurophysiology of Neuromodulatory Inputs to Main Olfactory Bulb), and (2) feedback inputs arising from olfactory-related structures, in particular those arising from POC. Feedback projections to MOB from POC arises predominantly from glutamatergic, pyramidal neurons in layers II and III of piriform cortex, as well as other POC structures (for review, see Ennis, M. *et al.*, 2007). These projections massively target GCs, where they form asymmetrical synapses on the cell bodies, basal dendrites, and spines of GCs (Price, J. L. and Powell, T. P. S., 1970a). Activation of these feedback projections produces a negative field potential recorded in the GCL (Walsh, R. R., 1959; Nakashima, M. *et al.*, 1978), as expected if excitatory currents are flowing into GCs. Similar stimulation elicits IPSPs in mitral cells (Yamamoto, C. *et al.*, 1963; Nicoll, R. A., 1971; Mori, K. and Takagi, S. F., 1978), due to excitation of GCs, followed by GABA release onto mitral cells, that is, dendrodendritic inhibition (Halász, N. and Shepherd, G. M., 1983). The transmitter of these feedback projections to GCs is glutamate. Activation of POC input to MOB excites GCs as measured by voltage-sensitive dye and field potential recordings *in vitro*. This excitation is mediated by glutamate acting at both AMPA and NMDA receptors (Laaris, N. *et al.*, 2007). A major function of these projections is to provide an inhibitory regulation of the firing rate and excitability state of mitral/tufted cells. Activation of these inputs modifies (inhibits) odor responses in MOB (Kerr, D. I. B. and Hagbarth, K. E., 1955).

4.37.9 Oscillations and Synchrony in Main Olfactory Bulb

4.37.9.1 Oscillations

Oscillations and synchronous activity are characteristic features of spontaneous and odor-driven activity in the MOB and other olfactory structures. A detailed consideration of oscillations in the MOB is beyond the scope of the present review and has recently been reviewed elsewhere (Lledo, P. M. *et al.*, 2005; Gelperin, A., 2006; Kepecs, A. *et al.*, 2006). Prominent oscillatory activity is present in MOB field potentials, which primarily reflect synchronized subthreshold membrane potentials arising from large

neuronal populations, or in membrane potential or spike activity of individual MOB neurons. Sensory input to the olfactory system occurs as a result of respiration which by its very nature is cyclical or rhythmical. Relatively low-frequency oscillations synchronized to respiration, in the presence and absence of odors, occur in the theta band (2–12 Hz). Higher frequency oscillations, superimposed on the theta, occur in the beta (15–40 Hz) and gamma (30–80 Hz) bands. Beta frequency oscillations are relatively poorly understood, but have recently been discussed elsewhere (Neville, K. R. and Haberly, L. B., 2003; Martin, C. *et al.*, 2004; Fletcher, M. L. *et al.*, 2005). Lower frequency oscillatory (0.1–0.03 Hz) and synchronous activity has been observed in MOB slices under certain conditions (Puopolo, M. and Belluzzi, O., 2001).

4.37.9.1.1 Theta rhythm

In the MOB, 2 to 12 Hz oscillations are referred to as theta principally because they occupy a highly overlapping frequency band with hippocampal theta oscillations (Kay, L. M. and Laurent, G., 1999; Kay, L. M., 2003; 2005). As noted above, theta frequency oscillations are driven by, and thus synchronized or phase-locked with, respiration. Therefore, the specific theta frequency varies with the sniffing frequency (i.e., the pattern of airflow across the nasal epithelium) and includes components related to low-frequency (1–3 Hz) passive sniffing as well as a higher frequency component (5–12 Hz) characteristic of active investigative sniffing (Adrian, E. D., 1950; Welker, W. I., 1964; Macrides, F. *et al.*, 1982; Eeckman, F. H. and Freeman, W. J., 1990; Kay, L. M. and Laurent, G., 1999; Kay, L. M., 2003). As noted earlier, mitral cell subthreshold membrane potential oscillations and/or spike activity is synchronized with certain phases of the respiratory cycle. Manipulations that impair or disrupt respiratory-driven input to the MOB disrupts mitral cell activity and decouples it from the respiratory cycle (Philpot, B. D. *et al.*, 1997). Thus, theta oscillations appear to be driven by the pattern of respiratory-driven input to MOB. However, intrinsic membrane and synaptic properties enhance the ability of some MOB neurons to entrain or synchronize with rhythmical respiratory-driven activity. For example, ET cells in MOB slices intrinsically generate spike bursts at frequencies ranging from ~1 to 8 Hz, and readily entrain to patterned sensory input over this same range. Thus, the properties of these cells may allow them to discharge in synchrony with the respiratory cycle.

4.37.9.1.2 Gamma rhythm

Gamma oscillations arise during odorant or direct ON stimulation and appear to be generated intrinsically in the MOB. Thus, they persist when centrifugal input to the bulb is severed or impaired (Gray, C. M. and Skinner, J. E., 1988; Neville, K. R. and Haberly, L. B., 2003). As with theta oscillations, odor- or ON-evoked gamma oscillations are apparent in the EEG, MOB field potentials, and subthreshold membrane potential and spiking activity of mitral cells (Adrian, E. D., 1950; Eeckman, F. H. and Freeman, W. J., 1990; Kay, L. M. and Freeman, W. J., 1998; Kashiwadani, H. *et al.*, 1999; Debarbieux, F. *et al.*, 2003; Friedman, D. and Strowbridge, B. W., 2003; Neville, K. R. and Haberly, L. B., 2003; Lagier, S. *et al.*, 2004; Martin, C. *et al.*, 2004; Fletcher, M. L. *et al.*, 2005). Current source-density analyses indicate that the gamma oscillations in MOB field potentials primarily reflect synaptic currents flowing in GCs (Neville, K. R. and Haberly, L. B., 2003). Multisite recordings demonstrated that the phase of gamma oscillation may vary in different parts of the MOB, especially at low odor concentrations. Gamma oscillations are thought to arise primarily as a result of rhythmic or reverbatory interactions between mitral/tufted cells and inhibitory interneurons, chiefly GCs (Friedman, D. and Strowbridge, B. W., 2003; Neville, K. R. and Haberly, L. B., 2003; Lagier, S. *et al.*, 2004). Mitral cell activity is synchronized with γ-frequency MOB field potentials and also with GABAergic synaptic input (Lagier, S. *et al.*, 2004). The duration of mitral cell to GC synaptic input occurs over a time course of approximately one-half cycle of the gamma oscillation. Gamma oscillatory activity in mitral cells is suppressed or attenuated by glutamate receptor antagonists (decreasing mitral/tufted cell excitation of GCs) or by GABA$_A$ receptor antagonists (blocking GC-mediated inhibition of mitral/tufted cells) (Friedman, D. and Strowbridge, B. W., 2003; Lagier, S. *et al.*, 2004). In GABA$_A$ $\beta 3$ receptor subunit-deficient mice, functional expression of GC GABA$_A$ receptors was almost eliminated (Nusser *et al.*, 2001). In these mice amplitudes of mitral/tufted cell miniature IPSPs, theta-frequency oscillations, and γ-frequency oscillations were increased, and discrimination of closely related mixtures of alcohols after training was poor relative to normal mice. Gamma oscillations also appear to involve electrical synapses

as they are suppressed by gap junction inhibitors (Friedman, D. and Strowbridge, B. W., 2003). Consistent with this, gamma oscillations are disrupted in connexin36-knockout mice (Hormuzdi, S. G. *et al.*, 2001). Intrinsic membrane properties of mitral cells may facilitate gamma oscillatory activity and spiking. When near spike threshold, mitral cells exhibit intrinsically generated subthreshold membrane potential oscillations in the γ-frequency range (Chen, W. R. and Shepherd, G. M., 1997; Desmaisons, D. *et al.*, 1999).

4.37.9.2 Synchrony

Synchrony can be defined as a temporal coincidence between two or more events that have a low probability of occurring by random or chance. Neurons involved in the detection and processing of odors show temporally correlated or synchronized activity, a feature that is expected to have functional relevance for understanding the population code in the olfactory system (Laurent, G., 2002).

4.37.9.2.1 Synchrony among JG cells

JG neurons associated with the same glomerulus exhibit highly synchronous spontaneous activity. Simultaneous recordings of ET–PG or ET–SA cell pairs demonstrated that spikes in ET cells drive synchronous activity in PG and SA cells, but only if the dendrites of both cells ramified in the same glomerulus (Figure 14) (Hayar, A. *et al.*, 2004b). This intraglomerular synchronous activity is driven by glutamatergic input from ET cells to PG/SA cells and is abolished by AMPA receptor antagonists. Similar studies revealed that spontaneous spikes and subthreshold membrane potential activity (i.e., EPSPs, IPSPs) among ET cells of the same glomerulus is highly synchronous (Figure 14) (Hayar, A. *et al.*, 2004b; 2005). Interestingly, the frequency of spontaneous inhibitory input to ET cells, which primarily originates from PG cells, occurs around 50 Hz (γ frequency) and is similar to the frequency of ON-evoked γ-oscillations in mitral cells (mean 45 Hz, Lagier, S. *et al.*, 2004). Although spontaneous synaptic input can synchronize ET cells, synchrony among these cells persists when ionotropic glutamate and GABA receptors are blocked (Hayar, A. *et al.*, 2004b; 2005). Additional electrophysiological results demonstrated that ET cells of the same glomerulus are electrotonically coupled via gap junctions with a relatively small conductance (0.1 nS). Spontaneous synchrony among ET cells is

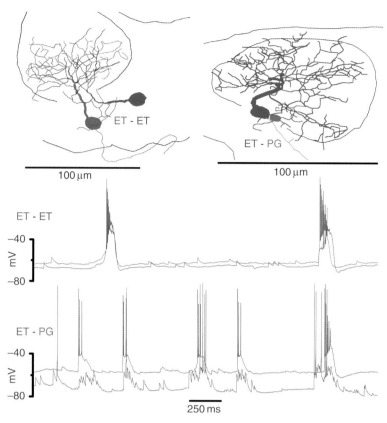

Figure 14 Synchronous activity among juxtaglomerular (JG) cells associated with the same glomerulus. Upper panel shows reconstructions of the recorded cells (left, two external tufted (ET) cells; right, an ET cell and a periglomerular (PG) cell). Middle panel: Dual whole-cell current clamp recording from two ET cells (red and blue traces, respectively) showing correlation of spike bursts and membrane potentials. Lower panel: Dual whole-cell current clamp recordings show that EPSP bursts and spikes in a PG cell (red trace) are synchronous with spike bursts in an ET cell (blue trace).

abolished by the gap junction blocker carbenoxolone (Hayar, A. *et al.*, 2005). These results are consistent with anatomical evidence for gap junctions between mitral/tufted cell apical dendrites (See Section 4.37.9.2.2). Unlike mitral cells, which exhibit a narrow window of spike-to-spike correlation (see below; Schoppa, N. E. and Westbrook, G. L., 2002), ET cells exhibit a broader window of burst-to-burst correlation. The narrow window of correlation among mitral cells is due to the fact that mitral cell spikes induced fast EPSPs in other mitral cells (Schoppa, N. E. and Westbrook, G. L., 2002). In contrast, ET cells seem to communicate mainly via slow inward currents produced by gap junctions with relatively low conductance. Therefore, one potential function of gap junctions in ET cells is to filter fast spiking activity and propagate slow membrane potential oscillations, which are driven mainly by persistent Na$^+$ current (Hayar, A. *et al.*, 2004a).

4.37.9.2.2 Synchrony among mitral cells

Synchronous activity in the mammalian MOB occurs among mitral cells that send an apical dendrite to receive input from a single glomerulus. By using simultaneous recordings from pairs of mitral cells, it has been shown that pairs with apical dendrites in the same glomerulus display highly synchronized synaptic and/or spike activity, whereas pairs associated with different glomeruli do not. As reviewed earlier, mitral cells associated with the same glomerulus exhibit synchronous AMPA receptor-mediated LLDs that spontaneously occur at a frequency of ~1 Hz (Carlson, G. C. *et al.*, 2000). LLDs (Figure 7) are generated by network activity in the glomeruli and could be elicited by ON or antidromic stimulation, but never by depolarization of single mitral cells. Subsequent studies demonstrated that slow (2 Hz) oscillations and spikes elicited by single ON shocks or NMDA application were synchronous among mitral cells of the same glomerulus; cells associated with distant glomeruli did not exhibit synchrony, while those associated with adjacent glomeruli exhibited temporally correlated but not synchronous oscillations (Schoppa, N. E. and Westbrook, G. L., 2001; 2002). The synchronous activity among mitral cells was abolished by AMPA receptor antagonists and dampened by mGluR antagonists (Schoppa, N. E. and Westbrook, G. L., 2001; 2002). These and later studies indicated that mitral cell apical dendrites in the same glomerulus exhibit electrotonic coupling (Schoppa, N. E. and Westbrook, G. L., 2002; Christie, J. M. et al., 2005). In mouse MOB slices, current injections into one cell elicited correlated spikes in a second mitral cell if the dendrites extended into the same glomerulus. Such synchrony, however, was absent in slices from connexin36-knockout mice (Christie, J. M. et al., 2005). Connexins are a member of gap junction proteins involved in electrical signaling between cells. The preceding finding is consistent with the presence of high levels of connexin36 in MOB neurons, and specifically in mitral/tufted dendrites (Condorelli, D. F. et al., 1998; 2000; Belluardo, N. et al., 2000; Rash, J. E. et al., 2000; Teubner, B. et al., 2000; Zhang, C. and Restrepo, D., 2003; Christie, J. M. et al., 2005). Other studies have also reported gap junctions in, or between, mitral/tufted cell dendrites (Paternostro, M. A. et al., 1995; Miragall, F. et al., 1996; Kosaka, T. and Kosaka, K., 2003; 2004). In slices from wildtype, but not connexin36-knockout mice, a spike elicited in one mitral cell evoked an AMPA receptor-mediated potential (D_{AMPA}) in the second mitral cell (Christie, J. M. et al., 2005). Correlated spiking in connexin36-knockout mice could be reinstated by boosting glutamate levels with uptake inhibitors. Taken together, these findings indicate that electrical coupling is necessary to drive D_{AMPA}, which in turn, drives correlated spiking among mitral cells. Thus, the apical dendrites of mitral cells generate a unique form of electrochemical transmission. The dependence of correlated spiking on AMPA receptors is a distinct feature of mitral–mitral cell synchrony.

GABAergic inhibition is also thought to play an important role in synchronizing mitral cell activity. As noted earlier, γ-frequency oscillation and spiking is a prominent feature of mitral cell odor-evoked activity. Spontaneous synchronous IPSPs can reset the phase of subthreshold membrane oscillations and synchronize the firing of multiple mitral cells (Desmaisons, D. et al., 1999). Correlated IPSPs might also contribute to the odor-elicited high-frequency oscillations in the olfactory bulb of adult rats (Fletcher, M. L. et al., 2005). The frequency of ON-evoked local field potential oscillations (mean 45 Hz, Lagier, S. et al., 2004) recorded in the MCL is attributed to rhythmic dendrodendritic granule to mitral cell inhibition. Unlike glomerular mechanisms, GABAergic inhibition appears to synchronize mitral cells associated with different glomeruli. Mitral cell associated with different glomeruli can be synchronized by rhythmic stimulation of the ON at theta frequencies (Schoppa, N. E., 2006). Such synchrony does not depend upon electrotonic coupling, but appears to be driven by synchronous IPSPs from GCs, such that synchronous spikes in mitral cells occur during the recovery from the synchronous IPSP. One possible function of spike synchrony among mitral cells of different odorant receptor specificities is to provide a mechanism for downstream cortical neurons to decode information about different odorant receptors (Schoppa, N. E., 2006). As would be required by such a mechanism, labeling studies suggest that each pyramidal cell in the anterior piriform cortex receives anatomical connections from mitral cells with different odorant receptor specificities (Zou, Z. et al., 2001). Assuming that pyramidal cells in the olfactory cortex have synaptic integration windows comparable to other pyramidal neurons (e.g., ~7 ms), synchrony among mitral cells would promote the summation of EPSPs and the likelihood that cortical neurons fire action potentials in response to input from activated mitral cells of different odorant receptor specificities.

4.37.9.3 Oscillations, Synchrony and Odor Coding

Temporal firing patterns, including rhythmic oscillations and neuronal synchronization, are thought to be important in sensory information processing (Alonso, J. M. et al., 1996; Konig, P. et al., 1996; Roy, S. A. and Alloway, K. D., 2001) including odor coding (for review, see Friedrich, R. W., 2002). Rhythmic activity and synchrony are thought to coordinate intra- and interstructural communication. Olfactory information is thought to be encoded, at least in part, by network oscillations (Wehr, M. and Laurent, G., 1996; Kauer, J. S., 1998; Kay, L. M. and Laurent, G., 1999). In the first stage of olfactory processing, network interactions within each glomerulus act to synchronize the discharge of mitral cells associated with that glomerulus to the theta pattern of rhythmical sensory input. By temporally binding the discharge of mitral cells, such synchrony would maximize the transfer of sensory input onto the output neurons. This, in turn, would facilitate faithful transfer of glomerular sensory input to higher order olfactory structures (i.e., POC). In other words, temporal summation of EPSPs produced by synchronous spikes from the glomerular ensemble of mitral/tufted cells would increase the likelihood of spike initiation in post-synaptic POC neurons. Glomerular synchrony may be a particularly important amplification mechanism when odor concentrations are insufficient to activate all mitral cells of a particular glomerulus. Odors typically activate multiple glomeruli, but the mechanism that drives the sensory input (respiration) is common to all glomeruli. Does this imply that oscillations and synchrony should be phase-locked across responsive glomeruli? This does not seem to be the case as the widespread distribution of ORNs and their odorant response kinetics produce temporal variations in odor response properties in different glomeruli. Thus, different glomeruli activated by the same odorant exhibit temporally distinct response profiles (Spors, H. et al., 2006). It is likely, therefore, that at least initially there will be temporal differences in the phase of oscillatory neuronal activity and synchrony among neurons in different responsive glomeruli. In addition, both glomerular odor activity patterns and discharge of mitral cells change within and across sniffs (Friedrich, R. W., 2002; Spors, H. et al., 2006), and as noted above, the phase of gamma oscillations may not always be uniform throughout the MOB. Neurons in POC appear to receive convergent input from multiple glomeruli (Zou, Z. et al., 2001; Illig, K. R. and Haberly, L. B., 2003). Temporal variations in activity patterns from different glomeruli, via mitral/tufted cells, may be an important element in odor recognition and discrimination.

4.37.10 Neurophysiology of Neuromodulatory Inputs to Main Olfactory Bulb

MOB receives extrinsic inputs from cholinergic, noradrenergic, and serotonergic groups in the basal forebrain and brainstem. As these inputs innervate neurons in multiple layers of MOB, and therefore modulate multiple neuronal subtypes, their physiological effects are collectively reviewed here.

4.37.10.1 Cholinergic Inputs to MOB

In the mouse, about 3.5% of all neurons that project to the bulb originate in the nucleus of the horizontal limb of the diagonal band (NDB); far fewer originate in the vertical limb of DB (Carson, K. A., 1984; Shipley, M. T., and Adamek, G. D., 1984). At least two distinct transmitter-specific populations of NDB neurons project to the MOB (Zaborszky, L., et al., 1986). About 20% of the NDB neurons that project to the bulb are cholinergic; most of these cells are concentrated in the rostromedial portion of the horizontal limb of NDB. Many NDB–MOB projection neurons are GABAergic and occur mainly in the caudo-lateral aspect of NDB (Zaborszky, L., et al., 1986). Choline acetyltransferase, the biosynthetic enzyme for ACh synthesis, is located in axons distributed across most layers of the MOB, except the ONL; cholinergic fibers are especially heavy in the GL and IPL (Ennis, M. et al., 2007). As reviewed elsewhere, muscarinic and nicotinic receptors are found in most layers of MOB (Ennis, M. et al., 2007).

Electrical activation of NBD has been reported to depress (Nickell, W. T. and Shipley, M. T., 1988) or to increase (Kunze, W. A. A. et al., 1991; 1992) mitral cell activity indirectly via primary effects on GABAergic GCs. NDB stimulation also reduced the field potential in the MOB caused by stimulation of the anterior commissure (Nickell, W. T. and Shipley, M. T., 1993), an effect mediated by presynaptic inhibition of anterior commissure terminals via muscarinic receptors. One interpretation of these results is that cholinergic input to MOB may function to modulate interhemispheric transmission of olfactory information. In this regard, it is noteworthy that anterior commissural fibers are required for access and recall of olfactory memories between the two hemispheres. Infusion of ACh into MOB was reported to reduce paired-pulse depression of LOT-evoked field potentials recorded in the GCL. This effect was attributed to muscarinic receptor-mediated inhibition of GABA release from GCs (Elaagouby, A. et al., 1991). In slice preparations, nicotinic but not muscarinic receptor agonists directly excited mitral cells, and this effect appeared to be due to an inward current with a reversal potential of -5 to $+10\,\text{mV}$ (Castillo, P. E. et al., 1999). In slices, muscarinic receptor agonists inhibited GCs (Castillo, P. E. et al., 1999), and paradoxically, also appeared to increase GABA release from these cells. The same study reported that in the GL only bipolar PG cells were sensitive to nicotine (Castillo, P. E.

et al., 1999). The morphological identity of these cells is unclear. Behaviorally, muscarinic receptor antagonists impair discrimination among closely related odors (Fletcher, M. L. and Wilson, D. A., 2002).

4.37.10.2 Noradrenergic Input to MOB

A significant modulatory input to the bulb is from the pontine nucleus locus coeruleus (LC). In the rat, all LC neurons contain the neurotransmitter, NE; LC contains the largest population of NE neurons in the brain. It has been estimated that up to 40% of LC neurons (400–600 of a total of 1600 LC neurons) project to the bulb in the rat (Shipley, M. T. *et al.*, 1985). A subset of LC neurons projecting to MOB contain NPY (Bouna, S. *et al.*, 1994). LC axons project mainly to the subglomerular layers of the bulb, particularly in the IPL and GCL (McLean, J. H., *et al.*, 1989). The EPL and MCL are moderately innervated, while the GL is nearly devoid of NE input. NE receptors occur in multiple layers of the MOB and are expressed by multiple cell types, in general consistent with the pattern of NE fiber innervation (for review, see Ennis, M. *et al.*, 2007).

While NE clearly plays significant roles in olfactory function, the effects of NE at the cellular and network levels are somewhat discrepant. For example, LC stimulation was reported to have no effect on LOT-evoked field potentials recorded in the GCL (Perez, H. *et al.*, 1987). A subsequent study reported that LC stimulation initially decreased and then subsequently increased paired-pulse depression of GC field potential responses to LOT stimulation (Okutani, F. *et al.*, 1998). These effects were attributed to activation of β receptors. Another field potential study reported that NE infusion into MOB, acting at $\alpha 1$ receptors, increased the depolarization of GC dendrites elicited by LOT stimulation. Mitral cell responses to antidromic shocks were not affected, suggesting that NE excites GCs (Mouly, A. M. *et al.*, 1995). In neonatal animals, β receptor stimulation in MOB decreased LOT-evoked, paired-pulse inhibition of GC field potentials (Wilson, D. A. and Leon, M. 1988). It is unclear if this was mediated by presynaptic inhibition of transmitter release from MCs and/or increased excitability of GCs.

Cellular recording studies are also somewhat discrepant. In the rabbit and cat, Salmoiraghi *et al.* (1964) and McLennan (1971) found that iontophoretically applied NE inhibited mitral cells. This effect was blocked by the $GABA_A$ receptor antagonist, bicuculline. In the isolated turtle bulb (Jahr and Nicoll,

1982), mitral cell spike rate increased and IPSPs decreased following both application of NE. In dissociated MOB cultures, NE decreased mitral to granule dendrodendritic synaptic transmission by acting presynaptically at $\alpha 2$ receptors to decrease Ca^{2+} currents in both granule and mitral cells (Trombley, P. Q., 1992; Trombley, P. Q. and Shepherd, G. M., 1992).

NE release or NE agonists have more consistent effects on ON-evoked responses in mitral cells. NE release, evoked by selective chemical activation of LC *in vivo*, enhanced the response of mitral cells in response to weak (i.e. perithreshold) but not strong (i.e. suprathreshold) stimulation of the olfactory epithelium (Jiang, M. R. *et al.*, 1996). Interestingly, NE release from LC axon terminals is facilitated and suppressed by activation of presynaptic nicotinic and muscarinic cholinergic receptors, respectively (El-Etri, M. M. *et al.*, 1999). Consistent with these *in vivo* findings, application of NE or $\alpha 1$ receptor agonists, but not $\alpha 2$ or β-receptor agonists, selectively increased mitral cell responses to perithreshold intensity ON stimulation in rat MOB slices (Figure 15) (Ciombor, K. J. *et al.*, 1999). Noradrenergic agonists were without effect on ON-evoked field potentials recorded in the GL, or on ON-evoked postsynaptic currents in mitral cells (Hayar, A. *et al.*, 2001). This suggests that NE-evoked modulation of ON-evoked mitral cell spiking is mediated by postsynaptic actions on bulb neurons. In voltage-clamp recordings, NE or $\alpha 1$ agonists directly evoked an inward current in mitral cells that appeared to be due to closure of K^+ channels (Figure 16). In current clamp recordings from bistable mitral cells, $\alpha 1$ agonists shifted the membrane potential from the downstate ($-52\,mV$) toward the upstate ($-40\,mV$) and significantly increased spike generation in response to perithreshold ON input. Taken together, these findings suggest that NE release directly alters mitral cell excitability in a manner that could increase their sensitivity to weak ON input, perhaps to improve the detection of weak odorants.

NE inputs to the bulb are critical to olfactory function. Olfactory cues increase the discharge of LC neurons in behaving animals (Aston-Jones, G. and Bloom, F. E., 1981) and trigger rapid increases in NE levels in the MOB (Chanse, N. T. and Kopin, I. J., 1968; Rosser, A. E. and Keverne, E. B., 1985; Brennan, P. *et al.*, 1990). LC projections to the main and accessory olfactory bulb are pivotal to the formation of specific olfactory memories, pheromonal

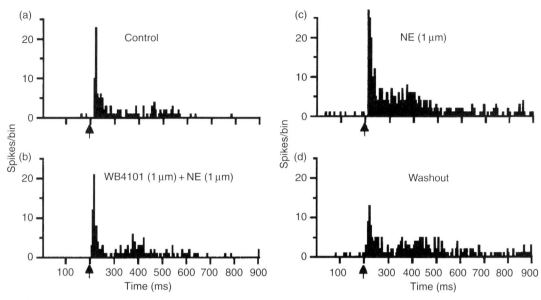

Figure 15 The α1 receptor antagonist WB-4101 prevents norepinephrine (NE)-induced facilitation of mitral cell excitatory responses to perithreshold intensity olfactory nerve (ON) stimulation. (a–d) Peri-stimulus time histograms (PSTHs) showing responses of a mitral cell to perithreshold intensity ON shocks (12 μA, at arrows). (a and b) In the presence of WB-4101 (1 μM), superfusion of NE (1 μM) did not alter (103% of control) the early excitation elicited by ON stimulation. (c) After washout of WB-4101, NE substantially increased (184% of control) the early excitatory response component. (d) The facilitation of the early excitation recovered by 53 min after washout of NE. All PSTHs were generated for 50 consecutive ON shocks. Reprinted from Ciombor, K. J., Ennis, M., and Shipley, M. T. 1999. Norepinephrine increases rat mitral cell excitatory responses to weak olfactory nerve input via alpha-1 receptors *in vitro*. Neuroscience 90, 595–606, with permission from Elsevier, Ltd.

regulation of pregnancy and post-partum maternal behavior (Kaba, H. and Keverne, E. B., 1988; Brennan, P. *et al.*, 1990; Kendrick, K. M. *et al.*, 1992). NE plays an important role in the so-called Bruce effect in mice: when impregnated female mice are exposed to the odor of a strange male, they abort; if exposed to the odor of the impregnating male, they do not abort (Kaba, H. and Keverne, E. B., 1989). Systemic administration of adrenergic receptor antagonists or 6-OHDA lesions, selectively destroying only the NE inputs to the MOB, causes the female to abort when presented with the odor of the impregnating male (Rosser, A. E. and Keverne, E. B., 1985; Kaba, H. and Keverne, E. B., 1988; Brennan, P. *et al.*, 1990). Finally, NE has been shown to play a critical role in olfactory learning in young animals. In neonatal rats, NE release via tactile stimulation leads to a preference for an odor associatively paired with this stimulation (Sullivan, R. M. *et al.*, 1989). The conditioned preference is associated with odor-specific metabolic changes in the bulb (Coopersmith, R. and Leon, M., 1984). Following the conditioning, there is an increased inhibition of mitral cells by the odor (Sullivan, R. M. *et al.*, 1989). Such conditioning is

abolished by eliminating NE input to the bulb or via β-receptor antagonists (Sullivan, R. M. *et al.*, 1989; Wilson, D. A. and Sullivan, R. M., 1991; Sullivan, R. M. *et al.*, 1992; 2000; Moriceau, S. and Sullivan, R. M., 2004). Recent studies from McLean's laboratory demonstrate that this β-receptor-dependent neonatal learning involves activation of intracellular cAMP and cAMP-response element-binding protein (CREP) pathways (Yuan, Q. *et al.*, 2003; McLean, J. H. and Harley, C. W., 2004).

4.37.10.3   Serotonergic (5-HT) Input to MOB

The midbrain dorsal and median raphe provides strong serotonergic inputs to the MOB. In the rat, about 1000 dorsal and 300 median raphe neurons project to the bulb (McLean, J. H. and Shipley, M. T., 1987a; 1987b). 5-HT fibers are present in all layers of MOB, but with varying densities. Input to the GL is especially dense, while the EPL contains very low density. The MCL, IPL, and GCL have a fairly heavy and uniform innervation, but not as dense as that of the GL. Thick serotonergic fibers preferentially innervate the glomeruli of MOB, while

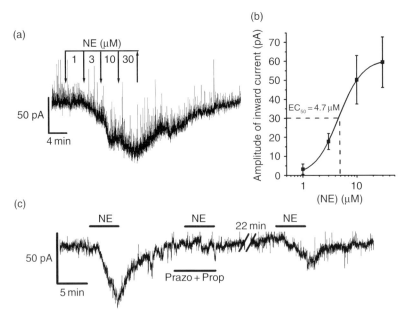

Figure 16 Effect of norepinephrine (NE) in voltage-clamp recordings from mitral cells. (a) Inward currents evoked by NE at different concentrations (1, 3, 10, and 30 μM added cumulatively, 4 min at each concentration). (b) A sigmoidal curve was fitted to the NE concentration–response data obtained from four cells. The holding potential was −60 mV. (c) The response to a second application of NE (30 μM) was blocked in the presence of the alpha 1 receptor antagonist prazosin (Prazo; 1 μM) and the beta receptor antagonist propranolol (Prop; 10 μM). The NE-induced inward current recovered partially after washout of the antagonists (about 30 min). Reprinted from Hayar, A., Heyward, P. M., Heinbockel, T., Shipley, M. T., and Ennis, M. 2001. Direct excitation of mitral cells via activation of alpha1-noradrenergic receptors in rat olfactory bulb slices. J. Neurophysiol. 86, 2173–2182, with permission from The American Physiological Society.

thinner serotonergic axons preferentially innervate inframitral layers (McLean, J. H. and Shipley, M. T., 1987a; 1987b). In neocortex, thick axons arise from the median raphe and thin axons arise from dorsal raphe (McLean, J. H. and Shipley, M. T., 1987a; 1987b), and the same segregation occurs in MOB. In agreement with the 5-HT fiber distribution, 5-HT receptors are localized in most layers of the MOB (for review, see Ennis, M. et al., 2007).

In the GL, it was recently reported that 5-HT depolarized 34% of JG cells *in vitro* via activation of $5HT_{2C}$ receptors (Hardy, A. et al., 2005). The 5-HT-induced depolarization was due to activation of a nonselective cation current with a reversal potential of −44 mV. The heterogeneous electrophysiological properties of 5-HT-responsive JG cells suggested that several types of JG cells could be targeted by 5-HT centrifugal fibers. A subset of mitral cells was also depolarized by 5-HT acting at $5-HT_{2A}$ receptors. By contrast with these results, another subset of mitral cells was hyperpolarized by 5-HT, an action that was indirectly mediated by GCs as it was blocked by $GABA_A$ receptor antagonists (Hardy, A. et al., 2005). This effect of 5-HT was also thought to be mediated

by $5-HT_{2A}$ receptors. Behavioral studies indicate that lesion of serotonergic fibers reversed conditioned olfactory learning (Morizumi, T. et al., 1994) and also induced glomerular atrophy. Behavioral work on neonates by McLean (McLean, J. H. et al., 1996) showed that 5-HT depletion or $5-HT_2$ receptor antagonism compromised olfactory learning and that 5-HT release or $5-HT_2$ receptor activation promoted odor conditioning. 5-HT release also appears to facilitate β receptor-mediated, NE-induced olfactory learning by facilitating cAMP-mediated mechanisms (McLean, J. H. and Harley, C. W., 2004).

Acknowledgments

We acknowledge the grant support for portions of this chapter that refer to research of the authors: PHS Grants DC03195, DC06356, DC07123, and RR020146.

References

Adrian, E. D. 1950. The electrical activity of the olfactory bulb. Electroencephalogr. Clin. Neurophysiol. 2, 377–388.

Allison, A. 1953. The morphology of the olfactory system in the vertebrates. Biol. Rev. Camb. Philos. Soc. 28, 195–244.

Alonso, J. M., Usrey, W. M., and Reid, R. C. 1996. Precisely correlated firing in cells of the lateral geniculate nucleus. Nature 383, 815–819.

Alonso, J. R., Arévalo, R., Garcia-Ojeda, E., Porteros, A., Briñón, J. G., and Aijón, J. 1995. NADPH-diaphorase active and calbindin D-28k-immunoreactive neurons and fibers in the olfactory bulb of the hedgehog (Erinaceus europaeus). J. Comp. Neurol. 351, 307–327.

Alonso, J. R., Arévalo, R., Porteros, A., Briñón, J. G., Lara, J., and Aijón, J. 1993. Calbindin D-28K and NADPH-diaphorase activity are localized in different populations of periglomerular cells in the rat olfactory bulb. J. Chem. Neuroanat. 6, 1–6.

Alreja, M. and Aghajanian, G. K. 1995. Use of the whole-cell patch-clamp method in studies on the role of cAMP in regulating the spontaneous firing of locus coeruleus neurons. J. Neurosci. Methods 59, 67–75.

Altman, J. 1969. Autoradiographic and histological studies of postnatal neurogenesis. IV. Cell proliferation and migration in the anterior forebrain, with special reference to persisting neurogenesis in the olfactory bulb. J. Comp. Neurol. 137, 433–457.

Aroniadou-Anderjaska, V., Ennis, M., and Shipley, M. T. 1997. Glomerular synaptic responses to olfactory nerve input in rat olfactory bulb slices. Neuroscience 79, 425–434.

Aroniadou-Anderjaska, V., Ennis, M., and Shipley, M. T. 1999a. Current-source density analysis in the rat olfactory bulb: laminar distribution of kainate/AMPA and NMDA receptor-mediated currents. J. Neurophysiol. 81, 15–28.

Aroniadou-Anderjaska, V., Ennis, M., and Shipley, M. T. 1999b. Dendrodendritic recurrent excitation in mitral cells of the rat olfactory bulb. J. Neurophysiol. 82, 489–494.

Aroniadou-Anderjaska, V., Zhou, F. M., Priest, C. A., Ennis, M., and Shipley, M. T. 2000. Tonic and synaptically evoked presynaptic inhibition of sensory input to the rat olfactory bulb via GABA(B) heteroreceptors. J. Neurophysiol. 84, 1194–1203.

Aston-Jones, G. and Bloom, F. E. 1981. Norepinephrine-containing locus coeruleus neurons in behaving rats exhibit pronounced responses to non-noxious environmental stimuli. J. Neurosci. 1, 887–900.

Aungst, J. L., Heyward, P. M., Puche, A. C., Karnup, S. V., Hayar, A., Szabo, G., and Shipley, M. T. 2003. Center-surround inhibition among olfactory bulb glomeruli. Nature 426, 623–629.

Bailey, M. S. and Shipley, M. T. 1993. Astrocyte subtypes in the rat olfactory bulb: morphological heterogeneity and differential laminar distribution. J. Comp. Neurol. 328, 501–526.

Baker, H. 1986. Species differences in the distribution of substance P and tyrosine hydroxylase immunoreactivity in the olfactory bulb. J. Comp. Neurol. 252, 206–226.

Baker, H., Liu, N., Chun, H. S., Saino, S., Berlin, R., Volpe, B., and Son, J. H. 2001. Phenotypic differentiation during migration of dopaminergic progenitor cells to the olfactory bulb. J. Neurosci. 21, 8505–8513.

Balu, R., Larimer, P., and Strowbridge, B. W. 2004. Phasic stimuli evoke precisely timed spikes in intermittently discharging mitral cells. J. Neurophysiol. 92, 743–753.

Banay-Schwartz, M., Lajtha, A., and Palkovits, M. 1989a. Changes with aging in the levels of amino acids in rat CNS structural elements: I glutamate and related amino acids. Neurochem. Res. 14, 555–562.

Banay-Schwartz, M., Lajtha, A., and Palkovits, M. 1989b. Changes with aging in the levels of amino acids in rat CNS structural elements: II. Taurine and small neutral amino acids. Neurochem. Res. 14, 563–570.

Bardoni, R., Magherini, P. C., and Belluzzi, O. 1995. Sodium current in periglomerular cells of frog olfactory bulb in vitro. Brain Res. 703, 19–25.

Bardoni, R., Magherini, P. C., and Belluzzi, O. 1996. Excitatory synapses in the glomerular triad of frog olfactory bulb in vitro. Neuroreport 7, 1851–1855.

Bayer, S. A. 1983. 3H-Thymidine-radiographic studies of neurogenesis in the rat olfactory bulb. Exp. Brain Res. 50, 329–340.

Belluardo, N., Mudò, G., Trovato-Salinaro, A., Le Gurun, S., Charollais, A., Serre-Beinier, V., Amato, G., Haefliger, J.-A., Meda, P., and Condorelli, D. F. 2000. Expression of connexin36 in the adult and developing rat brain. Brain Res. 865, 121–138.

Belluscio, L., Lodovichi, C., Feinstein, P., Mombaerts, P., and Katz, L. C. 2002. Odorant receptors instruct functional circuitry in the mouse olfactory bulb. Nature 419, 296–300.

Belluzzi, O., Benedusi, M., Ackman, J., and LoTurco, J. 2003. Electrophysiological identification of new neurons in the olfactory bulb. J. Neurosci. 23, 10411–10418.

Belluzzi, O., Puopolo, M., Benedusi, M., and Kratskin, I. 2004. Selective neuroinhibitory effects of taurine in slices of rat main olfactory bulb. Neuroscience 124, 929–944.

Berkowicz, D. A. and Trombley, P. Q. 2000. Dopaminergic modulation at the olfactory nerve synapse. Brain Res. 855, 90–99.

Biffo, S., Grillo, M., and Margolis, F. L. 1990. Cellular localization of carnosine-like and anserine-like immunoreactivities in rodent and avian central nervous system. Neuroscience 35, 637–651.

Bischofberger, J. and Jonas, P. 1997. Action potential propagation into the presynaptic dendrites of rat mitral cells. J. Physiol. 504, 359–365.

Blakemore, L. J. and Trombley, P. Q. 2003. Kinetic variability of AMPA receptors among olfactory bulb neurons in culture. Neuroreport 14, 965–970.

Blanes, T. 1898. Sobre algunos puntos dudosos de la estructura del bulbo olfactorio. Revista Trimestral micrográgica 3, 99–127.

Bogan, N., Brecha, N., Gall, C., and Karten, H. J. 1982. Distribution of enkephalin-like immunoreactivity in the rat main olfactory bulb. Neuroscience 7, 895–906.

Bokil, H., Laaris, N., Ennis, M., and Keller, A. 2001. Ephaptic interactions in the mammalian olfactory system. J. Neurosci. 21, RC173.

Bonino, M., Cantino, D., and Sassoe-Pognetto, M. 1999. Cellular and subcellular localization of gamma-aminobutyric acid B receptors in the rat olfactory bulb. Neurosci. Lett. 274, 195–198.

Bouna, S., Gysling, K., Calas, A., and Araneda, S. 1994. Some noradrenergic neurons of locus ceruleus-olfactory pathways contain neuropeptide-Y. Brain Res. Bull. 34, 413–417.

Bowery, N. G., Hudson, A. L., and Price, G. W. 1987. GABAA and GABAB receptor site distribution in the rat central nervous system. Neuroscience 20, 365–383.

Brennan, P., Kaba, H., and Keverne, E. B. 1990. Olfactory recognition: a simple memory system. Science 250, 1223–1226.

Brumberg, J. C., Nowak, L. G., and McCormick, D. A. 2000. Ionic mechanisms underlying repetitive high-frequency burst firing in supragranular cortical neurons. J. Neurosci. 20, 4829–4843.

Burd, G. D., Davis, B. J., Macrides, F., Grillo, M., and Margolis, F. 1982. Carnosine in primary afferents of the olfactory system: an autoradiographic and biochemical study. J. Neurosci. 2, 244–255.

Cadetti, L. and Belluzzi, O. 2001. Hyperpolarisation-activated current in glomerular cells of the rat olfactory bulb. Neuroreport 12, 3117–3120.

Cajal, S. R. Y. 1890. The origins and terminations of the olfactory, optic, and acoustic nerves in vertebrates. In: *Studies on the Cerebral Cortex [Limbic Structures]*. (Translated by L.M. Kraft (1955)), pp. 1–27 The Year Book Publishers.

Cajal, S. R. Y. 1911a. Olfactory apparatus: olfactory mucosa and olfactory bulb or first-order olfactory center, In: Histology of the Nervous System, Vol. II. (Translated by N. Swanson and L. Swanson (1995)), pp. 532–554. Oxford University Press.

Cajal, R. S. Y. 1911b. *Histologie du Systeme Nerveux de l'Hommes et des Vertebres*. Paris: Maloine.

Camps, M., Kelly, P. H., and Palacios, J. M. 1990. Autoradiographic localization of D1 and D2 receptors in the brain of several mammalian species. J. Neural Transm. 80, 105–127.

Cang, J. and Isaacson, J. S. 2003. *In vivo* whole-cell recording of odor-evoked synaptic transmission in the rat olfactory bulb. J. Neurosci. 23, 4108–4116.

Carleton, A., Petreanu, L. T., Lansford, R., Alvarez-Buylla, A., and Lledo, P.-M. 2003. Becoming a new neuron in the adult olfactory bulb. Nat. Neurosci. 6, 507–518.

Carlson, G. C., Shipley, M. T., and Keller, A. 2000. Long-lasting depolarizations in mitral cells of the rat olfactory bulb. J. Neurosci. 20, 2011–2021.

Carlson, G. C., Slawecki, M. L., Lancaster, E., and Keller, A. 1997. Distribution and activation of intracellular Ca^{2+} stores in cultured olfactory bulb neurons. J. Neurophysiol. 78, 2176–2185.

Carson, K. A. 1984. Quantitative localization of neurons projecting to the mouse main olfactory bulb. Brain Res. Bull. 12, 629–634.

Castillo, P. E., Carleton, A., Vincent, J. D., and Lledo, P. M. 1999. Multiple and opposing roles of cholinergic transmission in the main olfactory bulb. J. Neurosci. 19, 9180–9191.

Chalansonnet, M. and Chaput, M. A. 1998. Olfactory bulb output cell temporal response patterns to increasing odor concentrations in freely breathing rats. Chem. Senses 23, 1–9.

Chanse, N. T. and Kopin, I. J. 1968. Stimulus-induced release of substances from olfactory bulb using the push–pull cannula. Nature 217, 466–498.

Chaput, M. 1983. Effects of olfactory peduncle sectioning on the single unit responses of olfactory bulb neurons to odor presentation in awake rabbits. Chem. Senses 8, 161–177.

Chaput, M. and Holley, A. 1979. Spontaneous activity of olfactory bulb neurons in awake rabbits, with some observations on the effects of pentobarbital anaesthesia. J. Physiol. (Paris) 75, 939–948.

Charpak, S., Mertz, J., Beaurepaire, E., Moreaux, L., and Delaney, K. 2001. Odor-evoked calcium signals in dendrites of rat mitral cells. Proc. Natl. Acad. Sci. U. S. A. 98, 1230–1234.

Chen, W. R. and Shepherd, G. M. 1997. Membrane and synaptic properties of mitral cells in slices of rat olfactory bulb. Brain Res. 745, 189–196.

Chen, W. R., Midtgaard, J., and Shepherd, G. M. 1997. Forward and backward propagation of dendritic impulses and their synaptic control in mitral cells. Science 278, 463–467.

Chen, W. R., Xiong, W., and Shepherd, G. M. 2000. Analysis of relations between NMDA receptors and GABA release at olfactory bulb reciprocal synapses. Neuron 25, 625–633.

Christie, J. M. and Westbrook, G. L. 2003. Regulation of backpropagating action potential in mitral cell lateral dendrites by A-type potassium currents. J. Neurophysiol. 89, 2466–2472.

Christie, J. M. and Westbrook, G. L. 2006. Lateral excitation within the olfactory bulb. J. Neurosci. 26, 2269–2277.

Christie, J. M., Bark, C., Hormuzdi, S. G., Helbig, I., Monyer, H., and Westbrook, G. L. 2005. Connexin36 mediates spike synchrony in olfactory bulb glomeruli. Neuron 46, 761–772.

Christie, J. M., Schoppa, N. E., and Westbrook, G. L. 2001. Tufted cell dendrodendritic inhibition in the olfactory bulb is dependent on NMDA receptor activity. J. Neurophysiol. 85, 169–173.

Chu, D. C. M., Albin, R. L., Young, A. B., and Penney, J. B. 1990. Distribution and kinetics of GABAB binding sites in rat central nervous system: a quantitative autoradiographic study. Neuroscience 34, 341–357.

Ciombor, K. J., Ennis, M., and Shipley, M. T. 1999. Norepinephrine increases rat mitral cell excitatory responses to weak olfactory nerve input via alpha-1 receptors *in vitro*. Neuroscience 90, 595–606.

Collins, G. G. 1974. The rates of synthesis, uptake and disappearance of [14C]-taurine in eight areas of the rat central nervous system. Brain Res. 76, 447–459.

Condorelli, D. F., Belluardo, N., Trovato-Salimaro, A., and Mudò 2000. Expression of Cx36 in mammalian neurons. Brain Res. Rev. 32, 72–85.

Condorelli, D. F., Parenti, R., Spinella, F., Salinaro, A. T., Belluardo, N., Cardile, V., and Cicirata, F. 1998. Cloning of a new gap junction gene (Cx36) highly expressed in mammalian neurons. Eur. J. Neurosci. 10, 1202–1208.

Coopersmith, R. and Leon, M. 1984. Enhanced neural response to familiar olfactory cues. Science 225, 849–851.

Coronas, V., Srivastava, L. K., Liang, J. J., Jourdan, F., and Moyse, E. 1997. Identification and localization of dopamine receptor subtypes in rat olfactory mucosa and bulb: a combined in situ hybridization and ligand binding radioautographic approach. J. Chem. Neuroanat. 12, 243–257.

Davila, N. G., Blakemore, L. J., and Trombley, P. Q. 2003. Dopamine modulates synaptic transmission between rat olfactory bulb neurons in culture. J. Neurophysiol. 90, 395–404.

Davis, B. J. 1991. NADPH-diaphorase activity in the olfactory system of the hamster and rat. J. Comp. Neurol. 314, 493–511.

Davis, B. J. and Macrides, F. 1983. Tyrosine hydroxylase immunoreactive neurons and fibers in the olfactory system of the hamster. J. Comp. Neurol. 214, 427–440.

Davis, B. J., Burd, G. D., and Macrides, F. 1982. Localization of methionine-enkephalin, substance P, and somatostatin immunoreactivities in the main olfactory bulb of the hamster. J. Comp. Neurol. 204, 377–383.

Debarbieux, F., Audinat, E., and Charpak, S. 2003. Action potential propagation in dendrites of rat mitral cells *in vivo*. J. Neurosci. 23, 5553–5560.

Desmaisons, D., Vincent, J-D., and Lledo, P-M. 1999. Control of action potential timing by intrinsic subthreshold oscillations in olfactory bulb neurons. J. Neurosci. 19, 10727–10737.

De Saint Jan, D. and Westbrook, G. L. 2005. Detecting activity in olfactory bulb glomeruli with astrocyte recording. J. Neurosci. 25, 2917–2924.

Didier, A., Ottersen, O. P., and Storm-Mathisen, J. 1994. Differential subcellular distribution of glutamate and taurine in primary olfactory neurons. Neuroreport 6, 145–148.

Dietz, S. B. and Murthy, V. N. 2005. Contrasting short-term plasticity at two sides of the mitral-granule reciprocal synapse in the mammalian olfactory bulb. J. Physiol. 569, 475–488.

DiFrancesco, D. 1993. Pacemaker mechanisms in cardiac tissue. Annu. Rev. Physiol. 55, 455–472.

Djurisic, M., Antic, S., Chen, W. R., and Zecevic, D. 2004. Voltage imaging from dendrites of mitral cells: EPSP attenuation and spike trigger zones. J. Neurosci. 24, 6703–6714.

Doty, R. L. and Risser, J. M. 1989. Influence of the D-2 dopamine receptor agonist quinpirole on the odor detection performance of rats before and after spiperone administration. Psychopharmacology 98, 310–315.

Doucette, R. 1989. Development of the nerve fiber layer in the olfactory bulb of mouse embryos. J. Comp. Neurol. 285, 514–527.

Duchamp-Viret, P., Chaput, M. A., and Duchamp, A. 1999. Odor response properties of rat olfactory receptor neurons. Science 284, 2171–2174.

Eeckman, F. H. and Freeman, W. J. 1990. Correlations between unit firing and EEG in the rat olfactory system. Brain Res. 528, 238–244.

Egger, V., Svoboda, K., and Mainen, Z. F. 2003. Mechanisms of lateral inhibition in the olfactory bulb: efficiency and modulation of spike-evoked calcium influx into granule cells. J. Neurosci. 23, 7551–7558.

Egger, V., Svoboda, K., and Mainen, Z. F. 2005. Dendrodendritic synaptic signals in olfactory bulb granule cells: local spine boost and global low-threshold spike. J. Neurosci. 25, 3521–3530.

Elaagouby, A., Ravel, N., and Gervais, R. 1991. Cholinergic modulation of excitability in the rat olfactory bulb: effect of local application of cholinergic agents on evoked field potentials. Neuroscience 45, 653–662.

El-Etri, M. M., Griff, E. R., Ennis, M., and Shipley, M. T. 1999. Evidence for cholinergic regulation of basal norepinephrine in the rat olfactory bulb. Neuroscience 93, 611–617.

Ennis, M., Hamilton, K. A., and Hayar, A. 2007. Neurochemistry of the main olfactory system. In: Sensory Neurochemistry In: Handbook of Neurochemistry and Molecular Neurobiology 3rd Edition (ed. A. Lajtha), Vol. 20, Sensory Neurochemistry (ed D. A. Johnson) pp. 137–204. Spinger.

Ennis, M., Zhou, F. M., Ciombor, K. J., Aroniadou-Anderjaska, V., Hayar, A., Borrelli, E., Zimmer, L. A., Margolis, F., and Shipley, M. T. 2001. Dopamine D2 receptor-mediated presynaptic inhibition of olfactory nerve terminals. J. Neurophysiol. 86, 2986–2997.

Ennis, M., Zhu, M., Heinbockel, T., and Hayar, A. 2006. Olfactory nerve-evoked metabotropic glutamate receptor-mediated responses in olfactory bulb mitral cells. J. Neurophysiol. 95, 2233–2241.

Ennis, M., Zimmer, L. A., and Shipley, M. T. 1996. Olfactory nerve stimulation activates rat mitral cells via NMDA and non- NMDA receptors in vitro. Neuroreport 7, 989–992.

Ezeh, P. I., Wellis, D. P., and Scott, J. W. 1993. Organization of inhibition in the rat olfactory bulb external plexiform layer. J. Neurophysiol. 70, 263–274.

Fadool, D. A., Tucker, K., Perkins, R., Fasciani, G., Thompson, R. N., Parsons, A. D., Overton, J. M., Koni, P. A., Flavell, R. A., and Kaczmarek, L. K. 2004. Kv1.3 channel gene-targeted deletion produces "Super-Smeller Mice" with altered glomeruli, interacting scaffolding proteins, and biophysics. Neuron 41, 389–404.

Ferriero, D. and Margolis, F. L. 1975. Denervation in the primary olfactory pathway of mice. II. Effects on carnosine and other amine compounds. Brain Res. 94, 75–86.

Fletcher, M. L. and Wilson, D. A. 2002. Experience modifies olfactory acuity: acetylcholine-dependent learning decreases behavioral generalization between similar odorants. J. Neurosci. 22, RC201.

Fletcher, M. L. and Wilson, D. A. 2003. Olfactory bulb mitral-tufted cell plasticity: odorant-specific tuning reflects previous odorant exposure. J. Neurosci. 23, 6946–6955.

Fletcher, M. L., Smith, A. M., Best, A. R., and Wilson, D. A. 2005. High-frequency oscillations are not necessary for simple olfactory discriminations in young rats. J. Neurosci. 25, 792–798.

Frazier, L. L. and Brunjes, P. C. 1988. Unilateral odor deprivation: early postnatal changes in olfactory bulb cell density and number. J. Comp. Neurol. 269, 355–370.

Friedman, D. and Strowbridge, B. W. 2000. Functional role of NMDA autoreceptors in olfactory mitral cells. J. Neurophysiol. 84, 39–50.

Friedman, D. and Strowbridge, B. W. 2003. Both electrical and chemical synapses mediate fast network oscillations in the olfactory bulb. J. Neurophysiol. 89, 2601–2610.

Friedrich, R. W. 2002. Real time odor representations. Trends Neurosci. 25, 487–489.

Frosch, M. P. and Dichter, M. A. 1982. Physiology and pharmacology of olfactory bulb neurons in dissociated cell culture. Brain Res. 290, 321–332.

Gall, C., Seroogy, K. B., and Brecha, N. 1986. Distribution of VIP- and NPY-like immunoreactivities in rat main olfactory bulb. Brain Res. 374, 389–394.

Gall, C. M., Hendry, S. H. C., Seroogy, K. B., Jones, E. G., and Haycock, J. W. 1987. Evidence for coexistence of GABA and dopamine in neurons of the rat olfactory bulb. J. Comp. Neurol. 266, 307–318.

Geiling, H. and Schild, D. 1996. Glutamate-mediated release of Ca^{2+} in mitral cells of the olfactory bulb. J. Neurophysiol. 76, 563–570.

Gelperin, A. 2006. Olfactory computations and network oscillation. J. Neurosci. 26, 1663–1668.

Getchell, T. V. and Shepherd, G. M. 1975. Synaptic actions on mitral and tufted cells elicited by olfactory nerve volleys in the rabbit. J. Physiol. 251, 497–522.

Gheusi, G., Cremer, H., McLean, H., Chazal, G., Vincent, J.-D., and Lledo, P.-M. 2000. Importance of newly generated neurons in the adult olfactory bulb for odor discrimination. Proc. Natl. Acad. Sci. U. S. A. 97, 1823–1828.

Giustetto, M., Bovolin, P., Fasolo, A., Bonino, M., Cantino, D., and Sassoe-Pognetto, M. 1997. Glutamate receptors in the olfactory bulb synaptic circuitry: heterogeneity and synaptic localization of N-methyl-D-aspartate receptor subunit 1 and AMPA receptor subunit 1. Neuroscience 76, 787–798.

Golgi, C. 1875. Sulla Fina Structrura del Bulbi Olfattori. Rev Sper Feniatr 1, 405–425.

Gray, C. M. and Skinner, J. E. 1988. Centrifugal regulation of olfactory bulb of the waking rabbit as revealed by reversible cryogenic blockade. Exp. Brain Res. 69, 378–386.

Greer, C. A. 1987. Golgi analyses of dendritic organization among denervated olfactory bulb granule cells. J. Comp. Neurol. 256, 442–452.

Halabisky, B. and Strowbridge, B. W. 2003. γ-Frequency excitatory input to granule cells facilitates dendrodendritic inhibition in the rat olfactory bulb. J. Neurophysiol. 90, 644–654.

Halabisky, B., Friedman, D., Radojicic, M., and Strowbridge, B. W. 2000. Calcium influx through NMDA receptors directly evokes GABA release in olfactory bulb granule cells. J. Neurosci. 20, 5124–5134.

Halász, N. and Shepherd, G. M. 1983. Neurochemistry of the vertebrate olfactory bulb. Neuroscience 10, 579–619.

Halász, N., Hokfelt, T., Norman, A. W., and Goldstein, M. 1985. Tyrosine hydroxylase and 28K-vitamin D-dependent calcium binding protein are localized in different subpopulations of periglomerular cells of the rat olfactory bulb. Neurosci. Lett. 61, 103–107.

Halász, N., Ljungdahl, A., Hokfelt, T., Johansson, O., Goldstein, M., Park, D., and Biberfeld, P. 1977. Transmitter histochemistry of the rat olfactory bulb. I. Immunohistochemical localization of monoamine synthesizing enzymes. Support for intrabulbar, peri-glomerular dopamine neurons. Brain Res. 126, 455–474.

Hamilton, K. A. and Coppola, D. M. 2003. Distribution of GluR1 is altered in the olfactory bulb following neonatal naris occlusion. J. Neurobiol. 54, 326–336.

Hamilton, K. A., Heinbockel, T., Ennis, M., Szabó, G., Erdélyi, F., and Hayar, A. 2005. Properties of external plexiform layer interneurons in mouse olfactory bulb slices. Neuroscience 133, 819–829.

Hardy, A., Palouzier-Paulignan, B., Duchamp, A., Royet, J. P., and Duchamp-Viret, P. 2005. 5-hydroxytryptamine action in the rat olfactory bulb: In vitro electrophysiological patch-clamp recordings of juxtaglomerular and mitral cells. Neuroscience 131, 717–731.

Hayar, A., Heyward, P. M., Heinbockel, T., Shipley, M. T., and Ennis, M. 2001. Direct excitation of mitral cells via activation of alpha1-noradrenergic receptors in rat olfactory bulb slices. J. Neurophysiol. 86, 2173–2182.

Hayar, A., Karnup, S., Ennis, M., and Shipley, M. T. 2004b. External tufted cells: a major excitatory element that coordinates glomerular activity. J. Neurosci. 24, 6676–6685.

Hayar, A., Karnup, S., Shipley, M. T., and Ennis, M. 2004a. Olfactory bulb glomeruli: external tufted cells intrinsically burst at theta frequency and are entrained by patterned olfactory input. J. Neurosci. 24, 1190–1199.

Hayar, A., Shipley, M. T., and Ennis, M. 2005. Olfactory bulb external tufted cells are synchronized by multiple intraglomerular mechanisms. J. Neurosci. 25, 8197–8208.

Heinbockel, T., Heyward, P., Conquet, F., and Ennis, M. 2004. Regulation of main olfactory bulb mitral cell excitability by metabotropic glutamate receptor mGluR1. J. Neurophysiol. 92, 3085–3096.

Heinbockel, T., Laaris, N., and Ennis, M. 2007. Metabotropic glutamate receptors in the main olfactory bulb drive granule cell-mediated inhibition. J. Neurophysiol. 97, 858–870.

Heyward, P. M., Ennis, M., Keller, A., and Shipley, M. T. 2001. Membrane bistability in olfactory bulb mitral cells. J. Neurosci. 21, 5311–5320.

Hinds, J. W. 1968. Autoradiographic study of histogenesis in the mouse olfactory bulb. I. Time of origin of neurons and neuroglia. J. Comp. Neurol. 134, 287–304.

Hirata, Y. 1964. Some observations on the fine structure of the synapses in the olfactory bulb of the mouse, with particular reference to the atypical synaptic configurations. Arch. Histol. Jpn. 24, 293–302.

Hormuzdi, S. G., Pais, I., LeBleu, F. E. N., Towers, S. K., Rozov, A., Buhl, E. H., Whittington, M. A., and Monyer, H. 2001. Impaired electrical signaling disrupts gamma frequency oscillations in connexin 36-deficient mice. Neuron 31, 487–495.

Horning, M. S., Blakemore, L. J., and Trombley, P. Q. 2000. Endogenous mechanisms of neuroprotection: role of zinc, copper, and carnosine. Brain Res. 852, 56–61.

Hsia, A. Y., Vincent, J. D., and Lledo, P. M. 1999. Dopamine depresses synaptic inputs into the olfactory bulb. J. Neurophysiol. 82, 1082–1085.

Illig, K. R. and Haberly, L. B. 2003. Odor-evoked activity is spatially distributed in piriform cortex. J. Comp. Neurol. 457, 361–373.

Inagaki, S., Sakanaka, M., Shiosaka, S., Senba, E., Takatsuki, K., Takagi, H., Kawai, Y., Minagawa, H., and Tohyama, M. 1982. Ontogeny of substance P-containing neuron system of the rat: immunohistochemical analysis – I. Forebrain and upper brain stem. Neuroscience 7, 251–277.

Isaacson, J. S. 1999. Glutamate spillover mediates excitatory transmission in the rat olfactory bulb. Neuron 23, 377–384.

Isaacson, J. S. 2001. Mechanisms governing dendritic gamma-aminobutyric acid (GABA) release in the rat olfactory bulb. Proc. Natl. Acad. Sci. U.S.A. 98, 337–342.

Isaacson, J. S. and Murphy, G. J. 2001. Glutamate-mediated extrasynaptic inhibition: direct coupling of NMDA receptors to Ca^{2+}-activated K^+ channels. Neuron 31, 1027–1034.

Isaacson, J. S. and Strowbridge, B. W. 1998. Olfactory reciprocal synapses: dendritic signaling in the CNS. Neuron 20, 749–761.

Isaacson, J. S. and Vitten, H. 2003. GABAB receptors inhibit dendrodendritic transmission in the rat olfactory bulb. J. Neurosci. 23, 2032–2039.

Jackowski, A., Parnavelas, J. G., and Lieberman, A. R. 1978. The reciprocal synapse in the external plexiform layer of the mammalian olfactory bulb. Brain Res. 159, 17–28.

Jahr, C. E. and Nicoll, R. A. 1982. Noradrenergic modulation of dendrodendritic inhibition in the olfactory bulb. Nature 297, 227–229.

Jardemark, K., Nilsson, M., Muyderman, H., and Jacobson, I. 1997. Ca2+ ion permeability properties of (RS)α-amino-3-hydroxy-5-methyl-4-isoxazolepropionate (AMPA) receptors in isolated interneurons from the olfactory bulb of the rat. J. Neurophysiol. 77, 702–708.

Jiang, M. R., Griff, E. R., Ennis, M., Zimmer, L. A., and Shipley, M. T. 1996. Activation of locus coeruleus enhances the responses of olfactory bulb mitral cells to weak olfactory nerve input. J. Neurosci. 16, 6319–6329.

Kaba, H. and Keverne, E. B. 1988. The effect of microinfusions of drugs into the accessory olfactory block to pregnancy. Neuroscience 25, 1007–1011.

Kamisaki, Y., Wada, K., Nakamoto, K., and Itoh, T. 1996. Effects of taurine on GABA release from synaptosomes of rat olfactory bulb. Amino Acids 10, 49–57.

Karnup, S. V., Hayar, A., Shipley, M. T., and Kurnikova, M. G. 2006. Spontaneous field potentials in the glomeruli of the olfactory bulb: the leading role of juxtaglomerular cells. Neuroscience 142, 203–221.

Kashiwadani, H., Sasaki, Y. F., Uchida, N., and Mori, K. 1999. Synchronized oscillatory discharges of mitral/tufted cells with different molecular receptive ranges in the rabbit olfactory bulb. J. Neurophysiol. 182, 1786–1792.

Kasowski, H. J., Kim, H., and Greer, C. A. 1999. Compartmental organization of the olfactory bulb glomerulus. J. Comp. Neurol. 407, 261–274.

Kauer, J. S. 1998. Olfactory processing: a time and place for everything. Curr. Biol. 8, R282–R283.

Kay, L. M. 2003. A challenge to chaotic intinerancy from brain dynamics. Chaos 13, 1057–1066.

Kay, L. M. 2005. Theta oscillations and sensorimotor performance. Proc. Natl. Acad. Sci. U. S. A. 102, 3863–3868.

Kay, L. M. and Freeman, W. J. 1998. Bidirectional processing in the olfactory-limbic axis during olfactory behavior. Behav. Neurosci. 112, 541–553.

Kay, L. M. and Laurent, G. 1999. Odor- and context-dependent modulation of mitral cell activity in behaving rats. Nat. Neurosci. 2, 1003–1009.

Keller, A., Yagodin, S., Aroniadou-Anderjaska, V., Zimmer, L. A., Ennis, M., Sheppard, N. F., Jr, and Shipley, M. T. 1998. Functional organization of rat olfactory bulb glomeruli revealed by optical imaging. J. Neurosci. 18, 2602–2612.

Kendrick, K. M., Levy, F., and Keverne, E. B. 1992. Changes in the sensory processing of olfactory signals induced by birth in sheep. Science 256, 833–836.

Kepecs, A., Uchida, N., and Mainen, Z. F. 2006. The sniff as a unit of olfactory processing. Chem. Senses 31, 167–179.

Kerr, D. I. B. and Hagbarth, K. E. 1955. An investigation of the olfactory centrifugal system. J. Neurophysiol. 18, 362–374.

Konig, P., Engel, A. K., and Singer, W. 1996. Integrator or coincidence detector? The role of the cortical neuron revisited. Trends Neurosci. 19, 130–137.

Kosaka, K., Aika, Y., Toida, K., Heizmann, C. W., Hunziker, W., Jacobowitz, D. M., Nagatsu, I., Streit, P., Visser, T. J., and Kosaka, T. 1995. Chemically defined neuron groups and their subpopulations in the glomerular layer of the rat main olfactory bulb. Neurosci. Res. 23, 73–88.

Kosaka, K., Hama, K., Nagatsu, I., Wu, J. Y., and Kosaka, T. 1988. Possible coexistence of amino acid (gamma-aminobutyric acid), amine (dopamine) and peptide (substance P); neurons containing immunoreactivities for glutamic acid decarboxylase, tyrosine hydroxylase and substance P in the hamster main olfactory bulb. Exp. Brain Res. 71, 633–642.

Kosaka, K., Hama, K., Nagatsu, I., Wu, J. Y., Ottersen, O. P., Storm-Mathisen, J., and Kosaka, T. 1987. Postnatal development of neurons containing both catecholaminergic and GABAergic traits in the rat main olfactory bulb. Brain Res. 403, 355–360.

Kosaka, K., Toida, K., Aika, Y., and Kosaka, T. 1998. How simple is the organization of the olfactory glomerulus?: the heterogeneity of so-called periglomerular cells. Neurosci. Res. 30, 101–110.

Kosaka, K., Toida, K., Margolis, F. L., and Kosaka, T. 1997. Chemically defined neuron groups and their subpopulations in the glomerular layer of the rat main olfactory bulb – II. Prominent differences in the intraglomerular dendritic arborization and their relationship to olfactory nerve terminals. Neuroscience 76, 775–786.

Kosaka, T. and Kosaka, K. 2003. Neuronal gap junctions in the rat main olfactory bulb, with special reference to intraglomerular gap junction. Neurosci. Res. 45, 189–209.

Kosaka, T. and Kosaka, K. 2004. Neuronal gap junctions between intraglomerular mitral/tufted cell dendrites in the mouse main olfactory bulb. Neurosci. Res. 49, 373–378.

Kosaka, T., Hataguchi, Y., Hama, K., Nagatsu, I., and Wu, J. Y. 1985. Coexistence of immunoreactivities for glutamate decarboxylase and tyrosine hydroxylase in some neurons in the periglomerular region of the rat main olfactory bulb: possible coexistence of gamma-aminobutyric acid (GABA) and dopamine. Brain Res. 343, 166–171.

Kosaka, T., Kosaka, K., Hama, K., Wu, J.-Y., and Nagatsy, I. 1987a. Differential effect of functional olfactory deprivation on the GABAergic and catecholaminergic traits in the rat main olfactory bulb. Brain Res. 413, 197–203.

Kosaka, T., Kosaka, K., Hataguchi, Y., Nagatsu, I., Wu, J. Y., Ottersen, O. P., Storm-Mathisen, J., and Hama, K. 1987b. Catecholaminergic neurons containing GABA-like and/or glutamic acid decarboxylase-like immunoreactivities in various brain regions of the rat. Exp. Brain Res. 66, 191–210.

Kosaka, T., Kosaka, K., Heizmann, C. W., Nagatsu, I., Wu, J. Y., Yanaihara, N., and Hama, K. 1987c. An aspect of the organization of the GABAergic system in the rat main olfactory bulb: laminar distribution of immunohistochemically defined subpopulations of GABAergic neurons. Brain Res. 411, 373–378.

Koster, N. L., Norman, A. B., Richtand, N. M., Nickell, W. T., Puche, A. C., Pixley, S. K., and Shipley, M. T. 1999. Olfactory receptor neurons express D2 dopamine receptors. J. Comp. Neurol. 411, 666–673.

Kratskin, I. and Belluzzi, O. 2002. Immunohistochemical localization of the taurine-synthesizing enzyme in the rat olfactory bulb. Chem. Senses 27, 46.

Kratskin, I. and Belluzzi, O. 2003. Anatomy and neurochemistry of the olfactory bulb. In: Handbook of Olfaction and Gustation, 2nd Edition (ed. R. L. Doty), pp. 139–164. Marcel Dekker.

Kratskin, I. L., Rio, J. P., Kenigfest, N. B., Doty, R. L., and Repérant, J. 2000. A light and electron microscopic study of taurine-like immunoreactivity in the main olfactory bulb of frogs. J. Chem. Neuroanat. 18, 87–101.

Kunze, W. A. A., Shafton, A. D., Kemm, R. E., and McKenzie, J. S. 1991. Effect of stimulating the nucleus of the horizontal limb of the diagonal band on single unit activity in the olfactory bulb. Neuroscience 40, 21–27.

Kunze, W. A. A., Shafton, A. D., Kemm, R. E., and McKenzie, J. S. 1992. Intracellular responses of olfactory bulb granule cells to stimulating the horizontal diagonal band nucleus. Neuroscience 48, 363–369.

Laaris, N., Heinbockel, T., and Ennis, M. 2007. Complementary Postsynaptic activity patterns elicited in olfactory bulb by stimulation mitral/tufted an centrifugal fiber inputs to granule cells. J. Neurophysiol. 97, 296–306.

Lagier, S., Carleton, A., and Lledo, P. M. 2004. Interplay between local GABAergic interneurons and relay neurons generates gamma oscillations in the rat olfactory bulb. J. Neurosci. 24, 4382–4392.

Laurent, G. 2002. Olfactory network dynamics and the coding of multidimensional signals. Nat. Rev. Neurosci. 3, 884–895.

Levey, A. I., Hersch, S. M., Rye, D. B., Sunahara, R. K., Niznik, H. B., Kitt, C. A., Price, D. L., Maggio, R., Brann, M. R., and Ciliax, B. J. 1993. Localization of D1 and D2 dopamine receptors in brain with subtype-specific antibodies. Proc. Natl. Acad. Sci. U. S. A. 90, 8861–8865.

Liu, C. J., Grandes, P., Matute, C., Cuenod, M., and Streit, P. 1989. Glutamate-like immunoreactivity revealed in rat olfactory bulb, hippocampus and cerebellum by monoclonal antibody and sensitive staining method. Histochemistry 90, 427–445.

Liu, W. L. and Shipley, M. T. 1994. Intrabulbar associational system in the rat olfactory bulb comprises cholecystokinin-containing tufted cells that synapse onto the dendrites of GABAergic granule cells. J. Comp. Neurol. 346, 541–558.

Lledo, P. M., Gheusi, G., and Vincent, J. D. 2005. Information processing in the mammalian olfactory system. Physiol. Rev. 85, 281–317.

Lodovichi, C., Belluscio, L., and Katz, L. C. 2003. Functional topography of connections linking mirror-symmetric maps in the mouse olfactory bulb. Neuron 38, 265–276.

Lois, C. and Alvarez-Buylla, A. 1994. Long-distance neuronal migration in the adult mammalian brain. Science 264, 1145–1148.

López-Mascaraque, L., de Carlos, J. A., and Valverde, F. 1990. Structure of the olfactory bulb of the hedgehog (Erinaceus europaeus): a Golgi study of the intrinsic organization of the superficial layers. J. Comp. Neurol. 301, 243–261.

López-Mascaraque, L., Villalba, R. M., and de Carlos, J. A. 1989. Vasoactive intestinal polypeptide-immunoreactive neurons in the main olfactory bulb of the hedgehog (Erinaceus europaeus). Neurosci. Lett. 98, 19–21.

Lowe, G. 2002. Inhibition of backpropagating action potentials in mitral cell secondary dendrites. J. Neurophysiol. 88, 64–85.

Luo, M. and Katz, L. 2001. Response correlation maps of neurons in the mammalian olfactory bulb. Neuron 32, 1165–1179.

Luskin, M. B. 1993. Restricted proliferation and migration of postnatally generated neurons derived from the forebrain subventricular zone. Neuron 11, 173–189.

Lüthi, A. and McCormick, D. A. 1998. H-current: properties of a neuronal and network pacemaker. Neuron 21, 9–12.

MacLeod, N. K. and Straughan, D. W. 1979. Responses of olfactory bulb neurones to the dipeptide carnosine. Exp. Brain Res. 34, 183–188.

Macrides, F. and Schneider, S. P. 1982. Laminar organization of mitral and tufted cells in the main olfactory bulb of the adult hamster. J. Comp. Neurol. 208, 419–430.

Macrides, F., Eichenbaum, H. B., and Forbes, W. B. 1982. Temporal relationship between sniffing and the limbic theta rhythm during odor discrimination and reversal learning. J. Neurosci. 2, 1705–1717.

Maher, B. J. and Westbrook, G. L. 2005. SK channel regulation of dendritic excitability and dendrodendritic inhibition in the olfactory bulb. J. Neurophysiol. 94, 3743–3750.

Mansour, A., Meador-Woodruff, J. H., Bunzow, J. R., Civelli, O., Akil, H., and Watson, S. J. 1990. Localization of dopamine D2 receptor mRNA and D1 and D2 receptor binding in the rat brain and pituitary: an in situ hybridization-receptor autoradiographic analysis. J. Neurosci. 10, 2587–2600.

Margeta-Mitrovic, M., Mitrovic, I., Riley, R. C., Jan, L. Y., and Basbaum, A. I. 1999. Immunohistochemical localization of GABA(B) receptors in the rat central nervous system. J. Comp. Neurol. 405, 299–321.

Margolis, F. L. 1974. Carnosine in the primary olfactory pathway. Science 184, 909–911.

Margolis, F. L. 1980. Carnosine: an olfactory neuropeptide. In: The Role of Peptides in Neuronal Function (eds. J. L. Barker and T. G. J. Smith), pp. 545–572. Dekker.

Margolis, F. L. and Grillo, M. 1984. Carnosine, homocarnosine and anserine in vertebrate retinas. Neurochem. Int. 6, 207–209.

Margolis, F. L., Grillo, M., Grannot-Reisfeld, N., and Farbman, A. I. 1983. Purification, characterization and immunocytochemical localization of mouse kidney carnosinase. Biochim. Biophys. Acta 744, 237–248.

Margolis, F. L, Grillo, M., Hempstead, J., and Morgan, J. I. 1987. Monoclonal antibodies to mammalian carnosine synthetase. J. Neurochem. 48, 593–600.

Margolis, F. L, Grillo, M., Kawano, T., and Farbman, A. I. 1985. Carnosine synthesis in olfactory tissue during ontogeny: Effect of exogenous b-alanine. J. Neurochem. 44, 1459–1464.

Margrie, T. W. and Schaefer, A. T. 2003. Theta oscillation coupled spike latencies yield computational vigour in a mammalian sensory system. J. Physiol. 546, 363–374.

Margrie, T. W., Sakmann, B., and Urban, N. N. 2001. Action potential propagation in mitral cell lateral dendrites is decremental and controls recurrent and lateral inhibition in the mammalian olfactory bulb. Proc. Natl. Acad. Sci. U. S. A. 98, 319–324.

Martin, C., Gervais, R., Hugues, E., Messaoudi, B., and Ravel, N. 2004. Learning modulation of odor-induced oscillatory responses in the rat olfactory bulb: a correlate of odor recognition? Neurosci. 24, 389–397.

Matsutani, S., Senba, E., and Tohyama, M. 1988. Neuropeptide- and neurotransmitter-related immunoreactivities in the developing rat olfactory bulb. J. Comp. Neurol. 272, 331–342.

McLean, J. H. and Harley, C. W. 2004. Olfactory learning in the rat pup: a model that may permit visualization of a mammalian memory trace. Neuroreport 15, 1691–1697.

McLean, J. H. and Shipley, M. T. 1987a. Serotonergic afferents to the rat olfactory bulb: I Origins and laminar specificity of serotonergic inputs in the adult rat. J. Neurosci. 7, 3016–3028.

McLean, J. H. and Shipley, M. T. 1987b. Serotonergic afferents to the rat olfactory bulb: II Changes in fiber distribution during development. J. Neurosci. 7, 3029–3039.

McLean, J. H. and Shipley, M. T. 1988. Postmitotic, postmigrational expression of tyrosine hydroxylase in olfactory bulb dopaminergic neurons. J. Neurosci. 8, 3658–3669.

McLean, J. H., Darby-King, A., and Hodge, E. 1996. 5-HT2 receptor involvement in conditioned olfactory learning in the neonatal rat pup. Behav. Neurosci. 110, 1426–1434.

McLean, J. H., Shipley, M. T., Nickell, W. T., Aston-Jones, G., and Reyher, C. K. 1989. Chemoanatomical organization of the noradrenergic input from locus coeruleus to the olfactory bulb of the adult rat. J. Comp. Neurol. 285, 339–349.

McLennan, H. 1971. The Pharmacology of inhibition of mitral cells in the olfactory bulb. Brain. Res. 29, 177–184.

McQuiston, A. R. and Katz, L. C. 2001. Electrophysiology of interneurons in the glomerular layer of the rat olfactory bulb. J. Neurophysiol. 86, 1899–1907.

Meisami, E. 1989. A proposed relationship between increases in the number of olfactory receptor neurons, convergence ratio and sensitivity in the developing rat. Brain Res. Dev. Brain. Res. 46, 9–19.

Merchenthaler, I., Csernus, V., Csontos, C., Petrusz, P., and Mess, B. 1988. New data on the immunocytochemical localization of thyrotropin-releasing hormone in the rat central nervous system. Am. J. Anat. 181, 359–376.

Miragall, F., Simburger, E., and Dermietzel, R. 1996. Mitral and tufted cells of the mouse olfactory bulb possess gap junctions and express connecin43 mRNA. Neurosci. Lett. 216, 199–202.

Mombaerts, P., Wang, F., Dulac, C., Chao, S. K., Nemes, A., Mendelsohn, M., Edmondson, J., and Axel, R. 1996. Visualizing an olfactory sensory map. Cell 87, 675–686.

Montague, A. A. and Greer, C. A. 1999. Differential distribution of ionotropic glutamate receptor subunits in the rat olfactory bulb. J. Comp. Neurol. 405, 233–246.

Mori, K. 1987. Membrane and synaptic properties of identified neurons in the olfactory bulb. Prog. Neurobiol. 29, 275–320.

Mori, K. and Takagi, S. F. 1978. Activation and inhibition of olfactory bulb neurones by anterior commissure volleys in the rabbit. J. Physiol. (Lond.) 279, 589–604.

Mori, K., Kishi, K., and Ojima, H. 1983. Distribution of dendrites of mitral, displaced mitral, tufted, and granule cells in the rabbit olfactory bulb. J. Comp. Neurol. 219, 339–355.

Mori, K., Takahashi, Y. K., Igarashi, K. M., and Yamaguchi, M. 2006. Maps of odorant molecular features in the mammalian olfactory bulb. Physiol. Rev. 86, 409–433.

Moriceau, S. and Sullivan, R. M. 2004. Unique neural circuitry for neonatal olfactory learning. J. Neurosci. 24, 1182–1189.

Morizumi, T., Tsukatani, H., and Miwa, T. 1994. Olfactory disturbance induced by deafferentation of serotonergic fibers in the olfactory bulb. Neuroscience 61, 733–738.

Mouly, A. M., Elaagouby, A., and Ravel, N. 1995. A study of the effects of noradrenaline in the rat olfactory bulb using evoked field potential response. Brain Res. 681, 47–57.

Mugnaini, E., Oertel, W. H., and Wouterlood, F. F. 1984. Immunocytochemical localization of GABA neurons and dopamine neurons in the rat main and accessory olfactory bulbs. Neurosci. Lett. 47, 221–226.

Murphy, G. J., Darcy, D. P., and Isaacson, J. S. 2005. Intraglomerular inhibition: signaling mechanisms of an olfactory microcircuit. Nat. Neurosci. 8, 354–364.

Murphy, G. J., Glickfeld, L. L., Balsen, Z., and Isaacson, J. S. 2004. Sensory neuron signaling to the brain: properties of transmitter release from olfactory nerve terminals. J. Neurosci. 24, 3023–3030.

Mutoh, H., Yuan, Q., and Knöpfel, T. 2005. Long-term depression at olfactory nerve synapses. J. Neurosci. 25, 4252–4259.

Nagayama, S., Takahashi, Y. K., Yoshihara, Y., and Mori, K. 2004. Mitral and tufted cells differ in the decoding manner of odor maps in the rat olfactory bulb. J. Neurophysiol. 91, 2532–2540.

Nakajima, T., Okamura, M., Ogawa, K., and Taniguchi, K. 1996. Immunohistochemical and enzyme histochemical characteristics of short axon cells in the olfactory bulb of the golden hamster. J. Vet. Med. Sci. 58, 903–908.

Nakashima, M., Mori, K., and Takagi, S. F. 1978. Centrifugal influence on olfactory bulb activity in the rabbit. Brain Res. 154, 301–316.

Neville, K. R. and Haberly, L. B. 2003. Beta and gamma oscillations in the olfactory system of the urethane-anesthetized rat. J. Neurophysiol. 90, 3921–3930.

Nickell, W. T. and Shipley, M. T. 1988. Neurophysiology of magnocellular forebrain inputs to the olfactory bulb in the rat: frequency potentiation of field potentials and inhibition of output neurons. J. Neurosci. 8, 4492–4502.

Nickell, W. T. and Shipley, M. T. 1992. Neurophysiology of the olfactory bulb. In: Science of Olfaction (eds. M. J. Serby and K. L. Chobor), pp. 172–212. Springer.

Nickell, W. T. and Shipley, M. T. 1993. Evidence for presynaptic inhibition of the olfactory commissural pathway by cholinergic agonists and stimulation of the nucleus of the diagonal band. J. Neurosci. 13, 650–659.

Nickell, W. T., Norman, A. B., Wyatt, L. M., and Shipley, M. T. 1991. Olfactory bulb DA receptors may be located on terminals of the olfactory nerve. Neuroreport 2, 9–12.

Nicoll, R. A. 1971. Pharmacological evidence for GABA as the transmitter in granule cell inhibition in the olfactory bulb. Brain Res. 35, 137–149.

Nicoll, R. A. and Alger, B. E. 1980. Peptides as putative excitatory neurotransmitters: carnosine, enkephalin, substance P and TRH. Proc. R. Soc. Lond. 210, 133–149.

Nowycky, M. C., Halasz, N., and Shepherd, G. M. 1983. Evoked field potential analysis of dopaminergic mechanisms in the isolated turtle olfactory bulb. Neuroscience 8, 717–722.

Nusser, Z., Kay, L. M., Laurent, G., Homanics, G. E., and Mody, I. 2001. Disruption of GABAA receptors on GABAergic interneurons leads to increased oscillatory power in the olfactory bulb network. J. Neurophysiol. 86, 2823–2833.

Ohm, T. G., Müller, H., Ulfig, N., and Braak, E. 1990. Glutamic-acid-decarboxylase- and parvalbumin-like-immunoreactive structures in the olfactory bulb of the human adult. J. Comp. Neurol. 291, 1–8.

Okutani, F., Kaba, H., Takahashi, S., and Seto, K. 1998. The biphasic effect of locus coeruleus noradrenergic activation on dendrodendritic inhibition in the rat olfactory bulb. Brain Res. 783, 272–279.

Orona, E., Rainer, E. C., and Scott, J. W. 1984. Dendritic and axonal organization of mitral and tufted cells in the rat olfactory bulb. J. Comp. Neurol. 226, 346–356.

Orona, E., Scott, J. W., and Rainer, E. C. 1983. Different granule cell populations innervate superficial and deep regions of the external plexiform layer in rat olfactory bulb. J. Comp. Neurol. 217, 227–237.

Palouzier-Paulignan, B., Duchamp-Viret, P., Hardy, A. B., and Duchamp, A. 2002. GABA(B) receptor-mediated inhibition of mitral/tufted cell activity in the rat olfactory bulb: a whole-cell patch-clamp study in vitro. Neuroscience 111, 241–250.

Pape, H. C. 1996. Queer current and pacemaker: the hyperpolarization-activated cation current in neurons. Annu. Rev. Physiol. 58, 299–327.

Paternostro, M. A., Reyher, C. K. H., and Brunjes, P. C. 1995. Intracellular injections of Lucifer Yellow into lightly fixed mitral cells reveal neuronal dye-coupling in the developing rat olfactory bulb. Dev. Brain Res. 84, 1–10.

Perez, H., Hernandez, A., and Almli, C. R. 1987. Locus coeruleus stimulation modulates olfactory bulb evoked potentials. Brain Res. Bull. 18, 767–770.

Petralia, R. S. and Wenthold, R. J. 1992. Light and electron immunocytochemical localization of AMPA-selective glutamate receptors in the rat brain. J. Comp. Neurol. 318, 329–354.

Philpot, B. D., Foster, T. C., and Brunjes, P. C. 1997. Mitral/tufted cell activity is attenuated and becomes uncoupled from respiration following naris closure. J. Neurobiol. 33, 374–386.

Pignatelli, A., Kobayashi, K., Okano, H., and Belluzzi, O. 2005. Functional properties of dopaminergic neurones in the mouse olfactory bulb. J. Physiol. 564, 501–514.

Pinato, G. and Midtgaard, J. 2003. Regulation of granule cell excitability by a low-threshold calcium spike in turtle olfactory bulb. J. Neurophysiol. 90, 3341–3351.

Pinato, G. and Midtgaard, J. 2005. Dendritic sodium spikelets and low-threshold calcium spikes in turtle olfactory bulb granule cells. J. Neurophysiol. 93, 1285–1294.

Pinching, A. J. and Powell, T. P. 1971a. The neuron types of the glomerular layer of the olfactory bulb. J. Cell Sci. 9, 305–345.

Pinching, A. J. and Powell, T. P. 1971b. The neuropil of the glomeruli of the olfactory bulb. J. Cell Sci. 9, 347–377.

Pinching, A. J. and Powell, T. P. 1971c. The neuropil of the periglomerular region of the olfactory bulb. J. Cell Sci. 9, 379–409.

Pinching, A. J. and Powell, T. P. 1972b. Experimental studies on the axons intrinsic to the glomerular layer of the olfactory bulb. J. Cell Sci. 10, 637–655.

Pinching, A. J. and Powell, T. P. 1972c. The termination of centrifugal fibres in the glomerular layer of the olfactory bulb. J. Cell Sci. 10, 621–635.

Potter, S. M., Zheng, C., Koos, D. S., Feinstein, P., Fraser, S. E., and Mombaerts, P. 2001. Structure and emergence of specific olfactory glomeruli in the mouse. J. Neurosci. 21, 9713–9723.

Pow, D. V., Sullivan, R., Reye, P., and Hermanussen, S. 2002. Localization of taurine transporters, taurine, and 3H taurine accumulation in the rat retina, pituitary, and brain. Glia 37, 153–168.

Pressler, R. T. and Strowbridge, B. W. 2006. Blanes cells mediate persistent feedforward inhibition onto granule cells in the olfactory bulb. Neuron 49, 889–904.

Price, J. L. and Powell, T. P. S. 1970a. An electron-microscopic study of the termination of the afferent fibres to the olfactory bulb from the cerebral hemisphere. J. Cell Sci. 7, 157–187.

Price, J. L. and Powell, T. P. S. 1970b. The mitral and short axon cells of the olfactory bulb. J. Cell Sci. 7, 631–651.

Price, J. L. and Powell, T. P. S. 1970c. The morphology of the granule cells of the olfactory bulb. J. Cell Sci. 9, 91–123.

Price, J. L. and Powell, T. P. S. 1970d. The synaptology of the granule cells of the olfactory bulb. J. Cell Sci. 7, 125–155.

Puopolo, M. and Belluzzi, O. 1996. Sodium current in periglomerular cells of rat olfactory bulb in vitro. Neuroreport 7, 1846–1850.

Puopolo, M. and Belluzzi, O. 1998. Functional heterogeneity of periglomerular cells in the rat olfactory bulb. Eur. J. Neurosci. 10, 1073–1083.

Puopolo, M. and Belluzzi, O. 2001. NMDA-dependent, network-driven oscillatory activity induced by bicuculline or removal of Mg2+ in rat olfactory bulb neurons. Eur. J. Neurosci. 13, 92–102.

Puopolo, M., Bean, B. P., and Raviola, E. 2005. Spontaneous activity of isolated dopaminergic periglomerular cells of the main olfactory bulb. J. Neurophysiol. 94, 3618–3627.

Rall, W., Shepherd, G. M., Reese, T. S., and Brightman, M. W. 1966. Dendrodendritic synaptic pathway for inhibition in the olfactory bulb. Exp. Neurol. 14, 44–56.

Rash, J. E., Staines, W. A., Yasmura, T., Patel, D., Furman, D. S., Stelmack, G. L., and Nagy, J. I. 2000. Immunogold evidence that the neuronal gap junctions in adult rat brain and spinal cord contain connexin-36 but not connexin-32 or connexin-43. Proc. Natl. Acad. Sci. U. S. A. 97, 7573–7578.

Ressler, K. J., Sullivan, S. L., and Buck, L. B. 1993. A zonal organization of odorant receptor gene expression in the olfactory epithelium. Cell 73, 597–609.

Ressler, K. J., Sullivan, S. L., and Buck, L. B. 1994. Information coding in the olfactory system: evidence for a stereotyped and highly organized epitope map in the olfactory bulb. Cell 79, 1245–1255.

Reyher, C. K., Lubke, J., Larsen, W. J., Hendrix, G. M., Shipley, M. T., and Baumgarten, H. G. 1991. Olfactory bulb granule cell aggregates: morphological evidence for interperikaryal electrotonic coupling via gap junctions. J. Neurosci. 11, 1485–1495.

Ribak, C. E., Vaughn, J. E., Saito, K., Barber, R., and Roberts, E. 1977. Glutamate decarboxylase localization in neurons of the olfactory bulb. Brain Res. 126, 1–18.

Rochefort, C., Gheusi, G., Vincent, J.-D., and Lledo, P.-M. 2002. Enriched odor exposure increases the number of newborn neurons in the adult olfactory bulb and improves odor memory. J. Neurosci. 22, 2679–2689.

Rochel, S. and Margolis, F. L. 1982. Carnosine release from olfactory bulb synaptosomes is calcium-dependent and depolarization-stimulated. J. Neurochem. 38, 505–1514.

Ross, C. D., Godfrey, D. A., and Parli, J. A. 1995. Amino acid concentrations and selected enzyme activities in rat auditory, olfactory, and visual systems. Neurochem. Res. 20, 1483–1490.

Rosser, A. E. and Keverne, E. B. 1985. The importance of central noradrenergic neurones in the formation of an olfactory memory in the prevention of pregnancy block. Neuroscience 15, 1141–1147.

Roy, S. A. and Alloway, K. D. 2001. Coincidence detection or temporal integration? What the neurons in somatosensory cortex are doing. J. Neurosci. 21, 2462–2473.

Salin, P. A., Lledo, P. M., Vincent, J. D., and Charpak, S. 2001. Dendritic glutamate autoreceptors modulate signal processing in rat mitral cells. J. Neurophysiol. 85, 1275–1282.

Sallaz, M. and Jourdan, F. 1992. Apomorphine disrupts odour-induced patterns of glomerular activation in the olfactory bulb. Neuroreport 3, 833–836.

Salmoiraghi, G. C., Bloom, F. E., and Costa, E. 1964. Aderenergic mechanisms in rabbit olfactory bulb. Am. J. Physiol. 207, 1417–1424.

Sanides-Kohlrausch, C. and Wahle, P. 1990a. Morphology of neuropeptide Y-immunoreactive neurons in the cat olfactory bulb and olfactory peduncle: postnatal development and species comparison. J. Comp. Neurol. 291, 468–489.

Sanides-Kohlrausch, C. and Wahle, P. 1990b. VIP- and PHI-immunoreactivity in olfactory centers of the adult cat. J. Comp. Neurol. 294, 325–339.

Sassoe-Pognetto, M., Cantino, D., and Panzanelli, P. 1993. Presynaptic co-localization of carnosine and glutamate in olfactory neurons. Neuroreport 5, 7–10.

Schneider, S. P. and Macrides, F. 1978. Laminar distributions of interneurons in the main olfactory bulb of the adult hamster. Brain Res. Bull. 3, 73–82.

Schneider, S. P. and Scott, J. W. 1983. Orthodromic response properties of rat olfactory mitral and tufted cells correlate with their projection patterns. J. Neurophysiol. 50, 358–378.

Schoenfeld, T. A. and Macrides, F. 1984. Topographic organization of connections between the main olfactory bulb and pars externa of the anterior olfactory nucleus in the hamster. J. Comp. Neurol. 227, 121–135.

Schoenfeld, T. A., Marchand, J. E., and Macrides, F. 1985. Topographic organization of tufted cell axonal projections in the hamster main olfactory bulb: an intrabulbar associational system. J. Comp. Neurol. 235, 503–518.

Schoppa, N. E. 2006. Synchronization of olfactory bulb mitral cells by precisely timed inhibitory inputs. Neuron 49, 271–283.

Schoppa, N. E. and Westbrook, G. L. 1997. Modulation of mEPSCs in olfactory bulb mitral cells by metabotropic glutamate receptors. J. Neurophysiol. 78, 1468–1475.

Schoppa, N. E. and Westbrook, G. L. 1999. Regulation of synaptic timing in the olfactory bulb by an A-type potassium current. Nat. Neurosci. 2, 1106–1113.

Schoppa, N. E. and Westbrook, G. L. 2001. Glomerulus-specific synchronization of mitral cells in the olfactory bulb. Neuron 31, 639–651.

Schoppa, N. E. and Westbrook, G. L. 2002. AMPA receptors drive correlated spiking in olfactory bulb glomeruli. Nat. Neurosci. 5, 1194–1202.

Schoppa, N. E., Kinzie, J. M., Sahara, Y., Segerson, T. P., and Westbrook, G. L. 1998. Dendrodendritic inhibition in the olfactory bulb is driven by NMDA receptors. J. Neurosci. 18, 6790–6802.

Scott, J. W. 1986. The olfactory bulb and central pathways. Experientia 42, 223–232.

Scott, J. W., McDonald, J. K., and Pemberton, J. L. 1987. Short axon cells of the rat olfactory bulb display NADPH-diaphorase activity, neuropeptide Y-like immunoreactivity, and somatostatin-like immunoreactivity. J. Comp. Neurol. 260, 378–391.

Seroogy, K. B., Brecha, N., and Gall, C. 1985. Distribution of cholecystokinin-like immunoreactivity in the rat main olfactory bulb. J Comp. Neurol. 239, 373–383.

Shepherd, G. M. 1972. Synaptic organization of the mammalian olfactory bulb. Physiol. Rev. 52, 864–917.

Shepherd, G. M. and Greer, C. A. 1998. The olfactory bulb. In: The Synaptic Organization of the Brain (ed. G. M. Shepherd), pp. 159–203. Oxford University Press.

Shipley, M. T. and Adamek, G. D. 1984. The connections of the mouse olfactory bulb: a study using orthograde and retrograde transport of wheat germ agglutinin conjugated to horseradish peroxidase. Brain Res. Bull. 12, 669–688.

Shipley, M. T., Halloran, F. J., and de la Torre, J. 1985. Surprisingly rich projection from locus coeruleus to the olfactory bulb in the rat. Brain Res. 329, 294–299.

Shults, C. W., Quirion, R., Chronwall, B., Chase, T. N., and O'Donohue, T. L. 1984. A comparison of the anatomical distribution of substance P and substance P receptors in the rat central nervous system. Peptides 5, 1097–1128.

Siklos, L., Rickmann, M., Joo, F., Freeman, W. J., and Wolff, J. R. 1995. Chloride is preferentially accumulated in a subpopulation of dendrites and periglomerular cells of the main olfactory bulb in adult rats. Neuroscience 64, 165–172.

Smith, C. and Lushkin, M. B. 1998. Cell cycle length of olfactory bulb neuronal progenitors in the rostral migratory stream. Dev Dyn. 213, 220–227.

Smith, T..C. and Jahr, C. E. 2002. Self-inhibition of olfactory bulb neurons. Nat. Neurosci. 5, 760–766.

Spors, H., Wachowiak, M., Cohen, L. B., and Friedrich, R. W. 2006. Temporal dynamics and latency patterns of receptor neuron input to the olfactory bulb. J. Neurosci. 26, 1247–1259.

Sullivan, R. M., Stackenwalt, G., Nasr, F., Lemon, C., and Wilson, D. A. 2000. Association of an odor with activation of olfactory bulb noradrenergic beta-receptors or locus coeruleus stimulation is sufficient to produce learned approach responses to that odor in neonatal rats. Behav. Neurosci. 114, 957–962.

Sullivan, R. M., Wilson, D. A., and Leon, M. 1989. Norepinephrine and learning-induced plasticity in infant rat olfactory system. J. Neurosci. 9, 3998–4006.

Sullivan, R. M., Zyzak, D. R., Skierkowski, P., and Wilson, D. A. 1992. The role of olfactory bulb norepinephrine in early olfactory learning. Brain Res. Dev. 70, 279–282.

Teubner, B., Degen, J., Söhl, G., Güldenagel, M., Bukauskas, F. F., Trexler, E. B., Verselis, V. K., De Zeeuw, C. I., Lee, C. G., Kozak, C. A., Petrasch-Parwez, E., Dermietzel, R., and Willecke, K. 2000. Functional expression of the murine connexin 36 gene coding for a neuron-specific gap junctional protein. J. Membr. Biol. 176, 249–263.

Toida, K., Kosaka, K., Aika, Y., and Kosaka, T. 2000. Chemically defined neuron groups and their subpopulations in the glomerular layer of the rat main olfactory bulb – IV. Intraglomerular synapses of tyrosine hydroxylase-immunoreactive neurons. Neuroscience 101, 11–17.

Toida, K., Kosaka, K., Heizmann, C. W., and Kosaka, T. 1996. Electron microscopic serial-sectioning/reconstruction study of parvalbumin-containing neurons in the external plexiform layer of the rat olfactory bulb. Neuroscience 72, 449–466.

Toida, K., Kosaka, K., Heizmann, C. W., and Kosaka, T. 1998. Chemically defined neuron groups and their subpopulations in the glomerular layer of the rat main olfactory bulb: III Structural features of calbindin D28K-immunoreactive neurons. J. Comp. Neurol. 392, 179–198.

Treloar, H. B., Feinstein, P., Mombaerts, P., and Greer, C. A. 2002. Specificity of glomerular targeting by olfactory sensory axons. J. Neurosci. 22, 2469–2477.

Trombley, P. Q. 1992. Norepinephrine inhibits calcium currents and EPSPs via a G-protein-coupled mechanism in olfactory bulb neurons. J. Neurosci. 12, 3992–3998.

Trombley, P. Q. and Shepherd, G. M. 1992. Noradrenergic inhibition of synaptic transmission between mitral and granule cells in mammalian olfactory bulb cultures. J. Neurosci. 12, 3985–3991.

Trombley, P. Q. and Shepherd, G. M. 1996. Differential modulation zinc and copper of amino acid receptors from rat olfactory bulb neurons. J. Neurophysiol. 76, 2536–2546.

Trombley, P. Q. and Westbrook, G. L. 1992. L-AP4 inhibits calcium currents and synaptic transmission via a G-protein-coupled glutamate receptor. J. Neurosci. 12, 2043–2050.

Trombley, P. Q., Horning, M. S., and Blakemore, L. J. 1998. Carnosine modulates zinc and copper effects on amino acid receptors and synaptic transmission. Neuroreport 9, 3503–3507.

Tsuruo, Y., Hokfelt, T., and Visser, T. J. 1988. Thyrotropin-releasing hormone (TRH)-immunoreactive neuron populations in the rat olfactory bulb. Brain Res. 447, 183–187.

Urban, N. N. and Sakmann, B. 2002. Reciprocal intraglomerular excitation and intra- and interglomerular lateral inhibition between mouse olfactory bulb mitral cells. J. Physiol. (Lond.) 542, 355–367.

Van Gehuchten, L. E. and Martin, A. 1891. Le bulbe olfactif chez quelques mammiferes. Cellule 205–237.

Vassar, R., Chao, S. K., Sitcheran, R., Nunez, J. M., Vosshall, L. B., and Axel, R. 1994. Topographic organization of sensory projections to the olfactory bulb. Cell 79, 981–991.

Villalba, R. M., Rodrigo, J., Alvarez, F. J., Achaval, M., and Martinez-Murillo, R. 1989. Localization of C-PON immunoreactivity in the rat main olfactory bulb. Demonstration that the population of neurons containing endogenous C-PON display NADPH-diaphorase activity. Neuroscience 33, 373–382.

Wachowiak, M. and Cohen, L. B. 1999. Presynaptic inhibition of primary olfactory afferents mediated by different mechanisms in lobster and turtle. J. Neurosci. 19, 8808–8817.

Wachowiak, M., McGann, J. P., Heyward, P. M., Shao, Z., Puche, A. C., and Shipley, M. T. 2005. Inhibition [corrected] of olfactory receptor neuron input to olfactory bulb glomeruli mediated by suppression of presynaptic calcium influx. J. Neurophysiol. 94, 2700–2712.

Wahle, P., Sanides-Kohlrausch, S., Meyer, G., and Lubke, J. 1990. Substance P- and opioid immunoreactive structures in olfactory centers of the cat: adult pattern and postnatal development. J. Comp. Neurol. 302, 349–369.

Walsh, R. R. 1959. Olfactory bulb potentials evoked by electrical stimulation of the contralateral bulb. Am. J. Physiol. 196, 327–329.

Wang, F., Nemes, A., Mendelsohn, M., and Axel, R. 1998. Odorant receptors govern the formation of a precise topographic map. Cell 93, 47–60.

Warden, M. K. and Young, W. S. 1988. Distribution of cells containing mRNAs encoding substance P and neurokinin B in the rat central nervous system. J. Comp. Neurol. 272, 90–113.

Wehr, M. and Laurent, G. 1996. Odour encoding by temporal sequences of firing in oscillating neural assemblies. Nature 384, 162–166.

Welker, W. I. 1964. Analysis of sniffing of the albino rat. Behavior 22, 223–244.

Wellis, D. P. and Scott, J. W. 1990. Intracellular responses of identified rat olfactory bulb interneurons to electrical and odor stimulation. J. Neurophysiol. 64, 932–947.

Wellis, D. P., Scott, J. W., and Harrison, T. A. 1989. Discrimination among odorants by single neurons of the rat olfactory bulb. J. Neurophysiol. 61, 1161–1177.

Wichterle, H., Turnbull, D. H., Nery, S., Fishell, G., and Alvareq-Buylla, A. 2001. In utero fate mapping reveals distinct migratory pathways and fates of neurons born in the mammalian basal forebrain. Development 128, 3759–3771.

Wilson, D. A. and Leon, M. 1988. Noradrenergic modulation of olfactory bulb excitability in the postnatal rat. Dev. Brain Res. 42, 69–75.

Wilson, D. A. and Sullivan, R. M. 1991. Olfactory associative conditioning in infant rats with brain stimulation as reward: II. Norepinephrine mediates a specific component of the bulb response to reward. Behav. Neurosci. 105, 843–849.

Wilson, D. A. and Sullivan, R. M. 1995. The D2 antagonist spiperone mimics the effects of olfactory deprivation on mitral/tufted cell odor response patterns. J. Neurosci. 15, 5574–5581.

Woolf, T. B., Shepherd, G. M., and Greer, C. A. 1991. Serial reconstruction of granule cell spines in the mammalian olfactory bulb. Synapse 7, 181–191.

Xiong, W. and Chen, W. R. 2002. Dynamic gating of spike propagation in the mitral cell lateral dendrites. Neuron 34, 115–126.

Yamamoto, C., Yamamoto, T., and Iwama, K. 1963. The inhibitory systems in the olfactory bulb studied by intracellular recording. J. Neurophysiol. 26, 403–415.

Yu, G.-Z., Kaba, H., Saito, H., and Seto, K. 1993. Heterogeneous characteristics of mitral cells in the rat olfactory bulb. Brain Res. Bull. 31, 701–706.

Yuan, Q. and Knopfel, T. 2006. Olfactory nerve stimulation-evoked mGluR1 slow potentials, oscillations and calcium signaling in mouse olfactory bulb mitral cells. J. Neurophysiol. 95, 3097–3104.

Yuan, Q., Harley, C. W., and McLean, J. H. 2003. Mitral cell $\alpha 1$ and 5-HT2A receptor colocalization and cAMP coregulation: a new model of norepinephrine-induced learning in the olfactory bulb. Learn Mem. 10, 5–15.

Zaborszky, L., Carlsen, J., Brashear, H. R., and Heimer, L. 1986. Cholinergic and GABAergic afferents to the olfactory bulb in the rat with special emphasis on the projection neurons in the nucleus of the horizontal limb of the diagonal band. J. Comp. Neurol. 243, 488–509.

Zelles, T., Boyd, J. D., Hardy, A. B., and Delaney, K. R. 2006. Branch-specific Ca2+ influx from Na+-dependent dendritic spikes in olfactory granule cells. J. Neurosci. 26, 30–40.

Zhang, C. and Restrepo, D. 2003. Heterogeneous expression of connexin 36 in the olfactory epithelium and glomerular layer of the olfactory bulb. J. Comp. Neurol. 459, 426–439.

Zou, Z., Horowitz, L. F., Montmayeur, J. P., Snapper, S., and Buck, L. B. 2001. Genetic tracing reveals a stereotyped sensory map in the olfactory cortex. Nature 414, 173–179.

Further Reading

Duchamp-Viret, P., Coronas, V., Delaleu, J.-C., Moyse, E., and Duchamp, A. 1997. Dopaminergic modulation of mitral cell activity in the frog olfactory bulb: a combined radioligand binding-electrophysiological study. Neuroscience 79, 203–216.

Duchamp-Viret, P., Delaleu, J. C., and Duchamp, A. 2000. GABA$_B$-mediated action in the frog olfactory bulb makes odor responses more salient. Neuroscience 97, 771–777.

Kaba, H., Rosser, A., and Keverne, B. 1989. Neural basis of olfactory memory in the context of pregnancy block. Neuroscience 32, 657–662.

Persohn, E., Malherbe, P., and Richards, J. G. 1992. Comparative molecular neuroanatomy of cloned GABAA receptor subunits in the rat CNS. J. Comp. Neurol. 326, 193–216.

Pinching, A. J. and Powell, T. P. 1972a. A study of terminal degeneration in the olfactory bulb of the rat. J. Cell Sci. 10, 585–619.

4.38 Olfactory Cortex

D A Wilson, University of Oklahoma, Norman, OK, USA

Glossary

odorant The physical stimulus that gives rise to the perception of an odor. The same odorant may give rise to different odor percepts, depending on context, expectation, and past experience.

paleocortex An evolutionarily old cortical structure with generally three lamina, as opposed to the six-layered evolutionarily more recent neocortex. Piriform cortex is an example of paleocortex.

perceptual learning An enhancement in fine sensory discrimination based on past experience. Perceptual learning has been demonstrated in most sensory modalities including vision, audition, somatosensation, and olfaction.

thalamocortical system An organization typical of most sensory systems, wherein a thalamic nucleus receives input from the sensory peripheral circuits and transmits this information to the sensory neocortex, Olfaction includes both a nonthalamocortical system of olfactory bulb projections directly to the piriform cortex, and also a thalamocortical system of mediodorsal thalamic nucleus input to the orbitofrontal cortex.

4.38.1 Introduction

Sensory receptors are generally highly analytical. Your retina and photoreceptors pixilate this printed page into tiny points of light within a relatively narrow range of wavelengths. The tips of your fingers have separate receptor populations for rough surfaces and smooth surfaces, and for hot and cold. As noted in earlier chapters, your nose contains several hundred different olfactory receptors responsive to different submolecular features of volatile compounds, huge subsets of which are activated when inhaling chemically rich everyday odors such as cheese or wine. In contrast, sensory cortex is often the epitome of synthetic. In visual sensory cortex, pixilated images get combined with past experience to allow to one see and recognize objects. In somatosensory cortex, smooth gets put together with cold when we run our finger tips over an ice cube. And in the olfactory sensory cortex, the 655 different volatiles rising over a cup of coffee, so carefully dissected by the olfactory periphery, get merged into the only stimulus capable of getting me out of bed this morning – the single percept, coffee.

This chapter will review our current understanding of the structure and function of olfactory sensory cortex. Olfaction is unique among mammalian sensory systems in that second-order neurons project directly to the primary sensory cortex, without an intervening thalamic relay. However, there is also a thalamocortical olfactory pathway, highly interconnected with primary olfactory cortex and only just beginning to be fully explored. While our understanding of peripheral mechanisms of olfactory coding has grown exponentially in the past decade or two, it is increasingly apparent that to understand olfactory perception we must understand what olfactory sensory cortex does. For example, knowing that there are different opsin proteins in the cones of your retina differentially sensitive to light wavelength is not sufficient to explain the visual perception of a

rainbow or an orchid. Similarly, our perception of, and ability to discriminate between, odors depends not only on events happening at the receptor, but very critically on events within sensory cortex. The experience- and context-dependent nature of olfactory perception, in fact, make very clear that knowing which receptors and olfactory bulb glomeruli are activated by a particular stimulus is insufficient to predict the resulting percept. We will review what we currently understand about the contribution of olfactory cortex to odor coding and the mechanisms underlying these contributions.

4.38.1.1 Functions of Sensory Cortex

Sensory cortical circuits have at least three primary functions (see Figure 1). The first is to elaborate upon the representation of afferent input begun by more peripheral structures. This cortical elaboration may occur through processes such as decorrelation of patterned afferent activity to enhance discrimination of similar patterns, convergence and synthesis of diverse inputs to promote emergence of wholistic

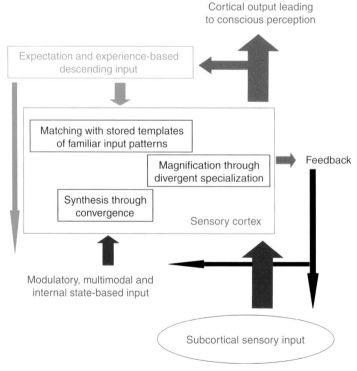

Figure 1 A schematic representation of some of the basic functions ascribed to primary and earlier higher-order sensory cortices. In thalamocortical sensory systems, sensory cortex plays a central role in constructing sensory images from the feature analyzed input it receives, which are ultimately important for conscious perception. Sensory cortex also plays an important role in feedback processes, underlying attention and expectation-induced changes in sensory processing.

percepts, and comparison of afferent activity patterns with stored templates to promote recognition. Furthermore, sensory cortices often embellish this emerging representation with information from other sensory modalities, memory, and emotional and internal state. The second function of sensory cortices is to use the information obtained through these convergent processes to provide feedback to peripheral stages in order to selectively modify gain or acuity, just as sensory cortex itself is under the influence of higher-order cortices. Finally, sensory cortical efferents provide downstream structures with highly processed representations of sensory stimuli and their associated external and internal context to drive behavioral and cognitive outcomes. In effect, sensory cortex can be the first stage of conscious perception.

In performing these functions, sensory cortices utilize at least three basic tools. The first tool is convergence. Stimulus features extracted by peripheral circuits begin to be merged through anatomical convergence in the sensory cortex. Hierarchical processing in the visual system is a classic example, wherein spots of light extracted by center-surround receptive fields in the retina and visual thalamus are merged through anatomical convergence into bars and contrast edges driving visual cortex neurons. Convergence can occur both within a sensory modality, such as the hierarchical visual processing just mentioned, and on a wider scale including disparate regions of sensory cortex as well as potentially including multimodal interactions.

The second tool used by sensory cortices, and directly related to convergence, is synaptic plasticity. Experience-dependent synaptic plasticity is critical for both developmental emergence of local cortical circuits and fine-tuning of those circuits throughout life. This plasticity can allow increased processing power to be devoted to familiar or learned important stimuli by adjusting receptive fields and cortical maps. It can also help to associate stimuli and stimulus features which frequently co-occur to form templates or internal representations of perceptual objects. An advantage of object representation is an increased ability to recognize familiar objects despite degraded inputs or background interference. Finally, synaptic plasticity can allow on-line adjustment of stimulus processing through experience-dependent short-term modulation of synaptic strength.

A third sensory cortical tool is magnification. In nearly all thalamocortical sensory systems, pathways display anatomical organizations laid out as functional sensory maps. Retinotopic, tonotopic, and somatotopic maps are all examples of sensory maps which are expressed within the corresponding sensory cortex. Within each of these cortical maps, cortical space is not divided evenly across sensory space, but rather there are regions of disproportionate investment in specific areas of sensory space. The fovea, which covers a physically very small region of the retina, commands a relatively huge physical proportion of the visual cortex. Similarly, the small range of auditory frequencies corresponding to biologically important sounds (e.g., human voice in humans, echolocating sounds in bats) dominate cortical space within audition. Furthermore, the somatosensory cortex devotes a large region to inputs from the fingers, which account for a relatively small proportion of the human body surface. Thus, cortical processing power, as indexed by cortical volume, is differentially devoted to some regions of sensory world over others, in effect magnifying those stimuli which fall on those receptors. Beyond this magnification in cortical maps, additional cortical space can also be specialized for important stimuli, outside the primary sensory pathway. For example, the fusiform face region in the primate visual system is largely devoted to processing biologically important face stimuli. Similar specializations are seen in other systems, for example the Doppler-shift compensation area in bat echolocating auditory cortex which allows calculation of relative speed of approach toward prey targets.

As we review the structure and physiology of olfactory cortex, the functions and tools used by other sensory systems will be used as guideposts to determine how different (or similar) olfaction is to the other senses. In the following section, we will review the anatomical organization of olfactory cortex, local circuit connectivity, and afferent and efferent connections. We will then review current synaptic and sensory physiological data, followed by an abstraction of general principles of olfactory cortical function.

4.38.2 Cortical Neurocircuitry

The olfactory cortex consists of a collection of laminar structures arranged along the ventrolateral surface of the mammalian brain which receive direct input from olfactory bulb mitral and/or tufted cells (see Figure 2). These target structures include the anterior olfactory cortex, piriform cortex, olfactory

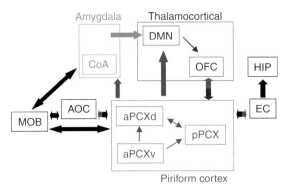

Figure 2 A schematic representation of the major olfactory cortical areas and their interrelationships. Additional details can be found in Cleland T. A. and Linster C. (2003) and Neville K. R. and Haberly L. (2004). This review focuses on the piriform and orbitofrontal cortices. Almost nothing is known regarding the sensory physiology of the anterior olfactory cortex (AOC). MOB, main olfactory bulb; AOC, anterior olfactory cortex; aPXDd, dorsal anterior piriform cortex; aPCXv, ventral anterior piriform cortex; pPCX, posterior piriform cortex; CoA, cortical nucleus of the amygdala; DMN, dorsomedial nucleus of the thalamus; OFC, orbitofrontal cortex; EC, entorhinal cortex; HIP, hippocampus.

tubercle, and the cortical nucleus of the amygdala (Haberly, L. B., 2001). Each of these structures is trilaminar paleocortex, with olfactory bulb afferent input terminating in superficial layer I, a pyramidal cell dense layer II, and a layer III composed of association fiber axons, deep pyramidal cell somata, and interneurons. The olfactory cortex does not express a columnar organization, but rather receives afferent input as a wave of activity spreading caudally through its arrayed dendritic trees. Each of the olfactory cortical structures is interconnected and also project back to the olfactory bulb.

In addition to the paleocortical areas, the olfactory system also includes a thalamocortical pathway, with a projection from the piriform cortex to the dorsomedial nucleus of the thalamus, which in turn projects to the orbitofrontal cortex. The dorsomedial nucleus of the thalamus also receives input from the amygdala and other limbic areas. There are also direct, reciprocal connections between the orbitofrontal cortex and the piriform cortex (Neville, K. R. and Haberly, L., 2004). Orbitofrontal cortex is located anterior in the mammalian brain and juxtaposes piriform cortex along the rhinal fissure in rodents. The orbitofrontal cortex is neocortical in structure and thus displays a six-layer, presumably columnar organization.

While the anatomy of all of these cortical areas dealing with olfaction has been described to some extent in the literature, the piriform cortex has received the most attention. Thus, this review will emphasize piriform cortical structure and function. This emphasis is not meant to imply dominance of the piriform cortex in olfactory processing and perception, but rather, unfortunately, reflects a search where the light is currently brightest. That is, while the piriform cortex, and increasingly orbitofrontal cortex, appear to perform computations of critical importance for the rich experience that is olfactory perception, we simply do not have sufficient data at present to state definitively how areas such as the anterior olfactory cortex and olfactory tubercle contribute.

4.38.2.1 Cortical–Cortical Connectivity

The piriform cortex can be cytoarchitecturally divided into at least four major subregions. In rodents, the anterior piriform cortex is defined as extending caudally from the anterior olfactory cortex to the caudal end of the lateral olfactory tract. Immediately caudal to the anterior piriform cortex lies the posterior piriform cortex. Mitral cell axons no longer lie in a dense tract, but instead diffusely spread in a thin layer across the posterior piriform cortical surface. In addition to this anterior–posterior division, the anterior piriform can be further subdivided into a dorsal and ventral component. Each of these four subdivisions expresses differences in cortical–cortical connectivity and, as more research is performed, increasingly demonstrates differences in sensory physiology (see Section 4.38.3).

The piriform cortex receives direct input from the olfactory bulb, as do all other olfactory cortical structures (see Figure 3). Single mitral/tufted cell axons terminate in small patches in piriform cortex. Mitral cells conveying input from specific olfactory receptor types terminate in overlapping patches in anterior piriform cortex, and even more broadly in posterior piriform (Zou, Z. *et al.*, 2001). The primary target of olfactory bulb input to cortical areas is pyramidal cell apical dendrites.

Pyramidal cells within the piriform cortex can be divided into at least three subgroups, deep pyramidal cells with somata in layer III, pyramidal cells with somata in layer II, and semilunar pyramidal cells with somata in superficial layer II (layer IIa). Each of these neuron classes has spiny apical dendrites extending into superficial layer I. Mitral/tufted cell afferent

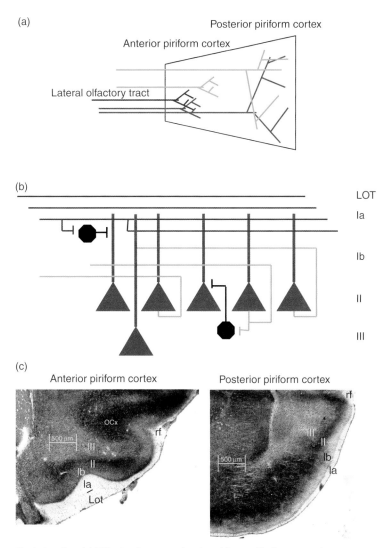

Figure 3 Piriform cortical circuitry. (a) Mitral cells conveying input from olfactory receptor neuron expressing specific olfactory receptor genes terminate in patches in the anterior piriform cortex that overlap with termination zones of mitral cells conveying information from other receptors. Mitral cell termination in the posterior piriform is much more diffuse with information from different receptors broadly overlapping. (b) A schematic representation of piriform circuitry emphasizing lamination of afferent (blue) and association (green) fiber input into trilaminar cortical sheet. Black cells represent local feedforward and feedback inhibitory interneurons. (c) Coronal sections through the anterior and posterior piriform cortex stained with Timm stain for heavy metals which selectively stain intracortical association fiber axons brown and counterstained with cresyl violet. Notice the proportional differences in association fiber input to neurons in dorsal anterior piriform and posterior piriform cortex relative to ventral anterior piriform cortex. rf, rhinal fissure; Ocx, orbital cortex; LOT, lateral olfactory tract.

input to the piriform terminates on the distal most regions of the apical dendrites in cortical layer Ia. Semilunar pyramidal cells may receive largely exclusive excitatory input from mitral/tufted cells, as semilunar cells possess large dendritic spines within layer Ia, but not within layer Ib. Axons of pyramidal cells form the intracortical association fiber system, generally synapsing on spines of both the apical and

basal dendrites of other piriform cortical neurons. Individual pyramidal cells have diffuse projections throughout the ipsilateral piriform cortex, and extending into the orbitofrontal cortex and other olfactory cortical areas, the cortical nucleus of the amygdala, as well as the ipsilateral olfactory bulb. Other ipsilateral projects include the dorsomedial nucleus of the thalamus, the hypothalamus, and entorhinal cortex. These neurons also send axons to the contralateral piriform via the anterior commissure. Semilunar cells also project widely throughout the piriform cortex but do not send feedback to the olfactory bulb. The piriform cortex receives input from a number of areas other than the olfactory bulb, including the entorhinal cortex, orbitofrontal cortex, and amygdala. More complete recent descriptions of local and extrinsic circuitry can be found in Cleland T. A. and Linster C. (2003) and Neville K. R. and Haberly L. (2004).

Olfactory areas also receive input from a variety of modulatory systems, including the noradrenergic nucleus locus coeruleus, the cholinergic horizontal limb of the diagonal band, and serotonergic raphe nucleus. These systems have been shown to play important roles in olfactory cortical function and plasticity (see Section 4.38.3).

In addition to pyramidal cells, there are a variety of classes of inhibitory interneurons. Interneurons in layer I appear to receive input from the lateral olfactory tract and in turn synapse onto apical dendritic shafts of piriform cortical pyramidal cells in a feedforward inhibitory circuit. Interneurons with somata in layer III receive input from pyramidal cell axon collaterals and feedback onto neighboring pyramidal cell somata and basal dendritic shafts. However, extensive variations from these canonical patterns exist (Neville, K. R. and Haberly, L., 2004).

4.38.2.2 Piriform Cortex Synaptic Physiology

Mitral/tufted cell axons terminating on distal apical dendrites of piriform cortical pyramidal neurons release glutamate as a neurotransmitter (Jung, M. W. *et al.*, 1990). Both AMPA and NMDA receptors are located postsynaptically on pyramidal cell apical dendrites. Mitral/tufted cell presynaptic terminals express metabotropic glutamate receptors, as well as receptors for modulatory inputs such as norepinephrine. Activation of NMDA receptors appears to be regulated in part by apical dendritic inhibition mediated by layer I gamma-aminobutyric acid (GABA)ergic interneurons. Layer III GABAergic interneurons synapse heavily on pyramidal cell somata and induce current shunting inhibition. Thus, these two forms of inhibition can be both anatomically and functionally distinct.

Activation of the lateral olfactory tract evokes an excitatory postsynaptic potential (EPSP) in the distal half of the pyramidal cell which, due to the electrotonic distance, can be reduced by 75% at the cell body (Ketchum, K. L. and Haberly, L. B., 1993). Depolarization of pyramidal cells to spike threshold activates the glutamatergic association fiber system, and thus within approximately 5 m, the afferent-evoked EPSP is followed by an association fiber-evoked EPSP. These EPSPs are due to activation of synapses in the proximal half of the apical dendrite; thus, they decay much less than lateral olfactory tract-evoked EPSPs. The association synapse-mediated EPSPs, however, coincide with feedback inhibitory portrynaptic potentials (IPSPs). Following this initial rapid sequence of synaptic events, a later, prolonged K^+-mediated IPSP is apparent. This sequence of events can initiate oscillatory (gamma frequency) activity within the piriform following a single afferent volley as association fiber activity circulates throughout the piriform. Single-unit activity in the piriform has been shown to occur in phase with the gamma frequency oscillation, as would be expected given its generation mechanism (Eeckman, F. H. and Freeman, W. J., 1990).

Modulatory inputs to the piriform cortex appear to have differential effects on the afferent and intrinsic fiber pathways. Both acetylcholine and norepinephrine have direct effects on piriform cortical neurons, but interestingly they also have presynaptic effects on synaptic transmission. Both acetylcholine and norepinephrine preferentially depress association fiber excitatory synapses, while having minimal impact on lateral olfactory tract afferent input (Hasselmo, M. E. *et al.*, 1990; Hasselmo, M. E. and Bower, J. M., 1992).

Both lateral olfactory tract and association fiber synapses express use-dependent plasticity that may be involved in piriform cortical odor information processing. As described below, lateral olfactory tract synapses express a short-term depression following intense electrical or odor-evoked activation that recovers within 2 min (Best, A. R. and Wilson, D. A., 2004). This depression is composed of two phases, a short (<15 s) phase that is consistent with neurotransmitter depletion and a longer phase (<2 min) that can be blocked by antagonists of presynaptic group III metabotropic glutamate receptors (Best, A. R. and Wilson, D. A., 2004).

Lateral olfactory tract and the association fiber synapses also express long-term potentiation, though potentiation of the lateral olfactory tract synapses appear minimal at best (Neville, K. R. and Haberly, L., 2004). The association fiber synapses display robust (20–30%), NMDA receptor-dependent, long-term potentiation. Theta burst stimulation paradigms, which involve short gamma (40–100 Hz) frequency bursts of pulses with the burst delivered at theta (5–10 Hz) frequency appear most effective at inducing long-term potentiation in association fiber synapses. These stimulation parameters roughly match what would occur during active sniffing at theta frequency.

This physiology and pharmacology suggests a particularly important role for association fibers in piriform cortical function and plasticity. Association fibers are electrotonically closer to the somata, are more plastic, and more strongly influenced by modulatory inputs than is the primary cortical afferent. The differential roles of these two excitatory cortical pathways in information processing by olfactory cortex are discussed below.

4.38.2.3 Olfactory Thalamocortical Pathway Functional Anatomy

In addition to the direct olfactory bulb input to the primary olfactory cortical areas, the olfactory pathway also includes a thalamic representation and a thalamocortical projection. The piriform cortex projects to the mediodorsal thalamic nucleus which in turn projects to the orbitofrontal cortex (and other prefrontal regions). The mediodorsal thalamus also receives input from the cortical and basal nuclei of the amygdala, potentially enriching the olfactory input to orbitofrontal cortex with emotional context (Schoenbaum, G. *et al.*, 2003a). The orbitofrontal cortex also receives direct input from piriform cortex, in addition to its thalamic input, leading to a convergent triangulation between piriform, mediodorsal thalamus, and orbitofrontal cortex. As in the piriform cortex, orbitofrontal neurons respond not only to specific odors, but also to odor-contextual cues (Schoenbaum, G. and Eichenbaum, H., 1995; Rolls, E. T., 2004). Orbitofrontal cortex neurons also respond to specific odor-taste compounds, potentially contributing to taste and odor hedonics.

The specific functions of thalamocortical olfactory pathways in odor perception are not known. Lesions of the dorsomedial thalamus in rats impair learning and memory of discriminative odor cues, although this impairment appears to not be related to odor discrimination per se, but rather may represent a memory disruption (Zhang, Y. *et al.*, 1998).

Single neurons in the orbitofrontal cortex of both rodents and primates encode both odor quality and odor associations. In primates, both the identity of an odor being inhaled and its reinforcement value and taste associations can be extracted from orbitofrontal cortex neuron spike trains (Rolls, E. T., 2004). Odor quality is encoded very sparsely in orbitofrontal cortex, and there is diversity in responses with, for example, some neurons encoding odor quality independently of reinforcement associations and some encoding the reinforcement associations. Similar results have been seen in rat orbitofrontal cortex and have led to the suggestion that the orbitofrontal cortex integrates sensory representations with associated reward value to help guide motivated behavior – a conclusion supported by lesion studies (Schoenbaum, G. *et al.*, 2003b). Given the interconnectivity between the orbitofrontal and piriform cortices, however, the specific role of thalamocortical inputs versus cortical–cortical inputs to orbitofrontal cortex in this higher-order sensory representation is unknown. It is interesting to speculate that the orbitofrontal cortex may serve in part in a magnification role as discussed above, providing dedicated additional processing support for hedonically important odors beyond that performed in the piriform cortex.

4.38.3 Cortical Sensory Physiology

Despite the wealth of information regarding olfactory cortical circuitry and synaptic physiology, our understanding of olfactory cortical sensory physiology lags far behind most other sensory systems. As more knowledge is gained as to the nature of sensory information being fed into the cortex, and as more people turn centrally to understand how the outcome of peripheral coding is acted upon by cortical circuits, this lag should be rapidly reduced. This section will review our current state of understanding regarding cortical coding of odor intensity, quality, familiarity, and context. Sensory physiological and computational work suggests that the piriform cortex is critically involved in synthetic coding of complex odorants and in some cases their nonolfactory contexts, gating olfactory information flow relative to experience and behavioral state, and creating templates of familiar and learned odorants which facilitates recognition and discrimination.

4.38.3.1 Odorant Intensity

Piriform cortical representation of odor intensity has been sorely understudied. Intensity codes in other systems include a variety of options. Firing rate of individual neurons can be correlated with stimulus intensity. Similarly, the number or type of activated neurons can change with intensity. For example, there may be neurons with differing thresholds that are preferentially activated by high or low intensities. Finally, temporal patterning of individual or ensemble activity may be correlated with stimulus intensity.

Recently, Sugai and colleagues have examined the spatial extent of piriform cortical activation by odors of varying intensity (Sugai, T. *et al.*, 2005). Using a combination of intrinsic signal imaging and single-unit recording in the dorsal anterior piriform cortex, these authors report a rostral to caudal spread of activity as odor concentration increases. Thus, single units in caudal regions of the dorsal anterior piriform cortex had higher-odor thresholds than more rostral cells. Interestingly, these results correspond with findings that superficial tufted cells in the olfactory bulb project more heavily to rostral portions of the piriform cortex and are more responsive to olfactory nerve input compared to internal tufted or mitral cells which have higher thresholds and project more caudally. Thus, the differential sensitivity and projection patterns of cortical afferents could produce a rostral to caudal gradient of odor sensitivity across the piriform cortex.

It should also be noted that that there is a rostral to caudal gradient of cortical feedback to the olfactory bulb, wherein more caudal olfactory cortical neurons terminate on deep granule cells and more rostral cortical neurons terminate on more superficial granule cells. Deep granule cells are believed to synapse preferentially with mitral and internal tufted cells, while superficial granule cells may preferentially synapse with middle or external tufted cells. Thus, the rostral to caudal sensitivity gradient in cortex could be a part of a closed feedback loop to specific populations of olfactory bulb output neurons. Tufted and mitral cells also display differential phase locking to the respiratory cycle, with tufted cells firing earlier in the respiratory cycle than mitral cells, potentially creating rostral–caudal temporal gradients in cortical activity relative to intensity as well.

One important issue that the olfactory cortex must deal with, however, is that the afferent input to the cortex may change with stimulus intensity. Olfactory receptors, olfactory glomerular layer spatial patterns, and mitral/tufted output are all modified by changes in stimulus intensity. Some odorants change in perceptual quality as stimulus intensity increases;

however, most odorants are qualitatively stable over many orders of magnitude of concentration (Laing, D. G. *et al.*, 2003). Therefore, cortical circuits described below may be particularly important for maintaining odorant-specific patterns of cortical output across concentration range, thus maintaining perceptual quality despite intensity-induced changes in afferent input.

4.38.3.2 Odorant Quality

Cortical coding of odorant quality ultimately involves two levels of information. At one level, cortical odorant quality encoding may reflect the structural make-up of the inhaled stimulus – essentially a reflection of the sum of the odorant features extracted by more peripheral circuits. However, odor perception must involve coding beyond a simple reflection of convergent input. For example, structurally similar molecules (similar collections of features) can smell quite different, and similar smells may derive from structurally dissimilar molecules. Furthermore, the perceived smell of a banana is relatively stable, regardless of the background odorants inhaled at the same time whose components should mix with those of the banana at the receptor sheet. Finally, the integral hedonic quality of that banana smell perception is dependent on current internal state (e.g., hunger vs. nausea) and expectations. Thus, a full account of cortical odorant quality coding (and ultimately odor perception) must go beyond simple description of responses to isolated monomolecular stimuli and begin to describe such higher-order perceptual quality phenomenon as mixture synthesis, sensory gating, figure-ground separation, and the effects of expectation.

In fact, an examination of odorant receptive fields of piriform cortical neurons (see Figure 4) suggests superficial similarity to those of olfactory bulb mitral/tufted cells. Individual cortical neurons respond to multiple odorants (Wilson, D. A., 1998; 2000; Litaudon, P. *et al.*, 2003) and demonstrate differential responsiveness to odorant features such as carbon chain length (Wilson, D. A. and Stevenson, R. J., 2003). Selectivity for simple odorants is increased in neurons in posterior piriform compared to anterior piriform cortex (Litaudon, P. *et al.*, 2003), and in fact appears to increase as one moves from piriform to orbitofrontal cortex (Tanabe, T. *et al.*, 1975), suggesting a possible refinement of odorant identity coding in a hierarchical manner.

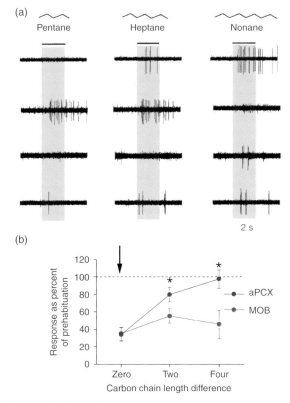

Figure 4 Examples of anterior piriform cortex single-unit responses to odorant stimuli. (a) Responses of four different anterior piriform cortical neurons to second stimuli presented during the red-shaded region. Note that, similar to reports of olfactory receptor neurons and mitral/tufted cells, piriform cortical neurons respond differentially to odorants varying along a specific dimension such as carbon chain length. (b) However, using a cross-adaptation procedure to determine odorant discrimination abilities of these neurons revealed that cortical neurons were significantly better at discriminating between odorants within their receptive fields than mitral/tufted cells of the main olfactory bulb (MOB). aPCX, anterior piriform cortex.

In contrast to a possible intensity-dependent spatial activation of the piriform cortex described above, initial attempts to describe odorant quality-dependent spatial patterns of piriform cortical activation were largely unsuccessful. Studies utilizing 2-deoxyglucose metabolic activity imaging, intrinsic signal imaging, voltage sensitive dye imaging, and immediate early gene immunohistochemistry (Litaudon, P. *et al.*, 1997; Illig, K. R. and Haberly, L. B., 2003; Sugai, T. *et al.*, 2005) have all failed to detect precise monomolecular odorant quality-specific patterns of piriform cortex activation comparable to those seen in the olfactory bulb.

Illig K. R. and Haberly L. B. (2003) reported (see Figure 5) that monomolecular odorant stimulation evoked rostrocaudal bands of c-fos labeling of pyramidal cells in anterior piriform cortex. However, these bands appeared to correspond primarily with the anatomical subdivisions of the anterior piriform more than demonstrate an odorant-specific pattern. Thus for example, most odorants more strongly activated the dorsal anterior piriform and a narrow band of ventrally located cells than the intermediate regions of the anterior piriform cortex. The authors suggested that there may be subtle variations in c-fos labeling that could be odor specific, but that there were no clear odorant-specific patches evident.

Recent work by the Buck laboratory using a more diverse odorant set and more quantitative analyses of odorant-evoked Fos-staining patterns suggests there may in fact be regional specificity in response to odorants within the anterior piriform cortex (Zou, Z. *et al.*, 2005). Structurally similar odorants evoke odorant-specific spatial patterns of neural activation within the anterior piriform cortex. Activated neurons are dispersed within activated regions; thus individual neighboring neurons do not necessarily have similar response properties. Similarly, mapping of single-unit odorant responses and cell locations shows no strong spatial clustering of neurons responsive to molecularly and perceptually dissimilar odorants in anterior piriform cortex (Kadohisa, M. and Wilson, D. A., 2006), though this has not yet been thoroughly investigated.

If the piriform cortex processes odors as synthetic objects built through experience as discussed elsewhere, then perhaps strong spatial clustering or mapping of activity evoked by particular odorants would not be expected. For example, although inferotemporal visual cortical neurons responding to a particular visual object lie within a particular cortical column, neighboring columns may contain neurons responding to completely different visual objects (Tanaka, K., 2000; see Figure 6). As experience throughout life shapes the tuning properties of these cortical neurons, objects become encoded by randomly scattered cortical neurons, with neurons maximal responsive to ball objects, located next to neurons responsive to tree objects and star-shaped objects. Similarly, we predict that the location of piriform cortical neurons responding to odor objects will not conform to a consistent organized map, but rather show discontinuities. This effect may be more pronounced in posterior piriform than in anterior

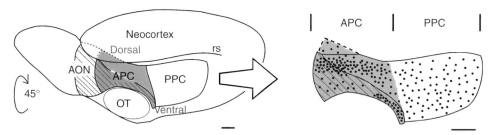

Figure 5 Odorant-evoked neural activity in the piriform cortex as revealed by c-fos immunohistochemistry (Illig, K. R. and Haberly, L. B., 2003). (Left) Image of the rat brain viewed from the side with the subregions of the piriform cortex displayed. (Right) Patterning of activated neurons in response to odorant stimulation. No distinct odorant-specific differences were noticed in activation patterns in piriform, although distinct subregion differences in odorant responsiveness were noted. Recent work by Zou Z. *et al.* (2005), using similar techniques, suggests that odorant-specific patterning may be present in the anterior piriform. Both studies clearly demonstrate that odorant-evoked activity is dispersed, with individual neighboring neurons responsive to different odorants. It must be noted, however, that the techniques used would not reveal neurons with rapidly adapting responses (Wilson, D. A., 1998). APC, anterior piriform cortex; PPC, posterior piriform cortex. Image adapted from original image provided by Kurt Illig.

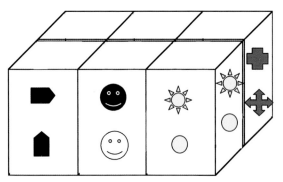

Figure 6 Schematic representation of the spatial organization of visual object encoding in the primate visual inferotemporal cortex. Although primary visual cortex is highly and predictably organized, in inferotemporal cortex where neurons have receptive fields for visual objects, the organization is less predictable. Neurons within a single column may have receptive fields for similar objects, perhaps at different views or angles, but neurons in neighboring columns may have receptive fields for completely different objects. Such a dispersed organization may also be expected to occur in piriform cortex for odor objects. Adapted from Tanaka, K. 2000. Mechanisms of visual object recognition studied in monkeys. Spat. Vis. 13, 147–163.

piriform as association fibers come to dominate excitatory control.

While much of sensory physiology in the olfactory system has focused on monomolecular odorants, most odorants experienced in nature are complex mixtures. However, as noted above, while composed of often hundreds of different components, these natural mixtures are perceived as individual, indivisible odor objects. For at least 20 years, the piriform

cortex has been seen as a possible site of synthetic odor quality processing in olfaction. Mitral/tufted cells, conveying information about specific odorant features (Fletcher, M. L. and Wilson, D. A., 2003; Linda, Y. *et al.*, 2005), converge onto individual piriform cortical pyramidal cells through overlapping, patchy axonal termination zones (Zou, Z. *et al.*, 2001). Thus, through simple coincidence detection, single cortical neurons could transmit information concerning the occurrence of specific combinations of odorant features. However, simple coincidence detection is insufficient to account for complex odor processing such as occurs in odorant figure-ground separation (Kadohisa, M. and Wilson, D. A., 2006) (see Section 4.38.4.3). In the situation where odorants are experienced against a background (nearly all situations), those odorants would be merged with the background based on their coincident inhalation resulting in context-variant odor perception of stable objects. This does not generally occur, and thus a separate mechanism, beyond simple afferent coincidence detection, must be invoked.

Intracortical association fibers and synaptic plasticity may solve the problem of stable odor representations regardless of background, and often regardless of variations in stimulus make-up or intensity. As noted above, single piriform cortical neurons have axon collaterals broadly dispersed throughout the piriform cortex (Johnson, D. M. *et al.*, 2000). Unique combinations of mitral cell-mediated odorant feature input should activate a unique ensemble of piriform cortical pyramidal neurons, based on termination patterns of those mitral cell axons. In some cases, individual cortical neurons will receive convergent input from mitral cells conveying information regarding two or more odorant features, and in other cases a cortical neuron may receive input regarding a single odorant feature, depending on the pattern of mitral cell–cortical connectivity. Cortical pyramidal cells sufficiently activated by this input in turn activate additional cortical neurons via the intracortical association fibers, thus greatly expanding the combinatorial activation of single cortical neurons.

Specifically, imagine an odorant composed of features A, B, and C. Each of these features is recognized by individual mitral cells which convey processed feature information to the cortex. Through simple anatomical convergence and coincidence detection, individual cortical neurons receive partial combinatorial information (AB, AC, and BC), depending on the initial state of afferent connectivity for each cortical neuron (it should be noted that odor experience early in development can modify overall mitral cell termination patterns in piriform cortex) (Wilson, D. A. et al., 2000).

Pyramidal neurons activated by these partial combinations can then, via intracortical association fibers, create more complete combinations on other cortical neurons, resulting in an ensemble of neurons driven by the complete ABC mixture. This process can occur passively on each new presentation of ABC through coincidence detection. However, given the subtle variations inherent in naturally occurring complex mixtures (with exceptions for tightly controlled pheromonal mixtures), including variation in mixture component ratios, variation in overall intensity, occasional missing or extra components, each new presentation of that mixture may evoke a different ensemble of cortical neuron activation, and in turn a different percept. Plasticity of intracortical association fiber synapses, and creation of stored templates of previously experienced patterns of input, could allow perceptual stability of complex mixtures over time.

In effect, the piriform cortex has been hypothesized to function as a form of auto-associative, content-addressable memory. Circuits using content-addressable memory allow an input pattern to be entered, and then the entire memory array circuit is searched to determine whether that pattern exists in the stored memory, and in some cases identifies other items associated with that pattern. This process is distinct from common computer random access memory wherein the input is an address or storage site and the output is the contents of that site. Thus, content-addressable memory allows familiar input

patterns to be recognized, completed and elaborated upon, even in the face of degraded input. As input patterns become more divergent, they activate different memory templates, and thus different odorant representations. In contrast, two well-learned representations can involve very similar inputs and yet evoke distinctly separate stored representation.

Cortical content-addressable memory can rely on associative synaptic plasticity such as long-term potentiation described above. As combinations of odorant features co-occur, association fibers conveying information about these features and their combinations are strengthened through associative plasticity (Haberly, L. B., 2001). On subsequent presentation of the experienced, familiar pattern, circuits containing these now strengthened synapses are more easily activated – signaling a matched template. If the input pattern is degraded or slightly modified, the template pattern can be completed because of the enhanced synaptic strength (Hasselmo, M. E. and Barkai, E., 1995). In fact, with experience, anterior piriform cortical neurons learn to distinguish binary mixtures from their components (Wilson, D. A., 2003). Through this process, odor experience and familiarization come to have strong effects on odor discrimination and perceptual stability (Wilson, D. A. and Stevenson, R. J., 2003). In fact, pharmacological disruption of normal synaptic plasticity within the piriform cortex can disrupt experience-induced enhancement in cortical odorant discrimination. Furthermore, when content-addressable memory is combined with selective cortical afferent adaptation, the piriform can begin to perform more complex functions such as figure-ground separation (see below).

There is a distinct difference in the relative contribution of the afferent lateral olfactory tract and the intrinsic association fiber system to the anterior and posterior regions of the piriform cortex. The anterior piriform cortex is dominated by excitatory input from the lateral olfactory tract, as evidenced both anatomically (see Figure 3) and physiologically. In contrast, the posterior piriform cortex is dominated by association fiber excitatory activity from both within the piriform cortex and from other cortical areas. For example, local field potential oscillations recorded in the anterior piriform cortex are similar in frequency characteristics and strongly coherent with the olfactory bulb, while the oscillations in the posterior piriform are more similar to, and strongly coherent with, the entorhinal cortex. In light of the association fiber plasticity described above, this difference suggests that odor quality representations by ensembles of neurons within these two areas may also differ substantially.

In the only comparison of anterior and posterior piriform single-unit activity, Litaudon and colleagues reported that cells in posterior piriform showed less spontaneous activity and responded to fewer odors than cells in the anterior piriform cortex (Litaudon, P. *et al.*, 2003). Posterior cortical cells were more diverse in their activity phase relationship to the respiratory cycle than were anterior cells, again suggesting a reduced reliance on lateral olfactory tract-driven activity and more on association fiber activity. Precise comparisons of odorant receptive fields or responses to complex odor mixtures and learned odor objects were not made, although the reduced odor responsiveness in the posterior hints at a more narrow tuning.

We are currently examining the effects of familiarity with odorant mixtures on responses to mixtures and their components in the anterior and posterior piriform cortex, with the hypothesis that posterior neurons may become highly selective for familiar mixtures. In fact, in several studies of piriform cortical responses to novel binary mixtures, cells responding selectively to the mixture, without also responding to one or both of the components, are extremely rare (Kadohisa, M. and Wilson, D. A., 2006). It is hypothesized that as odor mixtures become more familiar, piriform cortical neurons, perhaps especially posterior cortical neurons, will enhance their responsiveness to the familiar mixtures independent of their existing responsiveness to the components alone. This could result in stimulus response patterns similar to that observed in visual inferotemporal cortex, wherein single units may respond vigorously to a visual object (e.g., a face) yet not to components of the object if presented in isolation (Figure 7).

The preceding discussion of odorant quality coding has focused on the structural aspects of odorant molecules as the discriminant variable. Ultimately, of course, odor quality encoding moves away from molecular structure toward perceptual objects, where similar molecular structures may have distinctly different odor qualities, and different molecular structures can evoke similar odor qualities. Recent human cortical fMRI imaging works confirms that perceptual odor quality is a cortical phenomenon. Using a creative cross-habituation paradigm, Gottfried and colleagues suggest that the human anterior piriform cortex encodes information about

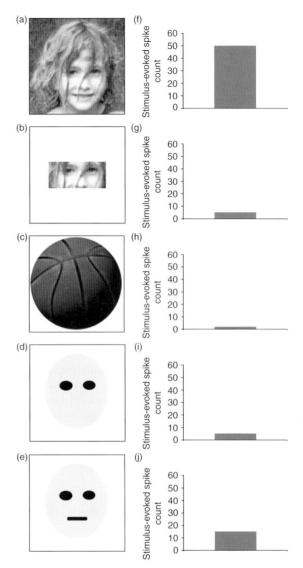

Figure 7 Representation of single-unit response properties in primate visual inferotemporal cortex. Individual inferotemporal neurons have receptive fields for visual objects and are often much less responsive or nonresponsive to individual features of those objects. These receptive fields emerge with experience and are believed to be dependent on plasticity of intracortical association fiber synapses. Similar response properties are predicted for familiar complex odor objects and their components in piriform cortex. Adapted from Tanaka, K. 2000. Mechanisms of visual object recognition studied in monkeys. Spat. Vis. 13, 147–163.

the structural conformation of odorants (e.g., aldehydes cross-adapt with aldehydes regardless of their perceptual quality), while the posterior piriform cortex encodes the perceptual conformation of odors (e.g., fruity odors cross-adapt with fruity odors regardless of their structural make-up). These data, if confirmed, are an excellent example of the emergence of odor quality through synthesis within the piriform cortex. Similarly, presentation of the odorant isovaleric acid can produce the perception either of cheese or of an unpleasant percept of vomit or body odor. When subjects were informed they were smelling cheese, the orbitofrontal cortex was activated. However, when the same molecular stimulus was labeled as body odor, the orbitofrontal cortex was not activated (DeAraujo, I. E. *et al.*, 2005). This strongly suggests that perceptual odor quality is an emergent property of the cortical synthesis of odorant input with multimodal, internal state, and expectation information to create a unique perceptual object.

A final point regarding encoding of odor quality in the piriform cortex involves multimodal, hedonic, and context-dependent components of the representation. As discussed in more detail below, cortical odor representations – similar to their perceptual outcome – can be multifaceted. The smell of an orange can include taste components (e.g., sweet), hedonic components (which may vary depending on recent experience, e.g., sensory specific satiety), and even visual imagery components (e.g., visual imagery of the color or of the fruit). Olfactory cortical areas, including both the piriform cortex and the orbitofrontal cortex, can serve as sites of multimodal and hedonic convergence (Schoenbaum, G. and Eichenbaum, H., 1995; Rolls, E. T., 2004), which may allow for the richness of odor quality perception.

In summary, cortical coding of odor quality appears to go far beyond a simple read-out of peripheral odorant feature extraction, allowing for a synthetic process wherein cortical odor coding and odor perception are shaped by past experience. Experience-dependent cortical synthetic processing enhances discriminability and perceptual constancy of familiar odors and adds multimodal complexity to odor quality representations. The cortical circuitry underlying these phenomena is characteristic of an auto-associative, content-addressable memory system, relying on synaptic plasticity of the intrinsic association fiber system to store templates of familiar patterns of afferent input. In accord with this view, disruption of cortical synaptic plasticity disrupts experience-dependent cortical odorant quality discrimination. Given this view of cortical odor representation, it is hypothesized, and generally supported by the data, that odor quality will be encoded by dispersed ensembles of cortical neurons, with

minimal clustering of cells sharing similar tuning characteristics. It must also be emphasized that quality discrimination of familiar odorants occurs very rapidly in both freely breathing animals and humans, general within a single sniff. The memory-based process described here could facilitate this rapid processing.

4.38.3.3 Sensory Gating

In addition to serving as the primary site of synthesis and recognition of perceptual objects, sensory cortex is also a critical component of the circuitry underlying attention and sensory gating. Cortical processing resources can be differentially allocated over perceptual space and time depending on internal state and arousal, recent stimulation history, and cognitively directed attention. In olfaction, evidence has been obtained for a direct role of the anterior piriform cortex in sensory gating related to both internal state and recent stimulation history.

In thalamocortical sensory systems, the sensory-evoked activity of thalamic neurons is modulated by arousal and the sleep–wake cycle. For example, in the auditory system of restrained, unanesthetized guinea pigs, tone-evoked responses of auditory thalamic neurons decrease in magnitude and become more selective during slow-wave sleep compared to waking (Edeline, J. M. *et al.*, 2000). These changes appear to occur within the thalamus, as subthalamic activity is either not modulated or variably modulated by sleep–wake cycles. The decrease in thalamic excitability appears to be mediated by a membrane hyperpolarization during slow-wave sleep, presumably driven by changes in modulatory input. The decreased thalamic response during slow-wave sleep still allows accurate throughput of frequency-specific information to the cortex, however at a much reduced level (Edeline, J. M. *et al.*, 2000).

In the olfactory system, a similar state-dependent modulation of information transfer occurs, though it appears in the anterior piriform cortex. In urethane anesthetized rats, which display electrographic evidence of spontaneous slow-wave and fast-wave neocortical EEG transitions, anterior piriform cortical neurons decrease their responsiveness to odor input during the slow-wave state (comparable to slow-wave sleep) relative to responsiveness during the fast-wave state (perhaps comparable to awake or fast-wave sleep states) (Murakami, M. *et al.*, 2005). Electrical stimulation of the brainstem which shifts activity between these two states similarly modifies cortical odorant responsiveness. Olfactory bulb evoked activity is not reliably modified in a state-dependent manner, suggesting a piriform cortical origin for these changes. In fact, similar to the mechanism of thalamic modulation, anterior piriform cortical neurons show a hyperpolarization during the slow-wave state that can account for the reduced sensory responsiveness (Murakami, M. *et al.*, 2005). Whether similar changes occur in olfactory thalamus is not known.

A second form of sensory gating is based on recent stimulation history and is commonly associated with adaptation or habituation to background or currently nonsignificant stimuli. In thalamocortical systems, precise mechanisms of sensory habituation have not been determined. In general, however, adaptation to prolonged or repeated stimuli tends to be most pronounced in the sensory cortex compared to subcortical areas. Two specific mechanisms for cortical adaptation have been suggested. First, similar to the state-dependent gating mechanism described above, in the visual system, there is evidence for a sensory-dependent hyperpolarization in visual cortical neurons following prolonged stimulation (Carandini, M. and Ferster, D., 1997). As with state-dependent gating, this hyperpolarization could reduce excitability and sensory-evoked responses in cortical neurons. A second proposed mechanism in thalamocortical systems is a stimulation-dependent synaptic depression of thalamic inputs to visual sensory cortex (Chung, S. *et al.*, 2002). Synaptic depression has also been proposed as a mechanism of adaptation of brainstem-mediated acoustic startle reflexes.

In the olfactory system, neurons in the anterior piriform cortex express significantly more rapid and complete adaptation to prolonged or repeated odor stimulation than their subcortical afferents, mitral/tufted cells in the olfactory bulb (Wilson, D. A., 1998). In awake animals performing a simple odor task, piriform neurons rapidly reduce their odor responses to a short burst of a few spikes (McCollum, J. *et al.*, 1991). In anesthetized animals, the adaptation is odorant specific, expressed in both evoked firing and subthreshold postsynaptic potentials, and recovers within 2 min following a 50 s odorant exposure. There is no corresponding hyperpolarization to account for the decreased odor responsiveness (Wilson, D. A., 1998). However, intracellularly recorded postsynaptic potentials in response to both odorant stimulation and electrical stimulation of the lateral olfactory tract are transiently depressed during

and after prolonged odorant exposure (Wilson, D. A., 1998). These results suggest that synaptic depression of cortical afferents may underlie short-term cortical odorant adaptation.

To test this hypothesis, the synaptic depression evoked by odorant exposure was modeled in vitro using patterned electrical stimulation of the lateral olfactory tract in slices of anterior piriform cortex. Electrical stimulation of the lateral olfactory tract in vitro, in a manner meant to mimic odorant-evoked mitral cell activity, induced a short-term synaptic depression, with a magnitude and duration similar to that of odorant exposure-induced cortical odor adaptation (Best, A. R. and Wilson, D. A., 2004). This synaptic depression could be blocked by the metabotropic glutamate receptor group II/III (mGluRIII) antagonist CPPG. mGluRIII receptors are expressed on the presynaptic terminals of mitral cell axons, and when activated, reduce neurotransmitter release. *In vivo*, bilateral infusion of CPPG into the anterior piriform cortex prevents adaptation of both cortical and simple behavioral responses to odorants (Best, A. R. and Wilson, D. A., 2004; Best, A. R. *et al.*, 2005). These studies are the most complete mechanistic description of sensory adaptation in any mammalian sensory system and demonstrate the power of taking advantage of the relatively simple structure of the olfactory system to address questions of this nature.

Rapid cortical adaptation in the olfactory system may not only be important for reducing responses to persistent stimuli, but also may play an important role in olfactory figure-ground separation. The piriform cortex can function as a high-pass filter, responding most strongly to new stimuli and much less so to stable stimuli. As odorants are encountered against a background, the piriform cortex can respond as if these new odorants were in isolation. In contrast, under similar conditions, mitral/tufted cells respond to the actual inhaled stimulus – a mixture of background and new additive. Rapid cortical adaptation thus can allow the piriform cortex to selective read olfactory bulb output, ignoring temporally stable components (background) while remaining responsive to novel stimuli occurring against this background (Kadohisa, M. and Wilson, D. A., 2006).

Cortical adaptation may also play a role in modulation of cortical responsiveness to food stimuli based on recent feeding patterns. Sensory-specific satiety is a decrease in behavioral responsiveness to a particular food stimulus (e.g., banana's) following prolonged feeding with that stimulus. The behavioral satiety can be highly stimulus specific (Rolls, E. T., 2004). In primates, behavioral sensory-specific satiety is associated with a corresponding decrease in responsiveness of orbitofrontal cortex neurons to that stimulus.

Together, these descriptions of state-dependent and experience-dependent sensory gating suggest a role for anterior piriform cortex that combines both thalamic and cortical components of sensory gating in thalamocortical systems. The state-dependent process and mechanism mediated by thalamic neurons in other sensory systems is expressed in anterior piriform cortical neurons in olfaction. Similarly, the experience-dependent adaptation mediated by cortical neurons in other sensory systems is expressed in anterior piriform cortical neurons in olfaction. The separation of these two function in thalamocortical systems may be in response to differences in the nature of the stimulus representation or other information processing demands in these nonolfactory sensory systems.

A third form of sensory gating – cognitively directed attention – has not been explored to any major extent in olfaction. Attention has been likened to a searchlight, wherein differential processing power can be devoted to some portions of a sensory scene compared to other portions of the same scene (Crick, F., 1984). This may be most easily imagined as occurring in the visual or auditory systems. An example might include switching auditory attention between different, simultaneously occurring conversations. It is not clear whether a similar phenomenon exists in olfaction, although it is anecdotally possible to adapt to a background odor and then redirect attention toward that odor and begin to perceive it again.

In thalamocortical sensory systems, the thalamic reticular nucleus may play an important role in the attentional search light. The thalamic reticular nucleus is a thin lamina of neurons along the superficial edge of the thalamus and is organized both topographically and according to modality. For example, a subset of reticular thalamic neurons receive input from visual cortex and project to visual lateral geniculate nucleus neurons in a retinotopic manner. The reticular thalamus receives excitatory input from both sensory thalamus and sensory cortex, and projects back to sensory thalamus. Importantly, this feedback to sensory thalamus is inhibitory. Thus, the reticular thalamus is a group of inhibitory neurons, receiving cortical feedback (as well as

state-dependent modulatory inputs) and capable of gating the flow of thalamic information to the sensory cortex. The reticular thalamus thus could play a role in attentional changes in sensory coding, through enhancing processing of important information while suppressing processing of unimportant or unexpected information.

Although data on olfactory-directed attention and scene analysis are limited at best, it is interesting to note a potential olfactory circuit analogy to the reticular thalamic circuit described above. Olfactory bulb granule cells monitor mitral cell output via dendrodendritic synapses and receive axosomatic and axodendritic synapses from cortical feedback fibers. While several hypotheses exist regarding the role of mitral–granule cell reciprocal connections in lateral and feedback inhibition, these hypotheses largely overlook the impact of huge excitatory input to granule cells from olfactory cortical areas. Olfactory cortical feedback to the bulb could provide a higher-order modulation of granule cell excitability, resulting in top-down control of olfactory processing of sensory afferent activity. This cortical modulation of granule cell excitability could be evidenced by changes in gamma frequency local field potential oscillations during periods of strong descending input to the bulb, just as has been found in freely behaving animals performing well-learned olfactory behaviors, as described above. Thus, the olfactory cortex, through descending control based on attention and expectancy, could modulate how the olfactory bulb processes odors (Kay, L. M. and Laurent, G., 1999).

4.38.3.4 Odorant Familiarity and Meaning

In thalamocortical sensory systems, individual neuron receptive fields and cortical maps are shaped by experience. Familiar or learned, important stimuli have enhanced representations at the single cell and circuit level, while less familiar or unimportant stimuli have reduced representations. Where it has been examined, these learned changes in cortical sensory representations involve modifications of intracortical association fiber connectivity, potentially including both excitatory and inhibitory synapses, and are mediated by phenomena such as long-term potentiation and long-term depression. The synaptic and local circuit plasticity induced by experience can lead to changes in sensory acuity (perceptual learning) as well as memory of the learned stimuli (associative learning). Experience also appears to be important in creating synthetic receptive fields, for example in higher-order infero-temporal visual cortex.

Experience also shapes olfactory cortical circuit function (Wilson, D. A. *et al.*, 2004). As noted above, simple odor exposure can result in cortical afferent synaptic depression and cortical odor adaptation. However, several forms of experience-dependent plasticity contributing to odor discrimination and odor memory have also been described in the piriform cortex. This plasticity includes synaptic and circuit changes necessary to synthesize multiple odor features into unique objects distinct from those components as a process of perceptual learning, as well as circuit and membrane biophysical changes important for associative memory of learned odors and the rules involved in learning new odors.

As noted above, familiar odors are more easily identified and discriminated from similar odors than are novel odors (Wilson, D. A. and Stevenson, R. J., 2003). This experience-induced increase in olfactory acuity may be due in part to changes in feature encoding by olfactory bulb circuitry (Fletcher, M. L. and Wilson, D. A., 2003), but also appears to be critically dependent on changes within olfactory cortex. For example, single units in the anterior piriform cortex are unable to discriminate between single odorants and binary mixtures made of those components as determined with a simple habituation/cross-habituation paradigm. Thus, exposure to a novel binary mixture for 10 s is sufficient to cause adaptation of anterior piriform cortical neuron responses to that mixture and causes cross-adaptation to the components of that mixture (Wilson, D. A., 2003). This suggests that these neurons do not discriminate between this novel (never smelled by the animal before) mixture and its components. However, simply extending the duration of exposure to the mixture to 50 s is sufficient to reduce cross-adaptation between the mixture and its components, suggesting that as the mixture becomes more familiar, it becomes more distinctive from similar odors. Olfactory bulb mitral/tufted cells do not appear to be capable of discriminating mixtures from their components in similar conditions. This supports the hypothesis that mitral/tufted cells encode for odorant features, rather than odors as a whole. These cells will respond to any odorant containing a particular feature, and adaptation to that feature will reduce responsiveness to all odors sharing it.

The familiarization effect on cortical odorant discrimination appears to involve plasticity within cortical circuits as noted above in the discussion of content-addressable memory. Application of the muscarinic acetylcholine receptor antagonist scopolamine directly onto the anterior piriform cortex during initial odorant exposure prevents experience-dependent enhancement in cortical odor discrimination (Wilson, D. A. and Stevenson, R. J., 2003). Cholinergic input to the piriform cortex comes primarily from the horizontal limb of the diagonal band and has been shown to modulate both synaptic transmission of piriform cortical associative fiber synapses and plasticity of those synapses. This selective modulation of association fiber pathways might be expected to strongly influence the content-addressable memory function hypothesized to be involved in odor representations, thus impairing odor recognition and discrimination (Hasselmo, M. E. and Barkai, E., 1995).

In addition to perceptual learning, odor experience can also result in associative memories wherein odors acquire predictive value or meaning. Associative olfactory conditioning modifies c-fos expression and 2-deoxyglucose uptake in piriform cortex in response to the conditioned odor, with differential effects on anterior and posterior regions (Datiche, F. *et al.*, 2001; Roth, T. L. and Sullivan, R. M., 2005). Associative conditioning also modifies the effectiveness of olfactory bulb electrically evoked responses in the piriform cortex, again most prominently in the posterior regions. Using high-speed voltage-sensitive dye imaging across a wide region of the piriform cortex, Litaudon P. *et al* (1997) reported that the most significant changes in piriform cortical responsiveness to olfactory bulb input following conditioning were in the longer-latency components of the evoked response, suggesting a primary effect on association fiber synapses. These results further argue that cortical representations of familiar odorants, stored by changes in association connections, may include nonolfactory components.

In fact, activity of neurons in both the piriform cortex and orbitofrontal cortex is influenced by a variety of nonolfactory stimuli in awake animals, including visuospatial or expectancy cues in a well-learned odor discrimination task. Schoenbaum and colleagues (Schoenbaum, G. and Eichenbaum, H., 1995) found no significant difference between piriform cortical neuron single-unit activity and orbitofrontal cortical neuron activity in an odor learning task, with cells in both areas responding to such apparently nonolfactory phases of the task as approach to the odor sampling port and water reward consumption.

In a series of well-designed experiments, Barkai and colleagues have found that the piriform cortex is also modified by odor rule learning in a form of metaplasticity (Saar, D. and Barkai, E., 2003). Rats trained in a series of simple odor discrimination tasks (go, no-go discrimination) dramatically improve their acquisition over repeated odor pairs. For example, if training on an odor A versus odor B task is followed by similar C versus D and then E versus F discriminations, acquisition on the E versus F or subsequent pairings can often occur on a single trial. This has been described as odor set learning, odor rule learning, or learning to learn. Barkai has found that such odor rule learning is associated with heightened excitability of piriform cortical (Saar, D. and Barkai, E., 2003) and hippocampal neurons, as well as changes in dendritic spine number on piriform cortical pyramidal cell dendrites. The membrane excitability changes in piriform cortex are acetylcholine dependent and work through cholinergic reduction of a K-mediated after-hyperpolarization (Saar, D. and Barkai, E., 2003). This enhanced excitability and connectivity may facilitate learning of subsequent odor pairs, as the change would be expected to be a whole-cell phenomenon and thus not odor specific.

4.38.3.5 Nonolfactory Context

As noted above, activity of olfactory cortical neurons is modulated by a variety of nonolfactory stimuli. This modulation could reflect a direct multisensory input to olfactory cortex. For example, the orbitofrontal cortex receives strong input from gustatory and orosensory areas (Rolls, E. T., 2004), and the piriform cortex receives strong input from the multisensory entorhinal cortex (Neville, K. R. and Haberly, L., 2004). The orbitofrontal and piriform cortices are also reciprocally interconnected; thus, there are multiple opportunities for multisensory convergence in either structure. In addition, nonolfactory modulation of olfactory cortical activity could reflect input from nonsensory-specific areas responsive to changes in arousal or behavioral state. For example, the olfactory system is richly innervated by noradrenergic fibers from the nucleus locus coeruleus. Activation of the locus coeruleus can modify piriform cortical single-unit excitability and activity phase relationship to the respiratory cycle (Bouret, S. and Sara, S. J., 2002), either through direct modulation of cortical synaptic physiology or through modulation of olfactory bulb afferent activity (Jiang, M. et al., 1996). Similarly, cholinergic input to the olfactory cortex can

modulate piriform cortical synaptic physiology. Both of these systems have direct effects on cortical pyramidal cells, as well as modulatory effects on association fiber synaptic function. They can also both modulate synaptic plasticity in the piriform cortex as noted above, as well as olfactory bulb local circuits (Sullivan, R. M. and Wilson, D. A., 1994).

Modulatory inputs such as from the locus coeruleus and horizontal limb of the diagonal band can signal changes in arousal and internal behavioral state. In doing so, they can differentially influence response properties of olfactory cortical circuits depending on context and expectations. For example, the response seen in the olfactory bulb and piriform cortex as an animal performs a well-learned task of approaching an odor port may reflect either a nonspecific change in cortical arousal and attention signaled by the locus coeruleus or cholinergic inputs. Alternatively, it may reflect other, more specific top-down activation, for example from the entorhinal and higher cortex (Kay, L. M. et al., 1996). Dissecting the differential roles of these two forces will be an important future avenue of research for understanding the effects of context and expectancy in cortical odor processing.

4.38.4 Summary of Olfactory Cortical Information Processing and Unresolved Issues

4.38.4.1 Odor Discrimination

There is increasing evidence that odor discrimination, identification, and adaptation all require olfactory cortical involvement. In humans, lesions of temporal lobe structures that include olfactory cortex, diseases that affect temporal lobe cortical memory areas, and some developmental disorders have all been shown to impair odor identification and discrimination in the absence of impairment of odor detection thresholds (Wilson, D. A. and Stevenson, R. J., 2003). The thesis presented here and elsewhere (Haberly, L. B., 2001; Wilson, D. A. and Stevenson, R. J., 2003) is that olfactory cortex synthesizes odorant features extracted at the periphery and refined by the olfactory bulb into unique perceptual odor objects. This synthesis is an active process involving synaptic plasticity and memory of familiar input patterns. Experience with complex odors results in plasticity of intracortical association fiber synapses within the olfactory cortex, leading to the formation of a content-addressable memory for that odor amongst an ensemble of activated cortical neurons. Subsequent activation of the cortical circuit with the same, or similar input, reactivates the complete ensemble, leading to odor recognition. As two similar inputs are repeatedly, separately activated, they may come to diverge, enhancing odor discrimination. Thus, experience with odors and the resulting cortical plasticity enhances odor perceptual acuity (olfactory perceptual learning). These general mechanisms are hypothesized to be similar to those involved in visual object recognition and the emergence of visual object receptive fields in inferotemporal cortex. Further examination of odor quality coding in the olfactory cortex, using multiple single-unit ensemble recording, is clearly warranted. In addition, it is now clear that odor quality discriminations can be made within a single sniff in both humans and rodents. A closer analysis of the temporal dimensions of odor quality coding in olfactory cortex should also prove fruitful.

Odor objects may be multimodal, including for example gustatory or other integral components. The influence of stimulus color on odor perception (Gottfried, J. A. and Dolan, R. J., 2003) and the inclusion of gustatory descriptors for purely olfactory stimuli (Stevenson, R. J. et al., 1995) are good examples of this process. The multimodal inputs to olfactory cortex, combined with the synthetic processing just summarized, could easily contribute to this multimodal perception. In fact, associative experience with odors and tastes can facilitate the merger of their perceptual properties.

As in other sensory systems, olfaction may include specialist subregions for specific classes of odorants or odor functions. The orbitofrontal cortex, for example, while heavily interconnected with the piriform cortex, may be particularly important for food odors and other hedonically charged odorants. The different subregions of the piriform cortex may also have specialized roles in olfaction, with for example the posterior piriform more involved in higher-order or multimodal odor representations, though this is a topic in need of further research. In fact, the sensory physiology of large regions of olfactory cortex have not been examined at all.

Finally, it must be emphasized that odorant molecular structure is not synonymous with odor quality percept. Expectations, past experience, internal state, and stimulus intensity can all affect how a given inhaled molecular stimulus is perceived. A major point of this review, and new human imaging evidence (DeAraujo, I. E. et al., 2005), suggests that the transform from molecular features to perceptual object is a cortical event.

4.38.4.2 Sensory Gating

Sensory gating takes several forms including modulation of sensation and perception due to changes in arousal, recent stimulus exposure, and selective attention. Olfactory cortex appears to play a major role in the first two sensory gating phenomena, and may be expected to also be involved in the latter. Spontaneous alternation between fast-wave and slow-wave cortical states, roughly comparable to alert and sleep states respectively, produces corresponding shifts in the state of piriform cortical membrane depolarization. Fast-wave cortical states are associated with depolarized piriform cortical neurons and robust responsiveness to odorant stimulation, while slow-wave states are associated with membrane hyperpolarization and reduced sensory responsiveness (Murakami, M. et al., 2005). In this way, the piriform cortex can gate sensory throughput to higher cortical areas based on arousal level. Similarly, the piriform cortex is a critical component of odor habituation. Activity-dependent presynaptic depression of cortical afferents can account for both cortical and simple behavioral adaptation to odors (Best, A. R. and Wilson, D. A., 2004; Best, A. R. et al., 2005). Prevention of cortical short-term adaptation prevents adaptation of simple behavioral responses, demonstrating that adaptation of receptors in the nose cannot account for short-term behavioral habituation.

Cortical mechanisms of selective attention in olfaction have not been examined. A hypothesis is proposed that olfactory cortical feedback to the olfactory bulb granule cell layer may play a role similar to cortical input to the reticular thalamus in other sensory systems (Crick, F., 1984). Another possibility is that short-term cortical adaptation may be reversible, for example through noradrenergic input (Best, A. R. and Wilson, D. A., 2004), thus allowing an arousal or attentional shift to enhance cortical responsiveness to previously adapted odors. This area requires further attention.

4.38.4.3 Background Segmentation

A cortical and perceptual function that merges odor discrimination and sensory gating is figure-ground separation. In the natural world, odorants are invariable experienced against backgrounds, yet the perceived quality of the foreground odorants can remain constant despite background changes. If odor discrimination and identity were dependent solely on identification of inhaled features, then odorants experienced against backgrounds would change in quality dependent on the identity of the background. Formation of odorant objects in cortical memory, combined with odor-specific cortical adaptation to background odorants, solves the problem of olfactory figure-ground separation. In fact, recent evidence suggests that piriform cortical neurons respond to odorants against a preexisting background odorant as if the background were not present (Kadohisa, M. and Wilson, D. A., 2006). Olfactory bulb mitral/tufted cells, in contrast, respond to the same situation as if a binary mixture of the background and foreground odorants were present (as they are). Thus, cortical plasticity may be critical for this important ability of the olfactory system to deal with variation in stimulus complexity and analysis of olfactory scenes.

4.38.4.4 Corticobulbar Feedback and Laterality

As in all sensory systems, the olfactory system includes a rich feedback circuitry. In thalamocortical sensory systems, cortical feedback plays an important role in experience and expectation modulation of sensory acuity and sensitivity, as well as selective attention. In olfaction, the olfactory bulb receives strong input from all olfactory cortical structures, and olfactory cortical structures are strongly interconnected and receive descending input from higher-order cortex (Neville, K. R. and Haberly, L. B., 2003). These feedback pathways are activated by past experience and expectancy and may shape odor sensitivity and acuity. This is another area in need of future research.

Similarly, many sensory and higher-order systems express some lateralization of function. The olfactory cortex includes extensive commissural connectivity, and piriform cortical neurons are differentially sensitive to unilateral and bilateral odorant stimulation. Human imaging work suggests there may be some lateralization of function in olfaction (Zatorre, R. J. *et al.*, 1992), though this needs to be further examined. The causes and consequences of lateralized cortical function in olfaction should be examined.

In summary, as in all other sensory systems, olfactory cortical structures underlie the critical transition from peripheral analytical processing to synthetic, global processing that leads to the rich perceptual experience that defines the sense of smell. The unique nature of the stimulus in olfaction, however, creates unique problems for processing this information. Though substantial progress has been made in the past 10 years toward

understanding the role of olfactory cortex in smell, the field is poised to make major strides by the time this volume is revised.

References

Best, A. R. and Wilson, D. A. 2004. Coordinate synaptic mechanisms contributing to olfactory cortical adaptation. J. Neurosci. 24, 652–660.

Best, A. R., Thompson, J. V., Fletcher, M. L., and Wilson, D. A. 2005. Cortical metabotropic glutamate receptors contribute to habituation of a simple odor-evoked behavior. J. Neurosci. 25, 2513–2517.

Bouret, S and Sara, S. J. 2002. Locus coeruleus activation modulates firing rate and temporal organization of odour-induced single-cell responses in rat piriform cortex. Eur. J. Neurosci. 16, 2371–2382.

Carandini, M. and Ferster, D. 1997. A tonic hyperpolarization underlying contrast adaptation in cat visual cortex. Science 276, 949–952.

Chung, S., Li, X., and Nelson, S. B. 2002. Short-term depression at thalamocortical synapses contributes to rapid adaptation of cortical sensory responses in vivo. Neuron 34, 437–446.

Cleland, T. A. and Linster, C. 2003. Central Olfactory Structures. In: Handbook of Olfaction and Gustation, 2nd edn. (*ed*. R. L. Doty), pp. 165–180. Dekker.

Crick, F. 1984. Function of the thalamic reticular complex: the searchlight hypothesis. Proc. Natl. Acad. Sci. U. S. A. 81, 4586–4590.

Datiche, F., Roullet, F., and Cattarelli, M. 2001. Expression of Fos in the piriform cortex after acquisition of olfactory learning: an immunohistochemical study in the rat. Brain Res. Bull. 55, 95–99.

DeAraujo, I. E., Rolls, E. T., Velazco, M. I., Margot, C., and Cayeux, I. 2005. Cognitive modulation of olfactory processing. Neuron 46, 671–679.

Edeline, J. M., Manunta, Y., and Hennevin, E. 2000. Auditory thalamus neurons during sleep: changes in frequency selectivity, threshold, and receptive field size. J. Neurophysiol. 84, 934–952.

Eeckman, F. H. and Freeman, W. J. 1990. Correlations between unit firing and EEG in the rat olfactory system. Brain Res. 528, 238–244.

Fletcher, M. L. and Wilson, D. A. 2003. Olfactory bulb mitral-tufted cell plasticity: odorant-specific tuning reflects previous odorant exposure. J. Neurosci. 23, 6946–6955.

Gottfried, J. A. and Dolan, R. J. 2003. The nose smells what the eye sees: crossmodal visual facilitation of human olfactory perception. Neuron 39, 375–386.

Haberly, L. B. 2001. Parallel-distributed processing in olfactory cortex: new insights from morphological and physiological analysis of neuronal circuitry. Chem. Senses 26, 551–576.

Hasselmo, M. E. and Barkai, E. 1995. Cholinergic modulation of activity-dependent synaptic plasticity in the piriform cortex and associative memory function in a network biophysical simulation. J. Neurosci. 15, 6592–6604.

Hasselmo, M. E. and Bower, J. M. 1992. Cholinergic suppression specific to intrinsic not afferent fiber synapses in rat piriform (olfactory) cortex. J. Neurophysiol. 67, 1222–1229.

Hasselmo, M. E., Wilson, M. A., Anderson, B. P., and Bower, J. M. 1990. Associative memory function in piriform (olfactory) cortex: computational modeling and neuropharmacology. Cold Spring Harb. Symp. Quant. Biol. 55, 599–610.

Illig, K. R. and Haberly, L. B. 2003. Odor-evoked activity is spatially distributed in piriform cortex. J. Comp. Neurol. 457, 361–373.

Jiang, M., Griff, E. R., Ennis, M., Zimmer, L. A., and Shipley, M. T. 1996. Activation of locus coeruleus enhances the responses of olfactory bulb mitral cells to weak olfactory nerve input. J. Neurosci. 16, 6319–6329.

Johnson, D. M., Illig, K. R., Behan, M., and Haberly, L. B. 2000. New features of connectivity in piriform cortex visualized by intracellular injection of pyramidal cells suggest that "primary" olfactory cortex functions like "association" cortex in other sensory systems. J. Neurosci. 20, 6974–6982.

Jung, M. W., Larson, J., and Lynch, G. 1990. Role of NMDA and non-NMDA receptors in synaptic transmission in rat piriform cortex. Exp. Brain. Res. 82, 451–455.

Kadohisa, M. and Wilson, D. A. 2006. Olfactory cortical adaptation facilies detection of odors against background. J. Neurophysiol. 95, 1888–1896.

Kay, L. M. and Laurent, G. 1999. Odor- and context-dependent modulation of mitral cell activity in behaving rats. Nat. Neurosci. 2, 1003–1009.

Kay, L. M., Lancaster, L. R., and Freeman, W. J. 1996. Reafference and attractors in the olfactory system during odor recognition. Int. J. Neural Syst. 7, 489–495.

Ketchum, K. L. and Haberly, L. B. 1993. Membrane currents evoked by afferent fiber stimulation in rat piriform cortex. I. Current source-density analysis. J. Neurophysiol. 69, 248–260.

Laing, D. G., Legha, P. K., Jinks, A. L., and Hutchinson, I. 2003. Relationship between molecular structure, concentration and odor qualities of oxygenated aliphatic molecules. Chem. Senses 28, 57–69.

Lin da, Y., Zhang, S. Z., Block, E., and Katz, L. C. 2005. Encoding social signals in the mouse main olfactory bulb. Nature 434, 470–477.

Litaudon, P., Amat, C., Bertrand, B., Vigouroux, M., and Buonviso, N. 2003. Piriform cortex functional heterogeneity revealed by cellular responses to odours. Eur. J. Neurosci. 17, 2457–2461.

Litaudon, P., Mouly, A. M., Sullivan, R., Gervais, R., and Cattarelli, M. 1997. Learning-induced changes in rat piriform cortex activity mapped using multisite recording with voltage sensitive dye. Eur. J. Neurosci. 9, 1593–1602.

McCollum, J., Larson, J., Otto, T., Schottler, F., Granger, R., and Lynch, G. 1991. Short-latency single-unit processing in olfactory cortex. J. Cogn. Neurosci. 3, 293–299.

Murakami, M., Kashiwadani, H., Kirino, Y., and Mori, K. 2005. State-dependent sensory gating in olfactory cortex. Neuron 46, 285–296.

Neville, K. R. and Haberly, L. B. 2003. Beta and gamma oscillations in the olfactory system of the urethane-anesthetized rat. J. Neurophysiol. 90, 3921–3930.

Neville, K. R. and Haberly, L. 2004. Olfactory Cortex. In: The Synaptic Organization of the Brain, 5th edn. (ed. G. M. Shepherd), pp. 415–454. Oxford University Press.

Rolls, E. T. 2004. The functions of the orbitofrontal cortex. Brain Cogn. 55, 11–29.

Roth, T. L. and Sullivan, R. M. 2005. Memory of early maltreatment: neonatal behavioral and neural correlates of maternal maltreatment within the context of classical conditioning. Biol. Psychiatry 57, 823–831.

Saar, D. and Barkai, E. 2003. Long-term modifications in intrinsic neuronal properties and rule learning in rats. Eur. J. Neurosci. 17, 2727–2734.

Schoenbaum, G. and Eichenbaum, H. 1995. Information coding in the rodent prefrontal cortex. I. Single-neuron activity in orbitofrontal cortex compared with that in pyriform cortex. J. Neurophysiol. 74, 733–750.

Schoenbaum, G., Setlow, B., and Ramus, S. J. 2003a. A systems approach to orbitofrontal cortex function: recordings in rat orbitofrontal cortex reveal interactions with different learning systems. Behav. Brain Res. 146, 19–29.

Schoenbaum, G., Setlow, B., Saddoris, M. P., and Gallagher, M. 2003b. Encoding predicted outcome and acquired value in orbitofrontal cortex during cue sampling depends upon input from basolateral amygdala. Neuron 39, 855–867.

Stevenson, R. J., Prescott, J., and Boakes, R. A. 1995. The acquisition of taste properties by odors. Learn. Motiv. 26, 433–455.

Sugai, T., Miyazawa, T., Fukuda, M., Yoshimura, H., and Onoda, N. 2005. Odor-concentration coding in the guinea-pig piriform cortex. Neuroscience 130, 769–781.

Sullivan, R. M. and Wilson, D. A. 1994. The locus coeruleus, norepinephrine, and memory in newborns. Brain Res. Bull. 35, 467–472.

Tanabe, T., Iino, M., and Takagi, S. F. 1975. Discrimination of odors in olfactory bulb, pyriform-amygdaloid areas, and orbitofrontal cortex of the monkey. J. Neurophysiol. 38, 1284–1296.

Tanaka, K. 2000. Mechanisms of visual object recognition studied in monkeys. Spat. Vis. 13, 147–163.

Wilson, D. A. 1998. Habituation of odor responses in the rat anterior piriform cortex. J. Neurophysiol. 79, 1425–1440.

Wilson, D. A. 2000. Comparison of odor receptive field plasticity in the rat olfactory bulb and anterior piriform cortex. J. Neurophysiol. 84, 3036–3042.

Wilson, D. A. 2003. Rapid, experience-induced enhancement in odorant discrimination by anterior piriform cortex neurons. J. Neurophysiol. 90, 65–72.

Wilson, D. A. and Stevenson, R. J. 2003. Olfactory perceptual learning: the critical role of memory in odor discrimination. Neurosci. Biobehav. Rev. 27, 307–328.

Wilson, D. A., Best, A. R., and Brunjes, P. C. 2000. Trans-neuronal modification of anterior piriform cortical circuitry in the rat. Brain Res. 853, 317–322.

Wilson, D. A., Best, A. R., and Sullivan, R. M. 2004. Plasticity in the olfactory system: lessons for the neurobiology of memory. Neuroscientist 10, 513–524.

Zatorre, R. J., Jones-Gotman, M., Evans, A. C., and Meyer, E. 1992. Functional localization and lateralization of human olfactory cortex. Nature 360, 339–340.

Zhang, Y., Burk, J. A., Glode, B. M., and Mair, R. G. 1998. Effects of thalamic and olfactory cortical lesions on continuous olfactory delayed nonmatching-to-sample and olfactory discrimination in rats (Rattus norvegicus). Behav. Neurosci. 112, 39–53.

Zou, Z., Fusheng, L., and Buck, L. B. 2005. Odor maps in the olfactory cortex. Proc. Natl. Acad. Sci. U. S. A. 102, 7724–7729.

Zou, Z., Horowitz, L. F., Montmayeur, J. P., Snapper, S., and Buck, L. B. 2001. Genetic tracing reveals a stereotyped sensory map in the olfactory cortex. Nature 414, 173–179.

4.39 Modeling of Olfactory Processing

C Linster and T A Cleland, Cornell University, Ithaca, NY, USA

Glossary

associative memory In this context, the term refers to a type of neural network in which associations between different elements of a stimulus are stored by strengthening connections between these elements. The most important feature of an associative memory is that it will restore a previously stored pattern when presented with an incomplete or distorted version of this pattern.

attractor In dynamical systems analysis, an attractor is any state of a system in which it will remain if otherwise undisturbed. The attractor of a simple pendulum is the vertical position; adding magnets or other disturbances to the system may move the attractor to a new position, or generate additional attractors. Attractors need not be single points; cyclic attractors are common.

Hebbian learning A type of plasticity (learning) rule in which interactions between simultaneously active elements, such as neurons, are strengthened incrementally.

molecular receptive range Equivalent to receptive field for different chemical ligands. In the olfactory system, a neuron's molecular receptive range describes the ensemble of chemicals to which it is responsive.

olfactory bulb Cortical telencephalic structure receiving direct afferent input from olfactory sensory neurons located in the nasal cavity, as well as efferent input from multiple regions of the brain.

signal-to-noise ratio Engineering term quantifying the proportion of the signal of interest (signal) in the total information stream that is transmitted (signal + noise). Low signal-to-noise ratios render it difficult to detect or interpret the signal, as it is buried in noise.

4.39.1 Introduction

In natural environments, airborne chemical stimuli are distributed unpredictably in time and space, and odorants from innumerable sources intermix freely. The olfactory system must be able to detect potential signals of interest within these chemically noisy environments, correctly extract these signals from a complex and changing odor background to form stimulus representations, compare these constructed representations to those of previously experienced odors, differentiate relevant from irrelevant stimuli, and cue an appropriate response. Many of the neural circuit elements comprising the olfactory system have been proposed to contribute to these processes in particular ways; in particular, multiple feedback and feed-forward interactions among olfactory structures, as well as between olfactory and nonolfactory areas, are thought to contribute to the filtering and construction of olfactory representations. Computational models of olfactory processing have been increasingly utilized to describe and interpret these complex and interrelated phenomena.

Primary olfactory sensory neurons (OSNs) number in the millions in rodents. Their axons are highly convergent, targeting specific, discrete neuropilar synaptic regions within the input layer of the olfactory bulb called glomeruli. In hamsters, for example, between 1300 and 4700 OSNs presumably expressing the same odorant receptor complement converge upon each glomerulus (Schoenfeld, T. A. and Knott, T. K., 2004). These large populations of redundant OSNs and their correspondingly high convergence ratios have been proposed to yield advantages such as improved stimulus sensitivity, an improved signal-to-noise ratio, and an increased range of tuning to different odorant concentrations (van Drongelen, W. et al., 1978; Duchamp-Viret, P. et al, 1989; Meisami, E., 1989; Cleland, T. A. and Linster, C., 1999). The molecular receptive ranges, or chemical receptive fields, of these odorant receptors overlap substantially, such that the identity of odorants is not associated with the activation of a specific receptor, but rather is represented by a distributed, combinatorial code (Adrian, E. D., 1953; Moulton, D. G., 1967; Stewart, W. B. et al., 1979; Kauer, J. S., 1991), now recognized as a pattern of activation across many receptors. Owing to the specific convergence of OSN axons, these odor-specific activity patterns can be most clearly observed in imaging studies of olfactory bulb glomeruli (Friedrich, R. W. and Korsching, S. I., 1997; Rubin, B. D. and Katz, L. C., 1999; Meister, M. and Bonhoeffer, T., 2001; Wachowiak, M. et al., 2002). These overlapping representations underlie two critical properties of the olfactory system that a labeled-line solution would not. First, the number of unique odor representations is not limited to the number of different receptor types (roughly 1000 in mice) (Mombaerts, P., 1996), but can be estimated as m^n, where n denotes the number of receptor types and m denotes the number of recognizable states that each sensor can assume, ultimately limited by the signal-to-noise ratio of the system. Even if only two receptor states, active and inactive, are recognized, this enables roughly 2^{1000} potential odor stimuli to be discriminated in mice. Second, the fact that structurally and perceptually similar odorant molecules will activate correspondingly overlapping sets of olfactory receptors (Linster, C. et al., 2001; Cleland, T. A., et al., 2002; Linster, C. et al., 2002) establishes a basis for the recognition of stimulus similarity in the olfactory system. This is a prerequisite for basic postsensory cognitive processes such as generalization (Shepard, R. N. and Chang, J. J., 1963; Shepard, R. N., 1987; Cleland, T. A. et al., 2002) and a tolerance for variance among repeated stimulus samples that a labeled-line system would have no clear means of generating.

Distributed patterns of activity in response to chemical stimuli are transmitted to the olfactory bulb via OSN axons that terminate in the glomeruli of its input layer. The olfactory bulb is believed to filter and transform these incoming sensory data, performing normalization, contrast enhancement, and similar operations before conveying the processed olfactory information to several different secondary olfactory structures via mitral cell axon collaterals (Cleland, T. A. and Linster, C., 2003). Notably, the bulb constitutes the last common stage at which olfactory sensory representations can be processed before the signal diverges dramatically into these multiple secondary structures. It is clear from recent investigations that the perceptual qualities of odorants can be predicted, to a limited degree, from the patterns of activation that they evoke at the olfactory bulb input layer (Linster, C. and Hasselmo, M. E., 1999; Linster, C. et al., 2001; Cleland, T. A. et al., 2002; Linster, C. et al., 2002). However, several aspects of odor perception, for example, changes in perception and discrimination capacity due to odor intensity or prior experience, cannot be predicted solely by this first-order representation as reflected in glomerular activation patterns. Furthermore, the olfactory bulb receives substantial centrifugal projections from both cortical and neuromodulatory centers, and its activity is strongly regulated by learning and experience. It is therefore safe to assume that the olfactory bulb plays an important role in processing incoming sensory information. Indeed, many models of olfactory bulb signal processing have been developed, which are grouped here into studies of (1) filtering and contrast enhancement, (2) mechanisms underlying oscillations and spike synchronization, and (3) odor segmentation and associative memory function. In addition, a number of detailed biophysical models of bulbar neurons have been constructed, in many cases, to address how their intrinsic properties underlie and interact with network properties.

4.39.2 Filtering and Contrast Enhancement

The high convergence ratio between OSNs expressing a particular odorant receptor and their target glomeruli in the olfactory bulb (Figure 1(a)) is believed to improve the signal-to-noise ratio during odorant detection (van Drongelen, W. et al., 1978), potentially overcoming the limiting noise inherent in

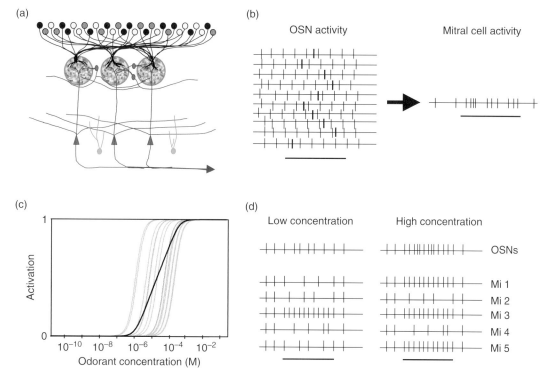

Figure 1 Glomerular computations and convergence. (a) Simplified structure of the olfactory bulb as represented in most computational models. Large populations of olfactory sensory neurons (OSNs) expressing the same receptor (particular shades of gray) converge onto common glomeruli in the olfactory bulb, within which they synapse onto the dendrites of periglomerular (blue) and mitral (red) cells. Mitral cells also excite granule cells (green) via their lateral dendrites, and granule cells in turn inhibit mitral cells; functionally, this mediates lateral inhibition between mitral cells. Periglomerular cells are primarily associated with one or a small number of glomeruli and project axons to a few other glomeruli; these have also been suggested to mediate lateral inhibition between mitral cells. A number of other cell types have also been characterized in the bulb, such as short-axon and tufted cells that are rarely included in computational models; furthermore, established heterogeneity within the cell types modeled has usually been neglected. (b) The high convergence of OSN axons onto mitral cells within each glomerulus significantly improves the signal-to-noise ratio in mitral cells as compared with OSNs. Even if a weak odorant stimulus were to evoke only a single additional spike in each OSN above its basal rate of spiking (OSN activity; bold lines), this high convergence ratio can generate a robust odor responses in a postsynaptic mitral cell (Mitral cell activity), hence enabling mitral cells to reliably represent odor stimuli at lower concentrations than can OSNs (van Drongelen, W. *et al.*, 1978). Horizontal bars depict the time of odorant presentation. (c) While OSN concentration–activation curves are steep, enabling accurate representations of ligand–receptor binding over a range of roughly one log unit concentration, the considerably broader curves observed in glomeruli can be explained if the convergent OSN population is nonuniform in spare receptor capacity or other determinants of intracellular gain (Cleland, T. A. and Linster, C., 1999). If OSNs with identical odorant selectivity but exhibiting different functional spare receptor capacities (gray sigmoids) project onto a single glomerulus, the summed concentration–activation curve of that glomerulus can span several log units of concentration (black sigmoid). (d) Frequency-to-spatial transformation by local glomerular circuits would activate different populations of mitral cells within each glomerulus as a function of the average firing rate of the convergent OSNs (Anton, P. S. *et al.*, 1991). Low concentration odorants (left panel) evoke weak activity in a given receptor-specific OSN subpopulation, generating increased activity in only one mitral cell. In contrast, a high concentration odor stimulus (right panel) evokes greater activity in the OSN population and hence generates measurable excitatory responses in three mitral cells. Horizontal bars depict the time of odorant presentation.

the transduction mechanisms of individual OSNs (Lowe, G. and Gold, G. H., 1995). In principle, this property improves the coding capacity of the olfactory system, increasing the number of different odor-specific patterns that can be discriminated, as well as improving the maximum reliable sensitivity of the system (Figure 1(b)). Another modeling study has suggested that this same property can be employed to increase the range of odor ligand concentrations that can be represented by olfactory bulb glomeruli without saturation (Figure 1(c)) (Cleland, T. A. and Linster, C., 1999). This hypothesis provides a possible answer to the conundrum of how collective concentration–response curves measured by glomerular imaging can be substantially broader than those measured in individual OSN recordings (Friedrich, R. W. and Korsching, S. I., 1997; Bozza, T. *et al.*, 2002), enabling preservation of the ratios of activation levels among glomeruli across broader concentration ranges and hence potentially facilitating the recognition of odor quality across changes in concentration. Another approach has been suggested by Anton P. S. *et al.* (1991). Noting that the activity of each glomerulus is sampled and conveyed centrally by a number of mitral cells (on the order of 50 in hamsters) (Schoenfeld, T. A. and Knott, T. K., 2004), and that mitral cell firing frequencies do not scale monotonically with concentration as do those of OSNs, these authors proposed that the synaptic circuitry within each glomerulus could compute a frequency-to-spatial transformation on the incoming information. That is, the number of responding mitral cells within a glomerulus, but not their firing rate, would depend on the firing rates of the sensory neurons projecting to that glomerulus in response to odor stimulation (Figure 1(d)).

Contrast enhancement is a common property of sensory systems that narrows, or sharpens, sensory representations by specifically inhibiting neurons on the periphery of the representation, hence enhancing the contrast between signal and background. A number of computational models have investigated the contrast enhancement potential of olfactory bulb circuitry (Figure 2(a)), most of which, by analogy with the retina, have investigated the potential role of lateral inhibitory projections. Classically, bulb models have emphasized lateral inhibition mediated by mitral cell lateral dendrites (Rall, W. and Shepherd, G. M., 1968; Shepherd, G. M. and Brayton, R. K., 1979; Schild, D., 1988; Urban, N. N., 2002; Davison, A. P. *et al.*, 2003). These lateral dendrites form reciprocal synapses with inhibitory granule cell spines in the external plexiform layer of the bulb, forming a network through which mitral cells inhibit one another as well as themselves (Isaacson, J. S. and Strowbridge, B. W., 1998), although the region receiving this inhibition is not clearly localized (Luo, M. and Katz, L. C., 2001; Debarbieux, F. *et al.*, 2003; Djurisic, M. *et al.*, 2004). Subsequent models proposed that contrast enhancement was instead mediated by lateral inhibition mediated by the relatively superficial periglomerular (PG) cells (Linster, C. and Gervais, R., 1996; Linster, C. and Hasselmo, M., 1997) (Figures 1(a) and 2(b)). This hypothesis offered the substantial advantage that lateral

inhibition could be delivered onto mitral cells in the glomerular region of their apical dendrites, a location better capable of preventing spike initiation in these cells. In a model based on this hypothesis, Linster C. and Hasselmo M. (1997) further showed that if the activation of PG cells is modulated by cholinergic inputs from the horizontal limb of the diagonal band, a relatively stable number of active mitral cells can be maintained independent of the intensity of olfactory input or the set of OSNs activated by the odorant. In this model, granule cells served instead to modulate the gain of mitral cell activity and were necessary in order to obtain stable average firing rates. Indeed, subsequent studies have shown that PG cells are appropriately modulated by acetylcholine (Castillo, P. E. *et al.*, 1999), and some of the behavioral predictions from these models have been tested in rats (Linster, C. and Cleland, T. A., 2002; Linster, C. *et al.*, 2003; Mandairon, N. *et al.*, 2006).

Contrast enhancement, the effects of which have been observed in the olfactory bulb (Yokoi, M. *et al.*, 1995), can be functionally defined as a process of competition between neurons proportional to the similarity of the information that they mediate. Simplified models of the olfactory system, based on one-dimensional odor subspaces, have been able to implement contrast enhancement using lateral inhibition (Linster, C. and Gervais, R., 1996; Linster, C. and Hasselmo, M., 1997; Linster, C. and Smith, B. H., 1997; Linster, C. and Hasselmo, M., 1999; Linster, C. and Cleland, T. A., 2001; Cleland, T. A. and Linster, C., 2002) and have been effective at interpreting behavioral and physiological data derived from single monotonically varying odorant series (Yokoi, M. *et al.*, 1995; Linster, C. and Hasselmo, M. E., 1999; Cleland, T. A. *et al.*, 2002; Cleland, T. A. and Narla, V. A., 2003). However, to escape this limitation and model more realistic, high-dimensional odor spaces (Hudson, R., 1999; Korsching, S. I., 2001; Alkasab, T. K. *et al.*, 2002), subsequent models have relied upon networks constructed so that the strength of PG-mediated inhibition is effectively proportional to receptive field similarity rather than the physical proximity of glomeruli. Networks based on this assumption have been shown to best reproduce calcium imaging data obtained from honeybee OSNs and projection neurons (analogous to mitral cells), while networks based on decremental lateral inhibition performed comparably to networks based on random inhibitory projections (Linster, C. *et al.*, 2005). Another such model has successfully reproduced mixture processing properties

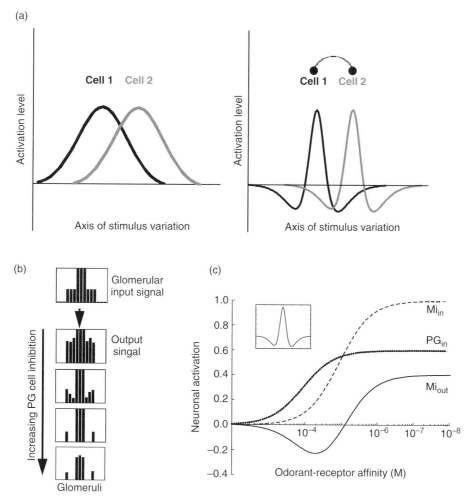

Figure 2 Contrast enhancement. (a) Contrast enhancement is a phenomenon observed in most sensory systems by which marginally activated neurons are excluded from a stimulus-specific ensemble by inhibition, hence sharpening the sensory representation and differentiating it from other, similar representations. In the absence of contrast enhancement, the tuning curves of cells 1 and 2 substantially overlap (left panel). If contrast enhancement is enabled, e.g., by the addition of lateral inhibition such that each cell inhibits the other, the two cells both become more narrowly tuned, and their receptive fields may no longer overlap (right panel). (b) Computational models of the olfactory bulb have proposed that inhibitory periglomerular (PG) cells could mediate contrast enhancement in the glomerular layer of the olfactory bulb. In these models, PG cells receive direct sensory input within a given glomerulus and inhibit mitral cells in neighboring glomeruli via axonal projections (Figure 1(a)). If an odor stimulus activates a range of neighboring glomeruli (Glomerular input signal), increasing the level of PG cell-mediated inhibition would lead to concomitantly sharpened representations among mitral cells, the output layer of the olfactory bulb. With strong contrast enhancement, only the most strongly activated mitral cells are activated, while weakly activated mitral cells are inhibited (Linster, C. and Gervais, R., 1996; Linster, C. and Hasselmo, M. E., 1997). (c) Nontopographical contrast enhancement based on local glomerular computations. In this model, contrast enhancement is generated within the stimulus–response profile of each individual glomerulus, resulting in sharpening of the odor-evoked representation across the glomerular layer. Within each glomerulus, mitral (Mi_{in}) and periglomerular (PG_{in}) cells have the same tuning curves but different response properties, and both receive parallel input from olfactory sensory neurons (OSNs). Mitral cell output (Mi_{out}) is additionally shaped by PG-mediated dendrodendritic inhibition, such that only the most strongly excited mitral cells are activated, while more weakly excited mitral cells exhibit net inhibitory responses (Cleland, T. A. and Sethupathy, P., 2006). The shape of the Mi_{out} curve generates the familiar on-center/inhibitory-surround function of contrast enhancement (inset).
(b) Reproduced from Linster, C. and Gervais, R. 1996. Investigation of the role of interneurons and their modulation by centrifugal fibers in a neural model of the olfactory bulb. J. Comput. Neurosci. 3(3), 225–246, with permission from springer.

measured in rats (Wiltrout, C. *et al.*, 2003; Linster, C. and Cleland, T. A., 2004). Finally, nontopographical models take an entirely different approach to olfactory contrast enhancement, relying solely on intraglomerular computations and broad feedback inhibition to effect contrast enhancement via a 'winner-take-most' algorithm (Figure 2(c)) (Cleland, T. A. and Sethupathy, P., 2006; Cleland, T. A. *et al.*, 2007). This approach does not rely on odorant feature similarities being embedded into bulbar odor maps *a priori*, as it is independent of the physical location of glomeruli and of specific lateral inhibitory projections. Rather, the maps of stimulus feature similarity that underlie contrast enhancement are naturally inherited from the cross-activation patterns of OSNs by odorants.

4.39.3 Mechanisms Underlying Oscillations and Spike Synchronization

While recent models have begun to favor glomerular layer mechanisms for contrast enhancement, granule cell activity also clearly shapes mitral cell response patterns, and hence the presumptive odor representations that emerge from the olfactory bulb. Specifically, several computational models of the olfactory bulb have suggested that the temporal pattern of spiking among mitral cells may play a role in odor representation (Schild, D., 1988; Meredith, M., 1992; White, J. *et al.*, 1992; 1998; White, J. and Kauer, J. S., 2001). While it is clear that temporal response patterns in mitral cells do change as a function of odor identity (Figure 3(a)), there is as yet no broadly accepted theory of how these response patterns may contribute to the representation of odorant stimuli.

One of the most widely studied features of olfactory bulb processing has been the dynamic oscillatory activity patterns observed in the bulb in response to odor stimulation. Models of these phenomena have traditionally studied these responses as coupled oscillators, attributing the dynamics to reciprocal feedback interactions between mitral cell secondary dendrites and granule cells (Figure 3(b)) (Freeman, W. J., 1979; 1987; Li, Z. and Hopfield, J. J., 1989; Grobler, T. and Erdi, P., 1991; Erdi, P. *et al.*, 1993; Freeman, W. J., 1994; Ermentrout, G. B. and Kleinfeld, D., 2001). Some researchers have proposed that odor quality may be represented in dynamic attractors formed in the olfactory bulb (Freeman, W. J., 1979; 1987; Li, Z. and Hopfield, J. J., 1989; Erdi, P. *et al.*, 1993; Freeman, W. J., 1994; Fukai, T., 1996; Hoshino, O.

(a)

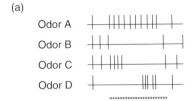

(b)

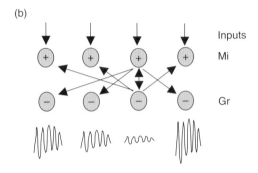

(c)

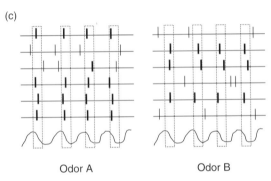

Figure 3 Oscillations and synchrony. (a) Any given mitral cell in the olfactory bulb may respond to different odorant stimuli (A, B, C, and D) with a variety of temporally complex spike patterns including interwoven excitatory and inhibitory phases. It has been proposed that these temporal patterns may contribute to odor representations in the vertebrate olfactory bulb and the analogous insect antennal lobe (Laurent, G., 1999; Laurent, G. *et al.*, 2001). Horizontal bar depicts the time of odorant presentation. (b) The reciprocal synaptic interactions between mitral (Mi) and granule (Gr) cells have often been simulated as a system of coupled oscillators driven by external inputs. In such models, variance of stimulus amplitudes across these inputs generates a map of field oscillations with variable amplitudes and fixed phase lags across the olfactory bulb. For clarity, the output from only a single column is depicted. (c) Field oscillatory dynamics are believed to reflect and/or influence spike timing in mitral cells, potentially resulting in odor-specific populations of mitral cells based on spike synchrony rather than overall activity. While the overall activity patterns evoked by odors A and B are very similar, selection for spikes relatively synchronized with one another and with the oscillatory field potential reveals two clearly odor-specific subpopulations (bold spikes within dotted boxes).

et al., 1998; Breakspear, M., 2001). However, it is increasingly clear that the dynamics of the olfactory bulb are tightly coupled with those of the piriform cortex, and that both depend on mutual feedback between the two structures (Gray, C. M. and Skinner, J. E., 1988; Neville, K. R. and Haberly, L. B., 2003; Martin, C. *et al.*, 2004); combined bulb–cortex models have suggested possible roles for these interactions (Fukai, T., 1996; Li, Z. and Hertz, J., 2000). Other aspects of piriform cortical dynamics have also been modeled (Wilson, M. and Bower, J. M., 1992; Liljenstrom, H. and Hasselmo, M. E., 1995; Claverol, E. T. *et al.*, 2002; Xu, D. and Principe, J. C., 2004), albeit with less focus on their functional role (but see Granger, R. and Lynch, G., 1991).

Recently, models of these field oscillatory properties have begun to emphasize their relationship to the regulation of spike timing in mitral cells (Davison, A. P. *et al.*, 2003; Margrie, T. W. and Schaefer, A. T., 2003), suggesting that patterns of spike synchronization among mitral cells responding to the same sensory input are important contributors to the odorant representation at this level (Figure 3(c)). This phenomenon has been extensively modeled in the insect antennal lobe, a functionally comparable analog of the olfactory bulb. In these models, the regulation of spike synchronization among secondary neurons can mediate stimulus salience and contrast enhancement between similar odorants (Linster, C. and Cleland, T. A., 2001; Cleland, T. A. and Linster, C., 2002) and can dramatically influence the readout of information at the next level of processing (Sivan, E. and Kopell, N., 2004).

4.39.4 Odor Segmentation and Associative Memory Function

Odor segmentation is the general term for the problem of how the olfactory system is able to segregate and identify different odorants that are encountered simultaneously. As most odors comprise multiple separate odorant molecules, it is far from clear how the olfactory system can parse the multitude of odorant stimuli present at any given time and attribute each to appropriately separate sources. One approach has been to hypothesize that odors emitted from different sources can be segregated by OB circuitry based on their differential fluctuations in time (Fort, J. C. and Rospars, J. P., 1992; Hendin, O. *et al.*, 1998; Hopfield, J. J., 1999). Odor segmentation could thereafter be performed in the OB using source-separation

algorithms dependent upon associative memory function. Generally, such models hypothesize that associative memories for patterns of olfactory bulb activity evoked by known odorants become embedded in bulbar circuitry and can then be used to recognize these patterns when they recur, even in degraded form. Specifically, a model by Hendin O. *et al.* (1998) illustrates how, if the glomerular layer feeds into a mitral–granule cell layer for which appropriate dynamics for an associative memory function have been implemented, each odor can be separately represented in successive inhalation cycles when multiple (known) odors are presented at the same time.

Olfactory associative memory functions have been more commonly attributed to the piriform cortex, one of the targets of mitral cell axons projecting from the olfactory bulb (Figure 4). Specifically, the piriform cortex has been proposed to mediate the associative memory functions necessary for odor-context learning (Haberly, L. B. and Bower, J. M., 1989; Hasselmo, M. E. *et al.*, 1990; Haberly, L. B., 2001) and hierarchical clustering (Ambros-Ingerson, J. *et al.*, 1990). Indeed, the extensive intrinsic feedback network in this cortex and its integration with afferent inputs closely resembles the structure of traditional theoretical associative memory networks as first described by Marr D. (1971). Several laboratories have constructed models of piriform cortex implementing these associative memory functions (Haberly, L. B. and Bower, J. M., 1989; Ambros-Ingerson, J. *et al.*, 1990; Hasselmo *et al.*, 1990; Barkai, E. and Hasselmo, M. E., 1994; Hasselmo, M. E. *et al.*, 1997) as well as exploring the role of cholinergic modulation in their regulation (Hasselmo, M. E. *et al.*, 1990; 1997; Patil, M. M. and Hasselmo, M. E., 1999; Linster, C. and Hasselmo, M. E., 2001; Linster, C. *et al.*, 2003).

4.39.5 Detailed Biophysical Models of OB Neurons

Most models of the olfactory system to date have emphasized network-level interactions and the properties of olfactory bulb and piriform cortical circuitry, using simplified cell models conducive to these larger-scale simulations. However, several relatively detailed biophysical models of olfactory neurons – particularly OSNs and mitral cells – have also been constructed. Several detailed OSN models have illustrated how ligand–receptor binding and nonlinear transduction processes can underlie several experimentally observed response properties of these cells (Malaka, R. *et al.*, 1995; Rospars, J. P. *et al.*, 1996;

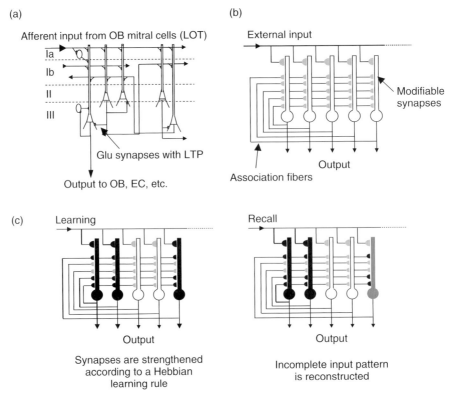

(a)

Afferent input from OB mitral cells (LOT)

Ia
Ib
II
III

Glu synapses with LTP

Output to OB, EC, etc.

(b)

External input

Modifiable
synapses

Output

Association fibers

(c) Learning

Output

Synapses are strengthened
according to a Hebbian
learning rule

Recall

Output

Incomplete input pattern
is reconstructed

Figure 4 Associative memory function in olfactory cortex. (a) Piriform cortex (PC) exhibits the fundamental anatomical features necessary for the implementation of associative memory: extrinsic input from the olfactory bulb (OB) to each pyramidal cell via the lateral olfactory tract (LOT) and extensive intrinsic excitatory connections among pyramidal cells. These intrinsic associative connections are subject to synaptic plasticity, also crucial for associative memory function, and exhibit long-term potentiation (LTP). Additionally, several classes of local inhibitory interneurons have been described in the piriform cortex. Piriform pyramidal neurons project back to the OB, as well as to other structures such as the entorhinal cortex (EC). (b) Schematic representation of an associative memory network. The critical features are (1) external inputs to associative neurons, (2) all-to-all excitatory connections among these neurons, and (3) a learning rule that modifies the strengths of these connections when external inputs are being learned by the network. (c) Learning and recall in an associative memory network. Learning (left panel): An olfactory stimulus activates a subset of pyramidal cells (three dark cells) via distributed afferent projections from the OB to the PC. During learning, excitatory synaptic connections between pyramidal cells activated by that odorant are strengthened via a Hebbian plasticity rule (dark semicircles denote strengthened synapses). Recall (right panel): After learning of a pattern, if a noisy or degraded example of that pattern is presented to the system, the stored odorant pattern can be reconstructed due to the previously strengthened connections. Dark semicircles denote strengthened synapses. Owing to the previously strengthened synapses, activation of the two dark cells secondarily activates a third cell (gray), reconstructing the previously learned pattern.

Vermeulen, A. *et al.*, 1996; 1997; Kaissling, K. E., 1998; Lansky, P. and Rospars, J. P., 1998; Vermeulen, A. and Rospars, J. P., 1998; Cleland, T. A. and Linster, C., 1999; Rospars, J. P. *et al.*, 2000; Kaissling, K. E., 2001; Kaissling, K. E. and Rospars, J. P., 2004). Compartmental models of mitral cells have been used to elucidate intrinsic cellular phenomena, such as the localization of spike initiation in mitral cells (Shen, G. Y. *et al.*, 1999; Chen, W. R. *et al.*, 2002), or long-term potentiation at the OSN-mitral synapse (Ennis, M. *et al.*, 1998). Other compartmental models have focused on bridging the gap between cellular

and systems properties in both vertebrate and insect systems (Bhalla, U. S. and Bower, J. M., 1993; Davison, A. P. *et al.*, 2000; Bazhenov, M. *et al.*, 2001; Davison, A. P. *et al.*, 2003; Cleland, T. A. and Sethupathy, P., 2006; Rubin, D. B. and Cleland, T. A., 2006; Cleland, T. A. *et al.*, 2007).

4.39.6 Conclusions

Computational models have an established and growing role within systems neuroscience. As our understanding of neural processing and interactions becomes more sophisticated, computer models of these systems are increasingly necessary in order to understand and interpret experimental results. Models serve as proofs of concept, tests of sufficiency, and as quantitative embodiments of working hypotheses. In the olfactory system in particular, computational modeling will no doubt be essential to understanding the integration of the many factors influencing the construction and transformation of odor representations.

References

Adrian, E. D. 1953. Sensory messages and sensation; the response of the olfactory organ to different smells. Acta Physiol. Scand. 29(1), 5–14.

Alkasab, T. K., White, J., and Kauer, J. S. 2002. A computational system for simulating and analyzing arrays of biological and artificial chemical sensors. Chem. Senses 27(3), 261–275.

Ambros-Ingerson, J., Granger, R., and Lynch, G. 1990. Simulation of paleocortex performs hierarchical clustering. Science 247(4948), 1344–1348.

Anton, P. S., Lynch, G., and Granger, R. 1991. Computation of frequency-to-spatial transform by olfactory bulb glomeruli. Biol. Cybern. 65(5), 407–414.

Barkai, E. and Hasselmo, M. E. 1994. Modulation of the input/output function of rat piriform cortex pyramidal cells. J. Neurophysiol. 72(2), 644–658.

Bazhenov, M., Stopfer, M., Rabinovich, M., Abarbanel, H. D., Sejnowski, T. J., and Laurent, G. 2001. Model of cellular and network mechanisms for odor-evoked temporal patterning in the locust antennal lobe. Neuron 30(2), 569–581.

Bhalla, U. S. and Bower, J. M. 1993. Exploring parameter space in detailed single neuron models: simulations of the mitral and granule cells of the olfactory bulb. J. Neurophysiol. 69(6), 1948–1965.

Bozza, T., Feinstein, P., Zheng, C., and Mombaerts, P. 2002. Odorant receptor expression defines functional units in the mouse olfactory system. J. Neurosci. 22(8), 3033–3043.

Breakspear, M. 2001. Perception of odors by a nonlinear model of the olfactory bulb. Int. J. Neural Syst. 11(2), 101–124.

Castillo, P. E., Carleton, A., Vincent, J. D., and Lledo, P. M. 1999. Multiple and opposing roles of cholinergic transmission in the main olfactory bulb. J. Neurosci. 19(21), 9180–9191.

Chen, W. R., Shen, G. Y., Shepherd, G. M., Hines, M. L., and Midtgaard, J. 2002. Multiple modes of action potential initiation and propagation in mitral cell primary dendrite. J. Neurophysiol. 88(5), 2755–2764.

Claverol, E. T., Brown, A. D., and Chad, J. E. 2002. A large-scale simulation of the piriform cortex by a cell automaton-based network model. IEEE Trans. Biomed. Eng. 49(9), 921–935.

Cleland, T. A. and Linster, C. 1999. Concentration tuning mediated by spare receptor capacity in olfactory sensory neurons: a theoretical study. Neural Comput. 11(7), 1673–1690.

Cleland, T. A. and Linster, C. 2002. How synchronization properties among second-order sensory neurons can mediate stimulus salience. Behav. Neurosci. 116(2), 212–221.

Cleland, T. A. and Linster, C. 2003. Central Olfactory Processing. In: Handbook of Olfaction and Gustation, 2nd Edn, (ed. R. L. Doty), pp. 165–180. Dekker.

Cleland, T. A. and Narla, V. A. 2003. Intensity modulation of olfactory acuity. Behav. Neurosci. 117(6), 1434–1440.

Cleland, T. A. and Sethupathy, P. 2006. Non-topographical contrast enhancement in the olfactory bulb BMC Neurosci. 7, 7.

Cleland, T. A, Johnson, B. A., Leon, M., and Linster, C. 2007. Relational representation in the olfactory system. Proc. Natl. Acad. Sci. U.S.A. 104(6), 1953–1958.

Cleland, T. A., Morse, A., Yue, E. L., and Linster, C. 2002. Behavioral models of odor similarity. Behav. Neurosci. 116(2), 222–231.

Davison, A. P., Feng, J., and Brown, D. 2000. A reduced compartmental model of the mitral cell for use in network models of the olfactory bulb. Brain Res. Bull. 51(5), 393–399.

Davison, A. P., Feng, J., and Brown, D. 2003. Dendrodendritic inhibition and simulated odor responses in a detailed olfactory bulb network model. J. Neurophysiol. 90(3), 1921–1935.

Debarbieux, F., Audinat, E., and Charpak, S. 2003. Action potential propagation in dendrites of rat mitral cells *in vivo*. J. Neurosci. 23(13), 5553–5560.

Djurisic, M., Antic, S., Chen, W. R., and Zecevic, D. 2004. Voltage imaging from dendrites of mitral cells: EPSP attenuation and spike trigger zones. J. Neurosci. 24(30), 6703–6714.

van Drongelen, W., Holley, A., and Doving, K. B. 1978. Convergence in the olfactory system: quantitative aspects of odour sensitivity. J. Theor. Biol. 71(1), 39–48.

Duchamp-Viret, P., Duchamp, A., and Vigouroux, M. 1989. Amplifying role of convergence in olfactory system a comparative study of receptor cell and second-order neuron sensitivities. J. Neurophysiol. 61(5), 1085–1094.

Ennis, M., Linster, C., Aroniadou-Anderjaska, V., Ciombor, K., and Shipley, M. T. 1998. Glutamate and synaptic plasticity at mammalian primary olfactory synapses. Ann. N. Y. Acad. Sci. 855, 457–466.

Erdi, P., Grobler, T., Barna, G., and Kaski, K. 1993. Dynamics of the olfactory bulb: bifurcations, learning, and memory. Biol. Cybern. 69(1), 57–66.

Ermentrout, G. B. and Kleinfeld, D. 2001. Traveling electrical waves in cortex: insights from phase dynamics and speculation on a computational role. Neuron 29(1), 33–44.

Fort, J. C. and Rospars, J. P. 1992. Modelling of the qualitative discrimination of odours in the first two layers of olfactory system by Jutten and Herault algorithm. C. R. Acad. Sci. III 315(9), 331–336.

Freeman, W. J. 1979. Nonlinear dynamics of paleocortex manifested in the olfactory EEG. Biol. Cybern. 35(1), 21–37.

Freeman, W. J. 1987. Simulation of chaotic EEG patterns with a dynamic model of the olfactory system. Biol. Cybern. 56(2–3), 139–150.

Freeman, W. J. 1994. Characterization of state transitions in spatially distributed, chaotic, nonlinear, dynamical systems in cerebral cortex. Integr. Physiol. Behav. Sci. 29(3), 294–306.

Friedrich, R. W. and Korsching, S. I. 1997. Combinatorial and chemotopic odorant coding in the zebrafish olfactory bulb visualized by optical imaging. Neuron 18(5), 737–752.

Fukai, T. 1996. Bulbocortical interplay in olfactory information processing via synchronous oscillations. Biol. Cybern. 74(4), 309–317.

Granger, R. and Lynch, G. 1991. Higher olfactory processes: perceptual learning and memory. Curr. Opin. Neurobiol. 1(2), 209–214.

Gray, C. M. and Skinner, J. E. 1988. Centrifugal regulation of neuronal activity in the olfactory bulb of the waking rabbit as revealed by reversible cryogenic blockade. Exp. Brain Res. 69(2), 378–386.

Grobler, T. and Erdi, P. 1991. Dynamic phenomena in the olfactory bulb. Acta Biochim. Biophys. Hung. 26(1–4), 61–65.

Haberly, L. B. 2001. Parallel-distributed processing in olfactory cortex: new insights from morphological and physiological analysis of neuronal circuitry. Chem. Senses 26(5), 551–576.

Haberly, L. B. and Bower, J. M. 1989. Olfactory cortex: model circuit for study of associative memory? Trends Neurosci. 12(7), 258–264.

Hasselmo, M. E., Linster, C., Patil, M., Ma, D., and Cekic, M. 1997. Noradrenergic suppression of synaptic transmission may influence cortical signal-to-noise ratio. J. Neurophysiol. 77(6), 3326–3339.

Hasselmo, M. E., Wilson, M. A., Anderson, B. P., and Bower, J. M. 1990. Associative memory function in piriform (olfactory) cortex: computational modeling and neuropharmacology. Cold Spring Harb. Symp. Quant. Biol. 55, 599–610.

Hendin, O., Horn, D., and Tsodyks, M. V. 1998. Associative memory and segmentation in an oscillatory neural model of the olfactory bulb. J. Comput. Neurosci. 5(2), 157–169.

Hopfield, J. J. 1999. Odor space and olfactory processing: collective algorithms and neural implementation. Proc. Natl. Acad. Sci. U. S. A. 96(22), 12506–12511.

Hoshino, O., Kashimori, Y., and Kambara, T. 1998. An olfactory recognition model based on spatio-temporal encoding of odor quality in the olfactory bulb. Biol. Cybern. 79(2), 109–120.

Hudson, R. 1999. From molecule to mind: the role of experience in shaping olfactory function. J. Comp. Physiol. [A], 185(4), 297–304.

Isaacson, J. S. and Strowbridge, B. W. 1998. Olfactory reciprocal synapses: dendritic signaling in the CNS. Neuron 20(4), 749–761.

Kaissling, K. E. 1998. A quantitative model of odor deactivation based on the redox shift of the pheromone-binding protein in moth antennae. Ann. N.Y. Acad. Sci. 855, 320–322.

Kaissling, K. E. 2001. Olfactory perireceptor and receptor events in moths: a kinetic model. Chem. Senses 26(2), 125–150.

Kaissling, K. E. and Rospars, J. P. 2004. Dose–response relationships in an olfactory flux detector model revisited. Chem. Senses 29(6), 529–531.

Kauer, J. S. 1991. Contributions of topography and parallel processing to odor coding in the vertebrate olfactory pathway. Trends Neurosci. 14(2), 79–85.

Korsching, S. I. 2001. Odor maps in the brain: spatial aspects of odor representation in sensory surface and olfactory bulb. Cell. Mol. Life Sci. 58(4), 520–530.

Lansky, P. and Rospars, J. P. 1998. Odorant concentration and receptor potential in olfactory sensory neurons. Biosystems 48(1–3), 131–138.

Laurent, G. 1999. A systems perspective on early olfactory coding. Science 286(5440), 723–728.

Laurent, G., Stopfer, M., Friedrich, R. W., Rabinovich, M. I., Volkovskii, A., and Abarbanel, H. D. 2001. Odor encoding as an active, dynamical process: experiments, computation, and theory. Annu. Rev. Neurosci. 24, 263–297.

Li, Z. and Hertz, J. 2000. Odour recognition and segmentation by a model olfactory bulb and cortex. Network 11(1), 83–102.

Li, Z. and Hopfield, J. J. 1989. Modeling the olfactory bulb and its neural oscillatory processings. Biol. Cybern. 61(5), 379–392.

Liljenstrom, H. and Hasselmo, M. E. 1995. Cholinergic modulation of cortical oscillatory dynamics. J. Neurophysiol. 74(1), 288–297.

Linster, C. and Cleland, T. A. 2001. How spike synchronization among olfactory neurons can contribute to sensory discrimination. J. Comput. Neurosci. 10(2), 187–193.

Linster, C. and Cleland, T. A. 2002. Cholinergic modulation of sensory representations in the olfactory bulb. Neural Netw. 15(4–6), 709–717.

Linster, C. and Cleland, T. A. 2004. Configurational and elemental odor mixture perception can arise from local inhibition. J. Comput. Neurosci. 16(1), 39–47.

Linster, C. and Gervais, R. 1996. Investigation of the role of interneurons and their modulation by centrifugal fibers in a neural model of the olfactory bulb. J. Comput. Neurosci. 3(3), 225–246.

Linster, C. and Hasselmo, M. 1997. Modulation of inhibition in a model of olfactory bulb reduces overlap in the neural representation of olfactory stimuli. Behav. Brain Res. 84(1–2), 117–127.

Linster, C. and Hasselmo, M. E. 1999. Behavioral responses to aliphatic aldehydes can be predicted from known electrophysiological responses of mitral cells in the olfactory bulb. Physiol. Behav. 66(3), 497–502.

Linster, C. and Hasselmo, M. E. 2001. Neuromodulation and the functional dynamics of piriform cortex. Chem Senses 26(5), 585–594.

Linster, C. and Smith, B. H. 1997. A computational model of the response of honey bee antennal lobe circuitry to odor mixtures: overshadowing, blocking and unblocking can arise from lateral inhibition. Behav. Brain Res. 87(1), 1–14.

Linster, C., Johnson, B. A., Morse, A., Yue, E., and Leon, M. 2002. Spontaneous versus reinforced olfactory discriminations. J. Neurosci. 22(16), 6842–6845.

Linster, C., Johnson, B. A., Yue, E., Morse, A., Xu, Z., Hingco, E. E., Choi, Y., Choi, M., Messiha, A., and Leon, M. 2001. Perceptual correlates of neural representations evoked by odorant enantiomers. J. Neurosci. 21(24), 9837–9843.

Linster, C., Maloney, M., Patil, M., and Hasselmo, M. E. 2003. Enhanced cholinergic suppression of previously strengthened synapses enables the formation of self-organized representations in olfactory cortex. Neurobiol. Learn. Mem. 80(3), 302–314.

Linster, C., Sachse, S., and Galizia, G. 2005. Computational modeling suggests that response properties rather than spatial position determine connectivity between olfactory glomeruli. J. Neurophysiol. 93(6), 3410–3417.

Lowe, G. and Gold, G. H. 1995. Olfactory transduction is intrinsically noisy. Proc. Natl. Acad. Sci. U.S.A. 92(17), 7864–7868.

Luo, M. and Katz, L. C. 2001. Response correlation maps of neurons in the mammalian olfactory bulb. Neuron 32(6), 1165–1179.

Malaka, R., Ragg, T., and Hammer, M. 1995. Kinetic models of odor transduction implemented as artificial neural networks. Simulations of complex response properties of honeybee olfactory neurons. Biol. Cybern. 73(3), 195–207.

Mandairon, N., Ferretti, C. J., Stack, C. M., Rubin, D. B., Cleland, T. A., and Linster, C. 2006. Cholinergic modulation in the olfactory bulb influences spontaneous olfactory discrimination in adult rats. Eur. J. Neurosci. 24(11), 3234–3244.

Margrie, T. W. and Schaefer, A. T. 2003. Theta oscillation coupled spike latencies yield computational vigour in a mammalian sensory system. J. Physiol. 546(Pt 2), 363–374.

Marr, D. 1971. Simple memory: a theory for archicortex. Philos. Trans. R. Soc. Lond. B Biol. Sci. 262(841), 23–81.

Martin, C., Gervais, R., Hugues, E., Messaoudi, B., and Ravel, N. 2004. Learning modulation of odor-induced oscillatory responses in the rat olfactory bulb: a correlate of odor recognition? J. Neurosci. 24(2), 389–397.

Meisami, E. 1989. A proposed relationship between increases in the number of olfactory receptor neurons, convergence ratio and sensitivity in the developing rat. Brain Res. Dev. Brain Res. 46(1), 9–19.

Meister, M. and Bonhoeffer, T. 2001. Tuning and topography in an odor map on the rat olfactory bulb. J. Neurosci. 21(4), 1351–1360.

Meredith, M. 1992. Neural circuit computation: complex patterns in the olfactory bulb. Brain Res. Bull. 29(1), 111–117.

Mombaerts, P. 1996. Targeting olfaction. Curr. Opin. Neurobiol. 6(4), 481–486.

Moulton, D. G. 1967. Spatio-temporal patterning of response in the olfactory system. In: Olfaction and Taste II (ed. T. Hayashi), pp. 109–116. Pergamon.

Neville, K. R. and Haberly, L. B. 2003. Beta and gamma oscillations in the olfactory system of the urethane-anesthetized rat. J. Neurophysiol. 90(6), 3921–3930.

Patil, M. M. and Hasselmo, M. E. 1999. Modulation of inhibitory synaptic potentials in the piriform cortex. J. Neurophysiol. 81(5), 2103–2118.

Rall, W. and Shepherd, G. M. 1968. Theoretical reconstruction of field potentials and dendrodendritic synaptic interactions in olfactory bulb. J. Neurophysiol. 31(6), 884–915.

Rospars, J. P., Lansky, P., Duchamp-Viret, P., and Duchamp, A. 2000. Spiking frequency versus odorant concentration in olfactory receptor neurons. Biosystems 58(1–3), 133–141.

Rospars, J. P., Lansky, P., Tuckwell, H. C., and Vermeulen, A. 1996. Coding of odor intensity in a steady-state deterministic model of an olfactory receptor neuron. J. Comput. Neurosci. 3(1), 51–72.

Rubin, D. B. and Cleland, T. A. 2006. Dynamical mechanisms of odor processing in olfactory bulb mitral cells. J. Neurophysiol. 96(2), 555–568.

Rubin, B. D. and Katz, L. C. 1999. Optical imaging of odorant representations in the mammalian olfactory bulb. Neuron 23(3), 499–511.

Schild, D. 1988. Principles of odor coding and a neural network for odor discrimination. Biophys. J. 54(6), 1001–1011.

Schoenfeld, T. A. and Knott, T. K. 2004. Evidence for the disproportionate mapping of olfactory airspace onto the main olfactory bulb of the hamster. J. Comp. Neurol. 476(2), 186–201.

Shen, G. Y., Chen, W. R., Midtgaard, J., Shepherd, G. M., and Hines, M. L. 1999. Computational analysis of action potential initiation in mitral cell soma and dendrites based on dual patch recordings. J. Neurophysiol. 82(6), 3006–3020.

Shepard, R. N. 1987. Toward a universal law of generalization for psychological science. Science 237(4820), 1317–1323.

Shepard, R. N. and Chang, J. J. 1963. Stimulus generalization in the learning of classifications. J. Exp. Psychol. 65, 94–102.

Shepherd, G. M. and Brayton, R. K. 1979. Computer simulation of a dendrodendritic synaptic circuit for self- and lateral-inhibition in the olfactory bulb. Brain Res. 175(2), 377–382.

Sivan, E. and Kopell, N. 2004. Mechanism and circuitry for clustering and fine discrimination of odors in insects. Proc. Natl. Acad. Sci. U. S. A. 101(51), 17861–17866.

Stewart, W. B., Kauer, J. S., and Shepherd, G. M. 1979. Functional organization of rat olfactory bulb analysed by the 2-deoxyglucose method. J. Comp. Neurol. 185(4), 715–734.

Urban, N. N. 2002. Lateral inhibition in the olfactory bulb and in olfaction. Physiol. Behav. 77(4–5), 607–612.

Vermeulen, A. and Rospars, J. P. 1998. Dendritic integration in olfactory sensory neurons: a steady-state analysis of how the neuron structure and neuron environment influence the coding of odor intensity. J. Comput. Neurosci. 5(3), 243–266.

Vermeulen, A., Lansky, P., Tuckwell, H., and Rospars, J. P. 1997. Coding of odour intensity in a sensory neuron. Biosystems 40(1–2), 203–210.

Vermeulen, A., Rospars, J. P., Lansky, P., and Tuckwell, H. C. 1996. Coding of stimulus intensity in an olfactory receptor neuron: role of neuron spatial extent and passive dendritic backpropagation of action potentials. Bull. Math. Biol. 58(3), 493–512.

Wachowiak, M., Cohen, L. B., and Zochowski, M. R. 2002. Distributed and concentration-invariant spatial representations of odorants by receptor neuron input to the turtle olfactory bulb. J. Neurophysiol. 87(2), 1035–1045.

White, J. and Kauer, J. S. 2001. Exploring olfactory population coding using an artificial olfactory system. Prog. Brain Res. 130, 191–203.

White, J., Dickinson, T. A., Walt, D. R., and Kauer, J. S. 1998. An olfactory neuronal network for vapor recognition in an artificial nose. Biol. Cybern. 78(4), 245–251.

White, J., Hamilton, K. A., Neff, S. R., and Kauer, J. S. 1992. Emergent properties of odor information coding in a representational model of the salamander olfactory bulb. J. Neurosci. 12(5), 1772–1780.

Wilson, M. and Bower, J. M. 1992. Cortical oscillations and temporal interactions in a computer simulation of piriform cortex. J. Neurophysiol. 67(4), 981–995.

Wiltrout, C., Dogra, S., and Linster, C. 2003. Configurational and nonconfigurational interactions between odorants in binary mixtures. Behav. Neurosci. 117(2), 236–245.

Xu, D. and Principe, J. C. 2004. Dynamical analysis of neural oscillators in an olfactory cortex model. IEEE Trans. Neural Netw. 15(5), 1053–1062.

Yokoi, M., Mori, K., and Nakanishi, S. 1995. Refinement of odor molecule tuning by dendrodendritic synaptic inhibition in the olfactory bulb. Proc. Natl. Acad. Sci. U. S. A. 92(8), 3371–3375.

4.40 Understanding Olfactory Coding via an Analysis of Odorant-Evoked Glomerular Response Maps

B A Johnson and M Leon, University of California, Irvine, CA, USA

Glossary

glomeruli Ball-like regions in which homologous olfactory receptor neurons synapse with second-order neurons in the olfactory bulb.

neuropil A dense network of neural and glial processes.

combinatorial coding The sensory processing of odorant chemicals by the binding of their multiple chemical features to multiple odorant receptors.

2-deoxyglucose A glucose analog that is incompletely metabolized in cells and that can be used for mapping neural activity.

hydrophobic A property that describes the degree of resistance of a molecule to mix with water.

tuning Describes the relationship between systematic changes in the physical properties of sensory stimuli and the level of neural response in a sensory system.

aliphatic Possessing a nonaromatic hydrocarbon structure.

enantiomers Molecules that are non-superimposable mirror images of one another.

4.40.1 Introduction

Buck L. B. and Axel R. (1991) reported the discovery of rat odorant receptor genes and began a series of studies that clarified the foundation of odor perception. Most olfactory sensory neurons in the nose appear to express only one of these genes, and while homologous sensory neurons expressing the same odorant receptor gene are distributed in fairly broad zones stretching from anterior to posterior within the main olfactory epithelium (Ressler, K. J. *et al.*, 1993), they converge in their projections to the brain and connect to a small number of glomeruli in the main olfactory bulb (Vassar, R. *et al.*, 1994). Glomeruli are roughly spherical masses of neuropil that contain synapses between sensory neuron axons and dendrites of projection neurons and interneurons. The typical projection from homologous sensory neurons involves one glomerulus or small grouping of glomeruli on the lateral aspect and a similar projection on the medial aspect of the bulb (Mombaerts, P. *et al.*, 1996). All sensory neuron synapses in a glomerulus appear to be from cells expressing the same receptor (Treloar, H. B. *et al.*, 2002).

This orderly projection pattern, which is consistent across different animals, suggests that one can get a read-out of odorant receptor activation from the pattern of odorant-evoked activity in the glomerular layer. This prediction now has been tested using a variety of activity-mapping methods. In our own work, we use the uptake of radiolabeled 2-deoxyglucose (2DG) as a metabolic marker, as this is the only method currently able to give a quantitative measure of relative activity throughout the entire structure at a spatial resolution sufficient to detect activity in a single glomerulus.

4.40.2 Combinatorial Coding

Even pure, simple odorant molecules tend to activate multiple locations in the olfactory bulb, and each area of activation tends to encompass many glomeruli (Johnson, B. A. *et al.*, 1998; 1999). These findings suggest that odor perception probably arises from a combination of activated odorant receptors. Similar conclusions have come from the study of sensory neurons expressing distinct receptor genes (Malnic, B. *et al.*, 1999), from the study of glomerular responses using other methods (Uchida, N. *et al.*, 2000; Wachowiak, M. and Cohen, L. B., 2001), and from the study of action potentials in mitral cells (Imamura, K. *et al.*, 1992). Indeed, there are far more unique odor perceptions than there are odorant receptors, a situation that logically necessitates a combinatorial code.

4.40.3   Molecular Feature Detection

Most receptors appear not to interact with the entirety of their ligand molecules. Rather, parts of the ligand molecule can be modified with minimal impact on receptor binding and activation, while other parts of the ligand molecule are more critical. Indeed, one can think of receptors as detectors of these critical molecular features. We have designed a large number of experiments using sets of odorants differing systematically in chemical structure to explore the detection of molecular features by odorant receptors. In one experiment, each of two ethyl esters stimulated 2DG uptake in a part of the bulb that was not activated by either of two isoamyl esters (Johnson, B. A. *et al.*, 1998). Conversely, both isoamyl esters activated a location not stimulated by either ethyl ester. The activity patterns were consistent with the recognition of the ethyl and isoamyl groups as distinct features by different glomerular modules in the bulb. All four esters stimulated a distinct, overlapping module, consistent with the overall similarity in their core structure.

We tested a set of five odorants sharing a straight-chained hydrocarbon structure but differing in oxygen-containing functional groups to determine whether functional groups would be recognized as distinct molecular features (Johnson, B. A. and Leon, M., 2000a). While all of these odorants stimulated a cluster of glomeruli in the posterior bulb, consistent with their shared hydrocarbon structure, they activated distinct anterior modules, consistent with the recognition of functional groups as distinct features. When we used carboxylic acid odorants that differed greatly in hydrocarbon structure, anterior clusters of activated glomeruli overlapped, consistent with the

shared acid functional group, but the posterior patterns differed greatly, consistent with the recognition of different hydrocarbon structures as distinct features (Johnson, B. A. and Leon, M., 2000b). As we increased our sample of the odorant world to include more and more complex molecules, the data showed that the recognition of functional groups as molecular features is not entirely independent of hydrocarbon structure, but rather, that the two types of features interact to determine which glomeruli are activated (Johnson, B. A. *et al.*, 2005a; 2005b).

4.40.4   Chemotopic Progressions

Chemotopy refers to any systematic spatial representation of odorant chemistry. We have been particularly interested in determining whether the olfactory system uses this kind of spatial organization in odor coding. Indeed, glomeruli of related specificity are arrayed systematically within response modules. Uptake of 2DG in a medial, acid-sensitive glomerular module shifted ventrally with increasing carbon number in a series of straight-chained carboxylic acids (Johnson, B. A. *et al.*, 1999). In the corresponding lateral module, uptake shifted both rostrally and ventrally. Separate responses in the posterior, lateral, and the medial bulb also shifted ventrally with increasing carbon number. By using a series of carboxylic acids of the same carbon number, but with different hydrocarbon structures, we found that the ventral shift in the medial acid-responsive module appeared to be due specifically to changes in molecular length (Johnson, B. A. and Leon, M., 2000b). More recently, we found ventral progressions of activity with increasing odorant carbon number using aldehydes, esters, acetates, primary and secondary alcohols, ketones, and alkanes in a number of different glomerular regions, suggesting that such chemotopic progressions represent a fundamental organizational principle in the olfactory system (Johnson, B. A. *et al.*, 2004; Ho *et al.*, 2006) (Figure 1).

Why are the chemotopic progressions with increasing carbon number always in the ventral direction? One possibility is that the chemotopic progressions are related to the spatial distributions of the odorants themselves across the olfactory epithelium. Mozell and co-workers have shown that odorants partition across the epithelium chromatographically (Mozell, M. M., 1964; Hornung, D. E. and Mozell, M. M., 1977; Mozell, M. M. *et al.*, 1987). Molecules that are more water soluble (hydrophilic)

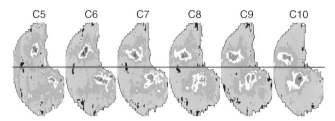

Figure 1   Chemotopic progression of glomerular responses in the ventral and rostral direction with increasing carbon number in straight-chained aldehydes. The maps are in a ventral-centered orientation such that the red line indicates the ventral extremity of the bulb. In each map, rostral is to the left. Warmer colors indicate higher uptake of the metabolic marker, 2-deoxyglucose.

absorb more readily into the olfactory mucosa, which would prevent them from reaching the more peripheral and ventral parts of the rat epithelium (Schoenfeld, T. A. and Knott, T. K., 2004). The less water-soluble (more hydrophobic) odorant molecules would be freer to distribute across the epithelium to reach the more peripheral and ventral zones. Longer odorants within any homologous series of straight-chained chemicals would be more hydrophobic and therefore would be expected to reach the peripheral and ventral epithelium. Because the more peripheral and ventral regions of the epithelium project to more ventral bulbar targets (Schoenfeld, T. A. et al., 1994), this topography could explain the ventral chemotopic progressions within many glomerular modules in the olfactory bulb (Johnson, B. A. et al., 2004).

Local bulbar anatomy suggests that chemotopic progressions in the glomerular layer might have important functional consequences. Reciprocal synapses between mitral cell projection neurons and inhibitory interneurons in the glomerular layer and in the external plexiform layer are expected to create local lateral inhibitory networks. Strong activity in mitral cells associated with a given glomerulus would likely suppress activity in less strongly stimulated mitral cells associated with neighboring glomeruli (Mori, K. and Shepherd, G. M., 1994; Aungst, J. L. et al., 2003). This pattern of activity would be predicted to tune individual mitral cells to a more narrow range of molecules than would excite individual sensory neurons. Indeed, pharmacological blockade of inhibition in the bulb widens the molecular receptive range of mitral cells (Yokoi, M. et al., 1995). Chemotopic progressions ensure that the glomeruli that are nearest neighbors within a module would have the most similar specificity within a homologous series, thereby ensuring that tuning involves the most similar odorant molecules.

4.40.5   Predictive Value of Activity Maps

The value of any scientific observation is perhaps best judged by whether it allows one to predict the outcome of future experiments. Maps of odorant-evoked 2DG uptake across the glomerular layer of the olfactory bulb have been predictive in two senses: they have predicted aspects of the patterns that would be evoked by previously untested odorants, and they have predicted relative similarities in odors perceived by rats. The original associations between functional group-related molecular features and activity in particular anterior glomerular modules (Johnson, B. A. and Leon, M., 2000a) successfully predicted that other simple aliphatic odorant chemicals possessing the same features would also stimulate those modules (Johnson, B. A. et al., 2002; 2004). Aspects of glomerular activity patterns related to hydrocarbon structure also have had predictive value (Johnson, B. A. et al., 2002). The ventral shifts in activity that we observed with increasing carbon number for aliphatic, oxygen-containing odorants (Johnson, B. A. et al., 1999; 2004) led us to predict that even larger straight-chained molecules might stimulate an extremely ventral portion of the bulb that we had not seen activated previously. We tested this prediction using very long alkanes such as pentadecane, which indeed activated the extreme ventral aspect (Ho et al., 2006).

As part of our exploration of the relationships between odorant chemistry and bulbar activity patterns, we mapped responses to three pairs of odorant enantiomers: carvone, limonene, and terpinen-4-ol. The patterns evoked by the two carvones were more distinct than those for the other two enantiomer pairs (Linster, C. et al., 2001). We therefore predicted that the odors of the carvone enantiomers might be more readily distinguished than those of the limonenes and terpinen-4-ols. Indeed, a cross-habituation assay revealed that rats spontaneously distinguished the carvones, but treated limonene and terpinen-4-ol enantiomers as being similar in odor (Linster, C. *et al.*, 2001). The quantitative differences in activity patterns across homologous series of acids and alkanes also correlated well with behavioral measures of odor discrimination (Cleland, T. A. *et al.*, 2002; Ho *et al.*, 2006).

4.40.6 Summary

As predicted from the projection patterns of olfactory sensory neurons expressing distinct odorant receptor genes, different odorants generate different spatial activity patterns in the olfactory bulb. Mapping activity patterns in response to odorants differing systematically in structure has shown that receptors likely extract information about odorant chemistry, especially regarding functional groups and hydrocarbon structure. The bulb responses are organized in space according to the relative specificity of the input, such that increasingly long straight-chained odorants optimally stimulate increasingly ventral glomeruli. Finally, the overall pattern of glomerular activation is related to the odor perceived by rats.

Acknowledgments

This work has been supported by grants DC6516, DC3545, and DC6391 from NIDCD.

References

Aungst, J. L., Heyward, P. M., Puche, A. C., Karnup, S. V., Hayar, A., Szabo, G., and Shipley, M. T. 2003. Centre-surround inhibition among olfactory bulb glomeruli. Nature 426, 623–629.

Buck, L. B. and Axel, R. 1991. A novel multigene family may encode odorant receptors: a molecular basis for odor recognition. Cell 65, 175–187.

Cleland, T. A., Morse, A., Yue, E. L., and Linster, C. 2002. Behavioral models of odor similarity. Behav. Neurosci. 116, 222–231.

Ho, S. L., Johnson, B. A., and Leon, M. 2006. Long hydrocarbon chains serve as unique molecular features recognized by ventral glomeruli of the rat olfactory bulb. J. Comp. Neurol. 498, 16–30.

Hornung, D. E. and Mozell, M. M. 1977. Factors influencing the differential sorption of odorant molecules across the olfactory mucosa. J. Gen. Physiol. 69, 343–361.

Imamura, K., Mataga, N., and Mori, K. 1992. Coding of odor molecules by mitral/tufted cells in rabbit olfactory bulb. I. Aliphatic compounds. J. Neurophysiol. 68, 1986–2002.

Johnson, B. A. and Leon, M. 2000a. Modular representations of odorants in the glomerular layer of the rat olfactory bulb and the effects of stimulus concentration. J. Comp. Neurol. 422, 496–509.

Johnson, B. A. and Leon, M. 2000b. Odorant molecular length: one aspect of the olfactory code. J. Comp. Neurol. 426, 330–338.

Johnson, B. A., Farahbod, H., and Leon, M. 2005b. Interactions between odorant functional group and hydrocarbon structure influence activity in glomerular response modules in the rat olfactory bulb. J. Comp. Neurol. 483, 205–216.

Johnson, B. A., Farahbod, H., Saber, S., and Leon, M. 2005a. Effects of functional group position on spatial representations of aliphatic odorants in the rat olfactory bulb. J. Comp. Neurol. 483, 192–204.

Johnson, B. A., Farahbod, H., Xu, Z., Saber, S., and Leon, M. 2004. Local and global chemotopic organization: general features of the glomerular representations of aliphatic odorants differing in carbon number. J. Comp. Neurol. 480, 234–249.

Johnson, B. A., Ho, S. L., Xu, Z., Yihan, J. S., Yip, S., Hingco, E. E., and Leon, M. 2002. Functional mapping of the rat olfactory bulb using diverse odorants reveals modular responses to functional groups and hydrocarbon structural features. J. Comp. Neurol. 449, 180–194.

Johnson, B. A., Woo, C. C., Hingco, E. E., Pham, K. L., and Leon, M. 1999. Multidimensional chemotopic responses to n-aliphatic acid odorants in the rat olfactory bulb. J. Comp. Neurol. 409, 529–548.

Johnson, B. A., Woo, C. C., and Leon, M. 1998. Spatial coding of odorant features in the glomerular layer of the rat olfactory bulb. J. Comp. Neurol. 393, 457–471.

Linster, C., Johnson, B. A., Yue, E., Morse, A., Xu, Z., Hingco, E., Choi, Y., Choi, M., Messiha, A., and Leon, M. 2001. Perceptual correlates of neural representations evoked by odorant enantiomers. J. Neurosci. 21, 9837–9843.

Malnic, B., Hirono, J., Sato, T., and Buck, L. B. 1999. Combinatorial receptor codes for odors. Cell 96, 713–723.

Mombaerts, P., Wang, F., Dulac, C., Chao, S. K., Nemes, A., Mendelsohn, M., Edmonson, J., and Axel, R. 1996. Visualizing an olfactory sensory map. Cell 87, 675–686.

Mori, K. and Shepherd, G. M. 1994. Emerging principles of molecular signal processing by mitral/tufted cells in the olfactory bulb. Semin. Cell Biol. 5, 65–74.

Mozell, M. M. 1964. Evidence for sorption as a mechanism of the olfactory analysis of vapors. Nature 203, 1181–1182.

Mozell, M. M., Sheehe, P. R., Hornung, D. E., Kent, P. F., Youngentob, S. L., and Murphy, S. J. 1987. "Imposed" and "inherent" mucosal activity patterns. J. Gen. Physiol. 90, 625–650.

Ressler, K. J., Sullivan, S. L., and Buck, L. B. 1993. A zonal organization of odorant receptor gene expression in the olfactory epithelium. Cell 73, 597–609.

Schoenfeld, T. A. and Knott, T. K. 2004. Evidence for the disproportionate mapping of olfactory airspace onto the main olfactory bulb of the hamster. J. Comp. Neurol. 476, 186–201.

Schoenfeld, T. A., Clancy, A. N., Forbes, W. B., and Macrides, F. 1994. The spatial organization of the peripheral olfactory system of the hamster. I. Receptor neuron projections to the main olfactory bulb. Brain Res. Bull. 34, 183–210.

Treloar, H. B., Feinstein, P., Mombaerts, P., and Greer, C. A. 2002. Specificity of glomerular targeting by olfactory sensory axons. J. Neurosci. 22, 2469–2477.

Uchida, N., Takahashi, Y. K., Tanifuji, M., and Mori, K. 2000. Odor maps in the mammalian olfactory bulb: domain organization and odorant structural features. Nat. Neurosci. 3, 1035–1043.

Vassar, R., Chao, S. K., Sitcheran, R., Nuñez, J. M., Vosshall, L. B., and Axel, R. 1994. Topographic organization of sensory projections to the olfactory bulb. Cell 79, 981–991.

Wachowiak, M. and Cohen, L. B. 2001. Representation of odorants by receptor neuron input to the mouse olfactory bulb. Neuron 32, 723–735.

Yokoi, M., Mori, K., and Nakanishi, S. 1995. Refinement of odor molecule tuning by dendrodendritic synaptic inhibition in the olfactory bulb. Proc. Natl. Acad. Sci. U. S. A. 92, 3371–3375.

Further Reading

Arctander, S. 1994. Perfume and Flavor Chemicals (Aroma Chemicals). Allured Publishing Company.

Dravnieks, A. 1985. Atlas of Odor Character Profiles. ASTM Data Series DS 61.

Johnson, B. A. and Leon, M. 2007. Chemotopic odorant coding in a mammalian olfactory system. J. Comp. Neurol. 503, 1–34.

Takahashi, Y. K., Kurosaki, M., Hirono, S., and Mori, K. 2004. Topographic representation of odorant molecular features in the rat olfactory bulb. J. Neurophysiol. 92, 2413–2427.

4.41 Insect Olfaction

G Galizia, Universität Konstanz, Konstanz, Germany

Glossary

2-DG labeling A technique used to label active cells. Radioactive 2-DG ($[^{14}C]$2-deoxyglucose) is injected into the animal. Physiologically active cells accumulate 2-DG because it is taken up as a glucose analogue but accumulates because it cannot be metabolized. Cells that are particularly active accumulate more 2-DG and are therefore apparent in an autoradiograph. For each activity map, a single animal has to be killed.

acetylcholine (ACh) A neurotransmitter. In the insect brain, ACh is the major excitatory transmitter between neurons. The major excitatory neurotransmitter at the insect neuromuscular junction is glutamate.

antennal lobe (AL) The first neuropil in the insect brain to receive olfactory information, comparable to the mammalian olfactory bulb. It is located in the deutocerebrum.

antennal mechanosensory and motor center (AMMC) Neuropil situated in the deutocerebrum dorsally to the insect's AL (also called dorsal lobe). Among others, it receives the axonal terminals of mechanoreceptor cells on the antenna and contains the somata of antennal motor neurons.

antennocerebralis tract (ACT) Axonal tract between the AL and the protocerebrum. There are several ACTs in each brain, traveling in different routes.

arista Feather-like appendix on the *Drosophila* antenna involved in forming a recipient structure for vibrations. Sensory sensilla are also located on the arista.

associative memory See *Pavlovian conditioning*.

biogenic amines Neurotransmitters that have an amine group, comprising catecholamines (epinephrine, dopamine, and norepinephrine = noradrenalin) and indoles (tryptophan and serotonin). Many of them are involved in modulatory functions in the insect brain.

blocking Term from learning psychology. Assume that an association of a stimulus A with a reward is learned, and subsequently stimulus A is trained together with a stimulus B (i.e., AB). When learning about B is reduced (even though B was trained as a compound in AB), then A has blocked B. Blocking is generally explained using cognitive concepts of attention or expectation.

boutons Presynaptic structures in neurons.

calcium-sensitive dye A dye that changes its light properties as a function of calcium concentration. Most are fluorescent substances. These dyes are often used as reporter molecules for neural activity. Examples are calcium green, FURA, and fluo-3.

calyx, calyces See *mushroom body*.

cameleon See *G-CaMP*.

classical conditioning See *Pavlovian conditioning*.

combinatorial code A coding logic whereby information is not stored in a single channel (see *labeled line*), but rather across several channels. Different combinations of active channels indicate different information. Take, for example, 10 channels: simultaneous activity in channels 1, 3, and 5 constitutes a different combination (and thus coded information) than activity in channels 1, 3, and 6. A combinatorial code greatly amplifies the coding capacity of a given number of channels.

conditioned stimulus (CS) See *Pavlovian conditioning*.

deutocerebrum See *supraesophageal ganglia*.

differential training In Pavlovian conditioning, when two stimuli are used as CS, with one being rewarded and the other not, or one being punished and the other not.

dorsal lobe (DL) Neuropil in the deutocerebrum located dorsally to the AL. Contains the AMMC.

G-CaMP A genetically engineered fluorescent protein that changes its fluorescent properties with changing Ca^{2+} concentration. G-CaMP can be used to monitor cell activity when it is expressed in a cell. Another such reporter protein is cameleon. The two differ in spectral properties, Ca^{2+} affinity, and binding kinetics.

gamma-aminobutyric acid (GABA)
Neurotransmitter that in most cases binds to receptors that have an inhibitory effect on the recipient cell. GABA is the most important inhibitory transmitter in the brain.

glomerulus Named after the spherical arrangement of dense neuropil in the AL which is apparent in microscopic sections, the glomeruli form the functional units within the AL. Generally, they collect all input from a particular type of olfactory sensory neurons (OSNs).

habituation A term from learning psychology. An animal habituates to a repeatedly presented stimulus and ignores its presence. Habituation is different from sensory adaptation or fatigue. Habituation is a nonassociative learning phenomenon: a strong stimulus of a different or even the same modality can instantly dishabituate an animal, showing that sensory information about the stimulus is not reduced.

hemimetabolous insects Insects that go through several developmental stages (instar larvae) that resemble the adult stage. Metamorphosis is gradual, without a pupal stage. For example, cockroaches and locusts. See also *holometabolous insects*.

heterogeneous local neuron (heteroLN) In this chapter, local neurons in the AL that branch in many glomeruli but have a clear polarity in that one or at most a few glomeruli are densely innervated, while most glomeruli are innervated only sparsely.

holometabolous insects Insects that undergo complete developmental metamorphosis, with a larval stage, followed by a pupal stage, followed by the adult. The larva is very different from the adult and generally occupies a different ecological niche. The best-known example is the butterfly, with the caterpillar larva. Bees, flies, and moths are holometabolous.

homogeneous LN In this chapter, local neurons in the AL that branch in many glomeruli without a clear polarity.

ionotropic receptor Ligand-gated ion channels that mediate fast excitatory or inhibitory neurotransmission.

Kenyon cell (KC) Intrinsic cell of the mushroom body in insects, first described by Kenyon F. C. in 1896.

labeled line Coding principle whereby a single channel carries information independently from the activity status of other channels, as opposed to a combinatorial code. In a binary system, four channels can carry 4 bits of information in a labeled line system, while four channels can carry 16 bits of information in a combinatorial code.

latent inhibition Term from learning psychology, which refers to the effect of unreinforced preexposure with a CS that leads to a reduced learning.

local neuron (LN) In this chapter used for neurons that are local to the AL, i.e., their entire branching arbors are confined within the AL.

mating disruption A technique in pest control aimed at modifying pest behavior with semiochemicals. For example, when large quantities of female pheromone are released in the environment, males are less efficient in finding their females, and thus natural mating patterns are disrupted.

metabotropic receptor A type of neurotransmitter receptor that, upon binding with a neurotransmitter, activates an intracellular second messenger cascade (e.g., involving cAMP), as opposed to opening an ion channel.

molecular receptive range (MRR) The chemosensory response properties of a chemoreceptor or of a chemosensory neuron, i.e., the range of molecules for which the receptor is receptive.

multielectrode array A technique to measure many neurons at the same time with multiple electrodes that penetrate the brain. Each electrode can record extracellular activity from one or a few cells in its vicinity. Different cells are sorted by the spike shapes that they produce. Some multielectrode arrays are made of silicon and others of isolated wires.

mushroom body (MB) A structure in the insect protocerebrum. The intrinsic cells of the MBs are the KCs. Their somata and dendrites form the calyx, and their axons form the α-lobe, β-lobe, and

γ-lobe. The name comes from the shape of this structure, reminiscent of a mushroom.

negative patterning Term from learning psychology. In negative patterning, two single stimuli are reinforced, but their compound is not (i.e., A^+, B^+, AB^-). Therefore, while A and B each have predictive power for the reward, their mixture has not. Solving a negative patterning problem requires configural association and cannot be solved with elemental learning alone (see also *positive patterning*).

neuropeptides Short-chain peptides in the brain, e.g., enkephalins and endorphins, that function as intercellular messengers. There is a large variety of peptides, which are often species-specific and have dedicated receptors in the recipient cells. Peptides are often released as cotransmitters and modulate the affect of a transmitter at the recipient cell.

neuropil A dense feltwork of neural processes (dendrites, axons, and neurites) and glial processes. Generally, in insects, the cell bodies of the brain are arranged peripherally and the neuropil areas contain only very few cell bodies.

nitric oxide (NO) A gas that is also produced by cells and functions as a messenger between cells.

olfactory binding protein (OBP) A protein in the sensillar lymph of olfactory sensilla that participates in the olfactory process. Its function is as yet unclear. Among the current hypotheses for its function are that it might help dissolve the odors in the aqueous lymph or that might remove odor from the receptor after it has interacted.

olfactory bulb (OB) The vertebrate counterpart to the insect AL.

olfactory receptor (OR) A seven-transmembrane protein in OSNs that interacts with the odorant molecules and initiates the intracellular transduction cascade.

olfactory sensilla In insects, ORs are housed in sensilla, which are cuticular structures with pores on the insect's surface. Sensilla vary greatly in shape and size, including hair-like protrusions, flat plates, or funnel-like indentations.

olfactory sensory neuron (OSN) The primary olfactory neurons. OSNs express ORs on their dendrites and express the transduction cascade. Upon contact with the appropriate stimulus, they send action potentials (APs) along their axons into the ALs in the brain. Often referred to as OR neurons, or sometimes even receptors. The latter term

is now avoided in order to prevent confusion with the OR proteins.

Pavlovian conditioning Also called classical conditioning. A neutral stimulus (the CS) is repeatedly presented immediately before a reward (or punishment, the unconditioned stimulus, US), so that the CS has predictive value for the US. The animal learns the coincidence of the two events forming an associative memory. Before learning, only the US leads to a response (the unconditioned response, UR). After learning, also the CS leads to a response (the conditioned response, CR). In Pavlov's famous dog experiment, the bell is the CS, the food is the US, and salivation is the UR and the CR. In bee PER (proboscis extension reflex) conditioning, the CS is an odor stimulus, the US is nectar (or sugar water), and the CR is the extension of the bee's proboscis (PER).

pheromone Chemical substance used for communication within a species. Sexual pheromones are used for sexual attraction and/or arousal. Other pheromones include alarm pheromones and trail pheromones in ants. The first pheromone to be isolated biochemically was Bombycol, the sexual pheromone of the moth *Bombix mori*, found by Butenand A. in the 1950s. Chemical messengers are also used between species: those advantageous for the receiver are called kairomones, those advantageous for the emitter are called allomones, and those that benefit both are called synomones.

Picrotoxin A pharmacological substance that blocks many ionotropic chloride channels. Generally, these channels are GABAergic receptors, and Picrotoxin is used to block these receptors.

positive patterning Term from learning psychology. In positive patterning, two single stimuli are not reinforced, but their compound is (i.e., A^-, B^-, AB^+). Therefore, neither A alone nor B alone has predictive power for the reward, but only their mixture. Unlike negative patterning, positive patterning can be solved with elemental learning mechanisms, without the need of configural association.

proboscis extension reflex (PER) The response to an appetitive stimulus. For example, when a honeybee is presented a drop of sugar water, it reflexively extends its proboscis.

projection neuron (PN) In this chapter, a neuron that projects out from the AL.

protocerebrum See *supraesophageal ganglia*.

subesophageal ganglia (SEG) The insect brain consists of the supraesophageal ganglia, which are joined to the SEG by two connectives. Together, these structures form a ring around the esophagus. The SEG consists of a fusion of several pairs of ganglia. Often referred to as SOG (based on British spelling: suboesophageal ganglia).

supraesophageal ganglia The insect brain consists of a fusion of several ganglia arranged above the esophagus. The anterior part is the protocerebrum, which contains the MBs, the central complex, and the optic lobes. The posterior part contains the deutocerebrum (AL and DL) and the smaller tritocerebrum.

synaptopHluorin An artificially created fluorescent protein that changes its fluorescent properties with changing pH value and is localized in synaptic vesicles. synaptopHluorin can be used to monitor synaptic transmitter release when it is expressed in a cell.

tritocerebrum See *supraesophageal ganglia*.

unconditioned stimulus (US) See *Pavlovian conditioning*.

voltage-sensitive dye (VSD) A dye that intercalates into cell membranes and changes its light properties with the membrane potential. VSDs can be used to measure membrane potentials of cells with light.

α-**lobe** See *mushroom body*.

γ-**lobe** See *mushroom body*.

4.41.1 Introduction

In order to understand how the insect brain processes information, it is necessary to understand the anatomy and physiology of the neurons involved in this task. In a very simplified picture, information by olfactory sensory neurons (OSNs) is processed first in the antennal lobes (ALs), and then in the mushroom bodies (MBs) and the lateral protocerebrums (LPs). In this chapter, these three areas are treated sequentially, and information about morphology, physiology, and function is collected for each. At the end, olfactory behavior in insects is discussed.

It is useful to separate at least three transformation steps involved in olfactory coding, which correspond to transformations between four multidimensional spaces. First is the external multidimensional space, that is, the odor world, which has dimensionality and axes defined by the chemical and physical properties of the odorant molecules. Some of these axes are easy to define, like carbon chain length for simple hydrocarbons or electronegativity for functional group residues. Molecular resonance frequencies, which are exploited in spectroscopic techniques, also define a classification system (Turin, L., 1996). However, any arrangement of molecules in such a multidimensional space is a matter of definition, rather than reflecting a naturally dictated order. The second one is a biologically defined multidimensional space, where each dimension is given by the response properties of OSNs. The mapping of the physical–chemical space onto the receptive space is the first step in olfactory coding. It consists of interactions with olfactory binding proteins (OBPs) and other extracellular proteins, the olfactory receptor proteins (ORs), and intracellular second messenger pathways. The third multidimensional space consists of activity patterns across neurons in the AL and its several additional transformations along processing steps within the brain, which are mediated by the neural networks within the AL or the MBs. These steps enhance information that is relevant to the animal and filter less-relevant content, increase signal over noise, optimize sensitivity, and/or reformat the coding scheme in a way that makes it easier and more reliable to decode, so that information is better stored and/or extracted. This information consists of any combination of (1) the identity of an odor stimulus (an identification task), (2) its concentration (a measurement along a continuous variable), (3) the identity of odor components in a mixture (an analytical task), (4) the identity of a combination of odor components in a mixture (a synthetic task), and (5) the analysis of odor plume sequences (a temporal task). Finally, the last and biologically most relevant olfactory space is the perceptual space, which results in behavioral decisions. Thus, the first, physical odor space (of unknown order and dimensionality) is mapped onto a primary, receptive odor space (with a clearly defined dimensionality dictated by the number of ORs), then transformed into several physiological spaces (which might work sequentially, in parallel, or both), which in turn is the basis for the perceptual odor space.

4.41.2 Cellular Organization of the Antennal Lobe

The AL is the first neuropil in the insect brain that processes olfactory information. It is evolutionarily shared within the protostome clade and present in all insects, with the exception of some anosmic species where it was secondarily lost (Strausfeld, N. J. and Hildebrand, J. G., 1999). Both in its structure and in its function, it is remarkably similar and analogous to the vertebrate OB, although it evolved independently (Strausfeld, N. J. and Hildebrand, J. G., 1999). There is one major difference in design, which reflects a structural difference between the nervous tissue in insects and that in vertebrates: In the insect brain, most neuronal cell bodies lie at the periphery of the brain, and they extend neurites that branch within the brain, forming axons and dendrites. In the AL, neurons have their somata in cell clusters around the AL itself and branch in olfactory glomeruli, which are the interaction sites of OSNs, local neurons (LNs), projection neurons (PNs), and others. Very few synapses are found outside the glomerulus (Gascuel, J. and Masson, C., 1991; Boeckh, J. and Tolbert, L. P., 1993). In vertebrates, the bulb is organized in a more columnar fashion, with cellular interactions within several layers beneath the olfactory glomeruli, and LN networks separated in these layers. To a first approximation, therefore, the insect glomerulus is functionally comparable to a vertebrate glomerular column, rather than to the glomerulus alone. However, this analogy remains an approximation, because the OB is not columnar beneath the glomerulus. For example, mitral cell somata can be displaced laterally with respect to the glomerulus innervated by their dendritic tuft (Shepherd, G. M. and Greer, C. A., 1998).

The considerable similarity to vertebrates should not prevent us to recognize that within insects the diversity of AL organization is large (Schachtner, J. *et al.*, 2005). Insects have an evolutionary history of over 400 000 000 years, and most modern insect orders were already present since 250 000 000 years, which allows for considerable divergent evolution (Grimaldi, D. A. and Engel, M. S., 2004). As a minor example, cell somata are arranged in distinct clusters around the AL in most insects, but the distribution of cell bodies for functionally different neurons (e.g., LNs vs. PNs) is not conserved among species. In the following, I will mention the different organizational principles found in those species that have been studied in more detail than others. For hemimetabolous insects, most results are from locusts, among which *Locusta migratoria* (Orthoptera), and cockroaches, notably *Periplaneta americana* (Blattodea). In holometabolous insects, the best studied are the fruit fly, *Drosophila melanogaster* (Diptera), a variety of moths, in particular *Manduca sexta* (Lepidoptera), and the honeybee, *Apis mellifera* (Hymenoptera). Where useful, I will follow the sequence bee, moth, fly, cockroach, and locust for clarity.

4.41.2.1 Antennal Lobe Glomeruli

A striking similarity across animal phyla is the glomerular organization of the primary neuropil of the olfactory system, be it the OB of vertebrates or the AL in insects (Hildebrand, J. G. and Shepherd, G. M., 1997). This unique structural principle might reflect a basic property of olfactory coding, possibly because there is no physical property common to all odors that can be mapped in a small number of dimensions. Such an oligodimensional representation is possible for spatial position of light stimuli, or frequency coding for sounds. Color coding, however, does not make use of the unidimensional property of photon wavelength but rather creates artificial dimensions with different color-sensitive photoreceptors and subsequent color-opponency channels, which can be very few (two such channels in humans and honeybees). Olfaction is intrinsically multidimensional, and a glomerular morphology may be a structural organization that helps convert a multidimensional set of inputs into a topological map. The resulting glomerular organizing principle would be functionally dictated. Alternatively, a glomerular organization may just be the most parsimonious organization of many coalescing axons with equal response properties, resulting from the convergence of OSNs onto one point in the central nervous system.

In insects, the number, shape, and arrangement of glomeruli vary greatly among species: an adult *Drosophila* fruit fly has less than 50 glomeruli (Stocker, R. F., 1994; Laissue, P. P. *et al.*, 1999), a moth (*Manduca*) has ~60 (Sanes, J. R. and Hildebrand, J. G., 1976b; Rospars, J. P. and Hildebrand, J. G., 1992), a cockroach has ~125 (Ernst, K. D. *et al.*, 1977), a worker honeybee has ~160 (Flanagan, D. and Mercer, A. R., 1989a; Galizia, C. G. *et al.*, 1999b), and some ant species have over 200 (Rospars, J. P., 1988) or even 400 glomeruli (Ozaki, M., personal communication). In many species, glomeruli are arranged around a central area of the AL, the coarse neuropil, while in others, generally smaller animals, glomeruli are more densely packed, filling most of the AL. Glomeruli are the functional units within the AL, but nevertheless, there is a substructure within them. In bees the outer cap and the core of each glomerulus receive input from different neuron types: OSNs innervate the outer cap, and within this cap, the antenna is topologically arranged. Many LNs innervate only the core of those glomeruli that they sparsely innervate. In honeybees, serotonergic neurons innervate only the cap and dopaminergic neurons only the core. Consequently, each glomerulus is concentrically organized, and the cap may serve a function different from the core. Furthermore, in immunostainings for protein kinase C (PKC), small circular regions of dense staining are visible within honeybee glomeruli (Müller, U., personal communication; see also Grünbaum, L. and Müller, U., 1998). In mass-fills of OSNs similar dense knots can be seen (own observations). Whether these two substructures coincide is not known, and their functional significance, if any, also remains obscure. A subcompartmentalization of glomeruli has been shown for *Drosophila* (Laissue, P. P. *et al.*, 1999) and is also known from vertebrates (Kasowski, H. J. *et al.*, 1999). Morphologically and most likely functionally, there is an inner life to a glomerulus that remains to be elucidated: possibly, glomeruli are not the smallest unit of the olfactory code.

The term glomeruli is also used in locusts, even though their arrangement differs from that in most other species (Ignell, R. *et al.*, 2001). There are about 1000 glomeruli, with each receptor neuron axon innervating several of them, and each projection neuron branching in many of them (Ernst, K. D. *et al.*, 1977; Anton, S. and Hansson, B. S., 1996). It is clear that there is no morphological homology between the locust glomeruli and those of other insects. Nevertheless, anatomically they are dense knots of neuropil and therefore referred to as glomeruli. Furthermore, even if the AL is structured differently, basic mechanisms of olfactory coding may still be the same in locusts and other insects. For example, application of the chloride channel blocker Picrotoxin has similar effects on the temporal response structure in PNs in bees (Stopfer, M. *et al.*, 1997) and in locusts (MacLeod, K. and Laurent, G., 1996). Therefore, a comparative approach, and a thorough comparison of results from all species, is needed.

The most complete ultrastructural analysis of synaptic connections between neurons in the AL has been done in cockroaches. Briefly, almost any possible synaptic contact is realized (Malun, D. 1991a; 1991b; Distler, P. G. and Boeckh, J., 1996; 1997a; Distler, P. G. *et al.*, 1998b). That is, OSNs synapse onto LNs and onto PNs, and their presynaptic terminals themselves receive input from AL neurons. LNs synapse onto OSN terminals, other LNs, and PNs. PNs get input from OSNs and LNs, and they synapse onto other cells in the AL. No synapses from PNs onto OSNs have been found. The presence of LN input to OSNs indicates that the AL network already controls magnitude and maybe timing of afferent input activity, as shown in crustaceans and vertebrates (Wachowiak, M. *et al.*, 2002).

4.41.2.2 Glomerular Atlas of the Antennal Lobe

The low numbers of glomeruli in insects, and their characteristic arrangements and sizes, have allowed mapping them, and creating 3D-Atlases (Rospars, J. P., 1988). Initially, these were morphological descriptions based on serial sections, but in recent years confocal reconstructions of whole ALs have been used. Computer-based atlases are increasingly available over the Internet and can be used for relating findings to identified glomeruli, thus greatly enhancing the comparability of results among research groups. Unfortunately, the electronic format used by different laboratories and for different species still needs unification. The quest for electronic brain atlases in general (Toga, A. W. and Thompson, P. M., 2001) will soon lead to more uniform standards among insect researches, not only for the AL, but also for the entire brain (Rein, K. *et al.*, 2002; Brandt, R. *et al.*, 2005).

The honeybee AL has been mapped, and glomeruli were named according to the antennal tract that innervates them (T1–T4) and a number (i.e., T1-1, T1-2, or T3-33) (Flanagan, D. and Mercer, A. R., 1989a). An electronic version based on confocal sections was published later (Galizia, C. G. *et al.*, 1999b) and is available for download over the Internet (see Relevant Websites section).

Atlases are also available for different moth species (Rospars, J. P. and Hildebrand, J. G., 1992; Berg, B. G. *et al.*, 2002; Huetteroth, W. and Schachtner, J., 2005; Ignell, R. *et al.*, 2005; Masante-Roca, I. *et al.*, 2005; Skiri, H. T. *et al.*, 2005).

The *Drosophila* AL was first mapped based on serial sections (Stocker, R. F. *et al.*, 1990) and later based on confocal sections (Laissue, P. P. *et al.*, 1999). Each of the 43 glomeruli mapped so far is named with a letter or two indicating the AL area and a number, e.g., DM2 for dorsomedial-2. This atlas is also freely available for downloading (see Relevant Websites section). Glomeruli that are identified later can be easily included into this nomenclature (Couto, A. *et al.*, 2005).

Currently, there is no glomerular atlas for cockroaches. It is unlikely that there will ever be an atlas for locusts, given the ambiguity of glomerular borders and the high number of glomeruli.

4.41.2.3 Sensory Neuron Axons

4.41.2.3.1 Antennal nerves

Chemosensory neurons are present on the surface of many areas of the insect body, including legs, wings, genitalia, and head appendices. All (nonanosmic) insects have olfactory sensilla that contain OSNs on their antennae, and some, notably dipterans, also on the maxillary palps. Structure and function of OSNs and the diversity of olfactory sensilla are treated in other chapters. OSN axons project to the AL via the antennal tract(s). In many species, there is more than one antennal tract. In honeybees, there are four tracts, T1–T4, that innervate different glomeruli (Suzuki, H., 1975). The honeybee T1 and T3 innervate ~70 glomeruli each, and T2 and T4 innervate seven glomeruli each (Arnold, G. *et al.*, 1983; Flanagan, D. and Mercer, A. R., 1989a; Galizia, C. G. *et al.*, 1999b). Since each glomerulus has a distinct functional response spectrum (see below), the antennal tract segregation may have a functional significance. In honeybees, the morphology of the innervating OSN terminals differs between glomeruli innervated by the four antennal tracts (Mobbs, P. G., 1982). Furthermore, this segregation corresponds to distinct groups of PNs that leave the AL following different tracts (see below) (Abel, R. *et al.*, 2001; Kirschner, S. *et al.*, 2006).

Antennal nerves contain other axons too: mechanosensory and gustatory axons from the antenna travel along the AL, bypassing it toward the antennal mechanosensory and motor center (AMMC) in the dorsal lobe (DL) (Suzuki, H., 1975; Gewecke, M., 1979; Mobbs, P. G., 1982; Kloppenburg, P. *et al.*, 1997). In honeybees, these are tracts T5 and T6. Efferent axons for antennal muscles, originating in motor neurons in the DL, use these tracts to enter the antenna. Finally, axons from modulatory neurons, with their unpaired cell bodies in the subesophageal ganglion, also enter the antenna. These neurons most likely use biogenic amines as transmitters. While their function has not yet been elucidated directly,

a modulatory effect of biogenic amines on OSNs has been shown (Pophof, B., 2002).

4.41.2.3.2 Olfactory sensory neuron numbers

The number of OSNs differs greatly among species, and within some species among sexes. Male honeybees (drones) have ~300 000 OSNs, while (female) worker bees have ~65 000, the difference being probably due to the male's sensitivity for the queen's sexual pheromone (Esslen, J. and Kaissling, K. E., 1976). In *Manduca*, sexual dimorphism is reflected in OSN response properties, and not their numbers: both males and females have ~300 000 OSNs on each antenna (Sanes, J. R. and Hildebrand, J. G., 1976c; Oland, L. A. and Tolbert, L. P., 1988). *Drosophila* has ~1200 OSNs in both sexes (Stocker, R. F. *et al.*, 1990). Adult cockroaches have ~150 000 OSNs (Ernst, K. D. *et al.*, 1977). As juveniles, cockroaches have less OSNs. As they grow, they increase the number of antennal segments and sensilla with each nymphal instar (Schafer, R. and Sanchez, T. V., 1973). The number of OSNs in locusts is between 50 000 (Ernst, K. D. *et al.*, 1977) and 105 000 (Anton, S. *et al.*, 2002) and also increases over developmental stages (Ochieng, S. A. *et al.*, 1998; Chapman, R. F., 2002).

4.41.2.3.3 Projection of olfactory sensory neurons into the antennal lobe

Generally, one OSN axon innervates a single ipsilateral glomerulus in the AL, where it branches and where it forms many synapses. It is unknown how many synapses a single axon forms within the glomerulus, but given the often extensive branching of these axons, the number is likely to be high. In most cases, branching is strongest in the outermost layer of the innervated glomerulus, but in some species (e.g., cockroaches) or some subpopulations of glomeruli (e.g., T4 in bees), OSN axons invade the entire glomerulus. In bees, there is a concentric topological arrangement of OSN input to glomeruli, with OSNs from the distal antennal segments innervating the outer layer of the glomerulus, and more proximal OSNs occupying more central layers (Pareto, A., 1972). These layers are, however, strongly overlapping and may be dictated by developmental constraints rather than by functional necessity. Antennotopic projections have also been found in *Manduca* (Christensen, T. A. *et al.*, 1995) and *P. americana* (Hösl, M., 1990). Synapses are formed onto both LNs and PNs and are often dyadic, that is, one presynaptic element with two postsynaptic elements, or even more complex (Malun, D. 1991a; Distler, P. G. and Boeckh, J., 1996; Sun, X.-J. *et al.*, 1997). In flies, most axons innervate a bilateral pair of homologous glomeruli, and OSNs form an axonal commissure between the two ALs (Strausfeld, N. J., 1976; Stocker, R. F., 1994). Each bilaterally innervated glomerulus receives equal input from both antennae (Vosshall, L. B. *et al.*, 2000). Some glomeruli, however, are innervated only unilaterally; in *D. melanogaster*, these are V, VL1, VP1, as well as VP2 and VP3 that collect all input from sensilla located on the *Drosophila* arista (Stocker, R. F. *et al.*, 1983; Stocker, R. F., 2001).

In locusts, most OSN axons innervate several glomeruli ipsilaterally (Ernst, K. D. *et al.*, 1977), but some innervate only one glomerulus (Anton, S. *et al.*, 2002).

4.41.2.3.4 Targeting mechanisms of olfactory sensory neurons

Targeting of OSNs onto glomeruli is best understood in *Drosophila*. After the ORs were characterized (Clyne, P. J. *et al.*, 1999; Gao, Q. and Chess, A., 1999; Vosshall, L. B. *et al.*, 1999), the mapping of their target areas has been intensely pursued (Couto, A. *et al.*, 2005; Fishilevich, E. and Vosshall, L. B., 2005). Indeed, the nomenclature of olfactory glomeruli by spatial location (e.g., DM2 for glomerulus 2 in the dorsomedial area) might eventually be used synonymously with the name of the OR gene expressed in the OSNs innervating it (e.g., dOr22a for DM2). The current rule is that each OSN expresses only one OR and that OSNs expressing the same OR converge on a single glomerulus in each hemisphere (Vosshall, L. B. *et al.*, 2000), but there are exceptions: in transgenic animals, a few axons go astray to other glomeruli (Bhalerao, S. *et al.*, 2003). Some OSNs express more than one OR. For example, dOr22a and dOr22b are coexpressed in ab3a (for antennal basiconic sensillum 3, neuron a), and their axons converge onto glomerulus DM2 in the AL, but dOr22b does not seem to contribute to the olfactory response of those neurons (Dobritsa, A. A. *et al.*, 2003). In pb2a OSNs (palp basiconic type 2, neuron a), receptors dOr33c and dOr85e are coexpressed, and in this case, both contribute to these neurons' molecular response profiles (MRRs) (Goldman, A. L. *et al.*, 2005). Other OSN axon populations innervate two glomeruli even though expressing the same OR. For example, OSNs expressing dOr67d innervate the two glomeruli DA1 and VA6 (Fishilevich, E. and Vosshall, L. B., 2005). An almost complete mapping of dOrs and their target glomeruli is now available for adult *Drosophila* (Couto, A. *et al.*, 2005; Fishilevich, E.

and Vosshall, L. B., 2005). From this it is clear that the general rule one OSN – one receptor – one glomerulus is valid in most cases, but not followed strictly, allowing for additional degrees of freedom in functionally organizing chemosensory responses. In addition, most but not all OSNs express an additional OR-like receptor, which is dOr83b in *Drosophila* (Larsson, M. C. *et al.*, 2004), and which has an ortholog receptor in all other insects investigated so far, including mosquitoes, bees, moths, and beetles (Krieger, J. *et al.*, 2003; Pitts, R. J. *et al.*, 2004; Jones, W. D. *et al.*, 2005). OR expression in *Drosophila* OSNs is discussed in more Chapter Phototransduction in Microvillar Photoreceptors of Drosophila and Other Invertebrates.

In mammals, ORs are important for OSN axon targeting to the appropriate glomerulus (Wang, F. *et al.*, 1998; Feinstein, P. and Mombaerts, P., 2004; Feinstein, P. *et al.*, 2004). In insects, the innervated glomerulus is independent of the OR that is expressed (Dobritsa, A. A. *et al.*, 2003). Other factors are required for correct OSN targeting in *Drosophila*, including the Src homology domain 2 (SH2)/SH3 adapter Dock (Ang, L. H. *et al.*, 2003), the serine/threonine kinase Pak (Ang, L. H. *et al.*, 2003), the cell surface proteins *N*-cadherin (Hummel, T. and Zipursky, S. L., 2004), the immunoglobulin Dscam (Hummel, T. *et al.*, 2003), and the POU domain transcriptional factor Acj6 (Komiyama, T. *et al.*, 2004). Many of these factors are only expressed in subpopulations, suggesting some sort of combinatorial mechanism. For example, Dscam disrupts OSN targeting in many OSN axons, but not in those expressing the receptors dOr22a, and only partially in some others (Hummel, T. *et al.*, 2003). All OSNs in the antenna express Acj6, but not all OSNs in the maxillary palps do. Furthermore, precise OSN targeting is not only regulated by intrinsic transcriptional control but also by axon–axon interaction (Komiyama, T. *et al.*, 2004) and interactions with glial cells (Tolbert, L. P. *et al.*, 2004). Glomerular identity appears to be determined before OSN ingrowth, with the participation of PNs forming protoglomeruli (Jefferis, G. *et al.*, 2004).

4.41.2.3.5 Olfactory sensory neuron odor responses

Responses in OSN axons to odorants originate from the response properties of the ORs expressed in that axon, and other interactions at the periphery including OBPs. Response properties measured at the periphery are covered in another chapter.

Functional response properties of olfactory glomeruli have been measured in several species using optical imaging techniques, notably flies (Ng, M. *et al.*, 2002; Fiala, A. *et al.*, 2002; Wang, J. W. *et al.*, 2003), bees (Galizia, C. G. and Menzel, R., 2001; Fiala, A. *et al.*, 2002; Sachse, S. and Galizia, C. G., 2002; 2003), moths (Galizia, C. G. *et al.*, 2000a; Carlsson, M. A. *et al.*, 2002; Skiri, H. T. *et al.*, 2004), and ants (Galizia, C. G. *et al.*, 1999a). In all cases, glomeruli had characteristic odor-response profiles, confirming that a single or at most very few ORs dictate their response properties. Consequently, also in species other than *Drosophila*, OSNs expressing the same OR are likely to converge generally onto a single glomerulus. Additional support comes from the observation that in the recently sequenced genome of the honeybee, *A. mellifera*, the number of OR genes is close to 160 (Robertson, H. M. and Wanner, K. W., 2006). This number matches closely the number of glomeruli found in this species. Thus, molecular and optophysiological data have superseded earlier models of AL innervation based on single-cell electrophysiology that assumed OSNs with equal response profiles to innervate several glomeruli, and each glomerulus to receive input from several OSN types (Boeckh, J. and Tolbert, L. P., 1993). For locusts, where individual OSN axons innervate several glomeruli, no data about OSN targeting are as yet available.

OSNs differ in how broad their response spectrum is. Narrowly tuned OSNs form a labeled line: each OSN responds only to a single substance, and thus when that OSN is active, animals can behave assuming that the substance is in the environment. Such a system is very specialized, and in its extreme form, the number of odors that can be coded corresponds to the number of OSNs with different response profile (which would be ~43 in the adult *Drosophila*). On the other hand, broadly tuned OSNs create a combinatorial code. Each odor activates many OSN types, and each OSN can be activated by many odors. Thus, the information carried by individual glomeruli becomes ambiguous. It is only by analyzing the pattern of active glomeruli that the animal can identify an odor. Experiments show that many OSNs respond to a variety of substances at high concentration but may be exquisitely sensitive to a few molecules at a concentration several orders of magnitude lower (Røstelien, T. *et al.*, 2000; Stensmyr, M. C. *et al.*, 2003; Pelz, D. *et al.*, 2006). Are they highly sensitive labeled lines, or rather units in a combinatorial code? Animals are exposed to odorants over a large range of

concentrations in their natural environment. A honeybee, for example, can smell an odor source over large distances (low-concentration task) but also recognize it when sitting on a flower (high-concentration task). The olfactory system should therefore be capable of differentiating, for a given OSN, between the occurrence of a low-concentration best ligand and a high-concentration secondary ligand. This cannot be done on the basis of a single OSN response itself and must be done by a pattern analysis across several OSNs. Even for sexual pheromones, the labeled line at OSN level is converted into a combinatorial system for the next processing step. Pheromones consist of blends of several substances, and extraction of the correct pheromone blend information is done by a combinatorial analysis of the respective glomerular activities (Christensen, T. A. and Hildebrand, J. G., 2002; Shirokova, E. *et al.*, 2005). Some species, such as *Helicoverpa armigera* and *Helicoverpa assulta*, have pheromone blends consisting of the same substances, but in different relative concentrations, so that relative activity becomes important information (Berg, B. G. *et al.*, 2005).

Most OSN responses are temporally structured: some are inhibited by odors and show rebound excitation at the end of the stimulus, some fire for a long time irrespective of stimulus duration, others only fire for a very short time, and stop even if the stimulus continues (de Bruyne, M. *et al.*, 2001). Some but not all of these effects are due to adaptation of the OSNs (tonic versus phasic response properties). The consequence for the central nervous system is that it gets a temporally complex input from the OSNs. This is in addition to the intrinsic temporal complexity that air-borne odors always have (Murlis, J. *et al.*, 1992). The question as to whether these temporal structures are used for odor analysis and whether the periphery rather than the AL network may cause part of the temporal patterns observed within the AL is discussed below.

OSN responses can also be inhibitory. Because the background activity differs among OSN types, an inhibitory ligand might become evident only when an OSN is also activated by another ligand, resulting in nonlinear mixture interactions (de Bruyne, M. *et al.*, 2001).

4.41.2.3.6 *Olfactory sensory neuron transmitters*

The OSN transmitter is likely to be acetylcholine (ACh), which is the primary excitatory transmitter in the insect brain. The evidence, however, is not conclusive yet, and mostly based on the detection of the ACh-synthesizing enzyme choline acetyltransferase (ChAT) (Kreissl, S. and Bicker, G., 1989; Bicker, G., 1999a). ACh is present in *Manduca* OSNs (Sanes, J. R. and Hildebrand, J. G., 1976a; Stengl, M. *et al.*, 1990). Pressure application of ACh leads to activity (depolarization and hyperpolarization) in moth AL neurons, mediated by nicotinic receptors (Waldrop, B. and Hildebrand, J. G., 1989). In *Drosophila*, a ChAT/lacZ tranformant labels OSNs (Yasuyama, K. *et al.*, 1995), but immunoreactivity for ACh is low (Yasuyama, K. and Salvaterra, P. M., 1999). In all species studied so far, however, the presence of ACh, ChAT, and α-bungarotoxin binding sites was heterogeneous among glomeruli, leaving the possibility that there might be another, as yet unknown, transmitter (or cotransmitter) in OSNs (Kreissl, S. and Bicker, G., 1989; Homberg, U. *et al.*, 1995; Homberg, U. and Müller, U., 1999) or that glomeruli differ in receptor and/or vesicle density.

In moths, OSN axons contain nitric oxide (NO) synthase and release NO when activated by odors (Gibson, N. J. and Nighorn, A., 2000). NO signaling stimulates glial migration, and blocking NO signaling in development disrupts proper glomerular development in the AL (see Section 4.41.2.7) (Gibson, N. J. *et al.*, 2001). NO is also present in honeybee ALs, but it is unclear whether NO is produced in OSNs or other cells (Hosler, J. S. *et al.*, 2000; Bicker, G., 2001). The NO/cGMP system is involved in olfactory habituation in honeybees (Müller, U. and Hildebrandt, H., 2002), and blocking NO activity in adult honeybees disrupts olfactory discrimination in a similar way as when chloride channels are blocked (Hosler, J. S. *et al.*, 2000). In cockroaches and locusts, NO is not synthesized in OSNs but is synthesized in LNs (Elphick, M. *et al.*, 1995; Seidel, C. and Bicker, G. 1997; Ott, S. R. and Elphick, M. R., 2002).

4.41.2.4 Local Neurons

LNs are defined as neurons whose branching patterns are limited to the AL alone. Their somata are located close to the AL, and their neurites branch among olfactory glomeruli, and within them. Many different morphologies have been described, and a variety of transmitters and expressed peptides are known. Synaptic connections among LNs and with other neurons have been analyzed in few species only. Descriptions from different species vary in approach and technique, making a thorough comparison

difficult. It is clear that we are lacking considerable knowledge about LNs in insects.

4.41.2.4.1 Morphology of local neurons

Morphologically, LNs fall into two major classes. One type innervates most if not all glomeruli uniformly. These neurons have been named symmetrical LNs (Matsumoto, S. G. and Hildebrand, J. G., 1981; Ernst, K. D. and Boeckh, J., 1983; Anton, S. and Homberg, U., 1999) or, in bees, homogeneous LNs (homoLNs) (Flanagan, D. and Mercer, A. R., 1989b). Another type innervates only a small number of glomeruli. In bees, these heterogeneous LNs (heteroLNs) have one dominant glomerulus with dense innervation, and several glomeruli with very sparse innervation (Flanagan, D. and Mercer, A. R., 1989b; Sun, X.-J. *et al.*, 1993a). They branch only in the core of the sparsely innervated glomeruli but branch in the core and the outer cap of the densely innervated glomeruli (Fonta, C. *et al.*, 1993; Abel, R. *et al.*, 2001). In moths, all asymmetric neurons described so far innervate a small group of glomeruli sparsely (asymmetric LNs) without a single, dominant glomerulus (Matsumoto, S. G. and Hildebrand, J. G., 1981). Symmetric and asymmetric LNs together account for the vast majority of LNs in the AL, with heteroLNs (asymmetric) dominating in bees (Flanagan, D. and Mercer, A. R., 1989b; Fonta, C. *et al.*, 1993; Sun, X.-J. *et al.*, 1993a), and symmetric LNs dominating in moths (Christensen, T. A. *et al.*, 1993). Other morphological types are LNs that innervate some dorsal glomeruli of the AL and areas of the AMMC (which makes their definition as LNs questionable), so far published for bees (Flanagan, D. and Mercer, A. R., 1989b). A thin connection between the two ALs has been described in bees (Arnold, G. *et al.*, 1985; Mobbs, P. G., 1985), which appears to contain about 30 GABAergic fibers in immunohistochemical sections (Schäfer, S. and Bicker, G., 1986). These neurons are multiglomerular (Fonta, C. *et al.*, 1993) and may be related to bilateral comparisons, regulation, and inhibition of AL activity.

In *Drosophila*, two GAL4 lines that label two distinct sets of LNs have been described. Both consist of homogeneously branching LNs, but density, morphology, and location of branches within glomeruli differ (Ito, K., personal communication). The functional differences in these two lines are also unknown. A third group of LNs is cholinergic and mediates excitatory interactions within the AL (Shang, Y, *et al.*, 2007). LNs that bilaterally connect the two ALs are also found in *Drosophila* and at least some of them innervate only glomerular subsets and not the entire AL (Stocker, R. F. *et al.*, 1990; Stocker, R. F., 1994).

4.41.2.4.2 Local neuron numbers

Counts of LNs in insects are difficult, and rarely better than a gross estimate. There are up to 4000 LNs in bees (see below), ~360 LNs in moths (*M. sexta*) (Homberg, U. *et al.*, 1988), and ~100 (GABAergic) LNs in *Drosophila* (Ng, M. *et al.*, 2002). The estimates for other species are ~300 LNs both in cockroaches and in locusts (Anton, S. and Homberg, U., 1999).

Honeybees have by far the highest number, with up to 4000 LNs. This count is based on about 7350 neurons in each deutocerebrum, that is, the AL and DL taken together, of which 4750 are directly associated with the AL (Witthöft, W., 1967). The vast majority of these are either LNs or PNs. Assuming 800 PNs (see below), this leaves almost 4000 LNs.

4.41.2.4.3 Transmitters used by local neurons

Most LNs are GABAergic, with GABA-like immunoreactivity shown in bees (Schäfer, S. and Bicker, G., 1986), moths (Hoskins, S. G. *et al.*, 1986), *Drosophila* (Jackson, F. R. *et al.*, 1990; Buchner, E., 1991), cockroaches (Malun, D., 1991b; Distler, P. G. *et al.*, 1998b), and locusts (Leitch, B. and Laurent, G., 1996; Ignell, R. *et al.*, 2001). LNs get synaptic input also from GABAergic LNs. In *Drosophila*, GABAergic LNs expressing the GABA biosynthetic enzyme glutamic acid decarboxylase also express GABA receptors, both metabotropic and ionotropic (Ito, K., personal communication).

In bees, ~750 LNs stain for GABA (Schäfer, S. and Bicker, G., 1986), which forms a rather low proportion of GABAergic interneurons as compared with other insects, where most LNs appear to be GABAergic. Knowledge about LNs that are nonGABAergic is limited, with the exception of a population of about 35 LNs in honeybees that stain for histamine as a transmitter (Bornhauser, B. C. and Meyer, E. P., 1997). Application of histamine blocks activity in the honeybee AL (Sachse, S. and Galizia, C. G., 2002), suggesting that it acts as an inhibitory transmitter. The function of histaminergic LNs with regard to odor coding and processing is still unknown and needs further investigation. Since OSNs in flies express G-protein-coupled glutamate receptors, there probably is a population of glutamatergic LNs

that synapse onto (at least some) OSN terminals (Ramaekers, A. *et al.*, 2001).

A pharmacological analysis is difficult in insects, because receptor affinities are not well characterized for most artificial antagonists and agonists. Furthermore, the long evolutionary history of insects results in considerable diversity. For example, the GABA$_A$ antagonist bicucculine modifies AL activity in moths (Christensen, T. A. *et al.*, 1998b), but not in bees, flies (own observations), and locusts (MacLeod, K. and Laurent, G., 1996). The chloride channel blocker Picrotoxin has both a washable and a nonwashable effect in bees (Sachse, S. and Galizia, C. G., 2002), but only a nonwashable effect in moths (Waldrop, B. *et al.*, 1987). The GABA$_B$ blocker 2H-saclofen is effective in moths (Lei, H personal communication), and the GABA$_B$ blocker CGP54626 is effective in fruit flies (Wilson, R. I. and Laurent, G., 2005).

LNs differ in their expression of neuropeptides. The available data show that many, often small populations of LNs, express characteristic peptides, including but by far not limited to allatotropin, allatostatins, tachykinins, FMRFamide, and other RFamide peptides. The patterns of peptide antiserum stainings differ widely among the species (Homberg, U. *et al.*, 1990; Nässel, D. R., 1993; Davis, N. T. *et al.*, 1996; Homberg, U. and Müller, U., 1999; Nässel, D. R., 2000; Homberg, U., 2002; Nassel, D. R. 2002; Schachtner, J. *et al.*, 2004; Iwano, M. and Kanzaki, R., 2005). Some neuropeptides are coexpressed with GABA while others are not. In locusts, locustatachikinin is found in a population of about 20 GABAergic neurons (Ignell, R. *et al.*, 2001). Myotropic neuropeptides of the orcokinin family are found in cockroach and locust LNs, but not in moths, bees, and flies (Hofer, S. *et al.*, 2005).

4.41.2.4.4 Physiology of local neurons

LNs have action potentials (APs) in honeybees (Sun, X.-J. *et al.*, 1993a; Galizia, C. G. and Kimmerle, B., 2004), moths (Christensen, T. A. *et al.*, 1993), flies (Wilson, R. I. *et al.*, 2004), and cockroaches (Ernst, K. D. and Boeckh, J., 1983), but in locusts LNs are nonspiking (Laurent, G. and Davidowitz, H., 1994).

In species where LNs spike, they can be tonically active in the resting animal and increase or decrease their firing upon stimulation with odors. In bees and moths, most intracellular recordings from LNs show multiple spike amplitudes. This might indicate that LNs with complex branching patterns have multiple spike initiation zones or that LNs are electrically coupled. Alternatively, their fine processes might lead to artificial electrical connections created by the penetrating sharp electrode, in which case multiple spike heights would be an experimental artifact (Galizia, C. G. and Kimmerle, B., 2004). HeteroLNs in honeybees have distinct odor-response profiles comparable to those of uniglomerular PNs. Generally, the odors they respond to correspond to the odors expected from the honeybee functional atlas, that is, LNs respond to those odors associated with its dominant glomerulus, suggesting that their morphologically dominant glomerulus is their input site (Galizia, C. G. and Kimmerle, B., 2004). In contrast, homoLNs studied in *Drosophila* have very broad response profiles, responding to most odors, with activity spread across their entire arborization (Ng, M. *et al.*, 2002; Wilson, R. I. *et al.*, 2004). Furthermore, *Drosophila* is the only species, so far, where both inhibitory and excitatory LNs have been characterized (Shang, Y. *et al.* 2007).

LNs influence the responses in PNs in at least two ways: first, with a very fast inhibition, mediated by GABA acting on ionotropic channels, and second, on a slower timescale, which leads to continuing inhibitions even after stimulus offset (Christensen, T. A. *et al.*, 1998a). Studies with pharmacological antagonists have shown that the fast GABA-mediated currents cause synchrony among PN spikes, and odor-evoked oscillations in the 20–30 Hz range (MacLeod, K. and Laurent, G., 1996). Furthermore, LNs also mediate a disinhibitory connection between OSNs and PNs, which leads to an excitatory response in PNs. In this sequential arrangement OSN–LN–LN–PN, a tonically active LN tonically inhibits a PN; upon OSN activity, another LN becomes active, inhibits this LN, and thereby releases the PN from its inhibition, resulting in PN activity (Av-ron, E. and Rospars, J. P., 1995).

Across insects, the available data are insufficient to draw firm conclusions about LN response profiles. The function that LNs play for olfactory coding is discussed below (Section 4.41.3.2).

4.41.2.5 Projection Neurons

PNs connect the AL to other brain areas. Uniglomerular neurons branch in a single glomerulus within the AL and have axons that lead from the AL to the MBs and to the LP. Multiglomerular neurons branch in several if not all glomeruli and generally have axons that do not innervate the MBs, but a variety of other areas in the protocerebrum, both lateral (close to and overlapping with the LP) and more central (surrounding the α-lobe of the MB). PNs have both input and output synapses within olfactory glomeruli. It is unclear, though, whether all PNs have output synapses within the AL or only some of them.

4.41.2.5.1 Projection neuron morphology and tracts

Nomenclature of PN tracts is inconsistent across insect species, reflecting both their morphological diversity and a lack of knowledge about the homology of the tracts involved. Morphologically, in most species there is a strong antennocerebralis tract (ACT) that travels along the brain midline (mACT (medial ACT) in bees and iACT (inner ACT) in most other species), and a generally smaller tract that travels close to the optic lobes (lACT (lateral ACT) in bees, oACT (outer ACT) in moth, and mACT in flies). In addition, there are several minor tracts that sometimes share parts with the lACT/oACT, and whose axons target different areas (multiglomerular ACTs (mlACTs) in bees and mACTs in most other species). I will therefore present different insects sequentially. Branching patterns of PNs in the MBs and the LP are discussed in their respective sections below.

In bees, uniglomerular PNs form two tracts, the lACT and the mACT (Mobbs, P. G., 1982). The mACT travels along the brain midline toward the MBs where collaterals enter into the MB calyces, and continues to end branching in the LP. PNs in the mACT originate from glomeruli of the T2, T3, or T4 group (Bicker, G. *et al.*, 1993; Abel, R. *et al.*, 2001; Kirschner, S. 2006) and have their somata in the ventrolateral and dorsal soma cluster of the AL (Bicker, G. *et al.*, 1993). The lACT leaves the AL dorsally and then travels on the lateral, or outer side, close to the optic lobes. Collaterals enter the LP, and the tract continues and also innervates the MB calyces. lACT PNs innervate T1 glomeruli and their somata lie in the anteroventral and lateral part of the AL (Schäfer, S. *et al.*, 1988). Thus, uniglomerular PNs separate in tracts according to their innervated glomerulus, and the separation corresponds to separate antennal tracts (T1–T4). The molecular mechanisms for this separation are unknown, but some insight comes from flies (see below). Uniglomerular PNs in honeybees have small branches at their exit points of the AL, which may form connections between cells outside their innervated glomerulus, but whose computational function is not yet understood (Müller, D. *et al.*, 2002; Galizia, C. G. and Kimmerle, B., 2004). The initial portion of the lACT is also shared by multiglomerular fibers that innervate only the LP and the ring neuropil of the α-lobe (Abel, R. *et al.*, 2001). Most multiglomerular PNs, however, use the mlACT, which in fact consists of several small tracts. Their axons also innervate the LP and the neuropil around the α-lobe. Within the AL, they branch sparsely within glomeruli (Fonta, C. *et al.*, 1993; Abel, R. *et al.*, 2001). The number of glomeruli that each multiPN innervates within the AL varies, ranging from a few to many or possibly all.

In *Manduca*, tracts are iACT, mACT, and oACT (Homberg, U. *et al.*, 1988). Most neurons in the iACT are uniglomerular, while most neurons in mACT and oACT are multiglomerular. Axons in the iACT and oACT innervate MB calyces and the LP, while axons in the mACT innervate a variety of foci in the LP but not the MBs (Homberg, U. *et al.*, 1988).

Fly tracts are also named iACT, mACT, and oACT (and, as in other species, these may be subdivided further), but these do not correspond to the moth nomenclature. The inner tract (traveling along the brain midline) has the strongest input to the MB calyces. The medial tract has some axons that innervate the MB calyces, while most axons target only the LP. Axons in the oACT do not innervate the MB calyces. All tracts innervate the LP (Stocker, R. F. *et al.*, 1990; Stocker, R. F., 1994; Strausfeld, N. J. *et al.*, 2003). Thus, *Drosophila*'s iACT and mACT may correspond to the honeybee's mACT and lACT, or the moth's iACT and oACT, respectively. As in bees, glomerular identity and tract are probably related in flies: the GH146 line labels PNs with somata in the anterodorsal cluster (adPN) and lateral cluster (lPN), which have axons that use the iACT and send their dendrites to about 30–35 stereotypical glomeruli (Jefferis, G. *et al.*, 2001; Marin, E. C. *et al.*, 2002; Wong, A. M. *et al.*, 2002; Jefferis, G. S. *et al.*, 2007; Lin, H. H. *et al.*, 2007). The tract used by PNs innervating the remaining glomeruli is unknown. GH146 also labels at least four PNs that use the mACT and at least one PN that uses the oACT. The POU domain transcription factors, Acj6 and Drifter, are expressed in adPNs and lPNs, respectively, and are required for their dendritic targeting (Komiyama, T. *et al.*, 2003). Interestingly, these genes are also necessary for OSN targeting (see Section 4.41.2.3.4).

PN tracts in cockroaches follow an only partially similar nomenclature: the iACT travels along the brain midline, the oACT travels along the optic lobes, and intermediate tracts have been named ACT tracts II, III, and IV (Malun, D. *et al.*, 1993). PNs in iACT are uniglomerular and innervate both

the MB calyces and the LP. Pheromone-sensitive PNs in male cockroaches only use the iACT (Malun, D. *et al.*, 1993). PNs in tract II are also uniglomerular and innervate both MB and LP. PNs in the other tracts are multiglomerular and innervate the LP only (tracts III and oACT) or several brain areas including the MBs (tract IV, which originates from the same root in the AL as the oACT) (Malun, D. *et al.*, 1993).

PNs in locusts have a different morphology from what has been described so far. No uniglomerular PNs have been reported yet; rather, within the AL, PNs branch in a limited number of glomeruli (Laurent, G. *et al.*, 1996; Anton, S. *et al.*, 2002). This is reminiscent of the multiple glomeruli innervated by individual OSNs, but the two do not form corresponding groups, that is, PNs innervate groups of glomeruli that do not correspond to groups innervated by an OSN (Anton, S. *et al.*, 2002). In other words, glomerular groups in locusts cannot be functionally analogous to single glomeruli in other insects. In locusts, only a single ACT close to the brain midline connects the ALs to the MB (Leitch, B. and Laurent, G., 1996), but minor tracts that connect the AL to other areas do exist (Ignell, R. *et al.*, 2001).

4.41.2.5.2 Projection neuron numbers

In honeybees, the lACT and mACT probably house about 400 fibers each, and the mlACT much less, giving a total of less than 1000 (Rybak, J., 1994). This figure may represent an upper limit, considering that the number of PNs has also been estimated at 500 (Bicker, G. *et al.*, 1993) or 800 PNs (Hammer, M., 1997).

Manduca has $\sim$400 uniglomerular PNs in the iACT and $\sim$340 in the oACT, and $\sim$122 multiglomerular PNs in the mACT (Homberg, U. *et al.*, 1988).

In *Drosophila*, the GAL4 line that is most often used for investigating properties of PNs is GH146 (Heimbeck, G. *et al.*, 2001). This line labels about 80–90 PNs, of which 85 are uniglomerular and travel in the iACT, with somata in the anterodorsal (40–50) and lateral cluster (30–35), and 6–7 are multiglomerular, do not use iACT, do not innervate the MB, and have their somata in the ventral cluster (Jefferis, G. *et al.*, 2001; Wong, A. M. *et al.*, 2002). The total number of PNs in *Drosophila* is estimated to be 150–200 (Stocker, R. F. *et al.*, 1997). The proportion of multiglomerular PNs with respect to uniglomerular PNs is below 10% in *Drosophila*.

An estimate of cockroach PN number is 250 (Ernst, K. D. *et al.*, 1977). The number of PNs in locusts is ~830 (Leitch, B. and Laurent, G., 1996), based on counting axon profiles in the main ACT tract.

4.41.2.5.3 Projection neuron transmitters

Most uniglomerular PNs have ACh as transmitter. In bees, mACT PNs are AchE-positive, showing that they use ACh (Kreissl, S. and Bicker, G., 1989). Furthermore, Kenyon cells (KCs) express nicotinic ACh receptors and are activated by ACh (Bicker, G. and Kreissl, S., 1994). Fibers in the honeybee lACT are not cholinergic; they are taurin-immunoreactive (Bicker, G. et al., 1993). No responses to taurin application could yet be measured in cultured KCs (Grünewald, B., personal communication). In moths, ~67 somata associated with the AL stain for AChE, and the oACT leading to the LP and the MB is stained (Homberg, U. et al., 1995). The situation in *Drosophila* is similar, with at least one tract strongly stained for AChE (Yasuyama, K. and Salvaterra, P. M., 1999). In locusts, the ACT tract stains for AChE (Homberg, U., 2002).

Multiglomerular PNs that are GABAergic have been shown in bees (Schäfer, S. and Bicker, G., 1986), moths (Hoskins, S. G. et al., 1986), and flies (Ito, K., personal communication). No other transmitter has so far been associated with multiglomerular PNs.

4.41.2.5.4 Projection neuron physiology

Responses in insect PNs have been measured at the single-cell level with electrophysiological techniques in a variety of species (Hansson, B. S. et al., 1991; Christensen, T. A. et al., 1998a; 1998b; Abel, R. et al., 2001; Müller, D. et al., 2002; Galizia, C. G. and Kimmerle, B., 2004; Wilson, R. I. and Laurent, G., 2005). Optical methods were successful in bees after loading PNs with calcium-sensitive dyes (Sachse, S. and Galizia, C. G., 2002; 2003) and in flies after genetically expressing activity-sensitive proteins (Fiala, A. et al., 2002; Ng, M. et al., 2002; Wang, J. W. et al., 2003; Yu, D. et al., 2004). PNs are spontaneously active in insects, including honeybees (Abel, R. et al., 2001; Galán, R. F. et al., 2005), moths (Christensen, T. A. et al., 1998b), *Drosophila* (Wilson, R. I. et al., 2004), cockroaches (Boeckh, J. et al., 1987), and locusts (Perez-Orive, J. et al., 2002). Against this background, responses to odors can be both excitatory and inhibitory. Their responses are shaped by the AL network (Christensen, T. A. et al., 1998a; Sachse, S. and Galizia, C. G., 2002; Wilson, R. I. and Laurent, G., 2005; Olsen, S. R. et al., 2007; Schlief, M. L. and Wilson, R. I. 2007; Shang, Y. et al., 2007).

Instantaneous response frequencies of excitatory responses can be several hundreds of Hertz and persist over the duration of a 2-s stimulus. In honeybees, lACT neurons tend to respond to many odors with excitation, whereas mACT neurons display more complex responses to different odors, including phases of excitation and inhibition (Abel, R. et al., 2001; Müller, D. et al., 2002). PN physiology is discussed below in more detail (see Section 4.41.3).

4.41.2.6 Feedback Neurons and Biogenic Amines

Neurons that branch in the AL and also in other brain areas have been found in many insect species. There are many different morphologies and biochemical profiles of these neurons, but part of the apparent variability among species might be due to limited knowledge of these cells. Many of these neurons stain with antibodies against biogenic amines (dopamine, serotonin, octopamine, and histamine) and are believed to have a modulatory function, including up- and downregulation, thresholding, motivational states, attention, and learning (Hammer, M. and Menzel, R., 1998; Bicker, G., 1999b; Homberg, U. and Müller, U., 1999). Other possible feedback neurons are the honeybee AL-1 neurons that originate in the MB α-lobe and project widely through the ALs (Rybak, J. and Menzel, R., 1993). Dye injections into single neurons have also revealed the existence of neurons with somata in the subesophageal ganglia (SEG), of which some innervate only a few glomeruli in the AL symmetrically on both sides, and send collaterals to the MB (Abel, R. et al., 2001).

I review evidence for biogenic amines (Bicker, G., 1999b; Monastirioti, M., 1999; Homberg, U., 2002) in the order dopamine, octopamine (Stevenson, P. A. and Sporhase-Eichmann, U., 1995; Roeder, T., 1999; Pflüger, H. J. and Stevenson, P. A., 2005), serotonin (Homberg, U., 2002), and histamine (Nässel, D. R., 1999).

Dopamine: Dopamine has been found in many insect ALs, but there is considerable variation in its presence and distribution (Homberg, U. and Müller, U., 1999). In honeybees, four large dopamine-immunoreactive cells innervate the glomeruli and central neuropil of each AL. Two of their somata lie in the dorsolateral cell-body rind of the AL and the other two close to the dorsal rind of the SEG (Schäfer, S. and Rehder, V., 1989). These neurons form a fine plexus both in the central neuropil of the AL and in the glomeruli and also branch in the DL, the protocerebrum, and the SEG. Within the glomeruli, they are restricted to the core (Schürmann, F. W. *et al.*, 1989). While there are dopaminergic neurons in moth brains, their precise morphology is not yet known (Mesce, K. A. *et al.*, 2001). No dopamine was found in fly ALs (Nässel, D. R. and Elekes, K., 1992; Friggi-Grelin, F. *et al.*, 2003). In cockroaches, there are dopaminergic centrifugal neurons, and in addition, some LNs are dopamine-positive (Distler, P. G., 1990). In locust ALs, no dopamine immunoreactivity was found (Wendt, B. and Homberg, U., 1992).

Octopamine: All honeybee glomeruli contain octopamine-immunoreactive fibers, deriving from very few profiles entering the AL from ventroposterior. Within the outer cap of the glomeruli, these fibers appear to be of considerably larger diameter than the very thin fibers in the glomerular centers (Kreissl, S. *et al.*, 1994). A neuron that most likely uses octopamine in honeybees is the ventral unpaired median cell of maxillary neuromere 1 (VUMmx1), which has an unpaired soma in the subesophageal ganglion. Apart from the AL, this neuron has branches in the subesophageal ganglion itself and in the protocerebrum (MB and LP). The input–output relationship in the innervated areas is unknown (Kreissl, S. *et al.*, 1994; Bicker, G., 1999b). Functionally, it represents (part of) the unconditioned stimulus (US) in an appetitive learning paradigm (Hammer, M., 1993). When a honeybee is presented an odor (conditioned stimulus, CS) and rewarded with sugar water (US), she will extend the proboscis to lick the sugar water (proboscis extension reflex (PER)). After the learning trial, odor alone elicits a PER (classical or Pavlovian conditioning). Replacing the US with a depolarization of VUMmx1 also leads to an association, and preventing VUMmx1 to spike by hyperpolarizing it during a sugar water stimulus impairs learning (Hammer, M., 1993). In honeybees, octopamine is also related to division of labor within the hive (Schulz, D. J. *et al.*, 2002).

Octopamine is also present in fine arborizations within the AL in moths (Homberg, U. and Müller, U., 1999). These arborizations probably derive from at least three VUM neurons that also innervate the MBs, somewhat reminiscent of VUMmx1 in honeybees (Dacks, A. M. *et al.*, 2005). However, the detailed anatomy of these neurons is quite different between bees and moths, leaving their homology in doubt (Dacks, A. M. *et al.*, 2005). In the moth's periphery, octopamine increases responses to pheromones, but not to general odors (Pophof, B., 2002). In flies, a cluster of neurons in the SEG (SM cells) has a similar morphology to the honeybee VUM neurons, also innervating the ALs (Monastirioti, M., 1999). Octopaminergic neurons in the cockroach AL branch similar as in honeybee (Sinakevitch, I. *et al.*, 2005). In locusts, octopamine is present in fine arborizations within the AL (Homberg, U. and Müller, U., 1999), probably from dorsal unpaired neurons (DUM) in the SEG (Bräunig, P., 1991; Stevenson, P. A. and Sporhase-Eichmann, U., 1995).

Serotonin: In honeybees, there are two paired neurons immunoreactive for serotonin (5-hydroxytryptamine (5HT)), one in each side of the body, called the deutocerebral giant (DCG) neurons, which innervate the ipsilateral AL, the DL, and the SEG and descend into the ventral nerve cord. Their somata lie in the dorsolateral cell body cluster of the AL (Rehder, V. *et al.*, 1987). Within the glomeruli, serotonergic fibers are restricted to the outer cap (Schürmann, F. W. and Klemm, N., 1984). Two serotonergic neurons in the moth *M. sexta* have been well characterized (Kent, K. S. *et al.*, 1987; Sun, X.-J. *et al.*, 1993b). These neurons have their soma close to the AL and branch in the core of all glomeruli within the contralateral AL, without forming overlap with OSN afferent fibers that innervate the outer cap. They receive input in the superior protocerebrum, the MB, and the lateral accessory lobes. This anatomy suggests a role as feedback neurons, or as centrifugal modulators of AL responses. In flies, there are two serotonergic neurons associated with each AL that branch in olfactory glomeruli (Nässel, D. R., 1988). A pair of serotonergic neurons is also found in cockroaches, where each neuron has a soma in the AL, extending a neurite to the protocerebrum including the MB and then back to the AL where it branches and makes output synapses in most if not all glomeruli (Salecker, I. and Distler, P., 1990). In locusts, there are three serotonin-positive cells in each AL, innervating all glomeruli. These neurons are multiglomerular PNs that project toward the MB, without entering them. One has an axon in the iACT, and the other in an intermediate ACT (dmACT) (Ignell, R. *et al.*, 2001).

Histamine: In the honeybee AL, there are ~35 LNs that are immunoreactive for histamine (Bornhauser, B. C. and Meyer, E. P., 1997). Histamine acts as an inhibitory transmitter in this species (Sachse, S. and Galizia, C. G., 2002). In moths, histamine is not present in LNs, but it is present in two centrifugal neurons that innervate both ALs each and that also arborize in the AMMC and the SEG (Homberg, U. and Hildebrand, J. G., 1991). There are no histaminergic neurites in *Drosophila* ALs (Nässel, D. R., 1999). Histamine is found in cockroach LNs, with a few somata around the ALs, and possibly a tract connecting to the AL ventrally (Pirvola, U. *et al.*, 1988). In the locust, there is one histamine-positive neuron in each AL that projects to the LP (Ignell, R. *et al.*, 2001).

4.41.2.7 Glial Cells

Glial cells have been analyzed in several insect species, including bees (Hähnlein, I. and Bicker, G., 1996), *Drosophila* (Jhaveri, D. *et al.*, 2000; Jhaveri, D. and Rodrigues, V., 2002; Jhaveri, D. *et al.*, 2004), cockroaches (Prillinger, L., 1981), and locust ALs (Hähnlein, I. *et al.*, 1996). Glial cells are certainly best known in moths (Kretzschmar, D. and Pflugfelder, G. O., 2002; Oland, L. A. and Tolbert, L. P., 2003; Tolbert, L. P. *et al.*, 2004). The cells are prominent on the outside of the AL and form thin processes that digitate between glomeruli well into the AL, thus separating them from each other. Glial cells are invisible in light microscopy and difficult to discern in electron microscopy. In species where glomeruli are arranged around a central, nonglomerular neuropil (coarse neuropil, e.g., honeybees, moths) glial cells do not form a border between glomeruli and the central neuropil. It should be noted that there is no universal glial cell marker in insects and that glial fibrillary acidic protein (GFAP) does not label insect glial cells.

Several different types of glial cells are associated with the AL. Along the antennal nerve, glial cells with long processes and multiple expansions enwrap axon fascicles. OSN axons travel as parallel bundles within the antennal nerve, but as they approach the AL, they reach a so-called sorting zone, where they form a dense and complex network, which results in a rearrangement of neighborhood relationships that ultimately leads to appropriate glomerular targeting. Glial cells are necessary for this rearrangement to occur (Oland, L. A. *et al.*, 1988; Rössler, W. *et al.*, 1999b). Within the AL, there are at least two glial-cell types that form borders around glomeruli: one has large cell bodies and branching, vellate arbors and the other has multiple, mostly unbranched processes with many lamellate expansions along their length that surround glomeruli as part of a multilamellar envelope (Oland, L. A. *et al.*, 1999). Glia in the neuropil is instrumental for protoglomerulus formation and proper development of the AL (Baumann, P. M. *et al.*, 1996). These mechanisms are mediated by

direct contact between the OSN axon growth cones and the glial cells themselves, as shown *in vitro* (Tucker, E. S. and Tolbert, L. P., 2003; Tucker, E. S. *et al.*, 2004).

In the adult, glial cells might help keeping neural activity separate between glomeruli, and thus increasing the discriminatory power of the system. Goriely and colleagues simulated the movement of K^+ ions within the AL in a mathematical diffusion model based on electron microscopic data and found that the glial envelope could efficiently reduce ion movement between glomeruli. Since K^+ accumulation in extracellular space causes cell depolarization, this barrier might prevent cross talk between adjacent glomeruli (Goriely, A. R. *et al.*, 2002). Glia might also form a sink for the gaseous messenger NO, given that NO is involved in glial cell physiology (Gibson, N. J. *et al.*, 2001).

4.41.2.8 Cellular Organization and Number Relationships

What is the numerical relationship of the different neuron types in the AL, and how many neurons of each type innervate each glomerulus? The numbers differ widely among insects. For example, a single honeybee glomerulus is innervated on average by almost 400 OSNs (60 000 OSNs onto 160 glomeruli), 1000 LNs (as a gross estimate, assuming a total of 4000 LNs, with each innervating an average of 40 glomeruli), and 5 PNs (800 PNs in 160 glomeruli). The fruit fly *D. melanogaster* has fewer OSNs, but since most OSNs project into the ALs of both sides, the relationship of OSNs and glomerulus is 50:1. Each glomerulus is innervated by three PNs, on average (Wong, A. M. *et al.*, 2002). These average considerations do not consider that OSNs of different classes vary greatly in number, and therefore, the convergence/divergence ratio might span a wide range in different glomeruli not only across species but also within a species.

4.41.3 Functional Organization in the Antennal Lobe

Is it possible to attribute functional tasks to each neuron type in the AL? In a greatly simplified picture, the answer is yes, as developed elsewhere (Sachse, S. and Galizia, C. G., 2005). An interpretation of the main cell types forming the AL circuitry could then be as follows (see below for more details).

1. Homogeneous LNs uniformly innervate the AL and accomplish a gain control function. They set the background activity close to the threshold, thus optimizing the system's sensitivity. They also protect the system from overloading and are ideal candidates for driving oscillations.

2. Heterogeneous or asymmetric LNs enhance the contrast of across-glomeruli patterns by unidirectional connections between input glomeruli and output glomeruli.

3. Excitatory LNs may distribute information accross glomarular boundaries or amplify responses within glomeruli (see above).

4. Multiglomerular PNs have access to the entire AL. They may respond to global activity and thus give information about stimulus timing (onset/offset) and concentration and would be ideal candidates for cellular learning in the AL.

5. Uniglomerular PNs extract the activity of single glomeruli and project to higher-order brain centers. The identity of an odor is encoded in the combinatorial activity pattern across their axons' identities.

4.41.3.1 Odor-Evoked Activity Across Glomeruli

When an animal is challenged with an odor, many different receptor neurons are activated. Their axons send that activity to the AL in the form of AP trains. Due to the convergence of functionally equivalent axons onto individual glomeruli, the result is a pattern of glomeruli that are activated to a varying degree. Therefore, activity patterns across glomeruli are not binary (on/off), but continuous (each glomerulus can be activated to varying degrees). The resulting combinatorial odor patterns are sufficient, in terms of the information contained in the patterns, to identify an odor, and therefore could represent the olfactory code. However, it is not known what information is really used by the brain, and therefore, the biological olfactory code remains to be deciphered.

Odor-evoked combinatorial patterns can be measured with imaging techniques, because this allows for measuring several combinatorial odor responses across glomeruli within individual animals. Unfortunately, the temporal resolution is not yet sufficient to resolve individual APs, and therefore, imaging studies are limited to measuring time-averages of neural activity.

Responses have been measured in *D. melanogaster* by using genetically encoded reporter proteins.

synaptopHluorin was used to measure synaptic vesicle release (Ng, M. *et al.*, 2002; Yu, D. *et al.*, 2004) and cameleon or G-CaMP for intracellular calcium (Fiala, A. *et al.*, 2002; Wang, J. W. *et al.*, 2003; Suh, G. S. *et al.*, 2004). These proteins were expressed under the control of specific promoters, which allow for a good reproducibility of the measured cells. For example, using the GAL4 line GH146, a reporter protein can be expressed in a population of brain cells that within the AL only consists of PNs. Similarly, expressing the reporter in GABAergic cells allows to exclusively measure inhibitory cells (mostly LNs, and a subpopulation of PNs). However, the relative contribution of different cell groups labeled by the same genetic line cannot be separated if they overlap spatially. Currently, the clearest separation can be expected from OSN populations, where reporter proteins can be driven to be expressed in all cells that also express ORs, or just a particular OR.

In species that do not have the molecular power of *Drosophila*, synthetic reporters can be used to stain the cells. Bath application of calcium-sensitive dyes has revealed odor responses in bees (Joerges, J. *et al.*, 1997; Galizia, C. G. and Menzel, R., 2001), ants (Galizia, C. G. *et al.*, 1999a), and moths (Galizia, C. G. *et al.*, 2000a; Carlsson, M. A. *et al.*, 2002; Hansson, B. S. *et al.*, 2003). Because the dye is applied in the bath, the exact proportion of measured cells cannot be determined and might include all neuron types as well as glial cells, making a cellular interpretation of the results difficult. In bees, however, the response magnitude (though not the time course) is a good estimate of receptor neuron response for each glomerulus (Galizia, C. G. and Vetter, R. S., 2004). In moths, measurements of released NO closely mirror the activity patterns measured with bath-applied calcium dyes (Collmann, C. *et al.*, 2004). Since in moths NO is produced only by OSNs (Gibson, N. J. and Nighorn, A., 2000), this suggests that also in this species the (bath-applied) calcium indicator response magnitude can be used to estimate OSN activity for each glomerulus. Time courses of these calcium responses do not correspond to electrophysiological measurements of OSNs: they rise continuously during the stimulus and decay slowly following the shape of a mathematical α-function (Stetter, M. *et al.*, 2001). Odor responses can also be measured with voltage-sensitive dyes (VSDs) (Galizia, C. G. *et al.*, 1997; 2000b). PNs can be measured without a contribution from other cell types if they are backfilled from the protocerebrum (Sachse, S. and Galizia, C. G., 2002; 2003). These selective stainings lead to calcium responses that are fast, with a steep rise at stimulus onset and a steep decay at stimulus offset. As in mammals (Xu, F. *et al.*, 2000; Leon, M. and Johnson, B. A., 2003), the spatial odor-response patterns have also been recorded using 2-DG labeling in fruit flies (Buchner, E. and Rodrigues, V., 1983; Rodrigues, V. and Buchner, E., 1984) and in *Calliphora* (Distler, P. G. *et al.*, 1998a). This brings the advantage of having a permanent preparation, but the disadvantage of having but one measurement per animal, and no dynamic information.

An important strength of insect preparations is that glomeruli can be identified among animals, given their small number and their characteristic arrangements. Using VSD in honeybees allows recognizing the glomerular layout very clearly in the staining pattern (Galizia, C. G. and Vetter, R. S., 2004). This shows that activity patterns are indeed glomerular, that is, that the boundaries of highly active regions coincide with the boundaries of individual glomeruli. With identifiable glomeruli, comparison and characterization is straightforward. Even in species where a morphological atlas of glomeruli is not yet available, or when identifying glomeruli is not possible, odor-evoked patterns are so stereotype among individuals that a direct quantification can often be achieved (Meijerink, J. *et al.*, 2003; Skiri, H. T. *et al.*, 2004).

A recent development in insect olfactory recordings is the use of multielectrode arrays, which allow the simultaneous extracellular recording of many neurons but do not allow identifying the neurons involved (Christensen, T. A. *et al.*, 2000).

4.41.3.2 Processing in the Antennal Lobe

The only olfactory information that is available to the brain must be present in OSN activity. However, not all information present in the OSNs is kept in PNs. The cellular network formed by OSNs, LNs, and PNs extracts information and creates a modified odor representation. The activity patterns across PNs are combinatorial and temporally fluctuating.

4.41.3.2.1 Contrast enhancement in the antennal lobe

In the AL network, an input layer (OSN) is transformed by internal connections (LNs) into an output layer (PNs). This resembles a neural network as it is often used in computational neurobiology (Getz, W. M., 1991; Linster, C. *et al.*, 1994; Getz, W. M. and Lutz, A., 1999; Brody, C. D. and Hopfield, J. J., 2003), and indeed, the AL has often been modeled as such

(Getz, W. M., 1991; Linster, C. *et al.*, 1994; Getz W. M. and Lutz, A., 1999; Rabinovich, M. I. *et al.*, 2000; Bazhenov, M. *et al.*, 2001). Since both input and output are organized in the same glomeruli, this network has the additional beauty of having equal dimensionality for input and output. Therefore, the question of what role the hidden layer plays can easily be addressed by comparing the glomerular activity of PNs with the glomerular activity in the OSNs. This approach has been taken in several studies, with surprisingly disparate results. In optical imaging studies in *Drosophila*, the difference between PNs and OSNs was so small that no apparent processing was deducible (Ng, M. *et al.*, 2002; Wang, J. W. *et al.*, 2003). In an electrophysiological study comparing the input and the output of glomerulus DM2, also in *Drosophila*, the response spectrum was apparently broader in PNs than in OSNs (Wilson, R. I. *et al.*, 2004). Comparing the input and the output for odor-concentration dose–response curves in honeybees showed that glomeruli with a low threshold for an odor have almost identical dose–response curves in PNs and in OSNs, while less-sensitive glomeruli have a shifted dose–response curve in PNs, showing that at higher concentrations LNs suppress these PN responses (Sachse, S. and Galizia, C. G., 2003). Thus, within 2 years, publications suggesting no change, response broadening, and response sharpening have been published. One difficulty arises from comparing activity across different neuron populations (OSNs and PNs). Conceptually, the quantitative interpretation of such a comparison is impossible, because the conversion rate of neural activity is unknown. In electrophysiological studies, an AP in an OSN has a different value from an AP in a PN, and therefore, comparing AP numbers *per se* is impossible. For example, one AP in an OSN might be sufficient to strongly depolarize the PNs in the innervated glomerulus (given the high convergence of many OSNs in that glomerulus and the many synapses made by each OSN axon), while even several APs in a PN might not represent a significant signal in the MBs (given the high spontaneous activity level in these neurons and the possible need for APs to synchronize with other APs). The same argument is valid when optical signals are measured: calcium measurements record changes due to several intracellular cascades of dendritic and presynaptic calcium, and the measurements of synaptic exocytosis are similarly affected. Furthermore, transfer functions among neurons are not linear, again clouding a quantitative comparison of different neuron populations.

Therefore, the question of what the population of LNs is computing within the AL has to be addressed with different techniques. Linster C. and coworkers (2005) created a computer model of the AL and ran experimental data from honeybees under different assumptions about the connectivity of the LN network, comparing three network types: one where LNs interconnect neighboring glomeruli, one with stochastic connections, and one based on the odor-response properties of glomeruli. This approach avoids comparing PN and OSN activity directly but rather compares different network architectures and finds the one that matches best with experimental data. The results show that, in honeybees, inhibitory connections are strongest among glomeruli that have similar odor-response profiles, and weakest among glomeruli that do not overlap in their odor-response profiles, irrespective of their spatial position (Linster, C. et al., 2005). This leads to an optimization of odor responses in PNs across odors, by reducing response overlap to similar odors, and supports the theory that the AL sharpens odor-response profiles.

Such a glomerulus-specific network is likely to be realized by heteroLNs, which densely innervate one glomerulus and diffusely innervate a limited number of other glomeruli. Because these neurons are not symmetrical, a glomerulus A that inhibits a glomerulus B does not need to receive inhibitory input from B. Physiological evidence for such a nonsymmetrical connectivity has been found in bees (Sachse, S. and Galizia, C. G., 2002). Interestingly, in behavioral experiments where odor similarity was studied, such an asymmetry has also been found: similarity from an odor A to B does not imply the same similarity from odor B to A (Guerrieri, F. et al., 2005).

4.41.3.2.2 Sensitivity optimization in the antennal lobe

Some OSNs are active even without a stimulus, but this activity is not temporally complex. On the other hand, many electrophysiological recordings show that PNs are spontaneously active with a pronounced temporal complexity. As a consequence, continuously changing glomerular activity patterns are seen across glomeruli (for example see Relevant Websites section) (Sachse, S. and Galizia, C. G., 2002; Galán, R. F. et al., 2005). If the response strength were normalized, an experimenter would not be able to distinguish individual events of spontaneous activity from odor-evoked responses. The spontaneous activity might result from a mechanism that increases the sensitivity to weak odors (Sachse, S. and Galizia, C. G.,

2005). In this model, PNs can be compared to a loaded spring. An inhibitory feedback loop of LNs keeps PNs at firing threshold, so that minimal olfactory stimuli will suffice to elicit an odor-evoked pattern. In order to maintain PNs at threshold, their depolarization must constantly be probed, which is apparent as spontaneous activity. Such a mechanism may seem to interfere with a reliable representation of odor concentration. However, measuring odor concentration is much more affected by odor adaptation at the periphery, a phenomenon known from all sensory systems. Due to the phasic response properties of OSNs, the olfactory system always measures concentration changes rather than absolute concentrations. In the adapted state, the loaded spring model of the AL therefore ensures that even small increases in odor concentration will lead to them being detected by the animal. The strength of the olfactory stimulus (i.e., the concentration change magnitude) remains reliably coded in the response magnitude change across olfactory glomeruli (Sachse, S. and Galizia, C. G., 2005).

4.41.3.2.3 Temporal activity structures

Odor responses are already temporally structured at the level of OSNs: some OSNs have phasic, some have tonic responses, some have activity that outlasts the odor stimulus, and some reduce activity upon olfactory stimulation (de Bruyne, M. et al., 2001). As a consequence, an olfactory stimulus that is temporally uniform leads to a pattern of activity that is temporally complex. In addition, in the natural environment of insects, odors themselves are temporally complex due to air turbulence (Murlis, J. et al., 1992; Justus, K. A. et al., 2005). This complicates the analysis of temporal patterns within the brain, suggesting that slow patterns might predominantly reflect stimulus variation (Vickers, N. J. et al., 2001).

Response properties on a fast temporal scale can only be measured with electrophysiological techniques. Single-cell recordings allow to fill neurons with a dye and then to identify the glomerulus that was innervated (Müller, D. et al., 2002; Galizia, C. G. and Kimmerle, B., 2004). Slow temporal structures consisting of sequences of bursts and inhibitory events are found in PNs in all insects studied so far, including moths (Christensen, T. A. et al., 1998b), locusts (Laurent, G., 1996), and honeybees (Abel, R. et al., 2001; Müller, U. and Hildebrandt, H., 2002). In locusts and honeybees, blocking chloride channels with Picrotoxin does not prevent these slow temporal patterns (MacLeod, K. and Laurent, G., 1996;

Stopfer, M. *et al.*, 1997) but modifies them (Sachse, S. and Galizia, C. G., 2002). This suggests that they may partially originate in the OSN input or from a Picrotoxin-insensitive inhibitory network.

Behavioral studies have shown that olfactory decisions are taken rapidly in rats (Uchida, N. and Mainen, Z. F., 2003; Abraham, N. M. *et al.*, 2004) and in bees (Ditzen, M. *et al.*, 2003). This time was ~200 ms for mice and ~690 ms for honeybees, which included the time needed for olfactory recognition as well as motor responses and physical displacement of the animal. The time needed for olfactory discrimination is not affected by the difficulty of the olfactory task in terms of odor similarity or concentration (Ditzen, M. *et al.*, 2003; Uchida, N. and Mainen, Z. F., 2003), or the difference in time, when present, is small (Abraham, N. M. *et al.*, 2004). Physiological studies of PN responses show that 200–300 ms in locusts (Stopfer, M. *et al.*, 2003) and 400 ms in bees (Galán, R. F. *et al.*, 2004) are needed to reach the most distinct odor classification in the AL irrespective of odor concentration. Therefore, the later phases of slow activity components cannot be behaviorally relevant for odor discrimination. However, these late activities might play a role in olfactory learning or other aspects of olfactory processing that might occur after discrimination has taken place. An argument in favor of this interpretation is that odor representation is ameliorated during the first 2 s after stimulus onset, leading to a clearer distinction of odors (Galizia, C. G. *et al.*, 2000b; Friedrich, R. W. and Laurent, G., 2001).

Fast temporal structures are evident in odor-evoked oscillations, which are found almost ubiquitously in olfactory systems. So far, within insects, only *Drosophila* seems not to respond with oscillation (Wilson, R. I. *et al.*, 2004), while cockroaches (27 Hz), locusts (22 Hz), bees (30 Hz), wasps (17 Hz) all do (Stopfer, M. *et al.*, 1999), and moths also (Heinbockel, T. *et al.*, 1998). Applying Picrotoxin abolishes these oscillations (MacLeod, K. and Laurent, G., 1996; Stopfer, M. *et al.*, 1997) and also modifies the spatial activity patterns across glomeruli (Sachse, S. and Galizia, C. G., 2002). Oscillations originate from many PNs firing APs simultaneously, due to a rhythmical inhibition from LNs that branch over large areas within the AL and the brain. Because individual PNs do not fire in every oscillation cycle, and because some APs also occur out of the rhythmic synchrony pattern, a model of olfactory coding has been proposed where odor identity would be encoded in sequences of changing PN ensembles (Laurent, G., 1999). However,

other interpretations for synchronized APs have also been proposed, where synchrony would be more related to odor concentration or intermittency rather than to odor quality (Christensen, T. A. *et al.*, 2000).

4.41.3.3 Combinatorial Odor Codes

Odors evoke combinatorial patterns of activated glomeruli, with each glomerulus participating in activity to many odors. These patterns are conserved among individuals (Galizia, C. G. *et al.*, 1999c; Wang, J. W. *et al.*, 2003), which is a consequence of the innate mapping of OSNs that express a given OR to individual glomeruli (Vosshall, L. B. *et al.*, 2000; Couto, A. *et al.*, 2005; Fishilevich, E. and Vosshall, L. B., 2005). As a corollary, patterns are also bilaterally symmetrical (Galizia, C. G. *et al.*, 1998).

MRRs are best described as (some more, some less) broadly tuned to a best molecule (Sachse, S. *et al.*, 1999). Importantly, there are no glomeruli for functional groups or other chemical parameters (e.g., aldehyde or C6 carbon chain). While some glomeruli preferentially respond to one functional group rather than another when tested for aldehydes, ketones, alcohols, carbon acids, or alkanes, they always respond also to other functional groups, even though the response is weaker. Furthermore, they always have a particular preferred range of molecule size to which they respond. Therefore, the response profile of individual glomeruli is not determined by particular features of the odorant (sometimes referred to as odotopes), such as ketone group or aldehyde, and the olfactory code is not a building set, where 1-heptanol would be coded in an alcohol glomerulus plus a C7 aliphatic chain glomerulus.

4.41.3.4 Coding Odor Concentration and Odor Mixtures

With increasing concentration, activity increases both in magnitude and in number of active glomeruli. As a corollary, with decreasing concentration, activity decreases and consists of a single glomerulus being active at very low concentration. This is a direct consequence of the receptors' MRRs that form the input: each receptor has a few substances to which it responds with a higher affinity than any other receptor, and at its lowest effective concentration, that substance will elicit a combinatorial pattern of activity in the AL that consists of that single glomerulus being active, and all other glomeruli being silent. At higher concentrations, many glomeruli are active, and every glomerulus takes part in activity patterns of several odors. Again, this is a direct consequence of the broadening MRR of receptors with increasing concentrations. PN responses in honeybees are qualitatively stable over a concentrations range of up to 4 log units, because higher odor concentrations increase total response intensity without changing the relative intensity across glomeruli; at the input level, however, the activity patterns are more affected by concentration differences (Sachse, S. and Galizia, C. G., 2003). Thus, the neural network within the AL contributes to concentration invariance.

Mixture coding is complex and far from understood in insects. Most naturally occurring odors consist of a large number of volatile compounds. Additionally, the turbulent nature of air leads to odor plumes being mixed in a chaotic fashion (Murlis, J. *et al.*, 1992). This creates a difficult task for the olfactory system: should it create a synthetic representation of the blend, or rather an analytical separation of the components, or both? Several studies have asked how many odors animals can differentiate in a mixture. However, results cannot be generalized: perfumers know that some odors do blend, while others do not. Therefore, the number of discriminable odors depends on the quality of the odors mixed. Electrophysiological studies have also shown different kinds of mixture interactions. The presence of an odor B can interfere with the normally strong response to an odor A, which is termed mixture suppression. Conversely, a neuron or glomerulus may respond to a binary mixture with a response that exceeds the summed responses to the single components, defined as synergism. An inhibitory network within the AL optimized for sharpening odor-response patterns should have the effect of creating stronger mixture interactions with similar odors than with odors that elicit nonoverlapping response patterns. Psychophysically, this implies that similar odors in a mixture compose a new odor, which should make it difficult for the olfactory system to extract the identity of the odor components (synthetic representation), while mixtures of dissimilar substances are represented as the sum of the optimized representation of each component (analytical representation) (Wilson, D. A. and Stevenson, R. J., 2003). Behavioral data in rats support this idea by showing that rats perceive binary mixtures composed of dissimilar odors as very similar to their components, whereas binary mixtures containing similar odors appeared dissimilar to the animal (Wiltrout, C. *et al.*, 2003). By increasing the number of components in an odor mixture, mixture interactions are likely to increase and further reduce the similarity to the single-component patterns (Deisig, N. *et al.*, 2003). With this coding strategy of odor mixtures, the olfactory system implements a logic that allows a unique representation of odor mixtures without saturating the olfactory code, at the expense of a loss in analytical capacity.

4.41.3.5 Coding of Special Odors

There are several chemosensory tasks that are special for olfactory systems. Among them are pheromones (which can be further subdivided into sexual and nonsexual pheromones), CO_2 concentration, humidity, and contact (nontaste) odors.

4.41.3.5.1 Sexual pheromones

The sexual pheromone system is the classic example for a labeled line system, with each OSN type, and consequently the corresponding glomerulus, representing a component of the pheromone blend. Glomeruli specific for sexual pheromones do not have a broad response spectrum in the natural environment of the species, because none of the other ligands that exist (and are known) occur in nature. For example, in *M. sexta*, the sexual pheromone is a blend of two chemicals, the main component (E,Z)-10,12-hexadecadienal and the secondary component (E,E,Z)-10,12,14-hexadecatrienal. A different and more stable molecule, (E,Z)-11,13-pentadecadienal, is a good mimic and is routinely used instead in physiological experiments (Christensen, T. A. and Hildebrand, J. G., 1997). However, because it does not occur in nature, this glomerular activity channel is uniquely activated by its natural ligand within a natural environment (Kaissling, K. E. *et al.*, 1989). Female moths do not have a macroglomerular complex, since they are not receptive to sexual pheromones. They do, however, have a group of large, female-specific glomeruli that respond to plant-derived odors and might also be involved in odor-driven navigation (Reisenman, C. E. *et al.*, 2004).

Sexual pheromones in *Drosophila* are not involved in long-distance navigation, but rather part of a close-encounter olfactory display, that is detected by contact chemoreceptors. The only candidate for a volatile pheromone is *cis*-vaccenyl acetate (Costa, R., 1989; Amrein, H., 2004). Its precise behavioral significance remains to be elucidated, and includes attractive, repellent, and sex specific effects. There are several OSNs sensitive to *cis*-vaccenyl acetate. Cells expressing dOr67d reside in T1 sensilla on the antenna and mediate sex recognition (Xu, P. *et al.*, 2005; Kurtovic, A., *et al.*, 2007), while cells expressing dOr65a reside in T3 sensilla and mediate generalization of courtship learning (Ejima, A., *et al.*, 2007).

The sexual pheromone of bees consists of several substances, with 9-keto-2(*E*)-decenoic acid (9-ODA) as main component, often referred to as queen's substance. Drones find queens using both olfactory and visual cues, that is, they follow her scent trail, and they identify a dark spot against the bright sky. Mating occurs while flying. The male AL is entirely different from the female, with a limited number of ordinary glomeruli, and four large macroglomeruli (Arnold, G. *et al.*, 1985). Macroglomeruli have complex substructures. In optical imaging experiments, small areas within macroglomeruli respond to many odors, suggesting a more complex arrangement than just separation of sexual pheromone processing from other odors (Sandoz, J. C. and Galizia, C. G., unpublished data). Queen's substance is also a social pheromone within the hive and prevents the emergence of new queens when it is present (Moritz, R. F. A. and Southwick, E. E., 1992).

4.41.3.5.2 Nonsexual pheromones

Nonsexual pheromones differ chemically from sexual pheromones in that they are often less complex molecules that may also be encountered as environmental odors. They sometimes also act as kairomones, as for the case of alarm pheromones that can elicit alarm in sympatric species (Hölldobler, B. and Wilson, E. O., 1990). Ants have a well-developed intraspecific communication system based on pheromones, which they use to communicate food sources, trails, alarm, and more (Hölldobler, B. and Wilson, E. O., 1990). In the ant *Camponotus rufipes*, activity elicited by environmental odors and by pheromonal substances overlap in the AL, suggesting that the two systems share their olfactory circuitry (Galizia, C. G. *et al.*, 1999a).

Important nonsexual pheromones for bees are citral and geraniol, both components of the Nasanov gland, a gland important for the social cohesion of the hive (Free, J. B., 1987): isoamyl acetate, which is the major component of the alarm pheromone associated with the sting apparatus (Free, J. B., 1987; Nuñez, J. *et al.*, 1997), and the repellent scent-marker 2-heptanone, which bees use to mark already visited flowers, thus avoiding revisiting a depleted nectar source (Giurfa, M. and Nunez, J. A., 1992; Giurfa, M., 1993). All of these substances elicit activity in the same region as environmental odors

(Joerges, J. *et al.*, 1997; Galizia, C. G. and Menzel, R., 2001). Isoamyl acetate elicits strong responses in several glomeruli, among which T1-48, a glomerulus that is also activated by orange, clove oil, limonene, and several plant extracts. It is unknown whether there is a parallel, more 'labeled line'-like pathway for nonsexual pheromones, reminiscent of the macroglomerular system for sexual pheromones. If there is, this system is certainly not in the T1 region, whose response profiles have already been well mapped but could be located in the T2, T3, or the T4 glomeruli.

Aggregation pheromones are also found, e.g., in male Colorado beetles (Dickens, J. C. *et al.*, 2002). Some aggregation pheromones are effectively sexual pheromones. For example, the cerambycid beetle *Neoclytus acuminatur* releases a species-specific aggregation pheromone that attracts both sexes (Lacey, E. S. *et al.*, 2004). As with sexual pheromones, aggregation pheromones are quite different for closely related beetles and may be involved in species segregation (Symonds, M. R. and Elgar, M. A., 2004). This is different in fruit flies, where aggregation pheromones are also found but do not form boundaries among species (Symonds, M. R. and Wertheim, B., 2005).

4.41.3.5.3 Carbon dioxide

Carbon dioxide detection is widespread among arthropods, with a wide variability in structure and function (Bogner, F. *et al.*, 1986; Kleineidam, C. and Tautz, J., 1996; Stange, G. and Stowe, S., 1999). CO_2-sensitive sensilla can be hairs, pegs, plugged, or open grooves and can be on the surface or located within a depression or a pit with a restricted opening (Keil, T. A., 1996; Stange, G. and Stowe, S., 1999). Similarly, the functional context for CO_2 detection differs. In hematophagous insects, including Ixodidae and Diptera, CO_2 sensitivity might be related to finding the host (Grant, A. J. *et al.*, 1995; Dekker, T. *et al.*, 2002). Insects that live in confined spaces, such as centipedes or beetle larvae, are capable of sensing CO_2. Similarly, social insects that live in hives (ants, bees, and termites) monitor CO_2 with tonic receptors and control its concentration in a manner reminiscent of homeostatic control (Lacher, V., 1964; Stange, G. and Stowe, S., 1999; Weidenmüller, A. *et al.*, 2002). For moths, in particular nocturnal moths feeding on nectar, CO_2 might be a component of the attractive flower odor, given that flowers that produce high-quality nectar also release considerable amounts of metabolic CO_2 (Guerenstein, P. G. *et al.*, 2004; Raguso, R. A., 2004; Thom, C. *et al.*, 2004).

Drosophila is repelled by CO_2, and a role of this gas as a component of a stress signal has been suggested (Suh, G. S. *et al.*, 2004). However, CO_2 is also produced by rotting fruit, and by fly aggregations on such fruit that might indicate good ovipositioning sites. In the AL, CO_2 activates the V glomerulus. It is innervated by OSNs located on the maxillary palps. These OSNs express two receptors: dGR21a (de Bruyne, M. *et al.*, 2001; Suh, G. S. *et al.*, 2004) and dGr6a (Jones, W. D., *et al.*, 2007) and do not coexpress the otherwise common coreceptor dOR83b (Larsson, M. C. *et al.*, 2004). The innate avoidance behavior to CO_2 is abolished when these OSNs are silenced, suggesting that there is no parallel, additional CO_2 channel (Suh, G. S. *et al.*, 2004). A homologous glomerulus is present in mosquitoes, where it is innervated by OSNs on the mouthparts (Distler, P. G. and Boeckh, J., 1997b). Similarly, CO_2-sensitive OSNs are located in a pit organ on the labial palps in moths, and their axons form glomeruli in the dorsal area of the AL (Kent, K. S. *et al.*, 1986). In locusts and cockroaches, these fibers project close to the AL but do not form a glomerulus within the AL (Ernst, K. D. *et al.*, 1977; Ignell, R. *et al.*, 2001).

4.41.3.5.4 Humidity

Hygroreceptors are often found on insect antennae (Tominaga, Y. and Yokohari, F., 1982; Yokohari, F., *et al.*, 1982; Yokohari, F., 1983; Nishikawa, M. *et al.*, 1995; Shanbhag, S. R. *et al.*, 1995; Nishino, H. *et al.*, 2003; Tichy, H. 2007). Bees are known to control the humidity in their hives, and to carry water into the hive when they need it (von Frisch, K., 1966). The homeostatic nature of this water detection task makes it likely that humidity be coded in a nonadapting labeled line system, possibly with a single glomerulus. The function of this glomerulus would then be limited to integrating the signals from all hygroreceptors on the antenna, rather than taking part in across-glomeruli processing. However, experimental data do not support this view; bees can associate vapor with a sucrose reward, and reliably respond to it, just as they learn appetitive odors (Menzel, R., personal communication). In intracellular recordings from mlACT neurons, some PNs showed responses to water vapor, which was increased by association with sugar water (Abel, R., 1997). Therefore, the information channel for water must overlap with the circuits responsible for associative learning. No localization of the representation of humidity in the AL is possible from the data currently available, apart from some intracellular recordings of honeybee PNs that across the odors tested

responded exclusively to vapor and branched in a glomerulus of the T2 tract (Müller, D., personal communication).

4.41.3.5.5 Other special odors

In ants, a sensillum that does not respond to nest mates but does respond to cuticular components of nonnest mates has recently been found (Ozaki, M. *et al.*, 2005). This sensillum contains about 200 OSNs. The projections of these OSNs are not known. Nestmate recognition is also found in honeybees, where guarding bees at the hive entrance attack approaching bees that do not have the nest's own cuticular odor (von Frisch, K., 1966). In Hymenoptera, this recognition relies on the mixture of hydro- carbons present on the animal's cuticle (Dani, F. R. *et al.*, 2005), which is often exploited by parasite species (Lenoir, A. *et al.*, 2001).

4.41.3.6 Effects of Memory

The AL is the first olfactory neuropil where substantial cellular interaction takes place. Some of that interaction is experience dependent. What, then, are the memory traces that can be found in the AL? Both nonassociative and associative effects have been found.

Nonassociative learning occurs as a consequence of odor exposure. When a locust is repeatedly exposed to an odor, odor responses in PNs change in a characteristic way: they have less APs, but these APs are more precisely timed, suggesting that odor coding is more efficient for the later puffs in the series (Stopfer, M. and Laurent, G., 1999). This sensory memory trace is already visible in the second and third puff and decays within 10–15 min after the last puff (Stopfer, M. and Laurent, G., 1999). A related effect was found in honeybees: even without an olfactory stimulus, PNs are spontaneously active, but there is no preferential pattern of activity. After a single odor exposure, however, a pattern corresponding to the experienced odor reoccurs repeatedly within the sequence of spontaneous events, showing that spontaneous activity within the AL network is temporarily biased toward this pattern (Galán, R. F. *et al.*, 2005).

Most studies investigating associative learning in the AL have looked at classical conditioning, in which an odor is associated with a punishment (e.g., electroshock and aversive learning) or a reward (e.g., sugar water and appetitive learning). In flies, after aversive conditioning, some glomeruli change their odor responses for a short time window, that is, 3 min after conditioning (Yu, D. *et al.*, 2004). These experiments were done by optically measuring synaptic vesicle release within the AL using the reporter protein synaptopHluorin in the PN line GH146. Thus, the measurements monitored the feedback activity in PNs within the AL, in a population of mixed multiglomerular and uniglomerular PNs comprising approximately 60% of the total PN population.

Application of sucrose (or water) to the honeybee antenna elicits a transient increase of PKA activity in the AL, but odor stimulation alone does not (Hildebrandt, H. and Müller, U., 1995a; 1995b). This effect is mediated by octopamine and reverts to baseline within 3 s (Hildebrandt, H. and Müller, U., 1995b). The sugar water stimulus of the appetitive learning paradigm (the US) is also mediated by octopamine: it can be replaced by octopamine injections restricted to the AL (octopamine is the putative transmitter of VUMmx1, the neuron that mediates the US in the bee brain) (Hammer, M. and Menzel, R., 1998). Blocking octopaminergic transmission by injecting an octopamine receptor antagonist (mianserin) or by injecting double-stranded octopamine receptor RNA into the AL interferes both with odor memory acquisition and with its recall, suggesting that octopamine is important for memory consolidation but also that octopaminergic neurons become reactivated during memory recall (Farooqui, T. *et al.*, 2003). In honeybees, single-trial conditioning leads to short-term memory (STM), while multiple conditioning leads to LTM. In the AL, multiple conditioning trials lead to an elevated PKA response mediated by the NO/cGMP system. Artificially increasing PKA activity in the AL after single-trial learning elicits long-term memory (Müller, U., 2000), suggesting that the PKA level, rather than its presence alone, is important for a switch between STM and long-term memory. Bath application of a calcium-sensitive dye revealed an effect of differential conditioning (one odor was rewarded, the other not) in the time window of 5–15 min after conditioning, showing that learning modifies the odor representation in the AL (Faber, T. *et al.*, 1999). Similarly, extracellular recordings from moths showed a net recruitment of AL neurons activated by the rewarded odor and a net loss of neurons activated by the unrewarded odor (Daly, K. C. *et al.*, 2004).

Other studies have disputed the role of the AL in olfactory associative memory. In *Drosophila*, KCs have been shown to be necessary and sufficient for short-term learning of odors (see below), leaving no space

for a memory trace in the AL (Gerber, B. *et al.*, 2004a). In honeybees, a subpopulation of PNs (the uniglomerular PNs from the lACT tract) have very stable odor responses that are not affected by single-odor training or differential appetitive training (Peele, P., *et al.*, 2006). These neurons might represent a learning-resistant channel for reliable transfer of odor-related information to higher-order brain centers.

Together, these studies show that odor learning in the AL is a complex process that might affect spike timing and/or relative activities to different odors. They also show that these effects only occur in very limited time windows and that not all cell populations are involved in these plastic processes.

Because the development of the olfactory system is not part of this review, developmental plasticity is also not treated, even though developmental experience does shape odor responses and AL morphology (Winnington, A. P. *et al.*, 1996; Sigg, D. *et al.*, 1997).

4.41.4 Cellular Organization of the Mushroom Bodies

Mushroom bodies (MBs) are multimodal structures and are involved in learning (Strausfeld, N. J. and Gilbert, C., 1992; Heisenberg, M., 2003; Davis, R. L., 2004). In most insects, they receive both olfactory and visual input (Strausfeld, N. J. *et al.*, 1998; Farris, S. M., 2005). MBs owe their names to massive peduncles with large cup-shaped protuberances, which are called the calyces.

4.41.4.1 Kenyon Cells

The intrinsic neurons in the MBs are the KCs. Their number differs widely among species: in honeybees, there are ~170 000 KCs in each hemisphere (Witthöft, W., 1967; Mobbs, P. G., 1982), and *Drosophila* counts give ~2500 KCs (Stocker, R. F., 1994). In adult cockroaches, the number is ~175 000 (Neder, R., 1959), but juveniles have much smaller numbers (Farris, S. M. and Strausfeld, N. J., 2001), and adult locusts have ~50 000 KCs (Farivar, S. S., 2005). KC somata are small and densely packed close to the MB calyces. Their axons are long and thin and form the peduncles. Most axons branch, so that the peduncle splits into two lobes: the vertical (α) lobe and the horizontal (β) lobe. In *Drosophila*, these lobes are accompanied by the α'- and the β'-lobes (Strausfeld, N. J. *et al.*, 2003). A third lobe, the γ-lobe, is morphologically distinct in *Drosophila*, but physically juxtaposed to the ventral aspect of the vertical lobe (the α-lobe) in honeybees (Strausfeld, N. J., 2002; Farris, S. M. *et al.*, 2004). The γ-lobe is formed by axons of clawed KCs, which have been found in *Drosophila*, bees, and cockroaches (they are sometimes referred to as class II or type 5 KCs) (Mobbs, P. G., 1982; Rybak, J. and Menzel, R., 1993; Strausfeld, N. J., 2002; Farris, S. M. *et al.*, 2004). Their somata are located outside the calyx, unlike all other KCs that have their somata inside the calyces. The name clawed derives from the claw-like dendritic shapes within the calyces. These KCs are the first population to occur in development. While in a simplified view of the KCs, the calyces are the input region and the lobes the output region of these cells, input synapses are also found in the lobes, allowing for local feedback within the lobes (Strausfeld, N. J. *et al.*, 2003).

The calyces are subdividable: in hymenoptera, these subdivisions are lip, collar, and basal ring, which correspond to different layers in the lobes. In honeybees, KC from lip, collar, and basal ring project into three separate bands in the α-lobe (or vertical lobe). Similarly, in *Drosophila*, subdivisions from the calyces can be traced to a layered structure in the α'/β'-lobe, and to a concentrically circular arrangement in the α/β-lobe (Tanaka, N. K. *et al.*, 2004). These layers partially correspond to KCs that differ in their morphology and/or in what transmitters they use, or in other molecular markers. Glutamate has been found in a KC subpopulation of bees (Bicker, G. *et al.*, 1988). Aspartate, glutamate, and taurine immunocytochemistry leads to different KC populations being labeled in *Drosophila* (Strausfeld, N. J. *et al.*, 2003) and in cockroaches (Sinakevitch, I. *et al.*, 2001). In addition, in *Drosophila*, KCs produce NO (Schürmann, F. W., 2000), but ACh and GABA are excluded as KC transmitters (Yusuyama, K. *et al.*, 2002). The complexity in KC morphology, pharmacology, and peptide expression is considerable in all species studied so far (Iwasaki, M. *et al.*, 1999; Strausfeld, N. J. and Li, Y., 1999a; 1999b; Strausfeld, N. J. *et al.*, 2000; Sinakevitch, I. *et al.*, 2001; Strausfeld, N. J., 2002; Strausfeld, N. J. *et al.*, 2003).

4.41.4.2 Projection Neuron Input

Olfactory input to the MB calyces occurs via uniglomerular PNs. In honeybees, chemosensory input is targeted at the lip and the basal ring of the calyces, while the intermediate area, the collar, receives input from the optic lobes. Each of these input areas is further subdivided into segregated zones, suggesting the existence of several parallel channels (Gronenberg, W., 2001). Similar arrangements are also found in other hymenoptera (Gronenberg, W., 1999; 2001).

In *Drosophila*, each PN axon travels over large areas of the calyx, forming synapses with many intrinsic KCs (Marin, E. C. *et al.*, 2002; Wong, A. M. *et al.*, 2002). This is reminiscent of the situation in the mammalian olfactory cortex (Zou, Z. *et al.*, 2005) and allows for a combinatorial readout of PN response patterns. PNs within the MB calyx region occupy concentric layers (Tanaka, N. K. *et al.*, 2004). PNs from identified glomeruli branch in a stereotype manner within the MB calyces, and PNs from different glomeruli have distinct but overlapping branching patterns (Marin, E. C. *et al.*, 2002; Wong, A. M. *et al.*, 2002). On average, three uniglomerular PNs innervate each glomerulus, and these have the same projection pattern in MB and LP, suggesting that they are not functionally distinct (Wong, A. M. *et al.*, 2002). PNs with similar axon projection patterns in the LP tend to receive input from neighboring glomeruli in *Drosophila*, suggesting that the organization of the LP mirrors some aspects of the organization in the AL (Marin, E. C. *et al.*, 2002; Tanaka, N. K. *et al.*, 2004), but other studies have not found this relationship (Wong, A. M. *et al.*, 2002). However, whether such spatial correlations are dictated by developmental constraints or are the result of functional neuropil optimization, or both, remains to be investigated.

Schematically, MB input can be described as a scaffold, with arrays of PN axons running perpendicular to arrays of KCs, and forming synapses with some, but not all KCs (Heisenberg, M., 2003). This is an ideal arrangement for a combinatorial readout across PNs. In *Drosophila*, PN axons have large boutons, 2–7 μm in diameter, throughout all calycal subdivisions, which is a considerable size when compared to KC somata of about 4 μm (Yusuyama, K. *et al.*, 2002). PN boutons are multisynaptic, with GABAergic input and KC output connections arranged to form glomerular structures (see below). The detailed geometrical layout of this structure has led to comparisons with the mammalian cerebellum, where the GABAergic circuit in MBs would correspond to Golgi cells in the cerebellum (Yusuyama, K. *et al.*, 2002), but it is unclear whether this similarity includes the cerebellar specialization for precise AP timing. Others have likened the vertebrate cerebellum to the central complex, rather than to the mushroom bodies (Svidersky, V. L. and Plotnikova, S. I., 2004).

Although not all PNs may be cholinergic (see above), PNs are the only cholinergic input to the MBs in *Drosophila* (Yasuyama, K. and Salvaterra, P. M., 1999).

4.41.4.3 Local Inhibitory and Modulatory Neurons; Mushroom Body Glomeruli

Dense GABAergic processes invade the MBs (Yusuyama, K. *et al.*, 2002). These synaptic arrangements form microglomeruli with very local computational capabilities. In *Drosophila*, each MB glomerulus comprises a large cholinergic bouton formed by a PN axon from the AL, which is surrounded by tiny vesicle-free KC dendrites and several GABAergic terminals (Yusuyama, K. *et al.*, 2002). GABAergic terminals contact both KC dendrites and PN axon terminals, suggesting that PN input is modulated both pre- and postsynaptically (Yusuyama, K. *et al.*, 2002). This electron microscopic arrangement is also found in bees (Ganeshina, O. and Menzel, R., 2001) and in locusts (Leitch, B. and Laurent, G., 1996). MB glomeruli have no glial sheath (Ganeshina, O. and Menzel, R., 2001; Yusuyama, K. *et al.*, 2002). KC dendritic spines label for F-actin in all insects studied so far, while the PN terminal boutons do not (Frambach, I. *et al.*, 2004). Physiological evidence for the functional role of this circuit is discussed below.

A second inhibitory circuit within the MBs consists of GABAergic feedback neurons from the MB lobes back onto their calyces (Bicker, G. *et al.*, 1985). These neurons are few in number, in honeybees ∼55. Each feedback neuron innervates a subcompartment in the calyx. Due to the arrangement of intrinsic KCs, each subcompartment in the calyx is connected to its specific, corresponding layer in the α-lobe (Grünewald, B., 1999b). These neurons are found in bees (Bicker, G. *et al.*, 1985; Schäfer, S. and Bicker, G., 1986), moths (Homberg, U. and Hildebrand, J. G., 1994), *Drosophila* (Yusuyama, K. *et al.*, 2002), cockroaches (Farris, S. M. and Strausfeld, N. J., 2001), and locusts (Leitch, B. and Laurent, G., 1996).

Octopaminergic cellular processes sparsely but uniformly innervate MB calyces (Strausfeld, N. J. *et al.*, 2003). In honeybees, these are likely to be from the VUMmx1 neuron that represents the CS during olfactory learning (Hammer, M., 1997). VUMmx1 branches in the AL, the MB, and the LP and has been discussed in the Section 4.41.2.6.

4.41.4.4 Output Lobe Circuitry

In comparison with the many studies published about input connections to the MB calyces, about the internal organization between calyces and lobes, and about the feedback from the lobes back onto the calyces, what has been published about the MB output to the LP and other brain areas is little. These target areas probably include premotor areas, so that the loop OSN–AL–MB–LP–motor neuron may form a complete behaviorally significant neural control loop, paralleled by the direct AL–LP loop that bypasses the MBs. In the *Drosophila* brain, at least three areas are MB output regions: the superior medial protocerebrum, the inferior medial protocerebrum, and the superior lateral protocerebrum (Ito, K. *et al.*, 1998; Tanaka, N. K. *et al.*, 2004). However, the precise delimitations of these areas are still being worked out.

Honeybee MB output neurons include unilateral neurons as well as bilaterally innervating neurons (Rybak, J. and Menzel, R., 1993). A prominent large neuron in honeybee brains is PE1, which is a single neuron in each brain hemisphere and connects the α-lobe to the LP and the ring neuropil around the α-lobe (Rybak, J. and Menzel, R., 1998; Brandt, R. *et al.*, 2005). Responses of this multimodal neuron are modified after olfactory learning (Mauelshagen, J., 1993).

4.41.5 Functional Organization of the Mushroom Bodies

4.41.5.1 Odor-Evoked Activity and Processing

There is a massive divergence between PNs and KCs. In honeybees, 800 PNs project onto 180 000 KCs, and this ratio is similar in other species. Patch-clamp recordings in locusts show that KC activity is rare as compared with PNs (Stopfer, M. *et al.*, 2003): while PNs had a probability of $p = 0.64$ to respond to any of the tested panel of odors, KCs responded with $p = 0.11$ (Perez-Orive, J. *et al.*, 2002). Furthermore, while PNs respond with trains of spikes, KCs often respond with single or very few spikes only. These results were confirmed in flies and honeybees using calcium imaging of PNs and KCs (Wang, Y. L. *et al.*, 2004; Szyszka, P. *et al.*, 2005). Thus, odor representation is sparse across KCs (Laurent, G., 2002), where sparse is used both as population sparseness (a low proportion of units active at any time) and as lifetime sparseness (few spikes in each neuron with narrow tuning) (Olshausen, B. A. and Field, D. J., 2004). The transformation between PNs and KCs appears useful because it may facilitate the readout mechanism (Laurent, G., 2002) and may allow more efficient learning due to the perpendicular array structure between PNs and KCs, as noted above (Heisenberg, M., 2003).

A comparison of PN activity in the ALs with PN activity at the synaptic boutons in the MBs and with KC activity showed that this sparsening occurs progressively: activity trains arriving at the MB terminals are inhibited presynaptically within the MBs by GABAergic glomerular microcircuits, so that only the first APs are likely to drive activity in KCs (Szyszka, P. *et al.*, 2005). This mechanism implies that only the first activity wave of every odor response is relevant for decoding within the MBs. In addition, the inhibitory feedback loop from the MB output lobes onto the calyces further sharpens that response (Szyszka, P. *et al.*, 2005). This feedback loop also locks activity into a global oscillatory rhythm (Perez-Orive, J. *et al.*, 2002), which in turn favors the extraction of synchronized APs from PNs (Perez-Orive, J. *et al.*, 2004). In this framework, only spikes that are in phase with the oscillatory activity carry relevant olfactory information, and extracting them would reduce noise and optimize coding (Laurent, G., 2002).

4.41.5.2 Olfactory Coding

These studies result in two distinct models (that need not be entirely exclusive). In one, only the initial firing pattern is read out for each odor puff (Szyszka, P. *et al.*, 2005), in the other only APs in synchrony with oscillations are relevant (Laurent, G., 2002). In both, most APs of PNs are not involved in olfactory coding within the MB calyces. Therefore, the question arises whether the remaining APs are wasted. Given that firing APs is among the most energy-costly activities of the brain (Attwell, D. and Laughlin, S. B., 2001), such a waste would appear as quite inefficient. However, the surplus spikes may not be wasted at all: PNs make synaptic output not only in the MB but also within the AL and in the LP, and this output is not filtered by the MB circuitry. Therefore, APs that are ignored within the MBs could still be an important part for olfactory coding within the AL and the LP.

The olfactory system has been compared with a support vector machine (Galán, R. F. *et al.*, 2004).

Assuming a simplified system that is based on a combinatorial readout without any time-dependent mechanisms, a recipient neuron that reads across PN activities can only perform a very limited classification, which statistically corresponds to a linear classification in a multidimensional space. However, the addition of the MB as an intermediate step changes this picture: with a relatively small number of PNs, combinatorially mapped onto a very large number of KCs, which then are combinatorially read out by very few recipient neurons, very complex pattern families can be extracted. The intermediate step of enormously expanding the number of units by a large number of KCs allows for the computation of highly nonlinear classification schemes across PNs, which is reminiscent of support vector machines (Huerta, R. *et al.*, 2004). By adding temporal complexity to the code, the theoretical capacity of the system increases even further (Laurent, G. *et al.*, 2001).

4.41.5.3 Effects of Memory

Early confirmation of the role that MBs play in olfactory memory came from cooling experiments in honeybees, where memory retrieval was impaired when the MBs were cooled (Erber, J. *et al.*, 1980). Similarly, *Drosophila* mutants where MB structure is altered (*MB deranged, mbd,* and *MB miniature, mbm*) have learning deficits (Heisenberg, M. *et al.*, 1985). When the MBs are chemically ablated by applying the DNA synthesis inhibitor hydroxyurea during the early proliferation phase of KCs, olfactory memory is impaired in *Drosophila* (de Belle, J. S. and Heisenberg, M., 1994). In honeybees, partial ablation of MBs does not impair easy learning tasks (Malun, D. *et al.*, 2002), but more complex tasks involving several odors are impaired (Komischke, B. *et al.*, 2005). Similarly, blocking synaptic activity or disrupting MB physiology also leads to memory deficits (Connolly, J. B. *et al.*, 1996; Dubnau, J. *et al.*, 2001; McGuire, S. E. *et al.*, 2001).

There are many olfactory tasks that do not require MBs, and indeed, fly mutants that lack MBs are remarkably normal: they court and copulate, feed, lay eggs, are alert, seem to be well oriented in space, and respond to odors (Heisenberg, M., 2003). Furthermore, MBs are necessary not only for olfactory tasks but also for spatial memory and navigation (Mizunami, M. *et al.*, 1998; Strausfeld, N. J. *et al.*, 1998; Kwon, H. W. *et al.*, 2004).

In honeybees, the VUMmx1 neuron that represents appetitive reinforcer also innervates the MB calyces, suggesting that coincidence detection for appetitive olfactory learning in honeybees occurs in the MB input region (Menzel, R. and Giurfa, M., 2001), in contrast to findings from *Drosophila* (see below). Optical imaging experiments show that a rewarded odor leads to increased calcium responses in the MB calyces as compared to before learning (Faber, T. and Menzel, R., 2001). Experience leads to morphological changes in honeybee MBs: worker bees that forage have KC dendrites with more branches than age-matched bees that do not forage, while the density of dendritic spines remained constant (Farris, S. M. *et al.*, 2001).

The best mechanistic analysis of olfactory memory traces in MBs comes from *Drosophila* (Waddell, S. and Quinn, W. G., 2001; Dubnau, J. *et al.*, 2003; Heisenberg, M., 2003; Davis, R. L., 2004). Olfactory memory can be subdivided into at least four phases: STM decays within an hour, middle-term memory (MTM) within 3 h, while anesthesia-resistant memory (ARM) and long-term memory (LTM) are two forms of LTM that differ in their training procedures: ARM occurs after massed training, and the protein synthesis-dependent LTM occurs after spaced training, that is, a training protocol where individual learning events occur with longer intervals in between (Tully, T. *et al.*, 1994).

Based on our current knowledge, KCs are the cells where this memory is located (Gerber, B. *et al.*, 2004a). A necessary second messenger in STM is cyclic AMP, and mutants for the genes *dunce* (*dnc*, which is a cAMP phosphodiesterase), DC0 (which is a PKA catalytic subunit), or CREB (cAMP response element binding protein) all are impaired in memory tasks. The *Drosophila* gene *rutabaga* (*rut*) codes for a Ca/CaM-dependent adenylyl cyclase (Levin, L. R. *et al.*, 1992). When *rut* is mutated, olfactory learning is impaired, but when *rut* is expressed in a mutant background in MB cells only, learning is restored (Zars, T. *et al.*, 2000). Several lines were used in these experiments, and the clawed KCs were the set of cells common to all of the effective lines, suggesting that clawed KCs are sufficient for short-term olfactory learning. A selective loss-of-function study of these cells showed that they are also necessary for olfactory learning (Connolly, J. B. *et al.*, 1996). The function of *rut* is only necessary in adult animals, but not during development (McGuire, S. E. *et al.*, 2003; Mao, Z. *et al.*, 2004). Furthermore, it acts presynaptically at the output synapses of KCs, and blocking synaptic release of these cells impairs retrieval, but not acquisition (Dubnau, J. *et al.*, 2001; Schwaerzel, M. *et al.*, 2002).

The gene *amnesiac* (*amn*) is strongly expressed in two large neurons, the two dorsal paired medial (DPM) neurons, and codes for a neuropeptide PACAP (pituitary adenylate cyclase activating peptide) that modulates *rut* activity in KCs. Disruption of *amn* leads to loss of MTM, as does silencing of DPM neurons. For the odors octanol and methylcyclohexanol, DPM neuron activity is necessary during storage and possibly consolidation, but not during acquisition and recall. However, memories for benzaldehyde need DPM neuron activity during acquisition (Keene, A. C. *et al.*, 2004). These data show that olfactory memory is not only highly complex in terms of memory phases and neural and genetic networks, but also diverse with respect to the odors used, adding tremendous complexity to the system.

The MBs are also involved in the two forms of long-term memory, ARM and LTM. These memory stages are not sequential: ARM is formed even in *amn* mutants that show no MTM and at least partially also in *rut* mutants that show no STM (Isabel, G. *et al.*, 2004). STM/MTM and ARM/LTM possibly rely on different KC populations, that is, those from the γ-lobe and from the α/β-lobe, respectively (Zars, T. *et al.*, 2000; Pascual, A. and Preat, T., 2001; Isabel, G. *et al.*, 2004). Nevertheless, LTM and ARM are not independent pathways: rather, LTM leads to an active erasure of ARM memory (Isabel, G. *et al.*, 2004). The cAMP response element CREB appears to be related to LTM (Yin, J. C. and Tully, T., 1996), although the scientific evidence has recently been questioned (Perazzona, B. *et al.*, 2004).

Associative learning occurs when a CS (which in this case is an odor) is paired with a US in a way that the CS acts as an efficient predictor for the US. A positive reward as US, such as food, elicits an appetitive response, while a negative US, such as an electric shock, leads to aversive conditioning. The sensory input for the US and the behavioral responses of these two paradigms are clearly distinct. Therefore, the neuronal pathways should be different. Indeed, in *Drosophila* an appetitive US is mediated by octaminergic neurons, while an aversive US is mediated by dopaminergic neurons (Schwaerzel, M. *et al.*, 2003). This is consistent with findings in honeybees, where appetitive learning is mediated by octopamine (Hammer, M. and Menzel, R., 1998).

4.41.6 Cellular and Functional Organization of the Lateral Protocerebrum

In insect olfaction, the LP is, unfortunately, the least studied area. A frequent belief is that this area is more related to innate behavior control, and less plastic than the circuitry in the MBs (Tanaka, N. K. *et al.*, 2004). Several lines of evidence support this model: in *Drosophila*, individual PNs branch in a highly stereotype and genetically predetermined way in the LP, while their branching pattern in the MBs is less predictable (Marin, E. C. *et al.*, 2002), and animals with ablated MBs but intact LP are not impaired in innate odor responses (de Belle, J. S. and Heisenberg, M., 1994). Because LP areas inherit the plasticity that comes from plastic MB output neurons such as the PE1 neuron (Mauelshagen, J., 1993) and the PCT neurons in honeybees (Grünewald, B., 1999a), local circuits involved in learning are more difficult to identify. Within the LP, there are many subregions that receive olfactory information, some from the AL only, and some from both the AL and the MB (Ito, K. *et al.*, 1998; Tanaka, N. K. *et al.*, 2004). Furthermore, the ring neuropil surrounding the vertical lobe is densely innervated by olfactory neurons.

4.41.7 The Larval System

This review focuses on the olfactory system of adult insects. The larval system has recently started to receive closer attention (Cobb, M., 1999; Heimbeck, G. *et al.*, 1999; Stocker, R. F., 2001; Scherer, S. *et al.*, 2003; Kreher, S. A. *et al.*, 2005; Marin, E. C. *et al.*, 2005). Most importantly, the larval system is much simpler than the adult, in terms of both cell numbers and organization, but shares important principles of the adult (Python, F. and Stocker, R. F., 2002a; 2002b). Larvae are suited for complex behavioral assays, including learning experiments (Scherer, S. *et al.*, 2003; Gerber, B. *et al.*, 2004b; Hendel, T. *et al.*, 2005). OSN afferents to the larval AL also form glomeruli that are interconnected by LNs, even though each glomerulus only receives input from a single OSN axon and is innervated only by a single PN (Kreher, S. A. *et al.*, 2005; Ramaekers, A. *et al.*, 2005). In *Drosophila* larvae, there are about 21 OSNs and about 25 expressed OR genes, of which only 13 are expressed in adults (Kreher, S. A. *et al.*, 2005; Ramaekers, A. *et al.*, 2005). That indicates that the different environment to which the larva is adapted has led to evolving a dedicated chemosensory system. This is further substantiated by the larger number of gustatory neurons in larvae than in adults. The larval mushroom bodies also share many of the characteristics with its adult counterpart (Marin, E. C. *et al.*,

2005). Given its numerical simplicity, it is easy to predict that we will learn a lot more about the larval olfactory system in the near future.

4.41.8 Olfactory Behavior

Theodosius Dobzhansky (1900–1975) stated, "Nothing in biology makes sense except in the light of evolution." For the individual animal, this can be paraphrased to behavior, and the statement becomes even more relevant for the scientific approach to studying olfaction, because it is only through observing behavior that it is possible to judge and investigate how an animal is making use of olfactory cues.

Olfactory behavior in insects has been studied in many species. An extensive review is beyond the scope of this chapter. Therefore, I limit myself to list some types of behavior that impose different constraints and/or requirements onto olfactory processing. Generally, insects are considered as more genetically predetermined and less plastic in their behavior than mammals. However, there are some olfactory behaviors that are exquisitely plastic, and even cognitive capabilities can be found in insects.

4.41.8.1 Odor Learning

Olfactory learning in insects has been reviewed extensively (Menzel, R. and Müller, U., 1996; Dubnau, J. *et al.*, 2003; Giurfa, M., 2003a; Heisenberg, M., 2003; Davis, R. L., 2004; Gerber, B. *et al.*, 2004a; Giurfa, M. and Malun, D., 2004), having been studied in many insects, notably honeybees (von Frisch, K., 1966; Menzel, R. and Müller, U., 1996), bumblebees (Plowright, R. C. and Laverty, T. M., 1984), moths (Fan, R. J. *et al.*, 1997; Hartlieb, E. *et al.*, 1999; Daly, K. C. and Smith, B. H., 2000), cockroaches (Sakura, M. *et al.*, 2002; Watanabe, H. *et al.*, 2003), and crickets (Matsumoto, Y. and Mizunami, M., 2004). No olfactory learning studies with locusts have yet been published, although non-olfactory learning capacities have been shown in *Schistocerca* (Dukas, R. and Bernays, E. A., 2000). Interestingly, also animals for which the ecological relevance of learning is not fully understood can be trained, and the most important example of this category is appetitive learning in the fruit fly, *Drosophila*, which can be trained both as adult (Borst, A., 1983; Schwaerzel, M. *et al.*, 2003) and as larva (Scherer, S. *et al.*, 2003; Hendel, T. *et al.*, 2005).

The most powerful approach to study olfactory learning relies on classical conditioning, by pairing a US with an odor as CS. Appetitive training is commonly used in bees, and sugar water (as nectar surrogate) is used as US. Most (but not all) studies in *Drosophila* use aversive learning with electroshocks as US. At a cellular level, appetitive and aversive pathways are different (see above) (Schwaerzel, M. *et al.*, 2003). Decay of memory over time appears as seamless. However, this is the result of overlapping distinct memory phases (ATM, MTM, ARM, and LTM, in *Drosophila*), which can be induced by different training protocols (massed versus spaced training, and/or number of trials), and which rely on different cellular mechanisms (transcription, translation only, and phosphorylation only, see above). For example, long-term memory depends on a distributed arrangement of repeated conditioning trials over time (spaced training), whereas the same number of conditioning trials induces STM if applied in rapid succession (i.e., massed training) (Tully, T. *et al.*, 1994).

Classical conditioning experiments in insects show all characteristics studied in psychological learning theory: the CS and the US have to coincide within a critical time interval, and the CS has to precede the US (Sutton, R. S. and Barto, A. G., 1981; Giurfa, M., 2003b). Protocols investigating blocking, latent inhibition, negative patterning, and positive patterning all work in honeybees, and complex training protocols reveal that honeybees have almost cognitive capacities (Menzel, R. and Giurfa, M., 1999; Giurfa, M., 2003a), including the capacity to discriminate between 'same' and 'different' and to transfer this concept between the visual and the olfactory system (Giurfa, M. *et al.*, 2001).

4.41.8.2 Attractive and Repulsive Odors

In appetitive training experiments in bees, not all odors are equally easy to learn, suggesting that some odors have an innate bias toward being attractive (Vareschi, E., 1971). Similarly, in *Drosophila*, some odors are attractive and others repellent, but almost all odors are repellent when tested at very high concentrations (Borst, A., 1983).

In some cases, adding a component to a mixture can create a repellent effect. For example, the commercially used substance *N,N*-diethyltoluamide (DEET) acts as a repellent for many insects including mosquitoes (Rutledge, L. C. *et al.*, 1978). The mechanistic explanation for its activity is still missing. Its effect on many species would suggest that it acts physiologically on olfactory mechanisms rather than interfering with a behavioral pattern. In other cases, specialized communication channels can be disrupted by added odors, such as when additional substances are added to a sexual pheromone. For example, sympatric moth species often use similar substances in their sexual pheromone blends, and a small amount of a substance belonging to a sympatric species can already block the attractive action of a pheromone (Vickers, N. J. and Baker, T. C., 1997). Interestingly, adding such a behavioral antagonist to the environment has been tested as one means for mating disruption in pest control, but the effects are limited, suggesting that the plume structure from an individual source can be recognized even against the repellent signal from another source (Cardé, R. T. and Minks, A. K., 1995).

4.41.8.3 Social Communication – Pheromones

Social insects have a large number of olfactory-guided behaviors, many of which are controlled by pheromones. Honeybees use social pheromones such as aggregation pheromones and alarm pheromones, among others. Processing of these odors in the AL is treated above. The behavior elicited by social pheromones is a stereotypical and innate behavior. However, the behavior is not conditional on the odor only: individuals are receptive or not, depending on the circumstances. This is true both for the released pheromone itself, which can elicit different behaviors in different hive situations (e.g., for a stable hive, or when a hive is swarming), and for a chemical component (e.g., isoamyl acetate, which is a component of the alarm pheromone but does not elicit defensive behavior when it is the component of a flower odor). Whether these differences in elicited behavior are controlled within the olfactory system (e.g., by feedback neurons), or in multimodal brain centers, is unknown.

Social pheromones have been studied particularly well in ants (Hölldobler, B. and Wilson, E. O., 1990). Some pheromones have different behavioral effects with increasing distance from the releasing individual, either due to decreased odor concentration or because in a mixture different volatilities of the components lead to changing component ratios with increasing distances, thus creating unique odor fingerprints for distance. For example, the alarm secretion of the ant *Oecophylla longinoda* contains several substances that are active in concentric circles: 2-butyl-2-octenal in the inner circle elicits biting, 2-undecanone in the next elicits attraction and biting, 1-hexanol in the third elicits attraction, while the fourth concentric circle, 1-hexanol, only elicits attraction – all circles together having a radius of <10 cm (Hölldobler, B. and Wilson, E. O., 1990). Such a separation of mixture components is only possible because, physically, the aerodynamics of an ant's olfactory environment is different from that of a flying insect: in the air layer immediately above ground, turbulence is almost irrelevant, and odor diffusion becomes dominant. Tropotaxis is common, that is, ants follow odor concentration gradients, and they compare bilateral olfactory input from their antennae. The olfactory environment of a cockroach is also a substrate-bound still-air environment, and much behavior is related to active antennal movement. Spatial sampling of the olfactory environment is common in cockroaches (Boeckh, J. and Tolbert, L. P., 1993). However, even flying insects such as *Drosophila* are capable of using concentration differences between the two antennae for orientation (Borst, A., 1983), although the ecological relevance of this behavior has not yet been analyzed.

4.41.8.4 Sexual Communication – Pheromones

Sexual pheromone-related behavior (best studied in moths) is fundamentally different. These free-flying animals follow pheromone plumes that are transported downwind over large distances. The nature of the trail is inherently turbulent (Murlis, J. *et al.*, 1992). Odor concentration varies greatly, but the relative ratio of pheromone components does not. Consequently, the olfactory system is optimized to recognize a particular, genetically determined blend of substances, and male moth ALs have a stereotypical macroglomerular complex with glomeruli specifically dedicated to their species' pheromone components and to behavioral antagonists, which often are pheromones of closely related, sympatric species. Moths do not follow concentration gradients of pheromones. Rather, when they encounter an odor plume, they fly upwind, and when contact with the pheromone stops, they fly sideways (casting behavior) (Willis, M. A. and Baker, T. C., 1984; Cardé, R. T., 1996). In other words, the odor gives the information of when to fly, and the wind gives the information of where to fly. Therefore, the olfactory system is optimized for detecting short bursts of odors, with high frequencies (Christensen, T. A. and

Sorensen, P. W., 1996). These mechanisms are not universal: some moths, such as storage pests that live in enclosed environments, have lost the requirement for turbulent pheromone trails (Justus, K. A. *et al.*, 2005).

Some plants, notably orchids, have evolved the capacity to mimic sexual pheromones and attract insects to pseudocopulate with them, thus insuring pollination (Schiestl, F. P. *et al.*, 2000; Ayasse, M. *et al.*, 2003).

4.41.8.5 Navigation – Innate Odors

Navigation toward an odor source occurs not only in the context of sexual calling but also in other behavioral contexts, for example in food source finding. Innate odors that are used for navigation are likely to rely on genetically encoded, hardwired olfactory coding mechanisms. Alternative models are also possible, such as early olfactory imprinting, in particular for phytophagous insects where the feeding substrate of a larva might influence the olfactory preference of the ovipositioning adult female.

Female moths follow characteristic odor trails of their host plants for finding good oviposition sites. Possibly, the neural circuits responsible for this are related to those of the male moth finding the female pheromone trail, as suggested by cross-grafting of female antennae onto male moths and vice versa (Schneiderman, A. M. *et al.*, 1986; Rössler, W. *et al.*, 1999a). Plants produce the attractive odors. They are often part of their obligate metabolic processes, which means that they represent evolutionarily reliable signals. Plant odor preferences can cause sympatric speciation, as shown in *Rhagoletis* flies, where two hosts (apple and hawthorn) are visited by different fly populations that prefer the respective plant odor substances (Linn, C. *et al.*, 2003; Linn, C. E. *et al.*, 2004).

Some plants respond to insect challenge by releasing stress volatiles, which act as attractants for other insect species that parasitize on the phytophagous species, creating tritrophic interaction systems (Karban, R. and Baldwin, I. T., 1997; Pare, P. W. and Tumlinson, J. H., 1999). Olfactory orientation of the parasitoids themselves consists in being attracted to an innately attractive odor.

Mosquitoes find their blood meals following a combination of cues, including lactic acid, ammonia, and CO_2, which act synergistically (Kline, D. L., 1998; Geier, M. *et al.*, 1999a; 1999b; Costantini, C. *et al.*, 2001; Steib, B. M. *et al.*, 2001; Dekker, T. *et al.*, 2002; Dekker, T. *et al.*, 2005). Such synergistic effects are caused by mixture interactions, but it remains to be elucidated where within the olfactory pathway these interactions occur. Interestingly, humans that are infected with malaria gametocytes (the transmissible stage) are particularly attractive for the mosquito vector *Anopheles gambiae*, suggesting that the parasite might induce a change in the olfactory signature of the human host (Lacroix, R. *et al.*, 2005).

Insects feeding on carrion (e.g., blowflies) are also attracted by specific, innate odor cues, which often act synergistically with other sensory cues, such as heat or color. Some flowers have evolved appropriate displays to mimic these signals and use carrion-feeding insects as pollinators (Angioy, A. M. *et al.*, 2004; Raguso, R. A., 2004).

Navigation for innate odors includes trail pheromone navigation, used in many ant species (Hölldobler, B. and Wilson, E. O., 1990).

4.41.8.6 Navigation – Learned Odors

Not all odors used for navigation are innate. Many pollinating insects learn which flower is currently in blossom and use this information to improve their foraging success. When an odor trail is used for navigation in these instances, the brain cannot rely on innate circuitry. Whether the behavioral strategies and the neural circuitries for finding the odor source are comparable with those used for innate odors is currently unknown.

Odors are not the only cues used for navigation. Bees learn both olfactory and visual elements of a flower. However, optical resolution is low, and odors are the only long-range signals that are available for a bee. The optical resolution of a bee allows to see an inflorescence of 8 cm at a distance of ~90 cm and to recognize its color at a distance of ~30 cm, leaving olfaction alone for everything farther away than 90 cm (Galizia, C. G. *et al.*, 2005). Honeybees also use their topographical knowledge of the landscape for navigating (Giurfa, M. and Capaldi, E. A., 1999; Menzel, R. *et al.*, 2000). Scout bees communicate the location of a new crop using the waggle dance (von Frisch, K., 1966). But even during the waggle dance, follower bees experience the flower's odor on the dancer's body and might use that odor as an additional clue for navigation. This has led to a long controversy among bee researchers, with some researchers advocating odors as the main navigation clue opposed to those arguing for the dominant role of the bees' waggle dances (Wenner, A. M. and Johnson, D. L., 1967; Wenner, A. M., 2002). Recent

studies that monitored individual bees' flights after receiving information from recruiter bees have shown that the waggle dance is the dominant information used (Riley, J. R. *et al.*, 2005).

Some orchids mimic a rewarding flower. This situation is quite different from sexual pheromone mimicry, because flower odors consist in more complex mixtures than pheromones, and therefore, their reproduction by a deceiving plant may be biochemically more difficult. In at least one of these cases, the mimicry only involves visual parameters, while the long-range task of attracting pollinating bees is solved not by mimicry but by having a model inflorescence in close neighborhood (Galizia, C. G. *et al.*, 2005). In other words, these orchids hitchhike on their model's olfactory display and attract the pollinators at close range with their visual display only.

4.41.9 Limitations and Potential of Using Insects to Study Olfactory Coding

Some people argue that it becomes increasingly difficult to justify research in insect olfaction. The classical advantages of insect models are the limited number of neurons and even more so the possibility of identifying individual neurons, thus allowing for detailed analyses across animals. With the recent technological advances in mammals, both in recording techniques and in molecular and genetic manipulations, these advantages lose in importance: insects no longer are the simpler models for mammalian neurophysiological research.

However, insects remain important models on their own. The relative simplicity of their brains makes it possible to study many species in great detail and thus to understand not just one solution that evolution proposed for olfactory coding, but several solutions. This comparative approach is the main treasure that insect studies can contribute. However, it comes with a price, and the consequences of this price weaken the case for studying insect olfaction: because technical approaches and possibilities differ in different laboratories and species, our information is often insufficient for a thorough comparison among species. For example, as stated above, temporal aspects of odor responses in PNs are best studied in locusts, while across-PN patterns are best understood in honeybees and flies, with moths somewhere in between. This fragmentation of knowledge and research time severely limits the conclusions that can be drawn, in particular with respect to understanding which mechanisms are universal and which are specific specializations. The massive advance in possibilities provided by molecular biology in *Drosophila*, and the concomitant increase in laboratories involved in studying fly olfaction, will help by creating a well-understood insect model species. However, it is important to realize that *Drosophila*, just as any other insect species, is not a good model for insect olfaction. The diversity among insects is enormous, and it is this treasure that we need to unearth for our understanding of the basic mechanisms in olfaction.

References

Abel, R., Rybak, J., and Menzel, R. 2001. Structure and response patterns of olfactory interneurons in the honeybee, *Apis mellifera*. J. Comp. Neurol. 437, 363–383.

Abraham, N. M., Spors, H., Carleton, A., Margrie, T. W., Kuner, T., and Schaefer, A. T. 2004. Maintaining accuracy at the expense of speed; stimulus similarity defines odor discrimination time in mice. Neuron 44, 865–876.

Amrein, H. 2004. Pheromone perception and behavior in Drosophila. Curr. Opin. Neurobiol. 14, 435–442.

Ang, L. H., Kim, J., Stepensky, V., and Hing, H. 2003. Dock and Pak regulate olfactory axon pathfinding in Drosophila. Development 130, 1307–1316.

Angioy, A. M., Stensmyr, M. C., Urru, I., Puliafito, M., Collu, I., and Hansson, B. S. 2004. Function of the heater: the dead horse arum revisited. Proc. Biol. Sci. 271(Suppl. 3), S13–S15.

Anton, S. and Hansson, B. S. 1996. Antennal lobe interneurons in the desert locust *Schistocerca gregaria* (Forskal): processing of aggregation pheromones in adult males and females. J. Comp. Neurol. 370, 85–96.

Anton, S. and Homberg, U. 1999. Antennal Lobe Structure. In: Insect Olfaction (*ed.* B. S. Hansson), pp. 97–124. Springer.

Anton, S., Ignell, R., and Hansson, B. S. 2002. Developmental changes in the structure and function of the central olfactory system in gregarious and solitary desert locusts. Microsc. Res. Tech. 56, 281–291.

Arnold, G., Masson, C., and Budharugsa, S. 1983. Spatial organization of the sensory antennal system in the honeybee by using a cobalt ions marketing method. Apidologie 14, 127–135.

Arnold, G., Masson, C., and Budharugsa, S. 1985. Comparative study of the antennal lobes and their afferent pathway in the worker bee and the drone (*Apis mellifera*). Cell Tissue Res. 242, 593–605.

Attwell, D. and Laughlin, S. B. 2001. An energy budget for signaling in the grey matter of the brain. J. Cereb. Blood Flow Metab. 21, 1133–1145.

Av-ron, E. and Rospars, J. P. 1995. Modeling insect olfactory neuron signaling by a network utilizing disinhibition. Biosystems 36, 101–108.

Ayasse, M., Schiestl, F. P., Paulus, H. F., Ibarra, F., and Francke, W. 2003. Pollinator attraction in a sexually deceptive orchid by means of unconventional chemicals. Proc. Biol. Sci. 270, 517–522.

Baumann, P. M., Oland, L. A., and Tolbert, L. P. 1996. Glial cells stabilize axonal protoglomeruli in the developing olfactory lobe of the moth *Manduca sexta*. J. Comp. Neurol. 373, 118–128.

Bazhenov, M., Stopfer, M., Rabinovich, M., Abarbanel, H. D., Sejnowski, T. J., and Laurent, G. 2001. Model of cellular and network mechanisms for odor-evoked temporal patterning in the locust antennal lobe. Neuron 30, 569–581.

de Belle, J. S. and Heisenberg, M. 1994. Associative odor learning in Drosophila abolished by chemical ablation of mushroom bodies. Science 263, 692–695.

Berg, B. G., Almaas, T. J., Bjaalie, J. G., and Mustaparta, H. 2005. Projections of male-specific receptor neurons in the antennal lobe of the oriental tobacco budworm moth, *Helicoverpa assulta*: a unique glomerular organization among related species. J. Comp. Neurol. 486, 209–220.

Berg, B. G., Galizia, C. G., Brandt, R., and Mustaparta, H. 2002. Digital atlases of the antennal lobe in two species of tobacco budworm moths, the Oriental *Helicoverpa assulta* (male) and the American *Heliothis virescens* (male and female). J. Comp. Neurol. 446, 123–134.

Bhalerao, S., Sen, A., Stocker, R., and Rodrigues, V. 2003. Olfactory neurons expressing identified receptor genes project to subsets of glomeruli within the antennal lobe of *Drosophila melanogaster*. J. Neurobiol. 54, 577–592.

Bicker, G. 1999a. Histochemistry of classical neurotransmitters in antennal lobes and mushroom bodies of the honeybee. Microsc. Res. Tech. 45, 174–183.

Bicker, G. 1999b. Biogenic amines in the brain of the honeybee: cellular distribution, development, and behavioral functions. Microsc. Res. Tech. 44, 166–178.

Bicker, G. 2001. Sources and targets of nitric oxide signalling in insect nervous systems. Cell Tissue Res. 303, 137–146.

Bicker, G. and Kreissl, S. 1994. Calcium imaging reveals nicotinic acetylcholine receptors on cultured mushroom body neurons. J. Neurophysiol. 71, 808–810.

Bicker, G., Kreissl, S., and Hofbauer, A. 1993. Monoclonal antibody labels olfactory and visual pathways in Drosophila and Apis brains. J. Comp. Neurol. 335, 413–424.

Bicker, G., Schäfer, S., and Kingan, T. G. 1985. Mushroom body feedback interneurones in the honeybee show GABA-like immunoreactivity. Brain Res. 360, 394–397.

Bicker, G., Schäfer, S., Ottersen, O. P., and Storm-Mathisen, J. 1988. Glutamate-like immunoreactivity in identified neuronal populations of insect nervous systems. J. Neurosci. 8, 2108–2122.

Boeckh, J. and Tolbert, L. P. 1993. Synaptic organization and development of the antennal lobe in insects. Microsc. Res. Tech. 24, 260–280.

Boeckh, J., Ernst, K. D., and Selsam, P. 1987. Neurophysiology and neuroanatomy of the olfactory pathway in the cockroach. Ann. N. Y. Acad. Sci. 510, 39–43.

Bogner, F., Boppre, M., Ernst, K. D., and Boeckh, J. 1986. CO2 sensitive receptors on labial palps of rhodogastria moths (Lepidoptera, Arctiidae) – physiology, fine-structure and central projection. J. Comp. Physiol. [A] 158, 741–749.

Bornhauser, B. C. and Meyer, E. P. 1997. Histamine-like immunoreactivity in the visual system and brain of an orthopteran and a hymenopteran insect. Cell Tissue Res. 287, 211–221.

Borst, A. 1983. Computation of olfactory signals in *Drosophila melanogaster*. J. Comp. Physiol. [A] 152, 373–383.

Brandt, R., Rohlfing, T., Rybak, J., Krofczik, S., Maye, A., Westerhoff, M., Hege, H. C., and Menzel, R. 2005. Three-dimensional average-shape atlas of the honeybee brain and its applications. J. Comp. Neurol. 492, 1–19.

Bräunig, P. 1991. Suboesophageal DUM neurons innervate the principal neuropiles of the locust brain. Philos. Trans. R. Soc. Lond. B 332, 221–240.

Brody, C. D. and Hopfield, J. J. 2003. Simple networks for spike-timing-based computation, with application to olfactory processing. Neuron 37, 843–852.

de Bruyne, M., Foster, K., and Carlson, J. R. 2001. Odor coding in the Drosophila antenna. Neuron 30, 537–552.

Buchner, E. 1991. Genes expressed in the adult brain of Drosophila and effects of their mutations on behavior: a survey of transmitter- and second messenger-related genes. J. Neurogenet. 7, 153–192.

Buchner, E. and Rodrigues, V. 1983. Autoradiographic localization of [3H]choline uptake in the brain of Drosophila melanogaster. Neurosci. Lett. 42, 25–31.

Cardé, R. T. 1996. Odour plumes and odour-mediated flight in insects. Ciba Found. Symp. 200, 54–66; discussion 66–70.

Cardé, R. T. and Minks, A. K. 1995. Control of moth pests by mating disruption: successes and constraints. Annu. Rev. Entomol. 40, 559–585.

Carlsson, M. A., Galizia, C. G., and Hansson, B. S. 2002. Spatial representation of odours in the antennal lobe of the moth Spodoptera littoralis (Lepidoptera: Noctuidae). Chem. Senses 27, 231–244.

Chapman, R. F. 2002. Development of phenotypic differences in sensillum populations on the antennae of a grasshopper, Schistocerca americana. J. Morphol. 254, 186–194.

Christensen, T. A. and Hildebrand, J. G. 1997. Coincident stimulation with pheromone components improves temporal pattern resolution in central olfactory neurons. J. Neurophysiol. 77, 775–781.

Christensen, T. A. and Hildebrand, J. G. 2002. Pheromonal and host-odor processing in the insect antennal lobe: how different? Curr. Opin. Neurobiol. 12, 393–399.

Christensen, T. A. and Sorensen, P. W. 1996. Pheromones as tools for olfactory research. Introduction. Chem. Senses 21, 241–243.

Christensen, T. A., Harrow, I. D., Cuzzocrea, C., Randolph, P. W., and Hildebrand, J. G. 1995. Distinct projections of two populations of olfactory receptor axons in the antennal lobe of the sphinx moth Manduca sexta. Chem. Senses 20, 313–323.

Christensen, T. A., Pawlowski, V. M., Lei, H., and Hildebrand, J. G. 2000. Multi-unit recordings reveal context-dependent modulation of synchrony in odor-specific neural ensembles. Nat. Neurosci. 3, 927–931.

Christensen, T. A., Waldrop, B. R., Harrow, I. D., and Hildebrand, J. G. 1993. Local interneurons and information processing in the olfactory glomeruli of the moth Manduca sexta. J. Comp. Physiol. [A] 173, 385–399.

Christensen, T. A., Waldrop, B. R., and Hildebrand, J. G. 1998a. GABAergic mechanisms that shape the temporal response to odors in moth olfactory projection neurons. Ann. N. Y. Acad. Sci. 855, 475–481.

Christensen, T. A., Waldrop, B. R., and Hildebrand, J. G. 1998b. Multitasking in the olfactory system: context-dependent responses to odors reveal dual GABA-regulated coding mechanisms in single olfactory projection neurons. J. Neurosci. 18, 5999–6008.

Clyne, P. J., Warr, C. G., Freeman, M. R., Lessing, D., Kim, J., and Carlson, J. R. 1999. A novel family of divergent seven-transmembrane proteins: candidate odorant receptors in Drosophila. Neuron 22, 327–338.

Cobb, M. 1999. What and how do maggots smell? Biol. Rev. 74, 425–459.

Collmann, C., Carlsson, M. A., Hansson, B. S., and Nighorn, A. 2004. Odorant-evoked nitric oxide signals in the antennal lobe of Manduca sexta. J. Neurosci. 24, 6070–6077.

Connolly, J. B., Roberts, I. J., Armstrong, J. D., Kaiser, K., Forte, M., Tully, T., and O'Kane, C. J. 1996. Associative learning disrupted by impaired Gs signaling in Drosophila mushroom bodies. Science 274, 2104–2107.

Costa, R. 1989. Esterase-6 and the pheromonal effects of cis-vaccenyl acetate in Drosophila melanogaster. J. Evol. Biol. 2, 395–407.

Costantini, C., Birkett, M. A., Gibson, G., Ziesmann, J., Sagnon, N. F., Mohammed, H. A., Coluzzi, M., and Pickett, J. A. 2001. Electroantennogram and behavioural responses of the malaria vector Anopheles gambiae to human-specific sweat components. Med. Vet. Entomol. 15, 259–266.

Couto, A., Alenius, M., and Dickson, B. J. 2005. Molecular, anatomical, and functional organization of the Drosophila olfactory system. Curr. Biol. 15, 1535–1547.

Dacks, A. M., Christensen, T. A., Agricola, H. J., Wollweber, L., and Hildebrand, J. G. 2005. Octopamine-immunoreactive neurons in the brain and subesophageal ganglion of the hawkmoth Manduca sexta. J. Comp. Neurol. 488, 255–268.

Daly, K. C. and Smith, B. H. 2000. Associative olfactory learning in the moth Manduca sexta. J. Exp. Biol. 203, 2025–2038.

Daly, K. C., Christensen, T. A., Lei, H., Smith, B. H., and Hildebrand, J. G. 2004. Learning modulates the ensemble representations for odors in primary olfactory networks. Proc. Natl. Acad. Sci. U. S. A. 101, 10476–10481.

Dani, F. R., Jones, G. R., Corsi, S., Beard, R., Pradella, D., and Turillazzi, S. 2005. Nestmate recognition cues in the honey bee: differential importance of cuticular alkanes and alkenes. Chem. Senses 30, 477–489.

Davis, R. L. 2004. Olfactory learning. Neuron 44, 31–48.

Davis, N. T., Homberg, U., Teal, P. E. A., Altstein, M., Agricola, H. J., and Hildebrand, J. G. 1996. Neuroanatomy and immunocytochemistry of the median neuroendocrine cells of the subesophageal ganglion of the tobacco hawkmoth, Manduca sexta: immunoreactivities to PBAN and other neuropeptides. Microsc. Res. Tech. 35, 201–229.

Deisig, N., Lachnit, H., Sandoz, J. C., Lober, K., and Giurfa, M. 2003. A modified version of the unique cue theory accounts for olfactory compound processing in honeybees. Learn. Mem. 10, 199–208.

Dekker, T., Geier, M., and Cardé, R. T. 2005. Carbon dioxide instantly sensitizes female yellow fever mosquitoes to human skin odours. J. Exp. Biol. 208, 2963–2972.

Dekker, T., Steib, B., Cardé, R. T., and Geier, M. 2002. L-lactic acid: a human-signifying host cue for the anthropophilic mosquito Anopheles gambiae. Med. Vet. Entomol. 16, 91–98.

Dickens, J. C., Oliver, J. E., Hollister, B., Davis, J. C., and Klun, J. A. 2002. Breaking a paradigm: male-produced aggregation pheromone for the Colorado potato beetle. J. Exp. Biol. 205, 1925–1933.

Distler, P. G. 1990. Synaptic connections of dopamine-immunoreactive neurons in the antennal lobes of Periplaneta americana. Colocalization with GABA-like immunoreactivity. Histochemistry 93, 401–408.

Distler, P. G. and Boeckh, J. 1996. Synaptic connection between olfactory receptor cells and uniglomerular projection neurons in the antennal lobe of the American cockroach, Periplaneta americana. J. Comp. Neurol. 370, 35–46.

Distler, P. G. and Boeckh, J. 1997a. Synaptic connections between identified neuron types in the antennal lobe glomeruli of the cockroach, Periplaneta americana: II. Local multiglomerular interneurons. J. Comp. Neurol. 383, 529–540.

Distler, P. G. and Boeckh, J. 1997b. Central projections of the maxillary and antennal nerves in the mosquito Aedes aegypti. J. Exp. Biol. 200, 1873–1879.

Distler, P. G., Bausenwein, B., and Boeckh, J. 1998a. Localization of odor-induced neuronal activity in the antennal lobes of the blowfly *Calliphora vicina*: a [3H] 2-deoxyglucose labeling study. Brain Res. 805, 263–266.

Distler, P. G., Gruber, C., and Boeckh, J. 1998b. Synaptic connections between GABA-immunoreactive neurons and uniglomerular projection neurons within the antennal lobe of the cockroach, *Periplaneta americana*. Synapse 29, 1–13.

Ditzen, M., Evers, J. F., and Galizia, C. G. 2003. Odor similarity does not influence the time needed for odor processing. Chem. Senses 28, 781–789.

Dobritsa, A. A., van der Goes van Naters, W., Warr, C. G., Steinbrecht, R. A., and Carlson, J. R. 2003. Integrating the molecular and cellular basis of odor coding in the *Drosophila* antenna. Neuron 37, 827–841.

Dubnau, J., Chiang, A. S., and Tully, T. 2003. Neural substrates of memory: from synapse to system. J. Neurobiol. 54, 238–253.

Dubnau, J., Grady, L., Kitamoto, T., and Tully, T. 2001. Disruption of neurotransmission in Drosophila mushroom body blocks retrieval but not acquisition of memory. Nature 411, 476–480.

Dukas, R. and Bernays, E. A. 2000. Learning improves growth rate in grasshoppers. Proc. Natl. Acad. Sci. U. S. A. 97, 2637–2640.

Ejima, A., Smith, B. P., Lucas, C., van der Goes van Naters, W., Miller, C. J., Carlson, J. R., Kevine, J. D., and Griffith, L. C. 2007. Generalization of courtship learning in Drosophila is mediated by *cis*-vaccenyl acetate. Curr. Biol. 17, 599–605.

Elphick, M., Rayne, R., Riveros-Moreno, V. V., Moncada, S., and Shea, M. 1995. Nitric oxide synthesis in locust olfactory interneurones. J. Exp. Biol. 198, 821–829.

Erber, J., Masuhr, T., and Menzel, R. 1980. Localization of short-term-memory in the brain of the bee, *Apis mellifera*. Physiol. Entomol. 5, 343–358.

Ernst, K. D. and Boeckh, J. 1983. A neuroanatomical study on the organization of the central antennal pathways in insects. III. Neuroanatomical characterization of physiologically defined response types of deutocerebral neurons in *Periplaneta americana*. Cell Tissue Res. 229, 1–22.

Ernst, K. D., Boeckh, J., and Boeckh, V. 1977. A neuroanatomical study on the organization of the central antennal pathways in insects. Cell Tissue Res. 176, 285–306.

Esslen, J. and Kaissling, K. E. 1976. Number and distribution of sensilla on antennal flagellum of honeybee (*Apis mellifera* L.). Zoomorphologie 83, 227–251.

Faber, T. and Menzel, R. 2001. Visualizing mushroom body response to a conditioned odor in honeybees. Naturwissenschaften 88, 472–476.

Faber, T., Joerges, J., and Menzel, R. 1999. Associative learning modifies neural representations of odors in the insect brain. Nat. Neurosci. 2, 74–78.

Fan, R. J., Anderson, P., and Hansson, B. 1997. Behavioural analysis of olfactory conditioning in the moth *Spodoptera littoralis* (Boisd.) (Lepidoptera: noctuidae). J. Exp. Biol. 200 (Pt 23), 2969–2976.

Farivar, S. S. 2005. Cytoarchitecture of the Locust Olfactory System. PhD Thesis, http://resolver.caltech.edu/Caltech ETD:etd-04212005-143332.

Farooqui, T., Robinson, K., Vaessin, H., and Smith, B. H. 2003. Modulation of early olfactory processing by an octopaminergic reinforcement pathway in the honeybee. J. Neurosci. 23, 5370–5380.

Farris, S. M. 2005. Evolution of insect mushroom bodies: old clues, new insights. Arthropod Struct. Dev. 34, 211–234.

Farris, S. M. and Strausfeld, N. J. 2001. Development of laminar organization in the mushroom bodies of the cockroach: Kenyon cell proliferation, outgrowth, and maturation. J. Comp. Neurol. 439, 331–351.

Farris, S. M., Abrams, A. I., and Strausfeld, N. J. 2004. Development and morphology of class II Kenyon cells in the mushroom bodies of the honey bee, *Apis mellifera*. J. Comp. Neurol. 474, 325–339.

Farris, S. M., Robinson, G. E., and Fahrbach, S. E. 2001. Experience- and age-related outgrowth of intrinsic neurons in the mushroom bodies of the adult worker honeybee. J. Neurosci. 21, 6395–6404.

Feinstein, P. and Mombaerts, P. 2004. A contextual model for axonal sorting into glomeruli in the mouse olfactory system. Cell 117, 817–831.

Feinstein, P., Bozza, T., Rodriguez, I., Vassalli, A., and Mombaerts, P. 2004. Axon guidance of mouse olfactory sensory neurons by odorant receptors and the beta2 adrenergic receptor. Cell 117, 833–846.

Fiala, A., Spall, T., Diegelmann, S., Eisermann, B., Sachse, S., Devaud, J. M., Buchner, E., and Galizia, C. G. 2002. Genetically expressed cameleon in *Drosophila melanogaster* is used to visualize olfactory information in projection neurons. Curr. Biol. 12, 1877–1884.

Fishilevich, E. and Vosshall, L. B. 2005. Genetic and functional subdivision of the Drosophila antennal lobe. Curr. Biol. 15, 1548–1553.

Flanagan, D. and Mercer, A. R. 1989a. An atlas and 3-D reconstruction of the antennal lobes in the worker honey bee, *Apis mellifera* L. (Hymenoptera: Apidae). Int. J. Insect Morphol. Embryol. 18, 145–159.

Flanagan, D. and Mercer, A. R. 1989b. Morphology and response characteristics of neurones in the deutocerebrum of the brain in the honeybee *Apis mellifera*. J. Comp. Physiol. [A] 164, 483–494.

Fonta, C., Sun, X. J., and Masson, C. 1993. Morphology and spatial distribution of bee antennal lobe interneurones responsive to odours. Chem. Senses 18, 101–119.

Frambach, I., Rössler, W., Winkler, M., and Schurmann, F. W. 2004. F-actin at identified synapses in the mushroom body neuropil of the insect brain. J. Comp. Neurol. 475, 303–314.

Free, J. B. 1987. Pheromones of social bees. Comstock Pub. Associates.

Friedrich, R. W. and Laurent, G. 2001. Dynamic optimization of odor representations by slow temporal patterning of mitral cell activity. Science 291, 889–894.

Friggi-Grelin, F., Coulom, H., Meller, M., Gomez, D., Hirsh, J., and Birman, S. 2003. Targeted gene expression in Drosophila dopaminergic cells using regulatory sequences from tyrosine hydroxylase. J. Neurobiol. 54, 618–627.

von Frisch, K. 1966. The Dancing Bees: An Account of the Life and Senses of the Honey Bee, 2nd Edition. Methuen.

Galán, R. F., Sachse, S., Galizia, C. G., and Herz, A. V. 2004. Odor-driven attractor dynamics in the antennal lobe allow for simple and rapid olfactory pattern classification. Neural. Comput. 16, 999–1012.

Galán, R. F., Weidert, M., Menzel, R., Herz, A. V. M., and Galizia, C. G. 2006. Sensory memory for odors is encoded in spontaneous correlated activity between olfactory glomeruli. Neural Comput.,18, 10–25.

Galizia, C. G. and Kimmerle, B. 2004. Physiological and morphological characterization of honeybee olfactory neurons combining electrophysiology, calcium imaging and confocal microscopy. J. Comp. Physiol. [A] 190, 21–38.

Galizia, C. G. and Menzel, R. 2001. The role of glomeruli in the neural representation of odours: results from optical recording studies. J. Insect Physiol. 47, 115–130.

Galizia, C. G. and Vetter, R. S. 2004. Optical Methods for Analyzing Odor-Evoked Activity in the Insect Brain. In: Advances in Insect Sensory Neuroscience (*ed*. T. A. Christensen), pp. 349–392. CRC Press.

Galizia, C. G., Joerges, J., Küttner, A., Faber, T., and Menzel, R. 1997. A semi *in-vivo* preparation for optical recording of the insect brain. J. Neurosci. Methods 76, 61–69.

Galizia, C. G., Kunze, J., Gumbert, A., Borg-Karlson, A. K., Sachse, S., Markl, C., and Menzel, R. 2005. Relationship of visual and olfactory signal parameters in a food-deceptive flower mimicry system. Behav. Ecol. 16, 159–168.

Galizia, C. G., Küttner, A., Joerges, J., and Menzel, R. 2000b. Odour representation in honeybee olfactory glomeruli shows slow temporal dynamics: an optical recording study using a voltage-sensitive dye. J. Insect Physiol. 46, 877–886.

Galizia, C. G., McIlwrath, S. L., and Menzel, R. 1999b. A digital three-dimensional atlas of the honeybee antennal lobe based on optical sections acquired by confocal microscopy. Cell Tissue Res. 295, 383–394.

Galizia, C. G., Menzel, R., and Hölldobler, B. 1999a. Optical imaging of odor-evoked glomerular activity patterns in the antennal lobes of the ant *Camponotus rufipes*. Naturwissenschaften 86, 533–537.

Galizia, C. G., Nägler, K., Hölldobler, B., and Menzel, R. 1998. Odour coding is bilaterally symmetrical in the antennal lobes of honeybees (*Apis mellifera*). Eur. J. Neurosci. 10, 2964–2974.

Galizia, C. G., Sachse, S., and Mustaparta, H. 2000a. Calcium responses to pheromones and plant odours in the antennal lobe of the male and female moth *Heliothis virescens*. J. Comp. Physiol. [A] 186, 1049–1063.

Galizia, C. G., Sachse, S., Rappert, A., and Menzel, R. 1999c. The glomerular code for odor representation is species specific in the honeybee *Apis mellifera*. Nat. Neurosci. 2, 473–478.

Ganeshina, O. and Menzel, R. 2001. GABA-immunoreactive neurons in the mushroom bodies of the honeybee: an electron microscopic study. J. Comp. Neurol. 437, 335–349.

Gao, Q. and Chess, A. 1999. Identification of candidate *Drosophila* olfactory receptors from genomic DNA sequence. Genomics 60, 31–39.

Gascuel, J. and Masson, C. 1991. A quantitative ultrastructural study of the honeybee antennal lobe. Tissue and Cell 23, 341–355.

Geier, M., Bosch, O. J., and Boeckh, J. 1999a. Ammonia as an attractive component of host odour for the yellow fever mosquito, *Aedes aegypti*. Chem. Senses 24, 647–653.

Geier, M., Bosch, O. J., and Boeckh, J. 1999b. Influence of odour plume structure on upwind flight of mosquitoes towards hosts. J. Exp. Biol. 202, 1639–1648.

Gerber, B., Scherer, S., Neuser, K., Michels, B., Hendel, T., Stocker, R. F., and Heisenberg, M. 2004b. Visual learning in individually assayed Drosophila larvae. J. Exp. Biol. 207, 179–188.

Gerber, B., Tanimoto, H., and Heisenberg, M. 2004a. An engram found? Evaluating the evidence from fruit flies. Curr. Opin. Neurobiol. 14, 737–744.

Getz, W. M. 1991. A neural network for processing olfactory-like stimuli. Bull. Math. Biol. 53, 805–823.

Getz, W. M. and Lutz, A. 1999. A neural network model of general olfactory coding in the insect antennal lobe. Chem. Senses 24, 351–372.

Gewecke, M. 1979. Central projection of antennal afferents for the flight motor in *Locusta migratoria* (Orthoptera, Acrididae). Entomologia Generalis 5, 317–320.

Gibson, N. J. and Nighorn, A. 2000. Expression of nitric oxide synthase and soluble guanylyl cyclase in the developing olfactory system of *Manduca sexta*. J. Comp. Neurol. 422, 191–205.

Gibson, N. J., Rossler, W., Nighorn, A. J., Oland, L. A., Hildebrand, J. G., and Tolbert, L. P. 2001. Neuron-glia communication via nitric oxide is essential in establishing antennal-lobe structure in *Manduca sexta*. Dev. Biol. 240, 326–339.

Giurfa, M. 1993. The repellent scent-mark of the honeybee *Apis mellifera ligustica* and its role as communication cue during foraging. Insectes Sociaux 40, 59–67.

Giurfa, M. 2003a. Cognitive neuroethology: dissecting non-elemental learning in a honeybee brain. Curr. Opin. Neurobiol. 13, 726–735.

Giurfa, M. 2003b. The amazing mini-brain: lessons from a honey bee. Bee World 84, 5–18.

Giurfa, M. and Capaldi, E. A. 1999. Vectors, routes and maps: new discoveries about navigation in insects. Trends Neurosci. 22, 237–242.

Giurfa, M. and Malun, D. 2004. Associative mechanosensory conditioning of the proboscis extension reflex in honeybees. Learn. Mem. 11, 294–302.

Giurfa, M. and Nunez, J. A. 1992. Honeybees mark with scent and reject recently visited flowers. Oecologia 89, 113–117.

Giurfa, M., Zhang, S., Jenett, A., Menzel, R., and Srinivasan, M. V. 2001. The concepts of "sameness" and "difference" in an insect. Nature 410, 930–933.

Goldman, A. L., Van der Goes van Naters, W., Lessing, D., Warr, C. G., and Carlson, J. R. 2005. Coexpression of two functional odor receptors in one neuron. Neuron 45, 661–666.

Goriely, A. R., Secomb, T. W., and Tolbert, L. P. 2002. Effect of the glial envelope on extracellular K^+ diffusion in olfactory glomeruli. J. Neurophysiol. 87, 1712–1722.

Grant, A. J., Wigton, B. E., Aghajanian, J. G., and O'Connell, R. J. 1995. Electrophysiological responses of receptor neurons in mosquito maxillary palp sensilla to carbon dioxide. J. Comp. Physiol. [A] 177, 389–396.

Grimaldi, D. A. and Engel, M. S. 2004. Evolution of the Insects. Cambridge University Press.

Gronenberg, W. 1999. Modality-specific segregation of input to ant mushroom bodies. Brain Behav. Evol. 54, 85–95.

Gronenberg, W. 2001. Subdivisions of hymenopteran mushroom body calyces by their afferent supply. J. Comp. Neurol. 435, 474–489.

Grünbaum, L. and Müller, U. 1998. Induction of a specific olfactory memory leads to a long-lasting activation of protein kinase C in the antennal lobe of the honeybee. J. Neurosci. 18, 4384–4392.

Grünewald, B. 1999a. Physiological properties and response modulations of mushroom body feedback neurons during olfactory learning in the honeybee, *Apis mellifera*. J. Comp. Physiol. [A] 185, 565–576.

Grünewald, B. 1999b. Morphology of feedback neurons in the mushroom body of the honeybee, *Apis mellifera*. J. Comp. Neurol. 404, 114–126.

Guerenstein, P. G., Yepez, E. A., van Haren, J., Williams, D. G., and Hildebrand, J. G. 2004. Floral CO2 emission may indicate food abundance to nectar-feeding moths. Naturwissenschaften 91, 329–333.

Guerrieri, F., Schubert, M., Sandoz, J. C., and Giurfa, M. 2005. Perceptual and neural olfactory similarity in honeybees. PLoS Biol 3, e60.

Hähnlein, I. and Bicker, G. 1996. Morphology of neuroglia in the antennal lobes and mushroom bodies of the brain of the honeybee. J. Comp. Neurol. 367, 235–245.

Hähnlein, I., Hartig, W., and Bicker, G. 1996. Datura stramonium lectin staining of glial associated extracellular material in insect brains. J. Comp. Neurol. 376, 175–187.

Hammer, M. 1993. An identified neuron mediates the unconditioned stimulus in associative olfactory learning in honeybees. Nature 366, 59–63.

Hammer, M. 1997. The neural basis of associative reward learning in honeybees. Trends Neurosci. 20, 245–252.

Hammer, M. and Menzel, R. 1998. Multiple sites of associative odor learning as revealed by local brain microinjections of octopamine in honeybees. Learn. Mem. 5, 146–156.

Hansson, B. S., Carlsson, M. A., and Kalinova, B. 2003. Olfactory activation patterns in the antennal lobe of the sphinx moth, *Manduca sexta*. J. Comp. Physiol. [A] 189, 301–308.

Hansson, B. S., Christensen, T. A., and Hildebrand, J. G. 1991. Functionally distinct subdivisions of the macroglomerular complex in the antennal lobe of the male sphinx moth *Manduca sexta*. J. Comp. Neurol. 312, 264–278.

Hartlieb, E., Anderson, P., and Hansson, B. S. 1999. Appetitive learning of odours with different behavioural meaning in moths. Physiol. Behav. 67, 671–677.

Heimbeck, G., Bugnon, V., Gendre, N., Haberlin, C., and Stocker, R. F. 1999. Smell and taste perception in *Drosophila melanogaster* larva: toxin expression studies in chemosensory neurons. J. Neurosci. 19, 6599–6609.

Heimbeck, G., Bugnon, V., Gendre, N., Keller, A., and Stocker, R. F. 2001. A central neural circuit for experience-independent olfactory and courtship behavior in *Drosophila melanogaster*. Proc. Natl. Acad. Sci. U. S. A. 98, 15336–15341.

Heinbockel, T., Kloppenburg, P., and Hildebrand, J. G. 1998. Pheromone-evoked potentials and oscillations in the antennal lobes of the sphinx moth *Manduca sexta*. J. Comp. Physiol. [A] 182, 703–714.

Heisenberg, M. 2003. Mushroom body memoir: from maps to models. Nat. Rev. Neurosci. 4, 266–275.

Heisenberg, M., Borst, A., Wagner, S., and Byers, D. 1985. Drosophila mushroom body mutants are deficient in olfactory learning. J. Neurogenet. 2, 1–30.

Hendel, T., Michels, B., Neuser, K., Schipanski, A., Kaun, K., Sokolowski, M. B., Marohn, F., Michel, R., Heisenberg, M., and Gerber, B. 2005. The carrot, not the stick: appetitive rather than aversive gustatory stimuli support associative olfactory learning in individually assayed *Drosophila* larvae. J. Comp. Physiol. [A] 191, 265–279.

Hildebrand, J. G. and Shepherd, G. M. 1997. Mechanisms of olfactory discrimination: converging evidence for common principles across phyla. Annu. Rev. Neurosci. 20, 595–631.

Hildebrandt, H. and Müller, U. 1995a. PKA activity in the antennal lobe of honeybees is regulated by chemosensory stimulation in vivo. Brain Res. 679, 281–288.

Hildebrandt, H. and Müller, U. 1995b. Octopamine mediates rapid stimulation of protein kinase A in the antennal lobe of honeybees. J. Neurobiol. 27, 44–50.

Hofer, S., Dircksen, H., Tollback, P., and Homberg, U. 2005. Novel insect orcokinins: characterization and neuronal distribution in the brains of selected dicondylian insects. J. Comp. Neurol. 490, 57–71.

Hölldobler, B. and Wilson, E. O. 1990. The Ants. Belknap Press of Harvard University Press.

Homberg, U. 2002. Neurotransmitters and neuropeptides in the brain of the locust. Microsc. Res. Tech. 56, 189–209.

Homberg, U. and Hildebrand, J. G. 1991. Histamine immunoreactive neurons in the midbrain and subesophageal ganglion of the sphinx moth *Manduca sexta*. J. Comp. Neurol. 307, 647–657.

Homberg, U. and Hildebrand, J. G. 1994. Postembryonic development of gamma-aminobutyric acid-like immunoreactivity in the brain of the sphinx moth *Manduca-Sexta*. J. Comp. Neurol. 339, 132–149.

Homberg, U. and Müller, U. 1999. Neuroactive Substances in the Antennal Lobe. In: Insect Olfaction (*ed*. B. S. Hansson), pp. 181–206. Springer.

Homberg, U., Hoskins, S. G., and Hildebrand, J. G. 1995. Distribution of acetylcholinesterase activity in the deutocerebrum of the sphinx moth *Manduca sexta*. Cell Tissue Res. 279, 249–259.

Homberg, U., Kingan, T. G., and Hildebrand, J. G. 1990. Distribution of FMRFamide-like immunoreactivity in the brain

and suboesophageal ganglion of the sphinx moth *Manduca sexta* and colocalization with SCPB-, BPP-, and GABA-like immunoreactivity. Cell Tissue Res. 259, 401–419.

Homberg, U., Montague, R. A., and Hildebrand, J. G. 1988. Anatomy of antenno-cerebral pathways in the brain of the sphinx moth *Manduca sexta*. Cell Tissue Res. 254, 255–281.

Hoskins, S. G., Homberg, U., Kingan, T. G., Christensen, T. A., and Hildebrand, J. G. 1986. Immunocytochemistry of GABA in the antennal lobes of the sphinx moth *Manduca sexta*. Cell Tissue Res. 244, 243–252.

Hösl, M. 1990. Pheromone-sensitive neurons in the deutocerebrum of *Periplaneta americana* – Receptive fields on the antenna. J. Comp. Physiol. [A] 167, 321–327.

Hosler, J. S., Buxton, K. L., and Smith, B. H. 2000. Impairment of olfactory discrimination by blockade of GABA and nitric oxide activity in the honey bee antennal lobes. Behav. Neurosci. 114, 514–525.

Huerta, R., Nowotny, T., Garcia-Sanchez, M., Abarbanel, H. D., and Rabinovich, M. I. 2004. Learning classification in the olfactory system of insects. Neural Comput. 16, 1601–1640.

Huetteroth, W. and Schachtner, J. 2005. Standard three-dimensional glomeruli of the *Manduca sexta* antennal lobe: a tool to study both developmental and adult neuronal plasticity. Cell Tissue Res. 319, 513–524.

Hummel, T. and Zipursky, S. L. 2004. Afferent induction of olfactory glomeruli requires *N*-cadherin. Neuron 42, 77–88.

Hummel, T., Vasconcelos, M. L., Clemens, J. C., Fishilevich, Y., Vosshall, L. B., and Zipursky, S. L. 2003. Axonal targeting of olfactory receptor neurons in Drosophila is controlled by Dscam. Neuron 37, 221–231.

Ignell, R. 2001. Monoamines and neuropeptides in antennal lobe interneurons of the desert locust, Schistocerca gregana: an immunocytochemical study. Cell Tissue Res. 306, 143–156.

Ignell, R., Dekker, T., Ghaninia, M., and Hansson, B. 2005. The neuronal architecture of the mosquito deutocerebrum. J. Comp. Neurol. 493, 207–240.

Isabel, G., Pascual, A., and Preat, T. 2004. Exclusive consolidated memory phases in Drosophila. Science 304, 1024–1027.

Ito, K., Suzuki, K., Estes, P., Ramaswami, M., Yamamoto, D., and Strausfeld, N. J. 1998. The organization of extrinsic neurons and their implications in the functional roles of the mushroom bodies in *Drosophila melanogaster* Meigen. Learn. Mem. 5, 52–77.

Iwano, M. and Kanzaki, R. 2005. Immunocytochemical identification of neuroactive substances in the antennal lobe of the male silkworm moth *Bombyx mori*. Zoolog. Sci. 22, 199–211.

Iwasaki, M., Mizunami, M., Nishikawa, M., Itoh, T., and Tominaga, Y. 1999. Ultrastructural analyses of modular subunits in the mushroom bodies of the cockroach. J. Electron Microsc. 48, 55–62.

Jackson, F. R., Newby, L. M., and Kulkarni, S. J. 1990. Drosophila GABAergic systems: sequence and expression of glutamic acid decarboxylase. J. Neurochem. 54, 1068–1078.

Jefferis, G., Marin, E. C., Stocker, R. F., and Luo, L. Q. 2001. Target neuron prespecification in the olfactory map of Drosophila. Nature 414, 204–208.

Jefferis, G. S., Potter, C. J., Chan, A. M., Marin, E. C., Rohlfing, T., Maurer, C. R., Jr., and Luo, L. 2007. Comprehensive maps of Drosophila higher olfactory centers: spatially segregated fruit and pheromone representation. Cell 128, 1187–1203.

Jefferis, G., Vyas, R. M., Berdnik, D., Ramaekers, A., Stocker, R. F., Tanaka, N. K., Ito, K., and Luo, L. Q. 2004. Developmental origin of wiring specificity in the olfactory system of Drosophila. Development 131, 117–130.

Jhaveri, D. and Rodrigues, V. 2002. Sensory neurons of the Atonal lineage pioneer the formation of glomeruli within the adult Drosophila olfactory lobe. Development 129, 1251–1260.

Jhaveri, D., Saharan, S., Sen, A., and Rodrigues, V. 2004. Positioning sensory terminals in the olfactory lobe of Drosophila by Robo signaling. Development 131, 1903–1912.

Jhaveri, D., Sen, A., and Rodrigues, V. 2000. Mechanisms underlying olfactory neuronal connectivity in Drosophila – The atonal lineage organizes the periphery while sensory neurons and glia pattern the olfactory lobe. Dev. Biol. 226, 73–87.

Joerges, J., Küttner, A., Galizia, C. G., and Menzel, R. 1997. Representations of odours and odour mixtures visualized in the honeybee brain. Nature 387, 285–288.

Jones, W. D., Cayirlioglu, P., Kadow, I. G., and Vosshall, L. B. 2007. Two chemosensory receptors together mediated carbon dioxide detection in Drosophila. Nature 445, 86–90.

Jones, W. D., Nguyen, T. A., Kloss, B., Lee, K. J., and Vosshall, L. B. 2005. Functional conservation of an insect odorant receptor gene across 250 million years of evolution. Curr. Biol. 15, R119–121.

Justus, K. A., Cardé, R. T., and French, A. S. 2005. Dynamic properties of antennal responses to pheromone in two moth species. J. Neurophysiol. 93, 2233–2239.

Kaissling, K. E., Hildebrand, J. G., and Tumlinson, J. H. 1989. Pheromone receptor cells in the male moth Manduca sexta. Arch. Insect Biochem. and Physiol. 10, 273–279.

Karban, R. and Baldwin, I. T. 1997. Induced Responses to Herbivory. University of Chicago Press.

Kasowski, H. J., Kim, H., and Greer, C. A. 1999. Compartmental organization of the olfactory bulb glomerulus. J. Comp. Neurol. 407, 261–274.

Keene, A. C., Stratmann, M., Keller, A., Perrat, P. N., Vosshall, L. B., and Waddell, S. 2004. Diverse odor-conditioned memories require uniquely timed dorsal paired medial neuron output. Neuron 44, 521–533.

Keil, T. A. 1996. Sensilla on the maxillary palps of Helicoverpa armigera caterpillars: in search of the CO_2-receptor. Tissue Cell 28, 703–717.

Kent, K. S., Harrow, I. D., Quartararo, P., and Hildebrand, J. G. 1986. An accessory olfactory pathway in Lepidoptera: the labial pit organ and its central projections in Manduca sexta and certain other sphinx moths and silk moths. Cell Tissue Res. 245, 237–245.

Kent, K. S., Hoskins, S. G., and Hildebrand, J. G. 1987. A novel serotonin-immunoreactive neuron in the antennal lobe of the sphinx moth Manduca sexta persists throughout postembryonic life. J. Neurobiol. 18, 451–465.

Kirschner, S., Kleineidam, C. J., Zube, C., Rybak, J., Grubewakd, B., and Rossler, W. 2006. Dual olfactoty pathway in the honeybee, Apis mellifera. J. Comp Neurol. 499, 933–952.

Kleineidam, C. and Tautz, J. 1996. Perception of carbon dioxide and other "air-condition" parameters in the leaf cutting ant Atta cephalotes. Naturwissenschaften 83, 566–568.

Kline, D. L. 1998. Olfactory responses and field attraction of mosquitoes to volatiles from Limburger cheese and human foot odor. J. Vector Ecol. 23, 186–194.

Kloppenburg, P., Camazine, S. M., Sun, X. J., Randolph, P., and Hildebrand, J. G. 1997. Organization of the antennal motor system in the sphinx moth Manduca sexta. Cell Tissue Res. 287, 425–433.

Komischke, B., Sandoz, J. C., Malun, D., and Giurfa, M. 2005. Partial unilateral lesions of the mushroom bodies affect olfactory learning in honeybees Apis mellifera L. Eur. J. Neurosci. 21, 477–485.

Komiyama, T., Carlson, J. R., and Luo, L. 2004. Olfactory receptor neuron axon targeting: intrinsic transcriptional control and hierarchical interactions. Nat. Neurosci. 7, 819–825.

Komiyama, T., Johnson, W. A., Luo, L., and Jefferis, G. S. 2003. From lineage to wiring specificity. POU domain transcription factors control precise connections of Drosophila olfactory projection neurons. Cell 112, 157–167.

Kreher, S. A., Kwon, J. Y., and Carlson, J. R. 2005. The molecular basis of odor coding in the Drosophila larva. Neuron 46, 445–456.

Kreissl, S. and Bicker, G. 1989. Histochemistry of acetylcholinesterase and immunocytochemistry of an acetylcholine receptor-like antigen in the brain of the honeybee. J. Comp. Neurol. 286, 71–84.

Kreissl, S., Eichmuller, S., Bicker, G., Rapus, J., and Eckert, M. 1994. Octopamine-like immunoreactivity in the brain and subesophageal ganglion of the honeybee. J. Comp. Neurol. 348, 583–595.

Kretzschmar, D. and Pflugfelder, G. O. 2002. Glia in development, function, and neurodegeneration of the adult insect brain. Brain Res. Bull. 57, 121–131.

Krieger, J., Klink, O., Mohl, C., Raming, K., and Breer, H. 2003. A candidate olfactory receptor subtype highly conserved across different insect orders. J. Comp. Physiol. [A] 189, 519–526.

Kurtovic, A., Widmer, A., and Dickson, B. J. 2007. A single class of olfactory neurons mediates behavioural responses to a Drosophila sex pheromone. Nature 446, 542–546.

Kwon, H. W., Lent, D. D., and Strausfeld, N. J. 2004. Spatial learning in the restrained American cockroach Periplaneta americana. J. Exp. Biol. 207, 377–383.

Lacey, E. S., Ginzel, M. D., Millar, J. G., and Hanks, L. M. 2004. Male-produced aggregation pheromone of the cerambycid beetle Neoclytus acuminatus acuminatus. J. Chem. Ecol. 30, 1493–1507.

Lacher, V. 1964. Elektrophysiologische Untersuchungen an einzelnen Rezeptoren für Geruch, Kohlendioxyd, Luftfeuchtigkeit und Temperatur auf den Antennen der Arbeitsbiene und der Drohne (Apis mellifica L.) [Electrophysiological single-receptor analysis for odor, CO_2, humidity and temperature on the antennae of worker and drone honeybees]. Zeitschrift für vergleichende Physiologie 48, 587–623.

Lacroix, R., Mukabana, W. R., Gouagna, L. C., and Koella, J. C. 2005. Malaria infection increases attractiveness of humans to mosquitoes. PLoS Biol. 3, e298.

Laissue, P. P., Reiter, C., Hiesinger, P. R., Halter, S., Fischbach, K. F., and Stocker, R. F. 1999. Three-dimensional reconstruction of the antennal lobe in Drosophila melanogaster. J. Comp. Neurol. 405, 543–552.

Larsson, M. C., Domingos, A. I., Jones, W. D., Chiappe, M. E., Amrein, H., and Vosshall, L. B. 2004. Or83b encodes a broadly expressed odorant receptor essential for Drosophila olfaction. Neuron 43, 703–714.

Laurent, G. 1996. Dynamical representation of odors by oscillating and evolving neural assemblies. Trends Neurosci. 19, 489–496.

Laurent, G. 1999. A systems perspective on early olfactory coding. Science 286, 723–728.

Laurent, G. 2002. Olfactory network dynamics and the coding of multidimensional signals. Nat. Rev. Neurosci. 3, 884–895.

Laurent, G. and Davidowitz, H. 1994. Encoding of olfactory information with oscillating neural assemblies. Science 265, 1872–1875.

Laurent, G., Stopfer, M., Friedrich, R. W., Rabinovich, M. I., Volkovskii, A., and Abarbanel, H. D. 2001. Odor encoding as an active, dynamical process: experiments, computation, and theory. Annu. Rev. Neurosci. 24, 263–297.

Laurent, G., Wehr, M., and Davidowitz, H. 1996. Temporal representations of odors in an olfactory network. J. Neurosci. 16, 3837–3847.

Leitch, B. and Laurent, G. 1996. GABAergic synapses in the antennal lobe and mushroom body of the locust olfactory system. J. Comp. Neurol. 372, 487–514.

Lenoir, A., D'Ettorre, P., Errard, C., and Hefetz, A. 2001. Chemical ecology and social parasitism in ants. Annu. Rev. Entomol. 46, 573–599.

Leon, M. and Johnson, B. A. 2003. Olfactory coding in the mammalian olfactory bulb. Brain Res. Brain Res. Rev. 42, 23–32.

Levin, L. R., Han, P. L., Hwang, P. M., Feinstein, P. G., Davis, R. L., and Reed, R. R. 1992. The Drosophila learning and memory gene rutabaga encodes a Ca^{2+}/Calmodulin-responsive adenylyl cyclase. Cell 68, 479–489.

Lin, H. H., Lai, J. S., Chin, A. L., Chen, Y. C., and Chiang, A. S. 2007. A map of olfactory representation in the Drosophila mushroom body. Cell 128, 1205–1217.

Linn, C. E., Jr., Dambroski, H. R., Feder, J. L., Berlocher, S. H., Nojima, S., and Roelofs, W. L. 2004. Postzygotic isolating factor in sympatric speciation in Rhagoletis flies: reduced response of hybrids to parental host-fruit odors. Proc. Natl. Acad. Sci. U. S. A. 101, 17753–17758.

Linn, C., Jr., Feder, J. L., Nojima, S., Dambroski, H. R., Berlocher, S. H., and Roelofs, W. 2003. Fruit odor discrimination and sympatric host race formation in Rhagoletis. Proc. Natl. Acad. Sci. U. S. A. 100, 11490–11493.

Linster, C., Kerszberg, M., and Masson, C. 1994. How neurons may compute: the case of insect sexual pheromone discrimination. J. Comput. Neurosci. 1, 231–238.

Linster, C., Sachse, S., and Galizia, C. G. 2005. Computational modeling suggests that response properties rather than spatial position determine connectivity between olfactory glomeruli. J. Neurophysiol. 93, 3410–3417.

MacLeod, K. and Laurent, G. 1996. Distinct mechanisms for synchronization and temporal patterning of odor-encoding neural assemblies. Science 274, 976–979.

Malun, D. 1991a. Inventory and distribution of synapses of identified uniglomerular projection neurons in the antennal lobe of Periplaneta americana. J. Comp. Neurol. 305, 348–360.

Malun, D. 1991b. Synaptic relationships between GABA-immunoreactive neurons and an identified uniglomerular projection neuron in the antennal lobe of Periplaneta americana: a double-labeling electron microscopic study. Histochemistry 96, 197–207.

Malun, D., Giurfa, M., Galizia, C. G., Plath, N., Brandt, R., Gerber, B., and Eisermann, B. 2002. Hydroxyurea-induced partial mushroom body ablation does not affect acquisition and retention of olfactory differential conditioning in honeybees. J. Neurobiol. 53, 343–360.

Malun, D., Waldow, U., Kraus, D., and Boeckh, J. 1993. Connections between the deutocerebrum and the protocerebrum, and neuroanatomy of several classes of deutocerebral projection neurons in the brain of male Periplaneta americana. J. Comp. Neurol. 329, 143–162.

Mao, Z., Roman, G., Zong, L., and Davis, R. L. 2004. Pharmacogenetic rescue in time and space of the rutabaga memory impairment by using Gene-Switch. Proc. Natl. Acad. Sci. U. S. A. 101, 198–203.

Marin, E. C., Jefferis, G. S., Komiyama, T., Zhu, H., and Luo, L. 2002. Representation of the glomerular olfactory map in the Drosophila brain. Cell 109, 243–255.

Marin, E. C., Watts, R. J., Tanaka, N. K., Ito, K., and Luo, L. Q. 2005. Developmentally programmed remodeling of the Drosophila olfactory circuit. Development 132, 725–737.

Masante-Roca, I., Gadenne, C., and Anton, S. 2005. Three-dimensional antennal lobe atlas of male and female moths, Lobesia botrana (Lepidoptera: Tortricidae) and glomerular representation of plant volatiles in females. J. Exp. Biol. 208, 1147–1159.

Matsumoto, S. G. and Hildebrand, J. G. 1981. Olfactory mechanisms in the moth Manduca sexta – response characteristics and morphology of central neurons in the antennal lobes. Proc. R. Soc. Lond. B Biol. Sci. 213, 249–277.

Matsumoto, Y. and Mizunami, M. 2004. Context-dependent olfactory learning in an insect. Learn. Mem. 11, 288–293.

Mauelshagen, J. 1993. Neural correlates of olfactory learning paradigms in an identified neuron in the honeybee brain. J. Neurophysiol. 69, 609–625.

McGuire, S. E., Le, P. T., and Davis, R. L. 2001. The role of Drosophila mushroom body signaling in olfactory memory. Science 293, 1330–1333.

McGuire, S. E., Le, P. T., Osborn, A. J., Matsumoto, K., and Davis, R. L. 2003. Spatiotemporal rescue of memory dysfunction in Drosophila. Science 302, 1765–1768.

Meijerink, J., Carlsson, M. A., and Hansson, B. S. 2003. Spatial representation of odorant structure in the moth antennal lobe: a study of structure-response relationships at low doses. J. Comp. Neurol. 467, 11–21.

Menzel, R. and Giurfa, M. 1999. Cognition by a mini brain. Nature 400, 718–719.

Menzel, R. and Giurfa, M. 2001. Cognitive architecture of a mini-brain: the honeybee. Trends Cogn. Sci. 5, 62–71.

Menzel, R. and Müller, U. 1996. Learning and memory in honeybees: from behavior to neural substrates. Annu. Rev. Neurosci. 19, 379–404.

Menzel, R., Brandt, R., Gumbert, A., Komischke, B., and Kunze, J. 2000. Two spatial memories for honeybee navigation. Proc. R. Soc. Lond. B Biol. Sci. 267, 961–968.

Mesce, K. A., DeLorme, A. W., Brelje, T. C., and Klukas, K. A. 2001. Dopamine-synthesizing neurons include the putative H-cell homologue in the moth Manduca sexta. J. Comp. Neurol. 430, 501–517.

Mizunami, M., Weibrecht, J. M., and Strausfeld, N. J. 1998. Mushroom bodies of the cockroach: their participation in place memory. J. Comp. Neurol. 402, 520–537.

Mobbs, P. G. 1982. The brain of the honeybee Apis mellifera 1. The connections and spatial organization of the mushroom bodies. Philos. R. Soc. Lond. B Biol. Sci. 298, 309–354.

Mobbs, P. G. 1985. Brain Structure. In: Comprehensive Insect Physiology Biochemistry and Pharmacology. In: Nervous System: Structure and Motor Function, Vol. 5, (eds. G. A. Kerkut and L. I. Gilbert), pp. 299–370. Pergamon Press.

Monastirioti, M. 1999. Biogenic amine systems in the fruit fly Drosophila melanogaster. Microsc. Res. Tech. 45, 106–121.

Moritz, R. F. A. and Southwick, E. E. 1992. Bees as Superorganisms: An Evolutionary Reality. Springer-Verlag.

Müller, U. 2000. Prolonged activation of cAMP-dependent protein kinase during conditioning induces long-term memory in honeybees. Neuron 27, 159–168.

Müller, U. and Hildebrandt, H. 2002. Nitric oxide/cGMP-mediated protein kinase A activation in the antennal lobes plays an important role in appetitive reflex habituation in the honeybee. J. Neurosci. 22, 8739–8747.

Müller, D., Abel, R., Brandt, R., Zockler, M., and Menzel, R. 2002. Differential parallel processing of olfactory information in the honeybee, Apis mellifera L. J. Comp. Physiol. [A] 188, 359–370.

Murlis, J., Elkinton, J. S., and Cardé, R. T. 1992. Odor plumes and how insects use them. Annu. Rev. Entomol. 37, 505–532.

Nässel, D. R. 1988. Serotonin and serotonin-immunoreactive neurons in the nervous system of insects. Prog. Neurobiol. 30, 1–85.

Nässel, D. R. 1993. Neuropeptides in the insect brain: a review. Cell Tissue Res. 273, 1–29.

Nässel, D. R. 1999. Histamine in the brain of insects: a review. Microsc. Res. Tech. 44, 121–136.

Nässel, D. R. 2000. Functional roles of neuropeptides in the insect central nervous system. Naturwissenschaften 87, 439–449.

Nässel, D. R. and Elekes, K. 1992. Aminergic neurons in the brain of blowflies and *Drosophila* – Dopamine-immunoreactive and tyrosine hydroxylase-immunoreactive neurons and their relationship with putative histaminergic neurons. Cell Tissue Res. 267, 147–167.

Neder, R. 1959. Allometrisches Wachstum von Hirnteilen bei drei verschieden grossen Schabenarten. (Allometric growth of brain structures in three cockroaches of different size). Zool. Jahrb. Anat. 77, 411–464.

Ng, M., Roorda, R. D., Lima, S. Q., Zemelman, B. V., Morcillo, P., and Miesenbock, G. 2002. Transmission of olfactory information between three populations of neurons in the antennal lobe of the fly. Neuron 36, 463–474.

Nishikawa, M., Yokohari, F., and Ishibashi, T. 1995. Central projections of the antennal cold receptor neurons and hygroreceptor neurons of the cockroach *Periplaneta americana*. J. Comp. Neurol. 361, 165–176.

Nishino, H., Yamashita, S., Yamazaki, Y., Nishikawa, M., Yokohari, F., and Mizunami, M. 2003. Projection neurons originating from thermo- and hygrosensory glomeruli in the antennal lobe of the cockroach. J. Comp. Neurol. 455, 40–55.

Nuñez, J., Almeida, L., Balderrama, N., and Giurfa, M. 1997. Alarm pheromone induces stress analgesia via an opioid system in the honeybee. Physiol. Behav. 63, 75–80.

Ochieng, S. A., Hallberg, E., and Hansson, B. S. 1998. Fine structure and distribution of antennal sensilla of the desert locust, *Schistocerca gregaria* (Orthoptera: Acrididae). Cell Tissue Res. 291, 525–536.

Oland, L. A. and Tolbert, L. P. 1988. Effects of hydroxyurea parallel the effects of radiation in developing olfactory glomeruli in insects. J. Comp. Neurol. 278, 377–387.

Oland, L. A. and Tolbert, L. P. 2003. Key interactions between neurons and glial cells during neural development in insects. Annu. Rev. Entomol. 48, 89–110.

Oland, L. A., Marrero, H. G., and Burger, I. 1999. Glial cells in the developing and adult olfactory lobe of the moth *Manduca sexta*. Cell Tissue Res. 297, 527–545.

Oland, L. A., Tolbert, L. P., and Mossman, K. L. 1988. Radiation-induced reduction of the glial population during development disrupts the formation of olfactory glomeruli in an insect. J. Neurosci. 8, 353–367.

Olsen, S. R., Bhandawat, V., and Wilson, R. I. 2007. Excitatory interactions between olfactory processing channels in the Drosophila antennal lobe. Neuron 54, 89–103.

Olshausen, B. A. and Field, D. J. 2004. Sparse coding of sensory inputs. Curr. Opin. Neurobiol. 14, 481–487.

Ott, S. R. and Elphick, M. R. 2002. Nitric oxide synthase histochemistry in insect nervous systems: methanol/formalin fixation reveals the neuroarchitecture of formaldehyde-sensitive NADPH diaphorase in the cockroach Periplaneta americana. J. Comp. Neurol. 448, 165–185.

Ozaki, M., Wada-Katsumata, A., Fujikawa, K., Iwasaki, M., Yokohari, F., Satoji, Y., Nisimura, T., and Yamaoka, R. 2005. Ant nestmate and non-nestmate discrimination by a chemosensory sensillum. Science 309, 311–314.

Pare, P. W. and Tumlinson, J. H. 1999. Plant volatiles as a defense against insect herbivores. Plant Physiol. 121, 325–332.

Pareto, A. 1972. Die zentrale Verteilung der Fühlerrafferenz bei Arbeiterinnen der Honigbiene, *Apis mellifera* L. [Spatial distribution of sensory antennal fibres in the central nervous system of worker bees]. Z Zellforsch Mikrosk Anat 131, 109–140.

Pascual, A. and Preat, T. 2001. Localization of long-term memory within the Drosophila mushroom body. Science 294, 1115–1117.

Peele, P., Ditzen, M., Menzel, R., and Galzia, C. G. 2006. Appetitive odor learning does not change olfactory coding in a subpopulation of honeybee antennal lobe neurons. J. Comp. Physiol. A 192, 1083–1103.

Pelz, D., Roeske, C., and Galizia, C. G. 2004. Odour response spectrum of a population of genetically identified olfactory sensory neurons: characterizing the input to a single glomerulus in *Drosophila melanogaster*. *ECRO*, Dijon, France.

Perazzona, B., Isabel, G., Preat, T., and Davis, R. L. 2004. The role of cAMP response element-binding protein in Drosophila long-term memory. J. Neurosci. 24, 8823–8828.

Perez-Orive, J., Bazhenov, M., and Laurent, G. 2004. Intrinsic and circuit properties favor coincidence detection for decoding oscillatory input. J. Neurosci. 24, 6037–6047.

Perez-Orive, J., Mazor, O., Turner, G. C., Cassenaer, S., Wilson, R. I., and Laurent, G. 2002. Oscillations and sparsening of odor representations in the mushroom body. Science 297, 359–365.

Pflüger, H. J. and Stevenson, P. A. 2005. Evolutionary aspects of octopaminergic systems with emphasis on arthropods. Arthropod Struct. Dev. 34, 379–396.

Pirvola, U., Tuomisto, L., Yamatodani, A., and Panula, P. 1988. Distribution of histamine in the cockroach brain and visual system: an immunocytochemical and biochemical study. J. Comp. Neurol. 276, 514–526.

Pitts, R. J., Fox, A. N., and Zwiebel, L. J. 2004. A highly conserved candidate chemoreceptor expressed in both olfactory and gustatory tissues in the malaria vector *Anopheles gambiae*. Proc. Natl. Acad. Sci. U. S. A. 101, 5058–5063.

Plowright, R. C. and Laverty, T. M. 1984. The ecology and sociobiology of bumble bees. Annu. Rev. Entomol. 29, 175–199.

Pophof, B. 2002. Octopamine enhances moth olfactory responses to pheromones, but not those to general odorants. J. Comp. Physiol. [A] 188, 659–662.

Prillinger, L. 1981. Postembryonic development of the antennal lobes in *Periplaneta americana* L. Cell Tissue Res. 215, 563–575.

Python, F. and Stocker, R. F. 2002a. Adult-like complexity of the larval antennal lobe of *D. melanogaster* despite markedly low numbers of odorant receptor neurons. J. Comp. Neurol. 445, 374–387.

Python, F. and Stocker, R. F. 2002b. Immunoreactivity against choline acetyltransferase, gamma-aminobutyric acid, histamine, octopamine, and serotonin in the larval chemosensory system of *Dosophila melanogaster*. J. Comp. Neurol. 453, 157–167.

Rabinovich, M. I., Huerta, R., Volkovskii, A., Abarbanel, H. D., Stopfer, M., and Laurent, G. 2000. Dynamical coding of sensory information with competitive networks. J. Physiol. Paris 94, 465–471.

Raguso, R. A. 2004. Flowers as sensory billboards: progress towards an integrated understanding of floral advertisement. Curr. Opin. Plant Biol. 7, 434–440.

Ramaekers, A., Magnenat, E., Marin, E. C., Gendre, N., Jefferis, G. S., Luo, L., and Stocker, R. F. 2005. Glomerular maps without cellular redundancy at successive levels of the Drosophila larval olfactory circuit. Curr. Biol. 15, 982–992.

Ramaekers, A., Parmentier, M. L., Lasnier, C., Bockaert, J., and Grau, Y. 2001. Distribution of metabotropic glutamate receptor DmGlu A in *Drosophila melanogaster* central nervous system. J. Comp. Neurol. 438, 213–225.

Rehder, V., Bicker, G., and Hammer, M. 1987. Serotonin-immunoreactive neurons in the antennal lobes and suboesophageal ganglion of the honeybee. Cell Tissue Res. 247, 59–66.

Rein, K., Zockler, M., Mader, M. T., Grubel, C., and Heisenberg, M. 2002. The Drosophila standard brain. Curr. Biol. 12, 227–231.

Reisenman, C. E., Christensen, T. A., Francke, W., and Hildebrand, J. G. 2004. Enantioselectivity of projection neurons innervating identified olfactory glomeruli. J. Neurosci. 24, 2602–2611.

Riley, J. R., Greggers, U., Smith, A. D., Reynolds, D. R., and Menzel, R. 2005. The flight paths of honeybees recruited by the waggle dance. Nature 435, 205–207.

Robertson, H. M. and Wanner, K. W. 2006. The chemoreceptor superfamily in the honey bee, *Apis mellifera*: expansion of the odorant, but not gustatory, receptor family. Genome. Res. 16, 1395–1403.

Rodrigues, V. and Buchner, E. 1984. [3H]2-deoxyglucose mapping of odor-induced neuronal activity in the antennal lobes of *Drosophila melanogaster*. Brain Res. 324, 374–378.

Roeder, T. 1999. Octopamine in invertebrates. Prog. Neurobiol. 59, 533–561.

Rospars, J. P. 1988. Structure and development of the insect antennodeutocerebral system. Int. J. Insect Morphol. Embryol. 17, 243–294.

Rospars, J. P. and Hildebrand, J. G. 1992. Anatomical identification of glomeruli in the antennal lobes of the male sphinx moth *Manduca sexta*. Cell Tissue Res. 270, 205–227.

Rössler, W., Oland, L. A., Higgins, M. R., Hildebrand, J. G., and Tolbert, L. P. 1999b. Development of a glia-rich axon-sorting zone in the olfactory pathway of the moth *Manduca sexta*. J. Neurosci. 19, 9865–9877.

Rössler, W., Randolph, P. W., Tolbert, L. P., and Hildebrand, J. G. 1999a. Axons of olfactory receptor cells of transsexually grafted antennae induce development of sexually dimorphic glomeruli in *Manduca sexta*. J. Neurobiol. 38, 521–541.

Rostelien, T., Borg-Karlson, A. K., Fäldt, J., Jacobsen, U., and Mustaparta, H. 2000. The plant sesquiterpene germacrene D specifically activates a major type of antennal receptor neruons of the tobacco budworm moth *Heliothis virescens*. Chem. Senses 25, 141–148.

Rutledge, L. C., Moussa, M. A., Lowe, C. A., and Sofield, R. K. 1978. Comparative sensitivity of mosquito species and strains to the repellent diethyl toluamide. J. Med. Entomol. 14, 536–541.

Rybak, J. 1994. Die strukturelle Organisation der Pilzkîrper und synaptische Konnektivität protocerebraler Interneurone im Gehirn der Honigbiene, *Apis mellifera*. Eine licht- und elektronenmikroskopische Studie. Ph.D. thesis, Freie Universität Berlin.

Rybak, J. and Menzel, R. 1993. Anatomy of the mushroom bodies in the honey bee brain: the neuronal connections of the alpha-lobe. J. Comp. Neurol. 334, 444–465.

Rybak, J. and Menzel, R. 1998. Integrative properties of the Pe1 neuron, a unique mushroom body output neuron. Learn. Mem. 5, 133–145.

Sachse, S. and Galizia, C. G. 2002. Role of inhibition for temporal and spatial odor representation in olfactory output neurons: a calcium imaging study. J. Neurophysiol. 87, 1106–1117.

Sachse, S. and Galizia, C. G. 2003. The coding of odour-intensity in the honeybee antennal lobe: local computation optimizes odour representation. Eur. J. Neurosci. 18, 2119–2132.

Sachse, S. and Galizia, C. G. 2006. Topography and Dynamics of the Olfactory System. In Microcircuits: The Interface Between Neurons and Global Brain Function. Dahlem Workshop Report 93 (ed. S. Grillner), pp. 251–273. MIT Press.

Sachse, S., Rappert, A., and Galizia, C. G. 1999. The spatial representation of chemical structures in the antennal lobe of honeybees: steps towards the olfactory code. Eur. J. Neurosci. 11, 3970–3982.

Sakura, M., Okada, R., and Mizunami, M. 2002. Olfactory discrimination of structurally similar alcohols by cockroaches. J. Comp. Physiol. [A] 188, 787–797.

Salecker, I. and Distler, P. 1990. Serotonin-immunoreactive neurons in the antennal lobes of the American cockroach *Periplaneta americana*: light- and electron-microscopic observations. Histochemistry 94, 463–473.

Sanes, J. R. and Hildebrand, J. G. 1976a. Acetylcholine and its metabolic enzymes in developing antennae of moth, *Manduca sexta*. Dev. Biol. 52, 105–120.

Sanes, J. R. and Hildebrand, J. G. 1976b. Structure and development of antennae in a moth, *Manduca sexta*. Dev. Biol. 51, 282–299.

Sanes, J. R. and Hildebrand, J. G. 1976c. Origin and morphogenesis of sensory neurons in an insect antenna. Dev. Biol. 51, 300–319.

Schachtner, J., Schmidt, M., and Homberg, U. 2005. Organization and evolutionary trends of primary olfactory brain centers in Tetraconata (Crustacea + Hexapoda). Arthropod Struct. Dev. 34, 257–299.

Schachtner, J., Trosowski, B., D'Hanis, W., Stubner, S., and Homberg, U. 2004. Development and steroid regulation of RFamide inummoreactivity in antennal lobe neurons of the sphinx moth *Manduca sexta*. J. Exp. Biol. 207, 2389–2400.

Schäfer, S. and Bicker, G. 1986. Distribution of GABA-like immunoreactivity in the brain of the honeybee. J. Comp. Neurol. 246, 287–300.

Schäfer, S. and Rehder, V. 1989. Dopamine-like immunoreactivity in the brain and suboesophageal ganglion of the honeybee. J. Comp. Neurol. 280, 43–58.

Schafer, R. and Sanchez, T. V. 1973. Antennal sensory system of the cockroach, *Periplaneta americana*: postembryonic development and morphology of the sense organs. J. Comp. Neurol. 149, 335–354.

Schäfer, S., Bicker, G., Ottersen, O. P., and Storm-Mathisen, J. 1988. Taurine-like immunoreactivity in the brain of the honeybee. J. Comp. Neurol. 268, 60–70.

Scherer, S., Stocker, R. F., and Gerber, B. 2003. Olfactory learning in individually assayed *Drosophila* larvae. Learn. Mem. 10, 217–225.

Schief, M. L. and Wilson, R. I. 2007. Olfactory processing and behavior downstream from highly selective receptor neurons. Nat. Neurosci. 10, 623–630.

Schiestl, F. P., Ayasse, M., Paulus, H. F., Lofstedt, C., Hansson, B. S., Ibarra, F., and Francke, W. 2000. Sex pheromone mimicry in the early spider orchid (*Ophrys sphegodes*): patterns of hydrocarbons as the key mechanism for pollination by sexual deception. J. Comp. Physiol. [A] 186, 567–574.

Schneiderman, A. M., Hildebrand, J. G., Brennan, M. M., and Tumlinson, J. H. 1986. Transsexually grafted antennae alter pheromone directed behavior in a moth. Nature 323, 801–803.

Schulz, D. J., Barron, A. B., and Robinson, G. E. 2002. A role for octopamine in honey bee division of labor. Brain Behavior and Evolution 60, 350–359.

Schürmann, F. W. 2000. Acetylcholine, GABA, glutamate and NO as putative transmitters indicated by immunocytochemistry in the olfactory mushroom body system of the insect brain. Acta Biol. Hung. 51, 355–362.

Schürmann, F. W. and Klemm, N. 1984. Serotonin-immunoreactive neurons in the brain of the honeybee. J. Comp. Neurol. 225, 570–580.

Schürmann, F. W., Elekes, K., and Geffard, M. 1989. Dopamine-like immunoreactivity in the bee brain. Cell Tissue Res. 256, 399–410.

Schwaerzel, M., Heisenberg, M., and Zars, T. 2002. Extinction antagonizes olfactory memory at the subcellular level. Neuron 35, 951–960.

Schwaerzel, M., Monastirioti, M., Scholz, H., Friggi-Grelin, F., Birman, S., and Heisenberg, M. 2003. Dopamine and octopamine differentiate between aversive and appetitive olfactory memories in Drosophila. J. Neurosci. 23, 10495–10502.

Seidel, C. and Bicker, G. 1997. Colocalization of NADPH-diaphorase and GABA-immunoreactivity in the olfactory and visual system of the locust. Brain Res. 769, 273–280.

Shanbhag, S. R., Singh, K., and Singh, R. N. 1995. Fine-structure and primary sensory projections of sensilla located in the sacculus of the antenna of Drosophila melanogaster. Cell Tissue Res. 282, 237–249.

Shang, Y., Claridge-Chang, A., Sjulson, L., Oyoaert, M., and Miesenbock, G. 2007. Exciatory local circuits and their implications for olfactory processing in the fly antennal lobe. Cell 128, 601–612.

Shepherd, G. M. and Greer, C. A. 1998. Olfactory Bulb. In: The Synaptic Organization of the Brain, 4th ed. (ed. G. M. Shepherd), pp. 159–203. Oxford University Press.

Shirokova, E., Schmiedeberg, K., Bedner, P., Niessen, H., Willecke, K., Raguse, J. D., Meyerhof, W., and Krautwurst, D. 2005. Identification of specific ligands for orphan olfactory receptors. G protein-dependent agonism and antagonism of odorants. J. Biol. Chem. 280, 11807–11815.

Sigg, D., Thompson, C. M., and Mercer, A. R. 1997. Activity-dependent changes to the brain and behavior of the honey bee, Apis mellifera (L.). J. Neurosci. 17, 7148–7156.

Sinakevitch, I., Farris, S. M., and Strausfeld, N. J. 2001. Taurine-, aspartate- and glutamate-like immunoreactivity identifies chemically distinct subdivisions of Kenyon cells in the cockroach mushroom body. J. Comp. Neurol. 439, 352–367.

Sinakevitch, I., Niwa, M., and Strausfeld, N. J. 2005. Octopamine-like immunoreactivity in the honey bee and cockroach: comparable organization in the brain and subesophageal ganglion. J. Comp. Neurol. 488, 233–254.

Skiri, H. T., Galizia, C. G., and Mustaparta, H. 2004. Representation of primary plant odorants in the antennal lobe of the moth Heliothis virescens using calcium imaging. Chem. Senses 29, 253–267.

Skiri, H. T., Ro, H., Berg, B. G., and Mustaparta, H. 2005. Consistent organization of glomeruli in the antennal lobes of related species of heliothine moths. J. Comp. Neurol. 491, 367–380.

Stange, G. and Stowe, S. 1999. Carbon-dioxide sensing structures in terrestrial arthropods. Microsc. Res. Tech. 47, 416–427.

Steib, B. M., Geier, M., and Boeckh, J. 2001. The effect of lactic acid on odour-related host preference of yellow fever mosquitoes. Chem. Senses 26, 523–528.

Stengl, M., Homberg, U., and Hildebrand, J. G. 1990. Acetylcholinesterase activity in antennal receptor neurons of the sphinx moth Manduca sexta. Cell Tissue Res. 262, 245–252.

Stensmyr, M. C., Giordano, E., Balloi, A., Angioy, A. M., and Hansson, B. S. 2003. Novel natural ligands for Drosophila olfactory receptor neurones. J. Exp. Biol. 206, 715–724.

Stetter, M., Greve, H., Galizia, C. G., and Obermayer, K. 2001. Analysis of calcium imaging signals from the honeybee brain by nonlinear models. Neuroimage 13, 119–128.

Stevenson, P. A. and Sporhase-Eichmann, U. 1995. Localization of octopaminergic neurones in insects. Comp. Biochem. Physiol. A 110, 203–215.

Stocker, R. F. 1994. The organization of the chemosensory system in Drosophila melanogaster: a review. Cell Tissue Res. 275, 3–26.

Stocker, R. F. 2001. Drosophila as a focus in olfactory research: mapping of olfactory sensilla by fine structure, odor specificity, odorant receptor expression, and central connectivity. Microsc. Res. Tech. 55, 284–296.

Stocker, R. F., Heimbeck, G., Gendre, N., and de Belle, J. S. 1997. Neuroblast ablation in Drosophila P[GAL4] lines reveals origins of olfactory interneurons. J. Neurobiol. 32, 443–456.

Stocker, R. F., Lienhard, M. C., Borst, A., and Fischbach, K. F. 1990. Neuronal architecture of the antennal lobe in Drosophila melanogaster. Cell Tissue Res. 262, 9–34.

Stocker, R. F., Singh, R. N., Schorderet, M., and Siddiqi, O. 1983. Projection patterns of different types of antennal sensilla in the antennal glomeruli of Drosophila melanogaster. Cell Tissue Res. 232, 237–248.

Stopfer, M. and Laurent, G. 1999. Short-term memory in olfactory network dynamics. Nature 402, 664–668.

Stopfer, M., Bhagavan, S., Smith, B. H., and Laurent, G. 1997. Impaired odour discrimination on desynchronization of odour-encoding neural assemblies. Nature 390, 70–74.

Stopfer, M., Jayaraman, V., and Laurent, G. 2003. Intensity versus identity coding in an olfactory system. Neuron 39, 991–1004.

Stopfer, M., Wehr, M., MacLeod, K., and Laurent, G. 1999. Neural Dynamics, Oscillatory Synchronisation, and Odour Codes. In: Insect Olfaction (ed. B. S. Hansson), pp. 163–180. Springer.

Strausfeld, N. J. 1976. Atlas of an Insect Brain. Springer-Verlag.

Strausfeld, N. J. 2002. Organization of the honey bee mushroom body: representation of the calyx within the vertical and gamma lobes. J. Comp. Neurol. 450, 4–33.

Strausfeld, N. J. and Gilbert, C. 1992. Small-field neurons associated with oculomotor control in muscoid flies: cellular organization in the lobula plate. J. Comp. Neurol. 316, 56–71.

Strausfeld, N. J. and Hildebrand, J. G. 1999. Olfactory systems: common design, uncommon origins? Curr. Opin. Neurobiol. 9, 634–639.

Strausfeld, N. J. and Li, Y. 1999a. Representation of the calyces in the medial and vertical lobes of cockroach mushroom bodies. J. Comp. Neurol. 409, 626–646.

Strausfeld, N. J. and Li, Y. 1999b. Organization of olfactory and multimodal afferent neurons supplying the calyx and pedunculus of the cockroach mushroom bodies. J. Comp. Neurol. 409, 603–625.

Strausfeld, N. J., Hansen, L., Li, Y., Gomez, R. S., and Ito, K. 1998. Evolution, discovery, and interpretations of arthropod mushroom bodies. Learn. Mem. 5, 11–37.

Strausfeld, N. J., Homberg, U., and Kloppenburg, P. 2000. Parallel organization in honey bee mushroom bodies by peptidergic Kenyon cells. J. Comp. Neurol. 424, 179–195.

Strausfeld, N. J., Sinakevitch, I., and Vilinsky, I. 2003. The mushroom bodies of Drosophila melanogaster: an immunocytological and golgi study of Kenyon cell organization in the calyces and lobes. Microsc. Res. Tech. 62, 151–169.

Suh, G. S., Wong, A. M., Hergarden, A. C., Wang, J. W., Simon, A. F., Benzer, S., Axel, R., and Anderson, D. J. 2004. A single population of olfactory sensory neurons mediates an innate avoidance behaviour in Drosophila. Nature 431, 854–859.

Sun, X.-J., Fonta, C., and Masson, C. 1993a. Odour quality processing by bee antennal lobe interneurones. Chem. Senses 18, 355–377.

Sun, X.-J., Tolbert, L. P., and Hildebrand, J. G. 1993b. Ramification pattern and ultrastructural characteristics of the serotonin-immunoreactive neuron in the antennal lobe of the moth Manduca sexta: a laser scanning confocal and electron microscopic study. J. Comp. Neurol. 338, 5–16.

Sun, X.-J., Tolbert, L. P., and Hildebrand, J. G. 1997. Synaptic organization of the uniglomerular projection neurons of the antennal lobe of the moth *Manduca sexta*: a laser scanning confocal and electron microscopic study. J. Comp. Neurol. 379, 2–20.

Sutton, R. S. and Barto, A. G. 1981. Toward a modern theory of adaptive networks: expectation and prediction. Psychol. Rev. 88, 135–170.

Suzuki, H. 1975. Convergence of olfactory inputs from both antennae in the brain of the honeybee. J. Exp. Biol. 62, 11–26.

Sviderisky, V. L. and Plotnikova, S. I. 2004. On structural-functional organization of dragonfly mushroom bodies and some general considerations about purpose of these formations. J. Evol. Biochem. Physiol. 40, 608–624.

Symonds, M. R. and Elgar, M. A. 2004. The mode of pheromone evolution: evidence from bark beetles. Proc. Biol. Sci. 271, 839–846.

Symonds, M. R. and Wertheim, B. 2005. The mode of evolution of aggregation pheromones in Drosophila species. J. Evol. Biol. 18, 1253–1263.

Szyszka, P., Ditzen, M., Galkin, A., Galizia, C. G., and Menzel, R. 2005. Sparsening and temporal sharpening of olfactory representations in the honeybee mushroom bodies. J. Neurophysiol. 94, 3303–3313.

Tanaka, N. K., Awasaki, T., Shimada, T., and Ito, K. 2004. Integration of chemosensory pathways in the Drosophila second-order olfactory centers. Curr. Biol. 14, 449–457.

Thom, C., Guerenstein, P. G., Mechaber, W. L., and Hildebrand, J. G. 2004. Floral CO_2 reveals flower profitability to moths. J. Chem. Ecol. 30, 1285–1288.

Tichy, H. 2007. Humidity-dependent cold cells on the antenna of the stick insect. J. Neurophysiol. 97, 3851–3858.

Toga, A. W. and Thompson, P. M. 2001. Maps of the brain. Anat. Rec. 265, 37–53.

Tolbert, L. P., Oland, L. A., Tucker, E. S., Gibson, N. J., Higgins, M. R., and Lipscomb, B. W. 2004. Bidirectional influences between neurons and glial cells in the developing olfactory system. Prog. Neurobiol. 73, 73–105.

Tominaga, Y. and Yokohari, F. 1982. External structure of the sensillum capitulum, a hygro- and thermoreceptive sensillum of the cockroach, *Periplaneta americana*. Cell Tissue Res. 226, 309–318.

Tucker, E. S. and Tolbert, L. P. 2003. Reciprocal interactions between olfactory receptor axons and olfactory nerve glia cultured from the developing moth *Manduca sexta*. Dev. Biol. 260, 9–30.

Tucker, E. S., Oland, L. A., and Tolbert, L. P. 2004. In vitro analyses of interactions between olfactory receptor growth cones and glial cells that mediate axon sorting and glomerulus formation. J. Comp. Neurol. 472, 478–495.

Tully, T., Preat, T., Boynton, S. C., and Del Vecchio, M. 1994. Genetic dissection of consolidated memory in Drosophila. Cell 79, 35–47.

Turin, L. 1996. A spectroscopic mechanism for primary olfactory reception. Chem. Senses 21, 773–791.

Uchida, N. and Mainen, Z. F. 2003. Speed and accuracy of olfactory discrimination in the rat. Nat. Neurosci. 6, 1224–1229.

Vareschi, E. 1971. Duftunterscheidung bei der Honigbiene. Einzelzellableitungen und Verhaltensreaktionen. (Odor discrimination in honeybees. Single cell recordings and behavioral studies). ZverglPhysiol 75, 143–173.

Vickers, N. J. and Baker, T. C. 1997. Chemical communication in heliothine moths. VII. correlation between diminished responses to point-source plumes and single filaments similarly tainted with a behavioral antagonist. J. Comp. Physiol. [A] 180, 523–536.

Vickers, N. J., Christensen, T. A., Baker, T. C., and Hildebrand, J. G. 2001. Odour-plume dynamics influence the brain's olfactory code. Nature 410, 466–470.

Vosshall, L. B., Amrein, H., Morozov, P. S., Rzhetsky, A., and Axel, R. 1999. A spatial map of olfactory receptor expression in the *Drosophila* antenna. Cell 96, 725–736.

Vosshall, L. B., Wong, A. M., and Axel, R. 2000. An olfactory sensory map in the fly brain. Cell 102, 147–159.

Wachowiak, M., Cohen, L. B., and Ache, B. W. 2002. Presynaptic inhibition of olfactory receptor neurons in crustaceans. Microsc. Res. Tech. 58, 365–375.

Waddell, S. and Quinn, W. G. 2001. Flies, genes, and learning. Annu. Rev. Neurosci. 24, 1283–1309.

Waldrop, B. and Hildebrand, J. G. 1989. Physiology and pharmacology of acetylcholinergic responses of interneurons in the antennal lobes of the moth *Manduca sexta*. J. Comp. Physiol. [A] 164, 433–441.

Waldrop, B., Christensen, T. A., and Hildebrand, J. G. 1987. GABA-mediated synaptic inhibition of projection neurons in the antennal lobes of the sphinx moth, *Manduca sexta*. J. Comp. Physiol. [A] 161, 23–32.

Wang, Y. L., Guo, H. F., Pologruto, T. A., Hannan, F., Hakker, I., Svoboda, K., and Zhong, Y. 2004. Stereotyped odor-evoked activity in the mushroom body of Drosophila revealed by green fluorescent protein-based Ca^{2+} imaging. J. Neurosci. 24, 6507–6514.

Wang, F., Nemes, A., Mendelsohn, M., and Axel, R. 1998. Odorant receptors govern the formation of a precise topographic map. Cell 93, 47–60.

Wang, J. W., Wong, A. M., Flores, J., Vosshall, L. B., and Axel, R. 2003. Two-photon calcium imaging reveals an odor-evoked map of activity in the fly brain. Cell 112, 271–282.

Watanabe, H., Kobayashi, Y., Sakura, M., Matsumoto, Y., and Mizunami, M. 2003. Classical olfactory conditioning in the cockroach *Periplaneta americana*. Zoolog. Sci. 20, 1447–1454.

Weidenmüller, A., Kleineidam, C., and Tautz, J. 2002. Collective control of nest climate parameters in bumblebee colonies. Anim. Behav. 63, 1065–1071.

Wendt, B. and Homberg, U. 1992. Immunocytochemistry of dopamine in the brain of the locust *Schistocerca gregaria*. J. Comp. Neurol. 321, 387–403.

Wenner, A. M. 2002. The elusive honey bee dance "language" hypothesis. J. Ins. Behav. 15, 859–878.

Wenner, A. M. and Johnson, D. L. 1967. Honeybees: do they use direction and distance information provided by their dancers? Science 158, 1072–1077.

Willis, M. A. and Baker, T. C. 1984. Effects of intermittent and continuous pheromone stimulation on the flight behavior of the oriental fruit moth, *Grapholita molesta*. Physiol. Entomol. 9, 341–358.

Wilson, R. I. and Laurent, G. 2005. Role of GABAergic inhibition in shaping odor-evoked spatiotemporal patterns in the Drosophila antennal lobe. J. Neurosci. 25, 9069–9079.

Wilson, D. A. and Stevenson, R. J. 2003. The fundamental role of memory in olfactory perception. Trends Neurosci. 26, 243–247.

Wilson, R. I., Turner, G. C., and Laurent, G. 2004. Transformation of olfactory representations in the Drosophila antennal lobe. Science 303, 366–370.

Wiltrout, C., Dogra, S., and Linster, C. 2003. Configurational and nonconfigurational interactions between odorants in binary mixtures. Behav. Neurosci. 117, 236–245.

Winnington, A. P., Napper, R. M., and Mercer, A. R. 1996. Structural plasticity of identified glomeruli in the antennal lobes of the adult worker honey bee. J. Comp. Neurol. 365, 479–490.

Witthöft, W. 1967. Absolute Anzahl und Verteilung der Zellen im Hirn der Honigbiene [Absolute number and distribution of cells in the brain of honeybees]. Z. Morphol. Tiere 61, 160–184.

Wong, A. M., Wang, J. W., and Axel, R. 2002. Spatial representation of the glomerular map in the *Drosophila* protocerebrum. Cell 109, 229–241.

Xu, P., Atkinson, R., Jones, D. N., and Smith, D. P. 2005. Drosophila OBP LUSH is required for activity of pheromone-sensitive neurons. Neuron 45, 193–200.

Xu, F., Greer, C. A., and Shepherd, G. M. 2000. Odor maps in the olfactory bulb. J. Comp. Neurol. 422, 489–495.

Yokohari, F. 1983. The coelocapitular sensillum, an antennal hygro- and thermoreceptive sensillum of the honey bee, *Apis mellifera* L. Cell Tissue Res. 233, 355–365.

Yasuyama, K. and Salvaterra, P. M. 1999. Localization of choline acetyltransferase-expressing neurons in Drosophila nervous system. Microsc. Res. Tech. 45, 65–79.

Yasuyama, K., Kitamoto, T., and Salvaterra, P. M. 1995. Localization of choline acetyltransferase-expressing neurons in the larval visual system of *Drosophila melanogaster*. Cell Tissue Res. 282, 193–202.

Yin, J. C. and Tully, T. 1996. CREB and the formation of long-term memory. Curr. Opin. Neurobiol. 6, 264–268.

Yokohari, F., Tominaga, Y., and Tateda, H. 1982. Antennal hygroreceptors of the honey bee, *Apis mellifera* L. Cell Tissue Res. 226, 63–73.

Yu, D., Ponomarev, A., and Davis, R. L. 2004. Altered representation of the spatial code for odors after olfactory classical conditioning; memory trace formation by synaptic recruitment. Neuron 42, 437–449.

Zars, T., Fischer, M., Schulz, R., and Heisenberg, M. 2000. Localization of a short-term memory in Drosophila. Science 288, 672–675.

Zou, Z., Li, F., and Buck, L. B. 2005. Odor maps in the olfactory cortex. Proc. Natl. Acad. Sci. U. S. A. 102, 7724–7729.

Further Reading

Yasuyama, K., Meinertzhagen, I. A., and Schurmann, F. W. 2002. Synaptic organization of the mushroom body calyx in *Drosophila melanogaster*. J. Comp. Neurol. 445, 211–226.

Relevant Websites

http://www.flybrain.org – An online database of *Drosophila* nervous system.

http://neuro.uni-konstanz.de – Research group in insect olfaction.

4.42 Odor Plumes and Animal Orientation

M A Willis, Case Western Reserve University, Cleveland, OH, USA

Glossary

aesthetasc sensilla Peg-shaped sensillae on the antennules, or first antenna, of crustaceans that detect chemicals dissolved in the surrounding water. Each aesthetasc can contain dendrites from a few, to more than 100 olfactory receptor neurons.

boundary layer The layer of any fluid defined by the velocity gradient that begins at zero flow at the surface of an object and extends upward until the average ambient flow velocity is reached.

concha Several fine, narrow, and curled boney shelves that project into the nasal passages of mammals. Also known as turbinates.

epithelium A thin layer of cells that covers the outer surfaces and lines the inner surfaces of our bodies. A sensory epithelium is such an outer layer that lines a sensory structure and carries the sensory receptor cells embedded in it.

fluid dynamics The field of engineering involved in the description, characterization, and manipulation of the behavior of fluids, particularly water and air.

neuromast a sensory structure in the skin of fish that detects changes in the direction, or intensity of water movement.

The distribution of chemicals resulting from the interaction of flowing fluid with a source of volatile chemicals is commonly called a plume, which has a discontinuous patchy structure (Figure 1). The internal structure of odor plumes changes predictably with both distance downstream from the source and distance from centerline across the plume, and behavioral experiments indicate that odor-tracking animals alter their tracking behavior in response to these changes. In most organisms tested, odor-tracking behavior requires the intermittent olfactory stimulation generated by the patchy structure of the plume to generate upstream movement and successful location of the source. Neurons at all levels of the olfactory system are sensitive to intermittent stimulation, and some circuits appear to preserve the temporal activation patterns experienced

(a)

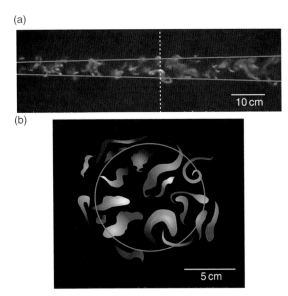

(b)

Figure 1 Two perspectives on instantaneous views of smoke visualizations of wind-borne odor plumes. (a) A single field from a video recording of a TiCl₄ smoke plume in a laboratory wind tunnel viewed from above. The time-averaged plume boundaries are represented by red lines running from the upwind to downwind side of the image (i.e., left to right). In this image, the wind is blowing from left to right at 75 cm s⁻¹, and the source is a 0.7 cm filter paper disk to which a drop of TiCl₄ has been applied. Since this is the same source used in insect odor-tracking experiments, according to the laws of fluid dynamics, the structure of the plume should be similar to the odor plume experienced by the insects. (b) A schematic approximation illustrating the tracker's eye view of the smoke plume visualization depicted in (a) above. The red circle represents an extension of the time-averaged plume boundaries in (a) onto this cross section. White dashed line is approximately 1 m downwind from the source. (a) Reproduced from Webb, B., Harrison, R. R., and Willis, M. A. 2004. Sensorimotor control of navigation in arthropod and artificial systems. Arthropod Struct. Dev. 33, 301–330, with permission from Elsevier.

in a plume and relay them to higher processing centers in the brain. Recent studies using simulation models and odor-guided robots are leading to tests of biological hypotheses that would be difficult or impossible to address with living organisms, and are also leading to the first man-made plume-tracking systems.

4.42.1 Ubiquity of the Use of Olfaction to Locate Distant Resources

Virtually every object on our planet, biotic or abiotic, sheds molecules into the environment. If these molecules are in high enough concentrations, they can be detected and processed by the olfactory systems of animals as they move through the environment, seeking opportunities to forage (Arbas, E. A. *et al.*, 1993; Weissburg, M. J. and Zimmer-Faust, R. K., 1994; Basil, J. A. *et al.*, 2000; Raguso, R. and Willis, M., 2002; Ferner and Weissburg, M. J., 2005; Raguso, R. and Willis, M., 2005), mate (Cardé, R. T. and Minks, A. K., 1997 and citations therein), locate home and appropriate territories (Wolf, H. and Wehner, R., 2000), and avoid predators (Susswein, A. J. *et al.*, 1982). The behavior of two fluids, air and water, dominates the distribution of these molecules, and to a large degree determines how living organisms are able to use this molecular information to adapt their behavior, increasing their probability of survival and reproduction (Vickers, N. J., 2000; Weissburg, M. J., 2000).

Flow-borne odor plumes provide information important for aquatic, marine, and terrestrial organisms that walk, fly, and swim to locate resources. Despite these seemingly different environments and modes of locomotion, the paths generated by these animals often appear remarkably similar (Figures 2 and 3). Whether this similarity in tracking trajectories stems from a similarity in underlying physiological and behavioral mechanisms is almost entirely unknown.

4.42.2 Distribution of Odorants in the Environment

Our understanding of how volatile chemicals are distributed in the environment and the effects that this has on the behavior of animals tracking odor plumes has advanced considerably in recent years. This has resulted from behavioral biologists working together with fluid dynamics engineers to study the fine structure of odor plumes in different and dynamically changing environments (Wright, R. H., 1958, Figure 1; Murlis, J. and Jones, C. D., 1981; Murlis, J. *et al.*, 1990; Weissburg, M. J. and Zimmer-Faust, R. K., 1993; Griffiths, N. and Brady, J., 1995; Murlis, J. *et al.*, 2000; Webster, D. R. and Weissburg, M. J., 2001). Odor plume structure has significant effects on the tracking behavior of almost every animal in which this has been tested (Vickers, N. J., 2000; Weissburg, M. J., 2000).

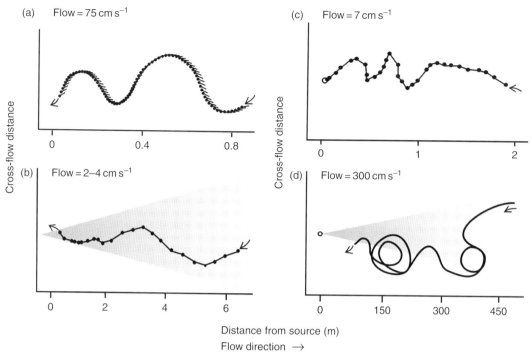

Figure 2 Movement paths of animals tracking plumes while supported by the fluid through which they are tracking, as viewed from above. (a) Flight path of a male moth, *Manduca sexta*, tracking an airborne plume of female sex-attractant pheromone in a laboratory wind tunnel. (b) Swimming path of the Antarctic fish, *Trematomus bernachii*, while tracking a water-borne plume of prey extract recorded in its natural environment. (c) Swimming path of the cephalopod mollusk, *Nautilus pompilus*, tracking a water-borne plume of prey extract recorded in a laboratory flume. (d) Observer's approximation of the flight path of a Turkey vulture, *Cathartes aura*, tracking an airborne plume of ethyl mercaptan, a volatile chemical associated with dead animals. In all images, the fluid flows from left to right, and the flow velocities for each example are presented in each panel. (a) Reproduced from Webb, B., Harrison, R. R., and Willis, M. A. 2004. Sensorimotor control of navigation in arthropod and artificial systems. Arthropod Struct. Dev. 33, 301–330, with permission from Elsevier. (b) Reproduced from Polar Biology, Vol. 21, 1999, pp. 151–154, Olfactory search tracks in the Antarctic fish. *Trematomous bernacchi*, Montgomery, J. C., Diebel, C., Halstead, M. B. D., and Downer, J., figure 1, with kind permission of Springer Science and Business Media and the authors. (c) Reproduced from Basil, J. A., Hanlon, R. T., Sheikh, S. I., and Atema, J. 2000. Three-dimentional odor tracking by *Nautilus pompilius*. J. Exp. Biol. 203, 1409–1414, with permission from the Company of Biologists Ltd. (d) Image was provided by Willis, unpublished data.

4.42.2.1 Diffusion versus Turbulent Diffusion

The distribution of volatile odorants in fluids is determined primarily by the processes of diffusion and turbulence (Murlis, J. *et al.*, 1990; Vogel, S., 1994; Weissburg, M. J., 2000). For animals larger than approximately 1 mm, turbulence is usually the more important of these two processes (Weissburg, M. J., 2000). According to the laws of fluid dynamics, the behavior of air and water are governed by similar constraints, and their behavior can be predicted depending on viscosity, flow velocity, and density (Vogel, S., 1994). Turbulence can be described as velocity variations in the flow itself or can result from velocity differences caused when flowing fluid interacts with fixed objects in the flow. From a fluid dynamical perspective, plumes are typically described by a statistical distribution that has been depicted as a smooth concentration gradient that diminishes both with distance downstream from the source and across the flow from the plume's centerline to the edges, both laterally and vertically (Bossert, W. H. and Wilson, E. O., 1963). However, a snapshot of an odor plume reveals a discontinuous structure composed of patches of odor separated by clean air (Figure 1).

The olfactory sensory cells and central processing circuitry of animals do not typically respond to distributions of odorants averaged over time. Rather, they respond to changes in sensory stimuli and initiate behavioral responses in tens or hundreds of

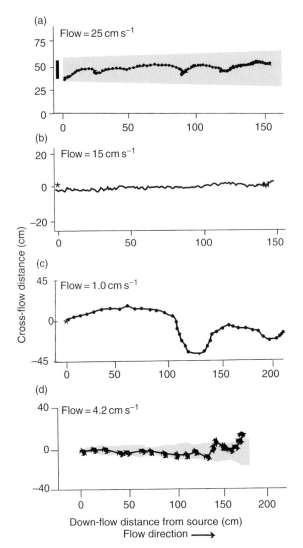

Figure 3 Movement paths of substrate-borne animals tracking air- and water-borne odor plumes as viewed from above. (a) Walking path of a male cockroach, *Periplaneta americana*, tracking an airborne plume of female sex attractant issuing from source 14.3 cm across in a laboratory wind tunnel. (a) Reproduced from Willis, M. A. and Avondet, J. L. 2005. Odor-modulated orientation in walking male cockroaches, *Periplaneta americana* (L.), and the effects of odor plumes of different structure. J. Exp. Biol. 208, 721–735, with permission from the Company of Biologists Ltd. (b) Crawling path of a predatory marine snail, *Busycon carica*, in a plume of prey odorant in a laboratory flume. (b) Reproduced from Ferner, M. C. and Weissburg, M. J. 2005. Slow-moving predatory gastropods track prey odors in fast and turbulent flow. J. Exp. Biol. 208, 809–819, with permission of the Company of Biologists Ltd. (c) Walking path of a lobster, *Homarus americanus*, tracking a water-borne plume of prey odorants in a laboratory flume. (c) Reproduced from J. Chem. Ecol., Vol. 17, 1991, pp. 1293–1307, Chemical orientation of lobsters, *Homarus americanus*, in turbulent odor plumes, Moore, P. A., Scholtz, N., and Atema, J., figure 5, with kind permission of Springer Science and Business Media. (d) Walking path of a blue crab, *Callinectes sapidus*, tracking prey odor in its natural environment. In all images, the fluid flows from left to right, and the flow velocities for each example are presented in each panel. (d) Reproduced from Zimmer-Faust, R. K., Finelli, C. M., Pentcheff, N. D., and Wethey, D. S. 1995. Odor plumes and animal navigation in turbulent water flow: A field study. Biol. Bull. 188, 111–116, with permission from Marine Biological Laboratory and the author.

milliseconds. High-frequency sampling of plumes performed in water (Moore, P. A. and Atema, J., 1991) and air (Vickers, N. J. *et al.*, 2001) reveals the structure underlying time-averaged plume statistics. Any tracker moving across the plume will have an increased probability of encountering filaments of higher concentration as it moves toward the plume centerline, across the flow, from any direction

(Figure 1(b)). Likewise, in both air (Murlis, J. and Jones, C. D., 1981; Murlis, J. *et al.*, 1990; 2000) and water (Webster, D. R. and Weissburg, M. J., 2001), a tracker will have an increased probability of encountering higher concentration plume filaments as it moves into the flow toward the source. The lateral and longitudinal decrease in concentration (i.e., molecules/volume of fluid) results primarily from turbulent mixing of the filaments within the plume, not diffusion. Behavioral experiments suggest that plume-tracking animals use these changes in plume structure to adapt their behavior. Considering the speed of most animal locomotion and fluid flow in most of earth's environments (i.e., tens of centimeters to meters/second) together with the small distances/time generated by diffusion (*c.* 9.5 cm in 1 h for a typical insect sex pheromone in air; Loudon, C., 2003), it is easy to understand that turbulence dominates the distribution of the stimulus and thus the structure of the tracking behavior observed from odor-tracking animals.

4.42.2.2 Environmental Effects on Odorant Distribution

Although we can measure the differences in plumes in different environments (Weissburg, M. J. and Zimmer-Faust, R. K., 1994; Griffiths, N. and Brady, J., 1995; Murlis, J. *et al.*, 2000), the primary significance of these differences stems from the fact that plume-tracking animals show dramatically different behavior and success at source location in different turbulent flows. Even though many organisms are superbly adapted to track plumes to their sources, ambient flow and turbulence conditions can exist in natural environments that make the task almost impossible to perform (Elkinton, J. S. *et al.*, 1987; Weissburg, M. J. and Zimmer-Faust, R. K., 1993; Griffiths, N. and Brady, J., 1995).

4.42.3 Plume Characteristics Important to Plume-Tracking Organisms

Behaviors that are initiated or modulated by olfactory information can be difficult to study because the primary stimulus is out of the control of the researchers almost as soon as it evaporates from the dispenser. However, even with this major difficulty, it is possible to define three broad categories of information known to be detected by plume-tracking organisms: (1) temporal/spatial structure of molecules distributed in the plume, (2) quality of the molecules in the plume, and (3) quantity of molecules in the plume. Experimental manipulation of variables in each of these three categories results in significant changes in the structure of the paths taken by odor-tracking organisms as they orient and move upstream or upwind toward the source (Belanger, J. H. and Willis, M. A., 1996).

4.42.3.1 Temporal/Spatial Structure

The patchy structure of odor plumes has been known to researchers studying olfactory orientation for some time (Wright, R. H., 1958), but its dramatic effect on orientation and tracking behavior has been fully appreciated only recently (Cardé, R. T. and Minks, A. K., 1997). The intermittent stimulation resulting from an odor plume is necessary for most animals tested to orient to odor, except for the fruit fly *Drosophila melanogaster* (Budick, S. A. and Dickinson, M. H., 2006), one species of moth (*Cadra cautella*; Justus, K. A. and Cardé, R. T., 2002), and small organisms in a detectable gradient (e.g., microorganisms, copepods; Weissburg, M. J., 2000). Recent results show that the temporal/spatial structure of an odor plume has a more dramatic effect on the structure of tracking behavior than any other known variable (Mafra-Neto, A. and Cardé, R. T., 1994; Vickers, N. J. and Baker, T. C., 1994). For example, it has been observed that sources emitting similar levels of chemicals but manipulated to generate structurally different plumes elicit characteristically different behavior from males approaching from downwind (Charlton, R. E. *et al.*, 1993; Mafra-Neto, A. and Cardé, R. T., 1994; Vickers N. J. and Baker T. C., 1994; Willis, M. A. *et al.*, 1994). A seminal study by Vickers N. J. and Baker T. C. (1992) showed that moths alter their tracking performance in response to both the timing of encounters with structural elements of the plume and the concentration detected from those elements. Their study used a puffing device to experimentally control the interval between puffs while they varied the concentration of the source independently. By altering these variables independently, Vickers N. J. and Baker T. C. (1992) demonstrated that decreasing the number of pheromone puffs encountered enabled their moths to track upwind to high concentrations that they normally could not.

Dogs challenged to track trails of odor deposited on the ground were more successful (over 90%) tracking a path of discontinuous patches (i.e., human foot steps) than a continuous trail of odor (Hepper, P. G. and Wells, D. L., 2005). So the combination of discrete patches of odor with no odor in between, and the active sniffing behavior performed while tracking, results in an intermittent odor signal that appears to be required for dogs to track substrate-borne trails. In addition, recent measurements of respiratory air flow in dogs during the tracking of wind-borne odors, combined with modeling of the flow through the upper respiratory tract, suggest that a nearly continuous inflow of odor-bearing air over the olfactory epithelium is probable, even while running (Steen, J. B. *et al.*, 1996). Thus, in this situation, some if not all of the temporal/spatial structure of the odor might be preserved to be presented to the olfactory epithelium.

What effect does the sniffing behavior of vertebrate animals have on the structure of the plume? One possibility is that the increased flow velocity interacting with the concha in the nasal passages would result in increased turbulence and a completely mixed sample arriving at the olfactory epithelium. This could effectively smooth out the dramatic peaks and valleys in the odor signal caused by the filamentous structure of the plume. Whether this actually happens, and how it might affect the trackers' behavior, awaits further study.

Crustaceans also sniff as a way to enhance or alter their olfactory inputs (Schmidt, B. C. and Ache, B. W., 1979). When responding to odor, they move the olfactory part of their second antennae (i.e., antennules) through the water very rapidly (Schmidt, B. C. and Ache, B. W., 1979). Flicking behavior, as this is known, has been shown to decrease the boundary layer around the antennules and force the odor filaments into the array of odor-detecting sensilla (Koehl, M. A. R., 2006). The effect of the shallower boundary layer is to decrease the distance that the odor molecules must diffuse to reach the receptors and thus enhances the extraction of odor information from the environment. To date, the effects of flicking behavior on the perceived structure of an odor plume have not been measured directly, and how this could affect the animal's tracking behavior has not been determined.

4.42.3.2 Quality

In moths, the best-known example, detection of the number and type of molecules in an odorant plume is determined by the combination of peripheral sensory structures and central processing. This has been demonstrated in behavioral experiments by independently manipulating the structure of the plume and the chemical compounds making up that structure (Vickers, N. J. and Baker, T. C., 1992). Two sensory cells that detect the two components of the pheromone required for behavioral activity are housed a few microns apart in the same sensory hair on the antenna. Their spatial coincidence means that they are perfectly situated to detect temporal coincidences of odorant components in the pheromone. This arrangement of sensory structures, in combination with central coincidence detecting interneurons (Christensen, T. C. and Hildebrand, J. G., 1988), is capable of resolving spatial/temporal separation of structural elements in the plume in the millisecond (i.e., millimeter) range. At the wind speeds at which these experiments were performed, this resulted in male moths being able to discriminate between odorant encounters approximately 3 m apart (Baker, T. C. *et al.*, 1998). Thus, in a fundamental way, the ability of these animals to use blends of odorant compounds in a meaningful way may *require* the patchiness of the distribution and detection that characterizes plumes. The components of a behaviorally relevant blend of compounds must be detected nearly simultaneously to activate a behavioral response (Vickers, N. J. and Baker, T. C., 1992; Baker, T. C. *et al.*, 1998; Vickers, N. J., 2006).

4.42.3.3 Quantity

Concentration of chemicals in the plume or plume filaments is known to change predictably with distance from the source. As the individual filaments expand as a result of turbulence and, as the period from evaporation from the source to detection gets longer, the same amount of chemicals is spread over a larger space, that is, dilution, resulting in lower concentration. This will also result in the ratio of the peak concentration to the mean decreasing with distance from the source. Thus, the peak concentration will be lower, and the time from odor onset to peak concentration detected upon contacting a filament of odor will be longer as the plume moves farther and farther downwind. It is clear that animals modulate their behavior according to their distance from the source and, when tested independently, they modulate their behavior according to changes in concentration. Thus, plume-tracking animals appear to modulate their behavior according to both changes in odorant concentration and the spatial/temporal structure of those concentration differences, and these responses can operate independently (Vickers, N. J. and Baker, T. C., 1992).

4.42.4 Tracking Odor Plumes Typically Requires Information from Multiple Senses

In most known cases, plume-tracking animals require information from multiple sensory systems to successfully perform this task. The environmental cue that the tracking organism uses to orient its steering is the direction of the fluid flow, not a nonexistent odor gradient. The preferred direction of orientation with respect to the flow, if there is one, is usually modulated by the presence or absence of the odor.

4.42.4.1 Orientation to Flow

The direction of flow may be extracted from the environment directly by the deflection of a mechanosensory structure, for example, antennae and wind-sensitive hairs in arthropods (Arbas, E. A., 1986), superficial neuromast sensors in the lateral line system in fish (Baker, C. F. and Montgomery, J. C., 1999), feathers or wind-sensitive breast feathers in birds (Gewecke, M. and Woike, M., 1978), or facial whiskers in mammals (Dehnhardt, G. *et al.*, 1998). When the animal is in contact with a solid substrate, these sensors will provide the speed and direction of the flow. However, if the animal is suspended in the flow, like a swimming fish or flying bird or insect, the information provided by these sensors will be the speed of the organism with respect to the flow. Flying or swimming animals almost always detect the flow direction indirectly by determining how they are moving with respect to the nonmoving parts of their environment (e.g., the ground or trees). The single well-documented contradiction to this statement is the Mexican blind cave fish (Baker, C. F. and Montgomery, J. C., 1999). However, these results suggest that the lateral line system could provide flow direction to any fish.

All well-described examples of indirect detection of flow direction involve animals using visually detected movement with respect to an assumed stationary environment. In the case of flying moths, it is thought that they are able to compute the wind speed and direction by comparing their intended course to their realized course, with any difference between the two assumed to be caused by wind-induced drift (Marsh, D. *et al.*, 1978). This method of using visual information to steer with respect to the wind direction and speed while tracking an odor plume was first demonstrated by Kennedy J. S. (1940) in female mosquitoes tracking human breath. Kennedy and his co-workers went on to demonstrate that using visually perceived wind information to steer and control flight speed (known as optomotor anemotaxis) was ubiquitous in flying insects, whether or not they were tracking odor (Kennedy, J. S. and Marsh, D., 1974; Marsh, D. *et al.*, 1978; David, C. T., 1986). The use of visually detected flow information to control the speed and direction of movement has also been demonstrated clearly in male rainbow trout tracking female oviduct extract upstream (Emanuel, M. E. and Dodson, J. J., 1979).

4.42.4.2 Orientation to Odors

Animals that use odor cues to track plumes may use two hypothetical mechanisms to acquire chemical information on which to base their steering decisions: (1) spatial comparisons of odor inputs to bilaterally symmetrical sensors (i.e., antennae or nostrils), and (2) temporal comparisons of odor inputs sensed at different points in time (Frankel, G. S. and Gunn, D. L., 1961; Schöne, H., 1984; Dusenbery, D. B., 1992). It is possible that either of these two strategies may be more effective at tracking plumes generated under different environmental and flow conditions (Weissburg, M. J., 2000). These two hypotheses are typically referred to as spatial and temporal strategies respectively. Since the vast majority of odor-tracking animals have bilaterally symmetrical odor sensors, the possibility that both strategies could be operating at the same time, either synergistically or independently, has always been acknowledged (Frankel, G. S. and Gunn, D. L., 1961; Schöne, H., 1984). However, careful experiments necessary to resolve these questions have rarely been pursued (Martin, H., 1965; Hangartner, W., 1967; Bell, W. J. and Tobin, T. R., 1981; Borst, A. and Heisenberg, M., 1982; Johnsen, P. B. and Teeter, J. H., 1985).

4.42.4.2.1 *Spatial comparisons of odor concentration (bilaterally symmetrical sensors)*

Animals walking slowly through slower flows experience relatively less turbulent plumes because the surfaces they walk on damp its vertical component (Weissburg, M. J., 2000). Slow movement relative to the slowly moving plume may allow such plume trackers to steer to remain in the plume by using a spatial sampling strategy to experience the odor edges defined by the rapid drop of concentration at the lateral margins of the plume (Figure 1(b)). Chemical concentration differences that would inform these steering decisions could be sensed instantaneously using bilateral sensors, or because of the relatively low encounter rate with odor structures in the plume, the same sensors could enable time averaging of concentration in the central nervous system. If an odor-tracking organism were using spatial comparisons to control its maneuvering, it would be possible, knowing the plume structure, to predict when and in what direction it would turn. Recent observations in plume-tracking cockroaches suggest that many, but perhaps not all, of the turns in their walking tracks may result from bilateral comparisons (Figure 4). Recent analyses indicate that knowledge

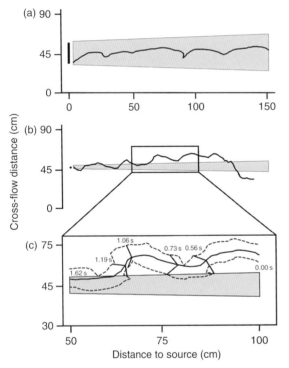

Figure 4 Movement paths of male cockroaches as they track female sex-attractant pheromone plumes of different structures. (a) Walking path of a male cockroach tracking a turbulent plume issuing from a 14.3 cm wide source in a laboratory wind tunnel. Note that the cockroach never approaches the lateral pheromone-clean air boundary of the plume. (b) Walking path of a male cockroach tracking a turbulent plume issuing from a source 0.7 cm wide (i.e., point-source plume). Note the regular sinusoidal zigzagging turning executed by this male as he tracked the plume upwind toward the source. (c) Detail of (b) above showing the position and orientation of the animal's antennae during a brief sequence of turning. In all images, the fluid flows from left to right, and the flow velocities for each example are presented in each panel. Reproduced from Willis, M. A. and Avondet, J. L. 2005. Odor-modulated orientation in walking male cockroaches, *Periplaneta americana* (L.), and the effects of odor plumes of different structure. J. Exp. Biol. 208, 721–735, with permission of the Company of Biologists Ltd.

of the plume structure should allow prediction of sensor spacing appropriate to support bilateral concentration comparisons (Webster, D. R. *et al.*, 2001).

4.42.4.2.2 *Temporal comparisons of odor concentration (sensors small relative to the scale of the plume, memory required)*

In contrast, animals swimming or flying are moving relatively more rapidly, through faster turbulent flows and thus experience an odor plume at a much faster rate (which makes it essentially a continuous series of pheromone-clean air edges). These animals may not have time to extract information about changes in the plume fast enough to support adaptive maneuvering using a spatial strategy. To a system focused on the timing of the odor inputs and not where it is, the edge of the plume is just another odor on–off in the sequence (Webster, D. R. *et al.*, 2001).

The inability, in most cases, to determine when and where an organism detects an odor plume, both in relation to the plume and in relation to the animals' own preceding and subsequent behavior, has made it almost impossible to determine which of the existing hypothetical mechanisms may underlie the behavior we observe. In only one case has the physiological response of the olfactory sensors to plume structure been measured during a plume-tracking performance (Vickers, N. J. and Baker, T. C., 1994). Even in this example, the olfactory responses were measured from a surgically detached antenna carried by the odor-tracking individual, not from the antennae of the individual actually performing the tracking behavior.

4.42.4.3 Importance of Active Sampling Behaviors to Odor Detection and Processing

Loudon C. and Koehl M. A. R. (2000) have shown that flows induced by pheromone-activated wing fanning behavior in *Bombyx mori* males may enhance the ability of the males to respond to pheromone during plume-tracking behavior. Measurements of air flow over the male moth's feathery antenna showed rhythmically varying velocities that matched the moths' wing beat frequency. Such a rhythmically varying flow would cause intermittent encounters between the antennae and any odor plume near enough to be drawn into the flow generated by the beating wings. When the wings of *B. mori* males were removed, their success at locating the females decreased (Ishida, H. *et al.*, 1996). Similar active sampling behaviors are generated by lobsters, crayfish, and other crustaceans when they are tracking odors as they rapidly flex their antennules driving their chemosensors through the plume in a behavior known as flicking (Schmidt, B. C. and Ache, B. W., 1979; Koehl, M. A. R., 2006).

4.42.5 Man-Made Odor Tracking Systems

Although our knowledge and understanding of odor plumes and the neural and behavioral responses of plume-tracking organisms is far from complete, enough is known that some have begun to apply this knowledge to man-made odor-tracking systems. Some of the first robotic odor-tracking systems were conceived and fabricated by Russell R. A. (1999) to provide the chemical communication capabilities of social insects (e.g., ants and bees) to interacting robots. The idea was to use odor tracking in a manner similar to ants and termites, as a way to mark trails through the environment, to guide the foraging behavior of other members of a robot team. Russell's first attempts included fabricating the entire system, from the robot to the odor detectors (Russell, R. A., 1999). These early robots were able to track a volatile chemical trail deposited on the substrate for some time and distance. However, in both the robotic system and its insect model, the source of the odor (i.e., the odor trail) was always very close to the tracker. These initial experiments then led to the development of wind direction sensors and tracking algorithms that mimicked the pheromone-tracking behavior of male silk worm moths, *B. mori* (Russell, R. A. *et al.*, 1995). These new sensory and control systems performed well enough to guide a mobile robot around obstacles in a laboratory environment and successfully locate a source of volatile chemicals (Russell, R. A. *et al.*, 1995). Similar experiments were being conducted in other laboratories with similar results (Ishida, H. *et al.*, 1994).

To date, the most significant hindrances to the development of odor-tracking robots are the chemical detectors; the robots' olfactory sensors. None of the currently available off-the-shelf man-made odor detectors approach the temporal response or exquisite sensitivity of their biological precursors. In one series of experiments, this problem was overcome by surgically removing the biological odor detectors (i.e., the antennae of male moths) and mounting them on a mobile robot (Kuwana, Y. *et al.*, 1995). Although these experiments demonstrated the

efficacy of interfacing living (i.e., dying) tissue with simple electronics, and generated predictable odor-activated steering responses, the relatively short life of the antenna after removal from the insect makes this alternative impractical.

Another approach has been to fabricate robotic odor-tracking systems explicitly as a means to test biological hypotheses about how animals track odor plumes (Grasso, F. W. *et al.*, 2000; Edwards, S. *et al.*, 2005). These robotic studies usually form part of a multipronged research program aimed at working iteratively between robotic or simulation experiments and animal studies. In these studies, simulation and mobile robot experiments are developed using new and existing biological knowledge, and their results then are used to inform the design and interpretation of the subsequent behavioral or neurophysiological studies. Results from some of these robotic experiments (Grasso, F. W. *et al.*, 2000; Edwards, S. *et al.*, 2005) have caused the reexamination of our previous understanding of odor tracking.

4.42.6 Conclusions

If we could know when the organisms detected an odor filament together with the preceding and subsequent behavior, we could begin to understand how the animals are using the information to control their behavior. Do they modify their tracking behavior based on each encounter with odor as suggested by recent experiments (Mafra-Neto, A. and Cardé, R. T., 1994; Vickers, N. J. and Baker, T. C., 1994), or do they continuously modulate an odor-activated internal program as proposed earlier (Kennedy, J. S., 1983). Or does the trackers' response change in a context-dependent manner? Do some animals have alternative tracking algorithms that are employed in different environmental conditions (i.e., different plumes structures) or does a single simple tracking algorithm yield apparently different but adaptive behavior as the animal actively interacts with different plume structures generated in different environmental conditions? Much is known about this interesting biological phenomenon, but many lacunae in our knowledge at all levels of organization, environmental, neural, and behavioral, remain to be filled before a complete understanding is revealed.

References

Arbas, E. A. 1986. Control of hindlimb posture by wind-sensitive hairs and antennae during locust flight. J. Comp. Physiol. A 159, 849–857.

Arbas, E. A., Willis, M. A., and Kanzaki, R. 1993. Organization of Goal-Oriented Locomotion: Pheromone-Modulated Flight Behavior of Moths. In: Biological Neural Networks in Invertebrate Neuroethology and Robotics (eds. R. D. Beer, R. E. Ritzmann, and T. McKenna). pp. 159–198. Academic Press.

Baker, C. F. and Montgomery, J. C. 1999. The sensory basis for rheotaxis in the blind Mexican cave fish, Astynax fasciatus. J. Comp. Physiol. A 184, 519–527.

Baker, T. C., Fadamiro, H. Y., and Cosse, A. A. 1998. Moths use fine tuning for odour resolution. Nature 393, 530.

Basil, J. A., Hanlon, R. T., Sheikh, S. I., and Atema, J. 2000. Three-dimentional odor tracking by Nautilus pompilius. J. Exp. Biol. 203, 1409–1414.

Belanger, J. H. and Willis, M. A., 1996. Adaptive control of odor-guided locomotion: behavioral flexibility as an antidote to environmental unpredictability. Adapt. Behav. 4, 217–253.

Bell, W. J. and Tobin, T. R. 1981. Orientation to sex pheromone in the American cockroach: analysis of chemo-orientation mechanisms. J. Insect Physiol. 27, 501–508.

Borst, A. and Heisenberg, M. 1982. Osmotropotaxis in Drosophila melanogaster. J. Comp. Physiol. 147, 479–484.

Bossert, W. H. and Wilson, E. O. 1963. The analysis of olfactory communication among animals. J. Theor. Biol. 48, 443–469.

Budick, S. A. and Dickinson, M. H. 2006. Free flight responses of Drosophila melanogaster to attractive odors. J. Exp. Biol. 209, 3001–3017.

Cardé, R. T. and Minks, A. K. 1997. Pheromone Research: New Directions. Chapman & Hall.

Charlton, R. E., Kanno, H., Collins, R. D., and Cardé, R. T. 1993. Influence of pheromone concentration and ambient temperature on the flight of the gypsy moth, Lymantria dispar (L), in a sustained-flight wind-tunnel. Physiol. Entomol. 18, 349–362.

Christensen, T. C. and Hildebrand, J. G. 1988. Frequency coding in central olfactory neurons in the sphinx moth, Manduca sexta. Chem. Senses 13, 23–30.

David, C. T. 1986. Mechanisms of Directional Flight in Wind. In: Mechanisms in Insect Olfaction (eds. T. L. Payne, M. C. Birch, and C. E. J. Kennedy), pp. 49–58. Clarendon.

Dehnhardt, G., Mauck, B., and Bleckmann, H. 1998. Seal whiskers detect water movements. Nature, 394, 235–236.

Dusenberry, D. B. 1992. Sensory Ecology., W.H. Freeman.

Edwards, S., Rutkowski, A., Quinn, R., and Willis, M. 2005. Moth-inspired plume tracking strategies in three-dimensions. Proceedings of the 2005 IEEE International Conference on Robotics and Automation, 1681–1686.

Elkinton, J. S., Schal, C., Ono, T., and Cardé, R. T. 1987. Pheromone puff trajectory and upwind flight of male gypsy moth in a forest. Physiol. Entomol. 12, 399–406.

Emanuel, M. E. and Dodson, J. J. 1979. Modification of the rheotropic behavior of male rainbow trout (Salmo gairderi) by ovarian fluid. J. Fish. Res. Board Can. 36, 63–68.

Ferner, M. C. and Weissburg, M. J. 2005. Slow-moving predatory gastropods track prey odors in fast and turbulent flow. J. Exp. Biol. 208, 809–819.

Frankel, G. S. and Gunn, D. L. 1961. The Orientation of Animals. Dover.

Gewecke, M. and Woike, M. 1978. Breast feathers as an air-current sense organ for the control of flight behaviour in a songbird (Carduelis spinus). Zietschrift für Tierpsychology 47, 293–298.

Grasso, F. W., Consi, T. R., Mountain, D. C., and Atema, J. 2000. Biomimetic robot lobster performs chemo-orientation in turbulence using a pair of spatially separated sensors: Progress and challenges. Robot. Auton. Syst. 30, 115–131.

Griffiths, N. and Brady, J. 1995. Wind structure in relation to odour plumes in tsetse fly habitats. Physiol. Entomol. 20, 286–292.

Hangartner, W. 1967. Spezifität und inaktivierung des spurpheromons von Lasius fuliginosus Latr. und orientierung der arbeiterinnen im duftfeld. Zeitschrift für Vergleichende Physiologie 57, 103–136.

Hepper, P. G. and Wells, D. L. 2005. How many footsteps do dogs need to determine the direction of an odor trail? Chem. Senses 30, 291–298.

Ishida, H., Hayashi, K., Takakusaki, M., Nakamoto, T., Moriizumi, T., and Kanzaki, R. 1996. Odour-source localization system mimicking behaviour of silkworm moth. Sens. Actuators A 99, 225–230.

Ishida, H., Suetsugu, T., Nakamoto, T., and Moriizumi, T. 1994. Study of autonomous mobile sensing system for localization of odor source using gas sensors and anemotactic sensors. Sensors and Actuators A 45, 154–157.

Johnsen, P. B. and Teeter, J. H. 1985. Behavioral responses of Bonnethead sharks (Sphyrna tiburo) to controlled olfactory stimulation. Mar. Behav. Physiol. 11, 283–291.

Justus, K. A. and Cardé, R. T. 2002. Flight behaviour of males of two moths, Cadra cautella and Pectinophora gossypiella, in homogeneous clouds of pheromone. Physiol. Entomol. 27, 67–75.

Kennedy, J. S. 1940. The visual responses of flying mosquitoes. Proceedings of the Zoological Society of London A 109, 221–242.

Kennedy, J. S. 1983. Zigzagging and casting as a response to wind-borne odor: a review. Physiol. Entomol. 8, 109–120.

Kennedy, J. S. and Marsh, D. 1974. Pheromone regulated anemotaxis in flying moths. Science 184, 999–1001.

Koehl, M. A. R. 2006. The fluid mechanics of arthropod sniffing in turbulent odor plumes. Chem. Senses 31, 93–105.

Kuwana, Y., Shimoyama, I., and Miura, H. 1995. Steering control of a mobile robot using insect antennae. IEEE/RSJ International Conference on Intelligent Robots and Systems, 2, 530–535.

Loudon, C. 2003. The Biomechanical Design of an Insect Antenna as an Odor Capture Device. In: Insect Pheromone Biochemistry and Molecular Biology (eds. G. J. Blomquist and R. G. Vogt), pp. 609–630. Elsevier Academic Press.

Loudon, C. and Koehl, M. A. R. 2000. Sniffing by a silkworm moth: wing fanning enhances air penetration through and pheromone interception by antennae. J. Exp. Biol. 203, 2977–2990.

Mafra-Neto, A. and Cardé, R. T. 1994. Fine-scale structure of pheromone plumes modulates upwind orientation of flying male moths. Nature 369, 142.

Marsh, D., Kennedy, J. S., and Ludlow, A. R. 1978. An analysis of anemotactic zigzagging flight in male moths stimulated by pheromone. Physiol. Entomol. 3, 221–240.

Martin, H. 1965. Osmotropotaxis in the honeybee. Nature 208, 59–63.

Moore, P. A. and Atema, J. 1991. Spatial information in the three-dimensional fine structure of an aquatic odor plume. Biol. Bull. 181, 408–418.

Moore, P. A., Scholtz, N., and Atema, J. 1991. Chemical orientation of lobsters, Homarus americanus, in turbulent odor plumes. J. Chem. Ecol. 17, 1293–1307.

Montgomery, J. C., Diebel, C., Halstead, M. D. B., and Downer, J. 1999. Olfactory search tracks in the Antarctic fish Trematomus bernachii. Polar Biol. 21, 151–154.

Murlis, J. and Jones, C. D. 1981. Fine-scale structure of odour plumes in relation to insect orientation to distant pheromone and other attractant sources. Physiol. Entomol. 6, 71–86.

Murlis, J., Willis, M. A., and Cardé, R. T. 1990. Odor Signals: Patterns in Time and Space. In: Proceedings of the X International Symposium on Olfaction and Taste, Oslo (ed. K. Døving).

Murlis, J., Willis, M. A., and Cardé, R. T. 2000. Spatial and temporal structure of pheromone plumes in fields and forests. Physiol. Entomol. 25, 211–222.

Raguso, R. and Willis, M. 2002. Synergy between visual and olfactory cues in nectar feeding by naïve hawkmoths. Anim. Behav. 64, 685–695.

Raguso, R. and Willis, M. 2005. Synergy between visual and olfactory cues in nectar feeding by wild hawkmoths, *Manduca sexta*. Anim. Behav. 69, 407–418.

Russell, R. A. 1999. Odour Detection by Mobile Robots. World Scientific Publishing.

Russell, R. A., Thiel, D., Deveza, R., and Mackay-Sim, A. 1995. A robotic system to locate hazardous chemical leaks. IEEE International Conference on Robotics and Automation, 1, 556–561.

Schmidt, B. C. and Ache, B. W. 1979. Olfaction: response enhancement by flicking in a decapod crustacean. Science 205, 204–206.

Schöne, H. 1984. Spatial Orientation. Princeton University Press.

Steen, J. B., Mohus, I., Kvesetberg, T., and Walløe, L. 1996. Olfaction in bird dogs during hunting. Acta Physiol. Scand. 157, 115–229.

Susswein, A. J., Cappell, M. S., and Bennett, M. V. L. 1982. Distance chemoreception in *Navanax inermis*. Marine Behav. Physiol. 8, 231–241.

Vickers, N. J. 2000. Mechanisms of animal navigation in odor plumes. Biol. Bull. 198, 203–212.

Vickers, N. J. 2006. Winging it: moth flight behavior and responses of olfactory neurons are shaped by pheromone plume dynamics. Chem. Senses 31, 155–166.

Vickers, N. J. and Baker, T. C. 1992. Male *Heliothis virescens* maintain upwind flight in response to experimentally pulsed filaments of their sex pheromone (Lepidoptera: Noctuidae). J. Insect Behav. 5, 669–687.

Vickers, N. J. and Baker, T. C. 1994. Reiterative responses to single strands of odor promote sustained upwind flight and odor source location by moths. Proc. Natl. Acad. Sci. U. S. A. 91, 5756–5760.

Vickers, N. J., Christensen, T. A., Baker, T. C, and Hildebrand, J. G. 2001. Odour-plume dynamics influence the brain's olfactory code. Nature 410, 466–470.

Vogel, S. 1994. Life in Moving Fluids, Princeton University Press.

Webb, B., Harrison, R. R., and Willis, M. A. 2004. Sensorimotor control of navigation in arthropod and artificial systems. Arthropod Struct. Dev. 33, 301–330.

Webster, D. R. and Weissburg, M. J. 2001. Chemosensory guidance cues in a turbulent odor plume. Limn. Oceanogr. 46, 1034–1047.

Webster, D. R., Rahman, S., and Dasi, L. P. 2001. On the usefulness of bilateral comparison to tracking turbulent chemical plumes. Limnol. Oceanogr. 46, 1048–1053.

Weissburg, M. J. 2000. The fluid dynamical context of chemosensory behavior. Biol. Bull. 198, 188–200.

Weissburg, M. J. and Zimmer-Faust, R. K. 1993. Life and death in moving fluids: Hydrodynamic effects on chemosensory-mediated predation. Ecology 74, 1428–1443.

Weissburg, M. J. and Zimmer-Faust, R. K. 1994. Odor plumes and how blue crabs use them in finding prey. J. Exp. Biol. 197, 349–375.

Willis, M. A. and Avondet, J. L. 2005. Odor-modulated orientation in walking male cockroaches, *Periplaneta americana* (L.), and the effects of odor plumes of different structure. J. Exp. Biol. 208, 721–735.

Willis, M. A., David, C. T., Murlis, J., and Cardé, R. T. 1994. Effects of pheromone plume structure and visual stimuli on the pheromone-modulated upwind flight of male gypsy moths (*Lymantria dispar* L.), in a forest. J. Insect Behav. 7, 385–409.

Wolf, H. and Wehner, R. 2000. Pinpointing food sources: Olfactory and anemotactic orientation in desert ants, *Cataglyphis fortis*. J. Exp. Biol. 203, 857–868.

Wright, R. H. 1958. The olfactory guidance of flying insects. The Can. Entomol. 90, 81–89.

Zimmer-Faust, R. K., Finelli, C. M., Pentcheff, N. D., and Wethey, D. S. 1995. Odor plumes and animal navigation in turbulent water flow: A field study. Biol. Bull. 188, 111–116.

4.43 Accessory Olfactory System

F Zufall and T Leinders-Zufall, University of Saarland School of Medicine, Homburg, Germany

A C Puche, University of Maryland School of Medicine, Baltimore, MD, USA

Glossary

aEPL Accessory external plexiform layer.

aGCL Accessory granule layer.

aGL Accessory glomerular layer.

aMCL Accessory mitral cell layer.

AOB Accessory olfactory bulb.

AOC Accessory olfactory cortex.

aONL Accessory olfactory nerve layer.

AONm Anterior olfactory nucleus, medial division.

AP Posterior amygdaloid nucleus.
APVH Anterior periventricular hypothalamus.
BAOT Bed nucleus of the accessory olfactory tract.
Bar Barrington's nucleus.
BST Bed nucleus of the stria terminalis.
CG Central (periaqueductal) gray.
CNGA2 Cyclic nucleotide-gated channel type A2.
DAG Diacylglycerol.
DB Nucleus of the diagonal band.
DR Dorsal raphe nucleus.
DT Dendritic tip.
EVG Electrovomeronasogram.
GC-D Guanylyl cyclase type D.
GG Grüneberg ganglion.
LC Locus coeruleus.
LOT Lateral olfactory tract.
LPO Lateral preoptic area.
MC Mitral cell.
Me Medial amygdaloid nucleus.
MHC Major histocompatibility complex.
MPO Medial preoptic area.
NSE Nonsensory epithelium.

P/Amb Nucleus ambiguous and periabigual area.
PDE4A Phosphodiesterase type 4A.
PG Periglomerular.
Pit Pituitary.
PMCo Posteromedial cortical amygdaloid nucleus.
RVL Rostroventrolateral medulla.
SAG 1-Stearoyl-2-arachidonoyl-*sn*-glycerol.
SC Sustentacular cell.
SCis Lumbosacral spinal cord.
SO Supraoptic nucleus.
SOM Septal organ of Masera.
TRP Transient receptor potential.
TRPC2 Canonical transient receptor potential channel type 2.
VB Vomer bone.
VMH Ventromedial hypothalamic nucleus.
VN Vomeronasal nerve.
VNO Vomeronasal organ.
VSN Vomeronasal sensory neuron.
V1R Vomeronasal receptor family type 1.
V2R Vomeronasal receptor family type 2.
wt Wild type.

4.43.1 Introduction

The majority of vertebrates have dual independent olfactory systems; possessing an accessory olfactory system (AOS) in addition to the main olfactory system. The AOS has attracted a great deal of attention in recent years because of a growing recognition of the essential role in chemical communication and the regulation of social behaviors this system plays. Key elements of the AOS (Figure 1) include

1. the vomeronasal organ (VNO), a chemoreceptive structure situated at the base of the nasal septum which houses the vomeronasal sensory neurons (VSNs);
2. the accessory olfactory bulb (AOB), a structure of the forebrain that receives input from the VNO serving as the first processing center of vomeronasal information; and
3. higher olfactory centers that receive direct or indirect information from the AOB.

These higher brain structures include, in mammals, the medial amygdala and parts of the cortical and posterior amygdala, the bed nucleus of the accessory olfactory tract (BAOT), the bed nucleus of the stria terminalis (BST), and the medial hypothalamus. The focus of this chapter will be on physiology and function of the VNO and AOB. Whenever possible, we attempt to link genetic information with anatomical, physiological, and behavioral approaches. Because the mouse has become the principal model system for a genetically driven, systems-oriented approach, the focus of this chapter will be on rodent AOS function. However, it is important to realize that the precise role of the AOS must be viewed in close relationship to species-specific adaptations. Indeed, the initial discovery of pheromonal–behavioral signaling through specific neural pathways was pioneered in invertebrate biology, and many of the fundamental concepts of VNO–AOB function have developed from research in nonmammalian species.

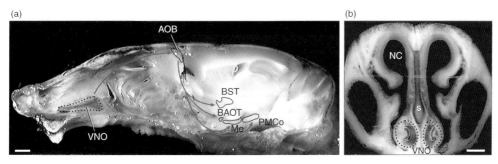

Figure 1   Anatomy of the accessory olfactory system. (a) Perisaggital view of the mouse head with a superimposed diagram representing the accessory olfactory pathway. Sensory inputs in the accessory olfactory system are detected by sensory neurons in the vomenonasal organ (VNO), a bilateral tubular structure in the anterior region of the nasal cavity (see (b)). An extensive network of blood sinuses is located at the lateral external wall of the organ. Together with a band of cavernous erectile tissue, it provides the hardware for a vomeronasal pump that aids in stimulus delivery. Axons of the vomeronasal sensory neurons (VSNs) project to the accessory olfactory bulb (AOB), which in turn transmits sensory information to the bed nucleus of the stria terminalis (BST), the bed nucleus of the accessory olfactory tract (BAOT), and the amygdala, that is, the medial amygdaloid nucleus (Me) as well as the posteromedial cortical amydaloid nucleus (PMCo). The information is then transmitted from the amygdala to specific nuclei of the hypothalamus. These tertiary projections to the hypothalamic region are also known as the neuroendocrine hypothalamus. This area controls the release of hormones by the pituitary and therefore modulates the endocrine status of the animal. As a reference for the location of the VNO, the septum was partially removed to reveal the turbinates of the main olfactory organ. Scale bar = 1.5 mm. (b) Transmitted light image of a coronal section through a postnatal day 7 mouse head showing the location of the bilateral VNO (dashed circles) at the base of the nasal septum (S). NC, nasal cavity. Scale bar = 200 μm.

4.43.2   A Brief Historical Perspective

The VNO was discovered by Ludvig Jacobson (and hence often referred to as Jacobson's organ) and was originally described in a Danish publication (Jacobson, L., 1813). This work has recently been translated into English and is thus accessible to a wider audience (Doving, K. B. and Trotier, D., 1998; Jacobson, L. *et al.*, 1998). Jacobson described the organ in a variety of domesticated mammals, some wild carnivores, ungulates, and marine mammals. He assumed that the organ was secretory in nature but suspected that it could also function as a sensory organ.

Balogh C. (1860), Klein E. (1881), Piana G. P. (1882), and Retzius G. (1894) observed similarities in the morphology of sensory neurons in the main olfactory epithelium and VNO, and proposed a common function between the cells. During the following years, the presence of the VNO was confirmed in most mammals (von Mihalkovics, V., 1899; Perlman, S. M., 1934). A review by Doving K. B. and Trotier D. (1998) has reproduced the excellent quality of cross sections of the mouse VNO originally published by von Mihalkovics V. (1899).

The existence of the AOB in mammals was recognized by von Gudden B. (1870). McCotter R. E. (1912) was the first to clearly demonstrate that the AOB is a brain structure dedicated to the reception of vomeronasal nerve (VN) fibers, and that it gives off fibers that join the lateral olfactory tract. He suggested that the olfactory system should be divided into two distinct subsystems, namely an ordinary and an AOS.

Beginning in the 1930s and extending to the 1950s, Lord Adrian began to use electrophysiological methods to record extracellular activity from the VNs of the rabbit. He was unable to show a reaction to chemical stimuli in the VNO, but he noted considerable rhythmic activity in response to localized pressure (Adrian, E. D., 1955).

During the 1950s and 1960s, an intense period of pheromone research began. Following the seminal work on insect pheromones (Karlson, P. and Lüscher, M., 1959), behavioral endocrinologists had discovered several pheromonal effects in mice and a consensus emerged that these effects require a functional olfactory system. Winans S. S. and Scalia F. (1970) suggested that some of these effects might depend on a functional AOS. This idea eventually led to the dual olfactory hypothesis (reviewed by Halpern, M., 1987), which proposed that distinct parallel pathways could be traced from the periphery of the main and AOSs into the telencephalon and diencephalon of all animals possessing these two chemoreceptive systems. As pointed out by Halpern M. (1987), a main corollary of this hypothesis is that each system should be involved in distinct behavioral domains. During the 1970s and 1980s, research concentrated on testing this hypothesis and its corollary by using a combination of anatomical, behavioral, and physiological approaches. This work has been summarized in a number of excellent reviews (Burghardt, G. M., 1970; Madison, D. W., 1977; Keverne, B., 1979; Wysocki, C. J., 1979; Johnston, R. E., 1983; Keverne, E. B., 1983; Meredith, M., 1983; Halpern, M., 1987; Wysocki, C. J. and Meredith, M., 1987).

The 1990s saw an explosion of interest in AOS function triggered by the application of molecular biological and genetic approaches, together with the use of more sophisticated electrophysiological recording techniques. A significant finding was the segregation of the mouse and rat vomeronasal system into two major subsystems, each expressing distinct classes of signal transduction-related molecules (Halpern, M. *et al.*, 1995; Berghard, A. *et al.*, 1996; Jia, C. and Halpern, M., 1996). This was followed by the identification of the first (Dulac, C. and Axel, R., 1995) and second families of vomeronasal receptor genes (Herrada, G. and Dulac, C., 1997; Ryba, N. J. and Tirindelli, R., 1997; Matsunami, H. and Buck, L. B., 1997), and the cloning of the ion channel TRPC2 (canonical transient receptor potential channel type 2) (Liman, E. R. *et al.*, 1999). The first patch clamp recordings from single VSNs were carried out by Trotier D. *et al* (1993), Taniguchi, M. *et al.* (1995), and Liman E. R. and Corey D. P. (1996). In parallel, gene-targeting methods were developed by Mombaerts P. *et al.* (1996), which subsequently led to an analysis of the axonal projection patterns of specific subsets of VSNs (Belluscio, L. *et al.*, 1999; Rodriguez, I. *et al.*, 1999).

Since 2000, research activities aimed at understanding AOS function have further increased. The year 2000 saw the development of reduced mouse VNO preparations that allowed for simultaneous recording of stimulus-induced activity from hundreds of VSNs (Holy, T. E. *et al.*, 2000; Leinders-Zufall, T. *et al.*, 2000) leading to the identification of the first prospective pheromones acting on mammalian VSNs (Leinders-Zufall, T. *et al.*, 2000). Next came the targeted deletion of specific signal transduction genes, which enabled, for the first time in the AOS, a correlation between gene function, cellular activity, and behavioral changes (Del Punta, K. *et al.*, 2002a; Leypold, B. G. *et al.*, 2002; Stowers, L. *et al.*, 2002). Progress in understanding VSN signal transduction mechanisms (Lucas, P. *et al.*, 2003) was quickly followed by the finding that a subset of VSNs function as detectors of peptide ligands of the major histocompatibility complex (MHC), and that these peptides influence selective pregnancy termination (the Bruce effect) (Leinders-Zufall, T. *et al.*, 2004). In parallel, the first recordings from AOB neurons in freely behaving mice were undertaken, making it now possible to correlate neuronal activity in the AOB with specific behavioral changes (Luo, M. *et al.*, 2003). Several reviews have summarized these most recent aspects of AOS research (Dulac, C. and Torello, A. T., 2003; Brennan, P. A. and Keverne, E. B., 2004; Luo, M. and Katz, L. C., 2004; Mombaerts, P., 2004; Bigiani, A. *et al.*, 2005; Boehm, T. and Zufall, F., 2006). Halpern M. and Martínez-Marcos A. (2003) have provided the most comprehensive review on structure and function of the vomeronasal system published thus far.

4.43.3 The Vomeronasal Organ

4.43.3.1 Basic Anatomical Organization

The structure of the VNO and its neuroepithelium has been described in representatives from virtually every major vertebrate class at the light microscopic and ultrastructural level, demonstrating a remarkable conservation between different species. Many of these original studies have been summarized by Halpern M. (1987) and Halpern M. and Martínez-Marcos A. (2003) and the reader is referred to these extensive reviews. In mammals, the VNO is a bilateral, cigar-shaped organ that is located at the base of the nasal septum, anterior and ventral to the main olfactory epithelium (Figures 1 and 2). Each VNO is encased in a bony and/or cartilaginous capsule known as vomer bone capsule. Chemosensory stimuli gain access to the VNO through its single rostral opening which, in rodents, opens into the nasal cavity. The main structural elements of the VNO are best seen in coronal sections (Figures 2(b)–2(d) and 2(h)) and include a central lumen, a medially located crescent-shaped sensory epithelium, and a laterally located nonsensory epithelium together with an extensive network of blood sinuses surrounded by a band of cavernous tissue. These latter components provide the hardware for a vomeronasal pump that aids in stimulus delivery (Meredith, M. and O'Connell, R. J., 1979; Meredith, M. *et al.*, 1980) (see below).

The VNO sensory epithelium consists primarily of three cell types (Ciges, M. *et al.*, 1977; Vaccarezza, O. L. *et al.*, 1981; Garrosa, M. and Coca, S., 1991):

1. elongated glia-like supporting cells,
2. bipolar sensory cells known as VSNs, and
3. a sparse population of basal stem cells.

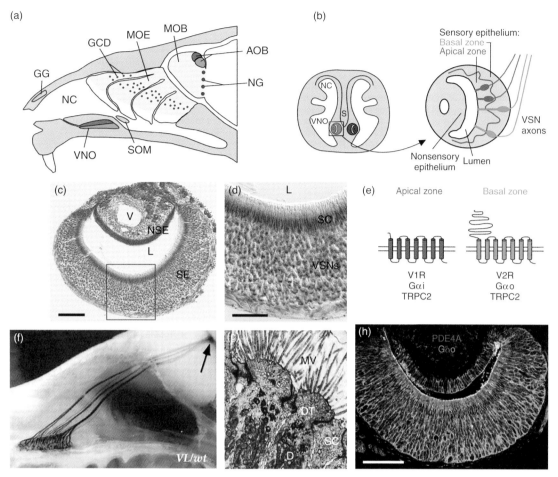

Figure 2 Organization of the sense of smell. (a) Midsagittal view of the rodent nasal cavity (NC) and forebrain. Vomeronasal sensory neurons (VSNs) in the vomeronasal organ (VNO), which is located at the base of the nasal septum (S), project to the anterior (red) or posterior (green) accessory olfactory bulb (AOB). Sensory neurons in the main olfactory epithelium (MOE) project their axons to glomeruli in the main olfactory bulb (MOB). These MOE neurons employ cyclic nucleotide-gated channels for sensory transduction. The septal organ of Masera (SOM) is an isolated island of olfactory sensory neurons that project axons to glomeruli in the ventromedial olfactory bulb. The Grüeneberg ganglion (GG) projects axons to specific glomeruli in the caudal region of the MOB. GCD (guanylyl cyclase D-expressing) neurons are located in the MOE, but project to a limited number of necklace glomeruli (NG) in the caudal MOB. (b–d) Coronal sectioning through the nasal cavity reveals the typical crescent-shaped organization of the VNO sensory epithelium (SE). The region delimited by the black box in (c) is shown at higher magnification in (d). L, lumen; NSE, nonsensory epithelium; SC, sustentacular cell; V, vein. Original sections in (c) and (d) are Nissl stains. Scale bars = 100 μm (c) and 50 μm (d). (e) The sensory epithelium is segregated into two distinct zones, both of which express a unique set of transduction-related molecules: (1) an apical zone (red VSNs) that expresses the G protein $G\alpha_{i2}$ as well as members of the V1R family of vomeronasal receptors, and (2) a basal zone that characteristically contains VSNs (green) that express $G\alpha_o$ and members of the V2R receptor family. Both neuron types express TRPC2. (f) Distribution and axonal projection pattern of a subpopulation of gene-targeted VSNs, showing that axon bundles (vomeronasal nerves) terminate in the AOB (arrow). To visualize this topography, a targeted mutation was generated that led to co expression of taulacZ in all VSNs expressing a distinct V1R receptor. (g) Electron micrograph of a transverse section through two mouse VSNs showing their dendritic process (D), which enlarges into a dendritic tip or knob (DT). From the knob emanate various microvilli (MV). (h) Visualization by confocal microscopy of the two principal expression zones of the VNO sensory epithelium using double-label immunohistochemistry with antibodies to phosphodiesterase type 4A (PDE4A, red) and $G\alpha_o$ (green). Scale bar = 100 μm. (a–e) Reprinted from Zufall, F., Ukhanov, K., Lucas, P., Liman, E., and Leinders-Zufall, T. 2005. Neurobiology of TRPC2: from gene to behavior. Pflügers Arch. 451, 61–71, with kind permission of Springer Science and Business Media. (f) Reprinted from Rodriguez, I., Feinstein, P., and Mombaerts, P. 1999. Variable patterns of axonal projections of sensory neurons in the mouse vomeronasal system. Cell 97, 199–208, with permission from Elsevier. (g) Reprinted from Ciges, M., Labella, T., Gayoso, M., Sanchez, G. 1977. Ultrastructure of the organ of Jacobson and comparative study with olfactory mucosa. Acta Otolaryngol. 83, 47–58, Taylor & Francis. (h) Reprinted with permission from Leinders-Zufall, T., Brennan, P., Widmayer, P., Prasanth Chandramani, S., Maul-Pavicic, A., Jager, M., Li, X. H., Breer, H., Zufall, F., and Boehm, T. 2004. MHC class I peptides as chemosensory signals in the vomeronasal organ. Science 306, 1033–1037, AAAS.

VSNs make up the majority of the cells in the sensory epithelium and are responsible for the detection of chemostimuli. VSNs extend a dendritic process to the luminal surface where it enlarges into an apical dendritic knob (Figure 2(g)). Unlike olfactory sensory neurons (OSNs), the dendritic knob of VSNs lacks cilia instead containing up to 100 microvilli (Figure 2(g)). In rat, VSN microvilli are 3–6 μm long and approximately 100 nm in diameter (Vaccarezza, O. L. *et al.*, 1981). The primary chemotransduction events are thought to take place in these microvilli. The most prominent feature of the dendritic knob is an extensive network of cisterns that appear as a series of small cylindrical rods without any specific orientation (Vaccarezza, O. L. *et al.*, 1981). Cisterns have not been detected in the microvilli. By contrast, the apical VSN dendrite has an extensive network of mitochondria and endoplasmic reticulum (Ciges, M. *et al.*, 1977; Vaccarezza, O. L. *et al.*, 1981). The mitochondria typically appear radially oriented in the dendrite and are generally parallel to the dendritic plasmalemma. The soma is characterized by its clear round nucleus and the extensive presence of rough endoplasmic reticulum. Particularly at the apical pole of the soma, where the dendrite begins to emerge, there is a complex stack of Golgi apparatus, rough endoplasmic reticulum, and a multivesicular body. Thus, organelles that could serve as intracellular Ca^{2+} stores are spatially compartmentalized in VSNs but, as far as we know, are not present in the microvilli. A single unmyelinated, axon emerges from the basal aspect of the soma of each VSN and exits the epithelium through the basal lamina. VSN axons coalesce to form the relatively long VNs that course along the septum, pass through the cribriform plate, and end in the glomeruli of the AOB, where they synapse onto second-order mitral cells (Halpern, M., 1987 and references therein). This axonal projection pattern is best observed in mutant mouse strains in which specific subpopulations of VSNs coexpress histological markers such as β-galactosidase or green fluorescent protein (Belluscio, L. *et al.*, 1999; Rodriguez, I. *et al.*, 1999) (Figure 2(f)).

The evolution of the VNO has been reviewed by Bertmar G. (1981) and Eisthen H. L. (1992; 1997). In general, the presence of a discrete VNO appears to be a derived character of tetrapods and did not arise as a terrestrial adaptation. VSNs are thought to first appear in amphibians, although a recent study using transgenic zebrafish in combination with VNO-specific markers showed the existence of a VNO-like subsystem of microvillous sensory neurons within the olfactory epithelium of fish (Sato, Y. *et al.*, 2005). Many reptiles possess a well-developed VNO; the VNO of the garter snake has become one of the best-studied model systems for understanding VNO function (Halpern, M. and Martinez-Marcos, A., 2003). The AOS has been lost independently in several taxa, including crocodilians, birds, some bats, and some primates (Eisthen, H. L., 1992). Among primates, New World monkeys possess VNOs whereas adult Old World monkeys do not appear to have well developed VNOs (Halpern, M. and Martinez-Marcos, A., 2003).

Do humans possess a functional VNO? While this question has been hotly debated for many years, a consensus is emerging (Meisami, E. and Bhatnagar, K. P., 1998; Trotier, D. *et al.*, 2000; Meredith, M., 2001; Dulac, C. and Torello, A. T., 2003; Halpern, M. and Martinez-Marcos, A., 2003; Wysocki, C. J. and Preti, G., 2004). At the anatomical level, bipolar receptor cells can be found within the VNO of the developing human fetus, but are absent in the adult (Boehm, N. and Gasser, B., 1993; Boehm, N. *et al.*, 1994; Wysocki, C. J. and Preti, G., 2004). More importantly, there appear to be no connections of the VNO with the adult brain, and in adult humans an anatomically distinct AOB cannot be located (Meisami, E. and Bhatnagar, K. P., 1998). Thus, any putative pheromonal responses by humans are likely mediated by the main olfactory system rather than by an AOS (Wysocki, C. J. and Preti, G., 2004). Consistent with a vestigial AOS in humans, most of the genes encoding vomeronasal classes of receptors are pseudogenes in humans (Dulac, C. and Torello, A. T., 2003).

4.43.3.2 Zonal Segregation of the Vomeronasal System

The VNO sensory epithelium of many mammals is not homogeneous, but it can be partitioned into at least two distinct zones or layers containing apical (superficial) and basal (deep) VSN populations, respectively (Halpern, M. *et al.*, 1998; Tirindelli, R. *et al.*, 1998; Dulac, C., 2000) (Figures 2(b), 2(e), and 2(h)). The two subsets of sensory neurons project their axons to segregated zones of the AOB (von Campenhausen, H. *et al.*, 1997; Mori, K. *et al.*, 2000; Halpern, M. and Martinez-Marcos, A., 2003). As will be summarized in subsequent sections of this chapter, it is becoming increasingly clear that this spatial segregation has important consequences for the handling of structurally and functionally distinct sets of chemostimuli. The zonal separation was first recognized in the AOS of opossum, mouse, and rat by localization of the G protein subunits $G\alpha_{i2}$ and $G\alpha_o$ to two distinct zones: $G\alpha_{i2}$-positive VSNs define an apical zone whereas $G\alpha_o$-positive VSNs define a basal zone of the VNO epithelium (Halpern, M. et al., 1995; Berghard, A. et al., 1996; Jia, C. and Halpern, M., 1996) (Figures 2(e) and 2(h)).

As reviewed in detail by Halpern M. and Martínez-Marcos A. (2003), numerous other markers (often with unknown functions) are expressed in the VNO and AOB in a heterogeneous manner. For example, an axonal surface glycoprotein, olfactory cell adhesion molecule (OCAM), is expressed exclusively in apical VSNs (von Campenhausen, H. et al., 1997; Yoshihara, Y. et al., 1997). Likewise, the cAMP-specific phosphodiesterase type 4A (PDE4A) is found specifically in apical VSNs (Lau, Y. E. and Cherry, J. A., 2000) (Figure 2(h)). The axon guidance molecule neuropilin-2, a coreceptor for a subfamily of secreted guidance cues known as class 3 semaphorins, is also detected in VSNs of the apical layer (Walz, A. et al., 2002). By contrast, olfactory marker protein (OMP), a 19 kDa cytoplasmic protein of unknown function, is usually present in all mature mammalian VSNs (Halpern, M. and Martinez-Marcos, A., 2003) (Figure 3(a)). Maybe most interestingly, the partitioning of the AOS is maintained at the level of different vomeronasal receptors, summarized in the next section.

4.43.3.3 A Genetic Basis for Stimulus Detection

Differential screening of cDNA libraries constructed from single rat VSNs led to the isolation of the first family of vomeronasal, G-protein-coupled receptor genes, the V1Rs (vomeronasal receptor family type 1) (Dulac, C. and Axel, R., 1995). These are completely distinct from the olfactory receptor superfamily. In the mouse genome, 293 V1R sequences have been identified of which 137 have uninterrupted full-length open reading frames. These are divided into 12 subfamilies that are considerably divergent from each other (Del Punta, K. et al., 2000; Rodriguez, I. et al., 2002). V1R genes are expressed in VSNs of the apical zone of the sensory epithelium (Figure 2(e)) and each individual receptor gene is expressed by only a small subset of VSNs (Dulac, C. and Axel, R., 1995) (Figures 3(b) and 3(c)). Evidence for a functional role of V1Rs in vomeronasal chemodetection comes from the observation that a mouse line lacking

a cluster of V1Rs fails to respond to urinary chemicals with known pheromonal effects (6-hydroxy-6-methyl-3-heptanone, n-pentylacetate, and isobutylamine) (Del Punta, K. et al., 2002a) (Figures 3(d) and 3(e)). In an alternative approach, green fluorescent protein-tagged VSNs that express the V1rb2 gene have been shown to respond to the urinary compound 2-heptanone and this response was absent when V1rb2 gene was deleted (Boschat, C. et al., 2002). This result strongly implies that V1rb2 is a receptor for 2-heptanone.

A second large family of G-protein-coupled receptors expressed in the VNO is formed by the V2R (vomeronasal receptor family type 2) genes, which are expressed in VSNs of the basal zone (Herrada, G. and Dulac, C., 1997; Matsunami, H. and Buck, L. B., 1997; Ryba, N. J. and Tirindelli, R., 1997). With their large extracellular, N-terminal domain (Figure 2(e)), V2Rs resemble the calcium-sensing and metabotropic glutamate receptors rather than V1Rs and olfactory receptors. In the mouse, the V2R gene repertoire consists of 208 members of which only 61 exhibit an intact open reading frame (Yang, H. et al., 2005). Although each V2R gene is generally expressed by only a small subset of VSNs, one subfamily of V2R genes is coexpressed broadly across basal VSNs (Martini, S. et al., 2001). It still remains to be seen whether this results in the formation of functional homo- or heterodimeric receptor proteins (Bigiani, A. et al., 2005).

VSNs in the basal layer express yet another multigene family representing nonclassical class Ib genes of the MHC (Ishii, T. et al., 2003; Loconto, J. et al., 2003). These seem to form multimolecular complexes with V2Rs and have been proposed to function as escort molecules in the transport of V2Rs to the cell surface (Loconto, J. et al., 2003). These findings are intriguing in light of the evidence that subsets of basal VSNs are tuned to respond to class I MHC peptide ligands (Leinders-Zufall, T. et al., 2004). However, a recent structural analysis of MHC class 1b proteins found no evidence for bound peptides, suggesting that class Ib molecules are unlikely to provide specific recognition of MHC peptide ligands in VSNs (Olson, R. et al., 2005).

4.43.3.4 Molecular Physiology of Neuronal Sensing

Like OSNs, VSNs have evolved distinct ionic mechanisms that enable them to generate specific neural signals in response to chemostimulation. Recent years

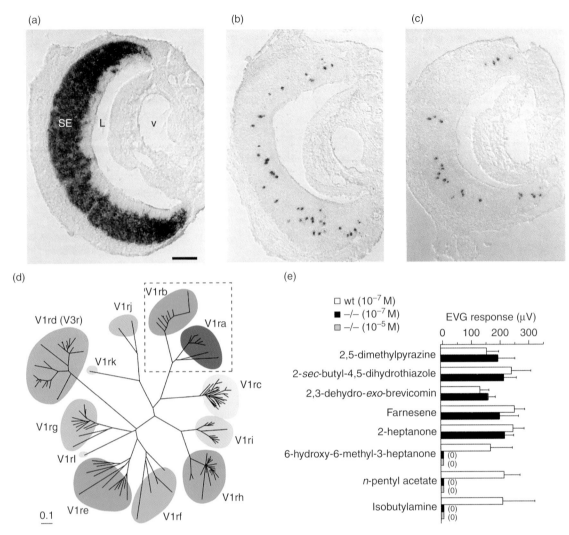

Figure 3 (a–c) Expression of Vr1 receptor RNA is restricted to a subset of vomeronasal sensory neurons (VSNs). Coronal sections of the vomeronasal organ (VNO) dissected from adult male rats were annealed with digoxigenin-labeled antisense RNA probes for olfactory marker protein (a). *In situ* hybridization to coronal sections of a dissected VNO using digoxigenin-labeled probes for rat Vr1 receptors VN1 (b) and VN3 (c). VNO cDNA clones 1 and 3 label 2.7% and 1.1% of cells in the rat sensory epithelium, respectively. L, lumen; SE, sensory epithelium; V, blood vessel. The two above-mentioned rat V1r genes have been predicted to belong to the V1ra group of vomeronasal receptors based on homology with the mouse V1r genes (Del Punta, K. *et al.*, 2000). (d) Unrooted tree representing the 137 mouse V1r genes, with 12 families readily distinguishable. Dashed lines enclose the V1ra and V1rb families of which most members were deleted from the germ line in the study shown in (e). (e) Histogram of electrovomeronasogram (EVG) responses (mean ± SD) from wild type (wt) and $\Delta V1rab\Delta$ mutant mice. Both male and female mutant mice do not respond to 6-hydroxy-6-methyl-3-heptanone, *n*-pentyl acetate, and isobutylamine, but are otherwise normal in their responses to other tested compounds. This chemosensory deficiency supports a role of V1r receptors as pheromone receptors. Stimulus concentration, 10^{-7} M (white and black bars) or 10^{-5} M (gray bars). (a–c) Reprinted from Dulac, C. and Axel, R. 1995. A novel family of genes encoding putative pheromone receptors in mammals. Cell 83, 195–208, with permission from Elsevier. (d, e) reprinted from Del Punta, K., Leinders-Zufall, T., Rodriguez, I., Jukam, D., Wysocki, C. J., Ogawa, S., Zufall, F., and Mombaerts, P. 2002a. Deficient pheromone responses in mice lacking a cluster of vomeronasal receptor genes. Nature 419, 70–74, with permission from Nature Publishing Group.

have delivered significant progress in our ability to record physiological responses in VSNs and to analyze the ionic and molecular mechanisms underlying such responses.

Patch clamp recordings using individual VSNs from mouse, rat, frog, snake, turtle, or lizard – either freshly dissociated or in acute VNO tissue slices – have established that VSNs exhibit resting potentials, ranging from −60 to −80 mV or even more negative (Trotier, D. *et al.*, 1993; Liman, E. R. and Corey, D. P. 1996; Trotier, D. and Doving, K. B., 1996; Inamura, K. and Kashiwayanagi, M., 2000; Taniguchi, M. *et al.*, 2000; Fadool, D. A. *et al.*, 2001; Fieni, F. *et al.*, 2003; Lucas, P. *et al.*, 2003; Leinders-Zufall, T. *et al.*, 2004; Labra, A. *et al.*, 2005). An electrogenic Na$^+$, K$^+$-ATPase appears to be essential in setting this resting potential (Trotier, D. and Doving, K. B., 1996). VSNs express a variety of voltage-dependent currents, including rapid, tetrodotoxin-sensitive Na$^+$ currents (Trotier, D. *et al.*, 1993; Liman, E. R. and Corey, D. P. 1996; Fieni, F. *et al.*, 2003), voltage- and Ca^{2+}-dependent K$^+$ currents (Trotier, D. *et al.*, 1993; Liman, E. R. and Corey, D. P., 1996; Inamura, K., *et al.*, 1997; Fadool, D. A. *et al.*, 2001; Labra, A. *et al.*, 2005), several types of voltage-activated Ca^{2+} currents (Trotier, D. *et al.*, 1993; Liman, E. R. and Corey, D. P., 1996; Fieni, F. *et al.*, 2003), and hyperpolarization-activated inward rectifying currents (Trotier, D. *et al.*, 1993; Liman, E. R. and Corey, D. P., 1996). A common feature of most VSNs analyzed thus far seems to be their high input resistance. Combined with a low threshold for action potential generation, this means that only a few picoampere of inward current are sufficient to evoke repetitive spiking in these neurons (Liman, E. R. and Corey, D. P., 1996; Inamura, K., *et al.*, 1997; Trotier, D., 1998; Taniguchi, M. *et al.*, 2000), an ability that has important consequences for the generation of ligand-induced responses in these cells. The biophysical features of VSNs differ considerably from those observed in supporting cells of the vomeronasal epithelium (Ghiaroni, V. *et al.*, 2003).

How do VSNs respond to chemical stimulation? Several laboratories used extracellular or whole-cell current clamp recordings to demonstrate that VSNs, at rest, maintain a relatively low spiking activity and that chemostimulation produces a transient increase in the rate of action potential firing (Trotier, D., 1998; Inamura, K. *et al.*, 1999; Holy, T. E. *et al.*, 2000; Leinders-Zufall, T. *et al.*, 2000; Lucas, P. *et al.*, 2003; Leinders-Zufall, T. *et al.*, 2004) (Figures 4(a) and 4(b)). This excitation is caused by the generation of a depolarizing receptor potential sufficient in size to activate voltage-gated Na$^+$ channels (Lucas, P. *et al.*, 2003) (Figure 4(c)). The ionic properties of this receptor potential are not yet fully understood, but it is clear that the receptor potential depends critically on the generation of a transient inward current in the distal dendrites and microvilli of VSNs which is associated with an increase in membrane conductance (Inamura, K. and Kashiwayanagi, M.,

2000; Taniguchi, M. *et al.*, 2000; Lucas, P. *et al.*, 2003) (Figure 4(d)). The primary electrical events in VSN dendrites can also be monitored by placing micropipettes on VSN microvilli using a semi-intact VNO preparation and recording local field potentials that are known as an electrovomeronasogram (EVG) (Figures 4(e) and 4(f)). This technique has been very useful in the search for molecular cues acting on VSNs (Taniguchi, M. *et al.*, 1998; Leinders-Zufall, T. *et al.*, 2000; 2004) (Figure 4(f)), as well as for the phenotypic characterization of VSN responses in animals with targeted mutations (Del Punta, K. *et al.*, 2000; Leypold, B. G. *et al.*, 2002) (Figures 3(e) and 5(f)). Finally, it is important to note that chemostimulation also causes a transient rise in the intracellular Ca^{2+} concentration of VSNs (Leinders-Zufall, T. *et al.*, 2000; Cinelli, A. R. *et al.*, 2002; Spehr, M. *et al.*, 2002; Leinders-Zufall, T. *et al.*, 2004) (Figures 4(g) and 4(h)). This stimulus-evoked Ca^{2+} response is used increasingly to identify molecular cues recognized by the VSNs and to determine VNO coding mechanisms (Figure 4(i) and see below).

Although these basic response properties have now been observed independently by several investigators, other experiments suggest that there are also differences in the properties of VSN sensory responses amongst species. For example, VSNs from turtle and lizard, respectively, have been shown to exhibit both inward and outward currents in response to chemostimulation (Fadool, D. A. *et al.*, 2001; Labra, A. *et al.*, 2005). In particular, the properties of the voltage-activated K$^+$ currents differ between VSNs from male and female animals (Fadool, D. A. *et al.*, 2001). Thus, a modicum of caution needs to be used for extrapolating cellular mechanisms in response to chemostimulation across species.

4.43.3.5 Signal Transduction Mechanisms

How do VSNs transduce a sensory stimulus into a change in membrane conductance? Increasing evidence indicates that this process depends on a G-protein-coupled second messenger cascade that ultimately leads to the activation of specific ion channels. Here, we will summarize the key results that led to our current understanding of this signaling cascade (Halpern, M. and Martinez-Marcos, A., 2003; Bigiani, A. *et al.*, 2005; Zufall, F. *et al.*, 2005).

An important development during the mid 1990s was the demonstration that VSNs and OSNs differ in their chemosensory signal transduction pathways. Several molecules that are essential for transduction

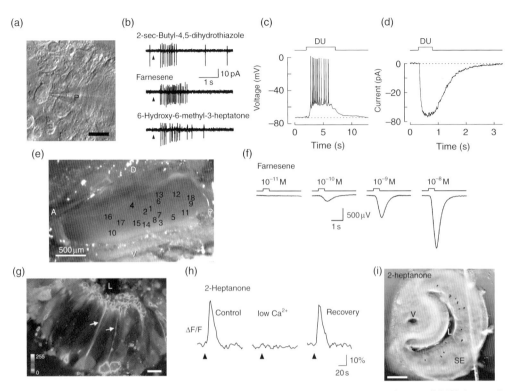

Figure 4 (a) Infrared–differential interference contrast micrograph showing a vomeronasal organ (VNO) tissue slice. Individual vomeronasal sensory neuron (VSN) somata are clearly visible. P, patch electrode. Scale bar = 10 µm. (b) Trains of extracellularly recorded, action-potential-driven, capacitive currents produced by bath application of 2-*sec*-butyl-4,5-dihydrothiazole, farnesene, or 6-hydroxy-6-methyl-3-heptanone (all at 10^{-7} M). Each response was from a different vomeronasal sensory neuron (VSN) using the approach shown in (a). Arrowheads, start of bath application (duration, 1–3 s). (c) Focal stimulation of a current-clamped VSN with dilute urine (DU) produces a depolarizing receptor potential leading to robust action potential discharges. (d) Under voltage clamp, 500-ms stimulation generates a rapidly activating and then deactivating inward current (the sensory current). Holding potential, −70 mV. (e, f) Image and electrovomeronasogram (EVG) responses of the microvillus surface of the vomeronasal sensory epithelium. Numbers in the image indicate the various sites at which field potentials (EVG responses) were recorded. 500-ms pulses of increasing concentrations of farnesene produced dose-dependent EVG responses. Ligands were focally applied to the microvillus surface of the sensory epithelium by a multibarreled stimulation pipette. A, anterior; D, dorsal; V, ventral; P, posterior. (g, h) Pheromone-induced Ca^{2+} elevations in single VSNs. (g) Confocal fluorescence image acquired at rest (pseudocolor scale) showing subcellular structures of VSNs such as single dendrites. Scale bar = 10 µm. (h) Waveform of somatic Ca^{2+} transient of a single VSN evoked by bath-applied 2-heptanone (10^{-7} M). The response was reversible abolished by lowering the Ca^{2+} concentration in the bath solution. (i) Spatial representation of functional, pheromone-induced activity in VNO. Reconstructed VSN response map ($\Delta F/F$ Ca^{2+} images digitally superimposed onto transmitted light image of the same slice) for 10^{-7} M 2-heptanone, showing that VSNs activated by this ligand (black dots) were interspersed within the apical zone of the sensory epithelium (SE). L, lumen; V, vein. Scale bar = 100 µm. (a, b, f, g, h, i) Reprinted from Leinders-Zufall, T., Lane, A. P., Puche, A. C., Ma, W., Novotny, M. V., Shipley, M. T., and Zufall, F. 2000. Ultrasensitive pheromone detection by mammalian vomeronasal neurons. Nature 405, 792–796, with permission from Nature Publishing Group. (c, d) Reprinted from Lucas, P., Ukhanov, K., Leinders-Zufall, T., and Zufall F. 2003. A diacylglycerol-gated cation channel in vomeronasal neuron dendrites is impaired in *TRPC2* mutant mice: mechanism of pheromone transduction. Neuron 40, 551–561, with permission from Elsevier. (e) Reprinted from Zufall, F., Kelliher, K. R., and Leinders-Zufall, T. 2002. Pheromone detection by mammalian vomeronasal neurons. Microsc. Res. Tech. 58, 251–260, with permission of Wiley.

in OSNs – the G protein G_{olf}, adenylyl cyclase type III, and the cyclic nucleotide-gated channel type A2 (CNGA2) subunit – are not present in the VNO sensory epithelium (Berghard, A. *et al.*, 1996; Wu, Y. *et al.*, 1996). Unlike OSNs, VSNs do not respond to cyclic nucleotides with a conductance change or

Ca^{2+} rise (Liman, E. R. and Corey, D. P., 1996; Cinelli, A. R. *et al.*, 2002; but see also Taniguchi, M. *et al.*, 1996). The focus of interest therefore shifted toward the other major family of ion channels involved in sensory signal transduction, the transient receptor potential (TRP) channels. An important advance came with the cloning of *TRPC2* from bovine, mouse, and rat (Wissenbach, U. *et al.*, 1998; Liman, E. R. *et al.*, 1999; Vannier, B. *et al.*, 1999; Hofmann, T. *et al.*, 2000). Because *TRPC2* encodes a protein that is expressed in all VSNs and specifically localized to the sensory microvilli of these cells (Figures 5(a) and 5(b)), it was proposed to participate in VSN sensory transduction (Liman, E. R. *et al.*,

1999; Menco, B. P. *et al.*, 2001). Many TRP channels, especially those of the classical TRPC subfamily, are known to depend on signaling pathways involving phospholipase C (PLC), an enzyme that hydrolyzes phosphatidylinositol-4,5-bisphosphate (PIP$_2$) into inositol 1,4,5-trisphosphate (Ins(1,4,5)P$_3$) and diacylglycerol (DAG) (Montell, C. *et al.*, 2002). Therefore, the localization of TRPC2 to VSN microvilli

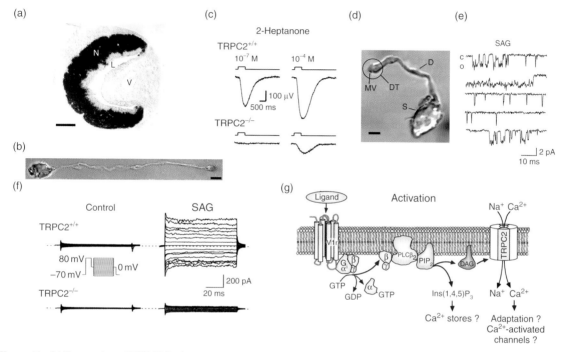

Figure 5 (a) Expression of TRPC2 in the vomeronasal organ (VNO). Labeling of a section of VNO from adult rat with a digoxigenin antisense probe directed against TRPC2 reveals strong expression (dark reaction product) of the TRPC2 mRNA in all VSNs (N). Scale bar = 100 μm. (b) In a singly dissociated vomeronasal sensory neuron (VSN), TRPC2 immunoreactivity (red) is clearly seen in the tuft of microvilli at the distal end of the dendrite. Scale bar = 5 μm. (c) Strongly diminished responses are observed in TRPC2$^{-/-}$ VSNs to 2-heptanone. Representative field potential responses from wild-type (TRPC2$^{/}$) (upper) and TRPC2$^{-/-}$ mice (lower). Responses were induced by 500 ms pulses of 2-heptanone at 10^{-7} or 10^{-4} M, respectively. (d) Micrograph showing an isolated mouse VSN. D, dendrite; MV, microvilli; S, soma. Plasma membrane inside out patches were taken from the encircled area denoted as dendritic tip (DT) in experiments shown in (e). Scale bar = 3 μm. (e) The endogenous diacylglycerol analog 1-stearoyl-2-arachidonoyl-*sn*-glycerol (SAG) activates a 42-pS channel. Control recordings before SAG application showed no activity or only an occasional brief channel opening. Continuous application of SAG (10 μM) evoked single-channel events. (f) TRPC2$^{-/-}$ VSNs display a striking defect in the activation of the diacylglycerol-gated channel. Representative families of whole-cell currents to a series of depolarizing and hyperpolarizing voltage steps (as indicated in the figure) were recorded from an isolated wild-type (upper) and TRPC2$^{-/-}$ VSN (lower). VSNs were exposed successively to extracellular solution (control) and solution containing SAG (50 μM). Voltage-gated channels were blocked in these experiments. Dotted line, zero current level. (g) Schematic representation of pathways that might regulate the activity of the TRPC2 channel in the DT of a mammalian VSN. The model refers to VSNs of the apical zone of the VNO. Whether this applies also to VSNs of the basal zone remains to be investigated. (a, b) Reprinted from Liman, E. R., Corey, D. P., and Dulac, C. 1999. TRP2: a candidate transduction channel for mammalian pheromone sensory signaling. Proc. Natl. Acad. Sci. U. S. A. 96, 5791–5796, with permission from the publisher. (c) Reprinted from Leypold, B. G., Yu, C. R., Leinders-Zufall, T., Kim, M. M., Zufall, F., and Axel, R. 2002. Altered sexual and social behaviors in trp2 mutant mice. Proc. Natl. Acad. Sci. U. S. A. 99, 6376–6381, with permission from the publisher. (d–f) Reprinted from Lucas, P., Ukhanov, K., Leinders-Zufall, T., and Zufall F. 2003. A diacylglycerol-gated cation channel in vomeronasal neuron dendrites is impaired in TRPC2 mutant mice: mechanism of pheromone transduction. Neuron 40, 551–561, with permission from Elsevier. (g) Reprinted from Zufall, F., Ukhanov, K., Lucas, P., Liman, E., and Leinders-Zufall, T. 2005. Neurobiology of TRPC2: from gene to behavior. Pflügers Arch. 451, 61–71, with kind permission of Springer Science and Business Media.

suggested the use of a phosphoinositide signaling pathway for VSN transduction (Liman, E. R. *et al.*, 1999). Several studies have now provided evidence that PLC is indeed critical for the generation of sensory responses in VSNs (Inamura, K. *et al.*, 1997; Holy, T. E. *et al.*, 2000; Lucas, P. *et al.*, 2003).

Investigations of mice with targeted deletions in *TRPC2* have established an essential role of this gene in the detection of pheromonal stimuli by the VNO (Leypold, B. G. *et al.*, 2002; Stowers, L. *et al.*, 2002). In *TRPC2$^{-/-}$* mice, pheromone-evoked EVG responses were either absent or strongly diminished, depending on stimulus concentration (Leypold, B. G. *et al.*, 2002) (Figure 5(c)). Likewise, action potential recordings showed that VSNs from *TRPC2$^{-/-}$* mice are unable to respond to cues present in dilute urine (Stowers, L. *et al.*, 2002). Furthermore, *TRPC2$^{-/-}$* mice showed striking defects in social behaviors (Leypold, B. G. *et al.*, 2002; Stowers, L. *et al.*, 2002), results that will be discussed in Section 4.43.5.

What kind of ion channel is formed by TRPC2? Although this question is still being investigated, evidence indicates that TRPC2 is required for a Ca^{2+}-permeable cation channel that is gated by the lipid messenger DAG or one of its endogenous analogs (Lucas, P. *et al.*, 2003). Excised patch recordings identified a 42-pS DAG-activated cation channel present in high density in the plasma membrane of dendritic knobs of VSNs and the activation of this channel is strongly defective in *TRPC2$^{-/-}$* VSNs (Figures 5(d)–5(f)). Furthermore, the functional properties of this channel are consistent with those of the pheromone-evoked conductance (Lucas, P. *et al.*, 2003). Independent evidence that endogenously produced DAG may be a critical second messenger in the sensory response of VSNs comes from pharmacological blockade of DAG kinase, an enzyme involved in termination of DAG signaling (Lucas, P. *et al.*, 2003; Zufall, F. *et al.*, 2005). Thus a direct link between stimulus-induced PLC activity, gating of the TRPC2 channel by DAG, and the sensory current has been established (Figure 5(g)).

Several other second messengers such as Ca^{2+} or $Ins(1,4,5)P_3$ are likely to be involved in VSN signal transduction (Cinelli, A. R. *et al.*, 2002; Lucas, P. *et al.*, 2003). TRPC2 seems to form protein–protein interactions with the type III $Ins(1,4,5)P_3$ receptor (Brann, J. H. *et al.*, 2002) and biochemical studies in several species have reported increased $Ins(1,4,5)P_3$ levels in response to sensory stimulation of the VNO (Luo, Y. *et al.*, 1994; Kroner, C. *et al.*, 1996; Wekesa, K. S. and Anholt, R. R., 1997; Krieger, J. and Breer, H., 1999; Sasaki, K. *et al.*,

1999; Cinelli, A. R. *et al.*, 2002). Furthermore, flash photolysis of $Ins(1,4,5)P_3$ has been shown to excite frog VSNs (Gjerstad, J. *et al.*, 2003). One possibility is that $Ins(1,4,5)P_3$-induced Ca^{2+} release from intracellular stores (Lucas, P. *et al.*, 2003) leads to the opening of Ca^{2+}-dependent channels. These could be Ca^{2+}-activated cation channels, such as those found in hamster VSNs (Liman, E. R., 2003), or other Ca^{2+}-dependent channels. Such a mechanism could be involved in the primary sensory response, or it could mediate further amplification of this response. Even less clear than the roles of $Ins(1,4,5)P_3$ and Ca^{2+} are the proposed roles for arachidonic acid (Spehr, M. *et al.*, 2002) and cAMP (Halpern, M. and Martinez-Marcos, A., 2003) in VSN transduction. More information will be required to assess the precise function of these messenger molecules.

4.43.3.6 The Search for Sensory Stimuli

Understanding the functional role of the AOS requires identification of the sensory cues that are detected by the VSNs. It has been known since the 1970s that the mammalian VNO plays an essential role in the detection of stimuli of social nature, including some pheromone-like signals (Meredith, M. *et al.*, 1983; Halpern, M., 1987; Johnston, A. N. *et al.*, 1993); however, identification of the molecular nature of these stimuli proved surprisingly difficult. The advent of large-scale neurophysiological recording techniques in the VNO and AOB has enabled a more systematic search for vomeronasal stimuli. Surprisingly, the long held notion that VSNs detect only nonvolatile molecules appears an oversimplification of the systems function. VSNs show responses to a wider range of stimuli, including molecules that would not normally be categorized as pheromones. Additionally, it is now clear that many social recognition signals including some pheromones can also be detected by OSNs in the main olfactory system and that the mammalian main and AOSs detect, in part, overlapping sets of stimuli. Putatively, information about the same stimuli derived from each system is integrated in higher brain centers (Meredith, M., 1998; Boehm, T. and Zufall, F., 2006; Boehm, U. *et al.*, 2005; Shepherd, G. M., 2006; Spehr, M. *et al.*, 2006).

4.43.3.6.1 Stimulus access

Abundant evidence has shown that animals must actively enlarge the openings to the VNO and activate a pumping or suction mechanism to transport chemical stimuli into the lumen of the VNO (Halpern, M., 1987; Wysocki, C. J. and Meredith, M., 1987). Work by Meredith and coworkers in hamster has established that the large blood vessels located within the VNO are capable of dilation and constriction and thus serve as a pumping mechanism that is activated in situations of novelty (Meredith, M., 1994). This vascular pump is under the control of sympathetic nerve fibers and interruption of these efferent nerves causes behavioral deficits similar to those produced by lesioning of the VNs (Meredith, M. and O'Connell, R. J., 1979; Meredith, M. et al., 1980; Meredith, M. and O'Connell, J. M., 1988). There is now excellent evidence that activation of the VNO by appropriate stimuli can occur following direct physical contact of the nose with a stimulus source (Luo, M. et al., 2003; Leinders-Zufall, T. et al., 2004). Recordings from single neurons in the AOB of mice that were engaged in natural behaviors showed increased neuronal firing after physical contact with a conspecific (Luo, M. et al., 2003) (see also Section 4.43.4). For nonvolatile cues such as peptide ligands, such physical contact is indeed required to reach sensory neurons in the nose in an *in vivo* situation (Leinders-Zufall, T. et al., 2004; Spehr, M. et al.,2006). Volatile cues dissolved in bodily secretions will also reach the VNO following direct physical contact between the nose and a stimulus source. However, whether the VNO can detect such volatiles in the absence of direct physical contact remains debated (Luo, M. et al., 2003; Trinh, K. and Storm, D. R., 2003). In at least one example, clear activation of the mouse AOB in vivo was observed in response to volatile stimuli that were delivered via the airstream (Xu, F. et al., 2005).

4.43.3.6.2 Complex stimuli

To evoke sensory responses in the AOS, numerous researchers have tested complex natural communication stimuli such as urine, soiled bedding, vaginal secretion, and seminal fluid. Functional responses to these stimuli were monitored using a variety of techniques including immediate early gene expression in VNO and AOB (Halem, H. A. et al., 1999; Kumar, A. et al., 1999; Halem, H. A., et al., 2001) or medial amygdala (Meredith, M. and Westberry, J. M., 2004), biochemical analysis of second messenger formation in VNO tissue (Wekesa, K. S. and Anholt, R. R., 1997), electrophysiological recording or Ca^{2+} imaging from single neurons in the VNO (Inamura, K. et al., 1999; Holy, T. E. et al., 2000; Spehr, M. et al., 2002; Leinders-Zufall, T. et al., 2004) or AOB (Luo, M. et al., 2003), and functional magnetic resonance

imaging (fMRI) of AOB (Xu, F. et al., 2005). Although the details of the findings in these studies differ in many cases, taken together the evidence clearly shows that complex stimuli contain specific molecular cues that are recognized by VSNs. Particularly interesting data from single cell VSN recordings suggest that VSNs respond to cues present in urine that carry information about strain, species, and sex (Inamura, K. et al., 1999; Holy, T. E. et al., 2000). Recordings from single AOB neurons of freely behaving mice led to similar conclusions regarding the information content of vomeronasal stimuli (Luo, M. et al., 2003). These latter authors also showed for the first time neuronal activity in response to active social investigation of anogenital or head/face regions of conspecifics of different genetic strains (Luo, M. et al., 2003).

4.43.3.6.3 Urinary volatiles

Mouse urine contains hundreds of volatile compounds (Schwende, F. J. et al., 1986; Novotny, M. V. et al., 1999a) some of which have been associated with distinct endocrine or behavioral responses in conspecifics and are therefore considered prospective pheromones (see Section 4.43.5). A direct demonstration that at least some of these molecules directly activate VSNs came from a combination of EVG recording and Ca^{2+} imaging in VNO slices (Leinders-Zufall, T. et al., 2000). The putative mouse pheromones 2,5-dimethylpyrazine, 2-sec-butyl-4,5-dihydrothiazole, 2,3-dehydro-exo-brevicomin, a mixture of E,E-α-farnesene and E-β-farnesene (referred to as farnesene), 2-heptanone, and 6-hydroxy-6-methyl-3-heptanone were potent sensory cues in the mouse VNO (Table 1) (Leinders-Zufall, T. et al., 2000). VSNs responded to these compounds at nanomolar to subnanomolar concentrations, several orders of magnitude more sensitive than previously suspected. Activated VSNs were located in the apical, V1R-expressing layer of the VNO (Leinders-Zufall, T. et al., 2000) (Figure 4(i)). Several other urinary volatiles including n-pentyl acetate and isobutylamine (Del Punta, K. et al., 2002a; Trinh, K. and Storm, D. R., 2003) were subsequently identified as stimuli for mouse VSNs. Genetic ablation of a group of V1R genes showed an essential role for these receptors in the detection of 6-hydroxy-6-methyl-3-heptanone, n-pentyl acetate, isobutylamine, and 2-heptanone (Boschat, C. et al., 2002; Del Punta, K. et al., 2002a). Further evidence for the activity of these molecules derives from functional imaging in the AOB of intact mice showing that 2-heptanone activates the AOS (Xu, F. et al., 2005).

Table 1 Structure, origin, and function of some urinary volatiles acting in the VNO

Name	Chemical structure	Origin	Possible chemosignaling function in female mice	Detection threshold EVG response (M)
2,5-Dimethylpyrazine		Female urine	Puberty delay[a]	10^{-8}–10^{-7}
2-sec-Butyl-4,5-dihydrothiazole		Male bladder, urine	Estrus synchronization[b], puberty acceleration[c]	10^{-10}–10^{-9}
2,3-Dehydro-exo-brevicomin		Male bladder, urine	Estrus synchronization[b], puberty acceleration[c]	10^{-10}–10^{-9}
α- and β-farnesenes		Male preputial gland	Puberty acceleration[c]	10^{-11}–10^{-10}
2-Heptanone		Female or male urine	Estrus extension[d]	10^{-11}–10^{-10}
6-Hydroxy-6-methyl-3-heptanone		Male bladder urine	Puberty acceleration[e]	10^{-8}–10^{-7}

[a]Novotny M. et al. (1986).
[b]Jemiolo B. et al. (1986).
[c]Novotny M. V. et al. (1999b).
[e]Jemiolo B. et al. (1989).
[f]Novotny M. V. et al. (1999a).
EVG, electrovomeronasogram; VNO, vomeronasal organ.
Reprinted from Leinders-Zufall T., Lane A. P., Puche A. C., Ma W., Novotny M. V., Shipley M. T., and Zufall F. 2000. Ultrasensitive pheromone detection by mammalian vomeronasal neurons. Nature 405, 792–796, with permission from Nature Publishing Group.

4.43.3.6.4 General odors

Early studies indicated that VSNs are activated by a fraction of odorants without known pheromonal functions (Müller, W., 1971; Tucker, D., 1971). More recent investigations have confirmed and extended these findings (Sam, M. et al., 2001; Trinh, K. and Storm, D. R., 2003). Functional imaging studies of the AOB used a mixture of six of the compounds reported by Sam M. et al. (2001) – methyl anisole, patchone, indole, helional, butyrophenone, and fenchone – to demonstrate that they stimulate the mouse AOB in vivo (Xu, F. et al., 2005). Sam M. et al. (2001) suggested that these odorants could be released from plants or other animal species that may signal the presence of a predator or indicate the suitability of a particular site for feeding or nesting.

4.43.3.6.5 Major histocompatibility complex peptide ligands

A particularly surprising finding was the discovery that subsets of mouse VSNs are tuned to recognize members of a large family of immune system molecules known as the MHC peptide ligands (Leinders-Zufall, T. et al., 2004). Together with the evidence presented above, this demonstrates that the mouse VNO can detect both volatile and nonvolatile cues. As pointed out by Boehm T. and Zufall F. (2006), MHC peptides are not considered pheromones according to the classical definition by Karlson P. and Lüscher M. (1959). Hence, the detection of MHC peptide ligands by VSNs shows that the VNO also plays important roles in the sensing of nonpheromonal chemical signals of social nature.

Table 2 Structure of peptide ligands acting in the VNO

MHC peptide ligands	Other specific peptides
AAPDNRETF[a]	SIPSKDALLK (sodefrin)[b]
SYFPEITHI[a]	SILSKDAQLK (silefrin)[c]
SYIPSAEKI[a]	1
	ADQKTNHEADLKNPDPQEVQRALARILC ALGELDKLVKDQ 40
	41 ANAGQQEFKLPKDFTGRSK CRSLGRIK 67 (exocrine gland-secreting peptide 1)[d]

[a]Leinders-Zufall T. et al. (2004).
[b]Kikuyama S. et al. (1995).
[c]Yamamoto K. et al. (2000).
[d]Kimoto H. et al. (2005).
MHC, major histocompatibility complex; VNO, vomeronasal organ.

Immune system researchers have established a clear structural relationship between MHC molecules and their peptide ligands, which are usually nine amino acids long (Table 2; summarized in Boehm T. and Zufall F. 2006). MHC/peptide complexes present at the cell surface carry information about the genetic makeup of cells. They are shed from the cell surface and their fragments appear in bodily fluids such as serum, saliva, sweat, and urine (Singh, P. B. et al., 1987). These truncated MHC molecules are thought to have a reduced affinity to their peptide ligands and are thus likely to release them into the extracellular space. The finding that the structural complexity of MHC peptide ligands can be sensed by neurons in the VNO suggests that this information is used for the sensory evaluation of the genotype of a conspecific, or in other words for the detection of genetic individuality (Boehm, T. and Zufall, F., 2006). Experiments in behaving mice strongly support this notion (Leinders-Zufall, T. et al., 2004) (see Section 4.43.5). VSNs can sense MHC peptides at low concentrations, with detection thresholds near 10^{-12} M. Mapping studies in VNO slices combined with the use of immunocytochemical markers showed that MHC peptides are detected by VSNs in the basal layer (Leinders-Zufall, T. et al., 2004). This indicates that VSNs of the apical and basal zones detect structurally and functionally distinct sets of molecular cues.

4.43.3.6.6 Other specific peptides

It is worth noting that a number of specific peptide pheromones have been identified previously in organisms as diverse as bacteria, fungi, mollusks, arthropods, and a variety of vertebrates (Altstein, M., 2004). Among vertebrates, maybe the best example comes from work in closely related newt species where the decapeptides sodefrin (SIPSKDALLK) and silefrin (SILSKDAQLK) have been shown to attract conspecific females (Kikuyama, S. et al., 1995; Yamamoto, K. et al., 2000) (Table 2), most likely via activation of their VNOs (Toyoda, F. and Kikuyama, S., 2000). More recently, in mice a male-specific 7 kDa peptide secreted from the extraorbital lacrimal gland has been shown to stimulate VSNs (Kimoto, H. et al., 2005). This peptide (Table 2), which has been called exocrine gland-secreting peptide 1 (ESP1), seems to be encoded by a gene from a new multigene family consisting of at least 23 members (Kimoto, H. et al., 2005). Although the behavioral role of ESP1 is not yet clear, these results together with those of Leinders-Zufall T. et al. (2004) indicate that social communication in mammals involves the detection of extended families of peptidergic chemosignals by the VNO.

4.43.3.6.7 Proteins

It has long been thought that VSNs can detect chemical signals of proteinaceous nature. The best evidence for protein detection by the VNO comes from work using snakes (Jiang, X. C. et al., 1990; summarized in Halpern, M. and Martinez-Marcos, A., 2003) where a 20-kDa chemoattractant purified from earthworm shock secretions (known as ES20) has been shown to induce excitatory responses in VNO and AOB. This protein has been used extensively to characterize second messenger formation and signal transduction events in snake VSNs (Halpern, M. and Martinez-Marcos, A., 2003). Collectively, this work has shown that prey detection is a major function of the snake VNO. In a terrestrial salamander, a 22 kDa protein with structural homology to members of the interleukin-6 cytokine family has been shown to affect female receptivity but it is not yet clear whether this molecule is detected by the VNO (Rollmann, S. M. et al., 1999).

Work in rodents also suggested an involvement of proteins in VNO sensing. Mouse urine contains a class of highly polymorphic proteins known as major urinary proteins (MUPs) that are implicated in individuality recognition (Hurst, J. L. et al., 2001). It seems clear that MUPs bind urinary volatiles (Novotny, M. V. et al., 1999a; Zidek, L. et al., 1999) and release them from scent marks (Hurst, J. L. and Beynon, R. J., 2004). However, whether MUPs devoid of their volatile ligands function as VSN stimuli as suggested by Mucignat-Caretta C. *et al.* (1995) and Cavaggioni A. and Mucignat-Caretta C. (2000) remains on open question (see Novotny, M. V. *et al.*, 1999a). Essentially, the same can be said about a protein isolated from hamster vaginal fluid known as aphrodisin (Singer, A. G. *et al.*, 1986). Like MUPs, aphrodisin belongs to the lipocalin superfamily and it remains unclear whether it functions as a carrier for small hydrophobic chemicals or as a pheromone in its own right (Briand, L. *et al.*, 2004a; 2004b).

4.43.3.7 Sensory Coding Strategies in the Vomeronasal Organ

In the main olfactory system, our current understanding is that OSNs respond to chemical epitopes that may be present on different odorant molecules, and that the tuning of main olfactory neurons is broad; that is, a single OSN responds to related chemical epitopes with different affinities. By contrast, mouse VSNs – at least those located in the apical layer of the VNO – show much narrower tuning properties. Specifically, VSN tuning curves to putative pheromone molecules do not seem to broaden with increasing concentrations of ligand and an increase in stimulus strength does not recruit activation of more neurons (Leinders-Zufall, T. *et al.*, 2000). Populations of VSNs also exhibit strong selectivity for either male or female urine, although the molecules underlying this response are unknown (Holy, T. E. *et al.*, 2000). Thus, if OSNs in the MOE are hypothesized to be epitope sensors, VSNs of the apical zone might be considered molecule detectors for single putative pheromones. However, not all VSNs in apical and basal zones may employ the same strategies for encoding socially relevant chemosignals. For example, work using the MHC peptide ligands as sensory stimuli has clearly shown that a given basal VSN can detect multiple peptides exhibiting distinct chemical structures (Leinders-Zufall, T. *et al.*, 2004). Furthermore, the vast structural repertoire of the family of MHC peptide ligands argues against a one cell–one receptor–one ligand strategy but rather for the existence of some kind of combinatorial coding mechanism in peptide-detecting VSNs. Thus, mammalian VSNs may employ a variety of coding strategies. Similar conclusions were reached by analyzing sensory responses to complex stimuli from lizard VSNs (Labra, A. *et al.*, 2005).

4.43.4 The Accessory Olfactory Bulb

4.43.4.1 Inputs from the Vomeronasal Organ

VSNs within the VNO project axons that traverse along the nasal septum, through the cribriform plate, along the medial surface of the main olfactory bulb (MOB), and terminate exclusively in the AOB (Figure 2(f)). In rodents, the AOB is located at a dorsocaudal position relative to the MOB (Figures 1(a), 2(a), and 2(f)). The fascicles of axons comprising this VN are distinct from those fascicles innervating the MOB (Halpern, M., 1987). These axons pass through the nerve fiber layer (aONL) of the AOB and converge to glomeruli in the accessory glomerular layer (aGL) (Figure 6). Unlike the MOB where glomeruli are large (50–100 μm diameter) and organized into a distinct layer one to two glomeruli thick segregated by numerous juxtaglomerular cells, AOB glomeruli are smaller (10–50 μm diameter), numerous, poorly separated by juxtaglomerular cells, and distributed in a loosely organized layer that is many glomeruli thick. Early studies using cell surface antigens showed that the projections from the VNO segregated into two zones, an anterior and a posterior zone (Mori, K. *et al.*, 1985; Mori, K., 1987; Schwarting, G. A. and Crandall, J. E., 1991; Takami, S. *et al.*, 1992b; Osada, T. *et al.*, 1994). Recent studies demonstrated that these anterior and posterior zones receive input from distinct classes of VSNs (Halpern, M. *et al.*, 1995; Rodriguez, I. *et al.*, 1999; Del Punta, K. *et al.*, 2002b). VSNs located apically in the VNO sensory epithelium expressing the V1R receptor family and the G protein $G\alpha_{i2}$ (described above) project to the anterior aGL domain, while VSNs located basal expressing the V2R receptor family and the $G\alpha_o$ G protein project axons into the posterior aGL domain.

In the main olfactory system, OSNs expressing the same odor receptor gene project to only two or a few glomeruli in each bulb (Rodriguez, I. *et al.*, 1999; Del Punta, K. *et al.*, 2002b). However, unlike the high degree of axon convergence in the MOB, VSNs expressing the same receptor gene project to multiple glomeruli in either the anterior or posterior domain of the AOB (six to 20 glomeruli per gene have been reported; Belluscio, L. *et al.*, 1999; Rodriguez, I. *et al.*, 1999; Del Punta, K. *et al.*, 2002b). The distribution of these multiple glomeruli within each domain exhibits a high degree of variability compared to the convergence and position of glomeruli in the MOB, but there is a general similarity across animals.

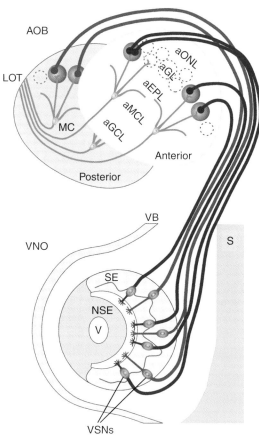

Figure 6 Diagram showing the mosaic projection of vomeronasal sensory neurons (VSNs). The sensory epithelium (SE) is segregated into two distinct zones, both of which express a unique set of transduction-related molecules: (1) an apical zone (blue VSNs) that expresses the G protein $G\alpha_{i2}$ as well as members of the V1R family of vomeronasal receptors, and (2) a basal zone that characteristically contains VSNs (purple) that express $G\alpha_o$ and members of the V2R receptor family. Unlike the main olfactory bulb, VSNs expressing the same receptors project to multiple glomeruli in the AOB, and AOB mitral cells (MC) innervate multiple glomeruli. The AOB can be further divided into mainly five layers depending on the cell types present, that is, the accessory olfactory nerve layer (aONL), accessory glomerular layer (aGL), the accessory external plexiform layer (aEPL), the accessory mitral cell layer (aMCL), and the accessory granule cell layer (aGCL). LOT, lateral olfactory tract; NSE, nonsensory epithelium; S, septum; V, blood vessel; VB, vomer bone.

4.43.4.2 Functional Organization of the Accessory Olfactory Bulb

With respect to circuit organization, the AOB appears quite similar to the MOB; however, there are points of significant differences in the arrangement of circuits in the AOB. Within each glomerulus, VSN axons synapse onto the dendrites of mitral, tufted, and juxtaglomerular neurons. Electrophysiological studies in tissue slices demonstrated that, similar to OSNs in the main olfactory system, glutamate is the primary VSN neurotransmitter (Dudley, C. A. and Moss, R. L., 1995; Jia, C. et al., 1999). Glutamatergic transmission from the VSN to postsynaptic target neurons is heavily dependent on alpha-amino-3-hydroxy-5-methyl-4-isoxazolepropionic acid (AMPA)/kainate receptors for fast postsynaptic responses, and on N-methyl-D-aspartate (NMDA) receptors for slower components of the response (Dudley, C. A. and Moss, R. L., 1995; Jia, C. et al., 1999). AOB mitral cell responses in the awake animal consist of a gradual elevation in firing rate over 2–5 s following contact with a stimulus, for example, anogenital investigation (Luo, M. et al., 2003). The slowly evolving elevation of firing rate is in sharp contrast to other sensory systems such as vision, auditory or somatosensory where the precise timing of stimulus-triggered responses are highly conserved. However, the slow evolution of activity in the AOB mitral cell is consistent with the nature of odor access and receptor kinetics discussed above. The relatively slow characteristics of olfactory responses could allow for significant temporal evolution of local inhibitory feedback and cortical feedback during the detection event, as postulated by Laurent and coworkers for the main olfactory system (for a review, see Laurent, G., 2002).

VSNs expressing the same receptor gene project to the AOB in a complex, multiglomerular pattern, but how is this complex glomerular projection pattern read by the AOB? The major output neuron of the AOB, the mitral cell, has a strikingly different structure to mitral cells of the MOB. It has been known for nearly a century that AOB mitral cells possess multiple apical dendrites, upto five, that each ramify within a different glomerulus (Cajal, R. S., 1911; Takami, S. and Graziadei, P. P., 1990; 1991) (Figure 7(a)). It was thought for many years that this multiglomerular organization of mitral cell dendrites allowed for sampling from different glomeruli and thus presumably different olfactory specificity; a cross-modality model. A logical prediction of this idea is that AOB mitral cells would be broadly tuned, capable of responding and integrating different kinds of vomeronasal input. A blend of pheromonal compounds could thus be integrated in the context of relative intensity of input by concurrent activation of glomeruli with different receptor identity. A difficulty with such a model is the requirement that mitral cell dendritic connectivity between different glomeruli precisely match

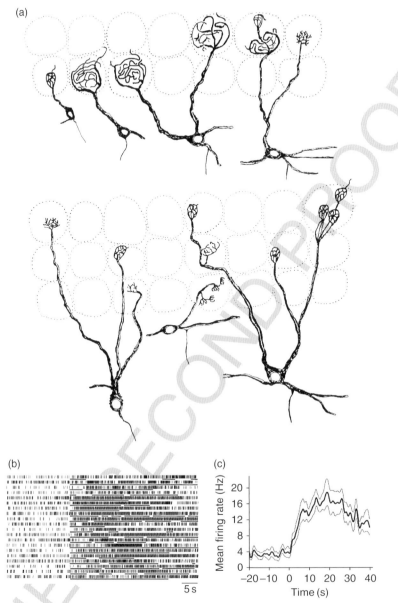

Figure 7 (a) Schematic drawing of accessory olfactory bulb (AOB) mitral and tufted cells. These cells have between one and five apical dendrites that terminate in different glomeruli. (b, c) Response kinetics of AOB neurons in the awake mouse. (b) Raster plots of the activation patterns of 14 AOB cells. Plots are aligned with the initiation of the first direct contact. Black bars beneath rasters represent periods of direct contact between the snout of test animals and the body of stimulus animals. (c) Plot of the mean firing rate (thick black line) from the AOB cells shown in (b). The thin gray lines represent ±1 SEM. The activity of the AOB cells gradually increased following the initiation of active investigation, which is unlike the main olfactory bulb (MOB) neurons that respond invariably to odor ligands with rapid and abrupt changes in firing frequency. (a) Reprinted from Takami, S. and Graziadei, P. P. 1991. Light microscopy Golgi study of mitral/tufted cells in the accessory olfactory bulb of the adult rat. J. Comp. Neurol. 311, 65–83, with permission from Wiley. (b, c) reprinted with permission from Luo, M., Fee, M. S., and Katz, L. C. 2003. Encoding pheromonal signals in the accessory olfactory bulb of behaving mice. Science 299, 1196–1201, AAAS.

biologically relevant pheromonal blends, a formidable problem of wiring specificity.

However, it was recently shown that the divergent pattern of VSN projections into the AOB could be rendered convergent due to the multiple glomerular projections of the mitral cell (Del Punta, K. *et al.*, 2002b). This study suggested that at least some of the mitral cell multiple processes, but not all, innervate glomeruli that receive axons from the same population of VSNs (Del Punta, K. *et al.*, 2002b).

The convergence of VSNs with the same receptor gene and presumably the same receptive field to a single mitral cell challenges the notion that AOB mitral cells would show a cross-modal response by integrating different sensory stimuli. Instead, AOB mitral cell response profiles may closely match those of the VSNs, modified by spatiotemporal circuit processing in the AOB; a labeled-line model. Because mouse VSNs, at least those of the apical layer, exhibit a narrow receptive field tuned to a single putative pheromone even at a wide range of concentrations (as discussed in the preceding section), a labeled line predicts that AOB mitral cells should thus exhibit similar narrow response profiles and that increasing concentrations of ligand will not recruit additional mitral cells. Elegant work by Luo M. and Katz L. C. (2004) in awake, behaving mice tested these predictions. Single unit recordings from the mitral cell layer of the awake, behaving mouse showed a high degree of selectivity for sex and strain of animal used as the stimulus (Luo, M. *et al.*, 2003; Luo, M. and Katz, L. C., 2004) (Figures 7(b) and 7(c)). The high selectivity of AOB mitral cell physiological responses supports the notion of a labeled line from VSN to AOB mitral cell.

What then is the benefit of a divergent VSN projection that is subsequently rendered, at least in part, convergent at the AOB mitral cell? One possibility is that the formation of multiple glomeruli of the same receptor type may be important for enhancing signal to noise via glomerular level lateral inhibition. Multiple small glomeruli of the same type are adjacent to a greater number of other receptor type glomeruli compared to a single large glomerulus such as present in the MOB. In the MOB an extensive lateral interglomerular network allows for integration between numerous different receptor type glomeruli (Aungst, J. L. *et al.*, 2003); however, there does not appear to be such a long-range interglomerular network in the AOB (unpublished observations). The ensemble of multiple glomeruli of the same receptor type with only short interglomerular neuronal connections would still have the potential for extensive and diverse lateral inhibition compared to a single glomerulus of larger size. Luo M. and Katz L. C. (2004) observed inhibition that does not require preceding excitation, suggesting the presence of selective lateral inhibition. The authors attribute this to selective lateral inhibition at the level of granule–mitral–granule cell connections (discussed below), but the observation is also consistent with glomerular lateral inhibition.

Together, the recent studies on VNO and AOB support the notion that VNO–AOB follows a labeled line neural encoding strategy rather than a cross modality, and that AOB mitral cells could act as functional modules. Further evidence for the role of AOB mitral cells as functional units comes from recent cellular physiology experiments (Lowe, G., 2003; Urban, N. N. and Castro, J. B., 2005). Calcium imaging and paired dendritic recordings were used to show action potentials in AOB mitral cells back propagate without attenuation through the entire dendritic tree (Ma, J. and Lowe, G., 2004). This active nondecremental back propagation could assist in coordinating activity of all the mitral cells projecting to the same glomerulus. In the MOB mitral cells projecting to the same glomerulus show synchronous membrane oscillations that are likely to depend on intraglomerular coupling between the cells (Carlson, G. C., *et al.*, 2000; Schoppa, N. E. and Westbrook, G. L., 2001; 2002). If a similar coupling exists in the AOB the nonattenuating back propagation of action potentials would contribute to binding all the mitral cells of that glomerulus into a single responsive unit.

4.43.4.2.1 Interneurons

The AOB contains a sparse population of juxtaglomerular cells located between glomeruli, although relatively little is known about these cells. Golgi studies show the dendrites of these AOB juxtaglomerular cells ramify in the aGL (Cajal, R. S., 1911) and can be categorized into periglomerular (PG) neurons and tufted cells. Immunohistochemistry shows the presence of larger glutamate containing cells, presumably the tufted cells, and smaller γ-aminobutyric acid (GABA)-containing cells, presumably the inhibitory PG neurons (Takami, S. *et al.*, 1992a; Quaglino, E. *et al.*, 1999). The role of these interneurons is presumed to be similar to the MOB, although no studies have yet examined these cells in detail. Any role of these cells is, however, likely to be local to one or neighboring glomeruli as there is no indication of the extensive interglomerular network present in the MOB (unpublished observations).

AOB granule cells are a numerous population of GABAergic inhibitory neurons located in the deep layers of the AOB. Unlike the MOB the mitral and granule cell layers are less distinct in the AOB, with mitral and granule cells intermingled through the broad mitral cell layer, and dispersed granule cells present throughout the deeper granule cell layer. In general, MOB olfactory bulb granule cells are thought to play two roles in the olfactory circuit; feedback inhibition of mitral cells and lateral inhibition between mitral cells. AOB mitral cells have lateral dendrites that run for hundreds of microns, significantly shorter than the multimillimeter lengths of MOB mitral cell lateral dendrites (Cajal, R. S., 1911; Takami, S. and Graziadei, P. P., 1990; 1991). However, as the AOB is smaller than the MOB the fraction of AOB within reach of the mitral lateral dendrites is very high. AOB granule cell dendrodendritic connections with mitral cells can be formed with the lateral dendrites or with the basal portions of the apical dendrites which, given the considerable distances that these lateral/apical dendrites can project, still allows for either feedback or lateral inhibition between diverse AOB mitral cells. Shepherd and co-workers using field potential and whole-cell recordings in AOB slices showed activation of the VN results in a sequence of neural responses in the AOB that is consistent with VN stimulation of mitral cells, followed by mitral cell dendrodendritic excitation of granule cells and subsequent reciprocal inhibition of mitral cells by granule cell GABA release (Jia, C. *et al.*, 1999). This work demonstrated the basic reciprocal mitral–granule–mitral inhibition circuit follows the same principles as the MOB. Mitral–granule–mitral lateral inhibition is also present in the AOB. Mitral cell inhibitory responses can occur without a preceding excitatory response in that cell, consistent with lateral inhibition from a distant excited mitral cell (Luo, M. *et al.*, 2003).

There have been numerous studies examining the patterns of neural activation using increased expression of c-fos (an immediate early gene) as an activity marker following stimulation with different strain/sex mouse urine/odor (for review Halpern, M. and Martinez-Marcos, A., 2003 and reference therein). These stimuli evoke c-fos activity in cells across all layers of the bulb, particularly in the granule and mitral layers. The dispersed location of mitral cells together with the inherent difficulties of attributing c-fos staining in nuclei to a particular cell type, make the mapping of AOB mitral cell activity problematic. However, these studies clearly show extensive c-fos activity in the granule cell layer which is consistent with the notion that there is extensive feedback and lateral inhibition shaping AOB output.

4.43.4.2.2 Subdivisions

The anterior and posterior domains of the AOB receive input from different populations of VSNs that express different families of receptor genes; thus, it is seems likely that different stimuli may elicit

responses from only one of the two AOB domains. Although c-fos responses to natural stimuli are generally dispersed throughout the AOB there are reports of preferential activity to a species/sex difference in the rostral domain (Matsuoka, M. *et al.*, 1999). Optical imaging and field potential recordings of AOB slices showed electrical stimulation of the anterior aONL elicited activity from the anterior internal plexiform layer. Stimulation of parts of the posterior aONL suggest this region is further subdivided into a rostral two-thirds and caudal one-third AOB domain (Sugai, T. *et al.*, 2000). Additional evidence for these further divisions comes from recent genetic tract tracing showing a VSN–VR population projecting into a restricted middle part of the bulb, putatively into the more rostral part of the posterior division (Mombaerts, unpublished observation).

4.43.4.3 Outputs of the Accesory Olfactory Bulb

Track tracing studies from the 1960s and 1970s demonstrated that the central connections of the AOB and MOB to higher order olfactory structures are essentially nonoverlapping. These observations lead to the hypothesis that the AOB and MOB pathways would serve different functional roles (Winans, S. S. and Scalia, F., 1970; Raisman, G., 1972; Scalia, F. and Winans, S. S., 1975). Subsequent research over the past 30 years has clarified many behavioral roles that are specific to vomeronasal function (several example effects are discussed in Section 4.43.5). The AOB has direct projections (Figure 8) to the medial (Me) and posteromedial cortical amygdaloid nucleus (PMCo), the BST, and the BAOT (Scalia, F. and Winans, S. S., 1975; de Olmos, J. *et al.*, 1978; Shipley, M. T. and Adamek, G. D., 1984). Together these structures can be called the accessory olfactory cortex (AOC). A recent examination of higher cortical projections from the anterior and posterior domains of the mouse AOB concluded both domains project into the same brain regions (Sugai, T. *et al.*, 2000). However, mitral cells in the opossum anterior AOB appear to project into deep cell layers of the medial amygdaloid complex (corresponding to the ventral division of the medial amygdala in rats) while the posterior AOB domain mitral cells project only into the superficial layers (Martinez-Marcos, A. and Halpern, M., 1999). Opossum mitral cell projections from the different domains of the AOB to the other AOC structures were essentially overlapping (Martinez-Marcos, A. and Halpern, M., 1999). These data show the

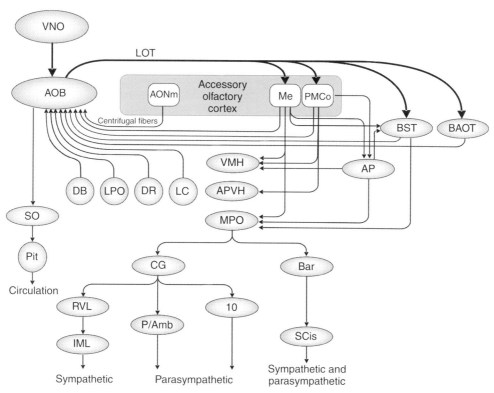

Figure 8 Diagrammatic representation of the connections of the accessory olfactory bulb (AOB) with cortical and subcortical structures. The major outputs of the AOB along the lateral olfactory tract (LOT) are shown at the upper part of the diagram. The major centrifugal inputs from the cortical regions back to the olfactory bulb are shown centrally along with higher-order connections. Emphasis is placed on possible circuits that mediate output responses such as autonomic, behavioral, or hormonal changes. 10, dorsal motor nucleus of the vagus; AOB, accessory olfactory bulb; AONm, anterior olfactory nucleus, medial division; AP, posterior amygdaloid nucleus; APVH, anterior periventricular hypothalamus; BAOT, bed nucleus of the lateral olfactory tract; Bar, Barrington's nucleus; BST, bed nucleus of the stria terminalis; CG, central (periaqueductal) gray; DB, nucleus of the diagonal band; DR, dorsal raphe nucleus; LC, locus coeruleus; LPO, lateral preoptic area; Me, medial amygdaloid nucleus; MPO, medial preoptic area; P/Amb, nucleus ambiguous and periambigual area; PMCo, posteromedial cortical amygdaloid nucleus; Pit, pituitary; RVL, rostroventrolateral medulla; SCis, lumbosacral spinal cord; SO, supraoptic nucleus; VMH, ventromedial hypothalamic nucleus; VNO, vomeronasal organ.

divergence into segregated pathways in the AOB generally, with some species differences, converge to the same higher brain regions.

The structures comprising the AOC are complex in organization and innervate many cortical and subcortical structures throughout the central nervous system (CNS). For example, the medial amygdala is a complex heterogeneous structure that is partitioned into at least four regions on the basis of cytoarchitecture, neurotransmitter content, hormone receptor expression, connections and putative functions (for a review, see Paxinos, G., 2004). The amygdala is a neural structure thought to be heavily involved in emotive states including behaviors such as mating and aggression behaviors in rat. The Me and PMCo have projections to other amygdaloid nuclei, notably the posterior amygdaloid nucleus (AP) (Canteras, N.

S. *et al.*, 1992), and to the medial preoptic area and the hypothalamus (for a review, see Shiosaka, S. *et al.*, 1982). The AP appears to receive convergent input from both the Me and PMCo and projects heavily upon some of the same structures targeted by Me and PMCo, namely the medial preoptic area and the ventromedial hypothalamic nucleus. Some of these secondary olfactory connections strongly influence sexual drive. The posterodorsal part of Me (MePD) contains neurons that project to four cell groups that are known to be sexually dimorphic and differ in their roles in reproduction. The medial preoptic area (MPO) is one of the sexually dimorphic targets of MePD and lesions of MPO decreases male copulatory behavior (for review Simerly, R. B. *et al.*, 1989). These higher-order connections of the AOB are highly consistent with roles for VNO function in mating, species recognition, and aggression. Indeed, studies of c-fos expression following mating in the AOB and higher brain regions by many investigators clearly establishes a link for vomeronasal activity in this behavior (Brennan, P. A. *et al.*, 1992; Fernandez-Fewell, G. D. and Meredith, M., 1994; Halem, H. A. *et al.*, 1999; Fewell, G. D. and Meredith, M., 2002).

4.43.4.4 Centrifugal Afferents to the Accessory Olfactory Bulb

The AOB receives heavy centrifugal inputs from higher brain regions (Figure 8), although from fewer sites than the MOB. The major olfactory cortex afferents to AOB are from the BST, the nucleus of the accessory olfactory tract, the Me, and the PMCo (de Olmos, J. *et al.*, 1978; Shipley, M. T. and Adamek, G. D., 1984). A restricted part of the medial division of AON sends a dense projection to the granule cell layer of AOB (Rizvi, T. A. *et al.*, 1992), but all other divisions of AON lack connections with AOB. Several of these structures (e.g., the amygdala) receive input from the AOB and thus are situated to provide a potential rapid feedback circuit with the AOB.

There are also extensive nonolfactory modulatory inputs to the AOB arising from the nucleus of the diagonal band (DB), lateral preoptic area, median and dorsal raphe, and the locus coeruleus (LC). The terminal distribution of these common inputs differs in AOB and MOB, especially with respect to the cholinergic inputs from the DB and serotonergic inputs from the LC. The cholinergic–serotonergic inputs to AOB are mainly to the granule cell layer and internal plexiform layer and only sparsely to the glomerular layer, whereas in the MOB glomeruli are heavily innervated (McLean, J. H. and Shipley, M. T., 1987; Le Jeune, H. and Jourdan, F., 1991). These fibers are thought to target PG cells in the MOB (McLean, J. H. and Shipley, M. T., 1987), as the AOB has fewer PG cells this may account for the paucity of serotonin and acetylcholine innervation of the AOB glomerular layer. The roles of these extensive centrifugal fiber tracts in circuit processing in the AOB (and the MOB) are largely unknown.

4.43.5 Mechanisms of Behavior

Following the pioneering work of Powers J. B. and Winans S. S. (1975), a vast number of studies have been published on the role of the VNO and AOS in behavioral responses using a variety of animal models

such as hamsters, voles, mice, rats, farm animals, and reptiles. This work has been summarized previously and the reader is referred to several extensive reviews (Meredith, M., 1983; Wysocki, C. J. and Meredith, M., 1987; Halpern, M., 1987; Johnston, R. E., 2000; Halpern, M. and Martinez-Marcos, A., 2003). Here, we will focus on only a few selected functions of the AOS for which the evidence seems particularly well established. Most recently, it has become clear that the mammalian main and AOSs detect, at least in part, overlapping sets of sensory stimuli (see Spehr, M. *et al.*, 2006) and that both systems play important roles in the regulation of social behaviors and the sensing of pheromone-like chemical signals. Consequently, it has been suggested that the main olfactory and vomeronasal systems should be viewed as complementary rather than separate pathways for chemical communication (Brennan, P. A. and Keverne, E. B., 2004; Shepherd, G. M., 2006). Current work is focusing on a dissection of the different nasal subsystems in behavioral responses using a combination of gene-targeting methods and more classical lesioning approaches. However, this goal is proving more difficult than previously thought because of the growing complexity of the function of the mammalian nose in chemosensory communication and the fact that each method exhibits specific limitations. For example, surgical removal of the VNO (VNX) also leads to removal of the terminal nerve system (Wysocki, C. J. and Meredith, M., 1987; Wysocki, C. J. and Lepri, J. J., 1991) and lesioning of the MOE by application of zinc sulphate is often incomplete. Similarly, the use of gene-targeted animals can involve a number of limitations. For example, deletion of the ion channel TRPC2 causes a dramatic loss in VSN sensitivity rather than a total loss of function (Leypold, B. G. *et al.*, 2002) and deletion of the CNGA2 subunit as a method to disrupt signaling in the main olfactory epithelium is likely to cause other deficits in the CNS (Mandiyan, V S. *et al.*, 2005). Furthermore, the role of these signal transduction molecules in additional olfactory subsystems (Figure 2(a)) such as the guanylyl cyclase type (GC-D) cell system (Zufall, F. and Munger, S. D., 2001), the septal organ of Masera (Tian, H. and Ma, M., 2004), and the Grueneberg ganglion (Fuss, S. H. *et al.*, 2005; Storan, M. J. and Key, B., 2006) is not yet clear. Thus, we must be careful in the interpretation of behavioral results based on the use of either method and in most cases, only a combination of methodological approaches, together with parallel analysis at the cellular level, will provide definitive answers. Table 3 shows a comparison of behavioral impairments caused by surgical VNO ablation or genetic manipulation of VSNs.

Table 3 Behavioral impairments, relative to wild type, caused by surgical ablation of VNO or genetic manipulation of VSNs

Phenotype	VNO ablation[a]	Trp2[−/−b,c]	V1Rab[−/−d]	β2m[−/−e]	Gα$_{i2}^{−/−f}$
Male sexual behavior directed to females	Reduced, fewer mounts and longer latency, experience-dependent	Normal	Reduced drive to initiate sexual behavior	Not reported	Normal
Male sexual behavior directed to males	Not reported	Increased	Normal/reduced	Normal	Normal
Male ultrasonic vocalizations to females	Eliminated/reduced, experience-dependent	Normal	Normal	Not reported	Not reported
Male ultrasonic vocalizations to males	Eliminated/reduced, experience-dependent	Increased	Not reported	Not reported	Not reported
Male aggression to intruder male	Eliminated	Eliminated	Normal	Eliminated	Reduced, depends on genetic background
Maternal aggression to intruder male	Eliminated	Eliminated	Eliminated	Not reported	Eliminated/reduced

[a]Lepri J. J. and Wysocki C. J. (1987).
[b]Leypold B. G. *et al.* (2002).
[c]Stowers L. *et al.* (2002).
[d]Del Punta K. *et al.* (2002a).
[e]Loconto J. *et al.* (2003).
[f]Norlin E. M. *et al.* (2003).
VNO, vomeronasal organ; VSN, vomeronasal sensory neuron.
Reprinted from Brennan, P. A. and Keverne, E. B. 2004. Something in the air? New insights into mammalian pheromones. Curr. Biol. R81–R89, with permission from Elsevier.

4.43.5.1 Detection of Heterospecific Signals

Work in the AOS of snakes has provided the best example that the VNO can be involved in the initiation of behavioral responses to both social and nonsocial chemical cues. The role of the snake VNO in prey recognition and response to prey odors has been especially well documented (Halpern, M., 1987; Halpern, M. and Martinez-Marcos, A., 2003; Baxi, K. N. *et al.*, 2006). Snakes with lesioned VNs stop following prey odor trails and eventually stop eating prey (Halpern, M., 1987). As pointed out by Baxi K. N. *et al.* (2006), heterospecific stimuli such as those involved in foraging cannot be considered pheromones.

4.43.5.2 Male Aggression to Intruder Males

In rodents, one of the best examples for a role of the VNO in the regulation of social behaviors comes from work on intermale aggression in mice using a resident–intruder assay. In this assay, a male intruder is added to the home cage of a singly housed test mouse. After initial olfactory exploration, a wild-type resident mouse will initiate vigorous attacks against the intruder. The involvement of the VNO in this type of aggression is clearly demonstrated by several studies using surgical VNO ablations (Clancy, A. N. *et al.*, 1984; Wysocki, C. J. and Lepri, J. J., 1991; Halpern, M. and Martinez-Marcos, A., 2003). An excellent correspondence (Table 3) exists between these studies and more recent work that employed mice with a targeted mutation in the cation channel gene TRPC2, also showing absence or strong reduction of male–male aggression (Leypold, B. G. *et al.*, 2002; Stowers, L. *et al.*, 2002). Consistent with these findings, TRPC2-deficient mice fail to establish territorial dominance relationships (Leypold, B. G. *et al.*, 2002). Together, these results demonstrate the importance of the VNO in sensing male chemo-signals and initiating an appropriate aggressive response.

4.43.5.3 Maternal Aggression to Intruder Males

The correspondence between the behavioral effects of the TRPC2 deletion and surgical VNO ablation is also very good in the context of maternal aggression (Table 3). Female mice are usually not aggressive toward intruders, but lactating females vigorously attack intruder males (Gandelman, R. *et al.*, 1972). Ablation of the VNO eliminates the display of maternal aggression (Bean, N. J. and Wysocki, C. J., 1989; Wysocki, C. J. and Lepri, J. J., 1991) and this aggression is also absent in $TRPC2^{-/-}$ females (Leypold, B. G. *et al.*, 2002). Similarly, maternal aggression is substantially reduced in mice lacking a cluster V1R receptor genes ($V1rab^{-/-}$, Del Punta, K. *et al.*, 2002a) as well as in $G\alpha_{i2}$ mutant mice (Norlin, E. M. *et al.*, 2003). Thus, it seems clear that the loss of specific sensory inputs in the VNO causes a loss of the display of maternal aggression in mice. Interestingly, although the $V1rab^{-/-}$ mice fail to initiate maternal aggression, there are no differences in male–male aggression compared with control mice (Del Punta, K. *et al.*, 2002a). This suggests that different male chemosignals might be responsible for triggering aggression in males and females (Brennan, P. A. and Keverne, E. B., 2004).

4.43.5.4 Male Mating Behavior

Although there is abundant evidence for an involvement of the rodent VNO in male sexual behavior (Wysocki, C. J. and Meredith, M., 1987; Halpern, M. and Martinez-Marcos, A., 2003) its exact role remains unclear, possibly due to complex interactions with other chemosensory systems, especially the main olfactory system (Brennan, P. A. and Keverne, E. B., 2004). Initial studies were done in hamster where approximately one-third of male hamsters with VN cuts developed a severe mating deficit (Powers, J. B. and Winans, S. S., 1975). Subsequently, Meredith M. (1986) showed that sexually inexperienced male hamsters are severely defective in male–female mating behavior when the VNO is removed, but that prior sexual experience can mitigate the lesion effects. Results in mice are less clear. Although several studies reported the importance of an intact VNO for the normal display of sexual behavior in male mice (Clancy, A. N. *et al.*, 1984; Bean, N. J. and Wysocki, C. J., 1989; Wysocki, C. J. and Lepri, J. J., 1991), a more recent study found that the rates of mounting and intromission and the

timing of ejaculation are equivalent in sexually naive control and VNX mice (Pankevich, D. E. *et al.*, 2004).

In $TRPC2^{-/-}$ mice, sexual behavior toward females appears to be unaffected. Instead, $TRPC2^{-/-}$ males show abnormally high levels of mounting behavior toward other males (Leypold, B. G. *et al.*, 2002; Stowers, L. *et al.*, 2002), a phenotype not previously observed in surgical lesioning studies (Pankevich, D. E. *et al.*, 2004). It has been argued that the VNO plays an essential role in gender discrimination (Stowers, L. *et al.*, 2002), but this is currently debated (Pankevich, D. E. *et al.*, 2004).

4.43.5.5 Male Ultrasonic Vocalizations

Male mice emit 70 kHz precopulatory vocalizations in the presence of female mice or their chemical cues (Nyby, J. *et al.*, 1979; Wysocki, C. J. and Lepri, J. J., 1991). Deafferentation of the VNO abolishes vocalizing in sexually naïve males and reduces it in males that have had previous sexual experience (Wysocki, C. J. and Lepri, J. J., 1991). However, ultrasonic vocalizations persist in $TRPC2^{-/-}$ males and are not suppressed in the presence of male chemosignals (Stowers, L. *et al.*, 2002; Brennan, P. A. and Keverne, E. B., 2004). The discrepancies between the effects of surgical and genetic lesions might be explained, in part, by the result that the loss of TRPC2 does not cause a complete loss of function in the VNO (Leypold, B. G. *et al.*, 2002).

4.43.5.6 Timing of Puberty, Cyclicity, and Ovulation in Females

A vast literature exists on the role of the VNO in altering the course of puberty and modulating female cyclicity and ovulation in a variety of species (Meredith, M. *et al.*, 1983; Halpern, M., 1987; Halpern, M. and Martinez-Marcos, A., 2003). Among these effects, puberty acceleration (Vandenbergh, J. G., 1983), caused by exposure to male chemosignals during female development, represents one of the best examples for a clear dependence on an intact VNO (Kaneko, N. *et al.*, 1980; Lomas, D. E. and Keverne, E. B., 1982; Sanchez-Criado, J. E., 1982). Novotny and collaborators have identified several volatile urine constituents that are capable of accelerating puberty in female mice (Novotny, M. V. *et al.*, 1999a; 1999b). Other evidence suggests that the delay of the onset of puberty, observed in crowded female house mice, depends also upon a functional VNO (Wysocki, C. J. and Lepri, J. J., 1991). Further work using animals with specific genetic deletions will be required to examine some of these effects.

4.43.5.7 Hormonal Responses

Some of the VNO-dependent functions summarized above depend on the coordination of neuroendocrine functions and the release of specific hormones. Significant efforts have been undertaken to establish a direct link between VNO stimulation and hormonal function (Keverne, E. B., 1983; Meredith, M. *et al.*, 1983). It is important to note that the level of reproductive hormones can also be influenced by the detection of chemosignals via the main olfactory system (Meredith, M., 1998) and current work is focusing on establishing neural connections between chemosensory neurons of the main and AOSs and neural circuits involved in hormone regulation (Boehm, U. *et al.*, 2005; Yoon, H. *et al.*, 2005; Shepherd, G. M., 2006). With respect to VNO stimulation, some of the best evidence comes from measurement of the gonadotropin luteinizing hormone (LH), which itself depends on the release of gonadotropin releasing hormone (GnRH, also known as luteinizing hormone releasing hormone (LHRH)) (Wysocki, C. J. and Meredith, M., 1987; Meredith, M., 1998; Halpern, M. and Martinez-Marcos, A., 2003). Female rats, when exposed to males or their odors, show a surge in LH and this effect is absent after removal of the VNO or lesioning of the AOB (Beltramino, C. and Taleisnik, S., 1983). Electrical stimulation of the AOB or medial (vomeronasal) amygdala produces a similar LH surge, whereas stimulation of the MOB or lateral (olfactory) amygdala does not elicit a facilitatory but rather an inhibitory effect (Beltramino, C. and Taleisnik, S., 1978; 1979).

Male mice exhibit a plasma testosterone surge after an encounter with strange females or their odors (Macrides, F. *et al.*, 1975). Removal of the VNO blocks this testosterone surge after exposure to an anesthetized female (Wysocki, C. J. *et al.*, 1983). Similarly, reflexive release of LH following exposure to female urine odor is blocked in VNO-lesioned male mice, although exposure to females did cause LH responses in these mice (Coquelin, A. *et al.*, 1984). These results stress the existence of multiple convergent pathways involved in LH regulation.

4.43.5.8 Pregnancy Block and Individual Recognition in Females: The Bruce Effect

One of the best-known examples of olfactory imprinting in adult vertebrates is the selective pregnancy block (or Bruce effect; Bruce, H. M., 1959) in the mouse, which depends on the formation and maintenance of a long-lasting odor recognition memory by the vomeronasal system (Brennan, P. *et al.*, 1990; Brennan, P. A. and Keverne, E. B., 1997; Halpern, M. and Martinez-Marcos, A., 2003). The Bruce effect occurs when a recently mated female mouse aborts if exposed to the urine odor of a male genetically different from the stud male (Figure 9). This unfamiliar male odor results in pregnancy block and return to estrus in the recently mated female. Keverne and colleagues suggest that the dendrodendritic synapse between granule cells and mitral cells in AOB may be critical for odor memory of the mate in the female (Keverne, E. B. and de la Riva, C., 1982; Rosser, A. E. and Keverne, E. B., 1985; Kaba, H. *et al.*, 1989). The Bruce effect can be abolished if noradrenergic centrifugal input to the female AOB is blocked immediately after mating, presumably before olfactory memories of the mate are formed (Keverne, E. B. and de la Riva, C., 1982; Rosser, A. E. and Keverne, E. B., 1985). Thus, noradrenalin appears to be important in strengthening the memory of the odor of the mating partner. Noradrenergic innervation of the AOB derives from LC projections primarily to the accessory granule layer (aGCL) and accessory mitral cell layer (aMCL) with lower density in the accessory external plexiform layer (aEPL) and aGL (McLean, J. H. *et al.*, 1989). Based on these light microscopic studies it was suggested that the major target of the noradrenalin input are granule cells (McLean, J. H. *et al.*, 1989). Thus, noradrenalin exciting granule cells and enhancing inhibition of a subset of mitral cells for several hours following mating may facilitate this selective odor memory.

The MHC peptide ligands are the first identified vomeronasal stimuli that can mediate the Bruce effect. To show that MHC peptides provide a crucial signal among the plethora of other cues present in male urine, peptides characteristic of a disparate MHC molecule were synthesized and added to the urine of a familiar male. Compared to the addition of cognate peptides, which had no effect, disparate peptides caused pregnancy block by otherwise familiar urine (Leinders-Zufall, T. *et al.*, 2004) (Figure 9). These results indicate that MHC peptides are recognized as natural odors and convey genotypic information in the context of a distinct social behavior (Boehm, T. and Zufall, F., 2006).

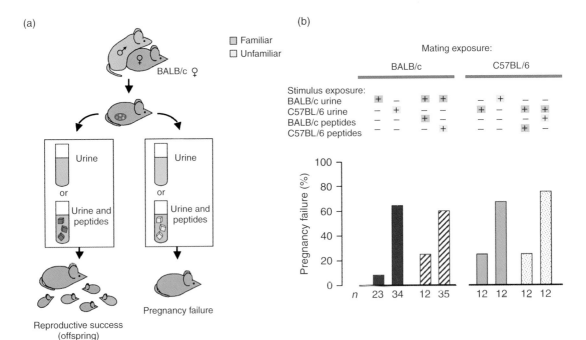

Figure 9 (a) Processing of major histocompatibility complex (MHC) peptide ligands via the accessory system provides a crucial signal in the pregnancy block paradigm of female mice (the Bruce effect), where disparate but not cognate peptides caused pregnancy block when added to otherwise familiar urine. (b) Percent pregnancy failure in female BALB/c (H-2d haplotype) mice mated with either BALB/c (H-2d haplotype) or C57BL/6 (H-2b haplotype) males. Exposure to different male urine types, which, in some cases, were supplemented with MHC class I peptides specific for BALB/c or C57BL/6 haplotypes is indicated. Reprinted with permission from Leinders-Zufall, T., Brennan, P., Widmayer, P., Chandramani, S. P., Maul-Pavicic, A., Jager, M., Li, X. H., Breer, H., Zufall, F., and Boehm, T. 2004. MHC class I peptides as chemosensory signals in the vomeronasal organ. Science 306, 1033–1037, AAAS.

In summary, it is now clear that the AOS is involved in a wide variety of functions including complex mechanisms such as those mediating the sensing of genome composition of a conspecific. The upcoming years, undoubtedly, will provide further exciting results with respect to the role of the VNO and AOS in chemical communication.

References

Adrian, E. D. 1955. Synchronised activity in the vomero-nasal nerves with a note on the function of the organ of Jacobsen. Pflugers Arch. 260, 188–192.

Altstein, M. 2004. Peptide pheromones: an overview. Peptides 25, 1373–1376.

Aungst, J. L., Heyward, P. M., Puche, A. C., Karnup, S. V., Hayar, A., Szabo, G., and Shipley, M. T. 2003. Centre-surround inhibition among olfactory bulb glomeruli. Nature 426, 623–629.

Balogh, C. 1860. Über das Jacobson'sche Organ des Schafes. Sitzungsbericht Akad. Wiss. Wien 42, 449–549.

Baxi, K. N., Dorries, K. M., and Eisthen, H. L. 2006. Is the vomeronasal system really specialized for detecting pheromones? Trends Neurosci. 29, 1–7.

Bean, N. J. and Wysocki, C. J. 1989. Vomeronasal organ removal and female mouse aggression: the role of experience. Physiol. Behav. 45, 875–882.

Belluscio, L., Koentges, G., Axel, R., and Dulac, C. 1999. A map of pheromone receptor activation in the mammalian brain. Cell 97, 209–220.

Beltramino, C. and Taleisnik, S. 1978. Facilitatory and inhibitory effects of electrochemical stimulation of the amygdala on the release of luteinizing hormone. Brain Res. 144, 95–107.

Beltramino, C. and Taleisnik, S. 1979. Effect of electrochemical stimulation in the olfactory bulbs on the release of gonadotropin hormones in rats. Neuroendocrinology 28, 320–328.

Beltramino, C. and Taleisnik, S. 1983. Release of LH in the female rat by olfactory stimuli. Effect of the removal of the vomeronasal organs or lesioning of the accessory olfactory bulbs. Neuroendocrinology 36, 53–58.

Berghard, A., Buck, L. B., and Liman, E. R. 1996. Evidence for distinct signaling mechanisms in two mammalian olfactory sense organs. Proc. Natl. Acad. Sci. U. S. A. 93, 2365–2369.

Bertmar, G. 1981. Evolution of vomeronasal organs in vertebrates. Evolution 35, 359–366.

Bigiani, A., Mucignat-Caretta, C., Montani, G., and Tirindelli, R. 2005. Pheromone reception in mammals. Rev. Physiol. Biochem. Pharmacol. 154, 1–35.

Boehm, N. and Gasser, B. 1993. Sensory receptor-like cells in the human foetal vomeronasal organ. Neuroreport 4, 867–870.

Boehm, T. and Zufall, F. 2006. MHC peptides and the sensory evaluation of genotype. Trends Neurosci. 29, 100–107.

Boehm, N., Roos, J., and Gasser, B. 1994. Luteinizing hormone-releasing hormone (LHRH)-expressing cells in the nasal septum of human fetuses. Brain Res. Dev. Brain Res. 82, 175–180.

Boehm, U., Zou, Z., and Buck, L. B. 2005. Feedback loops link odor and pheromone signaling with reproduction. Cell 123, 683–695.

Boschat, C., Pelofi, C., Randin, O., Roppolo, D., Luscher, C., Broillet, M. C., and Rodriguez, I. 2002. Pheromone detection mediated by a V1r vomeronasal receptor. Nat. Neurosci. 5, 1261–1262.

Brann, J. H., Dennis, J. C., Morrison, E. E., and Fadool, D. A. 2002. Type-specific inositol 1,4,5-trisphosphate receptor localization in the vomeronasal organ and its interaction with a transient receptor potential channel, TRPC2. J. Neurochem. 83, 1452–1460.

Brennan, P. A. and Keverne, E. B. 1997. Neural mechanisms of mammalian olfactory learning. Prog. Neurobiol. 51, 457–481.

Brennan, P. A. and Keverne, E. B. 2004. Something in the air? New insights into mammalian pheromones. Curr. Biol. 14, R81–R89.

Brennan, P. A., Hancock, D., and Keverne, E. B. 1992. The expression of the immediate-early genes c-fos, egr-1 and c-jun in the accessory olfactory bulb during the formation of an olfactory memory in mice. Neuroscience 49, 277–284.

Brennan, P., Kaba, H., and Keverne, E. B. 1990. Olfactory recognition: a simple memory system. Science 250, 1223–1226.

Briand, L., Blon, F., Trotier, D., and Pernollet, J. C. 2004a. Natural ligands of hamster aphrodisin. Chem. Senses 29, 425–430.

Briand, L., Trotier, D., and Pernollet, J. C. 2004b. Aphrodisin, an aphrodisiac lipocalin secreted in hamster vaginal secretions. Peptides 25, 1545–1552.

Bruce, H. M. 1959. An exteroceptive block to pregnancy in the mouse. Nature 184, 105.

Burghardt, G. M. 1970. Chemical Reception in Reptiles. In: Communication by Chemical Signals (eds. J. W. Johnson, D. R. Moulton, and A. Turk A), pp. 241–308. Appleton-Century-Crofts.

von Campenhausen, H., Yoshihara, Y., and Mori, K. 1997. OCAM reveals segregated mitral/tufted cell pathways in developing accessory olfactory bulb. Neuroreport 8, 2607–2612.

Cajal, R. S. 1911. Histologie du Systeme Neurveux de l'Hommes et des Vertebres. Maloine.

Canteras, N. S., Simerly, R. B., and Swanson, L. W. 1992. Connections of the posterior nucleus of the amygdala. J. Comp. Neurol. 324, 143–179.

Carlson, G. C., Shipley, M. T., and Keller, A. 2000. Long-lasting depolarizations in mitral cells of the rat olfactory bulb. J. Neurosci. 20, 2011–2021.

Cavaggioni, A. and Mucignat-Caretta, C. 2000. Major urinary proteins, alpha(2U)-globulins and aphrodisin. Biochim. Biophys. Acta 1482, 218–228.

Ciges, M., Labella, T., Gayoso, M., and Sanchez, G. 1977. Ultrastructure of the organ of Jacobson and comparative study with olfactory mucosa. Acta Otolaryngol. 83, 47–58.

Cinelli, A. R., Wang, D., Chen, P., Liu, W., and Halpern, M. 2002. Calcium transients in the garter snake vomeronasal organ. J. Neurophysiol. 87, 1449–1472.

Clancy, A. N., Coquelin, A., Macrides, F., Gorski, R. A., and Noble, E. P. 1984. Sexual behavior and aggression in male mice: involvement of the vomeronasal system. J. Neurosci. 4, 2222–2229.

Coquelin, A., Clancy, A. N., Macrides, F., Noble, E. P., and Gorski, R. A. 1984. Pheromonally induced release of luteinizing hormone in male mice: involvement of the vomeronasal system. J. Neurosci. 4, 2230–2236.

Del Punta, K., Leinders-Zufall, T., Rodriguez, I., Jukam, D., Wysocki, C. J., Ogawa, S., Zufall, F., and Mombaerts, P. 2002a. Deficient pheromone responses in mice lacking a cluster of vomeronasal receptor genes. Nature 419, 70–74.

Del Punta, K., Puche, A., Adams, N., Rodriguez, I., and Mombaerts, P. 2002b. A divergent pattern of sensory axonal projections is rendered convergent by second-order neurons in the accessory olfactory bulb. Neuron 35, 1057.

Del Punta, K., Rothman, A., Rodriguez, I., and Mombaerts, P. 2000. Sequence diversity and genomic organization of vomeronasal receptor genes in the mouse. Genome Res. 10, 1958–1967.

Doving, K. B. and Trotier, D. 1998. Structure and function of the vomeronasal organ. J. Exp. Biol. 201, 2913–2925.

Dudley, C. A. and Moss, R. L. 1995. Electrophysiological evidence for glutamate as a vomeronasal receptor cell neurotransmitter. Brain Res. 675, 208–214.

Dulac, C. 2000. Sensory coding of pheromone signals in mammals. Curr. Opin. Neurobiol. 10, 511–518.

Dulac, C. and Axel, R. 1995. A novel family of genes encoding putative pheromone receptors in mammals. Cell 83, 195–206.

Dulac, C. and Torello, A. T. 2003. Molecular detection of pheromone signals in mammals: from genes to behaviour. Nat. Rev. Neurosci. 4, 551–562.

Eisthen, H. L. 1992. Phylogeny of the vomeronasal system and of receptor cell types in the olfactory and vomeronasal epithelia of vertebrates. Microsc. Res. Tech. 23, 1–21.

Eisthen, H. L. 1997. Evolution of vertebrate olfactory systems. Brain Behav. Evol. 50, 222–233.

Fadool, D. A., Wachowiak, M., and Brann, J. H. 2001. Patch-clamp analysis of voltage-activated and chemically activated currents in the vomeronasal organ of *Sternotherus odoratus* (stinkpot/musk turtle). J. Exp. Biol. 204, 4199–4212.

Fernandez-Fewell, G. D. and Meredith, M. 1994. c-fos expression in vomeronasal pathways of mated or pheromone-stimulated male golden hamsters: contributions from vomeronasal sensory input and expression related to mating performance. J. Neurosci. 14, 3643–3654.

Fewell, G. D. and Meredith, M. 2002. Experience facilitates vomeronasal and olfactory influence on Fos expression in medial preoptic area during pheromone exposure or mating in male hamsters. Brain Res. 941, 91–106.

Fieni, F., Ghiaroni, V., Tirindelli, R., Pietra, P., and Bigiani, A. 2003. Apical and basal neurones isolated from the mouse vomeronasal organ differ for voltage-dependent currents. J. Physiol. 552, 425–436.

Fuss, S. H., Omura, M., and Mombaerts, P. 2005. The Grueneberg ganglion of the mouse projects axons to glomeruli in the olfactory bulb. Eur. J. Neurosci. 22, 2649–2654.

Gandelman, R., Zarrow, M. X., and Denenberg, V. H. 1972. Reproductive and maternal performance in the mouse following removal of the olfactory bulbs. J. Reprod. Fertil. 28, 453–456.

Garrosa, M. and Coca, S. 1991. Postnatal development of the vomeronasal epithelium in the rat: an ultrastructural study. J. Morphol. 208, 257–269.

Ghiaroni, V., Fieni, F., Tirindelli, R., Pietra, P., and Bigiani, A. 2003. Ion conductances in supporting cells isolated from the mouse vomeronasal organ. J. Neurophysiol. 89, 118–127.

Gjerstad, J., Valen, E. C., Trotier, D., and Doving, K. 2003. Photolysis of caged inositol 1,4,5-trisphosphate induces action potentials in frog vomeronasal microvillar receptor neurones. Neuroscience 119, 193–200.

von Gudden, B. 1870. Experimentaluntersuchung über das peripherische und centrale Nervensystem. Archiv. Psychol. 2, 699–720.

Halem, H. A., Cherry, J. A., and Baum, M. J. 1999. Vomeronasal neuroepithelium and forebrain Fos responses to male pheromones in male and female mice. J. Neurobiol. 39, 249–263.

Halem, H. A., Cherry, J. A., and Baum, M. J. 2001. Central forebrain Fos responses to familiar male odours are attenuated in recently mated female mice. Eur. J. Neurosci. 13, 389–399.

Halpern, M. 1987. The organization and function of the vomeronasal system. Annu. Rev. Neurosci. 10, 325–362.

Halpern, M. and Martinez-Marcos, A. 2003. Structure and function of the vomeronasal system: an update. Prog. Neurobiol. 70, 245–318.

Halpern, M., Jia, C., and Shapiro, L. S. 1998. Segregated pathways in the vomeronasal system. Microsc. Res. Tech. 41, 519–529.

Halpern, M., Shapiro, L. S., and Jia, C. 1995. Differential localization of G proteins in the opossum vomeronasal system. Brain Res. 677, 157–161.

Herrada, G. and Dulac, C. 1997. A novel family of putative pheromone receptors in mammals with a topographically organized and sexually dimorphic distribution. Cell 90, 763–773.

Hofmann, T., Schaefer, M., Schultz, G., and Gudermann, T. 2000. Cloning, expression and subcellular localization of two novel splice variants of mouse transient receptor potential channel 2. Biochem. J. 351, 115–122.

Holy, T. E., Dulac, C., and Meister, M. 2000. Responses of vomeronasal neurons to natural stimuli. Science 289, 1569–1572.

Hurst, J. L. and Beynon, R. J. 2004. Scent wars: the chemobiology of competitive signalling in mice. Bioessays 26, 1288–1298.

Hurst, J. L., Payne, C. E., Nevison, C. M., Marie, A. D., Humphries, R. E., Robertson, D. H., Cavaggioni, A., and Beynon, R. J. 2001. Individual recognition in mice mediated by major urinary proteins. Nature 414, 631–634.

Inamura, K. and Kashiwayanagi, M. 2000. Inward current responses to urinary substances in rat vomeronasal sensory neurons. Eur. J. Neurosci. 12, 3529–3536.

Inamura, K., Kashiwayanagi, M., and Kurihara, K. 1997. Inositol-1,4,5-trisphosphate induces responses in receptor neurons in rat vomeronasal sensory slices. Chem. Senses 22, 93–103.

Inamura, K., Matsumoto, Y., Kashiwayanagi, M., and Kurihara, K. 1999. Laminar distribution of pheromone-receptive neurons in rat vomeronasal epithelium. J. Physiol. 517(Pt 3), 731–739.

Ishii, T., Hirota, J., and Mombaerts, P. 2003. Combinatorial coexpression of neural and immune multigene families in mouse vomeronasal sensory neurons. Curr. Biol. 13, 394–400.

Jacobson, L. 1813. Anatoisk beskrievelse over et nyt organ i huusdyrenes naese. Vet. Selsk. Skrift 2, 209–246.

Jacobson, L., Trotier, D., and Doving, K. B. 1998. Anatomical description of a new organ in the nose of domesticated animals by Ludvig Jacobson (1813) [classical article]. Chem. Senses 23, 743–754.

Jemiolo, B., Andreolini, F., Xie, T. M., Wiesler, D., and Novotny, M. 1989. Puberty-affecting synthetic analogs of urinary chemosignals in the house mouse, *Mus domesticus*. Physiol. Behav. 46, 293–298.

Jemiolo, B., Harvey, S., and Novotny, M. 1986. Promotion of the Whitten effect in female mice by synthetic analogs of male urinary constituents. Proc. Natl. Acad. Sci. U. S. A. 83, 4576–4579.

Jia, C. and Halpern, M. 1996. Subclasses of vomeronasal receptor neurons: differential expression of G proteins (Gi alpha 2 and G(o alpha)) and segregated projections to the accessory olfactory bulb. Brain Res. 719, 117–128.

Jia, C., Chen, W. R., and Shepherd, G. M. 1999. Synaptic organization and neurotransmitters in the rat accessory olfactory bulb. J. Neurophysiol. 81, 345–355.

Jiang, X. C., Inouchi, J., Wang, D., and Halpern, M. 1990. Purification and characterization of a chemoattractant from electric shock-induced earthworm secretion, its receptor binding, and signal transduction through the vomeronasal system of garter snakes. J. Biol. Chem. 265, 8736–8744.

Johnston, R. E. 1983. Chemical Signals and Reproductive Behavior. In: Pheromones and Reproduction in Mammals (ed. J. G. Vandenbergh), pp. 71–79. Academic Press.

Johnston, R. E. 2000. Chemical Communication and Pheromones: The Types of Chemical Signals and the Role of the Vomeronasal System. In: The Neurobiology of Taste and Smell (eds. T. E. Finger, W. L. Silver, and D. Restrepo), pp. 101–127. Wiley-Liss.

Johnston, A. N., Rogers, L. J., and Johnston, G. A. 1993. Glutamate and imprinting memory: the role of glutamate receptors in the encoding of imprinting memory. Behav. Brain Res. 54, 137–143.

Kaba, H., Rosser, A., and Keverne, B. 1989. Neural basis of olfactory memory in the context of pregnancy block. Neuroscience 32, 657–662

Kaneko, N., Debski, E. A., Wilson, M. C., and Whitten, W. K. 1980. Puberty acceleration in mice. II. Evidence that the vomeronasal organ is a receptor for the primer pheromone in male mouse urine. Biol. Reprod. 22, 873–878.

Karlson, P. and Lüscher, M. 1959. Pheromones: a new term for a class of biologically active substances. Nature 183, 55–56.

Keverne, B. 1979. The Dual Olfactory Projections and Their Significance for Behaviour. In: Chemical Ecology: Odour Communication in Animals (ed. F. J. Ritter), pp. 75–83. Elsevier/North-Holland Biomed Press.

Keverne, E. B. 1983. Pheromonal influences on the endocrine regulation of reproduction. Trends Neurosci. 6, 381–384.

Keverne, E. B. and de la Riva, C. 1982. Pheromones in mice: reciprocal interaction between the nose and brain. Nature 296, 148–150.

Kikuyama, S., Toyoda, F., Ohmiya, Y., Matsuda, K., Tanaka, S., and Hayashi, H. 1995. Sodefrin: a female-attracting peptide pheromone in newt cloacal glands. Science 267, 1643–1645.

Kimoto, H., Haga, S., Sato, K., and Touhara, K. 2005. Sex-specific peptides from exocrine glands stimulate mouse vomeronasal sensory neurons. Nature 437, 898–901.

Klein, E. 1881. The organ of Jacobson in the rabbit. Q. J. Microsc. Sci. 21, 549–555.

Krieger, J. and Breer, H. 1999. Olfactory reception in invertebrates. Science 286, 720–723.

Kroner, C., Breer, H., Singer, A. G., and O'Connell, R. J. 1996. Pheromone-induced second messenger signaling in the hamster vomeronasal organ. Neuroreport 7, 2989–2992.

Kumar, A., Dudley, C. A., and Moss, R. L. 1999. Functional dichotomy within the vomeronasal system: distinct zones of neuronal activity in the accessory olfactory bulb correlate with sex-specific behaviors. J. Neurosci. 19, RC32.

Labra, A., Brann, J. H., and Fadool, D. A. 2005. Heterogeneity of voltage- and chemosignal-activated response profiles in vomeronasal sensory neurons. J. Neurophysiol. 94, 2535–2548.

Lau, Y. E. and Cherry, J. A. 2000. Distribution of PDE4A and G(o) alpha immunoreactivity in the accessory olfactory system of the mouse. Neuroreport 11, 27–32.

Laurent, G. 2002. Olfactory network dynamics and the coding of multidimensional signals. Nat. Rev. Neurosci. 3, 884–895.

Le Jeune, H. and Jourdan, F. 1991. Postnatal development of cholinergic markers in the rat olfactory bulb: a histochemical and immunocytochemical study. J. Comp. Neurol. 314, 383–395.

Leinders-Zufall, T., Brennan, P., Widmayer, P., Chandramani, S. P., Maul-Pavicic, A., Jäger, M., Li, X. H., Breer, H., Zufall, F., and Boehm, T. 2004. MHC class I peptides as chemosensory signals in the vomeronasal organ. Science 306, 1033–1037.

Leinders-Zufall, T., Lane, A. P., Puche, A. C., Ma, W., Novotny, M. V., Shipley, M. T., and Zufall, F. 2000. Ultrasensitive pheromone detection by mammalian vomeronasal neurons. Nature 405, 792–796.

Lepri, J. J. and Wysocki, C. J. 1987. Removal of the vomeronasal organ disrupts the activation of reproduction in female voles. Physiol. Behav. 40, 349–355.

Leypold, B. G., Yu, C. R., Leinders-Zufall, T., Kim, M. M., Zufall, F., and Axel, R. 2002. Altered sexual and social behaviors in trp2 mutant mice. Proc. Natl. Acad. Sci. U. S. A. 99, 6376–6381.

Liman, E. R. 2003. Regulation by voltage and adenine nucleotides of a Ca^{2+}-activated cation channel from hamster vomeronasal sensory neurons. J. Physiol. 548, 777–787.

Liman, E. R. and Corey, D. P. 1996. Electrophysiological characterization of chemosensory neurons from the mouse vomeronasal organ. J. Neurosci. 16, 4625–4637.

Liman, E. R., Corey, D. P., and Dulac, C. 1999. TRP2: a candidate transduction channel for mammalian pheromone sensory signaling. Proc. Natl. Acad. Sci. U. S. A. 96, 5791–5796.

Loconto, J., Papes, F., Chang, E., Stowers, L., Jones, E. P., Takada, T., Kumanovics, A., Fischer, L. K., and Dulac, C. 2003. Functional expression of murine V2R pheromone receptors involves selective association with the M10 and M1 families of MHC class Ib molecules. Cell 112, 607–618.

Lomas, D. E. and Keverne, E. B. 1982. Role of the vomeronasal organ and prolactin in the acceleration of puberty in female mice. J. Reprod. Fertil. 66, 101–107.

Lowe, G. 2003. Electrical signaling in the olfactory bulb. Curr. Opin. Neurobiol. 13, 476–481.

Lucas, P., Ukhanov, K., Leinders-Zufall, T., and Zufall, F. 2003. A diacylglycerol-gated cation channel in vomeronasal neuron dendrites is impaired in TRPC2 mutant mice: mechanism of pheromone transduction. Neuron 40, 551–561.

Luo, M. and Katz, L. C. 2004. Encoding pheromonal signals in the mammalian vomeronasal system. Curr. Opin. Neurobiol. 14, 428–434.

Luo, M., Fee, M. S., and Katz, L. C. 2003. Encoding pheromonal signals in the accessory olfactory bulb of behaving mice. Science 299, 1196–1201.

Luo, Y., Lu, S., Chen, P., Wang, D., and Halpern, M. 1994. Identification of chemoattractant receptors and G-proteins in the vomeronasal system of garter snakes. J. Biol. Chem. 269, 16867–16877.

Ma, J. and Lowe, G. 2004. Action potential backpropagation and multiglomerular signaling in the rat vomeronasal system. J. Neurosci. 24, 9341–9352.

Macrides, F., Bartke, A., and Dalterio, S. 1975. Strange females increase plasma testosterone levels in male mice. Science 189, 1104–1106.

Madison, D. W. 1977. Chemical Communication in Amphibians and Reptiles. In: Chemical Signals in Vertebrates (eds. D. Muller-Schwarze and M. M. Mozell), pp. 135–168. Plenum.

Mandiyan, V. S., Coats, J. K., and Shah, N. M. 2005. Deficits in sexual and aggressive behaviors in Cnga2 mutant mice. Nat. Neurosci. 8, 1660–1662.

Martinez-Marcos, A. and Halpern, M. 1999. Differential projections from the anterior and posterior divisions of the accessory olfactory bulb to the medial amygdala in the opossum, Monodelphis domestica. Eur. J. Neurosci. 11, 3789–3799.

Martini, S., Silvotti, L., Shirazi, A., Ryba, N. J., and Tirindelli, R. 2001. Co-expression of putative pheromone receptors in the sensory neurons of the vomeronasal organ. J. Neurosci. 21, 843–848.

Matsunami, H. and Buck, L. B. 1997. A multigene family encoding a diverse array of putative pheromone receptors in mammals. Cell 90, 775–784.

Matsuoka, M., Yokosuka, M., Mori, Y., and Ichikawa, M. 1999. Specific expression pattern of Fos in the accessory olfactory bulb of male mice after exposure to soiled bedding of females. Neurosci. Res. 35, 189–195.

McCotter, R. E. 1912. The connection of the vomeronasal nerves with the accessory olfactory bulb in the opossum and other mammals. Anat. Rec. 6, 299–318.

McLean, J. H. and Shipley, M. T. 1987. Serotonergic afferents to the rat olfactory bulb: II. Changes in fiber distribution during development. J. Neurosci. 7, 3029–3039.

McLean, J. H., Shipley, M. T., Nickell, W. T., Aston-Jones, G., and Reyher, C. K. 1989. Chemoanatomical organization of the noradrenergic input from locus coeruleus to the olfactory bulb of the adult rat. J. Comp. Neurol. 285, 339–349.

von Mihalkovics, V. 1899. Nasenhöhle und Jacobsonsches Organ. Anat. Embryol. (Berl.) 11, 1–108.

Meisami, E. and Bhatnagar, K. P. 1998. Structure and diversity in mammalian accessory olfactory bulb. Microsc. Res. Tech. 43, 476–499.

Menco, B. P., Carr, V. M., Ezeh, P. I., Liman, E. R., and Yankova, M. P. 2001. Ultrastructural localization of G-proteins and the channel protein TRP2 to microvilli of rat vomeronasal receptor cells. J. Comp. Neurol. 438, 468–489.

Meredith, M. 1983. Sensory Physiology of Pheromone Communication. In: Pheromones and Reproduction in Mammals (ed. J. G. Vandenbergh), pp. 199–252. Academic Press.

Meredith, M. 1986. Vomeronasal organ removal before sexual experience impairs male hamster mating behavior. Physiol. Behav. 36, 737–743.

Meredith, M. 1994. Chronic recording of vomeronasal pump activation in awake behaving hamsters. Physiol. Behav. 56, 345–354.

Meredith, M. 1998. Vomeronasal, olfactory, hormonal convergence in the brain. Cooperation or coincidence? Ann. N. Y. Acad. Sci. 855, 349–361.

Meredith, M. 2001. Human vomeronasal organ function: a critical review of best and worst cases. Chem. Senses 26, 433–445.

Meredith, M. and O'Connell, R. J. 1979. Efferent control of stimulus access to the hamster vomeronasal organ. J. Physiol. 286, 301–316.

Meredith, M. and O'Connell, J. M. 1988. HRP uptake by olfactory and vomeronasal receptor neurons: use as an indicator of incomplete lesions and relevance for non-volatile chemoreception. Chem. Senses 13, 487–515.

Meredith, M. and Westberry, J. M. 2004. Distinctive responses in the medial amygdala to same-species and different-species pheromones. J. Neurosci. 24, 5719–5725.

Meredith, M., Graziadei, P. P., Graziadei, G. A., Rashotte, M. E., and Smith, J. C. 1983. Olfactory function after bulbectomy. Science 222, 1254–1255.

Meredith, M., Marques, D. M., O'Connell, R. O., and Stern, F. L. 1980. Vomeronasal pump: significance for male hamster sexual behavior. Science 207, 1224–1226.

Mombaerts, P. 2004. Genes and ligands for odorant, vomeronasal and taste receptors. Nat. Rev. Neurosci. 5, 263–278.

Mombaerts, P., Wang, F., Dulac, C., Chao, S. K., Nemes, A., Mendelsohm, M., Edmondson, J., and Axel, R. 1996. Visualizing an olfactory sensory map. Cell 87, 675–686.

Montell, C., Birnbaumer, L., and Flockerzi, V. 2002. The TRP channels, a remarkably functional family. Cell 108, 595–598.

Mori, K. 1987. Monoclonal antibodies (2C5 and 4C9) against lactoseries carbohydrates identify subsets of olfactory and vomeronasal receptor cells and their axons in the rabbit. Brain Res. 408, 215–221.

Mori, K., von Campenhausen, H., and Yoshihara, Y. 2000. Zonal organization of the mammalian main and accessory olfactory systems. Philos. Trans. R. Soc. Lond. B Biol. Sci. 355, 1801–1812.

Mori, K., Fujita, S. C., Imamura, K., and Obata, K. 1985. Immunohistochemical study of subclasses of olfactory nerve fibers and their projections to the olfactory bulb in the rabbit. J. Comp. Neurol. 242, 214–229.

Mucignat-Caretta, C., Caretta, A., and Cavaggioni, A. 1995. Acceleration of puberty onset in female mice by male urinary proteins. J. Physiol. 486(Pt 2), 517–522.

Müller, W. 1971. Vergleichende elektrophysiologische Untersuchungen an den Sinnesepithelien des Jacobsonschen Organs und der Nase von Amphibien (Rana), Reptilien (Lacerta) und Säugetieren (Mus). Z. vergl. Physiologie 72, 370–385.

Norlin, E. M., Gussing, F., and Berghard, A. 2003. Vomeronasal phenotype and behavioral alterations in G alpha i2 mutant mice. Curr. Biol. 13, 1214–1219.

Novotny, M., Jemiolo, B., Harvey, S., Wiesler, D., and Marchlewska-Koj, A. 1986. Adrenal-mediated endogenous metabolites inhibit puberty in female mice. Science 231, 722–725.

Novotny, M. V., Jemiolo, B., Wiesler, D., Ma, W., Harvey, S., Xu, F., Xie, T. M., and Carmack, M. 1999a. A unique urinary constituent, 6-hydroxy-6-methyl-3-heptanone, is a pheromone that accelerates puberty in female mice. Chem. Biol. 6, 377–383.

Novotny, M. V., Ma, W., Wiesler, D., and Zidek, L. 1999b. Positive identification of the puberty-accelerating pheromone of the house mouse: the volatile ligands associating with the major urinary protein. Proc. Biol. Sci. 266, 2017–2022.

Nyby, J., Wysocki, C. J., Whitney, G., Dizinno, G., and Schneider, J. 1979. Elicitation of male mouse (*Mus musculus*) ultrasonic vocalizations I: urinary cues. J. Comp. Physiol. Psychol. 93, 957–975.

de Olmos, J., Hardy, H., and Heimer, L. 1978. The afferent connections of the main and the accessory olfactory bulb formations in the rat: an experimental HRP-study. J. Comp. Neurol. 181, 213–244.

Olson, R., Huey-Tubman, K. E., Dulac, C., and Bjorkman, P. J. 2005. Structure of a pheromone receptor-associated MHC molecule with an open and empty groove. PLoS Biol. 3, e257.

Osada, T., Kito, K., Ookata, K., Graziadei, P. P., Ikai, A., and Ichikawa, M. 1994. Monoclonal antibody (VOM2) specific for the luminal surface of the rat vomeronasal sensory epithelium. Neurosci. Lett. 170, 47–50.

Pankevich, D. E., Baum, M. J., and Cherry, J. A. 2004. Olfactory sex discrimination persists, whereas the preference for urinary odorants from estrous females disappears in male mice after vomeronasal organ removal. J. Neurosci. 24, 9451–9457.

Paxions, G. 2004. The Rat Nervous System, 3rd edn. London Academic Press.

Perlman, S. M. 1934. Jacobson's organ (organon vomeronasale, Jacobsoni): its anatomy, gross, microscopic and comparative, with some observations as well on its function. Ann. Otol. Rhinol. Larangyol. 43, 739–768.

Piana, G. P. 1882. Contribuzioni alla conoscenza della strutture e della funzione dell' organo di Jacobson. Bologna 1880. Deusch Zeitsch Tiermedizin 7, 325–335.

Powers, J. B. and Winans, S. S. 1975. Vomeronasal organ: critical role in mediating sexual behavior of the male hamster. Science 187, 961–963.

Quaglino, E., Giustetto, M., Panzanelli, P., Cantino, D., Fasolo, A., and Sassoe-Pognetto, M. 1999. Immunocytochemical localization of glutamate and gamma-aminobutyric acid in the accessory olfactory bulb of the rat. J. Comp. Neurol. 408, 61–72.

Raisman, G. 1972. An experimental study of the projection of the amygdala to the accessory olfactory bulb and its relationship to the concept of a dual olfactory system. Exp. Brain Res. 14, 395–408.

Retzius, G. 1894. Die Riechzellen der Ophidier in der Riechschleimhaut und im Jacobson'schen Organ. Biol. Untersuch. Neue Folge 6, 48–51.

Rizvi, T. A., Ennis, M., and Shipley, M. T. 1992. Reciprocal connections between the medial preoptic area and the midbrain periaqueductal gray in rat: a WGA-HRP and PHA-L study. J. Comp. Neurol. 315, 1–15.

Rodriguez, I., Del Punta, K., Rothman, A., Ishii, T., and Mombaerts, P. 2002. Multiple new and isolated families within the mouse superfamily of V1r vomeronasal receptors. Nat. Neurosci. 5, 134–140.

Rodriguez, I., Feinstein, P., and Mombaerts, P. 1999. Variable patterns of axonal projections of sensory neurons in the mouse vomeronasal system. Cell 97, 199–208.

Rollmann, S. M., Houck, L. D., and Feldhoff, R. C. 1999. Proteinaceous pheromone affecting female receptivity in a terrestrial salamander. Science 285, 1907–1909.

Rosser, A. E. and Keverne, E. B. 1985. The importance of central noradrenergic neurones in the formation of an olfactory memory in the prevention of pregnancy block. Neuroscience 15, 1141–1147.

Ryba, N. J. and Tirindelli, R. 1997. A new multigene family of putative pheromone receptors. Neuron 19, 371–379.

Sam, M., Vora, S., Malnic, B., Ma, W., Novotny, M. V., and Buck, L. B. 2001. Odorants may arouse instinctive behaviours. Nature 412, 142.

Sanchez-Criado, J. E. 1982. Involvement of the Vomeronasal System in the Reproductive Physiology of the Rat. In: Olfaction and Endocrine Regulation (*ed*. W. Breipohl), pp. 209–217. IRL Press.

Sasaki, K., Okamoto, K., Inamura, K., Tokumitsu, Y., and Kashiwayanagi, M. 1999. Inositol-1,4,5-trisphosphate accumulation induced by urinary pheromones in female rat vomeronasal epithelium. Brain Res. 823, 161–168.

Sato, Y., Miyasaka, N., and Yoshihara, Y. 2005. Mutually exclusive glomerular innervation by two distinct types of olfactory sensory neurons revealed in transgenic zebrafish. J. Neurosci. 25, 4889–4897.

Scalia, F. and Winans, S. S. 1975. The differential projections of the olfactory bulb and accessory olfactory bulb in mammals. J. Comp. Neurol. 161, 31–55.

Schoppa, N. E. and Westbrook, G. L. 2001. Glomerulus-specific synchronization of mitral cells in the olfactory bulb. Neuron 31, 639–651.

Schoppa, N. E. and Westbrook, G. L. 2002. AMPA autoreceptors drive correlated spiking in olfactory bulb glomeruli. Nat. Neurosci. 5, 1194–1202.

Schwarting, G. A. and Crandall, J. E. 1991. Subsets of olfactory and vomeronasal sensory epithelial cells and axons revealed by monoclonal antibodies to carbohydrate antigens. Brain Res. 547, 239–248.

Schwende, F. J., Wiesler, D., Jorgenson, J. W., Carmack, M., and Novotny, M. 1986. Urinary volatile constituents of the house mouse, *Mus musculus*, and their endocrine dependency. J. Chem. Ecol. 12, 277–296.

Shepherd, G. M. 2006. Smells, brains and hormones. Nature 439, 149–151.

Shiosaka, S., Sakanaka, M., Inagaki, S., Senba, E., Hara, Y., Takatsuki, K., Takagi, H., Kawai, Y., and Tohyama, M. 1982. Putative Neurotransmitters in the Amygdaloid Complex With

Special Reference to Peptidergic Pathways. In: Chemical Neuroanatomy (*ed*. P. C. Emson), pp. 359–389. Raven Press.

Shipley, M. T. and Adamek, G. D. 1984. The connections of the mouse olfactory bulb: a study using orthograde and retrograde transport of wheat germ agglutinin conjugated to horseradish peroxidase. Brain Res. Bull. 12, 669–688.

Simerly, R. B., Young, B. J., Capozza, M. A., and Swanson, L. W. 1989. Estrogen differentially regulates neuropeptide gene expression in a sexually dimorphic olfactory pathway. Proc. Natl. Acad. Sci. U. S. A. 86, 4766–4770.

Singer, A. G., Macrides, F., Clancy, A. N., and Agosta, W. C. 1986. Purification and analysis of a proteinaceous aphrodisiac pheromone from hamster vaginal discharge. J. Biol. Chem. 261, 13323–13326.

Singh, P. B., Brown, R. E., and Roser, B. 1987. MHC antigens in urine as olfactory recognition cues. Nature 327, 161–164.

Spehr, M., Hatt, H., and Wetzel, C. H. 2002. Arachidonic acid plays a role in rat vomeronasal signal transduction. J. Neurosci. 22, 8429–8437.

Spehr, M., Kelliher, K. R., Li, X. H., Boehm, T., Leinders-Zufall, T., and Zufall, F. 2006. Essential role of the main olfactory system in social recognition of major histocompatibility complex peptide ligands. J. Neurosci. 26, 1961–1970.

Storan, M. J. and Key, B. 2006. Septal organ of Gruneberg is part of the olfactory system. J. Comp. Neurol. 494, 834–844.

Stowers, L., Holy, T. E., Meister, M., Dulac, C., and Koentges, G. 2002. Loss of sex discrimination and male-male aggression in mice deficient for TRP2. Science 295, 1493–1500.

Sugai, T., Sugitani, M., and Onoda, N. 2000. Novel subdivisions of the rat accessory olfactory bulb revealed by the combined method with lectin histochemistry, electrophysiological and optical recordings. Neuroscience 95, 23–32.

Takami, S. and Graziadei, P. P. 1990. Morphological complexity of the glomerulus in the rat accessory olfactory bulb – a Golgi study. Brain Res. 510, 339–342.

Takami, S. and Graziadei, P. P. 1991. Light microscopic Golgi study of mitral/tufted cells in the accessory olfactory bulb of the adult rat. J. Comp. Neurol. 311, 65–83.

Takami, S., Fernandez, G. D., and Graziadei, P. P. 1992a. The morphology of GABA-immunoreactive neurons in the accessory olfactory bulb of rats. Brain Res. 588, 317–323.

Takami, S., Graziadei, P. P., and Ichikawa, M. 1992b. The differential staining patterns of two lectins in the accessory olfactory bulb of the rat. Brain Res. 598, 337–342.

Taniguchi, M., Kashiwayanagi, M., and Kurihara, K. 1995. Intracellular injection of inositol 1,4,5-trisphosphate increases a conductance in membranes of turtle vomeronasal receptor neurons in the slice preparation. Neurosci. Lett. 188, 5–8.

Taniguchi, M., Kashiwayanagi, M., and Kurihara, K. 1996. Intracellular dialysis of cyclic nucleotides induces inward currents in turtle vomeronasal receptor neurons. J. Neurosci. 16, 1239–1246.

Taniguchi, M., Wang, D., and Halpern, M. 1998. The characteristics of the electrovomeronasogram: its loss following vomeronasal axotomy in the garter snake. Chem. Senses 23, 653–659.

Taniguchi, M., Wang, D., and Halpern, M. 2000. Chemosensitive conductance and inositol 1,4,5-trisphosphate-induced conductance in snake vomeronasal receptor neurons. Chem. Senses 25, 67–76.

Tian, H. and Ma, M. 2004. Molecular organization of the olfactory septal organ. J. Neurosci. 24, 8383–8390.

Tirindelli, R., Mucignat-Caretta, C., and Ryba, N. J. 1998. Molecular aspects of pheromonal communication via the vomeronasal organ of mammals. Trends Neurosci. 21, 482–486.

Toyoda, F. and Kikuyama, S. 2000. Hormonal influence on the olfactory response to a female-attracting pheromone, sodefrin, in the newt, *Cynops pyrrhogaster*. Comp. Biochem. Physiol. B Biochem. Mol. Biol. 126, 239–245.

Trinh, K. and Storm, D. R. 2003. Vomeronasal organ detects odorants in absence of signaling through main olfactory epithelium. Nat. Neurosci. 6, 519–525.

Trotier, D. 1998. Electrophysiological properties of frog olfactory supporting cells. Chem. Senses 23, 363–369.

Trotier, D. and Doving, K. B. 1996. Functional role of receptor neurons in encoding olfactory information. J. Neurobiol. 30, 58–66.

Trotier, D., Doving, K. B., and Rosin, J. F. 1993. Voltage-dependent currents in microvillar receptor cells of the frog vomeronasal organ. Eur. J. Neurosci. 5, 995–1002.

Trotier, D., Eloit, C., Wassef, M., Talmain, G., Bensimon, J. L., Doving, K. B., and Ferrand, J. 2000. The vomeronasal cavity in adult humans. Chem. Senses 25, 369–380.

Tucker, D. 1971. Nonolfactory Responses from the Nasal Cavity: Jacobson's Organ and the Trigeminal System. In: Handbook of Sensory Physiology (*ed*. L. M. Beidler), pp. 151–181. Springer Verlag.

Urban, N. N. and Castro, J. B. 2005. Tuft calcium spikes in accessory olfactory bulb mitral cells. J. Neurosci. 25, 5024–5028.

Vaccarezza, O. L., Sepich, L. N., and Tramezzani, J. H. 1981. The vomeronasal organ of the rat. J. Anat. 132, 167–185.

Vandenbergh, J. G. 1983. Social Factors Controlling Puberty in the Female Mouse. In: Hormones and Behaviours in Higher Vertebrates (*eds*. J. Balthazart, J. Prive, and R. Gilles), pp. 342–349. Springer Verlag.

Vannier, B., Peyton, M., Boulay, G., Brown, D., Qin, N., Jiang, M., Zhu, X., and Birnbaumer, L. 1999. Mouse trp2, the homologue of the human trpc2 pseudogene, encodes mTrp2, a store depletion-activated capacitative Ca^{2+} entry channel. Proc. Natl. Acad. Sci. U. S. A. 96, 2060–2064.

Walz, A., Rodriguez, I., and Mombaerts, P. 2002. Aberrant sensory innervation of the olfactory bulb in neuropilin-2 mutant mice. J. Neurosci. 22, 4025–4035.

Wekesa, K. S. and Anholt, R. R. 1997. Pheromone regulated production of inositol-(1, 4, 5)-trisphosphate in the mammalian vomeronasal organ. Endocrinology 138, 3497–3504.

Winans, S. S. and Scalia, F. 1970. Amygdaloid nucleus: new afferent input from the vomeronasal organ. Science 170, 330–332.

Wissenbach, U., Schroth, G., Philipp, S., and Flockerzi, V. 1998. Structure and mRNA expression of a bovine trp homologue related to mammalian trp2 transcripts. FEBS Lett. 429, 61–66.

Wu, Y., Tirindelli, R., and Ryba, N. J. 1996. Evidence for different chemosensory signal transduction pathways in olfactory and vomeronasal neurons. Biochem. Biophys. Res. Commun. 220, 900–904.

Wysocki, C. J. 1979. Neurobehavioral evidence for the involvement of the vomeronasal system in mammalian reproduction. Neurosci. Biobehav. Rev. 3, 301–341.

Wysocki, C. J. and Lepri, J. J. 1991. Consequences of removing the vomeronasal organ. J. Steroid Biochem. Mol. Biol. 39, 661–669.

Wysocki, C. J. and Meredith, M. 1987. The Vomeronasal System. In: Neurobiology of Taste and Smell (*eds*. T. E. Finger and W. L. Silver), pp. 125–150. Wiley and Sons.

Wysocki, C. J. and Preti, G. 2004. Facts, fallacies, fears, and frustrations with human pheromones. Anat. Rec. A Discov. Mol. Cell. Evol. Biol. 281, 1201–1211.

Wysocki, C. J., Katz, Y., and Bernhard, R. 1983. Male vomeronasal organ mediates female-induced testosterone surges in mice. Biol. Reprod. 28, 917–922.

Xu, F., Schaefer, M., Kida, I., Schafer, J., Liu, N., Rothman, D. L., Hyder, F., Restrepo, D., and Shepherd, G. M. 2005. Simultaneous activation of mouse main and accessory olfactory bulbs by odors or pheromones. J. Comp. Neurol. 489, 491–500.

Yamamoto, K., Kawai, Y., Hayashi, T., Ohe, Y., Hayashi, H., Toyoda, F., Kawahara, G., Iwata, T., and Kikuyama, S. 2000. Silefrin, a sodefrin-like pheromone in the abdominal gland of the sword-tailed newt, *Cynops ensicauda*. FEBS Lett. 472, 267–270.

Yang, H., Shi, P., Zhang, Y. P., and Zhang, J. 2005. Composition and evolution of the V2r vomeronasal receptor gene repertoire in mice and rats. Genomics 86, 306–315.

Yoon, H., Enquist, L. W., and Dulac, C. 2005. Olfactory inputs to hypothalamic neurons controlling reproduction and fertility. Cell 123, 669–682.

Yoshihara, Y., Kawasaki, M., Tamada, A., Fujita, H., Hayashi, H., Kagamiyama, H., and Mori, K. 1997. OCAM: A new member of the neural cell adhesion molecule family related to zone-to-zone projection of olfactory and vomeronasal axons. J. Neurosci. 17, 5830–5842.

Zidek, L., Stone, M. J., Lato, S. M., Pagel, M. D., Miao, Z., Ellington, A. D., and Novotny, M. V. 1999. NMR mapping of the recombinant mouse major urinary protein I binding site occupied by the pheromone 2-sec-butyl-4,5-dihydrothiazole. Biochemistry 38, 9850–9861.

Zufall, F. and Munger, S. D. 2001. From odor and pheromone transduction to the organization of the sense of smell. Trends Neurosci. 24, 191–193.

Zufall, F., Ukhanov, K., Lucas, P., Liman, E., and Leinders-Zufall, T. 2005. Neurobiology of TRPC2: from gene to behavior. Pflugers. Arch. 451, 61–71.

4.44 Genomics of Vomeronasal Receptors

I Rodriguez, University of Geneva, Geneva, Switzerland

Glossary

vomeronasal organ (VNO) Sensory olfactory structure, physically separate from the main olfactory system and present in mammals, amphibians, and reptiles. Its function is often linked to pheromone perception.

V1r receptors Pheromone receptors expressed by vomeronasal sensory neurons present in the apical part of the vomeronasal sensory neuroepithelium in rodents.

V2r receptors Putative pheromone receptors expressed by vomeronasal sensory neurons present in the basal part of the vomeronasal sensory neuroepithelium in rodents.

4.44.1 Introduction

The survival of multicellular organisms depends on their ability to perceive and recognize the outside and their own inside worlds. The main tool selected by evolution to achieve these basic tasks is based on the expression of multiple and different receptors, in specific cells. These cells then either directly respond to the stimuli, or instead transmit the information to others, which process the signals. To achieve this general need for receptor variability, various strategies have been selected by different systems and species. The vertebrate immune system, for example, bases its T-cell receptor (TCR) and Ig remarkable diversity on genomic recombination between a relatively limited number of coding segments. Another inventive strategy, used by many life forms including viruses, achieves variability by random mutation of messenger RNA transcripts. A very different option, which is based neither on genome recombination nor on transcript alteration, is used by vertebrate chemosensory systems to face the outside world: they benefit from receptor gene repertoires of amazing size and diversity already present in the genomes, which directly translate into an exceptional variability of chemoreceptors. Thus, odorant, taste, and vomeronasal receptor genes represent a substantial part of the coding fraction of mammalian genomes.

Vomeronasal receptor genes will be discussed here. Although some of their characteristics may be shared by odorant and taste receptor genes, they should not be assumed to be so, since their likely direct involvement in pheromone perception involved specific evolutionary pressures, pressures probably different from those which affected other chemosensory receptor genes. Moreover, the signal transduction cascades of sensory cells expressing odorant or vomeronasal receptors, but also their targets in the brain, two processes which directly involve the expressed chemoreceptors, are dissimilar.

In mammals, pheromones are mostly (but not only) perceived by the vomeronasal organ. Two large putative seven transmembrane receptor families, the V1r and V2r superfamilies (Dulac, C. and Axel, R., 1995; Herrada, G. and Dulac, C., 1997; Matsunami, H. and Buck, L. B., 1997; Ryba, N. J. and Tirindelli, R., 1997), are almost exclusively expressed by vomeronasal sensory neurons.

4.44.2 V1r Receptors

4.44.2.1 On the Size of the Repertoires

All mammalian genomes apparently contain V1r genes. V1r repertoire sizes can be relatively limited, such as in dogs, for example, with only 8–10 potentially functional V1r genes (Rodriguez, I., 2005; Young, J. M., 2005), or particularly expanded such as in mice, where V1rs number about 160 (Rodriguez, I. *et al.*, 2002; Zhang, X. *et al.*, 2004) (Figure 1). Other species, such as cows or opossums, possess V1r repertoires of intermediate size (Grus, W. *et al.*, 2005). A few potentially functional V1r

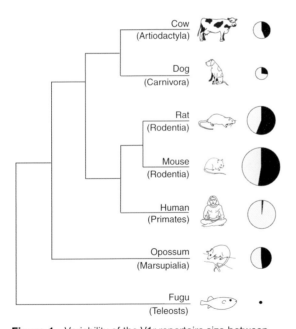

Figure 1 Variability of the V1r repertoire size between different vertebrate species. Pies represent the complete V1r repertoires corresponding to each species, and their surface the number of V1r genes. Black and gray indicate potentially functional V1r genes and pseudogenes, respectively. Adapted from Rodriguez, I. 2005. Remarkable diversity of mammalian pheromone receptor repertoires. Proc. Natl. Acad. Sci. U.S.A. 102, 6639–6640.

genes (among many pseudogenes; Giorgi, D. *et al.*, 2000; Kouros-Mehr, H. *et al.*, 2001) have been reported in primates, including humans (Rodriguez, I. *et al.*, 2000; Rodriguez, I. and Mombaerts, P., 2002). This may appear surprising since catarrhine primates only possess a vestigial and thus likely nonfunctional vomeronasal organ after birth (Maier, W., 1997) (a situation thought to be a result of trichromatic vision acquisition; Liman, E. R. and Innan, H., 2003; Zhang, J. and Webb, D. M., 2003; Gilad, Y. *et al.*, 2004). Possibly explaining the presence of V1r genes without functional vomeronasal system, expression of V1r genes has been reported outside the vomeronasal organ, in the main olfactory epithelium, in some species (Rodriguez, I. *et al.*, 2000; Wakabayashi, Y. *et al.*, 2002; Karunadasa, D. K. *et al.*, 2006). Thus, despite our strong taste for clear-cut expression profiles, the exclusive association between V1r expression and vomeronasal function is incorrect. This parallels the wrongly assumed functional dichotomy between the main olfactory and the vomeronasal systems, supposedly exclusively responsible for odor and pheromone perception, respectively. The origin of V1rs dates back apparently over 450 million years ago, since V1rs have also been identified in teleost species (Pfister, P. and Rodriguez, I., 2005). In these species, however, and unlike all mammals that express numerous V1rs, a very limited number of V1r genes are present in the genome (Pfister, P. *et al.*, 2007; Shi, P. and Zhang, J., 2007). These observations argue both for the presence of V1rs in other vertebrate phyla such as reptiles, and for a specific advantage in non-teleost species to maintain large and diversified V1r repertoires.

In addition to functional genes, a sizeable number of V1r pseudogenes are found in most species. Functional V1r genes are indeed sometimes less numerous than pseudogenic sequences, even in orders with well-developed vomeronasal systems such as rodents (Giorgi, D. and Rouquier, S., 2002; Rodriguez, I. *et al.*, 2002; Grus, W. E. and Zhang, J., 2004; Young, J. M. *et al.*, 2005) (Figure 1). Based on the sequence of their products, V1r genes can be divided into multiple, highly divergent families (12–15 in rodents), which only share a few common, highly conserved amino acids. Intrafamily homologies typically range from 40% to 98%, while interfamily identities can be as low as 15% (Rodriguez, I. *et al.*, 2002).

But does size matter? The very closely related and numerous V1r genes in a given family may each play a specific role and be responsible for the recognition of a narrow set of agonists; alternatively, these could represent a very redundant set of receptors with very little evolutionary pressure to stay intact at the individual level, a view possibly supported by the large number of pseudogenes found in each family. Genetic experiments involving the partial deletion of defined V1r families, and the analysis of the potential resulting behavioral deficits have not yet settled this issue.

4.44.2.2 Remarkably Variable Repertoires

The recent sequencing of multiple mammalian genomes allowed the comparison of complete V1r repertoires between different species. These comparisons led to the surprising observation that V1r genes are not only highly variable inside the repertoire of a given species, but that repertoires of different species are also remarkably different (Lane, R. P. *et al.*, 2004). Thus, the parallel exploration of the cow, dog, opossum, and mouse V1r repertoires revealed species-specific expansions and absence of some V1r families (Grus, W. *et al.*, 2005). More unexpected, a similar observation was made when analyzing the mouse and rat V1r repertoires. Although both of these species are closely related and mostly express common V1r families, each also possesses their own V1r receptor families (Grus, W. E. and Zhang, J., 2004; Lane, R. P. *et al.*, 2004; Shi, P. *et al.*, 2005). V1r repertoires are thus characterized by rapid gene turnover and lineage-specific phylogenetic clustering, via gene duplications, deletions, and gene conversions.

A closer look at the evolutionary forces acting on these genes revealed that the diversification of the V1r repertoires, both inside a given species and between species, is at least partly the result of positive Darwinian selective pressure (i.e., the fixation of mutations that alter the translated sequence were favored over those which did not) (Mundy, N. I. and Cook, S., 2003; Lane, R. P. *et al.*, 2004; Shi, P. *et al.*, 2005). Positive selection has been observed in a few genes, evolving rapidly and often involved in the recognition of foreign molecules or cells, and usually related to the immune or fertilization processes. This latter function involves the recognition of same-species gametes, or the nonrecognition of gametes from different species: variability is there thought to provide postmating species barriers. One may suggest that V1r gene variability could similarly provide reproduction barriers, but in this case premating barriers, allowing only individuals possessing a given set of V1r receptors to adequately respond to the pheromonal blends of their own species. An attractive hypothesis would be to consider V1r diversity or novelty as potentially instrumental during speciation

events, by their contribution to premating reproductive isolation. In favor of such possible role in speciation, V1r receptors are known to be narrowly tuned (Leinders-Zufall, T. *et al.*, 2000), unlike odorant receptors, and are therefore likely to dramatically alter perception and thus social behavior if lost or modified. In addition, in support of this view, a link between pheromone perception and speciation has been observed in a moth species (*Ostrinia*), in which changes in pheromone signaling and perception apparently allowed isolation and possibly speciation (Roelofs, W. L. *et al.*, 2002).

4.44.2.3 Clustered Gene Families

V1r families are often organized in clusters in the genome, and are present on virtually all autosomes in the few species where this was analyzed. These clusters result from multiple duplication events, possibly facilitated by Line1 repetitive elements, elements which are found at unusual high density in genomic regions harboring V1r genes (Lane, R. P. *et al.*, 2002; Lane, R. P. *et al.*, 2004), and whose retrotransposition apparently correlates with V1r duplications (Lane, R. P. *et al.*, 2004). Although V1rs are multiexonic genes (usually 2 or 3 exons), V1r coding sequences are entirely included in single exons (often the last one), similarly to many other G-coupled receptor genes. As previously mentioned, V1r families are organized, or better said found, in clusters; each of these contain multiple V1r units, whose transcription orientation is apparently random. Evidence for cluster-mediated regulatory mechanisms has recently been reported by us (Roppolo, D. *et al.*, 2007). Each V1r gene contains a relatively large segment of about 600 bp, located mostly upstream its transcription start site and overlapping slightly with the 5′-untranslated region, which is remarkably conserved (over 80%) between members of a given V1r family but not between different V1r family members (Lane, R. P. *et al.*, 2002; Lane, R. P. *et al.*, 2004). This observation contrasts with the poor promoter sequence conservation reported for odorant receptor genes. The role of these ultraconserved elements, largely more conserved than what one would expect from conventional promoters, is unknown.

4.44.2.4 Monogenic and Monoallelic V1r Expression

At the base of our current understanding of pheromone coding is the belief that each vomeronasal sensory neuron expresses a single (or a few) V1r receptor genes (Dulac, C. and Axel, R., 1995), from a single, randomly chosen allele (Rodriguez, I. *et al.*, 1999). The perception of blends would thus involve a more central integration of multiple highly specific signals. Although monoallelic expression is experimentally well supported, the monogenic assumption still needs stronger experimental evidence. What is clear is that neurons expressing V1rs are not the same as those expressing V2rs. The first ones are expressed by sensory neurons located in the apical part of the pseudostratified vomeronasal epithelium, while the second ones are found in the more basal strata. The joined patterns of expression indicate that all mature sensory neurons express either V1r or V2r genes.

4.44.2.5 V1rs as Chemodetectors and Guidance Molecules

A remarkable dual role is apparently played by V1r receptors. First, not surprisingly, those expressed at the vomeronasal dendritic endings in contact with the outside world, play a role in chemodetection. This is supported by

1. a genetic approach in the mouse involving the deletion of an entire V1r cluster, and the observation of a correlated deficiency to perceive pheromones *in vitro*, and of inadequate behavior responses to specific interindividual interactions (Del Punta, K. *et al.*, 2000; Del Punta, K. *et al.*, 2002), and
2. by the identification of 2-heptanone, a molecule present in mouse urine, which plays a pheromonal role in this species, as an agonist for the V1rb2 receptor (Boschat, C. *et al.*, 2002).

Second, the expression of V1rs is instrumental in the elaboration of an adequate topographical map in the accessory olfactory bulb. Genetic experiments have shown that vomeronasal sensory neurons expressing a given V1r gene project their axons toward 10–20 neuropil-rich spherical structures, called glomeruli, in the caudal part of the olfactory bulb termed the accessory olfactory bulb (Belluscio, L. *et al.*, 1999; Rodriguez, I. *et al.*, 1999). Sensory neurons expressing other V1r receptors project to different glomeruli, leading to a complex topographical map of a few hundred glomeruli. The prevention of expression of a given V1r in a sensory neuron population leads to an altered axonal projection map, no longer convergent, which innervates glomeruli corresponding other receptors. This very particular dual role is also shared by odorant receptors, receptors that can

even substitute for V1rs, in a vomeronasal context (Rodriguez, I. *et al.*, 1999). This and other observations in the olfactory system (Feinstein, P. *et al.*, 2004) suggest that very unrelated seven transmembrane receptors, in different systems, may also play a role in axonal targeting.

4.44.3 V2r Receptors

Three research groups independently identified V2rs in 1998 (Herrada, G. and Dulac, C., 1997; Matsunami, H. and Buck, L. B., 1997; Ryba, N. J. and Tirindelli, R., 1997). Our understanding of V2r biology, both in terms of role and expression in different species, lags behind the little we know about V1rs, likely for the simple reason that their identification postdates the one of V1rs. Unlike these latter, V2rs share strong homologies with well-known seven transmembrane G-coupled receptors, with long extracellular N-termini, including calcium-sensing and glutamate receptors.

4.44.3.1 Genomic Parallels and Divergences with V1rs

V2r genes have been identified in vertebrates, and are apparently restricted, as V1rs, to this subphylum. Numerous V2rs are found in frogs (Hagino-Yamagishi, K. *et al.*, 2004), and in teleost species (Cao, Y. *et al.*, 1998; Naito, T. *et al.*, 1998; Speca, D. J. *et al.*, 1999; Hashiguchi, Y. and Nishida, M., 2005), such as *Danio rerio*, for example, whose genome contains over 50 V2rs, (contrasting with the extremely limited number of V1rs in this species). In mammals such as mice and rats, 61 and 57 V2r genes with intact open reading frames have been reported, respectively (Yang, H. *et al.*, 2005). These numbers represent a strikingly low proportion (between 29% and 34%) of the complete, pseudogene-including V2r repertoire of these two species. The chromosomal distribution of these genes is wide, since in the mouse they are located on at least 12 chromosomes, including the X chromosome. V2rs tend, like V1rs, to be clustered in the genome, with no apparent specific organization within these clusters.

The exact number of different expressed receptors corresponding to the known V2r genes is difficult to evaluate, since the large extracellular part of V2r receptors is encoded by multiple exons that are often alternatively spliced, and which therefore lead to a corresponding expansion of the V2r repertoires variability. In rodents, V2rs have been classified into three families, among which one, family C, is particularly distant from the others, a situation likely reflecting an ancient divergence with the other two families, predating the separation between teleosts and tetrapods (Yang, H. *et al.*, 2005). Repertoire comparisons between mouse and rat indicate that similarly to what is observed for V1rs, V2rs form species-specific clades, with the result that only a few rat–mouse orthologs can be identified. A more detailed analysis of the evolution of rodent V2rs suggests that most species-specific clades resulted from recent gene duplications/losses, with a minor role played by gene conversion events. Moreover, similarly to what was observed for V1rs, positive selection was detected for specific amino acids in V2rs (Yang, H. *et al.*, 2005), with the particularity that most of them were located on the N-terminal extracellular region of the receptor, a segment thought to mediate ligand/receptor interactions.

4.44.3.2 Immune Multigene Families and V2r-Expressing Sensory Neurons

Vomeronasal receptors are thought to interact with G proteins, respectively, the Gi2a and Goa subunits for V1rs and V2rs. An unexpected complex and non-random coexpression, and interaction, was recently reported between two unlikely partners: V2r receptors and major histocompatibility complex (MHC) molecules (Ishii, T. *et al.*, 2003; Loconto, J. *et al.*, 2003). Members of the nonclassical MHC class I M10 and M1 families (6 and 3 members, respectively), are exclusively expressed by vomeronasal sensory neurons. Genes encoding for these receptors are clustered on chromosome 17 in the mouse, and are part of the H2-M locus, the most distal segment encoding class I genes.

These MHC receptors are involved, via a still unknown mechanism, in the adequate export of at least some V2rs to the cellular membrane (Loconto, J. *et al.*, 2003). Reports suggest however, in addition, a more direct involvement of MHC molecules in pheromone perception. For example, MHC-diassortative pregnancy block and preferences have been demonstrated in multiple species and conditions. Moreover, some vomeronasal sensory neurons respond to class I, MHC-binding peptides (Leinders-Zufall, T. *et al.*, 2004). The presence of specific MHC molecules in the vomeronasal organ forces us to consider them as prime candidates for such interactions. Thus, it is possible that not only do MHC type I function as

facilitators of V2r cell surface expression, but that they may be able to detect peptides (although crystal structure data does not support this hypothesis (Olson, R. *et al.*, 2005)) via the use of their binding grooves, a crucial tool used by their classical counterparts.

4.44.4 Perspective

4.44.4.1 A Genomic Trail of Speciation Events

A similar strategy has thus been selected by evolution to provide vertebrate pheromone-sensing systems with two large and diverse repertoires of chemoreceptors: rapid gene birth and death. We look today at a battlefield with hundreds of pseudogenic sequences, often more numerous than potentially functional ones. This represents an amazing source of information, and because of the speed at which the repertoires evolved, a unique recording of speciation events, including those that occurred relatively recently. We are now lacking data relative to the other side of the coin: the pheromones themselves. These are today, in vertebrates, still remarkably poorly described (insects got somehow more attention for once). Even in rodents, which are largely the best-studied mammals in the pheromone field, <10 compounds have been reported to mediate pheromonal interactions, a situation largely due to the tedious biochemical purifications involved in the identification of these chemicals. Thus, unless we benefit from a biochemical technical revolution allowing large-scale approaches similar to the one which makes today a quick survey of complete genomes an easy task, a global understanding of vertebrate pheromone perception at the peripheral level will be incomplete, for a long time.

References

Belluscio, L., Koentges, G., Axel, R., and Dulac, C. 1999. A map of pheromone receptor activation in the mammalian brain. Cell 97, 209–220.

Boschat, C., Pelofi, C., Randin, O., Roppolo, D., Luscher, C., Broillet, M. C., and Rodriguez, I. 2002. Pheromone detection mediated by a V1r vomeronasal receptor. Nat. Neurosci. 5, 1261–1262.

Cao, Y., Oh, B. C., and Stryer, L. 1998. Cloning and localization of two multigene receptor families in goldfish olfactory epithelium. Proc. Natl. Acad. Sci. U. S. A. 95, 11987–11992.

Del Punta, K., Leinders-Zufall, T., Rodriguez, I., Jukam, D., Wysocki, C. J., Ogawa, S., Zufall, F., and Mombaerts, P. 2002. Deficient pheromone responses in mice lacking a cluster of vomeronasal receptor genes. Nature 419, 70–74.

Del Punta, K., Rothman, A., Rodriguez, I., and Mombaerts, P. 2000. Sequence diversity and genomic organization of vomeronasal receptor genes in the mouse. Genome Res. 10, 1958–1967.

Dulac, C. and Axel, R. 1995. A novel family of genes encoding putative pheromone receptors in mammals. Cell 83, 195–206.

Feinstein, P., Bozza, T., Rodriguez, I., Vassalli, A., and Mombaerts, P. 2004. Axon guidance of mouse olfactory sensory neurons by odorant receptors and the beta2 adrenergic receptor. Cell 117, 833–846.

Gilad, Y., Wiebe, V., Przeworski, M., Lancet, D., and Pääbo, S. 2004. Loss of olfactory receptor genes coincides with the acquisition of full trichromatic vision in primates. PLoS Biol. 2, E5.

Giorgi, D. and Rouquier, S. 2002. Identification of V1R-like putative pheromone receptor sequences in non-human primates. Characterization of V1R pseudogenes in marmoset, a primate species that possesses an intact vomeronasal organ. Chem. Senses 27, 529–537.

Giorgi, D., Friedman, C., Trask, B. J., and Rouquier, S. 2000. Characterization of nonfunctional V1R-like pheromone receptor sequences in human. Genome Res. 10, 1979–1985.

Grus, W. E. and Zhang, J. 2004. Rapid turnover and species-specificity of vomeronasal pheromone receptor genes in mice and rats. Gene 340, 303–312.

Grus, W., Shi, P., Zhang, Y., and Zhang, J., 2005. Dramatic variation of the vomeronasal pheromone gene repertoire among five orders of placental and marsupial mammals. PNAS 102, 5767–5772.

Hagino-Yamagishi, K., Moriya, K., Kubo, H., Wakabayashi, Y., Isobe, N., Saito, S., Ichikawa, M., and Yazaki, K. 2004. Expression of vomeronasal receptor genes in Xenopus laevis. J. Comp. Neurol. 472, 246–256.

Herrada, G. and Dulac, C. 1997. A novel family of putative pheromone receptors in mammals with a topographically organized and sexually dimorphic distribution. Cell 90, 763–773.

Hashiguchi, Y. and Nishida, M. 2005. Evolution of vomeronasal-type odorant receptor genes in the zebrafish genome. Gene 362, 19–28.

Ishii, T., Hirota, J., and Mombaerts, P. 2003. Combinatorial coexpression of neural and immune multigene families in mouse vomeronasal sensory neurons. Curr. Biol. 13, 394–400.

Karunadasa, D. K., Chapman, C., and Bicknell, R. J. 2006. Expression of pheromone receptor gene families during olfactory development in the mouse: expression of a V1 receptor in the main olfactory epithelium. Eur. J. Neurosci. 23, 2563–2572.

Kouros-Mehr, H., Pintchovski, S., Melnyk, J., Chen, Y. J., Friedman, C., Trask, B., and Shizuya, H. 2001. Identification of non-functional human VNO receptor genes provides evidence for vestigiality of the human VNO. Chem. Senses 26, 1167–1174.

Lane, R. P., Cutforth, T., Axel, R., Hood, L., and Trask, B. J. 2002. Sequence analysis of mouse vomeronasal receptor gene clusters reveals common promoter motifs and a history of recent expansion. Proc. Natl. Acad. Sci. U. S. A. 99, 291–296.

Lane, R. P., Young, J., Newman, T., and Trask, B. J. 2004. Species specificity in rodent pheromone receptor repertoires. Genome Res. 14, 603–608.

Leinders-Zufall, T., Brennan, P., Widmayer, P. S. P. C., Maul-Pavicic, A., Jager, M., Li, X. H., Breer, H., Zufall, F., and Boehm, T. 2004. MHC class I peptides as chemosensory signals in the vomeronasal organ. Science 306, 1033–1037.

Leinders-Zufall, T., Lane, A. P., Puche, A. C., Ma, W., Novotny, M. V., Shipley, M. T., and Zufall, F. 2000.

Ultrasensitive pheromone detection by mammalian vomeronasal neurons. Nature 405, 792–796.

Liman, E. R. and Innan, H. 2003. Relaxed selective pressure on an essential component of pheromone transduction in primate evolution. Proc. Natl. Acad. Sci. U. S. A. 100, 3328–3332.

Loconto, J., Papes, F., Chang, E., Stowers, L., Jones, E. P., Takada, T., Kumanovics, A., Fischer Lindahl, K., and Dulac, C. 2003. Functional expression of murine V2R pheromone receptors involves selective association with the M10 and M1 families of MHC class Ib molecules. Cell 112, 607–618.

Maier, W. 1997. The nasopalatine duct and the nasal floor cartilages in catarrhine primates. Z. Morphol. Anthropol. 81, 289–300.

Matsunami, H. and Buck, L. B. 1997. A multigene family encoding a diverse array of putative pheromone receptors in mammals. Cell 90, 775–784.

Mundy, N. I. and Cook, S. 2003. Positive selection during the diversification of class I vomeronasal receptor-like (V1RL) genes, putative pheromone receptor genes, in human and primate evolution. Mol. Biol. Evol. 20, 1805–1810.

Naito, T., Saito, Y., Yamamoto, J., Nozaki, Y., Tomura, K., Hazama, M., Nakanishi, S., and Brenner, S. 1998. Putative pheromone receptors related to the Ca^{2+}-sensing receptor in Fugu. Proc. Natl. Acad. Sci. U. S. A. 95, 5178–5181.

Olson, R., Huey-Tubman, K. E., Dulac, C., and Bjorkman, P. J. 2005. Structure of a pheromone receptor-associated MHC molecule with an open and empty groove. PLoS Biol. 3, e257.

Pfister, P. and Rodriguez, I. 2005. Olfactory expression of a single and highly variable V1r pheromone receptor-like gene in fish species. Proc. Natl. Acad. Sci. U. S. A. 102(15), 5489–5494.

Pfister, P., Randall, J., Montoya-Burgos, J. I., and Rodriguez, I. 2007. Divergent evolution among teleost V1r receptor genes. PLoS ONE 2, e379.

Rodriguez, I. 2005. Remarkable diversity of mammalian pheromone receptor repertoires. Proc. Natl. Acad. Sci. U. S. A. 102, 6639–6640.

Rodriguez, I. and Mombaerts, P. 2002. Novel human vomeronasal receptor-like genes reveal species-specific families. Curr. Biol. 12, R409–R411.

Rodriguez, I., Del Punta, K., Rothman, A., Ishii, T., and Mombaerts, P. 2002. Multiple new and isolated families within the mouse superfamily of V1r vomeronasal receptors. Nat. Neurosci. 5, 134–140.

Rodriguez, I., Feinstein, P., and Mombaerts, P. 1999. Variable patterns of axonal projections of sensory neurons in the mouse vomeronasal system. Cell 97, 199–208.

Rodriguez, I., Greer, C. A., Mok, M. Y., and Mombaerts, P. 2000. A putative pheromone receptor gene expressed in human olfactory mucosa. Nat. Genet. 26, 18–19.

Roelofs, W. L., Liu, W., Hao, G., Jiao, H., Rooney, A. P., and Linn, C. E., Jr. 2002. Evolution of moth sex pheromones via ancestral genes. Proc. Natl. Acad. Sci. U. S. A. 99, 13621–13626.

Roppolo, D., Vollery, S., Kan, C., Lüscher, C., Broillet, M. C., and Rodriguez, I. 2007. Gene cluster lock after pheromone receptor gene choice. EMBO J. (in press).

Ryba, N. J. and Tirindelli, R. 1997. A new multigene family of putative pheromone receptors. Neuron 19, 371–379.

Shi, P. and Zhang, J. 2007. Comparative genomic analysis identifies an evolutionary shift of vomeronasal receptor gene repertoires in the vertebrate transition from water to land. Genome Res. 17(2), 166–174.

Shi, P., Bielawski, J. P., Yang, H., and Zhang, Y. P. 2005. Adaptive diversification of vomeronasal receptor 1 genes in rodents. J. Mol. Evol. 60, 566–576.

Speca, D. J., Lin, D. M., Sorensen, P. W., Isacoff, E. Y., Ngai, J., and Dittman, A. H. 1999. Functional identification of a goldfish odorant receptor. Neuron 23, 487–498.

Wakabayashi, Y., Mori, Y., Ichikawa, M., Yazaki, K., and Hagino-Yamagishi, K. 2002. A putative pheromone receptor gene is expressed in two distinct olfactory organs in goats. Chem. Senses 27, 207–213.

Yang, H., Shi, P., Zhang, Y. P., and Zhang, J. 2005. Composition and evolution of the V2r vomeronasal receptor gene repertoire in mice and rats. Genomics 86, 306–315.

Young, J. M., Kambere, M., Trask, B. J., and Lane, R. P. 2005. Divergent V1R repertoires in five species: Amplification in rodents, decimation in primates, and a surprisingly small repertoire in dogs. Genome Res. 15, 231–240.

Zhang, J. and Webb, D. M. 2003. Evolutionary deterioration of the vomeronasal pheromone transduction pathway in catarrhine primates. Proc. Natl. Acad. Sci. U. S. A. 100, 8337–8341.

Zhang, X., Rodriguez, I., Mombaerts, P., and Firestein, S. 2004. Odorant and vomeronasal receptor genes in two mouse genome assemblies. Genomics 83, 802–811.

4.45 Human Olfactory Psychophysics

Brad Johnson, Rehan M Khan, and Noam Sobel, UC Berkeley, Berkeley, CA, USA

Published by Elsevier Inc.

Glossary

2AFC Two-alternative forced-choice psychophysical paradigm in which a subject must choose which of two alternatives (one target and one distracter) is the target.

3AFC Three-alternative forced-choice psychophysical paradigm in which a subject must choose which of three alternatives (one target and two distracters) is the target.

adaptation The temporary diminution of response to successive repetitions of a stimulus.

adaptive staircase method A method for presenting stimuli to ascertain a detection threshold in which the choice of successive stimuli depends on prior responses.

anosmia A condition in which a subject cannot detect any odorant.

anosmia, specific An inability to detect a specific odorant in the absence of a general anosmia.

ascending method of limits A method for presenting stimuli in sequence from low to high concentration to determine detection threshold.

detection threshold The minimal concentration of an odorant sufficient for reliable detection.

JND Just noticeable difference. The smallest increment in the intensity of a stimulus that can be discerned.

magnitude estimation The assignment of a numerical value to the perceived intensity of a stimulus

MDS Multidimensional scaling. A technique for inferring a coordinate space in which objects whose pairwise distances are known can be placed.

method of constant stimuli A procedure for determining absolute or discrimination thresholds wherein a standard stimulus is presented repeatedly against distractors drawn one at a time from a set spanning a range of intensities and subjects make a forced-choice decision at each trial.

odorant Airborne molecule perceived as an odor.

olfactometer A device for the precise delivery of odorants to a subject.

osmic Able to smell, in contradistinction to anosmic.

power law A mathematical function that describes the relationship between a stimulus magnitude and its corresponding perceived magnitude.

psychometric function A function that relates a behavioral response (e.g., percent correct identification) to a stimuli to some parameter (usually a magnitude) of the stimulus.

psychophysics The precise measurement of perceptual phenomena in relation to their physical stimuli.

pure olfactant An odorant that activates only the olfactory and not the trigeminal system

signal detection theory A theory that describes how well a signal can be detected embedded within noise as a function of the signal strength and the detector's criterion or bias.

trigeminal odorant An odorant that stimulates the trigeminal system.

UPSIT University of Pennsylvania smell identification test. A standardized multiple-choice smell identification test that is used to identify olfactory function as a function of age.

vibrational theory A controversial theory of how molecules interact with olfactory receptors according to which receptors use a form of electron spectroscopy to measure the vibrational spectra of molecules.

Weber–Fechner law A law which states that the perceived magnitude of a stimulus is proportional to the logarithm of the intensity of the stimulus.

Weber fraction The minimal increment in a stimulus necessary to perceive a change. It is a constant proportion of the original comparator stimulus.

4.45.1 Introduction

The field of psychophysics deals with describing the relationship between physical stimuli and their resultant perceptual phenomena within a rigorous, quantifiable framework. The field of neuroscience is then tasked with elucidating the genetic, molecular, cellular, and system-level physiological mechanisms that underlie this psychophysical framework. This pattern whereby psychophysics guides physiological research has been central to our understanding of the physiology of vision and audition. For example, psychophysical studies in vision demonstrated that all perceived colors can result from the mixing of just three colors, and that these perceived colors are largely determined by the wavelength of the impinging light. This psychophysical trichromatic model of color perception was the impetus that guided physiologists in their discovery of the three receptor types in the retina that provide the basis for color perception (De Valois, R. L. and De Valois, K. K. 1993).

Similarly, olfactory psychophysics has provided a broad basis for understanding olfaction. Often, however, what is observed about olfactory psychophysics is not the great mass of knowledge, but rather one prominent lack: the core question in olfactory psychophysics, namely, which particular properties of airborne molecules give rise to particular odor qualities, remains unanswered. The reason for this lacuna is not, as we shall see in this chapter, want of industry, but rather the peculiar complexities of both the stimulus and the perceptual response. Sometimes characterized as the most primitive sense, olfaction has also proved itself to be the most stubbornly inaccessible. Undoubtedly, this is in part because the stimulus is more complex than that in other distal senses. In vision, three spatial dimensions, one of intensity and another of wavelength, do much to constrain the stimulus space. In audition a similar number of dimensions are sufficient to characterize the basic properties of the stimulus. In olfaction, the number of stimulus features necessary to characterize the countless discriminable odor molecules may well number in the hundreds, and so olfaction may well be orders of magnitude more intrinsically complex than the other distal senses. Notwithstanding all this, a considerable body of psychophysics exists and provides important constraints on efforts to understand olfactory physiology.

Olfactory perception can be described as a hierarchical pyramid (Savic, I. *et al.*, 2000) building up from a base of olfactory detection, followed by olfactory discrimination, and culminating in olfactory identification. We consider these three levels of processing as hierarchical because of their pattern of exclusivity. One can have olfactory detection without olfactory discrimination and identification, but not vice versa. Likewise, one can have detection and discrimination between two odorants, without odorant identification. Finally, olfactory identification requires olfactory detection, implicitly entails olfactory discrimination, and represents the highest level of olfactory processing whereby an olfactory percept is matched to an olfactory memory and given a semantic label. In this chapter, we will use this hierarchy of olfactory processing as our organizational guide and describe what the psychophysical study of human olfaction has taught us regarding each of these three processing stages.

It is noteworthy that the above functional hierarchy is loosely related to a neuroanatomical structural hierarchy (Savic, I. *et al.*, 2000; Royet, J. P. *et al.*, 2001; Sabri, M. *et al.*, 2005). The first phase in olfactory processing consists of sniffing odorants (reviewed in Mainland, J. and Sobel, N. 2005) that then cross a mucous membrane (reviewed in Lewis, J. E. and Dahl, A. R. 1995) in order to interact with (in humans) ~14 million olfactory receptors of ~400 different types (Gilad, Y. and Lancet, D. 2003) that line the olfactory epithelium. Same receptor types then converge via the olfactory nerve onto unique locations in the olfactory bulb called glomeruli. Following some processing at the level of the olfactory bulb, information is conveyed via the olfactory tract to olfactory cortex, mostly inhabiting the ventral portion of the temporal lobe (reviewed in Price, J. L. 1990). From there, olfactory information is relayed throughout the brain to multiple sites, most notably the limbic system, the orbitofrontal gyri, and insular cortex (reviewed in Sobel, N. *et al.*, 2003).

The process of olfactory detection is largely linked to events at the level of the epithelium in that it reflects a nonspecific increase is system activation beyond noise levels. The process of olfactory discrimination is linked to both events at the epithelium and events in the bulb, in that it reflects specific rather than nonspecific patterns of activity. Discrimination is

also linked to cortex in that it calls upon short-term memory (Wilson, D. A. and Stevenson, R. J. 2003a; 2003b). Finally, the process of olfactory identification is linked to events in the epithelium and bulb, and strongly linked to events in cortex, in that it reflects linking specific patterns of odorant-induced activity to previously stored information such as semantic representations (Haberly, L. B. and Bower, J. M. 1989; Stevenson, R. J. and Boakes, R. A. 2003). Some human lesion studies support this loose link between the structural and functional hierarchy of olfaction. For example, temporal lobe lesions hamper olfactory discrimination and identification, but not detection (Zatorre, R. J. and Jones-Gotman, M. 1991). This link, however, should be treated as a loose framework only, as even the basic process of olfactory detection is undoubtedly subserved by cortical mechanisms of attention and noise suppression.

Although in this chapter we will focus on basic psychophysical mechanisms, it is noteworthy that these basic mechanisms subserve an extraordinary human sense (Shepherd, G. M., 2004). Humans have superb olfaction. For example, it may not surprise many that dogs can recognize humans by their odor (Schoon, G. A. A. and Debruin, J. C., 1994), but it will probably surprise most that reciprocally, humans can identify dogs by their odor (Wells, D. L. and Hepper, P. G., 2000). Similarly, the dominance of olfactory cues in the interaction between a rabbit and its pups is common knowledge (Schaal, B. *et al.*, 2003), but not all are aware that human mothers can identify their babies by smell (Porter, R. H. *et al.*, 1983), and human babies can identify the smell of their breast-feeding mother by 6 days after birth (Macfarelane, A. 1975; Schaal, B. *et al.*, 1980). Finally, that a pig can locate a truffle by its smell (Ackerman, D., 1990) is appreciated by all who like to eat it (the truffle), but that humans can spatially localize an odorant in a laboratory setting (Békésy, G. V. 1964; Porter, J. *et al.*, 2005) (Figure 1), or track an odor trail in a field setting (Porter, J. *et al.*, 2007), is appreciated by only an esoteric few. These are only but a few of the extraordinary olfactory abilities of humans.

Finally, despite the temptation of digressing to relevant data from physiology and functional imaging, as well as data from nonhuman animals, in this chapter we will limit such excursions to the absolute minimum. Here, we concentrate on human olfactory psychophysics only, but we encourage the interested reader to pursue the relevant physiology and imaging literature.

Figure 1 Scent tracking in dogs and humans. The left panel consists of an image published in the 1978 National Geographic Survey on Smell. A dead pheasant was dragged along a field as denoted by the yellow line. A dog with light-emitting diodes on its collar was then released at the far end of the field. The time-lapsed image reveals the dog's tracking behavior. The right panel consists of data from ongoing experiments in our laboratory. The odorant track was made of a mixture of common odorants. All sensory input other than olfaction was prevented by opaque goggles, earmuffs, and thick gloves. A backpack was equipped with monitors that measured subject's nasal respiration in real time and monitored subject location with RTK-GPS offering 1 mm resolution. Although slow, human subjects were surprisingly good at tracking scent trails.

4.45.2 Olfactory Detection

A basic idea of psychophysics is that there is a minimum stimulus quantity that is needed for conscious perception of a sensory event (Herbart, J. F., 1824–1825). In olfaction, estimates of this minimal quantity vary across odorants and even for the same odorant have varied tremendously across laboratories (Laffort, P. 1963; Patte, F. *et al.*, 1975; Van Gemert, L. J. and Nettenbreijer, A. H., 1977; Fazzalari, F. A., 1978; Van Gemert, L. J. 1984). For example, Amoore J. E. and Hautala E. (1983) report 10^6-fold range in detection threshold across laboratories for some odorants. This state of affairs, understatedly described by Devos M. and Patte F. (1990) as disappointing, is mostly the result of several clearly understood extraneous sources of variance having to do with methods of stimulus generation and delivery. Thus, any reading of the olfactory psychophysics literature depends on an understanding of these extraneous sources, which we will first review.

4.45.2.1 Method of Odorant Delivery

4.45.2.1.1 Olfactometers

In an otherwise excellent review of olfactory psychophysics, Engen T. (1973) noted that "[a] great deal has been done recently in instrumentation of the presentation of odorants, enough surely to satisfy those who feel psychophysical research starts with the apparatus." Yet 30 years after this perhaps too optimistic appraisal, we find that there is still no uniform standard device for the delivery of odorants. Odorant delivery methods vary widely across laboratories, and even in those that expend the most effort (and funds) toward odorant generation, there is still no olfactory device on a par with the visual projector or auditory amplifier.

The two most commonly used methods of odorant delivery are either sniffing the headspace over an odorant-filled jar or delivery through a dedicated apparatus usually referred to as an olfactometer, a term coined by Hans Zwaardemaker, one of the first to build such a device (Zwaardemaker, H., 1889). Olfactometers range from simple devices such as squeezable bottles fitted with nose cones, to slightly more complex devices using solenoid valves to control odorized airflow, to sophisticated devices using computer-controlled mass flow controllers (MFCs) that allow for precise control over not only odorant concentration, but also timing, temperature, and humidity (reviewed in Prah, D. *et al.*, 1995). See Figure 2 for an example of such an olfactometer and associated imaging and psychophysics uses.

Regardless of complexity, most olfactometers odorize the airflow by flowing a carrier (clean air or nitrogen) either over an odorant within a canister of some sort or through an odorant-infused sponge-type media, or bubbling the carrier through liquid odorant in a canister. The latter enables obtaining higher overall concentrations of odorized airflows, but carries the risk of generating aerosols within the odorized airflow. Thus, flowing carrier over, rather than through the odorant, generates a far more stable stimulus and is the recommended approach whenever it is sufficient for generating a dynamic range of concentrations. Most current olfactometers are air-dilution olfactometers, which means that they obtain the intended odorant concentration by diluting an odorized airflow with a clean airflow. The ratio of dilution will determine the end concentration. This stands in contrast to simpler olfactometers in which the concentration of the resultant airflow is set by using different concentrations of odorant liquid within the odorant canisters. High-end air-dilution olfactometers also enable mixing different odorized airflows with each other in order to obtain air-dilution odorant mixtures. MFC-equipped olfactometers offer the highest level of control over airflow parameters, enable odorant delivery onset times within the millisecond range, have very high repeatability and stability, and when properly calibrated with analytical devices such as photoionization or flame-ionization detectors, offer accurate control over odorant concentration. Although all the above attributes of high-end

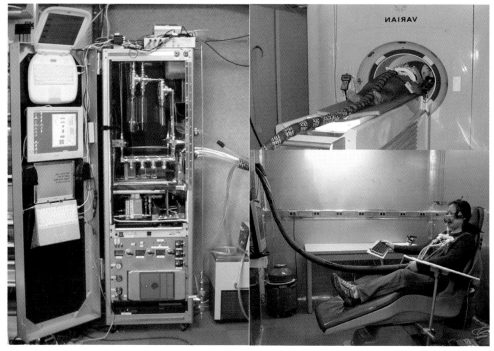

Figure 2 An air-dilution olfactometer. The olfactometer (left panel) is about $6 \times 2 \times 2$ feet and weighs about 400 lbs. Using 13 MFCs and 18 solenoid valves, it delivers odorants to subjects in either the psychophysics laboratory (bottom right panel) or the fMRI scanner (top right panel) while simultaneously measuring airflow in each nostril separately. Although the olfactometer controls odorant temperature, humidity, flow rate, identity (one of five odorants, or any binary mixture thereof), intensity (verified by photoionization), and temporal resolution to the order of $\sim$2 ms, it is still a far cry from a visual projector or auditory amplifier in its accuracy and versatility.

olfactometers render them the likely candidate for generating odorant stimuli when measuring absolute olfactory detection thresholds, these devices also suffer from typical drawbacks that hamper such efforts. Most prominent of these drawbacks is contamination. Within an olfactometer, odorized air passes through tubing before it reaches the sampling port, which may be in the form of a nasal mask, or tubes (canulas) that are inserted into the nostrils. Odorant molecules from ongoing experimental events tend to stick to these tubes, and then present themselves at unwanted times. Such contamination can be minimized by using nonstick tubing materials such as Teflon, glass, and stainless steel, but the contamination cannot be altogether eliminated. Considering the outstanding sensitivity of the human olfactory system, such contamination renders olfactometers a problematic device for the determination of absolute detection thresholds. Olfactometers remain, however, the preferred method of stimulus delivery for other olfactory tasks such as suprathreshold magnitude estimation, olfactory discrimination, identification, and so on.

4.45.2.1.2 *To sniff or not to sniff*

Olfaction is strongly dependent on a sensory-motor component, namely sniffing (Mainland, J. and Sobel, N. 2005). When using an olfactometer, one can eliminate sniffing by injecting the odorant into the nostrils of an otherwise passive subject. Furthermore, the subject can be trained in the behavior of velopharyngeal closure, whereby the soft palate elevates and lateral pharyngeal walls move medially to meet the soft palate to separate oral and nasal cavities, a method developed to reduce sniff-related artifacts that may be present in various methods of recording neural activity (Kobal, G. and Hummel, C., 1988). However, in most psychophysical studies, preventing sniffing significantly alters the behavior under investigation. Sniffs transport approximately 5–10% of the airflow entering the external nares to the epithelium (Keyhani, K. *et al.*, 1997). Sniff airflow velocity influences olfactory detection thresholds (Le Magnen, J., 1945; Laing, D. G., 1983; Sobel, N. *et al.*, 2000a) and olfactory intensity estimates (Rehn, T., 1978; Teghtsoonian, R. and Teghtsoonian, M., 1984; Youngentob, S. L. *et al.*, 1986; Hornung, D. E. *et al.*, 1997). Moreover, the somatosensory stimulation associated with airflow in the nose is represented by neural activity in both human olfactory bulb (Hughes, J. R. *et al.*, 1969) and olfactory cortex (Sobel, N. *et al.*, 1998), and it is necessary for olfactory perception. For example, when an odorant is injected

intravenously into the bloodstream reaching the olfactory epithelium, it is not perceived as an odor. However, if the injection is simultaneously accompanied by a sniff of odorless air, or by a puff of odorless nitrogen into the nose, it is perceived as odor (Engen, T., 1973). In fact, a sniff may be sufficient to generate a rudimentary olfactory percept even in the absence of odorants. For example, when an early-type olfactometer that injected puffs of air into the nose (blast olfactometer) was tested, subjects responded to the blast pressure as an odor, even when odorant was not present (Wenzel, B. M. 1949; 1955). Similarly, when humans try to imagine an odor, they spontaneously sniff (Bensafi, M. *et al.*, 2003b), and if this spontaneous sniffing is prevented, the vividness of their olfactory imagery is reduced (Bensafi, M. *et al.*, 2005). All of this goes to show that sniffing is an integral and essential part of olfaction, and thus, when measuring olfactory behavior, sniffing should be enabled and measured whenever possible, even if it is avoidable through cannulated olfactometry.

4.45.2.1.3 *Jars*

An alternative to olfactometers is odor delivery by jars. Here the odorant is presented to the subject by smelling the headspace over a glass jar containing odorant at a given concentration, or by smelling the resultant airflow from a squeeze bottle similarly containing a fixed concentration of odorant. One can point to several disadvantages of this method in comparison with an olfactometer. In a high-end olfactometer equipped with an analytical device such as photoionization or flame ionization one has a dependable measure of odorant concentration in the vapor smelled by subjects. In contrast, when using jars, one can only estimate odor concentration based on the ideal gas law and Henry's law that assume proportionality of concentration in the headspace above a solution to the molar fraction within the solution. Unfortunately, due to various parameters such as interactions of the odorant with the diluents (e.g., mineral oil), the headspace concentration over a jar does not always match the theoretical expected concentration. Stone H. (1963) analytically measured the concentration of propionic acid within a mineral oil-based dilution series in jars and found that the actual headspace concentration deviated from the expected concentration, as deduced from the ideal gas law and Henry's law, by as much as two orders of magnitude (Stone, H., 1963; Haring, H. G. *et al.*, 1972). This limitation may be overcome by somehow incorporating an analytical sampling protocol from jar headspace during jar delivery. However, besides the technical impracticality of such a solution, it would still not assure a controlled stimulus concentration. This is because factors such as variable distance of jar from nose across trials and across subjects, as well as variable sniffing behavior (Pfaffmann, C. 1951; Cain, W. S. and Engen, T. 1969), will all influence end concentrations. Jars have additional shortcomings in comparison with an olfactometer in the context of threshold testing. Whereas an air-dilution olfactometer can produce a very large number of discrete odorant concentrations, when using jars, one must have a separate jar for each concentration. This either limits the number of concentrations one can test, or requires use of many jars, a cumbersome arrangement indeed. In our experiments, we typically use ~25 jars in the process of a single odorant threshold determination. All that said, despite these and other drawbacks, the issue of olfactometer contamination renders jars the preferred method in our view for testing absolute detection threshold. Even having settled on jars another complication still looms large. When using jars, comparisons across groups (e.g., patients vs. healthy controls) or individuals within a laboratory are possible, but comparisons across laboratories are limited. Different laboratories use different type jars, different ratios in their dilution series, different diluents, and different methods for presentation of jar to nose. A solution that may minimize this variance may be found in two threshold test kits that have recently been made commercially available. One kit uses jars for presentations (The Smell Threshold Test, Sensonics) and the other uses odorant-pens (Sniffin' Sticks, Burghart Medical Technology) (Hummel, T. *et al.*, 1997; Kobal, G. *et al.*, 2000). Although these kits promise some uniformity across laboratories, they are each limited to one odorant, have rather widely spaced concentration intervals, and are expensive. An alternative to the commercial solution is for olfactory psychophysicists to finally agree on standards they will all use. The foundation for such an effort can be found in the work of groups such as Walker and colleagues (Walker, J. C. *et al.*, 2003) or Cometto-Muniz and their colleagues (Abraham, M. H. *et al.*, 2002; Cometto-Muniz, J. E. *et al.*, 2002; 2003), who have meticulously measured detection thresholds for many odorants (Cometto-Muniz, J. E. and Cain, W. S. 1990; Cometto-Muniz, J. E. *et al.*, 1998a; 1998b; 2001; 2002; 2003), and more critically, have described

the relationship between liquid-phase and vapor-phase concentrations for many odorants (Cometto-Muniz, J. E. *et al.*, 2003).

4.45.2.2 The Assumed Threshold Model

Regardless of whether delivering odorants with olfactometers or jars, determination of detection threshold depends on the application of a particular sampling methodology and statistical criterion. In general, one assumes that detection around the threshold concentration conforms to a psychometric function, typically an S-shaped or ogive function wherein stimulus detectability increases continuously from chance to perfect accuracy over a range of stimulus concentrations. Such psychometric functions are prevalent across sensory modalities, and a typical example is shown in Figure 3. In the figure, stimulus intensity (concentration) is plotted on the abscissa and the observer's performance in a two-alternative forced-choice (2AFC) procedure is plotted on the ordinate. In this context, the observer's threshold is defined as some level of detectability, typically midway between chance and perfect performance, corresponding to 75% performance in the figure. Different procedures will define threshold at different levels of performance, accounting for some of the variance in results reported in the literature.

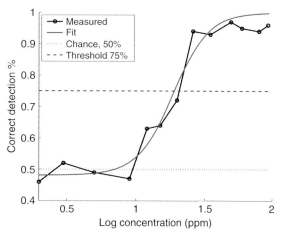

Figure 3 Detection threshold. Detection performance (percent correct in a two-alternative forced-choice (2AFC) task) is plotted as a function of odorant concentration. Raw data (black curve) are fitted with a Weibull psychometric function (red curve) which is used to determine the threshold detection value (value of the red curve at 75% correct performance; in this case ~1.3 log[ppm]). The use of a fitted curve allows more precise determination of thresholds than the raw data, which often has a fixed arbitrary step size.

A second source of variance is observer bias. An observer's response in the context of a threshold test is the product of (at least) two components, a sensory component reflecting the neural activity induced by the stimulus and a psychological component or bias reflecting the observer's predisposition to give one reply or another. This bias is usually different across observers and can also be modified by changing the task structure. For example, bias towards one response or another can change dramatically as a function of task contingencies, that is, the cost associated with making an error versus the benefit of making a correct selection. Methods of signal detection theory have been developed to control the influence of psychological bias on deduced detection thresholds. It is important to note, however, that complete disentanglement of these processes may be impossible, because psychological bias may influence not only the report of a sensory process, but also the sensory process itself. This is particularly true for olfaction that is typified by robust centrifugal projections from central brain structures to the periphery (Luskin, M. B. and Price, J. L., 1983). For example, the expectation of odor modifies patterns of neural activity in the olfactory bulb (Kay, L. M. and Laurent, G., 1999) and olfactory cortex (Zelano, C. *et al.*, 2005); thus, the bias of expectation will modify the sensory process of detection. With these issues in mind, we review methods for determining detection thresholds below.

4.45.2.3   Measuring the Threshold Model

The most widely used methods for measuring olfactory thresholds are the ascending method of limits (AML) (Cometto-Muniz, J. E. and Cain, W. S., 1998) and the adaptive staircase method (Dixon, W. J. and Massey, F. J., 1957; Cornsweet, T. N., 1962). A third method called the method of constant stimuli, which has been used in other sensory modalities, has been rarely used in olfaction, as the relatively large number of trials required by this method combined with the relatively large interstimulus intervals (ISIs) in olfaction make it prohibitive in most situations (Cain, W. S. *et al.*, 1988). In the AML, a sequence of concentrations of the odorant around the probable threshold concentration are prepared and trials are presented in sequence starting from the lowest and proceeding to the highest. Each trial consists of a 2AFC in which subjects are presented with the target odorant mixed in diluent and with just the diluent and are forced to indicate which of the two contained

the odorant. If the subject is correct, another trial at the same concentration is initiated. If the subject is incorrect, the next highest concentration is used. The procedure is terminated when a predetermined number (typically five) (Cometto-Muniz, J. E. and Cain, W. S., 1998) of successive correct trials is achieved. Detection threshold is typically defined as the last concentration used.

The adaptive staircase is similar to the AML in that it too samples a predetermined series of concentrations with a 2AFC per trial procedure but differs in how it samples these concentrations and in its stopping rule. The staircase may be started at any trial, typically one is randomly chosen above the expected threshold, and then a rule is applied after each trial which determines the next concentration to use. In a typical staircase, a one-up two-down rule is used which means that after one incorrect trial the next highest concentration is used, and after two successive correct trials, the next lowest concentration is used. After a single correct trial, the same concentration is sampled again. Typical stopping rules for staircases include a predetermined number of sampling reversals, or a predetermined number of total trials (Linschoten, M. R. *et al.*, 2001). The data from the staircase are used to fit a psychometric function, typically an S-shaped function such as a logistic, cumulative normal, or Weibull function (see Figure 3). Threshold is defined as the concentration derived from the fitted function corresponding to performance halfway between chance and perfect performance, which is 75% correct in the case of a 2AFC. It should be noted that this definition of threshold differs from that used in the AML, which defines threshold as 100% performance, which explains why thresholds using AML are typically higher than those using staircases.

A variant of the adaptive staircase is the maximum likelihood adaptive staircase method (MLASM) in which a psychometric function is fit to all the data collected thus far after each trial, the threshold is estimated from this fitted function, and a concentration close to it is chosen for the next trial (Taylor, M. M. and Creelman, C. D., 1967). A number of procedures of MLASM are available (Linschoten, M. R. *et al.*, 2001). The method described by Linschoten M. R. *et al.* (2001) uses a logistic-function psychometric model and uses a sufficiently small error estimate on the estimated threshold as a stopping criterion. Their method compares favorably in terms of expected stopping time and accuracy of the estimated threshold with a one-up two-down staircase method, and both methods outperform AML. It should be noted that this result is partly due to the different definitions of threshold employed, and because the AML method uses only a few data points rather than the whole set to estimate threshold. Indeed, when the same logistic function is fitted to the data collected from AML, i.e., when all the data collected are used to estimate the threshold rather than just the last few data points, the thresholds derived from the fit agree well with those derived from the staircase and MLASM methods (Linschoten, M. R. *et al.*, 2001). These results suggest that the method of sampling data around the threshold is less important than how this data are analyzed, i.e., when a proper psychometric function is fitted to the data and a common definition of threshold is applied (75% correct), all three methods are about equally useful. Indeed, using the AML procedure, while having the advantage of providing a threshold without recourse to computation, has little justification given the artificially large error of the method, and the trivial effort involved in significantly improving it. When a psychometric function is fit to all the data, the most significant factor in the precision of the estimated threshold is not the method of sampling concentration steps, but the step size between them (Linschoten, M. R. *et al.*, 2001). Considering all of the above, we recommend using the MLASM method and further recommend using the convenient shareware software available for this purpose called MLpest.

4.45.2.4 Measuring Near-Threshold Performance

Although determining a threshold concentration for detection or recognition is the most common question in research, a somewhat different procedure is needed to answer questions about near-threshold detection performance. In questions of absolute detectability, such as for example, whether a subject can truly detect the presence of a compound he or she appears to have no percept for, one might wish to use a procedure that explicitly incorporates a decision behavior model. The simplest and most widely used such procedure is derived from signal detection theory (Tanner, W. P. and Swets, J. A., 1954; Green, D. M. and Swets, J. A., 1966). In this model, the subject is presented with a large number of trials consisting of a single sample that either contains just the diluent or the diluent plus an odorant at some fixed concentration. The subject must indicate

on every trial whether an odorant was present or not (yes or no). This results in four kinds of stimulus-response situations, which are conventionally called hits (stimulus present, yes), misses (stimulus present, no), correct rejections (stimulus absent, no), and false alarms (stimulus absent, yes). In this situation, different subjects may have different decision criteria or biases: a conservative subject may only indicate that the odorant is present only when he or she has high certainty resulting in relatively more misses and fewer false alarms, whereas a less conservative subject will indicate the presence of the odorant with significantly less certainty, resulting in relatively more false alarms and fewer misses. Signal detection theory provides a model of subject behavior that explicitly includes terms for the discriminability of the odorant from the diluent (termed d-prime), and for the subject bias or degree of conservatism (termed beta). These parameters are separable and subject to different manipulations: changes in the concentration of the olfactant will change d-prime, whereas differences in the rewards and penalties associated with correct and incorrect responses will change beta. Thus, subject bias may be factored out when discriminability is the issue, or conversely may be focused upon when it is the question of interest.

An example of the use of signal detection is found in a study by Bremner E. A. *et al.* (2003) in which eight putative nondetectors of the steroidal compound androstenone were tested on 74 yes/no detection trials. Although these putative nondetectors were determined to be insensitive to androstenone in a screening study, d-prime analysis of their yes/no trials revealed that seven of eight putative nondetectors could in fact detect androstenone. A second example from Sobel N. *et al.* (1999) found that eight subjects who reported no odor percept for a variant of the steroidal compound androstodienone were able to detect it in a signal detection experiment.

4.45.2.5 Nonolfactory Detection of Airborne Chemicals

The perception of smell is the result of odorant transduction at several types of nerve-endings, which will often occur in concert (Bojsen-Moller, F., 1975). An odorant that enters the nasal passage can stimulate both cranial nerves I (olfactory) and V (trigeminal), and most known odorants stimulate both. However, sensitivity and specificity are greater in the olfactory than the trigeminal nerve, and the latter conveys primarily nasal irritation (Cain, W. S.

and Murphy, C. L., 1980; Hummel, T., 2000). In this respect, the trigeminal system may serve as a non-specific chemical warning device, rather than assist in olfactory discrimination and identification. There is some debate in the literature as to which odorants are pure olfactants, that is, stimulate the olfactory nerve only. There is, however, general agreement that vanillin is such an odorant (Doty, R. L. *et al.*, 1978; Doty, R. L., 1995). The odorants hydrogen sulfide, phenyl ethyl alcohol (PEA), and decanoic acid are also considered to have minimal trigeminal stimulation, if any (Doty, R. L. *et al.*, 1978; Livermore, A. *et al.*, 1992; Doty, R. L., 1995). By contrast, carbon dioxide can be considered a pure trigeminal in that it stimulates the trigeminal but not olfactory nerve (Hummel, T. *et al.*, 2003). Although most odorants will stimulate both systems, trigeminal thresholds are often several orders of magnitude higher in concentration than olfactory thresholds for the same odorants (Cometto-Muniz, J. E. and Cain, W. S., 1990). Thus, degree of trigeminal activation will have little influence on results of detection threshold studies, but may have significant impact on the results of studies using suprathreshold concentrations.

4.45.2.6 Adaptation

Any consideration of reported olfactory detection thresholds must consider the phenomenon of adaptation. Adaptation, in all sensory systems, is the waning of response with stimulus repetition. Adaptation can be viewed as either a limitation or advantage in sensory processing. It is a limitation in that the adapted sense may be unable to register the occurrence of an event that closely followed a previous, adapting event. By contrast, adaptation functions as gain control within sensory systems. This enables the system to detect changes under conditions of vastly differing background levels. Olfactory adaptation may result from mechanisms at every level of the olfactory process, starting from molecular mechanisms at the level of olfactory receptors, and on to cellular mechanisms at the level of olfactory bulb and cortex (Sobel, N. *et al.*, 2000b; Poellinger, A. *et al.*, 2001). Olfactory adaptation was first demonstrated by Zwaardemaker H. (1895) who found that during constant olfactory stimulation, detection threshold decreases linearly (Pfaffmann, C. 1951). Thus, in the context of determining absolute olfactory detection thresholds, one must consider adaptation in the experimental design. Specifically, an early trial within the threshold test may influence

the outcome of the next trial. To minimize this, one must take the following two steps. Considering that adaptation is increased following exposure to higher concentration odorants (Berglund, B. *et al.*, 1971), one must use ascending concentrations within an olfactory threshold test. Pangborn R. M. *et al.* (1964) tested detection threshold for 2-heptanone three times, once in an ascending concentration order, once in a descending concentration order, and once in random order. Detection thresholds were lowest when tested with ascending concentrations and highest when tested with descending concentrations. A second step is to employ a sufficiently long ISI so as to minimize adaptation effects by allowing full recovery. Although most all published determinations of detection threshold employ ISIs in the range of 30–45 s, Koster E. P. (1971) reports that even ISIs of as long as 120 s do not allow for complete recovery at near-threshold levels. Effects on olfactory detection threshold can also result from the inverse of olfactory adaptation, namely olfactory facilitation. Facilitation consists of lower detection thresholds for one odorant after exposure to another (Engen, T. and Bosack, T. N., 1969). Interestingly, a determining factor on whether or not one odorant will facilitate the detection of another appears to be odorant solubility in water. When a significantly less soluble odorant is presented first, it facilitates the detectability of an ensuing more soluble odorant. By contrast, when a more soluble odorant is presented first, it reduces the detectability of an ensuing less soluble odorant, a phenomenon termed cross-adaptation. For example, using the long- and short-chain homologous alcohols C_3 (high solubility), C_4 (high solubility), and C_7 (low solubility), Corbit T. E. and Engen T. (1971) found that using C_7 as the adapting stimulus facilitated detection of C_3 and C_4, whereas using C_4 as the adapting stimulus inhibited detection of C_3. They hypothesized that these effects reflected the relative rates at which the odorants could cross the olfactory mucosa in order to then bind to receptors. Specifically, they argued that the low-solubility odorant changed the mucosal environment such that the higher-solubility odorants could then cross the mucosa more rapidly.

4.45.2.7 Reported Absolute Thresholds

Considering the above described sources of variance, namely method of odorant delivery, sniffing behavior, statistical analysis, nonolfactory effects, and odorant order effects (adaptation and/or facilitation), it is not surprising that reports on absolute detection thresholds varied so widely across laboratories (Amoore, J. E. and Hautala, E., 1983). However, even data collected within laboratories initially suggested unusually high populational variance in olfactory detection thresholds. For example, detection threshold spans of as much as 5 (Brown, K. S. *et al.*, 1968) and 16 (Yoshida, M., 1984) log units within a homogeneous young and healthy population have been reported. Furthermore, the magnitude of this populational variance appears odorant dependent, significantly greater for some odorants than for others (Punter, P. H., 1983). In a critical study, Stevens and colleagues suggested that much of this variation may be explained by individual variation over time (Stevens, J. C. *et al.*, 1988). They tested three subjects on detection of three odorants over a period of 30 days. Every test consisted of an ascending staircase 2AFC paradigm, with five consecutive correct as criteria (plus an additional five correct for the next higher concentration). They found that the variability of each subject over time was similar to that seen in large groups at one point in time. In other words, they suggested that the dynamic properties of individual olfactory detection thresholds may underlie much of the apparent populational variance. However, when they pooled all their data for each of the three subjects, they found that the subjects were nearly identical. Taken together, these results suggest that the across-test variability reflected the poor power of the single test. Had the single test consisted of the 150 trials that contributed to each point in the pooled data, subjects' performance may have appeared stable. With this issue in mind, Walker J. C. *et al.* (2003) conducted repeated threshold testing using a 75 trial yes/no paradigm that took about 2.3 h to complete. Using such strict criteria, they failed to replicate the variability reported by Stevens J. C. *et al.* and concluded that thresholds are stable when meticulously determined. As concluded in this respect by Cain W. S. and Gent J. F. (1991), there is "little solace to those who may wish to measure quick and dirty olfactory thresholds".

In contrast to all of the above unintended sources of variance, olfactory thresholds are also influenced by various known factors. Principal amongst these is age. Olfactory detection thresholds are typically elevated two- to 15-fold in the healthy elderly in comparison with young adults (Schiffman, S. 1979; Cain, W. S. and Stevens, J. C. 1989; Doty, R. L. *et al.*, 1989; Cain, W. S. and Gent, J. F., 1991; Serby, M. *et al.*,

1991). This reduction in sensitivity is in fact clearly apparent by middle age (Cain, W. S. and Gent, J. F., 1991). In addition to age, detection thresholds are increased by exposure to environmental toxins, diseases that insult the peripheral components of the olfactory system (e.g., rhinitis), developmental diseases that insult the olfactory bulb (e.g., Kalmans), neurodegenerative diseases that may insult both peripheral and central components of the olfactory system (e.g., Parkinson's disease and Alzheimer's disease), and self-injurious behaviors such as excessive smoking or drinking of alcohol. Although reports of gender differences in olfactory detection thresholds have been mixed, a preponderance of evidence points to lower thresholds in women. Early studies by Bailey and colleagues suggested lower thresholds in men than in women (Bailey, E. H. S. and Nichols, E. L., 1884; Bailey, E. H. S. and Powell, L. M., 1885); however, later studies by Toulouse E. and Vaschide N. (1899) suggested the opposite. Still other studies found no difference between men and women in olfactory detection thresholds (Mesolella, V., 1934; Kloek, J., 1961; Matzker, J., 1965; Amoore, J. E. and Venstrom, D., 1966; Venstrom, D. and Amoore, J. E., 1968; Griffiths, N. M. and Patterson, R. L., 1970; Koelega, H. S. 1970; Koelega, H. S. and Koster, E. P., 1974; Dorries, K. M. *et al.*, 1989; Zatorre, R. J. and Jones-Gotman, M. 1990; Stevens, D. A. and O'Connell, R. J., 1991; Segal, N. *et al.*, 1995). In efforts to account for this variability, it has been suggested that a sex difference may be limited to specific odorants (Koelega, H. S. and Koster, E. P., 1974), or apparent only at particular points along the menstrual cycle. This too, however, has been a point of contention, with some studies suggesting lower (better) thresholds at time of ovulation (Le Magnen, J., 1952; Vierling, J. S. and Rock, J., 1967; Mair, R. G. *et al.*, 1978; Doty, R. L. *et al.*, 1981; 1982; Narita, S. *et al.*, 1992; Pause, B. M. *et al.*, 1996; 1999), others at the follicular phase (Henkin, R. I., 1974), others around menses (Koster, E. P., 1968; Doty R. L., 1976), or at menstruation (Le Magnen, J., 1952; Schneider, R. A. and Wolf, S., 1955; Good, P. R. *et al.*, 1976; Mair, R. G. *et al.*, 1978; Moriyama, M. and Kurahashi, T., 2000). Still other studies failed to find any differences across the menstrual cycle in olfactory detection thresholds (Amoore, J. E. *et al.*, 1975b; Herberhold, C. *et al.*, 1982; Filsinger, E. E. and Monte, W. C., 1986; Hummel, T. *et al.*, 1991; Kanamura, S. and Takashima, Y., 1991). Similarly, early pregnancy, which influences various aspects of olfactory perception, does not influence olfactory detection threshold (Kolble, N. *et al.*, 2001). Whereas the bulk of the literature taken together may only imply lower thresholds in women, one recent study may have finally decided this issue. Dalton P. and colleagues (2002) measured detection threshold for benzaldehyde every 2 days, for 60 days. They found that whereas men's detection thresholds maintained a constant mean, women's thresholds decreased dramatically with time (Figure 4, middle panel). They both replicated this finding with an additional odorant (citralva) and demonstrated that it does not generalize to taste thresholds. In other words, whereas men and women had equal detection thresholds at baseline, they did not after training. This convincing study suggests that given the undertrained state of olfaction in most humans, differences between men and women are obscured (as reflected in the bulk of the literature). But once the full potential of the olfactory system is realized through training, women significantly outperform men in olfactory detection thresholds.

Training the olfactory system can serve not only to reduce detection threshold values from one concentration to a lower one, but can in fact also serve to enable detection of a previously completely undetectable odorant. The term specific anosmia is used to describe a state wherein a person with otherwise healthy olfaction is completely unable to detect a particular odorant, at any concentration. Although specific anosmia has been described for several odorants (Amoore, J. E., 1967; Amoore, J. E. *et al.*, 1975a; Amoore, J. E., 1977a; 1977b), it has been primarily studied with the odorant androstenone. Estimates of populational prevalence for androstenone anosmia range from ~2% to ~30% (review in Bremner, E. A. *et al.*, 2003). Considering that androstenone anosmia has a genetic component (Beets, M. G. J. and Theimer, E. T., 1970; Polak, E. H., 1973; Amoore, J. E., 1977a; 1977b), it has been considered as a possible psychophysical tool for probing the relationship between genetic makeup and specific olfactory abilities. One of the chief researchers of this phenomenon is Charles Wysocki. Wysocki, himself a nondetector of androstenone, noticed that after months of research and exposure, he had developed the ability to detect androstenone. In an effort to further explore this serendipitous observation, Wyscoki and colleagues systematically exposed androstenone nondetectors to androstenone (Wysocki, C. J. *et al.*, 1989). They found that repeated exposure induced detectability, which by 6 weeks of exposure was

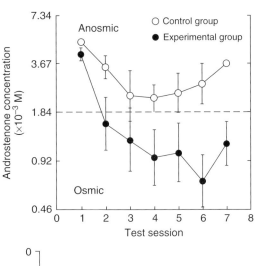

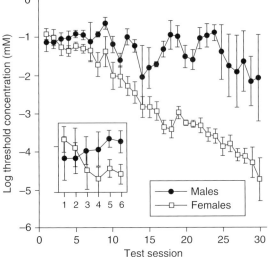

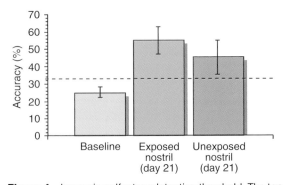

Figure 4 Improving olfactory detection threshold. The top panel depicts reductions in detection thresholds for androstenone during 6 weeks of exposure (Wysocki, C. J. *et al.*, 1989). The middle panel depicts reductions in detection threshold for benzaldehyde induced by 60 days of exposure in women but not men (Dalton, P. *et al.*, 2002). The bottom panel depicts detection of androstenone in the nostril that was exposed to androstenone vs. the unexposed nostril, following 21 days of exposure (Mainland, J. D. *et al.*, 2002).

significantly better than chance, and dramatically different from baseline (Figure 4, top panel).

What mechanisms underlie the plasticity evident in this and other studies where olfactory detection improved through practice? Was the improved performance the result of events in the nose or events in the brain? This question was addressed in experiments on rodents and humans. Yee K. K. and Wysocki C. J. (2001) surgically disconnected the epithelium from bulb in androstenone-anosmic mice. They then systematically exposed the now disconnected epithelium to androstenone over 10 days. When the epithelium-to-bulb connection regenerated 45–50 days later, they tested the mice for androstenone detection. Androstenone detection had improved significantly, leading the authors to suggest that plasticity underlying acquired olfactory abilities was peripheral, in the epithelium. Mainland J. D. *et al.* (2002) addressed the same question in humans. They repeatedly exposed only one nostril of androstenone nondetectors to androstenone, and then tested the unexposed nostril for detection. Following exposure, both the exposed and the naïve nostrils could detect androstenone, effectively doubling their detection accuracy (Figure 4, bottom panel). Since the two olfactory epithelia are not neurally connected at the peripheral level, Mainland J. D. *et al.* concluded that learning occurred via a central brain mechanism, such as olfactory bulb or cortex, which shared information from both nostrils. In contrast, Wang L. *et al.* (2004) conducted a study that led them to the opposite conclusion. These authors repeatedly measured olfactory evoked potentials (EOG) from the epithelium in the nose and olfactory event-related potentials (OERP) from the scalp concomitantly with an androstenone exposure paradigm. They found that increased androstenone detection was associated with changes in both the EOG and OERP. They concluded that a modified EOG necessarily implies plasticity and change at the level of the epithelium. A conclusion consistent with all the data is that exposure may induce changes at both the peripheral and central levels. Repeated exposure may indeed lead to increased expression of receptors at the epithelial level (Yee, K. K. and Wysocki, C. J., 2001), as well as to an increased ability of the brain to make sense of what was a previously senseless message (Mainland, J. D. *et al.*, 2002).

With all the confusion and disagreement regarding human olfactory detection thresholds, there is one fact that can be stated without controversy:

human olfactory detection thresholds are exceedingly low, or in other words, the human olfactory system is a first-rate chemical detector. For example, the odorant ethyl mercaptan that is often added as a warning sign to propane can be detected at concentrations far below 1 part per billion (ppb), typically as low as 0.2 ppb (Whisman, M. *et al.*, 1978). This is equivalent to approximately three drops of odorant within an Olympic-size swimming pool – given two pools, a human could detect which pool contained the three drops of odorant! Finally, a report by the Japan Sanitation Center (Nagata, Y. and Takeuchi, N., 1990) suggests humans can detect isoamyl mercaptan at 0.77 parts per trillion (ppt)! This, to our knowledge, is the lowest reported human detection threshold (Figure 5).

How does this sensitivity compare with other mammals? The few direct within-experiment comparisons of olfaction in humans and other species have focused on tasks other than absolute detection threshold (Slotnick, B. M. and Ptak, J. E., 1977). Matthias Laska and colleagues have amassed detection thresholds across several species and have compared their data to that obtained by others from humans. In one such study, they measured detection threshold for the fox anal gland odorant 2,4,5-tri-methylthiazoline (TMT) in rats, squirrel monkeys, and spider monkeys (Laska, M. *et al.*, 2005a). They found that whereas monkeys had thresholds comparable to those of previously found in humans (in the single ppb range) (Figure 6), rats detected TMT in

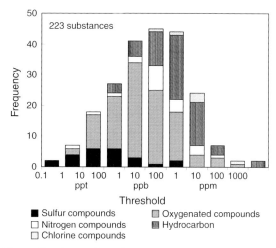

Figure 5 Human detection thresholds. Distribution of detection thresholds for different types of odorants. Humans can detect compounds containing sulfur in the ppt range (Nagata, Y. and Takeuchi, N., 1990).

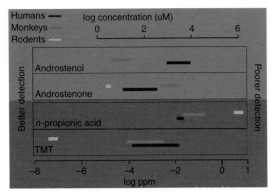

Figure 6 Detection thresholds across species. The data are amassed from studies by Laska and colleagues and are for detection of the fox odor 2,4,5-trimethylthiazoline (TMT), n-propionic acid, and the two steroidal compounds androstenol and androstenone. The two shades of gray distinguish the units in which results were reported, dark gray in log concentration of the vapor, and light gray in log concentration of the odorant liquid. The extent of the lines reflects the reported variance across studies. The important point illustrated is that each species excels at detecting particular odorants. For example, humans outperform rats and monkeys at detecting n-propionic acid. That said, one must keep in mind the limitation of comparing across studies that used different methods of delivery and statistical criteria.

the subpart per trillion (ppt) range (between 0.04 and 0.1 ppt). This is the lowest olfactory detection threshold ever reported for any animal.

This dramatic difference between rats on one hand and primates (including humans) and monkeys on the other seemingly supports the qualitative distinction considering rats and dogs as macrosmatic and primates and monkeys as microsmatic. However, Laska has suggested that this distinction may have been significantly influenced by the behavioral relevance of the odorants tested. Squirrel monkeys and pigtail macaques are better than rats at detecting a host of odorants that are not behaviorally significant to rats. For example, both squirrel monkeys and pigtail macaques outperformed rats by at least an order of magnitude at detecting n-propionic acid and n-pentanoic acid (Laska, M. et al., 2004). Thus, one would like to see a comparison of detection thresholds for an odorant that is behaviorally significant to humans. Unlike the case of rats and TMT, however, humans are not faced with ongoing olfactory detection of predators, other than other humans. Indeed, although not in a predatory context, a host of steroidal compounds have been implicated in olfactory communication between humans (Monti-Bloch, L. et al., 1998; Jacob, S. et al., 2002;

Bensafi, M. et al., 2003a; 2004a; 2004b). Laska and colleagues measured detection thresholds for such compounds in monkeys (Laska, M. et al., 2005b). In comparing their results to those obtained by others, they concluded that androstenone is best detected by mice, less so by monkeys, and even less by humans and pigs. In contrast, androstenol is better detected by humans than by monkeys (no data for mice and pigs). Summing up the efforts of Laska and colleagues, rodents outperform monkeys for some odorants, but monkeys outperform rodents for others. Similarly, monkeys outperform humans for some odorants, but humans outperform monkeys for others – these results are illustrated in Figure 6. Thus, any cross-species comparison of olfactory detection threshold must consider the behavioral relevance of the odorants tested. Furthermore, it is noteworthy that reported detection thresholds in nonhuman animals were obtained after hundreds, sometimes thousands of training trials. By contrast, most human thresholds were obtained after minimal training, if any. Considering the previously noted malleability of olfactory detection, it is tempting to consider the possibility that extensive training may induce/reveal detection thresholds in humans that rival those of macrosmatic mammals.

4.45.3 Olfactory Discrimination

One step up from olfactory detection in the hierarchy of olfaction is olfactory discrimination. Discrimination can be between two concentrations of the same odorant, or between two different odorants, at either the same or different concentrations. In order to discriminate between two odorants (smelled in succession), one must be able to detect them, form a neural representation of the first, and maintain this representation in short-term memory (Wilson, D. A. and Stevenson, R. J. 2003a; 2003b). The nature of this representation, however, is more akin to a sensory memory (Cowan, N., 1995) than a semantic or linguistically mediated memory, and discrimination is generally understood to occur without accessing long-term memory associations with the odorants, or semantic representations of the odorant, processes more properly associated with olfactory identification. The study of olfactory discrimination is susceptible to all the unwanted sources of variance described for the study of detection, namely methods of odorant delivery, sniffing behavior, statistical deduction, and experimental arrangement.

4.45.3.1 Discriminating between Different Concentrations

The simplest case of olfactory discrimination is that between two concentrations of the same odorant. Closely linked to the question of the minimal difference in concentration that can be discriminated is the nature of the relationship between odorant concentration and perception. In general, as the concentration of an odorant increases so does its perceived intensity (it is noteworthy that in some odorants, shifts in concentration are accompanied by shifts in odor quality as well) (Gross-Isseroff, R. and Lancet, D., 1988). The function that relates stimulus magnitude (concentration) to perceived magnitude (intensity) is a kind of psychometric function that can be measured. In this section, we will first describe how this psychometric function is estimated, and then describe how it is related to the issue of intensity discrimination.

The basic method for relating stimulus to perceived magnitude is referred to as magnitude estimation. Typically, many subjects are presented with many trials of an odorant stimulus, each trial at a different concentration. Subjects are asked to rate the intensity of each trial on a linear scale. Rating intensity can be achieved by assigning a number to the stimulus on a Likert scale (typically between 1 for low intensity and 9 for high intensity), or alternatively, by marking a line on a visual-analog scale (VAS). A VAS consists of a horizontal line presented on either paper or monitor, which is also accompanied by verbal descriptors at two or more points along the line. A typical VAS for intensity estimations may have the descriptor "barely detectable" at one end, and "the strongest smell I have ever experienced" at the other. On every trial, subjects cross the line at any location between these two extremes in a manner that reflects their perception. Using number assignment, Stevens S. S. (1957) and others found that the function which best describes the relation of odor concentration to perceived intensity was a simple power function defined as:

$$R = kS^b$$

Where R is perceived intensity, S is stimulus concentration, k is a constant for the logarithmic intercept, and b is a constant for the logarithmic slope. In logarithmic terms, the function is:

$$LogR = b(LogS) + Logk$$

Figure 7 contains typical data from such an experiment. Power functions describe the relationship between perceived magnitude and stimulus magnitude in all sensory modalities including olfaction, and in all cases the slope is less than 1, so that successive increases in stimulus magnitude produce successively smaller and smaller increases in perceived magnitude. One implication of this saturating or compressed response is that doubling of stimulus magnitude will produce less than a doubling in its perceived intensity. Furthermore, the degree of this compression (i.e., the slope parameter) appears to be odorant specific. Cain W. S. (1969) demonstrated that this compression reflected an odorant's solubility in water. As odorant solubility increased, the degree of compression decreased. This, like the previously described effect on facilitation, may be the result of the higher rate at which more soluble odorants can cross the olfactory mucosa in order to bind to olfactory receptors (Mozell, M. M. and Jagodowicz, M., 1973). This predictable link between an odorant's solubility and the slope of it s power function forms, in our view, one of the most successful implementations of olfactory psychophysics in that is fulfils the previously mentioned goal of defining rules that link the physical stimulus to the resultant percept.

With an understanding of how odorant concentration is related to perceived intensity, we turn to the question of how discriminable different concentrations are from one another. The minimal perceptible change

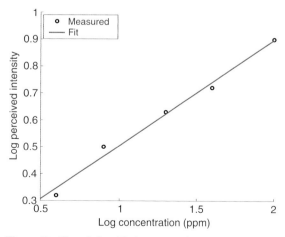

Figure 7 The relationship between odorant concentration and perceived intensity. As odorant concentration increases, so does perceived intensity. When expressing concentration and perception in log units, the relationship can be described by a power function, with a slope that is typically <1.

in stimulus magnitude is known as the just noticeable difference (JND) and follows the law of JND, which states that the minimum increase in a stimulus required for a subject to take notice (symbolized as ΔS) is in constant proportion to the initial stimulus (Fechner, G. T. 1860). This proportion, called the Weber fraction:

$$\Delta S = kS$$

where ΔS is the increment in S (JND), S is the standard stimulus, and k is Weber's fraction, is measured empirically by the method of constant stimuli. This equation implies the Weber–Fechner law:

$$R = k \log(S/S_a)$$

where R is the perceived magnitude and S_a is the absolute threshold for detection of the stimulus. Of course, this law applies only to detectable or suprathreshold stimuli.

In a typical JND experiment, a subject performs many trials in which they smell in rapid succession first a standard and then a comparison stimulus. Different comparison stimuli are used, some equal to the standard, and others differing by fixed proportions. The subject rates the two stimuli as either same or different. Typical Weber fractions for human vision and audition are in the 10% range (Mueller, C. G., 1951). In other words, reliable discrimination is achieved for a 10% increment from the standard. Initial experiments using the method of constant stimuli in olfaction suggested that olfaction indeed follows Weber's law, but that it is characterized by relatively high Weber fractions of between 25% and 33% (Gamble, R. C. 1899). This value was not only significantly higher than human auditory and visual Weber fractions, but also orders of magnitude higher than olfactory Weber fractions obtained in macrosmatic mammals. Slotnick B. M. and Ptak J. E. (1977) used a modified AML procedure to determine Weber fractions for the odorant amyl acetate in rats and humans within the same study. Similar to previous reports, they found human Weber fractions of ∼30%, but rat Weber fractions of ∼4%. However, in a landmark study, Cain W. S. (1977) found that the high Weber fractions reported for humans may have reflected fluctuations in the stimulus. Once the stimulus was carefully controlled, Cain revealed human olfactory Weber fractions of as low as 7% for some odorants (*n*-butyl alcohol), thus grouping olfaction together with vision and audition in its relative sensitivity, and reducing the implied magnitude of difference between humans and macrosmatic mammals in this respect.

4.45.3.2 Discriminating between Different Odorants

In the preceding section, the discriminability of two kinds of odor was parameterized in terms of the only difference between them, namely the concentration of the odorant. Concentration is a numerical scale, which permits us to define a metric on the difference in stimuli, and then relate the difference in perception to this stimulus difference. In comparing the discriminability of different odorants, that is, different species of molecules, there is no obvious metric that defines the extent of differences between them. For example, how different is methane from ammonia? How can we define this as a number? The question of how to define a metric of stimulus similarity against which to gauge discriminability is crucial to understanding olfactory discrimination. A variety of studies have used a variety of different chemical properties of odorants to define metrics of similarity to see how they affect discriminability. The ways in which molecules can change, however, number in the hundreds, and a systematic study of all of them with respect to discriminability, is close to impossible.

Beyond the essential question of how to define stimulus similarity, there are further issues involved in how to quantify discrimination. Discrimination between different odorants can be considered as an absolute binary decision, in which case it can be probed using 2AFC with same or different as answers, or three-alternative forced-choice (3AFC) where subjects select the odorant that is different from the two others (more than two others can also be used). Matthias Laska and colleagues have used such 3AFC paradigms to test discriminability of odorant molecules as a function of various aspects of their physical structure. They found that the greater the difference in carbon chain length between any two odorants, the easier they were to discriminate (Laska, M. and Freyer, D., 1997). Furthermore, they found that humans can discriminate aliphatic odorants equal in number of carbons, but different in oxygen moiety (Laska, M. *et al.*, 2000), and differences in structures that share a benzene ring rather than a linear and unbranched carbon chain (Laska, M., 2002). Laska and colleagues argued that such discrimination data can be used as a tool to probe the dimensionality of odor quality. Specifically, they assume that easier discrimination reflects greater perceptual difference. Thus, because discrimination became easier as the difference in carbon chain length increased, they conclude that carbon chain length reflects a dimension of olfactory perception. Unfortunately, it is generally impossible to vary only a single parameter of variance across a molecule. For example, increasing odorant carbon chain length will decrease odorant solubility, which leads to a diminution in perceived intensity (Cain, W. S., 1969). Thus, when subjects are able to discriminate odorants of different carbon chain length, it is impossible to say whether the ability is due to differences in intensity, solubility, some other attribute, or chain length *per se*.

One class of molecular changes that avoids most of these confounding problems is that of enantiomeric pairs. Enantiomers are mirror-image or chiral molecular structures that under most circumstances will not differ in solubility or many other chemical properties. Humans can discriminate many enantiomers by smell (Jones, F. N. and Velasquez, V., 1974; Jones, F. N. and Elliot, D., 1975). However, some enantiomeric pairs are not easily discriminated (Theimer, E. T. *et al.*, 1977; Laska, M. and Teubner, P. 1999), and still others are easily discriminated whereby one has a distinct odor and the other has no clear olfactory percept at all (Laska, M. and Teubner, P., 1999). Although these experiments reduce the potential sources of variance underlying discrimination, they still fail to link a particular molecular aspect of odorants to a particular olfactory percept or quality.

The overwhelming impression following review of the human olfactory discrimination literature is that humans can discriminate nearly any molecular change in odorant structure. An alternative strategy to probing the olfactory system through discrimination studies is to focus not on the incredible achievements of olfactory discrimination, but rather on those rare instances where the olfactory system fails to discriminate changes in odorant structure. Keller A. and Vosshall L. B. (2004) used discrimination studies to probe what has been termed the vibrational theory of olfaction. Specifically, the most widely accepted theory of olfaction holds that the molecular structure, or shape, of an odorant is the determining factor in its interaction with a receptor and hence in creating a percept. According to this theory, different receptors bind to particular structural elements of the odorant molecule, termed odotopes, creating a combinatorial pattern of neural activity across receptors that is the

basis for the odor percept. A controversial alternative theory of olfaction suggests that olfactory receptors act as biological spectroscopes, encoding the vibrational spectrum of odorant molecules (Turin, L. 1996; 2002). Whereas the vibrational theory predicts that the shape of a molecule does not determine its percept, the simplest version of odotope theory predicts that the vibrational spectrum of a molecule does not determine its percept. These different predictions suggest a simple test of the two theories, namely locating a pair of molecules that are isoteric (same shape), yet whose vibrational spectra differ. The necessary pair of molecules can be made by isotope substitution; for example, substituting deuterium for hydrogen in one version of a molecule will change its vibrational spectrum but leave its shape unchanged. A simple version of odotope theory predicts these two molecules will smell the same, while the vibrational theory predicts they will smell different. In support of the vibrational theory, Luca Turin suggested that deuterated and undeuterated acetophenone smell quite different (Turin, L. 1996; 2002). To further address this, Keller and Vosshall systematically tested discriminability of these two odorants using several methods. They used a number-based similarity scale, and two variants of 3AFC, a triangle-test where subjects were asked to identify the one odorant jar of three that was different (two were indeed the same, either deuterated or undeuterated), and a duo-trio test where two stimuli were presented, and subjects were asked to match one to a third reference. Overall, subjects were unable to discriminate deuterated from undeuterated acetophenone, and Keller and Vosshall concluded that they found no psychophysical support for the vibrational theory.

Given the difficulties of determining how the myriad possible factors that vary across molecules might change perceptual qualities, that is, the difficulties in determining a similarity metric in the stimulus space, one possible strategy is to directly probe the similarity metric in the perceptual space. This is typically done by presenting pairs of stimuli and asking subjects to rate their similarity on a Likert scale or VAS (e.g., the question "how similar are these two odors?", with a line ranging from "not at all" to "identical"). A shortcoming of such scales is that they typical asymptote at both floor and ceiling, making them relatively insensitive for odors that are very similar or dissimilar. This limitation can in part be addressed using performance-based measures of similarity scaling (Wise, P. *et al.*, 2000). For example, in both vision and audition, more difficult discriminations, namely discriminations between more similar stimuli, take longer. Thus, one can measure time-to-discrimination between various olfactory stimuli and assume that stimuli that take longer to discriminate are more similar in terms of olfactory quality (Wise, P. M. and Cain, W. S., 2000).

4.45.3.3 Speed of Olfactory Discrimination

Olfaction is not typically thought of as a rapid process. Indeed, when faced with olfactory discriminations, humans will typically take several sniffs, a sniffing bout, before they settle upon their discrimination judgment. Laing D. G. (1983) compared performance obtained in such natural sniffing bouts to that obtained in a single sniff. He found that for olfactory detection and intensity discrimination, one sniff produced results as good as those from a sniffing bout. In fact, a sniff as short as 400 ms was sufficient for maximal intensity discrimination accuracy. In a second study (Laing, D. G. 1986), subjects were trained to limit sniff duration to the sound of a buzzer. Sniffs lasting 420 ms were sufficient for 90% discrimination accuracy across different odorants. However, when odorant concentration was lowered to near-threshold, there was a time-accuracy trade-off in that shorter sniffs impaired discrimination accuracy. Whereas this experiment by Laing emphasized speed over accuracy, Wise and Cain emphasized accuracy over speed (Wise, P. M. and Cain, W. S., 2000). They compared discrimination times for perceptually similar (difficult comparison) and dissimilar (easy comparison) odorant pairs and found that latency decreased as accuracy increased. In other words, easy comparisons were made faster than difficult comparisons. These seemingly contradictory results (Figure 8) are simply two sides of the same coin, namely the speed-accuracy trade-off, which is affected by bias. If subjects are instructed to be fast, high speed will reflect inaccuracy. If subjects are instructed to be accurate, high speed will reflect accuracy. These contingencies are easily controlled in human experiments through appropriate instructions, but can be a cause for confusion in interpreting animal experiments (see Uchida, N. and Mainen, Z. F., 2003 vs. Abraham, M. H. *et al.*, 2004, compared in Khan, R. M. and Sobel, N., 2004).

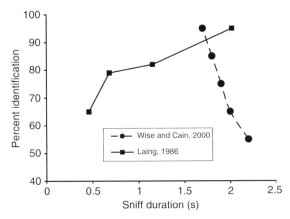

Figure 8 Time-accuracy trade-off. Behavior is modified by bias. Thus, when Laing D. G. (1986) stressed speed, the more subjects rushed, the less correct discriminations they made. In contrast, Wise and Cain stressed accuracy. Thus, when discriminations were easy and subjects were very confident, they answered at once. When discriminations were difficult, they took their time.

Finally, Johnson B. N. *et al.* (2003) used measurements of sniffing as an indirect performance-based measure to assess the latency-to-discrimination of differences in odorant intensity. Specifically, sniff vigor (flow rate, volume, and duration) is inversely proportional to odorant concentration (Laing, D. G. 1983; Warren, D. W. *et al.*, 1994; Sobel, N. *et al.*, 2001; Walker, J. C. *et al.*, 2001). Higher intensity odorants are sampled with reduced vigor, and lower intensity odorants are sampled with increased vigor. Johnson *et al.*, used careful recordings of sniff airflow combined with tight temporal control over odorant delivery. They found that sniffs start off with an extremely stereotyped structure, but are then modulated in accordance with odorant concentration by as early as 160 ms following sniff onset for an odorant that may have a trigeminal component, and 240 ms for a pure olfactant (Figure 9). These values are the lowest we are aware of for olfactory discrimination in humans and suggest that contrary to the common notion, olfaction is fast. Here a nonverbal measure of discrimination suggested that the olfactory system discriminated between two stimuli in less than 200 ms. Considering that odorant transduction alone is expected to account for at least ~150 ms of this latency (Firestein, S. *et al.*, 1990), this result suggests that the cognitive process of olfactory discrimination was extremely rapid.

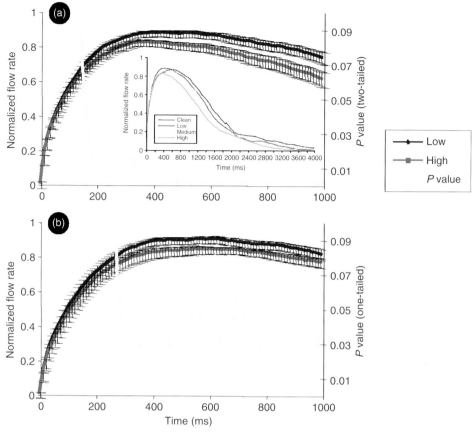

Figure 9 Speed of olfactory discrimination. Sniff magnitude serves as a nonverbal performance-based measure of discrimination. The inset depicts the full mean sniff of 28 subjects sniffing clean air, or one of three intensities of odor. Sniffs clearly separate as a function of odorant concentration. To probe the point of separation, only the first second of the sniff of 28 subjects for high and low concentration of propionic acid (a) and phenyl ethyl alcohol (PEA) (b) are depicted. The yellow line reflects a *t*-test between both traces. Sniff velocity drops when sniffing high as compared with low-concentration propionic acid by 160 ms after sniff onset and 240 ms for PEA.

4.45.4 Olfactory Identification

One step up from olfactory discrimination in the hierarchy of olfaction is olfactory identification. Whereas discrimination calls upon working memory to enable comparison of one odor with another, identification calls upon long-term memory in that in its explicit form, it consists of linking an odorant with a previously stored semantic representation. The study of olfactory identification is methodologically simpler than the study of detection and discrimination because the process of interest is slower and is mostly irrelevant at near-threshold concentrations. Despite this simpler methodological setting, olfactory identification remains the least understood perceptual process in olfaction. A basic goal of the study of olfactory psychophysics is a characterization of the structure of the perceptual space and its relation to

the structure of the stimulus space. In vision, the laws of optics and the nature of light provide powerful constraints on the perceptual space: spatial location in the world is mapped onto spatial location in the retina which is subsequently preserved in subcortical and cortical processing, photon density maps onto brightness, and photon wavelength maps onto perceived color. Similarly in audition, characteristics of vibrations in the air can be mapped onto perceived qualities such as loudness, frequency, timbre, and so on. In olfaction, although there is agreement about the basic stimulus, namely relatively small molecules, there is surprisingly little consensus as to which aspects of the stimulus are connected with which aspects of perceptual experience. Indeed, there is not a perfumist or scientist who could predict the smell of a novel chemical structure, nor the chemical structure of a novel smell. This is unequivocally the major question in the field of olfaction in general and olfactory perception in particular. Regarding this lacuna, Amoore, J. E. asked in 1970 "[h]ow can something *right in front of your nose* still remain a mystery while men are flying to the moon?" (Amoore, J. E., 1970). Now 35 years later, rovers are exploring mars, yet this question remains unanswered.

4.45.4.1 Language and Olfactory Identification

Free identification of odors is a surprisingly difficult task. Unprompted (without the use of multiple choices) odor identification accuracy of generally familiar odors is around 50% (Sumner, D., 1962; Lawless, H. T. and Engen, T., 1977; Cain, W. S., 1979; Engen, T., 1987; de Wijk, R. A. and Cain, W. S., 1994), although it can exceed 85% in experts (Bensafi, M. *et al.*, 2002b). Although odor familiarity can increase identification scores (Homewood, J. and Stevens, R. J., 2001), and discrimination sores (Savic, I. and Berglund, H. 2000), it does not always enable unprompted identification. Often subjects will acknowledge that the odor is familiar, but will be unable to name it. This situation has been named the tip-of-the-nose phenomenon (Lawless, H. T. and Engen, T., 1977), analogous to the tip-of-the-tongue phenomenon in word recall (Brown, R. and McNeill, D., 1966). Richardson J. T. and Zucco G. M. (1989) suggested that this phenomenon may result from subjects' tendency to use a small, impoverished set of verbal labels to encode odor memories, rather than precise and differentiating labels. When trying to recall a label at re-exposure to an odorant, the odor may be recognized, but may not activate a sufficiently precise differentiating label, leading to recognition without identification. To probe this issue, Cain W. S. (1979) trained participants to identify odors before testing them. Teaching subjects labels for odors over five sessions improved identification performance from 60% to 94%. In view of these results, Schab F. R. (1991) suggested that odor identification consists of two components, a perceptual process of odor encoding, and a verbal-semantic process that consists of the activation of potential labels in semantic memory, and the selection of the most appropriate label from those activated. On this account, the poor unprompted identification performance commonly observed might be explained by a limitation in odor encoding, or in the verbal-semantic treatment, or both (Schab, F. R., 1991; de Wijk, R. A., Schab, F. R. and Cain, W. S., 1995). Evidence for such a dual-component system can be seen in the influence of verbal

context on olfactory perception. If the same odorant is presented within a positive versus negative verbal context, such as natural versus artificial, it is perceived as more pleasant (Herz, R. S., 2003). The magnitude of this effect can be significant. For example, Herz R. S. and von Clef J. (2001) presented subjects with a mixture of isovaleric acid and butyric acid labeled as either parmesan cheese or vomit. Using a 9-point pleasantness scale (9 = most pleasant), the same odorant was rated at 5.06 in the former condition and 1.68 in the latter. This result suggests an interesting interaction whereby odor perception is a poor generator of verbal labels (poorer than in other senses), yet verbal labels are a viable modifier of odor percepts (seemingly more so than in other senses).

Although the tip-of-the-nose phenomenon intuitively suggests that poor naming of odorants reflects failures at the level of verbal-semantic processing rather than perceptual processing, some lines of evidence suggest otherwise. For example, odor naming ability is highly correlated with odor discrimination ability, despite the latter's independence of semantic processes (de Wijk, R. A. and Cain, W. S., 1994). Jonsson, F. U. *et al.* (2005) further argued that the tip-of-the-nose phenomenon may not be as prevalent as one might think. These authors dissociated the tip-of-the-nose phenomenon from what they called a feeling of knowing, a sense of general familiarity that they argue reflects poor perceptual encoding of the odor. They presented subjects with 30 common odorants and collected subjective assessments from subjects regarding their experience during attempted identification. These subjective assessments were classified as either reflecting tip-of-the-nose phenomenon or feeling of knowing. They found that in failed identification, 4% of trials were accompanied by strong tip-of-the-nose phenomenon and 24% were accompanied by a strong feeling of knowing. From this, the authors argued that poor perceptual encoding accounts for a greater proportion of failed odor naming than does poor verbal-semantic processing.

The universally poor ability to freely recall odor names has shaped standard tests of olfactory identification, which generally eschew free recall in favor of forced multiple choice (Doty, R. L. *et al.*, 1984a; 1984b; Cain, W. S. *et al.*, 1988; Kobal, G. *et al.*, 1996; Hummel, T. *et al.*, 1997; Nordin, S. *et al.*, 1998). The most widely used of these tests is the University of Pennsylvania smell identification test (UPSIT) (Doty, R. L. *et al.*, 1984a; 1984b). This test consists of 40 microencapsulated odorants that subjects have to scratch and sniff. For each odorant, their task is to choose among four alternative labels the one that best describes each odor (e.g., for the smell of pizza, the labels gasoline, pizza, peanuts, and lilac are proposed). The UPSIT may have some shortcomings as a scientific tool in that it was designed with clinical testing in mind. That said, it remains the only real standard test in olfaction and has proved enormously valuable in characterizing olfactory identification in both health and disease. The UPSIT and similar tests have revealed that olfactory identification is generally better in women than in men (Doty, R. L. *et al.*, 1985), that odor identification performance declines with age, usually following an inverted U-curve, best in young-adults and poorest in children and old-adults (Doty *et al.*, 1984c; Doty, R. L., 1989), and that olfactory identification declines in many diseases, to the extent that it may be diagnostic for some.

4.45.4.2 Identification of an Odor Rather than Its Parts

The nose is commonly thought of as a kind of chemical analyzer. As reviewed elsewhere in this book, specific chemicals are identified by specific receptors in a lock-and-key-type interaction based on receptor configuration and chemical functional groups. Furthermore, these individual chemicals are then represented in specific spatial patterns of activity on the surface of the olfactory bulb. In other words, the olfactory system is studied as a type of analytical tool, such as a chromatograph, which parses and stores information such that subcomponents may be identified from the original stimulus. Some behavioral researchers, however, hold a different view of olfactory processing. They believe that olfaction functions as a synthetic rather than an analytic process (Livermore, A. and Laing, D. G., 1998a; 1998b; Wilson, D. A. and Stevenson, R. J. 2003a; 2003b). By this, they mean that the olfactory system does not reduce the input into recognizable subcomponents but rather that it is the amalgam input that is recognized as a whole. The main line of evidence for this view is from the so-called gestalt aspect of olfaction evidenced in the perception of odor mixtures.

At the simplest level, human subjects are poor at determining whether an odor is a pure odorant (that we will refer to here as a simple odorant) or odorant mixture (that we will refer to here as a complex odorant) (Moskowitz, H. R. and Barbe, C. D., 1977). Furthermore, extensive characterization of odor mixture perception by Laing and colleagues has

repeatedly found that humans cannot correctly identify the individual constituents of a mixture containing more than four or five constituents. In one study (Laing, D. G. and Francis, G. W., 1989), subjects were initially familiarized with seven odorants and their common names. Subjects were then asked to identify which of these seven components were in mixtures presented by an air-dilution olfactometer. Identification accuracy decreased as the number of components in the mixture increased. Performance was only marginally better than chance for four components and at chance for five components (Figure 10). As the actual number of components in the mixtures increased, the number of estimated components increased as well, but lagged behind the true number. Such performance is unlikely to be the result of inadequate experience as expert perfumers revealed very similar results (Livermore, A. and Laing, D. G., 1996). However, component identification is better for familiar than unfamiliar odors (Rabin, M. D., 1988).

Subjects' ability to identify mixture components when constituents were themselves complex odorants was at the same level of performance as when constituents were simple odorants (Livermore, A. and Laing, D. G., 1998a; 1998b). This critical result suggests that odorants were represented as irreducible objects that used equal neural processing capacity, whether they were simple or complex. This irreducibility was further evidenced in that subjects are worse at distinguishing the components of complex odors they had been trained to identify as a single odor, yet when presented with only some of

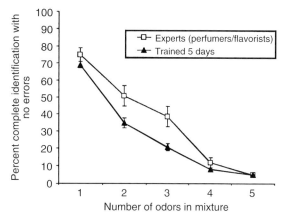

Figure 10 Odor mixture perception. Results from Laing and colleagues demonstrate that both expert perfumers and laypeople alike can identify up to four components of an odorant mixture.

the constituents of a learned odor object, they reported greater perception of qualities normally associated with the other constituents they had previously associated (Stevenson, R. J., 2001a; 2001b). This suggests that the mixed odorant is being represented as a single object, and when a portion of the object is presented the whole object is recalled.

These studies suggested that odorant qualities are blending together to decrease identification of components. To test this, Livermore A. and Laing D. G. (1998a; 1998b) tested mixture component identification using two sets of eight odorants chosen by a panel of expert perfumers as a set of bad blending odorants and a set of good blending odorants. As they predicted, component identification accuracy decreased with mixture complexity regardless of odorant group (good or bad blenders). Subjects were unable to identify more than three or four components in a mixture. However, subjects were significantly better at identifying components for the bad blending odorants than the good blending odorants.

Taken together, these experiments have been presented by some as evidence that olfaction is a synthetic rather than an analytic sense, which is to say that olfaction deals primarily with odor objects rather than odor molecules (Wilson, D. A. and Stevenson, R. J. 2003a; 2003b). It is clear, however, that olfaction, like all other senses, is both analytic and synthetic, and that these terms should be understood not as alternative construals but as descriptive of different levels of sensory processing. Consider an analogous situation in auditory processing. The sensory apparatus (the hair cells on the cochlea) is laid out to encode sound frequency in a linear fashion, and so the analytic basis set for audition is the set of pure tones within the audible range. As it is easy to generate pure tones (e.g., with tuning forks), we have the advantage of being able to probe the system with its analytic atoms. In olfaction, however, the basis set of the system is comprised of the few hundred different olfactory receptor types that are expressed in a particular organism. In general, each receptor responds to a range of molecules, and each molecule reacts with a number of receptors, so that individual molecular species are not the analytic atoms of the system. The set of atoms (or tuning forks) for olfaction would be a set of stimuli each of which reacted with only a single type of olfactory receptor. It is unlikely that many such olfactory stimuli exist. Continuing the comparison to audition, even an odorant composed of a single molecular type is thus analogous to a complex sound, not a pure tone. It is

little surprise, on this view, that identification of complex and simple odors in mixtures is equal. Thus, to say that olfaction deals with olfactory objects rather than olfactory molecules is somewhat to have missed the point: molecules are not the analytic unit for olfaction but are themselves olfactory objects. Reverting to the analog in audition, if a viola is played simultaneously with a violin, a flute, a cello, and a trumpet, the ability of the system to distinguish the components (the particular instruments) from the mixture will be no better than the case of mixtures of molecules, and both may reflect fundamental information processing constraints.

Thus, in both audition and olfaction, the system begins with an analytic basis set, which must then be processed, in the light of much top-down information such as familiarity and learning the use of verbal labels, and attention to perceptual objects. The problem of mixture analysis in olfaction is akin to auditory scene analysis (Alain, C. and Arnott, S. R., 2000; Haykin, S. and Chen, Z., 2005) and should not be understood as a problem peculiar to olfaction. Inasmuch as the labels analytic and synthetic are useful, it is to remind us that olfaction, like all sensory modalities, is a complex, multilayered process.

4.45.4.3 Nostril-Specific Odor Tuning

Almost all odors that humans encounter in their environment are mixtures. Using mixture perception, Sobel N. *et al.* (1999) set out to ask whether the olfactory percept is equal across nostrils. Nostrils can be divided according to two criteria. One is the simple anatomical criterion, namely a left and right nostril. A second criterion is nasal airflow. In mammals, the rate of airflow is usually higher in one nostril than in the other. This occurs because a bilateral highly vascularized structure, the nasal turbinate, swells with blood flow in either one nostril or the other, increasing the resistance to airflow in one nostril in comparison to the other (Principato, J. J. and Ozenberger, J. M., 1970; Bojsen-Moller, F. and Fahrenkrug, J., 1971; Hasegawa, M. and Kern, E. B., 1977) (Figure 11(a)). The nostril with higher airflow, left or right, alternates on an ultradian rhythm of uncertain periodicity (Gilbert, A. N. and Rosenwasser, A. M., 1987; Mirza, N. et *al.*, 1997). Thus, nostrils can be designated either left and right or high flow (HF) and low flow (LF). However, in studies where one nostril was occluded in order to test the other, results regarding asymmetry in olfactory performance were mixed.

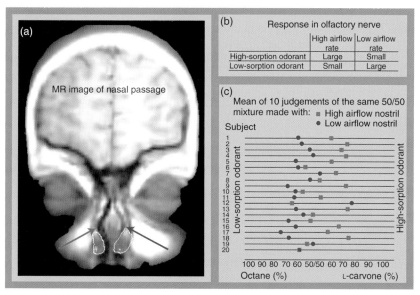

Figure 11 Nostril-specific odor tuning. (a) Magnetic resonance (MR) image of the nasal passage, which appears dark. The swollen (*) and relaxed (#) turbinates, outlined in white, result in an occluded right nostril (red arrow) and a clearer left nostril (green arrow). (b) The interaction between airflow rate and odorant sorption, which brings about a response in the olfactory nerve. (c) On each of ten trials, subjects were asked to smell an identical mixture of 50% octane and 50% L-carvone using either the left or right nostril. They were then given each individual odorant component to smell separately and judged the composition of the mixture by marking the line. Using the HF rate nostril (green), the average judgment was that the mixture consisted of 55% L-carvone and 45% octane. Using the LF rate nostril (red), the judgment was that it consisted of 61% octane and 39% L-carvone ($t(19) = 3.74$, $P = 0.001$).

Although some advantages have been suggested for right versus left nostril (Toulouse, E. and Vaschide, N., 1899; Koelega, H. S., 1979; Youngentob, S. L. *et al.*, 1982; Zatorre, R. J. and Jones-Gotman, M. 1990; Broman, D. A. *et al.*, 2001), and left versus right nostril (Homewood, J. and Stevens, R. J., 2001), no differences were found for HF versus LF nostrils (Eccles, R. *et al.*, 1989; Frye, R. E., 1995). This was paradoxical in view of the previously noted link between nasal airflow velocity and olfactory performance. Because increased velocity improves detection (Le Magnen, J., 1945; Laing, D. G. and Wilcox, M. E., 1983), and olfactory intensity estimates (Rehn, T., 1978; Teghtsoonian, R. and Teghtsoonian, M., 1984; Youngentob, S. L. *et al.*, 1986; Hornung, D. E. *et al.*, 1997), one would expect better olfactory performance in the HF nostril versus the LF nostril. Yet, no such differences were observed (Eccles, R. *et al.*, 1989; Frye, R. E., 1995). This paradox was resolved by Sobel N. *et al.* (2000a) who found that when humans are restricted to sniffing through one nostril only, if it is the LF nostril, they spontaneously increase sniff duration in that nostril so as to equate volume with the HF nostril. In other words, the olfactomotor system generates a compensatory sniff that accounts for the lower airflow in that nostril.

Critically, Sobel N. *et al.*, found that once activation of this compensatory sniffing was prevented, olfactory performance was poorer in the LF nostril compared to the HF nostril (Sobel, N. *et al.*, 2000). Thus, the valid method for comparison of olfactory performance in the HF nostril versus the LF nostril is to assure that both sniff for the same duration (as they would have had one of them not been occluded for testing). This can be achieved by training subjects to match their sniff to either an auditory tone, or on-monitor measurement of nasal airflow (Mainland, J. D. *et al.*, 2005). With this in mind, Sobel N. *et al.* (1999) set out to test the hypothesis that the HF nostril and LF nostril would each be better tuned to a different aspect of an odor mixture. The framework for this hypothesis was set by the pioneering work of Mozell and colleagues, who demonstrated that odorants absorb in, and cross, the olfactory mucosa at different rates (Mozell, M. M. and Jagodowicz, M., 1973; Mozell, M. *et al.*, 1991). They established this phenomenon by measuring the relative sorption rates of 15 odorants across the mucosa of the bullfrog. Mozell and colleagues later found that the effect of an odorant on the magnitude of response in the olfactory nerve of the frog results from an interaction between the particular sorption rate of that odorant and the flow rate at which it flows across the olfactory mucosa: a high-sorption rate odorant will induce a large response when delivered at a high airflow and a smaller response when delivered at a lower airflow. In contrast, a low-sorption rate odorant will induce a small response when delivered at a high airflow and a larger response when delivered at a lower airflow. A theoretical explanation to this is as follows: when a bolus of low-sorption odorant enters the nostril at a slow flow rate, there is a weak vector along the epithelium (slow flow) and a weak vector across the mucus (low sorption). This results in an even distribution of odorant molecules along the epithelium whereby a large epithelial surface area is affected, and the resulting response is large. In turn, when the same bolus of low-sorption odorant is flown rapidly across the mucosa, there is a strong vector along the epithelium (fast flow) and a weak vector across the mucus (low sorption). This results in a proportion of the molecules never sorbing before they are cleared into the respiratory system, an uneven distribution of odorant molecules along the epithelium whereby only the posterior end of the epithelial surface area is affected, and the resulting response is small. In contrast, when a bolus of high-sorption odorant enters the nostril at a slow flow rate, there is a weak vector along the epithelium (slow flow) and a strong vector across the mucus (high sorption). This results in an uneven distribution of odorant molecules along the epithelium whereby the full bolus saturates the anterior portion of the epithelium, a small epithelial surface area is affected, and the resulting response is small. In turn, when the same bolus of high-sorption odorant is flown rapidly across the mucosa, there is a strong vector along the epithelium (fast flow) and a strong vector across the mucus (high sorption) (Figure 11(b)). This results in an even distribution of odorant molecules along the epithelium whereby a large epithelial surface area is affected, and the resulting response is large.

The foregoing findings suggest that particular airflows will optimize perception for particular odorants. High airflows will optimize perception of higher-sorption rate odorants and low airflows will optimize perception of lower-sorption rate odorants. Considering these findings in the frog, one may predict that the difference in airflow rate between the nostrils in the human will result in a different olfactory percept in each nostril as a function of the interaction between airflow and odorant sorption rates. Accordingly, we hypothesized that the high-

airflow nostril is better tuned to high-sorption rate odorants and the low-airflow nostril is better tuned to low-sorption rate odorants.

To test this, 20 subjects performed a task in which on each trial an olfactometer produced an equally proportioned mixture of the low-sorption rate odorant octane and the high-sorption rate odorant L-carvone (Figure 11(c)). Although subjects were deceived and told that the mixtures would be slightly different on every trial, they were actually identical mixtures. The subject then (1) took a monorhinal fixed-duration sniff of the mixture with either the HF rate nostril or LF rate nostril, (2) smelled each component odorant individually, and (3) judged the composition of the mixture on a proportion scale (ranging from 100% octane, to 50/50 octane/L-carvone to 100% L-carvone, as illustrated at bottom of Figure 11(c)). Each subject performed 20 trials with each nostril. All experimental components were counter-balanced to prevent confounds of cross-adaptation and learning.

Although the mixture was always the same, we predicted that when smelling the mixture with the high-airflow nostril, the high-sorption rate odorant would seem more prominent in the mixture, and when smelling the same mixture with the low-airflow nostril, the low-sorption rate odorant would seem more prominent. As predicted, 17 of the 20 subjects judged the same mixture to have a higher L-carvone content (high-sorption-rate odorant) when using the high-airflow nostril and a higher octane content (low-sorption-rate odorant) when using the low-airflow nostril (binomial, $P < .002$) (Figure 11(c)). This finding demonstrated that the olfactory content obtained from each nostril in a given sniff was different and related to sniff airflow. Each nostril was slightly better tuned to odorants that optimally sorb to the mucosa at the current flow rate in that nostril. This finding points to a fundamental commonality across distal sensory systems. In both vision and audition, the two sense organs simultaneously provide a slightly disparate image to the brain. In vision, this disparity is computationally manipulated to provide an additional dimension – depth. In audition, this disparity is computationally manipulated to provide an additional dimension – spatial location of sounds. We expect that the disparity of input across nostrils is similarly computationally manipulated to provide an additional dimension in olfaction. The nature of this olfactory dimension awaits characterization.

4.45.4.4 Identifying Odor Characteristics

Whereas odor identification can prove difficult, identifying particular odor characteristics or attributes is readily achieved. Even when they cannot identify an odor by name, subjects can generate descriptors for an odorant. Wise P. *et al.* (2000) have pointed to some limitations in assessing such behavior, such as (1) such judgments are difficult to verify, (2) different people may use widely different words to describe the same odors. This difference in language use across subjects can make it difficult to assess both interrater and intrarater agreement. Notwithstanding these issues, the analysis of subjective characteristics of odors provides a number of insights into how olfaction may be organized.

The most consistently agreed upon and meaningful characteristic of odors is their pleasantness, or valence as it is sometimes described. Whereas explicit identification, particularly of unfamiliar odors, can be relatively difficult, people can quite readily make judgments of odor pleasantness. Indeed, the primary perceptual aspect subjects use in odorant discrimination studies is odorant pleasantness (Schiffman, S. S. 1974; Godinot, N. and Sicard, G., 1995), and odor-grouping experiments show that pleasantness is the most salient dimension of olfaction (Berglund, B. *et al.*, 1973; Schiffman, S. *et al.*, 1977). Pleasant and unpleasant odors are evaluated at different speeds (Bensafi, M. *et al.*, 2002a) and act differentially on peripheral nervous responses (Jäncke, L. and Kaufmann, N., 1994; Brauchli, P. *et al.*, 1995; AlaouiIsmaili, O. *et al.*, 1997; AlaouiIsmaili, O. *et al.*, 1997; Bensafi, M., 2002b; 2002c; 2002d). Studies of startle reflexes suggest that unpleasant and pleasant smells might be mediated by at least two distinct fundamental emotional systems: an aversive system (associated with negative affects) and an appetitive one (associated with positive affects). According to Lang P. J. *et al.* (1990), activation of the aversive system might increase the magnitude of the startle reflex, while activation of the appetitive system might decrease the magnitude of the startle reflex. Odors have been used to produce positive and negative affective states during which unpleasant odors activate the aversive system and enhance startle reflexes (Miltner, W. *et al.*, 1994; Ehrlichman, H. *et al.*, 1995; 1997), whilst pleasant odors activate the appetitive system and attenuate startle reflexes (Ehrlichman, H. *et al.*, 1997). Startle in the presence of unpleasant odors may be a defensive behavior present early in life: in newborns, differential facial

reactions have been found for pleasant and unpleasant odors (Steiner, J. E., 1979; Soussignan, R. *et al.*, 1997). The primacy of pleasantness judgments is also supported by electrophysiological recordings (Kobal, G. *et al.*, 1992; Pause, B. M. and Krauel, K. 2000; Masago, R. *et al.*, 2001) and functional neuroimaging studies (Zald, D. H. and Pardo, J. V., 1997; Fulbright, R. K. *et al.*, 1998; Crespo-Facorro, B. *et al.*, 2001; Henkin, R. I. and Levy, L. M., 2001; Gottfried, J. A. *et al.*, 2002a; 2002b).

Perceptually, odor valence is tightly linked with odorant intensity. In fact, the inverse correlation between odor intensity and valance is so pronounced that Henion K. E. (1971) suggested odor pleasantness and intensity are a single dimension. Henion based this claim on a study of a single odorant (*n*-amyl acetate). More comprehensive studies on the relationship between odor intensity and pleasantness, however, revealed a complex interaction whereby some odorants display a positive correlation (increased intensity leads to increased pleasantness), some an inverse correlation (increased intensity leads to decreased pleasantness), and yet others a variable inverted U-shaped correlation between pleasantness and intensity (Henion, K. E., 1971; Doty, R. L., 1975; Moskowitz, H. R. *et al.*, 1976; Distel, H. *et al.*, 1999) (Figure 12). The psychophysical data suggesting

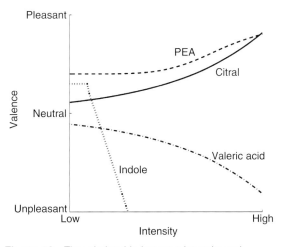

Figure 12 The relationship between intensity and valence. Different odors behave differently. Citral is pleasant, and as you increase its intensity, it becomes more pleasant. Valeric acid is unpleasant, and as you increase its intensity, it becomes more unpleasant. Phenyl ethyl alcohol (PEA) is pleasant, and this valence does not change much across a wide range of intensities, although it eventually does increase. Indole is pleasant at very low intensities, but even a relatively small increase in intensity shifts its percept to one that is extremely unpleasant.

intensity and valence are indeed dissociable constructs received support from neuroimaging studies that found olfactory intensity and valence are subserved by dissociable neural substrates (Gottfried, J. A. *et al.*, 2002b; Anderson, A. K. *et al.*, 2003; Anderson, A. K. and Sobel, N., 2003). Studies of olfactory psychophysics must consider this relationship in their design. For example, if one wants to probe intensity coding, one must keep valence constant, and vice versa.

4.45.4.5 Odor Space

In vision and audition, perceptual experience organizes itself, both through language and, directly, into an ordered space: perceptual events have a distinct relationship to one another that most people acknowledge. In vision, these organizational principles are so basic and ingrained, that they may scarcely need acknowledging: objects exist in a three-dimensional space with all the spatial relational operators, they have colors that can be ordered into a color wheel, so that colors themselves have a kind of spatial relationship. These basic facts arise from basic facts about the world: wavelengths of light, which map onto colors, themselves have an ordering, as does the nature of physical space. In audition, the sensory receptors are organized in the cochlea to map sound frequency onto a linear spatial gradient, and so within auditory perception, the experience of frequency has a linear ordering, namely pitch. That such a latent organization in the olfactory sensory apparatus may exist, and that it may map onto some intrinsic organizational principle within the set of olfactant molecules, has fueled numerous efforts to understand the olfactory odor space. These efforts are of two basic kinds: efforts to understand chemical features of molecules that may map onto perceptual differences and efforts to understand the structure of olfactory perceptual space itself, without reference to chemical properties. These different efforts are, of course, complementary, and the ultimate goal of each is to move from one to the other, so that specific changes in a perceptual space can be related to specific changes in molecular features. We will refer to these two kinds of efforts as stimulus-based maps and perception-based maps. We will first discuss the methods of acquiring feature information from these maps and subsequently discuss mathematical techniques usefully for reconstructing the maps from the information.

4.45.4.5.1 Perception-based maps

Attempts to assess the perceptual structure of olfaction have assumed there is an implicit set of features that are extracted by the nervous system from all odorants, and that these features can therefore be combined to form an olfactory space. The advantages of producing such maps is that from them unknown features of an odorant may be inferred from partial information, they can be used to investigate possible molecular features that correspond to underlying perceptual features, and they provide information about the possible organization of the olfactory nervous system.

All maps are connected to the idea of similarity: objects closer together in the perceptual space are more similar to each other than are objects further apart. Efforts to derive perceptual spaces can thus proceed in either of two ways: they can measure similarity between odorants directly and use this data to derive a map, or they can measure some other features of the objects and infer similarity from them indirectly and thus build a map. These two approaches can be thought of as explicit and implicit measures, depending on whether they measure or derive odorant similarity.

4.45.4.5.2 Explicit similarity methods

One approach to assessing similarity is having subjects group samples of odorants into categories. Odorants placed within the same category are judged to be more similar than odorants that span different categories. Subjects perform either a task where they place each odorant into one of the predefined categories to which it is most similar (Linneaus, C., 1756), or a task where they organize the odorants into any number of categories the subject deems appropriate (MacRae, A. W. *et al.*, 1990; 1992; Stevens, D. A. and O'Connell, R. J., 1996). The first method imposes a structure which may bias subjects and constrains the odor space (Lawless, H. T., 1989), but has the advantage of ensuring consistency of categories across subjects. Unrestricted sorting of odorants permits subjects greater freedom in expressing their perception, but there is greater intersubject variability, and the inconsistency of categories across subjects introduces further difficulties in combining subject data. However, this method appears reliable within subjects (MacRae, A. W. *et al.*, 1990; 1992), and certain differences in sorting might reflect true differences in perception by special subpopulations of subjects (Stevens, D. A. and O'Connell, R. J., 1996).

A second method is to present pairs of odorants and ask subjects to rate the similarity between them. The results from several such studies found considerable variability in judgments between subjects (Dimmick, F. L., 1927; Yoshida, M., 1964a; 1964b) but, as in the unrestricted sorting experiments, these differences may reflect true differences in perception (Dimmick, F. L., 1927; Gregson, R. M., 1972; Davis, R. G., 1979). Analysis of direct similarity scores suggest that in making judgments, subjects generally rely on a hedonic factor and several other factors which may vary across subjects (Yoshida, M., 1972a; 1972b; Moskowitz, H. R. and Gerbers, C. L., 1974; Schiffman, S. S. 1984). Part of this variability may result from a lack of specific instructions for similarity criteria (Wise, P. *et al.*, 2000). While useful for a relatively small set of odorants, the task becomes burdensome as the number of odorants increases, as the number of possible pairwise comparisons increases as the square of the number of odorants.

4.45.4.5.3 *Implicit similarity methods*

A second approach that is feasible for large sets of odorants is to instruct subjects to rate or rank odorants one at a time against a series of descriptors (e.g., verbal labels) (Dravnieks, A. *et al.*, 1978; Dravnieks, A., 1982; Laing, D. G. and Wilcox, M. E., 1983) and then inferring the similarity of pairs of odorants from their responses. This method assumes that similar odorants will have similar feature values, and thus will cluster together in the feature space. Dravnieks A. (1982) demonstrated that responses on a large set of semantic descriptors were stable across 150 subjects from 15 different locations across the United States. The use of a large set of descriptors reduces the concern that the choice of labels will limit the diversity of the method (Thiboud, M., 1991).

There are three large databases of odor category rankings available as commercial products – a strikingly small number considering the importance of such data and the uncertain reliability of these existing products. The largest and most comprehensive is the multivolume/compact-disc set by Steffen Actander. The Sigma-Aldrich Flavors & Fragrances Catalog contains a handful of odorant descriptors for each of 800 + chemicals. The most popular reference for obtaining odorant profiles is the Atlas of odor character profiles, which ranks 160 chemicals against each of the 146 descriptors (Dravnieks, A., 1985).

Once a full set of similarity data (either implicit or explicit) is available for a set of odorants, the organization of the result into a coherent odor space can be

achieved through a variety of statistical methods. A common method for map reconstruction is with multidimensional scaling (MDS) (Kruskal, J. B. and Wish, M. 1978). We will only briefly introduce this method and leave the interested reader to seek out the provided references. The first step of MDS is to create an N by N distance matrix, where there are N odorants and the (i, j)th cell of the matrix contains the distance (the inverse of similarity) between the ith and jth odorants. The goal of MDS is to assign each odorant to a point in a metric space (typically Euclidean) so that the distance between odorants in this coordinate system preserves the distance structure in the original distance matrix (Figure 13). MDS is an iterative optimization algorithm that is often performed many times for high-dimensional spaces to reduce the chances of finding local minima. MDS is essential a data-compression technique, and there are many studies that set out to validate the technique for reducing olfactory perception data. The dimensionality of the constructed space must be specified as a prior constraint.

Using similarity ratings, several researchers found a primary pleasant–unpleasant dimension (Yoshida, M., 1972a; 1972b; Berglund, B. *et al.*, 1973;

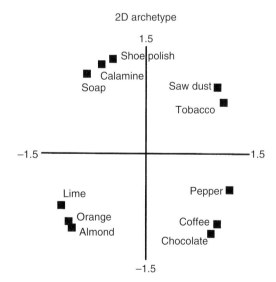

Figure 13 A two-dimensional multidimensional scaling (MDS) derived odor space for some common odors. The map is constructed from the similarity judgments of 110 subjects, so that the distances between odors in the MDS space is inversely proportional to perceived similarity. Although the axes are not meaningfully labeled, the odors cluster into sensible groups from Dawes, P. J. D., Dawes, M. T., *et al.* 2004. The smell map: commonality of odour perception confirmed. Clin. Otolaryngol. 29, 648–654.

Moskowitz, H. R. and Gerbers, C. L., 1974; Yoshida, M., 1975; Schiffman, S. *et al.*, 1977) and a secondary sweet-pungent dimension (Yoshida, M., 1964a; 1964b). The inability to associate more dimensions with perceptual qualities could be due to too few odorants or high subject variability that obscured everything except the most salient dimensions (e.g., hedonic value). Another weakness can be high between-subject variability for the underlying similarity ratings (Berglund, B. *et al.*, 1973). Cunningham M. and Crady C. A. (1971) used bipolar categories (e.g., pleasant or unpleasant or sour or sweet) instead of similarity ratings to probe the olfactory perceptual map and then used MDS to determine the most salient dimensions and found similar results in the salient dimensions.

Carrie S. *et al.* (1999) collected similarity reports for 20 normal and four hyposmic subjects on 11 common odors (i.e., coffee, chocolate, soap, and so on) and found very good agreement for the MDS maps of individual subjects. ALSCAL NMDS was used to compute a subject-weighted archetypal perception map. Individual subjects fit the archetypal map well, while the hyposmic did not. Dawes P. J. D. *et al.* (2004) repeated this experiment but used 110 new observers and found an almost identical MDS solution for the salient dimensions. However, the first three dimensions only accounted for 60% of response variance, suggesting either high variance between subjects, or that the odor space is in fact fairly high dimensional. Another source of this variance could be the known influence of olfactory sensitivity on perceptual quality (Stevens, D. A. and O'Connell, R. J., 1991).

Odor maps are also useful tools for understanding differences in perception across subjects. Stevens D. A. and O'Connell R. J. (1996) examined the perceptual map differences for three groups with differing sensitivity to the odorant pemenone. There are three typical perceptions of pemenone: putrid, nonputrid, and nondetection. Subjects responding in each of these ways were classified as osmic, allosmic, or anosmic, respectively. One hypothesis for these perceptual differences is differences in odorant receptor expression on the olfactory epithelium. Over 20 subjects from each group sorted 15 odorants (five putrid, five floral/fruity, and five wood/vegetable, note these are a priori groups created by the researchers and not based on the subjects) into any number of groups (e.g., free sorting). A similarity metric was created for each pair of odorants that was based on the percentage of subjects placing the pair in the same group. MDS was then used to project the distance matrix onto a three-dimensional space. The MDS created three clusters that reflected the three groups of odorants selected by the researchers (putrid, floral/fruity, and woody/vegetable). Additionally, the location of pemenone differed for each group of subjects. Pemenone was near the putrid cluster for osmics, near the vegetable/woody cluster for allosmics, and near the floral/fruity cluster for anosmics.

Perhaps the most extensive investigation of perceptual maps using MDS is by Chee-Ruiter and Madany-Mamlouk working in the laboratory of Jim Bower (Mamlouk, A. M., 2002; Mamlouk, A. M. *et al.*, 2003; Mamlouk, A. M. and Martinetz, T., 2004). Using the odor quality data in the Aldrich Flavor and Fragrances Catalog, Mamlouk computed the subdimensional distance (a weighted cross-entropy function) to estimate dissimilarity between the 851 odorants. The MDS solution suggested approximately 32 dimensions within the dissimilarity measurements, many more than typically presented in the MDS literature reviewed previously. When the solution was projected to a two-dimensional Euclidean space for visualization, distortions became apparent. Odorants that were not similar in the 32-dimensional solution became similar when collapsed, and some similarities were lost in the projection. In an interesting addition to the analysis, the number of nitrogen and sulfur atoms for each odorant emerged as the third dimension after the two-dimensional projection. The map shows that odorants containing nitrogen or sulfur were grouped together – supporting an eco-olfaction hypothesis in which metabolically relevant odorants are perceived similarly.

Altogether, efforts to understand the structure of the perceptual space reveal that a low dimensional description may be inadequate to capture the complexity of olfactory perception. Ultimately, the state of models of olfactory space is a half-finished project: the completion consists of connecting a space that represents meaningful features of perception with a physiological receptor space that represents the response of olfactory receptor cells to a given odor. This, of course, is no small task but is in fact tantamount to understanding a large part of all processing in the olfactory system. A fuller appreciation of the meaning of olfactory space models will emerge as progress towards this linkage bears fruit in future work.

4.45.4.6 Final Word

In this chapter, we have seen two things: first, that the human olfactory sensory system is an incredibly subtle molecular detector and discriminator; and second, that humans are an animal model particularly well suited for the study of what remains the biggest mystery in olfaction, namely the link between odorant structure and odorant percept. Whereas many other mammals exhibit olfactory detection and discrimination equal or better to that of humans, no other animal can convey its perception through language. Thus, for example, if one wanted to ask what properties of an odorant make it smell more like almonds, one could study nonhuman animals and devise complex behavioral tasks where one could only hope that decisions the animal would make (e.g., error rate or reaction time) might imply to what extent a given stimulus smells like almonds. However, any interpretation of animal behavior in this context is prone to error, as one can never be sure what stimulus features an animal is using in its decision process. This concern is increased due to the correlation between various olfactory features. For example, when studying intensity perception, how can we be sure that an animal is not basing its decisions in a task as a function of odorant valence rather than intensity, because these two dimensions of olfaction are tightly correlated. In contrast, when studying humans, the solution to such constraints is relatively easy. One can simply ask subjects: "which odorant smells more like almonds?". The power afforded by such a simple approach enables probing the more complex and essential questions of the qualities and similarities of odors. For these questions, the human model truly shines. With humans, not only can we learn much more about the percept because of language, but the datum of language use in connection with odors itself can be used to construct odor spaces that correlate with behavior. From the one side, the rules of biochemistry and physics govern how odorants react with olfactory receptors; from the other, there are rules of mind or perception. How they meet is the substance of psychophysics, and although the first bricks have been laid, the house remains to be built.

References

Abraham, M. H., Gola, J. M., et al. 2002. A model for odour thresholds. Chem. Senses 27(2), 95–104.

Abraham, N. M., Spors, H., et al. 2004. Maintaining accuracy at the expense of speed: stimulus similarity defines odor discrimination time in mice. Neuron 44(5), 865–876.

Ackerman, D. 1990. A Natural History of the Senses. Random House.

Alain, C. and Arnott, S. R. 2000. Selectively attending to auditory objects. Front. Biosci. 5, D202–212.

Alaouilsmaili, O., Robin, O., et al. 1997. Basic emotions evoked by odorants: Comparison between autonomic responses and self-evaluation. Physiol. Behav. 62(4), 713–720.

Alaouilsmaili, O., VernetMaury, E., et al. 1997. Odor hedonics: Connection with emotional response estimated by autonomic parameters. Chem. Senses 22(3), 237–248.

Amoore, J. E. 1967. Specific anosmia: a clue to the olfactory code. Nature 214(93), 1095–1098.

Amoore, J. E. 1970. The Molecular Basis of Odor, Charles C Thomas.

Amoore, J. E. 1977a. Specific anosmia and the concept of primary odors. Chem. Senses Flavor 2, 267–281.

Amoore, J. E. 1977b. Specific anosmias to 5-alpha-androst-16-en-3-one and omega-pentadecalactone: the urinous and musky primary odors. Chem. Senses Flavor 2, 401–425.

Amoore, J. E. and Hautala, E. 1983. Odor as an aid to chemical safety: odor thresholds compared with threshold limit values and volatilities for 214 industrial chemicals in air and water dilution. J. Appl. Toxicol. 3(6), 272–290.

Amoore, J. E. and Venstrom, D. 1966. Sensory analysis of odor qualities in terms of the stereochemical theory. J. Food Sci. 31, 118–128.

Amoore, J. E., Forrester, L. J., et al. 1975a. Specific anosmia to 1-pyrroline: the spermous primary odor. J. Chem. Ecol. 1(3), 299–310.

Amoore, J. E., Popplewell, J. R., et al. 1975b. Sensitivity of women to musk odor: no menstrual variation. J. Chem. Ecol. 1(3), 291–297.

Anderson, A. K. and Sobel, N. 2003. Dissociating intensity from valence as sensory inputs to emotion. Neuron 39(4), 581–583.

Anderson, A. K., Christoff, K., et al. 2003. Dissociated neural representations of intensity and valence in human olfaction. Nat. Neurosci. 6(2), 196–202.

Bailey, E. H. S. and Nichols, E. L. 1884. Preliminary notes on the delicacy of the special senses. NY Med. J. 40, 325.

Bailey, E. H. S. and Powell, L. M. 1885. Some special tests in regard to the delicacy of the sense of smell. Trans. Kans. Acad. Sci. 9, 100–101.

Beets, M. G. J. and Theimer, E. T. 1970. Odour Similarity between Structurally Unrelated Odorants. In: Taste and Smell in Vertebrates (eds. G. E. W. Wostenholme and J. Knight), pp. 313–323. J. & A. Churchill.

Békésy, G. V. 1964. Olfactory analogue to directional hearing. J. Appl. Physiol. 19, 369–373.

Bensafi, M., Brown, W. M., et al. 2003a. Sex-steroid derived compounds induce sex-specific effects on autonomic nervous system function in humans. Behav. Neurosci. 117(6), 1125–1134.

Bensafi, M., Brown, W. M., et al. 2004a. Sniffing human sex-steroid derived compounds modulates mood, memory and autonomic nervous system function in specific behavioral contexts. Behav. Brain Res. 152(1), 11–22.

Bensafi, M., Pierson, A., et al. 2002a. Modulation of visual event-related potentials by emotional olfactory stimuli. Neurophysiol. Clin. 32(6), 335–342.

Bensafi, M., Porter, J., et al. 2003b. Olfactomotor activity during imagery mimics that during perception. Nat. Neurosci. 6(11), 1142–1144.

Bensafi, M., Pouliot, S., et al. 2005. Odorant-specific patterns of sniffing during imagery distinguish "bad" and "good" olfactory imagers. Chem. Senses 30, 521–529.

Bensafi, M., Rouby, C., et al. 2002b. Do primacy, recency and experimental amnesia also exist for odor names? Curr. Psychol. Lett.: Behav. Brain Cogn. 8, 23–32.

Bensafi, M., Tsutsui, T., *et al.* 2004b. Sniffing a human sex-steroid derived compound affects mood and autonomic arousal in a dose-dependent manner. Psychoneuroendocrinology 29(10), 1290–1299.

Berglund, B., Berglund, U., *et al.* 1971. The effect of adaptation on odor detection. Percept. Psychophys. 9(5), 435–438.

Berglund, B., Berglund, U., *et al.* 1973. Multidimensional analysis of 21 odors. Scand. J. Psychol. 14(2), 131–137.

Bojsen-Moller, F. 1975. Demonstration of terminalis, olfactory, trigeminal and perivascular nerves in the rat nasal septum. J. Comp. Neurol. 159(2), 245–256.

Bojsen-Moller, F. and Fahrenkrug, J. 1971. Nasal swell-bodies and cyclic changes in the air passage of the rat and rabbit nose. J. Anat. 110(1), 25–37.

Brauchli, P., Ruegg, P. B., *et al.* 1995. Electrocortical and autonomic alteration by administration of a pleasant and an unpleasant odor. Chem. Senses 20(5), 505–515.

Bremner, E. A., Mainland, J. D., *et al.* 2003. The prevalence of androstenone anosmia. Chem. Senses 28(5), 423–432.

Broman, D. A., Olsson, M. J., *et al.* 2001. Lateralization of olfactory cognitive functions: effects of rhinal side of stimulation. Chem. Senses 26(9), 1187–1192.

Brown, R. and McNeill, D. 1966. The "tip-of-the-tongue" phenomenon. J. Verb. Learn. Verb. Behav. 5, 325–337.

Brown, K. S., Maclean, C. M., *et al.* 1968. Distribution of sensitivity to chemical odors in man. Hum. Biol. 40(4), 456–472.

Cain, W. S. 1969. Odor intensity. Differences in exponent of psychophysical function. Percept. Psychophys. 6(6A), 349–354.

Cain, W. S. 1977. Differential sensitivity for smell: "noise" at the nose. Science 195(4280), 796–798.

Cain, W. S. 1979. To know with the nose: keys to odor identification. Science 203(4379), 467–470.

Cain, W. S. and Engen, T. 1969. Olfactory Adaptation and the Scaling of Odor Intensity. International Symposium on Olfaction and Taste III. Academic Press.

Cain, W. S. and Gent, J. F. 1991. Olfactory sensitivity – reliability, generality, and association with aging. J. Exp. Psychol. Hum. Percept. Perform. 17(2), 382–391.

Cain, W. S. and Murphy, C. L. 1980. Interaction between chemoreceptive modalities of odour and irritation. Nature 284(5753), 255–257.

Cain, W. S. and Stevens, J. C. 1989. Uniformity of olfactory loss in aging. Ann. N. Y. Acad. Sci. 561, 29–38.

Cain, W. S., Gent, J. F., *et al.* 1988. Evaluation of olfactory dysfunction in the Connecticut Chemosensory Clinical Research Center. Laryngoscope 98(1), 83–88.

Carrie, S., Scannell, J. W., *et al.* 1999. The smell map: is there a commonality of odour perception? Clin. Otolaryngol. 24(3), 184–189.

Cometto-Muniz, J. E. and Cain, W. S. 1990. Thresholds for odor and nasal pungency. Physiol. Behav. 48(5), 719–725.

Cometto-Muniz, J. E. and Cain, W. S. 1998a. Trigeminal and olfactory sensitivity: comparison of modalities and methods of measurement. Int. Arch. Occup. Environ. Health 71(2), 105–110.

Cometto-Muniz, J. E., Cain, W. S., *et al.* 1998b. Nasal pungency and odor of homologous aldehydes and carboxylic acids. Exp. Brain Res. 118(2), 180–188.

Cometto-Muniz, J. E., Cain, W. S., *et al.* 2001. Ocular and nasal trigeminal detection of butyl acetate and toluene presented singly and in mixtures. Toxicol. Sci. 63(2), 233–244.

Cometto-Muniz, J. E., Cain, W. S., *et al.* 2002. Psychometric functions for the olfactory and trigeminal detectability of butyl acetate and toluene. J. Appl. Toxicol. 22(1), 25–30.

Cometto-Muniz, J. E., Cain, W. S., *et al.* 2003. Quantification of chemical vapors in chemosensory research. Chem Senses 28(6), 467–477.

Corbit, T. E. and Engen, T. 1971. Facilitation of olfactory detection. Percep. Psychophys. 10(6), 433–436.

Cornsweet, T. N. 1962. The staircase-method in psychophysics. Am. J. Psychol. 75, 485–491.

Cowan, N. 1995. Sensory memory and its role in information processing. Electroencephalogr. Clin. Neurophysiol. Suppl. 44, 21–31.

Crespo-Facorro, B., Paradiso, S., *et al.* 2001. Neural mechanisms of anhedonia in schizophrenia: a PET study of response to unpleasant and pleasant odors. JAMA 286(4), 427–435.

Cunningham, M. and Crady, C. A. 1971. Identification of olfactory dimensions by semantic differential technique. Psychon. Sci. 23(6), 387–388.

Dalton, P., Doolittle, N., *et al.* 2002. Gender-specific induction of enhanced sensitivity to odors. Nat. Neurosci. 5(3), 199–200.

Davis, R. G. 1979. Olfactory perceptual space model compared by quantitative methods. Chem. Senses Flavor 4, 21–33.

Dawes, P. J. D., Dawes, M. T., *et al.* 2004. The smell map: commonality of odour perception confirmed. Clin. Otolaryngol. 29, 648–654.

De Valois, R. L. and De Valois, K. K. 1993. A multi-stage color model. Vision Res. 33(8), 1053–1065.

de Wijk, R. A., Schab, F. R., and Cain, W. S. 1995. Odor identification. In: Odor Memory (*eds*. F. R. Schab and R. G. Crowder), pp. 21–37. Erlbaum.

Devos, M. and Patte, F. 1990. Standardized Human Olfactory Thresholds. Oxford University Press.

Dimmick, F. L. 1927. The investigation of the olfactory qualities. Psychol. Rev. 34(5), 321–335.

Distel, H., Ayabe-Kanamura, S., *et al.* 1999. Perception of everyday odors – correlation between intensity, familiarity and strength of hedonic judgement. Chem Senses 24(2), 191–199.

Dixon, W. J. and Massey, F. J. 1957. Introduction to Statistical Analysis. McGraw Hill.

Dorries, K. M., Schmidt, H. J., *et al.* 1989. Changes in sensitivity to the odor of androstenone during adolescence. Dev. Psychobiol. 22(5), 423–435.

Doty, R. L. 1975. An examination of relationships between the pleasantness, intensity and concentration of 10 odorous stimuli. Percept. Psychophysiol. 17, 492–496.

Doty, R. L. 1976. Reproductive Endocrine Influences upon Human Nasal Chemoreception: a Review. In: Mammalian Olfaction, Reproductive Processes and Behavior (*ed*. R. L. Doty), pp. 295–321. Academic Press.

Doty, R. L. 1989. Influence of age and age-related diseases on olfactory function. Ann. N. Y. Acad. Sci. 561, 76–86.

Doty, R. L. 1995. Intranasal Trigeminal Chemoreception: Anatomy, Physiology, and Psychophysics. In: Handbook of Olfaction and Gustation (*ed*. R. L. Doty), pp. 821–834. Marcel Dekker.

Doty, R. L., Brugger, W. E., *et al.* 1978. Intranasal trigeminal stimulation from odorous volatiles: psychometric responses from anosmic and normal humans. Physiol. Behav. 20(2), 175–185.

Doty, R. L., Hall, J. W., *et al.* 1982. Cyclical Changes in Olfactory and Auditory Sensitivity During the Menstrual Cycle: No Attenuation by Oral Contraceptive Medication. In: Olfaction and Endocrine Regulation (*ed*. W. Breipohl), IRL Press.

Doty, R. L., Riklan, M., *et al.* 1989. The olfactory and cognitive deficits of Parkinson's disease: evidence for independence. Ann. Neurol. 25(2), 166–171.

Doty, R. L., Shaman, P., and Dann, M. 1984a. Development of the University of Pennsylvania Smell Identification Test: a standardized microencapsulated test of olfactory function. Physiol. Behav. 32(3), 489–502.

Doty, R. L., Shaman, P., Kimmelman, C. P., and Dann, M. S. 1984b. University of Pennsylvania Smell Identification Test: a rapid quantitative olfactory function test for the clinic. Laryngoscope 94(2 Pt 1), 176–178.

Doty, R. L., Snyder, P. L., *et al.* 1981. Endocrine, cardiovascular, and psychological correlates of olfactory sensitivity changes during the human menstrual cycle. J. Comp. Physiol. Psychol. 95, 45–60.

Dravnieks, A. 1982. Odor quality: semantically generated multi-dimensional profiles are stable. Science 218, 799–801.

Dravnieks, A. 1985. Atlas of Odor Character Profiles. ASTM Press.

Dravnieks, A., Bock, F. C., *et al.* 1978. Comparison of odors directly and through odor profiling. Chem. Senses 3, 191–220.

Eccles, R., Jawad, M. S., *et al.* 1989. Olfactory and trigeminal thresholds and nasal resistance to airflow. Acta Otolaryngol. 108(3–4), 268–273.

Ehrlichman, H., Brown, S., *et al.* 1995. Startle reflex modulation during exposure to pleasant and unpleasant odors. Psychophysiology 32(2), 150–154.

Ehrlichman, H., Kuhl, S. B., *et al.* 1997. Startle reflex modulation by pleasant and unpleasant odors in a between-subjects design. Psychophysiology 34(6), 726–729.

Engen, T. 1973. Sense of smell. Annu. Rev. Psychol. 24, 187–206.

Engen, T. 1987. Remembering odors and their names. Am. Sci. 75, 497–503.

Engen, T. and Bosack, T. N. 1969. Facilitation in olfactory detection. J. Comp. Physiol. Psychol. 68(3), 320–326.

Fazzalari, F. A. 1978. Compilation of Odor and Taste Threshold Values Data. ASTM Data Series DS4 8A.

Fechner, G. T. 1860. Elemente der psychophysik. Breitkopf and Harterl.

Filsinger, E. E. and Monte, W. C. 1986. Sex history, menstrual cycle, and psychophysical ratings of alpha androstenon, a possible human sex phermone. J. Sex Res. 22, 243–248.

Firestein, S., Shepherd, G. M., *et al.* 1990. Time course of the membrane current underlying sensory transduction in salamander olfactory receptor neurones. J. Physiol. 430, 135–158.

Frye, R. E. 1995. Nasal Airway Dynamics and Olfactory Function. In: Handbook of Olfaction and Gustation, Vol. 1 (*ed.* R. L. Doty), pp. 471–491. Marcel Dekker.

Fulbright, R. K., Skudlarski, P., *et al.* 1998. Functional MR imaging of regional brain responses to pleasant and unpleasant odors. AJNR Am. J. Neuroradiol. 19(9), 1721–1726.

Gamble, R. C. 1899. The applicability of Weber's lab to smell. Am. J. Psychol. 10, 82–143.

Gilad, Y. and Lancet, D. 2003. Population differences in the human functional olfactory repertoire. Mol. Biol. Evol. 20(3), 307–314.

Gilbert, A. N. and Rosenwasser, A. M. 1987. Biological rhythmicity of nasal airway patency: a re-examination of the 'nasal cycle'. Acta Otolaryngol. 104(1–2), 180–186.

Godinot, N. and Sicard, G. 1995. Odor categorization by human-subjects – an experimental approach. Chem. Senses 20(1), 101.

Good, P. R., Geary, N., *et al.* 1976. The effect of estrogen on odor detection. Chem. Senses Flavor 2, 45–50.

Gottfried, J. A., Deichmann, R., *et al.* 2002a. Functional heterogeneity in human olfactory cortex: an event-related functional magnetic resonance imaging study. J. Neurosci. 22(24), 10819–10828.

Gottfried, J. A., O'Doherty, J., *et al.* 2002b. Appetitive and aversive olfactory learning in humans studied using event-related functional magnetic resonance imaging. J. Neurosci. 22(24), 10829–10837.

Green, D. M. and Swets, J. A. 1966. Signal Detection Theory and Psychophysics. Wiley.

Gregson, R. M. 1972. Odour similarities and their multidimensional metric representation. Multivar. Behav. Res. 4, 165–175.

Griffiths, N. M. and Patterson, R. L. 1970. Human olfactory responses to 5-alpha-androst-16-en-3-one-principal component of boar taint. J. Sci. Food Agric. 21(1), 4–6.

Gross-Isseroff, R. and Lancet, D. 1988. Concentration-dependent changes of perceived odor quality. Chem. Senses 13(2), 191–204.

Haberly, L. B. and Bower, J. M. 1989. Olfactory cortex: model circuit for study of associative memory? Trends Neurosci. 12(7), 258–264.

Haring, H. G., Boelens, H., *et al.* 1972. Olfactory studies on enantiomeric eremophilane sesquiterpenoids. J. Agric. Food Chem. 20(5), 1018–1021.

Hasegawa, M. and Kern, E. B. 1977. The human nasal cycle. Mayo Clin. Proc. 52(1), 28–34.

Haykin, S. and Chen, Z. 2005. The cocktail party problem. Neural Comput. 17(9), 1875–1902.

Henion, K. E. 1971. Odor pleasantness and intensity: a single dimension? J. Exp. Psychol. 90(2), 275–279.

Henkin, R. I. 1974. Sensory Changes During the Menstrual Cycle. In: Biorhythms and Human Reproduction (*eds.* M. Ferin, F. Halberg, R. M. Richard and R. L. Vandewiele), pp. 277–285. Wiley.

Henkin, R. I. and Levy, L. M. 2001. Lateralization of brain activation to imagination and smell of odors using functional magnetic resonance imaging (fMRI): left hemispheric localization of pleasant and right hemispheric localization of unpleasant odors. J. Comput. Assist. Tomogr. 25(4), 493–514.

Herbart, J. F. 1824–1825. Psychologie als wissenschaft, neu gegrundet auf erfahrung, metaphysik und mathematik, Unzer.

Herberhold, C., Genkin, H., *et al.* 1982. Olfactory Threshold and Hormone Levels During the Human Menstrual Cycle. In: Olfaction and Endocrine Regulation (*ed.* W. Breipohl), pp. 343–351. IRL Press.

Herz, R. S. 2003. The effect of verbal context on olfactory perception. J. Exp. Psychol. Gen. 132(4), 595–606.

Herz, R. S. and von Clef, J. 2001. The influence of verbal labeling on the perception of odors: evidence for olfactory illusions?. Perception 30(3), 381–391.

Homewood, J. and Stevens, R. J. 2001. Differences in naming accuracy of odors presented to the left and right nostrils. Biol. Psychol. 58(1), 65–73.

Hornung, D. E., Chin, C., *et al.* 1997. Effect of nasal dilators on perceived odor intensity. Chem. Senses 22(2), 177–180.

Hughes, J. R., Hendrix, D. E., *et al.* 1969. Correlations Between Electrophysiological Activity from the Human Olfactory Bulb and the Subjective Response to Odoriferous Stimuli. International Symposium on Olfaction and Taste III, Academic Press.

Hummel, T. 2000. Assessment of intranasal trigeminal function. Int. J. Psychophysiol. 36(2), 147–155.

Hummel, T., Gollisch, R., *et al.* 1991. Changes in olfactory perception during the menstrual cycle. Experientia 47(7), 712–715.

Hummel, T., Mohammadian, P., *et al.* 2003. Pain in the trigeminal system: irritation of the nasal mucosa using short- and long-lasting stimuli. Int. J. Psychophysiol. 47(2), 147–158.

Hummel, T., Sekinger, B., *et al.* 1997. 'Sniffing' Sticks: olfactory performance assessed by the combined testing of odor identification, odor discrimination and olfactory threshold. Chem. Senses 22, 39–52.

Jacob, S., McClintock, M. K., *et al.* 2002. Paternally inherited HLA alleles are associated with women's choice of male odor. Nat. Genet. 30(2), 175–179.

Jäncke, L. and Kaufmann, N. 1994. Facial EMG responses to odors in solitude and with an audience. Chem. Senses 19, 99–111.

Johnson, B. N., Mainland, J. D., *et al.* 2003. Rapid olfactory processing implicates subcortical control of an olfactomotor system. J. Neurophysiol 90, 1084–1094.

Jones, F. N. and Elliot, D. 1975. Individual and substance differences in discriminability of optical isomers. Chem. Senses Flavour 1(3), 317–321.

Jones, F. N. and Velasquez, V. 1974. Effect of repeated discriminations on the identifiability of the enantiomers of carvone. Percept. Mot. Skills 38(3), 1001–1002.

Jonsson, F. U., Tchekhova, A., *et al.* 2005. A metamemory perspective on odor naming and identification. Chem. Senses 30, 353–365.

Kanamura, S. and Takashima, Y. 1991. Effect of the menstrual cycle on olfactory sensitivity. Chem. Senses 16, 202–203.

Kay, L. M. and Laurent, G. 1999. Odor- and context-dependent modulation of mitral cell activity in behaving rats. Nat. Neurosci. 2(11), 1003–1009.

Keller, A. and Vosshall, L. B. 2004. A psychophysical test of the vibration theory of olfaction. Nat. Neurosci. 7(4), 337–338.

Keyhani, K., Scherer, P. W., *et al.* 1997. A numerical model of nasal odorant transport for the analysis of human olfaction. J. Theor. Biol. 186(3), 279–301.

Khan, R. M. and Sobel, N. 2004. Neural processing at the speed of smell. Neuron 44(5), 744–747.

Kloek, J. 1961. The smell of some steroid sex-hormones and their metabolites. Psychiatr. Neurol. Neurochir. 64, 309–344.

Kobal, G. and Hummel, C. 1988. Cerebral chemosensory evoked potentials elicited by chemical stimulation of the human olfactory and respiratory nasal mucosa. Electroencephalogr. Clin. Neurophysiol. 71(4), 241–250.

Kobal, G., Hummel, T., Sekinger, B., Barz, S., Roscher, S., and Wolf, S. 1996. "Sniffin' sticks": screening of olfactory performance. Rhinology 34(4), 222–226.

Kobal, G., Hummel, T., *et al.* 1992. Differences in human chemosensory evoked-potentials to olfactory and somatosensory chemical stimuli presented to left and right nostrils. Chem. Senses 17(3), 233–244.

Kobal, G., Klimek, L., *et al.* 2000. Multi-center investigation of 1036 subjects using a standardized method for the assessment of olfactory function combining tests of odor identification, odor discrimination, and olfactory thresholds. Eur. Arch. Otorhnolaryngol. 257, 205–211.

Koelega, H. S. 1970. Extraversion, sex, arousal and olfactory sensitivity. Acta Psychol. (Amst.) 34(1), 51–66.

Koelega, H. S. 1979. Olfaction and sensory asymmetry. Chem. Senses 4, 89–95.

Koelega, H. S. and Koster, E. P. 1974. Some experiments on sex differences in odor perception. Ann. N. Y. Acad. Sci. 237(0), 234–246.

Kolble, N., Hummel, T., *et al.* 2001. Gustatory and olfactory function in the first trimester of pregnancy. Eur. J. Obstet. Gynecol. Reprod. Biol. 99(2), 179–183.

Koster, E. P. 1968. Olfactory sensitivity and the menstrual cycle. Olfactologia 1, 57–64.

Koster, E. P. 1971. In: Adaptation and cross-adaptation in olfaction: an experimental study with olfactory stimuli at low levels of intensity. Utrecht University: p. 211.

Kruskal, J. B. and Wish, M. 1978. Multidemension Scaling, Sage Publications.

Laffort, P. 1963. Essai de standardisation des seuils olfactifs humains pour 192 corps pur. Arch. Sci. Physiol. 17, 75–105.

Laing, D. G. 1983. Natural sniffing gives optimum odour perception for humans. Perception 12(2), 99–117.

Laing, D. G. 1986. Identification of single dissimilar odors is achieved by humans with a single sniff. Physiol. Behav. 37(1), 163–170.

Laing, D. G. and Francis, G. W. 1989. The capacity of humans to identify odors in mixtures. Physiol. Behav. 46(5), 809–814.

Laing, D. G. and Wilcox, M. E. 1983. Perception of components in binary odour mixtures. Chem. Senses 7, 249–264.

Lang, P. J., Bradley, M. M., *et al.* 1990. Emotion, attention, and the startle reflex. Psychol. Rev. 97(3), 377–395.

Laska, M. 2002. Olfactory discrimination ability for aromatic odorants as a function of oxygen moiety. Chem. Senses 27(1), 23–29.

Laska, M. and Freyer, D. 1997. Olfactory discrimination ability for aliphatic esters in squirrel monkeys and humans. Chem. Senses 22(4), 457–465.

Laska, M. and Teubner, P. 1999. Olfactory discrimination ability of human subjects for ten pairs of enantiomers. Chem. Senses 24(2), 161–170.

Laska, M., Ayabe-Kanamura, S., *et al.* 2000. Olfactory discrimination ability for aliphatic odorants as a function of oxygen moiety. Chem. Senses 25(2), 189–197.

Laska, M., Fendt, M., *et al.* 2005a. Detecting danger – or just another odorant? Olfactory sensitivity for the fox odor component 2,4,5-trimethylthiazoline in four species of mammals. Physiol. Behav. 84(2), 211–215.

Laska, M., Wieser, A., *et al.* 2004. Olfactory sensitivity for carboxylic acids in spider monkeys and pigtail macaques. Chem. Senses 29(2), 101–109.

Laska, M., Wieser, A., *et al.* 2005b. Olfactory responsiveness to two odorous steroids in three species of nonhuman primates. Chem. Senses. 30(6), 505–511.

Lawless, H. T. 1989. Exploration of fragrance categories and ambiguous odors using multidimensional-scaling and cluster-analysis. Chem. Senses 14(3), 349–360.

Lawless, H. T. and Engen, T. 1977. Associations to odors: interference, mnemonics, and verbal labeling. J. Exp. Psychol. 3(1), 52–59.

Le Magnen, J. 1945. Etude des facteurs dynamiques de l'excitation olfactive. L'Année Psychologique. 6, 77–89.

Le Magnen, J. 1952. Les phénomènes olfacto-sexuels chez l'homme. Arch. Sci. Physiol. 6, 125–160.

Lewis, J. E. and Dahl, A. R. 1995. Olfactory Mucosa: Composition, Enzymatic Localization, and Metabolism. In: Handbook of Olfaction and Gustation (*ed.* R. L. Doty), pp. 33–52. Marcel Dekker.

Linneaus, C. 1756. Odores medicamentorum. Amoenitates Academicae 3, 183–201.

Linschoten, M. R., Harvey, L. O., Jr., *et al.* 2001. Fast and accurate measurement of taste and smell thresholds using a maximum-likelihood adaptive staircase procedure. Percept. Psychophys. 63(8), 1330–1347.

Livermore, A. and Laing, D. G. 1996. Influence of training and experience on the perception of multicomponent odor mixtures. J. Exp. Psychol. Hum. Percept. Perform. 22(2), 267–277.

Livermore, A. and Laing, D. G. 1998a. The influence of chemical complexity on the perception of multicomponent odor mixtures. Percept. Psychophys. 60(4), 650–661.

Livermore, A. and Laing, D. G. 1998b. The influence of odor type on the discrimination and identification of odorants in multicomponent odor mixtures. Physiol. Behav. 65(2), 311–320.

Livermore, A., Hummel, T., *et al.* 1992. Chemosensory event-related potentials in the investigation of interactions between the olfactory and the somatosensory (trigeminal) systems. Electroencephalogr. Clin. Neurophysiol. 83(3), 201–210.

Luskin, M. B. and Price, J. L. 1983. The topographic organization of associational fibers of the olfactory system in the rat, including centrifugal fibers to the olfactory bulb. J. Comp. Neurol. 216, 264–291.

Macfarelane, A. 1975. Olfaction in the development of social preferences in the human neonate. Ciba Found. Symp. 33, 103–117.

MacRae, A. W., Howgate, P., et al. 1990. Assessing the similarity of odors by sorting and by triadic comparison. Chem. Senses 15(6), 691–699.

MacRae, A. W., Rawcliffe, T., et al. 1992. Patterns of odor similarity among carbonyls and their mixtures. Chem. Senses 17(2), 119–125.

Mainland, J. and Sobel, N. 2006. The sniff is part of the olfactory percept. Chem. Senses 31(2), 181–196.

Mainland, J. D., Bremner, E. A., et al. 2002. Olfactory plasticity: one nostril knows what the other learns. Nature 419(6909), 802.

Mainland, J. D., Johnson, B. N., et al. 2005. Olfactory impairments in patients with unilateral cerebellar lesions are selective to inputs from the contralesion nostril. J. Neurosci. 25(27), 6362–6371.

Mair, R. G., Bouffard, J. A., et al. 1978. Olfactory sensitivity during the menstrual cycle. Sens. Processes 2(2), 90–98.

Mamlouk, A. M. 2002. Quantifying Olfactory Perception. Institute for Signal Processing.

Mamlouk, A. M. and Martinetz, T. 2004. On the dimensions of the olfactory perception space. Neurocomputing 58–60, 1019–1025.

Mamlouk, A. M., Chee-Ruiter, C., et al. 2003. Quantifying olfactory perception: mapping olfactory perception space by using multidimensional scaling and self-organizing maps. Neurocomputing 52–54, 591–597.

Masago, R., Shimomura, Y., et al. 2001. The effects of hedonic properties of odors and attentional modulation on the olfactory event-related potentials. J. Physiol. Anthropol. Appl. Hum. Sci. 20(1), 7–13.

Matzker, J. 1965. Riechen und Lebensalter-Riechen und Rauchen. Achiv. Ohren 185, 755–760.

Mesolella, V. 1934. L'ofatto nelle diverse eta. Arch. Ital. Otol. Rinol. Laring. 46, 43–62.

Miltner, W., Matjak, M., et al. 1994. Emotional qualities of odors and their influence on the startle reflex in humans. Psychophysiology 31(1), 107–110.

Mirza, N., Kroger, H., et al. 1997. Influence of age on the 'nasal cycle'. Laryngoscope 107(1), 62–66.

Monti-Bloch, L., Diaz-Sanchez, V., et al. 1998. Modulation of serum testosterone and autonomic function through stimulation of the male human vomeronasal organ (VNO) with pregna-4,20-diene-3,6-dione. J. Steroid Biochem. Mol. Biol. 65(1–6), 237–242.

Moriyama, M. and Kurahashi, T. 2000. Human olfactory contrast accompanies the menstrual cycle. Chem. Senses 25, 216.

Moskowitz, H. R. and Barbe, C. D. 1977. Profiling of odor components and their mixtures. Sens. Processes 1(3), 212–226.

Moskowitz, H. R. and Gerbers, C. L. 1974. Dimensional salience of odors. Ann. N. Y. Acad. Sci. 237(SEP27), 1–16.

Moskowitz, H. R., Dravnieks, A., et al. 1976. Odor intensity and pleasantness for a diverse set of odorants. Percept. Psychophysiol. 19, 122–128.

Mozell, M., Kent, P., et al. 1991. The effect of flow rate upon the magnitude of the olfactory response differs for different odorants. Chem. Senses 16(6), 631–649.

Mozell, M. M. and Jagodowicz, M. 1973. Chromatographic separation of odorants by the nose: retention times measured across in vivo olfactory mucosa. Science 181(106), 1247–1249.

Mueller, C. G. 1951. Frequency of seeing functions for intensity discrimination of various levels of adapting intensity. J. Gen. Physiol. 34(4), 463–474.

Nagata, Y. and Takeuchi, N. 1990. Measurement of odor threshold by triangle odor bag method. Bull. Jpn. Environ. Sanit. Cent. 17, 77–89.

Narita, S., Koama, T., et al. 1992. Variation of odor sensitivity and reaction time in menstrual cycle. Chem. Senses 17, 97.

Nordin, S., Bramerson, A., Liden, E., and Bende, M. 1998. The Scandinavian Odor-Identification Test: development, reliability, validity and normative data. Acta Otolaryngol. 118(2), 226–234.

Pangborn, R. M., Berg, H. W., et al. 1964. Influence of methodology on olfactory response. Percept. Motor Skills 18(1), 91–103.

Patte, F., Etcheto, M., et al. 1975. Selected and standardized values of suprathreshold odour intensities for 110 substances. Chem. Senses Flavour 1, 283–305.

Pause, B. M. and Krauel, K. 2000. Chemosensory event-related potentials (CSERP) as a key to the psychology of odors. Int. J. Psychophysiol. 36(2), 105–122.

Pause, B. M., Krauel, K., et al. 1999. Perception of HLA-Related Body Odors During the Course of the Menstrual Cycle. In: Advances in Chemical Signals in Vertebrates (eds. R. E. Johnston, D. Muller-Schwarze and P. W. Sorensen), pp. 201–207. Plenum.

Pause, B. M., Sojka, B., et al. 1996. Olfactory information processing during the course of the menstrual cycle. Biol. Psychol. 44(1), 31–54.

Pfaffmann, C. 1951. Taste and Smell. In: Handbook of Experimental Psychology (ed. S. W. Stevens). Wiley.

Poellinger, A., Thomas, R., et al. 2001. Activation and habituation in olfaction – an fMRI study. Neuroimage 13(4), 547–560.

Polak, E. H. 1973. Mutiple profile-multiple receptor site model for vertebrate olfaction. J. Theor. Biol. 40(3), 469–484.

Porter, J., Craven, B., Khan, R. M., Chang, S. J., Kang, I., Judkewitz, B., Volpe, J., Settles, G., and Sobel, N. 2007. Mechanisms of scent-tracking in humans. Nat Neurosci. 10(1), 27–29.

Porter, J., Anand, T., et al. 2005. Brain mechanisms for extracting spatial information from smell. Neuron 47, 581–592.

Porter, R. H., Cernoch, J. M., et al. 1983. Maternal recognition of neonates through olfactory cues. Physiol. Behav. 30(1), 151–154.

Prah, D., Sears, S., et al. 1995. Modern Approaches to Air Dilution Olfactometry. In: Handbook of Olfaction and Gustation, Vol. I (ed. R. L. Doty), pp. 227–257. Marcel Dekker, Inc.

Price, J. L. 1990. Olfactory System. In: The Human Nervous System (ed. G. Paxinos), pp. 979–1001. Academic Press.

Principato, J. J. and Ozenberger, J. M. 1970. Cyclical changes in nasal resistance. Arch. Otolaryngol. 91(1), 71–77.

Punter, P. H. 1983. Measurement of human olfactory thresholds for several groups of structurally related-compounds. Chem. Senses 7(3–4), 215–235.

Rabin, M. D. 1988. Experience facilitates olfactory discrimination and mixture component analysis by humans. Percept. Psychophys. 44(6), 532–540.

Rehn, T. 1978. Perceived odor intensity as a function of air flow through the nose. Sens. Processes 2(3), 198–205.

Richardson, J. T. and Zucco, G. M. 1989. Cognition and olfaction: a review. Psychol. Bull. 105(3), 352–360.

Royet, J. P., Hudry, J., et al. 2001. Functional neuroanatomy of different olfactory judgments. Neuroimage 13(5), 506–519.

Sabri, M., Radnovich, A. J., et al. 2005. Neural correlates of olfactory change detection. Neuroimage 25(3), 969–974.

Savic, I. and Berglund, H. 2000. Right-nostril dominance in discrimination of unfamiliar, but not familiar, odours. Chem. Senses 25(5), 517–523.

Savic, I., Gulyas, B., et al. 2000. Olfactory functions are mediated by parallel and hierarchical processing. Neuron 26(3), 735–745.

Schaal, B., Coureaud, G., et al. 2003. Chemical and behavioural characterization of the rabbit mammary pheromone. Nature 424(6944), 68–72.

Schaal, B., Montagner, H., *et al.* 1980. Olfactory stimulations in mother–child relations. Reprod. Nutr. Dev. 20(3B), 843–858.

Schab, F. R. 1991. Odor memory: taking stock. Psychol. Bull. 109(2), 242–251.

Schiffman, S. 1979. Changes in Taste and Smell with Age: Psychophysical Aspects. In: Sensory Systems and Communications in the Eldery: Aging, Vol. 10 (*eds*. J. M. Ordy and K. Brizzee), pp. 227–246. Raven Press.

Schiffman, S., Robinson, D. E., *et al.* 1977. Multidimensional-scaling of odorants – examination of psychological and physiochemical dimensions. Chem. Senses Flavour 2(3), 375–390.

Schiffman, S. S. 1974. Physicochemical correlates of olfactory quality. Science 185(146), 112–117.

Schiffman, S. S. 1984. Mathematical Approaches for Quantitative Design of Odorants and Tastants. In: Computers in Flavor and Fragrance Research (*eds*. C. Warren and J. Walradt), pp. 33–50. American Chemical Society.

Schneider, R. A. and Wolf, S. 1955. Olfactory perception thresholds for citral utilizing a new type olfactorium. 8, 337–342.

Schoon, G. A. A. and Debruin, J. C. 1994. The ability of dogs to recognize and cross-match human odors. Forensic Sci. Int. 69(2), 111–118.

Segal, N., Topolski, T. D., *et al.* 1995. Twin analysis of odor identification and perception. Physiol. Behav. 57, 605–609.

Serby, M., Larson, P., *et al.* 1991. The nature and course of olfactory deficits in Alzheimer's disease. Am. J. Psychiatr. 148(3), 357–360.

Shepherd, G. M. 2004. The human sense of smell: are we better than we think? PLoS Biol. 2(5), E146.

Slotnick, B. M. and Ptak, J. E. 1977. Olfactory intensity-difference thresholds in rats and humans. Physiol. Behav. 19(6), 795–802.

Sobel, N., Johnson, B. N., *et al.* 2003. Functional Neuroimaging of Human Olfaction. In: Handbook of Olfaction and Gustation (*ed*. R. L. Doty), pp. 251–273. Marcel Dekker, Inc.

Sobel, N., Khan, R. M., *et al.* 1999. The world smells different to each nostril. Nature 402(6757), 35.

Sobel, N., Khan, R. M., *et al.* 2000a. Sniffing longer rather than stronger to maintain olfactory detection threshold. Chem. Senses 25(1), 1–8.

Sobel, N., Prabhakaran, V., *et al.* 1998. Sniffing and smelling: separate subsystems in the human olfactory cortex. Nature 392(6673), 282–286.

Sobel, N., Prabhakaran, V., *et al.* 2000b. Time course of odorant-induced activation in the human primary olfactory cortex. J. Neurophysiol. 83(1), 537–551.

Sobel, N., Thomason, M. E., *et al.* 2001. An impairment in sniffing contributes to the olfactory impairment in Parkinson's disease. Proc. Natl. Acad. Sci. U. S. A. 98(7), 4154–4159.

Soussignan, R., Schaal, B., *et al.* 1997. Facial and autonomic responses to biological and artificial olfactory stimuli in human neonates: re-examining early hedonic discrimination of odors. Physiol. Behav. 62(4), 745–758.

Steiner, J. E. 1979. Human facial expressions in response to taste and smell stimulation. Adv. Child Dev. Behav. 13, 257–295.

Stevens, D. A. and O'Connell, R. J. 1991. Individual differences in thresholds and quality reports of human subjects to various odors. Chem. Senses 16(1), 57–67.

Stevens, D. A. and O'Connell, R. J. 1996. Semantic-free scaling of odor quality. Physiol. Behav. 60(1), 211–215.

Stevens, J. C., Cain, W. S., *et al.* 1988. Variability of olfactory thresholds. Chem. Senses 13(4), 643–653.

Stevens, S. S. 1957. On the psychophysical law. Psychol. Rev. 64(3), 153–181.

Stevenson, R. J. 2001a. The acquisition of odour qualities. Q. J. Exp. Psychol. Section a-Hum. Exp. Psychol. 54(2), 561–577.

Stevenson, R. J. 2001b. Associative learning and odor quality perception: how sniffing an odor mixture can alter the smell of its parts. Learn. Motiv. 32(2), 154–177.

Stevenson, R. J. and Boakes, R. A. 2003. A mnemonic theory of odor perception. Psychol. Rev. 110(2), 340–364.

Stone, H. 1963. Techniques for odor measurement – olfactometric vs. sniffing. J. Food Sci. 28(6), 719–725.

Sumner, D. 1962. On testing the sense of smell. Lancet 2, 895–897.

Tanner, W. P. and Swets, J. A. 1954. A decision-making theory of visual detection. Psychol. Rev. 61(6), 401–409.

Taylor, M. M. and Creelman, C. D. 1967. Pest – efficient estimates on probability functions. J. Acoust. Soc. Am. 41(4P1), 782–787.

Teghtsoonian, R. and Teghtsoonian, M. 1984. Testing a perceptual constancy model for odor strength: the effects of sniff pressure and resistance to sniffing. Perception 13(6), 743–752.

Theimer, E. T., Yoshida, T., and Klaiber, E. M. 1977. Olfaction and molecular shape. Chirality as a requisite for odor. J Agric Food Chem. 25(5), 1168–1177.

Thiboud, M. 1991. Empirical Classification of Odours. In: Perfumes: Art, Science, and Technology (*eds*. Müller P.M. and Lamparsky D.) pp. 253–286. Kluwer Academic Publishers.

Toulouse, E. and Vaschide, N. 1899. Mesure de l'odorat chez l'homme et chez la femme. Comp. Rend. Soc. Biol. 51, 381–383.

Turin, L. 1996. A spectroscopic mechanism for primary olfactory reception. Chem. Senses 21(6), 773–791.

Turin, L. 2002. A method for the calculation of odor character from molecular structure. J. Theor. Biol. 216(3), 367–385.

Uchida, N. and Mainen, Z. F. 2003. Speed and accuracy of olfactory discrimination in the rat. Nat. Neurosci. 6(11), 1224–1229.

Van Gemert, L. J. 1984. Compilations of odour threshold values in air. Supplement V. Report No 84.220 TNO-CIVO Food Analysis Institute.

Van Gemert, L. J. and Nettenbreijer, A. H. 1977. Compilation of odour threshold values in air and water. Central Institute for Nutrition and Food Research TNO and National Institute for Water Supply.

Venstrom, D. and Amoore, J. E. 1968. Olfactory threshold in relation to age, sex, or smoking. J. Food Sci. 33, 264–265.

Vierling, J. S. and Rock, J. 1967. Variations in olfactory sensitivity to exaltolide during the menstrual cycle. J. Appl. Physiol. 22, 311–315.

Walker, J. C., Hall, S. B., *et al.* 2003. Human odor detectability: new methodology used to determine threshold and variation. Chem. Senses 28(9), 817–826.

Walker, J. C., Kendal-Reed, M., *et al.* 2001. Human responses to propionic acid. II. Quantification of breathing responses and their relationship to perception. Chem. Senses 26(4), 351–358.

Wang, L., Chen, L., *et al.* 2004. Evidence for peripheral plasticity in human odour response. J. Physiol. 554, 236–244.

Warren, D. W., Walker, J. C., *et al.* 1994. Effects of odorants and irritants on respiratory behavior. Laryngoscope 104(5 Pt 1), 623–626.

Wells, D. L. and Hepper, P. G. 2000. The discrimination of dog odours by humans. Perception 29(1), 111–115.

Wenzel, B. M. 1949. Differential sensitivity in olfaction. J. Exp. Psychol. 39(2), 129–143.

Wenzel, B. M. 1955. Olfactometric method utilizing natural breathing in an odor-free environment. Science 121(3153), 802–803.

Whisman, M., Goetzinger, J., et al. 1978. Odorant evaluation: a study of ethanethiol and tetrahdrothiophene as warning agents in propane. Environ. Sci. Technol. 12, 1285–1288.

de Wijk, R. A. and Cain, W. S. 1994. Odor quality: discrimination versus free and cued identification. Percept. Psychophys. 56(1), 12–18.

Wilson, D. A. and Stevenson, R. J. 2003a. The fundamental role of memory in olfactory perception. Trends Neurosci. 26(5), 243–247.

Wilson, D. A. and Stevenson, R. J. 2003b. Olfactory perceptual learning: the critical role of memory in odor discrimination. Neurosci. Biobehav. Rev. 27(4), 307–328.

Wise, P., Olsson, M., et al. 2000. Quantification of odor quality. Chem. Senses 25(4), 429–443.

Wise, P. M. and Cain, W. S. 2000. Latency and accuracy of discriminations of odor quality between binary mixtures and their components. Chem. Senses 25(3), 247–265.

Wysocki, C. J., Dorries, K. M., et al. 1989. Ability to perceive androstenone can be acquired by ostensibly anosmic people. Proc. Natl. Acad. Sci. U. S. A. 86(20), 7976–7978.

Yee, K. K. and Wysocki, C. J. 2001. Odorant exposure increases olfactory sensitivity: olfactory epithelium is implicated. Physiol. Behav. 72(5), 705–711.

Yoshida, M. 1964a. Studies in psychometric classification of odors (4). Jpn. Psychol. Res. 6, 115–124.

Yoshida, M. 1964b. Studies in psychometric classification of odors (5). Jpn. Psychol. Res. 6, 145–154.

Yoshida, M. 1972a. Studies in psychometric classification of odors (6). Jpn. Psychol. Res. 14(2), 70–86.

Yoshida, M. 1972b. Studies in psychometric classification of odors (7). Jpn. Psychol. Res. 14(3), 101–108.

Yoshida, M. 1975. Psychometric classification of odors. Chem. Senses Flavour 1(4), 443–464.

Yoshida, M. 1984. Correlation analysis of detection threshold data for 'standard test' odors. Bull. Fac. Sci. Eng. 27, 343–353.

Youngentob, S. L., Kurtz, D. B., et al. 1982. Olfactory sensitivity: is there laterality? Chem. Senses 7(1), 11–21.

Youngentob, S. L. M., Stern, N. et al. 1986. Effect of airway resistance on perceived odor intensity. Am. J. Otolaryngol. 7(3), 187–193.

Zald, D. H. and Pardo, J. V. 1997. Emotion, olfaction, and the human amygdala: amygdala activation during aversive olfactory stimulation. Proc. Natl. Acad. Sci. U. S. A. 94(8), 4119–4124.

Zatorre, R. J. and Jones-Gotman, M. 1990. Right-nostril advantage for discrimination of odors. Percept. Psychophys. 47(6), 526–531.

Zatorre, R. J. and Jones-Gotman, M. 1991. Human olfactory discrimination after unilateral frontal or temporal lobectomy. Brain 114(Pt 1A), 71–84.

Zelano, C., Bensafi, M., et al. 2005. Attentional modulation in human primary olfactory cortex. Nat. Neurosci. 8(1), 114–120.

Zwaardemaker, H. 1889. On measurement of the sense of smell in clinical examination. Lancet i, 1300–1302.

Zwaardemaker, H. 1895. Die physiologie des geruchs. Wengelmann.

Relevant Websites

http://psych.colorado.edu – MLpest statistical thresholds.

http://shop.store.yahoo.com – Multivolume/compact-disc set of odor profiles by Steffen Actander.

http://www.sigmaaldrich.com – The Sigma-Aldrich Flavors & Fragrances Catalog.

http://www.smelltest.com – The Smell Threshold Test and the UPSIT, Sensonics;

http://www.burghart.net – Sniffin' Sticks, Burghart Medical Technology.

4.46 Disorders of Taste and Smell

R L Doty and K Saito, University of Pennsylvania, Philadelphia, PA, USA

S M Bromley, University of Pennsylvania, Philadelphia, PA, USA, UMDNJ-Robert Wood Johnson Medical School, Camden, NJ, USA

Glossary

accessory nerve The 11th cranial nerve (CN XI). This nerve controls specific muscles of the neck, namely the sternocleidomastoid muscle and the upper part of the trapezius muscle.

Addison's disease A rare endocrine disorder, also termed chronic adrenal insufficiency or hypocorticism, in which the body produces insufficient amounts of adrenal steroid hormones (glucocorticoids and often mineralocorticoids).

adrencorticotropic hormone A hormone, also termed ACTH or corticotrophin, produced and secreted by the pituitary gland. This hormone stimulates the cortex of the cortex of the adrenal gland and boosts the synthesis of corticosteroids, mainly glucocorticoids, but also adrenal sex steroids (androgens).

adrenocortical insufficiency The inability of the adrenal gland to produce adequate amounts of cortisol in response to stress. Also termed hypocortisolism.

agenesis The failure of an organ to develop during embryonic growth and development. Agenesis of key organs such as the heart and brain is usually fatal. However, for those organs which occur in pairs, the loss of one member (i.e., unilateral agenesis) is not.

ageusia The loss of taste functions of the tongue, particularly the inability to detect sweetness, sourness, bitterness, saltiness, and umami (the taste of monosodium glutamate).

albright hereditary osteodystrophy A disorder, also known as Martin–Albright syndrome, characterized by a lack of renal responsiveness to parathyroid hormone, resulting in low serum calcium, high serum phosphate, and high serum parathyroid hormone. Patients with this disorder have short stature, characteristically shortened fourth and fifth metacarpals, and rounded facies.

Alzheimer's disease The most common cause of dementia. Characterized by progressive cognitive deterioration, mainly loss of memory, together with declining activities of daily living and neuropsychiatric symptoms or behavioral changes. Neuronal loss or atrophy occurs largely in the temporoparietal cortex but also in the frontal cortex. Pathologically defined by abnormal deposition of amyloid plaques and neurofibrillary tangles.

amyotrophic lateral sclerosis (ALS; also known as Lou Gehrig's Disease) A progressive and ultimately fatal neurodegenerative disease caused by the gradual degeneration of CNS nerve cells that control voluntary muscle movement. Both the upper motor neurons and the lower motor neurons degenerate or die, ceasing to send messages to muscles. The muscles gradually weaken, waste away (atrophy), and have fasciculations (twitches) because of denervation. Eventually, the brain completely loses its ability to initiate and control voluntary movement.

aneurysm A localized, blood-filled dilation (bulge) of a blood vessel caused by disease or weakening of the vessel wall. Cerebral aneurysms most commonly occur in arteries at the base of the brain (the circle of Willis).

angiotensin-converting enzyme An enzyme that catalyzes the conversion of angiotensin I to angiotensin II, a potent vasoconstrictor, and participates in the inactivation of bradykinin, a potent vasodilator. These two actions make it an ideal target in the treatment of conditions such as high blood pressure, heart failure, diabetic nephropathy, and type 2 diabetes mellitus.

anorexia nervosa An eating disorder characterized by low body weight and body image distortion with an obsessive fear of gaining weight. Individuals with anorexia often control body weight by voluntary starvation, purging, vomiting, excessive exercise, or other weight control measures, such as diet pills or diuretic drugs.

anosmia The lack of the ability to smell. It can be either temporary or permanent.

anterior olfactory nucleus A nucleus located in the posterior part of the olfactory bulb which serves as a relay between some projection neurons of the bulb and other brain regions, including the contralateral olfactory bulb and cortex.

antihyperlipidemic drugs A diverse group of lipid-lowering drugs used in the treatment of hyperlipidemias. There are several classes of anti-hypolipidemic drugs which differ in their impact on the cholesterol profile and in adverse side effects. Examples include statins, fibrates, niacin, bile acid sequestrants (resins), and drugs that inhibit cholesterol absorption.

antihypertensive drugs A class of drugs used in medicine and pharmacology to treat hypertension (high blood pressure). There are many classes of antihypertensives which act by varying means to lower blood pressure.

ataxia A neurological sign and symptom consisting of unsteady and clumsy motion of the limbs and torso, due to a failure of the gross coordination of muscle movements, most evident on standing and walking.

auriculotemporal nerve A branch of the mandibular nerve that runs with the superficial temporal artery and vein, and provides sensory innervation to various regions on the side of the head.

autonomic nervous system That part of the peripheral nervous system that controls homeostatic maintenance of the body's internal state. Also termed the visceral nervous system.

Bardet–Biedl syndrome A genetic multi-system disorder that can cause chronic and end-stage renal failure in children. It is characterized by obesity, limb deformities, cognitive impairment, and genitourinary tract malformations. Despite its rarity, this syndrome has come to recent prominence because of the primary cilia dysfunction underlying its pathogenesis.

basal ganglia A group of brain nuclei, sometimes termed the basal nuclei, which are interconnected with the cerebral cortex, thalamus, and brainstem. Mammalian basal ganglia are associated with a variety of functions, including motor control, cognition, emotions, and learning, and are known to be compromised in Parkinson's disease.

bulemia nervosa A psychological eating disorder in which recurrent binge eating is followed by intentional purging. This purging is done to compensate for the excessive intake of food, usually to prevent weight gain. Purging can take the form of vomiting; inappropriate use of laxatives, enemas, diuretics, or other medication; or excessive physical exercise.

burning mouth syndrome A condition, also termed glossodynia, characterized by a burning or tingling sensation on the lips, tongue, or entire mouth. Typically, there are no visual signs like discoloration to aid in the diagnosis.

chorda tympani nerve A branch of the facial nerve (cranial nerve VII) that carries sensory fibers providing taste sensation from the anterior two-thirds of the tongue, as well as presynaptic parasympathetic fibers to the submandibular ganglion that provide secretomotor innervation to the submandibular and sublingual salivary glands.

cribriform plate A portion of the ethmoid bone separating the brain cavity from the nasal cavity and through which the olfactory nerves project.

cryptorchidism The absence from the scrotum of one or both testes.

Cushing's syndrome A rare endocrine disorder caused by high blood levels of cortisol, a hormone relesed from the adrenal gland in response to ACTH from the pituitary gland. Also termed hypercortisolsim or hyperadrenocorticism.

Down syndrome A genetic disorder, previously termed mongolism, caused by the presence of all or part of an extra 21st chromosome (trisomy 21). Mental retardation and physical growth changes are typical characteristics.

dysgenesis Abnormal development of a tissue or other body part.

dysosmia A disordered smell quality, such as the odor of roses smelling like feces.

dysgeusia A disordered taste quality, such as salt tasting bitter.

ethmoidal Nerve A branch of the trigeminal nerve. The anterior ethmoidal nerve provides somatosensation to the nasal cavity.

facial nerve The 7th cranial nerve (CN VII). It emerges from the brainstem between the pons and the medulla. This nerve controls the muscles of facial expression and innervates the taste buds on the anterior two-thirds of the tongue via the chorda tympani nerve branch. It also supplies preganglionic parasympathetic fibers to several head and neck ganglia.

frontal lobe An area in the brain of vertebrates located at the front of each cerebral hemisphere. The frontal lobes play a part in impulse control, judgment, language production, working memory, motor function, problem solving, sexual behavior, socialization, and spontaneity, and are critical for normal planning, coordinating, controlling, and executing of a range of behaviors.

frontal operculum The most posterior portion of the inferior frontal gyrus of the frontal lobe. One famous part of the operculum is Broca's area which plays an important role in conversation or speech production, reading, and writing.

glossalgia Same as glossadynia; see burning mouth syndrome.

glossopharyngeal nerve The 9th cranial nerve. This nerve exits the brainstem from the sides of the upper medulla, just rostral (closer to the nose) to the vagus nerve. This nerve supplies the taste buds on the posterior third of the tongue.

Huntington's disease A rare inherited neurological disorder caused by a trinucleotide repeat expansion in the Huntingtin (Htt) gene. This expansion produces an altered form of the Htt protein, mutant Huntingtin (mHtt), which results in neuronal cell death in select areas of the brain. Its most obvious symptoms are abnormal body movements called chorea and a lack of coordination, but mentation and some aspects of personality also can be affected.

hypogeusia Diminished taste sensitivity.

hypoglossal nerve The 12th cranial nerve (XII). This nerve supplies motor fibres to all of the muscles of the tongue, except the palatoglossus muscle which is innervated by the vagus nerve (X) and the accessory nerve (XI).

hypogonadotropic hypogonadism A disorder characterized by a failure of the function of the gonads (ovaries or testes), resulting in hormonal deficiency and infertility. Deficiency of the sex hormones can result in defective primary or secondary sexual development, or withdrawal effects (e.g., premature menopause) in adults.

hypoparathyroidism Decreased function of the parathyroid glands, leading to decreased levels of parathyroid hormone (PTH). The consequence, hypocalcemia, is a serious medical condition.

hyposmia Diminished smell function.

hypothyroidism A disease state caused by insufficient production of thyroid hormone by the thyroid gland.

iatrogenesis Literally means "brought forth by a healer" (*iatros* means healer in Greek). As such, it can refer to good or bad effects, but it is almost exclusively used to refer to a state of ill health or adverse effect or complication caused by or resulting from medical treatment.

insular cortex A brain structure that lies deep to the brain's lateral surface within the lateral sulcus, a groove which separates the temporal lobe and inferior parietal cortex. The latter overlying cortical areas are known as opercula (meaning "lids"), and parts of the frontal, temporal, and parietal lobes form opercula over the insular cortex (also termed insula).

internal capsule An area of brain white matter that separates the caudate nucleus and the thalamus from the lenticular nucleus. It consists of axonal fibers that run between the cerebral cortex and the pyramids of the medulla.

Kallmann syndrome A syndrome characterized by anosmia and hypogonadism (decreased functioning of the sex hormone-producing glands) caused by a deficiency of gonadotropin-releasing hormone (GnRH) in the hypothalamus. Also known as hypothalmic hypogonadism with anosmia, familial hypogonadism with anosmia, or hypogonadotropic hypogonadism with anosmia.

lingual nerve A branch of the mandibular nerve (CN V3), itself a branch of the trigeminal nerve. This nerve supplies sensory innervation to the mucous membrane of the anterior two-thirds of the tongue. It also carries non-trigeminal nerve fibers, including those responsible for taste sensation on the anterior tongue and both parasympathetic and sympathetic fibers to the submandibular ganglion, a structure suspended by two nerve filaments from the lingual nerve.

lymphedema A condition of localized fluid retention caused by a compromised lymphatic system. The lymphatic system (often referred to as the body's "second" circulatory system) collects and filters the interstitial fluid of the body.

magnetoencephalography An imaging technique used to measure the magnetic fields produced by electrical activity in the brain via extremely sensitive devices such as superconducting quantum Interference devices (SQUIDs). There are many uses for the MEG, including assisting surgeons in localizing a pathology, assisting researchers in determining the function of various parts of the brain, neurofeedback, and others.

meningioma Common benign tumors of the brain (95% of benign tumors) that can occasionally be malignant. They arise from the arachnoidal cap cells of the meninges and represent about 15% of all primary brain tumors.

midbrain The mesencephalon (or midbrain) is the middle of three vesicles that arise from the neural tube that forms the brain of developing animals.

Caudally the mesencephalon adjoins the pons (metencephalon) and rostrally it adjoins the diencephalons, a brain region containing the thalamus, hypothalamus, and other structures.

migraine A neurological disorder associated with an electrochemical change in the trigeminovascular regions of the brain. The most common symptom is an intense and disabling episodic headache. Migraine headaches are usually characterized by severe pain on one or both sides of the head as well as a multitude of potential neurological symptoms.

multiple sclerosis A chronic, inflammatory, demyelinating disease that affects the central nervous system (CNS). Symptoms can include sensory problems (most notably visual and balance problems), muscle weakness, depression, difficulties with coordination and speech, severe fatigue, cognitive impairment, overheating, and pain. MS causes impaired mobility and disability in more severe cases.

nucleus tractus solitarius The solitary nucleus and tract are structures in the brainstem that carry and receive visceral sensation and taste from the facial (VII), glossopharyngeal (IX), vagus (X) cranial nerves, as well as the cranial part of the accessory nerve (XI).

olfactory reference syndrome A psychiatric condition of excessive concern with odors, usually imagined. The disorder can be accompanied by shame, embarrassment, significant distress, avoidance behavior, and social isolation.

Parkinson's disease A degenerative disorder that often impairs motor skills and speech. It is classically characterized by muscle rigidity, tremor, a slowing of physical movement (bradykinesia) and, in extreme cases, a loss of physical movement (akinesia). The motor symptoms result from decreased dopamine within the basal ganglia, thereby reducing stimulation of the motor cortex. Secondary symptoms can include high level cognitive dysfunction and subtle language problems.

pons A central nervous system structure located on the brainstem above the medulla oblongata, below the midbrain, and anterior to the cerebellum. This structure relays sensory information between the cerebellum and cerebrum, and contains the pneumotaxic centers that regulate respiration. some theories pose that it has a role in dreaming.

postcentral gyrus The lateral postcentral gyrus is a prominent structure in the parietal lobe of the human brain and an important landmark. Often referred to as primary somatosensory cortex, as it received the bulk of the thalamocortical projection from the sensory input fields.

precentral gyrus The primary motor cortex (or M1) works in association with pre-motor areas to plan and execute movements. M1 contains large neurons known as Betz cells which send long axons down the spinal cord to synapse onto alpha motor neurons which connect to the muscles. Pre-motor areas are involved in planning actions (in concert with the basal ganglia) and refining movements based upon sensory input (this requires the cerebellum).

Primary olfactory cortex Cortical regions (e.g., entorhinocortex and pyriform cortex) that receive the initial projections from the olfactory bulb.

progressive supranuclear palsy A rare degenerative disorder involving the gradual deterioration and death of selected areas of the brain. The initial symptom in most cases is loss of balance and falls. Other common early symptoms are changes in personality or general slowing of movement. Later symptoms and signs are dementia (typically including loss of inhibition and ability to organize information), slurring of speech, difficulty swallowing, and difficulty moving the eyes, most specifically in the downward direction.

pseudohypoparathyroidism A condition caused by resistance to the parathyroid hormone. Patients have a low serum calcium and high phosphate, but the parathyroid hormone level is appropriately high.

schizophrenia A mental illness characterized by impairments in the perception or expression of reality, most commonly manifesting as auditory hallucinations, paranoid or bizarre delusions, or disorganized speech and thinking in the context of significant social or occupational dysfunction.

seasonal affective disorder An affective, or mood, disorder. Most sufferers experience normal mental health throughout most of the year, but experience depressive symptoms in the winter or summer.

sella turcica A saddle-shaped depression in the sphenoid bone at the base of the human skull. The seat of the saddle is known as the hypophyseal fossa which holds the pituitary gland.

striatum A subcortical part of the telencephalon. It is the major input station of the basal ganglia system. Anatomically, the striatum is comprised of the caudate and the putamen.

sulfhydryl group A functional group on a molecule composed of a sulfur and a hydrogen atom (–SH).

superior temporal gyrus The superior temporal gyrus is one of three (sometimes two) gyri in the temporal lobe of the human brain. A gyrus (plural gyri) is a bump or ridge on the surface of the brain.

temporal lobe These elements of the cerebrum lie at the sides of the brain, beneath the lateral or Sylvian fissure. Seen in profile, the human brain looks something like a boxing glove. The temporal lobes are where the thumbs would be. The temporal lobe is involved in sensory processing, as well as in semantics both in speech and vision. The temporal lobe contains the hippocampus and is therefore involved in memory formation as well.

thalamus A paired and symmetric deep portion of the brain which relays sensory information from sensory systems to various areas of the cortex. It constitutes the main part of the diencephalon.

trigeminal nerve The fifth cranial nerve (CN V). This nerve is responsible for sensation in the face, eyes, and oral cavity, and can respond to some chemicals, producing such sensations as cooling, sharpness, and warmth.

Turner syndrome This disorder occurs in 1 in 2500 female births. Instead of the normal XX sex chromosomes for a female, only one X chromo-some is present and fully functional. A normal female karyotype is labeled 46, XX; individuals with Turner syndrome are 45, X (also labeled 45, X0 or less commonly X(,)) though other genetic variants occur. In Turner syndrome, female sexual charac-teristics are present but generally underdeveloped.

uremia A toxic condition resulting from renal fail-ure, when kidney function is compromised and urea (a waste product normally excreted in the urine) is retained in the blood. Uremia can lead to distur-bances in the platelets and hypersomnia, among other problems.

vagus nerve The 10th cranial nerve (CN X). This nerve starts in the brainstem (within the medulla oblongata) and extends, through the jugular fora-men, down below the head, to the abdomen. The vagus nerve is responsible for such varied tasks as heart rate, gastrointestinal peristalsis, sweating, and quite a few muscle movements in the mouth, including speech (via the recurrent laryngeal nerve) and keeping the larynx open for breathing. It also receives some sensation from the outer ear, via the auricular branch (also known as Alderman's nerve) and part of the meninges. Taste buds at the back of the throat are innervated by this nerve.

vestibular nuclei The cranial nuclei for the vestib-ular nerve.

"...smell and taste are in fact but a single composite sense, whose laboratory is the mouth and its chimney the nose...." Jean Anthelme Brillat-Savarin, 1826

4.46.1 Introduction

The senses of taste and smell warn us of toxins, such as spoiled food, leaking natural gas, polluted air, and smoke. Moreover, as noted above by Brillat-Savarin, author of *Physiologie du Gout*, these senses work together in their production of flavor sensations. This is why many people mistake the olfaction-related experiences derived from eating (e.g., the sensations of chocolate, strawberry, lemon, etc.), which depend upon stimulation of the olfactory receptors via the retronasal route, as emanating from the taste buds. In fact, taste buds only mediate the sensations of sweet, sour, salty, bitter, and possibly metallic (iron salts), umami (sodium salts), and chalky (calcium salts). Like primary colors, smell and taste sensations blend to create variable hues of chemosensory experience, and when combined with the sensory input from the trigeminal system, ingested substances acquire an aroma, texture, tem-perature, and spiciness, allowing for the overall perception of flavor.

Damage to the olfactory and gustatory systems can profoundly influence a patient's quality of life. Persons without a sense of smell can no longer enjoy the nuances of perfume, wine, and fine food, in some cases leading to depression and, depending upon the individual, weight loss or weight gain. Taste dysfunc-tion, commonly presenting as an abnormal taste, may significantly affect food choices and adversely alter nutrient intake. Chemosensory dysfunction is parti-cularly disabling for those who depend upon these senses for their livelihood or safety, including plumbers, fire fighters, policemen, perfumers, and workers within the food and beverage industries. Surprisingly, disorders of taste and smell are common, being present in at least 1% of the population under the age of 65 years, and well over 50% of the population over the age of 65 years (Doty, R. L. *et al.*, 1984a; 1984b; Hoffman, H. J. *et al.*, 1998).

In this chapter, we review a number of human chemosensory disorders, present up-to-date practical techniques for their management and quantitative evaluation, and describe how they are classified, evaluated, and treated. This survey is not inclusive, and examples are presented only to show the relationship between specific disease states and chemosensory disturbances.

4.46.2   Anatomy of Human Taste and Smell Systems

Chemosensory dysfunction can arise from disorders that block delivery of stimuli to receptors, as well as from ones that damage neural processes at various levels of the sensory system. In the vast majority of cases of olfactory dysfunction, damage occurs to the olfactory membrane. The most typical site of taste dysfunction is generally less clear, although taste buds can be directly altered via injury to the lingual surfaces. Since the neural circuitry involved in olfactory and gustatory perception is complex, a basic overview of the anatomy of these systems is presented in this section to place clinical issues into perspective; more detailed descriptions of the biochemistry and physiology of these systems are found elsewhere in this volume.

4.46.2.1   Olfactory Pathways

Six to ten million olfactory receptor cells, collectively termed cranial nerve I (CN I), are located within the approximately $2\,cm^2$ pseudostratified columnar epithelium lining the region of the cribriform plate and sectors of the superior turbinate, middle turbinate, and superior septum. In the human, about 350 receptors are distributed across these millions of cells, localized to cilia extending from dendritic knobs into the overlying mucus (Rouquier, S. *et al.*, 2000). The axons of the receptor cells unite to form bundles of 30 to 50 fila ensheathed by glial cells that traverse the cribriform plate and pia matter to form the outermost layer of the olfactory

bulb. The axons of the olfactory receptors enter the glomeruli of the second layer of the olfactory bulb, where they form multiple synapses with the dendrites of the mitral and tufted cells, the bulb's primary projection neurons. The olfactory bulb is a layered structure that serves as the first relay system in the olfactory pathway. The bulb transforms the incoming signal, sharpening the information relative to the mix of chemicals involved and bodily state (e.g., hunger). In addition to sending afferents directly to the olfactory cortex via the lateral olfactory tract, the mitral and tufted cells send collaterals that synapse within the periglomerular and external plexiform layers of the bulb, where efferent influences from central structures can occur.

The primary olfactory cortex (POC), the direct recipient of the bulbar afferents, is comprised of the anterior olfactory nucleus (AON), the piriform cortex, the olfactory tubercle, the entorhinal area, the periamygdaloid cortex, and the corticomedial amygdala (Figure 1). While the majority of olfactory bulb neurons project to the piriform cortex, found within the medial portion of the temporal lobe, other fibers project directly to structures of the limbic system involved in memory and emotion (e.g., entorhinal cortex, periamygdaloid complex, and subsequently the hippocampus). Some projections occur, via the anterior commissure, from pyramidal cells of the AON to contralateral elements of the POC. Contralateral, as well as ipsilateral, activity is present, although generally the ipsilateral activity is more intense and rapid, as measured, for example, by magnetoencephalography (MEG) (Tonoike, M. *et al.*, 1998). Olfaction differs from the other senses in sending projections to the cortex that do not relay first within the thalamus. Thalamic connections do exist, however, between projections to and from the primary and secondary olfactory cortices.

4.46.2.2   Gustatory Pathways

Taste receptors are mostly found within taste buds located on lingual protuberances, termed papillae. Some taste buds are found on the epiglottis and soft palate. The receptor elements within the buds are innervated by branches of the facial (CN VII), glossopharyngeal (IX), and vagus (CN X) nerves, depending upon the bud's location within the oral cavity (Figure 2). Unlike CN I, these cranial nerves are mixed motor and sensory nerves. This becomes important when considering clinical syndromes that involve taste dysfunction, since variable degrees of

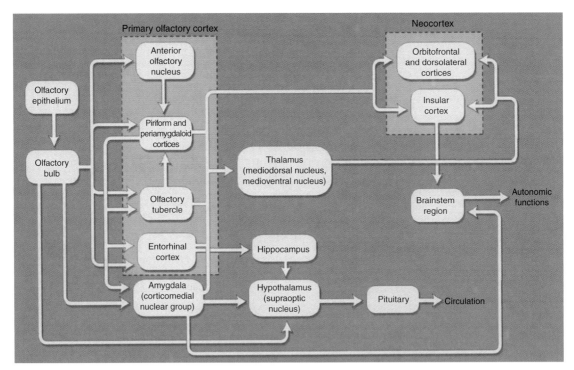

Figure 1 Primary afferent neural connections of the human olfactory system. Although olfactory fibers project from the amygdala to the feeding center of the hypothalamus, particularly the ventromedial and ventrolateral nuclei, it is unclear whether direct olfactohypothalamic routes are available in humans, as occurs in some other mammals. Copyright © 1996, Richard L. Doty.

either motor or sensory disturbance may be present depending upon where and to what extent the taste-mediating cranial nerve is injured.

All the peripheral gustatory fibers enter the brainstem and project to the nucleus of the tractus solitarius (NTS), which begins in the rosterolateral medulla and extends caudally along the ventral border of the vestibular nuclei. The major cortical projections from the NTS are to insular and perisylvian cortical regions, including the frontal operculum, superior temporal gyrus (opercular part), and inferior parts of the pre- and postcentral gyrus (Gloor, P., 1997; Faurion, A. *et al.*, 1998). A secondary cortical taste region is located in the caudomedial/caudolateral orbitofrontal cortex rostral to the primary taste cortex (Baylis, L. L. *et al.*, 1995). A number of such regions process touch and temperature, in addition to taste. The cells of the NTS also make viscerovisceral and viscerosomatic reflexive connections, via the interneurons of the reticular formation, with cranial nuclei that control (1) the muscles of facial expression; (2) taste-mediated behaviors, including chewing, licking, salivation, and swallowing; and (3) preabsorptive insulin release (Smith, D. V. and Shipley, M. T., 1992).

4.46.2.3 Trigeminal System

Free nerve endings from the trigeminal nerve (CN V) are found throughout the nasal and oral cavities. In the nose, they are responsible for somatosensory sensations induced by some odorants (e.g., stinging, sharpness, coolness, and warmth), whereas in the mouth, they contribute to the tactile, thermal, and spicy sensations essential for many flavor sensations. Anatomically, the anterior and lateral sectors of the nasal cavity are innervated by lateral and medial nasal branches of the ethmoidal nerve. The posterior region of the nasal cavity is innervated by the nasopalatine nerve. The gustatory fibers of the facial nerve (chorda tympani) travel in conjunction with the mandibular trigeminal fibers of the lingual nerve for a short distance near the tongue, making both of these nerves susceptible to orofacial causes of injury.

4.46.3 Symptoms of Chemosensory Dysfunction

The symptoms of chemosensory dysfunction can range from total loss of smell or taste perception

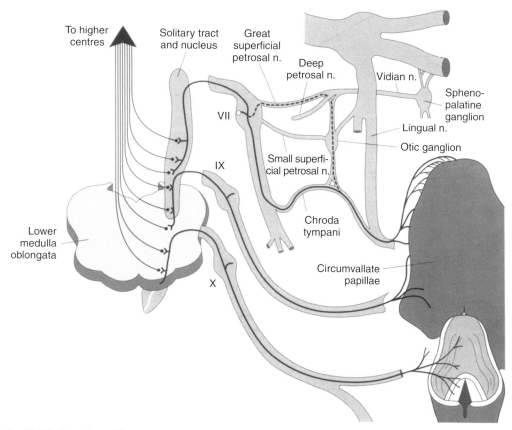

Figure 2   Distribution of cranial nerves to gustatory regions. CN VII fibers from the geniculate ganglion innervate taste buds on the anterior tongue and soft palate. CN IX fibers from cell bodies within the petrosal ganglion innervate taste bud on the lingual foliate and vallate papillae, as well as pharyngeal taste buds. CN X fibers from cell bodies in the nodose ganglion innervate taste buds on the epiglottis, larynx, and esophagus. Modified from Netter, F. H. 1964. The Ciba Collection of Medial Illustrations, Vol. 1, Nervous System. Ciba Pharmaceutical Corporation.

(termed anosmia and ageusia, respectively) to persistent and debilitating sensory disturbances, such as the presence of a strong and persistent salty taste (dysgeusia) or a bad smell (dysosmia). When such sensations appear without any apparent stimulus, they are termed phantogeusias or phantosmias, respectively. In most cases where a patient complains of chemosensory problems, some degree of smell or taste loss is apparent upon quantitative testing. In a few cases, however, disturbances in the perception of tastes or smells occur that cannot be objectively detected. A debilitating disorder, often associated with taste distortions that are not detectable by taste tests, is that of burning mouth syndrome (BMS; also known as glossodynia or glossalgia). In this syndrome, the patient experiences intense oral burning pain on the tongue or elsewhere within the oral cavity (Grushka, M. and Sessle, B. J., 1991).

4.46.4   Clinical Evaluation of Chemosensory Disorders

4.46.4.1   Medical History and Examination

The etiology of most cases of chemosensory dysfunction can be ascertained from carefully questioning the patient about the nature, timing, onset, duration, and pattern of symptoms, as well as a historical determination of antecedent events, such as head trauma, upper respiratory infections (URIs), toxic exposures, nasal or oral surgeries, and so on. A history of epistaxis, discharge (clear, purulent, or bloody), nasal obstruction, and somatic symptoms, including headache or irritation, may aid in establishing the probable etiology. Fluctuations in olfactory function usually reflect obstructive, rather than neural, factors. It is important to identify exacerbating or relieving foods or products, prior treatments and their efficacy and comorbid medical conditions (e.g., liver disease, kidney disease, hypothyroidism, diabetes, and vitamin deficiencies). Delayed puberty in association with anosmia, with or without midline craniofacial abnormalities, deafness, and renal anomalies, suggests the possibility of Kallmann's syndrome or one of its variants. Patients who report never having had a sense of smell typically lack normal olfactory bulbs or tracts upon appropriate magnetic resonance imaging (MRI) studies (Li, C. *et al.*, 2003). Medications being used prior to or at the time of the symptom should be identified, particularly when taste, per se, appears to be altered. Such widely used drugs as statins, antifungal agents, and angiotensin-converting enzyme (ACE) inhibitors commonly cause chemosensory disturbances, most notably alterations in taste perception. Although dysosmias or phantosmias usually reflect alterations at the level of olfactory epithelium, in rare cases they signify central tumors or lesions that can be associated with tremor or seizure activity (e.g., automatisms, occurrence of black-outs, auras, and déjà vu).

Problems with speech articulation, salivation, chewing, swallowing, oral pain or burning, dryness of the mouth, periodontal disease, foul breath odor, recent dental procedures or surgeries, recent radiation exposure, medications, and bruxism can aid in establishing the basis of a taste problem. Inquiry into the patient's diet and oral habits may reveal exposure to oral irritants. Questions about hearing, tinnitus, and balance are important since the vestibulocochlear nerve (CN VIII) travels in close proximity to the facial nerve and can be susceptible to similar etiologies of disturbance (e.g., cerebellopontine angle tumors). Considering the occasional effect of gastroesophageal reflux on taste, asking about stomach problems may also be relevant. Constitutional symptoms – such as fever, malaise, headache and body pains, and rashes – may also be important since such symptoms often accompany cancers or systemic inflammatory conditions (e.g., lupus).

4.46.4.2 Quantitative Chemosensory Testing

A common error made on the part of clinicians is to accept a patient's report of olfactory dysfunction and to fail to objectively verify the presence or magnitude of the problem. Unfortunately, many persons are inaccurate in assessing the magnitude of their dysfunction, particularly the elderly or those with dementia. As noted in the introduction, a large number of patients confuse the loss of flavor sensations derived from the

loss of olfactory function with taste, misleading themselves and their physicians as to the sensory system involved in their dysfunction (Figure 3). Some patients, particularly ones involved in litigation, attempt to malinger or overstate the nature of their dysfunction. Quantitative testing allows for a determination of the nature and degree of the chemosensory problem, the detection of malingering, and the tracking of changes in function over time and the efficacy of medical or surgical interventions.

Practical tests of smell function are now commercially available (for review, see Doty, R. L. 2001). The most widely used of such tests, the University of Pennsylvania smell identification test (UPSIT), has been administered to hundreds of thousands of patients and has been translated into a number of non-English languages. This test provides an indication of absolute smell loss, as well as the degree of smell loss relative to a patient's age and gender (i.e., percentile rank). The latter is critical, since smell ability decreases markedly in the later years (Figure 4). Electrophysiological tests, such as the odor event-related potential, can aid in the detection of malingering, although, unlike their auditory counterpart, they are presently unable to discern where in the olfactory pathway an anomaly exists. Although direct electrical recordings from the olfactory neuroepithelium can be made (termed the electro-olfactogram), such recordings can be misleading. Thus, anosmic patients lacking olfactory bulbs and tracts (e.g., those with Kallmann's syndrome), as well as patients with clearly determined psychophysical olfactory dysfunction (e.g., those with schizophrenia), commonly have clear electro-olfactograms.

Numerous taste tests are described in the literature that can be used to detect taste dysfunction, perhaps the most practical being that of electrogustometry. Testing of specific regions of the tongue is usually desired, since whole-mouth tests can be relatively insensitive to a number of gustatory alterations, given the redundancy of the innervation of the taste buds (for review, see Frank, M. E. *et al.*, 1995).

4.46.5 Causes of Chemosensory Dysfunction

Although there are reports of hundreds of disorders associated with chemosensory disturbance, a number of such reports are anecdotal and require empirical validation. Olfactory dysfunction can reflect, among other things, (1) altered airflow to the olfactory

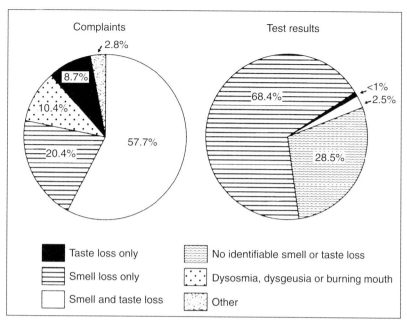

Figure 3 Distribution of primary chemosensory complaints (left diagram) and test results (right diagram) in 750 consecutive patients presenting to the University of Pennsylvania Smell and Taste Center for chemosensory evaluation. Note that the complaint categories of smell loss, taste loss, and smell and taste loss include some patients with secondary complaints of chemosensory distortion and burning mouth from Deems, D. A., Doty, R. L., Settle, R. G., Moore-Gillon, V., Shaman, P., Mester, A. F., *et al.*, 1991. Smell and taste disorders, a study of 750 patients from the University of Pennsylvania Smell and Taste Center. Arch. Otolaryngol. Head Neck Surg. 117, 519–528. Copyright © 1991, American Medical Association.

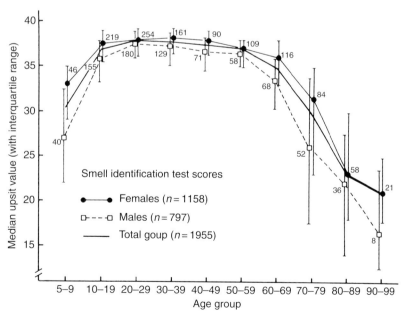

Figure 4 University of Pennsylvania smell identification test (UPSIT) scores as a function of age and gender. Numbers by data points indicate sample sizes modified from Doty, R. L., Shaman, P., Applebaum, S. L., Giberson, R., Siksorski, L., Rosenberg, *et al.* 1984a. Smell identification ability: changes with age. Science 226, 1441–1443; Doty, R. L., Shaman, P. and Dann, M. 1984b. Development of the University of Pennsylvania smell identification test: a standardized microencapsulated test of olfactory function. Physiol. Behav. 32, 489–502.

receptors (e.g., secondary to polyposis and excessive nasal engorgement), (2) damage to the olfactory membrane proper (e.g., from URIs, toxic chemical exposure, and occlusion of foramina through which the olfactory fila project), (3) damage to central nervous system (CNS) structures involved in olfactory transduction/processing (e.g., from strokes, epilepsy, and so on), and (4) systemic disturbances of metabolism or transduction (e.g., due to medications, hypothyroidism, diabetes, and liver or kidney disease). Approximately two-thirds of cases of chronic anosmia or hyposmia are due to prior URIs, rhinosinusitis, or head trauma (Deems, D. S. *et al.*, 1991). Mechanisms for the alteration of taste include (1) blockage of stimulus access to the taste buds (e.g., from damage to taste pores from caustic agents or burns, chronic lack of salivary proteins critical for taste bud maintenance), (2) the release of bad-tasting materials from dental appliances (e.g., bacterial infections, gingivitis, and purulent sialadenitis), (3) damage to one or more neural pathways or central brain regions (e.g., from epilepsy, tumors, strokes, viruses, occlusion of foramina through which nerves pass, and trauma), and (4) systemic disturbances of metabolism (e.g., medications, hypothyroidism, diabetes, and liver or kidney disease). Avitaminosis, particularly of the B group and C group, can result in pellagra, sprue, pernicious anemia, scurvy, or glossitis. Additionally, zinc and copper deficiencies are reportedly linked to taste disturbances (Mattes, R. D., 1995).

4.46.5.1 Upper Respiratory and Oral Infections

The most frequent cause of permanent smell loss in the adult human is an URI, such as associated with the common cold, influenza, pneumonia, or HIV+ (Deems, D. S. *et al.*, 1991). Often the respiratory illness is described as being more severe than usual, and in many cases dysosmia or phantosmia is present, at least initially. Most dysosmias subside over time, although some patients are left with a noticeable olfactory deficit. Such losses are most common in middle or older age, presumably reflecting cumulative insult and the challenge of epithelial regeneration in the face of reductions in the proliferation of basal cells and immature neurons. Evidence of damage to the olfactory epithelia from patients with post-URI anosmia includes extensive cicatrization, decreases in receptor cell number, absent or decreased numbers of cilia on remaining receptor

cells, and replacement of sensory epithelium with respiratory-like epithelium (Yamagishi, M. *et al.*, 1994).

The taste system is similarly susceptible to damage from bacteria, viruses, and other agents. Some cases of Bell's palsy, for example, reflect viral damage to the facial nerve (e.g., from *Herpes simplex* virus-1), producing unilateral facial weakness, gustatory loss, and in some cases damage to the taste-salivary reflex, which is dependent upon a relay from the nucleus of the solitary tract to the parasympathetic fibers innervating the salivary glands (Murakami, S. *et al.*, 1996). Involvement of taste in Bell's palsy, which may involve over half the cases, localizes the lesion to the neural pathways from the pons, along the nervus intermedius portion of the facial nerve, through the geniculate ganglion, to the point where the chorda tympani joins the facial nerve in the facial canal. A more severe form of Bell's palsy, Ramsy–Hunt syndrome, results from an active *Herpes* infection of the geniculate ganglion and involves pain and vesicles in the external auditory canal or soft palate. Interestingly, dental procedures can activate viruses that, in turn, lead to infections that sometimes influence olfactory function.

4.46.5.2 Allergies and Rhinosinusitis

A significant number of patients with allergic rhinitis exhibit greater olfactory dysfunction than their non-allergic counterparts. In one study, for example, 23.1% of those with allergic rhinitis had a threshold at or above the 2½ percentile of the controls. The smell loss was associated with clinical or radiographic evidence of rhinosinusitis or nasal polyps. Thus, hyposmia was present in 42.9% of those with associated rhinosinusitis. The presence of hyposmia was only 14.3% in those with no associated rhinosinusitis (Cowart, B. J. *et al.*, 1993).

In general, olfactory dysfunction is related to the severity of rhinosinusitis [e.g., in one study, the mean UPSIT scores were 35, 31, 26, and 23 for Kennedy stages I to IV of the disease, respectively (Downey, L. L. *et al.*, 1996)]. While the smell loss associated with rhinosinusitis has been traditionally viewed as conductive (Figure 1), at best only weak associations have been found between quantitative olfactory test scores and measures of nasal airway patency (Cowart, B. J. *et al.*, 1993). Moreover, medical (e.g., administration of topical or systemic steroids) or surgical (e.g., excision of polyps) treatments rarely return olfactory function to normal, suggesting that the severity of histopathological changes within the olfactory mucosa may be the defining factor. Support for this hypothesis comes from evidence that degree of epithelial inflammation is related to olfactory test scores in patients with chronic rhinosinusitis (Kern, R. C., 2000), and that olfactory epithelial biopsies from patients with nasal disease are less likely to yield olfactory-related tissue than those from controls (Feron, F. *et al.*, 1998). The same is true for anosmic vs. nonanosmic rhinosinusitis patients, the former of whom typically exhibit a more pathological epithelium (e.g., disordered arrangement of cells and more islands of respiratory-like epithelium) (Lee, S. H. *et al.*, 2000). While the taste system, per se, is not directly influenced by rhinosinusitis, unpleasant tasting discharge, e.g., from postnasal drip, is often present that is of significant concern to the patient.

4.46.5.3 Disorders Influencing Airway Patency

In addition to rhinosinusitis, other conditions that block or otherwise hinder airflow to the olfactory receptors can adversely influence the ability to smell. Children with choanal atresia, for example, suffer from smell loss. In this congenital disorder, the airway is obstructed by membranous or bony tissue. The limited data suggest that permanent olfactory dysfunction may occur in such patients, presumably reflecting the need for early exposure to odorants for normal olfactory system development. Thus, in one study, smell was tested before and after such repair in patients of relatively advanced age (8–31 years). Although one patient who had suffered from unilateral atresia had normal smell

function, three others who had suffered from bilateral atresia had permanent olfactory loss (Gross-Isseroff, R. *et al.*, 1989). A reversible smell loss occurs in children whose nasal airflow is compromised by hypertrophied adenoid tissue (Figure 5) (Ghorbanian, S. N. *et al.*, 1983).

4.46.5.4 Head Trauma

Head trauma likely accounts for 10–20% of all smell and taste disorders (Deems, D. S. *et al.*, 1991). While there are studies that suggest between 4 and 7% of head trauma patients lose their sense of smell, incidence rates as high as 60% in cases of severe head injury have been reported. However, such high incidences are usually found in centers specializing in chemosensory disturbances and therefore are influenced by referral patterns (Doty, R. L. *et al.*, 1997c; Ogawa, T. and Rutka, J., 1999). In general, taste dysfunction resulting from trauma is much less common than posttraumatic smell dysfunction, with solitary ageusia of one or more primary taste modalities, as measured by whole-mouth testing, occurring in less than 1% of persons with major head injury (Deems, D. S. *et al.*, 1991). Only rarely are trauma-related taste disturbances unaccompanied by smell impairment.

A number of mechanisms have been implicated in trauma-related smell impairment, including (1) shearing injury to the olfactory nerve filaments of the cribriform plate as a result of brain movement relative to the skull, (2) damage to the nasal and sinus

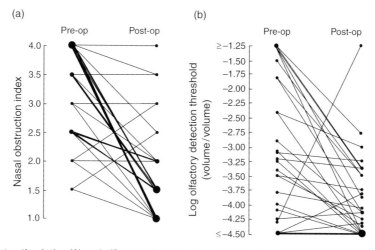

Figure 5 Nasal obstruction index (A) and olfactory detection threshold for phenyl ethyl alcohol (B) in 28 children before and after adenoidectomy. Each line joins preoperative (pre-op) and postoperative (post-op) values for an individual subject. Heavy lines denote confluence of several subjects who had the same preoperative and postoperative values (from Ghorbanian, S. N., Paradise, J. L., Doty, R. L. 1983. Odor perception in children in relation to nasal obstruction. Pediatrics 72, 510–516). Copyright © 1983 by the American Academy of Pediatrics.

architecture – with subsequent alteration of nasal airflow – without specific injury to the olfactory epithelium or neural projections, and (3) contusion or hemorrhage to central regions that mediate olfactory information (e.g., olfactory bulbs and tracts themselves, septal nuclei of the inferior frontal region, orbitofrontal territories, or the anteromedial temporal lobes). In addition to the severity of injury and loss of consciousness, there appear to be other factors associated with increased risk of developing anosmia after head trauma, including (1) anterior skull base fractures, (2) bilateral subfrontal lobe injury, (3) dural lacerations, and (4) cerebrospinal fluid leakage (Giasson, B. I. *et al.*, 2002). Occipital and lateral blows are more likely to result in anosmia than blows to the front of the head (Doty, R. L. *et al.*, 1997a; 1997b; 1997c). Compound fractures in the occipital, frontal, skull base, and mid-face are reported to be twice as likely as temporal and parietal fractures to result in olfactory alterations (Ogawa, T. and Rutka, J., 1999). In terms of facial fractures, nasozygomatic Le Fort fractures, fronto-orbital fractures, and pure La Fort fractures appear to most likely to be associated with smell disturbances (Horiguchi, T. *et al.*, 2003).

A strong correlate of head trauma-related anosmia is the severity of the injury (Doty, R. L. *et al.*, 1997a; 1997b; 1997c). Lower Glasgow coma scale ratings (GCS 3–8) are associated with a greater than 50% increase in the development of anosmia compared to moderate ratings (GCS 9–12) (Giasson, B. I. and Mushynski, W. E., 1996). Trauma-related loss of consciousness lasting 24 h or longer results in significantly more loss than that found in patients who have no loss of consciousness (Doty, R. L. *et al.*, 1997a; 1997b; 1997c).

The prognosis for posttraumatic ageusia is much better than for posttraumatic anosmia, and recovery from ageusia usually occurs in most patients over a period of a few weeks to months. The etiology of posttraumatic taste dysfunction varies. Head trauma that results in injury to the middle ear, a finding often encountered with temporal bone fractures, can result in both ipsilateral impairment of taste function and injury to the preganglionic parasympathetic fibers, thus preventing salivary secretion. Seven to 10% of temporal bone fractures reportedly result in facial nerve dysfunction (see Chang, C. Y. J. and Cass, S. P. 1999, for review). The lingual nerve, a branch of the mandibular division of the trigeminal nerve that subserves general somatic sensation and taste (due to coexisting chorda tympani fibers), can be damaged

from trauma in and around the mouth and tongue, resulting in numbness and taste loss over the anterior tongue ispilateral to the side of injury. Trauma can produce Vernet's syndrome – a syndrome consisting of loss of taste in posterior third of tongue, paralysis of the vocal cord, palate, pharynx, trapezius, and sternocleidomastoid muscles secondary to lesions of the jugular foramen that affect CNs IX, X, and XI.

Isolated damage to the auriculotemporal nerve (ATN) can produce posttraumatic gustatory neuralgia, a taste-initiated, episodic, paroxysmal lancinating facial pain (often electric shock-like) in the cutaneous distribution of this nerve (Scrivani, S. J. *et al.*, 1998). Posttraumatic abnormalities involving the ATN, such as sweating and flushing following a taste stimulus, are termed Frey's syndrome. Although typically provoked by tastants, this syndrome can be elicited in some cases by the smell of food or emotional excitement (Scrivani, S. J. *et al.*, 1998).

4.46.5.5 Iatrogenesis

Head trauma is not the only type of trauma that can influence chemosensory function. In the case of olfaction, for example, various nasal and other surgical operations have been associated with smell loss (e.g., anterior ethmoidectomy, posterior ethmoidectomy, polypectomy, middle turbinate medialization, uncinectomy, sphenoidectomy, and anterior craniotomy). Although there are scattered reports of adverse influences of septoplasty and rhinoplasty on olfactory function, these operations only rarely influence smell function.

Trauma-related damage to the glossopharyngeal nerve, a nerve which is relatively well protected in comparison with the facial nerve, can occur from tonsillectomy, bronchoscopy, or laryngoscopy, reflecting the proximity of the lingual branch of this nerve to the muscle layer of the palatine tonsillar bed (Arnhold-Schneider, M. and Bernemann, D., 1987). Damage to this nerve is usually accompanied by damage to the nearby vagus, accessory, or hypoglossal nerves. Vernet's syndrome can also arise after parotid gland surgery, temporomandibular joint (TMJ) surgery, carotid endarterectomy, orthognathic surgery, and oncologic surgery.

4.46.5.6 Tumors and Central Nervous System Lesions

Damage to the olfactory bulbs and tracts can occur from numerous central tumors, including meningiomas from the dura of the cribriform plate and surrounding regions (e.g., olfactory groove meningiomas, suprasellar ridge meningiomas), pituitary growths that extend above the sella turcica, and tumors within or around the third ventricle (for review, see Kern, R. C. and Mathog, R. H., 1991). Hyposmia or anosmia can be the principal or sole feature of an olfactory groove neoplasm. Unilateral and bilateral olfactory loss has been reported in frontal lobe glioma, suprasellar meningioma, and sphenoidal ridge meningioma, as well as in non-neoplastic space filling lesions, such as aneurysms of the anterior communicating bifurcation, large internal carotid aneurysms extending over the pituitary fossa, and hydrocephalus that forces the floor of the third ventricle downward (Graff-Radford, N. R. et al., 1997). Removal of anterior skull base tumors via classical bifrontal craniotomy is commonly associated with postoperative anosmia and other complications (Jung, T. M. et al., 1997).

Tumors growing anywhere along the path of the facial nerve – from the oral cavity, through the skull base, temporal bone and middle ear, to the cerebellopontine angle and the brainstem – can cause a mononeuropathy of the nerve and alterations in taste function. Nearly 10% of patients with surgically confirmed acoustic neuromas experience loss of taste as a primary complaint (Harner, S. G. and Laws, E. R., Jr., 1981). Lesions to brainstem structures that subserve taste can produce dysgeusia, although usually not without significant impairment of other cranial nerves or long tracts. Notable brainstem areas involved in taste that are susceptible to CNS lesions include the tractus solitarius and its nucleus (where injury produces ipsilateral deficits) and the pontine tegmentum (where injury produces bilateral deficits). Bilateral injury to the thalamus can result in ageusia, although unilateral lesions above the brainstem do not usually cause complete loss of function (likely due to the multiple areas involved in processing taste information). An isolated unilateral lesion to the thalamus or the parietal lobe may cause contralateral taste disturbance.

Gustatory perception may also be affected by cortical injury to areas of association cortex outside of the main gustatory processing regions. In a manner similar to that seen in the visual, auditory, tactile, and olfactory sensory systems, left unilateral neglect of taste sensations can result from contralateral injury to the right hemisphere (e.g., parietal lobe, thalamus, basal ganglia) (Bellas, D. N. et al., 1988). In fact, a specific syndrome of buccal hemineglect exists in which patients neglect mouth and food products in the left half of the mouth and fail to initiate chewing and swallowing when food is in this location (Andre, J. M. et al., 2000). This syndrome can result in choking or a socially embarrassing tendency to drool and regurgitate unnoticed food.

In a review of 15 reported cases, Onoda, K. and Ikeda, M. (1999) found that unilateral impairment of taste function occurs with injury to any one of several central locations, including the pons (eight cases), the thalamus (five cases), the midbrain (one case), and the internal capsule (one case). Based on the side of the respective central lesion and its effect on either unilateral or contralateral gustatory function, these authors surmised that the gustatory pathways ascend ipsilaterally from the medulla, cross over in their course from the pons to the midbrain, and synapse within the contralateral thalamus. Nonetheless, unilateral taste stimulation can reportedly evoke contralateral and bilateral cortical electrical responses (Kida, A., 1974), and one study suggests that taste information from both sides of the tongue must pass through the left insula for proper recognition to occur (Pritchard, T. C. et al., 1999). In this study of six stroke patients with unilateral lesions of the insular cortex and three patients with brain damage outside the insula, damage to the right insula caused ipsilateral taste deficits in recognition and intensity, whereas damage to the left insula caused an ipsilateral deficit in taste intensity but a bilateral deficit in taste recognition.

Before the advent of modern imaging, olfactory tests were used to help diagnose and localize cribriform plate meningiomas and other tumors impacting the olfactory nerve (Elsberg, C. A., 1935a; 1935b). Olfactory recognition thresholds were reported to be elevated on the side where pressure was exerted on one olfactory nerve. When both nerves were involved, thresholds were bilaterally elevated, with more dysfunction on the most affected side. Expanding lesions on the ventral surfaces of the frontal cerebral lobes, as well as for suprasellar meningiomas and aneurysms of the internal carotid artery or the anterior part Willis' circle, produced such changes. Thresholds were uninfluenced by pituitary adenomas confined to the sella turcica, but were heightened if growth occurred above this bony structure. Tumors in or near the midline of the cranial cavity produced long-lasting adaptation (e.g., tumors of the corpus callosum, parasagittal meningiomas, and infiltrating growths extending to the medial surface of a cerebral hemisphere). Decreased recognition threshold sensitivity sometimes is accompanied by generalized increased intracranial pressure

(Elsberg, C. A., 1935b). However, the findings of a recent study of ten patients with tumors localized to the left or right temporal or frontal lobes differed somewhat from these results (Daniels, C. *et al.*, 2001). Thus, patients with right-side tumors had odor discrimination deficits on both the left and the right. Those with left-side tumors exhibited attenuated function only when the odorant was presented to the left side, in accord with the notion that right hemisphere is more involved in olfactory processing than the left.

4.46.5.7 Drug Effects

Although both taste and smell can be altered by medications, drug-related taste disturbances occur much more frequently than drug-related olfactory disturbances. Over 250 medications reportedly alter the sense of taste, including antiproliferative drugs, antirheumatic drugs, antibiotic drugs, psychotrophic drugs, and drugs with sulfhydryl groups, such as penicillamine and captopril (Doty, R. L. and Bromley, S. M., 2004). Antihypertensive and antihyperlipidemic drugs, including ACE inhibitors, are commonly associated with taste disturbances, including hypogeusia and dysgeusia (Ackerman, B. H. and Kasbekar, N., 1997; Doty, R. L. *et al.*, 2003). Terbinafine, a commonly used antifungal medication, has been linked to taste disturbance that reportedly can last many months and, in some cases, for years (Doty, R. L. and Haxel, B. R. 2005). Among the medications used in the treatment of HIV and AIDS, the protease inhibitors have a taste themselves and appear to adversely modify the taste perception of other taste compounds (Schiffman, S. S. *et al.*, 1999). In a study of six psychotropic medications, including amytriptyline, clomipramine, desimpramine, imipramine, doxepin, and trifluoperazine, Schiffman, S. S. *et al.* (1998) found that these drugs do not only have a taste of their own, but alter the intensity of other tastants (such as salt and sugar).

Medication-related chemosensory changes can, in some instances, be reversed by discontinuance of the offending drug, by employing alternative medications, or by changing drug dosage. Unfortunately, many pharmacological agents appear to induce long-term alterations in taste that may take months to disappear even after drug discontinuance. Although the Physicians' Desk Reference (PDR) (2005) can be helpful in identifying medications that may result in drug-related chemosensory dysfunction, there is lack of uniformity and control in the acquisition of this information, and the relative influences on taste and smell can be confused.

4.46.5.8 Congenital and Inherited Disorders

The vast majority of idiopathic cases of apparent congenital anosmia exhibit agenesis or dysgenesis of the olfactory bulbs and stalks (Yousem, D. M. *et al.*, 1996). Several congenital olfactory syndromes are accompanied by endocrine dysfunction (e.g., Kallmann's syndrome), as discussed in the Section 4.46.5.9. Although most idiopathic congenital cases are presumably inherited, some may reflect early viral- or trauma-induced insults to the olfactory epithelium that produce regression of olfactory bulb/tract volume over time. Lygonis, C. S. (1969) reported total anosmia (without any apparent associated disorders) in members of four generations of a family living in an isolated island community and suggested that the pattern of inheritance was likely autosomal dominant. Singh, N., *et al.* (1970) described a familial anosmia loosely associated with premature baldness and vascular headaches, which appeared to be inherited as dominant with varying penetrance.

Recently, smell dysfunction has been reported in Bardet–Biedl syndrome (BBS), a rare pleiotropic genetic disorder associated with, among other things, retinal degeneration, obesity, limb anomalies, genitourinary abnormalities, and cognitive impairment (Iannaconne, A. *et al.*, 2005). Eight BBS genes have been cloned. One of these genes, the BBS4 gene, has been found in the olfactory neuroepithelium and plays a significant role in ciliary physiology. Interestingly, BBS4-knockout mice exhibit smell loss (Iannaconne, A. *et al.*, 2005).

Familial dysautonomia (Riley–Day syndrome) is an autosomal recessively inherited disease with sensory neuropathy that mostly affects Ashkenazi Jewish patients. This disorder is associated with defective autonomic control, insensitivity to pain, hyporeflexia, and impaired lacrimation. Most notably for our purposes, patients with familial dysautonomia exhibit taste impairment secondary to the lack of fungiform papillae and lingual taste buds (Gadoth, N. *et al*, 1997). Taste dysfunction secondary to the absence of these structures was also described in two Japanese siblings who had late-onset progressive ataxia, global thermoanalgesia, sensorineural deafness, and mild autonomic disturbances (Fukutake, T. *et al*, 1996). This particular disease entity may represent a variant of late-onset hereditary spinocerebellar degeneration. More recently, Uchiyama, T. *et al.* (2001) described an individual with an apparently new variant of Machado–Joseph disease that lacked fungiform papillae.

4.46.5.9 Endocrine and Metabolic Diseases

A number of endocrine disorders have been associated with smell dysfunction. In general, the mechanisms responsible for the smell loss – aside from obvious anatomical alterations in the primary olfactory pathways – are poorly understood.

4.46.5.9.1 Addison's disease

One study has reported that patients with untreated adrenocortical insufficiency exhibit hypersensitivity to several odorants (e.g., pyridine, thiophene, and nitrobenzene), as well as to the vapors above aqueous solutions of six tastants (NaCl, KCl, HCl, NaHCO$_3$, sucrose, and urea) (Henkin, R. I. and Bartter, F. C. 1966). No overlap of the distribution of scores from the patient and control groups was observed, which is a remarkable finding. The patients detected differences between water and the tastant vapors at 1/10 000 the concentration of that observed for normals. The increased sensitivity was not influenced by daily 20 mg injections of deoxycorticosterone acetate (DOCA) for periods up to 10 days. In contrast, a similar regimen of 20 mg of the carbohydrate-active steroid prednisolone returned sensitivity to normal within 24 h.

Unfortunately, no attempts have been made to replicate this early work using more sophisticated sensory testing. If valid, the effects may reflect general processes, as hypersensitivity was also noted for pure-tone auditory thresholds and filtered and nonfiltered speech discrimination tests (Henkin, R. I. and Daly, R. L. 1968).

4.46.5.9.2 Cushing's syndrome

Cushing's syndrome is characterized by hypercortisolism, a chronic excessive secretion of adrenal corticosteroids. Adrenocorticotropic hormone (ACTH) hypersecretion occurs in approximately 80% of cases; the remaining approximately 20% reflect ACTH-independent etiologies (e.g., cortisol-secreting tumors or adrenal gland adenomas and carcinomas) (Ferrante, M. A., 1999). According to Henkin R. I. (1975), these patients have decreased sensitivity to odors, tastes, and sounds. In accord with this notion is a report by Ezeh P. I. et al. (1992) that dogs given chronic dexamethasone exhibit increased detection thresholds for benzaldehyde and eugenol (Ezeh, P. I. et al., 1992).

4.46.5.9.3 Diabetes

A number of studies suggest that both taste and smell functions are altered by diabetes (for review, see Settle, R. G., 1991). In the case of smell, both threshold and suprathreshold deficits have been observed. One study found a correlation between smell identification scores and macrovascular disease in such patients (Weinstock, R. S. et al., 1993). A progressive loss of taste beginning with glucose and extending to other sweeteners, salty stimuli, and then all stimuli has been noted (Hardy, S. L. et al., 1981). In newly diagnosed noninsulin-dependent diabetics, quantifiable taste impairment exists that may be reversible, to some degree, with correction of the hyperglycemia (Perros, P. et al., 1996). A subgroup of approximately 10% of first-degree relatives of noninsulin-dependent diabetics that have no neuropathies evidence taste dysfunction specific to glucose, but not other sugars such as fructose (Settle, R. G., 1981). It is not known if these individuals are the ones who go on to later develop diabetes, but this would seem quite possible.

4.46.5.9.4 Hypothyroidism

One might expect hypothyroidism to influence taste and smell functions, since this disease produces alterations in other sensory systems. Thus, 36–83% of hypothyroid patients are said to experience somesthetic disturbances, and 25–45% auditory or visual dysfunction, with night blindness reportedly being a common feature of the disorder (Mattes, R. D. et al., 1986). In fact, dysgeusias appear to be disproportionately represented in hypothyroid patients (Deems, D. S. et al., 1991). McConnell, R. J. et al. (1975) reported that seven of 18 patients with untreated primary hypothyroidism (39%) were cognizant of some alteration in their sense of smell, with three (17%) experiencing dysosmia. Lewitt M. S. et al. (1989) noted that 11 of 16 (69%) hypothyroid patients complained of alterations in taste and smell, although no detection threshold differences between the patients and controls were found for the odorant phenyl ethyl alcohol. Pittman J. A. and Beschi R. J. (1967) reported no influences of hypothyroidism on taste thresholds. Although experimentally induced hypothyroidism in rats fails to alter detection sensitivity to either tastes or odors, taste preferences are significantly effected, suggesting that thyroid disease may influence qualitative sensations, rather than sensitivity, per se (Brosvic, G. M. et al., 1996).

4.46.5.9.5 Kallmann's syndrome

Kallmann's syndrome, which is more prevalent in men than in women, is defined as anosmia accompanied by idiopathic hypogonadotrophic hypogonadism (IHH).

This autosomal dominant disorder has incomplete expressivity and can be accompanied by cryptorchidism, deafness, midline craniofacial abnormalities, tooth agenesis, and renal anomalies. Relatives may be normal, may have anosmia unaccompanied by gonadal dysfunction, or may have anosmia and hypogonadism. Kallmann's syndrome patients typically have bilateral agenesis or dysgenesis of the olfactory bulbs and tracts, although lateralized variability may be present (Yousem, D. M. *et al.*, 1993).

The anosmia of Kallmann's syndrome is not reversed by either gonadotropin or gonadal steroid therapy, reflecting its anatomical basis. Given that testicular derangement can be largely reversed by hormones when given at the appropriate time, it is important to detect the endocrine problem as early as possible to prevent the possible development of irreversible atrophic changes and to minimize social consequences. Unfortunately, routine pediatric examinations do not test for olfactory function, and many individuals lacking such function, particularly children, do not seek help for this difficulty.

4.46.5.9.6 Kidney disease
Chronic renal failure is associated with alterations in smell and taste function, and in some circumstances, transplantation may improve the function. For example, Griep, M. I. *et al.* (1997) determined amyl acetate detection thresholds on 101 patients with chronic renal failure. Patients with cognitive deficits were excluded from analysis. No differences were noted for patients on peritoneal and hemodialysis, or for a subgroup of patients tested before and after hemodialysis. However, transplanted patients had better odor perception relative to matched dialysis controls. Significant negative correlations were noted between threshold sensitivity and serum concentrations of urea and phosphorus, as well as between such sensitivity and protein catabolic rates.

In patients younger than 55 years with uremia, a significant impairment of taste recognition was reported for salty, sweet, bitter, and sour agents relative to controls (Ciechanover, M. *et al.*, 1980). This impairment for sour and bitter, however, was found to show improvement immediately after dialysis. These uremic patients reportedly showed no signs of zinc deficiency.

4.46.5.9.7 Pseudohypoparathyroidism
Pseudohypoparathyroidism (PHP) is characterized by deficient responsiveness to, but not deficient production of, parathyroid hormone. Type Ia PHP exhibits generalized hormone resistance, a deficiency of the

alpha chain of the stimulatory guanine nucleotide-binding protein ($G_s\alpha$) of adenylyl cyclase, and a constellation of developmental and skeletal anomalies termed Albright hereditary osteodystrophy (AHO). Type Ia patients exhibit obesity, short stature, brachydactyly, round faces, and subcutaneous ossifications. In contrast, type Ib patients have a specific hormone resistance to parathyroid hormone and lack a deficiency in $G_s\alpha$ protein activity. They also do not have AHO.

Patients with type Ia PHP, but not type Ib PHP, were reported, in the 1980s, to have olfactory dysfunction (Weinstock, R. S. *et al.*, 1986). The problem was attributed to the $G_s\alpha$ protein deficiency, since G_s proteins are involved in the first stages of olfactory transduction (Weinstock, R. S. *et al.*, 1986). More recent work, however, found type 1b patients to have similar olfactory dysfunction, making it unlikely that the PHP-related deficit reflects a $G_s\alpha$ protein deficiency (Doty, R. L. *et al.*, 1997a). In this same study, patients with pseudopseudohypoparathyroidism (PPHP), a disorder found in some relatives of patients with type Ia PHP, were found to have relatively normal smell function. Like type Ia PHP, PPHP is accompanied by a $G_s\alpha$ protein deficiency and by AHO, but not by a generalized end-organ hormone insensitivity. These findings suggest that the smell loss of PHP type Ia patients is unlikely due to $G_s\alpha$ deficiency. More work is clearly needed to determine the basis of the olfactory deficits in some patients with PHP.

4.46.5.9.8 Turner's syndrome
Chromatin-negative gonadal dysgenesis (Turner's syndrome) results from a X-chromosomal monosomy (i.e., 45,X karyotype). This disorder is characterized by a feminized phenotype, a short and sometimes webbed neck, small stature, low-set ears, a high-arched palate, shield-like chest, and sexual infantilism. Congenital lymphedema, cardiac deficits, and renal deficits are also sometimes found. Such patients reportedly have elevated detection and recognition thresholds to pyridine, thiophene, and nitrobenzene (Henkin, R. I., 1967). Gonadal hormone therapy was found to reverse the chemosensory deficits. A genetic basis for the chemosensory disorder is suggested by the fact that the mothers of the patients exhibited similar olfactory abnormalities.

4.46.5.10 Psychiatric Disorders

Several psychiatric disorders are associated with altered chemosensation. Such disorders include severe stage anorexia nervosa, alcoholism with or without Wernicke–Korsakoff syndrome, seasonal affective disorder (SAD), attention-deficit/hyperactivity disorder (ADHD), depression, olfactory reference syndrome (ORS), schizophrenia, and chronic hallucinatory psychoses.

4.46.5.10.1 Alcoholism

Olfactory function is depressed in chronic alcoholics, particularly those whose alcoholism has led to vitamin deficiencies and the Wernicke–Korsakoff syndrome (Mair, R. G. et al., 1986). For example, DiTraglia, G. M. et al. (1991) found alcoholics to have lower UPSIT scores than nonalcoholic controls; thus, 32% of 37 alcoholics exhibited scores within the clinically impaired range, as compared with 5% of 21 controls. A significant correlation was found between UPSIT scores and cortical sulcal volumes. In a follow-up study of 23 alcoholic subjects, those who had remained abstinent ($n = 15$) had significantly higher UPSIT scores than those who had not ($n = 8$). Subsequent studies by this group found impairment on the UPSIT to be associated with elevated cortical and ventricular CSF volumes, as well as with reduced cortical and subcortical gray matter volumes (Shear, P. K. et al., 1992). Moreover, thalamic volume was found to be a significant unique predictor of UPSIT scores.

4.46.5.10.2 Anorexia and bulimia nervosa

Anorexia is associated with loss of olfactory function. Fedoroff, I. C. et al. (1995) administered the UPSIT and the phenyl ethyl alcohol threshold test to 55 eating-disordered hospitalized women. Very low-weight anorexics ($n = 11$) exhibited statistically significant deficits, although the magnitude of the losses was relative small, being approximately 3 UPSIT points and approximately 1.25 log vol/vol dilution units. The problem remained even after considerable weight has been regained at the time of discharge (although ideal body weight had not yet been achieved), suggesting the possibility that the effect may be long lasting.

4.46.5.10.3 Attention-deficit/hyperactivity disorder

Olfactory dysfunction has been associated with some forms of ADHDs. For example, Gansler, D. A. et al. (1998) found that patients diagnosed with ADHD/inattentive type (ADHD−) exhibited, among other things, odor identification deficits and deficits on a test of working memory, whereas those diagnosed with ADHD/hyperactive impulsive type (ADHD+)

exhibited relatively more dysfunction on the Wisconsin card sorting test. These authors suggested that such differences may be linked to dysregulation of separate frontal neurotransmitter systems and/or brain regions.

4.46.5.10.4 Depression

Although persons with olfactory dysfunction often are depressed, the reverse is rarely true (i.e., depression does not cause smell loss). Thus, Amsterdam, J. D. et al. (1987) administered the UPSIT to 51 unmedited patients with well-documented moderate-to-severe depression and to 51 age-, race-, and gender-matched controls. No differences were observed between the scores of the two groups, a finding that has been replicated by others using this same test (Kopala, L. C. et al., 1994).

Electrogustometric thresholds do not differ between depressed and nondepressed subjects (Scinska, A. et al., 2004). Nevertheless, clinically depressed patients sometimes complain of an unpleasant or diminished taste that is not readily explainable by medications or other causes (Miller, S. M. and Naylor, G. J., 1989). Paradoxically, pleasantness ratings are higher in depressed patients than in controls for high concentrations of sucrose (Amsterdam, J. D. et al., 1987; Kavaliers, M., 1989).

4.46.5.10.5 Olfactory reference syndrome

Patients with ORS believe that unpleasant or unhealthy smells emanate from their body (intrinsic hallucinations) or from some external source (extrinsic hallucinations). In cases where the hallucinations are minimal, patients complain about the odor but do not take steps to remove it. In other cases, the odor is so problematic that reasonable steps are taken to eliminate the odor (e.g., complaining to authorities or plugging up the chimney with newspapers). Compulsive washing behavior, changing clothes, and restriction of social activity is common in intrinsic hallucinatory cases. Some cases of ORC seem closely linked with an obsessive compulsive disorder and are amenable to treatment with serotonin uptake inhibitors (Dominguez, R. A. and Puig, A., 1997). ORS is generally viewed as distinct from schizophrenic or affective disorders (Pryse-Phillips, W., 1971; Bishop, E. R. 1980).

4.46.5.10.6 Seasonal affective disorder

There is some suggestion that SAD is associated with enhanced olfactory sensitivity. Thus, lower detection thresholds for the odorant phenyl ethyl alcohol have been reported in such patients relative to controls

(Postolache, T. T. *et al.*, 2002). This effect occurred regardless of whether the testing was done in the summer and winter, and no evidence of seasonal changes was found. However, a subsequent report, in which both UPSIT and threshold measures were determined for each side of the nose separately, found no differences between SAD patients and controls. Interestingly, however, depression scores were negatively correlated with right-, but not left-side UPSIT scores, suggesting the possibility of some laterality in affective state that influences smell function (Postolache, T. T. *et al.*, 1999).

4.46.5.10.7 Schizophrenia

Patients with schizophrenia exhibit moderate bilateral losses in olfactory function. In a comprehensive meta-analytic study, the median UPSIT score from 13 SZ studies was 32.5 (Moberg, P. J. *et al.*, 1999). The median effect size (i.e., the mean difference between patient and control groups expressed in SD units) for 17 SZ studies employing the UPSIT was 0.92. This deficit is clearly less, on average, than that seen in such neurodegenerative disorders as Alzheimer's disease (AD) and Parkinson's disease (PD) (see Section 4.46.5.11), where median effect sizes for eight AD and four PD studies have been reported as 2.01 and 1.66, respectively (Mesholam, R. I. *et al.*, 1998). In a recent study from our center, 23% of a group of 41 patients with SZ exhibited UPSIT scores indicative of microsmia or anosmia, in contrast to 94.4% of a group of 18 AD patients and 96.3% of a group of 54 PD patients (Moberg, P. J. *et al.*, unpublished data).

The olfactory decrement of SZ appears early in the disease process and, in fact, is found in many patients who may be prone to SZ, such as those with schizotypy or with family histories positive for significant mental illness (Park, S. and Schoppe, S., 1997; Kopala, L. C. *et al.*, 1998). This deficit appears to be more strongly related to neuropsychological measures of frontal lobe function (e.g., problems in executive function) than to measures of medial temporal lobe function (e.g., memory) (Pantelis, C. and Brewer, W. J., 1995); functional imaging studies report hypometabolism or decreased activation within the subtemporal and subfrontal lobes, particularly on the right (Buchsbaum, M. S. *et al.*, 1991). The olfactory bulbs and tracts are reduced in patients with SZ, as well as in their first-order relatives (Turetsky, B. I. *et al.*, 2003), although the significance of such reduction is not entirely clear. The smell loss of SZ correlates with negative, but not positive, symptoms of the disorder (Brewer, W. J. *et al.*, 1996). UPSIT

scores are inversely and significantly correlated with disease duration, suggesting that some progression of pathology may be occurring somewhere in the olfactory pathways (Moberg, P. J. *et al.*, 1997). This phenomenon appears to be independent of medication history and gender. Presently, odor identification ability appears to be the only neuropsychological marker known to correlate with disease duration in SZ.

4.46.5.11 Neurodegenerative Diseases

The ability to smell is compromised in a number of neurodegenerative diseases, including AD, Down's syndrome (DS), Huntington's disease (HD), idiopathic PD, multi-infarct dementia (MID), multiple sclerosis (MS), Pick's disease, restless leg syndrome, and the parkinsonism–dementia complex of Guam (PDC) (for recent review, see Doty, R. L., 2003). The degree of olfactory dysfunction differs, on average, amongst most of these disorders. For example, as discussed in more detail below, AD, PD, and PDC are accompanied by marked alterations in the ability to smell, whereas HD and MID are accompanied by more moderate alterations. Essential tremor, progressive supranuclear palsy (PSP), and 1-methyl-4-phenyl-l,2,3,6-tetrahydropyridine-induced parkinsonism (MPTP-P) are associated with no, or only minor, changes in the ability to smell, in spite of the fact that they share major clinical features with PD.

4.46.5.11.1 Alzheimer's disease

The smell dysfunction of AD, although prevalent and marked (present in 85–90% of patients), is usually not total. Indeed, many patients are unaware of their disorder until after being tested ((Doty, R. L. *et al.*, 1987). The loss affects both nasal chambers in the earliest disease stages, including some cases of questionable AD, and appears to increase in magnitude with time. In a meta-analysis of 11 olfactory/AD studies, effect sizes ranging from 0.98 to 12.15 (median = 2.17) were noted (Mesholam, R. I. *et al.*, 1998). The olfactory dysfunction is reflected by decreased odor-related activation of central structures (e.g., subfrontal temporal lobe), as measured by positron emission tomography (PET; Figure 6) (Kareken, D. A. *et al.*, 2001).

Olfactory dysfunction – particularly in conjunction with other risk factors – appears to be a predictor of subsequent development of AD in older persons. For example, Graves, A. B. *et al.* (1999) administered a 12-item smell identification test to 1604 nondemented community-dwelling senior citizens 65 years of age or

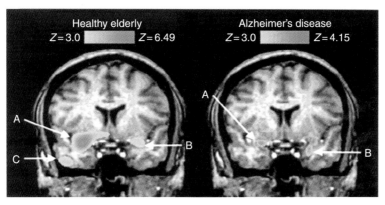

Figure 6 Left: Whole brain subtraction of olfactory stimulation minus control stimulation in healthy elderly control participants, coregistered on a T1-weighted reference magnetic resonance image ($n = 7$), accounting for one control participant whose right temporal lobe fell outside the scanner field of view. Right and left are reversed. Regions of interest are as follows: (a) Right frontotemporal junction (piriform area); (b) Left piriform area; (c) Right anterior-ventral temporal lobe (Brodmann 20). Right: Same whole-brain subtraction in patients with Alzheimer's disease. Regions of interest are as follows: A. 7 mm anterior to the frontotemporal junction (piriform) region in the control participants' image; B. Left amygdala-uncus (from Kareken, D. A., Doty, R. L., Moberg, P. J., Mosnik, D., Chen, S. H., Farlow, M. R., *et al.* 2001. Olfactory-evoked regional cerebral blood flow in Alzheimer's disease. Neuropsychology 15, 18–29). Copyright © 2001 by the American Psychological Association.

older. Over a subsequent 2 year time period, the identification test scores proved to be a better predictor of cognitive decline than global cognitive test scores. Anosmics who possessed at least one APOE-4 allele had 4.9 times the risk of having cognitive decline than normosmics not possessing this allele. This is in contrast to a 1.23 times greater risk for cognitive decline in normosmic individuals possessing at least one such APOE allele. Anosmic women who possessed at least one APOE-4 allele had an odds ratio of 9.71, compared to an odds ratio of 1.90 for women who were normosmic and possessed at least one allele. The corresponding odds ratios for men were 3.18 and 0.67, respectively. This study, along with similar longitudinal studies (e.g., Tabert, M. H. *et al.*, 2005), strongly suggests that olfactory testing is a useful aid in determining those at-risk individuals who are most likely to develop AD.

4.46.5.11.2 Amyotrophic lateral sclerosis
Amyotrophic lateral sclerosis (ALS), traditionally viewed as a motor neuron disease, is associated with olfactory dysfunction. The degree of dysfunction is less than that seen in AD or PD, with average UPSIT scores falling around 30. One study of 58 patients with motor neuron disease reported that only bulbar patients exhibit significant deficits (Hawkes, C. H. *et al.*, 1998). Another study of 37 ALS patients found that nearly half had UPSIT scores that fell within the normal range (Sajjadian, A. *et al.*, 1994). Nonetheless,

in the latter study, 75.7% of the patients scored below their individually matched controls. Only 11% of the ALS patients had total or near-total anosmia. Although no sex differences or laterality in the ALS-related test scores were observed, an age-related decrement was present. Interestingly, significant correlations were found between UPSIT scores and peripheral nerve conductance measures. One histological study of the olfactory bulbs from eight ALS patients revealed excessive lipofuscin deposition in all eight cases, suggesting subclinical neuronal damage (Hawkes, C. H. *et al.*, 1998).

4.46.5.11.3 Multiple sclerosis
Although rare, smell loss can be a presenting symptom of MS (Constantinescu, C. S. *et al.*, 1994). About a quarter of MS patients appear to exhibit olfactory dysfunction at any one time (Doty, R. L. *et al.*, 1984b). The demyelinating plaques appear to be the basis of the disorder, as strong inverse correlations have been found between odor identification test scores and the number of plaques within the subfrontal and subtemporal lobes (Doty, R. L. *et al.*, 1997b; Zorzon, M. *et al.*, 2000). No such relationship is present in other brain regions. Longitudinal studies suggest that olfactory function waxes and wanes in relation to the exacerbation and remission of plaques within subtemporal and subfrontal lobes (Doty, R. L. *et al.*, 1999). Evidence that MS alters taste function is limited,

although suprathreshold deficits have been observed for both tests of identification and perceived intensity (Catalanotto, F. A. *et al.*, 1984).

4.46.5.11.4 Parkinson's disease

Parkinson's disease, which is classically considered a motor disorder, is accompanied by smell loss with a prevalence rate greater than tremor, one of the cardinal signs of the disorder (Doty, R. L. *et al.*, 1988). As in AD, total anosmia is not the rule. Thus, only 38.3% of 81 patients evaluated in one study had UPSIT scores suggestive of anosmia, and only 13% of 38 patients who received an odor detection threshold test were unable to detect the highest odorant concentration presented (Doty, R. L. *et al.*, 1988). Moreover, all but one of 41 PD patients asked whether an odor was present on each of 40 UPSIT items answered affirmatively to 35 or more of the items, even though most were unable to identify the majority of the odors or felt that the perceived sensation did not correspond to the response alternatives. While olfactory loss is generally bilateral, there can be slight individual differences in the degree to which the left and right sides of the nose are involved. No association exists, however, between the side of relatively greater involvement and the side of hemiparkinsonism, as might be expected if asymmetrical damage to striatal dopamine systems were involved in the olfactory problem (Doty, R. L. *et al.*, 1992). The smell loss is unrelated to the magnitude of the motor symptoms, as well as a range of neuropsychological measures (Doty, R. L. *et al.*, 1989), although subtle variations among so-called benign vs. malignant forms may be present (Stern, M. B. *et al.*, 1994). Anti-PD medications (e.g., L-dopa, dopamine agonists, and anticholinergic compounds) have no influence on the smell deficit, which occurs as severely in nonmediated or never-medicated patients as in medicated ones (Doty, R. L. *et al.*, 1992). A correlation between striatal dopamine transport activity, an index of the health of a neuron, and smell function has been recently reported (Siderowf, A. *et al.*, 2005).

The PD-related smell loss appears in both familial and sporadic forms and may be a sign of the preclinical disease state (Markopoulou, K. *et al.*, 1997; Berendse, H. W. B., 2001). As in the case with AD, some asymptomatic first-degree relatives of patients with either familial or sporadic forms of PD appear to exhibit olfactory dysfunction (Montgomery, E. B. *et al.,*. 1999). Recently, Berendse, H. W. B. (2001) administered tests of odor detection, identification, and discrimination to 250 relatives of PD patients

(approximately 84% children, approximately 16% siblings, and one parent). In 25 hyposmic and 23 normosmic individuals sampled from this group, nigrostriatal dopaminergic function was assessed using single-photon emission computer tomography (SPECT) with $[^{125}I]\beta$-CIT as a dopamine transporter ligand. An abnormal reduction in striatal dopamine transporter binding was present in four of the 25 (16%) hyposmic relatives, two of whom subsequently developed clinical parkinsonism, and in none of the 23 normosmic relatives. Significant reductions in dopamine transporter binding were found only in hyposmic relatives of PD patients, suggesting olfactory dysfunction may precede clinical motor signs in PD. Such findings reiterate the point that olfactory testing, in conjunction with other measures, is likely useful in the early detection of PD.

Although the basis of the PD-related olfactory loss has been debated, Braak *et al.* present anatomical evidence suggesting that first CNS lesions of PD develop simultaneously within (a) the olfactory bulb and related portions of the AON and (b) the dorsal motor nucleus of the vagus nerve and adjoining intermediate reticular zone (Figure 7) (Del Tredici, K. *et al.*, 2002; Braak, H. *et al.*, 2003b). Cell types that succumb to PD-related development of inclusion bodies are primarily unmyelinated or poorly myelinated projection neurons with disproportionately long axons in relation to their somata (Braak, H. *et al.*, 2003b). It is noteworthy that dopamine-containing cells within the olfactory bulb are increased twofold in PD, at least in its later stages, despite the bulb's early involvement in the degenerative process and evidence of altered function (Huisman, E. *et al.*, 2004). The reason for this paradoxical finding is not known, although it may relate to the repopulation of periglomerular cells from progenitors within the anterior subventricular zone that migrate to the olfactory bulb via the rostral migratory stream (Saino-Saito, S. *et al.*, 2004).

4.46.5.11.5 Progressive supranuclear palsy

PSP, a disorder that accounts for approximately 4% of patients with parkinsonian symptoms, is commonly misdiagnosed as PD since it shares many motor features with PD. Unlike PD, however, tremor is rarely present in this disorder, and the parkinsonian features are less responsive to anti-PD medications (Jackson, J. A. *et al.*, 1983). Compared to PD, PSP is typically characterized by more frontal lobe dysfunction, more neuronal degeneration within the basal ganglia and upper brain

Preclinical PD

Clinical PD

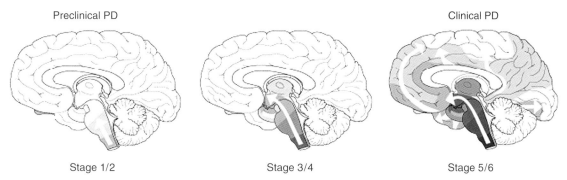

Stage 1/2   Stage 3/4   Stage 5/6

Figure 7 Parkinson's disease (PD)-related Lewy body pathology evolves in predictable stages. According to the staging system of Braak, Lewy bodies (LB) first form within in the olfactory bulb and dorsal motor nucleus of the vagal nerve (stage 1). In stage 2 and stage 3, LB pathology expands from these induction sites into additional brain stem nuclei (e.g., locus coeruleus and substantia nigra) and then into the amygdala. In stage 5 to stage 6, the pathology extends into the cerebral cortex. Clinical symptoms arise during stage 4 to stage 6 when the pathology involves significant regions of the substantia nigra and related brain areas (from Thal, D. R., Del, T. K., and Braak, H. 2004. Neurodegeneration in normal brain aging and disease 15. Sci. Aging Knowledge Environ. e26.).

stem, and less involvement of mesolimbic and meso-cortical dopamine systems (Jankovic, J., 1989). Conceivably because of this difference in distribution of pathology, about half of PSP patients have a normal sense of smell, with the remainder having slight-to-moderate dysfunction. In one study of 22 PSP patients, for example, the mean PSP UPSIT score was approximately 32, compared to a mean PD UPSIT score of approximately 19 and a mean control UPSIT score of approximately 36.

4.46.5.12 Epilepsy

It is widely recognized that olfactory and gustatory auras can occur secondary to seizure activity within the temporal lobes and, more rarely, within the orbitofrontal lobes (Acharya, V. *et al.*, 1998). Examples of taste sensations that have been reported in such cases include peculiar, rotten, sweet, like a cigarette, like rotten apples, and like vomitus (for review, see (West, S. E. and Doty, R. L., 1995). However, some of these tastes likely represent smell sensations that are miscategorized as tastes by both the patients and their physicians.

A number of early olfactory threshold studies reported heightened overall bilateral olfactory sensitivity in patients with epilepsy, particularly prior to an ictal event. More recent studies, however, have not reported such hypersensitivity, conceivably reflecting (1) mitigating influences of modern anti-seizure medications or (2) evaluation of more severe and largely intractable cases. For example, one study reported normal bilaterally determined *n*-butanol

odor detection thresholds in 18 epileptic patients relative to 17 controls (Eskenazi, B. *et al.*, 1986), a finding that was confirmed for unilaterally determined odor detection thresholds in 21 epileptic patients (Martinez, B. A. *et al.*, 1993). Similarly, another study of 16 patients with left foci and 14 patients with right foci found no detection threshold deficits on either side of the nose to the odorant phenyl ethyl alcohol (West, S. E. and Doty, R. L. 1995). In contrast to threshold studies, suprathreshold deficits have been reported in epileptic patients, including deficits in odor identification, discrimination, and both short- and long-term memory (e.g., Martinez, B. A. *et al.* (1993) . Prolonged odor event-related potential latencies have also been noted in patients with right-side foci after right-side odorant stimulation, and in patients with left-side foci after left-side odorant stimulation (Hummel, T. *et al.*, 1995).

Small, D. M., *et al.* (1997) observed that patients who had a right anterior temporal lobectomy had elevated citric acid recognition thresholds, but normal detection thresholds, relative to both controls and patients who had undergone left temporal lobectomy. Employing PET, these researchers noted that orally presented citric acid increased regional cerebral blood flow (rCBF) bilaterally in the caudolateral orbitofrontal cortex, but relatively more in the right anteromedial temporal lobe and in the right caudomedial orbitofrontal cortex. They speculated that (1) damage to the area corresponding to the rCBF increase in right anterior medial temporal lobe in normal subjects may account for elevation in taste recognition thresholds in lobectomized patients and (2) taste recognition requires further processing by structures located in the anteromedial temporal lobe.

4.46.5.13 Migraine

The limited evidence suggests that 20–50% of migraineurs experience olfactory auras, osmophobia (intolerance of odors), and/or taste abnormalities during migraine attacks (Diamond, S. *et al.*, 1985). Moreover, both odorants and tastants can trigger migraine attacks in susceptible individuals (Blau, J. N. and Solomon, F., 1985). Unfortunately, few empirical studies of chemosensory function have been made in migraine sufferers, and consensus is lacking as to the direction of the effects. Hirsch, A. R. (1992) reported that of 67 migraineurs, 12 were either hyposmic or anosmic, a percentage much higher than that expected in the general population. No controls were tested. More recently, Snyder, R. D. and Drummond, P. D. (1997) determined odor recognition thresholds to acetone and vanillin in 20 migraine suffers and 21 controls. Rather than decreased sensitivity, these authors found heightened sensitivity (i.e., lower thresholds) in the migraineurs for the odor of vanillin. Although this was not the case for the odor of acetone, those migraineurs who reported that odors generally seemed stronger during migraine attacks exhibited lower acetone thresholds than those exhibited by the others. Hedonic ratings for concentrated acetone did not differ between the migraineurs and the controls.

4.46.6 Treatment of Chemosensory Disorders

Meaningful treatments are available for some, but not all, patients whose olfactory dysfunction is conductive, including dysfunction resulting from blockage of airflow to the olfactory neuroepithelium or from inflammation within the olfactory membrane proper (for review, see Doty, R. L. and Mishra, A., 2001). Thus, surgical procedures that reduce nasal obstruction can improve olfactory function in some patients. Effective therapies for olfactory loss secondary to allergic rhinitis include allergy management, topical cromolyn, topical and systemic corticosteroid therapies, and surgical procedures to reduce inflammation or obstructions. A brief course of systemic steroid therapy can be useful in distinguishing between conductive and sensorineural olfactory loss, since those with conductive loss will often respond to treatment. However, longer term systemic steroid therapy is not advised.

Topical nasal steroids are often ineffectual in returning smell function because the steroid fails to reach the affected regions in the upper nasal passages. Increased efficacy presumably occurs when nasal steroid drops or sprays are administered in the head-down Moffett's position. In patients with obstructive inflammatory disease, swelling of the ostiomeatal complex can prevent drainage from the sinuses causing chronic sinusitis. Antibiotic therapy in combination with control of the allergic symptoms underlying the inflammation is effective in many of these cases. Resistant cases typically require surgery to improve drainage and clear infection (Bromley, S. M. and Doty, R. L., 2002).

It is an empirical question as to whether exercising the olfactory system by repeatedly exposing it to odorants improves olfactory function over time. Animal studies suggest that repeated presentations of an odor may alter detection thresholds (Doty, R. L. and Ferguson-Segall, M., 1989), as well as functional connectivity in higher brain regions (Wilson, D. A. *et al.*, 1985). Moreover, there is evidence, in humans, that repeated presentation of odorants increases sensitivity to them (Doty, R. L. *et al.*, 1981). A recently clinical study on this topic, which is unpublished, suggested that some benefit may accrue by exposing hyposmic subjects to a set of odorants in the morning and evening, although the degree of benefit was small and an atypically low spontaneous return of function was present in the control group (Hummel, T. *et al.*, 2005).

Currently, there are no treatment options for smell loss secondary to congenital or other malformations of the olfactory bulbs or stalks. In general, olfactory dysfunction due to sensorineural causes is difficult to manage, and the prognosis for patients suffering from long-standing anosmia due to upper respiratory illness or head trauma is relatively poor. Most patients who recover smell function subsequent to head trauma do so within 12 weeks of injury (Costanzo, R. M. *et al.*, 1995). Although there are no verified treatments for trauma-related olfactory loss, anti-inflammatory agents may minimize post-traumatic sequelae in some cases. Rat research suggests that application of nerve growth factor onto the olfactory epithelium may alleviate axotomy-induced degenerative changes in the olfactory receptor neurons, although it is not known whether this has any functional consequence or if such a procedure in humans would be efficacious (Yasuno, H. *et al.*, 2000). Rat studies also suggest that estrogens may offer some prophylaxis to neurotoxin-related damage to the olfactory receptor region (Dhong, H. J. *et al.*, 1999).

CNS tumors that impinge upon the olfactory bulbs and tracts, as well as sclerotic tissue within the medial temporal lobes, can be resected in some instances with a resultant improvement in olfaction. Although permanent sensory abnormalities can result from damage to a peripheral nerve supplying taste (e.g., lingual nerve and facial nerve), such damage is amenable to surgical repair, particularly in the first months after the injury (Robinson, P. P. *et al.*, 2000). However, the degree of return of function is highly variable and return of normal function is rare. In a series of 53 patients with prior lingual nerve injury who underwent direct reapposition of the nerve with epineurial sutures, Robinson, P. P. *et al.* (2000) found that 41% more patients could detect some taste solutions postoperatively.

A number of oral infections and inflammatory problems that alter taste function can be treated with appropriate antibiotic or anti-inflammatory medication. Even though a number of dysgeusias spontaneously remit with time (Deems, D. S. *et al.*, 1996), this is not true in all cases, and several interventions have been reported to aid in their resolution. First, those dysgeusias due to medications, such as salty or bitter tastes secondary to the use of drugs (e.g., ACE inhibitors and antifugals), commonly resolve following drug discontinuance, although such resolution may require months after cessation. Second, some pharmacological treatments are available for dysgeusias. Thus, low doses of anticonvulsants can eliminate dysgeusias due to central causes, such as partial seizures. Over-the-counter antioxidants or mineral supplements, such as alpha lipoic acid and zinc gluconate (140 mg day^{-1}), may mitigate some dysgeusias (Femiano, F. *et al.*, 2002; Heckmann, S. M. *et al.*, 2005), although zinc therapies are generally ineffectual for most chemosensory disturbances in the absence of frank zinc deficiencies such as occur in liver and kidney disease (Price, S., 1986). Tricyclic antidepressants (e.g., amitriptyline, desipramine, and nortriptyline) and benzodiazepines are reportedly useful as therapy in some cases. Specialized mouthwashes have been reported to provide symptomatic relief in some patients (e.g., chlorhexidine digluconate, dyclonine HCl) (Lang, N. P. *et al.*, 1988). In cases of dysgeusia or taste loss secondary to hypothyroidism, thyroxin replacement therapy is likely beneficial. Mucosal function and comfort can be provided to some patients with xerostomia and excessive dryness using artificial salivas (e.g., Xerolube) (Bromley, S. M. and Doty, R. L., 2003).

There is some rationale to controlling dysgeusias by applying a local anesthetic, such as a lidocaine or cancor sore solution, to selected sectors of the tongue. One basis for this approach comes from studies by Bartoshuk *et al.* (e.g., Yanagisawa, K. *et al.*, 1998). These authors found that anesthetizing one chorda tympani nerve increases the perceived intensity of bitter substances, such as quinine, applied to taste fields innervated by the contralateral glossopharyngeal nerve. In contrast, perceived intensity of NaCl applied to an area innervated by the ipsilateral glossopharyngeal nerve is decreased. When both chorda tympani nerves are anesthetized, the taste of quinine is intensified and the taste of NaCl diminished in areas innervated by the glossopharyngeal on both sides of the tongue. In about 40% of their subjects, a phantom dysgeusia, usually localized to the posterior tongue contralateral to the anesthesia, appeared in the absence of stimulation. This phantom taste was eliminated when the region of origin was anesthetized.

4.46.7 Summary

Possibly owing to the fact that, historically, most disorders of smell and taste have been difficult to diagnose and treat, physicians often downplay these senses in the routine neurological examination. This is unfortunate when one considers that such disorders are common and profoundly affect a patient's quality of life. The chemical senses largely determine the flavor of foods and beverages and provide an early warning for leaking natural gas, spoiled food, fire, and other adverse environmental situations. Importantly, olfactory disturbances can be an early sign of a number of serious diseases or anomalies, including AD, PD, epilepsy, MS, and schizophrenia. Although some patients initially present with a frank complaint of a smell disturbance, others are unaware of their dysfunction, pointing out the need for routine clinical quantitative assessment, which is now easily performed in the office.

In this chapter, we (1) reviewed the anatomy of the human taste and smell systems, (2) described the major symptoms experienced by patients suffering from chemosensory disorders, (3) described the major causes of chemosensory dysfunction, and (4) presented up-to-date practical techniques for their management and quantitative evaluation. It is clear from the material reviewed that chemosensory disorders are varied in nature and can result from a wide range of diseases and disturbances, including such common neurological disorders as AD and idiopathic PD. It is also apparent that while some types of chemosensory disturbances can be treated medically or surgically, in many cases treatment options are limited or nonexistent. Accurate assessment of the dysfunction is of value to both the patient and physician, allowing for a determination of the relative degree of dysfunction and whether the alteration is abnormal relative to a patient's age and gender.

References

Acharya, V., Acharya, J., and Luders, H. 1998. Olfactory epileptic auras. Neurology, 51, 56–61.

Ackerman, B. H. and Kasbekar, N. 1997. Disturbances of taste and smell induced by drugs. Pharmacotherapy, 17, 482–496.

Amsterdam, J. D., Settle, R. G., Doty, R. L., Abelman, E., and Winokur, A. 1987. Taste and smell perception in depression. Biol. Psychiatry, 22, 1481–1485.

Andre, J. M., Beis, J. M., Morin, N., and Paysant, J. 2000. Buccal hemineglect. Arch. Neurol. 57, 1734–1741.

Arnhold-Schneider, M. and Bernemann, D. 1987. Uber die Haufigkeit von Geschmacksstorungen nach Tonsillektomie. HNO, 35, 195–198.

Baylis, L. L., Rolls, E. T., and Baylis, G. C. 1995. Afferent connections of the caudolateral orbitofrontal cortex taste area of the primate. Neuroscience 64, 801–812.

Bellas, D. N., Novelly, R. A., Eskenazi, B., and Wasserstein, J. 1988. The nature of unilateral neglect in the olfactory sensory system. Neuropsychologia 26, 45–52.

Berendse, H. W. B. 2001. Subclinical dopaminergic dysfunction in asymptomatic Parkinson's disease patients' relatives with a decreased sense of smell. Ann. Neurol. 50, 34–41.

Bishop, E. R. 1980. An olfactory reference syndrome – monosymptomatic hypochondriasis. J. Clin. Psychiatry 41, 57–59.

Blau, J. N. and Solomon, F. 1985. Smell and other sensory disturbances in migraine. J. Neurol. 232, 275–276.

Braak, H., Del Tredici, K., Rub, U., De Vos, R. A., Jansen Steur, E. N., and Braak, E. 2003a. Staging of brain pathology related to sporadic Parkinson's disease. Neurobiol. Aging 24, 197–211.

Braak, H., Rub, U., Gai, W. P., and Del, T. K. 2003b. Idiopathic Parkinson's disease: possible routes by which vulnerable neuronal types may be subject to neuroinvasion by an unknown pathogen. J. Neural Transm. 110, 517–536.

Brewer, W. J., Edwards, J., Anderson, V., Robinson, T., and Pantelis, C. 1996. Neuropsychological, olfactory, and hygiene deficits in men with negative symptom schizophrenia. Biol. Psychiatry 40, 1021–1031.

Bromley, S. M. and Doty, R. L. 2002. Taste. In: Diseases of the Nervous System: Clinical Neuroscience and Therapeutic Principles (eds. A. K. Asbury, G. M. McKhann, W. I. McDonald, P. J. Goadsby, and J. C. McArthur), 3rd ed., pp. 610–620. Cambridge University Press.

Bromley, S. M. and Doty, R. L. 2003. Clinical disorders affecting taste: evaluation and management. In: Handbook of Olfaction and Gustation (ed. R. L. Doty), 2nd ed., pp. 935–957. Marcel Dekker.

Brosvic, G. M., Risser, J. M., Mackay-Sim, A., and Doty, R. L. 1996. Odor detection performance in hypothyroid and euthyroid rats. Physiol. Behav. 59, 117–121.

Buchsbaum, M. S., Kesslak, J. P., Lynch, G., Chui, H., Wu, J., Sicotte, N., et al. 1991. Temporal and hippocampal metabolic rate during an olfactory memory task assessed by positron emission tomography in patients with dementia of the Alzheimer type and controls. Preliminary studies. Arch. Gen. Psychiatry 48, 840–847.

Catalanotto, F. A., Dore-Duffy, P., Donaldson, J. O., Testa, M., Peterson, M., and Ostrom, K. M. 1984. Quality-specific taste changes in multiple sclerosis. Ann. Neurol. 16, 611–615.

Chang, C. Y. J. and Cass, S. P. 1999. Management of facial nerve injury due to temporal bone trauma. Amer. J. Otology. 20, 96–111.

Ciechanover, M., Peresecenschi, G., Aviram, A., and Steiner, J. E. 1980. Malrecognition of taste in uremia. Nephron 26, 20–22.

Constantinescu, C. S., Raps, E. C., Cohen, J. A., West, S. E., and Doty, R. L. 1994. Olfactory disturbances as the initial or most prominent symptom of multiple sclerosis. J. Neurol. Neurosurg. Psychiatry 57, 1011–1012.

Costanzo, R. M., DiNardo, L. J., and Zasler, N. D. 1995. Head injury and olfaction. In: Handbook of Olfaction and Gustation (ed. R. L. Doty), pp. 493–502. Marcel Dekker.

Cowart, B. J., Flynn-Rodden, K., McGeady, S. J., and Lowry, L. D. 1993. Hyposmia in allergic rhinitis. J. Allergy Clin. Immunol. 91, 747–751.

Daniels, C., Gottwald, B., Pause, B. M., Sojka, B., Mehdorn, H. M., and Ferstl, R. 2001. Olfactory event-related potentials in patients with brain tumors. Clin. Neurophysiol. 112, 1523–1530.

Deems, D. A., Doty, R. L., Settle, R. G., Moore-Gillon, V., Shaman, P., Mester, A. F., et al. 1991. Smell and taste disorders, a study of 750 patients from the University of Pennsylvania Smell and Taste Center. Arch. Otolaryngol. Head Neck Surg. 117, 519–528.

Deems, D. A., Yen, D. M., Kreshak, A., and Doty, R. L. 1996. Spontaneous resolution of dysgeusia. Arch. Otolaryngol. Head Neck Surg. 122, 961–963.

Del Tredici, K., Rub, U., De Vos, R. A., Bohl, J. R., and Braak, H. 2002. Where does Parkinson disease pathology begin in the brain? J. Neuropathol. Exp. Neurol. 61, 413–426.

Dhong, H. J., Chung, S. K., and Doty, R. L. 1999. Estrogen protects against 3-methylindole-induced olfactory loss. Brain Res. 824, 312–315.

Diamond, S., Freitag, F. G., Prager, J., and Gandi, S. 1985. Olfactory aura in migraine. N. Engl. J. Med. 312, 1390–1391.

DiTraglia, G. M., Press, D. S., Butters, N., Jernigan, T. L., Cermak, L. S., Velin, R. A., et al. 1991. Assessment of olfactory deficits in detoxified alcoholics. Alcohol 8, 109–115.

Dominguez, R. A. and Puig, A. 1997. Olfactory reference syndrome responds to clomipramine but not fluoxetine: a case report. J. Clin. Psychiatry 58, 497–498.

Doty, R. L. 2001. Olfaction. Annu. Rev. Psychol. 52, 423–452.

Doty, R. L. 2003. Odor perception in neurodegenerative diseases. In: Handbook of Olfaction and Gustation (ed. R. L. Doty), 2nd ed., pp. 479–502. Marcel Dekker.

Doty, R. L. and Bromley, S. M. 2004. Effects of drugs on olfaction and taste. Otolaryngol. Clin. North Am. 37, 1229–1254.

Doty, R. L. and Ferguson-Segall, M. 1989. Influence of adult castration on the olfactory sensitivity of the male rat: a signal detection analysis. Behav. Neurosci. 103, 691–694.

Doty, R. L. and Haxel, B. R. 2005. Objective assessment of terbinafine-induced taste loss. Laryngoscope 115, 2035–2037.

Doty, R. L. and Mishra, A. 2001. Olfaction and its alteration by nasal obstruction, rhinitis, and rhinosinusitis. Laryngoscope 111, 409–423.

Doty, R. L., Deems, D. A., and Stellar, S. 1988. Olfactory dysfunction in parkinsonism: a general deficit unrelated to neurologic signs, disease stage, or disease duration. Neurology 38, 1237–1244.

Doty, R. L., Fernandez, A. D., Levine, M. A., Moses, A., and McKeown, D. A. 1997a. Olfactory dysfunction in type I pseudohypoparathyroidism: dissociation from Gs alpha protein deficiency. J. Clin. Endocrinol. Metab. 82, 247–250.

Doty, R. L., Li, C., Mannon, L. J., and Yousem, D. M. 1997b. Olfactory dysfunction in multiple sclerosis. N. Engl. J. Med. 336, 1918–1919.

Doty, R. L., Li, C., Mannon, L. J., and Yousem, D. M. 1999. Olfactory dysfunction in multiple sclerosis: relation to longitudinal changes in plaque numbers in central olfactory structures. Neurology 53, 880–882.

Doty, R. L., Philip, S., Reddy, K., and Kerr, K. L. 2003. Influences of antihypertensive and antihyperlipidemic drugs on the senses of taste and smell: a review. J. Hypertens. 21, 1805–1813.

Doty, R. L., Reyes, P. F., and Gregor, T. 1987. Presence of both odor identification and detection deficits in Alzheimer's disease. Brain Res. Bull. 18, 597–600.

Doty, R. L., Riklan, M., Deems, D. A., Reynolds, C., and Stellar, S. 1989. The olfactory and cognitive deficits of Parkinson's disease: evidence for independence. Ann. Neurol. 25, 166–171.

Doty, R. L., Shaman, P., and Dann, M. 1984b. Development of the University of Pennsylvania smell identification test: a standardized microencapsulated test of olfactory function. Physiol. Behav. 32, 489–502.

Doty, R. L., Shaman, P., Applebaum, S. L., Giberson, R., Siksorski, L., Rosenberg, et al. 1984a. Smell identification ability: changes with age. Science 226, 1441–1443.

Doty, R. L., Snyder, P. J., Huggins, G. R., and Lowry, L. D. 1981. Endocrine, cardiovascular, and psychological correlated of olfactory sensitivity changes during the human menstrual cycle. J. Comp. Physiol. Psychol. 95, 45–60.

Doty, R. L., Stern, M. B., Pfeiffer, C., Gollomp, S. M., and Hurtig, H. I. 1992. Bilateral olfactory dysfunction in early stage treated and untreated idiopathic Parkinson's disease. J. Neurol. Neurosurg. Psychiatry 55, 138–142.

Doty, R. L., Yousem, D. M., Pham, L. T., Kreshak, A. A., Geckle, R., and Lee, W. W. 1997c. Olfactory dysfunction in patients with head trauma. Arch. Neurol. 54, 1131–1140.

Downey, L. L., Jacobs, J. B., and Lebowitz, R. A. 1996. Anosmia and chronic sinus disease. Otolaryngol. Head Neck Surg 115, 24–28.

Elsberg, C. A. 1935a. XI. The value of quantitative olfactory tests for the localization of supratentorial tumors of the brain. A preliminary report. Bull. Neurol. Inst. NY 4, 511–522.

Elsberg, C. A. 1935b. XII. The localization of tumors of the frontal lobe of the brain by quantitative olfactory tests. Bull. Neurol. Inst. NY 4, 535–543.

Eskenazi, B., Cain, W. S., Novelly, R. A., and Mattson, R. 1986. Odor perception in temporal lobe epilepsy patients with and without temporal lobectomy. Neuropsychologia 24, 553–562.

Ezeh, P. I., Myers, L. J., Hanrahan, L. A., Kemppainen, R. J., and Cummins, K. A. 1992. Effects of steroids on the olfactory function of the dog. Physiol. Behav. 51, 1183–1187.

Faurion, A., Cerf, B., Le, B. D., and Pillias, A. M. 1998. fMRI study of taste cortical areas in humans. Ann. NY Acad. Sci. 855, 535–545.

Fedoroff, I. C., Stoner, S. A., Andersen, A. E., Doty, R. L., and Rolls, B. J. 1995. Olfactory dysfunction in anorexia and bulimia nervosa. Int. J. Eating Disord 18, 71–77.

Femiano, F., Scully, C., and Gombos, F. 2002. Idiopathic dysgeusia; an open trial of alpha lipoic acid (ALA) therapy. Int. J. Oral Maxillofac. Surg. 31, 625–628.

Feron, F., Perry, C., McGrath, J. J., and Mackay, S. 1998. New techniques for biopsy and culture of human olfactory epithelial neurons. Arch. Otolaryngol. Head Neck Surg. 124, 861–866.

Ferrante, M. A. 1999. Endogenous metabolic disorders. In: Textbook of Clinical Neurology (eds. C. G. Goetz and E. J. Pappert), pp. 731–767. W.B. Saunders Company.

Frank, M. E., Hettinger, T. P., and Clive, J. M. 1995. Current trends in measuring taste. In: Handbook of Olfaction and Gustation (ed. R. L. Doty), pp. 669–688. Marcel Dekker.

Funkutake, T., Kita, KI., Sakakibara, R., Takagi, K., Tokumaru, Y., Kojima, J., Hattari, T., and Hirayana, K. 1996. Late-onset hereditary ataxia with global thermoanalgesia and absence of fungiform papillae on the tongue in a Japanese family. Brain 119, 1011–1021.

Gadoth, N., Mass, E., Gordon, C. R., et al. 1997. Taste and smell in familial dysautonomia. Dev. Med. Child. Neurol. 39, 393–397.

Gansler, D. A., Fucetola, R., Krengel, M., Stetson, S., Zimering, R., and Makary, C. 1998. Are there cognitive subtypes in adult attention deficit/hyperactivity disorder? J. Nerv. Ment. Dis. 186, 776–781.

Ghorbanian, S. N., Paradise, J. L., and Doty, R. L. 1983. Odor perception in children in relation to nasal obstruction. Pediatrics 72, 510–516.

Giasson, B. I. and Mushynski, W. E. 1996. Aberrant stress-induced phosphorylation of perikaryal neurofilaments. J. Biol. Chem. 271, 30404–30409.

Giasson, B. I., Sampathu, D. M., Wilson, C. A., Vogelsberg-Ragaglia, V., Mushynski, W. E., and Lee, V. M. 2002. The environmental toxin arsenite induces tau hyperphosphorylation. Biochemistry 41, 15376–15387.

Gloor, P. 1997. The Temporal Lobe and the Limbic System. Oxford University Press.

Graff-Radford, N. R., Lin, S. C., Brazis, P. W., Bolling, J. P., Liesegang, T. J., Lucas, J. A., et al. 1997. Tropicamide eyedrops cannot be used for reliable diagnosis of Alzheimer's disease. [see comments]. Mayo Clin. Proc. 72, 495–504.

Graves, A. B., Bowen, J. D., Rajaram, L., McCormick, W. C., McCurry, S. M., Schellenberg, G. D., et al. 1999. Impaired olfaction as a marker for cognitive decline: interaction with apolipoprotein E epsilon4 status. Neurology 53, 1480–1487.

Griep, M. I., Van der, N. P., Sennesael, J. J., Mets, T. F., Massart, D. L., and Verbeelen, D. L. 1997. Odour perception in chronic renal disease. Nephrol. Dial. Transplant. 12, 2093–2098.

Gross-Isseroff, R., Ophir, D., Marshak, G., Ganchrow, J. R., Beizer, M., and Lancet, D. 1989. Olfactory function following late repair of choanal atresia. Laryngoscope 99, 1165–1166.

Grushka, M. and Sessle, B. J. 1991. Burning mouth syndrome. [Review] [101 refs]. Dent. Clin. North Am. 35, 171–184.

Hardy, S. L., Brennand, C. P., and Wyse, B. W. 1981. Taste thresholds of individuals with diabetes mellitus and of control subjects. J. Am. Diet. Assoc. 79, 286–289.

Harner, S. G. and Laws, E. R., Jr. 1981. Diagnosis of acoustic neurinoma. Neurosurgery 9, 373–379.

Hawkes, C. H., Shephard, B. C., Geddes, J. F., Body, G. D., and Martin, J. E. 1998. Olfactory disorder in motor neuron disease. Exp. Neurol. 150, 248–253.

Heckmann, S. M., Hujoel, P., Habiger, S., Friess, W., Wichmann, M., Heckmann, J. G., et al. 2005. Zinc gluconate in the treatment of dysgeusia – a randomized clinical trial. J. Dent. Res. 84, 35–38.

Henkin, R. I. 1967. Abnormalities of taste and olfaction in patients with chromatin negative gonadal dysgenesis. J. Clin. Endocrinol. Metab. 27, 1436–1440.

Henkin, R. I. 1975. The Role of Adrenal Corticosteroids in Sensory Processes. In: Handbook of Physiology (eds. H. Blaschko, A. D. Smith, and G. Sayers), pp. 209–230. American Physiological Society.

Henkin, R. I. and Bartter, F. C. 1966. Studies on olfactory thresholds in normal man and in patients with adrenal cortical insufficiency: the role of adrenal cortical steroids and of serum sodium concentration. J. Clin. Invest. 45, 1631–1639.

Henkin, R. I. and Daly, R. L. 1968. Auditory detection and perception in normal man and in patients with adrenal cortical insufficiency: effect of adrenal cortical steroids. J. Clin. Invest. 47, 1269–1280.

Hirsch, A. R. 1992. Olfaction in migraineurs. Headache 32, 233–236.

Hoffman, H. J., Ishii, E. K., and Macturk, R. H. 1998. Age-related changes in the prevalence of smell/taste problems among the United States adult population. Results of the 1994 disability supplement to the National Health Interview Survey (NHIS). Ann. NY Acad. Sci. 855, 716–722.

Horiguchi, T., Uryu, K., Giasson, B. I., Ischiropoulos, H., LightFoot, R., Bellmann, C., et al. 2003. Nitration of tau protein is linked to neurodegeneration in tauopathies.[erratum appears in Am. J. Pathol. 2003 Dec; 163(6), 2645] . Am. J. Pathol. 163, 1021–1031.

Huisman, E., Uylings, H. B. M., and Hoogland, P. V. 2004. A 100% increase of 687 dopaminergie cells in the olfactory bulb may explain hyposmia in parkinson's disease. Mov. Disord. 19, 687–692.

Hummel, T., Pauli, E., Schuler, P., Kettenmann, B., Stefan, H., and Kobal, G. 1995. Chemosensory event-related potentials in patients with temporal lobe epilepsy. Epilepsia 36, 79–85.

Hummel, T., Risson, K., Muller, A., Reden, J., Weidenbecher, M., and Huttenbrink, K. B. 2005. "Olfactory training" in patients with olfactory loss. Presented at the Association for Chemoreception Sciences XXVII Annual Meeting. In Sarasota, Florida.

Iannaconne, A., Mykytyn, K., Persico, A. M., Searby, C. C., Baldi, A., Jablonski, M. M., et al. 2005. Clinical evidence of decreased olfaction in Bardet–Biedl syndrome caused by a deletion in the BBS4 gene. Am. J. Med. Genet. 132A, 343–346.

Jackson, J. A., Jankovic, J., and Ford, J. 1983. Progressive supranuclear palsy: Clinical features and response to treatment in 16 patients. Ann. Neurol. 13, 273–278.

Jankovic, J. 1989. Parkinsonism-plus syndromes. Mov. Disord. 4, S95–S119.

Jung, T. M., TerKonda, R. P., Haines, S. J., Strome, S., and Marentette, L. J. 1997. Outcome analysis of the transglabellar/subcranial approach for lesions of the anterior cranial fossa: a comparison with the classic craniotomy approach. Otolaryngol. Head Neck Surg. 116, 642–646.

Kareken, D. A., Doty, R. L., Moberg, P. J., Mosnik, D., Chen, S. H., Farlow, M. R., et al. 2001. Olfactory-evoked regional cerebral blood flow in Alzheimer's disease. Neuropsychology 15, 18–29.

Kavaliers, M. 1989. Beta-funaltrexamine disrupts the day-night rhythm of nociception in mice. Brain Res. Bull. 22, 783–785.

Kern, R. C. 2000. Chronic sinusitis and anosmia: pathologic changes in the olfactory mucosa. Laryngoscope 110, 1071–1077.

Kern, R. C. and Mathog, R. H. 1991. Neoplasms of the nose and paranasal sinuses. In: Smell and Taste in Health and Disease (eds. T. V. Getchell, R. L. Doty, L. M. Bartoshuk, and J. B. Snow, Jr.), pp. 599–620. Raven Press.

Kida, A. 1974. Relations between the taste evoked potential and the background brain wave. J. Nihon Univ. Med. Assoc., 34, 43–52.

Kopala, L. C., Good, K. P., and Honer, W. G. 1994. Olfactory hallucinations and olfactory identification ability in patients with schizophrenia and other psychiatric disorders. Schizophr. Res. 12(3), 205–211.

Kopala, L. C., Good, K. P., Torrey, E. F., and Honer, W. G. 1998. Olfactory function in monozygotic twins discordant for schizophrenia. Am. J. Psychiatry 155, 134–136.

Lang, N. P., Catalanotto, F. A., Knopfli, R. U., and Antczak, A. A. A. 1988. Quality-specific taste impairment following the application of chlorhexidine digluconate mouth rinses. J. Clin. Periodontol. 15, 43–48.

Lee, S. H., Lim, H. H., Lee, H. M., Park, H. J., and Choi, J. O. 2000. Olfactory mucosal findings in patients with persistent anosmia after endoscopic sinus surgery. Ann. Otol. Rhinol. Laryngol. 109, 720–725.

Lewitt, M. S., Laing, D. G., Panhuber, H., Corbett, A., and Carter, J. N. 1989. Sensory perception and hypothyroidism. Chem. Senses 14, 537–546.

Li, C., Doty, R. L., Kennedy, D. W., and Yousem, D. M. 2003. Evaluation of olfactory deficits by structural medical imaging. In: Handbook of Olfaction and Gustation (ed. R. L. Doty), 2nd edn., pp. 593–613. Marcel Dekker.

Lygonis, C. S. 1969. Familial absence of olfaction. Heredity 61, 413–416.

Mair, R. G., Doty, R. L., Kelly, K. M., Wilson, C. S., Langlais, P. J., McEntee, W. J., et al. 1986. Multimodal sensory discrimination deficits in Korsakoff's psychosis. Neuropsychologia 24, 831–839.

Markopoulou, K., Larsen, K. W., Wszolek, E. K., Denson, M. A., Lang, A. E., Pfeiffer, R. F., et al. 1997. Olfactory dysfunction in familial parkinsonism. Neurology 49, 1262–1267.

Martinez, B. A., Cain, W. S., de Wijk, R. A., Spencer, D. D., Novelly, R. A., and Sass, K. J. 1993. Olfactory functioning before and after temporal lobe resection for intractable seizures. Neuropsychology 7, 351–363.

Mattes, R. D. 1995. Nutritional implications of taste and smell. In: Handbook of Olfaction and Gustation (ed. R. L. Doty), pp. 731–744. Marcel Dekker.

Mattes, R. D., Heller, A. D., and Rivlin, R. S. 1986. Abnormalities in suprathreshold taste function in early hypothyroidism in humans. In: Clinical Measurement of Taste and Smell (eds. H. L. Meiselman, and R. S. Rivlin), pp. 467–486. Macmillan Publishing Company.

McConnell, R. J., Menendez, C. E., Smith, F. R., Henkin, R. I., and Rivlin, R. S. 1975. Defects of taste and smell in patients with hypothyroidism. Am. J. Med. 59, 354–364.

Mesholam, R. I., Moberg, P. J., Mahr, R. N., and Doty, R. L. 1998. Olfaction in neurodegenerative disease: a meta-analysis of olfactory functioning in Alzheimer's and Parkinson's diseases. Arch. Neurol. 55, 84–90.

Miller, S. M. and Naylor, G. J. 1989. Unpleasant taste: A neglected symptom in depression. J. Affect. Disord. 17, 291–293.

Moberg, P. J., Agrin, R., Gur, R. E., Gur, R. C., Turetsky, B. I., and Doty, R. L. 1999. Olfactory dysfunction in schizophrenia: a qualitative and quantitative review. [Review] [90 refs]. Neuropsychopharmacology 21, 325–340.

Moberg, P. J., Doty, R. L., Turetsky, B. I., Arnold, S. E., Mahr, R. N., Gur, R. C., et al. 1997. Olfactory identification deficits in schizophrenia: correlation with duration of illness. Am. J. Psychiatry 154, 1016–1018.

Montgomery, E. B., Jr., Baker, K. B., Lyons, K., and Koller, W. C. 1999. Abnormal performance on the PD test battery by asymptomatic first-degree relatives. Neurology 52, 757–762.

Murakami, S., Mizobuchi, M., Nakashiro, Y., Doi, T., Hato, N., and Yanagihara, N. 1996. Bell palsy and herpes simplex virus: identification of viral DNA in endoneurial fluid and muscle [see comments]. Ann. Int. Med. 124, 27–30.

Netter, F. H. 1964. The Ciba Collection of Medial Illustrations, Vol. 1, Nervous System. Ciba Pharmaceutical Corporation. .

Ogawa, T. and Rutka, J. 1999. Olfactory dysfunction in head injured workers. Acta Otolaryngol. 540, 50–57.

Onoda, K. and Ikeda, M. 1999. Gustatory disturbance due to cerebrovascular disorder. Laryngoscope 109, 123–128.

Pantelis, C. and Brewer, W. J. 1995. Neuropsychological and olfactory dysfunction in schizophrenia: relationship of frontal syndromes to syndromes of schizophrenia. Schizophr. Res. 17, 35–45.

Park, S. and Schoppe, S. 1997. Olfactory identification deficit in relation to schizotypy. Schizophr. Res. 26, 191–197.

Perros, P., MacFarlane, T. W., Counsell, C., and Frier, B. M. 1996. Altered taste sensation in newly-diagnosed NIDDM. Diabetes Care 19, 768–770.

Pittman, J. A. and Beschi, R. J. 1967. Taste thresholds in hyper- and hypothyroidism. J. Clin. Endocrinol. Metab. 27, 895–896.

Physicians' Desk Reference. 2005. (59th ed.) Montvale, N.J: Medical Economics Company, Inc.

Postolache, T. T., Doty, R. L., Wehr, T. A., Jimma, L. A., Han, L., Turner, E. H., et al. 1999. Monorhinal odor identification and depression scores in patients with seasonal affective disorder. J. Affect. Disord. 56, 27–35.

Postolache, T. T., Wehr, T. A., Doty, R. L., Sher, L., Turner, E. H., Bartko, J. J., et al. 2002. Patients with seasonal affective disorder have lower odor detection thresholds than control subjects. Arch. Gen. Psychiatry 59, 1119–1122.

Price, S. 1986. The role of zinc in taste and smell. In: Clinical Measurement of Taste and Smell (eds. H. L. Meiselman and R. S. Rivlin), pp. 443–445. MacMillan.

Pritchard, T. C., Macaluso, D. A., and Eslinger, P. J. 1999. Taste perception in patients with insular cortex lesions. Behav. Neurosci. 113, 663–671.

Pryse-Phillips, W. 1971. An olfactory reference syndrome. Acta Psychiatr. Scand. 47, 484–509.

Robinson, P. P., Loescher, A. R., and Smith, K. G. 2000. A prospective, quantitative study on the clinical outcome of lingual nerve repair. Br. J. Oral Maxillofac. Surg. 38, 255–263.

Rouquier, S., Blancher, A., and Giorgi, D. 2000. The olfactory receptor gene repertoire in primates and mouse: evidence for reduction of the functional fraction in primates. Proc. Natl Acad Sci. U.S.A. 97, 2870–2874.

Saino-Saito, S., Sasaki, H., Volpe, B. T., Kobayashi, K., Berlin, R., and Baker, H. 2004. Differentiation of the dopaminergic phenotype in the olfactory system of neonatal and adult mice 3. J. Comp. Neurol. 479, 389–398.

Sajjadian, A., Doty, R. L., Gutnick, D. N., Chirurgi, R. J., Sivak, M., and Perl, D. 1994. Olfactory dysfunction in amyotrophic lateral sclerosis. Neurodegeneration 3, 153–157.

Schiffman, S. S., Graham, B. G., Suggs, M. S., and Sattely-Miller, E. A. 1998. Effect of psychotropic drugs on taste responses in young and elderly persons. Ann. NY Acad. Sci. 855, 732–737.

Schiffman, S. S., Zervakis, J., Heffron, S., and Heald, A. E. 1999. Effect of protease inhibitors on the sense of taste. Nutrition 15, 767–772.

Scinska, A., Sienkiewicz-Jarosz, H., Kuran, W., Ryglewicz, D., Rogowski, A., Wrobel, E., et al. 2004. Depressive symptoms and taste reactivity in humans. Physiol. Behav. 82, 899–904.

Scrivani, S. J., Keith, D. A., Kulich, R., Mehta, N., and Maciewicz, R. J. 1998. Posttraumatic gustatory neuralgia: a clinical model of trigeminal neuropathic pain. J. Orofac. Pain 12, 287–292.

Settle, R. G. 1981. Suprathreshold glucose and fructose sensitivity in individuals with different family histories of non-insulin-dependent diabetes mellitus. Chem. Senses 6, 435–443.

Settle, R. G. 1991. The chemical senses in diabetes mellitus. In: Smell and Taste in Health and Disease (eds. T. V. Getchell, R. L. Doty, L. M. Bartoshuk, and J. B. Snow, Jr.), pp. 829–843. Raven Press.

Shear, P. K., Butters, N., Jernigan, T. L., DiTraglia, G. M., Irwin, M., Schuckit, M. A., et al. 1992. Olfactory loss in alcoholics: correlations with cortical and subcortical MRI indices. Alcohol 9, 247–255.

Siderowf, A., Newberg, A., Chou, K. L., Lloyd, M., Colcher, A., Hurtig, H. I., et al. 2005. [99mTc]TRODAT-1 SPECT imaging correlates with odor identification in early Parkinson disease. Neurology 64, 1716–1720.

Singh, N., Grewal, M. S., and Austin, J. H. 1970. Familial anosmia. Arch. Neurol. 22, 40–44.

Small, D. M., Jones-Gotman, M., Zatorre, R. J., Petrides, M., and Evans, A. C. 1997. A role for the right anterior temporal lobe in taste quality recognition. J. Neurosci. 17, 5136–5142.

Smith, D. V. and Shipley, M. T. 1992. Anatomy and physiology of taste and smell. J. Head Trauma Rehabil. 7, 1–14.

Snyder, R. D. and Drummond, P. D. 1997. Olfaction in migraine. [see comment]. Cephalalgia 17, 729–732.

Stern, M. B., Doty, R. L., Dotti, M., Corcoran, P., Crawford, D., McKeown, D. A., et al. 1994. Olfactory function in Parkinson's disease subtypes. Neurology 44, 266–268.

Tabert, M. H., Liu, X., Doty, R. L., Serby, M., Zamora, D., Pelton, G. H., et al. 2005. A 10-item smell identification scale related to risk for Alzheimer's disease. Ann. Neurol. 58, 155–160.

Thal, D. R., Del, T. K., and Braak, H. 2004. Neurodegeneration in normal brain aging and disease 15. Sci. Aging Knowledge Environ. e26.

Tonoike, M., Yamaguchi, M., Kaetsu, I., Kida, H., Seo, R., and Koizuka, I. 1998. Ipsilateral dominance of human olfactory activated centers estimated from event-related magnetic fields measured by 122-channel whole-head neuromagnetometer using odorant stimuli synchronized with respirations. Ann. NY Acad. Sci. 855, 579–590.

Turetsky, B. I., Moberg, P. J., Arnold, S. E., Doty, R. L., and Gur, R. E. 2003. Low olfactory bulb volume in first-degree relatives of patients with schizophrenia. Am. J. Psychiatry 160, 703–708.

Uchiyama, T., Fukutake, T., Arai, K., et al. 2001. Machado-Joseph disease associated with an absence of fungiform papillae on the tongue. Neurology 56, 558–560.

Weinstock, R. S., Wright, H. N., and Smith, D. U. 1993. Olfactory dysfunction in diabetes mellitus. Physiol. Behav. 53, 17–21.

Weinstock, R. S., Wright, H. N., Spiegel, A. M., Levine, M. A., and Moses, A. M. 1986. Olfactory dysfunction in humans with deficient guanine nucleotide-binding protein. Nature 322, 635–636.

West, S. E. and Doty, R. L. 1995. Influence of epilepsy and temporal lobe resection on olfactory function. Epilepsia 36, 531–542.

Wilson, D. A., Sullivan, R. M., and Leon, M. 1985. Odor familiarity alters mitral cell response in the olfactory bulb of neonatal rats. Brain Res. 354, 314–317.

Yamagishi, M., Fujiwara, M., and Nakamura, H. 1994. Olfactory mucosal findings and clinical course in patients with olfactory disorders following upper respiratory viral infection. Rhinology 32, 113–118.

Yanagisawa, K., Bartoshuk, L. M., Catalanotto, F. A., Karrer, T. A., and Kveton, J. F. 1998. Anesthesia of the chorda tympani nerve and taste phantoms. Physiol. Behav. 63, 329–335.

Yasuno, H., Fukazawa, K., Fukuoka, T., Kondo, E., Sakagami, M., and Noguchi, K. 2000. Nerve growth factor applied onto the olfactory epithelium alleviates degenerative changes of the olfactory receptor neurons following axotomy. Brain Res. 887, 53–62.

Yousem, D. M., Geckle, R. J., Bilker, W., McKeown, D. A., and Doty, R. L. 1996. MR evaluation of patients with congenital hyposmia or anosmia. AJR 166, 439–443.

Yousem, D. M., Turner, W. J., Li, C., Snyder, P. J., and Doty, R. L. 1993. Kallmann syndrome: MR evaluation of olfactory system. [see comments]. AJNR Am. J. Neuroradiol. 14, 839–843.

Zorzon, M., Ukmar, M., Bragadin, L. M., Zanier, F., Antonello, R. M., Cazzato, G., et al. 2000. Olfactory dysfunction and extent of white matter abnormalities in multiple sclerosis: a clinical and MR study. Mult. Scler. 6, 386–390.

Index to Volume 4

Cross-reference terms in *italics* are general cross-references, or refer to subentry terms within the main entry (the main entry is not repeated to save space). Readers are also advised to refer to the end of each article for additional cross-references - not all of these cross-references have been included in the index cross-references.

The index is arranged in set-out style with a maximum of three levels of heading. Major discussion of a subject is indicated by bold page numbers. Page numbers suffixed by *t* and *f* refer to Tables and Figures respectively. *vs.* indicates a comparison.

This index is in letter-by-letter order, whereby hyphens and spaces within index headings are ignored in the alphabetization. Prefixes and terms in parentheses are excluded from the initial alphabetization.